Stats

Modeling the World

THIRD EDITION

EDITION 3

Stats
Modeling the World

David E. Bock
Ithaca High School
Cornell University

Paul F. Velleman
Cornell University

Richard D. De Veaux
Williams College

Addison-Wesley

Boston San Francisco New York
London Toronto Sydney Tokyo Singapore Madrid
Mexico City Munich Paris Cape Town Hong Kong Montreal

Editor in Chief	Deirdre Lynch
Acquisitions Editor	Christopher Cummings
Senior Editor, AP and Electives	Andrea Sheehan
Assistant Editor	Christina Lepre
Editorial Assistant	Dana Jones
Senior Project Editor	Chere Bemelmans
Senior Managing Editor	Karen Wernholm
Senior Production Supervisor	Sheila Spinney
Cover Design	Barbara T. Atkinson
Digital Assets Manager	Marianne Groth
Media Producer	Christine Stavrou
Software Development	Edward Chappell (MathXL) and Marty Wright (TestGen)
Marketing Manager	Alex Gay
Marketing Coordinator	Kathleen DeChavez
Senior Author Support/ Technology Specialist	Joe Vetere
Senior Prepress Supervisor	Caroline Fell
Senior Manufacturing Buyer	Carol Melville
Senior Media Buyer	Ginny Michaud
Production Coordination, Composition, and Illustrations	Pre-Press PMG
Interior Design	The Davis Group, Inc.
Cover Photo	Pete McArthur

Library of Congress Cataloging-in-Publication Data
Bock, David E.
  Stats : modeling the world / David E. Bock, Paul F. Velleman, Richard D. De Veaux.— 3rd ed.
      p. cm.
  Includes index.
  ISBN 13: 978-0-13-135958-1 (high school student edition)   ISBN 13: 978-0-13-135959-8 (high school teacher's edition)
  ISBN 10: 0-13-135958-4 (high school student edition)   ISBN 10: 0-13-135959-2 (high school teachers's edition)
   1. Graphic calculators—Textbooks. I. Velleman, Paul F., 1949- II. De Veaux, Richard D. III. Title.
  QA276.12.B628 2010
  519.5—dc22

                                                                                    2008029019

For permission to use copyrighted material, grateful acknowledgement has been made to the copyright holders listed in Appendix D, which is hereby made part of this copyright page.

Many of the designations used by manufacturers and sellers to distinguish their products are claimed as trademarks. Where those designations appear in this book, and Addison-Wesley was aware of a trademark claim, the designations have been printed in initial caps or all caps. TI-Nspire and the TI-Nspire logo are trademarks of Texas Instruments, Inc.

4 5 6 7 8 9 10—CRK—12 11

Addison-Wesley
is an imprint of

ISBN 13: 978-0-13-135959-8
ISBN 10: 0-13-135959-2

www.PearsonSchool.com/Advanced

To Greg and Becca, great fun as kids and great friends as adults,
and especially to my wife and best friend, Joanna, for her
understanding, encouragement, and love
—*Dave*

To my sons, David and Zev, from whom I've learned so much,
and to my wife, Sue, for taking a chance on me
—*Paul*

To Sylvia, who has helped me in more ways than she'll ever know,
and to Nicholas, Scyrine, Frederick, and Alexandra,
who make me so proud in everything that they are and do
—*Dick*

Meet the Authors

David E. Bock taught mathematics at Ithaca High School for 35 years. He has taught Statistics at Ithaca High School, Tompkins-Cortland Community College, Ithaca College, and Cornell University. Dave has won numerous teaching awards, including the MAA's Edyth May Sliffe Award for Distinguished High School Mathematics Teaching (twice), Cornell University's Outstanding Educator Award (three times), and has been a finalist for New York State Teacher of the Year.

Dave holds degrees from the University at Albany in Mathematics (B.A.) and Statistics/Education (M.S.). Dave has been a reader and table leader for the AP Statistics exam, serves as a Statistics consultant to the College Board, and leads workshops and institutes for AP Statistics teachers. He has recently served as K–12 Education and Outreach Coordinator and a senior lecturer for the Mathematics Department at Cornell University. His understanding of how students learn informs much of this book's approach.

Dave relaxes by biking and hiking. He and his wife have enjoyed many days camping across Canada and through the Rockies. They have a son, a daughter, and three grandchildren.

Paul F. Velleman has an international reputation for innovative Statistics education. He is the author and designer of the multimedia statistics CD-ROM *ActivStats*, for which he was awarded the EDUCOM Medal for innovative uses of computers in teaching statistics, and the ICTCM Award for Innovation in Using Technology in College Mathematics. He also developed the award-winning statistics program, Data Desk, and the Internet site Data And Story Library (DASL) (http://dasl.datadesk.com), which provides data sets for teaching Statistics. Paul's understanding of using and teaching with technology informs much of this book's approach.

Paul has taught Statistics at Cornell University since 1975. He holds an A.B. from Dartmouth College in Mathematics and Social Science, and M.S. and Ph.D. degrees in Statistics from Princeton University, where he studied with John Tukey. His research often deals with statistical graphics and data analysis methods. Paul co-authored (with David Hoaglin) *ABCs of Exploratory Data Analysis*. Paul is a Fellow of the American Statistical Association and of the American Association for the Advancement of Science.

Out of class, Paul sings baritone in a barbershop quartet. He is the father of two boys.

Richard D. De Veaux is an internationally known educator and consultant. He has taught at the Wharton School and the Princeton University School of Engineering, where he won a "Lifetime Award for Dedication and Excellence in Teaching." Since 1994, he has been Professor of Statistics at Williams College. Dick has won both the Wilcoxon and Shewell awards from the American Society for Quality. He is a fellow of the American Statistical Association. Dick is also well known in industry, where for the past 20 years he has consulted for such companies as Hewlett-Packard, Alcoa, DuPont, Pillsbury, General Electric, and Chemical Bank. He has also sometimes been called the "Official Statistician for the Grateful Dead." His real-world experiences and anecdotes illustrate many of this book's chapters.

Dick holds degrees from Princeton University in Civil Engineering (B.S.E.) and Mathematics (A.B.) and from Stanford University in Dance Education (M.A.) and Statistics (Ph.D.), where he studied with Persi Diaconis. His research focuses on the analysis of large data sets and data mining in science and industry.

In his spare time he is an avid cyclist and swimmer. He also is the founder and bass for the "Diminished Faculty," an a cappella Doo-Wop quartet at Williams College. Dick is the father of four children.

Contents

Appendixes

*Indicates an optional chapter.

Preface

About the Book

We've been thrilled with the feedback we've received from teachers and students using *Stats: Modeling the World*, Second Edition. If there is a single hallmark of this book it is that students actually read it. We have reports from every level—from high school to graduate school—that students find our books easy and even enjoyable to read. We strive for a conversational, approachable style, and introduce anecdotes to maintain students' interest. And it works. Teachers report their amazement that students are voluntarily reading ahead of their assignments. Students write to tell us (to their amazement) that they actually enjoyed the book.

Stats: Modeling the World, Third Edition is written from the ground up with the understanding that Statistics is practiced with technology. This insight informs everything from our choice of forms for equations (favoring intuitive forms over calculation forms) to our extensive use of real data. Most important, it allows us to focus on teaching Statistical Thinking rather than calculation. The questions that motivate each of our hundreds of examples are not "how do you find the answer?" but "how do you think about the answer?"

Our Goal: Read This Book!

The best text in the world is of little value if students don't read it. Here are some of the ways we have made *Stats: Modeling the World*, Third Edition even more approachable:

- *Readability.* You'll see immediately that this book doesn't read like other Statistics texts. The style, both colloquial (with occasional humor) and informative, engages students to actually read the book to see what it says.
- *Informality.* Our informal diction doesn't mean that the subject matter is covered lightly or informally. We have tried to be precise and, wherever possible, to offer deeper explanations and justifications than those found in most introductory texts.
- *Focused lessons.* The chapters are shorter than in most other texts, to make it easier to focus on one topic at a time.
- *Consistency.* We've worked hard to avoid the "do what we say, not what we do" trap. From the very start we teach the importance of plotting data and checking assumptions and conditions, and we have been careful to model that behavior right through the rest of the book.
- *The need to read.* Students who plan just to skim the book may find our presentation a bit frustrating. The important concepts, definitions, and sample solutions don't sit in little boxes. This is a book that needs to be read, so we've tried to make the reading experience enjoyable.

New to the Third Edition

The third edition of *Stats: Modeling the World* continues and extends the successful innovations pioneered in our books, teaching Statistics and statistical thinking as it is practiced today. We've rewritten sections throughout the book to make them clearer and more interesting. We've introduced new up-to-the-minute motivating examples throughout. And, we've added a number of new features, each with the goal of making it even easier for students to put the concepts of Statistics together into a coherent whole.

FOR EXAMPLE

▶ *For Example.* In every chapter, you'll find approximately 4 new worked examples that illustrate how to apply new concepts and methods—**more than 100 new illustrative examples.** But these aren't isolated examples. We carry a discussion through the chapter with each *For Example*, picking up the story and moving it forward as students learn to apply each new concept.

STEP-BY-STEP EXAMPLE

▶ *Step-by-Step Worked Examples.* We've brought our innovative *Think/Show/Tell Step-by-Step* examples up-to-date with new examples and data.

 A S

▶ *ActivStats Pointers.* In the third edition, the *ActivStats* pointers have been revised for clarity and now indicate exactly what they are pointing to—activity, video, simulation, or animation—paralleling the book's discussions to enhance learning.

TI-nspire

▶ *TI-Nspire Activities.* We've created many demonstrations and investigations for TI-Nspire handhelds to enhance each chapter. They're on the DVD and at the book's Web site.

▶ *Exercises.* We've added **hundreds of new exercises,** including more single-concept exercises at the beginning of each set so students can be sure they have a clear understanding of each important topic before they're asked to tie them all together in more comprehensive exercises. Continuing exercises have been **updated with the most recent data.** Whenever possible, the data are on the DVD and the book's Web site so students can explore them further.

▶ *Data Sources.* Most of the data used in examples and exercises are from recent news stories, research articles, and other real-world sources. We've listed more of those sources in this edition.

▶ *Chapters 4 and 5* have been entirely rewritten and reorganized. We think you'll agree with our reviewers that the new organization—discussing displays and summaries for quantitative data in Chapter 4 and then expanding on those ideas to discuss comparisons across groups, outliers, and other more sophisticated topics in Chapter 5—provides a more exciting and interesting way to approach these fundamental topics.

▶ *Simulation.* We've improved the discussion of simulation in Chapter 11 so it could relate more easily to discussions of experimental design and probability. The simulations included in the *ActivStats* multimedia software on the book's DVD carry those ideas forward in a student-friendly fashion.

▶ *Teacher's Podcasts* (10 points in 10 minutes). Created and presented by the authors, these podcasts focus on key points in each chapter to help you with class preparation. These podcasts are available on the Instructor's Resource CD.

▶ *Video Lectures on DVD* featuring the textbook authors will help students review the high points of each chapter. Video presenters also work through examples from the text. The presentations feature the same student-friendly style and emphasis on critical thinking as the text.

Continuing Features

▶ *Think, Show, Tell.* The worked examples repeat the mantra of *Think, Show*, and *Tell* in every chapter. They emphasize the importance of thinking about a Statistics question (What do we know? What do we hope to learn? Are the assumptions and conditions satisfied?) and reporting our findings (the *Tell* step). The *Show* step contains the mechanics of calculating results and conveys our belief that it is only one part of the process. This rubric is highlighted in the *Step-by-Step* examples that guide the students through the process of analyzing the problem with the general explanation on the left and the worked-out problem on the right. The result is a better understanding of the concept, not just number crunching.

▶ *Just Checking.* Within each chapter, we ask students to pause and think about what they've just read. These questions are designed to be a quick check that they understand the material. Answers are at the end of the exercise sets in each chapter so students can easily check themselves.

▶ *TI Tips.* We emphasize sound understanding of formulas and methods, but want students to use technology for actual calculations. Easy-to-read "TI Tips" in the chapters show students how to use TI-83/84 Plus statistics functions. (Help using a TI-89 or TI-Nspire appears in Appendix B.) We do remind students that calculators are just for "Show"—they cannot Think about what to do nor Tell what it all means.

▶ *Math Boxes.* In many chapters we present the mathematical underpinnings of the statistical methods and concepts. By setting these proofs, derivations, and justifications apart from the narrative, we allow the student to continue to follow the logical development of the topic at hand, yet also refer to the underlying mathematics for greater depth.

▶ *What Can Go Wrong?* Each chapter still contains our innovative *What Can Go Wrong?* sections that highlight the most common errors people make and the misconceptions they have about Statistics. Our goals are to help students avoid these pitfalls, and to arm them with the tools to detect statistical errors and to debunk misuses of statistics, whether intentional or not. In this spirit, some of our exercises probe the understanding of such failures.

▶ *What Have We Learned?* These chapter-ending summaries are great study guides providing complete overviews that highlight the new concepts, define the new terms, and list the skills that the student should have acquired in the chapter.

▶ *Exercises.* Throughout, we've maintained the pairing of examples so that each odd-numbered exercise (with an answer in the back of the book) is followed by an even-numbered exercise on the same concept. Exercises are still ordered by level of difficulty.

REALITY CHECK ▶ *Reality Check.* We regularly remind students that Statistics is about understanding the world with data. Results that make no sense are probably wrong, no matter how carefully we think we did the calculations. Mistakes are often easy to spot with a little thought, so we ask students to stop for a reality check before interpreting their result.

NOTATION ALERT: ▶ *Notation Alert.* Throughout this book we emphasize the importance of clear communication, and proper notation is part of the vocabulary of Statistics. We've found that it helps students when we call attention to the letters and symbols statisticians use to mean very specific things.

ON THE COMPUTER

▶ *Connections.* Each chapter has a *Connections* section to link key terms and concepts with previous discussions and to point out continuing themes, helping students fit newly learned concepts into a growing understanding of Statistics.

▶ *On the Computer.* In the real world, Statistics is practiced with computers. We prefer not to choose a particular Statistics program. Instead, at the end of each chapter, we summarize what students can find in the most common packages, often with an annotated example. Computer output appearing in the book and in exercises is often generic, resembling all of the common packages to some degree.

Coverage

Textbooks are often defined more by what they choose not to cover than by what they do cover. We've been guided in the choice and order of topics by several fundamental principles. First, we have tried to ensure that each new topic fits into the growing structure of understanding that we hope students will build. Several topic orders can support this goal. We explain our reasons for the topic order of the chapters in the ancillary Printed Test Bank and Resource Guide.

GAISE Guidelines. We have worked to provide materials to help each class, in its own way, follow the guidelines of the GAISE (Guidelines for Assessment and Instruction in Statistics Education) project sponsored by the American Statistical Association. That report urges that Statistics education should

1. emphasize Statistical literacy and develop Statistical thinking,
2. use real data,
3. stress conceptual understanding rather than mere knowledge of procedures,
4. foster active learning,
5. use technology for developing concepts and analyzing data, and
6. make assessment a part of the learning process.

We also have been guided by the syllabus of the AP* Statistics course. We agree with the wisdom of those who designed that course in their selection of topics and their emphasis on Statistics as a practical discipline. *Stats: Modeling the World* provides complete discussions of all AP* topics and teaches students communication skills that lead to success on the AP* examination. A correlation of the text to the AP* Statistics course standards is available in the Printed Test Bank and Resource Guide, on the Instructor's Resource CD, and at www.phschool.com/advanced/correlations/statistics.html.

Mathematics

Mathematics traditionally appears in Statistics texts in several roles:

1. It can provide a concise, clear statement of important concepts.
2. It can describe calculations to be performed with data.
3. It can embody proofs of fundamental results.

Of these, we emphasize the first. Mathematics can make discussions of Statistics concepts, probability, and inference clear and concise. We have tried to be sensitive to those who are discouraged by equations by also providing verbal descriptions and numerical examples.

This book is not concerned with proving theorems about Statistics. Some of these theorems are quite interesting, and many are important. Often, though, their proofs are not enlightening to introductory Statistics students, and can distract the audience from the concepts we want them to understand. However, we have not shied away from the mathematics where we believed that it helped clarify without intimidating. You will find some important proofs, derivations, and justifications in Math Boxes that accompany the development of many topics.

Nor do we concentrate on calculations. Although statistics calculations are generally straightforward, they are also usually tedious. And, more to the point, they are often unnecessary. Today, virtually all statistics are calculated with technology, so there is little need for students to work by hand. The equations we use have been selected for their focus on understanding concepts and methods.

Technology and Data

To experience the real world of Statistics, it's best to explore real data sets using modern technology.

▶ *Technology.* We assume that you are using some form of technology in your Statistics course. That could be a calculator, a spreadsheet, or a statistics package. Rather than adopt any particular software, we discuss generic computer output. "TI-Tips"—included in most chapters—show students how to use statistics features of the TI-83/84 Plus series. The Companion DVD, included in the Teacher's Edition, may be purchased for students and includes *ActivStats* and the software package Data Desk. Also, in Appendix B, we offer guidance (by chapter) to help students get started on common software platforms (Excel, MINITAB, Data Desk, JMP, and SPSS), a TI-89 calculator, and a TI-Nspire.

▶ *Data.* Because we use technology for computing, we don't limit ourselves to small, artificial data sets. In addition to including some small data sets, we have built examples and exercises on real data with a moderate number of cases— usually more than you would want to enter by hand into a program or calculator. These data are included on the DVD as well as on the book's Web site, **www.aw.com/bock**.

ON THE DVD

The DVD holds a number of supporting materials, including *ActivStats*, the *Data Desk* statistics package, an Excel add-in (DDXL), all large data sets from the text formatted for the most popular technologies, and two additional chapters.

ActivStats (for Data Desk). The award-winning *ActivStats* multimedia program supports learning chapter by chapter. It complements the book with videos of real-word stories, worked examples, animated expositions of each of the major Statistics topics, and tools for performing simulations, visualizing inference, and learning to use statistics software. The new version of *ActivStats* includes

- improved navigation and a cleaner design that makes it easier to find and use tools such as the Index and Glossary
- more than **1000 homework exercises,** including many new exercises, plus answers to the "odd numbered" exercises. Many are from the text, providing the data already set up for calculations, and some are unique to *ActivStats*. Many exercises link to data files for each statistics package.
- **17** short **video clips,** many new and updated
- **70 animated activities**
- **117 teaching applets**
- more than **300 data sets**

Supplements

STUDENT SUPPLEMENTS

The following supplements are available for purchase:

Graphing Calculator Manual, by Patricia Humphrey (Georgia Southern University) and John Diehl (Hinsdale Central High School), is organized to follow the sequence of topics in the text, and is an easy to-follow, step-by-step guide on how to use the TI-83/84 Plus, TI-89, and TI-Nspire™ graphing calculators. It provides worked-out examples to help students fully understand and use the graphing calculator. (ISBN-13: 978-0-321-57094-9; ISBN-10: 0-321-57094-4)

Pearson Education AP* Test Prep Series: Statistics by Anne Carroll, Ruth Carver, Susan Peters, and Janice Ricks, is written specifically to complement *Stats: Modeling the World, Third Edition, AP* Edition*, and to help students prepare for the AP* Statistics exam. Students can review topics that are discussed in *Stats: Modeling the World, Third Edition AP* Edition*, and are likely to appear on the Advanced Placement Exam. The guide also contains test-taking strategies as well as practice tests. (ISBN 13: 978-0-13-135964-2; ISBN-10: 0-13-135964-9)

Statistics Study Card is a resource for students containing important formulas, definitions, and tables that correspond precisely to the De Veaux/Velleman/Bock Statistics series. This card can work as a reference for completing homework assignments or as an aid in studying. (ISBN-13: 978-0-321-46370-8; ISBN-10: 0-321-46370-6)

Graphing Calculator Tutorial for Statistics will guide students through the keystrokes needed to most efficiently use their graphing calculator. Although based on the TI-84 Plus Silver Edition, operating system 2.30, the keystrokes for this calculator are identical to those on the TI-84 Plus, and very similar to the TI-83 and TI-83 Plus. This tutorial should be helpful to students using any of these calculators, though there may be differences in some lessons. The tutorial is organized by topic. (ISBN-13: 978-0-321-41382-6; ISBN-10: 0-321-41382-2)

TEACHER SUPPLEMENTS

Most of the teacher supplements and resources for this book are available electronically. On adoption or to preview, please go to PearsonSchool.com/Advanced and click "Online Teacher Supplements." You will be required to complete a one-time registration subject to verification before being emailed access information to download materials.

The following supplements are available to qualified adopters:

Teacher's Edition contains answers to all exercises. Packaged with the Teacher's Edition is the Companion DVD and the Instructor's Resource CD. The Instructor's Resource CD includes the Teachers' Solutions Manual, Test Bank and Resource Guide, Audio Podcasts, PowerPoint slides, and Graphing Calculator Manual. (ISBN-13: 978-0-13-135959-8; ISBN-10: 0-13-135959-2)

Printed Test Bank and Resource Guide, by William Craine, contains chapter-by-chapter comments on the major concepts, tips on presenting topics (and what to avoid), teaching examples, suggested assignments, Web links and lists of other resources, as well as chapter quizzes, unit tests, investigative tasks, TI-Nspire activities, and suggestions for projects. An indispensable guide to help teachers prepare for class, the previous editions were soundly praised by new teachers of Statistics and seasoned veterans alike. The Printed Test Bank and Resource Guide is on the Instructor's Resource CD and available for download. (ISBN-13: 978-0-13-135960-4; ISBN-10: 0-13-135960-6)

Teacher's Solutions Manual, by William Craine, contains detailed solutions to all of the exercises. (ISBN-13: 978-0-13-136009-9; ISBN-10: 0-13-136009-4)

TestGen® CD enables teachers to build, edit, print, and administer tests using a computerized bank of questions developed to cover all the objectives of the text. TestGen is algorithmically based, allowing teachers to create multiple but equivalent versions of the same question or test with the click of a button. Teachers can also modify test bank questions or add new questions. Tests can be printed or administered online. (ISBN-13: 978-0-13-135961-1: ISBN-10: 0-13-135961-4)

PowerPoint Lecture Slides provide an outline to use in a lecture setting, presenting definitions, key concepts, and figures from the text. These slide are available on the Instructor's Resource CD and available for download. (ISBN-13: 978-0-321-57101-4; ISBN 10: 0-321-57101-0)

Technology Resources

Instructor's Resource CD, packaged with every new Teacher's Edition, includes the Teacher's Solutions Manual, Test Bank and Resource Guide (which includes a correlation to the AP* Statistics course standards), Audio Podcasts, PowerPoint slides, and Graphing Calculator Manual. A replacement CD is available for purchase. (ISBN-13: 978-0-13-136349-6; ISBN-10: 0-13-136349-2)

Companion DVD A multimedia program on DVD designed to support learning chapter by chapter comes with the Teacher's Edition. It may be purchased separately for individual students or as a lab version (per work station). (ISBN-13: 978-0-13-136608-4; ISBN-10: 0-13-136608-4) A replacement DVD is available for purchase. The DVD holds a number of supporting materials, including:

- *ActivStats® for Data Desk.* The award-winning *ActivStats* multimedia program supports learning chapter by chapter with the book. It complements the book with videos of real-word stories, worked examples, animated expositions of each of the major Statistics topics, and tools for performing simulations, visualizing inference, and learning to use statistics software. The new version of *ActivStats* includes 17 short video clips; 170 animated activities and teaching applets; 300 data sets; 1,000 homework exercises, many with links to Data Desk files; interactive graphs, simulations, activities for the TI-Nspire graphing calculator, visualization tools, and much more.
- *Data Desk* statistics package.
- *TI-Nspire activities.* These investigations and demonstrations for the TI-Nspire handheld illustrate and explore important concepts from each chapter.
- *DDXL,* an Excel add-in, adds sound statistics and statistical graphics capabilities to Excel. DDXL adds, among other capabilities, boxplots, histograms, statistical scatterplots, normal probability plots, and statistical inference procedures not available in Excel's Data Analysis pack.
- *Data.* Data for exercises marked ⓣ are available on the DVD and at www.aw.com/bock formatted for Data Desk, Excel, JMP, MINITAB, SPSS, and the TI calculators, and as text files suitable for these and virtually any other statistics software.
- *Additional Chapters.* Two additional chapters cover **Analysis of Variance** (Chapter 28) and **Multiple Regression** (Chapter 29). These topics point the way to further study in Statistics.

ActivStats® The award-winning ActivStats multimedia program supports learning chapter by chapter with the book. It is available as a standalone DVD, or in a lab version (per work station). It complements the book with videos of real-word stories, worked examples, animated expositions of each of the major Statistics topics, and tools for performing simulations, visualizing inference, and learning to use statistics software. The new version of ActivStats includes 17 short video clips; 170 animated activities and teaching applets; 300 data sets; 1,000 homework exercises, many with links to Data Desk files; interactive graphs, simulations, visualization tools, and much more. ActivStats (Mac and PC) is available in an all-in-one version for Data Desk, Excel, JMP, MINITAB, and SPSS. This DVD also includes Data Desk statistical software. For more information on options for purchasing ActivStats, contact Customer Service at 1-800-848-9500.

MathXL® for School is a powerful online homework, tutorial, and assessment system that accompanies Pearson textbooks in Statistics. With MathXL for School, teachers can create, edit, and assign online homework and tests using algorithmically generated exercises correlated at the objective level to the textbook. They can also create and assign their own online exercises and import TestGen tests for added flexibility. All student work is tracked in MathXL for School's online gradebook. Students can take chapter tests in MathXL for School and receive personalized study plans based on their test results. The study plan diagnoses weaknesses and links students directly to tutorial exercises for the objectives they need to study and retest. Students can also access supplemental animations directly from selected exercises. MathXL for School is available to qualified adopters. For more information, visit our Web site at www.MathXLforSchool.com, or contact your Pearson sales representative.

StatCrunch is a powerful online tool that provides an interactive environment for doing Statistics. StatCrunch can be used for both numerical and graphical data analysis, and uses interactive graphics to illustrate the connection between objects selected in a graph and the underlying data. StatCrunch may be purchased in a Registration Packet of 10 "redemptions." One redemption is for one student for 12 months beginning at the time of registration. Teacher access for StatCrunch adoptors or for those wishing to preview the product may be obtained by filling out the form at www.pearsonschool.com/access_request (ISBN-13: 978-0-13-136416-5; ISBN-10: 0-13-136416-2)

Video Lectures on DVD with Subtitles feature the textbook authors reviewing the high points of each chapter. The presentations continue the same student-friendly style and emphasis on critical thinking as the text. The DVD format makes it easy and convenient to watch the videos from a computer at home or on campus. (ISBN 13: 978-0-321-57103-8; ISBN-10: 0-321-57103-7)

Companion Web Site (www.aw.com/bock) provides additional resources for instructors and students.

Acknowledgments

Many people have contributed to this book in all three of its editions. This edition would have never seen the light of day without the assistance of the incredible team at Addison-Wesley. Our editor in chief, Deirdre Lynch, was central to the genesis, development, and realization of the book from day one. Chris Cummings, acquisitions editor, provided much needed support. Chere Bemelmans, senior project editor, kept us on task as much as humanly possible. Sheila Spinney, senior production supervisor, kept the cogs from getting into the wheels where they often wanted to wander. Christina Lepre, assistant editor, and Kathleen DeChavez, marketing assistant, were essential in managing all of the behind-the-scenes work that needed to be done. Christine Stavrou, media producer, put together a top-notch media package for this book. Barbara T. Atkinson, senior designer, and Geri Davis are responsible for the wonderful way the book looks. Carol Melville, manufacturing buyer, and Ginny Michaud, senior media buyer, worked miracles to get this book and DVD in your hands, and Greg Tobin, publisher, was supportive and good-humored throughout all aspects of the project. Special thanks go out to Pre-Press PMG, the compositor, for the wonderful work they did on this book, and in particular to Laura Hakala, senior project manager, for her close attention to detail. We'd also like to thank our accuracy checkers whose monumental task was to make sure we said what we thought we were saying. They are Jackie Miller, The Ohio State University; Douglas Cashing, St. Bonaventure University; Jared Derksen, Rancho Cucamonga High School; and Susan Blackwell, First Flight High School.

We extend our sincere thanks for the suggestions and contributions made by the following reviewers of this edition:

Allen Back, *Cornell University, New York*

Susan Blackwell, *First Flight High School, North Carolina*

Kevin Crowther, *Lake Orion High School, Michigan*

Sam Erickson, *North High School, Wisconsin*

Guillermo Leon, *Coral Reef High School, Florida*

Martha Lowther, *The Tatnall School, Delaware*

Karl Ronning, *Davis Senior High School, California*

Agatha Shaw, *Valencia Community College, Florida*

We extend our sincere thanks for the suggestions and contributions made by the following reviewers, focus group participants, and class-testers of the previous edition:

John Arko, *Glenbrook South High School, IL*

Kathleen Arthur, *Shaker High School, NY*

Beverly Beemer, *Ruben S. Ayala High School, CA*

Judy Bevington, *Santa Maria High School, CA*

Susan Blackwell, *First Flight High School, NC*

Gail Brooks, *McLennan Community College, TX*

Walter Brown, *Brackenridge High School, TX*

Darin Clifft, *Memphis University School, TN*

Bill Craine, *Ithaca High School, NY*

Sybil Coley, *Woodward Academy, GA*

Caroline DiTullio, *Summit High School, NJ*

Jared Derksen, *Rancho Cucamonga High School, CA*

Laura Estersohn, *Scarsdale High School, NY*

Laura Favata, *Niskayuna High School, NY*

David Ferris, *Noblesville High School, IN*

Linda Gann, *Sandra Day O'Connor High School, TX*

Randall Groth, *Illinois State University, IL*

Donnie Hallstone, *Green River Community College, WA*

Howard W. Hand, *St. Marks School of Texas, TX*

Bill Hayes, *Foothill High School, CA*

Miles Hercamp, *New Palestine High School, IN*

Michelle Hipke, *Glen Burnie Senior High School, MD*

Carol Huss, *Independence High School, NC*

Sam Jovell, *Niskayuna High School, NY*

Peter Kaczmar, *Lower Merion High School, PA*

John Kotmel, *Lansing High School, NY*

Beth Lazerick, *St. Andrews School, FL*

Michael Legacy, *Greenhill School, TX*

John Lieb, *The Roxbury Latin School, MA*

John Maceli, *Ithaca College, NY*

Jim Miller, *Alta High School, UT*

Timothy E. Mitchell, *King Philip Regional High School, MA*

Maxine Nesbitt, *Carmel High School, IN*

Elizabeth Ann Przybysz, *Dr. Phillips High School, FL*

Diana Podhrasky, *Hillcrest High School, TX*

Rochelle Robert, *Nassau Community College, NY*

Bruce Saathoff, *Centennial High School, CA*

Murray Siegel, *Sam Houston State University, TX*

Chris Sollars, *Alamo Heights High School, TX*

Darren Starnes, *The Webb Schools, CA*

PART I

Exploring and Understanding Data

Stats Starts Here[1]

"But where shall I begin?" asked Alice. "Begin at the beginning," the King said gravely, "and go on till you come to the end: then stop."

—Lewis Carroll,
*Alice's Adventures
in Wonderland*

Statistics gets no respect. People say things like "You can prove anything with Statistics." People will write off a claim based on data as "just a statistical trick." And Statistics courses don't have the reputation of being students' first choice for a fun elective.

But Statistics *is* fun. That's probably not what you heard on the street, but it's true. Statistics is about how to think clearly with data. A little practice thinking statistically is all it takes to start seeing the world more clearly and accurately.

So, What Is (Are?) Statistics?

Q: What is Statistics?
A: Statistics is a way of reasoning, along with a collection of tools and methods, designed to help us understand the world.
Q: What are statistics?
A: Statistics (plural) are particular calculations made from data.
Q: So what is data?
A: You mean, "what *are* data?" Data is the plural form. The singular is datum.
Q: OK, OK, so what are data?
A: Data are values along with their context.

It seems every time we turn around, someone is collecting data on us, from every purchase we make in the grocery store, to every click of our mouse as we surf the Web. The United Parcel Service (UPS) tracks every package it ships from one place to another around the world and stores these records in a giant database. You can access part of it if you send or receive a UPS package. The database is about 17 terabytes big—about the same size as a database that contained every book in the Library of Congress would be. (But, we suspect, not *quite* as interesting.) What can anyone hope to do with all these data?

Statistics plays a role in making sense of the complex world in which we live today. Statisticians assess the risk of genetically engineered foods or of a new drug being considered by the Food and Drug Administration (FDA). They predict the number of new cases of AIDS by regions of the country or the number of customers likely to respond to a sale at the mall. And statisticians help scientists and social scientists understand how unemployment is related to environmental controls, whether enriched early education af-

---

[1] This chapter might have been called "Introduction," but nobody reads the introduction, and we wanted you to read this. We feel safe admitting this here, in the footnote, because nobody reads footnotes either.

> The ads say, "Don't drink and drive; you don't want to be a statistic." But you can't be a statistic.
>   We say: "Don't be a datum."

fects later performance of school children, and whether vitamin C really prevents illness. Whenever there are data and a need for understanding the world, you need Statistics.

So our objectives in this book are to help you develop the insights to think clearly about the questions, use the tools to show what the data are saying, and acquire the skills to tell clearly what it all means.

FRAZZ reprinted by permission of United Feature Syndicate, Inc.

Statistics in a Word

> Statistics is about variation.
>   Data vary because we don't see everything and because even what we do see and measure, we measure imperfectly.
>   So, in a very basic way, Statistics is about the real, imperfect world in which we live.

It can be fun, and sometimes useful, to summarize a discipline in only a few words. So,

> Economics is about . . . *Money (and why it is good).*
> Psychology: *Why we think what we think (we think).*
> Biology: *Life.*
> Anthropology: *Who?*
> History: *What, where, and when?*
> Philosophy: *Why?*
> Engineering: *How?*
> Accounting: *How much?*
> In such a caricature, Statistics is about . . . **Variation.**

Data vary. People are different. We can't see everything, let alone measure it all. And even what we do measure, we measure imperfectly. So the data we wind up looking at and basing our decisions on provide, at best, an imperfect picture of the world. This fact lies at the heart of what Statistics is all about. How to make sense of it is a central challenge of Statistics.

So, How Will This Book Help?

A fair question. Most likely, this book will not turn out to be quite what you expected.

What's different?

Close your eyes and open the book to a page at random. Is there a graph or table on that page? Do that again, say, 10 times. We'll bet you saw data displayed in many ways, even near the back of the book and in the exercises.

We can better understand everything we do with data by making pictures. This book leads you through the entire process of thinking about a problem, finding and showing results, and telling others about what you have discovered. At each of these steps, we display data for better understanding and insight.

You looked at only a few randomly selected pages to get an impression of the entire book. We'll see soon that doing so was sound Statistics practice and reasoning.

Next, pick a chapter and read the first two sentences. (Go ahead; we'll wait.)

We'll bet you didn't see anything about Statistics. Why? Because the best way to understand Statistics is to see it at work. In this book, chapters usually start by presenting a story and posing questions. That's when Statistics really gets down to work.

There are three simple steps to doing Statistics right: *think, show,* and *tell:*

Think first. Know where you're headed and why. It will save you a lot of work.

Show is what most folks think Statistics is about. The *mechanics* of calculating statistics and making displays is important, but not the most important part of Statistics.

Tell what you've learned. Until you've explained your results so that someone else can understand your conclusions, the job is not done.

The best way to learn new skills is to take them out for a spin. In **For Example** boxes you'll see brief ways to apply new ideas and methods as you learn them. You'll also find more comprehensive worked examples called **Step-by-Steps.** These show you fully worked solutions side by side with commentary and discussion, modeling the way statisticians attack and solve problems. They illustrate how to think about the problem, what to show, and how to tell what it all means. These step-by-step examples will show you how to produce the kind of solutions instructors hope to see.

Sometimes, in the middle of the chapter, we've put a section called **Just Checking** . . . . There you'll find a few short questions you can answer without much calculation—a quick way to check to see if you've understood the basic ideas in the chapter. You'll find the answers at the end of the chapter's exercises.

MATH BOX

Knowing where the formulas and procedures of Statistics come from and why they work will help you understand the important concepts. We'll provide brief, clear explanations of the mathematics that supports many of the statistical methods in **Math Boxes** like this.

TI Tips   Do statistics on your calculator!

Although we'll show you all the formulas you need to understand the calculations, you will most often use a calculator or computer to perform the mechanics of a statistics problem. Your graphing calculator has a specialized program called a "statistics package." Each chapter contains **TI Tips** that teach you how to use it (and avoid doing most of the messy calculations).

"Get your facts first, and then you can distort them as much as you please. (Facts are stubborn, but statistics are more pliable.)"
—Mark Twain

From time to time, you'll see an icon like this in the margin to signal that the *ActivStats* multimedia materials on the available DVD in the back of the book have an activity that you might find helpful at this point. Typically, we've flagged simulations and interactive activities because they're the most fun and will probably help you see how things work best. The chapters in *ActivStats* are the same as those in the text—just look for the named activity in the corresponding chapter.

If you are using TI-Nspire™ technology, these margin icons will alert you to activities and demonstrations that can help you understand important ideas in the text. If you have the DVD that's available with this book, you'll find these there; if not, they're also available on the book's Web site www.aw.com/bock.

One of the interesting challenges of Statistics is that, unlike in some math and science courses, there can be more than one right answer. This is why two statisticians can testify honestly on opposite sides of a court case. And it's why some people think that you can prove anything with statistics. But that's not true. People make mistakes using statistics, sometimes on purpose in order to mislead others. Most of the unintentional mistakes people make, though, are avoidable. We're not talking about arithmetic. More often, the mistakes come from using a method in the wrong situation or misinterpreting the results. Each chapter has a section called **What Can Go Wrong?** to help you avoid some of the most common mistakes.

> **Time out.** From time to time, we'll take time out to discuss an interesting or important side issue. We indicate these by setting them apart like this.[2]

There are a number of statistics packages available for computers, and they differ widely in the details of how to use them and in how they present their results. But they all work from the same basic information and find the same results. Rather than adopt one package for this book, we present generic output and point out common features that you should look for. The **. . . on the Computer** section of most chapters (just before the exercises) holds this information. We also give a table of instructions to get you started on any of several commonly used packages, organized by chapters in Appendix B's Guide to Statistical Software.

At the end of each chapter, you'll see a brief summary of the important concepts you've covered in a section called **What Have We Learned?** That section includes a list of the **Terms** and a summary of the important **Skills** you've acquired in the chapter. You won't be able to learn the material from these summaries, but you can use them to check your knowledge of the important ideas in the chapter. If you have the skills, know the terms, and understand the concepts, you should be well prepared for the exam—and ready to use Statistics!

Beware: No one can learn Statistics just by reading or listening. The only way to learn it is to do it. So, of course, at the end of each chapter (except this one) you'll find **Exercises** designed to help you learn to use the Statistics you've just read about.

Some exercises are marked with an orange (T). You'll find the data for these exercises on the DVD in the back of the book or on the book's Web site at www.aw.com/bock.

---

[2] Or in a footnote.

We've paired up the exercises, putting similar ones together. So, if you're having trouble doing an exercise, you will find a similar one either just before or just after it. You'll find answers to the odd-numbered exercises at the back of the book. But these are only "answers" and not complete "solutions." Huh? What's the difference? The answers are sketches of the complete solutions. For most problems, your solution should follow the model of the Step-By-Step Examples. If your calculations match the numerical parts of the "answer" and your argument contains the elements shown in the answer, you're on the right track. Your complete solution should explain the context, show your reasoning and calculations, and state your conclusions. Don't fret too much if your numbers don't match the printed answers to every decimal place. Statistics is more about getting the reasoning correct—pay more attention to how you interpret a result than what the digit in the third decimal place was.

In the real world, problems don't come with chapters attached. So, in addition to the exercises at the ends of chapters, we've also collected a variety of problems at the end of each part of the text to make it more like the real world. This should help you to see whether you can sort out which methods to use when. If you can do that successfully, then you'll know you understand Statistics.

Onward!

It's only fair to warn you: You can't get there by just picking out the highlighted sentences and the summaries. This book is different. It's not about memorizing definitions and learning equations. It's deeper than that. And much more fun. But . . .

You have to read the book![3]

---

[3] So, turn the page.

CHAPTER 2

Data

"Data is king at Amazon. Clickstream and purchase data are the crown jewels at Amazon. They help us build features to personalize the Web site experience."

—Ronny Kohavi,
Director of Data Mining
and Personalization,
Amazon.com

Many years ago, most stores in small towns knew their customers personally. If you walked into the hobby shop, the owner might tell you about a new bridge that had come in for your Lionel train set. The tailor knew your dad's size, and the hairdresser knew how your mom liked her hair. There are still some stores like that around today, but we're increasingly likely to shop at large stores, by phone, or on the Internet. Even so, when you phone an 800 number to buy new running shoes, customer service representatives may call you by your first name or ask about the socks you bought 6 weeks ago. Or the company may send an e-mail in October offering new head warmers for winter running. This company has millions of customers, and you called without identifying yourself. How did the sales rep know who you are, where you live, and what you had bought?

The answer is data. Collecting data on their customers, transactions, and sales lets companies track their inventory and helps them predict what their customers prefer. These data can help them predict what their customers may buy in the future so they know how much of each item to stock. The store can use the data and what it learns from the data to improve customer service, mimicking the kind of personal attention a shopper had 50 years ago.

Amazon.com opened for business in July 1995, billing itself as "Earth's Biggest Bookstore." By 1997, Amazon had a catalog of more than 2.5 million book titles and had sold books to more than 1.5 million customers in 150 countries. In 2006, the company's revenue reached $10.7 billion. Amazon has expanded into selling a wide selection of merchandise, from $400,000 necklaces[1] to yak cheese from Tibet to the largest book in the world.

Amazon is constantly monitoring and evolving its Web site to serve its customers better and maximize sales performance. To decide which changes to make to the site, the company experiments, collecting data and analyzing what works best. When you visit the Amazon Web site, you may encounter a different look or different suggestions and offers. Amazon statisticians want to know whether you'll follow the links offered, purchase the items suggested, or even spend a

---

[1] Please get credit card approval before purchasing online.

longer time browsing the site. As Ronny Kohavi, director of Data Mining and Personalization, said, "Data trumps intuition. Instead of using our intuition, we experiment on the live site and let our customers tell us what works for them."

But What *Are* Data?

<table>
<tr><td>THE W'S:</td></tr>
<tr><td>WHO</td></tr>
<tr><td>WHAT</td></tr>
<tr><td>and in what units</td></tr>
<tr><td>WHEN</td></tr>
<tr><td>WHERE</td></tr>
<tr><td>WHY</td></tr>
<tr><td>HOW</td></tr>
</table>

We bet you thought you knew this instinctively. Think about it for a minute. What exactly *do* we mean by "data"?

Do data have to be numbers? The amount of your last purchase in dollars is numerical data, but some data record names or other labels. The names in Amazon.com's database are data, but not numerical.

Sometimes, data can have values that look like numerical values but are just numerals serving as labels. This can be confusing. For example, the ASIN (Amazon Standard Item Number) of a book, like 0321570448, may have a numerical value, but it's really just another name for *Stats: Modeling the World.*

Data values, no matter what kind, are useless without their context. Newspaper journalists know that the lead paragraph of a good story should establish the "Five W's": *Who, What, When, Where,* and (if possible) *Why.* Often we add *How* to the list as well. Answering these questions can provide the **context** for data values. The answers to the first two questions are essential. If you can't answer *Who* and *What,* you don't have **data,** and you don't have any useful information.

Data Tables

Here are some data Amazon might collect:

B000001OAA	10.99	Chris G.	902	15783947	15.98	Kansas	Illinois	Boston
Canada	Samuel P.	Orange County	N	B000068ZVQ	Bad Blood	Nashville	Katherine H.	N
Mammals	10783489	Ohio	N	Chicago	12837593	11.99	Massachusetts	16.99
312	Monique D.	10675489	413	B00000I5Y6	440	B000002BK9	Let Go	Y

A S   *Activity:* **What Is (Are) Data?** Do you really know what's data and what's just numbers?

Try to guess what they represent. Why is that hard? Because these data have no *context.* If we don't know *Who* they're about or *What* they measure, these values are meaningless. We can make the meaning clear if we organize the values into a **data table** such as this one:

Purchase Order	Name	Ship to State/Country	Price	Area Code	Previous CD Purchase	Gift?	ASIN	Artist
10675489	Katharine H.	Ohio	10.99	440	Nashville	N	B00000I5Y6	Kansas
10783489	Samuel P.	Illinois	16.99	312	Orange County	Y	B000002BK9	Boston
12837593	Chris G.	Massachusetts	15.98	413	Bad Blood	N	B000068ZVQ	Chicago
15783947	Monique D.	Canada	11.99	902	Let Go	N	B000001OAA	Mammals

Now we can see that these are four purchase records, relating to CD orders from Amazon. The column titles tell *What* has been recorded. The rows tell us *Who.* But be careful. Look at all the variables to see *Who* the variables are about. Even if people are involved, they may not be the *Who* of the data. For example, the *Who* here are the purchase orders (not the people who made the purchases).

A common place to find the *Who* of the table is the leftmost column. The other W's might have to come from the company's database administrator.[2]

Who

In general, the rows of a data table correspond to individual **cases** about *Whom* (or about which—if they're not people) we record some characteristics. These cases go by different names, depending on the situation. Individuals who answer a survey are referred to as *respondents.* People on whom we experiment are *subjects* or (in an attempt to acknowledge the importance of their role in the experiment) *participants,* but animals, plants, Web sites, and other inanimate subjects are often just called *experimental units.* In a database, rows are called *records*—in this example, purchase records. Perhaps the most generic term is **cases.** In the Amazon table, the cases are the individual CD orders.

Sometimes people just refer to data values as *observations,* without being clear about the *Who.* Be sure you know the *Who* of the data, or you may not know what the data say.

Often, the cases are a **sample** of cases selected from some larger **population** that we'd like to understand. Amazon certainly cares about its customers, but also wants to know how to attract all those other Internet users who may never have made a purchase from Amazon's site. To be able to generalize from the sample of cases to the larger population, we'll want the sample to be *representative* of that population—a kind of snapshot image of the larger world.

> **A S** *Activity:* **Consider the Context** . . . Can you tell who's *Who* and what's *What?* And *Why?* This activity offers real-world examples to help you practice identifying the context.

FOR EXAMPLE — Identifying the "Who"

In March 2007, *Consumer Reports* published an evaluation of large-screen, high-definition television sets (HDTVs). The magazine purchased and tested 98 different models from a variety of manufacturers.

Question: Describe the population of interest, the sample, and the *Who* of this study.

The magazine is interested in the performance of all HDTVs currently being offered for sale. It tested a sample of 98 sets, the "Who" for these data. Each HDTV set represents all similar sets offered by that manufacturer.

What and Why

The characteristics recorded about each individual are called **variables.** These are usually shown as the columns of a data table, and they should have a name that identifies *What* has been measured. Variables may seem simple, but to really understand your variables, you must *Think* about what you want to know.

Although area codes are numbers, do we use them that way? Is 610 twice 305? Of course it is, but is that the question? Why would we want to know whether Allentown, PA (area code 610), is twice Key West, FL (305)? Variables play different roles, and you can't tell a variable's role just by looking at it.

Some variables just tell us what group or category each individual belongs to. Are you male or female? Pierced or not? . . . What kinds of things can we learn about variables like these? A natural start is to *count* how many cases belong in each category. (Are you listening to music while reading this? We could count

---
[2] In database management, this kind of information is called "metadata."

It is wise to be careful. The *What* and *Why* of area codes are not as simple as they may first seem. When area codes were first introduced, AT&T was still the source of all telephone equipment, and phones had dials.

To reduce wear and tear on the dials, the area codes with the lowest digits (for which the dial would have to spin least) were assigned to the most populous regions—those with the most phone numbers and thus the area codes most likely to be dialed. New York City was assigned 212, Chicago 312, and Los Angeles 213, but rural upstate New York was given 607, Joliet was 815, and San Diego 619. For that reason, at one time the numerical value of an area code could be used to guess something about the population of its region. Now that phones have push-buttons, area codes have finally become just categories.

By international agreement, the International System of Units links together all systems of weights and measures. There are seven base units from which all other physical units are derived:

- Distance   Meter
- Mass   Kilogram
- Time   Second
- Electric current Ampere
- Temperature   °Kelvin
- Amount of substance Mole
- Intensity of light Candela

A S *Activity:* Recognize variables measured in a variety of ways. This activity shows examples of the many ways to measure data.

A S *Activities:* Variables. Several activities show you how to begin working with data in your statistics package.

the number of students in the class who were and the number who weren't.) We'll look for ways to compare and contrast the sizes of such categories.

Some variables have measurement **units.** Units tell how each value has been measured. But, more importantly, units such as yen, cubits, carats, angstroms, nanoseconds, miles per hour, or degrees Celsius tell us the *scale* of measurement. The units tell us how much of something we have or how far apart two values are. Without units, the values of a measured variable have no meaning. It does little good to be promised a raise of 5000 a year if you don't know whether it will be paid in euros, dollars, yen, or Estonian krooni.

What kinds of things can we learn about measured variables? We can do a lot more than just counting categories. We can look for patterns and trends. (How much did you pay for your last movie ticket? What is the range of ticket prices available in your town? How has the price of a ticket changed over the past 20 years?)

When a variable names categories and answers questions about how cases fall into those categories, we call it a **categorical variable.**[3] When a measured variable with units answers questions about the quantity of what is measured, we call it a **quantitative variable.** These types can help us decide what to do with a variable, but they are really more about what we hope to learn from a variable than about the variable itself. It's the questions we ask a variable (the *Why* of our analysis) that shape how we think about it and how we treat it.

Some variables can answer questions only about categories. If the values of a variable are words rather than numbers, it's a good bet that it is categorical. But some variables can answer both kinds of questions. Amazon could ask for your *Age* in years. That seems quantitative, and would be if the company wanted to know the average age of those customers who visit their site after 3 a.m. But suppose Amazon wants to decide which CD to offer you in a special deal—one by Raffi, Blink-182, Carly Simon, or Mantovani—and needs to be sure to have adequate supplies on hand to meet the demand. Then thinking of your age in one of the categories—child, teen, adult, or senior—might be more useful. If it isn't clear whether a variable is categorical or quantitative, think about *Why* you are looking at it and what you want it to tell you.

A typical course evaluation survey asks, "How valuable do you think this course will be to you?": 1 = Worthless; 2 = Slightly; 3 = Middling; 4 = Reasonably; 5 = Invaluable. Is *Educational Value* categorical or quantitative? Once again, we'll look to the *Why.* A teacher might just count the number of students who gave each response for her course, treating *Educational Value* as a categorical variable. When she wants to see whether the course is improving, she might treat the responses as the *amount* of perceived value—in effect, treating the variable as quantitative. But what are the units? There is certainly an *order* of perceived worth: Higher numbers indicate higher perceived worth. A course that averages 4.5 seems more valuable than one that averages 2, but we should be careful about treating *Educational Value* as

---

[3] You may also see it called a *qualitative variable.*

One tradition that hangs on in some quarters is to name variables with cryptic abbreviations written in uppercase letters. This can be traced back to the 1960s, when the very first statistics computer programs were controlled with instructions punched on cards. The earliest punch card equipment used only uppercase letters, and the earliest statistics programs limited variable names to six or eight characters, so variables were called things like PRSRF3. Modern programs do not have such restrictive limits, so there is no reason for variable names that you wouldn't use in an ordinary sentence.

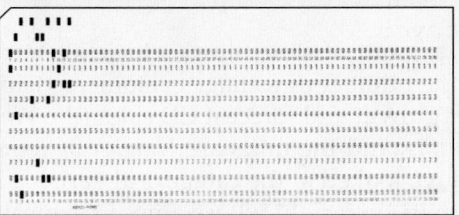

purely quantitative. To treat it as quantitative, she'll have to imagine that it has "educational value units" or some similar arbitrary construction. Because there are no natural units, she should be cautious. Variables like this that report order without natural units are often called "ordinal" variables. But saying "that's an ordinal variable" doesn't get you off the hook. You must still look to the *Why* of your study to decide whether to treat it as categorical or quantitative.

Identifying "What" and "Why" of HDTVs.

Recap: A *Consumer Reports* article about 98 HDTVs lists each set's manufacturer, cost, screen size, type (LCD, plasma, or rear projection), and overall performance score (0–100).

Question: Are these variables categorical or quantitative? Include units where appropriate, and describe the "Why" of this investigation.

The "what" of this article includes the following variables:

- manufacturer (categorical);
- cost (in dollars, quantitative);
- screen size (in inches, quantitative);
- type (categorical);
- performance score (quantitative).

The magazine hopes to help consumers pick a good HDTV set.

Counts Count

In Statistics, we often count things. When Amazon considers a special offer of free shipping to customers, it might first analyze how purchases are shipped. They'd probably start by counting the number of purchases shipped by ground transportation, by second-day air, and by overnight air. Counting is a natural way to summarize the categorical variable *Shipping Method*. So every time we see counts, does that mean the variable is categorical? Actually, no.

We also use counts to measure the amounts of things. How many songs are on your digital music player? How many classes are you taking this semester? To measure these quantities, we'd naturally count. The variables (*Songs, Classes*) would be quantitative, and we'd consider the units to be "number of . . ." or, generically, just "counts" for short.

So we use counts in two different ways. When we count the cases in each category of a categorical variable, the category labels are the *What* and the individuals counted are the *Who* of our data. The counts themselves are not the data, but are something we summarize about the data. Amazon counts the number of purchases in each category of the categorical variable *Shipping Method.* For this purpose (the *Why*), the *What* is shipping method and the *Who* is purchases.

Shipping Method	Number of Purchases
Ground	20,345
Second-day	7,890
Overnight	5,432

Other times our focus is on the amount of something, which we measure by counting. Amazon might record the number of teenage customers visiting their site each month to track customer growth and forecast CD sales (the *Why*). Now the *What* is *Teens,* the *Who* is *Months,* and the units are *Number of Teenage Customers. Teen* was a category when we looked at the categorical variable *Age.* But now it is a quantitative variable in its own right whose amount is measured by counting the number of customers.

Month	Number of Teenage Customers
January	123,456
February	234,567
March	345,678
April	456,789
May	. . .
. . .	. . .

Identifying Identifiers

What's your student ID number? It is numerical, but is it a quantitative variable? No, it doesn't have units. Is it categorical? Yes, but it is a special kind. Look at how many categories there are and at how many individuals are in each. There are as many categories as individuals and only one individual in each category. While it's easy to count the totals for each category, it's not very interesting. Amazon wants to know who you are when you sign in again and doesn't want to confuse you with some other customer. So it assigns you a unique identifier.

Identifier variables themselves don't tell us anything useful about the categories because we know there is exactly one individual in each. However, they are crucial in this age of large data sets. They make it possible to combine data from different sources, to protect confidentiality, and to provide unique labels. The variables *UPS Tracking Number, Social Security Number,* and Amazon's *ASIN* are all examples of identifier variables.

You'll want to recognize when a variable is playing the role of an identifier so you won't be tempted to analyze it. There's probably a list of unique ID numbers for students in a class (so they'll each get their own grade confidentially), but you might worry about the professor who keeps track of the average of these numbers from class to class. Even though this year's average ID number happens to be higher than last's, it doesn't mean that the students are better.

Where, When, and How

We must know *Who, What,* and *Why* to analyze data. Without knowing these three, we don't have enough to start. Of course, we'd always like to know more. The more we know about the data, the more we'll understand about the world.

If possible, we'd like to know the **When** and **Where** of data as well. Values recorded in 1803 may mean something different than similar values recorded last year. Values measured in Tanzania may differ in meaning from similar measurements made in Mexico.

How the data are collected can make the difference between insight and nonsense. As we'll see later, data that come from a voluntary survey on the Internet are almost always worthless. One primary concern of Statistics, to be discussed in Part III, is the design of sound methods for collecting data.

Throughout this book, whenever we introduce data, we'll provide a margin note listing the W's (and H) of the data. It's a habit we recommend. The first step of any data analysis is to know why you are examining the data (what you want to know), whom each row of your data table refers to, and what the variables (the columns of the table) record. These are the *Why,* the *Who,* and the *What.* Identifying them is a key part of the *Think* step of any analysis. Make sure you know all three before you proceed to *Show* or *Tell* anything about the data.

JUST CHECKING

In the 2003 Tour de France, Lance Armstrong averaged 40.94 kilometers per hour (km/h) for the entire course, making it the fastest Tour de France in its 100-year history. In 2004, he made history again by winning the race for an unprecedented sixth time. In 2005, he became the only 7-time winner and once again set a new record for the fastest average speed. You can find data on all the Tour de France races on the DVD. Here are the first three and last ten lines of the data set. Keep in mind that the entire data set has nearly 100 entries.

1. List as many of the W's as you can for this data set.

2. Classify each variable as categorical or quantitative; if quantitative, identify the units.

Year	Winner	Country of origin	Total time (h/min/s)	Avg. speed (km/h)	Stages	Total distance ridden (km)	Starting riders	Finishing riders
1903	Maurice Garin	France	94.33.00	25.3	6	2428	60	21
1904	Henri Cornet	France	96.05.00	24.3	6	2388	88	23
1905	Louis Trousselier	France	112.18.09	27.3	11	2975	60	24
⋮								
1999	Lance Armstrong	USA	91.32.16	40.30	20	3687	180	141
2000	Lance Armstrong	USA	92.33.08	39.56	21	3662	180	128
2001	Lance Armstrong	USA	86.17.28	40.02	20	3453	189	144
2002	Lance Armstrong	USA	82.05.12	39.93	20	3278	189	153
2003	Lance Armstrong	USA	83.41.12	40.94	20	3427	189	147
2004	Lance Armstrong	USA	83.36.02	40.53	20	3391	188	147
2005	Lance Armstrong	USA	86.15.02	41.65	21	3608	189	155
2006	Óscar Periero	Spain	89.40.27	40.78	20	3657	176	139
2007	Alberto Contador	Spain	91.00.26	38.97	20	3547	189	141
2008	Carlos Sastre	Spain	87.52.52	40.50	21	3559	199	145

There's a world of data on the Internet.   These days, one of the richest sources of data is the Internet. With a bit of practice, you can learn to find data on almost any subject. Many of the data sets we use in this book were found in this way. The Internet has both advantages and disadvantages as a source of data. Among the advantages are the fact that often you'll be able to find even more current data than those we present. The disadvantage is that references to Internet addresses can "break" as sites evolve, move, and die.

Our solution to these challenges is to offer the best advice we can to help you search for the data, wherever they may be residing. We usually point you to a Web site. We'll sometimes suggest search terms and offer other guidance.

Some words of caution, though: Data found on Internet sites may not be formatted in the best way for use in statistics software. Although you may see a data table in standard form, an attempt to copy the data may leave you with a single column of values. you may have to work in your favorite statistics or spreadsheet program to reformat the data into variables. You will also probably want to remove commas from large numbers and such extra symbols as money indicators ($, ¥, £); few statistics packages can handle these.

WHAT CAN GO WRONG?

▶ **Don't label a variable as categorical or quantitative without thinking about the question you want it to answer.** The same variable can sometimes take on different roles.

▶ **Just because your variable's values are numbers, don't assume that it's quantitative.** Categories are often given numerical labels. Don't let that fool you into thinking they have quantitative meaning. Look at the context.

▶ **Always be skeptical.** One reason to analyze data is to discover the truth. Even when you are told a context for the data, it may turn out that the truth is a bit (or even a lot) different. The context colors our interpretation of the data, so those who want to influence what you think may slant the context. A survey that seems to be about all students may in fact report just the opinions of those who visited a fan Web site. The question that respondents answered may have been posed in a way that influenced their responses.

TI Tips

Working with data

You'll need to be able to enter and edit data in your calculator. Here's how.

To enter data:
Hit the **STAT** button, and choose **EDIT** from the menu. You'll see a set of columns labeled **L1, L2,** and so on. Here is where you can enter, change, or delete a set of data.

Let's enter the heights (in inches) of the five starting players on a basketball team: 71, 75, 75, 76, and 80. Move the cursor to the space under **L1,** type in 71, and hit **ENTER** (or the down arrow). There's the first player. Now enter the data for the rest of the team.

To change a datum:
Suppose the 76" player grew since last season; his height should be listed as 78". Use the arrow keys to move the cursor onto the 76, then change the value and **ENTER** the correction.

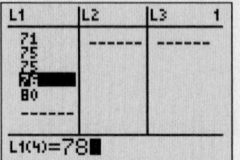

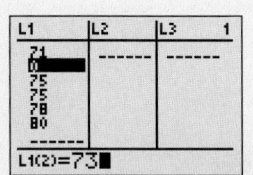

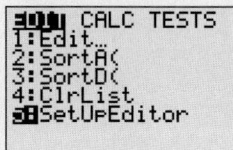

To add more data:
We want to include the sixth man, 73" tall. It would be easy to simply add this new datum to the end of the list. However, sometimes the order of the data matters, so let's place this datum in numerical order. Move the cursor to the desired position (atop the first 75). Hit **2ND INS**, then **ENTER** the 73 in the new space.

To delete a datum:
The 78" player just quit the team. Move the cursor there. Hit **DEL**. Bye.

To clear the datalist:
Finished playing basketball? Move the cursor atop the **L1**. Hit **CLEAR**, then **ENTER** (or down arrow). You should now have a blank datalist, ready for you to enter your next set of values.

Lost a datalist?
Oops! Is **L1** now missing entirely? Did you delete **L1** by mistake, instead of just *clearing* it? Easy problem to fix: buy a new calculator. No? OK, then simply go to the **STAT EDIT** menu, and run **SetUpEditor** to recreate all the lists.

WHAT HAVE WE LEARNED?

We've learned that data are information in a context.
- The W's help nail down the context: *Who, What, Why, Where, When*, and *hoW*.
- We must know at least the *Who, What*, and *Why* to be able to say anything useful based on the data. The *Who* are the *cases*. The *What* are the *variables*. A variable gives information about each of the cases. The *Why* helps us decide which way to treat the variables.

We treat variables in two basic ways: as *categorical* or *quantitative*.
- Categorical variables identify a category for each case. Usually, we think about the counts of cases that fall into each category. (An exception is an identifier variable that just names each case.)
- Quantitative variables record measurements or amounts of something; they must have *units*.
- Sometimes we treat a variable as categorical or quantitative depending on what we want to learn from it, which means that some variables can't be pigeonholed as one type or the other. That's an early hint that in Statistics we can't always pin things down precisely.

Terms

Context 8. The context ideally tells *Who* was measured, *What* was measured, *How* the data were collected, *Where* the data were collected, and *When* and *Why* the study was performed.

Data 8. Systematically recorded information, whether numbers or labels, together with its context.

Data table 8. An arrangement of data in which each row represents a case and each column represents a variable.

Case 9. A case is an individual about whom or which we have data.

Population 9. All the cases we wish we knew about.

Sample 9. The cases we actually examine in seeking to understand the much larger population.

Variable 9. A variable holds information about the same characteristic for many cases.

Units 10. A quantity or amount adopted as a standard of measurement, such as dollars, hours, or grams.

Categorical variable 10. A variable that names categories (whether with words or numerals) is called categorical.

Quantitative variable 10. A variable in which the numbers act as numerical values is called quantitative. Quantitative variables always have units.

Skills

▸ Be able to identify the *Who, What, When, Where, Why,* and *How* of data, or recognize when some of this information has not been provided.

▸ Be able to identify the cases and variables in any data set.

▸ Be able to identify the population from which a sample was chosen.

▸ Be able to classify a variable as categorical or quantitative, depending on its use.

▸ For any quantitative variable, be able to identify the units in which the variable has been measured (or note that they have not been provided).

▸ Be able to describe a variable in terms of its *Who, What, When, Where, Why,* and *How* (and be prepared to remark when that information is not provided).

DATA ON THE COMPUTER

Activity: Examine the Data. Take a look at your own data from your experiment (p. 12) and get comfortable with your statistics package as you find out about the experiment test results.

Most often we find statistics on a computer using a program, or *package,* designed for that purpose. There are many different statistics packages, but they all do essentially the same things. If you understand what the computer needs to know to do what you want and what it needs to show you in return, you can figure out the specific details of most packages pretty easily.

For example, to get your data into a computer statistics package, you need to tell the computer:

▸ Where to find the data. This usually means directing the computer to a file stored on your computer's disk or to data on a database. Or it might just mean that you have copied the data from a spreadsheet program or Internet site and it is currently on your computer's clipboard. Usually, the data should be in the form of a data table. Most computer statistics packages prefer the *delimiter* that marks the division between elements of a data table to be a *tab* character and the delimiter that marks the end of a case to be a *return* character.

▸ Where to put the data. (Usually this is handled automatically.)

▸ What to call the variables. Some data tables have variable names as the first row of the data, and often statistics packages can take the variable names from the first row automatically.

EXERCISES

1. **Voters.** A February 2007 Gallup Poll question asked, "In politics, as of today, do you consider yourself a Republican, a Democrat, or an Independent?" The possible responses were "Democrat", "Republican", "Independent", "Other", and "No Response". What kind of variable is the response?

2. **Mood.** A January 2007 Gallup Poll question asked, "In general, do you think things have gotten better or gotten worse in this country in the last five years?" Possible answers were "Better", "Worse", "No Change", "Don't Know", and "No Response". What kind of variable is the response?

3. **Medicine.** A pharmaceutical company conducts an experiment in which a subject takes 100 mg of a substance orally. The researchers measure how many minutes it takes for half of the substance to exit the bloodstream. What kind of variable is the company studying?

4. **Stress.** A medical researcher measures the increase in heart rate of patients under a stress test. What kind of variable is the researcher studying?

(Exercises 5–12) For each description of data, identify Who and What were investigated and the population of interest.

5. **The news.** Find a newspaper or magazine article in which some data are reported. For the data discussed in the article, answer the questions above. Include a copy of the article with your report.

6. **The Internet.** Find an Internet source that reports on a study and describes the data. Print out the description and answer the questions above.

7. **Bicycle safety.** Ian Walker, a psychologist at the University of Bath, wondered whether drivers treat bicycle riders differently when they wear helmets. He rigged his bicycle with an ultrasonic sensor that could measure how close each car was that passed him. He then rode on alternating days with and without a helmet. Out of 2500 cars passing him, he found that when he wore his helmet, motorists passed 3.35 inches closer to him, on average, than when his head was bare. [*NY Times*, Dec. 10, 2006]

8. **Investments.** Some companies offer 401(k) retirement plans to employees, permitting them to shift part of their before-tax salaries into investments such as mutual funds. Employers typically match 50% of the employees' contribution up to about 6% of salary. One company, concerned with what it believed was a low employee participation rate in its 401(k) plan, sampled 30 other companies with similar plans and asked for their 401(k) participation rates.

9. **Honesty.** Coffee stations in offices often just ask users to leave money in a tray to pay for their coffee, but many people cheat. Researchers at Newcastle University alternately taped two posters over the coffee station. During one week, it was a picture of flowers; during the other, it was a pair of staring eyes. They found that the average contribution was significantly higher when the eyes poster was up than when the flowers were there. Apparently, the mere feeling of being watched—even by eyes that were not real—was enough to encourage people to behave more honestly. [*NY Times*, Dec. 10, 2006]

10. **Movies.** Some motion pictures are profitable and others are not. Understandably, the movie industry would like to know what makes a movie successful. Data from 120 first-run movies released in 2005 suggest that longer movies actually make *less* profit.

11. **Fitness.** Are physically fit people less likely to die of cancer? An article in the May 2002 issue of *Medicine and Science in Sports and Exercise* reported results of a study that followed 25,892 men aged 30 to 87 for 10 years. The most physically fit men had a 55% lower risk of death from cancer than the least fit group.

12. **Molten iron.** The Cleveland Casting Plant is a large, highly automated producer of gray and nodular iron automotive castings for Ford Motor Company. The company is interested in keeping the pouring temperature of the molten iron (in degrees Fahrenheit) close to the specified value of 2550 degrees. Cleveland Casting measured the pouring temperature for 10 randomly selected crankshafts.

(Exercises 13–26) *For each description of data, identify the W's, name the variables, specify for each variable whether its use indicates that it should be treated as categorical or quantitative, and, for any quantitative variable, identify the units in which it was measured (or note that they were not provided).*

13. **Weighing bears.** Because of the difficulty of weighing a bear in the woods, researchers caught and measured 54 bears, recording their weight, neck size, length, and sex. They hoped to find a way to estimate weight from the other, more easily determined quantities.

14. **Schools.** The State Education Department requires local school districts to keep these records on all students: age, race or ethnicity, days absent, current grade level, standardized test scores in reading and mathematics, and any disabilities or special educational needs.

15. **Arby's menu.** A listing posted by the Arby's restaurant chain gives, for each of the sandwiches it sells, the type of meat in the sandwich, the number of calories, and the serving size in ounces. The data might be used to assess the nutritional value of the different sandwiches.

16. **Age and party.** The Gallup Poll conducted a representative telephone survey of 1180 American voters during the first quarter of 2007. Among the reported results were the voter's region (Northeast, South, etc.), age, party affiliation, and whether or not the person had voted in the 2006 midterm congressional election.

17. **Babies.** Medical researchers at a large city hospital investigating the impact of prenatal care on newborn health collected data from 882 births during 1998–2000. They kept track of the mother's age, the number of weeks the pregnancy lasted, the type of birth (cesarean, induced, natural), the level of prenatal care the mother had (none, minimal, adequate), the birth weight and sex of the baby, and whether the baby exhibited health problems (none, minor, major).

18. **Flowers.** In a study appearing in the journal *Science*, a research team reports that plants in southern England are flowering earlier in the spring. Records of the first flowering dates for 385 species over a period of 47 years show that flowering has advanced an average of 15 days per decade, an indication of climate warming, according to the authors.

19. **Herbal medicine.** Scientists at a major pharmaceutical firm conducted an experiment to study the effectiveness of an herbal compound to treat the common cold. They exposed each patient to a cold virus, then gave them either the herbal compound or a sugar solution known to have no effect on colds. Several days later they assessed each patient's condition, using a cold severity scale ranging from 0 to 5. They found no evidence of the benefits of the compound.

20. **Vineyards.** Business analysts hoping to provide information helpful to American grape growers compiled these data about vineyards: size (acres), number of years in existence, state, varieties of grapes grown, average case price, gross sales, and percent profit.

21. **Streams.** In performing research for an ecology class, students at a college in upstate New York collect data on streams each year. They record a number of biological, chemical, and physical variables, including the stream name, the substrate of the stream (limestone, shale, or mixed), the acidity of the water (pH), the temperature (°C), and the BCI (a numerical measure of biological diversity).

22. **Fuel economy.** The Environmental Protection Agency (EPA) tracks fuel economy of automobiles based on information from the manufacturers (Ford, Toyota, etc.). Among the data the agency collects are the manufacturer, vehicle type (car, SUV, etc.), weight, horsepower, and gas mileage (mpg) for city and highway driving.

23. **Refrigerators.** In 2006, *Consumer Reports* published an article evaluating refrigerators. It listed 41 models, giving the brand, cost, size (cu ft), type (such as top freezer), estimated annual energy cost, an overall rating (good, excellent, etc.), and the repair history for that brand (percentage requiring repairs over the past 5 years).

24. **Walking in circles.** People who get lost in the desert, mountains, or woods often seem to wander in circles rather than walk in straight lines. To see whether people naturally walk in circles in the absence of visual clues, researcher Andrea Axtell tested 32 people on a football field. One at a time, they stood at the center of one goal line, were blindfolded, and then tried to walk to the other goal line. She recorded each individual's sex, height, handedness, the number of yards each was able to walk before going out of bounds, and whether each wandered off course to the left or the right. No one made it all the way to the far end of the field without crossing one of the sidelines. [*STATS* No. 39, Winter 2004]

25. **Horse race 2008.** The Kentucky Derby is a horse race that has been run every year since 1875 at Churchill Downs, Louisville, Kentucky. The race started as a 1.5-mile race, but in 1896, it was shortened to 1.25 miles because experts felt that 3-year-old horses shouldn't run such a long race that early in the season. (It has been run in May every year but one—1901—when it took place on April 29). Here are the data for the first four and several recent races.

Date	Winner	Margin (lengths)	Jockey	Winner's Payoff ($)	Duration (min:sec)	Track Condition
May 17, 1875	Aristides	2	O. Lewis	2850	2:37.75	Fast
May 15, 1876	Vagrant	2	B. Swim	2950	2:38.25	Fast
May 22, 1877	Baden-Baden	2	W. Walker	3300	2:38.00	Fast
May 21, 1878	Day Star	1	J. Carter	4050	2:37.25	Dusty
......						
May 1, 2004	Smarty Jones	2 3/4	S. Elliott	854800	2:04.06	Sloppy
May 7, 2005	Giacomo	1/2	M. Smith	5854800	2:02.75	Fast
May 6, 2006	Barbaro	6 1/2	E. Prado	1453200	2:01.36	Fast
May 5, 2007	Street Sense	2 1/4	C. Borel	1450000	2:02.17	Fast
May 3, 2008	Big Brown	4 3/4	K. Desormeaux	1451800	2:01.82	Fast

26. **Indy 2008.** The 2.5-mile Indianapolis Motor Speedway has been the home to a race on Memorial Day nearly every year since 1911. Even during the first race, there were controversies. Ralph Mulford was given the checkered flag first but took three extra laps just to make sure he'd completed 500 miles. When he finished, another driver, Ray Harroun, was being presented with the winner's trophy, and Mulford's protests were ignored. Harroun averaged 74.6 mph for the 500 miles. In 2008, the winner, Scott Dixon, averaged 143.567 mph.

Here are the data for the first five races and five recent Indianapolis 500 races. Included also are the pole winners (the winners of the trial races, when each driver drives alone to determine the position on race day).

Year	Winner	Pole Position	Average Speed (mph)	Pole Winner	Average Pole Speed (mph)
1911	Ray Harroun	28	74.602	Lewis Strang	.
1912	Joe Dawson	7	78.719	Gil Anderson	.
1913	Jules Goux	7	75.933	Caleb Bragg	.
1914	René Thomas	15	82.474	Jean Chassagne	.
1915	Ralph DePalma	2	89.840	Howard Wilcox	98.580
...					
2004	Buddy Rice	1	138.518	Buddy Rice	220.024
2005	Dan Wheldon	16	157.603	Tony Kanaan	224.308
2006	Sam Hornish Jr.	1	157.085	Sam Hornish Jr.	228.985
2007	Dario Franchitti	3	151.744	Hélio Castroneves	225.817
2008	Scott Dixon	1	143.567	Scott Dixon	221.514

JUST CHECKING
Answers

1. Who—Tour de France races; What—year, winner, country of origin, total time, average speed, stages, total distance ridden, starting riders, finishing riders; How—official statistics at race; Where—France (for the most part); When—1903 to 2008; Why—not specified (To see progress in speeds of cycling racing?)

2.

Variable	Type	Units
Year	Quantitative or Categorical	Years
Winner	Categorical	
Country of Origin	Categorical	
Total Time	Quantitative	Hours/minutes/seconds
Average Speed	Quantitative	Kilometers per hour
Stages	Quantitative	Counts (stages)
Total Distance	Quantitative	Kilometers
Starting Riders	Quantitative	Counts (riders)
Finishing Riders	Quantitative	Counts (riders)

Displaying and Describing Categorical Data

WHO	People on the *Titanic*
WHAT	Survival status, age, sex, ticket class
WHEN	April 14, 1912
WHERE	North Atlantic
HOW	A variety of sources and Internet sites
WHY	Historical interest

What happened on the *Titanic* at 11:40 on the night of April 14, 1912, is well known. Frederick Fleet's cry of "Iceberg, right ahead" and the three accompanying pulls of the crow's nest bell signaled the beginning of a nightmare that has become legend. By 2:15 a.m., the *Titanic*, thought by many to be unsinkable, had sunk, leaving more than 1500 passengers and crew members on board to meet their icy fate.

Here are some data about the passengers and crew aboard the *Titanic*. Each case (row) of the data table represents a person on board the ship. The variables are the person's *Survival* status (Dead or Alive), *Age* (Adult or Child), *Sex* (Male or Female), and ticket *Class* (First, Second, Third, or Crew).

The problem with a data table like this—and in fact with all data tables—is that you can't *see* what's going on. And seeing is just what we want to do. We need ways to show the data so that we can see patterns, relationships, trends, and exceptions.

A S *Video:* **The Incident** tells the story of the *Titanic*, and includes rare film footage.

Survival	Age	Sex	Class
Dead	Adult	Male	Third
Dead	Adult	Male	Crew
Dead	Adult	Male	Third
Dead	Adult	Male	Crew
Dead	Adult	Male	Crew
Dead	Adult	Male	Crew
Alive	Adult	Female	First
Dead	Adult	Male	Third
Dead	Adult	Male	Crew

Table 3.1

Part of a data table showing four variables for nine people aboard the *Titanic*.

The Three Rules of Data Analysis

FIGURE 3.1 *A Picture to Tell a Story*

Florence Nightingale (1820–1910), a founder of modern nursing, was also a pioneer in health management, statistics, and epidemiology. She was the first female member of the British Statistical Society and was granted honorary membership in the newly formed American Statistical Association.

To argue forcefully for better hospital conditions for soldiers, she and her colleague, Dr. William Farr, invented this display, which showed that in the Crimean War, far more soldiers died of illness and infection than of battle wounds. Her campaign succeeded in improving hospital conditions and nursing for soldiers.

Florence Nightingale went on to apply statistical methods to a variety of important health issues and published more than 200 books, reports, and pamphlets during her long and illustrious career.

So, what should we do with data like these? There are three things you should always do first with data:

1. **Make a picture.** A display of your data will reveal things you are not likely to see in a table of numbers and will help you to *Think* clearly about the patterns and relationships that may be hiding in your data.
2. **Make a picture.** A well-designed display will *Show* the important features and patterns in your data. A picture will also show you the things you did not expect to see: the extraordinary (possibly wrong) data values or unexpected patterns.
3. **Make a picture.** The best way to *Tell* others about your data is with a well-chosen picture.

These are the three rules of data analysis. There are pictures of data throughout the book, and new kinds keep showing up. These days, technology makes drawing pictures of data easy, so there is no reason not to follow the three rules.

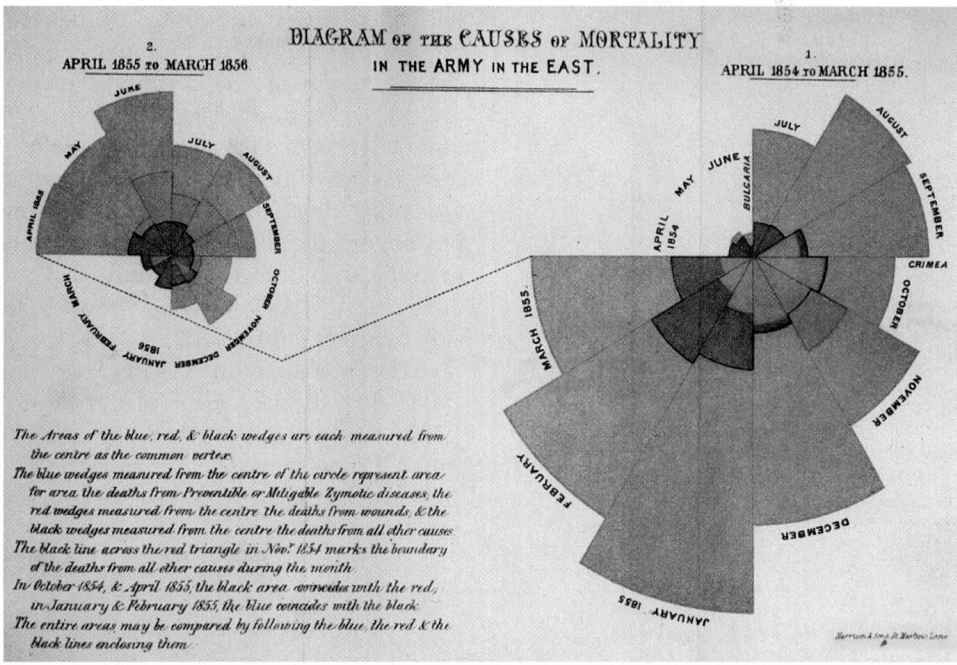

Frequency Tables: Making Piles

Class	Count
First	325
Second	285
Third	706
Crew	885

Table 3.2

A frequency table of the *Titanic* passengers.

To make a picture of data, the first thing we have to do is to make piles. Making piles is the beginning of understanding about data. We pile together things that seem to go together, so we can see how the cases distribute across different categories. For categorical data, piling is easy. We just count the number of cases corresponding to each category and pile them up.

One way to put all 2201 people on the *Titanic* into piles is by ticket *Class,* counting up how many had each kind of ticket. We can organize these counts into a **frequency table,** which records the totals and the category names.

Even when we have thousands of cases, a variable like ticket *Class,* with only a few categories, has a frequency table that's easy to read. A frequency table with dozens or hundreds of categories would be much harder to read. We use the names of the categories to label each row in the frequency table. For ticket *Class,* these are "First," "Second," "Third," and "Crew."

Class	%
First	14.77
Second	12.95
Third	32.08
Crew	40.21

Table 3.3

A relative frequency table for the same data.

Counts are useful, but sometimes we want to know the fraction or **proportion** of the data in each category, so we divide the counts by the total number of cases. Usually we multiply by 100 to express these proportions as **percentages**. A **relative frequency table** displays the *percentages*, rather than the counts, of the values in each category. Both types of tables show how the cases are distributed across the categories. In this way, they describe the **distribution** of a categorical variable because they name the possible categories and tell how frequently each occurs.

The Area Principle

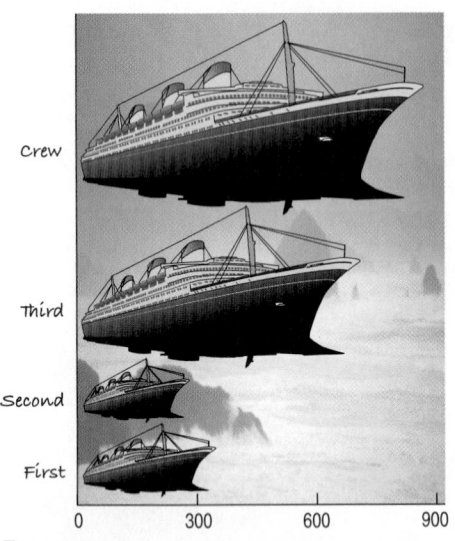

FIGURE 3.2

How many people were in each class on the Titanic? From this display, it looks as though the service must have been great, since most aboard were crew members. Although the length of each ship here corresponds to the correct number, the impression is all wrong. In fact, only about 40% were crew.

Now that we have the frequency table, we're ready to follow the three rules of data analysis and make a picture of the data. But a bad picture can distort our understanding rather than help it. Here's a graph of the *Titanic* data. What impression do you get about who was aboard the ship?

It sure looks like most of the people on the *Titanic* were crew members, with a few passengers along for the ride. That doesn't seem right. What's wrong? The lengths of the ships *do* match the totals in the table. (You can check the scale at the bottom.) However, experience and psychological tests show that our eyes tend to be more impressed by the *area* than by other aspects of each ship image. So, even though the *length* of each ship matches up with one of the totals, it's the associated *area* in the image that we notice. Since there were about 3 times as many crew as second-class passengers, the ship depicting the number of crew is about 3 times longer than the ship depicting second-class passengers, but it occupies about 9 times the area. As you can see from the frequency table (Table 3.2), that just isn't a correct impression.

The best data displays observe a fundamental principle of graphing data called the **area principle**. The area principle says that the area occupied by a part of the graph should correspond to the magnitude of the value it represents. Violations of the area principle are a common way to lie (or, since most mistakes are unintentional, we should say err) with Statistics.

Bar Charts

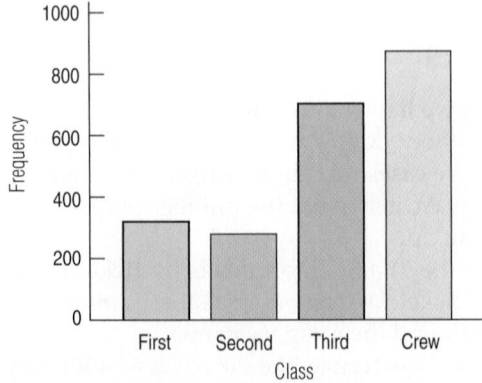

FIGURE 3.3 *People on the* Titanic *by Ticket Class*

With the area principle satisfied, we can see the true distribution more clearly.

Here's a chart that obeys the area principle. It's not as visually entertaining as the ships, but it does give an *accurate* visual impression of the distribution. The height of each bar shows the count for its category. The bars are the same width, so their heights determine their areas, and the areas are proportional to the counts in each class. Now it's easy to see that the majority of people on board were *not* crew, as the ships picture led us to believe. We can also see that there were about 3 times as many crew as second-class passengers. And there were more than twice as many third-class passengers as either first- or second-class passengers, something you may have missed in the frequency table. Bar charts make these kinds of comparisons easy and natural.

A **bar chart** displays the distribution of a categorical variable, showing the counts for each category next to each other for easy comparison. Bar charts should have small spaces between the bars to indicate that these are freestanding bars that could be rearranged into any order. The bars are lined up along a common base.

Usually they stick up like this 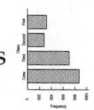 but sometimes they run

sideways like this

If we really want to draw attention to the relative *proportion* of passengers falling into each of these classes, we could replace the counts with percentages and use a **relative frequency bar chart.**

For some reason, some computer programs give the name "bar chart" to any graph that uses bars. And others use different names according to whether the bars are horizontal or vertical. Don't be misled. "Bar chart" is the term for a *display of counts of a categorical variable* with bars.

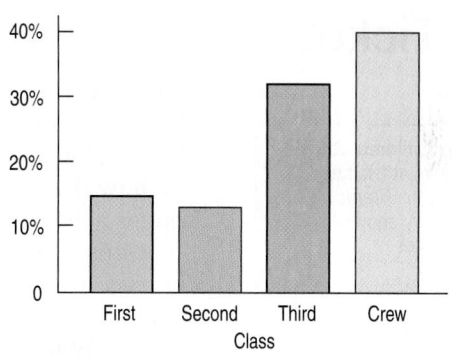

FIGURE 3.4

The relative frequency bar chart looks the same as the bar chart (Figure 3.3) but shows the proportion of people in each category rather than the counts.

Pie Charts

Another common display that shows how a whole group breaks into several categories is a pie chart. **Pie charts** show the whole group of cases as a circle. They slice the circle into pieces whose sizes are proportional to the fraction of the whole in each category.

Pie charts give a quick impression of how a whole group is partitioned into smaller groups. Because we're used to cutting up pies into 2, 4, or 8 pieces, pie charts are good for seeing relative frequencies near 1/2, 1/4, or 1/8. For example, you may be able to tell that the pink slice, representing the second-class passengers, is very close to 1/8 of the total. It's harder to see that there were about twice as many third-class as first-class passengers. Which category had the most passengers? Were there more crew or more third-class passengers? Comparisons such as these are easier in a bar chart.

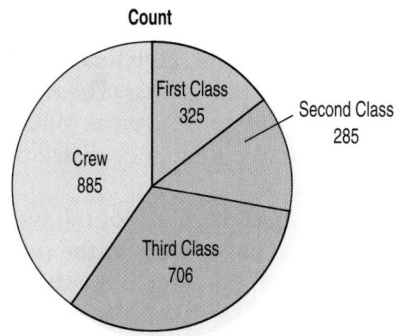

FIGURE 3.5 *Number of Titanic passengers in each class*

Think before you draw. Our first rule of data analysis is *Make a picture.* But what kind of picture? We don't have a lot of options—yet. There's more to Statistics than pie charts and bar charts, and knowing when to use each type of graph is a critical first step in data analysis. That decision depends in part on what type of data we have.

It's important to check that the data are appropriate for whatever method of analysis you choose. Before you make a bar chart or a pie chart, always check the

Categorical Data Condition: The data are counts or percentages of individuals in categories.

If you want to make a relative frequency bar chart or a pie chart, you'll need to also make sure that the categories don't overlap so that no individual is counted twice. If the categories do overlap, you can still make a bar chart, but the percentages won't add up to 100%. For the *Titanic* data, either kind of display is appropriate because the categories don't overlap.

Throughout this course, you'll see that doing Statistics right means selecting the proper methods. That means you have to *Think* about the situation at hand. An important first step, then, is to check that the type of analysis you plan is appropriate. The Categorical Data Condition is just the first of many such checks.

Contingency Tables: Children and First-Class Ticket Holders First?

A S   *Activity:* **Children at Risk.** This activity looks at the fates of children aboard the *Titanic*; the subsequent activity shows how to make such tables on a computer.

We know how many tickets of each class were sold on the *Titanic,* and we know that only about 32% of all those aboard the *Titanic* survived. After looking at the distribution of each variable by itself, it's natural and more interesting to ask how they relate. Was there a relationship between the kind of ticket a passenger held and the passenger's chances of making it into the lifeboat? To answer this question, we need to look at the two categorical variables *Class* and *Survival* together.

To look at two categorical variables together, we often arrange the counts in a two-way table. Here is a two-way table of those aboard the *Titanic,* classified according to the class of ticket and whether the ticket holder survived or didn't. Because the table shows how the individuals are distributed along each variable, contingent on the value of the other variable, such a table is called a **contingency table.**

Contingency table of ticket *Class* and *Survival*. The bottom line of "Totals" is the same as the previous frequency table.
Table 3.4

		First	Second	Third	Crew	Total
Survival	**Alive**	203	118	178	212	**711**
	Dead	122	167	528	673	**1490**
	Total	325	285	706	885	**2201**

*(Column group heading: **Class**)*

The margins of the table, both on the right and at the bottom, give totals. The bottom line of the table is just the frequency distribution of ticket *Class.* The right column of the table is the frequency distribution of the variable *Survival.* When presented like this, in the margins of a contingency table, the frequency distribution of one of the variables is called its **marginal distribution.**

Each **cell** of the table gives the count for a combination of values of the two variables. If you look down the column for second-class passengers to the first cell, you can see that 118 second-class passengers survived. Looking at the third-class passengers, you can see that more third-class passengers (178) survived. Were second-class passengers more likely to survive? Questions like this are easier to address by using percentages. The 118 survivors in second class were 41.4% of the total 285 second-class passengers, while the 178 surviving third-class passengers were only 25.2% of that class's total.

We know that 118 second-class passengers survived. We could display this number as a percentage—but as a percentage of what? The total number of passengers? (118 is 5.4% of the total: 2201.) The number of second-class passengers?

A bell-shaped artifact from the Titanic.

(118 is 41.4% of the 285 second-class passengers.) The number of survivors? (118 is 16.6% of the 711 survivors.) All of these are possibilities, and all are potentially useful or interesting. You'll probably wind up calculating (or letting your technology calculate) lots of percentages. Most statistics programs offer a choice of total percent, row percent, or column percent for contingency tables. Unfortunately, they often put them all together with several numbers in each cell of the table. The resulting table holds lots of information, but it can be hard to understand:

Another contingency table of ticket Class. This time we see not only the counts for each combination of *Class* and *Survival* (in bold) but the percentages these counts represent. For each count, there are three choices for the percentage: by row, by column, and by table total. There's probably too much information here for this table to be useful.
Table 3.5

		Class				
		First	Second	Third	Crew	Total
Alive	Count	203	118	178	212	711
	% of Row	28.6%	16.6%	25.0%	29.8%	100%
	% of Column	62.5%	41.4%	25.2%	24.0%	32.3%
	% of Table	9.2%	5.4%	8.1%	9.6%	32.3%
Dead	Count	122	167	528	673	1490
	% of Row	8.2%	11.2%	35.4%	45.2%	100%
	% of Column	37.5%	58.6%	74.8%	76.0%	67.7%
	% of Table	5.6%	7.6%	24.0%	30.6%	67.7%
Total	Count	325	285	706	885	2201
	%of Row	14.8%	12.9%	32.1%	40.2%	100%
	% of Column	100%	100%	100%	100%	100%
	% of Table	14.8%	12.9%	32.1%	40.2%	100%

To simplify the table, let's first pull out the percent of table values:

A contingency table of *Class* by *Survival* with only the table percentages
Table 3.6

	Class				
Survival	First	Second	Third	Crew	Total
Alive	9.2%	5.4%	8.1%	9.6%	32.3%
Dead	5.6%	7.6%	24.0%	30.6%	67.7%
Total	14.8%	12.9%	32.1%	40.2%	100%

These percentages tell us what percent of *all* passengers belong to each combination of column and row category. For example, we see that although 8.1% of the people aboard the *Titanic* were surviving third-class ticket holders, only 5.4% were surviving second-class ticket holders. Is this fact useful? Comparing these percentages, you might think that the chances of surviving were better in third class than in second. But be careful. There were many more third-class than second-class passengers on the *Titanic*, so there were more third-class survivors. That group is a larger percentage of the passengers, but is that really what we want to know?

> **Percent of what?**   The English language can be tricky when we talk about percentages. If you're asked "What percent *of the survivors* were in second class?" it's pretty clear that we're interested only in survivors. It's as if we're restricting the *Who* in the question to the survivors, so we should look at the number of second-class passengers among all the survivors—in other words, the row percent.
> But if you're asked "What percent were second-class passengers who survived?" you have a different question. Be careful; here, the *Who* is everyone on board, so 2201 should be the denominator, and the answer is the table percent.

And if you're asked "What percent of the second-class passengers survived?" you have a third question. Now the *Who* is the second-class passengers, so the denominator is the 285 second-class passengers, and the answer is the column percent.

Always be sure to ask "percent of what?" That will help you to know the *Who* and whether we want *row, column,* or *table* percentages.

FOR EXAMPLE   Finding marginal distributions

In January 2007, a Gallup poll asked 1008 Americans age 18 and over whether they planned to watch the upcoming Super Bowl. The pollster also asked those who planned to watch whether they were looking forward more to seeing the football game or the commercials. The results are summarized in the table:

Question: What's the marginal distribution of the responses?

To determine the percentages for the three responses, divide the count for each response by the total number of people polled:

		Sex		
		Male	**Female**	**Total**
Response	Game	279	200	479
	Commercials	81	156	237
	Won't watch	132	160	292
	Total	**492**	**516**	**1008**

$$\frac{479}{1008} = 47.5\% \quad \frac{237}{1008} = 23.5\% \quad \frac{292}{1008} = 29.0\%$$

According to the poll, 47.5% of American adults were looking forward to watching the Super Bowl game, 23.5% were looking forward to watching the commercials, and 29% didn't plan to watch at all.

Conditional Distributions

The more interesting questions are *contingent.* We'd like to know, for example, what percentage of *second-class passengers* survived and how that compares with the survival rate for third-class passengers.

It's more interesting to ask whether the chance of surviving the *Titanic* sinking *depended* on ticket class. We can look at this question in two ways. First, we could ask how the distribution of ticket *Class* changes between survivors and nonsurvivors. To do that, we look at the *row percentages:*

The conditional distribution of ticket *Class* conditioned on each value of *Survival: Alive* and *Dead.*
Table 3.7

		Class				
		First	**Second**	**Third**	**Crew**	**Total**
Survival	**Alive**	203 28.6%	118 16.6%	178 25.0%	212 29.8%	711 100%
	Dead	122 8.2%	167 11.2%	528 35.4%	673 45.2%	1490 100%

By focusing on each row separately, we see the distribution of class under the *condition* of surviving or not. The sum of the percentages in each row is 100%, and we divide that up by ticket class. In effect, we temporarily restrict the *Who* first to survivors and make a pie chart for them. Then we refocus the *Who* on the nonsurvivors and make their pie chart. These pie charts show the distribution of ticket classes *for each row* of the table: survivors and nonsurvivors. The distributions we create this way are called **conditional distributions,** because they show the distribution of one variable for just those cases that satisfy a condition on another variable.

FIGURE 3.6

Pie charts of the conditional distributions of ticket Class *for the survivors and nonsurvivors, separately. Do the distributions appear to be the same? We're primarily concerned with percentages here, so pie charts are a reasonable choice.*

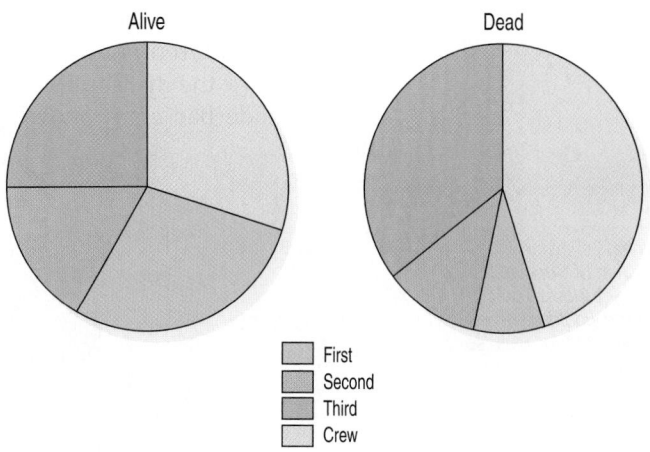

Alive   Dead

☐ First
☐ Second
☐ Third
☐ Crew

FOR EXAMPLE   Finding conditional distributions

Recap: The table shows results of a poll asking adults whether they were looking forward to the Super Bowl game, looking forward to the commercials, or didn't plan to watch.

Question: How do the conditional distributions of interest in the commercials differ for men and women?

		Sex		
		Male	**Female**	**Total**
Response	**Game**	279	200	479
	Commercials	81	156	237
	Won't watch	132	160	292
	Total	492	516	1008

Look at the group of people who responded "Commercials" and determine what percent of them were male and female:

$$\frac{81}{237} = 34.2\% \quad \frac{156}{237} = 65.8\%$$

Women make up a sizable majority of the adult Americans who look forward to seeing Super Bowl commercials more than the game itself. Nearly 66% of people who voiced a preference for the commercials were women, and only 34% were men.

But we can also turn the question around. We can look at the distribution of *Survival* for each category of ticket *Class*. To do this, we look at the *column percentages*. Those show us whether the chance of surviving was roughly the same *for each of the four classes*. Now the percentages in each column add to 100%, because we've restricted the *Who*, in turn, to each of the four ticket classes:

A contingency table of *Class* **by** *Survival* **with only counts and column percentages. Each column represents the conditional distribution of** *Survival* **for a given category of ticket** *Class***.**

Table 3.8

			Class				
			First	**Second**	**Third**	**Crew**	**Total**
Survival	**Alive**	**Count** **% of Column**	203 62.5%	118 41.4%	178 25.2%	212 24.0%	711 32.3%
	Dead	**Count** **% of Column**	122 37.5%	167 58.6%	528 74.8%	673 76.0%	1490 67.7%
	Total	**Count**	325 100%	285 100%	706 100%	885 100%	2201 100%

Looking at how the percentages change across each row, it sure looks like ticket class mattered in whether a passenger survived. To make it more vivid, we could show the distribution of *Survival* for each ticket class in a display. Here's a side-by-side bar chart showing percentages of surviving and not for each category:

FIGURE 3.7

Side-by-side bar chart showing the conditional distribution of Survival for each category of ticket Class. The corresponding pie charts would have only two categories in each of four pies, so bar charts seem the better alternative.

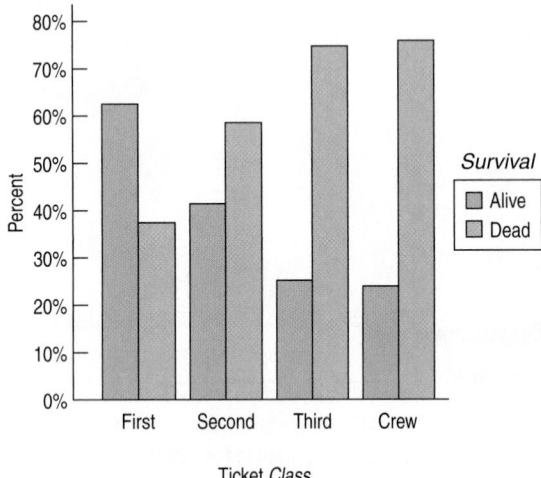

These bar charts are simple because, for the variable *Survival,* we have only two alternatives: Alive and Dead. When we have only two categories, we really need to know only the percentage of one of them. Knowing the percentage that survived tells us the percentage that died. We can use this fact to simplify the display even more by dropping one category. Here are the percentages of dying *across the classes* displayed in one chart:

FIGURE 3.8

Bar chart showing just nonsurvivor percentages for each value of ticket Class. Because we have only two values, the second bar doesn't add any information. Compare this chart to the side-by-side bar chart shown earlier.

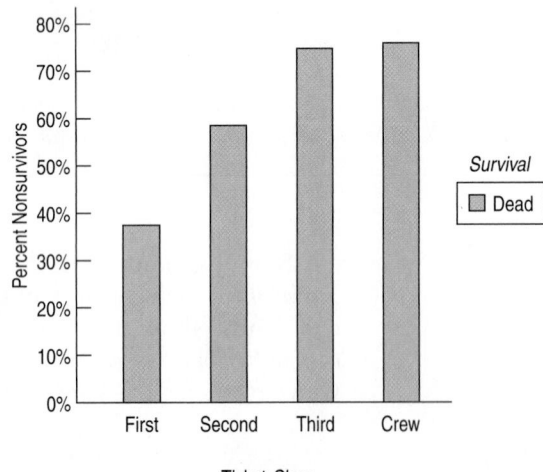

TI-*nspire*

Conditional distributions and association. Explore the *Titanic* data to see which passengers were most likely to survive.

Now it's easy to compare the risks. Among first-class passengers, 37.5% perished, compared to 58.6% for second-class ticket holders, 74.8% for those in third class, and 76.0% for crew members.

If the risk had been about the same across the ticket classes, we would have said that survival was *independent* of class. But it's not. The differences we see among these conditional distributions suggest that survival may have depended on ticket class. You may find it useful to consider conditioning on each variable in a contingency table in order to explore the dependence between them.

It is interesting to know that *Class* and *Survival* are associated. That's an important part of the *Titanic* story. And we know how important this is because the margins show us the actual numbers of people involved.

Variables can be associated in many ways and to different degrees. The best way to tell whether two variables are associated is to ask whether they are *not*.[1] In a contingency table, when the distribution of *one* variable is the same for all categories of another, we say that the variables are **independent**. That tells us there's no association between these variables. We'll see a way to check for independence formally later in the book. For now, we'll just compare the distributions.

FOR EXAMPLE   Looking for associations between variables

Recap: The table shows results of a poll asking adults whether they were looking forward to the Super Bowl game, looking forward to the commercials, or didn't plan to watch.

Question: Does it seem that there's an association between interest in Super Bowl TV coverage and a person's sex?

		Sex		
		Male	Female	Total
Response	Game	279	200	479
	Commercials	81	156	237
	Won't watch	132	160	292
	Total	492	516	1008

First find the distribution of the three responses for the men (the column percentages):

$$\frac{279}{492} = 56.7\% \qquad \frac{81}{492} = 16.5\% \qquad \frac{132}{492} = 26.8\%$$

Then do the same for the women who were polled, and display the two distributions with a side-by-side bar chart:

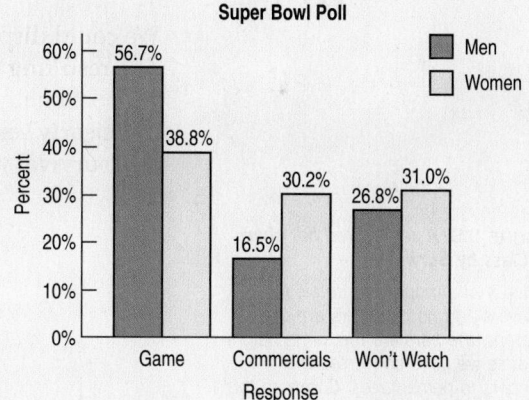

Based on this poll it appears that women were only slightly less interested than men in watching the Super Bowl telecast: 31% of the women said they didn't plan to watch, compared to just under 27% of men. Among those who planned to watch, however, there appears to be an association between the viewer's sex and what the viewer is most looking forward to. While more women are interested in the game (39%) than the commercials (30%), the margin among men is much wider: 57% of men said they were looking forward to seeing the game, compared to only 16.5% who cited the commercials.

---

[1] This kind of "backwards" reasoning shows up surprisingly often in science—and in Statistics. We'll see it again.

JUST CHECKING

A Statistics class reports the following data on Sex and Eye Color for students in the class:

		Eye Color			
		Blue	Brown	Green/Hazel/Other	Total
Sex	Males	6	20	6	32
	Females	4	16	12	32
	Total	10	36	18	64

1. What percent of females are brown-eyed?
2. What percent of brown-eyed students are female?
3. What percent of students are brown-eyed females?
4. What's the distribution of Eye Color?

5. What's the conditional distribution of Eye Color for the males?
6. Compare the percent who are female among the blue-eyed students to the percent of all students who are female.
7. Does it seem that Eye Color and Sex are independent? Explain.

Segmented Bar Charts

We could display the *Titanic* information by dividing up bars rather than circles. The resulting **segmented bar chart** treats each bar as the "whole" and divides it proportionally into segments corresponding to the percentage in each group. We can clearly see that the distributions of ticket *Class* are different, indicating again that survival was not independent of ticket *Class*.

FIGURE 3.9 *A segmented bar chart for Class by Survival*

Notice that although the totals for survivors and nonsurvivors are quite different, the bars are the same height because we have converted the numbers to percentages. Compare this display with the side-by-side pie charts of the same data in Figure 3.6.

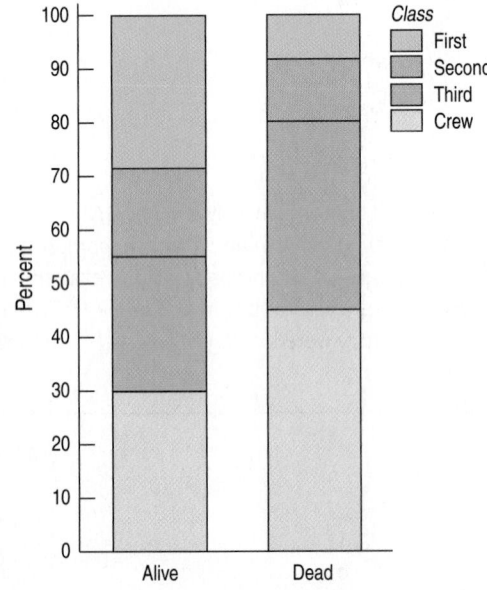

STEP-BY-STEP EXAMPLE   **Examining Contingency Tables**

Medical researchers followed 6272 Swedish men for 30 years to see if there was any association between the amount of fish in their diet and prostate cancer ("Fatty Fish Consumption and Risk of Prostate Cancer," *Lancet,* June 2001). Their results are summarized in this table:

We asked for a picture of a man eating fish. This is what we got.

	Prostate Cancer	
	No	Yes
Never/seldom	110	14
Small part of diet	2420	201
Moderate part	2769	209
Large part	507	42

(Fish Consumption)

Table 3.9

Question: Is there an association between fish consumption and prostate cancer?

THINK

Plan Be sure to state what the problem is about.

I want to know if there is an association between fish consumption and prostate cancer.

Variables Identify the variables and report the W's.

The individuals are 6272 Swedish men followed by medical researchers for 30 years. The variables record their fish consumption and whether or not they were diagnosed with prostate cancer.

Be sure to check the appropriate condition.

✔ **Categorical Data Condition:** I have counts for both fish consumption and cancer diagnosis. The categories of diet do not overlap, and the diagnoses do not overlap. It's okay to draw pie charts or bar charts.

SHOW

Mechanics It's a good idea to check the marginal distributions first before looking at the two variables together.

Fish Consumption	Prostate Cancer		
	No	Yes	Total
Never/seldom	110	14	124 (2.0%)
Small part of diet	2420	201	2621 (41.8%)
Moderate part	2769	209	2978 (47.5%)
Large part	507	42	549 (8.8%)
Total	5806 (92.6%)	466 (7.4%)	6272 (100%)

Two categories of the diet are quite small, with only 2.0% Never/Seldom eating fish and 8.8% in the "Large part" category. Overall, 7.4% of the men in this study had prostate cancer.

Then, make appropriate displays to see whether there is a difference in the relative proportions. These pie charts compare fish consumption for men who have prostate cancer to fish consumption for men who don't.

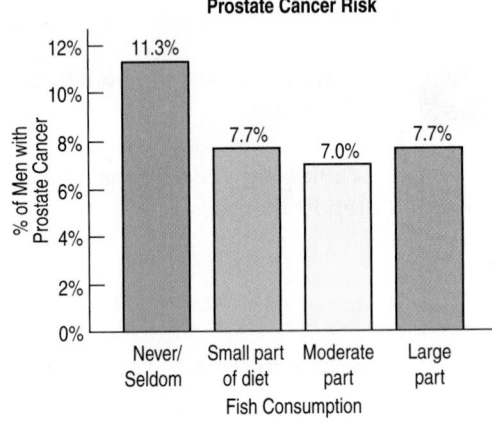

It's hard to see much difference in the pie charts. So, I made a display of the row percentages. Because there are only two alternatives, I chose to display the risk of prostate cancer for each group:

Both pie charts and bar charts can be used to compare conditional distributions. Here we compare prostate cancer rates based on differences in fish consumption.

 **Conclusion** Interpret the patterns in the table and displays in context. If you can, discuss possible real-world consequences. Be careful not to overstate what you see. The results may not generalize to other situations.

Overall, there is a 7.4% rate of prostate cancer among men in this study. Most of the men (89.3%) ate fish either as a moderate or small part of their diet. From the pie charts, it's hard to see a difference in cancer rates among the groups. But in the bar chart, it looks like the cancer rate for those who never/seldom ate fish may be somewhat higher.

However, only 124 of the 6272 men in the study fell into this category, and only 14 of them developed prostate cancer. More study would probably be needed before we would recommend that men change their diets.[2]

---

[2] The original study actually used pairs of twins, which enabled the researchers to discern that the risk of cancer for those who never ate fish actually *was* substantially greater. Using pairs is a special way of gathering data. We'll discuss such study design issues and how to analyze the data in the later chapters.

This study is an example of looking at a sample of data to learn something about a larger population. We care about more than these particular 6272 Swedish men. We hope that learning about their experiences will tell us something about the value of eating fish in general. That raises the interesting question of what population we think this sample might represent. Do we hope to learn about all Swedish men? About all men? About the value of eating fish for all adult humans? [3] Often, it can be hard to decide just which population our findings may tell us about, but that also is how researchers decide what to look into in future studies.

WHAT CAN GO WRONG?

▶ **Don't violate the area principle.** This is probably the most common mistake in a graphical display. It is often made in the cause of artistic presentation. Here, for example, are two displays of the pie chart of the *Titanic* passengers by class:

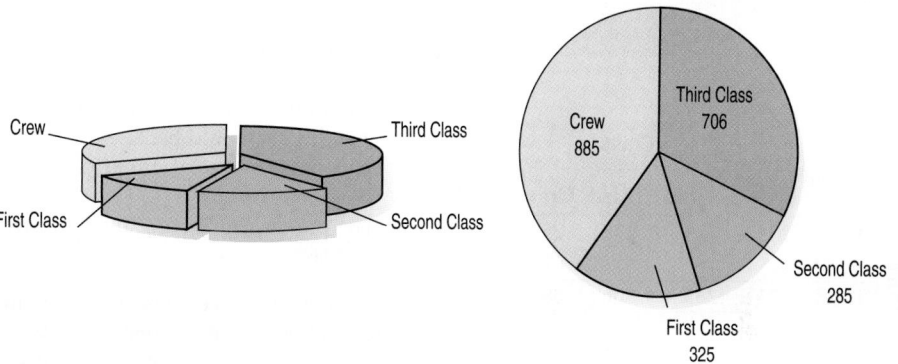

The one on the left looks pretty, doesn't it? But showing the pie on a slant violates the area principle and makes it much more difficult to compare fractions of the whole made up of each class—the principal feature that a pie chart ought to show.

▶ **Keep it honest.** Here's a pie chart that displays data on the percentage of high school students who engage in specified dangerous behaviors as reported by the Centers for Disease Control. What's wrong with this plot?

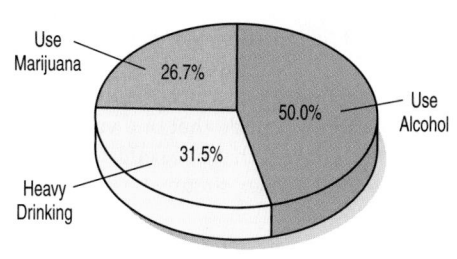

Try adding up the percentages. Or look at the 50% slice. Does it look right? Then think: What are these percentages of? Is there a "whole" that has been sliced up? In a pie chart, the proportions shown by each slice of the pie must add up to 100% and each individual must fall into only one category. Of course, showing the pie on a slant makes it even harder to detect the error.

(continued)

---

[3] Probably not, since we're looking only at prostate cancer risk.

Here's another. This bar chart shows the number of airline passengers searched in security screening, by year:

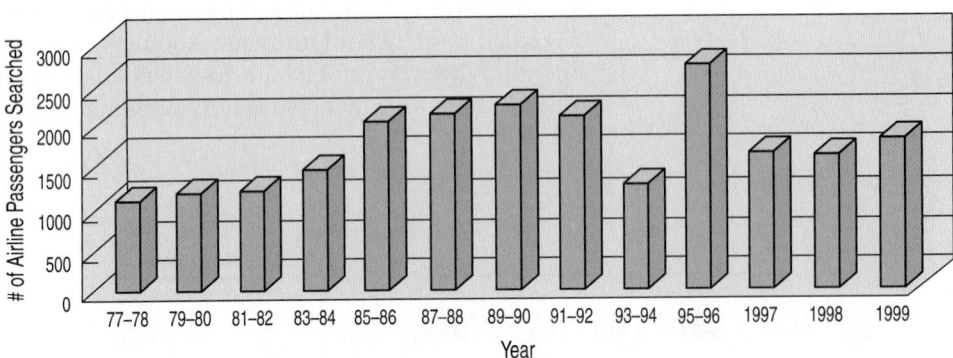

Looks like things didn't change much in the final years of the 20th century—until you read the bar labels and see that the last three bars represent single years while all the others are for *pairs* of years. Of course, the false depth makes it harder to see the problem.

▶ **Don't confuse similar-sounding percentages.** These percentages sound similar but are different:

	Class				
Survival	**First**	**Second**	**Third**	**Crew**	**Total**
Alive	203	118	178	212	**711**
Dead	122	167	528	673	**1490**
Total	325	285	706	885	**2201**

  ▶ The percentage of the passengers who were both in first class and survived: This would be 203/2201, or 9.4%.
  ▶ The percentage of the first-class passengers who survived: This is 203/325, or 62.5%.
  ▶ The percentage of the survivors who were in first class: This is 203/711, or 28.6%.

In each instance, pay attention to the *Who* implicitly defined by the phrase. Often there is a restriction to a smaller group (all aboard the *Titanic*, those in first class, and those who survived, respectively) before a percentage is found. Your discussion of results must make these differences clear.

▶ **Don't forget to look at the variables separately, too.** When you make a contingency table or display a conditional distribution, be sure you also examine the marginal distributions. It's important to know how many cases are in each category.

▶ **Be sure to use enough individuals.** When you consider percentages, take care that they are based on a large enough number of individuals. Take care not to make a report such as this one:

> *We found that 66.67% of the rats improved their performance with training. The other rat died.*

▶ **Don't overstate your case.** Independence is an important concept, but it is rare for two variables to be *entirely* independent. We can't conclude that one variable has no effect whatsoever on another. Usually, all we know is that little effect was observed in our study. Other studies of other groups under other circumstances could find different results.

Entering Centerville

Established	1793
Population	7943
Elevation	710
Average	3482

SIMPSON'S PARADOX

▶ **Don't use unfair or silly averages.** Sometimes averages can be misleading. Sometimes they just don't make sense at all. Be careful when averaging different variables that the quantities you're averaging are comparable. The Centerville sign says it all.

When using averages of proportions across several different groups, it's important to make sure that the groups really are comparable.

It's easy to make up an example showing that averaging across very different values or groups can give absurd results. Here's how that might work: Suppose there are two pilots, Moe and Jill. Moe argues that he's the better pilot of the two, since he managed to land 83% of his last 120 flights on time compared with Jill's 78%. But let's look at the data a little more closely. Here are the results for each of their last 120 flights, broken down by the time of day they flew:

Table 3.10

On-time flights by *Time of Day* and *Pilot*. Look at the percentages within each *Time of Day* category. Who has a better on-time record during the day? At night? Who is better overall?

		Time of Day		
		Day	**Night**	**Overall**
Pilot	**Moe**	90 out of 100 90%	10 out of 20 50%	100 out of 120 83%
	Jill	19 out of 20 95%	75 out of 100 75%	94 out of 120 78%

One famous example of Simpson's paradox arose during an investigation of admission rates for men and women at the University of California at Berkeley's graduate schools. As reported in an article in *Science,* about 45% of male applicants were admitted, but only about 30% of female applicants got in. It looked like a clear case of discrimination. However, when the data were broken down by school (Engineering, Law, Medicine, etc.), it turned out that, within each school, the women were admitted at nearly the same or, in some cases, much *higher* rates than the men. How could this be? Women applied in large numbers to schools with very low admission rates (Law and Medicine, for example, admitted fewer than 10%). Men tended to apply to Engineering and Science. Those schools have admission rates above 50%. When the *average* was taken, the women had a much lower *overall* rate, but the average didn't really make sense.

Look at the daytime and nighttime flights separately. For day flights, Jill had a 95% on-time rate and Moe only a 90% rate. At night, Jill was on time 75% of the time and Moe only 50%. So Moe is better "overall," but Jill is better both during the day and at night. How can this be?

What's going on here is a problem known as **Simpson's paradox,** named for the statistician who discovered it in the 1960s. It comes up rarely in real life, but there have been several well-publicized cases. As we can see from the pilot example, the problem is *unfair averaging* over different groups. Jill has mostly night flights, which are more difficult, so her *overall average* is heavily influenced by her nighttime average. Moe, on the other hand, benefits from flying mostly during the day, with its higher on-time percentage. With their very different patterns of flying conditions, taking an overall average is misleading. It's not a fair comparison.

The moral of Simpson's paradox is to be careful when you average across different levels of a second variable. It's always better to compare percentages or other averages *within* each level of the other variable. The overall average may be misleading.

CONNECTIONS

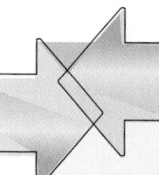

All of the methods of this chapter work with *categorical variables.* You must know the *Who* of the data to know who is counted in each category and the *What* of the variable to know where the categories come from.

WHAT HAVE WE LEARNED?

We've learned that we can summarize categorical data by counting the number of cases in each category, sometimes expressing the resulting distribution as percents. We can display the distribution in a bar chart or a pie chart. When we want to see how two categorical variables are related, we put the counts (and/or percentages) in a two-way table called a contingency table.

▸ We look at the marginal distribution of each variable (found in the margins of the table).
▸ We also look at the conditional distribution of a variable within each category of the other variable.
▸ We can display these conditional and marginal distributions by using bar charts or pie charts.
▸ If the conditional distributions of one variable are (roughly) the same for every category of the other, the variables are independent.

Terms

Frequency table
(Relative frequency table)

21. A frequency table lists the categories in a categorical variable and gives the count (or percentage of observations for each category.

Distribution

22. The distribution of a variable gives
▸ the possible values of the variable and
▸ the relative frequency of each value.

Area principle

22. In a statistical display, each data value should be represented by the same amount of area.

Bar chart
(Relative frequency bar chart)

22. Bar charts show a bar whose area represents the count (or percentage) of observations for each category of a categorical variable.

Pie chart

23. Pie charts show how a "whole" divides into categories by showing a wedge of a circle whose area corresponds to the proportion in each category.

Categorical data condition

24. The methods in this chapter are appropriate for displaying and describing categorical data. Be careful not to use them with quantitative data.

Contingency table

24. A contingency table displays counts and, sometimes, percentages of individuals falling into named categories on two or more variables. The table categorizes the individuals on all variables at once to reveal possible patterns in one variable that may be contingent on the category of the other.

Marginal distribution

24. In a contingency table, the distribution of either variable alone is called the marginal distribution. The counts or percentages are the totals found in the margins (last row or column) of the table.

Conditional distribution

26. The distribution of a variable restricting the *Who* to consider only a smaller group of individuals is called a conditional distribution.

Independence

29. Variables are said to be independent if the conditional distribution of one variable is the same for each category of the other. We'll show how to check for independence in a later chapter.

Segmented bar chart

30. A segmented bar chart displays the conditional distribution of a categorical variable within each category of another variable.

Simpson's paradox

34. When averages are taken across different groups, they can appear to contradict the overall averages. This is known as "Simpson's paradox."

Skills

▸ Be able to recognize when a variable is categorical and choose an appropriate display for it.
▸ Understand how to examine the association between categorical variables by comparing conditional and marginal percentages.

▸ Be able to summarize the distribution of a categorical variable with a frequency table.
▸ Be able to display the distribution of a categorical variable with a bar chart or pie chart.
▸ Know how to make and examine a contingency table.

▸ Know how to make and examine displays of the conditional distributions of one variable for two or more groups.

▸ Be able to describe the distribution of a categorical variable in terms of its possible values and relative frequencies.

▸ Know how to describe any anomalies or extraordinary features revealed by the display of a variable.

▸ Be able to describe and discuss patterns found in a contingency table and associated displays of conditional distributions.

DISPLAYING CATEGORICAL DATA ON THE COMPUTER

Although every package makes a slightly different bar chart, they all have similar features:

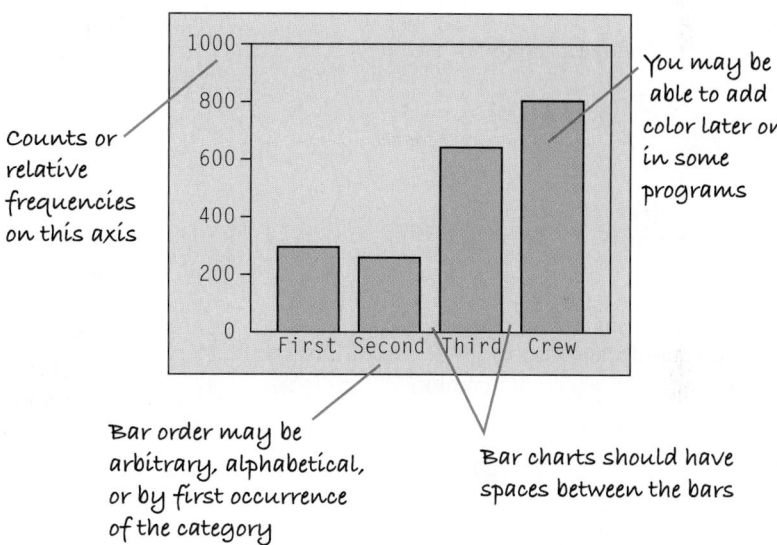

Counts or relative frequencies on this axis

You may be able to add color later on in some programs

Bar order may be arbitrary, alphabetical, or by first occurrence of the category

Bar charts should have spaces between the bars

Sometimes the count or a percentage is printed above or on top of each bar to give some additional information. You may find that your statistics package sorts category names in annoying orders by default. For example, many packages sort categories alphabetically or by the order the categories are seen in the data set. Often, neither of these is the best choice.

EXERCISES

1. **Graphs in the news.**   Find a bar graph of categorical data from a newspaper, a magazine, or the Internet.
   a) Is the graph clearly labeled?
   b) Does it violate the area principle?
   c) Does the accompanying article tell the W's of the variable?
   d) Do you think the article correctly interprets the data? Explain.

2. **Graphs in the news II.**   Find a pie chart of categorical data from a newspaper, a magazine, or the Internet.
   a) Is the graph clearly labeled?
   b) Does it violate the area principle?
   c) Does the accompanying article tell the W's of the variable?
   d) Do you think the article correctly interprets the data? Explain.

3. **Tables in the news.** Find a frequency table of categorical data from a newspaper, a magazine, or the Internet.
   a) Is it clearly labeled?
   b) Does it display percentages or counts?
   c) Does the accompanying article tell the W's of the variable?
   d) Do you think the article correctly interprets the data? Explain.

4. **Tables in the news II.** Find a contingency table of categorical data from a newspaper, a magazine, or the Internet.
   a) Is it clearly labeled?
   b) Does it display percentages or counts?
   c) Does the accompanying article tell the W's of the variables?
   d) Do you think the article correctly interprets the data? Explain.

5. **Movie genres.** The pie chart summarizes the genres of 120 first-run movies released in 2005.
   a) Is this an appropriate display for the genres? Why/why not?
   b) Which genre was least common?

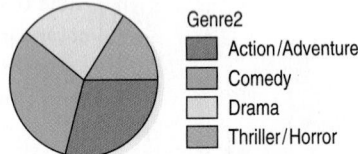

Genre2
- ■ Action/Adventure
- ■ Comedy
- ☐ Drama
- ■ Thriller/Horror

6. **Movie ratings.** The pie chart shows the ratings assigned to 120 first-run movies released in 2005.
   a) Is this an appropriate display for these data? Explain.
   b) Which was the most common rating?

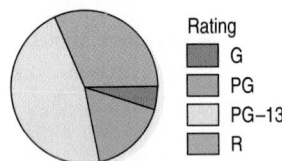

Rating
- ■ G
- ■ PG
- ☐ PG–13
- ■ R

7. **Genres again.** Here is a bar chart summarizing the 2005 movie genres, as seen in the pie chart in Exercise 5.
   a) Which genre was most common?
   b) Is it easier to see that in the pie chart or the bar chart? Explain.

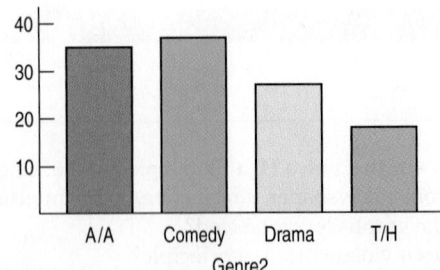

8. **Ratings again.** Here is a bar chart summarizing the 2005 movie ratings, as seen in the pie chart in Exercise 6.
   a) Which was the least common rating?
   b) An editorial claimed that there's been a growth in PG-13 rated films that, according to the writer, "have too much sex and violence," at the expense of G-rated films that offer "good, clean fun." The writer offered the bar chart below as evidence to support his claim. Does the bar chart support his claim? Explain.

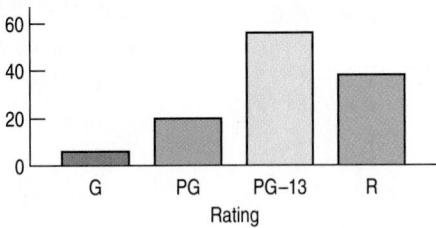

9. **Magnet schools.** An article in the Winter 2003 issue of *Chance* magazine reported on the Houston Independent School District's magnet schools programs. Of the 1755 qualified applicants, 931 were accepted, 298 were waitlisted, and 526 were turned away for lack of space. Find the relative frequency distribution of the decisions made, and write a sentence describing it.

10. **Magnet schools again.** The *Chance* article about the Houston magnet schools program described in Exercise 9 also indicated that 517 applicants were black or Hispanic, 292 Asian, and 946 white. Summarize the relative frequency distribution of ethnicity with a sentence or two (in the proper context, of course).

11. **Causes of death 2004.** The Centers for Disease Control and Prevention (www.cdc.gov) lists causes of death in the United States during 2004:

Cause of Death	Percent
Heart disease	27.2
Cancer	23.1
Circulatory diseases and stroke	6.3
Respiratory diseases	5.1
Accidents	4.7

   a) Is it reasonable to conclude that heart or respiratory diseases were the cause of approximately 33% of U.S. deaths in 2004?
   b) What percent of deaths were from causes not listed here?
   c) Create an appropriate display for these data.

12. **Plane crashes.** An investigation compiled information about recent nonmilitary plane crashes (www.planecrashinfo.com). The causes, to the extent that they could be determined, are summarized in the table.

Cause	Percent
Pilot error	40
Other human error	5
Weather	6
Mechanical failure	14
Sabotage	6

   a) Is it reasonable to conclude that the weather or mechanical failures caused only about 20% of recent plane crashes?
   b) In what percent of crashes were the causes not determined?
   c) Create an appropriate display for these data.

13. Oil spills 2006. Data from the International Tanker Owners Pollution Federation Limited (www.itopf.com) give the cause of spillage for 312 large oil tanker accidents from 1974–2006. Here are displays.
a) Write a brief report interpreting what the displays show.
b) Is a pie chart an appropriate display for these data? Why or why not?

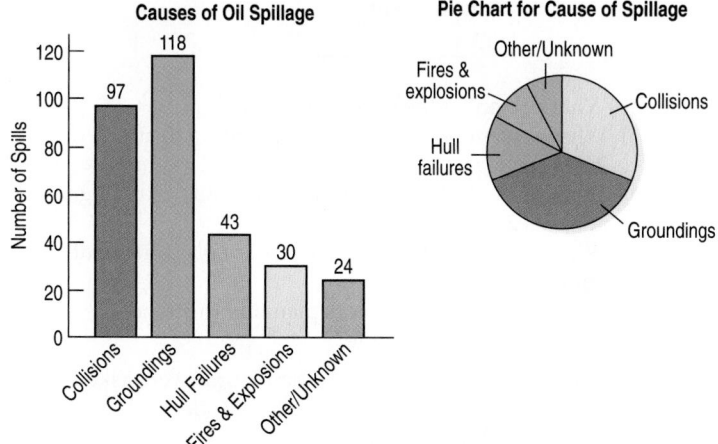

14. Winter Olympics 2006. Twenty-six countries won medals in the 2006 Winter Olympics. The table lists them, along with the total number of medals each won:

Country	Medals	Country	Medals
Germany	29	Finland	9
United States	25	Czech Republic	4
Canada	24	Estonia	3
Austria	23	Croatia	3
Russia	22	Australia	2
Norway	19	Poland	2
Sweden	14	Ukraine	2
Switzerland	14	Japan	1
South Korea	11	Belarus	1
Italy	11	Bulgaria	1
China	11	Great Britain	1
France	9	Slovakia	1
Netherlands	9	Latvia	1

a) Try to make a display of these data. What problems do you encounter?
b) Can you find a way to organize the data so that the graph is more successful?

15. Global Warming. The Pew Research Center for the People and the Press (http://people-press.org) has asked a representative sample of U.S. adults about global warming, repeating the question over time. In January 2007, the responses reflected an increased belief that global warming is real and due to human activity. Here's a display of the percentages of respondents choosing each of the major alternatives offered:

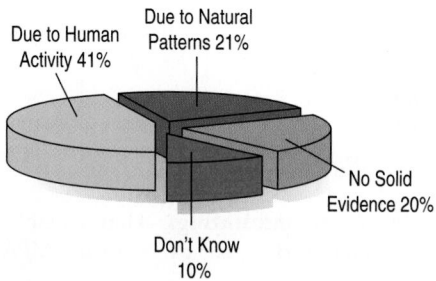

List the errors in this display.

16. Modalities. A survey of athletic trainers (Scott F. Nadler, Michael Prybicien, Gerard A. Malanga, and Dan Sicher. "Complications from Therapeutic Modalities: Results of a National Survey of Athletic Trainers." *Archives of Physical Medical Rehabilitation* 84 [June 2003]) asked what modalities (treatment methods such as ice, whirlpool, ultrasound, or exercise) they commonly use to treat injuries. Respondents were each asked to list three modalities. The article included the following figure reporting the modalities used:

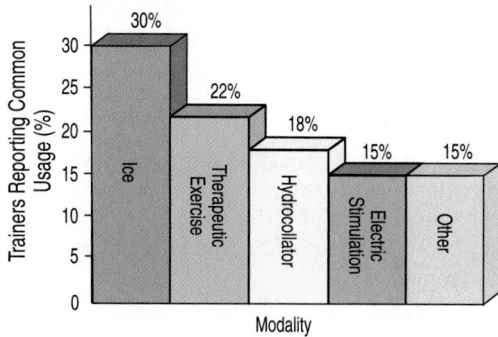

a) What problems do you see with the graph?
b) Consider the percentages for the named modalities. Do you see anything odd about them?

17. Teen smokers. The organization Monitoring the Future (www.monitoringthefuture.org) asked 2048 eighth graders who said they smoked cigarettes what brands they preferred. The table below shows brand preferences for two regions of the country. Write a few sentences describing the similarities and differences in brand preferences among eighth graders in the two regions listed.

Brand preference	South	West
Marlboro	58.4%	58.0%
Newport	22.5%	10.1%
Camel	3.3%	9.5%
Other (over 20 brands)	9.1%	9.5%
No usual brand	6.7%	12.9%

18. Handguns. In an effort to reduce the number of gun-related homicides, some cities have run buyback programs in which the police offer cash (often $50) to anyone who turns in an operating handgun. *Chance* magazine looked at results from a four-year period in Milwaukee. The table on the next page shows what types of guns were turned in and what types were used in homicides during a four-year period. Write a few sentences comparing the two distributions.

Caliber of gun	Buyback	Homicide
Small (.22, .25, .32)	76.4%	20.3%
Medium (.357, .38, 9 mm)	19.3%	54.7%
Large (.40, .44, .45)	2.1%	10.8%
Other	2.2%	14.2%

19. Movies by Genre and Rating.   Here's a table that classifies movies released in 2005 by genre and MPAA rating:

	G	PG	PG-13	R	Total
Action/Adventure	66.7	25	30.4	23.7	**29.2**
Comedy	33.3	60.0	35.7	10.5	**31.7**
Drama	0	15.0	14.3	44.7	**23.3**
Thriller/Horror	0	0	19.6	21.1	**15.8**
Total	100%	100%	100%	100%	**100%**

a) The table gives column percents. How could you tell that from the table itself?
b) What percentage of these movies were comedies?
c) What percentage of the PG-rated movies were comedies?
d) Which of the following can you learn from this table? Give the answer if you can find it from the table.
   i)   The percentage of PG-13 movies that were comedies
   ii) The percentage of dramas that were R-rated
   iii) The percentage of dramas that were G-rated
   iv) The percentage of 2005 movies that were PG-rated comedies

20. The Last Picture Show.   Here's another table showing information about 120 movies released in 2005. This table gives percentages of the table total:

	G	PG	PG-13	R	Total
Action/Adventure	3.33%	4.17	14.2	7.50	**29.2**
Comedy	1.67	10	16.7	3.33	**31.7**
Drama	0	2.50	6.67	14.2	**23.3**
Thriller/Horror	0	0	9.17	6.67	**15.8**
Total	5	16.7	46.7	31.7	**100%**

a) How can you tell that this table holds table percentages (rather than row or column percentages)?
b) What was the most common genre/rating combination in 2005 movies?
c) How many of these movies were PG-rated comedies?
d) How many were G-rated?
e) An editorial about the movies noted, "More than three-quarters of the movies made today can be seen only by patrons 13 years old or older." Does this table support that assertion? Explain.

21. Seniors.   Prior to graduation, a high school class was surveyed about its plans. The following table displays the results for white and minority students (the "Minority" group included African-American, Asian, Hispanic, and Native American students):

		Seniors	
		White	Minority
Plans	**4-year college**	198	44
	2-year college	36	6
	Military	4	1
	Employment	14	3
	Other	16	3

a) What percent of the seniors are white?
b) What percent of the seniors are planning to attend a 2-year college?
c) What percent of the seniors are white and planning to attend a 2-year college?
d) What percent of the white seniors are planning to attend a 2-year college?
e) What percent of the seniors planning to attend a 2-year college are white?

22. Politics.   Students in an Intro Stats course were asked to describe their politics as "Liberal," "Moderate," or "Conservative." Here are the results:

		Politics			
		L	M	C	Total
Sex	**Female**	35	36	6	77
	Male	50	44	21	115
	Total	85	80	27	192

a) What percent of the class is male?
b) What percent of the class considers themselves to be "Conservative"?
c) What percent of the males in the class consider themselves to be "Conservative"?
d) What percent of all students in the class are males who consider themselves to be "Conservative"?

23. More about seniors.   Look again at the table of post-graduation plans for the senior class in Exercise 21.
a) Find the conditional distributions (percentages) of plans for the white students.
b) Find the conditional distributions (percentages) of plans for the minority students.
c) Create a graph comparing the plans of white and minority students.
d) Do you see any important differences in the post-graduation plans of white and minority students? Write a brief summary of what these data show, including comparisons of conditional distributions.

24. Politics revisited.   Look again at the table of political views for the Intro Stats students in Exercise 22.
a) Find the conditional distributions (percentages) of political views for the females.
b) Find the conditional distributions (percentages) of political views for the males.
c) Make a graphical display that compares the two distributions.
d) Do the variables *Politics* and *Sex* appear to be independent? Explain.

25. Magnet schools revisited. The *Chance* magazine article described in Exercise 9 further examined the impact of an applicant's ethnicity on the likelihood of admission to the Houston Independent School District's magnet schools programs. Those data are summarized in the table below:

	Admission Decision			
Ethnicity	Accepted	Wait-listed	Turned away	Total
Black/Hispanic	485	0	32	**517**
Asian	110	49	133	**292**
White	336	251	359	**946**
Total	**931**	**300**	**524**	**1755**

a) What percent of all applicants were Asian?
b) What percent of the students accepted were Asian?
c) What percent of Asians were accepted?
d) What percent of all students were accepted?

26. More politics. Look once more at the table summarizing the political views of Intro Stats students in Exercise 22.
a) Produce a graphical display comparing the conditional distributions of males and females among the three categories of politics.
b) Comment briefly on what you see from the display in a.

27. Back to school. Examine the table about ethnicity and acceptance for the Houston Independent School District's magnet schools program, shown in Exercise 25. Does it appear that the admissions decisions are made independent of the applicant's ethnicity? Explain.

28. Cars. A survey of autos parked in student and staff lots at a large university classified the brands by country of origin, as seen in the table.

	Driver	
Origin	Student	Staff
American	107	105
European	33	12
Asian	55	47

a) What percent of all the cars surveyed were foreign?
b) What percent of the American cars were owned by students?
c) What percent of the students owned American cars?
d) What is the marginal distribution of origin?
e) What are the conditional distributions of origin by driver classification?
f) Do you think that the origin of the car is independent of the type of driver? Explain.

29. Weather forecasts. Just how accurate are the weather forecasts we hear every day? The following table compares the daily forecast with a city's actual weather for a year:

	Actual Weather	
Forecast	Rain	No rain
Rain	27	63
No rain	7	268

a) On what percent of days did it actually rain?
b) On what percent of days was rain predicted?
c) What percent of the time was the forecast correct?
d) Do you see evidence of an association between the type of weather and the ability of forecasters to make an accurate prediction? Write a brief explanation, including an appropriate graph.

30. Twins. In 2000, the *Journal of the American Medical Association (JAMA)* published a study that examined pregnancies that resulted in the birth of twins. Births were classified as preterm with intervention (induced labor or cesarean), preterm without procedures, or term/post-term. Researchers also classified the pregnancies by the level of prenatal medical care the mother received (inadequate, adequate, or intensive). The data, from the years 1995–1997, are summarized in the table below. Figures are in thousands of births. (*JAMA* 284 [2000]:335–341)

TWIN BIRTHS 1995–1997 (IN THOUSANDS)				
Level of Prenatal Care	Preterm (induced or cesarean)	Preterm (without procedures)	Term or post-term	Total
Intensive	18	15	28	**61**
Adequate	46	43	65	**154**
Inadequate	12	13	38	**63**
Total	**76**	**71**	**131**	**278**

a) What percent of these mothers received inadequate medical care during their pregnancies?
b) What percent of all twin births were preterm?
c) Among the mothers who received inadequate medical care, what percent of the twin births were preterm?
d) Create an appropriate graph comparing the outcomes of these pregnancies by the level of medical care the mother received.
e) Write a few sentences describing the association between these two variables.

31. Blood pressure. A company held a blood pressure screening clinic for its employees. The results are summarized in the table below by age group and blood pressure level:

	Age		
Blood Pressure	Under 30	30–49	Over 50
Low	27	37	31
Normal	48	91	93
High	23	51	73

a) Find the marginal distribution of blood pressure level.
b) Find the conditional distribution of blood pressure level within each age group.
c) Compare these distributions with a segmented bar graph.
d) Write a brief description of the association between age and blood pressure among these employees.
e) Does this prove that people's blood pressure increases as they age? Explain.

32. Obesity and exercise. The Centers for Disease Control and Prevention (CDC) has estimated that 19.8% of Americans over 15 years old are obese. The CDC conducts a survey on obesity and various behaviors. Here is a table on self-reported exercise classified by body mass index (BMI):

	Body Mass Index		
	Normal (%)	Overweight (%)	Obese (%)
Inactive	23.8	26.0	35.6
Irregularly active	27.8	28.7	28.1
Regular, not intense	31.6	31.1	27.2
Regular, intense	16.8	14.2	9.1

(Physical Activity)

a) Are these percentages column percentages, row percentages, or table percentages?
b) Use graphical displays to show different percentages of physical activities for the three BMI groups.
c) Do these data prove that lack of exercise causes obesity? Explain.

33. Anorexia. Hearing anecdotal reports that some patients undergoing treatment for the eating disorder anorexia seemed to be responding positively to the antidepressant Prozac, medical researchers conducted an experiment to investigate. They found 93 women being treated for anorexia who volunteered to participate. For one year, 49 randomly selected patients were treated with Prozac and the other 44 were given an inert substance called a placebo. At the end of the year, patients were diagnosed as healthy or relapsed, as summarized in the table:

	Prozac	Placebo	Total
Healthy	35	32	67
Relapse	14	12	26
Total	49	44	93

Do these results provide evidence that Prozac might be helpful in treating anorexia? Explain.

34. Antidepressants and bone fractures. For a period of five years, physicians at McGill University Health Center followed more than 5000 adults over the age of 50. The researchers were investigating whether people taking a certain class of antidepressants (SSRIs) might be at greater risk of bone fractures. Their observations are summarized in the table:

	Taking SSRI	No SSRI	Total
Experienced fractures	14	244	258
No fractures	123	4627	4750
Total	137	4871	5008

Do these results suggest there's an association between taking SSRI antidepressants and experiencing bone fractures? Explain.

35. Drivers' licenses 2005. The following table shows the number of licensed U.S. drivers by age and by sex (www.dot.gov):

Age	Male Drivers (number)	Female Drivers (number)	Total
19 and under	4,777,694	4,553,946	9,331,640
20–24	8,611,161	8,398,879	17,010,040
25–29	8,879,476	8,666,701	17,546,177
30–34	9,262,713	8,997,662	18,260,375
35–39	9,848,050	9,576,301	19,424,351
40–44	10,617,456	10,484,149	21,101,605
45–49	10,492,876	10,482,479	20,975,355
50–54	9,420,619	9,475,882	18,896,501
55–59	8,218,264	8,265,775	16,484,039
60–64	6,103,732	6,147,569	12,251,361
65–69	4,571,157	4,643,913	9,215,070
70–74	3,617,908	3,761,039	7,378,947
75–79	2,890,155	3,192,408	6,082,563
80–84	1,907,743	2,222,412	4,130,155
85 and over	1,170,817	1,406,271	2,577,088
Total	100,389,881	100,275,386	200,665,267

a) What percent of total drivers are under 20?
b) What percent of total drivers are male?
c) Write a few sentences comparing the number of male and female licensed drivers in each age group.
d) Do a driver's age and sex appear to be independent? Explain?

36. Tattoos. A study by the University of Texas Southwestern Medical Center examined 626 people to see if an increased risk of contracting hepatitis C was associated with having a tattoo. If the subject had a tattoo, researchers asked whether it had been done in a commercial tattoo parlor or elsewhere. Write a brief description of the association between tattooing and hepatitis C, including an appropriate graphical display.

	Tattoo done in commercial parlor	Tattoo done elsewhere	No tattoo
Has hepatitis C	17	8	18
No hepatitis C	35	53	495

37. Hospitals. Most patients who undergo surgery make routine recoveries and are discharged as planned. Others suffer excessive bleeding, infection, or other postsurgical complications and have their discharges from the hospital delayed. Suppose your city has a large hospital and a small hospital, each performing major and minor surgeries. You collect data to see how many surgical patients have their discharges delayed by postsurgical complications, and you find the results shown in the following table.

Discharge Delayed		
	Large hospital	**Small hospital**
Major surgery	120 of 800	10 of 50
Minor surgery	10 of 200	20 of 250

a) Overall, for what percent of patients was discharge delayed?
b) Were the percentages different for major and minor surgery?
c) Overall, what were the discharge delay rates at each hospital?
d) What were the delay rates at each hospital for each kind of surgery?
e) The small hospital advertises that it has a lower rate of postsurgical complications. Do you agree?
f) Explain, in your own words, why this confusion occurs.

38. Delivery service. A company must decide which of two delivery services it will contract with. During a recent trial period, the company shipped numerous packages with each service and kept track of how often deliveries did not arrive on time. Here are the data:

Delivery Service	Type of Service	Number of Deliveries	Number of Late Packages
Pack Rats	Regular	400	12
	Overnight	100	16
Boxes R Us	Regular	100	2
	Overnight	400	28

a) Compare the two services' overall percentage of late deliveries.
b) On the basis of the results in part a, the company has decided to hire Pack Rats. Do you agree that Pack Rats delivers on time more often? Explain.
c) The results here are an instance of what phenomenon?

39. Graduate admissions. A 1975 article in the magazine *Science* examined the graduate admissions process at Berkeley for evidence of sex discrimination. The table below shows the number of applicants accepted to each of four graduate programs:

		Males accepted (of applicants)	Females accepted (of applicants)
Program	1	511 of 825	89 of 108
	2	352 of 560	17 of 25
	3	137 of 407	132 of 375
	4	22 of 373	24 of 341
	Total	**1022 of 2165**	**262 of 849**

a) What percent of total applicants were admitted?
b) Overall, was a higher percentage of males or females admitted?
c) Compare the percentage of males and females admitted in each program.
d) Which of the comparisons you made do you consider to be the most valid? Why?

40. Be a Simpson! Can you design a Simpson's paradox? Two companies are vying for a city's "Best Local Employer" award, to be given to the company most committed to hiring local residents. Although both employers hired 300 new people in the past year, Company A brags that it deserves the award because 70% of its new jobs went to local residents, compared to only 60% for Company B. Company B concedes that those percentages are correct, but points out that most of its new jobs were full-time, while most of Company A's were part-time. Not only that, says Company B, but a higher percentage of its full-time jobs went to local residents than did Company A's, and the same was true for part-time jobs. Thus, Company B argues, it's a better local employer than Company A.

Show how it's possible for Company B to fill a higher percentage of both full-time and part-time jobs with local residents, even though Company A hired more local residents overall.

JUST CHECKING
Answers

1. 50.0%
2. 44.4%
3. 25.0%
4. 15.6% Blue, 56.3% Brown, 28.1% Green/Hazel/Other
5. 18.8% Blue, 62.5% Brown, 18.8% Green/Hazel/Other
6. 40% of the blue-eyed students are female, while 50% of all students are female.
7. Since blue-eyed students appear less likely to be female, it seems that *Sex* and *Eye Color* may not be independent. (But the numbers are small.)

Displaying and Summarizing Quantitative Data

Tsunamis are potentially destructive waves that can occur when the sea floor is suddenly and abruptly deformed. They are most often caused by earthquakes beneath the sea that shift the earth's crust, displacing a large mass of water.

The tsunami of December 26, 2004, with epicenter off the west coast of Sumatra, was caused by an earthquake of magnitude 9.0 on the Richter scale. It killed an estimated 297,248 people, making it the most disastrous tsunami on record. But was the earthquake that caused it truly extraordinary, or did it just happen at an unlucky place and time? The U.S. National Geophysical Data Center[1] has information on more than 2400 tsunamis dating back to 2000 B.C.E., and we have estimates of the magnitude of the underlying earthquake for 1240 of them. What can we learn from these data?

Histograms

WHO	1240 earthquakes known to have caused tsunamis for which we have data or good estimates
WHAT	Magnitude (Richter scale [2]), depth (m), date, location, and other variables
WHEN	From 2000 B.C.E. to the present
WHERE	All over the earth

Let's start with a picture. For categorical variables, it is easy to draw the distribution because each category is a natural "pile." But for quantitative variables, there's no obvious way to choose piles. So, usually, we slice up all the possible values into equal-width bins. We then count the number of cases that fall into each bin. The bins, together with these counts, give the **distribution** of the quantitative variable and provide the building blocks for the histogram. By representing the counts as bars and plotting them against the bin values, the **histogram** displays the distribution at a glance.

---

[1] www.ngdc.noaa.gov

[2] Technically, Richter scale values are in units of log dyne-cm. But the Richter scale is so common now that usually the units are assumed. The U.S. Geological Survey gives the background details of Richter scale measurements on its Web site www.usgs.gov/.

For example, here are the *Magnitudes* (on the Richter scale) of the 1240 earthquakes in the NGDC data:

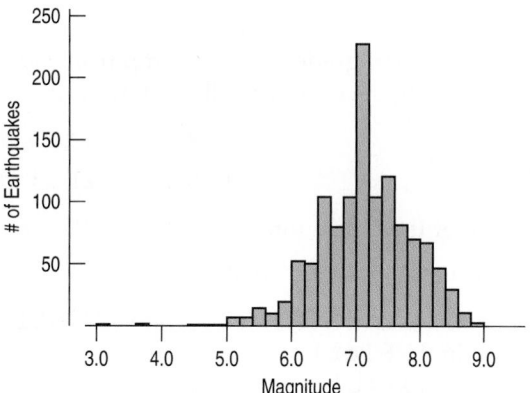

FIGURE 4.1

A histogram of earthquake magnitudes shows the number of earthquakes with magnitudes (in Richter scale units) in each bin.

One surprising feature of the earthquake magnitudes is the spike around magnitude 7.0. Only one other bin holds even half that many earthquakes. These values include historical data for which the magnitudes were estimated by experts and not measured by modern seismographs. Perhaps the experts thought 7 was a typical and reasonable value for a tsunami-causing earthquake when they lacked detailed information. That would explain the overabundance of magnitudes right at 7.0 rather than spread out near that value.

Like a bar chart, a histogram plots the bin counts as the heights of bars. In this histogram of earthquake magnitudes, each bin has a width of 0.2, so, for example, the height of the tallest bar says that there were about 230 earthquakes with magnitudes between 7.0 and 7.2. In this way, the histogram displays the entire distribution of earthquake magnitudes.

Does the distribution look as you expected? It is often a good idea to *imagine* what the distribution might look like before you make the display. That way you'll be less likely to be fooled by errors in the data or when you accidentally graph the wrong variable.

From the histogram, we can see that these earthquakes typically have magnitudes around 7. Most are between 5.5 and 8.5, and some are as small as 3 and as big as 9. Now we can answer the question about the Sumatra tsunami. With a value of 9.0 it's clear that the earthquake that caused it was an extraordinarily powerful earthquake—one of the largest on record.[3]

The bar charts of categorical variables we saw in Chapter 3 had spaces between the bars to separate the counts of different categories. But in a histogram, the bins slice up *all the values* of the quantitative variable, so any spaces in a histogram are actual **gaps** in the data, indicating a region where there are no values.

Sometimes it is useful to make a **relative frequency histogram,** replacing the counts on the vertical axis with the *percentage* of the total number of cases falling in each bin. Of course, the shape of the histogram is exactly the same; only the vertical scale is different.

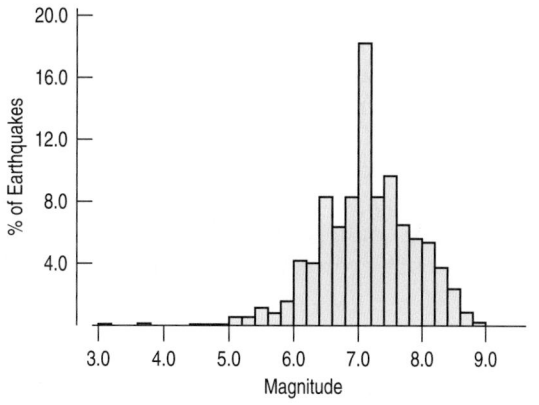

FIGURE 4.2

A relative frequency histogram looks just like a frequency histogram except for the labels on the y-axis, which now show the percentage of earthquakes in each bin.

[3] Some experts now estimate the magnitude at between 9.1 and 9.3.

TI Tips

Making a histogram

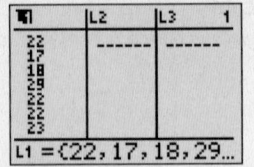

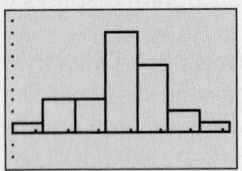

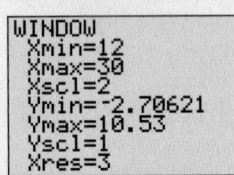

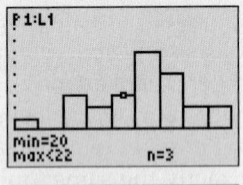

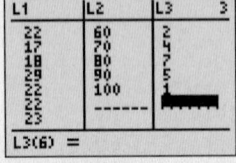

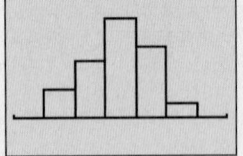

Your calculator can create histograms. First you need some data. For an agility test, fourth-grade children jump from side to side across a set of parallel lines, counting the number of lines they clear in 30 seconds. Here are their scores:

22, 17, 18, 29, 22, 22, 23, 24, 23, 17, 21, 25, 20

12, 19, 28, 24, 22, 21, 25, 26, 25, 16, 27, 22

Enter these data into L1.

Now set up the calculator's plot:

- Go to 2nd STATPLOT, choose Plot1, then ENTER.
- In the Plot1 screen choose On, select the little histogram icon, then specify Xlist:L1 and Freq:1.
- Be sure to turn off any other graphs the calculator may be set up for. Just hit the Y= button, and deactivate any functions seen there.

All set? To create your preliminary plot go to ZOOM, select 9:ZoomStat, and then ENTER.

You now see the calculator's initial attempt to create a histogram of these data. Not bad. We can see that the distribution is roughly symmetric. But it's hard to tell exactly what this histogram shows, right? Let's fix it up a bit.

- Under WINDOW, let's reset the bins to convenient, sensible values. Try Xmin=12, Xmax=30 and Xscl=2. That specifies the range of values along the x-axis and makes each bar span two lines.
- Hit GRAPH (not ZoomStat—this time we want control of the scale!).

There. We still see rough symmetry, but also see that one of the scores was much lower than the others. Note that you can now find out exactly what the bars indicate by activating TRACE and then moving across the histogram using the arrow keys. For each bar the calculator will indicate the interval of values and the number of data values in that bin. We see that 3 kids had agility scores of 20 or 21.

Play around with the WINDOW settings. A different Ymax will make the bars appear shorter or taller. What happens if you set the bar width (Xscl) smaller? Or larger? You don't want to lump lots of values into just a few bins or make so many bins that the overall shape of the histogram is not clear. Choosing the best bar width takes practice.

Finally, suppose the data are given as a frequency table. Consider a set of test scores, with two grades in the 60s, four in the 70s, seven in the 80s, five in the 90s, and one 100. Enter the group cutoffs 60, 70, 80, 90, 100 in L2 and the corresponding frequencies 2, 4, 7, 5, 1 in L3. When you set up the histogram STATPLOT, specify Xlist:L2 and Freq:L3. Can you specify the WINDOW settings to make this histogram look the way you want it? (By the way, if you get a DIM MISMATCH error, it means you can't count. Look at L2 and L3; you'll see the two lists don't have the same number of entries. Fix the problem by correcting the data you entered.)

Stem-and-Leaf Displays

Histograms provide an easy-to-understand summary of the distribution of a quantitative variable, but they don't show the data values themselves. Here's a histogram of the pulse rates of 24 women, taken by a researcher at a health clinic:

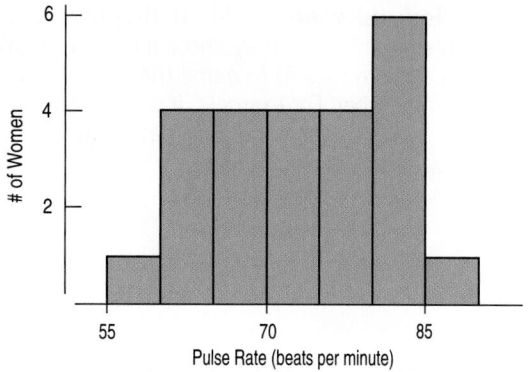

FIGURE 4.3
The pulse rates of 24 women at a health clinic

The story seems pretty clear. We can see the entire span of the data and can easily see what a typical pulse rate might be. But is that all there is to these data?

A **stem-and-leaf display** is like a histogram, but it shows the individual values. It's also easier to make by hand. Here's a stem-and-leaf display of the same data:

```
8 | 8
8 | 000044
7 | 6666
7 | 2222
6 | 8888
6 | 0444
5 | 6
```
Pulse Rate
(8|8 means 88 beats/min)

The Stem-and-Leaf display was devised by John W. Tukey, one of the greatest statisticians of the 20th century. It is called a "Stemplot" in some texts and computer programs, but we prefer Tukey's original name for it.

A S *Activity:* **Stem-and-Leaf Displays.** As you might expect of something called "stem-and leaf," these displays grow as you consider each data value.

Turn the stem-and-leaf on its side (or turn your head to the right) and squint at it. It should look roughly like the histogram of the same data. Does it? Well, it's backwards because now the higher values are on the left, but other than that, it has the same shape.[4]

What does the line at the top of the display that says 8 | 8 mean? It stands for a pulse of 88 beats per minute (bpm). We've taken the tens place of the number and made that the "stem." Then we sliced off the ones place and made it a "leaf." The next line down is 8 | 000044. That shows that there were four pulse rates of 80 and two of 84 bpm.

Stem-and-leaf displays are especially useful when you make them by hand for batches of fewer than a few hundred data values. They are a quick way to display—and even to record—numbers. Because the leaves show the individual values, we can sometimes see even more in the data than the distribution's shape. Take another look at all the leaves of the pulse data. See anything

---

[4] You could make the stem-and-leaf with the higher values on the bottom. Usually, though, higher on the top makes sense.

unusual? At a glance you can see that they are all even. With a bit more thought you can see that they are all multiples of 4—something you couldn't possibly see from a histogram. How do you think the nurse took these pulses? Counting beats for a full minute or counting for only 15 seconds and multiplying by 4?

How do stem-and-leaf displays work?   Stem-and-leaf displays work like histograms, but they show more information. They use part of the number itself (called the stem) to name the bins. To make the "bars," they use the next digit of the number. For example, if we had a test score of 83, we could write it 8|3, where 8 serves as the stem and 3 as the leaf. Then, to display the scores 83, 76, and 88 together, we would write

```
8 | 38
7 | 6
```

For the pulse data, we have

```
8 | 0000448
7 | 22226666
6 | 04448888
5 | 6
   Pulse Rate
(5|6 means 56 beats/min)
```

This display is OK, but a little crowded. A histogram might split each line into two bars. With a stem-and-leaf, we can do the same by putting the leaves 0–4 on one line and 5–9 on another, as we saw above:

```
8 | 8
8 | 000044
7 | 6666
7 | 2222
6 | 8888
6 | 0444
5 | 6
   Pulse Rate
(8|8 means 88 beats/min)
```

For numbers with three or more digits, you'll often decide to truncate (or round) the number to two places, using the first digit as the stem and the second as the leaf. So, if you had 432, 540, 571, and 638, you might display them as shown below with an indication that 6|3 means 630–639.

```
6 | 3
5 | 47
4 | 3
```

When you make a stem-and-leaf by hand, make sure to give each digit the same width, in order to preserve the area principle. (That can lead to some fat 1's and thin 8's—but it makes the display honest.)

Dotplots

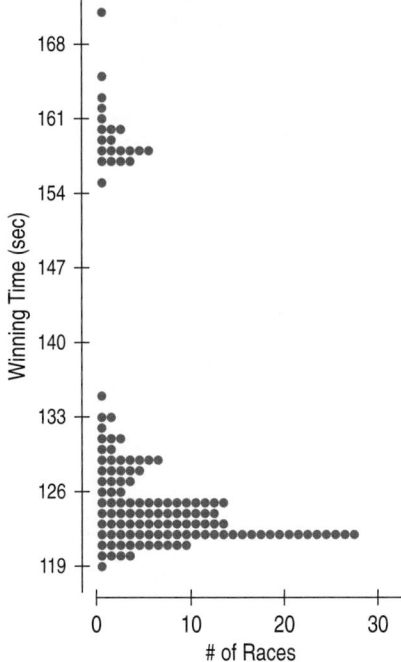

A **dotplot** is a simple display. It just places a dot along an axis for each case in the data. It's like a stem-and-leaf display, but with dots instead of digits for all the leaves. Dotplots are a great way to display a small data set (especially if you forget how to write the digits from 0 to 9). Here's a dotplot of the time (in seconds) that the winning horse took to win the Kentucky Derby in each race between the first Derby in 1875 and the 2008 Derby.

Dotplots show basic facts about the distribution. We can find the slowest and quickest races by finding times for the topmost and bottommost dots. It's also clear that there are two clusters of points, one just below 160 seconds and the other at about 122 seconds. Something strange happened to the Derby times. Once we know to look for it, we can find out that in 1896 the distance of the Derby race was changed from 1.5 miles to the current 1.25 miles. That explains the two clusters of winning times.

Some dotplots stretch out horizontally, with the counts on the vertical axis, like a histogram. Others, such as the one shown here, run vertically, like a stem-and-leaf display. Some dotplots place points next to each other when they would otherwise overlap. Others just place them on top of one another. Newspapers sometimes offer dotplots with the dots made up of little pictures.

FIGURE 4.4

A dotplot of Kentucky Derby winning times plots each race as its own dot, showing the bimodal distribution.

Think Before You Draw, Again

Suddenly, we face a lot more options when it's time to invoke our first rule of data analysis and make a picture. You'll need to *Think* carefully to decide which type of graph to make. In the previous chapter you learned to check the Categorical Data Condition before making a pie chart or a bar chart. Now, before making a stem-and-leaf display, a histogram, or a dotplot, you need to check the

Quantitative Data Condition: The data are values of a quantitative variable whose units are known.

Although a bar chart and a histogram may look somewhat similar, they're not the same display. You can't display categorical data in a histogram or quantitative data in a bar chart. Always check the condition that confirms what type of data you have before proceeding with your display.

Step back from a histogram or stem-and-leaf display. What can you say about the distribution? When you describe a distribution, you should always tell about three things: its **shape, center,** and **spread.**

The Shape of a Distribution

1. *Does the histogram have a single, central hump or several separated humps?* These humps are called **modes.**[5] The earthquake magnitudes have a single mode

---

[5] Well, technically, it's the value on the horizontal axis of the histogram that is the mode, but anyone asked to point to the mode would point to the hump.

The **mode** is sometimes defined as the single value that appears most often. That definition is fine for categorical variables because all we need to do is count the number of cases for each category. For quantitative variables, the mode is more ambiguous. What is the mode of the Kentucky Derby times? Well, seven races were timed at 122.2 seconds—more than any other race time. Should that be the mode? Probably not. For quantitative data, it makes more sense to use the term "mode" in the more general sense of the peak of the histogram rather than as a single summary value. In this sense, the important feature of the Kentucky Derby races is that there are two distinct modes, representing the two different versions of the race and warning us to consider those two versions separately.

at just about 7. A histogram with one peak, such as the earthquake magnitudes, is dubbed **unimodal;** histograms with two peaks are **bimodal,** and those with three or more are called **multimodal**.[6] For example, here's a bimodal histogram.

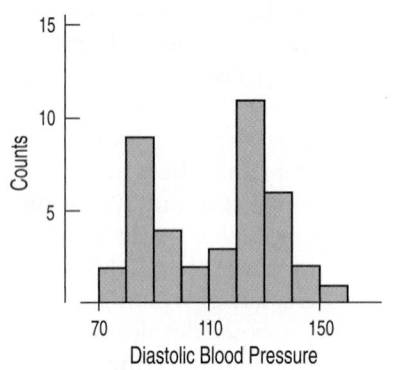

FIGURE 4.5

A bimodal histogram has two apparent peaks.

A histogram that doesn't appear to have any mode and in which all the bars are approximately the same height is called **uniform.**

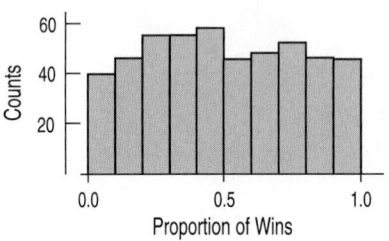

FIGURE 4.6

In a uniform histogram, the bars are all about the same height. The histogram doesn't appear to have a mode.

You've heard of pie à la mode. Is there a connection between pie and the mode of a distribution? Actually, there is! The mode of a distribution is a *popular* value near which a lot of the data values gather. And "à la mode" means "in style"—*not* "with ice cream." That just happened to be a *popular* way to have pie in Paris around 1900.

2. *Is the histogram **symmetric?*** Can you fold it along a vertical line through the middle and have the edges match pretty closely, or are more of the values on one side?

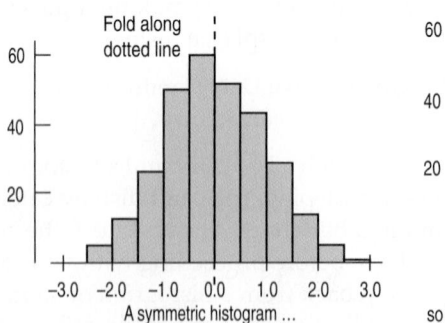

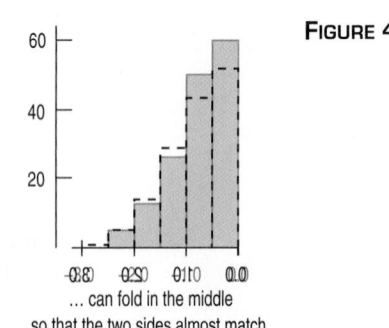

FIGURE 4.7

The (usually) thinner ends of a distribution are called the **tails.** If one tail stretches out farther than the other, the histogram is said to be **skewed** to the side of the longer tail.

---

[6] Apparently, statisticians don't like to count past two.

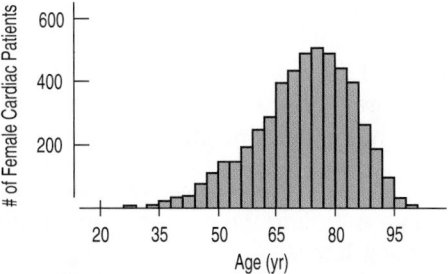

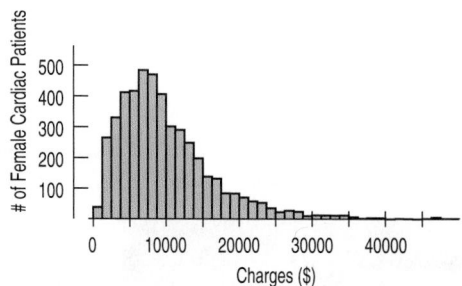

FIGURE 4.8

Two skewed histograms showing data on two variables for all female heart attack patients in New York state in one year. The blue one (age in years) is skewed to the left. The purple one (charges in $) is skewed to the right.

3. *Do any unusual features stick out?* Often such features tell us something interesting or exciting about the data. You should always mention any stragglers, or **outliers,** that stand off away from the body of the distribution. If you're collecting data on nose lengths and Pinocchio is in the group, you'd probably notice him, and you'd certainly want to mention it.

Outliers can affect almost every method we discuss in this course. So we'll always be on the lookout for them. An outlier can be the most informative part of your data. Or it might just be an error. But don't throw it away without comment. Treat it specially and discuss it when you tell about your data. Or find the error and fix it if you can. Be sure to look for outliers. Always.

In the next chapter you'll learn a handy rule of thumb for deciding when a point might be considered an outlier.

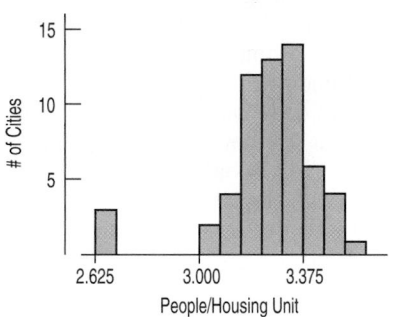

FIGURE 4.9

A histogram with outliers. There are three cities in the leftmost bar.

FOR EXAMPLE Describing histograms

A credit card company wants to see how much customers in a particular segment of their market use their credit card. They have provided you with data[7] on the amount spent by 500 selected customers during a 3-month period and have asked you to summarize the expenditures. Of course, you begin by making a histogram.

Question: Describe the shape of this distribution.

The distribution of expenditures is unimodal and skewed to the high end. There is an extraordinarily large value at about $7000, and some of the expenditures are negative.

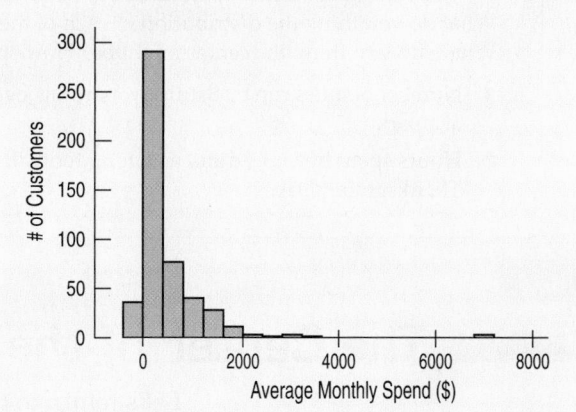

---

[7] These data are real, but cannot be further identified for obvious privacy reasons.

Are there any gaps in the distribution? The Kentucky Derby data that we saw in the dotplot on page 49 has a large gap between two groups of times, one near 120 seconds and one near 160. Gaps help us see multiple modes and encourage us to notice when the data may come from different sources or contain more than one group.

Toto, I've a feeling we're not in math class anymore . . . When Dorothy and her dog Toto land in Oz, everything is more vivid and colorful, but also more dangerous and exciting. Dorothy has new choices to make. She can't always rely on the old definitions, and the yellow brick road has many branches. You may be coming to a similar realization about Statistics.

When we summarize data, our goal is usually more than just developing a detailed knowledge of the data we have at hand. Scientists generally don't care about the particular guinea pigs they've treated, but rather about what their reactions say about how animals (and, perhaps, humans) would respond.

When you look at data, you want to know what the data say about the world, so you'd like to know whether the patterns you see in histograms and summary statistics generalize to other individuals and situations. You'll want to calculate summary statistics accurately, but then you'll also want to think about what they may say beyond just describing the data. And your knowledge about the world matters when you think about the overall meaning of your analysis.

It may surprise you that many of the most important concepts in Statistics are not defined as precisely as most concepts in mathematics. That's done on purpose, to leave room for judgment.

Because we want to see broader patterns rather than focus on the details of the data set we're looking at, we deliberately leave some statistical concepts a bit vague. Whether a histogram is symmetric or skewed, whether it has one or more modes, whether a point is far enough from the rest of the data to be considered an outlier—these are all somewhat vague concepts. And they all require judgment. You may be used to finding a single correct and precise answer, but in Statistics, there may be more than one interpretation. That may make you a little uncomfortable at first, but soon you'll see that this room for judgment brings you enormous power and responsibility. It means that using your own knowledge and judgment and supporting your findings with statistical evidence and justifications entitles you to your own opinions about what you see.

JUST CHECKING

It's often a good idea to think about what the distribution of a data set might look like before we collect the data. What do you think the distribution of each of the following data sets will look like? Be sure to discuss its shape. Where do you think the center might be? How spread out do you think the values will be?

1. Number of miles run by Saturday morning joggers at a park.

2. Hours spent by U.S. adults watching football on Thanksgiving Day.

3. Amount of winnings of all people playing a particular state's lottery last week.

4. Ages of the faculty members at your school.

5. Last digit of phone numbers on your campus.

The Center of the Distribution: The Median

Let's return to the tsunami earthquakes. But this time, let's look at just 25 years of data: 176 earthquakes that occurred from 1981 through 2005. These should be more accurately measured than prehistoric quakes because seismographs were in wide use. Try to put your finger on the histogram at the value you think is

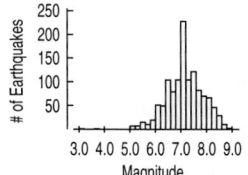

typical. (Read the value from the horizontal axis and remember it.) When we think of a typical value, we usually look for the **center** of the distribution. Where do you think the center of this distribution is? For a unimodal, symmetric distribution such as these earthquake data, it's easy. We'd all agree on the center of symmetry, where we would fold the histogram to match the two sides. But when the distribution is skewed or possibly multimodal, it's not immediately clear what we even mean by the center.

One reasonable choice of typical value is the value that is literally in the middle, with half the values below it and half above it.

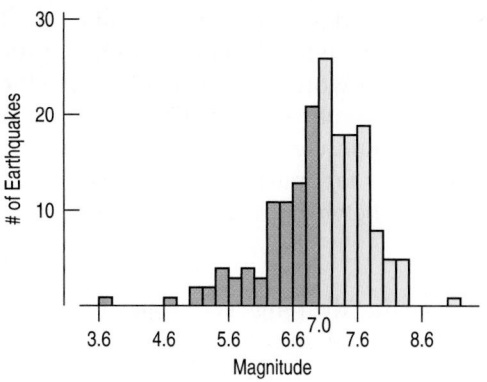

FIGURE 4.10 *Tsunami-causing earthquakes (1981–2005)*

The median splits the histogram into two halves of equal area.

Histograms follow the area principle, and each half of the data has about 88 earthquakes, so each colored region has the same area in the display. The middle value that divides the histogram into two equal areas is called the **median**.

The median has the same units as the data. Be sure to include the units whenever you discuss the median.

For the recent tsunamis, there are 176 earthquakes, so the median is found at the $(176 + 1)/2 = 88.5$th place in the sorted data. That ".5" just says to average the two values on either side: the 88th and the 89th. The median earthquake magnitude is 7.0.

NOTATION ALERT:

We always use n to indicate the number of values. Some people even say, "How big is the n?" when they mean the number of data values.

How do medians work? Finding the median of a batch of n numbers is easy as long as you remember to order the values first. If n is odd, the median is the middle value. Counting in from the ends, we find this value in the $\dfrac{n + 1}{2}$ position.

When n is even, there are two middle values. So, in this case, the median is the average of the two values in positions $\dfrac{n}{2}$ and $\dfrac{n}{2} + 1$.

Here are two examples:

Suppose the batch has these values: 14.1, 3.2, 25.3, 2.8, −17.5, 13.9, 45.8. First we order the values: −17.5, 2.8, 3.2, 13.9, 14.1, 25.3, 45.8.

Since there are 7 values, the median is the $(7 + 1)/2 = 4$th value, counting from the top or bottom: 13.9. Notice that 3 values are lower, 3 higher.

Suppose we had the same batch with another value at 35.7. Then the ordered values are −17.5, 2.8, 3.2, 13.9, 14.1, 25.3, 35.7, 45.8.

The median is the average of the 8/2 or 4th, and the (8/2) + 1, or 5th, values. So the median is $(13.9 + 14.1)/2 = 14.0$. Four data values are lower, and four higher.

The median is one way to find the center of the data. But there are many others. We'll look at an even more important measure later in this chapter.

Knowing the median, we could say that a typical tsunami-causing earthquake, worldwide, was about 7.0 on the Richter scale. How much does that really say? How well does the median describe the data? After all, not every earthquake has a Richter scale value of 7.0. Whenever we find the center of data, the next step is always to ask how well it actually summarizes the data.

Spread: Home on the Range

> Statistics pays close attention to what we *don't* know as well as what we do know. Understanding how spread out the data are is a first step in understanding what a summary *cannot* tell us about the data. It's the beginning of telling us what we don't know.

If every earthquake that caused a tsunami registered 7.0 on the Richter scale, then knowing the median would tell us everything about the distribution of earthquake magnitudes. The more the data vary, however, the less the median alone can tell us. So we need to measure how much the data values vary around the center. In other words, how spread out are they? When we describe a distribution numerically, we always report a measure of its **spread** along with its center.

How should we measure the spread? We could simply look at the extent of the data. How far apart are the two extremes? The **range** of the data is defined as the *difference* between the maximum and minimum values:

$$Range = max - min.$$

Notice that the range is a *single number, not* an interval of values, as you might think from its use in common speech. The maximum magnitude of these earthquakes is 9.0 and the minimum is 3.7, so the *range* is $9.0 - 3.7 = 5.3$.

The range has the disadvantage that a single extreme value can make it very large, giving a value that doesn't really represent the data overall.

Spread: The Interquartile Range

A better way to describe the spread of a variable might be to ignore the extremes and concentrate on the middle of the data. We could, for example, find the range of just the middle half of the data. What do we mean by the middle half? Divide the data in half at the median. Now divide both halves in half again, cutting the data into four quarters. We call these new dividing points **quartiles**. One quarter of the data lies below the **lower quartile,** and one quarter of the data lies above the **upper quartile,** so half the data lies between them. The quartiles border the middle half of the data.

> **How do quartiles work?**   A simple way to find the quartiles is to start by splitting the batch into two halves at the median. (When n is odd, some statisticians include the median in both halves; others omit it.) The lower quartile is the median of the lower half, and the upper quartile is the median of the upper half.
>
> Here are our two examples again.
>
> The ordered values of the first batch were −17.5, 2.8, 3.2, 13.9, 14.1, 25.3, and 45.8, with a median of 13.9. Excluding the median, the two halves of the list are −17.5, 2.8, 3.2 and 14.1, 25.3, 45.8.
>
> Each half has 3 values, so the median of each is the middle one. The lower quartile is 2.8, and the upper quartile is 25.3.
>
> The second batch of data had the ordered values −17.5, 2.8, 3.2, 13.9, 14.1, 25.3, 35.7, and 45.8.
>
> Here n is even, so the two halves of 4 values are −17.5, 2.8, 3.2, 13.9 and 14.1, 25.3, 35.7, 45.8.
>
> Now the lower quartile is $(2.8 + 3.2)/2 = 3.0$, and the upper quartile is $(25.3 + 35.7)/2 = 30.5$.

The difference between the quartiles tells us how much territory the middle half of the data covers and is called the **interquartile range.** It's commonly abbreviated IQR (and pronounced "eye-cue-are," not "ikker"):

$$IQR = upper\ quartile - lower\ quartile.$$

For the earthquakes, there are 88 values below the median and 88 values above the median. The midpoint of the lower half is the average of the 44th and 45th values in the ordered data; that turns out to be 6.6. In the upper half we average the 132nd and 133rd values, finding a magnitude of 7.6 as the third quartile. The *difference* between the quartiles gives the IQR:

$$IQR = 7.6 - 6.6 = 1.0.$$

Now we know that the middle half of the earthquake magnitudes extends across a (interquartile) range of 1.0 Richter scale units. This seems like a reasonable summary of the spread of the distribution, as we can see from this histogram:

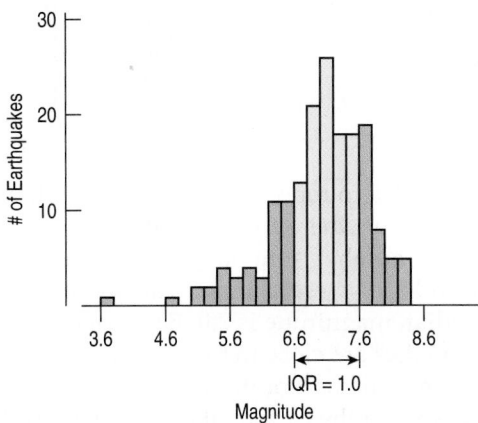

FIGURE 4.11

The quartiles bound the middle 50% of the values of the distribution. This gives a visual indication of the spread of the data. Here we see that the IQR is 1.0 Richter scale units.

The IQR is almost always a reasonable summary of the spread of a distribution. Even if the distribution itself is skewed or has some outliers, the IQR should provide useful information. The one exception is when the data are strongly bimodal. For example, remember the dotplot of winning times in the Kentucky Derby (page 49)? Because the race distance was changed, we really have data on two different races, and they shouldn't be summarized together.

So, what is a quartile anyway? Finding the quartiles sounds easy, but surprisingly, the quartiles are not well-defined. It's not always clear how to find a value such that exactly one quarter of the data lies above or below that value. We offered a simple rule for Finding Quartiles in the box on page 54: Find the median of each half of the data split by the median. When n is odd, we (and your TI calculator) omit the median from each of the halves. Some other texts include the median in both halves before finding the quartiles. Both methods are commonly used. If you are willing to do a bit more calculating, there are several other methods that locate a quartile somewhere between adjacent data values. We know of at least six different rules for finding quartiles. Remarkably, each one is in use in some software package or calculator.

So don't worry too much about getting the "exact" value for a quartile. All of the methods agree pretty closely when the data set is large. When the data set is small, different rules will disagree more, but in that case there's little need to summarize the data anyway.

Remember, Statistics is about understanding the world, not about calculating the right number. The "answer" to a statistical question is a sentence about the issue raised in the question.

The lower and upper quartiles are also known as the 25th and 75th **percentiles** of the data, respectively, since the lower quartile falls above 25% of the data and the upper quartile falls above 75% of the data. If we count this way, the median is the 50th percentile. We could, of course, define and calculate any percentile that we want. For example, the 10th percentile would be the number that falls above the lowest 10% of the data values.

5-Number Summary

The **5-number summary** of a distribution reports its median, quartiles, and extremes (maximum and minimum). The 5-number summary for the recent tsunami earthquake *Magnitudes* looks like this:

Max	9.0
Q3	7.6
Median	7.0
Q1	6.6
Min	3.7

It's good idea to report the number of data values and the identity of the cases (the *Who*). Here there are 176 earthquakes.

The 5-number summary provides a good overview of the distribution of magnitudes of these tsunami-causing earthquakes. For a start, we can see that the median magnitude is 7.0. Because the IQR is only 7.6 − 6.6 = 1, we see that many quakes are close to the median magnitude. Indeed, the quartiles show us that the middle half of these earthquakes had magnitudes between 6.6 and 7.6. One quarter of the earthquakes had magnitudes above 7.6, although one tsunami was caused by a quake measuring only 3.7 on the Richter scale.

STEP-BY-STEP EXAMPLE   **Shape, Center, and Spread: Flight Cancellations**

The U.S. Bureau of Transportation Statistics (www.bts.gov) reports data on airline flights. Let's look at data giving the percentage of flights cancelled each month between 1995 and 2005.

Question: How often are flights cancelled?

WHO	Months
WHAT	Percentage of flights cancelled at U.S. airports
WHEN	1995–2005
WHERE	United States

Variable: Identify the *variable*, and decide how you wish to display it.

To identify a variable, report the W's.

I want to learn about the monthly percentage of flight cancellations at U.S airports.

I have data from the U.S. Bureau of Transportation Statistics giving the percentage of flights cancelled at U.S. airports each month between 1995 and 2005.

Select an appropriate display based on the nature of the data and what you want to know.

✔ **Quantitative Data Condition:** Percentages are quantitative. A histogram and numerical summaries would be appropriate.

Mechanics: We usually make histograms with a computer or graphing calculator.

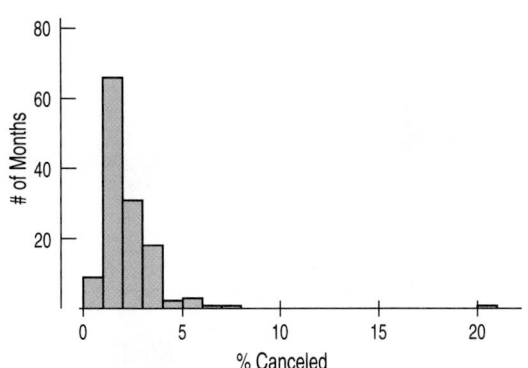

The histogram shows a distribution skewed to the high end and one extreme outlier, a month in which more than 20% of flights were cancelled.

In most months, fewer than 5% of flights are cancelled and usually only about 2% or 3%. That seems reasonable.

REALITY CHECK ▷ It's always a good idea to think about what you expect to see so that you can check whether the histogram looks like what you expected.

With 132 cases, we probably have more data than you'd choose to work with by hand. The results given here are from technology.

Count	132
Max	20.240
Q3	2.615
Median	1.755
Q1	1.445
Min	0.770
IQR	1.170

Interpretation: Describe the shape, center, and spread of the distribution. Report on the symmetry, number of modes, and any gaps or outliers. You should also mention any concerns you may have about the data.

The distribution of cancellations is skewed to the right, and this makes sense: The values can't fall below 0%, but can increase almost arbitrarily due to bad weather or other events.

The median is 1.76% and the IQR is 1.17%. The low IQR indicates that in most months the cancellation rate is close to the median. In fact, it's between 1.4% and 2.6% in the middle 50% of all months, and in only 1/4 of the months were more than 2.6% of flights cancelled.

There is one extraordinary value: 20.2%. Looking it up, I find that the extraordinary month was September 2001. The attacks of September 11 shut down air travel for several days, accounting for this outlier.

Summarizing Symmetric Distributions: The Mean

NOTATION ALERT:

In Algebra you used letters to represent values in a problem, but it didn't matter what letter you picked. You could call the width of a rectangle X or you could call it w (or *Fred*, for that matter). But in Statistics, the notation is part of the vocabulary. For example, in Statistics n is always the number of data values. Always.

We have already begun to point out such special notation conventions: n, Q1, and Q3. Think of them as part of the terminology you need to learn in this course.

Here's another one: Whenever we put a bar over a symbol, it means "find the mean."

Medians do a good job of summarizing the center of a distribution, even when the shape is skewed or when there is an outlier, as with the flight cancellations. But when we have symmetric data, there's another alternative. You probably already know how to average values. In fact, to find the median when n is even, we said you should average the two middle values, and you didn't even flinch.

The earthquake magnitudes are pretty close to symmetric, so we can also summarize their center with a mean. The mean tsunami earthquake magnitude is 6.96—about what we might expect from the histogram. You already know how to average values, but this is a good place to introduce notation that we'll use throughout the book. We use the Greek capital letter sigma, Σ, to mean "sum" (sigma is "S" in Greek), and we'll write:

$$\bar{y} = \frac{Total}{n} = \frac{\sum y}{n}.$$

The formula says to add up all the values of the variable and divide that sum by the number of data values, n—just as you've always done.[8]

Once we've averaged the data, you'd expect the result to be called the *average*, but that would be too easy. Informally, we speak of the "average person" but we don't add up people and divide by the number of people. A median is also a kind of average. To make this distinction, the value we calculated is called the mean, $\bar{y}$, and pronounced "y-bar."

---

[8] You may also see the variable called x and the equation written $\bar{x} = \frac{Total}{n} = \frac{\sum x}{n}$. Don't let that throw you. You are free to name the variable anything you want, but we'll generally use y for variables like this that we want to summarize, model, or predict. (Later we'll talk about variables that are used to explain, model, or predict y. We'll call them x.)

The **mean** feels like the center because it is the point where the histogram balances:

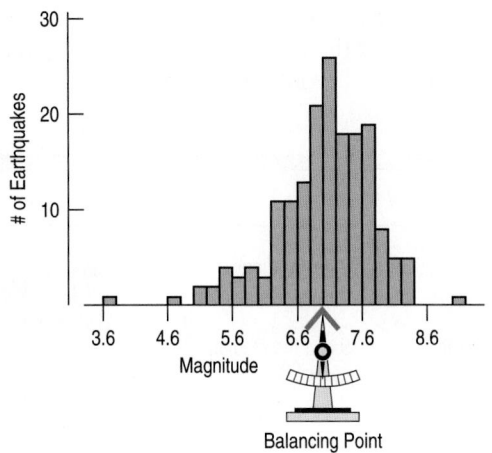

FIGURE 4.12

The mean is located at the balancing point of the histogram.

In everyday language, sometimes "average" *does* mean what we want it to mean. We don't talk about your grade point mean or a baseball player's batting mean or the Dow Jones Industrial mean. So we'll continue to say "average" when that seems most natural. When we do, though, you may assume that what we mean is the mean.

Mean or Median?

Using the center of balance makes sense when the data are symmetric. But data are not always this well behaved. If the distribution is skewed or has outliers, the center is not so well defined and the mean may not be what we want. For example, the mean of the flight cancellations doesn't give a very good idea of the typical percentage of cancellations.

TI-*nspire*

Mean, median, and outliers. Drag data points around to explore how outliers affect the mean and median.

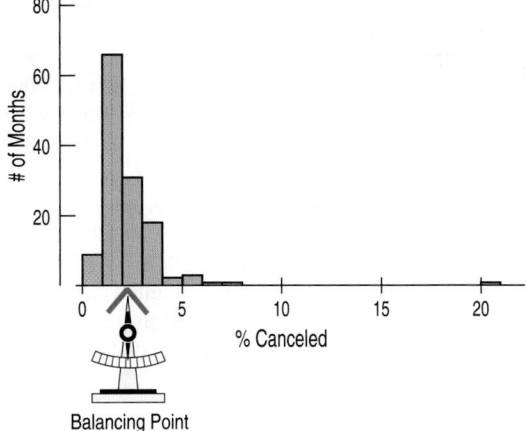

FIGURE 4.13

The median splits the area of the histogram in half at 1.755%. Because the distribution is skewed to the right, the mean (2.28%) is higher than the median. The points at the right have pulled the mean toward them away from the median.

A S *Activity*: **The Center of a Distribution.** Compare measures of center by dragging points up and down and seeing the consequences. Another activity shows how to find summaries with your statistics package.

The mean is 2.28%, but nearly 70% of months had cancellation rates below that, so the mean doesn't feel like a good overall summary. Why is the balancing point so high? The large outlying value pulls it to the right. For data like these, the median is a better summary of the center.

Because the median considers only the order of the values, it is **resistant** to values that are extraordinarily large or small; it simply notes that they are one of the "big ones" or the "small ones" and ignores their distance from the center.

For the tsunami earthquake magnitudes, it doesn't seem to make much difference—the mean is 6.96; the median is 7.0. When the data are symmetric, the mean and median will be close, but when the data are skewed, the median is likely to be a better choice. So, why not just use the median? Well, for one, the median can go overboard. It's not just resistant to occasional outliers, but can be unaffected by changes in up to half the data values. By contrast, the mean includes input from each data value and gives each one equal weight. It's also easier to work with, so when the distribution is unimodal and symmetric, we'll use the mean.

Of course, to choose between mean and median, we'll start by looking at the data. If the histogram is symmetric and there are no outliers, we'll prefer the mean. However, if the histogram is skewed or has outliers, we're usually better off with the median. If you're not sure, report both and discuss why they might differ.

FOR EXAMPLE   Describing center

Recap: You want to summarize the expenditures of 500 credit card company customers, and have looked at a histogram.

Question: You have found the mean expenditure to be $478.19 and the median to be $216.28. Which is the more appropriate measure of center, and why?

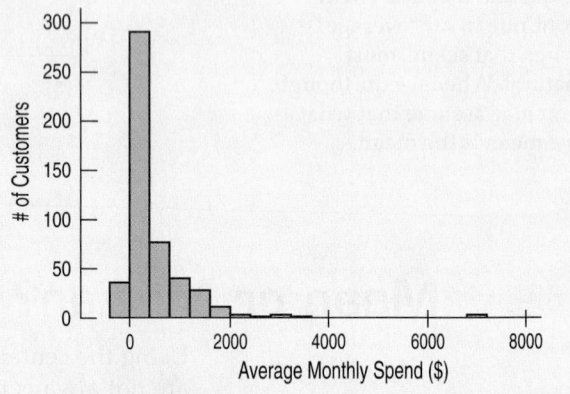

Because the distribution of expenditures is skewed, the median is the more appropriate measure of center. Unlike the mean, it's not affected by the large outlying value or by the skewness. Half of these credit card customers had average monthly expenditures less than $216.28 and half more.

When to expect skewness   Even without making a histogram, we can expect some variables to be skewed. When values of a quantitative variable are bounded on one side but not the other, the distribution may be skewed. For example, incomes and waiting times can't be less than zero, so they are often skewed to the right. Amounts of things (dollars, employees) are often skewed to the right for the same reason. If a test is too easy, the distribution will be skewed to the left because many scores will bump against 100%. And combinations of things are often skewed. In the case of the cancelled flights, flights are more likely to be cancelled in January (due to snowstorms) and in August (thunderstorms). Combining values across months leads to a skewed distribution.

What About Spread? The Standard Deviation

A S   Activity: **The Spread of a Distribution.** What happens to measures of spread when data values change may not be quite what you expect.

The IQR is always a reasonable summary of spread, but because it uses only the two quartiles of the data, it ignores much of the information about how individual values vary. A more powerful approach uses the **standard deviation,** which takes into account how far *each* value is from the mean. Like the mean, the standard deviation is appropriate only for symmetric data.

One way to think about spread is to examine how far each data value is from the mean. This difference is called a *deviation*. We could just average the deviations, but the positive and negative differences always cancel each other out. So the average deviation is always zero—not very helpful.

To keep them from canceling out, we *square* each deviation. Squaring always gives a positive value, so the sum won't be zero. That's great. Squaring also emphasizes larger differences—a feature that turns out to be both good and bad.

When we add up these squared deviations and find their average (almost), we call the result the **variance:**

$$s^2 = \frac{\sum (y - \bar{y})^2}{n - 1}.$$

Why almost? It *would* be a mean if we divided the sum by n. Instead, we divide by $n - 1$. Why? The simplest explanation is "to drive you crazy." But there are good technical reasons, some of which we'll see later.

The variance will play an important role later in this book, but it has a problem as a measure of spread. Whatever the units of the original data are, the variance is in *squared* units. We want measures of spread to have the same units as the data. And we probably don't want to talk about squared dollars or mpg^2. So, to get back to the original units, we take the square root of s^2. The result, s, is the **standard deviation.**

Putting it all together, the standard deviation of the data is found by the following formula:

$$s = \sqrt{\frac{\sum (y - \bar{y})^2}{n - 1}}.$$

WHO	52 adults
WHAT	Resting heart rates
UNITS	Beats per minute

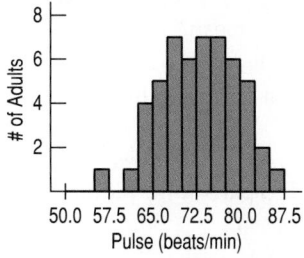

You will almost always rely on a calculator or computer to do the calculating.

Understanding what the standard deviation really means will take some time, and we'll revisit the concept in later chapters. For now, have a look at this histogram of resting pulse rates. The distribution is roughly symmetric, so it's okay to choose the mean and standard deviation as our summaries of center and spread. The mean pulse rate is 72.7 beats per minute, and we can see that's a typical heart rate. We also see that some heart rates are higher and some lower—but how much? Well, the standard deviation of 6.5 beats per minute indicates that, on average, we might expect people's heart rates to differ from the mean rate by about 6.5 beats per minute. Looking at the histogram, we can see that 6.5 beats above or below the mean appears to be a typical deviation.

How does standard deviation work? To find the standard deviation, start with the mean, $\bar{y}$. Then find the *deviations* by taking $\bar{y}$ from each value: $(y - \bar{y})$ Square each deviation: $(y - \bar{y})^2$.

Now you're nearly home. Just add these up and divide by $n - 1$. That gives you the variance, s^2. To find the standard deviation, s, take the square root. Here we go:

Suppose the batch of values is 14, 13, 20, 22, 18, 19, and 13.

The mean is $\bar{y} = 17$. So the deviations are found by subtracting 17 from each value:

Original Values	Deviations	Squared Deviations
14	$14 - 17 = -3$	$(-3)^2 = 9$
13	$13 - 17 = -4$	$(-4)^2 = 16$
20	$20 - 17 = 3$	9
22	$22 - 17 = 5$	25
18	$18 - 17 = 1$	1
19	$19 - 17 = 2$	4
13	$13 - 17 = -4$	16

Add up the squared deviations: $9 + 16 + 9 + 25 + 1 + 4 + 16 = 80$.

Now divide by $n - 1$: $80/6 = 13.33$.

Finally, take the square root: $s = \sqrt{13.33} = 3.65$

Thinking About Variation

Why do banks favor a single line that feeds several teller windows rather than separate lines for each teller? The average waiting time is the same. But the time you can expect to wait is less variable when there is a single line, and people prefer consistency.

Statistics is about variation, so spread is an important fundamental concept in Statistics. Measures of spread help us to be precise about what we *don't* know. If many data values are scattered far from the center, the IQR and the standard deviation will be large. If the data values are close to the center, then these measures of spread will be small. If all our data values were exactly the same, we'd have no question about summarizing the center, and all measures of spread would be zero—and we wouldn't need Statistics. You might think this would be a big plus, but it would make for a boring world. Fortunately (at least for Statistics), data do vary.

Measures of spread tell how well other summaries describe the data. That's why we always (always!) report a spread along with any summary of the center.

JUST CHECKING

6. The U.S. Census Bureau reports the median family income in its summary of census data. Why do you suppose they use the median instead of the mean? What might be the disadvantages of reporting the mean?

7. You've just bought a new car that claims to get a highway fuel efficiency of 31 miles per gallon. Of course, your mileage will "vary." If you had to guess, would you expect the IQR of gas mileage attained by all cars like yours to be 30 mpg, 3 mpg, or 0.3 mpg? Why?

8. A company selling a new MP3 player advertises that the player has a mean lifetime of 5 years. If you were in charge of quality control at the factory, would you prefer that the standard deviation of lifespans of the players you produce be 2 years or 2 months? Why?

What to *Tell* About a Quantitative Variable

What should you *Tell* about a quantitative variable?

▶ Start by making a histogram or stem-and-leaf display, and discuss the shape of the distribution.

▶ Next, discuss the center *and* spread.
  ▸ We always pair the median with the IQR and the mean with the standard deviation. It's not useful to report one without the other. Reporting a center without a spread is dangerous. You may think you know more than you do about the distribution. Reporting only the spread leaves us wondering where we are.
  ▸ If the shape is skewed, report the median and IQR. You may want to include the mean and standard deviation as well, but you should point out why the mean and median differ.
  ▸ If the shape is symmetric, report the mean and standard deviation and possibly the median and IQR as well. For unimodal symmetric data, the IQR is usually a bit larger than the standard deviation. If that's not true of your data set, look again to make sure that the distribution isn't skewed and there are no outliers.

How "Accurate" Should We Be?

Don't think you should report means and standard deviations to a zillion decimal places; such implied accuracy is really meaningless. Although there is no ironclad rule, statisticians commonly report summary statistics to one or two decimal places more than the original data have.

▷ Also, discuss any unusual features.

　▷ If there are multiple modes, try to understand why. If you can identify a reason for separate modes (for example, women and men typically have heart attacks at different ages), it may be a good idea to split the data into separate groups.

　▷ If there are any clear outliers, you should point them out. If you are reporting the mean and standard deviation, report them with the outliers present and with the outliers removed. The differences may be revealing. (Of course, the median and IQR won't be affected very much by the outliers.)

STEP-BY-STEP EXAMPLE	Summarizing a distribution

One of the authors owned a 1989 Nissan Maxima for 8 years. Being a statistician, he recorded the car's fuel efficiency (in mpg) each time he filled the tank. He wanted to know what fuel efficiency to expect as "ordinary" for his car. (Hey, he's a statistician. What would you expect?[9]) Knowing this, he was able to predict when he'd need to fill the tank again and to notice if the fuel efficiency suddenly got worse, which could be a sign of trouble.

Question: How would you describe the distribution of *Fuel efficiency* for this car?

Plan State what you want to find out.

Variable Identify the variable and report the W's.

Be sure to check the appropriate condition.

I want to summarize the distribution of Nissan Maxima fuel efficiency.

The data are the fuel efficiency values in miles per gallon for the first 100 fill-ups of a 1989 Nissan Maxima between 1989 and 1992.

✔ **Quantitative Data Condition:** The fuel efficiencies are quantitative with units of miles per gallon. Histograms and boxplots are appropriate displays for displaying the distribution. Numerical summaries are appropriate as well.

---

[9] He also recorded the time of day, temperature, price of gas, and phase of the moon. (OK, maybe not phase of the moon.) His data are on the DVD.

Mechanics Make a histogram and boxplot. Based on the shape, choose appropriate numerical summaries.

[A histogram with x-axis "Fuel efficiency (mpg)" ranging from 12 to 27+ and y-axis "# of Fill-ups" ranging from 0 to 25.]

REALITY CHECK A value of 22 mpg seems reasonable for such a car. The spread is reasonable, although the range looks a bit large.

A histogram of the data shows a fairly symmetric distribution with a low outlier.

Count	100
Mean	22.4 mpg
StdDev	2.45
Q1	20.8
Median	22.0
Q3	24.0
IQR	3.2

The mean and median are close, so the outlier doesn't seem to be a problem. I can use the mean and standard deviation.

Conclusion Summarize and interpret your findings in context. Be sure to discuss the distribution's shape, center, spread, and unusual features (if any).

The distribution of mileage is unimodal and roughly symmetric with a mean of 22.4 mpg. There is a low outlier that should be investigated, but it does not influence the mean very much. The standard deviation suggests that from tankful to tankful, I can expect the car's fuel economy to differ from the mean by an average of about 2.45 mpg.

Are my statistics "right"?   When you calculate a mean, the computation is clear: You sum all the values and divide by the sample size. You may round your answer less or more than someone else (we recommend one more decimal place than the data), but all books and technologies agree on how to find the mean. Some statistics, however, are more problematic. For example we've already pointed out that methods of finding quartiles differ.

Differences in numeric results can also arise from decisions in the middle of calculations. For example, if you round off your value for the mean before you calculate the sum of squared deviations, your standard deviation probably won't agree with a computer program that calculates using many decimal places. (We do recommend that you do calculations using as many digits as you can to minimize this effect.)

Don't be overly concerned with these discrepancies, especially if the differences are small. They don't mean that your answer is "wrong," and they usually won't change any conclusion you might draw about the data. Sometimes (in footnotes and in the answers in the back of the book) we'll note alternative results, but we could never list all the possible values, so we'll rely on your common sense to focus on the meaning rather than on the digits. Remember: Answers are sentences!

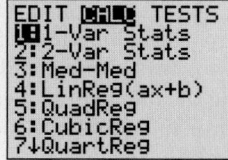
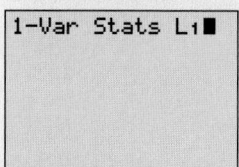
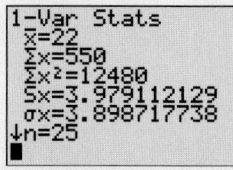
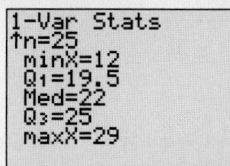

TI Tips

Calculating the statistics

Your calculator can easily find all the numerical summaries of data. To try it out, you simply need a set of values in one of your datalists. We'll illustrate using the boys' agility test results from this chapter's earlier TI Tips (still in L1), but you can use any data currently stored in your calculator.

- Under the STAT CALC menu, select 1-Var Stats and hit ENTER.
- Specify the location of your data, creating a command like 1-Var Stats L1.
- Hit ENTER again.

Voilà! Everything you wanted to know, and more. Among all of the information shown, you are primarily interested in these statistics: $\bar{x}$ (the mean), Sx (the standard deviation), n (the count), and—scrolling down—minX (the smallest datum), Q_1 (the first quartile), Med (the median), Q_3 (the third quartile), and maxX (the largest datum).

Sorry, but the TI doesn't explicitly tell you the range or the IQR. Just subtract: IQR $= Q_3 - Q_1 = 25 - 19.5 = 5.5$. What's the range?

By the way, if the data come as a frequency table with the values stored in, say, L4 and the corresponding frequencies in L5, all you have to do is ask for 1-Var Stats L4,L5.

WHAT CAN GO WRONG?

A data display should tell a story about the data. To do that, it must speak in a clear language, making plain what variable is displayed, what any axis shows, and what the values of the data are. And it must be consistent in those decisions.

A display of quantitative data can go wrong in many ways. The most common failures arise from only a few basic errors:

▶ **Don't make a histogram of a categorical variable.** Just because the variable contains numbers doesn't mean that it's quantitative. Here's a histogram of the insurance policy numbers of some workers. It's not very informative because the policy numbers are just labels. A histogram or stem-and-leaf display of a categorical variable makes no sense. A bar chart or pie chart would be more appropriate.

▶ **Don't look for shape, center, and spread of a bar chart.** A bar chart showing the sizes of the piles displays the distribution of a categorical variable, but the bars could be arranged in any order left to right. Concepts like symmetry, center, and spread make sense only for quantitative variables.

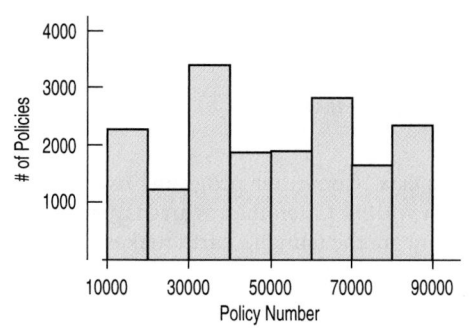

FIGURE 4.14

It's not appropriate to display these data with a histogram.

(continued)

▶ **Don't use bars in every display—save them for histograms and bar charts.** In a bar chart, the bars indicate how many cases of a categorical variable are piled in each category. Bars in a histogram indicate the number of cases piled in each interval of a quantitative variable. In both bar charts and histograms, the bars represent counts of data values. Some people create other displays that use bars to represent individual data values. Beware: Such graphs are neither bar charts nor histograms. For example, a student was asked to make a histogram from data showing the number of juvenile bald eagles seen during each of the 13 weeks in the winter of 2003–2004 at a site in Rock Island, IL. Instead, he made this plot:

FIGURE 4.15

This isn't a histogram or a bar chart. It's an ill-conceived graph that uses bars to represent individual data values (number of eagles sighted) week by week.

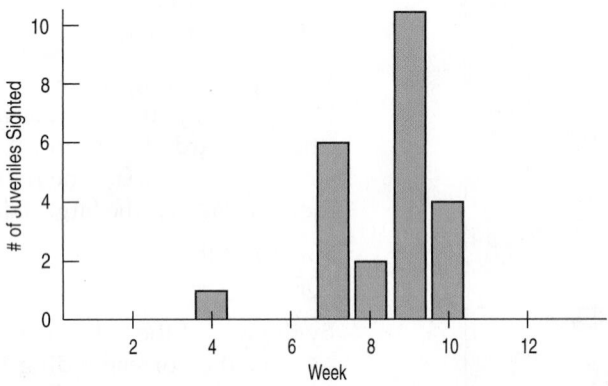

Look carefully. That's not a histogram. A histogram shows *What* we've measured along the horizontal axis and counts of the associated *Who*'s represented as bar heights. This student has it backwards: He used bars to show counts of birds for each week.[10] We need counts of weeks. A correct histogram should have a tall bar at "0" to show there were many weeks when no eagles were seen, like this:

FIGURE 4.16

A histogram of the eagle-sighting data shows the number of weeks in which different counts of eagles occurred. This display shows the distribution of juvenile-eagle sightings.

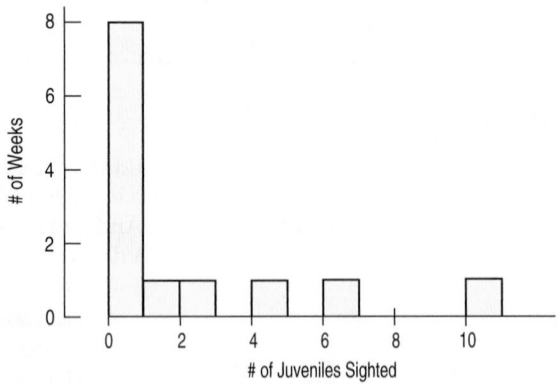

▶ **Choose a bin width appropriate to the data.** Computer programs usually do a pretty good job of choosing histogram bin widths. Often there's an easy way to adjust the width, sometimes interactively. Here are the tsunami earthquakes with two (rather extreme) choices for the bin size:

---

[10] Edward Tufte, in his book *The Visual Display of Quantitative Information*, proposes that graphs should have a high data-to-ink ratio. That is, we shouldn't waste a lot of ink to display a single number when a dot would do the job.

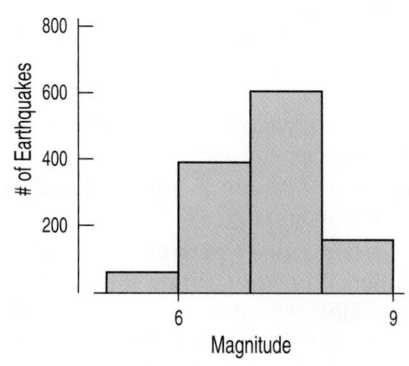

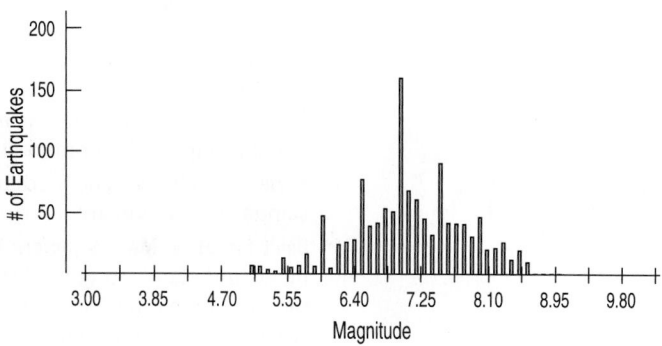

The task of summarizing a quantitative variable is relatively simple, and there is a simple path to follow. However, you need to watch out for certain features of the data that make summarizing them with a number dangerous. Here's some advice:

▶ **Don't forget to do a reality check.** Don't let the computer or calculator do your thinking for you. Make sure the calculated summaries make sense. For example, does the mean look like it is in the center of the histogram? Think about the spread: An IQR of 50 mpg would clearly be wrong for gas mileage. And no measure of spread can be negative. The standard deviation can take the value 0, but only in the very unusual case that all the data values equal the same number. If you see an IQR or standard deviation equal to 0, it's probably a sign that something's wrong with the data.

▶ **Don't forget to sort the values before finding the median or percentiles.** It seems obvious, but when you work by hand, it's easy to forget to sort the data first before counting in to find medians, quartiles, or other percentiles. Don't report that the median of the five values 194, 5, 1, 17, and 893 is 1 just because 1 is the middle number.

▶ **Don't worry about small differences when using different methods.** Finding the 10th percentile or the lower quartile in a data set sounds easy enough. But it turns out that the definitions are not exactly clear. If you compare different statistics packages or calculators, you may find that they give slightly different answers for the same data. These differences, though, are unlikely to be important in interpreting the data, the quartiles, or the IQR, so don't let them worry you.

▶ **Don't compute numerical summaries of a categorical variable.** Neither the mean zip code nor the standard deviation of social security numbers is meaningful. If the variable is categorical, you should instead report summaries such as percentages of individuals in each category. It is easy to make this mistake when using technology to do the summaries for you. After all, the computer doesn't care what the numbers mean.

▶ **Don't report too many decimal places.** Statistical programs and calculators often report a ridiculous number of digits. A general rule for numerical summaries is to report one or two more digits than the number of digits in the data. For example, earlier we saw a dotplot of the Kentucky Derby race times. The mean and standard deviation of those times could be reported as:

$$\bar{y} = 130.63401639344262 \, \text{sec} \qquad s = 13.66448201942662 \, \text{sec}$$

But we knew the race times only to the nearest quarter second, so the extra digits are meaningless.

▶ **Don't round in the middle of a calculation.** Don't *report* too many decimal places, but it's best not to do any rounding until the end of your calculations. Even though you might report the mean of the earthquakes as 7.08, it's really 7.08339. Use the more precise number in your calculations if you're finding the standard deviation by hand—or be prepared to see small differences in your final result.

Gold Card Customers—Regions National Banks		
Month	April 2007	May 2007
Average Zip Code	45,034.34	38,743.34

(continued)

▶ **Watch out for multiple modes.** The summaries of the Kentucky Derby times are meaningless for another reason. As we saw in the dotplot, the Derby was initially a longer race. It would make much more sense to report that the old 1.5 mile Derby had a mean time of 159.6 seconds, while the current Derby has a mean time of 124.6 seconds. If the distribution has multiple modes, consider separating the data into different groups and summarizing each group separately.

▶ **Beware of outliers.** The median and IQR are resistant to outliers, but the mean and standard deviation are not. To help spot outliers . . .

▶ **Don't forget to: Make a picture (make a picture, make a picture).** The sensitivity of the mean and standard deviation to outliers is one reason you should always make a picture of the data. Summarizing a variable with its mean and standard deviation when you have not looked at a histogram or dotplot to check for outliers or skewness invites disaster. You may find yourself drawing absurd or dangerously wrong conclusions about the data. And, of course, you should demand no less of others. Don't accept a mean and standard deviation blindly without some evidence that the variable they summarize is unimodal, symmetric, and free of outliers.

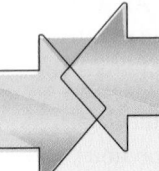

CONNECTIONS

Distributions of quantitative variables, like those of categorical variables, show the possible values and their relative frequencies. A histogram shows the distribution of values in a quantitative variable with adjacent bars. Don't confuse histograms with bar charts, which display categorical variables. For categorical data, the mode is the category with the biggest count. For quantitative data, modes are peaks in the histogram.

The shape of the distribution of a quantitative variable is an important concept in most of the subsequent chapters. We will be especially interested in distributions that are unimodal and symmetric.

In addition to their shape, we summarize distributions with center and spread, usually pairing a measure of center with a measure of spread: median with IQR and mean with standard deviation. We favor the mean and standard deviation when the shape is unimodal and symmetric, but choose the median and IQR for skewed distributions or when there are outliers we can't otherwise set aside.

WHAT HAVE WE LEARNED?

We've learned how to make a picture of quantitative data to help us see the story the data have to *Tell*.

▶ We can display the distribution of quantitative data with a *histogram*, a *stem-and-leaf* display, or a *dotplot*.

▶ We *Tell* what we see about the distribution by talking about *shape*, *center*, *spread*, and any *unusual features*.

We've learned how to summarize distributions of quantitative variables numerically.

▶ Measures of center for a distribution include the median and the mean.

We write the formula for the mean as $\bar{y} = \dfrac{\sum y}{n}$.

▶ Measures of spread include the range, IQR, and standard deviation.

The standard deviation is computed as $s = \sqrt{\dfrac{\sum (y - \bar{y})^2}{n - 1}}$.

The median and IQR are not usually given as formulas.

► We'll report the median and IQR when the distribution is skewed. If it's symmetric, we'll summarize the distribution with the mean and standard deviation (and possibly the median and IQR as well). Always pair the median with the IQR and the mean with the standard deviation.

We've learned to *Think* about the type of variable we're summarizing.

► All the methods of this chapter assume that the data are quantitative.
► The **Quantitative Data Condition** serves as a check that the data are, in fact, quantitative. One good way to be sure is to know the measurement units. You'll want those as part of the *Think* step of your answers.

Terms

Distribution	44. The distribution of a quantitative variable slices up all the possible values of the variable into equal-width bins and gives the number of values (or counts) falling into each bin.
Histogram (relative frequency histogram)	45. A histogram uses adjacent bars to show the distribution of a quantitative variable. Each bar represents the frequency (or relative frequency) of values falling in each bin.
Gap	45. A region of the distribution where there are no values.
Stem-and-leaf display	47. A stem-and-leaf display shows quantitative data values in a way that sketches the distribution of the data. It's best described in detail by example.
Dotplot	49. A dotplot graphs a dot for each case against a single axis.
Shape	49. To describe the shape of a distribution, look for
	► single vs. multiple modes.
	► symmetry vs. skewness.
	► outliers and gaps.
Center	52, 58. The place in the distribution of a variable that you'd point to if you wanted to attempt the impossible by summarizing the entire distribution with a single number. Measures of center include the mean and median.
Spread	54, 61. A numerical summary of how tightly the values are clustered around the center. Measures of spread include the IQR and standard deviation.
Mode	49. A hump or local high point in the shape of the distribution of a variable. The apparent location of modes can change as the scale of a histogram is changed.
Unimodal (Bimodal)	50. Having one mode. This is a useful term for describing the shape of a histogram when it's generally mound-shaped. Distributions with two modes are called **bimodal.** Those with more than two are **multimodal.**
Uniform	50. A distribution that's roughly flat is said to be uniform.
Symmetric	50. A distribution is symmetric if the two halves on either side of the center look approximately like mirror images of each other.
Tails	50. The tails of a distribution are the parts that typically trail off on either side. Distributions can be characterized as having long tails (if they straggle off for some distance) or short tails (if they don't).
Skewed	50. A distribution is skewed if it's not symmetric and one tail stretches out farther than the other. Distributions are said to be **skewed left** when the longer tail stretches to the left, and **skewed right** when it goes to the right.
Outliers	51. Outliers are extreme values that don't appear to belong with the rest of the data. They may be unusual values that deserve further investigation, or they may be just mistakes; there's no obvious way to tell. Don't delete outliers automatically—you have to think about them. Outliers can affect many statistical analyses, so you should always be alert for them.
Median	52. The median is the middle value, with half of the data above and half below it. If n is even, it is the average of the two middle values. It is usually paired with the IQR.
Range	54. The difference between the lowest and highest values in a data set. $Range = max - min$.
Quartile	54. The lower quartile (Q1) is the value with a quarter of the data below it. The upper quartile (Q3) has three quarters of the data below it. The median and quartiles divide data into four parts with equal numbers of data values.

Interquartile range (IQR)

55. The IQR is the difference between the first and third quartiles. $IQR = Q3 - Q1$. It is usually reported along with the median.

Percentile

55. The ith percentile is the number that falls above i% of the data.

5-Number Summary

56. The 5-number summary of a distribution reports the minimum value, Q1, the median, Q3, and the maximum value.

Mean

58. The mean is found by summing all the data values and dividing by the count:

$$\bar{y} = \frac{Total}{n} = \frac{\sum y}{n}.$$

It is usually paired with the standard deviation.

Resistant

59. A calculated summary is said to be resistant if outliers have only a small effect on it.

Variance

61. The variance is the sum of squared deviations from the mean, divided by the count minus 1:

$$s^2 = \frac{\sum (y - \bar{y})^2}{n - 1}.$$

It is useful in calculations later in the book.

Standard deviation

61. The standard deviation is the square root of the variance:

$$s = \sqrt{\frac{\sum (y - \bar{y})^2}{n - 1}}$$

It is usually reported along with the mean.

Skills

- ▸ Be able to identify an appropriate display for any quantitative variable.
- ▸ Be able to guess the shape of the distribution of a variable by knowing something about the data.
- ▸ Be able to select a suitable measure of center and a suitable measure of spread for a variable based on information about its distribution.
- ▸ Know the basic properties of the median: The median divides the data into the half of the data values that are below the median and the half that are above.
- ▸ Know the basic properties of the mean: The mean is the point at which the histogram balances.
- ▸ Know that the standard deviation summarizes how spread out all the data are around the mean.
- ▸ Understand that the median and IQR resist the effects of outliers, while the mean and standard deviation do not.
- ▸ Understand that in a skewed distribution, the mean is pulled in the direction of the skewness (toward the longer tail) relative to the median.

- ▸ Know how to display the distribution of a quantitative variable with a stem-and-leaf display (drawn by hand for smaller data sets), a dotplot, or a histogram (made by computer for larger data sets).
- ▸ Know how to compute the mean and median of a set of data.
- ▸ Know how to compute the standard deviation and IQR of a set of data.

- ▸ Be able to describe the distribution of a quantitative variable in terms of its shape, center, and spread.
- ▸ Be able to describe any anomalies or extraordinary features revealed by the display of a variable.
- ▸ Know how to describe summary measures in a sentence. In particular, know that the common measures of center and spread have the same units as the variable that they summarize, and should be described in those units.
- ▸ Be able to describe the distribution of a quantitative variable with a description of the shape of the distribution, a numerical measure of center, and a numerical measure of spread. Be sure to note any unusual features, such as outliers, too.

DISPLAYING AND SUMMARIZING QUANTITATIVE VARIABLES ON THE COMPUTER

Almost any program that displays data can make a histogram, but some will do a better job of determining where the bars should start and how they should partition the span of the data.

The vertical scale may be counts or proportions. Sometimes it isn't clear which. But the shape of the histogram is the same either way.

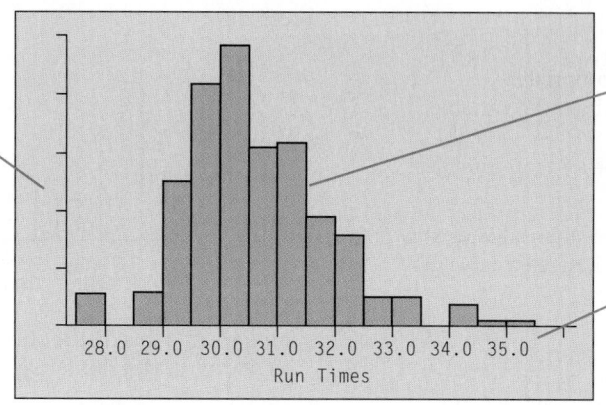

Most packages choose the number of bars for you automatically. Often you can adjust that choice.

The axis should be clearly labeled so you can tell what "pile" each bar represents. You should be able to tell the lower and upper bounds of each bar.

Many statistics packages offer a prepackaged collection of summary measures. The result might look like this:

Variable: Weight
N = 234
Mean = 143.3   Median = 139
St. Dev = 11.1   IQR = 14

Alternatively, a package might make a table for several variables and summary measures:

A S *Case Study:* **Describing Distribution Shapes.** Who's safer in a crash—passengers or the driver? Investigate with your statistics package.

Variable	N	mean	median	stdev	IQR
Weight	234	143.3	139	11.1	14
Height	234	68.3	68.1	4.3	5
Score	234	86	88	9	5

It is usually easy to read the results and identify each computed summary. You should be able to read the summary statistics produced by any computer package.

Packages often provide many more summary statistics than you need. Of course, some of these may not be appropriate when the data are skewed or have outliers. It is your responsibility to check a histogram or stem-and-leaf display and decide which summary statistics to use.

It is common for packages to report summary statistics to many decimal places of "accuracy." Of course, it is rare data that have such accuracy in the original measurements. The ability to calculate to six or seven digits beyond the decimal point doesn't mean that those digits have any meaning. Generally it's a good idea to round these values, allowing perhaps one more digit of precision than was given in the original data.

Displays and summaries of quantitative variables are among the simplest things you can do in most statistics packages.

EXERCISES

1. **Histogram.**   Find a histogram that shows the distribution of a variable in a newspaper, a magazine, or the Internet.
   a) Does the article identify the W's?
   b) Discuss whether the display is appropriate.
   c) Discuss what the display reveals about the variable and its distribution.
   d) Does the article accurately describe and interpret the data? Explain.

2. **Not a histogram.**   Find a graph other than a histogram that shows the distribution of a quantitative variable in a newspaper, a magazine, or the Internet.
   a) Does the article identify the W's?
   b) Discuss whether the display is appropriate for the data.
   c) Discuss what the display reveals about the variable and its distribution.
   d) Does the article accurately describe and interpret the data? Explain.

3. **In the news.**   Find an article in a newspaper, a magazine, or the Internet that discusses an "average."
   a) Does the article discuss the W's for the data?
   b) What are the units of the variable?
   c) Is the average used the median or the mean? How can you tell?
   d) Is the choice of median or mean appropriate for the situation? Explain.

4. **In the news II.**   Find an article in a newspaper, a magazine, or the Internet that discusses a measure of spread.
   a) Does the article discuss the W's for the data?
   b) What are the units of the variable?
   c) Does the article use the range, IQR, or standard deviation?
   d) Is the choice of measure of spread appropriate for the situation? Explain.

5. **Thinking about shape.**   Would you expect distributions of these variables to be uniform, unimodal, or bimodal? Symmetric or skewed? Explain why.
   a) The number of speeding tickets each student in the senior class of a college has ever had.
   b) Players' scores (number of strokes) at the U.S. Open golf tournament in a given year.
   c) Weights of female babies born in a particular hospital over the course of a year.
   d) The length of the average hair on the heads of students in a large class.

6. **More shapes.**   Would you expect distributions of these variables to be uniform, unimodal, or bimodal? Symmetric or skewed? Explain why.
   a) Ages of people at a Little League game.
   b) Number of siblings of people in your class.
   c) Pulse rates of college-age males.
   d) Number of times each face of a die shows in 100 tosses.

7. **Sugar in cereals.**   The histogram displays the sugar content (as a percent of weight) of 49 brands of breakfast cereals.

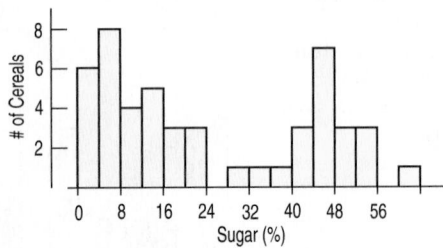

   a) Describe this distribution.
   b) What do you think might account for this shape?

8. **Singers.**   The display shows the heights of some of the singers in a chorus, collected so that the singers could be positioned on stage with shorter ones in front and taller ones in back.

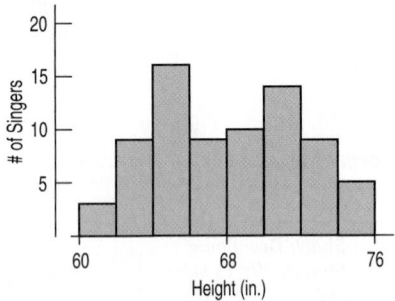

   a) Describe the distribution.
   b) Can you account for the features you see here?

9. **Vineyards.**   The histogram shows the sizes (in acres) of 36 vineyards in the Finger Lakes region of New York.

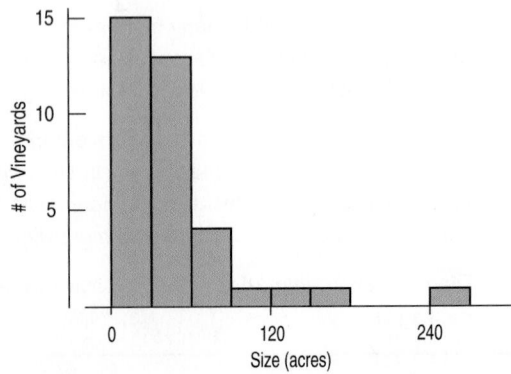

   a) Approximately what percentage of these vineyards are under 60 acres?
   b) Write a brief description of this distribution (shape, center, spread, unusual features).

10. Run times. One of the authors collected the times (in minutes) it took him to run 4 miles on various courses during a 10-year period. Here is a histogram of the times.

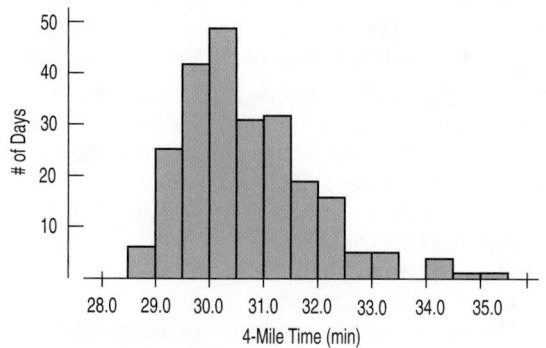

Describe the distribution and summarize the important features. What is it about running that might account for the shape you see?

11. Heart attack stays. The histogram shows the lengths of hospital stays (in days) for all the female patients admitted to hospitals in New York during one year with a primary diagnosis of acute myocardial infarction (heart attack).

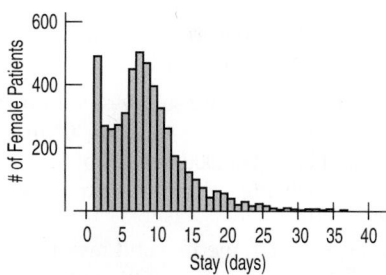

a) From the histogram, would you expect the mean or median to be larger? Explain.
b) Write a few sentences describing this distribution (shape, center, spread, unusual features).
c) Which summary statistics would you choose to summarize the center and spread in these data? Why?

12. E-mails. A university teacher saved every e-mail received from students in a large Introductory Statistics class during an entire term. He then counted, for each student who had sent him at least one e-mail, how many e-mails each student had sent.

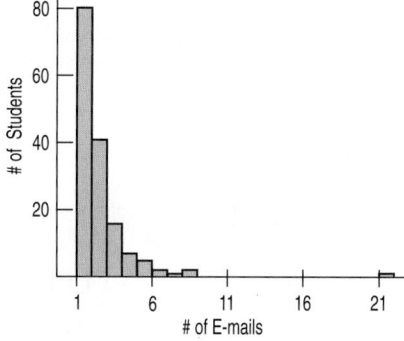

a) From the histogram, would you expect the mean or the median to be larger? Explain.
b) Write a few sentences describing this distribution (shape, center, spread, unusual features).

c) Which summary statistics would you choose to summarize the center and spread in these data? Why?

13. Super Bowl points. How many points do football teams score in the Super Bowl? Here are the total numbers of points scored by both teams in each of the first 42 Super Bowl games:

45, 47, 23, 30, 29, 27, 21, 31, 22, 38, 46, 37, 66, 50, 37, 47, 44, 47, 54, 56, 59, 52, 36, 65, 39, 61, 69, 43, 75, 44, 56, 55, 53, 39, 41, 37, 69, 61, 45, 31, 46, 31

a) Find the median.
b) Find the quartiles.
c) Write a description based on the 5-number summary.

14. Super Bowl wins. In the Super Bowl, by how many points does the winning team outscore the losers? Here are the winning margins for the first 42 Super Bowl games:

25, 19, 9, 16, 3, 21, 7, 17, 10, 4, 18, 17, 4, 12, 17, 5, 10, 29, 22, 36, 19, 32, 4, 45, 1, 13, 35, 17, 23, 10, 14, 7, 15, 7, 27, 3, 27, 3, 3, 11, 12, 3

a) Find the median.
b) Find the quartiles.
c) Write a description based on the 5-number summary.

15. Standard deviation I. For each lettered part, a through c, examine the two given sets of numbers. Without doing any calculations, decide which set has the larger standard deviation and explain why. Then check by finding the standard deviations *by hand*.

	Set 1	Set 2
a)	3, 5, 6, 7, 9	2, 4, 6, 8, 10
b)	10, 14, 15, 16, 20	10, 11, 15, 19, 20
c)	2, 6, 6, 9, 11, 14	82, 86, 86, 89, 91, 94

16. Standard deviation II. For each lettered part, a through c, examine the two given sets of numbers. Without doing any calculations, decide which set has the larger standard deviation and explain why. Then check by finding the standard deviations *by hand*.

	Set 1	Set 2
a)	4, 7, 7, 7, 10	4, 6, 7, 8, 10
b)	100, 140, 150, 160, 200	10, 50, 60, 70, 110
c)	10, 16, 18, 20, 22, 28	48, 56, 58, 60, 62, 70

17. Pizza prices. The histogram shows the distribution of the prices of plain pizza slices (in $) for 156 weeks in Dallas, TX.

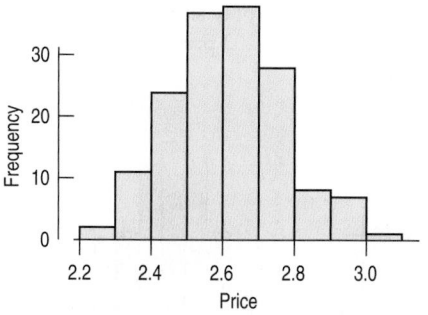

Which summary statistics would you choose to summarize the center and spread in these data? Why?

18. Neck size.   The histogram shows the neck sizes (in inches) of 250 men recruited for a health study in Utah.

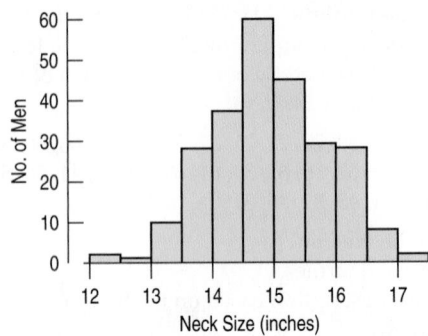

Which summary statistics would you choose to summarize the center and spread in these data? Why?

19. Pizza prices again.   Look again at the histogram of the pizza prices in Exercise 17.
a) Is the mean closer to $2.40, $2.60, or $2.80? Why?
b) Is the standard deviation closer to $0.15, $0.50, or $1.00? Explain.

20. Neck sizes again.   Look again at the histogram of men's neck sizes in Exercise 18.
a) Is the mean closer to 14, 15, or 16 inches? Why?
b) Is the standard deviation closer to 1 inch, 3 inches, or 5 inches? Explain.

21. Movie lengths.   The histogram shows the running times in minutes of 122 feature films released in 2005.

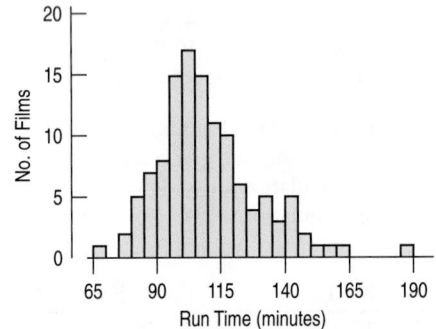

a) You plan to see a movie this weekend. Based on these movies, how long do you expect a typical movie to run?
b) Would you be surprised to find that your movie ran for $2\frac{1}{2}$ hours (150 minutes)?
c) Which would you expect to be higher: the mean or the median run time for all movies? Why?

22. Golf drives.   The display shows the average drive distance (in yards) for 202 professional golfers on the men's PGA tour.

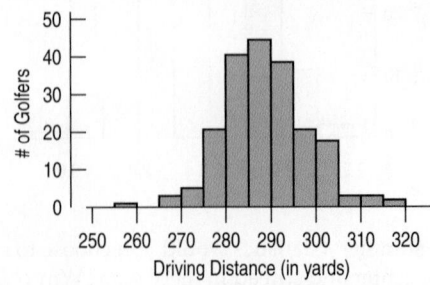

a) Describe this distribution.
b) Approximately what proportion of professional male golfers drive, on average, less than 280 yards?
c) Estimate the mean by examining the histogram.
d) Do you expect the mean to be smaller than, approximately equal to, or larger than the median? Why?

23. Movie lengths II.   Exercise 21 looked at the running times of movies released in 2005. The standard deviation of these running times is 19.6 minutes, and the quartiles are $Q_1 = 97$ minutes and $Q_3 = 119$ minutes.
a) Write a sentence or two describing the spread in running times based on
  i) the quartiles.
  ii) the standard deviation.
b) Do you have any concerns about using either of these descriptions of spread? Explain.

24. Golf drives II.   Exercise 22 looked at distances PGA golfers can hit the ball. The standard deviation of these average drive distances is 9.3 yards, and the quartiles are $Q_1 = 282$ yards and $Q_3 = 294$ yards.
a) Write a sentence or two describing the spread in distances based on
  i) the quartiles.
  ii) the standard deviation.
b) Do you have any concerns about using either of these descriptions of spread? Explain.

25. Mistake.   A clerk entering salary data into a company spreadsheet accidentally put an extra "0" in the boss's salary, listing it as $2,000,000 instead of $200,000. Explain how this error will affect these summary statistics for the company payroll:
a) measures of center: median and mean.
b) measures of spread: range, IQR, and standard deviation.

26. Cold weather.   A meteorologist preparing a talk about global warming compiled a list of weekly low temperatures (in degrees Fahrenheit) he observed at his southern Florida home last year. The coldest temperature for any week was 36°F, but he inadvertently recorded the Celsius value of 2°. Assuming that he correctly listed all the other temperatures, explain how this error will affect these summary statistics:
a) measures of center: mean and median.
b) measures of spread: range, IQR, and standard deviation.

27. Movie budgets.   The histogram shows the budgets (in millions of dollars) of major release movies in 2005.

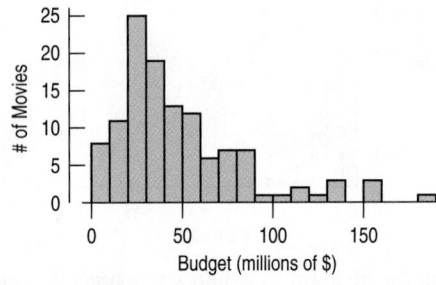

An industry publication reports that the average movie costs $35 million to make, but a watchdog group concerned with rising ticket prices says that the average cost is $46.8 million. What statistic do you think each group is using? Explain.

28. **Sick days.** During contract negotiations, a company seeks to change the number of sick days employees may take, saying that the annual "average" is 7 days of absence per employee. The union negotiators counter that the "average" employee misses only 3 days of work each year. Explain how both sides might be correct, identifying the measure of center you think each side is using and why the difference might exist.

29. **Payroll.** A small warehouse employs a supervisor at $1200 a week, an inventory manager at $700 a week, six stock boys at $400 a week, and four drivers at $500 a week.
    a) Find the mean and median wage.
    b) How many employees earn more than the mean wage?
    c) Which measure of center best describes a typical wage at this company: the mean or the median?
    d) Which measure of spread would best describe the payroll: the range, the IQR, or the standard deviation? Why?

30. **Singers.** The frequency table shows the heights (in inches) of 130 members of a choir.

Height	Count	Height	Count
60	2	69	5
61	6	70	11
62	9	71	8
63	7	72	9
64	5	73	4
65	20	74	2
66	18	75	4
67	7	76	1
68	12		

    a) Find the median and IQR.
    b) Find the mean and standard deviation.
    c) Display these data with a histogram.
    d) Write a few sentences describing the distribution.

31. **Gasoline.** In March 2006, 16 gas stations in Grand Junction, CO, posted these prices for a gallon of regular gasoline:

2.22	2.21	2.45	2.24
2.27	2.28	2.27	2.23
2.26	2.46	2.29	2.32
2.36	2.38	2.33	2.27

    a) Make a stem-and-leaf display of these gas prices. Use split stems; for example, use two 2.2 stems—one for prices between $2.20 and $2.24 and the other for prices from $2.25 to $2.29.
    b) Describe the shape, center, and spread of this distribution.
    c) What unusual feature do you see?

32. **The Great One.** During his 20 seasons in the NHL, Wayne Gretzky scored 50% more points than anyone who ever played professional hockey. He accomplished this amazing feat while playing in 280 fewer games than Gordie Howe, the previous record holder. Here are the number of games Gretzky played during each season:

79, 80, 80, 80, 74, 80, 80, 79, 64, 78, 73, 78, 74, 45, 81, 48, 80, 82, 82, 70

    a) Create a stem-and-leaf display for these data, using split stems.
    b) Describe the shape of the distribution.
    c) Describe the center and spread of this distribution.
    d) What unusual feature do you see? What might explain this?

33. **States.** The stem-and-leaf display shows populations of the 50 states and Washington, DC, in millions of people, according to the 2000 census.

```
3 | 4
2 |
2 | 1
1 | 69
1 | 0122
0 | 5555666667888
0 | 111111111111122222333333344444
```
State Populations (1| 2 means 12 million)

    a) What measures of center and spread are most appropriate?
    b) Without doing any calculations, which must be larger: the median or the mean? Explain how you know.
    c) From the stem-and-leaf display, find the median and the interquartile range.
    d) Write a few sentences describing this distribution.

34. **Wayne Gretzky.** In Exercise 32, you examined the number of games played by hockey great Wayne Gretzky during his 20-year career in the NHL.
    a) Would you use the median or the mean to describe the center of this distribution? Why?
    b) Find the median.
    c) Without actually finding the mean, would you expect it to be higher or lower than the median? Explain.

35. **Home runs.** The stem-and-leaf display shows the number of home runs hit by Mark McGwire during the 1986–2001 seasons. Describe the distribution, mentioning its shape and any unusual features.

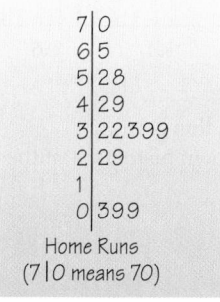

```
7 | 0
6 | 5
5 | 28
4 | 29
3 | 22399
2 | 29
1 |
0 | 399
```
Home Runs
(7|0 means 70)

36. Bird species. The Cornell Lab of Ornithology holds an annual Christmas Bird Count (www.birdsource.org), in which bird watchers at various locations around the country see how many different species of birds they can spot. Here are some of the counts reported from sites in Texas during the 1999 event:

228	178	186	162	206	166	163
183	181	206	177	175	167	162
160	160	157	156	153	153	152

a) Create a stem-and-leaf display of these data.
b) Write a brief description of the distribution. Be sure to discuss the overall shape as well as any unusual features.

37. Hurricanes 2006. The data below give the number of hurricanes classified as major hurricanes in the Atlantic Ocean each year from 1944 through 2006, as reported by *NOAA* (www.nhc.noaa.gov):

3, 2, 1, 2, 4, 3, 7, 2, 3, 3, 2, 5, 2, 2, 4, 2, 2, 6, 0, 2, 5, 1, 3, 1, 0, 3, 2, 1, 0, 1, 2, 3, 2, 1, 2, 2, 2, 3, 1, 1, 1, 3, 0, 1, 3, 2, 1, 2, 1, 1, 0, 5, 6, 1, 3, 5, 3, 3, 2, 3, 6, 7, 2

a) Create a dotplot of these data.
b) Describe the distribution.

38. Horsepower. Create a stem-and-leaf display for these horsepowers of autos reviewed by *Consumer Reports* one year, and describe the distribution:

155	103	130	80	65
142	125	129	71	69
125	115	138	68	78
150	133	135	90	97
68	105	88	115	110
95	85	109	115	71
97	110	65	90	
75	120	80	70	

39. Home runs again. Students were asked to make a histogram of the number of home runs hit by Mark McGwire from 1986 to 2001 (see Exercise 35). One student submitted the following display:

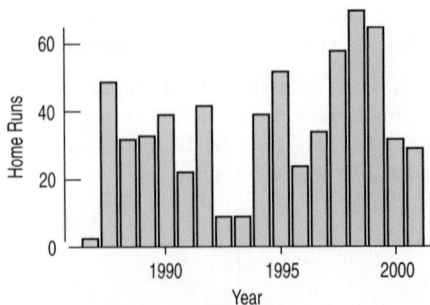

a) Comment on this graph.
b) Create your own histogram of the data.

40. Return of the birds. Students were given the assignment to make a histogram of the data on bird counts reported in Exercise 36. One student submitted the following display:

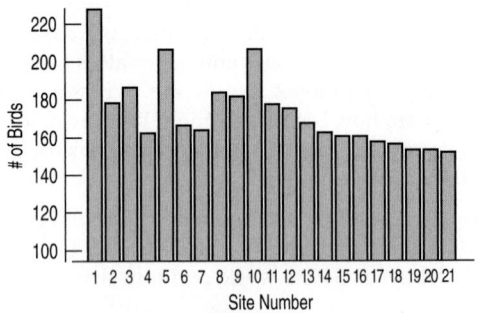

a) Comment on this graph.
b) Create your own histogram of the data.

41. Acid rain. Two researchers measured the pH (a scale on which a value of 7 is neutral and values below 7 are acidic) of water collected from rain and snow over a 6-month period in Allegheny County, PA. Describe their data with a graph and a few sentences:

4.57	5.62	4.12	5.29	4.64	4.31	4.30	4.39	4.45
5.67	4.39	4.52	4.26	4.26	4.40	5.78	4.73	4.56
5.08	4.41	4.12	5.51	4.82	4.63	4.29	4.60	

42. Marijuana 2003. In 2003 the Council of Europe published a report entitled *The European School Survey Project on Alcohol and Other Drugs* (www.espad.org). Among other issues, the survey investigated the percentages of 16-year-olds who had used marijuana. Shown here are the results for 20 European countries. Create an appropriate graph of these data, and describe the distribution.

Country	Percentage	Country	Percentage
Austria	21%	Italy	27%
Belgium	32%	Latvia	16%
Bulgaria	21%	Lithuania	13%
Croatia	22%	Malta	10%
Cyprus	4%	Netherlands	28%
Czech		Norway	9%
Republic	44%	Poland	18%
Denmark	23%	Portugal	15%
Estonia	23%	Romania	3%
Faroe		Russia	22%
Islands	9%	Slovak	
Finland	11%	Republic	27%
France	22%	Slovenia	28%
Germany	27%	Sweden	7%
Greece	6%	Switzerland	40%
Greenland	27%	Turkey	4%
Hungary	16%	Ukraine	21%
Iceland	13%	United	
Ireland	39%	Kingdom	38%
Isle of Man	39%		

43. Final grades. A professor (of something other than Statistics!) distributed the following histogram to show the distribution of grades on his 200-point final exam. Comment on the display.

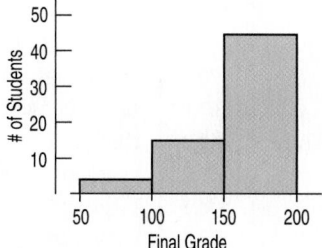

44. Final grades revisited. After receiving many complaints about his final-grade histogram from students currently taking a Statistics course, the professor from Exercise 43 distributed the following revised histogram:

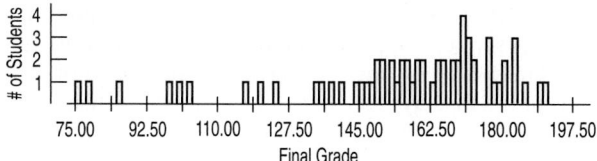

a) Comment on this display.
b) Describe the distribution of grades.

45. Zip codes. Holes-R-Us, an Internet company that sells piercing jewelry, keeps transaction records on its sales. At a recent sales meeting, one of the staff presented a histogram of the zip codes of the last 500 customers, so that the staff might understand where sales are coming from. Comment on the usefulness and appropriateness of the display.

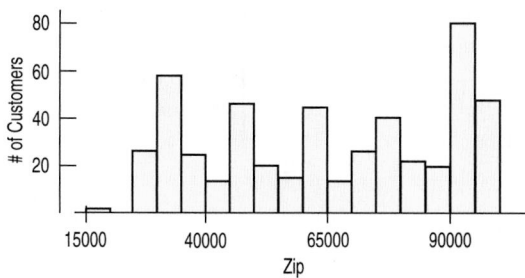

46. Zip codes revisited. Here are some summary statistics to go with the histogram of the zip codes of 500 customers from the Holes-R-Us Internet Jewelry Salon that we saw in Exercise 45:

Count	500
Mean	64,970.0
StdDev	23,523.0
Median	64,871
IQR	44,183
Q1	46,050
Q3	90,233

What can these statistics tell you about the company's sales?

47. Math scores 2005. The National Center for Education Statistics (http://nces.ed.gov/nationsreportcard/) reported 2005 average mathematics achievement scores for eighth graders in all 50 states:

State	Score	State	Score
Alabama	225	Montana	241
Alaska	236	Nebraska	238
Arizona	230	Nevada	230
Arkansas	236	New Hampshire	246
California	230	New Jersey	244
Colorado	239	New Mexico	224
Connecticut	242	New York	238
Delaware	240	North Carolina	241
Florida	239	North Dakota	243
Georgia	234	Ohio	242
Hawaii	230	Oklahoma	234
Idaho	242	Oregon	238
Illinois	233	Pennsylvania	241
Indiana	240	Rhode Island	233
Iowa	240	South Carolina	238
Kansas	246	South Dakota	242
Kentucky	231	Tennessee	232
Louisiana	230	Texas	242
Maine	241	Utah	239
Maryland	238	Vermont	244
Massachusetts	247	Virginia	240
Michigan	238	Washington	242
Minnesota	246	West Virginia	231
Mississippi	227	Wisconsin	241
Missouri	235	Wyoming	243

a) Find the median, the IQR, the mean, and the standard deviation of these state averages.
b) Which summary statistics would you report for these data? Why?
c) Write a brief summary of the performance of eighth graders nationwide.

48. Boomtowns. In 2006, *Inc.* magazine (www.inc.com) listed its choice of "boomtowns" in the United States—larger cities that are growing rapidly. Here is the magazine's top 20, along with their job growth percentages:

City	1-Year Job Growth (%)
Las Vegas, NV	7.5
Fort Lauderdale, FL	4.2
Orlando, FL	4.5
West Palm Beach-Boca Raton, FL	3.4
San Bernadino-Riverside, CA	1.9
Phoenix, AZ	4.4
Northern Virginia, VA	3.1
Washington, DC-Arlington-Alexandria, VA	3.2
Tampa-St. Petersburg, FL	2.6
Camden-Burlington counties, NJ	2.6

(*continued*)

City	1-Year Job Growth (%)
Jacksonville, FL	2.6
Charlotte, NC	3.3
Raleigh-Cary, NC	2.8
Richmond, VA	2.9
Salt Lake City, UT	3.3
Putnam-Rockland-Westchester counties, New York	2.3
Santa Ana-Anaheim-Irvine, CA	1.7
Miami-Miami Beach, FL	2.2
Sacramento, CA	1.5
San Diego, CA	1.4

a) Make a suitable display of the growth rates.
b) Summarize the typical growth rate among these cities with a median and mean. Why do they differ?
c) Given what you know about the distribution, which of the measures in b) does the better job of summarizing the growth rates? Why?
d) Summarize the spread of the growth rate distribution with a standard deviation and with an IQR.
e) Given what you know about the distribution, which of the measures in d) does the better job of summarizing the growth rates? Why?
f) Suppose we subtract from each of the preceding growth rates the predicted U.S. average growth rate of 1.20%, so that we can look at how much these growth rates exceed the U.S. rate. How would this change the values of the summary statistics you calculated above? (*Hint:* You need not recompute any of the summary statistics from scratch.)
g) If we were to omit Las Vegas from the data, how would you expect the mean, median, standard deviation, and IQR to change? Explain your expectations for each.
h) Write a brief report about all of these growth rates.

T 49. **Gasoline usage 2004.**   The California Energy Commission (www.energy.ca.gov/gasoline/) collects data on the amount of gasoline sold in each state. The following data show the per capita (gallons used per person) consumption in the year 2004. Using appropriate graphical displays and summary statistics, write a report on the gasoline use by state in the year 2004.

State	Gallons per Capita	State	Gallons per Capita
Alabama	529.4	Hawaii	358.7
Alaska	461.7	Idaho	454.8
Arizona	381.9	Illinois	408.3
Arkansas	512.0	Indiana	491.7
California	414.4	Iowa	555.1
Colorado	435.7	Kansas	511.8
Connecticut	435.7	Kentucky	526.6
Delaware	541.6	Louisiana	507.8
Florida	496.0	Maine	576.3
Georgia	537.1	Maryland	447.5

State	Gallons per Capita	State	Gallons per Capita
Massachusetts	458.5	Oklahoma	614.2
Michigan	482.0	Oregon	418.4
Minnesota	527.7	Pennsylvania	386.8
Mississippi	558.5	Rhode Island	454.6
Missouri	550.5	South Carolina	578.6
Montana	544.4	South Dakota	564.4
Nebraska	470.1	Tennessee	552.5
Nevada	367.9	Texas	532.7
New Hampshire	544.4	Utah	460.6
New Jersey	488.2	Vermont	545.5
New Mexico	508.8	Virginia	526.9
New York	293.4	Washington	423.6
North Carolina	505.0	West Virginia	426.7
North Dakota	553.7	Wisconsin	449.8
Ohio	451.1	Wyoming	615.0

T 50. **Prisons 2005.**   A report from the U.S. Department of Justice (www.ojp.usdoj.gov/bjs/) reported the percent changes in federal prison populations in 21 northeastern and midwestern states during 2005. Using appropriate graphical displays and summary statistics, write a report on the changes in prison populations.

State	Percent Change	State	Percent Change
Connecticut	−0.3	Iowa	2.5
Maine	0.0	Kansas	1.1
Massachusetts	5.5	Michigan	1.4
New Hampshire	3.3	Minnesota	6.0
New Jersey	2.2	Missouri	−0.8
New York	−1.6	Nebraska	7.9
Pennsylvania	3.5	North Dakota	4.4
Rhode Island	6.5	Ohio	2.3
Vermont	5.6	South Dakota	11.9
Illinois	2.0	Wisconsin	−1.0
Indiana	1.9		

JUST CHECKING
Answers

(Thoughts will vary.)

1. Roughly symmetric, slightly skewed to the right. Center around 3 miles? Few over 10 miles.

2. Bimodal. Center between 1 and 2 hours? Many people watch no football; others watch most of one or more games. Probably only a few values over 5 hours.

3. Strongly skewed to the right, with almost everyone at $0; a few small prizes, with the winner an outlier.

4. Fairly symmetric, somewhat uniform, perhaps slightly skewed to the right. Center in the 40s? Few ages below 25 or above 70.

5. Uniform, symmetric. Center near 5. Roughly equal counts for each digit 0–9.

6. Incomes are probably skewed to the right and not symmetric, making the median the more appropriate measure of center. The mean will be influenced by the high end of family incomes and not reflect the "typical" family income as well as the median would. It will give the impression that the typical income is higher than it is.

7. An IQR of 30 mpg would mean that only 50% of the cars get gas mileages in an interval 30 mpg wide. Fuel economy doesn't vary that much. 3 mpg is reasonable. It seems plausible that 50% of the cars will be within about 3 mpg of each other. An IQR of 0.3 mpg would mean that the gas mileage of half the cars varies little from the estimate. It's unlikely that cars, drivers, and driving conditions are that consistent.

8. We'd prefer a standard deviation of 2 months. Making a consistent product is important for quality. Customers want to be able to count on the MP3 player lasting somewhere close to 5 years, and a standard deviation of 2 years would mean that lifespans were highly variable.

Understanding and Comparing Distributions

WHO	Days during 1989
WHAT	Average daily wind speed (mph), Average barometric pressure (mb), Average daily temperature (deg Celsius)
WHEN	1989
WHERE	Hopkins Forest, in Western Massachusetts
WHY	Long-term observations to study ecology and climate

The Hopkins Memorial Forest is a 2500-acre reserve in Massachusetts, New York, and Vermont managed by the Williams College Center for Environmental Studies (CES). As part of their mission, CES monitors forest resources and conditions over the long term. They post daily measurements at their Web site.[1] You can go there, download, and analyze data for any range of days. We'll focus for now on 1989. As we'll see, some interesting things happened that year.

One of the variables measured in the forest is wind speed. Three remote anemometers generate far too much data to report, so, as summaries, you'll find the minimum, maximum, and average wind speed (in mph) for each day.

Wind is caused as air flows from areas of high pressure to areas of low pressure. Centers of low pressure often accompany storms, so both high winds and low pressure are associated with some of the fiercest storms. Wind speeds can vary greatly during a day and from day to day, but if we step back a bit farther, we can see patterns. By modeling these patterns, we can understand things about *Average Wind Speed* that we may not have known.

In Chapter 3 we looked at the association between two categorical variables using contingency tables and displays. Here we'll explore different ways of examining the relationship between two variables when one is quantitative, and the other is categorical and indicates groups to compare. We are given wind speed averages for each day of 1989. But we can collect the days together into different size groups and compare the wind speeds among them. If we consider *Time* as a categorical variable in this way, we'll gain enormous flexibility for our analysis and for our understanding. We'll discover new insights as we change the granularity of the grouping variable—from viewing the whole year's data at one glance, to comparing seasons, to looking for patterns across months, and, finally, to looking at the data day by day.

---

[1] www.williams.edu/CES/hopkins.htm

The Big Picture

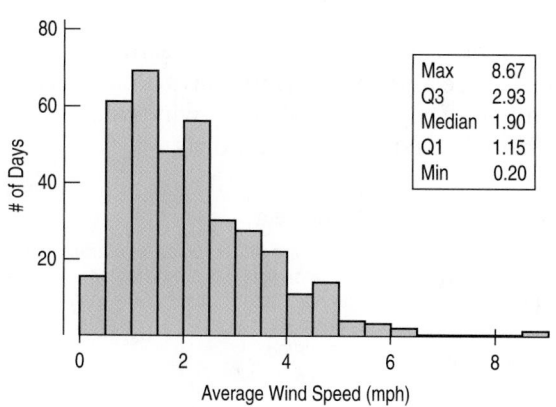

Max	8.67
Q3	2.93
Median	1.90
Q1	1.15
Min	0.20

Let's start with the "big picture." Here's a histogram and 5-number summary of the *Average Wind Speed* for every day in 1989. Because of the skewness, we'll report the median and IQR. We can see that the distribution of *Average Wind Speed* is unimodal and skewed to the right. Median daily wind speed is about 1.90 mph, and on half of the days, the average wind speed is between 1.15 and 2.93 mph. We also see a rather windy 8.67-mph day. Was that unusually windy or just the windiest day of the year? To answer that, we'll need to work with the summaries a bit more.

FIGURE 5.1

A histogram of daily Average Wind Speed *for 1989. It is unimodal and skewed to the right, with a possible high outlier.*

Boxplots and 5-Number Summaries

Once we have a 5-number summary of a (quantitative) variable, we can display that information in a **boxplot.** To make a boxplot of the average wind speeds, follow these steps:

1. Draw a single vertical axis spanning the extent of the data.[2] Draw short horizontal lines at the lower and upper quartiles and at the median. Then connect them with vertical lines to form a box. The box can have any width that looks OK.[3]
2. To help us construct the boxplot, we erect "fences" around the main part of the data. We place the upper fence 1.5 IQRs above the upper quartile and the lower fence 1.5 IQRs below the lower quartile. For the wind speed data, we compute

$$Upper\ fence = Q3 + 1.5\ IQR = 2.93 + 1.5 \times 1.78 = 5.60\ mph$$

and

$$Lower\ fence = Q1 - 1.5\ IQR = 1.15 - 1.5 \times 1.78 = -1.52\ mph$$

The fences are just for construction and are not part of the display. We show them here with dotted lines for illustration. You should never include them in your boxplot.

3. We use the fences to grow "whiskers." Draw lines from the ends of the box up and down to *the most extreme data values found within the fences.* If a data value falls outside one of the fences, we do *not* connect it with a whisker.
4. Finally, we add the **outliers** by displaying any data values beyond the fences with special symbols. (We often use a different symbol for "**far outliers**"—data values farther than 3 IQRs from the quartiles.)

What does a boxplot show? The center of a boxplot is (remarkably enough) a box that shows the middle half of the data, between the quartiles. The height of the box is equal to the IQR. If the median is roughly centered between the quartiles, then the middle half of the data is roughly symmetric. If the median is not centered, the distribution is skewed. The whiskers show skewness as well if they are not roughly the same length. Any outliers are displayed individually, both to keep them out of the way for judging skewness and to encourage you to give them special attention. They may be mistakes, or they may be the most interesting cases in your data.

A S **Boxplots.** Watch a boxplot under construction.

TI-*nspire*

Boxplots and dotplots. Drag data points around to explore what a boxplot shows (and doesn't).

---

[2] The axis could also run horizontally.

[3] Some computer programs draw wider boxes for larger data sets. That can be useful when comparing groups.

The prominent statistician John W. Tukey, the originator of the boxplot, was asked by one of the authors why the outlier nomination rule cut at 1.5 IQRs beyond each quartile. He answered that the reason was that 1 IQR would be too small and 2 IQRs would be too large. That works for us.

For the Hopkins Forest data, the central box contains each day whose *Average Wind Speed* is between 1.15 and 2.93 miles per hour (see Figure 5.2). From the shape of the box, it looks like the central part of the distribution of wind speeds is roughly symmetric, but the longer upper whisker indicates that the distribution stretches out at the upper end. We also see a few very windy days. Boxplots are particularly good at pointing out outliers. These extraordinarily windy days may deserve more attention. We'll give them that extra attention shortly.

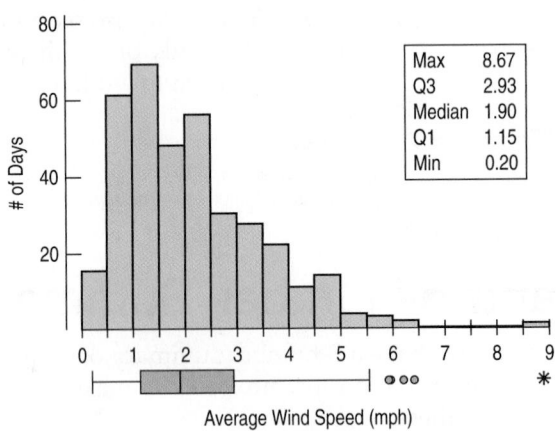

FIGURE 5.2

By turning the boxplot and putting it on the same scale as the histogram, we can compare both displays of the daily wind speeds and see how each represents the distribution.

Max	8.67
Q3	2.93
Median	1.90
Q1	1.15
Min	0.20

A S *Activity:* **Playing with Summaries.** See how different summary measures behave as you place and drag values, and see how sensitive some statistics are to individual data values.

Comparing Groups with Histograms

TI-*nspire*

Histograms and boxplots. See that the shape of a distribution is not always evident in a boxplot.

It is almost always more interesting to compare groups. Is it windier in the winter or the summer? Are any months particularly windy? Are weekends a special problem? Let's split the year into two groups: April through September (Spring/Summer) and October through March (Fall/Winter). To compare the groups, we create two histograms, being careful to use the same scale. Here are displays of the average daily wind speed for Spring/Summer (on the left) and Fall/Winter (on the right):

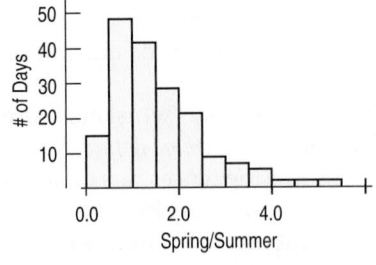
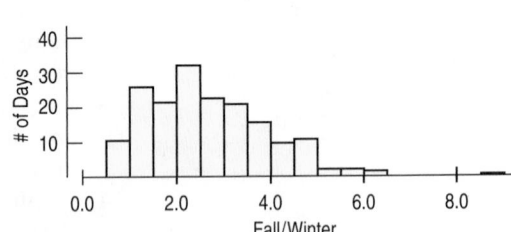

FIGURE 5.3

Histograms of Average Wind Speed *for days in Spring/Summer (left) and Fall/Winter (right) show very different patterns.*

The shapes, centers, and spreads of these two distributions are strikingly different. During spring and summer (histogram on the left), the distribution is skewed to the right. A typical day during these warmer months has an average wind speed of only 1 to 2 mph, and few have average speeds above 3 mph. In the colder months (histogram on the right), however, the shape is less strongly skewed and more spread out. The typical wind speed is higher, and days with average wind speeds above 3 mph are not unusual. There are several noticeable high values.

Summaries for *Average Wind Speed* by Season				
Group	Mean	StdDev	Median	IQR
Fall/Winter	2.71	1.36	2.47	1.87
Spring/Summer	1.56	1.01	1.34	1.32

Comparing groups with stem-and-leaf displays

In 2004 the infant death rate in the United States was 6.8 deaths per 1000 live births. The Kaiser Family Foundation collected data from all 50 states and the District of Columbia, allowing us to look at different regions of the country. Since there are only 51 data values, a back-to-back stem-and-leaf plot is an effective display. Here's one comparing infant death rates in the Northeast and Midwest to those in the South and West. In this display the stems run down the middle of the plot, with the leaves for the two regions to the left or right. Be careful when you read the values on the left: 4 | 11 | means a rate of 11.4 deaths per 1000 live birth for one of the southern or western states.

Question: How do infant death rates compare for these regions?

In general, infant death rates were generally higher for states in the South and West than in the Northeast and Midwest. The distribution for the northeastern and midwestern states is roughly uniform, varying from a low of 4.8 to a high of 8.1 deaths per 1000 live births. Ten southern and western states had higher infant death rates than any in the Northeast or Midwest, with one state over 11. Rates varied more widely in the South and West, where the distribution is skewed to the right and possibly bimodal. We should investigate further to see which states represent the cluster of high death rates.

Infant Death Rates (by state) 2004

South and West		North and Midwest
4	11	
3 0	10	
0 0	9	
0 4 1 6 9 5 8	8	1 0
0 5 0 3	7	5 8 0 7 4 1
4 1 0 4 9 1 1 6 4	6	3 1 5 4 4
6 3 6 2	5	8 4 0 6
	4	8 8 9 7
	3	

(4 |11| means 11.4 deaths per 1000 live births)

Comparing Groups with Boxplots

Are some months windier than others? Even residents may not have a good idea of which parts of the year are the most windy. (Do you know for your hometown?) We're not interested just in the centers, but also in the spreads. Are wind speeds equally variable from month to month, or do some months show more variation?

Earlier, we compared histograms of the wind speeds for two halves of the year. To look for seasonal trends, though, we'll group the daily observations by month. Histograms or stem-and-leaf displays are a fine way to look at one distribution or two. But it would be hard to see patterns by comparing 12 histograms. Boxplots offer an ideal balance of information and simplicity, hiding the details while displaying the overall summary information. So we often plot them side by side for groups or categories we wish to compare.

By placing boxplots side by side, we can easily see which groups have higher medians, which have the greater IQRs, where the central 50% of the data is located in each group, and which have the greater overall range. And, when the boxes are in an order, we can get a general idea of patterns in both the centers and the spreads. Equally important, we can see past any outliers in making these comparisons because they've been displayed separately.

Here are boxplots of the *Average Daily Wind Speed* by month:

FIGURE 5.4

Boxplots of the average daily wind speed for each month show seasonal patterns in both the centers and spreads.

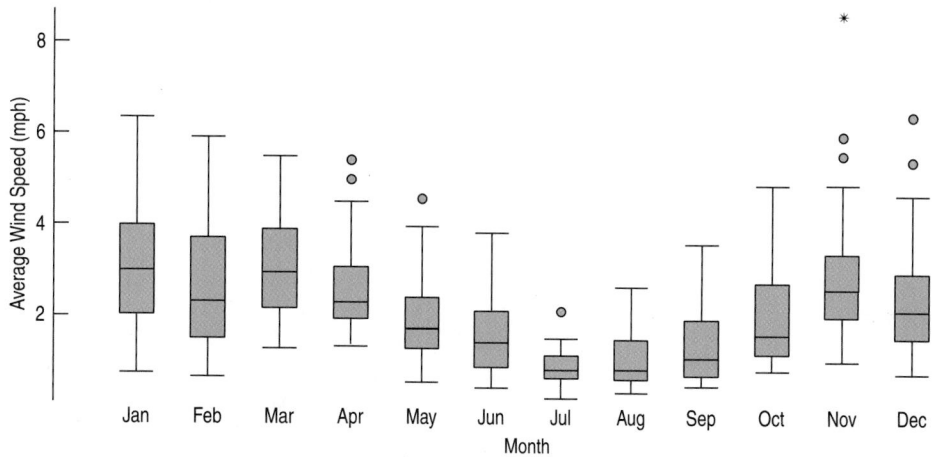

Here we see that wind speeds tend to decrease in the summer. The months in which the winds are both strongest and most variable are November through March. And there was one remarkably windy day in November.

When we looked at a boxplot of wind speeds for the entire year, there were only 5 outliers. Now, when we group the days by *Month*, the boxplots display more days as outliers and call out one in November as a far outlier. The boxplots show different outliers than before because some days that seemed ordinary when placed against the entire year's data looked like outliers for the month that they're in. That windy day in July certainly wouldn't stand out in November or December, but for July, it was remarkable.

FOR EXAMPLE | Comparing distributions

Roller coasters[4] are a thrill ride in many amusement parks worldwide. And thrill seekers want a coaster that goes fast. There are two main types of roller coasters: those with wooden tracks and those with steel tracks. Do they typically run at different speeds? Here are boxplots:

Question: Compare the speeds of wood and steel roller coasters.

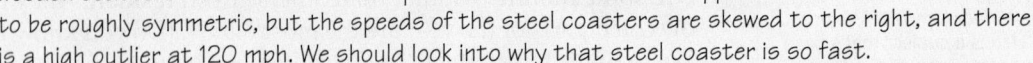

Overall, wooden-track roller coasters are slower than steel-track coasters. In fact, the fastest half of the steel coasters are faster than three quarters of the wooden coasters. Although the IQRs of the two groups are similar, the range of speeds among steel coasters is larger than the range for wooden coasters. The distribution of speeds of wooden coasters appears to be roughly symmetric, but the speeds of the steel coasters are skewed to the right, and there is a high outlier at 120 mph. We should look into why that steel coaster is so fast.

STEP-BY-STEP EXAMPLE | **Comparing Groups**

Of course, we can compare groups even when they are not in any particular order. Most scientific studies compare two or more groups. It is almost always a good idea to start an analysis of data from such studies by comparing boxplots for the groups. Here's an example:

For her class project, a student compared the efficiency of various coffee containers. For her study, she decided to try 4 different containers and to test each of them 8 different times. Each time, she heated water to 180°F, poured it into a container, and sealed it. (We'll learn the details of how to set up experiments in Chapter 13.) After 30 minutes, she measured the temperature again and recorded the difference in temperature. Because these are temperature differences, smaller differences mean that the liquid stayed hot—just what we would want in a coffee mug.

Question: What can we say about the effectiveness of these four mugs?

---

[4] See the Roller Coaster Data Base at www.rcdb.com.

Plan State what you want to find out.

Variables Identify the *variables* and report the W's.

Be sure to check the appropriate condition.

I want to compare the effectiveness of the different mugs in maintaining temperature. I have 8 measurements of Temperature Change for each of the mugs.

✔ **Quantitative Data Condition:** The Temperature Changes are quantitative, with units of °F. Boxplots are appropriate displays for comparing the groups. Numerical summaries of each group are appropriate as well.

Mechanics Report the 5-number summaries of the four groups. Including the IQR is a good idea as well.

	Min	Q1	Median	Q3	Max	IQR
CUPPS	6°F	6	8.25	14.25	18.50	8.25
Nissan	0	1	2	4.50	7	3.50
SIGG	9	11.50	14.25	21.75	24.50	10.25
Starbucks	6	6.50	8.50	14.25	17.50	7.75

Make a picture. Because we want to compare the distributions for four groups, boxplots are an appropriate choice.

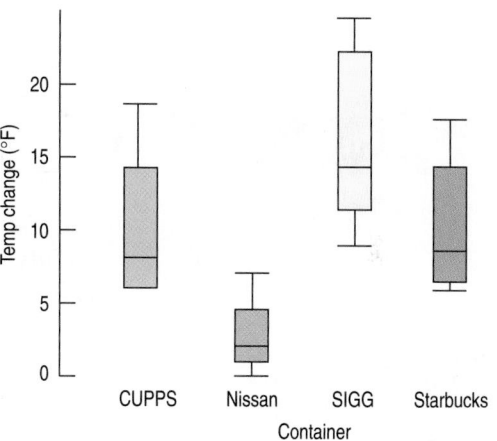

Conclusion Interpret what the boxplots and summaries say about the ability of these mugs to retain heat. Compare the shapes, centers, and spreads, and note any outliers.

The individual distributions of temperature changes are all slightly skewed to the high end. The Nissan cup does the best job of keeping liquids hot, with a median loss of only 2°F, and the SIGG cup does the worst, typically losing 14°F. The difference is large enough to be important: A coffee drinker would be likely to notice a 14° drop in temperature. And the mugs are clearly different: 75% of the Nissan tests showed less heat loss than any of the other mugs in the study. The IQR of results for the Nissan cup is also the smallest of these test cups, indicating that it is a consistent performer.

JUST CHECKING

The Bureau of Transportation Statistics of the U.S. Department of Transportation collects and publishes statistics on airline travel (www.transtats.bts.gov). Here are three displays of the % of flights arriving late each month from 1995 through 2005:

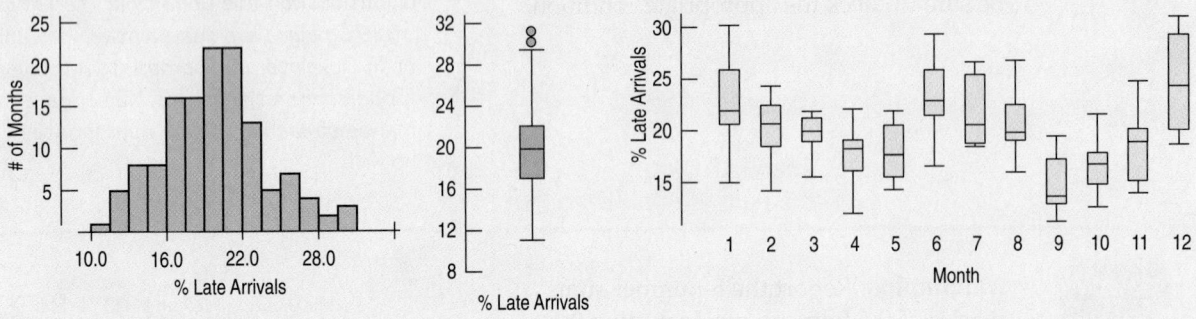

1. Describe what the histogram says about late arrivals.

2. What does the boxplot of late arrivals suggest that you can't see in the histogram?

3. Describe the patterns shown in the boxplots by month. At what time of year are flights least likely to be late? Can you suggest reasons for this pattern?

TI Tips | **Comparing groups with boxplots**

In the last chapter we looked at the performances of fourth-grade students on an agility test. Now let's make comparative boxplots for the boys' scores and the girls' scores:

Boys: 22, 17, 18, 29, 22, 22, 23, 24, 23, 17, 21

Girls: 25, 20, 12, 19, 28, 24, 22, 21, 25, 26, 25, 16, 27, 22

Enter these data in L1 (*Boys*) and L2 (*Girls*).

Set up STATPLOT's Plot1 to make a boxplot of the boys' data:

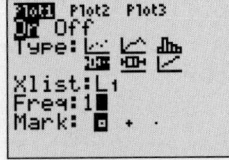

- Turn the plot On;
- Choose the first boxplot icon (you want your plot to indicate outliers);
- Specify Xlist:L1 and Freq:1, and select the Mark you want the calculator to use for displaying any outliers.

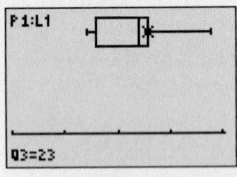

Use ZoomStat to display the boxplot for *Boys*. You can now TRACE to see the statistics in the five-number summary. Try it!

As you did for the boys, set up Plot2 to display the girls' data. This time when you use ZoomStat with both plots turned on, the display shows the parallel boxplots. See the outlier?

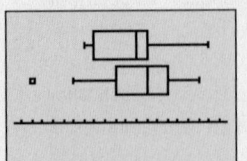

This is a great opportunity to practice your "Tell" skills. How do these fourth graders compare in terms of agility?

Outliers

When we looked at boxplots for the *Average Wind Speed* by *Month*, we noticed that several days stood out as possible outliers and that one very windy day in November seemed truly remarkable. What should we do with such outliers?

Cases that stand out from the rest of the data almost always deserve our attention. An outlier is a value that doesn't fit with the rest of the data, but exactly how different it should be to be treated specially is a judgment call. Boxplots provide a rule of thumb to highlight these unusual points, but that rule doesn't tell you what to do with them.

So, what *should* we do with outliers? The first thing to do is to try to understand them in the context of the data. A good place to start is with a histogram. Histograms show us more detail about a distribution than a boxplot can, so they give us a better idea of how the outlier fits (or doesn't fit) in with the rest of the data.

A histogram of the *Average Wind Speed* in November shows a slightly skewed main body of data and that very windy day clearly set apart from the other days. When considering whether a case is an outlier, we often look at the gap between that case and the rest of the data. A large gap suggests that the case really is quite different. But a case that just happens to be the largest or smallest value at the end of a possibly stretched-out tail may be best thought of as just . . . the largest or smallest value. After all, *some* case has to be the largest or smallest.

Some outliers are simply unbelievable. If a class survey includes a student who claims to be 170 inches tall (about 14 feet, or 4.3 meters), you can be pretty sure that's an error.

Once you've identified likely outliers, you should always investigate them. Some outliers are just errors. A decimal point may have been misplaced, digits transposed, or digits repeated or omitted. The units may be wrong. (Was that outlying height reported in centimeters rather than in inches [170 cm = 65 in.]?) Or a number may just have been transcribed incorrectly, perhaps copying an adjacent value on the original data sheet. If you can identify the correct value, then you should certainly fix it. One important reason to look into outliers is to correct errors in your data.

Many outliers are not wrong; they're just different. Such cases often repay the effort to understand them. You can learn more from the extraordinary cases than from summaries of the overall data set.

What about that windy November day? Was it really that windy, or could there have been a problem with the anemometers? A quick Internet search for weather on November 21, 1989, finds that there was a severe storm:

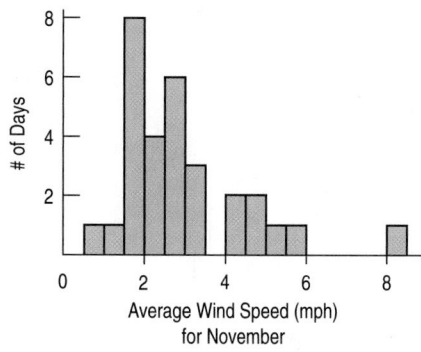

FIGURE 5.5

The Average Wind Speed *in November is slightly skewed with a high outlier.*

WIND, SNOW, COLD GIVE N.E. A TASTE OF WINTER

Published on November 22, 1989
Author: Andrew Dabilis, Globe Staff

An intense storm roared like the Montreal Express through New England yesterday, bringing frigid winds of up to 55 m.p.h., 2 feet of snow in some parts of Vermont and a preview of winter after weeks of mild weather. Residents throughout the region awoke yesterday to an icy vortex that lifted an airplane off the runway in Newark and made driving dangerous in New England because of rapidly shifting winds that seemed to come from all directions.

When we have outliers, we need to decide what to *Tell* about the data. If we can correct an error, we'll just summarize the corrected data (and note the correction). But if we see no way to correct an outlying value, or if we confirm that it is correct, our best path is to report summaries and analyses with *and* without the outlier. In this way a reader can judge for him- or herself what influence the outlier has and decide what to think about the data.

There are two things we should *never* do with outliers. The first is to silently leave an outlier in place and proceed as if nothing were unusual. Analyses of data with outliers are very likely to be influenced by those outliers—sometimes to a large and misleading degree. The other is to drop an outlier from the analysis without comment just because it's unusual. If you want to exclude an outlier, you must discuss your decision and, to the extent you can, justify your decision.

> **A S**   *Case Study:* **Are passengers or drivers safer in a crash?** Practice the skills of this chapter by comparing these two groups.

FOR EXAMPLE   Checking out the outliers

Recap: We've looked at the speeds of roller coasters and found a difference between steel- and wooden-track coasters. We also noticed an extraordinary value.

Question: The fastest coaster in this collection turns out to be the "Top Thrill Dragster" at Cedar Point amusement park. What might make this roller coaster unusual? You'll have to do some research, but that's often what happens with outliers.

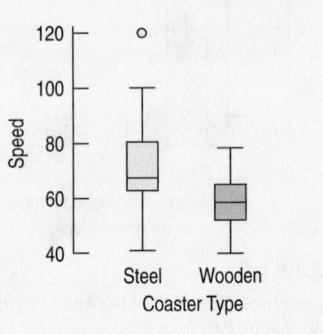

The Top Thrill Dragster is easy to find in an Internet search. We learn that it is a "hydraulic launch" coaster. That is, it doesn't get its remarkable speed just from gravity, but rather from a kick-start by a hydraulic piston. That could make it different from the other roller coasters.

(You might also discover that it is no longer the fastest roller coaster in the world.)

Timeplots: Order, Please!

The Hopkins Forest wind speeds are reported as daily averages. Previously, we grouped the days into months or seasons, but we could look at the wind speed values day by day. Whenever we have data measured over time, it is a good idea to look for patterns by plotting the data in time order. Here are the daily average wind speeds plotted over time:

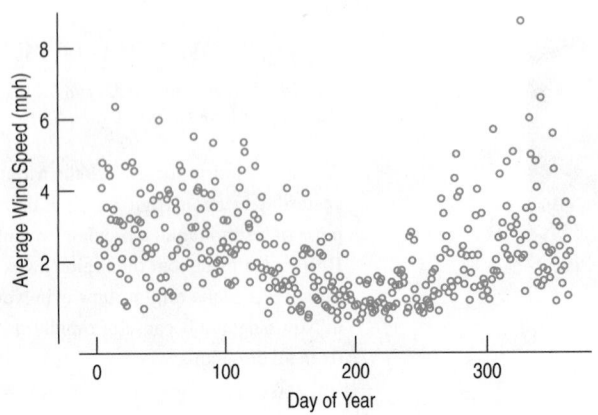

FIGURE 5.6

A timeplot of Average Wind Speed *shows the overall pattern and changes in variation.*

A display of values against time is sometimes called a **timeplot**. This timeplot reflects the pattern that we saw when we plotted the wind speeds by month. But without the arbitrary divisions between months, we can see a calm period during the summer, starting around day 200 (the middle of July), when the wind is relatively mild and doesn't vary greatly from day to day. We can also see that the wind becomes both more variable and stronger during the early and late parts of the year.

Looking into the Future

It is always tempting to try to extend what we see in a timeplot into the future. Sometimes that makes sense. Most likely, the Hopkins Forest climate follows regular seasonal patterns. It's probably safe to predict a less windy June next year and a windier November. But we certainly wouldn't predict another storm on November 21.

Other patterns are riskier to extend into the future. If a stock has been rising, will it continue to go up? No stock has ever increased in value indefinitely, and no stock analyst has consistently been able to forecast when a stock's value will turn around. Stock prices, unemployment rates, and other economic, social, or psychological concepts are much harder to predict than physical quantities. The path a ball will follow when thrown from a certain height at a given speed and direction is well understood. The path interest rates will take is much less clear. Unless we have strong (nonstatistical) reasons for doing otherwise, we should resist the temptation to think that any trend we see will continue, even into the near future.

Statistical models often tempt those who use them to think beyond the data. We'll pay close attention later in this book to understanding when, how, and how much we can justify doing that.

Re-expressing Data: A First Look

RE-EXPRESSING TO IMPROVE SYMMETRY

When the data are skewed, it can be hard to summarize them simply with a center and spread, and hard to decide whether the most extreme values are outliers or just part of the stretched-out tail. How can we say anything useful about such data? The secret is to *re-express* the data by applying a simple function to each value.

Many relationships and "laws" in the sciences and social sciences include functions such as logarithms, square roots, and reciprocals. Similar relationships often show up in data. Here's a simple example:

In 1980 large companies' chief executive officers (CEOs) made, on average, about 42 times what workers earned. In the next two decades, CEO compensation soared when compared to the average worker. By 2000 that multiple had jumped[5]

---

[5] *Sources:* United for a Fair Economy, *Business Week* annual CEO pay surveys, Bureau of Labor Statistics, "Average Weekly Earnings of Production Workers, Total Private Sector." Series ID: EEU00500004.

to 525. What does the distribution of the compensation of Fortune 500 companies' CEOs look like? Here's a histogram and boxplot for 2005 compensation:

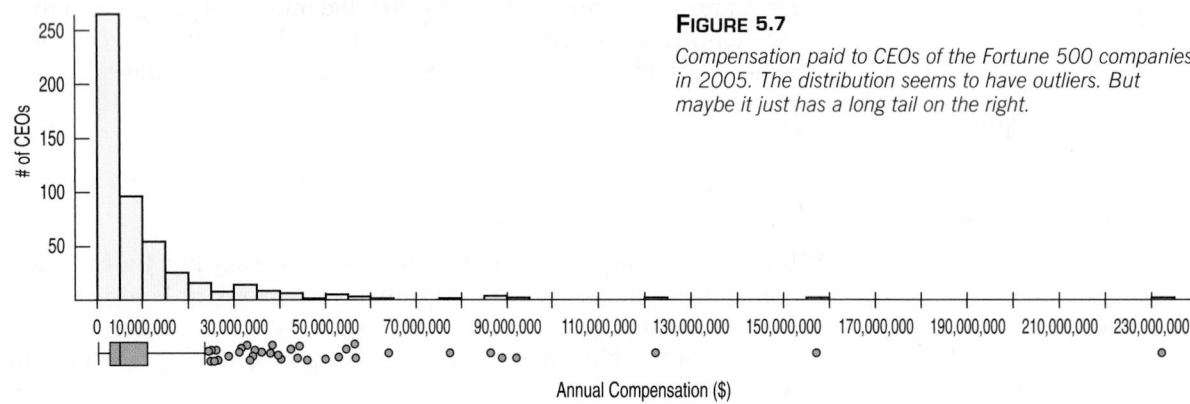

FIGURE 5.7

Compensation paid to CEOs of the Fortune 500 companies in 2005. The distribution seems to have outliers. But maybe it just has a long tail on the right.

We have 500 CEOs and about 48 possible histogram bins, most of which are empty—but don't miss the tiny bars straggling out to the right. The boxplot indicates that some CEOs received extraordinarily high compensations, while the majority received relatively "little." But look at the values of the bins. The first bin, with about half the CEOs, covers incomes from $0 to $5,000,000. Imagine receiving a salary survey with these categories:

> What is your income?
> a) $0 to $5,000,000
> b) $5,000,001 to $10,000,000
> c) $10,000,001 to $15,000,000
> d) More than $15,000,000

The reason that the histogram seems to leave so much of the area blank is that the salaries are spread all along the axis from about $15,000,000 to $240,000,000. After $50,000,000 there are so few for each bin that it's very hard to see the tiny bars. What we *can* see from this histogram and boxplot is that this distribution is highly skewed to the right.

It can be hard to decide what we mean by the "center" of a skewed distribution, so it's hard to pick a typical value to summarize the distribution. What would you say was a typical CEO total compensation? The mean value is $10,307,000, while the median is "only" $4,700,000. Each tells us something different about the data.

One approach is to **re-express,** or **transform,** the data by applying a simple function to make the skewed distribution more symmetric. For example, we could take the square root or logarithm of each compensation value. Taking logs works pretty well for the CEO compensations, as you can see:

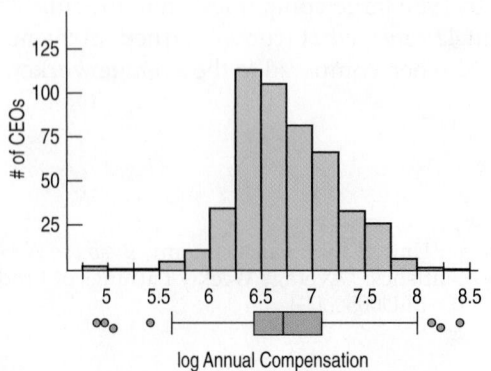

FIGURE 5.8

The logarithms of 2005 CEO compensations are much more nearly symmetric.

The histogram of the logs of the total CEO compensations is much more nearly symmetric, so we can see that a typical log compensation is between 6, which corresponds to $1,000,000, and 7, corresponding to $10,000,000. And it's easier to talk about a typical value for the logs. The mean log compensation is 6.73, while the median is 6.67. (That's $5,370,317 and $4,677,351, respectively.) Notice that nearly all the values are between 6.0 and 8.0—in other words, between $1,000,000 and $100,000,000 a year, but who's counting?

Against the background of a generally symmetric main body of data, it's easier to decide whether the largest compensations are outliers. In fact, the three most highly compensated CEOs are identified as outliers by the boxplot rule of thumb even after this re-expression. It's perhaps impressive to be an outlier CEO in annual compensation. It's even more impressive to be an outlier in the log scale!

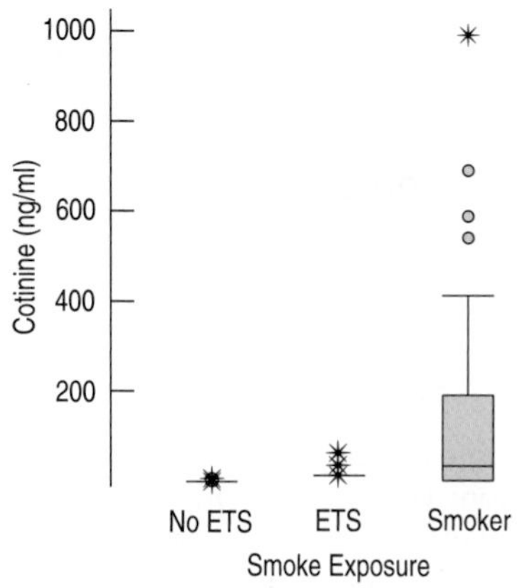

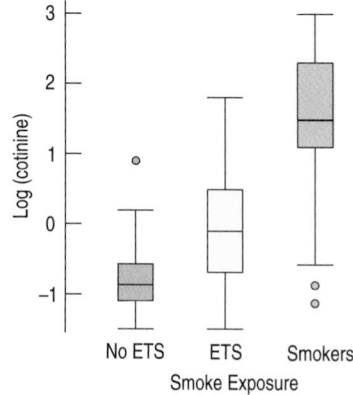

FIGURE 5.9

Cotinine levels (nanograms per milliliter) for three groups with different exposures to tobacco smoke. Can you compare the ETS (exposed to smoke) and No-ETS groups?

> **Dealing with logarithms** You have probably learned about logs in math courses and seen them in psychology or science classes. In this book, we use them only for making data behave better. Base 10 logs are the easiest to understand, but natural logs are often used as well. (Either one is fine.) You can think of base 10 logs as roughly one less than the number of digits you need to write the number. So 100, which is the smallest number to require 3 digits, has a $\log_{10}$ of 2. And 1000 has a $\log_{10}$ of 3. The $\log_{10}$ of 500 is between 2 and 3, but you'd need a calculator to find that it's approximately 2.7. All salaries of "six figures" have $\log_{10}$ between 5 and 6. Logs are incredibly useful for making skewed data more symmetric. But don't worry—nobody does logs without technology and neither should you. Often, remaking a histogram or other display of the data is as easy as pushing another button.

RE-EXPRESSING TO EQUALIZE SPREAD ACROSS GROUPS

Researchers measured the concentration (nanograms per milliliter) of cotinine in the blood of three groups of people: nonsmokers who have not been exposed to smoke, nonsmokers who have been exposed to smoke (ETS), and smokers. Cotinine is left in the blood when the body metabolizes nicotine, so this measure gives a direct measurement of the effect of passive smoke exposure. The boxplots of the cotinine levels of the three groups tell us that the smokers have higher cotinine levels, but if we want to compare the levels of the passive smokers to those of the nonsmokers, we're in trouble, because on this scale, the cotinine levels for both nonsmoking groups are too low to be seen.

Re-expressing can help alleviate the problem of comparing groups that have very different spreads. For measurements like the cotinine data, whose values can't be negative and whose distributions are skewed to the high end, a good first guess at a re-expression is the logarithm.

After taking logs, we can compare the groups and see that the nonsmokers exposed to environmental smoke (the ETS group) do show increased levels of (log) cotinine, although not the high levels found in the blood of smokers.

Notice that the same re-expression has also improved the symmetry of the cotinine distribution for smokers and pulled in most of the apparent outliers in all of the groups. It is not unusual for a re-expression that improves one aspect of data to improve others as well. We'll talk about other ways to re-express data as the need arises throughout the book. We'll explore some common re-expressions more thoroughly in Chapter 10.

FIGURE 5.10

Blood cotinine levels after taking logs. What a difference a log makes!

WHAT CAN GO WRONG?

▶ **Avoid inconsistent scales.** Parts of displays should be mutually consistent—no fair changing scales in the middle or plotting two variables on different scales but on the same display. When comparing two groups, be sure to compare them on the same scale.

▶ **Label clearly.** Variables should be identified clearly and axes labeled so a reader knows what the plot displays.

Here's a remarkable example of a plot gone wrong. It illustrated a news story about rising college costs. It uses time-plots, but it gives a misleading impression. First think about the story you're being told by this display. Then try to figure out what has gone wrong.

What's wrong? Just about everything.

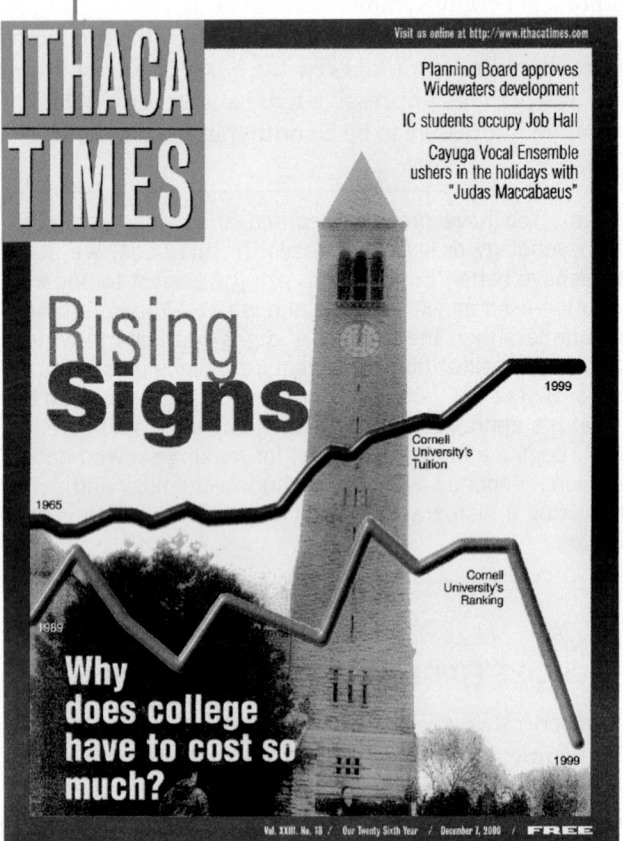

- The horizontal scales are inconsistent. Both lines show trends over time, but exactly for what years? The tuition sequence starts in 1965, but rankings are graphed from 1989. Plotting them on the same (invisible) scale makes it seem that they're for the same years.

- The vertical axis isn't labeled. That hides the fact that it's inconsistent. Does it graph dollars (of tuition) or ranking (of Cornell University)?

This display violates three of the rules. And it's even worse than that: It violates a rule that we didn't even bother to mention.

- The two inconsistent scales for the vertical axis don't point in the same direction! The line for Cornell's rank shows that it has "plummeted" from 15th place to 6th place in academic rank. Most of us think that's an *improvement*, but that's not the message of this graph.

▶ **Beware of outliers.** If the data have outliers and you can correct them, you should do so. If they are clearly wrong or impossible, you should remove them and report on them. Otherwise, consider summarizing the data both with and without the outliers.

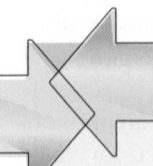

CONNECTIONS

We discussed the value of summarizing a distribution with shape, center, and spread in Chapter 4, and we developed several ways to measure these attributes. Now we've seen the value of comparing distributions for different groups and of looking at patterns in a quantitative variable measured over time. Although it can be interesting to summarize a single variable for a single group, it is almost always more interesting to compare groups and look for patterns across several groups and over time. We'll continue to make comparisons like these throughout the rest of our work.

WHAT HAVE WE LEARNED?

▸ We've learned the value of comparing groups and looking for patterns among groups and over time.
▸ We've seen that boxplots are very effective for comparing groups graphically. When we compare groups, we discuss their shape, center, and spreads, and any unusual features.
▸ We've experienced the value of identifying and investigating outliers. And we've seen that when we group data in different ways, it can allow different cases to emerge as possible outliers.
▸ We've graphed data that have been measured over time against a time axis and looked for long-term trends.

Terms

Boxplot
81. A boxplot displays the 5-number summary as a central box with whiskers that extend to the non-outlying data values. Boxplots are particularly effective for comparing groups and for displaying outliers.

Outlier
81, 87. Any point more than 1.5 IQR from either end of the box in a boxplot is nominated as an outlier.

Far Outlier
81. If a point is more than 3.0 IQR from either end of the box in a boxplot, it is nominated as a *far outlier*.

Comparing distributions
82. When comparing the distributions of several groups using histograms or stem-and-leaf displays, consider their:
▸ Shape
▸ Center
▸ Spread

Comparing boxplots
83. When comparing groups with boxplots:
▸ Compare the shapes. Do the boxes look symmetric or skewed? Are there differences between groups?
▸ Compare the medians. Which group has the higher center? Is there any pattern to the medians?
▸ Compare the IQRs. Which group is more spread out? Is there any pattern to how the IQRs change?
▸ Using the IQRs as a background measure of variation, do the medians seem to be different, or do they just vary much as you'd expect from the overall variation?
▸ Check for possible outliers. Identify them if you can and discuss why they might be unusual. Of course, correct them if you find that they are errors.

Timeplot
88. A timeplot displays data that change over time. Often, successive values are connected with lines to show trends more clearly. Sometimes a smooth curve is added to the plot to help show long-term patterns and trends.

Skills

▸ Be able to select a suitable display for comparing groups. Understand that histograms show distributions well, but are difficult to use when comparing more than two or three groups. Boxplots are more effective for comparing several groups, in part because they show much less information about the distribution of each group.

▸ Understand that how you group data can affect what kinds of patterns and relationships you are likely to see. Know how to select groupings to show the information that is important for your analysis.

▸ Be aware of the effects of skewness and outliers on measures of center and spread. Know how to select appropriate measures for comparing groups based on their displayed distributions.

▸ Understand that outliers can emerge at different groupings of data and that, whatever their source, they deserve special attention.

▸ Recognize when it is appropriate to make a timeplot.

▸ Know how to make side-by-side histograms on comparable scales to compare the distributions of two groups.

▸ Know how to make side-by-side boxplots to compare the distributions of two or more groups.

▸ Know how to describe differences among groups in terms of patterns and changes in their center, spread, shape, and unusual values.

▸ Know how to make a timeplot of data that have been measured over time.

▸ Know how to compare the distributions of two or more groups by comparing their shapes, centers, and spreads. Be prepared to explain your choice of measures of center and spread for comparing the groups.

▸ Be able to describe trends and patterns in the centers and spreads of groups—especially if there is a natural order to the groups, such as a time order.

▸ Be prepared to discuss patterns in a timeplot in terms of both the general trend of the data and the changes in how spread out the pattern is.

▸ Be cautious about assuming that trends over time will continue into the future.

▸ Be able to describe the distribution of a quantitative variable in terms of its shape, center, and spread.

▸ Be able to describe any anomalies or extraordinary features revealed by the display of a variable.

▸ Know how to compare the distributions of two or more groups by comparing their shapes, centers, and spreads.

▸ Know how to describe patterns over time shown in a timeplot.

▸ Be able to discuss any outliers in the data, noting how they deviate from the overall pattern of the data.

COMPARING DISTRIBUTIONS ON THE COMPUTER

Most programs for displaying and analyzing data can display plots to compare the distributions of different groups. Typically these are boxplots displayed side-by-side.

Side-by-side boxplots should be on the same y-axis scale so they can be compared.

Some programs offer a graphical way to assess how much the medians differ by drawing a band around the median or by "notching" the boxes.

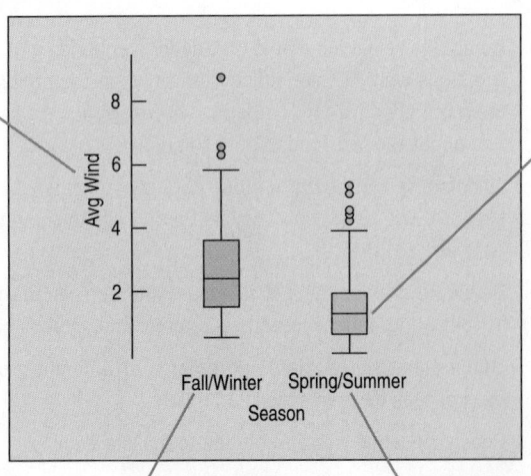

Boxes are typically labeled with a group name. Often they are placed in alphabetical order by group name—not the most useful order.

EXERCISES

1. **In the news.**   Find an article in a newspaper, magazine, or the Internet that compares two or more groups of data.
   a) Does the article discuss the W's?
   b) Is the chosen display appropriate? Explain.
   c) Discuss what the display reveals about the groups.
   d) Does the article accurately describe and interpret the data? Explain.

2. **In the news.**   Find an article in a newspaper, magazine, or the Internet that shows a time plot.
   a) Does the article discuss the W's?
   b) Is the timeplot appropriate for the data? Explain.
   c) Discuss what the timeplot reveals about the variable.
   d) Does the article accurately describe and interpret the data? Explain.

3. **Time on the Internet.**   Find data on the Internet (or elsewhere) that give results recorded over time. Make an appropriate display and discuss what it shows.

4. **Groups on the Internet.**   Find data on the Internet (or elsewhere) for two or more groups. Make appropriate displays to compare the groups, and interpret what you find.

5. **Pizza prices.**   A company that sells frozen pizza to stores in four markets in the United States (Denver, Baltimore, Dallas, and Chicago) wants to examine the prices that the stores charge for pizza slices. Here are boxplots comparing data from a sample of stores in each market:

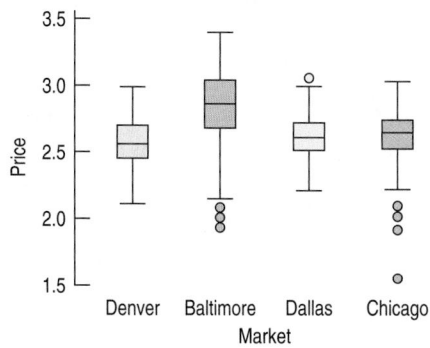

   a) Do prices appear to be the same in the four markets? Explain.
   b) Does the presence of any outliers affect your overall conclusions about prices in the four markets?

6. **Costs.**   To help travelers know what to expect, researchers collected the prices of commodities in 16 cities throughout the world. Here are boxplots comparing the prices of a ride on public transportation, a newspaper, and a cup of coffee in the 16 cities (prices are all in $US).

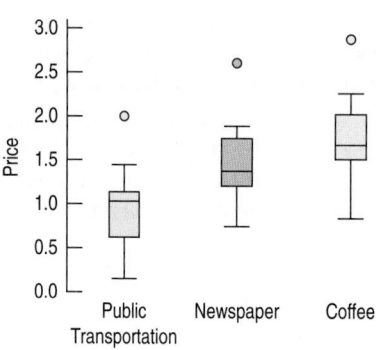

   a) On average, which commodity is the most expensive?
   b) Is a newspaper always more expensive than a ride on public transportation? Explain.
   c) Does the presence of outliers affect your conclusions in a) or b)?

7. **Still rockin'.**   Crowd Management Strategies monitors accidents at rock concerts. In their database, they list the names and other variables of victims whose deaths were attributed to "crowd crush" at rock concerts. Here are the histogram and boxplot of the victims' ages for data from 1999 to 2000:

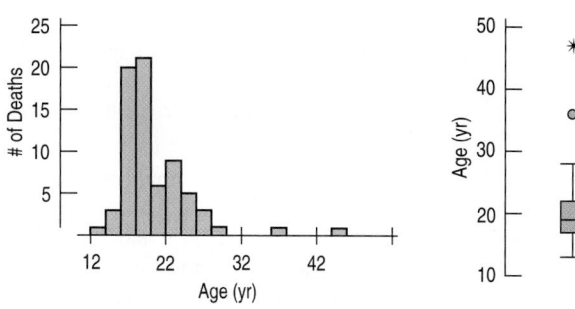

   a) What features of the distribution can you see in both the histogram and the boxplot?
   b) What features of the distribution can you see in the histogram that you could not see in the boxplot?
   c) What summary statistic would you choose to summarize the center of this distribution? Why?
   d) What summary statistic would you choose to summarize the spread of this distribution? Why?

8. **Slalom times.**   The Men's Combined skiing event consists of a downhill and a slalom. Here are two displays of the slalom times in the Men's Combined at the 2006 Winter Olympics:

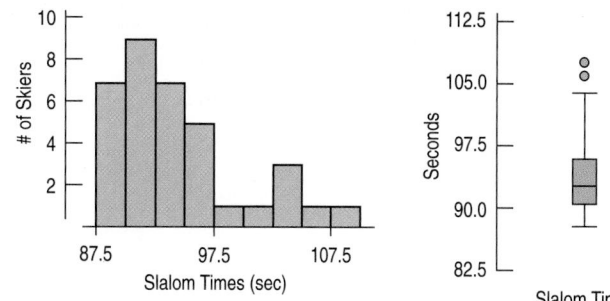

a) What features of the distribution can you see in both the histogram and the boxplot?
b) What features of the distribution can you see in the histogram that you could not see in the boxplot?
c) What summary statistic would you choose to summarize the center of this distribution? Why?
d) What summary statistic would you choose to summarize the spread of this distribution? Why?

9. Cereals. Sugar is a major ingredient in many breakfast cereals. The histogram displays the sugar content as a percentage of weight for 49 brands of cereal. The boxplot compares sugar content for adult and children's cereals.

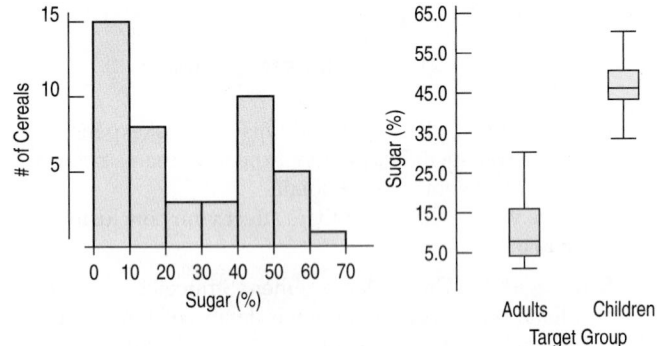

a) What is the range of the sugar contents of these cereals.
b) Describe the shape of the distribution.
c) What aspect of breakfast cereals might account for this shape?
d) Are all children's cereals higher in sugar than adult cereals?
e) Which group of cereals varies more in sugar content? Explain.

10. Tendon transfers. People with spinal cord injuries may lose function in some, but not all, of their muscles. The ability to push oneself up is particularly important for shifting position when seated and for transferring into and out of wheelchairs. Surgeons compared two operations to restore the ability to push up in children. The histogram shows scores rating pushing strength two years after surgery and boxplots compare results for the two surgical methods. (Mulcahey, Lutz, Kozen, Betz, "Prospective Evaluation of Biceps to Triceps and Deltoid to Triceps for Elbow Extension in Tetraplegia," *Journal of Hand Surgery*, 28, 6, 2003)

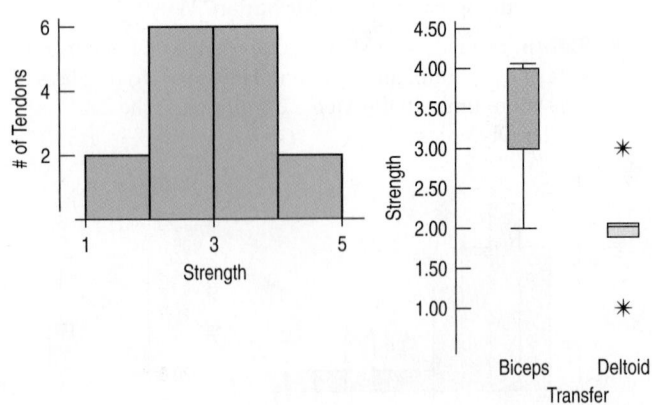

a) Describe the shape of this distribution.
b) What is the range of the strength scores?
c) What fact about results of the two procedures is hidden in the histogram?
d) Which method had the higher (better) median score?
e) Was that method always best?
f) Which method produced the most consistent results? Explain.

11. Population growth. Here is a "back-to-back" stem-and-leaf display that shows two data sets at once—one going to the left, one to the right. The display compares the percent change in population for two regions of the United States (based on census figures for 1990 and 2000). The fastest growing states were Nevada at 66% and Arizona at 40%. To show the distributions better, this display breaks each stem into two lines, putting leaves 0–4 on one stem and leaves 5–9 on the other.

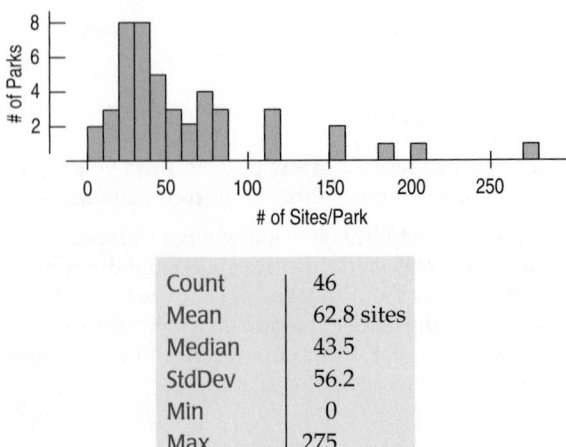

a) Use the data displayed in the stem-and-leaf display to construct comparative boxplots.
b) Write a few sentences describing the difference in growth rates for the two regions of the United States.

12. Camp sites. Shown below are the histogram and summary statistics for the number of camp sites at public parks in Vermont.

Count	46
Mean	62.8 sites
Median	43.5
StdDev	56.2
Min	0
Max	275
Q1	28
Q3	78

a) Which statistics would you use to identify the center and spread of this distribution? Why?

b) How many parks would you classify as outliers? Explain.

c) Create a boxplot for these data.

d) Write a few sentences describing the distribution.

13. **Hospital stays.** The U.S. National Center for Health Statistics compiles data on the length of stay by patients in short-term hospitals and publishes its findings in *Vital and Health Statistics*. Data from a sample of 39 male patients and 35 female patients on length of stay (in days) are displayed in the histograms below.

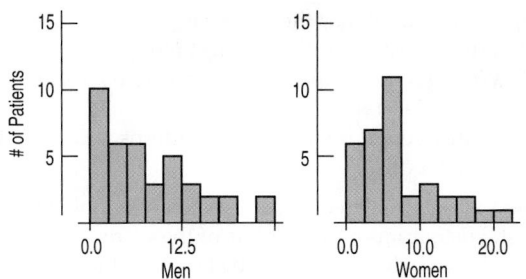

a) What would you suggest be changed about these histograms to make them easier to compare?

b) Describe these distributions by writing a few sentences comparing the duration of hospitalization for men and women.

c) Can you suggest a reason for the peak in women's length of stay?

14. **Deaths 2003.** A National Vital Statistics Report (www.cdc.gov/nchs/) indicated that nearly 300,000 black Americans died in 2003, compared with just over 2 million white Americans. Here are histograms displaying the distributions of their ages at death:

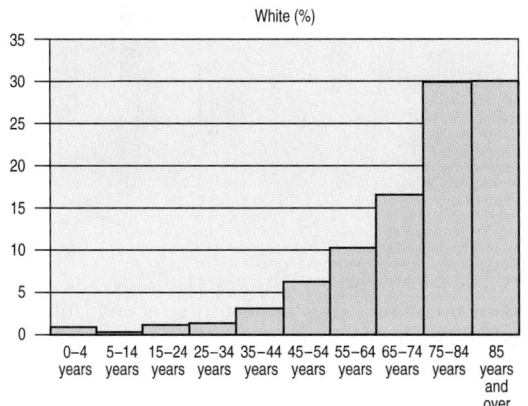

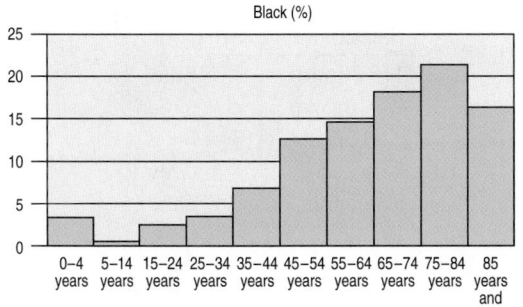

a) Describe the overall shapes of these distributions.

b) How do the distributions differ?

c) Look carefully at the bar definitions. Where do these plots violate the rules for statistical graphs?

15. **Women's basketball.** Here are boxplots of the points scored during the first 10 games of the season for both Scyrine and Alexandra:

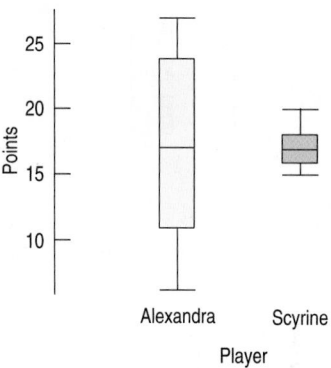

a) Summarize the similarities and differences in their performance so far.

b) The coach can take only one player to the state championship. Which one should she take? Why?

16. **Gas prices.** Here are boxplots of weekly gas prices at a service station in the midwestern United States (prices in $ per gallon):

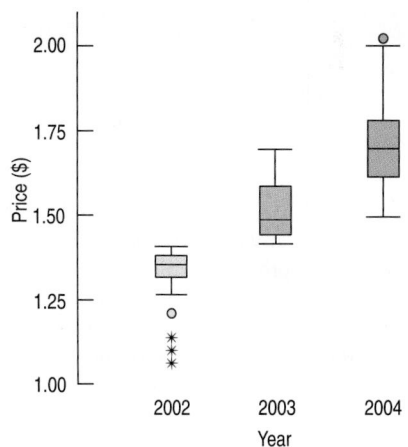

a) Compare the distribution of prices over the three years.

b) In which year were the prices least stable? Explain.

17. **Marriage age.** In 1975, did men and women marry at the same age? Here are boxplots of the age at first marriage for a sample of U.S. citizens then. Write a brief report discussing what these data show.

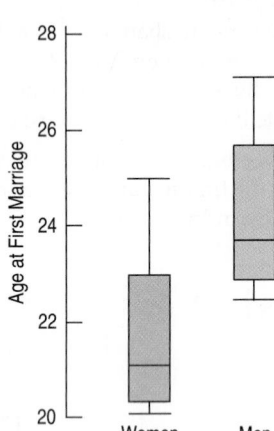

18. Fuel economy. Describe what these boxplots tell you about the relationship between the number of cylinders a car's engine has and the car's fuel economy (mpg):

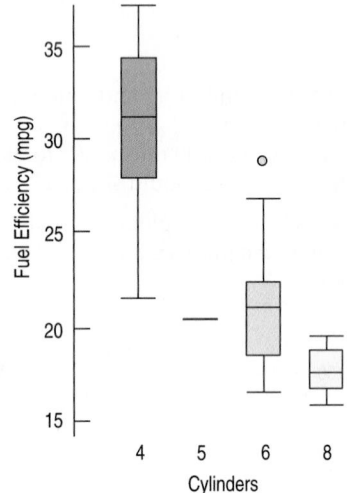

19. Fuel economy II. The Environmental Protection Agency provides fuel economy and pollution information on over 2000 car models. Here is a boxplot of *Combined Fuel Economy* (using an average of driving conditions) in *miles per gallon* by vehicle *Type* (car, van, or SUV). Summarize what you see about the fuel economies of the three vehicle types.

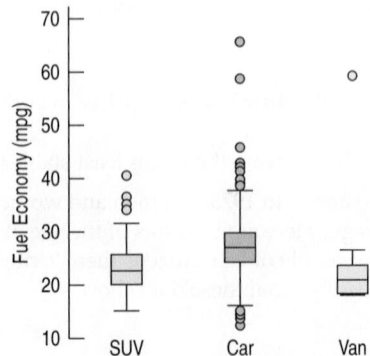

20. Ozone. Ozone levels (in parts per billion, ppb) were recorded at sites in New Jersey monthly between 1926 and 1971. Here are boxplots of the data for each month (over the 46 years), lined up in order (January = 1):

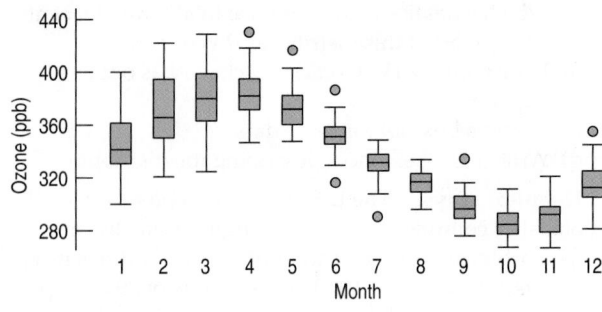

a) In what month was the highest ozone level ever recorded?
b) Which month has the largest IQR?
c) Which month has the smallest range?
d) Write a brief comparison of the ozone levels in January and June.
e) Write a report on the annual patterns you see in the ozone levels.

21. Test scores. Three Statistics classes all took the same test. Histograms and boxplots of the scores for each class are shown below. Match each class with the corresponding boxplot.

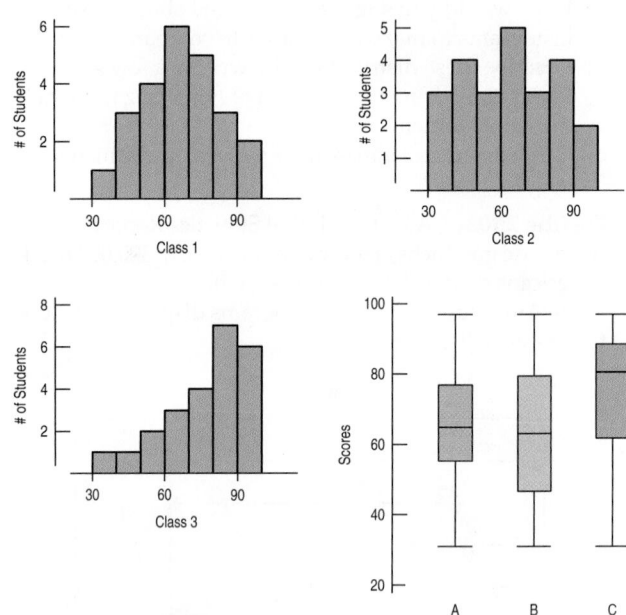

22. Eye and hair color. A survey of 1021 school-age children was conducted by randomly selecting children from several large urban elementary schools. Two of the questions concerned eye and hair color. In the survey, the following codes were used:

Hair Color	Eye Color
1 = Blond	1 = Blue
2 = Brown	2 = Green
3 = Black	3 = Brown
4 = Red	4 = Grey
5 = Other	5 = Other

The Statistics students analyzing the data were asked to study the relationship between eye and hair color. They produced this plot:

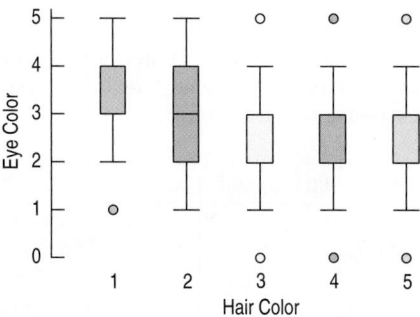

Is their graph appropriate? If so, summarize the findings. If not, explain why not.

23. Graduation? A survey of major universities asked what percentage of incoming freshmen usually graduate "on time" in 4 years. Use the summary statistics given to answer the questions that follow.

	% on Time
Count	48
Mean	68.35
Median	69.90
StdDev	10.20
Min	43.20
Max	87.40
Range	44.20
25th %tile	59.15
75th %tile	74.75

a) Would you describe this distribution as symmetric or skewed? Explain.
b) Are there any outliers? Explain.
c) Create a boxplot of these data.
d) Write a few sentences about the graduation rates.

T 24. Vineyards. Here are summary statistics for the sizes (in acres) of Finger Lakes vineyards:

Count	36
Mean	46.50 acres
StdDev	47.76
Median	33.50
IQR	36.50
Min	6
Q1	18.50
Q3	55
Max	250

a) Would you describe this distribution as symmetric or skewed? Explain.
b) Are there any outliers? Explain.
c) Create a boxplot of these data.
d) Write a few sentences about the sizes of the vineyards.

25. Caffeine. A student study of the effects of caffeine asked volunteers to take a memory test 2 hours after drinking soda. Some drank caffeine-free cola, some drank regular cola (with caffeine), and others drank a mixture of the two (getting a half-dose of caffeine). Here are the 5-number summaries for each group's scores (number of items recalled correctly) on the memory test:

	n	Min	Q1	Median	Q3	Max
No caffeine	15	16	20	21	24	26
Low caffeine	15	16	18	21	24	27
High caffeine	15	12	17	19	22	24

a) Describe the W's for these data.
b) Name the variables and classify each as categorical or quantitative.
c) Create parallel boxplots to display these results as best you can with this information.
d) Write a few sentences comparing the performances of the three groups.

26. SAT scores. Here are the summary statistics for Verbal SAT scores for a high school graduating class:

	n	Mean	Median	SD	Min	Max	Q1	Q3
Male	80	590	600	97.2	310	800	515	650
Female	82	602	625	102.0	360	770	530	680

a) Create parallel boxplots comparing the scores of boys and girls as best you can from the information given.
b) Write a brief report on these results. Be sure to discuss the shape, center, and spread of the scores.

T 27. Derby speeds 2007. How fast do horses run? Kentucky Derby winners top 30 miles per hour, as shown in this graph. The graph shows the percentage of Derby winners that have run *slower* than each given speed. Note that few have won running less than 33 miles per hour, but about 86% of the winning horses have run less than 37 miles per hour. (A cumulative frequency graph like this is called an "ogive.")

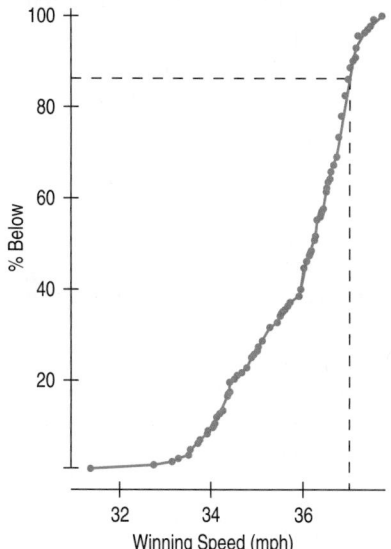

a) Estimate the median winning speed.
b) Estimate the quartiles.
c) Estimate the range and the IQR.
d) Create a boxplot of these speeds.
e) Write a few sentences about the speeds of the Kentucky Derby winners.

28. Cholesterol.   The Framingham Heart Study recorded the cholesterol levels of more than 1400 men. Here is an ogive of the distribution of these cholesterol measures. (An ogive shows the percentage of cases at or below a certain value.) Construct a boxplot for these data, and write a few sentences describing the distribution.

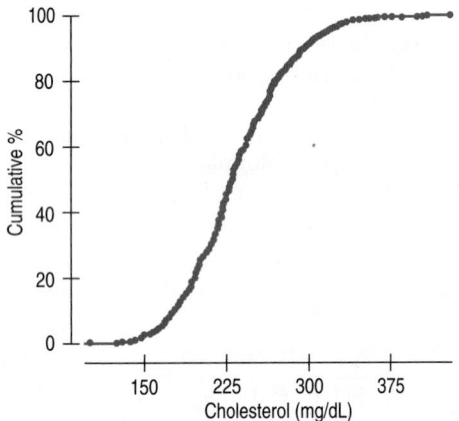

29. Reading scores.   A class of fourth graders takes a diagnostic reading test, and the scores are reported by reading grade level. The 5-number summaries for the 14 boys and 11 girls are shown:

Boys: 2.0 3.9 4.3 4.9 6.0

Girls: 2.8 3.8 4.5 5.2 5.9

a) Which group had the highest score?
b) Which group had the greater range?
c) Which group had the greater interquartile range?
d) Which group's scores appear to be more skewed? Explain.
e) Which group generally did better on the test? Explain.
f) If the mean reading level for boys was 4.2 and for girls was 4.6, what is the overall mean for the class?

30. Rainmakers?   In an experiment to determine whether seeding clouds with silver iodide increases rainfall, 52 clouds were randomly assigned to be seeded or not. The amount of rain they generated was then measured (in acre-feet). Here are the summary statistics:

	n	Mean	Median	SD	IQR	Q1	Q3
Unseeded	26	164.59	44.20	278.43	138.60	24.40	163
Seeded	26	441.98	221.60	650.79	337.60	92.40	430

a) Which of the summary statistics are most appropriate for describing these distributions. Why?
b) Do you see any evidence that seeding clouds may be effective? Explain.

31. Industrial experiment.   Engineers at a computer production plant tested two methods for accuracy in drilling holes into a PC board. They tested how fast they could set the drilling machine by running 10 boards at each of two different speeds. To assess the results, they measured the distance (in inches) from the center of a target on the board to the center of the hole. The data and summary statistics are shown in the table:

Distance (in.)	Speed	Distance (in.)	Speed
0.000101	Fast	0.000098	Slow
0.000102	Fast	0.000096	Slow
0.000100	Fast	0.000097	Slow
0.000102	Fast	0.000095	Slow
0.000101	Fast	0.000094	Slow
0.000103	Fast	0.000098	Slow
0.000104	Fast	0.000096	Slow
0.000102	Fast	0.975600	Slow
0.000102	Fast	0.000097	Slow
0.000100	Fast	0.000096	Slow
Mean	0.000102	Mean	0.097647
StdDev	0.000001	StdDev	0.308481

Write a report summarizing the findings of the experiment. Include appropriate visual and verbal displays of the distributions, and make a recommendation to the engineers if they are most interested in the accuracy of the method.

32. Cholesterol.   A study examining the health risks of smoking measured the cholesterol levels of people who had smoked for at least 25 years and people of similar ages who had smoked for no more than 5 years and then stopped. Create appropriate graphical displays for both groups, and write a brief report comparing their cholesterol levels. Here are the data:

Smokers				Ex-Smokers		
225	211	209	284	250	134	300
258	216	196	288	249	213	310
250	200	209	280	175	174	328
225	256	243	200	160	188	321
213	246	225	237	213	257	292
232	267	232	216	200	271	227
216	243	200	155	238	163	263
216	271	230	309	192	242	249
183	280	217	305	242	267	243
287	217	246	351	217	267	218
200	280	209		217	183	228

33. MPG.   A consumer organization compared gas mileage figures for several models of cars made in the United States with autos manufactured in other countries. The data are shown in the table:

Gas Mileage (mpg)	Country	Gas Mileage (mpg)	Country
16.9	U.S.	26.8	U.S.
15.5	U.S.	33.5	U.S.
19.2	U.S.	34.2	U.S.
18.5	U.S.	16.2	Other
30.0	U.S.	20.3	Other
30.9	U.S.	31.5	Other
20.6	U.S.	30.5	Other
20.8	U.S.	21.5	Other
18.6	U.S.	31.9	Other
18.1	U.S.	37.3	Other
17.0	U.S.	27.5	Other
17.6	U.S.	27.2	Other
16.5	U.S.	34.1	Other
18.2	U.S.	35.1	Other
26.5	U.S.	29.5	Other
21.9	U.S.	31.8	Other
27.4	U.S.	22.0	Other
28.4	U.S.	17.0	Other
28.8	U.S.	21.6	Other

a) Create graphical displays for these two groups.
b) Write a few sentences comparing the distributions.

34. Baseball. American League baseball teams play their games with the designated hitter rule, meaning that pitchers do not bat. The League believes that replacing the pitcher, typically a weak hitter, with another player in the batting order produces more runs and generates more interest among fans. Following are the average number of runs scored in American League and National League stadiums for the first half of the 2001 season:

Average Runs	League	Average Runs	League
11.1	American	14.0	National
10.8	American	11.6	National
10.8	American	10.4	National
10.3	American	10.9	National
10.3	American	10.2	National
10.1	American	9.5	National
10.0	American	9.5	National
9.5	American	9.5	National
9.4	American	9.5	National
9.3	American	9.1	National
9.2	American	8.8	National
9.2	American	8.4	National
9.0	American	8.3	National
8.3	American	8.2	National
		8.1	National
		7.9	National

a) Create an appropriate graphical display of these data.
b) Write a few sentences comparing the average number of runs scored per game in the two leagues. (Remember: shape, center, spread, unusual features!)

c) Coors Field in Denver stands a mile above sea level, an altitude far greater than that of any other major league ball park. Some believe that the thinner air makes it harder for pitchers to throw curveballs and easier for batters to hit the ball a long way. Do you see any evidence that the 14 runs scored per game there is unusually high? Explain.

35. Fruit Flies. Researchers tracked a population of 1,203,646 fruit flies, counting how many died each day for 171 days. Here are three timeplots offering different views of these data. One shows the number of flies alive on each day, one the number who died that day, and the third the mortality rate—the fraction of the number alive who died. On the last day studied, the last 2 flies died, for a mortality rate of 1.0.

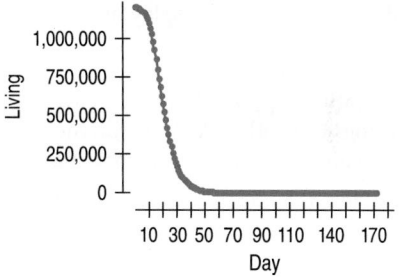

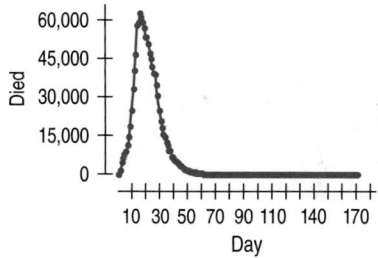

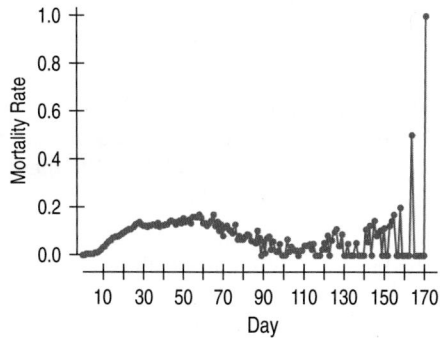

a) On approximately what day did the most flies die?
b) On what day during the first 100 days did the largest *proportion* of flies die?
c) When did the number of fruit flies alive stop changing very much from day to day?

36. Drunk driving 2005. Accidents involving drunk drivers account for about 40% of all deaths on the nation's highways. The table tracks the number of alcohol-related fatalities for 24 years. (www.madd.org)

Year	Deaths (thousands)	Year	Deaths (thousands)
1982	26.2	1994	17.3
1983	24.6	1995	17.7
1984	24.8	1996	17.7
1985	23.2	1997	16.7
1986	25.0	1998	16.7
1987	24.1	1999	16.6
1988	23.8	2000	17.4
1989	22.4	2001	17.4
1990	22.6	2002	17.5
1991	20.2	2003	17.1
1992	18.3	2004	16.9
1993	17.9	2005	16.9

a) Create a stem-and-leaf display or a histogram of these data.
b) Create a timeplot.
c) Using features apparent in the stem-and-leaf display (or histogram) and the timeplot, write a few sentences about deaths caused by drunk driving.

37. Assets. Here is a histogram of the assets (in millions of dollars) of 79 companies chosen from the *Forbes* list of the nation's top corporations:

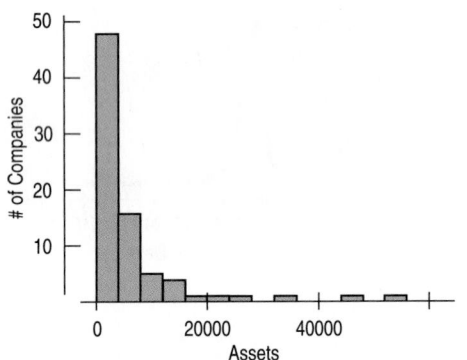

a) What aspect of this distribution makes it difficult to summarize, or to discuss, center and spread?
b) What would you suggest doing with these data if we want to understand them better?

38. Music library. Students were asked how many songs they had in their digital music libraries. Here's a display of the responses:

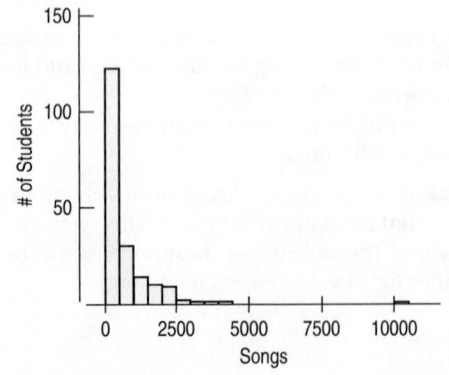

a) What aspect of this distribution makes it difficult to summarize, or to discuss, center and spread?
b) What would you suggest doing with these data if we want to understand them better?

39. Assets again. Here are the same data you saw in Exercise 37 after re-expressions as the square root of assets and the logarithm of assets:

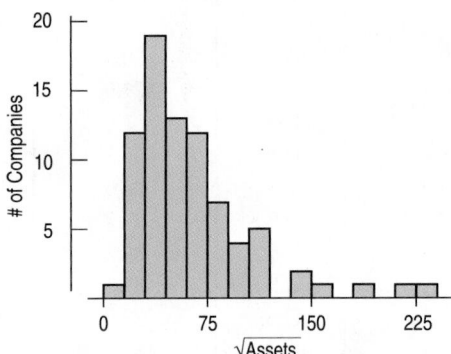

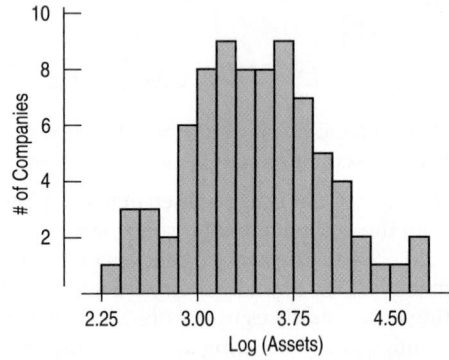

a) Which re-expression do you prefer? Why?
b) In the square root re-expression, what does the value 50 actually indicate about the company's assets?
c) In the logarithm re-expression, what does the value 3 actually indicate about the company's assets?

40. Rainmakers. The table lists the amount of rainfall (in acre-feet) from the 26 clouds seeded with silver iodide discussed in Exercise 30:

2745	703	302	242	119	40	7
1697	489	274	200	118	32	4
1656	430	274	198	115	31	
978	334	255	129	92	17	

a) Why is acre-feet a good way to measure the amount of precipitation produced by cloud seeding?
b) Plot these data, and describe the distribution.
c) Create a re-expression of these data that produces a more advantageous distribution.
d) Explain what your re-expressed scale means.

41. Stereograms. Stereograms appear to be composed entirely of random dots. However, they contain separate images that a viewer can "fuse" into a three-dimensional (3D) image by staring at the dots while defocusing the eyes. An experiment was performed to determine whether knowledge of the embedded image affected the time required for subjects to fuse the images. One group of subjects (group NV) received no information or just verbal information about the shape of the embedded object. A second group (group VV) received both verbal information and visual information (specifically, a drawing of the object). The experimenters measured how many seconds it took for the subject to report that he or she saw the 3D image.

a) What two variables are discussed in this description?

b) For each variable, is it quantitative or categorical? If quantitative, what are the units?

c) The boxplots compare the fusion times for the two treatment groups. Write a few sentences comparing these distributions. What does the experiment show?

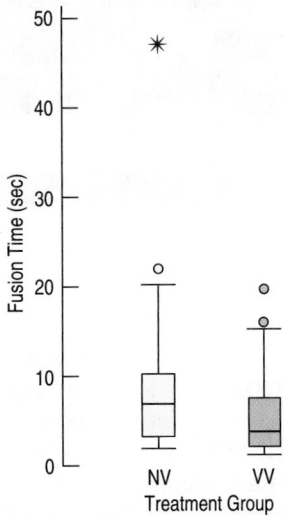

T **42. Stereograms, revisited.** Because of the skewness of the distributions of fusion times described in Exercise 41, we might consider a re-expression. Here are the boxplots of the *log* of fusion times. Is it better to analyze the original fusion times or the log fusion times? Explain.

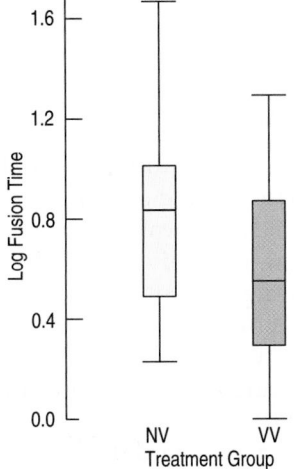

The Standard Deviation as a Ruler and the Normal Model

The women's heptathlon in the Olympics consists of seven track and field events: the 200-m and 800-m runs, 100-m high hurdles, shot put, javelin, high jump, and long jump. To determine who should get the gold medal, somehow the performances in all seven events have to be combined into one score. How can performances in such different events be compared? They don't even have the same units; the races are recorded in minutes and seconds and the throwing and jumping events in meters. In the 2004 Olympics, Austra Skujytė of Lithuania put the shot 16.4 meters, about 3 meters farther than the average of all contestants. Carolina Klüft won the long jump with a 6.78-m jump, about a meter better than the average. Which performance deserves more points? Even though both events are measured in meters, it's not clear how to compare them. The solution to the problem of how to compare scores turns out to be a useful method for comparing all sorts of values whether they have the same units or not.

The Standard Deviation as a Ruler

Grading on a Curve

If you score 79% on an exam, what grade should you get? One teaching philosophy looks only at the raw percentage, 79, and bases the grade on that alone. Another looks at your *relative* performance and bases the grade on how you did compared with the rest of the class. Teachers and students still debate which method is better.

The trick in comparing very different-looking values is to use standard deviations. The standard deviation tells us how the whole collection of values varies, so it's a natural ruler for comparing an individual value to the group. Over and over during this course, we will ask questions such as "How far is this value from the mean?" or "How different are these two statistics?" The answer in every case will be to measure the distance or difference in standard deviations.

The concept of the standard deviation as a ruler is not special to this course. You'll find statistical distances measured in standard deviations throughout Statistics, up to the most advanced levels.[1] This approach is one of the basic tools of statistical thinking.

---

[1] Other measures of spread could be used as well, but the standard deviation is the most common measure, and it is almost always used as the ruler.

In order to compare the two events, let's start with a picture. This time we'll use stem-and-leaf displays so we can see the individual distances.

Long Jump		Shot Put	
Stem	Leaf	Stem	Leaf
67	8	16	4
66		15	
65	1	15	
64	2	14	56778
63	0566	14	24
62	11235	13	5789
61	0569	13	012234
60	2223	12	55
59	0278	12	0144
58	4	11	59
57	0	11	23

FIGURE 6.1
Stem-and-leaf displays for both the long jump and the shot put in the 2004 Olympic Heptathlon. Carolina Klüft (green scores) won the long jump, and Austra Skujyté (red scores) won the shot put. Which heptathlete did better for both events combined?

The two winning performances on the top of each stem-and-leaf display appear to be about the same distance from the center of the pack. But look again carefully. What do we mean by the *same distance*? The two displays have different scales. Each line in the stem-and-leaf for the shot put represents half a meter, but for the long jump each line is only a tenth of a meter. It's only because our eyes naturally adjust the scales and use the standard deviation as the ruler that we see each as being about the same distance from the center of the data. How can we make this hunch more precise? Let's see how many standard deviations each performance is from the mean.

Klüft's 6.78-m long jump is 0.62 meters longer than the mean jump of 6.16 m. How many *standard deviations* better than the mean is that? The standard deviation for this event was 0.23 m, so her jump was $(6.78 - 6.16)/0.23 = 0.62/0.23 = 2.70$ *standard deviations better* than the mean. Skujyté's winning shot put was $16.40 - 13.29 = 3.11$ meters longer than the mean shot put distance, and that's $3.11/1.24 = 2.51$ standard deviations better than the mean. That's a great performance but not quite as impressive as Klüft's long jump, which was farther above the mean, as measured in *standard deviations*.

	Event	
	Long Jump	Shot Put
Mean (all contestants)	6.16 m	13.29 m
SD	0.23 m	1.24 m
n	26	28
Klüft	6.78 m	14.77 m
Skujyté	6.30 m	16.40 m

Standardizing with z-Scores

To compare these athletes' performances, we determined how many standard deviations from the event's mean each was.

Expressing the distance in standard deviations *standardizes* the performances. To standardize a value, we simply subtract the mean performance in that event and then divide this difference by the standard deviation. We can write the calculation as

$$z = \frac{y - \bar{y}}{s}.$$

These values are called **standardized values,** and are commonly denoted with the letter z. Usually, we just call them **z-scores.**

Standardized values have *no units*. z-scores measure the distance of each data value from the mean in standard deviations. A z-score of 2 tells us that a data value is 2 standard deviations above the mean. It doesn't matter whether the original variable was measured in inches, dollars, or seconds. Data values below the mean have negative z-scores, so a z-score of -1.6 means that the data value was 1.6 standard deviations below the mean. Of course, regardless of the direction, the farther a data value is from the mean, the more unusual it is, so a z-score of -1.3

is more extraordinary than a z-score of 1.2. Looking at the z-scores, we can see that even though both were winning scores, Klüft's long jump with a z-score of 2.70 is slightly more impressive than Skujyté's shot put with a z-score of 2.51.

FOR EXAMPLE | Standardizing skiing times

The men's combined skiing event in the winter Olympics consists of two races: a downhill and a slalom. Times for the two events are added together, and the skier with the lowest total time wins. In the 2006 Winter Olympics, the mean slalom time was 94.2714 seconds with a standard deviation of 5.2844 seconds. The mean downhill time was 101.807 seconds with a standard deviation of 1.8356 seconds. Ted Ligety of the United States, who won the gold medal with a combined time of 189.35 seconds, skied the slalom in 87.93 seconds and the downhill in 101.42 seconds.

Question: On which race did he do better compared with the competition?

For the slalom, Ligety's z-score is found by subtracting the mean time from his time and then dividing by the standard deviation:

$$z_{Slalom} = \frac{87.93 - 94.2714}{5.2844} = -1.2$$

Similarly, his z-score for the downhill is:

$$z_{Downhill} = \frac{101.42 - 101.807}{1.8356} = -0.21$$

The z-scores show that Ligety's time in the slalom is farther below the mean than his time in the downhill. His performance in the slalom was more remarkable.

By using the standard deviation as a ruler to measure statistical distance from the mean, we can compare values that are measured on different variables, with different scales, with different units, or for different individuals. To determine the winner of the heptathlon, the judges must combine performances on seven very different events. Because they want the score to be absolute, and *not* dependent on the particular athletes in each Olympics, they use predetermined tables, but they could combine scores by standardizing each, and then adding the z-scores together to reach a total score. The only trick is that they'd have to switch the sign of the z-score for running events, because unlike throwing and jumping, it's better to have a running time below the mean (with a negative z-score).

To combine the scores Skujyté and Klüft earned in the long jump and the shot put, we standardize both events as shown in the table. That gives Klüft her 2.70 z-score in the long jump and a 1.19 in the shot put, for a total of 3.89. Skujyté's shot put gave her a 2.51, but her long jump was only 0.61 SDs above the mean, so her total is 3.12.

Is this the result we wanted? Yes. Each won one event, but Klüft's shot put was second best, while Skujyté's long jump was seventh. The z-scores measure how far each result is from the event mean in standard deviation units. And because they are both in standard deviation units, we can combine them. Not coincidentally, Klüft went on to win the gold medal for the entire seven-event heptathlon, while Skujyté got the silver.

		Event	
		Long Jump	Shot Put
	Mean	6.16 m	13.29 m
	SD	0.23 m	1.24 m
Klüft	Performance	6.78 m	14.77 m
	z-score	$\frac{6.78 - 6.16}{0.23} = 2.70$	$\frac{14.77 - 13.29}{1.24} = 1.19$
	Total z-score	2.70 + 1.19 = 3.89	
Skujyté	Performance	6.30 m	16.40 m
	z-score	$\frac{6.30 - 6.16}{0.23} = 0.61$	$\frac{16.40 - 13.29}{1.24} = 2.51$
	Total z-score	0.61 + 2.51 = 3.12	

Combining z-scores

In the 2006 winter Olympics men's combined event, Ivica Kostelić of Croatia skied the slalom in 89.44 seconds and the downhill in 100.44 seconds. He thus beat Ted Ligety in the downhill, but not in the slalom. Maybe he should have won the gold medal.

Question: Considered in terms of standardized scores, which skier did better?

Kostelić's z-scores are:

$$z_{Slalom} = \frac{89.44 - 94.2714}{5.2844} = -0.91 \quad \text{and} \quad z_{Downhill} = \frac{100.44 - 101.807}{1.8356} = -0.74$$

The sum of his z-scores is approximately −1.65. Ligety's z-score sum is only about −1.41. Because the standard deviation of the downhill times is so much smaller, Kostelić's better performance there means that he would have won the event if standardized scores were used.

When we standardize data to get a z-score, we do two things. First, we shift the data by subtracting the mean. Then, we rescale the values by dividing by their standard deviation. We often shift and rescale data. What happens to a grade distribution if *everyone* gets a five-point bonus? Everyone's grade goes up, but does the shape change? (*Hint:* Has anyone's distance from the mean changed?) If we switch from feet to meters, what happens to the distribution of heights of students in your class? Even though your intuition probably tells you the answers to these questions, we need to look at exactly how shifting and rescaling work.

JUST CHECKING

1. Your Statistics teacher has announced that the lower of your two tests will be dropped. You got a 90 on test 1 and an 80 on test 2. You're all set to drop the 80 until she announces that she grades "on a curve." She standardized the scores in order to decide which is the lower one. If the mean on the first test was 88 with a standard deviation of 4 and the mean on the second was 75 with a standard deviation of 5,

　a) Which one will be dropped?
　b) Does this seem "fair"?

Shifting Data

Since the 1960s, the Centers for Disease Control's National Center for Health Statistics has been collecting health and nutritional information on people of all ages and backgrounds. A recent survey, the National Health and Nutrition Examination Survey (NHANES) 2001–2002,[2] measured a wide variety of variables, including body measurements, cardiovascular fitness, blood chemistry, and demographic information on more than 11,000 individuals.

[2] www.cdc.gov/nchs/nhanes.htm

WHO	80 male participants of the NHANES survey between the ages of 19 and 24 who measured between 68 and 70 inches tall
WHAT	Their weights
UNIT	Kilograms
WHEN	2001–2002
WHERE	United States
WHY	To study nutrition, and health issues and trends
HOW	National survey

Included in this group were 80 men between 19 and 24 years old of average height (between 5'8" and 5'10" tall). Here are a histogram and boxplot of their weights:

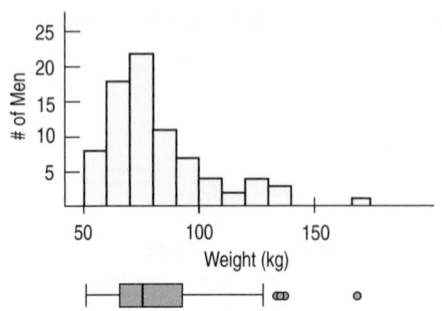

FIGURE 6.2

Histogram and boxplot for the men's weights. The shape is skewed to the right with several high outliers.

A S *Activity:* **Changing the Baseline.** What happens when we shift data? Do measures of center and spread change?

Their mean weight is 82.36 kg. For this age and height group, the National Institutes of Health recommends a maximum healthy weight of 74 kg, but we can see that some of the men are heavier than the recommended weight. To compare their weights to the recommended maximum, we could subtract 74 kg from each of their weights. What would that do to the center, shape, and spread of the histogram? Here's the picture:

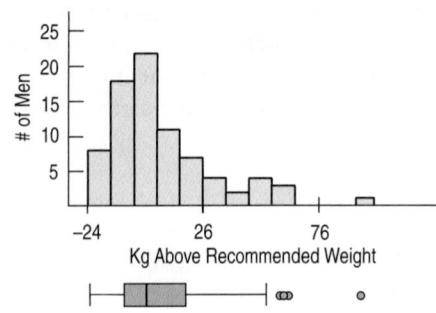

FIGURE 6.3

Subtracting 74 kilograms shifts the entire histogram down but leaves the spread and the shape exactly the same.

Doctors' height and weight charts sometimes give ideal weights for various heights that include 2-inch heels. If the mean height of adult women is 66 inches including 2-inch heels, what is the mean height of women without shoes? Each woman is shorter by 2 inches when barefoot, so the mean is decreased by 2 inches, to 64 inches.

On average, they weigh 82.36 kg, so on average they're 8.36 kg overweight. And, after subtracting 74 from each weight, the mean of the new distribution is $82.36 - 74 = 8.36$ kg. In fact, when we **shift** the data by adding (or subtracting) a constant to each value, all measures of position (center, percentiles, min, max) will increase (or decrease) by the same constant.

What about the spread? What does adding or subtracting a constant value do to the spread of the distribution? Look at the two histograms again. Adding or subtracting a constant changes each data value equally, so the entire distribution just shifts. Its shape doesn't change and neither does the spread. None of the measures of spread we've discussed—not the range, not the IQR, not the standard deviation—changes.

> *Adding (or subtracting) a constant to every data value adds (or subtracts) the same constant to measures of position, but leaves measures of spread unchanged.*

Rescaling Data

Not everyone thinks naturally in metric units. Suppose we want to look at the weights in pounds instead. We'd have to **rescale** the data. Because there are about 2.2 pounds in every kilogram, we'd convert the weights by multiplying each value by 2.2. Multiplying or dividing each value by a constant changes the measurement units. Here are histograms of the two weight distributions, plotted on the same scale, so you can see the effect of multiplying:

FIGURE 6.4

Men's weights in both kilograms and pounds. How do the distributions and numerical summaries change?

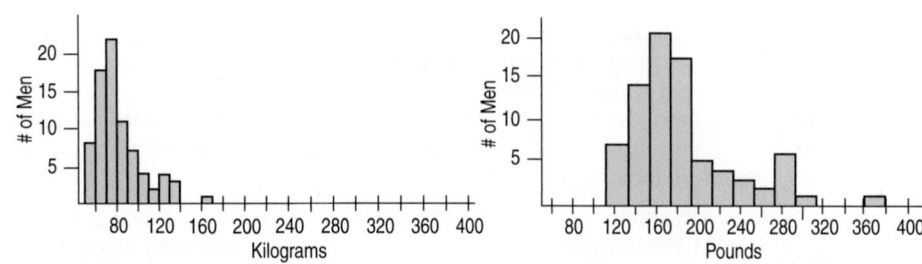

A S *Simulation:* **Changing the Units.** Change the center and spread values for a distribution and watch the summaries change (or not, as the case may be).

What happens to the shape of the distribution? Although the histograms don't look exactly alike, we see that the shape really hasn't changed: Both are uni-modal and skewed to the right.

What happens to the mean? Not too surprisingly, it gets multiplied by 2.2 as well. The men weigh 82.36 kg on average, which is 181.19 pounds. As the box-plots and 5-number summaries show, all measures of position act the same way. They all get multiplied by this same constant.

What happens to the spread? Take a look at the boxplots. The spread in pounds (on the right) is larger. How much larger? If you guessed 2.2 times, you've figured out how measures of spread get rescaled.

FIGURE 6.5

The boxplots (drawn on the same scale) show the weights measured in kilograms (on the left) and pounds (on the right). Because 1 kg is 2.2 lb, all the points in the right box are 2.2 times larger than the corresponding points in the left box. So each meas-ure of position and spread is 2.2 times as large when measured in pounds rather than kilograms.

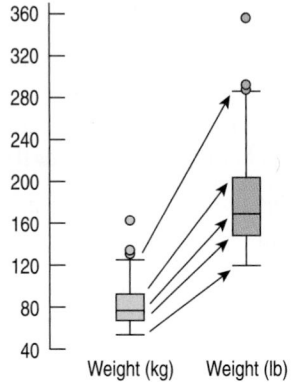

	Weight (kg)	Weight (lb)
Min	54.3	119.46
Q1	67.3	148.06
Median	76.85	169.07
Q3	92.3	203.06
Max	161.5	355.30
IQR	25	55
SD	22.27	48.99

When we multiply (or divide) all the data values by any constant, all measures of position (such as the mean, median, and percentiles) and measures of spread (such as the range, the IQR, and the standard deviation) are multiplied (or divided) by that same constant.

FOR EXAMPLE Rescaling the slalom

Recap: The times in the men's combined event at the winter Olympics are reported in minutes and seconds. Previously, we converted these to seconds and found the mean and standard deviation of the slalom times to be 94.2714 seconds and 5.2844 seconds, respectively.

Question: Suppose instead that we had reported the times in minutes—that is, that each individual time was divided by 60. What would the resulting mean and standard deviation be?

Dividing all the times by 60 would divide both the mean and the standard deviation by 60:

Mean = 94.2714/60 = 1.5712 minutes;   SD = 5.2844/60 = 0.0881 minutes.

JUST CHECKING

2. In 1995 the Educational Testing Service (ETS) adjusted the scores of SAT tests. Before ETS recentered the SAT Verbal test, the mean of all test scores was 450.

   a) How would adding 50 points to each score affect the mean?

   b) The standard deviation was 100 points. What would the standard deviation be after adding 50 points?

   c) Suppose we drew boxplots of test takers' scores a year before and a year after the recentering. How would the boxplots of the two years differ?

3. A company manufactures wheels for in-line skates. The diameters of the wheels have a mean of 3 inches and a standard deviation of 0.1 inches. Because so many of their customers use the metric system, the company decided to report their production statistics in millimeters (1 inch = 25.4 mm). They report that the standard deviation is now 2.54 mm. A corporate executive is worried about this increase in variation. Should he be concerned? Explain.

Back to *z*-scores

A S *Activity:* **Standardizing.**
What if we both shift and rescale? The result is so nice that we give it a name.

Standardizing data into *z*-scores is just shifting them by the mean and rescaling them by the standard deviation. Now we can see how standardizing affects the distribution. When we subtract the mean of the data from every data value, we shift the mean to zero. As we have seen, such a shift doesn't change the standard deviation.

When we *divide* each of these shifted values by *s*, however, the standard deviation should be divided by *s* as well. Since the standard deviation was *s* to start with, the new standard deviation becomes 1.

How, then, does standardizing affect the distribution of a variable? Let's consider the three aspects of a distribution: the shape, center, and spread.

> *z*-scores have mean 0 and standard deviation 1.

▶ *Standardizing into z-scores does not change the **shape** of the distribution of a variable.*

▶ *Standardizing into z-scores changes the **center** by making the mean 0.*

▶ *Standardizing into z-scores changes the **spread** by making the standard deviation 1.*

STEP-BY-STEP EXAMPLE   **Working with Standardized Variables**

Many colleges and universities require applicants to submit scores on standardized tests such as the SAT Writing, Math, and Critical Reading (Verbal) tests. The college your little sister wants to apply to says that while there is no minimum score required, the middle 50% of their students have combined SAT scores between 1530 and 1850. You'd feel confident if you knew her score was in their top 25%, but unfortunately she took the ACT test, an alternative standardized test.

Question: How high does her ACT need to be to make it into the top quarter of equivalent SAT scores?

To answer that question you'll have to standardize all the scores, so you'll need to know the mean and standard deviations of scores for some group on both tests. The college doesn't report the mean or standard deviation for their applicants on either test, so we'll use the group of all test takers nationally. For college-bound seniors, the average combined SAT score is about 1500 and the standard deviation is about 250 points. For the same group, the ACT average is 20.8 with a standard deviation of 4.8.

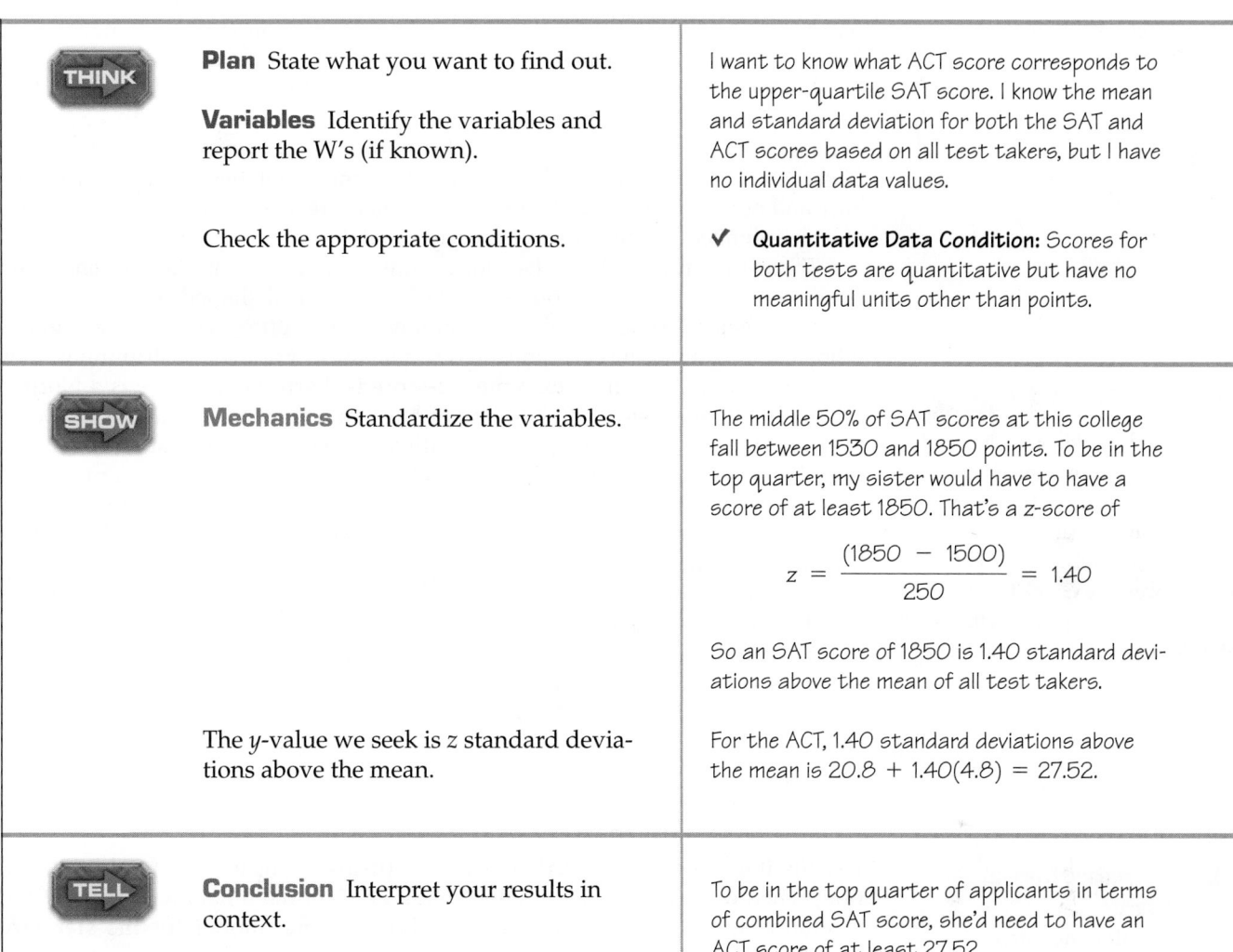

THINK	**Plan** State what you want to find out.	I want to know what ACT score corresponds to the upper-quartile SAT score. I know the mean and standard deviation for both the SAT and ACT scores based on all test takers, but I have no individual data values.
	Variables Identify the variables and report the W's (if known).	
	Check the appropriate conditions.	✔ **Quantitative Data Condition:** Scores for both tests are quantitative but have no meaningful units other than points.
SHOW	**Mechanics** Standardize the variables.	The middle 50% of SAT scores at this college fall between 1530 and 1850 points. To be in the top quarter, my sister would have to have a score of at least 1850. That's a z-score of $$z = \frac{(1850 - 1500)}{250} = 1.40$$ So an SAT score of 1850 is 1.40 standard deviations above the mean of all test takers.
	The *y*-value we seek is *z* standard deviations above the mean.	For the ACT, 1.40 standard deviations above the mean is $20.8 + 1.40(4.8) = 27.52$.
TELL	**Conclusion** Interpret your results in context.	To be in the top quarter of applicants in terms of combined SAT score, she'd need to have an ACT score of at least 27.52.

When Is a *z*-score BIG?

A *z*-score gives us an indication of how unusual a value is because it tells us how far it is from the mean. If the data value sits right at the mean, it's not very far at all and its *z*-score is 0. A *z*-score of 1 tells us that the data value is 1 standard deviation above the mean, while a *z*-score of −1 tells us that the value is 1 standard deviation below the mean. How far from 0 does a *z*-score have to be to be interesting or unusual? There is no universal standard, but the larger the score is (negative or positive), the more unusual it is. We know that 50% of the data lie between the quartiles. For symmetric data, the standard deviation is usually a bit smaller than the IQR, and it's not uncommon for at least half of the data to have *z*-scores between −1 and 1. But no matter what the shape of the distribution, a *z*-score of 3 (plus or minus) or more is rare, and a *z*-score of 6 or 7 shouts out for attention.

To say more about how big we expect a *z*-score to be, we need to *model* the data's distribution. A model will let us say much more precisely how often we'd be likely to see *z*-scores of different sizes. Of course, like all models of the real world, the model will be wrong—wrong in the sense that it can't match

Is Normal Normal?
Don't be misled. The name "Normal" doesn't mean that these are the *usual* shapes for histograms. The name follows a tradition of positive thinking in Mathematics and Statistics in which functions, equations, and relationships that are easy to work with or have other nice properties are called "normal", "common", "regular", "natural", or similar terms. It's as if by calling them ordinary, we could make them actually occur more often and simplify our lives.

reality exactly. But it can still be useful. Like a physical model, it's something we can look at and manipulate in order to learn more about the real world.

Models help our understanding in many ways. Just as a model of an airplane in a wind tunnel can give insights even though it doesn't show every rivet,[3] models of data give us summaries that we can learn from and use, even though they don't fit each data value exactly. It's important to remember that they're only *models* of reality and not reality itself. But without models, what we can learn about the world at large is limited to only what we can say about the data we have at hand.

There is no universal standard for z-scores, but there is a model that shows up over and over in Statistics. You may have heard of "bell-shaped curves." Statisticians call them Normal models. **Normal models** are appropriate for distributions whose shapes are unimodal and roughly symmetric. For these distributions, they provide a measure of how extreme a z-score is. Fortunately, there is a Normal model for every possible combination of mean and standard deviation. We write $N(\mu, \sigma)$ to represent a Normal model with a mean of μ and a standard deviation of σ. Why the Greek? Well, *this* mean and standard deviation are not numerical summaries of data. They are part of the model. They don't come from the data. Rather, they are numbers that we choose to help specify the model. Such numbers are called **parameters** of the model.

We don't want to confuse the parameters with summaries of the data such as $\bar{y}$ and s, so we use special symbols. In Statistics, we almost always use Greek letters for parameters. By contrast, summaries of data are called **statistics** and are usually written with Latin letters.

If we model data with a Normal model and standardize them using the corresponding μ and σ, we still call the standardized value a **z-score**, and we write

$$z = \frac{y - \mu}{\sigma}.$$

Usually it's easier to standardize data first (using its mean and standard deviation). Then we need only the model $N(0,1)$. The Normal model with mean 0 and standard deviation 1 is called the **standard Normal model** (or the **standard Normal distribution**).

But be careful. You shouldn't use a Normal model for just any data set. Remember that standardizing won't change the shape of the distribution. If the distribution is not unimodal and symmetric to begin with, standardizing won't make it Normal.

When we use the Normal model, we assume that the distribution of the data is, well, Normal. Practically speaking, there's no way to check whether this **Normality Assumption** is true. In fact, it almost certainly is not true. Real data don't behave like mathematical models. Models are idealized; real data are real. The good news, however, is that to use a Normal model, it's sufficient to check the following condition:

Nearly Normal Condition. The shape of the data's distribution is unimodal and symmetric. Check this by making a histogram (or a Normal probability plot, which we'll explain later).

Don't model data with a Normal model without checking whether the condition is satisfied.

All models make **assumptions.** Whenever we model—and we'll do that often—we'll be careful to point out the assumptions that we're making. And, what's even more important, we'll check the associated **conditions** in the data to make sure that those assumptions are reasonable.

---

[3] In fact, the model is useful *because* it doesn't have every rivet. It is because models offer a simpler view of reality that they are so useful as we try to understand reality.

The 68–95–99.7 Rule

One in a Million

These magic 68, 95, 99.7 values come from the Normal model. As a model, it can give us corresponding values for any *z*-score. For example, it tells us that fewer than 1 out of a million values have *z*-scores smaller than −5.0 or larger than +5.0. So if someone tells you you're "one in a million," they must really admire your *z*-score.

TI-*nspire*

The 68–95–99.7 Rule. See it work for yourself.

Normal models give us an idea of how extreme a value is by telling us how likely it is to find one that far from the mean. We'll soon show how to find these numbers precisely—but one simple rule is usually all we need.

It turns out that in a Normal model, about 68% of the values fall within 1 standard deviation of the mean, about 95% of the values fall within 2 standard deviations of the mean, and about 99.7%—almost all—of the values fall within 3 standard deviations of the mean. These facts are summarized in a rule that we call (let's see . . .) the **68–95–99.7 Rule**.[4]

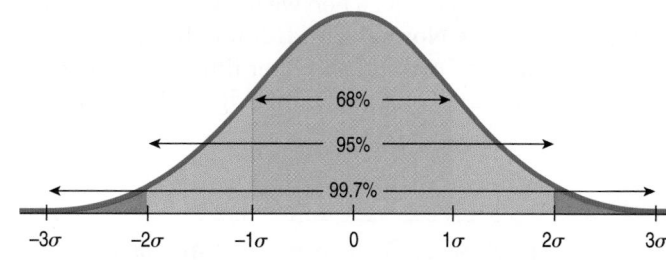

FIGURE 6.6

Reaching out one, two, and three standard deviations on a Normal model gives the 68−95−99.7 Rule, seen as proportions of the area under the curve.

FOR EXAMPLE Using the 68–95–99.7 Rule

Question: In the 2006 Winter Olympics men's combined event, Jean-Baptiste Grange of France skied the slalom in 88.46 seconds—about 1 standard deviation faster than the mean. If a Normal model is useful in describing slalom times, about how many of the 35 skiers finishing the event would you expect skied the slalom *faster* than Jean-Baptiste?

From the 68–95–99.7 Rule, we expect 68% of the skiers to be within one standard deviation of the mean. Of the remaining 32%, we expect half on the high end and half on the low end. 16% of 35 is 5.6, so, conservatively, we'd expect about 5 skiers to do better than Jean-Baptiste.

JUST CHECKING

4. As a group, the Dutch are among the tallest people in the world. The average Dutch man is 184 cm tall—just over 6 feet (and the average Dutch woman is 170.8 cm tall—just over 5′7″). If a Normal model is appropriate and the standard deviation for men is about 8 cm, what percentage of all Dutch men will be over 2 meters (6′6″) tall?

5. Suppose it takes you 20 minutes, on average, to drive to school, with a standard deviation of 2 minutes. Suppose a Normal model is appropriate for the distributions of driving times.

   a) How often will you arrive at school in less than 22 minutes?

   b) How often will it take you more than 24 minutes?

   c) Do you think the distribution of your driving times is unimodal and symmetric?

   d) What does this say about the accuracy of your predictions? Explain.

---

[4] This rule is also called the "Empirical Rule" because it originally came from observation. The rule was first published by Abraham de Moivre in 1733, 75 years before the Normal model was discovered. Maybe it should be called "de Moivre's Rule," but that wouldn't help us remember the important numbers, 68, 95, and 99.7.

The First Three Rules for Working with Normal Models

1. Make a picture.
2. Make a picture.
3. Make a picture.

Although we're thinking about models, not histograms of data, the three rules don't change. To help you think clearly, a simple hand-drawn sketch is all you need. Even experienced statisticians sketch pictures to help them think about Normal models. You should too.

Of course, when we have data, we'll also need to make a histogram to check the **Nearly Normal Condition** to be sure we can use the Normal model to model the data's distribution. Other times, we may be told that a Normal model is appropriate based on prior knowledge of the situation or on theoretical considerations.

How to Sketch a Normal Curve That Looks Normal   To sketch a good Normal curve, you need to remember only three things:

▶ The Normal curve is bell-shaped and symmetric around its mean. Start at the middle, and sketch to the right and left from there.

▶ Even though the Normal model extends forever on either side, you need to draw it only for 3 standard deviations. After that, there's so little left that it isn't worth sketching.

▶ The place where the bell shape changes from curving downward to curving back up—the *inflection point*—is exactly one standard deviation away from the mean.

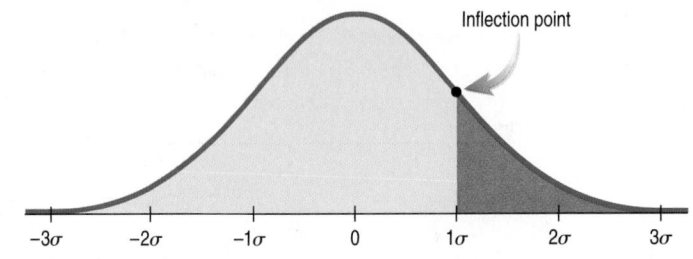

STEP-BY-STEP EXAMPLE   **Working with the 68–95–99.7 Rule**

The SAT Reasoning Test has three parts: Writing, Math, and Critical Reading (Verbal). Each part has a distribution that is roughly unimodal and symmetric and is designed to have an overall mean of about 500 and a standard deviation of 100 for all test takers. In any one year, the mean and standard deviation may differ from these target values by a small amount, but they are a good overall approximation.

Question: Suppose you earned a 600 on one part of your SAT. Where do you stand among all students who took that test?

You could calculate your *z*-score and find out that it's $z = (600 - 500)/100 = 1.0$, but what does that tell you about your percentile? You'll need the Normal model and the 68–95–99.7 Rule to answer that question.

	**Plan** State what you want to know.	*I want to see how my SAT score compares with the scores of all other students. To do that, I'll need to model the distribution.*
	Variables Identify the variable and report the W's.	*Let y = my SAT score. Scores are quantitative but have no meaningful units other than points.*
	Be sure to check the appropriate conditions.	✔ **Nearly Normal Condition:** *If I had data, I would check the histogram. I have no data, but I am told that the SAT scores are roughly unimodal and symmetric.*
	Specify the parameters of your model.	*I will model SAT score with a N(500, 100) model.*
SHOW	**Mechanics** Make a picture of this Normal model. (A simple sketch is all you need.)	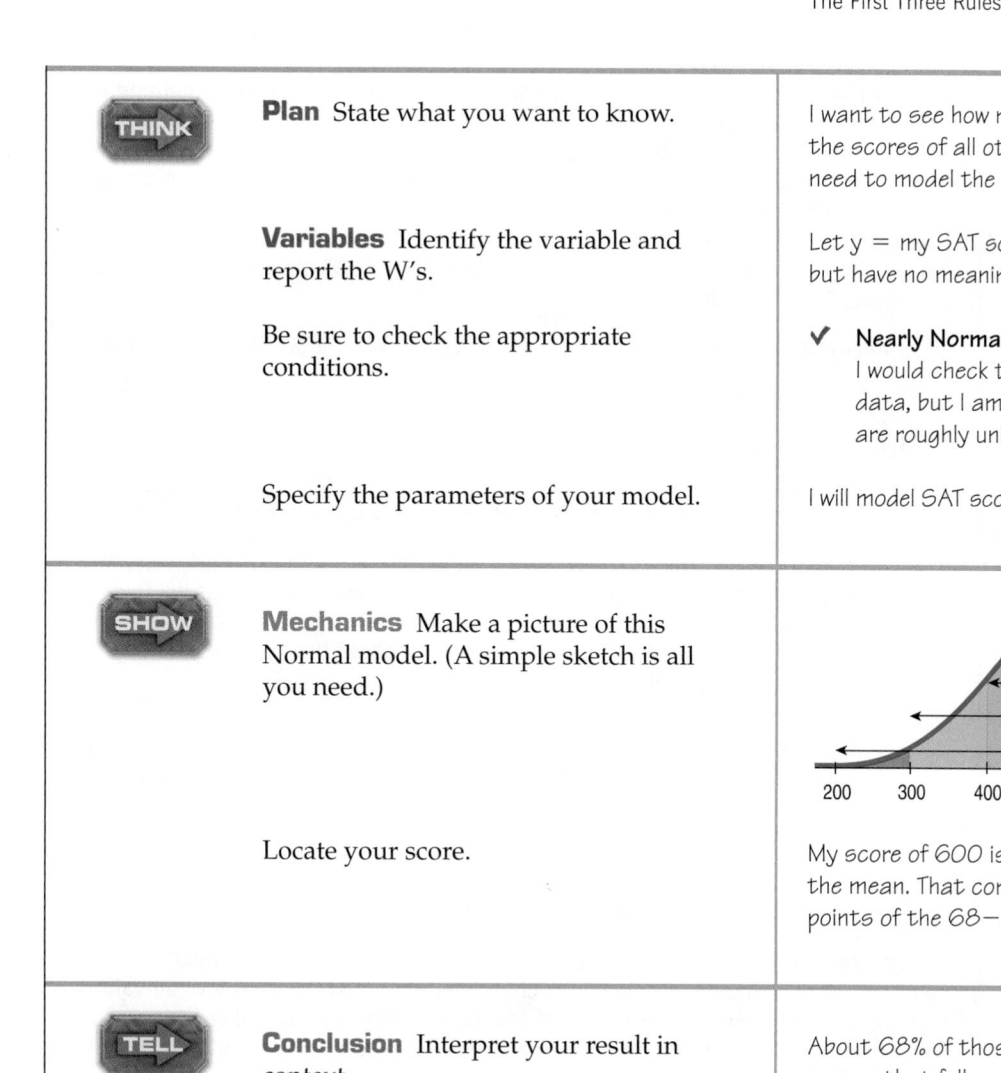
	Locate your score.	*My score of 600 is 1 standard deviation above the mean. That corresponds to one of the points of the 68–95–99.7 Rule.*
TELL	**Conclusion** Interpret your result in context.	*About 68% of those who took the test had scores that fell no more than 1 standard deviation from the mean, so 100% − 68% = 32% of all students had scores more than 1 standard deviation away. Only half of those were on the high side, so about 16% (half of 32%) of the test scores were better than mine. My score of 600 is higher than about 84% of all scores on this test.*

The bounds of SAT scoring at 200 and 800 can also be explained by the 68–95–99.7 Rule. Since 200 and 800 are three standard deviations from 500, it hardly pays to extend the scoring any farther on either side. We'd get more information only on 100 − 99.7 = 0.3% of students.

The Worst-Case Scenario* Suppose we encounter an observation that's 5 standard deviations above the mean. Should we be surprised? We've just seen that when a Normal model is appropriate, such a value is exceptionally rare. After all, 99.7% of all the values should be within 3 standard deviations of the mean, so anything farther away would be unusual indeed.

But our handy 68–95–99.7 Rule applies only to Normal models, and the Normal is such a *nice* shape. What if we're dealing with a distribution that's strongly skewed (like the CEO salaries), or one that is uniform or bimodal or something really strange? A Normal model has 68% of its observations within one standard deviation of the mean, but a bimodal distribution could even be entirely empty in the middle. In that case could we still say anything at all about an observation 5 standard deviations above the mean?

Remarkably, even with really weird distributions, the worst case can't get all that bad. A Russian mathematician named Pafnuty Tchebycheff[5] answered the question by proving this theorem:

In any distribution, at least $1 - \dfrac{1}{k^2}$ of the values must lie within $\pm k$ standard deviations of the mean.

What does that mean?

▶ For $k = 1$, $1 - \dfrac{1}{1^2} = 0$; if the distribution is far from Normal, it's possible that none of the values are within 1 standard deviation of the mean. We should be really cautious about saying anything about 68% unless we think a Normal model is justified. (Tchebycheff's theorem really is about the worst case; it tells us nothing about the middle; only about the extremes.)

▶ For $k = 2$, $1 - \dfrac{1}{2^2} = \dfrac{3}{4}$; no matter how strange the shape of the distribution, at least 75% of the values must be within 2 standard deviations of the mean. Normal models may expect 95% in that 2-standard-deviation interval, but even in a worst-case scenario it can never go lower than 75%.

▶ For $k = 3$, $1 - \dfrac{1}{3^2} = \dfrac{8}{9}$; in any distribution, at least 89% of the values lie within 3 standard deviations of the mean.

What we see is that values beyond 3 standard deviations from the mean are uncommon, Normal model or not. Tchebycheff tells us that at least 96% of all values must be within 5 standard deviations of the mean. While we can't always apply the 68–95–99.7 Rule, we can be sure that the observation we encountered 5 standard deviations above the mean is unusual.

Finding Normal Percentiles

A S *Activity:* **Your Pulse z-Score.** Is your pulse rate high or low? Find its z-score with the Normal Model Tool.

An SAT score of 600 is easy to assess, because we can think of it as one standard deviation above the mean. If your score was 680, though, where do you stand among the rest of the people tested? Your z-score is 1.80, so you're somewhere between 1 and 2 standard deviations above the mean. We figured out that no more than 16% of people score better than 600. By the same logic, no more than 2.5% of people score better than 700. Can we be more specific than "between 16% and 2.5%"?

When the value doesn't fall exactly 1, 2, or 3 standard deviations from the mean, we can look it up in a table of **Normal percentiles** or use technology.[6] Either way, we first convert our data to z-scores before using the table. Your SAT score of 680 has a z-score of $(680 - 500)/100 = 1.80$.

A S *Activity:* **The Normal Table.** Table Z just sits there, but this version of the Normal table changes so it always Makes a Picture that fits.

[5] He may have made the worst case for deviations clear, but the English spelling of his name is not. You'll find his first name spelled Pavnutii or Pavnuty and his last name spelled Chebsheff, Cebysev, and other creative versions.
[6] See Table Z in Appendix G, if you're curious. But your calculator (and any statistics computer package) does this, too—and more easily!

FIGURE 6.7

A table of Normal percentiles (Table Z in Appendix G) lets us find the percentage of individuals in a Standard Normal distribution falling below any specified z-score value.

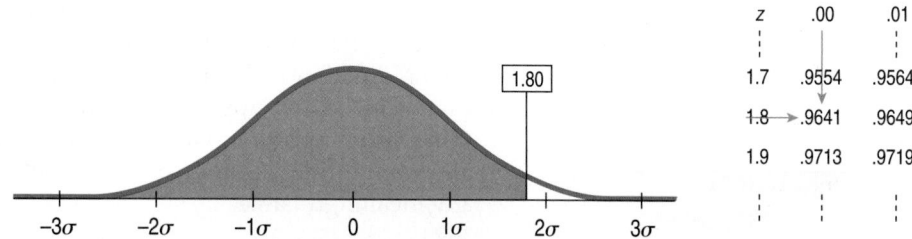

z	.00	.01
⋮	⋮	⋮
1.7	.9554	.9564
1.8 →	.9641	.9649
1.9	.9713	.9719
⋮	⋮	⋮

In the piece of the table shown, we find your z-score by looking down the left column for the first two digits, 1.8, and across the top row for the third digit, 0. The table gives the percentile as 0.9641. That means that 96.4% of the z-scores are less than 1.80. Only 3.6% of people, then, scored better than 680 on the SAT.

Most of the time, though, you'll do this with your calculator.

TI-*nspire*

Normal percentiles. Explore the relationship between z-scores and areas in a Normal model.

TI Tips — Finding Normal percentages

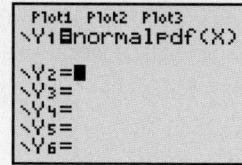

Your calculator knows the Normal model. Have a look under 2nd DISTR. There you will see three "norm" functions, normalpdf(, normalcdf(, and invNorm(. Let's play with the first two.

- normalpdf(calculates *y*-values for graphing a Normal curve. You probably won't use this very often, if at all. If you want to try it, graph Y1=normalpdf(X) in a graphing WINDOW with Xmin=-4, Xmax=4, Ymin=-0.1, and Ymax=0.5.

- normalcdf(finds the proportion of area under the curve between two z-score cut points, by specifying normalcdf(zLeft,zRight). Do make friends with this function; you will use it often!

Example 1

The Normal model shown shades the region between *z* = −0.5 and *z* = 1.0.

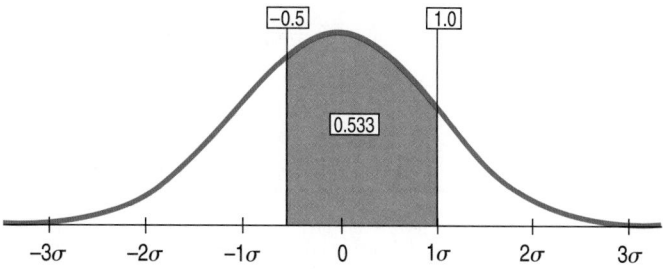

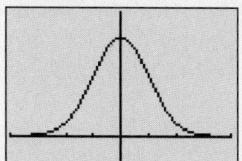

To find the shaded area:

Under 2nd DISTR select normalcdf(; hit ENTER.
Specify the cut points: normalcdf(-.5,1.0) and hit ENTER again.

There's the area. Approximately 53% of a Normal model lies between half a standard deviation below and one standard deviation above the mean.

Example 2

In the example in the text we used Table Z to determine the fraction of SAT scores above your score of 680. Now let's do it again, this time using your TI.

First we need z-scores for the cut points:

- Since 680 is 1.8 standard deviations above the mean, your z-score is 1.8; that's the left cut point.

• Theoretically the standard Normal model extends rightward forever, but you can't tell the calculator to use infinity as the right cut point. Recall that for a Normal model almost all the area lies within ±3 standard deviations of the mean, so any upper cut point beyond, say, $z = 5$ does not cut off anything very important. We suggest you always use 99 (or −99) when you really want infinity as your cut point—it's easy to remember and way beyond any meaningful area.

Now you're ready. Use the command `normalcdf(1.8,99)`.

There you are! The Normal model estimates that approximately 3.6% of SAT scores are higher than 680.

STEP-BY-STEP EXAMPLE | **Working with Normal Models Part I**

The Normal model is our first model for data. It's the first in a series of modeling situations where we step away from the data at hand to make more general statements about the world. We'll become more practiced in thinking about and learning the details of models as we progress through the book. To give you some practice in thinking about the Normal model, here are several problems that ask you to find percentiles in detail.

Question: What proportion of SAT scores fall between 450 and 600?

Plan State the problem.

Variables Name the variable.

Check the appropriate conditions and specify which Normal model to use.

I want to know the proportion of SAT scores between 450 and 600.

Let y = SAT score.

✔ **Nearly Normal Condition:** We are told that SAT scores are nearly Normal.

I'll model SAT scores with a $N(500, 100)$ model, using the mean and standard deviation specified for them.

Mechanics Make a picture of this Normal model. Locate the desired values and shade the region of interest.

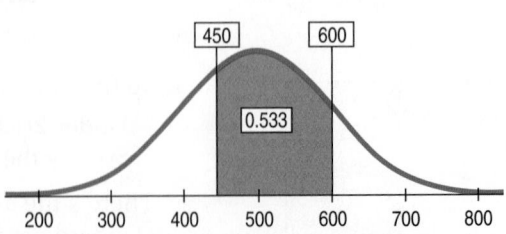

Find z-scores for the cut points 450 and 600. Use technology to find the desired proportions, represented by the area under the curve. (This was Example 1 in the TI Tips—take another look.)

Standardizing the two scores, I find that

$$z = \frac{(y - \mu)}{\sigma} = \frac{(600 - 500)}{100} = 1.00$$

and

$$z = \frac{(450 - 500)}{100} = -0.50$$

	So,
	Area $(450 < y < 600)$ = Area $(-0.5 < z < 1.0)$ = 0.5328
(If you use a table, then you need to subtract the two areas to find the area *between* the cut points.)	(**OR**: From Table Z, the area $(z < 1.0)$ = 0.8413 and area $(z < -0.5)$ = 0.3085, so the proportion of z-scores between them is $0.8413 - 0.3085 = 0.5328$, or 53.28%.)
TELL **Conclusion** Interpret your result in context.	The Normal model estimates that about 53.3% of SAT scores fall between 450 and 600.

From Percentiles to Scores: *z* in Reverse

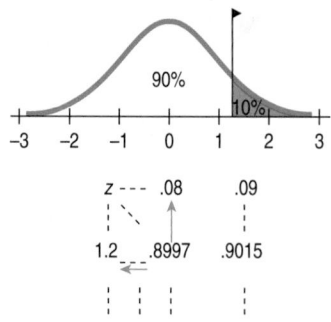

Finding areas from z-scores is the simplest way to work with the Normal model. But sometimes we start with areas and are asked to work backward to find the corresponding z-score or even the original data value. For instance, what z-score cuts off the top 10% in a Normal model?

Make a picture like the one shown, shading the rightmost 10% of the area. Notice that this is the 90th percentile. Look in Table Z for an area of 0.900. The exact area is not there, but 0.8997 is pretty close. That shows up in the table with 1.2 in the left margin and .08 in the top margin. The z-score for the 90th percentile, then, is approximately z = 1.28.

Computers and calculators will determine the cut point more precisely (and more easily).

TI Tips   Finding Normal cutpoints

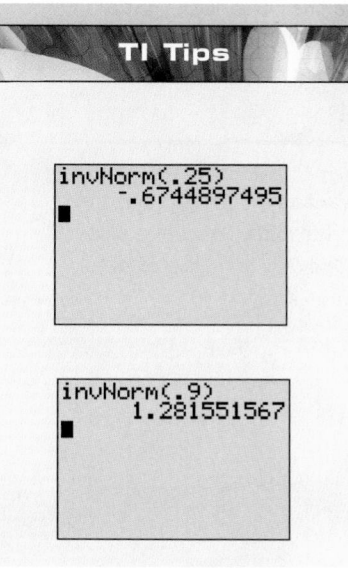

To find the z-score at the 25th percentile, go to **2nd DISTR** again. This time we'll use the third of the "norm" functions, **invNorm(**.

Just specify the desired percentile with the command **invNorm(.25)** and hit **ENTER**. The calculator says that the cut point for the leftmost 25% of a Normal model is approximately z = −0.674.

One more example: What z-score cuts off the highest 10% of a Normal model? That's easily done—just remember to specify the *percentile*. Since we want the cut point for the *highest* 10%, we know that the other 90% must be *below* that z-score. The cut point, then, must stand at the 90th percentile, so specify **invNorm(.90)**.

Only 10% of the area in a Normal model is more than about 1.28 standard deviations above the mean.

 **STEP-BY-STEP EXAMPLE** | **Working with Normal Models Part II**

Question: Suppose a college says it admits only people with SAT Verbal test scores among the top 10%. How high a score does it take to be eligible?

 THINK

Plan State the problem.

How high an SAT Verbal score do I need to be in the top 10% of all test takers?

Variable Define the variable.

Let y = my SAT score.

Check to see if a Normal model is appropriate, and specify which Normal model to use.

✔ **Nearly Normal Condition:** I am told that SAT scores are nearly Normal. I'll model them with N(500, 100).

 SHOW

Mechanics Make a picture of this Normal model. Locate the desired percentile approximately by shading the rightmost 10% of the area.

90%   10%

200 300 400 500 600 700 800

The college takes the top 10%, so its cutoff score is the 90th percentile. Find the corresponding z-score using your calculator as shown in the TI Tips. (**OR**: Use Table Z as shown on p. 119.)

The cut point is z = 1.28.

Convert the z-score back to the original units.

A z-score of 1.28 is 1.28 standard deviations above the mean. Since the SD is 100, that's 128 SAT points. The cutoff is 128 points above the mean of 500, or 628.

 TELL

Conclusion Interpret your results in the proper context.

Because the school wants SAT Verbal scores in the top 10%, the cutoff is 628. (Actually, since SAT scores are reported only in multiples of 10, I'd have to score at least a 630.)

TI-*nspire*

Normal models. Watch the Normal model react as you change the mean and standard deviation.

STEP-BY-STEP EXAMPLE | **More Working with Normal Models**

Working with Normal percentiles can be a little tricky, depending on how the problem is stated. Here are a few more worked examples of the kind you're likely to see.

A cereal manufacturer has a machine that fills the boxes. Boxes are labeled "16 ounces," so the company wants to have that much cereal in each box, but since no packaging process is perfect, there will be minor variations. If the machine is set at exactly 16 ounces and the Normal model applies (or at least the distribution is roughly symmetric), then about half of the boxes will be underweight, making consumers unhappy and exposing the company to bad publicity and possible lawsuits. To prevent underweight boxes, the manufacturer has to set the mean a little higher than 16.0 ounces.

Based on their experience with the packaging machine, the company believes that the amount of cereal in the boxes fits a Normal model with a standard deviation of 0.2 ounces. The manufacturer decides to set the machine to put an average of 16.3 ounces in each box. Let's use that model to answer a series of questions about these cereal boxes.

Question 1: What fraction of the boxes will be underweight?

Plan State the problem.

What proportion of boxes weigh less than 16 ounces?

Variable Name the variable.

Let y = weight of cereal in a box.

Check to see if a Normal model is appropriate.

✔ **Nearly Normal Condition:** I have no data, so I cannot make a histogram, but I am told that the company believes the distribution of weights from the machine is Normal.

Specify which Normal model to use.

I'll use a $N(16.3, 0.2)$ model.

Mechanics Make a picture of this Normal model. Locate the value you're interested in on the picture, label it, and shade the appropriate region.

REALITY CHECK ▶ Estimate from the picture the percentage of boxes that are underweight. (This will be useful later to check that your answer makes sense.) It looks like a low percentage. Less than 20% for sure.

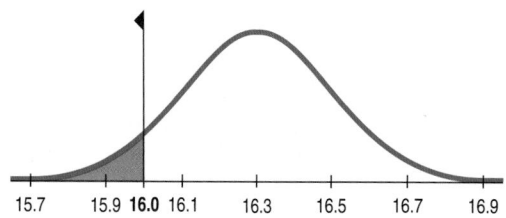

I want to know what fraction of the boxes will weigh less than 16 ounces.

Convert your cutoff value into a z-score.

$$z = \frac{y - \mu}{\sigma} = \frac{16 - 16.3}{0.2} = -1.50$$

Find the area with your calculator (or use the Normal table).

Area $(y < 16)$ = Area$(z < -1.50)$ = 0.0668

 **Conclusion** State your conclusion, and check that it's consistent with your earlier guess. It's below 20%—seems okay.

I estimate that approximately 6.7% of the boxes will contain less than 16 ounces of cereal.

Question 2: The company's lawyers say that 6.7% is too high. They insist that no more than 4% of the boxes can be underweight. So the company needs to set the machine to put a little more cereal in each box. What mean setting do they need?

 **Plan** State the problem.

What mean weight will reduce the proportion of underweight boxes to 4%?

Variable Name the variable.

Let y = weight of cereal in a box.

Check to see if a Normal model is appropriate.

✔ **Nearly Normal Condition:** *I am told that a Normal model applies.*

Specify which Normal model to use. This time you are not given a value for the mean!

I don't know μ, the mean amount of cereal. The standard deviation for this machine is 0.2 ounces. The model is N(μ, 0.2).

 **REALITY CHECK** We found out earlier that setting the machine to $\mu = 16.3$ ounces made 6.7% of the boxes too light. We'll need to raise the mean a bit to reduce this fraction.

No more than 4% of the boxes can be below 16 ounces.

 **Mechanics** Make a picture of this Normal model. Center it at μ (since you don't know the mean), and shade the region below 16 ounces.

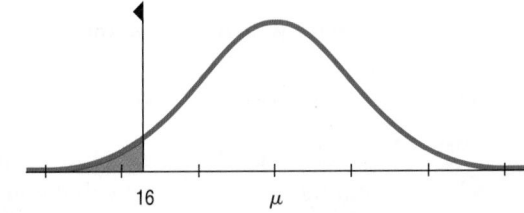

Using your calculator (or the Normal table), find the z-score that cuts off the lowest 4%.

The z-score that has 0.04 area to the left of it is z = −1.75.

Use this information to find μ. It's located 1.75 standard deviations to the right of 16. Since σ is 0.2, that's 1.75×0.2, or 0.35 ounces more than 16.

For 16 to be 1.75 standard deviations below the mean, the mean must be

$$16 + 1.75\,(0.2) = 16.35 \text{ ounces.}$$

Conclusion Interpret your result in context.
(This makes sense; we knew it would have to be just a bit higher than 16.3.)

The company must set the machine to average 16.35 ounces of cereal per box.

Question 3: The company president vetoes that plan, saying the company should give away less free cereal, not more. Her goal is to set the machine no higher than 16.2 ounces and still have only 4% underweight boxes. The only way to accomplish this is to reduce the standard deviation. What standard deviation must the company achieve, and what does that mean about the machine?

Plan State the problem.

What standard deviation will allow the mean to be 16.2 ounces and still have only 4% of boxes underweight?

Variable Name the variable.

Let y = weight of cereal in a box.

Check conditions to be sure that a Normal model is appropriate.

✓ **Nearly Normal Condition:** The company believes that the weights are described by a Normal model.

Specify which Normal model to use. This time you don't know σ.

I know the mean, but not the standard deviation, so my model is N(16.2, σ).

REALITY CHECK We know the new standard deviation must be less than 0.2 ounces.

Mechanics Make a picture of this Normal model. Center it at 16.2, and shade the area you're interested in. We want 4% of the area to the left of 16 ounces.

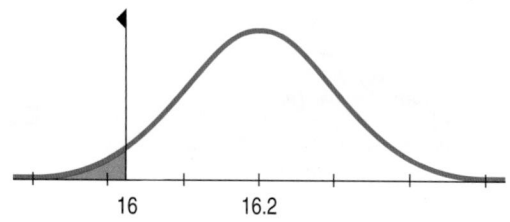

Find the z-score that cuts off the lowest 4%.

I know that the z-score with 4% below it is z = −1.75.

Solve for σ. (We need 16 to be 1.75 σ's below 16.2, so 1.75 σ must be 0.2 ounces. You could just start with that equation.)

$$z = \frac{y - \mu}{\sigma}$$

$$-1.75 = \frac{16 - 16.2}{\sigma}$$

$$1.75\,\sigma = 0.2$$

$$\sigma = 0.114$$

Conclusion Interpret your result in context.

As we expected, the standard deviation is lower than before—actually, quite a bit lower.

The company must get the machine to box cereal with a standard deviation of only 0.114 ounces. This means the machine must be more consistent (by nearly a factor of 2) in filling the boxes.

Are You Normal? Find Out with a Normal Probability Plot

In the examples we've worked through, we've assumed that the underlying data distribution was roughly unimodal and symmetric, so that using a Normal model makes sense. When you have data, you must *check* to see whether a Normal model is reasonable. How? Make a picture, of course! Drawing a histogram of the data and looking at the shape is one good way to see if a Normal model might work.

There's a more specialized graphical display that can help you to decide whether the Normal model is appropriate: the **Normal probability plot.** If the distribution of the data is roughly Normal, the plot is roughly a diagonal straight line. Deviations from a straight line indicate that the distribution is not Normal. This plot is usually able to show deviations from Normality more clearly than the corresponding histogram, but it's usually easier to understand *how* a distribution fails to be Normal by looking at its histogram.

Some data on a car's fuel efficiency provide an example of data that are nearly Normal. The overall pattern of the Normal probability plot is straight. The two trailing low values correspond to the values in the histogram that trail off the low end. They're not quite in line with the rest of the data set. The Normal probability plot shows us that they're a bit lower than we'd expect of the lowest two values in a Normal model.

TI-*nspire*

Normal probability plots and histograms. See how a normal probability plot responds as you change the shape of a distribution.

FIGURE 6.9

Histogram and Normal probability plot for gas mileage (mpg) recorded by one of the authors over the 8 years he owned a 1989 Nissan Maxima. The vertical axes are the same, so each dot on the probability plot would fall into the bar on the histogram immediately to its left.

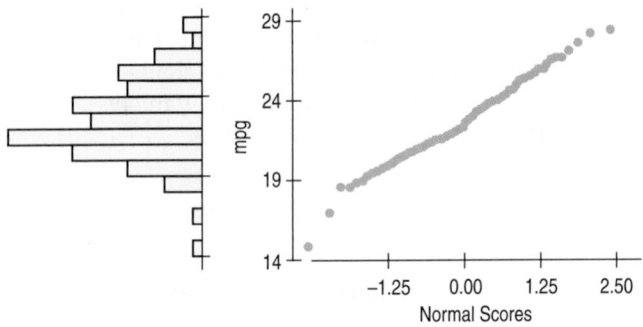

By contrast, the Normal probability plot of the men's *Weight*s from the NHANES Study is far from straight. The weights are skewed to the high end, and the plot is curved. We'd conclude from these pictures that approximations using the 68–95–99.7 Rule for these data would not be very accurate.

FIGURE 6.10

Histogram and Normal probability plot for men's weights. Note how a skewed distribution corresponds to a bent probability plot.

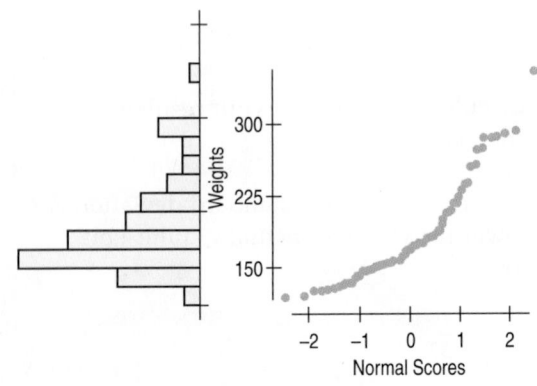

TI Tips

Creating a Normal probability plot

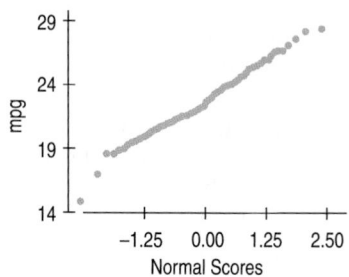

Let's make a Normal probability plot with the calculator. Here are the boys' agility test scores we looked at in Chapter 5; enter them in **L1**:

$$22, 17, 18, 29, 22, 23, 24, 23, 17, 21$$

Now you can create the plot:

- Turn a **STATPLOT On**.
- Tell it to make a Normal probability plot by choosing the last of the icons.
- Specify your datalist and which axis you want the data on. (We'll use **Y** so the plot looks like the others we showed you.)
- Specify the **Mark** you want the plot to use.
- Now **ZoomStat** does the rest.

The plot doesn't look very straight. Normality is certainly questionable here.

(Not that it matters in making this decision, but that vertical line is the y-axis. Points to the left have negative z-scores and points to the right have positive z-scores.)

How Does a Normal Probability Plot Work?

Why does the Normal probability plot work like that? We looked at 100 fuel efficiency measures for the author's Nissan car. The smallest of these has a z-score of -3.16. The Normal model can tell us what value to expect for the smallest z-score in a batch of 100 if a Normal model were appropriate. That turns out to be -2.58. So our first data value is smaller than we would expect from the Normal.

We can continue this and ask a similar question for each value. For example, the 14th-smallest fuel efficiency has a z-score of almost exactly -1, and that's just what we should expect (well, -1.1 to be exact). A Normal probability plot takes each data value and plots it against the z-score you'd expect that point to have if the distribution were perfectly Normal.[7]

When the values match up well, the line is straight. If one or two points are surprising from the Normal's point of view, they don't line up. When the entire distribution is skewed or different from the Normal in some other way, the values don't match up very well at all and the plot bends.

It turns out to be tricky to find the values we expect. They're called *Normal scores*, but you can't easily look them up in the tables. That's why probability plots are best made with technology and not by hand.

The best advice on using Normal probability plots is to see whether they are straight. If so, then your data look like data from a Normal model. If not, make a histogram to understand how they differ from the model.

A S *Activity:* **Assessing Normality.** This activity guides you through the process of checking the Nearly Normal condition using your statistics package.

---

[7] Sometimes the Normal probability plot switches the two axes, putting the data on the x-axis and the z-scores on the y-axis.

WHAT CAN GO WRONG?

▶ **Don't use a Normal model when the distribution is not unimodal and symmetric.** Normal models are so easy and useful that it is tempting to use them even when they don't describe the data very well. That can lead to wrong conclusions. Don't use a Normal model without first checking the **Nearly Normal Condition.** Look at a picture of the data to check that it is unimodal and symmetric. A histogram, or a Normal probability plot, can help you tell whether a Normal model is appropriate.

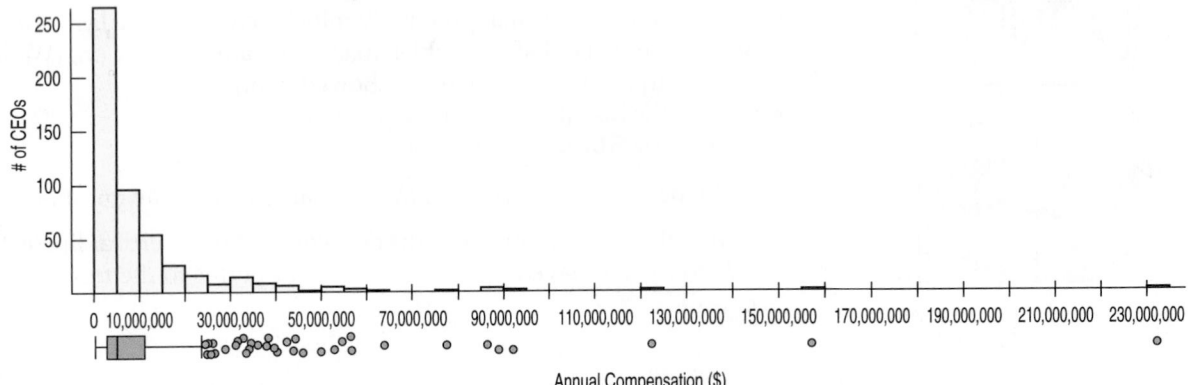

The CEOs (p. 90) had a mean total compensation of $10,307,311.87 with a standard deviation of $17,964,615.16. Using the Normal model rule, we should expect about 68% of the CEOs to have compensations between −$7,657,303.29 and $28,271,927.03. In fact, more than 90% of the CEOs have annual compensations in this range. What went wrong? The distribution is skewed, not symmetric. Using the 68–95–99.7 Rule for data like these will lead to silly results.

▶ **Don't use the mean and standard deviation when outliers are present.** Both means and standard deviations can be distorted by outliers, and no model based on distorted values will do a good job. A z-score calculated from a distribution with outliers may be misleading. It's always a good idea to check for outliers. How? Make a picture.

▶ **Don't round your results in the middle of a calculation.** We *reported* the mean of the heptathletes' long jump as 6.16 meters. More precisely, it was 6.16153846153846 meters.

You should use all the precision available in the data for all the intermediate steps of a calculation. Using the more precise value for the mean (and also carrying 15 digits for the SD), the z-score calculation for Klüft's long jump comes out to

$$z = \frac{6.78 - 6.16153846153846}{0.2297597407326585} = 2.691775053755667700$$

We'd report that as 2.692, as opposed to the rounded-off value of 2.70 we got earlier from the table.

▶ **Don't worry about minor differences in results.** Because various calculators and programs may carry different precision in calculations, your answers may differ slightly from those we show in the text and in the Step-By-Steps, or even from the values given in the answers in the back of the book. Those differences aren't anything to worry about. They're not the main story Statistics tries to tell.

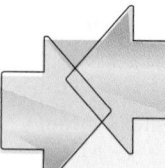

CONNECTIONS

Changing the center and spread of a variable is equivalent to changing its *units*. Indeed, the only part of the data's context changed by standardizing is the units. All other aspects of the context do not depend on the choice or modification of measurement units. This fact points out an important distinction between the numbers the data provide for calculation and the meaning of the variables and the relationships among them. Standardizing can make the numbers easier to work with, but it does not alter the meaning.

Another way to look at this is to note that standardizing may change the center and spread values, but it does not affect the *shape* of a distribution. A histogram or boxplot of standardized values looks just the same as the histogram or boxplot of the original values except, perhaps, for the numbers on the axes.

When we summarized *shape, center,* and *spread* for histograms, we compared them to unimodal, symmetric shapes. You couldn't ask for a nicer example than the Normal model. And if the shape *is* like a Normal, we'll use the the mean and standard deviation to standardize the values.

WHAT HAVE WE LEARNED?

We've learned that the story data can tell may be easier to understand after shifting or rescaling the data.

▸ Shifting data by adding or subtracting the same amount from each value affects measures of center and position but not measures of spread.
▸ Rescaling data by multiplying or dividing every value by a constant, changes all the summary statistics—center, position, and spread.

We've learned the power of standardizing data.

▸ Standardizing uses the standard deviation as a ruler to measure distance from the mean, creating z-scores.
▸ Using these z-scores, we can compare apples and oranges—values from different distributions or values based on different units.
▸ And a z-score can identify unusual or surprising values among data.

We've learned that the 68–95–99.7 Rule can be a useful rule of thumb for understanding distributions.

▸ For data that are unimodal and symmetric, about 68% fall within 1 SD of the mean, 95% fall within 2 SDs of the mean, and 99.7% fall within 3 SDs of the mean (see p. 130).

Again we've seen the importance of *Thinking* about whether a method will work.

▸ **Normality Assumption:** We sometimes work with Normal tables (Table Z). Those tables are based on the Normal model.
▸ Data can't be exactly Normal, so we check the **Nearly Normal Condition** by making a histogram (is it unimodal, symmetric, and free of outliers?) or a Normal probability plot (is it straight enough?). (See p. 125.)

Terms

Standardizing	105. We standardize to eliminate units. Standardized values can be compared and combined even if the original variables had different units and magnitudes.
Standardized value	105. A value found by subtracting the mean and dividing by the standard deviation.

Shifting	107. Adding a constant to each data value adds the same constant to the mean, the median, and the quartiles, but does not change the standard deviation or IQR.
Rescaling	108. Multiplying each data value by a constant multiplies both the measures of position (mean, median, and quartiles) and the measures of spread (standard deviation and IQR) by that constant.
Normal model	112. A useful family of models for unimodal, symmetric distributions.
Parameter	112. A numerically valued attribute of a model. For example, the values of μ and σ in a $N(\mu, \sigma)$ model are parameters.
Statistic	112. A value calculated from data to summarize aspects of the data. For example, the mean, $\bar{y}$ and standard deviation, s, are statistics.
z-score	105. A z-score tells how many standard deviations a value is from the mean; z-scores have a mean of 0 and a standard deviation of 1. When working with data, use the statistics $\bar{y}$ and s:

$$z = \frac{y - \bar{y}}{s}.$$

112. When working with models, use the parameters μ and σ:

$$z = \frac{y - \mu}{\sigma}.$$

Standard Normal model	112. A Normal model, $N(\mu, \sigma)$ with mean $\mu = 0$ and standard deviation $\sigma = 1$. Also called the **standard Normal distribution.**
Nearly Normal Condition	112. A distribution is nearly Normal if it is unimodal and symmetric. We can check by looking at a histogram or a Normal probability plot.
68–95–99.7 Rule	113. In a Normal model, about 68% of values fall within 1 standard deviation of the mean, about 95% fall within 2 standard deviations of the mean, and about 99.7% fall within 3 standard deviations of the mean.
Normal percentile	116. The Normal percentile corresponding to a z-score gives the percentage of values in a standard Normal distribution found at that z-score or below.
Normal probability plot	124. A display to help assess whether a distribution of data is approximately Normal. If the plot is nearly straight, the data satisfy the **Nearly Normal Condition.**

Skills

- ▸ Understand how adding (subtracting) a constant or multiplying (dividing) by a constant changes the center and/or spread of a variable.
- ▸ Recognize when standardization can be used to compare values.
- ▸ Understand that standardizing uses the standard deviation as a ruler.
- ▸ Recognize when a Normal model is appropriate.

- ▸ Know how to calculate the z-score of an observation.
- ▸ Know how to compare values of two different variables using their z-scores.
- ▸ Be able to use Normal models and the 68–95–99.7 Rule to estimate the percentage of observations falling within 1, 2, or 3 standard deviations of the mean.
- ▸ Know how to find the percentage of observations falling below any value in a Normal model using a Normal table or appropriate technology.
- ▸ Know how to check whether a variable satisfies the **Nearly Normal Condition** by making a Normal probability plot or a histogram.

- ▸ Know what z-scores mean.
- ▸ Be able to explain how extraordinary a standardized value may be by using a Normal model.

NORMAL PLOTS ON THE COMPUTER

The best way to tell whether your data can be modeled well by a Normal model is to make a picture or two. We've already talked about making histograms. Normal probability plots are almost never made by hand because the values of the Normal scores are tricky to find. But most statistics software make Normal plots, though various packages call the same plot by different names and array the information differently.

EXERCISES

1. **Shipments.** A company selling clothing on the Internet reports that the packages it ships have a median weight of 68 ounces and an IQR of 40 ounces.
   a) The company plans to include a sales flyer weighing 4 ounces in each package. What will the new median and IQR be?
   b) If the company recorded the shipping weights of these new packages in pounds instead of ounces, what would the median and IQR be? (1 lb. = 16 oz.)

2. **Hotline.** A company's customer service hotline handles many calls relating to orders, refunds, and other issues. The company's records indicate that the median length of calls to the hotline is 4.4 minutes with an IQR of 2.3 minutes.
   a) If the company were to describe the duration of these calls in seconds instead of minutes, what would the median and IQR be?
   b) In an effort to speed up the customer service process, the company decides to streamline the series of push-button menus customers must navigate, cutting the time by 24 seconds. What will the median and IQR of the length of hotline calls become?

3. **Payroll.** Here are the summary statistics for the weekly payroll of a small company: lowest salary = \$300, mean salary = \$700, median = \$500, range = \$1200, IQR = \$600, first quartile = \$350, standard deviation = \$400.
   a) Do you think the distribution of salaries is symmetric, skewed to the left, or skewed to the right? Explain why.
   b) Between what two values are the middle 50% of the salaries found?
   c) Suppose business has been good and the company gives every employee a \$50 raise. Tell the new value of each of the summary statistics.
   d) Instead, suppose the company gives each employee a 10% raise. Tell the new value of each of the summary statistics.

4. **Hams.** A specialty foods company sells "gourmet hams" by mail order. The hams vary in size from 4.15 to 7.45 pounds, with a mean weight of 6 pounds and standard deviation of 0.65 pounds. The quartiles and median weights are 5.6, 6.2, and 6.55 pounds.
   a) Find the range and the IQR of the weights.
   b) Do you think the distribution of the weights is symmetric or skewed? If skewed, which way? Why?

   c) If these weights were expressed in ounces (1 pound = 16 ounces) what would the mean, standard deviation, quartiles, median, IQR, and range be?
   d) When the company ships these hams, the box and packing materials add 30 ounces. What are the mean, standard deviation, quartiles, median, IQR, and range of weights of boxes shipped (in ounces)?
   e) One customer made a special order of a 10-pound ham. Which of the summary statistics of part d might *not* change if that data value were added to the distribution?

5. **SAT or ACT?** Each year thousands of high school students take either the SAT or the ACT, standardized tests used in the college admissions process. Combined SAT Math and Verbal scores go as high as 1600, while the maximum ACT composite score is 36. Since the two exams use very different scales, comparisons of performance are difficult. A convenient rule of thumb is $SAT = 40 \times ACT + 150$; that is, multiply an ACT score by 40 and add 150 points to estimate the equivalent SAT score. An admissions officer reported the following statistics about the ACT scores of 2355 students who applied to her college one year. Find the summaries of equivalent SAT scores.

   Lowest score = 19   Mean = 27   Standard deviation = 3
   Q3 = 30   Median = 28   IQR = 6

6. **Cold U?** A high school senior uses the Internet to get information on February temperatures in the town where he'll be going to college. He finds a Web site with some statistics, but they are given in degrees Celsius. The conversion formula is $°F = 9/5 \, °C + 32$. Determine the Fahrenheit equivalents for the summary information below.

   Maximum temperature = 11°C   Range = 33°
   Mean = 1°   Standard deviation = 7°
   Median = 2°   IQR = 16°

7. **Stats test.** Suppose your Statistics professor reports test grades as z-scores, and you got a score of 2.20 on an exam. Write a sentence explaining what that means.

8. **Checkup.** One of the authors has an adopted grandson whose birth family members are very short. After examining him at his 2-year checkup, the boy's pediatrician said that the z-score for his height relative to American 2-year-olds was −1.88. Write a sentence explaining what that means.

9. **Stats test, part II.**   The mean score on the Stats exam was 75 points with a standard deviation of 5 points, and Gregor's z-score was −2. How many points did he score?

10. **Mensa.**   People with z-scores above 2.5 on an IQ test are sometimes classified as geniuses. If IQ scores have a mean of 100 and a standard deviation of 16 points, what IQ score do you need to be considered a genius?

11. **Temperatures.**   A town's January high temperatures average 36°F with a standard deviation of 10°, while in July the mean high temperature is 74° and the standard deviation is 8°. In which month is it more unusual to have a day with a high temperature of 55°? Explain.

12. **Placement exams.**   An incoming freshman took her college's placement exams in French and mathematics. In French, she scored 82 and in math 86. The overall results on the French exam had a mean of 72 and a standard deviation of 8, while the mean math score was 68, with a standard deviation of 12. On which exam did she do better compared with the other freshmen?

13. **Combining test scores.**   The first Stats exam had a mean of 65 and a standard deviation of 10 points; the second had a mean of 80 and a standard deviation of 5 points. Derrick scored an 80 on both tests. Julie scored a 70 on the first test and a 90 on the second. They both totaled 160 points on the two exams, but Julie claims that her total is better. Explain.

14. **Combining scores again.**   The first Stat exam had a mean of 80 and a standard deviation of 4 points; the second had a mean of 70 and a standard deviation of 15 points. Reginald scored an 80 on the first test and an 85 on the second. Sara scored an 88 on the first but only a 65 on the second. Although Reginald's total score is higher, Sara feels she should get the higher grade. Explain her point of view.

15. **Final exams.**   Anna, a language major, took final exams in both French and Spanish and scored 83 on each. Her roommate Megan, also taking both courses, scored 77 on the French exam and 95 on the Spanish exam. Overall, student scores on the French exam had a mean of 81 and a standard deviation of 5, and the Spanish scores had a mean of 74 and a standard deviation of 15.
    a) To qualify for language honors, a major must maintain at least an 85 average for all language courses taken. So far, which student qualifies?
    b) Which student's overall performance was better?

16. **MP3s.**   Two companies market new batteries targeted at owners of personal music players. DuraTunes claims a mean battery life of 11 hours, while RockReady advertises 12 hours.
    a) Explain why you would also like to know the standard deviations of the battery lifespans before deciding which brand to buy.
    b) Suppose those standard deviations are 2 hours for DuraTunes and 1.5 hours for RockReady. You are headed for 8 hours at the beach. Which battery is most likely to last all day? Explain.
    c) If your beach trip is all weekend, and you probably will have the music on for 16 hours, which battery is most likely to last? Explain.

17. **Cattle.**   The Virginia Cooperative Extension reports that the mean weight of yearling Angus steers is 1152 pounds. Suppose that weights of all such animals can be described by a Normal model with a standard deviation of 84 pounds.
    a) How many standard deviations from the mean would a steer weighing 1000 pounds be?
    b) Which would be more unusual, a steer weighing 1000 pounds or one weighing 1250 pounds?

18. **Car speeds.**   John Beale of Stanford, CA, recorded the speeds of cars driving past his house, where the speed limit read 20 mph. The mean of 100 readings was 23.84 mph, with a standard deviation of 3.56 mph. (He actually recorded every car for a two-month period. These are 100 representative readings.)
    a) How many standard deviations from the mean would a car going under the speed limit be?
    b) Which would be more unusual, a car traveling 34 mph or one going 10 mph?

19. **More cattle.**   Recall that the beef cattle described in Exercise 17 had a mean weight of 1152 pounds, with a standard deviation of 84 pounds.
    a) Cattle buyers hope that yearling Angus steers will weigh at least 1000 pounds. To see how much over (or under) that goal the cattle are, we could subtract 1000 pounds from all the weights. What would the new mean and standard deviation be?
    b) Suppose such cattle sell at auction for 40 cents a pound. Find the mean and standard deviation of the sale prices for all the steers.

20. **Car speeds again.**   For the car speed data of Exercise 18, recall that the mean speed recorded was 23.84 mph, with a standard deviation of 3.56 mph. To see how many cars are speeding, John subtracts 20 mph from all speeds.
    a) What is the mean speed now? What is the new standard deviation?
    b) His friend in Berlin wants to study the speeds, so John converts all the original miles-per-hour readings to kilometers per hour by multiplying all speeds by 1.609 (km per mile). What is the mean now? What is the new standard deviation?

21. **Cattle, part III.**   Suppose the auctioneer in Exercise 19 sold a herd of cattle whose minimum weight was 980 pounds, median was 1140 pounds, standard deviation 84 pounds, and IQR 102 pounds. They sold for 40 cents a pound, and the auctioneer took a $20 commission on each animal. Then, for example, a steer weighing 1100 pounds would net the owner 0.40 (1100) − 20 = $420. Find the minimum, median, standard deviation, and IQR of the net sale prices.

22. **Caught speeding.**   Suppose police set up radar surveillance on the Stanford street described in Exercise 18. They handed out a large number of tickets to speeders going a mean of 28 mph, with a standard deviation of 2.4 mph, a maximum of 33 mph, and an IQR of 3.2 mph. Local law prescribes fines of $100, plus $10 per mile per hour over the 20 mph speed limit. For example, a driver convicted of going 25 mph would be fined 100 + 10(5) = $150. Find the mean, maximum, standard deviation, and IQR of all the potential fines.

23. Professors. A friend tells you about a recent study dealing with the number of years of teaching experience among current college professors. He remembers the mean but can't recall whether the standard deviation was 6 months, 6 years, or 16 years. Tell him which one it must have been, and why.

24. Rock concerts. A popular band on tour played a series of concerts in large venues. They always drew a large crowd, averaging 21,359 fans. While the band did not announce (and probably never calculated) the standard deviation, which of these values do you think is most likely to be correct: 20, 200, 2000, or 20,000 fans? Explain your choice.

25. Guzzlers? Environmental Protection Agency (EPA) fuel economy estimates for automobile models tested recently predicted a mean of 24.8 mpg and a standard deviation of 6.2 mpg for highway driving. Assume that a Normal model can be applied.
a) Draw the model for auto fuel economy. Clearly label it, showing what the 68–95–99.7 Rule predicts.
b) In what interval would you expect the central 68% of autos to be found?
c) About what percent of autos should get more than 31 mpg?
d) About what percent of cars should get between 31 and 37.2 mpg?
e) Describe the gas mileage of the worst 2.5% of all cars.

26. IQ. Some IQ tests are standardized to a Normal model, with a mean of 100 and a standard deviation of 16.
a) Draw the model for these IQ scores. Clearly label it, showing what the 68–95–99.7 Rule predicts.
b) In what interval would you expect the central 95% of IQ scores to be found?
c) About what percent of people should have IQ scores above 116?
d) About what percent of people should have IQ scores between 68 and 84?
e) About what percent of people should have IQ scores above 132?

27. Small steer. In Exercise 17 we suggested the model $N(1152, 84)$ for weights in pounds of yearling Angus steers. What weight would you consider to be unusually low for such an animal? Explain.

28. High IQ. Exercise 26 proposes modeling IQ scores with $N(100, 16)$. What IQ would you consider to be unusually high? Explain.

29. Trees. A forester measured 27 of the trees in a large woods that is up for sale. He found a mean diameter of 10.4 inches and a standard deviation of 4.7 inches. Suppose that these trees provide an accurate description of the whole forest and that a Normal model applies.
a) Draw the Normal model for tree diameters.
b) What size would you expect the central 95% of all trees to be?
c) About what percent of the trees should be less than an inch in diameter?
d) About what percent of the trees should be between 5.7 and 10.4 inches in diameter?
e) About what percent of the trees should be over 15 inches in diameter?

30. Rivets. A company that manufactures rivets believes the shear strength (in pounds) is modeled by $N(800, 50)$.
a) Draw and label the Normal model.
b) Would it be safe to use these rivets in a situation requiring a shear strength of 750 pounds? Explain.
c) About what percent of these rivets would you expect to fall below 900 pounds?
d) Rivets are used in a variety of applications with varying shear strength requirements. What is the maximum shear strength for which you would feel comfortable approving this company's rivets? Explain your reasoning.

31. Trees, part II. Later on, the forester in Exercise 29 shows you a histogram of the tree diameters he used in analyzing the woods that was for sale. Do you think he was justified in using a Normal model? Explain, citing some specific concerns.

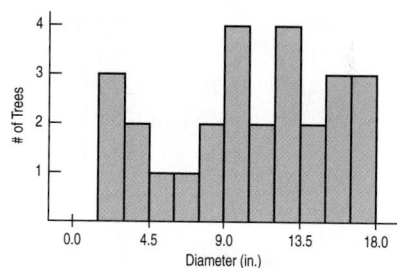

T 32. Car speeds, the picture. For the car speed data of Exercise 18, here is the histogram, boxplot, and Normal probability plot of the 100 readings. Do you think it is appropriate to apply a Normal model here? Explain.

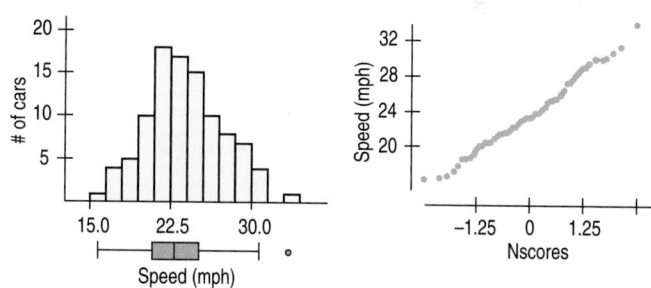

T 33. Winter Olympics 2006 downhill. Fifty-three men qualified for the men's alpine downhill race in Torino. The gold medal winner finished in 1 minute, 48.8 seconds. All competitors' times (in seconds) are found in the following list:

108.80	109.52	109.82	109.88	109.93	110.00
110.04	110.12	110.29	110.33	110.35	110.44
110.45	110.64	110.68	110.70	110.72	110.84
110.88	110.88	110.90	110.91	110.98	111.37
111.48	111.51	111.55	111.70	111.72	111.93
112.17	112.55	112.87	112.90	113.34	114.07
114.65	114.70	115.01	115.03	115.73	116.10
116.58	116.81	117.45	117.54	117.56	117.69
118.77	119.24	119.41	119.79	120.93	

a) The mean time was 113.02 seconds, with a standard deviation of 3.24 seconds. If the Normal model is appropriate, what percent of times will be less than 109.78 seconds?
b) What is the actual percent of times less than 109.78 seconds?
c) Why do you think the two percentages don't agree?
d) Create a histogram of these times. What do you see?

T 34. Check the model. The mean of the 100 car speeds in Exercise 20 was 23.84 mph, with a standard deviation of 3.56 mph.
a) Using a Normal model, what values should border the middle 95% of all car speeds?
b) Here are some summary statistics.

Percentile		Speed
100%	**Max**	34.060
97.5%		30.976
90.0%		28.978
75.0%	**Q3**	25.785
50.0%	**Median**	23.525
25.0%	**Q1**	21.547
10.0%		19.163
2.5%		16.638
0.0%	**Min**	16.270

From your answer in part a, how well does the model do in predicting those percentiles? Are you surprised? Explain.

T 35. Receivers. NFL data from the 2006 football season reported the number of yards gained by each of the league's 167 wide receivers:

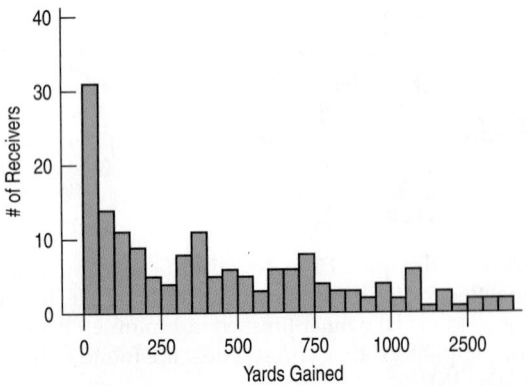

The mean is 435 yards, with a standard deviation of 384 yards.
a) According to the Normal model, what percent of receivers would you expect to gain fewer yards than 2 standard deviations below the mean number of yards?
b) For these data, what does that mean?
c) Explain the problem in using a Normal model here.

36. Customer database. A large philanthropic organization keeps records on the people who have contributed to their cause. In addition to keeping records of past giving, the organization buys demographic data on neighbor-

hoods from the U.S. Census Bureau. Eighteen of these variables concern the ethnicity of the neighborhood of the donor. Here are a histogram and summary statistics for the percentage of whites in the neighborhoods of 500 donors:

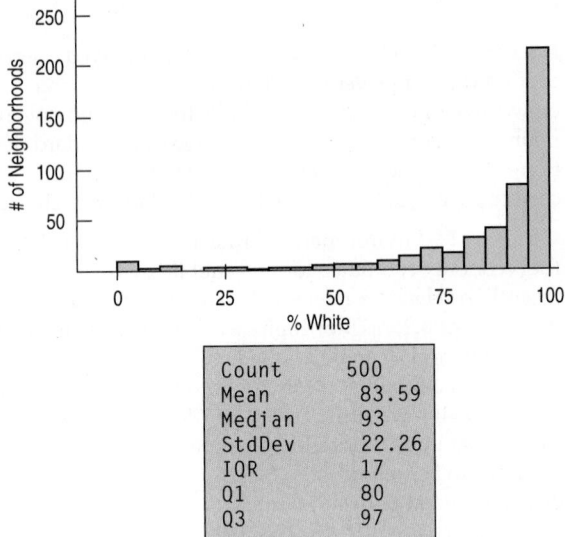

Count	500
Mean	83.59
Median	93
StdDev	22.26
IQR	17
Q1	80
Q3	97

a) Which is a better summary of the percentage of white residents in the neighborhoods, the mean or the median? Explain.
b) Which is a better summary of the spread, the IQR or the standard deviation? Explain.
c) From a Normal model, about what percentage of neighborhoods should have a percent white within one standard deviation of the mean?
d) What percentage of neighborhoods actually have a percent white within one standard deviation of the mean?
e) Explain the discrepancy between parts c and d.

37. Normal cattle. Using $N(1152, 84)$, the Normal model for weights of Angus steers in Exercise 17, what percent of steers weigh
a) over 1250 pounds?
b) under 1200 pounds?
c) between 1000 and 1100 pounds?

38. IQs revisited. Based on the Normal model $N(100, 16)$ describing IQ scores, what percent of people's IQs would you expect to be
a) over 80?
b) under 90?
c) between 112 and 132?

39. More cattle. Based on the model $N(1152, 84)$ describing Angus steer weights, what are the cutoff values for
a) the highest 10% of the weights?
b) the lowest 20% of the weights?
c) the middle 40% of the weights?

40. More IQs. In the Normal model $N(100, 16)$, what cutoff value bounds
a) the highest 5% of all IQs?
b) the lowest 30% of the IQs?
c) the middle 80% of the IQs?

41. Cattle, finis. Consider the Angus weights model $N(1152, 84)$ one last time.
a) What weight represents the 40th percentile?
b) What weight represents the 99th percentile?
c) What's the IQR of the weights of these Angus steers?

42. IQ, finis. Consider the IQ model $N(100, 16)$ one last time.
a) What IQ represents the 15th percentile?
b) What IQ represents the 98th percentile?
c) What's the IQR of the IQs?

43. Cholesterol. Assume the cholesterol levels of adult American women can be described by a Normal model with a mean of 188 mg/dL and a standard deviation of 24.
a) Draw and label the Normal model.
b) What percent of adult women do you expect to have cholesterol levels over 200 mg/dL?
c) What percent of adult women do you expect to have cholesterol levels between 150 and 170 mg/dL?
d) Estimate the IQR of the cholesterol levels.
e) Above what value are the highest 15% of women's cholesterol levels?

44. Tires. A tire manufacturer believes that the treadlife of its snow tires can be described by a Normal model with a mean of 32,000 miles and standard deviation of 2500 miles.
a) If you buy a set of these tires, would it be reasonable for you to hope they'll last 40,000 miles? Explain.
b) Approximately what fraction of these tires can be expected to last less than 30,000 miles?
c) Approximately what fraction of these tires can be expected to last between 30,000 and 35,000 miles?
d) Estimate the IQR of the treadlives.
e) In planning a marketing strategy, a local tire dealer wants to offer a refund to any customer whose tires fail to last a certain number of miles. However, the dealer does not want to take too big a risk. If the dealer is willing to give refunds to no more than 1 of every 25 customers, for what mileage can he guarantee these tires to last?

45. Kindergarten. Companies that design furniture for elementary school classrooms produce a variety of sizes for kids of different ages. Suppose the heights of kindergarten children can be described by a Normal model with a mean of 38.2 inches and standard deviation of 1.8 inches.
a) What fraction of kindergarten kids should the company expect to be less than 3 feet tall?
b) In what height interval should the company expect to find the middle 80% of kindergarteners?
c) At least how tall are the biggest 10% of kindergarteners?

46. Body temperatures. Most people think that the "normal" adult body temperature is 98.6°F. That figure, based on a 19th-century study, has recently been challenged.

In a 1992 article in the *Journal of the American Medical Association*, researchers reported that a more accurate figure may be 98.2°F. Furthermore, the standard deviation appeared to be around 0.7°F. Assume that a Normal model is appropriate.
a) In what interval would you expect most people's body temperatures to be? Explain.
b) What fraction of people would be expected to have body temperatures above 98.6°F?
c) Below what body temperature are the coolest 20% of all people?

47. Eggs. Hens usually begin laying eggs when they are about 6 months old. Young hens tend to lay smaller eggs, often weighing less than the desired minimum weight of 54 grams.
a) The average weight of the eggs produced by the young hens is 50.9 grams, and only 28% of their eggs exceed the desired minimum weight. If a Normal model is appropriate, what would the standard deviation of the egg weights be?
b) By the time these hens have reached the age of 1 year, the eggs they produce average 67.1 grams, and 98% of them are above the minimum weight. What is the standard deviation for the appropriate Normal model for these older hens?
c) Are egg sizes more consistent for the younger hens or the older ones? Explain.

48. Tomatoes. Agricultural scientists are working on developing an improved variety of Roma tomatoes. Marketing research indicates that customers are likely to bypass Romas that weigh less than 70 grams. The current variety of Roma plants produces fruit that averages 74 grams, but 11% of the tomatoes are too small. It is reasonable to assume that a Normal model applies.
a) What is the standard deviation of the weights of Romas now being grown?
b) Scientists hope to reduce the frequency of undersized tomatoes to no more than 4%. One way to accomplish this is to raise the average size of the fruit. If the standard deviation remains the same, what target mean should they have as a goal?
c) The researchers produce a new variety with a mean weight of 75 grams, which meets the 4% goal. What is the standard deviation of the weights of these new Romas?
d) Based on their standard deviations, compare the tomatoes produced by the two varieties.

JUST CHECKING

Answers

1. **a)** On the first test, the mean is 88 and the SD is 4, so $z = (90 - 88)/4 = 0.5$. On the second test, the mean is 75 and the SD is 5, so $z = (80 - 75)/5 = 1.0$. The first test has the lower z-score, so it is the one that will be dropped.

   b) No. The second test is 1 standard deviation above the mean, farther away than the first test, so it's the better score relative to the class.

2. **a)** The mean would increase to 500.

   b) The standard deviation is still 100 points.

   c) The two boxplots would look nearly identical (the shape of the distribution would remain the same), but the later one would be shifted 50 points higher.

3. The standard deviation is now 2.54 millimeters, which is the same as 0.1 inches. Nothing has changed. The standard deviation has "increased" only because we're reporting it in millimeters now, not inches.

4. The mean is 184 centimeters, with a standard deviation of 8 centimeters. 2 meters is 200 centimeters, which is 2 standard deviations above the mean. We expect 5% of the men to be more than 2 standard deviations below or above the mean, so half of those, 2.5%, are likely to be above 2 meters.

5. **a)** We know that 68% of the time we'll be within 1 standard deviation (2 min) of 20. So 32% of the time we'll arrive in less than 18 or more than 22 minutes. Half of those times (16%) will be greater than 22 minutes, so 84% will be less than 22 minutes.

   b) 24 minutes is 2 standard deviations above the mean. Because of the 95% rule, we know 2.5% of the times will be more than 24 minutes.

   c) Traffic incidents may occasionally increase the time it takes to get to school, so the driving times may be skewed to the right, and there may be outliers.

   d) If so, the Normal model would not be appropriate and the percentages we predict would not be accurate.

REVIEW OF PART I

Exploring and Understanding Data

Quick Review

It's time to put it all together. Real data don't come tagged with instructions for use. So let's step back and look at how the key concepts and skills we've seen work together. This brief list and the review exercises that follow should help you check your understanding of Statistics so far.

▶ We treat data two ways: as categorical and as quantitative.

▶ To describe categorical data:

- Make a picture. Bar graphs work well for comparing counts in categories.
- Summarize the distribution with a table of counts or relative frequencies (percents) in each category.
- Pie charts and segmented bar charts display divisions of a whole.
- Compare distributions with plots side by side.
- Look for associations between variables by comparing marginal and conditional distributions.

▶ To describe quantitative data:

- Make a picture. Use histograms, boxplots, stem-and-leaf displays, or dotplots. Stem-and-leafs are great when working by hand and good for small data sets. Histograms are a good way to see the distribution. Boxplots are best for comparing several distributions.
- Describe distributions in terms of their shape, center, and spread, and note any unusual features such as gaps or outliers.
- The shape of most distributions you'll see will likely be uniform, unimodal, or bimodal. It may be multimodal. If it is unimodal, then it may be symmetric or skewed.
- A 5-number summary makes a good numerical description of a distribution: min, Q1, median, Q3, and max.

- If the distribution is skewed, be sure to include the median and interquartile range (IQR) when you describe its center and spread.
- A distribution that is severely skewed may benefit from re-expressing the data. If it is skewed to the high end, taking logs often works well.
- If the distribution is unimodal and symmetric, describe its center and spread with the mean and standard deviation.
- Use the standard deviation as a ruler to tell how unusual an observed value may be, or to compare or combine measurements made on different scales.
- Shifting a distribution by adding or subtracting a constant affects measures of position but not measures of spread. Rescaling by multiplying or dividing by a constant affects both.
- When a distribution is roughly unimodal and symmetric, a Normal model may be useful. For Normal models, the 68–95–99.7 Rule is a good rule of thumb.
- If the Normal model fits well (check a histogram or Normal probability plot), then Normal percentile tables or functions found in most statistics technology can provide more detailed values.

Need more help with some of this? It never hurts to reread sections of the chapters! And in the following pages we offer you more opportunities[1] to review these concepts and skills.

The exercises that follow use the concepts and skills you've learned in the first six chapters. To be more realistic and more useful for your review, they don't tell you which of the concepts or methods you need. But neither will the exam.

---

[1] If you doubted that we are teachers, this should convince you. Only a teacher would call additional homework exercises "opportunities."

REVIEW EXERCISES

1. Bananas. Here are the prices (in cents per pound) of bananas reported from 15 markets surveyed by the U.S. Department of Agriculture.

51	52	45
48	53	52
50	49	52
48	43	46
45	42	50

a) Display these data with an appropriate graph.
b) Report appropriate summary statistics.
c) Write a few sentences about this distribution.

2. Prenatal care. Results of a 1996 American Medical Association report about the infant mortality rate for twins carried for the full term of a normal pregnancy are shown on the next page, broken down by the level of prenatal care the mother had received.

Full-Term Pregnancies, Level of Prenatal Care	Infant Mortality Rate Among Twins (deaths per thousand live births)
Intensive	5.4
Adequate	3.9
Inadequate	6.1
Overall	5.1

a) Is the overall rate the average of the other three rates? Should it be? Explain.

b) Do these results indicate that adequate prenatal care is important for pregnant women? Explain.

c) Do these results suggest that a woman pregnant with twins should be wary of seeking too much medical care? Explain.

3. Singers.   The boxplots shown display the heights (in inches) of 130 members of a choir.

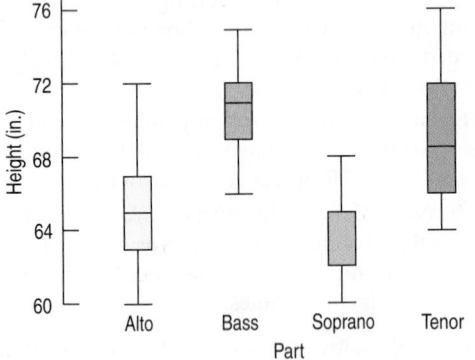

a) It appears that the median height for sopranos is missing, but actually the median and the upper quartile are equal. How could that happen?

b) Write a few sentences describing what you see.

4. Dialysis.   In a study of dialysis, researchers found that "of the three patients who were currently on dialysis, 67% had developed blindness and 33% had their toes amputated." What kind of display might be appropriate for these data? Explain.

5. Beanstalks.   Beanstalk Clubs are social clubs for very tall people. To join, a man must be over 6′2″ tall, and a woman over 5′10″. The National Health Survey suggests that heights of adults may be Normally distributed, with mean heights of 69.1″ for men and 64.0″ for women. The respective standard deviations are 2.8″ and 2.5″.

a) You are probably not surprised to learn that men are generally taller than women, but what does the greater standard deviation for men's heights indicate?

b) Who are more likely to qualify for Beanstalk membership, men or women? Explain.

6. Bread.   Clarksburg Bakery is trying to predict how many loaves to bake. In the last 100 days, they have sold between 95 and 140 loaves per day. Here is a histogram of the number of loaves they sold for the last 100 days.

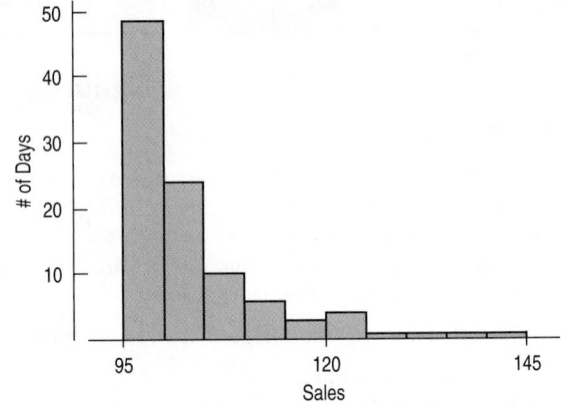

a) Describe the distribution.

b) Which should be larger, the mean number of sales or the median? Explain.

c) Here are the summary statistics for Clarksburg Bakery's bread sales. Use these statistics and the histogram above to create a boxplot. You may approximate the values of any outliers.

Summary of Sales	
Median	100
Min	95
Max	140
25th %tile	97
75th %tile	105.5

d) For these data, the mean was 103 loaves sold per day, with a standard deviation of 9 loaves. Do these statistics suggest that Clarksburg Bakery should expect to sell between 94 and 112 loaves on about 68% of the days? Explain.

7. State University.   Public relations staff at State U. collected data on people's opinions of various colleges and universities in their state. They phoned 850 local residents. After identifying themselves, the callers asked the survey participants their ages, whether they had attended college, and whether they had a favorable opinion of the university. The official report to the university's directors claimed that, in general, people had very favorable opinions about their university.

a) Identify the W's of these data.

b) Identify the variables, classify each as categorical or quantitative, and specify units if relevant.

c) Are you confident about the report's conclusion? Explain.

8. Acid rain.   Based on long-term investigation, researchers have suggested that the acidity (pH) of rainfall in the Shenandoah Mountains can be described by the Normal model N(4.9, 0.6).

a) Draw and carefully label the model..

b) What percent of storms produce rainfall with pH over 6?

c) What percent of storms produce rainfall with pH under 4?

d) The lower the pH, the more acidic the rain. What is the pH level for the most acidic 20% of all storms?

e) What is the pH level for the least acidic 5% of all storms?

f) What is the IQR for the pH of rainfall?

9. Fraud detection. A credit card bank is investigating the incidence of fraudulent card use. The bank suspects that the type of product bought may provide clues to the fraud. To examine this situation, the bank looks at the Standard Industrial Code (SIC) of the business related to the transaction. This is a code that was used by the U.S. Census Bureau and Statistics Canada to identify the type of every registered business in North America.[2] For example, 1011 designates Meat and Meat Products (except Poultry), 1012 is Poultry Products, 1021 is Fish Products, 1031 is Canned and Preserved Fruits and Vegetables, and 1032 is Frozen Fruits and Vegetables.

A company intern produces the following histogram of the SIC codes for 1536 transactions:

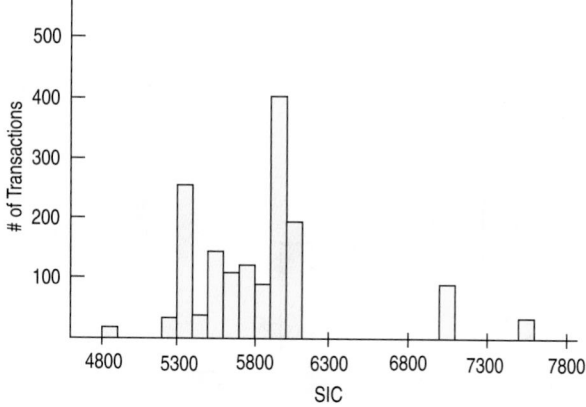

He also reports that the mean SIC is 5823.13 with a standard deviation of 488.17.

a) Comment on any problems you see with the use of the mean and standard deviation as summary statistics.

b) How well do you think the Normal model will work on these data? Explain.

10. Streams. As part of the course work, a class at an upstate NY college collects data on streams each year. Students record a number of biological, chemical, and physical variables, including the stream name, the substrate of the stream (*limestone, shale,* or *mixed*), the pH, the temperature (°C), and the BCI, a measure of biological diversity.

Group	Count	%
Limestone	77	44.8
Mixed	26	15.1
Shale	69	40.1

[2] Since 1997 the SIC has been replaced by the NAICS, a code of six letters.

a) Name each variable, indicating whether it is categorical or quantitative, and giving the units if available.

b) These streams have been classified according to their substrate—the composition of soil and rock over which they flow—as summarized in the table. What kind of graph might be used to display these data?

11. Cramming. One Thursday, researchers gave students enrolled in a section of basic Spanish a set of 50 new vocabulary words to memorize. On Friday the students took a vocabulary test. When they returned to class the following Monday, they were retested—without advance warning. Both sets of test scores for the 28 students are shown below.

Fri	Mon	Fri	Mon
42	36	50	47
44	44	34	34
45	46	38	31
48	38	43	40
44	40	39	41
43	38	46	32
41	37	37	36
35	31	40	31
43	32	41	32
48	37	48	39
43	41	37	31
45	32	36	41
47	44		

a) Create a graphical display to compare the two distributions of scores.

b) Write a few sentences about the scores reported on Friday and Monday.

c) Create a graphical display showing the distribution of the *changes* in student scores.

d) Describe the distribution of changes.

12. Computers and Internet. A U.S. Census Bureau report (August 2000, *Current Population Survey*) found that 51.0% of homes had a personal computer and 41.5% had access to the Internet. A newspaper concluded that 92.5% of homes had either a computer or access to the Internet. Do you agree? Explain.

13. Let's play cards. You pick a card from a deck (see description in Chapter 11) and record its denomination (7, say) and its suit (maybe spades).

a) Is the variable *suit* categorical or quantitative?

b) Name a game you might be playing for which you would consider the variable *denomination* to be categorical. Explain.

c) Name a game you might be playing for which you would consider the variable *denomination* to be quantitative. Explain.

14. Accidents. In 2001, Progressive Insurance asked customers who had been involved in auto accidents how far they were from home when the accident happened. The data are summarized in the table.

Miles from Home	% of Accidents
Less than 1	23
1 to 5	29
6 to 10	17
11 to 15	8
16 to 20	6
Over 20	17

a) Create an appropriate graph of these data.
b) Do these data indicate that driving near home is particularly dangerous? Explain.

15. Hard water. In an investigation of environmental causes of disease, data were collected on the annual mortality rate (deaths per 100,000) for males in 61 large towns in England and Wales. In addition, the water hardness was recorded as the calcium concentration (parts per million, ppm) in the drinking water.
a) What are the variables in this study? For each, indicate whether it is quantitative or categorical and what the units are.
b) Here are histograms of calcium concentration and mortality. Describe the distributions of the two variables.

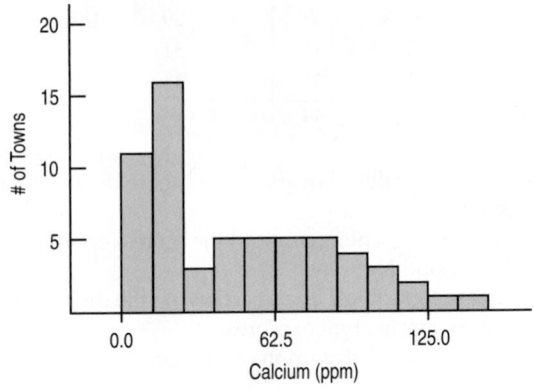

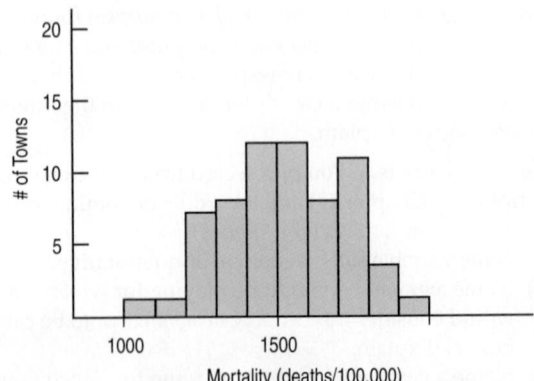

16. Hard water II. The data set from England and Wales also notes for each town whether it was south or north of Derby. Here are some summary statistics and a comparative boxplot for the two regions.

Summary of Mortality				
Group	Count	Mean	Median	StdDev
North	34	1631.59	1631	138.470
South	27	1388.85	1369	151.114

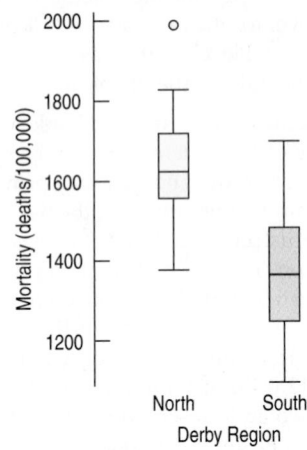

a) What is the overall mean mortality rate for the two regions?
b) Do you see evidence of a difference in mortality rates? Explain.

17. Seasons. Average daily temperatures in January and July for 60 large U.S. cities are graphed in the histograms below.

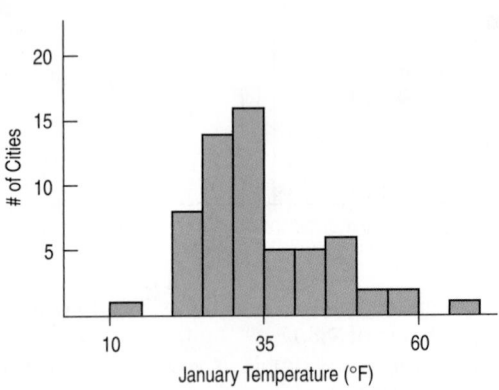

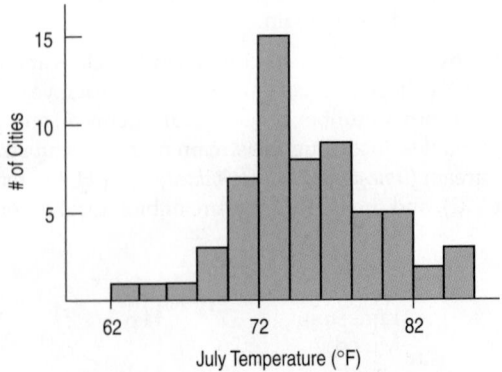

a) What aspect of these histograms makes it difficult to compare the distributions?

b) What differences do you see between the distributions of January and July average temperatures?

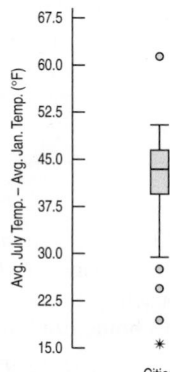

c) Differences in temperatures (July–January) for each of the cities are displayed in the boxplot above. Write a few sentences describing what you see.

18. Old Faithful. It is a common belief that Yellowstone's most famous geyser erupts once an hour at very predictable intervals. The histogram below shows the time gaps (in minutes) between 222 successive eruptions. Describe this distribution.

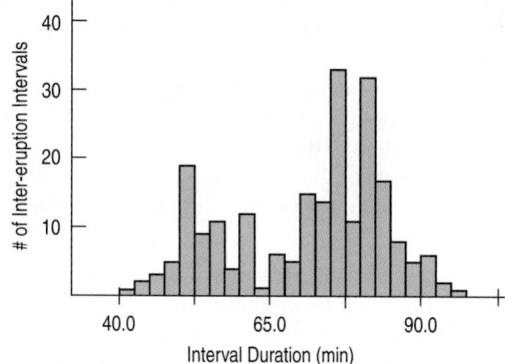

19. Old Faithful? Does the duration of an eruption have an effect on the length of time that elapses before the next eruption?

a) The histogram below shows the duration (in minutes) of those 222 eruptions. Describe this distribution.

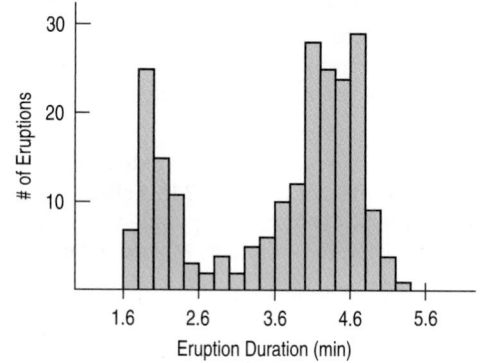

b) Explain why it is not appropriate to find summary statistics for this distribution.

c) Let's classify the eruptions as "long" or "short," depending upon whether or not they last at least 3 minutes. Describe what you see in the comparative boxplots.

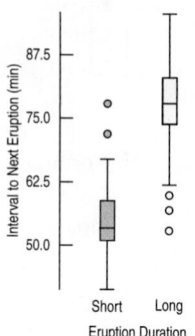

20. Teen drivers. In its *Traffic Safety Facts 2005*, the National Highway Traffic Safety Administration reported that 6.3% of licensed drivers were between the ages of 15 and 20, yet this age group was behind the wheel in 15.9% of all fatal crashes. Use these statistics to explain the concept of independence.

T 21. Liberty's nose. Is the Statue of Liberty's nose too long? Her nose measures, 4'6", but she is a large statue, after all. Her arm is 42 feet long. That means her arm is $42/45 = 9.3$ times as long as her nose. Is that a reasonable ratio? Shown in the table are arm and nose lengths of 18 girls in a Statistics class, and the ratio of arm-to-nose length for each.

Arm (cm)	Nose (cm)	Arm/Nose Ratio
73.8	5.0	14.8
74.0	4.5	16.4
69.5	4.5	15.4
62.5	4.7	13.3
68.6	4.4	15.6
64.5	4.8	13.4
68.2	4.8	14.2
63.5	4.4	14.4
63.5	5.4	11.8
67.0	4.6	14.6
67.4	4.4	15.3
70.7	4.3	16.4
69.4	4.1	16.9
71.7	4.5	15.9
69.0	4.4	15.7
69.8	4.5	15.5
71.0	4.8	14.8
71.3	4.7	15.2

a) Make an appropriate plot and describe the distribution of the ratios.

b) Summarize the ratios numerically, choosing appropriate measures of center and spread.

c) Is the ratio of 9.3 for the Statue of Liberty unrealistically low? Explain.

T **22. Winter Olympics 2006 speed skating.** The top 25 women's 500-m speed skating times are listed in the table below:

Skater	Country	Time
Svetlana Zhurova	Russia	76.57
Wang Manli	China	76.78
Hui Ren	China	76.87
Tomomi Okazaki	Japan	76.92
Lee Sang-Hwa	South Korea	77.04
Jenny Wolf	Germany	77.25
Wang Beixing	China	77.27
Sayuri Osuga	Japan	77.39
Sayuri Yoshii	Japan	77.43
Chiara Simionato	Italy	77.68
Jennifer Rodriguez	United States	77.70
Annette Gerritsen	Netherlands	78.09
Xing Aihua	China	78.35
Sanne van der Star	Netherlands	78.59
Yukari Watanabe	Japan	78.65
Shannon Rempel	Canada	78.85
Amy Sannes	United States	78.89
Choi Seung-Yong	South Korea	79.02
Judith Hesse	Germany	79.03
Kim You-Lim	South Korea	79.25
Kerry Simpson	Canada	79.34
Krisy Myers	Canada	79.43
Elli Ochowicz	United States	79.48
Pamela Zoellner	Germany	79.56
Lee Bo-Ra	South Korea	79.73

a) The mean finishing time was 78.21 seconds, with a standard deviation of 1.03 second. If the Normal model is appropriate, what percent of the times should be within 0.5 second of 78.21?

b) What percent of the times actually fall within this interval?

c) Explain the discrepancy between a and b.

23. Sample. A study in South Africa focusing on the impact of health insurance identified 1590 children at birth and then sought to conduct follow-up health studies 5 years later. Only 416 of the original group participated in the 5-year follow-up study. This made researchers concerned that the follow-up group might not accurately resemble the total group in terms of health insurance. The table in the next column summarizes the two groups by race and by presence of medical insurance when the child was born. Carefully explain how this study demonstrates Simpson's paradox. (*Birth to Ten Study*, Medical Research Council, South Africa)

		Number (%) Insured	
		Follow-up	**Not traced**
Race	**Black**	36 of 404 (8.9%)	91 of 1048 (8.7%)
	White	10 of 12 (83.3%)	104 of 126 (82.5%)
	Overall	46 of 416 (11.1%)	195 of 1174 (16.6%)

24. Sluggers. Roger Maris's 1961 home run record stood until Mark McGwire hit 70 in 1998. Listed below are the home run totals for each season McGwire played. Also listed are Babe Ruth's home run totals.

McGwire: 3*, 49, 32, 33, 39, 22, 42, 9*, 9*, 39, 52, 58, 70, 65, 32*, 29*

Ruth: 54, 59, 35, 41, 46, 25, 47, 60, 54, 46, 49, 46, 41, 34, 22

a) Find the 5-number summary for McGwire's career.

b) Do any of his seasons appear to be outliers? Explain.

c) McGwire played in only 18 games at the end of his first big league season, and missed major portions of some other seasons because of injuries to his back and knees. Those seasons might not be representative of his abilities. They are marked with asterisks in the list above. Omit these values and make parallel boxplots comparing McGwire's career to Babe Ruth's.

d) Write a few sentences comparing the two sluggers.

e) Create a side-by-side stem-and-leaf display comparing the careers of the two players.

f) What aspects of the distributions are apparent in the stem-and-leaf displays that did not clearly show in the boxplots?

25. Be quick! Avoiding an accident when driving can depend on reaction time. That time, measured from the moment the driver first sees the danger until he or she steps on the brake pedal, is thought to follow a Normal model with a mean of 1.5 seconds and a standard deviation of 0.18 seconds.

a) Use the 68–95–99.7 Rule to draw the Normal model.

b) Write a few sentences describing driver reaction times.

c) What percent of drivers have a reaction time less than 1.25 seconds?

d) What percent of drivers have reaction times between 1.6 and 1.8 seconds?

e) What is the interquartile range of reaction times?

f) Describe the reaction times of the slowest 1/3 of all drivers.

26. Music and memory. Is it a good idea to listen to music when studying for a big test? In a study conducted by some Statistics students, 62 people were randomly assigned to listen to rap music, Mozart, or no music while attempting to memorize objects pictured on a page. They were then asked to list all the objects they could remember. Here are the 5-number summaries for each group:

	n	Min	Q1	Median	Q3	Max
Rap	29	5	8	10	12	25
Mozart	20	4	7	10	12	27
None	13	8	9.5	13	17	24

a) Describe the W's for these data: *Who, What, Where, Why, When, How.*
b) Name the variables and classify each as categorical or quantitative.
c) Create parallel boxplots as best you can from these summary statistics to display these results.
d) Write a few sentences comparing the performances of the three groups.

Ⓣ **27. Mail.** Here are the number of pieces of mail received at a school office for 36 days.

123	70	90	151	115	97
80	78	72	100	128	130
52	103	138	66	135	76
112	92	93	143	100	88
118	118	106	110	75	60
95	131	59	115	105	85

a) Plot these data.
b) Find appropriate summary statistics.
c) Write a brief description of the school's mail deliveries.
d) What percent of the days actually lie within one standard deviation of the mean? Comment.

Ⓣ **28. Birth order.** Is your birth order related to your choice of major? A Statistics professor at a large university polled his students to find out what their majors were and what position they held in the family birth order. The results are summarized in the table.
a) What percent of these students are oldest or only children?
b) What percent of Humanities majors are oldest children?
c) What percent of oldest children are Humanities students?
d) What percent of the students are oldest children majoring in the Humanities?

	Birth Order*				
	1	**2**	**3**	**4+**	**Total**
Math/Science	34	14	6	3	57
Agriculture	52	27	5	9	93
Humanities	15	17	8	3	43
Other	12	11	1	6	30
Total	113	69	20	21	223

Major (row label at left)

* 1 = oldest or only child

29. Herbal medicine. Researchers for the Herbal Medicine Council collected information on people's experiences with a new herbal remedy for colds. They went to a store that sold natural health products. There they asked 100 customers whether they had taken the cold remedy and, if so, to rate its effectiveness (on a scale from 1 to 10) in curing their symptoms. The Council concluded that this product was highly effective in treating the common cold.
a) Identify the W's of these data.
b) Identify the variables, classify each as categorical or quantitative, and specify units if relevant.
c) Are you confident about the Council's conclusion? Explain.

Ⓣ **30. Birth order revisited.** Consider again the data on birth order and college majors in Exercise 28.
a) What is the marginal distribution of majors?
b) What is the conditional distribution of majors for the oldest children?
c) What is the conditional distribution of majors for the children born second?
d) Do you think that college major appears to be independent of birth order? Explain.

31. Engines. One measure of the size of an automobile engine is its "displacement," the total volume (in liters or cubic inches) of its cylinders. Summary statistics for several models of new cars are shown. These displacements were measured in cubic inches.

Summary of Displacement	
Count	38
Mean	177.29
Median	148.5
StdDev	88.88
Range	275
25th %tile	105
75th %tile	231

a) How many cars were measured?
b) Why might the mean be so much larger than the median?
c) Describe the center and spread of this distribution with appropriate statistics.
d) Your neighbor is bragging about the 227-cubic-inch engine he bought in his new car. Is that engine unusually large? Explain.
e) Are there any engines in this data set that you would consider to be outliers? Explain.
f) Is it reasonable to expect that about 68% of car engines measure between 88 and 266 cubic inches? (That's 177.289 ± 88.8767.) Explain.
g) We can convert all the data from cubic inches to cubic centimeters (cc) by multiplying by 16.4. For example, a 200-cubic-inch engine has a displacement of 3280 cc. How would such a conversion affect each of the summary statistics?

32. Engines, again. Horsepower is another measure commonly used to describe auto engines. Here are the summary statistics and histogram displaying horsepowers of the same group of 38 cars discussed in Exercise 31.

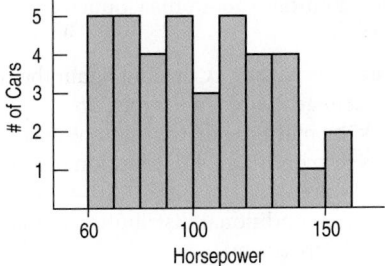

Summary of Horsepower

Count	38
Mean	101.7
Median	100
StdDev	26.4
Range	90
25th %tile	78
75th %tile	125

a) Describe the shape, center, and spread of this distribution.
b) What is the interquartile range?
c) Are any of these engines outliers in terms of horsepower? Explain.
d) Do you think the 68–95–99.7 Rule applies to the horsepower of auto engines? Explain.
e) From the histogram, make a rough estimate of the percentage of these engines whose horsepower is within one standard deviation of the mean.
f) A fuel additive boasts in its advertising that it can "add 10 horsepower to any car." Assuming that is true, what would happen to each of these summary statistics if this additive were used in all the cars?

33. **Age and party 2007.**   The Pew Research Center conducts surveys regularly asking respondents which political party they identify with. Among their results is the following table relating preferred political party and age. (http://people-press.org/reports/)

		Party			
		Republican	**Democrat**	**Others**	Total
Age	**18–29**	2636	2738	4765	**10139**
	30–49	6871	6442	8160	**21473**
	50–64	3896	4286	4806	**12988**
	65+	3131	3718	2934	**9784**
	Total	**16535**	**17183**	**20666**	**54384**

a) What percent of people surveyed were Republicans?
b) Do you think this might be a reasonable estimate of the percentage of all voters who are Republicans? Explain.
c) What percent of people surveyed were under 30 or over 65?
d) What percent of people were classified as "Other" and under the age of 30?

e) What percent of the people classified as "Other" were under 30?
f) What percent of people under 30 were classified as "Other"?

34. **Pay.**   According to the *2006 National Occupational Employment and Wage Estimates for Management Occupations,* the mean hourly wage for Chief Executives was $69.52 and the median hourly wage was "over $70.00." By contrast, for General and Operations Managers, the mean hourly wage was $47.73 and the median was $40.97. Are these wage distributions likely to be symmetric, skewed left, or skewed right? Explain.

35. **Age and party II.**   Consider again the Pew Research Center results on age and political party in Exercise 33.
a) What is the marginal distribution of party affiliation?
b) Create segmented bar graphs displaying the conditional distribution of party affiliation for each age group.
c) Summarize these poll results in a few sentences that might appear in a newspaper article about party affiliation in the United States.
d) Do you think party affiliation is independent of the voter's age? Explain.

36. **Bike safety 2003.**   The Bicycle Helmet Safety Institute website includes a report on the number of bicycle fatalities per year in the United States. The table below shows the counts for the years 1994–2003.

Year	Bicycle fatalities
1994	796
1995	828
1996	761
1997	811
1998	757
1999	750
2000	689
2001	729
2002	663
2003	619

a) What are the W's for these data?
b) Display the data in a stem-and-leaf display.
c) Display the data in a timeplot.
d) What is apparent in the stem-and-leaf display that is hard to see in the timeplot?
e) What is apparent in the timeplot that is hard to see in the stem-and-leaf display?
f) Write a few sentences about bicycle fatalities in the United States.

37. **Some assembly required.**   A company that markets build-it-yourself furniture sells a computer desk that is advertised with the claim "less than an hour to assemble." However, through postpurchase surveys the company has learned that only 25% of its customers succeeded in building the desk in under an hour. The mean time was 1.29 hours. The company assumes that consumer assembly time follows a Normal model.

a) Find the standard deviation of the assembly time model.

b) One way the company could solve this problem would be to change the advertising claim. What assembly time should the company quote in order that 60% of customers succeed in finishing the desk by then?

c) Wishing to maintain the "less than an hour" claim, the company hopes that revising the instructions and labeling the parts more clearly can improve the 1-hour success rate to 60%. If the standard deviation stays the same, what new lower mean time does the company need to achieve?

d) Months later, another postpurchase survey shows that new instructions and part labeling did lower the mean assembly time, but only to 55 minutes. Nonetheless, the company did achieve the 60%-in-an-hour goal, too. How was that possible?

T **38. Profits.** Here is a stem-and-leaf display showing profits as a percent of sales for 29 of the *Forbes* 500 largest U.S. corporations. The stems are split; each stem represents a span of 5%, from a loss of 9% to a profit of 25%.

```
-0 | 99
-0 | 1 2 3 4
 0 | 1 1 1 1 2 3 4 4 4
 0 | 5 5 5 5 6 7 9
 1 | 0 0 1 1 3
 1 |
 2 | 2
 2 | 5
```
Profits (% of sales)
(−0|3 means a loss of 3%)

a) Find the 5-number summary.

b) Draw a boxplot for these data.

c) Find the mean and standard deviation.

d) Describe the distribution of profits for these corporations.

PART II

Exploring Relationships Between Variables

Scatterplots, Association, and Correlation

WHO	Years 1970–2005
WHAT	Mean error in the position of Atlantic hurricanes as predicted 72 hours ahead by the NHC
UNITS	nautical miles
WHEN	1970–2005
WHERE	Atlantic and Gulf of Mexico
WHY	The NHC wants to improve prediction models

Hurricane Katrina killed 1,836 people[1] and caused well over 100 billion dollars in damage—the most ever recorded. Much of the damage caused by Katrina was due to its almost perfectly deadly aim at New Orleans.

Where will a hurricane go? People want to know if a hurricane is coming their way, and the National Hurricane Center (NHC) of the National Oceanic and Atmospheric Administration (NOAA) tries to predict the path a hurricane will take. But hurricanes tend to wander around aimlessly and are pushed by fronts and other weather phenomena in their area, so they are notoriously difficult to predict. Even relatively small changes in a hurricane's track can make big differences in the damage it causes.

To improve hurricane prediction, NOAA[2] relies on sophisticated computer models, and has been working for decades to improve them. How well are they doing? Have predictions improved in recent years? Has the improvement been consistent? Here's a timeplot of the mean error, in nautical miles, of the NHC's 72-hour predictions of Atlantic hurricanes since 1970:

> **Look, Ma, no origin!**
> Scatterplots usually don't—and shouldn't—show the origin, because often neither variable has values near 0. The display should focus on the part of the coordinate plane that actually contains the data. In our example about hurricanes, none of the prediction errors or years were anywhere near 0, so the computer drew the scatterplot with axes that don't quite meet.

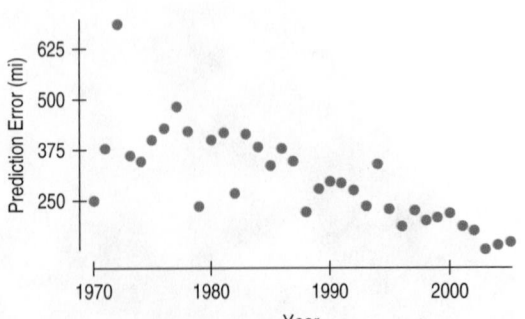

FIGURE 7.1

A scatterplot of the average error in nautical miles of the predicted position of Atlantic hurricanes for predictions made by the National Hurricane Center of NOAA, plotted against the Year in which the predictions were made.

---

[1] In addition, 705 are still listed as missing.
[2] www.nhc.noaa.gov

Clearly, predictions have improved. The plot shows a fairly steady decline in the average error, from almost 500 nautical miles in the late 1970s to about 150 nautical miles in 2005. We can also see a few years when predictions were unusually good and that 1972 was a really bad year for predicting hurricane tracks.

This timeplot is an example of a more general kind of display called a **scatterplot.** Scatterplots may be the most common displays for data. By just looking at them, you can see patterns, trends, relationships, and even the occasional extraordinary value sitting apart from the others. As the great philosopher Yogi Berra[3] once said, "You can observe a lot by watching."[4] Scatterplots are the best way to start observing the relationship between two *quantitative* variables.

Relationships between variables are often at the heart of what we'd like to learn from data:

▷ Are grades actually higher now than they used to be?

▷ Do people tend to reach puberty at a younger age than in previous generations?

▷ Does applying magnets to parts of the body relieve pain? If so, are stronger magnets more effective?

▷ Do students learn better with more use of computer technology?

Questions such as these relate two quantitative variables and ask whether there is an **association** between them. Scatterplots are the ideal way to *picture* such associations.

Looking at Scatterplots

How would you describe the association of hurricane *Prediction Error* and *Year*? Everyone looks at scatterplots. But, if asked, many people would find it hard to say what to look for in a scatterplot. What do *you* see? Try to describe the scatterplot of *Prediction Error* against *Year*.

You might say that the **direction** of the association is important. Over time, the NHC's prediction errors have decreased. A pattern like this that runs from the

upper left to the lower right is said to be **negative.** A pattern running

the other way is called **positive.**

The second thing to look for in a scatterplot is its **form.** If there is a straight line relationship, it will appear as a cloud or swarm of points stretched out in a generally consistent, straight form. For example, the scatterplot of *Prediction Error* vs. *Year* has such an underlying **linear** form, although some points stray away from it.

Scatterplots can reveal many kinds of patterns. Often they will not be straight, but straight line patterns are both the most common and the most useful for statistics.

If the relationship isn't straight, but curves gently, while still increasing or

decreasing steadily, , we can often find ways to make it more nearly

straight. But if it curves sharply—up and then down, for example —there is much less we can say about it with the methods of this book.

---

[3] Hall of Fame catcher and manager of the New York Mets and Yankees.
[4] But then he also said "I really didn't say everything I said." So we can't really be sure.

Look for **Strength:** how much scatter?

The third feature to look for in a scatterplot is how strong the relationship is.

At one extreme, do the points appear tightly clustered in a single stream (whether straight, curved, or bending all over the place)? Or, at the other extreme, does the swarm of points seem to form a vague cloud through which we can

barely discern any trend or pattern? The *Prediction error* vs. *Year* plot shows moderate scatter around a generally straight form. This indicates that the linear trend of improving prediction is pretty consistent and moderately strong.

Finally, always look for the unexpected. Often the most interesting thing to see in a scatterplot is something you never thought to look for. One example of such a surprise is an **outlier** standing away from the overall pattern of the scatterplot. Such a point is almost always interesting and always deserves special attention. In the scatterplot of prediction errors, the year 1972 stands out as a year with very high prediction errors. An Internet search shows that it was a relatively quiet hurricane season. However, it included the very unusual—and deadly—Hurricane Agnes, which combined with another low-pressure center to ravage the northeastern United States, killing 122 and causing 1.3 billion 1972 dollars in damage. Possibly, Agnes was also unusually difficult to predict.

Look for **Unusual Features:** Are there outliers or subgroups?

You should also look for clusters or subgroups that stand away from the rest of the plot or that show a trend in a different direction. Deviating groups should raise questions about why they are different. They may be a clue that you should split the data into subgroups instead of looking at them all together.

FOR EXAMPLE   Describing the scatterplot of hurricane winds and pressure

Hurricanes develop low pressure at their centers. This pulls in moist air, pumps up their rotation, and generates high winds. Standard sea-level pressure is around 1013 millibars (mb), or 29.9 inches of mercury. Hurricane Katrina had a central pressure of 920 mb and sustained winds of 110 knots.

Here's a scatterplot of *Maximum Wind Speed* (kts) vs. *Central Pressure* (mb) for 163 hurricanes that have hit the United States since 1851.

Question: Describe what this plot shows.

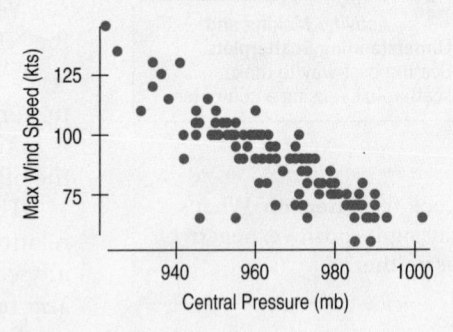

The scatterplot shows a negative direction: in general, lower central pressure is found in hurricanes that have higher maximum wind speeds. This association is linear and moderately strong.

Roles for Variables

Which variable should go on the *x*-axis and which on the *y*-axis? What we want to know about the relationship can tell us how to make the plot. We often have questions such as:

▶ Do baseball teams that score more runs sell more tickets to their games?

▶ Do older houses sell for less than newer ones of comparable size and quality?

▶ Do students who score higher on their SAT tests have higher grade point averages in college?

▶ Can we estimate a person's percent body fat more simply by just measuring waist or wrist size?

In these examples, the two variables play different roles. We'll call the variable of interest the **response variable** and the other the **explanatory** or **predictor variable**.[5] We'll continue our practice of naming the variable of interest y. Naturally we'll plot it on the y-axis and place the explanatory variable on the x-axis. Sometimes, we'll call them the x- and y-**variables**. When you make a scatterplot, you can assume that those who view it will think this way, so choose which variables to assign to which axes carefully.

The roles that we choose for variables are more about how we *think* about them than about the variables themselves. Just placing a variable on the x-axis doesn't necessarily mean that it explains or predicts *anything*. And the variable on the y-axis may not respond to it in any way. We plotted prediction error on the y-axis against year on the x-axis because the National Hurricane Center is interested in how their predictions have changed over time. Could we have plotted them the other way? In this case, it's hard to imagine reversing the roles—knowing the prediction error and wanting to guess in what year it happened. But for some scatterplots, it can make sense to use either choice, so you have to think about how the choice of role helps to answer the question you have.

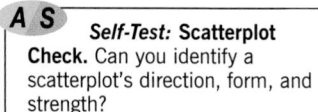

A S *Self-Test:* **Scatterplot Check.** Can you identify a scatterplot's direction, form, and strength?

TI Tips

Creating a scatterplot

Let's use your calculator to make a scatterplot. First you need some data. It's okay to just enter the data in any two lists, but let's get fancy. When you are handling lots of data and several variables (as you will be soon), remembering what you stored in **L1, L2**, and so on can become confusing. You can—and should—give your variables meaningful names. To see how, let's store some data that you will use several times in this chapter and the next. They show the change in tuition costs at Arizona State University during the 1990s.

Naming the Lists

- Go into **STAT Edit**, place the cursor on one of the list names (**L1**, say), and use the arrow key to move to the right across all the lists until you encounter a blank column.
- Type **YR** to name this first variable, then hit **ENTER**.
- Often when we work with years it makes sense to use values like "90" (or even "0") rather than big numbers like "1990." For these data enter the years 1990 through 2000 as 0, 1, 2, . . . , 10.
- Now go to the next blank column, name this variable **TUIT**, and enter these values: 6546, 6996, 6996, 7350, 7500, 7978, 8377, 8710, 9110, 9411, 9800.

---

[5] The x- and y-variables have sometimes been referred to as the *independent* and *dependent* variables, respectively. The idea was that the y-variable depended on the x-variable and the x-variable acted independently to make y respond. These names, however, conflict with other uses of the same terms in Statistics.

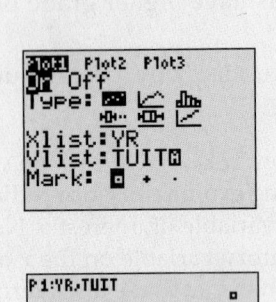

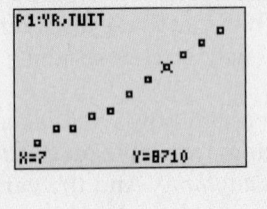

Making the Scatterplot
- Set up the **STATPLOT** by choosing the scatterplot icon (the first option).
- Identify which lists you want as **Xlist** and **Ylist**. If the data are in **L1** and **L2**, that's easy to do—but your data are stored in lists with special names. To specify your **Xlist**, go to **2nd LIST NAMES**, scroll down the list of variables until you find **YR**, then hit **ENTER**.
- Use **LIST NAMES** again to specify **Ylist: TUIT**.
- Pick a symbol for displaying the points.
- Now **ZoomStat** to see your scatterplot. (Didn't work? **ERR: DIM MISMATCH** means you don't have the same number of *x*'s and *y*'s. Go to **STAT Edit** and look carefully at your two datalists. You can easily fix the problem once you find it.)
- Notice that if you **TRACE** the scatterplot the calculator will tell you the *x*- and *y*-value at each point.

What can you Tell about the trend in tuition costs at ASU? (Remember: direction, form, and strength!)

Correlation

WHO	Students
WHAT	Height (inches), weight (pounds)
WHERE	Ithaca, NY
WHY	Data for class
HOW	Survey

Data collected from students in Statistics classes included their *Height* (in inches) and *Weight* (in pounds). It's no great surprise to discover that there is a positive association between the two. As you might suspect, taller students tend to weigh more. (If we had reversed the roles and chosen height as the explanatory variable, we might say that heavier students tend to be taller.)[6] And the form of the scatterplot is fairly straight as well, although there seems to be a high outlier, as the plot shows.

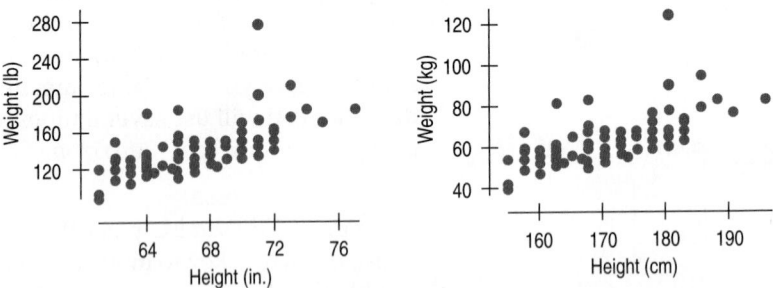

FIGURE 7.2 *Weight vs. Height of Statistics students.*

Plotting Weight *vs.* Height *in different units doesn't change the shape of the pattern.*

 Activity: Correlation. Here's a good example of how correlation works to summarize the strength of a linear relationship and disregard scaling.

The pattern in the scatterplots looks straight and is clearly a positive association, but how strong is it? If you had to put a number (say, between 0 and 1) on the strength, what would it be? Whatever measure you use shouldn't depend on the choice of units for the variables. After all, if we measure heights and weights in centimeters and kilograms instead, it doesn't change the direction, form, or strength, so it shouldn't change the number.

---

[6] The son of one of the authors, when told (as he often was) that he was tall for his age, used to point out that, actually, he was young for his height.

Since the units shouldn't matter to our measure of strength, we can remove them by standardizing each variable. Now, for each point, instead of the values (x, y) we'll have the standardized coordinates (z_x, z_y). Remember that to standardize values, we subtract the mean of each variable and then divide by its standard deviation:

$$(z_x, z_y) = \left(\frac{x - \bar{x}}{s_x}, \frac{y - \bar{y}}{s_y} \right).$$

Because standardizing makes the means of both variables 0, the center of the new scatterplot is at the origin. The scales on both axes are now standard deviation units.

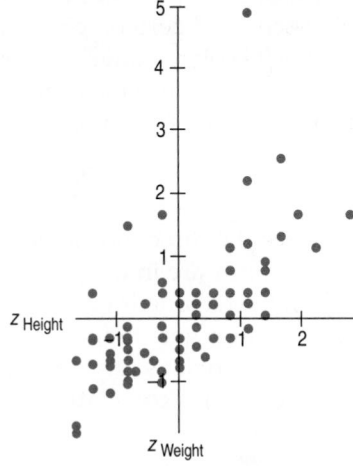

FIGURE 7.3

A scatterplot of standardized heights and weights.

Standardizing shouldn't affect the appearance of the plot. Does the plot of z-scores (Figure 7.3) look like the previous plots? Well, no. The underlying linear pattern seems steeper in the standardized plot. That's because the scales of the axes are now the same, so the length of one standard deviation is the same vertically and horizontally. When we worked in the original units, we were free to make the plot as tall and thin

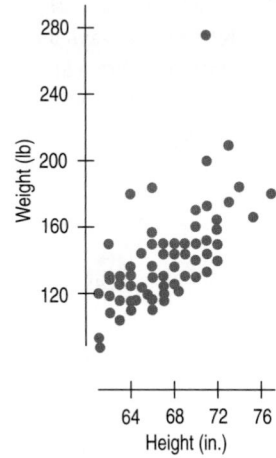

or as squat and wide

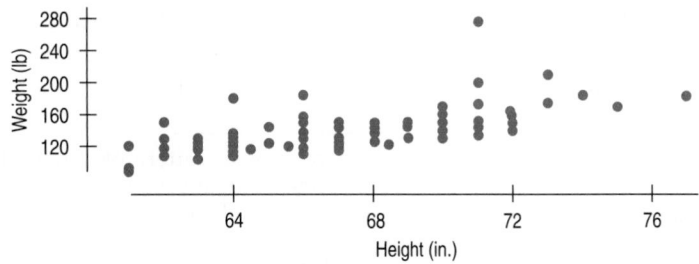

as we wanted to, but that can change the impression the plot gives. By contrast,

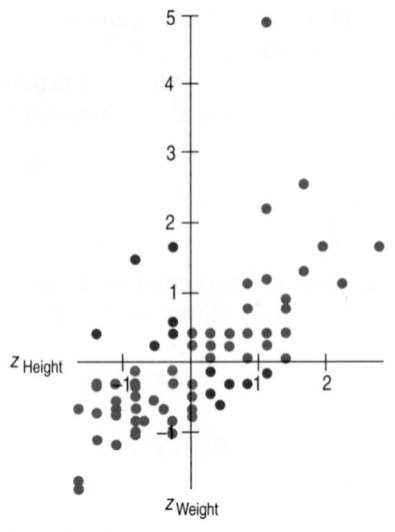

FIGURE 7.4

In this scatterplot of z-scores, points are colored according to how they affect the association: green for positive, red for negative, and blue for neutral.

A S   *Activity: Correlation and Relationship Strength.* What does a correlation of 0.8 look like? How about 0.3?

NOTATION ALERT

The letter r is always used for correlation, so you can't use it for anything else in Statistics. Whenever you see an r, it's safe to assume it's a correlation.

---

equal scaling gives a neutral way of drawing the scatterplot and a fairer impression of the strength of the association.[7]

Which points in the scatterplot of the z-scores give the impression of a positive association? In a positive association, y tends to increase as x increases. So, the points in the upper right and lower left (colored green) strengthen that impression. For these points, z_x and z_y have the same sign, so the product $z_x z_y$ is positive. Points far from the origin (which make the association look more positive) have bigger products.

The red points in the upper left and lower right quadrants tend to weaken the positive association (or support a negative association). For these points, z_x and z_y have opposite signs. So the product $z_x z_y$ for these points is negative. Points far from the origin (which make the association look more negative) have a negative product even larger in magnitude.

Points with z-scores of zero on either variable don't vote either way, because $z_x z_y = 0$. They're colored blue.

To turn these products into a measure of the strength of the association, just add up the $z_x z_y$ products for every point in the scatterplot:

$$\sum z_x z_y.$$

This summarizes the direction *and* strength of the association for all the points. If most of the points are in the green quadrants, the sum will tend to be positive. If most are in the red quadrants, it will tend to be negative.

But the *size* of this sum gets bigger the more data we have. To adjust for this, the natural (for statisticians anyway) thing to do is to divide the sum by $n - 1$.[8] The ratio is the famous **correlation coefficient:**

$$r = \frac{\sum z_x z_y}{n - 1}.$$

For the students' heights and weights, the correlation is 0.644. There are a number of alternative formulas for the correlation coefficient, but this form using z-scores is best for understanding what correlation means.

Correlation Conditions

Correlation measures the strength of the *linear* association between two *quantitative* variables. Before you use correlation, you must check several *conditions:*

▶ **Quantitative Variables Condition:** Are both variables quantitative? Correlation applies only to quantitative variables. Don't apply correlation to categorical data masquerading as quantitative. Check that you know the variables' units and what they measure.

▶ **Straight Enough Condition:** Is the form of the scatterplot straight enough that a linear relationship makes sense? Sure, you can *calculate* a correlation coefficient for any pair of variables. But correlation measures the strength only

A S   *Simulation: Correlation and Linearity.* How much does straightness matter?

---

[7] When we draw a scatterplot, what often looks best is to make the length of the x-axis slightly larger than the length of the y-axis. This is an aesthetic choice, probably related to the Golden Ratio of the Greeks.

[8] Yes, the same $n - 1$ as in the standard deviation calculation. And we offer the same promise to explain it later.

of the *linear* association, and will be misleading if the relationship is not linear. What is "straight enough"? How non-straight would the scatterplot have to be to fail the condition? This is a judgment call that you just have to think about. Do you think that the underlying relationship is curved? If so, then summarizing its strength with a correlation would be misleading.

 *Case Study:* **Mortality and Education.** Is the mortality rate lower in cities with higher education levels?

▶ **Outlier Condition:** Outliers can distort the correlation dramatically. An outlier can make an otherwise weak correlation look big or hide a strong correlation. It can even give an otherwise positive association a negative correlation coefficient (and vice versa). When you see an outlier, it's often a good idea to report the correlation with and without that point.

Each of these conditions is easy to check with a scatterplot. Many correlations are reported without supporting data or plots. Nevertheless, you should still think about the conditions. And you should be cautious in interpreting (or accepting others' interpretations of) the correlation when you can't check the conditions for yourself.

FOR EXAMPLE   Correlating wind speed and pressure

Recap: We looked at the scatterplot displaying hurricane wind speeds and central pressures.

The correlation coefficient for these wind speeds and pressures is $r = -0.879$.

Question: Check the conditions for using correlation. If you feel they are satisfied, interpret this correlation.

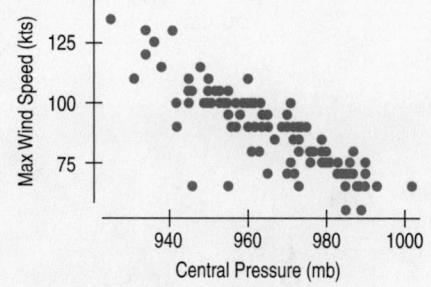

▶ Quantitative Variables Condition: Both wind speed and central pressure are quantitative variables, measured (respectively) in knots and millibars.

▶ Straight Enough Condition: The pattern in the scatterplot is quite straight.

▶ Outlier Condition: A few hurricanes seem to straggle away from the main pattern, but they don't appear to be extreme enough to be called outliers. It may be worthwhile to check on them, however.

The conditions for using correlation are satisfied. The correlation coefficient of $r = -0.879$ indicates quite a strong negative linear association between the wind speeds of hurricanes and their central pressures.

✓ JUST CHECKING

Your Statistics teacher tells you that the correlation between the scores (points out of 50) on Exam 1 and Exam 2 was 0.75.

1. Before answering any questions about the correlation, what would you like to see? Why?

2. If she adds 10 points to each Exam 1 score, how will this change the correlation?

3. If she standardizes scores on each exam, how will this affect the correlation?

4. In general, if someone did poorly on Exam 1, are they likely to have done poorly or well on Exam 2? Explain.

5. If someone did poorly on Exam 1, can you be sure that they did poorly on Exam 2 as well? Explain.

 **STEP-BY-STEP EXAMPLE** | **Looking at Association**

When your blood pressure is measured, it is reported as two values: systolic blood pressure and diastolic blood pressure.

Questions: How are these variables related to each other? Do they tend to be both high or both low? How strongly associated are they?

Plan State what you are trying to investigate.

I'll examine the relationship between two measures of blood pressure.

Variables Identify the two quantitative variables whose relationship we wish to examine. Report the W's, and be sure both variables are recorded for the same individuals.

The variables are systolic and diastolic blood pressure (SBP and DBP), recorded in millimeters of mercury (mm Hg) for each of 1406 participants in the Framingham Heart Study, a famous health study in Framingham, MA.[9]

Plot Make the scatterplot. Use a computer program or graphing calculator if you can.

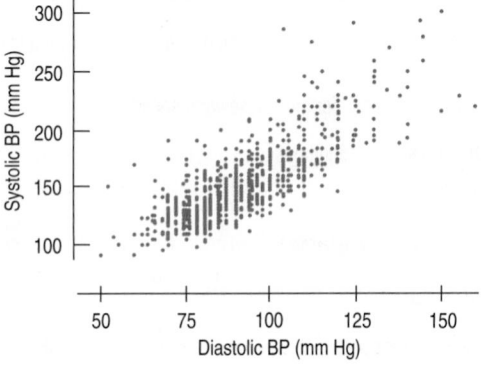

Check the conditions.

✔ **Quantitative Variables Condition:** Both SBP and DBP are quantitative and measured in mm Hg.

✔ **Straight Enough Condition:** The scatterplot looks straight.

✔ **Outlier Condition:** There are a few straggling points, but none far enough from the body of the data to be called outliers.

REALITY CHECK Looks like a strong positive linear association. We shouldn't be surprised if the correlation coefficient is positive and fairly large.

I have two quantitative variables that satisfy the conditions, so correlation is a suitable measure of association.

 **Mechanics** We usually calculate correlations with technology. Here we have 1406 cases, so we'd never try it by hand.

The correlation coefficient is r = 0.792.

[9] www.nhlbi.nih.gov/about/framingham

Conclusion Describe the direction, form, and strength you see in the plot, along with any unusual points or features. Be sure to state your interpretations in the proper context.

The scatterplot shows a positive direction, with higher SBP going with higher DBP. The plot is generally straight, with a moderate amount of scatter. The correlation of 0.792 is consistent with what I saw in the scatterplot. A few cases stand out with unusually high SBP compared with their DBP. It seems far less common for the DBP to be high by itself.

TI Tips — Finding the correlation

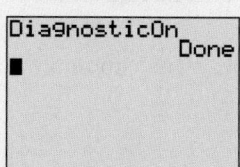

Now let's use the calculator to find a correlation. Unfortunately, the statistics package on your TI calculator does not automatically do that. Correlations are one of the most important things we might want to do, so here's how to fix that, once and for all.

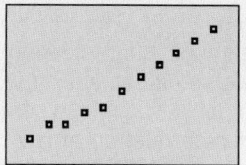

- Hit **2nd CATALOG** (on the zero key). You now see a list of everything the calculator knows how to do. Impressive, huh?
- Scroll down until you find **DiagnosticOn**. Hit **ENTER**. Again. It should say **Done**.

Now and forevermore (or perhaps until you change batteries) your calculator will find correlations.

Finding the Correlation

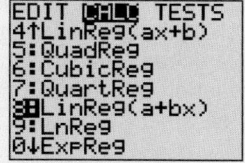

- *Always* check the conditions first. Look at the scatterplot for the Arizona State tuition data again. Does this association look linear? Are there outliers? This plot looks fine, but remember that correlation can be used to describe the strength of *linear* associations only, and outliers can distort the results. Eyeballing the scatterplot is an essential first step. (You should be getting used to checking on assumptions and conditions before jumping into a statistical procedure—it's always important.)
- Under the **STAT CALC** menu, select **8:LinReg(a+bx)** and hit **ENTER**.
- Now specify x and y by importing the names of your variables from the **LIST NAMES** menu. First name your x-variable followed by a comma, then your y-variable, creating the command

$$\text{LinReg(a+bx)} \mathord{\text{\tiny L}}\text{YR}, \mathord{\text{\tiny L}}\text{TUIT}$$

Wow! A lot of stuff happened. If you suspect all those other numbers are important, too, you'll really enjoy the next chapter. But for now, it's the value of **r** you care about. What does this correlation, $r = 0.993$, say about the trend in tuition costs?

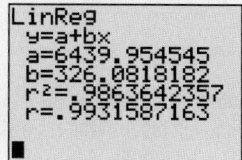

Correlation Properties

A S *Activity:* **Construct Scatterplots with a Given Correlation.** Try to make a scatterplot that has a given correlation. How close can you get?

Here's a useful list of facts about the correlation coefficient:

▶ The sign of a correlation coefficient gives the direction of the association.

▶ Correlation is always between −1 and +1. Correlation *can* be exactly equal to −1.0 or +1.0, but these values are unusual in real data because they mean that all the data points fall *exactly* on a single straight line.

▶ Correlation treats *x* and *y* symmetrically. The correlation of *x* with *y* is the same as the correlation of *y* with *x*.

▶ Correlation has no units. This fact can be especially appropriate when the data's units are somewhat vague to begin with (IQ score, personality index, socialization, and so on). Correlation is sometimes given as a percentage, but you probably shouldn't do that because it suggests a percentage of *something*—and correlation, lacking units, has no "something" of which to be a percentage.

▶ Correlation is not affected by changes in the center or scale of either variable. Changing the units or baseline of either variable has no effect on the correlation coefficient. Correlation depends only on the z-scores, and they are unaffected by changes in center or scale.

▶ Correlation measures the strength of the *linear* association between the two variables. Variables can be strongly associated but still have a small correlation if the association isn't linear.

▶ Correlation is sensitive to outliers. A single outlying value can make a small correlation large or make a large one small.

Height and Weight, Again

We could have measured the students' weights in stones. In the now outdated UK system of measures, a stone is a measure equal to 14 pounds. And we could have measured heights in hands. Hands are still commonly used to measure the heights of horses. A hand is 4 inches. But no matter what *units* we use to measure the two variables, the *correlation* stays the same.

TI-*nspire*

Correlation and Scatterplots. See how the correlation changes as you drag data points around in a scatterplot.

How strong is strong? You'll often see correlations characterized as "weak," "moderate," or "strong," but be careful. There's no agreement on what those terms mean. The same numerical correlation might be strong in one context and weak in another. You might be thrilled to discover a correlation of 0.7 between the new summary of the economy you've come up with and stock market prices, but you'd consider it a design failure if you found a correlation of "only" 0.7 between two tests intended to measure the same skill. Deliberately vague terms like "weak," "moderate," or "strong" that describe a linear association can be useful additions to the numerical summary that correlation provides. But be sure to include the correlation and show a scatterplot, so others can judge for themselves.

FOR EXAMPLE   Changing scales

Recap: We found a correlation of $r = -0.879$ between hurricane wind speeds in knots and their central pressures in millibars.

Question: Suppose we wanted to consider the wind speeds in miles per hour (1 mile per hour = 0.869 knots) and central pressures in inches of mercury (1 inch of mercury = 33.86 millibars). How would that conversion affect the conditions, the value of *r*, and our interpretation of the correlation coefficient?

Not at all! Correlation is based on standardized values (z-scores), so the conditions, the value of r, and the proper interpretation are all unaffected by changes in units.

Warning: Correlation ≠ Causation

Whenever we have a strong correlation, it's tempting to try to explain it by imagining that the predictor variable has *caused* the response to change. Humans are like that; we tend to see causes and effects in everything.

Sometimes this tendency can be amusing. A scatterplot of the human population (*y*) of Oldenburg, Germany, in the beginning of the 1930s plotted against the number of storks nesting in the town (*x*) shows a tempting pattern.

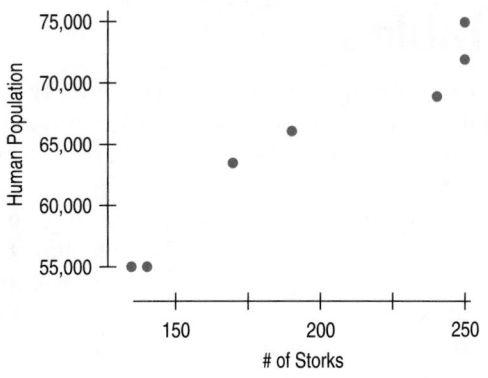

FIGURE 7.5
The number of storks in Oldenburg, Germany, plotted against the population of the town for 7 years in the 1930s. The association is clear. How about the causation? (Ornithologishe Monatsberichte, 44, no. 2)

Anyone who has seen the beginning of the movie *Dumbo* remembers Mrs. Jumbo anxiously waiting for the stork to bring her new baby. Even though you know it's silly, you can't help but think for a minute that this plot shows that storks are the culprits. The two variables are obviously related to each other (the correlation is 0.97!), but that doesn't prove that storks bring babies.

It turns out that storks nest on house chimneys. More people means more houses, more nesting sites, and so more storks. The causation is actually in the *opposite* direction, but you can't tell from the scatterplot or correlation. You need additional information—not just the data—to determine the real mechanism.

A scatterplot of the damage (in dollars) caused to a house by fire would show a strong correlation with the number of firefighters at the scene. Surely the damage doesn't cause firefighters. And firefighters do seem to cause damage, spraying water all around and chopping holes. Does that mean we shouldn't call the fire department? Of course not. There is an underlying variable that leads to both more damage and more firefighters: the size of the blaze.

A hidden variable that stands behind a relationship and determines it by simultaneously affecting the other two variables is called a **lurking variable.** You can often debunk claims made about data by finding a lurking variable behind the scenes.

Scatterplots and correlation coefficients *never* prove causation. That's one reason it took so long for the U.S. Surgeon General to get warning labels on cigarettes. Although there was plenty of evidence that increased smoking was *associated* with increased levels of lung cancer, it took years to provide evidence that smoking actually *causes* lung cancer.

Does cancer cause smoking?   Even if the correlation of two variables is due to a causal relationship, the correlation itself cannot tell us what causes what.

Sir Ronald Aylmer Fisher (1890–1962) was one of the greatest statisticians of the 20th century. Fisher testified in court (in testimony paid for by the tobacco companies) that a causal relationship might underlie the correlation of smoking and cancer:

"Is it possible, then, that lung cancer . . . is one of the causes of smoking cigarettes? I don't think it can be excluded . . . the pre-cancerous condition is one involving a certain amount of slight chronic inflammation . . . .

A slight cause of irritation . . . is commonly accompanied by pulling out a cigarette, and getting a little compensation for life's minor ills in that way. And . . . is not unlikely to be associated with smoking more frequently."

Ironically, the proof that smoking indeed is the cause of many cancers came from experiments conducted following the principles of experiment design and analysis that Fisher himself developed—and that we'll see in Chapter 13.

Correlation Tables

It is common in some fields to compute the correlations between every pair of variables in a collection of variables and arrange these correlations in a table. The rows and columns of the table name the variables, and the cells hold the correlations.

Correlation tables are compact and give a lot of summary information at a glance. They can be an efficient way to start to look at a large data set, but a dangerous one. By presenting all of these correlations without any checks for linearity and outliers, the correlation table risks showing truly small correlations that have been inflated by outliers, truly large correlations that are hidden by outliers, and correlations of any size that may be meaningless because the underlying form is not linear.

	Assets	Sales	Market Value	Profits	Cash Flow	Employees
Assets	1.000					
Sales	0.746	1.000				
Market Value	0.682	0.879	1.000			
Profits	0.602	0.814	0.968	1.000		
Cash Flow	0.641	0.855	0.970	0.989	1.000	
Employees	0.594	0.924	0.818	0.762	0.787	1.000

Table 7.1

A correlation table of data reported by *Forbes* magazine for large companies. From this table, can you be sure that the variables are linearly associated and free from outliers?

The diagonal cells of a correlation table always show correlations of exactly 1. (Can you see why?) Correlation tables are commonly offered by statistics packages on computers. These same packages often offer simple ways to make all the scatterplots that go with these correlations.

Straightening Scatterplots

Correlation is a suitable measure of strength for straight relationships only. When a scatterplot shows a bent form that consistently increases or decreases, we can often straighten the form of the plot by re-expressing one or both variables.

Some camera lenses have an adjustable aperture, the hole that lets the light in. The size of the aperture is expressed in a mysterious number called the f/stop. Each increase of one f/stop number corresponds to a halving of the light that is allowed to come through. The f/stops of one digital camera are

f/stop: 2.8 4 5.6 8 11 16 22 32

When you halve the shutter speed, you cut down the light, so you have to open the aperture one notch. We could experiment to find the best f/stop value for each shutter speed. A table of recommended shutter speeds and f/stops for a camera lists the relationship like this:

Shutter speed:	1/1000	1/500	1/250	1/125	1/60	1/30	1/15	1/8
f/stop:	2.8	4	5.6	8	11	16	22	32

The correlation of these shutter speeds and f/stops is 0.979. That sounds pretty high. You might assume that there must be a strong linear relationship. But when we check the scatterplot (we *always* check the scatterplot), it shows that something is not quite right:

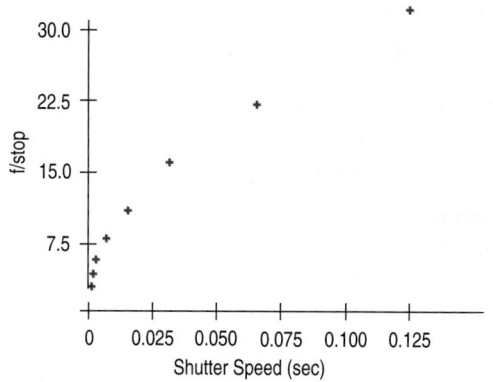

FIGURE 7.6

A scatterplot of f/stop *vs.* Shutter Speed *shows a bent relationship.*

We can see that the f/stop is not *linearly* related to the shutter speed. Can we find a transformation of f/stop that straightens out the line? What if we look at the *square* of the f/stop against the shutter speed?

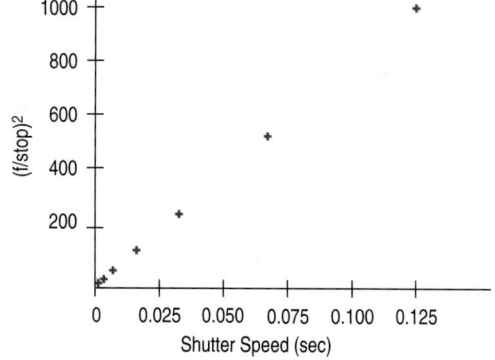

FIGURE 7.7

Re-expressing f/stop *by squaring straightens the plot.*

The second plot looks much more nearly straight. In fact, the correlation is now 0.998, but the increase in correlation is not important. (The original value of 0.979 should please almost anyone who sought a large correlation.) What is important is that the *form* of the plot is now straight, so the correlation is now an appropriate measure of association.[10]

We can often find transformations that straighten a scatterplot's form. Here, we found the square. Chapter 10 discusses simple ways to find a good re-expression.

---

[10] Sometimes we can do a "reality check" on our choice of re-expression. In this case, a bit of research reveals that f/stops are related to the diameter of the open shutter. Since the amount of light that enters is determined by the *area* of the open shutter, which is related to the diameter by squaring, the square re-expression seems reasonable. Not all re-expressions have such nice explanations, but it's a good idea to think about them.

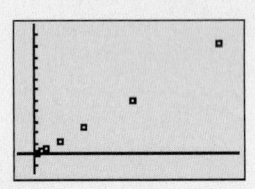

TI Tips

Straightening a curve

Let's straighten the f/stop scatterplot with your calculator.

- Enter the data in two lists, *shutterspeed* in L1 and *f/stop* in L2.
- Set up a STAT PLOT to create a scatterplot with Xlist:L1 and Ylist:L2.
- Hit ZoomStat. See the curve?

We want to find the squares of all the f/stops and save those re-expressed values in another datalist. That's easy to do.

- Create the command to square all the values in L2 and STOre those results in L3, then hit ENTER.

Now make the new scatterplot.

- Go back to STAT PLOT and change the setup. Xlist is still L1, but this time specify Ylist:L3.
- ZoomStat again.

You now see the straightened plot for these data. On deck: drawing the best line through those points!

WHAT CAN GO WRONG?

▶ **Don't say "correlation" when you mean "association."** How often have you heard the word "correlation"? Chances are pretty good that when you've heard the term, it's been misused. When people want to sound scientific, they often say "correlation" when talking about the relationship between two variables. It's one of the most widely misused Statistics terms, and given how often statistics are misused, that's saying a lot. One of the problems is that many people use the specific term *correlation* when they really mean the more general term *association*. "Association" is a deliberately vague term describing the relationship between two variables.

"Correlation" is a precise term that measures the strength and direction of the linear relationship between quantitative variables.

▶ **Don't correlate categorical variables.** People who misuse the term "correlation" to mean "association" often fail to notice whether the variables they discuss are quantitative. Be sure to check the **Quantitative Variables Condition.**

▶ **Don't confuse correlation with causation.** One of the most common mistakes people make in interpreting statistics occurs when they observe a high correlation between two variables and jump to the perhaps tempting conclusion that one thing must be causing the other. Scatterplots and correlations *never* demonstrate causation. At best, these statistical tools can only reveal an association between variables, and that's a far cry from establishing cause and effect. While it's true that some associations may be causal, the nature and direction of the causation can be very hard to establish, and there's always the risk of overlooking lurking variables.

▶ **Make sure the association is linear.** Not all associations between quantitative variables are linear. Correlation can miss even a strong nonlinear association. A student project evaluating the quality of brownies baked at different temperatures reports a correlation of −0.05 between judges' scores and baking temperature. That seems to say there is no relationship—until we look at the scatterplot:

Did you know that there's a strong correlation between playing an instrument and drinking coffee? No? One reason might be that the statement doesn't make sense. Correlation is a statistic that's valid only for *quantitative* variables.

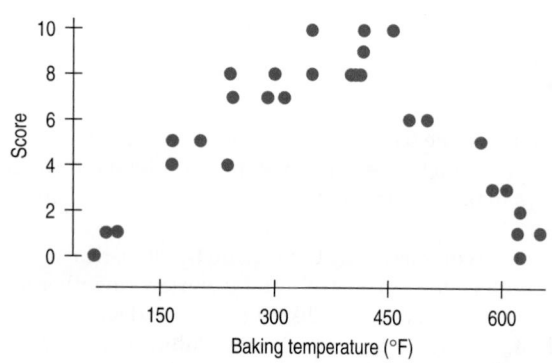

FIGURE 7.8

The relationship between brownie taste Score *and* Baking Temperature *is strong, but not at all linear.*

There is a strong association, but the relationship is not linear. Don't forget to check the Straight Enough Condition.

▶ **Don't assume the relationship is linear just because the correlation coefficient is high.** Recall that the correlation of f/stops and shutter speeds is 0.979 and yet the relationship is clearly not straight. Although the relationship must be straight for the correlation to be an appropriate measure, a high correlation is no guarantee of straightness. Nor is it safe to use correlation to judge the best re-expression. It's always important to look at the scatterplot.

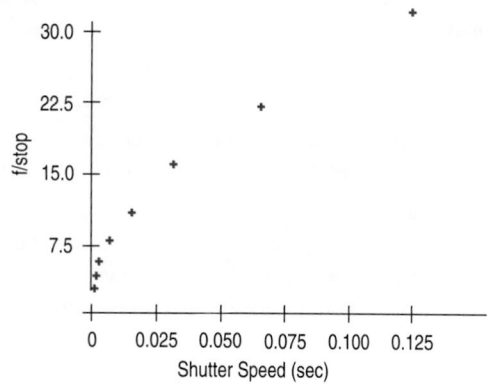

FIGURE 7.9

A scatterplot of f/stop *vs.* Shutter Speed *shows a bent relationship even though the correlation is* r = 0.979.

▶ **Beware of outliers.** You can't interpret a correlation coefficient safely without a background check for outliers. Here's a silly example:

The relationship between IQ and shoe size among comedians shows a surprisingly strong positive correlation of 0.50. To check assumptions, we look at the scatterplot:

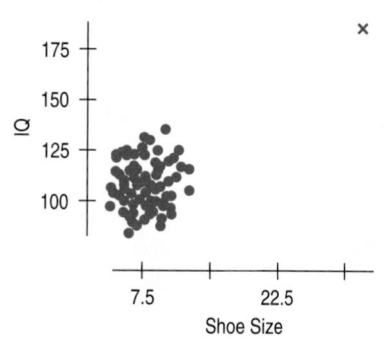

FIGURE 7.10

A scatterplot of IQ *vs.* Shoe Size. *From this "study," what is the relationship between the two? The correlation is 0.50. Who does that point (the green x) in the upper right-hand corner belong to?*

The outlier is Bozo the Clown, known for his large shoes, and widely acknowledged to be a comic "genius." Without Bozo, the correlation is near zero.

Even a single outlier can dominate the correlation value. That's why you need to check the Outlier Condition.

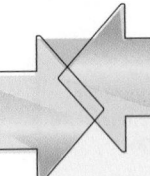

CONNECTIONS

Scatterplots are the basic tool for examining the relationship between two quantitative variables. We start with a picture when we want to understand the distribution of a single variable, and we always make a scatterplot to begin to understand the relationship between two quantitative variables.

We used z-scores as a way to measure the statistical distance of data values from their means. Now we've seen the z-scores of x and y working together to build the correlation coefficient. Correlation is a summary statistic like the mean and standard deviation—only it summarizes the strength of a linear relationship. And we interpret it as we did z-scores, using the standard deviations as our rulers in both x and y.

WHAT HAVE WE LEARNED?

In recent chapters we learned how to listen to the story told by data from a single variable. Now we've turned our attention to the more complicated (and more interesting) story we can discover in the association between two quantitative variables.

We've learned to begin our investigation by looking at a scatterplot. We're interested in the *direction* of the association, the *form* it takes, and its *strength*.

We've learned that, although not every relationship is linear, when the scatterplot is straight enough, the *correlation coefficient* is a useful numerical summary.

▸ The sign of the correlation tells us the direction of the association.
▸ The magnitude of the correlation tells us the *strength* of a linear association. Strong associations have correlations near -1 or $+1$ and very weak associations near 0.
▸ Correlation has no units, so shifting or scaling the data, standardizing, or even swapping the variables has no effect on the numerical value.

Once again we've learned that doing Statistics right means we have to *Think* about whether our choice of methods is appropriate.

▸ The correlation coefficient is appropriate only if the underlying relationship is linear.
▸ We'll check the **Straight Enough Condition** by looking at a scatterplot.
▸ And, as always, we'll watch out for outliers!

Finally, we've learned not to make the mistake of assuming that a high correlation or strong association is evidence of a cause-and-effect relationship. Beware of lurking variables!

A S *Simulation: Correlation, Center, and Scale.* If you have any lingering doubts that shifting and rescaling the data won't change the correlation, watch nothing happen right before your eyes!

Terms

Scatterplots
: 147. A scatterplot shows the relationship between two quantitative variables measured on the same cases.

Association
: ▸ 147. **Direction:** A positive direction or association means that, in general, as one variable increases, so does the other. When increases in one variable generally correspond to decreases in the other, the association is negative.
: ▸ 147. **Form:** The form we care about most is straight, but you should certainly describe other patterns you see in scatterplots.
: ▸ 148. **Strength:** A scatterplot is said to show a strong association if there is little scatter around the underlying relationship.

Outlier
: 148. A point that does not fit the overall pattern seen in the scatterplot.

Response variable, Explanatory variable, *x*-variable, *y*-variable	149. In a scatterplot, you must choose a role for each variable. Assign to the *y*-axis the response variable that you hope to predict or explain. Assign to the *x*-axis the explanatory or predictor variable that accounts for, explains, predicts, or is otherwise responsible for the *y*-variable.
Correlation Coefficient	152. The correlation coefficient is a numerical measure of the direction and strength of a linear association.

$$r = \frac{\sum z_x z_y}{n - 1}.$$

Lurking variable	157. A variable other than *x* and *y* that simultaneously affects both variables, accounting for the correlation between the two.

Skills

▸ Recognize when interest in the pattern of a possible relationship between two quantitative variables suggests making a scatterplot.

▸ Know how to identify the roles of the variables and that you should place the response variable on the *y*-axis and the explanatory variable on the *x*-axis.

▸ Know the conditions for correlation and how to check them.

▸ Know that correlations are between −1 and +1, and that each extreme indicates a perfect linear association.

▸ Understand how the magnitude of the correlation reflects the strength of a linear association as viewed in a scatterplot.

▸ Know that correlation has no units.

▸ Know that the correlation coefficient is not changed by changing the center or scale of either variable.

▸ Understand that causation cannot be demonstrated by a scatterplot or correlation.

▸ Know how to make a scatterplot by hand (for a small set of data) or with technology.

▸ Know how to compute the correlation of two variables.

▸ Know how to read a correlation table produced by a statistics program.

▸ Be able to describe the direction, form, and strength of a scatterplot.

▸ Be prepared to identify and describe points that deviate from the overall pattern.

▸ Be able to use correlation as part of the description of a scatterplot.

▸ Be alert to misinterpretations of correlation.

▸ Understand that finding a correlation between two variables does not indicate a causal relationship between them. Beware the dangers of suggesting causal relationships when describing correlations.

SCATTERPLOTS AND CORRELATION ON THE COMPUTER

Statistics packages generally make it easy to look at a scatterplot to check whether the correlation is appropriate. Some packages make this easier than others.

Many packages allow you to modify or enhance a scatterplot, altering the axis labels, the axis numbering, the plot symbols, or the colors used. Some options, such as color and symbol choice, can be used to display additional information on the scatterplot.

EXERCISES

1. **Association.** Suppose you were to collect data for each pair of variables. You want to make a scatterplot. Which variable would you use as the explanatory variable and which as the response variable? Why? What would you expect to see in the scatterplot? Discuss the likely direction, form, and strength.
   a) Apples: weight in grams, weight in ounces
   b) Apples: circumference (inches), weight (ounces)
   c) College freshmen: shoe size, grade point average
   d) Gasoline: number of miles you drove since filling up, gallons remaining in your tank

2. **Association.** Suppose you were to collect data for each pair of variables. You want to make a scatterplot. Which variable would you use as the explanatory variable and which as the response variable? Why? What would you expect to see in the scatterplot? Discuss the likely direction, form, and strength.
   a) T-shirts at a store: price each, number sold
   b) Scuba diving: depth, water pressure
   c) Scuba diving: depth, visibility
   d) All elementary school students: weight, score on a reading test

3. **Association.** Suppose you were to collect data for each pair of variables. You want to make a scatterplot. Which variable would you use as the explanatory variable and which as the response variable? Why? What would you expect to see in the scatterplot? Discuss the likely direction, form, and strength.
   a) When climbing mountains: altitude, temperature
   b) For each week: ice cream cone sales, air-conditioner sales
   c) People: age, grip strength
   d) Drivers: blood alcohol level, reaction time

4. **Association.** Suppose you were to collect data for each pair of variables. You want to make a scatterplot. Which variable would you use as the explanatory variable and which as the response variable? Why? What would you expect to see in the scatterplot? Discuss the likely direction, form, and strength.
   a) Long-distance calls: time (minutes), cost
   b) Lightning strikes: distance from lightning, time delay of the thunder
   c) A streetlight: its apparent brightness, your distance from it
   d) Cars: weight of car, age of owner

5. **Scatterplots.** Which of the scatterplots at the top of the next column show
   a) little or no association?
   b) a negative association?
   c) a linear association?
   d) a moderately strong association?
   e) a very strong association?

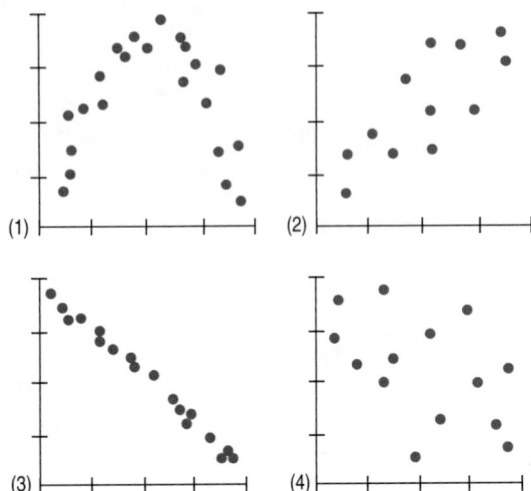

6. **Scatterplots.** Which of the scatterplots below show
   a) little or no association?
   b) a negative association?
   c) a linear association?
   d) a moderately strong association?
   e) a very strong association?

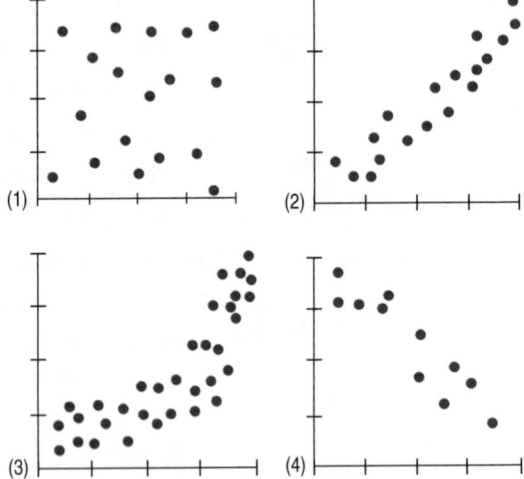

7. **Performance IQ scores vs. brain size.** A study examined brain size (measured as pixels counted in a digitized magnetic resonance image [MRI] of a cross section of the brain) and IQ (4 Performance scales of the Weschler IQ test) for college students. The scatterplot shows the Performance IQ scores vs. the brain size. Comment on the association between brain size and IQ as seen in the scatterplot on the next page.

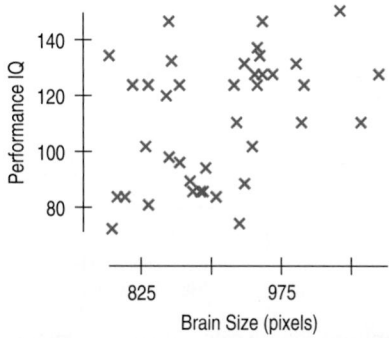

8. Kentucky Derby 2006. The fastest horse in Kentucky Derby history was Secretariat in 1973. The scatterplot shows speed (in miles per hour) of the winning horses each year.

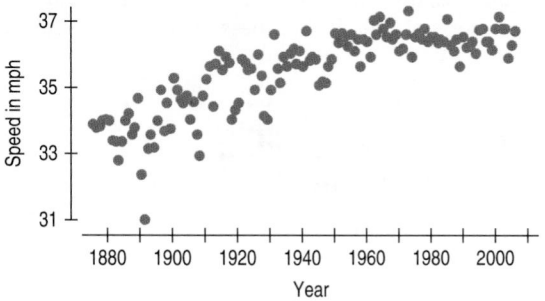

What do you see? In most sporting events, performances have improved and continue to improve, so surely we anticipate a positive direction. But what of the form? Has the performance increased at the same rate throughout the last 130 years?

9. Firing pottery. A ceramics factory can fire eight large batches of pottery a day. Sometimes a few of the pieces break in the process. In order to understand the problem better, the factory records the number of broken pieces in each batch for 3 days and then creates the scatterplot shown.

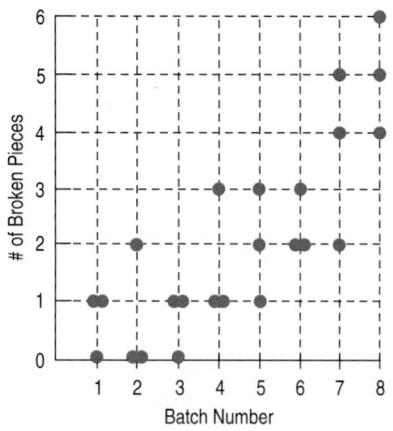

a) Make a histogram showing the distribution of the number of broken pieces in the 24 batches of pottery examined.
b) Describe the distribution as shown in the histogram. What feature of the problem is more apparent in the histogram than in the scatterplot?
c) What aspect of the company's problem is more apparent in the scatterplot?

10. Coffee sales. Owners of a new coffee shop tracked sales for the first 20 days and displayed the data in a scatterplot (by day).

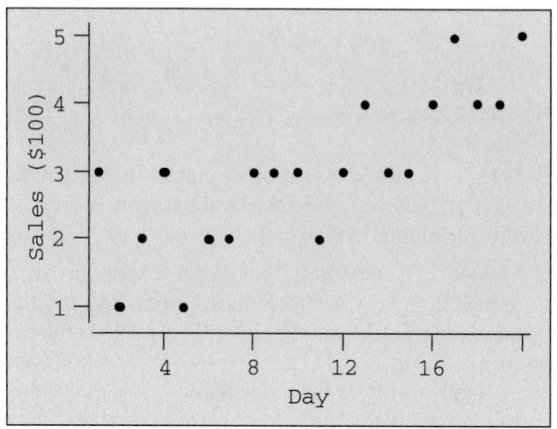

a) Make a histogram of the daily sales since the shop has been in business.
b) State one fact that is obvious from the scatterplot, but not from the histogram.
c) State one fact that is obvious from the histogram, but not from the scatterplot.

11. Matching. Here are several scatterplots. The calculated correlations are −0.923, −0.487, 0.006, and 0.777. Which is which?

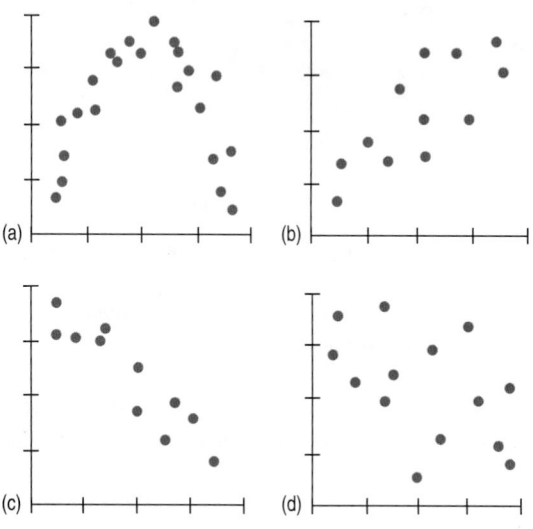

12. Matching. Here and on the next page are several scatterplots. The calculated correlations are −0.977, −0.021, 0.736, and 0.951. Which is which?

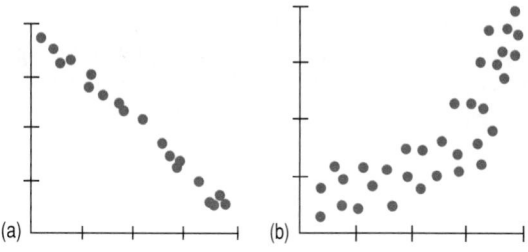

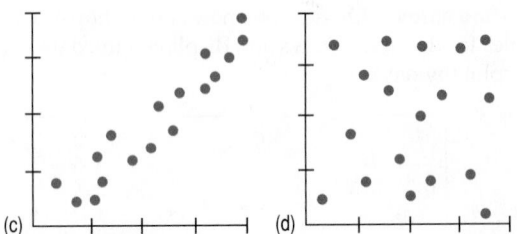

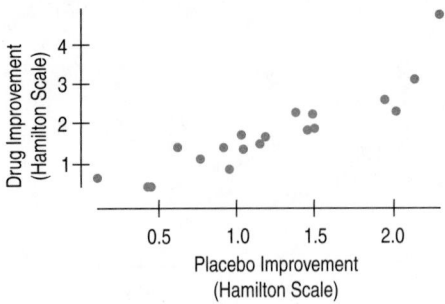

13. Politics.   A candidate for office claims that "there is a correlation between television watching and crime." Criticize this statement on statistical grounds.

14. Car thefts.   The National Insurance Crime Bureau reports that Honda Accords, Honda Civics, and Toyota Camrys are the cars most frequently reported stolen, while Ford Tauruses, Pontiac Vibes, and Buick LeSabres are stolen least often. Is it reasonable to say that there's a correlation between the type of car you own and the risk that it will be stolen?

T 15. Roller coasters.   Roller coasters get all their speed by dropping down a steep initial incline, so it makes sense that the height of that drop might be related to the speed of the coaster. Here's a scatterplot of top *Speed* and largest *Drop* for 75 roller coasters around the world.

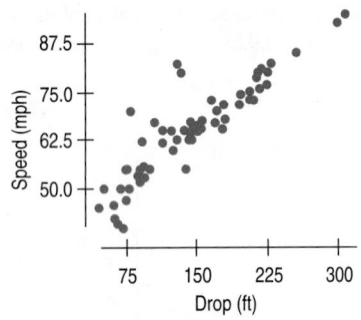

a) Does the scatterplot indicate that it is appropriate to calculate the correlation? Explain.
b) In fact, the correlation of *Speed* and *Drop* is 0.91. Describe the association.

T 16. Antidepressants.   A study compared the effectiveness of several antidepressants by examining the experiments in which they had passed the FDA requirements. Each of those experiments compared the active drug with a placebo, an inert pill given to some of the subjects. In each experiment some patients treated with the placebo had improved, a phenomenon called the *placebo effect*. Patients' depression levels were evaluated on the Hamilton Depression Rating Scale, where larger numbers indicate greater improvement. (The Hamilton scale is a widely accepted standard that was used in each of the independently run studies.) The scatterplot at the top of the next column compares mean improvement levels for the antidepressants and placebos for several experiments.

a) Is it appropriate to calculate the correlation? Explain.
b) The correlation is 0.898. Explain what we have learned about the results of these experiments.

T 17. Hard water.   In a study of streams in the Adirondack Mountains, the following relationship was found between the water's pH and its hardness (measured in grains):

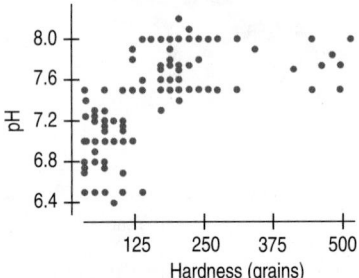

Is it appropriate to summarize the strength of association with a correlation? Explain.

18. Traffic headaches.   A study of traffic delays in 68 U.S. cities found the following relationship between total delays (in total hours lost) and mean highway speed:

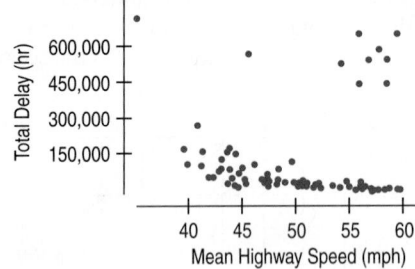

Is it appropriate to summarize the strength of association with a correlation? Explain.

19. Cold nights.   Is there an association between time of year and the nighttime temperature in North Dakota? A researcher assigned the numbers 1–365 to the days January 1–December 31 and recorded the temperature at 2:00 a.m. for each. What might you expect the correlation between *DayNumber* and *Temperature* to be? Explain.

20. Association. A researcher investigating the association between two variables collected some data and was surprised when he calculated the correlation. He had expected to find a fairly strong association, yet the correlation was near 0. Discouraged, he didn't bother making a scatterplot. Explain to him how the scatterplot could still reveal the strong association he anticipated.

21. Prediction units. The errors in predicting hurricane tracks (examined in this chapter) were given in nautical miles. An ordinary mile is 0.86898 nautical miles. Most people living on the Gulf Coast of the United States would prefer to know the prediction errors in miles rather than nautical miles. Explain why converting the errors to miles would not change the correlation between *Prediction Error* and *Year*.

22. More predictions. Hurricane Katrina's hurricane force winds extended 120 miles from its center. Katrina was a big storm, and that affects how we think about the prediction errors. Suppose we add 120 miles to each error to get an idea of how far from the predicted track we might still find damaging winds. Explain what would happen to the correlation between *Prediction Error* and *Year,* and why.

23. Correlation errors. Your Economics instructor assigns your class to investigate factors associated with the gross domestic product (*GDP*) of nations. Each student examines a different factor (such as *Life Expectancy, Literacy Rate,* etc.) for a few countries and reports to the class. Apparently, some of your classmates do not understand Statistics very well because you know several of their conclusions are incorrect. Explain the mistakes in their statements below.
a) "My very low correlation of −0.772 shows that there is almost no association between *GDP* and *Infant Mortality Rate.*"
b) "There was a correlation of 0.44 between *GDP* and *Continent.*"

24. More correlation errors. Students in the Economics class discussed in Exercise 23 also wrote these conclusions. Explain the mistakes they made.
a) "There was a very strong correlation of 1.22 between *Life Expectancy* and *GDP.*"
b) "The correlation between *Literacy Rate* and *GDP* was 0.83. This shows that countries wanting to increase their standard of living should invest heavily in education."

25. Height and reading. A researcher studies children in elementary school and finds a strong positive linear association between height and reading scores.
a) Does this mean that taller children are generally better readers?
b) What might explain the strong correlation?

26. Cellular telephones and life expectancy. A survey of the world's nations in 2004 shows a strong positive correlation between percentage of the country using cell phones and life expectancy in years at birth.
a) Does this mean that cell phones are good for your health?
b) What might explain the strong correlation?

27. Correlation conclusions I. The correlation between *Age* and *Income* as measured on 100 people is $r = 0.75$. Explain whether or not each of these possible conclusions is justified:
a) When *Age* increases, *Income* increases as well.
b) The form of the relationship between *Age* and *Income* is straight.
c) There are no outliers in the scatterplot of *Income* vs. *Age.*
d) Whether we measure *Age* in years or months, the correlation will still be 0.75.

28. Correlation conclusions II. The correlation between *Fuel Efficiency* (as measured by miles per gallon) and *Price* of 150 cars at a large dealership is $r = -0.34$. Explain whether or not each of these possible conclusions is justified:
a) The more you pay, the lower the fuel efficiency of your car will be.
b) The form of the relationship between *Fuel Efficiency* and *Price* is moderately straight.
c) There are several outliers that explain the low correlation.
d) If we measure *Fuel Efficiency* in kilometers per liter instead of miles per gallon, the correlation will increase.

29. Baldness and heart disease. Medical researchers followed 1435 middle-aged men for a period of 5 years, measuring the amount of *Baldness* present (none = 1, little = 2, some = 3, much = 4, extreme = 5) and presence of *Heart Disease* (No = 0, Yes = 1). They found a correlation of 0.089 between the two variables. Comment on their conclusion that this shows that baldness is not a possible cause of heart disease.

30. Sample survey. A polling organization is checking its database to see if the two data sources it used sampled the same zip codes. The variable *Datasource* = 1 if the data source is MetroMedia, 2 if the data source is DataQwest, and 3 if it's RollingPoll. The organization finds that the correlation between five-digit zip code and *Datasource* is −0.0229. It concludes that the correlation is low enough to state that there is no dependency between *Zip Code* and *Source of Data*. Comment.

Ⓣ **31. Income and housing.** The Office of Federal Housing Enterprise Oversight (www.ofheo.gov) collects data on various aspects of housing costs around the United States. Here is a scatterplot of the *Housing Cost Index* versus the *Median Family Income* for each of the 50 states. The correlation is 0.65.

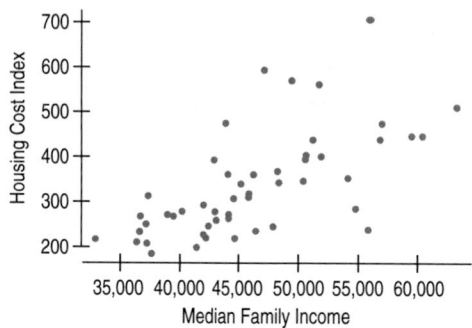

a) Describe the relationship between the *Housing Cost Index* and the *Median Family Income* by state.

b) If we standardized both variables, what would the correlation coefficient between the standardized variables be?

c) If we had measured *Median Family Income* in thousands of dollars instead of dollars, how would the correlation change?

d) Washington, DC, has a Housing Cost Index of 548 and a median income of about $45,000. If we were to include DC in the data set, how would that affect the correlation coefficient?

e) Do these data provide proof that by raising the median income in a state, the Housing Cost Index will rise as a result? Explain.

32. Interest rates and mortgages. Since 1980, average mortgage interest rates have fluctuated from a low of under 6% to a high of over 14%. Is there a relationship between the amount of money people borrow and the interest rate that's offered? Here is a scatterplot of *Total Mortgages* in the United States (in millions of 2005 dollars) versus *Interest Rate* at various times over the past 26 years. The correlation is −0.84.

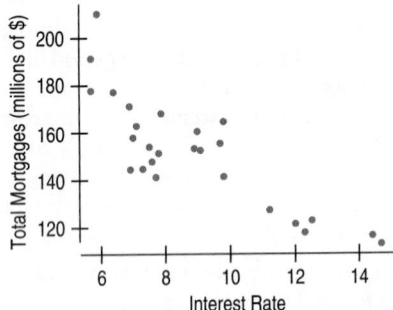

a) Describe the relationship between *Total Mortgages* and *Interest Rate*.

b) If we standardized both variables, what would the correlation coefficient between the standardized variables be?

c) If we were to measure *Total Mortgages* in thousands of dollars instead of millions of dollars, how would the correlation coefficient change?

d) Suppose in another year, interest rates were 11% and mortgages totaled $250 million. How would including that year with these data affect the correlation coefficient?

e) Do these data provide proof that if mortgage rates are lowered, people will take out more mortgages? Explain.

33. Fuel economy 2007. Here are advertised horsepower ratings and expected gas mileage for several 2007 vehicles. (http://www.kbb.com/KBB/ReviewsAndRatings)

Vehicle	Horsepower	Highway Gas Mileage (mpg)
Audi A4	200	32
BMW 328	230	30
Buick LaCrosse	200	30
Chevy Cobalt	148	32
Chevy TrailBlazer	291	22
Ford Expedition	300	20
GMC Yukon	295	21
Honda Civic	140	40
Honda Accord	166	34
Hyundai Elantra	138	36
Lexus IS 350	306	28
Lincoln Navigator	300	18
Mazda Tribute	212	25
Toyota Camry	158	34
Volkswagen Beetle	150	30

a) Make a scatterplot for these data.

b) Describe the direction, form, and strength of the plot.

c) Find the correlation between horsepower and miles per gallon.

d) Write a few sentences telling what the plot says about fuel economy.

34. Drug abuse. A survey was conducted in the United States and 10 countries of Western Europe to determine the percentage of teenagers who had used marijuana and other drugs. The results are summarized in the table.

	Percent Who Have Used	
Country	**Marijuana**	**Other Drugs**
Czech Rep.	22	4
Denmark	17	3
England	40	21
Finland	5	1
Ireland	37	16
Italy	19	8
No. Ireland	23	14
Norway	6	3
Portugal	7	3
Scotland	53	31
USA	34	24

a) Create a scatterplot.

b) What is the correlation between the percent of teens who have used marijuana and the percent who have used other drugs?

c) Write a brief description of the association.

d) Do these results confirm that marijuana is a "gateway drug," that is, that marijuana use leads to the use of other drugs? Explain.

35. Burgers. Fast food is often considered unhealthy because much of it is high in both fat and sodium. But are the two related? Here are the fat and sodium contents of several brands of burgers. Analyze the association between fat content and sodium.

Fat (g)	19	31	34	35	39	39	43
Sodium (mg)	920	1500	1310	860	1180	940	1260

36. Burgers II. In the previous exercise you analyzed the association between the amounts of fat and sodium in fast food hamburgers. What about fat and calories? Here are data for the same burgers:

Fat (g)	19	31	34	35	39	39	43
Calories	410	580	590	570	640	680	660

37. Attendance 2006. American League baseball games are played under the designated hitter rule, meaning that pitchers, often weak hitters, do not come to bat. Baseball owners believe that the designated hitter rule means more runs scored, which in turn means higher attendance. Is there evidence that more fans attend games if the teams score more runs? Data collected from American League games during the 2006 season indicate a correlation of 0.667 between runs scored and the number of people at the game. (http://mlb.mlb.com)

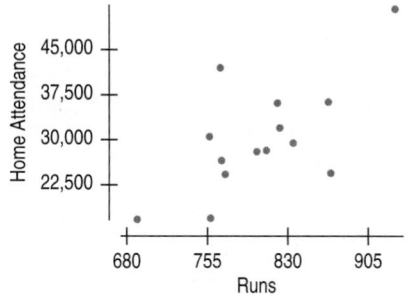

a) Does the scatterplot indicate that it's appropriate to calculate a correlation? Explain.
b) Describe the association between attendance and runs scored.
c) Does this association prove that the owners are right that more fans will come to games if the teams score more runs?

38. Second inning 2006. Perhaps fans are just more interested in teams that win. The displays below are based on American League teams for the 2006 season. (http://espn.go.com) Are the teams that win necessarily those which score the most runs?

CORRELATION			
	Wins	Runs	Attend
Wins	1.000		
Runs	0.605	1.000	
Attend	0.697	0.667	1.000

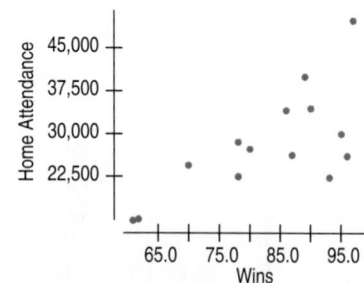

a) Do winning teams generally enjoy greater attendance at their home games? Describe the association.
b) Is attendance more strongly associated with winning or scoring runs? Explain.
c) How strongly is scoring more runs associated with winning more games?

39. Thrills. People who responded to a July 2004 Discovery Channel poll named the 10 best roller coasters in the United States. The table below shows the length of the initial drop (in feet) and the duration of the ride (in seconds). What do these data indicate about the height of a roller coaster and the length of the ride you can expect?

Roller Coaster	State	Drop (ft)	Duration (sec)
Incredible Hulk	FL	105	135
Millennium Force	OH	300	105
Goliath	CA	255	180
Nitro	NJ	215	240
Magnum XL-2000	OH	195	120
The Beast	OH	141	65
Son of Beast	OH	214	140
Thunderbolt	PA	95	90
Ghost Rider	CA	108	160
Raven	IN	86	90

40. Vehicle weights. The Minnesota Department of Transportation hoped that they could measure the weights of big trucks without actually stopping the vehicles by using a newly developed "weight-in-motion" scale. To see if the new device was accurate, they conducted a calibration test. They weighed several stopped trucks (static weight) and assumed that this weight was correct. Then they weighed the trucks again while they were moving to see how well the new scale could estimate the actual weight. Their data are given in the table on the next page.

WEIGHTS (1000S OF LBS)	
Weight-in-Motion	Static Weight
26.0	27.9
29.9	29.1
39.5	38.0
25.1	27.0
31.6	30.3
36.2	34.5
25.1	27.8
31.0	29.6
35.6	33.1
40.2	35.5

a) Make a scatterplot for these data.
b) Describe the direction, form, and strength of the plot.
c) Write a few sentences telling what the plot says about the data. (*Note:* The sentences should be about weighing trucks, not about scatterplots.)
d) Find the correlation.
e) If the trucks were weighed in kilograms, how would this change the correlation? (1 kilogram = 2.2 pounds)
f) Do any points deviate from the overall pattern? What does the plot say about a possible recalibration of the weight-in-motion scale?

41. Planets (more or less). On August 24, 2006, the International Astronomical Union voted that Pluto is not a planet. Some members of the public have been reluctant to accept that decision. Let's look at some of the data. (We'll see more in the next chapter.) Is there any pattern to the locations of the planets? The table shows the average distance of each of the traditional nine planets from the sun.

Planet	Position Number	Distance from Sun (million miles)
Mercury	1	36
Venus	2	67
Earth	3	93
Mars	4	142
Jupiter	5	484
Saturn	6	887
Uranus	7	1784
Neptune	8	2796
Pluto	9	3666

a) Make a scatterplot and describe the association. (Remember: direction, form, and strength!)
b) Why would you not want to talk about the correlation between a planet's *Position* and *Distance* from the sun?
c) Make a scatterplot showing the logarithm of *Distance* vs. *Position*. What is better about this scatterplot?

42. Flights. The number of flights by U.S. Airlines has grown rapidly. Here are the number of flights flown in each year from 1995 to 2005.
a) Find the correlation of *Flights* with *Year*.
b) Make a scatterplot and describe the trend.
c) Note two reasons that the correlation you found in (a) is not a suitable summary of the strength of the association. Can you account for these violations of the conditions?

Year	Flights
1995	5,327,435
1996	5,351,983
1997	5,411,843
1998	5,384,721
1999	5,527,884
2000	5,683,047
2001	5,967,780
2002	5,271,359
2003	6,488,539
2004	7,129,270
2005	7,140,596

JUST CHECKING
Answers

1. We know the scores are quantitative. We should check to see if the Straight Enough Condition and the Outlier Condition are satisfied by looking at a scatterplot of the two scores.
2. It won't change.
3. It won't change.
4. They are likely to have done poorly. The positive correlation means that low scores on Exam 1 are associated with low scores on Exam 2 (and similarly for high scores).
5. No. The general association is positive, but individual performances may vary.

Linear Regression

WHO	Items on the Burger King menu
WHAT	Protein content and total fat content
UNITS	Grams of protein Grams of fat
HOW	Supplied by BK on request or at their Web site

A S *Video:* **Manatees and Motorboats.** Are motorboats killing more manatees in Florida? Here's the story on video.

A S *Activity:* **Linear Equations.** For a quick review of linear equations, view this activity and play with the interactive tool.

The Whopper™ has been Burger King's signature sandwich since 1957. One Double Whopper with cheese provides 53 grams of protein—all the protein you need in a day. It also supplies 1020 calories and 65 grams of fat. The Daily Value (based on a 2000-calorie diet) for fat is 65 grams. So after a Double Whopper you'll want the rest of your calories that day to be fat-free.[1]

Of course, the Whopper isn't the only item Burger King sells. How are fat and protein related on the entire BK menu? The scatterplot of the *Fat* (in grams) versus the *Protein* (in grams) for foods sold at Burger King shows a positive, moderately strong, linear relationship.

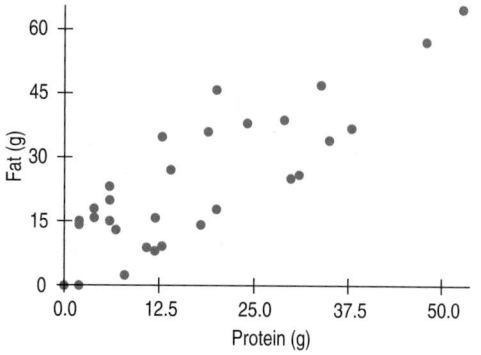

FIGURE 8.1

Total Fat *versus* Protein *for 30 items on the BK menu. The Double Whopper is in the upper right corner. It's extreme, but is it out of line?*

If you want 25 grams of protein in your lunch, how much fat should you expect to consume at Burger King? The correlation between *Fat* and *Protein* is 0.83, a sign that the linear association seen in the scatterplot is fairly strong. But *strength* of the relationship is only part of the picture. The correlation says, "The linear association between these two variables is fairly strong," but it doesn't tell us *what the line is.*

---

[1] Sorry about the fries.

> *"Statisticians, like artists, have the bad habit of falling in love with their models."*
>
> —George Box, famous statistician

> **A S** **Activity: Residuals.** Residuals are the basis for fitting lines to scatterplots. See how they work.

Now we *can* say more. We can **model** the relationship with a line and give its equation. The equation will let us predict the fat content for any Burger King food, given its amount of protein.

We met our first model in Chapter 6. We saw there that we can specify a Normal model with two parameters: its mean (μ) and standard deviation (σ).

For the Burger King foods, we'd choose a linear model to describe the relationship between *Protein* and *Fat*. The **linear model** is just an equation of a straight line through the data. Of course, no line can go through all the points, but a linear model can summarize the general pattern with only a couple of parameters. Like all models of the real world, the line will be wrong—wrong in the sense that it can't match reality *exactly*. But it can help us understand how the variables are associated.

Residuals

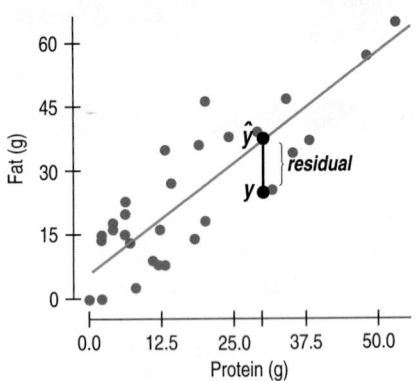

Not only can't we draw a line through all the points, the best line might not even hit *any* of the points. Then how can it be the "best" line? We want to find the line that somehow comes *closer* to all the points than any other line. Some of the points will be above the line and some below. For example, the line might suggest that a BK Broiler chicken sandwich with 30 grams of protein should have 36 grams of fat when, in fact, it actually has only 25 grams of fat. We call the estimate made from a model the **predicted value,** and write it as $\hat{y}$ (called *y-hat*) to distinguish it from the true value y (called, uh, y). The difference between the observed value and its associated predicted value is called the **residual.** The residual value tells us how far off the model's prediction is at that point. The BK Broiler chicken residual would be $y - \hat{y} = 25 - 36 = -11$ g of fat.

> *residual = observed value − predicted value*
> A *negative* residual means the predicted value is too big—an overestimate. And a *positive* residual shows that the model makes an underestimate. These may seem backwards until you think about them.

To find the residuals, we always subtract the predicted value from the observed one. The negative residual tells us that the actual fat content of the BK Broiler chicken is about 11 grams *less* than the model predicts for a typical Burger King menu item with 30 grams of protein.

Our challenge now is how to find the right line.

"Best Fit" Means Least Squares

> **A S** **Activity: The Least Squares Criterion.** Does your sense of "best fit" look like the least squares line?

When we draw a line through a scatterplot, some residuals are positive and some negative. We can't assess how well the line fits by adding up all the residuals—the positive and negative ones would just cancel each other out. We faced the same issue when we calculated a standard deviation to measure spread. And we deal with it the same way here: by squaring the residuals. Squaring makes them all positive. Now we can add them up. Squaring also emphasizes the large residuals. After all, points near the line are consistent with the model, but we're more concerned about points far from the line. When we add all the squared residuals together, that sum indicates how well the line we drew fits the data—the smaller the sum, the better the fit. A different line will produce a different sum, maybe bigger, maybe smaller. The **line of best fit** is the line for which the sum of the squared residuals is smallest, the **least squares** line.

> ## Who's on First
> In 1805, Legendre was the first to publish the "least squares" solution to the problem of fitting a line to data when the points don't all fall exactly on the line. The main challenge was how to distribute the errors "fairly." After considerable thought, he decided to minimize the sum of the squares of what we now call the residuals. When Legendre published his paper, though, Gauss claimed he had been using the method since 1795. Gauss later referred to the "least squares" solution as "*our* method" (*principium nostrum*), which certainly didn't help his relationship with Legendre.

You might think that finding this line would be pretty hard. Surprisingly, it's not, although it was an exciting mathematical discovery when Legendre published it in 1805 (see margin note on previous page).

Correlation and the Line

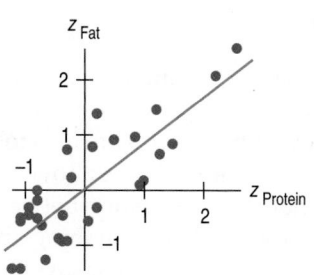

FIGURE 8.2

The Burger King scatterplot in z-scores.

If you suspect that what we know about correlation can lead us to the equation of the linear model, you're headed in the right direction. It turns out that it's not a very big step. In Chapter 7 we learned a lot about how correlation worked by looking at a scatterplot of the standardized variables. Here's a scatterplot of z_y (standardized *Fat*) vs. z_x (standardized *Protein*).

What line would you choose to model the relationship of the standardized values? Let's start at the center of the scatterplot. How much protein and fat does a *typical* Burger King food item provide? If it has average protein content, $\bar{x}$, what about its fat content? If you guessed that its fat content should be about average, $\bar{y}$, as well, then you've discovered the first property of the line we're looking for. The line must go through the point $(\bar{x}, \bar{y})$. In the plot of z-scores, then, the line passes through the origin $(0, 0)$.

You might recall that the equation for a line that passes through the origin can be written with just a slope and no intercept:

$$y = mx.$$

The coordinates of our standardized points aren't written (x, y); their coordinates are z-scores: (z_x, z_y). We'll need to change our equation to show that. And we'll need to indicate that the point on the line corresponding to a particular z_x is $\hat{z}_y$, the model's estimate of the actual value of z_y. So our equation becomes

$$\hat{z}_y = mz_x.$$

Many lines with different slopes pass through the origin. Which one fits our data the best? That is, which slope determines the line that minimizes the sum of the squared residuals? It turns out that the best choice for m is the correlation coefficient itself, r! (You must really wonder where that stunning assertion comes from. Check the Math Box.)

Wow! This line has an equation that's about as simple as we could possibly hope for:

$$\hat{z}_y = rz_x.$$

Great. It's simple, but what does it tell us? It says that in moving one standard deviation from the mean in x, we can expect to move about r standard deviations away from the mean in y. Now that we're thinking about least squares lines, the correlation is more than just a vague measure of strength of association. It's a great way to think about what the model tells us.

Let's be more specific. For the sandwiches, the correlation is 0.83. If we standardize both protein and fat, we can write

$$\hat{z}_{Fat} = 0.83z_{Protein}.$$

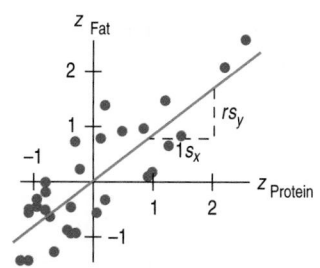

FIGURE 8.3

Standardized fat vs. standardized protein with the regression line. Each one standard deviation change in protein results in a predicted change of r standard deviations in fat.

This model tells us that for every standard deviation above (or below) the mean a sandwich is in protein, we'll predict that its fat content is 0.83 standard deviations above (or below) the mean fat content. A double hamburger has 31 grams of protein, about 1 SD above the mean. Putting 1.0 in for $z_{Protein}$ in the model gives a $\hat{z}_{Fat}$ value of 0.83. If you trust the model, you'd expect the fat content to be about 0.83 fat SDs above the mean fat level. Moving one standard deviation away from the mean in x moves our estimate r standard deviations away from the mean in y.

If $r = 0$, there's no linear relationship. The line is horizontal, and no matter how many standard deviations you move in x, the predicted value for y doesn't change. On the other hand, if $r = 1.0$ or -1.0, there's a perfect linear association. In that case, moving any number of standard deviations in x moves exactly the same number of standard deviations in y. In general, moving any number of standard deviations in x moves r times that number of standard deviations in y.

How Big Can Predicted Values Get?

Sir Francis Galton was the first to speak of "regression," although others had fit lines to data by the same method.

Suppose you were told that a new male student was about to join the class, and you were asked to guess his height in inches. What would be your guess? A safe guess would be the mean height of male students. Now suppose you are also told that this student has a grade point average (*GPA*) of 3.9—about 2 SDs above the mean *GPA*. Would that change your guess? Probably not. The correlation between *GPA* and *height* is near 0, so knowing the *GPA* value doesn't tell you anything and doesn't move your guess. (And the equation tells us that as well, since it says that we should move 0×2 SDs from the mean.)

On the other hand, suppose you were told that, measured in centimeters, the student's height was 2 SDs above the mean. There's a perfect correlation between *height in inches* and *height in centimeters*, so you'd know he's 2 SDs above mean height in inches as well. (The equation would tell us to move 1.0×2 SDs from the mean.)

What if you're told that the student is 2 SDs above the mean in *shoe size*? Would you still guess that he's of average *height*? You might guess that he's taller than average, since there's a positive correlation between *height* and *shoe size*. But would you guess that he's 2 SDs above the mean? When there was no correlation, we didn't move away from the mean at all. With a perfect correlation, we moved our guess the full 2 SDs. Any correlation between these extremes should lead us to move somewhere between 0 and 2 SDs above the mean. (To be exact, the equation tells us to move $r \times 2$ standard deviations away from the mean.)

Notice that if x is 2 SDs above its mean, we won't ever guess more than 2 SDs away for y, since r can't be bigger than 1.0.[2] So, each predicted y tends to be closer to its mean (in standard deviations) than its corresponding x was. This property of the linear model is called **regression to the mean,** and the line is called the **regression line.**

> **The First Regression**
>
> Sir Francis Galton related the heights of sons to the heights of their fathers with a regression line. The slope of his line was less than 1. That is, sons of tall fathers were tall, but not as much above the average height as their fathers had been above their mean. Sons of short fathers were short, but generally not as far from their mean as their fathers. Galton interpreted the slope correctly as indicating a "regression" toward the mean height—and "regression" stuck as a description of the method he had used to find the line.

JUST CHECKING

A scatterplot of house *Price* (in thousands of dollars) vs. house *Size* (in thousands of square feet) for houses sold recently in Saratoga, NY shows a relationship that is straight, with only moderate scatter and no outliers. The correlation between house *Price* and house *Size* is 0.77.

1. You go to an open house and find that the house is 1 standard deviation above the mean in size. What would you guess about its price?

2. You read an ad for a house priced 2 standard deviations below the mean. What would you guess about its size?

3. A friend tells you about a house whose size in square meters (he's European) is 1.5 standard deviations above the mean. What would you guess about its size in square feet?

---

[2] In the last chapter we asserted that correlations max out at 1, but we never actually *proved* that. Here's yet another reason to check out the Math Box on the next page.

MATH BOX

Where does the equation of the line of best fit come from? To write the equation of any line, we need to know a point on the line and the slope. The point is easy. Consider the BK menu example. Since it is logical to predict that a sandwich with average protein will contain average fat, the line passes through the point $(\bar{x}, \bar{y})$.[3]

To think about the slope, we look once again at the z-scores. We need to remember a few things:

1. The mean of any set of z-scores is 0. This tells us that the line that best fits the z-scores passes through the origin (0,0).
2. The standard deviation of a set of z-scores is 1, so the variance is also 1. This means that
$$\frac{\sum(z_y - \bar{z}_y)^2}{n-1} = \frac{\sum(z_y - 0)^2}{n-1} = \frac{\sum z_y^2}{n-1} = 1, \text{ a fact that will be important soon.}$$
3. The correlation is $r = \dfrac{\sum z_x z_y}{n-1}$, also important soon.

Ready? Remember that our objective is to find the slope of the best fit line. Because it passes through the origin, its equation will be of the form $\hat{z}_y = m z_x$. We want to find the value for m that will minimize the sum of the squared residuals. Actually we'll divide that sum by $n-1$ and minimize this "mean squared residual," or *MSR*. Here goes:

Minimize:
$$MSR = \frac{\sum(z_y - \hat{z}_y)^2}{n-1}$$

Since $\hat{z}_y = m z_x$:
$$MSR = \frac{\sum(z_y - m z_x)^2}{n-1}$$

Square the binomial:
$$= \frac{\sum(z_y^2 - 2m z_x z_y + m^2 z_x^2)}{n-1}$$

Rewrite the summation:
$$= \frac{\sum z_y^2}{n-1} - 2m\frac{\sum z_x z_y}{n-1} + m^2\frac{\sum z_x^2}{n-1}$$

4. Substitute from (2) and (3):
$$= 1 - 2mr + m^2$$

Wow! That simplified nicely! And as a bonus, the last expression is quadratic. Remember parabolas from algebra class? A parabola in the form $y = ax^2 + bx + c$ reaches its minimum at its turning point, which occurs when $x = \dfrac{-b}{2a}$. We can minimize the mean of squared residuals by choosing $m = \dfrac{-(-2r)}{2(1)} = r$.

Wow, again! The slope of the best fit line for z-scores is the correlation, r. This stunning fact immediately leads us to two important additional results, listed below. As you read on in the text, we explain them in the context of our continuing discussion of Burger King foods.

• A slope of r for z-scores means that for every increase of 1 standard deviation in z_x there is an increase of r standard deviations in $\hat{z}_y$. "Over one, up r," as you probably said in algebra class. Translate that back to the original x and y values: "Over one standard deviation in x, up r standard deviations in $\hat{y}$."

That's it! In x- and y-values, the slope of the regression line is $b = \dfrac{r s_y}{s_x}$.

---
[3] It's actually not hard to prove this too.

• We know choosing $m = r$ minimizes the sum of the squared residuals, but how small does that sum get? Equation (4) told us that the mean of the squared residuals is $1 - 2mr + m^2$. When $m = r$, $1 - 2mr + m^2 = 1 - 2r^2 + r^2 = 1 - r^2$. This is the variability *not* explained by the regression line. Since the variance in z_y was 1 (Equation 2), the percentage of variability in y that is explained by x is r^2. This important fact will help us assess the strength of our models.

And there's still another bonus. Because r^2 is the percent of variability explained by our model, r^2 is at most 100%. If $r^2 \leq 1$, then $-1 \leq r \leq 1$, proving that correlations are always between -1 and $+1$. (Told you so!)

The Regression Line in Real Units

Why Is Correlation "r"?
In his original paper on correlation, Galton used r for the "index of correlation" that we now call the correlation coefficient. He calculated it from the regression of y on x or of x on y after standardizing the variables, just as we have done. It's fairly clear from the text that he used r to stand for (standardized) regression.

A S *Simulation: Interpreting Equations.* This demonstrates how to use and interpret linear equations.

Protein	Fat
$\bar{x} = 17.2$ g	$\bar{y} = 23.5$ g
$s_x = 14.0$ g	$s_y = 16.4$ g
$r = 0.83$	

Slope
$$b_1 = \frac{rs_y}{s_x}$$

Intercept
$$b_0 = \bar{y} - b_1\bar{x}$$

When you read the Burger King menu, you probably don't think in z-scores. But you might want to know the fat content in grams for a specific amount of protein in grams.

How much fat should we predict for a double hamburger with 31 grams of protein? The mean protein content is near 17 grams and the standard deviation is 14, so that item is 1 SD above the mean. Since $r = 0.83$, we predict the fat content will be 0.83 SDs above the mean fat content. Great. How much fat is that? Well, the mean fat content is 23.5 grams and the standard deviation of fat content is 16.4, so we predict that the double hamburger will have $23.5 + 0.83 \times 16.4 = 37.11$ grams of fat.

We can always convert both x and y to z-scores, find the correlation, use $\hat{z}_y = rz_x$, and then convert $\hat{z}_y$ back to its original units so that we can understand the prediction. But can't we do this more simply?

Yes. Let's write the equation of the line for protein and fat—that is, the actual x and y values rather than their z-scores. In Algebra class you may have once seen lines written in the form $y = mx + b$. Statisticians do exactly the same thing, but with different notation:

$$\hat{y} = b_0 + b_1 x.$$

In this equation, b_0 is the **y-intercept**, the value of y where the line crosses the y-axis, and b_1 is the **slope**.[4]

First we find the slope, using the formula we developed in the Math Box.[5] Remember? We know that our model predicts that for each increase of one standard deviation in protein we'll see an increase of about 0.83 standard deviations in fat.

In other words, the slope of the line in original units is

$$b_1 = \frac{rs_y}{s_x} = \frac{0.83 \times 16.4 \text{ g fat}}{14 \text{ g protein}} = 0.97 \text{ grams of fat per gram of protein.}$$

Next, how do we find the y-intercept, b_0? Remember that the line has to go through the mean-mean point $(\bar{x}, \bar{y})$. In other words, the model predicts $\bar{y}$ to be the value that corresponds to $\bar{x}$. We can put the means into the equation and write $\bar{y} = b_0 + b_1\bar{x}$.

Solving for b_0, we see that the intercept is just $b_0 = \bar{y} - b_1\bar{x}$.

---

[4] We changed from $mx + b$ to $b_0 + b_1x$ for a reason—not just to be difficult. Eventually we'll want to add more x's to the model to make it more realistic and we don't want to use up the entire alphabet. What would we use after m? The next letter is n, and that one's already taken. o? See our point? Sometimes subscripts are the best approach.
[5] Several important results popped up in that Math Box. Check it out!

For the Burger King foods, that comes out to

$$b_0 = 23.5 \text{ g fat} - 0.97 \frac{\text{g fat}}{\text{g protein}} \times 17.2 \text{ g protein} = 6.8 \text{ g fat}.$$

Putting this back into the regression equation gives

$$\widehat{fat} = 6.8 + 0.97 \, protein.$$

> **Units of *y* per unit of *x***
> Get into the habit of identifying the units by writing down "*y*-units per *x*-unit," with the unit names put in place. You'll find it'll really help you to Tell about the line in context.

What does this mean? The slope, 0.97, says that an additional gram of protein is associated with an additional 0.97 grams of fat, on average. Less formally, we might say that Burger King sandwiches pack about 0.97 grams of fat per gram of protein. Slopes are always expressed in *y*-units per *x*-unit. They tell how the *y*-variable changes (in its units) for a one-unit change in the *x*-variable. When you see a phrase like "students per teacher" or "kilobytes per second" think slope.

Changing the units of the variables doesn't change the *correlation*, but for the *slope*, units do matter. We may know that age and height in children are positively correlated, but the *value* of the slope depends on the units. If children grow an average of 3 inches per year, that's the same as 0.21 millimeters per day. For the slope, it matters whether you express age in days or years and whether you measure height in inches or millimeters. How you choose to express *x* and *y*—what units you use—affects the slope directly. Why? We know changing units doesn't change the correlation, but does change the standard deviations. The slope introduces the units into the equation by multiplying the correlation by the ratio of s_y to s_x. The units of the **slope** are always the units of *y* per unit of *x*.

How about the **intercept** of the BK regression line, 6.8? Algebraically, that's the value the line takes when *x* is zero. Here, our model predicts that even a BK item with no protein would have, on average, about 6.8 grams of fat. Is that reasonable? Well, the apple pie, with 2 grams of protein, has 14 grams of fat, so it's not impossible. But often 0 is not a plausible value for *x* (the year 0, a baby born weighing 0 grams, ...). Then the intercept serves only as a starting value for our predictions and we don't interpret it as a meaningful predicted value.

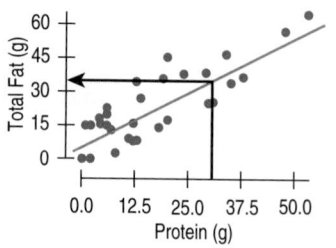

FIGURE 8.4

Burger King menu items in their natural units with the regression line.

FOR EXAMPLE   A regression model for hurricanes

In Chapter 7 we looked at the relationship between the central pressure and maximum wind speed of Atlantic hurricanes. We saw that the scatterplot was straight enough, and then found a correlation of −0.879, but we had no model to describe how these two important variables are related or to allow us to predict wind speed from pressure. Since the conditions we need to check for regression are the same ones we checked before, we can use technology to find the regression model. It looks like this:

$$\widehat{MaxWindSpeed} = 955.27 - 0.897 \, CentralPressure$$

Question: Interpret this model. What does the slope mean in this context? Does the intercept have a meaningful interpretation?

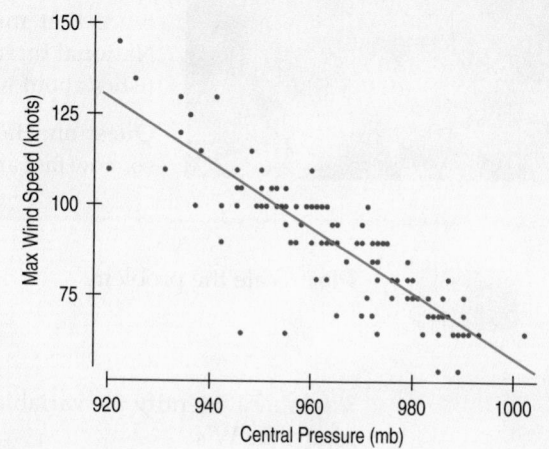

The negative slope says that as *CentralPressure* falls, *MaxWindSpeed* increases. That makes sense from our general understanding of how hurricanes work: Low central pressure pulls in moist air, driving the rotation and the resulting destructive winds. The slope's value says that, on average, the maximum wind speed increases by about 0.897 knots for every 1-millibar drop in central pressure.

It's not meaningful, however, to interpret the intercept as the wind speed predicted for a central pressure of 0—that would be a vacuum. Instead, it is merely a starting value for the model.

With the estimated linear model, $\widehat{fat} = 6.8 + 0.97\ protein$, it's easy to predict fat content for any menu item we want. For example, for the BK Broiler chicken sandwich with 30 grams of *protein*, we can plug in 30 grams for the amount of *protein* and see that the *predicted fat* content is $6.8 + 0.97(30) = 35.9$ grams of fat. Because the BK Broiler chicken sandwich actually has 25 grams of fat, its residual is

$$fat - \widehat{fat} = 25 - 35.9 = -10.9\ \text{g}.$$

To use a regression model, we should check the same conditions for regressions as we did for correlation: the **Quantitative Variables Condition,** the **Straight Enough Condition,** and the **Outlier Condition.**

JUST CHECKING

Let's look again at the relationship between house *Price* (in thousands of dollars) and house *Size* (in thousands of square feet) in Saratoga. The regression model is

$$\widehat{Price} = -3.117 + 94.454\ Size.$$

4. What does the slope of 94.454 mean?

5. What are the units of the slope?

6. Your house is 2000 sq ft bigger than your neighbor's house. How much more do you expect it to be worth?

7. Is the *y*-intercept of −3.117 meaningful? Explain.

STEP-BY-STEP EXAMPLE | **Calculating a Regression Equation**

Wildfires are an ongoing source of concern shared by several government agencies. In 2004, the Bureau of Land Management, Bureau of Indian Affairs, Fish and Wildlife Service, National Park Service, and USDA Forest Service spent a combined total of $890,233,000 on fire suppression, down from nearly twice that much in 2002. These government agencies join together in the National Interagency Fire Center, whose Web site (www.nifc.gov) reports statistics about wildfires.

Question: Has the annual number of wildfires been changing, on average? If so, how fast and in what way?

Plan State the problem.

I want to know how the number of wildfires in the continental United States has changed in the past two decades.

Variables Identify the variables and report the W's.

I have data giving the number of wildfires for each year (in thousands of fires) from 1982 to 2005.

Check the appropriate assumptions and conditions.

✔ **Quantitative Variables Condition:** Both the number of fires and the year are quantitative.

Just as we did for correlation, check the conditions for a regression by making a picture. Never fit a regression without looking at the scatterplot first.

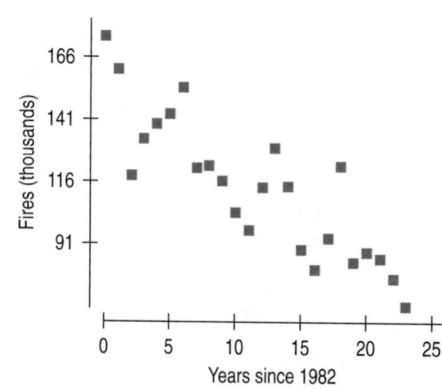

Note: It's common (and usually simpler) not to use four-digit numbers to identify years. Here we have chosen to number the years beginning in 1982, so 1982 is represented as year 0 and 2005 as year 23.

✓ **Straight Enough Condition:** The scatterplot shows a strong linear relationship with a negative association.

✓ **Outlier Condition:** No outliers are evident in the scatterplot.

Because these conditions are satisfied, it is OK to model the relationship with a regression line.

Mechanics Find the equation of the regression line. Summary statistics give the building blocks of the calculation.

(We generally report summary statistics to one more digit of accuracy than the data. We do the same for intercept and predicted values, but for slopes we usually report an additional digit. Remember, though, not to round off until you finish computing an answer.)[6]

Find the slope, b_1.

Find the intercept, b_0.

Write the equation of the model, using meaningful variable names.

Year:

$$\bar{x} = 11.5 \text{ (representing 1993.5)}$$
$$s_x = 7.07 \text{ years}$$

Fires:

$$\bar{y} = 114.098 \text{ fires}$$
$$s_y = 28.342 \text{ fires}$$

Correlation:

$$r = -0.862$$
$$b_1 = \frac{rs_y}{s_x} = \frac{-0.862(28.342)}{7.07}$$
$$= -3.4556 \text{ fires per year}$$
$$b_0 = \bar{y} - b_1\bar{x} = 114.098 - (-3.4556)11.5$$
$$= 153.837$$

So the least squares line is

$$\hat{y} = 153.837 - 3.4556x, \text{ or}$$
$$\widehat{Fires} = 153.837 - 3.4556 \text{ year}$$

---

[6] We warned you in Chapter 6 that we'll round in the intermediate steps of a calculation to show the steps more clearly. If you repeat these calculations yourself on a calculator or statistics program, you may get somewhat different results. When calculated with more precision, the intercept is 153,809 and the slope is −3.453.

Conclusion Interpret what you have found in the context of the question. Discuss in terms of the variables and their units.

> *During the period from 1982 to 2005, the annual number of fires declined at an average rate of about 3,456 (3.456 thousand) fires per year. For prediction, the model uses a base estimation of 153,837 fires in 1982.*

A S *Activity:* **Find a Regression Equation.** Now that we've done it by hand, try it with technology using the statistics package paired with your version of *ActivStats.*

Residuals Revisited

Why *e* for "Residual"?
The flip answer is that *r* is already taken, but the truth is that *e* stands for "error." No, that doesn't mean it's a mistake. Statisticians often refer to variability not explained by a model as error.

The linear model we are using assumes that the relationship between the two variables is a perfect straight line. The residuals are the part of the data that *hasn't* been modeled. We can write

$$Data = Model + Residual$$

or, equivalently,

$$Residual = Data - Model.$$

Or, in symbols,

$$e = y - \hat{y}.$$

When we want to know how well the model fits, we can ask instead what the model missed. To see that, we look at the residuals.

FOR EXAMPLE   Katrina's residual

Recap: The linear model relating hurricanes' wind speeds to their central pressures was

$$\widehat{MaxWindSpeed} = 955.27 - 0.897CentralPressure$$

Let's use this model to make predictions and see how those predictions do.

Question: Hurricane Katrina had a central pressure measured at 920 millibars. What does our regression model predict for her maximum wind speed? How good is that prediction, given that Katrina's actual wind speed was measured at 110 knots?

Substituting 920 for the central pressure in the regression model equation gives

$$\widehat{MaxWindSpeed} = 955.27 - 0.897(920) = 130.03$$

The regression model predicts a maximum wind speed of 130 knots for Hurricane Katrina.

The residual for this prediction is the observed value minus the predicted value:

$$110 - 130 = -20kts.$$

In the case of Hurricane Katrina, the model predicts a wind speed 20 knots higher than was actually observed.

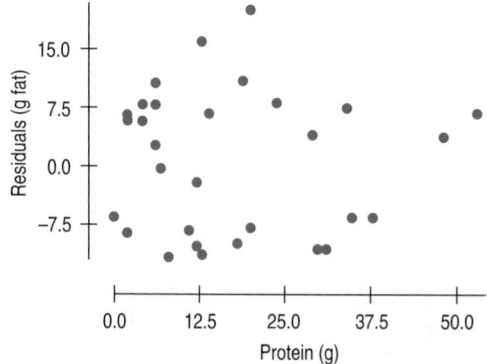

FIGURE 8.5

The residuals for the BK menu regression look appropriately boring.

Residuals help us to see whether the model makes sense. When a regression model is appropriate, it should model the underlying relationship. Nothing interesting should be left behind. So after we fit a regression model, we usually plot the residuals in the hope of finding . . . nothing.

A scatterplot of the residuals versus the x-values should be the most boring scatterplot you've ever seen. It shouldn't have any interesting features, like a direction or shape. It should stretch horizontally, with about the same amount of scatter throughout. It should show no bends, and it should have no outliers. If you see any of these features, find out what the regression model missed.

Most computer statistics packages plot the residuals against the predicted values $\hat{y}$, rather than against x. When the slope is negative, the two versions are mirror images. When the slope is positive, they're virtually identical except for the axis labels. Since all we care about is the patterns (or, better, lack of patterns) in the plot, it really doesn't matter which way we plot the residuals.

JUST CHECKING

Our linear model for Saratoga homes uses the *Size* (in thousands of square feet) to estimate the *Price* (in thousands of dollars): $\widehat{Price} = -3.117 + 94.454 Size$. Suppose you're thinking of buying a home there.

8. Would you prefer to find a home with a negative or a positive residual? Explain.

9. You plan to look for a home of about 3000 square feet. How much should you expect to have to pay?

10. You find a nice home that size selling for $300,000. What's the residual?

The Residual Standard Deviation

If the residuals show no interesting pattern when we plot them against x, we can look at how big they are. After all, we're trying to make them as small as possible. Since their mean is always zero, though, it's only sensible to look at how much they vary. The standard deviation of the residuals, s_e, gives us a measure of how much the points spread around the regression line. Of course, for this summary to make sense, the residuals should all share the same underlying spread, so we check to make sure that the residual plot has about the same amount of scatter throughout.

This gives us a new assumption: the **Equal Variance Assumption.** The associated condition to check is the **Does the Plot Thicken? Condition.** We check to make sure that the spread is about the same all along the line. We can check that either in the original scatterplot of y against x or in the scatterplot of residuals.

We estimate the standard deviation of the residuals in almost the way you'd expect:

$$s_e = \sqrt{\frac{\Sigma e^2}{n-2}}$$

We don't need to subtract the mean because the mean of the residuals $\bar{e} = 0$.

For the Burger King foods, the standard deviation of the residuals is 9.2 grams of fat. That looks about right in the scatterplot of residuals. The residual for the BK Broiler chicken was -11 grams, just over one standard deviation.

Why $n - 2$ rather than $n - 1$? We used $n - 1$ for s when we estimated the mean. Now we're estimating both a slope and an intercept. Looks like a pattern—and it is. We subtract one more for each parameter we estimate.

It's a good idea to make a histogram of the residuals. If we see a unimodal, symmetric histogram, then we can apply the 68–95–99.7 Rule to see how well the regression model describes the data. In particular, we know that 95% of the residuals should be no larger in size than $2s_e$. The Burger King residuals look like this:

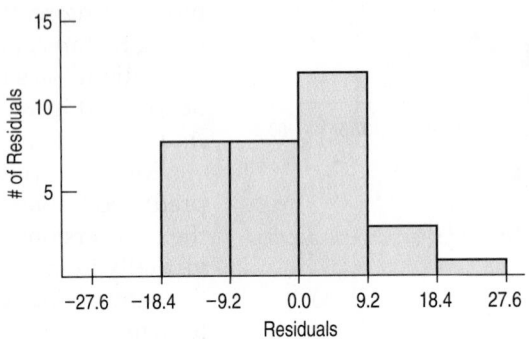

Sure enough, almost all are less than 2(9.2), or 18.4, g of fat in size.

R^2—The Variation Accounted For

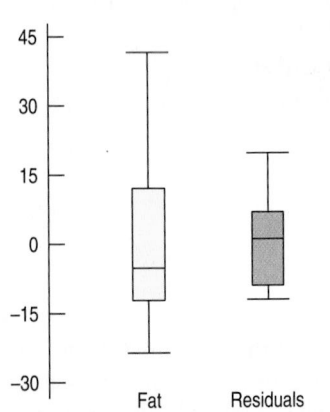

FIGURE 8.6

Compare the variability of total Fat with the residuals from the regression. The means have been subtracted to make it easier to compare spreads. The variation left in the residuals is unaccounted for by the model, but it's less than the variation in the original data.

The variation in the residuals is the key to assessing how well the model fits. Let's compare the variation of the response variable with the variation of the residuals. The total *Fat* has a standard deviation of 16.4 grams. The standard deviation of the residuals is 9.2 grams. If the correlation were 1.0 and the model predicted the *Fat* values perfectly, the residuals would all be zero and have no variation. We couldn't possibly do any better than that.

On the other hand, if the correlation were zero, the model would simply predict 23.5 grams of *Fat* (the mean) for all menu items. The residuals from that prediction would just be the observed *Fat* values minus their mean. These residuals would have the same variability as the original data because, as we know, just subtracting the mean doesn't change the spread.

How well does the BK regression model do? Look at the boxplots. The variation in the residuals is smaller than in the data, but certainly bigger than zero. That's nice to know, but how much of the variation is still left in the residuals? If you had to put a number between 0% and 100% on the fraction of the variation left in the residuals, what would you say?

All regression models fall somewhere between the two extremes of zero correlation and perfect correlation. We'd like to gauge where our model falls. As we showed in the Math Box,[7] the squared correlation, r^2, gives the fraction of the data's variation accounted for by the model, and $1 - r^2$ is the fraction of the original variation left in the residuals. For the Burger King model, $r^2 = 0.83^2 = 0.69$, and $1 - r^2$ is 0.31, so 31% of the variability in total *Fat* has been left in the residuals. How close was that to your guess?

All regression analyses include this statistic, although by tradition, it is written with a capital letter, R^2, and pronounced "R-squared." An R^2 of 0 means that none of the variance in the data is in the model; all of it is still in the residuals. It would be hard to imagine using that model for anything.

TI-*nspire*

Understanding R^2. Watch the unexplained variability decrease as you drag points closer to the regression line.

---

[7] Have you looked yet? Please do.

Because R^2 is a fraction of a whole, it is often given as a percentage.[8] For the Burger King data, R^2 is 69%. When interpreting a regression model, you need to *Tell* what R^2 means. According to our linear model, 69% of the variability in the fat content of Burger King sandwiches is accounted for by variation in the protein content.

> Is a correlation of 0.80 twice as strong as a correlation of 0.40? Not if you think in terms of R^2. A correlation of 0.80 means an R^2 of $0.80^2 = 64\%$. A correlation of 0.40 means an R^2 of $0.40^2 = 16\%$—only a quarter as much of the variability accounted for. A correlation of 0.80 gives an R^2 *four* times as strong as a correlation of 0.40 and accounts for four times as much of the variability.

> **How can we see that R^2 is really the fraction of variance accounted for by the model?** It's a simple calculation. The variance of the fat content of the Burger King foods is $16.4^2 = 268.42$. If we treat the residuals as data, the variance of the residuals is 83.195.[9] As a fraction, that's $83.195/268.42 = 0.31$, or 31%. That's the fraction of the variance that is not accounted for by the model. The fraction that is accounted for is $100\% - 31\% = 69\%$, just the value we got for R^2.

FOR EXAMPLE   Interpreting R^2

Recap: Our regression model that predicts maximum wind speed in hurricanes based on the storm's central pressure has $R^2 = 77.3\%$.

Question: What does that say about our regression model?

An R^2 of 77.3% indicates that 77.3% of the variation in maximum wind speed can be accounted for by the hurricane's central pressure. Other factors, such as temperature and whether the storm is over water or land, may explain some of the remaining variation.

JUST CHECKING

Back to our regression of house *Price* (in thousands of $) on house *Size* (in thousands of square feet). The R^2 value is reported as 59.5%, and the standard deviation of the residualsis 53.79.

11. What does the R^2 value mean about the relationship of *Price* and *Size*?

12. Is the correlation of *Price* and *Size* positive or negative? How do you know?

13. If we measure house *Size* in square meters instead, would R^2 change? Would the slope of the line change? Explain.

14. You find that your house in Saratoga is worth $100,000 more than the regression model predicts. Should you be very surprised (as well as pleased)?

How Big Should R^2 Be?

R^2 is always between 0% and 100%. But what's a "good" R^2 value? The answer depends on the kind of data you are analyzing and on what you want to do with it. Just as with correlation, there is no value for R^2 that automatically determines

[8] By contrast, we usually give correlation coefficients as decimal values between -1.0 and 1.0.
[9] This isn't quite the same as squaring the s_e that we discussed on the previous page, but it's very close. We'll deal with the distinction in Chapter 27.

that the regression is "good." Data from scientific experiments often have R^2 in the 80% to 90% range and even higher. Data from observational studies and surveys, though, often show relatively weak associations because it's so difficult to measure responses reliably. An R^2 of 50% to 30% or even lower might be taken as evidence of a useful regression. The standard deviation of the residuals can give us more information about the usefulness of the regression by telling us how much scatter there is around the line.

As we've seen, an R^2 of 100% is a perfect fit, with no scatter around the line. The s_e would be zero. All of the variance is accounted for by the model and none is left in the residuals at all. This sounds great, but it's too good to be true for real data.[10]

Along with the slope and intercept for a regression, you should always report R^2 so that readers can judge for themselves how successful the regression is at fitting the data. Statistics is about variation, and R^2 measures the success of the regression model in terms of the fraction of the variation of y accounted for by the regression. R^2 is the first part of a regression that many people look at because, along with the scatterplot, it tells whether the regression model is even worth thinking about.

Regression Assumptions and Conditions

The linear regression model is perhaps the most widely used model in all of Statistics. It has everything we could want in a model: two easily estimated parameters, a meaningful measure of how well the model fits the data, and the ability to predict new values. It even provides a self-check in plots of the residuals to help us avoid silly mistakes.

Like all models, though, linear models don't apply all the time, so we'd better think about whether they're reasonable. It makes no sense to make a scatterplot of categorical variables, and even less to perform a regression on them. Always check the **Quantitative Variables Condition** to be sure a regression is appropriate.

The linear model makes several assumptions. First, and foremost, is the **Linearity Assumption**—that the relationship between the variables is, in fact, linear. You can't verify an assumption, but you can check the associated condition. A quick look at the scatterplot will help you check the **Straight Enough Condition.** You don't need a *perfectly* straight plot, but it must be straight enough for the linear model to make sense. If you try to model a curved relationship with a straight line, you'll usually get exactly what you deserve.

If the scatterplot is not straight enough, stop here. You can't use a linear model for *any* two variables, even if they are related. They must have a *linear* association, or the model won't mean a thing.

For the standard deviation of the residuals to summarize the scatter, all the residuals should share the same spread, so we need the **Equal Variance Assumption.** The **Does the Plot Thicken? Condition** checks for changing spread in the scatterplot.

Check the **Outlier Condition.** Outlying points can dramatically change a regression model. Outliers can even change the sign of the slope, misleading us about the underlying relationship between the variables. We'll see examples in the next chapter.

Even though we've checked the conditions in the scatterplot of the data, a scatterplot of the residuals can sometimes help us see any violations even more

---

[10] If you see an R^2 of 100%, it's a good idea to figure out what happened. You may have discovered a new law of Physics, but it's much more likely that you accidentally regressed two variables that measure the same thing.

clearly. And examining the residuals is the best way to look for additional patterns and interesting quirks in the data.

A Tale of Two Regressions

Regression slopes may not behave exactly the way you'd expect at first. Our regression model for the Burger King sandwiches was $\widehat{fat} = 6.8 + 0.97\ protein$. That equation allowed us to estimate that a sandwich with 30 grams of protein would have 35.9 grams of fat. Suppose, though, that we knew the fat content and wanted to predict the amount of protein. It might seem natural to think that by solving our equation for *protein* we'd get a model for predicting *protein* from *fat*. But that doesn't work.

Our original model is $\hat{y} = b_0 + b_1 x$, but the new one needs to evaluate an $\hat{x}$ based on a value of y. There's no y in our original model, only $\hat{y}$, and that makes all the difference. Our model doesn't fit the BK data values perfectly, and the least squares criterion focuses on the vertical errors the model makes in using to model y—not on horizontal errors related to x.

A quick look at the equations reveals why. Simply solving our equation for x would give a new line whose slope is the reciprocal of ours. To model y in terms of x, our slope is $b_1 = \frac{rs_y}{s_x}$. To model x in terms of y, we'd need to use the slope $b_1 = \frac{rs_x}{s_y}$. Notice that it's *not* the reciprocal of ours.

If we want to predict *protein* from *fat*, we need to create that model. The slope is $b_1 = \frac{(0.83)(14.0)}{16.4} = 0.709$ grams of protein per gram of fat. The equation turns out to be $\widehat{protein} = 0.55 + 0.709\ fat$, so we'd predict that a sandwich with 35.9 grams of fat should have 26.0 grams of protein—not the 30 grams that we used in the first equation.

Moral of the story: *Think*. (Where have you heard *that* before?) Decide which variable you want to use (x) to predict values for the other (y). Then find the model that does that. If, later, you want to make predictions in the other direction, you'll need to start over and create the other model from scratch.

Protein	Fat
$\bar{x} = 17.2$ g	$\bar{y} = 23.5$ g
$s_x = 14.0$ g	$s_y = 16.4$ g
$r = 0.83$	

STEP-BY-STEP EXAMPLE   **Regression**

Even if you hit the fast food joints for lunch, you should have a good breakfast. Nutritionists, concerned about "empty calories" in breakfast cereals, recorded facts about 77 cereals, including their *Calories* per serving and *Sugar* content (in grams).

Question: How are calories and sugar content related in breakfast cereals?

Plan State the problem and determine the role of the variables.	I am interested in the relationship between sugar content and calories in cereals. I'll use *Sugar* to estimate *Calories*.
Variables Name the variables and report the W's.	✔ **Quantitative Variables Condition:** I have two quantitative variables, *Calories* and *Sugar* content per serving, measured on 77 breakfast cereals. The units of measurement are calories and grams of sugar, respectively.

Check the conditions for a regression by making a picture. Never fit a regression without looking at the scatterplot first.

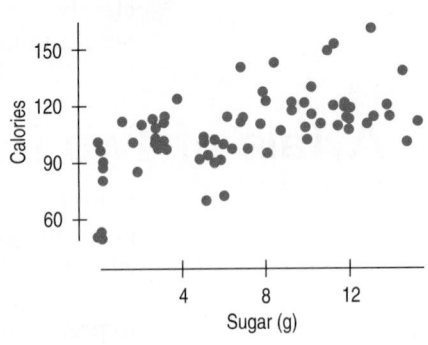

✔ **Outlier Condition:** There are no obvious outliers or groups.

✔ The **Straight Enough Condition** is satisfied; I will fit a regression model to these data.

✔ The **Does the Plot Thicken? Condition** is satisfied. The spread around the line looks about the same throughout.

Mechanics If there are no clear violations of the conditions, fit a straight line model of the form $\hat{y} = b_0 + b_1 x$ to the data. Summary statistics give the building blocks of the calculation.

Calories
$\bar{y} = 107.0$ calories
$s_y = 19.5$ calories

Sugar
$\bar{x} = 7.0$ grams
$s_x = 4.4$ grams

Correlation
$r = 0.564$

Find the slope.

$$b_1 = \frac{rs_y}{s_x} = \frac{0.564(19.5)}{4.4}$$

$= 2.50$ calories per gram of sugar.

Find the intercept.

$b_0 = \bar{y} - b_1\bar{x} = 107 - 2.50(7) = 89.5$ calories.

So the least squares line is

$$\hat{y} = 89.5 + 2.50\,x \text{ or}$$
$$\widehat{Calories} = 89.5 + 2.50\ Sugar.$$

Write the equation, using meaningful variable names.

Squaring the correlation gives

State the value of R^2.

$$R^2 = 0.564^2 = 0.318 \text{ or } 31.8\%.$$

Conclusion Describe what the model says in words and numbers. Be sure to use the names of the variables and their units.

The key to interpreting a regression model is to start with the phrase "b_1 y-units per x-unit," substituting the estimated value of the slope for b_1 and the names of the

The scatterplot shows a positive, linear relationship and no outliers. The slope of the least squares regression line suggests that cereals have about 2.50 *Calories* more per additional gram of *Sugar*.

respective units. The intercept is then a starting or base value.

The intercept predicts that sugar-free cereals would average about 89.5 calories.

R^2 gives the fraction of the variability of y accounted for by the linear regression model.

The R^2 says that 31.8% of the variability in Calories is accounted for by variation in Sugar content.

Find the standard deviation of the residuals, s_e, and compare it to the original s_y.

$s_e = 16.2$ calories. That's smaller than the original SD of 19.5, but still fairly large.

THINK AGAIN

Check Again Even though we looked at the scatterplot *before* fitting a regression model, a plot of the residuals is essential to any regression analysis because it is the best check for additional patterns and interesting quirks in the data.

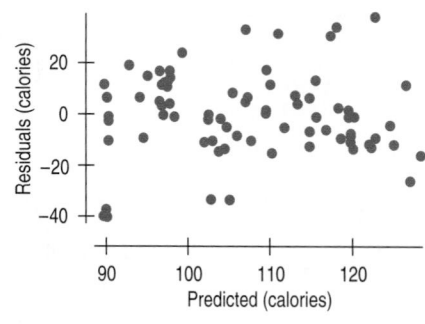

The residuals show a horizontal direction, a shapeless form, and roughly equal scatter for all predicted values. The linear model appears to be appropriate.

TI-*nspire*

Residuals plots. See how the residuals plot changes as you drag points around in a scatterplot.

TI Tips   **Regression lines and residuals plots**

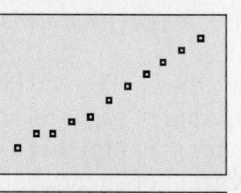

By now you will not be surprised to learn that your calculator can do it all: scatterplot, regression line, and residuals plot. Let's try it using the Arizona State tuition data from the last chapter. (TI Tips, p. 149) You should still have that saved in lists named **YR** and **TUIT**. First, recreate the scatterplot.

1. Find the equation of the regression line.
Actually, you already found the line when you used the calculator to get the correlation. But this time we'll be a little fancier so that we can display the line on our scatterplot. We want to tell the calculator to do the regression and save the equation of the model as a graphing variable.

- Under **STAT CALC** choose **LinReg(a+bx)**.
- Specify that x and y are **YR** and **TUIT**, as before, but . . .
- Now add a comma and one more specification. Press **VARS**, go to the **Y-VARS** menu, choose **1:Function**, and finally(!) choose **Y1**.
- Hit **ENTER**.

There's the equation. The calculator tells you that the regression line is $\widehat{tuit} = 6440 + 326\ year$. Can you explain what the slope and y-intercept mean?

2. Add the line to the plot.
When you entered this command, the calculator automatically saved the equation as **Y1**. Just hit **GRAPH** to see the line drawn across your scatterplot.

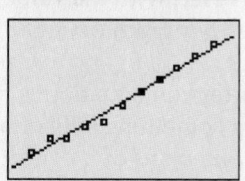

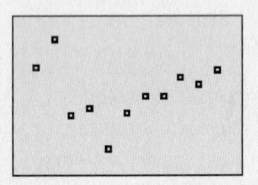

3. Check the residuals.

Remember, you are not finished until you check to see if a linear model is appropriate. That means you need to see if the residuals appear to be randomly distributed. To do that, you need to look at the residuals plot.

This is made easy by the fact that the calculator has already placed the residuals in a list named RESID. Want to see them? Go to STAT EDIT and look through the lists. (If RESID is not already there, go to the first blank list and import the name RESID from your LIST NAMES menu. The residuals should appear.) Every time you have the calculator compute a regression analysis, it will automatically save this list of residuals for you.

4. Now create the residuals plot.

- Set up STAT PLOT Plot2 as a scatterplot with Xlist: YR and Ylist: RESID.
- Before you try to see the plot, go to the Y= screen. By moving the cursor around and hitting ENTER in the appropriate places you can turn off the regression line and Plot1, and turn on Plot2.
- ZoomStat will now graph the residuals plot.

Uh-oh! See the curve? The residuals are high at both ends, low in the middle. Looks like a linear model may not be appropriate after all. Notice that the residuals plot makes the curvature much clearer than the original scatterplot did.

Moral: Always check the residuals plot!

So a linear model might not be appropriate here. What now? The next two chapters provide techniques for dealing with data like these.

Reality Check: Is the Regression Reasonable?

Adjective, Noun, or Verb

You may see the term *regression* used in different ways. There are many ways to fit a line to data, but the term "regression line" or "regression" without any other qualifiers always means least squares. People also use *regression* as a verb when they speak of *regressing* a *y*-variable on an *x*-variable to mean fitting a linear model.

Statistics don't come out of nowhere. They are based on data. The results of a statistical analysis should reinforce your common sense, not fly in its face. If the results are surprising, then either you've learned something new about the world or your analysis is wrong.

Whenever you perform a regression, think about the coefficients and ask whether they make sense. Is a slope of 2.5 calories per gram of sugar reasonable? That's hard to say right off. We know from the summary statistics that a typical cereal has about 100 calories and 7 grams of sugar. A gram of sugar contributes some calories (actually, 4, but you don't need to know that), so calories should go up with increasing sugar. The direction of the slope seems right.

To see if the *size* of the slope is reasonable, a useful trick is to consider its order of magnitude. We'll start by asking if deflating the slope by a factor of 10 seems reasonable. Is 0.25 calories per gram of sugar enough? The 7 grams of sugar found in the average cereal would contribute less than 2 calories. That seems too small.

Now let's try inflating the slope by a factor of 10. Is 25 calories per gram reasonable? Then the average cereal would have 175 calories from sugar alone. The average cereal has only 100 calories per serving, though, so that slope seems too big.

We have tried inflating the slope by a factor of 10 and deflating it by 10 and found both to be unreasonable. So, like Goldilocks, we're left with the value in the middle that's just right. And an increase of 2.5 calories per gram of sugar is certainly *plausible*.

The small effort of asking yourself whether the regression equation is plausible is repaid whenever you catch errors or avoid saying something silly or absurd about the data. It's too easy to take something that comes out of a computer at face value and assume that it makes sense.

Always be skeptical and ask yourself if the answer is reasonable.

WHAT CAN GO WRONG?

There are many ways in which data that appear at first to be good candidates for regression analysis may be unsuitable. And there are ways that people use regression that can lead them astray. Here's an overview of the most common problems. We'll discuss them at length in the next chapter.

▶ **Don't fit a straight line to a nonlinear relationship.** Linear regression is suited only to relationships that are, well, *linear*. Fortunately, we can often improve the linearity easily by using re-expression. We'll come back to that topic in Chapter 10.

▶ **Beware of extraordinary points.** Data points can be extraordinary in a regression in two ways: They can have *y*-values that stand off from the linear pattern suggested by the bulk of the data, or extreme *x*-values. Both kinds of extraordinary points require attention.

▶ **Don't extrapolate beyond the data.** A linear model will often do a reasonable job of summarizing a relationship in the narrow range of observed *x*-values. Once we have a working model for the relationship, it's tempting to use it. But beware of predicting *y*-values for *x*-values that lie outside the range of the original data. The model may no longer hold there, so such *extrapolations* too far from the data are dangerous.

▶ **Don't infer that *x* causes *y* just because there is a good linear model for their relationship.** When two variables are strongly correlated, it is often tempting to assume a causal relationship between them. Putting a regression line on a scatterplot tempts us even further, but it doesn't make the assumption of causation any more valid. For example, our regression model predicting hurricane wind speeds from the central pressure was reasonably successful, but the relationship is very complex. It is reasonable to say that low central pressure at the eye is responsible for the high winds because it draws moist, warm air into the center of the storm, where it swirls around, generating the winds. But as is often the case, things aren't quite that simple. The winds themselves also contribute to lowering the pressure at the center of the storm as it becomes a hurricane. Understanding causation requires far more work than just finding a correlation or modeling a relationship.

R^2 does not mean that protein accounts for 69% of the fat in a BK food item. It is the *variation* in fat content that is accounted for by the linear model.

▶ **Don't choose a model based on R^2 alone.** Although R^2 measures the *strength* of the linear association, a high R^2 does not demonstrate the *appropriateness* of the regression. A single outlier, or data that separate into two groups rather than a single cloud of points, can make R^2 seem quite large when, in fact, the linear regression model is simply inappropriate. Conversely, a low R^2 value may be due to a single outlier as well. It may be that most of the data fall roughly along a straight line, with the exception of a single point. Always look at the scatterplot.

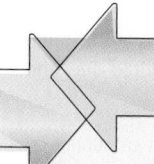

CONNECTIONS

We've talked about the importance of models before, but have seen only the Normal model as an example. The linear model is one of the most important models in Statistics. Chapter 7 talked about the assignment of variables to the *y*- and *x*-axes. That didn't matter to correlation, but it does matter to regression because *y* is predicted by *x* in the regression model.

The connection of R^2 to correlation is obvious, although it may not be immediately clear that just by squaring the correlation we can learn the fraction of the variability of *y* accounted for by a regression on *x*. We'll return to this in subsequent chapters.

We made a big fuss about knowing the units of your quantitative variables. We didn't need units for correlation, but without the units we can't define the slope of a regression. A regression makes no sense if you don't know the *Who,* the *What,* and the *Units* of both your variables.

We've summed squared deviations before when we computed the standard deviation and variance. That's not coincidental. They are closely connected to regression.

When we first talked about models, we noted that deviations away from a model were often interesting. Now we have a formal definition of these deviations as residuals.

WHAT HAVE WE LEARNED?

We've learned that when the relationship between quantitative variables is fairly straight, a linear model can help summarize that relationship and give us insights about it:

▸ The regression (best fit) line doesn't pass through all the points, but it is the best compromise in the sense that the sum of squares of the residuals is the smallest possible.

We've learned several things the correlation, *r*, tells us about the regression:

▸ The slope of the line is based on the correlation, adjusted for the units of x and y:

$$b_1 = \frac{rs_y}{s_x}$$

We've learned to interpret that slope in context:

▸ For each SD of x that we are away from the x mean, we expect to be r SDs of y away from the y mean.

▸ Because r is always between -1 and $+1$, each predicted y is fewer SDs away from its mean than the corresponding x was, a phenomenon called regression to the mean.

▸ The square of the correlation coefficient, R^2, gives us the fraction of the variation of the response accounted for by the regression model. The remaining $1 - R^2$ of the variation is left in the residuals.

The residuals also reveal how well the model works:

▸ If a plot of residuals against predicted values shows a pattern, we should re-examine the data to see why.

▸ The standard deviation of the residuals, s_e, quantifies the amount of scatter around the line.

Of course, the linear model makes no sense unless the **Linearity Assumption** is satisfied. We check the **Straight Enough Condition** and **Outlier Condition** with a scatterplot, as we did for correlation, and also with a plot of residuals against either the x or the predicted values. For the standard deviation of the residuals to make sense as a summary, we have to make the **Equal Variance Assumption**. We check it by looking at both the original scatterplot and the residual plot for the **Does the Plot Thicken? Condition**.

Terms

Model
172. An equation or formula that simplifies and represents reality.

Linear model
172. A linear model is an equation of a line. To interpret a linear model, we need to know the variables (along with their W's) and their units.

Predicted value
172. The value of $\hat{y}$ found for a given x-value in the data. A predicted value is found by substituting the x-value in the regression equation. The predicted values are the values on the fitted line; the points $(x, \hat{y})$ all lie exactly on the fitted line.

Residuals
172. Residuals are the differences between data values and the corresponding values predicted by the regression model—or, more generally, values predicted by any model.

$$\text{Residual} = \text{observed value} - \text{predicted value} = e = y - \hat{y}$$

Least squares
172. The least squares criterion specifies the unique line that minimizes the variance of the residuals or, equivalently, the sum of the squared residuals.

Regression to the mean
174. Because the correlation is always less than 1.0 in magnitude, each predicted $\hat{y}$ tends to be fewer standard deviations from its mean than its corresponding x was from its mean. This is called regression to the mean.

Regression line
Line of best fit
174. The particular linear equation

$$\hat{y} = b_0 + b_1 x$$

that satisfies the least squares criterion is called the least squares regression line. Casually, we often just call it the regression line, or the line of best fit.

Slope   176. The slope, b_1, gives a value in "y-units *per* x-unit." Changes of one unit in x are associated with changes of b_1 units in predicted values of y. The slope can be found by

$$b_1 = \frac{r s_y}{s_x}.$$

Intercept   176. The intercept, b_0, gives a starting value in y-units. It's the $\hat{y}$-value when x is 0. You can find it from $b_0 = \bar{y} - b_1 \bar{x}$.

s_e 181. The standard deviation of the residuals is found by $s_e = \sqrt{\dfrac{\Sigma e^2}{n - 2}}$. When the assumptions and conditions are met, the residuals can be well described by using this standard deviation and the 68–95–99.7 Rule.

R^2 ▸   182. R^2 is the square of the correlation between y and x.

   ▸ R^2 gives the fraction of the variability of y accounted for by the least squares linear regression on x.

   ▸ R^2 is an overall measure of how successful the regression is in linearly relating y to x.

Skills

   ▸   Be able to identify response (y) and explanatory (x) variables in context.

   ▸   Understand how a linear equation summarizes the relationship between two variables.

   ▸   Recognize when a regression should be used to summarize a linear relationship between two quantitative variables.

   ▸   Be able to judge whether the slope of a regression makes sense.

   ▸   Know how to examine your data for violations of the **Straight Enough Condition** that would make it inappropriate to compute a regression.

   ▸   Understand that the least squares slope is easily affected by extreme values.

   ▸   Know that residuals are the differences between the data values and the corresponding values predicted by the line and that the *least squares criterion* finds the line that minimizes the sum of the squared residuals.

   ▸   Know how to use a plot of residuals against predicted values to check the **Straight Enough Condition**, the **Does the Plot Thicken? Condition,** and the **Outlier Condition.**

   ▸   Understand that the standard deviation of the residuals, s_e, measures variability around the line. A large s_e means the points are widely scattered; a small s_e means they lie close to the line.

SHOW   ▸   Know how to find a regression equation from the summary statistics for each variable and the correlation between the variables.

   ▸   Know how to find a regression equation using your statistics software and how to find the slope and intercept values in the regression output table.

   ▸   Know how to use regression to predict a value of y for a given x.

   ▸   Know how to compute the residual for each data value and how to display the residuals.

TELL   ▸   Be able to write a sentence explaining what a linear equation says about the relationship between y and x, basing it on the fact that the slope is given in *y-units per x-unit.*

   ▸   Understand how the correlation coefficient and the regression slope are related. Know how R^2 describes how much of the variation in y is accounted for by its linear relationship with x.

   ▸   Be able to describe a prediction made from a regression equation, relating the predicted value to the specified x-value.

   ▸   Be able to write a sentence interpreting s_e as representing typical errors in predictions—the amounts by which actual y-values differ from the $\hat{y}$'s estimated by the model.

REGRESSION ON THE COMPUTER

All statistics packages make a table of results for a regression. These tables may differ slightly from one package to another, but all are essentially the same—and all include much more than we need to know for now. Every computer regression table includes a section that looks something like this:

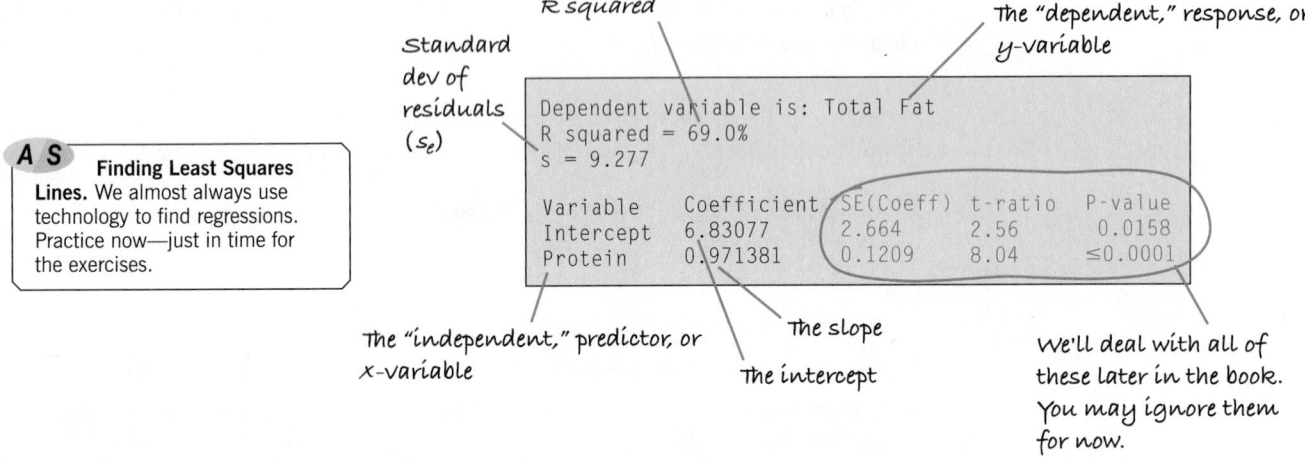

A S **Finding Least Squares Lines.** We almost always use technology to find regressions. Practice now—just in time for the exercises.

The slope and intercept coefficient are given in a table such as this one. Usually the slope is labeled with the name of the x-variable, and the intercept is labeled "Intercept" or "Constant." So the regression equation shown here is

$$\widehat{Fat} = 6.83077 + 0.97138Protein.$$

It is not unusual for statistics packages to give many more digits of the estimated slope and intercept than could possibly be estimated from the data. (The original data were reported to the nearest gram.) Ordinarily, you should round most of the reported numbers to one digit more than the precision of the data, and the slope to two. We will learn about the other numbers in the regression table later in the book. For now, all you need to be able to do is find the coefficients, the s_e, and the R^2 value.

EXERCISES

1. **Cereals.** For many people, breakfast cereal is an important source of fiber in their diets. Cereals also contain potassium, a mineral shown to be associated with maintaining a healthy blood pressure. An analysis of the amount of fiber (in grams) and the potassium content (in milligrams) in servings of 77 breakfast cereals produced the regression model $\widehat{Potassium} = 38 + 27Fiber$. If your cereal provides 9 grams of fiber per serving, how much potassium does the model estimate you will get?

2. **Horsepower.** In Chapter 7's Exercise 33 we examined the relationship between the fuel economy (mpg) and horsepower for 15 models of cars. Further analysis produces the regression model $\widehat{mpg} = 46.87 - 0.084HP$. If the car you are thinking of buying has a 200-horsepower engine, what does this model suggest your gas mileage would be?

3. **More cereal.** Exercise 1 describes a regression model that estimates a cereal's potassium content from the amount of fiber it contains. In this context, what does it mean to say that a cereal has a negative residual?

4. **Horsepower, again.** Exercise 2 describes a regression model that uses a car's horsepower to estimate its fuel economy. In this context, what does it mean to say that a certain car has a positive residual?

5. **Another bowl.** In Exercise 1, the regression model $\widehat{Potassium} = 38 + 27Fiber$ relates fiber (in grams) and potassium content (in milligrams) in servings of breakfast cereals. Explain what the slope means.

6. **More horsepower.** In Exercise 2, the regression model $\widehat{mpg} = 46.87 - 0.084HP$ relates cars' horsepower to their fuel economy (in mpg). Explain what the slope means.

7. Cereal again. The correlation between a cereal's fiber and potassium contents is $r = 0.903$. What fraction of the variability in potassium is accounted for by the amount of fiber that servings contain?

8. Another car. The correlation between a car's horse-power and its fuel economy (in mpg) is $r = -0.869$. What fraction of the variability in fuel economy is accounted for by the horsepower?

9. Last bowl! For Exercise 1's regression model predicting potassium content (in milligrams) from the amount of fiber (in grams) in breakfast cereals, $s_e = 30.77$. Explain in this context what that means.

10. Last tank! For Exercise 2's regression model predicting fuel economy (in mpg) from the car's horsepower, $s_e = 3.287$. Explain in this context what that means.

11. Residuals. Tell what each of the residual plots below indicates about the appropriateness of the linear model that was fit to the data.

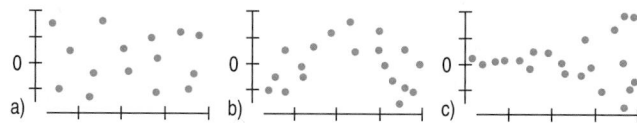

12. Residuals. Tell what each of the residual plots below indicates about the appropriateness of the linear model that was fit to the data.

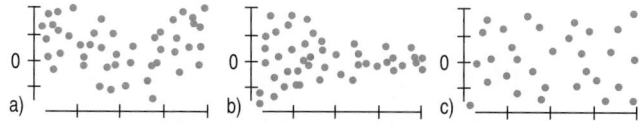

13. What slope? If you create a regression model for predicting the *Weight* of a car (in pounds) from its *Length* (in feet), is the slope most likely to be 3, 30, 300, or 3000? Explain.

14. What slope? If you create a regression model for estimating the *Height* of a pine tree (in feet) based on the *Circumference* of its trunk (in inches), is the slope most likely to be 0.1, 1, 10, or 100? Explain.

15. Real estate. A random sample of records of sales of homes from Feb. 15 to Apr. 30, 1993, from the files maintained by the Albuquerque Board of Realtors gives the *Price* and *Size* (in square feet) of 117 homes. A regression to predict *Price* (in thousands of dollars) from *Size* has an R-squared of 71.4%. The residuals plot indicated that a linear model is appropriate.
a) What are the variables and units in this regression?
b) What units does the slope have?
c) Do you think the slope is positive or negative? Explain.

16. Roller coaster. People who responded to a July 2004 Discovery Channel poll named the 10 best roller coasters in the United States. A table in the last chapter's exercises shows the length of the initial drop (in feet) and the duration of the ride (in seconds). A regression to predict *Duration* from *Drop* has $R^2 = 12.4\%$.
a) What are the variables and units in this regression?
b) What units does the slope have?
c) Do you think the slope is positive or negative? Explain.

17. Real estate again. The regression of *Price* on *Size* of homes in Albuquerque had $R^2 = 71.4\%$, as described in Exercise 15. Write a sentence (in context, of course) summarizing what the R^2 says about this regression.

18. Coasters again. Exercise 16 examined the association between the *Duration* of a roller coaster ride and the height of its initial *Drop*, reporting that $R^2 = 12.4\%$. Write a sentence (in context, of course) summarizing what the R^2 says about this regression.

19. Real estate redux. The regression of *Price* on *Size* of homes in Albuquerque had $R^2 = 71.4\%$, as described in Exercise 15.
a) What is the correlation between *Size* and *Price*? Explain why you chose the sign (+ or −) you did.
b) What would you predict about the *Price* of a home 1 standard deviation above average in *Size*?
c) What would you predict about the *Price* of a home 2 standard deviations below average in *Size*?

20. Another ride. The regression of *Duration* of a roller coaster ride on the height of its initial *Drop*, described in Exercise 16, had $R^2 = 12.4\%$.
a) What is the correlation between *Drop* and *Duration*? Explain why you chose the sign (+ or −) you did.
b) What would you predict about the *Duration* of the ride on a coaster whose initial *Drop* was 1 standard deviation below the mean *Drop*?
c) What would you predict about the *Duration* of the ride on a coaster whose initial *Drop* was 3 standard deviations above the mean *Drop*?

21. More real estate. Consider the Albuquerque home sales from Exercise 15 again. The regression analysis gives the model $\widehat{Price} = 47.82 + 0.061\ Size$.
a) Explain what the slope of the line says about housing prices and house size.
b) What price would you predict for a 3000-square-foot house in this market?
c) A real estate agent shows a potential buyer a 1200-square-foot home, saying that the asking price is $6000 less than what one would expect to pay for a house of this size. What is the asking price, and what is the $6000 called?

22. Last ride. Consider the roller coasters described in Exercise 16 again. The regression analysis gives the model $\widehat{Duration} = 91.033 + 0.242\ Drop$.
a) Explain what the slope of the line says about how long a roller coaster ride may last and the height of the coaster.
b) A new roller coaster advertises an initial drop of 200 feet. How long would you predict the rides last?
c) Another coaster with a 150-foot initial drop advertises a 2-minute ride. Is this longer or shorter than you'd expect? By how much? What's that called?

23. Misinterpretations. A Biology student who created a regression model to use a bird's *Height* when perched for predicting its *Wingspan* made these two statements. Assuming the calculations were done correctly, explain what is wrong with each interpretation.
a) My R^2 of 93% shows that this linear model is appropriate.
b) A bird 10 inches tall will have a wingspan of 17 inches.

24. **More misinterpretations.**   A Sociology student investigated the association between a country's *Literacy Rate* and *Life Expectancy,* then drew the conclusions listed below. Explain why each statement is incorrect. (Assume that all the calculations were done properly.)
    a) The *Literacy Rate* determines 64% of the *Life Expectancy* for a country.
    b) The slope of the line shows that an increase of 5% in *Literacy Rate* will produce a 2-year improvement in *Life Expectancy.*

25. **ESP.**   People who claim to "have ESP" participate in a screening test in which they have to guess which of several images someone is thinking of. You and a friend both took the test. You scored 2 standard deviations above the mean, and your friend scored 1 standard deviation below the mean. The researchers offer everyone the opportunity to take a retest.
    a) Should you choose to take this retest? Explain.
    b) Now explain to your friend what his decision should be and why.

26. **SI jinx.**   Players in any sport who are having great seasons, turning in performances that are much better than anyone might have anticipated, often are pictured on the cover of *Sports Illustrated*. Frequently, their performances then falter somewhat, leading some athletes to believe in a "*Sports Illustrated* jinx." Similarly, it is common for phenomenal rookies to have less stellar second seasons—the so-called "sophomore slump." While fans, athletes, and analysts have proposed many theories about what leads to such declines, a statistician might offer a simpler (statistical) explanation. Explain.

27. **Cigarettes.**   Is the nicotine content of a cigarette related to the "tars"? A collection of data (in milligrams) on 29 cigarettes produced the scatterplot, residuals plot, and regression analysis shown:

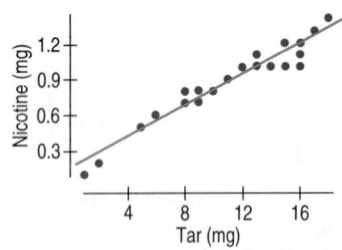

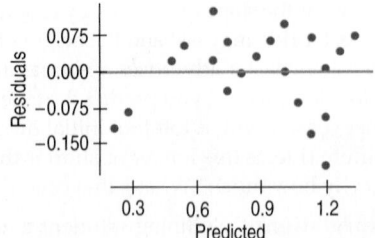

Dependent variable is: nicotine
R squared = 92.4%

Variable	Coefficient
Constant	0.154030
Tar	0.065052

a) Do you think a linear model is appropriate here? Explain.
b) Explain the meaning of R^2 in this context.

28. **Attendance 2006.**   In the previous chapter you looked at the relationship between the number of wins by American League baseball teams and the average attendance at their home games for the 2006 season. Here are the scatterplot, the residuals plot, and part of the regression analysis:

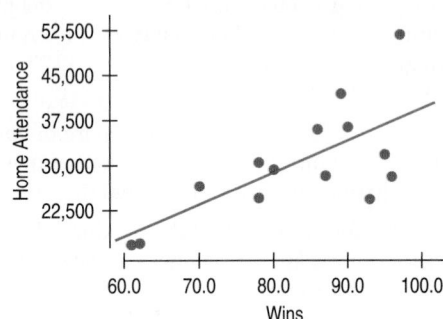

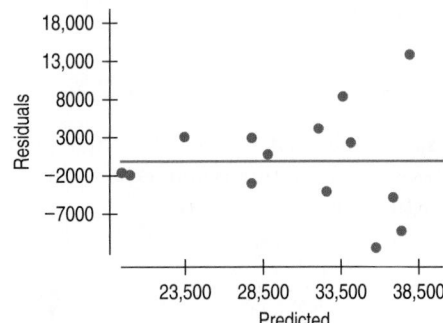

Dependent variable is: Home Attendance
R squared = 48.5%

Variable	Coefficient
Constant	−14364.5
Wins	538.915

a) Do you think a linear model is appropriate here? Explain.
b) Interpret the meaning of R^2 in this context.
c) Do the residuals show any pattern worth remarking on?
d) The point in the upper right of the plots is the New York Yankees. What can you say about the residual for the Yankees?

29. **Another cigarette.**   Consider again the regression of *Nicotine* content on *Tar* (both in milligrams) for the cigarettes examined in Exercise 27.
    a) What is the correlation between *Tar* and *Nicotine*?
    b) What would you predict about the average *Nicotine* content of cigarettes that are 2 standard deviations below average in *Tar* content?
    c) If a cigarette is 1 standard deviation above average in *Nicotine* content, what do you suspect is true about its *Tar* content?

30. **Second inning 2006.**   Consider again the regression of *Average Attendance* on *Wins* for the baseball teams examined in Exercise 28.

a) What is the correlation between *Wins* and *Average Attendance*?

b) What would you predict about the *Average Attendance* for a team that is 2 standard deviations above average in *Wins*?

c) If a team is 1 standard deviation below average in attendance, what would you predict about the number of games the team has won?

T **31. Last cigarette.** Take another look at the regression analysis of tar and nicotine content of the cigarettes in Exercise 27.

a) Write the equation of the regression line.

b) Estimate the *Nicotine* content of cigarettes with 4 milligrams of *Tar*.

c) Interpret the meaning of the slope of the regression line in this context.

d) What does the *y*-intercept mean?

e) If a new brand of cigarette contains 7 milligrams of tar and a nicotine level whose residual is –0.5 mg, what is the nicotine content?

T **32. Last inning 2006.** Refer again to the regression analysis for average attendance and games won by American League baseball teams, seen in Exercise 28.

a) Write the equation of the regression line.

b) Estimate the *Average Attendance* for a team with 50 *Wins*.

c) Interpret the meaning of the slope of the regression line in this context.

d) In general, what would a negative residual mean in this context?

e) The St. Louis Cardinals, the 2006 World Champions, are not included in these data because they are a National League team. During the 2006 regular season, the Cardinals won 83 games and averaged 42,588 fans at their home games. Calculate the residual for this team, and explain what it means.

T **33. Income and housing revisited.** In Chapter 7, Exercise 31, we learned that the Office of Federal Housing Enterprise Oversight (OFHEO) collects data on various aspects of housing costs around the United States. Here's a scatterplot (by state) of the *Housing Cost Index* (HCI) versus the *Median Family Income* (MFI) for the 50 states. The correlation is $r = 0.65$. The mean HCI is 338.2, with a standard deviation of 116.55. The mean MFI is $46,234, with a standard deviation of $7072.47.

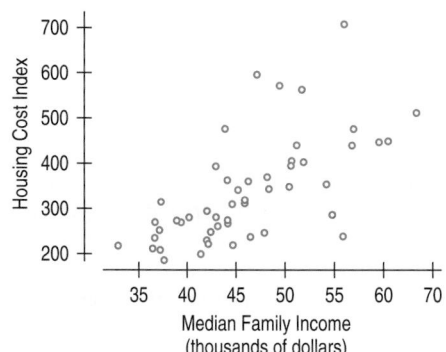

a) Is a regression analysis appropriate? Explain.

b) What is the equation that predicts Housing Cost Index from median family income?

c) For a state with MFI = $44,993, what would be the predicted HCI?

d) Washington, DC, has an MFI of $44,993 and an HCI of 548.02. How far off is the prediction in b) from the actual HCI?

e) If we standardized both variables, what would be the regression equation that predicts standardized HCI from standardized MFI?

f) If we standardized both variables, what would be the regression equation that predicts standardized MFI from standardized HCI?

34. Interest rates and mortgages again. In Chapter 7, Exercise 32, we saw a plot of total mortgages in the United States (in millions of 2005 dollars) versus the interest rate at various times over the past 26 years. The correlation is $r = -0.84$. The mean mortgage amount is $151.9 million and the mean interest rate is 8.88%. The standard deviations are $23.86 million for mortgage amounts and 2.58% for the interest rates.

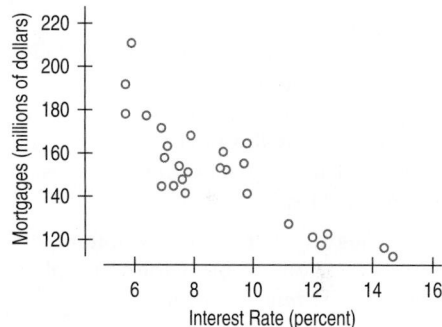

a) Is a regression model appropriate for predicting mortgage amount from interest rates? Explain.

b) What is the equation that predicts mortgage amount from interest rates?

c) What would you predict the mortgage amount would be if the interest rates climbed to 20%?

d) Do you have any reservations about your prediction in part c?

e) If we standardized both variables, what would be the regression equation that predicts standardized mortgage amount from standardized interest rates?

f) If we standardized both variables, what would be the regression equation that predicts standardized interest rates from standardized mortgage amount?

35. Online clothes. An online clothing retailer keeps track of its customers' purchases. For those customers who signed up for the company's credit card, the company also has information on the customer's *Age* and *Income*. A random sample of 500 of these customers shows the following scatterplot of *Total Yearly Purchases* by *Age*:

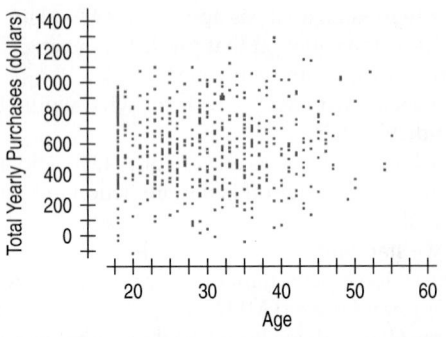

The correlation between *Total Yearly Purchases* and *Age* is $r = 0.037$. Summary statistics for the two variables are:

	Mean	SD
Age	29.67 yrs	8.51 yrs
Total Yearly Purchase	$572.52	$253.62

a) What is the linear regression equation for predicting *Total Yearly Purchase* from *Age*?
b) Do the assumptions and conditions for regression appear to be met?
c) What is the predicted average *Total Yearly Purchase* for an 18-year-old? For a 50-year-old?
d) What percent of the variability in *Total Yearly Purchases* is accounted for by this model?
e) Do you think the regression might be a useful one for the company? Explain.

36. **Online clothes II.**   For the online clothing retailer discussed in the previous problem, the scatterplot of *Total Yearly Purchases* by *Income* shows

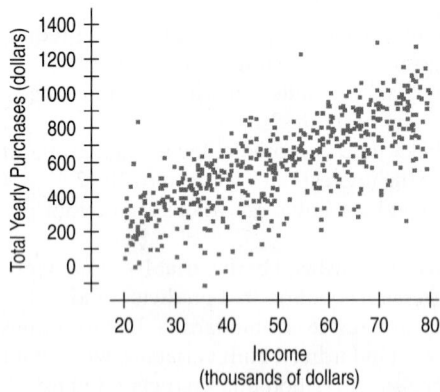

The correlation between *Total Yearly Purchases* and *Income* is 0.722. Summary statistics for the two variables are:

	Mean	SD
Income	$50,343.40	$16,952.50
Total Yearly Purchase	$572.52	$253.62

a) What is the linear regression equation for predicting *Total Yearly Purchase* from *Income*?
b) Do the assumptions and conditions for regression appear to be met?

c) What is the predicted average *Total Yearly Purchase* for someone with a yearly *Income* of $20,000? For someone with an annual *Income* of $80,000?
d) What percent of the variability in *Total Yearly Purchases* is accounted for by this model?
e) Do you think the regression might be a useful one for the company? Comment.

T 37. **SAT scores.**   The SAT is a test often used as part of an application to college. SAT scores are between 200 and 800, but have no units. Tests are given in both Math and Verbal areas. Doing the SAT-Math problems also involves the ability to read and understand the questions, but can a person's verbal score be used to predict the math score? Verbal and math SAT scores of a high school graduating class are displayed in the scatterplot, with the regression line added.

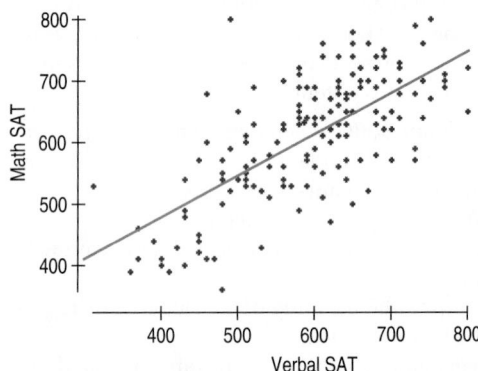

a) Describe the relationship.
b) Are there any students whose scores do not seem to fit the overall pattern?
c) For these data, $r = 0.685$. Interpret this statistic.
d) These verbal scores averaged 596.3, with a standard deviation of 99.5, and the math scores averaged 612.2, with a standard deviation of 96.1. Write the equation of the regression line.
e) Interpret the slope of this line.
f) Predict the math score of a student with a verbal score of 500.
g) Every year some student scores a perfect 1600. Based on this model, what would be that student's Math score residual?

38. **Success in college.**   Colleges use SAT scores in the admissions process because they believe these scores provide some insight into how a high school student will perform at the college level. Suppose the entering freshmen at a certain college have mean combined *SAT Scores* of 1833, with a standard deviation of 123. In the first semester these students attained a mean *GPA* of 2.66, with a standard deviation of 0.56. A scatterplot showed the association to be reasonably linear, and the correlation between *SAT* score and *GPA* was 0.47.
a) Write the equation of the regression line.
b) Explain what the *y*-intercept of the regression line indicates.
c) Interpret the slope of the regression line.
d) Predict the GPA of a freshman who scored a combined 2100.

e) Based upon these statistics, how effective do you think SAT scores would be in predicting academic success during the first semester of the freshman year at this college? Explain.

f) As a student, would you rather have a positive or a negative residual in this context? Explain.

39. SAT, take 2. Suppose we wanted to use SAT math scores to estimate verbal scores based on the information in Exercise 37.

a) What is the correlation?

b) Write the equation of the line of regression predicting verbal scores from math scores.

c) In general, what would a positive residual mean in this context?

d) A person tells you her math score was 500. Predict her verbal score.

e) Using that predicted verbal score and the equation you created in Exercise 37, predict her math score.

f) Why doesn't the result in part e) come out to 500?

40. Success, part 2. Based on the statistics for college freshmen given in Exercise 38, what SAT score might be expected among freshmen who attained a first-semester GPA of 3.0?

41. Used cars 2007. Classified ads in the *Ithaca Journal* offered several used Toyota Corollas for sale. Listed below are the ages of the cars and the advertised prices.

Age (yr)	Price Advertised ($)
1	13,990
1	13,495
3	12,999
4	9500
4	10,495
5	8995
5	9495
6	6999
7	6950
7	7850
8	6999
8	5995
10	4950
10	4495
13	2850

a) Make a scatterplot for these data.

b) Describe the association between *Age* and *Price* of a used Corolla.

c) Do you think a linear model is appropriate?

d) Computer software says that $R^2 = 94.4\%$. What is the correlation between *Age* and *Price*?

e) Explain the meaning of R^2 in this context.

f) Why doesn't this model explain 100% of the variability in the price of a used Corolla?

42. Drug abuse. In the exercises of the last chapter you examined results of a survey conducted in the United States and 10 countries of Western Europe to determine the

percentage of teenagers who had used marijuana and other drugs. Below is the scatterplot. Summary statistics showed that the mean percent that had used marijuana was 23.9%, with a standard deviation of 15.6%. An average of 11.6% of teens had used other drugs, with a standard deviation of 10.2%.

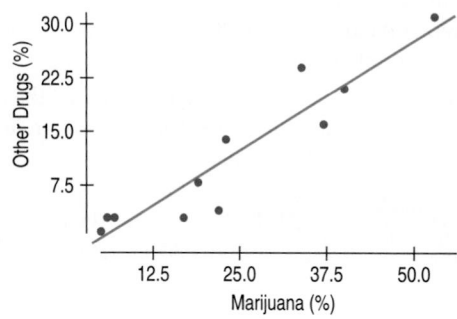

a) Do you think a linear model is appropriate? Explain.

b) For this regression, R^2 is 87.3%. Interpret this statistic in this context.

c) Write the equation you would use to estimate the percentage of teens who use other drugs from the percentage who have used marijuana.

d) Explain in context what the slope of this line means.

e) Do these results confirm that marijuana is a "gateway drug," that is, that marijuana use leads to the use of other drugs?

43. More used cars 2007. Use the advertised prices for Toyota Corollas given in Exercise 41 to create a linear model for the relationship between a car's *Age* and its *Price*.

a) Find the equation of the regression line.

b) Explain the meaning of the slope of the line.

c) Explain the meaning of the *y*-intercept of the line.

d) If you want to sell a 7-year-old Corolla, what price seems appropriate?

e) You have a chance to buy one of two cars. They are about the same age and appear to be in equally good condition. Would you rather buy the one with a positive residual or the one with a negative residual? Explain.

f) You see a "For Sale" sign on a 10-year-old Corolla stating the asking price as $3500. What is the residual?

g) Would this regression model be useful in establishing a fair price for a 20-year-old car? Explain.

44. Birthrates 2005. The table shows the number of live births per 1000 women aged 15–44 years in the United States, starting in 1965. (National Center for Health Statistics, www.cdc.gov/nchs/)

Year	1965	1970	1975	1980	1985	1990	1995	2000	2005
Rate	19.4	18.4	14.8	15.9	15.6	16.4	14.8	14.4	14.0

a) Make a scatterplot and describe the general trend in *Birthrates*. (Enter *Year* as years since 1900: 65, 70, 75, etc.)

b) Find the equation of the regression line.

c) Check to see if the line is an appropriate model. Explain.

d) Interpret the slope of the line.
e) The table gives rates only at 5-year intervals. Estimate what the rate was in 1978.
f) In 1978 the birthrate was actually 15.0. How close did your model come?
g) Predict what the *Birthrate* will be in 2010. Comment on your faith in this prediction.
h) Predict the *Birthrate* for 2025. Comment on your faith in this prediction.

45. Burgers. In the last chapter, you examined the association between the amounts of *Fat* and *Calories* in fast-food hamburgers. Here are the data:

Fat (g)	19	31	34	35	39	39	43
Calories	410	580	590	570	640	680	660

a) Create a scatterplot of *Calories* vs. *Fat*.
b) Interpret the value of R^2 in this context.
c) Write the equation of the line of regression.
d) Use the residuals plot to explain whether your linear model is appropriate.
e) Explain the meaning of the *y*-intercept of the line.
f) Explain the meaning of the slope of the line.
g) A new burger containing 28 grams of fat is introduced. According to this model, its residual for calories is +33. How many calories does the burger have?

46. Chicken. Chicken sandwiches are often advertised as a healthier alternative to beef because many are lower in fat. Tests on 11 brands of fast-food chicken sandwiches produced the following summary statistics and scatterplot from a graphing calculator:

	Fat (g)	Calories
Mean	20.6	472.7
St. Dev.	9.8	144.2
Correlation	0.947	

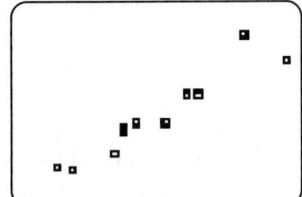

a) Do you think a linear model is appropriate in this situation?
b) Describe the strength of this association.
c) Write the equation of the regression line to estimate calories from the fat content.
d) Explain the meaning of the slope.
e) Explain the meaning of the *y*-intercept.
f) What does it mean if a certain sandwich has a negative residual?

47. A second helping of burgers. In Exercise 45 you created a model that can estimate the number of *Calories* in a burger when the *Fat* content is known.
a) Explain why you cannot use that model to estimate the fat content of a burger with 600 calories.
b) Using an appropriate model, estimate the fat content of a burger with 600 calories.

48. A second helping of chicken. In Exercise 46 you created a model to estimate the number of *Calories* in a chicken sandwich when you know the *Fat*.
a) Explain why you cannot use that model to estimate the fat content of a 400-calorie sandwich.
b) Make that estimate using an appropriate model.

49. Body fat. It is difficult to determine a person's body fat percentage accurately without immersing him or her in water. Researchers hoping to find ways to make a good estimate immersed 20 male subjects, then measured their waists and recorded their weights.

Waist (in.)	Weight (lb)	Body Fat (%)	Waist (in.)	Weight (lb)	Body Fat (%)
32	175	6	33	188	10
36	181	21	40	240	20
38	200	15	36	175	22
33	159	6	32	168	9
39	196	22	44	246	38
40	192	31	33	160	10
41	205	32	41	215	27
35	173	21	34	159	12
38	187	25	34	146	10
38	188	30	44	219	28

a) Create a model to predict *%Body Fat* from *Weight*.
b) Do you think a linear model is appropriate? Explain.
c) Interpret the slope of your model.
d) Is your model likely to make reliable estimates? Explain.
e) What is the residual for a person who weighs 190 pounds and has 21% body fat?

50. Body fat again. Would a model that uses the person's *Waist* size be able to predict the *%Body Fat* more accurately than one that uses *Weight*? Using the data in Exercise 49, create and analyze that model.

51. Heptathlon 2004. We discussed the women's 2004 Olympic heptathlon in Chapter 6. The table on the next page shows the results from the high jump, 800-meter run, and long jump for the 26 women who successfully completed all three events in the 2004 Olympics.

Name	Country	High Jump (m)	800-m (sec)	Long Jump (m)
Carolina Klüft	SWE	1.91	134.15	6.51
Austra Skujyté	LIT	1.76	135.92	6.30
Kelly Sotherton	GBR	1.85	132.27	6.51
Shelia Burrell	USA	1.70	135.32	6.25
Yelena Prokhorova	RUS	1.79	131.31	6.21
Sonja Kesselschlaeger	GER	1.76	135.21	6.42
Marie Collonville	FRA	1.85	133.62	6.19
Natalya Dobrynska	UKR	1.82	137.01	6.23
Margaret Simpson	GHA	1.79	137.72	6.02
Svetlana Sokolova	RUS	1.70	133.23	5.84
J. J. Shobha	IND	1.67	137.28	6.36
Claudia Tonn	GER	1.82	130.77	6.35
Naide Gomes	POR	1.85	140.05	6.10
Michelle Perry	USA	1.70	133.69	6.02
Aryiro Strataki	GRE	1.79	137.90	5.97
Karin Ruckstuhl	NED	1.85	133.95	5.90
Karin Ertl	GER	1.73	138.68	6.03
Kylie Wheeler	AUS	1.79	137.65	6.36
Janice Josephs	RSA	1.70	138.47	6.21
Tiffany Lott Hogan	USA	1.67	145.10	6.15
Magdalena Szczepanska	POL	1.76	133.08	5.98
Irina Naumenko	KAZ	1.79	134.57	6.16
Yuliya Akulenko	UKR	1.73	142.58	6.02
Soma Biswas	IND	1.70	132.27	5.92
Marsha Mark-Baird	TRI	1.70	141.21	6.22
Michaela Hejnova	CZE	1.70	145.68	5.70

Let's examine the association among these events. Perform a regression to predict high-jump performance from the 800-meter results.
a) What is the regression equation? What does the slope mean?
b) What percent of the variability in high jumps can be accounted for by differences in 800-m times?
c) Do good high jumpers tend to be fast runners? (Be careful—low times are good for running events and high distances are good for jumps.)
d) What does the residuals plot reveal about the model?
e) Do you think this is a useful model? Would you use it to predict high-jump performance? (Compare the residual standard deviation to the standard deviation of the high jumps.)

T 52. **Heptathlon 2004 again.** We saw the data for the women's 2004 Olympic heptathlon in Exercise 51. Are the two jumping events associated? Perform a regression of the long-jump results on the high-jump results.

a) What is the regression equation? What does the slope mean?
b) What percentage of the variability in long jumps can be accounted for by high-jump performances?
c) Do good high jumpers tend to be good long jumpers?
d) What does the residuals plot reveal about the model?
e) Do you think this is a useful model? Would you use it to predict long-jump performance? (Compare the residual standard deviation to the standard deviation of the long jumps.)

53. **Least squares.** Consider the four points (10,10), (20,50), (40,20), and (50,80). The least squares line is $\hat{y} = 7.0 + 1.1x$. Explain what "least squares" means, using these data as a specific example.

54. **Least squares.** Consider the four points (200,1950), (400,1650), (600,1800), and (800,1600). The least squares line is $\hat{y} = 1975 - 0.45x$. Explain what "least squares" means, using these data as a specific example.

JUST CHECKING
Answers

1. You should expect the price to be 0.77 standard deviations above the mean.

2. You should expect the size to be $2(0.77) = 1.54$ standard deviations below the mean.

3. The home is 1.5 standard deviations above the mean in size no matter how size is measured.

4. An increase in home size of 1000 square feet is associated with an increase in price of $94,454, on average.

5. Units are thousands of dollars per thousand square feet.

6. About $188,908, on average

7. No. Even if it were positive, no one wants a house with 0 square feet!

8. Negative; that indicates it's priced lower than a typical home of its size.

9. $280,245

10. $19,755 (positive!)

11. Differences in the size of houses account for about 59.5% of the variation in the house prices.

12. It's positive. The correlation and the slope have the same sign.

13. R^2 would not change, but the slope would. Slope depends on the units used but correlation doesn't.

14. No, the standard deviation of the residuals is 53.79 thousand dollars. We shouldn't be surprised by any residual smaller than 2 standard deviations, and a residual of $100,000 is less than 2(53,790).

CHAPTER 9

Regression Wisdom

A S   *Activity:* **Construct a Plot with a Given Slope.** How's your feel for regression lines? Can you make a scatterplot that has a specified slope?

Regression may be the most widely used Statistics method. It is used every day throughout the world to predict customer loyalty, numbers of admissions at hospitals, sales of automobiles, and many other things. Because regression is so widely used, it's also widely abused and misinterpreted. This chapter presents examples of regressions in which things are not quite as simple as they may have seemed at first, and shows how you can still use regression to discover what the data have to say.

Getting the "Bends": When the Residuals Aren't Straight

No regression analysis is complete without a display of the residuals to check that the linear model is reasonable. Because the residuals are what is "left over" after the model describes the relationship, they often reveal subtleties that were not clear from a plot of the original data. Sometimes these are additional details that help confirm or refine our understanding. Sometimes they reveal violations of the regression conditions that require our attention.

The fundamental assumption in working with a linear model is that the relationship you are modeling is, in fact, linear. That sounds obvious, but when you fit a regression, you can't take it for granted. Often it's hard to tell from the scatterplot you looked at before you fit the regression model. Sometimes you can't see a bend in the relationship until you plot the residuals.

> We can't *know* whether the **Linearity Assumption** is true, but we can see if it's *plausible* by checking the **Straight Enough Condition.**

Jessica Meir and Paul Ponganis study emperor penguins at the Scripps Institution of Oceanography's Center for Marine Biotechnology and Biomedicine at the University of California at San Diego. Says Jessica:

> *Emperor penguins are the most accomplished divers among birds, making routine dives of 5–12 minutes, with the longest recorded dive over 27 minutes. These birds can also dive to depths of over 500 meters! Since air-breathing animals like penguins must hold their breath while submerged, the duration of any given dive depends on how much oxygen is in the bird's body at the beginning of the dive, how quickly that oxygen gets used,*

and the lowest level of oxygen the bird can tolerate. The rate of oxygen depletion is primarily determined by the penguin's heart rate. Consequently, studies of heart rates during dives can help us understand how these animals regulate their oxygen consumption in order to make such impressive dives.

The researchers equip emperor penguins with devices that record their heart rates during dives. Here's a scatterplot of the *Dive Heart Rate* (beats per minute) and the *Duration* (minutes) of dives by these high-tech penguins.

The scatterplot looks fairly linear with a moderately strong negative association ($R^2 = 71.5\%$). The linear regression equation

$$\widehat{DiveHeartRate} = 96.9 - 5.47\,Duration$$

says that for longer dives, the average *Dive Heart Rate* is lower by about 5.47 beats per dive minute, starting from a value of 96.9 beats per minute.

The scatterplot of the residuals against *Duration* holds a surprise. The Linearity Assumption says we should not see a pattern, but instead there's a bend, starting high on the left, dropping down in the middle of the plot, and rising again at the right. Graphs of residuals often reveal patterns such as this that were easy to miss in the original scatterplot.

Now looking back at the original scatterplot, you may see that the scatter of points isn't really straight. There's a slight bend to that plot, but the bend is much easier to see in the residuals. Even though it means rechecking the Straight Enough Condition *after* you find the regression, it's always a good idea to check your scatterplot of the residuals for bends that you might have overlooked in the original scatterplot.

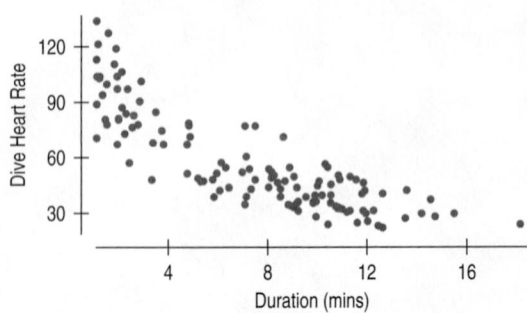

FIGURE 9.1

The scatterplot of Dive Heart Rate *in beats per minute (bpm) vs.* Duration *(minutes) shows a strong, roughly linear, negative association.*

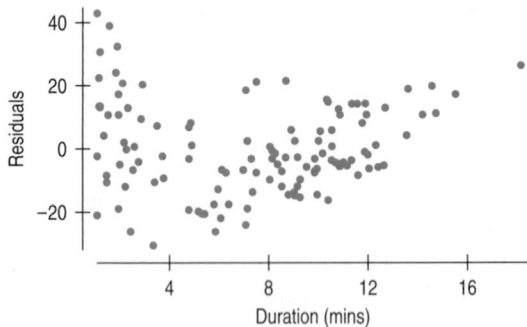

FIGURE 9.2

Plotting the residuals against Duration *reveals a bend. It was also in the original scatterplot, but here it's easier to see.*

Sifting Residuals for Groups

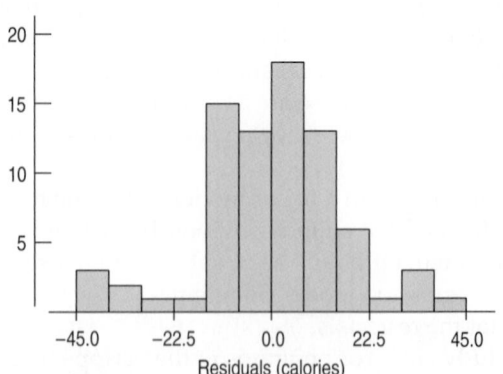

FIGURE 9.3

A histogram of the regression residuals shows small modes both above and below the central large mode. These may be worth a second look.

In the Step-By-Step analysis in Chapter 8 to predict *Calories* from *Sugar* content in breakfast cereals, we examined a scatterplot of the residuals. Our first impression was that it had no particular structure—a conclusion that supported using the regression model. But let's look again.

Here's a histogram of the residuals. How would you describe its shape? It looks like there might be small modes on both sides of the central body of the data. One group of cereals seems to stand out as having large negative residuals, with fewer calories than we might have predicted, and another stands out with large positive residuals. The calories in these cereals were underestimated by the model. Whenever we suspect multiple modes, we ask whether they are somehow different.

On the next page is the residual plot, with the points in those modes marked. Now we can see that those two groups stand away from the central pattern in the scatterplot. The high-residual cereals are Just Right Fruit & Nut; Muesli Raisins, Dates & Almonds; Peaches & Pecans; Mueslix Crispy Blend; and Nutri-Grain Almond Raisin. Do these cereals seem to have something in common? They all present themselves as "healthy." This might be surprising, but in fact, "healthy" cereals

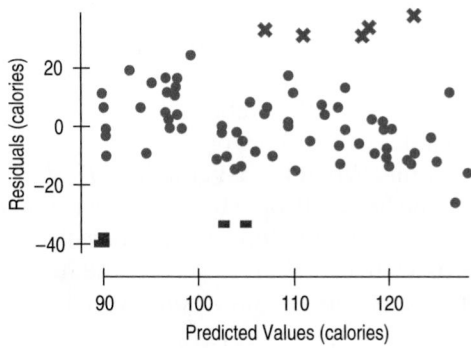

FIGURE 9.4

A scatterplot of the residuals vs. predicted values for the cereal regression. The green "x" points are cereals whose calorie content is higher than the linear model predicts. The red "–" points show cereals with fewer calories than the model predicts. Is there something special about these cereals?

often contain more fat, and therefore more calories, than we might expect from looking at their sugar content alone.

The low-residual cereals are Puffed Rice, Puffed Wheat, three bran cereals, and Golden Crisps. You might not have grouped these cereals together before. What they have in common is a low calorie count *relative to their sugar content*—even though their sugar contents are quite different.

These observations may not lead us to question the overall linear model, but they do help us understand that other factors may be part of the story. An examination of residuals often leads us to discover groups of observations that are different from the rest.

When we discover that there is more than one group in a regression, we may decide to analyze the groups separately, using a different model for each group. Or we can stick with the original model and simply note that there are groups that are a little different. Either way, the model will be wrong, but useful, so it will improve our understanding of the data.

Subsets

> Here's an important unstated condition for fitting models: **All the data must come from the same population.**

Cereal manufacturers aim cereals at different segments of the market. Supermarkets and cereal manufacturers try to attract different customers by placing different types of cereals on certain shelves. Cereals for kids tend to be on the "kid's shelf," at their eye level. Toddlers wouldn't be likely to grab a box from this shelf and beg, "Mom, can we please get this All-Bran with Extra Fiber?"

Should we take this extra information into account in our analysis? Figure 9.5 shows a scatterplot of *Calories* and *Sugar,* colored according to the shelf on which the cereals were found and with a separate regression line fit for each. The top shelf is clearly different. We might want to report two regressions, one for the top shelf and one for the bottom two shelves.[1]

FIGURE 9.5

Calories *and* Sugar *colored according to the shelf on which the cereal was found in a supermarket, with regression lines fit for each shelf individually. Do these data appear homogeneous? That is, do all the cereals seem to be from the same population of cereals? Or are there different kinds of cereals that we might want to consider separately?*

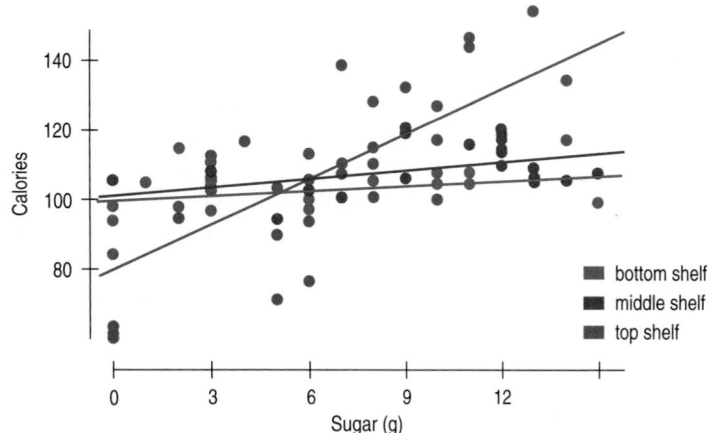

Extrapolation: Reaching Beyond the Data

Linear models give a predicted value for each case in the data. Put a new *x*-value into the equation, and it gives a predicted value, $\hat{y}$, to go with it. But when the new *x*-value lies far from the data we used to build the regression, how trustworthy is the prediction?

---

[1] More complex models can take into account both sugar content and shelf information. This kind of *multiple regression* model is a natural extension of the model we're using here. You can learn about such models in Chapter 29 on the DVD.

"Prediction is difficult, especially about the future."

—Niels Bohr, Danish physicist

The simple answer is that the farther the new *x*-value is from $\bar{x}$, the less trust we should place in the predicted value. Once we venture into new *x* territory, such a prediction is called an **extrapolation.** Extrapolations are dubious because they require the very questionable assumption that nothing about the relationship between *x* and *y* changes even at extreme values of *x* and beyond.

Extrapolations can get us into deep trouble. When the *x*-variable is *Time*, extrapolation becomes an attempt to peer into the future. People have always wanted to see into the future, and it doesn't take a crystal ball to foresee that they always will. In the past, seers, oracles, and wizards were called on to predict the future. Today mediums, fortune-tellers, and Tarot card readers still find many customers.

Those with a more scientific outlook may use a linear model as their digital crystal ball. Linear models are based on the *x*-values of the data at hand and cannot be trusted beyond that span. Some physical phenomena do exhibit a kind of "inertia" that allows us to guess that current systematic behavior will continue, but regularity can't be counted on in phenomena such as stock prices, sales figures, hurricane tracks, or public opinion.

Extrapolating from current trends is so tempting that even professional forecasters make this mistake, and sometimes the errors are striking. In the mid-1970s, oil prices surged and long lines at gas stations were common. In 1970, oil cost about $17 a barrel (in 2005 dollars)—about what it had cost for 20 years or so. But then, within just a few years, the price surged to over $40. In 1975, a survey of 15 top econometric forecasting models (built by groups that included Nobel prize–winning economists) found predictions for 1985 oil prices that ranged from $300 to over $700 a barrel (in 2005 dollars). How close were these forecasts?

Here's a scatterplot of oil prices from 1972 to 1981 (in 2005 dollars).

When the Data Are Years. . .

. . . we usually don't enter them as four-digit numbers. Here we used 0 for 1970, 10 for 1980, and so on. Or we may simply enter two digits, using 82 for 1982, for instance. Rescaling years like this often makes calculations easier and equations simpler. We recommend you do it, too. But be careful: If 1982 is 82, then 2004 is 104 (not 4), right?

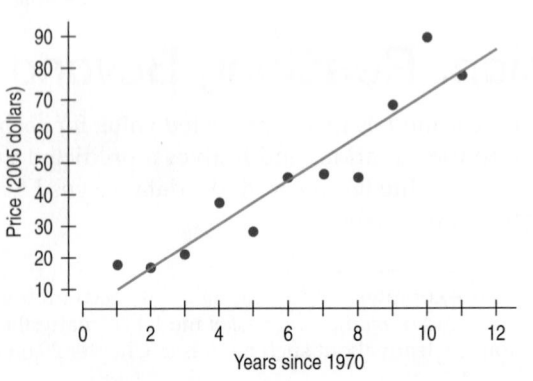

FIGURE 9.6

The scatterplot shows an average increase in the price of a barrel of oil of over $7 per year from 1971 to 1982.

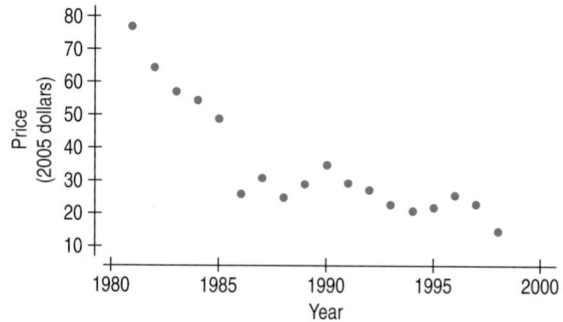

FIGURE 9.7

This scatterplot of oil prices from 1981 to 1998 shows a fairly constant decrease of about $3 per barrel per year.

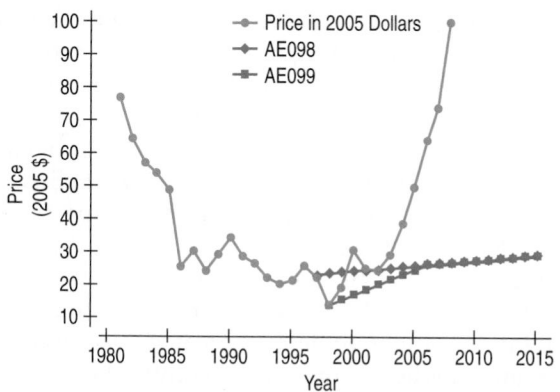

FIGURE 9.8

Here are the EIA forecasts with the actual prices from 1981 to 2008. Neither forecast predicted the sharp run-up in the past few years.

The regression model

$$\widehat{Price} = -0.85 + 7.39 \, Years \; since \; 1970$$

says that prices had been going up 7.39 dollars per year, or nearly $74 in 10 years. If you assume that they would *keep going up*, it's not hard to imagine almost any price you want.

So, how did the forecasters do? Well, in the period from 1982 to 1998 oil prices didn't exactly continue that steady increase. In fact, they went down so much that by 1998, prices (adjusted for inflation) were the lowest they'd been since before World War II.

Not one of the experts' models predicted that.

Of course, these decreases clearly couldn't continue, or oil would be free by now. The Energy Information Administration offered two *different* 20-year forecasts for oil prices after 1998, and both called for relatively modest increases in oil prices. So, how accurate have *these* forecasts been? Here's a timeplot of the EIA's predictions and the actual prices (in 2005 dollars).

Oops! They seemed to have missed the sharp run-up in oil prices in the past few years.

Where do you think oil prices will go in the next decade? *Your* guess may be as good as anyone's!

Of course, knowing that extrapolation is dangerous doesn't stop people. The temptation to see into the future is hard to resist. So our more realistic advice is this:

If you must extrapolate into the future, at least don't believe that the prediction will come true.

Outliers, Leverage, and Influence

The outcome of the 2000 U.S. presidential election was determined in Florida amid much controversy. The main race was between George W. Bush and Al Gore, but two minor candidates played a significant role. To the political right of the main party candidates was Pat Buchanan, while to the political left was Ralph Nader. Generally, Nader earned more votes than Buchanan throughout the state. We would expect counties with larger vote totals to give more votes to each candidate. Here's a regression relating *Buchanan's* vote totals by county in the state of Florida to *Nader's*:

Dependent variable is: Buchanan
R-squared = 42.8%

Variable	Coefficient
Intercept	50.3
Nader	0.14

The regression model,

$$\widehat{Buchanan} = 50.3 + 0.14 \, Nader,$$

says that, in each county, Buchanan received about 0.14 times (or 14% of) the vote Nader received, starting from a base of 50.3 votes.

This seems like a reasonable regression, with an R^2 of almost 43%. But we've violated all three Rules of Data Analysis by going straight to the regression table without making a picture.

Here's a scatterplot that shows the vote for Buchanan in each county of Florida plotted against the vote for Nader. The striking **outlier** is Palm Beach County.

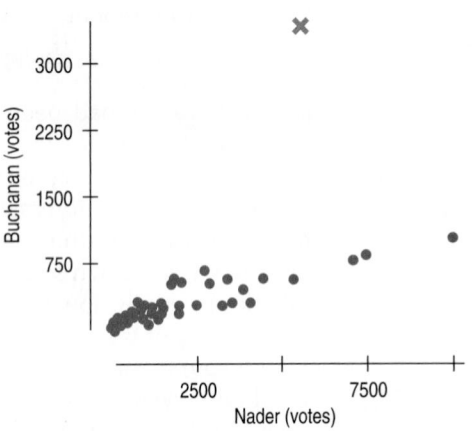

FIGURE 9.9

Votes received by Buchanan against votes for Nader in all Florida counties in the presidential election of 2000. The red "x" point is Palm Beach County, home of the "butterfly ballot."

The so-called "butterfly ballot," used only in Palm Beach County, was a source of controversy. It has been claimed that the format of this ballot confused voters so that some who intended to vote for the Democrat, Al Gore, punched the wrong hole next to his name and, as a result, voted for Buchanan.

The scatterplot shows a strong, positive, linear association, and one striking point. With Palm Beach removed from the regression, the R^2 jumps from 42.8% to 82.1% and the slope of the line changes to 0.1, suggesting that Buchanan received only about 10% of the vote that Nader received. With more than 82% of the variability of the Buchanan vote accounted for, the model when Palm Beach is omitted certainly fits better. Palm Beach County now stands out, not as a Buchanan stronghold, but rather as a clear violation of the model that begs for explanation.

One of the great values of models is that, by establishing an idealized behavior, they help us to see when and how data values are unusual. In regression, a point can stand out in two different ways. First, a data value can have a large residual, as Palm Beach County does in this example. Because they seem to be different from the other cases, points whose residuals are large always deserve special attention.

FIGURE 9.10

The red line shows the effect that one unusual point can have on a regression.

A data point can also be unusual if its x-value is far from the mean of the x-values. Such a point is said to have high **leverage**. The physical image of a lever is exactly right. We know the line must pass through $(\bar{x}, \bar{y})$, so you can picture that point as the fulcrum of the lever. Just as sitting farther from the hinge on a see-saw gives you more leverage to pull it your way, points with values far from $\bar{x}$ pull more strongly on the regression line.

A point with high leverage has the potential to change the regression line. But it doesn't always use that potential. If the point lines up with the pattern of the other points, then including it doesn't change our estimate of the line. By sitting so far from $\bar{x}$, though, it may strengthen the relationship, inflating the correlation and R^2. How can you tell if a high-leverage point actually changes the model? Just fit the linear model twice, both with and without the point in question. We say that a point is **influential** if omitting it from the analysis gives a very different model.[2]

Influence depends on both leverage and residual; a case with high leverage whose y-value sits right on the line fit to the rest of the data is not influential.

A S **Activity: Leverage.** You may be surprised to see how sensitive to a single influential point a regression line is.

---

[2] Some textbooks use the term *influential point* for any observation that influences the slope, intercept, or R^2. We'll reserve the term for points that influence the slope.

TI-*nspire*

Influential points. Try to make the regression line's slope change dramatically by dragging a point around in the scatterplot.

"For whoever knows the ways of Nature will more easily notice her deviations; and, on the other hand, whoever knows her deviations will more accurately describe her ways."

—Francis Bacon
(1561–1626)

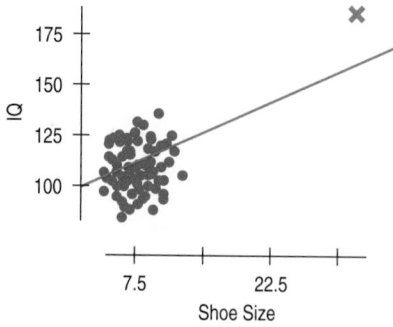

FIGURE 9.11

Bozo's extraordinarily large shoes give his data point high leverage in the regression. Wherever Bozo's IQ falls, the regression line will follow.

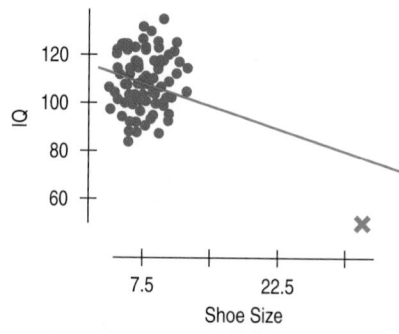

FIGURE 9.12

If Bozo's IQ were low, the regression slope would change from positive to negative. A single influential point can change a regression model drastically.

Removing that case won't change the slope, even if it does affect R^2. A case with modest leverage but a very large residual (such as Palm Beach County) can be influential. Of course, if a point has enough leverage, it can pull the line right to it. Then it's highly influential, but its residual is small. The only way to be sure is to fit both regressions.

Unusual points in a regression often tell us more about the data and the model than any other points. We face a challenge: The best way to identify unusual points is against the background of a model, but good models are free of the influence of unusual points. (That insight's at least 400 years old. See the sidebar.) Don't give in to the temptation to simply delete points that don't fit the line. You can take points out and discuss what the model looks like with and without them, but arbitrarily deleting points can give a false sense of how well the model fits the data. Your goal should be understanding the data, not making R^2 as big as you can.

In 2000, George W. Bush won Florida (and thus the presidency) by only a few hundred votes, so Palm Beach County's residual is big enough to be meaningful. It's the rare unusual point that determines a presidency, but all are worth examining and trying to understand.

A point with so much influence that it pulls the regression line close to it can make its residual deceptively small. Influential points like that can have a shocking effect on the regression. Here's a plot of *IQ* against *Shoe Size*, again from the fanciful study of intelligence and foot size in comedians we saw in Chapter 7. The linear regression output shows

Dependent variable is: IQ
R-squared = 24.8%

Variable	Coefficient
Intercept	93.3265
Shoe size	2.08318

Although this is a silly example, it illustrates an important and common potential problem: Almost all of the variance accounted for (R^2 = 24.8%) is due to *one* point, namely, Bozo. Without Bozo, there is little correlation between *Shoe Size* and *IQ*. Look what happens to the regression when we take him out:

Dependent variable is: IQ
R-squared = 0.7%

Variable	Coefficient
Intercept	105.458
Shoe size	−0.460194

The R^2 value is now 0.7%—a very weak linear relationship (as one might expect!). One single point exhibits a great influence on the regression analysis.

What would have happened if Bozo hadn't shown his comic genius on IQ tests? Suppose his measured *IQ* had been only 50. The slope of the line would then drop from 0.96 IQ points/shoe size to −0.69 IQ points/shoe size. No matter where Bozo's *IQ* is, the line tends to follow it because his *Shoe Size*, being so far from the mean *Shoe Size*, makes this a high-leverage point.

Even though this example is far fetched, similar situations occur all the time in real life. For example, a regression of sales against floor space for hardware stores that looked primarily at small-town businesses could be dominated in a similar way if The Home Depot were included.

> **Warning:** Influential points can hide in plots of residuals. Points with high leverage pull the line close to them, so they often have small residuals. You'll see influential points more easily in scatterplots of the original data or by finding a regression model with and without the points.

JUST CHECKING

Each of these scatterplots shows an unusual point. For each, tell whether the point is a high-leverage point, would have a large residual, or is influential.

1.

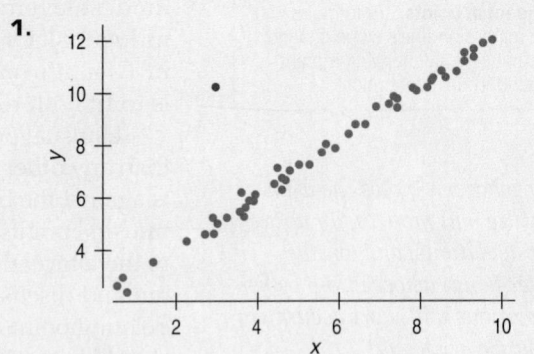

2.

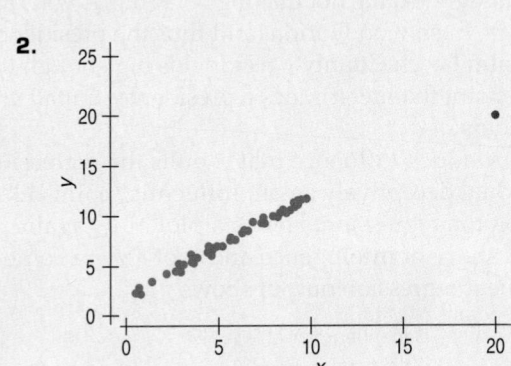

3.

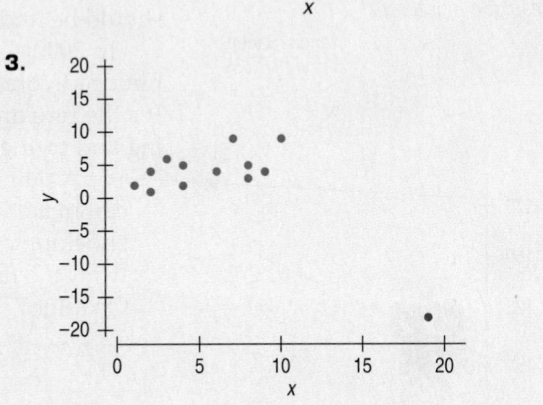

Lurking Variables and Causation

> One common way to interpret a regression slope is to say that "a change of 1 unit in x results in a change of b_1 units in y." This way of saying things encourages causal thinking. Beware.

In Chapter 7, we tried to make it clear that no matter how strong the correlation is between two variables, there's no simple way to show that one variable causes the other. Putting a regression line through a cloud of points just increases the temptation to think and to say that the x-variable *causes* the y-variable. Just to make sure, let's repeat the point again: No matter how strong the association, no matter how large the R^2 value, no matter how straight the line, there is no way to conclude from a regression alone that one variable *causes* the other. There's always the possibility that some third variable is driving both of the variables you have observed. With observational data, as opposed to data from a designed experiment, there is no way to be sure that a **lurking variable** is not the cause of any apparent association.

Here's an example: The scatterplot shows the *Life Expectancy* (average of men and women, in years) for each of 41 countries of the world, plotted against the square root of the number of *Doctors* per person in the country. (The square root is here to make the relationship satisfy the Straight Enough Condition, as we saw back in Chapter 7.)

The strong positive association ($R^2 = 62.4\%$) seems to confirm our expectation that more *Doctors* per person improves healthcare, leading to longer lifetimes and a greater *Life Expectancy*. The strength of the association would *seem* to argue that we should send more doctors to developing countries to increase life expectancy.

That conclusion is about the consequences of a change. Would sending more doctors increase life expectancy? Specifically, do doctors *cause* greater life expectancy? Perhaps, but these are observed data, so there may be another explanation for the association.

On the next page, the similar-looking scatterplot's x-variable is the square root of the number of *Televisions* per person in each country. The positive association in this scatterplot is even *stronger* than the association in the previous plot

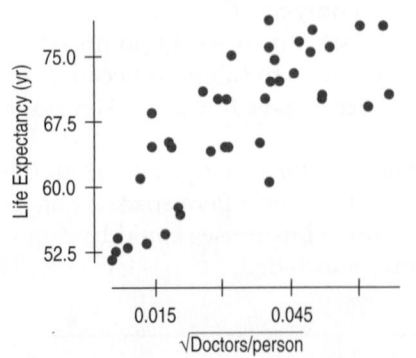

FIGURE 9.13

The relationship between Life Expectancy *(years) and availability of* Doctors *(measured as* √doctors per person*) for countries of the world is strong, positive, and linear.*

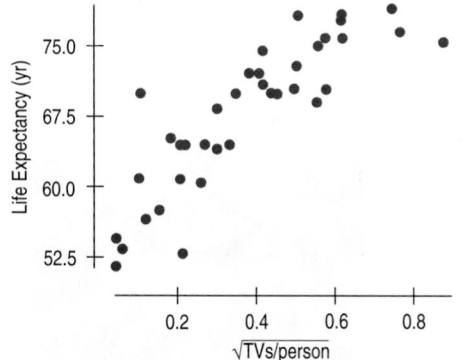

FIGURE 9.14

To increase life expectancy, don't send doctors, send TVs; they're cheaper and more fun. Or maybe that's not the right interpretation of this scatterplot of life expectancy against availability of TVs (as $\sqrt{TVs\,per\,person}$).

($R^2 = 72.3\%$). We can fit the linear model, and quite possibly use the number of TVs as a way to predict life expectancy. Should we conclude that increasing the number of TVs actually extends lifetimes? If so, we should send TVs instead of doctors to developing countries. Not only is the correlation with life expectancy higher, but TVs are much cheaper than doctors.

What's wrong with this reasoning? Maybe we were a bit hasty earlier when we concluded that doctors *cause* longer lives. Maybe there's a lurking variable here. Countries with higher standards of living have both longer life expectancies *and* more doctors (and more TVs). Could higher living standards cause changes in the other variables? If so, then improving living standards might be expected to prolong lives, increase the number of doctors, and increase the number of TVs.

From this example, you can see how easy it is to fall into the trap of mistakenly inferring causality from a regression. For all we know, doctors (or TVs!) *do* increase life expectancy. But we can't tell that from data like these, no matter how much we'd like to. Resist the temptation to conclude that x causes y from a regression, no matter how obvious that conclusion seems to you.

Working with Summary Values

Scatterplots of statistics summarized over groups tend to show less variability than we would see if we measured the same variable on individuals. This is because the summary statistics themselves vary less than the data on the individuals do—a fact we will make more specific in coming chapters.

In Chapter 7 we looked at the heights and weights of individual students. There we saw a correlation of 0.644, so R^2 is 41.5%.

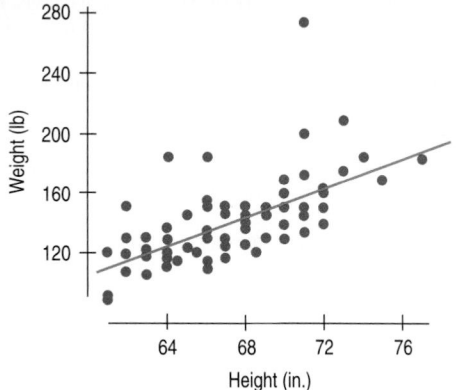

FIGURE 9.15

Weight (lb) against Height (in.) for a sample of men. There's a strong, positive, linear association.

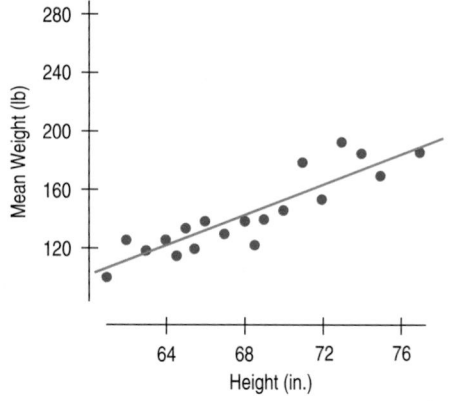

FIGURE 9.16

Mean Weight (lb) shows a stronger linear association with Height than do the weights of individuals. Means vary less than individual values.

Suppose, instead of data on individuals, we knew only the mean weight for each height value. The scatterplot of mean weight by height would show less scatter. And the R^2 would increase to 80.1%.

Scatterplots of summary statistics show less scatter than the baseline data on individuals and can give a false impression of how well a line summarizes the data. There's no simple correction for this phenomenon. Once we're given summary data, there's no simple way to get the original values back.

In the life expectancy and TVs example, we have no good measure of exposure to doctors or to TV on an individual basis. But if we did, we should expect the scatterplot to show more variability and the corresponding R^2 to be smaller. The bottom line is that you should be a bit suspicious of conclusions based on regressions of summary data. They may look better than they really are.

FOR EXAMPLE Using several of these methods together

Motorcycles designed to run off-road, often known as dirt bikes, are specialized vehicles.

We have data on 104 dirt bikes available for sale in 2005. Some cost as little as $3000, while others are substantially more expensive. Let's investigate how the size and type of engine contribute to the cost of a dirt bike. As always, we start with a scatterplot.

Here's a scatterplot of the manufacturer's suggested retail price (*MSRP*) in dollars against the engine *Displacement*, along with a regression analysis:

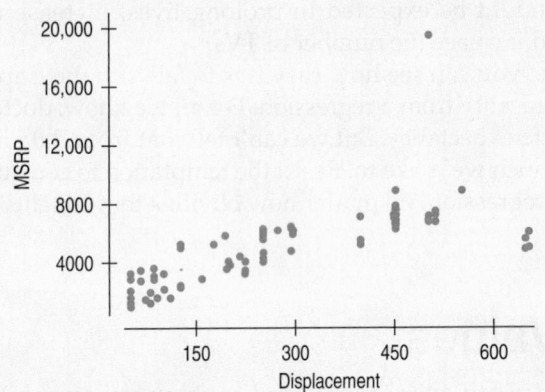

Dependent variable is: MSRP
R-squared = 49.9% s = 1737

Variable	Coefficient
Intercept	2273.67
Displacement	10.0297

Question: What do you see in the scatterplot?

There is a strong positive association between the engine displacement of dirt bikes and the manufacturer's suggested retail price. One of the dirt bikes is an outlier; its price is more than double that of any other bike.

The outlier is the Husqvarna TE 510 Centennial. Most of its components are handmade exclusively for this model, including extensive use of carbon fiber throughout. That may explain its $19,500 price tag! Clearly, the TE 510 is not like the other bikes. We'll set it aside for now and look at the data for the remaining dirt bikes.

Question: What effect will removing this outlier have on the regression? Describe how the slope, R^2, and s_e will change.

The TE 510 was an influential point, tilting the regression line upward. With that point removed, the regression slope will get smaller. With that dirt bike omitted, the pattern becomes more consistent, so the value of R^2 should get larger and the standard deviation of the residuals, s_e, should get smaller.

With the outlier omitted, here's the new regression and a scatterplot of the residuals:

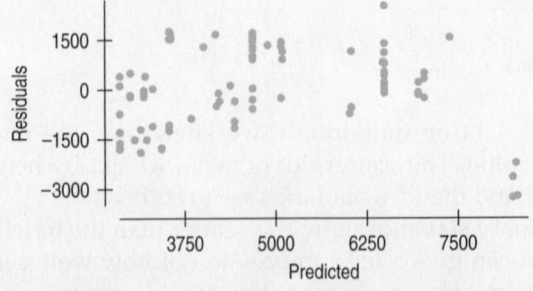

Dependent variable is: MSRP
R-squared = 61.3% s = 1237

Variable	Coefficient
Intercept	2411.02
Displacement	9.05450

Question: What do you see in the residuals plot?

The points at the far right don't fit well with the other dirt bikes. Overall, there appears to be a bend in the relationship, so a linear model may not be appropriate.

Let's try a re-expression. Here's a scatterplot showing *MSRP* against the cube root of *Displacement* to make the relationship closer to straight. (Since displacement is measured in cubic centimeters, its cube root has the simple units of centimeters.) In addition, we've colored the plot according to the cooling method used in the bike's engine: liquid or air. Each group is shown with its own regression line, as we did for the cereals on different shelves.

Question: What does this plot say about dirt bikes?

There appears to be a positive, linear relationship between MSRP and the cube root of Displacement. In general, the larger the engine a bike has, the higher the suggested price. Liquid-cooled dirt bikes, however, typically cost more than air-cooled bikes with comparable displacement. A few liquid-cooled bikes appear to be much less expensive than we might expect, given their engine displacements.

[Jiang Lu, Joseph B. Kadane, and Peter Boatwright, "The Dirt on Bikes: An Illustration of CART Models for Brand Differentiation," provides data on 2005-model bikes.]

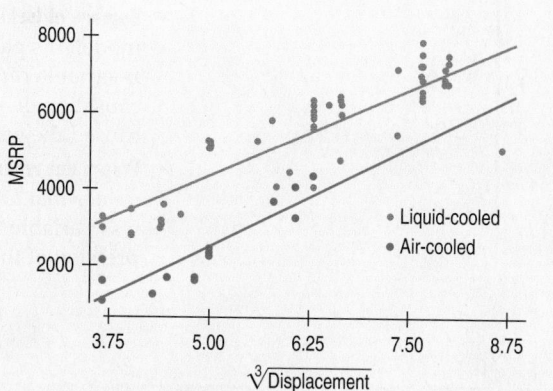

WHAT CAN GO WRONG?

This entire chapter has held warnings about things that can go wrong in a regression analysis. So let's just recap. When you make a linear model:

▶ **Make sure the relationship is straight.** Check the Straight Enough Condition. Always examine the residuals for evidence that the Linearity Assumption has failed. It's often easier to see deviations from a straight line in the residuals plot than in the scatterplot of the original data. Pay special attention to the most extreme residuals because they may have something to add to the story told by the linear model.

▶ **Be on guard for different groups in your regression.** Check for evidence that the data consist of separate subsets. If you find subsets that behave differently, consider fitting a different linear model to each subset.

▶ **Beware of extrapolating.** Beware of extrapolation beyond the x-values that were used to fit the model. Although it's common to use linear models to extrapolate, the practice is dangerous.

▶ **Beware especially of extrapolating into the future!** Be especially cautious about extrapolating into the future with linear models. To predict the future, you must assume that future changes will continue at the same rate you've observed in the past. Predicting the future is particularly tempting and particularly dangerous.

▶ **Look for unusual points.** Unusual points always deserve attention and may well reveal more about your data than the rest of the points combined. Always look for them and try to understand why they stand apart. A scatterplot of the data is a good way to see high-leverage and influential points. A scatterplot of the residuals against the predicted values is a good tool for finding points with large residuals.

▶ **Beware of high-leverage points and especially of those that are influential.** Influential points can alter the regression model a great deal. The resulting model may say more about one or two points than about the overall relationship.

▶ **Consider comparing two regressions.** To see the impact of outliers on a regression, it's often wise to run two regressions, one with and one without the extraordinary points, and then to discuss the differences.

▶ **Treat unusual points honestly.** If you remove enough carefully selected points, you can always get a regression with a high R^2 eventually. But it won't give you much understanding. Some variables are not related in a way that's simple enough for a linear model to fit very well. When that happens, report the failure and stop.

(continued)

▶ **Beware of lurking variables.** Think about lurking variables before interpreting a linear model. It's particularly tempting to explain a strong regression by thinking that the x-variable *causes* the y-variable. A linear model alone can never demonstrate such causation, in part because it cannot eliminate the chance that a lurking variable has caused the variation in both x and y.

▶ **Watch out when dealing with data that are summaries.** Be cautious in working with data values that are themselves summaries, such as means or medians. Such statistics are less variable than the data on which they are based, so they tend to inflate the impression of the strength of a relationship.

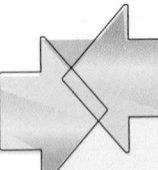

CONNECTIONS

We are always alert to things that can go wrong if we use statistics without thinking carefully. Regression opens new vistas of potential problems. But each one relates to issues we've thought about before.

It is always important that our data be from a single homogeneous group and not made up of disparate groups. We looked for multiple modes in single variables. Now we check scatterplots for evidence of subgroups in our data. As with modes, it's often best to split the data and analyze the groups separately.

Our concern with unusual points and their potential influence also harks back to our earlier concern with outliers in histograms and boxplots—and for many of the same reasons. As we've seen here, regression offers such points new scope for mischief.

The risks of interpreting linear models as causal or predictive arose in Chapters 7 and 8. And they're important enough to mention again in later chapters.

WHAT HAVE WE LEARNED?

We've learned that there are many ways in which a data set may be unsuitable for a regression analysis.

▶ Watch out for more than one group hiding in your regression analysis. If you find subsets of the data that behave differently, consider fitting a different regression model to each subset.

▶ The **Straight Enough Condition** says that the relationship should be reasonably straight to fit a regression. Somewhat paradoxically, sometimes it's easier to see that the relationship is not straight *after* fitting the regression by examining the residuals. The same is true of outliers.

▶ The **Outlier Condition** actually means two things: Points with large residuals or high leverage (especially both) can influence the regression model significantly. It's a good idea to perform the regression analysis with and without such points to see their impact.

And we've learned that even a good regression doesn't mean we should believe that the model says more than it really does.

▶ Extrapolation far from $\bar{x}$ can lead to silly and useless predictions.

▶ Even an R^2 near 100% doesn't indicate that x causes y (or the other way around). Watch out for lurking variables that may affect both x and y.

▶ Be careful when you interpret regressions based on *summaries* of the data sets. These regressions tend to look stronger than the regression based on all the individual data.

Terms

Extrapolation
203. Although linear models provide an easy way to predict values of y for a given value of x, it is unsafe to predict for values of x far from the ones used to find the linear model equation. Such extrapolation may pretend to see into the future, but the predictions should not be trusted.

Outlier

205. Any data point that stands away from the others can be called an outlier. In regression, outliers can be extraordinary in two ways: by having a large residual or by having high leverage.

Leverage

206. Data points whose x-values are far from the mean of x are said to exert leverage on a linear model. High-leverage points pull the line close to them, and so they can have a large effect on the line, sometimes completely determining the slope and intercept. With high enough leverage, their residuals can be deceptively small.

Influential point

206. If omitting a point from the data results in a very different regression model, then that point is called an influential point.

Lurking variable

208. A variable that is not explicitly part of a model but affects the way the variables in the model appear to be related is called a lurking variable. Because we can never be certain that observational data are not hiding a lurking variable that influences both x and y, it is never safe to conclude that a linear model demonstrates a causal relationship, no matter how strong the linear association.

Skills

- ▸ Understand that we cannot fit linear models or use linear regression if the underlying relationship between the variables is not itself linear.

- ▸ Understand that data used to find a model must be homogeneous. Look for subgroups in data before you find a regression, and analyze each separately.

- ▸ Know the danger of extrapolating beyond the range of the x-values used to find the linear model, especially when the extrapolation tries to predict into the future.

- ▸ Understand that points can be unusual by having a large residual or by having high leverage.

- ▸ Understand that an influential point can change the slope and intercept of the regression line.

- ▸ Look for lurking variables whenever you consider the association between two variables. Understand that a strong association does not mean that the variables are causally related.

- ▸ Know how to display residuals from a linear model by making a scatterplot of residuals against predicted values or against the x-variable, and know what patterns to look for in the picture.

- ▸ Know how to look for high-leverage and influential points by examining a scatterplot of the data and how to look for points with large residuals by examining a scatterplot of the residuals against the predicted values or against the x-variable. Understand how fitting a regression line with and without influential points can add to your understanding of the regression model.

- ▸ Know how to look for high-leverage points by examining the distribution of the x-values or by recognizing them in a scatterplot of the data, and understand how they can affect a linear model.

- ▸ Include diagnostic information such as plots of residuals and leverages as part of your report of a regression.

- ▸ Report any high-leverage points.

- ▸ Report any outliers. Consider reporting analyses with and without outliers, to assess their influence on the regression.

- ▸ Include appropriate cautions about extrapolation when reporting predictions from a linear model.

- ▸ Discuss possible lurking variables.

REGRESSION DIAGNOSIS ON THE COMPUTER

Most statistics technology offers simple ways to check whether your data satisfy the conditions for regression. We have already seen that these programs can make a simple scatterplot. They can also help us check the conditions by plotting residuals.

EXERCISES

1. Marriage age 2003. Is there evidence that the age at which women get married has changed over the past 100 years? The scatterplot shows the trend in age at first marriage for American women (www.census.gov).

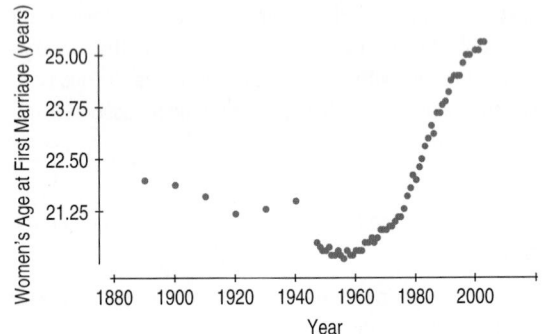

a) Is there a clear pattern? Describe the trend.
b) Is the association strong?
c) Is the correlation high? Explain.
d) Is a linear model appropriate? Explain.

2. Smoking 2004. The Centers for Disease Control and Prevention track cigarette smoking in the United States. How has the percentage of people who smoke changed since the danger became clear during the last half of the 20th century? The scatterplot shows percentages of smokers among men 18–24 years of age, as estimated by surveys, from 1965 through 2004 (www.cdc.gov/nchs/).

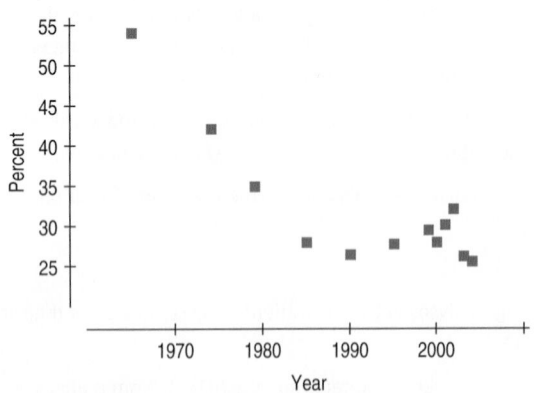

a) Is there a clear pattern? Describe the trend.
b) Is the association strong?
c) Is a linear model appropriate? Explain.

3. Human Development Index. The United Nations Development Programme (UNDP) uses the Human Development Index (HDI) in an attempt to summarize in one number the progress in health, education, and economics of a country. In 2006, the HDI was as high as 0.965 for Norway and as low as 0.331 for Niger. The gross domestic product per capita (GDPPC), by contrast, is often used to summarize the *overall* economic strength of a country. Is the HDI related to the GDPPC? Here is a scatterplot of *HDI* against *GDPPC*.

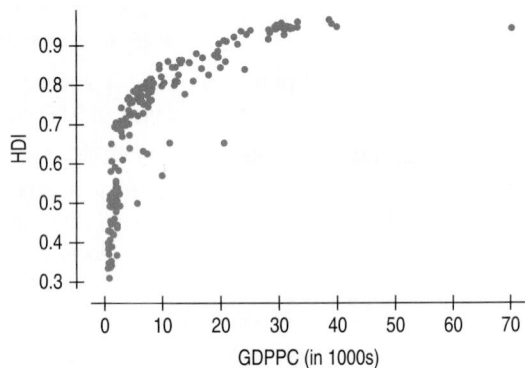

a) Explain why fitting a linear model to these data might be misleading.
b) If you fit a linear model to the data, what do you think a scatterplot of residuals versus predicted *HDI* will look like?
c) There is an outlier (Luxembourg) with a *GDPPC* of around $70,000. Will setting this point aside improve the model substantially? Explain.

4. HDI Revisited. The United Nations Development Programme (UNDP) uses the Human Development Index (HDI) in an attempt to summarize in one number the progress in health, education, and economics of a country. The number of cell phone subscribers per 1000 people is positively associated with economic progress in a country. Can the number of cell phone subscribers be used to predict the HDI? Here is a scatterplot of HDI against cell phone subscribers:

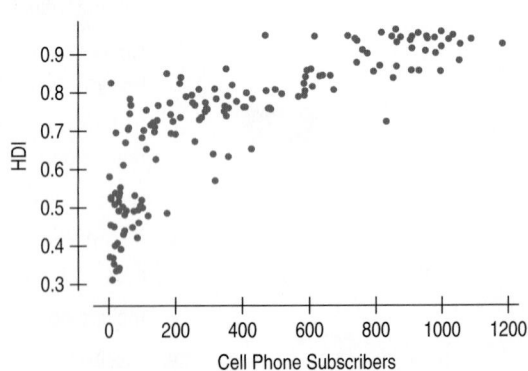

a) Explain why fitting a linear model to these data might be misleading.
b) If you fit a linear model to the data, what do you think a scatterplot of residuals versus predicted *HDI* will look like?

5. Good model? In justifying his choice of a model, a student wrote, "I know this is the correct model because $R^2 = 99.4\%$."
a) Is this reasoning correct? Explain.
b) Does this model allow the student to make accurate predictions? Explain.

6. Bad model?   A student who has created a linear model is disappointed to find that her R^2 value is a very low 13%.
   a) Does this mean that a linear model is not appropriate? Explain.
   b) Does this model allow the student to make accurate predictions? Explain.

7. Movie Dramas.   Here's a scatterplot of the production budgets (in millions of dollars) vs. the running time (in minutes) for major release movies in 2005. Dramas are plotted in red and all other genres are plotted in black. A separate least squares regression line has been fitted to each group. For the following questions, just examine the plot:

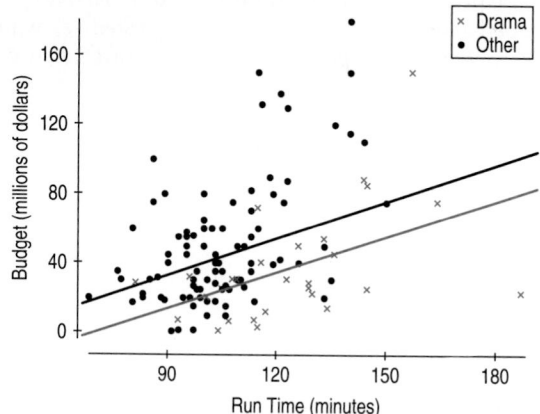

   a) What are the units for the slopes of these lines?
   b) In what way are dramas and other movies similar with respect to this relationship?
   c) In what way are dramas different from other genres of movies with respect to this relationship?

8. Movie Ratings.   Does the cost of making a movie depend on its audience? Here's a scatterplot of the same data we examined in Exercise 7. Movies with an R rating are colored purple, those with a PG-13 rating are red, and those with a PG rating are green. Regression lines have been found for each group. (The black points are G-rated, but there were too few to fit a line reliably.)

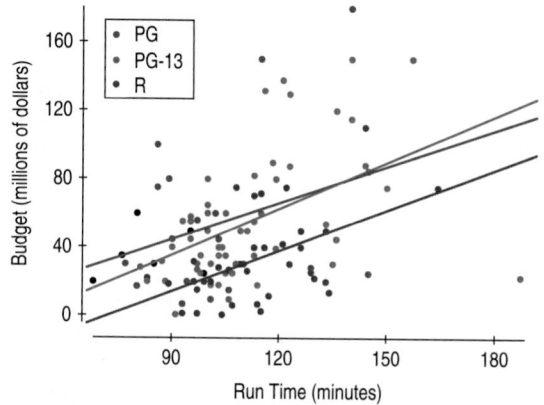

   a) In what ways is the relationship between run times and budgets similar for the three ratings groups?
   b) How do the costs of R-rated movies differ from those of PG-13 and PG rated movies? Discuss both the slopes and the intercepts.

   c) The film *King Kong,* with a run time of 187 minutes, is the red point sitting at the lower right. If it were omitted from this analysis, how might that change your conclusions about PG-13 movies?

9. Oakland passengers.   The scatterplot below shows the number of passengers departing from Oakland (CA) airport month by month since the start of 1997. Time is shown as years since 1990, with fractional years used to represent each month. (Thus, June of 1997 is 7.5—halfway through the 7th year after 1990.) www.oaklandairport.com

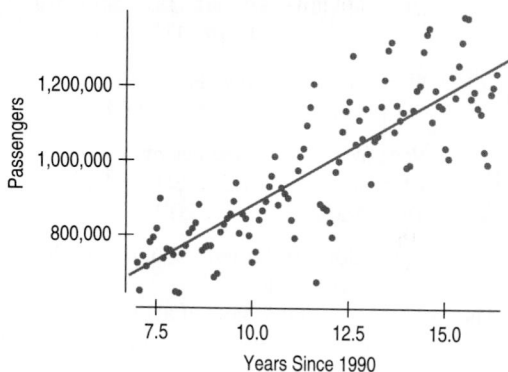

Here's a regression and the residuals plot:

Dependent variable is: Passengers
R-squared = 71.1%   s = 104330

Variable	Coefficient
Constant	282584
Year-1990	59704.4

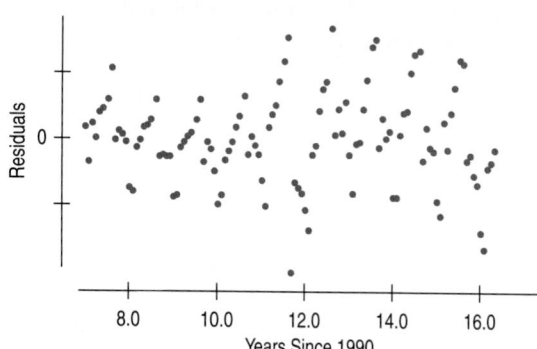

   a) Interpret the slope and intercept of the model.
   b) What does the value of R^2 say about the model?
   c) Interpret s_e in this context.
   d) Would you use this model to predict the numbers of passengers in 2010 ($YearsSince1990 = 20$)? Explain.
   e) There's a point near the middle of this time span with a large negative residual. Can you explain this outlier?

10. Tracking hurricanes.   In a previous chapter, we saw data on the errors (in nautical miles) made by the National Hurricane Center in predicting the path of hurricanes. The scatterplot on the next page shows the trend in the 24-hour tracking errors since 1970 (www.nhc.noaa.gov).

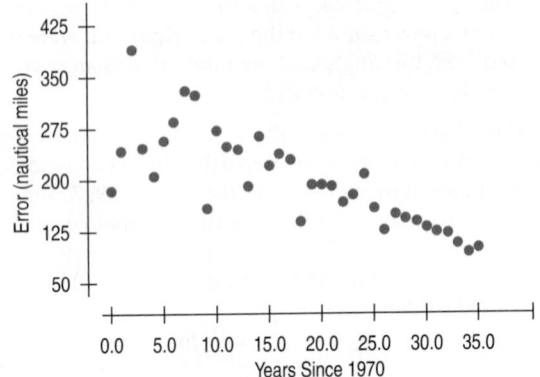

Dependent variable is: Error
R-squared = 63.0%   s = 42.87

Variable	Coefficient
Intercept	292.089
Years-1970	−5.22924

a) Interpret the slope and intercept of the model.
b) Interpret s_e in this context.
c) The Center had a stated goal of achieving an average tracking error of 125 nautical miles in 2009. Will they make it? Why do you think so?
d) What if their goal were an average tracking error of 90 nautical miles?
e) What cautions would you state about your conclusion?

11. **Unusual points.**   Each of the four scatterplots that follow shows a cluster of points and one "stray" point. For each, answer these questions:
   1) In what way is the point unusual? Does it have high leverage, a large residual, or both?
   2) Do you think that point is an influential point?
   3) If that point were removed, would the correlation become stronger or weaker? Explain.
   4) If that point were removed, would the slope of the regression line increase or decrease? Explain.

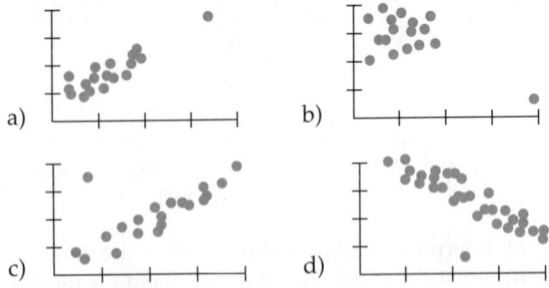

12. **More unusual points.**   Each of the following scatterplots shows a cluster of points and one "stray" point. For each, answer these questions:
   1) In what way is the point unusual? Does it have high leverage, a large residual, or both?
   2) Do you think that point is an influential point?
   3) If that point were removed, would the correlation become stronger or weaker? Explain.
   4) If that point were removed, would the slope of the regression line increase or decrease? Explain.

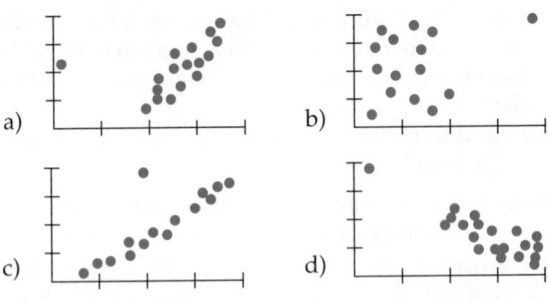

13. **The extra point.**   The scatterplot shows five blue data points at the left. Not surprisingly, the correlation for these points is $r = 0$. Suppose *one* additional data point is added at one of the five positions suggested below in green. Match each point (a–e) with the correct new correlation from the list given.
   1) −0.90   4) 0.05
   2) −0.40   5) 0.75
   3) 0.00

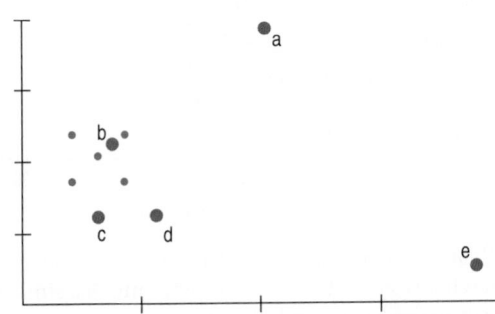

14. **The extra point revisited.**   The original five points in Exercise 13 produce a regression line with slope 0. Match each of the green points (a–e) with the slope of the line after that one point is added:
   1) −0.45   4) 0.05
   2) −0.30   5) 0.85
   3) 0.00

15. **What's the cause?**   Suppose a researcher studying health issues measures blood pressure and the percentage of body fat for several adult males and finds a strong positive association. Describe three different possible cause-and-effect relationships that might be present.

16. **What's the effect?**   A researcher studying violent behavior in elementary school children asks the children's parents how much time each child spends playing computer games and has their teachers rate each child on the level of aggressiveness they display while playing with other children. Suppose that the researcher finds a moderately strong positive correlation. Describe three different possible cause-and-effect explanations for this relationship.

17. **Reading.**   To measure progress in reading ability, students at an elementary school take a reading comprehension test every year. Scores are measured in "grade-level" units; that is, a score of 4.2 means that a student is reading at slightly above the expected level for a fourth grader. The school principal prepares a report to parents that includes a graph showing the mean reading score for

each grade. In his comments he points out that the strong positive trend demonstrates the success of the school's reading program.

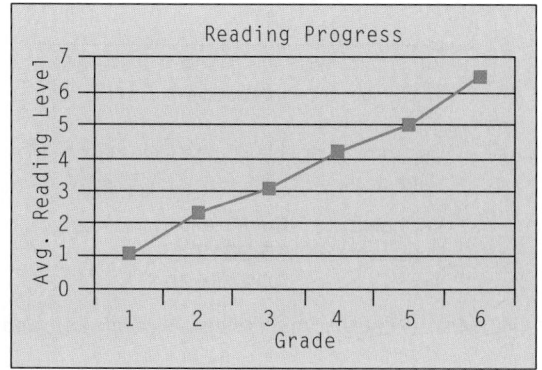

a) Does this graph indicate that students are making satisfactory progress in reading? Explain.
b) What would you estimate the correlation between *Grade* and *Average Reading Level* to be?
c) If, instead of this plot showing average reading levels, the principal had produced a scatterplot of the reading levels of all the individual students, would you expect the correlation to be the same, higher, or lower? Explain.
d) Although the principal did not do a regression analysis, someone as statistically astute as you might do that. (But don't bother.) What value of the slope of that line would you view as demonstrating acceptable progress in reading comprehension? Explain.

18. Grades. A college admissions officer, defending the college's use of SAT scores in the admissions process, produced the graph below. It shows the mean GPAs for last year's freshmen, grouped by SAT scores. How strong is the evidence that *SAT Score* is a good predictor of *GPA*? What concerns you about the graph, the statistical methodology or the conclusions reached?

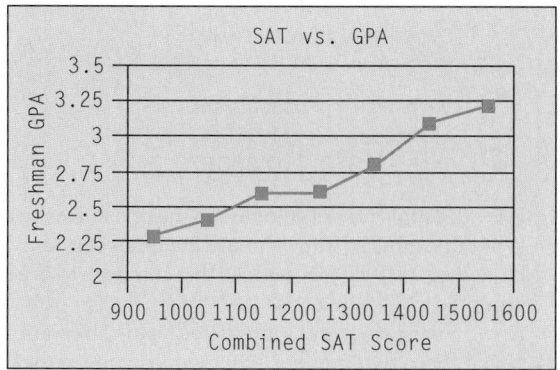

19. Heating. After keeping track of his heating expenses for several winters, a homeowner believes he can estimate the monthly cost from the average daily Fahrenheit temperature by using the model $\widehat{Cost} = 133 - 2.13\ Temp$. Here is the residuals plot for his data:

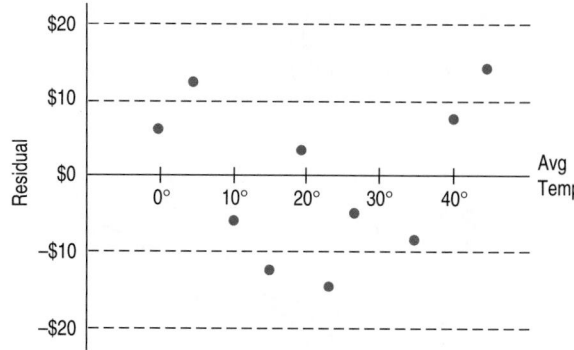

a) Interpret the slope of the line in this context.
b) Interpret the *y*-intercept of the line in this context.
c) During months when the temperature stays around freezing, would you expect cost predictions based on this model to be accurate, too low, or too high? Explain.
d) What heating cost does the model predict for a month that averages 10°?
e) During one of the months on which the model was based, the temperature did average 10°. What were the actual heating costs for that month?
f) Should the homeowner use this model? Explain.
g) Would this model be more successful if the temperature were expressed in degrees Celsius? Explain.

20. Speed. How does the speed at which you drive affect your fuel economy? To find out, researchers drove a compact car for 200 miles at speeds ranging from 35 to 75 miles per hour. From their data, they created the model $\widehat{Fuel\,Efficiency} = 32 - 0.1\ Speed$ and created this residual plot:

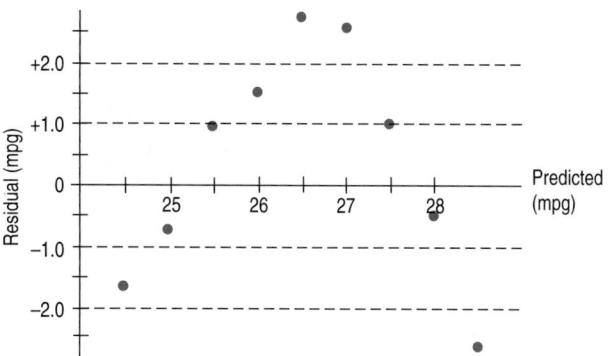

a) Interpret the slope of this line in context.
b) Explain why it's silly to attach any meaning to the *y*-intercept.
c) When this model predicts high *Fuel Efficiency*, what can you say about those predictions?
d) What *Fuel Efficiency* does the model predict when the car is driven at 50 mph?
e) What was the actual *Fuel Efficiency* when the car was driven at 45 mph?
f) Do you think there appears to be a strong association between *Speed* and *Fuel Efficiency*? Explain.
g) Do you think this is the appropriate model for that association? Explain.

21. Interest rates. Here's a plot showing the federal rate on 3-month Treasury bills from 1950 to 1980, and a regression model fit to the relationship between the *Rate* (in %) and *Years since 1950* (www.gpoaccess.gov/eop/).

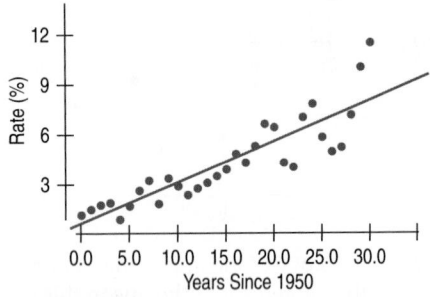

Dependent variable is: Rate
R-squared = 77.4%   s = 1.239

Variable	Coefficient
Intercept	0.640282
Year − 1950	0.247637

a) What is the correlation between *Rate* and *Year*?
b) Interpret the slope and intercept.
c) What does this model predict for the interest rate in the year 2000?
d) Would you expect this prediction to have been accurate? Explain.

22. Ages of couples 2003. The graph shows the ages of both men and women at first marriage (www.census.gov).

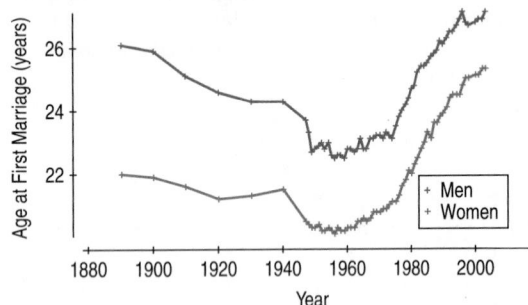

Clearly, the pattern for men is similar to the pattern for women. But are the two lines getting closer together?

Here's a timeplot showing the *difference* in average age (men's age − women's age) at first marriage, the regression analysis, and the associated residuals plot.

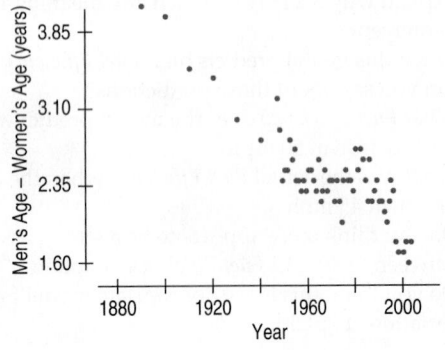

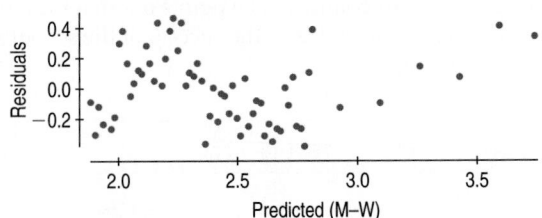

Dependent variable is: Age Difference
R-squared = 75.1%   s = 0.2333

Variable	Coefficient
Constant	35.0617
Year	−0.016565

a) What is the correlation between *Age Difference* and *Year*?
b) Interpret the slope of this line.
c) Predict the average age difference in 2015.
d) Describe reasons why you might not place much faith in that prediction.

23. Interest rates revisited. In Exercise 21 you investigated the federal rate on 3-month Treasury bills between 1950 and 1980. The scatterplot below shows that the trend changed dramatically after 1980.

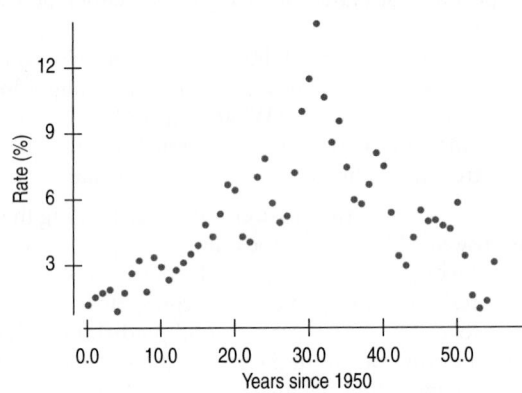

Here's a regression model for the data since 1980.

Dependent variable is: Rate
R-squared = 74.5%   s = 1.630

Variable	Coefficient
Intercept	21.0688
Year − 1950	−0.356578

a) How does this model compare to the one in Exercise 21?
b) What does this model estimate the interest rate to have been in 2000? How does this compare to the rate you predicted in Exercise 21?
c) Do you trust this newer predicted value? Explain.
d) Given these two models, what would you predict the interest rate on 3-month Treasury bills will be in 2020?

24. Ages of couples, again. Has the trend of decreasing difference in age at first marriage seen in Exercise 22 gotten stronger recently? The scatterplot and residual plot for the data from 1975 through 2003, along with a regression for just those years, are on the next page.

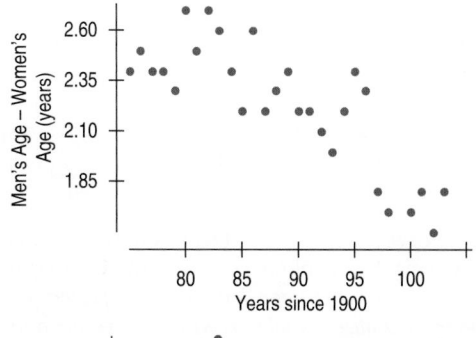

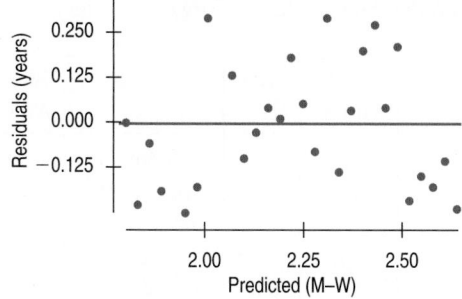

Dependent variable is: Men − Women
R-Squared = 65.6% s = 0.1869

Variable	Coefficient
Intercept	4.88424
Year	−0.029959

a) Why is R^2 higher for the first model (in Exercise 22)?
b) Is this linear model appropriate for the post-1975 data? Explain.
c) What does the slope say about marriage ages?
d) Explain why it's not reasonable to interpret the y-intercept.

25. **Gestation.**   For women, pregnancy lasts about 9 months. In other species of animals, the length of time from conception to birth varies. Is there any evidence that the gestation period is related to the animal's lifespan? The first scatterplot shows *Gestation Period* (in days) vs. *Life Expectancy* (in years) for 18 species of mammals. The highlighted point at the far right represents humans.

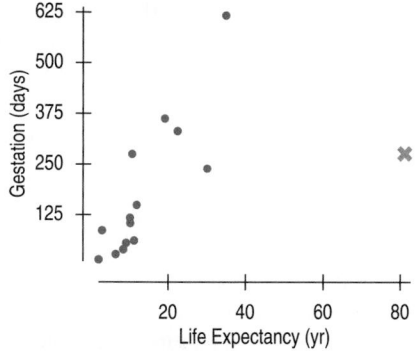

a) For these data, r = 0.54, not a very strong relationship. Do you think the association would be stronger or weaker if humans were removed? Explain.
b) Is there reasonable justification for removing humans from the data set? Explain.

c) Here are the scatterplot and regression analysis for the 17 nonhuman species. Comment on the strength of the association.

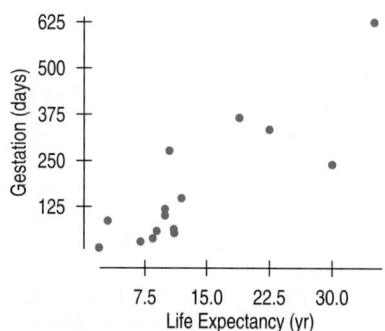

Dependent variable is: Gestation
R-Squared = 72.2%

Variable	Coefficient
Constant	−39.5172
LifExp	15.4980

d) Interpret the slope of the line.
e) Some species of monkeys have a life expectancy of about 20 years. Estimate the expected gestation period of one of these monkeys.

Ⓣ 26. **Swim the lake 2006.**   People swam across Lake Ontario 42 times between 1974 and 2006 (www.soloswims.com). We might be interested in whether they are getting any faster or slower. Here are the regression of the crossing *Times* (minutes) against the *Year* of the crossing and the residuals plot:

Dependent variable is: Time
R-Squared = 1.3% s = 443.8

Variable	Coefficient
Intercept	−8950.40
Year	5.14171

a) What does the R^2 mean for this regression?
b) Are the swimmers getting faster or slower? Explain.
c) The outlier seen in the residuals plot is a crossing by Vicki Keith in 1987 in which she swam a round trip, north to south, and then back again. Clearly, this swim doesn't belong with the others. Would removing it change the model a lot? Explain.

27. **Elephants and hippos.**   We removed humans from the scatterplot in Exercise 25 because our species was an outlier in life expectancy. The resulting scatterplot (next page) shows two points that now may be of concern. The point in the upper right corner of this scatterplot is for elephants, and the other point at the far right is for hippos.

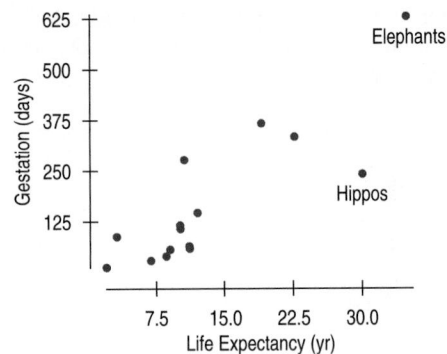

a) By removing one of these points, we could make the association appear to be stronger. Which point? Explain.
b) Would the slope of the line increase or decrease?
c) Should we just keep removing animals to increase the strength of the model? Explain.
d) If we remove elephants from the scatterplot, the slope of the regression line becomes 11.6 days per year. Do you think elephants were an influential point? Explain.

28. Another swim 2006. In Exercise 26 we saw that Vicki Keith's round-trip swim of Lake Ontario was an obvious outlier among the other one-way times. Here is the new regression after this unusual point is removed:

Dependent variable is: Time
R-Squared = 4.1%   s = 292.6

Variable	Coefficient
Intercept	−11048.7
Year	6.17091

a) In this new model, the value of s_e is much smaller. Explain what that means in this context.
b) Now would you be willing to say that the Lake Ontario swimmers are getting faster (or slower)?

29. Marriage age 2003 revisited. Suppose you wanted to predict the trend in marriage age for American women into the early part of this century.
a) How could you use the data graphed in Exercise 1 to get a good prediction? Marriage ages in selected years starting in 1900 are listed below. Use all or part of these data to create an appropriate model for predicting the average age at which women will first marry in 2010.

1900–1950 (10-yr intervals): 21.9, 21.6, 21.2, 21.3, 21.5, 20.3
1955–2000 (5-yr intervals): 20.2, 20.2, 20.6, 20.8, 21.1, 22.0, 23.3, 23.9, 24.5, 25.1

b) How much faith do you place in this prediction? Explain.
c) Do you think your model would produce an accurate prediction about your grandchildren, say, 50 years from now? Explain.

30. Unwed births. The National Center for Health Statistics reported the data below, showing the percentage of all births that are to unmarried women for selected years

between 1980 and 1998. Create a model that describes this trend. Justify decisions you make about how to best use these data.

Year	1980	1985	1990	1991	1992	1993	1994	1995	1996	1997	1998
%	18.4	22.0	28.0	29.5	30.1	31.0	32.6	32.2	32.4	32.4	32.8

31. Life Expectancy 2004. Data from the World Bank for 26 Western Hemisphere countries can be used to examine the association between female *Life Expectancy* and the average *Number of Children* women give birth to (http://devdata.worldbank.org/data-query/).

Country	Births/Woman	Life Exp.	Country	Births/Woman	Life Exp.
Argentina	2.3	74.6	Guatemala	4.4	67.6
Bahamas	2.3	70.5	Honduras	3.6	68.2
Barbados	1.7	75.4	Jamaica	2.4	70.8
Belize	3.0	71.9	Mexico	2.2	75.1
Bolivia	3.7	64.5	Nicaragua	3.2	70.1
Brazil	2.3	70.9	Panama	2.6	75.1
Canada	1.5	79.8	Paraguay	3.7	71.2
Chile	2.0	78.0	Peru	2.8	70.4
Colombia	2.4	72.6	Puerto Rico	1.9	77.5
Costa Rica	24.9	78.7	United States	2.0	77.4
Dominican Republic	2.8	67.8	Uruguay	2.1	75.2
Ecuador	2.7	74.5	Venezuela	2.7	73.7
El Salvador	2.8	71.1	Virgin Islands	2.2	78.6

a) Create a scatterplot relating these two variables, and describe the association.
b) Are there any countries that do not seem to fit the overall pattern?
c) Find the correlation, and interpret the value of R^2.
d) Find the equation of the regression line.
e) Is the line an appropriate model? Describe what you see in the residuals plot.
f) Interpret the slope and the y-intercept of the line.
g) If government leaders wanted to increase life expectancy in their country, should they encourage women to have fewer children? Explain.

32. Tour de France 2007. We met the Tour de France data set in Chapter 2 (in Just Checking). One hundred years ago, the fastest rider finished the course at an average speed of about 25.3 kph (around 15.8 mph). In 2005, Lance Armstrong averaged 41.65 kph (25.88 mph) for the fastest average winning speed in history.
a) Make a scatterplot of *Avg Speed* against *Year*. Describe the relationship of *Avg Speed* by *Year*, being careful to point out any unusual features in the plot.
b) Find the regression equation of *Avg Speed* on *Year*.
c) Are the conditions for regression met? Comment.

33. Inflation 2006. The Consumer Price Index (CPI) tracks the prices of consumer goods in the United States, as shown in the table on the next page (ftp://ftp.bis.gov). It indicates, for example, that the average item costing $17.70 in 1926 cost $201.60 in the year 2006.

Year	CPI	Year	CPI
1914	10.0	1962	30.2
1918	15.1	1966	32.4
1922	16.8	1970	38.8
1926	17.7	1974	49.3
1930	16.7	1978	65.2
1934	13.4	1982	96.5
1938	14.1	1986	109.6
1942	16.3	1990	130.7
1946	19.5	1994	148.2
1950	24.1	1998	163.0
1954	26.9	2002	179.9
1958	28.9	2006	201.6

a) Make a scatterplot showing the trend in consumer prices. Describe what you see.

b) Be an economic forecaster: Project increases in the cost of living over the next decade. Justify decisions you make in creating your model.

T 34. Second stage 2007. Look once more at the data from the Tour de France. In Exercise 32 we looked at the whole history of the race, but now let's consider just the post–World War II era.

a) Find the regression of *Avg Speed* by *Year* only for years from 1947 to the present. Are the conditions for regression met?

b) Interpret the slope.

c) In 1979 Bernard Hinault averaged 39.8 kph, while in 2005 Lance Armstrong averaged 41.65 kph. Which was the more remarkable performance and why?

JUST CHECKING
Answers

1. Not high leverage, not influential, large residual
2. High leverage, not influential, small residual
3. High leverage, influential, not large residual

CHAPTER 10

Re-expressing Data: Get It Straight!

A S **Activity: Re-expressing Data.** Should you re-express data? Actually, you already do.

Scan through any Physics book. Most equations have powers, reciprocals, or logs.

How fast can you go on a bicycle? If you measure your speed, you probably do it in miles per hour or kilometers per hour. In a 12-mile-long time trial in the 2005 Tour de France, Dave Zabriskie *averaged* nearly 35 mph (54.7 kph), beating Lance Armstrong by 2 seconds. You probably realize that's a tough act to follow. It's fast. You can tell that at a glance because you have no trouble thinking in terms of distance covered per time.

OK, then, if you averaged 12.5 mph (20.1 kph) for a mile *run*, would *that* be fast? Would it be fast for a 100-m dash? Even if you run the mile often, you probably have to stop and calculate. Running a mile in under 5 minutes (12 mph) is fast. A mile at 16 mph would be a world record (that's a 3-minute, 45-second mile). There's no single *natural* way to measure speed. Sometimes we use time over distance; other times we use the *reciprocal*, distance over time. Neither one is *correct*. We're just used to thinking that way in each case.

So, how does this insight help us understand data? All quantitative data come to us measured in some way, with units specified. But maybe those units aren't the best choice. It's not that meters are better (or worse) than fathoms or leagues. What we're talking about is re-expressing the data another way by applying a function, such as a square root, log, or reciprocal. You already use some of them, even though you may not know it. For example, the Richter scale of earthquake strength (logs), the decibel scale for sound intensity (logs), the f/stop scale for camera aperture openings (squares), and the gauges of shotguns (square roots) all include simple functions of this sort.

Why bother? As with speeds, some expressions of the data may be easier to think about. And some may be much easier to analyze with statistical methods. We've seen that symmetric distributions are easier to summarize and straight scatterplots are easier to model with regressions. We often look to re-express our data if doing so makes them more suitable for our methods.

Straight to the Point

We know from common sense and from physics that heavier cars need more fuel, but exactly how does a car's weight affect its fuel efficiency? Here are the scatterplot of *Weight* (in pounds) and *Fuel Efficiency* (in miles per gallon) for 38 cars, and the residuals plot:

FIGURE 10.1

Fuel Efficiency *(mpg)* vs. Weight *for 38 cars as reported by* Consumer Reports. *The scatterplot shows a negative direction, roughly linear shape, and strong relationship. However, the residuals from a regression of* Fuel Efficiency *on* Weight *reveal a bent shape when plotted against the predicted values. Looking back at the original scatterplot, you may be able to see the bend.*

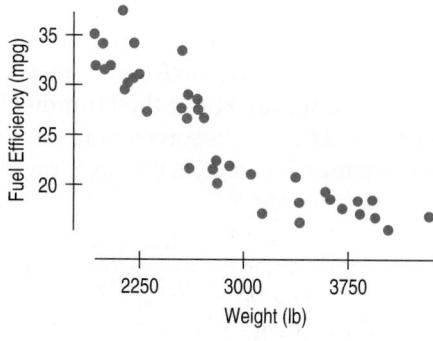 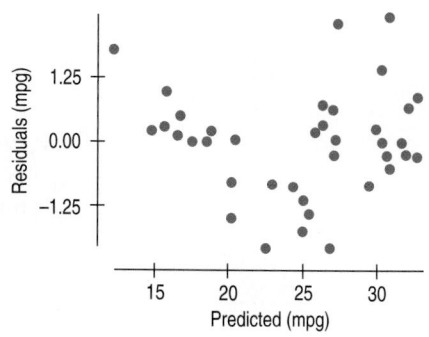

Hmm . . . . Even though R^2 is 81.6%, the residuals don't show the random scatter we were hoping for. The shape is clearly bent. Looking back at the first scatterplot, you can probably see the slight bending. Think about the regression line through the points. How heavy would a car have to be to have a predicted gas mileage of 0? It looks like the *Fuel Efficiency* would go negative at about 6000 pounds. A Hummer H2 weighs about 6400 pounds. The H2 is hardly known for fuel efficiency, but it does get more than the *minus* 5 mpg this regression predicts. Extrapolation is always dangerous, but it's more dangerous the more the model is wrong, because wrong models tend to do even worse the farther you get from the middle of the data.

The bend in the relationship between *Fuel Efficiency* and *Weight* is the kind of failure to satisfy the conditions for an analysis that we can repair by re-expressing the data. Instead of looking at miles per gallon, we could take the reciprocal and work with gallons per hundred miles.[1]

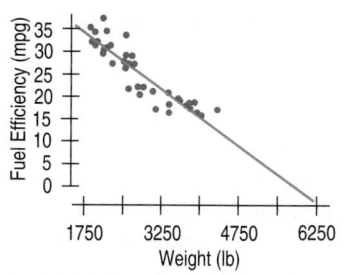

FIGURE 10.2

Extrapolating the regression line gives an absurd answer for vehicles that weigh as little as 6000 pounds.

> **"Gallons per hundred miles—what an absurd way to measure fuel efficiency! Who would ever do it that way?"** Not all re-expressions are easy to understand, but in this case the answer is "Everyone except U.S. drivers." Most of the world measures fuel efficiency in liters per 100 kilometers (L/100 km). This is the same reciprocal form (fuel amount per distance driven) and differs from gallons per 100 miles only by a constant multiple of about 2.38. It has been suggested that most of the world says, "I've got to go 100 km; how much gas do I need?" But Americans say, "I've got 10 gallons in the tank. How far can I drive?" In much the same way, re-expressions "think" about the data differently but don't change what they mean.

FIGURE 10.3

The reciprocal (1/y) is measured in gallons per mile. Gallons per 100 miles gives more meaningful numbers. The reciprocal is more nearly linear against Weight *than the original variable, but the re-expression changes the direction of the relationship. The residuals from the regression of* Fuel Consumption *(gal/100 mi) on* Weight *show less of a pattern than before.*

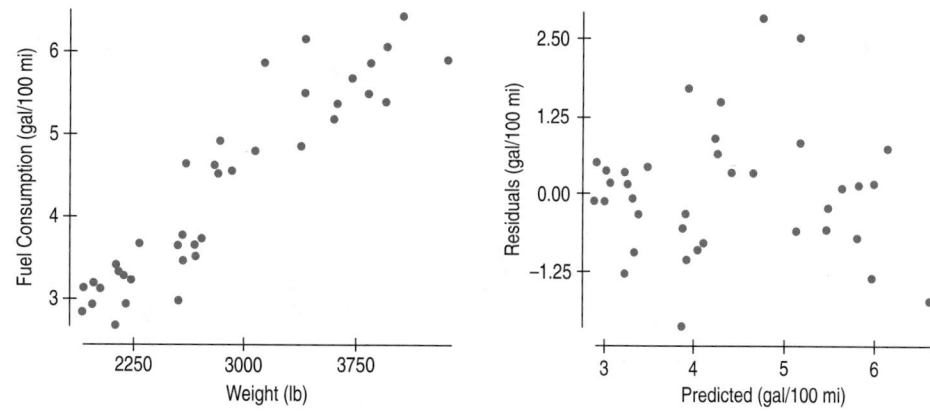

---

[1] Multiplying by 100 to get gallons per 100 miles simply makes the numbers easier to think about: You might have a good idea of how many gallons your car needs to drive 100 miles, but probably a much poorer sense of how much gas you need to go just 1 mile.

The direction of the association is positive now, since we're measuring gas consumption and heavier cars consume more gas per mile. The relationship is much straighter, as we can see from a scatterplot of the regression residuals.

This is more the kind of boring residuals plot (no direction, no particular shape, no outliers, no bends) that we hope to see, so we have reason to think that the Straight Enough Condition is now satisfied. Now here's the payoff: What does the reciprocal model say about the Hummer? The regression line fit to *Fuel Consumption vs. Weight* predicts somewhere near 9.7 for a car weighing 6400 pounds. What does this mean? It means the car is predicted to use 9.7 gallons for every 100 miles, or in other words,

$$\frac{100 \ miles}{9.7 \ gallons} = 10.3 \ mpg.$$

That's a much more reasonable prediction and very close to the reported value of 11.0 miles per gallon (of course, *your* mileage may vary . . .).

Goals of Re-expression

We re-express data for several reasons. Each of these goals helps make the data more suitable for analysis by our methods.

GOAL 1

Make the distribution of a variable (as seen in its histogram, for example) more symmetric. It's easier to summarize the center of a symmetric distribution, and for nearly symmetric distributions, we can use the mean and standard deviation. If the distribution is unimodal, then the resulting distribution may be closer to the Normal model, allowing us to use the 68–95–99.7 Rule.

Here are a histogram, quite skewed, showing the *Assets* of 77 companies selected from the Forbes 500 list (in $100,000) and the more symmetric histogram after taking logs.

WHO	77 large companies
WHAT	Assets, sales, and market sector
UNITS	$100,000
HOW	Public records
WHEN	1986
WHY	By *Forbes* magazine in reporting on the Forbes 500 for that year

A S *Simulation:* **Re-expression in Action.** Slide the re-expression power and watch the histogram change.

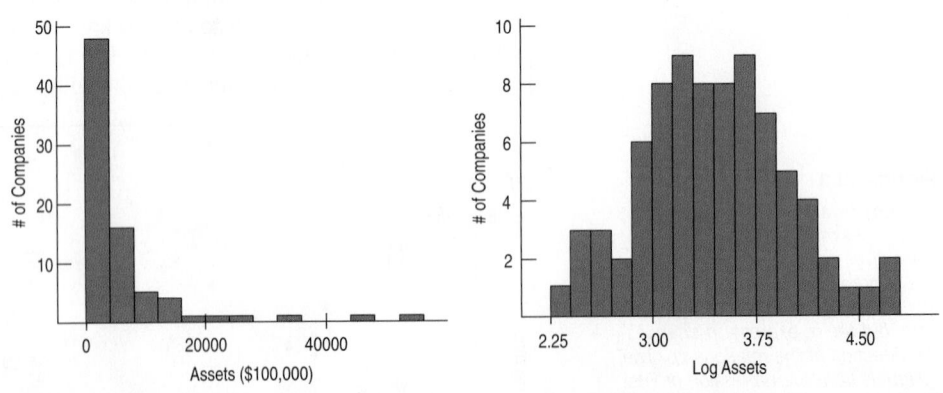

FIGURE 10.4

The distribution of the Assets *of large companies is skewed to the right. Data on wealth often look like this. Taking logs makes the distribution more nearly symmetric.*

GOAL 2

Make the spread of several groups (as seen in side-by-side boxplots) more alike, even if their centers differ. Groups that share a common spread are easier to compare. We'll see methods later in the book that can be applied only to groups with a common standard deviation. We saw an example of re-expression for comparing groups with boxplots in Chapter 5.

Here are the *Assets* of these companies by *Market Sector:*

FIGURE 10.5

Assets *of large companies by* **Market Sector.** *It's hard to compare centers or spreads, and there seem to be a number of high outliers.*

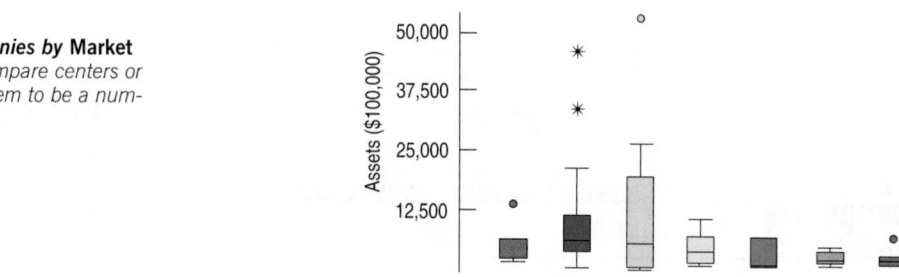

Taking logs makes the individual boxplots more symmetric and gives them spreads that are more nearly equal.

FIGURE 10.6

After re-expressing by logs, it's much easier to compare across market sectors. The boxplots are more nearly symmetric, most have similar spreads, and the companies that seemed to be outliers before are no longer extraordinary. Two new outliers have appeared in the finance sector. They are the only companies in that sector that are not banks. Perhaps they don't belong there.

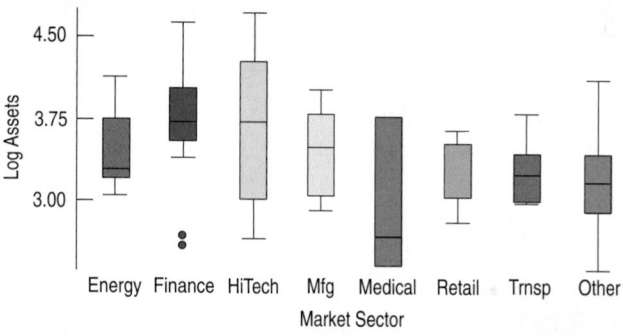

Doing this makes it easier to compare assets across market sectors. It can also reveal problems in the data. Some companies that looked like outliers on the high end turned out to be more typical. But two companies in the finance sector now stick out. Unlike the rest of the companies in that sector, they are not banks. They may have been placed in the wrong sector, but we couldn't see that in the original data.

GOAL 3

Make the form of a scatterplot more nearly linear. Linear scatterplots are easier to model. We saw an example of scatterplot straightening in Chapter 7. The greater value of re-expression to straighten a relationship is that we can fit a linear model once the relationship is straight.

Here are *Assets* of the companies plotted against the logarithm of *Sales*, clearly bent. Taking logs makes things much more linear.

FIGURE 10.7

Assets vs. *log* Sales *shows a positive association (bigger sales go with bigger assets) but a bent shape. Note also that the points go from tightly bunched at the left to widely scattered at the right; the plot "thickens." In the second plot, log* Assets *vs. log* Sales *shows a clean, positive, linear association. And the variability at each value of x is about the same.*

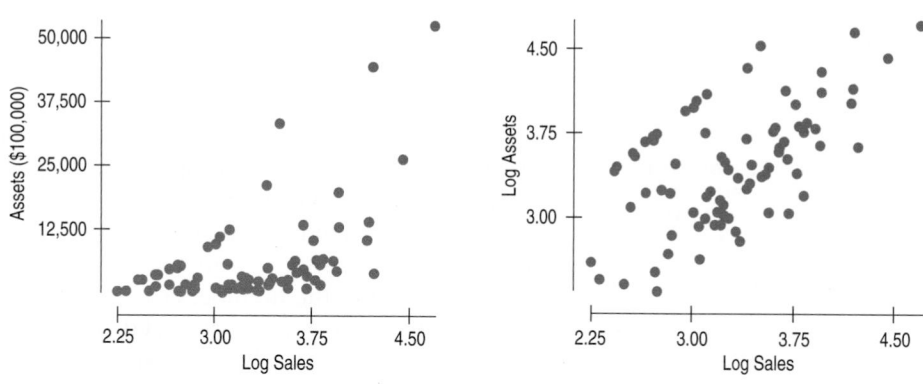

GOAL 4

Make the scatter in a scatterplot spread out evenly rather than thickening at one end. Having an even scatter is a condition of many methods of Statistics, as we'll see in later chapters. This goal is closely related to Goal 2, but it often comes along with Goal 3. Indeed, a glance back at the scatterplot (Figure 10.7) shows that the plot for *Assets* is much more spread out on the right than on the left, while the plot for log *Assets* has roughly the same variation in log *Assets* for any *x*-value.

FOR EXAMPLE | Recognizing when a re-expression can help

In Chapter 9, we saw the awesome ability of emperor penguins to slow their heart rates while diving. Here are three displays relating to the diving heart rates:

(The boxplots show the diving heart rates for each of the 9 penguins whose dives were tracked. The names are those given by the researchers; EP = emperor penguin.)

Question: What features of each of these displays suggest that a re-expression might be helpful?

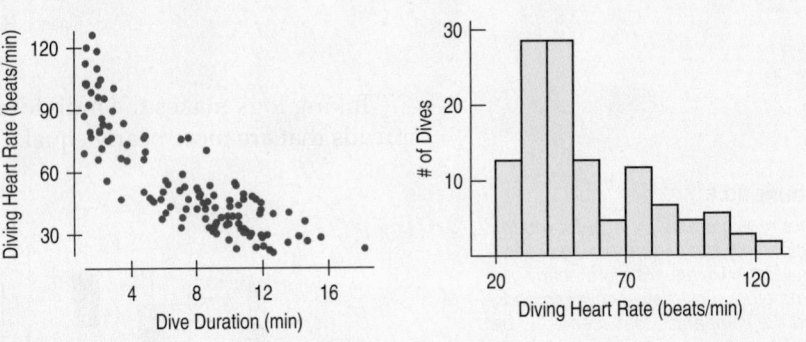

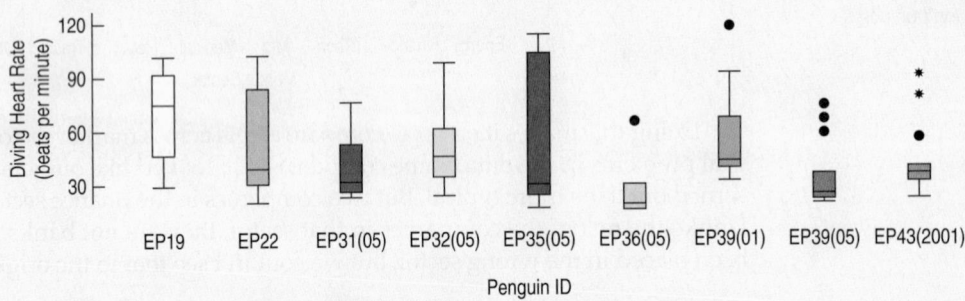

The scatterplot shows a curved relationship, concave upward, between the duration of the dives and penguins' heart rates. Re-expressing either variable may help to straighten the pattern.

The histogram of heart rates is skewed to the high end. Re-expression often helps to make skewed distributions more nearly symmetric.

The boxplots each show skewness to the high end as well. The medians are low in the boxes, and several show high outliers.

The Ladder of Powers

A S *Activity:* **Re-expression in Action** Here's the animated version of the Ladder of Powers. Slide the power and watch the change.

How can we pick a re-expression to use? Some kinds of data favor certain re-expressions. But even starting from a suggested one, it's always a good idea to look around a bit. Fortunately, the re-expressions line up in order, so it's easy to slide up and down to find the best one. The trick is to choose our re-expressions from a simple family that includes the most common ways to re-express data. More important, the members of the family line up in order, so that the farther you move away from the original data (the "1" position), the greater is the effect on the data. This fact lets you search systematically for a re-expression that works, stepping a bit farther from "1" or taking a step back toward "1" as you see the results.

Where to start? It turns out that certain kinds of data are more likely to be helped by particular re-expressions. Knowing that gives you a good place to start your search for a re-expression. We call this collection of re-expressions the **Ladder of Powers.**

Power	Name	Comment
2	The square of the data values, y^2.	Try this for unimodal distributions that are skewed to the left.
1	The raw data—no change at all. This is "home base." The farther you step from here up or down the ladder, the greater the effect.	Data that can take on both positive and negative values with no bounds are less likely to benefit from re-expression.
1/2	The square root of the data values, $\sqrt{y}$.	Counts often benefit from a square root re-expression. For counted data, start here.
"0"	Although mathematicians define the "0-th" power differently,[2] for us the place is held by the logarithm. You may feel uneasy about logarithms. Don't worry; the computer or calculator does the work.[3]	Measurements that cannot be negative, and especially values that grow by percentage increases such as salaries or populations, often benefit from a log re-expression. When in doubt, start here. If your data have zeros, try adding a small constant to all values before finding the logs.
−1/2	The (negative) reciprocal square root, $-1/\sqrt{y}$.	An uncommon re-expression, but sometimes useful. Changing the sign to take the *negative* of the reciprocal square root preserves the direction of relationships, making things a bit simpler.
−1	The (negative) reciprocal, $-1/y$.	Ratios of two quantities (miles per hour, for example) often benefit from a reciprocal. (You have about a 50–50 chance that the original ratio was taken in the "wrong" order for simple statistical analysis and would benefit from re-expression.) Often, the reciprocal will have simple units (hours per mile). Change the sign if you want to preserve the direction of relationships. If your data have zeros, try adding a small constant to all values before finding the reciprocal.

JUST CHECKING

1. You want to model the relationship between the number of birds counted at a nesting site and the temperature (in degrees Celsius). The scatterplot of counts vs. temperature shows an upwardly curving pattern, with more birds spotted at higher temperatures. What transformation (if any) of the bird counts might you start with?

2. You want to model the relationship between prices for various items in Paris and in Hong Kong. The scatterplot of Hong Kong prices vs. Parisian prices shows a generally straight pattern with a small amount of scatter. What transformation (if any) of the Hong Kong prices might you start with?

3. You want to model the population growth of the United States over the past 200 years. The scatterplot shows a strongly upwardly curved pattern. What transformation (if any) of the population might you start with?

[2] You may remember that for any nonzero number y, $y^0 = 1$. This is not a very exciting transformation for data; every data value would be the same. We use the logarithm in its place.

[3] Your calculator or software package probably gives you a choice between "base 10" logarithms and "natural (base e)" logarithms. Don't worry about that. It doesn't matter at all which you use; they have exactly the same effect on the data. If you want to choose, base 10 logarithms can be a bit easier to interpret.

Scientific laws often include simple re-expressions. For example, in Psychology, Fechner's Law states that sensation increases as the logarithm of stimulus intensity ($S = k \log R$).

The Ladder of Powers orders the effects that the re-expressions have on data. If you try, say, taking the square roots of all the values in a variable and it helps, but not enough, then move farther down the ladder to the logarithm or reciprocal root. Those re-expressions will have a similar, but even stronger, effect on your data. If you go too far, you can always back up. But don't forget—when you take a negative power, the *direction* of the relationship will change. That's OK. You can always change the sign of the response variable if you want to keep the same direction. With modern technology, finding a suitable re-expression is no harder than the push of a button.

FOR EXAMPLE — Trying a re-expression

Recap: We've seen curvature in the relationship between emperor penguins' diving heart rates and the duration of the dive. Let's start the process of finding a good re-expression. Heart rate is in beats per minute; maybe heart "speed" in minutes per beat would be a better choice. Here are the corresponding displays for this reciprocal re-expression (as we often do, we've changed the sign to preserve the order of the data values):

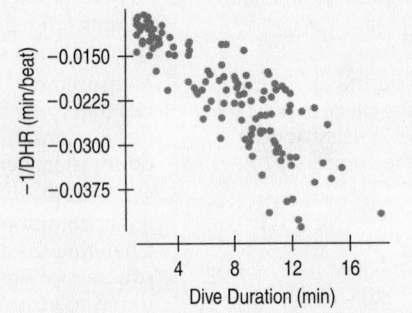 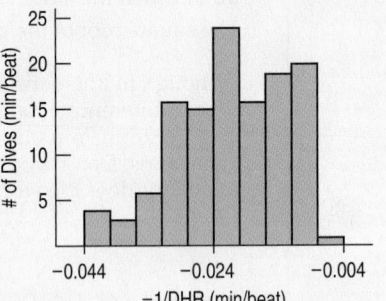

Question: Were the re-expressions successful?

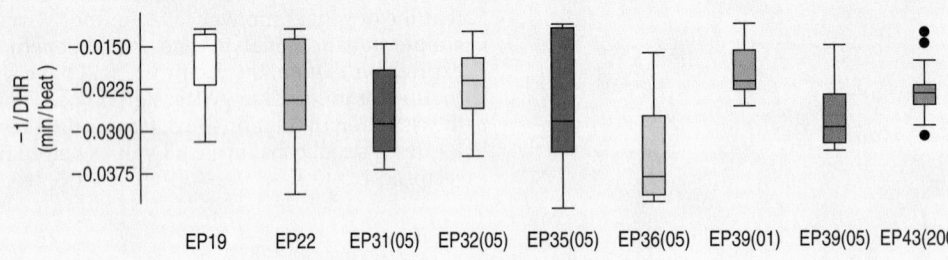

The scatterplot bends less than before, but now may be slightly concave downward. The histogram is now slightly skewed to the low end. Most of the boxplots have no outliers. These boxplots seem better than the ones for the raw heart rates.

Overall, it looks like I may have moved a bit "too far" on the ladder of powers. Halfway between "1" (the original data) and "−1" (the reciprocal) is "0", which represents the logarithm. I'd try that for comparison.

STEP-BY-STEP EXAMPLE — Re-expressing to Straighten a Scatterplot

Standard (monofilament) fishing line comes in a range of strengths, usually expressed as "test pounds." Five-pound test line, for example, can be expected to withstand a pull of up to five pounds without breaking. The convention in selling fishing line is that the price of a spool doesn't vary with strength. Instead, the length of line on the spool varies. Higher test pound line is thicker, though, so spools of fishing line hold about the same amount of material. Some spools hold line that is thinner and longer, some fatter and shorter. Let's look at the *Length* and *Strength* of spools of monofilament line manufactured by the same company and sold for the same price at one store.

Questions: How are the *Length* on the spool and the *Strength* related? And what re-expression will straighten the relationship?

Plan State the problem.

I want to fit a linear model for the length and strength of monofilament fishing line.

Variables Identify the variables and report the W's.

I have the *length* and "pound test" strength of monofilament fishing line sold by a single vendor at a particular store. Each case is a different strength of line, but all spools of line sell for the same price.

Let *Length* = length (in yards) of fishing line on the spool

Strength = the test strength (in pounds).

Plot Check that even if there is a curve, the overall pattern does not reach a minimum or maximum and then turn around and go back. An up-and-down curve can't be fixed by re-expression.

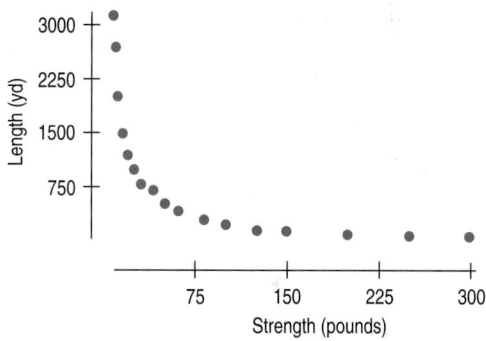

The plot shows a negative direction and an association that has little scatter but is not straight.

Mechanics Try a re-expression.

The lesson of the Ladder of Powers is that if we're moving in the right direction but have not had sufficient effect, we should go farther along the ladder. This example shows improvement, but is still not straight.

(Because *Length* is an amount of something and cannot be negative, we probably should have started with logs. This plot is here in part to illustrate how the Ladder of Powers works.)

Here's a plot of the square root of *Length* against *Strength*:

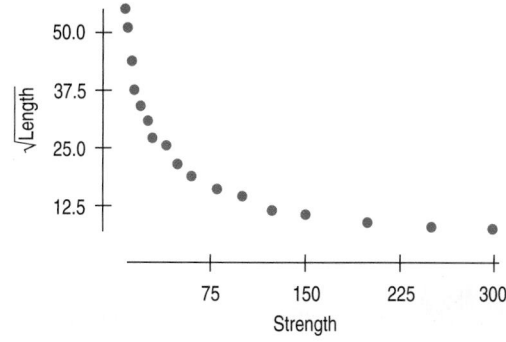

The plot is less bent, but still not straight.

Stepping from the 1/2 power to the "0" power, we try the logarithm of *Length* against *Strength*.

The scatterplot of the logarithm of *Length* against *Strength* is even less bent:

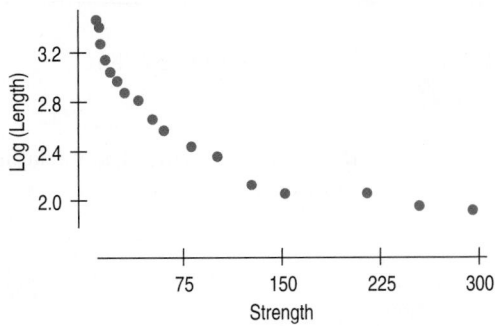

The straightness is improving, so we know we're moving in the right direction. But since the plot of the logarithms is not yet straight, we know we haven't gone far enough. To keep the direction consistent, change the sign and re-express to $-1/Length$.

This is much better, but still not straight, so I'll take another step to the "−1" power, or reciprocal.

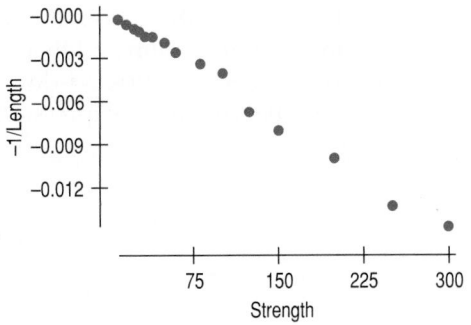

Maybe now I moved too far along the ladder.

We may have to choose between two adjacent re-expressions. For most data analyses, it really doesn't matter which we choose.

A half-step back is the −1/2 power: the reciprocal square root.

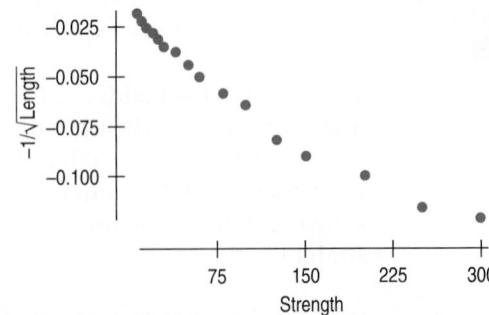

Conclusion Specify your choice of re-expression. If there's some natural interpretation (as for gallons per 100 miles), give that.

It's hard to choose between the last two alternatives. Either of the last two choices is good enough. I'll choose the −1/2 power.

Now that the re-expressed data satisfy the Straight Enough Condition, we can fit a linear model by least squares. We find that

$$\frac{-1}{\sqrt{\widehat{Length}}} = -0.023 - 0.000373\ Strength.$$

We can use this model to predict the length of a spool of, say, 35-pound test line:

$$\frac{-1}{\sqrt{\widehat{Length}}} = -0.023 - 0.000373 \times 35 = -0.036$$

We could leave the result in these units ($-1/\sqrt{\text{yards}}$). Sometimes the new units may be as meaningful as the original, but here we want to transform the predicted value back into yards. Fortunately, each of the re-expressions in the Ladder of Powers can be reversed.

To reverse the process, we first take the reciprocal: $\sqrt{\widehat{Length}} = -1/(-0.036) = 27.778$. Then squaring gets us back to the original units:

$$\widehat{Length} = 27.778^2 = 771.6\ yards.$$

This may be the most painful part of the re-expression. Getting back to the original units can sometimes be a little work. Nevertheless, it's worth the effort to always consider re-expression. Re-expressions extend the reach of all of your Statistics tools by helping more data to satisfy the conditions they require. Just think how much more useful this course just became!

FOR EXAMPLE   Comparing re-expressions

Recap: We've concluded that in trying to straighten the relationship between *Diving Heart Rate* and *Dive Duration* for emperor penguins, using the reciprocal re-expression goes a bit "too far" on the ladder of powers. Now we try the logarithm. Here are the resulting displays:

Questions: Comment on these displays. Now that we've looked at the original data (rung 1 on the Ladder), the reciprocal (rung -1), and the logarithm (rung 0), which re-expression of *Diving Heart Rate* would you choose?

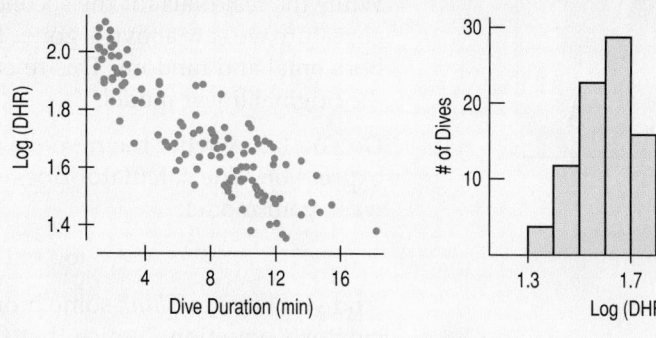

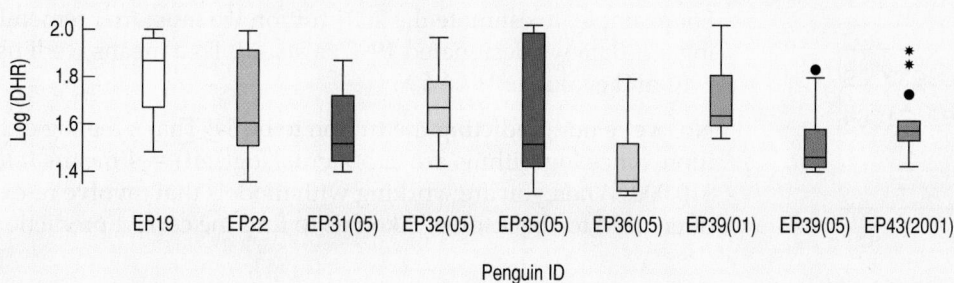

The scatterplot is now more linear and the histogram is symmetric. The boxplots are still a bit skewed to the high end, but less so than for the original Diving Heart Rate values. We don't expect real data to cooperate perfectly, and the logarithm seems like the best compromise re-expression, improving several different aspects of the data.

Re-expressing data to achieve linearity

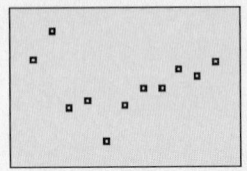

Let's revisit the Arizona State tuition data. Recall that back in Chapter 8 when we tried to fit a linear model to the yearly tuition costs, the residuals plot showed a distinct curve. Residuals are high (positive) at the left, low in the middle of the decade, and high again at the right.

This curved pattern indicates that data re-expression may be in order. If you have no clue what re-expression to try, the Ladder of Powers may help. We just used that approach in the fishing line example. Here, though, we can play a hunch. It is reasonable to suspect that tuition increases at a relatively consistent percentage year by year. This suggests that using the logarithm of tuition may help.

- Tell the calculator to find the logs of the tuitions, and store them as a new list. Remember that you must import the name **TUIT** from the **LIST NAMES** menu. The command is **log(ʟTUIT) STO L1**.
- Check the scatterplot for the re-expressed data by changing your **STATPLOT** specifications to **Xlist:YR** and **Ylist:L1**. (Don't forget to use **9:ZoomStat** to resize the window properly.)

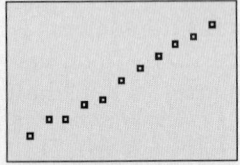

The new scatterplot looks quite linear, but it's really the residuals plot that will tell the story. Remember that the TI automatically finds and stores the residuals whenever you ask it to calculate a regression.

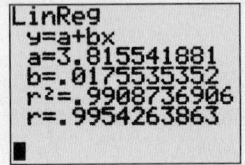

- Perform the regression for the logarithm of *tuition* vs. *year* with the command **LinReg(a+bx)ʟYR,L1,Y1**. That both creates the residuals and reports details about the model (storing the equation for later use).
- Now that the residuals are stored in **RESID**, set up a new scatterplot, this time specifying **Xlist:YR** and **Ylist:RESID**.

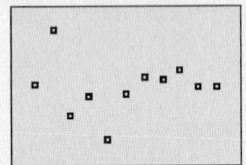

While the residuals for the second and fifth years are comparatively large, the curvature we saw above is gone. The pattern in these residuals seem essentially horizontal and random. This re-expressed model is probably more useful than the original linear model.

Do you know what the model's equation is? Remember, it involves a log re-expression. The calculator does not indicate that; be sure to *Think* when you write your model!

$$\log \widehat{tuit} = 3.816 + 0.018\,yr$$

And you have to *Think* some more when you make an estimate using the calculator's equation. Notice that this model does not actually predict tuition; rather, it predicts the *logarithm* of the tuition.

For example, to estimate the 2001 tuition we must first remember that in entering our data we designated 1990 as year 0. That means we'll use 11 for the year 2001 and evaluate **Y1(11)**.

No, we're not predicting the tuition to be $4! That's the log of the estimated tuition. Since logarithms are exponents, $\log(\widehat{tuit}) = 4$ means $\widehat{tuit} = 10^4$, or about $10,000. When you are working with models that involve re-expressions, you'll often need to "backsolve" like this to find the correct predictions.

Plan B: Attack of the Logarithms

The Ladder of Powers is often successful at finding an effective re-expression. Sometimes, though, the curvature is more stubborn, and we're not satisfied with the residual plots. What then?

When none of the data values is zero or negative, logarithms can be a helpful ally in the search for a useful model. Try taking the logs of both the *x*- and *y*-variables. Then re-express the data using some combination of *x* or log(*x*) vs. *y* or log(*y*). You may find that one of these works pretty well.

Model Name	*x*-axis	*y*-axis	Comment
Exponential	*x*	log(*y*)	This model is the "0" power in the ladder approach, useful for values that grow by percentage increases.
Logarithmic	log(*x*)	*y*	A wide range of *x*-values, or a scatterplot descending rapidly at the left but leveling off toward the right, may benefit from trying this model.
Power	log(*x*)	log(*y*)	The Goldilocks model: When one of the ladder's powers is too big and the next is too small, this one may be just right.

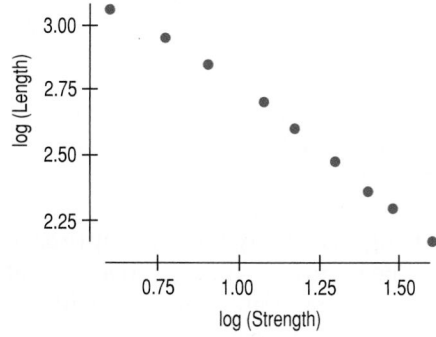

FIGURE 10.8

Plotting log (Length) *against log* (Strength) *gives a straighter shape.*

When we tried to model the relationship between the length of fishing line and its strength, we were torn between the "−1" power and the "−1/2" power. The first showed slight upward curvature, and the second downward. Maybe there's a better power between those values.

The scatterplot shows what happens when we graph the logarithm of *Length* against the logarithm of *Strength*. Technology reveals that the equation of our log–log model is

$$\widehat{\log(Length)} = 4.49 - 1.08\log(Strength).$$

It's interesting that the slope of this line (−1.08) is a power[4] we didn't try. After all, the ladder can't have every imaginable rung.

A warning, though! Don't expect to be able to straighten every curved scatterplot you find. It may be that there just isn't a very effective re-expression to be had. You'll certainly encounter situations when nothing seems to work the way you wish it would. Don't set your sights too high—you won't find a perfect model. Keep in mind: We seek a *useful* model, not perfection (or even "the best").

TI Tips   Using logarithmic re-expressions

In Chapter 7 we looked at data showing the relationship between the *f*/stop of a camera's lens and its shutter speed. Let's use the attack of the logarithms to model this situation.

Shutter speed:	1/1000	1/500	1/250	1/125	1/60	1/30	1/15	1/8
***f*/stop:**	2.8	4	5.6	8	11	16	22	32

• Enter these data into your calculator, shutter *speed* in L1 and *f*/stop in L2.
• Create the scatterplot with Xlist:L1 and Ylist:L2. See the curve?

---

[4] For logarithms, $-1.08\log(Strength) = \log(Strength^{-1.08})$.

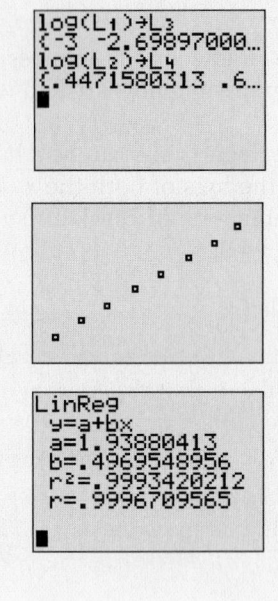

- Find the logarithms of each variable's values. Keep track of where you store everything so you don't get confused! We put log(*speed*) in **L3** and log(*f/stop*) in **L4**.
- Make three scatterplots:
  - *f/stop* vs. log(*speed*) using **Xlist:L3** and **Ylist:L2**
  - log(*f/stop*) vs. *speed* using **Xlist:L1** and **Ylist:L4**
  - log(*f/stop*) vs. log(*speed*) using **Xlist:L3** and **Ylist:L4**
- Pick your favorite. We liked log(*f/stop*) vs. log(*speed*) a lot! It appears to be very straight. (Don't be misled—this is a situation governed by the laws of Physics. Real data are not so cooperative. Don't expect to achieve this level of perfection often!)
- Remember that before you check the residuals plot, you first have to calculate the regression. In this situation all the errors in the residuals are just round-off errors in the original *f/stops*.
- Use your regression to write the equation of the model. Remember: The calculator does not know there were logarithms involved. You have to Think about that to be sure you write your model correctly.[5]

$$\log(\widehat{f/stop}) = 1.94 + 0.497\log(speed)$$

Why Not Just Use a Curve?

When a clearly curved pattern shows up in the scatterplot, why not just fit a curve to the data? We saw earlier that the association between the *Weight* of a car and its *Fuel Efficiency* was not a straight line. Instead of trying to find a way to straighten the plot, why not find a curve that seems to describe the pattern well?

We can find "curves of best fit" using essentially the same approach that led us to linear models. You won't be surprised, though, to learn that the mathematics and the calculations are considerably more difficult for curved models. Many calculators and computer packages do have the ability to fit curves to data, but this approach has many drawbacks.

Straight lines are easy to understand. We know how to think about the slope and the *y*-intercept, for example. We often want some of the other benefits mentioned earlier, such as making the spread around the model more nearly the same everywhere. In later chapters you will learn more advanced statistical methods for analyzing linear associations.

We give all of that up when we fit a model that is not linear. For many reasons, then, it is usually better to re-express the data to straighten the plot.

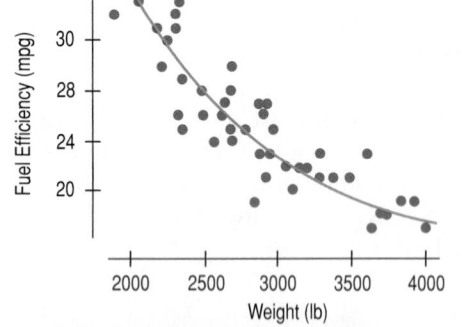

TI Tips | Some shortcuts to avoid

Your calculator offers many regression options in the **STAT CALC** menu. There are three that automate fitting simple re-expressions of *y* or *x*:

- **9:LnReg**—fits a logarithmic model ($\hat{y} = a + b\ln x$)

---

[5] See the slope, 0.497? Just about 0.5. That's because the actual relationship involves the square root of shutter speeds. Technically the *f/stop* listed as 2.8 should be $2\sqrt{2} \approx 2.8284$. Rounding off to 2.8 makes sense for photographers, but it's what led to the minor errors you saw in the residuals plot.

- `0:ExpReg`—fits an exponential model ($\hat{y} = ab^x$)
- `A:PwrReg`—fits a power model ($\hat{y} = ax^b$)

In addition, the calculator offers two other functions:

- `5:QuadReg`—fits a quadratic model ($\hat{y} = ax^2 + bx + c$)
- `6:CubicReg`—fits a cubic model ($\hat{y} = ax^3 + bx^2 + cx + d$)

These two models have a form we haven't seen, with several x-terms. Because x, x^2, and x^3 are likely to be highly correlated with each other, the quadratic and cubic models are almost sure to be unreliable to fit, difficult to understand, and dangerous to use for predictions even slightly outside the range of the data. We recommend that you be very wary of models of this type.

Let's try out one of the calculator shortcuts; we'll use the Arizona State tuition data. (For the last time, we promise!) This time, instead of re-expressing *tuition* to straighten the scatterplot, we'll have the calculator do more of the work.

Which model should you use? You could always just play hit-and-miss, but knowing something about the data can save a lot of time. If tuition increases by a consistent percentage each year, then the growth is exponential.

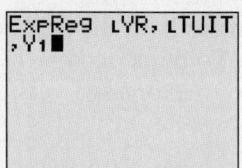

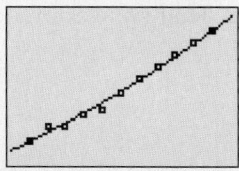

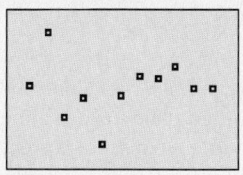

- Choose the exponential model, and specify your variables by importing `YR` and `TUIT` from the list names menu. And, because you'll want to graph the curve later, save its equation by adding `Y1` (from `VARS`, `Y-VARS`, `Function`) to create the command `ExpReg ᴌYR, ᴌTUIT, Y1`.
- Set up the scatterplot. `ZoomStat` should show you the curve too.
- Graph the residuals plot.

This all looks very good. R^2 is high, the curve appears to fit the points quite well, and the residuals plot is acceptably random.

The equation of the model is $\widehat{tuit} = 6539.46(1.041^{year})$.

Notice that this is the same residuals plot we saw when we re-expressed the data and fit a line to the logarithm of *tuition*. That's because what the calculator just did is mathematically the very same thing. This new equation may look different, but it is equivalent to our earlier model $\log \widehat{tuit} = 3.816 + 0.018\ year$.

Not easy to see that, is it? Here's how it works:

Initially we used a logarithmic re-expression to create a linear model:

$$\log \hat{y} = a + bx$$

Rewrite that equation in exponential form:

$$\hat{y} = 10^{a+bx}$$

Simplify, using the laws of exponents:

$$\hat{y} = 10^a(10^b)^x$$

Let $10^a = a$ and $10^b = b$ (different a and b!)

$$\hat{y} = ab^x$$

See? Your linear model created by logarithmic re-expression is the same as the calculator model created by `ExpReg`.

Three of the special TI functions correspond to a simple regression model involving re-expression. The calculator presents the results in an equation of a different form, but it doesn't actually fit that equation. Instead it is just doing the re-expression for you automatically.

Here are the equivalent models for the two approaches.

Type of Model	Re-expression Equation	Calculator's Curve	
		Command	Equation
Logarithmic	$\hat{y} = a + b\log x$	LnReg	$\hat{y} = a + b\ln x$
Exponential	$\log \hat{y} = a + bx$	ExpReg	$\hat{y} = ab^x$
Power	$\log \hat{y} = a + b\log x$	PwrReg	$\hat{y} = ax^b$

Be careful. It may look like the calculator is fitting these equations to the data by minimizing the sum of squared residuals, but it isn't really doing that. It handles the residuals differently, and the difference matters. If you use a statistics program to fit an "exponential model," it will probably fit the exponential form of the equation and give you a different answer. So think of these TI functions as just shortcuts for fitting linear regressions to re-expressed versions of your data.

You've seen two ways to handle bent relationships:

- straighten the data, then fit a line, or
- use the calculator shortcut to create a curve.

Note that the calculator does not have a shortcut for every model you might want to use—models involving square roots or reciprocals, for instance. And remember: The calculator may be quick, but there are real advantages to finding *linear* models by actually re-expressing the data. That's the approach we strongly recommend you use.

WHAT CAN GO WRONG?

Occam's Razor

If you think that simpler explanations and simpler models are more likely to give a true picture of the way things work, then you should look for opportunities to re-express your data and simplify your analyses.

The general principle that simpler explanations are likely to be the better ones is known as Occam's Razor, after the English philosopher and theologian William of Occam (1284–1347).

▶ **Don't expect your model to be perfect.** In Chapter 6 we quoted statistician George Box: "All models are wrong, but some are useful." Be aware that the real world is a messy place and data can be uncooperative. Don't expect to find one elusive re-expression that magically irons out every kink in your scatterplot and produces perfect residuals. You aren't looking for the Right Model, because that mythical creature doesn't exist. Find a useful model and use it wisely.

▶ **Don't stray too far from the ladder.** It's wise not to stray too far from the powers that we suggest in the Ladder of Powers. Taking the y-values to an extremely high power may artificially inflate R^2, but it won't give a useful or meaningful model, so it doesn't really simplify anything. It's better to stick to powers between 2 and −2. Even in that range, you should prefer the simpler powers in the ladder to those in the cracks. A square root is easier to understand than the 0.413 power. That simplicity may compensate for a slightly less straight relationship.

▶ **Don't choose a model based on R^2 alone.** You've tried re-expressing your data to straighten a curved relationship and found a model with a high R^2. Beware: That doesn't mean the pattern is straight now. On the next page is a plot of a relationship with an R^2 of 98.3%.

The R^2 is about as high as we could ask for, but if you look closely, you'll see that there's a consistent bend. Plotting the residuals from the least squares line makes the bend much easier to see.

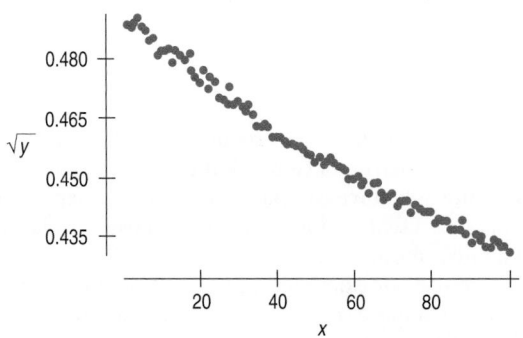

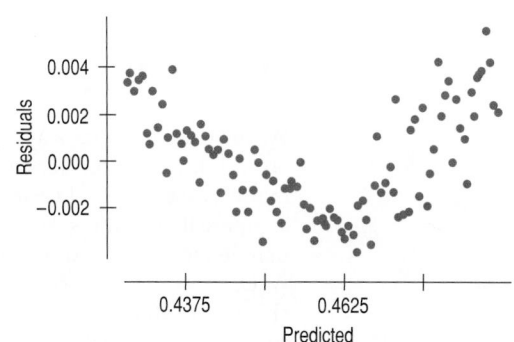

Remember the basic rule of data analysis: *Make a picture*. Before you fit a line, always look at the pattern in the scatterplot. After you fit the line, check for linearity again by plotting the residuals.

▶ **Beware of multiple modes.** Re-expression can often make a skewed unimodal histogram more nearly symmetric, but it cannot pull separate modes together. A suitable re-expression may, however, make the separation of the modes clearer, simplifying their interpretation and making it easier to separate them to analyze individually.

▶ **Watch out for scatterplots that turn around.** Re-expression can straighten many bent relationships but not those that go up and then down or down and then up. You should refuse to analyze such data with methods that require a linear form.

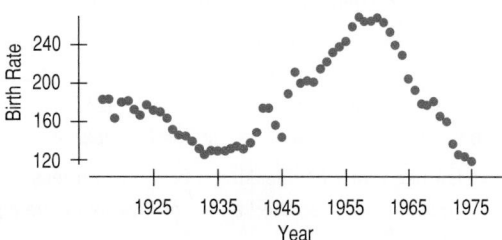

FIGURE 10.9

The shape of the scatterplot of Birth Rates *(births per 100,000 women) in the United States shows an oscillation that cannot be straightened by re-expressing the data.*

▶ **Watch out for negative data values.** It's impossible to re-express negative values by any power that is not a whole number on the Ladder of Powers or to re-express values that are zero for negative powers. Most statistics programs will just mark the result of trying to re-express such values "missing" if they can't be re-expressed. But that might mean that when you try a re-expression, you inadvertently lose a bunch of data values. The effect of that loss may be surprising and may substantially change your analysis. Because you are likely to be working with a computer package or calculator, take special care that you do not lose otherwise good data values when you choose a re-expression.

One possible cure for zeros and small negative values is to add a constant ($\frac{1}{2}$ and $\frac{1}{6}$ are often used) to bring all the data values above zero.

▶ **Watch for data far from 1.** Data values that are all very far from 1 may not be much affected by re-expression unless the range is very large. Re-expressing numbers between 1 and 100 will have a much greater effect than re-expressing numbers between 100,001 and 100,100. When all your data values are large (for example, working with years), consider subtracting a constant to bring them back near 1. (For example, consider "years since 1950" as an alternative variable for re-expression. Unless your data start at 1950, then avoid creating a zero by using "years since 1949.")

CONNECTIONS

We have seen several ways to model or summarize data. Each requires that the data have a particular simple structure. We seek symmetry for summaries of center and spread and to use a Normal model. We seek equal variation across groups when we compare groups with boxplots or want to compare their centers. We seek linear shape in a scatterplot so that we can use correlation to summarize the scatter and regression to fit a linear model.

Data do often satisfy the requirements to use Statistics methods. But often they do not. Our choice is to stop with just displays, to use much more complex methods, or to re-express the data so that we can use the simpler methods we have developed.

In this fundamental sense, this chapter connects to everything we have done thus far and to all of the methods we will introduce throughout the rest of the book. Re-expression greatly extends the reach and applicability of all of these methods.

WHAT HAVE WE LEARNED?

We've learned that when the conditions for regression are not met, a simple re-expression of the data may help. There are several reasons to consider a re-expression:

▸ To make the distribution of a variable more symmetric (as we saw in Chapter 5)
▸ To make the spread across different groups more similar
▸ To make the form of a scatterplot straighter
▸ To make the scatter around the line in a scatterplot more consistent

We've learned that when seeking a useful re-expression, taking logs is often a good, simple starting point. To search further, the Ladder of Powers or the log–log approach can help us find a good re-expression.

We've come to understand that our models won't be perfect, but that re-expression can lead us to a useful model.

Terms

Re-expression   224. We re-express data by taking the logarithm, the square root, the reciprocal, or some other mathematical operation on all values of a variable.

Ladder of Powers   226. The Ladder of Powers places in order the effects that many re-expressions have on the data.

Skills

▸ Recognize when a well-chosen re-expression may help you improve and simplify your analysis.

▸ Understand the value of re-expressing data to improve symmetry, to make the scatter around a line more constant, or to make a scatterplot more linear.

▸ Recognize when the pattern of the data indicates that no re-expression can improve the structure of the data.

▸ Know how to re-express data with powers and how to find an effective re-expression for your data using your statistics software or calculator.

▸ Be able to reverse any of the common re-expressions to put a predicted value or residual back into the original units.

▸ Be able to describe a summary or display of a re-expressed variable, making clear how it was re-expressed and giving its re-expressed units.

▸ Be able to describe a regression model fit to re-expressed data in terms of the re-expressed variables.

RE-EXPRESSION ON THE COMPUTER

Computers and calculators make it easy to re-express data. Most statistics packages offer a way to re-express and compute with variables. Some packages permit you to specify the power of a re-expression with a slider or other moveable control, possibly while watching the consequences of the re-expression on a plot or analysis. This, of course, is a very effective way to find a good re-expression.

EXERCISES

1. Residuals. Suppose you have fit a linear model to some data and now take a look at the residuals. For each of the following possible residuals plots, tell whether you would try a re-expression and, if so, why.

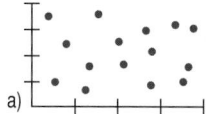 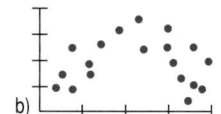 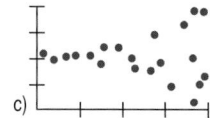

a) b) c)

2. Residuals. Suppose you have fit a linear model to some data and now take a look at the residuals. For each of the following possible residuals plots, tell whether you would try a re-expression and, if so, why.

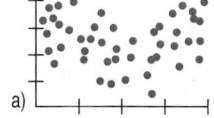 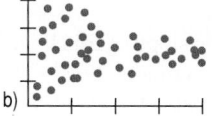 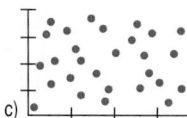

a) b) c)

3. Airline passengers revisited. In Chapter 9, Exercise 9, we created a linear model describing the trend in the number of passengers departing from the Oakland (CA) airport each month since the start of 1997. Here's the residual plot, but with lines added to show the order of the values in time:

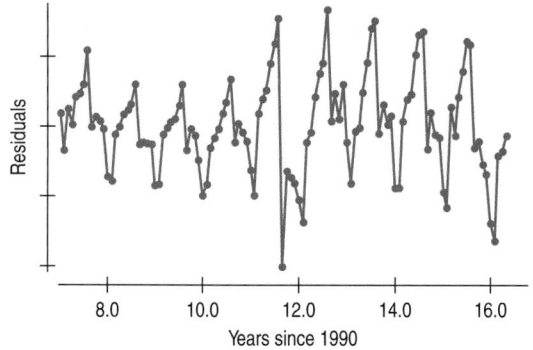

a) Can you account for the pattern shown here?
b) Would a re-expression help us deal with this pattern? Explain.

4. Hopkins winds, revisited. In Chapter 5, we examined the wind speeds in the Hopkins forest over the course of a year. Here's the scatterplot we saw then:

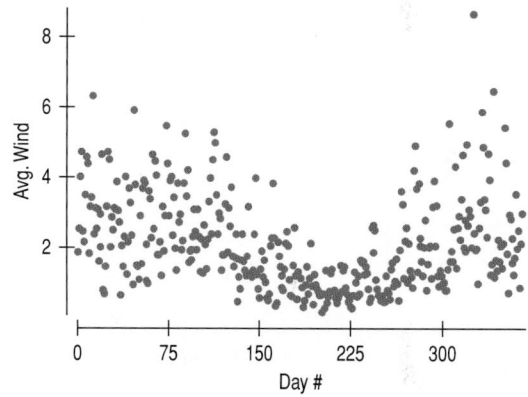

a) Describe the pattern you see here.
b) Should we try re-expressing either variable to make this plot straighter? Explain.

5. Models. For each of the models listed below, predict y when $x = 2$.
a) $\ln \hat{y} = 1.2 + 0.8x$
b) $\sqrt{\hat{y}} = 1.2 + 0.8x$
c) $\dfrac{1}{\hat{y}} = 1.2 + 0.8x$
d) $\hat{y} = 1.2 + 0.8 \ln x$
e) $\log \hat{y} = 1.2 + 0.8 \log x$

6. More models. For each of the models listed below, predict y when $x = 2$.
a) $\hat{y} = 1.2 + 0.8 \log x$
b) $\log \hat{y} = 1.2 + 0.8x$
c) $\ln \hat{y} = 1.2 + 0.8 \ln x$
d) $\hat{y}^2 = 1.2 + 0.8x$
e) $\dfrac{1}{\sqrt{\hat{y}}} = 1.2 + 0.8x$

7. Gas mileage. As the example in the chapter indicates, one of the important factors determining a car's *Fuel Efficiency* is its *Weight*. Let's examine this relationship again, for 11 cars.
a) Describe the association between these variables shown in the scatterplot on the next page.

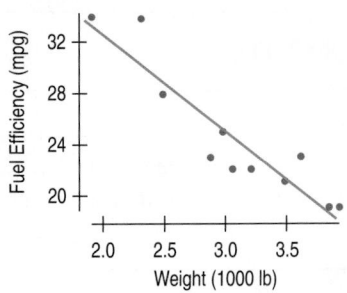

b) Here is the regression analysis for the linear model. What does the slope of the line say about this relationship?

Dependent variable is: Fuel Efficiency
R-squared = 85.9%

Variable	Coefficient
Intercept	47.9636
Weight	−7.65184

c) Do you think this linear model is appropriate? Use the residuals plot to explain your decision.

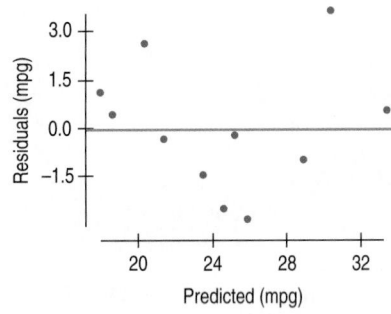

T 8. **Crowdedness.** In a *Chance* magazine article (Summer 2005), Danielle Vasilescu and Howard Wainer used data from the United Nations Center for Human Settlements to investigate aspects of living conditions for several countries. Among the variables they looked at were the country's per capita gross domestic product (*GDP*, in $) and *Crowdedness*, defined as the average number of persons per room living in homes there. This scatterplot displays these data for 56 countries:

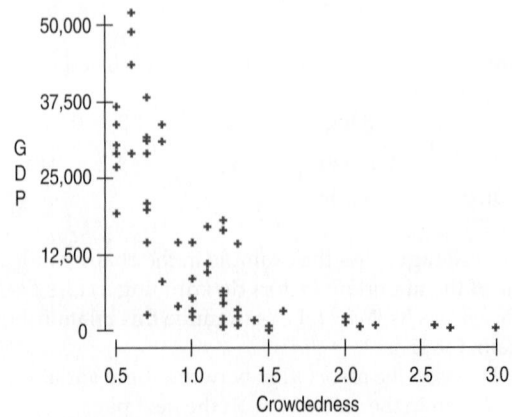

a) Explain why you should re-express these data before trying to fit a model.

b) What re-expression of *GDP* would you try as a starting point?

9. **Gas mileage revisited.** Let's try the re-expressed variable *Fuel Consumption* (gal/100 mi) to examine the fuel efficiency of the 11 cars in Exercise 7. Here are the revised regression analysis and residuals plot:

Dependent variable is: Fuel Consumption
R-squared = 89.2%

Variable	Coefficient
Intercept	0.624932
Weight	1.17791

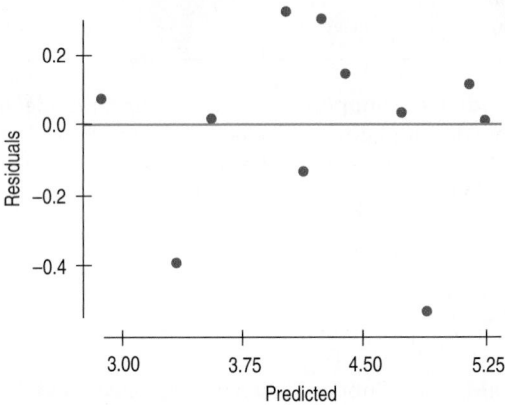

a) Explain why this model appears to be better than the linear model.

b) Using the regression analysis above, write an equation of this model.

c) Interpret the slope of this line.

d) Based on this model, how many miles per gallon would you expect a 3500-pound car to get?

10. **Crowdedness again.** In Exercise 8 we looked at United Nations data about a country's *GDP* and the average number of people per room (*Crowdedness*) in housing there. For a re-expression, a student tried the reciprocal −10000/*GDP*, representing the number of people per $10,000 of gross domestic product. Here are the results, plotted against *Crowdedness*:

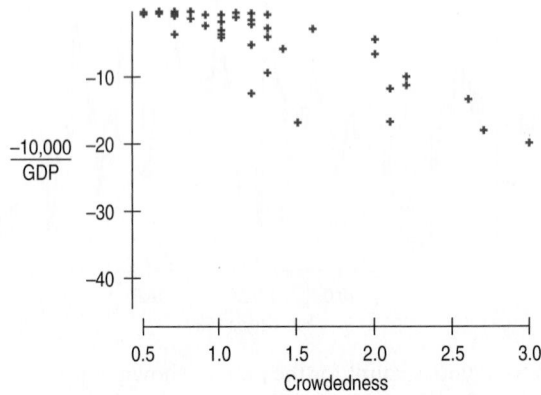

a) Is this a useful re-expression? Explain.

b) What re-expression would you suggest this student try next?

11. GDP. The scatterplot shows the gross domestic product (GDP) of the United States in billions of dollars plotted against years since 1950.

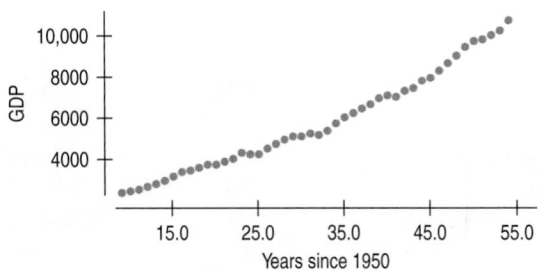

A linear model fit to the relationship looks like this:

Dependent variable is: GDP
R-squared = 97.2% s = 406.6

Variable	Coefficient
Intercept	240.171
Year−1950	177.689

a) Does the value 97.2% suggest that this is a good model? Explain.
b) Here's a scatterplot of the residuals. Now do you think this is a good model for these data? Explain?

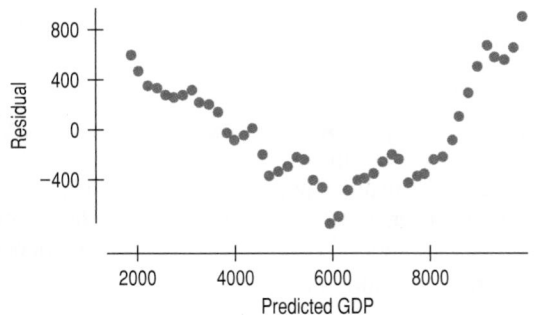

T 12. Treasury Bills. The 3-month Treasury bill interest rate is watched by investors and economists. Here's a scatterplot of the 3-month Treasury bill rate since 1950:

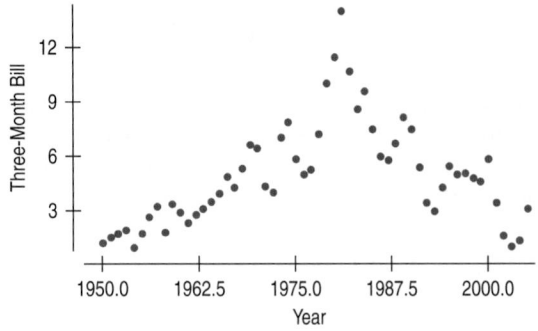

Clearly, the relationship is not linear. Can it be made nearly linear with a re-expression? If so, which one would you suggest? If not, why not?

13. Better GDP model? Consider again the post-1950 trend in U.S. GDP we examined in Exercise 11. Here are a regression and residual plot when we use the log of GDP in the model. Is this a better model for GDP? Explain.

Dependent variable is: LogGDP
R-squared = 99.4% s = 0.0150

Variable	Coefficient
Intercept	3.29092
Year−1950	0.013881

T 14. Pressure. Scientist Robert Boyle examined the relationship between the volume in which a gas is contained and the pressure in its container. He used a cylindrical container with a moveable top that could be raised or lowered to change the volume. He measured the *Height* in inches by counting equally spaced marks on the cylinder, and measured the *Pressure* in inches of mercury (as in a barometer). Some of his data are listed in the table. Create an appropriate model.

Height	48	44	40	36	32	28
Pressure	29.1	31.9	35.3	39.3	44.2	50.3
Height	24	20	18	16	14	12
Pressure	58.8	70.7	77.9	87.9	100.4	117.6

T 15. Brakes. The table below shows stopping distances in feet for a car tested 3 times at each of 5 speeds. We hope to create a model that predicts *Stopping Distance* from the *Speed* of the car.

Speed (mph)	Stopping Distances (ft)
20	64, 62, 59
30	114, 118, 105
40	153, 171, 165
50	231, 203, 238
60	317, 321, 276

a) Explain why a linear model is not appropriate.
b) Re-express the data to straighten the scatterplot.
c) Create an appropriate model.
d) Estimate the stopping distance for a car traveling 55 mph.
e) Estimate the stopping distance for a car traveling 70 mph.
f) How much confidence do you place in these predictions? Why?

T 16. Pendulum. A student experimenting with a pendulum counted the number of full swings the pendulum made in 20 seconds for various lengths of string. Her data are shown on the next page.

Length (in.)	6.5	9	11.5	14.5	18	21	24	27	30	37.5
Number of Swings	22	20	17	16	14	13	13	12	11	10

a) Explain why a linear model is not appropriate for using the *Length* of a pendulum to predict the *Number of Swings* in 20 seconds.
b) Re-express the data to straighten the scatterplot.
c) Create an appropriate model.
d) Estimate the number of swings for a pendulum with a 4-inch string.
e) Estimate the number of swings for a pendulum with a 48-inch string.
f) How much confidence do you place in these predictions? Why?

17. Baseball salaries 2005. Ballplayers have been signing ever larger contracts. The highest salaries (in millions of dollars per season) for some notable players are given in the following table.

Player	Year	Salary (million $)
Nolan Ryan	1980	1.0
George Foster	1982	2.0
Kirby Puckett	1990	3.0
Jose Canseco	1990	4.7
Roger Clemens	1991	5.3
Ken Griffey, Jr.	1996	8.5
Albert Belle	1997	11.0
Pedro Martinez	1998	12.5
Mike Piazza	1999	12.5
Mo Vaughn	1999	13.3
Kevin Brown	1999	15.0
Carlos Delgado	2001	17.0
Alex Rodriguez	2001	22.0
Manny Ramirez	2004	22.5
Alex Rodriguez	2005	26.0

a) Examine a scatterplot of the data. Does it look straight?
b) Find the regression of *Salary* vs. *Year* and plot the residuals. Do they look straight?
c) Re-express the data, if necessary, to straighten the relationship.
d) What model would you report for the trend in salaries?

18. Planet distances and years 2006. At a meeting of the International Astronomical Union (IAU) in Prague in 2006, Pluto was determined not to be a planet, but rather the largest member of the Kuiper belt of icy objects. Let's examine some facts. Here is a table of the 9 sun-orbiting objects formerly known as planets:

Planet	Position Number	Distance from Sun (million miles)	Length of Year (Earth years)
Mercury	1	36	0.24
Venus	2	67	0.61
Earth	3	93	1.00
Mars	4	142	1.88
Jupiter	5	484	11.86
Saturn	6	887	29.46
Uranus	7	1784	84.07
Neptune	8	2796	164.82
Pluto	9	3707	247.68

a) Plot the *Length* of the year against the *Distance* from the sun. Describe the shape of your plot.
b) Re-express one or both variables to straighten the plot. Use the re-expressed data to create a model describing the length of a planet's year based on its distance from the sun.
c) Comment on how well your model fits the data.

19. Planet distances and order 2006. Let's look again at the pattern in the locations of the planets in our solar system seen in the table in Exercise 18.
a) Re-express the distances to create a model for the *Distance* from the sun based on the planet's *Position*.
b) Based on this model, would you agree with the International Astronomical Union that Pluto is not a planet? Explain.

20. Planets 2006, part 3. The asteroid belt between Mars and Jupiter may be the remnants of a failed planet. If so, then Jupiter is really in position 6, Saturn is in 7, and so on. Repeat Exercise 19, using this revised method of numbering the positions. Which method seems to work better?

21. Eris: Planets 2006, part 4. In July 2005, astronomers Mike Brown, Chad Trujillo, and David Rabinowitz announced the discovery of a sun-orbiting object, since named Eris,[6] that is 5% larger than Pluto. Eris orbits the sun once every 560 earth years at an average distance of about 6300 million miles from the sun. Based on its *Position*, how does Eris's *Distance* from the sun (re-expressed to logs) compare with the prediction made by your model of Exercise 19?

22. Models and laws: Planets 2006 part 5. The model you found in Exercise 18 is a relationship noted in the 17th century by Kepler as his Third Law of Planetary Motion. It was subsequently explained as a consequence of Newton's Law of Gravitation. The models

---

[6] Eris is the Greek goddess of warfare and strife who caused a quarrel among the other goddesses that led to the Trojan war. In the astronomical world, Eris stirred up trouble when the question of its proper designation led to the raucous meeting of the IAU in Prague where IAU members voted to demote Pluto and Eris to dwarf-planet status—http://www.gps.caltech.edu/~mbrown/planetlila/#paper.

for Exercises 19–21 relate to what is sometimes called the Titius-Bode "law," a pattern noticed in the 18th century but lacking any scientific explanation.

Compare how well the re-expressed data are described by their respective linear models. What aspect of the model of Exercise 18 suggests that we have found a physical law? In the future, we may learn enough about a planetary system around another star to tell whether the Titius-Bode pattern applies there. If you discovered that another planetary system followed the same pattern, how would it change your opinion about whether this is a real natural "law"? What would you think if the next system we find does not follow this pattern?

23. Logs (not logarithms). The value of a log is based on the number of board feet of lumber the log may contain. (A board foot is the equivalent of a piece of wood 1 inch thick, 12 inches wide, and 1 foot long. For example, a 2" × 4" piece that is 12 feet long contains 8 board feet.) To estimate the amount of lumber in a log, buyers measure the diameter inside the bark at the smaller end. Then they look in a table based on the Doyle Log Scale. The table below shows the estimates for logs 16 feet long.

Diameter of Log	8"	12"	16"	20"	24"	28"
Board Feet	16	64	144	256	400	576

a) What model does this scale use?
b) How much lumber would you estimate that a log 10 inches in diameter contains?
c) What does this model suggest about logs 36 inches in diameter?

24. Weightlifting 2004. Listed below are the gold medal-winning men's weight-lifting performances at the 2004 Olympics.

Weight Class (kg)	Winner (country)	Weight Lifted (kg)
56	Halil Mutlu (Turkey)	295.0
62	Zhiyong Shi (China)	325.0
69	Guozheng Zhang (China)	347.5
77	Taner Sagir (Turkey)	375.0
85	George Asanidze (Georgia)	382.5
94	Milen Dobrev (Bulgaria)	407.5
105	Dmitry Berestov (Russia)	425.0

a) Create a linear model for the *Weight Lifted* in each *Weight Class.*
b) Check the residuals plot. Is your linear model appropriate?
c) Create a better model.
d) Explain why you think your new model is better.
e) Based on your model, which of the medalists turned in the most surprising performance? Explain.

25. Life expectancy. The data in the next column list the *Life Expectancy* for white males in the United States every decade during the last century (1 = 1900 to 1910, 2 = 1911

to 1920, etc.). Create a model to predict future increases in life expectancy. (National Vital Statistics Report)

Decade	1	2	3	4	5	6	7	8	9	10
Life exp.	48.6	54.4	59.7	62.1	66.5	67.4	68.0	70.7	72.7	74.9

26. Lifting more weight 2004. In Exercise 24 you examined the winning weight-lifting performances for the 2004 Olympics. One of the competitors turned in a performance that appears not to fit the model you created.
a) Consider that competitor to be an outlier. Eliminate that data point and re-create your model.
b) Using this revised model, how much would you have expected the outlier competitor to lift?
c) Explain the meaning of the residual from your new model for that competitor.

27. Slower is cheaper? Researchers studying how a car's *Fuel Efficiency* varies with its *Speed* drove a compact car 200 miles at various speeds on a test track. Their data are shown in the table.

Speed (mph)	35	40	45	50	55	60	65	70	75
Fuel Eff. (mpg)	25.9	27.7	28.5	29.5	29.2	27.4	26.4	24.2	22.8

Create a linear model for this relationship and report any concerns you may have about the model.

28. Orange production. The table below shows that as the number of oranges on a tree increases, the fruit tends to get smaller. Create a model for this relationship, and express any concerns you may have.

Number of Oranges/Tree	Average Weight/Fruit (lb)
50	0.60
100	0.58
150	0.56
200	0.55
250	0.53
300	0.52
350	0.50
400	0.49
450	0.48
500	0.46
600	0.44
700	0.42
800	0.40
900	0.38

29. Years to live 2003. Insurance companies and other organizations use actuarial tables to estimate the remaining lifespans of their customers. On the next page are the estimated additional years of life for black males in the United States, according to a 2003 National Vital Statistics Report. (www.cdc.gov/nchs/deaths.htm)

Age	10	20	30	40	50	60	70	80	90	100
Years Left	60.3	50.7	41.8	32.9	24.8	17.9	12.1	7.9	5.0	3.0

a) Find a re-expression to create an appropriate model.
b) Predict the lifespan of an 18-year-old black man.
c) Are you satisfied that your model has accounted for the relationship between *Years Left* and *Age?* Explain.

 30. **Tree growth.** A 1996 study examined the growth of grapefruit trees in Texas, determining the average trunk *Diameter* (in inches) for trees of varying *Ages:*

Age (yr)	2	4	6	8	10	12	14	16	18	20
Diameter (in.)	2.1	3.9	5.2	6.2	6.9	7.6	8.3	9.1	10.0	11.4

a) Fit a linear model to these data. What concerns do you have about the model?
b) If data had been given for individual trees instead of averages, would you expect the fit to be stronger, less strong, or about the same? Explain.

JUST CHECKING
Answers

1. Counts are often best transformed by using the square root.
2. None. The relationship is already straight.
3. Even though, technically, the population values are counts, you should probably try a stronger transformation like log(population) because populations grow in proportion to their size.

PART
II

REVIEW OF PART II

Exploring Relationships Between Variables

Quick Review

You have now survived your second major unit of Statistics. Here's a brief summary of the key concepts and skills:

▶ We treat data two ways: as categorical and as quantitative.
▶ To explore relationships in categorical data, check out Chapter 3.
▶ To explore relationships in quantitative data:
  • Make a picture. Use a scatterplot. Put the explanatory variable on the x-axis and the response variable on the y-axis.
  • Describe the association between two quantitative variables in terms of direction, form, and strength.
  • The amount of scatter determines the strength of the association.
  • If, as one variable increases so does the other, the association is positive. If one increases as the other decreases, it's negative.
  • If the form of the association is linear, calculate a correlation to measure its strength numerically, and do a regression analysis to model it.
  • Correlations closer to −1 or +1 indicate stronger linear associations. Correlations near 0 indicate weak linear relationships, but other forms of association may still be present.
  • The line of best fit is also called the least squares regression line because it minimizes the sum of the squared residuals.
  • The regression line predicts values of the response variable from values of the explanatory variable.

• A residual is the difference between the true value of the response variable and the value predicted by the regression model.
• The slope of the line is a rate of change, best described in "y-units" per "x-unit."
• R^2 gives the fraction of the variation in the response variable that is accounted for by the model.
• The standard deviation of the residuals measures the amount of scatter around the line.
• Outliers and influential points can distort any of our models.
• If you see a pattern (a curve) in the residuals plot, your chosen model is not appropriate; use a different model. You may, for example, straighten the relationship by re-expressing one of the variables.
• To straighten bent relationships, re-express the data using logarithms or a power (squares, square roots, reciprocals, etc.).
• Always remember that an association is not necessarily an indication that one of the variables causes the other.

Need more help with some of this? Try rereading some sections of Chapters 7 through 10. And go on to the next page for more opportunities to review these concepts and skills.

"One must learn by doing the thing; though you think you know it, you have no certainty until you try."
—Sophocles (495–406 BCE)

REVIEW EXERCISES

1. College. Every year *US News and World Report* publishes a special issue on many U.S. colleges and universities. The scatterplots below have *Student/Faculty Ratio* (number of students per faculty member) for the colleges and universities on the *y*-axes plotted against 4 other variables. The correct correlations for these scatterplots appear in this list. Match them.

$$-0.98 \quad -0.71 \quad -0.51 \quad 0.09 \quad 0.23 \quad 0.69$$

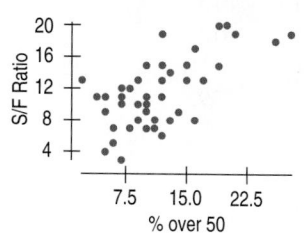
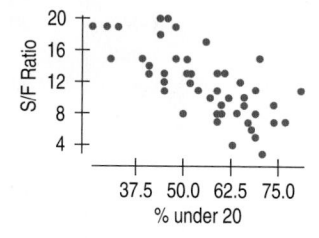
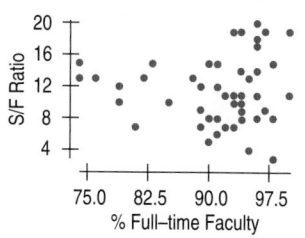
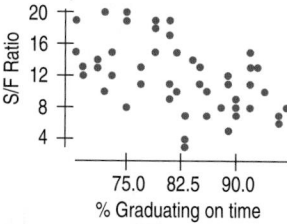

2. Togetherness. Are good grades in high school associated with family togetherness? A random sample of 142 high school students was asked how many meals per week their families ate together. Their responses produced a mean of 3.78 meals per week, with a standard deviation of 2.2. Researchers then matched these responses against the students' grade point averages (GPAs). The scatterplot appeared to be reasonably linear, so they created a line of regression. No apparent pattern emerged in the residuals plot. The equation of the line was $\widehat{GPA} = 2.73 + 0.11\ Meals$.
a) Interpret the *y*-intercept in this context.
b) Interpret the slope in this context.
c) What was the mean GPA for these students?
d) If a student in this study had a negative residual, what did that mean?
e) Upon hearing of this study, a counselor recommended that parents who want to improve the grades their children get should get the family to eat together more often. Do you agree with this interpretation? Explain.

3. Vineyards. Here are the scatterplot and regression analysis for *Case Prices* of 36 wines from vineyards in the Finger Lakes region of New York State and the *Ages* of the vineyards.

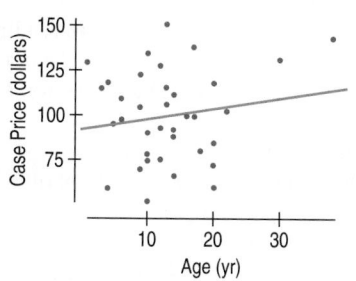

Dependent variable is: Case Price
R-squared = 2.7%

Variable	Coefficient
Constant	92.7650
Age	0.567284

a) Does it appear that vineyards in business longer get higher prices for their wines? Explain.
b) What does this analysis tell us about vineyards in the rest of the world?
c) Write the regression equation.
d) Explain why that equation is essentially useless.

4. Vineyards again. Instead of *Age,* perhaps the *Size* of the vineyard (in acres) is associated with the price of the wines. Look at the scatterplot:

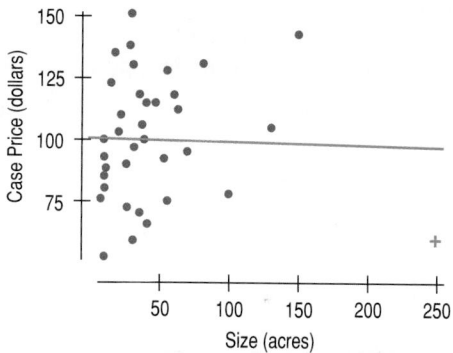

a) Do you see any evidence of an association?
b) What concern do you have about this scatterplot?
c) If the red "+" data point is removed, would the correlation become stronger or weaker? Explain.
d) If the red "+" data point is removed, would the slope of the line increase or decrease? Explain.

5. More twins 2004? As the table shows, the number of twins born in the United States has been increasing. (www.cdc.gov/nchs/births.htm)

Year	Twin Births	Year	Twin Births
1980	68,339	1993	96,445
1981	70,049	1994	97,064
1982	71,631	1995	96,736
1983	72,287	1996	100,750
1984	72,949	1997	104,137
1985	77,102	1998	110,670
1986	79,485	1999	114,307
1987	81,778	2000	118,916
1988	85,315	2001	121,246
1989	90,118	2002	125,134
1990	93,865	2003	128,665
1991	94,779	2004	132,219
1992	95,372		

a) Find the equation of the regression line for predicting the number of twin births.
b) Explain in this context what the slope means.
c) Predict the number of twin births in the United States for the year 2010. Comment on your faith in that prediction.
d) Comment on the residuals plot.

6. **Dow Jones 2006.** The Dow Jones stock index measures the performance of the stocks of America's largest companies (http://finance.yahoo.com). A regression of the Dow prices on years 1972–2006 looks like this:

Dependent variable is: Dow Index
R-squared = 83.5%   s = 1577

Variable	Coefficient
Intercept	−2294.01
Year since 1970	341.095

a) What is the correlation between *Dow Index* and *Year?*
b) Write the regression equation.
c) Explain in this context what the equation says.
d) Here's a scatterplot of the residuals. Which assumption(s) of the regression analysis appear to be violated?

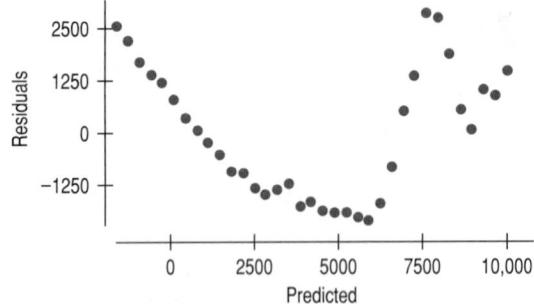

7. **Acid rain.** Biologists studying the effects of acid rain on wildlife collected data from 163 streams in the Adirondack Mountains. They recorded the *pH* (acidity) of the water and the *BCI*, a measure of biological diversity, and they calculated $R^2 = 27\%$. Here's a scatterplot of *BCI* against *pH:*

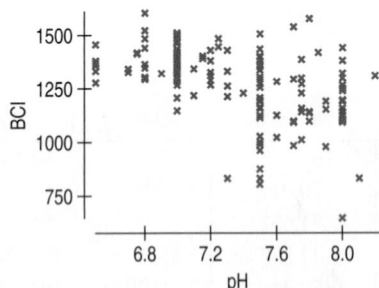

a) What is the correlation between *pH* and *BCI?*
b) Describe the association between these two variables.
c) If a stream has average *pH*, what would you predict about the *BCI?*
d) In a stream where the *pH* is 3 standard deviations above average, what would you predict about the *BCI?*

8. **Manatees 2005.** Marine biologists warn that the growing number of powerboats registered in Florida threatens the existence of manatees. The data below come from the

Florida Fish and Wildlife Conservation Commission (www.floridamarine.org) and the National Marine Manufacturers Association (www.nmma.org/facts).

Year	Manatees Killed	Powerboat Registrations (in 1000s)
1982	13	447
1983	21	460
1984	24	481
1985	16	498
1986	24	513
1987	20	512
1988	15	527
1989	34	559
1990	33	585
1992	33	614
1993	39	646
1994	43	675
1995	50	711
1996	47	719
1997	53	716
1998	38	716
1999	35	716
2000	49	735
2001	81	860
2002	95	923
2003	73	940
2004	69	946
2005	79	974

a) In this context, which is the explanatory variable?
b) Make a scatterplot of these data and describe the association you see.
c) Find the correlation between *Boat Registrations* and *Manatee Deaths.*
d) Interpret the value of R^2.
e) Does your analysis prove that powerboats are killing manatees?

9. **A manatee model 2005.** Continue your analysis of the manatee situation from the previous exercise.
a) Create a linear model of the association between *Manatee Deaths* and *Powerboat Registrations.*
b) Interpret the slope of your model.
c) Interpret the *y*-intercept of your model.
d) How accurately did your model predict the high number of manatee deaths in 2005?
e) Which is better for the manatees, positive residuals or negative residuals? Explain.
f) What does your model suggest about the future for the manatee?

10. **Grades.** A Statistics instructor created a linear regression equation to predict students' final exam scores from their midterm exam scores. The regression equation was $\widehat{Fin} = 10 + 0.9\,Mid$.
a) If Susan scored a 70 on the midterm, what did the instructor predict for her score on the final?

b) Susan got an 80 on the final. How big is her residual?

c) If the standard deviation of the final was 12 points and the standard deviation of the midterm was 10 points, what is the correlation between the two tests?

d) How many points would someone need to score on the midterm to have a predicted final score of 100?

e) Suppose someone scored 100 on the final. Explain why you can't estimate this student's midterm score from the information given.

f) One of the students in the class scored 100 on the midterm but got overconfident, slacked off, and scored only 15 on the final exam. What is the residual for this student?

g) No other student in the class "achieved" such a dramatic turnaround. If the instructor decides not to include this student's scores when constructing a new regression model, will the R^2 value of the regression increase, decrease, or remain the same? Explain.

h) Will the slope of the new line increase or decrease?

11. Traffic.   Highway planners investigated the relationship between traffic *Density* (number of automobiles per mile) and the average *Speed* of the traffic on a moderately large city thoroughfare. The data were collected at the same location at 10 different times over a span of 3 months. They found a mean traffic *Density* of 68.6 cars per mile (cpm) with standard deviation of 27.07 cpm. Overall, the cars' average *Speed* was 26.38 mph, with standard deviation of 9.68 mph. These researchers found the regression line for these data to be $\widehat{Speed} = 50.55 - 0.352\,Density$.

a) What is the value of the correlation coefficient between *Speed* and *Density*?

b) What percent of the variation in average *Speed* is explained by traffic *Density*?

c) Predict the average *Speed* of traffic on the thoroughfare when the traffic *Density* is 50 cpm.

d) What is the value of the residual for a traffic *Density* of 56 cpm with an observed *Speed* of 32.5 mph?

e) The data set initially included the point *Density* = 125 cpm, *Speed* = 55 mph. This point was considered an outlier and was not included in the analysis. Will the slope increase, decrease, or remain the same if we redo the analysis and include this point?

f) Will the correlation become stronger, weaker, or remain the same if we redo the analysis and include this point (125,55)?

g) A European member of the research team measured the *Speed* of the cars in kilometers per hour (1 km ≈ 0.62 miles) and the traffic *Density* in cars per kilometer. Find the value of his calculated correlation between speed and density.

12. Cramming.   One Thursday, researchers gave students enrolled in a section of basic Spanish a set of 50 new vocabulary words to memorize. On Friday the students took a vocabulary test. When they returned to class the following Monday, they were retested—without advance warning. Here are the test scores for the 25 students.

Fri.	Mon.	Fri.	Mon.	Fri.	Mon.
42	36	48	37	39	41
44	44	43	41	46	32
45	46	45	32	37	36
48	38	47	44	40	31
44	40	50	47	41	32
43	38	34	34	48	39
41	37	38	31	37	31
35	31	43	40	36	41
43	32				

a) What is the correlation between *Friday* and *Monday* scores?

b) What does a scatterplot show about the association between the scores?

c) What does it mean for a student to have a positive residual?

d) What would you predict about a student whose *Friday* score was one standard deviation below average?

e) Write the equation of the regression line.

f) Predict the *Monday* score of a student who earned a 40 on Friday.

13. Correlations.   What factor most explains differences in *Fuel Efficiency* among cars? Below is a correlation matrix exploring that relationship for the car's *Weight, Horsepower*, engine *size* (*Displacement*), and number of *Cylinders*.

	MPG	Weight	Horse-power	Displace-ment	Cylinders
MPG	1.000				
Weight	−0.903	1.000			
Horsepower	−0.871	0.917	1.000		
Displacement	−0.786	0.951	0.872	1.000	
Cylinders	−0.806	0.917	0.864	0.940	1.000

a) Which factor seems most strongly associated with *Fuel Efficiency*?

b) What does the negative correlation indicate?

c) Explain the meaning of R^2 for that relationship.

14. Autos revisited.   Look again at the correlation table for cars in the previous exercise.

a) Which two variables in the table exhibit the strongest association?

b) Is that strong association necessarily cause-and-effect? Offer at least two explanations why that association might be so strong.

c) Engine displacements for U.S.-made cars are often measured in cubic inches. For many foreign cars, the units are either cubic centimeters or liters. How would changing from cubic inches to liters affect the calculated correlations involving *Displacement*?

d) What would you predict about the *Fuel Efficiency* of a car whose engine *Displacement* is one standard deviation above the mean?

15. Cars, one more time!   Can we predict the *Horsepower* of the engine that manufacturers will put in a car by knowing the *Weight* of the car? Here are the regression analysis and residuals plot:

Dependent variable is: Horsepower
R-squared = 84.1%

Variable	Coefficient
Intercept	3.49834
Weight	34.3144

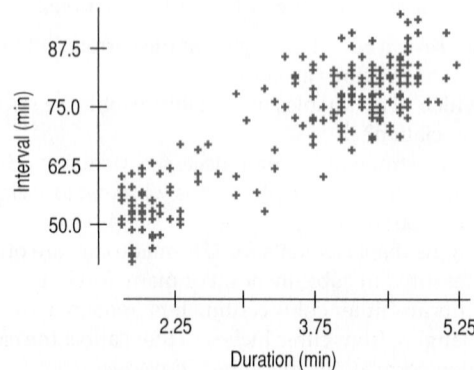

a) Write the equation of the regression line.
b) Do you think the car's *Weight* is measured in pounds or thousands of pounds? Explain.
c) Do you think this linear model is appropriate? Explain.
d) The highest point in the residuals plot, representing a residual of 22.5 horsepower, is for a Chevy weighing 2595 pounds. How much horsepower does this car have?

16. Colorblind. Although some women are colorblind, this condition is found primarily in men. Why is it wrong to say there's a strong correlation between *Sex* and *Colorblindness?*

17. Old Faithful. There is evidence that eruptions of Old Faithful can best be predicted by knowing the duration of the previous eruption.
a) Describe what you see in the scatterplot of *Intervals* between eruptions vs. *Duration* of the previous eruption.

b) Write the equation of the line of best fit. Here's the regression analysis:

Dependent variable is: Interval
R-squared = 77.0%

Variable	Coefficient
Intercept	33.9668
Duration	10.3582

c) Carefully explain what the slope of the line means in this context.
d) How accurate do you expect predictions based on this model to be? Cite statistical evidence.
e) If you just witnessed an eruption that lasted 4 minutes, how long do you predict you'll have to wait to see the next eruption?
f) So you waited, and the next eruption came in 79 minutes. Use this as an example to define a residual.

18. Which croc? The ranges inhabited by the Indian gharial crocodile and the Australian saltwater crocodile overlap in Bangladesh. Suppose a very large crocodile skeleton is found there, and we wish to determine the species of the animal. Wildlife scientists have measured the lengths of the heads and the complete bodies of several crocs (in centimeters) of each species, creating the regression analyses below:

Indian Crocodile		Australian Crocodile	
Dependent variable is: IBody		Dependent variable is: ABody	
R-squared = 97.2%		R-squared = 98.0%	
Variable	**Coefficient**	**Variable**	**Coefficient**
Intercept	−69.3693	Intercept	−20.2245
IHead	7.40004	AHead	7.71726

a) Do the associations between the sizes of the heads and bodies of the two species appear to be strong? Explain.
b) In what ways are the two relationships similar? Explain.
c) What is different about the two models? What does that mean?
d) The crocodile skeleton found had a head length of 62 cm and a body length of 380 cm. Which species do you think it was? Explain why.

19. How old is that tree? One can determine how old a tree is by counting its rings, but that requires cutting the tree down. Can we estimate the tree's age simply from its diameter? A forester measured 27 trees of the same species that had been cut down, and counted the rings to determine the ages of the trees.

Diameter (in.)	Age (yr)	Diameter (in.)	Age (yr)
1.8	4	10.3	23
1.8	5	14.3	25
2.2	8	13.2	28
4.4	8	9.9	29
6.6	8	13.2	30
4.4	10	15.4	30
7.7	10	17.6	33
10.8	12	14.3	34
7.7	13	15.4	35
5.5	14	11.0	38
9.9	16	15.4	38
10.1	18	16.5	40
12.1	20	16.5	42
12.8	22		

a) Find the correlation between *Diameter* and *Age*. Does this suggest that a linear model may be appropriate? Explain.

b) Create a scatterplot and describe the association.

c) Create the linear model.

d) Check the residuals. Explain why a linear model is probably not appropriate.

e) If you used this model, would it generally overestimate or underestimate the ages of very large trees? Explain.

T 20. Improving trees. In the last exercise you saw that the linear model had some deficiencies. Let's create a better model.

a) Perhaps the cross-sectional area of a tree would be a better predictor of its age. Since area is measured in square units, try re-expressing the data by squaring the diameters. Does the scatterplot look better?

b) Create a model that predicts *Age* from the square of the *Diameter*.

c) Check the residuals plot for this new model. Is this model more appropriate? Why?

d) Estimate the age of a tree 18 inches in diameter.

21. New homes. A real estate agent collects data to develop a model that will use the *Size* of a new home (in square feet) to predict its *Sale Price* (in thousands of dollars). Which of these is most likely to be the slope of the regression line: 0.008, 0.08, 0.8, or 8? Explain.

T 22. Smoking and pregnancy 2003. The organization Kids Count monitors issues related to children. The table shows a 50-state average of the percent of expectant mothers who smoked cigarettes during their pregnancies.

Year	% Smoking While Pregnant	Year	% Smoking While Pregnant
1990	19.2	1997	14.9
1991	18.7	1998	14.8
1992	17.9	1999	14.1
1993	16.8	2000	14.0
1994	16.0	2001	13.8
1995	15.4	2002	13.3
1996	15.3	2003	12.7

a) Create a scatterplot and describe the trend you see.

b) Find the correlation.

c) How is the value of the correlation affected by the fact that the data are averages rather than percentages for each of the 50 states?

d) Write a linear model and interpret the slope in context.

T 23. No smoking? The downward trend in smoking you saw in the last exercise is good news for the health of babies, but will it ever stop?

a) Explain why you can't use the linear model you created in Exercise 22 to see when smoking during pregnancy will cease altogether.

b) Create a model that could estimate the year in which the level of smoking would be 0%.

c) Comment on the reliability of such a prediction.

24. Tips. It's commonly believed that people use tips to reward good service. A researcher for the hospitality industry examined tips and ratings of service quality from 2645 dining parties at 21 different restaurants. The correlation between ratings of service and tip percentages was 0.11. (M. Lynn and M. McCall, "Gratitude and Gratuity." *Journal of Socio-Economics* 29: 203–214)

a) Describe the relationship between *Quality of Service* and *Tip Size*.

b) Find and interpret the value of R^2 in this context.

T 25. US Cities. Data from 50 large U.S. cities show the mean *January Temperature* and the *Latitude*. Describe what you see in the scatterplot.

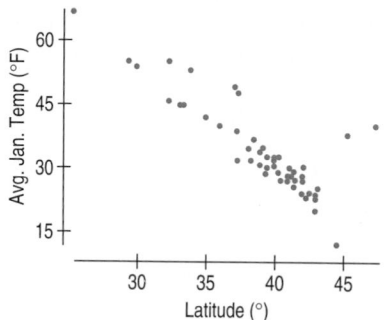

T 26. Correlations. The study of U.S. cities in Exercise 25 found the mean *January Temperature* (degrees Fahrenheit), *Altitude* (feet above sea level), and *Latitude* (degrees north of the equator) for 55 cities. Here's the correlation matrix:

	Jan. Temp	Latitude	Altitude
Jan. Temp	1.000		
Latitude	−0.848	1.000	
Altitude	−0.369	0.184	1.000

a) Which seems to be more useful in predicting *January Temperature*—*Altitude* or *Latitude*? Explain.

b) If the *Temperature* were measured in degrees Celsius, what would be the correlation between *Temperature* and *Latitude*?

c) If the *Temperature* were measured in degrees Celsius and the *Altitude* in meters, what would be the correlation? Explain.

d) What would you predict about the January *Temperatures* in a city whose *Altitude* is two standard deviations higher than the average *Altitude*?

T 27. Winter in the city. Summary statistics for the data relating the latitude and average January temperature for 55 large U.S. cities are given below.

Variable	Mean	StdDev
Latitude	39.02	5.42
JanTemp	26.44	13.49

Correlation = −0.848

a) What percent of the variation in January *Temperatures* can be explained by variation in *Latitude*?

b) What is indicated by the fact that the correlation is negative?

c) Write the equation of the line of regression for predicting January *Temperature* from *Latitude*.

d) Explain what the slope of the line means.

e) Do you think the *y*-intercept is meaningful? Explain.

f) The latitude of Denver is 40° N. Predict the mean January temperature there.

g) What does it mean if the residual for a city is positive?

28. **Depression.**   The September 1998 issue of the *American Psychologist* published an article by Kraut et al. that reported on an experiment examining "the social and psychological impact of the Internet on 169 people in 73 households during their first 1 to 2 years online." In the experiment, 73 households were offered free Internet access for 1 or 2 years in return for allowing their time and activity online to be tracked. The members of the households who participated in the study were also given a battery of tests at the beginning and again at the end of the study. The conclusion of the study made news headlines: Those who spent more time online tended to be more depressed at the end of the experiment. Although the paper reports a more complex model, the basic result can be summarized in the following regression of *Depression* (at the end of the study, in "depression scale units") vs. *Internet Use* (in mean hours per week):

Dependent variable is: Depression
R-squared = 4.6%
s = 0.4563

Variable	Coefficient
Intercept	0.5655
Internet use	0.0199

The news reports about this study clearly concluded that using the Internet causes depression. Discuss whether such a conclusion can be drawn from this regression. If so, discuss the supporting evidence. If not, say why not.

29. **Jumps 2004.**   How are Olympic performances in various events related? The plot shows winning long-jump and high-jump distances, in inches, for the Summer Olympics from 1912 through 2004.

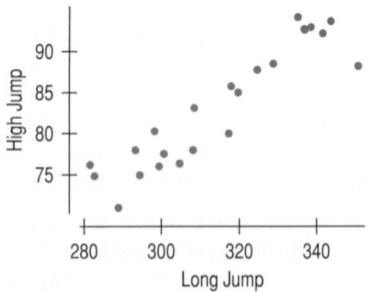

a) Describe the association.

b) Do long-jump performances somehow influence the high-jumpers? How do you account for the relationship you see?

c) The correlation for the given scatterplot is 0.925, but at the Olympics these jumps are actually measured in meters rather than inches. Does that make the actual correlation higher or lower?

d) What would you predict about the long jump in a year when the high-jumper jumped one standard deviation better than the average high jump?

30. **Modeling jumps.**   Here are the summary statistics for the Olympic long jumps and high jumps displayed in the scatterplot above:

Event	Mean	StdDev
Long Jump	316.04	20.85
High Jump	83.85	7.46

Correlation = 0.925

a) Write the equation of the line of regression for estimating *High Jump* from *Long Jump*.

b) Interpret the slope of the line.

c) In a year when the long jump is 350 inches, what high jump would you predict?

d) Why can't you use this line to estimate the long jump for a year when you know the high jump was 85 inches?

e) Write the equation of the line you need to make that prediction.

31. **French.**   Consider the association between a student's score on a French vocabulary test and the weight of the student. What direction and strength of correlation would you expect in each of the following situations? Explain.

a) The students are all in third grade.

b) The students are in third through twelfth grades in the same school district.

c) The students are in tenth grade in France.

d) The students are in third through twelfth grades in France.

32. **Twins.**   Twins are often born after a pregnancy that lasts less than 9 months. The graph from the *Journal of the American Medical Association (JAMA)* shows the rate of preterm twin births in the United States over the past 20 years. In this study, *JAMA* categorized mothers by the level of prenatal medical care they received: inadequate, adequate, or intensive.

a) Describe the overall trend in preterm twin births.

b) Describe any differences you see in this trend, depending on the level of prenatal medical care the mother received.

c) Should expectant mothers be advised to cut back on the level of medical care they seek in the hope of avoiding preterm births? Explain.

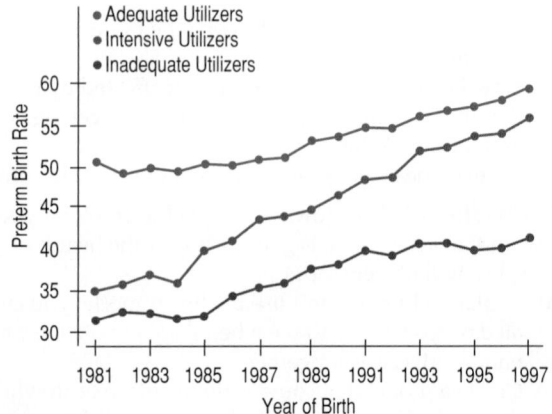

Preterm Birth Rate *per 100 live twin births among U.S. twins by intensive, adequate, and less than adequate prenatal care utilization, 1981–1997. (JAMA 284[2000]: 335–341)*

T 33. Lunchtime. Create and interpret a model for the toddlers' lunchtime data presented in Chapter 7. The table and graph show the number of minutes the kids stayed at the table and the number of calories they consumed.

Calories	Time	Calories	Time
472	21.4	450	42.4
498	30.8	410	43.1
465	37.7	504	29.2
456	33.5	437	31.3
423	32.8	489	28.6
437	39.5	436	32.9
508	22.8	480	30.6
431	34.1	439	35.1
479	33.9	444	33.0
454	43.8	408	43.7

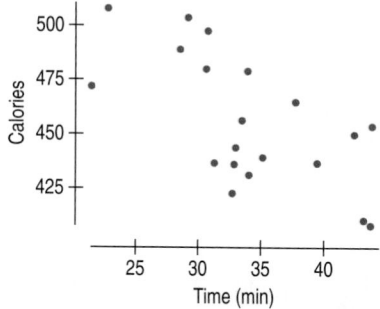

34. Gasoline. Since clean-air regulations have dictated the use of unleaded gasoline, the supply of leaded gas in New York state has diminished. The table below was given on the August 2001 New York State Math B exam, a statewide achievement test for high school students.

Year	1984	1988	1992	1996	2000
Gallons (1000's)	150	124	104	76	50

a) Create a linear model to predict the number of gallons that will be available in 2005.
b) The exam then asked students to estimate the year when leaded gasoline will first become unavailable, expecting them to use the model from part a to answer the question. Explain why that method is incorrect.
c) Create a model that *would* be appropriate for that task, and make the estimate.
d) The "wrong" answer from the other model is fairly accurate in this case. *Why?*

T 35. Tobacco and alcohol. Are people who use tobacco products more likely to consume alcohol? Here are data on household spending (in pounds) taken by the British Government on 11 regions in Great Britain. Do tobacco and alcohol spending appear to be related? What questions do you have about these data? What conclusions can you draw?

Region	Alcohol	Tobacco
North	6.47	4.03
Yorkshire	6.13	3.76
Northeast	6.19	3.77
East Midlands	4.89	3.34
West Midlands	5.63	3.47
East Anglia	4.52	2.92
Southeast	5.89	3.20
Southwest	4.79	2.71
Wales	5.27	3.53
Scotland	6.08	4.51
Northern Ireland	4.02	4.56

T 36. Football weights. The Sears Cup was established in 1993 to honor institutions that maintain a broad-based athletic program, achieving success in many sports, both men's and women's. Since its Division III inception in 1995, the cup has been won by Williams College in every year except one. Their football team has a 85.3% winning record under their current coach. Why does the football team win so much? Is it because they're heavier than their opponents? The table shows the average team weights for selected years from 1973 to 1993.

Year	Weight (lb)	Year	Weight (lb)
1973	185.5	1983	192.0
1975	182.4	1987	196.9
1977	182.1	1989	202.9
1979	191.1	1991	206.0
1981	189.4	1993	198.7

a) Fit a straight line to the relationship between *Weight* and *Year*.
b) Does a straight line seem reasonable?
c) Predict the average weight of the team for the year 2003. Does this seem reasonable?
d) What about the prediction for the year 2103? Explain.
e) What about the prediction for the year 3003? Explain.

37. Models. Find the predicted value of *y*, using each model for $x = 10$.
a) $\hat{y} = 2 + 0.8\ln x$
b) $\log \hat{y} = 5 - 0.23x$
c) $\frac{1}{\sqrt{\hat{y}}} = 17.1 - 1.66x$

T 38. Williams vs. Texas. Here are the average weights of the football team for the University of Texas for various years in the 20th century.

Year	1905	1919	1932	1945	1955	1965
Weight (lb)	164	163	181	192	195	199

a) Fit a straight line to the relationship of *Weight* by *Year* for Texas football players.
b) According to these models, in what year will the predicted weight of the Williams College team from

Exercise 36 first be more than the weight of the University of Texas team?

c) Do you believe this? Explain.

39. **Vehicle weights.** The Minnesota Department of Transportation hoped that they could measure the weights of big trucks without actually stopping the vehicles by using a newly developed "weigh-in-motion" scale. After installation of the scale, a study was conducted to find out whether the scale's readings correspond to the true weights of the trucks being monitored. In Exercise 40 of Chapter 7, you examined the scatterplot for the data they collected, finding the association to be approximately linear with $R^2 = 93\%$. Their regression equation is $\widehat{Wt} = 10.85 + 0.64\ Scale$, where both the scale reading and the predicted weight of the truck are measured in thousands of pounds.

a) Estimate the weight of a truck if this scale read 31,200 pounds.

b) If that truck actually weighed 32,120 pounds, what was the residual?

c) If the scale reads 35,590 pounds, and the truck has a residual of −2440 pounds, how much does it actually weigh?

d) In general, do you expect estimates made using this equation to be reasonably accurate? Explain.

e) If the police plan to use this scale to issue tickets to trucks that appear to be overloaded, will negative or positive residuals be a greater problem? Explain.

40. **Profit.** How are a company's profits related to its sales? Let's examine data from 71 large U.S. corporations. All amounts are in millions of dollars.

a) Histograms of *Profits* and *Sales* and histograms of the logarithms of *Profits* and *Sales* are on the next page. Why are the re-expressed data better for regression?

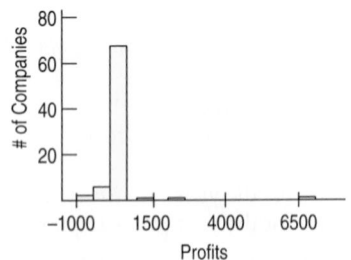

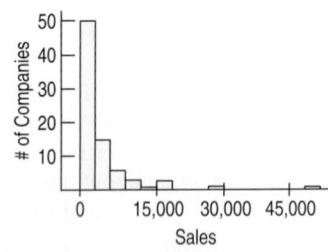

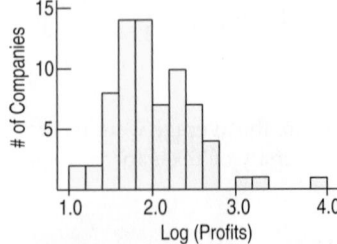

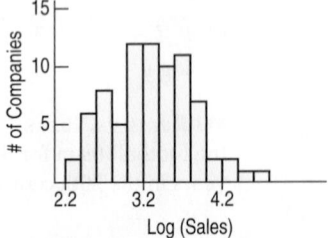

b) Here are the scatterplot and residuals plot for the regression of logarithm of *Profits* vs. log of *Sales*. Do you think this model is appropriate? Explain.

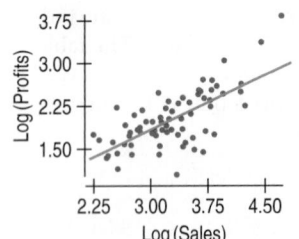

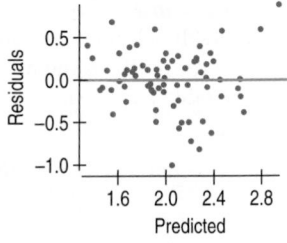

c) Here's the regression analysis. Write the equation.

Dependent variable is: Log Profit
R-squared = 48.1%

Variable	Coefficient
Intercept	−0.106259
LogSales	0.647798

d) Use your equation to estimate profits earned by a company with sales of 2.5 billion dollars. (That's 2500 million.)

41. **Down the drain.** Most water tanks have a drain plug so that the tank may be emptied when it's to be moved or repaired. How long it takes a certain size of tank to drain depends on the size of the plug, as shown in the table. Create a model.

Plug Dia (in.)	$\frac{3}{8}$	$\frac{1}{2}$	$\frac{3}{4}$	1	$1\frac{1}{4}$	$1\frac{1}{2}$	2
Drain Time (min.)	140	80	35	20	13	10	5

42. **Chips.** A start-up company has developed an improved electronic chip for use in laboratory equipment. The company needs to project the manufacturing cost, so it develops a spreadsheet model that takes into account the purchase of production equipment, overhead, raw materials, depreciation, maintenance, and other business costs. The spreadsheet estimates the cost of producing 10,000 to 200,000 chips per year, as seen in the table. Develop a regression model to predict *Costs* based on the *Level* of production.

Chips Produced (1000s)	Cost per Chip ($)	Chips Produced (1000s)	Cost per Chip ($)
10	146.10	90	47.22
20	105.80	100	44.31
30	85.75	120	42.88
40	77.02	140	39.05
50	66.10	160	37.47
60	63.92	180	35.09
70	58.80	200	34.04
80	50.91		

PART III

Gathering Data

CHAPTER
11

Understanding Randomness

We all know what it means for something to be random. Or do we? Many children's games rely on chance outcomes. Rolling dice, spinning spinners, and shuffling cards all select at random. Adult games use randomness as well, from card games to lotteries to Bingo. What's the most important aspect of the randomness in these games? It must be fair.

What is it about random selection that makes it seem fair? It's really two things. First, nobody can guess the outcome before it happens. Second, when we want things to be fair, usually some underlying set of outcomes will be equally likely (although in many games, some combinations of outcomes are more likely than others).

Randomness is not always what we might think of as "at random." Random outcomes have a lot of structure, especially when viewed in the long run. You can't predict how a fair coin will land on any single toss, but you're pretty confident that if you flipped it thousands of times you'd see about 50% heads. As we will see, randomness is an essential tool of Statistics. Statisticians don't think of randomness as the annoying tendency of things to be unpredictable or haphazard. Statisticians use randomness as a tool. In fact, without deliberately applying randomness, we couldn't do most of Statistics, and this book would stop right about here.[1]

But truly random values are surprisingly hard to get. Just to see how fair humans are at selecting, pick a number at random from the top of the next page. Go ahead. Turn the page, look at the numbers quickly, and pick a number at random.

Ready?

Go.

---

[1] Don't get your hopes up.

1 2 3 4

It's Not Easy Being Random

> *"The generation of random numbers is too important to be left to chance."*
>
> —Robert R. Coveyou,
> Oak Ridge National
> Laboratory

Did you pick 3? If so, you've got company. Almost 75% of all people pick the number 3. About 20% pick either 2 or 4. If you picked 1, well, consider yourself a little different. Only about 5% choose 1. Psychologists have proposed reasons for this phenomenon, but for us, it simply serves as a lesson that we've got to find a better way to choose things at random.

So how should we generate **random numbers?** It's surprisingly difficult to get random values even when they're equally likely. Computers have become a popular way to generate random numbers. Even though they often do much better than humans, computers can't generate truly random numbers either. Computers follow programs. Start a computer from the same place, and it will always follow exactly the same path. So numbers generated by a computer program are not truly random. Technically, "random" numbers generated this way are *pseudorandom* numbers. Pseudorandom values are generated in a fixed sequence, and because computers can represent only a finite number of distinct values, the sequence of pseudorandom numbers must eventually repeat itself. Fortunately, pseudorandom values are good enough for most purposes because they are virtually indistinguishable from truly random numbers.

A S **Activity: Random Behavior.** *ActivStats'* Random Experiment Tool lets you experiment with truly random outcomes. We'll use it a lot in the coming chapters.

A S **Activity: Truly Random Values on the Internet.** This activity will take you to an Internet site (www.random.org) that generates all the truly random numbers you could want.

There *are* ways to generate random numbers so that they are both equally likely and truly random. In the past, entire books of carefully generated random numbers were published. The books never made the best-seller lists and probably didn't make for great reading, but they were quite valuable to those who needed truly random values.[2] Today, we have a choice. We can use these books or find genuinely random digits from several Internet sites. The sites use methods like timing the decay of a radioactive element or even the random changes of lava

---

[2] You'll find a table of random digits of this kind in the back of this book.

lamps to generate truly random digits.[3] In either case, a string of random digits might look like this:

```
22177263043874100925370862705819976227258497959070328250001108963
32175358226438002922546449437606423890437665572041073541860024508
89064273086456814121982266538858732858016990278431103804200067664
87405226398245305199020270444649843220009462386785779026390002954
88870033199331475083312651923214139086086744963835289689749105333
69441827131689194060221812813047510193215463038704814076766366740
60702049165089136328553513613610437942934284869094628881431793360
77063565133105632105089936242728722505353955136459910155328128202
```

You probably have more interesting things to download than a few million random digits, but we'll discuss ways to use such random digits to apply randomness to real situations soon. The best ways we know to generate data that give a fair and accurate picture of the world rely on randomness, and the ways in which we draw conclusions from those data depend on the randomness, too.

An ordinary deck of playing cards, like the ones used in bridge and many other card games, consists of 52 cards. There are numbered cards (2 through 10), and face cards (Jack, Queen, King, Ace) whose value depends on the game you are playing. Each card is also marked by one of four suits (clubs, diamonds, hearts, or spades) whose significance is also game-specific.

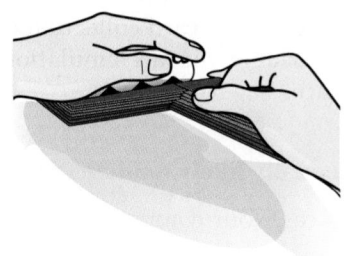

> **Aren't you done shuffling yet?** Even something as common as card shuffling may not be as random as you might think. If you shuffle cards by the usual method in which you split the deck in half and try to let cards fall roughly alternately from each half, you're doing a "riffle shuffle."
> How many times should you shuffle cards to make the deck random? A surprising fact was discovered by statisticians Persi Diaconis, Ronald Graham, and W. M. Kantor. It takes seven riffle shuffles. Fewer than seven leaves order in the deck, but after that, more shuffling does little good. Most people, though, don't shuffle that many times.
> When computers were first used to generate hands in bridge tournaments, some professional bridge players complained that the computer was making too many "weird" hands—hands with 10 cards of one suit, for example. Suddenly these hands were appearing more often than players were used to when cards were shuffled by hand. The players assumed that the computer was doing something wrong. But it turns out that it's humans who hadn't been shuffling enough to make the decks really random and have those "weird" hands appear as often as they should.

Practical Randomness

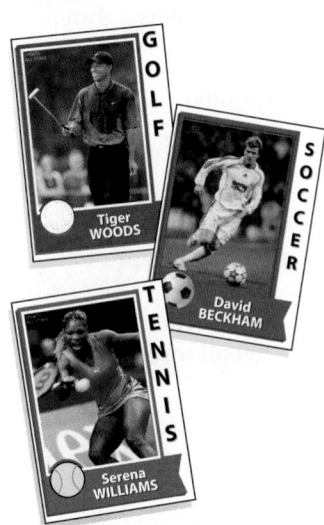

Suppose a cereal manufacturer puts pictures of famous athletes on cards in boxes of cereal in the hope of boosting sales. The manufacturer announces that 20% of the boxes contain a picture of Tiger Woods, 30% a picture of David Beckham, and the rest a picture of Serena Williams. You want all three pictures. How many boxes of cereal do you expect to have to buy in order to get the complete set?

How can we answer questions like this? Well, one way is to buy hundreds of boxes of cereal to see what might happen. But let's not. Instead, we'll consider using a random model. Why random? When we pick a box of cereal off the shelf, we don't know what picture is inside. We'll assume that the pictures are randomly placed in the boxes and that the boxes are distributed randomly to stores around the country. Why a model? Because we won't actually buy the cereal boxes. We can't afford all those boxes and we don't want to waste food. So we need an imitation of the real process that we can manipulate and control. In short, we're going to **simulate** reality.

[3] For example, www.random.org or www.randomnumbers.info.

A Simulation

The question we've asked is how many boxes do you expect to buy to get a complete card collection. But we can't answer our question by completing a card collection just once. We want to understand the *typical* number of boxes to open, how that number varies, and, often, the shape of the distribution. So we'll have to do this over and over. We call each time we obtain a simulated answer to our question a **trial.**

For the sports cards, a trial's outcome is the number of boxes. We'll need at least 3 boxes to get one of each card, but with really bad luck, you could empty the shelves of several supermarkets before finding the card you need to get all 3. So, the possible outcomes of a trial are 3, 4, 5, or lots more. But we can't simply pick one of those numbers at random, because they're not equally likely. We'd be surprised if we only needed 3 boxes to get all the cards, but we'd probably be even more surprised to find that it took exactly 7,359 boxes. In fact, the reason we're doing the simulation is that it's hard to guess how many boxes we'd expect to open.

BUILDING A SIMULATION

We know how to find equally likely random digits. How can we get from there to simulating the trial outcomes? We know the relative frequencies of the cards: 20% Tiger, 30% Beckham, and 50% Serena. So, we can interpret the digits 0 and 1 as finding Tiger; 2, 3, and 4 as finding Beckham; and 5 through 9 as finding Serena to simulate opening one box. Opening one box is the basic building block, called a **component** of our simulation. But the component's outcome isn't the result we want. We need to observe a sequence of components until our card collection is complete. The *trial's* outcome is called the **response variable**; for this simulation that's the *number* of components (boxes) in the sequence.

Let's look at the steps for making a simulation:

Specify how to model a component outcome using equally likely random digits:

1. **Identify the component to be repeated.** In this case, our component is the opening of a box of cereal.
2. **Explain how you will model the component's outcome.** The digits from 0 to 9 are equally likely to occur. Because 20% of the boxes contain Tiger's picture, we'll use 2 of the 10 digits to represent that outcome. Three of the 10 digits can model the 30% of boxes with David Beckham cards, and the remaining 5 digits can represent the 50% of boxes with Serena. One possible assignment of the digits, then, is

0, 1 Tiger   2, 3, 4 Beckham   5, 6, 7, 8, 9 Serena.

Specify how to simulate trials:

3. **Explain how you will combine the components to model a trial.** We pretend to open boxes (repeat components) until our collection is complete. We do this by looking at each random digit and indicating what picture it represents. We continue until we've found all three.
4. **State clearly what the response variable is.** What are we interested in? We want to find out the number of boxes it might take to get all three pictures.

Put it all together to run the simulation:

5. **Run several trials.** For example, consider the third line of random digits shown earlier (p. 257):

8906427308645681412198226653885873285801699027843110380420067664.

Let's see what happened.

The first random digit, 8, means you get Serena's picture. So the first component's outcome is Serena. The second digit, 9, means Serena's picture is also in the next box. Continuing to interpret the random digits, we get Tiger's picture (0) in the third, Serena's (6) again in the fourth, and finally Beckham (4) on the fifth box. Since we've now found all three pictures, we've finished one trial of our simulation. This trial's outcome is 5 boxes.

Now we keep going, running more trials by looking at the rest of our line of random digits:

89064 2730 8645681 41219 822665388587328580 169902 78431 1038 042006 7664.

It's best to create a chart to keep track of what happens:

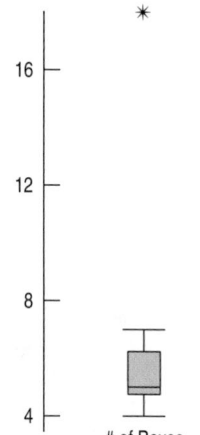

16

12

8

4

# of Boxes

Trial Number	Component Outcomes	Trial Outcomes: y = Number of boxes
1	89064 = Serena, Serena, Tiger, Serena, Beckham	5
2	2730 = Beckham, Serena, Beckham, Tiger	4
3	8645681 = Serena, Serena, Beckham, . . . , Tiger	7
4	41219 = Beckham, Tiger, Beckham, Tiger, Serena	5
5	822665388587328580 = Serena, Beckham, . . . , Tiger	18
6	169902 = Tiger, Serena, Serena, Serena, Tiger, Beckham	6
7	78431 = Serena, Serena, Beckham, Beckham, Tiger	5
8	1038 = Tiger, Tiger, Beckham, Serena	4
9	042006 = Tiger, Beckham, Beckham, Tiger, Tiger, Serena	6
10	7664 . . . = Serena, Serena, Serena, Beckham . . .	?

Analyze the response variable:

6. **Collect and summarize the results of all the trials.** You know how to summarize and display a response variable. You'll certainly want to report the shape, center, and spread, and depending on the question asked, you may want to include more.

7. **State your conclusion,** as always, in the context of the question you wanted to answer. Based on this simulation, we estimate that customers hoping to complete their card collection will need to open a median of 5 boxes, but it could take a lot more.

If you fear that these may not be accurate estimates because we ran only nine trials, you are absolutely correct. The more trials the better, and nine is woefully inadequate. Twenty trials is probably a reasonable minimum if you are doing this by hand. Even better, use a computer and run a few hundred trials.

A S   *Activity:* **Bigger Samples Are Better.** The random simulation tool can generate lots of outcomes with a single click, so you can see more of the long run with less effort.

FOR EXAMPLE   Simulating a dice game

The game of 21 can be played with an ordinary 6-sided die. Competitors each roll the die repeatedly, trying to get the highest total less than or equal to 21. If your total exceeds 21, you lose.
    Suppose your opponent has rolled an 18. Your task is to try to beat him by getting more than 18 points without going over 21. How many rolls do you expect to make, and what are your chances of winning?

Question: How will you simulate the components?

A component is one roll of the die. I'll simulate each roll by looking at a random digit from a table or an Internet site. The digits 1 through 6 will represent the results on the die; I'll ignore digits 7–9 and 0.

(continued)

For Example (*continued*)

Question: How will you combine components to model a trial? What's the response variable?

I'll add components until my total is greater than 18, counting the number of rolls. If my total is greater than 21, it is a loss; if not, it is a win. There are two response variables. I'll count the number of times I roll the die, and I'll keep track of whether I win or lose.

Question: How would you use these random digits to run trials? Show your method clearly for two trials.

91129 58757 69274 92380 82464 33089

I've marked the discarded digits in color.

Trial #1:	9	1	1	2	9	5	8	7	5	7	6
Total:		1	2	4		9			14		20

Outcomes: 6 rolls, won

Trial #2:	9	2	7	4	9	2	3	8	0	8	2	4	6
Total:		2		6		8	11				13	17	23

Outcomes: 7 rolls, lost

Question: Suppose you run 30 trials, getting the outcomes tallied here. What is your conclusion?

Based on my simulation, when competing against an opponent who has a score of 18, I expect my turn to usually last 5 or 6 rolls, and I should win about 70% of the time.

Number of rolls		Result	
4	///	Won	ⅣⅣ ⅣⅣ ⅣⅣ ⅣⅣ /
5	ⅣⅣ ⅣⅣ	Lost	ⅣⅣ ////
6	ⅣⅣ ⅣⅣ /		
7	ⅣⅣ		
8	/		

JUST CHECKING

The baseball World Series consists of up to seven games. The first team to win four games wins the series. The first two are played at one team's home ballpark, the next three at the other team's park, and the final two (if needed) are played back at the first park. Records over the past century show that there is a home field advantage; the home team has about a 55% chance of winning. Does the current system of alternating ballparks even out the home field advantage? How often will the team that begins at home win the series?

Let's set up the simulation:

1. What is the component to be repeated?

2. How will you model each component from equally likely random digits?

3. How will you model a trial by combining components?

4. What is the response variable?

5. How will you analyze the response variable?

STEP-BY-STEP EXAMPLE | **Simulation**

Fifty-seven students participated in a lottery for a particularly desirable dorm room—a triple with a fireplace and private bath in the tower. Twenty of the participants were members of the same varsity team. When all three winners were members of the team, the other students cried foul.

Question: Could an all-team outcome reasonably be expected to happen if everyone had a fair shot at the room?

Plan State the problem. Identify the important parts of your simulation.

I'll use a simulation to investigate whether it's unlikely that three varsity athletes would get the great room in the dorm if the lottery were fair.

Components Identify the components.

A component is the selection of a student.

Outcomes State how you will model each component using equally likely random digits. You can't just use the digits from 0 to 9 because the outcomes you are simulating are not multiples of 10%.

I'll look at two-digit random numbers.

Let 00–19 represent the 20 varsity applicants.

Let 20–56 represent the other 37 applicants.

There are 20 and 37 students in the two groups. This time you must use *pairs* of random digits (and ignore some of them) to represent the 57 students.

Skip 57–99. If I get a number in this range, I'll throw it away and go back for another two-digit random number.

Trial Explain how you will combine the components to simulate a trial. In each of these trials, you can't choose the same student twice, so you'll need to ignore a random number if it comes up a second or third time. Be sure to mention this in describing your simulation.

Each trial consists of identifying pairs of digits as V (varsity) or N (nonvarsity) until 3 people are chosen, ignoring out-of-range or repeated numbers (X)—I can't put the same person in the room twice.

Response Variable Define your response variable.

The response variable is whether or not all three selected students are on the varsity team.

Mechanics Run several trials. Carefully record the random numbers, indicating

1) the corresponding component outcomes (here, Varsity, Nonvarsity, or ignored number) and
2) the value of the response variable.

Trial Number	Component Outcomes	All Varsity?
1	74 02 94 39 02 77 55 X V X N X X N	No
2	18 63 33 25 V X N N	No
3	05 45 88 91 56 V N X X N	No
4	39 09 07 N V V	No
5	65 39 45 95 43 X N N X N	No
6	98 95 11 68 77 12 17 X X V X X V V	Yes
7	26 19 89 93 77 27 N V X X X N	No

(continued)

8	23 52 37 N N N	No
9	16 50 83 44 V N X N	No
10	74 17 46 85 09 X V N X V	No

Analyze Summarize the results across all trials to answer the initial question.

"All varsity" occurred once, or 10% of the time.

Conclusion Describe what the simulation shows, and interpret your results in the context of the real world.

In my simulation of "fair" room draws, the three people chosen were all varsity team members only 10% of the time. While this result could happen by chance, it is not particularly likely. I'm suspicious, but I'd need many more trials and a smaller frequency of the all-varsity outcome before I would make an accusation of unfairness.

TI Tips

Generating random numbers

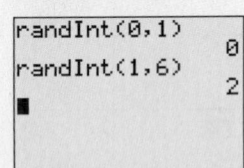

Instead of using coins, dice, cards, or tables of random numbers, you may decide to use your calculator for simulations. There are several random number generators offered in the MATH PRB menu.

5:randInt(is of particular importance. This command will produce any number of random integers in a specified range.

Here are some examples showing how to use randInt for simulations:

- randInt(0,1) randomly chooses a 0 or a 1. This is an effective simulation of a coin toss. You could let 0 represent tails and 1 represent heads.
- randInt(1,6) produces a random integer from 1 to 6, a good way to simulate rolling a die.
- randInt(1,6,2) simulates rolling *two* dice. To do several rolls in a row, just hit ENTER repeatedly.

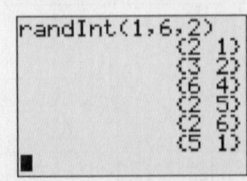

- randInt(0,9,5) produces five random integers that might represent the pictures in the cereal boxes. Our run gave us two Tigers (0, 1), no Beckhams (2, 3, 4), and three Serenas (5–9).

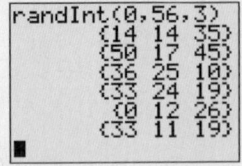

- randInt(0,56,3) produces three random integers between 0 and 56, a nice way to simulate the dorm room lottery. The window shows 6 trials, but we would skip the first one because one student was chosen twice. In none of the remaining 5 trials did three athletes (0–19) win.

WHAT CAN GO WRONG?

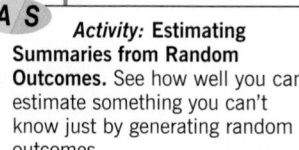

Activity: Estimating Summaries from Random Outcomes. See how well you can estimate something you can't know just by generating random outcomes.

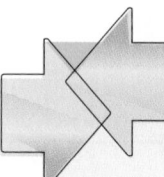

TI-*nspire*

Simulations. Improve your predictions by running thousands of trials.

▶ **Don't overstate your case.** Let's face it: In some sense, a simulation is *always* wrong. After all, it's not the real thing. We didn't buy any cereal or run a room draw. So beware of confusing what *really* happens with what a simulation suggests *might* happen. Never forget that future results will not match your simulated results exactly.

▶ **Model outcome chances accurately.** A common mistake in constructing a simulation is to adopt a strategy that may appear to produce the right kind of results, but that does not accurately model the situation. For example, in our room draw, we could have gotten 0, 1, 2, or 3 team members. Why not just see how often these digits occur in random digits from 0 to 9, ignoring the digits 4 and up?

3 2 1 7 9 0 0 5 9 7 3 7 9 2 5 2 4 1 3 8

3 2 1 x x 0 0 x x x 3 x x 2 x 2 x 1 3 x

This "simulation" makes it seem fairly likely that three team members would be chosen. There's a big problem with this approach, though: The digits 0, 1, 2, and 3 occur with equal frequency among random digits, making each outcome appear to happen 25% of the time. In fact, the selection of 0, 1, 2, or all 3 team members are not all equally likely outcomes. In our correct simulation, we estimated that all 3 would be chosen only about 10% of the time. If your simulation overlooks important aspects of the real situation, your model will not be accurate.

▶ **Run enough trials.** Simulation is cheap and fairly easy to do. Don't try to draw conclusions based on 5 or 10 trials (even though we did for illustration purposes here). We'll make precise how many trials to use in later chapters. For now, err on the side of large numbers of trials.

CONNECTIONS

Simulations often generate many outcomes of a response variable, and we are often interested in the distribution of these responses. The tools we use to display and summarize the distribution of any real variable are appropriate for displaying and summarizing randomly generated responses as well.

Make histograms, boxplots, and Normal probability plots of the response variables from simulations, and summarize them with measures of center and spread. Be especially careful to report the variation of your response variable.

Don't forget to think about your analyses. Simulations can hide subtle errors. A careful analysis of the responses can save you from erroneous conclusions based on a faulty simulation.

You may be less likely to find an outlier in simulated responses, but if you find one, you should certainly determine how it happened.

WHAT HAVE WE LEARNED?

We've learned to harness the power of randomness. We've learned that a simulation model can help us investigate a question for which many outcomes are possible, we can't (or don't want to) collect data, and a mathematical answer is hard to calculate. We've learned how to base our simulation on random values generated by a computer, generated by a randomizing device such as a die or spinner, or found on the Internet. Like all models, simulations can provide us with useful insights about the real world.

Terms

Random	255. An outcome is random if we know the possible values it can have, but not which particular value it takes.
Generating random numbers	256. Random numbers are hard to generate. Nevertheless, several Internet sites offer an unlimited supply of equally likely random values.
Simulation	258. A simulation models a real-world situation by using random-digit outcomes to mimic the uncertainty of a response variable of interest.
Simulation component	258. A component uses equally likely random digits to model simple random occurrences whose outcomes may not be equally likely.
Trial	258. The sequence of several components representing events that we are pretending will take place.
Response variable	258. Values of the response variable record the results of each trial with respect to what we were interested in.

Skills

▸ Be able to recognize random outcomes in a real-world situation.

▸ Be able to recognize when a simulation might usefully model random behavior in the real world.

▸ Know how to perform a simulation either by generating random numbers on a computer or calculator, or by using some other source of random values, such as dice, a spinner, or a table of random numbers.

▸ Be able to describe a simulation so that others can repeat it.

▸ Be able to discuss the results of a simulation study and draw conclusions about the question being investigated.

SIMULATION ON THE COMPUTER

Simulations are best done with the help of technology simply because more trials makes a better simulation, and computers are fast. There are special computer programs designed for simulation, and most statistics packages and calculators can at least generate random numbers to support a simulation.

All technology-generated random numbers are *pseudorandom*. The random numbers available on the Internet may technically be better, but the differences won't matter for any simulation of modest size. Pseudorandom numbers generate the next random value from the previous one by a specified algorithm. But they have to start somewhere. This starting point is called the "seed." Most programs let you set the seed. There's usually little reason to do this, but if you wish to, go ahead. If you reset the seed to the same value, the programs will generate the same sequence of "random" numbers.

> **A S** *Activity:* **Creating Random Values.** Learn to use your statistics package to generate random outcomes.

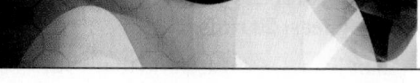

EXERCISES

1. **Coin toss.** Is a coin flip random? Why or why not?

2. **Casino.** A casino claims that its electronic "video roulette" machine is truly random. What should that claim mean?

3. **The lottery.** Many states run lotteries, giving away millions of dollars if you match a certain set of winning numbers. How are those numbers determined? Do you think this method guarantees randomness? Explain.

4. **Games.** Many kinds of games people play rely on randomness. Cite three different methods commonly used in the attempt to achieve this randomness, and discuss the effectiveness of each.

5. **Birth defects.** The American College of Obstetricians and Gynecologists says that out of every 100 babies born in the United States, 3 have some kind of major birth defect. How would you assign random numbers to conduct a simulation based on this statistic?

6. **Colorblind.** By some estimates, about 10% of all males have some color perception defect, most commonly red–green colorblindness. How would you assign random numbers to conduct a simulation based on this statistic?

7. **Geography.** An elementary school teacher with 25 students plans to have each of them make a poster about two different states. The teacher first numbers the states (in alphabetical order, from 1-Alabama to 50-Wyoming), then uses a random number table to decide which states each kid gets. Here are the random digits:

    45921 01710 22892 37076

    a) Which two state numbers does the first student get?
    b) Which two state numbers go to the second student?

8. **Get rich.** Your state's BigBucks Lottery prize has reached $100,000,000, and you decide to play. You have to pick five numbers between 1 and 60, and you'll win if your numbers match those drawn by the state. You decide to pick your "lucky" numbers using a random number table. Which numbers do you play, based on these random digits?

    43680 98750 13092 76561 58712

9. **Play the lottery.** Some people play state-run lotteries by always playing the same favorite "lucky" number. Assuming that the lottery is truly random, is this strategy better, worse, or the same as choosing different numbers for each play? Explain.

10. **Play it again, Sam.** In Exercise 8 you imagined playing the lottery by using random digits to decide what numbers to play. Is this a particularly good or bad strategy? Explain.

11. **Bad simulations.** Explain why each of the following simulations fails to model the real situation properly:
    a) Use a random integer from 0 through 9 to represent the number of heads when 9 coins are tossed.
    b) A basketball player takes a foul shot. Look at a random digit, using an odd digit to represent a good shot and an even digit to represent a miss.
    c) Use random digits from 1 through 13 to represent the denominations of the cards in a five-card poker hand.

12. **More bad simulations.** Explain why each of the following simulations fails to model the real situation:
    a) Use random numbers 2 through 12 to represent the sum of the faces when two dice are rolled.
    b) Use a random integer from 0 through 5 to represent the number of boys in a family of 5 children.
    c) Simulate a baseball player's performance at bat by letting 0 = an out, 1 = a single, 2 = a double, 3 = a triple, and 4 = a home run.

13. **Wrong conclusion.** A Statistics student properly simulated the length of checkout lines in a grocery store and then reported, "The average length of the line will be 3.2 people." What's wrong with this conclusion?

14. **Another wrong conclusion.** After simulating the spread of a disease, a researcher wrote, "24% of the people contracted the disease." What should the correct conclusion be?

15. **Election.** You're pretty sure that your candidate for class president has about 55% of the votes in the entire school. But you're worried that only 100 students will show up to vote. How often will the underdog (the one with 45% support) win? To find out, you set up a simulation.
    a) Describe how you will simulate a component.
    b) Describe how you will simulate a trial.
    c) Describe the response variable.

16. **Two pair or three of a kind?** When drawing five cards randomly from a deck, which is more likely, two pairs or three of a kind? A pair is exactly two of the same denomination. Three of a kind is exactly 3 of the same denomination. (Don't count three 8's as a pair—that's 3 of a kind. And don't count 4 of the same kind as two pair—that's 4 of a kind, a very special hand.) How could you simulate 5-card hands? Be careful; once you've picked the 8 of spades, you can't get it again in that hand.
    a) Describe how you will simulate a component.
    b) Describe how you will simulate a trial.
    c) Describe the response variable.

17. **Cereal.** In the chapter's example, 20% of the cereal boxes contained a picture of Tiger Woods, 30% David Beckham, and the rest Serena Williams. Suppose you buy five boxes of cereal. Estimate the probability that you end up with a complete set of the pictures. Your simulation should have at least 20 runs.

18. **Cereal, again.** Suppose you really want the Tiger Woods picture. How many boxes of cereal do you need to buy to be pretty sure of getting at least one? Your simulation should use at least 10 trials.

19. Multiple choice. You take a quiz with 6 multiple choice questions. After you studied, you estimated that you would have about an 80% chance of getting any individual question right. What are your chances of getting them all right? Use at least 20 trials.

20. Lucky guessing? A friend of yours who took the multiple choice quiz in Exercise 19 got all 6 questions right, but now claims to have guessed blindly on every question. If each question offered 4 possible answers, do you believe her? Explain, basing your argument on a simulation involving at least 10 trials.

21. Beat the lottery. Many states run lotteries to raise money. A Web site advertises that it knows "how to increase YOUR chances of Winning the Lottery." They offer several systems and criticize others as foolish. One system is called *Lucky Numbers*. People who play the *Lucky Numbers* system just pick a "lucky" number to play, but maybe some numbers are luckier than others. Let's use a simulation to see how well this system works.

To make the situation manageable, simulate a simple lottery in which a single digit from 0 to 9 is selected as the winning number. Pick a single value to bet, such as 1, and keep playing it over and over. You'll want to run at least 100 trials. (If you can program the simulations on a computer, run several hundred. Or generalize the questions to a lottery that chooses two- or three-digit numbers—for which you'll need thousands of trials.)
a) What proportion of the time do you expect to win?
b) Would you expect better results if you picked a "luckier" number, such as 7? (Try it if you don't know.) Explain.

22. Random is as random does. The "beat the lottery" Web site discussed in Exercise 21 suggests that because lottery numbers are random, it is better to select your bet randomly. For the same simple lottery in Exercise 21 (random values from 0 to 9), generate each bet by choosing a separate random value between 0 and 9. Play many games. What proportion of the time do you win?

23. It evens out in the end. The "beat the lottery" Web site of Exercise 21 notes that in the long run we expect each value to turn up about the same number of times. That leads to their recommended strategy. First, watch the lottery for a while, recording the winners. Then bet the value that has turned up the least, because it will need to turn up more often to even things out. If there is more than one "rarest" value, just take the lowest one (since it doesn't matter). Simulating the simplified lottery described in Exercise 21, play many games with this system. What proportion of the time do you win?

24. Play the winner? Another strategy for beating the lottery is the reverse of the system described in Exercise 23. Simulate the simplified lottery described in Exercise 21. Each time, bet the number that just turned up. The Web site suggests that this method should do worse. Does it? Play many games and see.

25. Driving test. You are about to take the road test for your driver's license. You hear that only 34% of candidates pass the test the first time, but the percentage rises to 72% on subsequent retests. Estimate the average number of tests drivers take in order to get a license. Your simulation should use at least 20 runs.

26. Still learning? As in Exercise 25, assume that your chance of passing the driver's test is 34% the first time and 72% for subsequent retests. Estimate the percentage of those tested who still do not have a driver's license after two attempts.

27. Basketball strategy. Late in a basketball game, the team that is behind often fouls someone in an attempt to get the ball back. Usually the opposing player will get to shoot foul shots "one and one," meaning he gets a shot, and then a second shot only if he makes the first one. Suppose the opposing player has made 72% of his foul shots this season. Estimate the number of points he will score in a one-and-one situation.

28. Blood donors. A person with type O-positive blood can receive blood only from other type O donors. About 44% of the U.S. population has type O blood. At a blood drive, how many potential donors do you expect to examine in order to get three units of type O blood?

29. Free groceries. To attract shoppers, a supermarket runs a weekly contest that involves "scratch-off" cards. With each purchase, customers get a card with a black spot obscuring a message. When the spot is scratched away, most of the cards simply say, "Sorry—please try again." But during the week, 100 customers will get cards that make them eligible for a drawing for free groceries. Ten of the cards say they may be worth $200, 10 others say $100, 20 may be worth $50, and the rest could be worth $20. To register those cards, customers write their names on them and put them in a barrel at the front of the store. At the end of the week the store manager draws cards at random, awarding the lucky customers free groceries in the amount specified on their card. The drawings continue until the store has given away more than $500 of free groceries. Estimate the average number of winners each week.

30. Find the ace. A new electronics store holds a contest to attract shoppers. Once an hour someone in the store is chosen at random to play the Music Game. Here's how it works: An ace and four other cards are shuffled and placed face down on a table. The customer gets to turn cards over one at a time, looking for the ace. The person wins $100 worth of free CDs or DVDs if the ace is the first card, $50 if it is the second card, and $20, $10, or $5 if it is the third, fourth, or fifth card chosen. What is the average dollar amount of music the store will give away?

31. The family. Many couples want to have both a boy and a girl. If they decide to continue to have children until they have one child of each sex, what would the average family size be? Assume that boys and girls are equally likely.

32. A bigger family. Suppose a couple will continue having children until they have at least two children of each sex (two boys *and* two girls). How many children might they expect to have?

33. Dice game. You are playing a children's game in which the number of spaces you get to move is determined by the rolling of a die. You must land exactly on the final space in order to win. If you are 10 spaces away, how many turns might it take you to win?

34. Parcheesi. You are three spaces from a win in Parcheesi. On each turn, you will roll two dice. To win, you must roll a total of 3 or roll a 3 on one of the dice. How many turns might you expect this to take?

35. The hot hand. A basketball player with a 65% shooting percentage has just made 6 shots in a row. The announcer says this player "is hot tonight! She's in the zone!" Assume the player takes about 20 shots per game. Is it unusual for her to make 6 or more shots in a row during a game?

36. The World Series. The World Series ends when a team wins 4 games. Suppose that sports analysts consider one team a bit stronger, with a 55% chance to win any individual game. Estimate the likelihood that the underdog wins the series.

37. Teammates. Four couples at a dinner party play a board game after the meal. They decide to play as teams of two and to select the teams randomly. All eight people write their names on slips of paper. The slips are thoroughly mixed, then drawn two at a time. How likely is it that every person will be teamed with someone other than the person he or she came to the party with?

38. Second team. Suppose the couples in Exercise 37 choose the teams by having one member of each couple write their names on the cards and the other people each pick a card at random. How likely is it that every person will be teamed with someone other than the person he or she came with?

39. Job discrimination? A company with a large sales staff announces openings for three positions as regional managers. Twenty-two of the current salespersons apply, 12 men and 10 women. After the interviews, when the company announces the newly appointed managers, all three positions go to women. The men complain of job discrimination. Do they have a case? Simulate a random selection of three people from the applicant pool, and make a decision about the likelihood that a fair process would result in hiring all women.

40. Cell phones. A proud legislator claims that your state's new law against talking on a cell phone while driving has reduced cell phone use to less than 12% of all drivers. While waiting for your bus the next morning, you notice that 4 of the 10 people who drive by are using their cell phones. Does this cast doubt on the legislator's figure of 12%? Use a simulation to estimate the likelihood of seeing at least 4 of 10 randomly selected drivers talking on their cell phones if the actual rate of usage is 12%. Explain your conclusion clearly.

JUST CHECKING
Answers

1. The component is one game.
2. I'll generate random numbers and assign numbers from 00 to 54 to the home team's winning and from 55 to 99 to the visitors' winning.
3. I'll generate components until one team wins 4 games. I'll record which team wins the series.
4. The response is who wins the series.
5. I'll calculate the proportion of wins by the team that starts at home.

Sample Surveys

In 2007, Pew Research conducted a survey to assess Americans' knowledge of current events. They asked a random sample of 1,502 U.S. adults 23 factual questions about topics currently in the news.[1] Pew also asked respondents where they got their news. Those who frequented major newspaper Web sites or who are regular viewers of the *Daily Show* or *Colbert Report* scored best on knowledge of current events.[2] Even among those viewers, only 54% responded correctly to 15 or more of the questions. Pew claimed that this was close to the true percentage responding correctly that they would have found if they had asked all U.S. adults who got their news from those sources. That step from a small sample to the entire population is impossible without understanding Statistics. To make business decisions, to do science, to choose wise investments, or to understand what voters think they'll do the next election, we need to stretch beyond the data at hand to the world at large.

To make that stretch, we need three ideas. You'll find the first one natural. The second may be more surprising. The third is one of the strange but true facts that often confuse those who don't know Statistics.

Idea 1: Examine a Part of the Whole

The first idea is to draw a sample. We'd like to know about an entire **population** of individuals, but examining all of them is usually impractical, if not impossible. So we settle for examining a smaller group of individuals—a **sample**—selected from the population.

You do this every day. For example, suppose you wonder how the vegetable soup you're cooking for dinner tonight is going to go over with your friends. To decide whether it meets your standards, you only need to try a small amount. You might taste just a spoonful or two. You certainly don't have to consume the whole

---

[1] For example, two of the questions were "Who is the vice-president of the United States?" and "What party controls Congress?"
[2] The lowest scores came from those whose main source of news was network morning shows or *Fox News*.

pot. You trust that the taste will *represent* the flavor of the entire pot. The idea behind your tasting is that a small sample, if selected properly, can represent the entire population.

It's hard to go a day without hearing about the latest opinion poll. These polls are examples of **sample surveys,** designed to ask questions of a small group of people in the hope of learning something about the entire population. Most likely, you've never been selected to be part of one of these national opinion polls. That's true of most people. So how can the pollsters claim that a sample is representative of the entire population? The answer is that professional pollsters work quite hard to ensure that the "taste"—the sample that they take—represents the population. If not, the sample can give misleading information about the population.

Bias

In 1936, a young pollster named George Gallup used a subsample of only 3000 of the 2.4 million responses that the Literary Digest received to reproduce the wrong prediction of Landon's victory over Roosevelt. He then used an entirely different sample of 50,000 and predicted that Roosevelt would get 56% of the vote to Landon's 44%. His sample was apparently much more representative of the actual voting populace. The Gallup Organization went on to become one of the leading polling companies.

A S **Video: The *Literary Digest* Poll and the Election of 1936.** Hear the story of one of the most famous polling failures in history.

Selecting a sample to represent the population fairly is more difficult than it sounds. Polls or surveys most often fail because they use a sampling method that tends to over- or underrepresent parts of the population. The method may overlook subgroups that are harder to find (such as the homeless or those who use only cell phones) or favor others (such as Internet users who like to respond to online surveys). Sampling methods that, by their nature, tend to over- or underemphasize some characteristics of the population are said to be **biased.** Bias is the bane of sampling—the one thing above all to avoid. Conclusions based on samples drawn with biased methods are inherently flawed. There is usually no way to fix bias after the sample is drawn and no way to salvage useful information from it.

Here's a famous example of a really dismal failure. By the beginning of the 20th century, it was common for newspapers to ask readers to return "straw" ballots on a variety of topics. (Today's Internet surveys are the same idea, gone electronic.) The earliest known example of such a straw vote in the United States dates back to 1824.

During the period from 1916 to 1936, the magazine *Literary Digest* regularly surveyed public opinion and forecast election results correctly. During the 1936 presidential campaign between Alf Landon and Franklin Delano Roosevelt, it mailed more than 10 million ballots and got back an astonishing 2.4 million. (Polls were still a relatively novel idea, and many people thought it was important to send back their opinions.) The results were clear: Alf Landon would be the next president by a landslide, 57% to 43%. You remember President Landon? No? In fact, Landon carried only two states. Roosevelt won, 62% to 37%, and, perhaps coincidentally, the *Digest* went bankrupt soon afterward.

What went wrong? One problem was that the *Digest*'s sample wasn't representative. Where would *you* find 10 million names and addresses to sample? The *Digest* used the phone book, as many surveys do.[3] But in 1936, at the height of the Great Depression, telephones were a real luxury, so they sampled more rich than poor voters. The campaign of 1936 focused on the economy, and those who were less well off were more likely to vote for the Democrat. So the *Digest*'s sample was hopelessly biased.

How do modern polls get their samples to *represent* the entire population? You might think that they'd handpick individuals to sample with care and precision.

---

[3] Today phone numbers are computer-generated to make sure that unlisted numbers are included. But even now, cell phones and VOIP Internet phones are often not included.

But in fact, they do something quite different: They select individuals to sample *at random*. The importance of deliberately using randomness is one of the great insights of Statistics.

Idea 2: Randomize

Think back to the soup sample. Suppose you add some salt to the pot. If you sample it from the top before stirring, you'll get the misleading idea that the whole pot is salty. If you sample from the bottom, you'll get an equally misleading idea that the whole pot is bland. By stirring, you *randomize* the amount of salt throughout the pot, making each taste more typical of the whole pot.

Not only does randomization protect you against factors that you know are in the data, it can also help protect against factors that you didn't even know were there. Suppose, while you weren't looking, a friend added a handful of peas to the soup. If they're down at the bottom of the pot, and you don't randomize the soup by stirring, your test spoonful won't have any peas. By stirring in the salt, you *also* randomize the peas throughout the pot, making your sample taste more typical of the overall pot *even though you didn't know the peas were there.* So randomizing protects us even in this case.

How do we "stir" people in a survey? We select them at random. **Randomizing** protects us from the influences of *all* the features of our population by making sure that, *on average,* the sample looks like the rest of the population.

> **Why not match the sample to the population?** Rather than randomizing, we could try to design our sample so that the people we choose are typical in terms of every characteristic we can think of. We might want the income levels of those we sample to match the population. How about age? Political affiliation? Marital status? Having children? Living in the suburbs? We can't possibly think of all the things that might be important. Even if we could, we wouldn't be able to match our sample to the population for all these characteristics.

A S *Activity:* **Sampling from Some Real Populations.** Draw random samples to see how closely they resemble each other and the population.

FOR EXAMPLE   Is a random sample representative?

Here are summary statistics comparing two samples of 8000 drawn at random from a company's database of 3.5 million customers:

Age (yr)	White (%)	Female (%)	# of Children	Income Bracket (1–7)	Wealth Bracket (1–9)	Homeowner? (% Yes)
61.4	85.12	56.2	1.54	3.91	5.29	71.36
61.2	84.44	56.4	1.51	3.88	5.33	72.30

Question: Do you think these samples are representative of the population? Explain.

The two samples look very similar with respect to these seven variables. It appears that randomizing has automatically matched them pretty closely. We can reasonably assume that since the two samples don't differ too much from each other, they don't differ much from the rest of the population either.

Idea 3: It's the Sample Size

How large a random sample do we need for the sample to be reasonably representative of the population? Most people think that we need a large percentage, or *fraction,* of the population, but it turns out that what matters is the

A friend who knows that you are taking Statistics asks your advice on her study. What can you possibly say that will be helpful? Just say, "If you could just get a larger sample, it would probably improve your study." Even though a larger sample might not be worth the cost, it will almost always make the results more precise.

number of individuals *in the sample*, not the size of the population. A random sample of 100 students in a college represents the student body just about as well as a random sample of 100 voters represents the entire electorate of the United States. This is the *third* idea and probably the most surprising one in designing surveys.

How can it be that only the size of the sample, and not the population, matters? Well, let's return one last time to that pot of soup. If you're cooking for a banquet rather than just for a few people, your pot will be bigger, but do you need a bigger spoon to decide how the soup tastes? Of course not. The same-size spoonful is probably enough to make a decision about the entire pot, no matter how large the pot. The *fraction* of the population that you've sampled doesn't matter.[4] It's the **sample size** itself that's important.

How big a sample do you need? That depends on what you're estimating. To get an idea of what's really in the soup, you'll need a large enough taste to get a *representative* sample from the pot. For a survey that tries to find the proportion of the population falling into a category, you'll usually need several hundred respondents to say anything precise enough to be useful.[5]

> **What do the pollsters do?**   How do professional polling agencies do their work? The most common polling method today is to contact respondents by telephone. Computers generate random telephone numbers, so pollsters can even call some people with unlisted phone numbers. The person who answers the phone is invited to respond to the survey—if that person qualifies. (For example, only if it's an adult who lives at that address.) If the person answering doesn't qualify, the caller will ask for an appropriate alternative. In phrasing questions, pollsters often list alternative responses (such as candidates' names) in different orders to avoid biases that might favor the first name on the list.
>
> Do these methods work? The Pew Research Center for the People and the Press, reporting on one survey, says that
>
> *Across five days of interviewing, surveys today are able to make some kind of contact with the vast majority of households (76%), and there is no decline in this contact rate over the past seven years. But because of busy schedules, skepticism and outright refusals, interviews were completed in just 38% of households that were reached using standard polling procedures.*
>
> Nevertheless, studies indicate that those actually sampled can give a good snapshot of larger populations from which the surveyed households were drawn.

Does a Census Make Sense?

Why bother determining the right sample size? Wouldn't it be better to just include everyone and "sample" the entire population? Such a special sample is called a **census.** Although a census would appear to provide the best possible information about the population, there are a number of reasons why it might not.

First, it can be difficult to complete a census. Some individuals in the population will be hard (and expensive) to locate. Or a census might just be impractical. If you were a taste tester for the Hostess™ Company, you probably wouldn't want to census *all* the Twinkies on the production line. Not only might this be life-endangering, but you wouldn't have any left to sell.

---

[4] Well, that's not exactly true. If the population is small enough and the sample is more than 10% of the whole population, it *can* matter. It doesn't matter whenever, as usual, our sample is a very small fraction of the population.

[5] Chapter 19 gives the details behind this statement and shows how to decide on a sample size for a survey.

Second, populations rarely stand still. In populations of people, babies are born and folks die or leave the country. In opinion surveys, events may cause a shift in opinion during the survey. A census takes longer to complete and the population changes while you work. A sample surveyed in just a few days may give more accurate information.

Third, taking a census can be more complex than sampling. For example, the U.S. Census records too many college students. Many are counted once with their families and are then counted a second time in a report filed by their schools.

> **The undercount.** It's particularly difficult to compile a complete census of a population as large, complex, and spread out as the U.S. population. The U.S. Census is known to miss some residents. On occasion, the undercount has been striking. For example, there have been blocks in inner cities in which the number of residents recorded by the Census was smaller than the number of electric meters for which bills were being paid. What makes the problem particularly important is that some groups have a higher probability of being missed than others—undocumented immigrants, the homeless, the poor. The Census Bureau proposed the use of random sampling to estimate the number of residents missed by the ordinary census. Unfortunately, the resulting debate has become more political than statistical.

Populations and Parameters

> Any quantity that we calculate from data could be called a "statistic." But in practice, we usually use a statistic to estimate a population parameter.

> **A S** *Activity:* **Statistics and Parameters.** Explore the difference between statistics and parameters.

> Remember: Population model parameters are not just unknown—usually they are *unknowable*. We have to settle for sample statistics.

A study found that teens were less likely to "buckle up." The National Center for Chronic Disease Prevention and Health Promotion reports that 21.7% of U.S. teens never or rarely wear seatbelts. We're sure they didn't take a census, so what *does* the 21.7% mean? We can't know what percentage of teenagers wear seatbelts. Reality is just too complex. But we can simplify the question by building a model.

Models use mathematics to represent reality. Parameters are the key numbers in those models. A parameter used in a model for a population is sometimes called (redundantly) a **population parameter.**

But let's not forget about the data. We use summaries of the data to estimate the population parameters. As we know, any summary found from the data is a **statistic.** Sometimes you'll see the (also redundant) term **sample statistic.**[6]

We've already met two parameters in Chapter 6: the mean, μ, and the standard deviation, σ. We'll try to keep denoting population model parameters with Greek letters and the corresponding statistics with Latin letters. Usually, but not always, the letter used for the statistic and the parameter correspond in a natural way. So the standard deviation of the data is s, and the corresponding parameter is σ (Greek for s). In Chapter 7, we used r to denote the sample correlation. The corresponding correlation in a model for the population would be called ρ (rho). In Chapter 8, b_1 represented the slope of a linear regression estimated from the data. But when we think about a (linear) *model* for the population, we denote the slope parameter β_1 (beta).

Get the pattern? Good. Now it breaks down. We denote the mean of a population model with μ (because μ is the Greek letter for m). It might make sense to denote the sample mean with m, but long-standing convention is to put a bar over anything when we average it, so we write $\bar{y}$. What about proportions? Suppose we want to talk about the proportion of teens who don't wear seatbelts. If we use p to denote the proportion from the data, what is the corresponding model parameter? By all rights it should be π. But statements like $\pi = 0.25$ might be confusing because π has been equal to 3.1415926 . . . for so long, and it's worked so *well*. So, once again we violate the rule. We'll use p for the population model

---

[6] Where else besides a sample *could* a statistic come from?

parameter and $\hat{p}$ for the proportion from the data (since, like $\hat{y}$ in regression, it's an estimated value).

Here's a table summarizing the notation:

Name	Statistic	Parameter
Mean	$\bar{y}$	μ (mu, pronounced "meeoo," not "moo")
Standard deviation	s	σ (sigma)
Correlation	r	ρ (rho)
Regression coefficient	b	β (beta, pronounced "baytah"[7])
Proportion	$\hat{p}$	p (pronounced "pee"[8])

We draw samples because we can't work with the entire population, but we want the statistics we compute from a sample to reflect the corresponding parameters accurately. A sample that does this is said to be **representative.** A biased sampling methodology tends to over- or underestimate the parameter of interest.

JUST CHECKING

1. Various claims are often made for surveys. Why is each of the following claims not correct?
  a) It is always better to take a census than to draw a sample.
  b) Stopping students on their way out of the cafeteria is a good way to sample if we want to know about the quality of the food there.
  c) We drew a sample of 100 from the 3000 students in a school. To get the same level of precision for a town of 30,000 residents, we'll need a sample of 1000.
  d) A poll taken at a statistics support Web site garnered 12,357 responses. The majority said they enjoy doing statistics homework. With a sample size that large, we can be pretty sure that most Statistics students feel this way, too.
  e) The true percentage of all Statistics students who enjoy the homework is called a "population statistic."

Simple Random Samples

How would you select a representative sample? Most people would say that every individual in the population should have an equal chance to be selected, and certainly that seems fair. But it's not sufficient. There are many ways to give everyone an equal chance that still wouldn't give a representative sample. Consider, for example, a school that has equal numbers of males and females. We could sample like this: Flip a coin. If it comes up heads, select 100 female students at random. If it comes up tails, select 100 males at random. Everyone has an equal chance of selection, but every sample is of only a single sex—hardly representative.

We need to do better. Suppose we insist that every possible *sample* of the size we plan to draw has an equal chance to be selected. This ensures that situations like the one just described are not likely to occur and still guarantees that each person has an equal chance of being selected. What's different is that with this method, each *combination* of people has an equal chance of being selected as well. A sample drawn in this way is called a **Simple Random Sample,** usually abbreviated **SRS.** An SRS is the standard against which we measure other sampling methods, and the sampling method on which the theory of working with sampled data is based.

To select a sample at random, we first need to define where the sample will come from. The **sampling frame** is a list of individuals from which the sample is drawn.

---

[7] If you're from the United States. If you're British or Canadian, it's "beetah."
[8] Just in case you weren't sure.

For example, to draw a random sample of students at a college, we might obtain a list of all registered full-time students and sample from that list. In defining the sampling frame, we must deal with the details of defining the population. Are part-time students included? How about those who are attending school elsewhere and transferring credits back to the college?

Once we have a sampling frame, the easiest way to choose an SRS is to assign a random number to each individual in the sampling frame. We then select only those whose random numbers satisfy some rule.[9] Let's look at some ways to do this.

FOR EXAMPLE Using random numbers to get an SRS

There are 80 students enrolled in an introductory Statistics class; you are to select a sample of 5.

Question: How can you select an SRS of 5 students using these random digits found on the Internet: 05166 29305 77482?

First I'll number the students from 00 to 79. Taking the random numbers two digits at a time gives me 05, 16, 62, 93, 05, 77, and 48. I'll ignore 93 because the students were numbered only up to 79. And, so as not to pick the same person twice, I'll skip the repeated number 05. My simple random sample consists of students with the numbers 05, 16, 62, 77, and 48.

Error Okay, Bias Bad!
Sampling variability is sometimes referred to as *sampling error*, making it sound like it's some kind of mistake. It's not. We understand that samples will vary, so "sampling error" is to be expected. It's *bias* we must strive to avoid. Bias means our sampling method distorts our view of the population, and that will surely lead to mistakes.

▶ We can be more efficient when we're choosing a larger sample from a sampling frame stored in a data file. First we assign a random number with several digits (say, from 0 to 10,000) to each individual. Then we arrange the random numbers in numerical order, keeping each name with its number. Choosing the first *n* names from this re-arranged list will give us a random sample of that size.

▶ Often the sampling frame is so large that it would be too tedious to number everyone consecutively. If our intended sample size is approximately 10% of the sampling frame, we can assign each individual a single random digit 0 to 9. Then we select only those with a specific random digit, say, 5.

Samples drawn at random generally differ one from another. Each draw of random numbers selects *different* people for our sample. These differences lead to different values for the variables we measure. We call these sample-to-sample differences **sampling variability.** Surprisingly, sampling variability isn't a problem; it's an opportunity. In future chapters we'll investigate what the variation in a sample can tell us about its population.

Stratified Sampling

Simple random sampling is not the only fair way to sample. More complicated designs may save time or money or help avoid sampling problems. All statistical sampling designs have in common the idea that chance, rather than human choice, is used to select the sample.

Designs that are used to sample from large populations—especially populations residing across large areas—are often more complicated than simple random samples. Sometimes the population is first sliced into homogeneous groups, called **strata,** before the sample is selected. Then simple random sampling is used within each stratum before the results are combined. This common sampling design is called **stratified random sampling.**

Why would we want to complicate things? Here's an example. Suppose we want to learn how students feel about funding for the football team at a large

---

[9] Chapter 11 presented ways of finding and working with random numbers.

university. The campus is 60% men and 40% women, and we suspect that men and women have different views on the funding. If we use simple random sampling to select 100 people for the survey, we could end up with 70 men and 30 women or 35 men and 65 women. Our resulting estimates of the level of support for the football funding could vary widely. To help reduce this sampling variability, we can decide to force a representative balance, selecting 60 men at random and 40 women at random. This would guarantee that the proportions of men and women within our sample match the proportions in the population, and that should make such samples more accurate in representing population opinion.

You can imagine the importance of stratifying by race, income, age, and other characteristics, depending on the questions in the survey. Samples taken within a stratum vary less, so our estimates can be more precise. This reduced sampling variability is the most important benefit of stratifying.

Stratified sampling can also help us notice important differences among groups. As we saw in Chapter 3, if we unthinkingly combine group data, we risk reaching the wrong conclusion, becoming victims of Simpson's paradox.

FOR EXAMPLE   Stratifying the sample

Recap: You're trying to find out what freshmen think of the food served on campus. Food Services believes that men and women typically have different opinions about the importance of the salad bar.

Question: How should you adjust your sampling strategy to allow for this difference?

I will stratify my sample by drawing an SRS of men and a separate SRS of women—assuming that the data from the registrar include information about each person's sex.

Cluster and Multistage Sampling

Suppose we wanted to assess the reading level of this textbook based on the length of the sentences. Simple random sampling could be awkward; we'd have to number each sentence, then find, for example, the 576th sentence or the 2482nd sentence, and so on. Doesn't sound like much fun, does it?

It would be much easier to pick a few *pages* at random and count the lengths of the sentences on those pages. That works if we believe that each page is representative of the entire book in terms of reading level. Splitting the population into representative **clusters** can make sampling more practical. Then we could simply select one or a few clusters at random and perform a census within each of them. This sampling design is called **cluster sampling.** If each cluster represents the full population fairly, cluster sampling will be unbiased.

FOR EXAMPLE   Cluster sampling

Recap: In trying to find out what freshmen think about the food served on campus, you've considered both an SRS and a stratified sample. Now you have run into a problem: It's simply too difficult and time consuming to track down the individuals whose names were chosen for your sample. Fortunately, freshmen at your school are all housed in 10 freshman dorms.

Questions: How could you use this fact to draw a cluster sample? How might that alleviate the problem? What concerns do you have?

To draw a cluster sample, I would select one or two dorms at random and then try to contact everyone in each selected dorm. I could save time by simply knocking on doors on a given evening and interviewing people. I'd have to assume that freshmen were assigned to dorms pretty much at random and that the people I'm able to contact are representative of everyone in the dorm.

What's the difference between cluster sampling and stratified sampling? We stratify to ensure that our sample represents different groups in the population, and we sample randomly within each stratum. Strata are internally homogeneous, but differ from one another. By contrast, clusters are internally heterogeneous, each resembling the overall population. We select clusters to make sampling more practical or affordable.

> **Stratified vs. cluster sampling.** Boston cream pie consists of a layer of yellow cake, a layer of pastry creme, another cake layer, and then a chocolate frosting. Suppose you are a professional taster (yes, there really are such people) whose job is to check your company's pies for quality. You'd need to eat small samples of randomly selected pies, tasting all three components: the cake, the creme, and the frosting.
>
> One approach is to cut a thin vertical slice out of the pie. Such a slice will be a lot like the entire pie, so by eating that slice, you'll learn about the whole pie. This vertical slice containing all the different ingredients in the pie would be a *cluster* sample.
>
> Another approach is to sample in *strata:* Select some tastes of the cake at random, some tastes of creme at random, and some bits of frosting at random. You'll end up with a reliable judgment of the pie's quality.
>
> Many populations you might want to learn about are like this Boston cream pie. You can think of the subpopulations of interest as horizontal strata, like the layers of pie. Cluster samples slice vertically across the layers to obtain clusters, each of which is representative of the entire population. Stratified samples represent the population by drawing some from each layer, reducing variability in the results that could arise because of the differences among the layers.

Strata or Clusters?
We may split a population into strata or clusters. What's the difference? We create strata by dividing the population into groups of similar individuals so that each stratum is different from the others. By contrast, since clusters each represent the entire population, they all look pretty much alike.

Sometimes we use a variety of sampling methods together. In trying to assess the reading level of this book, we might worry that it starts out easy and then gets harder as the concepts become more difficult. If so, we'd want to avoid samples that selected heavily from early or from late chapters. To guarantee a fair mix of chapters, we could randomly choose one chapter from each of the seven parts of the book and then randomly select a few pages from each of those chapters. If, altogether, that made too many sentences, we might select a few sentences at random from each of the chosen pages. So, what is our sampling strategy? First we stratify by the part of the book and randomly choose a chapter to represent each stratum. Within each selected chapter, we choose pages as clusters. Finally, we consider an SRS of sentences within each cluster. Sampling schemes that combine several methods are called **multistage samples.** Most surveys conducted by professional polling organizations use some combination of stratified and cluster sampling as well as simple random samples.

FOR EXAMPLE Multistage sampling

Recap: Having learned that freshmen are housed in separate dorms allowed you to sample their attitudes about the campus food by going to dorms chosen at random, but you're still concerned about possible differences in opinions between men and women. It turns out that these freshmen dorms house the sexes on alternate floors.

Question: How can you design a sampling plan that uses this fact to your advantage?

Now I can stratify my sample by sex. I would first choose one or two dorms at random and then select some dorm floors at random from among those that house men and, separately, from among those that house women. I could then treat each floor as a cluster and interview everyone on that floor.

Systematic Samples

Some samples select individuals systematically. For example, you might survey every 10th person on an alphabetical list of students. To make it random, you still must start the systematic selection from a randomly selected individual. When the order of the list is not associated in any way with the responses sought, **systematic sampling** can give a representative sample. Systematic sampling can be much less expensive than true random sampling. When you use a systematic sample, you should justify the assumption that the systematic method is not associated with any of the measured variables.

Think about the reading-level sampling example again. Suppose we have chosen a chapter of the book at random, then three pages at random from that chapter, and now we want to select a sample of 10 sentences from the 73 sentences found on those pages. Instead of numbering each sentence so we can pick a simple random sample, it would be easier to sample systematically. A quick calculation shows $73/10 = 7.3$, so we can get our sample by just picking every seventh sentence on the page. But where should you start? At random, of course. We've accounted for $10 \times 7 = 70$ of the sentences, so we'll throw the extra 3 into the starting group and choose a sentence at random from the first 10. Then we pick every seventh sentence after that and record its length.

JUST CHECKING

2. We need to survey a random sample of the 300 passengers on a flight from San Francisco to Tokyo. Name each sampling method described below.
   a) Pick every 10th passenger as people board the plane.
   b) From the boarding list, randomly choose 5 people flying first class and 25 of the other passengers.
   c) Randomly generate 30 seat numbers and survey the passengers who sit there.
   d) Randomly select a seat position (right window, right center, right aisle, etc.) and survey all the passengers sitting in those seats.

STEP-BY-STEP EXAMPLE | Sampling

The assignment says, "Conduct your own sample survey to find out how many hours per week students at your school spend watching TV during the school year." Let's see how we might do this step by step. (Remember, though—actually collecting the data from your sample can be difficult and time consuming.)

Question: How would you design this survey?

Plan State what you want to know.

I wanted to design a study to find out how many hours of TV students at my school watch.

Population and Parameter Identify the W's of the study. The *Why* determines the population and the associated sampling frame. The *What* identifies the parameter of interest and the variables measured. The *Who* is the sample we actually draw. The *How, When,* and *Where* are given by the sampling plan.

The population studied was students at our school. I obtained a list of all students currently enrolled and used it as the sampling frame. The parameter of interest was the number of TV hours watched per week during the school year, which I attempted to measure by asking students how much TV they watched during the previous week.

Often, thinking about the *Why* will help us see whether the sampling frame and plan are adequate to learn about the population.

Sampling Plan Specify the sampling method and the sample size, *n*. Specify how the sample was actually drawn. What is the sampling frame? How was the randomization performed?

A good description should be complete enough to allow someone to replicate the procedure, drawing another sample from the same population in the same manner.

I decided against stratifying by class or sex because I didn't think TV watching would differ much between males and females or across classes. I selected a simple random sample of students from the list. I obtained an alphabetical list of students, assigned each a random digit between 0 and 9, and then selected all students who were assigned a "4." This method generated a sample of 212 students from the population of 2133 students.

Sampling Practice Specify *When, Where,* and *How* the sampling was performed. Specify any other details of your survey, such as how respondents were contacted, what incentives were offered to encourage them to respond, how nonrespondents were treated, and so on.

The survey was taken over the period Oct. 15 to Oct. 25. Surveys were sent to selected students by e-mail, with the request that they respond by e-mail as well. Students who could not be reached by e-mail were handed the survey in person.

Summary and Conclusion This report should include a discussion of all the elements. In addition, it's good practice to discuss any special circumstances. Professional polling organizations report the *When* of their samples but will also note, for example, any important news that might have changed respondents' opinions during the sampling process. In this survey, perhaps, a major news story or sporting event might change students' TV viewing behavior.

The question you ask also matters. It's better to be specific ("How many hours did you watch TV last week?") than to ask a general question ("How many hours of TV do you usually watch in a week?").

During the period Oct. 15 to Oct. 25, 212 students were randomly selected, using a simple random sample from a list of all students currently enrolled. The survey they received asked the following question: "How many hours did you spend watching television last week?"

Of the 212 students surveyed, 110 responded. It's possible that the nonrespondents differ in the number of TV hours watched from those who responded, but I was unable to follow up on them due to limited time and funds. The 110 respondents reported an average 3.62 hours of TV watching per week. The median was only 2 hours per week. A histogram of the data shows that the distribution is highly right-skewed, indicating that the median might be a more appropriate summary of the typical TV watching of the students.

The report should show a display of the data, provide and interpret the statistics from the sample, and state the conclusions that you reached about the population.

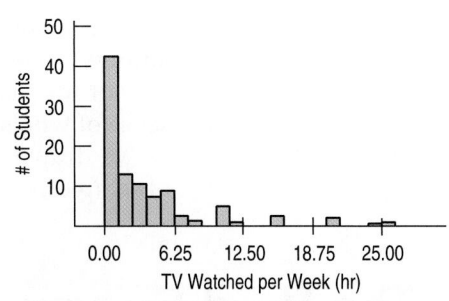

Most of the students (90%) watch between 0 and 10 hours per week, while 30% reported watching less than 1 hour per week. A few watch much more. About 3% reported watching more than 20 hours per week.

Defining the "Who": You Can't Always Get What You Want

> The population is determined by the *Why* of the study. Unfortunately, the sample is just those we can reach to obtain responses—the *Who* of the study. This difference could undermine even a well-designed study.

Before you start a survey, think first about the population you want to study. You may find that it's not the well-defined group you thought it was. Who, exactly, is a student, for example? Even if the population seems well defined, it may not be a practical group from which to draw a sample. For example, election polls want to sample from all those who will vote in the next election—a population that is impossible to identify before Election Day.

Next, you must specify the sampling frame. (Do you have a list of students to sample from? How about a list of registered voters?) Usually, the sampling frame is not the group you *really* want to know about. (All those registered to vote are not equally likely to show up.) The sampling frame limits what your survey can find out.

Then there's your target sample. These are the individuals for whom you *intend* to measure responses. You're not likely to get responses from all of them. ("I know it's dinnertime, but I'm sure you wouldn't mind answering a few questions. It'll only take 20 minutes or so. Oh, you're busy?") Nonresponse is a problem in many surveys.

Finally, there's your sample—the actual respondents. These are the individuals about whom you *do* get data and can draw conclusions. Unfortunately, they might not be representative of the sampling frame or the population.

CALVIN AND HOBBES © 1993 Watterson. Reprinted with permission of Universal Press Syndicate. All rights reserved.

At each step, the group we can study may be constrained further. The *Who* keeps changing, and each constraint can introduce biases. A careful study should address the question of how well each group matches the population of interest. One of the main benefits of simple random sampling is that it never loses its sense of who's *Who*. The *Who* in an SRS is the population of interest from which we've drawn a representative sample. That's not always true for other kinds of samples.

The Valid Survey

It isn't sufficient to just draw a sample and start asking questions. We'll want our survey to be *valid*. A valid survey yields the information we are seeking about the population we are interested in. Before setting out to survey, ask yourself:

- ▶ What do I want to know?
- ▶ Am I asking the right respondents?
- ▶ Am I asking the right questions?
- ▶ What would I do with the answers if I had them; would they address the things I want to know?

These questions may sound obvious, but there are a number of pitfalls to avoid.

Know what you want to know. Before considering a survey, understand what you hope to learn and about whom you hope to learn it. Far too often, people decide to perform a survey without any clear idea of what they hope to learn.

Use the right frame. A valid survey obtains responses from the appropriate respondents. Be sure you have a suitable *sampling frame*. Have you identified the population of interest and sampled from it appropriately? A company might survey customers who returned warranty registration cards, a readily available sampling frame. But if the company wants to know how to make their product more attractive, the most important population is the customers who rejected their product in favor of one from a competitor.

Tune your instrument. It is often tempting to ask questions you don't really need, but beware—longer questionnaires yield fewer responses and thus a greater chance of nonresponse bias.

Ask specific rather than general questions. People are not very good at estimating their typical behavior, so it is better to ask "How many hours did you sleep last night?" than "How much do you usually sleep?" Sure, some responses will include some unusual events (My dog was sick; I was up all night.), but overall you'll get better data.

Ask for quantitative results when possible. "How many magazines did you read last week?" is better than "How much do you read: A lot, A moderate amount, A little, or None at all?"

Be careful in phrasing questions. A respondent may not understand the question—or may understand the question differently than the researcher intended it. ("Does anyone in your family belong to a union?" Do you mean just me, my spouse, and my children? Or does "family" include my father, my siblings, and my second cousin once removed? What about my grandfather, who is staying with us? I think he once belonged to the Autoworkers Union.) Respondents are unlikely (or may not have the opportunity) to ask for clarification. A question like "Do you approve of the recent actions of the Secretary of Labor?" is likely not to measure what you want if many respondents don't know who the Secretary of Labor is or what actions he or she recently made.

Respondents may even lie or shade their responses if they feel embarrassed by the question ("Did you have too much to drink last night?"), are intimidated or insulted by the question ("Could you understand our new *Instructions for Dummies* manual, or was it too difficult for you?"), or if they want to avoid offending the interviewer ("Would you hire a man with a tattoo?" asked by a tattooed interviewer). Also, be careful to avoid phrases that have double or regional meanings. "How often do you go to town?" might be interpreted differently by different people and cultures.

Even subtle differences in phrasing can make a difference. In January 2006, the *New York Times* asked half of the 1229 U.S. adults in their sample the following question:

> After 9/11, President Bush authorized government wiretaps on some phone calls in the U.S. without getting court warrants, saying this was necessary to reduce the threat of terrorism. Do you approve or disapprove of this?

They found that 53% of respondents approved. But when they asked the other half of their sample a question with only slightly different phrasing,

> After 9/11, George W. Bush authorized government wiretaps on some phone calls in the U.S. without getting court warrants. Do you approve or disapprove of this?

only 46% approved.

Be careful in phrasing answers. It's often a good idea to offer choices rather than inviting a free response. Open-ended answers can be difficult to analyze. "How did you like the movie?" may start an interesting debate, but it may be better to give a range of possible responses. Be sure to phrase them in a neutral way. When asking "Do you support higher school taxes?" positive responses could be worded "Yes," "Yes, it is important for our children," or "Yes, our future depends on it." But those are not equivalent answers.

THE WIZARD OF ID   parker and hart

The best way to protect a survey from such unanticipated measurement errors is to perform a pilot survey. A **pilot** is a trial run of the survey you eventually plan to give to a larger group, using a draft of your survey questions administered to a small sample drawn from the same sampling frame you intend to use. By analyzing the results from this smaller survey, you can often discover ways to improve your instrument.

WHAT CAN GO WRONG?—OR, HOW TO SAMPLE BADLY

Bad sample designs yield worthless data. Many of the most convenient forms of sampling can be seriously biased. And there is no way to correct for the bias from a bad sample. So it's wise to pay attention to sample design—and to beware of reports based on poor samples.

SAMPLE BADLY WITH VOLUNTEERS

One of the most common dangerous sampling methods is a voluntary response sample. In a **voluntary response sample,** a large group of individuals is invited to respond, and all who do respond are counted. This method is used by call-in shows, 900 numbers, Internet polls, and letters written to members of Congress. Voluntary response samples are almost always biased, and so conclusions drawn from them are almost always wrong.

It's often hard to define the sampling frame of a voluntary response study. Practically, the frames are groups such as Internet users who frequent a particular Web site or those who happen to be watching a particular TV show at the moment. But those sampling frames don't correspond to interesting populations.

Even within the sampling frame, voluntary response samples are often biased toward those with strong opinions or those who are strongly motivated. People with very negative opinions tend to respond more often than those with equally strong positive opinions. The sample is not representative, even though every individual in the population may have been offered the chance to respond. The resulting **voluntary response bias** invalidates the survey.

A S *Activity:* **Sources of Sampling Bias.** Here's a narrated exploration of sampling bias.

> **If you had it to do over again, would you have children?** Ann Landers, the advice columnist, asked parents this question. The overwhelming majority—70% of the more than 10,000 people who wrote in—said no, kids weren't worth it. A more carefully designed survey later showed that about 90% of parents actually are happy with their decision to have children. What accounts for the striking difference in these two results? What parents do you think are most likely to respond to the original question?

FOR EXAMPLE | Bias in sampling

Recap: You're trying to find out what freshmen think of the food served on campus, and have thought of a variety of sampling methods, all time consuming. A friend suggests that you set up a "Tell Us What You Think" Web site and invite freshmen to visit the site to complete a questionnaire.

Question: What's wrong with this idea?

Letting each freshman decide whether to participate makes this a voluntary response survey. Students who were dissatisfied might be more likely to go to the Web site to record their complaints, and this could give me a biased view of the opinions of all freshmen.

SAMPLE BADLY, BUT CONVENIENTLY

Do you use the Internet?
Click here ○ for yes
Click here ○ for no

Another sampling method that doesn't work is convenience sampling. As the name suggests, in **convenience sampling** we simply include the individuals who are convenient for us to sample. Unfortunately, this group may not be representative of the population. A recent survey of 437 potential home buyers in Orange County, California, found, among other things, that

All but 2 percent of the buyers have at least one computer at home, and 62 percent have two or more. Of those with a computer, 99 percent are connected to the Internet (Jennifer Hieger, "Portrait of Homebuyer Household: 2 Kids and a PC," Orange County Register, 27 July 2001).

Later in the article, we learn that the survey was conducted via the Internet! That was a convenient way to collect data and surely easier than drawing a simple random sample, but perhaps home builders shouldn't conclude from this study that *every* family has a computer and an Internet connection.

Many surveys conducted at shopping malls suffer from the same problem. People in shopping malls are not necessarily representative of the population of interest. Mall shoppers tend to be more affluent and include a larger percentage of teenagers and retirees than the population at large. To make matters worse, survey interviewers tend to select individuals who look "safe," or easy to interview.

> Internet convenience surveys are worthless. As voluntary response surveys, they have no well-defined sampling frame (all those who use the Internet and visit their site?) and thus report no useful information. Do not believe them.

FOR EXAMPLE | Bias in sampling

Recap: To try to gauge freshman opinion about the food served on campus, Food Services suggests that you just stand outside a school cafeteria at lunchtime and stop people to ask them questions.

Questions: What's wrong with this sampling strategy?

This would be a convenience sample, and it's likely to be biased. I would miss people who use the cafeteria for dinner, but not for lunch, and I'd never hear from anyone who hates the food so much that they have stopped coming to the school cafeterias.

SAMPLE FROM A BAD SAMPLING FRAME

An SRS from an incomplete sampling frame introduces bias because the individuals included may differ from the ones not in the frame. People in prison, homeless people, students, and long-term travelers are all likely to be missed. In telephone surveys, people who have only cell phones or who use VOIP Internet phones are often missing from the sampling frame.

UNDERCOVERAGE

Many survey designs suffer from **undercoverage,** in which some portion of the population is not sampled at all or has a smaller representation in the sample than it has in the population. Undercoverage can arise for a number of reasons, but it's always a potential source of bias.

Telephone surveys are usually conducted when you are likely to be home, interrupting your dinner. If you eat out often, you may be less likely to be surveyed, a possible source of undercoverage.

WHAT *Else* CAN GO WRONG?

▶ **Watch out for nonrespondents.** A common and serious potential source of bias for most surveys is **nonresponse bias.** No survey succeeds in getting responses from everyone. The problem is that those who don't respond may differ from those who do. And they may differ on just the variables we care about. The lack of response will

(continued)

bias the results. Rather than sending out a large number of surveys for which the response rate will be low, it is often better to design a smaller randomized survey for which you have the resources to ensure a high response rate. One of the problems with nonresponse bias is that it's usually impossible to tell what the nonrespondents might have said.

Remember the *Literary Digest* Survey?   It turns out that they were wrong on two counts. First, their list of 10 million people was not representative. There was a selection bias in their sampling frame. There was also a nonresponse bias. We know this because the *Digest* also surveyed a *systematic* sample in Chicago, sending the same question used in the larger survey to every third registered voter. They *still* got a result in favor of Landon, even though Chicago voted overwhelmingly for Roosevelt in the election. This suggests that the Roosevelt supporters were less likely to respond to the *Digest* survey. There's a modern version of this problem: It's been suggested that those who screen their calls with caller ID or an answering machine, and so might not talk to a pollster, may differ in wealth or political views from those who just answer the phone.

> **Work hard to avoid influencing responses.** Response bias[10] refers to anything in the survey design that influences the responses. Response biases include the tendency of respondents to tailor their responses to try to please the interviewer, the natural unwillingness of respondents to reveal personal facts or admit to illegal or unapproved behavior and the ways in which the wording of the questions can influence responses.

HOW TO THINK ABOUT BIASES

> **Look for biases in any survey you encounter.** If you design one of your own, ask someone else to help look for biases that may not be obvious to you. And do this *before* you collect your data. There's no way to recover from a biased sampling method or a survey that asks biased questions. Sorry, it just can't be done.

A bigger sample size for a biased study just gives you a bigger useless study. A really big sample gives you a really big useless study. (Think of the 2.4 million *Literary Digest* responses.)

> **Spend your time and resources reducing biases.** No other use of resources is as worthwhile as reducing the biases.

> **If you can, pilot-test your survey.** Administer the survey in the exact form that you intend to use it to a small sample drawn from the population you intend to sample. Look for misunderstandings, misinterpretation, confusion, or other possible biases. Then refine your survey instrument.

> **Always report your sampling methods in detail.** Others may be able to detect biases where you did not expect to find them.

A S *Video:* **Biased Question Wording.** Watch a hapless interviewer make every mistake in the book.

A Short Survey
Given the fact that those who understand Statistics are smarter and better looking than those who don't, don't you think it is important to take a course in Statistics?

A S *Activity:* **Can a Large Sample Protect Against Bias?** Explore how we can learn about the population from large or repeated samples.

A researcher distributed a survey to an organization before some economizing changes were made. She asked how people felt about a proposed cutback in secretarial and administrative support on a seven-point scale from Very Happy to Very Unhappy.

But virtually all respondents were very unhappy about the cutbacks, so the results weren't particularly useful. If she had pretested the question, she might have chosen a scale that ran from Unhappy to Outraged.

---

[10] Response bias is not the opposite of nonresponse bias. (We don't make these terms up; we just try to explain them.)

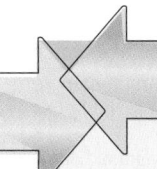

CONNECTIONS

With this chapter, we take our first formal steps to relate our sample data to a larger population. Some of these ideas have been lurking in the background as we sought patterns and summaries for data. Even when we only worked with the data at hand, we often thought about implications for a larger population of individuals.

Notice the ongoing central importance of models. We've seen models in several ways in previous chapters. Here we recognize the value of a model for a population. The parameters of such a model are values we will often want to estimate using statistics such as those we've been calculating. The connections to summary statistics for center, spread, correlation, and slope are obvious.

We now have a specific application for random numbers. The idea of applying randomness deliberately showed up in Chapter 11 for simulation. Now we need randomization to get good-quality data from the real world.

WHAT HAVE WE LEARNED?

We've learned that a representative sample can offer us important insights about populations. It's the size of the sample—and not its fraction of the larger population—that determines the precision of the statistics it yields.

We've learned several ways to draw samples, all based on the power of randomness to make them representative of the population of interest:

▸ A Simple Random Sample (SRS) is our standard. Every possible group of n individuals has an equal chance of being our sample. That's what makes it *simple*.
▸ Stratified samples can reduce sampling variability by identifying homogeneous subgroups and then randomly sampling within each.
▸ Cluster samples randomly select among heterogeneous subgroups that each resemble the population at large, making our sampling tasks more manageable.
▸ Systematic samples can work in some situations and are often the least expensive method of sampling. But we still want to start them randomly.
▸ Multistage samples combine several random sampling methods.

We've learned that bias can destroy our ability to gain insights from our sample:

▸ Nonresponse bias can arise when sampled individuals will not or cannot respond.
▸ Response bias arises when respondents' answers might be affected by external influences, such as question wording or interviewer behavior.

We've learned that bias can also arise from poor sampling methods:

▸ Voluntary response samples are almost always biased and should be avoided and distrusted.
▸ Convenience samples are likely to be flawed for similar reasons.
▸ Even with a reasonable design, sample frames may not be representative. Undercoverage occurs when individuals from a subgroup of the population are selected less often than they should be.

Finally, we've learned to look for biases in any survey we find and to be sure to report our methods whenever we perform a survey so that others can evaluate the fairness and accuracy of our results.

Terms

Population	268.	The entire group of individuals or instances about whom we hope to learn.
Sample	268.	A (representative) subset of a population, examined in hope of learning about the population.

Sample survey	269. A study that asks questions of a sample drawn from some population in the hope of learning something about the entire population. Polls taken to assess voter preferences are common sample surveys.
Bias	269. Any systematic failure of a sampling method to represent its population is bias. Biased sampling methods tend to over- or underestimate parameters. It is almost impossible to recover from bias, so efforts to avoid it are well spent. Common errors include
	▸ relying on voluntary response.
	▸ undercoverage of the population.
	▸ nonresponse bias.
	▸ response bias.
Randomization	270. The best defense against bias is randomization, in which each individual is given a fair, random chance of selection.
Sample size	271. The number of individuals in a sample. The sample size determines how well the sample represents the population, not the fraction of the population sampled.
Census	271. A sample that consists of the entire population is called a census.
Population parameter	272. A numerically valued attribute of a model for a population. We rarely expect to know the true value of a population parameter, but we do hope to estimate it from sampled data. For example, the mean income of all employed people in the country is a population parameter.
Statistic, sample statistic	272. Statistics are values calculated for sampled data. Those that correspond to, and thus estimate, a population parameter, are of particular interest. For example, the mean income of all employed people in a representative sample can provide a good estimate of the corresponding population parameter. The term "sample statistic" is sometimes used, usually to parallel the corresponding term "population parameter."
Representative	273. A sample is said to be representative if the statistics computed from it accurately reflect the corresponding population parameters.
Simple random sample (SRS)	273. A simple random sample of sample size n is a sample in which each set of n elements in the population has an equal chance of selection.
Sampling frame	273. A list of individuals from whom the sample is drawn is called the sampling frame. Individuals who may be in the population of interest, but who are not in the sampling frame, cannot be included in any sample.
Sampling variability	274. The natural tendency of randomly drawn samples to differ, one from another. Sometimes, unfortunately, called *sampling error*, sampling variability is no error at all, but just the natural result of random sampling.
Stratified random sample	274. A sampling design in which the population is divided into several subpopulations, or **strata,** and random samples are then drawn from each stratum. If the strata are homogeneous, but are different from each other, a stratified sample may yield more consistent results than an SRS.
Cluster sample	275. A sampling design in which entire groups, or **clusters,** are chosen at random. Cluster sampling is usually selected as a matter of convenience, practicality, or cost. Each cluster should be representative of the population, so all the clusters should be heterogeneous and similar to each other.
Multistage sample	276. Sampling schemes that combine several sampling methods are called multistage samples. For example, a national polling service may stratify the country by geographical regions, select a random sample of cities from each region, and then interview a cluster of residents in each city.
Systematic sample	277. A sample drawn by selecting individuals systematically from a sampling frame. When there is no relationship between the order of the sampling frame and the variables of interest, a systematic sample can be representative.
Pilot	281. A small trial run of a survey to check whether questions are clear. A pilot study can reduce errors due to ambiguous questions.
Voluntary response bias	282. Bias introduced to a sample when individuals can choose on their own whether to participate in the sample. Samples based on voluntary response are always invalid and cannot be recovered, no matter how large the sample size.

Convenience sample	282. A convenience sample consists of the individuals who are conveniently available. Convenience samples often fail to be representative because every individual in the population is not equally convenient to sample.
Undercoverage	283. A sampling scheme that biases the sample in a way that gives a part of the population less representation than it has in the population suffers from undercoverage.
Nonresponse bias	283. Bias introduced when a large fraction of those sampled fails to respond. Those who do respond are likely to not represent the entire population. Voluntary response bias is a form of nonresponse bias, but nonresponse may occur for other reasons. For example, those who are at work during the day won't respond to a telephone survey conducted only during working hours.
Response bias	284. Anything in a survey design that influences responses falls under the heading of response bias. One typical response bias arises from the wording of questions, which may suggest a favored response. Voters, for example, are more likely to express support of "the president" than support of the particular person holding that office at the moment.

Skills

> ► Know the basic concepts and terminology of sampling (see the preceding list).
>
> ► Recognize population parameters in descriptions of populations and samples.
>
> ► Understand the value of randomization as a defense against bias.
>
> ► Understand the value of sampling to estimate population parameters from statistics calculated on representative samples drawn from the population.
>
> ► Understand that the size of the sample (not the fraction of the population) determines the precision of estimates.

> ► Know how to draw a simple random sample from a master list of a population, using a computer or a table of random numbers.

> ► Know what to report about a sample as part of your account of a statistical analysis.
>
> ► Report possible sources of bias in sampling methods. Recognize voluntary response and nonresponse as sources of bias in a sample survey.

SAMPLING ON THE COMPUTER

Computer-generated pseudorandom numbers are usually good enough for drawing random samples. But there is little reason not to use the truly random values available on the Internet.

Here's a convenient way to draw an SRS of a specified size using a computer-based sampling frame. The sampling frame can be a list of names or of identification numbers arrayed, for example, as a column in a spreadsheet, statistics program, or database:

1. Generate random numbers of enough digits so that each exceeds the size of the sampling frame list by several digits. This makes duplication unlikely.
2. Assign the random numbers arbitrarily to individuals in the sampling frame list. For example, put them in an adjacent column.
3. Sort the list of random numbers, carrying along the sampling frame list.
4. Now the first *n* values in the sorted sampling frame column are an SRS of *n* values from the entire sampling frame.

EXERCISES

1. **Roper.** Through their *Roper Reports Worldwide*, GfK Roper conducts a global consumer survey to help multinational companies understand different consumer attitudes throughout the world. Within 30 countries, the researchers interview 1000 people aged 13–65. Their samples are designed so that they get 500 males and 500 females in each country. (www.gfkamerica.com)
   a) Are they using a simple random sample? Explain.
   b) What kind of design do you think they are using?

2. **Student Center Survey.** For their class project, a group of Statistics students decide to survey the student body to assess opinions about the proposed new student center. Their sample of 200 contained 50 first-year students, 50 sophomores, 50 juniors, and 50 seniors.
   a) Do you think the group was using an SRS? Why?
   b) What sampling design do you think they used?

3. **Emoticons.** The Web site www.gamefaqs.com asked, as their question of the day to which visitors to the site were invited to respond, *"Do you ever use emoticons when you type online?"* Of the 87,262 respondents, 27% said that they did not use emoticons. ;-(
   a) What kind of sample was this?
   b) How much confidence would you place in using 27% as an estimate of the fraction of people who use emoticons?

4. **Drug tests.** Major League Baseball tests players to see whether they are using performance-enhancing drugs. Officials select a team at random, and a drug-testing crew shows up unannounced to test all 40 players on the team. Each testing day can be considered a study of drug use in Major League Baseball.
   a) What kind of sample is this?
   b) Is that choice appropriate?

5. **Gallup.** At its Web site (www.gallup.com) the Gallup Poll publishes results of a new survey each day. Scroll down to the end, and you'll find a statement that includes words such as these:

   Results are based on telephone interviews with 1,008 national adults, aged 18 and older, conducted April 2–5, 2007. . . . In addition to sampling error, question wording and practical difficulties in conducting surveys can introduce error or bias into the findings of public opinion polls.

   a) For this survey, identify the population of interest.
   b) Gallup performs its surveys by phoning numbers generated at random by a computer program. What is the sampling frame?
   c) What problems, if any, would you be concerned about in matching the sampling frame with the population?

6. **Gallup World.** At its Web site (www.gallupworldpoll.com) the Gallup World Poll describes their methods. After one report they explained:

   Results are based on face-to-face interviews with randomly selected national samples of approximately 1,000 adults,

aged 15 and older, who live permanently in each of the 21 sub-Saharan African nations surveyed. Those countries include Angola (areas where land mines might be expected were excluded), Benin, Botswana, Burkina Faso, Cameroon, Ethiopia, Ghana, Kenya, Madagascar (areas where interviewers had to walk more than 20 kilometers from a road were excluded), Mali, Mozambique, Niger, Nigeria, Senegal, Sierra Leone, South Africa, Tanzania, Togo, Uganda (the area of activity of the Lord's Resistance Army was excluded from the survey), Zambia, and Zimbabwe. . . . In all countries except Angola, Madagascar, and Uganda, the sample is representative of the entire population.

   a) Gallup is interested in sub-Saharan Africa. What kind of survey design are they using?
   b) Some of the countries surveyed have large populations. (Nigeria is estimated to have about 130 million people.) Some are quite small. (Togo's population is estimated at 5.4 million.) Nonetheless, Gallup sampled 1000 adults in each country. How does this affect the precision of its estimates for these countries?

7–14. What did they do? *For the following reports about statistical studies, identify the following items (if possible). If you can't tell, then say so—this often happens when we read about a survey.*

   a) The population
   b) The population parameter of interest
   c) The sampling frame
   d) The sample
   e) The sampling method, including whether or not randomization was employed
   f) Any potential sources of bias you can detect and any problems you see in generalizing to the population of interest

7. Consumers Union asked all subscribers whether they had used alternative medical treatments and, if so, whether they had benefited from them. For almost all of the treatments, approximately 20% of those responding reported cures or substantial improvement in their condition.

8. A question posted on the Lycos Web site on 18 June 2000 asked visitors to the site to say whether they thought that marijuana should be legally available for medicinal purposes. (www.lycos.com)

9. Researchers waited outside a bar they had randomly selected from a list of such establishments. They stopped every 10th person who came out of the bar and asked whether he or she thought drinking and driving was a serious problem.

10. Hoping to learn what issues may resonate with voters in the coming election, the campaign director for a mayoral candidate selects one block from each of the city's election districts. Staff members go there and interview all the residents they can find.

11. The Environmental Protection Agency took soil samples at 16 locations near a former industrial waste dump and checked each for evidence of toxic chemicals. They found no elevated levels of any harmful substances.

12. State police set up a roadblock to estimate the percentage of cars with up-to-date registration, insurance, and safety inspection stickers. They usually find problems with about 10% of the cars they stop.

13. A company packaging snack foods maintains quality control by randomly selecting 10 cases from each day's production and weighing the bags. Then they open one bag from each case and inspect the contents.

14. Dairy inspectors visit farms unannounced and take samples of the milk to test for contamination. If the milk is found to contain dirt, antibiotics, or other foreign matter, the milk will be destroyed and the farm reinspected until purity is restored.

15. **Mistaken poll.** A local TV station conducted a "PulsePoll" about the upcoming mayoral election. Evening news viewers were invited to phone in their votes, with the results to be announced on the late-night news. Based on the phone calls, the station predicted that Amabo would win the election with 52% of the vote. They were wrong: Amabo lost, getting only 46% of the vote. Do you think the station's faulty prediction is more likely to be a result of bias or sampling error? Explain.

16. **Another mistaken poll.** Prior to the mayoral election discussed in Exercise 15, the newspaper also conducted a poll. The paper surveyed a random sample of registered voters stratified by political party, age, sex, and area of residence. This poll predicted that Amabo would win the election with 52% of the vote. The newspaper was wrong: Amabo lost, getting only 46% of the vote. Do you think the newspaper's faulty prediction is more likely to be a result of bias or sampling error? Explain.

17. **Parent opinion, part 1.** In a large city school system with 20 elementary schools, the school board is considering the adoption of a new policy that would require elementary students to pass a test in order to be promoted to the next grade. The PTA wants to find out whether parents agree with this plan. Listed below are some of the ideas proposed for gathering data. For each, indicate what kind of sampling strategy is involved and what (if any) biases might result.
    a) Put a big ad in the newspaper asking people to log their opinions on the PTA Web site.
    b) Randomly select one of the elementary schools and contact every parent by phone.
    c) Send a survey home with every student, and ask parents to fill it out and return it the next day.
    d) Randomly select 20 parents from each elementary school. Send them a survey, and follow up with a phone call if they do not return the survey within a week.

18. **Parent opinion, part 2.** Let's revisit the school system described in Exercise 17. Four new sampling strategies have been proposed to help the PTA determine whether parents favor requiring elementary students to pass a test in order to be promoted to the next grade. For each, indi-

cate what kind of sampling strategy is involved and what (if any) biases might result.
    a) Run a poll on the local TV news, asking people to dial one of two phone numbers to indicate whether they favor or oppose the plan.
    b) Hold a PTA meeting at each of the 20 elementary schools, and tally the opinions expressed by those who attend the meetings.
    c) Randomly select one class at each elementary school and contact each of those parents.
    d) Go through the district's enrollment records, selecting every 40th parent. PTA volunteers will go to those homes to interview the people chosen.

19. **Churches.** For your political science class, you'd like to take a survey from a sample of all the Catholic Church members in your city. A list of churches shows 17 Catholic churches within the city limits. Rather than try to obtain a list of all members of all these churches, you decide to pick 3 churches at random. For those churches, you'll ask to get a list of all current members and contact 100 members at random.
    a) What kind of design have you used?
    b) What could go wrong with your design?

20. **Playground.** Some people have been complaining that the children's playground at a municipal park is too small and is in need of repair. Managers of the park decide to survey city residents to see if they believe the playground should be rebuilt. They hand out questionnaires to parents who bring children to the park. Describe possible biases in this sample.

21. **Roller coasters.** An amusement park has opened a new roller coaster. It is so popular that people are waiting for up to 3 hours for a 2-minute ride. Concerned about how patrons (who paid a large amount to enter the park and ride on the rides) feel about this, they survey every 10th person on the line for the roller coaster, starting from a randomly selected individual.
    a) What kind of sample is this?
    b) What is the sampling frame?
    c) Is it likely to be representative?

22. **Playground, act two.** The survey described in Exercise 20 asked,

    Many people believe this playground is too small and in need of repair. Do you think the playground should be repaired and expanded even if that means raising the entrance fee to the park?

    Describe two ways this question may lead to response bias.

23. **Wording the survey.** Two members of the PTA committee in Exercises 17 and 18 have proposed different questions to ask in seeking parents' opinions.

    Question 1: Should elementary school–age children have to pass high-stakes tests in order to remain with their classmates?
    Question 2: Should schools and students be held accountable for meeting yearly learning goals by testing students before they advance to the next grade?

    a) Do you think responses to these two questions might differ? How? What kind of bias is this?
    b) Propose a question with more neutral wording that might better assess parental opinion.

24. Banning ephedra. An online poll at a Web site asked:

A nationwide ban of the diet supplement ephedra went into effect recently. The herbal stimulant has been linked to 155 deaths and many more heart attacks and strokes. Ephedra manufacturer NVE Pharmaceuticals, claiming that the FDA lacked proof that ephedra is dangerous if used as directed, was denied a temporary restraining order on the ban yesterday by a federal judge. Do you think that ephedra should continue to be banned nationwide?

65% of 17,303 respondents said "yes." Comment on each of the following statements about this poll:
a) With a sample size that large, we can be pretty certain we know the true proportion of Americans who think ephedra should be banned.
b) The wording of the question is clearly very biased.
c) The sampling frame is all Internet users.
d) Results of this voluntary response survey can't be reliably generalized to any population of interest.

25. Survey questions. Examine each of the following questions for possible bias. If you think the question is biased, indicate how and propose a better question.
a) Should companies that pollute the environment be compelled to pay the costs of cleanup?
b) Given that 18-year-olds are old enough to vote and to serve in the military, is it fair to set the drinking age at 21?

26. More survey questions. Examine each of the following questions for possible bias. If you think the question is biased, indicate how and propose a better question.
a) Do you think high school students should be required to wear uniforms?
b) Given humanity's great tradition of exploration, do you favor continued funding for space flights?

27. Phone surveys. Anytime we conduct a survey, we must take care to avoid undercoverage. Suppose we plan to select 500 names from the city phone book, call their homes between noon and 4 p.m., and interview whoever answers, anticipating contacts with at least 200 people.
a) Why is it difficult to use a simple random sample here?
b) Describe a more convenient, but still random, sampling strategy.
c) What kinds of households are likely to be included in the eventual sample of opinion? Excluded?
d) Suppose, instead, that we continue calling each number, perhaps in the morning or evening, until an adult is contacted and interviewed. How does this improve the sampling design?
e) Random-digit dialing machines can generate the phone calls for us. How would this improve our design? Is anyone still excluded?

28. Cell phone survey. What about drawing a random sample only from cell phone exchanges? Discuss the advantages and disadvantages of such a sampling method compared with surveying randomly generated telephone numbers from non–cell phone exchanges. Do you think these advantages and disadvantages have changed over time? How do you expect they'll change in the future?

29. Arm length. How long is your arm compared with your hand size? Put your right thumb at your left shoulder bone, stretch your hand open wide, and extend your hand down your arm. Put your thumb at the place where your little finger is, and extend down the arm again. Repeat this a third time. Now your little finger will probably have reached the back of your left hand. If the fourth hand width goes past the end of your middle finger, turn your hand sideways and count finger widths to get there.
a) How many hand and finger widths is your arm?
b) Suppose you repeat your measurement 10 times and average your results. What parameter would this average estimate? What is the population?
c) Suppose you now collect arm lengths measured in this way from 9 friends and average these 10 measurements. What is the population now? What parameter would this average estimate?
d) Do you think these 10 arm lengths are likely to be representative of the population of arm lengths in your community? In the country? Why or why not?

30. Fuel economy. Occasionally, when I fill my car with gas, I figure out how many miles per gallon my car got. I wrote down those results after 6 fill-ups in the past few months. Overall, it appears my car gets 28.8 miles per gallon.
a) What statistic have I calculated?
b) What is the parameter I'm trying to estimate?
c) How might my results be biased?
d) When the Environmental Protection Agency (EPA) checks a car like mine to predict its fuel economy, what parameter is it trying to estimate?

31. Accounting. Between quarterly audits, a company likes to check on its accounting procedures to address any problems before they become serious. The accounting staff processes payments on about 120 orders each day. The next day, the supervisor rechecks 10 of the transactions to be sure they were processed properly.
a) Propose a sampling strategy for the supervisor.
b) How would you modify that strategy if the company makes both wholesale and retail sales, requiring different bookkeeping procedures?

32. Happy workers? A manufacturing company employs 14 project managers, 48 foremen, and 377 laborers. In an effort to keep informed about any possible sources of employee discontent, management wants to conduct job satisfaction interviews with a sample of employees every month.
a) Do you see any potential danger in the company's plan? Explain.
b) Propose a sampling strategy that uses a simple random sample.
c) Why do you think a simple random sample might not provide the representative opinion the company seeks?
d) Propose a better sampling strategy.
e) Listed below are the last names of the project managers. Use random numbers to select two people to be interviewed. Explain your method carefully.

Barrett	Bowman	Chen
DeLara	DeRoos	Grigorov
Maceli	Mulvaney	Pagliarulo
Rosica	Smithson	Tadros
Williams	Yamamoto	

33. Quality control. Sammy's Salsa, a small local company, produces 20 cases of salsa a day. Each case contains 12 jars and is imprinted with a code indicating the date and batch number. To help maintain consistency, at the end of each day, Sammy selects three jars of salsa, weighs the contents, and tastes the product. Help Sammy select the sample jars. Today's cases are coded 07N61 through 07N80.
a) Carefully explain your sampling strategy.
b) Show how to use random numbers to pick 3 jars.
c) Did you use a simple random sample? Explain.

34. A fish story. Concerned about reports of discolored scales on fish caught downstream from a newly sited chemical plant, scientists set up a field station in a shoreline public park. For one week they asked fishermen there to bring any fish they caught to the field station for a brief inspection. At the end of the week, the scientists said that 18% of the 234 fish that were submitted for inspection displayed the discoloration. From this information, can the researchers estimate what proportion of fish in the river have discolored scales? Explain.

35. Sampling methods. Consider each of these situations. Do you think the proposed sampling method is appropriate? Explain.
a) We want to know what percentage of local doctors accept Medicaid patients. We call the offices of 50 doctors randomly selected from local Yellow Page listings.
b) We want to know what percentage of local businesses anticipate hiring additional employees in the upcoming month. We randomly select a page in the Yellow Pages and call every business listed there.

36. More sampling methods. Consider each of these situations. Do you think the proposed sampling method is appropriate? Explain.
a) We want to know if there is neighborhood support to turn a vacant lot into a playground. We spend a Saturday afternoon going door-to-door in the neighborhood, asking people to sign a petition.
b) We want to know if students at our college are satisfied with the selection of food available on campus. We go to the largest cafeteria and interview every 10th person in line.

JUST CHECKING
Answers

1. a) It can be hard to reach all members of a population, and it can take so long that circumstances change, affecting the responses. A well-designed sample is often a better choice.
b) This sample is probably biased—students who didn't like the food at the cafeteria might not choose to eat there.
c) No, only the sample size matters, not the fraction of the overall population.
d) Students who frequent this Web site might be more enthusiastic about Statistics than the overall population of Statistics students. A large sample cannot compensate for bias.
e) It's the population "parameter." "Statistics" describe samples.

2. a) systematic
b) stratified
c) simple
d) cluster

Experiments and Observational Studies

CHAPTER 13

W ho gets good grades? And, more importantly, why? Is there something schools and parents could do to help weaker students improve their grades? Some people think they have an answer: music! No, not your iPod, but an instrument. In a study conducted at Mission Viejo High School, in California, researchers compared the scholastic performance of music students with that of non-music students. Guess what? The music students had a much higher overall grade point average than the non-music students, 3.59 to 2.91. Not only that: A whopping 16% of the music students had all A's compared with only 5% of the non-music students.

As a result of this study and others, many parent groups and educators pressed for expanded music programs in the nation's schools. They argued that the work ethic, discipline, and feeling of accomplishment fostered by learning to play an instrument also enhance a person's ability to succeed in school. They thought that involving more students in music would raise academic performance. What do you think? Does this study provide solid evidence? Or are there other possible explanations for the difference in grades? Is there any way to really prove such a conjecture?

Observational Studies

This research tried to show an association between music education and grades. But it wasn't a survey. Nor did it assign students to get music education. Instead, it simply observed students "in the wild," recording the choices they made and the outcome. Such studies are called **observational studies.** In observational studies, researchers don't *assign* choices; they simply observe them. In addition, this was a **retrospective study,** because researchers first identified subjects who studied music and then collected data on their past grades.

What's wrong with concluding that music education causes good grades? One high school during one academic year may not be representative of the whole United States. That's true, but the real problem is that the claim that music study *caused* higher grades depends on there being *no other differences* between the groups that could account for the differences in grades, and studying music was not the *only* difference between the two groups of students.

We can think of lots of lurking variables that might cause the groups to perform differently. Students who study music may have better work habits to start with, and this makes them successful in both music and course work. Music students may have more parental support (someone had to pay for all those lessons), and that support may have enhanced their academic performance, too. Maybe they came from wealthier homes and had other advantages. Or it could be that smarter kids just like to play musical instruments.

> For rare illnesses, it's not practical to draw a large enough sample to see many ill respondents, so the only option remaining is to develop retrospective data. For example, researchers can interview those who have become ill. The likely causes of both legionnaires' disease and HIV were initially identified from such retrospective studies of the small populations who were initially infected. But to confirm the causes, researchers needed laboratory-based experiments.

Observational studies are valuable for discovering trends and possible relationships. They are used widely in public health and marketing. Observational studies that try to discover variables related to rare outcomes, such as specific diseases, are often retrospective. They first identify people with the disease and then look into their history and heritage in search of things that may be related to their condition. But retrospective studies have a restricted view of the world because they are usually restricted to a small part of the entire population. And because retrospective records are based on historical data, they can have errors. (Do you recall *exactly* what you ate even yesterday? How about last Wednesday?)

A somewhat better approach is to observe individuals over time, recording the variables of interest and ultimately seeing how things turn out. For example, we might start by selecting young students who have not begun music lessons. We could then track their academic performance over several years, comparing those who later choose to study music with those who do not. Identifying subjects in advance and collecting data as events unfold would make this a **prospective study.**

Although an observational study may identify important variables related to the outcome we are interested in, there is no guarantee that we have found the right or the most important related variables. Students who choose to study an instrument might still differ from the others in some important way that we failed to observe. It may be this difference—whether we know what it is or not—rather than music itself that leads to better grades. It's just not possible for observational studies, whether prospective or retrospective, to demonstrate a causal relationship.

FOR EXAMPLE   Designing an observational study

In early 2007, a larger-than-usual number of cats and dogs developed kidney failure; many died. Initially, researchers didn't know why, so they used an observational study to investigate.

Question: Suppose you were called on to plan a study seeking the cause of this problem. Would your design be retrospective or prospective? Explain why.

I would use a retrospective observational study. Even though the incidence of disease was higher than usual, it was still rare. Surveying all pets would have been impractical. Instead, it makes sense to locate some who were sick and ask about their diets, exposure to toxins, and other possible causes.

Randomized, Comparative Experiments

Experimental design was advanced in the 19th century by work in psychophysics by Gustav Fechner (1801–1887), the founder of experimental psychology. Fechner designed ingenious experiments that exhibited many of the features of modern designed experiments. Fechner was careful to control for the effects of factors that might affect his results. For example, in his 1860 book Elemente der Psychophysik *he cautioned readers to group experiment trials together to minimize the possible effects of time of day and fatigue.*

An Experiment:

Manipulates the factor levels to create treatments. *Randomly assigns* subjects to these treatment levels. *Compares* the responses of the subject groups across treatment levels.

"He that leaves nothing to chance will do few things ill, but he will do very few things."

—Lord Halifax
(1633–1695)

Is it *ever* possible to get convincing evidence of a cause-and-effect relationship? Well, yes it is, but we would have to take a different approach. We could take a group of third graders, randomly assign half to take music lessons, and forbid the other half to do so. Then we could compare their grades several years later. This kind of study design is called an **experiment.**

An experiment requires a **random assignment** of subjects to treatments. Only an experiment can justify a claim like "Music lessons cause higher grades." Questions such as "Does taking vitamin C reduce the chance of getting a cold?" and "Does working with computers improve performance in Statistics class?" and "Is this drug a safe and effective treatment for that disease?" require a designed experiment to establish cause and effect.

Experiments study the relationship between two or more variables. An experimenter must identify at least one explanatory variable, called a **factor,** to manipulate and at least one **response variable** to measure. What distinguishes an experiment from other types of investigation is that the experimenter actively and deliberately manipulates the factors to control the details of the possible treatments, and assigns the subjects to those treatments *at random.* The experimenter then observes the response variable and *compares* responses for different groups of subjects who have been treated differently. For example, we might design an experiment to see whether the amount of sleep and exercise you get affects your performance.

The individuals on whom or which we experiment are known by a variety of terms. Humans who are experimented on are commonly called **subjects** or **participants.** Other individuals (rats, days, petri dishes of bacteria) are commonly referred to by the more generic term **experimental unit.** When we recruit subjects for our sleep deprivation experiment by advertising in Statistics class, we'll probably have better luck if we invite them to be participants than if we advertise that we need experimental units.

The specific values that the experimenter chooses for a factor are called the **levels** of the factor. We might assign our participants to sleep for 4, 6, or 8 hours. Often there are several factors at a variety of levels. (Our subjects will also be assigned to a treadmill for 0 or 30 minutes.) The combination of specific levels from all the factors that an experimental unit receives is known as its **treatment.** (Our subjects could have any one of six different treatments—three sleep levels, each at two exercise levels.)

How should we assign our participants to these treatments? Some students prefer 4 hours of sleep, while others need 8. Some exercise regularly; others are couch potatoes. Should we let the students choose the treatments they'd prefer? No. That would not be a good idea. To have any hope of drawing a fair conclusion, we must assign our participants to their treatments *at random.*

It may be obvious to you that we shouldn't let the students choose the treatment they'd prefer, but the need for random assignment is a lesson that was once hard for some to accept. For example, physicians might naturally prefer to assign patients to the therapy that they think best rather than have a random element such as a coin flip determine the treatment. But we've known for more than a century that for the results of an experiment to be valid, we must use deliberate randomization.

The Women's Health Initiative is a major 15-year research program funded by the National Institutes of Health to address the most common causes of death, disability, and poor quality of life in older women. It consists of both an observational study with more than 93,000 participants and several randomized comparative experiments. The goals of this study include

▶ giving reliable estimates of the extent to which known risk factors predict heart disease, cancers, and fractures;

No drug can be sold in the United States without first showing, in a suitably designed experiment approved by the Food and Drug Administration (FDA), that it's safe and effective. The small print on the booklet that comes with many prescription drugs usually describes the outcomes of that experiment.

▶ identifying "new" risk factors for these and other diseases in women;
▶ comparing risk factors, presence of disease at the start of the study, and new occurrences of disease during the study across all study components; and
▶ creating a future resource to identify biological indicators of disease, especially substances and factors found in blood.

That is, the study seeks to identify possible risk factors and assess how serious they might be. It seeks to build up data that might be checked retrospectively as the women in the study continue to be followed. There would be no way to find out these things with an experiment because the task includes identifying new risk factors. If we don't know those risk factors, we could never control them as factors in an experiment.

By contrast, one of the clinical trials (randomized experiments) that received much press attention randomly assigned postmenopausal women to take either hormone replacement therapy or an inactive pill. The results published in 2002 and 2004 concluded that hormone replacement with estrogen carried increased risks of stroke.

FOR EXAMPLE   Determining the treatments and response variable

Recap: In 2007, deaths of a large number of pet dogs and cats were ultimately traced to contamination of some brands of pet food. The manufacturer now claims that the food is safe, but before it can be released, it must be tested.

Question: In an experiment to test whether the food is now safe for dogs to eat,[1] what would be the treatments and what would be the response variable?

The treatments would be ordinary-size portions of two dog foods: the new one from the company (the *test food*) and one that I was certain was safe (perhaps prepared in my kitchen or laboratory). The response would be a veterinarian's assessment of the health of the test animals.

The Four Principles of Experimental Design

A S *Video:* **An Industrial Experiment.** Manufacturers often use designed experiments to help them perfect new products. Watch this video about one such experiment.

1. **Control.** We control sources of variation other than the factors we are testing by making conditions as similar as possible for all treatment groups. For human subjects, we try to treat them alike. However, there is always a question of degree and practicality. Controlling extraneous sources of variation reduces the variability of the responses, making it easier to detect differences among the treatment groups.

   Making generalizations from the experiment to other levels of the controlled factor can be risky. For example, suppose we test two laundry detergents and carefully control the water temperature at 180°F. This would reduce the variation in our results due to water temperature, but what could we say about the detergents' performance in cold water? Not much. It would be hard to justify extrapolating the results to other temperatures.

   Although we control both experimental factors and other sources of variation, we think of them very differently. We control a factor by assigning subjects to different factor levels because we want to see how the response will change at those different levels. We control other sources of variation to *prevent* them from changing and affecting the response variable.

---

[1] It may disturb you (as it does us) to think of deliberately putting dogs at risk in this experiment, but in fact that is what is done. The risk is borne by a small number of dogs so that the far larger population of dogs can be kept safe.

The deep insight that experiments should use random assignment is quite an old one. It can be attributed to the American philosopher and scientist C. S. Peirce in his experiments with J. Jastrow, published in 1885.

A S *Activity:* **The Three Rules of Experimental Design.** Watch an animated discussion of three rules of design.

A S *Activity:* **Perform an Experiment.** How well can you read pie charts and bar charts? Find out as you serve as the subject in your own experiment.

2. **Randomize.** As in sample surveys, **randomization** allows us to equalize the effects of unknown or uncontrollable sources of variation. It does not eliminate the effects of these sources, but it should spread them out across the treatment levels so that we can see past them. If experimental units were not assigned to treatments at random, we would not be able to use the powerful methods of Statistics to draw conclusions from an experiment. Assigning subjects to treatments at random reduces bias due to uncontrolled sources of variation. Randomization protects us even from effects we didn't know about. There's an adage that says "control what you can, and randomize the rest."

3. **Replicate.** Two kinds of replication show up in comparative experiments. First, we should apply each treatment to a number of subjects. Only with such replication can we estimate the variability of responses. If we have not assessed the variation, the experiment is not complete. The outcome of an experiment on a single subject is an anecdote, not data.

A second kind of replication shows up when the experimental units are not a representative sample from the population of interest. We may believe that what is true of the students in Psych 101 who volunteered for the sleep experiment is true of all humans, but we'll feel more confident if our results for the experiment are *replicated* in another part of the country, with people of different ages, and at different times of the year. **Replication** of an entire experiment with the controlled sources of variation at different levels is an essential step in science.

4. **Block.** The ability of randomizing to equalize variation across treatment groups works best in the long run. For example, if we're allocating players to two 6-player soccer teams from a pool of 12 children, we might do so at random to equalize the talent. But what if there were two 12-year-olds and ten 6-year-olds in the group? Randomizing may place both 12-year-olds on the same team. In the long run, if we did this over and over, it would all equalize. But wouldn't it be better to assign one 12-year-old to each group (at random) and five 6-year-olds to each team (at random)? By doing this, we would improve fairness in the short run. This approach makes the division more fair by recognizing the variation in *age* and allocating the players at random *within* each age level. When we do this, we call the variable *age* a **blocking variable.** The levels of *age* are called blocks.

Sometimes, attributes of the experimental units that we are not studying and that we can't control may nevertheless affect the outcomes of an experiment. If we group similar individuals together and then randomize within each of these **blocks,** we can remove much of the variability due to the difference among the blocks. Blocking is an important compromise between randomization and control. However, unlike the first three principles, blocking is not *required* in an experimental design.

Control, randomize, and replicate

Recap: We're planning an experiment to see whether the new pet food is safe for dogs to eat. We'll feed some animals the new food and others a food known to be safe, comparing their health after a period of time.

Questions: In this experiment, how will you implement the principles of control, randomization, and replication?

I'd control the portion sizes eaten by the dogs. To reduce possible variability from factors other than the food, I'd standardize other aspects of their environments—housing the dogs in similar pens and ensuring that each got the same amount of water, exercise, play, and sleep time, for example. I might restrict the experiment to a single breed of dog and to adult dogs to further minimize variation.

To equalize traits, pre-existing conditions, and other unknown influences, I would assign dogs to the two feed treatments randomly.

I would replicate by assigning more than one dog to each treatment to allow for variability among individual dogs. If I had the time and funding, I might replicate the entire experiment using, for example, a different breed of dog.

Diagrams

An experiment is carried out over time with specific actions occurring in a specified order. A diagram of the procedure can help in thinking about experiments.[2]

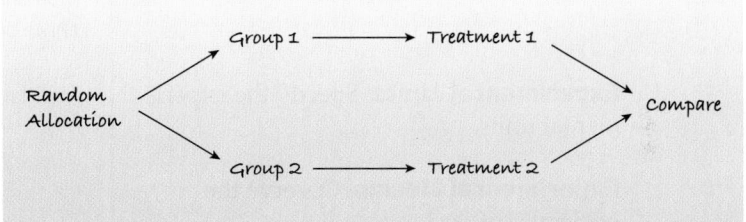

The diagram emphasizes the random allocation of subjects to treatment groups, the separate treatments applied to these groups, and the ultimate comparison of results. It's best to specify the responses that will be compared. A good way to start comparing results for the treatment groups is with boxplots.

STEP-BY-STEP EXAMPLE Designing an Experiment

An ad for OptiGro plant fertilizer claims that with this product you will grow "juicier, tastier" tomatoes. You'd like to test this claim, and wonder whether you might be able to get by with half the specified dose. How can you set up an experiment to check out the claim?

Of course, you'll have to get some tomatoes, try growing some plants with the product and some without, and see what happens. But you'll need a clearer plan than that. How should you design your experiment?

[2] Diagrams of this sort were introduced by David Moore in his textbooks and are widely used.

A completely randomized experiment is the ideal simple design, just as a *simple random sample* is the ideal simple sample—and for many of the same reasons.

Let's work through the design, step by step. We'll design the simplest kind of experiment, a **completely randomized experiment in one factor**. Since this is a *design* for an experiment, most of the steps are part of the *Think* stage. The statements in the right column are the kinds of things you would need to say in *proposing* an experiment. You'd need to include them in the "methods" section of a report once the experiment is run.

Question: How would you design an experiment to test OptiGro fertilizer?

THINK

Plan State what you want to know.

I want to know whether tomato plants grown with OptiGro yield juicier, tastier tomatoes than plants raised in otherwise similar circumstances but without the fertilizer.

Response Specify the response variable.

I'll evaluate the juiciness and taste of the tomatoes by asking a panel of judges to rate them on a scale from 1 to 7 in juiciness and in taste.

Treatments Specify the factor levels and the treatments.

The factor is fertilizer, specifically OptiGro fertilizer. I'll grow tomatoes at three different factor levels: some with no fertilizer, some with half the specified amount of OptiGro, and some with the full dose of OptiGro. These are the three treatments.

Experimental Units Specify the experimental units.

I'll obtain 24 tomato plants of the same variety from a local garden store.

Experimental Design Observe the principles of design:

　Control any sources of variability you know of and can control.

I'll locate the farm plots near each other so that the plants get similar amounts of sun and rain and experience similar temperatures. I will weed the plots equally and otherwise treat the plants alike.

　Replicate results by placing more than one plant in each treatment group.

I'll use 8 plants in each treatment group.

　Randomly assign experimental units to treatments, to equalize the effects of unknown or uncontrollable sources of variation.

Describe how the randomization will be accomplished.

To randomly divide the plants into three groups, first I'll label the plants with numbers 00–23. I'll look at pairs of digits across a random number table. The first 8 plants identified (ignoring numbers 24–99 and any repeats) will go in Group 1, the next 8 in Group 2, and the remaining plants in Group 3.

Make a Picture A diagram of your design can help you think about it clearly.

Specify any other experiment details. You must give enough details so that another experimenter could exactly replicate your experiment. It's generally better to include details that might seem irrelevant than to leave out matters that could turn out to make a difference.

Specify how to measure the response.

I will grow the plants until the tomatoes are mature, as judged by reaching a standard color.

I'll harvest the tomatoes when ripe and store them for evaluation.

I'll set up a numerical scale of juiciness and one of tastiness for the taste testers. Several people will taste slices of tomato and rate them.

SHOW Once you collect the data, you'll need to display them and compare the results for the three treatment groups.

I will display the results with side-by-side box-plots to compare the three treatment groups.

I will compare the means of the groups.

TELL To answer the initial question, we ask whether the differences we observe in the means of the three groups are meaningful.

Because this is a randomized experiment, we can attribute significant differences to the treatments. To do this properly, we'll need methods from what is called "statistical inference," the subject of the rest of this book.

If the differences in taste and juiciness among the groups are greater than I would expect by knowing the usual variation among tomatoes, I may be able to conclude that these differences can be attributed to treatment with the fertilizer.

Does the Difference Make a Difference?

Activity: **Graph the Data.**
Do you think there's a significant difference in your perception of pie charts and bar charts? Explore the data from your plot perception experiment.

If the differences among the treatment groups are big enough, we'll attribute the differences to the treatments, but how can we decide whether the differences are big enough?

Would we expect the group means to be identical? Not really. Even if the treatment made no difference whatever, there would still be some variation. We assigned the tomato plants to treatments at random. But a different random assignment would have led to different results. Even a repeat of the *same* treatment on a different randomly assigned set of plants would lead to a different mean. The real question is whether the differences we observed are about as big as we might get just from the randomization alone, or whether they're bigger than that. If we decide that they're bigger, we'll attribute the differences to the treatments. In that case we say the differences are **statistically significant.**

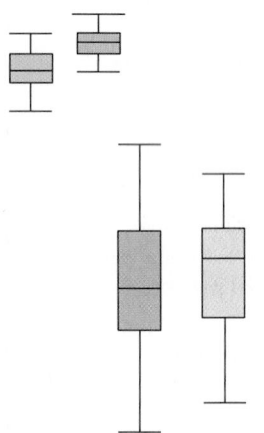

FIGURE 13.1

The boxplots in both pairs have centers the same distance apart, but when the spreads are large, the observed difference may be just from random fluctuation.

How will we decide if something is different enough to be considered statistically significant? Later chapters will offer methods to help answer that question, but to get some intuition, think about deciding whether a coin is fair. If we flip a fair coin 100 times, we expect, *on average*, to get 50 heads. Suppose we get 54 heads out of 100. That doesn't seem very surprising. It's well within the bounds of ordinary random fluctuations. What if we'd seen 94 heads? That's clearly outside the bounds. We'd be pretty sure that the coin flips were not random. But what about 74 heads? Is that far enough from 50% to arouse our suspicions? That's the sort of question we need to ask of our experiment results.

In Statistics terminology, 94 heads would be a statistically significant difference from 50, and 54 heads would not. Whether 74 is *statistically significant* or not would depend on the chance of getting 74 heads in 100 flips of a fair coin and on our tolerance for believing that rare events can happen to us.

Back at the tomato stand, we ask whether the differences we see among the treatment groups are the kind of differences we'd expect from randomization. A good way to get a feeling for that is to look at how much our results vary among plants that get the *same* treatment. Boxplots of our results by treatment group can give us a general idea.

For example, Figure 13.1 shows two pairs of boxplots whose centers differ by exactly the same amount. In the upper set, that difference appears to be larger than we'd expect just by chance. Why? Because the variation is quite small *within* treatment groups, so the larger difference *between* the groups is unlikely to be just from the randomization. In the bottom pair, that same difference between the centers looks less impressive. There the variation *within* each group swamps the difference *between* the two medians. We'd say the difference is statistically significant in the upper pair and not statistically significant in the lower pair.

In later chapters we'll see statistical tests that quantify this intuition. For now, the important point is that a difference is statistically significant if we don't believe that it's likely to have occurred only by chance.

 **JUST CHECKING**

1. At one time, a method called "gastric freezing" was used to treat people with peptic ulcers. An inflatable bladder was inserted down the esophagus and into the stomach, and then a cold liquid was pumped into the bladder. Now you can find the following notice on the Internet site of a major insurance company:

   [Our company] does not cover gastric freezing (intragastric hypothermia) for chronic peptic ulcer disease. . . .

   Gastric freezing for chronic peptic ulcer disease is a non-surgical treatment which was popular about 20 years ago but now is seldom performed. It has been abandoned due to a high complication rate, only temporary improvement experienced by patients, and a lack of effectiveness when tested by double-blind, controlled clinical trials.

   What did that "controlled clinical trial" (experiment) probably look like? (Don't worry about "double-blind"; we'll get to that soon.)

   a) What was the factor in this experiment?
   b) What was the response variable?
   c) What were the treatments?
   d) How did researchers decide which subjects received which treatment?
   e) Were the results statistically significant?

Experiments and Samples

Both experiments and sample surveys use randomization to get unbiased data. But they do so in different ways and for different purposes. Sample surveys try to estimate population parameters, so the sample needs to be as representative of the population as possible. By contrast, experiments try to assess the effects of treatments. Experimental units are not always drawn randomly from the population. For example, a medical experiment may deal only with local patients who have the disease under study. The randomization is in the assignment of their therapy. We want a sample to exhibit the diversity and variability of the population, but for an experiment the more homogeneous the subjects the more easily we'll spot differences in the effects of the treatments.

Experiments are rarely performed on random samples from a population. Don't describe the subjects in an experiment as a random sample unless they really are. More likely, the randomization was in assigning subjects to treatments.

Unless the experimental units are chosen from the population at random, you should be cautious about generalizing experiment results to larger populations until the experiment has been repeated under different circumstances. Results become more persuasive if they remain the same in completely different settings, such as in a different season, in a different country, or for a different species, to name a few.

Even without choosing experimental units from a population at random, experiments can draw stronger conclusions than surveys. By looking only at the differences across treatment groups, experiments cancel out many sources of bias. For example, the entire pool of subjects may be biased and not representative of the population. (College students may need more sleep, on average, than the general population.) When we assign subjects randomly to treatment groups, all the groups are still biased, but *in the same way*. When we consider the differences in their responses, these biases cancel out, allowing us to see the *differences* due to treatment effects more clearly.

Control Treatments

A S *Activity:* **Control Groups in Experiments**. Is a control group really necessary?

Suppose you wanted to test a $300 piece of software designed to shorten download times. You could just try it on several files and record the download times, but you probably want to *compare* the speed with what would happen *without* the software installed. Such a baseline measurement is called a **control** treatment, and the experimental units to whom it is applied are called a **control group.**

This is a use of the word "control" in an entirely different context. Previously, we controlled extraneous sources of variation by keeping them constant. Here, we use a control treatment as another *level* of the factor in order to compare the treatment results to a situation in which "nothing happens." That's what we did in the tomato experiment when we used no fertilizer on the 8 tomatoes in Group 1.

Blinding

Humans are notoriously susceptible to errors in judgment.[3] All of us. When we know what treatment was assigned, it's difficult not to let that knowledge influence our assessment of the response, even when we try to be careful.

Suppose you were trying to advise your school on which brand of cola to stock in the school's vending machines. You set up an experiment to see which of the three competing brands students prefer (or whether they can tell the difference at all). But people have brand loyalties. You probably prefer one brand already. So if you knew which brand you were tasting, it might influence your rating. To avoid this problem, it would be better to disguise the brands as much as possible. This strategy is called **blinding** the participants to the treatment.[4]

But it isn't just the subjects who should be blind. Experimenters themselves often subconsciously behave in ways that favor what they believe. Even technicians may treat plants or test animals differently if, for example, they expect them to die. An animal that starts doing a little better than others by showing an increased appetite may get fed a bit more than the experimental protocol specifies.

---

[3] For example, here we are in Chapter 13 and you're still reading the footnotes.
[4] C. S. Peirce, in the same 1885 work in which he introduced randomization, also recommended blinding.

Blinding by Misleading

Social science experiments can sometimes blind subjects by misleading them about the purpose of a study. One of the authors participated as an undergraduate volunteer in a (now infamous) psychology experiment using such a blinding method. The subjects were told that the experiment was about three-dimensional spatial perception and were assigned to draw a model of a horse. While they were busy drawing, a loud noise and then groaning were heard coming from the room next door. The *real* purpose of the experiment was to see how people reacted to the apparent disaster. The experimenters wanted to see whether the social pressure of being in groups made people react to the disaster differently. Subjects had been randomly assigned to draw either in groups or alone; that was the treatment. The experimenter had no interest in how well the subjects could draw the horse, but the subjects were blinded to the treatment because they were misled.

People are so good at picking up subtle cues about treatments that the best (in fact, the *only*) defense against such biases in experiments on human subjects is to keep *anyone* who could affect the outcome or the measurement of the response from knowing which subjects have been assigned to which treatments. So, not only should your cola-tasting subjects be blinded, but also *you,* as the experimenter, shouldn't know which drink is which, either—at least until you're ready to analyze the results.

There are two main classes of individuals who can affect the outcome of the experiment:

▶ those who could influence the results (the subjects, treatment administrators, or technicians)

▶ those who evaluate the results (judges, treating physicians, etc.)

When all the individuals in either one of these classes are blinded, an experiment is said to be **single-blind.** When everyone in *both* classes is blinded, we call the experiment **double-blind.** Even if several individuals in one class are blinded—for example, both the patients and the technicians who administer the treatment—the study would still be just single-blind. If only some of the individuals in a class are blind—for example, if subjects are not told of their treatment, but the administering technician is not blind—there is a substantial risk that subjects can discern their treatment from subtle cues in the technician's behavior or that the technician might inadvertently treat subjects differently. Such experiments cannot be considered truly blind.

In our tomato experiment, we certainly don't want the people judging the taste to know which tomatoes got the fertilizer. That makes the experiment single-blind. We might also not want the people caring for the tomatoes to know which ones were being fertilized, in case they might treat them differently in other ways, too. We can accomplish this double-blinding by having some fake fertilizer for them to put on the other plants. Read on.

FOR EXAMPLE Blinding

Recap: In our experiment to see if the new pet food is now safe, we're feeding one group of dogs the new food and another group a food we know to be safe. Our response variable is the health of the animals as assessed by a veterinarian.

Questions: Should the vet be blinded? Why or why not? How would you do this? (Extra credit: Can this experiment be double-blind? Would that mean that the test animals wouldn't know what they were eating?)

Whenever the response variable involves judgment, it is a good idea to blind the evaluator to the treatments. The veterinarian should not be told which dogs ate which foods.

Extra credit: There is a need for double-blinding. In this case, the workers who care for and feed the animals should not be aware of which dogs are receiving which food. We'll need to make the "safe" food look as much like the "test" food as possible.

Placebos

 A S **Activity: Blinded Experiments.** This narrated account of blinding isn't a placebo!

Often, simply applying *any* treatment can induce an improvement. Every parent knows the medicinal value of a kiss to make a toddler's scrape or bump stop hurting. Some of the improvement seen with a treatment—even an effective treatment—can be due simply to the act of treating. To separate these two effects, we can use a control treatment that mimics the treatment itself.

The placebo effect is stronger when placebo treatments are administered with authority or by a figure who appears to be an authority. "Doctors" in white coats generate a stronger effect than salespeople in polyester suits. But the placebo effect is not reduced much even when subjects know that the effect exists. People often suspect that they've gotten the placebo if nothing at all happens. So, recently, drug manufacturers have gone so far in making placebos realistic that they cause the same side effects as the drug being tested! Such "active placebos" usually induce a stronger placebo effect. When those side effects include loss of appetite or hair, the practice may raise ethical questions.

A "fake" treatment that looks just like the treatments being tested is called a **placebo.** Placebos are the best way to blind subjects from knowing whether they are receiving the treatment or not. One common version of a placebo in drug testing is a "sugar pill." Especially when psychological attitude can affect the results, control group subjects treated with a placebo may show an improvement.

The fact is that subjects treated with a placebo sometimes improve. It's not unusual for 20% or more of subjects given a placebo to report reduction in pain, improved movement, or greater alertness, or even to demonstrate improved health or performance. This **placebo effect** highlights both the importance of effective blinding and the importance of comparing treatments with a control. Placebo controls are so effective that you should use them as an essential tool for blinding whenever possible.

The best experiments are usually

- randomized.
- comparative.
- double-blind.
- placebo-controlled.

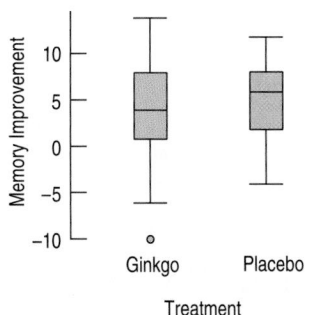

Does ginkgo biloba improve memory? Researchers investigated the purported memory-enhancing effect of ginkgo biloba tree extract (P. R. Solomon, F. Adams, A. Silver, J. Zimmer, R. De Veaux, "Ginkgo for Memory Enhancement. A Randomized Controlled Trial." *JAMA* 288 [2002]: 835–840). In a randomized, comparative, double-blind, placebo-controlled study, they administered treatments to 230 elderly community members. One group received Ginkoba™ according to the manufacturer's instructions. The other received a similar-looking placebo. Thirteen different tests of memory were administered before and after treatment. The placebo group showed greater improvement on 7 of the tests, the treatment group on the other 6. None showed any significant differences. Here are boxplots of one measure.

By permission of John L. Hart FLP and Creators Syndicate, Inc.

Blocking

We wanted to use 18 tomato plants of the same variety for our experiment, but suppose the garden store had only 12 plants left. So we drove down to the nursery and bought 6 more plants of that variety. We worry that the tomato plants from the two stores are different somehow, and, in fact, they don't really look the same.

How can we design the experiment so that the differences between the stores don't mess up our attempts to see differences among fertilizer levels? We can't measure the effect of a store the same way as we can the fertilizer because we can't assign it as we would a factor in the experiment. You can't tell a tomato what store to come from.

Because stores may vary in the care they give plants or in the sources of their seeds, the plants from either store are likely to be more like each other than they are like the plants from the other store. When groups of experimental units are similar, it's often a good idea to gather them together into **blocks.** By blocking, we isolate the variability attributable to the differences between the blocks, so that we can see the differences caused by the treatments more clearly. Here, we would define the plants from each store to be a block. The randomization is introduced when we randomly assign treatments within each block.

In a completely randomized design, each of the 18 plants would have an equal chance to land in each of the three treatment groups. But we realize that the store may have an effect. To isolate the store effect, we block on store by assigning the plants from each store to treatments at random. So we now have six treatment groups, three for each block. Within each block, we'll randomly assign the same number of plants to each of the three treatments. The experiment is still fair because each treatment is still applied (at random) to the same number of plants and to the same proportion from each store: 4 from store A and 2 from store B. Because the randomization occurs only within the blocks (plants from one store cannot be assigned to treatment groups for the other), we call this a **randomized block design.**

In effect, we conduct two parallel experiments, one for tomatoes from each store, and then combine the results. The picture tells the story:

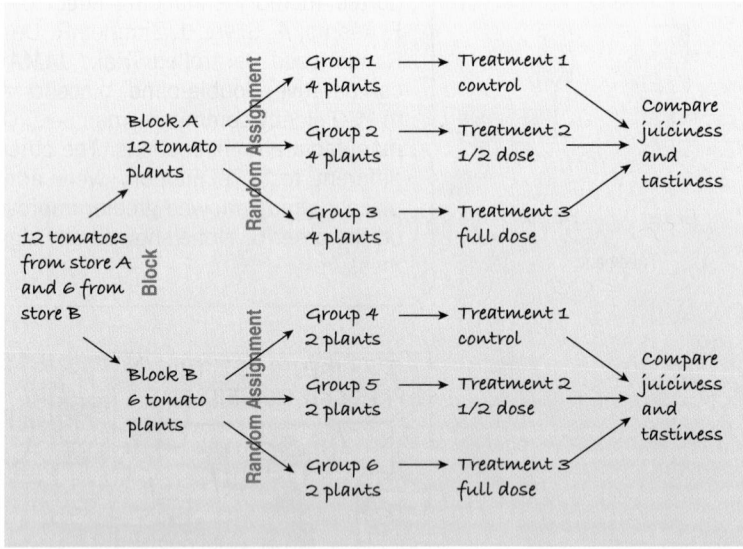

In a retrospective or prospective study, subjects are sometimes paired because they are similar in ways *not* under study. **Matching** subjects in this way can reduce variation in much the same way as blocking. For example, a retrospective study of music education and grades might match each student who studies an instrument with someone of the same sex who is similar in family income but didn't study an instrument. When we compare grades of music students with those of non-music students, the matching would reduce the variation due to income and sex differences.

Blocking is the same idea for experiments as stratifying is for sampling. Both methods group together subjects that are similar and randomize within those groups as a way to remove unwanted variation. (But be careful to keep the terms straight. Don't say that we "stratify" an experiment or "block" a sample.) We use blocks to reduce variability so we can see the effects of the factors; we're not usually interested in studying the effects of the blocks themselves.

Blocking

Recap: In 2007, pet food contamination put cats at risk, as well as dogs. Our experiment should probably test the safety of the new food on both animals.

Questions: Why shouldn't we randomly assign a mix of cats and dogs to the two treatment groups? What would you recommend instead?

Dogs and cats might respond differently to the foods, and that variability could obscure my results. Blocking by species can remove that superfluous variation. I'd randomize cats to the two treatments (test food and safe food) separately from the dogs. I'd measure their responses separately and look at the results afterward.

JUST CHECKING

2. Recall the experiment about gastric freezing, an old method for treating peptic ulcers that you read about in the first Just Checking. Doctors would insert an inflatable bladder down the patient's esophagus and into the stomach and then pump in a cold liquid. A major insurance company now states that it doesn't cover this treatment because "double-blind, controlled clinical trials" failed to demonstrate that gastric freezing was effective.
a) What does it mean that the experiment was double-blind?
b) Why would you recommend a placebo control?
c) Suppose that researchers suspected that the effectiveness of the gastric freezing treatment might depend on whether a patient had recently developed the peptic ulcer or had been suffering from the condition for a long time. How might the researchers have designed the experiment?

Adding More Factors

There are two kinds of gardeners. Some water frequently, making sure that the plants are never dry. Others let Mother Nature take her course and leave the watering to her. The makers of OptiGro want to ensure that their product will work under a wide variety of watering conditions. Maybe we should include the amount of watering as part of our experiment. Can we study a second factor at the same time and still learn as much about fertilizer?

We now have two factors (fertilizer at three levels and irrigation at two levels). We combine them in all possible ways to yield six treatments:

	No Fertilizer	Half Fertilizer	Full Fertilizer
No Added Water	1	2	3
Daily Watering	4	5	6

If we allocate the original 12 plants, the experiment now assigns 2 plants to each of these six treatments at random. This experiment is a **completely randomized two-factor experiment** because any plant could end up assigned at random to any of the six treatments (and we have two factors).

It's often important to include several factors in the same experiment in order to see what happens when the factor levels are applied in different *combinations*. A common misconception is that applying several factors at once makes it difficult to separate the effects of the individual factors. You may hear people say that experiments should always be run "one factor at a time." In fact, just the opposite

> **Think Like a Statistician**
> With two factors, we can account for more of the variation. That lets us see the underlying patterns more clearly.

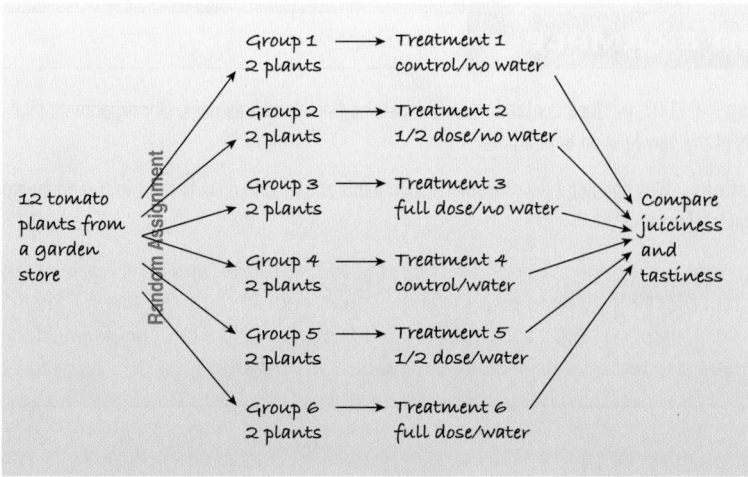

is true: Experiments with more than one factor are both more efficient and provide more information than one-at-a-time experiments. There are many ways to design efficient multifactor experiments. You can take a whole course on the design and analysis of such experiments.

Confounding

Professor Stephen Ceci of Cornell University performed an experiment to investigate the effect of a teacher's classroom style on student evaluations. He taught a class in developmental psychology during two successive terms to a total of 472 students in two very similar classes. He kept everything about his teaching identical (same text, same syllabus, same office hours, etc.) and modified only his style in class. During the fall term, he maintained a subdued demeanor. During the spring term, he used expansive gestures and lectured with more enthusiasm, varying his vocal pitch and using more hand gestures. He administered a standard student evaluation form at the end of each term.

The students in the fall term class rated him only an average teacher. Those in the spring term class rated him an excellent teacher, praising his knowledge and accessibility, and even the quality of the textbook. On the question "How much did you learn in the course?" the average response changed from 2.93 to 4.05 on a 5-point scale.[5]

How much of the difference he observed was due to his difference in manner, and how much might have been due to the season of the year? Fall term in Ithaca, NY (home of Cornell University), starts out colorful and pleasantly warm but ends cold and bleak. Spring term starts out bitter and snowy and ends with blooming flowers and singing birds. Might students' overall happiness have been affected by the season and reflected in their evaluations?

Unfortunately, there's no way to tell. Nothing in the data enables us to tease apart these two effects, because all the students who experienced the subdued manner did so during the fall term and all who experienced the expansive manner did so during the spring. When the levels of one factor are associated with the levels of another factor, we say that these two factors are **confounded.**

In some experiments, such as this one, it's just not possible to avoid some confounding. Professor Ceci could have randomly assigned students to one of two classes during the same term, but then we might question whether mornings or

_____

[5] But the two classes performed almost identically well on the final exam.

afternoons were better, or whether he really delivered the same class the second time (after practicing on the first class). Or he could have had another professor deliver the second class, but that would have raised more serious issues about differences in the two professors and concern over more serious confounding.

FOR EXAMPLE   Confounding

Recap: After many dogs and cats suffered health problems caused by contaminated foods, we're trying to find out whether a newly formulated pet food is safe. Our experiment will feed some animals the new food and others a food known to be safe, and a veterinarian will check the response.

Question: Why would it be a bad design to feed the test food to some dogs and the safe food to cats?

This would create confounding. We would not be able to tell whether any differences in animals' health were attributable to the food they had eaten or to differences in how the two species responded.

> **A two-factor example**   Confounding can also arise from a badly designed multifactor experiment. Here's a classic. A credit card bank wanted to test the sensitivity of the market to two factors: the annual fee charged for a card and the annual percentage rate charged. Not wanting to scrimp on sample size, the bank selected 100,000 people at random from a mailing list. It sent out 50,000 offers with a low rate and no fee and 50,000 offers with a higher rate and a $50 annual fee. Guess what happened? That's right—people preferred the low-rate, no-fee card. No surprise. In fact, they signed up for that card at over twice the rate as the other offer. And because of the large sample size, the bank was able to estimate the difference precisely. But the question the bank really wanted to answer was "how much of the change was due to the rate, and how much was due to the fee?" unfortunately, there's simply no way to separate out the two effects. If the bank had sent out all four possible different treatments—low rate with no fee, low rate with $50 fee, high rate with no fee, and high rate with $50 fee—each to 25,000 people, it could have learned about both factors and could have also seen what happens when the two factors occur in combination.

Lurking or Confounding?

Confounding may remind you of the problem of lurking variables we discussed back in Chapters 7 and 9. Confounding variables and lurking variables are alike in that they interfere with our ability to interpret our analyses simply. Each can mislead us, but there are important differences in both how and where the confusion may arise.

A lurking variable creates an association between two other variables that tempts us to think that one may cause the other. This can happen in a regression analysis or an observational study when a lurking variable influences both the explanatory and response variables. Recall that countries with more TV sets per capita tend to have longer life expectancies. We shouldn't conclude it's the TVs "causing" longer life. We suspect instead that a generally higher standard of living may mean that people can afford more TVs and get better health care, too. Our data revealed an association between TVs and life expectancy, but economic conditions were a likely lurking variable. A lurking variable, then, is usually thought of as a variable associated with both y and x that makes it appear that x may be causing y.

Confounding can arise in experiments when some other variable associated with a factor has an effect on the response variable. However, in a designed experiment, the experimenter *assigns* treatments (at random) to subjects rather than just observing them. A confounding variable can't be thought of as causing that assignment. Professor Ceci's choice of teaching styles was not caused by the weather, but because he used one style in the fall and the other in spring, he was unable to tell how much of his students' reactions were attributable to his teaching and how much to the weather. A confounding variable, then, is associated in a noncausal way with a factor and affects the response. Because of the confounding, we find that we can't tell whether any effect we see was caused by our factor or by the confounding variable—or even by both working together.

Both confounding and lurking variables are outside influences that make it harder to understand the relationship we are modeling. However, the nature of the causation is different in the two situations. In regression and observational studies, we can only observe associations between variables. Although we can't demonstrate a causal relationship, we often imagine whether *x* could cause *y*. We can be misled by a lurking variable that influences both. In a designed experiment, we often hope to show that the factor causes a response. Here we can be misled by a confounding variable that's associated with the factor and causes or contributes to the differences we observe in the response.

It's worth noting that the role of blinding in an experiment is to combat a possible source of confounding. There's a risk that knowledge about the treatments could lead the subjects or those interacting with them to behave differently or could influence judgments made by the people evaluating the responses. That means we won't know whether the treatments really do produce different results or if we're being fooled by these confounding influences.

WHAT CAN GO WRONG?

▶ **Don't give up just because you can't run an experiment.** Sometimes we can't run an experiment because we can't identify or control the factors. Sometimes it would simply be unethical to run the experiment. (Consider randomly assigning students to take—and be graded in—a Statistics course deliberately taught to be boring and difficult or one that had an unlimited budget to use multimedia, real-world examples, and field trips to make the subject more interesting.) If we can't perform an experiment, often an observational study is a good choice.

▶ **Beware of confounding.** Use randomization whenever possible to ensure that the factors not in your experiment are not confounded with your treatment levels. Be alert to confounding that cannot be avoided, and report it along with your results.

▶ **Bad things can happen even to good experiments.** Protect yourself by recording additional information. An experiment in which the air conditioning failed for 2 weeks, affecting the results, was saved by recording the temperature (although that was not originally one of the factors) and estimating the effect the higher temperature had on the response.[6]

It's generally good practice to collect as much information as possible about your experimental units and the circumstances of the experiment. For example, in the tomato experiment, it would be wise to record details of the weather (temperature, rainfall, sunlight) that might affect the plants and any facts available about their

[6] R. D. DeVeaux and M. Szelewski, "Optimizing Automatic Splitless Injection Parameters for Gas Chromatographic Environmental Analysis." *Journal of Chromatographic Science* 27, no. 9 (1989): 513–518.

growing situation. (Is one side of the field in shade sooner than the other as the day proceeds? Is one area lower and a bit wetter?) Sometimes we can use this extra information during the analysis to reduce biases.

▶ **Don't spend your entire budget on the first run.** Just as it's a good idea to pretest a survey, it's always wise to try a small pilot experiment before running the full-scale experiment. You may learn, for example, how to choose factor levels more effectively, about effects you forgot to control, and about unanticipated confoundings.

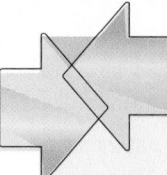

CONNECTIONS

The fundamental role of randomization in experiments clearly points back to our discussions of randomization, to our experiments with simulations, and to our use of randomization in sampling. The similarities and differences between experiments and samples are important to keep in mind and can make each concept clearer.

If you think that blocking in an experiment resembles stratifying in a sample, you're quite right. Both are ways of removing variation we can identify to help us see past the variation in the data.

Experiments compare groups of subjects that have been treated differently. Graphics such as boxplots that help us compare groups are closely related to these ideas. Think about what we look for in a boxplot to tell whether two groups look really different, and you'll be thinking about the same issues as experiment designers.

Generally, we're going to consider how different the mean responses are for different treatment groups. And we're going to judge whether those differences are large by using standard deviations as rulers. (That's why we needed to replicate results for each treatment; we need to be able to estimate those standard deviations.) The discussion in Chapter 6 introduced this fundamental statistical thought, and it's going to keep coming back over and over again. Statistics is about variation.

We'll see a number of ways to analyze results from experiments in subsequent chapters.

WHAT HAVE WE LEARNED?

We've learned to recognize sample surveys, observational studies, and randomized comparative experiments. We know that these methods collect data in different ways and lead us to different conclusions.

We've learned to identify retrospective and prospective observational studies and understand the advantages and disadvantages of each.

We've learned that only well-designed experiments can allow us to reach cause-and-effect conclusions. We manipulate levels of treatments to see if the factor we have identified produces changes in our response variable.

We've learned the principles of experimental design:

▶ We want to be sure that variation in the response variable can be attributed to our factor, so we identify and control as many other sources of variability as possible.
▶ Because there are many possible sources of variability that we cannot identify, we try to equalize those by randomly assigning experimental units to treatments.
▶ We replicate the experiment on as many subjects as possible.
▶ We consider blocking to reduce variability from sources we recognize but cannot control.

We've learned the value of having a control group and of using blinding and placebo controls.

Finally, we've learned to recognize the problems posed by confounding variables in experiments and lurking variables in observational studies.

Terms

Observational study	292. A study based on data in which no manipulation of factors has been employed.
Retrospective study	292. An observational study in which subjects are selected and then their previous conditions or behaviors are determined. Retrospective studies need not be based on random samples and they usually focus on estimating differences between groups or associations between variables.
Prospective study	293. An observational study in which subjects are followed to observe future outcomes. Because no treatments are deliberately applied, a prospective study is not an experiment. Nevertheless, prospective studies typically focus on estimating differences among groups that might appear as the groups are followed during the course of the study.
Experiment	294. An experiment *manipulates* factor levels to create treatments, *randomly assigns* subjects to these treatment levels, and then *compares* the responses of the subject groups across treatment levels.
Random assignment	294. To be valid, an experiment must assign experimental units to treatment groups at random. This is called random assignment.
Factor	294. A variable whose levels are manipulated by the experimenter. Experiments attempt to discover the effects that differences in factor levels may have on the responses of the experimental units.
Response	294. A variable whose values are compared across different treatments. In a randomized experiment, large response differences can be attributed to the effect of differences in treatment level.
Experimental units	294. Individuals on whom an experiment is performed. Usually called **subjects** or **participants** when they are human.
Level	294. The specific values that the experimenter chooses for a factor are called the levels of the factor.
Treatment	294. The process, intervention, or other controlled circumstance applied to randomly assigned experimental units. Treatments are the different levels of a single factor or are made up of combinations of levels of two or more factors.
Principles of experimental design	▸ 295. **Control** aspects of the experiment that we know may have an effect on the response, but that are not the factors being studied.
	▸ 296. **Randomize** subjects to treatments to even out effects that we cannot control.
	▸ 296. **Replicate** over as many subjects as possible. Results for a single subject are just anecdotes. If, as often happens, the subjects of the experiment are not a representative sample from the population of interest, replicate the entire study with a different group of subjects, preferably from a different part of the population.
	▸ 296. **Block** to reduce the effects of identifiable attributes of the subjects that cannot be controlled.
Statistically significant	299. When an observed difference is too large for us to believe that it is likely to have occurred naturally, we consider the difference to be statistically significant. Subsequent chapters will show specific calculations and give rules, but the principle remains the same.
Control group	301. The experimental units assigned to a baseline treatment level, typically either the default treatment, which is well understood, or a null, placebo treatment. Their responses provide a basis for comparison.
Blinding	301. Any individual associated with an experiment who is not aware of how subjects have been allocated to treatment groups is said to be blinded.
Single-blind Double-blind	302. There are two main classes of individuals who can affect the outcome of an experiment:
	▸ those who could *influence the results* (the subjects, treatment administrators, or technicians).
	▸ those who *evaluate the results* (judges, treating physicians, etc.).
	When every individual in *either* of these classes is blinded, an experiment is said to be single-blind. When everyone in *both* classes is blinded, we call the experiment double-blind.
Placebo	303. A treatment known to have no effect, administered so that all groups experience the same conditions. Many subjects respond to such a treatment (a response known as a placebo effect). Only by comparing with a placebo can we be sure that the observed effect of a treatment is not due simply to the placebo effect.
Placebo effect	303. The tendency of many human subjects (often 20% or more of experiment subjects) to show a response even when administered a placebo.

Blocking   303. When groups of experimental units are similar, it is often a good idea to gather them together into blocks. By blocking, we isolate the variability attributable to the differences between the blocks so that we can see the differences caused by the treatments more clearly.

Matching   304. In a retrospective or prospective study, subjects who are similar in ways not under study may be matched and then compared with each other on the variables of interest. Matching, like blocking, reduces unwanted variation.

Designs   298, 305. In **a completely randomized design,** all experimental units have an equal chance of receiving any treatment.

304. In **a randomized block design,** the randomization occurs only within blocks.

Confounding   306. When the levels of one factor are associated with the levels of another factor in such a way that their effects cannot be separated, we say that these two factors are confounded.

Skills

- Recognize when an observational study would be appropriate.

- Be able to identify observational studies as retrospective or prospective, and understand the strengths and weaknesses of each method.

- Know the four basic principles of sound experimental design—control, randomize, replicate, and block—and be able to explain each.

- Be able to recognize the factors, the treatments, and the response variable in a description of a designed experiment.

- Understand the essential importance of randomization in assigning treatments to experimental units.

- Understand the importance of replication to move from anecdotes to general conclusions.

- Understand the value of blocking so that variability due to differences in attributes of the subjects can be removed.

- Understand the importance of a control group and the need for a placebo treatment in some studies.

- Understand the importance of blinding and double-blinding in studies on human subjects, and be able to identify blinding and the need for blinding in experiments.

- Understand the value of a placebo in experiments with human participants.

- Be able to design a completely randomized experiment to test the effect of a single factor.

- Be able to design an experiment in which blocking is used to reduce variation.

- Know how to use graphical displays to compare responses for different treatment groups. Understand that you should *never* proceed with any other analysis of a designed experiment without first looking at boxplots or other graphical displays.

- Know how to report the results of an observational study. Identify the subjects, how the data were gathered, and any potential biases or flaws you may be aware of. Identify the factors known and those that might have been revealed by the study.

- Know how to compare the responses in different treatment groups to assess whether the differences are larger than could be reasonably expected from ordinary sampling variability.

- Know how to report the results of an experiment. Tell who the subjects are and how their assignment to treatments was determined. Report how and in what measurement units the response variable was measured.

- Understand that your description of an experiment should be sufficient for another researcher to replicate the study with the same methods.

- Be able to report on the statistical significance of the result in terms of whether the observed group-to-group differences are larger than could be expected from ordinary sampling variation.

EXPERIMENTS ON THE COMPUTER

Most experiments are analyzed with a statistics package. You should almost always display the results of a comparative experiment with side-by-side boxplots. You may also want to display the means and standard deviations of the treatment groups in a table.

The analyses offered by statistics packages for comparative randomized experiments fall under the general heading of Analysis of Variance, usually abbreviated ANOVA. These analyses are beyond the scope of this chapter.

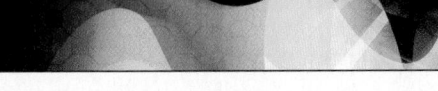

EXERCISES

1. **Standardized test scores.** For his Statistics class experiment, researcher J. Gilbert decided to study how parents' income affects children's performance on standardized tests like the SAT. He proposed to collect information from a random sample of test takers and examine the relationship between parental income and SAT score.
   a) Is this an experiment? If not, what kind of study is it?
   b) If there is relationship between parental income and SAT score, why can't we conclude that differences in score are caused by differences in parental income?

2. **Heart attacks and height.** Researchers who examined health records of thousands of males found that men who died of myocardial infarction (heart attack) tended to be shorter than men who did not.
   a) Is this an experiment? If not, what kind of study is it?
   b) Is it correct to conclude that shorter men are at higher risk for heart attack? Explain.

3. **MS and vitamin D.** Multiple sclerosis (MS) is an autoimmune disease that strikes more often the farther people live from the equator. Could vitamin D—which most people get from the sun's ultraviolet rays—be a factor? Researchers compared vitamin D levels in blood samples from 150 U.S. military personnel who have developed MS with blood samples of nearly 300 who have not. The samples were taken, on average, five years before the disease was diagnosed. Those with the highest blood vitamin D levels had a 62% lower risk of MS than those with the lowest levels. (The link was only in whites, not in blacks or Hispanics.)
   a) What kind of study was this?
   b) Is that an appropriate choice for investigating this problem? Explain.
   c) Who were the subjects?
   d) What were the variables?

4. **Super Bowl commercials.** When spending large amounts to purchase advertising time, companies want to know what audience they'll reach. In January 2007, a poll asked 1008 American adults whether they planned to watch the upcoming Super Bowl. Men and women were asked separately whether they were looking forward more to the football game or to watching the commercials. Among the men, 16% were planning to watch and were looking forward primarily to the commercials. Among women, 30% were looking forward primarily to the commercials.
   a) Was this a stratified sample or a blocked experiment? Explain.
   b) Was the design of the study appropriate for the advertisers' questions?

5. **Menopause.** Researchers studied the herb black cohosh as a treatment for hot flashes caused by menopause. They randomly assigned 351 women aged 45 to 55 who reported at least two hot flashes a day to one of five groups: (1) black cohosh, (2) a multiherb supplement with black cohosh, (3) the multiherb supplement plus advice to consume more soy foods, (4) estrogen replacement therapy, or (5) receive a placebo. After a year, only the women given estrogen replacement therapy had symptoms different from those of the placebo group. [*Annals of Internal Medicine* 145:12, 869–897]
   a) What kind of study was this?
   b) Is that an appropriate choice for this problem?
   c) Who were the subjects?
   d) Identify the treatment and response variables.

6. **Honesty.** Coffee stations in offices often just ask users to leave money in a tray to pay for their coffee, but many people cheat. Researchers at Newcastle University replaced the picture of flowers on the wall behind the coffee station with a picture of staring eyes. They found that the average contribution increased significantly above the well-established standard when people felt they were being watched, even though the eyes were patently not real. (NY *Times* 12/10/06)
   a) Was this a survey, an observational study, or an experiment? How can we tell?
   b) Identify the variables.
   c) What does "increased significantly" mean in a statistical sense?

7–20. What's the design? Read each brief report of statistical research, and identify
   a) whether it was an observational study or an experiment.

If it was an observational study, identify (if possible)
   b) whether it was retrospective or prospective.
   c) the subjects studied and how they were selected.

d) the parameter of interest.

e) the nature and scope of the conclusion the study can reach.

If it was an experiment, identify (if possible)

b) the subjects studied.

c) the factor(s) in the experiment and the number of levels for each.

d) the number of treatments.

e) the response variable measured.

f) the design (completely randomized, blocked, or matched).

g) whether it was blind (or double-blind).

h) the nature and scope of the conclusion the experiment can reach.

7. Over a 4-month period, among 30 people with bipolar disorder, patients who were given a high dose (10 g/day) of omega-3 fats from fish oil improved more than those given a placebo. (*Archives of General Psychiatry* 56 [1999]: 407)

8. Among a group of disabled women aged 65 and older who were tracked for several years, those who had a vitamin B$_{12}$ deficiency were twice as likely to suffer severe depression as those who did not. (*American Journal of Psychiatry* 157 [2000]: 715)

9. In a test of roughly 200 men and women, those with moderately high blood pressure (averaging 164/89 mm Hg) did worse on tests of memory and reaction time than those with normal blood pressure. (*Hypertension* 36 [2000]: 1079)

10. Is diet or exercise effective in combating insomnia? Some believe that cutting out desserts can help alleviate the problem, while others recommend exercise. Forty volunteers suffering from insomnia agreed to participate in a month-long test. Half were randomly assigned to a special no-desserts diet; the others continued desserts as usual. Half of the people in each of these groups were randomly assigned to an exercise program, while the others did not exercise. Those who ate no desserts and engaged in exercise showed the most improvement.

11. After menopause, some women take supplemental estrogen. There is some concern that if these women also drink alcohol, their estrogen levels will rise too high. Twelve volunteers who were receiving supplemental estrogen were randomly divided into two groups, as were 12 other volunteers not on estrogen. In each case, one group drank an alcoholic beverage, the other a nonalcoholic beverage. An hour later, everyone's estrogen level was checked. Only those on supplemental estrogen who drank alcohol showed a marked increase.

12. Researchers have linked an increase in the incidence of breast cancer in Italy to dioxin released by an industrial accident in 1976. The study identified 981 women who lived near the site of the accident and were under age 40 at the time. Fifteen of the women had developed breast cancer at an unusually young average age of 45. Medical records showed that they had heightened concentrations of dioxin in their blood and that each tenfold increase in dioxin level was associated with a doubling of the risk of breast cancer. (*Science News,* Aug. 3, 2002)

13. In 2002 the journal *Science* reported that a study of women in Finland indicated that having sons shortened the lifespans of mothers by about 34 weeks per son, but that daughters helped to lengthen the mothers' lives. The data came from church records from the period 1640 to 1870.

14. Scientists at a major pharmaceutical firm investigated the effectiveness of an herbal compound to treat the common cold. They exposed each subject to a cold virus, then gave him or her either the herbal compound or a sugar solution known to have no effect on colds. Several days later they assessed the patient's condition, using a cold severity scale ranging from 0 to 5. They found no evidence of benefits associated with the compound.

15. The May 4, 2000, issue of *Science News* reported that, contrary to popular belief, depressed individuals cry no more often in response to sad situations than nondepressed people. Researchers studied 23 men and 48 women with major depression and 9 men and 24 women with no depression. They showed the subjects a sad film about a boy whose father has died, noting whether or not the subjects cried. Women cried more often than men, but there were no significant differences between the depressed and nondepressed groups.

16. Some people who race greyhounds give the dogs large doses of vitamin C in the belief that the dogs will run faster. Investigators at the University of Florida tried three different diets in random order on each of five racing greyhounds. They were surprised to find that when the dogs ate high amounts of vitamin C they ran more slowly. (*Science News,* July 20, 2002)

17. Some people claim they can get relief from migraine headache pain by drinking a large glass of ice water. Researchers plan to enlist several people who suffer from migraines in a test. When a participant experiences a migraine headache, he or she will take a pill that may be a standard pain reliever or a placebo. Half of each group will also drink ice water. Participants will then report the level of pain relief they experience.

18. A dog food company wants to compare a new lower-calorie food with their standard dog food to see if it's effective in helping inactive dogs maintain a healthy weight. They have found several dog owners willing to participate in the trial. The dogs have been classified as small, medium, or large breeds, and the company will supply some owners of each size of dog with one of the two foods. The owners have agreed not to feed their dogs anything else for a period of 6 months, after which the dogs' weights will be checked.

19. Athletes who had suffered hamstring injuries were randomly assigned to one of two exercise programs. Those who engaged in static stretching returned to sports activity in a mean of 15.2 days faster than those assigned to a program of agility and trunk stabilization exercises. (*Journal of Orthopaedic & Sports Physical Therapy* 34 [March 2004]: 3)

20. Pew Research compared respondents to an ordinary 5-day telephone survey with respondents to a 4-month-long rigorous survey designed to generate the highest

possible response rate. They were especially interested in identifying any variables for which those who responded to the ordinary survey were different from those who could be reached only by the rigorous survey.

21. Omega-3. Exercise 7 describes an experiment that showed that high doses of omega-3 fats might be of benefit to people with bipolar disorder. The experiment involved a control group of subjects who received a placebo. Why didn't the experimenters just give everyone the omega-3 fats to see if they improved?

22. Insomnia. Exercise 10 describes an experiment showing that exercise helped people sleep better. The experiment involved other groups of subjects who didn't exercise. Why didn't the experimenters just have everyone exercise and see if their ability to sleep improved?

23. Omega-3 revisited. Exercises 7 and 21 describe an experiment investigating a dietary approach to treating bipolar disorder. Researchers randomly assigned 30 subjects to two treatment groups, one group taking a high dose of omega-3 fats and the other a placebo.
a) Why was it important to randomize in assigning the subjects to the two groups?
b) What would be the advantages and disadvantages of using 100 subjects instead of 30?

24. Insomnia again. Exercises 10 and 22 describe an experiment investigating the effectiveness of exercise in combating insomnia. Researchers randomly assigned half of the 40 volunteers to an exercise program.
a) Why was it important to randomize in deciding who would exercise?
b) What would be the advantages and disadvantages of using 100 subjects instead of 40?

25. Omega-3, finis. Exercises 7, 21, and 23 describe an experiment investigating the effectiveness of omega-3 fats in treating bipolar disorder. Suppose some of the 30 subjects were very active people who walked a lot or got vigorous exercise several times a week, while others tended to be more sedentary, working office jobs and watching a lot of TV. Why might researchers choose to block the subjects by activity level before randomly assigning them to the omega-3 and placebo groups?

26. Insomnia, at last. Exercises 10, 22, and 24 describe an experiment investigating the effectiveness of exercise in combating insomnia. Suppose some of the 40 subjects had maintained a healthy weight, but others were quite overweight. Why might researchers choose to block the subjects by weight level before randomly assigning some of each group to the exercise program?

27. Tomatoes. Describe a strategy to randomly split the 24 tomato plants into the three groups for the chapter's completely randomized single factor test of OptiGro fertilizer.

28. Tomatoes II. The chapter also described a completely randomized two-factor experiment testing OptiGro fertilizer in conjunction with two different routines for watering the plants. Describe a strategy to randomly assign the 24 tomato plants to the six treatments.

29. Shoes. A running-shoe manufacturer wants to test the effect of its new sprinting shoe on 100-meter dash times.

The company sponsors 5 athletes who are running the 100-meter dash in the 2004 Summer Olympic games. To test the shoe, it has all 5 runners run the 100-meter dash with a competitor's shoe and then again with their new shoe. The company uses the difference in times as the response variable.
a) Suggest some improvements to the design.
b) Why might the shoe manufacturer not be able to generalize the results they find to all runners?

30. Swimsuits. A swimsuit manufacturer wants to test the speed of its newly designed suit. The company designs an experiment by having 6 randomly selected Olympic swimmers swim as fast as they can with their old swimsuit first and then swim the same event again with the new, expensive swimsuit. The company will use the difference in times as the response variable. Criticize the experiment and point out some of the problems with generalizing the results.

31. Hamstrings. Exercise 19 discussed an experiment to see if the time it took athletes with hamstring injuries to be able to return to sports was different depending on which of two exercise programs they engaged in.
a) Explain why it was important to assign the athletes to the two different treatments randomly.
b) There was no control group consisting of athletes who did not participate in a special exercise program. Explain the advantage of including such a group.
c) How might blinding have been used?
d) One group returned to sports activity in a mean of 37.4 days ($SD = 27.6$ days) and the other in a mean of 22.2 days ($SD = 8.3$ days). Do you think this difference is statistically significant? Explain.

32. Diet and blood pressure. An experiment that showed that subjects fed the DASH diet were able to lower their blood pressure by an average of 6.7 points compared to a group fed a "control diet." All meals were prepared by dieticians.
a) Why were the subjects randomly assigned to the diets instead of letting people pick what they wanted to eat?
b) Why were the meals prepared by dieticians?
c) Why did the researchers need the control group? If the DASH diet group's blood pressure was lower at the end of the experiment than at the beginning, wouldn't that prove the effectiveness of that diet?
d) What additional information would you want to know in order to decide whether an average reduction in blood pressure of 6.7 points was statistically significant?

33. Mozart. Will listening to a Mozart piano sonata make you smarter? In a 1995 study published in the journal *Psychological Science*, Rauscher, Shaw, and Ky reported that when students were given a spatial reasoning section of a standard IQ test, those who listened to Mozart for 10 minutes improved their scores more than those who simply sat quietly.
a) These researchers said the differences were statistically significant. Explain what that means in context.
b) Steele, Bass, and Crook tried to replicate the original study. In their study, also published in *Psychological Science* (1999), the subjects were 125 college students

who participated in the experiment for course credit. Subjects first took the test. Then they were assigned to one of three groups: listening to a Mozart piano sonata, listening to music by Philip Glass, and sitting for 10 minutes in silence. Three days after the treatments, they were retested. Draw a diagram displaying the design of this experiment.

c) These boxplots show the differences in score before and after treatment for the three groups. Did the Mozart group show improvement?

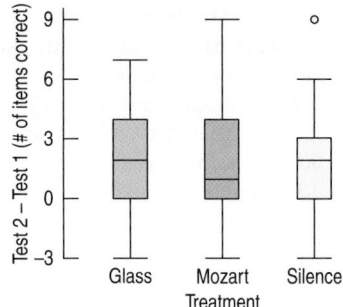

d) Do you think the results prove that listening to Mozart is beneficial? Explain.

34. Full moon. It's a common belief that people behave strangely when there's a full moon and that as a result police and emergency rooms are busier than usual. Design a way you could find out whether there is any merit to this belief. Will you use an observational study or an experiment? Why?

35. Wine. A 2001 Danish study published in the *Archives of Internal Medicine* casts significant doubt on suggestions that adults who drink wine have higher levels of "good" cholesterol and fewer heart attacks. These researchers followed a group of individuals born at a Copenhagen hospital between 1959 and 1961 for 40 years. Their study found that in this group the adults who drank wine were richer and better educated than those who did not.
a) What kind of study was this?
b) It is generally true that people with high levels of education and high socioeconomic status are healthier than others. How does this call into question the supposed health benefits of wine?
c) Can studies such as these prove causation (that wine helps prevent heart attacks, that drinking wine makes one richer, that being rich helps prevent heart attacks, etc.)? Explain.

36. Swimming. Recently, a group of adults who swim regularly for exercise were evaluated for depression. It turned out that these swimmers were less likely to be depressed than the general population. The researchers said the difference was statistically significant.
a) What does "statistically significant" mean in this context?
b) Is this an experiment or an observational study? Explain.
c) News reports claimed this study proved that swimming can prevent depression. Explain why this conclusion is not justified by the study. Include an example of a possible lurking variable.

d) But perhaps it is true. We wonder if exercise can ward off depression, and whether anaerobic exercise (like weight training) is as effective as aerobic exercise (like swimming). We find 120 volunteers not currently engaged in a regular program of exercise. Design an appropriate experiment.

37. Dowsing. Before drilling for water, many rural homeowners hire a dowser (a person who claims to be able to sense the presence of underground water using a forked stick.) Suppose we wish to set up an experiment to test one dowser's ability. We get 20 identical containers, fill some with water, and ask him to tell which ones they are.
a) How will we randomize this procedure?
b) The dowser correctly identifies the contents of 12 out of 20 containers. Do you think this level of success is statistically significant? Explain.
c) How many correct identifications (out of 20) would the dowser have to make to convince you that the forked-stick trick works? Explain.

38. Healing. A medical researcher suspects that giving post-surgical patients large doses of vitamin E will speed their recovery times by helping their incisions heal more quickly. Design an experiment to test this conjecture. Be sure to identify the factors, levels, treatments, response variable, and the role of randomization.

39. Reading. Some schools teach reading using phonics (the sounds made by letters) and others using whole language (word recognition). Suppose a school district wants to know which method works better. Suggest a design for an appropriate experiment.

40. Gas mileage. Do cars get better gas mileage with premium instead of regular unleaded gasoline? It might be possible to test some engines in a laboratory, but we'd rather use real cars and real drivers in real day-to-day driving, so we get 20 volunteers. Design the experiment.

41. Weekend deaths. A study published in the *New England Journal of Medicine* (Aug. 2001) suggests that it's dangerous to enter a hospital on a weekend. During a 10-year period, researchers tracked over 4 million emergency admissions to hospitals in Ontario, Canada. Their findings revealed that patients admitted on weekends had a much higher risk of death than those who went on weekdays.
a) The researchers said the difference in death rates was "statistically significant." Explain in this context what that means.
b) What kind of study was this? Explain.
c) If you think you're quite ill on a Saturday, should you wait until Monday to seek medical help? Explain.
d) Suggest some possible explanations for this troubling finding.

42. Shingles. A research doctor has discovered a new ointment that she believes will be more effective than the current medication in the treatment of shingles (a painful skin rash). Eight patients have volunteered to participate in the initial trials of this ointment. You are the statistician hired as a consultant to help design a completely randomized experiment.
a) Describe how you will conduct this experiment.
b) Suppose the eight patients' last names start with the letters A to H. Using the random numbers listed below,

show which patients you will assign to each treatment. Explain your randomization procedure clearly.

41098   18329   78458   31685   55259

c) Can you make this experiment double-blind? How?
d) The initial experiment revealed that males and females may respond differently to the ointment. Further testing of the drug's effectiveness is now planned, and many patients have volunteered. What changes in your first design, if any, would you make for this second stage of testing?

43. Beetles.   Hoping to learn how to control crop damage by a certain species of beetle, a researcher plans to test two different pesticides in small plots of corn. A few days after application of the chemicals, he'll check the number of beetle larvae found on each plant. The researcher wants to know whether either pesticide works and whether there is a significant difference in effectiveness between them. Design an appropriate experiment.

44. SAT Prep.   Can special study courses actually help raise SAT scores? One organization says that the 30 students they tutored achieved an average gain of 60 points when they retook the test.
a) Explain why this does not necessarily prove that the special course caused the scores to go up.
b) Propose a design for an experiment that could test the effectiveness of the tutorial course.
c) Suppose you suspect that the tutorial course might be more helpful for students whose initial scores were particularly low. How would this affect your proposed design?

45. Safety switch.   An industrial machine requires an emergency shutoff switch that must be designed so that it can be easily operated with either hand. Design an experiment to find out whether workers will be able to deactivate the machine as quickly with their left hands as with their right hands. Be sure to explain the role of randomization in your design.

46. Washing clothes.   A consumer group wants to test the effectiveness of a new "organic" laundry detergent and make recommendations to customers about how to best use the product. They intentionally get grass stains on 30 white T-shirts in order to see how well the detergent will clean them. They want to try the detergent in cold water and in hot water on both the "regular" and "delicates" wash cycles. Design an appropriate experiment, indicating the number of factors, levels, and treatments. Explain the role of randomization in your experiment.

47. Skydiving, anyone?   A humor piece published in the *British Medical Journal* ("Parachute use to prevent death and major trauma related to gravitational challenge: systematic review of randomized control trials," Gordon, Smith, and Pell, *BMJ*, 2003:327) notes that we can't tell for sure whether parachutes are safe and effective because there has never been a properly randomized, double-blind, placebo-controlled study of parachute effectiveness in skydiving. (Yes, this is the sort of thing statisticians find funny . . . .) Suppose you were designing such a study:
a) What is the factor in this experiment?
b) What experimental units would you propose?[7]
c) What would serve as a placebo for this study?
d) What would the treatments be?
e) What would the response variable be?
f) What sources of variability would you control?
g) How would you randomize this "experiment"?
h) How would you make the experiment double-blind?

JUST CHECKING
Answers

1. a) The factor was type of treatment for peptic ulcer.
   b) The response variable could be a measure of relief from gastric ulcer pain or an evaluation by a physician of the state of the disease.
   c) Treatments would be gastric freezing and some alternative control treatment.
   d) Treatments should be assigned randomly.
   e) No. The Web site reports "lack of effectiveness," indicating that no large differences in patient healing were noted.

2. a) Neither the patients who received the treatment nor the doctor who evaluated them afterward knew what treatment they had received.
   b) The placebo is needed to accomplish blinding. The best alternative would be using body-temperature liquid rather than the freezing liquid.
   c) The researchers should block the subjects by the length of time they had had the ulcer, then randomly assign subjects in each block to the freezing and placebo groups.

---

[7] Don't include your Statistics instructor!

PART III

REVIEW OF PART III

Gathering Data

QUICK REVIEW

Before you can make a boxplot, calculate a mean, describe a distribution, or fit a line, you must have meaningful data to work with. Getting good data is essential to any investigation. No amount of clever analysis can make up for badly collected data. Here's a brief summary of the key concepts and skills:

▶ The way you gather data depends both on what you want to discover and on what is practical.

▶ To get some insight into what might happen in a real situation, model it with a **simulation** using random numbers.

▶ To answer questions about a target population, collect information from a sample with a **survey** or poll.

- Choose the sample randomly. Random sampling designs include simple, stratified, systematic, cluster, and multistage.
- A simple random sample draws without restriction from the entire target population.
- When there are subgroups within the population that may respond differently, use a stratified sample.
- Avoid bias, a systematic distortion of the results. Sample designs that allow undercoverage or response bias and designs such as voluntary response or convenience samples don't faithfully represent the population.
- Samples will naturally vary one from another. This sample-to-sample variation is called sampling error. Each sample only approximates the target population.

▶ **Observational studies** collect information from a sample drawn from a target population.

- Retrospective studies examine existing data. Prospective studies identify subjects in advance, then follow them to collect data as the data are created, perhaps over many years.
- Observational studies can spot associations between variables but cannot establish cause and effect. It's impossible to eliminate the possibility of lurking or confounding variables.

▶ To see how different treatments influence a response variable, design an **experiment.**

- Assign subjects to treatments randomly. If you don't assign treatments randomly, your experiment is not likely to yield valid results.
- Control known sources of variation as much as possible. Reduce variation that cannot be controlled by using blocking, if possible.
- Replicate the experiment, assigning several subjects to each treatment level.
- If possible, replicate the entire experiment with an entirely different collection of subjects.
- A well-designed experiment can provide evidence that changes in the factors cause changes in the response variable.

Now for more opportunities to review these concepts and skills . . .

REVIEW EXERCISES

1–18. What design? *Analyze the design of each research example reported. Is it a sample survey, an observational study, or an experiment? If a sample, what are the population, the parameter of interest, and the sampling procedure? If an observational study, was it retrospective or prospective? If an experiment, describe the factors, treatments, randomization, response variable, and any blocking, matching, or blinding that may be present. In each, what kind of conclusions can be reached?*

1. Researchers identified 242 children in the Cleveland area who had been born prematurely (at about 29 weeks). They examined these children at age 8 and again at age 20, comparing them to another group of 233 children not born prematurely. Their report, published in the *New England Journal of Medicine,* said the "preemies" engaged in significantly less risky behavior than the others. Differences

showed up in the use of alcohol and marijuana, conviction of crimes, and teenage pregnancy.

2. The journal *Circulation* reported that among 1900 people who had heart attacks, those who drank an average of 19 cups of tea a week were 44% more likely than nondrinkers to survive at least 3 years after the attack.

3. Researchers at the Purina Pet Institute studied Labrador retrievers for evidence of a relationship between diet and longevity. At 8 weeks of age, 2 puppies of the same sex and weight were randomly assigned to one of two groups—a total of 48 dogs in all. One group was allowed to eat all they wanted, while the other group was fed a diet about 25% lower in calories. The median lifespan of dogs fed the restricted diet was 22 months longer than that of other dogs. (*Science News* 161, no. 19)

4. The radioactive gas radon, found in some homes, poses a health risk to residents. To assess the level of contamination in their area, a county health department wants to test a few homes. If the risk seems high, they will publicize the results to emphasize the need for home testing. Officials plan to use the local property tax list to randomly choose 25 homes from various areas of the county.

5. Almost 90,000 women participated in a 16-year study of the role of the vitamin folate in preventing colon cancer. Some of the women had family histories of colon cancer in close relatives. In this at-risk group, the incidence of colon cancer was cut in half among those who maintained a high folate intake. No such difference was observed in those with no family-based risk. (*Science News,* Feb. 9, 2002)

6. In the journal *Science,* a research team reported that plants in southern England are flowering earlier in the spring. Records of the first flowering dates for 385 species over a period of 47 years indicate that flowering has advanced an average of 15 days per decade, an indication of climate warming, according to the authors.

7. Fireworks manufacturers face a dilemma. They must be sure that the rockets work properly, but test firing a rocket essentially destroys it. On the other hand, not testing the product leaves open the danger that they sell a bunch of duds, leading to unhappy customers and loss of future sales. The solution, of course, is to test a few of the rockets produced each day, assuming that if those tested work properly, the others are ready for sale.

8. Can makeup damage fetal development? Many cosmetics contain a class of chemicals called phthalates. Studies that exposed some laboratory animals to these chemicals found a heightened incidence of damage to male reproductive systems. Since traces of phthalates are found in the urine of women who use beauty products, there is growing concern that they may present a risk to male fetuses. (*Science News,* July 20, 2002)

9. Can long-term exposure to strong electromagnetic fields cause cancer? Researchers in Italy tracked down 13 years of medical records for people living near Vatican Radio's powerful broadcast antennas. A disproportionate share of the leukemia cases occurred among men and children who lived within 6 kilometers of the antennas. (*Science News,* July 20, 2002)

10. Some doctors have expressed concern that men who have vasectomies seemed more likely to develop prostate cancer. Medical researchers used a national cancer registry to identify 923 men who had had prostate cancer and 1224 men of similar ages who had not. Roughly one quarter of the men in each group had undergone a vasectomy, many more than 25 years before the study. The study's authors concluded that there is strong evidence that having the operation presents no long-term risk for developing prostate cancer. (*Science News,* July 20, 2002)

11. Researchers investigating appetite control as a means of losing weight found that female rats ate less and lost weight after injections of the hormone leptin, while male rats responded better to insulin. (*Science News,* July 20, 2002)

12. An artisan wants to create pottery that has the appearance of age. He prepares several samples of clay with four different glazes and test fires them in a kiln at three different temperature settings.

13. Tests of gene therapy on laboratory rats have raised hopes of stopping the degeneration of tissue that characterizes chronic heart failure. Researchers at the University of California, San Diego, used hamsters with cardiac disease, randomly assigning 30 to receive the gene therapy and leaving the other 28 untreated. Five weeks after treatment the gene therapy group's heart muscles stabilized, while those of the untreated hamsters continued to weaken. (*Science News,* July 27, 2002)

14. Researchers at the University of Bristol (England) investigated reasons why different species of birds begin to sing at different times in the morning. They captured and examined birds of 57 species at seven different sites. They measured the diameter of the birds' eyes and also recorded the time of day at which each species began to sing. These researchers reported a strong relationship between eye diameter and time of singing, saying that birds with bigger eyes tended to sing earlier. (*Science News,* 161, no. 16 [2002])

15. An orange-juice processing plant will accept a shipment of fruit only after several hundred oranges selected from various locations within the truck are carefully inspected. If too many show signs of unsuitability for juice (bruised, rotten, unripe, etc.), the whole truckload is rejected.

16. A soft-drink manufacturer must be sure the bottle caps on the soda are fully sealed and will not come off easily. Inspectors pull a few bottles off the production line at regular intervals and test the caps. If they detect any problems, they will stop the bottling process to adjust or repair the machine that caps the bottles.

17. Physically fit people seem less likely to die of cancer. A report in the May 2002 issue of *Medicine and Science in Sports and Exercise* followed 25,892 men aged 30 to 87 for 10 years. The most physically fit men had a 55% lower risk of death from cancer than the least fit group.

18. Does the use of computer software in Introductory Statistics classes lead to better understanding of the concepts? A professor teaching two sections of Statistics decides to investigate. She teaches both sections using the same lectures and assignments, but gives one class statistics software to help them with their homework. The classes take the same final exam, and graders do not know which students used computers during the semester. The professor is also concerned that students who have had calculus may perform differently from those who have not, so she plans to compare software vs. no-software scores separately for these two groups of students.

19. **Point spread.** When taking bets on sporting events, bookmakers often include a "point spread" that awards the weaker team extra points. In theory this makes the outcome of the bet a toss-up. Suppose a gambler places a $10 bet and picks the winners of five games. If he's right about fewer than three of the games, he loses. If he gets three, four, or all five correct, he's paid $10, $20, and $50, respectively. Estimate the amount such a bettor might expect to lose over many weeks of gambling.

20. The lottery. Many people spend a lot of money trying to win huge jackpots in state lotteries. Let's play a simplified version using only the numbers from 1 to 20. You bet on three numbers. The state picks five winning numbers. If your three are all among the winners, you are rich!
a) Simulate repeated plays. How long did it take you to win?
b) In real lotteries, there are many more choices (often 54) and you must match all five winning numbers. Explain how these changes affect your chances of hitting the jackpot.

21. Everyday randomness. Aside from casinos, lotteries, and games, there are other situations you encounter in which something is described as "random" in some way. Give three different examples. Describe how randomness is (or is not) achieved in each.

22. Cell phone risks. Researchers at the Washington University School of Medicine randomly placed 480 rats into one of three chambers containing radio antennas. One group was exposed to digital cell phone radio waves, the second to analog cell phone waves, and the third group to no radio waves. Two years later the rats were examined for signs of brain tumors. In June 2002 the scientists said that differences among the three groups were not statistically significant.
a) Is this a study or an experiment? Explain.
b) Explain in this context what "not statistically significant" means.
c) Comment on the fact that this research was funded by Motorola, a manufacturer of cell phones.

23. Tips. In restaurants, servers rely on tips as a major source of income. Does serving candy after the meal produce larger tips? To find out, two waiters determined randomly whether or not to give candy to 92 dining parties. They recorded the sizes of the tips and reported that guests getting candy tipped an average of 17.8% of the bill, compared with an average tip of only 15.1% from those who got no candy. ("Sweetening the Till: The Use of Candy to Increase Restaurant Tipping." *Journal of Applied Social Psychology* 32, no. 2 [2002]: 300–309)
a) Was this an experiment or an observational study? Explain.
b) Is it reasonable to conclude that the candy caused guests to tip more? Explain.
c) The researchers said the difference was statistically significant. Explain in this context what that means.

24. Tips, take 2. In another experiment to see if getting candy after a meal would induce customers to leave a bigger tip, a waitress randomly decided what to do with 80 dining parties. Some parties received no candy, some just one piece, and some two pieces. Others initially got just one piece of candy, and then the waitress suggested that they take another piece. She recorded the tips received, finding that, in general, the more candy, the higher the tip, but the highest tips (23%) came from the parties who got one piece and then were offered more. ("Sweetening the Till: The Use of Candy to Increase Restaurant Tipping." *Journal of Applied Social Psychology* 32, no. 2 [2002]: 300–309)

a) Diagram this experiment.
b) How many factors are there? How many levels?
c) How many treatments are there?
d) What is the response variable?
e) Did this experiment involve blinding? Double blinding?
f) In what way might the waitress, perhaps unintentionally, have biased the results?

25. Cloning. In September 1998, *USA Weekend* magazine asked, "Should humans be cloned?" Readers were invited to vote "Yes" or "No" by calling one of two different 900 numbers. Based on 38,023 responses, the magazine reported that "9 out of 10 readers oppose cloning."
a) Explain why you think the conclusion is not justified. Describe the types of bias that may be present.
b) Reword the question in a way that you think might create a more positive response.

26. Laundry. An experiment to test a new laundry detergent, SparkleKleen, is being conducted by a consumer advocate group. They would like to compare its performance with that of a laboratory standard detergent they have used in previous experiments. They can stain 16 swatches of cloth with 2 tsp of a common staining compound and then use a well-calibrated optical scanner to detect the amount of the stain left after washing. To save time in the experiment, several suggestions have been made. Comment on the possible merits and drawbacks of each one.
a) Since data for the laboratory standard detergent are already available from previous experiments, for this experiment wash all 16 swatches with SparkleKleen, and compare the results with the previous data.
b) Use both detergents with eight separate runs each, but to save time, use only a 10-second wash time with very hot water.
c) To ease bookkeeping, first run all of the standard detergent washes on eight swatches, then run all of the SparkleKleen washes on the other eight swatches.
d) Rather than run the experiment, use data from the company that produced SparkleKleen, and compare them with past data from the standard detergent.

27. When to stop? You play a game that involves rolling a die. You can roll as many times as you want, and your score is the total for all the rolls. But ... if you roll a 6 your score is 0 and your turn is over. What might be a good strategy for a game like this?
a) One of your opponents decides to roll 4 times, then stop (hoping not to get the dreaded 6 before then). Use a simulation to estimate his average score.
b) Another opponent decides to roll until she gets at least 12 points, then stop. Use a simulation to estimate her average score.
c) Propose another strategy that you would use to play this game. Using your strategy, simulate several turns. Do you think you would beat the two opponents?

28. Rivets. A company that manufactures rivets believes the shear strength of the rivets they manufacture follows a Normal model with a mean breaking strength of 950 pounds and a standard deviation of 40 pounds.

a) What percentage of rivets selected at random will break when tested under a 900-pound load?
b) You're trying to improve the rivets and want to examine some that fail. Use a simulation to estimate how many rivets you might need to test in order to find three that fail at 900 pounds (or below).

29. Homecoming. A college Statistics class conducted a survey concerning community attitudes about the college's large homecoming celebration. That survey drew its sample in the following manner: Telephone numbers were generated at random by selecting one of the local telephone exchanges (first three digits) at random and then generating a random four-digit number to follow the exchange. If a person answered the phone and the call was to a residence, then that person was taken to be the subject for interview. (Undergraduate students and those under voting age were excluded, as was anyone who could not speak English.) Calls were placed until a sample of 200 eligible respondents had been reached.
a) Did every telephone number that could occur in that community have an equal chance of being generated?
b) Did this method of generating telephone numbers result in a simple random sample (SRS) of local residences? Explain.
c) Did this method generate an SRS of local voters? Explain.
d) Is this method unbiased in generating samples of households? Explain.

30. Youthful appearance. *Readers' Digest* reported results of several surveys that asked graduate students to examine photographs of men and women and try to guess their ages. Researchers compared these guesses with the number of times the people in the pictures reported having sexual intercourse. It turned out that those who had been more sexually active were judged as looking younger, and that the difference was described as "statistically significant." Psychologist David Weeks, who compiled the research, speculated that lovemaking boosts hormones that "reduce fatty tissue and increase lean muscle, giving a more youthful appearance."
a) What does "statistically significant" mean in this context?
b) Explain in statistical terms why you might be skeptical about Dr. Weeks's conclusion. Propose an alternative explanation for these results.

31. Smoking and Alzheimer's. Medical studies indicate that smokers are less likely to develop Alzheimer's disease than people who never smoked.
a) Does this prove that smoking may offer some protection against Alzheimer's? Explain.
b) Offer an alternative explanation for this association.
c) How would you conduct a study to investigate this?

32. Antacids. A researcher wants to compare the performance of three types of antacid in volunteers suffering from acid reflux disease. Because men and women may react differently to this medication, the subjects are split into two groups, by sex. Subjects in each group are randomly assigned to take one of the antacids or to take a sugar pill made to look the same. The subjects will rate their level of discomfort 30 minutes after eating.
a) What kind of design is this?
b) The experiment uses volunteers rather than a random sample of all people suffering from acid reflux disease. Does this make the results invalid? Explain.
c) How may the use of the placebo confound this experiment? Explain.

33. Sex and violence. Does the content of a television program affect viewers' memory of the products advertised in commercials? Design an experiment to compare the ability of viewers to recall brand names of items featured in commercials during programs with violent content, sexual content, or neutral content.

34. Pubs. In England, a Leeds University researcher said that the local watering hole's welcoming atmosphere helps men get rid of the stresses of modern life and is vital for their psychological well-being. Author of the report, Dr. Colin Gill, said rather than complain, women should encourage men to "pop out for a swift half." "Pub-time allows men to bond with friends and colleagues," he said. "Men need break-out time as much as women and are mentally healthier for it." Gill added that men might feel unfulfilled or empty if they had not been to the pub for a week. The report, commissioned by alcohol-free beer brand Kaliber, surveyed 900 men on their reasons for going to the pub. More than 40% said they went for the conversation, with relaxation and a friendly atmosphere being the other most common reasons. Only 1 in 10 listed alcohol as the overriding reason.

Let's examine this news story from a statistical perspective.
a) What are the W's: *Who, What, When, Where, Why?*
b) What population does the researcher think the study applies to?
c) What is the most important thing about the selection process that the article does *not* tell us?
d) How do *you* think the 900 respondents were selected? (Name a method of drawing a sample that is likely to have been used.)
e) Do you think the report that only 10% of respondents listed alcohol as an important reason for going to the pub might be a biased result? Why?

35. Age and party. The Gallup Poll conducted a representative telephone survey during the first quarter of 1999. Among its reported results was the following table concerning the preferred political party affiliation of respondents and their ages:

		Party			
		Republican	Democratic	Independent	Total
Age	18–29	241	351	409	1001
	30–49	299	330	370	999
	50–64	282	341	375	998
	65+	279	382	343	1004
	Total	1101	1404	1497	**4002**

a) What sampling strategy do you think the pollsters used? Explain.
b) What percentage of the people surveyed were Democrats?
c) Do you think this is a good estimate of the percentage of voters in the United States who are registered Democrats? Why or why not?
d) In creating this sample design, what question do you think the pollsters were trying to answer?

36. Bias? Political analyst Michael Barone has written that "conservatives are more likely than others to refuse to respond to polls, particularly those polls taken by media outlets that conservatives consider biased" (*The Weekly Standard,* March 10, 1997). The Pew Research Foundation tested this assertion by asking the same questions in a national survey run by standard methods and in a more rigorous survey that was a true SRS with careful follow-up to encourage participation. The response rate in the "standard survey" was 42%. The response rate in the "rigorous survey" was 71%.
a) What kind of bias does Barone claim may exist in polls?
b) What is the population for these surveys?
c) On the question of political position, the Pew researchers report the following table:

	Standard Survey	Rigorous Survey
Conservative	37%	35%
Moderate	40%	41%
Liberal	19%	20%

What makes you think these results are incomplete?
d) The Pew researchers report that differences between opinions expressed on the two surveys were not statistically significant. Explain what "not statistically significant" means in this context.

37. Save the grapes. Vineyard owners have problems with birds that like to eat the ripening grapes. Some vineyards use scarecrows to try to keep birds away. Others use netting that covers the plants. Owners really would like to know if either method works and, if so, which one is better. One owner has offered to let you use his vineyard this year for an experiment. Propose a design. Carefully indicate how you would set up the experiment, specifying the factor(s) and response variable.

38. Bats. It's generally believed that baseball players can hit the ball farther with aluminum bats than with the traditional wooden ones. Is that true? And, if so, how much farther? Players on your local high school baseball team have agreed to help you find out. Design an appropriate experiment.

39. Knees. Research reported in the spring of 2002 cast doubt on the effectiveness of arthroscopic knee surgery for patients with arthritis. Patients suffering from arthritis pain who volunteered to participate in the study were randomly divided into groups. One group received arthroscopic knee surgery. The other group underwent "placebo surgery" during which incisions were made in their knees, but no surgery was actually performed. Follow-up evaluations over a period of 2 years found that differences in the amount of pain relief experienced by the two groups were not statistically significant. (*NEJM* 347:81–88 July 11, 2002)
a) Why did the researchers feel it was necessary to have some of the patients undergo "placebo surgery"?
b) Because patients had to consent to participate in this experiment, the subjects were essentially self-selected—a kind of voluntary response group. Explain why that does not invalidate the findings of the experiment.
c) What does "statistically significant" mean in this context?

40. NBA draft lottery. Professional basketball teams hold a "draft" each year in which they get to pick the best available college and high school players. In an effort to promote competition, teams with the worst records get to pick first, theoretically allowing them to add better players. To combat the fear that teams with no chance to make the playoffs might try to get better draft picks by intentionally losing late-season games, the NBA's Board of Governors adopted a weighted lottery system in 1990. Under this system, the 11 teams that did not make the playoffs were eligible for the lottery. The NBA prepared 66 cards, each naming one of the teams. The team with the worst win-loss record was named on 11 of the cards, the second-worst team on 10 cards, and so on, with the team having the best record among the nonplayoff clubs getting only one chance at having the first pick. The cards were mixed, then drawn randomly to determine the order in which the teams could draft players. (Since 1995, 13 teams have been involved in the lottery, using a complicated system with 14 numbered Ping-Pong balls drawn in groups of four.) Suppose there are two exceptional players available in this year's draft and your favorite team had the third-worst record. Use a simulation to find out how likely it is that your team gets to pick first or second. Describe your simulation carefully.

41. Security. There are 20 first-class passengers and 120 coach passengers scheduled on a flight. In addition to the usual security screening, 10% of the passengers will be subjected to a more complete search.
a) Describe a sampling strategy to randomly select those to be searched.
b) Here is the first-class passenger list and a set of random digits. Select two passengers to be searched, carefully demonstrating your process.

65436 71127 04879 41516 20451 02227 94769 23593

Bergman	Cox	Fontana	Perl
Bowman	DeLara	Forester	Rabkin
Burkhauser	Delli-Bovi	Frongillo	Roufaiel
Castillo	Dugan	Furnas	Swafford
Clancy	Febo	LePage	Testut

c) Explain how you would use a random number table to select the coach passengers to be searched.

42. **Profiling?** Among the 20 first-class passengers on the flight described in Exercise 41, there were four businessmen from the Middle East. Two of them were the two passengers selected to be searched. They complained of profiling, but the airline claims that the selection was random. What do you think? Support your conclusion with a simulation.

43. **Par 4.** In theory, a golfer playing a par-4 hole tees off, hitting the ball in the fairway, then hits an approach shot onto the green. The first putt (usually long) probably won't go in, but the second putt (usually much shorter) should. Sounds simple enough, but how many strokes might it really take? Use a simulation to estimate a pretty good golfer's score based on these assumptions:
    - The tee shot hits the fairway 70% of the time.
    - A first approach shot lands on the green 80% of the time from the fairway, but only 40% of the time otherwise.
    - Subsequent approach shots land on the green 90% of the time.
    - The first putt goes in 20% of the time, and subsequent putts go in 90% of the time.

44. **The back nine.** Use simulations to estimate more golf scores, similar to the procedure in Exercise 43.
    a) On a par 3, the golfer hopes the tee shot lands on the green. Assume that the tee shot behaves like the first approach shot described in Exercise 43.
    b) On a par 5, the second shot will reach the green 10% of the time and hit the fairway 60% of the time. If it does not hit the green, the golfer must play an approach shot as described in Exercise 43.
    c) Create a list of assumptions that describe your golfing ability, and then simulate your score on a few holes. Explain your simulation clearly.

Randomness and Probability

CHAPTER
14

From Randomness to Probability

E arly humans saw a world filled with random events. To help them make sense of the chaos around them, they sought out seers, consulted oracles, and read tea leaves. As science developed, we learned to recognize some events as predictable. We can now forecast the change of seasons, tell when eclipses will occur precisely, and even make a reasonably good guess at how warm it will be tomorrow. But many other events are still essentially random. Will the stock market go up or down today? When will the next car pass this corner? And we now know from quantum mechanics that the universe is in some sense random at the most fundamental levels of subatomic particles.

But we have also learned to understand randomness. The surprising fact is that in the long run, even truly random phenomena settle down in a way that's consistent and predictable. It's this property of random phenomena that makes the next steps we're about to take in Statistics possible.

Dealing with Random Phenomena

Every day you drive through the intersection at College and Main. Even though it may seem that the light is never green when you get there, you know this can't really be true. In fact, if you try really hard, you can recall just sailing through the green light once in a while.

What's random here? The light itself is governed by a timer. Its pattern isn't haphazard. In fact, the light may even be red at precisely the same times each day. It's the pattern of *your driving* that is random. No, we're certainly not insinuating that you can't keep the car on the road. At the precision level of the 30 seconds or so that the light spends being red or green, the time you arrive at the light *is random*. Even if you try to leave your house at exactly the same time every day, whether the light is red or green as *you* reach the intersection is a **random phenomenon.**[1]

---

[1] If you somehow managed to leave your house at *precisely* the same time every day and there was *no* variation in the time it took you to get to the light, then there wouldn't be any randomness, but that's not very realistic.

Is the color of the light completely unpredictable? When you stop to think about it, it's clear that you do expect some kind of *regularity* in your long-run experience. Some *fraction* of the time, the light will be green as you get to the intersection. How can you figure out what that fraction is?

You might record what happens at the intersection each day and graph the *accumulated percentage* of green lights like this:

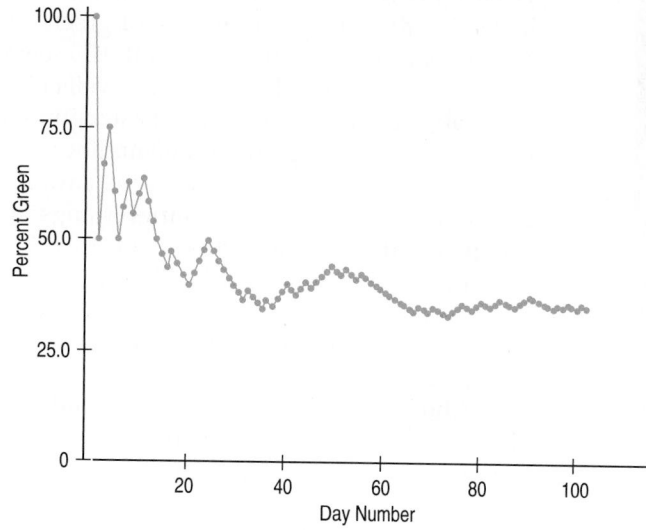

FIGURE 14.1

The overall percentage of times the light is green settles down as you see more outcomes.

Day	Light	% Green
1	Green	100
2	Red	50
3	Green	66.7
4	Green	75
5	Red	60
6	Red	50
⋮	⋮	⋮

The first day you recorded the light, it was green. Then on the next five days, it was red, then green again, then green, red, and red. If you plot the percentage of green lights against days, the graph would start at 100% (because the first time, the light was green, so 1 out of 1, for 100%). Then the next day it was red, so the accumulated percentage dropped to 50% (1 out of 2). The third day it was green again (2 out of 3, or 67% green), then green (3 out of 4, or 75%), then red twice in a row (3 out of 5, for 60% green, and then 3 out of 6, for 50%), and so on. As you collect a new data value for each day, each new outcome becomes a smaller and smaller fraction of the accumulated experience, so, in the long run, the graph settles down. As it settles down, you can see that, in fact, the light is green about 35% of the time.

When talking about random phenomena such as this, it helps to define our terms. You aren't interested in the traffic light *all* the time. You pull up to the intersection only once a day, so you care about the color of the light only at these particular times.[2] In general, each occasion upon which we observe a random phenomenon is called a **trial**. At each trial, we note the value of the random phenomenon, and call that the trial's **outcome**. (If this language reminds you of Chapter 11, that's *not* unintentional.)

For the traffic light, there are really three possible outcomes: red, yellow, or green. Often we're more interested in a combination of outcomes rather than in the individual ones. When you see the light turn yellow, what do *you* do? If you race through the intersection, then you treat the yellow more like a green light. If you step on the brakes, you treat it more like a red light. Either way, you might want to group the yellow with one or the other. When we combine outcomes like that, the resulting combination is an **event**.[3] We sometimes talk about the collection of *all possible outcomes* and call that event the **sample space**.[4] We'll denote the sample

> A phenomenon consists of trials. Each trial has an outcome. Outcomes combine to make events.

---

[2] Even though the randomness here comes from the uncertainty in our arrival time, we can think of the light itself as showing a color at random.

[3] Each individual outcome is also an event.

[4] Mathematicians like to use the term "space" as a fancy name for a set. Sort of like referring to that closet colleges call a dorm room as "living space." But remember that it's really just the set of all outcomes.

space **S**. (Some books are even fancier and use the Greek letter Ω.) For the traffic light, **S** = {red, green, yellow}.

The Law of Large Numbers

"For even the most stupid of men . . . is convinced that the more observations have been made, the less danger there is of wandering from one's goal."

—Jacob Bernoulli, 1713, discoverer of the LLN

What's the *probability* of a green light at College and Main? Based on the graph, it looks like the relative frequency of green lights settles down to about 35%, so saying that the probability is about 0.35 seems like a reasonable answer. But do random phenomena always behave well enough for this to make sense? Perhaps the relative frequency of an event can bounce back and forth between two values forever, never settling on just one number.

Fortunately, a principle called the **Law of Large Numbers** (LLN) gives us the guarantee we need. It simplifies things if we assume that the events are **independent.** Informally, this means that the outcome of one trial doesn't affect the outcomes of the others. (We'll see a formal definition of independent events in the next chapter.) The LLN says that as the number of independent trials increases, the long-run *relative frequency* of repeated events gets closer and closer to a single value.

Although the LLN wasn't proven until the 18th century, everyone expects the kind of long-run regularity that the Law describes from everyday experience. In fact, the first person to prove the LLN, Jacob Bernoulli, thought it was pretty obvious, too, as his remark quoted in the margin shows.[5]

Because the LLN guarantees that relative frequencies settle down in the long run, we can now officially give a name to the value that they approach. We call it the **probability** of the event. If the relative frequency of green lights at that intersection settles down to 35% in the long run, we say that the probability of encountering a green light is 0.35, and we write $P(\text{green}) = 0.35$. Because this definition is based on repeatedly observing the event's outcome, this definition of probability is often called **empirical probability.**

> **Empirical Probability**
> For any event **A**,
> $$P(\mathbf{A}) = \frac{\#\text{ times }\mathbf{A}\text{ occurs}}{\text{total }\#\text{ of trials}}$$
> in the long run.

The Nonexistent Law of Averages

Don't let yourself think that there's a Law of Averages that promises short-term compensation for recent deviations from expected behavior. A belief in such a "Law" can lead to money lost in gambling and to poor business decisions.

"Slump? I ain't in no slump. I just ain't hittin'."

—Yogi Berra

Even though the LLN seems natural, it is often misunderstood because the idea of the *long run* is hard to grasp. Many people believe, for example, that an outcome of a random event that hasn't occurred in many trials is "due" to occur. Many gamblers bet on numbers that haven't been seen for a while, mistakenly believing that they're likely to come up sooner. A common term for this is the "Law of Averages." After all, we know that in the long run, the relative frequency will settle down to the probability of that outcome, so now we have some "catching up" to do, right?

Wrong. The Law of Large Numbers says nothing about short-run behavior. Relative frequencies even out *only in the long run.* And, according to the LLN, the long run is *really* long (*infinitely* long, in fact).

The so-called Law of Averages doesn't exist at all. But you'll hear people talk about it as if it does. Is a good hitter in baseball who has struck out the last six times *due* for a hit his next time up? If you've been doing particularly well in weekly quizzes in Statistics class, are you *due* for a bad grade? No. This isn't the way random phenomena work. There is *no* Law of Averages for short runs.

The lesson of the LLN is that sequences of random events don't compensate in the *short* run and don't need to do so to get back to the right long-run probability.

---

[5] In case you were wondering, Jacob's reputation was that he was every bit as nasty as this quotation suggests. He and his brother, who was also a mathematician, fought publicly over who had accomplished the most.

If the probability of an outcome doesn't change and the events are independent, the probability of any outcome in another trial is *always* what it was, no matter what has happened in other trials.

Coins, Keno, and the Law of Averages You've just flipped a fair coin and seen six heads in a row. Does the coin "owe" you some tails? Suppose you spend that coin and your friend gets it in change. When she starts flipping the coin, should she expect a run of tails? Of course not. Each flip is a new event. The coin can't "remember" what it did in the past, so it can't "owe" any particular outcomes in the future.

Just to see how this works in practice, we ran a simulation of 100,000 flips of a fair coin. We collected 100,000 random numbers, letting the numbers 0 to 4 represent heads and the numbers 5 to 9 represent tails. In our 100,000 "flips," there were 2981 streaks of at least 5 heads. The "Law of Averages" suggests that the next flip after a run of 5 heads should be tails more often to even things out. Actually, the next flip was heads more often than tails: 1550 times to 1431 times. That's 51.9% heads. You can perform a similar simulation easily on a computer. Try it!

Of course, sometimes an apparent drift from what we expect means that the probabilities are, in fact, *not* what we thought. If you get 10 heads in a row, maybe the coin has heads on both sides!

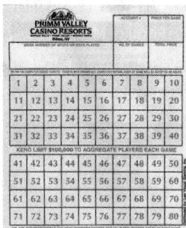

Keno is a simple casino game in which numbers from 1 to 80 are chosen. The numbers, as in most lottery games, are supposed to be equally likely. Payoffs are made depending on how many of those numbers you match on your card. A group of graduate students from a Statistics department decided to take a field trip to Reno. They (*very* discreetly) wrote down the outcomes of the games for a couple of days, then drove back to test whether the numbers were, in fact, equally likely. It turned out that some numbers were *more likely* to come up than others. Rather than bet on the Law of Averages and put their money on the numbers that were "due," the students put their faith in the LLN—and all their (and their friends') money on the numbers that had come up before. After they pocketed more than $50,000, they were escorted off the premises and invited never to show their faces in that casino again.

TI-*nspire*

The Law of Large Numbers. Watch the relative frequency of a random event approach the true probability *in the long run*.

JUST CHECKING

1. One common proposal for beating the lottery is to note which numbers have come up lately, eliminate those from consideration, and bet on numbers that have not come up for a long time. Proponents of this method argue that in the long run, every number should be selected equally often, so those that haven't come up are due. Explain why this is faulty reasoning.

Modeling Probability

A S *Activity:* **What Is Probability?** The best way to get a feel for probabilities is to experiment with them. We'll use this random-outcomes tool many more times.

Probability was first studied extensively by a group of French mathematicians who were interested in games of chance.[6] Rather than *experiment* with the games (and risk losing their money), they developed mathematical models of **theoretical probability.** To make things simple (as we usually do when we build models), they started by looking at games in which the different outcomes were equally likely. Fortunately, many games of chance are like that. Any of 52 cards is equally

---

[6] Ok, gambling.

likely to be the next one dealt from a well-shuffled deck. Each face of a die is equally likely to land up (or at least it *should be*).

It's easy to find probabilities for events that are made up of several *equally likely* outcomes. We just count all the outcomes that the event contains. The probability of the event is the number of outcomes in the event divided by the total number of possible outcomes. We can write

$$P(\mathbf{A}) = \frac{\text{\# outcomes in } \mathbf{A}}{\text{\# of possible outcomes}}.$$

For example, the probability of drawing a face card (JQK) from a deck is

$$P(\text{face card}) = \frac{\text{\# face cards}}{\text{\# cards}} = \frac{12}{52} = \frac{3}{13}.$$

> **Is that all there is to it?** Finding the probability of any event when the outcomes are equally likely is straightforward, but not necessarily easy. It gets hard when the number of outcomes in the event (and in the sample space) gets big. Think about flipping two coins. The sample space is **S** = {HH,HT,TH,TT} and each outcome is equally likely. So, what's the probability of getting *exactly* one head and one tail? Let's call that event **A**. Well, there are two outcomes in the event **A** = {HT,TH} out of the 4 possible equally likely ones in **S**, so $P(\mathbf{A}) = \frac{2}{4}$, or $\frac{1}{2}$.
>
> OK, now flip 100 coins. What's the probability of exactly 67 heads? Well, first, how many outcomes are in the sample space? **S** = {HHHHHHHHHHH...H, HH...T,...} Hmm. A lot. In fact, there are 1,267,650,600,228,229,401,496,703,205,376 different outcomes possible when flipping 100 coins. To answer the question, we'd still have to figure out how many ways there are to get 67 heads. That's coming in Chapter 17; stay tuned!

Don't get trapped into thinking that random events are always equally likely. The chance of winning a lottery—especially lotteries with very large payoffs—is small. Regardless, people continue to buy tickets. In an attempt to understand why, an interviewer asked someone who had just purchased a lottery ticket, "What do you think your chances are of winning the lottery?" The reply was, "Oh, about 50–50." The shocked interviewer asked, "How do you get that?" to which the response was, "Well, the way I figure it, either I win or I don't!"

The moral of this story is that events are *not* always equally likely.

Personal Probability

What's the probability that your grade in this Statistics course will be an **A**? You may be able to come up with a number that seems reasonable. Of course, no matter how confident or depressed you feel about your chances for success, your probability should be between 0 and 1. How did you come up with this probability? Is it an empirical probability? Not unless you plan on taking the course over and over (and over . . .), calculating the proportion of times you get an **A**. And, unless you assume the outcomes are equally likely, it will be hard to find the theoretical probability. But people use probability in a third sense as well.

We use the language of probability in everyday speech to express a degree of uncertainty *without* basing it on long-run relative frequencies or mathematical models. Your personal assessment of your chances of getting an **A** expresses your uncertainty about the outcome. That uncertainty may be based on how comfortable you're feeling in the course or on your midterm grade, but it can't be based on long-run behavior. We call this third kind of probability a subjective or **personal probability.**

Although personal probabilities may be based on experience, they're not based either on long-run relative frequencies or on equally likely events. So they don't display the kind of consistency that we'll need probabilities to have. For that reason, we'll stick to formally defined probabilities. You should be alert to the difference.

The First Three Rules for Working with Probability

1. Make a picture.
2. Make a picture.
3. Make a picture.

We're dealing with probabilities now, not data, but the three rules don't change. The most common kind of picture to make is called a Venn diagram. We'll use Venn diagrams throughout the rest of this chapter. Even experienced statisticians make Venn diagrams to help them think about probabilities of compound and overlapping events. You should, too.

John Venn (1834–1923) created the Venn diagram. His book on probability, The Logic of Chance, *was "strikingly original and considerably influenced the development of the theory of Statistics," according to John Maynard Keynes, one of the luminaries of Economics.*

Formal Probability

Surprising Probabilities

We've been careful to discuss probabilities only for situations in which the outcomes were finite, or even countably infinite. But if the outcomes can take on *any* numerical value at all (we say they are *continuous*), things can get surprising. For example, what is the probability that a randomly selected child will be *exactly* 3 feet tall? Well, if we mean 3.00000 . . . feet, the answer is zero. No randomly selected child—even one whose height would be recorded as 3 feet, will be *exactly* 3 feet tall (to an infinite number of decimal places). But, if you've grown taller than 3 feet, there must have been a time in your life when you actually *were* exactly 3 feet tall, even if only for a second. So this is an outcome with probability 0 that not only has happened—it has happened to *you*.

We've seen another example of this already in Chapter 6 when we worked with the Normal model. We said that the probability of any *specific* value—say, $z = 0.5$—is zero. The model gives a probability for any *interval* of values, such as $0.49 < z < 0.51$. The probability is smaller if we ask for $0.499 < z < 0.501$, and smaller still for $0.49999999 < z < 0.50000001$. Well, you get the idea. Continuous probabilities are useful for the mathematics behind much of what we'll do, but it's easier to deal with probabilities for countable outcomes.

For some people, the phrase "50/50" means something vague like "I don't know" or "whatever." But when we discuss probabilities of outcomes, it takes on the precise meaning of *equally likely*. Speaking vaguely about probabilities will get us into trouble, so whenever we talk about probabilities, we'll need to be precise.[7] And to do that, we'll need to develop some formal rules[8] about how probability works.

1. If the probability is 0, the event can't occur, and likewise if it has probability 1, it *always* occurs. Even if you think an event is very unlikely, its probability can't be negative, and even if you're sure it will happen, its probability can't be greater than 1. So we require that

   A probability is a number between 0 and 1.

   For any event A, $0 \leq P(A) \leq 1$.

---

[7] And to be precise, we will be talking only about sample spaces where we can enumerate all the outcomes. Mathematicians call this a countable number of outcomes.

[8] Actually, in mathematical terms, these are axioms—statements that we assume to be true of probability. We'll derive other rules from these in the next chapter.

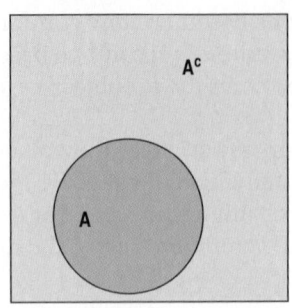

*The set **A** and its complement **A**^{C}. Together, they make up the entire sample space **S**.*

We write $P($**A** or **B**$)$ as $P($**A** $\cup$ **B**$)$. The symbol $\cup$ means "union," representing the outcomes in event **A** *or* event **B** (or both). The symbol $\cap$ means "intersection," representing outcomes that are in both event **A** *and* event **B**. We write $P($**A** *and* **B**$)$ as $P($**A** $\cap$ **B**$)$.

2. If a random phenomenon has only one possible outcome, it's not very interesting (or very random). So we need to distribute the probabilities among all the outcomes a trial can have. How can we do that so that it makes sense? For example, consider what you're doing as you read this book. The possible outcomes might be

A: You read to the end of this chapter before stopping.
B: You finish this section but stop reading before the end of the chapter.
C: You bail out before the end of this section.

When we assign probabilities to these outcomes, the first thing to be sure of is that we distribute all of the available probability. Something always occurs, so the probability of the entire sample space is 1.

Making this more formal gives the **Probability Assignment Rule.**

The set of all possible outcomes of a trial must have probability 1.

$$P(S) = 1$$

3. Suppose the probability that you get to class on time is 0.8. What's the probability that you don't get to class on time? Yes, it's 0.2. The set of outcomes that are *not* in the event **A** is called the **complement** of **A,** and is denoted **A**^{C}. This leads to the **Complement Rule:**

The probability of an event occurring is 1 minus the probability that it doesn't occur.

$$P(A) = 1 - P(A^C)$$

FOR EXAMPLE   Applying the Complement Rule

Recap: We opened the chapter by looking at the traffic light at the corner of College and Main, observing that when we arrive at that intersection, the light is green about 35% of the time.

Question: If $P($green$) = 0.35$, what's the probability the light isn't green when you get to College and Main?

"Not green" is the complement of "green," so $P($not green$) = 1 - P($green$)$
$$= 1 - 0.35 = 0.65$$

There's a 65% chance I won't have a green light.

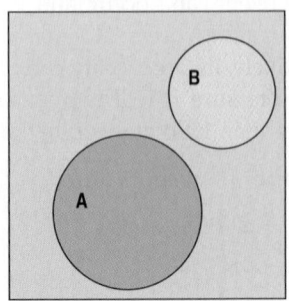

*Two disjoint sets, **A** and **B**.*

4. Suppose the probability that (**A**) a randomly selected student is a sophomore is 0.20, and the probability that (**B**) he or she is a junior is 0.30. What is the probability that the student is *either* a sophomore *or* a junior, written $P($**A** $\cup$ **B**$)$? If you guessed 0.50, you've deduced the **Addition Rule,** which says that you can add the probabilities of events that are disjoint. To see whether two events are disjoint, we take them apart into their component outcomes and check whether they have any outcomes in common. **Disjoint** (or **mutually exclusive**) events have no outcomes in common. The **Addition Rule** states,

For two disjoint events A and B, the probability that one *or* the other occurs is the sum of the probabilities of the two events.

$$P(A \cup B) = P(A) + P(B), \text{ provided that A and B are disjoint.}$$

FOR EXAMPLE   Applying the Addition Rule

Recap: When you get to the light at College and Main, it's either red, green, or yellow. We know that $P(\text{green}) = 0.35$.

Question: Suppose we find out that $P(\text{yellow})$ is about 0.04. What's the probability the light is red?

To find the probability that the light is green or yellow, I can use the Addition Rule because these are disjoint events: The light can't be both green and yellow at the same time.

$$P(\text{green} \cup \text{yellow}) = 0.35 + 0.04 = 0.39$$

Red is the only remaining alternative, and the probabilities must add up to 1, so

$$P(\text{red}) = P(\text{not (green} \cup \text{yellow}))$$
$$= 1 - P(\text{green} \cup \text{yellow})$$
$$= 1 - 0.39 = 0.61$$

"Baseball is 90% mental. The other half is physical."

—Yogi Berra

A S
Activity: **Addition Rule for Disjoint Events.** Experiment with disjoint events to explore the Addition Rule.

Because sample space outcomes are disjoint, we have an easy way to check whether the probabilities we've assigned to the possible outcomes are **legitimate.** The Probability Assignment Rule tells us that the sum of the probabilities of all possible outcomes must be exactly 1. No more, no less. For example, if we were told that the probabilities of selecting at random a freshman, sophomore, junior, or senior from all the undergraduates at a school were 0.25, 0.23, 0.22, and 0.20, respectively, we would know that something was wrong. These "probabilities" sum to only 0.90, so this is not a legitimate probability assignment. Either a value is wrong, or we just missed some possible outcomes, like "pre-freshman" or "postgraduate" categories that soak up the remaining 0.10. Similarly, a claim that the probabilities were 0.26, 0.27, 0.29, and 0.30 would be wrong because these "probabilities" sum to more than 1.

But be careful: The Addition Rule doesn't work for events that aren't disjoint. If the probability of owning an MP3 player is 0.50 and the probability of owning a computer is 0.90, the probability of owning either an MP3 player or a computer may be pretty high, but it is *not* 1.40! Why can't you add probabilities like this? Because these events are not disjoint. You *can* own both. In the next chapter, we'll see how to add probabilities for events like these, but we'll need another rule.

5. Suppose your job requires you to fly from Atlanta to Houston every Monday morning. The airline's Web site reports that this flight is on time 85% of the time. What's the chance that it will be on time two weeks in a row? That's the same as asking for the probability that your flight is on time this week *and* it's on time again next week. For independent events, the answer is very simple. Remember that independence means that the outcome of one event doesn't influence the outcome of the other. What happens with your flight this week doesn't influence whether it will be on time next week, so it's reasonable to assume that those events are independent. The **Multiplication Rule** says that for independent events, to find the probability that both events occur, we just multiply the probabilities together. Formally,

For two independent events A and B, the probability that both A *and* B occur is the product of the probabilities of the two events.

$$P(A \cap B) = P(A) \times P(B), \text{ provided that}$$
A and B are independent.

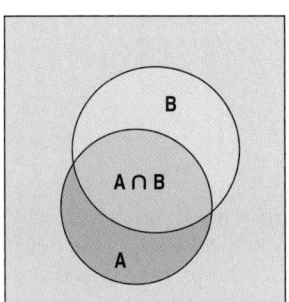

Two sets **A** *and* **B** *that are not disjoint. The event* ($\mathbf{A} \cap \mathbf{B}$) *is their intersection.*

Activity: **Multiplication Rule for Independent Events.** Experiment with independent random events to explore the Multiplication Rule.

Activity: **Probabilities of Compound Events.** The Random tool also lets you experiment with Compound random events to see if they are independent.

This rule can be extended to more than two independent events. What's the chance of your flight being on time for a month—four Mondays in a row? We can multiply the probabilities of it happening each week:

$$0.85 \times 0.85 \times 0.85 \times 0.85 = 0.522$$

or just over 50–50. Of course, to calculate this probability, we have used the assumption that the four events are independent.

Many Statistics methods require an **Independence Assumption,** but *assuming* independence doesn't make it true. Always *Think* about whether that assumption is reasonable before using the Multiplication Rule.

FOR EXAMPLE Applying the Multiplication Rule (and others)

Recap: We've determined that the probability that we encounter a green light at the corner of College and Main is 0.35, a yellow light 0.04, and a red light 0.61. Let's think about your morning commute in the week ahead.

Question: What's the probability you find the light red both Monday and Tuesday?

Because the color of the light I see on Monday doesn't influence the color I'll see on Tuesday, these are independent events; I can use the Multiplication Rule:

$$P(\text{red Monday} \cap \text{red Tuesday}) = P(\text{Red}) \times P(\text{red})$$
$$= (0.61)(0.61)$$
$$= 0.3721$$

There's about a 37% chance I'll hit red lights both Monday and Tuesday mornings.

Question: What's the probability you don't encounter a red light until Wednesday?

For that to happen, I'd have to see green or yellow on Monday, green or yellow on Tuesday, and then red on Wednesday. I can simplify this by thinking of it as not red on Monday and Tuesday and then red on Wednesday.

$$P(\text{not red}) = 1 - P(\text{red}) = 1 - 0.61 = 0.39, \text{ so}$$

$$P(\text{not red Monday} \cap \text{not red Tuesday} \cap \text{red Wednesday}) = P(\text{not red}) \times P(\text{not red}) \times P(\text{red})$$
$$= (0.39)(0.39)(0.61)$$
$$= 0.092781$$

There's about a 9% chance that this week I'll hit my first red light there on Wednesday morning.

Question: What's the probability that you'll have to stop *at least once* during the week?

Having to stop at least once means that I have to stop for the light either 1, 2, 3, 4, or 5 times next week. It's easier to think about the complement: never having to stop at a red light. Having to stop at least once means that I didn't make it through the week with no red lights.

$P(\text{having to stop at the light at least once in 5 days})$
$$= 1 - P(\text{no red lights for 5 days in a row})$$
$$= 1 - P(\text{not red} \cap \text{not red} \cap \text{not red} \cap \text{not red} \cap \text{not red})$$
$$= 1 - (0.39)(0.39)(0.39)(0.39)(0.39)$$
$$= 1 - 0.0090$$
$$= 0.991$$

There's over a 99% chance I'll hit at least one red light sometime this week.

Note that the phrase "at least" is often a tip-off to think about the complement. Something that happens *at least once* <u>does</u> happen. Happening at least once is the complement of not happening at all, and that's easier to find.

> In informal English, you may see "some" used to mean "at least one." "What's the probability that some of the eggs in that carton are broken?" means at least one.

JUST CHECKING

2. Opinion polling organizations contact their respondents by telephone. Random telephone numbers are generated, and interviewers try to contact those households. In the 1990s this method could reach about 69% of U.S. households. According to the Pew Research Center for the People and the Press, by 2003 the contact rate had risen to 76%. We can reasonably assume each household's response to be independent of the others. What's the probability that . . .

   a) the interviewer successfully contacts the next household on her list?

   b) the interviewer successfully contacts both of the next two households on her list?

   c) the interviewer's first successful contact is the third household on the list?

   d) the interviewer makes at least one successful contact among the next five households on the list?

STEP-BY-STEP EXAMPLE   **Probability**

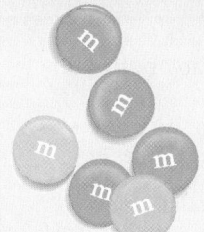

The five rules we've seen can be used in a number of different combinations to answer a surprising number of questions. Let's try one to see how we might go about it.

In 2001, Masterfoods, the manufacturers of M&M's® milk chocolate candies, decided to add another color to the standard color lineup of brown, yellow, red, orange, blue, and green. To decide which color to add, they surveyed people in nearly every country of the world and asked them to vote among purple, pink, and teal. The global winner was purple!

In the United States, 42% of those who voted said purple, 37% said teal, and only 19% said pink. But in Japan the percentages were 38% pink, 36% teal, and only 16% purple. Let's use Japan's percentages to ask some questions:

1. What's the probability that a Japanese M&M's survey respondent selected at random preferred either pink or teal?
2. If we pick two respondents at random, what's the probability that they both selected purple?
3. If we pick three respondents at random, what's the probability that *at least one* preferred purple?

The probability of an event is its long-term relative frequency. It can be determined in several ways: by looking at many replications of an event, by deducing it from equally likely events, or by using some other information. Here, we are told the relative frequencies of the three responses.

Make sure the probabilities are legitimate. Here, they're not. Either there was a mistake, or the other voters must have chosen a color other than the three given. A check of the reports from other countries shows a similar deficit, so probably we're seeing those who had no preference or who wrote in another color.

The M&M's Web site reports the proportions of Japanese votes by color. These give the probability of selecting a voter who preferred each of the colors:

$$P(\text{pink}) = 0.38$$
$$P(\text{teal}) = 0.36$$
$$P(\text{purple}) = 0.16$$

Each is between 0 and 1, but they don't all add up to 1. The remaining 10% of the voters must have not expressed a preference or written in another color. I'll put them together into "no preference" and add $P(\text{no preference}) = 0.10$.

With this addition, I have a legitimate assignment of probabilities.

Question 1. What's the probability that a Japanese M&M's survey respondent selected at random preferred either pink or teal?

THINK	**Plan** Decide which rules to use and check the conditions they require.	The events "Pink" and "Teal" are individual outcomes (a respondent can't choose both colors), so they are disjoint. I can apply the Addition Rule.
SHOW	**Mechanics** Show your work.	$P(\text{pink} \cup \text{teal}) = P(\text{pink}) + P(\text{teal})$ $= 0.38 + 0.36 = 0.74$
TELL	**Conclusion** Interpret your results in the proper context.	The probability that the respondent said pink or teal is 0.74.

Question 2. If we pick two respondents at random, what's the probability that they both said purple?

THINK	**Plan** The word "both" suggests we want $P(A \text{ and } B)$, which calls for the Multiplication Rule. Think about the assumption.	✓ **Independence Assumption:** It's unlikely that the choice made by one random respondent affected the choice of the other, so the events seem to be independent. I can use the Multiplication Rule.
SHOW	**Mechanics** Show your work. For both respondents to pick purple, each one has to pick purple.	$P(\text{both purple})$ $= P(\text{first respondent picks purple} \cap \text{second respondent picks purple})$ $= P(\text{first respondent picks purple}) \times P(\text{second respondent picks purple})$ $= 0.16 \times 0.16 = 0.0256$
TELL	**Conclusion** Interpret your results in the proper context.	The probability that both respondents pick purple is 0.0256.

Question 3. If we pick three respondents at random, what's the probability that at least one preferred purple?

 **Plan** The phrase "at least . . ." often flags a question best answered by looking at the complement, and that's the best approach here. The complement of "At least one preferred purple" is "None of them preferred purple."

Think about the assumption.

P(at least one picked purple)
$$= P(\{\text{none picked purple}\}^C)$$
$$= 1 - P(\text{none picked purple}).$$
$$= 1 - P(\text{not purple} \cap \text{not purple} \cap \text{not purple}).$$

✔ **Independence Assumption:** These are independent events because they are choices by three random respondents. I can use the Multiplication Rule.

 **Mechanics** First we find P(not purple) with the Complement Rule.

Next we calculate P(none picked purple) by using the Multiplication Rule.

Then we can use the Complement Rule to get the probability we want.

P(not purple) $= 1 - P$(purple)
$$= 1 - 0.16 = 0.84$$

P(at least one picked purple)
$$= 1 - P(\text{none picked purple})$$
$$= 1 - P(\text{not purple} \cap \text{not purple} \cap \text{not purple})$$
$$= 1 - (0.84)(0.84)(0.84)$$
$$= 1 - 0.5927$$
$$= 0.4073$$

 **Conclusion** Interpret your results in the proper context.

There's about a 40.7% chance that at least one of the respondents picked purple.

WHAT CAN GO WRONG?

▶ **Beware of probabilities that don't add up to 1.** To be a legitimate probability assignment, the sum of the probabilities for all possible outcomes must total 1. If the sum is less than 1, you may need to add another category ("other") and assign the remaining probability to that outcome. If the sum is more than 1, check that the outcomes are disjoint. If they're not, then you can't assign probabilities by just counting relative frequencies.

▶ **Don't add probabilities of events if they're not disjoint.** Events must be disjoint to use the Addition Rule. The probability of being under 80 *or* a female is not the probability of being under 80 *plus* the probability of being female. That sum may be more than 1.

▶ **Don't multiply probabilities of events if they're not independent.** The probability of selecting a student at random who is over 6'10" tall *and* on the basketball team is *not* the probability the student is over 6'10" tall *times* the probability he's on the basketball team. Knowing that the student is over 6'10" changes the probability of his being on the basketball team. You can't multiply these probabilities. The multiplication of probabilities of events that are not independent is one of the most common errors people make in dealing with probabilities.

▶ **Don't confuse disjoint and independent.** Disjoint events *can't* be independent. If **A** = {you get an A in this class} and **B** = {you get a B in this class}, **A** and **B** are disjoint. Are they independent? If you find out that **A** is true, does that change the probability of **B**? You bet it does! So they can't be independent. we'll return to this issue in the next chapter.

CONNECTIONS

We saw in the previous three chapters that randomness plays a critical role in gathering data. That fact alone makes it important that we understand how random events behave. The rules and concepts of probability give us a language to talk and think about random phenomena. From here on, randomness will be fundamental to how we think about data, and probabilities will show up in every chapter.

We began thinking about independence back in Chapter 3 when we looked at contingency tables and asked whether the distribution of one variable was the same for each category of another. Then, in Chapter 12, we saw that independence was fundamental to drawing a Simple Random Sample. For computing compound probabilities, we again ask about independence. And we'll continue to think about independence throughout the rest of the book.

Our interest in probability extends back to the start of the book. We've talked about "relative frequencies" often. But—let's be honest—that's just a casual term for probability. For example, you can now rephrase the 68–95–99.7 Rule to talk about the *probability* that a random value selected from a Normal model will fall within 1, 2, or 3 standard deviations of the mean.

Why not just say "probability" from the start? Well, we didn't need any of the formal rules of this chapter (or the next one), so there was no point to weighing down the discussion with those rules. And "relative frequency" is the right intuitive way to think about probability in this course, so you've been thinking right all along.

Keep it up.

WHAT HAVE WE LEARNED?

We've learned that probability is based on long-run relative frequencies. We've thought about the Law of Large Numbers and noted that it speaks only of long-run behavior. Because the long run is a very long time, we need to be careful not to misinterpret the Law of Large Numbers. Even when we've observed a string of heads, we shouldn't expect extra tails in subsequent coin flips.

Also, we've learned some basic rules for combining probabilities of outcomes to find probabilities of more complex events. These include

▸ the Probability Assignment Rule,
▸ the Complement Rule,
▸ the Addition Rule for disjoint events, and
▸ the Multiplication Rule for independent events.

Terms

Random phenomenon 324. A phenomenon is random if we know what outcomes could happen, but not which particular values will happen.

Trial 325. A single attempt or realization of a random phenomenon.

Outcome 325. The outcome of a trial is the value measured, observed, or reported for an individual instance of that trial.

Event 325. A collection of outcomes. Usually, we identify events so that we can attach probabilities to them. We denote events with bold capital letters such as **A**, **B**, or **C**.

Sample Space 325. The collection of all possible outcome values. The sample space has a probability of 1.

Law of Large Numbers 326. The Law of Large Numbers states that the long-run *relative frequency* of repeated independent events gets closer and closer to the *true* relative frequency as the number of trials increases.

Independence (informally) 326. Two events are *independent* if learning that one event occurs does not change the probability that the other event occurs.

Probability	326. The probability of an event is a number between 0 and 1 that reports the likelihood of that event's occurrence. We write $P(\mathbf{A})$ for the probability of the event **A**.
Empirical probability	326. When the probability comes from the long-run relative frequency of the event's occurrence, it is an **empirical probability.**
Theoretical probability	327. When the probability comes from a model (such as equally likely outcomes), it is called a **theoretical probability.**
Personal probability	328. When the probability is subjective and represents your personal degree of belief, it is called a **personal probability.**
The Probability Assignment Rule	330. The probability of the entire sample space must be 1. $P(\mathbf{S}) = 1$.
Complement Rule	330. The probability of an event occurring is 1 minus the probability that it doesn't occur. $$P(\mathbf{A}) = 1 - P(\mathbf{A}^{\mathbf{C}})$$
Disjoint (Mutually exclusive)	330. Two events are disjoint if they share no outcomes in common. If **A** and **B** are disjoint, then knowing that **A** occurs tells us that **B** cannot occur. Disjoint events are also called "mutually exclusive."
Addition Rule	330. If **A** and **B** are disjoint events, then the probability of **A** *or* **B** is $$P(\mathbf{A} \cup \mathbf{B}) = P(\mathbf{A}) + P(\mathbf{B}).$$
Legitimate probability assignment	331. An assignment of probabilities to outcomes is legitimate if ▸ each probability is between 0 and 1 (inclusive). ▸ the sum of the probabilities is 1.
Multiplication Rule	331. If **A** and **B** are independent events, then the probability of **A** *and* **B** is $$P(\mathbf{A} \cap \mathbf{B}) = P(\mathbf{A}) \times P(\mathbf{B}).$$
Independence Assumption	332. We often require events to be independent. (So you should think about whether this assumption is reasonable.)

Skills

- ▸ Understand that random phenomena are unpredictable in the short term but show long-run regularity.

- ▸ Be able to recognize random outcomes in a real-world situation.

- ▸ Know that the relative frequency of a random event settles down to a value called the (empirical) probability. Know that this is guaranteed for independent events by the Law of Large Numbers.

- ▸ Know the basic definitions and rules of probability.

- ▸ Recognize when events are disjoint and when events are independent. Understand the difference and that disjoint events cannot be independent.

- ▸ Be able to use the facts about probability to determine whether an assignment of probabilities is legitimate. Each probability must be a number between 0 and 1, and the sum of the probabilities assigned to all possible outcomes must be 1.

- ▸ Know how and when to apply the Addition Rule. Know that events must be disjoint for the Addition Rule to apply.

- ▸ Know how and when to apply the Multiplication Rule. Know that events must be independent for the Multiplication Rule to apply. Be able to use the Multiplication Rule to find probabilities for combinations of independent events.

- ▸ Know how to use the Complement Rule to make calculating probabilities simpler. Recognize that probabilities of "at least. . ." are likely to be simplified in this way.

- ▸ Be able to use statements about probability in describing a random phenomenon. You will need this skill soon for making statements about statistical inference.

- ▸ Know and be able to use the terms "sample space", "disjoint events", and "independent events" correctly.

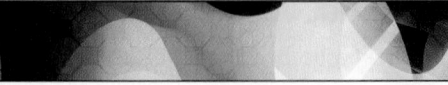

1. **Sample spaces.**   For each of the following, list the sample space and tell whether you think the events are equally likely:
   a) Toss 2 coins; record the order of heads and tails.
   b) A family has 3 children; record the number of boys.
   c) Flip a coin until you get a head or 3 consecutive tails; record each flip.
   d) Roll two dice; record the larger number.

2. **Sample spaces.**   For each of the following, list the sample space and tell whether you think the events are equally likely:
   a) Roll two dice; record the sum of the numbers.
   b) A family has 3 children; record each child's sex in order of birth.
   c) Toss four coins; record the number of tails.
   d) Toss a coin 10 times; record the length of the longest run of heads.

3. **Roulette.**   A casino claims that its roulette wheel is truly random. What should that claim mean?

4. **Rain.**   The weather reporter on TV makes predictions such as a 25% chance of rain. What do you think is the meaning of such a phrase?

5. **Winter.**   Comment on the following quotation:

   "What I think is our best determination is it will be a colder than normal winter," said Pamela Naber Knox, a Wisconsin state climatologist. "I'm basing that on a couple of different things. First, in looking at the past few winters, there has been a lack of really cold weather. Even though we are not supposed to use the law of averages, we are due." (Associated Press, fall 1992, quoted by Schaeffer et al.)

6. **Snow.**   After an unusually dry autumn, a radio announcer is heard to say, "Watch out! We'll pay for these sunny days later on this winter." Explain what he's trying to say, and comment on the validity of his reasoning.

7. **Cold streak.**   A batter who had failed to get a hit in seven consecutive times at bat then hits a game-winning home run. When talking to reporters afterward, he says he was very confident that last time at bat because he knew he was "due for a hit." Comment on his reasoning.

8. **Crash.**   Commercial airplanes have an excellent safety record. Nevertheless, there are crashes occasionally, with the loss of many lives. In the weeks following a crash, airlines often report a drop in the number of passengers, probably because people are afraid to risk flying.
   a) A travel agent suggests that since the law of averages makes it highly unlikely to have two plane crashes within a few weeks of each other, flying soon after a crash is the safest time. What do you think?
   b) If the airline industry proudly announces that it has set a new record for the longest period of safe flights, would you be reluctant to fly? Are the airlines due to have a crash?

9. **Fire insurance.**   Insurance companies collect annual payments from homeowners in exchange for paying to rebuild houses that burn down.
   a) Why should you be reluctant to accept a $300 payment from your neighbor to replace his house should it burn down during the coming year?
   b) Why can the insurance company make that offer?

10. **Jackpot.**   On January 20, 2000, the International Gaming Technology company issued a press release:

    (LAS VEGAS, Nev.)—Cynthia Jay was smiling ear to ear as she walked into the news conference at The Desert Inn Resort in Las Vegas today, and well she should. Last night, the 37-year-old cocktail waitress won the world's largest slot jackpot—$34,959,458—on a Megabucks machine. She said she had played $27 in the machine when the jackpot hit. Nevada Megabucks has produced 49 major winners in its 14-year history. The top jackpot builds from a base amount of $7 million and can be won with a 3-coin ($3) bet.

    a) How can the Desert Inn afford to give away millions of dollars on a $3 bet?
    b) Why did the company issue a press release? Wouldn't most businesses want to keep such a huge loss quiet?

11. **Spinner.**   The plastic arrow on a spinner for a child's game stops rotating to point at a color that will determine what happens next. Which of the following probability assignments are possible?

	Probabilities of . . .			
	Red	**Yellow**	**Green**	**Blue**
a)	0.25	0.25	0.25	0.25
b)	0.10	0.20	0.30	0.40
c)	0.20	0.30	0.40	0.50
d)	0	0	1.00	0
e)	0.10	0.20	1.20	−1.50

12. **Scratch off.**   Many stores run "secret sales": Shoppers receive cards that determine how large a discount they get, but the percentage is revealed by scratching off that black stuff (what *is* that?) only after the purchase has been totaled at the cash register. The store is required to reveal (in the fine print) the distribution of discounts available. Which of these probability assignments are legitimate?

	Probabilities of . . .			
	10% off	**20% off**	**30% off**	**50% off**
a)	0.20	0.20	0.20	0.20
b)	0.50	0.30	0.20	0.10
c)	0.80	0.10	0.05	0.05
d)	0.75	0.25	0.25	−0.25
e)	1.00	0	0	0

13. Vehicles. Suppose that 46% of families living in a certain county own a car and 18% own an SUV. The Addition Rule might suggest, then, that 64% of families own either a car or an SUV. What's wrong with that reasoning?

14. Homes. Funding for many schools comes from taxes based on assessed values of local properties. People's homes are assessed higher if they have extra features such as garages and swimming pools. Assessment records in a certain school district indicate that 37% of the homes have garages and 3% have swimming pools. The Addition Rule might suggest, then, that 40% of residences have a garage or a pool. What's wrong with that reasoning?

15. Speeders. Traffic checks on a certain section of highway suggest that 60% of drivers are speeding there. Since $0.6 \times 0.6 = 0.36$, the Multiplication Rule might suggest that there's a 36% chance that two vehicles in a row are both speeding. What's wrong with that reasoning?

16. Lefties. Although it's hard to be definitive in classifying people as right- or left-handed, some studies suggest that about 14% of people are left-handed. Since $0.14 \times 0.14 = 0.0196$, the Multiplication Rule might suggest that there's about a 2% chance that a brother and a sister are both lefties. What's wrong with that reasoning?

17. College admissions. For high school students graduating in 2007, college admissions to the nation's most selective schools were the most competitive in memory. (*The New York Times*, "A Great Year for Ivy League Schools, but Not So Good for Applicants to Them," April 4, 2007). Harvard accepted about 9% of its applicants, Stanford 10%, and Penn 16%. Jorge has applied to all three. Assuming that he's a typical applicant, he figures that his chances of getting into both Harvard and Stanford must be about 0.9%.
a) How has he arrived at this conclusion?
b) What additional assumption is he making?
c) Do you agree with his conclusion?

18. College admissions II. In Exercise 17, we saw that in 2007 Harvard accepted about 9% of its applicants, Stanford 10%, and Penn 16%. Jorge has applied to all three. He figures that his chances of getting into at least one of the three must be about 35%.
a) How has he arrived at this conclusion?
b) What assumption is he making?
c) Do you agree with his conclusion?

19. Car repairs. A consumer organization estimates that over a 1-year period 17% of cars will need to be repaired once, 7% will need repairs twice, and 4% will require three or more repairs. What is the probability that a car chosen at random will need
a) no repairs?
b) no more than one repair?
c) some repairs?

20. Stats projects. In a large Introductory Statistics lecture hall, the professor reports that 55% of the students enrolled have never taken a Calculus course, 32% have taken only one semester of Calculus, and the rest have taken two or more semesters of Calculus. The professor randomly assigns students to groups of three to work on a project for the course. What is the probability that the first groupmate you meet has studied
a) two or more semesters of Calculus?
b) some Calculus?
c) no more than one semester of Calculus?

21. More repairs. Consider again the auto repair rates described in Exercise 19. If you own two cars, what is the probability that
a) neither will need repair?
b) both will need repair?
c) at least one car will need repair?

22. Another project. You are assigned to be part of a group of three students from the Intro Stats class described in Exercise 20. What is the probability that of your other two groupmates,
a) neither has studied Calculus?
b) both have studied at least one semester of Calculus?
c) at least one has had more than one semester of Calculus?

23. Repairs, again. You used the Multiplication Rule to calculate repair probabilities for your cars in Exercise 21.
a) What must be true about your cars in order to make that approach valid?
b) Do you think this assumption is reasonable? Explain.

24. Final project. You used the Multiplication Rule to calculate probabilities about the Calculus background of your Statistics groupmates in Exercise 22.
a) What must be true about the groups in order to make that approach valid?
b) Do you think this assumption is reasonable? Explain.

25. Energy 2007. A Gallup poll in March 2007 asked 1005 U.S. adults whether increasing domestic energy production or protecting the environment should be given a higher priority. Here are the results:

Response	Number
Increase production	342
Protect environment	583
Equally important	30
No opinion	50
Total	**1005**

If we select a person at random from this sample of 1005 adults,
a) what is the probability that the person responded "Increase production"?
b) what is the probability that the person responded "Equally important" or had no opinion?

26. Failing fathers?   A Pew Research poll in 2007 asked 2020 U.S. adults whether fathers today were doing as good a job of fathering as fathers of 20–30 years ago. Here's how they responded:

Response	Number
Better	424
Same	566
Worse	950
No Opinion	80
Total	**2020**

If we select a respondent at random from this sample of 2020 adults,
a) what is the probability that the selected person responded "Worse"?
b) what is the probability that the person responded the "Same" or "Better"?

27. More energy.   Exercise 25 shows the results of a Gallup Poll about energy. Suppose we select three people at random from this sample.
a) What is the probability that all three responded "Protect the environment"?
b) What is the probability that none responded "Equally important"?
c) What assumption did you make in computing these probabilities?
d) Explain why you think that assumption is reasonable.

28. Fathers revisited.   Consider again the results of the poll about fathering discussed in Exercise 26. If we select two people at random from this sample,
a) what is the probability that both think fathers are better today?
b) what is the probability that neither thinks fathers are better today?
c) what is the probability that one person thinks fathers are better today and the other doesn't?
d) What assumption did you make in computing these probabilities?
e) Explain why you think that assumption is reasonable.

29. Polling.   As mentioned in the chapter, opinion-polling organizations contact their respondents by sampling random telephone numbers. Although interviewers now can reach about 76% of U.S. households, the percentage of those contacted who agree to cooperate with the survey has fallen from 58% in 1997 to only 38% in 2003 (Pew Research Center for the People and the Press). Each household, of course, is independent of the others.
a) What is the probability that the next household on the list will be contacted but will refuse to cooperate?
b) What is the probability (in 2003) of failing to contact a household or of contacting the household but not getting them to agree to the interview?
c) Show another way to calculate the probability in part b.

30. Polling, part II.   According to Pew Research, the contact rate (probability of contacting a selected household) was 69% in 1997 and 76% in 2003. However, the cooperation rate (probability of someone at the contacted household agreeing to be interviewed) was 58% in 1997 and dropped to 38% in 2003.

a) What is the probability (in 2003) of obtaining an interview with the next household on the sample list? (To obtain an interview, an interviewer must both contact the household and then get agreement for the interview.)
b) Was it more likely to obtain an interview from a randomly selected household in 1997 or in 2003?

31. M&M's.   The Masterfoods company says that before the introduction of purple, yellow candies made up 20% of their plain M&M's, red another 20%, and orange, blue, and green each made up 10%. The rest were brown.
a) If you pick an M&M at random, what is the probability that
   1) it is brown?
   2) it is yellow or orange?
   3) it is not green?
   4) it is striped?
b) If you pick three M&M's in a row, what is the probability that
   1) they are all brown?
   2) the third one is the first one that's red?
   3) none are yellow?
   4) at least one is green?

32. Blood.   The American Red Cross says that about 45% of the U.S. population has Type O blood, 40% Type A, 11% Type B, and the rest Type AB.
a) Someone volunteers to give blood. What is the probability that this donor
   1) has Type AB blood?
   2) has Type A or Type B?
   3) is not Type O?
b) Among four potential donors, what is the probability that
   1) all are Type O?
   2) no one is Type AB?
   3) they are not all Type A?
   4) at least one person is Type B?

33. Disjoint or independent?   In Exercise 31 you calculated probabilities of getting various M&M's. Some of your answers depended on the assumption that the outcomes described were *disjoint*; that is, they could not both happen at the same time. Other answers depended on the assumption that the events were *independent*; that is, the occurrence of one of them doesn't affect the probability of the other. Do you understand the difference between disjoint and independent?
a) If you draw one M&M, are the events of getting a red one and getting an orange one disjoint, independent, or neither?
b) If you draw two M&M's one after the other, are the events of getting a red on the first and a red on the second disjoint, independent, or neither?
c) Can disjoint events ever be independent? Explain.

34. Disjoint or independent?   In Exercise 32 you calculated probabilities involving various blood types. Some of your answers depended on the assumption that the outcomes described were *disjoint*; that is, they could not both happen at the same time. Other answers depended on the assumption that the events were *independent*; that is, the occurrence of one of them doesn't affect the probability of

the other. Do you understand the difference between disjoint and independent?
a) If you examine one person, are the events that the person is Type A and that the person is Type B disjoint, independent, or neither?
b) If you examine two people, are the events that the first is Type A and the second Type B disjoint, independent, or neither?
c) Can disjoint events ever be independent? Explain.

35. **Dice.** You roll a fair die three times. What is the probability that
a) you roll all 6's?
b) you roll all odd numbers?
c) none of your rolls gets a number divisible by 3?
d) you roll at least one 5?
e) the numbers you roll are not all 5's?

36. **Slot machine.** A slot machine has three wheels that spin independently. Each has 10 equally likely symbols: 4 bars, 3 lemons, 2 cherries, and a bell. If you play, what is the probability that
a) you get 3 lemons?
b) you get no fruit symbols?
c) you get 3 bells (the jackpot)?
d) you get no bells?
e) you get at least one bar (an automatic loser)?

37. **Champion bowler.** A certain bowler can bowl a strike 70% of the time. What's the probability that she
a) goes three consecutive frames without a strike?
b) makes her first strike in the third frame?
c) has at least one strike in the first three frames?
d) bowls a perfect game (12 consecutive strikes)?

38. **The train.** To get to work, a commuter must cross train tracks. The time the train arrives varies slightly from day to day, but the commuter estimates he'll get stopped on about 15% of work days. During a certain 5-day work week, what is the probability that he
a) gets stopped on Monday and again on Tuesday?
b) gets stopped for the first time on Thursday?
c) gets stopped every day?
d) gets stopped at least once during the week?

39. **Voters.** Suppose that in your city 37% of the voters are registered as Democrats, 29% as Republicans, and 11% as members of other parties (Liberal, Right to Life, Green, etc.). Voters not aligned with any official party are termed "Independent." You are conducting a poll by calling registered voters at random. In your first three calls, what is the probability you talk to
a) all Republicans?
b) no Democrats?
c) at least one Independent?

40. **Religion.** Census reports for a city indicate that 62% of residents classify themselves as Christian, 12% as Jewish, and 16% as members of other religions (Muslims, Buddhists, etc.). The remaining residents classify themselves as nonreligious. A polling organization seeking information about public opinions wants to be sure to talk with people holding a variety of religious views, and makes random phone calls. Among the first four people they call, what is the probability they reach

a) all Christians?
b) no Jews?
c) at least one person who is nonreligious?

41. **Tires.** You bought a new set of four tires from a manufacturer who just announced a recall because 2% of those tires are defective. What is the probability that at least one of yours is defective?

42. **Pepsi.** For a sales promotion, the manufacturer places winning symbols under the caps of 10% of all Pepsi bottles. You buy a six-pack. What is the probability that you win something?

43. **9/11?** On September 11, 2002, the first anniversary of the terrorist attack on the World Trade Center, the New York State Lottery's daily number came up 9–1–1. An interesting coincidence or a cosmic sign?

a) What is the probability that the winning three numbers match the date on any given day?
b) What is the probability that a whole year passes without this happening?
c) What is the probability that the date and winning lottery number match at least once during any year?
d) If every one of the 50 states has a three-digit lottery, what is the probability that at least one of them will come up 9–1–1 on September 11?

44. **Red cards.** You shuffle a deck of cards and then start turning them over one at a time. The first one is red. So is the second. And the third. In fact, you are surprised to get 10 red cards in a row. You start thinking, "The next one is due to be black!"
a) Are you correct in thinking that there's a higher probability that the next card will be black than red? Explain.
b) Is this an example of the Law of Large Numbers? Explain.

JUST CHECKING
Answers

1. The LLN works only in the long run, not in the short run. The random methods for selecting lottery numbers have no memory of previous picks, so there is no change in the probability that a certain number will come up.

2. a) 0.76
b) $0.76(0.76) = 0.5776$
c) $(1 - 0.76)^2(0.76) = 0.043776$
d) $1 - (1 - 0.76)^5 = 0.9992$

Probability Rules!

Pull a bill from your wallet or pocket without looking at it. An outcome of this trial is the bill you select. The sample space is all the bills in circulation: **S** = {$1 bill, $2 bill, $5 bill, $10 bill, $20 bill, $50 bill, $100 bill}.[1] These are *all* the possible outcomes. (In spite of what you may have seen in bank robbery movies, there are no $500 or $1000 bills.)

We can combine the outcomes in different ways to make many different events. For example, the event **A** = {$1, $5, $10} represents selecting a $1, $5, or $10 bill. The event **B** = {a bill that does not have a president on it} is the collection of outcomes (Don't look! Can you name them?): {$10 (Hamilton), $100 (Franklin)}. The event **C** = {enough money to pay for a $12 meal with one bill} is the set of outcomes {$20, $50, $100}.

Notice that these outcomes are not equally likely. You'd no doubt be more surprised (and pleased) to pull out a $100 bill than a $1 bill—it's not very likely, though. You probably carry many more $1 than $100 bills, but without information about the probability of each outcome, we can't calculate the probability of an event.

The probability of the event **C** (getting a bill worth more than $12) is *not* 3/7. There are 7 possible outcomes, and 3 of them exceed $12, but they are not *equally likely*. (Remember the probability that your lottery ticket will win rather than lose still isn't 1/2.)

The General Addition Rule

Now look at the bill in your hand. There are images of famous buildings in the center of the backs of all but two bills in circulation. The $1 bill has the word ONE in the center, and the $2 bill shows the signing of the Declaration of Independence.

---

[1] Well, technically, the sample space is all the bills in your pocket. You may be quite sure there isn't a $100 bill in there, but *we* don't know that, so humor us that it's at least *possible* that any legal bill could be there.

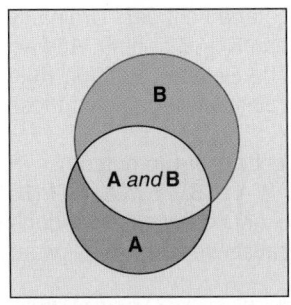

*Events **A** and **B** and their intersection.*

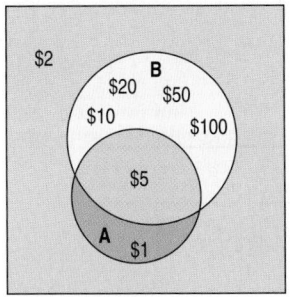

*Denominations of bills that are odd (**A**) or that have a building on the reverse side (**B**). The two sets both include the $5 bill, and both exclude the $2 bill.*

What's the probability of randomly selecting **A** = {a bill with an odd-numbered value} *or* **B** = {a bill with a building on the reverse}? We know **A** = {$1, $5} and **B** = {$5, $10, $20, $50, $100}. But *P*(**A** *or* **B**) is not simply the sum *P*(**A**) + *P*(**B**), because the events **A** and **B** are not disjoint. The $5 bill is in both sets. So what can we do? We'll need a new probability rule.

As the diagrams show, we can't use the Addition Rule and add the two probabilities because the events are not disjoint; they overlap. There's an outcome (the $5 bill) in the *intersection* of **A** and **B**. The Venn diagram represents the sample space. Notice that the $2 bill has neither a building nor an odd denomination, so it sits outside both circles.

The $5 bill plays a crucial role here because it is both odd *and* has a building on the reverse. It's in both **A** and **B,** which places it in the *intersection* of the two circles. The reason we can't simply add the probabilities of **A** and **B** is that we'd count the $5 bill twice.

If we did add the two probabilities, we could compensate by *subtracting* out the probability of that $5 bill. So,

P(odd number value or building)
    = *P*(odd number value) + *P*(building) − *P*(odd number value *and* building)
    = *P*($1, $5) + *P*($5, $10, $20, $50, $100) − *P*($5).

This method works in general. We add the probabilities of two events and then subtract out the probability of their intersection. This approach gives us the **General Addition Rule,** which does not require disjoint events:

$$P(A \cup B) = P(A) + P(B) - P(A \cap B).$$

FOR EXAMPLE   Using the General Addition Rule

A survey of college students found that 56% live in a campus residence hall, 62% participate in a campus meal program, and 42% do both.

Question: What's the probability that a randomly selected student either lives or eats on campus?

Let **L** = {student lives on campus} and **M** = {student has a campus meal plan}.

P(a student either lives or eats on campus) = $P(L \cup M)$
                                           = $P(L) + P(M) - P(L \cap M)$
                                           = $0.56 + 0.62 - 0.42$
                                           = 0.76

There's a 76% chance that a randomly selected college student either lives or eats on campus.

Would you like dessert or coffee?   Natural language can be ambiguous. In this question, is the answer one of the two alternatives, or simply "yes"? Must you decide between them, or may you have both? That kind of ambiguity can confuse our probabilities.

Suppose we had been asked a different question: What is the probability that the bill we draw has *either* an odd value *or* a building but *not both*? Which bills are we talking about now? The set we're interested in would be {$1, $10, $20, $50, $100}. We don't include the $5 bill in the set because it has both characteristics.

Why isn't this the same answer as before? The problem is that when we say the word "or," we usually mean *either* one *or* both. We don't usually mean the *exclusive* version of "or" as in, "Would you like the steak *or* the vegetarian entrée?" Ordinarily when we ask for the probability that **A** *or* **B** occurs, we mean **A** *or* **B** or both. And we know *that* probability is $P(\mathbf{A}) + P(\mathbf{B}) - P(\mathbf{A}\ and\ \mathbf{B})$. The General Addition Rule subtracts the probability of the outcomes in **A** *and* **B** because we've counted those outcomes *twice*. But they're still there.

If we really mean **A** or **B** but NOT both, we have to get rid of the outcomes in {**A** *and* **B**}. So $P(\mathbf{A}\ or\ \mathbf{B}\ but\ not\ \text{both}) = P(\mathbf{A} \cup \mathbf{B}) - P(\mathbf{A} \cap \mathbf{B}) = P(\mathbf{A}) + P(\mathbf{B}) - 2 \times P(\mathbf{A} \cap \mathbf{B})$. Now we've subtracted $P(\mathbf{A} \cap \mathbf{B})$ twice—once because we don't want to double-count these events and a second time because we really didn't want to count them at all.

Confused? *Make a picture*. It's almost always easier to think about such situations by looking at a Venn diagram.

FOR EXAMPLE   Using Venn diagrams

Recap: We return to our survey of college students: 56% live on campus, 62% have a campus meal program, and 42% do both.

Questions: Based on a Venn diagram, what is the probability that a randomly selected student

a) lives off campus and doesn't have a meal program?
b) lives in a residence hall but doesn't have a meal program?

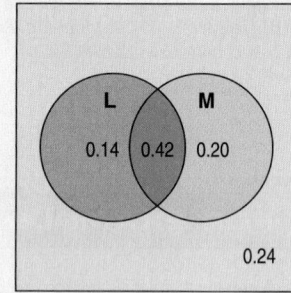

Let **L** = {student lives on campus} and **M** = {student has a campus meal plan}. In the Venn diagram, the intersection of the circles is $P(\mathbf{L} \cap \mathbf{M}) = 0.42$. Since $P(\mathbf{L}) = 0.56$, $P(\mathbf{L} \cap \mathbf{M}^C) = 0.56 - 0.42 = 0.14$. Also, $P(\mathbf{L}^C \cap \mathbf{M}) = 0.62 - 0.42 = 0.20$. Now, $0.14 + 0.42 + 0.20 = 0.76$, leaving $1 - 0.76 = 0.24$ for the region outside both circles.

Now . . . $P(\text{off campus and no meal program}) = P(\mathbf{L}^C \cap \mathbf{M}^C) = 0.24$

$P(\text{on campus and no meal program}) = P(\mathbf{L} \cap \mathbf{M}^C) = 0.14$

JUST CHECKING

1. Back in Chapter 1 we suggested that you sample some pages of this book at random to see whether they held a graph or other data display. We actually did just that. We drew a representative sample and found the following:

48% of pages had some kind of data display,

27% of pages had an equation, and

7% of pages had both a data display and an equation.

a) Display these results in a Venn diagram.
b) What is the probability that a randomly selected sample page had neither a data display nor an equation?
c) What is the probability that a randomly selected sample page had a data display but no equation?

Using the General Addition Rule

Police report that 78% of drivers stopped on suspicion of drunk driving are given a breath test, 36% a blood test, and 22% both tests.

Question: What is the probability that a randomly selected DWI suspect is given
1. a test?
2. a blood test or a breath test, but not both?
3. neither test?

THINK

Plan Define the events we're interested in. There are no conditions to check; the General Addition Rule works for any events!

Plot Make a picture, and use the given probabilities to find the probability for each region.

The blue region represents **A** but not **B**. The green intersection region represents **A** *and* **B**. Note that since $P(\mathbf{A}) = 0.78$ and $P(\mathbf{A} \cap \mathbf{B}) = 0.22$, the probability of **A** but not **B** must be $0.78 - 0.22 = 0.56$.

The yellow region is **B** but not **A**.

The gray region outside both circles represents the outcome neither **A** nor **B**. All the probabilities must total 1, so you can determine the probability of that region by subtraction.

Now, figure out what you want to know. The probabilities can come from the diagram or a formula. Sometimes translating the words to equations is the trickiest step.

Let **A** = {suspect is given a breath test}.
Let **B** = {suspect is given a blood test}.

I know that $P(\mathbf{A}) = 0.78$
$$P(\mathbf{B}) = 0.36$$
$$P(\mathbf{A} \cap \mathbf{B}) = 0.22$$
So $P(\mathbf{A} \cap \mathbf{B}^{C}) = 0.78 - 0.22 = 0.56$
$$P(\mathbf{B} \cap \mathbf{A}^{C}) = 0.36 - 0.22 = 0.14$$
$$P(\mathbf{A}^{C} \cap \mathbf{B}^{C}) = 1 - (0.56 + 0.22 + 0.14)$$
$$= 0.08$$

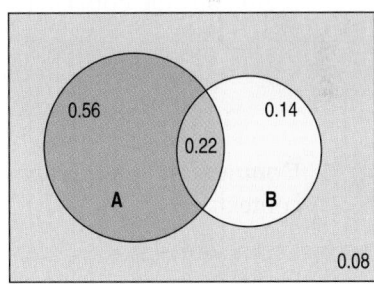

Question 1. What is the probability that the suspect is given a test?

SHOW

Mechanics The probability the suspect is given a test is $P(\mathbf{A} \cup \mathbf{B})$. We can use the General Addition Rule, or we can add the probabilities seen in the diagram.

$$P(\mathbf{A} \cup \mathbf{B}) = P(\mathbf{A}) + P(\mathbf{B}) - P(\mathbf{A} \cap \mathbf{B})$$
$$= 0.78 + 0.36 - 0.22$$
$$= 0.92$$

OR

$$P(\mathbf{A} \cup \mathbf{B}) = 0.56 + 0.22 + 0.14 = 0.92$$

TELL

Conclusion Don't forget to interpret your result in context.

92% of all suspects are given a test.

Question 2. What is the probability that the suspect gets either a blood test or a breath test but NOT both?

SHOW	**Mechanics** We can use the rule, or just add the appropriate probabilities seen in the Venn diagram.	$P(\textbf{A} \text{ or } \textbf{B} \text{ but NOT both}) = P(\textbf{A} \cup \textbf{B}) - P(\textbf{A} \cap \textbf{B})$ $= 0.92 - 0.22 = 0.70$ OR $P(\textbf{A} \text{ or } \textbf{B} \text{ but NOT both}) = P(\textbf{A} \cap \textbf{B}^C) + P(\textbf{B} \cap \textbf{A}^C)$ $= 0.56 + 0.14 = 0.70$
TELL	**Conclusion** Interpret your result in context.	70% of the suspects get exactly one of the tests.

Question 3. What is the probability that the suspect gets neither test?

SHOW	**Mechanics** Getting neither test is the complement of getting one or the other. Use the Complement Rule or just notice that "neither test" is represented by the region outside both circles.	$P(\text{neither test}) = 1 - P(\text{either test})$ $= 1 - P(\textbf{A} \cup \textbf{B})$ $= 1 - 0.92 = 0.08$ OR $P(\textbf{A}^C \cap \textbf{B}^C) = 0.08$
TELL	**Conclusion** Interpret your result in context.	Only 8% of the suspects get no test.

It Depends . . .

Two psychologists surveyed 478 children in grades 4, 5, and 6 in elementary schools in Michigan. They stratified their sample, drawing roughly 1/3 from rural, 1/3 from suburban, and 1/3 from urban schools. Among other questions, they asked the students whether their primary goal was to get good grades, to be popular, or to be good at sports. One question of interest was whether boys and girls at this age had similar goals.

Here's a *contingency table* giving counts of the students by their goals and sex:

		Goals			
		Grades	**Popular**	**Sports**	**Total**
Sex	**Boy**	117	50	60	**227**
	Girl	130	91	30	**251**
	Total	**247**	**141**	**90**	**478**

Table 15.1

The distribution of goals for boys and girls.

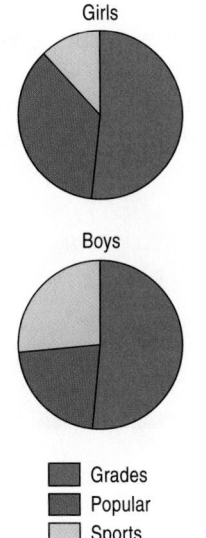

Girls

Boys

■ Grades
■ Popular
□ Sports

FIGURE 15.1

The distribution of goals for boys and girls.

We looked at contingency tables and graphed *conditional distributions* back in Chapter 3. The pie charts show the *relative frequencies* with which boys and girls named the three goals. It's only a short step from these relative frequencies to probabilities.

Let's focus on this study and make the sample space just the set of these 478 students. If we select a student at random from this study, the probability we select a girl is just the corresponding relative frequency (since we're equally likely to select any of the 478 students). There are 251 girls in the data out of a total of 478, giving a probability of

$$P(\text{girl}) = 251/478 = 0.525$$

The same method works for more complicated events like intersections. For example, what's the probability of selecting a girl whose goal is to be popular? Well, 91 girls named popularity as their goal, so the probability is

$$P(\text{girl} \cap \text{popular}) = 91/478 = 0.190$$

The probability of selecting a student whose goal is to excel at sports is

$$P(\text{sports}) = 90/478 = 0.188$$

What if we are given the information that the selected student is a girl? Would that change the probability that the selected student's goal is sports? You bet it would! The pie charts show that girls are much less likely to say their goal is to excel at sports than are boys. When we restrict our focus to girls, we look only at the girls' row of the table. Of the 251 girls, only 30 of them said their goal was to excel at sports.

We write the probability that a selected student wants to excel at sports *given that we have selected a girl* as

$$P(\text{sports} \mid \text{girl}) = 30/251 = 0.120$$

For boys, we look at the conditional distribution of goals given "boy" shown in the top row of the table. There, of the 227 boys, 60 said their goal was to excel at sports. So, $P(\text{sports} \mid \text{boy}) = 60/227 = 0.264$, more than twice the girls' probability.

In general, when we want the probability of an event from a *conditional* distribution, we write $P(\mathbf{B} \mid \mathbf{A})$ and pronounce it "the probability of **B** *given* **A**." A probability that takes into account a given *condition* such as this is called a **conditional probability.**

Let's look at what we did. We worked with the counts, but we could work with the probabilities just as well. There were 30 students who both were girls and had sports as their goal, and there are 251 girls. So we found the probability to be 30/251. To find the probability of the event **B** *given* the event **A**, we restrict our attention to the outcomes in **A**. We then find in what fraction of *those* outcomes **B** also occurred. Formally, we write:

$$P(\mathbf{B} \mid \mathbf{A}) = \frac{P(\mathbf{A} \cap \mathbf{B})}{P(\mathbf{A})}.$$

Thinking this through, we can see that it's just what we've been doing, but now with probabilities rather than with counts. Look back at the girls for whom sports was the goal. How did we calculate $P(\text{sports} \mid \text{girl})$?

The rule says to use probabilities. It says to find $P(\mathbf{A} \cap \mathbf{B})/P(\mathbf{A})$. The result is the same whether we use counts or probabilities because the total number in the sample cancels out:

$$\frac{P(\text{sports} \cap \text{girl})}{P(\text{girl})} = \frac{30/478}{251/478} = \frac{30}{251}.$$

A S   *Activity:* **Birthweights and Smoking.** Does smoking increase the chance of having a baby with low birth weight?

NOTATION ALERT:

$P(\mathbf{B} \mid \mathbf{A})$ is the conditional probability of **B** *given* **A**.

To use the formula for conditional probability, we're supposed to insist on one restriction. The formula doesn't work if $P(\mathbf{A})$ is 0. After all, we can't be "given" the fact that $\mathbf{A}$ was true if the probability of $\mathbf{A}$ is 0!

Let's take our rule out for a spin. What's the probability that we have selected a girl *given* that the selected student's goal is popularity? Applying the rule, we get

$$P(\text{girl} \mid \text{popular}) = \frac{P(\text{girl} \cap \text{popular})}{P(\text{popular})}$$

$$= \frac{91/478}{141/478} = \frac{91}{141}.$$

FOR EXAMPLE Finding a conditional probability

Recap: Our survey found that 56% of college students live on campus, 62% have a campus meal program, and 42% do both.

Question: While dining in a campus facility open only to students with meal plans, you meet someone interesting. What is the probability that your new acquaintance lives on campus?

Let **L** = {student lives on campus} and **M** = {student has a campus meal plan}.

$P(\text{student lives on campus given that the student has a meal plan}) = P(\mathbf{L} \mid \mathbf{M})$

$$= \frac{P(\mathbf{L} \cap \mathbf{M})}{P(\mathbf{M})}$$

$$= \frac{0.42}{0.62}$$

$$\approx 0.677$$

There's a probability of about 0.677 that a student with a meal plan lives on campus.

The General Multiplication Rule

Remember the Multiplication Rule for the probability of **A** *and* **B**? It said

$$P(\mathbf{A} \cap \mathbf{B}) = P(\mathbf{A}) \times P(\mathbf{B}) \text{ when } \mathbf{A} \text{ and } \mathbf{B} \text{ are independent.}$$

Now we can write a more general rule that doesn't require independence. In fact, we've *already* written it down. We just need to rearrange the equation a bit.

The equation in the definition for conditional probability contains the probability of **A** *and* **B**. Rewriting the equation gives

$$P(\mathbf{A} \cap \mathbf{B}) = P(\mathbf{A}) \times P(\mathbf{B} \mid \mathbf{A}).$$

This is a **General Multiplication Rule** for compound events that does not require the events to be independent. Better than that, it even makes sense. The probability that two events, **A** and **B**, *both* occur is the probability that event **A** occurs multiplied by the probability that event **B** *also* occurs—that is, by the probability that event **B** occurs *given* that event **A** occurs.

Of course, there's nothing special about which set we call **A** and which one we call **B**. We should be able to state this the other way around. And indeed we can. It is equally true that

$$P(\mathbf{A} \cap \mathbf{B}) = P(\mathbf{B}) \times P(\mathbf{A} \mid \mathbf{B}).$$

Independence

If we had to pick one idea in this chapter that you should understand and remember, it's the definition and meaning of independence. We'll need this idea in every one of the chapters that follow.

 A S *Activity:* **Independence.** Are *Smoking and Low Birthweight* independent?

In earlier chapters we said informally that two events were independent if learning that one occurred didn't change what you thought about the other occurring. Now we can be more formal. Events **A** and **B** are independent if (and only if) the probability of **A** is the same when we are given that **B** has occurred. That is, $P(\mathbf{A}) = P(\mathbf{A}\,|\,\mathbf{B})$.

Although sometimes your intuition is enough, now that we have the formal rule, use it whenever you can.

Let's return to the question of just what it means for events to be independent. We've said informally that what we mean by independence is that the outcome of one event does not influence the probability of the other. With our new notation for conditional probabilities, we can write a formal definition: Events **A** and **B** are **independent** whenever

$$P(\mathbf{B}\,|\,\mathbf{A}) = P(\mathbf{B}).$$

Now we can see that the Multiplication Rule for independent events we saw in Chapter 14 is just a special case of the General Multiplication Rule. The general rule says

$$P(\mathbf{A} \cap \mathbf{B}) = P(\mathbf{A}) \times P(\mathbf{B}\,|\,\mathbf{A}).$$

whether the events are independent or not. But when events **A** and **B** are independent, we can write $P(\mathbf{B})$ for $P(\mathbf{B}\,|\,\mathbf{A})$ and we get back our simple rule:

$$P(\mathbf{A} \cap \mathbf{B}) = P(\mathbf{A}) \times P(\mathbf{B}).$$

Sometimes people use this statement as the definition of independent events, but we find the other definition more intuitive. Either way, the idea is that for independent events, the probability of one doesn't change when the other occurs.

Is the probability of having good grades as a goal independent of the sex of the responding student? Looks like it might be. We need to check whether

$$P(\text{grades}\,|\,\text{girl}) = P(\text{grades})$$

$$\frac{130}{251} = 0.52 \overset{?}{=} \frac{247}{478} = 0.52$$

To two decimal place accuracy, it looks like we can consider choosing good grades as a goal to be independent of sex.

On the other hand, $P(\text{sports})$ is 90/478, or about 18.8%, but $P(\text{sports}\,|\,\text{boy})$ is $60/227 = 26.4\%$. Because these probabilities aren't equal, we can be pretty sure that choosing success in sports as a goal is not independent of the student's sex.

FOR EXAMPLE   Checking for independence

Recap: Our survey told us that 56% of college students live on campus, 62% have a campus meal program, and 42% do both.

Question: Are living on campus and having a meal plan independent? Are they disjoint?

Let **L** = {student lives on campus} and **M** = {student has a campus meal plan}. If these events are independent, then knowing that a student lives on campus doesn't affect the probability that he or she has a meal plan. I'll check to see if $P(\mathbf{M}\,|\,\mathbf{L}) = P(\mathbf{M})$:

$$P(\mathbf{M}\,|\,\mathbf{L}) = \frac{P(\mathbf{L} \cap \mathbf{M})}{P(\mathbf{L})}$$

$$= \frac{0.42}{0.56}$$

$$= 0.75, \qquad \text{but } P(\mathbf{M}) = 0.62.$$

Because $0.75 \neq 0.62$, the events are not independent; students who live on campus are more likely to have meal plans. Living on campus and having a meal plan are not disjoint either; in fact, 42% of college students do both.

Independent ≠ Disjoint

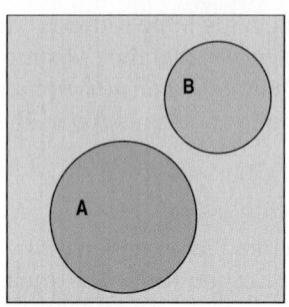

FIGURE 15.2

*Because these events are mutually exclusive, learning that **A** happened tells us that **B** didn't. The probability of **B** has changed from whatever it was to zero. So the disjoint events **A** and **B** are not independent.*

Are disjoint events independent? These concepts seem to have similar ideas of separation and distinctness about them, but in fact disjoint events *cannot* be independent.[2] Let's see why. Consider the two disjoint events {you get an A in this course} and {you get a B in this course}. They're disjoint because they have no outcomes in common. Suppose you learn that you *did* get an A in the course. Now what is the probability that you got a B? You can't get both grades, so it must be 0.

Think about what that means. Knowing that the first event (getting an A) occurred changed your probability for the second event (down to 0). So these events aren't independent.

Mutually exclusive events can't be independent. They have no outcomes in common, so if one occurs, the other doesn't. A common error is to treat disjoint events as if they were independent and apply the Multiplication Rule for independent events. Don't make that mistake.

JUST CHECKING

2. The American Association for Public Opinion Research (AAPOR) is an association of about 1600 individuals who share an interest in public opinion and survey research. They report that typically as few as 10% of random phone calls result in a completed interview. Reasons are varied, but some of the most common include no answer, refusal to cooperate, and failure to complete the call.

Which of the following events are independent, which are disjoint, and which are neither independent nor disjoint?

a) **A** = Your telephone number is randomly selected. **B** = You're not at home at dinnertime when they call.

b) **A** = As a selected subject, you complete the interview. **B** = As a selected subject, you refuse to cooperate.

c) **A** = You are not at home when they call at 11 a.m. **B** = You are employed full-time.

Depending on Independence

A S *Video: Is There a Hot Hand in Basketball?* Most coaches and fans believe that basketball players sometimes get "hot" and make more of their shots. What do the conditional probabilities say?

A S *Activity: Hot Hand Simulation.* Can you tell the difference between real and simulated sequences of basketball shot hits and misses?

It's much easier to think about independent events than to deal with conditional probabilities. It seems that most people's natural intuition for probabilities breaks down when it comes to conditional probabilities. Someone may estimate the probability of a compound event by multiplying the probabilities of its component events together without asking seriously whether those probabilities are independent.

For example, experts have assured us that the probability of a major commercial nuclear plant failure is so small that we should not expect such a failure to occur even in a span of hundreds of years. After only a few decades of commercial nuclear power, however, the world has seen two failures (Chernobyl and Three Mile Island). How could the estimates have been so wrong?

---

[2] Well, technically two disjoint events *can* be independent, but only if the probability of one of the events is 0. For practical purposes, though, we can ignore this case. After all, as statisticians we don't anticipate having data about things that never happen.

One simple part of the failure calculation is to test a particular valve and determine that valves such as this one fail only once in, say, 100 years of normal use. For a coolant failure to occur, several valves must fail. So we need the compound probability, *P*(valve 1 fails *and* valve 2 fails *and* . . .). A simple risk assessment might multiply the small probability of one valve failure together as many times as needed.

But if the valves all came from the same manufacturer, a flaw in one might be found in the others. And maybe when the first fails, it puts additional pressure on the next one in line. In either case, the events aren't independent and so we can't simply multiply the probabilities together.

Whenever you see probabilities multiplied together, stop and ask whether you think they are really independent.

Tables and Conditional Probability

One of the easiest ways to think about conditional probabilities is with contingency tables. We did that earlier in the chapter when we began our discussion. But sometimes we're given probabilities without a table. You can often construct a simple table to correspond to the probabilities.

For instance, in the drunk driving example, we were told that 78% of suspect drivers get a breath test, 36% a blood test, and 22% both. That's enough information. Translating percentages to probabilities, what we know looks like this:

		Breath Test		
		Yes	**No**	Total
Blood Test	**Yes**	0.22		**0.36**
	No			
	Total	**0.78**		**1.00**

Notice that the 0.78 and 0.36 are *marginal* probabilities and so they go into the *margins*. The 0.22 is the probability of getting both tests—a breath test *and* a blood test—so that's a *joint* probability. Those belong in the interior of the table.

Because the cells of the table show disjoint events, the probabilities always add to the marginal totals going across rows or down columns. So, filling in the rest of the table is quick:

		Breath Test		
		Yes	**No**	Total
Blood Test	**Yes**	0.22	0.14	**0.36**
	No	0.56	0.08	**0.64**
	Total	**0.78**	**0.22**	**1.00**

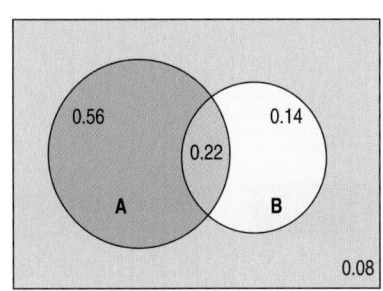

Compare this with the Venn diagram. Notice which entries in the table match up with the sets in this diagram. Whether a Venn diagram or a table is better to use will depend on what you are given and the questions you're being asked. Try both.

STEP-BY-STEP EXAMPLE | Are the Events Disjoint? Independent?

Let's take another look at the drunk driving situation. Police report that 78% of drivers are given a breath test, 36% a blood test, and 22% both tests.

Questions: 1. Are giving a DWI suspect a blood test and a breath test mutually exclusive?
2. Are giving the two tests independent?

Plan Define the events we're interested in.

State the given probabilities.

Let **A** = {suspect is given a breath test}
Let **B** = {suspect is given a blood test}.

I know that $P(A) = 0.78$
$P(B) = 0.36$
$P(A \cap B) = 0.22$

Question 1. Are giving a DWI suspect a blood test and a breath test mutually exclusive?

Mechanics Disjoint events cannot *both* happen at the same time, so check to see if $P(A \cap B) = 0$.

$P(A \cap B) = 0.22$. Since some suspects are given both tests, $P(A \cap B) \neq 0$. The events are not mutually exclusive.

Conclusion State your conclusion in context.

22% of all suspects get both tests, so a breath test and a blood test are not disjoint events.

Question 2. Are the two tests independent?

Plan Make a table.

		Breath Test		
		Yes	No	Total
Blood Test	Yes	0.22	0.14	0.36
	No	0.56	0.08	0.64
	Total	0.78	0.22	1.00

Mechanics Does getting a breath test change the probability of getting a blood test? That is, does $P(B \mid A) = P(B)$?

Because the two probabilities are *not* the same, the events are not independent.

$$P(B \mid A) = \frac{P(A \cap B)}{P(A)} = \frac{0.22}{0.78} \approx 0.28$$
$$P(B) = 0.36$$
$$P(B \mid A) \neq P(B)$$

TELL **Conclusion** Interpret your results in context.	Overall, 36% of the drivers get blood tests, but only 28% of those who get a breath test do. Since suspects who get a breath test are less likely to have a blood test, the two events are not independent.

JUST CHECKING

3. Remember our sample of pages in this book from the earlier Just Checking . . . ?

48% of pages had a data display.

27% of pages had an equation, and

7% of pages had both a data display and an equation.

a) Make a contingency table for the variables *display* and *equation*.
b) What is the probability that a randomly selected sample page with an equation also had a data display?
c) Are having an equation and having a data display disjoint events?
d) Are having an equation and having a data display independent events?

Drawing Without Replacement

Room draw is a process for assigning dormitory rooms to students who live on campus. Sometimes, when students have equal priority, they are randomly assigned to the currently available dorm rooms. When it's time for you and your friend to draw, there are 12 rooms left. Three are in Gold Hall, a very desirable dorm with spacious wood-paneled rooms. Four are in Silver Hall, centrally located but not quite as desirable. And five are in Wood Hall, a new dorm with cramped rooms, located half a mile from the center of campus on the edge of the woods.

You get to draw first, and then your friend will draw. Naturally, you would both like to score rooms in Gold. What are your chances? In particular, what's the chance that you *both* can get rooms in Gold?

When you go first, the chance that *you* will draw one of the Gold rooms is 3/12. Suppose you do. Now, with you clutching your prized room assignment, what chance does your friend have? At this point there are only 11 rooms left and just 2 left in Gold, so your friend's chance is now 2/11.

Using our notation, we write

$$P(\text{friend draws Gold} \mid \text{you draw Gold}) = 2/11.$$

The reason the denominator changes is that we draw these rooms *without replacement.* That is, once one is drawn, it doesn't go back into the pool.

We often sample without replacement. When we draw from a very large population, the change in the denominator is too small to worry about. But when there's a small population to draw from, as in this case, we need to take note and adjust the probabilities.

What are the chances that *both* of you will luck out? Well, now we've calculated the two probabilities we need for the General Multiplication Rule, so we can write:

$$P(\text{you draw Gold} \cap \text{friend draws Gold})$$
$$= P(\text{you draw Gold}) \times P(\text{friend draws Gold} \mid \text{you draw Gold})$$
$$= 3/12 \times 2/11 = 1/22 = 0.045$$

In this instance, it doesn't matter who went first, or even if the rooms were drawn simultaneously. Even if the room draw was accomplished by shuffling cards containing the names of the dormitories and then dealing them out to 12 applicants (rather than by each student drawing a room in turn), we can still *think* of the calculation as having taken place in two steps:

Diagramming conditional probabilities leads to a more general way of helping us think with pictures—one that works for calculating conditional probabilities even when they involve different variables.

Tree Diagrams

For men, binge drinking is defined as having five or more drinks in a row, and for women as having four or more drinks in a row. (The difference is because of the average difference in weight.) According to a study by the Harvard School of Public Health (H. Wechsler, G. W. Dowdall, A. Davenport, and W. DeJong, "Binge Drinking on Campus: Results of a National Study"), 44% of college students engage in binge drinking, 37% drink moderately, and 19% abstain entirely. Another study, published in the *American Journal of Health Behavior,* finds that among binge drinkers aged 21 to 34, 17% have been involved in an alcohol-related automobile accident, while among non-bingers of the same age, only 9% have been involved in such accidents.

What's the probability that a randomly selected college student will be a binge drinker who has had an alcohol-related car accident?

To start, we see that the probability of selecting a binge drinker is about 44%. To find the probability of selecting someone who is both a binge drinker and a driver with an alcohol-related accident, we would need to pull out the General Multiplication Rule and multiply the probability of one of the events by the conditional probability of the other given the first.

Or we *could* make a picture. Which would you prefer?

We thought so.

The kind of picture that helps us think through this kind of reasoning is called a **tree diagram,** because it shows sequences of events, like those we had in room draw, as paths that look like branches of a tree. It is a good idea to make a tree diagram almost any time you plan to use the General Multiplication Rule. The number of different paths we can take can get large, so we usually draw the tree starting from the left and growing vine-like across the page, although sometimes you'll see them drawn from the bottom up or top down.

"Why," said the Dodo, "the best way to explain it is to do it."
—Lewis Carroll

The first branch of our tree separates students according to their drinking habits. We label each branch of the tree with a possible outcome and its corresponding probability.

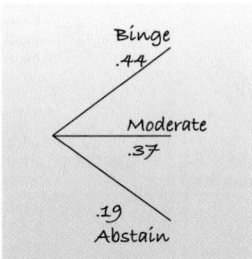

FIGURE 15.3

We can diagram the three outcomes of drinking and indicate their respective probabilities with a simple tree diagram.

Notice that we cover all possible outcomes with the branches. The probabilities add up to one. But we're also interested in car accidents. The probability of having an alcohol-related accident *depends* on one's drinking behavior. Because the probabilities are *conditional*, we draw the alternatives separately on each branch of the tree:

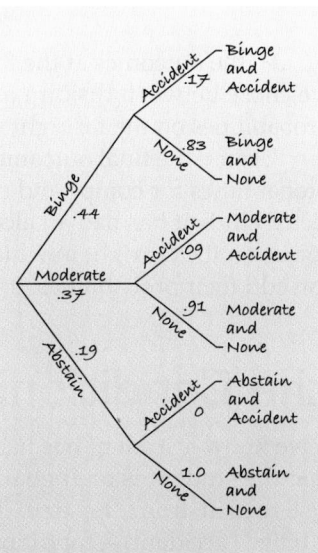

FIGURE 15.4

Extending the tree diagram, we can show both drinking and accident outcomes. The accident probabilities are conditional on the drinking outcomes, and they change depending on which branch we follow. Because we are concerned only with alcohol-related accidents, the conditional probability P(accident | abstinence) must be 0.

On each of the second set of branches, we write the possible outcomes associated with having an alcohol-related car accident (having an accident or not) and the associated probability. These probabilities are different because they are *conditional* depending on the student's drinking behavior. (It shouldn't be too surprising that those who binge drink have a higher probability of alcohol-related accidents.) The probabilities add up to one, because given the outcome on the first branch, these outcomes cover all the possibilities. Looking back at the General Multiplication Rule, we can see how the tree depicts the calculation. To find the probability that a randomly selected student will be a binge drinker who has had an alcohol-related car accident, we follow the top branches. The probability of selecting a binger is 0.44. The conditional probability of an accident *given* binge drinking is 0.17. The General Multiplication Rule tells us that to find the *joint* probability of being a binge drinker and having an accident, we multiply these two probabilities together:

$$P(\text{binge} \cap \text{accident}) = P(\text{binge}) \times P(\text{accident} \,|\, \text{binge})$$
$$= 0.44 \times 0.17 = 0.075$$

And we can do the same for each combination of outcomes:

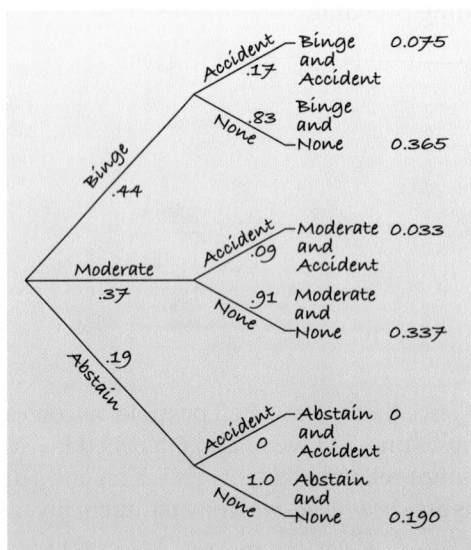

FIGURE 15.5

We can find the probabilities of compound events by multiplying the probabilities along the branch of the tree that leads to the event, just the way the General Multiplication Rule specifies.

The probability of abstaining and having an alcohol-related accident is, of course, zero.

All the outcomes at the far right are disjoint because at each branch of the tree we chose between disjoint alternatives. And they are *all* the possibilities, so the probabilities on the far right must add up to one. Always check!

Because the final outcomes are disjoint, we can add up their probabilities to get probabilities for compound events. For example, what's the probability that a selected student has had an alcohol-related car accident? We simply find *all* the outcomes on the far right in which an accident has happened. There are three and we can add their probabilities: 0.075 + 0.033 + 0 = 0.108—almost an 11% chance.

Reversing the Conditioning

If we know a student has had an alcohol-related accident, what's the probability that the student is a binge drinker? That's an interesting question, but we can't just read it from the tree. The tree gives us $P(\text{accident} \mid \text{binge})$, but we want $P(\text{binge} \mid \text{accident})$—conditioning in the other direction. The two probabilities are definitely *not* the same. We have reversed the conditioning.

We may not have the conditional probability we want, but we do know everything we need to know to find it. To find a conditional probability, we need the probability that both events happen divided by the probability that the given event occurs. We have already found the probability of an alcohol-related accident: 0.075 + 0.033 + 0 = 0.108.

The joint probability that a student is both a binge drinker and someone who's had an alcohol-related accident is found at the top branch: 0.075. We've restricted the *Who* of the problem to the students with alcohol-related accidents, so we divide the two to find the conditional probability:

$$P(\text{binge} \mid \text{accident}) = \frac{P(\text{binge} \cap \text{accident})}{P(\text{accident})}$$

$$= \frac{0.075}{0.108} = 0.694$$

The chance that a student who has an alcohol-related car accident is a binge drinker is more than 69%! As we said, reversing the conditioning is rarely intuitive, but tree diagrams help us keep track of the calculation when there aren't too many alternatives to consider.

STEP-BY-STEP EXAMPLE | **Reversing the Conditioning**

When the authors were in college, there were only three requirements for graduation that were the same for all students: You had to be able to tread water for 2 minutes, you had to learn a foreign language, and you had to be free of tuberculosis. For the last requirement, all freshmen had to take a TB screening test that consisted of a nurse jabbing what looked like a corncob holder into your forearm. You were then expected to report back in 48 hours to have it checked. If you were healthy and TB-free, your arm was supposed to look as though you'd never had the test.

Sometime during the 48 hours, one of us had a reaction. When he finally saw the nurse, his arm was about 50% bigger than normal and a very unhealthy red. Did he have TB? The nurse had said that the test was about 99% effective, so it seemed that the chances must be pretty high that he had TB. How high do you think the chances were? Go ahead and guess. Guess low.

We'll call **TB** the event of actually having TB and + the event of testing positive. To start a tree, we need to know $P(\textbf{TB})$, the probability of having TB.[3] We also need to know the conditional probabilities $P(+|\textbf{TB})$ and $P(+|\textbf{TB}^C)$. Diagnostic tests can make two kinds of errors. They can give a positive result for a healthy person (a *false positive*) or a negative result for a sick person (a *false negative*). Being 99% accurate usually means a false-positive rate of 1%. That is, someone who doesn't have the disease has a 1% chance of testing positive anyway. We can write $P(+|\textbf{TB}^C) = 0.01$.

Since a false negative is more serious (because a sick person might not get treatment), tests are usually constructed to have a lower false-negative rate. We don't know exactly, but let's assume a 0.1% false-negative rate. So only 0.1% of sick people test negative. We can write $P(-|\textbf{TB}) = 0.001$.

Plan Define the events we're interested in and their probabilities.

Figure out what you want to know in terms of the events. Use the notation of conditional probability to write the event whose probability you want to find.

Let **TB** = {having TB} and **TB**^{C} = {no TB}
+ = {testing positive} and
− = {testing negative}

I know that $P(+|\textbf{TB}^C) = 0.01$ and $P(-|\textbf{TB}) = 0.001$. I also know that $P(\textbf{TB}) = 0.00005$.

I'm interested in the probability that the author had TB given that he tested positive: $P(\textbf{TB}|+)$.

Plot Draw the tree diagram. When probabilities are very small like these are, be careful to keep all the significant digits.

To finish the tree we need $P(\textbf{TB}^C)$, $P(-|\textbf{TB}^C)$, and $P(-|\textbf{TB})$. We can find each of these from the Complement Rule:

$$P(\textbf{TB}^C) = 1 - P(\textbf{TB}) = 0.99995$$
$$P(-|\textbf{TB}^C) = 1 - P(+|\textbf{TB}^C)$$
$$= 1 - 0.01 = 0.99 \text{ and}$$
$$P(+|\textbf{TB}) = 1 - P(-|\textbf{TB})$$
$$= 1 - 0.01 = 0.999$$

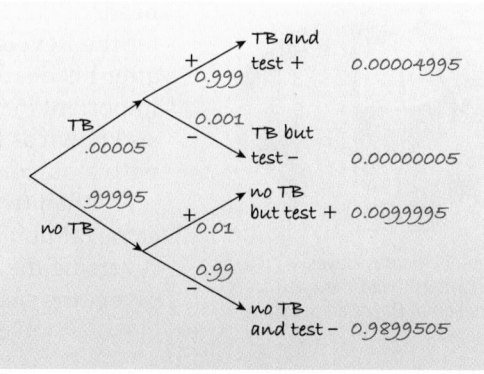

---

[3] This isn't given, so we looked it up. Although TB is a matter of serious concern to public health officials, it is a fairly uncommon disease, with an incidence of about 5 cases per 100,000 in the United States (see *http://www.cdc.gov/tb/default.htm*).

Mechanics Multiply along the branches to find the probabilities of the four possible outcomes. Check your work by seeing if they total 1.

(Check: $0.00004995 + 0.00000005 + 0.0099995 + 0.98995050 = 1$)

Add up the probabilities corresponding to the condition of interest—in this case, testing positive. We can add because the tree shows disjoint events.

$$P(+) = P(TB \cap +) + P(TB^c \cap +)$$
$$P = 0.00004995 + 0.0099995$$
$$= 0.01004945$$

Divide the probability of both events occuring (here, having TB and a positive test) by the probability of satisfying the condition (testing positive).

$$P(TB \mid +) = \frac{P(TB \cap +)}{P(+)}$$
$$= \frac{0.00004995}{0.01004945}$$
$$= 0.00497$$

Conclusion Interpret your result in context.

The chance of having TB after you test positive is less than 0.5%.

When we reverse the order of conditioning, we change the *Who* we are concerned with. With events of low probability, the result can be surprising. That's the reason patients who test positive for HIV, for example, are always told to seek medical counseling. They may have only a small chance of actually being infected. That's why global drug or disease testing can have unexpected consequences if people interpret *testing* positive as *being* positive.

Bayes's Rule

The Reverend Thomas Bayes is credited posthumously with the rule that is the foundation of Bayesian Statistics.

When we have $P(A \mid B)$ but want the *reverse* probability $P(B \mid A)$, we need to find $P(A \cap B)$ and $P(A)$. A tree is often a convenient way of finding these probabilities. It can work even when we have more than two possible events, as we saw in the binge-drinking example. Instead of using the tree, we *could* write the calculation algebraically, showing exactly how we found the quantities that we needed: $P(A \cap B)$ and $P(A)$. The result is a formula known as Bayes's Rule, after the Reverend Thomas Bayes (1702?–1761), who was credited with the rule after his death, when he could no longer defend himself. Bayes's Rule is quite important in Statistics and is the foundation of an approach to Statistical analysis known as Bayesian Statistics. Although the simple rule deals with two alternative outcomes, the rule can be extended to the situation in which there are more than two branches to the first split of the tree. The principle remains the same (although the math gets more difficult). Bayes's Rule is just a formula[4] for reversing the probability from the conditional probability that you're originally given, the same feat we accomplished with our tree diagram.

---

[4] Bayes's Rule for two events says that $P(B \mid A) = \dfrac{P(A \mid B)P(B)}{P(A \mid B)P(B) + P(A \mid B^C)P(B^C)}$.

Masochists may wish to try it with the TB testing probabilities. (It's easier to just draw the tree, isn't it?)

FOR EXAMPLE   Reversing the conditioning

A recent Maryland highway safety study found that in 77% of all accidents the driver was wearing a seatbelt. Accident reports indicated that 92% of those drivers escaped serious injury (defined as hospitalization or death), but only 63% of the non-belted drivers were so fortunate.

Question: What's the probability that a driver who was seriously injured wasn't wearing a seatbelt?

Let B = the driver was wearing a seatbelt, and NB = no belt.

Let I = serious injury or death, and OK = not seriously injured.

I know $P(B) = 0.77$, so $P(NB) = 1 - 0.77 = 0.23$.

Also, $P(OK \mid B) = 0.92$, so $P(I \mid B) = 0.08$
and $P(OK \mid B) = 0.63$, so $P(I \mid NB) = 0.37$

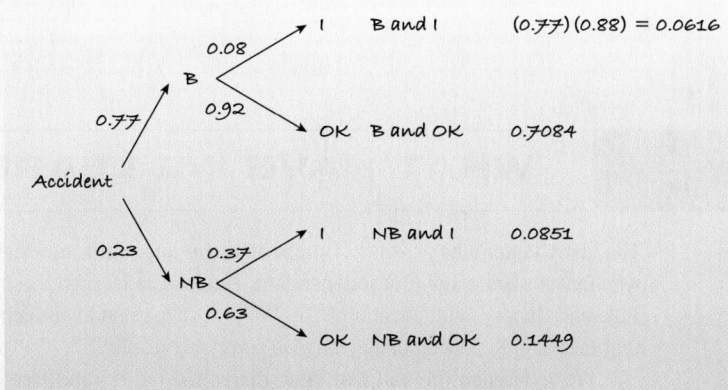

$$P(NB \mid I) = \frac{P(NB \text{ and } I)}{P(I)} = \frac{0.0851}{0.0616 + 0.0851} = 0.58$$

Even though only 23% of drivers weren't wearing seatbelts, they accounted for 58% of all the deaths and serious injuries.

Just some advice from your friends, the authors: *Please buckle up!* (We want you to finish this course.)

WHAT CAN GO WRONG?

▶ **Don't use a simple probability rule where a general rule is appropriate.** Don't assume independence without reason to believe it. Don't assume that outcomes are disjoint without checking that they are. Remember that the general rules always apply, even when outcomes are in fact independent or disjoint.

▶ **Don't find probabilities for samples drawn without replacement as if they had been drawn with replacement.** Remember to adjust the denominator of your probabilities. This warning applies only when we draw from small populations or draw a large fraction of a finite population. When the population is very large relative to the sample size, the adjustments make very little difference, and we ignore them.

▶ **Don't reverse conditioning naively.** As we have seen, the probability of **A** *given* **B** may not, and, in general does not, resemble the probability of **B** *given* **A**. The true probability may be counterintuitive.

▶ **Don't confuse "disjoint" with "independent."** Disjoint events *cannot* happen at the same time. When one happens, you know the other did not, so $P(\mathbf{B} \mid \mathbf{A}) = 0$. Independent events *must* be able to happen at the same time. When one happens, you know it has no effect on the other, so $P(\mathbf{B} \mid \mathbf{A}) = P(\mathbf{B})$.

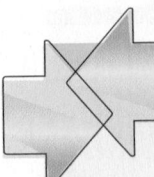

CONNECTIONS

This chapter shows the unintuitive side of probability. If you've been thinking, "My mind doesn't work this way," you're probably right. Humans don't seem to find conditional and compound probabilities natural and often have trouble with them. Even statisticians make mistakes with conditional probability.

Our central connection is to the guiding principle that Statistics is about understanding the world. The events discussed in this chapter are close to the kinds of real-world situations in which understanding probabilities matters. The methods and concepts of this chapter are the tools you need to understand the part of the real world that deals with the outcomes of complex, uncertain events.

WHAT HAVE WE LEARNED?

The last chapter's basic rules of probability are important, but they work only in special cases—when events are disjoint or independent. Now we've learned the more versatile General Addition Rule and General Multiplication Rule. We've also learned about conditional probabilities, and seen that reversing the conditioning can give surprising results.

We've learned the value of Venn diagrams, tables, and tree diagrams to help organize our thinking about probabilities.

Most important, we've learned to think clearly about independence. We've seen how to use conditional probability to determine whether two events are independent and to work with events that are not independent. A sound understanding of independence will be important throughout the rest of this book.

Terms

General Addition Rule 343. For any two events, **A** and **B**, the probability of **A** *or* **B** is

$$P(\mathbf{A} \cup \mathbf{B}) = P(\mathbf{A}) + P(\mathbf{B}) - P(\mathbf{A} \cap \mathbf{B}).$$

Conditional probability 347. $P(\mathbf{B} \mid \mathbf{A}) = \dfrac{P(\mathbf{A} \cap \mathbf{B})}{P(\mathbf{A})}$

$P(\mathbf{B} \mid \mathbf{A})$ is read "the probability of **B** *given* **A**."

General Multiplication Rule 348. For any two events, **A** and **B**, the probability of **A** and **B** is

$$P(\mathbf{A} \cap \mathbf{B}) = P(\mathbf{A}) \times P(\mathbf{B} \mid \mathbf{A}).$$

Independence (used formally) 349. Events **A** and **B** are independent when $P(\mathbf{B} \mid \mathbf{A}) = P(\mathbf{B})$.

Tree diagram 354. A display of conditional events or probabilities that is helpful in thinking through conditioning.

Skills

▸ Understand the concept of conditional probability as redefining the *Who* of concern, according to the information about the event that is *given*.

▸ Understand the concept of independence.

▸ Know how and when to apply the General Addition Rule.

▸ Know how to find probabilities for compound events as fractions of counts of occurrences in a two-way table.

▸ Know how and when to apply the General Multiplication Rule.

▸ Know how to make and use a tree diagram to understand conditional probabilities and reverse conditioning.

▸ Be able to make a clear statement about a conditional probability that makes clear how the condition affects the probability.

▸ Avoid making statements that assume independence of events when there is no clear evidence that they are in fact independent.

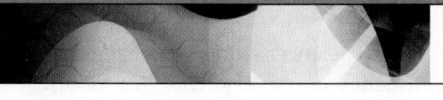

EXERCISES

1. Homes. Real estate ads suggest that 64% of homes for sale have garages, 21% have swimming pools, and 17% have both features. What is the probability that a home for sale has
a) a pool or a garage?
b) neither a pool nor a garage?
c) a pool but no garage?

2. Travel. Suppose the probability that a U.S. resident has traveled to Canada is 0.18, to Mexico is 0.09, and to both countries is 0.04. What's the probability that an American chosen at random has
a) traveled to Canada but not Mexico?
b) traveled to either Canada or Mexico?
c) not traveled to either country?

3. Amenities. A check of dorm rooms on a large college campus revealed that 38% had refrigerators, 52% had TVs, and 21% had both a TV and a refrigerator. What's the probability that a randomly selected dorm room has
a) a TV but no refrigerator?
b) a TV or a refrigerator, but not both?
c) neither a TV nor a refrigerator?

4. Workers. Employment data at a large company reveal that 72% of the workers are married, that 44% are college graduates, and that half of the college grads are married. What's the probability that a randomly chosen worker
a) is neither married nor a college graduate?
b) is married but not a college graduate?
c) is married or a college graduate?

5. Global survey. The marketing research organization GfK Custom Research North America conducts a yearly survey on consumer attitudes worldwide. They collect demographic information on the roughly 1500 respondents from each country that they survey. Here is a table showing the number of people with various levels of education in five countries:

Educational Level by Country						
	Post-graduate	College	Some high school	Primary or less	No answer	Total
China	7	315	671	506	3	1502
France	69	388	766	309	7	1539
India	161	514	622	227	11	1535
U.K.	58	207	1240	32	20	1557
USA	84	486	896	87	4	1557
Total	379	1910	4195	1161	45	7690

If we select someone at random from this survey,
a) what is the probability that the person is from the United States?
b) what is the probability that the person completed his or her education before college?
c) what is the probability that the person is from France *or* did some post-graduate study?
d) what is the probability that the person is from France *and* finished only primary school or less?

6. Birth order. A survey of students in a large Introductory Statistics class asked about their birth order (1 = oldest or only child) and which college of the university they were enrolled in. Here are the results:

		Birth Order		
		1 or only	2 or more	Total
College	Arts & Sciences	34	23	57
	Agriculture	52	41	93
	Human Ecology	15	28	43
	Other	12	18	30
	Total	113	110	223

Suppose we select a student at random from this class. What is the probability that the person is
a) a Human Ecology student?
b) a firstborn student?
c) firstborn *and* a Human Ecology student?
d) firstborn *or* a Human Ecology student?

7. Cards. You draw a card at random from a standard deck of 52 cards. Find each of the following conditional probabilities:
a) The card is a heart, given that it is red.
b) The card is red, given that it is a heart.
c) The card is an ace, given that it is red.
d) The card is a queen, given that it is a face card.

8. Pets. In its monthly report, the local animal shelter states that it currently has 24 dogs and 18 cats available for adoption. Eight of the dogs and 6 of the cats are male. Find each of the following conditional probabilities if an animal is selected at random:
a) The pet is male, given that it is a cat.
b) The pet is a cat, given that it is female.
c) The pet is female, given that it is a dog.

9. Health. The probabilities that an adult American man has high blood pressure and/or high cholesterol are shown in the table.

	Blood Pressure	
Cholesterol	**High**	**OK**
High	0.11	0.21
OK	0.16	0.52

What's the probability that
a) a man has both conditions?
b) a man has high blood pressure?
c) a man with high blood pressure has high cholesterol?
d) a man has high blood pressure if it's known that he has high cholesterol?

10. Death penalty. The table shows the political affiliations of American voters and their positions on the death penalty.

	Death Penalty	
Party	**Favor**	**Oppose**
Republican	0.26	0.04
Democrat	0.12	0.24
Other	0.24	0.10

a) What's the probability that
  i) a randomly chosen voter favors the death penalty?
  ii) a Republican favors the death penalty?
  iii) a voter who favors the death penalty is a Democrat?
b) A candidate thinks she has a good chance of gaining the votes of anyone who is a Republican or in favor of the death penalty. What portion of the voters is that?

11. Global survey, take 2. Look again at the table summarizing the Roper survey in Exercise 5.
a) If we select a respondent at random, what's the probability we choose a person from the United States who has done post-graduate study?
b) Among the respondents who have done post-graduate study, what's the probability the person is from the United States?
c) What's the probability that a respondent from the United States has done post-graduate study?
d) What's the probability that a respondent from China has only a primary-level education?
e) What's the probability that a respondent with only a primary-level education is from China?

12. Birth order, take 2. Look again at the data about birth order of Intro Stats students and their choices of colleges shown in Exercise 6.
a) If we select a student at random, what's the probability the person is an Arts and Sciences student who is a second child (or more)?
b) Among the Arts and Sciences students, what's the probability a student was a second child (or more)?
c) Among second children (or more), what's the probability the student is enrolled in Arts and Sciences?
d) What's the probability that a first or only child is enrolled in the Agriculture College?
e) What is the probability that an Agriculture student is a first or only child?

13. Sick kids. Seventy percent of kids who visit a doctor have a fever, and 30% of kids with a fever have sore throats. What's the probability that a kid who goes to the doctor has a fever and a sore throat?

14. Sick cars. Twenty percent of cars that are inspected have faulty pollution control systems. The cost of repairing a pollution control system exceeds $100 about 40% of the time. When a driver takes her car in for inspection, what's the probability that she will end up paying more than $100 to repair the pollution control system?

15. Cards. You are dealt a hand of three cards, one at a time. Find the probability of each of the following.
a) The first heart you get is the third card dealt.
b) Your cards are all red (that is, all diamonds or hearts).
c) You get no spades.
d) You have at least one ace.

16. Another hand. You pick three cards at random from a deck. Find the probability of each event described below.
a) You get no aces.
b) You get all hearts.
c) The third card is your first red card.
d) You have at least one diamond.

17. Batteries. A junk box in your room contains a dozen old batteries, five of which are totally dead. You start picking batteries one at a time and testing them. Find the probability of each outcome.
a) The first two you choose are both good.
b) At least one of the first three works.
c) The first four you pick all work.
d) You have to pick 5 batteries to find one that works.

18. Shirts. The soccer team's shirts have arrived in a big box, and people just start grabbing them, looking for the right size. The box contains 4 medium, 10 large, and 6 extra-large shirts. You want a medium for you and one for your sister. Find the probability of each event described.
a) The first two you grab are the wrong sizes.
b) The first medium shirt you find is the third one you check.
c) The first four shirts you pick are all extra-large.
d) At least one of the first four shirts you check is a medium.

19. Eligibility. A university requires its biology majors to take a course called BioResearch. The prerequisite for this course is that students must have taken either a Statistics course or a computer course. By the time they are juniors, 52% of the Biology majors have taken Statistics, 23% have had a computer course, and 7% have done both.
a) What percent of the junior Biology majors are ineligible for BioResearch?
b) What's the probability that a junior Biology major who has taken Statistics has also taken a computer course?
c) Are taking these two courses disjoint events? Explain.
d) Are taking these two courses independent events? Explain.

20. Benefits. Fifty-six percent of all American workers have a workplace retirement plan, 68% have health insurance, and 49% have both benefits. We select a worker at random.
a) What's the probability he has neither employer-sponsored health insurance nor a retirement plan?
b) What's the probability he has health insurance if he has a retirement plan?
c) Are having health insurance and a retirement plan independent events? Explain.
d) Are having these two benefits mutually exclusive? Explain.

21. For sale. In the real-estate ads described in Exercise 1, 64% of homes for sale have garages, 21% have swimming pools, and 17% have both features.
a) If a home for sale has a garage, what's the probability that it has a pool too?
b) Are having a garage and a pool independent events? Explain.
c) Are having a garage and a pool mutually exclusive? Explain.

22. On the road again. According to Exercise 2, the probability that a U.S. resident has traveled to Canada is 0.18, to Mexico is 0.09, and to both countries is 0.04.
a) What's the probability that someone who has traveled to Mexico has visited Canada too?
b) Are traveling to Mexico and to Canada disjoint events? Explain.
c) Are traveling to Mexico and to Canada independent events? Explain.

23. Cards. If you draw a card at random from a well-shuffled deck, is getting an ace independent of the suit? Explain.

24. Pets again. The local animal shelter in Exercise 8 reported that it currently has 24 dogs and 18 cats available for adoption; 8 of the dogs and 6 of the cats are male. Are the species and sex of the animals independent? Explain.

25. Unsafe food. Early in 2007 *Consumer Reports* published the results of an extensive investigation of broiler chickens purchased from food stores in 23 states. Tests for bacteria in the meat showed that 81% of the chickens were contaminated with campylobacter, 15% with salmonella, and 13% with both.
a) What's the probability that a tested chicken was not contaminated with either kind of bacteria?
b) Are contamination with the two kinds of bacteria disjoint? Explain.
c) Are contamination with the two kinds of bacteria independent? Explain.

26. Birth order, finis. In Exercises 6 and 12 we looked at the birth orders and college choices of some Intro Stats students. For these students:
a) Are enrolling in Agriculture and Human Ecology disjoint? Explain.
b) Are enrolling in Agriculture and Human Ecology independent? Explain.
c) Are being firstborn and enrolling in Human Ecology disjoint? Explain.
d) Are being firstborn and enrolling in Human Ecology independent? Explain.

27. Men's health, again. Given the table of probabilities from Exercise 9, are high blood pressure and high cholesterol independent? Explain.

		Blood Pressure	
		High	**OK**
Cholesterol	**High**	0.11	0.21
	OK	0.16	0.52

28. Politics. Given the table of probabilities from Exercise 10, are party affiliation and position on the death penalty independent? Explain.

		Death Penalty	
		Favor	**Oppose**
Party	**Republican**	0.26	0.04
	Democrat	0.12	0.24
	Other	0.24	0.10

29. Phone service. According to estimates from the federal government's 2003 National Health Interview Survey, based on face-to-face interviews in 16,677 households, approximately 58.2% of U.S. adults have both a landline in their residence and a cell phone, 2.8% have only cell phone service but no landline, and 1.6% have no telephone service at all.

a) Polling agencies won't phone cell phone numbers because customers object to paying for such calls. What proportion of U.S. households can be reached by a landline call?

b) Are having a cell phone and having a landline independent? Explain.

30. **Snoring.** After surveying 995 adults, 81.5% of whom were over 30, the National Sleep Foundation reported that 36.8% of all the adults snored. 32% of the respondents were snorers over the age of 30.

a) What percent of the respondents were under 30 and did not snore?

b) Is snoring independent of age? Explain.

31. **Montana.** A 1992 poll conducted by the University of Montana classified respondents by sex and political party, as shown in the table. Is party affiliation independent of the respondents' sex? Explain.

	Democrat	Republican	Independent
Male	36	45	24
Female	48	33	16

32. **Cars.** A random survey of autos parked in student and staff lots at a large university classified the brands by country of origin, as seen in the table. Is country of origin independent of type of driver?

		Driver	
		Student	Staff
Origin	**American**	107	105
	European	33	12
	Asian	55	47

33. **Luggage.** Leah is flying from Boston to Denver with a connection in Chicago. The probability her first flight leaves on time is 0.15. If the flight is on time, the probability that her luggage will make the connecting flight in Chicago is 0.95, but if the first flight is delayed, the probability that the luggage will make it is only 0.65.

a) Are the first flight leaving on time and the luggage making the connection independent events? Explain.

b) What is the probability that her luggage arrives in Denver with her?

34. **Graduation.** A private college report contains these statistics:

70% of incoming freshmen attended public schools.
75% of public school students who enroll as freshmen eventually graduate.
90% of other freshmen eventually graduate.

a) Is there any evidence that a freshman's chances to graduate may depend upon what kind of high school the student attended? Explain.

b) What percent of freshmen eventually graduate?

35. **Late luggage.** Remember Leah (Exercise 33)? Suppose you pick her up at the Denver airport, and her luggage is not there. What is the probability that Leah's first flight was delayed?

36. **Graduation, part II.** What percent of students who graduate from the college in Exercise 34 attended a public high school?

37. **Absenteeism.** A company's records indicate that on any given day about 1% of their day-shift employees and 2% of the night-shift employees will miss work. Sixty percent of the employees work the day shift.

a) Is absenteeism independent of shift worked? Explain.

b) What percent of employees are absent on any given day?

38. **Lungs and smoke.** Suppose that 23% of adults smoke cigarettes. It's known that 57% of smokers and 13% of nonsmokers develop a certain lung condition by age 60.

a) Explain how these statistics indicate that lung condition and smoking are not independent.

b) What's the probability that a randomly selected 60-year-old has this lung condition?

39. **Absenteeism, part II.** At the company described in Exercise 37, what percent of the absent employees are on the night shift?

40. **Lungs and smoke again.** Based on the statistics in Exercise 38, what's the probability that someone with the lung condition was a smoker?

41. **Drunks.** Police often set up sobriety checkpoints—roadblocks where drivers are asked a few brief questions to allow the officer to judge whether or not the person may have been drinking. If the officer does not suspect a problem, drivers are released to go on their way. Otherwise, drivers are detained for a Breathalyzer test that will determine whether or not they will be arrested. The police say that based on the brief initial stop, trained officers can make the right decision 80% of the time. Suppose the police operate a sobriety checkpoint after 9 p.m. on a Saturday night, a time when national traffic safety experts suspect that about 12% of drivers have been drinking.

a) You are stopped at the checkpoint and, of course, have not been drinking. What's the probability that you are detained for further testing?

b) What's the probability that any given driver will be detained?

c) What's the probability that a driver who is detained has actually been drinking?

d) What's the probability that a driver who was released had actually been drinking?

42. **No-shows.** An airline offers discounted "advance-purchase" fares to customers who buy tickets more than 30 days before travel and charges "regular" fares for tickets purchased during those last 30 days. The company has noticed that 60% of its customers take advantage of the advance-purchase fares. The "no-show" rate among people who paid regular fares is 30%, but only 5% of customers with advance-purchase tickets are no-shows.

a) What percent of all ticket holders are no-shows?

b) What's the probability that a customer who didn't show had an advance-purchase ticket?

c) Is being a no-show independent of the type of ticket a passenger holds? Explain.

43. Dishwashers. Dan's Diner employs three dishwashers. Al washes 40% of the dishes and breaks only 1% of those he handles. Betty and Chuck each wash 30% of the dishes, and Betty breaks only 1% of hers, but Chuck breaks 3% of the dishes he washes. (He, of course, will need a new job soon. . . .) You go to Dan's for supper one night and hear a dish break at the sink. What's the probability that Chuck is on the job?

44. Parts. A company manufacturing electronic components for home entertainment systems buys electrical connectors from three suppliers. The company prefers to use supplier A because only 1% of those connectors prove to be defective, but supplier A can deliver only 70% of the connectors needed. The company must also purchase connectors from two other suppliers, 20% from supplier B and the rest from supplier C. The rates of defective connectors from B and C are 2% and 4%, respectively. You buy one of these components, and when you try to use it you find that the connector is defective. What's the probability that your component came from supplier A?

45. HIV testing. In July 2005 the journal *Annals of Internal Medicine* published a report on the reliability of HIV testing. Results of a large study suggested that among people with HIV, 99.7% of tests conducted were (correctly) positive, while for people without HIV 98.5% of the tests were (correctly) negative. A clinic serving an at-risk population offers free HIV testing, believing that 15% of the patients may actually carry HIV. What's the probability that a patient testing negative is truly free of HIV?

46. Polygraphs. Lie detectors are controversial instruments, barred from use as evidence in many courts. Nonetheless, many employers use lie detector screening as part of their hiring process in the hope that they can avoid hiring people who might be dishonest. There has been some research, but no agreement, about the reliability of polygraph tests. Based on this research, suppose that a polygraph can detect 65% of lies, but incorrectly identifies 15% of true statements as lies.

A certain company believes that 95% of its job applicants are trustworthy. The company gives everyone a polygraph test, asking, "Have you ever stolen anything from your place of work?" Naturally, all the applicants answer "No," but the polygraph identifies some of those answers as lies, making the person ineligible for a job. What's the probability that a job applicant rejected under suspicion of dishonesty was actually trustworthy?

**JUST CHECKING
Answers**

1. a)

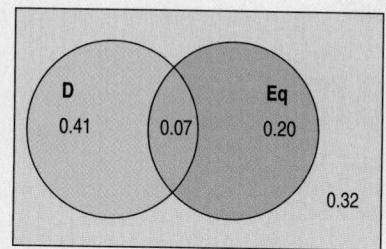

b) 0.32
c) 0.41

2. a) Independent
b) Disjoint
c) Neither

3. a)

		Equation		
		Yes	No	Total
Display	Yes	0.07	0.41	**0.48**
	No	0.20	0.32	**0.52**
	Total	**0.27**	**0.73**	**1.00**

b) $P(D \mid Eq) = P(D \text{ and } Eq)/P(Eq) = 0.07/0.27 = 0.259$
c) No, pages can (and 7% do) have both.
d) To be independent, we'd need $P(D \mid Eq) = P(D)$. $P(D \mid Eq) = 0.259$, but $P(D) = 0.48$. Overall, 48% of pages have data displays, but only about 26% of pages with equations do. They do not appear to be independent.

CHAPTER 16

Random Variables

What Is an Actuary?
Actuaries are the daring people who put a price on risk, estimating the likelihood and costs of rare events, so they can be insured. That takes financial, statistical, and business skills. It also makes them invaluable to many businesses. Actuaries are rather rare themselves; only about 19,000 work in North America. Perhaps because of this, they are well paid. If you're enjoying this course, you may want to look into a career as an actuary. Contact the Society of Actuaries or the Casualty Actuarial Society (who, despite what you may think, did not pay for this blurb).

Insurance companies make bets. They bet that you're going to live a long life. You bet that you're going to die sooner. Both you and the insurance company want the company to stay in business, so it's important to find a "fair price" for your bet. Of course, the right price for *you* depends on many factors, and nobody can predict exactly how long you'll live. But when the company averages over enough customers, it can make reasonably accurate estimates of the amount it can expect to collect on a policy before it has to pay its benefit.

Here's a simple example. An insurance company offers a "death and disability" policy that pays $10,000 when you die or $5000 if you are permanently disabled. It charges a premium of only $50 a year for this benefit. Is the company likely to make a profit selling such a plan? To answer this question, the company needs to know the *probability* that its clients will die or be disabled in any year. From actuarial information like this, the company can calculate the expected value of this policy.

Expected Value: Center

NOTATION ALERT:

The most common letters for random variables are X, Y, and Z. But be cautious: If you see any capital letter, it just might denote a random variable.

We'll want to build a probability model in order to answer the questions about the insurance company's risk. First we need to define a few terms. The amount the company pays out on an individual policy is called a **random variable** because its numeric value is based on the outcome of a random event. We use a capital letter, like X, to denote a random variable. We'll denote a particular value that it can have by the corresponding lowercase letter, in this case x. For the insurance company, x can be $10,000 (if you die that year), $5000 (if you are disabled), or $0 (if neither occurs). Because we can list all the outcomes, we might formally call this random variable a **discrete** random variable. Otherwise, we'd call it a **continuous** random variable. The collection of all the possible values and the probabilities that they occur is called the **probability model** for the random variable.

A S *Activity:* **Random Variables.** Learn more about random variables from this animated tour.

Suppose, for example, that the death rate in any year is 1 out of every 1000 people, and that another 2 out of 1000 suffer some kind of disability. Then we can display the probability model for this insurance policy in a table like this:

Policyholder Outcome	Payout x	Probability $P(X = x)$
Death	10,000	$\dfrac{1}{1000}$
Disability	5000	$\dfrac{2}{1000}$
Neither	0	$\dfrac{997}{1000}$

To see what the insurance company can expect, imagine that it insures exactly 1000 people. Further imagine that, in perfect accordance with the probabilities, 1 of the policyholders dies, 2 are disabled, and the remaining 997 survive the year unscathed. The company would pay $10,000 to one client and $5000 to each of 2 clients. That's a total of $20,000, or an average of 20000/1000 = $20 per policy. Since it is charging people $50 for the policy, the company expects to make a profit of $30 per customer. Not bad!

We can't predict what *will* happen during any given year, but we can say what we *expect* to happen. To do this, we (or, rather, the insurance company) need the probability model. The expected value of a policy is a parameter of this model. In fact, it's the mean. We'll signify this with the notation μ (for population mean) or $E(X)$ for expected value. This isn't an average of some data values, so we won't estimate it. Instead, we assume that the probabilities are known and simply calculate the expected value from them.

How did we come up with $20 as the expected value of a policy payout? Here's the calculation. As we've seen, it often simplifies probability calculations to think about some (convenient) number of outcomes. For example, we could imagine that we have exactly 1000 clients. Of those, exactly 1 died and 2 were disabled, corresponding to what the probabilities would say.

$$\mu = E(X) = \frac{10{,}000(1) + 5000(2) + 0(997)}{1000}$$

So our expected payout comes to $20,000, or $20 per policy.

Instead of writing the expected value as one big fraction, we can rewrite it as separate terms with a common denominator of 1000.

$$\mu = E(X)$$
$$= \$10{,}000\left(\frac{1}{1000}\right) + \$5000\left(\frac{2}{1000}\right) + \$0\left(\frac{997}{1000}\right)$$
$$= \$20.$$

How convenient! See the probabilities? For each policy, there's a 1/1000 chance that we'll have to pay $10,000 for a death and a 2/1000 chance that we'll have to pay $5000 for a disability. Of course, there's a 997/1000 chance that we won't have to pay anything.

Take a good look at the expression now. It's easy to calculate the **expected value** of a (discrete) random variable—just multiply each possible value by the probability that it occurs, and find the sum:

$$\mu = E(X) = \sum xP(x).$$

Be sure that every possible outcome is included in the sum. And verify that you have a valid probability model to start with—the probabilities should each be between 0 and 1 and should sum to one.

FOR EXAMPLE   Love and expected values

On Valentine's Day the *Quiet Nook* restaurant offers a *Lucky Lovers Special* that could save couples money on their romantic dinners. When the waiter brings the check, he'll also bring the four aces from a deck of cards. He'll shuffle them and lay them out face down on the table. The couple will then get to turn one card over. If it's a black ace, they'll owe the full amount, but if it's the ace of hearts, the waiter will give them a $20 Lucky Lovers discount. If they first turn over the ace of diamonds (hey—at least it's red!), they'll then get to turn over one of the remaining cards, earning a $10 discount for finding the ace of hearts this time.

Question: Based on a probability model for the size of the Lucky Lovers discounts the restaurant will award, what's the expected discount for a couple?

Let X = the Lucky Lovers discount. The probabilities of the three outcomes are:

$$P(X = 20) = P(A\heartsuit) = \frac{1}{4}$$

$$P(X = 10) = P(A\diamondsuit, \text{then } A\heartsuit) = P(A\diamondsuit) \cdot P(A\heartsuit | A\diamondsuit)$$
$$= \frac{1}{4} \cdot \frac{1}{3} = \frac{1}{12}$$

$$P(X = 0) = P(X \neq 20 \text{ or } 10) = 1 - \left(\frac{1}{4} + \frac{1}{12}\right) = \frac{2}{3}.$$

My probability model is:

Outcome	A♥	A♦, then A♥	Black Ace
x	20	10	0
$P(X = x)$	$\frac{1}{4}$	$\frac{1}{12}$	$\frac{2}{3}$

$$E(X) = 20 \cdot \frac{1}{4} + 10 \cdot \frac{1}{12} + 0 \cdot \frac{2}{3} = \frac{70}{12} \approx 5.83$$

Couples dining at the Quiet Nook can expect an average discount of $5.83.

 **JUST CHECKING**

1. One of the authors took his minivan in for repair recently because the air conditioner was cutting out intermittently. The mechanic identified the problem as dirt in a control unit. He said that in about 75% of such cases, drawing down and then recharging the coolant a couple of times cleans up the problem—and costs only $60. If that fails, then the control unit must be replaced at an additional cost of $100 for parts and $40 for labor.

   a) Define the random variable and construct the probability model.
   b) What is the expected value of the cost of this repair?
   c) What does that mean in this context?

Oh—in case you were wondering—the $60 fix worked!

First Center, Now Spread . . .

Of course, this expected value (or mean) is not what actually happens to any *particular* policyholder. No individual policy actually costs the company $20. We are dealing with random events, so some policyholders receive big payouts, others nothing. Because the insurance company must anticipate this variability, it needs to know the *standard deviation* of the random variable.

For data, we calculated the **standard deviation** by first computing the deviation from the mean and squaring it. We do that with (discrete) random variables as well. First, we find the deviation of each payout from the mean (expected value):

Policyholder Outcome	Payout x	Probability $P(X = x)$	Deviation $(x - \mu)$
Death	10,000	$\frac{1}{1000}$	$(10{,}000 - 20) = 9980$
Disability	5000	$\frac{2}{1000}$	$(5000 - 20) = 4980$
Neither	0	$\frac{997}{1000}$	$(0 - 20) = -20$

Next, we square each deviation. The **variance** is the expected value of those squared deviations, so we multiply each by the appropriate probability and sum those products. That gives us the variance of X. Here's what it looks like:

$$Var(X) = 9980^2 \left(\frac{1}{1000} \right) + 4980^2 \left(\frac{2}{1000} \right) + (-20)^2 \left(\frac{997}{1000} \right) = 149{,}600.$$

Finally, we take the square root to get the standard deviation:

$$SD(X) = \sqrt{149{,}600} \approx \$386.78.$$

The insurance company can expect an average payout of $20 per policy, with a standard deviation of $386.78.

Think about that. The company charges $50 for each policy and expects to pay out $20 per policy. Sounds like an easy way to make $30. In fact, most of the time (probability 997/1000) the company pockets the entire $50. But would you consider selling your roommate such a policy? The problem is that occasionally the company loses big. With probability 1/1000, it will pay out $10,000, and with probability 2/1000, it will pay out $5000. That may be more risk than you're willing to take on. The standard deviation of $386.78 gives an indication that it's no sure thing. That's a pretty big spread (and risk) for an average profit of $30.

Here are the formulas for what we just did. Because these are parameters of our probability model, the variance and standard deviation can also be written as σ^2 and σ. You should recognize both kinds of notation.

$$\sigma^2 = Var(X) = \sum (x - \mu)^2 P(x)$$
$$\sigma = SD(X) = \sqrt{Var(X)}$$

Finding the standard deviation

Recap: Here's the probability model for the Lucky Lovers restaurant discount.

Outcome	A♥	A♦, then A♥	Black Ace
x	20	10	0
$P(X = x)$	$\dfrac{1}{4}$	$\dfrac{1}{12}$	$\dfrac{2}{3}$

We found that couples can expect an average discount of $\mu = \$5.83$.

Question: What's the standard deviation of the discounts?

First find the variance: $Var(X) = \sum (x - \mu)^2 \cdot P(x)$

$$= (20 - 5.83)^2 \cdot \frac{1}{4} + (10 - 5.83)^2 \cdot \frac{1}{12} + (0 - 5.83)^2 \cdot \frac{2}{3}$$

$$\approx 74.306.$$

So, $SD(X) = \sqrt{74.306} \approx \$8.62$

Couples can expect the Lucky Lovers discounts to average \$5.83, with a standard deviation of \$8.62.

STEP-BY-STEP EXAMPLE

Expected Values and Standard Deviations for Discrete Random Variables

As the head of inventory for Knowway computer company, you were thrilled that you had managed to ship 2 computers to your biggest client the day the order arrived. You are horrified, though, to find out that someone had restocked refurbished computers in with the new computers in your storeroom. The shipped computers were selected randomly from the 15 computers in stock, but 4 of those were actually refurbished.

If your client gets 2 new computers, things are fine. If the client gets one refurbished computer, it will be sent back at your expense—\$100—and you can replace it. However, if both computers are refurbished, the client will cancel the order this month and you'll lose a total of \$1000.

Question: What's the expected value and the standard deviation of the company's loss?

 **Plan** State the problem.	I want to find the company's expected loss for shipping refurbished computers and the standard deviation.
Variable Define the random variable.	Let $X =$ amount of loss.

Plot Make a picture. This is another job for tree diagrams.

If you prefer calculation to drawing, find $P(\mathbf{NN})$ and $P(\mathbf{RR})$, then use the Complement Rule to find $P(\mathbf{NR}\ or\ \mathbf{RN})$.

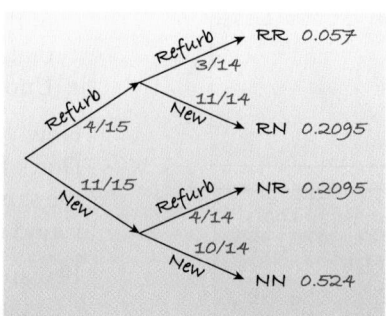

Model List the possible values of the random variable, and determine the probability model.

Outcome	x	$P(X = x)$
Two refurbs	1000	$P(\mathbf{RR}) = 0.057$
One refurb	100	$P(\mathbf{NR} \cup \mathbf{RN}) = 0.2095$
		$+\ 0.2095 = 0.419$
New/new	0	$P(\mathbf{NN}) = 0.524$

Mechanics Find the expected value.

$$E(X) = 0(0.524) + 100(0.419) + 1000(0.057)$$
$$= \$98.90$$

Find the variance.

$$Var(X) = (0 - 98.90)^2(0.524)$$
$$+ (100 - 98.90)^2(0.419)$$
$$+ (1000 - 98.90)^2(0.057)$$
$$= 51,408.79$$

Find the standard deviation.

$$SD(X) = \sqrt{51,408.79} = \$226.735$$

Conclusion Interpret your results in context.

I expect this mistake to cost the firm $98.90, with a standard deviation of $226.74. The large standard deviation reflects the fact that there's a pretty large range of possible losses.

REALITY CHECK Both numbers seem reasonable. The expected value of $98.90 is between the extremes of $0 and $1000, and there's great variability in the outcome values.

TI Tips   Finding the mean and SD of a random variable

You can easily calculate means and standard deviations for a random variable with your TI. Let's do the Knowway computer example.

- Enter the values of the variable in a list, say, **L1**: 0, 100, 1000.
- Enter the probability model in another list, say, **L2**. Notice that you can enter the probabilities as fractions. For example, multiplying along the top branches

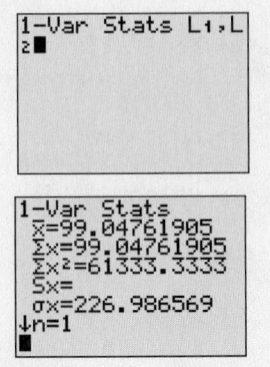

of the tree gives the probability of a $1000 loss to be $\frac{4}{15} \cdot \frac{3}{14}$. When you enter that, the TI will automatically calculate the probability as a decimal!

- Under the **STAT CALC** menu, ask for **1-Var Stats L1,L2**.

Now you see the mean and standard deviation (along with some other things). Don't fret that the calculator's mean and standard deviation aren't precisely the same as the ones we found. Such minor differences can arise whenever we round off probabilities to do the work by hand.

Beware: Although the calculator knows enough to call the standard deviation σ, it uses $\bar{x}$ where it should say μ. Make sure you don't make that mistake!

More About Means and Variances

Our insurance company expected to pay out an average of $20 per policy, with a standard deviation of about $387. If we take the $50 premium into account, we see the company makes a profit of $50 - 20 = \$30$ per policy. Suppose the company lowers the premium by $5 to $45. It's pretty clear that the expected profit also drops an average of $5 per policy, to $45 - 20 = \$25$.

What about the standard deviation? We know that adding or subtracting a constant from data shifts the mean but doesn't change the variance or standard deviation. The same is true of random variables.[1]

$$E(X \pm c) = E(X) \pm c \qquad Var(X \pm c) = Var(X).$$

FOR EXAMPLE   Adding a constant

Recap: We've determined that couples dining at the *Quiet Nook* can expect Lucky Lovers discounts averaging $5.83 with a standard deviation of $8.62. Suppose that for several weeks the restaurant has also been distributing coupons worth $5 off any one meal (one discount per table).

Question: If every couple dining there on Valentine's Day brings a coupon, what will be the mean and standard deviation of the total discounts they'll receive?

Let D = total discount (Lucky Lovers plus the coupon); then $D = X + 5$.

$$E(D) = E(X + 5) = E(X) + 5 = 5.83 + 5 = \$10.83$$
$$Var(D) = Var(X + 5) = Var(X) = 8.62^2$$
$$SD(D) = \sqrt{Var(X)} = \$8.62$$

Couples with the coupon can expect total discounts averaging $10.83. The standard deviation is still $8.62.

Back to insurance . . . What if the company decides to double all the payouts—that is, pay $20,000 for death and $10,000 for disability? This would double the average payout per policy and also increase the variability in payouts. We have seen that multiplying or dividing all data values by a constant changes both the mean and the standard deviation by the same factor. Variance, being the square of standard deviation, changes by the square of the constant. The same is true of random variables. In general, multiplying each value of a random variable by a

---

[1] The rules in this section are true for both discrete *and* continuous random variables.

constant multiplies the mean by that constant and the variance by the *square* of the constant.

$$E(aX) = aE(X) \qquad Var(aX) = a^2Var(X)$$

FOR EXAMPLE | Double the love

Recap: On Valentine's Day at the *Quiet Nook*, couples may get a Lucky Lovers discount averaging $5.83 with a standard deviation of $8.62. When two couples dine together on a single check, the restaurant doubles the discount offer—$40 for the ace of hearts on the first card and $20 on the second.

Question: What are the mean and standard deviation of discounts for such foursomes?

$$E(2X) = 2E(X) = 2(5.83) = \$11.66$$
$$Var(2x) = 2^2Var(x) = 2^2 \cdot 8.62^2 = 297.2176$$
$$SD(2X) = \sqrt{297.2176} = \$17.24$$

If the restaurant doubles the discount offer, two couples dining together can expect to save an average of $11.66 with a standard deviation of $17.24.

This insurance company sells policies to more than just one person. How can we figure means and variances for a collection of customers? For example, how can the company find the total expected value (and standard deviation) of policies taken over all policyholders? Consider a simple case: just two customers, Mr. Ecks and Ms. Wye. With an expected payout of $20 on each policy, we might predict a total of $20 + $20 = $40 to be paid out on the two policies. Nothing surprising there. The expected value of the sum is the sum of the expected values.

$$E(X + Y) = E(X) + E(Y).$$

The variability is another matter. Is the risk of insuring two people the same as the risk of insuring one person for twice as much? We wouldn't expect both clients to die or become disabled in the same year. Because we've spread the risk, the standard deviation should be smaller. Indeed, this is the fundamental principle behind insurance. By spreading the risk among many policies, a company can keep the standard deviation quite small and predict costs more accurately.

But how much smaller is the standard deviation of the sum? It turns out that, if the random variables are independent, there is a simple Addition Rule for variances: *The variance of the sum of two independent random variables is the sum of their individual variances.*

For Mr. Ecks and Ms. Wye, the insurance company can expect their outcomes to be independent, so (using X for Mr. Ecks's payout and Y for Ms. Wye's)

$$Var(X + Y) = Var(X) + Var(Y)$$
$$= 149{,}600 + 149{,}600$$
$$= 299{,}200.$$

If they had insured only Mr. Ecks for twice as much, there would only be one outcome rather than two *independent* outcomes, so the variance would have been

$$Var(2X) = 2^2Var(X) = 4 \times 149{,}600 = 598{,}400, \text{ or}$$

twice as big as with two independent policies.

Of course, variances are in squared units. The company would prefer to know standard deviations, which are in dollars. The standard deviation of the payout for two independent policies is $\sqrt{299{,}200} = \$546.99$. But the standard deviation of the payout for a single policy of twice the size is $\sqrt{598,400} = \$773.56$, or about 40% more.

If the company has two customers, then, it will have an expected annual total payout of \$40 with a standard deviation of about \$547.

FOR EXAMPLE | Adding the discounts

Recap: The Valentine's Day Lucky Lovers discount for couples averages \$5.83 with a standard deviation of \$8.62. We've seen that if the restaurant doubles the discount offer for two couples dining together on a single check, they can expect to save \$11.66 with a standard deviation of \$17.24. Some couples decide instead to get separate checks and pool their two discounts.

Question: You and your amour go to this restaurant with another couple and agree to share any benefit from this promotion. Does it matter whether you pay separately or together?

Let X_1 and X_2 represent the two separate discounts, and T the total; then $T = X_1 + X_2$.

$$E(T) = E(X_1 + X_2) = E(X_1) + E(X_2) = 5.83 + 5.83 = \$11.66,$$

so the expected saving is the same either way.

The cards are reshuffled for each couple's turn, so the discounts couples receive are independent. It's okay to add the variances:

$$Var(T) = Var(X_1 + X_2) = Var(X_1) + Var(X_2) = 8.62^2 + 8.62^2 = 148.6088$$
$$SD(T) = \sqrt{148.6088} = \$12.19$$

When two couples get separate checks, there's less variation in their total discount. The standard deviation is \$12.19, compared to \$17.24 for couples who play for the double discount on a single check. It does, therefore, matter whether they pay separately or together.

Pythagorean Theorem of Statistics

We often use the standard deviation to measure variability, but when we add independent random variables, we use their variances. Think of the Pythagorean Theorem. In a right triangle (only), the square of the length of the hypotenuse is the sum of the squares of the lengths of the other two sides:

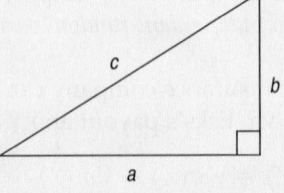

$$c^2 = a^2 + b^2.$$

For independent random variables (only), the square of the standard deviation of their sum is the sum of the squares of their standard deviations:

$$SD^2(X + Y) = SD^2(X) + SD^2(Y).$$

It's simpler to write this with *variances*:

For independent random variables, X and Y,
$Var(X + Y) = Var(X) + Var(Y)$.

In general,

▶ *The mean of the sum of two random variables is the sum of the means.*
▶ *The mean of the difference of two random variables is the difference of the means.*
▶ *If the random variables are independent, the variance of their sum or difference is always the sum of the variances.*

$$E(X \pm Y) = E(X) \pm E(Y) \qquad Var(X \pm Y) = Var(X) + Var(Y)$$

Wait a minute! Is that third part correct? Do we always *add* variances? Yes. Think about the two insurance policies. Suppose we want to know the mean and standard deviation of the *difference* in payouts to the two clients. Since each policy has an expected payout of \$20, the expected difference is $20 - 20 = \$0$. If we also subtract variances, we get \$0, too, and that surely doesn't make sense. Note that if the outcomes for the two clients are independent, the difference in payouts could range from $\$10,000 - \$0 = \$10,000$ to $\$0 - \$10,000 = -\$10,000$, a spread of \$20,000. The variability in differences increases as much as the variability in sums. If the company has two customers, the difference in payouts has a mean of \$0 and a standard deviation of about \$547 (again).

FOR EXAMPLE   Working with differences

Recap: The Lucky Lovers discount at the *Quiet Nook* averages $5.83 with a standard deviation of $8.62. Just up the street, the *Wise Fool* restaurant has a competing Lottery of Love promotion. There a couple can select a specially prepared chocolate from a large bowl and unwrap it to learn the size of their discount. The restaurant's manager says the discounts vary with an average of $10.00 and a standard deviation of $15.00.

Question: How much more can you expect to save at the *Wise Fool*? With what standard deviation?

Let W = discount at the Wise Fool, X = the discount at the Quiet Nook, and D = the difference: $D = W - X$. These are different promotions at separate restaurants, so the outcomes are independent.

$$E(W - X) = E(W) - E(X) = 10.00 - 5.83 = \$4.17$$
$$
\begin{aligned}
SD(W - X) &= \sqrt{Var(W - X)} \\
&= \sqrt{Var(W) + Var(X)} \\
&= \sqrt{15^2 + 8.62^2} \\
&\approx \$17.30
\end{aligned}
$$

Discounts at the Wise Fool will average $4.17 more than at the Quiet Nook, with a standard deviation of $17.30.

For random variables, does $X + X + X = 3X$?   Maybe, but be careful. As we've just seen, insuring one person for $30,000 is not the same risk as insuring three people for $10,000 each. When each instance represents a different outcome for the same random variable, it's easy to fall into the trap of writing all of them with the same symbol. Don't make this common mistake. Make sure you write each instance as a *different* random variable. Just because each random variable describes a similar situation doesn't mean that each random outcome will be the same.

These are *random* variables, not the variables you saw in Algebra. Being random, they take on different values each time they're evaluated. So what you really mean is $X_1 + X_2 + X_3$. Written this way, it's clear that the sum shouldn't necessarily equal 3 times *anything*.

FOR EXAMPLE   Summing a series of outcomes

Recap: The *Quiet Nook*'s Lucky Lovers promotion offers couples discounts averaging $5.83 with a standard deviation of $8.62. The restaurant owner is planning to serve 40 couples on Valentine's Day.

Question: What's the expected total of the discounts the owner will give? With what standard deviation?

Let $X_1, X_2, X_3, \ldots, X_{40}$ represent the discounts to the 40 couples, and T the total of all the discounts. Then:

$$
\begin{aligned}
T &= X_1 + X_2 + X_3 + \cdots + X_{40} \\
E(T) &= E(X_1 + X_2 + X_3 + \cdots + X_{40}) \\
&= E(X_1) + E(X_2) + E(X_3) + \cdots + E(X_{40}) \\
&= 5.83 + 5.83 + 5.83 + \cdots + 5.83 \\
&= \$233.20
\end{aligned}
$$

Reshuffling cards between couples makes the discounts independent, so:

$$
\begin{aligned}
SD(T) &= \sqrt{Var(X_1 + X_2 + X_3 + \cdots + X_{40})} \\
&= \sqrt{Var(X_1) + Var(X_2) + Var(X_3) + \cdots + Var(X_{40})} \\
&= \sqrt{8.62^2 + 8.62^2 + 8.62^2 + \cdots + 8.62^2} \\
&\approx \$54.52
\end{aligned}
$$

The restaurant owner can expect the 40 couples to win discounts totaling $233.20, with a standard deviation of $54.52.

JUST CHECKING

2. Suppose the time it takes a customer to get and pay for seats at the ticket window of a baseball park is a random variable with a mean of 100 seconds and a standard deviation of 50 seconds. When you get there, you find only two people in line in front of you.
   a) How long do you expect to wait for your turn to get tickets?
   b) What's the standard deviation of your wait time?
   c) What assumption did you make about the two customers in finding the standard deviation?

STEP-BY-STEP EXAMPLE | **Hitting the Road: Means and Variances**

You're planning to spend next year wandering through the mountains of Kyrgyzstan. You plan to sell your used SUV so you can purchase an off-road Honda motor scooter when you get there. Used SUVs of the year and mileage of yours are selling for a mean of $6940 with a standard deviation of $250. Your research shows that scooters in Kyrgyzstan are going for about 65,000 Kyrgyzstan som with a standard deviation of 500 som. One U.S. dollar is worth about 38.5 Kyrgyzstan som (38 som and 50 tylyn).

Question: How much cash can you expect to pocket after you sell your SUV and buy the scooter?

 **THINK**	**Plan** State the problem.	I want to model how much money I'd have (in som) after selling my SUV and buying the scooter.
	Variables Define the random variables.	Let A = sale price of my SUV (in dollars), B = price of a scooter (in som), and D = profit (in som)
	Write an appropriate equation. Think about the assumptions.	$D = 38.5A - B$ ✔ **Independence Assumption:** The prices are independent.
SHOW	**Mechanics** Find the expected value, using the appropriate rules.	$E(D) = E(38.5A - B)$ $\quad = 38.5E(A) - E(B)$ $\quad = 38.5(6,940) - (65,000)$ $E(D) = 202,190$ som

Find the variance, using the appropriate rules. Be sure to check the assumptions first!	Since sale and purchase prices are independent, $$Var(D) = Var(38.5A - B)$$ $$= Var(38.5A) + Var(B)$$ $$= (38.5)^2 Var(A) + Var(B)$$ $$= 1482.25(250)^2 + (500)^2$$ $$Var(D) = 92{,}890{,}625$$
Find the standard deviation.	$$SD(D) = \sqrt{92{,}890{,}625} = 9637.98 \text{ som}$$

 **Conclusion** Interpret your results in context. (Here that means talking about dollars.)

REALITY CHECK ➤ Given the initial cost estimates, the mean and standard deviation seem reasonable.

I can expect to clear about 202,190 som ($5252) with a standard deviation of 9638 som ($250).

Continuous Random Variables

A S *Activity:* **Numeric Outcomes.** You've seen how to simulate discrete random outcomes. There's a tool for simulating continuous outcomes, too.

A S *Activity:* **Means of Random Variables.** Experiment with continuous random variables to learn how their expected values behave.

A company manufactures small stereo systems. At the end of the production line, the stereos are packaged and prepared for shipping. Stage 1 of this process is called "packing." Workers must collect all the system components (a main unit, two speakers, a power cord, an antenna, and some wires), put each in plastic bags, and then place everything inside a protective styrofoam form. The packed form then moves on to Stage 2, called "boxing." There, workers place the form and a packet of instructions in a cardboard box, close it, then seal and label the box for shipping.

The company says that times required for the packing stage can be described by a Normal model with a mean of 9 minutes and standard deviation of 1.5 minutes. The times for the boxing stage can also be modeled as Normal, with a mean of 6 minutes and standard deviation of 1 minute.

This is a common way to model events. Do our rules for random variables apply here? What's different? We no longer have a list of discrete outcomes, with their associated probabilities. Instead, we have **continuous random variables** that can take on any value. Now any single value won't have a probability. We saw this back in Chapter 6 when we first saw the Normal model (although we didn't talk then about "random variables" or "probability"). We know that the probability that $z = 1.5$ doesn't make sense, but we *can* talk about the probability that z lies *between* 0.5 and 1.5. For a Normal random variable, the probability that it falls within an interval is just the area under the Normal curve over that interval.

Some continuous random variables have Normal models; others may be skewed, uniform, or bimodal. Regardless of shape, all continuous random variables have means (which we also call *expected values*) and variances. In this book we won't worry about how to calculate them, but we can still work with models for continuous random variables when we're given these parameters.

The good news is that nearly everything we've said about how discrete random variables behave is true of continuous random variables, as well. When two independent continuous random variables have Normal models, so does their sum or difference. This simple fact is a special property of Normal models and is very important. It allows us to apply our knowledge of Normal probabilities to questions about the sum or difference of independent random variables.

STEP·BY·STEP EXAMPLE | **Packaging Stereos**

Consider the company that manufactures and ships small stereo systems that we just discussed.

Recall that times required to pack the stereos can be described by a Normal model with a mean of 9 minutes and standard deviation of 1.5 minutes. The times for the boxing stage can also be modeled as Normal, with a mean of 6 minutes and standard deviation of 1 minute.

Questions:

1. What is the probability that packing two consecutive systems takes over 20 minutes?
2. What percentage of the stereo systems take longer to pack than to box?

Question 1: What is the probability that packing two consecutive systems takes over 20 minutes?

Plan State the problem.

I want to estimate the probability that packing two consecutive systems takes over 20 minutes.

Let P_1 = time for packing the first system
P_2 = time for packing the second
T = total time to pack two systems

Variables Define your random variables.

Write an appropriate equation.

$T = P_1 + P_2$

Think about the assumptions. Sums of independent Normal random variables follow a Normal model. Such simplicity isn't true in general.

✔ **Normal Model Assumption:** We are told that both random variables follow Normal models.

✔ **Independence Assumption:** We can reasonably assume that the two packing times are independent.

Mechanics Find the expected value.

$$E(T) = E(P_1 + P_2)$$
$$= E(P_1) + E(P_2)$$
$$= 9 + 9 = 18 \text{ minutes}$$

For sums of independent random variables, variances add. (We don't need the variables to be Normal for this to be true—just independent.)

Since the times are independent,
$$Var(T) = Var(P_1 + P_2)$$
$$= Var(P_1) + Var(P_2)$$
$$= 1.5^2 + 1.5^2$$
$$Var(T) = 4.50$$

Find the standard deviation.

$$SD(T) = \sqrt{4.50} \approx 2.12 \text{ minutes}$$

Now we use the fact that both random variables follow Normal models to say that their sum is also Normal.

I'll model T with $N(18, 2.12)$.

Sketch a picture of the Normal model for the total time, shading the region representing over 20 minutes.

Find the z-score for 20 minutes.

Use technology or Table Z to find the probability.

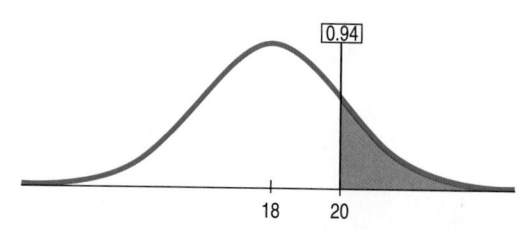

$$z = \frac{20 - 18}{2.12} = 0.94$$

$$P(T > 20) = P(z > 0.94) = 0.1736$$

Conclusion Interpret your result in context.

There's a little more than a 17% chance that it will take a total of over 20 minutes to pack two consecutive stereo systems.

Question 2: What percentage of the stereo systems take longer to pack than to box?

Plan State the question.

I want to estimate the percentage of the stereo systems that take longer to pack than to box.

Variables Define your random variables.

Let P = time for packing a system

B = time for boxing a system

D = difference in times to pack and box a system

Write an appropriate equation.

$D = P - B$

What are we trying to find? Notice that we can tell which of two quantities is greater by subtracting and asking whether the difference is positive or negative.

The probability that it takes longer to pack than to box a system is the probability that the difference $P - B$ is greater than zero.

Don't forget to think about the assumptions.

✔ **Normal Model Assumption:** We are told that both random variables follow Normal models.

✔ **Independence Assumption:** We can assume that the times it takes to pack and to box a system are independent.

Mechanics Find the expected value.

$$E(D) = E(P - B)$$
$$= E(P) - E(B)$$
$$= 9 - 6 = 3 \text{ minutes}$$

For the difference of independent random variables, variances add.	Since the times are independent, $$Var(D) = Var(P - B)$$ $$= Var(P) + Var(B)$$ $$= 1.5^2 + 1^2$$ $$Var(D) = 3.25$$
Find the standard deviation.	$$SD(D) = \sqrt{3.25} \approx 1.80 \text{ minutes}$$
State what model you will use.	I'll model D with $N(3, 1.80)$
Sketch a picture of the Normal model for the difference in times, and shade the region representing a difference greater than zero.	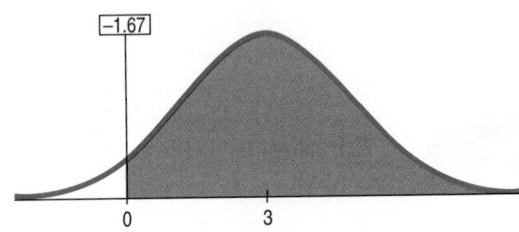
Find the z-score for 0 minutes, then use Table Z or technology to find the probability.	$$z = \frac{0 - 3}{1.80} = -1.67$$ $$P(D > 0) = P(z > -1.67) = 0.9525$$

	**Conclusion** Interpret your result in context.	About 95% of all the stereo systems will require more time for packing than for boxing.

WHAT CAN GO WRONG?

▶ **Probability models are still just models.** Models can be useful, but they are not reality. Think about the assumptions behind your models. Are your dice really perfectly fair? (They are probably pretty close.) But when you hear that the probability of a nuclear accident is 1/10,000,000 per year, is that likely to be a precise value? Question probabilities as you would data.

▶ **If the model is wrong, so is everything else.** Before you try to find the mean or standard deviation of a random variable, check to make sure the probability model is reasonable. As a start, the probabilities in your model should add up to 1. If not, you may have calculated a probability incorrectly or left out a value of the random variable. For instance, in the insurance example, the description mentions only death and disability. Good health is by far the most likely outcome, not to mention the best for both you and the insurance company (who gets to keep your money). Don't overlook that.

▶ **Don't assume everything's Normal.** Just because a random variable is continuous or you happen to know a mean and standard deviation doesn't mean that a Normal model will be useful. You must *Think* about whether the **Normality Assumption** is justified. Using a Normal model when it really does not apply will lead to wrong answers and misleading conclusions.

To find the expected value of the sum or difference of random variables, we simply add or subtract means. Center is easy; spread is trickier. Watch out for some common traps.

▶ **Watch out for variables that aren't independent.** You can add expected values of *any* two random variables, but you can only add variances of independent random variables. Suppose a survey includes questions about the number of hours of sleep people get each night and also the number of hours they are awake each day. From their answers, we find the mean and standard deviation of hours asleep and hours awake. The expected total must be 24 hours; after all, people are either asleep or awake.[2] The means still add just fine. Since all the totals are exactly 24 hours, however, the standard deviation of the total will be 0. We can't add variances here because the number of hours you're awake depends on the number of hours you're asleep. Be sure to check for independence before adding variances.

▶ **Don't forget: Variances of independent random variables add. Standard deviations don't.**

▶ **Don't forget: Variances of independent random variables add, even when you're looking at the difference between them.**

▶ **Don't write independent instances of a random variable with notation that looks like they are the same variables.** Make sure you write each instance as a different random variable. Just because each random variable describes a similar situation doesn't mean that each random outcome will be the same. These are *random* variables, not the variables you saw in Algebra. Write $X_1 + X_2 + X_3$ rather than $X + X + X$.

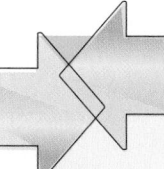

CONNECTIONS

We've seen means, variances, and standard deviations of data. We know that they estimate parameters of models for these data. Now we're looking at the probability models directly. We have only parameters because there are no data to summarize.

It should be no surprise that expected values and standard deviations adjust to shifts and changes of units in the same way as the corresponding data summaries. The fact that we can add variances of independent random quantities is fundamental and will explain why a number of statistical methods work the way they do.

WHAT HAVE WE LEARNED?

We've learned to work with random variables. We can use the probability model for a discrete random variable to find its expected value and its standard deviation.

We've learned that the mean of the sum or difference of two random variables, discrete or continuous, is just the sum or difference of their means. And we've learned the Pythagorean Theorem of Statistics: *For independent random variables*, the variance of their sum or difference is always the *sum* of their variances.

Finally, we've learned that Normal models are once again special. Sums or differences of Normally distributed random variables also follow Normal models.

---

[2] Although some students do manage to attain a state of consciousness somewhere between sleeping and wakefulness during Statistics class.

Terms

Random variable	366. A random variable assumes any of several different numeric values as a result of some random event. Random variables are denoted by a capital letter such as X.
Discrete random variable	366. A random variable that can take one of a finite number[3] of distinct outcomes is called a discrete random variable.
Continuous random variable	366, 367. A random variable that can take any numeric value within a range of values is called a continuous random variable. The range may be infinite or bounded at either or both ends.
Probability model	366. The probability model is a function that associates a probability P with each value of a discrete random variable X, denoted $P(X = x)$, or with any interval of values of a continuous random variable.
Expected value	367. The expected value of a random variable is its theoretical long-run average value, the center of its model. Denoted μ or $E(X)$, it is found (if the random variable is discrete) by summing the products of variable values and probabilities:

$$\mu = E(X) = \sum xP(x).$$

Variance	369. The variance of a random variable is the expected value of the squared deviation from the mean. For discrete random variables, it can be calculated as:

$$\sigma^2 = Var(X) = \sum (x - \mu)^2 P(x).$$

Standard deviation	369. The standard deviation of a random variable describes the spread in the model, and is the square root of the variance:

$$\sigma = SD(X) = \sqrt{Var(X)}.$$

Changing a random variable by a constant:	372. $E(X \pm c) = E(X) \pm c$ $\qquad$ $Var(X \pm c) = Var(X)$
	373. $\qquad E(aX) = aE(X)$ $\qquad\qquad$ $Var(aX) = a^2 Var(X)$
Adding or subtracting random variables:	373. $\qquad\qquad\qquad\qquad E(X \pm Y) = E(X) \pm E(Y)$
	374. *If X and Y are independent,* $Var(X \pm Y) = Var(X) + Var(Y)$.
	374. $\qquad\qquad\qquad$ (The Pythagorean Theorem of Statistics)

Skills

- ▸ Be able to recognize random variables.
- ▸ Understand that random variables must be independent in order to determine the variability of their sum or difference by adding variances.

- ▸ Be able to find the probability model for a discrete random variable.
- ▸ Know how to find the mean (expected value) and the variance of a random variable.
- ▸ Always use the proper notation for these population parameters: μ or $E(X)$ for the mean, and σ, $SD(X)$, σ^2, or $Var(X)$ when discussing variability.
- ▸ Know how to determine the new mean and standard deviation after adding a constant, multiplying by a constant, or adding or subtracting two independent random variables.

- ▸ Be able to interpret the meaning of the expected value and standard deviation of a random variable in the proper context.

---

[3] Technically, there could be an infinite number of outcomes, as long as they're *countable*. Essentially that means we can imagine listing them all in order, like the counting numbers 1, 2, 3, 4, 5, . . .

RANDOM VARIABLES ON THE COMPUTER

Statistics packages deal with data, not with random variables. Nevertheless, the calculations needed to find means and standard deviations of random variables are little more than weighted means. Most packages can manage that, but then they are just being overblown calculators. For technological assistance with these calculations, we recommend you pull out your calculator.

EXERCISES

1. **Expected value.** Find the expected value of each random variable:

   a)
x	10	20	30
$P(X=x)$	0.3	0.5	0.2

   b)
x	2	4	6	8
$P(X=x)$	0.3	0.4	0.2	0.1

2. **Expected value.** Find the expected value of each random variable:

   a)
x	0	1	2
$P(X=x)$	0.2	0.4	0.4

   b)
x	100	200	300	400
$P(X=x)$	0.1	0.2	0.5	0.2

3. **Pick a card, any card.** You draw a card from a deck. If you get a red card, you win nothing. If you get a spade, you win $5. For any club, you win $10 plus an extra $20 for the ace of clubs.
   a) Create a probability model for the amount you win.
   b) Find the expected amount you'll win.
   c) What would you be willing to pay to play this game?

4. **You bet!** You roll a die. If it comes up a 6, you win $100. If not, you get to roll again. If you get a 6 the second time, you win $50. If not, you lose.
   a) Create a probability model for the amount you win.
   b) Find the expected amount you'll win.
   c) What would you be willing to pay to play this game?

5. **Kids.** A couple plans to have children until they get a girl, but they agree that they will not have more than three children even if all are boys. (Assume boys and girls are equally likely.)
   a) Create a probability model for the number of children they might have.
   b) Find the expected number of children.
   c) Find the expected number of boys they'll have.

6. **Carnival.** A carnival game offers a $100 cash prize for anyone who can break a balloon by throwing a dart at it. It costs $5 to play, and you're willing to spend up to $20 trying to win. You estimate that you have about a 10% chance of hitting the balloon on any throw.

   a) Create a probability model for this carnival game.
   b) Find the expected number of darts you'll throw.
   c) Find your expected winnings.

7. **Software.** A small software company bids on two contracts. It anticipates a profit of $50,000 if it gets the larger contract and a profit of $20,000 on the smaller contract. The company estimates there's a 30% chance it will get the larger contract and a 60% chance it will get the smaller contract. Assuming the contracts will be awarded independently, what's the expected profit?

8. **Racehorse.** A man buys a racehorse for $20,000 and enters it in two races. He plans to sell the horse afterward, hoping to make a profit. If the horse wins both races, its value will jump to $100,000. If it wins one of the races, it will be worth $50,000. If it loses both races, it will be worth only $10,000. The man believes there's a 20% chance that the horse will win the first race and a 30% chance it will win the second one. Assuming that the two races are independent events, find the man's expected profit.

9. **Variation 1.** Find the standard deviations of the random variables in Exercise 1.

10. **Variation 2.** Find the standard deviations of the random variables in Exercise 2.

11. **Pick another card.** Find the standard deviation of the amount you might win drawing a card in Exercise 3.

12. **The die.** Find the standard deviation of the amount you might win rolling a die in Exercise 4.

13. **Kids again.** Find the standard deviation of the number of children the couple in Exercise 5 may have.

14. **Darts.** Find the standard deviation of your winnings throwing darts in Exercise 6.

15. **Repairs.** The probability model below describes the number of repair calls that an appliance repair shop may receive during an hour.

Repair Calls	0	1	2	3
Probability	0.1	0.3	0.4	0.2

   a) How many calls should the shop expect per hour?
   b) What is the standard deviation?

16. Red lights. A commuter must pass through five traffic lights on her way to work and will have to stop at each one that is red. She estimates the probability model for the number of red lights she hits, as shown below.

X = # of red	0	1	2	3	4	5
$P(X = x)$	0.05	0.25	0.35	0.15	0.15	0.05

a) How many red lights should she expect to hit each day?
b) What's the standard deviation?

17. Defects. A consumer organization inspecting new cars found that many had appearance defects (dents, scratches, paint chips, etc.). While none had more than three of these defects, 7% had three, 11% two, and 21% one defect. Find the expected number of appearance defects in a new car and the standard deviation.

18. Insurance. An insurance policy costs $100 and will pay policyholders $10,000 if they suffer a major injury (resulting in hospitalization) or $3000 if they suffer a minor injury (resulting in lost time from work). The company estimates that each year 1 in every 2000 policyholders may have a major injury, and 1 in 500 a minor injury only.
a) Create a probability model for the profit on a policy.
b) What's the company's expected profit on this policy?
c) What's the standard deviation?

19. Cancelled flights. Mary is deciding whether to book the cheaper flight home from college after her final exams, but she's unsure when her last exam will be. She thinks there is only a 20% chance that the exam will be scheduled after the last day she can get a seat on the cheaper flight. If it is and she has to cancel the flight, she will lose $150. If she can take the cheaper flight, she will save $100.
a) If she books the cheaper flight, what can she expect to gain, on average?
b) What is the standard deviation?

20. Day trading. An option to buy a stock is priced at $200. If the stock closes above 30 on May 15, the option will be worth $1000. If it closes below 20, the option will be worth nothing, and if it closes between 20 and 30 (inclusively), the option will be worth $200. A trader thinks there is a 50% chance that the stock will close in the 20–30 range, a 20% chance that it will close above 30, and a 30% chance that it will fall below 20 on May 15.
a) Should she buy the stock option?
b) How much does she expect to gain?
c) What is the standard deviation of her gain?

21. Contest. You play two games against the same opponent. The probability you win the first game is 0.4. If you win the first game, the probability you also win the second is 0.2. If you lose the first game, the probability that you win the second is 0.3.
a) Are the two games independent? Explain.
b) What's the probability you lose both games?
c) What's the probability you win both games?
d) Let random variable X be the number of games you win. Find the probability model for X.
e) What are the expected value and standard deviation?

22. Contracts. Your company bids for two contracts. You believe the probability you get contract #1 is 0.8. If you get contract #1, the probability you also get contract #2 will be 0.2, and if you do not get #1, the probability you get #2 will be 0.3.
a) Are the two contracts independent? Explain.
b) Find the probability you get both contracts.
c) Find the probability you get no contract.
d) Let X be the number of contracts you get. Find the probability model for X.
e) Find the expected value and standard deviation.

23. Batteries. In a group of 10 batteries, 3 are dead. You choose 2 batteries at random.
a) Create a probability model for the number of good batteries you get.
b) What's the expected number of good ones you get?
c) What's the standard deviation?

24. Kittens. In a litter of seven kittens, three are female. You pick two kittens at random.
a) Create a probability model for the number of male kittens you get.
b) What's the expected number of males?
c) What's the standard deviation?

25. Random variables. Given independent random variables with means and standard deviations as shown, find the mean and standard deviation of:
a) $3X$
b) $Y + 6$
c) $X + Y$
d) $X - Y$
e) $X_1 + X_2$

	Mean	SD
X	10	2
Y	20	5

26. Random variables. Given independent random variables with means and standard deviations as shown, find the mean and standard deviation of:
a) $X - 20$
b) $0.5Y$
c) $X + Y$
d) $X - Y$
e) $Y_1 + Y_2$

	Mean	SD
X	80	12
Y	12	3

27. Random variables. Given independent random variables with means and standard deviations as shown, find the mean and standard deviation of:
a) $0.8Y$
b) $2X - 100$
c) $X + 2Y$
d) $3X - Y$
e) $Y_1 + Y_2$

	Mean	SD
X	120	12
Y	300	16

28. Random variables. Given independent random variables with means and standard deviations as shown, find the mean and standard deviation of:
a) $2Y + 20$
b) $3X$
c) $0.25X + Y$
d) $X - 5Y$
e) $X_1 + X_2 + X_3$

	Mean	SD
X	80	12
Y	12	3

29. Eggs. A grocery supplier believes that in a dozen eggs, the mean number of broken ones is 0.6 with a standard

deviation of 0.5 eggs. You buy 3 dozen eggs without checking them.
a) How many broken eggs do you expect to get?
b) What's the standard deviation?
c) What assumptions did you have to make about the eggs in order to answer this question?

30. **Garden.** A company selling vegetable seeds in packets of 20 estimates that the mean number of seeds that will actually grow is 18, with a standard deviation of 1.2 seeds. You buy 5 different seed packets.
a) How many bad seeds do you expect to get?
b) What's the standard deviation?
c) What assumptions did you make about the seeds? Do you think that assumption is warranted? Explain.

31. **Repair calls.** Find the mean and standard deviation of the number of repair calls the appliance shop in Exercise 15 should expect during an 8-hour day.

32. **Stop!** Find the mean and standard deviation of the number of red lights the commuter in Exercise 16 should expect to hit on her way to work during a 5-day work week.

33. **Tickets.** A delivery company's trucks occasionally get parking tickets, and based on past experience, the company plans that the trucks will average 1.3 tickets a month, with a standard deviation of 0.7 tickets.
a) If they have 18 trucks, what are the mean and standard deviation of the total number of parking tickets the company will have to pay this month?
b) What assumption did you make in answering?

34. **Donations.** Organizers of a televised fundraiser know from past experience that most people donate small amounts ($10–$25), some donate larger amounts ($50–$100), and a few people make very generous donations of $250, $500, or more. Historically, pledges average about $32 with a standard deviation of $54.
a) If 120 people call in pledges, what are the mean and standard deviation of the total amount raised?
b) What assumption did you make in answering this question?

35. **Fire!** An insurance company estimates that it should make an annual profit of $150 on each homeowner's policy written, with a standard deviation of $6000.
a) Why is the standard deviation so large?
b) If it writes only two of these policies, what are the mean and standard deviation of the annual profit?
c) If it writes 10,000 of these policies, what are the mean and standard deviation of the annual profit?
d) Is the company likely to be profitable? Explain.
e) What assumptions underlie your analysis? Can you think of circumstances under which those assumptions might be violated? Explain.

36. **Casino.** A casino knows that people play the slot machines in hopes of hitting the jackpot but that most of them lose their dollar. Suppose a certain machine pays out an average of $0.92, with a standard deviation of $120.
a) Why is the standard deviation so large?
b) If you play 5 times, what are the mean and standard deviation of the casino's profit?

c) If gamblers play this machine 1000 times in a day, what are the mean and standard deviation of the casino's profit?
d) Is the casino likely to be profitable? Explain.

37. **Cereal.** The amount of cereal that can be poured into a small bowl varies with a mean of 1.5 ounces and a standard deviation of 0.3 ounces. A large bowl holds a mean of 2.5 ounces with a standard deviation of 0.4 ounces. You open a new box of cereal and pour one large and one small bowl.
a) How much more cereal do you expect to be in the large bowl?
b) What's the standard deviation of this difference?
c) If the difference follows a Normal model, what's the probability the small bowl contains more cereal than the large one?
d) What are the mean and standard deviation of the total amount of cereal in the two bowls?
e) If the total follows a Normal model, what's the probability you poured out more than 4.5 ounces of cereal in the two bowls together?
f) The amount of cereal the manufacturer puts in the boxes is a random variable with a mean of 16.3 ounces and a standard deviation of 0.2 ounces. Find the expected amount of cereal left in the box and the standard deviation.

38. **Pets.** The American Veterinary Association claims that the annual cost of medical care for dogs averages $100, with a standard deviation of $30, and for cats averages $120, with a standard deviation of $35.
a) What's the expected difference in the cost of medical care for dogs and cats?
b) What's the standard deviation of that difference?
c) If the costs can be described by Normal models, what's the probability that medical expenses are higher for someone's dog than for her cat?
d) What concerns do you have?

39. **More cereal.** In Exercise 37 we poured a large and a small bowl of cereal from a box. Suppose the amount of cereal that the manufacturer puts in the boxes is a random variable with mean 16.2 ounces and standard deviation 0.1 ounces.
a) Find the expected amount of cereal left in the box.
b) What's the standard deviation?
c) If the weight of the remaining cereal can be described by a Normal model, what's the probability that the box still contains more than 13 ounces?

40. **More pets.** You're thinking about getting two dogs and a cat. Assume that annual veterinary expenses are independent and have a Normal model with the means and standard deviations described in Exercise 38.
a) Define appropriate variables and express the total annual veterinary costs you may have.
b) Describe the model for this total cost. Be sure to specify its name, expected value, and standard deviation.
c) What's the probability that your total expenses will exceed $400?

41. **Medley.** In the 4 × 100 medley relay event, four swimmers swim 100 yards, each using a different stroke. A

college team preparing for the conference championship looks at the times their swimmers have posted and creates a model based on the following assumptions:

- The swimmers' performances are independent.
- Each swimmer's times follow a Normal model.
- The means and standard deviations of the times (in seconds) are as shown:

Swimmer	Mean	SD
1 (backstroke)	50.72	0.24
2 (breaststroke)	55.51	0.22
3 (butterfly)	49.43	0.25
4 (freestyle)	44.91	0.21

a) What are the mean and standard deviation for the relay team's total time in this event?
b) The team's best time so far this season was 3:19.48. (That's 199.48 seconds.) Do you think the team is likely to swim faster than this at the conference championship? Explain.

42. Bikes. Bicycles arrive at a bike shop in boxes. Before they can be sold, they must be unpacked, assembled, and tuned (lubricated, adjusted, etc.). Based on past experience, the shop manager makes the following assumptions about how long this may take:

- The times for each setup phase are independent.
- The times for each phase follow a Normal model.
- The means and standard deviations of the times (in minutes) are as shown:

Phase	Mean	SD
Unpacking	3.5	0.7
Assembly	21.8	2.4
Tuning	12.3	2.7

a) What are the mean and standard deviation for the total bicycle setup time?
b) A customer decides to buy a bike like one of the display models but wants a different color. The shop has one, still in the box. The manager says they can have it ready in half an hour. Do you think the bike will be set up and ready to go as promised? Explain.

43. Farmers' market. A farmer has 100 lb of apples and 50 lb of potatoes for sale. The market price for apples (per pound) each day is a random variable with a mean of 0.5 dollars and a standard deviation of 0.2 dollars. Similarly, for a pound of potatoes, the mean price is 0.3 dollars and the standard deviation is 0.1 dollars. It also costs him 2 dollars to bring all the apples and potatoes to the market. The market is busy with eager shoppers, so we can assume that he'll be able to sell all of each type of produce at that day's price.

a) Define your random variables, and use them to express the farmer's net income.
b) Find the mean.
c) Find the standard deviation of the net income.
d) Do you need to make any assumptions in calculating the mean? How about the standard deviation?

44. Bike sale. The bicycle shop in Exercise 42 will be offering 2 specially priced children's models at a sidewalk sale. The basic model will sell for $120 and the deluxe model for $150. Past experience indicates that sales of the basic model will have a mean of 5.4 bikes with a standard deviation of 1.2, and sales of the deluxe model will have a mean of 3.2 bikes with a standard deviation of 0.8 bikes. The cost of setting up for the sidewalk sale is $200.

a) Define random variables and use them to express the bicycle shop's net income.
b) What's the mean of the net income?
c) What's the standard deviation of the net income?
d) Do you need to make any assumptions in calculating the mean? How about the standard deviation?

45. Coffee and doughnuts. At a certain coffee shop, all the customers buy a cup of coffee; some also buy a doughnut. The shop owner believes that the number of cups he sells each day is normally distributed with a mean of 320 cups and a standard deviation of 20 cups. He also believes that the number of doughnuts he sells each day is independent of the coffee sales and is normally distributed with a mean of 150 doughnuts and a standard deviation of 12.

a) The shop is open every day but Sunday. Assuming day-to-day sales are independent, what's the probability he'll sell over 2000 cups of coffee in a week?
b) If he makes a profit of 50 cents on each cup of coffee and 40 cents on each doughnut, can he reasonably expect to have a day's profit of over $300? Explain.
c) What's the probability that on any given day he'll sell a doughnut to more than half of his coffee customers?

46. Weightlifting. The Atlas BodyBuilding Company (ABC) sells "starter sets" of barbells that consist of one bar, two 20-pound weights, and four 5-pound weights. The bars weigh an average of 10 pounds with a standard deviation of 0.25 pounds. The weights average the specified amounts, but the standard deviations are 0.2 pounds for the 20-pounders and 0.1 pounds for the 5-pounders. We can assume that all the weights are normally distributed.

a) ABC ships these starter sets to customers in two boxes: The bar goes in one box and the six weights go in another. What's the probability that the total weight in that second box exceeds 60.5 pounds? Define your variables clearly and state any assumptions you make.
b) It costs ABC $0.40 per pound to ship the box containing the weights. Because it's an odd-shaped package, though, shipping the bar costs $0.50 a pound plus a $6.00 surcharge. Find the mean and standard deviation of the company's total cost for shipping a starter set.
c) Suppose a customer puts a 20-pound weight at one end of the bar and the four 5-pound weights at the other end. Although he expects the two ends to weigh the same, they might differ slightly. What's the probability the difference is more than a quarter of a pound?

JUST CHECKING
Answers

1. a)

Outcome	X = cost	Probability
Recharging works	$60	0.75
Replace control unit	$200	0.25

b) 60(0.75) + 200(0.25) = $95

c) Car owners with this problem will spend an average of $95 to get it fixed.

2. a) 100 + 100 = 200 seconds

b) $\sqrt{50^2 + 50^2} = 70.7$ seconds

c) The times for the two customers are independent.

Probability Models

Suppose a cereal manufacturer puts pictures of famous athletes on cards in boxes of cereal, in the hope of increasing sales. The manufacturer announces that 20% of the boxes contain a picture of Tiger Woods, 30% a picture of David Beckham, and the rest a picture of Serena Williams.

Sound familiar? In Chapter 11 we simulated to find the number of boxes we'd need to open to get one of each card. That's a fairly complex question and one well suited for simulation. But many important questions can be answered more directly by using simple probability models.

Searching for Tiger

You're a huge Tiger Woods fan. You don't care about completing the whole sports card collection, but you've just *got* to have the Tiger Woods picture. How many boxes do you expect you'll have to open before you find him? This isn't the same question that we asked before, but this situation is simple enough for a probability model.

We'll keep the assumption that pictures are distributed at random and we'll trust the manufacturer's claim that 20% of the cards are Tiger. So, when you open the box, the probability that you succeed in finding Tiger is 0.20. Now we'll call the act of opening *each* box a trial, and note that:

▶ There are only two possible outcomes (called *success* and *failure*) on each trial. Either you get Tiger's picture (success), or you don't (failure).

▶ In advance, the probability of success, denoted p, is the same on every trial. Here $p = 0.20$ for each box.

▶ As we proceed, the trials are independent. Finding Tiger in the first box does not change what might happen when you reach for the next box.

Situations like this occur often, and are called **Bernoulli trials.** Common examples of Bernoulli trials include tossing a coin, looking for defective products rolling off an assembly line, or even shooting free throws in a basketball game. Just as we found equally likely random digits to be the building blocks for our simulation, we can use Bernoulli trials to build a wide variety of useful probability models.

Daniel Bernoulli (1700–1782) was the nephew of Jacob, whom you saw in Chapter 14. He was the first to work out the mathematics for what we now call Bernoulli trials.

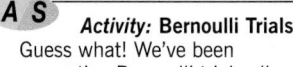
Activity: Bernoulli Trials. Guess what! We've been generating Bernoulli trials all along. Look at the Random Simulation Tool in a new way.

Back to Tiger. We want to know how many boxes we'll need to open to find his card. Let's call this random variable Y = # boxes, and build a probability model for it. What's the probability you find his picture in the first box of cereal? It's 20%, of course. We could write $P(Y = 1) = 0.20$.

How about the probability that you don't find Tiger until the second box? Well, that means you fail on the first trial and then succeed on the second. With the probability of success 20%, the probability of failure, denoted q, is $1 - 0.2 = 80\%$. Since the trials are independent, the probability of getting your first success on the second trial is $P(Y = 2) = (0.8)(0.2) = 0.16$.

Of course, you could have a run of bad luck. Maybe you won't find Tiger until the fifth box of cereal. What are the chances of that? You'd have to fail 4 straight times and then succeed, so $P(Y = 5) = (0.8)^4(0.2) = 0.08192$.

How many boxes might you expect to have to open? We could reason that since Tiger's picture is in 20% of the boxes, or 1 in 5, we expect to find his picture, on average, in the fifth box; that is, $E(Y) = \frac{1}{0.2} = 5$ boxes. That's correct, but not easy to prove.

The Geometric Model

Geometric probabilities. See what happens to a geometric model as you change the probability of success.

We want to model how long it will take to achieve the first success in a series of Bernoulli trials. The model that tells us this probability is called the **Geometric probability model.** Geometric models are completely specified by one parameter, p, the probability of success, and are denoted Geom(p). Since achieving the first success on trial number x requires first experiencing $x - 1$ failures, the probabilities are easily expressed by a formula.

NOTATION ALERT:

Now we have two more reserved letters. Whenever we deal with Bernoulli trials, p represents the probability of success, and q the probability of failure. (Of course, $q = 1 - p$.)

GEOMETRIC PROBABILITY MODEL FOR BERNOULLI TRIALS: Geom(p)

p = probability of success (and $q = 1 - p$ = probability of failure)
X = number of trials until the first success occurs

$$P(X = x) = q^{x-1}p$$

Expected value: $E(X) = \mu = \dfrac{1}{p}$

Standard deviation: $\sigma = \sqrt{\dfrac{q}{p^2}}$

FOR EXAMPLE Spam and the Geometric model

Postini is a global company specializing in communications security. The company monitors over 1 billion Internet messages per day and recently reported that 91% of e-mails are spam!

Let's assume that your e-mail is typical—91% spam. We'll also assume you aren't using a spam filter, so every message gets dumped in your inbox. And, since spam comes from many different sources, we'll consider your messages to be independent.

Questions: Overnight your inbox collects e-mail. When you first check your e-mail in the morning, about how many spam e-mails should you expect to have to wade through and discard before you find a real message? What's the probability that the 4th message in your inbox is the first one that isn't spam?

There are two outcomes: a real message (success) and spam (failure). Since 91% of e-mails are spam, the probability of success $p = 1 - 0.91 = 0.09$.

Let X = the number of e-mails I'll check until I find a real message. I assume that the messages arrive independently and in a random order. I can use the model Geom(0.09).

$$E(X) = \frac{1}{p} = \frac{1}{0.09} = 11.1$$

$$P(X = 4) = (0.91)^3(0.09) = 0.0678$$

On average, I expect to have to check just over 11 e-mails before I find a real message. There's slightly less than a 7% chance that my first real message will be the 4th one I check.

Note that the probability calculation isn't new. It's simply Chapter 14's Multiplication Rule used to find $P(\text{spam} \cap \text{spam} \cap \text{spam} \cap \text{real})$.

MATH BOX

We want to find the mean (expected value) of random variable X, using a geometric model with probability of success p.

First, write the probabilities:

x	1	2	3	4	$\cdots$
$P(X = x)$	p	qp	q^2p	q^3p	$\cdots$

The expected value is: $E(X) = 1p + 2qp + 3q^2p + 4q^3p + \cdots$

Let $p = 1 - q$: $\qquad = (1 - q) + 2q(1 - q) + 3q^2(1 - q) + 4q^3(1 - q) + \cdots$

Simplify: $\qquad = 1 - q + 2q - 2q^2 + 3q^2 - 3q^3 + 4q^3 - 4q^4 + \cdots$

That's an infinite geometric series, with first term 1 and common ratio q:

$\qquad = 1 + q + q^2 + q^3 + \cdots$

$\qquad = \dfrac{1}{1 - q}$

So, finally . . . $\qquad E(X) = \dfrac{1}{p}$.

Independence

One of the important requirements for Bernoulli trials is that the trials be independent. Sometimes that's a reasonable assumption—when tossing a coin or rolling a die, for example. But that becomes a problem when (often!) we're looking at situations involving samples chosen without replacement. We said that whether we find a Tiger Woods card in one box has no effect on the probabilities in other boxes. This is *almost* true. Technically, if exactly 20% of the boxes have Tiger Woods cards, then when you find one, you've reduced the number of remaining Tiger Woods cards. If you knew there were 2 Tiger Woods cards hiding in the 10 boxes of cereal on the market shelf, then finding one in the first box you try would clearly change your chances of finding Tiger in the next box. With a few million boxes of cereal, though, the difference is hardly worth mentioning.

If we had an infinite number of boxes, there wouldn't be a problem. It's selecting from a finite population that causes the probabilities to change, making the trials not independent. Obviously, taking 2 out of 10 boxes changes the probability. Taking even a few hundred out of millions, though, makes very little difference. Fortunately, we have a rule of thumb for the in-between cases. It turns out that if we look at less than 10% of the population, we can pretend that the trials are independent and still calculate probabilities that are quite accurate.

> **The 10% Condition:** Bernoulli trials must be independent. If that assumption is violated, it is still okay to proceed as long as the sample is smaller than 10% of the population.

STEP-BY-STEP EXAMPLE | **Working with a Geometric Model**

People with O-negative blood are called "universal donors" because O-negative blood can be given to anyone else, regardless of the recipient's blood type. Only about 6% of people have O-negative blood.

Questions:
1. If donors line up at random for a blood drive, how many do you expect to examine before you find someone who has O-negative blood?
2. What's the probability that the first O-negative donor found is one of the first four people in line?

 **THINK**	**Plan** State the questions.	I want to estimate how many people I'll need to check to find an O-negative donor, and the probability that 1 of the first 4 people is O-negative.

	Check to see that these are Bernoulli trials.	✔ There are two outcomes:
		    success = O-negative
		    failure = other blood types
		✔ The probability of success for each person is $p = 0.06$, because they lined up randomly.
		✔ **10% Condition:** Trials aren't independent because the population is finite, but the donors lined up are fewer than 10% of all possible donors.

	Variable Define the random variable.	Let X = number of donors until one is O-negative.
	Model Specify the model.	I can model X with Geom(0.06).

Mechanics Find the mean.	$$E(X) = \frac{1}{0.06} \approx 16.7$$	

Mechanics Find the mean.

Calculate the probability of success on one of the first four trials. That's the probability that $X = 1, 2, 3,$ *or* 4.

$$E(X) = \frac{1}{0.06} \approx 16.7$$

$$P(X \leq 4) = P(X = 1) + P(X = 2) +$$
$$P(X = 3) + P(X = 4)$$
$$= (0.06) + (0.94)(0.06) +$$
$$(0.94)^2(0.06) + (0.94)^3(0.06)$$
$$\approx 0.2193$$

Conclusion Interpret your results in context.

Blood drives such as this one expect to examine an average of 16.7 people to find a universal donor. About 22% of the time there will be one within the first 4 people in line.

TI TIPS Finding geometric probabilities

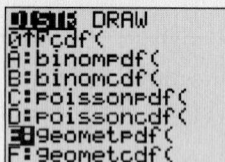

Your TI knows the geometric model. Just as you saw back in Chapter 6 with the Normal model, commands to calculate probability distributions are found in the 2nd DISTR menu. Have a look. After many others (Don't drop the course yet!) you'll see two Geometric probability functions at the bottom of the list.

- **geometpdf(.**

  The "pdf" stands for "probability density function." This command allows you to find the probability of any *individual* outcome. You need only specify *p*, which defines the Geometric model, and *x*, which indicates the number of trials until you get a success. The format is geometpdf(p, x).

  For example, suppose we want to know the probability that we find our first Tiger Woods picture in the fifth box of cereal. Since Tiger is in 20% of the boxes, we use $p = 0.2$ and $x = 5$, entering the command geometpdf(.2, 5). The calculator says there's about an 8% chance.

- **geometcdf(.**

  This is the "cumulative density function," meaning that it finds the sum of the probabilities of several possible outcomes. In general, the command geometcdf(p, x) calculates the probability of finding the first success *on or before* the *x*th trial.

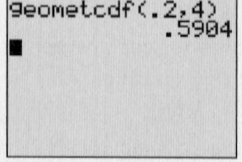

  Let's find the probability of getting a Tiger Woods picture by the time we open the fourth box of cereal—in other words, the probability our first success comes on the first box, or the second, or the third, or the fourth. Again we specify $p = 0.2$, and now use $x = 4$. The command geometcdf(.2, 4) calculates all the probabilities and adds them. There's about a 59% chance that our quest for a Tiger Woods photo will succeed by the time we open the fourth box.

The Binomial Model

We can use the Bernoulli trials to answer other questions. Suppose you buy 5 boxes of cereal. What's the probability you get *exactly* 2 pictures of Tiger Woods? Before, we asked how long it would take until our first success. Now we want to find the probability of getting 2 successes among the 5 trials. We are still talking about Bernoulli trials, but we're asking a different question.

This time we're interested in the *number of successes* in the 5 trials, so we'll call it X = number of successes. We want to find $P(X = 2)$. This is an example of a **Binomial probability.** It takes two parameters to define this **Binomial model:** the number of trials, n, and the probability of success, p. We denote this model Binom(n, p). Here, $n = 5$ trials, and $p = 0.2$, the probability of finding a Tiger Woods card in any trial.

Exactly 2 successes in 5 trials means 2 successes and 3 failures. It seems logical that the probability should be $(0.2)^2(0.8)^3$. Too bad! It's not that easy. That calculation would give you the probability of finding Tiger in the first 2 boxes and not in the next 3—*in that order*. But you could find Tiger in the third and fifth boxes and still have 2 successes. The probability of those outcomes in that particular order is $(0.8)(0.8)(0.2)(0.8)(0.2)$. That's also $(0.2)^2(0.8)^3$. In fact, the probability will always be the same, no matter what order the successes and failures occur in. Anytime we get 2 successes in 5 trials, no matter what the order, the probability will be $(0.2)^2(0.8)^3$. We just need to take account of all the possible orders in which the outcomes can occur.

Fortunately, these possible orders are *disjoint*. (For example, if your two successes came on the first two trials, they couldn't come on the last two.) So we can use the Addition Rule and add up the probabilities for all the possible orderings. Since the probabilities are all the same, we only need to know how many orders are possible. For small numbers, we can just make a tree diagram and count the branches. For larger numbers this isn't practical, so we let the computer or calculator do the work.

Each different order in which we can have k successes in n trials is called a "combination." The total number of ways that can happen is written $\binom{n}{k}$ or $_nC_k$ and pronounced "n choose k."

$$\binom{n}{k} = \frac{n!}{k!(n-k)!} \text{ where } n! \text{ (pronounced "}n \text{ factorial") } = n \times (n-1) \times \cdots \times 1$$

For 2 successes in 5 trials,

$$\binom{5}{2} = \frac{5!}{2!(5-2)!} = \frac{5 \times 4 \times 3 \times 2 \times 1}{2 \times 1 \times 3 \times 2 \times 1} = \frac{5 \times 4}{2 \times 1} = 10.$$

So there are 10 ways to get 2 Tiger pictures in 5 boxes, and the probability of each is $(0.2)^2(0.8)^3$. Now we can find what we wanted:

$$P(\#\text{success} = 2) = 10(0.2)^2(0.8)^3 = 0.2048$$

In general, the probability of exactly k successes in n trials is $\binom{n}{k} p^k q^{n-k}$.

Using this formula, we could find the expected value by adding up $xP(X = x)$ for all values, but it would be a long, hard way to get an answer that you already know intuitively. What's the expected value? If we have 5 boxes, and Tiger's picture is in 20% of them, then we would expect to have $5(0.2) = 1$ success. If we had 100 trials with probability of success 0.2, how many successes would you expect? Can you think of any reason not to say 20? It seems so simple that most people wouldn't even stop to think about it. You just multiply the probability of success by n. In other words, $E(X) = np$. Not fully convinced? We prove it in the next Math Box.

The standard deviation is less obvious; you can't just rely on your intuition. Fortunately, the formula for the standard deviation also boils down to something simple: $SD(X) = \sqrt{npq}$. (If you're curious about where that comes from, it's in the Math Box too!) In 100 boxes of cereal, we expect to find 20 Tiger Woods cards, with a standard deviation of $\sqrt{100 \times 0.8 \times 0.2} = 4$ pictures.

Time to summarize. A Binomial probability model describes the number of successes in a specified number of trials. It takes two parameters to specify this model: the number of trials n and the probability of success p.

BINOMIAL PROBABILITY MODEL FOR BERNOULLI TRIALS: Binom(n, p)

n = number of trials
p = probability of success (and $q = 1 - p$ = probability of failure)
X = number of successes in n trials

$$P(X = x) = {}_nC_x\, p^x q^{n-x}, where\ {}_nC_x = \frac{n!}{x!(n-x)!}$$

Mean: $\mu = np$
Standard Deviation: $\sigma = \sqrt{npq}$

MATH BOX

To derive the formulas for the mean and standard deviation of a Binomial model we start with the most basic situation.

Consider a single Bernoulli trial with probability of success p. Let's find the mean and variance of the number of successes.

Here's the probability model for the number of successes:

x	0	1
$P(X = x)$	q	p

Find the expected value:

$E(X) = 0q + 1p$
$E(X) = p$

And now the variance:

$Var(X) = (0 - p)^2 q + (1 - p)^2 p$
$\qquad = p^2 q + q^2 p$
$\qquad = pq(p + q)$
$\qquad = pq(1)$
$Var(X) = pq$

What happens when there is more than one trial, though? A Binomial model simply counts the number of successes in a series of n independent Bernoulli trials. That makes it easy to find the mean and standard deviation of a binomial random variable, Y.

$\text{Let } Y = X_1 + X_2 + X_3 + \cdots + X_n$
$E(Y) = E(X_1 + X_2 + X_3 + \cdots + X_n)$
$\qquad = E(X_1) + E(X_2) + E(X_3) + \cdots + E(X_n)$
$\qquad = p + p + p + \cdots + p \text{ (There are } n \text{ terms.)}$

So, as we thought, the mean is $E(Y) = np$.

And since the trials are independent, the variances add:

$Var(Y) = Var(X_1 + X_2 + X_3 + \cdots + X_n)$
$\qquad = Var(X_1) + Var(X_2) + Var(X_3) + \cdots + Var(X_n)$
$\qquad = pq + pq + pq + \cdots + pq \text{ (Again, } n \text{ terms.)}$
$Var(Y) = npq$

Voilà! The standard deviation is $SD(Y) = \sqrt{npq}$.

FOR EXAMPLE · Spam and the Binomial model

Recap: The communications monitoring company *Postini* has reported that 91% of e-mail messages are spam. Suppose your inbox contains 25 messages.

Questions: What are the mean and standard deviation of the number of real messages you should expect to find in your inbox? What's the probability that you'll find only 1 or 2 real messages?

I assume that messages arrive independently and at random, with the probability of success (a real message) $p = 1 - 0.91 = 0.09$. Let X = the number of real messages among 25. I can use the model Binom(25, 0.09).

$$E(X) = np = 25(0.09) = 2.25$$
$$SD(X) = \sqrt{npq} = \sqrt{25(0.09)(0.91)} = 1.43$$
$$P(X = 1 \text{ or } 2) = P(X = 1) + P(X = 2)$$
$$= \binom{25}{1}(0.09)^1(0.91)^{24} + \binom{25}{2}(0.09)^2(0.91)^{23}$$
$$= 0.2340 + 0.2777$$
$$= 0.5117$$

Among 25 e-mail messages, I expect to find an average of 2.25 that aren't spam, with a standard deviation of 1.43 messages. There's just over a 50% chance that 1 or 2 of my 25 e-mails will be real messages.

STEP-BY-STEP EXAMPLE · Working with a Binomial Model

Suppose 20 donors come to a blood drive. Recall that 6% of people are "universal donors."

Questions:
1. What are the mean and standard deviation of the number of universal donors among them?
2. What is the probability that there are 2 or 3 universal donors?

Plan State the question.

I want to know the mean and standard deviation of the number of universal donors among 20 people, and the probability that there are 2 or 3 of them.

Check to see that these are Bernoulli trials.

✔ There are two outcomes:

success = O-negative
failure = other blood types

✔ $p = 0.06$, because people have lined up at random.

✔ **10% Condition:** Trials are not independent, because the population is finite, but fewer than 10% of all possible donors are lined up.

Variable Define the random variable.

Let X = number of O-negative donors among $n = 20$ people.

Model Specify the model.

I can model X with Binom(20, 0.06).

<table>
<tr><td>
SHOW</td><td>**Mechanics** Find the expected value and standard deviation.</td><td>$E(X) = np = 20(0.06) = 1.2$
$SD(X) = \sqrt{npq} = \sqrt{20(0.06)(0.94)} \approx 1.06$

$P(X = 2 \text{ or } 3) = P(X = 2) + P(X = 3)$
$\qquad = \binom{20}{2}(0.06)^2(0.94)^{18}$
$\qquad\quad + \binom{20}{3}(0.06)^3(0.94)^{17}$
$\qquad \approx 0.2246 + 0.0860$
$\qquad = 0.3106$</td></tr>
<tr><td>TELL</td><td>**Conclusion** Interpret your results in context.</td><td>In groups of 20 randomly selected blood donors, I expect to find an average of 1.2 universal donors, with a standard deviation of 1.06. About 31% of the time, I'd find 2 or 3 universal donors among the 20 people.</td></tr>
</table>

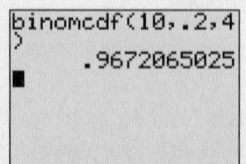 **TI Tips**

Finding binomial probabilities

Remember how the calculator handles Geometric probabilities? Well, the commands for finding Binomial probabilities are essentially the same. Again you'll find them in the **2nd DISTR** menu.

```
DISTR DRAW
0:Fcdf(
A↑binompdf(
B:binomcdf(
C:poissonpdf(
D:poissoncdf(
E:geometpdf(
F:geometcdf(
```

- **binompdf(**

  This probability density function allows you to find the probability of an *individual* outcome. You need to define the Binomial model by specifying n and p, and then indicate the desired number of successes, x. The format is **binompdf(n,p,X)**.

```
binompdf(5,.2,2)
           .2048
■
```

  For example, recall that Tiger Woods' picture is in 20% of the cereal boxes. Suppose that we want to know the probability of finding Tiger exactly twice among 5 boxes of cereal. We use $n = 5$, $p = 0.2$, and $x = 2$, entering the command **binompdf(5,.2,2)**. There's about a 20% chance of getting two pictures of Tiger Woods in five boxes of cereal.

- **binomcdf(**

  Need to add several Binomial probabilities? To find the total probability of getting x or fewer successes among the n trials use the cumulative Binomial density function **binomcdf(n,p,X)**.

```
binomcdf(10,.2,4
)
        .9672065025
■
```

  For example, suppose we have ten boxes of cereal and wonder about the probability of finding up to 4 pictures of Tiger. That's the probability of 0, 1, 2, 3 or 4 successes, so we specify the command **binomcdf(10,.2,4)**. Pretty likely!

```
1-binomcdf(10,.2
,3)
        .1208738816
■
```

  Of course "up to 4" allows for the possibility that we end up with none. What's the probability we get at least 4 pictures of Tiger in 10 boxes? Well, "at least 4" means "not 3 or fewer." That's the complement of 0, 1, 2, or 3 successes. Have your TI evaluate **1−binomcdf(10,.2,3)**. There's about a 12% chance we'll find at least 4 pictures of Tiger in 10 boxes of cereal.

The Normal Model to the Rescue!

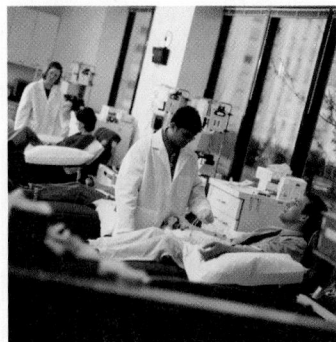

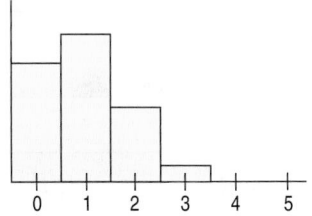

A S *Activity: Normal Approximation.* Binomial probabilities can be hard to calculate. With the Simulation Tool you'll see how well the Normal model can approximate the Binomial—a much easier method.

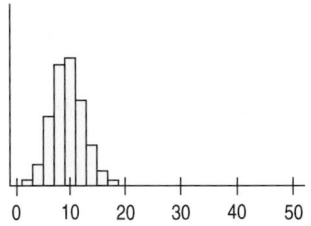

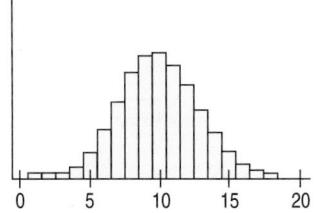

Suppose the Tennessee Red Cross anticipates the need for at least 1850 units of O-negative blood this year. It estimates that it will collect blood from 32,000 donors. How great is the risk that the Tennessee Red Cross will fall short of meeting its need? We've just learned how to calculate such probabilities. We can use the Binomial model with $n = 32,000$ and $p = 0.06$. The probability of getting *exactly* 1850 units of O-negative blood from 32,000 donors is $\binom{32000}{1850} \times 0.06^{1850} \times 0.94^{30150}$. No calculator on earth can calculate that first term (it has more than 100,000 digits).[1] And that's just the beginning. The problem said *at least* 1850, so we have to do it again for 1851, for 1852, and all the way up to 32,000. No thanks.

When we're dealing with a large number of trials like this, making direct calculations of the probabilities becomes tedious (or outright impossible). Here an old friend—the Normal model—comes to the rescue.

The Binomial model has mean $np = 1920$ and standard deviation $\sqrt{npq} \approx 42.48$. We could try approximating its distribution with a Normal model, using the same mean and standard deviation. Remarkably enough, that turns out to be a very good approximation. (We'll see why in the next chapter.) With that approximation, we can find the *probability*:

$$P(X < 1850) = P\left(z < \frac{1850 - 1920}{42.48}\right) \approx P(z < -1.65) \approx 0.05$$

There seems to be about a 5% chance that this Red Cross chapter will run short of O-negative blood.

Can we always use a Normal model to make estimates of Binomial probabilities? No. Consider the Tiger Woods situation—pictures in 20% of the cereal boxes. If we buy five boxes, the actual Binomial probabilities that we get 0, 1, 2, 3, 4, or 5 pictures of Tiger are 33%, 41%, 20%, 5%, 1%, and 0.03%, respectively. The first histogram shows that this probability model is skewed. That makes it clear that we should not try to estimate these probabilities by using a Normal model.

Now suppose we open 50 boxes of this cereal and count the number of Tiger Woods pictures we find. The second histogram shows this probability model. It is centered at $np = 50(0.2) = 10$ pictures, as expected, and it appears to be fairly symmetric around that center. Let's have a closer look.

The third histogram again shows Binom(50, 0.2), this time magnified somewhat and centered at the expected value of 10 pictures of Tiger. It looks close to Normal, for sure. With this larger sample size, it appears that a Normal model might be a useful approximation.

A Normal model, then, is a close enough approximation only for a large enough number of trials. And what we mean by "large enough" depends on the probability of success. We'd need a larger sample if the probability of success were very low (or very high). It turns out that a Normal model works pretty well if we expect to see at least 10 successes and 10 failures. That is, we check the **Success/Failure Condition.**

The Success/Failure Condition: A Binomial model is approximately Normal if we expect at least 10 successes and 10 failures:

$$np \geq 10 \text{ and } nq \geq 10.$$

---

[1] If your calculator *can* find Binom(32000,0.06), then it's smart enough to use an approximation. Read on to see how you can, too.

MATH BOX

It's easy to see where the magic number 10 comes from. You just need to remember how Normal models work. The problem is that a Normal model extends infinitely in both directions. But a Binomial model must have between 0 and n successes, so if we use a Normal to approximate a Binomial, we have to cut off its tails. That's not very important if the center of the Normal model is so far from 0 and n that the lost tails have only a negligible area. More than three standard deviations should do it, because a Normal model has little probability past that.

So the mean needs to be at least 3 standard deviations from 0 and at least 3 standard deviations from n. Let's look at the 0 end.

We require:	$\mu - 3\sigma > 0$
Or in other words:	$\mu > 3\sigma$
For a Binomial, that's:	$np > 3\sqrt{npq}$
Squaring yields:	$n^2p^2 > 9npq$
Now simplify:	$np > 9q$
Since $q \le 1$, we can require:	$np > 9$

For simplicity, we usually require that np (and nq for the other tail) be at least 10 to use the Normal approximation, the Success/Failure Condition.[2]

FOR EXAMPLE   Spam and the Normal approximation to the Binomial

Recap: The communications monitoring company *Postini* has reported that 91% of e-mail messages are spam. Recently, you installed a spam filter. You observe that over the past week it okayed only 151 of 1422 e-mails you received, classifying the rest as junk. Should you worry that the filtering is too aggressive?

Question: What's the probability that no more than 151 of 1422 e-mails is a real message?

I assume that messages arrive randomly and independently, with a probability of success (a real message) $p = 0.09$. The model Binom(1422, 0.09) applies, but will be hard to work with. Checking conditions for the Normal approximation, I see that:

✔ These messages represent less than 10% of all e-mail traffic.

✔ I expect $np = (1422)(0.09) = 127.98$ real messages and $nq = (1422)(0.91) = 1294.02$ spam messages, both far greater than 10.

It's okay to approximate this binomial probability by using a Normal model.

$\mu = np = 1422(0.09) = 127.98$

$\sigma = \sqrt{npq} = \sqrt{1422(0.09)(0.91)} \approx 10.79$

$P(x \le 151) = P\left(z \le \dfrac{151 - 127.98}{10.79}\right)$

$= P(z \le 2.13)$

$= 0.9834$

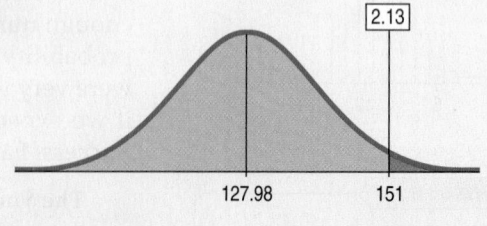

Among my 1422 e-mails, there's over a 98% chance that no more than 151 of them were real messages, so the filter may be working properly.

---

[2] Looking at the final step, we see that we need $np > 9$ in the worst case, when q (or p) is near 1, making the Binomial model quite skewed. When q and p are near 0.5—say between 0.4 and 0.6—the Binomial model is nearly symmetric and $np > 5$ ought to be safe enough. Although we'll always check for 10 expected successes and failures, keep in mind that for values of p near 0.5, we can be somewhat more forgiving.

Continuous Random Variables

There's a problem with approximating a Binomial model with a Normal model. The Binomial is discrete, giving probabilities for specific counts, but the Normal models a **continuous** random variable that can take on *any value*. For continuous random variables, we can no longer list all the possible outcomes and their probabilities, as we could for discrete random variables.[3]

   As we saw in the previous chapter, models for continuous random variables give probabilities for *intervals* of values. So, when we use the Normal model, we no longer calculate the probability that the random variable equals a *particular* value, but only that it lies *between* two values. We won't calculate the probability of getting exactly 1850 units of blood, but we have no problem approximating the probability of getting 1850 *or more*, which was, after all, what we really wanted.[4]

JUST CHECKING

As we noted a few chapters ago, the Pew Research Center (www.pewresearch.org) reports that they are actually able to contact only 76% of the randomly selected households drawn for a telephone survey.

1. Explain why these phone calls can be considered Bernoulli trials.

2. Which of the models of this chapter (Geometric, Binomial, Normal) would you use to model the number of successful contacts from a list of 1000 sampled households? Explain.

3. Pew further reports that even after they contacted a household, only 38% agree to be interviewed, so the probability of getting a completed interview for a randomly selected household is only 0.29. Which of the models of this chapter would you use to model the number of households Pew has to call before they get the first completed interview?

WHAT CAN GO WRONG?

▶ **Be sure you have Bernoulli trials.** Be sure to check the requirements first: two possible outcomes per trial ("success" and "failure"), a constant probability of success, and independence. Remember to check the 10% Condition when sampling without replacement.

▶ **Don't confuse Geometric and Binomial models.** Both involve Bernoulli trials, but the issues are different. If you are repeating trials until your first success, that's a Geometric probability. You don't know in advance how many trials you'll need—theoretically, it could take forever. If you are counting the number of successes in a specified number of trials, that's a Binomial probability.

▶ **Don't use the Normal approximation with small *n*.** To use a Normal approximation in place of a Binomial model, there must be at least 10 expected successes and 10 expected failures.

---

[3] In fact, some people use an adjustment called the "continuity correction" to help with this problem. It's related to the suggestion we make in the next footnote and is discussed in more advanced textbooks.

[4] If we really had been interested in a single value, we might have approximated it by finding the probability of getting between 1849.5 and 1850.5 units of blood.

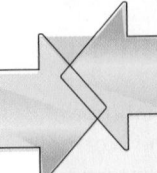

CONNECTIONS

This chapter builds on what we know about random variables. We now have two more probability models to join the Normal model.

There are a number of "forward" connections from this chapter. We'll see the **10% Condition** and the **Success/Failure Condition** often. And the facts about the Binomial distribution can help explain how proportions behave, as we'll see in the next chapter.

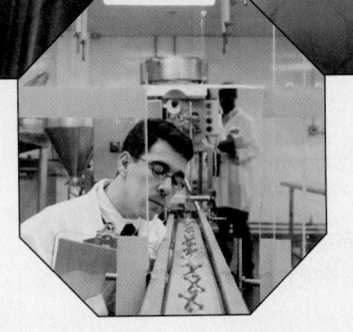

WHAT HAVE WE LEARNED?

We've learned that Bernoulli trials show up in lots of places. Depending on the random variable of interest, we can use one of three models to estimate probabilities for Bernoulli trials:

▸ a Geometric model when we're interested in the number of Bernoulli trials until the next success;
▸ a Binomial model when we're interested in the number of successes in a certain number of Bernoulli trials;
▸ a Normal model to approximate a Binomial model when we expect at least 10 successes and 10 failures.

Terms

Bernoulli trials, if . . .
388. 1. there are two possible outcomes.
2. the probability of success is constant.
3. the trials are independent.

Geometric probability model
389. A Geometric model is appropriate for a random variable that counts the number of Bernoulli trials until the first success.

Binomial probability model
393. A Binomial model is appropriate for a random variable that counts the number of successes in a fixed number of Bernoulli trials.

10% Condition
391. When sampling without replacement, trials are not independent. It's still okay to proceed as long as the sample is smaller than 10% of the population.

Success/Failure Condition
397. For a Normal model to be a good approximation of a Binomial model, we must expect at least 10 successes and 10 failures. That is, $np \geq 10$ and $nq \geq 10$.

Skills

▸ Know how to tell if a situation involves Bernoulli trials.
▸ Be able to choose whether to use a Geometric or a Binomial model for a random variable involving Bernoulli trials.

▸ Know the appropriate conditions for using a Geometric, Binomial, or Normal model.
▸ Know how to find the expected value of a Geometric model.

▸ Be able to calculate Geometric probabilities.
▸ Know how to find the mean and standard deviation of a Binomial model.
▸ Be able to calculate Binomial probabilities, perhaps approximating with a Normal model.

▸ Be able to interpret means, standard deviations, and probabilities in the Bernoulli trial context.

THE BINOMIAL AND THE GEOMETRIC ON THE COMPUTER

Most statistics packages offer functions that compute Binomial probabilities, and many offer functions for Geometric probabilities as well. Some technology solutions automatically use the Normal approximation for the Binomial when the exact calculations become unmanageable.

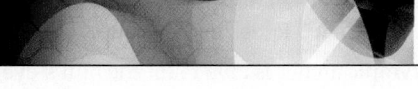

EXERCISES

1. **Bernoulli.**   Do these situations involve Bernoulli trials? Explain.
   a) We roll 50 dice to find the distribution of the number of spots on the faces.
   b) How likely is it that in a group of 120 the majority may have Type A blood, given that Type A is found in 43% of the population?
   c) We deal 7 cards from a deck and get all hearts. How likely is that?
   d) We wish to predict the outcome of a vote on the school budget, and poll 500 of the 3000 likely voters to see how many favor the proposed budget.
   e) A company realizes that about 10% of its packages are not being sealed properly. In a case of 24, is it likely that more than 3 are unsealed?

2. **Bernoulli 2.**   Do these situations involve Bernoulli trials? Explain.
   a) You are rolling 5 dice and need to get at least two 6's to win the game.
   b) We record the distribution of eye colors found in a group of 500 people.
   c) A manufacturer recalls a doll because about 3% have buttons that are not properly attached. Customers return 37 of these dolls to the local toy store. Is the manufacturer likely to find any dangerous buttons?
   d) A city council of 11 Republicans and 8 Democrats picks a committee of 4 at random. What's the probability they choose all Democrats?
   e) A 2002 Rutgers University study found that 74% of high school students have cheated on a test at least once. Your local high school principal conducts a survey in homerooms and gets responses that admit to cheating from 322 of the 481 students.

3. **Simulating the model.**   Think about the Tiger Woods picture search again. You are opening boxes of cereal one at a time looking for his picture, which is in 20% of the boxes. You want to know how many boxes you might have to open in order to find Tiger.
   a) Describe how you would simulate the search for Tiger using random numbers.
   b) Run at least 30 trials.
   c) Based on your simulation, estimate the probabilities that you might find your first picture of Tiger in the first box, the second, etc.

   d) Calculate the actual probability model.
   e) Compare the distribution of outcomes in your simulation to the probability model.

4. **Simulation II.**   You are one space short of winning a child's board game and must roll a 1 on a die to claim victory. You want to know how many rolls it might take.
   a) Describe how you would simulate rolling the die until you get a 1.
   b) Run at least 30 trials.
   c) Based on your simulation, estimate the probabilities that you might win on the first roll, the second, the third, etc.
   d) Calculate the actual probability model.
   e) Compare the distribution of outcomes in your simulation to the probability model.

5. **Tiger again.**   Let's take one last look at the Tiger Woods picture search. You know his picture is in 20% of the cereal boxes. You buy five boxes to see how many pictures of Tiger you might get.
   a) Describe how you would simulate the number of pictures of Tiger you might find in five boxes of cereal.
   b) Run at least 30 trials.
   c) Based on your simulation, estimate the probabilities that you get no pictures of Tiger, 1 picture, 2 pictures, etc.
   d) Find the actual probability model.
   e) Compare the distribution of outcomes in your simulation to the probability model.

6. **Seatbelts.**   Suppose 75% of all drivers always wear their seatbelts. Let's investigate how many of the drivers might be belted among five cars waiting at a traffic light.
   a) Describe how you would simulate the number of seatbelt-wearing drivers among the five cars.
   b) Run at least 30 trials.
   c) Based on your simulation, estimate the probabilities there are no belted drivers, exactly one, two, etc.
   d) Find the actual probability model.
   e) Compare the distribution of outcomes in your simulation to the probability model.

7. **On time.**   A Department of Transportation report about air travel found that, nationwide, 76% of all flights are on time. Suppose you are at the airport and your flight is one of 50 scheduled to take off in the next two hours. Can you consider these departures to be Bernoulli trials? Explain.

8. **Lost luggage.** A Department of Transportation report about air travel found that airlines misplace about 5 bags per 1000 passengers. Suppose you are traveling with a group of people who have checked 22 pieces of luggage on your flight. Can you consider the fate of these bags to be Bernoulli trials? Explain.

9. **Hoops.** A basketball player has made 80% of his foul shots during the season. Assuming the shots are independent, find the probability that in tonight's game he
   a) misses for the first time on his fifth attempt.
   b) makes his first basket on his fourth shot.
   c) makes his first basket on one of his first 3 shots.

10. **Chips.** Suppose a computer chip manufacturer rejects 2% of the chips produced because they fail presale testing.
    a) What's the probability that the fifth chip you test is the first bad one you find?
    b) What's the probability you find a bad one within the first 10 you examine?

11. **More hoops.** For the basketball player in Exercise 9, what's the expected number of shots until he misses?

12. **Chips ahoy.** For the computer chips described in Exercise 10, how many do you expect to test before finding a bad one?

13. **Customer center operator.** Raaj works at the customer service call center of a major credit card bank. Cardholders call for a variety of reasons, but regardless of their reason for calling, if they hold a platinum card, Raaj is instructed to offer them a double-miles promotion. About 10% of all cardholders hold platinum cards, and about 50% of those will take the double-miles promotion. On average, how many calls will Raaj have to take before finding the first cardholder to take the double-miles promotion?

14. **Cold calls.** Justine works for an organization committed to raising money for Alzheimer's research. From past experience, the organization knows that about 20% of all potential donors will agree to give something if contacted by phone. They also know that of all people donating, about 5% will give $100 or more. On average, how many potential donors will she have to contact until she gets her first $100 donor?

15. **Blood.** Only 4% of people have Type AB blood.
    a) On average, how many donors must be checked to find someone with Type AB blood?
    b) What's the probability that there is a Type AB donor among the first 5 people checked?
    c) What's the probability that the first Type AB donor will be found among the first 6 people?
    d) What's the probability that we won't find a Type AB donor before the 10th person?

16. **Colorblindness.** About 8% of males are colorblind. A researcher needs some colorblind subjects for an experiment and begins checking potential subjects.
    a) On average, how many men should the researcher expect to check to find one who is colorblind?
    b) What's the probability that she won't find anyone colorblind among the first 4 men she checks?

    c) What's the probability that the first colorblind man found will be the sixth person checked?
    d) What's the probability that she finds someone who is colorblind before checking the 10th man?

17. **Lefties.** Assume that 13% of people are left-handed. If we select 5 people at random, find the probability of each outcome described below.
    a) The first lefty is the fifth person chosen.
    b) There are some lefties among the 5 people.
    c) The first lefty is the second or third person.
    d) There are exactly 3 lefties in the group.
    e) There are at least 3 lefties in the group.
    f) There are no more than 3 lefties in the group.

18. **Arrows.** An Olympic archer is able to hit the bull's-eye 80% of the time. Assume each shot is independent of the others. If she shoots 6 arrows, what's the probability of each of the following results?
    a) Her first bull's-eye comes on the third arrow.
    b) She misses the bull's-eye at least once.
    c) Her first bull's-eye comes on the fourth or fifth arrow.
    d) She gets exactly 4 bull's-eyes.
    e) She gets at least 4 bull's-eyes.
    f) She gets at most 4 bull's-eyes.

19. **Lefties redux.** Consider our group of 5 people from Exercise 17.
    a) How many lefties do you expect?
    b) With what standard deviation?
    c) If we keep picking people until we find a lefty, how long do you expect it will take?

20. **More arrows.** Consider our archer from Exercise 18.
    a) How many bull's-eyes do you expect her to get?
    b) With what standard deviation?
    c) If she keeps shooting arrows until she hits the bull's-eye, how long do you expect it will take?

21. **Still more lefties.** Suppose we choose 12 people instead of the 5 chosen in Exercise 17.
    a) Find the mean and standard deviation of the number of right-handers in the group.
    b) What's the probability that
       i) they're not all right-handed?
       ii) there are no more than 10 righties?
       iii) there are exactly 6 of each?
       iv) the majority is right-handed?

22. **Still more arrows.** Suppose our archer from Exercise 18 shoots 10 arrows.
    a) Find the mean and standard deviation of the number of bull's-eyes she may get.
    b) What's the probability that
       i) she never misses?
       ii) there are no more than 8 bull's-eyes?
       iii) there are exactly 8 bull's-eyes?
       iv) she hits the bull's-eye more often than she misses?

23. **Vision.** It is generally believed that nearsightedness affects about 12% of all children. A school district tests the vision of 169 incoming kindergarten children. How many would you expect to be nearsighted? With what standard deviation?

24. International students. At a certain college, 6% of all students come from outside the United States. Incoming students there are assigned at random to freshman dorms, where students live in residential clusters of 40 freshmen sharing a common lounge area. How many international students would you expect to find in a typical cluster? With what standard deviation?

25. Tennis, anyone? A certain tennis player makes a successful first serve 70% of the time. Assume that each serve is independent of the others. If she serves 6 times, what's the probability she gets
a) all 6 serves in?
b) exactly 4 serves in?
c) at least 4 serves in?
d) no more than 4 serves in?

26. Frogs. A wildlife biologist examines frogs for a genetic trait he suspects may be linked to sensitivity to industrial toxins in the environment. Previous research had established that this trait is usually found in 1 of every 8 frogs. He collects and examines a dozen frogs. If the frequency of the trait has not changed, what's the probability he finds the trait in
a) none of the 12 frogs?
b) at least 2 frogs?
c) 3 or 4 frogs?
d) no more than 4 frogs?

27. And more tennis. Suppose the tennis player in Exercise 25 serves 80 times in a match.
a) What are the mean and standard deviation of the number of good first serves expected?
b) Verify that you can use a Normal model to approximate the distribution of the number of good first serves.
c) Use the 68–95–99.7 Rule to describe this distribution.
d) What's the probability she makes at least 65 first serves?

28. More arrows. The archer in Exercise 18 will be shooting 200 arrows in a large competition.
a) What are the mean and standard deviation of the number of bull's-eyes she might get?
b) Is a Normal model appropriate here? Explain.
c) Use the 68–95–99.7 Rule to describe the distribution of the number of bull's-eyes she may get.
d) Would you be surprised if she made only 140 bull's-eyes? Explain.

29. Apples. An orchard owner knows that he'll have to use about 6% of the apples he harvests for cider because they will have bruises or blemishes. He expects a tree to produce about 300 apples.
a) Describe an appropriate model for the number of cider apples that may come from that tree. Justify your model.
b) Find the probability there will be no more than a dozen cider apples.
c) Is it likely there will be more than 50 cider apples? Explain.

30. Frogs, part II. Based on concerns raised by his preliminary research, the biologist in Exercise 26 decides to collect and examine 150 frogs.

a) Assuming the frequency of the trait is still 1 in 8, determine the mean and standard deviation of the number of frogs with the trait he should expect to find in his sample.
b) Verify that he can use a Normal model to approximate the distribution of the number of frogs with the trait.
c) He found the trait in 22 of his frogs. Do you think this proves that the trait has become more common? Explain.

31. Lefties again. A lecture hall has 200 seats with folding arm tablets, 30 of which are designed for left-handers. The typical size of classes that meet there is 188, and we can assume that about 13% of students are left-handed. What's the probability that a right-handed student in one of these classes is forced to use a lefty arm tablet?

32. No-shows. An airline, believing that 5% of passengers fail to show up for flights, overbooks (sells more tickets than there are seats). Suppose a plane will hold 265 passengers, and the airline sells 275 tickets. What's the probability the airline will not have enough seats, so someone gets bumped?

33. Annoying phone calls. A newly hired telemarketer is told he will probably make a sale on about 12% of his phone calls. The first week he called 200 people, but only made 10 sales. Should he suspect he was misled about the true success rate? Explain.

34. The euro. Shortly after the introduction of the euro coin in Belgium, newspapers around the world published articles claiming the coin is biased. The stories were based on reports that someone had spun the coin 250 times and gotten 140 heads—that's 56% heads. Do you think this is evidence that spinning a euro is unfair? Explain.

35. Seatbelts II. Police estimate that 80% of drivers now wear their seatbelts. They set up a safety roadblock, stopping cars to check for seatbelt use.
a) How many cars do they expect to stop before finding a driver whose seatbelt is not buckled?
b) What's the probability that the first unbelted driver is in the 6th car stopped?
c) What's the probability that the first 10 drivers are all wearing their seatbelts?
d) If they stop 30 cars during the first hour, find the mean and standard deviation of the number of drivers expected to be wearing seatbelts.
e) If they stop 120 cars during this safety check, what's the probability they find at least 20 drivers not wearing their seatbelts?

36. Rickets. Vitamin D is essential for strong, healthy bones. Our bodies produce vitamin D naturally when sunlight falls upon the skin, or it can be taken as a dietary supplement. Although the bone disease rickets was largely eliminated in England during the 1950s, some people there are concerned that this generation of children is at increased risk because they are more likely to watch TV or play computer games than spend time outdoors. Recent research indicated that about 20% of British children are deficient in vitamin D. Suppose doctors test a group of elementary school children.

a) What's the probability that the first vitamin D–deficient child is the 8th one tested?

b) What's the probability that the first 10 children tested are all okay?

c) How many kids do they expect to test before finding one who has this vitamin deficiency?

d) They will test 50 students at the third-grade level. Find the mean and standard deviation of the number who may be deficient in vitamin D.

e) If they test 320 children at this school, what's the probability that no more than 50 of them have the vitamin deficiency?

37. **ESP.** Scientists wish to test the mind-reading ability of a person who claims to "have ESP." They use five cards with different and distinctive symbols (square, circle, triangle, line, squiggle). Someone picks a card at random and thinks about the symbol. The "mind reader" must correctly identify which symbol was on the card. If the test consists of 100 trials, how many would this person need to get right in order to convince you that ESP may actually exist? Explain.

38. **True-False.** A true-false test consists of 50 questions. How many does a student have to get right to convince you that he is not merely guessing? Explain.

39. **Hot hand.** A basketball player who ordinarily makes about 55% of his free throw shots has made 4 in a row. Is this evidence that he has a "hot hand" tonight? That is, is this streak so unusual that it means the probability he makes a shot must have changed? Explain.

40. **New bow.** Our archer in Exercise 18 purchases a new bow, hoping that it will improve her success rate to more than 80% bull's-eyes. She is delighted when she first tests her new bow and hits 6 consecutive bull's-eyes. Do you think this is compelling evidence that the new bow is better? In other words, is a streak like this unusual for her? Explain.

41. **Hotter hand.** Our basketball player in Exercise 39 has new sneakers, which he thinks improve his game. Over his past 40 shots, he's made 32—much better than the 55% he usually shoots. Do you think his chances of making a shot really increased? In other words, is making at least 32 of 40 shots really unusual for him? (Do you think it's his sneakers?)

42. **New bow, again.** The archer in Exercise 40 continues shooting arrows, ending up with 45 bull's-eyes in 50 shots. Now are you convinced that the new bow is better? Explain.

JUST CHECKING
Answers

1. There are two outcomes (contact, no contact), the probability of contact is 0.76, and random calls should be independent.

2. Binomial, with $n = 1000$ and $p = 0.76$. For actual calculations, we could approximate using a Normal model with $\mu = np = 1000(0.76) = 760$ and

$$\sigma = \sqrt{npq} = \sqrt{1000(0.76)(0.24)} \approx 13.5.$$

3. Geometric, with $p = 0.29$.

REVIEW OF PART IV

Randomness and Probability

Quick Review

Here's a brief summary of the key concepts and skills in probability and probability modeling:

▶ The Law of Large Numbers says that the more times we try something, the closer the results will come to theoretical perfection.
  • Don't mistakenly misinterpret the Law of Large Numbers as the "Law of Averages." There's no such thing.

▶ Basic rules of probability can handle most situations:
  • To find the probability that an event OR another event happens, add their probabilities and subtract the probability that both happen.
  • To find the probability that an event AND another independent event both happen, multiply probabilities.
  • Conditional probabilities tell you how likely one event is to happen, knowing that another event has happened.
  • Mutually exclusive events (also called "disjoint") cannot both happen at the same time.
  • Two events are independent if the occurrence of one doesn't change the probability that the other happens.

▶ A probability model for a random variable describes the theoretical distribution of outcomes.
  • The mean of a random variable is its expected value.
  • For sums or differences of independent random variables, variances add.
  • To estimate probabilities involving quantitative variables, you may be able to use a Normal model—but only if the distribution of the variable is unimodal and symmetric.
  • To estimate the probability you'll get your first success on a certain trial, use a Geometric model.
  • To estimate the probability you'll get a certain number of successes in a specified number of independent trials, use a Binomial model.

Ready? Here are some opportunities to check your understanding of these ideas.

REVIEW EXERCISES

1. **Quality control.** A consumer organization estimates that 29% of new cars have a cosmetic defect, such as a scratch or a dent, when they are delivered to car dealers. This same organization believes that 7% have a functional defect—something that does not work properly—and that 2% of new cars have both kinds of problems.
   a) If you buy a new car, what's the probability that it has some kind of defect?
   b) What's the probability it has a cosmetic defect but no functional defect?
   c) If you notice a dent on a new car, what's the probability it has a functional defect?
   d) Are the two kinds of defects disjoint events? Explain.
   e) Do you think the two kinds of defects are independent events? Explain.

2. **Workers.** A company's human resources officer reports a breakdown of employees by job type and sex shown in the table.

		Sex	
		Male	**Female**
Job Type	**Management**	7	6
	Supervision	8	12
	Production	45	72

   a) What's the probability that a worker selected at random is
      i) female?
      ii) female or a production worker?
      iii) female, if the person works in production?
      iv) a production worker, if the person is female?
   b) Do these data suggest that job type is independent of being male or female? Explain.

3. **Airfares.** Each year a company must send 3 officials to a meeting in China and 5 officials to a meeting in France. Airline ticket prices vary from time to time, but the company purchases all tickets for a country at the same price. Past experience has shown that tickets to China have a mean price of $1000, with a standard deviation of $150, while the mean airfare to France is $500, with a standard deviation of $100.
   a) Define random variables and use them to express the total amount the company will have to spend to send these delegations to the two meetings.
   b) Find the mean and standard deviation of this total cost.
   c) Find the mean and standard deviation of the difference in price of a ticket to China and a ticket to France.
   d) Do you need to make any assumptions in calculating these means? How about the standard deviations?

4. Bipolar. Psychiatrists estimate that about 1 in 100 adults suffers from bipolar disorder. What's the probability that in a city of 10,000 there are more than 200 people with this condition? Be sure to verify that a Normal model can be used here.

5. A game. To play a game, you must pay $5 for each play. There is a 10% chance you will win $5, a 40% chance you will win $7, and a 50% chance you will win only $3.
a) What are the mean and standard deviation of your net winnings?
b) You play twice. Assuming the plays are independent events, what are the mean and standard deviation of your total winnings?

6. Emergency switch. Safety engineers must determine whether industrial workers can operate a machine's emergency shutoff device. Among a group of test subjects, 66% were successful with their left hands, 82% with their right hands, and 51% with either hand.
a) What percent of these workers could not operate the switch with either hand?
b) Are success with right and left hands independent events? Explain.
c) Are success with right and left hands mutually exclusive? Explain.

7. Twins. In the United States, the probability of having twins (usually about 1 in 90 births) rises to about 1 in 10 for women who have been taking the fertility drug Clomid. Among a group of 10 pregnant women, what's the probability that
a) at least one will have twins if none were taking a fertility drug?
b) at least one will have twins if all were taking Clomid?
c) at least one will have twins if half were taking Clomid?

8. Deductible. A car owner may buy insurance that will pay the full price of repairing the car after an at-fault accident, or save $12 a year by getting a policy with a $500 deductible. Her insurance company says that about 0.5% of drivers in her area have an at-fault auto accident during any given year. Based on this information, should she buy the policy with the deductible or not? How does the value of her car influence this decision?

9. More twins. A group of 5 women became pregnant while undergoing fertility treatments with the drug Clomid, discussed in Exercise 7. What's the probability that
a) none will have twins?
b) exactly 1 will have twins?
c) at least 3 will have twins?

10. At fault. The car insurance company in Exercise 8 believes that about 0.5% of drivers have an at-fault accident during a given year. Suppose the company insures 1355 drivers in that city.
a) What are the mean and standard deviation of the number who may have at-fault accidents?
b) Can you describe the distribution of these accidents with a Normal model? Explain.

11. Twins, part III. At a large fertility clinic, 152 women became pregnant while taking Clomid. (See Exercise 7.)
a) What are the mean and standard deviation of the number of twin births we might expect?
b) Can we use a Normal model in this situation? Explain.
c) What's the probability that no more than 10 of the women have twins?

12. Child's play. In a board game you determine the number of spaces you may move by spinning a spinner and rolling a die. The spinner has three regions: Half of the spinner is marked "5," and the other half is equally divided between "10"and "20." The six faces of the die show 0, 0, 1, 2, 3, and 4 spots. When it's your turn, you spin and roll, adding the numbers together to determine how far you may move.
a) Create a probability model for the outcome on the spinner.
b) Find the mean and standard deviation of the spinner results.
c) Create a probability model for the outcome on the die.
d) Find the mean and standard deviation of the die results.
e) Find the mean and standard deviation of the number of spaces you get to move.

13. Language. Neurological research has shown that in about 80% of people, language abilities reside in the brain's left side. Another 10% display right-brain language centers, and the remaining 10% have two-sided language control. (The latter two groups are mainly left-handers; *Science News*, 161 no. 24 [2002].)
a) Assume that a freshman composition class contains 25 randomly selected people. What's the probability that no more than 15 of them have left-brain language control?
b) In a randomly chosen group of 5 of these students, what's the probability that no one has two-sided language control?
c) In the entire freshman class of 1200 students, how many would you expect to find of each type?
d) What are the mean and standard deviation of the number of these freshmen who might be right-brained in language abilities?
e) If an assumption of Normality is justified, use the 68–95–99.7 Rule to describe how many students in the freshman class might have right-brain language control.

14. Play again. If you land in a "penalty zone" on the game board described in Exercise 12, your move will be determined by subtracting the roll of the die from the result on the spinner. Now what are the mean and standard deviation of the number of spots you may move?

15. Beanstalks. In some cities tall people who want to meet and socialize with other tall people can join Beanstalk Clubs. To qualify, a man must be over 6'2" tall, and a woman over 5'10". According to the National Health Survey, heights of adults may have a Normal model with mean heights of 69.1" for men and 64.0" for women. The respective standard deviations are 2.8" and 2.5".

a) You're probably not surprised to learn that men are generally taller than women, but what does the greater standard deviation for men's heights indicate?

b) Are men or women more likely to qualify for Beanstalk membership?

c) Beanstalk members believe that height is an important factor when people select their spouses. To investigate, we select at random a married man and, independently, a married woman. Define two random variables, and use them to express how many inches taller the man is than the woman.

d) What's the mean of this difference?

e) What's the standard deviation of this difference?

f) What's the probability that the man is taller than the woman (that the difference in heights is greater than 0)?

g) Suppose a survey of married couples reveals that 92% of the husbands were taller than their wives. Based on your answer to part f, do you believe that people's choice of spouses is independent of height? Explain.

16. Stocks. Since the stock market began in 1872, stock prices have risen in about 73% of the years. Assuming that market performance is independent from year to year, what's the probability that

a) the market will rise for 3 consecutive years?

b) the market will rise 3 years out of the next 5?

c) the market will fall during at least 1 of the next 5 years?

d) the market will rise during a majority of years over the next decade?

17. Multiple choice. A multiple choice test has 50 questions, with 4 answer choices each. You must get at least 30 correct to pass the test, and the questions are very difficult.

a) Are you likely to be able to pass by guessing on every question? Explain.

b) Suppose, after studying for a while, you believe you have raised your chances of getting each question right to 70%. How likely are you to pass now?

c) Assuming you are operating at the 70% level and the instructor arranges questions randomly, what's the probability that the third question is the first one you get right?

18. Stock strategy. Many investment advisors argue that after stocks have declined in value for 2 consecutive years, people should invest heavily because the market rarely declines 3 years in a row.

a) Since the stock market began in 1872, there have been two consecutive losing years eight times. In six of those cases, the market rose during the following year. Does this confirm the advice?

b) Overall, stocks have risen in value during 95 of the 130 years since the market began in 1872. How is this fact relevant in assessing the statistical reasoning of the advisors?

19. Insurance. A 65-year-old woman takes out a $10,000 term life insurance policy. The company charges an annual premium of $500. Estimate the company's expected profit on such policies if mortality tables indicate that only 2.6% of women age 65 die within a year.

20. Teen smoking. The Centers for Disease Control say that about 30% of high-school students smoke tobacco (down from a high of 38% in 1997). Suppose you randomly select high-school students to survey them on their attitudes toward scenes of smoking in the movies. What's the probability that

a) none of the first 4 students you interview is a smoker?

b) the first smoker is the sixth person you choose?

c) there are no more than 2 smokers among 10 people you choose?

21. Passing stats. Molly's college offers two sections of Statistics 101. From what she has heard about the two professors listed, Molly estimates that her chances of passing the course are 0.80 if she gets Professor Scedastic and 0.60 if she gets Professor Kurtosis. The registrar uses a lottery to randomly assign the 120 enrolled students based on the number of available seats in each class. There are 70 seats in Professor Scedastic's class and 50 in Professor Kurtosis's class.

a) What's the probability that Molly will pass Statistics?

b) At the end of the semester, we find out that Molly failed. What's the probability that she got Professor Kurtosis?

22. Teen smoking II. Suppose that, as reported by the Centers for Disease Control, about 30% of high school students smoke tobacco. You randomly select 120 high school students to survey them on their attitudes toward scenes of smoking in the movies.

a) What's the expected number of smokers?

b) What's the standard deviation of the number of smokers?

c) The number of smokers among 120 randomly selected students will vary from group to group. Explain why that number can be described with a Normal model.

d) Using the 68–95–99.7 Rule, create and interpret a model for the number of smokers among your group of 120 students.

23. Random variables. Given independent random variables with means and standard deviations as shown, find the mean and standard deviation of each of these variables:

a) $X + 50$

b) $10Y$

c) $X + 0.5Y$

d) $X - Y$

e) $X_1 + X_2$

	Mean	SD
X	50	8
Y	100	6

24. Merger. Explain why the facts you know about variances of independent random variables might encourage two small insurance companies to merge. (*Hint:* Think about the expected amount and potential variability in payouts for the separate and the merged companies.)

25. Youth survey. According to a recent Gallup survey, 93% of teens use the Internet, but there are differences in how teen boys and girls say they use computers. The telephone poll found that 77% of boys had played computer games in the past week, compared with 65% of girls. On the other hand, 76% of girls said they had e-mailed friends in the past week, compared with only 65% of boys.

a) For boys, the cited percentages are 77% playing computer games and 65% using e-mail. That total is 142%, so there is obviously a mistake in the report. No? Explain.

b) Based on these results, do you think playing games and using e-mail are mutually exclusive? Explain.

c) Do you think whether a child e-mails friends is independent of being a boy or a girl? Explain.

d) Suppose that in fact 93% of the teens in your area do use the Internet. You want to interview a few who do not, so you start contacting teenagers at random. What is the probability that it takes you 5 interviews until you find the first person who does not use the Internet?

26. **Meals.** A college student on a seven-day meal plan reports that the amount of money he spends daily on food varies with a mean of $13.50 and a standard deviation of $7.

a) What are the mean and standard deviation of the amount he might spend in two consecutive days?

b) What assumption did you make in order to find that standard deviation? Are there any reasons you might question that assumption?

c) Estimate his average weekly food costs, and the standard deviation.

d) Do you think it likely he might spend less than $50 in a week? Explain, including any assumptions you make in your analysis.

27. **Travel to Kyrgyzstan.** Your pocket copy of *Kyrgyzstan on 4237 ± 360 Som a Day* claims that you can expect to spend about 4237 som each day with a standard deviation of 360 som. How well can you estimate your expenses for the trip?

a) Your budget allows you to spend 90,000 som. To the nearest day, how long can you afford to stay in Kyrgyzstan, on average?

b) What's the standard deviation of your expenses for a trip of that duration?

c) You doubt that your total expenses will exceed your expectations by more than two standard deviations. How much extra money should you bring? On average, how much of a "cushion" will you have per day?

28. **Picking melons.** Two stores sell watermelons. At the first store the melons weigh an average of 22 pounds, with a standard deviation of 2.5 pounds. At the second store the melons are smaller, with a mean of 18 pounds and a standard deviation of 2 pounds. You select a melon at random at each store.

a) What's the mean difference in weights of the melons?

b) What's the standard deviation of the difference in weights?

c) If a Normal model can be used to describe the difference in weights, what's the probability that the melon you got at the first store is heavier?

29. **Home, sweet home.** According to the 2000 Census, 66% of U.S. households own the home they live in. A mayoral candidate conducts a survey of 820 randomly selected homes in your city and finds only 523 owned by the current residents. The candidate then attacks the incumbent mayor, saying that there is an unusually low level of homeownership in the city. Do you agree? Explain.

30. **Buying melons.** The first store in Exercise 28 sells watermelons for 32 cents a pound. The second store is having a sale on watermelons—only 25 cents a pound. Find the mean and standard deviation of the difference in the price you may pay for melons randomly selected at each store.

31. **Who's the boss?** The 2000 Census revealed that 26% of all firms in the United States are owned by women. You call some firms doing business locally, assuming that the national percentage is true in your area.

a) What's the probability that the first 3 you call are all owned by women?

b) What's the probability that none of your first 4 calls finds a firm that is owned by a woman?

c) Suppose none of your first 5 calls found a firm owned by a woman. What's the probability that your next call does?

32. **Jerseys.** A Statistics professor comes home to find that all four of his children got white team shirts from soccer camp this year. He concludes that this year, unlike other years, the camp must not be using a variety of colors. But then he finds out that in each child's age group there are 4 teams, only 1 of which wears white shirts. Each child just happened to get on the white team at random.

a) Why was he so surprised? If each age group uses the same 4 colors, what's the probability that all four kids would get the same-color shirt?

b) What's the probability that all 4 would get white shirts?

c) We lied. Actually, in the oldest child's group there are 6 teams instead of the 4 teams in each of the other three groups. How does this change the probability you calculated in part b?

33. **When to stop?** In Exercise 27 of the Review Exercises for Part III, we posed this question:

You play a game that involves rolling a die. You can roll as many times as you want, and your score is the total for all the rolls. But . . . if you roll a 6, your score is 0 and your turn is over. What might be a good strategy for a game like this?

You attempted to devise a good strategy by simulating several plays to see what might happen. Let's try calculating a strategy.

a) On what roll would you expect to get a 6 for the first time?

b) So, roll *one time less* than that. Assuming all those rolls were not 6's, what's your expected score?

c) What's the probability that you can roll that many times without getting a 6?

34. Plan B.   Here's another attempt at developing a good strategy for the dice game in Exercise 33. Instead of stopping after a certain number of rolls, you could decide to stop when your score reaches a certain number of points.

a) How many points would you expect a roll to *add* to your score?

b) In terms of your current score, how many points would you expect a roll to *subtract* from your score?

c) Based on your answers in parts a and b, at what score will another roll "break even"?

d) Describe the strategy this result suggests.

35. Technology on campus.   Every 5 years the Conference Board of the Mathematical Sciences surveys college math departments. In 2000 the board reported that 51% of all undergraduates taking Calculus I were in classes that used graphing calculators and 31% were in classes that used computer assignments. Suppose that 16% used both calculators and computers.

a) What percent used neither kind of technology?

b) What percent used calculators but not computers?

c) What percent of the calculator users had computer assignments?

d) Based on this survey, do calculator and computer use appear to be independent events? Explain.

36. Dogs.   A census by the county dog control officer found that 18% of homes kept one dog as a pet, 4% had two dogs, and 1% had three or more. If a salesman visits two homes selected at random, what's the probability he encounters

a) no dogs?

b) some dogs?

c) dogs in each home?

d) more than one dog in each home?

37. Socks.   In your sock drawer you have 4 blue socks, 5 grey socks, and 3 black ones. Half asleep one morning, you grab 2 socks at random and put them on. Find the probability you end up wearing

a) 2 blue socks.

b) no grey socks.

c) at least 1 black sock.

d) a green sock.

e) matching socks.

38. Coins.   A coin is to be tossed 36 times.

a) What are the mean and standard deviation of the number of heads?

b) Suppose the resulting number of heads is unusual, two standard deviations above the mean. How many "extra" heads were observed?

c) If the coin were tossed 100 times, would you still consider the same number of extra heads unusual? Explain.

d) In the 100 tosses, how many extra heads would you need to observe in order to say the results were unusual?

e) Explain how these results refute the "Law of Averages" but confirm the Law of Large Numbers.

39. The Drake equation.   In 1961 astronomer Frank Drake developed an equation to try to estimate the number of extraterrestrial civilizations in our galaxy that might be able to communicate with us via radio transmissions. Now largely accepted by the scientific community, the Drake equation has helped spur efforts by radio astronomers to search for extraterrestrial intelligence. Here is the equation:

$$N_C = N \cdot f_p \cdot n_e \cdot f_l \cdot f_i \cdot f_c \cdot f_L$$

OK, it looks a little messy, but here's what it means:

Factor	What It Represents	Possible Value
N	Number of stars in the Milky Way Galaxy	200–400 billion
f_p	Probability that a star has planets	20%–50%
n_e	Number of planets in a solar system capable of sustaining earth-type life	1? 2?
f_l	Probability that life develops on a planet with a suitable environment	1%–100%
f_i	Probability that life evolves intelligence	50%?
f_c	Probability that intelligent life develops radio communication	10%–20%
f_L	Fraction of the planet's life for which the civilization survives	$\frac{1}{1,000,000}$?
N_c	Number of extraterrestrial civilizations in our galaxy with which we could communicate	?

So, how many ETs are out there? That depends; values chosen for the many factors in the equation depend on ever-evolving scientific knowledge and one's personal guesses. But now, some questions.

a) What quantity is calculated by the first product, $N \cdot f_p$?

b) What quantity is calculated by the product, $N \cdot f_p \cdot n_e \cdot f_l$?

c) What probability is calculated by the product $f_l \cdot f_i$?

d) Which of the factors in the formula are conditional probabilities? Restate each in a way that makes the condition clear.

Note: A quick Internet search will find you a site where you can play with the Drake equation yourself.

40. Recalls.   In a car rental company's fleet, 70% of the cars are American brands, 20% are Japanese, and the rest are German. The company notes that manufacturers' recalls seem to affect 2% of the American cars, but only 1% of the others.

a) What's the probability that a randomly chosen car is recalled?

b) What's the probability that a recalled car is American?

41. Pregnant? Suppose that 70% of the women who suspect they may be pregnant and purchase an in-home pregnancy test are actually pregnant. Further suppose that the test is 98% accurate. What's the probability that a woman whose test indicates that she is pregnant actually is?

42. Door prize. You are among 100 people attending a charity fundraiser at which a large-screen TV will be given away as a door prize. To determine who wins, 99 white balls and 1 red ball have been placed in a box and thoroughly mixed. The guests will line up and, one at a time, pick a ball from the box. Whoever gets the red ball wins the TV, but if the ball is white, it is returned to the box. If none of the 100 guests gets the red ball, the TV will be auctioned off for additional benefit of the charity.

a) What's the probability that the first person in line wins the TV?
b) You are the third person in line. What's the probability that you win the TV?
c) What's the probability that the charity gets to sell the TV because no one wins?
d) Suppose you get to pick your spot in line. Where would you want to be in order to maximize your chances of winning?
e) After hearing some protest about the plan, the organizers decide to award the prize by not returning the white balls to the box, thus ensuring that 1 of the 100 people will draw the red ball and win the TV. Now what position in line would you choose in order to maximize your chances?

PART V

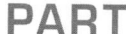

From the Data at Hand to the World at Large

Sampling Distribution Models

WHO	U.S. adults
WHAT	Belief in ghosts
WHEN	November 2005
WHERE	United States
WHY	Public attitudes

In November 2005 the Harris Poll asked 889 U.S. adults, "Do you believe in ghosts?" 40% said they did. At almost the same time, CBS News polled 808 U.S. adults and asked the same question. 48% of their respondents professed a belief in ghosts. Why the difference? This seems like a simple enough question. Should we be surprised to find that we could get proportions this different from properly selected random samples drawn from the same population? You're probably used to seeing that observations vary, but how much variability among polls should we expect to see?

Why do sample proportions vary at all? How can surveys conducted at essentially the same time by organizations asking the same questions get different results? The answer is at the heart of Statistics. The proportions vary from sample to sample because the samples are composed of different people.

It's actually pretty easy to predict how much a proportion will vary under circumstances like this. Understanding the variability of our estimates will let us actually use that variability to better understand the world.

The Central Limit Theorem for Sample Proportions

We've talked about *Think, Show,* and *Tell.* Now we have to add *Imagine.* In order to understand the CBS poll, we want to imagine the results from all the random samples of size 808 that CBS News didn't take. What would the histogram of all the sample proportions look like?

For people's belief in ghosts, where do you expect the center of that histogram to be? Of course, we don't *know* the answer to that (and probably never will). But we know that it will be at the true proportion in the population, and we can call that p. (See the Notation Alert.) For the sake of discussion here, let's suppose that 45% of all American adults believe in ghosts, so we'll use $p = 0.45$.

How about the *shape* of the histogram? We don't have to just imagine. We can simulate a bunch of random samples that we didn't really draw. Here's a histogram of the proportions saying they believe in ghosts for 2000 simulated independent samples of 808 adults when the true proportion is $p = 0.45$.

NOTATION ALERT:

The letter p is our choice for the *parameter* of the model for proportions. It violates our "Greek letters for parameters" rule, but if we stuck to that, our natural choice would be π. We could use π to be perfectly consistent, but then we'd have to write statements like $\pi = 0.46$. That just seems a bit weird to us. After all, we've known that $\pi = 3.1415926\ldots$ since the Greeks, and it's a hard habit to break.

So, we'll use p for the model parameter (the probability of a success) and $\hat{p}$ for the observed proportion in a sample. We'll also use q for the probability of a failure ($q = 1 - p$) and $\hat{q}$ for its observed value.

But be careful. We've already used capital P for a general probability. And we'll soon see another use of P in the next chapter! There are a lot of p's in this course; you'll need to think clearly about the context to keep them straight.

Pierre-Simon Laplace, 1749–1827.

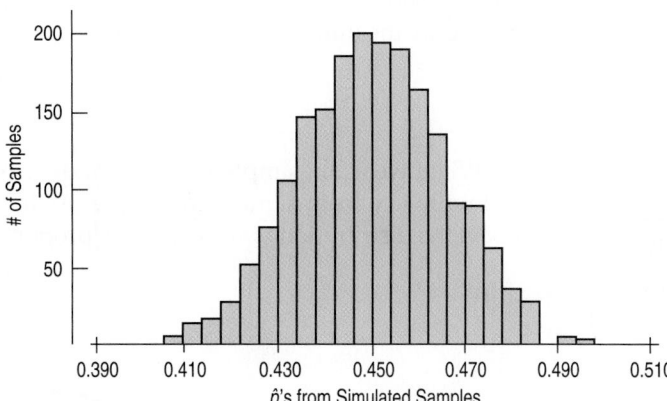

FIGURE 18.1

A histogram of sample proportions for 2000 simulated samples of 808 adults drawn from a population with p = 0.45. The sample proportions vary, but their distribution is centered at the true proportion, p.

It should be no surprise that we don't get the same proportion for each sample we draw, even though the underlying true value is the same for the population. Each $\hat{p}$ comes from a different simulated sample. The histogram above is a simulation of what we'd get if we could see *all the proportions from all possible samples.* That distribution has a special name. It is called the **sampling distribution** of the proportions.[1]

Does it surprise you that the histogram is unimodal? Symmetric? That it is centered at p? You probably don't find any of this shocking. Does the shape remind you of any model that we've discussed? It's an amazing and fortunate fact that a Normal model is just the right one for the histogram of sample proportions.

As we'll see in a few pages, this fact was proved in 1810 by the great French mathematician Pierre-Simon Laplace as part of a more general result. There is no reason you should guess that the Normal model would be the one we need here,[2] and, indeed, the importance of Laplace's result was not immediately understood by his contemporaries. But (unlike Laplace's contemporaries in 1810) we know how useful the Normal model can be.

Modeling how sample proportions vary from sample to sample is one of the most powerful ideas we'll see in this course. A **sampling distribution model** for how a sample proportion varies from sample to sample allows us to quantify that variation and to talk about how likely it is that we'd observe a sample proportion in any particular interval.

To use a Normal model, we need to specify two parameters: its mean and standard deviation. The center of the histogram is naturally at p, so we'll put μ, the mean of the Normal, at p.

What about the standard deviation? Usually the mean gives us no information about the standard deviation. Suppose we told you that a batch of bike helmets had a mean diameter of 26 centimeters and asked what the standard deviation was. If you said, "I have no idea," you'd be exactly right. There's no information about σ from knowing the value of μ.

But there's a special fact about proportions. With proportions we get something for free. Once we know the mean, p, we automatically also know the standard deviation. We saw in the last chapter that for a Binomial model the standard deviation of the *number* of successes is $\sqrt{npq}$. Now we want the standard deviation

---

[1] A word of caution. Until now we've been plotting the *distribution of the sample*, a display of the actual data that were collected in that one sample. But now we've plotted the *sampling distribution*; a display of summary statistics ($\hat{p}$'s, for example) for many different samples. "Sample distribution" and "sampling distribution" sound a lot alike, but they refer to very different things. (Sorry about that—we didn't make up the terms. It's just the way it is.) And the distinction is critical. Whenever you read or write something about one of these, think very carefully about what the words signify.

[2] Well, the fact that we spent most of Chapter 6 on the Normal model might have been a hint.

of the *proportion* of successes, $\hat{p}$. The sample proportion $\hat{p}$ is the number of successes divided by the number of trials, n, so the standard deviation is also divided by n:

$$\sigma(\hat{p}) = SD(\hat{p}) = \frac{\sqrt{npq}}{n} = \sqrt{\frac{pq}{n}}.$$

A S *Simulation:* **Simulating Sampling Distributions.** Watch the Normal model appear from random proportions.

When we draw simple random samples of n individuals, the proportions we find will vary from sample to sample. As long as n is reasonably large,[3] we can model the distribution of these sample proportions with a probability model that is

$$N\left(p, \sqrt{\frac{pq}{n}}\right).$$

FIGURE 18.2

A Normal model centered at p with a standard deviation of $\sqrt{\dfrac{pq}{n}}$ is a good model for a collection of proportions found for many random samples of size n from a population with success probability p.

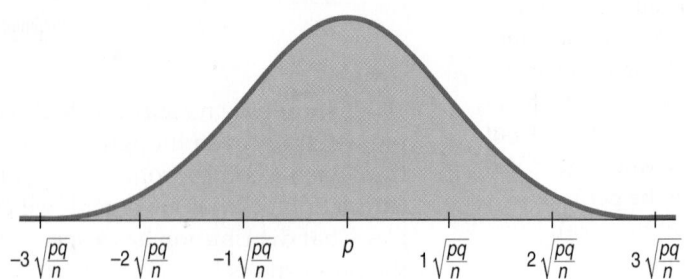

NOTATION ALERT:

In Chapter 8 we introduced $\hat{y}$ as the predicted value for y. The "hat" here plays a similar role. It indicates that $\hat{p}$—the observed proportion in our data—is our *estimate* of the parameter p.

Although we'll never know the true proportion of adults who believe in ghosts, we're supposing it to be 45%. Once we put the center at $p = 0.45$, the standard deviation for the CBS poll is

$$SD(\hat{p}) = \sqrt{\frac{pq}{n}} = \sqrt{\frac{(0.45)(0.55)}{808}} = 0.0175, \text{ or } 1.75\%.$$

Here's a picture of the Normal model for our simulation histogram:

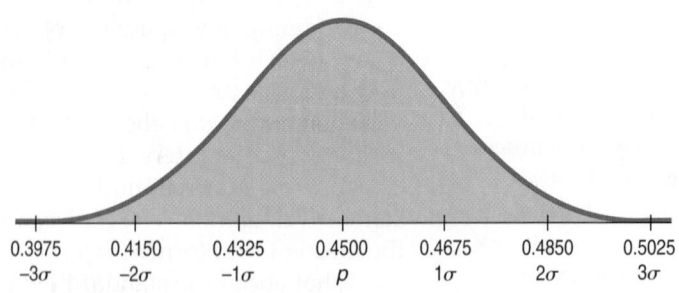

FIGURE 18.3

Using 0.45 for p gives this Normal model for Figure 18.1's histogram of the sample proportions of adults believing in ghosts (n = 808).

A S *Simulation:* **The Standard Deviation of a Proportion.** Do you believe this formula for standard deviation? Don't just take our word for it—convince yourself with an experiment.

Because we have a Normal model, we can use the 68–95–99.7 Rule or look up other probabilities using a table or technology. For example, we know that 95% of Normally distributed values are within two standard deviations of the mean, so we should not be surprised if 95% of various polls gave results that were near 45% but varied above and below that by no more than two standard deviations. Since $2 \times 1.75\% = 3.5\%$,[4] we see that the CBS poll estimating belief in ghosts at 48% is *consistent* with our guess of 45%. This is what we mean by **sampling error.** It's not really an *error* at all, but just *variability* you'd expect to see from one sample to another. A better term would be **sampling variability.**

---

[3] For smaller n, we can just use a Binomial model.
[4] The standard deviation is 1.75%. Remember that the standard deviation always has the same units as the data. Here our units are %. But that can be confusing, because the standard deviation is not 1.75% of anything. It is 1.75 percentage points. If that's confusing, try writing the units as "percentage points" instead of %.

How Good Is the Normal Model?

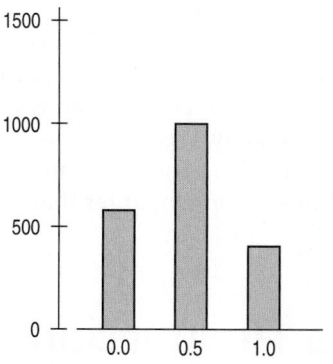

FIGURE 18.4

Proportions from samples of size 2 can take on only three possible values. A Normal model does not work well.

Stop and think for a minute about what we've just said. It's a remarkable claim. We've said that if we draw repeated random samples of the same size, n, from some population and measure the proportion, $\hat{p}$, we see in each sample, then the collection of these proportions will pile up around the underlying population proportion, p, and that a histogram of the sample proportions can be modeled well by a Normal model.

There must be a catch. Suppose the samples were of size 2, for example. Then the only possible proportion values would be 0, 0.5, and 1. There's no way the histogram could ever look like a Normal model with only three possible values for the variable.

Well, there *is* a catch. The claim is only approximately true. (But, that's OK. After all, models are only supposed to be approximately true.) And the model becomes a better and better representation of the distribution of the sample proportions as the sample size gets bigger.[5] Samples of size 1 or 2 just aren't going to work very well. But the distributions of proportions of many larger samples do have histograms that are remarkably close to a Normal model.

Assumptions and Conditions

To use a model, we usually must make some assumptions. To use the sampling distribution model for sample proportions, we need two assumptions:

The Independence Assumption: The sampled values must be independent of each other.

The Sample Size Assumption: The sample size, n, must be large enough.

Of course, assumptions are hard—often impossible—to check. That's why we *assume* them. But, as we saw in Chapter 8, we should check to see whether the assumptions are reasonable. To think about the Independence Assumption, we often wonder whether there is any reason to think that the data values might affect each other. Fortunately, we can often check *conditions* that provide information about the assumptions. Check these conditions before using the Normal to model the distribution of sample proportions:

Randomization Condition: If your data come from an experiment, subjects should have been randomly assigned to treatments. If you have a survey, your sample should be a simple random sample of the population. If some other sampling design was used, be sure the sampling method was not biased and that the data are representative of the population.

10% Condition: The sample size, n, must be no larger than 10% of the population. For national polls, the total population is usually very large, so the sample is a small fraction of the population.

Success/Failure Condition: The sample size has to be big enough so that we expect at least 10 successes and at least 10 failures. When np and nq are at least 10, we have enough data for sound conclusions. For the CBS survey, a "success" might be believing in ghosts. With $p = 0.45$, we expect $808 \times 0.45 = 364$ successes and $808 \times 0.55 = 444$ failures. Both are at least 10, so we certainly expect enough successes and enough failures for the condition to be satisfied.

> The terms "success" and "failure" for the outcomes that have probability p and q are common in Statistics. But they are completely arbitrary labels. When we say that a disease occurs with probability p, we certainly don't mean that getting sick is a "success" in the ordinary sense of the word.

[5] Formally, we say the claim is true in the limit as n grows.

These last two conditions seem to conflict with each other. The **Success/ Failure Condition** wants sufficient data. How much depends on p. If p is near 0.5, we need a sample of only 20 or so. If p is only 0.01, however, we'd need 1000. But the **10% Condition** says that a sample should be no larger than 10% of the population. If you're thinking, "Wouldn't a larger sample be better?" you're right of course. It's just that if the sample were more than 10% of the population, we'd need to use different methods to analyze the data. Fortunately, this isn't usually a problem in practice. Often, as in polls that sample from all U.S. adults or industrial samples from a day's production, the populations are much larger than 10 times the sample size.

A Sampling Distribution Model for a Proportion

We've simulated repeated samples and looked at a histogram of the sample proportions. We modeled that histogram with a Normal model. Why do we bother to model it? Because this model will give us insight into how much the sample proportion can vary from sample to sample. We've simulated many of the other random samples we might have gotten. The model is an attempt to show the distribution from *all* the random samples. But how do we know that a Normal model will really work? Is this just an observation based on some simulations that *might* be approximately true some of the time?

It turns out that this model can be justified theoretically and that the larger the sample size, the better the model works. That's the result Laplace proved. We won't bother you with the math because, in this instance, it really wouldn't help your understanding.[6] Nevertheless, the fact that we can think of the sample proportion as a random variable taking on a different value in each random sample, and then say something this specific about the distribution of those values, is a fundamental insight—one that we will use in each of the next four chapters.

We have changed our point of view in a very important way. No longer is a proportion something we just compute for a set of data. We now see it as a random variable quantity that has a probability distribution, and thanks to Laplace we have a model for that distribution. We call that the **sampling distribution model** for the proportion, and we'll make good use of it.

A S *Simulation: Simulate the* **Sampling Distribution Model of a Proportion.** You probably don't want to work through the formal mathematical proof; a simulation is far more convincing!

We have now answered the question raised at the start of the chapter. To know how variable a sample proportion is, we need to know the proportion and the size of the sample. That's all.

THE SAMPLING DISTRIBUTION MODEL FOR A PROPORTION

Provided that the sampled values are independent and the sample size is large enough, the sampling distribution of $\hat{p}$ is modeled by a Normal model with mean $\mu(\hat{p}) = p$ and standard deviation $SD(\hat{p}) = \sqrt{\dfrac{pq}{n}}$.

Without the sampling distribution model, the rest of Statistics just wouldn't exist. Sampling models are what makes Statistics work. They inform us about the amount of variation we should expect when we sample. Suppose we spin a coin 100 times in order to decide whether it's fair or not. If we get 52 heads, we're probably not surprised. Although we'd expect 50 heads, 52 doesn't seem particularly unusual for a fair coin. But we would be surprised to see 90 heads; that might really make us doubt that the coin is fair. How about 64 heads? Harder to say. That's a case where we need the sampling distribution model. The sampling model quantifies the variability, telling us how surprising any sample proportion is. And

---

[6] The proof is pretty technical. We're not sure it helps *our* understanding all that much either.

it enables us to make informed decisions about how precise our estimate of the true proportion might be. That's exactly what we'll be doing for the rest of this book.

Sampling distribution models act as a bridge from the real world of data to the imaginary model of the statistic and enable us to say something about the population when all we have is data from the real world. This is the huge leap of Statistics. Rather than thinking about the sample proportion as a fixed quantity calculated from our data, we now think of it as a random variable—our value is just one of many we might have seen had we chosen a different random sample. By imagining what *might* happen if we were to draw many, many samples from the same population, we can learn a lot about how close the statistics computed from our one particular sample may be to the corresponding population parameters they estimate. That's the path to the *margin of error* you hear about in polls and surveys. We'll see how to determine that in the next chapter.

FOR EXAMPLE   Using the sampling distribution model for proportions

---

The Centers for Disease Control and Prevention report that 22% of 18-year-old women in the United States have a body mass index (BMI)[7] of 25 or more—a value considered by the National Heart Lung and Blood Institute to be associated with increased health risk.

As part of a routine health check at a large college, the physical education department usually requires students to come in to be measured and weighed. This year, the department decided to try out a self-report system. It asked 200 randomly selected female students to report their heights and weights (from which their BMIs could be calculated). Only 31 of these students had BMIs greater than 25.

Question: Is this proportion of high-BMI students unusually small?

First, check the conditions:

✔ Randomization Condition: The department drew a random sample, so the respondents should be independent and randomly selected from the population.

✔ 10% Condition: 200 respondents is less than 10% of all the female students at a "large college."

✔ Success/Failure Condition: The department expected $np = 200(0.22) = 44$ "successes" and $nq = 200(0.78) = 156$ "failures," both at least 10.

It's okay to use a Normal model to describe the sampling distribution of the proportion of respondents with BMIs above 25.

The phys ed department observed $\hat{p} = \dfrac{31}{200} = 0.155$.

The department expected $E(\hat{p}) = p = 0.22$, with $SD(\hat{p}) = \sqrt{\dfrac{pq}{n}} = \sqrt{\dfrac{(0.22)(0.78)}{200}} = 0.029$,

so $z = \dfrac{\hat{p} - p}{SD(\hat{p})} = \dfrac{0.155 - 0.22}{0.029} = -2.24$.

By the 68–95–99.7 Rule, I know that values more than 2 standard deviations below the mean of a Normal model show up less than 2.5% of the time. Perhaps women at this college differ from the general population, or self-reporting may not provide accurate heights and weights.

---

[7] BMI = weight in kg/(height in m)^2.

418 CHAPTER 18 Sampling Distribution Models

JUST CHECKING

1. You want to poll a random sample of 100 students on campus to see if they are in favor of the proposed location for the new student center. Of course, you'll get just one number, your sample proportion, $\hat{p}$. But if you imagined all the possible samples of 100 students you could draw and imagined the histogram of all the sample proportions from these samples, what shape would it have?

2. Where would the center of that histogram be?

3. If you think that about half the students are in favor of the plan, what would the standard deviation of the sample proportions be?

STEP-BY-STEP EXAMPLE | **Working with Sampling Distribution Models for Proportions**

Suppose that about 13% of the population is left-handed.[8] A 200-seat school auditorium has been built with 15 "lefty seats," seats that have the built-in desk on the left rather than the right arm of the chair. (For the right-handed readers among you, have you ever tried to take notes in a chair with the desk on the left side?)

Question: In a class of 90 students, what's the probability that there will not be enough seats for the left-handed students?

Plan State what we want to know.

I want to find the probability that in a group of 90 students, more than 15 will be left-handed. Since 15 out of 90 is 16.7%, I need the probability of finding more than 16.7% left-handed students out of a sample of 90 if the proportion of lefties is 13%.

Model Think about the assumptions and check the conditions.

You might be able to think of cases where the **Independence Assumption** is not plausible—for example, if the students are all related, or if they were selected for being left- or right-handed. But for a random sample, the assumption of independence seems reasonable.

✓ **Independence Assumption:** *It is reasonable to assume that the probability that one student is left-handed is not changed by the fact that another student is right- or left-handed.*

✓ **Randomization Condition:** *The 90 students in the class can be thought of as a random sample of students.*

✓ **10% Condition:** *90 is surely less than 10% of the population of all students. (Even if the school itself is small, I'm thinking of the population of all possible students who could have gone to the school.)*

✓ **Success/Failure Condition:**

$$np = 90(0.13) = 11.7 \geq 10$$
$$nq = 90(0.87) = 78.3 \geq 10$$

---

[8] Actually, it's quite difficult to get an accurate estimate of the proportion of lefties in the population. Estimates range from 8% to 15%.

State the parameters and the sampling distribution model.	The population proportion is $p = 0.13$. The conditions are satisfied, so I'll model the sampling distribution of $\hat{p}$ with a Normal model with mean 0.13 and a standard deviation of $$SD(\hat{p}) = \sqrt{\frac{pq}{n}} = \sqrt{\frac{(0.13)(0.87)}{90}} \approx 0.035$$ My model for $\hat{p}$ is $N(0.13, 0.035)$.
 **Plot** Make a picture. Sketch the model and shade the area we're interested in, in this case the area to the right of 16.7%. **Mechanics** Use the standard deviation as a ruler to find the z-score of the cutoff proportion. We see that 16.7% lefties would be just over one standard deviation above the mean. Find the resulting probability from a table of Normal probabilities, a computer program, or a calculator.	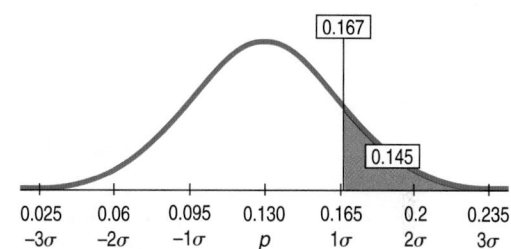 $$z = \frac{\hat{p} - p}{SD(\hat{p})} = \frac{0.167 - 0.13}{0.035} = 1.06$$ $$P(\hat{p} > 0.167) = P(z > 1.06) = 0.1446$$
 **Conclusion** Interpret the probability in the context of the question.	There is about a 14.5% chance that there will not be enough seats for the left-handed students in the class.

What About Quantitative Data?

Proportions summarize categorical variables. And the Normal sampling distribution model looks like it is going to be very useful. But can we do something similar with quantitative data?

Of course we can (or we wouldn't have asked). Even more remarkable, not only can we use all of the same concepts, but almost the same model, too.

What are the concepts? We know that when we sample at random or randomize an experiment, the results we get will vary from sample-to-sample and from experiment-to-experiment. The Normal model seems an incredibly simple way to summarize all that variation. Could something that simple work for means? We won't keep you in suspense. It turns out that means also have a sampling distribution that we can model with a Normal model. And it turns out that Laplace's theoretical result applies to means, too. As we did with proportions, we can get some insight from a simulation.

Simulating the Sampling Distribution of a Mean

Here's a simple simulation. Let's start with one fair die. If we toss this die 10,000 times, what should the histogram of the numbers on the face of the die look like? Here are the results of a simulated 10,000 tosses:

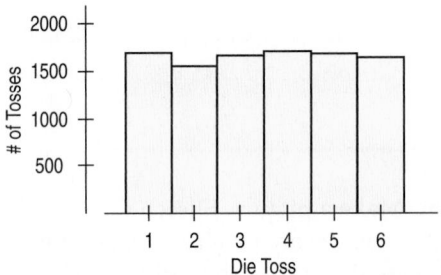

Now let's toss a *pair* of dice and record the average of the two. If we repeat this (or at least simulate repeating it) 10,000 times, recording the average of each pair, what will the histogram of these 10,000 averages look like? Before you look, think a minute. Is getting an average of 1 on *two* dice as likely as getting an average of 3 or 3.5?

Let's see:

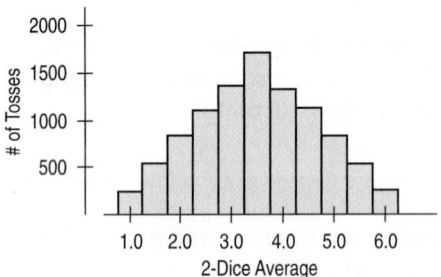

We're much more likely to get an average near 3.5 than we are to get one near 1 or 6. Without calculating those probabilities exactly, it's fairly easy to see that the *only* way to get an average of 1 is to get two 1's. To get a total of 7 (for an average of 3.5), though, there are many more possibilities. This distribution even has a name: the *triangular* distribution.

What if we average 3 dice? We'll simulate 10,000 tosses of 3 dice and take their average:

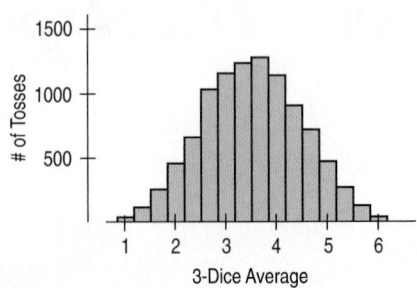

What's happening? First notice that it's getting harder to have averages near the ends. Getting an average of 1 or 6 with 3 dice requires all three to come up 1 or 6, respectively. That's less likely than for 2 dice to come up both 1 or both 6. The distribution is being pushed toward the middle. But what's happening to the shape? (This distribution doesn't have a name, as far as we know.)

Let's continue this simulation to see what happens with larger samples. Here's a histogram of the averages for 10,000 tosses of 5 dice:

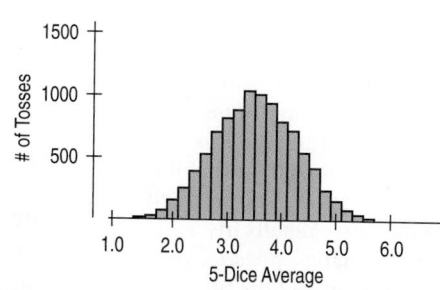

The pattern is becoming clearer. Two things continue to happen. The first fact we knew already from the Law of Large Numbers. It says that as the sample size (number of dice) gets larger, each sample average is more likely to be closer to the population mean. So, we see the shape continuing to tighten around 3.5. But the shape of the distribution is the surprising part. It's becoming bell-shaped. And not just bell-shaped; it's approaching the Normal model.

Are you convinced? Let's skip ahead and try 20 dice. The histogram of averages for 10,000 throws of 20 dice looks like this:

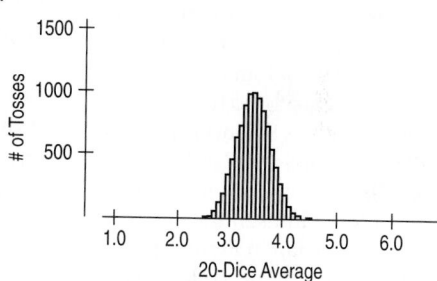

Now we see the Normal shape again (and notice how much smaller the spread is). But can we count on this happening for situations other than dice throws? What kinds of sample means have sampling distributions that we can model with a Normal model? It turns out that Normal models work well amazingly often.

The Fundamental Theorem of Statistics

The dice simulation may look like a special situation, but it turns out that what we saw with dice is true for means of repeated samples for almost every situation. When we looked at the sampling distribution of a proportion, we had to check only a few conditions. For means, the result is even more remarkable. *There are almost no conditions at all.*

Let's say that again: The sampling distribution of *any* mean becomes more nearly Normal as the sample size grows. All we need is for the observations to be independent and collected with randomization. We don't even care about the shape of the population distribution![9] This surprising fact is the result Laplace proved in a fairly general form in 1810. At the time, Laplace's theorem caused quite a stir (at least in mathematics circles) because it is so unintuitive. Laplace's result is called the **Central Limit Theorem**[10] (CLT).

"The theory of probabilities is at bottom nothing but common sense reduced to calculus."

—Laplace, in *Théorie analytique des probabilités*, 1812

---

[9] OK, one technical condition. The data must come from a population with a finite variance. You probably can't imagine a population with an infinite variance, but statisticians can construct such things, so we have to discuss them in footnotes like this. It really makes no difference in how you think about the important stuff, so you can just forget we mentioned it.

[10] The word "central" in the name of the theorem means "fundamental." It doesn't refer to the center of a distribution.

Why should the Normal model show up again for the sampling distribution of means as well as proportions? We're not going to try to persuade you that it is obvious, clear, simple, or straightforward. In fact, the CLT is surprising and a bit weird. Not only does the distribution of means of many random samples get closer and closer to a Normal model as the sample size grows, *this is true regardless of the shape of the population distribution!* Even if we sample from a skewed or bimodal population, the Central Limit Theorem tells us that means of repeated random samples will tend to follow a Normal model as the sample size grows. Of course, you won't be surprised to learn that it works better and faster the closer the population distribution is to a Normal model. And it works better for larger samples. If the data come from a population that's exactly Normal to start with, then the observations themselves are Normal. If we take samples of size 1, their "means" are just the observations—so, of course, they have Normal sampling distribution. But now suppose the population distribution is very skewed (like the CEO data from Chapter 5, for example). The CLT works, although it may take a sample size of dozens or even hundreds of observations for the Normal model to work well.

For example, think about a really bimodal population, one that consists of only 0's and 1's. The CLT says that even means of samples from this population will follow a Normal sampling distribution model. But wait. Suppose we have a categorical variable and we assign a 1 to each individual in the category and a 0 to each individual not in the category. And then we find the mean of these 0's and 1's. That's the same as counting the number of individuals who are in the category and dividing by n. That mean will be . . . the *sample proportion*, $\hat{p}$, of individuals who are in the category (a "success"). So maybe it wasn't so surprising after all that proportions, like means, have Normal sampling distribution models; they are actually just a special case of Laplace's remarkable theorem. Of course, for such an extremely bimodal population, we'll need a reasonably large sample size—and that's where the special conditions for proportions come in.

> ## THE CENTRAL LIMIT THEOREM (CLT)
> The mean of a random sample is a random variable whose sampling distribution can be approximated by a Normal model. The larger the sample, the better the approximation will be.

Assumptions and Conditions

The CLT requires essentially the same assumptions as we saw for modelling proportions:

> **Independence Assumption:** The sampled values must be independent of each other.
>
> **Sample Size Assumption:** The sample size must be sufficiently large.

We can't check these directly, but we can think about whether the **Independence Assumption** is plausible. We can also check some related conditions:

> **Randomization Condition:** The data values must be sampled randomly, or the concept of a sampling distribution makes no sense.
>
> **10% Condition:** When the sample is drawn without replacement (as is usually the case), the sample size, n, should be no more than 10% of the population.
>
> **Large Enough Sample Condition:** Although the CLT tells us that a Normal model is useful in thinking about the behavior of sample means when the

sample size is large enough, it doesn't tell us how large a sample we need. The truth is, it depends; there's no one-size-fits-all rule. If the population is unimodal and symmetric, even a fairly small sample is okay. If the population is strongly skewed, like the compensation for CEOs we looked at in Chapter 5, it can take a pretty large sample to allow use of a Normal model to describe the distribution of sample means. For now you'll just need to think about your sample size in the context of what you know about the population, and then tell whether you believe the **Large Enough Sample Condition** has been met.

But Which Normal?

Activity: **The Standard Deviation of Means.** Experiment to see how the variability of the mean changes with the sample size.

The CLT says that the sampling distribution of any mean or proportion is approximately Normal. But which Normal model? We know that any Normal is specified by its mean and standard deviation. For proportions, the sampling distribution is centered at the population proportion. For means, it's centered at the population mean. What else would we expect?

What about the standard deviations, though? We noticed in our dice simulation that the histograms got narrower as we averaged more and more dice together. This shouldn't be surprising. Means vary less than the individual observations. Think about it for a minute. Which would be more surprising, having *one* person in your Statistics class who is over 6'9" tall or having the *mean* of 100 students taking the course be over 6'9"? The first event is fairly rare.[11] You may have seen somebody this tall in one of your classes sometime. But finding a class of 100 whose mean height is over 6'9" tall just won't happen. Why? Because *means have smaller standard deviations than individuals.*

How much smaller? Well, we have good news and bad news. The good news is that the standard deviation of $\bar{y}$ falls as the sample size grows. The bad news is that it doesn't drop as fast as we might like. It only goes down by the *square root* of the sample size. Why? The Math Box will show you that the Normal model for the sampling distribution of the mean has a standard deviation equal to

$$SD(\bar{y}) = \frac{\sigma}{\sqrt{n}}$$

where σ is the standard deviation of the population. To emphasize that this is a standard deviation *parameter* of the sampling distribution model for the sample mean, $\bar{y}$, we write $SD(\bar{y})$ or $\sigma(\bar{y})$.

Activity: **The Sampling Distribution of the Mean.** The CLT tells us what to expect. In this activity you can work with the CLT or simulate it if you prefer.

> **THE SAMPLING DISTRIBUTION MODEL FOR A MEAN (CLT)**
> When a random sample is drawn from any population with mean μ and standard deviation σ, its sample mean, $\bar{y}$, has a sampling distribution with the same *mean* μ but whose *standard deviation* is $\frac{\sigma}{\sqrt{n}}$ (and we write $\sigma(\bar{y}) = SD(\bar{y}) = \frac{\sigma}{\sqrt{n}}$). No matter what population the random sample comes from, the *shape* of the sampling distribution is approximately Normal as long as the sample size is large enough. The larger the sample used, the more closely the Normal approximates the sampling distribution for the mean.

---

[11] If students are a random sample of adults, fewer than 1 out of 10,000 should be taller than 6'9". Why might college students not really be a random sample with respect to height? Even if they're not a perfectly random sample, a college student over 6'9" tall is still rare.

MATH BOX

We know that $\bar{y}$ is a sum divided by n:

$$\bar{y} = \frac{y_1 + y_2 + y_3 + \cdots + y_n}{n}.$$

As we saw in Chapter 16, when a random variable is divided by a constant its variance is divided by the *square* of the constant:

$$Var(\bar{y}) = \frac{Var(y_1 + y_2 + y_3 + \cdots + y_n)}{n^2}.$$

To get our sample, we draw the y's randomly, ensuring they are independent. For independent random variables, variances add:

$$Var(\bar{y}) = \frac{Var(y_1) + Var(y_2) + Var(y_3) + \cdots + Var(y_n)}{n^2}.$$

All n of the y's were drawn from our population, so they all have the same variance, σ^2:

$$Var(\bar{y}) = \frac{\sigma^2 + \sigma^2 + \sigma^2 + \cdots + \sigma^2}{n^2} = \frac{n\sigma^2}{n^2} = \frac{\sigma^2}{n}.$$

The standard deviation of $\bar{y}$ is the square root of this variance:

$$SD(\bar{y}) = \sqrt{\frac{\sigma^2}{n}} = \frac{\sigma}{\sqrt{n}}.$$

We now have two closely related sampling distribution models that we can use when the appropriate assumptions and conditions are met. Which one we use depends on which kind of data we have:

▶ When we have categorical data, we calculate a sample proportion, $\hat{p}$; the sampling distribution of this random variable has a Normal model with a mean at the true proportion ("Greek letter") p and a standard deviation of $SD(\hat{p}) = \sqrt{\frac{pq}{n}} = \frac{\sqrt{pq}}{\sqrt{n}}$. We'll use this model in Chapters 19 through 22.

▶ When we have quantitative data, we calculate a sample mean, $\bar{y}$; the sampling distribution of this random variable has a Normal model with a mean at the true mean, μ, and a standard deviation of $SD(\bar{y}) = \frac{\sigma}{\sqrt{n}}$. We'll use this model in Chapters 23, 24, and 25.

The means of these models are easy to remember, so all you need to be careful about is the standard deviations. Remember that these are standard deviations of the *statistics* $\hat{p}$ and $\bar{y}$. They both have a square root of n in the denominator. That tells us that the larger the sample, the less either statistic will vary. The only difference is in the numerator. If you just start by writing $SD(\bar{y})$ for quantitative data and $SD(\hat{p})$ for categorical data, you'll be able to remember which formula to use.

Using the CLT for means

Recap: A college physical education department asked a random sample of 200 female students to self-report their heights and weights, but the percentage of students with body mass indexes over 25 seemed suspiciously low. One possible explanation may be that the respondents "shaded" their weights down a bit. The CDC reports that the mean weight of 18-year-old women is 143.74 lb, with a standard deviation of 51.54 lb, but these 200 randomly selected women reported a mean weight of only 140 lb.

Question: Based on the Central Limit Theorem and the 68–95–99.7 Rule, does the mean weight in this sample seem exceptionally low, or might this just be random sample-to-sample variation?

The conditions check out okay:

✔ Randomization Condition: The women were a random sample and their weights can be assumed to be independent.

✔ 10% Condition: They sampled fewer than 10% of all women at the college.

✔ Large Enough Sample Condition: The distribution of college women's weights is likely to be unimodal and reasonably symmetric, so the CLT applies to means of even small samples; 200 values is plenty.

The sampling model for sample means is approximately Normal with $E(\bar{y}) = 143.7$ and $SD(\bar{y}) = \dfrac{\sigma}{\sqrt{n}} = \dfrac{51.54}{\sqrt{200}} = 3.64$. The expected distribution of sample means is:

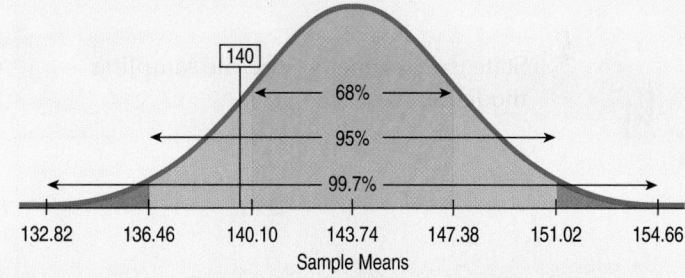

The 68–95–99.7 Rule suggests that although the reported mean weight of 140 pounds is somewhat lower than expected, it does not appear to be unusual. Such variability is not all that extraordinary for samples of this size.

STEP-BY-STEP EXAMPLE   Working with the Sampling Distribution Model for the Mean

The Centers for Disease Control and Prevention reports that the mean weight of adult men in the United States is 190 lb with a standard deviation of 59 lb.[12]

Question: An elevator in our building has a weight limit of 10 persons or 2500 lb. What's the probability that if 10 men get on the elevator, they will overload its weight limit?

Plan State what we want to know.

Asking the probability that the total weight of a sample of 10 men exceeds 2500 pounds is equivalent to asking the probability that their mean weight is greater than 250 pounds.

---

[12] Cynthia L. Ogden, Cheryl D. Fryar, Margaret D. Carroll, and Katherine M. Flegal, *Mean Body Weight, Height, and Body Mass Index, United States 1960–2002, Advance Data from Vital and Health Statistics Number 347*, Oct. 27, 2004. https//www.cdc.gov/nchs

Model Think about the assumptions and check the conditions.

✔ **Independence Assumption:** It's reasonable to think that the weights of 10 randomly sampled men will be independent of each other. (But there could be exceptions— for example, if they were all from the same family or if the elevator were in a building with a diet clinic!)

✔ **Randomization Condition:** I'll assume that the 10 men getting on the elevator are a random sample from the population.

✔ **10% Condition:** 10 men is surely less than 10% of the population of possible elevator riders.

Note that if the sample were larger we'd be less concerned about the shape of the distribution of all weights.

✔ **Large Enough Sample Condition:** I suspect the distribution of population weights is roughly unimodal and symmetric, so my sample of 10 men seems large enough.

State the parameters and the sampling model.

The mean for all weights is $\mu = 190$ and the standard deviation is $\sigma = 59$ pounds. Since the conditions are satisfied, the CLT says that the sampling distribution of $\bar{y}$ has a Normal model with mean 190 and standard deviation

$$SD(\bar{y}) = \frac{\sigma}{\sqrt{n}} = \frac{59}{\sqrt{10}} \approx 18.66$$

Plot Make a picture. Sketch the model and shade the area we're interested in. Here the mean weight of 250 pounds appears to be far out on the right tail of the curve.

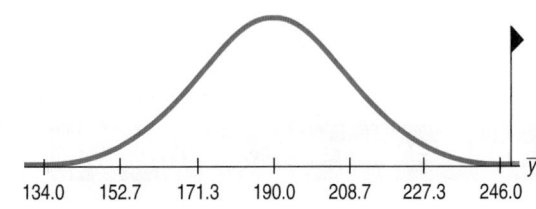

134.0   152.7   171.3   190.0   208.7   227.3   246.0 $\bar{y}$

Mechanics Use the standard deviation as a ruler to find the z-score of the cutoff mean weight. We see that an average of 250 pounds is more than 3 standard deviations above the mean.

$$z = \frac{\bar{y} - \mu}{SD(\bar{y})} = \frac{250 - 190}{18.66} = 3.21$$

Find the resulting probability from a table of Normal probabilities such as Table Z, a computer program, or a calculator.

$$P(\bar{y} > 250) = P(z > 3.21) = 0.0007$$

Conclusion Interpret your result in the proper context, being careful to relate it to the original question.

The chance that a random collection of 10 men will exceed the elevator's weight limit is only 0.0007. So, if they are a random sample, it is quite unlikely that 10 people will exceed the total weight allowed on the elevator.

About Variation

"The n's justify the means."
—Apocryphal
statistical saying

Means vary less than individual data values. That makes sense. If the same test is given to many sections of a large course and the class average is, say, 80%, some students may score 95% because individual scores vary a lot. But we'd be shocked (and pleased!) if the *average* score of the students in any section was 95%. Averages are much less variable. Not only do group averages vary less than individual values, but common sense suggests that averages should be more consistent for larger groups. The Central Limit Theorem confirms this hunch; the fact that $SD(\bar{y}) = \dfrac{\sigma}{\sqrt{n}}$ has n in the denominator shows that the variability of sample means decreases as the sample size increases. There's a catch, though. The standard deviation of the sampling distribution declines only with the square root of the sample size and not, for example, with $1/n$.

The mean of a random sample of 4 has half $\left(\dfrac{1}{\sqrt{4}} = \dfrac{1}{2}\right)$ the standard deviation of an individual data value. To cut the standard deviation in half again, we'd need a sample of 16, and a sample of 64 to halve it once more.

If only we had a much larger sample, we could get the standard deviation of the sampling distribution *really* under control so that the sample mean could tell us still more about the unknown population mean, but larger samples cost more and take longer to survey. And while we're gathering all that extra data, the population itself may change, or a news story may alter opinions. There are practical limits to most sample sizes. As we shall see, that nasty square root limits how much we can make a sample tell about the population. This is an example of something that's known as the Law of Diminishing Returns.

A Billion Dollar Misunderstanding? In the late 1990s the Bill and Melinda Gates Foundation began funding an effort to encourage the breakup of large schools into smaller schools. Why? It had been noticed that smaller schools were more common among the best-performing schools than one would expect. In time, the Annenberg Foundation, the Carnegie Corporation, the Center for Collaborative Education, the Center for School Change, Harvard's Change Leadership Group, the Open Society Institute, Pew Charitable Trusts, and the U.S. Department of Education's Smaller Learning Communities Program all supported the effort. Well over a billion dollars was spent to make schools smaller.

But was it all based on a misunderstanding of sampling distributions? Statisticians Howard Wainer and Harris Zwerling[13] looked at the mean test scores of schools in Pennsylvania. They found that indeed 12% of the top-scoring 50 schools were from the smallest 3% of Pennsylvania schools—substantially more than the 3% we'd naively expect. But then they looked at the *bottom* 50. There they found that 18% were small schools! The explanation? Mean test scores are, well, means. We are looking at a rough real-world simulation in which each school is a trial. Even if all Pennsylvania schools were equivalent, we'd expect their mean scores to vary. How much? The CLT tells us that means of test scores vary according to $\dfrac{\sigma}{\sqrt{n}}$. Smaller schools have (by definition) smaller n's, so the sampling distributions of their mean scores naturally have larger standard deviations. It's natural, then, that small schools have both higher and lower mean scores.

---

[13] Wainer, H. and Zwerling, H., "Legal and empirical evidence that smaller schools do not improve student achievement," *The Phi Delta Kappan* 2006 87:300–303. Discussed in Howard Wainer, "The Most Dangerous Equation," *American Scientist*, May–June 2007, pp. 249–256; also at www.Americanscientist.org.

On October 26, 2005, *The Seattle Times* reported:

[T]he Gates Foundation announced last week it is moving away from its emphasis on converting large high schools into smaller ones and instead giving grants to specially selected school districts with a track record of academic improvement and effective leadership. Education leaders at the Foundation said they concluded that improving classroom instruction and mobilizing the resources of an entire district were more important first steps to improving high schools than breaking down the size.

The Real World and the Model World

Be careful. We have been slipping smoothly between the real world, in which we draw random samples of data, and a magical mathematical model world, in which we describe how the sample means and proportions we observe in the real world behave as random variables in all the random samples that we might have drawn. Now we have *two* distributions to deal with. The first is the real-world distribution of the sample, which we might display with a histogram (for quantitative data) or with a bar chart or table (for categorical data). The second is the math world *sampling distribution model* of the statistic, a Normal model based on the Central Limit Theorem. Don't confuse the two.

For example, don't mistakenly think the CLT says that the *data* are Normally distributed as long as the sample is large enough. In fact, as samples get larger, we expect the distribution of the data to look more and more like the population from which they are drawn—skewed, bimodal, whatever—but not necessarily Normal. You can collect a sample of CEO salaries for the next 1000 years,[14] but the histogram will never look Normal. It will be skewed to the right. The Central Limit Theorem doesn't talk about the distribution of the data from the sample. It talks about the sample *means* and sample *proportions* of many different random samples drawn from the same population. Of course, the CLT does require that the sample be big enough when the population shape is not unimodal and symmetric, but the fact that, even then, a Normal model is useful is still a very surprising and powerful result.

JUST CHECKING

4. Human gestation times have a mean of about 266 days, with a standard deviation of about 16 days. If we record the gestation times of a sample of 100 women, do we know that a histogram of the times will be well modeled by a Normal model?

5. Suppose we look at the *average* gestation times for a sample of 100 women. If we imagined all the possible random samples of 100 women we could take and looked at the histogram of all the sample means, what shape would it have?

6. Where would the center of that histogram be?

7. What would be the standard deviation of that histogram?

---

[14] Don't forget to adjust for inflation.

Sampling Distribution Models

Let's summarize what we've learned about sampling distributions. At the heart is the idea that *the statistic itself is a random variable*. We can't know what our statistic will be because it comes from a random sample. It's just one instance of something that happened for our particular random sample. A different random sample would have given a different result. This sample-to-sample variability is what generates the sampling distribution. The sampling distribution shows us the distribution of possible values that the statistic could have had.

We could simulate that distribution by pretending to take lots of samples. Fortunately, for the mean and the proportion, the CLT tells us that we can model their sampling distribution directly with a Normal model.

The two basic truths about sampling distributions are:

1. Sampling distributions arise because samples vary. Each random sample will contain different cases and, so, a different value of the statistic.
2. Although we can always simulate a sampling distribution, the Central Limit Theorem saves us the trouble for means and proportions.

Here's a picture showing the process going into the sampling distribution model:

> **A S** *Simulation:* **The CLT for Real Data.** Why settle for a picture when you can see it in action?

FIGURE 18.5

We start with a population model, which can have any shape. It can even be bimodal or skewed (as this one is). We label the mean of this model μ and its standard deviation, σ.

We draw one real sample (solid line) of size n and show its histogram and summary statistics. We imagine (or simulate) drawing many other samples (dotted lines), which have their own histograms and summary statistics.

We (imagine) gathering all the means into a histogram.

The CLT tells us we can model the shape of this histogram with a Normal model. The mean of this Normal is μ, and the standard deviation is $SD(\bar{y}) = \dfrac{\sigma}{\sqrt{n}}$.

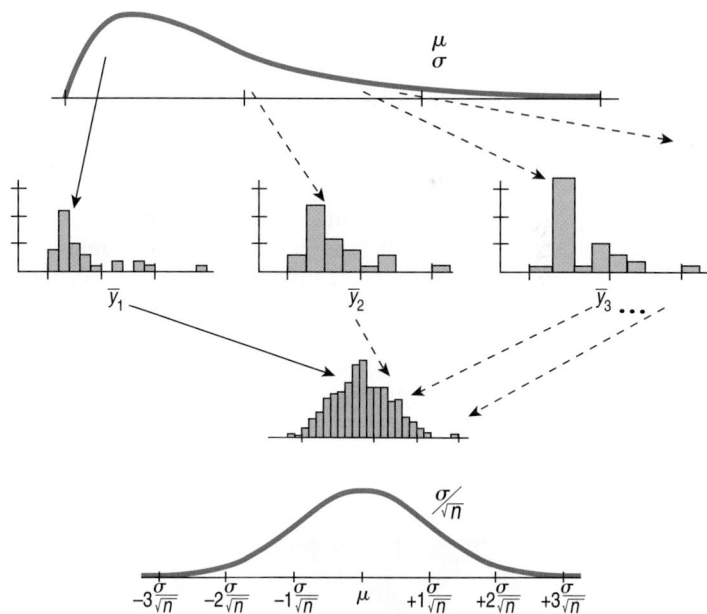

WHAT CAN GO WRONG?

▶ **Don't confuse the sampling distribution with the distribution of the sample.** When you take a sample, you always look at the distribution of the values, usually with a histogram, and you may calculate summary statistics. Examining the distribution of the sample data is wise. But that's not the sampling distribution. The sampling distribution is an imaginary collection of all the values that a statistic *might* have taken for all possible random samples—the one you got and the ones that you didn't get. We use the sampling distribution model to make statements about how the statistic varies.

(continued)

▶ **Beware of observations that are not independent.** The CLT depends crucially on the assumption of independence. If our elevator riders are related, are all from the same school (for example, an elementary school), or in some other way aren't a random sample, then the statements we try to make about the mean are going to be wrong. Unfortunately, this isn't something you can check in your data. You have to think about how the data were gathered. Good sampling practice and well-designed randomized experiments ensure independence.

▶ **Watch out for small samples from skewed populations.** The CLT assures us that the sampling distribution model is Normal if n is large enough. If the population is nearly Normal, even small samples (like our 10 elevator riders) work. If the population is very skewed, then n will have to be large before the Normal model will work well. If we sampled 15 or even 20 CEOs and used $\bar{y}$ to make a statement about the mean of all CEOs' compensation, we'd likely get into trouble because the underlying data distribution is so skewed. Unfortunately, there's no good rule of thumb.[15] It just depends on how skewed the data distribution is. Always plot the data to check.

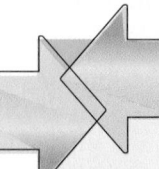

CONNECTIONS

The concept of a sampling distribution connects to almost everything we have done. The fundamental connection is to the deliberate application of randomness in random sampling and randomized comparative experiments. If we didn't employ randomness to generate unbiased data, then repeating the data collection would just get the same data values again (with perhaps a few new measurement or recording errors). The distribution of statistic values arises directly because different random samples and randomized experiments would generate different statistic values.

The connection to the Normal distribution is obvious. We first introduced the Normal model before because it was "nice." As a unimodal, symmetric distribution with 99.7% of its area within three standard deviations of the mean, the Normal model is easy to work with. Now we see that the Normal holds a special place among distributions because we can use it to model the sampling distributions of the mean and the proportion.

We use simulation to understand sampling distributions. In fact, some important sampling distributions were discovered first by simulation.

WHAT HAVE WE LEARNED?

Way back in Chapter 1 we said that Statistics is about variation. We know that no sample fully and exactly describes the population; sample proportions and means will vary from sample to sample. That's sampling error (or, better, sampling variability). We know it will always be present—indeed, the world would be a boring place if variability didn't exist. You might think that sampling variability would prevent us from learning anything reliable about a population by looking at a sample, but that's just not so. The fortunate fact is that sampling variability is not just unavoidable—it's predictable!

---

[15] For proportions, of course, there is a rule: the **Success/Failure Condition.** That works for proportions because the standard deviation of a proportion is linked to its mean.

We've learned how the Central Limit Theorem describes the behavior of sample proportions—shape, center, and spread—as long as certain assumptions and conditions are met. The sample must be independent, random, and large enough that we expect at least 10 successes and failures. Then:

▸ The sampling distribution (the imagined histogram of the proportions from all possible samples) is shaped like a Normal model.

▸ The mean of the sampling model is the true proportion in the population.

▸ The standard deviation of the sample proportions is $\sqrt{\dfrac{pq}{n}}$.

And we've learned to describe the behavior of sample means as well, based on this amazing result known as the Central Limit Theorem—the Fundamental Theorem of Statistics. Again the sample must be independent and random—no surprise there—and needs to be larger if our data come from a population that's not roughly unimodal and symmetric. Then:

▸ Regardless of the shape of the original population, the shape of the distribution of the means of all possible samples can be described by a Normal model, provided the samples are large enough.

▸ The center of the sampling model will be the true mean of the population from which we took the sample.

▸ The standard deviation of the sample means is the population's standard deviation divided by the square root of the sample size, $\dfrac{\sigma}{\sqrt{n}}$.

Terms

Sampling distribution model

413. Different random samples give different values for a statistic. The sampling distribution model shows the behavior of the statistic over all the possible samples for the same size n.

Sampling variability
Sampling error

414. The variability we expect to see from one random sample to another. It is sometimes called sampling error, but sampling variability is the better term.

Sampling distribution model for a proportion

416. If assumptions of independence and random sampling are met, and we expect at least 10 successes and 10 failures, then the sampling distribution of a proportion is modeled by a Normal model with a mean equal to the true proportion value, p, and a standard deviation equal to $\sqrt{\dfrac{pq}{n}}$.

Central Limit Theorem

421. The Central Limit Theorem (CLT) states that the sampling distribution model of the sample mean (and proportion) from a random sample is approximately Normal for large n, *regardless of the distribution of the population, as long as the observations are independent.*

Sampling distribution model for a mean

423. If assumptions of independence and random sampling are met, and the sample size is large enough, the sampling distribution of the sample mean is modeled by a Normal model with a mean equal to the population mean, μ, and a standard deviation equal to $\dfrac{\sigma}{\sqrt{n}}$.

Skills

▸ Understand that the variability of a statistic (as measured by the standard deviation of its sampling distribution) depends on the size of the sample. Statistics based on larger samples are less variable.

▸ Understand that the Central Limit Theorem gives the sampling distribution model of the mean for sufficiently large samples regardless of the underlying population.

▸ Be able to demonstrate a sampling distribution by simulation.

▸ Be able to use a sampling distribution model to make simple statements about the distribution of a proportion or mean under repeated sampling.

▸ Be able to interpret a sampling distribution model as describing the values taken by a statistic in all possible realizations of a sample or randomized experiment under the same conditions.

EXERCISES

1. **Send money.** When they send out their fundraising letter, a philanthropic organization typically gets a return from about 5% of the people on their mailing list. To see what the response rate might be for future appeals, they did a simulation using samples of size 20, 50, 100, and 200. For each sample size, they simulated 1000 mailings with success rate $p = 0.05$ and constructed the histogram of the 1000 sample proportions, shown below. Explain how these histograms demonstrate what the Central Limit Theorem says about the sampling distribution model for sample proportions. Be sure to talk about shape, center, and spread.

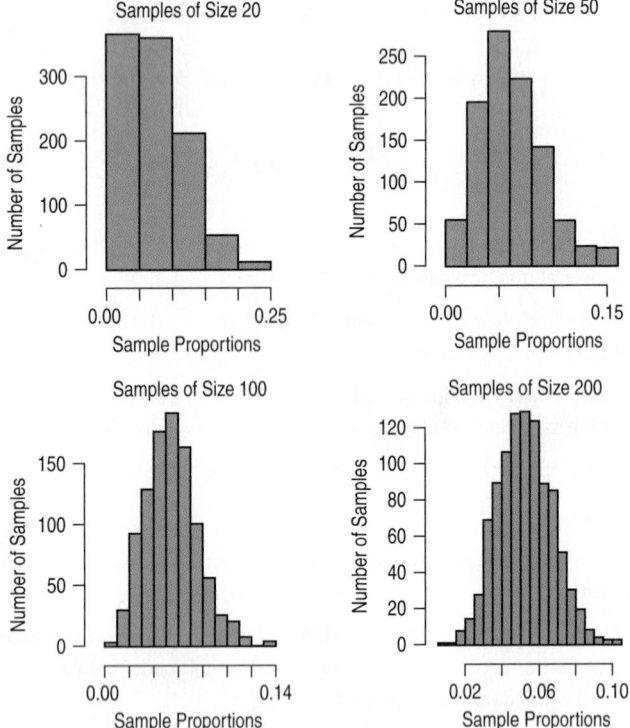

2. **Character recognition.** An automatic character recognition device can successfully read about 85% of handwritten credit card applications. To estimate what might happen when this device reads a stack of applications, the company did a simulation using samples of size 20, 50, 75, and 100. For each sample size, they simulated 1000 samples with success rate $p = 0.85$ and constructed the histogram of the 1000 sample proportions, shown here. Explain how these histograms demonstrate what the Central Limit Theorem says about the sampling distribution model for sample proportions. Be sure to talk about shape, center, and spread.

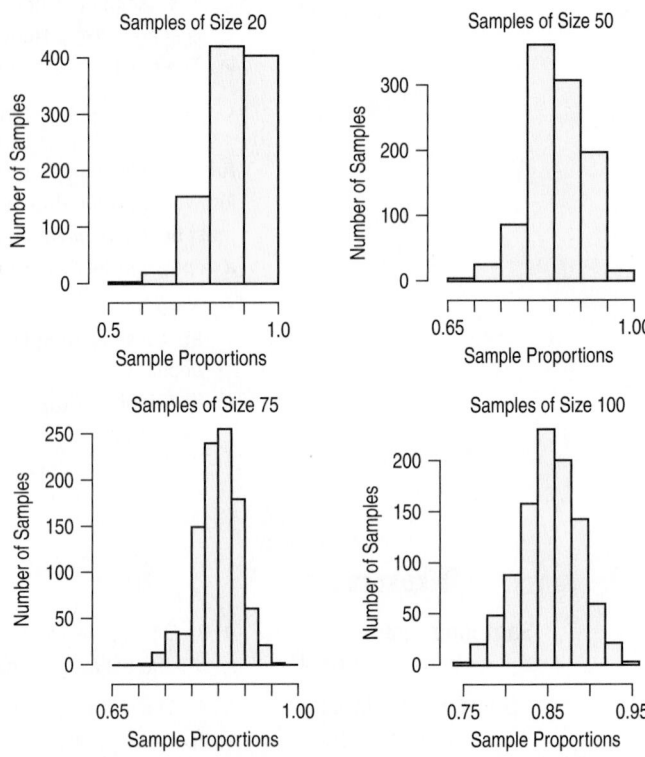

3. **Send money, again.** The philanthropic organization in Exercise 1 expects about a 5% success rate when they send fundraising letters to the people on their mailing list. In Exercise 1 you looked at the histograms showing distributions of sample proportions from 1000 simulated mailings for samples of size 20, 50, 100, and 200. The sample statistics from each simulation were as follows:

n	mean	st. dev.
20	0.0497	0.0479
50	0.0516	0.0309
100	0.0497	0.0215
200	0.0501	0.0152

a) According to the Central Limit Theorem, what should the theoretical mean and standard deviations be for these sample sizes?
b) How close are those theoretical values to what was observed in these simulations?
c) Looking at the histograms in Exercise 1, at what sample size would you be comfortable using the Normal model as an approximation for the sampling distribution?
d) What does the Success/Failure Condition say about the choice you made in part c?

4. Character recognition, again. The automatic character recognition device discussed in Exercise 2 successfully reads about 85% of handwritten credit card applications. In Exercise 2 you looked at the histograms showing distributions of sample proportions from 1000 simulated samples of size 20, 50, 75, and 100. The sample statistics from each simulation were as follows:

n	mean	st. dev.
20	0.8481	0.0803
50	0.8507	0.0509
75	0.8481	0.0406
100	0.8488	0.0354

a) According to the Central Limit Theorem, what should the theoretical mean and standard deviations be for these sample sizes?

b) How close are those theoretical values to what was observed in these simulations?

c) Looking at the histograms in Exercise 2, at what sample size would you be comfortable using the Normal model as an approximation for the sampling distribution?

d) What does the Success/Failure Condition say about the choice you made in part c?

5. Coin tosses. In a large class of introductory Statistics students, the professor has each person toss a coin 16 times and calculate the proportion of his or her tosses that were heads. The students then report their results, and the professor plots a histogram of these several proportions.

a) What shape would you expect this histogram to be? Why?

b) Where do you expect the histogram to be centered?

c) How much variability would you expect among these proportions?

d) Explain why a Normal model should not be used here.

6. M&M's. The candy company claims that 10% of the M&M's it produces are green. Suppose that the candies are packaged at random in small bags containing about 50 M&M's. A class of elementary school students learning about percents opens several bags, counts the various colors of the candies, and calculates the proportion that are green.

a) If we plot a histogram showing the proportions of green candies in the various bags, what shape would you expect it to have?

b) Can that histogram be approximated by a Normal model? Explain.

c) Where should the center of the histogram be?

d) What should the standard deviation of the proportion be?

7. More coins. Suppose the class in Exercise 5 repeats the coin-tossing experiment.

a) The students toss the coins 25 times each. Use the 68–95–99.7 Rule to describe the sampling distribution model.

b) Confirm that you can use a Normal model here.

c) They increase the number of tosses to 64 each. Draw and label the appropriate sampling distribution model. Check the appropriate conditions to justify your model.

d) Explain how the sampling distribution model changes as the number of tosses increases.

8. Bigger bag. Suppose the class in Exercise 6 buys bigger bags of candy, with 200 M&M's each. Again the students calculate the proportion of green candies they find.

a) Explain why it's appropriate to use a Normal model to describe the distribution of the proportion of green M&M's they might expect.

b) Use the 68–95–99.7 Rule to describe how this proportion might vary from bag to bag.

c) How would this model change if the bags contained even more candies?

9. Just (un)lucky? One of the students in the introductory Statistics class in Exercise 7 claims to have tossed her coin 200 times and found only 42% heads. What do you think of this claim? Explain.

10. Too many green ones? In a really large bag of M&M's, the students in Exercise 8 found 500 candies, and 12% of them were green. Is this an unusually large proportion of green M&M's? Explain.

11. Speeding. State police believe that 70% of the drivers traveling on a major interstate highway exceed the speed limit. They plan to set up a radar trap and check the speeds of 80 cars.

a) Using the 68–95–99.7 Rule, draw and label the distribution of the proportion of these cars the police will observe speeding.

b) Do you think the appropriate conditions necessary for your analysis are met? Explain.

12. Smoking. Public health statistics indicate that 26.4% of American adults smoke cigarettes. Using the 68–95–99.7 Rule, describe the sampling distribution model for the proportion of smokers among a randomly selected group of 50 adults. Be sure to discuss your assumptions and conditions.

13. Vision. It is generally believed that nearsightedness affects about 12% of all children. A school district has registered 170 incoming kindergarten children.

a) Can you apply the Central Limit Theorem to describe the sampling distribution model for the sample proportion of children who are nearsighted? Check the conditions and discuss any assumptions you need to make.

b) Sketch and clearly label the sampling model, based on the 68–95–99.7 Rule.

c) How many of the incoming students might the school expect to be nearsighted? Explain.

14. Mortgages. In early 2007 the Mortgage Lenders Association reported that homeowners, hit hard by rising interest rates on adjustable-rate mortgages, were defaulting in record numbers. The foreclosure rate of 1.6% meant that millions of families were losing their homes. Suppose a large bank holds 1731 adjustable-rate mortgages.

a) Can you apply the Central Limit Theorem to describe the sampling distribution model for the sample proportion of foreclosures? Check the conditions and discuss any assumptions you need to make.

b) Sketch and clearly label the sampling model, based on the 68–95–99.7 Rule.

c) How many of these homeowners might the bank expect will default on their mortgages? Explain.

15. **Loans.**   Based on past experience, a bank believes that 7% of the people who receive loans will not make payments on time. The bank has recently approved 200 loans.
    a) What are the mean and standard deviation of the proportion of clients in this group who may not make timely payments?
    b) What assumptions underlie your model? Are the conditions met? Explain.
    c) What's the probability that over 10% of these clients will not make timely payments?

16. **Contacts.**   Assume that 30% of students at a university wear contact lenses.
    a) We randomly pick 100 students. Let $\hat{p}$ represent the proportion of students in this sample who wear contacts. What's the appropriate model for the distribution of $\hat{p}$? Specify the name of the distribution, the mean, and the standard deviation. Be sure to verify that the conditions are met.
    b) What's the approximate probability that more than one third of this sample wear contacts?

17. **Back to school?**   Best known for its testing program, ACT, Inc., also compiles data on a variety of issues in education. In 2004 the company reported that the national college freshman-to-sophomore retention rate held steady at 74% over the previous four years. Consider random samples of 400 freshmen who took the ACT. Use the 68–95–99.7 Rule to describe the sampling distribution model for the percentage of those students we expect to return to that school for their sophomore years. Do you think the appropriate conditions are met?

18. **Binge drinking.**   As we learned in Chapter 15, a national study found that 44% of college students engage in binge drinking (5 drinks at a sitting for men, 4 for women). Use the 68–95–99.7 Rule to describe the sampling distribution model for the proportion of students in a randomly selected group of 200 college students who engage in binge drinking. Do you think the appropriate conditions are met?

19. **Back to school, again.**   Based on the 74% national retention rate described in Exercise 17, does a college where 522 of the 603 freshman returned the next year as sophomores have a right to brag that it has an unusually high retention rate? Explain.

20. **Binge sample.**   After hearing of the national result that 44% of students engage in binge drinking (5 drinks at a sitting for men, 4 for women), a professor surveyed a random sample of 244 students at his college and found that 96 of them admitted to binge drinking in the past week. Should he be surprised at this result? Explain.

21. **Polling.**   Just before a referendum on a school budget, a local newspaper polls 400 voters in an attempt to predict whether the budget will pass. Suppose that the budget actually has the support of 52% of the voters. What's the probability the newspaper's sample will lead them to predict defeat? Be sure to verify that the assumptions and conditions necessary for your analysis are met.

22. **Seeds.**   Information on a packet of seeds claims that the germination rate is 92%. What's the probability that more than 95% of the 160 seeds in the packet will germinate? Be sure to discuss your assumptions and check the conditions that support your model.

23. **Apples.**   When a truckload of apples arrives at a packing plant, a random sample of 150 is selected and examined for bruises, discoloration, and other defects. The whole truckload will be rejected if more than 5% of the sample is unsatisfactory. Suppose that in fact 8% of the apples on the truck do not meet the desired standard. What's the probability that the shipment will be accepted anyway?

24. **Genetic defect.**   It's believed that 4% of children have a gene that may be linked to juvenile diabetes. Researchers hoping to track 20 of these children for several years test 732 newborns for the presence of this gene. What's the probability that they find enough subjects for their study?

25. **Nonsmokers.**   While some nonsmokers do not mind being seated in a smoking section of a restaurant, about 60% of the customers demand a smoke-free area. A new restaurant with 120 seats is being planned. How many seats should be in the nonsmoking area in order to be very sure of having enough seating there? Comment on the assumptions and conditions that support your model, and explain what "very sure" means to you.

26. **Meals.**   A restauranteur anticipates serving about 180 people on a Friday evening, and believes that about 20% of the patrons will order the chef's steak special. How many of those meals should he plan on serving in order to be pretty sure of having enough steaks on hand to meet customer demand? Justify your answer, including an explanation of what "pretty sure" means to you.

27. **Sampling.**   A sample is chosen randomly from a population that can be described by a Normal model.
    a) What's the sampling distribution model for the sample mean? Describe shape, center, and spread.
    b) If we choose a larger sample, what's the effect on this sampling distribution model?

28. **Sampling, part II.**   A sample is chosen randomly from a population that was strongly skewed to the left.
    a) Describe the sampling distribution model for the sample mean if the sample size is small.
    b) If we make the sample larger, what happens to the sampling distribution model's shape, center, and spread?
    c) As we make the sample larger, what happens to the expected distribution of the data in the sample?

29. **Waist size.**   A study measured the *Waist Size* of 250 men, finding a mean of 36.33 inches and a standard deviation of 4.02 inches. Here is a histogram of these measurements

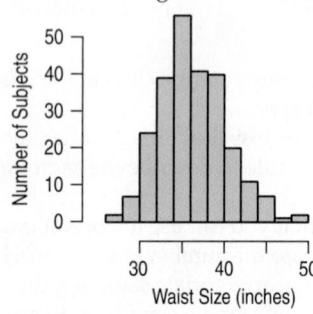

a) Describe the histogram of *Waist Size*.

b) To explore how the mean might vary from sample to sample, they simulated by drawing many samples of size 2, 5, 10, and 20, with replacement, from the 250 measurements. Here are histograms of the sample means for each simulation. Explain how these histograms demonstrate what the Central Limit Theorem says about the sampling distribution model for sample means.

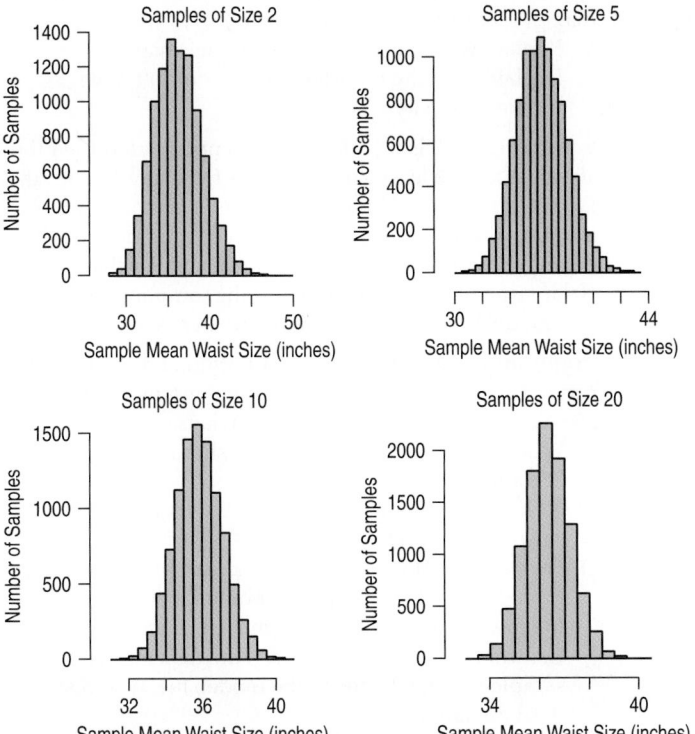

30. **CEO compensation.** In Chapter 5 we saw the distribution of the total compensation of the chief executive officers (CEOs) of the 800 largest U.S. companies (the Fortune 800). The average compensation (in thousands of dollars) is 10,307.31 and the standard deviation is 17,964.62. Here is a histogram of their annual compensations (in $1000):

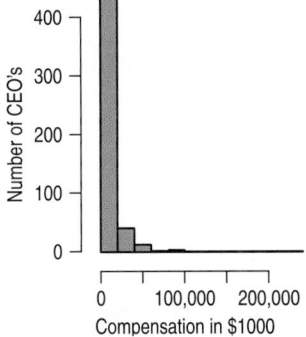

a) Describe the histogram of *Total Compensation*.
A research organization simulated sample means by drawing samples of 30, 50, 100, and 200, with replacement, from the 800 CEOs. The histograms show the distributions of means for many samples of each size.

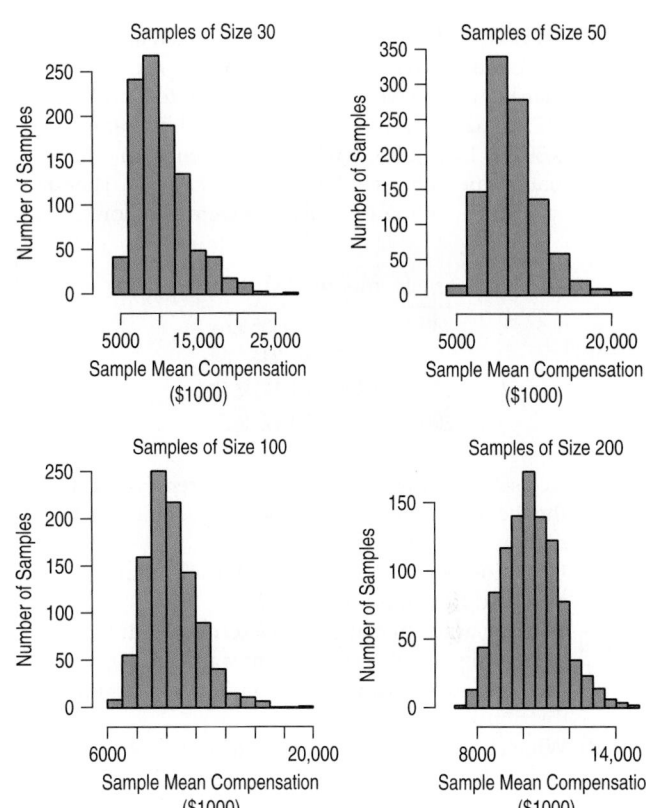

b) Explain how these histograms demonstrate what the Central Limit Theorem says about the sampling distribution model for sample means. Be sure to talk about shape, center, and spread.
c) Comment on the "rule of thumb" that "With a sample size of at least 30, the sampling distribution of the mean is Normal"?

31. **Waist size revisited.** Researchers measured the *Waist Sizes* of 250 men in a study on body fat. The true mean and standard deviation of the *Waist Sizes* for the 250 men are 36.33 in and 4.019 inches, respectively. In Exercise 29 you looked at the histograms of simulations that drew samples of sizes 2, 5, 10, and 20 (with replacement). The summary statistics for these simulations were as follows:

n	mean	st. dev.
2	36.314	2.855
5	36.314	1.805
10	36.341	1.276
20	36.339	0.895

a) According to the Central Limit Theorem, what should the theoretical mean and standard deviation be for each of these sample sizes?
b) How close are the theoretical values to what was observed in the simulation?
c) Looking at the histograms in Exercise 29, at what sample size would you be comfortable using the Normal model as an approximation for the sampling distribution?
d) What about the shape of the distribution of *Waist Size* explains your choice of sample size in part c?

32. CEOs revisited. In Exercise 30 you looked at the annual compensation for 800 CEOs, for which the true mean and standard deviation were (in thousands of dollars) 10,307.31 and 17,964.62, respectively. A simulation drew samples of sizes 30, 50, 100, and 200 (with replacement) from the total annual compensations of the Fortune 800 CEOs. The summary statistics for these simulations were as follows:

n	mean	st. dev.
30	10,251.73	3359.64
50	10,343.93	2483.84
100	10,329.94	1779.18
200	10,340.37	1230.79

a) According to the Central Limit Theorem, what should the theoretical mean and standard deviation be for each of these sample sizes?
b) How close are the theoretical values to what was observed from the simulation?
c) Looking at the histograms in Exercise 30, at what sample size would you be comfortable using the Normal model as an approximation for the sampling distribution?
d) What about the shape of the distribution of *Total Compensation* explains your answer in part c?

33. GPAs. A college's data about the incoming freshmen indicates that the mean of their high school GPAs was 3.4, with a standard deviation of 0.35; the distribution was roughly mound-shaped and only slightly skewed. The students are randomly assigned to freshman writing seminars in groups of 25. What might the mean GPA of one of these seminar groups be? Describe the appropriate sampling distribution model—shape, center, and spread—with attention to assumptions and conditions. Make a sketch using the 68–95–99.7 Rule.

34. Home values. Assessment records indicate that the value of homes in a small city is skewed right, with a mean of $140,000 and standard deviation of $60,000. To check the accuracy of the assessment data, officials plan to conduct a detailed appraisal of 100 homes selected at random. Using the 68–95–99.7 Rule, draw and label an appropriate sampling model for the mean value of the homes selected.

T 35. Lucky Spot? A reporter working on a story about the New York lottery contacted one of the authors of this book, wanting help analyzing data to see if some ticket sales outlets were more likely to produce winners. His data for each of the 966 New York lottery outlets are graphed below; the scatterplot shows the ratio *TotalPaid/TotalSales* vs. *TotalSales* for the state's "instant winner" games for all of 2007.

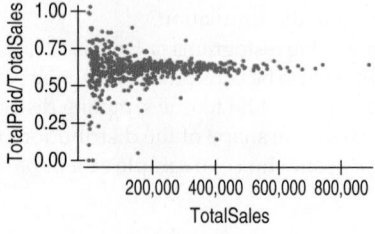

The reporter thinks that by identifying the outlets with the highest fraction of bets paid out, players might be able to increase their chances of winning. (Typically—but not always—instant winners are paid immediately (instantly) at the store at which they are purchased. However, the fact that tickets may be scratched off and then cashed in at any outlet may account for some outlets paying out more than they take in. The few with very low payouts may be on interstate highways where players may purchase cards but then leave.)
a) Explain why the plot has this funnel shape.
b) Explain why the reporter's idea wouldn't have worked anyway.

36. Safe cities. Allstate Insurance Company identified the 10 safest and 10 least-safe U.S. cities from among the 200 largest cities in the United States, based on the mean number of years drivers went between automobile accidents. The cities on both lists were all smaller than the 10 largest cities. Using facts about the sampling distribution model of the mean, explain why this is not surprising.

37. Pregnancy. Assume that the duration of human pregnancies can be described by a Normal model with mean 266 days and standard deviation 16 days.
a) What percentage of pregnancies should last between 270 and 280 days?
b) At least how many days should the longest 25% of all pregnancies last?
c) Suppose a certain obstetrician is currently providing prenatal care to 60 pregnant women. Let $\bar{y}$ represent the mean length of their pregnancies. According to the Central Limit Theorem, what's the distribution of this sample mean, $\bar{y}$? Specify the model, mean, and standard deviation.
d) What's the probability that the mean duration of these patients' pregnancies will be less than 260 days?

38. Rainfall. Statistics from Cornell's Northeast Regional Climate Center indicate that Ithaca, NY, gets an average of 35.4" of rain each year, with a standard deviation of 4.2". Assume that a Normal model applies.
a) During what percentage of years does Ithaca get more than 40" of rain?
b) Less than how much rain falls in the driest 20% of all years?
c) A Cornell University student is in Ithaca for 4 years. Let $\bar{y}$ represent the mean amount of rain for those 4 years. Describe the sampling distribution model of this sample mean, $\bar{y}$.
d) What's the probability that those 4 years average less than 30" of rain?

39. Pregnant again. The duration of human pregnancies may not actually follow the Normal model described in Exercise 37.
a) Explain why it may be somewhat skewed to the left.
b) If the correct model is in fact skewed, does that change your answers to parts a, b, and c of Exercise 37? Explain why or why not for each.

40. At work. Some business analysts estimate that the length of time people work at a job has a mean of 6.2 years and a standard deviation of 4.5 years.

a) Explain why you suspect this distribution may be skewed to the right.

b) Explain why you could estimate the probability that 100 people selected at random had worked for their employers an average of 10 years or more, but you could not estimate the probability that an individual had done so.

41. Dice and dollars. You roll a die, winning nothing if the number of spots is odd, $1 for a 2 or a 4, and $10 for a 6.

a) Find the expected value and standard deviation of your prospective winnings.

b) You play twice. Find the mean and standard deviation of your total winnings.

c) You play 40 times. What's the probability that you win at least $100?

42. New game. You pay $10 and roll a die. If you get a 6, you win $50. If not, you get to roll again. If you get a 6 this time, you get your $10 back.

a) Create a probability model for this game.

b) Find the expected value and standard deviation of your prospective winnings.

c) You play this game five times. Find the expected value and standard deviation of your average winnings.

d) 100 people play this game. What's the probability the person running the game makes a profit?

43. AP Stats 2006. The College Board reported the score distribution shown in the table for all students who took the 2006 AP Statistics exam.

Score	Percent of Students
5	12.6
4	22.2
3	25.3
2	18.3
1	21.6

a) Find the mean and standard deviation of the scores.

b) If we select a random sample of 40 AP Statistics students, would you expect their scores to follow a Normal model? Explain.

c) Consider the mean scores of random samples of 40 AP Statistics students. Describe the sampling model for these means (shape, center, and spread).

44. Museum membership. A museum offers several levels of membership, as shown in the table.

Member Category	Amount of Donation ($)	Percent of Members
Individual	50	41
Family	100	37
Sponsor	250	14
Patron	500	7
Benefactor	1000	1

a) Find the mean and standard deviation of the donations.

b) During their annual membership drive, they hope to sign up 50 new members each day. Would you expect the distribution of the donations for a day to follow a Normal model? Explain.

c) Consider the mean donation of the 50 new members each day. Describe the sampling model for these means (shape, center, and spread).

45. AP Stats 2006, again. An AP Statistics teacher had 63 students preparing to take the AP exam discussed in Exercise 43. Though they were obviously not a random sample, he considered his students to be "typical" of all the national students. What's the probability that his students will achieve an average score of at least 3?

46. Joining the museum. One of the museum's phone volunteers sets a personal goal of getting an average donation of at least $100 from the new members she enrolls during the membership drive. If she gets 80 new members and they can be considered a random sample of all the museum's members, what is the probability that she can achieve her goal?

47. Pollution. Carbon monoxide (CO) emissions for a certain kind of car vary with mean 2.9 g/mi and standard deviation 0.4 g/mi. A company has 80 of these cars in its fleet. Let $\bar{y}$ represent the mean CO level for the company's fleet.

a) What's the approximate model for the distribution of $\bar{y}$? Explain.

b) Estimate the probability that $\bar{y}$ is between 3.0 and 3.1 g/mi.

c) There is only a 5% chance that the fleet's mean CO level is greater than what value?

48. Potato chips. The weight of potato chips in a medium-size bag is stated to be 10 ounces. The amount that the packaging machine puts in these bags is believed to have a Normal model with mean 10.2 ounces and standard deviation 0.12 ounces.

a) What fraction of all bags sold are underweight?

b) Some of the chips are sold in "bargain packs" of 3 bags. What's the probability that none of the 3 is underweight?

c) What's the probability that the mean weight of the 3 bags is below the stated amount?

d) What's the probability that the mean weight of a 24-bag case of potato chips is below 10 ounces?

49. Tips. A waiter believes the distribution of his tips has a model that is slightly skewed to the right, with a mean of $9.60 and a standard deviation of $5.40.

a) Explain why you cannot determine the probability that a given party will tip him at least $20.

b) Can you estimate the probability that the next 4 parties will tip an average of at least $15? Explain.

c) Is it likely that his 10 parties today will tip an average of at least $15? Explain.

50. Groceries. A grocery store's receipts show that Sunday customer purchases have a skewed distribution with a mean of $32 and a standard deviation of $20.

a) Explain why you cannot determine the probability that the next Sunday customer will spend at least $40.

b) Can you estimate the probability that the next 10 Sunday customers will spend an average of at least $40? Explain.

c) Is it likely that the next 50 Sunday customers will spend an average of at least $40? Explain.

51. More tips. The waiter in Exercise 49 usually waits on about 40 parties over a weekend of work.
a) Estimate the probability that he will earn at least $500 in tips.
b) How much does he earn on the best 10% of such weekends?

52. More groceries. Suppose the store in Exercise 50 had 312 customers this Sunday.
a) Estimate the probability that the store's revenues were at least $10,000.
b) If, on a typical Sunday, the store serves 312 customers, how much does the store take in on the worst 10% of such days?

53. IQs. Suppose that IQs of East State University's students can be described by a Normal model with mean 130 and standard deviation 8 points. Also suppose that IQs of students from West State University can be described by a Normal model with mean 120 and standard deviation 10.
a) We select a student at random from East State. Find the probability that this student's IQ is at least 125 points.
b) We select a student at random from each school. Find the probability that the East State student's IQ is at least 5 points higher than the West State student's IQ.
c) We select 3 West State students at random. Find the probability that this group's average IQ is at least 125 points.
d) We also select 3 East State students at random. What's the probability that their average IQ is at least 5 points higher than the average for the 3 West Staters?

54. Milk. Although most of us buy milk by the quart or gallon, farmers measure daily production in pounds. Ayrshire cows average 47 pounds of milk a day, with a standard deviation of 6 pounds. For Jersey cows, the mean daily production is 43 pounds, with a standard

deviation of 5 pounds. Assume that Normal models describe milk production for these breeds.
a) We select an Ayrshire at random. What's the probability that she averages more than 50 pounds of milk a day?
b) What's the probability that a randomly selected Ayrshire gives more milk than a randomly selected Jersey?
c) A farmer has 20 Jerseys. What's the probability that the average production for this small herd exceeds 45 pounds of milk a day?
d) A neighboring farmer has 10 Ayrshires. What's the probability that his herd average is at least 5 pounds higher than the average for part c's Jersey herd?

JUST CHECKING
Answers

1. A Normal model (approximately).
2. At the actual proportion of all students who are in favor.
3. $SD(\hat{p}) = \sqrt{\dfrac{(0.5)(0.5)}{100}} = 0.05$
4. No, this is a histogram of individuals. It may or may not be Normal, but we can't tell from the information provided.
5. A Normal model (approximately).
6. 266 days
7. $\dfrac{16}{\sqrt{100}} = 1.6$ days

Confidence Intervals for Proportions

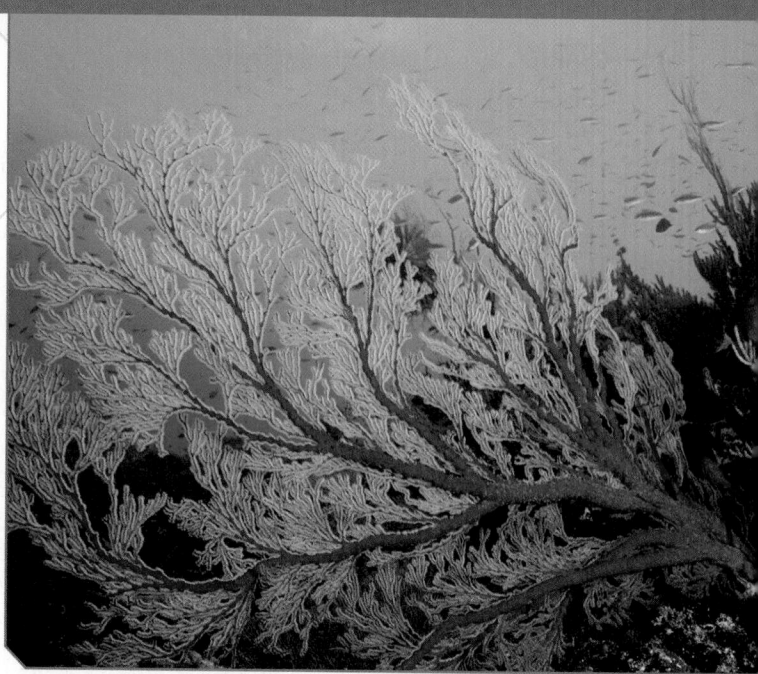

WHO	Sea fans
WHAT	Percent infected
WHEN	June 2000
WHERE	Las Redes Reef, Akumal, Mexico, 40 feet deep
WHY	Research

Coral reef communities are home to one quarter of all marine plants and animals worldwide. These reefs support large fisheries by providing breeding grounds and safe havens for young fish of many species. Coral reefs are seawalls that protect shorelines against tides, storm surges, and hurricanes, and are sand "factories" that produce the limestone and sand of which beaches are made. Beyond the beach, these reefs are major tourist attractions for snorkelers and divers, driving a tourist industry worth tens of billions of dollars.

But marine scientists say that 10% of the world's reef systems have been destroyed in recent times. At current rates of loss, 70% of the reefs could be gone in 40 years. Pollution, global warming, outright destruction of reefs, and increasing acidification of the oceans are all likely factors in this loss.

Dr. Drew Harvell's lab studies corals and the diseases that affect them. They sampled sea fans[1] at 19 randomly selected reefs along the Yucatan peninsula and diagnosed whether the animals were affected by the disease *aspergillosis*.[2] In specimens collected at a depth of 40 feet at the Las Redes Reef in Akumal, Mexico, these scientists found that 54 of 104 sea fans sampled were infected with that disease.

Of course, we care about much more than these particular 104 sea fans. We care about the health of coral reef communities throughout the Caribbean. What can this study tell us about the prevalence of the disease among sea fans?

We have a sample proportion, which we write as $\hat{p}$, of 54/104, or 51.9%. Our first guess might be that this observed proportion is close to the population proportion, p. But we also know that because of natural sampling variability, if the researchers had drawn a second sample of 104 sea fans at roughly the same time, the proportion infected from that sample probably wouldn't have been exactly 51.9%.

---

[1] That's a sea fan in the picture. Although they look like trees, they are actually colonies of genetically identical animals.
[2] K. M. Mullen, C. D. Harvell, A. P. Alker, D. Dube, E. Jordán-Dahlgren, J. R. Ward, and L. E. Petes, "Host range and resistance to aspergillosis in three sea fan species from the Yucatan," *Marine Biology* (2006), Springer-Verlag.

What *can* we say about the population proportion, p? To start to answer this question, think about how different the sample proportion might have been if we'd taken another random sample from the same population. But wait. Remember—we aren't actually going to take more samples. We just want to *imagine* how the sample proportions might vary from sample to sample. In other words, we want to know about the *sampling distribution* of the sample proportion of infected sea fans.

A Confidence Interval

Let's look at our model for the sampling distribution. What do we know about it? We know it's approximately Normal (under certain assumptions, which we should be careful to check) and that its mean is the proportion of all infected sea fans on the Las Redes Reef. Is the infected proportion of *all* sea fans 51.9%? No, that's just $\hat{p}$, our estimate. We don't know the proportion, p, of all the infected sea fans; that's what we're trying to find out. We do know, though, that the sampling distribution model of $\hat{p}$ is centered at p, and we know that the standard deviation of the sampling distribution is $\sqrt{\dfrac{pq}{n}}$.

Now we have a problem: Since we don't know p, we can't find the true standard deviation of the sampling distribution model. We do know the observed proportion, $\hat{p}$, so, of course we just use what we know, and we estimate. That may not seem like a big deal, but it gets a special name. Whenever we estimate the standard deviation of a sampling distribution, we call it a **standard error**.[3] For a sample proportion, $\hat{p}$, the standard error is

$$SE(\hat{p}) = \sqrt{\frac{\hat{p}\hat{q}}{n}}.$$

For the sea fans, then:

$$SE(\hat{p}) = \sqrt{\frac{\hat{p}\hat{q}}{n}} = \sqrt{\frac{(0.519)(0.481)}{104}} = 0.049 = 4.9\%.$$

Now we know that the sampling model for $\hat{p}$ should look like this:

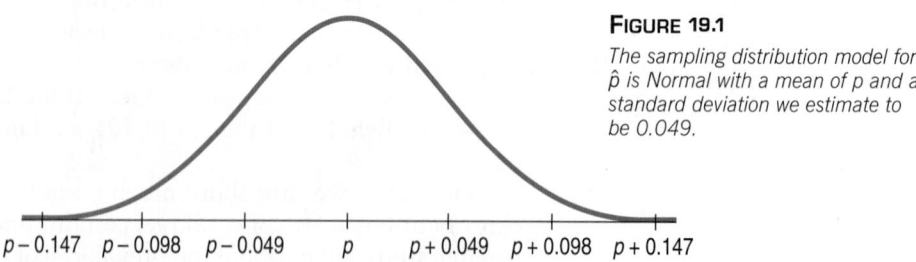

FIGURE 19.1

The sampling distribution model for $\hat{p}$ is Normal with a mean of p and a standard deviation we estimate to be 0.049.

$p - 0.147$ $p - 0.098$ $p - 0.049$ p $p + 0.049$ $p + 0.098$ $p + 0.147$

Great. What does that tell us? Well, because it's Normal, it says that about 68% of all samples of 104 sea fans will have $\hat{p}$'s within 1 *SE*, 0.049, of p. And about 95% of all these samples will be within $p \pm 2$ *SE*s. But where is *our* sample proportion in this picture? And what value does p have? We still don't know!

We do know that for 95% of random samples, $\hat{p}$ will be no more than 2 *SE*s away from p. So let's look at this from $\hat{p}$'s point of view. If I'm $\hat{p}$, there's a 95%

---

[3] This isn't such a great name because it isn't standard and nobody made an error. But it's much shorter and more convenient than saying, "the estimated standard deviation of the sampling distribution of the sample statistic."

chance that p is no more than 2 *SE*s away from me. If I reach out 2 *SE*s, or 2×0.049, away from me on both sides, I'm 95% sure that p will be within my grasp. Now I've got him! Probably. Of course, even if my interval does catch p, I still don't know its true value. The best I can do is an interval, and even then I can't be positive it contains p.

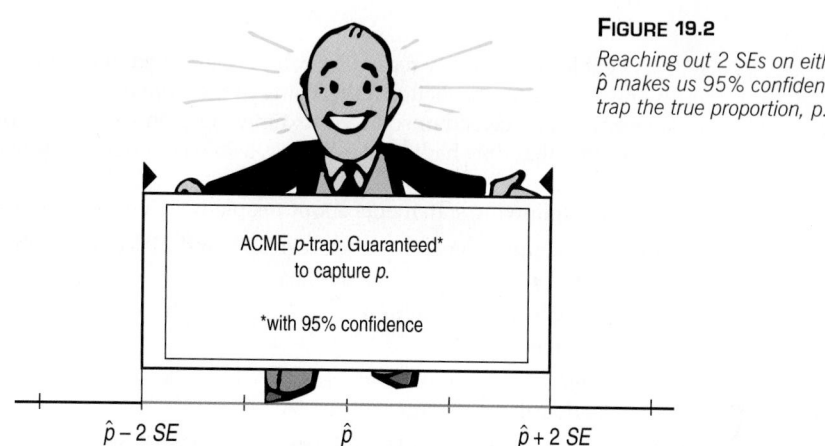

FIGURE 19.2

Reaching out 2 SEs on either side of $\hat{p}$ makes us 95% confident that we'll trap the true proportion, p.

$\hat{p} - 2\ SE$ $\hat{p}$ $\hat{p} + 2\ SE$

So what can we really say about p? Here's a list of things we'd like to be able to say, in order of strongest to weakest and the reasons we can't say most of them:

1. **"51.9% of *all* sea fans on the Las Redes Reef are infected."** It would be nice to be able to make absolute statements about population values with certainty, but we just don't have enough information to do that. There's no way to be sure that the population proportion is the same as the sample proportion; in fact, it almost certainly isn't. Observations vary. Another sample would yield a different sample proportion.
2. **"It is *probably* true that 51.9% of all sea fans on the Las Redes Reef are infected."** No. In fact, we can be pretty sure that whatever the true proportion is, it's not exactly 51.900%. So the statement is not true.
3. **"We don't know exactly what proportion of sea fans on the Las Redes Reef is infected, but we *know* that it's within the interval 51.9% $\pm$ 2 $\times$ 4.9%. That is, it's between 42.1% and 61.7%."** This is getting closer, but we still can't be certain. We can't know *for sure* that the true proportion is in this interval—or in any particular interval.
4. **"We don't know exactly what proportion of sea fans on the Las Redes Reef is infected, but the interval from 42.1% to 61.7% *probably* contains the true proportion."** We've now fudged twice—first by giving an interval and second by admitting that we only think the interval "probably" contains the true value. And this statement is true.

That last statement may be true, but it's a bit wishy-washy. We can tighten it up a bit by quantifying what we mean by "probably." We saw that 95% of the time when we reach out 2 *SE*s from $\hat{p}$ we capture p, so we can be 95% confident that this is one of those times. After putting a number on the probability that this interval covers the true proportion, we've given our best guess of where the parameter is and how certain we are that it's within some range.

5. **"We are 95% confident that between 42.1% and 61.7% of Las Redes sea fans are infected."** Statements like these are called **confidence intervals.** They're the best we can do.

Each confidence interval discussed in the book has a name. You'll see many different kinds of confidence intervals in the following chapters. Some will be

"Far better an approximate answer to the right question, . . . than an exact answer to the wrong question."
—John W. Tukey

about more than *one* sample, some will be about statistics other than *proportions*, and some will use models other than the Normal. The interval calculated and interpreted here is sometimes called a **one-proportion z-interval.**[4]

JUST CHECKING

A Pew Research study regarding cell phones asked questions about cell phone experience. One growing concern is unsolicited advertising in the form of text messages. Pew asked cell phone owners, "Have you ever received unsolicited text messages on your cell phone from advertisers?" and 17% reported that they had. Pew estimates a 95% confidence interval to be 0.17 ± 0.04, or between 13% and 21%.

Are the following statements about people who have cell phones correct? Explain.

1. In Pew's sample, somewhere between 13% and 21% of respondents reported that they had received unsolicited advertising text messages.

2. We can be 95% confident that 17% of U.S. cell phone owners have received unsolicited advertising text messages.

3. We are 95% confident that between 13% and 21% of all U.S. cell phone owners have received unsolicited advertising text messages.

4. We know that between 13% and 21% of all U.S. cell phone owners have received unsolicited advertising text messages.

5. 95% of all U.S. cell phone owners have received unsolicited advertising text messages.

What Does "95% Confidence" Really Mean?

What do we mean when we say we have 95% confidence that our interval contains the true proportion? Formally, what we mean is that "95% of samples of this size will produce confidence intervals that capture the true proportion." This is correct, but a little long winded, so we sometimes say, "we are 95% confident that the true proportion lies in our interval." Our uncertainty is about whether the particular sample we have at hand is one of the successful ones or one of the 5% that fail to produce an interval that captures the true value.

Back in Chapter 18 we saw that proportions vary from sample to sample. If other researchers select their own samples of sea fans, they'll also find some infected by the disease, but each person's sample proportion will almost certainly differ from ours. When they each try to estimate the true rate of infection in the entire population, they'll center *their* confidence intervals at the proportions they observed in their own samples. Each of us will end up with a different interval.

Our interval guessed the true proportion of infected sea fans to be between about 42% and 62%. Another researcher whose sample contained more infected fans than ours did might guess between 46% and 66%. Still another who happened to collect fewer infected fans might estimate the true proportion to be between 23% and 43%. And so on. Every possible sample would produce yet another confidence interval. Although wide intervals like these can't pin down the actual rate of infection very precisely, we expect that most of them should be winners, capturing the true value. Nonetheless, some will be duds, missing the population proportion entirely.

On the next page you'll see confidence intervals produced by simulating 20 different random samples. The red dots are the proportions of infected fans in

---

[4] In fact, this confidence interval is so standard for a single proportion that you may see it simply called a "confidence interval for the proportion."

each sample, and the blue segments show the confidence intervals found for each. The green line represents the true rate of infection in the population, so you can see that most of the intervals caught it—but a few missed. (And notice again that it is the *intervals* that vary from sample to sample; the green line doesn't move.)

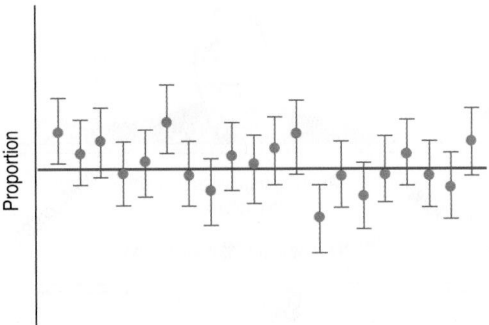

The horizontal green line shows the true percentage of all sea fans that are infected. Most of the 20 simulated samples produced confidence intervals that captured the true value, but a few missed.

Of course, there's a huge number of possible samples that *could* be drawn, each with its own sample proportion. These are just some of them. Each sample proportion can be used to make a confidence interval. That's a large pile of possible confidence intervals, and ours is just one of those in the pile. Did *our* confidence interval "work"? We can never be sure, because we'll never know the true proportion of all the sea fans that are infected. However, the Central Limit Theorem assures us that 95% of the intervals in the pile are winners, covering the true value, and only 5% are duds. *That's* why we're 95% confident that our interval is a winner!

FOR EXAMPLE   Polls and margin of error

On January 30–31, 2007, Fox News/Opinion Dynamics polled 900 registered voters nationwide.[5] When asked, "Do you believe global warming exists?" 82% said "Yes". Fox reported their margin of error to be ±3%.

Question: It is standard among pollsters to use a 95% confidence level unless otherwise stated. Given that, what does Fox News mean by claiming a margin of error of ±3% in this context?

If this polling were done repeatedly, 95% of all random samples would yield estimates that come within ±3% of the true proportion of all registered voters who believe that global warming exists.

Margin of Error: Certainty vs. Precision

We've just claimed that with a certain confidence we've captured the true proportion of all infected sea fans. Our confidence interval had the form

$$\hat{p} \pm 2 \, SE(\hat{p}).$$

The extent of the interval on either side of $\hat{p}$ is called the **margin of error** (*ME*). We'll want to use the same approach for many other situations besides estimating proportions. In general, confidence intervals look like this:

$$Estimate \pm ME.$$

---

[5] www.foxnews.com, "Fox News Poll: Most Americans Believe in Global Warming," Feb 7, 2007.

The margin of error for our 95% confidence interval was 2 *SE*. What if we wanted to be more confident? To be more confident, we'll need to capture *p* more often, and to do that we'll need to make the interval wider. For example, if we want to be 99.7% confident, the margin of error will have to be 3 *SE*.

FIGURE 19.3

Reaching out 3 SEs on either side of p̂ makes us 99.7% confident we'll trap the true proportion p. Compare with Figure 19.2.

$\hat{p} - 3\ SE$ $\hat{p}$ $\hat{p} + 3\ SE$

A S *Activity:* **Balancing Precision and Certainty.** What percent of parents expect their kids to pay for college with a student loan? Investigate the balance between the precision and the certainty of a confidence interval.

The more confident we want to be, the larger the margin of error must be. We can be 100% confident that the proportion of infected sea fans is between 0% and 100%, but this isn't likely to be very useful. On the other hand, we could give a confidence interval from 51.8% to 52.0%, but we can't be very confident about a precise statement like this. Every confidence interval is a balance between certainty and precision.

The tension between certainty and precision is always there. Fortunately, in most cases we can be both sufficiently certain and sufficiently precise to make useful statements. There is no simple answer to the conflict. You must choose a confidence level yourself. The data can't do it for you. The choice of confidence level is somewhat arbitrary. The most commonly chosen confidence levels are 90%, 95%, and 99%, but any percentage can be used. (In practice, though, using something like 92.9% or 97.2% is likely to make people think you're up to something.)

FOR EXAMPLE   Finding the margin of error (Take 1)

Recap: A January 2007 Fox poll of 900 registered voters reported a margin of error of ±3%. It is a convention among pollsters to use a 95% confidence level and to report the "worst case" margin of error, based on *p* = 0.5.

Question: How did Fox calculate their margin of error?

Assuming *p* = 0.5, for random samples of *n* = 900, $SD(\hat{p}) = \sqrt{\dfrac{pq}{n}} = \sqrt{\dfrac{(0.5)(0.5)}{900}} = 0.0167$

For a 95% confidence level, ME = 2(0.0167) = 0.033, so Fox's margin of error is just a bit over ±3%.

Critical Values

> **NOTATION ALERT:**
>
> We'll put an asterisk on a letter to indicate a critical value, so z^* is always a critical value from a Normal model.

In our sea fans example we used $2SE$ to give us a 95% confidence interval. To change the confidence level, we'd need to change the *number* of SEs so that the size of the margin of error corresponds to the new level. This number of SEs is called the **critical value**. Here it's based on the Normal model, so we denote it z^*. For any confidence level, we can find the corresponding critical value from a computer, a calculator, or a Normal probability table, such as Table Z.

For a 95% confidence interval, you'll find the precise critical value is $z^* = 1.96$. That is, 95% of a Normal model is found within ± 1.96 standard deviations of the mean. We've been using $z^* = 2$ from the 68–95–99.7 Rule because it's easy to remember.

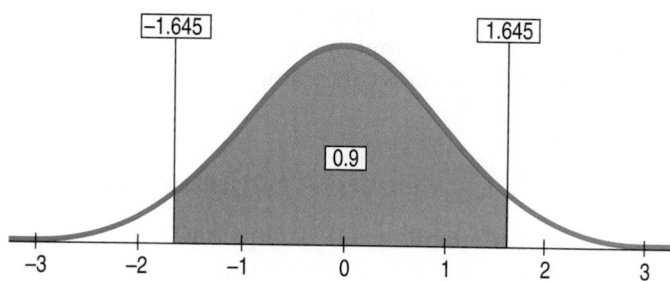

FIGURE 19.4

For a 90% confidence interval, the critical value is 1.645, because, for a Normal model, 90% of the values are within 1.645 standard deviations from the mean.

FOR EXAMPLE   Finding the margin of error (Take 2)

Recap: In January 2007 a Fox News poll of 900 registered voters found that 82% of the respondents believed that global warming exists. Fox reported a 95% confidence interval with a margin of error of $\pm 3\%$.

Questions: Using the critical value of z and the standard error based on the observed proportion, what would be the margin of error for a 90% confidence interval? What's good and bad about this change?

With $n = 900$ and $\hat{p} = 0.82$, $SE(\hat{p}) = \sqrt{\dfrac{\hat{p}\hat{q}}{n}} = \sqrt{\dfrac{(0.82)(0.18)}{900}} = 0.0128$

For a 90% confidence level, $z^* = 1.645$, so $ME = 1.645(0.0128) = 0.021$

Now the margin of error is only about $\pm 2\%$, producing a narrower interval. That makes for a more precise estimate of voter belief, but provides less certainty that the interval actually contains the true proportion of voters believing in global warming.

 **JUST CHECKING**

Think some more about the 95% confidence interval Fox News created for the proportion of registered voters who believe that global warming exists.

6. If Fox wanted to be 98% confident, would their confidence interval need to be wider or narrower?

7. Fox's margin of error was about $\pm 3\%$. If they reduced it to $\pm 2\%$, would their level of confidence be higher or lower?

8. If Fox News had polled more people, would the interval's margin of error have been larger or smaller?

Assumptions and Conditions

We've just made some pretty sweeping statements about sea fans. Those statements were possible because we used a Normal model for the sampling distribution. But is that model appropriate?

As we've seen, all statistical models make assumptions. Different models make different assumptions. If those assumptions are not true, the model might be inappropriate and our conclusions based on it may be wrong. Because the confidence interval is built on the Normal model for the sampling distribution, the assumptions and conditions are the same as those we discussed in Chapter 18. But, because they are so important, we'll go over them again.

We can never be certain that an assumption is true, but we can decide intelligently whether it is reasonable. When we have data, we can often decide whether an assumption is plausible by checking a related condition. However, we want to make a statement about the world at large, not just about the data we collected. So the assumptions we make are not just about how our data look, but about how representative they are.

A S *Activity:* Assumptions and **Conditions.** Here's an animated review of the assumptions and conditions.

INDEPENDENCE ASSUMPTION

Independence Assumption: We first need to *Think* about whether the independence assumption is plausible. We often look for reasons to suspect that it fails. We wonder whether there is any reason to believe that the data values somehow affect each other. (For example, might the disease in sea fans be contagious?) Whether you decide that the **Independence Assumption** is plausible depends on your knowledge of the situation. It's not one you can check by looking at the data.

However, now that we have data, there are two conditions that we can check:

Randomization Condition: Were the data sampled at random or generated from a properly randomized experiment? Proper randomization can help ensure independence.

10% Condition: Samples are almost always drawn without replacement. Usually, of course, we'd like to have as large a sample as we can. But when the population itself is small we have another concern. When we sample from small populations, the probability of success may be different for the last few individuals we draw than it was for the first few. For example, if most of the women have already been sampled, the chance of drawing a woman from the remaining population is lower. If the sample exceeds 10% of the population, the probability of a success changes so much during the sampling that our Normal model may no longer be appropriate. But if less than 10% of the population is sampled, the effect on independence is negligible.

SAMPLE SIZE ASSUMPTION

The model we use for inference is based on the Central Limit Theorem. The **Sample Size Assumption** addresses the question of whether the sample is large enough to make the sampling model for the sample proportions approximately Normal. It turns out that we need more data as the proportion gets closer and closer to either extreme (0 or 1). We can check this assumption with the:

Success/Failure Condition: We must expect at least 10 "successes" and at least 10 "failures." Recall that by tradition we arbitrarily label one alternative (usually the outcome being counted) as a "success" even if it's something bad (like a sick sea fan). The other alternative is, of course, then a "failure."

ONE-PROPORTION *z*-INTERVAL

When the conditions are met, we are ready to find the confidence interval for the population proportion, *p*. The confidence interval is $\hat{p} \pm z^* \times SE(\hat{p})$ where the standard deviation of the proportion is estimated by $SE(\hat{p}) = \sqrt{\dfrac{\hat{p}\hat{q}}{n}}$.

STEP-BY-STEP EXAMPLE | A Confidence Interval for a Proportion

In May 2006, the Gallup Poll[6] asked 510 randomly sampled adults the question "Generally speaking, do you believe the death penalty is applied fairly or unfairly in this country today?" Of these, 60% answered "Fairly," 35% said "Unfairly," and 4% said they didn't know.

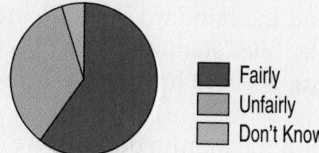

WHO	Adults in the United States
WHAT	Response to a question about the death penalty
WHEN	May 2006
WHERE	United States
HOW	510 adults were randomly sampled and asked by the Gallup Poll
WHY	Public opinion research

Question: From this survey, what can we conclude about the opinions of *all* adults?

To answer this question, we'll build a confidence interval for the proportion of all U.S. adults who believe the death penalty is applied fairly. There are four steps to building a confidence interval for proportions: Plan, Model, Mechanics, and Conclusion.

Plan State the problem and the W's.

Identify the *parameter* you wish to estimate.

Identify the *population* about which you wish to make statements.

Choose and state a confidence level.

Model Think about the assumptions and check the conditions.

I want to find an interval that is likely, with 95% confidence, to contain the true proportion, *p*, of U.S. adults who think the death penalty is applied fairly. I have a random sample of 510 U.S. adults.

✓ **Independence Assumption:** Gallup phoned a random sample of U.S. adults. It is very unlikely that any of their respondents influenced each other.

✓ **Randomization Condition:** Gallup drew a random sample from all U.S. adults. I don't have details of their randomization but assume that I can trust it.

✓ **10% Condition:** Although sampling was necessarily without replacement, there are many more U.S. adults than were sampled. The sample is certainly less than 10% of the population.

[6] www.gallup.com

✔ **Success/Failure Condition:**
$n\hat{p} = 510(60\%) = 306 \geq 10$ and
$n\hat{q} = 510(40\%) = 204 \geq 10$,
so the sample appears to be large enough
to use the Normal model.

State the sampling distribution model for the statistic.

Choose your method.

The conditions are satisfied, so I can use a Normal model to find a **one-proportion z-interval.**

 **SHOW**

Mechanics Construct the confidence interval.

First find the standard error. (Remember: It's called the "standard error" because we don't know p and have to use $\hat{p}$ instead.)

Next find the margin of error. We could informally use 2 for our critical value, but 1.96 is more accurate.

Write the confidence interval (CI).

REALITY CHECK The CI is centered at the sample proportion and about as wide as we might expect for a sample of 500.

$n = 510$, $\hat{p} = 0.60$, so

$$SE(\hat{p}) = \sqrt{\frac{\hat{p}\hat{q}}{n}} = \sqrt{\frac{(0.60)(0.40)}{510}} = 0.022$$

Because the sampling model is Normal, for a 95% confidence interval, the critical value $z^* = 1.96$.

The margin of error is

$$ME = z^* \times SE(\hat{p}) = 1.96(0.022) = 0.043$$

So the 95% confidence interval is

$$0.60 \pm 0.043 \text{ or } (0.557, 0.643)$$

 **TELL**

Conclusion Interpret the confidence interval in the proper context. We're 95% confident that our interval captured the true proportion.

I am 95% confident that between 55.7% and 64.3% of all U.S. adults think that the death penalty is applied fairly.

TI Tips

Finding confidence intervals

```
EDIT CALC TESTS
7↑ZInterval…
8:TInterval…
9:2-SampZInt…
0:2-SampTInt…
A:1-PropZInt…
B:2-PropZInt…
C↓χ²-Test…
```

```
1-PropZInt
 x:54
 n:104
 C-Level:.95
 Calculate
```

It will come as no surprise that your TI can calculate a confidence interval for a population proportion. Remember the sea fans? Of 104 sea fans, 54 were diseased. To find the resulting confidence interval, we first take a look at a whole new menu.

- Under **STAT** go to the **TESTS** menu. Quite a list! Commands are found here for the inference procedures you will learn through the coming chapters.
- We're using a Normal model to find a confidence interval for a proportion based on one sample. Scroll down the list and select **A:1-PropZInt**.
- Enter the number of successes observed and the sample size.
- Specify a confidence level and then **Calculate**.

And there it is! Note that the TI calculates the sample proportion for you, but the important result is the interval itself, 42% to 62%. The calculator did the easy part—just Show. Tell is harder. It's your job to interpret that interval correctly.

Beware: You may run into a problem. When you enter the value of x, you need a *count*, not a percentage. Suppose the marine scientists had reported that 52% of the 104 sea fans were infected. You can enter x: .52*104, and the calculator will evaluate that as 54.08. Wrong. Unless you fix that result, you'll get an error message. Think about it—the number of infected sea fans must have been a whole number, evidently 54. When the scientists reported the results, they rounded off the actual percentage (54 ÷ 104 = 51.923%) to 52%. Simply change the value of x to 54 and you should be able to Calculate the correct interval.

CHOOSING YOUR SAMPLE SIZE

The question of how large a sample to take is an important step in planning any study. We weren't ready to make that calculation when we first looked at study design in Chapter 12, but now we can—and we always should.

Suppose a candidate is planning a poll and wants to estimate voter support within 3% with 95% confidence. How large a sample does she need?

Let's look at the margin of error:

$$ME = z^* \sqrt{\frac{\hat{p}\hat{q}}{n}}$$

$$0.03 = 1.96 \sqrt{\frac{\hat{p}\hat{q}}{n}}.$$

We want to find n, the sample size. To find n we need a value for $\hat{p}$. We don't know $\hat{p}$ because we don't have a sample yet, but we can probably guess a value. The worst case—the value that makes $\hat{p}\hat{q}$ (and therefore n) largest—is 0.50, so if we use that value for $\hat{p}$, we'll certainly be safe. Our candidate probably expects to be near 50% anyway.

Our equation, then, is

$$0.03 = 1.96 \sqrt{\frac{(0.5)(0.5)}{n}}.$$

To solve for n, we first multiply both sides of the equation by $\sqrt{n}$ and then divide by 0.03:

$$0.03\sqrt{n} = 1.96\sqrt{(0.5)(0.5)}$$

$$\sqrt{n} = \frac{1.96\sqrt{(0.5)(0.5)}}{0.03} \approx 32.67$$

Notice that evaluating this expression tells us the *square root* of the sample size. We need to square that result to find n:

$$n \approx (32.67)^2 \approx 1067.1$$

To be safe, we round up and conclude that we need at least 1068 respondents to keep the margin of error as small as 3% with a confidence level of 95%.

What do I use instead of $\hat{p}$?
Often we have an estimate of the population proportion based on experience or perhaps a previous study. If so, use that value as $\hat{p}$ in calculating what size sample you need. If not, the cautious approach is to use $p = 0.5$ in the sample size calculation; that will determine the largest sample necessary regardless of the true proportion.

Recap: The Fox News poll which estimated that 82% of all voters believed global warming exists had a margin of error of ±3%. Suppose an environmental group planning a follow-up survey of voters' opinions on global warming wants to determine a 95% confidence interval with a margin of error of no more than ±2%.

Question: How large a sample do they need? Use the Fox News estimate as the basis for your calculation.

$$ME = z^* \sqrt{\frac{\hat{p}\hat{q}}{n}}$$

$$0.02 = 1.96 \sqrt{\frac{(0.82)(0.18)}{n}}$$

$$\sqrt{n} = \frac{1.96\sqrt{(0.82)(0.18)}}{0.02} \approx 37.65$$

$$n = 37.65^2 = 1{,}417.55$$

The environmental group's survey will need about 1,418 respondents.

Public opinion polls often sample 1000 people, which gives an ME of 3% when $p = 0.5$. But businesses and nonprofit organizations typically use much larger samples to estimate the proportion who will accept a direct mail offer. Why? Because that proportion is very low—often far below 5%. An ME of 3% wouldn't be precise enough. An ME like 0.1% would be more useful, and that requires a very large sample size.

Unfortunately, bigger samples cost more money and more effort. Because the standard error declines only with the *square root* of the sample size, to cut the standard error (and thus the ME) in half, we must *quadruple* the sample size.

Generally a margin of error of 5% or less is acceptable, but different circumstances call for different standards. For a pilot study, a margin of error of 10% may be fine, so a sample of 100 will do quite well. In a close election, a polling organization might want to get the margin of error down to 2%. Drawing a large sample to get a smaller ME, however, can run into trouble. It takes time to survey 2400 people, and a survey that extends over a week or more may be trying to hit a target that moves during the time of the survey. An important event can change public opinion in the middle of the survey process.

Keep in mind that the sample size for a survey is the number of respondents, not the number of people to whom questionnaires were sent or whose phone numbers were dialed. And keep in mind that a low response rate turns any study essentially into a voluntary response study, which is of little value for inferring population values. It's almost always better to spend resources on increasing the response rate than on surveying a larger group. A full or nearly full response by a modest-size sample can yield useful results.

Surveys are not the only place where proportions pop up. Banks sample huge mailing lists to estimate what proportion of people will accept a credit card offer. Even pilot studies may mail offers to over 50,000 customers. Most don't respond; that doesn't make the sample smaller—they simply said "No thanks". Those who do respond want the card. To the bank, the response rate[7] is $\hat{p}$. With a typical success rate around 0.5%, the bank needs a very small margin of error—often as low as 0.1%—to make a sound business decision. That calls for a large sample, and the bank must take care in estimating the size needed. For our election poll calculation we used $p = 0.5$, both because it's safe and because we honestly believed p to be near 0.5. If the bank used 0.5, they'd get an absurd answer. Instead, they base their calculation on a proportion closer to the one they expect to find.

---

[7] In marketing studies every mailing yields a response—"yes" or "no"—and "response rate" means the proportion of customers who accept an offer. That's not the way we use the term for survey response.

FOR EXAMPLE Sample size revisited

A credit card company is about to send out a mailing to test the market for a new credit card. From that sample, they want to estimate the true proportion of people who will sign up for the card nationwide. A pilot study suggests that about 0.5% of the people receiving the offer will accept it.

Question: To be within a tenth of a percentage point (0.001) of the true rate with 95% confidence, how big does the test mailing have to be?

Using the estimate $\hat{p} = 0.5\%$: $ME = 0.001 = z^* \sqrt{\dfrac{\hat{p}\hat{q}}{n}} = 1.96 \sqrt{\dfrac{(0.005)(0.995)}{n}}$

$$(0.001)^2 = 1.96^2 \frac{(0.005)(0.995)}{n} \Rightarrow n = \frac{1.96^2(0.005)(0.995)}{(0.001)^2}$$

$$= 19{,}111.96 \text{ or } 19{,}112$$

That's a lot, but it's actually a reasonable size for a trial mailing such as this. Note, however, that if they had assumed 0.50 for the value of p, they would have found

$$ME = 0.001 = z^* \sqrt{\frac{pq}{n}} = 1.96 \sqrt{\frac{(0.5)(0.5)}{n}}$$

$$(0.001)^2 = 1.96^2 \frac{(0.5)(0.5)}{n} \Rightarrow n = \frac{1.96^2(0.5)(0.5)}{(0.001)^2} = 960{,}400.$$

Quite a different (and unreasonable) result.

WHAT CAN GO WRONG?

Confidence intervals are powerful tools. Not only do they tell what we know about the parameter value, but—more important—they also tell what we *don't* know. In order to use confidence intervals effectively, you must be clear about what you say about them.

DON'T MISSTATE WHAT THE INTERVAL MEANS

▶ **Don't suggest that the parameter varies.** A statement like "There is a 95% chance that the true proportion is between 42.7% and 51.3%" sounds as though you think the population proportion wanders around and sometimes happens to fall between 42.7% and 51.3%. When you interpret a confidence interval, make it clear that *you* know that the population parameter is fixed and that it is the interval that varies from sample to sample.

▶ **Don't claim that other samples will agree with yours.** Keep in mind that the confidence interval makes a statement about the true population proportion. An interpretation such as "In 95% of samples of U.S. adults, the proportion who think marijuana should be decriminalized will be between 42.7% and 51.3%" is just wrong. The interval isn't about sample proportions but about the population proportion.

▶ **Don't be certain about the parameter.** Saying "Between 42.1% and 61.7% of sea fans are infected" asserts that the population proportion cannot be outside that interval. Of course, we can't be absolutely certain of that. (Just pretty sure.)

▶ **Don't forget: It's about the parameter.** Don't say, "I'm 95% confident that $\hat{p}$ is between 42.1% and 61.7%." Of course you are—in fact, we calculated that $\hat{p} = 51.9\%$ of the

(continued)

Confidence intervals are based on random samples, so the interval is random, too. The CLT tells us that 95% of the random samples will yield intervals that capture the true value. That's what we mean by being 95% confident.

Technically, we should say, "I am 95% confident that the interval from 42.1% to 61.7% captures the true proportion of infected sea fans." That formal phrasing emphasizes that *our confidence (and our uncertainty) is about the interval, not the true proportion.* But you may choose a more casual phrasing like "I am 95% confident that between 42.1% and 61.7% of the Las Redes fans are infected." Because you've made it clear that the uncertainty is yours and you didn't suggest that the randomness is in the true proportion, this is OK. Keep in mind that it's the interval that's random and is the focus of both our confidence and doubt.

fans in our sample were infected. So we already *know* the sample proportion. The confidence interval is about the (unknown) population parameter, *p*.

▶ **Don't claim to know too much.** Don't say, "I'm 95% confident that between 42.1% and 61.7% of all the sea fans in the world are infected." You didn't sample from all 500 species of sea fans found in coral reefs around the world. Just those of this type on the Las Redes Reef.

▶ **Do take responsibility.** Confidence intervals are about *uncertainty. You* are the one who is uncertain, not the parameter. You have to accept the responsibility and consequences of the fact that not all the intervals you compute will capture the true value. In fact, about 5% of the 95% confidence intervals you find will fail to capture the true value of the parameter. You *can* say, "I am 95% confident that between 42.1% and 61.7% of the sea fans on the Las Redes Reef are infected."[8]

▶ **Do treat the whole interval equally.** Although a confidence interval is a set of plausible values for the parameter, don't think that the values in the middle of a confidence interval are somehow "more plausible" than the values near the edges. Your interval provides no information about where in your current interval (if at all) the parameter value is most likely to be hiding.

MARGIN OF ERROR TOO LARGE TO BE USEFUL

We know we can't be exact, but how precise do we need to be? A confidence interval that says that the percentage of infected sea fans is between 10% and 90% wouldn't be of much use. Most likely, you have some sense of how large a margin of error you can tolerate. What can you do?

One way to make the margin of error smaller is to reduce your level of confidence. But that may not be a useful solution. It's a rare study that reports confidence levels lower than 80%. Levels of 95% or 99% are more common.

The time to think about whether your margin of error is small enough to be useful is when you design your study. Don't wait until you compute your confidence interval. To get a narrower interval without giving up confidence, you need to have less variability in your sample proportion. How can you do that? Choose a larger sample.

VIOLATIONS OF ASSUMPTIONS

Confidence intervals and margins of error are often reported along with poll results and other analyses. But it's easy to misuse them and wise to be aware of the ways things can go wrong.

▶ **Watch out for biased sampling.** Don't forget about the potential sources of bias in surveys that we discussed in Chapter 12. Just because we have more statistical machinery now doesn't mean we can forget what we've already learned. A questionnaire that finds that 85% of people enjoy filling out surveys still suffers from nonresponse bias even though now we're able to put confidence intervals around this (biased) estimate.

▶ **Think about independence.** The assumption that the values in our sample are mutually independent is one that we usually cannot check. It always pays to think about it, though. For example, the disease affecting the sea fans might be contagious, so that fans growing near a diseased fan are more likely themselves to be diseased. Such contagion would violate the Independence Assumption and could severely affect our sample proportion. It could be that the proportion of infected sea fans on the entire reef is actually quite small, and the researchers just happened to find an infected area. To avoid this, the researchers should be careful to sample sites far enough apart to make contagion unlikely.

[8] When we are being very careful we say, "95% of samples of this size will produce confidence intervals that capture the true proportion of infected sea fans on the Las Redes Reef."

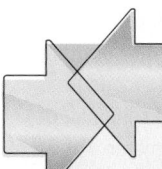

CONNECTIONS

Now we can see a practical application of sampling distributions. To find a confidence interval, we lay out an interval measured in standard deviations. We're using the standard deviation as a ruler again. But now the standard deviation we need is the standard deviation of the sampling distribution. That's the one that tells how much the proportion varies. (And when we estimate it from the data, we call it a standard error.)

WHAT HAVE WE LEARNED?

The first 10 chapters of the book explored graphical and numerical ways of summarizing and presenting sample data. We've learned (at last!) to use the sample we have at hand to say something about the *world at large*. This process, called statistical inference, is based on our understanding of sampling models and will be our focus for the rest of the book.

As our first step in statistical inference, we've learned to use our sample to make a *confidence interval* that estimates what proportion of a population has a certain characteristic.

We've learned that:

▶ Our best estimate of the true population proportion is the proportion we observed in the sample, so we center our confidence interval there.

▶ Samples don't represent the population perfectly, so we create our interval with a *margin of error.*

▶ This method successfully captures the true population proportion most of the time, providing us with a level of confidence in our interval.

▶ The higher the level of confidence we want, the *wider* our confidence interval becomes.

▶ The larger the sample size we have, the *narrower* our confidence interval can be.

▶ When designing a study, we can calculate the sample size we'll need to be able to reach conclusions that have a desired degree of precision and level of confidence.

▶ There are important assumptions and conditions we must check before using this (or any) statistical inference procedure.

We've learned to interpret a confidence interval by *Telling* what we believe is true in the entire population from which we took our random sample. Of course, we can't be *certain.* We've learned not to overstate or misinterpret what the confidence interval says.

Terms

Standard error
440. When we estimate the standard deviation of a sampling distribution using statistics found from the data, the estimate is called a standard error.

$$SE(\hat{p}) = \sqrt{\frac{\hat{p}\hat{q}}{n}}$$

Confidence interval
441. A level C confidence interval for a model parameter is an interval of values usually of the form

$$estimate \pm margin\ of\ error$$

found from data in such a way that C% of all random samples will yield intervals that capture the true parameter value.

One-proportion z-interval
442–444. A confidence interval for the true value of a proportion. The confidence interval is

$$\hat{p} \pm z^*SE(\hat{p}),$$

where z^* is a critical value from the Standard Normal model corresponding to the specified confidence level.

Margin of error 443. In a confidence interval, the extent of the interval on either side of the observed statistic value is called the margin of error. A margin of error is typically the product of a critical value from the sampling distribution and a standard error from the data. A small margin of error corresponds to a confidence interval that pins down the parameter precisely. A large margin of error corresponds to a confidence interval that gives relatively little information about the estimated parameter. For a proportion,

$$ME = z^* \sqrt{\frac{\hat{p}\hat{q}}{n}}$$

Critical value 445. The number of standard errors to move away from the mean of the sampling distribution to correspond to the specified level of confidence. The critical value, denoted z^*, is usually found from a table or with technology.

Skills

▸ Understand confidence intervals as a balance between the precision and the certainty of a statement about a model parameter.

▸ Understand that the margin of error of a confidence interval for a proportion changes with the sample size and the level of confidence.

▸ Know how to examine your data for violations of conditions that would make inference about a population proportion unwise or invalid.

▸ Be able to construct a one-proportion z-interval.

▸ Be able to interpret a one-proportion z-interval in a simple sentence or two. Write such an interpretation so that it does not state or suggest that the parameter of interest is itself random, but rather that the bounds of the confidence interval are the random quantities about which we state our degree of confidence.

CONFIDENCE INTERVALS FOR PROPORTIONS ON THE COMPUTER

Confidence intervals for proportions are so easy and natural that many statistics packages don't offer special commands for them. Most statistics programs want the "raw data" for computations. For proportions, the raw data are the "success" and "failure" status for each case. Usually, these are given as 1 or 0, but they might be category names like "yes" and "no." Often we just know the proportion of successes, $\hat{p}$, and the total count, n. Computer packages don't usually deal with summary data like this easily, but the statistics routines found on many graphing calculators allow you to create confidence intervals from summaries of the data—usually all you need to enter are the number of successes and the sample size.

In some programs you can reconstruct variables of 0's and 1's with the given proportions. But even when you have (or can reconstruct) the raw data values, you may not get exactly the same margin of error from a computer package as you would find working by hand. The reason is that some packages make approximations or use other methods. The result is very close but not exactly the same. Fortunately, Statistics means never having to say you're certain, so the approximate result is good enough.

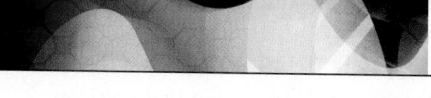

EXERCISES

1. **Margin of error.** A TV newscaster reports the results of a poll of voters, and then says, "The margin of error is plus or minus 4%." Explain carefully what that means.

2. **Margin of error.** A medical researcher estimates the percentage of children exposed to lead-base paint, adding that he believes his estimate has a margin of error of about 3%. Explain what the margin of error means.

3. **Conditions.** For each situation described below, identify the population and the sample, explain what p and $\hat{p}$ represent, and tell whether the methods of this chapter can be used to create a confidence interval.
   a) Police set up an auto checkpoint at which drivers are stopped and their cars inspected for safety problems. They find that 14 of the 134 cars stopped have at least one safety violation. They want to estimate the percentage of all cars that may be unsafe.
   b) A TV talk show asks viewers to register their opinions on prayer in schools by logging on to a Web site. Of the 602 people who voted, 488 favored prayer in schools. We want to estimate the level of support among the general public.
   c) A school is considering requiring students to wear uniforms. The PTA surveys parent opinion by sending a questionnaire home with all 1245 students; 380 surveys are returned, with 228 families in favor of the change.
   d) A college admits 1632 freshmen one year, and four years later 1388 of them graduate on time. The college wants to estimate the percentage of all their freshman enrollees who graduate on time.

4. **More conditions.** Consider each situation described. Identify the population and the sample, explain what p and $\hat{p}$ represent, and tell whether the methods of this chapter can be used to create a confidence interval.
   a) A consumer group hoping to assess customer experiences with auto dealers surveys 167 people who recently bought new cars; 3% of them expressed dissatisfaction with the salesperson.
   b) What percent of college students have cell phones? 2883 students were asked as they entered a football stadium, and 243 said they had phones with them.
   c) 240 potato plants in a field in Maine are randomly checked, and only 7 show signs of blight. How severe is the blight problem for the U.S. potato industry?
   d) 12 of the 309 employees of a small company suffered an injury on the job last year. What can the company expect in future years?

5. **Conclusions.** A catalog sales company promises to deliver orders placed on the Internet within 3 days. Follow-up calls to a few randomly selected customers show that a 95% confidence interval for the proportion of all orders that arrive on time is 88% ± 6%. What does this mean? Are these conclusions correct? Explain.
   a) Between 82% and 94% of all orders arrive on time.

   b) 95% of all random samples of customers will show that 88% of orders arrive on time.
   c) 95% of all random samples of customers will show that 82% to 94% of orders arrive on time.
   d) We are 95% sure that between 82% and 94% of the orders placed by the sampled customers arrived on time.
   e) On 95% of the days, between 82% and 94% of the orders will arrive on time.

6. **More conclusions.** In January 2002, two students made worldwide headlines by spinning a Belgian euro 250 times and getting 140 heads—that's 56%. That makes the 90% confidence interval (51%, 61%). What does this mean? Are these conclusions correct? Explain.
   a) Between 51% and 61% of all euros are unfair.
   b) We are 90% sure that in this experiment this euro landed heads on between 51% and 61% of the spins.
   c) We are 90% sure that spun euros will land heads between 51% and 61% of the time.
   d) If you spin a euro many times, you can be 90% sure of getting between 51% and 61% heads.
   e) 90% of all spun euros will land heads between 51% and 61% of the time.

7. **Confidence intervals.** Several factors are involved in the creation of a confidence interval. Among them are the sample size, the level of confidence, and the margin of error. Which statements are true?
   a) For a given sample size, higher confidence means a smaller margin of error.
   b) For a specified confidence level, larger samples provide smaller margins of error.
   c) For a fixed margin of error, larger samples provide greater confidence.
   d) For a given confidence level, halving the margin of error requires a sample twice as large.

8. **Confidence intervals, again.** Several factors are involved in the creation of a confidence interval. Among them are the sample size, the level of confidence, and the margin of error. Which statements are true?
   a) For a given sample size, reducing the margin of error will mean lower confidence.
   b) For a certain confidence level, you can get a smaller margin of error by selecting a bigger sample.
   c) For a fixed margin of error, smaller samples will mean lower confidence.
   d) For a given confidence level, a sample 9 times as large will make a margin of error one third as big.

9. **Cars.** What fraction of cars is made in Japan? The computer output below summarizes the results of a random sample of 50 autos. Explain carefully what it tells you.

   z-Interval for proportion
   With 90.00% confidence,
   0.29938661 < p(japan) < 0.46984416

10. Parole. A study of 902 decisions made by the Nebraska Board of Parole produced the following computer output. Assuming these cases are representative of all cases that may come before the Board, what can you conclude?

```
z-Interval for proportion
With 95.00% confidence,
0.56100658 < p(parole) < 0.62524619
```

11. Contaminated chicken. In January 2007 *Consumer Reports* published their study of bacterial contamination of chicken sold in the United States. They purchased 525 broiler chickens from various kinds of food stores in 23 states and tested them for types of bacteria that cause food-borne illnesses. Laboratory results indicated that 83% of these chickens were infected with *Campylobacter*.
a) Construct a 95% confidence interval.
b) Explain what your confidence interval says about chicken sold in the United States.
c) A spokesperson for the U.S. Department of Agriculture dismissed the *Consumer Reports* finding, saying, "That's 500 samples out of 9 billion chickens slaughtered a year. . . . With the small numbers they [tested], I don't know that one would want to change one's buying habits." Is this criticism valid? Explain.

12. Contaminated chicken, second course. The January 2007 *Consumer Reports* study described in Exercise 11 also found that 15% of the 525 broiler chickens tested were infected with *Salmonella*.
a) Are the conditions for creating a confidence interval satisfied? Explain.
b) Construct a 95% confidence interval.
c) Explain what your confidence interval says about chicken sold in the United States.

13. Baseball fans. In a poll taken in March of 2007, Gallup asked 1006 national adults whether they were baseball fans. 36% said they were. A year previously, 37% of a similar-size sample had reported being baseball fans.
a) Find the margin of error for the 2007 poll if we want 90% confidence in our estimate of the percent of national adults who are baseball fans.
b) Explain what that margin of error means.
c) If we wanted to be 99% confident, would the margin of error be larger or smaller? Explain.
d) Find that margin of error.
e) In general, if all other aspects of the situation remain the same, will smaller margins of error produce greater or less confidence in the interval?
f) Do you think there's been a change from 2006 to 2007 in the real proportion of national adults who are baseball fans? Explain.

14. Cloning 2007. A May 2007 Gallup Poll found that only 11% of a random sample of 1003 adults approved of attempts to clone a human.
a) Find the margin of error for this poll if we want 95% confidence in our estimate of the percent of American adults who approve of cloning humans.
b) Explain what that margin of error means.
c) If we only need to be 90% confident, will the margin of error be larger or smaller? Explain.
d) Find that margin of error.

e) In general, if all other aspects of the situation remain the same, would smaller samples produce smaller or larger margins of error?

15. Contributions, please. The Paralyzed Veterans of America is a philanthropic organization that relies on contributions. They send free mailing labels and greeting cards to potential donors on their list and ask for a voluntary contribution. To test a new campaign, they recently sent letters to a random sample of 100,000 potential donors and received 4781 donations.
a) Give a 95% confidence interval for the true proportion of their entire mailing list who may donate.
b) A staff member thinks that the true rate is 5%. Given the confidence interval you found, do you find that percentage plausible?

16. Take the offer. First USA, a major credit card company, is planning a new offer for their current cardholders. The offer will give double airline miles on purchases for the next 6 months if the cardholder goes online and registers for the offer. To test the effectiveness of the campaign, First USA recently sent out offers to a random sample of 50,000 cardholders. Of those, 1184 registered.
a) Give a 95% confidence interval for the true proportion of those cardholders who will register for the offer.
b) If the acceptance rate is only 2% or less, the campaign won't be worth the expense. Given the confidence interval you found, what would you say?

17. Teenage drivers. An insurance company checks police records on 582 accidents selected at random and notes that teenagers were at the wheel in 91 of them.
a) Create a 95% confidence interval for the percentage of all auto accidents that involve teenage drivers.
b) Explain what your interval means.
c) Explain what "95% confidence" means.
d) A politician urging tighter restrictions on drivers' licenses issued to teens says, "In one of every five auto accidents, a teenager is behind the wheel." Does your confidence interval support or contradict this statement? Explain.

18. Junk mail. Direct mail advertisers send solicitations (a.k.a. "junk mail") to thousands of potential customers in the hope that some will buy the company's product. The acceptance rate is usually quite low. Suppose a company wants to test the response to a new flyer, and sends it to 1000 people randomly selected from their mailing list of over 200,000 people. They get orders from 123 of the recipients.
a) Create a 90% confidence interval for the percentage of people the company contacts who may buy something.
b) Explain what this interval means.
c) Explain what "90% confidence" means.
d) The company must decide whether to now do a mass mailing. The mailing won't be cost-effective unless it produces at least a 5% return. What does your confidence interval suggest? Explain.

19. Safe food. Some food retailers propose subjecting food to a low level of radiation in order to improve safety, but sale of such "irradiated" food is opposed by many people. Suppose a grocer wants to find out what his customers think. He has cashiers distribute surveys at checkout and ask customers to fill them out and drop them in a box near the front door. He gets responses from 122 customers, of whom 78 oppose the radiation treatments. What can the grocer conclude about the opinions of all his customers?

20. **Local news.** The mayor of a small city has suggested that the state locate a new prison there, arguing that the construction project and resulting jobs will be good for the local economy. A total of 183 residents show up for a public hearing on the proposal, and a show of hands finds only 31 in favor of the prison project. What can the city council conclude about public support for the mayor's initiative?

21. **Death penalty, again.** In the survey on the death penalty you read about in the chapter, the Gallup Poll actually split the sample at random, asking 510 respondents the question quoted earlier, "Generally speaking, do you believe the death penalty is applied fairly or unfairly in this country today?" The other 510 were asked "Generally speaking, do you believe the death penalty is applied unfairly or fairly in this country today?" Seems like the same question, but sometimes the order of the choices matters. Suppose that for the second way of phrasing it, only 54% said they thought the death penalty was fairly applied.
    a) What kind of bias may be present here?
    b) If we combine them, considering the overall group to be one larger random sample of 1020 respondents, what is a 95% confidence interval for the proportion of the general public that thinks the death penalty is being fairly applied?
    c) How does the margin of error based on this pooled sample compare with the margins of error from the separate groups? Why?

22. **Gambling.** A city ballot includes a local initiative that would legalize gambling. The issue is hotly contested, and two groups decide to conduct polls to predict the outcome. The local newspaper finds that 53% of 1200 randomly selected voters plan to vote "yes," while a college Statistics class finds 54% of 450 randomly selected voters in support. Both groups will create 95% confidence intervals.
    a) Without finding the confidence intervals, explain which one will have the larger margin of error.
    b) Find both confidence intervals.
    c) Which group concludes that the outcome is too close to call? Why?

23. **Rickets.** Vitamin D, whether ingested as a dietary supplement or produced naturally when sunlight falls on the skin, is essential for strong, healthy bones. The bone disease rickets was largely eliminated in England during the 1950s, but now there is concern that a generation of children more likely to watch TV or play computer games than spend time outdoors is at increased risk. A recent study of 2700 children randomly selected from all parts of England found 20% of them deficient in vitamin D.
    a) Find a 98% confidence interval.
    b) Explain carefully what your interval means.
    c) Explain what "98% confidence" means.

24. **Pregnancy.** In 1998 a San Diego reproductive clinic reported 49 live births to 207 women under the age of 40 who had previously been unable to conceive.

a) Find a 90% confidence interval for the success rate at this clinic.
b) Interpret your interval in this context.
c) Explain what "90% confidence" means.
d) Do these data refute the clinic's claim of a 25% success rate? Explain.

25. **Payments.** In a May 2007 Experian/Gallup Personal Credit Index poll of 1008 U.S. adults aged 18 and over, 8% of respondents said they were very uncomfortable with their ability to make their monthly payments on their current debt during the next three months. A more detailed poll surveyed 1288 adults, reporting similar overall results and also noting differences among four age groups: 18–29, 30–49, 50–64, and 65+.
    a) Do you expect the 95% confidence interval for the true proportion of all 18- to 29-year-olds who are worried to be wider or narrower than the 95% confidence interval for the true proportion of all U.S. consumers? Explain.
    b) Do you expect this second poll's overall margin of error to be larger or smaller than the Experian/Gallup poll's? Explain.

26. **Back to campus again.** In 2004 ACT, Inc., reported that 74% of 1644 randomly selected college freshmen returned to college the next year. The study was stratified by type of college—public or private. The retention rates were 71.9% among 505 students enrolled in public colleges and 74.9% among 1139 students enrolled in private colleges.
    a) Will the 95% confidence interval for the true national retention rate in private colleges be wider or narrower than the 95% confidence interval for the retention rate in public colleges? Explain.
    b) Do you expect the margin of error for the overall retention rate to be larger or smaller? Explain.

27. **Deer ticks.** Wildlife biologists inspect 153 deer taken by hunters and find 32 of them carrying ticks that test positive for Lyme disease.
    a) Create a 90% confidence interval for the percentage of deer that may carry such ticks.
    b) If the scientists want to cut the margin of error in half, how many deer must they inspect?
    c) What concerns do you have about this sample?

28. **Pregnancy, II.** The San Diego reproductive clinic in Exercise 24 wants to publish updated information on its success rate.
    a) The clinic wants to cut the stated margin of error in half. How many patients' results must be used?
    b) Do you have any concerns about this sample? Explain.

29. **Graduation.** It's believed that as many as 25% of adults over 50 never graduated from high school. We wish to see if this percentage is the same among the 25 to 30 age group.
    a) How many of this younger age group must we survey in order to estimate the proportion of non-grads to within 6% with 90% confidence?
    b) Suppose we want to cut the margin of error to 4%. What's the necessary sample size?
    c) What sample size would produce a margin of error of 3%?

30. **Hiring.** In preparing a report on the economy, we need to estimate the percentage of businesses that plan to hire additional employees in the next 60 days.
    a) How many randomly selected employers must we contact in order to create an estimate in which we are 98% confident with a margin of error of 5%?
    b) Suppose we want to reduce the margin of error to 3%. What sample size will suffice?
    c) Why might it not be worth the effort to try to get an interval with a margin of error of only 1%?

31. **Graduation, again.** As in Exercise 29, we hope to estimate the percentage of adults aged 25 to 30 who never graduated from high school. What sample size would allow us to increase our confidence level to 95% while reducing the margin of error to only 2%?

32. **Better hiring info.** Editors of the business report in Exercise 30 are willing to accept a margin of error of 4% but want 99% confidence. How many randomly selected employers will they need to contact?

33. **Pilot study.** A state's environmental agency worries that many cars may be violating clean air emissions standards. The agency hopes to check a sample of vehicles in order to estimate that percentage with a margin of error of 3% and 90% confidence. To gauge the size of the problem, the agency first picks 60 cars and finds 9 with faulty emissions systems. How many should be sampled for a full investigation?

34. **Another pilot study.** During routine screening, a doctor notices that 22% of her adult patients show higher than normal levels of glucose in their blood—a possible warning signal for diabetes. Hearing this, some medical researchers decide to conduct a large-scale study, hoping to estimate the proportion to within 4% with 98% confidence. How many randomly selected adults must they test?

35. **Approval rating.** A newspaper reports that the governor's approval rating stands at 65%. The article adds that the poll is based on a random sample of 972 adults and has a margin of error of 2.5%. What level of confidence did the pollsters use?

36. **Amendment.** A TV news reporter says that a proposed constitutional amendment is likely to win approval in the upcoming election because a poll of 1505 likely voters indicated that 52% would vote in favor. The reporter goes on to say that the margin of error for this poll was 3%.
    a) Explain why the poll is actually inconclusive.
    b) What confidence level did the pollsters use?

JUST CHECKING
Answers

1. No. We know that in the sample 17% said "yes"; there's no need for a margin of error.
2. No, we are 95% confident that the percentage falls in some interval, not exactly on a particular value.
3. Yes. That's what the confidence interval means.
4. No. We don't know for sure that's true; we are only 95% confident.
5. No. That's our level of confidence, not the proportion of people receiving unsolicited text messages. The sample suggests the proportion is much lower.
6. Wider.
7. Lower.
8. Smaller.

Testing Hypotheses About Proportions

*"Half the money I spend on
advertising is wasted; the
trouble is I don't know which
half."*

—John Wanamaker
(attributed)

I ngots are huge pieces of metal, often weighing more than 20,000 pounds, made in a giant mold. They must be cast in one large piece for use in fabricating large structural parts for cars and planes. If they crack while being made, the crack can propagate into the zone required for the part, compromising its integrity. Airplane manufacturers insist that metal for their planes be defect-free, so the ingot must be made over if any cracking is detected.

Even though the metal from the cracked ingot is recycled, the scrap cost runs into the tens of thousands of dollars. Metal manufacturers would like to avoid cracking if at all possible. But the casting process is complicated and not everything is completely under control. In one plant, only about 80% of the ingots have been free of cracks. In an attempt to reduce the cracking proportion, the plant engineers and chemists recently tried out some changes in the casting process. Since then, 400 ingots have been cast and only 17% of them have cracked. Should management declare victory? Has the cracking rate really decreased, or was 17% just due to luck?

We can treat the 400 ingots cast with the new method as a random sample. We know that each random sample will have a somewhat different proportion of cracked ingots. Is the 17% we observe merely a result of natural sampling variability, or is this lower cracking rate strong enough evidence to assure management that the true cracking rate now is really below 20%?

People want answers to questions like these all the time. Has the president's approval rating changed since last month? Has teenage smoking decreased in the past five years? Is the global temperature increasing? Did the Super Bowl ad we bought actually increase sales? To answer such questions, we test *hypotheses* about models.

Hypotheses

How can we state and test a hypothesis about ingot cracking? Hypotheses are working models that we adopt temporarily. To test whether the changes made by the engineers have *improved* the cracking rate, we assume that they have in fact

Hypothesis n.;
pl. {Hypotheses}.
    A supposition; a
proposition or principle
which is supposed or taken
for granted, in order to draw
a conclusion or inference
for proof of the point in
question; something not
proved, but assumed for the
purpose of argument.
—*Webster's Unabridged
Dictionary, 1913*

NOTATION ALERT:

Capital H is the standard letter for hypotheses. H_0 always labels the null hypothesis, and H_A labels the alternative hypothesis.

To remind us that the parameter value comes from the null hypothesis, it is sometimes written as p_0 and the standard deviation as

$$SD(\hat{p}) = \sqrt{\frac{p_0 q_0}{n}}.$$

made no difference and that any apparent improvement is just random fluctuation (sampling error). So, our starting hypothesis, called the **null hypothesis,** is that the proportion of cracks is still 20%.

The null hypothesis, which we denote H_0, specifies a population model parameter of interest and proposes a value for that parameter. We usually write down the null hypothesis in the form H_0: *parameter = hypothesized value*. This is a concise way to specify the two things we need most: the identity of the parameter we hope to learn about and a specific hypothesized value for that parameter. (We need a hypothesized value so we can compare our observed statistic value to it.)

Which value to use is often obvious from the *Who* and *What* of the data. But sometimes it takes a bit of thinking to translate the question we hope to answer into a hypothesis about a parameter. For the ingots we can write H_0: $p = 0.20$.

The alternative hypothesis, which we denote H_A, contains the values of the parameter that we consider plausible if we reject the null hypothesis. In the ingots example, our null hypothesis is that $p = 0.20$. What's the alternative? Management is interested in *reducing* the cracking rate, so their alternative is H_A: $p < 0.20$.

What would convince you that the cracking rate had actually gone down? If you observed a cracking rate *much lower* than 20% in your sample, you'd likely be convinced. If only 3 out of the next 400 ingots crack (for a rate of 0.75%), most folks would conclude that the changes helped. But if the sample cracking rate is only slightly lower than 20%, you should be skeptical. After all, observed proportions do vary, so we wouldn't be surprised to see some difference. How much smaller must the cracking rate be before we *are* convinced that it has changed? Whenever we ask about the size of a statistical difference, we naturally think of using the standard deviation as a ruler. So let's start by finding the standard deviation of the sample cracking rate.

Since the company changed the process, 400 new ingots have been cast. The sample size of 400 is big enough to satisfy the **Success/Failure Condition.** (We expect $0.20 \times 400 = 80$ ingots to crack.) We have no reason to think the ingots are not independent, so the Normal sampling distribution model should work well. The standard deviation of the sampling model is

$$SD(\hat{p}) = \sqrt{\frac{pq}{n}} = \sqrt{\frac{(0.20)(0.80)}{400}} = 0.02$$

> **Why is this a standard deviation and not a standard error?** Because we haven't estimated anything. When we assume that the null hypothesis is true, it gives us a value for the model parameter p. With proportions, if we know p, then we also automatically know its standard deviation. And because we find the standard deviation from the model parameter, this is a standard deviation and not a standard error. When we found a confidence interval for p, we could not assume that we knew its value, so we estimated the standard deviation from the sample value $\hat{p}$.

Now we know both parameters of the Normal sampling distribution model: $p = 0.20$ and $SD(\hat{p}) = 0.02$, so we can find out how likely it would be to see the observed value of $\hat{p} = 17\%$. Since we are using a Normal model, we find the z-score:

$$z = \frac{0.17 - 0.20}{0.02} = -1.5$$

Then we ask, "How likely is it to observe a value at least 1.5 standard deviations below the mean of a Normal model?" The answer (from a calculator, computer program, or the Normal table) is about 0.067. This is the probability of observing a cracking rate of 17% or less in a sample of 400 if the null hypothesis is true.

Management now must decide whether an event that would happen 6.7% of the time by chance is strong enough evidence to conclude that the true cracking proportion has decreased.

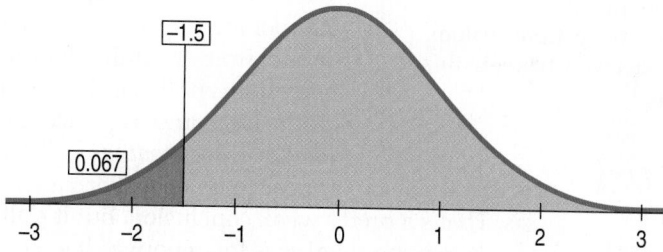

FIGURE 20.1

How likely is a z-score of −1.5 (or lower)? This is what it looks like. The red area is 0.067 of the total area under the curve.

A Trial as a Hypothesis Test

Does the reasoning of hypothesis tests seem backward? That could be because we usually prefer to think about getting things right rather than getting them wrong. You have seen this reasoning before in a different context. This is the logic of jury trials.

Let's suppose a defendant has been accused of robbery. In British common law and those systems derived from it (including U.S. law), the null hypothesis is that the defendant is innocent. Instructions to juries are quite explicit about this.

The evidence takes the form of facts that seem to contradict the presumption of innocence. For us, this means collecting data. In the trial, the prosecutor presents evidence. ("If the defendant were innocent, wouldn't it be remarkable that the police found him at the scene of the crime with a bag full of money in his hand, a mask on his face, and a getaway car parked outside?")

The next step is to judge the evidence. Evaluating the evidence is the responsibility of the jury in a trial, but it falls on your shoulders in hypothesis testing. The jury considers the evidence in light of the *presumption* of innocence and judges whether the evidence against the defendant would be plausible *if the defendant were in fact innocent.*

Like the jury, you ask, "Could these data plausibly have happened by chance if the null hypothesis were true?" If they are very unlikely to have occurred, then the evidence raises a reasonable doubt about the null hypothesis.

Ultimately, you must make a decision. The standard of "beyond a reasonable doubt" is wonderfully ambiguous because it leaves the jury to decide the degree to which the evidence contradicts the hypothesis of innocence. Juries don't explicitly use probability to help them decide whether to reject that hypothesis. But when you ask the same question of your null hypothesis, you have the advantage of being able to quantify exactly how surprising the evidence would be were the null hypothesis true.

How unlikely is unlikely? Some people set rigid standards, like 1 time out of 20 (0.05) or 1 time out of 100 (0.01). But if *you* have to make the decision, you must judge for yourself in each situation whether the probability of observing your data is small enough to constitute "reasonable doubt."

A S *Activity:* **The Reasoning of Hypothesis Testing.** Our reasoning is based on a rule of logic that dates back to ancient scholars. Here's a modern discussion of it.

P-Values

The fundamental step in our reasoning is the question "Are the data surprising, given the null hypothesis?" And the key calculation is to determine exactly how likely the data we observed would be were the null hypothesis a true model of the world. So we need a *probability.* Specifically, we want to find the probability of seeing data like these (or something even less likely) *given* that the null hypothesis is true. Statisticians are so thrilled with their ability to measure precisely

Beyond a Reasonable Doubt

We ask whether the data were unlikely beyond a reasonable doubt. We've just calculated that probability. The probability that the observed statistic value (or an even more extreme value) could occur if the null model were true—in this case, 0.067—is the P-value.

NOTATION ALERT:

We have many P's to keep straight. We use an uppercase P for probabilities, as in $P(A)$, and for the special probability we care about in hypothesis testing, the P-value.

We use lowercase p to denote our model's underlying proportion parameter and $\hat{p}$ to denote our observed proportion statistic.

how surprised they are that they give this probability a special name. It's called a **P-value.**[1]

When the P-value is high, we haven't seen anything unlikely or surprising at all. Events that have a high probability of happening happen often. The data are thus consistent with the model from the null hypothesis, and we have no reason to reject the null hypothesis. But we realize that many other similar hypotheses could also account for the data we've seen, so *we haven't proven that the null hypothesis is true*. The most we can say is that it doesn't appear to be false. Formally, we "fail to reject" the null hypothesis. That's a pretty weak conclusion, but it's all we're entitled to.

When the P-value *is* low enough, it says that it's very unlikely we'd observe data like these if our null hypothesis were true. We started with a model. Now that model tells us that the data we have are unlikely to have happened. The model and data are at odds with each other, so we have to make a choice. Either the null hypothesis is correct and we've just seen something remarkable, or the null hypothesis is wrong, and we were wrong to use it as the basis for computing our P-value. Perhaps another model is correct, and the data really aren't that remarkable after all. If you believe in data more than in assumptions, then, given that choice, you should reject the null hypothesis.

What to Do with an "Innocent" Defendant

"If the People fail to satisfy their burden of proof, you must find the defendant not guilty."

—NY state jury instructions

If the evidence is not strong enough to reject the defendant's presumption of innocence, what verdict does the jury return? They say "not guilty." Notice that they do not say that the defendant is innocent. All they say is that they have not seen sufficient evidence to convict, to reject innocence. The defendant may, in fact, be innocent, but the jury has no way to be sure.

Said statistically, the jury's null hypothesis is H_0: innocent defendant. If the evidence is too unlikely given this assumption, the jury rejects the null hypothesis and finds the defendant guilty. But—and this is an important distinction—if there is *insufficient evidence* to convict the defendant, the jury does not decide that H_0 is true and declare the defendant innocent. Juries can only *fail to reject* the null hypothesis and declare the defendant "not guilty."

In the same way, if the data are not particularly unlikely under the assumption that the null hypothesis is true, then the most we can do is to "fail to reject" our null hypothesis. We never declare the null hypothesis to be true (or "accept" the null), because we simply do not know whether it's true or not. (After all, more evidence may come along later.)

In the trial, the burden of proof is on the prosecution. In a hypothesis test, the burden of proof is on the unusual claim. The null hypothesis is the ordinary state of affairs, so it's the alternative to the null hypothesis that we consider unusual and for which we must marshal evidence.

Imagine a clinical trial testing the effectiveness of a new headache remedy. In Chapter 13 we saw the value of comparing such treatments to a placebo. The null hypothesis, then, is that the new treatment is no more effective than the placebo. This is important, because some patients will improve even when administered the placebo treatment. If we use only six people to test the drug, the results are likely *not to be clear* and we'll be unable to reject the hypothesis. Does this mean the drug doesn't work? Of course not. It simply means that we don't have enough

Don't "Accept" the Null Hypothesis

Every child knows that he (or she) is at the "center of the universe," so it's natural to suppose that the sun revolves around the earth. The fact that the sun appears to rise in the east every morning and set in the west every evening is *consistent* with this hypothesis and *seems* to lend support to it, but it certainly doesn't prove it, as we all eventually come to understand.

---

[1] You'd think if they were so excited, they'd give it a better name, but "P-value" is about as excited as statisticians get.

evidence to reject our assumption. That's why we don't start by assuming that the drug *is more effective*. If we were to do that, then we could test just a few people, find that the results aren't clear, and claim that since we've been unable to reject our original assumption the drug must be effective. The FDA is unlikely to be impressed by that argument.

JUST CHECKING

1. A research team wants to know if aspirin helps to thin blood. The null hypothesis says that it doesn't. They test 12 patients, observe the proportion with thinner blood, and get a P-value of 0.32. They proclaim that aspirin doesn't work. What would you say?

2. An allergy drug has been tested and found to give relief to 75% of the patients in a large clinical trial. Now the scientists want to see if the new, improved version works even better. What would the null hypothesis be?

3. The new drug is tested and the P-value is 0.0001. What would you conclude about the new drug?

The Reasoning of Hypothesis Testing

"The null hypothesis is never proved or established, but is possibly disproved, in the course of experimentation. Every experiment may be said to exist only in order to give the facts a chance of disproving the null hypothesis."

—Sir Ronald Fisher, *The Design of Experiments*

Hypothesis tests follow a carefully structured path. To avoid getting lost as we navigate down it, we divide that path into four distinct sections.

1. HYPOTHESES

First we state the null hypothesis. That's usually the skeptical claim that nothing's different. Are we considering a (New! Improved!) possibly better method? The null hypothesis says, "Oh yeah? Convince me!" To convert a skeptic, we must pile up enough evidence against the null hypothesis that we can reasonably reject it.

In statistical hypothesis testing, hypotheses are almost always about model parameters. To assess how unlikely our data may be, we need a null model. The null hypothesis specifies a particular parameter value to use in our model. In the usual shorthand, we write H_0: *parameter = hypothesized value*. The **alternative hypothesis**, H_A, contains the values of the parameter we consider plausible when we reject the null.

> Some folks pronounce the hypothesis labels "Ho!" and "Ha!" (but it makes them seem overexcited). We prefer to pronounce H_0 "H naught" (as in "all is for naught").

FOR EXAMPLE | Writing hypotheses

A large city's Department of Motor Vehicles claimed that 80% of candidates pass driving tests, but a newspaper reporter's survey of 90 randomly selected local teens who had taken the test found only 61 who passed.

Question: Does this finding suggest that the passing rate for teenagers is lower than the DMV reported? Write appropriate hypotheses.

I'll assume that the passing rate for teenagers is the same as the DMV's overall rate of 80%, unless there's strong evidence that it's lower.

$$H_O: p = 0.80$$
$$H_A: p < 0.80$$

2. MODEL

To plan a statistical hypothesis test, specify the *model* you will use to test the null hypothesis and the parameter of interest. Of course, all models require assumptions, so you will need to state them and check any corresponding conditions.

Your Model step should end with a statement such as

Because the conditions are satisfied, I can model the sampling distribution of the proportion with a Normal model.

Watch out, though. Your Model step could end with

Because the conditions are not satisfied, I can't proceed with the test. (If that's the case, stop and reconsider.)

Each test in the book has a name that you should include in your report. We'll see many tests in the chapters that follow. Some will be about more than one sample, some will involve statistics other than proportions, and some will use models other than the Normal (and so will not use z-scores). The test about proportions is called a **one-proportion z-test.**[2]

ONE-PROPORTION z-TEST

The conditions for the one-proportion z-test are the same as for the one-proportion z-interval. We test the hypothesis $H_0: p = p_0$ using the statistic $z = \dfrac{(\hat{p} - p_0)}{SD(\hat{p})}$. We use the hypothesized proportion to find the standard deviation, $SD(\hat{p}) = \sqrt{\dfrac{p_0 q_0}{n}}$.

When the conditions are met and the null hypothesis is true, this statistic follows the standard Normal model, so we can use that model to obtain a P-value.

When the Conditions Fail . . .
You might proceed with caution, explicitly stating your concerns. Or you may need to do the analysis with and without an outlier, or on different subgroups, or after re-expressing the response variable. Or you may not be able to proceed at all.

A S *Activity:* **Was the Observed Outcome Unlikely?** Complete the test you started in the first activity for this chapter. The narration explains the steps of the hypothesis test.

FOR EXAMPLE   Checking the conditions

Recap: A large city's DMV claimed that 80% of candidates pass driving tests. A reporter has results from a survey of 90 randomly selected local teens who had taken the test.

Question: Are the conditions for inference satisfied?

✔ The 90 teens surveyed were a random sample of local teenage driving candidates.

✔ 90 is fewer than 10% of the teenagers who take driving tests in a large city.

✔ We expect $np_0 = 90(0.80) = 72$ successes and $nq_0 = 90(0.20) = 18$ failures. Both are at least 10.

The conditions are satisfied, so it's okay to use a Normal model and perform a one-proportion z-test.

---

3. MECHANICS

Conditional Probability
Did you notice that a P-value is a conditional probability? It's the probability that the observed results could have happened *if the null hypothesis is true.*

Under "Mechanics," we place the actual calculation of our test statistic from the data. Different tests we encounter will have different formulas and different test statistics. Usually, the mechanics are handled by a statistics program or calculator, but it's good to have the formulas recorded for reference and to know what's

---

[2] It's also called the "one-sample test for a proportion."

being computed. The ultimate goal of the calculation is to obtain a P-value—the probability that the observed statistic value (or an even more extreme value) occur if the null model is correct. If the P-value is small enough, we'll reject the null hypothesis.

FOR EXAMPLE   Finding a P-value

Recap: A large city's DMV claimed that 80% of candidates pass driving tests, but a survey of 90 randomly selected local teens who had taken the test found only 61 who passed.

Question: What's the P-value for the one-proportion z-test?

I have $n = 90$, $x = 61$, and a hypothesized $p = 0.80$.

$$\hat{p} = \frac{61}{90} \approx 0.678$$

$$SD(\hat{p}) = \sqrt{\frac{p_0 q_0}{n}} = \sqrt{\frac{(0.8)(0.2)}{90}} \approx 0.042$$

$$z = \frac{\hat{p} - p_0}{SD(\hat{p})} = \frac{0.678 - 0.800}{0.042} \approx -2.90$$

$$\text{P-value} = P(z < -2.90) = 0.002$$

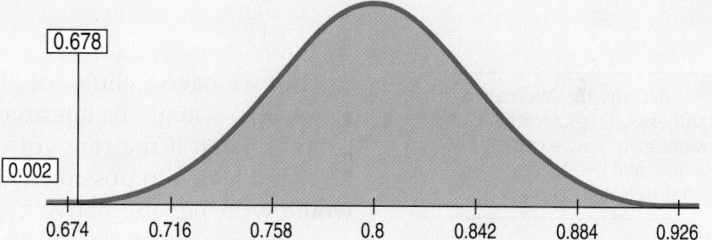

---

4. CONCLUSION

The conclusion in a hypothesis test is always a statement about the null hypothesis. The conclusion must state either that we reject or that we fail to reject the null hypothesis. And, as always, the conclusion should be stated in context.

FOR EXAMPLE   Stating the conclusion

Recap: A large city's DMV claimed that 80% of candidates pass driving tests. Data from a reporter's survey of randomly selected local teens who had taken the test produced a P-value of 0.002.

Question: What can the reporter conclude? And how might the reporter explain what the P-value means for the newspaper story?

Because the P-value of 0.002 is very low, I reject the null hypothesis. These survey data provide strong evidence that the passing rate for teenagers taking the driving test is lower than 80%.

If the passing rate for teenage driving candidates were actually 80%, we'd expect to see success rates this low in only about 1 in 500 samples (0.2%). This seems quite unlikely, casting doubt that the DMV's stated success rate applies to teens.

---

*". . . They make things
admirably plain,
But one hard question will
remain:
If one hypothesis you lose,
Another in its place you
choose . . ."*

—James Russell Lowell,
*Credidimus Jovem
Regnare*

Your conclusion about the null hypothesis should never be the end of a testing procedure. Often there are actions to take or policies to change. In our ingot example, management must decide whether to continue the changes proposed by the engineers. The decision always includes the practical consideration of whether the new method is worth the cost. Suppose management decides to reject the null hypothesis of 20% cracking in favor of the alternative that the percentage has been reduced. They must still evaluate how much the cracking rate has been reduced and how much it cost to accomplish the reduction. The *size of the effect* is always a concern when we test hypotheses. A good way to look at the effect size is to examine a confidence interval.

> **How much does it cost?** Formal tests of a null hypothesis base the decision of whether to reject the null hypothesis solely on the size of the P-value. But in real life, we want to evaluate the costs of our decisions as well. How much would you be willing to pay for a faster computer? Shouldn't your decision depend on how much faster? And on how much more it costs? Costs are not just monetary either. Would you use the same standard of proof for testing the safety of an airplane as for the speed of your new computer?

Alternative Alternatives

Tests on the ingot data can be viewed in two different ways. We know the old cracking rate is 20%, so the null hypothesis is

$$H_0: p = 0.20$$

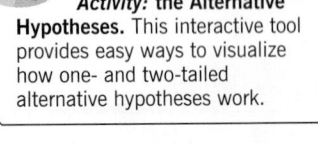

A S *Activity:* **the Alternative Hypotheses.** This interactive tool provides easy ways to visualize how one- and two-tailed alternative hypotheses work.

But we have a choice of alternative hypotheses. A metallurgist working for the company might be interested in *any* change in the cracking rate due to the new process. Even if the rate got worse, she might learn something useful from it. She's interested in possible changes on both sides of the null hypothesis. So she would write her alternative hypothesis as

$$H_A: p \neq 0.20$$

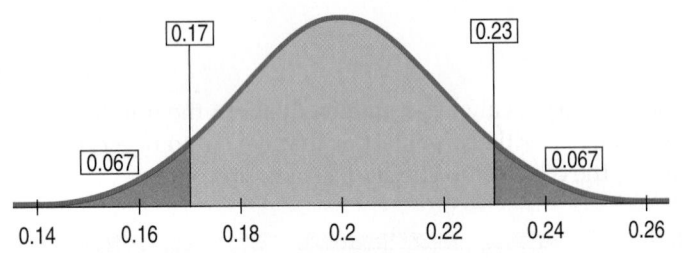

An alternative hypothesis such as this is known as a **two-sided alternative**,[3] because we are equally interested in deviations on either side of the null hypothesis value. For two-sided alternatives, the P-value is the probability of deviating in *either* direction from the null hypothesis value.

But management is really interested only in *lowering* the cracking rate below 20%. The scientific value of knowing how to *increase* the cracking rate may not appeal to them. The only alternative of interest to them is that the cracking rate *decreases*. They would write their alternative hypothesis as

$$H_A: p < 0.20$$

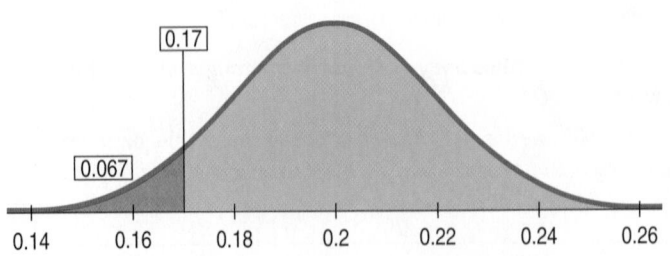

An alternative hypothesis that focuses on deviations from the null hypothesis value in only one direction is called a **one-sided alternative.**

For a hypothesis test with a one-sided alternative, the P-value is the probability of deviating *only in the direction of the alternative* away from the null hypothesis value. For the same data, the one-sided P-value is half the two-sided P-value. So, a one-sided test will reject the null hypothesis more often. If you aren't sure which to use, a two-sided test is always more conservative. Be sure you can justify the choice of a one-sided test from the *Why* of the situation.

---

[3] It is also called a **two-tailed alternative,** because the probabilities we care about are found in both tails of the sampling distribution.

STEP-BY-STEP EXAMPLE | Testing a Hypothesis

Anyone who plays or watches sports has heard of the "home field advantage." Teams tend to win more often when they play at home. Or do they?

If there were no home field advantage, the home teams would win about half of all games played. In the 2007 Major League Baseball season, there were 2431 regular-season games. (Tied at the end of the regular season, the Colorado Rockies and San Diego Padres played an extra game to determine who won the Wild Card playoff spot.) It turns out that the home team won 1319 of the 2431 games, or 54.26% of the time.

Question: Could this deviation from 50% be explained just from natural sampling variability, or is it evidence to suggest that there really is a home field advantage, at least in professional baseball?

Plan State what we want to know.

Define the variables and discuss the W's.

Hypotheses The null hypothesis makes the claim of no difference from the baseline. Here, that means no home field advantage.

We are interested only in a home field *advantage*, so the alternative hypothesis is one-sided.

Model Think about the assumptions and check the appropriate conditions.

I want to know whether the home team in professional baseball is more likely to win. The data are all 2431 games from the 2007 Major League Baseball season. The variable is whether or not the home team won. The parameter of interest is the proportion of home team wins. If there's no advantage, I'd expect that proportion to be 0.50.

$$H_O: p = 0.50$$
$$H_A: p > 0.50$$

✓ **Independence Assumption:** Generally, the outcome of one game has no effect on the outcome of another game. But this may not be strictly true. For example, if a key player is injured, the probability that the team will win in the next couple of games may decrease slightly, but independence is still roughly true. The data come from one entire season, but I expect other seasons to be similar.

✓ **Randomization Condition:** I have results for all 2431 games of the 2007 season. But I'm not just interested in 2007, and those games, while not randomly selected, should be a reasonable representative sample of all Major League Baseball games in the recent past and near future.

✓ **10% Condition:** We are interested in home field advantage for Major League Baseball for all seasons. While not a random sample, these 2431 games are fewer than 10% of all games played over the years.

✓ **Success/Failure Condition:** Both $np_O = 2431(0.50) = 1215.5$ and $nq_O = 2431(0.50) = 1215.5$ are at least 10.

Activity: Practice with Testing Hypotheses About Proportions. Here's an interactive tool that makes it easy to see what's going on in a hypothesis test.

Specify the sampling distribution model. State what test you plan to use.	Because the conditions are satisfied, I'll use a Normal model for the sampling distribution of the proportion and do a **one-proportion z-test.**

 **Mechanics** The null model gives us the mean, and (because we are working with proportions) the mean gives us the standard deviation.

The null model is a Normal distribution with a mean of 0.50 and a standard deviation of

$$SD(\hat{p}) = \sqrt{\frac{p_0 q_0}{n}} = \sqrt{\frac{(0.5)(1-0.5)}{2431}}$$
$$= 0.01014$$

Next, we find the z-score for the observed proportion, to find out how many standard deviations it is from the hypothesized proportion.

The observed proportion, $\hat{p}$, is 0.5426.

So the z-value is

$$z = \frac{0.5426 - 0.5}{0.01014} = 4.20$$

The sample proportion lies 4.20 standard deviations above the mean.

From the z-score, we can find the P-value, which tells us the probability of observing a value that extreme (or more).

The probability of observing a value 4.20 or more standard deviations above the mean of a Normal model can be found by computer, calculator, or table to be < 0.001.

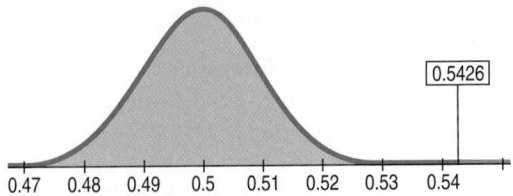

The corresponding P-value is < 0.001.

 **Conclusion** State your conclusion about the parameter—in context, of course!

The P-value of < 0.001 says that if the true proportion of home team wins were 0.50, then an observed value of 0.5426 (or larger) would occur less than 1 time in 1000. With a P-value so small, I reject H_0. I have evidence that the true proportion of home team wins is not 50%. It appears there is a home field advantage.

Ok, but how *big* is the home field advantage? Measuring the size of the effect involves a confidence interval. (Use your calculator.)

 Testing a hypothesis

By now probably nothing surprises you about your calculator. Of course it can help you with the mechanics of a hypothesis test. But that's not much. It cannot write the correct hypotheses, check the appropriate conditions, interpret the results, or state a conclusion. You have to do the tough stuff!

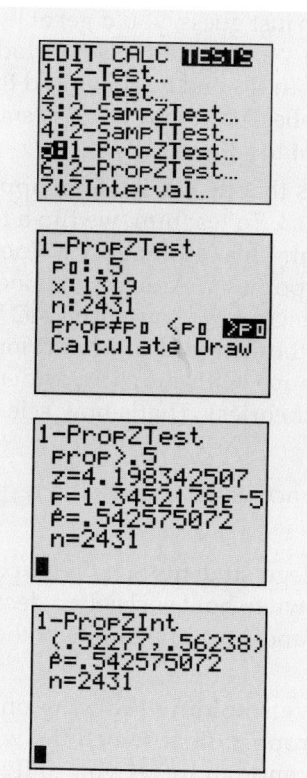

Let's do the mechanics of the Step-By-Step example about home field advantage in baseball. We hypothesized that home teams would win 50% of all games, but during this 2431-game season they actually won 54.26% of the time.

- Go to the **STAT TESTS** menu. Scroll down the list and select **5:1-Prop ZTest**.
- Specify the hypothesized proportion **P0**.
- Enter **x**, the observed number of wins: **1319**.
- Specify the sample size.
- Since this is a one-tail upper tail test, indicate that you want to see if the observed proportion is significantly greater than what was hypothesized.
- **Calculate** the result.

Ok, the rest is up to you. The calculator reports a z-score of 4.20 and a P-value (in scientific notation) of 1.35×10^{-5}, or about 0.00001. Such a small P-value indicates that the high percentage of home team wins is highly unlikely to be sampling error. State your conclusion in the appropriate context.

And how big is the advantage for the home team? In the last chapter you learned to create a 95% confidence interval. Try it here.

Looks like we can be 95% confident that in major league baseball games the home team wins between 52.3% and 56.2% of the time. Over a full season, the low end of this interval, 52.3% of the 81 home games, is nearly 2 extra victories, on average. The upper end, 56.2%, is 5 extra wins.

P-Values and Decisions: What to Tell About a Hypothesis Test

MORE

Hypothesis tests are particularly useful when we must make a decision. Is the defendant guilty or not? Should we choose print advertising or television? Questions like these cannot always be answered with the margins of error of confidence intervals. The absolute nature of the hypothesis test decision, however, makes some people (including the authors) uneasy. If possible, it's often a good idea to report a confidence interval for the parameter of interest as well.

How small should the P-value be in order for you to reject the null hypothesis? A jury needs enough evidence to show the defendant guilty "beyond a reasonable doubt." How does that translate to P-values? The answer is that it's highly context-dependent. When we're screening for a disease and want to be sure we treat all those who are sick, we may be willing to reject the null hypothesis of no disease with a P-value as large as 0.10. We would rather treat the occasional healthy person than fail to treat someone who was really sick. But a long-standing hypothesis, believed by many to be true, needs stronger evidence (and a correspondingly small P-value) to reject it.

See if you require the same P-value to reject each of the following null hypotheses:

▶ A renowned musicologist claims that she can distinguish between the works of Mozart and Haydn simply by hearing a randomly selected 20 seconds of music from any work by either composer. What's the null hypothesis? If she's just guessing, she'll get 50% of the pieces correct, on average. So our null hypothesis is that p is 50%. If she's for real, she'll get more than 50% correct. Now, we present her with 10 pieces of Mozart or Haydn chosen at random. She gets 9 out of 10 correct. It turns out that the P-value associated with

"Extraordinary claims require extraordinary proof."

—Carl Sagan

that result is 0.011. (In other words, if you tried to just guess, you'd get at least 9 out of 10 correct only about 1% of the time.) What would *you* conclude? Most people would probably reject the null hypothesis and be convinced that she has some ability to do as she claims. Why? Because the P-value is small and we don't have any particular reason to doubt the alternative.

▶ On the other hand, imagine a student who bets that he can make a flipped coin land the way he wants just by thinking hard. To test him, we flip a fair coin 10 times. Suppose he gets 9 out of 10 right. This also has a P-value of 0.011. Are you willing now to reject this null hypothesis? Are you convinced that he's not just lucky? What amount of evidence *would* convince you? We require more evidence if rejecting the null hypothesis would contradict long-standing beliefs or other scientific results. Of course, with sufficient evidence we would revise our opinions (and scientific theories). That's how science makes progress.

Another factor in choosing a P-value is the importance of the issue being tested. Consider the following two tests:

▶ A researcher claims that the proportion of college students who hold part-time jobs now is higher than the proportion known to hold such jobs a decade ago. You might be willing to believe the claim (and reject the null hypothesis of no change) with a P-value of 10%.

▶ An engineer claims that the proportion of rivets holding the wing on an airplane that are likely to fail is below the proportion at which the wing would fall off. What P-value would be small enough to get you to fly on that plane?

Your conclusion about any null hypothesis should be accompanied by the P-value of the test. Don't just declare the null hypothesis rejected or not rejected. Report the P-value to show the strength of the evidence against the hypothesis and the effect size. This will let each reader decide whether or not to reject the null hypothesis and whether or not to consider the result important if it is statistically significant.

To complete your analysis, follow your test with a confidence interval for the parameter of interest, to report the size of the effect.

A S *Activity:* **Hypothesis Tests for Proportions.** You've probably noticed that the tools for confidence intervals and for hypothesis tests are similar. See how tests and intervals for proportions are related—and an important way in which they differ.

JUST CHECKING

4. A bank is testing a new method for getting delinquent customers to pay their past-due credit card bills. The standard way was to send a letter (costing about $0.40) asking the customer to pay. That worked 30% of the time. They want to test a new method that involves sending a DVD to customers encouraging them to contact the bank and set up a payment plan. Developing and sending the video costs about $10.00 per customer. What is the parameter of interest? What are the null and alternative hypotheses?

5. The bank sets up an experiment to test the effectiveness of the DVD. They mail it out to several randomly selected delinquent customers and keep track of how many actually do contact the bank to arrange payments. The bank's statistician calculates a P-value of 0.003. What does this P-value suggest about the DVD?

6. The statistician tells the bank's management that the results are clear and that they should switch to the DVD method. Do you agree? What else might you want to know?

STEP-BY-STEP EXAMPLE | Tests and Intervals

Advances in medical care such as prenatal ultrasound examination now make it possible to determine a child's sex early in a pregnancy. There is a fear that in some cultures some parents may use this technology to select the sex of their children. A study from Punjab, India (E. E. Booth, M. Verma, and R. S. Beri, "Fetal Sex Determination in Infants in Punjab, India: Correlations and Implications," *BMJ* 309 [12 November 1994]: 1259–1261), reports that, in 1993, in one hospital, 56.9% of the 550 live births that year were boys. It's a medical fact that male babies are slightly more common than female babies. The study's authors report a baseline for this region of 51.7% male live births.

Question: Is there evidence that the proportion of male births has changed?

Plan State what we want to know.

Define the variables and discuss the W's.

Hypotheses The null hypothesis makes the claim of no difference from the baseline.

Before seeing the data, we were interested in any change in male births, so the alternative hypothesis is two-sided.

Model Think about the assumptions and check the appropriate conditions.

For testing proportions, the conditions are the same ones we had for making confidence intervals, except that we check the **Success/Failure Condition** with the *hypothesized* proportions rather than with the *observed* proportions.

Specify the sampling distribution model.

Tell what test you plan to use.

I want to know whether the proportion of male births has changed from the established baseline of 51.7%. The data are the recorded sexes of the 550 live births from a hospital in Punjab, India, in 1993, collected for a study on fetal sex determination. The parameter of interest, p, is the proportion of male births:

$$H_0: p = 0.517$$
$$H_A: p \neq 0.517$$

✔ **Independence Assumption:** There is no reason to think that the sex of one baby can affect the sex of other babies, so births can reasonably be assumed to be independent with regard to the sex of the child.

✔ **Randomization Condition:** The 550 live births are not a random sample, so I must be cautious about any general conclusions. I hope that this is a representative year, and I think that the births at this hospital may be typical of this area of India.

✔ **10% Condition:** I would like to be able to make statements about births at similar hospitals in India. These 550 births are fewer than 10% of all of those births.

✔ **Success/Failure Condition:** Both $np_0 = 550(0.517) = 284.35$ and $nq_0 = 550(0.483) = 265.65$ are greater than 10; I expect the births of at least 10 boys and at least 10 girls, so the sample is large enough.

The conditions are satisfied, so I can use a Normal model and perform a **one-proportion z-test.**

Mechanics The null model gives us the mean, and (because we are working with proportions) the mean gives us the standard deviation.

We find the z-score for the observed proportion to find out how many standard deviations it is from the hypothesized proportion.

Make a picture. Sketch a Normal model centered at $p_0 = 0.517$. Shade the region to the right of the observed proportion, and because this is a two-tail test, also shade the corresponding region in the other tail.

From the z-score, we can find the P-value, which tells us the probability of observing a value that extreme (or more). Use technology or a table (see p. 473.).

Because this is a two-tail test, the P-value is the probability of observing an outcome more than 2.44 standard deviations from the mean of a Normal model *in either direction*. We must therefore *double* the probability we find in the upper tail.

The null model is a Normal distribution with a mean of 0.517 and a standard deviation of

$$SD(\hat{p}) = \sqrt{\frac{p_0 q_0}{n}} = \sqrt{\frac{(0.517)(1 - 0.517)}{550}}$$
$$= 0.0213$$

The observed proportion, $\hat{p}$, is 0.569, so

$$z = \frac{\hat{p} - p_0}{SD(\hat{p})} = \frac{0.569 - 0.517}{0.0213} = 2.44$$

The sample proportion lies 2.44 standard deviations above the mean.

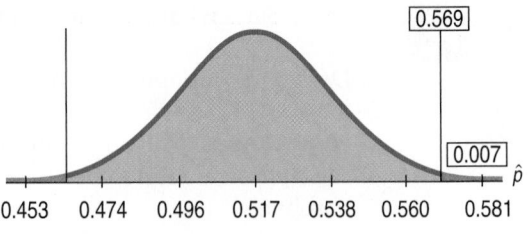

$$P = 2P(z > 2.44) = 2(0.0073) = 0.0146$$

Conclusion State your conclusion in context.

This P-value is roughly 1 time in 70. That's clearly significant, but don't jump to other conclusions. We can't be sure how this deviation came about. For instance, we don't know whether this hospital is typical, or whether the time period studied was selected at random.

The P-value of 0.0146 says that if the true proportion of male babies were still at 51.7%, then an observed proportion as different as 56.9% male babies would occur at random only about 15 times in 1000. With a P-value this small, I reject H_0. This is strong evidence that the birth ratio of boys to girls is not equal to its natural level. It appears that the proportion of boys may have increased.

How big an increase are we talking about? Let's find a confidence interval for the proportion of male births.

Model Check the conditions.

The conditions are identical to those for the hypothesis test, with one difference. Now we are not given a hypothesized proportion, p_0, so we must instead work with the observed proportion $\hat{p}$.

✓ **Success/Failure Condition:** Both $n\hat{p} = 550(0.569) = 313$ and $n\hat{q} = 237$ are at least 10.

Specify the sampling distribution model.	The conditions are satisfied, so I can model the sampling distribution of the proportion with a Normal model and find a **one-proportion z-interval.**
Tell what method you plan to use.	

Mechanics We can't find the sampling model standard deviation from the null model proportion. (In fact, we've just rejected it.) Instead, we find the standard error of $\hat{p}$ from the *observed* proportions. Other than that substitution, the calculation looks the same as for the hypothesis test.

With this large a sample size, the difference is negligible, but in smaller samples, it could make a bigger difference.

$$SE(\hat{p}) = \sqrt{\frac{\hat{p}\hat{q}}{n}} = \sqrt{\frac{(0.569)(1 - 0.569)}{550}}$$
$$= 0.0211$$

The sampling model is Normal, so for a 95% confidence interval, the critical value $z^* = 1.96$.

The margin of error is

$$ME = z^* \times SE(\hat{p}) = 1.96(0.0211) = 0.041$$

So the 95% confidence interval is

$$0.569 \pm 0.041 \text{ or } (0.528, 0.610).$$

Conclusion Confidence intervals help us think about the size of the effect. Here we can see that the change from the baseline of 51.7% male births might be quite substantial.

We are 95% confident that the true proportion of male births is between 52.8% and 61.0%.

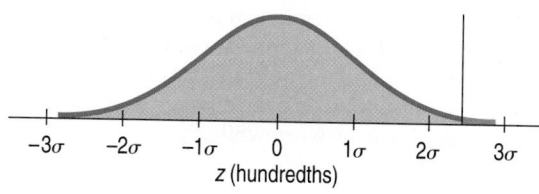

Here's a portion of a Normal table that gives the probability we needed for the hypothesis test. At $z = 2.44$, the table gives the percentile as 0.9927. The upper-tail probability (shaded red) is, therefore, $1 - 0.9927 = 0.0073$; so, for our two-sided test, the P-value is $2(0.0073) = 0.0146$.

z	0.00	0.01	0.02	0.03	0.04	0.05
1.9	0.9713	0.9719	0.9726	0.9732	0.9738	0.9744
2.0	0.9772	0.9778	0.9783	0.9788	0.9793	0.9798
2.1	0.9821	0.9826	0.9830	0.9834	0.9838	0.9842
2.2	0.9861	0.9864	0.9868	0.9871	0.9875	0.9878
2.3	0.9893	0.9896	0.9898	0.9901	0.9904	0.9906
2.4	0.9918	0.9920	0.9922	0.9925	0.9927	0.9929
2.5	0.9938	0.9940	0.9941	0.9943	0.9945	0.9946
2.6	0.9953	0.9955	0.9956	0.9957	0.9959	0.9960

WHAT CAN GO WRONG?

Hypothesis tests are so widely used—and so widely misused—that we've devoted all of the next chapter to discussing the pitfalls involved, but there are a few issues that we can talk about already.

▶ **Don't base your null hypotheses on what you see in the data.** You are not allowed to look at the data first and then adjust your null hypothesis so that it will be rejected. When your sample value turns out to be $\hat{p} = 51.8\%$, with a standard deviation of 1%, don't form a null hypothesis like $H_0: p = 49.8\%$, knowing that you can reject it. You should always *Think* about the situation you are investigating and make your null hypothesis describe the "nothing interesting" or "nothing has changed" scenario. No peeking at the data!

▶ **Don't base your alternative hypothesis on the data, either.** Again, you need to *Think* about the situation. Are you interested only in knowing whether something has *increased*? Then write a one-sided (upper-tail) alternative. Or would you be equally interested in a change in either direction? Then you want a two-sided alternative. You should decide whether to do a one- or two-sided test based on what results would be of interest to you, not what you see in the data.

▶ **Don't make your null hypothesis what you want to show to be true.** Remember, the null hypothesis is the status quo, the nothing-is-strange-here position a skeptic would take. You wonder whether the data cast doubt on that. You can reject the null hypothesis, but you can never "accept" or "prove" the null.

▶ **Don't forget to check the conditions.** The reasoning of inference depends on randomization. No amount of care in calculating a test result can recover from biased sampling. The probabilities we compute depend on the independence assumption. And our sample must be large enough to justify our use of a Normal model.

▶ **Don't accept the null hypothesis.** You may not have found enough evidence to reject it, but you surely have *not* proven it's true!

▶ **If you fail to reject the null hypothesis, don't think that a bigger sample would be more likely to lead to rejection.** If the results you looked at were "almost" significant, it's enticing to think that because you would have rejected the null had these same observations come from a larger sample, then a larger sample would surely lead to rejection. Don't be misled. Remember, each sample is different, and a larger sample won't necessarily duplicate your current observations. Indeed, the Central Limit Theorem tells us that statistics will vary *less* in larger samples. We should therefore expect such results to be less extreme. Maybe they'd be statistically significant but maybe (perhaps even probably) not. Even if you fail to reject the null hypothesis, it's a good idea to examine a confidence interval. If none of the plausible parameter values in the interval would matter to you (for example, because none would be *practically* significant), then even a larger study with a correspondingly smaller standard error is unlikely to be worthwhile.

Don't We Want to Reject the Null?

Often the folks who collect the data or perform the experiment hope to reject the null. (They hope the new drug is better than the placebo, or new ad campaign is better than the old one.) But when we practice Statistics, we can't allow that hope to affect our decision. The essential attitude for a hypothesis tester is skepticism. Until we become convinced otherwise, we cling to the null's assertion that there's nothing unusual, no effect, no difference, etc. As in a jury trial, the burden of proof rests with the alternative hypothesis—innocent until proven guilty. When you test a hypothesis, you must act as judge and jury, but you are not the prosecutor.

CONNECTIONS

Hypothesis tests and confidence intervals share many of the same concepts. Both rely on sampling distribution models, and because the models are the same and require the same assumptions, both check the same conditions. They also calculate many of the same statistics. Like confidence intervals, hypothesis tests use the standard deviation of the sampling distribution as a ruler, as we first saw in Chapter 6.

For testing, we find ourselves looking once again at z-scores, and we compute the P-value by finding the distance of our test statistic from the center of the null model. P-values are conditional probabilities. They give the probability of observing the result we have seen (or one even more extreme) *given* that the null hypothesis is true.

The Standard Normal model is here again as our connection between z-score values and probabilities.

WHAT HAVE WE LEARNED?

We've learned to use what we see in a random sample to test a particular hypothesis about the world. This is our second step in statistical inference, complementing our use of confidence intervals.

We've learned that testing a hypothesis involves proposing a model, then seeing whether the data we observe are consistent with that model or are so unusual that we must reject it. We do this by finding a P-value—the probability that data like ours could have occurred if the model is correct.

We've learned that:

▶ We start with a null hypothesis specifying the parameter of a model we'll test using our data.
▶ Our alternative hypothesis can be one- or two-sided, depending on what we want to learn.
▶ We must check the appropriate assumptions and conditions before proceeding with our test.
▶ If the data are out of line with the null hypothesis model, the P-value will be small and we will reject the null hypothesis.
▶ If the data are consistent with the null hypothesis model, the P-value will be large and we will not reject the null hypothesis.
▶ We must always state our conclusion in the context of the original question.

And we've learned that confidence intervals and hypothesis tests go hand in hand in helping us think about models. A hypothesis test makes a yes/no decision about the plausibility of a parameter value. The confidence interval shows us the range of plausible values for the parameter.

Terms

Null hypothesis

460. The claim being assessed in a hypothesis test is called the null hypothesis. Usually, the null hypothesis is a statement of "no change from the traditional value," "no effect," "no difference," or "no relationship." For a claim to be a testable null hypothesis, it must specify a value for some population parameter that can form the basis for assuming a sampling distribution for a test statistic.

Alternative hypothesis

460. The alternative hypothesis proposes what we should conclude if we find the null hypothesis to be unlikely.

Two-sided alternative (Two-tailed alternative)

466. An alternative hypothesis is two-sided ($H_A: p \neq p_0$) when we are interested in deviations in *either* direction away from the hypothesized parameter value.

One-sided alternative (One-tailed alternative)

466. An alternative hypothesis is one-sided (e.g., $H_A: p > p_0$ or $H_A: p < p_0$) when we are interested in deviations in *only one* direction away from the hypothesized parameter value.

P-value

461. The probability of observing a value for a test statistic at least as far from the hypothesized value as the statistic value actually observed if the null hypothesis is true. A small P-value indicates either that the observation is improbable or that the probability calculation was based on incorrect assumptions. The assumed truth of the null hypothesis is the assumption under suspicion.

One-proportion z-test

464. A test of the null hypothesis that the proportion of a single sample equals a specified value ($H_0: p = p_0$) by referring the statistic $z = \dfrac{\hat{p} - p_0}{SD(\hat{p})}$ to a Standard Normal model.

Skills

▶ Be able to state the null and alternative hypotheses for a one-proportion z-test.
▶ Know the conditions that must be true for a one-proportion z-test to be appropriate, and know how to examine your data for violations of those conditions.
▶ Be able to identify and use the alternative hypothesis when testing hypotheses. Understand how to choose between a one-sided and two-sided alternative hypothesis, and be able to explain your choice.

▶ Be able to perform a one-proportion z-test.
▶ Be able to write a sentence interpreting the results of a one-proportion z-test.
▶ Know how to interpret the meaning of a P-value in nontechnical language, making clear that the probability claim is made about computed values under the assumption that the null model is true and not about the population parameter of interest.

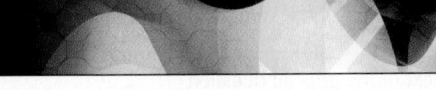

HYPOTHESIS TESTS FOR PROPORTIONS ON THE COMPUTER

Hypothesis tests for proportions are so easy and natural that many statistics packages don't offer special commands for them. Most statistics programs want to know the "success" and "failure" status for each case. Usually these are given as 1 or 0, but they might be category names like "yes" and "no." Often we just know the proportion of successes, $\hat{p}$, and the total count, n. Computer packages don't usually deal naturally with summary data like this, but the statistics routines found on many graphing calculators do. These calculators allow you to test hypotheses from summaries of the data—usually, all you need to enter are the number of successes and the sample size.

EXERCISES

1. **Hypotheses.** Write the null and alternative hypotheses you would use to test each of the following situations:
   a) A governor is concerned about his "negatives"—the percentage of state residents who express disapproval of his job performance. His political committee pays for a series of TV ads, hoping that they can keep the negatives below 30%. They will use follow-up polling to assess the ads' effectiveness.
   b) Is a coin fair?
   c) Only about 20% of people who try to quit smoking succeed. Sellers of a motivational tape claim that listening to the recorded messages can help people quit.

2. **More hypotheses.** Write the null and alternative hypotheses you would use to test each situation.
   a) In the 1950s only about 40% of high school graduates went on to college. Has the percentage changed?
   b) 20% of cars of a certain model have needed costly transmission work after being driven between 50,000 and 100,000 miles. The manufacturer hopes that a redesign of a transmission component has solved this problem.
   c) We field-test a new-flavor soft drink, planning to market it only if we are sure that over 60% of the people like the flavor.

3. **Negatives.** After the political ad campaign described in Exercise 1a, pollsters check the governor's negatives. They test the hypothesis that the ads produced no change against the alternative that the negatives are now below 30% and find a P-value of 0.22. Which conclusion is appropriate? Explain.
   a) There's a 22% chance that the ads worked.
   b) There's a 78% chance that the ads worked.
   c) There's a 22% chance that their poll is correct.
   d) There's a 22% chance that natural sampling variation could produce poll results like these if there's really no change in public opinion.

4. **Dice.** The seller of a loaded die claims that it will favor the outcome 6. We don't believe that claim, and roll the die 200 times to test an appropriate hypothesis. Our

P-value turns out to be 0.03. Which conclusion is appropriate? Explain.
   a) There's a 3% chance that the die is fair.
   b) There's a 97% chance that the die is fair.
   c) There's a 3% chance that a loaded die could randomly produce the results we observed, so it's reasonable to conclude that the die is fair.
   d) There's a 3% chance that a fair die could randomly produce the results we observed, so it's reasonable to conclude that the die is loaded.

5. **Relief.** A company's old antacid formula provided relief for 70% of the people who used it. The company tests a new formula to see if it is better and gets a P-value of 0.27. Is it reasonable to conclude that the new formula and the old one are equally effective? Explain.

6. **Cars.** A survey investigating whether the proportion of today's high school seniors who own their own cars is higher than it was a decade ago finds a P-value of 0.017. Is it reasonable to conclude that more high-schoolers have cars? Explain.

7. **He cheats!** A friend of yours claims that when he tosses a coin he can control the outcome. You are skeptical and want him to prove it. He tosses the coin, and you call heads; it's tails. You try again and lose again.
   a) Do two losses in a row convince you that he really can control the toss? Explain.
   b) You try a third time, and again you lose. What's the probability of losing three tosses in a row if the process is fair?
   c) Would three losses in a row convince you that your friend cheats? Explain.
   d) How many times in a row would you have to lose in order to be pretty sure that this friend really can control the toss? Justify your answer by calculating a probability and explaining what it means.

8. **Candy.** Someone hands you a box of a dozen chocolate-covered candies, telling you that half are vanilla creams and the other half peanut butter. You pick candies at random and discover the first three you eat are all vanilla.

a) If there really were 6 vanilla and 6 peanut butter candies in the box, what is the probability that you would have picked three vanillas in a row?
b) Do you think there really might have been 6 of each? Explain.
c) Would you continue to believe that half are vanilla if the fourth one you try is also vanilla? Explain.

9. **Cell phones.** Many people have trouble setting up all the features of their cell phones, so a company has developed what it hopes will be easier instructions. The goal is to have at least 96% of customers succeed. The company tests the new system on 200 people, of whom 188 were successful. Is this strong evidence that the new system fails to meet the company's goal? A student's test of this hypothesis is shown. How many mistakes can you find?

$H_0: \hat{p} = 0.96$
$H_A: \hat{p} \neq 0.96$
SRS, $0.96(200) > 10$
$\frac{188}{200} = 0.94; \quad SD(\hat{p}) = \sqrt{\frac{(0.94)(0.06)}{200}} = 0.017$
$z = \frac{0.96 - 0.94}{0.017} = 1.18$
$P = P(z > 1.18) = 0.12$
There is strong evidence the new instructions don't work.

10. **Got milk?** In November 2001, the *Ag Globe Trotter* newsletter reported that 90% of adults drink milk. A regional farmers' organization planning a new marketing campaign across its multicounty area polls a random sample of 750 adults living there. In this sample, 657 people said that they drink milk. Do these responses provide strong evidence that the 90% figure is not accurate for this region? Correct the mistakes you find in a student's attempt to test an appropriate hypothesis.

$H_0: \hat{p} = 0.9$
$H_A: \hat{p} < 0.9$
SRS, $750 > 10$
$\frac{657}{750} = 0.876; \quad SD(\hat{p}) = \sqrt{\frac{(0.88)(0.12)}{750}} = 0.012$
$z = \frac{0.876 - 0.90}{0.012} = -2$
$P = P(z > -2) = 0.977$
There is more than a 97% chance that the stated percentage is correct for this region.

11. **Dowsing.** In a rural area, only about 30% of the wells that are drilled find adequate water at a depth of 100 feet or less. A local man claims to be able to find water by "dowsing"—using a forked stick to indicate where the well should be drilled. You check with 80 of his customers and find that 27 have wells less than 100 feet deep. What do you conclude about his claim?
a) Write appropriate hypotheses.
b) Check the necessary assumptions.
c) Perform the mechanics of the test. What is the P-value?
d) Explain carefully what the P-value means in context.
e) What's your conclusion?

12. **Abnormalities.** In the 1980s it was generally believed that congenital abnormalities affected about 5% of the nation's children. Some people believe that the increase in the number of chemicals in the environment has led to an increase in the incidence of abnormalities. A recent study examined 384 children and found that 46 of them showed signs of an abnormality. Is this strong evidence that the risk has increased?
a) Write appropriate hypotheses.
b) Check the necessary assumptions.
c) Perform the mechanics of the test. What is the P-value?
d) Explain carefully what the P-value means in context.
e) What's your conclusion?
f) Do environmental chemicals cause congenital abnormalities?

13. **Absentees.** The National Center for Education Statistics monitors many aspects of elementary and secondary education nationwide. Their 1996 numbers are often used as a baseline to assess changes. In 1996 34% of students had not been absent from school even once during the previous month. In the 2000 survey, responses from 8302 students showed that this figure had slipped to 33%. Officials would, of course, be concerned if student attendance were declining. Do these figures give evidence of a change in student attendance?
a) Write appropriate hypotheses.
b) Check the assumptions and conditions.
c) Perform the test and find the P-value.
d) State your conclusion.
e) Do you think this difference is meaningful? Explain.

14. **Educated mothers.** The National Center for Education Statistics monitors many aspects of elementary and secondary education nationwide. Their 1996 numbers are often used as a baseline to assess changes. In 1996, 31% of students reported that their mothers had graduated from college. In 2000, responses from 8368 students found that this figure had grown to 32%. Is this evidence of a change in education level among mothers?
a) Write appropriate hypotheses.
b) Check the assumptions and conditions.
c) Perform the test and find the P-value.
d) State your conclusion.
e) Do you think this difference is meaningful? Explain.

15. **Contributions, please, part II.** In Exercise 19.15 you learned that the Paralyzed Veterans of America is a philanthropic organization that relies on contributions. They send free mailing labels and greeting cards to potential donors on their list and ask for a voluntary contribution. To test a new campaign, the organization recently sent letters to a random sample of 100,000 potential donors and received 4781 donations. They've had a contribution rate of 5% in past campaigns, but a staff member worries that the rate will be lower if they run this campaign as currently designed.
a) What are the hypotheses?
b) Are the assumptions and conditions for inference met?
c) Do you think the rate would drop? Explain.

16. **Take the offer, part II.**   In Exercise 19.16 you learned that First USA, a major credit card company, is planning a new offer for their current cardholders. First USA will give double airline miles on purchases for the next 6 months if the cardholder goes online and registers for this offer. To test the effectiveness of this campaign, the company recently sent out offers to a random sample of 50,000 cardholders. Of those, 1184 registered. A staff member suspects that the success rate for the full campaign will be comparable to the standard 2% rate that they are used to seeing in similar campaigns. What do you predict?
    a) What are the hypotheses?
    b) Are the assumptions and conditions for inference met?
    c) Do you think the rate would change if they use this fundraising campaign? Explain.

17. **Law School.**   According to the Law School Admission Council, in the fall of 2006, 63% of law school applicants were accepted to some law school.[4] The training program *LSATisfaction* claims that 163 of the 240 students trained in 2006 were admitted to law school. You can safely consider these trainees to be representative of the population of law school applicants. Has *LSATisfaction* demonstrated a real improvement over the national average?
    a) What are the hypotheses?
    b) Check the conditions and find the P-value.
    c) Would you recommend this program based on what you see here? Explain.

18. **Med School.**   According to the Association of American Medical Colleges, only 46% of medical school applicants were admitted to a medical school in the fall of 2006.[5] Upon hearing this, the trustees of Striving College expressed concern that only 77 of the 180 students in their class of 2006 who applied to medical school were admitted. The college president assured the trustees that this was just the kind of year-to-year fluctuation in fortunes that is to be expected and that, in fact, the school's success rate was consistent with the national average. Who is right?
    a) What are the hypotheses?
    b) Check the conditions and find the P-value.
    c) Are the trustees right to be concerned, or is the president correct? Explain.

19. **Pollution.**   A company with a fleet of 150 cars found that the emissions systems of 7 out of the 22 they tested failed to meet pollution control guidelines. Is this strong evidence that more than 20% of the fleet might be out of compliance? Test an appropriate hypothesis and state your conclusion. Be sure the appropriate assumptions and conditions are satisfied before you proceed.

20. **Scratch and dent.**   An appliance manufacturer stockpiles washers and dryers in a large warehouse for shipment to retail stores. Sometimes in handling them the appliances get damaged. Even though the damage may be minor, the company must sell those machines at drastically reduced prices. The company goal is to keep the level of damaged machines below 2%. One day an inspector randomly checks 60 washers and finds that 5 of them have scratches or dents. Is this strong evidence that the warehouse is failing to meet the company goal? Test an appropriate hypothesis and state your conclusion. Be sure the appropriate assumptions and conditions are satisfied before you proceed.

21. **Twins.**   In 2001 a national vital statistics report indicated that about 3% of all births produced twins. Is the rate of twin births the same among very young mothers? Data from a large city hospital found that only 7 sets of twins were born to 469 teenage girls. Test an appropriate hypothesis and state your conclusion. Be sure the appropriate assumptions and conditions are satisfied before you proceed.

22. **Football 2006.**   During the 2006 season, the home team won 136 of the 240 regular-season National Football League games. Is this strong evidence of a home field advantage in professional football? Test an appropriate hypothesis and state your conclusion. Be sure the appropriate assumptions and conditions are satisfied before you proceed.

23. **WebZine.**   A magazine is considering the launch of an online edition. The magazine plans to go ahead only if it's convinced that more than 25% of current readers would subscribe. The magazine contacted a simple random sample of 500 current subscribers, and 137 of those surveyed expressed interest. What should the company do? Test an appropriate hypothesis and state your conclusion. Be sure the appropriate assumptions and conditions are satisfied before you proceed.

24. **Seeds.**   A garden center wants to store leftover packets of vegetable seeds for sale the following spring, but the center is concerned that the seeds may not germinate at the same rate a year later. The manager finds a packet of last year's green bean seeds and plants them as a test. Although the packet claims a germination rate of 92%, only 171 of 200 test seeds sprout. Is this evidence that the seeds have lost viability during a year in storage? Test an appropriate hypothesis and state your conclusion. Be sure the appropriate assumptions and conditions are satisfied before you proceed.

25. **Women executives.**   A company is criticized because only 13 of 43 people in executive-level positions are women. The company explains that although this proportion is lower than it might wish, it's not surprising given that only 40% of all its employees are women. What do you think? Test an appropriate hypothesis and state your conclusion. Be sure the appropriate assumptions and conditions are satisfied before you proceed.

26. **Jury.**   Census data for a certain county show that 19% of the adult residents are Hispanic. Suppose 72 people are called for jury duty and only 9 of them are Hispanic. Does this apparent underrepresentation of Hispanics call into question the fairness of the jury selection system? Explain.

---

[4] As reported by the Cornell office of career services in their *Class of 2006 Postgraduate Report.*
[5] *Ibid.*

27. Dropouts. Some people are concerned that new tougher standards and high-stakes tests adopted in many states have driven up the high school dropout rate. The National Center for Education Statistics reported that the high school dropout rate for the year 2004 was 10.3%. One school district whose dropout rate has always been very close to the national average reports that 210 of their 1782 high school students dropped out last year. Is this evidence that their dropout rate may be increasing? Explain.

28. Acid rain. A study of the effects of acid rain on trees in the Hopkins Forest shows that 25 of 100 trees sampled exhibited some sort of damage from acid rain. This rate seemed to be higher than the 15% quoted in a recent *Environmetrics* article on the average proportion of damaged trees in the Northeast. Does the sample suggest that trees in the Hopkins Forest are more susceptible than trees from the rest of the region? Comment, and write up your own conclusions based on an appropriate confidence interval as well as a hypothesis test. Include any assumptions you made about the data.

29. Lost luggage. An airline's public relations department says that the airline rarely loses passengers' luggage. It further claims that on those occasions when luggage is lost, 90% is recovered and delivered to its owner within 24 hours. A consumer group that surveyed a large number of air travelers found that only 103 of 122 people who lost luggage on that airline were reunited with the missing items by the next day. Does this cast doubt on the airline's claim? Explain.

30. TV ads. A start-up company is about to market a new computer printer. It decides to gamble by running commercials during the Super Bowl. The company hopes that name recognition will be worth the high cost of the ads. The goal of the company is that over 40% of the public recognize its brand name and associate it with computer equipment. The day after the game, a pollster contacts 420 randomly chosen adults and finds that 181 of them know that this company manufactures printers. Would you recommend that the company continue to advertise during Super Bowls? Explain.

31. John Wayne. Like a lot of other Americans, John Wayne died of cancer. But is there more to this story? In 1955 Wayne was in Utah shooting the film *The Conqueror*. Across the state line, in Nevada, the United States military was testing atomic bombs. Radioactive fallout from those tests drifted across the filming location. A total of 46 of the 220 people working on the film eventually died of cancer. Cancer experts estimate that one would expect only about 30 cancer deaths in a group this size.
a) Is the death rate among the movie crew unusually high?
b) Does this prove that exposure to radiation increases the risk of cancer?

32. AP Stats. The College Board reported that 60% of all students who took the 2006 AP Statistics exam earned scores of 3 or higher. One teacher wondered if the performance of her school was different. She believed that year's students to be typical of those who will take AP Stats at that school and was pleased when 65% of her 54 students achieved scores of 3 or better. Can she claim that her school is different? Explain.

JUST CHECKING
Answers

1. You can't conclude that the null hypothesis is true. You can conclude only that the experiment was unable to reject the null hypothesis. They were unable, on the basis of 12 patients, to show that aspirin was effective.

2. The null hypothesis is $H_0: p = 0.75$.

3. With a P-value of 0.0001, this is very strong evidence against the null hypothesis. We can reject H_0 and conclude that the improved version of the drug gives relief to a higher proportion of patients.

4. The parameter of interest is the proportion, p, of all delinquent customers who will pay their bills. $H_0: p = 0.30$ and $H_A: p > 0.30$.

5. The very low P-value leads us to reject the null hypothesis. There is strong evidence that the DVD is more effective in getting people to start paying their debts than just sending a letter had been.

6. All we know is that there is strong evidence to suggest that $p > 0.30$. We don't know how much higher than 30% the new proportion is. We'd like to see a confidence interval to see if the new method is worth the cost.

CHAPTER
21

More About Tests and Intervals

WHO	Florida motorcycle riders aged 20 and younger involved in motorcycle accidents
WHAT	% wearing helmets
WHEN	2001–2003
WHERE	Florida
WHY	Assessment of injury rates commissioned by the National Highway Traffic Safety Administration (NHTSA)

In 2000 Florida changed its motorcycle helmet law. No longer are riders 21 and older required to wear helmets. Under the new law, those under 21 still must wear helmets, but a report by the Preusser Group (www .preussergroup. com) suggests that helmet use may have declined in this group, too.

It isn't practical to survey young motorcycle riders. (For example, how can you construct a sampling frame? If you contacted licensed riders, would they admit to riding illegally without a helmet?) The researchers adopted a different strategy. Police reports of motorcycle accidents record whether the rider wore a helmet and give the rider's age. Before the change in the helmet law, 60% of youths involved in a motorcycle accident had been wearing their helmets. The Preusser study looked at accident reports during 2001–2003, the three years following the law change, considering these riders to be a representative sample of the larger population. They observed 781 young riders who were involved in accidents. Of these, 396 (or 50.7%) were wearing helmets. Is this evidence of a decline in helmet-wearing, or just the natural fluctuation of such statistics?

Zero In on the Null

Null hypotheses have special requirements. In order to perform a statistical test of the hypothesis, the null must be a statement about the value of a parameter for a model. We use this value to compute the probability that the observed sample statistic—or something even farther from the null value—might occur.

How do we choose the null hypothesis? The appropriate null arises directly from the context of the problem. It is dictated, not by the data, but by the situation. One good way to identify both the null and alternative hypotheses is to think about the *Why* of the situation. Typical null hypotheses might be that the proportion of patients recovering after receiving a new drug is the same as we would expect of patients receiving a placebo or that the mean strength attained by athletes training with new equipment is the same as with the old equipment. The alternative hypotheses would be that the new drug cures a higher proportion of patients or that the new equipment results in a greater mean strength.

480

To write a null hypothesis, you can't just choose any parameter value you like. The null must relate to the question at hand. Even though the null usually means no difference or no change, you can't automatically interpret "null" to mean zero. A claim that "nobody" wears a motorcycle helmet would be absurd. The null hypothesis for the Florida study could be that the true rate of helmet use remained the same among young riders after the law changed. You need to find the value for the parameter in the null hypothesis from the context of the problem.

There is a temptation to state your *claim* as the null hypothesis. As we have seen, however, you cannot prove a null hypothesis true any more than you can prove a defendant innocent. So, it makes more sense to use what you want to show as the *alternative*. This way, if you reject the null, you are left with what you want to show.

FOR EXAMPLE Writing hypotheses

The diabetes drug Avandia® was approved to treat Type 2 diabetes in 1999. But in 2007 an article in the *New England Journal of Medicine* (*NEJM*)[1] raised concerns that the drug might carry an increased risk of heart attack. This study combined results from a number of other separate studies to obtain an overall sample of 4485 diabetes patients taking Avandia. People with Type 2 diabetes are known to have about a 20.2% chance of suffering a heart attack within a seven-year period. According to the article's author, Dr. Steven E. Nissen,[2] the risk found in the *NEJM* study was equivalent to a 28.9% chance of heart attack over seven years. The FDA is the government agency responsible for relabeling Avandia to warn of the risk if it is judged to be unsafe. Although the statistical methods they use are more sophisticated, we can get an idea of their reasoning with the tools we have learned.

Question: What null hypothesis and alternative hypothesis about seven-year heart attack risk would you test? Explain.

$$H_O: p = 0.202$$
$$H_A: p > 0.202$$

The parameter of interest is the proportion of diabetes patients suffering a heart attack in seven years. The FDA is concerned only with whether Avandia increases the seven-year risk of heart attacks above the baseline value of 20.2%, so a one-sided upper-tail test is appropriate.

One-sided or two? In the 1930s, a series of experiments was performed at Duke University in an attempt to see whether humans were capable of extrasensory perception, or ESP. Psychologist Karl Zener designed a set of cards with 5 symbols, later made infamous in the movie *Ghostbusters*:

In the experiment, the "sender" selects one of the 5 cards at random from a deck and then concentrates on it. The "receiver" tries to determine which card it is. If we let p be the proportion of correct responses, what's the null hypothesis? The null hypothesis is that ESP makes no difference. Without ESP, the receiver would just be guessing, and since there are 5 possible responses, there would be a 20% chance of guessing each card correctly. So, H_0 is $p = 0.20$. What's the alternative? It seems that it should be $p > 0.20$, a one-sided alternative. But some ESP researchers have expressed the claim that if the proportion guessed were much *lower* than expected, that would show an "interference" and should be considered evidence for ESP as well. So they argue for a two-sided alternative.

---

[1] Steven E. Nissen, M.D., and Kathy Wolski, M.P.H., "Effect of Rosiglitazone on the Risk of Myocardial Infarction and Death from Cardiovascular Causes," *NEJM* 2007; 356.
[2] Interview reported in the *New York Times* [May 26, 2007].

| STEP-BY-STEP EXAMPLE | Another One-Proportion z-Test |

Let's try to answer the question raised at the start of the chapter.

Question: Has helmet use in Florida declined among riders under the age of 21 subsequent to the change in the helmet laws?

Plan State the problem and discuss the variables and the W's.

Hypotheses The null hypothesis is established by the rate set before the change in the law. The study was concerned with safety, so they'll want to know of any decline in helmet use, making this a lower-tail test.

I want to know whether the rate of helmet wearing among Florida's motorcycle riders under the age of 21 remained at 60% after the law changed to allow older riders to go without helmets. I have data from accident records showing 396 of 781 young riders were wearing helmets.

$$H_O: p = 0.60$$
$$H_A: p < 0.60$$

Model Check the conditions.

The Risky Behavior Surveillance survey is in fact a complex, multistage sample, but it is randomized and great effort is taken to make it representative. It is safe to treat it as though it were a random sample.

✓ **Independence Assumption:** The data are for riders involved in accidents during a three-year period. Individuals are independent of one another.

✗ **Randomization Condition:** No randomization was applied, but we are considering these riders involved in accidents to be a representative sample of all riders. We should take care in generalizing our conclusions.

✓ **10% Condition:** These 781 riders are a small sample of a larger population of all young motorcycle riders.

✓ **Success/Failure Condition:** We'd expect $np = 781(0.6) = 468.6$ helmeted riders and $nq = 781(0.4) = 312.4$ non-helmeted. Both are at least 10.

Specify the sampling distribution model and name the test.

The conditions are satisfied, so I can use a Normal model and perform a **one-proportion z-test.**

Mechanics Find the standard deviation of the sampling model using the hypothesized proportion.

There were 396 helmet wearers among the 781 accident victims.

$$\hat{p} = \frac{396}{781} = 0.507$$

$$SD(\hat{p}) = \sqrt{\frac{p_0 q_0}{n}} = \sqrt{\frac{(0.60)(0.40)}{781}} = 0.0175$$

Find the z-score for the observed proportion.

$$z = \frac{\hat{p} - p_0}{SD(\hat{p})} = \frac{0.507 - 0.60}{0.0175} = -5.31$$

Make a picture. Sketch a Normal model centered at the hypothesized helmet rate of 60%. This is a lower-tail test, so shade the region to the left of the observed rate.

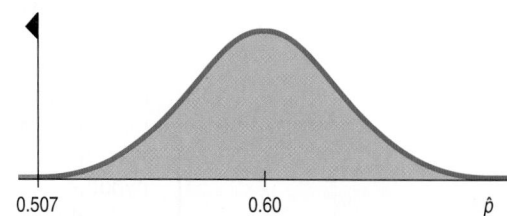

Given this *z*-score, the P-value is obviously very low.

The observed helmet rate is 5.31 standard deviations below the former rate. The corresponding P-value is less than 0.001.

Conclusion Link the P-value to your decision about the null hypothesis, and then state your conclusion in context.

The very small P-value says that if the true rate of helmet-wearing among riders under 21 were still 60%, the probability of observing a rate no higher than 50.7% in a sample like this is less than 1 chance in 1000, so I reject the null hypothesis. There is strong evidence that there has been a decline in helmet use among riders under 21.

How to Think About P-values

> ### Which Conditional?
> Suppose that as a political science major you are offered the chance to be a White House intern. There would be a very high probability that next summer you'd be in Washington, D.C. That is, *P*(Washington | Intern) would be high. But if we find a student in Washington, D.C., is it likely that he's a White House intern? Almost surely not; *P*(Intern | Washington) is low. You can't switch around conditional probabilities. The P-value is *P*(data|H$_0$). We might wish we could report *P*(H$_0$|data), but these two quantities are NOT the same.

A P-value actually is a conditional probability. It tells us the probability of getting results at least as unusual as the observed statistic, *given* that the null hypothesis is true. We can write P-value $= P$(observed statistic value [or even more extreme] | H$_0$).

Writing the P-value this way helps to make clear that the P-value is *not* the probability that the null hypothesis is true. It is a probability about the data. Let's say that again:

The P-value is not the probability that the null hypothesis is true.

The P-value is not even the conditional probability that the null hypothesis is true given the data. We would write that probability as *P*(H$_0$ | observed statistic value). This is a conditional probability but in reverse. It would be nice to know this, but it's impossible to calculate without making additional assumptions. As we saw in Chapter 15, reversing the order in a conditional probability is difficult, and the results can be counterintuitive.

We can find the P-value, *P*(observed statistic value | H$_0$), because H$_0$ gives the parameter values that we need to find the required probability. But there's no direct way to find *P*(H$_0$ | observed statistic value).[3] As tempting as it may be to say that a P-value of 0.03 means there's a 3% chance that the null hypothesis is true, that just isn't right. All we can say is that, given the null hypothesis, there's a 3% chance of observing the statistic value that we have actually observed (or one more unlike the null value).

---

[3] The approach to statistical inference known as Bayesian Statistics addresses the question in just this way, but it requires more advanced mathematics and more assumptions. See p. 358 for more about the founding father of this approach.

> **How guilty is the suspect?** We might like to know $P(H_0 \mid data)$, but when you think about it, we can't talk about the probability that the null hypothesis is true. The null is not a random event, so either it is true or it isn't. The data, however, are random in the sense that if we were to repeat a randomized experiment or draw another random sample, we'd get different data and expect to find a different statistic value. So we can talk about the probability of the data given the null hypothesis, and that's the P-value.
>
> But it does make sense that the smaller the P-value, the more confident we can be in declaring that we doubt the null hypothesis. Think again about the jury trial. Our null hypothesis is that the defendant is innocent. Then the evidence starts rolling in. A car the same color as his was parked in front of the bank. Well, there are lots of cars that color. The probability of that happening (given his innocence) is pretty high, so we're not persuaded that he's guilty. The bank's security camera showed the robber was male and about the dependant's height and weight. Hmmm. Could that be a coincidence? If he's innocent, then it's a little less likely that the car and description would *both* match, so our P-value goes down. We're starting to question his innocence a little. Witnesses said the robber wore a blue jacket just like the one the police found in a garbage can behind the defendant's house. Well, if he's innocent, then that doesn't seem very likely, does it? If he's really innocent, the probability that all of these could have happened is getting pretty low. Now our P-value may be small enough to be called "beyond a reasonable doubt" and lead to a conviction. Each new piece of evidence strains our skepticism a bit more. The more compelling the evidence—the more *unlikely* it would be were he innocent—the more convinced we become that he's guilty.
>
> But even though it may make *us* more confident in declaring him guilty, additional evidence does not make *him* any guiltier. Either he robbed the bank or he didn't. Additional evidence (like the teller picking him out of a police lineup) just makes us more confident that we did the right thing when we convicted him. The lower the P-value, the more comfortable we feel about our decision to reject the null hypothesis, but the null hypothesis doesn't get any more false.

> *"The wise man proportions his belief to the evidence."*
>
> —David Hume,
> "Enquiry Concerning Human Understanding," 1748

> *"You're so guilty now."*
>
> —Rearview Mirror

FOR EXAMPLE   Thinking about the P-value

Recap: A *New England Journal of Medicine* paper reported that the seven-year risk of heart attack in diabetes patients taking the drug Avandia was increased from the baseline of 20.2% to an estimated risk of 28.9% and said the P-value was 0.03.

Question: How should the P-value be interpreted in this context?

The P-value = $P(\hat{p} \geq 28.9\% \mid p = 20.2\%)$. That is, it's the probability of seeing such a high heart attack rate among the people studied if, in fact, taking Avandia really didn't increase the risk at all.

What to Do with a High P-value

Therapeutic touch (TT), taught in many schools of nursing, is a therapy in which the practitioner moves her hands near, but does not touch, a patient in an attempt to manipulate a "human energy field." Therapeutic touch practitioners believe that by adjusting this field they can promote healing. However, no instrument has ever detected a human energy field, and no experiment has ever shown that TT practitioners can detect such a field.

In 1998, the *Journal of the American Medical Association* published a paper reporting work by a then nine-year-old girl.[4] She had performed a simple experiment in

---

[4] L. Rosa, E. Rosa, L. Sarner, and S. Barrett, "A Close Look at Therapeutic Touch," *JAMA* 279(13) [1 April 1998]: 1005–1010.

A S   *Video:* **Is There Evidence for Therapeutic Touch?** This video shows the experiment and tells the story.

A S   *Activity:* **Testing Therapeutic Touch.** Perform the one-proportion *z*-test using *ActivStats* technology. The test in *ActivStats* is two-sided. Do you think this is the appropriate choice?

which she challenged 15 TT practitioners to detect whether her unseen hand was hovering over their left or right hand (selected by the flip of a coin).

The practitioners "warmed up" with a period during which they could see the experimenter's hand, and each said that they could detect the girl's human energy field. Then a screen was placed so that the practitioners could not see the girl's hand, and they attempted 10 trials each. Overall, of 150 trials, the TT practitioners were successful 70 times, for a success proportion of 46.7%. Is there evidence from this experiment that TT practitioners can successfully detect a "human energy field"?

When we see a small P-value, we could continue to believe the null hypothesis and conclude that we just witnessed a rare event. But instead, we trust the data and use it as evidence to reject the null hypothesis.

In the therapeutic touch example, the null hypothesis is that the practitioners are guessing, so we expect them to be right about half the time by chance. That's why we say $H_0: p = 0.5$. They claim that they can detect a "human energy field" and that their success rate should be well above chance, so our alternative is that they would do *better* than guessing. That's a one-sided alternative hypothesis: $H_A: p > 0.5$. With a one-sided hypothesis, our P-value is the probability the practitioners could achieve the observed number of successes or *more* even if they were just guessing.

If the practitioners had been highly successful, that would have been unusually lucky for guessing, so we would have seen a correspondingly low P-value. Since we don't believe in rare events, we would then have concluded that they weren't guessing.

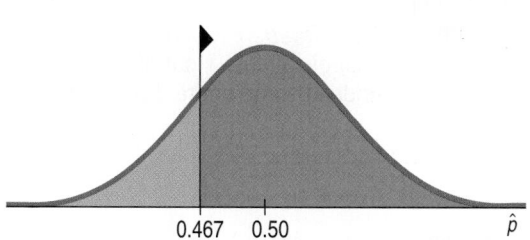

But that's not what happened. What we actually observed was that they did slightly *worse* than 50%, with a $\hat{p} = 0.467$ success rate.

As the figure shows, the probability of a success rate of 0.467 *or more* is even bigger than 0.5. In this case, it turns out to be 0.793. Obviously, we won't be rejecting the null hypothesis; for us to reject it, the P-value would have to be quite small. But a P-value of 0.788 seems so big it is almost awkward. With a success rate even lower than chance, we could have concluded right away that we have no evidence for rejecting H_0.

Big P-values just mean that what we've observed isn't surprising. That is, the results are in line with our assumption that the null hypothesis models the world, so we have no reason to reject it.

A big P-value doesn't prove that the null hypothesis is true, but it certainly offers no evidence that it's *not* true. When we see a large P-value, all we can say is that we "don't reject the null hypothesis."

FOR EXAMPLE   More about P-values

Recap: The question of whether the diabetes drug Avandia increased the risk of heart attack was raised by a study in the *New England Journal of Medicine*. This study estimated the seven-year risk of heart attack to be 28.9% and reported a P-value of 0.03 for a test of whether this risk was higher than the baseline seven-year risk of 20.2%. An earlier study (the ADOPT study) had estimated the seven-year risk to be 26.9% and reported a P-value of 0.27.

Question: Why did the researchers in the ADOPT study not express alarm about the increased risk they had seen?

A P-value of 0.27 means that a heart attack rate at least as high as the one they observed could be expected in 27% of similar experiments even if, in fact, there were no increased risk from taking Avandia. That's not remarkable enough to reject the null hypothesis. In other words, the ADOPT study wasn't convincing.

Alpha Levels

NOTATION ALERT:

The first Greek letter, α, is used in Statistics for the threshold value of a hypothesis test. You'll hear it referred to as the alpha level. Common values are 0.10, 0.05, 0.01, and 0.001.

Sir Ronald Fisher (1890–1962) was one of the founders of modern Statistics.

Sometimes we need to make a firm decision about whether or not to reject the null hypothesis. A jury must *decide* whether the evidence reaches the level of "beyond a reasonable doubt." A business must *select* a Web design. You need to decide which section of Statistics to enroll in.

When the P-value is small, it tells us that our data are rare, *given the null hypothesis.* As humans, we are suspicious of rare events. If the data are "rare enough," we just don't think that could have happened due to chance. Since the data *did* happen, something must be wrong. All we can do now is reject the null hypothesis.

But how rare is "rare"?

We can define "rare event" arbitrarily by setting a threshold for our P-value. If our P-value falls below that point, we'll reject the null hypothesis. We call such results **statistically significant.** The threshold is called an **alpha level.** Not surprisingly, it's labeled with the Greek letter α. Common α levels are 0.10, 0.05, and 0.01. You have the option—almost the *obligation*—to consider your alpha level carefully and choose an appropriate one for the situation. If you're assessing the safety of air bags, you'll want a low alpha level; even 0.01 might not be low enough. If you're just wondering whether folks prefer their pizza with or without pepperoni, you might be happy with $\alpha = 0.10$. It can be hard to justify your choice of α, though, so often we arbitrarily choose 0.05. Note, however: You must select the alpha level *before* you look at the data. Otherwise you can be accused of cheating by tuning your alpha level to suit the data.

> **Where did the value 0.05 come from?** In 1931, in a famous book called *The Design of Experiments*, Sir Ronald Fisher discussed the amount of evidence needed to reject a null hypothesis. He said that it was *situation dependent*, but remarked, somewhat casually, that for many scientific applications, 1 out of 20 *might be* a reasonable value. Since then, some people—indeed some entire disciplines—have treated the number 0.05 as sacrosanct.

The alpha level is also called the **significance level.** When we reject the null hypothesis, we say that the test is "significant at that level." For example, we might say that we reject the null hypothesis "at the 5% level of significance."

What can you say if the P-value does not fall below α?

When you have not found sufficient evidence to reject the null according to the standard you have established, you should say that "The data have failed to provide sufficient evidence to reject the null hypothesis." Don't say that you "accept the null hypothesis." You certainly haven't proven or established it; it was merely assumed to begin with. Say that you've failed to reject it.

Think again about the therapeutic touch example. The P-value was 0.788. This is so much larger than any reasonable alpha level that we can't reject H_0. For this test, we'd conclude, "We fail to reject the null hypothesis. There is insufficient evidence to conclude that the practitioners are performing better than they would if they were just guessing."

The automatic nature of the reject/fail-to-reject decision when we use an alpha level may make you uncomfortable. If your P-value falls just slightly above your alpha level, you're not allowed to reject the null. Yet a P-value just barely below the alpha level leads to rejection. If this bothers you, you're in good company. Many statisticians think it better to report the P-value than to base a decision on an arbitrary alpha level.

It Could Happen to You!
Of course, if the null hypothesis *is* true, no matter what alpha level you choose, you still have a probability α of rejecting the null hypothesis by mistake. This is the rare event we want to protect ourselves against. When we do reject the null hypothesis, no one ever thinks that *this* is one of those rare times. As statistician Stu Hunter notes, "*The statistician says 'rare events do happen—but not to me!'*"

> **It's in the stars** Some disciplines carry the idea further and code P-values by their size. In this scheme, a P-value between 0.05 and 0.01 gets highlighted by *. A P-value between 0.01 and 0.001 gets **, and a P-value less than 0.001 gets ***. This can be a convenient summary of the weight of evidence against the null hypothesis if it's not taken too literally. But we warn you against taking the distinctions too seriously and against making a black-and-white decision near the boundaries. The boundaries are a matter of tradition, not science; there is nothing special about 0.05. A P-value of 0.051 should be looked at very seriously and not casually thrown away just because it's larger than 0.05, and one that's 0.009 is not very different from one that's 0.011.

When you decide to declare a verdict, it's always a good idea to report the P-value as an indication of the strength of the evidence. Sometimes it's best to report that the conclusion is not yet clear and to suggest that more data be gathered. (In a trial, a jury may "hang" and be unable to return a verdict.) In these cases, the P-value is the best summary we have of what the data say or fail to say about the null hypothesis.

Significant vs. Important

Practical vs. Statistical Significance

A large insurance company mined its data and found a statistically significant ($P = 0.04$) difference between the mean value of policies sold in 2001 and 2002. The difference in the mean values was $9.83. Even though it was statistically significant, management did not see this as an important difference when a typical policy sold for more than $1000. On the other hand, even a clinically important improvement of 10% in cure rate with a new treatment is not likely to be statistically significant in a study of fewer than 225 patients. A small clinical trial would probably not be conclusive.

What do we mean when we say that a test is statistically significant? All we mean is that the test statistic had a P-value lower than our alpha level. Don't be lulled into thinking that statistical significance carries with it any sense of practical importance or impact.

For large samples, even small, unimportant ("insignificant") deviations from the null hypothesis can be statistically significant. On the other hand, if the sample is not large enough, even large financially or scientifically "significant" differences may not be statistically significant.

It's good practice to report the magnitude of the difference between the observed statistic value and the null hypothesis value (in the data units) along with the P-value on which we base statistical significance.

Confidence Intervals and Hypothesis Tests

For the motorcycle helmet example, a 95% confidence interval would give $0.507 \pm 1.96 \times 0.0179 = (0.472, 0.542)$, or 47.2% to 54.2%. If the previous rate of helmet compliance had been, say, 50%, we would not have been able to reject the null hypothesis because 50% is in the interval, so it's a plausible value. Indeed, *any* hypothesized value for the true proportion of helmet wearers in this interval is consistent with the data. Any value outside the confidence interval would make a null hypothesis that we would reject, but we'd feel more strongly about values far outside the interval.

Confidence intervals and hypothesis tests are built from the same calculations.[5] They have the same assumptions and conditions. As we have just seen, you can

[5] As we saw in Chapter 20, this is not *exactly* true for proportions. For a confidence interval, we estimate the standard deviation of $\hat{p}$ from $\hat{p}$ itself. Because we estimate it from the data, we have a *standard error*. For the corresponding hypothesis test, we use the model's standard deviation for $\hat{p}$, based on the null hypothesis value p_0. When $\hat{p}$ and p_0 are close, these calculations give similar results. When they differ, you're likely to reject H_0 (because the observed proportion is far from your hypothesized value). In that case, you're better off building your confidence interval with a standard error estimated from the data.

approximate a hypothesis test by examining the confidence interval. Just ask whether the null hypothesis value is consistent with a confidence interval for the parameter at the corresponding confidence level. Because confidence intervals are naturally two-sided, they correspond to two-sided tests. For example, a 95% confidence interval corresponds to a two-sided hypothesis test at $\alpha = 5\%$. In general, a confidence interval with a confidence level of $C\%$ corresponds to a two-sided hypothesis test with an α level of $100 - C\%$.

The relationship between confidence intervals and one-sided hypothesis tests is a little more complicated. For a one-sided test with $\alpha = 5\%$, the corresponding confidence interval has a confidence level of 90%—that's 5% in each tail. In general, a confidence interval with a confidence level of $C\%$ corresponds to a one-sided hypothesis test with an α level of $\frac{1}{2}(100 - C)\%$.

FOR EXAMPLE   Making a decision based on a confidence interval

Recap: The baseline seven-year risk of heart attacks for diabetics is 20.2%. In 2007 a *NEJM* study reported a 95% confidence interval equivalent to 20.8% to 40.0% for the risk among patients taking the diabetes drug Avandia.

Question: What did this confidence interval suggest to the FDA about the safety of the drug?

The FDA could be 95% confident that the interval from 20.8% to 40.0% included the true risk of heart attack for diabetes patients taking Avandia. Because the lower limit of this interval was higher than the baseline risk of 20.2%, there was evidence of an increased risk.

JUST CHECKING

1. An experiment to test the fairness of a roulette wheel gives a z-score of 0.62. What would you conclude?

2. In the last chapter we encountered a bank that wondered if it could get more customers to make payments on delinquent balances by sending them a DVD urging them to set up a payment plan. Well, the bank just got back the results on their test of this strategy. A 90% confidence interval for the success rate is (0.29, 0.45). Their old send-a-letter method had worked 30% of the time. Can you reject the null hypothesis that the proportion is still 30% at $\alpha = 0.05$? Explain.

3. Given the confidence interval the bank found in their trial of DVDs, what would you recommend that they do? Should they scrap the DVD strategy?

STEP-BY-STEP EXAMPLE   Wear that Seatbelt!

Teens are at the greatest risk of being killed or injured in traffic crashes. According to the National Highway Traffic Safety Administration, 65% of young people killed were not wearing a safety belt. In 2001, a total of 3322 teens were killed in motor vehicle crashes, an average of 9 teenagers a day. Because many of these deaths could easily be prevented by the use of safety belts, several states have begun "Click It or Ticket" campaigns in which increased enforcement and publicity have resulted in significantly higher seatbelt use. Overall use in Massachusetts quickly increased from 51% in 2002 to 64.8% in 2006, with a goal of surpassing the national average of 82%. Recently, a local newspaper reported that a roadblock resulted in 23 tickets to drivers who were unbelted out of 134 stopped for inspection.

Question: Does this provide evidence that the goal of over 82% compliance was met? Let's use a confidence interval to test this hypothesis.

Plan State the problem and discuss the variables and the W's.

Hypotheses The null hypothesis is that the compliance rate is only 82%. The alternative is that it is now higher. It's clearly a one-sided test, so if we use a confidence interval, we'll have to be careful about what level we use.

Model Think about the assumptions and check the conditions.

We are finding a confidence interval, so we work from the data rather than the null model.

State your method.

The data come from a local newspaper report that tells the number of tickets issued and number of drivers stopped at a recent road-block. I want to know whether the rate of compliance with the seatbelt law is greater than 82%.

$$H_0: p = 0.82$$
$$H_A: p > 0.82$$

✔ **Independence Assumption:** Drivers are not likely to influence one another when it comes to wearing a seatbelt.

✔ **Randomization Condition:** This wasn't a random sample, but I assume these drivers are representative of the driving public.

✔ **10% Condition:** The police stopped fewer than 10% of all drivers.

✔ **Success/Failure Condition:** There were 111 successes and 23 failures, both at least 10. The sample is large enough.

Under these conditions, the sampling model is Normal. I'll create a **one-proportion z-interval.**

Mechanics Write down the given information, and determine the sample proportion.

To use a confidence interval, we need a confidence level that corresponds to the alpha level of the test. If we use $\alpha = 0.05$, we should construct a 90% confidence interval, because this is a one-sided test.

That will leave 5% on *each* side of the observed proportion. Determine the standard error of the sample proportion and the margin of error. The critical value is $z^* = 1.645$.

The confidence interval is

estimate ± margin of error.

$n = 134$, so

$$\hat{p} = \frac{111}{134} = 0.828 \text{ and}$$

$$SE(\hat{p}) = \sqrt{\frac{\hat{p}\hat{q}}{n}} = \sqrt{\frac{(0.828)(0.172)}{134}} = 0.033$$

$$ME = z^* \times SE(\hat{p})$$
$$= 1.645(0.033) = 0.054$$

The 90% confidence interval is

$$0.828 \pm 0.054 \text{ or}$$
$$(0.774, 0.882).$$

Conclusion Link the confidence interval to your decision about the null hypothesis, and then state your conclusion in context.

I am 90% confident that between 77.4% and 88.2% of all drivers wear their seatbelts. Because the hypothesized rate of 82% is within this interval, I do not reject the null hypothesis. There is insufficient evidence to conclude that the campaign was truly effective and now more than 82% of all drivers are wearing seatbelts.

The upper limit of the confidence interval shows it's possible that the campaign is quite successful, but the small sample size makes the interval too wide to be very specific.

*A 95% Confidence Interval for Small Samples

When the **Success/Failure Condition** fails, all is not lost. A simple adjustment to the calculation lets us make a 95% confidence interval anyway.

All we do is add four *phony* observations—two to the successes, two to the failures. So instead of the proportion $\hat{p} = \dfrac{y}{n}$, we use the adjusted proportion $\widetilde{p} = \dfrac{y + 2}{n + 4}$ and, for convenience, we write $\widetilde{n} = n + 4$. We modify the interval by using these adjusted values for both the center of the interval *and* the margin of error. Now the adjusted interval is

$$\widetilde{p} \pm z^* \sqrt{\frac{\widetilde{p}(1 - \widetilde{p})}{\widetilde{n}}}.$$

This adjusted form gives better performance overall[6] and works much better for proportions near 0 or 1. It has the additional advantage that we no longer need to check the **Success/Failure Condition** that $n\hat{p}$ and $n\hat{q}$ are greater than 10.

FOR EXAMPLE   An Agresti-Coull "plus-four" interval

Surgeons examined their results to compare two methods for a surgical procedure used to alleviate pain on the outside of the wrist. A new method was compared with the traditional "freehand" method for the procedure. Of 45 operations using the "freehand" method, three were unsuccessful, for a failure rate of 6.7%. With only 3 failures, the data don't satisfy the **Success/Failure Condition**, so we can't use a standard confidence interval.

Question: What's the confidence interval using the "plus-four" method?

---

[6] By "better performance," we mean that a 95% confidence interval has more nearly a 95% chance of covering the true population proportion. Simulation studies have shown that our original, simpler confidence interval in fact is less likely than 95% to cover the true population proportion when the sample size is small or the proportion very close to 0 or 1. The original idea for this method can be attributed to E. B. Wilson. The simpler approach discussed here was proposed by Agresti and Coull (A. Agresti and B. A. Coull, "Approximate Is Better Than 'Exact' for Interval Estimation of Binomial Proportions," *The American Statistician*, 52[1998]: 119–129).

There were 42 successes and 3 failures. Adding 2 "pseudo-successes" and 2 "pseudo-failures," we find

$$\tilde{p} = \frac{3 + 2}{45 + 4} = 0.102$$

A 95% confidence interval is then

$$0.102 \pm 1.96 \sqrt{\frac{0.102(1 - 0.102)}{49}} = 0.102 \pm 0.085 \text{ or } (0.017, 0.187).$$

Notice that although the observed failure rate of 0.067 is contained in the interval, it is not at the center of the interval—something we haven't seen with any of the other confidence intervals we've considered.

Making Errors

A S   *Activity:* **Type I and Type II Errors.** View an animated exploration of Type I and Type II errors—a good backup for the reading in this section.

Nobody's perfect. Even with lots of evidence, we can still make the wrong decision. In fact, when we perform a hypothesis test, we can make mistakes in *two* ways:

I. The null hypothesis is true, but we mistakenly reject it.
II. The null hypothesis is false, but we fail to reject it.

These two types of errors are known as **Type I** and **Type II errors.** One way to keep the names straight is to remember that we start by assuming the null hypothesis is true, so a Type I error is the first kind of error we could make.

In medical disease testing, the null hypothesis is usually the assumption that a person is healthy. The alternative is that he or she has the disease we're testing for. So a Type I error is a *false positive:* A healthy person is diagnosed with the disease. A Type II error, in which an infected person is diagnosed as disease free, is a *false negative.* These errors have other names, depending on the particular discipline and context.

Some false-positive results mean no more than an unnecessary chest X-ray. But for a drug test or a disease like AIDS, a false-positive result that is not kept confidential could have serious consequences.

Which type of error is more serious depends on the situation. In the jury trial, a Type I error occurs if the jury convicts an innocent person. A Type II error occurs if the jury fails to convict a guilty person. Which seems more serious? In medical diagnosis, a false negative could mean that a sick patient goes untreated. A false positive might mean that the person must undergo further tests. In a Statistics final exam (with H_0: the student has learned only 60% of the material), a Type I error would be passing a student who in fact learned less than 60% of the material, while a Type II error would be failing a student who knew enough to pass. Which of these errors seems more serious? It depends on the situation, the cost, and your point of view.

Here's an illustration of the situations:

A S   *Activity:* **Hypothesis Tests Are Random.** Simulate hypothesis tests and watch Type I errors occur. When you conduct real hypothesis tests you'll never know, but simulation can tell you when you've made an error.

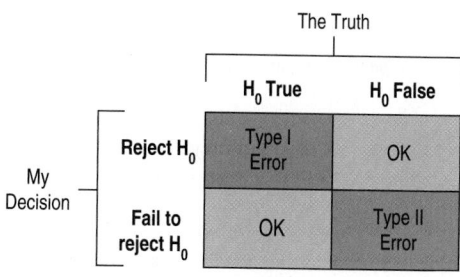

How often will a Type I error occur? It happens when the null hypothesis is true but we've had the bad luck to draw an unusual sample. To reject H_0, the P-value

NOTATION ALERT:

In Statistics, α is almost always saved for the alpha level. But β has already been used for the parameters of a linear model. Fortunately, it's usually clear whether we're talking about a Type II error probability or the slope or intercept of a regression model.

The null hypothesis specifies a single value for the parameter. So it's easy to calculate the probability of a Type I error. But the alternative gives a whole range of possible values, and we may want to find a β for several of them.

We have seen ways to find a sample size by specifying the margin of error. Choosing the sample size to achieve a specified β (for a particular alternative value) is sometimes more appropriate, but the calculation is more complex and lies beyond the scope of this book.

must fall below α. When H_0 is true, that happens *exactly* with probability α. So when you choose level α, you're setting the probability of a Type I error to α.

What if H_0 is not true? Then we can't possibly make a Type I error. You can't get a false positive from a sick person. A Type I error can happen only when H_0 is true.

When H_0 is false and we reject it, we have done the right thing. A test's ability to detect a false null hypothesis is called the **power** of the test. In a jury trial, power is the ability of the criminal justice system to convict people who are guilty—a good thing! We'll have a lot more to say about power soon.

When H_0 is false but we fail to reject it, we have made a Type II error. We assign the letter β to the probability of this mistake. What's the value of β? That's harder to assess than α because we don't know what the value of the parameter really is. When H_0 is true, it specifies a single parameter value. But when H_0 is false, we don't have a specific one; we have many possible values. We can compute the probability β for any parameter value in H_A. But which one should we choose?

One way to focus our attention is by thinking about the *effect size*. That is, we ask *"How big a difference would matter?"* Suppose a charity wants to test whether placing personalized address labels in the envelope along with a request for a donation increases the response rate above the baseline of 5%. If the minimum response that would pay for the address labels is 6%, they would calculate β for the alternative $p = 0.06$.

We could reduce β for *all* alternative parameter values by increasing α. By making it easier to reject the null, we'd be more likely to reject it whether it's true or not. So we'd reduce β, the chance that we fail to reject a false null—but we'd make more Type I errors. This tension between Type I and Type II errors is inevitable. In the political arena, think of the ongoing debate between those who favor provisions to reduce Type I errors in the courts (supporting Miranda rights, requiring warrants for wiretaps, providing legal representation for those who can't afford it) and those who advocate changes to reduce Type II errors (admitting into evidence confessions made when no lawyer is present, eavesdropping on conferences with lawyers, restricting paths of appeal, etc.).

The only way to reduce *both* types of error is to collect more evidence or, in statistical terms, to collect more data. Too often, studies fail because their sample sizes are too small to detect the change they are looking for.

FOR EXAMPLE **Thinking about errors**

Recap: A published study found the risk of heart attack to be increased in patients taking the diabetes drug Avandia. The issue of the *New England Journal of Medicine* (*NEJM*) in which that study appeared also included an editorial that said, in part, "A few events either way might have changed the findings for myocardial infarction[7] or for death from cardiovascular causes. In this setting, the possibility that the findings were due to chance cannot be excluded."

Question: What kind of error would the researchers have made if, in fact, their findings were due to chance? What could be the consequences of this error?

The null hypothesis said the risk didn't change, but the researchers rejected that model and claimed evidence of a higher risk. If these findings were just due to chance, they rejected a true null hypothesis—a Type I error.

If, in fact, Avandia carried no extra risk, then patients might be deprived of its benefits for no good reason.

---

[7] Doctorese for "heart attack."

Power

When we failed to reject the null hypothesis about TT practitioners, did we prove that they were just guessing? No, it could be that they actually *can* discern a human energy field but we just couldn't tell. For example, suppose they really have the ability to get 53% of the trials right but just happened to get only 47% in our experiment. Our confidence interval shows that with these data we wouldn't have rejected the null. And if we retained the null even though the true proportion was actually greater than 50%, we would have made a Type II error because we failed to detect their ability.

Remember, we can never prove a null hypothesis true. We can only fail to reject it. But when we fail to reject a null hypothesis, it's natural to wonder whether we looked hard enough. Might the null hypothesis actually be false and our test too weak to tell?

When the null hypothesis actually *is* false, we hope our test is strong enough to reject it. We'd like to know how likely we are to succeed. The power of the test gives us a way to think about that. The **power** of a test is the probability that it correctly rejects a false null hypothesis. When the power is high, we can be confident that we've looked hard enough. We know that β is the probability that a test *fails* to reject a false null hypothesis, so the power of the test is the probability that it *does* reject: $1 - \beta$.

Whenever a study fails to reject its null hypothesis, the test's power comes into question. Was the sample size big enough to detect an effect had there been one? Might we have missed an effect large enough to be interesting just because we failed to gather sufficient data or because there was too much variability in the data we could gather? The therapeutic touch experiment failed to reject the null hypothesis that the TT practitioners were just guessing. Might the problem be that the experiment simply lacked adequate power to detect their ability?

FOR EXAMPLE   Errors and power

Recap: The study of Avandia published in the *NEJM* combined results from 47 different trials—a method called *meta-analysis*. The drug's manufacturer, GlaxoSmithKline (GSK), issued a statement that pointed out, "Each study is designed differently and looks at unique questions: For example, individual studies vary in size and length, in the type of patients who participated, and in the outcomes they investigate." Nevertheless, by combining data from many studies, meta-analyses can achieve a much larger sample size.

Question: How could this larger sample size help?

If Avandia really did increase the seven-year heart attack rate, doctors needed to know. To overlook that would have been a Type II error (failing to detect a false null hypothesis), resulting in patients being put at greater risk. Increasing the sample size could increase the power of the analysis, making it more likely that researchers will detect the danger if there is one.

 Activity: The Power of a Test. Power is a concept that's much easier to understand when you can visualize what's happening.

When we calculate power, we imagine that the null hypothesis is false. The value of the power depends on how far the truth lies from the null hypothesis value. We call the distance between the null hypothesis value, p_0, and the truth, p, **the effect size.** The power depends directly on the effect size. It's easier to see larger effects, so the farther p_0 is from p, the greater the power. If the therapeutic touch practitioners were in fact able to detect human energy fields 90% of the time, it should be easy to see that they aren't guessing. With an effect size this large, we'd have a powerful test. If their true success rate were only 53%, however, we'd need a larger sample size to have a good chance of noticing that (and rejecting H_0).

How can we decide what power we need? Choice of power is more a financial or scientific decision than a statistical one because to calculate the power, we need to specify the "true" parameter value we're interested in. In other words, power is calculated for a particular effect size, and it changes depending on the size of the effect we want to detect. For example, do you think that health insurance companies should pay for therapeutic touch if practitioners could detect a human energy field only 53% of the time—just slightly better than chance? That doesn't seem clinically useful.[8] How about 75% of the time? No therapy works all the time, and insurers might be quite willing to pay for such a success rate. Let's take 75% as a reasonably interesting effect size (keeping in mind that 50% is the level of guessing). With 150 trials, the TT experiment would have been able to detect such an ability with a power of 99.99%. So power was not an issue in this study. There is only a very small chance that the study would have failed to detect a practitioner's ability, had it existed. The sample size was clearly big enough.

JUST CHECKING

4. Remember our bank that's sending out DVDs to try to get customers to make payments on delinquent loans? It is looking for evidence that the costlier DVD strategy produces a higher success rate than the letters it has been sending. Explain what a Type I error is in this context and what the consequences would be to the bank.

5. What's a Type II error in the bank experiment context, and what would the consequences be?

6. For the bank, which situation has higher power: a strategy that works really well, actually getting 60% of people to pay off their balances, or a strategy that barely increases the payoff rate to 32%? Explain briefly.

A Picture Worth $\dfrac{1}{P(z > 3.09)}$ Words

It makes intuitive sense that the larger the effect size, the easier it should be to see it. Obtaining a larger sample size decreases the probability of a Type II error, so it increases the power. It also makes sense that the more we're willing to accept a Type I error, the less likely we will be to make a Type II error.

FIGURE 21.1

The power of a test is the probability that it rejects a false null hypothesis. The upper figure shows the null hypothesis model. We'd reject the null in a one-sided test if we observed a value of $\hat{p}$ in the red region to the right of the critical value, p^*. The lower figure shows the true model. If the true value of p is greater than p_0, then we're more likely to observe a value that exceeds the critical value and make the correct decision to reject the null hypothesis. The power of the test is the purple region on the right of the lower figure. Of course, even drawing samples whose observed proportions are distributed around p, we'll sometimes get a value in the red region on the left and make a Type II error of failing to reject the null.

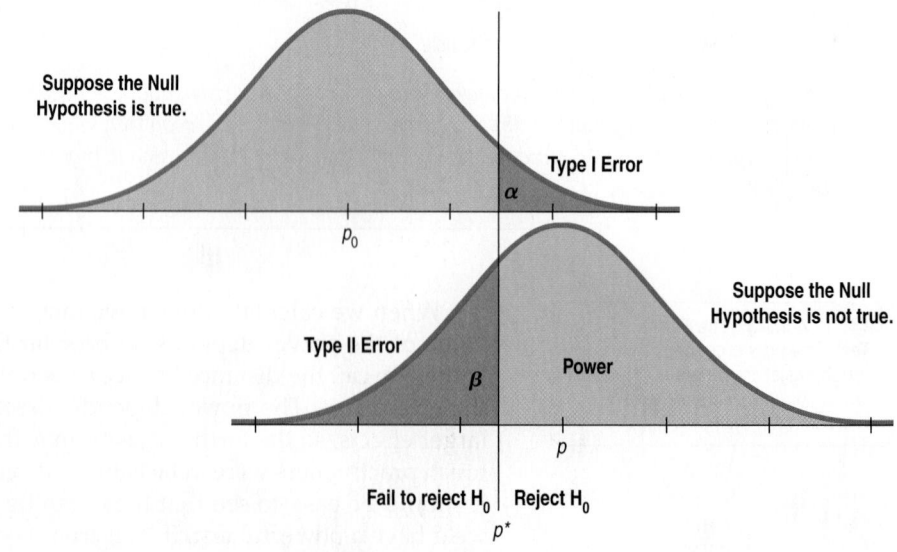

---

[8] On the other hand, a scientist might be interested in anything clearly different from the 50% guessing rate because that might suggest an entirely new physics at work. In fact, it could lead to a Nobel prize.

Fisher and $\alpha = 0.05$

Why did Sir Ronald Fisher suggest 0.05 as a criterion for testing hypotheses? It turns out that he had in mind small initial studies. Small studies have relatively little power. Fisher was concerned that they might make too many Type II errors—failing to discover an important effect—if too strict a criterion were used. Once a test failed to reject a null hypothesis, it was unlikely that researchers would return to that hypothesis to try again.

On the other hand, the increased risk of Type I errors arising from a generous criterion didn't concern him as much for exploratory studies because these are ordinarily followed by a replication or a larger study. The probability of a Type I error is α—in this case, 0.05. The probability that two independent studies would both make Type I errors is $0.05 \times 0.05 = 0.0025$, so Fisher was confident that Type I errors in initial studies were not a major concern.

The widespread use of the relatively generous 0.05 criterion even in large studies is most likely not what Fisher had in mind.

Figure 21.1 shows a good way to visualize the relationships among these concepts. Suppose we are testing $H_0: p = p_0$ against the alternative $H_A: p > p_0$. We'll reject the null if the observed proportion, $\hat{p}$, is big enough. By big enough, we mean $\hat{p} > p^*$ for some critical value, p^* (shown as the red region in the right tail of the upper curve). For example, we might be willing to believe the ability of therapeutic touch practitioners if they were successful in 65% of our trials. This is what the upper model shows. It's a picture of the sampling distribution model for the proportion if the null hypothesis were true. We'd make a Type I error whenever the sample gave us $\hat{p} > p^*$, because we would reject the (true) null hypothesis. And unusual samples like that would happen only with probability α.

In reality, though, the null hypothesis is rarely *exactly* true. The lower probability model supposes that H_0 is not true. In particular, it supposes that the true value is p, not p_0. (Perhaps the TT practitioner really can detect the human energy field 72% of the time.) It shows a distribution of possible observed $\hat{p}$ values around this true value. Because of sampling variability, sometimes $\hat{p} < p^*$ and we fail to reject the (false) null hypothesis. Suppose a TT practitioner with a true ability level of 72% is actually successful on fewer than 65% of our tests. Then we'd make a Type II error. The area under the curve to the left of p^* in the bottom model represents how often this happens. The probability is β. In this picture, β is less than half, so most of the time we *do* make the right decision. The *power* of the test—the probability that we make the right decision—is shown as the region to the right of p^*. It's $1 - \beta$.

We calculate p^* based on the upper model because p^* depends only on the null model and the alpha level. No matter what the true proportion, no matter whether the practitioners can detect a human energy field 90%, 53%, or 2% of the time, p^* doesn't change. After all, we don't *know* the truth, so we can't use it to determine the critical value. But we always reject H_0 when $\hat{p} > p^*$.

How often we correctly reject H_0 when it's *false* depends on the effect size. We can see from the picture that if the effect size were larger (the true proportion were farther above the hypothesized value), the bottom curve would shift to the right, making the power greater.

We can see several important relationships from this figure:

▶ Power $= 1 - \beta$.

▶ Reducing α to lower the chance of committing a Type I error will move the critical value, p^*, to the right (in this example). This will have the effect of increasing β, the probability of a Type II error, and correspondingly reducing the power.

▶ The larger the real difference between the hypothesized value, p_0, and the true population value, p, the smaller the chance of making a Type II error and the greater the power of the test. If the two proportions are very far apart, the two models will barely overlap, and we will not be likely to make any Type II errors at all—but then, we are unlikely to really need a formal hypothesis-testing procedure to see such an obvious difference. If the TT practitioners were successful almost all the time, we'd be able to see that with even a small experiment.

Reducing Both Type I and Type II Errors

Figure 21.1 seems to show that if we reduce Type I error, we automatically must increase Type II error. But there is a way to reduce both. Can you think of it?

If we can make both curves narrower, as shown in Figure 21.2, then both the probability of Type I errors and the probability of Type II errors will decrease, and the power of the test will increase.

FIGURE 21.2

Making the standard deviations smaller increases the power without changing the corresponding critical value. The means are just as far apart as in Figure 21.1, but the error rates are reduced.

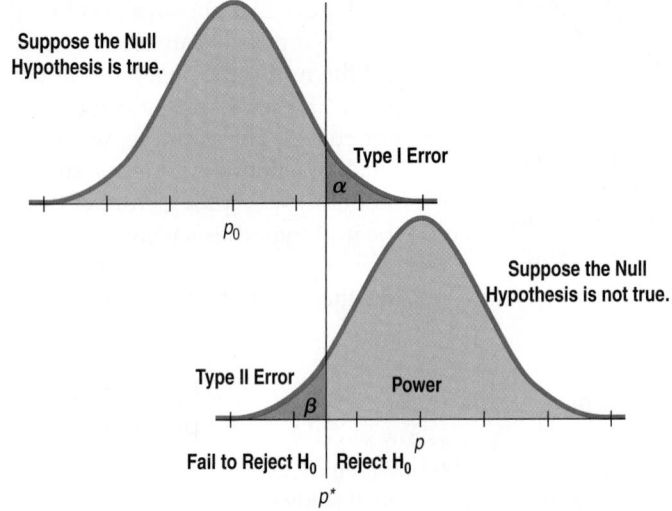

Suppose the Null Hypothesis is true.

Type I Error

α

p_0

Suppose the Null Hypothesis is not true.

Type II Error

β

Power

p

Fail to Reject H_0 | Reject H_0

p^*

TI-*nspire*

Errors and power. Explore the relationships among Type I and Type II errors, sample size, effect size, and the power of a test.

How can we accomplish that? The only way is to reduce the standard deviations by increasing the sample size. (Remember, these are pictures of sampling distribution models, not of data.) Increasing the sample size works regardless of the true population parameters. But recall the curse of diminishing returns. The standard deviation of the sampling distribution model decreases only as the *square root* of the sample size, so to halve the standard deviations we must *quadruple* the sample size.

FOR EXAMPLE Sample size, errors, and power

Recap: The meta-analysis of the risks of heart attacks in patients taking the diabetes drug Avandia combined results from 47 smaller studies. As GlaxoSmith-Kline (GSK), the drug's manufacturer, pointed out in their rebuttal, "Data from the ADOPT clinical trial did show a small increase in reports of myocardial infarction among the *Avandia*-treated group . . . however, the number of events is too small to reach a reliable conclusion about the role any of the medicines may have played in this finding."

Question: Why would this smaller study have been less likely to detect the difference in risk? What are the appropriate statistical concepts for comparing the smaller studies?

Smaller studies are subject to greater sampling variability; that is, the sampling distributions they estimate have a larger standard deviation for the sample proportion. That gives small studies less power: They'd be less able to discern whether an apparently higher risk was merely the result of chance variation or evidence of real danger. The FDA doesn't want to restrict the use of a drug that's safe and effective (Type I error), nor do they want patients to continue taking a medication that puts them at risk (Type II error). Larger sample sizes can reduce the risk of both kinds of error. Greater power (the probability of rejecting a false null hypothesis) means a better chance of spotting a genuinely higher risk of heart attacks.

WHAT CAN GO WRONG?

▸ **Don't interpret the P-value as the probability that H_0 is true.** The P-value is about the data, not the hypothesis. It's the probability of observing data this unusual, *given* that H_0 is true, not the other way around.

▸ **Don't believe too strongly in arbitrary alpha levels.** There's not really much difference between a P-value of 0.051 and a P-value of 0.049, but sometimes it's regarded as the difference between night (having to refrain from rejecting H_0) and day (being able to shout to the world that your results are "statistically significant"). It may just be better to report the P-value and a confidence interval and let the world decide along with you.

▶ **Don't confuse practical and statistical significance.** A large sample size can make it easy to discern even a trivial change from the null hypothesis value. On the other hand, an important difference can be missed if your test lacks sufficient power.

▶ **Don't forget that in spite of all your care, you might make a wrong decision.** We can never reduce the probability of a Type I error (α) or of a Type II error (β) to zero (but increasing the sample size helps).

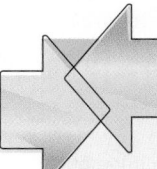

CONNECTIONS

All of the hypothesis tests we'll see boil down to the same question: "Is the difference between two quantities large?" We always measure "how large" by finding a ratio of this difference to the standard deviation of the sampling distribution of the statistic. Using the standard deviation as our ruler for inference is one of the core ideas of statistical thinking.

We've discussed the close relationship between hypothesis tests and confidence intervals. They are two sides of the same coin.

This chapter also has natural links to the discussion of probability, to the Normal model, and to the two previous chapters on inference.

WHAT HAVE WE LEARNED?

We've learned that there's a lot more to hypothesis testing than a simple yes/no decision.

▶ We've learned that the P-value can indicate evidence against the null hypothesis when it's small, but it does not tell us the probability that the null hypothesis is true.

▶ We've learned that the alpha level of the test establishes the level of proof we'll require. That determines the critical value of z that will lead us to reject the null hypothesis.

▶ We've also learned more about the connection between hypothesis tests and confidence intervals; they're really two ways of looking at the same question. The hypothesis test gives us the answer to a decision about a parameter; the confidence interval tells us the plausible values of that parameter.

We've learned about the two kinds of errors we might make, and we've seen why in the end we're never sure we've made the right decision.

▶ If the null hypothesis is really true and we reject it, that's a Type I error; the alpha level of the test is the probability that this could happen.

▶ If the null hypothesis is really false but we fail to reject it, that's a Type II error.

▶ The power of the test is the probability that we reject the null hypothesis when it's false. The larger the size of the effect we're testing for, the greater the power of the test to detect it.

▶ We've seen that tests with a greater likelihood of Type I error have more power and less chance of a Type II error. We can increase power while reducing the chances of both kinds of error by increasing the sample size.

Terms

Alpha level
486. The threshold P-value that determines when we reject a null hypothesis. If we observe a statistic whose P-value based on the null hypothesis is less than α, we reject that null hypothesis.

Statistically significant
486. When the P-value falls below the alpha level, we say that the test is "statistically significant" at that alpha level.

Significance level	486. The alpha level is also called the significance level, most often in a phrase such as a conclusion that a particular test is "significant at the 5% significance level."
Type I error	491. The error of rejecting a null hypothesis when in fact it is true (also called a "false positive"). The probability of a Type I error is α.
Type II error	491. The error of failing to reject a null hypothesis when in fact it is false (also called a "false negative"). The probability of a Type II error is commonly denoted β and depends on the effect size.
Power	492, 493. The probability that a hypothesis test will correctly reject a false null hypothesis is the power of the test. To find power, we must specify a particular alternative parameter value as the "true" value. For any specific value in the alternative, the power is $1 - \beta$.
Effect size	493. The difference between the null hypothesis value and true value of a model parameter is called the effect size.

Skills

▸ Understand that statistical significance does not measure the importance or magnitude of an effect. Recognize when others misinterpret statistical significance as proof of practical importance.

▸ Understand the close relationship between hypothesis tests and confidence intervals.

▸ Be able to identify and use the alternative hypothesis when testing hypotheses. Understand how to choose between a one-sided and two-sided alternative hypothesis, and know how to defend the choice of a one-sided alternative.

▸ Understand how the critical value for a test is related to the specified alpha level.

▸ Understand that the power of a test gives the probability that it correctly rejects a false null hypothesis when a specified alternative is true.

▸ Understand that the power of a test depends in part on the sample size. Larger sample sizes lead to greater power (and thus fewer Type II errors).

▸ Know how to complete a hypothesis test for a population proportion.

▸ Be able to interpret the meaning of a P-value in nontechnical language.

▸ Understand that the P-value of a test does not give the probability that the null hypothesis is correct.

▸ Know that we do not "accept" a null hypothesis if we cannot reject it but, rather, that we can only "fail to reject" the hypothesis for lack of evidence against it.

HYPOTHESIS TESTS ON THE COMPUTER

Reports about hypothesis tests generated by technologies don't follow a standard form. Most will name the test and provide the test statistic value, its standard deviation, and the P-value. But these elements may not be labeled clearly. For example, the expression "Prob > |z|" means the probability (the "Prob") of observing a test statistic whose magnitude (the absolute value tells us this) is larger than that of the one (the "z") found in the data (which, because it is written as "z," we know follows a Normal model). That is a fancy (and not very clear) way of saying P-value. In some packages, you can specify that the test be one-sided. Others might report three P-values, covering the ground for both one-sided tests and the two-sided test.

Sometimes a confidence interval and hypothesis test are automatically given together. The CI ought to be for the corresponding confidence level: $1 - \alpha$ for 2-tailed tests, $1 - 2\alpha$ for 1-tailed tests.

Often, the standard deviation of the statistic is called the "standard error," and usually that's appropriate because we've had to estimate its value from the data. That's not the case for proportions, however: We get the standard deviation for a proportion from the null hypothesis value. Nevertheless, you may see the standard deviation called a "standard error" even for tests with proportions.

It's common for statistics packages and calculators to report more digits of "precision" than could possibly have been found from the data. You can safely ignore them. Round values such as the standard deviation to one digit more than the number of digits reported in your data.

Here are the kind of results you might see. This is not from any program or calculator we know of, but it shows some of the things you might see in typical computer output.

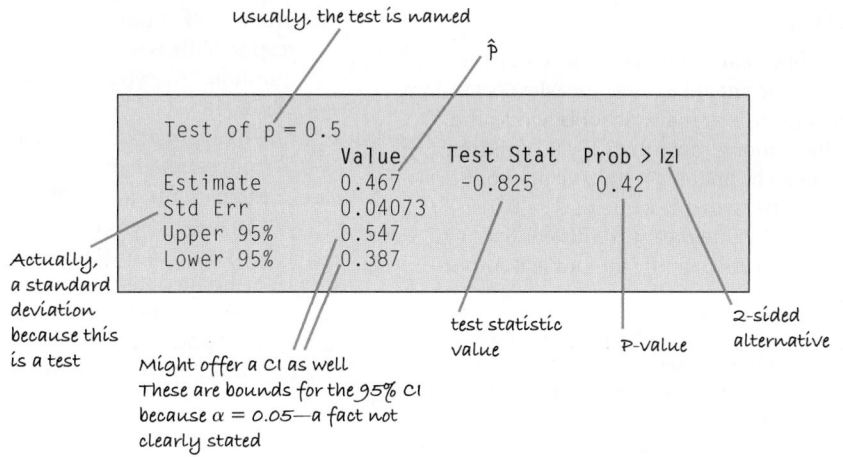

For information on hypothesis testing with particular statistics packages, see the table for Chapter 20 in Appendix B.

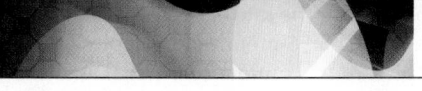

EXERCISES

1. **One sided or two?** In each of the following situations, is the alternative hypothesis one-sided or two-sided? What are the hypotheses?
   a) A business student conducts a taste test to see whether students prefer Diet Coke or Diet Pepsi.
   b) PepsiCo recently reformulated Diet Pepsi in an attempt to appeal to teenagers. They run a taste test to see if the new formula appeals to more teenagers than the standard formula.
   c) A budget override in a small town requires a two-thirds majority to pass. A local newspaper conducts a poll to see if there's evidence it will pass.
   d) One financial theory states that the stock market will go up or down with equal probability. A student collects data over several years to test the theory.

2. **Which alternative?** In each of the following situations, is the alternative hypothesis one-sided or two-sided? What are the hypotheses?
   a) A college dining service conducts a survey to see if students prefer plastic or metal cutlery.
   b) In recent years, 10% of college juniors have applied for study abroad. The dean's office conducts a survey to see if that's changed this year.

   c) A pharmaceutical company conducts a clinical trial to see if more patients who take a new drug experience headache relief than the 22% who claimed relief after taking the placebo.
   d) At a small computer peripherals company, only 60% of the hard drives produced passed all their performance tests the first time. Management recently invested a lot of resources into the production system and now conducts a test to see if it helped.

3. **P-value.** A medical researcher tested a new treatment for poison ivy against the traditional ointment. He concluded that the new treatment is more effective. Explain what the P-value of 0.047 means in this context.

4. **Another P-value.** Have harsher penalties and ad campaigns increased seat-belt use among drivers and passengers? Observations of commuter traffic failed to find evidence of a significant change compared with three years ago. Explain what the study's P-value of 0.17 means in this context.

5. **Alpha.** A researcher developing scanners to search for hidden weapons at airports has concluded that a new device is significantly better than the current scanner. He

made this decision based on a test using $\alpha = 0.05$. Would he have made the same decision at $\alpha = 0.10$? How about $\alpha = 0.01$? Explain.

6. **Alpha again.** Environmentalists concerned about the impact of high-frequency radio transmissions on birds found that there was no evidence of a higher mortality rate among hatchlings in nests near cell towers. They based this conclusion on a test using $\alpha = 0.05$. Would they have made the same decision at $\alpha = 0.10$? How about $\alpha = 0.01$? Explain.

7. **Significant?** Public health officials believe that 90% of children have been vaccinated against measles. A random survey of medical records at many schools across the country found that, among more than 13,000 children, only 89.4% had been vaccinated. A statistician would reject the 90% hypothesis with a P-value of $P = 0.011$.
   a) Explain what the P-value means in this context.
   b) The result is statistically significant, but is it important? Comment.

8. **Significant again?** A new reading program may reduce the number of elementary school students who read below grade level. The company that developed this program supplied materials and teacher training for a large-scale test involving nearly 8500 children in several different school districts. Statistical analysis of the results showed that the percentage of students who did not meet the grade-level goal was reduced from 15.9% to 15.1%. The hypothesis that the new reading program produced no improvement was rejected with a P-value of 0.023.
   a) Explain what the P-value means in this context.
   b) Even though this reading method has been shown to be significantly better, why might you not recommend that your local school adopt it?

9. **Success.** In August 2004, *Time* magazine reported the results of a random telephone poll commissioned by the Spike network. Of the 1302 men who responded, only 39 said that their most important measure of success was their work.
   a) Estimate the percentage of all American males who measure success primarily from their work. Use a 98% confidence interval. Check the conditions first.
   b) Some believe that few contemporary men judge their success primarily by their work. Suppose we wished to conduct a hypothesis test to see if the fraction has fallen below the 5% mark. What does your confidence interval indicate? Explain.
   c) What is the level of significance of this test? Explain.

10. **Is the Euro fair?** Soon after the Euro was introduced as currency in Europe, it was widely reported that someone had spun a Euro coin 250 times and gotten heads 140 times. We wish to test a hypothesis about the fairness of spinning the coin.
   a) Estimate the true proportion of heads. Use a 95% confidence interval. Don't forget to check the conditions.
   b) Does your confidence interval provide evidence that the coin is unfair when spun? Explain.
   c) What is the significance level of this test? Explain.

11. **Approval 2007.** In May 2007, George W. Bush's approval rating stood at 30% according to a CBS News/*New*

York Times national survey of 1125 randomly selected adults.
   a) Make a 95% confidence interval for his approval rating by all U.S. adults.
   b) Based on the confidence interval, test the null hypothesis that Bush's approval rating was no better than the 27% level established by Richard Nixon during the Watergate scandal.

12. **Superdads.** The Spike network commissioned a telephone poll of randomly sampled U.S. men. Of the 712 respondents who had children, 22% said "yes" to the question "Are you a stay-at-home dad?" (*Time*, August 23, 2004)
   a) To help market commercial time, Spike wants an accurate estimate of the true percentage of stay-at-home dads. Construct a 95% confidence interval.
   b) An advertiser of baby-carrying slings for dads will buy commercial time if at least 25% of men are stay-at-home dads. Use your confidence interval to test an appropriate hypothesis, and make a recommendation to the advertiser.
   c) Could Spike claim to the advertiser that it is possible that 25% of men with young children are stay-at-home dads? What is wrong with the reasoning?

13. **Dogs.** Canine hip dysplasia is a degenerative disease that causes pain in many dogs. Sometimes advanced warning signs appear in puppies as young as 6 months. A veterinarian checked 42 puppies whose owners brought them to a vaccination clinic, and she found 5 with early hip dysplasia. She considers this group to be a random sample of all puppies.
   a) Explain we cannot use this information to construct a confidence interval for the rate of occurrence of early hip dysplasia among all 6-month-old puppies.
   *b) Construct a "plus-four" confidence interval and interpret it in this context.

14. **Fans.** A survey of 81 randomly selected people standing in line to enter a football game found that 73 of them were home team fans.
   a) Explain why we cannot use this information to construct a confidence interval for the proportion of all people at the game who are fans of the home team.
   *b) Construct a "plus-four" confidence interval and interpret it in this context.

15. **Loans.** Before lending someone money, banks must decide whether they believe the applicant will repay the loan. One strategy used is a point system. Loan officers assess information about the applicant, totaling points they award for the person's income level, credit history, current debt burden, and so on. The higher the point total, the more convinced the bank is that it's safe to make the loan. Any applicant with a lower point total than a certain cutoff score is denied a loan.

We can think of this decision as a hypothesis test. Since the bank makes its profit from the interest collected on repaid loans, their null hypothesis is that the applicant will repay the loan and therefore should get the money. Only if the person's score falls below the minimum cutoff will the bank reject the null and deny

the loan. This system is reasonably reliable, but, of course, sometimes there are mistakes.
a) When a person defaults on a loan, which type of error did the bank make?
b) Which kind of error is it when the bank misses an opportunity to make a loan to someone who would have repaid it?
c) Suppose the bank decides to lower the cutoff score from 250 points to 200. Is that analogous to choosing a higher or lower value of α for a hypothesis test? Explain.
d) What impact does this change in the cutoff value have on the chance of each type of error?

16. **Spam.** Spam filters try to sort your e-mails, deciding which are real messages and which are unwanted. One method used is a point system. The filter reads each incoming e-mail and assigns points to the sender, the subject, key words in the message, and so on. The higher the point total, the more likely it is that the message is unwanted. The filter has a cutoff value for the point total; any message rated lower than that cutoff passes through to your inbox, and the rest, suspected to be spam, are diverted to the junk mailbox.

We can think of the filter's decision as a hypothesis test. The null hypothesis is that the e-mail is a real message and should go to your inbox. A higher point total provides evidence that the message may be spam; when there's sufficient evidence, the filter rejects the null, classifying the message as junk. This usually works pretty well, but, of course, sometimes the filter makes a mistake.
a) When the filter allows spam to slip through into your inbox, which kind of error is that?
b) Which kind of error is it when a real message gets classified as junk?
c) Some filters allow the user (that's you) to adjust the cutoff. Suppose your filter has a default cutoff of 50 points, but you reset it to 60. Is that analogous to choosing a higher or lower value of α for a hypothesis test? Explain.
d) What impact does this change in the cutoff value have on the chance of each type of error?

17. **Second loan.** Exercise 15 describes the loan score method a bank uses to decide which applicants it will lend money. Only if the total points awarded for various aspects of an applicant's financial condition fail to add up to a minimum cutoff score set by the bank will the loan be denied.
a) In this context, what is meant by the power of the test?
b) What could the bank do to increase the power?
c) What's the disadvantage of doing that?

18. **More spam.** Consider again the points-based spam filter described in Exercise 16. When the points assigned to various components of an e-mail exceed the cutoff value you've set, the filter rejects its null hypothesis (that the message is real) and diverts that e-mail to a junk mailbox.
a) In this context, what is meant by the power of the test?
b) What could you do to increase the filter's power?
c) What's the disadvantage of doing that?

19. **Homeowners 2005.** In 2005 the U.S. Census Bureau reported that 68.9% of American families owned their homes. Census data reveal that the ownership rate in one small city is much lower. The city council is debating a plan to offer tax breaks to first-time home buyers in order to encourage people to become homeowners. They decide to adopt the plan on a 2-year trial basis and use the data they collect to make a decision about continuing the tax breaks. Since this plan costs the city tax revenues, they will continue to use it only if there is strong evidence that the rate of home ownership is increasing.
a) In words, what will their hypotheses be?
b) What would a Type I error be?
c) What would a Type II error be?
d) For each type of error, tell who would be harmed.
e) What would the power of the test represent in this context?

20. **Alzheimer's.** Testing for Alzheimer's disease can be a long and expensive process, consisting of lengthy tests and medical diagnosis. Recently, a group of researchers (Solomon *et al.*, 1998) devised a 7-minute test to serve as a quick screen for the disease for use in the general population of senior citizens. A patient who tested positive would then go through the more expensive battery of tests and medical diagnosis. The authors reported a false positive rate of 4% and a false negative rate of 8%.
a) Put this in the context of a hypothesis test. What are the null and alternative hypotheses?
b) What would a Type I error mean?
c) What would a Type II error mean?
d) Which is worse here, a Type I or Type II error? Explain.
e) What is the power of this test?

21. **Testing cars.** A clean air standard requires that vehicle exhaust emissions not exceed specified limits for various pollutants. Many states require that cars be tested annually to be sure they meet these standards. Suppose state regulators double-check a random sample of cars that a suspect repair shop has certified as okay. They will revoke the shop's license if they find significant evidence that the shop is certifying vehicles that do not meet standards.
a) In this context, what is a Type I error?
b) In this context, what is a Type II error?
c) Which type of error would the shop's owner consider more serious?
d) Which type of error might environmentalists consider more serious?

22. **Quality control.** Production managers on an assembly line must monitor the output to be sure that the level of defective products remains small. They periodically inspect a random sample of the items produced. If they find a significant increase in the proportion of items that must be rejected, they will halt the assembly process until the problem can be identified and repaired.
a) In this context, what is a Type I error?
b) In this context, what is a Type II error?
c) Which type of error would the factory owner consider more serious?
d) Which type of error might customers consider more serious?

23. **Cars again.** As in Exercise 21, state regulators are checking up on repair shops to see if they are certifying vehicles that do not meet pollution standards.
    a) In this context, what is meant by the power of the test the regulators are conducting?
    b) Will the power be greater if they test 20 or 40 cars? Why?
    c) Will the power be greater if they use a 5% or a 10% level of significance? Why?
    d) Will the power be greater if the repair shop's inspectors are only a little out of compliance or a lot? Why?

24. **Production.** Consider again the task of the quality control inspectors in Exercise 22.
    a) In this context, what is meant by the power of the test the inspectors conduct?
    b) They are currently testing 5 items each hour. Someone has proposed that they test 10 instead. What are the advantages and disadvantages of such a change?
    c) Their test currently uses a 5% level of significance. What are the advantages and disadvantages of changing to an alpha level of 1%?
    d) Suppose that, as a day passes, one of the machines on the assembly line produces more and more items that are defective. How will this affect the power of the test?

25. **Equal opportunity?** A company is sued for job discrimination because only 19% of the newly hired candidates were minorities when 27% of all applicants were minorities. Is this strong evidence that the company's hiring practices are discriminatory?
    a) Is this a one-tailed or a two-tailed test? Why?
    b) In this context, what would a Type I error be?
    c) In this context, what would a Type II error be?
    d) In this context, what is meant by the power of the test?
    e) If the hypothesis is tested at the 5% level of significance instead of 1%, how will this affect the power of the test?
    f) The lawsuit is based on the hiring of 37 employees. Is the power of the test higher than, lower than, or the same as it would be if it were based on 87 hires?

26. **Stop signs.** Highway safety engineers test new road signs, hoping that increased reflectivity will make them more visible to drivers. Volunteers drive through a test course with several of the new- and old-style signs and rate which kind shows up the best.
    a) Is this a one-tailed or a two-tailed test? Why?
    b) In this context, what would a Type I error be?
    c) In this context, what would a Type II error be?
    d) In this context, what is meant by the power of the test?
    e) If the hypothesis is tested at the 1% level of significance instead of 5%, how will this affect the power of the test?
    f) The engineers hoped to base their decision on the reactions of 50 drivers, but time and budget constraints may force them to cut back to 20. How would this affect the power of the test? Explain.

27. **Dropouts.** A Statistics professor has observed that for several years about 13% of the students who initially enroll in his Introductory Statistics course withdraw before the end of the semester. A salesman suggests that he try a statistics software package that gets students more involved with computers, predicting that it will cut the dropout rate. The software is expensive, and the salesman offers to let the professor use it for a semester to see if the dropout rate goes down significantly. The professor will have to pay for the software only if he chooses to continue using it.
    a) Is this a one-tailed or two-tailed test? Explain.
    b) Write the null and alternative hypotheses.
    c) In this context, explain what would happen if the professor makes a Type I error.
    d) In this context, explain what would happen if the professor makes a Type II error.
    e) What is meant by the power of this test?

28. **Ads.** A company is willing to renew its advertising contract with a local radio station only if the station can prove that more than 20% of the residents of the city have heard the ad and recognize the company's product. The radio station conducts a random phone survey of 400 people.
    a) What are the hypotheses?
    b) The station plans to conduct this test using a 10% level of significance, but the company wants the significance level lowered to 5%. Why?
    c) What is meant by the power of this test?
    d) For which level of significance will the power of this test be higher? Why?
    e) They finally agree to use $\alpha = 0.05$, but the company proposes that the station call 600 people instead of the 400 initially proposed. Will that make the risk of Type II error higher or lower? Explain.

29. **Dropouts, part II.** Initially, 203 students signed up for the Stats course in Exercise 27. They used the software suggested by the salesman, and only 11 dropped out of the course.
    a) Should the professor spend the money for this software? Support your recommendation with an appropriate test.
    b) Explain what your P-value means in this context.

30. **Testing the ads.** The company in Exercise 28 contacts 600 people selected at random, and only 133 remember the ad.
    a) Should the company renew the contract? Support your recommendation with an appropriate test.
    b) Explain what your P-value means in this context.

31. **Two coins.** In a drawer are two coins. They look the same, but one coin produces heads 90% of the time when spun while the other one produces heads only 30% of the time. You select one of the coins. You are allowed to spin it *once* and then must decide whether the coin is the 90%- or the 30%-head coin. Your null hypothesis is that your coin produces 90% heads.
    a) What is the alternative hypothesis?
    b) Given that the outcome of your spin is tails, what would you decide? What if it were heads?
    c) How large is α in this case?
    d) How large is the power of this test? (*Hint:* How many possibilities are in the alternative hypothesis?)
    e) How could you lower the probability of a Type I error and increase the power of the test at the same time?

32. Faulty or not? You are in charge of shipping computers to customers. You learn that a faulty disk drive was put into some of the machines. There's a simple test you can perform, but it's not perfect. All but 4% of the time, a good disk drive passes the test, but unfortunately, 35% of the bad disk drives pass the test, too. You have to decide on the basis of one test whether the disk drive is good or bad. Make this a hypothesis test.

a) What are the null and alternative hypotheses?

b) Given that a computer fails the test, what would you decide? What if it passes the test?

c) How large is α for this test?

d) What is the power of this test? (*Hint:* How many possibilities are in the alternative hypothesis?)

33. Hoops. A basketball player with a poor foul-shot record practices intensively during the off-season. He tells the coach that he has raised his proficiency from 60% to 80%. Dubious, the coach asks him to take 10 shots, and is surprised when the player hits 9 out of 10. Did the player prove that he has improved?

a) Suppose the player really is no better than before—still a 60% shooter. What's the probability he can hit at least 9 of 10 shots anyway? (*Hint:* Use a Binomial model.)

b) If that is what happened, now the coach thinks the player has improved when he has not. Which type of error is that?

c) If the player really can hit 80% now, and it takes at least 9 out of 10 successful shots to convince the coach, what's the power of the test?

d) List two ways the coach and player could increase the power to detect any improvement.

34. Pottery. An artist experimenting with clay to create pottery with a special texture has been experiencing difficulty with these special pieces. About 40% break in the kiln during firing. Hoping to solve this problem, she buys some more expensive clay from another supplier. She plans to make and fire 10 pieces and will decide to use the new clay if at most one of them breaks.

a) Suppose the new, expensive clay really is no better than her usual clay. What's the probability that this test convinces her to use it anyway? (*Hint:* Use a Binomial model.)

b) If she decides to switch to the new clay and it is no better, what kind of error did she commit?

c) If the new clay really can reduce breakage to only 20%, what's the probability that her test will not detect the improvement?

d) How can she improve the power of her test? Offer at least two suggestions.

JUST CHECKING
Answers

1. With a z-score of 0.62, you can't reject the null hypothesis. The experiment shows no evidence that the wheel is not fair.

2. At $\alpha = 0.05$, you can't reject the null hypothesis because 0.30 is contained in the 90% confidence interval—it's plausible that sending the DVDs is no more effective than just sending letters.

3. The confidence interval is from 29% to 45%. The DVD strategy is more expensive and may not be worth it. We can't distinguish the success rate from 30% given the results of this experiment, but 45% would represent a large improvement. The bank should consider another trial, increasing their sample size to get a narrower confidence interval.

4. A Type I error would mean deciding that the DVD success rate is higher than 30% when it really isn't. They would adopt a more expensive method for collecting payments that's no better than the less expensive strategy.

5. A Type II error would mean deciding that there's not enough evidence to say that the DVD strategy works when in fact it does. The bank would fail to discover an effective method for increasing their revenue from delinquent accounts.

6. 60%; the larger the effect size, the greater the power. It's easier to detect an improvement to a 60% success rate than to a 32% rate.

CHAPTER

22

Comparing Two Proportions

WHO	6971 male drivers
WHAT	Seatbelt use
WHY	Highway safety
WHEN	2007
WHERE	Massachusetts

Do men take more risks than women? Psychologists have documented that in many situations, men choose riskier behavior than women do. But what is the effect of having a woman by their side? A recent seatbelt observation study in Massachusetts[1] found that, not surprisingly, male drivers wear seatbelts less often than women do. The study also noted that men's belt-wearing jumped more than 16 percentage points when they had a female passenger. Seatbelt use was recorded at 161 locations in Massachusetts, using random-sampling methods developed by the National Highway Traffic Safety Administration (NHTSA). Female drivers wore belts more than 70% of the time, regardless of the sex of their passengers. Of 4208 male drivers with female passengers, 2777 (66.0%) were belted. But among 2763 male drivers with male passengers only, 1363 (49.3%) wore seatbelts. This was only a random sample, but it suggests there may be a shift in men's risk-taking behavior when women are present. What would we estimate the true size of that gap to be?

Comparisons between two percentages are much more common than questions about isolated percentages. And they are more interesting. We often want to know how two groups differ, whether a treatment is better than a placebo control, or whether this year's results are better than last year's.

Another Ruler

We know the difference between the proportions of men wearing seatbelts seen in the *sample*. It's 16.7%. But what's the *true* difference for all men? We know that our estimate probably isn't exactly right. To say more, we need a new ruler—the standard deviation of the sampling distribution model for the difference in the proportions. Now we have two proportions, and each will vary from sample to sample. We are interested in the difference between them. So what is the correct standard deviation?

---

[1] Massachusetts Traffic Safety Research Program [June 2007].

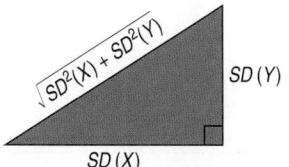

The answer comes to us from Chapter 16. Remember the Pythagorean Theorem of Statistics?

The variance of the sum or difference of two independent random variables is the sum of their variances.

This is such an important (and powerful) idea in Statistics that it's worth pausing a moment to review the reasoning. Here's some intuition about why variation increases even when we subtract two random quantities.

Grab a full box of cereal. The box claims to contain 16 ounces of cereal. We know that's not exact: There's some small variation from box to box. Now pour a bowl of cereal. Of course, your 2-ounce serving will not be exactly 2 ounces. There'll be some variation there, too. How much cereal would you guess was left in the box? Do you think your guess will be as close as your guess for the full box? *After* you pour your bowl, the amount of cereal in the box is still a random quantity (with a smaller mean than before), but it is even *more variable* because of the additional variation in the amount you poured.

According to our rule, the variance of the amount of cereal left in the box would now be the *sum* of the two *variances*.

We want a standard deviation, not a variance, but that's just a square root away. We can write symbolically what we've just said:

> For independent random variables, **variances add.**

$$Var(X - Y) = Var(X) + Var(Y), \text{so}$$

$$SD(X - Y) = \sqrt{SD^2(X) + SD^2(Y)} = \sqrt{Var(X) + Var(Y)}.$$

Be careful, though—this simple formula applies only when X and Y are independent. Just as the Pythagorean Theorem[2] works only for right triangles, our formula works only for independent random variables. Always check for independence before using it.

The Standard Deviation of the Difference Between Two Proportions

> Combining independent random quantities always *increases* the overall variation, so even for *differences* of independent random variables, **variances add.**

Fortunately, proportions observed in independent random samples *are* independent, so we can put the two proportions in for X and Y and add their variances. We just need to use careful notation to keep things straight.

When we have two samples, each can have a different size and proportion value, so we keep them straight with subscripts. Often we choose subscripts that remind us of the groups. For our example, we might use "$_M$" and "$_F$", but generically we'll just use "$_1$" and "$_2$". We will represent the two sample proportions as $\hat{p}_1$ and $\hat{p}_2$, and the two sample sizes as n_1 and n_2.

The standard deviations of the sample proportions are $SD(\hat{p}_1) = \sqrt{\dfrac{p_1 q_1}{n_1}}$ and

$SD(\hat{p}_2) = \sqrt{\dfrac{p_2 q_2}{n_2}}$, so the variance of the difference in the proportions is

$$Var(\hat{p}_1 - \hat{p}_2) = \left(\sqrt{\frac{p_1 q_1}{n_1}}\right)^2 + \left(\sqrt{\frac{p_2 q_2}{n_2}}\right)^2 = \frac{p_1 q_1}{n_1} + \frac{p_2 q_2}{n_2}.$$

The standard deviation is the square root of that variance:

$$SD(\hat{p}_1 - \hat{p}_2) = \sqrt{\frac{p_1 q_1}{n_1} + \frac{p_2 q_2}{n_2}}.$$

---

[2] If you don't remember the formula, don't rely on the Scarecrow's version from *The Wizard of Oz*. He may have a brain and have been awarded his Th.D. (Doctor of Thinkology), but he gets the formula wrong.

We usually don't know the true values of p_1 and p_2. When we have the sample proportions in hand from the data, we use them to estimate the variances. So the standard error is

$$SE(\hat{p}_1 - \hat{p}_2) = \sqrt{\frac{\hat{p}_1\hat{q}_1}{n_1} + \frac{\hat{p}_2\hat{q}_2}{n_2}}.$$

FOR EXAMPLE | Finding the standard error of a difference in proportions

A recent survey of 886 randomly selected teenagers (aged 12–17) found that more than half of them had online profiles.[3] Some researchers and privacy advocates are concerned about the possible access to personal information about teens in public places on the Internet. There appear to be differences between boys and girls in their online behavior. Among teens aged 15–17, 57% of the 248 boys had posted profiles, compared to 70% of the 256 girls. Let's start the process of estimating how large the true gender gap might be.

Question: What's the standard error of the difference in sample proportions?

Because the boys and girls were selected at random, it's reasonable to assume their behaviors are independent, so it's okay to use the Pythagorean Theorem of Statistics and add the variances:

$$SE(\hat{p}_{boys}) = \sqrt{\frac{0.57 \times 0.43}{248}} = 0.0314 \qquad SE(\hat{p}_{girls}) = \sqrt{\frac{0.70 \times 0.30}{256}} = 0.0286$$

$$SE(\hat{p}_{girls} - \hat{p}_{boys}) = \sqrt{0.0314^2 + 0.0286^2} = 0.0425$$

Assumptions and Conditions

Before we look at our example, we need to check assumptions and conditions.

INDEPENDENCE ASSUMPTIONS

Independence Assumption: Within each group, the data should be based on results for independent individuals. We can't check that for certain, but we *can* check the following:

Randomization Condition: The data in each group should be drawn independently and at random from a homogeneous population or generated by a randomized comparative experiment.

The 10% Condition: If the data are sampled without replacement, the sample should not exceed 10% of the population.

Because we are comparing two groups in this way, we need an additional Independence Assumption. In fact, this is the most important of these assumptions. If it is violated, these methods just won't work.

Independent Groups Assumption: The two groups we're comparing must also be independent *of each other.* Usually, the independence of the groups from each other is evident from the way the data were collected.

Why is the Independent Groups Assumption so important? If we compare husbands with their wives, or a group of subjects before and after some treatment, we can't just add the variances. Subjects' performance before a treatment might very well be related to their performance after the treatment. So the proportions are not independent and the Pythagorean-style variance formula does not hold. We'll see a way to compare a common kind of nonindependent samples in a later chapter.

---

[3] Princeton Survey Research Associates International for the Pew Internet & American Life Project.

SAMPLE SIZE CONDITION

Each of the groups must be big enough. As with individual proportions, we need larger groups to estimate proportions that are near 0% or 100%. We usually check the Success/Failure Condition for each group.

Success/Failure Condition: Both groups are big enough that at least 10 successes and at least 10 failures have been observed in each.

FOR EXAMPLE   Checking assumptions and conditions

Recap: Among randomly sampled teens aged 15–17, 57% of the 248 boys had posted online profiles, compared to 70% of the 256 girls.

Question: Can we use these results to make inferences about all 15–17-year-olds?

✔ **Randomization Condition:** The sample of boys and the sample of girls were both chosen randomly.

✔ **10% Condition:** 248 boys and 256 girls are each less than 10% of all teenage boys and girls.

✔ **Independent Groups Assumption:** Because the samples were selected at random, it's reasonable to believe the boys' online behaviors are independent of the girls' online behaviors.

✔ **Success/Failure Condition:** Among the boys, $248(0.57) = 141$ had online profiles and the other $248(0.43) = 107$ did not. For the girls, $256(0.70) = 179$ successes and $256(0.30) = 77$ failures. All counts are at least 10.

Because all the assumptions and conditions are satisfied, it's okay to proceed with inference for the difference in proportions.

(Note that when we find the *observed* counts of successes and failures, we round off to whole numbers. We're using the reported percentages to recover the actual counts.)

The Sampling Distribution

We're almost there. We just need one more fact about proportions. We already know that for large enough samples, each of our proportions has an approximately Normal sampling distribution. The same is true of their difference.

Why Normal?
In Chapter 16 we learned that sums and differences of independent Normal random variables also follow a Normal model. That's the reason we use a Normal model for the difference of two independent proportions.

> **THE SAMPLING DISTRIBUTION MODEL FOR A DIFFERENCE BETWEEN TWO INDEPENDENT PROPORTIONS**
> Provided that the sampled values are independent, the samples are independent, and the sample sizes are large enough, the sampling distribution of $\hat{p}_1 - \hat{p}_2$ is modeled by a Normal model with mean $\mu = p_1 - p_2$ and standard deviation
> $$SD(\hat{p}_1 - \hat{p}_2) = \sqrt{\frac{p_1 q_1}{n_1} + \frac{p_2 q_2}{n_2}}.$$

The sampling distribution model and the standard deviation give us all we need to find a margin of error for the difference in proportions—or at least they would if we knew the true proportions, p_1 and p_2. However, we don't know the true values, so we'll work with the observed proportions, $\hat{p}_1$ and $\hat{p}_2$, and use $SE(\hat{p}_1 - \hat{p}_2)$ to estimate the standard deviation. The rest is just like a one-proportion *z*-interval.

CHAPTER 22 Comparing Two Proportions

Activity: Compare Two Proportions. Does a preschool program help disadvantaged children later in life?

A TWO-PROPORTION z-INTERVAL

When the conditions are met, we are ready to find the confidence interval for the difference of two proportions, $p_1 - p_2$. The confidence interval is

$$(\hat{p}_1 - \hat{p}_2) \pm z^* \times SE(\hat{p}_1 - \hat{p}_2)$$

where we find the standard error of the difference,

$$SE(\hat{p}_1 - \hat{p}_2) = \sqrt{\frac{\hat{p}_1\hat{q}_1}{n_1} + \frac{\hat{p}_2\hat{q}_2}{n_2}},$$

from the observed proportions.

The critical value z^* depends on the particular confidence level, C, that we specify.

FOR EXAMPLE | Finding a two-proportion z-interval

Recap: Among randomly sampled teens aged 15–17, 57% of the 248 boys had posted online profiles, compared to 70% of the 256 girls. We calculated the standard error for the difference in sample proportions to be $SE(\hat{p}_{girls} - \hat{p}_{boys}) = 0.0425$ and found that the assumptions and conditions required for inference checked out okay.

Question: What does a confidence interval say about the difference in online behavior?

A 95% confidence interval for $p_{girls} - p_{boys}$ is $(\hat{p}_{girls} - \hat{p}_{boys}) \pm z^*SE(\hat{p}_{girls} - \hat{p}_{boys})$

$$(0.70 - 0.57) \pm 1.96(0.0425)$$

$$0.13 \pm 0.083$$

$$(4.7\%, 21.3\%)$$

We can be 95% confident that among teens aged 15–17, the proportion of girls who post online profiles is between 4.7 and 21.3 percentage points higher than the proportion of boys who do. It seems clear that teen girls are more likely to post profiles than are boys the same age.

STEP-BY-STEP EXAMPLE | A Two-Proportion z-Interval

Now we are ready to be more precise about the passenger-based gap in male drivers' seatbelt use. We'll estimate the difference with a confidence interval using a method called the **two-proportion z-interval** and follow the four confidence interval steps.

Question: How much difference is there in the proportion of male drivers who wear seatbelts when sitting next to a male passenger and the proportion who wear seatbelts when sitting next to a female passenger?

THINK AGAIN

Plan State what you want to know. Discuss the variables and the W's.

Identify the parameter you wish to estimate. (It usually doesn't matter in which direction we subtract, so, for convenience, we usually choose the direction with a positive difference.)

I want to know the true difference in the population proportion, p_M, of male drivers who wear seatbelts when sitting next to a man and p_F, the proportion who wear seatbelts when sitting next to a woman. The data are from a random sample of drivers in Massachusetts in 2007, observed according to procedures developed by the NHTSA. The parameter of interest is the difference $p_F - p_M$.

Choose and state a confidence level.	I will find a 95% confidence interval for this parameter.
Model Think about the assumptions and check the conditions.	✔ **Independence Assumption:** Driver behavior was independent from car to car.
	✔ **Randomization Condition:** The NHTSA methods are more complex than an SRS, but they result in a suitable random sample.
	✔ **10% Condition:** The samples include far fewer than 10% of all male drivers accompanied by male or by female passengers.
	✔ **Independent Groups Assumption:** There's no reason to believe that seatbelt use among drivers with male passengers and those with female passengers are not independent.
The Success/Failure Condition must hold for each group.	✔ **Success Failure Condition:** Among male drivers with female passengers, 2777 wore seatbelts and 1431 did not; of those driving with male passengers, 1363 wore seatbelts and 1400 did not. Each group contained far more than 10 successes and 10 failures.
State the sampling distribution model for the statistic. Choose your method.	Under these conditions, the sampling distribution of the difference between the sample proportions is approximately Normal, so I'll find a **two-proportion z-interval.**

Mechanics Construct the confidence interval.

As often happens, the key step in finding the confidence interval is estimating the standard deviation of the sampling distribution model of the statistic. Here the statistic is the difference in the proportions of men who wear seatbelts when they have a female passenger and the proportion who do so with a male passenger. Substitute the data values into the formula.

The sampling distribution is Normal, so the critical value for a 95% confidence interval, z^*, is 1.96. The margin of error is the critical value times the SE.

I know

$$n_F = 4208, n_M = 2763.$$

The observed sample proportions are

$$\hat{p}_F = \frac{2777}{4208} = 0.660, \hat{p}_M = \frac{1363}{2763} = 0.493$$

I'll estimate the SD of the difference with

$$SE(\hat{p}_F - \hat{p}_M) = \sqrt{\frac{\hat{p}_F\hat{q}_F}{n_F} + \frac{\hat{p}_M\hat{q}_M}{n_M}}$$

$$= \sqrt{\frac{(0.660)(0.340)}{4208} + \frac{(0.493)(0.507)}{2763}}$$

$$= 0.012$$

$$ME = z^* \times SE(\hat{p}_F - \hat{p}_M)$$
$$= 1.96(0.012) = 0.024$$

The confidence interval is the statistic ±ME.	The observed difference in proportions is $\hat{p}_F - \hat{p}_M = 0.660 - 0.493 = 0.167$, so the 95% confidence interval is $$0.167 \pm 0.024$$ or 14.3% to 19.1%
 **Conclusion** Interpret your confidence interval in the proper context. (Remember: We're 95% confident that our interval captured the true difference.)	I am 95% confident that the proportion of male drivers who wear seatbelts when driving next to a female passenger is between 14.3 and 19.1 percentage points higher than the proportion who wear seatbelts when driving next to a male passenger.

This is an interesting result—but be careful not to try to say too much! In Massachusetts, overall seatbelt use is lower than the national average, so we can't be certain that these results generalize to other states. And these were two different groups of men, so we can't say that, individually, men are more likely to buckle up when they have a woman passenger. You can probably think of several alternative explanations; we'll suggest just a couple. Perhaps age is a lurking variable: Maybe older men are more likely to wear seatbelts and also more likely to be driving with their wives. Or maybe men who don't wear seatbelts have trouble attracting women!

TI Tips

Finding a confidence interval

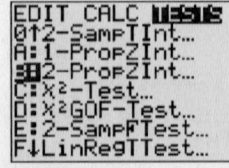

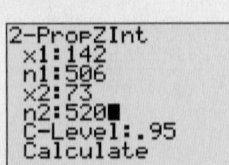

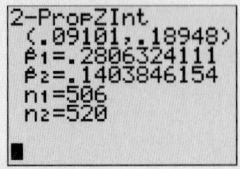

You can use a routine in the **STAT TESTS** menu to create confidence intervals for the difference of two proportions. Remember, the calculator can do only the mechanics—checking conditions and writing conclusions are still up to you.

A Gallup Poll asked whether the attribute "intelligent" described men in general. The poll revealed that 28% of 506 men thought it did, but only 14% of 520 women agreed. We want to estimate the true size of the gender gap by creating a 95% confidence interval.

- Go to the **STAT TESTS** menu. Scroll down the list and select **B:2-PropZInt**.
- Enter the observed number of males: **.28*506**. Remember that the actual number of males must be a whole number, so be sure to round off.
- Enter the sample size: **506** males.
- Repeat those entries for women: **.14*520** agreed, and the sample size was **520**.
- Specify the desired confidence level.
- **Calculate** the result.

And now explain what you see: We are 95% confident that the proportion of men who think the attribute "intelligent" describe males in general is between 9 and 19 percentage points higher than the proportion of women who think so.

JUST CHECKING

A public broadcasting station plans to launch a special appeal for additional contributions from current members. Unsure of the most effective way to contact people, they run an experiment. They randomly select two groups of current members. They send the same request for donations to everyone, but it goes to one group by e-mail and to the other group by regular mail. The station was successful in getting contributions from 26% of the members they e-mailed but only from 15% of those who received the request by regular mail. A 90% confidence interval estimated the difference in donation rates to be 11% ± 7%.

1. Interpret the confidence interval in this context.

2. Based on this confidence interval, what conclusion would we reach if we tested the hypothesis that there's no difference in the response rates to the two methods of fundraising? Explain.

Will I Snore When I'm 64?

WHO	Randomly selected U.S. adults over age 18
WHAT	Proportion who snore, categorized by age (less than 30, 30 or older)
WHEN	2001
WHERE	United States
WHY	To study sleep behaviors of U.S. adults

The National Sleep Foundation asked a random sample of 1010 U.S. adults questions about their sleep habits. The sample was selected in the fall of 2001 from random telephone numbers, stratified by region and sex, guaranteeing that an equal number of men and women were interviewed (2002 Sleep in America Poll, National Sleep Foundation, Washington, DC).

One of the questions asked about snoring. Of the 995 respondents, 37% of adults reported that they snored at least a few nights a week during the past year. Would you expect that percentage to be the same for all age groups? Split into two age categories, 26% of the 184 people under 30 snored, compared with 39% of the 811 in the older group. Is this difference of 13% real, or due only to natural fluctuations in the sample we've chosen?

The question calls for a hypothesis test. Now the parameter of interest is the true *difference* between the (reported) snoring rates of the two age groups.

What's the appropriate null hypothesis? That's easy here. We hypothesize that there is no difference in the proportions. This is such a natural null hypothesis that we rarely consider any other. But instead of writing $H_0: p_1 = p_2$, we usually express it in a slightly different way. To make it relate directly to the *difference*, we hypothesize that the difference in proportions is zero:

$$H_0: p_1 - p_2 = 0.$$

Everyone into the Pool

Our hypothesis is about a new parameter: the *difference* in proportions. We'll need a standard error for that. Wait—don't we know that already? Yes and no. We know that the standard error of the difference in proportions is

$$SE(\hat{p}_1 - \hat{p}_2) = \sqrt{\frac{\hat{p}_1\hat{q}_1}{n_1} + \frac{\hat{p}_2\hat{q}_2}{n_2}},$$

and we could just plug in the numbers, but we can do even better. The secret is that proportions and their standard deviations are linked. There are two proportions in the standard error formula—but look at the null hypothesis. It says that these proportions are equal. To do a hypothesis test, we *assume* that the null hypothesis is true. So there should be just a single value of $\hat{p}$ in the SE formula (and, of course, $\hat{q}$ is just $1 - \hat{p}$).

How would we do this for the snoring example? If the null hypothesis is true, then, among all adults, the two groups have the same proportion. Overall, we saw 48 + 318 = 366 snorers out of a total of 184 + 811 = 995 adults who responded to this question. The overall proportion of snorers was 366/995 = 0.3678.

Combining the counts like this to get an overall proportion is called **pooling.** Whenever we have data from different sources or different groups but we believe that they really came from the same underlying population, we pool them to get better estimates.

When we have counts for each group, we can find the pooled proportion as

$$\hat{p}_{\text{pooled}} = \frac{Success_1 + Success_2}{n_1 + n_2},$$

where $Success_1$ is the number of successes in group 1 and $Success_2$ is the number of successes in group 2. That's the overall proportion of success.

When we have only proportions and not the counts, as in the snoring example, we have to reconstruct the number of successes by multiplying the sample sizes by the proportions:

$$Success_1 = n_1\hat{p}_1 \quad \text{and} \quad Success_2 = n_2\hat{p}_2.$$

If these calculations don't come out to whole numbers, round them first. There must have been a whole number of successes, after all. (This is the *only* time you should round values in the middle of a calculation.)

We then put this pooled value into the formula, substituting it for *both* sample proportions in the standard error formula:

$$SE_{\text{pooled}}(\hat{p}_1 - \hat{p}_2) = \sqrt{\frac{\hat{p}_{\text{pooled}}\hat{q}_{\text{pooled}}}{n_1} + \frac{\hat{p}_{\text{pooled}}\hat{q}_{\text{pooled}}}{n_2}}$$

$$= \sqrt{\frac{0.3678 \times (1 - 0.3678)}{184} + \frac{0.3678 \times (1 - 0.3678)}{811}}.$$

This comes out to 0.039.

> When finding the number of successes, round the values to integers. For example, the 48 snorers among the 184 under-30 respondents are actually 26.1% of 184. We round back to the nearest whole number to find the count that could have yielded the rounded percent we were given.

Improving the Success/Failure Condition

The vaccine Gardasil® was introduced to prevent the strains of human papillomavirus (HPV) that are responsible for almost all cases of cervical cancer. In randomized placebo-controlled clinical trials,[4] only 1 case of HPV was diagnosed among 7897 women who received the vaccine, compared with 91 cases diagnosed among 7899 who received a placebo. The one observed HPV case ("success") doesn't meet the at-least-10-successes criterion. Surely, though, we should not refuse to test the effectiveness of the vaccine just because it failed so rarely; that would be absurd.

For that reason, in a two-proportion z-test, the proper Success/Failure test uses the *expected* frequencies, which we can find from the pooled proportion. In this case,

$$\hat{p}_{\text{pooled}} = \frac{91 + 1}{7899 + 7897} = 0.0058$$

$$n_1\hat{p}_{\text{pooled}} = 7899(0.0058) = 46$$

$$n_2\hat{p}_{\text{pooled}} = 7897(0.0058) = 46,$$

so we can proceed with the hypothesis test.

---

[4] *Quadrivalent Human Papillomavirus Vaccine: Recommendations of the Advisory Committee on Immunization Practices (ACIP)*, National Center for HIV/AIDS, Viral Hepatitis, STD and TB Prevention [May 2007].

Often it is easier just to check the observed numbers of successes and failures. If they are both greater than 10, you don't need to look further. But keep in mind that the correct test uses the expected frequencies rather than the observed ones.

Compared to What?

Naturally, we'll reject our null hypothesis if we see a large enough difference in the two proportions. How can we decide whether the difference we see, $\hat{p}_1 - \hat{p}_2$, is large? The answer is the same as always: We just compare it to its standard deviation.

Unlike previous hypothesis-testing situations, the null hypothesis doesn't provide a standard deviation, so we'll use a standard error (here, pooled). Since the sampling distribution is Normal, we can divide the observed difference by its standard error to get a z-score. The z-score will tell us how many standard errors the observed difference is away from 0. We can then use the 68–95–99.7 Rule to decide whether this is large, or some technology to get an exact P-value. The result is a **two-proportion z-test.**

A S **Activity: Test for a Difference Between Two Proportions.** Is premium-brand chicken less likely to be contaminated than store-brand chicken?

TWO-PROPORTION z-TEST

The conditions for the two-proportion z-test are the same as for the two-proportion z-interval. We are testing the hypothesis

$$H_0: p_1 - p_2 = 0.$$

Because we hypothesize that the proportions are equal, we pool the groups to find

$$\hat{p}_{\text{pooled}} = \frac{Success_1 + Success_2}{n_1 + n_2}$$

and use that pooled value to estimate the standard error:

$$SE_{\text{pooled}}(\hat{p}_1 - \hat{p}_2) = \sqrt{\frac{\hat{p}_{\text{pooled}}\,\hat{q}_{\text{pooled}}}{n_1} + \frac{\hat{p}_{\text{pooled}}\,\hat{q}_{\text{pooled}}}{n_2}}.$$

Now we find the test statistic,

$$z = \frac{(\hat{p}_1 - \hat{p}_2) - 0}{SE_{\text{pooled}}(\hat{p}_1 - \hat{p}_2)}.$$

When the conditions are met and the null hypothesis is true, this statistic follows the standard Normal model, so we can use that model to obtain a P-value.

STEP-BY-STEP EXAMPLE   **A Two-Proportion z-Test**

Question: Are the snoring rates of the two age groups really different?

THINK

Plan State what you want to know. Discuss the variables and the W's.

I want to know whether snoring rates differ for those under and over 30 years old. The data are from a random sample of 1010 U.S. adults surveyed in the 2002 Sleep in America Poll. Of these, 995 responded to the question about snoring, indicating whether or not they had snored at least a few nights a week in the past year.

Hypotheses The study simply broke down the responses by age, so there is no sense that either alternative was preferred. A two-sided alternative hypothesis is appropriate.

H_0: There is no difference in snoring rates in the two age groups:

$$p_{old} - p_{young} = 0.$$

H_A: The rates are different: $p_{old} - p_{young} \neq 0.$

Model Think about the assumptions and check the conditions.

✔ **Independence Assumption:** The National Sleep Foundation selected respondents at random, so they should be independent.

✔ **Randomization Condition:** The respondents were randomly selected by telephone number and stratified by sex and region.

✔ **10% Condition:** The number of adults surveyed in each age group is certainly far less than 10% of that population.

✔ **Independent Groups Assumption:** The two groups are independent of each other because the sample was selected at random.

✔ **Success/Failure Condition:** In the younger age group, 48 snored and 136 didn't. In the older group, 318 snored and 493 didn't. The observed numbers of both successes and failures are much more than 10 for both groups.[5]

State the null model.

Choose your method.

Because the conditions are satisfied, I'll use a Normal model and perform a **two-proportion z-test.**

 **Mechanics**

$$n_{young} = 184, y_{young} = 48, \hat{p}_{young} = 0.261$$
$$n_{old} = 811, \quad y_{old} = 318, \quad \hat{p}_{old} = 0.392$$

The hypothesis is that the proportions are equal, so pool the sample data.

$$\hat{p}_{pooled} = \frac{y_{old} + y_{young}}{n_{old} + n_{young}} = \frac{318 + 48}{811 + 184} = 0.3678$$

Use the pooled SE to estimate $SD(p_{old} - p_{young})$.

$$SE_{pooled}(\hat{p}_{old} - \hat{p}_{young})$$

$$= \sqrt{\frac{\hat{p}_{pooled}\hat{q}_{pooled}}{n_{old}} + \frac{\hat{p}_{pooled}\hat{q}_{pooled}}{n_{young}}}$$

$$= \sqrt{\frac{(0.3678)(0.6322)}{811} + \frac{(0.3678)(0.6322)}{184}}$$

$$\approx 0.039375$$

The observed difference in sample proportions is
$$\hat{p}_{old} - \hat{p}_{young} = 0.392 - 0.261 = 0.131$$

---

[5] This is one of those situations in which the traditional term "success" seems a bit weird. A success here could be that a person snores. "Success" and "failure" are arbitrary labels left over from studies of gambling games.

Make a picture. Sketch a Normal model centered at the hypothesized difference of 0. Shade the region to the right of the observed difference, and because this is a two-tailed test, also shade the corresponding region in the other tail.

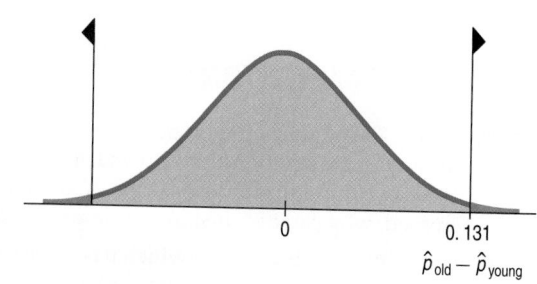

$$0 \qquad 0.131$$
$$\hat{p}_{old} - \hat{p}_{young}$$

Find the z-score for the observed difference in proportions, 0.131.

Find the P-value using Table Z or technology. Because this is a two-tailed test, we must *double* the probability we find in the upper tail.

$$z = \frac{(\hat{p}_{old} - \hat{p}_{young}) - 0}{SE_{pooled}(\hat{p}_{old} - \hat{p}_{young})} = \frac{0.131 - 0}{0.039375} = 3.33$$

$$P = 2P(z \geq 3.33) = 0.0008$$

Conclusion Link the P-value to your decision about the null hypothesis, and state your conclusion in context.

The P-value of 0.0008 says that if there really were no difference in (reported) snoring rates between the two age groups, then the difference observed in this study would happen only 8 times in 10,000. This is so small that I reject the null hypothesis of no difference and conclude that there is a difference in the rate of snoring between older adults and younger adults. It appears that older adults are more likely to snore.

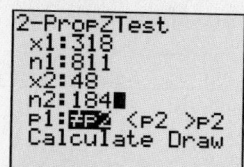

TI Tips

Testing the hypothesis

Yes, of course, there's a **STAT TESTS** routine to test a hypothesis about the difference of two proportions. Let's do the mechanics for the test about snoring. Of 811 people over 30 years old, 318 snored, while only 48 of the 184 people under 30 did.

- In the **STAT TESTS** menu select **6:2-PropZTest**.
- Enter the observed numbers of snorers and the sample sizes for both groups.
- Since this is a two-tailed test, indicate that you want to see if the proportions are unequal. When you choose this option, the calculator will automatically include both tails as it determines the P-value.
- **Calculate** the result. Don't worry; for this procedure the calculator will pool the proportions automatically.

Now it is up to you to interpret the result and state a conclusion. We see a z-score of 3.33 and the P-value is 0.0008. Such a small P-value indicates that the observed difference is unlikely to be sampling error. What does that mean about snoring and age? Here's a great opportunity to follow up with a confidence interval so you can Tell even more!

JUST CHECKING

3. A June 2004 public opinion poll asked 1000 randomly selected adults whether the United States should decrease the amount of immigration allowed; 49% of those responding said "yes." In June of 1995, a random sample of 1000 had found that 65% of adults thought immigration should be curtailed. To see if that percentage has decreased, why can't we just use a one-proportion z-test of $H_0: p = 0.65$ and see what the P-value for $\hat{p} = 0.49$ is?

4. For opinion polls like this, which has more variability: the percentage of respondents answering "yes" in either year or the difference in the percentages between the two years?

FOR EXAMPLE   Another 2-proportion z-test

Recap: One concern of the study on teens' online profiles was safety and privacy. In the random sample, girls were less likely than boys to say that they are easy to find online from their profiles. Only 19% (62 girls) of 325 teen girls with profiles say that they are easy to find, while 28% (75 boys) of the 268 boys with profiles say the same.

Question: Are these results evidence of a real difference between boys and girls? Perform a two-proportion z-test and discuss what you find.

$$H_O: p_{boys} - p_{girls} = 0$$
$$H_A: p_{boys} - p_{girls} \neq 0$$

✔ **Randomization Condition:** The sample of boys and the sample of girls were both chosen randomly.

✔ **10% Condition:** 268 boys and 325 girls are each less than 10% of all teenage boys and girls with online profiles.

✔ **Independent Groups Assumption:** Because the samples were selected at random, it's reasonable to believe the boys' perceptions are independent of the girls'.

✔ **Success/Failure Condition:** Among the girls, there were 62 "successes" and 263 failures, and among boys, 75 successes and 193 failures. These counts are at least 10 for each group.

Because all the assumptions and conditions are satisfied, it's okay to do a **two-proportion z-test:**

$$\hat{p}_{pooled} = \frac{75 + 62}{268 + 325} = 0.231$$

$$SE_{pooled}(\hat{p}_{boys} - \hat{p}_{girls}) = \sqrt{\frac{0.231 \times 0.769}{268} + \frac{0.231 \times 0.769}{325}} = 0.0348$$

$$z = \frac{(0.28 - 0.19) - 0}{0.0348} = 2.59$$

$$P(z > 2.59) = 0.0048$$

This is a two-tailed test, so the P-value = 2(0.0048) = 0.0096. Because this P-value is very small, I reject the null hypothesis. This study provides strong evidence that there really is a difference in the proportions of teen girls and boys who say they are easy to find online.

WHAT CAN GO WRONG?

▶ **Don't use two-sample proportion methods when the samples aren't independent.** These methods give wrong answers when this assumption of independence is violated. Good random sampling is usually the best insurance of independent groups. Make sure there is no relationship between the two groups. For example, you can't compare the proportion of respondents who own SUVs with the proportion of those same respondents who think the tax on gas should be eliminated. The responses are not independent because you've asked the same people. To use these methods to estimate or test the difference, you'd need to survey two different groups of people.

Alternatively, if you have a random sample, you can split your respondents according to their answers to one question and treat the two resulting groups as independent samples. So, you could test whether the proportion of SUV owners who favored eliminating the gas tax was the same as the corresponding proportion among non-SUV owners.

▶ **Don't apply inference methods where there was no randomization.** If the data do not come from representative random samples or from a properly randomized experiment, then the inference about the differences in proportions will be wrong.

▶ **Don't interpret a significant difference in proportions causally.** It turns out that people with higher incomes are more likely to snore. Does that mean money affects sleep patterns? Probably not. We have seen that older people are more likely to snore, and they are also likely to earn more. In a prospective or retrospective study, there is always the danger that other lurking variables not accounted for are the real reason for an observed difference. Be careful not to jump to conclusions about causality.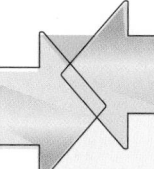

CONNECTIONS

In Chapter 3 we looked at contingency tables for two categorical variables. Differences in proportions are just 2×2 contingency tables. You'll often see data presented in this way. For example, the snoring data could be shown as

	18–29	30 and over	Total
Snore	48	318	**366**
Don't snore	136	493	**629**
Total	**184**	**811**	**995**

We tested whether the column percentages of snorers were the same for the two age groups.

This chapter gives the first examples we've seen of inference methods for a parameter other than a simple proportion. Although we have a different standard error, the step-by-step procedures are almost identical. In particular, once again we divide the statistic (the difference in proportions) by its standard error and get a z-score. You should feel right at home.

WHAT HAVE WE LEARNED?

In the last few chapters we began our exploration of statistical inference; we learned how to create confidence intervals and test hypotheses about a proportion. Now we've looked at inference for the difference in two proportions. In doing so, perhaps the most important thing we've learned is that the concepts and interpretations are essentially the same—only the mechanics have changed slightly.

We've learned that hypothesis tests and confidence intervals for the difference in two proportions are based on Normal models. Both require us to find the standard error of the difference in two proportions. We do that by adding the variances of the two sample proportions, assuming our two groups are independent. When we test a hypothesis that the two proportions are equal, we pool the sample data; for confidence intervals, we don't pool.

Terms

Variances of independent random variables add

506. The variance of a sum or difference of independent random variables is the sum of the variances of those variables.

Sampling distribution of the difference between two proportions

507. The sampling distribution of $\hat{p}_1 - \hat{p}_2$ is, under appropriate assumptions, modeled by a Normal model with mean $\mu = p_1 - p_2$ and standard deviation $SD(\hat{p}_1 - \hat{p}_2) = \sqrt{\dfrac{p_1 q_1}{n_1} + \dfrac{p_2 q_2}{n_2}}$.

Two-proportion z-interval

508. A two-proportion z-interval gives a confidence interval for the true difference in proportions, $p_1 - p_2$, in two independent groups.

The confidence interval is $(\hat{p}_1 - \hat{p}_2) \pm z^* \times SE(\hat{p}_1 - \hat{p}_2)$, where z^* is a critical value from the standard Normal model corresponding to the specified confidence level.

Pooling

512. When we have data from different sources that we believe are homogeneous, we can get a better estimate of the common proportion and its standard deviation. We can combine, or pool, the data into a single group for the purpose of estimating the common proportion. The resulting pooled standard error is based on more data and is thus more reliable (if the null hypothesis is true and the groups are truly homogeneous).

Two-proportion z-test

513. Test the null hypothesis $H_0: p_1 - p_2 = 0$ by referring the statistic

$$z = \frac{\hat{p}_1 - \hat{p}_2}{SE_{\text{pooled}}(\hat{p}_1 - \hat{p}_2)}$$

to a standard Normal model.

Skills

▸ Be able to state the null and alternative hypotheses for testing the difference between two population proportions.

▸ Know how to examine your data for violations of conditions that would make inference about the difference between two population proportions unwise or invalid.

▸ Understand that the formula for the standard error of the difference between two independent sample proportions is based on the principle that when finding the sum or difference of two independent random variables, their variances add.

▸ Know how to find a confidence interval for the difference between two proportions.

▸ Be able to perform a significance test of the natural null hypothesis that two population proportions are equal.

▸ Know how to write a sentence describing what is said about the difference between two population proportions by a confidence interval.

▸ Know how to write a sentence interpreting the results of a significance test of the null hypothesis that two population proportions are equal.

▸ Be able to interpret the meaning of a P-value in nontechnical language, making clear that the probability claim is made about computed values and not about the population parameter of interest.

▸ Know that we do not "accept" a null hypothesis if we fail to reject it.

INFERENCES FOR THE DIFFERENCE BETWEEN TWO PROPORTIONS ON THE COMPUTER

It is so common to test against the null hypothesis of no difference between the two true proportions that most statistics programs simply assume this null hypothesis. And most will automatically use the pooled standard deviation. If you wish to test a different null (say, that the true difference is 0.3), you may have to search for a way to do it.

Many statistics packages don't offer special commands for inference for differences between proportions. As with inference for single proportions, most statistics programs want the "success" and "failure" status for each case. Usually these are given as 1 or 0, but they might be category names like "yes" and "no." Often we just know the proportions of successes, $\hat{p}_1$ and $\hat{p}_2$, and the counts, n_1 and n_2. Computer packages don't usually deal with summary data like these easily. Calculators typically do a better job.

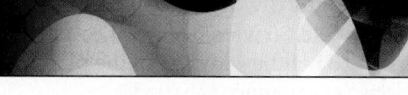

EXERCISES

1. **Online social networking.** The Parents & Teens 2006 Survey of 935 12- to 17-year-olds found that, among teens aged 15–17, girls were significantly more likely to have used social networking sites and online profiles. 70% of the girls surveyed had used an online social network, compared to 54% of the boys. What does it mean to say that the difference in proportions is "significant"?

2. **Science news.** In 2007 a Pew survey asked 1447 Internet users about their sources of news and information about science. Among those who had broadband access at home, 34% said they would turn to the Internet for most of their science news. The report on this survey claims that this is not significantly different from the percentage (33%) who said they ordinarily get their science news from television. What does it mean to say that the difference is not significant?

3. **Name recognition.** A political candidate runs a week-long series of TV ads designed to attract public attention to his campaign. Polls taken before and after the ad campaign show some increase in the proportion of voters who now recognize this candidate's name, with a P-value of 0.033. Is it reasonable to believe the ads may be effective?

4. **Origins.** In a 1993 Gallup poll, 47% of the respondents agreed with the statement *God created human beings pretty much in their present form at one time within the last 10,000 years or so.* When Gallup asked the same question in 2001, only 45% of those respondents agreed. Is it reasonable to conclude that there was a change in public opinion given that the P-value is 0.37? Explain.

5. **Revealing information.** 886 randomly sampled teens were asked which of several personal items of information they thought it okay to share with someone they had just met. 44% said it was okay to share their e-mail addresses, but only 29% said they would give out their cell phone numbers. A researcher claims that a two-proportion

z-test could tell whether there was a real difference among all teens. Explain why that test would not be appropriate for these data.

6. **Regulating access.** When a random sample of 935 parents were asked about rules in their homes, 77% said they had rules about the kinds of TV shows their children could watch. Among the 790 of those parents whose teenage children had Internet access, 85% had rules about the kinds of Internet sites their teens could visit. That looks like a difference, but can we tell? Explain why a two-sample z-test would not be appropriate here.

7. **Gender gap.** A presidential candidate fears he has a problem with women voters. His campaign staff plans to run a poll to assess the situation. They'll randomly sample 300 men and 300 women, asking if they have a favorable impression of the candidate. Obviously, the staff can't know this, but suppose the candidate has a positive image with 59% of males but with only 53% of females.
   a) What sampling design is his staff planning to use?
   b) What difference would you expect the poll to show?
   c) Of course, sampling error means the poll won't reflect the difference perfectly. What's the standard deviation for the difference in the proportions?
   d) Sketch a sampling model for the size difference in proportions of men and women with favorable impressions of this candidate that might appear in a poll like this.
   e) Could the campaign be misled by the poll, concluding that there really is no gender gap? Explain.

8. **Buy it again?** A consumer magazine plans to poll car owners to see if they are happy enough with their vehicles that they would purchase the same model again. They'll randomly select 450 owners of American-made cars and 450 owners of Japanese models. Obviously, the actual opinions of the entire population couldn't be

known, but suppose 76% of owners of American cars and 78% of owners of Japanese cars would purchase another.
a) What sampling design is the magazine planning to use?
b) What difference would you expect their poll to show?
c) Of course, sampling error means the poll won't reflect the difference perfectly. What's the standard deviation for the difference in the proportions?
d) Sketch a sampling model for the difference in proportions that might appear in a poll like this.
e) Could the magazine be misled by the poll, concluding that owners of American cars are much happier with their vehicles than owners of Japanese cars? Explain.

9. **Arthritis.** The Centers for Disease Control and Prevention reported a survey of randomly selected Americans age 65 and older, which found that 411 of 1012 men and 535 of 1062 women suffered from some form of arthritis.
a) Are the assumptions and conditions necessary for inference satisfied? Explain.
b) Create a 95% confidence interval for the difference in the proportions of senior men and women who have this disease.
c) Interpret your interval in this context.
d) Does this confidence interval suggest that arthritis is more likely to afflict women than men? Explain.

10. **Graduation.** In October 2000 the U.S. Department of Commerce reported the results of a large-scale survey on high school graduation. Researchers contacted more than 25,000 Americans aged 24 years to see if they had finished high school; 84.9% of the 12,460 males and 88.1% of the 12,678 females indicated that they had high school diplomas.
a) Are the assumptions and conditions necessary for inference satisfied? Explain.
b) Create a 95% confidence interval for the difference in graduation rates between males and females.
c) Interpret your confidence interval.
d) Does this provide strong evidence that girls are more likely than boys to complete high school? Explain.

11. **Pets.** Researchers at the National Cancer Institute released the results of a study that investigated the effect of weed-killing herbicides on house pets. They examined 827 dogs from homes where an herbicide was used on a regular basis, diagnosing malignant lymphoma in 473 of them. Of the 130 dogs from homes where no herbicides were used, only 19 were found to have lymphoma.
a) What's the standard error of the difference in the two proportions?
b) Construct a 95% confidence interval for this difference.
c) State an appropriate conclusion.

12. **Carpal tunnel.** The painful wrist condition called carpal tunnel syndrome can be treated with surgery or less invasive wrist splints. In September 2002, *Time* magazine reported on a study of 176 patients. Among the half that had surgery, 80% showed improvement after three months, but only 54% of those who used the wrist splints improved.
a) What's the standard error of the difference in the two proportions?
b) Construct a 95% confidence interval for this difference.
c) State an appropriate conclusion.

13. **Ear infections.** A new vaccine was recently tested to see if it could prevent the painful and recurrent ear infections that many infants suffer from. *The Lancet*, a medical journal, reported a study in which babies about a year old were randomly divided into two groups. One group received vaccinations; the other did not. During the following year, only 333 of 2455 vaccinated children had ear infections, compared to 499 of 2452 unvaccinated children in the control group.
a) Are the conditions for inference satisfied?
b) Find a 95% confidence interval for the difference in rates of ear infection.
c) Use your confidence interval to explain whether you think the vaccine is effective.

14. **Anorexia.** The *Journal of the American Medical Association* reported on an experiment intended to see if the drug Prozac® could be used as a treatment for the eating disorder anorexia nervosa. The subjects, women being treated for anorexia, were randomly divided into two groups. Of the 49 who received Prozac, 35 were deemed healthy a year later, compared to 32 of the 44 who got the placebo.
a) Are the conditions for inference satisfied?
b) Find a 95% confidence interval for the difference in outcomes.
c) Use your confidence interval to explain whether you think Prozac is effective.

15. **Another ear infection.** In Exercise 13 you used a confidence interval to examine the effectiveness of a vaccine against ear infections in babies. Suppose that instead you had conducted a hypothesis test. (Answer these questions *without* actually doing the test.)
a) What hypotheses would you test?
b) State a conclusion based on your confidence interval.
c) What alpha level did your test use?
d) If that conclusion is wrong, which type of error did you make?
e) What would be the consequences of such an error?

16. **Anorexia again.** In Exercise 14 you used a confidence interval to examine the effectiveness of Prozac in treating anorexia nervosa. Suppose that instead you had conducted a hypothesis test. (Answer these questions *without* actually doing the test.)
a) What hypotheses would you test?
b) State a conclusion based on your confidence interval.
c) What alpha level did your test use?
d) If that conclusion is wrong, which type of error did you make?
e) What would be the consequences of such an error?

17. **Teen smoking, part I.** A Vermont study published in December 2001 by the American Academy of Pediatrics examined parental influence on teenagers' decisions to smoke. A group of students who had never smoked were questioned about their parents' attitudes toward smoking. These students were questioned again two years later to see if they had started smoking. The researchers found that, among the 284 students who indicated that their parents disapproved of kids smoking, 54 had become established smokers. Among the 41 students who initially said their parents were lenient about smoking, 11 became

smokers. Do these data provide strong evidence that parental attitude influences teenagers' decisions about smoking?
a) What kind of design did the researchers use?
b) Write appropriate hypotheses.
c) Are the assumptions and conditions necessary for inference satisfied?
d) Test the hypothesis and state your conclusion.
e) Explain in this context what your P-value means.
f) If that conclusion is actually wrong, which type of error did you commit?

18. **Depression.** A study published in the *Archives of General Psychiatry* in March 2001 examined the impact of depression on a patient's ability to survive cardiac disease. Researchers identified 450 people with cardiac disease, evaluated them for depression, and followed the group for 4 years. Of the 361 patients with no depression, 67 died. Of the 89 patients with minor or major depression, 26 died. Among people who suffer from cardiac disease, are depressed patients more likely to die than non-depressed ones?
a) What kind of design was used to collect these data?
b) Write appropriate hypotheses.
c) Are the assumptions and conditions necessary for inference satisfied?
d) Test the hypothesis and state your conclusion.
e) Explain in this context what your P-value means.
f) If your conclusion is actually incorrect, which type of error did you commit?

19. **Teen smoking, part II.** Consider again the Vermont study discussed in Exercise 17.
a) Create a 95% confidence interval for the difference in the proportion of children who may smoke and have approving parents and those who may smoke and have disapproving parents.
b) Interpret your interval in this context.
c) Carefully explain what "95% confidence" means.

20. **Depression revisited.** Consider again the study of the association between depression and cardiac disease survivability in Exercise 18.
a) Create a 95% confidence interval for the difference in survival rates.
b) Interpret your interval in this context.
c) Carefully explain what "95% confidence" means.

21. **Pregnancy.** In 1998, a San Diego reproductive clinic reported 42 live births to 157 women under the age of 38, but only 7 live births for 89 clients aged 38 and older. Is this strong evidence of a difference in the effectiveness of the clinic's methods for older women?
a) Was this an experiment? Explain.
b) Test an appropriate hypothesis and state your conclusion in context.
c) If you concluded there was a difference, estimate that difference with a confidence interval and interpret your interval in context.

22. **Birthweight.** In 2003 the *Journal of the American Medical Association* reported a study examining the possible impact of air pollution caused by the 9/11 attack on New York's World Trade Center on the weight of babies.

Researchers found that 8% of 182 babies born to mothers who were exposed to heavy doses of soot and ash on September 11 were classified as having low birth weight. Only 4% of 2300 babies born in another New York City hospital whose mothers had not been near the site of the disaster were similarly classified. Does this indicate a possibility that air pollution might be linked to a significantly higher proportion of low-weight babies?
a) Was this an experiment? Explain.
b) Test an appropriate hypothesis and state your conclusion in context.
c) If you concluded there is a difference, estimate that difference with a confidence interval and interpret that interval in context.

23. **Politics and sex.** One month before the election, a poll of 630 randomly selected voters showed 54% planning to vote for a certain candidate. A week later it became known that he had had an extramarital affair, and a new poll showed only 51% of 1010 voters supporting him. Do these results indicate a decrease in voter support for his candidacy?
a) Test an appropriate hypothesis and state your conclusion.
b) If your conclusion turns out to be wrong, did you make a Type I or Type II error?
c) If you concluded there was a difference, estimate that difference with a confidence interval and interpret your interval in context.

24. **Shopping.** A survey of 430 randomly chosen adults found that 21% of the 222 men and 18% of the 208 women had purchased books online.
a) Is there evidence that men are more likely than women to make online purchases of books? Test an appropriate hypothesis and state your conclusion in context.
b) If your conclusion in fact proves to be wrong, did you make a Type I or Type II error?
c) Estimate this difference with a confidence interval.
d) Interpret your interval in context.

25. **Twins.** In 2001, one county reported that, among 3132 white women who had babies, 94 were multiple births. There were also 20 multiple births to 606 black women. Does this indicate any racial difference in the likelihood of multiple births?
a) Test an appropriate hypothesis and state your conclusion in context.
b) If your conclusion is incorrect, which type of error did you commit?

26. **Mammograms.** A 9-year study in Sweden compared 21,088 women who had mammograms with 21,195 who did not. Of the women who underwent screening, 63 died of breast cancer, compared to 66 deaths among the control group. (*The New York Times*, Dec 9, 2001)
a) Do these results support the effectiveness of regular mammograms in preventing deaths from breast cancer?
b) If your conclusion is incorrect, what kind of error have you committed?

27. **Pain.** Researchers comparing the effectiveness of two pain medications randomly selected a group of patients

who had been complaining of a certain kind of joint pain. They randomly divided these people into two groups, then administered the pain killers. Of the 112 people in the group who received medication A, 84 said this pain reliever was effective. Of the 108 people in the other group, 66 reported that pain reliever B was effective.

a) Write a 95% confidence interval for the percent of people who may get relief from this kind of joint pain by using medication A. Interpret your interval.

b) Write a 95% confidence interval for the percent of people who may get relief by using medication B. Interpret your interval.

c) Do the intervals for A and B overlap? What do you think this means about the comparative effectiveness of these medications?

d) Find a 95% confidence interval for the difference in the proportions of people who may find these medications effective. Interpret your interval.

e) Does this interval contain zero? What does that mean?

f) Why do the results in parts c and e seem contradictory? If we want to compare the effectiveness of these two pain relievers, which is the correct approach? Why?

28. **Gender gap.**   Candidates for political office realize that different levels of support among men and women may be a crucial factor in determining the outcome of an election. One candidate finds that 52% of 473 men polled say they will vote for him, but only 45% of the 522 women in the poll express support.

a) Write a 95% confidence interval for the percent of male voters who may vote for this candidate. Interpret your interval.

b) Write and interpret a 95% confidence interval for the percent of female voters who may vote for him.

c) Do the intervals for males and females overlap? What do you think this means about the gender gap?

d) Find a 95% confidence interval for the difference in the proportions of males and females who will vote for this candidate. Interpret your interval.

e) Does this interval contain zero? What does that mean?

f) Why do the results in parts c and e seem contradictory? If we want to see if there is a gender gap among voters with respect to this candidate, which is the correct approach? Why?

29. **Sensitive men.**   In August 2004, *Time* magazine, reporting on a survey of men's attitudes, noted that "Young men are more comfortable than older men talking about their problems." The survey reported that 80 of 129 surveyed 18- to 24-year-old men and 98 of 184 25- to 34-year-old men said they were comfortable. What do you think? Is *Time's* interpretation justified by these numbers?

30. **Retention rates.**   In 2004 the testing company ACT, Inc., reported on the percentage of first-year students at 4-year colleges who return for a second year. Their sample of 1139 students in private colleges showed a 74.9% retention rate, while the rate was 71.9% for the sample of 505 students at public colleges. Does this provide evidence that there's a difference in retention rates of first-year students at public and private colleges?

31. **Online activity checks.**   Are more parents checking up on their teen's online activities? A Pew survey in 2004 found that 33% of 868 randomly sampled teens said that their parents checked to see what Web sites they visited. In 2006 the same question posed to 811 teens found 41% reporting such checks. Do these results provide evidence that more parents are checking?

32. **Computer gaming.**   Who plays online or electronic games? A survey in 2006 found that 69% of 223 boys aged 12–14 said they "played computer or console games like Xbox or PlayStation . . . or games online." Of 248 boys aged 15–17, only 62% played these games. Is this evidence of a real age-based difference?

JUST CHECKING
Answers

1. We're 90% confident that if members are contacted by e-mail, the donation rate will be between 4 and 18 percentage points higher than if they received regular mail.

2. Since a difference of 0 is not in the confidence interval, we'd reject the null hypothesis. There is evidence that more members will donate if contacted by e-mail.

3. The proportion from the sample in 1995 has variability, too. If we do a one-proportion z-test, we won't take that variability into account and our P-value will be incorrect.

4. The difference in the proportions between the two years has more variability than either individual proportion. The variance of the difference is the sum of the two variances.

REVIEW OF PART V

From the Data at Hand to the World at Large

Quick Review

What do samples really tell us about the populations from which they are drawn? Are the results of an experiment meaningful, or are they just sampling error? Statistical inference based on our understanding of sampling models can help answer these questions. Here's a brief summary of the key concepts and skills:

▶ Sampling models describe the variability of sample statistics using a remarkable result called the Central Limit Theorem.

- When the number of trials is sufficiently large, proportions found in different samples vary according to an approximately Normal model.

- When samples are sufficiently large, the means of different samples vary, with an approximately Normal model.

- The variability of sample statistics decreases as sample size increases.

- Statistical inference procedures are based on the Central Limit Theorem.

- No inference procedure is valid unless the underlying assumptions are true. Always check the conditions before proceeding.

▶ A confidence interval uses a sample statistic (such as a proportion) to estimate a range of plausible values for the parameter of a population model.

- All confidence intervals involve an estimate of the parameter, a margin of error, and a level of confidence.

- For confidence intervals based on a given sample, the greater the margin of error, the higher the confidence.

- At a given level of confidence, the larger the sample, the smaller the margin of error.

▶ A hypothesis test proposes a model for the population, then examines the observed statistics to see if that model is plausible.

- A null hypothesis suggests a parameter value for the population model. Usually, we assume there is nothing interesting, unusual, or different about the sample results.

- The alternative hypothesis states what we will believe if the sample results turn out to be inconsistent with our null model.

- We compare the difference between the statistic and the hypothesized value with the standard deviation of the statistic. It's the sampling distribution of this ratio that gives us a P-value.

- The P-value of the test is the conditional probability that the null model could produce results at least as extreme as those observed in the sample or the experiment just as a result of sampling error.

- A low P-value indicates evidence against the null model. If it is sufficiently low, we reject the null model.

- A high P-value indicates that the sample results are not inconsistent with the null model, so we cannot reject it. However, this does not prove the null model is true.

- Sometimes we will mistakenly reject the null hypothesis even though it's actually true—that's called a Type I error. If we fail to reject a false null hypothesis, we commit a Type II error.

- The power of a test measures its ability to detect a false null hypothesis.

- You can lower the risk of a Type I error by requiring a higher standard of proof (lower P-value) before rejecting the null hypothesis. But this will raise the risk of a Type II error and decrease the power of the test.

- The only way to increase the power of a test while decreasing the chance of committing either error is to design a study based on a larger sample.

And now for some opportunities to review these concepts and skills . . .

REVIEW EXERCISES

1. Herbal cancer. A report in the *New England Journal of Medicine* (June 6, 2000) notes growing evidence that the herb *Aristolochia fangchi* can cause urinary tract cancer in those who take it. Suppose you are asked to design an experiment to study this claim. Imagine that you have data on urinary tract cancers in subjects who have used this herb and similar subjects who have not used it and that you can measure incidences of cancer and precancerous lesions in these subjects. State the null and alternative hypotheses you would use in your study.

2. Colorblind. Medical literature says that about 8% of males are colorblind. A university's introductory psychology course is taught in a large lecture hall. Among the students, there are 325 males. Each semester when the

professor discusses visual perception, he shows the class a test for colorblindness. The percentage of males who are colorblind varies from semester to semester.
a) Is the sampling distribution model for the sample proportion likely to be Normal? Explain.
b) What are the mean and standard deviation of this sampling distribution model?
c) Sketch the sampling model, using the 68–95–99.7 Rule.
d) Write a few sentences explaining what the model says about this professor's class.

3. **Birth days.** During a 2-month period in 2002, 72 babies were born at the Tompkins Community Hospital in upstate New York. The table shows how many babies were born on each day of the week.
a) If births are uniformly distributed across all days of the week, how many would you expect on each day?
b) Only 7 births occurred on a Monday. Does this indicate that women might be less likely to give birth on a Monday? Explain.
c) Are the 17 births on Tuesdays unusually high? Explain.
d) Can you think of any reasons why births may not occur completely at random?

Day	Births
Mon.	7
Tues.	17
Wed.	8
Thurs.	12
Fri.	9
Sat.	10
Sun.	9

4. **Polling 2004.** In the 2004 U.S. presidential election, the official results showed that George W. Bush received 50.7% of the vote and John Kerry received 48.3%. Ralph Nader, running as a third-party candidate, picked up only 0.4%. After the election, there was much discussion about exit polls, which had initially indicated a different result. Suppose you had taken a random sample of 1000 voters in an exit poll and asked them for whom they had voted.
a) Would you always get 507 votes for Bush and 483 for Kerry?
b) In 95% of such polls, your sample proportion of voters for Bush should be between what two values?
c) In 95% of such polls, your sample proportion of voters for Nader should be between what two numbers?
d) Would you expect the sample proportion of Nader votes to vary more, less, or about the same as the sample proportion of Bush votes? Why?

5. **Leaky gas tanks.** Nationwide, it is estimated that 40% of service stations have gas tanks that leak to some extent. A new program in California is designed to lessen the prevalence of these leaks. We want to assess the effectiveness of the program by seeing if the percentage of service stations whose tanks leak has decreased. To do this, we randomly sample 27 service stations in California and determine whether there is any evidence of leakage. In our sample, only 7 of the stations exhibit any leakage. Is there evidence that the new program is effective?
a) What are the null and alternative hypotheses?
b) Check the assumptions necessary for inference.
c) Test the null hypothesis.
d) What do you conclude (in plain English)?
e) If the program actually works, have you made an error? What kind?

f) What two things could you do to decrease the probability of making this kind of error?
g) What are the advantages and disadvantages of taking those two courses of action?

6. **Surgery and germs.** Joseph Lister (for whom Listerine is named!) was a British physician who was interested in the role of bacteria in human infections. He suspected that germs were involved in transmitting infection, so he tried using carbolic acid as an operating room disinfectant. In 75 amputations, he used carbolic acid 40 times. Of the 40 amputations using carbolic acid, 34 of the patients lived. Of the 35 amputations without carbolic acid, 19 patients lived. The question of interest is whether carbolic acid is effective in increasing the chances of surviving an amputation.
a) What kind of a study is this?
b) What do you conclude? Support your conclusion by testing an appropriate hypothesis.
c) What reservations do you have about the design of the study?

7. **Scrabble.** Using a computer to play many simulated games of Scrabble, researcher Charles Robinove found that the letter "A" occurred in 54% of the hands. This study had a margin of error of ±10%. (*Chance*, 15, no. 1 [2002])
a) Explain what the margin of error means in this context.
b) Why might the margin of error be so large?
c) Probability theory predicts that the letter "A" should appear in 63% of the hands. Does this make you concerned that the simulation might be faulty? Explain.

8. **Dice.** When one die is rolled, the number of spots showing has a mean of 3.5 and a standard deviation of 1.7. Suppose you roll 10 dice. What's the approximate probability that your total is between 30 and 40 (that is, the average for the 10 dice is between 3 and 4)? Specify the model you use and the assumptions and conditions that justify your approach.

9. **News sources.** In May of 2000, the Pew Research Foundation sampled 1593 respondents and asked how they obtain news. In Pew's report, 33% of respondents say that they now obtain news from the Internet at least once a week.
a) Pew reports a margin of error of ±3% for this result. Explain what the margin of error means.
b) Pew also asked about investment information, and 21% of respondents reported that the Internet is their main source of this information. When limited to the 780 respondents who identified themselves as investors, the percent who rely on the Internet rose to 28%. How would you expect the margin of error for this statistic to change in comparison with the margin of error for the percentage of all respondents?
c) When restricted to the 239 active traders in the sample, Pew reports that 45% rely on the Internet for investment information. Find a confidence interval for this statistic.
d) How does the margin of error for your confidence interval compare with the values in parts a and b? Explain why.

10. Death penalty 2006. In May of 2006, the Gallup Organization asked a random sample of 537 American adults this question:

> If you could choose between the following two approaches, which do you think is the better penalty for murder, the death penalty or life imprisonment, with absolutely no possibility of parole?

Of those polled, 47% chose the death penalty, the lowest percentage in the 21 years that Gallup has asked this question.
a) Create a 95% confidence interval for the percentage of all American adults who favor the death penalty.
b) Based on your confidence interval, is it clear that the death penalty no longer has majority support? Explain.
c) If pollsters wanted to follow up on this poll with another survey that could determine the level of support for the death penalty to within 2% with 98% confidence, how many people should they poll?

11. Bimodal. We are sampling randomly from a distribution known to be bimodal.
a) As our sample size increases, what's the expected shape of the sample's distribution?
b) What's the expected value of our sample's mean? Does the size of the sample matter?
c) How is the variability of sample means related to the standard deviation of the population? Does the size of the sample matter?
d) How is the shape of the sampling distribution model affected by the sample size?

12. Vitamin D. In July 2002 the *American Journal of Clinical Nutrition* reported that 42% of 1546 African-American women studied had vitamin D deficiency. The data came from a national nutrition study conducted by the Centers for Disease Control and Prevention in Atlanta.
a) Do these data meet the assumptions necessary for inference? What would you like to know that you don't?
b) Create a 95% confidence interval.
c) Interpret the interval in this context.
d) Explain in this context what "95% confidence" means.

13. Archery. A champion archer can generally hit the bull's-eye 80% of the time. Suppose she shoots 200 arrows during competition. Let $\hat{p}$ represent the percentage of bull's-eyes she gets (the sample proportion).
a) What are the mean and standard deviation of the sampling distribution model for $\hat{p}$?
b) Is a Normal model appropriate here? Explain.
c) Sketch the sampling model, using the 68–95–99.7 Rule.
d) What's the probability that she gets at least 85% bull's-eyes?

14. Free throws 2007. During the 2006–2007 NBA season, Kyle Korver led the league by making 191 of 209 free throws, for a success rate of 91.39%. But Matt Carroll was close behind, with 188 of 208 (90.39%).
a) Find a 95% confidence interval for the difference in their free throw percentages.
b) Based on your confidence interval, is it certain that Korver is better than Carroll at making free throws?

15. Twins. There is some indication in medical literature that doctors may have become more aggressive in inducing labor or doing preterm cesarean sections when a woman is carrying twins. Records at a large hospital show that, of the 43 sets of twins born in 1990, 20 were delivered before the 37th week of pregnancy. In 2000, 26 of 48 sets of twins were born preterm. Does this indicate an increase in the incidence of early births of twins? Test an appropriate hypothesis and state your conclusion.

16. Eclampsia. It's estimated that 50,000 pregnant women worldwide die each year of eclampsia, a condition involving elevated blood pressure and seizures. A research team from 175 hospitals in 33 countries investigated the effectiveness of magnesium sulfate in preventing the occurrence of eclampsia in at-risk patients. Results are summarized below. (*Lancet,* June 1, 2002)

	Total Subjects	Reported side effects	Developed eclampsia	Deaths
Magnesium sulfate (Treatment)	**4999**	1201	40	11
Placebo (Treatment)	**4993**	228	96	20

a) Write a 95% confidence interval for the increase in the proportion of women who may develop side effects from this treatment. Interpret your interval.
b) Is there evidence that the treatment may be effective in preventing the development of eclampsia? Test an appropriate hypothesis and state your conclusion.

17. Eclampsia. Refer again to the research summarized in Exercise 16. Is there any evidence that when eclampsia does occur, the magnesium sulfide treatment may help prevent the woman's death?
a) Write an appropriate hypothesis.
b) Check the assumptions and conditions.
c) Find the P-value of the test.
d) What do you conclude about the magnesium sulfide treatment?
e) If your conclusion is wrong, which type of error have you made?
f) Name two things you could do to increase the power of this test.
g) What are the advantages and disadvantages of those two options?

18. Eggs. The ISA Babcock Company supplies poultry farmers with hens, advertising that a mature B300 Layer produces eggs with a mean weight of 60.7 grams. Suppose that egg weights follow a Normal model with standard deviation 3.1 grams.
a) What fraction of the eggs produced by these hens weigh more than 62 grams?
b) What's the probability that a dozen randomly selected eggs average more than 62 grams?
c) Using the 68–95–99.7 Rule, sketch a model of the total weights of a dozen eggs.

19. Polling disclaimer. A newspaper article that reported the results of an election poll included the following explanation:

> *The Associated Press poll on the 2000 presidential campaign is based on telephone interviews with 798 randomly selected registered voters from all states except Alaska and Hawaii. The interviews were conducted June 21–25 by ICR of Media, Pa.*
>
> *The results were weighted to represent the population by demographic factors such as age, sex, region, and education.*
>
> *No more than 1 time in 20 should chance variations in the sample cause the results to vary by more than 4 percentage points from the answers that would be obtained if all Americans were polled.*
>
> *The margin of sampling error is larger for responses of subgroups, such as income categories or those in political parties. There are other sources of potential error in polls, including the wording and order of questions.*

a) Did they describe the 5 W's well?
b) What kind of sampling design could take into account the several demographic factors listed?
c) What was the margin of error of this poll?
d) What was the confidence level?
e) Why is the margin of error larger for subgroups?
f) Which kinds of potential bias did they caution readers about?

20. Enough eggs? One of the important issues for poultry farmers is the production rate—the percentage of days on which a given hen actually lays an egg. Ideally, that would be 100% (an egg every day), but realistically, hens tend to lay eggs on about 3 of every 4 days. ISA Babcock wants to advertise the production rate for the B300 Layer (see Exercise 18) as a 95% confidence interval with a margin of error of ±2%. How many hens must they collect data on?

21. Teen deaths. Traffic accidents are the leading cause of death among people aged 15 to 20. In May 2002, the National Highway Traffic Safety Administration reported that even though only 6.8% of licensed drivers are between 15 and 20 years old, they were involved in 14.3% of all fatal crashes. Insurance companies have long known that teenage boys were high risks, but what about teenage girls? One insurance company found that the driver was a teenage girl in 44 of the 388 fatal accidents they investigated. Is this strong evidence that the accident rate is lower for girls than for teens in general?
a) Test an appropriate hypothesis and state your conclusion.
b) Explain what your P-value means in this context.

22. Perfect pitch. A recent study of perfect pitch tested students in American music conservatories. It found that 7% of 1700 non-Asian and 32% of 1000 Asian students have perfect pitch. A test of the difference in proportions resulted in a P-value of < 0.0001.
a) What are the researchers' null and alternative hypotheses?
b) State your conclusion.

c) Explain in this context what the P-value means.
d) The researchers claimed that the data prove that genetic differences between the two populations cause a difference in the frequency of occurrence of perfect pitch. Do you agree? Why or why not?

23. Largemouth bass. Organizers of a fishing tournament believe that the lake holds a sizable population of largemouth bass. They assume that the weights of these fish have a model that is skewed to the right with a mean of 3.5 pounds and a standard deviation of 2.2 pounds.
a) Explain why a skewed model makes sense here.
b) Explain why you cannot determine the probability that a largemouth bass randomly selected ("caught") from the lake weighs over 3 pounds.
c) Each fisherman in the contest catches 5 fish each day. Can you determine the probability that someone's catch averages over 3 pounds? Explain.
d) The 12 fishermen competing each caught the limit of 5 fish. What's the probability that the total catch of 60 fish averaged more than 3 pounds?

24. Cheating. A Rutgers University study released in 2002 found that many high school students cheat on tests. The researchers surveyed a random sample of 4500 high school students nationwide; 74% of them said they had cheated at least once.
a) Create a 90% confidence interval for the level of cheating among high school students. Don't forget to check the appropriate conditions.
b) Interpret your interval.
c) Explain what "90% confidence" means.
d) Would a 95% confidence interval be wider or narrower? Explain without actually calculating the interval.

25. Language. Neurological research has shown that in about 80% of people language abilities reside in the brain's left side. Another 10% display right-brain language centers, and the remaining 10% have two-sided language control. (The latter two groups are mainly left-handers.) (*Science News*, 161, no. 24 [2002])
a) We select 60 people at random. Is it reasonable to use a Normal model to describe the possible distribution of the proportion of the group that has left-brain language control? Explain.
b) What's the probability that our group has at least 75% left-brainers?
c) If the group had consisted of 100 people, would that probability be higher, lower, or about the same? Explain why, without actually calculating the probability.
d) How large a group would almost certainly guarantee at least 75% left-brainers? Explain.

26. Cigarettes 2006. In 1999 the Centers for Disease Control and Prevention estimated that about 34.8% of high school students smoked cigarettes. They established a national health goal of reducing that figure to 16% by the year 2010. To that end, they hoped to achieve a reduction to 20% by 2006. In 2006 they released a research study in which 23% of a random sample of 1815 high school students said they were current smokers. Is this evidence that progress toward the goal is off track?
a) Write appropriate hypotheses.
b) Verify that the appropriate assumptions are satisfied.

c) Find the P-value of this test.

d) Explain what the P-value means in this context.

e) State an appropriate conclusion.

f) Of course, your conclusion may be incorrect. If so, which kind of error did you commit?

27. Crohn's disease. In 2002 the medical journal *The Lancet* reported that 335 of 573 patients suffering from Crohn's disease responded positively to injections of the arthritis-fighting drug infliximab.

a) Create a 95% confidence interval for the effectiveness of this drug.

b) Interpret your interval in context.

c) Explain carefully what "95% confidence" means in this context.

28. Teen smoking 2006. The Centers for Disease Control and Prevention say that about 23% of teenagers smoke tobacco (down from a high of 38% in 1997). A college has 522 students in its freshman class. Is it likely that more than 30% of them are smokers? Explain.

29. Alcohol abuse. Growing concern about binge drinking among college students has prompted one large state university to conduct a survey to assess the size of the problem on its campus. The university plans to randomly select students and ask how many have been drunk during the past week. If the school hopes to estimate the true proportion among all its students with 90% confidence and a margin of error of ±4%, how many students must be surveyed?

30. Errors. An auto parts company advertises that its special oil additive will make the engine "run smoother, cleaner, longer, with fewer repairs." An independent laboratory decides to test part of this claim. It arranges to use a taxicab company's fleet of cars. The cars are randomly divided into two groups. The company's mechanics will use the additive in one group of cars but not in the other. At the end of a year the laboratory will compare the percentage of cars in each group that required engine repairs.

a) What kind of a study is this?

b) Will they do a one-tailed or a two-tailed test?

c) Explain in this context what a Type I error would be.

d) Explain in this context what a Type II error would be.

e) Which type of error would the additive manufacturer consider more serious?

f) If the cabs with the additive do indeed run significantly better, can the company conclude it is an effect of the additive? Can they generalize this result and recommend the additive for all cars? Explain.

31. Preemies. Among 242 Cleveland-area children born prematurely at low birth weights between 1977 and 1979, only 74% graduated from high school. Among a comparison group of 233 children of normal birth weight, 83% were high school graduates. ("Outcomes in Young Adulthood for Very-Low-Birth-Weight Infants," *New England Journal of Medicine*, 346, no. 3 [2002])

a) Create a 95% confidence interval for the difference in graduation rates between children of normal and children of very low birth weights. Be sure to check the appropriate assumptions and conditions.

b) Does this provide evidence that premature birth may be a risk factor for not finishing high school? Use your confidence interval to test an appropriate hypothesis.

c) Suppose your conclusion is incorrect. Which type of error did you make?

32. Safety. Observers in Texas watched children at play in eight communities. Of the 814 children seen biking, roller skating, or skateboarding, only 14% wore a helmet.

a) Create and interpret a 95% confidence interval.

b) What concerns do you have about this study that might make your confidence interval unreliable?

c) Suppose we want to do this study again, picking various communities and locations at random, and hope to end up with a 98% confidence interval having a margin of error of ±4%. How many children must we observe?

33. Fried PCs. A computer company recently experienced a disastrous fire that ruined some of its inventory. Unfortunately, during the panic of the fire, some of the damaged computers were sent to another warehouse, where they were mixed with undamaged computers. The engineer responsible for quality control would like to check out each computer in order to decide whether it's undamaged or damaged. Each computer undergoes a series of 100 tests. The number of tests it fails will be used to make the decision. If it fails more than a certain number, it will be classified as damaged and then scrapped. From past history, the distribution of the number of tests failed is known for both undamaged and damaged computers. The probabilities associated with each outcome are listed in the table below:

Number of tests failed	0	1	2	3	4	5	>5
Undamaged (%)	80	13	2	4	1	0	0
Damaged (%)	0	10	70	5	4	1	10

The table indicates, for example, that 80% of the undamaged computers have no failures, while 70% of the damaged computers have 2 failures.

a) To the engineers, this is a hypothesis-testing situation. State the null and alternative hypotheses.

b) Someone suggests classifying a computer as damaged if it fails any of the tests. Discuss the advantages and disadvantages of this test plan.

c) What number of tests would a computer have to fail in order to be classified as damaged if the engineers want to have the probability of a Type I error equal to 5%?

d) What's the power of the test plan in part c?

e) A colleague points out that by increasing α just 2%, the power can be increased substantially. Explain.

34. Power. We are replicating an experiment. How will each of the following changes affect the power of our test? Indicate whether it will increase, decrease, or remain the same, assuming that all other aspects of the situation remain unchanged.

a) We increase the number of subjects from 40 to 100.

b) We require a higher standard of proof, changing from $\alpha = 0.05$ to $\alpha = 0.01$.

35. Approval 2007. Of all the post–World War II presidents, Richard Nixon had the highest *disapproval* rating near the end of his presidency. His disapproval rating peaked at 66% in July 1974, just before he resigned. In May 2007, George W. Bush's disapproval rating was 63%, according to a Gallup poll of 1000 voters. Pundits started discussing whether his rating was still discernibly better than Nixon's. What do you think?

36. Grade inflation. In 1996, 20% of the students at a major university had an overall grade point average of 3.5 or higher (on a scale of 4.0). In 2000, a random sample of 1100 student records found that 25% had a GPA of 3.5 or higher. Is this evidence of grade inflation?

37. Name recognition. An advertising agency won't sign an athlete to do product endorsements unless it is sure the person is known to more than 25% of its target audience. The agency always conducts a poll of 500 people to investigate the athlete's name recognition before offering a contract. Then it tests $H_0: p = 0.25$ against $H_A: p > 0.25$ at a 5% level of significance.
a) Why does the company use upper tail tests in this situation?
b) Explain what Type I and Type II errors would represent in this context, and describe the risk that each error poses to the company.
c) The company is thinking of changing its test to use a 10% level of significance. How would this change the company's exposure to each type of risk?

38. Name recognition, part II. The advertising company described in Exercise 37 is thinking about signing a WNBA star to an endorsement deal. In its poll, 27% of the respondents could identify her.
a) Fans who never took Statistics can't understand why the company did not offer this WNBA player an endorsement contract even though the 27% recognition rate in the poll is above the 25% threshold. Explain it to them.

b) Suppose that further polling reveals that this WNBA star really is known to about 30% of the target audience. Did the company initially commit a Type I or Type II error in not signing her?
c) Would the power of the company's test have been higher or lower if the player were more famous? Explain.

39. NIMBY. In March 2007, the Gallup Poll split a sample of 1003 randomly selected U.S. adults into two groups at random. Half ($n = 502$) of the respondents were asked,

"Overall, do you strongly favor, somewhat favor, somewhat oppose, or strongly oppose the use of nuclear energy as one of the ways to provide electricity for the U.S.?"

They found that 53% were either "somewhat" or "strongly" in favor. The other half ($n = 501$) were asked,

"Overall, would you strongly favor, somewhat favor, somewhat oppose, or strongly oppose the construction of a nuclear energy plant in your area as one of the ways to provide electricity for the U.S.?"

Only 40% were somewhat or strongly in favor. This difference is an example of the *NIMBY* (Not In My Back-Yard) phenomenon and is a serious concern to policy makers and planners. How large is the difference between the proportion of American adults who think nuclear energy is a good idea and the proportion who would be willing to have a nuclear plant in their area? Construct and interpret an appropriate confidence interval.

40. Dropouts. One study comparing various treatments for the eating disorder anorexia nervosa initially enlisted 198 subjects, but found overall that 105 failed to complete their assigned treatment programs. Construct and interpret an appropriate confidence interval. Discuss any reservations you have about this inference.

PART VI

Learning About the World

Inferences About Means

WHO	Vehicles on Triphammer Road
WHAT	Speed
UNITS	Miles per hour
WHEN	April 11, 2000, 1 p.m.
WHERE	A small town in the northeastern United States
WHY	Concern over impact on residential neighborhood

Motor vehicle crashes are the leading cause of death for people between 4 and 33 years old. In the year 2006, motor vehicle accidents claimed the lives of 43,300 people in the United States. This means that, on average, motor vehicle crashes resulted in 119 deaths each day, or 1 death every 12 minutes. Speeding is a contributing factor in 31% of all fatal accidents, according to the National Highway Traffic Safety Administration.

Triphammer Road is a busy street that passes through a residential neighborhood. Residents there are concerned that vehicles traveling on Triphammer often exceed the posted speed limit of 30 miles per hour. The local police sometimes place a radar speed detector by the side of the road; as a vehicle approaches, this detector displays the vehicle's speed to its driver.

The local residents are not convinced that such a passive method is helping the problem. They wish to persuade the village to add extra police patrols to encourage drivers to observe the speed limit. To help their case, a resident stood where he could see the detector and recorded the speed of vehicles passing it during a 15-minute period one day. When clusters of vehicles went by, he noted only the speed of the front vehicle. Here are his data and the histogram.

FIGURE 23.1

The speeds of cars on Triphammer Road seem to be unimodal and symmetric, at least at this scale.

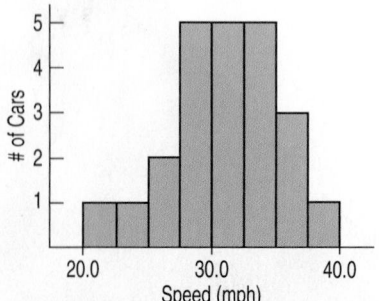

Speed		
29	29	24
34	34	34
34	32	36
28	31	31
30	27	34
29	37	36
38	29	21
31	26	

We're interested both in estimating the true mean speed and in testing whether it exceeds the posted speed limit. Although the sample of vehicles is a convenience sample, not a truly random sample, there's no compelling reason to believe that vehicles at one time of day are driving faster or slower than vehicles at another time of day,[1] so we can take the sample to be representative.

These data differ from data on proportions in one important way. Proportions are usually reported as summaries. After all, individual responses are just "success" and "failure" or "1" and "0." Quantitative data, though, usually report a value for each individual. When you have a value for each individual, you should remember the three rules of data analysis and plot the data, as we have done here.

We have quantitative data, so we summarize with means and standard deviations. Because we want to make inferences, we'll think about sampling distributions, too, and we already know most of the facts we need.

Getting Started

You've learned how to create confidence intervals and test hypotheses about proportions. We always center confidence intervals at our best guess of the unknown parameter. Then we add and subtract a margin of error. For proportions, that means $\hat{p} \pm ME$.

We found the margin of error as the product of the standard error, $SE(\hat{p})$, and a critical value, z^*, from the Normal table. So we had $\hat{p} \pm z^*SE(\hat{p})$.

We knew we could use z because the Central Limit Theorem told us (back in Chapter 18) that the sampling distribution model for proportions is Normal.

Now we want to do exactly the same thing for means, and fortunately, the Central Limit Theorem (still in Chapter 18) told us that the same Normal model works as the sampling distribution for means.

THE CENTRAL LIMIT THEOREM

When a random sample is drawn from any population with mean μ and standard deviation σ, its sample mean, $\bar{y}$, has a sampling distribution with the same *mean μ* but whose *standard deviation is* $\dfrac{\sigma}{\sqrt{n}}$ (and we write

$$\sigma(\bar{y}) = SD(\bar{y}) = \frac{\sigma}{\sqrt{n}}).$$

No matter what population the random sample comes from, the *shape* of the sampling distribution is approximately Normal as long as the sample size is large enough. The larger the sample used, the more closely the Normal approximates the sampling distribution for the mean.

FOR EXAMPLE   Using the CLT (as if we knew σ)

Based on weighing thousands of animals, the American Angus Association reports that mature Angus cows have a mean weight of 1309 pounds with a standard deviation of 157 pounds. This result was based on a very large sample of animals from many herds over a period of 15 years, so let's assume that these summaries are the population parameters and that the distribution of the weights was unimodal and reasonably symmetric.

Question: What does the CLT predict about the mean weight seen in random samples of 100 mature Angus cows?

(continued)

---

[1] Except, perhaps, at rush hour. But at that time, traffic is slowed. Our concern is with ordinary traffic during the day.

For Example (*continued*)

It's given that weights of all mature Angus cows have $\mu = 1309$ and $\sigma = 157$ pounds. Because $n = 100$ animals is a fairly large sample, I can apply the Central Limit Theorem. I expect the resulting sample means $\bar{y}$ will average 1309 pounds and have a standard deviation of $SD(\bar{y}) = \dfrac{\sigma}{\sqrt{n}} = \dfrac{157}{\sqrt{100}} = 15.7$ pounds.

The CLT also says that the distribution of sample means follows a Normal model, so the 68–95–99.7 Rule applies. I'd expect that

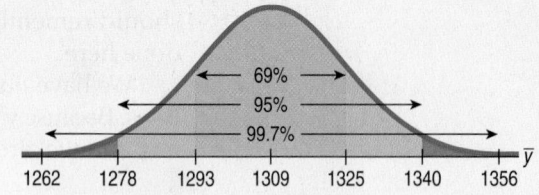

▶ in 68% of random samples of 100 mature Angus cows, the mean weight will be between $1309 - 15.7 = 1293.3$ and $1309 + 15.7 = 1324.7$ pounds;

▶ in 95% of such samples, $1277.6 \le \bar{y} \le 1340.4$ pounds;

▶ in 99.7% of such samples, $1261.9 \le \bar{y} \le 1356.1$ pounds.

The CLT says that all we need to model the sampling distribution of $\bar{y}$ is a random sample of quantitative data.

And the true population standard deviation, σ.

Uh oh. That could be a problem. How are we supposed to know σ? With proportions, we had a link between the proportion value and the standard deviation of the sample proportion: $SD(\hat{p}) = \sqrt{\dfrac{pq}{n}}$. And there was an obvious way to estimate the standard deviation from the data: $SE(\hat{p}) = \sqrt{\dfrac{\hat{p}\hat{q}}{n}}$. But for means, $SD(\bar{y}) = \dfrac{\sigma}{\sqrt{n}}$, so knowing $\bar{y}$ doesn't tell us anything about $SD(\bar{y})$. We know n, the sample size, but the population standard deviation, σ, could be *anything*. So what should we do? We do what any sensible person would do: We estimate the population parameter σ with s, the sample standard deviation based on the data. The resulting standard error is $SE(\bar{y}) = \dfrac{s}{\sqrt{n}}$.

A century ago, people used this standard error with the Normal model, assuming it would work. And for large sample sizes it *did* work pretty well. But they began to notice problems with smaller samples. The sample standard deviation, s, like any other statistic, varies from sample to sample. And this extra variation in the standard error was messing up the P-values and margins of error.

William S. Gosset is the man who first investigated this fact. He realized that not only do we need to allow for the extra variation with larger margins of error and P-values, but we even need a new sampling distribution model. In fact, we need a whole *family* of models, depending on the sample size, n. These models are unimodal, symmetric, bell-shaped models, but the smaller our sample, the more we must stretch out the tails. Gosset's work transformed Statistics, but most people who use his work don't even know his name.

> Because we estimate the standard deviation of the sampling distribution model from the data, it's a *standard error*. So we use the $SE(\bar{y})$ notation. Remember, though, that it's just the estimated standard deviation of the sampling distribution model for means.

A S *Activity: Estimating the Standard Error.* What's the average age at which people have heart attacks? A confidence interval gives a good answer, but we must estimate the standard deviation from the data to construct the interval.

Gosset's *t*

Gosset had a job that made him the envy of many. He was the quality control engineer for the Guinness Brewery in Dublin, Ireland. His job was to make sure that the stout (a thick, dark beer) leaving the brewery was of high enough quality to meet the demands of the brewery's many discerning customers. It's easy to imagine why a large sample with many observations might be undesirable when testing stout, not to mention dangerous to one's health. So Gosset often used small

To find the sampling distribution of $\frac{\bar{y}}{s/\sqrt{n}}$, Gosset simulated it by hand. He drew paper slips of small samples from a hat hundreds of times and computed the means and standard deviations with a mechanically cranked calculator. Today you could repeat in seconds on a computer the experiment that took him over a year. Gosset's work was so meticulous that not only did he get the shape of the new histogram approximately right, but he even figured out the exact formula for it from his sample. The formula was not confirmed mathematically until years later by Sir R. A. Fisher.

samples of 3 or 4. But he noticed that with samples of this size, his tests for quality weren't quite right. He knew this because when the batches that he rejected were sent back to the laboratory for more extensive testing, too often they turned out to be OK.

Gosset checked the stout's quality by performing hypothesis tests. He knew that the test would make some Type I errors and reject about 5% of the *good* batches of stout. However, the lab told him that he was in fact rejecting about 15% of the good batches. Gosset knew something was wrong, and it bugged him.

Gosset took time off to study the problem (and earn a graduate degree in the emerging field of Statistics). He figured out that when he used the standard error, $\frac{s}{\sqrt{n}}$, as an estimate of the standard deviation, the shape of the sampling model changed. He even figured out what the new model should be and called it a *t*-distribution.

The Guinness Company didn't give Gosset a lot of support for his work. In fact, it had a policy against publishing results. Gosset had to convince the company that he was not publishing an industrial secret, and (as part of getting permission to publish) he had to use a pseudonym. The pseudonym he chose was "Student," and ever since, the model he found has been known as **Student's t.**

Gosset's model is always bell-shaped, but the details change with different sample sizes. So the Student's *t*-models form a whole *family* of related distributions that depend on a parameter known as **degrees of freedom.** We often denote degrees of freedom as df and the model as t_{df}, with the degrees of freedom as a subscript.

A Confidence Interval for Means

To make confidence intervals or test hypotheses for means, we need to use Gosset's model. Which one? Well, for means, it turns out the right value for degrees of freedom is $df = n - 1$.

NOTATION ALERT:

Ever since Gosset, *t* has been reserved in Statistics for his distribution.

A PRACTICAL SAMPLING DISTRIBUTION MODEL FOR MEANS
When certain assumptions and conditions[2] are met, the standardized sample mean,

$$t = \frac{\bar{y} - \mu}{SE(\bar{y})},$$

follows a Student's *t*-model with $n - 1$ degrees of freedom. We estimate the standard deviation with

$$SE(\bar{y}) = \frac{s}{\sqrt{n}}.$$

When Gosset corrected the model for the extra uncertainty, the margin of error got bigger, as you might have guessed. When you use Gosset's model instead of the Normal model, your confidence intervals will be just a bit wider and your P-values just a bit larger. That's the correction you need. By using the *t*-model, you've compensated for the extra variability in precisely the right way.

---

[2] You can probably guess what they are. We'll see them in the next section.

When we found critical values from a Normal model, we called them $z*$. When we use a Student's t-model, we'll denote the critical values $t*$.

 Activity: Student's t in Practice. Use a statistics package to find a t-based confidence interval; that's how it's almost always done.

ONE-SAMPLE t-INTERVAL FOR THE MEAN

When the assumptions and conditions[3] are met, we are ready to find the confidence interval for the population mean, μ. The confidence interval is

$$\bar{y} \pm t^*_{n-1} \times SE(\bar{y}),$$

where the standard error of the mean is $SE(\bar{y}) = \dfrac{s}{\sqrt{n}}.$

The critical value t^*_{n-1} depends on the particular confidence level, C, that you specify and on the number of degrees of freedom, $n - 1$, which we get from the sample size.

FOR EXAMPLE A one-sample t-interval for the mean

In 2004, a team of researchers published a study of contaminants in farmed salmon.[4] Fish from many sources were analyzed for 14 organic contaminants. The study expressed concerns about the level of contaminants found. One of those was the insecticide mirex, which has been shown to be carcinogenic and is suspected to be toxic to the liver, kidneys, and endocrine system. One farm in particular produced salmon with very high levels of mirex. After those outliers are removed, summaries for the mirex concentrations (in parts per million) in the rest of the farmed salmon are:

$$n = 150 \qquad \bar{y} = 0.0913 \text{ ppm} \qquad s = 0.0495 \text{ ppm}.$$

Question: What does a 95% confidence interval say about mirex?

$df = 150 - 1 = 149$

$SE(\bar{y}) = \dfrac{s}{\sqrt{n}} = \dfrac{0.0495}{\sqrt{150}} = 0.0040$

$t^*_{149} \approx 1.977$ (from table T, using 140 df)

(actually, $t^*_{149} \approx 1.976$ from technology)

So the confidence interval for μ is $\bar{y} \pm t^*_{149} \times SE(\bar{y}) = 0.0913 \pm 1.977(0.0040)$

$$= 0.0913 \pm 0.0079$$
$$= (0.0834, 0.0992)$$

I'm 95% confident that the mean level of mirex concentration in farm-raised salmon is between 0.0834 and 0.0992 parts per million.

FIGURE 23.2

The t-model (solid curve) on 2 degrees of freedom has fatter tails than the Normal model (dashed curve). So the 68–95–99.7 Rule doesn't work for t-models with only a few degrees of freedom.

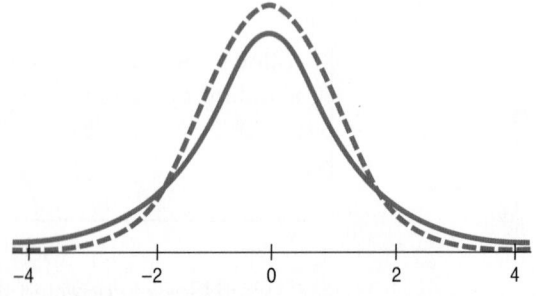

 Activity: Student's Distributions. Interact with Gosset's family of t-models. Watch the shape of the model change as you slide the degrees of freedom up and down.

---

[3] Yes, the same ones, and they're still coming in the next section.

[4] Ronald A. Hites, Jeffery A. Foran, David O. Carpenter, M. Coreen Hamilton, Barbara A. Knuth, and Steven J. Schwager, "Global Assessment of Organic Contaminants in Farmed Salmon," *Science* 9 January 2004: Vol. 303., no. 5655, pp. 226–229.

Student's *t*-models are unimodal, symmetric, and bell-shaped, just like the Normal. But *t*-models with only a few degrees of freedom have much fatter tails than the Normal. (That's what makes the margin of error bigger.) As the degrees of freedom increase, the *t*-models look more and more like the Normal. In fact, the *t*-model with infinite degrees of freedom is exactly Normal.[5] This is great news if you happen to have an infinite number of data values. Unfortunately, that's not practical. Fortunately, above a few hundred degrees of freedom it's very hard to tell the difference. Of course, in the rare situation that we *know* σ, it would be foolish not to use that information. And if we don't have to estimate σ, we can use the Normal model.

z or *t*?

If you know σ, use *z*. (That's rare!)
Whenever you use *s* to estimate σ, use *t*.

When σ is known   Administrators of a hospital were concerned about the prenatal care given to mothers in their part of the city. To study this, they examined the gestation times of babies born there. They drew a sample of 25 babies born in their hospital in the previous 6 months. Human gestation times for healthy pregnancies are thought to be well-modeled by a Normal with a mean of 280 days and a standard deviation of 14 days. The hospital administrators wanted to test the mean gestation time of their sample of babies against the known standard. For this test, they should use the established value for the standard deviation, 14 days, rather than estimating the standard deviation from their sample. Because they use the model parameter value for σ, they should base their test on the Normal model rather than Student's *t*.

TI Tips   **Finding *t*-model probabilities and critical values**

Finding Probabilities
You already know how to use your TI to find probabilities for Normal models using *z*-scores and `normalcdf`. What about *t*-models? Yes, the calculator can work with them, too.

You know from your experience with confidence intervals that $z = 1.645$ cuts off the upper 5% in a Normal model. Use the TI to check that. From the `DISTR` menu, enter `normalcdf(1.645,99)`. Only 0.04998? Close enough for statisticians!

We might wonder about the probability of observing a *t*-value greater than 1.645, but we can't find that. There's only one Normal model, but there are many *t*-models, depending on the number of degrees of freedom. We need to be more specific.

Let's find the probability of observing a *t*-value greater than 1.645 when there are 12 degrees of freedom. That we can do. Look in the `DISTR` menu again. See it? Yes, `tcdf`. That function works essentially like `normalcdf`, but after you enter the left and right cutoffs you must also specify the number of degrees of freedom. Try `tcdf(1.645,99,12)`.

The upper tail probability for t_{12} is 0.063, higher than the Normal model's 0.05. That should make sense to you—remember, *t*-models are a bit fatter in the tails, so more of the distribution lies beyond the 1.645 cutoff. (That means we'll have to go a little wider to make a 90% confidence interval.)

[5] Formally, in the limit as *n* goes to infinity.

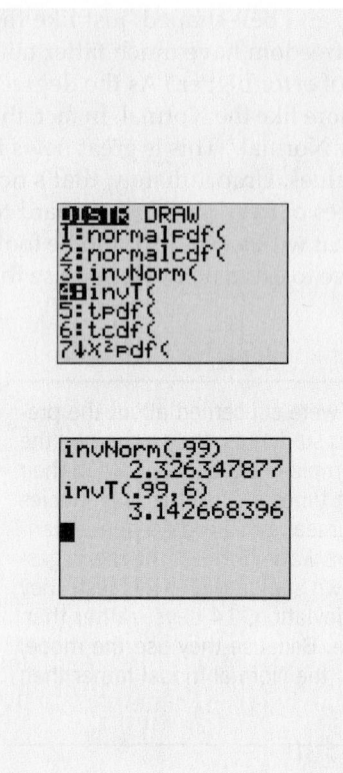

Check out what happens when there are more degrees of freedom, say, 25. The command `tcdf(1.645,99,25)` yields a probability of 0.056. That's closer to 0.05, for a good reason: *t*-models look more and more like the Normal model as the number of degrees of freedom increases.

Finding Critical Values

Your calculator can also determine the critical value of *t* that cuts off a specified percentage of the distribution, using `invT`. It works just like `invNorm`, but for *t* we also have to specify the number of degrees of freedom (of course).

Suppose we have 6 degrees of freedom and want to create a 98% confidence interval. A confidence level of 98% leaves 1% in each tail of our model, so we need to find the value of *t* corresponding to the 99th percentile. If a Normal model were appropriate, we'd use $z = 2.33$. (Try it: `invNorm(.99)`). Now think. How should the critical value for *t* compare?

If you thought, "It'll be larger, because *t*-models are more spread out," you're right. Check with your TI, remembering to specify our 6 degrees of freedom: `invT(.99,6)`. Were you surprised, though, that the critical value of *t* is so much larger?

So think once more. How would the critical value of *t* differ if there were 60 degrees of freedom instead of only 6? When you think you know, check it out on your TI.

Understanding *t*

Use your calculator to play around with `tcdf` and `invT` a bit. Try to develop a clear understanding of how *t*-models compare to the more familiar Normal model. That will help you as you learn to use *t*-models to make inferences about means.

Assumptions and Conditions

Gosset found the *t*-model by simulation. Years later, when Sir Ronald A. Fisher[6] showed mathematically that Gosset was right, he needed to make some assumptions to make it work. These are the assumptions we need to use the Student's *t*-models.

INDEPENDENCE ASSUMPTION

Independence Assumption: The data values should be independent. There's really no way to check independence of the data by looking at the sample, but we should think about whether the assumption is reasonable.

　　Randomization Condition: The data arise from a random sample or suitably randomized experiment. Randomly sampled data—and especially data from a Simple Random Sample—are ideal.

　　When a sample is drawn without replacement, technically we ought to confirm that we haven't sampled a large fraction of the population, which would threaten the independence of our selections. We check the

　　10% Condition: The sample is no more than 10% of the population.

　　In practice, though, we often don't mention the 10% Condition for means. Why not? When we made inferences about proportions, this condition was crucial

---

[6] We met Fisher back in Chapter 21. You can see his picture on page 486.

We Don't *Want* to Stop

We check conditions hoping that we can make a meaningful analysis of our data. The conditions serve as *disqualifiers*—we keep going unless there's a serious problem. If we find minor issues, we note them and express caution about our results. If the sample is not an SRS, but we believe it's representative of some populations, we limit our conclusions accordingly. If there are outliers, rather than stop, we perform the analysis both with and without them. If the sample looks bimodal, we try to analyze subgroups separately. Only when there's major trouble—like a strongly skewed small sample or an obviously nonrepresentative sample—are we unable to proceed at all.

because we usually had large samples. But for means our samples are generally smaller, so the independence problem arises only if we're sampling from a small population (and then there's a correction formula we could use—but let's not get into that here). And sometimes we're dealing with a randomized experiment; then there's no sampling at all.

NORMAL POPULATION ASSUMPTION

Student's *t*-models won't work for data that are badly skewed. How skewed is too skewed? Well, formally, we assume that the data are from a population that follows a Normal model. Practically speaking, there's no way to be certain this is true.

And it's almost certainly *not* true. Models are idealized; real data are, well, real—*never* Normal. The good news, however, is that even for small samples, it's sufficient to check the . . .

Nearly Normal Condition: The data come from a distribution that is unimodal and symmetric.

Check this condition by making a histogram or Normal probability plot. The importance of Normality for Student's *t* depends on the sample size. Just our luck: It matters most when it's hardest to check.[7]

For very small samples ($n < 15$ or so), the data should follow a Normal model pretty closely. Of course, with so little data, it's rather hard to tell. But if you do find outliers or strong skewness, don't use these methods.

For moderate sample sizes (n between 15 and 40 or so), the *t* methods will work well as long as the data are unimodal and reasonably symmetric. Make a histogram.

When the sample size is larger than 40 or 50, the *t* methods are safe to use unless the data are extremely skewed. Be sure to make a histogram. If you find outliers in the data, it's always a good idea to perform the analysis twice, once with and once without the outliers, even for large samples. They may well hold additional information about the data that deserves special attention. If you find multiple modes, you may well have different groups that should be analyzed and understood separately.

FOR EXAMPLE   Checking assumptions and conditions for Student's *t*

Recap: Researchers purchased whole farmed salmon from 51 farms in eight regions in six countries. The histogram shows the concentrations of the insecticide mirex in 150 farmed salmon.

Question: Are the assumptions and conditions for inference satisfied?

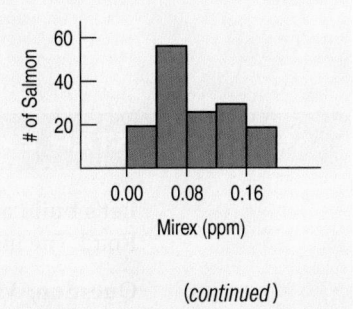

✔ **Independence Assumption:** The fish were raised in many different places, and samples were purchased independently from several sources.

✔ **Randomization Condition:** The fish were selected randomly from those available for sale.

(continued)

[7] There are formal tests of Normality, but they don't really help. When we have a small sample—just when we really care about checking Normality—these tests have very little power. So it doesn't make much sense to use them in deciding whether to perform a *t*-test. We don't recommend that you use them.

For Example *(continued)*

✔ **10% Conditions:** There's lots of fish in the sea (and at the fish farms); 150 is certainly far fewer than 10% of the population.

✔ **Nearly Normal Condition:** The histogram of the data is unimodal. Although it may be somewhat skewed to the right, this is not a concern with a sample size of 150.

It's okay to use these data for inference about farm-raised salmon.

JUST CHECKING

Every 10 years, the United States takes a census. The census tries to count every resident. There are two forms, known as the "short form," answered by most people, and the "long form," slogged through by about one in six or seven households chosen at random. According to the Census Bureau (www.census.gov), ". . . each estimate based on the long form responses has an associated confidence interval."

1. Why does the Census Bureau need a confidence interval for long-form information but not for the questions that appear on both the long and short forms?

2. Why must the Census Bureau base these confidence intervals on *t*-models?

The Census Bureau goes on to say, "These confidence intervals are wider . . . for geographic areas with smaller populations and for characteristics that occur less frequently in the area being examined (such as the proportion of people in poverty in a middle-income neighborhood)."

3. Why is this so? For example, why should a confidence interval for the mean amount families spend monthly on housing be wider for a sparsely populated area of farms in the Midwest than for a densely populated area of an urban center? How does the formula show this will happen?

To deal with this problem, the Census Bureau reports long-form data only for ". . . geographic areas from which about two hundred or more long forms were completed—which are large enough to produce good quality estimates. If smaller weighting areas had been used, the confidence intervals around the estimates would have been significantly wider, rendering many estimates less useful . . ."

4. Suppose the Census Bureau decided to report on areas from which only 50 long forms were completed. What effect would that have on a 95% confidence interval for, say, the mean cost of housing? Specifically, which values used in the formula for the margin of error would change? Which would change a lot and which would change only slightly?

5. Approximately how much wider would that confidence interval based on 50 forms be than the one based on 200 forms?

STEP-BY-STEP EXAMPLE A One-Sample *t*-Interval for the Mean

Let's build a 90% confidence interval for the mean speed of all vehicles traveling on Triphammer Road. The interval that we'll make is called the **one-sample *t*-interval**.

Question: What can we say about the mean speed of all cars on Triphammer Road?

Plan State what we want to know. Identify the parameter of interest.

Identify the variables and review the W's.

I want to find a 90% confidence interval for the mean speed, μ, of vehicles driving on Triphammer Road. I have data on the speeds of 23 cars there, sampled on April 11, 2000.

Make a picture. Check the distribution shape and look for skewness, multiple modes, and outliers.

Here's a histogram of the 23 observed speeds.

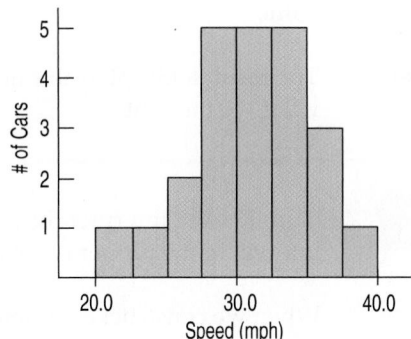

 **REALITY CHECK** The histogram centers around 30 mph, and the data lie between 20 and 40 mph. We'd expect a confidence interval to place the population mean within a few mph of 30.

Model Think about the assumptions and check the conditions.

Note that with this small sample we probably didn't need to check the 10% Condition.

On the other hand, doing so gives us a chance to think about what the population is.

✔ **Independence Assumption:** This is a convenience sample, but care was taken to select cars that were not driving near each other, so their speeds are plausibly independent.

✔ **Randomization Condition:** Not really met. This is a convenience sample, but I have reason to believe that it is representative.

✔ **10% Condition:** The cars I observed were fewer than 10% of all cars that travel Triphammer Road.

✔ **Nearly Normal Condition:** The histogram of the speeds is unimodal and symmetric.

State the sampling distribution model for the statistic.

Choose your method.

The conditions are satisfied, so I will use a Student's t-model with

$$(n - 1) = 22 \text{ degrees of freedom}$$

and find a **one-sample t-interval for the mean.**

Mechanics Construct the confidence interval.

Be sure to include the units along with the statistics.

Calculating from the data (see page 530):

$$n = 23 \text{ cars}$$
$$\bar{y} = 31.0 \text{ mph}$$
$$s = 4.25 \text{ mph.}$$

The standard error of $\bar{y}$ is

$$SE(\bar{y}) = \frac{s}{\sqrt{n}} = \frac{4.25}{\sqrt{23}} = 0.886 \text{ mph.}$$

The critical value we need to make a 90% interval comes from a Student's t table, a computer program, or a calculator. We have 23 − 1 = 22 degrees of freedom. The selected confidence level says that we want 90% of the probability to be caught in the middle, so we exclude 5% in *each* tail, for a total of 10%. The degrees

The 90% critical value is $t^*_{22} = 1.717$, so the margin of error is

$$ME = t^*_{22} \times SE(\bar{y})$$
$$= 1.717(0.886)$$
$$= 1.521 \text{ mph.}$$

The 90% confidence interval for the mean speed is 31.0 ± 1.5 mph.

of freedom and 5% tail probability are all we need to know to find the critical value.

 **REALITY CHECK** The result looks plausible and in line with what we thought.

TELL **Conclusion** Interpret the confidence interval in the proper context.

When we construct confidence intervals in this way, we expect 90% of them to cover the true mean and 10% to miss the true value. That's what "90% confident" means.

I am 90% confident that the interval from 29.5 mph to 32.5 mph contains the true mean speed of all vehicles on Triphammer Road.

Caveat: This was not a random sample of vehicles. It was a convenience sample taken at one time on one day. And the participants were not blinded. Drivers could see the police device, and some may have slowed down. I'm reluctant to extend this inference to other situations.

> **TI-*nspire***
>
> **Intervals for Means.** Generate confidence intervals from many samples to see how often they successfully capture the true mean.

Here's the part of the Student's *t* table that gives the critical value we needed for the Step-by-Step confidence interval. (See Table T in the back of the book.) To find a critical value, locate the row of the table corresponding to the degrees of freedom and the column corresponding to the probability you want. Our 90% confidence interval leaves 5% of the values on either side, so look for 0.05 at the top of the column or 90% at the bottom. The value in the table at that intersection is the critical value we need: 1.717.

> As degrees of freedom increase, the shape of Student's *t*-models changes more gradually. Table T at the back of the book includes degrees of freedom between 100 and 1000 selected so that you can pin down the P-value for just about any df. If your df's aren't listed, take the cautious approach by using the next lower value or use technology.

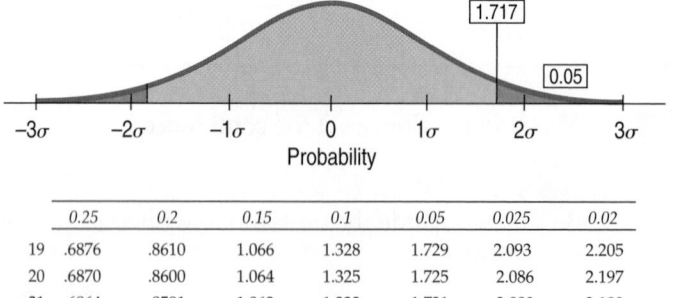

	0.25	0.2	0.15	0.1	0.05	0.025	0.02
19	.6876	.8610	1.066	1.328	1.729	2.093	2.205
20	.6870	.8600	1.064	1.325	1.725	2.086	2.197
21	.6864	.8591	1.063	1.323	1.721	2.080	2.189
22	.6858	.8583	1.061	1.321	1.717	2.074	2.183
23	.6853	.8575	1.060	1.319	1.714	2.069	2.177
24	.6848	.8569	1.059	1.318	1.711	2.064	2.172
25	.6844	.8562	1.058	1.316	1.708	2.060	2.167
26	.6840	.8557	1.058	1.315	1.706	2.056	2.162
27	.6837	.8551	1.057	1.314	1.703	2.052	2.158
C					80%	90%	95%

> **A S** *Activity:* **Building *t*-Intervals with the *t*-Table.** Interact with an animated version of Table T.

Of course, you can also create the confidence interval with computer software or a calculator.

| TI Tips | Finding a confidence interval for a mean |

Yes, your calculator can create a confidence interval for a mean. And it's so easy we'll do two!

Find a confidence interval given a set of data

- Type the speeds of the 23 Triphammer cars into **L1**. Go ahead; we'll wait.

| 29 | 34 | 34 | 28 | 30 | 29 | 38 | 31 | 29 | 34 | 32 | 31 |
| 27 | 37 | 29 | 26 | 24 | 34 | 36 | 31 | 34 | 36 | 21 | |

- Set up a **STATPLOT** to create a histogram of the data so you can check the nearly Normal condition. Looks okay—unimodal and roughly symmetric.
- Under **STAT TESTS** choose **8:TInterval**.
- Choose **Inpt:Data**, then specify that your data is **List:L1**.
- For these data the frequency is 1. (If your data have a frequency distribution stored in another list, you would specify that.)
- Choose the confidence level you want.
- **Calculate** the interval.

There's the 90% confidence interval. That was easy—but remember, the calculator only does the *Show*. Now you have to *Tell* what it means.

No data? Find a confidence interval given the sample's mean and standard deviation

Sometimes instead of the original data you just have the summary statistics. For instance, suppose a random sample of 53 lengths of fishing line had a mean strength of 83 pounds and standard deviation of 4 pounds. Let's make a 95% confidence interval for the mean strength of this kind of fishing line.

- Without the data you can't check the Nearly Normal Condition. But 53 is a moderately large sample, so assuming there were no outliers, it's okay to proceed. You need to say that.
- Go back to **STAT TESTS** and choose **8:TInterval** again. This time indicate that you wish to enter the summary statistics. To do that, select **Stats**, then hit **ENTER**.
- Specify the sample mean, standard deviation, and sample size.
- Choose a confidence level and **Calculate** the interval.
- If (repeat, IF . . .) strengths of fishing lines follow a Normal model, we are 95% confident that this kind of line has a mean strength between 81.9 and 84.1 pounds.

More Cautions About Interpreting Confidence Intervals

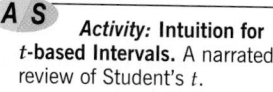

Activity: **Intuition for t-based Intervals.** A narrated review of Student's *t*.

Confidence intervals for means offer new tempting wrong interpretations. Here are some things you *shouldn't* say:

▶ ***Don't say,*** "90% of all the vehicles on Triphammer Road drive at a speed between 29.5 and 32.5 mph." The confidence interval is about the *mean* speed, not about the speeds of *individual* vehicles.

So What *Should* We Say?
Since 90% of random samples yield an interval that captures the true mean, we *should* say, "I am 90% confident that the interval from 29.5 to 32.5 mph contains the mean speed of all the vehicles on Triphammer Road." It's also okay to say something less formal: "I am 90% confident that the average speed of all vehicles on Triphammer Road is between 29.5 and 32.5 mph." Remember: *Our uncertainty is about the interval, not the true mean.* The interval varies randomly. The true mean speed is neither variable nor random—just unknown.

▶ *Don't say,* "We are 90% confident that *a randomly selected vehicle* will have a speed between 29.5 and 32.5 mph." This false interpretation is also about individual vehicles rather than about the *mean* of the speeds. We are 90% confident that the *mean* speed of all vehicles on Triphammer Road is between 29.5 and 32.5 mph.

▶ *Don't say,* "The mean speed of the vehicles is 31.0 mph 90% *of the time.*" That's about means, but still wrong. It implies that the true mean varies, when in fact it is the confidence interval that would have been different had we gotten a different sample.

▶ Finally, *don't say,* "90% *of all samples* will have mean speeds between 29.5 and 32.5 mph." That statement suggests that *this* interval somehow sets a standard for every other interval. In fact, this interval is no more (or less) likely to be correct than any other. You could say that 90% of all possible samples will produce intervals that actually do contain the true mean speed. (The problem is that, because we'll never know where the true mean speed really is, we can't know if our sample was one of those 90%.)

▶ *Do say,* "90% of intervals that could be found in this way would cover the true value." Or make it more personal and say, "I am 90% confident that the true mean speed is between 29.5 and 32.5 mph."

Make a Picture, Make a Picture, Make a Picture

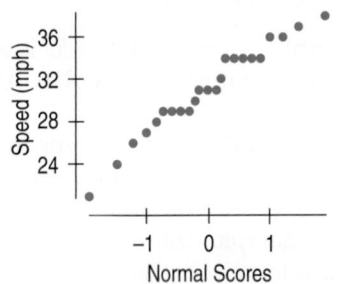

FIGURE 23.3
A Normal probability plot of speeds looks reasonably straight.

The only reasonable way to check the Nearly Normal Condition is with graphs of the data. Make a histogram of the data and verify that its distribution is unimodal and symmetric and that it has no outliers. You may also want to make a Normal probability plot to see that it's reasonably straight. You'll be able to spot deviations from the Normal model more easily with a Normal probability plot, but it's easier to understand the particular nature of the deviations from a histogram.

If you have a computer or graphing calculator doing the work, there's no excuse not to look at *both* displays as part of checking the Nearly Normal Condition.

A Test for the Mean

The residents along Triphammer Road have a more specific concern. It appears that the mean speed along the road is higher than it ought to be. To get the police to patrol more frequently, though, they'll need to show that the true mean speed is *in fact greater* than the 30 mph speed limit. This calls for a hypothesis test called the **one-sample *t*-test for the mean.**

You already know enough to construct this test. The test statistic looks just like the others we've seen. It compares the difference between the observed statistic and a hypothesized value to the standard error of the observed statistic. We already know that, for means, the appropriate probability model to use for P-values is Student's t with $n-1$ degrees of freedom.

We're ready to go:

ONE-SAMPLE *t*-TEST FOR THE MEAN

The assumptions and conditions for the one-sample *t*-test for the mean are the same as for the one-sample *t*-interval. We test the hypothesis $H_0: \mu = \mu_0$ using the statistic

$$t_{n-1} = \frac{\bar{y} - \mu_0}{SE(\bar{y})}.$$

The standard error of $\bar{y}$ is $SE(\bar{y}) = \frac{s}{\sqrt{n}}$.

When the conditions are met and the null hypothesis is true, this statistic follows a Student's *t*-model with $n - 1$ degrees of freedom. We use that model to obtain a P-value.

FOR EXAMPLE   A one-sample *t*-test for the mean

Recap: Researchers tested 150 farm-raised salmon for organic contaminants. They found the mean concentration of the carcinogenic insecticide mirex to be 0.0913 parts per million, with standard deviation 0.0495 ppm. As a safety recommendation to recreational fishers, the Environmental Protection Agency's (EPA) recommended "screening value" for mirex is 0.08 ppm.

Question: Are farmed salmon contaminated beyond the level permitted by the EPA? (We've already checked the conditions; see pages 537–8.)

$$H_0: \mu = 0.08$$
$$H_A: \mu > 0.08$$

These data satisfy the conditions for inference; I'll do a one-sample *t*-test for the mean:

$n = 150, df = 149$

$\bar{y} = 0.0913, s = 0.0495$

$SE(\bar{y}) = \dfrac{0.0495}{\sqrt{150}} = 0.0040$

$t_{149} = \dfrac{0.0913 - 0.08}{0.0040} = 2.825$

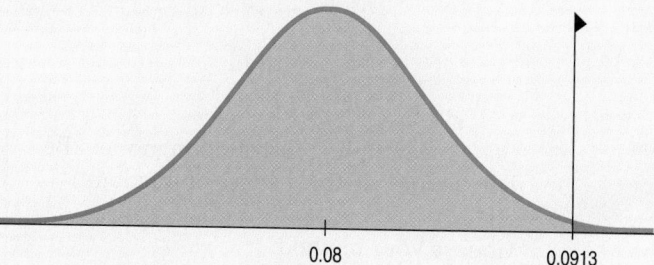

$P(t_{149} > 2.825) = 0.0027$ (from technology).

With a P-value that low, I reject the null hypothesis and conclude that, in farm-raised salmon, the mirex contamination level does exceed the EPA screening value.

STEP-BY-STEP EXAMPLE   A One-Sample *t*-Test for the Mean

Let's apply the one-sample *t*-test to the Triphammer Road car speeds. The speed limit is 30 mph, so we'll use that as the null hypothesis value.

Question: Does the mean speed of all cars exceed the posted speed limit?

Plan State what we want to know. Make clear what the population and parameter are.

Identify the variables and review the W's.

Hypotheses The null hypothesis is that the true mean speed is equal to the limit. Because we're interested in whether the vehicles are speeding, the alternative is one-sided.

Make a picture. Check the distribution for skewness, multiple modes, and outliers.

REALITY CHECK ▷ The histogram of the observed speeds is clustered around 30, so we'd be surprised to find that the mean was much higher than that. (The fact that 30 is within the confidence interval that we've just found confirms this suspicion.)

Model Think about the assumptions and check the conditions.

(We won't worry about the 10% Condition—it's a small sample.)

State the sampling distribution model. (Be sure to include the degrees of freedom.)

Choose your method.

I want to know whether the mean speed of vehicles on Triphammer Road exceeds the posted speed limit of 30 mph. I have a sample of 23 car speeds on April 11, 2000.

H_O: Mean speed, $\mu = 30$ mph
H_A: Mean speed, $\mu > 30$ mph

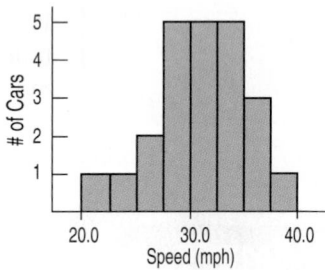

✔ **Independence Assumption:** These cars are a convenience sample, but they were selected so no two cars were driving near each other, so I am justified in believing that their speeds are independent.

✔ **Randomization Condition:** Although I have a convenience sample, I have reason to believe that it is a representative sample.

✔ **Nearly Normal Condition:** The histogram of the speeds is unimodal and reasonably symmetric.

The conditions are satisfied, so I'll use a Student's t-model with $(n - 1) = 22$ degrees of freedom to do a **one-sample t-test for the mean.**

Mechanics Be sure to include the units when you write down what you know from the data.

We use the null model to find the P-value. Make a picture of the t-model centered at $\mu = 30$. Since this is an upper-tail test, shade the region to the right of the observed mean speed.

From the data,

$$n = 23 \text{ cars}$$
$$\bar{y} = 31.0 \text{ mph}$$
$$s = 4.25 \text{ mph}$$
$$SE(\bar{y}) = \frac{s}{\sqrt{n}} = \frac{4.25}{\sqrt{23}} = 0.886 \text{ mph.}$$

The *t*-statistic calculation is just a standardized value, like *z*. We subtract the hypothesized mean and divide by the standard error.

The P-value is the probability of observing a sample mean as large as 31.0 (or larger) *if* the true mean were 30.0, as the null hypothesis states. We can find this P-value from a table, calculator, or computer program.

 We're not surprised that the difference isn't statistically significant.

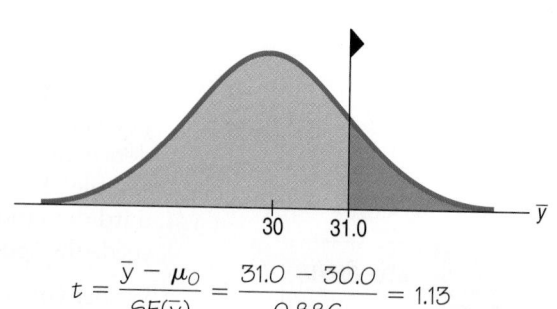

$$t = \frac{\bar{y} - \mu_0}{SE(\bar{y})} = \frac{31.0 - 30.0}{0.886} = 1.13$$

(The observed mean is 1.13 standard errors above the hypothesized value.)

$$\text{P-value} = P(t_{22} > 1.13) = 0.136$$

Conclusion Link the P-value to your decision about H_0, and state your conclusion in context.

Unfortunately for the residents, there is no course of action associated with failing to reject this particular null hypothesis.

The P-value of 0.136 says that if the true mean speed of vehicles on Triphammer Road were 30 mph, samples of 23 vehicles can be expected to have an observed mean of at least 31.0 mph 13.6% of the time. That P-value is not small enough for me to reject the hypothesis that the true mean is 30 mph at any reasonable alpha level. I conclude that there is not enough evidence to say the average speed is too high.

TI Tips

Testing a hypothesis about a mean

Testing a Hypothesis Given a Set of Data
Still have the Triphammer Road auto speeds in **L1**? Good. Let's use the TI to see if the mean is significantly higher than 30 mph (you've already checked the histogram to verify the nearly Normal condition, of course).

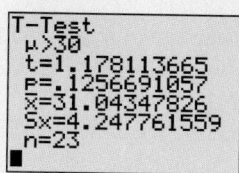

- Go to the **STAT TESTS** menu, and choose **2: T-Test**.
- Tell it you want to use the stored **Data**.
- Enter the mean of the null model, and indicate where the data are.
- Since this is an upper tail test, choose the $>\mu_0$ option.
- **Calculate**.

There's everything you need to know: the summary statistics, the calculated value of *t*, and the P-value of 0.126. (*t* and P differ slightly from the values in our worked example because when we did it by hand we rounded off the mean and standard deviation. No harm done.)

As always, the *Tell* is up to you.

Testing a Hypothesis Given the Sample's Mean and Standard Deviation

Don't have the actual data? Just summary statistics? No problem, assuming you can verify the necessary conditions. In the last TI Tips we created a confidence interval for the strength of fishing line. We had test results for a random sample of 53 lengths of line showing a mean strength of 83 pounds and a standard deviation of 4 pounds. Is there evidence that this kind of fishing line exceeds the "80-lb test" as labeled on the package?

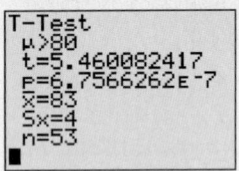

We bet you know what to do even without our help. Try it before you read on.

- Go back to 2: T-Test.
- You're entering Stats this time.
- Specify the hypothesized mean and the sample statistics.
- Choose the alternative being tested (upper tail here).
- Calculate.

The results of the calculator's mechanics show a large t and a really small P-value (0.0000007). We have very strong evidence that the mean breaking strength of this kind of fishing line is over the 80 pounds claimed by the manufacturer.

Significance and Importance

Recall that "statistically significant" does not mean "actually important" or "meaningful," even though it sort of sounds that way. In this example, it does seem that speeds may be a bit above 30 miles per hour. If so, it's possible that a larger sample would show statistical significance.

But would that be the right decision? The difference between 31 miles per hour and 30 miles per hour doesn't seem meaningful, and rejecting the null hypothesis wouldn't change that. Even with a statistically significant result, it would be hard to convince the police that vehicles on Triphammer Road were driving at dangerously fast speeds. It would probably also be difficult to persuade the town that spending more money to lower the average speed on Triphammer Road would be a good use of the town's resources. Looking at the confidence interval, we can say with 90% confidence that the mean speed is somewhere between 29.5 and 32.5 mph. Even in the worst case, if the mean speed is 32.5 mph, would this be a bad enough situation to convince the town to spend more money? Probably not. It's always a good idea when we test a hypothesis to also check the confidence interval and think about the likely values for the mean.

JUST CHECKING

In discussing estimates based on the long-form samples, the Census Bureau notes, "The disadvantage . . . is that . . . estimates of characteristics that are also reported on the short form will not match the [long-form estimates]."

The short-form estimates are values from a complete census, so they are the "true" values—something we don't usually have when we do inference.

6. Suppose we use long-form data to make 95% confidence intervals for the mean age of residents for each of 100 of the Census-defined areas. How many of these 100 intervals should we expect will fail to include the true mean age (as determined from the complete short-form Census data)?

7. Based only on the long-form sample, we might test the null hypothesis about the mean household income in a region. Would the power of the test increase or decrease if we used an area with more long forms?

Intervals and Tests

The 90% confidence interval for the mean speed was 31.0 mph ± 1.5, or (29.5 mph, 32.5 mph). If someone hypothesized that the mean speed was really 30 mph, how would you feel about it? How about 35 mph?

Because the confidence interval included the speed limit of 30 mph, it certainly looked like 30 mph might be a plausible value for the true mean speed of the vehicles on Triphammer Road. In fact, 30 mph gave a P-value of 0.136—too large to reject the null hypothesis. We should have seen this coming. The hypothesized mean of 30 mph lies *within the confidence interval*. It's one of the reasonable values for the mean.

Confidence intervals and significance tests are built from the same calculations. In fact, they are really complementary ways of looking at the same question. Here's the connection: The confidence interval contains all the null hypothesis values we can't reject with these data.

More precisely, a level C confidence interval contains *all* of the plausible null hypothesis values that would *not* be rejected by a two-sided hypothesis test at alpha level 1 − C. So a 95% confidence interval matches a 1 − 0.95 = 0.05 level two-sided test for these data.

Confidence intervals are naturally two-sided, so they match exactly with two-sided hypothesis tests. When, as in our example, the hypothesis is one-sided, the corresponding alpha level is $(1 - C)/2$.

> **Fail to reject**   Our 90% confidence interval was 29.5 to 32.5 mph. If any of these values had been the null hypothesis for the mean, then the corresponding hypothesis test at $\alpha = 0.05$ (because $\dfrac{1 - 0.90}{2} = 0.05$) would not have been able to reject the null. That is, the corresponding one-sided P-value for our observed mean of 31 mph would be greater than 0.05. So, we would not reject any hypothesized value between 29.5 and 32.5 mph.

Sample Size

A S   *Activity:* **The Real Effect of Small Sample Size.** We know that smaller sample sizes lead to wider confidence intervals, but is that just because they have fewer degrees of freedom?

How large a sample do we need? The simple answer is "more." But more data cost money, effort, and time, so how much is enough? Suppose your computer just took an hour to download a movie you wanted to watch. You're not happy. You hear about a program that claims to download movies in under a half hour. You're interested enough to spend $29.95 for it, but only if it really delivers. So you get the free evaluation copy and test it by downloading that movie 5 different times. Of course, the mean download time is not exactly 30 minutes as claimed. Observations vary. If the margin of error were 8 minutes, though, you'd probably be able to decide whether the software is worth the money. Doubling the sample size would require another 5 hours of testing and would reduce your margin of error to a bit under 6 minutes. You'll need to decide whether that's worth the effort.

As we make plans to collect data, we should have some idea of how small a margin of error we need to be able to draw a conclusion or detect a difference we want to see. If the size of the effect we're studying is large, then we may be able to tolerate a larger *ME*. If we need great precision, however, we'll want a smaller *ME*, and, of course, that means a larger sample size.

Armed with the *ME* and confidence level, we can find the sample size we'll need. Almost.

We know that for a mean, $ME = t^*_{n-1} \times SE(\bar{y})$ and that $SE(\bar{y}) = \dfrac{s}{\sqrt{n}}$, so we can determine the sample size by solving this equation for n:

$$ME = t^*_{n-1}\frac{s}{\sqrt{n}}.$$

The good news is that we have an equation; the bad news is that we won't know most of the values we need to solve it. When we thought about sample size for proportions back in Chapter 19, we ran into a similar problem. There we had to guess a working value for p to compute a sample size. Here, we need to know s. We don't know s until we get some data, but we want to calculate the sample size *before* collecting the data. We might be able to make a good guess, and that is often good enough for this purpose. If we have no idea what the standard deviation might be, or if the sample size really matters (for example, because each additional individual is very expensive to sample or experiment on), it might be a good idea to run a small *pilot study* to get some feeling for the standard deviation.

That's not all. Without knowing n, we don't know the degrees of freedom and we can't find the critical value, t^*_{n-1}. One common approach is to use the corresponding z^* value from the Normal model. If you've chosen a 95% confidence level, then just use 2, following the 68–95–99.7 Rule. If your estimated sample size is, say, 60 or more, it's probably okay—z^* was a good guess. If it's smaller than that, you may want to add a step, using z^* at first, finding n, and then replacing z^* with the corresponding t^*_{n-1} and calculating the sample size once more.

Sample size calculations are *never* exact. The margin of error you find *after* collecting the data won't match exactly the one you used to find n. The sample size formula depends on quantities that you won't have until you collect the data, but using it is an important first step. Before you collect data, it's always a good idea to know whether the sample size is large enough to give you a good chance of being able to tell you what you want to know.

FOR EXAMPLE   Finding sample size

A company claims its program will allow your computer to download movies quickly. We'll test the free evaluation copy by downloading a movie several times, hoping to estimate the mean download time with a margin of error of only 8 minutes. We think the standard deviation of download times is about 10 minutes.

Question: How many trial downloads must we run if we want 95% confidence in our estimate with a margin of error of only 8 minutes?
Using $z^* = 1.96$, solve

$$8 = 1.96\frac{10}{\sqrt{n}}$$

$$\sqrt{n} = \frac{1.96 \times 10}{8} = 2.45$$

$$n = (2.45)^2 = 6.0025$$

That's a small sample size, so I'll use $(6 - 1) = 5$ degrees of freedom[8] to substitute an appropriate t^* value. At 95%, $t^*_5 = 2.571$. Solving the equation one more time:

$$8 = 2.571\frac{10}{\sqrt{n}}$$

---

[8] Ordinarily we'd round the sample size *up*. But at this stage of the calculation, rounding *down* is the safer choice. Can you see why?

$$\sqrt{n} = \frac{2.571 \times 10}{8} \approx 3.214$$

$$n = (3.214)^2 \approx 10.33$$

To make sure the ME is no larger, I'll round *up*, which gives $n = 11$ runs. So, to get an ME of 8 minutes, I'll find the downloading times for 11 movies.

Degrees of Freedom

Some calculators offer an alternative button for standard deviation that divides by n instead of $n - 1$. Why don't you stick a wad of gum over the "n" button so you won't be tempted to use it? Use $n - 1$.

The number of degrees of freedom, $(n - 1)$, might have reminded you of the value we divide by to find the standard deviation of the data (since, in fact, it's the same number). When we introduced that formula, we promised to say a bit more about why we divide by $n - 1$ rather than by n. The reason is closely tied to the reasoning behind the t-distribution.

If only we knew the true population mean, μ, we would find the sample standard deviation as

$$s = \sqrt{\frac{\sum(y - \mu)^2}{n}} \qquad \text{(Equation 23.1)}[9]$$

We use $\bar{y}$ instead of μ, though, and that causes a problem. For any sample, the data values will generally be closer to their own sample mean than to the true population mean, μ. Why is that? Imagine that we take a random sample of 10 high school seniors. The mean SAT verbal score is 500 in the United States. But the sample mean, $\bar{y}$, for *these* 10 seniors won't be exactly 500. Are the 10 seniors' scores closer to 500 or $\bar{y}$? They'll always be closer to their own average $\bar{y}$. If we used $\sum(y - \bar{y})^2$ instead of $\sum(y - \mu)^2$ in Equation 23.1 to calculate s, our standard deviation estimate would be too small. How can we fix it? The amazing mathematical fact is that we an compensate for the smaller sum exactly by dividing by $n - 1$ instead of by n. So that's all the $n - 1$ is doing in the denominator of s. And we call $n - 1$ the degrees of freedom.

WHAT CAN GO WRONG?

The most fundamental issue you face is knowing when to use Student's t methods.

▶ **Don't confuse proportions and means.** When you treat your data as categorical, counting successes and summarizing with a sample proportion, make inferences using the Normal model methods you learned about in Chapters 19 through 22. When you treat your data as quantitative, summarizing with a sample mean, make your inferences using Student's t methods.

Student's t methods work only when the Normality Assumption is true. Naturally, many of the ways things can go wrong turn out to be different ways that the Normality

(continued)

---

[9] Statistics textbooks usually have equation numbers so they can talk about equations by name. We haven't needed equation numbers yet, but we admit it's useful here, so this is our first.

Assumption can fail. It's always a good idea to look for the most common kinds of failure. It turns out that you can even fix some of them.

▶ **Beware of multimodality.** The Nearly Normal Condition clearly fails if a histogram of the data has two or more modes. When you see this, look for the possibility that your data come from two groups. If so, your best bet is to try to separate the data into different groups. (Use the variables to help distinguish the modes, if possible. For example, if the modes seem to be composed mostly of men in one and women in the other, split the data according to sex.) Then you could analyze each group separately.

▶ **Beware of skewed data.** Make a Normal probability plot and a histogram of the data. If the data are very skewed, you might try re-expressing the variable. Re-expressing may yield a distribution that is unimodal and symmetric, more appropriate for Student's t inference methods for means. Re-expression cannot help if the sample distribution is not unimodal. Some people may object to re-expressing the data, but unless your sample is very large, you just can't use the methods of this chapter on skewed data.

▶ **Set outliers aside.** Student's t methods are built on the mean and standard deviation, so we should beware of outliers when using them. When you make a histogram to check the Nearly Normal Condition, be sure to check for outliers as well. If you find some, consider doing the analysis twice, both with the outliers excluded and with them included in the data, to get a sense of how much they affect the results.

The suggestion that you can perform an analysis with outliers removed may be controversial in some disciplines. Setting aside outliers is seen by some as "cheating." But an analysis of data with outliers left in place is *always* wrong. The outliers violate the Nearly Normal Condition and also the implicit assumption of a homogeneous population, so they invalidate inference procedures. An analysis of the non-outlying points, along with a separate discussion of the outliers, is often much more informative and can reveal important aspects of the data.

How can you tell whether there are outliers in your data? The "outlier nomination rule" of boxplots can offer some guidance, but it's just a rule of thumb and not an absolute definition. The best practical definition is that a value is an outlier if removing it substantially changes your conclusions about the data. You won't want a single value to determine your understanding of the world unless you are very, very sure that it is absolutely correct. Of course, when the outliers affect your conclusion, this can lead to the uncomfortable state of not really knowing what to conclude. Such situations call for you to use your knowledge of the real world and your understanding of the data you are working with.[10]

Of course, Normality issues aren't the only risks you face when doing inferences about means. Remember to *Think* about the usual suspects.

▶ **Watch out for bias.** Measurements of all kinds can be biased. If your observations differ from the true mean in a systematic way, your confidence interval may not capture the true mean. And there is no sample size that will save you. A bathroom scale that's 5 pounds off will be 5 pounds off even if you weigh yourself 100 times and take the average. We've seen several sources of bias in surveys, and measurements can be biased, too. Be sure to think about possible sources of bias in your measurements.

▶ **Make sure cases are independent.** Student's t methods also require the sampled values to be mutually independent. We check for random sampling and the 10% Condition. You should also think hard about whether there are likely violations of independence in the data collection method. If there are, be very cautious about using these methods.

▶ **Make sure that data are from an appropriately randomized sample.** Ideally, all data that we analyze are drawn from a simple random sample or generated by a randomized experiment. When they're not, be careful about making inferences from them. You

As tempting as it is to get rid of annoying values, you can't just throw away outliers and not discuss them. It isn't appropriate to lop off the highest or lowest values just to improve your results.

---

[10] An important reason for you to know Statistics rather than let someone else analyze your data.

may still compute a confidence interval correctly, or get the mechanics of the P-value right, but this might not save you from making a serious mistake in inference.

▶ **Interpret your confidence interval correctly.** Many statements that sound tempting are, in fact, misinterpretations of a confidence interval for a mean. You might want to have another look at some of the common mistakes, explained on pages 541–2. Keep in mind that a confidence interval is about the mean of the population, not about the means of samples, individuals in samples, or individuals in the population.

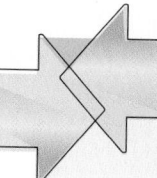

CONNECTIONS

The steps for finding a confidence interval or hypothesis test for means are just like the corresponding steps for proportions. Even the form of the calculations is similar. As the z-statistic did for proportions, the t-statistic tells us how many standard errors our sample mean is from the hypothesized mean. For means, though, we have to estimate the standard error separately. This added uncertainty changes the model for the sampling distribution from z to t.

As with all of our inference methods, the randomization applied in drawing a random sample or in randomizing a comparative experiment is what generates the sampling distribution. Randomization is what makes inference in this way possible at all.

The new concept of degrees of freedom connects back to the denominator of the sample standard deviation calculation, as shown earlier.

There's just no escaping histograms and Normal probability plots. The Nearly Normal Condition required to use Student's t can be checked best by making appropriate displays of the data. Back when we first used histograms, we looked at their shape and, in particular, checked whether they were unimodal and symmetric, and whether they showed any outliers. Those are just the features we check for here. The Normal probability plot zeros in on the Normal model a little more precisely.

WHAT HAVE WE LEARNED?

We first learned to create confidence intervals and test hypotheses about proportions. Now we've turned our attention to means, and learned that statistical inference for means relies on the same concepts; only the mechanics and our model have changed.

▶ We've learned that what we can say about a population mean is inferred from data, using the mean of a representative random sample.

▶ We've learned to describe the sampling distribution of sample means using a new model we select from the Student's t family based on our degrees of freedom.

▶ We've learned that our ruler for measuring the variability in sample means is the standard error $SE(\bar{y}) = \frac{s}{\sqrt{n}}$.

▶ We've learned to find the margin of error for a confidence interval using that ruler and critical values based on a Student's t-model.

▶ And we've also learned to use that ruler to test hypotheses about the population mean.

Above all, we've learned that the reasoning of inference, the need to verify that the appropriate assumptions are met, and the proper interpretation of confidence intervals and P-values all remain the same regardless of whether we are investigating means or proportions.

Terms

Student's t
Degrees of freedom (df)

533. A family of distributions indexed by its degrees of freedom. The t-models are unimodal symmetric, and bell shaped, but generally have fatter tails and a narrower center than the Normal model. As the degrees of freedom increase, t-distributions approach the Normal.

One-sample t-interval for the mean

534. A one-sample t-interval for the population mean is

$$\bar{y} \pm t^{*}_{n-1} \times SE(\bar{y}), \text{ where } SE(\bar{y}) = \frac{s}{\sqrt{n}}.$$

The critical value t^{*}_{n-1} depends on the particular confidence level, C, that you specify and on the number of degrees of freedom, $n - 1$.

One-sample t-test for the mean

543. The one-sample t-test for the mean tests the hypothesis $H_0: \mu = \mu_0$ using the statistic

$$t_{n-1} = \frac{\bar{y} - \mu_0}{SE(\bar{y})}.$$

The standard error of $\bar{y}$ is

$$SE(\bar{y}) = \frac{s}{\sqrt{n}}.$$

Skills

▸ Know the assumptions required for t-tests and t-based confidence intervals.

▸ Know how to examine your data for violations of conditions that would make inference about the population mean unwise or invalid.

▸ Understand that a confidence interval and a hypothesis test are essentially equivalent. You can do a two-tailed hypothesis test at level of significance α with a $1 - \alpha$ confidence interval, or a one-tailed test with a $1 - 2\alpha$ confidence interval.

▸ Be able to compute and interpret a t-test for the population mean using a statistics package or working from summary statistics for a sample.

▸ Be able to compute and interpret a t-based confidence interval for the population mean using a statistics package or working from summary statistics for a sample.

▸ Be able to explain the meaning of a confidence interval for a population mean. Make clear that the randomness associated with the confidence level is a statement about the interval bounds and not about the population parameter value.

▸ Understand that a 95% confidence interval does not trap 95% of the sample values.

▸ Be able to interpret the result of a test of a hypothesis about a population mean.

▸ Know that we do not "accept" a null hypothesis if we cannot reject it. We say that we fail to reject it.

▸ Understand that the P-value of a test does not give the probability that the null hypothesis is correct.

INFERENCE FOR MEANS ON THE COMPUTER

Statistics packages offer convenient ways to make histograms of the data. Even better for assessing near-Normality is a Normal probability plot. When you work on a computer, there is simply no excuse for skipping the step of plotting the data to check that it is nearly Normal. *Beware:* Statistics packages don't agree on whether to place the Normal scores on the x-axis (as we have done) or the y-axis. Read the axis labels.

Any standard statistics package can compute a hypothesis test. Here's what the package output might look like in general (although no package we know gives the results in exactly this form):[11]

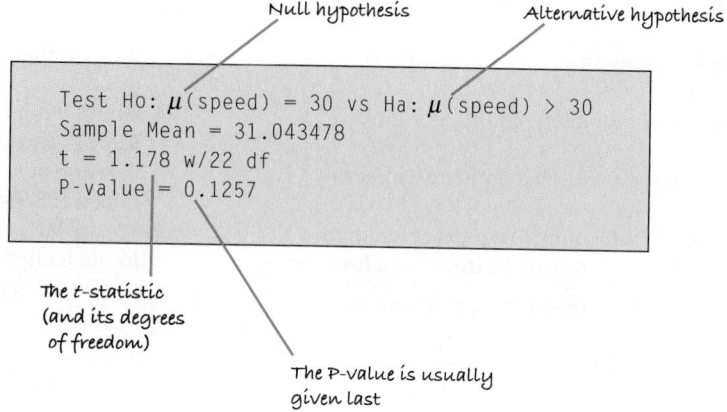

Null hypothesis   Alternative hypothesis

```
Test Ho: μ(speed) = 30 vs Ha: μ(speed) > 30
Sample Mean = 31.043478
t = 1.178 w/22 df
P-value = 0.1257
```

The *t*-statistic
(and its degrees
of freedom)

The P-value is usually
given last

A S *Activity:* **Student's *t* in Practice.** We almost always use technology to do inference with Student's *t*. Here's a chance to do that as you investigate several questions.

The package computes the sample mean and sample standard deviation of the variable and finds the P-value from the *t*-distribution based on the appropriate number of degrees of freedom. All modern statistics packages report P-values. The package may also provide additional information such as the sample mean, sample standard deviation, *t*-statistic value, and degrees of freedom. These are useful for interpreting the resulting P-value and telling the difference between a meaningful result and one that is merely statistically significant. Statistics packages that report the estimated standard deviation of the sampling distribution usually label it "standard error" or "SE."

Inference results are also sometimes reported in a table. You may have to read carefully to find the values you need. Often, test results and the corresponding confidence interval bounds are given together. And often you must read carefully to find the alternative hypotheses. Here's an example of that kind of output:

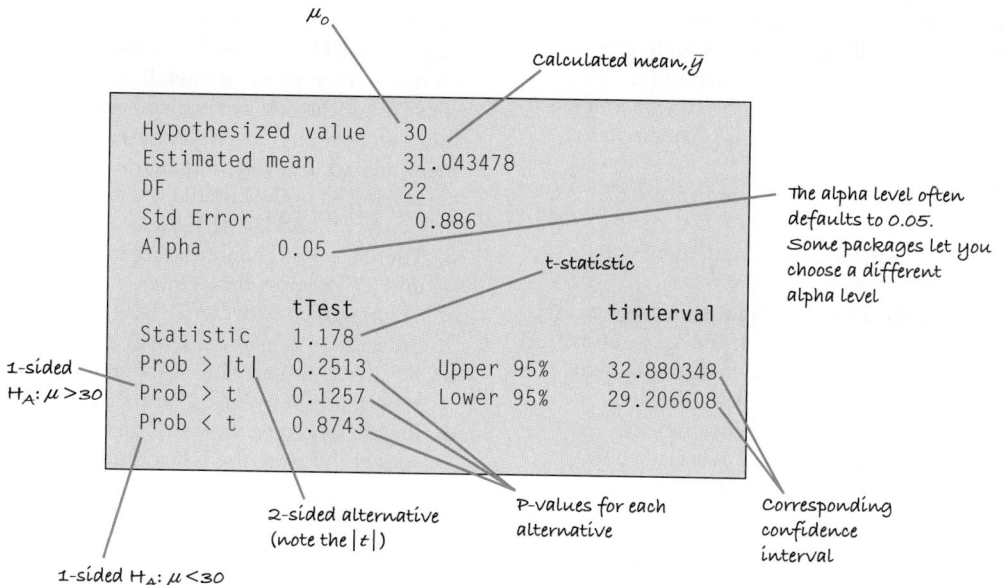

μ_0 Calculated mean, $\bar{y}$

```
Hypothesized value    30
Estimated mean        31.043478
DF                    22
Std Error              0.886
Alpha      0.05
                tTest                tinterval
Statistic       1.178
Prob > |t|      0.2513    Upper 95%    32.880348
Prob > t        0.1257    Lower 95%    29.206608
Prob < t        0.8743
```

The alpha level often
defaults to 0.05.
Some packages let you
choose a different
alpha level

t-statistic

1-sided
$H_A: \mu > 30$

1-sided $H_A: \mu < 30$

2-sided alternative
(note the |*t*|)

P-values for each
alternative

Corresponding
confidence
interval

The commands to do inference for means on common statistics programs and calculators are not always obvious. (By contrast, the resulting output is usually clearly labeled and easy to read.) The guides for each program can help you start navigating.

---

[11] Many statistics packages keep as many as 16 digits for all intermediate calculations. If we had kept as many, our results in the Step-By-Step section would have been closer to these.

EXERCISES

1. **t-models, part I.** Using the t tables, software, or a calculator, estimate
   a) the critical value of t for a 90% confidence interval with df = 17.
   b) the critical value of t for a 98% confidence interval with df = 88.
   c) the P-value for $t \geq 2.09$ with 4 degrees of freedom.
   d) the P-value for $|t| > 1.78$ with 22 degrees of freedom.

2. **t-models, part II.** Using the t tables, software, or a calculator, estimate
   a) the critical value of t for a 95% confidence interval with df = 7.
   b) the critical value of t for a 99% confidence interval with df = 102.
   c) the P-value for $t \leq 2.19$ with 41 degrees of freedom.
   d) the P-value for $|t| > 2.33$ with 12 degrees of freedom.

3. **t-models, part III.** Describe how the shape, center, and spread of t-models change as the number of degrees of freedom increases.

4. **t-models, part IV (last one!).** Describe how the critical value of t for a 95% confidence interval changes as the number of degrees of freedom increases.

5. **Cattle.** Livestock are given a special feed supplement to see if it will promote weight gain. Researchers report that the 77 cows studied gained an average of 56 pounds, and that a 95% confidence interval for the mean weight gain this supplement produces has a margin of error of ±11 pounds. Some students wrote the following conclusions. Did anyone interpret the interval correctly? Explain any misinterpretations.
   a) 95% of the cows studied gained between 45 and 67 pounds.
   b) We're 95% sure that a cow fed this supplement will gain between 45 and 67 pounds.
   c) We're 95% sure that the average weight gain among the cows in this study was between 45 and 67 pounds.
   d) The average weight gain of cows fed this supplement will be between 45 and 67 pounds 95% of the time.
   e) If this supplement is tested on another sample of cows, there is a 95% chance that their average weight gain will be between 45 and 67 pounds.

6. **Teachers.** Software analysis of the salaries of a random sample of 288 Nevada teachers produced the confidence interval shown below. Which conclusion is correct? What's wrong with the others?

   t-Interval for μ:
   with 90.00% Confidence,
   38944 < μ(TchPay) < 42893

   a) If we took many random samples of 288 Nevada teachers, about 9 out of 10 of them would produce this confidence interval.
   b) If we took many random samples of Nevada teachers, about 9 out of 10 of them would produce a confidence interval that contained the mean salary of all Nevada teachers.
   c) About 9 out of 10 Nevada teachers earn between $38,944 and $42,893.
   d) About 9 out of 10 of the teachers surveyed earn between $38,944 and $42,893.
   e) We are 90% confident that the average teacher salary in the United States is between $38,944 and $42,893.

7. **Meal plan.** After surveying students at Dartmouth College, a campus organization calculated that a 95% confidence interval for the mean cost of food for one term (of three in the Dartmouth trimester calendar) is ($1102, $1290). Now the organization is trying to write its report and is considering the following interpretations. Comment on each.
   a) 95% of all students pay between $1102 and $1290 for food.
   b) 95% of the sampled students paid between $1102 and $1290.
   c) We're 95% sure that students in this sample averaged between $1102 and $1290 for food.
   d) 95% of all samples of students will have average food costs between $1102 and $1290.
   e) We're 95% sure that the average amount all students pay is between $1102 and $1290.

8. **Snow.** Based on meteorological data for the past century, a local TV weather forecaster estimates that the region's average winter snowfall is 23", with a margin of error of ±2 inches. Assuming he used a 95% confidence interval, how should viewers interpret this news? Comment on each of these statements:
   a) During 95 of the last 100 winters, the region got between 21" and 25" of snow.
   b) There's a 95% chance the region will get between 21" and 25" of snow this winter.
   c) There will be between 21" and 25" of snow on the ground for 95% of the winter days.
   d) Residents can be 95% sure that the area's average snowfall is between 21" and 25".
   e) Residents can be 95% confident that the average snowfall during the last century was between 21" and 25" per winter.

9. **Pulse rates.** A medical researcher measured the pulse rates (beats per minute) of a sample of randomly selected adults and found the following Student's t-based confidence interval:

   With 95.00% Confidence,
   70.887604 < μ(Pulse) < 74.497011

   a) Explain carefully what the software output means.
   b) What's the margin of error for this interval?
   c) If the researcher had calculated a 99% confidence interval, would the margin of error be larger or smaller? Explain.

16. Parking II. Suppose that, for budget planning purposes, the city in Exercise 14 needs a better estimate of the mean daily income from parking fees.
   a) Someone suggests that the city use its data to create a 95% confidence interval instead of the 90% interval first created. How would this interval be better for the city? (You need not actually create the new interval.)
   b) How would the 95% interval be worse for the planners?
   c) How could they achieve an interval estimate that would better serve their planning needs?
   d) How many days' worth of data should they collect to have 95% confidence of estimating the true mean to within $3?

17. Speed of light. In 1882 Michelson measured the speed of light (usually denoted c as in Einstein's famous equation $E = mc^2$). His values are in km/sec and have 299,000 subtracted from them. He reported the results of 23 trials with a mean of 756.22 and a standard deviation of 107.12.
   a) Find a 95% confidence interval for the true speed of light from these statistics.
   b) State in words what this interval means. Keep in mind that the speed of light is a physical constant that, as far as we know, has a value that is true throughout the universe.
   c) What assumptions must you make in order to use your method?

T 18. Better light. After his first attempt to determine the speed of light (described in Exercise 17), Michelson conducted an "improved" experiment. In 1897 he reported results of 100 trials with a mean of 852.4 and a standard deviation of 79.0.
   a) What is the standard error of the mean for these data?
   b) Without computing it, how would you expect a 95% confidence interval for the second experiment to differ from the confidence interval for the first? Note at least three specific reasons why they might differ, and indicate the ways in which these differences would change the interval.
   c) According to Stigler (who reports these values), the true speed of light is 299,710.5 km/sec, corresponding to a value of 710.5 for Michelson's 1897 measurements. What does this indicate about Michelson's two experiments? Explain, using your confidence interval.

T 19. Departures. What are the chances your flight will leave on time? The U.S. Bureau of Transportation Statistics of the Department of Transportation publishes information about airline performance. Here are a histogram and summary statistics for the percentage of flights departing on time each month from 1995 thru 2006.

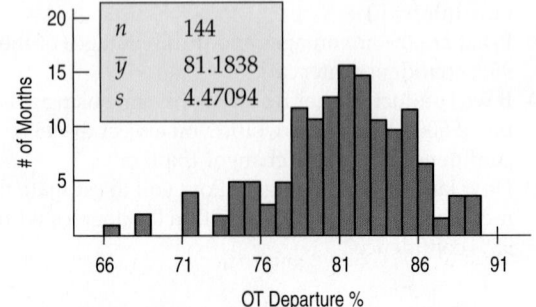

There is no evidence of a trend over time. (The correlation of On Time Departure% with time is $r = -0.016$.)
   a) Check the assumptions and conditions for inference.
   b) Find a 90% confidence interval for the true percentage of flights that depart on time.
   c) Interpret this interval for a traveler planning to fly.

T 20. Late arrivals. Will your flight get you to your destination on time? The U.S. Bureau of Transportation Statistics reported the percentage of flights that were late each month from 1995 through 2006. Here's a histogram, along with some summary statistics:

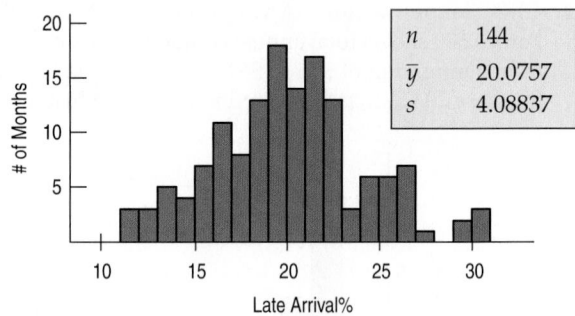

We can consider these data to be a representative sample of all months. There is no evidence of a time trend ($r = -0.07$).
   a) Check the assumptions and conditions for inference about the mean.
   b) Find a 99% confidence interval for the true percentage of flights that arrive late.
   c) Interpret this interval for a traveler planning to fly.

T 21. For Example, 2nd look. This chapter's For Examples looked at mirex contamination in farmed salmon. We first found a 95% confidence interval for the mean concentration to be 0.0834 to 0.0992 parts per million. Later we rejected the null hypothesis that the mean did not exceed the EPA's recommended safe level of 0.08 ppm based on a P-value of 0.0027. Explain how these two results are consistent. Your explanation should discuss the confidence level, the P-value, and the decision.

22. Hot Dogs. A nutrition lab tested 40 hot dogs to see if their mean sodium content was less than the 325 mg upper limit set by regulations for "reduced sodium" franks. The lab failed to reject the hypothesis that the hot dogs did not meet this requirement, with a P-value of 0.142. A 90% confidence interval estimated the mean sodium content for this kind of hot dog at 317.2 to 326.8 mg. Explain how these two results are consistent. Your explanation should discuss the confidence level, the P-value, and the decision.

23. Pizza. A researcher tests whether the mean cholesterol level among those who eat frozen pizza exceeds the value considered to indicate a health risk. She gets a P-value of 0.07. Explain in this context what the "7%" represents.

24. Golf balls. The United States Golf Association (USGA) sets performance standards for golf balls. For example, the initial velocity of the ball may not exceed 250 feet per second when measured by an apparatus approved by the USGA. Suppose a manufacturer introduces a new kind of ball and provides a sample for testing. Based on the mean speed in the test, the USGA comes up with a P-value of 0.34. Explain in this context what the "34%" represents.

25. **TV safety.** The manufacturer of a metal stand for home TV sets must be sure that its product will not fail under the weight of the TV. Since some larger sets weigh nearly 300 pounds, the company's safety inspectors have set a standard of ensuring that the stands can support an average of over 500 pounds. Their inspectors regularly subject a random sample of the stands to increasing weight until they fail. They test the hypothesis $H_0: \mu = 500$ against $H_A: \mu > 500$, using the level of significance $\alpha = 0.01$. If the sample of stands fail to pass this safety test, the inspectors will not certify the product for sale to the general public.

a) Is this an upper-tail or lower-tail test? In the context of the problem, why do you think this is important?
b) Explain what will happen if the inspectors commit a Type I error.
c) Explain what will happen if the inspectors commit a Type II error.

26. **Catheters.** During an angiogram, heart problems can be examined via a small tube (a catheter) threaded into the heart from a vein in the patient's leg. It's important that the company that manufactures the catheter maintain a diameter of 2.00 mm. (The standard deviation is quite small.) Each day, quality control personnel make several measurements to test $H_0: \mu = 2.00$ against $H_A: \mu \neq 2.00$ at a significance level of $\alpha = 0.05$. If they discover a problem, they will stop the manufacturing process until it is corrected.

a) Is this a one-sided or two-sided test? In the context of the problem, why do you think this is important?
b) Explain in this context what happens if the quality control people commit a Type I error.
c) Explain in this context what happens if the quality control people commit a Type II error.

27. **TV safety revisited.** The manufacturer of the metal TV stands in Exercise 25 is thinking of revising its safety test.

a) If the company's lawyers are worried about being sued for selling an unsafe product, should they increase or decrease the value of α? Explain.
b) In this context, what is meant by the power of the test?
c) If the company wants to increase the power of the test, what options does it have? Explain the advantages and disadvantages of each option.

28. **Catheters again.** The catheter company in Exercise 26 is reviewing its testing procedure.

a) Suppose the significance level is changed to $\alpha = 0.01$. Will the probability of a Type II error increase, decrease, or remain the same?
b) What is meant by the power of the test the company conducts?
c) Suppose the manufacturing process is slipping out of proper adjustment. As the actual mean diameter of the catheters produced gets farther and farther above the desired 2.00 mm, will the power of the quality control test increase, decrease, or remain the same?
d) What could they do to improve the power of the test?

29. **Marriage.** In 1960, census results indicated that the age at which American men first married had a mean of 23.3 years.

It is widely suspected that young people today are waiting longer to get married. We want to find out if the mean age of first marriage has increased during the past 40 years.

a) Write appropriate hypotheses.
b) We plan to test our hypothesis by selecting a random sample of 40 men who married for the first time last year. Do you think the necessary assumptions for inference are satisfied? Explain.
c) Describe the approximate sampling distribution model for the mean age in such samples.
d) The men in our sample married at an average age of 24.2 years, with a standard deviation of 5.3 years. What's the P-value for this result?
e) Explain (in context) what this P-value means.
f) What's your conclusion?

30. **Fuel economy.** A company with a large fleet of cars hopes to keep gasoline costs down and sets a goal of attaining a fleet average of at least 26 miles per gallon. To see if the goal is being met, they check the gasoline usage for 50 company trips chosen at random, finding a mean of 25.02 mpg and a standard deviation of 4.83 mpg. Is this strong evidence that they have failed to attain their fuel economy goal?

a) Write appropriate hypotheses.
b) Are the necessary assumptions to make inferences satisfied?
c) Describe the sampling distribution model of mean fuel economy for samples like this.
d) Find the P-value.
e) Explain what the P-value means in this context.
f) State an appropriate conclusion.

31. **Ruffles.** Students investigating the packaging of potato chips purchased 6 bags of Lay's Ruffles marked with a net weight of 28.3 grams. They carefully weighed the contents of each bag, recording the following weights (in grams): 29.3, 28.2, 29.1, 28.7, 28.9, 28.5.

a) Do these data satisfy the assumptions for inference? Explain.
b) Find the mean and standard deviation of the weights.
c) Create a 95% confidence interval for the mean weight of such bags of chips.
d) Explain in context what your interval means.
e) Comment on the company's stated net weight of 28.3 grams.

32. **Doritos.** Some students checked 6 bags of Doritos marked with a net weight of 28.3 grams. They carefully weighed the contents of each bag, recording the following weights (in grams): 29.2, 28.5, 28.7, 28.9, 29.1, 29.5.

a) Do these data satisfy the assumptions for inference? Explain.
b) Find the mean and standard deviation of the weights.
c) Create a 95% confidence interval for the mean weight of such bags of chips.
d) Explain in context what your interval means.
e) Comment on the company's stated net weight of 28.3 grams.

33. **Popcorn.** Yvon Hopps ran an experiment to test optimum power and time settings for microwave popcorn. His goal was to find a combination of power and time that would deliver high-quality popcorn with less than 10%

of the kernels left unpopped, on average. After experimenting with several bags, he determined that power 9 at 4 minutes was the best combination.

a) He concluded that this popping method achieved the 10% goal. If it really does not work that well, what kind of error did Hopps make?

b) To be sure that the method was successful, he popped 8 more bags of popcorn (selected at random) at this setting. All were of high quality, with the following percentages of uncooked popcorn: 7, 13.2, 10, 6, 7.8, 2.8, 2.2, 5.2. Does this provide evidence that he met his goal of an average of no more than 10% uncooked kernels? Explain.

34. Ski wax. Bjork Larsen was trying to decide whether to use a new racing wax for cross-country skis. He decided that the wax would be worth the price if he could average less than 55 seconds on a course he knew well, so he planned to test the wax by racing on the course 8 times.

a) Suppose that he eventually decides not to buy the wax, but it really would lower his average time to below 55 seconds. What kind of error would he have made?

b) His 8 race times were 56.3, 65.9, 50.5, 52.4, 46.5, 57.8, 52.2, and 43.2 seconds. Should he buy the wax? Explain.

35. Chips Ahoy. In 1998, as an advertising campaign, the Nabisco Company announced a "1000 Chips Challenge," claiming that every 18-ounce bag of their Chips Ahoy cookies contained at least 1000 chocolate chips. Dedicated Statistics students at the Air Force Academy (no kidding) purchased some randomly selected bags of cookies, and counted the chocolate chips. Some of their data are given below. (*Chance,* 12, no. 1[1999])

1219	1214	1087	1200	1419	1121	1325	1345
1244	1258	1356	1132	1191	1270	1295	1135

a) Check the assumptions and conditions for inference. Comment on any concerns you have.

b) Create a 95% confidence interval for the average number of chips in bags of Chips Ahoy cookies.

c) What does this evidence say about Nabisco's claim? Use your confidence interval to test an appropriate hypothesis and state your conclusion.

36. Yogurt. *Consumer Reports* tested 14 brands of vanilla yogurt and found these numbers of calories per serving:

160	200	220	230	120	180	140
130	170	190	80	120	100	170

a) Check the assumptions and conditions for inference.

b) Create a 95% confidence interval for the average calorie content of vanilla yogurt.

c) A diet guide claims that you will get 120 calories from a serving of vanilla yogurt. What does this evidence indicate? Use your confidence interval to test an appropriate hypothesis and state your conclusion.

37. Maze. Psychology experiments sometimes involve testing the ability of rats to navigate mazes. The mazes are classified according to difficulty, as measured by the mean length of time it takes rats to find the food at the

end. One researcher needs a maze that will take rats an average of about one minute to solve. He tests one maze on several rats, collecting the data shown.

Time (sec)	
38.4	57.6
46.2	55.5
62.5	49.5
38.0	40.9
62.8	44.3
33.9	93.8
50.4	47.9
35.0	69.2
52.8	46.2
60.1	56.3
55.1	

a) Plot the data. Do you think the conditions for inference are satisfied? Explain.

b) Test the hypothesis that the mean completion time for this maze is 60 seconds. What is your conclusion?

c) Eliminate the outlier, and test the hypothesis again. What is your conclusion?

d) Do you think this maze meets the "one-minute average" requirement? Explain.

38. Braking. A tire manufacturer is considering a newly designed tread pattern for its all-weather tires. Tests have indicated that these tires will provide better gas mileage and longer tread life. The last remaining test is for braking effectiveness. The company hopes the tire will allow a car traveling at 60 mph to come to a complete stop within an average of 125 feet after the brakes are applied. They will adopt the new tread pattern unless there is strong evidence that the tires do not meet this objective. The distances (in feet) for 10 stops on a test track were 129, 128, 130, 132, 135, 123, 102, 125, 128, and 130. Should the company adopt the new tread pattern? Test an appropriate hypothesis and state your conclusion. Explain how you dealt with the outlier and why you made the recommendation you did.

39. Driving distance. How far do professional golfers drive a ball? (For non-golfers, the drive is the shot hit from a tee at the start of a hole and is typically the longest shot.) Here's a histogram of the average driving distances of the 202 leading professional golfers in 2006 along with summary statistics.

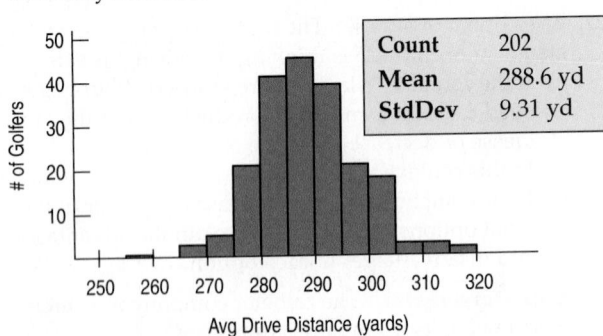

Count	202
Mean	288.6 yd
StdDev	9.31 yd

Avg Drive Distance (yards)

a) Find a 95% confidence interval for the mean drive distance.

b) Interpreting this interval raises some problems. Discuss.

c) The data are the mean driving distance for each golfer. Is that a concern in interpreting the interval? (*Hint:* Review the What Can Go Wrong warnings of Chapter 9. Chapter 9?! Yes, Chapter 9.)

40. Wind power. Should you generate electricity with your own personal wind turbine? That depends on whether you have enough wind on your site. To produce enough energy, your site should have an annual average wind speed above 8 miles per hour, according to the Wind Energy Association. One candidate site was monitored for a year, with wind speeds recorded every 6 hours. A total of 1114 readings of wind speed averaged 8.019 mph with a standard deviation of 3.813 mph. You've been asked to make a statistical report to help the landowner decide whether to place a wind turbine at this site.

a) Discuss the assumptions and conditions for using Student's t inference methods with these data. Here are some plots that may help you decide whether the methods can be used:

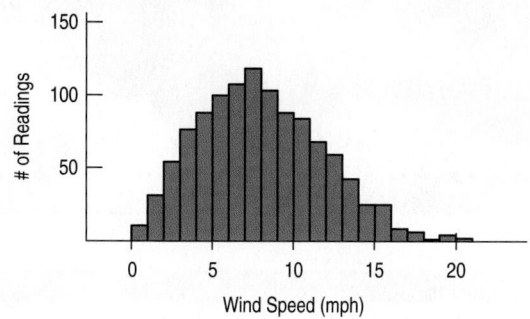

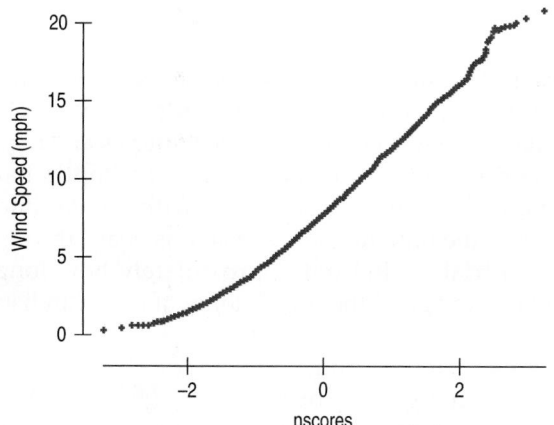

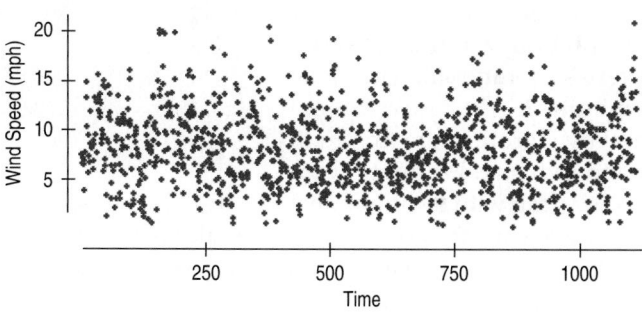

b) What would you tell the landowner about whether this site is suitable for a small wind turbine? Explain.

JUST CHECKING
Answers

1. Questions on the short form are answered by everyone in the population. This is a census, so means or proportions *are* the true population values. The long forms are given just to a sample of the population. When we estimate parameters from a sample, we use a confidence interval to take sample-to-sample variability into account.

2. They don't know the population standard deviation, so they must use the sample SD as an estimate. The additional uncertainty is taken into account by t-models.

3. The margin of error for a confidence interval for a mean depends, in part, on the standard error,

$$SE(\bar{y}) = \frac{s}{\sqrt{n}}.$$

Since n is in the denominator, smaller sample sizes lead to larger SEs and correspondingly wider intervals. Long forms returned by one in every six or seven households in a less populous area will be a smaller sample.

4. The critical values for t with fewer degrees of freedom would be slightly larger. The $\sqrt{n}$ part of the standard error changes a lot, making the SE much larger. Both would increase the margin of error.

5. The smaller sample is one fourth as large, so the confidence interval would be roughly twice as wide.

6. We expect 95% of such intervals to cover the true value, so 5 of the 100 intervals might be expected to miss.

7. The power would increase if we have a larger sample size.

Comparing Means

WHO	AA alkaline batteries
WHAT	Length of battery life while playing a CD continuously
UNITS	Minutes
WHY	Class project
WHEN	1998

A S **Video: Can Diet Prolong Life?** Watch a video that tells the story of an experiment. We'll analyze the data later in this chapter.

Should you buy generic rather than brand-name batteries? A Statistics student designed a study to test battery life. He wanted to know whether there was any real difference between brand-name batteries and a generic brand. To estimate the difference in mean lifetimes, he kept a battery-powered CD player[1] continuously playing the same CD, with the volume control fixed at 5, and measured the time until no more music was heard through the headphones. (He ran an initial trial to find out approximately how long that would take so that he didn't have to spend the first 3 hours of each run listening to the same CD.) For his trials he used six sets of AA alkaline batteries from two major battery manufacturers: a well-known brand name and a generic brand. He measured the time in minutes until the sound stopped. To account for changes in the CD player's performance over time, he randomized the run order by choosing sets of batteries at random. The table shows his data (times in minutes):

Brand Name	Generic
194.0	190.7
205.5	203.5
199.2	203.5
172.4	206.5
184.0	222.5
169.5	209.4

Studies that compare two groups are common throughout both science and industry. We might want to compare the effects of a new drug with the traditional therapy, the fuel efficiency of two car engine designs, or the sales of new products in two different test cities. In fact, battery manufacturers do research like this on their products and competitors' products themselves.

Plot the Data

The natural display for comparing two groups is boxplots of the data for the two groups, placed side by side. Although we can't make a confidence interval

---

[1] Once upon a time, not so very long ago, there were no iPods. At the turn of the century, people actually carried CDs around—and devices to play them. We bet you can find one in your parents' closet.

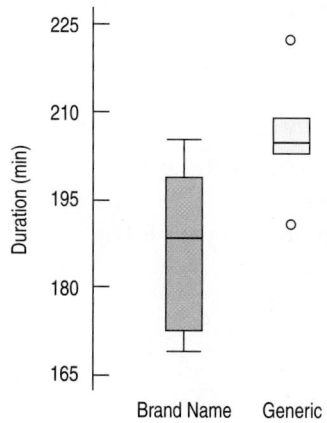

FIGURE 24.1

Boxplots comparing the brand-name and generic batteries suggest a difference in duration.

or test a hypothesis from the boxplots themselves, you should always start with boxplots when comparing groups. Let's look at the boxplots of the battery test data.

It sure looks like the generic batteries lasted longer. And we can see that they were also more consistent. But is the difference large enough to change our battery-buying behavior? Can we be confident that the difference is more than just random fluctuation? That's why we need statistical inference.

The boxplot for the generic data identifies two possible outliers. That's interesting, but with only six measurements in each group, the outlier nomination rule is not very reliable. Both of the extreme values are plausible results, and the range of the generic values is smaller than the range of the brand-name values, even with the outliers. So we're probably better off just leaving these values in the data.

Comparing Two Means

Comparing two means is not very different from comparing two proportions. In fact, it's not different in concept from any of the methods we've seen. Now, the population model parameter of interest is the difference between the *mean* battery lifetimes of the two brands, $\mu_1 - \mu_2$.

The rest is the same as before. The statistic of interest is the difference in the two observed means, $\bar{y}_1 - \bar{y}_2$. We'll start with this statistic to build our confidence interval, but we'll need to know its standard deviation and its sampling model. Then we can build confidence intervals and find P-values for hypothesis tests.

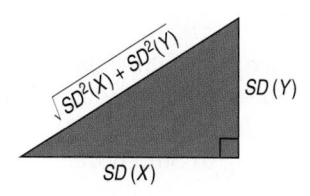

The Pythagorean
Theorem of Statistics

We know that, for independent random variables, the variance of their *difference* is the *sum* of their individual variances, $Var(Y - X) = Var(Y) + Var(X)$. To find the standard deviation of the difference between the two independent sample means, we add their variances and then take a square root:

$$SD(\bar{y}_1 - \bar{y}_2) = \sqrt{Var(\bar{y}_1) + Var(\bar{y}_2)}$$

$$= \sqrt{\left(\frac{\sigma_1}{\sqrt{n_1}}\right)^2 + \left(\frac{\sigma_2}{\sqrt{n_2}}\right)^2}$$

$$= \sqrt{\frac{\sigma_1^2}{n_1} + \frac{\sigma_2^2}{n_2}}.$$

Of course, we still don't know the true standard deviations of the two groups, σ_1 and σ_2, so as usual, we'll use the estimates, s_1 and s_2. Using the estimates gives us the *standard error*:

$$SE(\bar{y}_1 - \bar{y}_2) = \sqrt{\frac{s_1^2}{n_1} + \frac{s_2^2}{n_2}}.$$

We'll use the standard error to see how big the difference really is. Because we are working with means and estimating the standard error of their difference using the data, we shouldn't be surprised that the sampling model is a Student's *t*.

Finding the standard error of the difference in independent sample means

Can you tell how much you are eating from how full you are? Or do you need visual cues? Researchers[2] constructed a table with two ordinary 18 oz soup bowls and two identical-looking bowls that had been modified to slowly, imperceptibly, refill as they were emptied. They assigned experiment participants to the bowls randomly and served them tomato soup. Those eating from the ordinary bowls had their bowls refilled by ladle whenever they were one-quarter full. If people judge their portions by internal cues, they should eat about the same amount. How big a difference was there in the amount of soup consumed? The table summarizes their results.

	Ordinary bowl	Refilling bowl
n	27	27
$\bar{y}$	8.5 oz	14.7 oz
s	6.1 oz	8.4 oz

Question: How much variability do we expect in the difference between the two means? Find the standard error.

Participants were randomly assigned to bowls, so the two groups should be independent. It's okay to add variances.

$$SE(\bar{y}_{refill} - \bar{y}_{ordinary}) = \sqrt{\frac{s_r^2}{n_r} + \frac{s_o^2}{n_o}} = \sqrt{\frac{8.4^2}{27} + \frac{6.1^2}{27}} = 2.0 \ oz.$$

The confidence interval we build is called a **two-sample *t*-interval** (for the difference in means). The corresponding hypothesis test is called a **two-sample *t*-test.** The interval looks just like all the others we've seen—the statistic plus or minus an estimated margin of error:

$$(\bar{y}_1 - \bar{y}_2) \pm ME$$

$$\text{where } ME = t^* \times SE(\bar{y}_1 - \bar{y}_2).$$

Compare this formula with the one for the confidence interval for the difference of two proportions we saw in Chapter 22 (page 505). The formulas are almost the same. It's just that here we use a Student's *t*-model instead of a Normal model to find the appropriate critical *t**-value corresponding to our chosen confidence level.

What are we missing? Only the degrees of freedom for the Student's *t*-model. Unfortunately, *that* formula is strange.

The deep, dark secret is that the sampling model isn't *really* Student's *t*, but only something close. The trick is that by using a special, adjusted degrees-of-freedom value, we can make it so close to a Student's *t*-model that nobody can tell the difference. The adjustment formula is straightforward but doesn't help our understanding much, so we leave it to the computer or calculator. (If you are curious and really want to see the formula, look in the footnote.[3])

> **z or t?**
>
> If you know σ, use z. (That's rare!) Whenever you use s to estimate σ, use t.

---

[2] Brian Wansink, James E. Painter, and Jill North, "Bottomless Bowls: Why Visual Cues of Portion Size May Influence Intake," *Obesity Research*, Vol. 13, No. 1, January 2005.

[3]
$$df = \frac{\left(\frac{s_1^2}{n_1} + \frac{s_2^2}{n_2}\right)^2}{\frac{1}{n_1 - 1}\left(\frac{s_1^2}{n_1}\right)^2 + \frac{1}{n_2 - 1}\left(\frac{s_2^2}{n_2}\right)^2}$$

Are you sorry you looked? This formula usually doesn't even give a whole number. If you are using a table, you'll need a whole number, so round down to be safe. If you are using technology, it's even easier. The approximation formulas that computers and calculators use for the Student's *t*-distribution deal with degrees of freedom automatically.

> A SAMPLING DISTRIBUTION FOR THE DIFFERENCE BETWEEN TWO MEANS
>
> When the conditions are met, the sampling distribution of the standardized sample difference between the means of two independent groups,
>
> $$t = \frac{(\bar{y}_1 - \bar{y}_2) - (\mu_1 - \mu_2)}{SE(\bar{y}_1 - \bar{y}_2)},$$
>
> can be modeled by a Student's t-model with a number of degrees of freedom found with a special formula. We estimate the standard error with
>
> $$SE(\bar{y}_1 - \bar{y}_2) = \sqrt{\frac{s_1^2}{n_1} + \frac{s_2^2}{n_2}}.$$

Assumptions and Conditions

Now we've got everything we need. Before we can make a two-sample t-interval or perform a two-sample t-test, though, we have to check the assumptions and conditions.

INDEPENDENCE ASSUMPTION

Independence Assumption: The data in each group must be drawn independently and at random from a homogeneous population, or generated by a randomized comparative experiment. We can't expect that the data, taken as one big group, come from a homogeneous population, because that's what we're trying to test. But without randomization of some sort, there are no sampling distribution models and no inference. We can check two conditions:

Randomization Condition: Were the data collected with suitable randomization? For surveys, are they a representative random sample? For experiments, was the experiment randomized?

10% Condition: We usually don't check this condition for differences of means. We'll check it only if we have a very small population or an extremely large sample. We needn't worry about it at all for randomized experiments.

NORMAL POPULATION ASSUMPTION

As we did before with Student's t-models, we should check the assumption that the underlying populations are *each* Normally distributed. We check the . . .

Nearly Normal Condition: We must check this for *both* groups; a violation by either one violates the condition. As we saw for single sample means, the Normality Assumption matters most when sample sizes are small. For samples of $n < 15$ in either group, you should not use these methods if the histogram or Normal probability plot shows severe skewness. For n's closer to 40, a mildly skewed histogram is OK, but you should remark on any outliers you find and not work with severely skewed data. When both groups are bigger than 40, the Central Limit Theorem starts to kick in no matter how the data are distributed, so the Nearly Normal Condition for the data matters less. Even in large samples, however, you should still be on the lookout for outliers, extreme skewness, and multiple modes.

INDEPENDENT GROUPS ASSUMPTION

Independent Groups Assumption: To use the two-sample t methods, the two groups we are comparing must be independent of each other. In fact, this test is sometimes called the two *independent samples t*-test. No statistical test can verify this assumption. You have to think about how the data were collected. The assumption would be violated, for example, if one group consisted of husbands and the other group their wives. Whatever we measure on couples might naturally be related. Similarly, if we compared subjects' performances before some treatment with their performances afterward, we'd expect a relationship of each "before" measurement with its corresponding "after" measurement. In cases such as these, where the observational units in the two groups are related or matched, *the two-sample methods of this chapter can't be applied.* When this happens, we need a different procedure that we'll see in the next chapter.

FOR EXAMPLE   Checking assumptions and conditions

Recap: Researchers randomly assigned people to eat soup from one of two bowls: 27 got ordinary bowls that were refilled by ladle, and 27 others bowls that secretly refilled slowly as the people ate.

Question: Can the researchers use their data to make inferences about the role of visual cues in determining how much people eat?

✔ **Independence Assumption:** The amount consumed by one person should be independent of the amount consumed by others.

✔ **Randomization Condition:** Subjects were randomly assigned to the treatments.

✔ **Nearly Normal Condition:** The histograms for both groups look unimodal but somewhat skewed to the right. I believe both groups are large enough (27) to allow use of t-methods.

✔ **Independent Groups Assumption:** Randomization to treatment groups guarantees this.

It's okay to construct a two-sample t-interval for the difference in means.

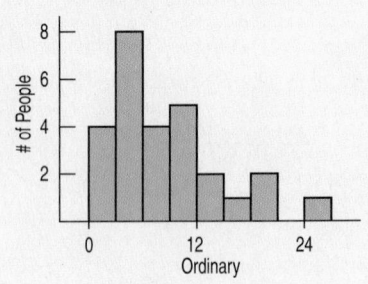

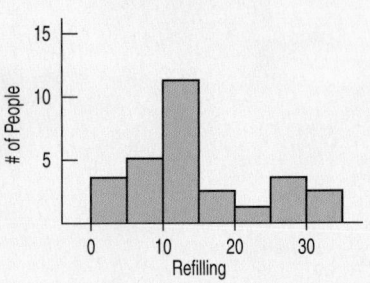

Note: When you check the Nearly Normal Condition it's important that you include the graphs you looked at (histograms or Normal probability plots).

TWO-SAMPLE *t*-INTERVAL FOR THE DIFFERENCE BETWEEN MEANS

When the conditions are met, we are ready to find the confidence interval for the difference between means of two independent groups, $\mu_1 - \mu_2$. The confidence interval is

$$(\bar{y}_1 - \bar{y}_2) \pm t^*_{df} \times SE(\bar{y}_1 - \bar{y}_2),$$

where the standard error of the difference of the means

$$SE(\bar{y}_1 - \bar{y}_2) = \sqrt{\frac{s_1^2}{n_1} + \frac{s_2^2}{n_2}}.$$

The critical value t^*_{df} depends on the particular confidence level, *C*, that you specify and on the number of degrees of freedom, which we get from the sample sizes and a special formula.

FOR EXAMPLE — Finding a confidence interval for the difference in sample means

Recap: Researchers studying the role of internal and visual cues in determining how much people eat conducted an experiment in which some people ate soup from bowls that secretly re-filled. The results are summarized in the table.

We've already checked the assumptions and conditions, and have found the standard error for the difference in means to be $SE(\bar{y}_{refill} - \bar{y}_{ordinary}) = 2.0$ oz.

	Ordinary bowl	Refilling bowl
n	27	27
$\bar{y}$	8.5 oz	14.7 oz
s	6.1 oz	8.4 oz

Question: What does a 95% confidence interval say about the difference in mean amounts eaten?

The observed difference in means is $\bar{y}_{refill} - \bar{y}_{ordinary} = (14.7 - 8.5) = 6.2\ oz$

$$df = 47.46 \quad t^*_{47.46} = 2.011\ (Table\ gives\ t^*_{45} = 2.014.)$$
$$ME = t^* \times SE(\bar{y}_{refill} - \bar{y}_{ordinary}) = 2.011(2.0) = 4.02\ oz$$

The 95% confidence interval for $\mu_{refill} - \mu_{ordinary}$ is 6.2 ± 4.02, or $(2.18, 10.22)$ oz.

I am 95% confident that people eating from a subtly refilling bowl will eat an average of between 2.18 and 10.22 more ounces of soup than those eating from an ordinary bowl.

STEP-BY-STEP EXAMPLE — A Two-Sample *t*-Interval

Judging from the boxplot, the generic batteries seem to have lasted about 20 minutes longer than the brand-name batteries. Before we change our buying habits, what should we expect to happen with the next batteries we buy?

Question: How much longer might the generic batteries last?

THINK

Plan State what we want to know.

Identify the *parameter* you wish to estimate. Here our parameter is the difference in the means, not the individual group means.

Identify the *population(s)* about which you wish to make statements. We hope to make decisions about purchasing batteries, so we're interested in all the AA batteries of these two brands.

Identify the variables and review the W's.

REALITY CHECK From the boxplots, it appears our confidence interval should be centered near a difference of 20 minutes. We don't have a lot of intuition about how far the interval should extend on either side of 20.

I have measurements of the lifetimes (in minutes) of 6 sets of generic and 6 sets of brand-name AA batteries from a randomized experiment. I want to find an interval that is likely, with 95% confidence, to contain the true difference $\mu_G - \mu_B$ between the mean lifetime of the generic AA batteries and the mean lifetime of the brand-name batteries.

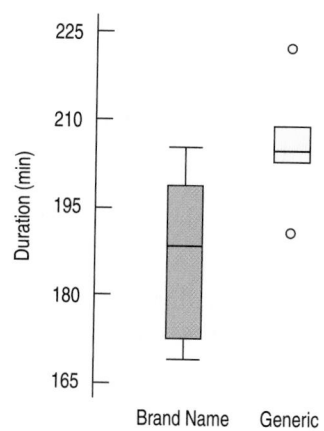

Model Think about the appropriate assumptions and check the conditions to be sure that a Student's *t*-model for the sampling distribution is appropriate.

For very small samples like these, we often don't worry about the 10% Condition.

✓ **Randomization Condition:** The batteries were selected at random from those available for sale. Not exactly an SRS, but a reasonably representative random sample.

✓ **Independence Assumption:** The batteries were packaged together, so they may not be independent. For example, a storage problem might affect all the batteries in the same pack. Repeating the study for several different packs of batteries would make the conclusions stronger.

✓ **Independent Groups Assumption:** Batteries manufactured by two different companies and purchased in separate packages should be independent.

✓ **Nearly Normal Condition:** The samples are small, but the histograms look unimodal and symmetric:

Make a picture. Boxplots are the display of choice for comparing groups, but now we want to check the *shape* of distribution of each group. Histograms or Normal probability plots do a better job there.

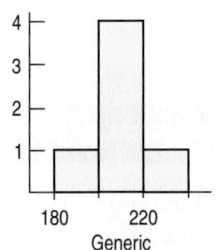 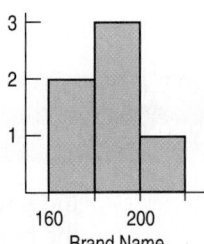

State the sampling distribution model for the statistic. Here the degrees of freedom will come from that messy approximation formula.

Under these conditions, it's okay to use a Student's t-model.

Specify your method.

I'll use a **two-sample t-interval.**

Mechanics Construct the confidence interval.

Be sure to include the units along with the statistics. Use meaningful subscripts to identify the groups.

Use the sample standard deviations to find the standard error of the sampling distribution.

We have three choices for degrees of freedom. The best alternative is to let the

I know $n_G = 6$ $\qquad$ $n_B = 6$

$\qquad \bar{y}_G = 206.0$ min $\qquad \bar{y}_B = 187.4$ min

$\qquad s_G = 10.3$ min $\qquad s_B = 14.6$ min

The groups are independent, so

$$SE(\bar{y}_G - \bar{y}_B) = \sqrt{SE^2(\bar{y}_G) + SE^2(\bar{y}_B)}$$

$$= \sqrt{\frac{s_G^2}{n_G} + \frac{s_B^2}{n_B}}$$

$$= \sqrt{\frac{10.3^2}{6} + \frac{14.6^2}{6}}$$

computer or calculator use the approximation formula for df. This gives a fractional degree of freedom (here df = 8.98), and technology can find a corresponding critical value. In this case, it is $t^* = 2.263$.

Or we could round the approximation formula's df value down to an integer so we can use a t table. That gives 8 df and a critical value $t^* = 2.306$.

The easy rule says to use only $6 - 1 = 5$ df. That gives a critical value $t^* = 2.571$. The corresponding confidence interval is about 14% wider—a high price to pay for a small savings in effort.

$$= \sqrt{\frac{106.09}{6} + \frac{213.16}{6}}$$
$$= \sqrt{53.208}$$
$$= 7.29 \text{ min.}$$

df (from technology[4]) = 8.98

The corresponding critical value for a 95% confidence level is $t^* = 2.263$.

So the margin of error is

$$ME = t^* \times SE(\bar{y}_G - \bar{y}_B)$$
$$= 2.263(7.29)$$
$$= 16.50 \text{ min.}$$

The 95% confidence interval is

$$(206.0 - 187.4) \pm 16.5 \text{ min.}$$
$$\text{or } 18.6 \pm 16.5 \text{ min.}$$
$$= (2.1, 35.1) \text{ min.}$$

Conclusion Interpret the confidence interval in the proper context.

Less formally, you could say, "I'm 95% confident that generic batteries last an average of 2.1 to 35.1 minutes longer than brand-name batteries."

I am 95% confident that the interval from 2.1 minutes to 35.1 minutes captures the mean amount of time by which generic batteries outlast brand-name batteries for this task. If generic batteries are cheaper, there seems little reason not to use them. If it is more trouble or costs more to buy them, then I'd consider whether the additional performance is worth it.

Another One Just Like the Other Ones?

A S *Activity:* **Find Two-Sample t-Intervals.** Who wants to deal with that ugly df formula? We usually find these intervals with a statistics package. Learn how here.

Yes. That's been our point all along. Once again we see a statistic plus or minus the margin of error. And the ME is just a critical value times the standard error. Just look out for that crazy degrees of freedom formula.

   Creating the confidence interval

If you have been successful using your TI to make confidence intervals for proportions and 1-sample means, then you can probably already use the 2-sample function just fine. But humor us while we do one. Please?

---

[4] If you try to find the degrees of freedom with that messy approximation formula (We dare you! It's in the footnote on page 562) using the values above, you'll get 8.99. The minor discrepancy is because we rounded the standard deviations to the nearest 10th.

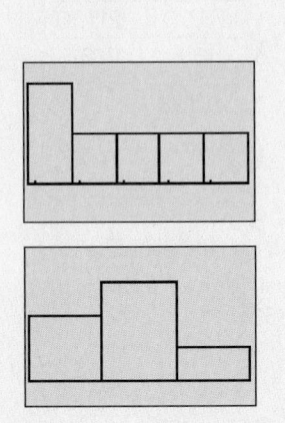

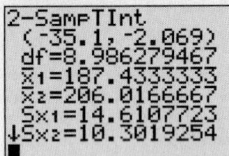

Find a confidence interval for the difference in means, given data from two independent samples.

- Let's do the batteries. Always think about whether the samples are independent. If not, stop right here. These procedures are appropriate only for independent groups.
- Enter the data into two lists.

| *NameBrand* in L1: | 194.0 | 205.5 | 199.2 | 172.4 | 184.0 | 169.5 |
| *Generic* in L2: | 190.7 | 203.5 | 203.5 | 206.5 | 222.5 | 209.4 |

- Make histograms of the data to check the Nearly Normal Condition. We see that L1's histogram doesn't look so good. But remember—this is a very small data set. The bars represent only one or two values each. It's not unusual for the histogram to look a little ragged. Try resetting the WINDOW to a range of 160 to 220 with XScl=20, and Ymax=4. Redraw the GRAPH. Looks better.
- It's your turn to try this. Check L2. Go on, do it.
- Under STAT TESTS choose 0:2-SampTint.
- Specify that you are using the Data in L1 and L2, specify 1 for both frequencies, and choose the confidence level you want.
- Pooled? We'll discuss this issue later in the chapter, but the easy advice is: Just Say No.
- To Calculate the interval, you need to scroll down one more line.

Now you have the 95% confidence interval. See df? The calculator did that messy degrees of freedom calculation for you. You have to love that!

Notice that the interval bounds are negative. That's because the TI is doing $\mu_1 - \mu_2$, and the generic batteries (L2) lasted longer. No harm done—you just need to be careful to interpret that result correctly when you *Tell* what the confidence interval means.

No data? Find a confidence interval using the sample statistics.

In many situations we don't have the original data, but must work with the summary statistics from the two groups. As we saw in the last chapter, you can still have your TI create the confidence interval with 0:2-SampTInt by choosing the Inpt:Stats option. Enter both means, standard deviations, and sample sizes, then Calculate. We show you the details in the next TI Tips.

 ## JUST CHECKING

Carpal tunnel syndrome (CTS) causes pain and tingling in the hand, sometimes bad enough to keep sufferers awake at night and restrict their daily activities. Researchers studied the effectiveness of two alternative surgical treatments for CTS (Mackenzie, Hainer, and Wheatley, *Annals of Plastic Surgery*, 2000). Patients were randomly assigned to have endoscopic or open-incision surgery. Four weeks later the endoscopic surgery patients demonstrated a mean pinch strength of 9.1 kg compared to 7.6 kg for the open-incision patients.

1. Why is the randomization of the patients into the two treatments important?

2. A 95% confidence interval for the difference in mean strength is about (0.04 kg, 2.96 kg). Explain what this interval means.

3. Why might we want to examine such a confidence interval in deciding between these two surgical procedures?

4. Why might you want to see the data before trusting the confidence interval?

Testing the Difference Between Two Means

If you bought a used camera in good condition from a friend, would you pay the same as you would if you bought the same item from a stranger? A researcher at Cornell University (J. J. Halpern, "The Transaction Index: A Method for Standardizing Comparisons of Transaction Characteristics Across Different Contexts," *Group Decision and Negotiation*, 6: 557–572) wanted to know how friendship might affect simple sales such as this. She randomly divided subjects into two groups and gave each group descriptions of items they might want to buy. One group was told to imagine buying from a friend whom they expected to see again. The other group was told to imagine buying from a stranger.

Here are the prices they offered for a used camera in good condition:

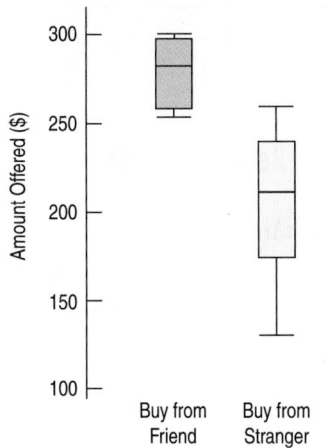

WHO	University students
WHAT	Prices offered for a used camera
UNITS	$
WHY	Study of the effects of friendship on transactions
WHEN	1990s
WHERE	U.C. Berkeley

PRICE OFFERED FOR A USED CAMERA ($)	
Buying from a Friend	Buying from a Stranger
275	260
300	250
260	175
300	130
255	200
275	225
290	240
300	

The researcher who designed this study had a specific concern. Previous theories had doubted that friendship had a measurable effect on pricing. She hoped to find an effect of friendship. This calls for a hypothesis test—in this case a **two-sample *t*-test for the difference between means.**[5]

A Test for the Difference Between Two Means

You already know enough to construct this test. The test statistic looks just like the others we've seen. It finds the difference between the observed group means and compares this with a hypothesized value for that difference. We'll call that hypothesized difference Δ_0 ("delta naught"). It's so common for that hypothesized difference to be zero that we often just assume $\Delta_0 = 0$. We then compare the difference in the means with the standard error of that difference. We already know that for a difference between independent means, we can find P-values from a Student's *t*-model on that same special number of degrees of freedom.

> TWO-SAMPLE *t*-TEST FOR THE DIFFERENCE BETWEEN MEANS
> The conditions for the two-sample *t*-test for the difference between the means of two independent groups are the same as for the two-sample *t*-interval. We test the hypothesis
> $$H_0: \mu_1 - \mu_2 = \Delta_0$$

[5] Because it is performed so often, this test is usually just called a "two-sample *t*-test."

Δ_0—delta naught—isn't so standard that you can assume everyone will understand it. We use it because it's the Greek letter (good for a parameter) "D" for "difference." You should say "delta naught" rather than "delta zero"—that's standard for parameters associated with null hypotheses.

where the hypothesized difference is almost always 0, using the statistic

$$t = \frac{(\bar{y}_1 - \bar{y}_2) - \Delta_0}{SE(\bar{y}_1 - \bar{y}_2)}.$$

The standard error of $\bar{y}_1 - \bar{y}_2$ is

$$SE(\bar{y}_1 - \bar{y}_2) = \sqrt{\frac{s_1^2}{n_1} + \frac{s_2^2}{n_2}}.$$

When the conditions are met and the null hypothesis is true, this statistic can be closely modeled by a Student's t-model with a number of degrees of freedom given by a special formula. We use that model to obtain a P-value.

STEP-BY-STEP EXAMPLE | **A Two-Sample t-Test for the Difference Between Two Means**

The usual null hypothesis is that there's no difference in means. That's just the right null hypothesis for the camera purchase prices.

Question: Is there a difference in the price people would offer a friend rather than a stranger?

THINK

Plan State what we want to know.

Identify the *parameter* you wish to estimate. Here our parameter is the difference in the means, not the individual group means.

Identify the variables and check the W's.

Hypotheses State the null and alternative hypotheses. The research claim is that friendship changes what people are willing to pay.[6] The natural null hypothesis is that friendship makes no difference.

We didn't start with any knowledge of whether friendship might increase or decrease the price, so we choose a two-sided alternative.

Model Think about the assumptions and check the conditions. (Note that, because this is a randomized experiment, we haven't sampled at all, so the 10% Condition does not apply.)

I want to know whether people are likely to offer a different amount for a used camera when buying from a friend than when buying from a stranger. I wonder whether the difference between mean amounts is zero. I have bid prices from 8 subjects buying from a friend and 7 buying from a stranger, found in a randomized experiment.

H_O: The difference in mean price offered to friends and the mean price offered to strangers is zero:

$$\mu_F - \mu_S = 0.$$

H_A: The difference in mean prices is not zero:

$$\mu_F - \mu_S \neq 0.$$

✔ **Randomization Condition:** The experiment was randomized. Subjects were assigned to treatment groups at random.

✔ **Independence Assumption:** This is an experiment, so there is no need for the subjects to be randomly selected from any

[6] This claim is a good example of what is called a "research hypothesis" in many social sciences. The only way to check it is to deny that it's true and see where the resulting null hypothesis leads us.

particular population. All we need to check is whether they were assigned randomly to treatment groups.

✔ **Independent Groups Assumption:** Randomizing the experiment gives independent groups.

✔ **Nearly Normal Condition:** Histograms of the two sets of prices are roughly unimodal and symmetric:

Make a picture. Boxplots are the display of choice for comparing groups, as seen on page 561. We also want to check the shapes of the distribution. Histograms or Normal probability plots do a better job for that.

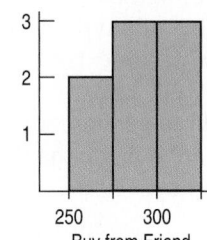
Buy from Friend

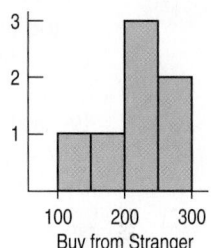
Buy from Stranger

State the sampling distribution model.

Specify your method.

The assumptions are reasonable and the conditions are okay, so I'll use a Student's t-model to perform a **two-sample t-test.**

Mechanics List the summary statistics. Be sure to use proper notation.

From the data:

$$n_F = 8 \qquad n_S = 7$$
$$\bar{y}_F = \$281.88 \qquad \bar{y}_S = \$211.43$$
$$s_F = \$18.31 \qquad s_S = \$46.43$$

Use the null model to find the P-value. First determine the standard error of the difference between sample means.

For independent groups,

$$SE(\bar{y}_F - \bar{y}_S) = \sqrt{SE^2(\bar{y}_F) + SE^2(\bar{y}_S)}$$

$$= \sqrt{\frac{s_F^2}{n_F} + \frac{s_S^2}{n_S}}$$

$$= \sqrt{\frac{18.31^2}{8} + \frac{46.43^2}{7}}$$

$$= 18.70$$

The observed difference is

$$(\bar{y}_F - \bar{y}_S) = 281.88 - 211.43 = \$70.45$$

Make a picture. Sketch the t-model centered at the hypothesized difference of zero. Because this is a two-tailed test, shade the region to the right of the observed difference and the corresponding region in the other tail.

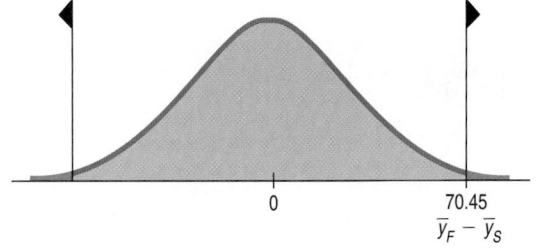

Find the *t*-value.	$t = \dfrac{(\bar{y}_F - \bar{y}_S) - (0)}{SE(\bar{y}_F - \bar{y}_S)} = \dfrac{70.45}{18.70} = 3.77$
A statistics program or graphing calculator finds the P-value using the fractional degrees of freedom from the approximation formula.	$df = 7.62$ (from technology) P-value $= 2P(t_{7.62} > 3.77) = 0.006$

 **Conclusion** Link the P-value to your decision about the null hypothesis, and state the conclusion in context.

Be cautious about generalizing to items whose prices are outside the range of those in this study.

If there were no difference in the mean prices, a difference this large would occur only 6 times in 1000. That's too rare to believe, so I reject the null hypothesis and conclude that people are likely to offer a friend more than they'd offer a stranger for a used camera (and possibly for other, similar items).

TI Tips

Testing a hypothesis about a difference in means

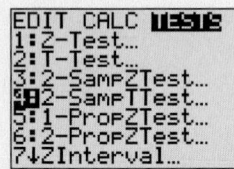

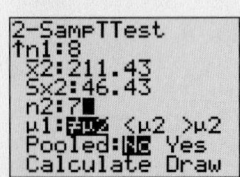

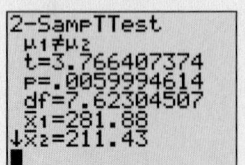

Now let's use the TI to do a hypothesis test for the difference of two means—independent, of course! (Have we said that enough times yet?)

Test a hypothesis when you know the sample statistics.
We'll demonstrate by using the statistics from the camera-pricing example. A sample of 8 people suggested they'd sell the camera to a friend for an average price of $281.88 with standard deviation $18.31. An independent sample of 7 other people would charge a stranger an average of $211.43 with standard deviation $46.43. Does this represent a significant difference in prices?

- From the **STAT TESTS** menu select **4:2-SampTTest**.
- Specify **Inpt:Stats**, and enter the appropriate sample statistics.
- You have to scroll down to complete the specifications. This is a two-tailed test, so choose alternative **≠μ2**.
- **Pooled**? Just say **No**. (We did promise to explain that and we will, coming up next.)
- Ready . . . set . . . **Calculate!**

The TI reports a calculated value of *t* = 3.77 and a P-value of 0.006. It's hard to tell who your real friends are.

By now we probably don't have to tell you how to do a 2-SampTTest starting with data in lists.
So we won't.

JUST CHECKING

Recall the experiment comparing patients 4 weeks after surgery for carpal tunnel syndrome. The patients who had endoscopic surgery demonstrated a mean pinch strength of 9.1 kg compared to 7.6 kg for the open-incision patients.

5. What hypotheses would you test?

6. The P-value of the test was less than 0.05. State a brief conclusion.

7. The study reports work on 36 "hands," but there were only 26 patients. In fact, 7 of the endoscopic surgery patients had both hands operated on, as did 3 of the open-incision group. Does this alter your thinking about any of the assumptions? Explain.

FOR EXAMPLE A two-sample *t*-test

Many office "coffee stations" collect voluntary payments for the food consumed. Researchers at the University of Newcastle upon Tyne performed an experiment to see whether the image of eyes watching would change employee behavior.[7] They alternated pictures (seen here) of eyes looking at the viewer with pictures of flowers each week on the cupboard behind the "honesty box." They measured the consumption of milk to approximate the amount of food consumed and recorded the contributions (in £) each week per liter of milk. The table summarizes their results.

Question: Do these results provide evidence that there really is a difference in honesty even when it's only photographs of eyes that are "watching"?

	Eyes	Flowers
n (# weeks)	5	5
$\bar{y}$	0.417 £/1	0.151 £/1
s	0.1811 £/1	0.067 £/1

$$H_O: \mu_{eyes} - \mu_{flowers} = 0$$
$$H_A: \mu_{eyes} - \mu_{flowers} \neq 0$$

✓ **Independence Assumption:** The amount paid by one person should be independent of the amount paid by others.

✓ **Randomization Condition:** This study was observational. Treatments alternated a week at a time and were applied to the same group of office workers.

✓ **Nearly Normal Condition:** I don't have the data to check, but it seems unlikely there would be outliers in either group. I could be more certain if I could see histograms for both groups.

✓ **Independent Groups Assumption:** The same workers were recorded each week, but week-to-week independence is plausible.

It's okay to do a two-sample *t*-test for the difference in means:

$$SE(\bar{y}_{eyes} - \bar{y}_{flowers}) = \sqrt{\frac{s^2_{eyes}}{n_{eyes}} + \frac{s^2_{flowers}}{n_{flowers}}} = \sqrt{\frac{0.1811^2}{5} + \frac{0.067^2}{5}} = 0.0864$$

$$df = 5.07$$

$$t_5 = \frac{(\bar{y}_{eyes} - \bar{y}_{flowers}) - 0}{SE(\bar{y}_{eyes} - \bar{y}_{flowers})} = \frac{0.417 - 0.151}{0.0864} = 3.08$$

$$P(|t_5| > 3.08) = 0.027$$

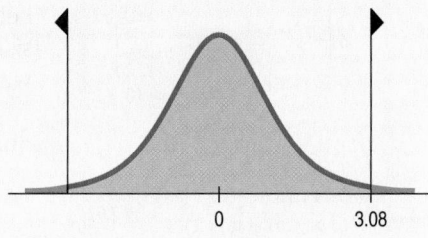

Assuming the data were free of outliers, the very low P-value leads me to reject the null hypothesis. This study provides evidence that people will leave higher average voluntary payments for food if pictures of eyes are "watching."

(Note: In Table T we can see that at 5 df, $t = 3.08$ lies between the critical values for $P = 0.02$ and $P = 0.05$, so we could report $P < 0.05$.)

---

[7] Melissa Bateson, Daniel Nettle, and Gilbert Roberts, "Cues of Being Watched Enhance Cooperation in a Real-World Setting," *Biol. Lett. doi*:10.1098/rsbl.2006.0509.

Back into the Pool

Remember that when we know a proportion, we know its standard deviation. When we tested the null hypothesis that two proportions were equal, that link meant we could assume their variances were equal as well. This led us to pool our data to estimate a standard error for the hypothesis test.

For means, there is also a pooled t-test. Like the two-proportions z-test, this test assumes that the variances in the two groups are equal. But be careful: Knowing the mean of some data doesn't tell you anything about their variance. And knowing that two means are equal doesn't say anything about whether their variances are equal. If we were willing to *assume* that their variances are equal, we could pool the data from two groups to estimate the common variance. We'd estimate this pooled variance from the data, so we'd still use a Student's t-model. This test is called a **pooled t-test (for the difference between means).**

Pooled t-tests have a couple of advantages. They often have a few more degrees of freedom than the corresponding two-sample test and a much simpler degrees of freedom formula. But these advantages come at a price: You have to pool the variances and think about another assumption. The assumption of equal variances is a strong one, is often not true, and is difficult to check. For these reasons, we recommend that you use a two-sample t-test instead.

The pooled t-test is the theoretically correct method only when we have a good reason to believe that the variances are equal. And (as we will see shortly) there are times when this makes sense. Keep in mind, however, that it's never wrong *not* to pool.

*The Pooled t-Test

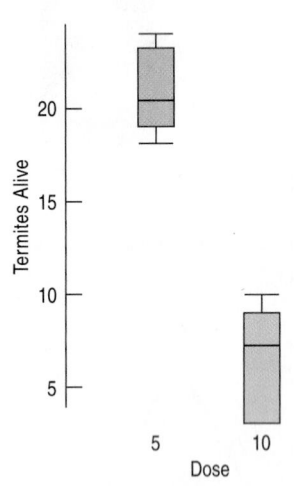

Termites cause billions of dollars of damage each year, to homes and other buildings, but some tropical trees seem to be able to resist termite attack. A researcher extracted a compound from the sap of one such tree and tested it by feeding it at two different concentrations to randomly assigned groups of 25 termites.[8] After 5 days, 8 groups fed the lower dose had an average of 20.875 termites alive, with a standard deviation of 2.23. But 6 groups fed the higher dose had an average of only 6.667 termites alive, with a standard deviation of 3.14. Is this a large enough difference to declare the sap compound effective in killing termites? In order to use the pooled t-test, we must make the **Equal Variance Assumption** that the variances of the two populations from which the samples have been drawn are equal. That is, $\sigma_1^2 = \sigma_2^2$. (Of course, we could think about the standard deviations being equal instead.) The corresponding **Similar Spreads Condition** really just consists of looking at the boxplots to check that the spreads are not wildly different. We were going to make boxplots anyway, so there's really nothing new here.

Once we decide to pool, we estimate the common variance by combining numbers we already have:

$$s_{pooled}^2 = \frac{(n_1 - 1)s_1^2 + (n_2 - 1)s_2^2}{(n_1 - 1) + (n_2 - 1)}.$$

$$s_{pooled}^2 = \frac{(8 - 1)2.23^2 + (6 - 1)3.14^2}{(8 - 1) + (6 - 1)} = 7.01$$

(If the two sample sizes are equal, this is just the average of the two variances.)

Now we just substitute this pooled variance in place of each of the variances in the standard error formula.

$$SE_{pooled}(\bar{y}_1 - \bar{y}_2) = \sqrt{\frac{s_{pooled}^2}{n_1} + \frac{s_{pooled}^2}{n_2}} = s_{pooled}\sqrt{\frac{1}{n_1} + \frac{1}{n_2}}.$$

$$SE_{pooled}(\bar{y}_1 - \bar{y}_2) = \sqrt{\frac{7.01}{8} + \frac{7.01}{6}} = 1.43$$

---

[8] Adam Messer, Kevin McCormick, Sunjaya, H. H. Hagedorm, Ferny Tumbel, and J. Meinwald, "Defensive role of tropical tree resins: antitermitic sesquiterpenes from Southeast Asian Dipterocarpaceae," *J Chem Ecology*, 16:122, pp. 3333–3352.

The formula for degrees of freedom for the Student's t-model is simpler, too. It was so complicated for the two-sample t that we stuck it in a footnote.[9] Now it's just df $= n_1 + n_2 - 2$.

Substitute the pooled-t estimate of the standard error and its degrees of freedom into the steps of the confidence interval or hypothesis test, and you'll be using the pooled-t method. For the termites, $\bar{y}_1 - \bar{y}_2 = 14.208$, giving a t-value $= 9.935$ with 12 df and a P-value ≤ 0.0001.

Of course, if you decide to use a pooled-t method, you must defend your assumption that the variances of the two groups are equal.

$$t = \frac{20.875 - 6.667}{1.43} = 9.935$$

POOLED t-TEST AND CONFIDENCE INTERVAL FOR MEANS

The conditions for the pooled t-test for the difference between the means of two independent groups (commonly called a "pooled t-test") are the same as for the two-sample t-test with the additional assumption that the variances of the two groups are the same. We test the hypothesis

$$H_0: \mu_1 - \mu_2 = \Delta_0$$

where the hypothesized difference, Δ_0, is almost always 0, using the statistic

$$t = \frac{(\bar{y}_1 - \bar{y}_2) - \Delta_0}{SE_{pooled}(\bar{y}_1 - \bar{y}_2)}.$$

The standard error of $\bar{y}_1 - \bar{y}_2$ is

$$SE_{pooled}(\bar{y}_1 - \bar{y}_2) = \sqrt{\frac{s^2_{pooled}}{n_1} + \frac{s^2_{pooled}}{n_2}} = s_{pooled}\sqrt{\frac{1}{n_1} + \frac{1}{n_2}},$$

where the pooled variance is

$$s^2_{pooled} = \frac{(n_1 - 1)s^2_1 + (n_2 - 1)s^2_2}{(n_1 - 1) + (n_2 - 1)}.$$

When the conditions are met and the null hypothesis is true, we can model this statistic's sampling distribution with a Student's t-model with $(n_1 - 1) + (n_2 - 1)$ degrees of freedom. We use that model to obtain a P-value for a test or a margin of error for a confidence interval.

The corresponding confidence interval is

$$(\bar{y}_1 - \bar{y}_2) \pm t^*_{df} \times SE_{pooled}(\bar{y}_1 - \bar{y}_2),$$

where the critical value t^* depends on the confidence level and is found with $(n_1 - 1) + (n_2 - 1)$ degrees of freedom.

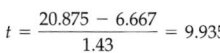

Activity: The Pooled t-Test. It's those hot dogs again. The same interactive tool can handle a pooled t-test, too. Take it for a spin here.

Is the Pool All Wet?

We're testing whether the means are equal, so we admit that we don't *know* whether they are equal. Doesn't it seem a bit much to just *assume* that the variances are equal? Well, yes—but there are some special cases to consider. So when *should* you use pooled-t methods rather than two-sample t methods?

Never.

What, never?

Well, hardly ever.

---

[9] But not this one. See page 562.

You see, when the variances of the two groups are in fact equal, the two methods give pretty much the same result. (For the termites, the two-sample t statistic is barely different—9.436 with 8 df—and the P-value is still < 0.001.) Pooled methods have a small advantage (slightly narrower confidence intervals, slightly more powerful tests) mostly because they usually have a few more degrees of freedom, but the advantage is slight.

When the variances are *not* equal, the pooled methods are just not valid and can give poor results. You have to use the two-sample methods instead.

As the sample sizes get bigger, the advantages that come from a few more degrees of freedom make less and less difference. So the advantage (such as it is) of the pooled method is greatest when the samples are small—just when it's hardest to check the conditions. And the difference in the degrees of freedom is greatest when the variances are not equal—just when you can't use the pooled method anyway. Our advice is to use the two-sample t methods to compare means.

Pooling may make sense in a randomized comparative experiment. We start by assigning our experimental units to treatments at random, as the experimenter did with the termites. We know that at the start of the experiment each treatment group is a random sample from the same population,[10] so each treatment group begins with the same population variance. In this case, assuming that the variances are equal after we apply the treatment is the same as assuming that the treatment doesn't change the variance. When we test whether the true means are equal, we may be willing to go a bit farther and say that the treatments made no difference *at all.* For example, we might suspect that the treatment is no different from the placebo offered as a control. Then it's not much of a stretch to assume that the variances have remained equal. It's still an assumption, and there are conditions that need to be checked (make the boxplots, make the boxplots, make the boxplots), but at least it's a plausible assumption.

This line of reasoning is important. The methods used to analyze comparative experiments *do* pool variances in exactly this way and defend the pooling with a version of this argument. The chapter on Analysis of Variance on the DVD introduces these methods.

> Because the advantages of pooling are small, and you are allowed to pool only rarely (when the Equal Variances Assumption is met), *don't.*
>
> It's never wrong *not* to pool.

WHAT CAN GO WRONG?

▶ **Watch out for paired data.** The Independent Groups Assumption deserves special attention. If the samples are not independent, you can't use these two-sample methods. This is probably the main thing that can go wrong when using these two-sample methods. The methods of this chapter can be used *only* if the observations in the two groups are *independent.* Matched-pairs designs in which the observations are deliberately related arise often and are important. The next chapter deals with them.

▶ **Look at the plots.** The usual (by now) cautions about checking for outliers and non-Normal distributions apply, of course. The simple defense is to make and examine boxplots. You may be surprised how often this simple step saves you from the wrong or even absurd conclusions that can be generated by a single undetected outlier. You don't want to conclude that two methods have very different means just because one observation is atypical.

---

[10] That is, the population of experimental subjects. Remember that to be valid, experiments do not need a representative sample drawn from a population because we are not trying to estimate a population model parameter.

Do what we say, not what we do . . .   Precision machines used in industry often have a bewildering number of parameters that have to be set, so experiments are performed in an attempt to try to find the best settings. Such was the case for a hole-punching machine used by a well-known computer manufacturer to make printed circuit boards. The data were analyzed by one of the authors, but because he was in a hurry, he didn't look at the boxplots first and just performed *t*-tests on the experimental factors. When he found extremely small P-values even for factors that made no sense, he plotted the data. Sure enough, there was one observation 1,000,000 times bigger than the others. It turns out that it had been recorded in microns (millionths of an inch), while all the rest were in inches.

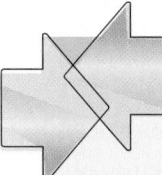

CONNECTIONS

The structure and reasoning of inference methods for comparing two means are very similar to what we used for comparing two proportions. Here we must estimate the standard errors independent of the means, so we use Student's *t*-models rather than the Normal.

We first learned about side-by-side boxplots in Chapter 5. There we made general statements about the shape, center, and spread of each group. When we compared groups, we asked whether their centers looked different compared to how spread out the distributions were. Here we've made that kind of thinking precise, with confidence intervals for the difference and tests of whether the means are the same.

We use Student's *t* as we did for single sample means, and for the same reasons: We are using standard errors from the data to estimate the standard deviation of the sample statistic. As before, to work with Student's *t*-models, we need to check the Nearly Normal Condition. Histograms and Normal probability plots are the best methods for such checks.

As always, we've decided whether a statistic is large by comparing it with its standard error. In this case, our statistic is the difference in means.

We pooled data to find a standard deviation when we tested the hypothesis of equal proportions. For that test, the assumption of equal variances was a consequence of the null hypothesis that the proportions were equal, so it didn't require an extra assumption. When two proportions are equal, so are their variances. But means don't have a linkage with their corresponding variances; so to use pooled-*t* methods, we must make the additional assumption of equal variances. When we can make this assumption, the pooled variance calculations are very similar to those for proportions, combining the squared deviations of each group from its own mean to find a common variance.

WHAT HAVE WE LEARNED?

Are the means of two groups the same? If not, how different are they? We've learned to use statistical inference to compare the means of two independent groups.

▸ We've seen that confidence intervals and hypothesis tests about the difference between two means, like those for an individual mean, use *t*-models.

▸ Once again we've seen the importance of checking assumptions that tell us whether our method will work.

▸ We've seen that, as when comparing proportions, finding the standard error for the difference in sample means depends on believing that our data come from independent groups. Unlike proportions, however, pooling is usually not the best choice here.

▸ And we've seen once again that we can add variances of independent random variables to find the standard deviation of the difference in two independent means.

▸ Finally, we've learned that the reasoning of statistical inference remains the same; only the mechanics change.

Terms

Two-sample t methods
562. Two-sample t methods allow us to draw conclusions about the difference between the means of two independent groups. The two-sample methods make relatively few assumptions about the underlying populations, so they are usually the method of choice for comparing two sample means. However, the Student's t-models are only approximations for their true sampling distribution. To make that approximation work well, the two-sample t methods have a special rule for estimating degrees of freedom.

Two-sample t-interval for the difference between means
564. A confidence interval for the difference between the means of two independent groups found as

$$(\bar{y}_1 - \bar{y}_2) \pm t^*_{df} \times SE(\bar{y}_1 - \bar{y}_2)$$

where

$$SE(\bar{y}_1 - \bar{y}_2) = \sqrt{\frac{s_1^2}{n_1} + \frac{s_2^2}{n_2}}$$

and the number of degrees of freedom is given by a special formula (see footnote 3 on page 562).

Two-sample t-test for the difference between means
569. A hypothesis test for the difference between the means of two independent groups. It tests the null hypothesis

$$H_0: \mu_1 - \mu_2 = \Delta_0,$$

where the hypothesized difference, Δ_0, is almost always 0, using the statistic

$$t_{df} = \frac{(\bar{y}_1 - \bar{y}_2) - \Delta_0}{SE(\bar{y}_1 - \bar{y}_2)},$$

with the number of degrees of freedom given by the special formula.

Pooling
574. Data from two or more populations may sometimes be combined, or *pooled*, to estimate a statistic (typically a pooled variance) when we are willing to assume that the estimated value is the same in both populations. The resulting larger sample size may lead to an estimate with lower sample variance. However, pooled estimates are appropriate only when the required assumptions are true.

Pooled-t methods
575. Pooled-t methods provide inferences about the difference between the means of two independent populations under the assumption that both populations have the same standard deviation. When the assumption is justified, pooled-t methods generally produce slightly narrower confidence intervals and more powerful significance tests than two-sample t methods. When the assumption is not justified, they generally produce worse results—sometimes substantially worse.

We recommend that you use two-sample t methods instead.

Skills

▸ Be able to recognize situations in which we want to do inference on the difference between the means of two independent groups.

▸ Know how to examine your data for violations of conditions that would make inference about the difference between two population means unwise or invalid.

▸ Be able to recognize when a pooled-t procedure might be appropriate and be able to explain why you decided to use a two-sample method anyway.

▸ Be able to perform a two-sample t-test using a statistics package or calculator (at least for finding the degrees of freedom).

▸ Be able to interpret a test of the null hypothesis that the means of two independent groups are equal. (If the test is a pooled t-test, your interpretation should include a defense of your assumption of equal variances.)

TWO-SAMPLE METHODS ON THE COMPUTER

Here's some typical computer package output with comments:

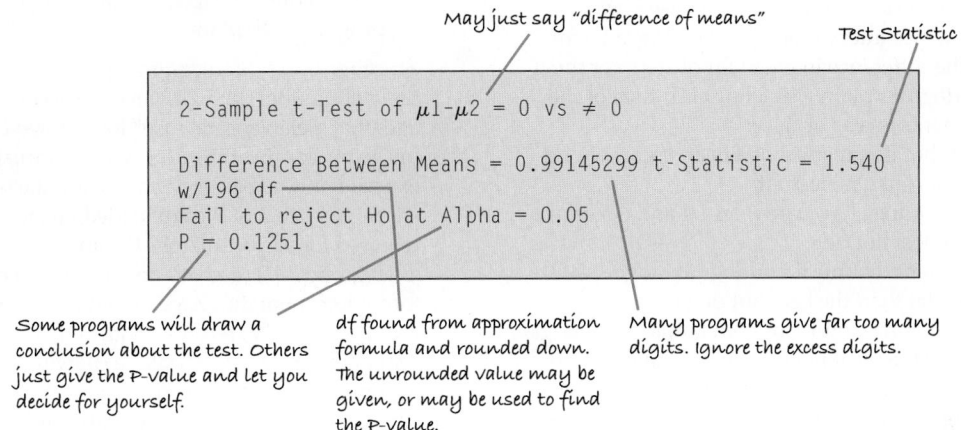

May just say "difference of means"

Test Statistic

2-Sample t-Test of $\mu1-\mu2$ = 0 vs ≠ 0

Difference Between Means = 0.99145299 t-Statistic = 1.540
w/196 df
Fail to reject Ho at Alpha = 0.05
P = 0.1251

Some programs will draw a conclusion about the test. Others just give the P-value and let you decide for yourself.

df found from approximation formula and rounded down. The unrounded value may be given, or may be used to find the P-value.

Many programs give far too many digits. Ignore the excess digits.

Most statistics packages compute the test statistic for you and report a P-value corresponding to that statistic. And, of course, statistics packages make it easy to examine the boxplots and histograms of the two groups, so you have no excuse for skipping this important check.

Some statistics software automatically tries to test whether the variances of the two groups are equal. Some automatically offer both the two-sample-t and pooled-t results. Ignore the test for the variances; it has little power in any situation in which its results could matter. If the pooled and two-sample methods differ in any important way, you should stick with the two-sample method. Most likely, the Equal Variance Assumption needed for the pooled method has failed.

The degrees of freedom approximation usually gives a fractional value. Most packages seem to round the approximate value down to the next smallest integer (although they may actually compute the P-value with the fractional value, gaining a tiny amount of power).

EXERCISES

1. **Dogs and calories.** In July 2007, *Consumer Reports* examined the calorie content of two kinds of hot dogs: meat (usually a mixture of pork, turkey, and chicken) and all beef. The researchers purchased samples of several different brands. The meat hot dogs averaged 111.7 calories, compared to 135.4 for the beef hot dogs. A test of the null hypothesis that there's no difference in mean calorie content yields a P-value of 0.124. Would a 95% confidence interval for $\mu_{Meat} - \mu_{Beef}$ include 0? Explain.

2. **Dogs and sodium.** The *Consumer Reports* article described in Exercise 1 also listed the sodium content (in mg) for the various hot dogs tested. A test of the null hypothesis that beef hot dogs and meat hot dogs don't differ in the mean amounts of sodium yields a P-value of 0.11. Would a 95% confidence interval for $\mu_{Meat} - \mu_{Beef}$ include 0? Explain.

3. **Dogs and fat.** The *Consumer Reports* article described in Exercise 1 also listed the fat content (in grams) for samples

of beef and meat hot dogs. The resulting 90% confidence interval for $\mu_{Meat} - \mu_{Beef}$ is (−6.5, −1.4).
   a) The endpoints of this confidence interval are negative numbers. What does that indicate?
   b) What does the fact that the confidence interval does not contain 0 indicate?
   c) If we use this confidence interval to test the hypothesis that $\mu_{Meat} - \mu_{Beef} = 0$, what's the corresponding alpha level?

4. **Washers.** In June 2007, *Consumer Reports* examined top-loading and front-loading washing machines, testing samples of several different brands of each type. One of the variables the article reported was "cycle time", the number of minutes it took each machine to wash a load of clothes. Among the machines rated good to excellent, the 98% confidence interval for the difference in mean cycle time ($\mu_{Top} - \mu_{Front}$) is (−40, −22).
   a) The endpoints of this confidence interval are negative numbers. What does that indicate?

b) What does the fact that the confidence interval does not contain 0 indicate?

c) If we use this confidence interval to test the hypothesis that $\mu_{Top} - \mu_{Front} = 0$, what's the corresponding alpha level?

5. **Dogs and fat, second helping.** In Exercise 3, we saw a 90% confidence interval of $(-6.5, -1.4)$ grams for $\mu_{Meat} - \mu_{Beef}$, the difference in mean fat content for meat vs. all-beef hot dogs. Explain why you think each of the following statements is true or false:

a) If I eat a meat hot dog instead of a beef dog, there's a 90% chance I'll consume less fat.

b) 90% of meat hot dogs have between 1.4 and 6.5 grams less fat than a beef hot dog.

c) I'm 90% confident that meat hot dogs average 1.4–6.5 grams less fat than the beef hot dogs.

d) If I were to get more samples of both kinds of hot dogs, 90% of the time the meat hot dogs would average 1.4–6.5 grams less fat than the beef hot dogs.

e) If I tested many samples, I'd expect about 90% of the resulting confidence intervals to include the true difference in mean fat content between the two kinds of hot dogs.

6. **Second load of wash.** In Exercise 4, we saw a 98% confidence interval of $(-40, -22)$ minutes for $\mu_{Top} - \mu_{Front}$, the difference in time it takes top-loading and front-loading washers to do a load of clothes. Explain why you think each of the following statements is true or false:

a) 98% of top loaders are 22 to 40 minutes faster than front loaders.

b) If I choose the laundromat's top loader, there's a 98% chance that my clothes will be done faster than if I had chosen the front loader.

c) If I tried more samples of both kinds of washing machines, in about 98% of these samples I'd expect the top loaders to be an average of 22 to 40 minutes faster.

d) If I tried more samples, I'd expect about 98% of the resulting confidence intervals to include the true difference in mean cycle time for the two types of washing machines.

e) I'm 98% confident that top loaders wash clothes an average of 22 to 40 minutes faster than front-loaders.

7. **Learning math.** The Core Plus Mathematics Project (CPMP) is an innovative approach to teaching Mathematics that engages students in group investigations and mathematical modeling. After field tests in 36 high schools over a three-year period, researchers compared the performances of CPMP students with those taught using a traditional curriculum. In one test, students had to solve applied Algebra problems using calculators. Scores for 320 CPMP students were compared to those of a control group of 273 students in a traditional Math program. Computer software was used to create a confidence interval for the difference in mean scores. (*Journal for Research in Mathematics Education*, 31, no. 3[2000])

Conf level: 95%   Variable: Mu(CPMP) – Mu(Ctrl)
Interval: (5.573, 11.427)

a) What's the margin of error for this confidence interval?

b) If we had created a 98% CI, would the margin of error be larger or smaller?

c) Explain what the calculated interval means in context.

d) Does this result suggest that students who learn Mathematics with CPMP will have significantly higher mean scores in Algebra than those in traditional programs? Explain.

8. **Stereograms.** Stereograms appear to be composed entirely of random dots. However, they contain separate images that a viewer can "fuse" into a three-dimensional (3D) image by staring at the dots while defocusing the eyes. An experiment was performed to determine whether knowledge of the form of the embedded image affected the time required for subjects to fuse the images. One group of subjects (group NV) received no information or just verbal information about the shape of the embedded object. A second group (group VV) received both verbal information and visual information (specifically, a drawing of the object). The experimenters measured how many seconds it took for the subject to report that he or she saw the 3D image.

2-Sample t-Interval for $\mu1 - \mu2$
Conf level = 90%   df = 70
μ(NV) – μ(VV) interval: (0.55, 5.47)

a) Interpret your interval in context.

b) Does it appear that viewing a picture of the image helps people "see" the 3D image in a stereogram?

c) What's the margin of error for this interval?

d) Explain what the 90% confidence level means.

e) Would you expect a 99% confidence level to be wider or narrower? Explain.

f) Might that change your conclusion in part b? Explain.

9. **CPMP, again.** During the study described in Exercise 7, students in both CPMP and traditional classes took another Algebra test that did not allow them to use calculators. The table below shows the results. Are the mean scores of the two groups significantly different?

Math Program	n	Mean	SD
CPMP	312	29.0	18.8
Traditional	265	38.4	16.2

Performance on Algebraic Symbolic Manipulation Without Use of Calculators

a) Write an appropriate hypothesis.

b) Do you think the assumptions for inference are satisfied? Explain.

c) Here is computer output for this hypothesis test. Explain what the P-value means in this context.

2-Sample t-Test of $\mu1 - \mu2 \neq 0$
t-Statistic = –6.451 w/574.8761 df
P < 0.0001

d) State a conclusion about the CPMP program.

10. **CPMP and word problems.** The study of the new CPMP Mathematics methodology described in Exercise 7 also tested students' abilities to solve word problems. This

table shows how the CPMP and traditional groups performed. What do you conclude?

Math Program	n	Mean	SD
CPMP	320	57.4	32.1
Traditional	273	53.9	28.5

11. **Commuting.** A man who moves to a new city sees that there are two routes he could take to work. A neighbor who has lived there a long time tells him Route A will average 5 minutes faster than Route B. The man decides to experiment. Each day he flips a coin to determine which way to go, driving each route 20 days. He finds that Route A takes an average of 40 minutes, with standard deviation 3 minutes, and Route B takes an average of 43 minutes, with standard deviation 2 minutes. Histograms of travel times for the routes are roughly symmetric and show no outliers.
   a) Find a 95% confidence interval for the difference in average commuting time for the two routes.
   b) Should the man believe the old-timer's claim that he can save an average of 5 minutes a day by always driving Route A? Explain.

12. **Pulse rates.** A researcher wanted to see whether there is a significant difference in resting pulse rates for men and women. The data she collected are displayed in the boxplots and summarized below.

	Sex	
	Male	**Female**
Count	28	24
Mean	72.75	72.625
Median	73	73
StdDev	5.37225	7.69987
Range	20	29
IQR	9	12.5

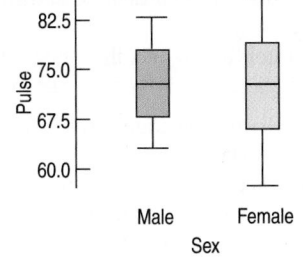

a) What do the boxplots suggest about differences between male and female pulse rates?
b) Is it appropriate to analyze these data using the methods of inference discussed in this chapter? Explain.
c) Create a 90% confidence interval for the difference in mean pulse rates.
d) Does the confidence interval confirm your answer to part a? Explain.

T 13. **Cereal.** The data below show the sugar content (as a percentage of weight) of several national brands of children's and adults' cereals. Create and interpret a 95% confidence interval for the difference in mean sugar content. Be sure to check the necessary assumptions and conditions.

Children's cereals: 40.3, 55, 45.7, 43.3, 50.3, 45.9, 53.5, 43, 44.2, 44, 47.4, 44, 33.6, 55.1, 48.8, 50.4, 37.8, 60.3, 46.6

Adults' cereals: 20, 30.2, 2.2, 7.5, 4.4, 22.2, 16.6, 14.5, 21.4, 3.3, 6.6, 7.8, 10.6, 16.2, 14.5, 4.1, 15.8, 4.1, 2.4, 3.5, 8.5, 10, 1, 4.4, 1.3, 8.1, 4.7, 18.4

T 14. **Egyptians.** Some archaeologists theorize that ancient Egyptians interbred with several different immigrant populations over thousands of years. To see if there is any indication of changes in body structure that might have resulted, they measured 30 skulls of male Egyptians dated from 4000 B.C.E and 30 others dated from 200 B.C.E. (A. Thomson and R. Randall-Maciver, *Ancient Races of the Thebaid,* Oxford: Oxford University Press, 1905)
   a) Are these data appropriate for inference? Explain.
   b) Create a 95% confidence interval for the difference in mean skull breadth between these two eras.
   c) Do these data provide evidence that the mean breadth of males' skulls changed over this period? Explain.

Maximum Skull Breadth (mm)			
4000 B.C.E.		**200 B.C.E.**	
131	131	141	131
125	135	141	129
131	132	135	136
119	139	133	131
136	132	131	139
138	126	140	144
139	135	139	141
125	134	140	130
131	128	138	133
134	130	132	138
129	138	134	131
134	128	135	136
126	127	133	132
132	131	136	135
141	124	134	141

T 15. **Reading.** An educator believes that new reading activities for elementary school children will improve reading comprehension scores. She randomly assigns third graders to an eight-week program in which some will use these activities and others will experience traditional teaching methods. At the end of the experiment, both groups take a reading comprehension exam. Their scores are shown in the back-to-back stem-and-leaf display. Do these results suggest that the new activities are better? Test an appropriate hypothesis and state your conclusion.

New Activities		Control
	1	07
4	2	068
3	3	377
96333	4	12222238
9876432	5	355
721	6	02
1	7	
	8	5

16. Streams. Researchers collected samples of water from streams in the Adirondack Mountains to investigate the effects of acid rain. They measured the pH (acidity) of the water and classified the streams with respect to the kind of substrate (type of rock over which they flow). A lower pH means the water is more acidic. Here is a plot of the pH of the streams by substrate (limestone, mixed, or shale):

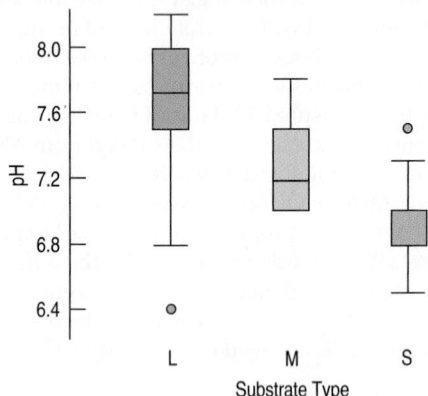

Here are selected parts of a software analysis comparing the pH of streams with limestone and shale substrates:

2-Sample t-Test of $\mu 1 - \mu 2$
Difference Between Means = 0.735
t-Statistic = 16.30 w/133 df
p ≤ 0.0001

a) State the null and alternative hypotheses for this test.
b) From the information you have, do the assumptions and conditions appear to be met?
c) What conclusion would you draw?

17. Baseball 2006. American League baseball teams play their games with the designated hitter rule, meaning that pitchers do not bat. The league believes that replacing the pitcher, traditionally a weak hitter, with another player in the batting order produces more runs and generates more interest among fans. Below are the average numbers of runs scored in American League and National League stadiums for the 2006 season.

American		National	
11.4	9.9	10.5	9.5
10.5	9.7	10.3	9.4
10.4	9.1	10.0	9.1
10.3	9.0	10.0	9.0
10.2	9.0	9.7	9.0
10.0	8.9	9.7	8.9
9.9	8.8	9.6	8.9
		9.5	7.9

a) Create an appropriate display of these data. What do you see?
b) With a 95% confidence interval, estimate the mean number of runs scored in American League games.
c) Coors Field, in Denver, stands a mile above sea level, an altitude far greater than that of any other National League ball park. Some believe that the thinner air makes it harder for pitchers to throw curve balls and easier for

batters to hit the ball a long way. Do you think the 10.5 runs scored per game at Coors is unusual? Explain.
d) Explain why you should not use two separate confidence intervals to decide whether the two leagues differ in average number of runs scored.

18. Handy. A factory hiring people to work on an assembly line gives job applicants a test of manual agility. This test counts how many strangely shaped pegs the applicant can fit into matching holes in a one-minute period. The table below summarizes the data by sex of the job applicant. Assume that all conditions necessary for inference are met.

	Male	Female
Number of subjects	50	50
Pegs placed:		
Mean	19.39	17.91
SD	2.52	3.39

a) Find 95% confidence intervals for the average number of pegs that males and females can each place.
b) Those intervals overlap. What does this suggest about any sex-based difference in manual agility?
c) Find a 95% confidence interval for the difference in the mean number of pegs that could be placed by men and women.
d) What does this interval suggest about any difference in manual agility between men and women?
e) The two results seem contradictory. Which method is correct: doing two-sample inference or doing one-sample inference twice?
f) Why don't the results agree?

19. Double header 2006. Do the data in Exercise 17 suggest that the American League's designated hitter rule may lead to more runs?
a) Using a 95% confidence interval, estimate the difference between the mean number of runs scored in American and National League games.
b) Interpret your interval.
c) Does that interval suggest that the two leagues may differ in average number of runs scored per game?

20. Hard water. In an investigation of environmental causes of disease, data were collected on the annual mortality rate (deaths per 100,000) for males in 61 large towns in England and Wales. In addition, the water hardness was recorded as the calcium concentration (parts per million, ppm) in the drinking water. The data set also notes, for each town, whether it was south or north of Derby. Is there a significant difference in mortality rates in the two regions? Here are the summary statistics.

Summary of:	mortality			
For categories in:	Derby			
Group	**Count**	**Mean**	**Median**	**StdDev**
North	34	1631.59	1631	138.470
South	27	1388.85	1369	151.114

a) Test appropriate hypotheses and state your conclusion.
b) On the next page, the boxplots of the two distributions show an outlier among the data north of Derby. What effect might that have had on your test?

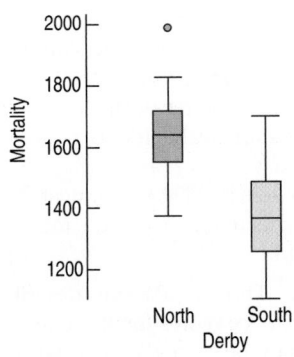

21. Job satisfaction. A company institutes an exercise break for its workers to see if this will improve job satisfaction, as measured by a questionnaire that assesses workers' satisfaction. Scores for 10 randomly selected workers before and after implementation of the exercise program are shown. The company wants to assess the effectiveness of the exercise program. Explain why you can't use the methods discussed in this chapter to do that. (Don't worry, we'll give you another chance to do this the right way.)

Worker	Job Satisfaction Index	
Number	**Before**	**After**
1	34	33
2	28	36
3	29	50
4	45	41
5	26	37
6	27	41
7	24	39
8	15	21
9	15	20
10	27	37

22. Summer school. Having done poorly on their math final exams in June, six students repeat the course in summer school, then take another exam in August. If we consider these students representative of all students who might attend this summer school in other years, do these results provide evidence that the program is worthwhile?

June	54	49	68	66	62	62
Aug.	50	65	74	64	68	72

23. Sex and violence. In June 2002, the *Journal of Applied Psychology* reported on a study that examined whether the content of TV shows influenced the ability of viewers to recall brand names of items featured in the commercials. The researchers randomly assigned volunteers to watch one of three programs, each containing the same nine commercials. One of the programs had violent content, another sexual content, and the third neutral content. After the shows ended, the subjects were asked to recall the brands of products that were advertised. Here are summaries of the results:

	Program Type		
	Violent	**Sexual**	**Neutral**
No. of subjects	108	108	108
Brands recalled			
Mean	2.08	1.71	3.17
SD	1.87	1.76	1.77

a) Do these results indicate that viewer memory for ads may differ depending on program content? A test of the hypothesis that there is no difference in ad memory between programs with sexual content and those with violent content has a P-value of 0.136. State your conclusion.

b) Is there evidence that viewer memory for ads may differ between programs with sexual content and those with neutral content? Test an appropriate hypothesis and state your conclusion.

24. Ad campaign. You are a consultant to the marketing department of a business preparing to launch an ad campaign for a new product. The company can afford to run ads during one TV show, and has decided not to sponsor a show with sexual content. You read the study described in Exercise 23, then use a computer to create a confidence interval for the difference in mean number of brand names remembered between the groups watching violent shows and those watching neutral shows.

    TWO-SAMPLE T
    95% CI FOR MUviol – MUneut: (–1.578, –0.602)

a) At the meeting of the marketing staff, you have to explain what this output means. What will you say?

b) What advice would you give the company about the upcoming ad campaign?

25. Sex and violence II. In the study described in Exercise 23, the researchers also contacted the subjects again, 24 hours later, and asked them to recall the brands advertised. Results are summarized below.

	Program Type		
	Violent	**Sexual**	**Neutral**
No. of subjects	101	106	103
Brands recalled			
Mean	3.02	2.72	4.65
SD	1.61	1.85	1.62

a) Is there a significant difference in viewers' abilities to remember brands advertised in shows with violent vs. neutral content?

b) Find a 95% confidence interval for the difference in mean number of brand names remembered between the groups watching shows with sexual content and those watching neutral content. Interpret your interval in this context.

26. Ad recall. In Exercises 23 and 25, we see the number of advertised brand names people recalled immediately after watching TV shows and 24 hours later. Strangely enough, it appears that they remembered more about the ads the next day. Should we conclude this is true in general about people's memory of TV ads?

a) Suppose one analyst conducts a two-sample hypothesis test to see if memory of brands advertised during violent TV shows is higher 24 hours later. If his P-value is 0.00013, what might he conclude?

b) Explain why his procedure was inappropriate. Which of the assumptions for inference was violated?

c) How might the design of this experiment have tainted the results?

d) Suggest a design that could compare immediate brand-name recall with recall one day later.

27. Hungry?   Researchers investigated how the size of a bowl affects how much ice cream people tend to scoop when serving themselves.[10] At an "ice cream social," people were randomly given either a 17 oz or a 34 oz bowl (both large enough that they would not be filled to capacity). They were then invited to scoop as much ice cream as they liked. Did the bowl size change the selected portion size? Here are the summaries:

	Small Bowl		Large Bowl
n	26	n	22
$\bar{y}$	5.07 oz	$\bar{y}$	6.58 oz
s	1.84 oz	s	2.91 oz

Test an appropriate hypothesis and state your conclusions. Assume any assumptions and conditions that you cannot test are sufficiently satisfied to proceed.

28. Thirsty?   Researchers randomly assigned participants either a tall, thin "highball" glass or a short, wide "tumbler," each of which held 355 ml. Participants were asked to pour a shot (1.5 oz = 44.3 ml) into their glass. Did the shape of the glass make a difference in how much liquid they poured?[11] Here are the summaries:

	highball		tumbler
n	99	n	99
$\bar{y}$	42.2 ml	$\bar{y}$	60.9 ml
s	16.2 ml	s	17.9 ml

Test an appropriate hypothesis and state your conclusions. Assume any assumptions and conditions that you cannot test are sufficiently satisfied to proceed.

29. Lower scores?   Newspaper headlines recently announced a decline in science scores among high school seniors. In 2000, a total of 15,109 seniors tested by The National Assessment in Education Program (NAEP)

---

[10] Brian Wansink, Koert van Ittersum, and James E. Painter, "Ice Cream Illusions: Bowls, Spoons, and Self-Served Portion Sizes," *Am J Prev Med* 2006.
[11] Brian Wansink and Koert van Ittersum, "Shape of Glass and Amount of Alcohol Poured: Comparative Study of Effect of Practice and Concentration," *BMJ* 2005;331;1512–1514.

scored a mean of 147 points. Four years earlier, 7537 seniors had averaged 150 points. The standard error of the difference in the mean scores for the two groups was 1.22.

a) Have the science scores declined significantly? Cite appropriate statistical evidence to support your conclusion.

b) The sample size in 2000 was almost double that in 1996. Does this make the results more convincing or less? Explain.

30. The Internet.   The NAEP report described in Exercise 29 compared science scores for students who had home Internet access to the scores of those who did not, as shown in the graph. They report that the differences are statistically significant.

a) Explain what "statistically significant" means in this context.

b) If their conclusion is incorrect, which type of error did the researchers commit?

c) Does this prove that using the Internet at home can improve a student's performance in science?

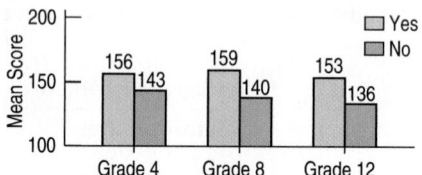

31. Running heats.   In Olympic running events, preliminary heats are determined by random draw, so we should expect that the abilities of runners in the various heats to be about the same, on average. Here are the times (in seconds) for the 400-m women's run in the 2004 Olympics in Athens for preliminary heats 2 and 5. Is there any evidence that the mean time to finish is different for randomized heats? Explain. Be sure to include a discussion of assumptions and conditions for your analysis.

Country	Name	Heat	Time
USA	HENNAGAN Monique	2	51.02
BUL	DIMITROVA Mariyana	2	51.29
CHA	NADJINA Kaltouma	2	51.50
JAM	DAVY Nadia	2	52.04
BRA	ALMIRAO Maria Laura	2	52.10
FIN	MYKKANEN Kirsi	2	52.53
CHN	BO Fanfang	2	56.01
BAH	WILLIAMS-DARLING Tonique	5	51.20
BLR	USOVICH Svetlana	5	51.37
UKR	YEFREMOVA Antonina	5	51.53
CMR	NGUIMGO Mireille	5	51.90
JAM	BECKFORD Allison	5	52.85
TOG	THIEBAUD-KANGNI Sandrine	5	52.87
SRI	DHARSHA K V Damayanthi	5	54.58

32. Swimming heats.   In Exercise 31 we looked at the times in two different heats for the 400-m women's run from the 2004 Olympics. Unlike track events, swimming heats are *not* determined at random. Instead, swimmers are seeded so that better swimmers are placed in later heats. Here are the times (in seconds) for the women's 400-m freestyle from heats 2 and 5. Do these results suggest that the mean times of seeded heats are not equal? Explain. Include a discussion of assumptions and conditions for your analysis.

Country	Name	Heat	Time
ARG	BIAGIOLI Cecilia Elizabeth	2	256.42
SLO	CARMAN Anja	2	257.79
CHI	KOBRICH Kristel	2	258.68
MKD	STOJANOVSKA Vesna	2	259.39
JAM	ATKINSON Janelle	2	260.00
NZL	LINTON Rebecca	2	261.58
KOR	HA Eun-Ju	2	261.65
UKR	BERESNYEVA Olga	2	266.30
FRA	MANAUDOU Laure	5	246.76
JPN	YAMADA Sachiko	5	249.10
ROM	PADURARU Simona	5	250.39
GER	STOCKBAUER Hannah	5	250.46
AUS	GRAHAM Elka	5	251.67
CHN	PANG Jiaying	5	251.81
CAN	REIMER Brittany	5	252.33
BRA	FERREIRA Monique	5	253.75

33. Tees. Does it matter what kind of tee a golfer places the ball on? The company that manufactures "Stinger" tees claims that the thinner shaft and smaller head will lessen drag, reducing spin and allowing the ball to travel farther. In August 2003, Golf Laboratories, Inc., compared the distance traveled by golf balls hit off regular wooden tees to those hit off Stinger tees. All the balls were struck by the same golf club using a robotic device set to swing the club head at approximately 95 miles per hour. Summary statistics from the test are shown in the table. Assume that 6 balls were hit off each tee and that the data were suitable for inference.

		Total Distance (yards)	Ball Velocity (mph)	Club Velocity (mph)
Regular tee	Avg.	227.17	127.00	96.17
	SD	2.14	0.89	0.41
Stinger tee	Avg.	241.00	128.83	96.17
	SD	2.76	0.41	0.52

Is there evidence that balls hit off the Stinger tees would have a higher initial velocity?

34. Golf again. Given the test results on golf tees described in Exercise 33, is there evidence that balls hit off Stinger tees would travel farther? Again, assume that 6 balls were hit off each tee and that the data were suitable for inference.

35. Crossing Ontario. Between 1954 and 2003, swimmers have crossed Lake Ontario 43 times. Both women and men have made the crossing. Here are some plots (we've

omitted a crossing by Vikki Keith, who swam a round trip—North to South to North—in 3390 minutes):

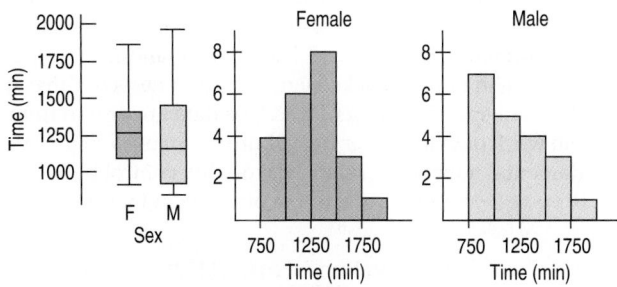

The summary statistics are:

	Summary of Time (min)		
Group	**Count**	**Mean**	**StdDev**
F	22	1271.59	261.111
M	20	1196.75	304.369

How much difference is there between the mean amount of time (in minutes) it would take female and male swimmers to swim the lake?
a) Construct and interpret a 95% confidence interval for the difference between female and male times.
b) Comment on the assumptions and conditions.

36. Music and memory. Is it a good idea to listen to music when studying for a big test? In a study conducted by some Statistics students, 62 people were randomly assigned to listen to rap music, music by Mozart, or no music while attempting to memorize objects pictured on a page. They were then asked to list all the objects they could remember. Here are summary statistics:

	Rap	Mozart	No Music
Count	29	20	13
Mean	10.72	10.00	12.77
SD	3.99	3.19	4.73

a) Does it appear that it is better to study while listening to Mozart than to rap music? Test an appropriate hypothesis and state your conclusion.
b) Create a 90% confidence interval for the mean difference in memory score between students who study to Mozart and those who listen to no music at all. Interpret your interval.

37. Rap. Using the results of the experiment described in Exercise 36, does it matter whether one listens to rap music while studying, or is it better to study without music at all?
a) Test an appropriate hypothesis and state your conclusion.
b) If you concluded there is a difference, estimate the size of that difference with a confidence interval and explain what your interval means.

38. Cuckoos. Cuckoos lay their eggs in the nests of other (host) birds. The eggs are then adopted and hatched by the host birds. But the potential host birds lay eggs of different sizes. Does the cuckoo change the size of her eggs for different foster species? The numbers in the table are lengths (in mm) of cuckoo eggs found in nests of three different species of other birds. The data are drawn from the work of O.M. Latter in 1902 and were used in a fundamental textbook on statistical quality control by L.H.C. Tippett (1902–1985), one of the pioneers in that field.

CUCKOO EGG LENGTH (MM)		
Foster Parent Species		
Sparrow	Robin	Wagtail
20.85	21.05	21.05
21.65	21.85	21.85
22.05	22.05	21.85
22.85	22.05	21.85
23.05	22.05	22.05
23.05	22.25	22.45
23.05	22.45	22.65
23.05	22.45	23.05
23.45	22.65	23.05
23.85	23.05	23.25
23.85	23.05	23.45
23.85	23.05	24.05
24.05	23.05	24.05
25.05	23.05	24.05
	23.25	24.85
	23.85	

Investigate the question of whether the mean length of cuckoo eggs is the same for different species, and state your conclusion.

JUST CHECKING
Answers

1. Randomization should balance unknown sources of variability in the two groups of patients and helps us believe the two groups are independent.

2. We can be 95% confident that after 4 weeks endoscopic surgery patients will have a mean pinch strength between 0.04 kg and 2.96 kg higher than open-incision patients.

3. The lower bound of this interval is close to 0, so the difference may not be great enough that patients could actually notice the difference. We may want to consider other issues such as cost or risk in making a recommendation about the two surgical procedures.

4. Without data, we can't check the Nearly Normal Condition.

5. H_0: Mean pinch strength is the same after both surgeries. ($\mu_E - \mu_O = 0$)

   H_A: Mean pinch strength is different after the two surgeries. ($\mu_E - \mu_O \neq 0$)

6. With a P-value this low, we reject the null hypothesis. We can conclude that mean pinch strength differs after 4 weeks in patients who undergo endoscopic surgery vs. patients who have open-incision surgery. Results suggest that the endoscopic surgery patients may be stronger, on average.

7. If some patients contributed two hands to the study, then the groups may not be internally independent. It is reasonable to assume that two hands from the same patient might respond in similar ways to similar treatments.

Paired Samples and Blocks

WHO	Olympic speed-skaters
WHAT	Time for women's 1500 m
UNITS	Seconds
WHEN	2006
WHERE	Torino, Italy
WHY	To see whether one lane is faster than the other

peed-skating races are run in pairs. Two skaters start at the same time, one on the inner lane and one on the outer lane. Halfway through the race, they cross over, switching lanes so that each will skate the same distance in each lane. Even though this seems fair, at the 2006 Olympics some fans thought there might have been an advantage to starting on the outside. After all, the winner, Cindy Klassen, started on the outside and skated a remarkable 1.47 seconds faster than the silver medalist.

Here are the data for the women's 1500-m race:

Inner Lane		Outer Lane	
Name	Time	Name	Time
OLTEAN Daniela	129.24	(no competitor)	
ZHANG Xiaolei	125.75	NEMOTO Nami	122.34
ABRAMOVA Yekaterina	121.63	LAMB Maria	122.12
REMPEL Shannon	122.24	NOH Seon Yeong	123.35
LEE Ju-Youn	120.85	TIMMER Marianne	120.45
ROKITA Anna Natalia	122.19	MARRA Adelia	123.07
YAKSHINA Valentina	122.15	OPITZ Lucille	122.75
BJELKEVIK Hedvig	122.16	HAUGLI Maren	121.22
ISHINO Eriko	121.85	WOJCICKA Katarzyna	119.96
RANEY Catherine	121.17	BJELKEVIK Annette	121.03
OTSU Hiromi	124.77	LOBYSHEVA Yekaterina	118.87
SIMIONATO Chiara	118.76	JI Jia	121.85
ANSCHUETZ THOMS Daniela	119.74	WANG Fei	120.13
BARYSHEVA Varvara	121.60	van DEUTEKOM Paulien	120.15
GROENEWOLD Renate	119.33	GROVES Kristina	116.74
RODRIGUEZ Jennifer	119.30	NESBITT Christine	119.15
FRIESINGER Anni	117.31	KLASSEN Cindy	115.27
WUST Ireen	116.90	TABATA Maki	120.77

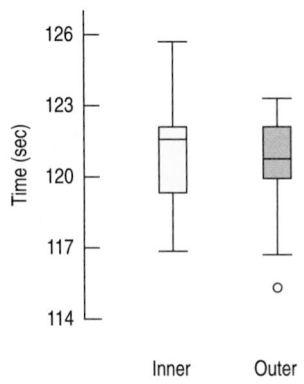

FIGURE 25.1

Using boxplots to compare times in the inner and outer lanes shows little because it ignores the fact that the skaters raced in pairs.

We can view this skating event as an experiment testing whether the lanes were equally fast. Skaters were assigned to lanes randomly. The boxplots of times recorded in the inner and outer lanes (look back a page) don't show much difference. But that's not the right way to compare these times. Conditions can change during the day. The data are recorded for races run two at a time, so the two groups are not independent.

Paired Data

Data such as these are called **paired.** We have the times for skaters in each lane for each race. The races are run in pairs, so they can't be independent. And since they're not independent, we can't use the two-sample t methods. Instead, we can focus on the *differences* in times for each racing pair.

Paired data arise in a number of ways. Perhaps the most common way is to compare subjects with themselves before and after a treatment. When pairs arise from an experiment, the pairing is a type of *blocking*. When they arise from an observational study, it is a form of *matching*.

FOR EXAMPLE   Identifying paired data

Do flexible schedules reduce the demand for resources? The Lake County, Illinois, Health Department experimented with a flexible four-day workweek. For a year, the department recorded the mileage driven by 11 field workers on an ordinary five-day workweek. Then it changed to a flexible four-day workweek and recorded mileage for another year.[1] The data are shown.

Question: Why are these data paired?

The mileage data are paired because each driver's mileage is measured before and after the change in schedule. I'd expect drivers who drove more than others before the schedule change to continue to drive more afterwards, so the two sets of mileages can't be considered independent.

Name	5-Day mileage	4-Day mileage
Jeff	2798	2914
Betty	7724	6112
Roger	7505	6177
Tom	838	1102
Aimee	4592	3281
Greg	8107	4997
Larry G.	1228	1695
Tad	8718	6606
Larry M.	1097	1063
Leslie	8089	6392
Lee	3807	3362

Pairing isn't a problem; it's an opportunity. If you know the data are paired, you can take advantage of that fact—in fact, you *must* take advantage of it. You *may not* use the two-sample and pooled methods of the previous chapter when the data are paired. Remember: Those methods rely on the Pythagorean Theorem of Statistics, and that requires the two samples be independent. Paired data aren't. There is no test to determine whether the data are paired. You must determine that from understanding how they were collected and what they mean (check the W's).

Once we recognize that the speed-skating data are matched pairs, it makes sense to consider the difference in times for each two-skater race. So we look at the *pairwise* differences:

---

[1] Charles S. Catlin, "Four-day Work Week Improves Environment," *Journal of Environmental Health*, Denver, 59:7.

A S *Activity:* **Differences in Means of Paired Groups.** Are married couples typically the same age, or do wives tend to be younger than their husbands, on average?

Skating Pair	Inner Time	Outer Time	Inner − Outer
1	129.24		•
2	125.75	122.34	3.41
3	121.63	122.12	−0.49
4	122.24	123.35	−1.11
5	120.85	120.45	0.40
6	122.19	123.07	−0.88
7	122.15	122.75	−0.60
8	122.16	121.22	0.94
9	121.85	119.96	1.89
10	121.17	121.03	0.14
11	124.77	118.87	5.90
12	118.76	121.85	−3.09
13	119.74	120.13	−0.39
14	121.60	120.15	1.45
15	119.33	116.74	2.59
16	119.30	119.15	0.15
17	117.31	115.27	2.04
18	116.90	120.77	−3.87

The first skater raced alone, so we'll omit that race. Because it is the *differences* we care about, we'll treat them as if *they* were the data, ignoring the original two columns. Now that we have only one column of values to consider, we can use a simple one-sample *t*-test. Mechanically, a **paired *t*-test** is just a one-sample *t*-test for the means of these pairwise differences. The sample size is the number of pairs.

So you've already seen the *Show.*

Assumptions and Conditions

PAIRED DATA ASSUMPTION

Paired Data Assumption: The data must be paired. You can't just decide to pair data when in fact the samples are independent. When you have two groups with the same number of observations, it may be tempting to match them up.

Don't, unless you are prepared to justify your claim that the data are paired.

On the other hand, be sure to recognize paired data when you have them. Remember, two-sample *t* methods aren't valid without independent groups, and paired groups aren't independent. Although this is a strictly required assumption, it is one that can be easy to check if you understand how the data were collected.

INDEPENDENCE ASSUMPTION

Independence Assumption: If the data are paired, the *groups* are not independent. For these methods, it's the *differences* that must be independent of each other. There's no reason to believe that the difference in speeds of one pair of races could affect the difference in speeds for another pair.

Randomization Condition: Randomness can arise in many ways. The pairs may be a random sample. In an experiment, the order of the two treatments may be randomly assigned, or the treatments may be randomly assigned to one member of each pair. In a before-and-after study, we may believe that the observed differences are a representative sample from a population of interest. If we have any doubts, we'll need to include a control group to be able to draw conclusions.

What we want to know usually focuses our attention on where the randomness should be.

In our example, skaters were assigned to the lanes at random.

10% Condition: We're thinking of the speed-skating data as an experiment testing the difference between lanes. The 10% Condition doesn't apply to randomized experiments, where no sampling takes place.

NORMAL POPULATION ASSUMPTION

We need to assume that the population of *differences* follows a Normal model. We don't need to check the individual groups.

Nearly Normal Condition: This condition can be checked with a histogram or Normal probability plot of the *differences*—but not of the individual groups. As with the one-sample *t*-methods, this assumption matters less the more pairs we have to consider. You may be pleasantly surprised when you check this condition. Even if your original measurements are skewed or bimodal, the *differences* may be nearly Normal. After all, the individual who was way out in the tail on an initial measurement is likely to still be out there on the second one, giving a perfectly ordinary difference.

FOR EXAMPLE   Checking assumptions and conditions

Recap: Field workers for a health department compared driving mileage on a five-day work schedule with mileage on a new four-day schedule. To see if the new schedule changed the amount of driving they did, we'll look at paired differences in mileages before and after.

Question: Is it okay to use these data to test whether the new schedule changed the amount of driving?

✓ **Paired Data Assumption:** The data are paired because each value is the mileage driven by the same person before and after a change in work schedule.

✓ **Independence Assumption:** The driving behavior of any individual worker is independent of the others, so the differences are mutually independent.

✓ **Randomization Condition:** The mileages are the sums of many individual trips, each of which experienced random events that arose while driving. Repeating the experiment in two new years would give randomly different values.

✓ **Nearly Normal Condition:** The histogram of the mileage differences is unimodal and symmetric:

Since the assumptions and conditions are satisfied, it's okay to use paired-t methods for these data.

Name	5-Day mileage	4-Day mileage	Difference
Jeff	2798	2914	−116
Betty	7724	6112	1612
Roger	7505	6177	1328
Tom	838	1102	−264
Aimee	4592	3281	1311
Greg	8107	4997	3110
Larry G.	1228	1695	−467
Tad	8718	6606	2112
Larry M.	1097	1063	34
Leslie	8089	6392	1697
Lee	3807	3362	445

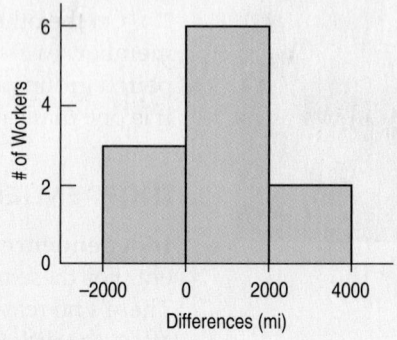

The steps in testing a hypothesis for paired differences are very much like the steps for a one-sample *t*-test for a mean.

THE PAIRED t-TEST

When the conditions are met, we are ready to test whether the mean of paired differences is significantly different from zero. We test the hypothesis

$$H_0: \mu_d = \Delta_0,$$

where the d's are the pairwise differences and Δ_0 is almost always 0.

We use the statistic

$$t_{n-1} = \frac{\bar{d} - \Delta_0}{SE(\bar{d})},$$

where $\bar{d}$ is the mean of the pairwise differences, n is the number of *pairs*, and

$$SE(\bar{d}) = \frac{s_d}{\sqrt{n}}.$$

$SE(\bar{d})$ is the ordinary standard error for the mean, applied to the differences.

When the conditions are met and the null hypothesis is true, we can model the sampling distribution of this statistic with a Student's t-model with $n - 1$ degrees of freedom, and use that model to obtain a P-value.

STEP-BY-STEP EXAMPLE   A Paired t-Test

Question: Was there a difference in speeds between the inner and outer speed-skating lanes at the 2006 Winter Olympics?

THINK

Plan State what we want to know.

Identify the *parameter* we wish to estimate. Here our parameter is the mean difference in race times.

Identify the variables and check the W's.

> I want to know whether there really was a difference in the speeds of the two lanes for speed skating at the 2006 Olympics. I have data for the women's 1500-m race.

Hypotheses State the null and alternative hypotheses.

Although fans suspected one lane was faster, we can't use the data we have to specify the direction of a test. We (and Olympic officials) would be interested in a difference in either direction, so we'd better test a two-sided alternative.

> H_0: Neither lane offered an advantage:
>
> $$\mu_d = 0.$$
>
> H_A: The mean difference is different from zero:
>
> $$\mu_d \neq 0.$$

REALITY CHECK The individual differences are all in seconds. We should expect the mean difference to be comparable in magnitude.

Model Think about the assumptions and check the conditions.

> ✔ **Independence Assumption:** Each race is independent of the others, so the differences are mutually independent.

State why you think the data are paired. Simply having the same number of individuals in each group and displaying them in side-by-side columns doesn't make them paired.

Think about what we hope to learn and where the randomization comes from. Here, the randomization comes from the racer pairings and lane assignments.

Make a picture—just one. Don't plot separate distributions of the two groups—that entirely misses the pairing. For paired data, it's the Normality of the *differences* that we care about. Treat those paired differences as you would a single variable, and check the Nearly Normal Condition with a histogram or a Normal probability plot.

✔ **Paired Data Assumption:** The data are paired because racers compete in pairs.

✔ **Randomization Condition:** Skaters are assigned to lanes at random. Repeating the experiment with different pairings and lane assignments would give randomly different values.

✔ **Nearly Normal Condition:** The histogram of the differences is unimodal and symmetric:

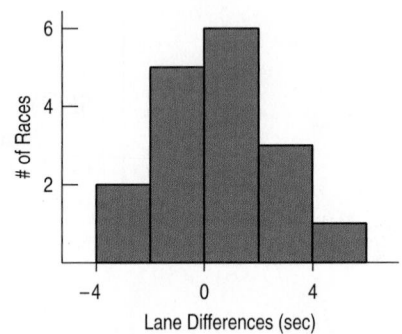

Specify the sampling distribution model.

Choose the method.

The conditions are met, so I'll use a Student's *t*-model with $(n - 1) = 16$ degrees of freedom, and perform a **paired *t*-test.**

Mechanics

n is the number of *pairs*—in this case, the number of races.

$\bar{d}$ is the mean difference.

s_d is the standard deviation of the differences.

Find the standard error and the *t*-score of the observed mean difference. There is nothing new in the mechanics of the paired-*t* methods. These are the mechanics of the *t*-test for a mean applied to the differences.

Make a picture. Sketch a *t*-model centered at the hypothesized mean of 0. Because this is a two-tail test, shade both the region to the right of the observed mean difference of 0.499 seconds and the corresponding region in the lower tail.

The data give

$$n = 17 \text{ pairs}$$
$$\bar{d} = 0.499 \text{ seconds}$$
$$s_d = 2.333 \text{ seconds.}$$

I estimate the standard deviation of $\bar{d}$ using

$$SE(\bar{d}) = \frac{s_d}{\sqrt{n}} = \frac{2.333}{\sqrt{17}} = 0.5658$$

So $t_{16} = \dfrac{\bar{d} - 0}{SE(\bar{d})} = \dfrac{0.499}{0.5658} = 0.882$

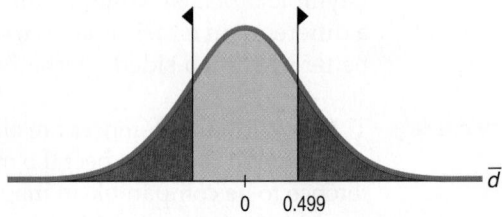

Find the P-value, using technology.

P-value $= 2P(t_{16} > 0.882) = 0.39$

 **REALITY CHECK** The mean difference is 0.499 seconds. That may not seem like much, but a smaller difference determined the Silver and Bronze medals. The standard error is about this big, so a *t*-value less than 1.0 isn't surprising. Nor is a large P-value.

TELL   **Conclusion** Link the P-value to your decision about H_0, and state your conclusion in context.

The P-value is large. Events that happen more than a third of the time are not remarkable. So, even though there is an observed difference between the lanes, I can't conclude that it isn't due simply to random chance. It appears the fans may have interpreted a random fluctuation in the data as favoring one lane. There's insufficient evidence to declare any lack of fairness.

FOR EXAMPLE   Doing a paired *t*-test

Recap: We want to test whether a change from a five-day workweek to a four-day workweek could change the amount driven by field workers of a health department. We've already confirmed that the assumptions and conditions for a paired *t*-test are met.

Question: Is there evidence that a four-day workweek would change how many miles workers drive?

H_0: The change in the health department workers' schedules didn't change the mean mileage driven; the mean difference is zero:

$$\mu_d = 0.$$

H_A: The mean difference is different from zero:

$$\mu_d \neq 0.$$

The conditions are met, so I'll use a Student's *t*-model with $(n - 1) = 10$ degrees of freedom and perform a **paired *t*-test.**

The data give

$$n = 11 \text{ pairs}$$
$$\bar{d} = 982 \text{ miles}$$
$$s_d = 1139.6 \text{ miles.}$$

$$SE(\bar{d}) = \frac{s_d}{\sqrt{n}} = \frac{1139.6}{\sqrt{11}} = 343.6$$

$$\text{So } t_{10} = \frac{\bar{d} - 0}{SE(\bar{d})} = \frac{982.0}{343.6} = 2.86$$

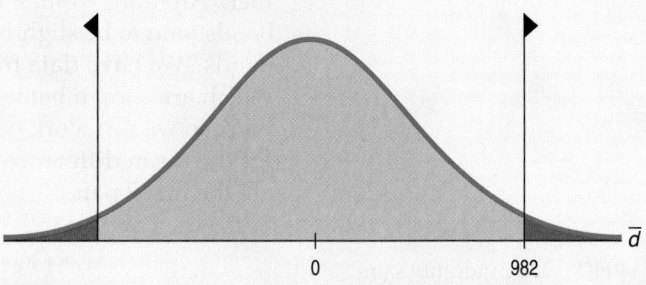

$$\text{P-value} = 2P(t_{10} > 2.86) = 0.017$$

The P-value is small, so I reject the null hypothesis and conclude that the change in workweek did lead to a change in average driving mileage. It appears that changing the work schedule may reduce the mileage driven by workers.

Note: We should propose a course of action, but it's hard to tell from the hypothesis test whether the reduction matters. Is the difference in mileage important in the sense of reducing air pollution or costs, or is it merely statistically significant? To help make that decision, we should look at a confidence interval. If the difference in mileage proves to be large in a practical sense, then we might recommend a change in schedule for the rest of the department.

```
L1-L2→L3
{-116 1612 1328…
■
```

```
L1    L2    L3    3
2798  2914  -116
7724  6112  1612
7505  6177  1328
838   1102  -264
4592  3281  1311
8107  4997  3110
1228  1695  -467
L3(1)= -116
```

```
T-Test
Inpt:Data Stats
μ0:0
List:L3
Freq:1■
μ:≠μ0 <μ0 >μ0
Calculate Draw
```

```
T-Test
μ≠0
t=2.85899122
p=.0169862463
x̄=982.8181818
Sx=1140.136116
n=11
■
```

Since the inference procedures for matched data are essentially just the one-sample *t* procedures, you already know what to do . . . once you have the list of paired differences, that is. That list is not hard to create.

Test a hypothesis about the mean of paired differences.

- Think: Are the samples independent or paired. Independent? Go back to the last chapter! Paired? Read on.
- Enter the driving data from page 588 into two lists, say *5-Day mileage* in L1, *4-Day mileage* in L2.
- Create a list of the differences. We want to take each value in L1, subtract the corresponding value in L2, and store the paired difference in L3. The command is L1-L2→L3. (The arrow is the STO button.) Now take a look at L3. See—it worked!
- Make a histogram of the differences, L3, to check the nearly Normal condition. Notice that we do not look at the histograms of the *5-day mileage* or the *4-day mileage*. Those are not the data that we care about now that we are using a paired procedure. Note also that the calculator's first histogram is not close to Normal. More work to do . . .
- As you have seen before, small samples often produce ragged histograms, and these may look very different after a change in bar width. Reset the WINDOW to Xmin=-3000, Xmax=4500, and Xscl=1500. The new histogram looks okay.
- Under STAT TESTS simply use 2:T-Test, as you've done before for hypothesis tests about a mean.
- Specify that the hypothesized difference is 0, you're using the Data in L3, and it's a two-tailed test.
- Calculate.

The small P-value shows strong evidence that on average the change in the workweek reduces the number of miles workers drive.

Confidence Intervals for Matched Pairs

In developed countries, the average age of women is generally higher than that of men. After all, women tend to live longer. But if we look at *married couples,* husbands tend to be slightly older than wives. How much older, on average, are husbands? We have data from a random sample of 200 British couples, the first 7 of which are shown below. Only 170 couples provided ages for both husband and wife, so we can work only with that many pairs. Let's form a confidence interval for the mean difference of husband's and wife's ages for these 170 couples. Here are the first 7 pairs:

WHO	170 randomly sampled couples
WHAT	Ages
UNITS	Years
WHEN	Recently
WHERE	Britain

Wife's Age	Husband's Age	Difference (husband − wife)
43	49	6
28	25	−3
30	40	10
57	52	−5
52	58	6
27	32	5
52	43	−9
⋮	⋮	⋮

Clearly, these data are paired. The survey selected *couples* at random, not individuals. We're interested in the mean age difference within couples. How would we construct a confidence interval for the true mean difference in ages?

PAIRED *t*-INTERVAL

When the conditions are met, we are ready to find the confidence interval for the mean of the paired differences. The confidence interval is

$$\bar{d} \pm t^*_{n-1} \times SE(\bar{d}),$$

where the standard error of the mean difference is $SE(\bar{d}) = \dfrac{s_d}{\sqrt{n}}$.

The critical value t^* from the Student's t-model depends on the particular confidence level, C, that you specify and on the degrees of freedom, $n - 1$, which is based on the number of pairs, n.

Making confidence intervals for matched pairs follows exactly the steps for a one-sample t-interval.

STEP-BY-STEP EXAMPLE   **A Paired *t*-Interval**

Question: How big a difference is there, on average, between the ages of husbands and wives?

Plan State what we want to know.

Identify the variables and check the W's.

Identify the parameter you wish to estimate. For a paired analysis, the parameter of interest is the mean of the differences. The population of interest is the population of differences.

Model Think about the assumptions and check the conditions.

I want to estimate the mean difference in age between husbands and wives. I have a random sample of 200 British couples, 170 of whom provided both ages.

✔ **Paired Data Assumption:** The data are paired because they are on members of married couples.

✔ **Independence Assumption:** The data are from a randomized survey, so couples should be independent of each other.

✔ **Randomization Condition:** These couples were randomly sampled.

✔ **10% Condition:** The sample is less than 10% of the population of married couples in Britain.

Make a picture. We focus on the differences, so a histogram or Normal probability plot is best here.

 The histogram shows husbands are often older than wives (because most of the differences are greater than 0). The mean difference seen here of about 2 years is reasonable.

State the sampling distribution model.

Choose your method.

✔ **Nearly Normal Condition:** The histogram of the husband – wife differences is unimodal and symmetric:

The conditions are met, so I can use a Student's t-model with $(n - 1) = 169$ degrees of freedom and find a **paired t-interval.**

 **Mechanics**

n is the number of *pairs*, here, the number of couples.

$\bar{d}$ is the mean difference.

s_d is the standard deviation of the differences.

Be sure to include the units along with the statistics.

The critical value we need to make a 95% interval comes from a Student's *t* table, a computer program, or a calculator.

This result makes sense. Our everyday experience confirms that an average age difference of about 2 years is reasonable.

$n = 170$ couples
$\bar{d} = 2.2$ years
$s_d = 4.1$ years

I estimate the standard error of $\bar{d}$ as

$$SE(\bar{d}) = \frac{s_d}{\sqrt{n}} = \frac{4.1}{\sqrt{170}} = 0.31 \text{ years.}$$

The df for the t-model is $n - 1 = 169$.

The 95% critical value for t_{169} (from the table) is 1.97.

The margin of error is

$$ME = t^*_{169} \times SE(\bar{d}) = 1.97(0.31) = 0.61$$

So the 95% confidence interval is

$$2.2 \pm 0.6 \text{ years,}$$

or an interval of $(1.6, 2.8)$ years.

TELL **Conclusion** Interpret the confidence interval in context.

I am 95% confident that British husbands are, on average, 1.6 to 2.8 years older than their wives.

TI Tips — Creating a confidence interval

Now let's get the TI to create a confidence interval for the mean of paired differences.

We'll demonstrate by using the statistics about the ages of the British married couples. (If we had all the data, we could enter that, of course. All 170 couples? Um, no thanks.) The husband in the sample were an average of 2.2 years older than their wives, with a standard deviation of 4.1 years. We've already seen that the data are paired and that a histogram of the differences satisfies the Nearly Normal Condition. (With a sample this large, we could proceed with inference even if we didn't have the actual data and were unable to make the histogram.)

- Once again, we treat the paired differences just like data from one sample. A confidence interval for the mean difference, then, like that for a mean, uses the STAT TESTS one-sample procedure 8:TInterval.
- Specify Inpt:Stats, and enter the statistics for the paired differences.
- Calculate.

Done. Finding the interval was the easy part. Now it's time for you to *Tell* what it means. Don't forget to talk about married couples in Britain.

```
TInterval
 Inpt:Data Stats
 x:2.2
 Sx:4.1
 n:170
 C-Level:.95
 Calculate
```

```
TInterval
 (1.5792,2.8208)
 x=2.2
 Sx=4.1
 n=170
```

Effect Size

When we examined the speed-skating times, we failed to reject the null hypothesis, so we couldn't be certain whether there really was a difference between the lanes. Maybe there wasn't any difference, or maybe whatever difference there might have been was just too small to matter at all. Were the fans right to be concerned?

We can't tell from the hypothesis test, but using the same summary statistics, we can find that the corresponding 95% confidence interval for the mean difference is $(-0.70 < \mu_d < 1.70)$ seconds.

A confidence interval is a good way to get a sense for the size of the effect we're trying to understand. That gives us a plausible range of values for the true mean difference in lane times. If differences of 1.7 seconds were too small to matter in 1500-m Olympic speed skating, we'd be pretty sure there was no need for concern.

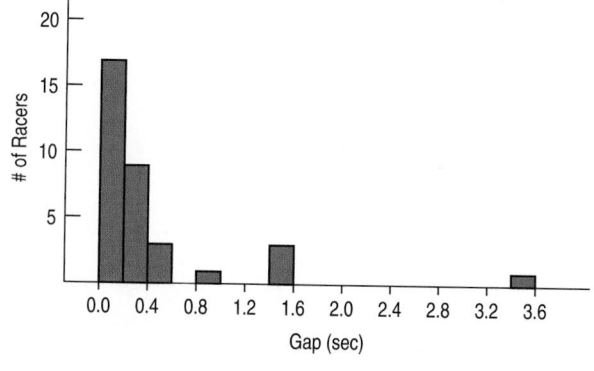

But in fact, except for the Gold − Silver gap, the successive gaps between each skater and the next-faster one were *all* less than the high end of this interval, and most were right around the middle of the interval.

So even though we were unable to discern a real difference, the confidence interval shows that the effects we're considering may be big enough to be important. We may want to continue this investigation by checking out other races on this ice and being alert for possible differences at other venues.

FOR EXAMPLE   Looking at effect size with a paired-*t* confidence interval

Recap: We know that, on average, the switch from a five-day workweek to a four-day workweek reduced the amount driven by field workers in that Illinois health department. However, finding that there is a significant difference doesn't necessarily mean that difference is meaningful or worthwhile. To assess the size of the effect, we need a confidence interval. We already know the assumptions and conditions are met.

Question: By how much, on average, might a change in workweek schedule reduce the amount driven by workers?

$$\bar{d} = 982 \text{ mi} \qquad SE(\bar{d}) = 343.6 \qquad t^*_{10} = 2.228 \text{ (for 95\%)}$$
$$ME = t^*_{10} \times SE(\bar{d}) = 2.228(343.6) = 765.54$$

So the 95% confidence interval for μ_d is 982 ± 765.54 or (216.46, 1747.54) fewer miles.

With 95% confidence, I estimate that by switching to a four-day workweek employees would drive an average of between 216 and 1748 fewer miles per year. With high gas prices, this could save a lot of money.

Blocking

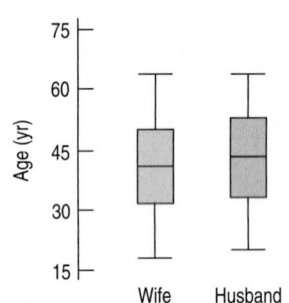

FIGURE 25.2

This display is worthless. It does no good to compare all the wives as a group with all the husbands. We care about the paired differences.

Because the sample of British husbands and wives includes both older and younger couples, there's a lot of variation in the ages of the men and in the ages of the women. In fact, that variation is so great that a boxplot of the two groups would show little difference. But that would be the wrong plot. It's the *difference* we care about. Pairing isolates the extra variation and allows us to focus on the individual differences. In Chapter 13 we saw how we could design an experiment with blocking to isolate the variability between identifiable groups of subjects, allowing us to better see variability among treatment groups due to their response to the treatment. A paired design is an example of blocking.

When we pair, we have roughly half the degrees of freedom of a two-sample test. You may see discussions that suggest that in "choosing" a paired analysis we "give up" these degrees of freedom. This isn't really true, though. If the data are paired, then there never were additional degrees of freedom, and we have no "choice." The fact of the pairing determines how many degrees of freedom are available.

Matching pairs generally removes so much extra variation that it more than compensates for having only half the degrees of freedom. Of course, inappropriate matching when the groups are in fact independent (say, by matching on the first letter of the last name of subjects) would cost degrees of freedom without the benefit of reducing the variance. When you design a study or experiment, you should consider using a paired design if possible.

JUST CHECKING

Think about each of the situations described below.

▶   Would you use a two-sample *t* or paired-*t* method (or neither)? Why?

▶   Would you perform a hypothesis test or find a confidence interval?

1. Random samples of 50 men and 50 women are asked to imagine buying a birthday present for their best friend. We want to estimate the difference in how much they are willing to spend.

2. Mothers of twins were surveyed and asked how often in the past month strangers had asked whether the twins were identical.

3. Are parents equally strict with boys and girls? In a random sample of families, researchers asked a brother and sister from each family to rate how strict their parents were.

4. Forty-eight overweight subjects are randomly assigned to either aerobic or stretching exercise programs. They are weighed at the beginning and at the end of the experiment to see how much weight they lost.

    a) We want to estimate the mean amount of weight lost by those doing aerobic exercise.

    b) We want to know which program is more effective at reducing weight.

5. Couples at a dance club were separated and each person was asked to rate the band. Do men or women like this band more?

WHAT CAN GO WRONG?

▶ **Don't use a two-sample *t*-test when you have paired data.** See the What Can Go Wrong? discussion in Chapter 24.

▶ **Don't use a paired-*t* method when the samples aren't paired.** Just because two groups have the same number of observations doesn't mean they can be paired, even if they are shown side by side in a table. We might have 25 men and 25 women in our study, but they might be completely independent of one another. If they were siblings or spouses, we might consider them paired. Remember that you cannot *choose* which method to use based on your preferences. If the data are from two independent samples, use two-sample *t* methods. If the data are from an experiment in which observations were paired, you must use a paired method. If the data are from an observational study, you must be able to defend your decision to use matched pairs or independent groups.

▶ **Don't forget outliers.** The outliers we care about now are in the differences. A subject who is extraordinary both before and after a treatment may still have a perfectly typical difference. But one outlying difference can completely distort your conclusions. Be sure to plot the differences (even if you also plot the data).

▶ **Don't look for the difference between the means of paired groups with side-by-side boxplots.** The point of the paired analysis is to remove extra variation. The boxplots of each group still contain that variation. Comparing them is likely to be misleading.

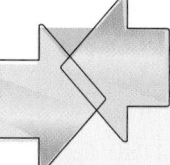

CONNECTIONS

The most important connection is to the concept of blocking that we first discussed when we considered designed experiments in Chapter 13. Pairing is a basic and very effective form of blocking.

Of course, the details of the mechanics for paired *t*-tests and intervals are identical to those for the one-sample *t*-methods. Everything we know about those methods applies here.

The connection to the two-sample and pooled methods of the previous chapter is that when the data are naturally paired, those methods are not appropriate because paired data fail the required condition of independence.

WHAT HAVE WE LEARNED?

When we looked at various ways to design experiments, back in Chapter 13, we saw that pairing can be a very effective strategy. Because pairing can help control variability between individual subjects, paired methods are usually more powerful than methods that compare independent groups. Now we've learned that analyzing data from matched pairs requires different inference procedures.

▶ We've learned that paired t-methods look at pairwise differences. Based on these differences, we test hypotheses and generate confidence intervals. These procedures are mechanically identical to the one-sample t-methods we saw in Chapter 23.

▶ We've also learned to *Think* about the design of the study that collected the data before we proceed with inference. We must be careful to recognize pairing when it is present but not assume it when it is not. Making the correct decision about whether to use independent t-procedures or paired t-methods is the first critical step in analyzing the data.

Terms

Paired data

588. Data are paired when the observations are collected in pairs or the observations in one group are naturally related to observations in the other. The simplest form of pairing is to measure each subject twice—often before and after a treatment is applied. More sophisticated forms of pairing in experiments are a form of blocking and arise in other contexts. Pairing in observational and survey data is a form of matching.

Paired t-test

591. A hypothesis test for the mean of the pairwise differences of two groups. It tests the null hypothesis

$$H_0: \mu_d = \Delta_0,$$

where the hypothesized difference is almost always 0, using the statistic

$$t = \frac{\bar{d} - \Delta_0}{SE(\bar{d})}$$

with $n - 1$ degrees of freedom, where $SE(\bar{d}) = \dfrac{s_d}{\sqrt{n}}$, and n is the number of pairs.

Paired-t confidence interval

595. A confidence interval for the mean of the pairwise differences between paired groups found as

$$\bar{d} \pm t^*_{n-1} \times SE(\bar{d}), \text{ where } SE(\bar{d}) = \frac{s_d}{\sqrt{n}} \text{ and } n \text{ is the number of pairs.}$$

Skills

▶ Be able to recognize whether a design that compares two groups is paired.

▶ Be able to find a paired confidence interval, recognizing that it is mechanically equivalent to doing a one-sample t-interval applied to the differences.

▶ Be able to perform a paired t-test, recognizing that it is mechanically equivalent to a one-sample t-test applied to the differences.

▶ Be able to interpret a paired t-test, recognizing that the hypothesis tested is about the mean of the differences between paired values rather than about the differences between the means of two independent groups.

▶ Be able to interpret a paired t-interval, recognizing that it gives an interval for the mean difference in the pairs.

PAIRED *t* ON THE COMPUTER

Most statistics programs can compute paired-*t* analyses. Some may want you to find the differences yourself and use the one-sample *t* methods. Those that perform the entire procedure will need to know the two variables to compare. The computer, of course, cannot verify that the variables are naturally paired. Most programs will check whether the two variables have the same number of observations, but some stop there, and that can cause trouble. Most programs will automatically omit any pair that is missing a value for either variable (as we did with the British couples). You must look carefully to see whether that has happened.

As we've seen with other inference results, some packages pack a lot of information into a simple table, but you must locate what you want for yourself. Here's a generic example with comments:

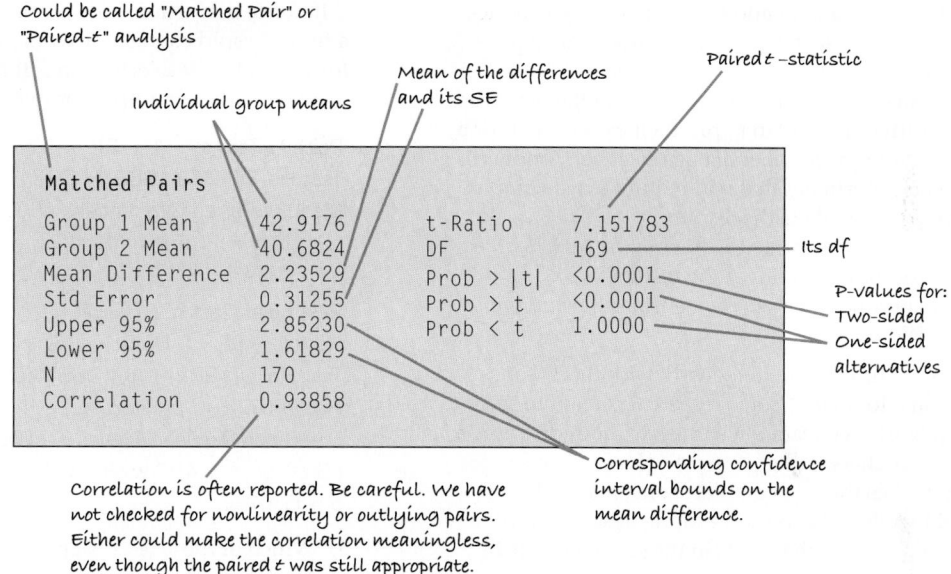

Could be called "Matched Pair" or "Paired-*t*" analysis

Individual group means

Mean of the differences and its SE

Paired *t* –statistic

```
Matched Pairs

Group 1 Mean      42.9176      t-Ratio      7.151783
Group 2 Mean      40.6824      DF           169
Mean Difference    2.23529     Prob > |t|   <0.0001
Std Error          0.31255     Prob > t     <0.0001
Upper 95%          2.85230     Prob < t      1.0000
Lower 95%          1.61829
N                 170
Correlation        0.93858
```

Its df

P-values for:
Two-sided
One-sided
alternatives

Correlation is often reported. Be careful. We have not checked for nonlinearity or outlying pairs. Either could make the correlation meaningless, even though the paired *t* was still appropriate.

Corresponding confidence interval bounds on the mean difference.

Other packages try to be more descriptive. It may be easier to find the results, but you may get less information from the output table.

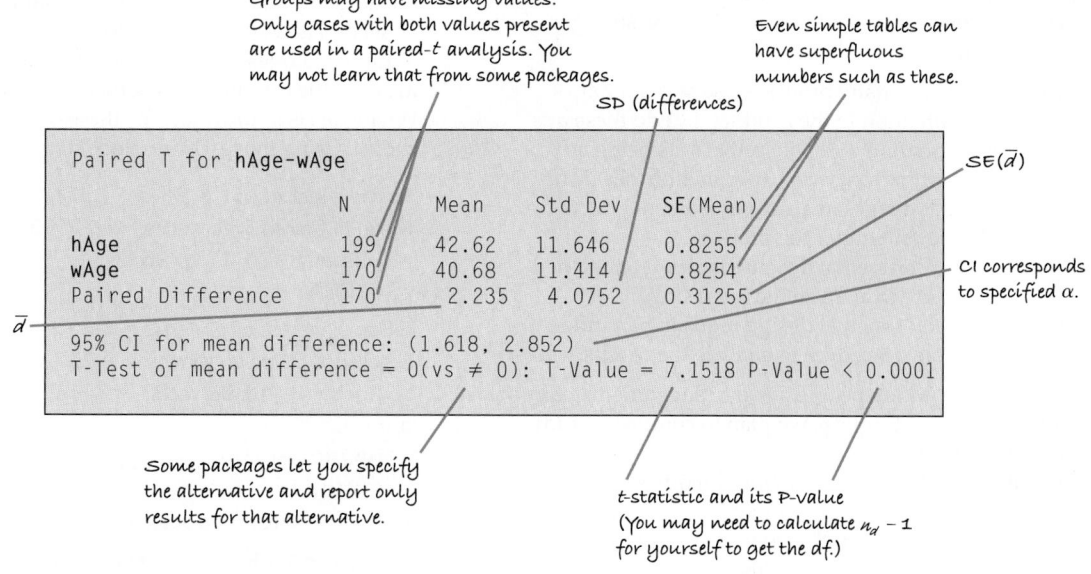

Groups may have missing values. Only cases with both values present are used in a paired-*t* analysis. You may not learn that from some packages.

SD (differences)

Even simple tables can have superfluous numbers such as these.

$SE(\bar{d})$

```
Paired T for hAge-wAge

                     N      Mean    Std Dev   SE(Mean)
hAge               199     42.62     11.646     0.8255
wAge               170     40.68     11.414     0.8254
Paired Difference  170      2.235     4.0752    0.31255

95% CI for mean difference: (1.618, 2.852)
T-Test of mean difference = 0(vs ≠ 0): T-Value = 7.1518 P-Value < 0.0001
```

$\bar{d}$

CI corresponds to specified α.

Some packages let you specify the alternative and report only results for that alternative.

t-statistic and its P-value (You may need to calculate $n_d - 1$ for yourself to get the df.)

Computers make it easy to examine the boxplots of the two groups and the histogram of the differences—both important steps. Some programs offer a scatterplot of the two variables. That can be helpful. In terms of the scatterplot, a paired t-test is about whether the points tend to be above or below the 45° line y = x. (Note, though, that pairing says nothing about whether the scatterplot should be straight. That doesn't matter for our t-methods.)

EXERCISES

1. **More eggs?** Can a food additive increase egg production? Agricultural researchers want to design an experiment to find out. They have 100 hens available. They have two kinds of feed: the regular feed and the new feed with the additive. They plan to run their experiment for a month, recording the number of eggs each hen produces.
   a) Design an experiment that will require a two-sample t procedure to analyze the results.
   b) Design an experiment that will require a matched-pairs t procedure to analyze the results.
   c) Which experiment would you consider the stronger design? Why?

2. **MTV.** Some students do homework with the TV on. (Anyone come to mind?) Some researchers want to see if people can work as effectively with as without distraction. The researchers will time some volunteers to see how long it takes them to complete some relatively easy crossword puzzles. During some of the trials, the room will be quiet; during other trials in the same room, a TV will be on, tuned to MTV.
   a) Design an experiment that will require a two-sample t procedure to analyze the results.
   b) Design an experiment that will require a matched-pairs t procedure to analyze the results.
   c) Which experiment would you consider the stronger design? Why?

3. **Sex sells?** Ads for many products use sexual images to try to attract attention to the product. But do these ads bring people's attention to the item that was being advertised? We want to design an experiment to see if the presence of sexual images in an advertisement affects people's ability to remember the product.
   a) Describe an experimental design requiring a matched-pairs t procedure to analyze the results.
   b) Describe an experimental design requiring an independent sample procedure to analyze the results.

4. **Freshman 15?** Many people believe that students gain weight as freshmen. Suppose we plan to conduct a study to see if this is true.
   a) Describe a study design that would require a matched-pairs t procedure to analyze the results.
   b) Describe a study design that would require a two-sample t procedure to analyze the results.

5. **Women.** Values for the labor force participation rate of women (LFPR) are published by the U.S. Bureau of Labor Statistics. We are interested in whether there was a difference between female participation in 1968 and 1972, a time of rapid change for women. We check LFPR values for 19 randomly selected cities for 1968 and 1972. Shown below is software output for two possible tests:

   Paired t-Test of $\mu(1 - 2)$
   Test Ho: $\mu(1972\text{-}1968) = 0$ vs Ha: $\mu(1972\text{-}1968) \neq 0$
   Mean of Paired Differences = 0.0337
   t-Statistic = 2.458 w/18 df
   p = 0.0244

   2-Sample t-Test of $\mu1 - \mu2$
   Ho: $\mu1 - \mu2 = 0$ Ha: $\mu1 - \mu2 \neq 0$
   Test Ho: $\mu(1972) - \mu(1968) = 0$ vs
   Ha: $\mu(1972) - \mu(1968) \neq 0$
   Difference Between Means = 0.0337
   t-Statistic = 1.496 w/35 df
   p = 0.1434

   a) Which of these tests is appropriate for these data? Explain.
   b) Using the test you selected, state your conclusion.

6. **Rain.** Simpson, Alsen, and Eden (*Technometrics* 1975) report the results of trials in which clouds were seeded and the amount of rainfall recorded. The authors report on 26 seeded and 26 unseeded clouds in order of the amount of rainfall, largest amount first. Here are two possible tests to study the question of whether cloud seeding works. Which test is appropriate for these data? Explain your choice. Using the test you select, state your conclusion.

   Paired t-Test of $\mu(1 - 2)$
   Mean of Paired Differences = −277.39615
   t-Statistic = −3.641 w/25 df
   p = 0.0012

   2-Sample t-Test of $\mu1 - \mu2$
   Difference Between Means = −277.4
   t-Statistic = −1.998 w/33 df
   p = 0.0538

   a) Which of these tests is appropriate for these data? Explain.
   b) Using the test you selected, state your conclusion.

7. **Friday the 13th, I.** In 1993 the *British Medical Journal* published an article titled, "Is Friday the 13th Bad for Your Health?" Researchers in Britain examined how Friday the 13th affects human behavior. One question was

whether people tend to stay at home more on Friday the 13th. The data below are the number of cars passing Junctions 9 and 10 on the M25 motorway for consecutive Fridays (the 6th and 13th) for five different periods.

Year	Month	6th	13th
1990	July	134,012	132,908
1991	September	133,732	131,843
1991	December	121,139	118,723
1992	March	124,631	120,249
1992	November	117,584	117,263

Here are summaries of two possible analyses:

Paired t-Test of mu(1 − 2) = 0 vs. mu(1 − 2) > 0
Mean of Paired Differences: 2022.4
t-Statistic = 2.9377 w/4 df
P = 0.0212

2-Sample t-Test of mu1 = mu2 vs. mu1 > mu2
Difference Between Means: 2022.4
t-Statistic = 0.4273 w/7.998 df
P = 0.3402

a) Which of the tests is appropriate for these data? Explain.
b) Using the test you selected, state your conclusion.
c) Are the assumptions and conditions for inference met?

8. Friday the 13th, II: The researchers in Exercise 7 also examined the number of people admitted to emergency rooms for vehicular accidents on 12 Friday evenings (6 each on the 6th and 13th).

Year	Month	6th	13th
1989	October	9	13
1990	July	6	12
1991	September	11	14
1991	December	11	10
1992	March	3	4
1992	November	5	12

Based on these data, is there evidence that more people are admitted, on average, on Friday the 13th? Here are two possible analyses of the data:

Paired t-Test of mu(1 − 2) = 0 vs. mu(1 − 2) < 0
Mean of Paired Differences = 3.333
t-Statistic = 2.7116 w/5 df
P = 0.0211

2-Sample t-Test of mu1 = mu2 vs. mu1 < mu2
Difference Between Means = 3.333
t-Statistic = 1.6644 w/9.940 df
P = 0.0636

a) Which of these tests is appropriate for these data? Explain.
b) Using the test you selected, state your conclusion.
c) Are the assumptions and conditions for inference met?

9. Online insurance I. After seeing countless commercials claiming one can get cheaper car insurance from an online company, a local insurance agent was concerned that he might lose some customers. To investigate, he randomly selected profiles (type of car, coverage, driving record, etc.) for 10 of his clients and checked online price quotes for their policies. The comparisons are shown in the table below. His statistical software produced the following summaries (where $PriceDiff = Local − Online$):

Variable	Count	Mean	StdDev
Local	10	799.200	229.281
Online	10	753.300	256.267
PriceDiff	10	45.9000	175.663

Local	Online	PriceDiff
568	391	177
872	602	270
451	488	−37
1229	903	326
605	677	−72
1021	1270	−249
783	703	80
844	789	55
907	1008	−101
712	702	10

At first, the insurance agent wondered whether there was some kind of mistake in this output. He thought the Pythagorean Theorem of Statistics should work for finding the standard deviation of the price differences—in other words, that $SD(Local − Online) = \sqrt{SD^2(Local) + SD^2(Online)}$. But when he checked, he found that $\sqrt{(229.281)^2 + (256.267)^2} = 343.864$, not 175.663 as given by the software. Tell him where his mistake is.

10. Windy, part I. To select the site for an electricity-generating wind turbine, wind speeds were recorded at several potential sites every 6 hours for a year. Two sites not far from each other looked good. Each had a mean wind speed high enough to qualify, but we should choose the site with a higher average daily wind speed. Because the sites are near each other and the wind speeds were recorded at the same times, we should view the speeds as paired. Here are the summaries of the speeds (in miles per hour):

Variable	Count	Mean	StdDev
site2	1114	7.452	3.586
site4	1114	7.248	3.421
site2 − site4	1114	0.204	2.551

Is there a mistake in this output? Why doesn't the Pythagorean Theorem of Statistics work here? In other words, shouldn't $SD(site2 − site4) = \sqrt{SD^2(site2) + SD^2(site4)}$? But $\sqrt{(3.586)^2 + (3.421)^2} = 4.956$, not 2.551 as given by the software. Explain why this happened.

11. Online insurance II. In Exercise 9, we saw summary statistics for 10 drivers' car insurance premiums quoted by a local agent and an online company. Here are displays for each company's quotes and for the difference (*Local – Online*):

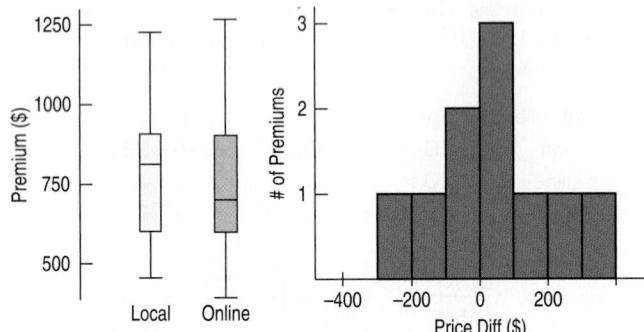

a) Which of the summaries would help you decide whether the online company offers cheaper insurance? Why?

b) The standard deviation of *PriceDiff* is quite a bit smaller than the standard deviation of prices quoted by either the local or online companies. Discuss why.

c) Using the information you have, discuss the assumptions and conditions for inference with these data.

12. Windy, part II. In Exercise 10, we saw summary statistics for wind speeds at two sites near each other, both being considered as locations for an electricity-generating wind turbine. The data, recorded every 6 hours for a year, showed each of the sites had a mean wind speed high enough to qualify, but how can we tell which site is best? Here are some displays:

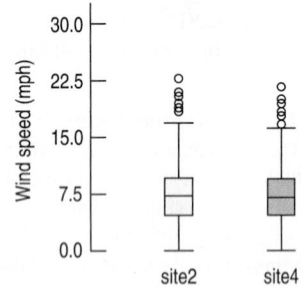

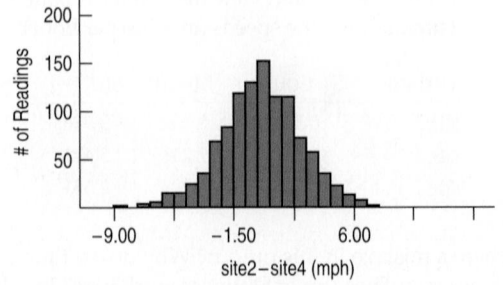

a) The boxplots show outliers for each site, yet the histogram shows none. Discuss why.

b) Which of the summaries would you use to select between these sites? Why?

c) Using the information you have, discuss the assumptions and conditions for paired *t* inference for these data. (*Hint:* Think hard about the independence assumption in particular.)

13. Online insurance 3. Exercises 9 and 11 give summaries and displays for car insurance premiums quoted by a local agent and an online company. Test an appropriate hypothesis to see if there is evidence that drivers might save money by switching to the online company.

14. Windy, part III. Exercises 10 and 12 give summaries and displays for two potential sites for a wind turbine. Test an appropriate hypothesis to see if there is evidence that either of these sites has a higher average wind speed.

15. Temperatures. The table below gives the average high temperatures in January and July for several European cities. Write a 90% confidence interval for the mean temperature difference between summer and winter in Europe. Be sure to check conditions for inference, and clearly explain what your interval means.

City	Mean High Temperatures (°F)	
	Jan.	**July**
Vienna	34	75
Copenhagen	36	72
Paris	42	76
Berlin	35	74
Athens	54	90
Rome	54	88
Amsterdam	40	69
Madrid	47	87
London	44	73
Edinburgh	43	65
Moscow	21	76
Belgrade	37	84

16. Marathons 2006. The table on the next page shows the winning times (in minutes) for men and women in the New York City Marathon between 1978 and 2006. Assuming that performances in the Big Apple resemble performances elsewhere, we can think of these data as a sample of performance in marathon competitions. Create a 90% confidence interval for the mean difference in winning times for male and female marathon competitors. (www.nycmarathon.org)

Year	Men	Women	Year	Men	Women
1978	132.2	152.5	1993	130.1	146.4
1979	131.7	147.6	1994	131.4	147.6
1980	129.7	145.7	1995	131.0	148.1
1981	128.2	145.5	1996	129.9	148.3
1982	129.5	147.2	1997	128.2	148.7
1983	129.0	147.0	1998	128.8	145.3
1984	134.9	149.5	1999	129.2	145.1
1985	131.6	148.6	2000	130.2	145.8
1986	131.1	148.1	2001	127.7	144.4
1987	131.0	150.3	2002	128.1	145.9
1988	128.3	148.1	2003	130.5	142.5
1989	128.0	145.5	2004	129.5	143.2
1990	132.7	150.8	2005	129.5	144.7
1991	129.5	147.5	2006	130.0	145.1
1992	129.5	144.7			

17. Push-ups. Every year the students at Gossett High School take a physical fitness test during their gym classes. One component of the test asks them to do as many push-ups as they can. Results for one class are shown below, separately for boys and girls. Assuming that students at Gossett are assigned to gym classes at random, create a 90% confidence interval for how many more push-ups boys can do than girls, on average, at that high school.

Boys	17	27	31	17	25	32	28	23	25	16	11	34
Girls	24	7	14	16	2	15	19	25	10	27	31	8

18. Brain waves. An experiment was performed to see whether sensory deprivation over an extended period of time has any effect on the alpha-wave patterns produced by the brain. To determine this, 20 subjects, inmates in a Canadian prison, were randomly split into two groups. Members of one group were placed in solitary confinement. Those in the other group were allowed to remain in their own cells. Seven days later, alpha-wave frequencies were measured for all subjects, as shown in the following table. (P. Gendreau et al., "Changes in EEG Alpha Frequency and Evoked Response Latency During Solitary Confinement," *Journal of Abnormal Psychology* 79 [1972]: 54–59)

Nonconfined	Confined
10.7	9.6
10.7	10.4
10.4	9.7
10.9	10.3
10.5	9.2
10.3	9.3
9.6	9.9
11.1	9.5
11.2	9.0
10.4	10.9

a) What are the null and alternative hypotheses? Be sure to define all the terms and symbols you use.
b) Are the assumptions necessary for inference met?
c) Perform the appropriate test, indicating the formula you used, the calculated value of the test statistic, the df, and the P-value.
d) State your conclusion.

19. Job satisfaction. (When you first read about this exercise break plan in Chapter 24, you did not have an inference method that would work. Try again now.) A company institutes an exercise break for its workers to see if it will improve job satisfaction, as measured by a questionnaire that assesses workers' satisfaction. Scores for 10 randomly selected workers before and after the implementation of the exercise program are shown in the table below.
a) Identify the procedure you would use to assess the effectiveness of the exercise program, and check to see if the conditions allow the use of that procedure.
b) Test an appropriate hypothesis and state your conclusion.
c) If your conclusion turns out to be incorrect, what kind of error did you commit?

Worker Number	Job Satisfaction Index	
	Before	After
1	34	33
2	28	36
3	29	50
4	45	41
5	26	37
6	27	41
7	24	39
8	15	21
9	15	20
10	27	37

20. Summer school. (When you first read about the summer school issue in Chapter 24 you did not have an inference method that would work. Try again now.) Having done poorly on their Math final exams in June, six students repeat the course in summer school and take another exam in August.

June	54	49	68	66	62	62
Aug.	50	65	74	64	68	72

a) If we consider these students to be representative of all students who might attend this summer school in other years, do these results provide evidence that the program is worthwhile?
b) This conclusion, of course, may be incorrect. If so, which type of error was made?

21. Yogurt. Is there a significant difference in calories between servings of strawberry and vanilla yogurt? Based on the data shown in the table, test an appropriate hypothesis and state your conclusion. Don't forget to check assumptions and conditions!

	Calories per Serving	
Brand	**Strawberry**	**Vanilla**
America's Choice	210	200
Breyer's Lowfat	220	220
Columbo	220	180
Dannon Light 'n Fit	120	120
Dannon Lowfat	210	230
Dannon la Crème	140	140
Great Value	180	80
La Yogurt	170	160
Mountain High	200	170
Stonyfield Farm	100	120
Yoplait Custard	190	190
Yoplait Light	100	100

T 22. Gasoline. Many drivers of cars that can run on regular gas actually buy premium in the belief that they will get better gas mileage. To test that belief, we use 10 cars from a company fleet in which all the cars run on regular gas. Each car is filled first with either regular or premium gasoline, decided by a coin toss, and the mileage for that tankful is recorded. Then the mileage is recorded again for the same cars for a tankful of the other kind of gasoline. We don't let the drivers know about this experiment. Here are the results (miles per gallon):

Car #	1	2	3	4	5	6	7	8	9	10
Regular	16	20	21	22	23	22	27	25	27	28
Premium	19	22	24	24	25	25	26	26	28	32

a) Is there evidence that cars get significantly better fuel economy with premium gasoline?
b) How big might that difference be? Check a 90% confidence interval.
c) Even if the difference is significant, why might the company choose to stick with regular gasoline?
d) Suppose you had done a "bad thing." (We're sure you didn't.) Suppose you had mistakenly treated these data as two independent samples instead of matched pairs. What would the significance test have found? Carefully explain why the results are so different.

T 23. Braking test. A tire manufacturer tested the braking performance of one of its tire models on a test track. The company tried the tires on 10 different cars, recording the stopping distance for each car on both wet and dry pavement. Results are shown in the table.

	Stopping Distance (ft)	
Car #	Dry Pavement	Wet Pavement
1	150	201
2	147	220
3	136	192
4	134	146
5	130	182
6	134	173
7	134	202
8	128	180
9	136	192
10	158	206

a) Write a 95% confidence interval for the mean dry pavement stopping distance. Be sure to check the appropriate assumptions and conditions, and explain what your interval means.
b) Write a 95% confidence interval for the mean increase in stopping distance on wet pavement. Be sure to check the appropriate assumptions and conditions, and explain what your interval means.

T 24. Braking test 2. For another test of the tires in Exercise 23, a car made repeated stops from 60 miles per hour. The test was run on both dry and wet pavement, with results as shown in the table. (Note that actual *braking distance*, which takes into account the driver's reaction time, is much longer, typically nearly 300 feet at 60 mph!)
a) Write a 95% confidence interval for the mean dry pavement stopping distance. Be sure to check the appropriate assumptions and conditions, and explain what your interval means.
b) Write a 95% confidence interval for the mean increase in stopping distance on wet pavement. Be sure to check the appropriate assumptions and conditions, and explain what your interval means.

Stopping Distance (ft)	
Dry Pavement	Wet Pavement
145	211
152	191
141	220
143	207
131	198
148	208
126	206
140	177
135	183
133	223

25. Tuition 2006. How much more do public colleges and universities charge out-of-state students for tuition per semester? A random sample of 19 public colleges and universities listed at www.collegeboard.com yielded the following data. Tuition figures per semester are rounded to the nearest hundred dollars.

Institution	Resident	Nonresident
Univ of Akron (OH)	4200	8800
Athens State (AL)	1900	3600
Ball State (IN)	3400	8600
Bloomsburg U (PA)	3200	7000
UC Irvine (CA)	3400	12700
Central State (OH)	2600	5700
Clarion U (PA)	3300	5900
Dakota State	2900	3400
Fairmont State (WV)	2200	4600
Johnson State (VT)	3400	7300
Lock Haven U (PA)	3200	6000
New College of Florida	1600	8300
Oakland U (MI)	3300	7700
U Pittsburgh	6100	10700
Savannah State (GA)	1600	5400
SE Louisiana	1700	4400
W Liberty State (WV)	2000	4800
W Texas College	800	1000
Worcester State (MA)	2800	5800

a) Create a 90% confidence interval for the mean difference in cost. Be sure to justify your procedure.
b) Interpret your interval in context.
c) A national magazine claims that public institutions charge state residents an average of $3500 less than out-of-staters for tuition each semester. What does your confidence interval indicate about this assertion?

26. Sex sells, part II. In Exercise 3 you considered the question of whether sexual images in ads affected people's abilities to remember the item being advertised. To investigate, a group of Statistics students cut ads out of magazines. They were careful to find two ads for each of 10 similar items, one with a sexual image and one without. They arranged the ads in random order and had 39 subjects look at them for one minute. Then they asked the subjects to list as many of the products as they could remember. Their data are shown in the table. Is there evidence that the sexual images mattered?

Subject Number	Ads Remembered Sexual Image	Ads Remembered No Sex	Subject Number	Ads Remembered Sexual Image	Ads Remembered No Sex
1	2	2	21	2	3
2	6	7	22	4	2
3	3	1	23	3	3
4	6	5	24	5	3
5	1	0	25	4	5
6	3	3	26	2	4
7	3	5	27	2	2
8	7	4	28	2	4
9	3	7	29	7	6
10	5	4	30	6	7
11	1	3	31	4	3
12	3	2	32	4	5
13	6	3	33	3	0
14	7	4	34	4	3
15	3	2	35	2	3
16	7	4	36	3	3
17	4	4	37	5	5
18	1	3	38	3	4
19	5	5	39	4	3
20	2	2			

27. Strikes. Advertisements for an instructional video claim that the techniques will improve the ability of Little League pitchers to throw strikes and that, after undergoing the training, players will be able to throw strikes on at least 60% of their pitches. To test this claim, we have 20 Little Leaguers throw 50 pitches each, and we record the number of strikes. After the players participate in the training program, we repeat the test. The table shows the number of strikes each player threw before and after the training.
a) Is there evidence that after training players can throw strikes more than 60% of the time?
b) Is there evidence that the training is effective in improving a player's ability to throw strikes?

Number of Strikes (out of 50)		Number of Strikes (out of 50)	
Before	After	Before	After
28	35	33	33
29	36	33	35
30	32	34	32
32	28	34	30
32	30	34	33
32	31	35	34
32	32	36	37
32	34	36	33
32	35	37	35
33	36	37	32

28. Freshman 15, revisited. In Exercise 4 you thought about how to design a study to see if it's true that students tend to gain weight during their first year in college. Well, Cornell Professor of Nutrition David Levitsky did just that. He recruited students from two large sections of an introductory health course. Although they were volunteers, they appeared to match the rest of the freshman class in terms of demographic variables such as sex and ethnicity. The students were weighed during the first week of the semester, then again 12 weeks later. Based on Professor Levitsky's data, estimate the mean weight gain in first-semester freshmen and comment on the "freshman 15." (Weights are in pounds.)

Subject Number	Initial Weight	Terminal Weight	Subject Number	Initial Weight	Terminal Weight
1	171	168	35	148	150
2	110	111	36	164	165
3	134	136	37	137	138
4	115	119	38	198	201
5	150	155	39	122	124
6	104	106	40	146	146
7	142	148	41	150	151
8	120	124	42	187	192
9	144	148	43	94	96
10	156	154	44	105	105
11	114	114	45	127	130
12	121	123	46	142	144
13	122	126	47	140	143
14	120	115	48	107	107
15	115	118	49	104	105
16	110	113	50	111	112
17	142	146	51	160	162
18	127	127	52	134	134
19	102	105	53	151	151
20	125	125	54	127	130
21	157	158	55	106	108
22	119	126	56	185	188
23	113	114	57	125	128
24	120	128	58	125	126
25	135	139	59	155	158
26	148	150	60	118	120
27	110	112	61	149	150
28	160	163	62	149	149
29	220	224	63	122	121
30	132	133	64	155	158
31	145	147	65	160	161
32	141	141	66	115	119
33	158	160	67	167	170
34	135	134	68	131	131

JUST CHECKING
Answers

1. These are independent groups sampled at random, so use a two-sample *t* confidence interval to estimate the size of the difference.

2. There is only one sample. Use a one-sample *t*-interval.

3. A brother and sister from the same family represent a matched pair. The question calls for a paired *t*-test.

4. **a)** A before-and-after study calls for paired *t*-methods. To estimate the loss, find a confidence interval for the before–after differences.

   b) The two treatment groups were assigned randomly, so they are independent. Use a two-sample *t*-test to assess whether the mean weight losses differ.

5. Sometimes it just isn't clear. Most likely, couples would discuss the band or even decide to go to the club because they both like a particular band. If we think that's likely, then these data are paired. But maybe not. If we asked them their opinions of, say, the decor or furnishings at the club, the fact that they were couples might not affect the independence of their answers.

REVIEW OF PART VI

Learning About the World

Quick Review

We continue to explore how to answer questions about the statistics we get from samples and experiments. In this part, those questions have been about means—means of one sample, two independent samples, or matched pairs. Here's a brief summary of the key concepts and skills:

▶ A confidence interval uses a sample statistic to estimate a range of possible values for a parameter of interest.

▶ A hypothesis test proposes a model, then examines the plausibility of that model by seeing how surprising our observed data would be if the model were true.

▶ Statistical inference procedures for proportions are based on the Central Limit Theorem. We can make inferences about a single proportion or the difference of two proportions using Normal models.

▶ Statistical inference procedures for means are also based on the Central Limit Theorem, but we don't usually know the population standard deviation. Student's *t*-models take into account the additional uncertainty of independently estimating the standard deviation.

- We can make inferences about one mean, the difference of two independent means, or the mean of paired differences using *t*-models.

- No inference procedure is valid unless the underlying assumptions are true. Always check the conditions before proceeding.

- Because *t*-models assume that samples are drawn from Normal populations, data in the sample should appear to be nearly Normal. Skewness and outliers are particularly problematic, especially for small samples.

- When there are two variables, you must think carefully about how the data were collected. You may use two-sample *t* procedures only if the groups are independent.

- Unless there is some obvious reason to suspect that two independent populations have the same standard deviation, you should not pool the variances. It is never wrong to use unpooled *t* procedures.

- If the two groups are somehow paired, the data are *not* from independent groups. You must use matched-pairs *t* procedures.

Now for some opportunities to review these concepts. Be careful. You have a lot of thinking to do. These review exercises mix questions about proportions and means. You have to determine which of our inference procedures is appropriate in each situation. Then you have to check the proper assumptions and conditions. Keeping track of those can be difficult, so first we summarize the many procedures with their corresponding assumptions and conditions on the next page. Look them over carefully . . . then, on to the Exercises!

Quick Guide to Inference

Think			Show				Tell?
Inference about?	One group or two?	Procedure	Model	Parameter	Estimate	SE	Chapter
Proportions	One sample	1-Proportion z-Interval	z	p	$\hat{p}$	$\sqrt{\dfrac{\hat{p}\hat{q}}{n}}$	19
		1-Proportion z-Test				$\sqrt{\dfrac{p_0 q_0}{n}}$	20, 21
	Two independent groups	2-Proportion z-Interval	z	$p_1 - p_2$	$\hat{p}_1 - \hat{p}_2$	$\sqrt{\dfrac{\hat{p}_1\hat{q}_1}{n_1} + \dfrac{\hat{p}_2\hat{q}_2}{n_2}}$	22
		2-Proportion z-Test				$\sqrt{\dfrac{\hat{p}\hat{q}}{n_1} + \dfrac{\hat{p}\hat{q}}{n_2}}, \ \hat{p} = \dfrac{y_1 + y_2}{n_1 + n_2}$	22
Means	One sample	t-Interval t-Test	t df $= n-1$	μ	$\bar{y}$	$\dfrac{s}{\sqrt{n}}$	23
	Two independent groups	2-Sample t-Test 2-Sample t-Interval	t df from technology	$\mu_1 - \mu_2$	$\bar{y}_1 - \bar{y}_2$	$\sqrt{\dfrac{s_1^2}{n_1} + \dfrac{s_2^2}{n_2}}$	24
	Matched pairs	Paired t-Test Paired t-Interval	t df $= n-1$	μ_d	$\bar{d}$	$\dfrac{s_d}{\sqrt{n}}$	25

Assumptions for Inference   And the Conditions That Support or Override Them

Proportions (z)
- **One sample**
  1. Individuals are independent.
  2. Sample is sufficiently large.

1. SRS and $n < 10\%$ of the population.
2. Successes and failures each ≥ 10.

- **Two groups**
  1. Groups are independent.
  2. Data in each group are independent.

1. (Think about how the data were collected.)
2. Both are SRSs and $n < 10\%$ of populations OR random allocation.

  3. Both groups are sufficiently large.

3. Successes and failures each ≥ 10 for both groups.

Means (t)
- **One sample** (df $= n - 1$)
  1. Individuals are independent.
  2. Population has a Normal model.

1. SRS and $n < 10\%$ of the population.
2. Histogram is unimodal and symmetric.*

- **Matched pairs** (df $= n - 1$)
  1. Data are matched.
  2. Individuals are independent.
  3. Population of differences is Normal.

1. (Think about the design.)
2. SRS and $n < 10\%$ OR random allocation.
3. Histogram of differences is unimodal and symmetric.*

- **Two independent groups** (df from technology)
  1. Groups are independent.
  2. Data in each group are independent.
  3. Both populations are Normal.

1. (Think about the design.)
2. SRSs and $n < 10\%$ OR random allocation.
3. Both histograms are unimodal and symmetric.*

(*less critical as n increases)

REVIEW EXERCISES

1. **Crawling.** A study published in 1993 found that babies born at different times of the year may develop the ability to crawl at different ages! The author of the study suggested that these differences may be related to the temperature at the time the infant is 6 months old. (Benson and Janette, *Infant Behavior and Development* [1993])

   a) The study found that 32 babies born in January crawled at an average age of 29.84 weeks, with a standard deviation of 7.08 weeks. Among 21 July babies, crawling ages averaged 33.64 weeks, with a standard deviation of 6.91 weeks. Is this difference significant?

   b) For 26 babies born in April the mean and standard deviation were 31.84 and 6.21 weeks, while for 44 October babies the mean and standard deviation of crawling ages were 33.35 and 7.29 weeks. Is this difference significant?

   c) Are these results consistent with the researcher's conjecture?

2. **Mazes and smells.** Can pleasant smells improve learning? Researchers timed 21 subjects as they tried to complete paper-and-pencil mazes. Each subject attempted a maze both with and without the presence of a floral aroma. Subjects were randomized with respect to whether they did the scented trial first or second. Is there any evidence that the floral scent improved the subjects' ability to complete the mazes? (A. R. Hirsch and L. H. Johnston, "Odors and Learning." Chicago: Smell and Taste Treatment and Research Foundation)

Time to Complete the Maze (sec)	
Unscented	**Scented**
25.7	30.2
41.9	56.7
51.9	42.4
32.2	34.4
64.7	44.8
31.4	42.9
40.1	42.7
43.2	24.8
33.9	25.1
40.4	59.2
58.0	42.2
61.5	48.4
44.6	32.0
35.3	48.1
37.2	33.7
39.4	42.6
77.4	54.9
52.8	64.5
63.6	43.1
56.6	52.8
58.9	44.3

3. **Women.** The U.S. Census Bureau reports that 26% of all U.S. businesses are owned by women. A Colorado consulting firm surveys a random sample of 410 businesses in the Denver area and finds that 115 of them have women owners. Should the firm conclude that its area is unusual? Test an appropriate hypothesis and state your conclusion.

4. **Drugs.** In a full-page ad that ran in many U.S. newspapers in August 2002, a Canadian discount pharmacy listed costs of drugs that could be ordered from a Web site in Canada. The table compares prices (in US$) for commonly prescribed drugs.

	Cost per 100 Pills		
Drug Name	**United States**	**Canada**	**Percent savings**
Cardizem	131	83	37
Celebrex	136	72	47
Cipro	374	219	41
Pravachol	370	166	55
Premarin	61	17	72
Prevacid	252	214	15
Prozac	263	112	57
Tamoxifen	349	50	86
Vioxx	243	134	45
Zantac	166	42	75
Zocor	365	200	45
Zoloft	216	105	51

   a) Give a 95% confidence interval for the average savings in dollars.

   b) Give a 95% confidence interval for the average savings in percent.

   c) Which analysis is more appropriate? Why?

   d) In small print the newspaper ad says, "Complete list of all 1500 drugs available on request." How does this comment affect your conclusions above?

5. **Pottery.** Archaeologists can use the chemical composition of clay found in pottery artifacts to determine whether different sites were populated by the same ancient people. They collected five samples of Romano–British pottery from each of two sites in Great Britain and measured the percentage of aluminum oxide in each. Based on these data, do you think the same people used these two kiln sites? Base your conclusion on a 95% confidence interval for the difference in aluminum oxide content of pottery made at the sites. (A. Tubb, A. J. Parker, and G. Nickless, "The Analysis of Romano–British Pottery by Atomic Absorption Spectrophotometry." *Archaeometry*, 22[1980]:153–171)

Ashley Rails	19.1	14.8	16.7	18.3	17.7
New Forest	20.8	18.0	18.0	15.8	18.3

6. **Streams.** Researchers in the Adirondack Mountains collect data on a random sample of streams each year. One of the variables recorded is the substrate of the streams—the type of soil and rock over which they flow. The researchers found that 69 of the 172 sampled streams had a substrate of shale. Construct a 95% confidence interval for the proportion of Adirondack streams with a shale substrate. Clearly interpret your interval in context.

7. **Gehrig.** Ever since Lou Gehrig developed amyotrophic lateral sclerosis (ALS), this deadly condition has been commonly known as Lou Gehrig's disease. Some believe that ALS is more likely to strike athletes or the very fit. Columbia University neurologist Lewis P. Rowland recorded personal histories of 431 patients he examined between 1992 and 2002. He diagnosed 280 as having ALS; 38% of them had been varsity athletes. The other 151 had other neurological disorders, and only 26% of them had been varsity athletes. (*Science News*, Sept. 28 [2002])
   a) Is there evidence that ALS is more common among athletes?
   b) What kind of study is this? How does that affect the inference you made in part a?

8. **Teen drinking.** A study of the health behavior of school-aged children asked a sample of 15-year-olds in several different countries if they had been drunk at least twice. The results are shown in the table, by gender. Give a 95% confidence interval for the difference in the rates for males and females. Be sure to check the assumptions that support your chosen procedure, and explain what your interval means. (*Health and Health Behavior Among Young People*. Copenhagen: World Health Organization, 2000)

	Percent of 15-Year-Olds Drunk at Least Twice	
Country	Female	Male
Denmark	63	71
Wales	63	72
Greenland	59	58
England	62	51
Finland	58	52
Scotland	56	53
No. Ireland	44	53
Slovakia	31	49
Austria	36	49
Canada	42	42
Sweden	40	40
Norway	41	37
Ireland	29	42
Germany	31	36
Latvia	23	47
Estonia	23	44
Hungary	22	43
Poland	21	39
USA	29	34
Czech Rep.	22	36
Belgium	22	36
Russia	25	32
Lithuania	20	32
France	20	29
Greece	21	24
Switzerland	16	25
Israel	10	18

9. **Babies.** The National Perinatal Statistics Unit of the Sydney Children's Hospital reports that the mean birth weight of all babies born in Australia in 1999 was 3361 grams—about 7.41 pounds. A Missouri hospital reports that the average weight of 112 babies born there last year was 7.68 pounds, with a standard deviation of 1.31 pounds. If we believe the Missouri babies fairly represent American newborns, is there any evidence that U.S. babies and Australian babies do not weigh the same amount at birth?

10. **Petitions.** To get a voter initiative on a state ballot, petitions that contain at least 250,000 valid voter signatures must be filed with the Elections Commission. The board then has 60 days to certify the petitions. A group wanting to create a statewide system of universal health insurance has just filed petitions with a total of 304,266 signatures. As a first step in the process, the Board selects an SRS of 2000 signatures and checks them against local voter lists. Only 1772 of them turn out to be valid.
   a) What percent of the sample signatures were valid?
   b) What percent of the petition signatures submitted must be valid in order to have the initiative certified by the Elections Commission?
   c) What will happen if the Elections Commission commits a Type I error?
   d) What will happen if the Elections Commission commits a Type II error?
   e) Does the sample provide evidence in support of certification? Explain.
   f) What could the Elections Commission do to increase the power of the test?

11. **Feeding fish.** In the midwestern United States, a large aquaculture industry raises largemouth bass. Researchers wanted to know whether the fish would grow better if fed a natural diet of fathead minnows or an artificial diet of food pellets. They stocked six ponds with bass fingerlings weighing about 8 grams. For one year, the fish in three of the ponds were fed minnows, and the others were fed the commercially prepared pellets. The fish were then harvested, weighed, and measured. The bass fed a natural food source had a higher average length (19.6 cm) and weight (95.9 g) than those fed the commercial fish food (17.3 cm and 72.0 g, respectively). The researchers reported P-values for differences in both measurements to be less than 0.001.
   a) Explain to someone who has not studied Statistics what the P-values mean here.
   b) What advice should the researchers give the people who raise largemouth bass?
   c) If that advice turns out to be incorrect, what type of error occurred?

12. **Risk.** A study of auto safety determined the number of driver deaths per million vehicle sales, classified by type of vehicle. The data on the next page are for 6 midsize

models and 6 SUVs. Wondering if there is evidence that drivers of SUVs are safer, we hope to create a 95% confidence interval for the difference in driver death rates for the two types of vehicles. Are these data appropriate for this inference? Explain. (Ross and Wenzel, *An Analysis of Traffic Deaths by Vehicle Type and Model*, March 2002)

Midsize	47	54	64	76	88	97
SUV	55	60	62	76	91	109

13. Age.   In a study of how depression may affect one's ability to survive a heart attack, the researchers reported the ages of the two groups they examined. The mean age of 2397 patients without cardiac disease was 69.8 years (SD = 8.7 years), while for the 450 patients with cardiac disease, the mean and standard deviation of the ages were 74.0 and 7.9, respectively.
   a) Create a 95% confidence interval for the difference in mean ages of the two groups.
   b) How might an age difference confound these research findings about the relationship between depression and ability to survive a heart attack?

14. Smoking.   In the depression and heart attack research described in Exercise 13, 32% of the diseased group were smokers, compared with only 23.7% of those free of heart disease.
   a) Create a 95% confidence interval for the difference in the proportions of smokers in the two groups.
   b) Is this evidence that the two groups in the study were different? Explain.
   c) Could this be a problem in analyzing the results of the study? Explain.

15. Computer use.   A Gallup telephone poll of 1240 teens conducted in 2001 found that boys were more likely than girls to play computer games, by a margin of 77% to 65%. Equal numbers of boys and girls were surveyed.
   a) What kind of sampling design was used?
   b) Give a 95% confidence interval for the difference in game playing by gender.
   c) Does your confidence interval suggest that among all teens a higher percentage of boys than girls play computer games?

16. Recruiting.   In September 2002, CNN reported on a method of grad student recruiting by the Haas School of Business at U.C.-Berkeley. The school notifies applicants by formal letter that they have been admitted, and also e-mails the accepted students a link to a Web site that greets them with personalized balloons, cheering, and applause. The director of admissions says this extra effort at recruiting has really worked well. The school accepts 500 applicants each year, and the percentage that actually choose to enroll at Berkeley increased from 52% the year before the Web greeting to 54% this year.
   a) Create a 95% confidence interval for the change in enrollment rates.

   b) Based on your confidence interval, are you convinced that this new form of recruiting has been effective? Explain.

17. Hearing.   Fitting someone for a hearing aid requires assessing the patient's hearing ability. In one method of assessment, the patient listens to a tape of 50 English words. The tape is played at low volume, and the patient is asked to repeat the words. The patient's hearing ability score is the number of words perceived correctly. Four tapes of equivalent difficulty are available so that each ear can be tested with more than one hearing aid. These lists were created to be equally difficult to perceive in silence, but hearing aids must work in the presence of background noise. Researchers had 24 subjects with normal hearing compare two of the tapes when a background noise was present, with the order of the tapes randomized. Is it reasonable to assume that the two lists are still equivalent for purposes of the hearing test when there is background noise? Base your decision on a confidence interval for the mean difference in the number of words people might misunderstand. (Faith Loven, *A Study of the Interlist Equivalency of the CID W-22 Word List Presented in Quiet and in Noise.* University of Iowa [1981])

Subject	List A	List B
1	24	26
2	32	24
3	20	22
4	14	18
5	32	24
6	22	30
7	20	22
8	26	28
9	26	30
10	38	16
11	30	18
12	16	34
13	36	32
14	32	34
15	38	32
16	14	18
17	26	20
18	14	20
19	38	40
20	20	26
21	14	14
22	18	14
23	22	30
24	34	42

18. Cesareans.   Some people fear that differences in insurance coverage can affect healthcare decisions. A survey of several randomly selected hospitals found that 16.6% of 223 recent births in Vermont involved cesarean deliveries, compared to 18.8% of 186 births in New Hampshire. Is this evidence that the rate of cesarean births in the two states is different?

19. Newspapers.   Who reads the newspaper more, men or women? Eurostat, an agency of the European Union (EU), conducts surveys on several aspects of daily life in EU countries. Recently, the agency asked samples of 1000 respondents in each of 14 European countries whether they read the newspaper on a daily basis. The table on the next page shows the data.

Country	% Reading a Newspaper Daily	
	Men	Women
Belgium	56.3	45.5
Denmark	76.8	70.3
Germany	79.9	76.8
Greece	22.5	17.2
Spain	46.2	24.8
Ireland	58.0	54.0
Italy	50.2	29.8
Luxembourg	71.0	67.0
Netherlands	71.3	63.0
Austria	78.2	74.1
Portugal	58.3	24.1
Finland	93.0	90.0
Sweden	89.0	88.0
UK	32.6	30.4

a) Examine the differences in the percentages for each country. Which of these countries seem to be outliers? What do they have in common?

b) After eliminating the outliers, is there evidence that in Europe men are more likely than women to read the newspaper?

20. Meals.   A college student is on a "meal program." His budget allows him to spend an average of $10 per day for the semester. He keeps track of his daily food expenses for 2 weeks; the data are given in the table. Is there strong evidence that he will overspend his food allowance? Explain.

Date	Cost ($)	Date	Cost ($)
7/29	15.20	8/5	8.55
7/30	23.20	8/6	20.05
7/31	3.20	8/7	14.95
8/1	9.80	8/8	23.45
8/2	19.53	8/9	6.75
8/3	6.25	8/10	0
8/4	0	8/11	9.01

21. Wall Street.   In September of 2000, the Harris Poll organization asked 1002 randomly sampled American adults whether they agreed or disagreed with the following statement:

Most people on Wall Street would be willing to break the law if they believed they could make a lot of money and get away with it.

Of those asked, 60% said they agreed with this statement. We know that if we could ask the entire population of American adults, we would not find that exactly 60% think that Wall Street workers would be willing to break the law to make money. Construct a 95% confidence interval for the true percentage of American adults who agree with the statement.

22. Teach for America.   Several programs attempt to address the shortage of qualified teachers by placing uncertified instructors in schools with acute needs—often in inner cities. A 1999–2000 study compared students taught by certified teachers to others taught by uncertified teachers in the same schools. Reading scores of the students of certified teachers averaged 35.62 points with standard deviation 9.31. The scores of students instructed by uncertified teachers had mean 32.48 points with standard deviation 9.43 points on the same test. There were 44 students in each group. The appropriate t procedure has 86 degrees of freedom. Is there evidence of lower scores with uncertified teachers? Discuss. (*The Effectiveness of "Teach for America" and Other Under-certified Teachers on Student Academic Achievement: A Case of Harmful Public Policy.* Education Policy Analysis Archives [2002])

23. Legionnaires' disease.   In 1974, the Bellevue-Stratford Hotel in Philadelphia was the scene of an outbreak of what later became known as legionnaires' disease. The cause of the disease was finally discovered to be bacteria that thrived in the air-conditioning units of the hotel. Owners of the Rip Van Winkle Motel, hearing about the Bellevue-Stratford, replaced their air-conditioning system. The following data are the bacteria counts in the air of eight rooms, before and after a new air-conditioning system was installed (measured in colonies per cubic foot of air). Has the new system succeeded in lowering the bacterial count? Base your analysis on a confidence interval. Be sure to list all your assumptions, methods, and conclusions.

Room Number	Before	After
121	11.8	10.1
163	8.2	7.2
125	7.1	3.8
264	14	12
233	10.8	8.3
218	10.1	10.5
324	14.6	12.1
325	14	13.7

24. Teach for America, Part II.   The study described in Exercise 22 also looked at scores in mathematics and language. Here are software outputs for the appropriate tests. Explain what they show.

Mathematics
T-TEST OF Mu(1) – Mu(2) = 0
Mu(Cert) – Mu(NoCert) = 4.53   t (86) = 2.95   p = 0.002

Language
T-TEST OF Mu(1) - Mu(2) = 0
Mu(Cert) – Mu(NoCert) = 2.13   t (84) = 1.71   p = 0.045

25. Bipolar kids.   The June 2002 *American Journal of Psychiatry* reported that researchers used medication and psychotherapy to treat children aged 7 to 16 who exhibit bipolar symptoms. After 2 years, symptoms had cleared up in only 26 of the 89 children involved in the study.

a) Write a 95% confidence interval; interpret it in context.
b) If researchers subsequently hope to produce an estimate of treatment effectiveness for bipolar disorder that has a margin of error of only 6%, how many patients should they study?

26. Online testing. The Educational Testing Service is now administering several of its standardized tests online—the CLEP and GMAT exams, for example. Since taking a test on a computer is different from taking a test with pencil and paper, one wonders if the scores will be the same. To investigate this question, researchers created two versions of an SAT-type test and got 20 volunteers to participate in an experiment. Each volunteer took both versions of the test, one with pencil and paper and the other online. Subjects were randomized with respect to the order in which they sat for the tests (online/paper) and which form they took (Test A, Test B) in which environment. The scores (out of a possible 20) are summarized in the table.

Subject	Paper	Online
	Test A	Test B
1	14	13
2	10	13
3	16	8
4	15	14
5	17	16
6	14	11
7	9	12
8	12	12
9	16	16
10	7	14
	Test B	Test A
11	8	13
12	11	13
13	15	17
14	11	13
15	13	14
16	9	9
17	15	9
18	14	15
19	16	12
20	8	10

a) Were the two forms (A/B) of the test equivalent in terms of difficulty? Test an appropriate hypothesis and state your conclusion.
b) Is there evidence that the testing environment (paper/online) matters? Test an appropriate hypothesis and state your conclusion.

27. Bread. Clarksburg Bakery is trying to predict how many loaves of bread to bake. In the last 100 days, the bakery has sold between 95 and 140 loaves per day. Here are a histogram and the summary statistics for the number of loaves sold for the last 100 days.

Summary of Sales

Mean	103
Median	100
SD	9.000
Min	95
Max	140
Q_1	97
Q_3	105.5

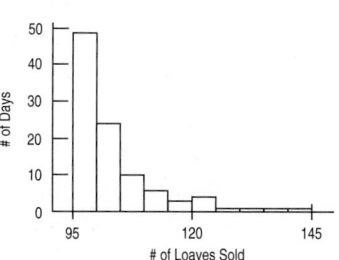

a) Can you use these data to estimate the number of loaves sold on the busiest 10% of all days? Explain.

b) Explain why you can use these data to construct a 95% confidence interval for the mean number of loaves sold per day.
c) Calculate a 95% confidence interval and carefully interpret what that confidence interval means.
d) If the bakery would have been satisfied with a confidence interval whose margin of error was twice as wide, how many days' data could they have used?
e) When the bakery opened, the owners estimated that they would sell an average of 100 loaves per day. Does your confidence interval provide strong evidence that this estimate was incorrect? Explain.

28. Irises. Can measurements of the petal length of flowers be of value when you need to determine the species of a certain flower? Here are the summary statistics from measurements of the petals of two species of irises. (R. A. Fisher, "The Use of Multiple Measurements in Axonomic Problems." *Annals of Eugenics 7* [1936]:179–188)

	Species	
	Versicolor	**Virginica**
Count	50	50
Mean	55.52	43.22
Median	55.50	44.00
SD	5.519	5.362
Min	45	30
Max	69	56
Lower Quartile	51	40
Upper Quartile	59	47

a) Make parallel boxplots of petal lengths for the two species.
b) Describe the differences seen in the boxplots.
c) Write a 95% confidence interval for the difference in petal length.
d) Explain what your interval means.
e) Based on your confidence interval, is there evidence of a difference in petal length? Explain.

29. Insulin and diet. A study published in the *Journal of the American Medical Association* examined people to see if they showed any signs of IRS (insulin resistance syndrome) involving major risk factors for Type 2 diabetes and heart disease. Among 102 subjects who consumed dairy products more than 35 times per week, 24 were identified with IRS. In comparison, IRS was identified in 85 of 190 individuals with the lowest dairy consumption, fewer than 10 times per week.
a) Is this strong evidence that IRS risk is different in people who frequently consume dairy products than in those who do not?
b) Does this indicate that diary consumption influences the development of IRS? Explain.

30. Speeding. A newspaper report in August 2002 raised the issue of racial bias in the issuance of speeding tickets. The following facts were noted:
• 16% of drivers registered in New Jersey are black.

- Of the 324 speeding tickets issued in one month on a 65-mph section of the New Jersey Turnpike, 25% went to black drivers.
  a) Is the percentage of speeding tickets issued to blacks unusually high compared to registrations?
  b) Does this suggest that racial profiling may be present?
  c) What other statistics would you like to know about this situation?

T **31. Rainmakers?** In an experiment to determine whether seeding clouds with silver iodide increases rainfall, 52 clouds were randomly assigned to be seeded or not. The amount of rain they generated was then measured (in acre-feet). Create a 95% confidence interval for the average amount of additional rain created by seeding clouds. Explain what your interval means.

	Unseeded Clouds	Seeded Clouds
Count	26	26
Mean	164.588	441.985
Median	44.200	221.600
SD	278.426	650.787
IntQRange	138.600	337.600
25 %ile	24.400	92.400
75 %ile	163	430

32. Fritos. As a project for an introductory Statistics course, students checked 6 bags of Fritos marked with a net weight of 35.4 grams. They carefully weighed the contents of each bag, recording the following weights (in grams): 35.5, 35.3, 35.1, 36.4, 35.4, 35.5. Is there evidence that the mean weight of bags of Fritos is less than advertised?
a) Write appropriate hypotheses.
b) Check the assumptions for inference.
c) Test your hypothesis using all 6 weights.
d) Retest your hypothesis with the one unusually high weight removed.
e) What would you conclude about the stated weight?

T **33. Color or text?** In an experiment, 32 volunteer subjects are briefly shown seven cards, each displaying the name of a color printed in a different color (example: red, blue, and so on). The subject is asked to perform one of two tasks: memorize the order of the words or memorize the order of the colors. Researchers record the number of cards remembered correctly. Then the cards are shuffled and the subject is asked to perform the other task. The table displays the results for each subject. Is there any evidence that either the color or the written word dominates perception?
a) What role does randomization play in this experiment?
b) Test appropriate hypotheses and state your conclusion.

Subject	Color	Word	Subject	Color	Word
1	4	7	17	4	3
2	1	4	18	7	4
3	5	6	19	4	3
4	1	6	20	0	6
5	6	4	21	3	3
6	4	5	22	3	5
7	7	3	23	7	3
8	2	5	24	3	7
9	7	5	25	5	6
10	4	3	26	3	4
11	2	0	27	3	5
12	5	4	28	1	4
13	6	7	29	2	3
14	3	6	30	5	3
15	4	6	31	3	4
16	4	7	32	6	7

34. And it means? Every statement about a confidence interval contains two parts: the level of confidence and the interval. Suppose that an insurance agent estimating the mean loss claimed by clients after home burglaries created the 95% confidence interval ($1644, $2391).
a) What's the margin of error for this estimate?
b) Carefully explain what the interval means.
c) Carefully explain what the confidence level means.

35. Batteries. We work for the "Watchdog for the Consumer" consumer advocacy group. We've been asked to look at a battery company that claims its batteries last an average of 100 hours under normal use. There have been several complaints that the batteries don't last that long, so we decide to test them. To do this, we select 16 batteries and run them until they die. They lasted a mean of 97 hours, with a standard deviation of 12 hours.
a) One of the editors of our newsletter (who does not know statistics) says that 97 hours is a lot less than the advertised 100 hours, so we should reject the company's claim. Explain to him the problem with doing that.
b) What are the null and alternative hypotheses?
c) What assumptions must we make in order to proceed with inference?
d) At a 5% level of significance, what do you conclude?
e) Suppose that, in fact, the average life of the company's batteries is only 98 hours. Has an error been made in part d? If so, what kind?

36. Hamsters. How large are hamster litters? Among 47 golden hamster litters recorded, there were an average of 7.72 baby hamsters, with a standard deviation of 2.5.
a) Create and interpret a 90% confidence interval.
b) Would a 98% confidence interval have a larger or smaller margin of error? Explain.
c) How many litters must be used to estimate the average litter size to within 1 baby hamster with 95% confidence?

PART VII

Inference When Variables Are Related

Comparing Counts

WHO	Executives of Fortune 400 companies
WHAT	Zodiac birth sign
WHY	Maybe the researcher was a Gemini and naturally curious?

A S *Activity:* **Children at Risk.** See how a contingency table helps us understand the different risks to which an incident exposed children.

oes your zodiac sign predict how successful you will be later in life? *Fortune* magazine collected the zodiac signs of 256 heads of the largest 400 companies. The table shows the number of births for each sign.

We can see some variation in the number of births per sign, and there *are* more Pisces, but is that enough to claim that successful people are more likely to be born under some signs than others?

Births	Sign
23	Aries
20	Taurus
18	Gemini
23	Cancer
20	Leo
19	Virgo
18	Libra
21	Scorpio
19	Sagittarius
22	Capricorn
24	Aquarius
29	Pisces

Birth totals by sign for 256 Fortune 400 executives.

Goodness-of-Fit

"All creatures have their determined time for giving birth and carrying fetus, only a man is born all year long, not in determined time, one in the seventh month, the other in the eighth, and so on till the beginning of the eleventh month."

—Aristotle

If births were distributed uniformly across the year, we would expect about 1/12 of them to occur under each sign of the zodiac. That suggests 256/12, or about 21.3 births per sign. How closely do the observed numbers of births per sign fit this simple "null" model?

A hypothesis test to address this question is called a test of **"goodness-of-fit."** The name suggests a certain badness-of-grammar, but it is quite standard. After all, we are asking whether the model that births are uniformly distributed over the signs fits the data good, . . . er, well. Goodness-of-fit involves testing a hypothesis. We have specified a model for the distribution and want to know whether it fits. There is no single parameter to estimate, so a confidence interval wouldn't make much sense.

If the question were about only one astrological sign (for example, "Are executives more likely to be Pisces?"[1]), we could use a one-proportion z-test and ask if

---

[1] A question actually asked us by someone who was undoubtedly a Pisces.

the true proportion of executives with that sign is equal to 1/12. However, here we have 12 hypothesized proportions, one for each sign. We need a test that considers all of them together and gives an overall idea of whether the observed distribution differs from the hypothesized one.

FOR EXAMPLE Finding expected counts

Birth month may not be related to success as a CEO, but what about on the ball field? It has been proposed by some researchers that children who are the older ones in their class at school naturally perform better in sports and that these children then get more coaching and encouragement. Could that make a difference in who makes it to the professional level in sports?

Baseball is a remarkable sport, in part because so much data are available. We have the birth dates of every one of the 16,804 players who ever played in a major league game. Since the effect we're suspecting may be due to relatively recent policies (and to keep the sample size moderate), we'll consider the birth months of the 1478 major league players born since 1975 and who have played through 2006. We can also look up the national demographic statistics to find what percentage of people were born in each month. Let's test whether the observed distribution of ballplayers' birth months shows just random fluctuations or whether it represents a real deviation from the national pattern.

Month	Ballplayer count	National birth %	Month	Ballplayer count	National birth %
1	137	8%	7	102	9%
2	121	7%	8	165	9%
3	116	8%	9	134	9%
4	121	8%	10	115	9%
5	126	8%	11	105	8%
6	114	8%	12	122	9%
			Total	1478	100%

Question: How can we find the expected counts?

There are 1478 players in this set of data. I'd expect 8% of them to have been born in January, and 1478(0.08) = 118.24. I won't round off, because expected "counts" needn't be integers. Multiplying 1478 by each of the birth percentages gives the expected counts shown in the table.

Month	Expected	Month	Expected
1	118.24	7	133.02
2	103.46	8	133.02
3	118.24	9	133.02
4	118.24	10	133.02
5	118.24	11	118.24
6	118.24	12	133.02

Assumptions and Conditions

These data are organized in tables as we saw in Chapter 3, and the assumptions and conditions reflect that. Rather than having an observation for each individual, we typically work with summary counts in categories. In our example, we don't see the birth signs of each of the 256 executives, only the totals for each sign.

Counted Data Condition: The data must be *counts* for the categories of a categorical variable. This might seem a simplistic, even silly condition. But many kinds of values can be assigned to categories, and it is unfortunately common to find the methods of this chapter applied incorrectly to proportions, percentages, or measurements just because they happen to be organized in a table. So check to be sure the values in each **cell** really are counts.

INDEPENDENCE ASSUMPTION

Independence Assumption: The counts in the cells should be independent of each other. The easiest case is when the individuals who are counted in the cells are sampled independently from some population. That's what we'd like to have if we want to draw conclusions about that population. Randomness can arise in other ways, though. For example, these Fortune 400 executives are not a random sample, but we might still think that their birth dates are randomly distributed throughout the year. If we want to generalize to a large population, we should check the Randomization Condition.

Randomization Condition: The individuals who have been counted should be a random sample from the population of interest.

SAMPLE SIZE ASSUMPTION

We must have enough data for the methods to work. We usually check the following:

Expected Cell Frequency Condition: We should expect to see at least 5 individuals in each cell.

The Expected Cell Frequency Condition sounds like—and is, in fact, quite similar to—the condition that np and nq be at least 10 when we tested proportions. In our astrology example, assuming equal births in each month leads us to expect 21.3 births per month, so the condition is easily met here.

FOR EXAMPLE | Checking assumptions and conditions

Recap: Are professional baseball players more likely to be born in some months than in others? We have observed and expected counts for the 1478 players born since 1975.

Question: Are the assumptions and conditions met for performing a goodness-of-fit test?

✓ **Counted Data Condition:** I have month-by-month counts of ballplayer births.

✓ **Independence Assumption:** These births were independent.

✓ **Randomization Condition:** Although they are not a random sample, we can take these players to be representative of players past and future.

✓ **Expected Cell Frequency Condition:** The expected counts range from 103.46 to 133.02, all much greater than 5.

✓ **10% Condition:** These 1478 players are less than 10% of the population of 16,804 players who have ever played (or will play) major league baseball.

It's okay to use these data for a goodness-of-fit test.

Calculations

NOTATION ALERT:

We compare the counts *observed* in each cell with the counts we *expect* to find. The usual notation uses O's and E's or abbreviations such as those we've used here. The method for finding the expected counts depends on the model.

We have observed a count in each category from the data, and have an expected count for each category from the hypothesized proportions. Are the differences just natural sampling variability, or are they so large that they indicate something important? It's natural to look at the *differences* between these observed and expected counts, denoted $(Obs - Exp)$. We'd like to think about the total of the differences, but just adding them won't work because some differences are positive, others negative. We've been in this predicament before—once when we looked at deviations from the mean and again when we dealt with residuals. In fact, these *are* residuals. They're just the differences between the observed data and the counts given by the (null) model. We handle these residuals in essentially the same way we did in regression: We square them. That gives us positive values and focuses attention on any cells with large differences from what we expected. Because the differences between observed and expected counts generally get larger the more data we have, we also need to get an idea of the *relative* sizes of the differences. To do that, we divide each squared difference by the expected count for that cell.

The test statistic, called the **chi-square** (or chi-squared) **statistic,** is found by adding up the sum of the squares of the deviations between the observed and expected counts divided by the expected counts:

$$\chi^2 = \sum_{all\ cells} \frac{(Obs - Exp)^2}{Exp}.$$

The chi-square statistic is denoted χ^2, where χ is the Greek letter chi (pronounced "ky" as in "sky"). It refers to a family of sampling distribution models we have not seen before called (remarkably enough) the **chi-square models.**

This family of models, like the Student's t-models, differ only in the number of degrees of freedom. The number of degrees of freedom for a goodness-of-fit test is $n - 1$. Here, however, n is *not* the sample size, but instead is the number of categories. For the zodiac example, we have 12 signs, so our χ^2 statistic has 11 degrees of freedom.

One-Sided or Two-Sided?

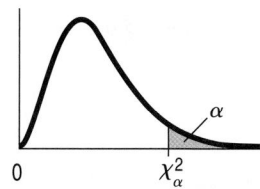

TI-*nspire*

The χ^2 **Models.** See what a χ^2 model looks like, and watch it change as you change the degrees as freedom.

The chi-square statistic is used only for testing hypotheses, not for constructing confidence intervals. If the observed counts don't match the expected, the statistic will be large. It can't be "too small." That would just mean that our model *really* fit the data well. So the chi-square test is always one-sided. If the calculated statistic value is large enough, we'll reject the null hypothesis. What could be simpler?

Even though its mechanics work like a one-sided test, the interpretation of a chi-square test is in some sense *many*-sided. With more than two proportions, there are many ways the null hypothesis can be wrong. By squaring the differences, we made all the deviations positive, whether our observed counts were higher or lower than expected. There's no direction to the rejection of the null model. All we know is that it doesn't fit.

FOR EXAMPLE Doing a goodness-of-fit test

Recap: We're looking at data on the birth months of major league baseball players. We've checked the assumptions and conditions for performing a χ^2 test.

Questions: What are the hypotheses, and what does the test show?

H_0: The distribution of birth months for major league ballplayers is the same as that for the general population.

H_A: The distribution of birth months for major league ballplayers differs from that of the rest of the population.

$df = 12 - 1 = 11$

$$\chi^2 = \sum \frac{(Obs - Exp)^2}{Exp}$$

$$= \frac{(137 - 118.24)^2}{118.24} + \frac{(121 - 103.46)^2}{103.46} + \cdots$$

$$= 26.48 \text{ (by technology)}$$

P-value $= P(\chi^2_{11} \geq 26.48) = 0.0055$ (by technology)

Because of the small P-value, I reject H_0; there's evidence that birth months of major league ballplayers have a different distribution from the rest of us.

 **STEP-BY-STEP EXAMPLE** | **A Chi-Square Test for Goodness-of-Fit**

We have counts of 256 executives in 12 zodiac sign categories. The natural null hypothesis is that birth dates of executives are divided equally among all the zodiac signs. The test statistic looks at how closely the observed data match this idealized situation.

Question: Are zodiac signs of CEOs distributed uniformly?

Plan State what you want to know.

Identify the variables and check the W's.

I want to know whether births of successful people are uniformly distributed across the signs of the zodiac. I have counts of 256 Fortune 400 executives, categorized by their birth sign.

Hypotheses State the null and alternative hypotheses. For χ^2 tests, it's usually easier to do that in words than in symbols.

H_O: Births are uniformly distributed over zodiac signs.[2]

H_A: Births are not uniformly distributed over zodiac signs.

Model Make a picture. The null hypothesis is that the frequencies are equal, so a bar chart (with a line at the hypothesized "equal" value) is a good display.

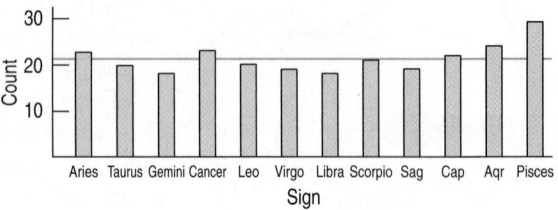

The bar chart shows some variation from sign to sign, and Pisces is the most frequent. But it is hard to tell whether the variation is more than I'd expect from random variation.

Think about the assumptions and check the conditions.

✔ **Counted Data Condition:** I have counts of the number of executives in 12 categories.

✔ **Independence Assumption:** The birth dates of executives should be independent of each other.

✔ **Randomization Condition:** This is a convenience sample of executives, but there's no reason to suspect bias.

✔ **Expected Cell Frequency Condition:** The null hypothesis expects that 1/12 of the 256 births, or 21.333, should occur in each sign. These expected values are all at least 5, so the condition is satisfied.

---

[2] It may seem that we have broken our rule of thumb that null hypotheses should specify parameter values. If you want to get formal about it, the null hypothesis is that

$$p_{Aries} = p_{Taurus} = \cdots = p_{Pisces}.$$

That is, we hypothesize that the true proportions of births of CEOs under each sign are equal. The role of the null hypothesis is to specify the model so that we can compute the test statistic. That's what this one does.

| Specify the sampling distribution model.

Name the test you will use. | The conditions are satisfied, so I'll use a χ^2 model with $12 - 1 = 11$ degrees of freedom and do a **chi-square goodness-of-fit test**. |

 **Mechanics** Each cell contributes an $\dfrac{(Obs - Exp)^2}{Exp}$ value to the chi-square sum. We add up these components for each zodiac sign. If you do it by hand, it can be helpful to arrange the calculation in a table. We show that after this Step-By-Step.

The P-value is the area in the upper tail of the χ^2 model above the computed χ^2 value.

The χ^2 models are skewed to the high end, and change shape depending on the degrees of freedom. The P-value considers only the right tail. Large χ^2 statistic values correspond to small P-values, which lead us to reject the null hypothesis.

The expected value for each zodiac sign is 21.333.

$$\chi^2 = \sum \frac{(Obs - Exp)^2}{Exp} = \frac{(23 - 21.333)^2}{21.333}$$
$$+ \frac{(20 - 21.333)^2}{21.333} + \cdots$$
$$= 5.094 \text{ for all 12 signs.}$$

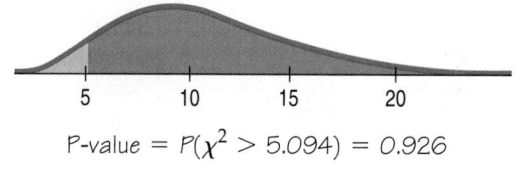

P-value $= P(\chi^2 > 5.094) = 0.926$

 **Conclusion** Link the P-value to your decision. Remember to state your conclusion in terms of what the data mean, rather than just making a statement about the distribution of counts.

The P-value of 0.926 says that if the zodiac signs of executives were in fact distributed uniformly, an observed chi-square value of 5.09 or higher would occur about 93% of the time. This certainly isn't unusual, so I fail to reject the null hypothesis, and conclude that these data show virtually no evidence of nonuniform distribution of zodiac signs among executives.

The Chi-Square Calculation

Activity: Calculating Standardized Residuals. Women were at risk, too. Standardized residuals help us understand the relative risks.

Let's make the chi-square procedure very clear. Here are the steps:

1. **Find the expected values.** These come from the null hypothesis model. Every model gives a hypothesized proportion for each cell. The expected value is the product of the total number of observations times this proportion.

   For our example, the null model hypothesizes *equal* proportions. With 12 signs, 1/12 of the 256 executives should be in each category. The expected number for each sign is 21.333.
2. **Compute the residuals.** Once you have expected values for each cell, find the residuals, *Observed − Expected*.
3. **Square the residuals.**
4. **Compute the components.** Now find the component, $\dfrac{(Observed - Expected)^2}{Expected}$, for each cell.

5. **Find the sum of the components.** That's the chi-square statistic.
6. **Find the degrees of freedom.** It's equal to the number of cells minus one. For the zodiac signs, that's $12 - 1 = 11$ degrees of freedom.
7. **Test the hypothesis.** Large chi-square values mean lots of deviation from the hypothesized model, so they give small P-values. Look up the critical value from a table of chi-square values, or use technology to find the P-value directly.

The steps of the chi-square calculations are often laid out in tables. Use one row for each category, and columns for observed counts, expected counts, residuals, squared residuals, and the contributions to the chi-square total like this:

Sign	Observed	Expected	Residual = $(Obs - Exp)$	$(Obs - Exp)^2$	Component = $\dfrac{(Obs - Exp)^2}{Exp}$
Aries	23	21.333	1.667	2.778889	0.130262
Taurus	20	21.333	−1.333	1.776889	0.083293
Gemini	18	21.333	−3.333	11.108889	0.520737
Cancer	23	21.333	1.667	2.778889	0.130262
Leo	20	21.333	−1.333	1.776889	0.083293
Virgo	19	21.333	−2.333	5.442889	0.255139
Libra	18	21.333	−3.333	11.108889	0.520737
Scorpio	21	21.333	−0.333	0.110889	0.005198
Sagittarius	19	21.333	−2.333	5.442889	0.255139
Capricorn	22	21.333	0.667	0.444889	0.020854
Aquarius	24	21.333	2.667	7.112889	0.333422
Pisces	29	21.333	7.667	58.782889	2.755491

$$\Sigma = 5.094$$

TI Tips

Testing goodness of fit

As always, the TI makes doing the mechanics of a goodness-of-fit test pretty easy, but it does take a little work to set it up. Let's use the zodiac data to run through the steps for a χ^2 GOF-Test.

- Enter the counts of executives born under each star sign in L1.

  Those counts were: 23 20 18 23 20 19 18 21 19 22 24 29

- Enter the expected percentages (or fractions, here 1/12) in L2. In this example they are all the same value, but that's not always the case.
- Convert the expected percentages to expected counts by multiplying each of them by the total number of observations. We use the calculator's summation command in the **LIST MATH** menu to find the total count for the data summarized in L1 and then multiply that sum by the percentages stored in L2 to produce the expected counts. The command is sum(L1)*L2→L2. (We don't ever need the percentages again, so we can replace them by storing the expected counts in L2 instead.)
- Choose D: χ^2 GOF-Test from the **STATS TESTS** menu.
- Specify the lists where you stored the observed and expected counts, and enter the number of degrees of freedom, here 11.

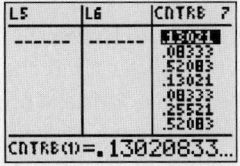

- Ready, set, `Calculate`...
- ... and there are the calculated value of χ^2 and your P-value.
- Notice, too, there's a list of values called `CNTRB`. You can scroll across them, or use `LIST NAMES` to display them as a data list (as seen on the next page). Those are the cell-by-cell components of the χ^2 calculation. We aren't very interested in them this time, because our data failed to provide evidence that the zodiac sign mattered. However, in a situation where we rejected the null hypothesis, we'd want to look at the components to see where the biggest effects occurred. You'll read more about doing that later in this chapter.

By hand?

If there are only a few cells, you may find that it's just as easy to write out the formula and then simply use the calculator to help you with the arithmetic. After you have found $\chi^2 = 5.09375$ you can use your TI to find the P-value, the probability of observing a χ^2 value at least as high as the one you calculated from your data. As you probably expect, that process is akin to `normalcdf` and `tcdf`. You'll find what you need in the `DISTR` menu at $8:\chi^2\text{cdf}$. Just specify the left and right boundaries and the number of degrees of freedom.

- Enter $\chi^2\text{cdf}(5.09375,999,11)$, as shown. (Why 999? Unlike t and z, chi-square values can get pretty big, especially when there are many cells. You may need to go a long way to the right to get to where the curve's tail becomes essentially meaningless. You can see what we mean by looking at Table C, showing chi-square values.)

And there's the P-value, a whopping 0.93! There's nothing at all unusual about these data. (So much for the zodiac's predictive power.)

How big is big? When we calculated χ^2 for the zodiac sign example, we got 5.094. That value would have been big for z or t, leading us to reject the null hypothesis. Not here. Were you surprised that $\chi^2 = 5.094$ had a huge P-value of 0.926? What *is* big for a χ^2 statistic, anyway?

Think about how χ^2 is calculated. In every cell, any deviation from the expected count contributes to the sum. Large deviations generally contribute more, but if there are a lot of cells, even small deviations can add up, making the χ^2 value larger. So the more cells there are, the higher the value of χ^2 has to get before it becomes noteworthy. For χ^2, then, the decision about how big is big depends on the number of degrees of freedom.

Unlike the Normal and t families, χ^2 models are skewed. Curves in the χ^2 family change both shape and center as the number of degrees of freedom grows. Here, for example, are the χ^2 curves for 5 and 9 degrees of freedom.

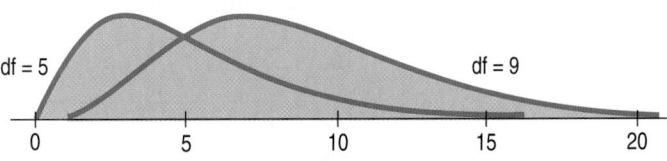

Notice that the value $\chi^2 = 10$ might seem somewhat extreme when there are 5 degrees of freedom, but appears to be rather ordinary for 9 degrees of freedom. Here are two simple facts to help you think about χ^2 models:

> ► The mode is at $\chi^2 = df - 2$. (Look back at the curves; their peaks are at 3 and 7, see?)
>
> ► The expected value (mean) of a χ^2 model is its number of degrees of freedom. That's a bit to the right of the mode—as we would expect for a skewed distribution.
>
> Our test for zodiac birthdays had 11 df, so the relevant χ^2 curve peaks at 9 and has a mean of 11. Knowing that, we might have easily guessed that the calculated χ^2 value of 5.094 wasn't going to be significant.

But I Believe the Model . . .

Goodness-of-fit tests are likely to be performed by people who have a theory of what the proportions *should* be in each category and who believe their theory to be true. Unfortunately, the only *null* hypothesis available for a goodness-of-fit test is that the theory is true. And as we know, the hypothesis-testing procedure allows us only to *reject* the null or *fail to reject* it. We can never confirm that a theory is in fact true, which is often what people want to do.

Unfortunately, they're stuck. At best, we can point out that the data are consistent with the proposed theory. But this doesn't *prove* the theory. The data *could* be consistent with the model even if the theory were wrong. In that case, we fail to reject the null hypothesis but can't conclude anything for sure about whether the theory is true.

And we can't fix the problem by turning things around. Suppose we try to make our favored hypothesis the alternative. Then it is impossible to pick a single null. For example, suppose, as a doubter of astrology, you want to prove that the distribution of executive births is uniform. If you choose uniform as the null hypothesis, you can only *fail* to reject it. So you'd like uniformity to be your alternative hypothesis. Which particular violation of equally distributed births would you choose as your null? The problem is that the model can be wrong in many, many ways. There's no way to frame a null hypothesis the other way around. There's just no way to prove that a favored model is true.

> **Why can't we prove the null?**   A biologist wanted to show that her inheritance theory about fruit flies is valid. It says that 10% of the flies should be type 1, 70% type 2, and 20% type 3. After her students collected data on 100 flies, she did a goodness-of-fit test and found a P-value of 0.07. She started celebrating, since her null hypothesis wasn't rejected—that is, until her students collected data on 100 more flies. With 200 flies, the P-value dropped to 0.02. Although she knew the answer was probably no, she asked the statistician somewhat hopefully if she could just ignore half the data and stick with the original 100. By this reasoning we could always "prove the null" just by not collecting much data. With only a little data, the chances are good that they'll be consistent with almost anything. But they also have little chance of disproving anything either. In this case, the test has no power. Don't let yourself be lured into this scientist's reasoning. With data, more is always better. But you can't ever prove that your null hypothesis is true.

Comparing Observed Distributions

Many colleges survey graduating classes to determine the plans of the graduates. We might wonder whether the plans of students are the same at different colleges. Here's a **two-way table** for Class of 2006 graduates from several colleges at one university. Each **cell** of the table shows how many students from a particular college made a certain choice.

WHO	Graduates from 4 colleges at an upstate New York university
WHAT	Post-graduation activities
WHEN	2006
WHY	Survey for general information

	Agriculture	Arts & Sciences	Engineering	Social Science	Total
Employed	379	305	243	125	**1052**
Grad School	186	238	202	96	**722**
Other	104	123	37	58	**322**
Total	**669**	**666**	**482**	**279**	**2096**

Table 26.1 Post-graduation activities of the class of 2006 for several colleges of a large university.

Because class sizes are so different, we see differences better by examining the proportions for each class rather than the counts:

A S *Video:* **The Incident.** You may have guessed which famous incident put women and children at risk. Here you can view the story complete with rare film footage.

	Agriculture	Arts & Sciences	Engineering	Social Science	Total
Employed	56.7%	45.8%	50.4%	44.8%	**50.2**
Grad School	27.8	35.7	41.9	34.4	**34.4**
Other	15.5	18.5	7.7	20.8	**15.4**
Total	**100**	**100**	**100**	**100**	**100**

Table 26.2 Activities of graduates as a percentage of respondents from each college.

We already know how to test whether *two* proportions are the same. For example, we could use a two-proportion z-test to see whether the proportion of students choosing graduate school is the same for Agriculture students as for Engineering students. But now we have more than two groups. We want to test whether the students' choices are the same across all four colleges. The z-test for two proportions generalizes to a **chi-square test of homogeneity.**

Chi-square again? It turns out that the mechanics of this test are *identical* to the chi-square test for goodness-of-fit that we just saw. (How similar can you get?) Why a different name, then? The goodness-of-fit test compared counts with a theoretical model. But here we're asking whether choices are the same among different groups, so we find the expected counts for each category directly from the data. As a result, we count the degrees of freedom slightly differently as well.

The term "homogeneity" means that things are the same. Here, we ask whether the post-graduation choices made by students are the *same* for these four colleges. The homogeneity test comes with a built-in null hypothesis: We hypothesize that the distribution does not change from group to group. The test looks for differences large enough to step beyond what we might expect from random sample-to-sample variation. It can reveal a large deviation in a single category or small, but persistent, differences over all the categories—or anything in between.

Assumptions and Conditions

The assumptions and conditions are the same as for the chi-square test for goodness-of-fit. The **Counted Data Condition** says that these data must be counts. You can't do a test of homogeneity on proportions, so we have to work with the counts of graduates given in the first table. Also, you can't do a chi-square test on measurements. For example, if we had recorded GPAs for these same groups, we wouldn't be able to determine whether the mean GPAs were different using this test.[3]

Often when we test for homogeneity, we aren't interested in some larger population, so we don't really need a random sample. (We would need one if we wanted to draw a more general conclusion—say, about the choices made by all members of the Class of '06.) Don't we need *some* randomness, though? Fortunately, the null hypothesis can be thought of as a model in which the counts in the table are distributed as if each student chose a plan randomly according to the overall proportions of the choices, regardless of the student's class. As long as we don't want to generalize, we don't have to check the **Randomization Condition** or the **10% Condition.**

We still must be sure we have enough data for this method to work. The **Expected Cell Frequency Condition** says that the expected count in each cell must be at least 5. We'll confirm that as we do the calculations.

Calculations

The null hypothesis says that the proportions of graduates choosing each alternative should be the same for all four colleges, so we can estimate those overall proportions by pooling our data from the four colleges together. Within each college, the expected proportion for each choice is just the overall proportion of all students making that choice. The expected counts are those proportions applied to the number of students in each graduating class.

For example, overall, 1052, or about 50.2%, of the 2096 students who responded to the survey were employed. If the distributions are homogeneous (as the null hypothesis asserts), then 50.2% of the 669 Agriculture school graduates (or about 335.8 students) should be employed. Similarly, 50.2% of the 482 Engineering grads (or about 241.96) should be employed.

Working in this way, we (or, more likely, the computer) can fill in expected values for each cell. Because these are theoretical values, they don't have to be integers. The expected values look like this:

	Agriculture	Arts & Sciences	Engineering	Social Science	Total
Employed	335.777	334.271	241.920	140.032	**1052**
Grad School	230.448	229.414	166.032	96.106	**722**
Other	102.776	102.315	74.048	42.862	**322**
Total	**669**	**666**	**482**	**279**	**2096**

Table 26.3 Expected values for the '06 graduates.

Now check the **Expected Cell Frequency Condition.** Indeed, there are at least 5 individuals expected in each cell.

Following the pattern of the goodness-of-fit test, we compute the component for each cell of the table. For the highlighted cell, employed students graduating from the Ag school, that's

$$\frac{(Obs - Exp)^2}{Exp} = \frac{(379 - 335.777)^2}{335.777} = 5.564$$

---

[3] To do that, you'd use a method called Analysis of Variance, discussed in a supplementary chapter on the DVD and in ActivStats.

Summing these components across all cells gives

$$\chi^2 = \sum_{all\ cells} \frac{(Obs - Exp)^2}{Exp} = 54.51$$

How about the degrees of freedom? We don't really need to calculate all the expected values in the table. We know there is a total of 1052 employed students, so once we find the expected values for three of the colleges, we can determine the expected number for the fourth by just subtracting. Similarly, we know how many students graduated from each college, so after filling in three rows, we can find the expected values for the remaining row by subtracting. To fill out the table, we need to know the counts in only $R - 1$ rows and $C - 1$ columns. So the table has $(R - 1)(C - 1)$ degrees of freedom.

In our example, we need to calculate only 2 choices in each column and counts for 3 of the 4 colleges, for a total of $2 \times 3 = 6$ degrees of freedom. We'll need the degrees of freedom to find a P-value for the chi-square statistic.

NOTATION ALERT:

For a contingency table, R represents the number of rows and C the number of columns.

STEP-BY-STEP EXAMPLE — A Chi-Square Test for Homogeneity

We have reports from four colleges on the post-graduation activities of their 2006 graduating classes.

Question: Are students' choices of post-graduation activities the same across all the colleges?

Plan State what you want to know.

Identify the variables and check the W's.

I want to know whether post-graduation choices are the same for students from each of four colleges. I have a table of counts classifying each college's Class of 2006 respondents according to their activities.

Hypotheses State the null and alternative hypotheses.

H_0: Students' post-graduation activities are distributed in the same way for all four colleges.

H_A: Students' plans do not have the same distribution.

Model Make a picture: A side-by-side bar chart shows the four distributions of post-graduation activities. Plot column percents to remove the effect of class size differences. A split bar chart would also be an appropriate choice.

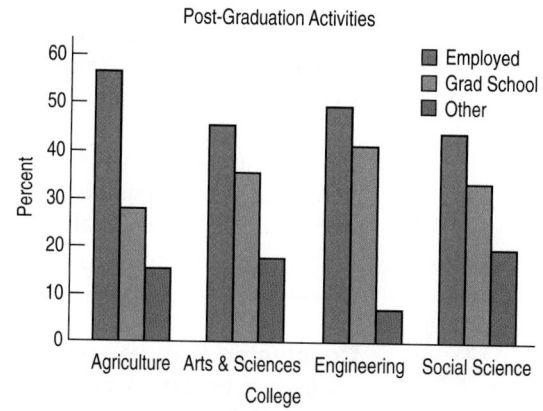

A side-by-side bar chart shows how the distributions of choices differ across the four colleges.

Think about the assumptions and check the conditions.

✔ **Counted Data Condition:** I have counts of the number of students in categories.

✔ **Independence Assumption:** Student plans should be largely independent of each other. The occasional friends who decide to join Teach for America together or couples who make grad school decisions together are too rare to affect this analysis.

✔ **Randomization Condition:** I don't want to draw inferences to other colleges or other classes, so there is no need to check for a random sample.

✔ **Expected Cell Frequency Condition:** The expected values (shown below) are all at least 5.

State the sampling distribution model and name the test you will use.

The conditions seem to be met, so I can use a χ^2 model with $(3 - 1) \times (4 - 1) = 6$ degrees of freedom and do a **chi-square test of homogeneity.**

SHOW

Mechanics Show the expected counts for each cell of the data table. You could make separate tables for the observed and expected counts, or put both counts in each cell as shown here. While observed counts must be whole numbers, expected counts rarely are—don't be tempted to round those off.

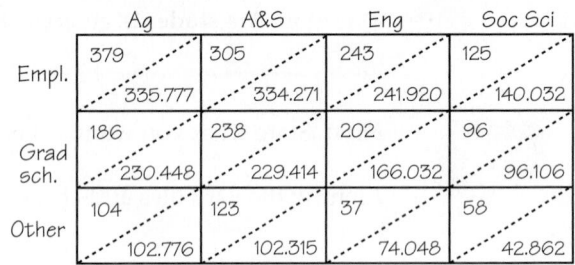

Calculate χ^2.

$$\chi^2 = \sum_{all\ cells} \frac{(Obs - Exp)^2}{Exp}$$
$$= \frac{(379 - 335.777)^2}{335.777} + \cdots$$
$$= 54.52$$

The shape of a χ^2 model depends on the degrees of freedom. A χ^2 model with 6 df is skewed to the high end.

The P-value considers only the right tail. Here, the calculated value of the χ^2 statistic is off the scale, so the P-value is quite small.

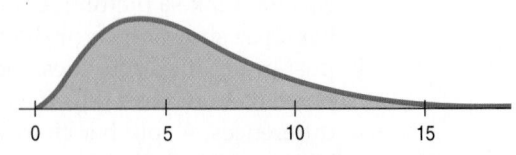

$$P\text{-value} = P(\chi^2 > 54.52) < 0.0001$$

TELL

Conclusion State your conclusion in the context of the data. You should specifically talk about whether the distributions for the groups appear to be different.

The P-value is very small, so I reject the null hypothesis and conclude that there's evidence that the post-graduation activities of students from these four colleges don't have the same distribution.

If you find that simply rejecting the hypothesis of homogeneity is a bit unsatisfying, you're in good company. Ok, so the post-graduation plans are different. What we'd really like to know is what the differences are, where they're the greatest, and where they're smallest. The test for homogeneity doesn't answer these interesting questions, but it does provide some evidence that can help us.

Examining the Residuals

Whenever we reject the null hypothesis, it's a good idea to examine residuals. (We don't need to do that when we fail to reject because when the χ^2 value is small, all of its components must have been small.) For chi-square tests, we want to compare residuals for cells that may have very different counts. So we're better off standardizing the residuals. We know the mean residual is zero,[4] but we need to know each residual's standard deviation. When we tested proportions, we saw a link between the expected proportion and its standard deviation. For counts, there's a similar link. To standardize a cell's residual, we just divide by the square root of its expected value:

$$c = \frac{(Obs - Exp)}{\sqrt{Exp}}.$$

Notice that these **standardized residuals** are just the square roots of the **components** we calculated for each cell, and their sign indicates whether we observed more cases than we expected, or fewer.

The standardized residuals give us a chance to think about the underlying patterns and to consider the ways in which the distribution of post-graduation plans may differ from college to college. Now that we've subtracted the mean (zero) and divided by their standard deviations, these are z-scores. If the null hypothesis were true, we could even appeal to the Central Limit Theorem, think of the Normal model, and use the 68–95–99.7 Rule to judge how extraordinary the large ones are.

Here are the standardized residuals for the Class of '06 data:

	Ag	A&S	Eng	Soc Sci
Employed	2.359	−1.601	0.069	−1.270
Grad School	−2.928	0.567	2.791	−0.011
Other	0.121	2.045	−4.305	2.312

Table 26.4

Standardized residuals can help show how the table differs from the null hypothesis pattern.

The column for Engineering students immediately attracts our attention. It holds both the largest positive and the largest negative standardized residuals. It looks like Engineering college graduates are more likely to go on to graduate work and very unlikely to take time off for "volunteering and travel, among other activities" (as the "Other" category is explained). By contrast, Ag school graduates seem to be readily employed and less likely to pursue graduate work immediately after college.

---

[4] Residual = observed − expected. Because the total of the expected values is set to be the same as the observed total, the residuals must sum to zero.

Looking at χ^2 residuals

Recap: Some people suggest that school children who are the older ones in their class naturally perform better in sports and therefore get more coaching and encouragement. To see if there's any evidence for this, we looked at major league baseball players born since 1975. A goodness-of-fit test found their birth months to have a distribution that's significantly different from the rest of us. The table shows the standardized residuals.

Question: What's different about the distribution of birth months among major league ballplayers?

Month	Residual	Month	Residual
1	1.73	7	−2.69
2	1.72	8	2.77
3	−0.21	9	0.08
4	0.25	10	−1.56
5	0.71	11	−1.22
6	−0.39	12	−0.96

It appears that, compared to the general population, fewer ballplayers than expected were born in July and more than expected in August. Either month would make them the younger kids in their grades in school, so these data don't offer support for the conjecture that being older is an advantage in terms of a career as a pro athlete.

JUST CHECKING

Tiny black potato flea beetles can damage potato plants in a vegetable garden. These pests chew holes in the leaves, causing the plants to wither or die. They can be killed with an insecticide, but a canola oil spray has been suggested as a non-chemical "natural" method of controlling the beetles. To conduct an experiment to test the effectiveness of the natural spray, we gather 500 beetles and place them in three Plexiglas® containers. Two hundred beetles go in the first container, where we spray them with the canola oil mixture. Another 200 beetles go in the second container; we spray them with the insecticide. The remaining 100 beetles in the last container serve as a control group; we simply spray them with water. Then we wait 6 hours and count the number of surviving beetles in each container.

1. Why do we need the control group?

2. What would our null hypothesis be?

3. After the experiment is over, we could summarize the results in a table as shown. How many degrees of freedom does our χ^2 test have?

4. Suppose that, all together, 125 beetles survived. (That's the first-row total.) What's the expected count in the first cell—survivors among those sprayed with the natural spray?

	Natural spray	Insecticide	Water	Total
Survived				
Died				
Total	200	200	100	500

5. If it turns out that only 40 of the beetles in the first container survived, what's the calculated component of χ^2 for that cell?

6. If the total calculated value of χ^2 for this table turns out to be around 10, would you expect the P-value of our test to be large or small? Explain.

Independence

A study from the University of Texas Southwestern Medical Center examined whether the risk of hepatitis C was related to whether people had tattoos and to where they got their tattoos. Hepatitis C causes about 10,000 deaths each year in the United States, but often lies undetected for years after infection.

The data from this study can be summarized in a two-way table, as follows:

WHO	Patients being treated for non–blood-related disorders
WHAT	Tattoo status and hepatitis C status
WHEN	1991, 1992
WHERE	Texas

	Hepatitis C	No Hepatitis C	Total
Tattoo, parlor	17	35	**52**
Tattoo, elsewhere	8	53	**61**
None	22	491	**513**
Total	**47**	**579**	**626**

Table 26.5

Counts of patients classified by their hepatitis C test status according to whether they had a tattoo from a tattoo parlor or from another source, or had no tattoo.

A S *Activity:* **Independence and Chi-Square.** This unusual simulation shows how independence arises (and fails) in contingency tables.

The only difference between the test for homogeneity and the test for independence is in what you . . .

THINK

These data differ from the kinds of data we've considered before in this chapter because they categorize subjects from a single group on two categorical variables rather than on only one. The categorical variables here are *Hepatitis C Status* ("Hepatitis C" or "No Hepatitis C") and *Tattoo Status* ("Parlor," "Elsewhere," "None"). We've seen counts classified by two categorical variables displayed like this in Chapter 3, so we know such tables are called contingency tables. **Contingency tables** categorize counts on two (or more) variables so that we can see whether the distribution of counts on one variable is contingent on the other.

The natural question to ask of these data is whether the chance of having hepatitis C is *independent* of tattoo status. Recall that for events **A** and **B** to be independent $P(\mathbf{A})$ must equal $P(\mathbf{A} \mid \mathbf{B})$. Here, this means the probability that a randomly selected patient has hepatitis C should not change when we learn the patient's tattoo status. We examined the question of independence in just this way back in Chapter 15, but we lacked a way to test it. The rules for independent events are much too precise and absolute to work well with real data. A **chi-square test for independence** is called for here.

If *Hepatitis Status* is independent of tattoos, we'd expect the proportion of people testing positive for hepatitis to be the same for the three levels of *Tattoo Status*. This sounds a lot like the test of homogeneity. In fact, the mechanics of the calculation are identical.

The difference is that now we have two categorical variables measured on a single population. For the homogeneity test, we had a single categorical variable measured independently on two or more populations. But now we ask a different question: "Are the variables independent?" rather than "Are the groups homogeneous?" These are subtle differences, but they are important when we state hypotheses and draw conclusions.

FOR EXAMPLE Which χ^2 test?

Many states and localities now collect data on traffic stops regarding the race of the driver. The initial concern was that Black drivers were being stopped more often (the "crime" ironically called "Driving While Black"). With more data in hand, attention has turned to other issues. For example, data from 2533 traffic stops in Cincinnati[5] report the race of the driver (Black, White, or Other) and whether the traffic stop resulted in a search of the vehicle.

Question: Which test would be appropriate to examine whether race is a factor in vehicle searches? What are the hypotheses?

		Race			
		Black	**White**	**Other**	**Total**
Search	**No**	787	594	27	1408
	Yes	813	293	19	1125
	Total	**1600**	**887**	**46**	**2533**

(continued)

[5] John E. Eck, Lin Liu, and Lisa Growette Bostaph, Police Vehicle Stops in Cincinnati, Oct. 1, 2003, available at http://www.cincinnati-oh.gov. Data for other localities can be found by searching from http://www.racialprofilinganalysis.neu.edu.

For Example (*continued*)

These data represent one group of traffic stops in Cincinnati, categorized on two variables, Race and Search. I'll do a chi-square test of independence.

H_O: Whether or not police search a vehicle is independent of the race of the driver.

H_A: Decisions to search vehicles are not independent of the driver's race.

Assumptions and Conditions

A S *Activity:* **Chi-Square Tables.** Work with *ActivStats'* interactive chi-square table to perform a hypothesis test.

Of course, we still need counts and enough data so that the expected values are at least 5 in each cell.

If we're interested in the independence of variables, we usually want to generalize from the data to some population. In that case, we'll need to check that the data are a representative random sample from, and fewer than 10% of, that population.

| STEP-BY-STEP EXAMPLE | A Chi-Square Test for Independence |

We have counts of 626 individuals categorized according to their "tattoo status" and their "hepatitis status."

Question: Are tattoo status and hepatitis status independent?

THINK

Plan State what you want to know.

Identify the variables and check the W's.

Hypotheses State the null and alternative hypotheses.

We perform a test of independence when we suspect the variables may not be independent. We are on the familiar ground of making a claim (in this case, that knowing *Tattoo Status* will change probabilities for *Hepatitis C Status*) and testing the null hypothesis that it is *not* true.

Model Make a picture. Because these are only two categories—Hepatitis C and No Hepatitis C—a simple bar chart of the distribution of tattoo sources for Hep C patients shows all the information.

I want to know whether the categorical variables *Tattoo Status* and *Hepatitis Status* are statistically independent. I have a contingency table of 626 Texas patients with an unrelated disease.

H_O: *Tattoo Status* and *Hepatitis Status* are independent.[6]

H_A: *Tattoo Status* and *Hepatitis Status* are not independent.

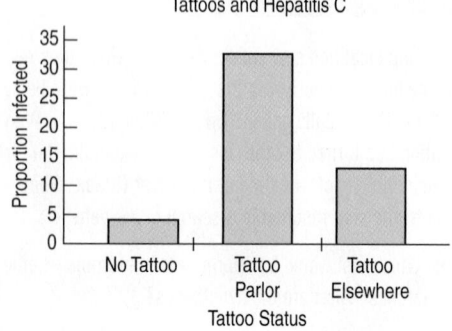

The bar chart suggests strong differences in Hepatitis C risk based on tattoo status.

---

[6] Once again, parameters are hard to express. The hypothesis of independence itself tells us how to find expected values for each cell of the contingency table. That's all we need.

Think about the assumptions and check the conditions.

✔ **Counted Data Condition:** I have counts of individuals categorized on two variables.

✔ **Independence Assumption:** The people in this study are likely to be independent of each other.

✔ **Randomization Condition:** These data are from a retrospective study of patients being treated for something unrelated to hepatitis. Although they are not an SRS, they were selected to avoid biases.

✔ **10% Condition:** These 626 patients are far fewer than 10% of all those with tattoos or hepatitis.

✘ **Expected Cell Frequency Condition:** The expected values do not meet the condition that all are at least 5.

This table shows both the observed and expected counts for each cell. The expected counts are calculated exactly as they were for a test of homogeneity; in the first cell, for example, we expect $\frac{52}{626}$ (that's 8.3%) of 47.

Warning: Be wary of proceeding when there are small expected counts, If we see expected counts that fall far short of 5, or if many cells violate the condition, we should not use χ^2. (We will soon discuss ways you can fix the problem.) If you do continue, always check the residuals to be sure those cells did not have a major influence on your result.

	Hepatitis C	No Hepatitis C	Total
Tattoo, parlor	17 3.904	35 48.096	52
Tattoo, elsewhere	8 4.580	53 56.420	61
None	22 38.516	491 474.484	513
Total	47	579	626

Although the Expected Cell Frequency Condition is not satisfied, the values are close to 5. I'll go ahead, but I'll check the residuals carefully. I'll use a χ^2 model with $(3-1) \times (2-1) = 2$ df and do a **chi-square test of independence**.

Specify the model.

Name the test you will use.

Mechanics Calculate χ^2.

The shape of a chi-square model depends on its degrees of freedom. With 2 df, the model looks quite different, as you can

$$\chi^2 = \sum_{all\ cells} \frac{(Obs - Exp)^2}{Exp}$$
$$= \frac{(17 - 3.094)^2}{3.094} + \cdots = 57.91$$

see here. We still care only about the right tail.

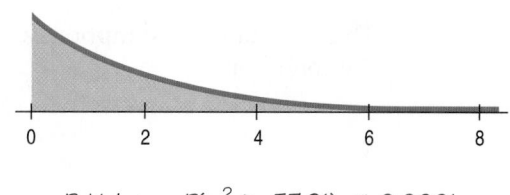

$$P\text{-}Value = P(\chi^2 > 57.91) < 0.0001$$

Conclusion Link the P-value to your decision. State your conclusion about the independence of the two variables.

(We should be wary of this conclusion because of the small expected counts. A complete solution must include the additional analysis, recalculation, and final conclusion discussed in the following section.)

The P-value is very small, so I reject the null hypothesis and conclude that *Hepatitis Status* is not independent of *Tattoo Status*. Because the Expected Cell Frequency Condition was violated, I need to check that the two cells with small expected counts did not influence this result too greatly.

FOR EXAMPLE Chi-square mechanics

Recap: We have data that allow us to investigate whether police searches of vehicles they stop are independent of the driver's race.

Questions: What are the degrees of freedom for this test? What is the expected frequency of searches for the Black drivers who were stopped? What's that cell's component in the χ^2 computation? And how is the standardized residual for that cell computed?

		Race			
		Black	White	Other	Total
Search	No	787	594	27	1408
	Yes	813	293	19	1125
	Total	1600	887	46	2533

This is a 2 × 3 contingency table, so $df = (2 - 1)(3 - 1) = 2$.

Overall, 1125 of 2533 vehicles were searched. If searches are conducted independent of race, then I'd expect $\frac{1125}{2533}$ of the 1600 Black drivers to have been searched: $\frac{1125}{2533} \times 1600 \approx 710.62$.

That cell's term in the χ^2 calculation is $\dfrac{(Obs - Exp)^2}{Exp} = \dfrac{(813 - 710.62)^2}{710.62} = 14.75$

The standardized residual for that cell is $\dfrac{Obs - Exp}{\sqrt{Exp}} = \dfrac{813 - 710.62}{\sqrt{710.62}} = 3.84$

Examine the Residuals

Each cell of the contingency table contributes a term to the chi-square sum. As we did earlier, we should examine the residuals because we have rejected the null hypothesis. In this instance, we have an additional concern that the cells with small expected frequencies not be the ones that make the chi-square statistic large.

Our interest in the data arises from the potential for improving public health. If patients with tattoos are more likely to test positive for hepatitis C, perhaps physicians should be advised to suggest blood tests for such patients.

The standardized residuals look like this:

	Hepatitis C	No Hepatitis C
Tattoo, parlor	6.628	−1.888
Tattoo, elsewhere	1.598	−0.455
None	−2.661	0.758

Table 26.6

Standardized residuals for the hepatitis and tattoos data. Are any of them particularly large in magnitude?

The chi-square value of 57.91 is the sum of the squares of these six values. The cell for people with tattoos obtained in a tattoo parlor who have hepatitis C is large and positive, indicating there are more people in that cell than the null hypothesis of independence would predict. Maybe tattoo parlors are a source of infection or maybe those who go to tattoo parlors also engage in risky behavior.

The second-largest component is a negative value for those with no tattoos who test positive for hepatitis C. A negative value says that there are fewer people in this cell than independence would expect. That is, those who have no tattoos are less likely to be infected with hepatitis C than we might expect if the two variables were independent.

What about the cells with small expected counts? The formula for the chi-square standardized residuals divides each residual by the square root of the expected frequency. Too small an expected frequency can arbitrarily inflate the residual and lead to an inflated chi-square statistic. Any expected count close to the arbitrary minimum of 5 calls for checking that cell's standardized residual to be sure it is not particularly large. In this case, the standardized residual for the "Hepatitis C and Tattoo, elsewhere" cell is not particularly large, but the standardized residual for the "Hepatitis C and Tattoo, parlor" cell is large.

We might choose not to report the results because of concern with the small expected frequency. Alternatively, we could include a warning along with our report of the results. Yet another approach is to combine categories to get a larger sample size and correspondingly larger expected frequencies, if there are some categories that can be appropriately combined. Here, we might naturally combine the two rows for tattoos, obtaining a 2 × 2 table:

	Hepatitis C	No Hepatitis C	Total
Tattoo	25	88	113
None	22	491	513
Total	47	579	626

Table 26.7

Combining the two tattoo categories gives a table with all expected counts greater than 5.

This table has expected values of at least 5 in every cell, and a chi-square value of 42.42 on 1 degree of freedom. The corresponding P-value is <0.0001.

We conclude that *Tattoo Status* and *Hepatitis C Status* are not independent. The data *suggest* that tattoo parlors may be a particular problem, but we haven't enough data to draw that conclusion.

FOR EXAMPLE Writing conclusions for χ^2 tests

Recap: We're looking at Cincinnati traffic stop data to see if police decisions about searching cars show evidence of racial bias. With 3 df, technology calculates $\chi^2 = 73.25$, a P-value less than 0.0001, and these standardized residuals:

Question: What's your conclusion?

		Race		
		Black	White	Other
Search	No	–3.43	4.55	0.28
	Yes	3.84	–5.09	–0.31

The very low P-value leads me to reject the null hypothesis.
There's strong evidence that police decisions to search cars at traffic stops are associated with the driver's race.

The largest residuals are for White drivers, who are searched less often than independence would predict. It appears that Black drivers' cars are searched more often.

TI Tips **Testing homogeneity or independence**

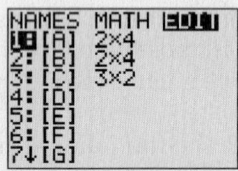

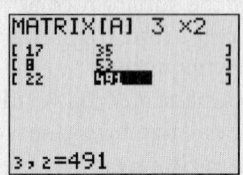

Yes, the TI will do chi-square tests of homogeneity and independence. Let's use the tattoo data. Here goes.

Test a hypothesis of homogeneity or independence

Stage 1: You need to enter the data as a matrix. A "matrix" is just a formal mathematical term for a table of numbers.

- Push the **MATRIX** button, and choose to **EDIT** matrix **[A]**.
- First specify the dimensions of the table, rows × columns.
- Enter the appropriate counts, one cell at a time. The calculator automatically asks for them row by row.

Stage 2: Do the test.

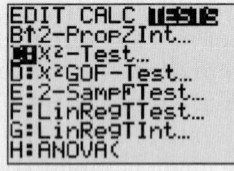

- In the **STAT TESTS** menu choose **C: χ^2-Test**.
- The TI now confirms that you have placed the observed frequencies in **[A]**. It also tells you that when it finds the expected frequencies it will store those in **[B]** for you. Now **Calculate** the mechanics of the test.

The TI reports a calculated value of $\chi^2 = 57.91$ and an exceptionally small P-value.

Stage 3: Check the expected counts.

- Go back to **MATRIX EDIT** and choose **[B]**.

Notice that two of the cells fail to meet the condition that expected counts be at least 5. This problem enters into our analysis and conclusions.

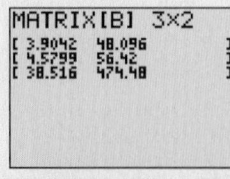

Stage 4: And now some bad news. There's no easy way to calculate the standardized residuals. Look at the two matrices, **[A]** and **[B]**. Large residuals will happen when the corresponding entries differ greatly, especially when the expected count in **[B]** is small (because you will divide by the square root of the entry in **[B]**). The first cell is a good candidate, so we show you the calculation of its standardized residual.

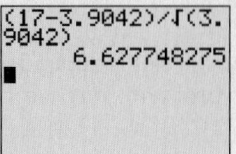

A residual of over 6 is pretty large—possibly an indication that you're more likely to get hepatitis in a tattoo parlor, but the expected count is smaller than 5. We're pretty sure that hepatitis status is not independent of having a tattoo, but we should be wary of saying anything more. Probably the best approach is to combine categories to get cells with expected counts above 5.

Chi-Square and Causation

Chi-square tests are common. Tests for independence are especially widespread. Unfortunately, many people interpret a small P-value as proof of causation. We know better. Just as correlation between quantitative variables does not demonstrate causation, a failure of independence between two categorical variables does not show a cause-and-effect relationship between them, nor should we say that one variable *depends* on the other.

The chi-square test for independence treats the two variables symmetrically. There is no way to differentiate the direction of any possible causation from one variable to the other. In our example, it is unlikely that having hepatitis causes one to crave a tattoo, but other examples are not so clear.

In this case it's easy to imagine that lurking variables are responsible for the observed lack of independence. Perhaps the lifestyles of some people include both tattoos and behaviors that put them at increased risk of hepatitis C, such as body piercings or even drug use. Even a small subpopulation of people with such a lifestyle among those with tattoos might be enough to create the observed result. After all, we observed only 25 patients with both tattoos and hepatitis.

In some sense, a failure of independence between two categorical variables is less impressive than a strong, consistent, linear association between quantitative variables. Two categorical variables can fail the test of independence in many ways, including ways that show no consistent pattern of failure. Examination of the chi-square standardized residuals can help you think about the underlying patterns.

JUST CHECKING

Which of the three chi-square tests—goodness-of-fit, homogeneity, or independence—would you use in each of the following situations?

7. A restaurant manager wonders whether customers who dine on Friday nights have the same preferences among the four "chef's special" entrées as those who dine on Saturday nights. One weekend he has the wait staff record which entrées were ordered each night. Assuming these customers to be typical of all weekend diners, he'll compare the distributions of meals chosen Friday and Saturday.

8. Company policy calls for parking spaces to be assigned to everyone at random, but you suspect that may not be so. There are three lots of equal size: lot A, next to the building; lot B, a bit farther away; and lot C, on the other side of the highway. You gather data about employees at middle management level and above to see how many were assigned parking in each lot.

9. Is a student's social life affected by where the student lives? A campus survey asked a random sample of students whether they lived in a dormitory, in off-campus housing, or at home, and whether they had been out on a date 0, 1–2, 3–4, or 5 or more times in the past two weeks.

WHAT CAN GO WRONG?

▶ **Don't use chi-square methods unless you have counts.** All three of the chi-square tests apply only to counts. Other kinds of data can be arrayed in two-way tables. Just because numbers are in a two-way table doesn't make them suitable for chi-square analysis. Data reported as proportions or percentages can be suitable for chi-square procedures, *but only after they are converted to counts.* If you try to do the calculations without first finding the counts, your results will be wrong.

(continued)

▶ **Beware large samples.** Beware *large* samples?! That's not the advice you're used to hearing. The chi-square tests, however, are unusual. You should be wary of chi-square tests performed on very large samples. No hypothesized distribution fits perfectly, no two groups are exactly homogeneous, and two variables are rarely perfectly independent. The degrees of freedom for chi-square tests don't grow with the sample size. With a sufficiently large sample size, a chi-square test can always reject the null hypothesis. But we have no measure of how far the data are from the null model. There are no confidence intervals to help us judge the effect size.

▶ **Don't say that one variable "depends" on the other just because they're not independent.** Dependence suggests a pattern and implies causation, but variables can fail to be independent in many different ways. When variables fail the test for independence, you might just say they are "associated."

CONNECTIONS

Chi-square methods relate naturally to inference methods for proportions. We can think of a test of homogeneity as stepping from a comparison of two proportions to a question of whether three or more proportions are equal. The standard deviations of the residuals in each cell are linked to the expected counts much like the standard deviations we found for proportions.

Independence is, of course, a fundamental concept in Statistics. But chi-square tests do not offer a general way to check on independence for all those times when we have had to assume it.

Stacked bar charts or side-by-side pie charts can help us think about patterns in two-way tables. A histogram or boxplot of the standardized residuals can help locate extraordinary values.

WHAT HAVE WE LEARNED?

We've learned how to test hypotheses about categorical variables. We use one of three related methods. All look at counts of data in categories, and all rely on chi-square models, a new family indexed by degrees of freedom.

▶ Goodness-of-fit tests compare the observed distribution of a single categorical variable to an expected distribution based on a theory or model.

▶ Tests of homogeneity compare the distribution of several groups for the same categorical variable.

▶ Tests of independence examine counts from a single group for evidence of an association between two categorical variables.

We've seen that, mechanically, these tests are almost identical. Although the tests appear to be one-sided, we've learned that conceptually they are many-sided, because there are many ways that a table of counts can deviate significantly from what we hypothesized. When that happens and we reject the null hypothesis, we've learned to examine standardized residuals in order to better understand patterns as in the table.

Terms

Chi-square model   621, 625. Chi-square models are skewed to the right. They are parameterized by their degrees of freedom and become less skewed with increasing degrees of freedom.

Cell   619, 626. A cell is one element of a table corresponding to a specific row and a specific column. Table cells can hold counts, percentages, or measurements on other variables. Or they can hold several values.

Chi-square statistic	621. The chi-square statistic can be used to test whether the observed counts in a frequency distribution or contingency table match the counts we would expect according to some model. It is calculated as $$\chi^2 = \sum_{all\ cells} \frac{(Obs - Exp)^2}{Exp}.$$ Chi-square statistics differ in how expected counts are found, depending on the question asked.
Chi-square test of goodness-of-fit	618, 622. A test of whether the distribution of counts in one categorical variable matches the distribution predicted by a model is called a test of goodness-of-fit. In a chi-square goodness-of-fit test, the expected counts come from the predicting model. The test finds a P-value from a chi-square model with $n-1$ degrees of freedom, where n is the number of categories in the categorical variable.
Chi-square test of homogeneity	627. A test comparing the distribution of counts for two or more groups on the same categorical variable is called a test of *homogeneity*. A chi-square test of homogeneity finds expected counts based on the overall frequencies, adjusted for the totals in each group under the (null hypothesis) assumption that the distributions are the same for each group. We find a P-value from a chi-square distribution with $(\#Rows - 1) \times (\#Cols - 1)$ degrees of freedom, where $\#Rows$ gives the number of categories and $\#Cols$ gives the number of independent groups.
Chi-square test of independence	633. A test of whether two categorical variables are independent examines the distribution of counts for one group of individuals classified according to both variables. A chi-square test of *independence* finds expected counts by assuming that knowing the marginal totals tells us the cell frequencies, assuming that there is no association between the variables. This turns out to be the same calculation as a test of homogeneity. We find a P-value from a chi-square distribution with $(\#Rows - 1) \times (\#Cols - 1)$ degrees of freedom, where $\#Rows$ gives the number of categories in one variable and $\#Cols$ gives the number of categories in the other.
Chi-square component	623, 628. The components of a chi-square calculation are $$\frac{(Observed - Expected)^2}{Expected},$$ found for each cell of the table.
Standardized residual	631. In each cell of a two-way table, a standardized residual is the square root of the chi-square component for that cell with the sign of the *Observed − Expected* difference: $$\frac{(Obs - Exp)}{\sqrt{Exp}}.$$ When we reject a chi-square test, an examination of the standardized residuals can sometimes reveal more about how the data deviate from the null model.
Two-way table	626, 633. Each *cell* of a two-way table shows counts of individuals. One way classifies a sample according to a categorical variable. The other way can classify different groups of individuals according to the same variable or classify the same individuals according to a different categorical variable.
Contingency table	633. A two-way table that classifies individuals according to two categorical variables is called a *contingency table*.

Skills

- ▸ Be able to recognize when a test of goodness-of-fit, a test of homogeneity, or a test of independence would be appropriate for a table of counts.
- ▸ Understand that the degrees of freedom for a chi-square test depend on the dimensions of the table and not on the sample size. Understand that this means that increasing the sample size increases the ability of chi-square procedures to reject the null hypothesis.

- ▸ Be able to display and interpret counts in a two-way table.
- ▸ Know how to use the chi-square tables to perform chi-square tests.

- ▸ Know how to compute a chi-square test using your statistics software or calculator.
- ▸ Be able to examine the standardized residuals to explain the nature of the deviations from the null hypothesis.

- ▸ Know how to interpret chi-square as a test of goodness-of-fit in a few sentences.
- ▸ Know how to interpret chi-square as a test of homogeneity in a few sentences.
- ▸ Know how to interpret chi-square as a test of independence in a few sentences.

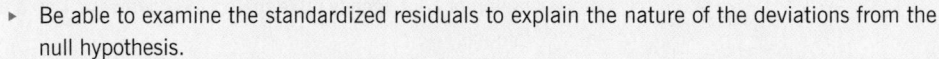

CHI-SQUARE ON THE COMPUTER

Most statistics packages associate chi-square tests with contingency tables. Often chi-square is available as an option only when you make a contingency table. This organization can make it hard to locate the chi-square test and may confuse the three different roles that the chi-square test can take. In particular, chi-square tests for goodness-of-fit may be hard to find or missing entirely. Chi-square tests for homogeneity are computationally the same as chi-square tests for independence, so you may have to perform the mechanics as if they were tests of independence and interpret them afterwards as tests of homogeneity.

Most statistics packages work with data on individuals rather than with the summary counts. If the only information you have is the table of counts, you may find it more difficult to get a statistics package to compute chi-square. Some packages offer a way to reconstruct the data from the summary counts so that they can then be passed back through the chi-square calculation, finding the cell counts again. Many packages offer chi-square standardized residuals (although they may be called something else).

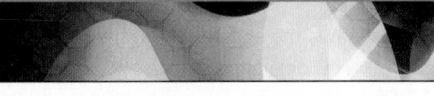

EXERCISES

1. **Which test?** For each of the following situations, state whether you'd use a chi-square goodness-of-fit test, a chi-square test of homogeneity, a chi-square test of independence, or some other statistical test:
   a) A brokerage firm wants to see whether the type of account a customer has (Silver, Gold, or Platinum) affects the type of trades that customer makes (in person, by phone, or on the Internet). It collects a random sample of trades made for its customers over the past year and performs a test.
   b) That brokerage firm also wants to know if the type of account affects the size of the account (in dollars). It performs a test to see if the mean size of the account is the same for the three account types.
   c) The academic research office at a large community college wants to see whether the distribution of courses chosen (Humanities, Social Science, or Science) is different for its residential and nonresidential students. It assembles last semester's data and performs a test.

2. **Which test again?** For each of the following situations, state whether you'd use a chi-square goodness-of-fit test,

a chi-square test of homogeneity, a chi-square test of independence, or some other statistical test:
   a) Is the quality of a car affected by what day it was built? A car manufacturer examines a random sample of the warranty claims filed over the past two years to test whether defects are randomly distributed across days of the work week.
   b) A medical researcher wants to know if blood cholesterol level is related to heart disease. She examines a database of 10,000 patients, testing whether the cholesterol level (in milligrams) is related to whether or not a person has heart disease.
   c) A student wants to find out whether political leaning (liberal, moderate, or conservative) is related to choice of major. He surveys 500 randomly chosen students and performs a test.

3. **Dice.** After getting trounced by your little brother in a children's game, you suspect the die he gave you to roll may be unfair. To check, you roll it 60 times, recording the number of times each face appears. Do these results cast doubt on the die's fairness?

a) If the die is fair, how many times would you expect each face to show?
b) To see if these results are unusual, will you test goodness-of-fit, homogeneity, or independence?
c) State your hypotheses.
d) Check the conditions.
e) How many degrees of freedom are there?
f) Find χ^2 and the P-value.
g) State your conclusion.

Face	Count
1	11
2	7
3	9
4	15
5	12
6	6

4. **M&M's.** As noted in an earlier chapter, the Masterfoods Company says that until very recently yellow candies made up 20% of its milk chocolate M&M's, red another 20%, and orange, blue, and green 10% each. The rest are brown. On his way home from work the day he was writing these exercises, one of the authors bought a bag of plain M&M's. He got 29 yellow ones, 23 red, 12 orange, 14 blue, 8 green, and 20 brown. Is this sample consistent with the company's stated proportions? Test an appropriate hypothesis and state your conclusion.
a) If the M&M's are packaged in the stated proportions, how many of each color should the author have expected to get in his bag?
b) To see if his bag was unusual, should he test goodness-of-fit, homogeneity, or independence?
c) State the hypotheses.
d) Check the conditions.
e) How many degrees of freedom are there?
f) Find χ^2 and the P-value.
g) State a conclusion.

5. **Nuts.** A company says its premium mixture of nuts contains 10% Brazil nuts, 20% cashews, 20% almonds, and 10% hazelnuts, and the rest are peanuts. You buy a large can and separate the various kinds of nuts. Upon weighing them, you find there are 112 grams of Brazil nuts, 183 grams of cashews, 207 grams of almonds, 71 grams of hazelnuts, and 446 grams of peanuts. You wonder whether your mix is significantly different from what the company advertises.
a) Explain why the chi-square goodness-of-fit test is not an appropriate way to find out.
b) What might you do instead of weighing the nuts in order to use a χ^2 test?

6. **Mileage.** A salesman who is on the road visiting clients thinks that, on average, he drives the same distance each day of the week. He keeps track of his mileage for several weeks and discovers that he averages 122 miles on Mondays, 203 miles on Tuesdays, 176 miles on Wednesdays, 181 miles on Thursdays, and 108 miles on Fridays. He wonders if this evidence contradicts his belief in a uniform distribution of miles across the days of the week. Explain why it is not appropriate to test his hypothesis using the chi-square goodness-of-fit test.

7. **NYPD and race.** Census data for New York City indicate that 29.2% of the under-18 population is white, 28.2% black, 31.5% Latino, 9.1% Asian, and 2% other ethnicities. The New York Civil Liberties Union points out that, of 26,181 police officers, 64.8% are white, 14.5% black, 19.1% Hispanic, and 1.4% Asian. Do the police officers reflect the ethnic composition of the city's youth? Test an appropriate hypothesis and state your conclusion.

8. **Violence against women 2005.** In its study *When Men Murder Women*, the Violence Policy Center (www.vpc.org) reported that 1857 women were murdered by men in 2005. Of these victims, a weapon could be identified for 1752 of them. Of those for whom a weapon could be identified, 966 were killed by guns, 390 by knives or other cutting instruments, 136 by other weapons, and 260 by personal attack (battery, strangulation, etc.). The FBI's Uniform Crime Report says that, among all murders nationwide, the weapon use rates were as follows: guns 63.4%, knives 13.1%, other weapons 16.8%, personal attack 6.7%. Is there evidence that violence against women involves different weapons than other violent attacks in the United States?

9. **Fruit flies.** Offspring of certain fruit flies may have yellow or ebony bodies and normal wings or short wings. Genetic theory predicts that these traits will appear in the ratio 9:3:3:1 (9 yellow, normal: 3 yellow, short: 3 ebony, normal: 1 ebony, short). A researcher checks 100 such flies and finds the distribution of the traits to be 59, 20, 11, and 10, respectively.
a) Are the results this researcher observed consistent with the theoretical distribution predicted by the genetic model?
b) If the researcher had examined 200 flies and counted exactly twice as many in each category—118, 40, 22, 20—what conclusion would he have reached?
c) Why is there a discrepancy between the two conclusions?

10. **Pi.** Many people know the mathematical constant π is approximately 3.14. But that's not exact. To be more precise, here are 20 decimal places: 3.14159265358979323846. Still not exact, though. In fact, the actual value is irrational, a decimal that goes on forever without any repeating pattern. But notice that there are no 0's and only one 7 in the 20 decimal places above. Does that pattern persist, or do all the digits show up with equal frequency? The table shows the number of times each digit appears in the first million digits. Test the hypothesis that the digits 0 through 9 are uniformly distributed in the decimal representation of π.

The first million digits of π	
Digit	Count
0	99,959
1	99,758
2	100,026
3	100,229
4	100,230
5	100,359
6	99,548
7	99,800
8	99,985
9	100,106

11. **Hurricane frequencies.** The National Hurricane Center provides data that list the numbers of large (category 3, 4, or 5) hurricanes that have struck the United States, by decade since 1851 (http://www.nhc.noaa.gov/Deadliest_Costliest.shtml). The data are on the next page.

Decade	Count	Decade	Count
1851–1860	6	1931–1940	8
1861–1870	1	1941–1950	10
1871–1880	7	1951–1960	9
1881–1890	5	1961–1970	6
1891–1900	8	1971–1980	4
1901–1910	4	1981–1990	4
1911–1920	7	1991–2000	5
1921–1930	5	2001–2006	7

Recently, there's been some concern that perhaps the number of large hurricanes has been increasing. The natural null hypothesis would be that the frequency of such hurricanes has remained constant.

a) With 96 large hurricanes observed over the 16 periods, what are the expected value(s) for each cell?

b) What kind of chi-square test would be appropriate?

c) State the null and alternative hypotheses.

d) How many degrees of freedom are there?

e) The value of χ^2 is 12.67. What's the P-value?

f) State your conclusion.

g) Look again at the definition of the last "decade". Does that alter your conclusion at all?

T **12. Lottery numbers.** The fairness of the South African lottery was recently challenged by one of the country's political parties. The lottery publishes historical statistics at its Website (http://www.nationallottery.co.za/lotto/statistics.aspx). Here is a table of the number of times each of the 49 numbers has been drawn in the main lottery and as the "bonus ball" number as of June 2007:

Number	Count	Bonus	Number	Count	Bonus
1	81	14	26	78	12
2	91	16	27	83	16
3	78	14	28	76	7
4	77	12	29	76	12
5	67	16	30	99	16
6	87	12	31	78	10
7	88	15	32	73	15
8	90	16	33	81	14
9	80	9	34	81	13
10	77	19	35	77	15
11	84	12	36	73	8
12	68	14	37	64	17
13	79	9	38	70	11
14	90	12	39	67	14
15	82	9	40	75	13
16	103	15	41	84	11
17	78	14	42	79	8
18	85	14	43	74	14
19	67	18	44	87	14
20	90	13	45	82	19
21	77	13	46	91	10
22	78	17	47	86	16
23	90	14	48	88	21
24	80	8	49	76	13
25	65	11			

We wonder if all the numbers are equally likely to be the "bonus ball".

a) What kind of test should we perform?

b) There are 655 bonus ball observations. What are the appropriate expected value(s) for the test?

c) State the null and alternative hypotheses.

d) How many degrees of freedom are there?

e) The value of χ^2 is 34.5. What's the P-value?

f) State your conclusion.

13. Childbirth, part 1. There is some concern that if a woman has an epidural to reduce pain during childbirth, the drug can get into the baby's bloodstream, making the baby sleepier and less willing to breastfeed. In December 2006, the *International Breastfeeding Journal* published results of a study conducted at Sydney University. Researchers followed up on 1178 births, noting whether the mother had an epidural and whether the baby was still nursing after 6 months. Here are their results:

		Epidural?		
		Yes	**No**	**Total**
Breastfeeding	**Yes**	206	498	**704**
@ 6 months?	**No**	190	284	**474**
	Total	**396**	**782**	**1178**

a) What kind of test would be appropriate?

b) State the null and alternative hypotheses.

14. Does your doctor know? A survey[7] of articles from the *New England Journal of Medicine* (*NEJM*) classified them according to the principal statistics methods used. The articles recorded were all non-editorial articles appearing during the indicated years. Let's just look at whether these articles used statistics at all.

	Publication Year			
	1978–79	**1989**	**2004–05**	**Total**
No stats	90	14	40	**144**
Stats	242	101	271	**614**
Total	**332**	**115**	**311**	**758**

Has there been a change in the use of Statistics?

a) What kind of test would be appropriate?

b) State the null and alternative hypotheses.

15. Childbirth, part 2. In Exercise 13, the table shows results of a study investigating whether aftereffects of epidurals administered during childbirth might interfere with successful breastfeeding. We're planning to do a chi-square test.

a) How many degrees of freedom are there?

b) The smallest expected count will be in the epidural/no breastfeeding cell. What is it?

c) Check the assumptions and conditions for inference.

---

[7] Suzanne S. Switzer and Nicholas J. Horton, "What Your Doctor Should Know about Statistics (but Perhaps Doesn't)" *Chance*, 20:1, 2007.

16. Does your doctor know? (part 2). The table in Exercise 14 shows whether *NEJM* medical articles during various time periods included statistics or not. We're planning to do a chi-square test.
   a) How many degrees of freedom are there?
   b) The smallest expected count will be in the 1989/No cell. What is it?
   c) Check the assumptions and conditions for inference.

17. Childbirth, part 3. In Exercises 13 and 15, we've begun to examine the possible impact of epidurals on successful breastfeeding.
   a) Calculate the component of chi-square for the epidural/no breastfeeding cell.
   b) For this test, $\chi^2 = 14.87$. What's the P-value?
   c) State your conclusion.

18. Does your doctor know? (part 3). In Exercises 14 and 16, we've begun to examine whether the use of statistics in *NEJM* medical articles has changed over time.
   a) Calculate the component of chi-square for the 1989/No cell.
   b) For this test, $\chi^2 = 25.28$. What's the P-value?
   c) State your conclusion.

19. Childbirth, part 4. In Exercises 13, 15, and 17, we've tested a hypothesis about the impact of epidurals on successful breastfeeding. The table shows the test's residuals.

		Epidural?	
		Yes	**No**
Breastfeeding	**Yes**	−1.99	1.42
at 6 months?	**No**	2.43	−1.73

   a) Show how the residual for the epidural/no breastfeeding cell was calculated.
   b) What can you conclude from the standardized residuals?

20. Does your doctor know? (part 4). In Exercises 14, 16, and 18, we've tested a hypothesis about whether the use of statistics in *NEJM* medical articles has changed over time. The table shows the test's residuals.

	1978–79	**1989**	**2004–05**
No stats	3.39	−1.68	−2.48
Stats	−1.64	0.81	1.20

   a) Show how the residual for the 1989/No cell was calculated.
   b) What can you conclude from the patterns in the standardized residuals?

21. Childbirth, part 5. In Exercises 13, 15, 17, and 19, we've looked at a study examining epidurals as one factor that might inhibit successful breastfeeding of newborn babies. Suppose a broader study included several additional issues, including whether the mother drank alcohol, whether this was a first child, and whether the parents occasionally supplemented breastfeeding with bottled formula. Why would it not be appropriate to use chi-square methods on the 2 × 8 table with yes/no columns for each potential factor?

22. Does your doctor know? (part 5). In Exercises 14, 16, 18, and 20, we considered data on articles in the *NEJM*. The original study listed 23 different Statistics methods. (The list read: *t*-tests, contingency tables, linear regression, ...) Why would it not be appropriate to use a chi-square test on the 23 × 3 table with a row for each method?

23. Titanic. Here is a table we first saw in Chapter 3 showing who survived the sinking of the *Titanic* based on whether they were crew members, or passengers booked in first-, second-, or third-class staterooms:

	Crew	**First**	**Second**	**Third**	**Total**
Alive	212	202	118	178	**710**
Dead	673	123	167	528	**1491**
Total	**885**	**325**	**285**	**706**	**2201**

   a) If we draw an individual at random, what's the probability that we will draw a member of the crew?
   b) What's the probability of randomly selecting a third-class passenger who survived?
   c) What's the probability of a randomly selected passenger surviving, given that the passenger was a first-class passenger?
   d) If someone's chances of surviving were the same regardless of their status on the ship, how many members of the crew would you expect to have lived?
   e) State the null and alternative hypotheses.
   f) Give the degrees of freedom for the test.
   g) The chi-square value for the table is 187.8, and the corresponding P-value is barely greater than 0. State your conclusions about the hypotheses.

24. NYPD and sex discrimination. The table below shows the rank attained by male and female officers in the New York City Police Department (NYPD). Do these data indicate that men and women are equitably represented at all levels of the department?

		Male	**Female**
	Officer	21,900	4,281
	Detective	4,058	806
Rank	**Sergeant**	3,898	415
	Lieutenant	1,333	89
	Captain	359	12
	Higher ranks	218	10

   a) What's the probability that a person selected at random from the NYPD is a female?
   b) What's the probability that a person selected at random from the NYPD is a detective?
   c) Assuming no bias in promotions, how many female detectives would you expect the NYPD to have?

d) To see if there is evidence of differences in ranks attained by males and females, will you test goodness-of-fit, homogeneity, or independence?
e) State the hypotheses.
f) Test the conditions.
g) How many degrees of freedom are there?
h) The chi-square value for the table is 290.1 and the P-value is less than 0.0001. State your conclusion about the hypotheses.

25. **Titanic again.** Examine and comment on this table of the standardized residuals for the chi-square test you looked at in Exercise 23.

	Crew	First	Second	Third
Alive	−4.35	9.49	2.72	−3.30
Dead	3.00	−6.55	−1.88	2.27

26. **NYPD again.** Examine and comment on this table of the standardized residuals for the chi-square test you looked at in Exercise 24.

	Male	Female
Officer	−2.34	5.57
Detective	−1.18	2.80
Sergeant	3.84	−9.14
Lieutenant	3.58	−8.52
Captain	2.46	−5.86
Higher ranks	1.74	−4.14

27. **Cranberry juice.** It's common folk wisdom that drinking cranberry juice can help prevent urinary tract infections in women. In 2001 the *British Medical Journal* reported the results of a Finnish study in which three groups of 50 women were monitored for these infections over 6 months. One group drank cranberry juice daily, another group drank a lactobacillus drink, and the third drank neither of those beverages, serving as a control group. In the control group, 18 women developed at least one infection, compared to 20 of those who consumed the lactobacillus drink and only 8 of those who drank cranberry juice. Does this study provide supporting evidence for the value of cranberry juice in warding off urinary tract infections?
a) Is this a survey, a retrospective study, a prospective study, or an experiment? Explain.
b) Will you test goodness-of-fit, homogeneity, or independence?
c) State the hypotheses.
d) Test the conditions.
e) How many degrees of freedom are there?
f) Find χ^2 and the P-value.
g) State your conclusion.
h) If you concluded that the groups are not the same, analyze the differences using the standardized residuals of your calculations.

28. **Cars.** A random survey of autos parked in the student lot and the staff lot at a large university classified the brands by country of origin, as seen in the table. Are there differences in the national origins of cars driven by students and staff?

		Driver	
		Student	Staff
Origin	American	107	105
	European	33	12
	Asian	55	47

a) Is this a test of independence or homogeneity?
b) Write appropriate hypotheses.
c) Check the necessary assumptions and conditions.
d) Find the P-value of your test.
e) State your conclusion and analysis.

29. **Montana.** A poll conducted by the University of Montana classified respondents by whether they were male or female and political party, as shown in the table. We wonder if there is evidence of an association between being male or female and party affiliation.

	Democrat	Republican	Independent
Male	36	45	24
Female	48	33	16

a) Is this a test of homogeneity or independence?
b) Write an appropriate hypothesis.
c) Are the conditions for inference satisfied?
d) Find the P-value for your test.
e) State a complete conclusion.

30. **Fish diet.** Medical researchers followed 6272 Swedish men for 30 years to see if there was any association between the amount of fish in their diet and prostate cancer. ("Fatty Fish Consumption and Risk of Prostate Cancer," *Lancet*, June 2001)

Fish Consumption	Total Subjects	Prostate Cancers
Never/seldom	124	14
Small part of diet	2621	201
Moderate part	2978	209
Large part	549	42

a) Is this a survey, a retrospective study, a prospective study, or an experiment? Explain.
b) Is this a test of homogeneity or independence?
c) Do you see evidence of an association between the amount of fish in a man's diet and his risk of developing prostate cancer?
d) Does this study prove that eating fish does not prevent prostate cancer? Explain.

31. Montana revisited. The poll described in Exercise 29 also investigated the respondents' party affiliations based on what area of the state they lived in. Test an appropriate hypothesis about this table and state your conclusions.

	Democrat	Republican	Independent
West	39	17	12
Northeast	15	30	12
Southeast	30	31	16

32. Working parents. In July 1991 and again in April 2001, the Gallup Poll asked random samples of 1015 adults about their opinions on working parents. The table summarizes responses to the question "Considering the needs of both parents and children, which of the following do you see as the ideal family in today's society?"

	1991	2001
Both work full time	142	131
One works full time, other part time	274	244
One works, other works at home	152	173
One works, other stays home for kids	396	416
No opinion	51	51

a) Is this a survey, a retrospective study, a prospective study, or an experiment? Explain.

b) Will you test goodness-of-fit, homogeneity, or independence?

c) Based on these results, do you think there was a change in people's attitudes during the 10 years between these polls?

33. Grades. Two different professors teach an introductory Statistics course. The table shows the distribution of final grades they reported. We wonder whether one of these professors is an "easier" grader.

	Prof. Alpha	Prof. Beta
A	3	9
B	11	12
C	14	8
D	9	2
F	3	1

a) Will you test goodness-of-fit, homogeneity, or independence?

b) Write appropriate null hypotheses.

c) Find the expected counts for each cell, and explain why the chi-square procedures are not appropriate.

34. Full moon. Some people believe that a full moon elicits unusual behavior in people. The table shows the number of arrests made in a small town during weeks of six full moons and six other randomly selected weeks in the same year. We wonder if there is evidence of a difference in the types of illegal activity that take place.

	Full Moon	Not Full
Violent (murder, assault, rape, etc.)	2	3
Property (burglary, vandalism, etc.)	17	21
Drugs/Alcohol	27	19
Domestic abuse	11	14
Other offenses	9	6

a) Will you test goodness-of-fit, homogeneity, or independence?

b) Write appropriate null hypotheses.

c) Find the expected counts for each cell, and explain why the chi-square procedures are not appropriate.

35. Grades again. In some situations where the expected cell counts are too small, as in the case of the grades given by Professors Alpha and Beta in Exercise 33, we can complete an analysis anyway. We can often proceed after combining cells in some way that makes sense and also produces a table in which the conditions are satisfied. Here we create a new table displaying the same data, but calling D's and F's "Below C":

	Prof. Alpha	Prof. Beta
A	3	9
B	11	12
C	14	8
Below C	12	3

a) Find the expected counts for each cell in this new table, and explain why a chi-square procedure is now appropriate.

b) With this change in the table, what has happened to the number of degrees of freedom?

c) Test your hypothesis about the two professors, and state an appropriate conclusion.

36. Full moon, next phase. In Exercise 34 you found that the expected cell counts failed to satisfy the conditions for inference.

a) Find a sensible way to combine some cells that will make the expected counts acceptable.

b) Test a hypothesis about the full moon and state your conclusion.

37. Racial steering. A subtle form of racial discrimination in housing is "racial steering." Racial steering occurs when real estate agents show prospective buyers only homes in neighborhoods already dominated by that family's race. This violates the Fair Housing Act of 1968. According to an article in *Chance* magazine (Vol. 14, no. 2 [2001]), tenants at a large apartment complex recently filed a lawsuit alleging racial steering. The complex is divided into two parts: Section A and Section B.

The plaintiffs claimed that white potential renters were steered to Section A, while African-Americans were steered to Section B. The table displays the data that were presented in court to show the locations of recently rented apartments. Do you think there is evidence of racial steering?

New Renters			
	White	Black	Total
Section A	87	8	95
Section B	83	34	117
Total	170	42	212

38. *Titanic*, **redux.** Newspaper headlines at the time, and traditional wisdom in the succeeding decades, have held that women and children escaped the *Titanic* in greater proportions than men. Here's a table with the relevant data. Do you think that survival was independent of whether the person was male or female? Explain.

	Female	Male	Total
Alive	343	367	710
Dead	127	1364	1491
Total	470	1731	2201

39. Steering revisited. You could have checked the data in Exercise 37 for evidence of racial steering using two-proportion z procedures.
a) Find the z-value for this approach, and show that when you square your z-value, you get the value of χ^2 you calculated in Exercise 37.
b) Show that the resulting P-values are the same.

40. Survival on the *Titanic*, one more time. In Exercise 38 you could have checked for a difference in the chances of survival for men and women using two-proportion z procedures.
a) Find the z-value for this approach.
b) Show that the square of your calculated value of z is the value of χ^2 you calculated in Exercise 38.
c) Show that the resulting P-values are the same.

41. Pregnancies. Most pregnancies result in live births, but some end in miscarriages or stillbirths. A June 2001 National Vital Statistics Report examined those outcomes in the United States during 1997, broken down by the age of the mother. The table shows counts consistent with that report. Is there evidence that the distribution of outcomes is not the same for these age groups?

		Live Births	Fetal Losses
Age of Mother	Under 20	49	13
	20–29	201	41
	30–34	88	21
	35 or over	49	21

42. Education by age. Use the survey results in the table to investigate differences in education level attained among different age groups in the United States.

		Age Group				
		25–34	35–44	45–54	55–64	≥65
Education Level	Not HS grad	27	50	52	71	101
	HS	82	19	88	83	59
	1–3 years college	43	56	26	20	20
	≥4 years college	48	75	34	26	20

JUST CHECKING
Answers

1. We need to know how well beetles can survive 6 hours in a Plexiglas® box so that we have a baseline to compare the treatments.
2. There's no difference in survival rate in the three groups.
3. $(2 - 1)(3 - 1) = 2\, df$
4. 50
5. 2
6. The mean value for a χ^2 with 2 df is 2, so 10 seems pretty large. The P-value is probably small.
7. This is a test of homogeneity. The clue is that the question asks whether the distributions are alike.
8. This is a test of goodness-of-fit. We want to test the model of equal assignment to all lots against what actually happened.
9. This is a test of independence. We have responses on two variables for the same individuals.

Inferences for Regression

WHO	250 male subjects
WHAT	Body fat and waist size
UNITS	% Body fat and inches
WHEN	1990s
WHERE	United States
WHY	Scientific research

Three percent of a man's body is essential fat. (For a woman, the percentage is closer to 12.5%.) As the name implies, essential fat is necessary for a normal, healthy body. Fat is stored in small amounts throughout your body. Too much body fat, however, can be dangerous to your health. For men between 18 and 39 years old, a healthy percent body fat ranges from 8% to 19%. (For women of the same age, it's 21% to 32%.)

Measuring body fat can be tedious and expensive. The "standard reference" measurement is by dual-energy X-ray absorptiometry (DEXA), which involves two low-dose X-ray generators and takes from 10 to 20 minutes.

How close can we get to a useable prediction of body fat from easily measurable variables such as *Height, Weight,* or *Waist* size? Here's a scatterplot of *%Body Fat* plotted against *Waist* size for a sample of 250 males of various ages.

FIGURE 27.1

Percent Body Fat vs. Waist size for 250 men of various ages. The scatterplot shows a strong, positive, linear relationship.

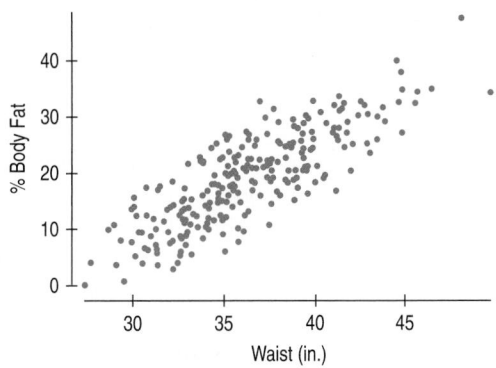

Back in Chapter 8 we modeled relationships like this by fitting a least squares line. The plot is clearly straight, so we can find that line. The equation of the least squares line for these data is

$$\widehat{\%Body\ Fat} = -42.7 + 1.7\ Waist.$$

The slope says that, on average, *%Body Fat* is greater by 1.7 percent for each additional inch around the waist.

How useful is this model? When we fit linear models before, we used them to describe the relationship between the variables and we interpreted the slope and intercept as descriptions of the data. Now we'd like to know what the regression model can tell us beyond the 250 men in this study. To do that, we'll want to make confidence intervals and test hypotheses about the slope and intercept of the regression line.

The Population and the Sample

When we found a confidence interval for a mean, we could imagine a single, true underlying value for the mean. When we tested whether two means or two proportions were equal, we imagined a true underlying difference. But what does it mean to do inference for regression? We know better than to think that even if we knew every population value, the data would line up perfectly on a straight line. After all, even in our sample, not all men who have 38-inch waists have the same *%Body Fat*. In fact, there's a whole distribution of *%Body Fat* for these men:

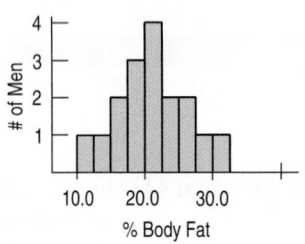

FIGURE 27.2

The distribution of %Body Fat for men with a Waist size of 38 inches is unimodal and symmetric.

This is true at each *Waist* size. In fact, we could depict the distribution of *%Body Fat* at different *Waist* sizes like this:

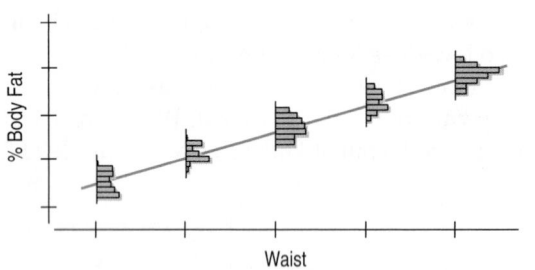

FIGURE 27.3

There's a distribution of %Body Fat for each value of Waist size. We'd like the means of these distributions to line up.

But we want to *model* the relationship between *%Body Fat* and *Waist* size for all men. To do that, we imagine an idealized regression line. The model assumes that the *means* of the distributions of *%Body Fat* for each *Waist* size fall along the line, even though the individuals are scattered around it. We know that this model is not a perfect description of how the variables are associated, but it may be useful for predicting *%Body Fat* and for understanding how it's related to *Waist* size.

If only we had all the values in the population, we could find the slope and intercept of this *idealized regression line* explicitly by using least squares. Following our usual conventions, we write the idealized line with Greek letters and consider the coefficients (the slope and intercept) to be *parameters:* β_0 is the intercept and β_1 is the slope. Corresponding to our fitted line of $\hat{y} = b_0 + b_1x$, we write

$$\mu_y = \beta_0 + \beta_1x.$$

Why μ_y instead of $\hat{y}$? Because this is a model. There is a distribution of *%Body Fat* for each *Waist* size. The model places the *means* of the distributions of *%Body Fat* for each *Waist* size on the same straight line.

NOTATION ALERT:

This time we used up only one Greek letter for two things. Lower-case Greek β (beta) is the natural choice to correspond to the *b*'s in the regression equation. We used β before for the probability of a Type II error, but there's little chance of confusion here.

Of course, not all the individual y's are at these means. (In fact, the line will miss most—and quite possibly all—of the plotted points.) Some individuals lie above and some below the line, so, like all models, this one makes **errors.** Lots of them. In fact, one at each point. These errors are random and, of course, can be positive or negative. They are model errors, so we use a Greek letter and denote them by ε.

When we put the errors into the equation, we can account for each individual y:

$$y = \beta_0 + \beta_1 x + \varepsilon.$$

This equation is now true for each data point (since there is an ε to soak up the deviation), so the model gives a value of y for any value of x.

For the body fat data, an idealized model such as this provides a summary of the relationship between *%Body Fat* and *Waist* size. Like all models, it simplifies the real situation. We know there is more to predicting body fat than waist size alone. But the advantage of a model is that the simplification might help us to think about the situation and assess how well *%Body Fat* can be predicted from simpler measurements.

We estimate the β's by finding a regression line, $\hat{y} = b_0 + b_1 x$, as we did in Chapter 8. The residuals, $e = y - \hat{y}$, are the sample-based versions of the errors, ε. We'll use them to help us assess the regression model.

We know that least squares regression will give reasonable estimates of the parameters of this model from a random sample of data. Our challenge is to account for our uncertainty in how well they do. For that, we need to make some assumptions about the model and the errors.

Assumptions and Conditions

A S   *Activity:* **Conditions for Regression Inference.** View an illustrated discussion of the conditions for regression inference.

Back in Chapter 8 when we fit lines to data, we needed to check only the Straight Enough Condition. Now, when we want to make inferences about the coefficients of the line, we'll have to make more assumptions. Fortunately, we can check conditions to help us judge whether these assumptions are reasonable for our data. And as we've done before, we'll make some checks *after* we find the regression equation.

Also, we need to be careful about the order in which we check conditions. If our initial assumptions are not true, it makes no sense to check the later ones. So now we number the assumptions to keep them in order.

1. Linearity Assumption

Check the scatterplot.
The shape must be linear or we can't use linear regression at all.

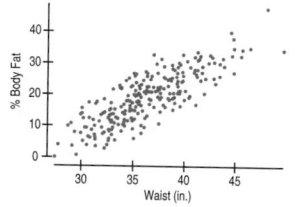

If the true relationship is far from linear and we use a straight line to fit the data, our entire analysis will be useless, so we always check this first.

The **Straight Enough Condition** is satisfied if a scatterplot looks straight. It's generally not a good idea to draw a line through the scatterplot when checking. That can fool your eyes into seeing the plot as more straight. Sometimes it's easier to see violations of the Straight Enough Condition by looking at a scatterplot of the residuals against x or against the predicted values, $\hat{y}$. That plot will have a horizontal direction and should have no pattern if the condition is satisfied.

If the scatterplot is straight enough, we can go on to some assumptions about the errors. If not, stop here, or consider re-expressing the data (see Chapter 10) to make the scatterplot more nearly linear. For the *%Body Fat* data, the scatterplot is beautifully linear. Of course, the data must be quantitative for this to make sense. Check the **Quantitative Data Condition.**

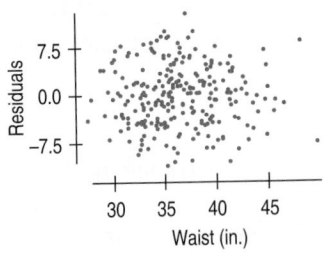

FIGURE 27.4

The residuals show only random scatter when plotted against Waist size.

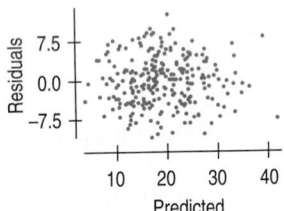

FIGURE 27.5

A scatterplot of residuals against predicted values can help check for plot thickening. Note that this plot looks identical to the plot of residuals against Waist size. For a regression of one response variable on one predictor, these plots differ only in the labels on the x-axis.

2. INDEPENDENCE ASSUMPTION

Independence Assumption: The errors in the true underlying regression model (the ε's) must be mutually independent. As usual, there's no way to be sure that the Independence Assumption is true.

Usually when we care about inference for the regression parameters, it's because we think our regression model might apply to a larger population. In such cases, we can check a **Randomization Condition** that the individuals are a representative sample from that population.

We can also check displays of the regression residuals for evidence of patterns, trends, or clumping, any of which would suggest a failure of independence. In the special case when the x-variable is related to time, a common violation of the Independence Assumption is for the errors to be correlated. (The error our model makes today may be similar to the one it made for yesterday.) This violation can be checked by plotting the residuals against the x-variable and looking for patterns.

The *%Body Fat* data were collected on a sample of men taken to be representative. The subjects were not related in any way, so we can be pretty sure that their measurements are independent. The residuals plot shows no pattern.

3. EQUAL VARIANCE ASSUMPTION

The variability of y should be about the same for all values of x. In Chapter 8 we looked at the standard deviation of the residuals (s_e) to measure the size of the scatter. Now we'll need this standard deviation to build confidence intervals and test hypotheses. The standard deviation of the residuals is the building block for the standard errors of all the regression parameters. But it makes sense only if the scatter of the residuals is the same everywhere. In effect, the standard deviation of the residuals "pools" information across all of the individual distributions at each x-value, and pooled estimates are appropriate only when they combine information for groups with the same variance.

Practically, what we can check is the **Does the Plot Thicken? Condition.** A scatterplot of y against x offers a visual check. Fortunately, we've already made one. Make sure the spread around the line is nearly constant. Be alert for a "fan" shape or other tendency for the variation to grow or shrink in one part of the scatterplot. Often it is better to look at the residuals plotted against the predicted values, $\hat{y}$. With the slope of the line removed, it's easier to see patterns left behind. For the body fat data, the spread of *%Body Fat* around the line is remarkably constant across *Waist* sizes from 30 inches to about 45 inches.

If the plot is straight enough, the data are independent, and the plot doesn't thicken, you can now move on to the final assumption.

4. NORMAL POPULATION ASSUMPTION

We assume the errors around the idealized regression line at each value of x follow a Normal model. We need this assumption so that we can use a Student's t-model for inference.

As we have at other times when we've used Student's t, we'll settle for the residuals satisfying the **Nearly Normal Condition** and the **Outlier Condition.** Look at a histogram or Normal probability plot of the residuals.[1]

---

[1] *This* is why we have to check the conditions in order. We have to check that the residuals are independent and that the variation is the same for all x's so that we can lump all the residuals together for a single check of the Nearly Normal Condition.

> **Check a histogram of the residuals.**
> The distribution of the residuals should be unimodal and symmetric.

The histogram of residuals in the *%Body Fat* regression certainly looks nearly Normal. As we have noted before, the Normality Assumption becomes less important as the sample size grows, because the model is about means and the Central Limit Theorem takes over.

If all four assumptions were true, the idealized regression model would look like this:

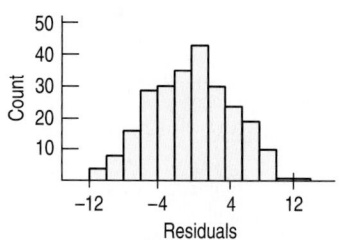

FIGURE 27.6

A histogram of the residuals is one way to check whether they are nearly Normal. Alternatively, we can look at a Normal probability plot.

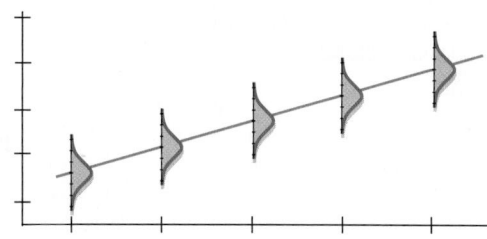

FIGURE 27.7

The regression model has a distribution of y-values for each x-value. These distributions follow a normal model with means lined up along the line and with the same standard deviations.

At each value of x there is a distribution of y-values that follows a Normal model, and each of these Normal models is centered on the line and has the same standard deviation. Of course, we don't expect the assumptions to be exactly true, and we know that all models are wrong, but the linear model is often close enough to be very useful.

FOR EXAMPLE   Checking assumptions and conditions

Look at the moon with binoculars or a telescope, and you'll see craters formed by thousands of impacts. The earth, being larger, has been hit even more often. Meteor Crater in Arizona was the first recognized impact crater and was identified as such only in the 1920s. With the help of satellite images, more and more craters have been identified; now more than 180 are known. These, of course, are only a small sample of all the impacts the earth has experienced: Only 29% of earth's surface is land, and many craters have been covered or eroded away. Astronomers have recognized a roughly 35 million-year cycle in the frequency of cratering, although the cause of this cycle is not fully understood. Here's a scatterplot of the known impact craters from the most recent 35 million years.[2] We've taken logs of both age (in millions of years ago) and diameter (km) to make the relationship simpler. (See Chapter 10.)

WHO	39 impact craters
WHAT	Diameter and age
UNITS	km and millions of years ago
WHEN	Past 35 million years
WHERE	Worldwide
WHY	Scientific research

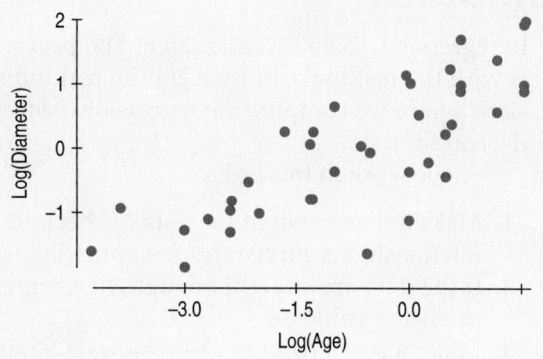

Question: Are the assumptions and conditions satisfied for fitting a linear regression model to these data?

✔ **Linearity Assumption:** The scatterplot satisfies the Straight Enough Condition.

✔ **Independence Assumption:** Sizes of impact craters are likely to be generally independent.

(continued)

---

[2] Data, pictures, and much more information at the Earth Impact Database found at http://www.unb.ca.

For Example (*continued*)

✔ **Randomization Condition:** These are the only known craters, and may differ from others that have disappeared or not yet been found. I'll need to be careful not to generalize my conclusions too broadly.

✔ **Does the Plot Thicken? Condition:** After fitting a linear model, I find the residuals shown.

Two points seem to give the impression that the residuals may be more variable for higher predicted values than for lower ones, but this doesn't seem to be a serious violation of the Equal Variance Assumption.

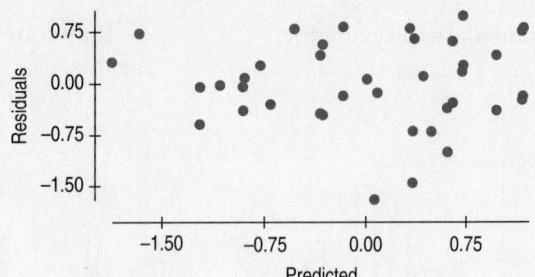

✔ **Nearly Normal Condition:** A Normal probability plot suggests a bit of skewness in the distribution of residuals, and the histogram confirms that.

There are no violations severe enough to stop my regression analysis, but I'll be cautious about my conclusions.

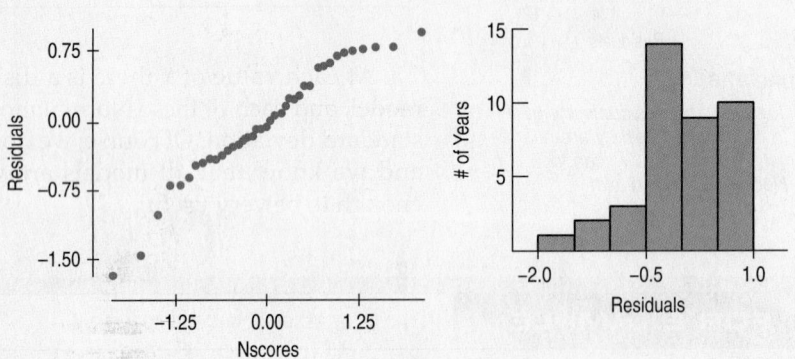

Which Come First: the Conditions or the Residuals?

> *"Truth will emerge more readily from error than from confusion."*
>
> —Francis Bacon (1561–1626)

In regression, there's a little catch. The best way to check many of the conditions is with the residuals, but we get the residuals only *after* we compute the regression. Before we compute the regression, however, we should check at least one of the conditions.

So we work in this order:

1. Make a scatterplot of the data to check the Straight Enough Condition. (If the relationship is curved, try re-expressing the data. Or stop.)
2. If the data are straight enough, fit a regression and find the residuals, e, and predicted values, $\hat{y}$.
3. Make a scatterplot of the residuals against x or the predicted values. This plot should have no pattern. Check in particular for any bend (which would suggest that the data weren't all that straight after all), for any thickening (or thinning), and, of course, for any outliers. (If there are outliers, and you can correct them or justify removing them, do so and go back to step 1, or consider performing two regressions—one with and one without the outliers.)
4. If the data are measured over time, plot the residuals against time to check for evidence of patterns that might suggest they are not independent.
5. If the scatterplots look OK, then make a histogram and Normal probability plot of the residuals to check the Nearly Normal Condition.
6. If all the conditions seem to be reasonably satisfied, go ahead with inference.

STEP-BY-STEP EXAMPLE | **Regression Inference**

If our data can jump through all these hoops, we're ready to do regression inference. Let's see how much more we can learn about body fat and waist size from a regression model.

Questions: What is the relationship between *%Body Fat* and *Waist* size in men?
What model best predicts body fat from waist size, and how well does it do the job?

Plan Specify the question of interest.

Name the variables and report the W's.

Identify the parameters you want to estimate.

Model Think about the assumptions and check the conditions.

Make pictures. For regression inference, you'll need a scatterplot, a residuals plot, and either a histogram or a Normal probability plot of the residuals.

(We've seen plots of the residuals already. See Figures 27.5 and 27.6.)

I have quantitative body measurements on 250 adult males from the BYU Human Performance Research Center. I want to understand the relationship between *%Body Fat* and *Waist* size.

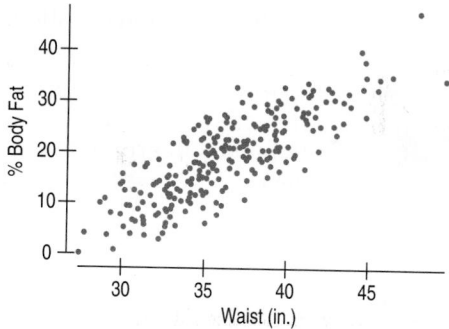

✓ **Straight Enough Condition:** There's no obvious bend in the original scatterplot of the data or in the plot of residuals against predicted values.

✓ **Independence Assumption:** These data are not collected over time, and there's no reason to think that the *%Body Fat* of one man influences the *%Body Fat* of another.

✓ **Does the Plot Thicken? Condition:** Neither the original scatterplot nor the residual scatterplot shows any changes in the spread about the line.

✓ **Nearly Normal Condition, Outlier Condition:** A histogram of the residuals is unimodal and symmetric. The Normal probability plot of the residuals is quite straight, indicating that the Normal model is reasonable for the errors.

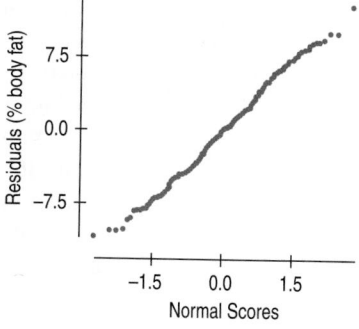

Choose your method.	Under these conditions a **regression model** is appropriate.

 **Mechanics** Let's just "push the button" and see what the regression looks like. The formula for the regression equation can be found in Chapter 8, and the standard error formulas will be shown a bit later, but regressions are almost always computed with a computer program or calculator. Write the regression equation.	Here's the computer output for this regression: Dependent variable is: %BF R-squared = 67.8% s = 4.713 with 250 − 2 = 248 degrees of freedom | Variable	Coeff	SE(Coeff)	t-ratio	P-value
Intercept	−42.734	2.717	−15.7	<0.0001
Waist	1.70	0.0743	22.9	<0.0001 | The estimated regression equation is $$\widehat{\%Body\ Fat} = -42.73 + 1.70\ Waist.$$

|  **Conclusion** Interpret your results in context.

More Interpretation We haven't worked it out in detail yet, but the output gives us numbers labeled as *t*-statistics and corresponding P-values, and we have a general idea of what those mean.

(Now it's time to learn more about regression inference so we can figure out what the rest of the output means.) | The R^2 for the regression is 67.8%. Waist size seems to account for about 2/3 of the %Body Fat variation in men. The slope of the regression says that %Body Fat increases by about 1.7 percentage points per inch of Waist size, on average.

The standard error of 0.07 for the slope is much smaller than the slope itself, so it looks like the estimate is reasonably precise. And there are a couple of t-ratios and P-values given. Because the P-values are small, it appears that some null hypotheses can be rejected. |

Intuition About Regression Inference

> **A S** *Simulation: Simulate the Sampling Distribution of a Regression Slope.* Draw samples repeatedly to see for yourself how slope can vary from sample to sample. This simulation experiment lets you build up a histogram to see the sampling distribution.

Wait a minute! We've just pulled a fast one. We've pushed the "regression button" on our computer or calculator but haven't discussed where the standard errors for the slope or intercept come from. We know that if we had collected similar data on a different random sample of men, the slope and intercept would be different. Each sample would have produced its own regression line, with slightly different b_0's and b_1's. This sample-to-sample variation is what generates the sampling distributions for the coefficients.

There's only one regression model; each sample regression is trying to estimate the same parameters, β_0 and β_1. We expect any sample to produce a b_1 whose expected value is the true slope, β_1. What about its standard deviation? What aspects of the data affect how much the slope (and intercept) vary from sample to sample?

> **Spread around the line.** Here are two situations in which we might do regression. Which situation would yield the more consistent slope? That is, if we were to sample over and over from the two underlying populations that these samples come from and compute all the slopes, which group of slopes would vary less?

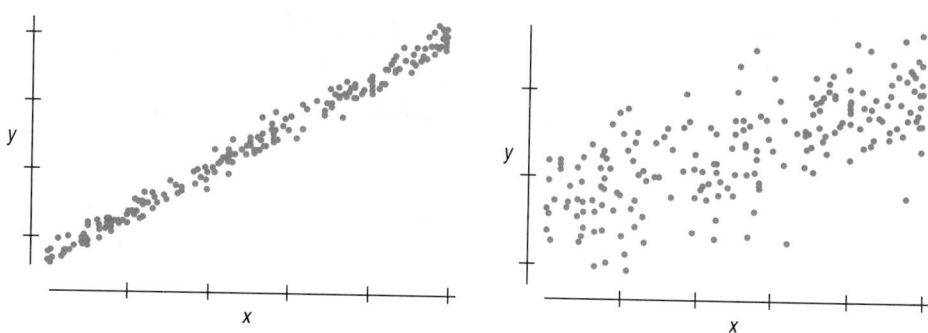

n − 2?

For standard deviation (in Chapter 4), we divided by $n-1$ because we didn't know the true mean and had to estimate it. Now it's later in the course and there's even more we don't know. Here we don't know *two* things: the slope and the intercept. If we knew them both, we'd divide by n and have n degrees of freedom. When we estimate both, however, we adjust by subtracting 2, so we divide by $n-2$ and (as we will see soon) have 2 fewer degrees of freedom.

Clearly, data like those in the left plot give more consistent slopes.

Less scatter around the line means the slope will be more consistent from sample to sample. The spread around the line is measured with the **residual standard deviation,** s_e. You can always find s_e in the regression output, often just labeled s. You're probably not going to calculate the residual standard deviation by hand. As we noted when we first saw this formula in Chapter 8, it looks a lot like the standard deviation of y, only now subtracting the predicted values rather than the mean and dividing by $n-2$ instead of $n-1$:

$$s_e = \sqrt{\frac{\sum (y-\hat{y})^2}{n-2}}.$$

The less scatter around the line, the smaller the residual standard deviation and the stronger the relationship between x and y.

Some people prefer to assess the strength of a regression by looking at s_e rather than R^2. After all, s_e has the same units as y, and because it's the standard deviation of the errors around the line, it tells you how close the data are to our model. By contrast, R^2 is the proportion of the variation of y accounted for by x. We say, why not look at both?

> **Spread of the x's:** Here are two more situations. Which of these would yield more consistent slopes?

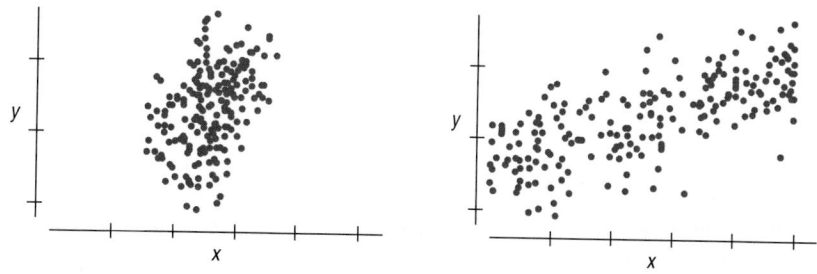

A plot like the one on the right has a broader range of x-values, so it gives a more stable base for the slope. We'd expect the slopes of samples from situations like that to vary less from sample to sample. A large standard deviation of x, s_x, provides a more stable regression.

> ▶ **Sample size.** Here we go again. What about these two?

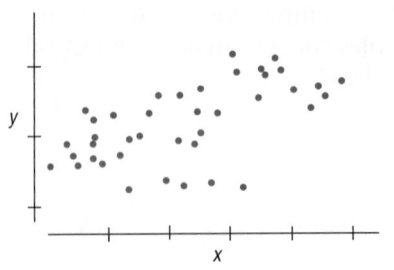 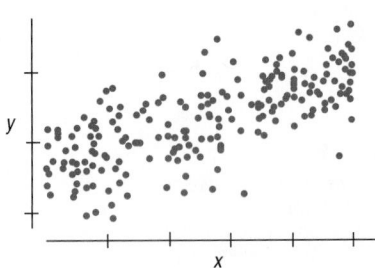

It shouldn't be a surprise that having a larger sample size, n, gives more consistent estimates from sample to sample.

Standard Error for the Slope

Three aspects of the scatterplot, then, affect the standard error of the regression slope:

- ▶ Spread around the line: s_e
- ▶ Spread of x values: s_x
- ▶ Sample size: n

These are in fact the *only* things that affect the standard error of the slope. Although you'll probably never have to calculate it by hand, the formula for the standard error is

$$SE(b_1) = \frac{s_e}{\sqrt{n-1}\, s_x}.$$

The error standard deviation, s_e, is in the *numerator*, since spread around the line *increases* the slope's standard error. The denominator has both a sample size term $\sqrt{n-1}$ and s_x, because increasing either of these *decreases* the slope's standard error.

We know the b_1's vary from sample to sample. As you'd expect, their sampling distribution model is centered at β_1, the slope of the idealized regression line. Now we can estimate its standard deviation with $SE(b_1)$. What about its shape? Here the Central Limit Theorem and "Wild Bill" Gosset come to the rescue again. When we standardize the slopes by subtracting the model mean and dividing by their standard error, we get a Student's t-model, this time with $n-2$ degrees of freedom:

$$\frac{b_1 - \beta_1}{SE(b_1)} \sim t_{n-2}.$$

A SAMPLING DISTRIBUTION FOR REGRESSION SLOPES

When the conditions are met, the standardized estimated regression slope,

$$t = \frac{b_1 - \beta_1}{SE(b_1)},$$

follows a Student's t-model with $n-2$ degrees of freedom. We estimate the standard error with

$$SE(b_1) = \frac{s_e}{\sqrt{n-1}\, s_x}, \text{ where } s_e = \sqrt{\frac{\sum (y - \hat{y})^2}{n-2}},$$

n is the number of data values, and s_x is the ordinary standard deviation of the x-values.

FOR EXAMPLE   Finding standard errors

Recap: Recent terrestrial impact craters seem to show a relationship between age and size that is linear when re-expressed using logarithms (see Chapter 10).

Here are summary statistics and regression output.

Variable	Count	Mean	StdDev
LogAge	39	−0.656310	1.57682
LogDiam	39	0.012600	1.04104

Dependent variable is: LogDiam
R-squared = 63.6%
s = 0.6362 with 39 − 2 = 37 degrees of freedom

Variable	Coefficient	Se(coeff)	t-ratio	P-value
Intercept	0.358262	0.1106	3.24	0.0025
LogAge	0.526674	0.0655	8.05	≤ 0.0001

Questions: How are the standard error of the slope and the t-ratio for the slope calculated? (And aren't you glad the software does this for you?)

$$SE(b_1) = \frac{s_e}{\sqrt{n-1} \times s_x} = \frac{0.6362}{\sqrt{39-1} \times 1.57682} = 0.0655$$

Assuming no linear association $(\beta_1 = 0)$, $t_{37} = \dfrac{b_1 - \beta_1}{SE(b_1)} = \dfrac{0.526674 - 0}{0.0655} = 8.05$

What About the Intercept?

The same reasoning applies for the intercept. We could write

$$\frac{b_0 - \beta_0}{SE(b_0)} \sim t_{n-2}$$

and use it to construct confidence intervals and test hypotheses, but often the value of the intercept isn't something we care about. The intercept usually isn't interesting. Most hypothesis tests and confidence intervals for regression are about the slope.

Regression Inference

TI-*nspire*

Regression Inference. How big must a slope be in order to be considered statistically significant? See for yourself by exploring the natural sample-to-sample variability in slopes.

Now that we have the standard error of the slope and its sampling distribution, we can test a hypothesis about it and make confidence intervals. The usual null hypothesis about the slope is that it's equal to 0. Why? Well, a slope of zero would say that y doesn't tend to change linearly when x changes—in other words, that there is no linear association between the two variables. If the slope were zero, there wouldn't be much left of our regression equation.

So a null hypothesis of a zero slope questions the entire claim of a linear relationship between the two variables—and often that's just what we want to know. In fact, every software package or calculator that does regression simply assumes that you want to test the null hypothesis that the slope is really zero.

What if the Slope Were 0?

If $b_1 = 0$, our prediction is $\hat{y} = b_0 + 0x$. The equation collapses to just $\hat{y} = b_0$. Now x is nowhere in sight, so y doesn't depend on x at all.

And b_0 would turn out to be $\bar{y}$. Why? We know that $b_0 = \bar{y} - b_1\bar{x}$, but when $b_1 = 0$, that becomes simply $b_0 = \bar{y}$. It turns out, then, that when the slope is 0, the equation is just $\hat{y} = \bar{y}$; at every value of x, we always predict the mean value for y.

To test $H_0: \beta_1 = 0$, we find

$$t_{n-2} = \frac{b_1 - 0}{SE(b_1)}.$$

This is just like every t-test we've seen: a difference between the statistic and its hypothesized value, divided by its standard error.

For our body fat data, the computer found the slope (1.7), its standard error (0.0743), and the ratio of the two: $\frac{1.7 - 0}{0.0743} = 22.9$ (see p. 656). Nearly 23 standard errors from the hypothesized value certainly seems big. The P-value (<0.0001) confirms that a t-ratio this large would be very unlikely to occur if the true slope were zero.

Maybe the standard null hypothesis isn't all that interesting here. Did you have any doubts that %Body Fat is related to Waist size? A more sensible use of these same values might be to make a confidence interval for the slope instead.

We can build a confidence interval in the usual way, as an estimate plus or minus a margin of error. As always, the margin of error is just the product of the standard error and a critical value. Here the critical value comes from the t-distribution with $n - 2$ degrees of freedom, so a 95% **confidence interval for β** is

$$b_1 \pm t^*_{n-2} \times SE(b_1).$$

For the body fat data, $t^*_{248} = 1.970$, so that comes to $1.7 \pm 1.97 \times 0.074$, or an interval from 1.55 to 1.85 %Body Fat per inch of Waist size.

FOR EXAMPLE Interpreting a regression model

Recap: On a log scale, there seems to be a linear relationship between the diameter and the age of recent terrestrial impact craters. We have regression output from statistics software:

Questions: What's the regression model, and what can it tell us?

Dependent variable is: LogDiam
R-squared = 63.6%
s = 0.6362 with 39 − 2 = 37 degrees of freedom

Variable	Coefficient	Se(coeff)	t-ratio	P-value
Intercept	0.358262	0.1106	3.24	0.0025
LogAge	0.526674	0.0655	8.05	≤0.0001

For terrestrial impact craters younger than 35 million years, the logarithm of Diameter grows linearly with the logarithm of Age: $\widehat{\log Diam} = 0.358 + 0.527 \log Age$. The P-value for each coefficient's t-statistic is very small, so I'm quite confident that neither coefficient is zero. Based on my model, I conclude that, on average, the older a crater is, the larger it tends to be. This model accounts for 63.6% of the variation in logDiam.

Although it is possible that impacts (and their craters) are getting smaller, it is more likely that I'm seeing the effects of age on craters. Small craters are probably more likely to erode or become buried or otherwise be difficult to find as they age. Larger craters may survive the huge expanses of geologic time more successfully.

JUST CHECKING

Researchers in Food Science studied how big people's mouths tend to be. They measured mouth volume by pouring water into the mouths of subjects who lay on their backs. Unless this is your idea of a good time, it would be helpful to have a model to estimate mouth volume more simply. Fortunately, mouth volume is related to height. (Mouth volume is measured in cubic centimeters and height in meters.)

The data were checked and deemed suitable for regression. Take a look at the computer output.

1. What does the *t*-ratio of 3.27 tell us about this relationship? How does the P-value help our understanding?

2. Would you say that measuring a person's height could reliably be used as a substitute for the wetter method of determining how big a person's mouth is? What numbers in the output helped you reach that conclusion?

3. What does the value of s_e add to this discussion?

Summary of	Mouth Volume
Mean	60.2704
StdDev	16.8777

Dependent variable is: Mouth Volume
R-squared = 15.3%
s = 15.66 with 61 − 2 = 59 degrees of freedom

Variable	Coefficient	SE(coeff)	t-ratio	P-value
Intercept	−44.7113	32.16	−1.39	0.1697
Height	61.3787	18.77	3.27	0.0018

Another Example

Every spring, Nenana, Alaska, hosts a contest in which participants try to guess the exact minute that a wooden tripod placed on the frozen Tanana River will fall through the breaking ice. The contest started in 1917 as a diversion for railroad engineers, with a jackpot of $800 for the closest guess. It has grown into an event in which hundreds of thousands of entrants enter their guesses on the Internet[3] and vie for as much as $300,000.

Because so much money and interest depends on the time of breakup, it has been recorded to the nearest minute with great accuracy ever since 1917. And because a standard measure of breakup has been used throughout this time, the data are consistent. An article in *Science*[4] used the data to investigate global warming—whether greenhouse gasses and other human actions have been making the planet warmer. Others might just want to make a good prediction of next year's breakup time.

Of course, we can't use regression to tell the *causes* of any change. But we can estimate the *rate* of change (if any) and use it to make better predictions.

Here are some of the data:

WHO	Years
WHAT	Year, day, and hour of ice breakup
UNITS	*x* is in years since 1900.
	y is in days after midnight Dec. 31.
WHEN	1917–present
WHERE	Nenana, Alaska
WHY	Wagering, but proposed to look at global warming

Year (since 1900)	Breakup Date (days after Jan. 1)	Year (since 1900)	Breakup Date (days after Jan. 1)
17	119.4792	30	127.7938
18	130.3979	31	129.3910
19	122.6063	32	121.4271
20	131.4479	33	127.8125
21	130.2792	34	119.5882
22	131.5556	35	134.5639
23	128.0833	36	120.5403
24	131.6319	37	131.8361
25	126.7722	38	125.8431
26	115.6688	39	118.5597
27	131.2375	40	110.6437
28	126.6840	41	122.0764
29	124.6535	⋮	⋮

---

[3] http://www.nenanaakiceclassic.com
[4] "Climate Change in Nontraditional Data Sets." *Science* 294 [26 October 2001]: 811.

STEP-BY-STEP EXAMPLE | A Regression Slope *t*-Test

The slope of the regression gives the change in Nenana ice breakup date per year.

Questions: Is there sufficient evidence to claim that ice breakup times are changing? If so, how rapid is the change?

Plan State what you want to know.

Identify the *parameter* you wish to estimate. Here our parameter is the slope.

Identify the variables and review the W's.

Hypotheses Write your null and alternative hypotheses.

Model Think about the assumptions and check the conditions.

Make pictures. Because the scatterplot seems straight enough, we can find and plot the residuals.

Usually, we check for suggestions that the Independence Assumption fails by plotting the residuals against the predicted values. Patterns and clusters in that plot raise our suspicions. But when the data are measured over time, it is always a good idea to plot residuals against time to look for trends and oscillations.

I wonder whether the date of ice breakup in Nenana has changed over time. The slope of that change might indicate climate change. I have the date of ice breakup annually since 1917, recorded as the number of days and fractions of a day until the ice breakup.

H_0: There is no change in the date of ice breakup: $\beta_1 = 0$

H_A: Yes, there is: $\beta_1 \neq 0$

✔ **Straight Enough Condition:** I have quantitative data with no obvious bend in the scatterplot.

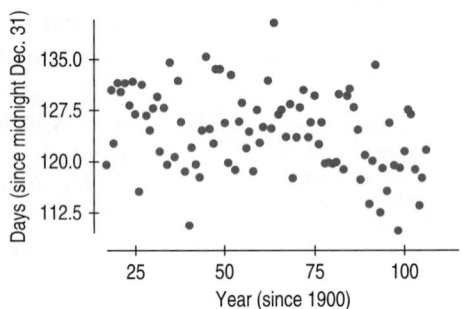

✔ **Independence Assumption:** These data are a time series, which raises my suspicions that they may not be independent. To check, here's a plot of the residuals against time, the x-variable of the regression:

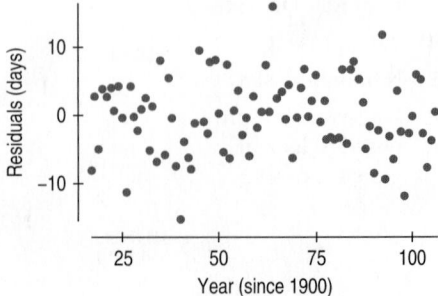

I see a hint that the data oscillate up and down, which suggests some failure of independence, but not so strongly that I can't proceed with the analysis. These data are not a random sample, so I'm reluctant to extend my conclusions beyond this river and these years.

✔ **Does the Plot Thicken? Condition:** The residuals plot shows no obvious trends in the spread.

✔ **Nearly Normal Condition, Outlier Condition:** A histogram of the residuals is unimodal and symmetric.

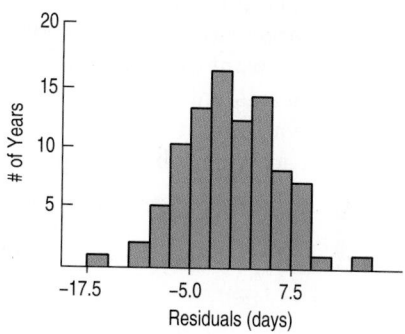

State the sampling distribution model.

Under these conditions, the sampling distribution of the regression slope can be modeled by a Student's t-model with $(n - 2) = 89$ degrees of freedom.

Choose your method.

I'll do a **regression slope t-test.**

Mechanics The regression equation can be found from the formulas in Chapter 8, but regressions are almost always found from a computer program or calculator.

The P-values given in the regression output table are from the Student's t-distribution on $(n - 2) = 89$ degrees of freedom. They are appropriate for two-sided alternatives.

Here's the computer output for this regression:

Dependent variable is: Breakup Date

R-squared = 11.3%
s = 5.673 with 91 − 2 = 89 degrees of freedom

Variable	Coeff	SE(Coeff)	t-ratio	P-value
Intercept	128.950	1.525	84.6	<0.0001
Year Since 1900	−0.07606	0.0226	−3.36	0.0012

The estimated regression equation is

$$\widehat{Date} = 128.95 - 0.076 \ YearSince1900.$$

Conclusion Link the P-value to your decision and state your conclusion in the proper context.

The P-value of 0.0012 means that the association we see in the data is unlikely to have occurred by chance. I reject the null hypothesis, and conclude that there is strong evidence that, on average, the ice breakup is occurring earlier each year. But the oscillation pattern in the residuals raises concerns.

 SHOW MORE	**Create a confidence interval for the true slope**	A 95% confidence interval for β_1 is $$b_1 \pm t^*_{89} \times SE(b_1)$$ $$-0.076 \pm (1.987)(0.0226)$$ or $(-0.12, -0.03)$ days per year.
 TELL MORE	**Interpret the interval** Simply rejecting the standard null hypothesis doesn't guarantee that the size of the effect is large enough to be important. Whether we want to know the breakup time to the nearest minute or are interested in global warming, a change measured in hours each year is big enough to be interesting.	I am 95% confident that the ice has been breaking up, on average, between 0.03 days (about 40 minutes) and 0.12 days (about 3 hours) earlier each year since 1900.

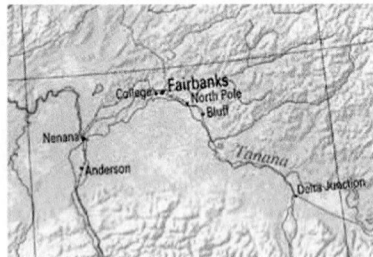

But is it global warming? So the ice is breaking up earlier. Temperatures are higher. Must be global warming, right?

Maybe.

An article challenging the original analysis of the Nenana data proposed a possible confounding variable. It noted that the city of Fairbanks is upstream from Nenana and suggested that the growth of Fairbanks could have warmed the river. So maybe it's not global warming.

Or maybe global warming is a lurking variable, leading more people to move to a now balmier Fairbanks and also leading to generally earlier ice breakup in Nenana.

Or maybe there's some other variable or combination of variables at work. We can't set up an experiment, so we may never really know.

Only one thing is for sure. When you try to explain an association by claiming cause and effect, you're bound to be on thin ice.[5]

 TI Tips   **Doing regression inference**

°F	Min
44	142.7
46	142.1
47	143.4
50	143.6
51	144.0
52	143.4
54	142.4
55	143.1
57	143.7
60	143.4
65	143.4

The TI will easily do almost everything you need for inference for regression: scatterplots, residual plots, histograms of residuals, and *t*-tests and confidence intervals for the slope of the regression line. OK, it won't tell you *SE(b)*, but it will give you enough information to easily figure it out for yourself. Not bad.

As an example we'll use data from *Chance* magazine (Vol. 12, No. 4, 1999), giving times and temperatures for 11 of the top performances in women's marathons during the 1990s. Let's examine the influence of temperature on the performance of elite runners in marathons.

---

[5] How *do* scientists sort out such messy situations? Even though they can't conduct an experiment, they *can* look for replications elsewhere. A number of studies of ice on other bodies of water have also shown earlier ice breakup times in recent years. That suggests they need an explanation that's more comprehensive than just Fairbanks and Nenana.

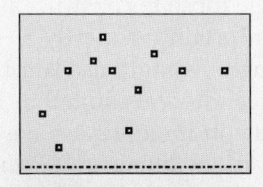

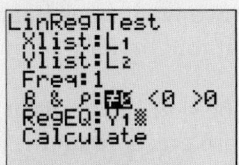

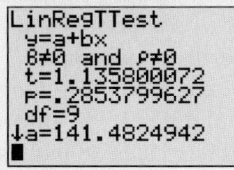

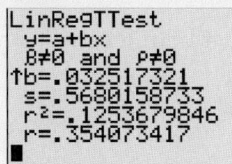

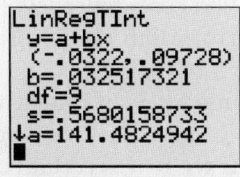

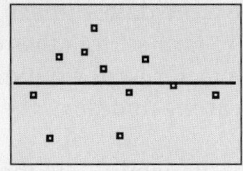

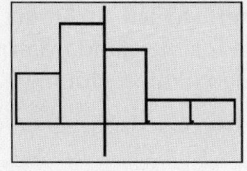

Test a Hypothesis About the Association
- Enter the temperatures (nearest degree Fahrenheit) in L1 and the runners' times (nearest tenth of a minute) in L2.
- Check the scatterplot. It's not obviously nonlinear, so go ahead.
- Under STAT TESTS choose LinRegTTest.
- Specify the two data lists (with Freq: 1).
- Choose the two-tailed option. (We are interested in whether higher temperatures enhance or interfere with a runner's performance.)
- Tell it to store the regression equation in Y1 (VARS, Y-VARS, Function . . . remember?), then Calculate.

The TI creates so much information you have to scroll down to see it all! Look what's there.

- The calculated value of t and the P-value.
- The coefficients of the regression equation, a and b.
- The value of s, our sample estimate of the common standard deviation of errors around the true line.
- The values of r^2 and r.

Wait, where's *SE(b)*? It's not there. No problem—if you need it, you can figure it out. Remember that the *t*-value is *b* divided by *SE(b)*. So *SE(b)* must be *b* divided by *t*. Here $SE(b) = 0.0325 \div 1.1358 = 0.0286$.

Create a Confidence Interval for the Slope
- Back to STAT TEST; this time you want LinRegTInt.
- The specifications for the data lists and the regression equation remain what you entered for the hypothesis test.
- Choose a confidence level, say 95%, and Calculate.

Checking Conditions
Beware!!! Before you try to interpret any of this, you must check the conditions to see if inference for regression is allowed.

- We already looked at the scatterplot; it was reasonably linear.
- To create the residuals plot, set up another scatterplot with RESID (from LIST NAMES) as your Ylist. OK, it looks fairly random.
- The residuals plot may show a slight hint of diminishing scatter, but with so few data values it's not very clear.
- The histogram of the residuals is unimodal and roughly symmetric.

What Does It All Mean?
Because the conditions check out okay, we can try to summarize what we have learned. With a P-value over 28%, it's quite possible that any perceived relationship could be just sampling error. The confidence interval suggests the slope could be positive or negative, so it's possible that as temperatures increase, women marathoners may run faster—or slower. Based on these 11 races there appears to be little evidence of a linear association between temperature and women's performances in the marathon.

*Standard Errors for Predicted Values

Once we have a useful regression, how can we indulge our natural desire to predict, without being irresponsible? We know how to compute predicted values of *y* for any value of *x*. We first did that in Chapter 8. This predicted value would be our best estimate, but it's still just an informed guess.

Now, however, we have standard errors. We can use those to construct a confidence interval for the predictions and to report our uncertainty honestly.

From our model of %*Body Fat* and *Waist* size, we might want to use *Waist* size to get a reasonable estimate of %*Body Fat*. A confidence interval can tell us how precise that prediction will be. The precision depends on the question we ask, however, and there are two questions: Do we want to know the mean %*Body Fat* for *all* men with a *Waist* size of, say, 38 inches? Or do we want to estimate the %*Body Fat* for a particular man with a 38-inch *Waist* without making him climb onto the X-ray table?

What's the difference between the two questions? The predicted %*Body Fat* is the same, but one question leads to an answer much more precise than the other. We can predict the *mean* %*Body Fat* for *all* men whose *Waist* size is 38 inches with a lot more precision than we can predict the %*Body Fat* of a *particular individual* whose *Waist* size happens to be 38 inches. Both are interesting questions.

We start with the same prediction in both cases. We are predicting the value for a new individual, one that was not part of the original data set. To emphasize this, we'll call his x-value "x sub new" and write it x_ν.[6] Here, x_ν is 38 inches. The regression equation predicts %*Body Fat* as $\hat{y}_\nu = b_0 + b_1 x_\nu$.

Now that we have the predicted value, we construct both intervals around this same number. Both intervals take the form

$$\hat{y}_\nu \pm t^*_{n-2} \times SE.$$

Even the t^* value is the same for both. It's the critical value (from Table T or technology) for $n - 2$ degrees of freedom and the specified confidence level. The intervals differ because they have different standard errors. Our choice of ruler depends on which interval we want.

The standard errors for prediction depend on the same kinds of things as the coefficients' standard errors. If there is more spread around the line, we'll be less certain when we try to predict the response. Of course, if we're less certain of the slope, we'll be less certain of our prediction. If we have more data, our estimate will be more precise. And there's one more piece: If we're farther from the center of our data, our prediction will be less precise. This last factor is new but makes intuitive sense: It's a lot easier to predict a data point near the middle of the data set than far from the center.

Each of these factors contributes uncertainty—that is, variability—to the estimate. Because the factors are independent of each other, we can add their variances to find the total variability. The resulting formula for the standard error of the predicted *mean* value explicitly takes into account each of the factors:

$$SE(\hat{\mu}_\nu) = \sqrt{SE^2(b_1) \cdot (x_\nu - \bar{x})^2 + \frac{s_e^2}{n}}.$$

Individual values vary more than means, so the standard error for a single predicted value has to be larger than the standard error for the mean. In fact, the standard error of a single predicted value has an *extra* source of variability: the variation of individuals around the predicted mean. That appears as the extra variance term, s_e^2, at the end under the square root:

$$SE(\hat{y}_\nu) = \sqrt{SE^2(b_1) \cdot (x_\nu - \bar{x})^2 + \frac{s_e^2}{n} + s_e^2}.$$

For the Nenana Ice Classic, someone who planned to place a bet would want to predict this year's breakup time. By contrast, scientists studying global warming are likely to be more interested in the mean breakup time. Unfortunately if you want to gamble, the variability is greater for predicting for a single year.

---

[6] Yes, this is a bilingual pun. The Greek letter ν is called "nu." Don't blame me; my co-author suggested this.

Keep in mind this distinction between the two kinds of confidence intervals: The narrower interval is a **confidence interval for the predicted mean value** at x_ν, and the wider interval is a **prediction interval for an individual** with that x-value.

FOR EXAMPLE   *Finding confidence intervals for predicted values

Let's use our analysis to create confidence intervals for predictions about *%Body Fat*. From the data and the regression output we know:

$$n = 250 \qquad \bar{x} = 36.3 \qquad s_e = 4.713 \qquad SE(b_1) = 0.074$$

Question 1: What's a 95% confidence interval for the mean *%Body Fat* for all men with 38-inch waists?

For $x_\nu = 38$ the regression model predicts

$$\hat{y}_\nu = -42.7 + 1.7(38) = 21.9\%.$$

The standard error is

$$SE(\hat{\mu}_\nu) = \sqrt{0.074^2(38 - 36.3)^2 + \frac{4.713^2}{250}} = 0.32\%.$$

With $250 - 2 = 248$ df, for 95% confidence $t^* = 1.97$.

Putting it all together, the 95% confidence interval is: $21.9\% \pm 1.97(0.32)$
$$21.9\% \pm 0.63\%, \text{ or } (21.27, 22.53)$$

I'm 95% confidence that the mean body fat level for all men with 38-inch waists is between 21.3% and 22.5% body fat.

Question 2: What's a 95% prediction interval for the *%Body Fat* of an individual man with a 38-inch waist?

The standard error is

$$SE(\hat{y}_\nu) = \sqrt{0.074^2(38 - 36.3)^2 + \frac{4.713^2}{250} + 4.713^2} = 4.72\%.$$

The prediction interval is: $21.9\% \pm 1.97(4.72)$
$$21.9\% \pm 9.3\%, \text{ or } (12.6, 31.2)$$

I'm 95% confident that a randomly selected man with a 38-inch waist will have between 12.6% and 31.2% body fat.

Notice how much wider this interval is than the first one. As we've known since Chapter 18, the mean is such less variable than a randomly selected individual value.

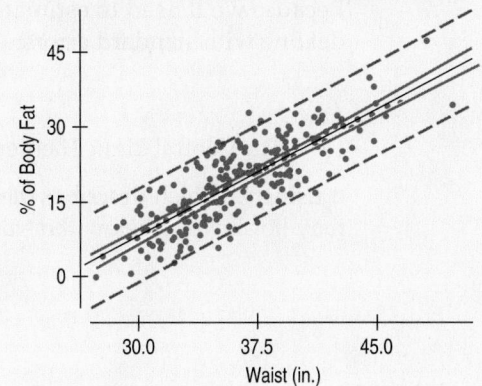

FIGURE 27.11

A scatterplot of %Body Fat vs. Waist size with a least squares regression line. The solid green lines near the regression line show the extent of the 95% confidence intervals for mean %Body Fat at each Waist size. The dashed red lines show the prediction intervals. Most of the points are contained within the prediction intervals, but not within the confidence intervals.

*MATH BOX

So where do those messy formulas for standard errors of predicted values come from? They're based on many of the ideas we've studied so far. Start with regression, add random variables, then throw in the Pythagorean Theorem, the Central Limit Theorem, and a dose of algebra. Mix well. . . .

We begin our quest with an equation of the regression line. Usually we write the line in the form $\hat{y} = b_0 + b_1 x$. Mathematicians call that the "slope-intercept" form; in your algebra class you wrote it as $y = mx + b$. In that algebra class you also learned another way to write equations of lines. When you know that a line with slope m passes through the point (x_1, y_1), the "point-slope" form of its equation is $y - y_1 = m(x - x_1)$.

We know the regression line passes through the mean-mean point $(\bar{x}, \bar{y})$ with slope b_1, so we can write its equation in point-slope form as $\hat{y} - \bar{y} = b_1(x - \bar{x})$. Solving for $\hat{y}$ yields $\hat{y} = b_1(x - \bar{x}) + \bar{y}$. This equation predicts the mean y-value for a specific x_ν:

$$\hat{\mu}_y = b_1(x_\nu - \bar{x}) + \bar{y}.$$

To create a confidence interval for the mean value we need to measure the variability in this prediction:

$$Var(\hat{\mu}_y) = Var(b_1(x_\nu - \bar{x}) + \bar{y}).$$

We now call on the Pythagorean Theorem of Statistics once more: the slope, b_1, and mean, $\bar{y}$, should be independent, so their variances add:

$$Var(\hat{\mu}_y) = Var(b_1(x_\nu - \bar{x})) + Var(\bar{y}).$$

The horizontal distance from our specific x-value to the mean, $x_\nu - \bar{x}$, is a constant:

$$Var(\hat{\mu}_y) = (Var(b_1))(x_\nu - \bar{x})^2 + Var(\bar{y}).$$

Let's write that equation in terms of standard deviations:

$$SD(\hat{\mu}_y) = \sqrt{(SD^2(b_1))(x_\nu - \bar{x})^2 + SD^2(\bar{y})}.$$

Because we'll need to estimate these standard deviations using samples statistics, we're really dealing with standard errors:

$$SE(\hat{\mu}_y) = \sqrt{(SE^2(b_1))(x_\nu - \bar{x})^2 + SE^2(\bar{y})}.$$

The Central Limit Theorem tells us that the standard deviation of $\bar{y}$ is $\frac{\sigma}{\sqrt{n}}$. Here we'll estimate σ using s_e, which describes the variability in how far the line we drew through our sample mean may lie above or below the true mean:

$$SE(\hat{\mu}_y) = \sqrt{(SE^2(b_1))(x_\nu - \bar{x})^2 + \left(\frac{s_e}{\sqrt{n}}\right)^2}$$

$$= \sqrt{(SE^2(b_1))(x_\nu - \bar{x})^2 + \frac{s_e^2}{n}}.$$

And there it is—the standard error we need to create a confidence interval for a predicted mean value.

When we try to predict an individual value of y, we must also worry about how far the true point may lie above or below the regression line. We represent that uncertainty by adding another term, e, to the original equation:

$$y = b_1(x_\nu - \bar{x}) + \bar{y} + e.$$

To make a long story short (and the equation a wee bit longer), that additional term simply adds one more standard error to the sum of the variances:

$$SE(\hat{y}) = \sqrt{(SE^2(b_1))(x_\nu - \bar{x}) + \frac{s_e^2}{n} + s_e^2}.$$

WHAT CAN GO WRONG?

In this chapter we've added inference to the regression explorations that we did in Chapters 8 and 9. Everything covered in those chapters that could go wrong with regression can still go wrong. It's probably a good time to review Chapter 9. Take your time; we'll wait.

With inference, we've put numbers on our estimates and predictions, but these numbers are only as good as the model. Here are the main things to watch out for:

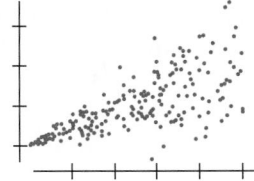

▶ **Don't fit a linear regression to data that aren't straight.** This is the most fundamental assumption. If the relationship between x and y isn't approximately linear, there's no sense in fitting a straight line to it.

▶ **Watch out for the plot thickening.** The common part of confidence and prediction intervals is the estimate of the error standard deviation, the spread around the line. If it changes with x, the estimate won't make sense. Imagine making a prediction interval for these data.

   When x is small, we can predict y precisely, but as x gets larger, it's much harder to pin y down. Unfortunately, if the spread changes, the single value of s_e won't pick that up. The prediction interval will use the average spread around the line, with the result that we'll be too pessimistic about our precision for low x-values and too optimistic for high x-values. A re-expression of y is often a good fix for changing spread.

▶ **Make sure the errors are Normal.** When we make a prediction interval for an individual, the Central Limit Theorem can't come to our rescue. For us to believe the prediction interval, the errors must be from the Normal model. Check the histogram and Normal probability plot of the residuals to see if this assumption looks reasonable.

▶ **Watch out for extrapolation.** It's tempting to think that because we have prediction *intervals*, they'll take care of all our uncertainty so we don't have to worry about extrapolating. Wrong. The interval is only as good as the model. The uncertainty our intervals predict is correct only if our model is true. There's no way to adjust for wrong models. That's why it's always dangerous to predict for x-values that lie far from the center of the data.

▶ **Watch out for influential points and outliers.** We always have to be on the lookout for a few points that have undue influence on our estimated model—and regression is certainly no exception.

▶ **Watch out for one-tailed tests.** Because tests of hypotheses about regression coefficients are usually two-tailed, software packages report two-tailed P-values. If you are using software to conduct a one-tailed test about slope, you'll need to divide the reported P-value in half.

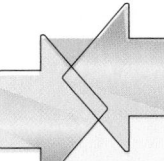

CONNECTIONS

Regression inference is connected to almost everything we've done so far. Scatterplots are essential for checking linearity and whether the plot thickens. Histograms and normal probability plots come into play to check the Nearly Normal condition. And we're still thinking about the same attributes of the data in these plots as we were back in the first part of the book.

Regression inference is also connected to just about every inference method we have seen for measured data. The assumption that the spread of data about the line is constant is essentially the same as the assumption of equal variances required for the pooled-t methods. Our use of all the residuals together to estimate their standard deviation is a form of pooling.

Inference for regression is closely related to inference for means, so your understanding of means transfers directly to your understanding of regression. Here's a table that displays the similarities:

	Means	Regression Slope
Parameter	μ	β_1
Statistic	$\bar{y}$	b_1
Population spread estimate	$s_y = \sqrt{\dfrac{\sum (y - \bar{y})^2}{n - 1}}$	$s_e = \sqrt{\dfrac{\sum (y - \hat{y})^2}{n - 2}}$
Standard error of the statistic	$SE(\bar{y}) = \dfrac{s_y}{\sqrt{n}}$	$SE(b_1) = \dfrac{s_e}{s_x \sqrt{n - 1}}$
Test statistic	$\dfrac{\bar{y} - \mu_0}{SE(\bar{y})} \sim t_{n-1}$	$\dfrac{b_1 - \beta_1}{SE(b_1)} \sim t_{n-2}$
Margin of error	$ME = t^*_{n-1} \times SE(\bar{y})$	$ME = t^*_{n-2} \times SE(b_1)$

WHAT HAVE WE LEARNED?

In Chapters 7, 8, and 9, we learned to examine the relationship between two quantitative variables in a scatterplot, to summarize its strength with correlation, and to fit linear relationships by least squares regression. And we saw that these methods are particularly powerful and effective for modeling, predicting, and understanding these relationships.

Now we have completed our study of inference methods by applying them to these regression models. We've found that the same methods we used for means—Student's t-models—work for regression in much the same way as they did for means. And we've seen that although this makes the mechanics familiar, there are new conditions to check and a need for care in describing the hypotheses we test and the confidence intervals we construct.

▶ We've learned that under certain assumptions, the sampling distribution for the slope of a regression line can be modeled by a Student's t-model with $n - 2$ degrees of freedom.

▶ We've learned to check four conditions to verify those assumptions before we proceed with inference. We've learned the importance of checking these conditions in order, and we've seen that most of the checks can be made by graphing the data and the residuals with the methods we learned in Chapters 4, 5, and 8.

▶ We've learned to use the appropriate t-model to test a hypothesis about the slope. If the slope of our regression line is significantly different from zero, we have strong evidence that there is an association between the two variables.

▶ We've also learned to create and interpret a confidence interval for the true slope.

▶ And we've been reminded yet again never to mistake the presence of an association for proof of causation.

Terms

Conditions for inference in regression (and checks for some of them)

▶ 651. **Straight Enough Condition** for linearity. (Check that the scatterplot of y against x has linear form and that the scatterplot of residuals against predicted values has no obvious pattern.)

▶ 652. **Independence Assumption.** (Think about the nature of the data. Check a residuals plot.)

▶ 652. **Does the Plot Thicken? Condition** for constant variance. (Check that the scatterplot shows consistent spread across the range of the x-variable, and that the residuals plot has constant variance, too. A common problem is increasing spread with increasing predicted values—the *plot thickens!*)

▸ 652. **Nearly Normal Condition** for Normality of the residuals. (Check a histogram of the residuals.)

Residual standard deviation

657. The spread of the data around the regression line is measured with the residual standard deviation, s_e:

$$s_e = \sqrt{\frac{\sum (y - \hat{y})^2}{n - 2}} = \sqrt{\frac{\sum e^2}{n - 2}}.$$

***t*-test for the regression slope**

658, 662. When the assumptions are satisfied, we can perform a test for the slope coefficient. We usually test the null hypothesis that the true value of the slope is zero against the alternative that it is not. A zero slope would indicate a complete absence of linear relationship between y and x.

To test $H_0: \beta_1 = 0$, we find

$$t = \frac{b_1 - 0}{SE(b_1)}$$

where

$$SE(b_1) = \frac{s_e}{\sqrt{n - 1}\, s_x}, \quad s_e = \sqrt{\frac{\sum (y - \hat{y})^2}{n - 2}},$$

n is the number of cases, and s_x is the standard deviation of the x-values. We find the P-value from the Student's t-model with $n - 2$ degrees of freedom.

Confidence interval for the regression slope (β)

660. When the assumptions are satisfied, we can find a confidence interval for the slope parameter from $b_1 \pm t^*_{n-2} \times SE(b_1)$. The critical value, t^*_{n-2}, depends on the confidence level specified and on Student's t-model with $n - 2$ degrees of freedom.

Skills

▸ Understand that the "true" regression line does not fit the population data perfectly, but rather is an idealized summary of that data.

▸ Know how to examine your data and a scatterplot of y vs. x for violations of assumptions that would make inference for regression unwise or invalid.

▸ Know how to examine displays of the residuals from a regression to double-check that the conditions required for regression have been met. In particular, know how to judge linearity and constant variance from a scatterplot of residuals against predicted values. Know how to judge Normality from a histogram and Normal probability plot.

▸ Remember to be especially careful to check for failures of the Independence Assumption when working with data recorded over time. To search for patterns, examine scatterplots both of x against time and of the residuals against time.

▸ Know how to test the standard hypothesis that the true regression slope is zero. Be able to state the null and alternative hypotheses. Know where to find the relevant numbers in standard computer regression output.

▸ Be able to find a confidence interval for the slope of a regression based on the values reported in a standard regression output table.

▸ Be able to summarize a regression in words. In particular, be able to state the meaning of the true regression slope, the standard error of the estimated slope, and the standard deviation of the errors.

▸ Be able to interpret the P-value of the t-statistic for the slope to test the standard null hypothesis.

▸ Be able to interpret a confidence interval for the slope of a regression.

REGRESSION ANALYSIS ON THE COMPUTER

All statistics packages make a table of results for a regression. These tables differ slightly from one package to another, but all are essentially the same. We've seen two examples of such tables already.

All packages offer analyses of the residuals. With some, you must request plots of the residuals as you request the regression. Others let you find the regression first and then analyze the residuals afterward. Either way, your analysis is not complete if you don't check the residuals with a histogram or Normal probability plot and a scatterplot of the residuals against x or the predicted values.

You should, of course, always look at the scatterplot of your two variables before computing a regression.

Regressions are almost always found with a computer or calculator. The calculations are too long to do conveniently by hand for data sets of any reasonable size. No matter how the regression is computed, the results are usually presented in a table that has a standard form. Here's a portion of a typical regression results table, along with annotations showing where the numbers come from:

> **A S**   **Activity: Regression on the Computer.** How fast is the universe expanding? And how old is it? A prominent astronomer used regression to astound the scientific community. Read the story, analyze the data, and interactively learn about each of the numbers in a typical computer regression output table.

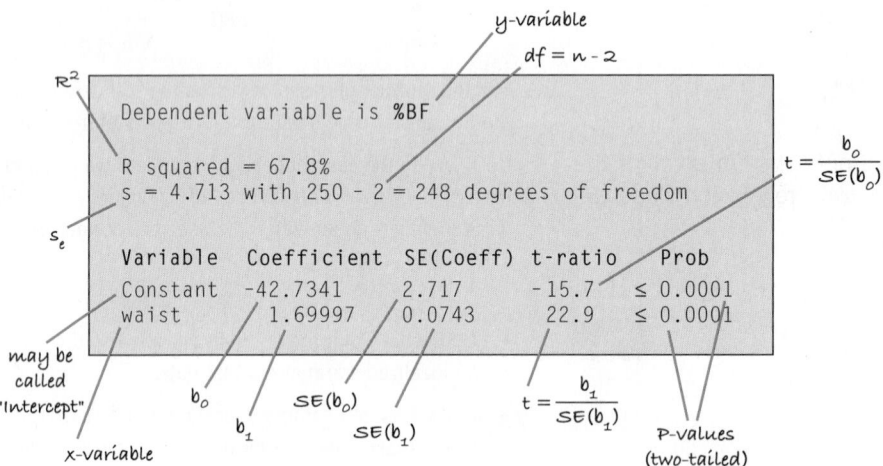

The regression table gives the coefficients (once you find them in the middle of all this other information), so we can see that the regression equation is

$$\widehat{\%BF} = -42.73 + 1.7\ Waist$$

and that the R^2 for the regression is 67.8%. (Is accounting for 68% of the variation in %Body Fat good enough to be useful? Is a prediction ME of more than 9% good enough? Health professionals might not be satisfied.)

The column of t-ratios gives the test statistics for the respective null hypotheses that the true values of the coefficients are zero. The corresponding P-values are also usually reported.

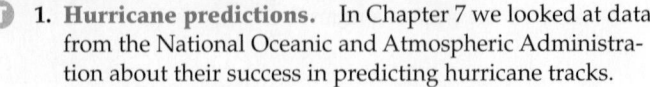

1. **Hurricane predictions.** In Chapter 7 we looked at data from the National Oceanic and Atmospheric Administration about their success in predicting hurricane tracks.

Here is a scatterplot of the error (in nautical miles) for predicting hurricane locations 72 hours in the future vs. the year in which the prediction (and the hurricane) occurred:

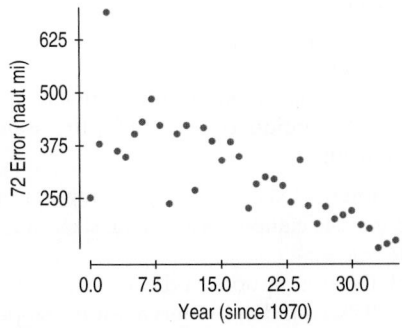

In Chapter 7 we could describe this relationship only in general terms. Now we can learn more. Here is the regression analysis:

Dependent variable is: 72Error
R squared = 58.5%
s = 75.38 with 36 − 2 = 34 degrees of freedom

Variable	Coefficient	SE(Coeff)	t-ratio	P-value
Intercept	453.223	24.61	18.4	≤0.0001
Year since 1970	−8.37084	1.209	−6.92	≤0.0001

a) Explain in context what the regression says.
b) State the hypothesis about the slope (both numerically and in words) that describes how hurricane prediction quality has changed.
c) Assuming that the assumptions for inference are satisfied, perform the hypothesis test and state your conclusion in context.
d) Explain what R-squared means in context.

2. Drug use. The *European School Study Project on Alcohol and Other Drugs*, published in 1995, investigated the use of marijuana and other drugs. Data from 11 countries are summarized in the following scatterplot and regression analysis. They show the association between the percentage of a country's ninth graders who report having smoked marijuana and who have used other drugs such as LSD, amphetamines, and cocaine.

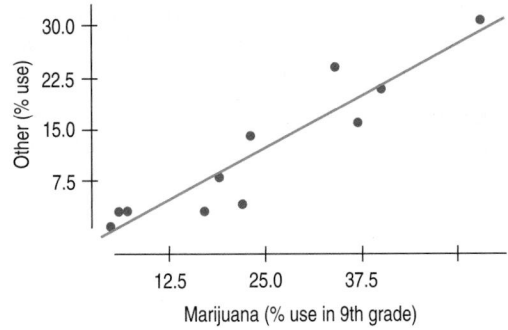

Dependent variable is: Other
R-squared = 87.3%
s = 3.853 with 11 − 2 = 9 degrees of freedom

Variable	Coefficient	SE(Coeff)	t-ratio	P-value
Intercept	−3.06780	2.204	−1.39	0.1974
Marijuana	0.615003	0.0784	7.85	<0.0001

a) Explain in context what the regression says.
b) State the hypothesis about the slope (both numerically and in words) that describes how use of marijuana is associated with other drugs.
c) Assuming that the assumptions for inference are satisfied, perform the hypothesis test and state your conclusion in context.
d) Explain what *R*-squared means in context.
e) Do these results indicate that marijuana use leads to the use of harder drugs? Explain.

3. Movie budgets. How does the cost of a movie depend on its length? Data on the cost (millions of dollars) and the running time (minutes) for major release films of 2005 are summarized in these plots and computer output:

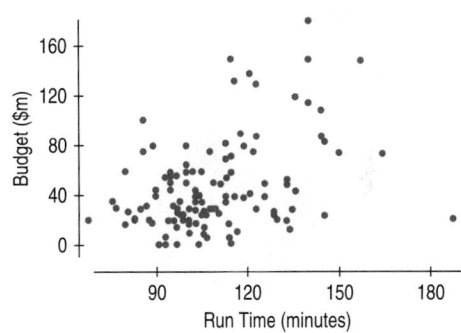

Dependent variable is: Budget($M)
R squared = 15.4%
s = 32.95 with 120 − 2 = 118 degrees of freedom

Variable	Coefficient	SE(Coeff)	t-ratio	P-value
Intercept	−31.3869	17.12	−1.83	0.0693
Run Time	0.714400	0.1541	4.64	≤0.0001

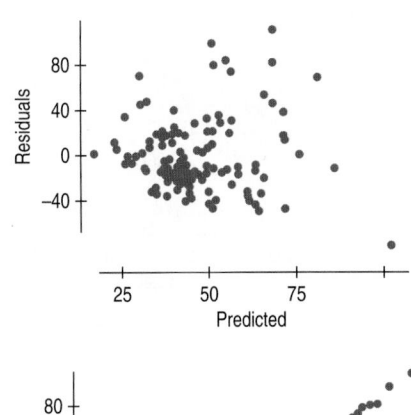

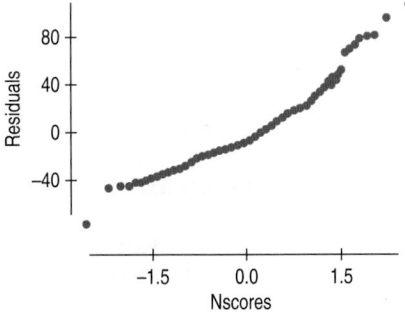

a) Explain in context what the regression says.
b) The intercept is negative. Discuss its value, taking note of the P-value.
c) The output reports $s = 32.95$. Explain what that means in this context.
d) What's the value of the standard error of the slope of the regression line?
e) Explain what that means in this context.

4. House prices. How does the price of a house depend on its size? Data from Saratoga, New York, on 1064 randomly selected houses that had been sold include data on price ($1000's) and size (1000's ft^2), producing the following graphs and computer output:

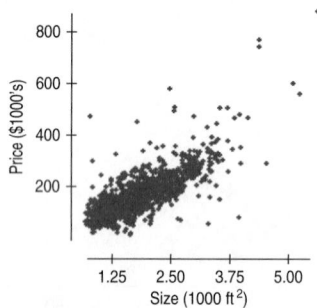

Dependent variable is: Price
R squared = 59.5%
$s = 53.79$ with $1064 - 2 = 1062$ degrees of freedom

Variable	Coefficient	SE(Coeff)	t-ratio	P-value
Intercept	-3.11686	4.688	-0.665	0.5063
Size	94.4539	2.393	39.5	≤0.0001

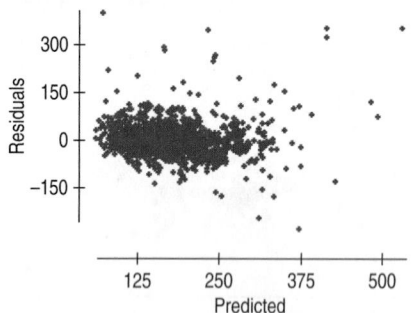

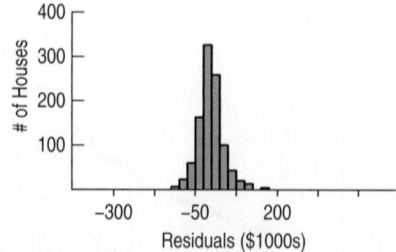

a) Explain in context what the regression says.
b) The intercept is negative. Discuss its value, taking note of its P-value.
c) The output reports $s = 53.79$. Explain what that means in this context.
d) What's the value of the standard error of the slope of the regression line?
e) Explain what that means in this context.

5. Movie budgets: the sequel. Exercise 3 shows computer output examining the association between the length of a movie and its cost.
a) Check the assumptions and conditions for inference.
b) Find a 95% confidence interval for the slope and interpret it in context.

6. Second home. Exercise 4 shows computer output examining the association between the sizes of houses and their sale prices.
a) Check the assumptions and conditions for inference.
b) Find a 95% confidence interval for the slope and interpret it in context.

7. Hot dogs. Healthy eating probably doesn't include hot dogs, but if you are going to have one, you'd probably hope it's low in both calories and sodium. In its July 2007 issue, *Consumer Reports* listed the number of calories and sodium content (in milligrams) for 13 brands of all-beef hot dogs it tested. Examine the association, assuming that the data satisfy the conditions for inference.

Dependent variable is: Sodium
R squared = 60.5%
$s = 59.66$ with $13 - 2 = 11$ degrees of freedom

Variable	Coefficient	SE(Coeff)	t-ratio	P-value
Constant	90.9783	77.69	1.17	0.2663
Calories	2.29959	0.5607	4.10	0.0018

a) State the appropriate hypotheses about the slope.
b) Test your hypotheses and state your conclusion in the proper context.

8. Cholesterol 2007. Does a person's cholesterol level tend to change with age? Data collected from 1406 adults aged 45 to 62 produced the regression analysis shown. Assuming that the data satisfy the conditions for inference, examine the association between age and cholesterol level.

Dependent variable is: Chol
$s = 46.16$

Variable	Coefficient	SE(Coeff)	t-ratio	P-value
Intercept	194.232	13.55	14.3	≤0.0001
Age	0.771639	0.2574	3.00	0.0056

a) State the appropriate hypothesis for the slope.
b) Test your hypothesis and state your conclusion in the proper context.

9. Second frank. Look again at Exercise 7's regression output for the calorie and sodium content of hot dogs.
a) The output reports $s = 59.66$. Explain what that means in this context.
b) What's the value of the standard error of the slope of the regression line?
c) Explain what that means in this context.

10. More cholesterol. Look again at Exercise 8's regression output for age and cholesterol level.
a) The output reports $s = 46.16$. Explain what that means in this context.
b) What's the value of the standard error of the slope of the regression line?
c) Explain what that means in this context.

11. Last dog. Based on the regression output seen in Exercise 7, create a 95% confidence interval for the slope of the regression line and interpret your interval in context.

12. Cholesterol, finis. Based on the regression output seen in Exercise 8, create a 95% confidence interval for the slope of the regression line and interpret it in context.

13. Marriage age 2003. The scatterplot suggests a decrease in the difference in ages at first marriage for men and women since 1975. We want to examine the regression to see if this decrease is significant.

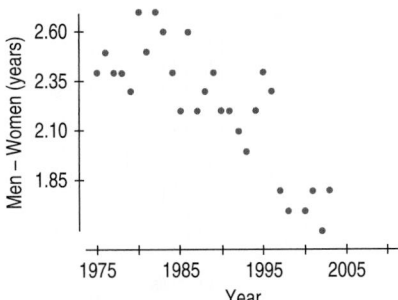

Dependent variable is: Men − Women
R squared = 65.6%
s = 0.1869 with 28 − 2 = 26 degrees of freedom

Variable	Coefficient	SE(Coeff)	t-ratio	P-value
Intercept	61.8067	8.468	7.30	≤0.0001
Year	−0.02996	0.0043	−7.04	≤0.0001

a) Write appropriate hypotheses.
b) Here are the residuals plot and a histogram of the residuals. Do you think the conditions for inference are satisfied? Explain.

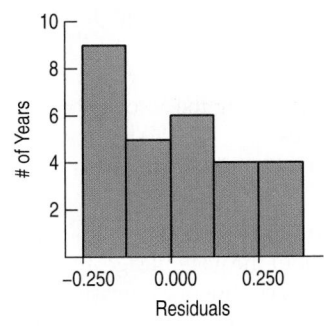

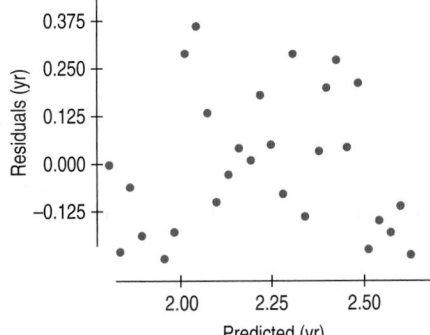

c) Test the hypothesis and state your conclusion about the trend in age at first marriage.

14. Used cars 2007. Classified ads in a newspaper offered several used Toyota Corollas for sale. Listed below are the ages of the cars and the advertised prices.

Age (yr)	Advertised Price ($)	Age (yr)	Advertised Price ($)
1	13990	7	6950
1	13495	7	7850
3	12999	8	6999
4	9500	8	5995
4	10495	10	4950
5	8995	10	4495
5	9495	13	2850
6	6999		

a) Make a scatterplot for these data.
b) Do you think a linear model is appropriate? Explain.
c) Find the equation of the regression line.
d) Check the residuals to see if the conditions for inference are met.

15. Marriage age 2003, again. Based on the analysis of marriage ages since 1975 given in Exercise 13, give a 95% confidence interval for the rate at which the age gap is closing. Explain what your confidence interval means.

16. Used cars 2007, again. Based on the analysis of used car prices you did for Exercise 14, create a 95% confidence interval for the slope of the regression line and explain what your interval means in context.

17. Fuel economy. A consumer organization has reported test data for 50 car models. We will examine the association between the weight of the car (in thousands of pounds) and the fuel efficiency (in miles per gallon). Here are the scatterplot, summary statistics, and regression analysis:

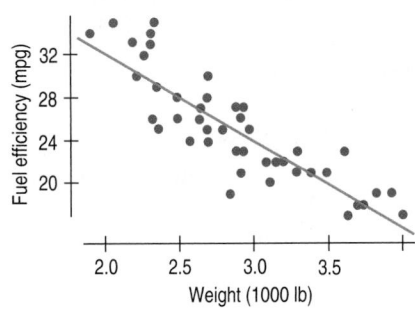

Variable	Count	Mean	StdDev
MPG	50	25.0200	4.83394
wt/1000	50	2.88780	0.511656

Dependent variable is: MPG
R-squared = 75.6%
s = 2.413 with 50 − 2 = 48 df

Variable	Coefficient	SE(Coeff)	t-ratio	P-value
Intercept	48.7393	1.976	24.7	≤0.0001
Weight	−8.21362	0.6738	−12.2	≤0.0001

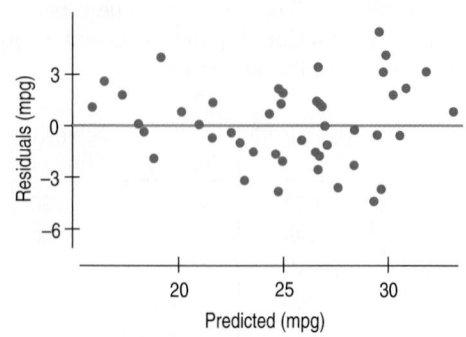

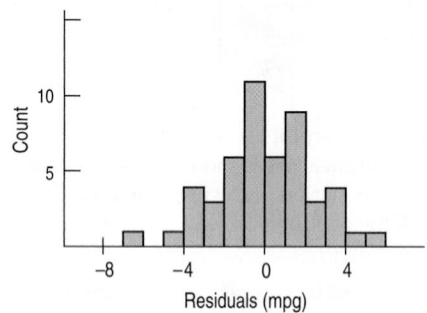

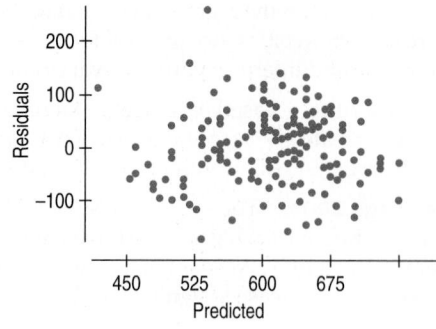

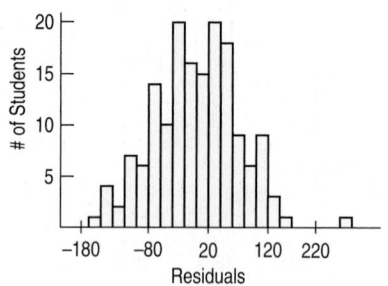

a) Is there strong evidence of an association between the weight of a car and its gas mileage? Write an appropriate hypothesis.
b) Are the assumptions for regression satisfied?
c) Test your hypothesis and state your conclusion.

18. SAT scores. How strong was the association between student scores on the Math and Verbal sections of the old SAT? Scores on each ranged from 200 to 800 and were widely used by college admissions offices. Here are summaries and plots of the scores for a graduating class at Ithaca High School:

Variable	Count	Mean	Median	StdDev	Range	IntQRange
Verbal	162	596.296	610	99.5199	490	140
Math	162	612.099	630	98.1343	440	150

Dependent variable is: Math
R-squared = 46.9%
s = 71.75 with 162 − 2 = 160 df

Variable	Coefficient	SE(Coeff)	t-ratio	P-value
Intercept	209.554	34.35	6.10	≤0.0001
Verbal	0.675075	0.0568	11.9	≤0.0001

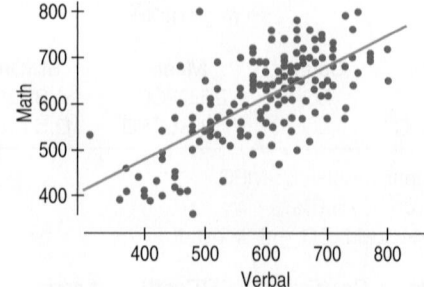

a) Is there evidence of an association between Math and Verbal scores? Write an appropriate hypothesis.
b) Discuss the assumptions for inference.
c) Test your hypothesis and state an appropriate conclusion.

19. Fuel economy, part II. Consider again the data in Exercise 17 about the gas mileage and weights of cars.
a) Create a 95% confidence interval for the slope of the regression line.
b) Explain in this context what your confidence interval means.

20. SATs, part II. Consider the high school SAT scores data from Exercise 18.
a) Find a 90% confidence interval for the slope of the true line describing the association between Math and Verbal scores.
b) Explain in this context what your confidence interval means.

21. *Fuel economy, part III. Consider again the data in Exercise 17 about the gas mileage and weights of cars.
a) Create a 95% confidence interval for the average fuel efficiency among cars weighing 2500 pounds, and explain what your interval means.
b) Create a 95% prediction interval for the gas mileage you might get driving your new 3450-pound SUV, and explain what that interval means.

22. *SATs again. Consider the high school SAT scores data from Exercise 18 once more.
a) Find a 90% confidence interval for the mean SAT-Math score for all students with an SAT-Verbal score of 500.
b) Find a 90% prediction interval for the Math score of the senior class president if you know she scored 710 on the Verbal section.

23. **Cereal.** A healthy cereal should be low in both calories and sodium. Data for 77 cereals were examined and judged acceptable for inference. The 77 cereals had between 50 and 160 calories per serving and between 0 and 320 mg of sodium per serving. Here's the regression analysis:

Dependent variable is: Sodium
R-squared = 9.0%
s = 80.49 with 77 − 2 = 75 degrees of freedom

Variable	Coefficient	SE(Coeff)	t-ratio	P-value
Intercept	21.4143	51.47	0.416	0.6786
Calories	1.29357	0.4738	2.73	0.0079

a) Is there an association between the number of calories and the sodium content of cereals? Explain.
b) Do you think this association is strong enough to be useful? Explain.

24. **Brain size.** Does your IQ depend on the size of your brain? A group of female college students took a test that measured their verbal IQs and also underwent an MRI scan to measure the size of their brains (in 1000s of pixels). The scatterplot and regression analysis are shown, and the assumptions for inference were satisfied.

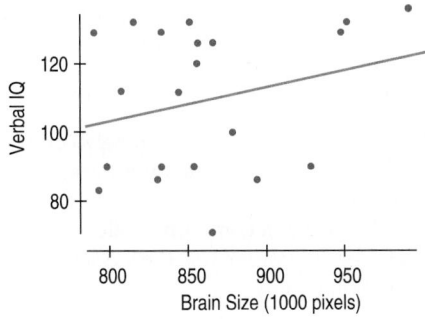

Dependent variable is: IQ_Verbal
R-squared = 6.5%

Variable	Coefficient	SE(Coeff)
Intercept	24.1835	76.38
Size	0.098842	0.0884

a) Test an appropriate hypothesis about the association between brain size and IQ.
b) State your conclusion about the strength of this association.

25. **Another bowl.** Further analysis of the data for the breakfast cereals in Exercise 23 looked for an association between *Fiber* content and *Calories* by attempting to construct a linear model. Here are several graphs. Which of the assumptions for inference are violated? Explain.

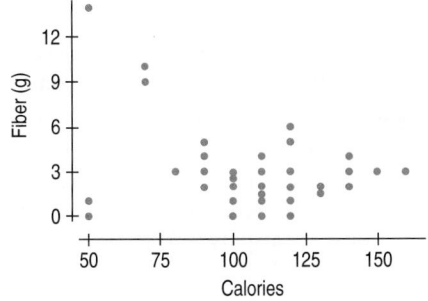

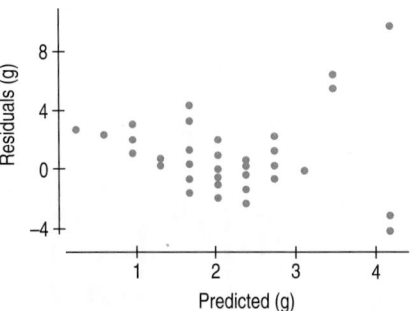

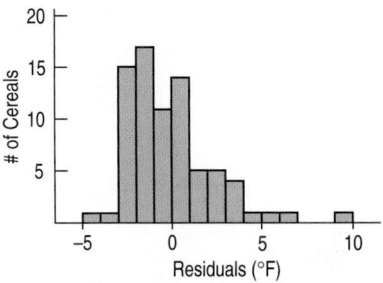

26. **Winter.** The output shows an attempt to model the association between average *January Temperature* (in degrees Fahrenheit) and *Latitude* (in degrees north of the equator) for 59 U.S. cities. Which of the assumptions for inference do you think are violated? Explain.

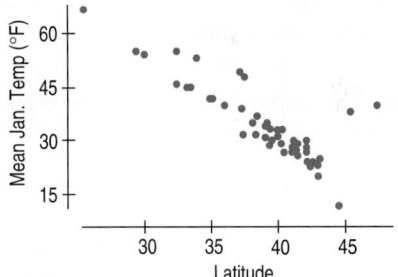

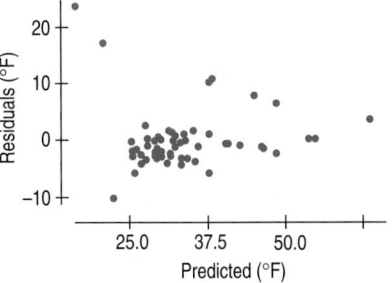

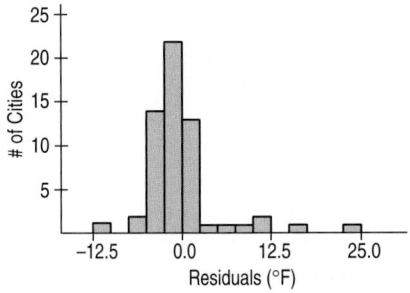

T **27. Acid rain.** Biologists studying the effects of acid rain on
wildlife collected data from 163 streams in the Adiron-
dack Mountains. They recorded the *pH* (acidity) of the
water and the *BCI*, a measure of biological diversity.
Here's a scatterplot of *BCI* against *pH*:

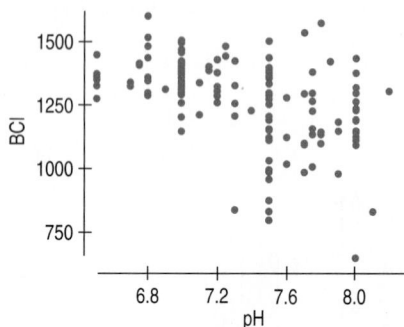

And here is part of the regression analysis:

Dependent variable is: BCI
R-squared = 27.1%
s = 140.4 with 163 − 2 = 161 degrees of freedom

Variable	Coefficient	SE(Coeff)
Intercept	2733.37	187.9
pH	−197.694	25.57

a) State the null and alternative hypotheses under
   investigation.
b) Assuming that the assumptions for regression inference are reasonable, find the t- and P-values.
c) State your conclusion.

T **28. El Niño.** Concern over the weather associated with El
Niño has increased interest in the possibility that the climate on earth is getting warmer. The most common theory relates an increase in atmospheric levels of carbon
dioxide (CO_2), a greenhouse gas, to increases in temperature. Here is part of a regression analysis of the mean
annual CO_2 concentration in the atmosphere, measured
in parts per million (ppm), at the top of Mauna Loa in
Hawaii and the mean annual air temperature over both
land and sea across the globe, in degrees Celsius. The
scatterplots and residuals plots indicated that the data
were appropriate for inference.

Dependent variable is: Temp
R-squared = 33.4%
s = 0.0809 with 37 − 2 = 35 degrees of freedom

Variable	Coefficient	SE(Coeff)
Intercept	15.3066	0.3139
CO2	0.004	0.0009

a) Write the equation of the regression line.
b) Is there evidence of an association between CO_2 level
   and global temperature?
c) Do you think predictions made by this regression will
   be very accurate? Explain.

29. Ozone. The Environmental Protection Agency is examining the relationship between the ozone level (in parts
per million) and the population (in millions) of U.S. cities.
Part of the regression analysis is shown.

Dependent variable is: Ozone
R-squared = 84.4%
s = 5.454 with 16 − 2 = 14 df

Variable	Coefficient	SE(Coeff)
Intercept	18.892	2.395
Pop	6.650	1.910

a) We suspect that the greater the population of a city,
   the higher its ozone level. Is the relationship significant? Assuming the conditions for inference are
   satisfied, test an appropriate hypothesis and state
   your conclusion in context.
b) Do you think that the population of a city is a useful
   predictor of ozone level? Use the values of both R^2
   and *s* in your explanation.

T **30. Sales and profits.** A business analyst was interested in
the relationship between a company's sales and its profits. She collected data (in millions of dollars) from a random sample of Fortune 500 companies and created the
regression analysis and summary statistics shown. The
assumptions for regression inference appeared to be
satisfied.

	Profits	Sales		Dependent variable is: Profits		
Count	79	79		R-squared = 66.2%	s = 466.2	
Mean	209.839	4178.29		**Variable**	**Coefficient**	**SE(Coeff)**
Variance	635,172	49,163,000		Intercept	−176.644	61.16
Std Dev	796.977	7011.63		Sales	0.092498	0.0075

a) Is there a significant association between sales and
   profits? Test an appropriate hypothesis and state your
   conclusion in context.
b) Do you think that a company's sales serve as a useful
   predictor of its profits? Use the values of both R^2 and *s*
   in your explanation.

31. Ozone, again. Consider again the relationship between
the population and ozone level of U.S. cities that you analyzed in Exercise 29.
a) Give a 90% confidence interval for the approximate increase in ozone level associated with each additional
   million city inhabitants.
*b) For the cities studied, the mean population was 1.7
   million people. The population of Boston is approximately 0.6 million people. Predict the mean ozone
   level for cities of that size with an interval in which
   you have 90% confidence.

T **32. More sales and profits.** Consider again the relationship between the sales and profits of Fortune 500 companies that you analyzed in Exercise 30.
a) Find a 95% confidence interval for the slope of the
   regression line. Interpret your interval in context.
*b) Last year the drug manufacturer Eli Lilly, Inc., re-
   ported gross sales of $9 billion (that's $9,000 million).
   Create a 95% prediction interval for the company's
   profits, and interpret your interval in context.

T **33. Start the car!** In October 2002, *Consumer Reports* listed
the price (in dollars) and power (in cold cranking amps)
of auto batteries. We want to know if more expensive batteries are generally better in terms of starting power. Here
are several software displays:

Dependent variable is: Power
R-squared = 25.2%
s = 116.0 with 33 − 2 = 31 degrees of freedom

Variable	Coefficient	SE(Coeff)	t-ratio	P-value
Intercept	384.594	93.55	4.11	0.0003
Cost	4.14649	1.282	3.23	0.0029

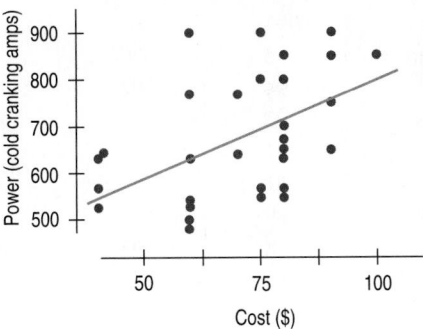

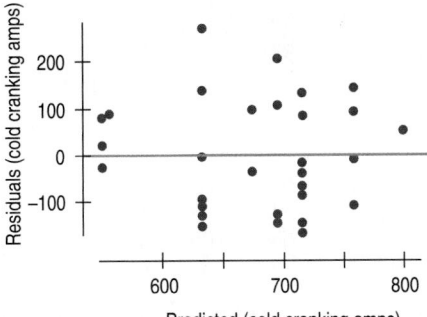

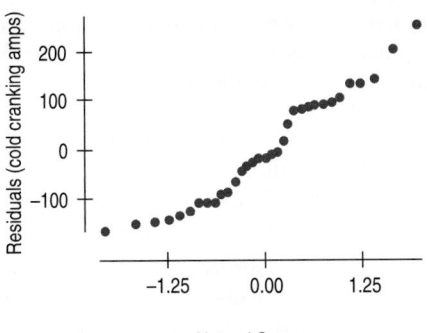

a) How many batteries were tested?
b) Are the conditions for inference satisfied? Explain.
c) Is there evidence of an association between the cost and cranking power of auto batteries? Test an appropriate hypothesis and state your conclusion.
d) Is the association strong? Explain.
e) What is the equation of the regression line?
f) Create a 90% confidence interval for the slope of the true line.
g) Interpret your interval in this context.

T 34. Crawling. Researchers at the University of Denver Infant Study Center wondered whether temperature might influence the age at which babies learn to crawl. Perhaps the extra clothing that babies wear in cold weather would restrict movement and delay the age at which they started crawling. Data were collected on 208 boys and 206 girls. Parents reported the month of the baby's birth and the age (in weeks) at which their child first crawled. The table gives the average *Temperature* (°F) when the babies were 6 months old and average *Crawling Age* (in weeks) for each month of the year. Make the plots and compute the analyses necessary to answer the following questions.

Birth Month	6-Month Temperature	Average Crawling Age
Jan.	66	29.84
Feb.	73	30.52
Mar.	72	29.70
April	63	31.84
May	52	28.58
June	39	31.44
July	33	33.64
Aug.	30	32.82
Sept.	33	33.83
Oct.	37	33.35
Nov.	48	33.38
Dec.	57	32.32

a) Would this association appear to be weaker, stronger, or the same if data had been plotted for individual babies instead of using monthly averages? Explain.
b) Is there evidence of an association between *Temperature* and *Crawling Age*? Test an appropriate hypothesis and state your conclusion. Don't forget to check the assumptions.
c) Create and interpret a 95% confidence interval for the slope of the true relationship.

T 35. Body fat. Do the data shown in the table below indicate an association between *Waist* size and *%Body Fat*?
a) Test an appropriate hypothesis and state your conclusion.
*b) Give a 95% confidence interval for the mean *%Body Fat* found in people with 40-inch *Waists*.

Waist (in.)	Weight (lb)	Body Fat (%)	Waist (in.)	Weight (lb)	Body Fat (%)
32	175	6	33	188	10
36	181	21	40	240	20
38	200	15	36	175	22
33	159	6	32	168	9
39	196	22	44	246	38
40	192	31	33	160	10
41	205	32	41	215	27
35	173	21	34	159	12
38	187	25	34	146	10
38	188	30	44	219	28

36. **Body fat, again.** Use the data from Exercise 35 to examine the association between *Weight* and *%Body Fat*.
    a) Find a 90% confidence interval for the slope of the regression line of *%Body Fat* on *Weight*.
    b) Interpret your interval in context.
    *c) Give a 95% prediction interval for the *%Body Fat* of an individual who weighs 165 pounds.

37. **Grades.** The data set below shows midterm scores from an Introductory Statistics course.

First Name	Midterm 1	Midterm 2	Homework
Timothy	82	30	61
Karen	96	68	72
Verena	57	82	69
Jonathan	89	92	84
Elizabeth	88	86	84
Patrick	93	81	71
Julia	90	83	79
Thomas	83	21	51
Marshall	59	62	58
Justin	89	57	79
Alexandra	83	86	78
Christopher	95	75	77
Justin	81	66	66
Miguel	86	63	74
Brian	81	86	76
Gregory	81	87	75
Kristina	98	96	84
Timothy	50	27	20
Jason	91	83	71
Whitney	87	89	85
Alexis	90	91	68
Nicholas	95	82	68
Amandeep	91	37	54
Irena	93	81	82
Yvon	88	66	82
Sara	99	90	77
Annie	89	92	68
Benjamin	87	62	72
David	92	66	78
Josef	62	43	56
Rebecca	93	87	80
Joshua	95	93	87
Ian	93	65	66
Katharine	92	98	77
Emily	91	95	83
Brian	92	80	82
Shad	61	58	65
Michael	55	65	51
Israel	76	88	67
Iris	63	62	67

First Name	Midterm 1	Midterm 2	Homework
Mark	89	66	72
Peter	91	42	66
Catherine	90	85	78
Christina	75	62	72
Enrique	75	46	72
Sarah	91	65	77
Thomas	84	70	70
Sonya	94	92	81
Michael	93	78	72
Wesley	91	58	66
Mark	91	61	79
Adam	89	86	62
Jared	98	92	83
Michael	96	51	83
Kathryn	95	95	87
Nicole	98	89	77
Wayne	89	79	44
Elizabeth	93	89	73
John	74	64	72
Valentin	97	96	80
David	94	90	88
Marc	81	89	62
Samuel	94	85	76
Brooke	92	90	86

a) Fit a model predicting the second midterm score from the first.
b) Comment on the model you found, including a discussion of the assumptions and conditions for regression. Is the coefficient for the slope statistically significant?
c) A student comments that because the P-value for the slope is very small, Midterm 2 is very well predicted from Midterm 1. So, he reasons, next term the professor can give just one midterm. What do you think?

38. **Grades?** The professor teaching the Introductory Statistics class discussed in Exercise 37 wonders whether performance on homework can accurately predict midterm scores.
    a) To investigate it, she fits a regression of the sum of the two midterms scores on homework scores. Fit the regression model.
    b) Comment on the model including a discussion of the assumptions and conditions for regression. Is the coefficient for the slope "statistically significant"?
    c) Do you think she can accurately judge a student's performance without giving the midterms? Explain.

39. **Strike two.** Remember the Little League instructional video discussed in Chapter 25? Ads claimed it would improve the performances of Little League pitchers. To test this claim, 20 Little Leaguers threw 50 pitches each,

and we recorded the number of strikes. After the players participated in the training program, we repeated the test. The table shows the number of strikes each player threw before and after the training. A test of paired differences failed to show that this training improves ability to throw strikes. Is there any evidence that the effectiveness of the video (*After – Before*) depends on the player's initial ability to throw strikes (*Before*)? Test an appropriate hypothesis and state your conclusion. Propose an explanation for what you find.

Number of Strikes (out of 50)			
Before	**After**	**Before**	**After**
28	35	33	33
29	36	33	35
30	32	34	32
32	28	34	30
32	30	34	33
32	31	35	34
32	32	36	37
32	34	36	33
32	35	37	35
33	36	37	32

T 40. All the efficiency money can buy. A sample of 84 model-2004 cars from an online information service was examined to see how fuel efficiency (as highway mpg) relates to the cost (Manufacturer's Suggested Retail Price in dollars) of cars. Here are displays and computer output:

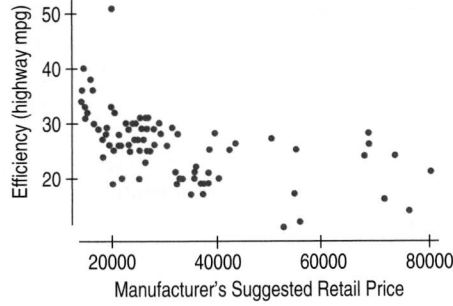

Dependent variable is: Highway MPG
R squared = 30.1%
s = 5.298 with 84 − 2 = 82 degrees of freedom

Variable	Coefficient	SE(Coeff)	t-ratio	P-value
Constant	33.0581	1.299	25.5	≤0.0001
MSRP	−2.16543e-4	0.0000	−5.95	≤0.0001

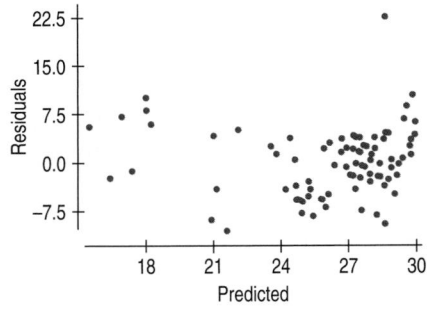

a) State what you want to know, identify the variables, and give the appropriate hypotheses.
b) Check the assumptions and conditions.
c) If the conditions are met, complete the analysis.

T 41. Education and mortality. The software output below is based on the mortality rate (deaths per 100,000 people) and the education level (average number of years in school) for 58 U.S. cities.

Variable	Count	Mean	StdDev
Mortality	58	942.501	61.8490
Education	58	11.0328	0.793480

Dependent variable is: Mortality
R-squared = 41.0%
s = 47.92 with 58 − 2 = 56 degrees of freedom

Variable	Coefficient	SE(Coeff)
Intercept	1493.26	88.48
Education	−49.9202	8.000

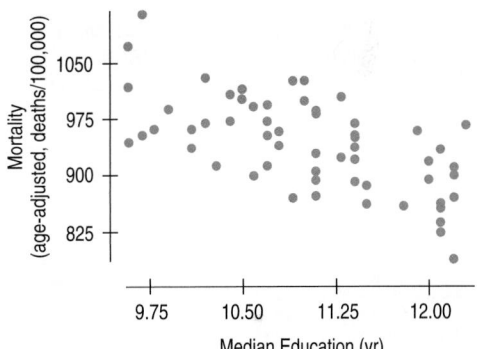

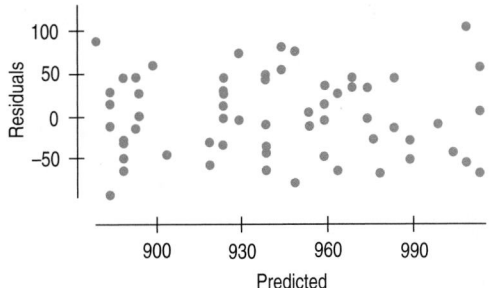

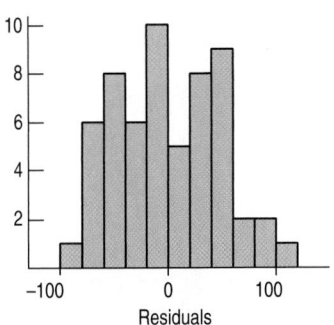

a) Comment on the assumptions for inference.
b) Is there evidence of a strong association between the level of *Education* in a city and the *Mortality* rate? Test an appropriate hypothesis and state your conclusion.

c) Can we conclude that getting more education is likely (on average) to prolong your life? Why or why not?
d) Find a 95% confidence interval for the slope of the true relationship.
e) Explain what your interval means.
*f) Find a 95% confidence interval for the average *Mortality* rate in cities where the adult population completed an average of 12 years of school.

42. Property assessments. The software outputs below provide information about the *Size* (in square feet) of 18 homes in Ithaca, New York, and the city's assessed *Value* of those homes.

Variable	Count	Mean	StdDev	Range
Size	18	2003.39	264.727	890
Value	18	60946.7	5527.62	19710

Dependent variable is: Value
R-squared = 32.5%
s = 4682 with 18 − 2 = 16 degrees of freedom

Variable	Coefficient	SE(Coeff)
Intercept	37108.8	8664
Size	11.8987	4.290

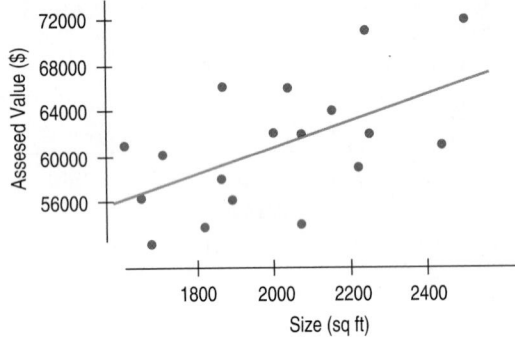

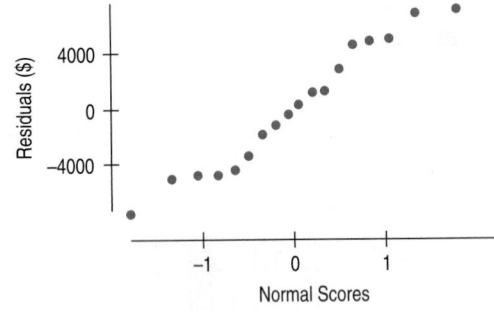
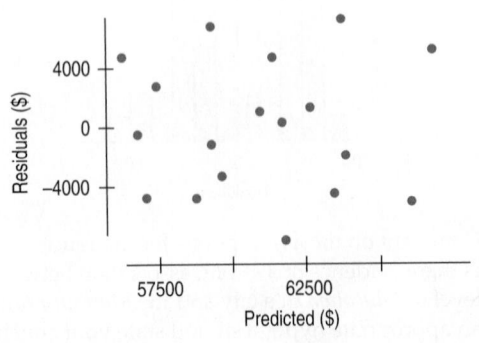

a) Explain why inference for linear regression is appropriate with these data.
b) Is there a significant association between the *Size* of a home and its assessed *Value*? Test an appropriate hypothesis and state your conclusion.
c) What percentage of the variability in assessed *Value* is explained by this regression?
d) Give a 90% confidence interval for the slope of the true regression line, and explain its meaning in the proper context.
e) From this analysis, can we conclude that adding a room to your house will increase its assessed *Value*? Why or why not?
*f) The owner of a home measuring 2100 square feet files an appeal, claiming that the $70,200 assessed *Value* is too high. Do you agree? Explain your reasoning.

REVIEW OF PART VII

Inference When Variables Are Related

Quick Review

With these last two chapters, you have added important analytical tools to your ways of looking at data. Here's a brief summary of those key concepts and skills, as well as an overview of statistical inference:

▶ Inferences about distributions of counts use chi-square models.

- To see if an observed distribution is consistent with a proposed model, use a goodness-of-fit test.
- To see if two or more observed distributions could have arisen from populations with the same model, use a test of homogeneity.

▶ Inference about association between two variables tests the hypothesis that it is plausible to consider the variables independent.

- If the variables are categorical, display the data in a contingency table and use a chi-square test of independence.
- If the variables are quantitative, display them with a scatterplot. You may use a linear regression t-test if there appears to be a linear association for which the residuals are random, consistent in terms of spread, and approximately Normal.

▶ You can now use statistical inference to answer questions about means, proportions, distributions, and associations.

- No inference procedure is valid unless the underlying assumptions are true. Always check the conditions before proceeding. Many of those checks should be made by examining a graph.
- You can make inferences about a single proportion or the difference of two proportions using Normal models.

- You can make inferences about one mean, the difference of two independent means, or the mean of paired differences using t-models.
- You can make inferences about distributions using chi-square models.
- You can make inferences about associations between categorical variables using chi-square models.
- You can make inferences about linear associations between quantitative variables using t-models.

If you look back at where we've been in this book, you'll see that statistical inference relies on almost everything we've seen. In Chapters 12 and 13 we learned techniques of collecting data using randomization—that's what makes inference possible at all. In Chapters 3, 4, and 7 we learned to plot our data and to look for the patterns and relationships we use to check the conditions that allow inference. In Chapters 3, 5, and 8 we learned about the summary statistics we use to do the mechanics of inference. We use our knowledge of randomness and probability from Chapters 11, 14, and 15 to help us think clearly about uncertainty, and the probability models of Chapters 6, 16, and 17 to measure our uncertainty precisely. Ultimately, the Central Limit Theorem of Chapter 18 makes all of inference possible.

Remember (have we said this often enough yet?): Never use any inference procedure without first checking the assumptions and conditions. On the next page we summarize the new types of inference procedures, the corresponding formulas, and the assumptions and conditions. You'll find complete summaries of all our inference procedures inside the back cover of the book. Have a look. Then you'll be ready for more opportunities to practice using these concepts and skills. . . .

Quick Guide to Inference

	Think			Show			Tell?
Inference about?	One group or two?	Procedure	Model	Parameter	Estimate	SE	Chapter
Distributions (one categorical variable)	One sample	Goodness-of-Fit	χ^2 $df = cells - 1$			$\sum \dfrac{(\text{obs} - \text{exp})^2}{\text{exp}}$	26
	Many independent groups	Homogeneity χ^2 Test					
Independence (two categorical variables)	One sample	Independence χ^2 Test	χ^2 $df = (r-1)(c-1)$				
Association (two quantitative variables)	One sample	Linear Regression t-Test or Confidence Interval for β	t $df = n - 2$	β_1	b_1	$\dfrac{s_e}{s_x \sqrt{n-1}}$ (compute with technology)	27
		Confidence Interval for μ_ν		μ_ν	$\hat{y}_\nu$	$\sqrt{SE^2(b_1)\cdot(x_\nu - \bar{x})^2 + \dfrac{s_e^2}{n}}$	
		Prediction Interval for y_ν		y_ν	$\hat{y}_\nu$	$\sqrt{SE^2(b_1)\cdot(x_\nu - \bar{x})^2 + \dfrac{s_e^2}{n} + s_e^2}$	

Assumptions for Inference   And the Conditions That Support or Override Them

Distributions/Association (χ^2)

- **Goodness-of-fit** (df = # of cells $-$ 1; one variable, one sample compared with population model)
  1. Data are counts.
  2. Data in sample are independent.
  3. Sample is sufficiently large.

  1. (Are they?)
  2. SRS and $n < 10\%$ of the population.
  3. All expected counts ≥ 5.

- **Homogeneity** [df = $(r-1)(c-1)$; samples from many populations compared on one variable]
  1. Data are counts.
  2. Data in groups are independent.
  3. Groups are sufficiently large.

  1. (Are they?)
  2. SRSs and $n < 10\%$ OR random allocation.
  3. All expected counts ≥ 5.

- **Independence** [df = $(r-1)(c-1)$; sample from one population classified on two variables]
  1. Data are counts.
  2. Data are independent.
  3. Sample is sufficiently large.

  1. (Are they?)
  2. SRSs and $n < 10\%$ of the population.
  3. All expected counts ≥ 5.

Regression (t, df = $n - 2$)

- **Association** between two quantitative variables ($\beta = 0$?)
  1. Form of relationship is linear.
  2. Errors are independent.
  3. Variability of errors is constant.
  4. Errors have a Normal model.

  1. Scatterplot looks approximately linear.
  2. No apparent pattern in residuals plot.
  3. Residuals plot has consistent spread.
  4. Histogram of residuals is approximately unimodal and symmetric or Normal probability plot reasonably straight.*

(*less critical as n increases)

REVIEW EXERCISES

1. Genetics. Two human traits controlled by a single gene are the ability to roll one's tongue and whether one's ear lobes are free or attached to the neck. Genetic theory says that people will have neither, one, or both of these traits in the ratio 1:3:3:9 (1 attached, noncurling; 3 attached, curling; 3 free, noncurling; 9 free, curling). An Introductory Biology class of 122 students collected the data shown. Are they consistent with the genetic theory? Test an appropriate hypothesis and state your conclusion.

	Trait			
	Attached, noncurling	Attached, curling	Free, noncurling	Free, curling
Count	10	22	31	59

T 2. Tableware. Nambe Mills manufactures plates, bowls, and other tableware made from an alloy of several metals. Each item must go through several steps, including polishing. To better understand the production process and its impact on pricing, the company checked the polishing time (in minutes) and the retail price (in US$) of these items. The regression analysis is shown below. The scatterplot showed a linear pattern, and residuals were deemed suitable for inference.

Dependent variable is: Price
R-squared = 84.5%
s = 20.50 with 59 − 2 = 57 degrees of freedom

Variable	Coefficient	SE(Coeff)
Intercept	−2.89054	5.730
Time	2.49244	0.1416

a) How many different products were included in this analysis?
b) What fraction of the variation in retail price is explained by the polishing time?
c) Create a 95% confidence interval for the slope of this relationship.
d) Interpret your interval in this context.

T 3. Hard water. In an investigation of environmental causes of disease, data were collected on the annual mortality rate (deaths per 100,000) for males in 61 large towns in England and Wales. In addition, the water hardness was recorded as the calcium concentration (parts per million, or ppm) in the drinking water. Here are the scatterplot and regression analysis of the relationship between mortality and calcium concentration.

Dependent variable is: mortality
R-squared = 43%
s = 143.0 with 61 − 2 = 59 degrees of freedom

Variable	Coefficient	SE(Coeff)
Intercept	1676	29.30
calcium	−3.23	0.48

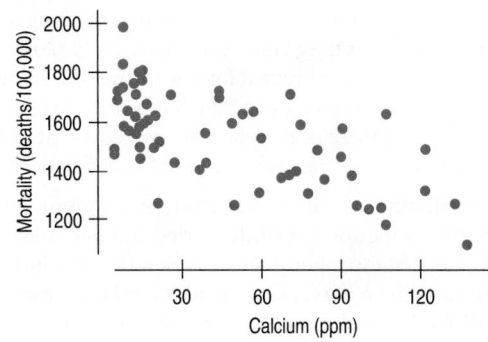

a) Is there an association between the hardness of the water and the mortality rate? Write the appropriate hypothesis.
b) Assuming the assumptions for regression inference are met, what do you conclude?
c) Create a 95% confidence interval for the slope of the true line relating calcium concentration and mortality.
d) Interpret your interval in context.

T 4. Mutual funds. In March 2002, *Consumer Reports* listed the rate of return for several large-cap mutual funds over the previous 3-year and 5-year periods. ("Large cap" refers to companies worth over $10 billion.)
a) Create a 95% confidence interval for the difference in rate of return for the 3- and 5-year periods covered by these data. Clearly explain what your interval means.
b) It's common for advertisements to carry the disclaimer "Past returns may not be indicative of future performance," but do these data indicate that there was an association between 3-year and 5-year rates of return?

Fund Name	Annualized Returns (%)	
	3-year	5-year
Ameristock	7.9	17.1
Clipper	14.1	18.2
Credit Suisse Strategic Value	5.5	11.5
Dodge & Cox Stock	15.2	15.7
Excelsior Value	13.1	16.4
Harbor Large Cap Value	6.3	11.5
ICAP Discretionary Equity	6.6	11.4
ICAP Equity	7.6	12.4
Neuberger Berman Focus	9.8	13.2
PBHG Large Cap Value	10.7	18.1
Pelican	7.7	12.1
Price Equity Income	6.1	10.9
USAA Cornerstone Strategy	2.5	4.9
Vanguard Equity Income	3.5	11.3
Vanguard Windsor	11.0	11.0

5. Resume fraud. In 2002 the Veritas Software company found out that its chief financial officer did not actually have the MBA he had listed on his resume. They fired him, and the value of the company's stock dropped 19%. Kroll, Inc., a firm that specializes in investigating such matters, said that they believe as many as 25% of background checks might reveal false information. How many such random checks would they have to do to estimate the true percentage of people who misrepresent their backgrounds to within ±5% with 98% confidence?

6. Paper airplanes. In preparation for a regional paper airplane competition, a student tried out her latest design. The distances her plane traveled (in feet) in 11 trial flights are given here. (The world record is an astounding 193.01 feet!) The data were 62, 52, 68, 23, 34, 45, 27, 42, 83, 56, and 40 feet. Here are some summaries:

Count	11
Mean	48.3636
Median	45
StdDev	18.0846
StdErr	5.45273
IntQRange	25
25th %tile	35.5000
75th %tile	60.5000

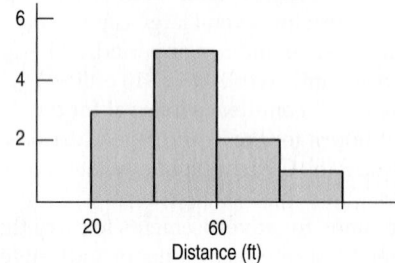

a) Construct a 95% confidence interval for the true distance.
b) Based on your confidence interval, is it plausible that the mean distance is 40 ft? Explain.
c) How would a 99% confidence interval for the true distance differ from your answer in part a? Explain briefly, without actually calculating a new interval.
d) How large a sample size would the student need to get a confidence interval half as wide as the one you got in part a, at the same confidence level?

7. Back to Montana. The respondents to the Montana poll described in Exercise 29 in Chapter 26 were also classified by income level: low (under $20,000), middle ($20,000–$35,000), or high (over $35,000). Is there any evidence that party enrollment there is associated with income? Test an appropriate hypothesis about this table, and state your conclusions.

	Democrat	Republican	Independent
Low	30	16	12
Middle	28	24	22
High	26	38	6

8. Wild horses. Large herds of wild horses can become a problem on some federal lands in the West. Researchers hoping to improve the management of these herds collected data to see if they could predict the number of foals that would be born based on the size of the current herd. Their attempt to model this herd growth is summarized in the output shown.

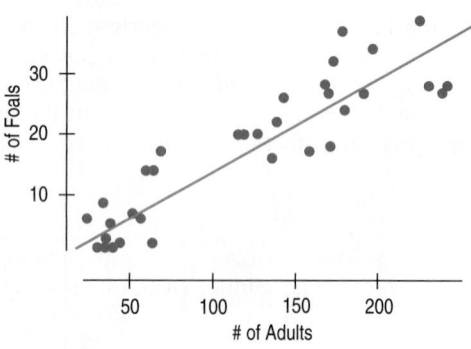

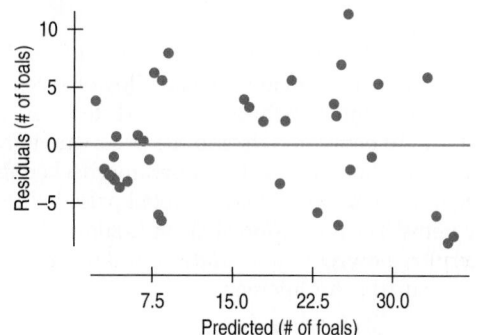

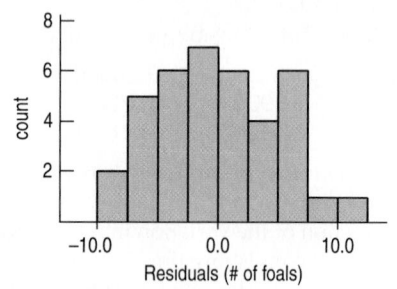

Variable	Count	Mean	StdDev
Adults	38	110.237	71.1809
Foals	38	15.3947	11.9945

Dependent variable is: Foals
R-squared = 83.5%
s = 4.941 with 38 − 2 = 36 degrees of freedom

Variable	Coefficient	SE(Coeff)	t-ratio	P-value
Intercept	−1.57835	1.492	−1.06	0.2970
Adults	0.153969	0.0114	13.5	≤ 0.0001

a) How many herds of wild horses were studied?
b) Are the conditions necessary for inference satisfied? Explain.
c) Create a 95% confidence interval for the slope of this relationship.
d) Explain in this context what that slope means.
e) Suppose that a new herd with 80 adult horses is located. Estimate, with a 90% prediction interval, the number of foals that may be born.

9. Lefties and music. In an experiment to see if left- and right-handed people have different abilities in music, subjects heard a tone and were then asked to identify which of several other tones matched the first. Of 76 right-handed subjects, 38 were successful in completing this test, compared with 33 of 53 lefties. Is this strong evidence of a difference in musical abilities based on handedness?

10. AP Statistics scores. In 2001, more than 41,000 Statistics students nationwide took the Advanced Placement Examination in Statistics. The national distribution of scores and the results at Ithaca High School are shown in the table.

Score	National Distribution	Ithaca High School Number of boys	Number of girls
5	11.5%	13	13
4	23.4%	21	15
3	24.9%	6	13
2	19.1%	7	3
1	21.1%	4	2

a) Is the distribution of scores at this high school significantly different from the national results?
b) Was there a significant different between the performances of boys and girls at this school?

11. Polling. How accurate are pollsters in predicting the outcomes of congressional elections? The table shows the actual number of Democratic party seats in the House of Representatives and the number predicted by the Gallup organization for nonpresidential election years between World War II and 1998.
a) Is there a significant difference between the number of seats predicted for the Democrats and the number they actually held? Test an appropriate hypothesis and state your conclusions.

Democratic Party Congressmen		
Year	Predicted	Actual
1946	190	188
1950	235	234
1954	232	232
1958	272	283
1962	259	258
1966	247	248
1970	260	255
1974	292	291
1978	277	277
1982	275	269
1986	264	258
1990	260	267
1994	201	204
1998	211	211

b) Is there a strong association between the pollsters' predictions and the outcomes of the elections? Test an appropriate hypothesis and state your conclusions.

12. Twins. In 2000 The *Journal of the American Medical Association* published a study that examined a sample of pregnancies that resulted in the birth of twins. Births were classified as preterm with intervention (induced labor or cesarean), preterm without such procedures, or term or postterm. Researchers also classified the pregnancies by the level of prenatal medical care the mother received (inadequate, adequate, or intensive). The data, from the years 1995–1997, are summarized in the table below. Figures are in thousands of births. (*JAMA* 284 [2000]: 335–341)

TWIN BIRTHS, 1995–1997 (IN THOUSANDS)	Preterm (induced or Cesarean)	Preterm (without procedures)	Term or postterm	Total
Intensive	18	15	28	61
Adequate	46	43	65	154
Inadequate	12	13	38	63
Total	76	71	131	278

Is there evidence of an association between the duration of the pregnancy and the level of care received by the mother?

13. Twins, again. After reading of the *JAMA* study in Exercise 12, a large city hospital examined their records of twin births for several years and found the data summarized in the table below. Is there evidence that the way the hospital deals with pregnancies involving twins may have changed?

	1990	1995	2000
Preterm (induced or cesarean)	11	13	19
Preterm (without procedures)	13	14	18
Term or postterm	27	26	32

14. Preemies. Do the effects of being born prematurely linger into adulthood? Researchers examined 242 Cleveland-area children born prematurely between 1977 and 1979, and compared them with 233 children of normal birth weight; 24 of the "preemies" and 12 of the other children were described as being of "subnormal height" as adults. Is this evidence that babies born with a very low birth weight are more likely to be smaller than normal adults? ("Outcomes in Young Adulthood for Very-Low-Birth-Weight Infants," *New England Journal of Medicine*, 346, no. 3 [January 2002])

T **15.** **LA rainfall.**   The Los Angeles Almanac Web site reports recent annual rainfall (in inches), as shown in the table.

a) Create a 90% confidence interval for the mean annual rainfall in LA.

b) If you wanted to estimate the mean annual rainfall with a margin of error of only 2 inches, how many years' data would you need?

c) Do these data suggest any change in annual rainfall as time passes? Check for an association between rainfall and year.

Year	Rain (in.)	Year	Rain (in.)
1980	8.96	1991	21.00
1981	10.71	1992	27.36
1982	31.28	1993	8.14
1983	10.43	1994	24.35
1984	12.82	1995	12.46
1985	17.86	1996	12.40
1986	7.66	1997	31.01
1987	12.48	1998	9.09
1988	8.08	1999	11.57
1989	7.35	2000	17.94
1990	11.99	2001	4.42

T **16.** **Age and party.**   The Gallup Poll conducted a representative telephone survey during the first quarter of 1999. Among the reported results was the following table concerning the preferred political party affiliation of respondents and their ages. Is there evidence of age-based differences in party affiliation in the United States?

	Republican	Democratic	Independent	Total
18–29	241	351	409	**1001**
30–49	299	330	370	**999**
50–64	282	341	375	**998**
65+	279	382	343	**1004**
Total	**1101**	**1404**	**1497**	**4002**

a) Will you conduct a test of homogeneity or independence? Why?

b) Test an appropriate hypothesis.

c) State your conclusion, including an analysis of differences you find (if any).

T **17.** **Eye and hair color.**   A survey of 1021 school-age children was conducted by randomly selecting children from several large urban elementary schools. Two of the questions concerned eye and hair color. In the survey, the following codes were used:

Hair Color	Eye Color
1 = Blond	1 = Blue
2 = Brown	2 = Green
3 = Black	3 = Brown
4 = Red	4 = Grey
5 = Other	5 = Other

The Statistics students analyzing the data were asked to study the relationship between eye and hair color.

a) One group of students produced the output shown below. What kind of analysis is this? What are the null and alternative hypotheses? Is the analysis appropriate? If so, summarize the findings, being sure to include any assumptions you've made and/or limitations to the analysis. If it's not an appropriate analysis, state explicitly why not.

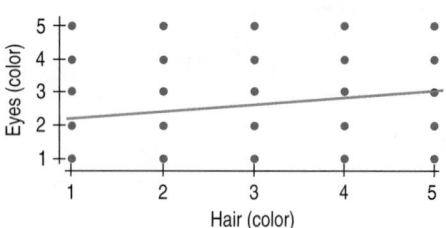

Dependent variable is: Eyes
R-squared = 3.7%
s = 1.112 with 1021 − 2 = 1019 degrees of freedom

Variable	Coefficient	SE(Coeff)	t-ratio	P-value
Intercept	1.99541	0.08346	23.9	≤0.0001
Hair	0.211809	0.03372	0.28	≤0.0001

b) A second group of students used the same data to produce the output shown below. The table displays counts and standardized residuals in each cell. What kind of analysis is this? What are the null and alternative hypotheses? Is the analysis appropriate? If so, summarize the findings, being sure to include any assumptions you've made and/or limitations to the analysis. If it's not an appropriate analysis, state explicitly why not.

		Eye Color				
		1	**2**	**3**	**4**	**5**
Hair Color	**1**	143	30	58	15	12
		7.67540	0.41799	−5.88169	−0.63925	−0.31451
	2	90	45	215	30	20
		−2.57141	0.29019	1.72235	0.49189	−0.08246
	3	28	15	190	10	10
		−5.39425	−2.34780	6.28154	−1.76376	−0.80382
	4	30	15	10	10	5
		2.06116	2.71589	−4.05540	2.37402	0.75993
	5	10	5	15	5	5
		−0.52195	0.33262	−0.94192	1.36326	2.07578

$$\sum \frac{(Observed - Expected)^2}{Expected} = 223.6 \quad \text{P-value} < 0.00001$$

18. Depression and the Internet. The September 1998 issue of the *American Psychologist* published an article reporting on an experiment examining "the social and psychological impact of the Internet on 169 people in 73 households during their first 1 to 2 years online." In the experiment, a sample of households was offered free Internet access for one or two years in return for allowing their time and activity online to be tracked. The members of the households who participated in the study were also given a battery of tests at the beginning and again at the end of the study. One of the tests measured the subjects' levels of depression on a 4-point scale, with higher numbers meaning the person was more depressed. Internet usage was measured in average number of hours per week. The regression analysis examines the association between the subjects' depression levels and the amounts of Internet use. The conditions for inference were satisfied.

Dependent variable is: Depression After
R-squared = 4.6%
s = 0.4563 with 162 − 2 = 160 degrees of freedom

Variable	Coefficient	SE(coeff)	t-ratio	Prob
Constant	0.565485	0.0399	14.2	≤0.0001
Intr_use	0.019948	0.0072	2.76	0.0064

a) Do these data indicate that there is an association between Internet use and depression? Test an appropriate hypothesis and state your conclusion clearly.
b) One conclusion of the study was that those who spent more time online tended to be more depressed at the end of the experiment. News headlines said that too much time on the Internet can lead to depression. Does the study support this conclusion? Explain.
c) As noted, the subjects' depression levels were tested at both the beginning and the end of this study; higher scores indicated the person was more depressed. Results are summarized in the table. Is there evidence that the depression level of the subjects changed during this study?

Depression Level
162 subjects

Variable	Mean	StdDev
DeprBfore	0.730370	0.487817
DeprAfter	0.611914	0.461932
Difference	−0.118457	0.552417

19. Pregnancy. In 1998 a San Diego reproductive clinic reported 42 live births to 157 women under the age of 38, but only 7 successes for 89 clients aged 38 and older. Is this evidence of a difference in the effectiveness of the clinic's methods for older women?
a) Test the appropriate hypotheses, using the two-proportion z-procedure.
b) Repeat the analysis, using an appropriate chi-square procedure.
c) Explain how the two results are equivalent.

20. Eating in front of the TV. Roper Reports asked a random sample of people in 30 countries whether they agreed with the statement "I like to nibble while reading or watching TV." Allowable responses were "Agree completely", "Agree somewhat", "Neither disagree nor agree", "Disagree somewhat", "Disagree completely", and "I Don't Know/No Response". Does a person's age influence their response? Here are data from 3792 respondents in the 2006 sample of five countries (China, India, France, United Kingdom, and United States) for three age groups (Teens, 30's (30–39) and Over 60):

	Agree Completely	Agree Somewhat	Neither Disagree Nor Agree	Disagree Somewhat	Disagree Completely
Teen	369	540	299	175	106
30's	272	522	325	229	170
60 +	93	207	153	154	178

a) Make an appropriate display of these data.
b) Does a person's age seem to affect their response to the question about nibbling?

21. Old Faithful. As you saw in an earlier chapter, Old Faithful isn't all that faithful. Eruptions do not occur at uniform intervals and may vary greatly. Can we improve our chances of predicting the time of the next eruption if we know how long the previous eruption lasted?
a) Describe what you see in this scatterplot.

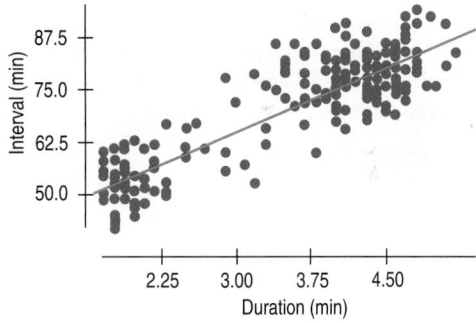

b) Write an appropriate hypothesis.
c) Here are a histogram of the residuals and the residuals plot. Do you think the assumptions for inference are met? Explain.

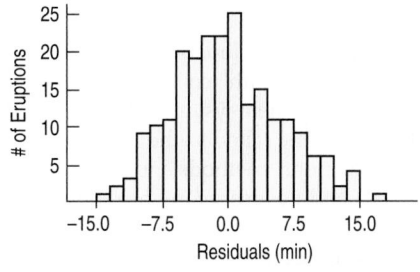

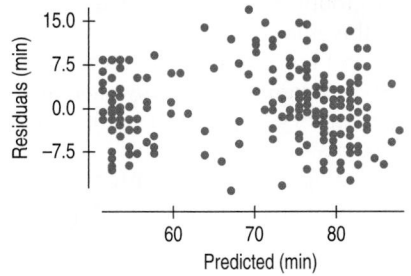

d) State a conclusion based on this regression analysis:

Dependent variable is: Interval
R-squared = 77.0%
s = 6.159 with 222 − 2 = 220 degrees of freedom

Variable	Coefficient	SE(Coeff)	t-ratio	P-value
Intercept	33.9668	1.428	23.8	≤0.0001
Duration	10.3582	0.3822	27.1	≤0.0001

Variable	Mean	StdDev
Duration	3.57613	1.08395
Interval	71.0090	12.7992

e) The second table shows the summary statistics for the two variables. Create a 95% confidence interval for the mean length of time that will elapse following a 2-minute eruption.

f) You arrive at Old Faithful just as an eruption ends. Witnesses say it lasted 4 minutes. Create a 95% prediction interval for the length of time you will wait to see the next eruption.

22. **Togetherness.** Are good grades in high school associated with family togetherness? A simple random sample of 142 high-school students was asked how many meals per week their families ate together. Their responses produced a mean of 3.78 meals per week, with a standard deviation of 2.2. Researchers then matched these responses against the students' grade point averages. The scatterplot appeared to be reasonably linear, so they went ahead with the regression analysis, seen below. No apparent pattern emerged in the residuals plot.

Dependent variable: GPA
R-squared = 11.0%
s = 0.6682 with 142 − 2 = 140 df

Variable	Coefficient	SE(Coeff)
Intercept	2.7288	0.1148
Meals/wk	0.1093	0.0263

a) Is there evidence of an association? Test an appropriate hypothesis and state your conclusion.

b) Do you think this association would be useful in predicting a student's grade point average? Explain.

c) Are your answers to parts a and b contradictory? Explain.

23. **Learning math.** Developers of a new math curriculum called "Accelerated Math" compared performances of students taught by their system with control groups of students in the same schools who were taught using traditional instructional methods and materials. Statistics about pretest and posttest scores are shown in the table. (J. Ysseldyke and S. Tardrew, *Differentiating Math Instruction*, Renaissance Learning, 2002)

a) Did the groups differ in average math score at the start of this study?

b) Did the group taught using the Accelerated Math program show a significant improvement in test scores?

c) Did the control group show a significant improvement in test scores?

d) Were gains significantly higher for the Accelerated Math group than for the control group?

		Instructional Method	
		Acc. math	Control
Number of students		231	245
Pretest	Mean	560.01	549.65
	St. Dev	84.29	74.68
Post-test	Mean	637.55	588.76
	St. Dev	82.9	83.24
Individual gain	Mean	77.53	39.11
	St. Dev.	78.01	66.25

24. **Pesticides.** A study published in 2002 in the journal *Environmental Health Perspectives* examined the gender ratios of children born to workers exposed to dioxin in Russian pesticide factories. The data covered the years from 1961 to 1988 in the city of Ufa, Bashkortostan, Russia. Of 227 children born to workers exposed to dioxin, only 40% were male. Overall in the city of Ufa, the proportion of males was 51.2%. Is this evidence that human exposure to dioxin may result in the birth of more girls? (An interesting note: It appeared that paternal exposure was most critical; 51% of babies born to mothers exposed to the chemical were boys.)

25. **Dairy sales.** Peninsula Creameries sells both cottage cheese and ice cream. The CEO recently noticed that in months when the company sells more cottage cheese, it seems to sell more ice cream as well. Two of his aides were assigned to test whether this is true or not. The first aide's plot and analysis of sales data for the past 12 months (in millions of pounds for cottage cheese and for ice cream) appear below.

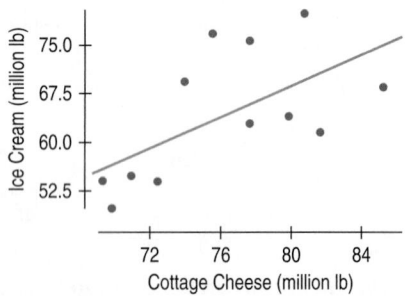

Dependent variable is: Ice cream
R-squared = 36.9%
s = 8.320 with 12 − 2 = 10 degrees of freedom

Variable	Coefficient	SE(Coeff)	t-ratio	P-value
Constant	−26.5306	37.68	−0.704	0.4975
Cottage C ...	1.19334	0.4936	2.42	0.0362

The other aide looked at the differences in sales of ice cream and cottage cheese for each month and created the following output:

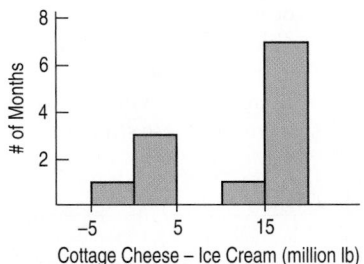

Cottage Cheese – Ice Cream (million lb)

Cottage Cheese – Ice Cream
Count   12
Mean   11.8000
Median   15.3500
StdDev   7.99386
IntQRange 14.3000
25th %tile   3.20000
75th %tile 17.5000

Test HO: $\mu(CC - IC) = 0$ vs Ha: $\mu(CC - IC) \neq 0$
Sample Mean = 11.800000 t-Statistic = 5.113 w/11 df
Prob = 0.0003
Lower 95% bound = 6.7209429
Upper 95% bound = 16.879057

a) Which analysis would you use to answer the CEO's question? Why?
b) What would you tell the CEO?
c) Which analysis would you use to test whether the company sells more cottage cheese or ice cream in a typical year? Why?
d) What would you tell the CEO about this other result?
e) What assumptions are you making in the analysis you chose in part a? What assumptions are you making in the analysis in part c?
f) Next month's cottage cheese sales are 82 million pounds. Ice cream sales are not yet available. How much ice cream do you predict Peninsula Creameries will sell?
g) Give a 95% confidence interval for the true slope of the regression equation of ice cream sales by cottage cheese sales.
h) Explain what your interval means.

26. **Infliximab.** In an article appearing in the journal *The Lancet* in 2002, medical researchers reported on the experimental use of the arthritis drug infliximab in treating Crohn's disease. In a trial, 573 patients were given initial 5-mg injections of the drug. Two weeks later, 335 had responded positively. These patients were then randomly assigned to three groups. Group I received continued injections of a placebo, Group II continued with 5 mg of infliximab, and Group III received 10 mg of the drug. After 30 weeks, 23 of 110 Group I patients were in remission, compared with 44 of 113 Group II and 50 of 112 Group III patients. Do these data indicate that continued treatment with infliximab is of value for Crohn's disease patients who exhibit a positive initial response to the drug?

27. **Weight loss.** A weight loss clinic advertises that its program of diet and exercise will allow clients to lose 10 pounds in one month. A local reporter investigating weight reduction gets permission to interview a randomly selected sample of clients who report the given weight losses during their first month in this program. Create a confidence interval to test the clinic's claim that the typical weight loss is 10 pounds.

Pounds Lost	
9.5	9.5
13	9
9	8
10	7.5
11	10
9	7
5	8
9	10.5
12.5	10.5
6	9

28. **Education vs. income.** The information below examines the median income and education level (years in school) for several U.S. cities.

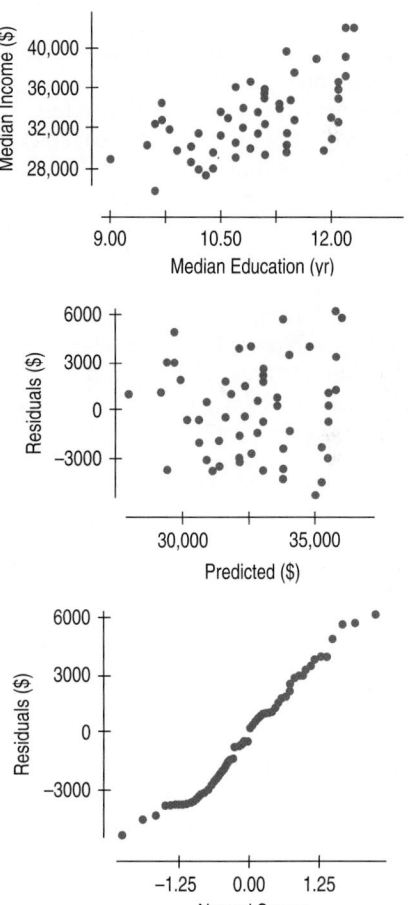

Variable	Count	Mean	StdDev
Education	57	10.9509	0.848344
Income	57	32742.6	3618.01

Dependent variable is: Income
R-squared = 32.9%
s = 2991 with 57 − 2 = 55 degrees of freedom

Variable	Coefficient	SE(Coeff)	t-ratio	P-value
Intercept	5970.05	5175	1.15	0.2537
Education	2444.79	471.2	5.19	≤0.0001

a) Do you think the assumptions for inference are met? Explain.
b) Does there appear to be an association between education and income levels in these cities?
c) Would this association appear to be weaker, stronger, or the same if data were plotted for individual people rather than for cities in aggregate? Explain.
d) Create and interpret a 95% confidence interval for the slope of the true line that describes the association between income and education.
e) Predict the median income for cities where residents spent an average of 11 years in school. Describe your estimate with a 90% confidence interval, and interpret that result.

T 29. **Diet.** Thirteen overweight women volunteered for a study to determine whether eating specially prepared crackers before a meal could help them lose weight. The subjects were randomly assigned to eat crackers with different types of fiber (bran fiber, gum fiber, both, and a control cracker). Unfortunately, some of the women developed uncomfortable bloating and upset stomachs. Researchers suspected that some of the crackers might be at fault. The contingency table of "Cracker" versus "Bloat" shows the relationship between the four different types of crackers and the reported bloating. The study was paid for by the manufacturers of the gum fiber. What would you recommend to them about the prospects for marketing their new diet cracker?

		Bloat	
		Little/None	Moderate/Severe
Cracker	Bran	11	2
	Gum	4	9
	Combo	7	6
	Control	8	4

T 30. **Cramming.** Students in two basic Spanish classes were required to learn 50 new vocabulary words. One group of 45 students received the list on Monday and studied the words all week. Statistics summarizing this group's scores on Friday's quiz are given. The other group of 25 students

did not get the vocabulary list until Thursday. They also took the quiz on Friday, after "cramming" Thursday night. Then, when they returned to class the following Monday, they were retested—without advance warning. Both sets of test scores for these students are shown.

Group 1

Fri.
Number of students = 45
Mean = 43.2 (of 50)
StDev = 3.4
Students passing (score ≥ 40) = 33%

Group 2

Fri.	Mon.	Fri.	Mon.
42	36	50	47
44	44	34	34
45	46	38	31
48	38	43	40
44	40	39	41
43	38	46	32
41	37	37	36
35	31	40	31
43	32	41	32
48	37	48	39
43	41	37	31
45	32	36	41
47	44		

a) On Friday, did the week-long study group have a mean score significantly higher than that of the overnight crammers?
b) Was there a significant difference in the percentages of students who passed the quiz on Friday?
c) Is there any evidence that when students cram for a test, their "learning" does not last for 3 days?
d) Use a 95% confidence interval to estimate the mean number of words that might be forgotten by crammers.
e) Is there any evidence that how much students forget depends on how much they "learned" to begin with?

Selected Formulas

$Range = Max - Min$

$IQR = Q3 - Q1$

Outlier Rule-of-Thumb: $y < Q1 - 1.5 \times IQR$ or $y > Q3 + 1.5 \times IQR$

$$\bar{y} = \frac{\Sigma y}{n}$$

$$s = \sqrt{\frac{\Sigma(y - \bar{y})^2}{n - 1}}$$

$$z = \frac{y - \mu}{\sigma} \text{ (model based)}$$

$$z = \frac{y - \bar{y}}{s} \text{ (data based)}$$

$$r = \frac{\Sigma z_x z_y}{n - 1}$$

$$\hat{y} = b_0 + b_1 x \qquad \text{where } b_1 = \frac{rs_y}{s_x} \text{ and } b_0 = \bar{y} - b_1 \bar{x}$$

$$P(\mathbf{A}) = 1 - P(\mathbf{A}^C)$$

$$P(\mathbf{A} \cup \mathbf{B}) = P(\mathbf{A}) + P(\mathbf{B}) - P(\mathbf{A} \cap \mathbf{B})$$

$$P(\mathbf{A} \cap \mathbf{B}) = P(\mathbf{A}) \times P(\mathbf{B}|\mathbf{A})$$

$$P(\mathbf{B}|\mathbf{A}) = \frac{P(\mathbf{A} \cap \mathbf{B})}{P(\mathbf{A})}$$

If $\mathbf{A}$ and $\mathbf{B}$ are independent, $P(\mathbf{B}|\mathbf{A}) = P(\mathbf{B})$

$E(X) = \mu = \Sigma x \cdot P(x)$ $\qquad\qquad$ $Var(X) = \sigma^2 = \Sigma(x - \mu)^2 P(x)$

$E(X \pm c) = E(X) \pm c$ $\qquad\qquad$ $Var(X \pm c) = Var(X)$

$E(aX) = aE(X)$ $\qquad\qquad$ $Var(aX) = a^2 Var(X)$

$E(X \pm Y) = E(X) \pm E(Y)$ $\qquad\qquad$ $Var(X \pm Y) = Var(X) + Var(Y),$
$\qquad\qquad\qquad\qquad\qquad\qquad\qquad$ if X and Y are independent

Geometric: $P(x) = q^{x-1}p$ $\qquad\qquad$ $\mu = \dfrac{1}{p}$ $\qquad\qquad$ $\sigma = \sqrt{\dfrac{q}{p^2}}$

Binomial: $P(x) = \dbinom{n}{x} p^x q^{n-x}$ $\qquad$ $\mu = np$ $\qquad\qquad$ $\sigma = \sqrt{npq}$

$\qquad\qquad\qquad$ $\hat{p} = \dfrac{x}{n}$ $\qquad\qquad\qquad\qquad$ $\mu(\hat{p}) = p$ $\qquad\qquad$ $SD(\hat{p}) = \sqrt{\dfrac{pq}{n}}$

Sampling distribution of $\bar{y}$:

(CLT) As n grows, the sampling distribution approaches the Normal model with

$$\mu(\bar{y}) = \mu_y \qquad SD(\bar{y}) = \frac{\sigma}{\sqrt{n}}$$

Inference:

Confidence interval for parameter = *statistic* ± *critical value* × *SD(statistic)*

$$\text{Test statistic} = \frac{Statistic - Parameter}{SD(statistic)}$$

Parameter	Statistic	SD(statistic)	SE(statistic)
p	$\hat{p}$	$\sqrt{\dfrac{pq}{n}}$	$\sqrt{\dfrac{\hat{p}\hat{q}}{n}}$
$p_1 - p_2$	$\hat{p}_1 - \hat{p}_2$	$\sqrt{\dfrac{p_1 q_1}{n_1} + \dfrac{p_2 q_2}{n_2}}$	$\sqrt{\dfrac{\hat{p}_1 \hat{q}_1}{n_1} + \dfrac{\hat{p}_2 \hat{q}_2}{n_2}}$
μ	$\bar{y}$	$\dfrac{\sigma}{\sqrt{n}}$	$\dfrac{s}{\sqrt{n}}$
$\mu_1 - \mu_2$	$\bar{y}_1 - \bar{y}_2$	$\sqrt{\dfrac{\sigma_1^2}{n_1} + \dfrac{\sigma_2^2}{n_2}}$	$\sqrt{\dfrac{s_1^2}{n_1} + \dfrac{s_2^2}{n_2}}$
μ_d	$\bar{d}$	$\dfrac{\sigma_d}{\sqrt{n}}$	$\dfrac{s_d}{\sqrt{n}}$
σ_ε	$s_e = \sqrt{\dfrac{\Sigma(y - \hat{y})^2}{n-2}}$		
β_1	b_1		$\dfrac{s_e}{s_x \sqrt{n-1}}$
$*\mu_\nu$	$\hat{y}_\nu$		$\sqrt{SE^2(b_1) \cdot (x_\nu - \bar{x})^2 + \dfrac{s_e^2}{n}}$
$*y_\nu$	$\hat{y}_\nu$		$\sqrt{SE^2(b_1) \cdot (x_\nu - \bar{x})^2 + \dfrac{s_e^2}{n} + s_e^2}$

Pooling: For testing difference between proportions: $\hat{p}_{pooled} = \dfrac{y_1 + y_2}{n_1 + n_2}$

For testing difference between means: $s_p = \sqrt{\dfrac{(n_1 - 1)s_1^2 + (n_2 - 1)s_2^2}{n_1 + n_2 - 2}}$

Substitute these pooled estimates in the respective SE formulas for both groups when assumptions and conditions are met.

Chi-square: $\chi^2 = \Sigma \dfrac{(obs - exp)^2}{exp}$

APPENDIX B

Guide to Statistical Software

Chapter 3. Displaying and Describing Categorical Data

DATA DESK

To make a bar chart or pie chart, select the variable.
In the **Plot** menu, choose **Bar Chart** or **Pie Chart.**
To make a frequency table, in the **Calc** menu choose
Frequency Table.

COMMENTS

These commands treat the data as categorical even if they are
numerals. If you select a quantitative variable by mistake, you'll
see an error message warning of too many categories.

EXCEL

First make a pivot table (Excel's name for a frequency table).
From the **Data** menu, choose **Pivot Table** and **Pivot Chart
Report.**
When you reach the Layout window, drag your variable to the
row area and drag your variable again to the data area. This
tells Excel to count the occurrences of each category.
Once you have an Excel pivot table, you can construct bar
charts and pie charts.
Click inside the Pivot Table.

Click the Pivot Table Chart Wizard button. Excel creates a bar
chart.
A longer path leads to a pie chart; see your Excel documentation.

COMMENTS

Excel uses the pivot table to specify the category names and
find counts within each category. If you already have that information, you can proceed directly to the Chart Wizard.

EXCEL 2007

To make a bar chart:

- Select the variable in Excel you want to work with.
- Choose the **Column** command from the Insert tab in the Ribbon.
- Select the appropriate chart from the drop-down dialog.

To change the bar chart into a pie chart:

- Right-click the chart and select **Change Chart Type...** from the
  menu. The Chart type dialog opens.
- Select a pie chart type.
- Click the **OK** button. Excel changes your bar chart into a pie
  chart.

JMP

JMP makes a bar chart and frequency table together.
From the **Analyze** menu, choose **Distribution.**
In the Distribution dialog, drag the name of the variable into the
empty variable window beside the label "Y, Columns"; click **OK.**
To make a pie chart, choose **Chart** from the **Graph** menu.
In the Chart dialog, select the variable name from the Columns

list, click on the button labeled "Statistics," and select "N" from
the drop-down menu.
Click the "**Categories, X, Levels**" button to assign the same vari-
able name to the X-axis.
Under Options, click on the **second** button—labeled "**Bar
Chart**"—and select "Pie" from the drop-down menu.

MINITAB

To make a bar chart, choose **Bar Chart** from the **Graph** menu.
Select "Counts of unique values" in the first menu, and select
"Simple" for the type of graph. Click **OK.**

In the Chart dialog, enter the name of the variable that you wish
to display in the box labeled "Categorical variables." Click **OK.**

A-3

To make a bar chart, open the **Chart Builder** from the **Graphs** menu.
Click the **Gallery** tab.
Choose **Bar Chart** from the list of chart types.
Drag the appropriate bar chart onto the canvas.

Drag a categorical variable onto the x-axis drop zone.
Click **OK.**

COMMENTS

A similar path makes a pie chart by choosing **Pie chart** from the list of chart types.

TI-NSPIRE

The TI-Nspire Handheld does not display plots for categorical variables.

TI-89

The TI-89 won't do displays for categorical variables.

Chapter 4. Displaying and Summarizing Quantitative Data

DATA DESK

To make a histogram:
▶ Select the variable to display.
▶ In the **Plot** menu, choose **Histogram.**

To calculate summaries:
▶ In the **Calc** menu, open the **summaries** submenu. **Options** offer separate tables, a single unified table, and other formats.

EXCEL

Excel cannot make histograms or dotplots without a third-party add-in.
To calculate summaries.
Click on an empty cell. Type an equal sign and choose "**Average**" from the popup list of functions that appears to the left of the text-editing box. Enter the data range in the box that says "**Number 1.**" Click the **OK** button.
To compute the standard deviation of a column of data directly, use the **STDEV** from the popup list of functions in the same way.

COMMENTS

Excel's Data Analysis add-in does offer something called a histogram, but it just makes a crude frequency table, and the Chart Wizard cannot then create a statistically appropriate histogram. The DDXL add-in provided on our DVD adds these and other capabilities to Excel.
Excel's STDEV function should not be used for data values larger in magnitude than 100,000 or for lists of more than a few thousand values. It is programmed with an unstable formula that can generate rounding errors when these limits are exceeded.

EXCEL 2007

In Excel 2007 there is another way to find some of the standard summary statistics. For example, to compute the mean:
▶ Click on an empty cell.
▶ Go to the Formulas tab in the Ribbon. Click on the drop down arrow next to "AutoSum" and choose "**Average**".
▶ Enter the data range in the formula displayed in the empty box you selected earlier.
▶ Press **Enter.** This computes the mean for the values in that range.

To compute the standard deviation:
▶ Click on an empty cell.
▶ Go to the Formulas tab in the Ribbon and click the drop down arrow next to "AutoSum" and select "**More functions...**"
▶ In the dialog window that opens, select "**STDEV**" from the list of functions and click **OK.** A new dialog window opens. Enter a range of fields into the text fields and click **OK.**
Excel 2007 computes the standard deviation for the values in that range and places it in the specified cell of the spreadsheet.

JMP

To make a histogram and find summary statistics:
▶ Choose **Distribution** from the **Analyze** menu.
▶ In the **Distribution** dialog, drag the name of the variable that you wish to analyze into the empty window beside the label "**Y, Columns.**"

▶ Click **OK.** JMP computes standard summary statistics along with displays of the variables.

MINITAB

To make a histogram:
- Choose **Histogram** from the **Graph** menu.
- Select "Simple" for the type of graph and click **OK.**
- Enter the name of the quantitative variable you wish to display in the box labeled "Graph variables." Click **OK.**

To calculate summary statistics:
- Choose **Basic statistics** from the **Stat** menu. From the **Basic Statistics** submenu, choose **Display Descriptive Statistics.**
- Assign variables from the variable list box to the Variables box. MINITAB makes a Descriptive Statistics table.

SPSS

To make a histogram in SPSS open the Chart Builder from the Graphs menu.
- Click the **Gallery** tab.
- Choose **Histogram** from the ist of chart types.
- Drag the histogram onto the canvas.
- Drag a scale variable to the y-axis drop zone.
- Click **OK.**

To calculate summary statistics:
- Choose **Explore** from the **Descriptive Statistics** submenu of the **Analyze** menu. In the Explore dialog, assign one or more variables from the source list to the Dependent List and click the **OK** button.

TI-NSPIRE

To plot a histogram using a named list, press ▲ several times so that the entire list is highlighted. Press ⒨, ③ for Data, and ④ for Quick Graph. Then press ⒨, ① for Plot Type, and ③ for Histogram.

To create the plot on a full page, press ⒢, and then ⑤ for Data & Statistics. Move the cursor to "Click to add variable," and then press ⓧ and select the list name. Then press ⒨, ① for Plot Type, and ③ for Histogram.

TI-89

To make a histogram:
- Select F2 **(Plots),** then 1: **Plot Setup.** Select a plot and press F1 to define it.
- Select plot type 4: **Histogram.** Use VAR-LINK to select the data list.
- Enter a number for the histogram bucket (bar) width.
- Press ENTER to complete the plot definition. Press F5 to display the histogram.
- Press ◆F2 to adjust the window appropriately, then press ◆F3 **(Graph).**

To calculate summary statistics:
- To compute summary statistics, press F4 **(Calc).** Input the name of the list using VAR-LINK. Press ENTER.
- Use the down arrow to scroll through the output.
- To create a boxplot, press F2 **(Plots)** then ENTER. Select a plot to define and press F1. Select either 3: **Box Plot** or 4: **Mod Box**

Plot (to identify outliers). Select the mark type of your choice (for outliers). Press ENTER to finish.
- Press F5 to display the graph.

COMMENTS

If the data are stored as a frequency table (say, with data values in list1 and frequencies in list2), change Use Freq and Categories to YES and use VAR-LINK to select list2 as the frequency variable on the plot definition screen.

If the data are stored as a frequency table (say, with data values in list1 and frequencies in list2), use VAR-LINK to select list2 as the frequency variable in 1-Var Stats.

For the plot, change Use Freq and Categories to YES and use VAR-LINK to select list2 as the frequency variable on the plot definition screen.

Chapter 5. Understanding and Comparing Distributions

There are two ways to organize data when we want to compare groups. Each group can be in its own variable (or list, on a calculator). In this form, the experiment comparing coffee cups would have four lists, one for each type of cup:

CUPPS	SIGG	Nissan	Starbucks
6	2	12	13
6	1.5	16	7
6	2	9	7
18.5	3	23	17.5
10	0	11	10
17.5	7	20.5	15.5
11	0.5	12.5	6
6.5	6	24.5	6

But there's another way to think about and organize the data. What is the variable of interest (the *What*) in this experiment? It's the number of degrees lost by the water in each cup. And the *Who* is each time she tested a cup. We could gather all the temperature values into one variable and put the names of the cups in a second variable listing the individual results, one on each row. Now the *Who* is clearer—it's an experimental run, one row of the table. Most statistics packages prefer data on groups organized in this way.

That's actually the way we've thought about the wind speed data in this chapter, treating wind speeds as one variable and the groups (whether seasons, months, or days) as a second variable.

Container	Temperature Difference		Container	Temperature Difference
CUPPS	6		SIGG	12
CUPPS	6		SIGG	16
CUPPS	6		SIGG	9
.			.	
.			.	
.			.	
Nissan	2		Starbucks	13
Nissan	1.5		Starbucks	7
Nissan	2		Starbucks	7
.			.	
.			.	
.			.	

DATA DESK

If the data are in separate variables, select the variables and choose **Boxplot side by side** from the **Plot** menu. The boxes will appear in the order in which the variables were selected. If the data are a single quantitative variable and a second variable holding group names, select the quantitative variable as

Y and the group variable as X. Then choose **Boxplot y by x** from the **Plot** menu. The boxes will appear in alphabetical order by group name.

Data Desk offers options for assessing whether any pair of medians differ.

EXCEL

Excel cannot make boxplots.

COMMENT

The DDXL add-on provided on the DVD adds the ability to make boxplots to Excel.

JMP

Choose **Fit y by x.** Assign a continuous response variable to **Y, Response** and a nominal group variable holding the group names to **X, Factor,** and click **OK.** JMP will offer (among other

things) dotplots of the data. Click the red triangle and, under **Display Options,** select Boxplots. *Note:* If the variables are of the wrong type, the display options might not offer boxplots.

MINITAB

Choose **Boxplot...** from the **Graph** menu. If your data are in the form of one quantitative variable and one group variable, choose

One Y and **with Groups.** If your data are in separate columns of the worksheet, choose **Multiple Y's.**

SPSS

To make a boxplot in SPSS, open the **Chart Builder** from the Graphs menu.
Click the **Gallery** tab.
Choose **Boxplot** from the list of chart types.
Drag a single or 2-D (side-by-side) boxplot onto the canvas.

Drag a scale variable to the y-axis drop zone.
To make side-by-side boxplots, drag a categorical variable to the x-axis drop zone.
Click **OK.**

TI-NSPIRE

To compute summary statistics using a named list, press (ᴳⁱ), (1) for Calculator, (menu), (6) for Statistics, (1) for Stat Calculations, and (1) for One-Variable Statistics. Complete the dialog boxes.
To create a box plot using a named list, press ▲ several times so that the entire list is highlighted. Press (menu), (3) for Data, and

(4) for Quick Graph. Then press (menu), (1) for Plot Type, and (2) for Box Plot.
To create the plot on a full page, press (ᴳⁱ), and then press (5) for Data & Statistics. Move the cursor to "Click to add variable," and then press (※) and select the list name. Then press (menu), (1) for Plot Type, and (2) for Box Plot.

TI-89

For the plot, change Use Freq and Categories to YES and use VAR-LINK to select list2 as the frequency variable on the plot definition screen.

To create a boxplot, press [F2] **(Plots),** then [ENTER]. Select a plot to define and press [F1]. Select either 3: **Box Plot** or 4: **Mod Box**

Plot (to identify outliers). Select the mark type of your choice (for outliers). Press [ENTER] to finish.
Press [F5] to display the graph.

Chapter 6. The Standard Deviation as a Ruler and the Normal Model

DATA DESK

To make a "Normal Probability Plot" in Data Desk,
▸ Select the Variable.
▸ Choose **Normal Prob Plot** from the **Plot** menu.

COMMENTS

Data Desk places the ordered data values on the vertical axis and the Normal scores on the horizontal axis.

EXCEL

Excel offers a "Normal probability plot" as part of the Regression command in the Data Analysis extension, but (as of this writing)

it is not a correct Normal probability plot and should not be used.

JMP

To make a "Normal Quantile Plot" in JMP,
▸ Make a histogram using **Distributions** from the **Analyze** menu.
▸ Click on the drop-down menu next to the variable name.
▸ Choose **Normal Quantile Plot** from the drop-down menu.
▸ JMP opens the plot next to the histogram.

COMMENTS

JMP places the ordered data on the vertical axis and the Normal scores on the horizontal axis. The vertical axis aligns with the histogram's axis, a useful feature.

MINITAB

To make a "Normal Probability Plot" in MINITAB,
▸ Choose **Probability Plot** from the **Graph** menu.
▸ Select "Single" for the type of plot. Click **OK.**
▸ Enter the name of the variable in the "Graph variables" box. Click **OK.**

COMMENTS

MINITAB places the ordered data on the horizontal axis and the Normal scores on the vertical axis.

SPSS

To make a Normal "P-P plot" in SPSS,
▸ Choose **P-P** from the **Graphs** menu.
▸ Select the variable to be displayed in the source list.
▸ Click the arrow button to move the variable into the target list.
▸ Click the **OK** button.

COMMENTS

SPSS places the ordered data on the horizontal axis and the Normal scores on the vertical axis. You may safely ignore the options in the P-P dialog.

TI-NSPIRE

To create a normal probability plot using a named list, press ▲ several times so that the entire list is highlighted. Press (menu), ③ for Data, and ④ for Quick Graph. Then press (menu), ① for Plot Type, and ④ for Normal Probability Plot.

To create the plot on a full page, press (🔓), and then ⑤ for Data & Statistics. Move the cursor to "Click to add variable," and then press (🔆) and select the list name. Then press (menu), ① for Plot Type, and ④ for Normal Probability Plot.

To compute the area under a normal curve, press (menu), ① for Calculator, (menu), ⑤ for Probability, ⑤ for Distributions, and ② for Normal Cdf. Complete the dialog box.

To compute the value for a given percentile, press (menu), ① for Calculator, (menu), ⑤ for Probability, ⑤ for Distributions, ③ for Inverse Normal. Complete the dialog box.

TI-89

▶ To create a "Normal Prob Plot", press [F2] and select choice 2: **Norm Prob Plot.** Select a plot number and use VAR-LINK to enter the data list. Select X or Y for the data axis. Press [ENTER] to calculate the z-scores.

▶ Press [F2] and select choice 1: **Plot Setup.** Turn off any undesired plots (either [F3] (Clear) or [F4] (√)). Press [F5] to display the plot.

▶ To find what percent of a Normal model lies between two z-scores, press [F5] (**Distr**). Then select 4: **Normal Cdf.** Enter the lower and upper z-scores, specify mean 0 and standard deviation 1, and press [ENTER].

▶ To find the z-score for a given percentile, press [F5] (**Distr**). Then arrow down to 2: **Inverse** press the right arrow to see the sub

menu and select 1: **Inverse Normal.** Enter the area to the left of the desired point, mean 0 and standard deviation 1, and press [ENTER].

COMMENTS

Normal models strictly go to infinity on either end, which is 1EE99 on the calculator. In practice, any "large" number will work. For example, the percentage of the Normal model over two standard deviations above the mean can use Lower Value 2 and Upper Value 99. To find area more than 2 standard deviations below the mean, use Lower Value −99, and Upper value −2.

Chapter 7. Scatterplots, Association, and Correlation

DATA DESK

To make a scatterplot of two variables, select one variable as Y and the other as X and choose **Scatterplot** from the **Plot** menu. Then find the correlation by choosing **Correlation** from the scatterplot's HyperView menu.

Alternatively, select the two variables and choose **Pearson Product-Moment** from the **Correlations** submenu of the **Calc** menu.

COMMENTS

We prefer that you look at the scatterplot first and then find the correlation. But if you've found the correlation first, click on the correlation value to drop down a menu that offers to make the scatterplot.

EXCEL

To make a Scatterplot with the Excel Chart Wizard:

▶ Click on the **Chart Wizard** Button in the menu bar. Excel opens the Chart Wizard's Chart Type Dialog window.

▶ Make sure the **Standard Types** tab is selected, and select **XY (Scatter)** from the choices offered.

▶ Specify the **scatterplot without lines** from the choices offered in the Chart subtype selections. The **Next** button takes you to the Chart Source Data dialog.

▶ If it is not already frontmost, click on the **Data Range** tab, and enter the data range in the space provided.

▶ By convention, we always represent variables in columns. The Chart Wizard refers to variables as Series. Be sure the **Column** option is selected.

▶ Excel places the leftmost column of those you select on the x-axis of the scatterplot. If the column you wish to see on the x-axis is not the leftmost column in your spreadsheet, click on the **Series** tab and edit the specification of the individual axis series.

▶ Click the **Next** button. The Chart Options dialog appears.

▶ Select the **Titles** tab. Here you specify the title of the chart and names of the variables displayed on each axis.

▶ Type the chart title in the **Chart title:** edit box.

▶ Type the x-axis variable name in the **Value (X) Axis:** edit box. Note that you must name the columns correctly here. Naming another variable will not alter the plot, only mislabel it.

▶ Type the y-axis variable name in the **Value (Y) Axis:** edit box.

▶ Click the **Next** button to open the chart location dialog.

▶ Select the **As new sheet:** option button.

▶ Click the **Finish** button.

Often, the resulting scatterplot will not be useful. By default, Excel includes the origin in the plot even when the data are far from zero. You can adjust the axis scales.

To change the scale of a plot axis in Excel:

▶ Double-click on the axis. The **Format Axis Dialog** appears.

▶ If the **scale tab** is not the frontmost, select it.

▶ Enter new minimum or new maximum values in the spaces provided. You can drag the dialog box over the scatterplot as a straightedge to help you read the maximum and minimum values on the axes.

▶ Click the **OK** button to view the rescaled scatterplot.

▶ Follow the same steps for the x-axis scale.

Compute a correlation in Excel with the **CORREL** function from the drop-down menu of functions. If CORREL is not on the menu, choose **More Functions** and find it among the statistical functions in the browser.

In the dialog that pops up, enter the range of cells holding one of the variables in the space provided.

Enter the range of cells for the other variable in the space provided.

EXCEL 2007

To make a scatterplot in Excel 2007:
▶ Select the columns of data to use in the scatterplot. You can select more than one column by holding down the control key while clicking.
▶ In the Insert tab, click on the **Scatter** button and select the **Scatter with only Markers** chart from the menu.

Unfortunately, the plot this creates is often statistically useless. To make the plot useful, we need to change the display:
▶ With the chart selected click on the **Gridlines** button in the Layout tab to cause the Chart Tools tab to appear.
▶ Within Primary Horizontal Gridlines, select **None**. This will remove the gridlines from the scatterplot.
▶ To change the axis scaling, click on the numbers of each axis of the chart, and click on the **Format Selection** button in the Layout tab.
▶ Select the **Fixed** option instead of the Auto option, and type a value more suited for the scatterplot. You can use the popup dialog window as a straightedge to approximate the appropriate values.

Excel 2007 automatically places the leftmost of the two columns you select on the x-axis, and the rightmost one on the y-axis. If that's not what you'd prefer for your plot, you'll want to switch them.

To switch the X and Y-variables:
▶ Click the chart to access the **Chart Tools** tabs.
▶ Click on the **Select Data** button in the Design tab.
▶ In the popup window's Legend Entries box, click on **Edit.**
▶ Highlight and delete everything in the Series X Values line, and select new data from the spreadsheet. (Note that selecting the column would inadvertently select the title of the column, which would not work well here.)
▶ Do the same with the Series Y Values line.
▶ Press **OK**, then press **OK** again.

JMP

To make a scatterplot and compute correlation, choose **Fit Y by X** from the **Analyze** menu.

In the Fit Y by X dialog, drag the Y variable into the **"Y, Response"** box, and drag the X variable into the **"X, Factor"** box. Click the **OK** button.

Once JMP has made the scatterplot, click on the red triangle next to the plot title to reveal a menu of options. Select **Density Ellipse** and select .95. JMP draws an ellipse around the data and reveals the **Correlation** tab. Click the blue triangle next to Correlation to reveal a table containing the correlation coefficient.

MINITAB

To make a scatterplot, choose **Scatterplot** from the **Graph** menu. Choose "Simple" for the type of graph. Click **OK**. Enter variable names for the *Y*-variable and *X*-variable into the table. Click **OK.**

To compute a correlation coefficient, choose **Basic Statistics** from the **Stat** menu. From the Basic Statistics submenu, choose **Correlation.** Specify the names of at least two quantitative variables in the "Variables" box. Click **OK** to compute the correlation table.

SPSS

To make a scatterplot in SPSS, open the Chart Builder from the Graphs menu. Then:
▶ Click the Gallery tab.
▶ Choose Scatterplot from the list of chart types.
▶ Drag the scatterplot onto the canvas.
▶ Drag a scale variable you want as the response variable to the y-axis drop zone.
▶ Drag a scale variable you want as the factor or predictor to the x-axis drop zone.
▶ Click OK.

To compute a correlation coefficient, choose **Correlate** from the **Analyze** menu. From the Correlate submenu, choose **Bivariate.** In the Bivariate Correlations dialog, use the arrow button to move variables between the source and target lists. Make sure the **Pearson** option is selected in the Correlation Coefficients field.

TI-NSPIRE

To create a scatterplot using named lists, press ▲ several times so that the first list is highlighted. Then press ⬇ ▶ so that the second list is highlighted. Press ⓜ, ③ for Data, and ④ for Quick Graph.

To create the plot on a full page, press ⓐ, then ⑤ for Data & Statistics. Move the cursor to "Click to add variable," and then press ⊙ and select the list name. Repeat for the other axis.

To find the correlation, press ⓐ, ① for Calculator, ⓜ, ⑥ for Statistics, ① for Stat Calculations, and ④ for Linear Regression. Complete the dialog boxes.

TI-89

To create a scatterplot, press [F2] (Plots). Select choice 1: **Plot Setup.** Select a plot to define and press [F1]. Select **Plot Type 1: Scatter.** Select a mark type. Specify the lists where the data are stored as Xlist and Ylist, using VAR-LINK. Press [ENTER] to finish. Press [F5] to display the plot.

To find the correlation, press [F4] (CALC), then arrow to 3: **Regressions,** press the right arrow, and select **1:LinReg(a+bx).**

Then specify the lists where the data are stored. You can also select a y-function to store the equation of the line.

COMMENTS

Notice that if you **TRACE** (press [F3]) the scatterplot, the calculator will tell you the x- and y-value at each point.

Chapter 8. Linear Regression

DATA DESK

Select the y-variable and the x-variable. In the **Plot** menu choose **Scatterplot.** from the scatterplot HyperView menu, choose **Add Regression Line** to display the line. from the HyperView menu, choose **Regression** to compute the regression.

COMMENTS

Alternatively, find the regression first with the **Regression** command in the **Calc** menu. Click on the x-variable's name to open a menu that offers the scatterplot.

EXCEL

Make a scatterplot of the data. With the scatterplot front-most, select **Add Trendline...** from the **Chart** menu. Click the **Options** tab and select **Display Equation on Chart.** Click **OK.**

COMMENTS

The computer section for Chapter 7 shows how to make a scatterplot. We don't repeat those steps here.

EXCEL 2007

▶ Click on a blank cell in the spreadsheet.
▶ Go to the **Formulas** tab in the Ribbon and click **More Functions → Statistical.**
▶ Choose the **CORREL** function from the drop-down menu of functions.
▶ In the dialog that pops up, enter the range of one of the variables in the space provided.
▶ Enter the range of the other variable in the space provided.
▶ Click **OK.**

COMMENTS

The correlation is computed in the selected cell. Correlations computed this way will update if any of the data values are changed. Before you interpret a correlation coefficient, always make a scatterplot to check for nonlinearity and outliers. If the variables are not linearly related, the correlation coefficient cannot be interpreted.

JMP

Choose **Fit Y by X** from the **Analyze** menu. Specify the y-variable in the Select Columns box and click the **"Y, Response"** button. Specify the x-variable and click the **"X, Factor"** button. Click **OK** to make a scatterplot. In the

scatterplot window, click on the red triangle beside the heading labeled "Bivariate Fit . . ." and choose **"Fit Line."** JMP draws the least squares regression line on the scatterplot and displays the results of the regression in tables below the plot.

MINITAB

Choose **Regression** from the **Stat** menu. From the Regression submenu, choose **Fitted Line Plot.** In the Fitted Line Plot dialog, click in the **Response Y** box, and assign the y-variable from the

Variable list. Click in the **Predictor X** box, and assign the x-variable from the Variable list. Make sure that the Type of Regression Model is set to Linear. Click the **OK** button.

SPSS

Choose **Interactive** from the **Graphs** menu. From the interactive Graphs submenu, choose **Scatterplot.** In the Create Scatterplot dialog, drag the y-variable into the **y-axis target,** and the

x-variable into the **x-axis target.** Click on the **Fit** tab. Choose **Regression** from the **Method** popup menu. Click the **OK** button.

TI-NSPIRE

To plot and find the equation of the regression line, first create a scatterplot. Using named lists, press ▲ several times so that the first list is highlighted. Then press ⌃ ▸ so that the second list is highlighted. Press (menu), ③ for Data, and ④ for Quick Graph. Then press (menu), ③ for Actions, ⑤ for Regression, and ② for Show Linear.

To find the equation of the regression line on a full page, press (ⓘ), ① for Calculator, (menu), ⑥ for Statistics, ① for Stat

Calculations, and ④ for Linear Regression. Complete the dialog boxes.

To see the plot on a full page, press (ⓘ), and then ⑤ for Data & Statistics. Move the cursor to "Click to add variable," and then press (⌘) and select the list name. Repeat for the other axis. Then press (menu), ③ for Actions, ⑤ for Regression, and ② for Show Linear.

TI-89

To find the equation of the regression line (and add the line to a scatterplot), choose **LinReg (a+bx)** from the **Calc Regressions** menu and tell it the list names and a function to store the equation. To make a residuals plot, define a **PLOT** as a scatterplot. Specify your explanatory datalist as Xlist. For Ylist, find the list name **resid** from VAR-LINK by arrowing to the **STATVARS** portion. then press ② **(r)** and locate the list. press [ENTER] to finish the plot definition and [F5] to display the plot.

COMMENTS

Each time you execute a **LinReg** command, the calculator automatically computes the residuals and stores them in a data list named RESID. If you don't want to see this (or any other calculator-generated list) anymore, press [F1] (Tools) and select choice 3: Setup Editor. Leaving the box for lists to display blank will reset the calculator to show only lists 1 through 6.

Chapter 9. Regression Wisdom

DATA DESK

Click on the **HyperView** menu on the **Regression** output table. A menu drops down to offer scatterplots of residuals against predicted values, Normal probability plots of residuals, or just the ability to save the residuals and predicted values.

Click on the name of a predictor in the regression table to be offered a scatterplot of the residuals against that predictor.

COMMENTS

If you change any of the variables in the regression analysis, Data Desk will offer to update the plots of residuals.

EXCEL

The Data Analysis add-in for Excel includes a Regression command. The dialog box it shows offers to make plots of residuals.

COMMENTS

Do not use the Normal probability plot offered in the regression dialog. It is not what it claims to be and is wrong.

JMP

From the **Analyze** menu, choose **Fit Y by X.** Select **Fit Line.** Under Linear Fit, Select **Plot Residuals.** You can also choose

to **Save Residuals.** Subsequently, from the **Distribution** menu, choose **Normal quantile plot** or **histogram** for the residuals.

MINITAB

From the **Stat** menu, choose **Regression.** From the **Regression** submenu, select **Regression** again. In the Regression dialog, enter the response variable name in the "Response" box and the predictor variable name in the "Predictor" box. To specify saved results, in the Regression dialog, click **Storage.** Check "Residu-

als" and "Fits." Click **OK.** To specify displays, in the Regression dialog, click **Graphs.** Under "Residual Plots," select "Individual plots" and check "Residuals versus fits." Click **OK.** Now back in the Regression dialog, click **OK.** Minitab computes the regression and the requested saved values and graphs.

SPSS

From the **Analyze** menu, choose **Regression.** From the Regression submenu, choose **Linear.** After assigning variables to their roles in the regression, click the "**Plots...**" button.

In the Plots dialog, you can specify a Normal probability plot of residuals and scatterplots of various versions of standardized residuals and predicted values.

COMMENTS

A plot of *ZRESID against *PRED will look most like the residual plots we've discussed. SPSS standardizes the residuals by dividing by their standard deviation. (There's no need to subtract their mean; it must be zero.) The standardization doesn't affect the scatterplot.

TI-NSPIRE

To create a residual plot, press (ⓘ), then ⑤ for Data & Statistics. Move the cursor to "Click to add variable," and then press (⌘)

and select the list name. For the other axis, select the variable name **stat.resid.**

TI-89

To make a residuals plot, define a Plot as a scatterplot. Specify your explanatory datalist as **Xlist.** For **Ylist,** find the list name resid from **VAR-LINK** by arrowing to the **STATVARS** portion. Then press ② **(r)** and locate the list. Press ENTER to finish the plot definition and F5 to display the plot.

COMMENTS

Each time you execute a **LinReg** command, the calculator automatically computes the residuals and stores them in a data list named **RESID.** If you don't want to see this (or any other calculator-generated list) anymore, press F1 (Tools) and select choice 3: Setup Editor. Leaving the box for lists to display blank will reset the calculator to show only lists 1 through 6.

Chapter 10. Re-expressing Data: Get It Straight!

DATA DESK

To re-express a variable in Data Desk, select the variable and Choose the function to re-express it from the **Manip > Transform** menu. Square root, log, reciprocal, and reciprocal root are immediately available. For others, make a derived variable and type the function. Data Desk makes a new derived variable that holds the re-expressed values. Any value changed in the original variable will immediately be re-expressed in the derived variable.

COMMENTS

Or choose **Manip > Transform > Dynamic > Box-Cox** to generate a continuously changeable variable and a slider that specifies the power. Set plots to **Automatic Update** in their HyperView menus and watch them change dynamically as you drag the slider.

EXCEL

To re-express a variable in Excel, use Excel's built-in functions as you would for any calculation. Changing a value in the original column will change the re-expressed value.

JMP

To re-express a variable in JMP, double-click to the right of the last column of data to create a new column. Name the new column and select it. Choose **Formula** from the **Cols** menu. In the Formula dialog, choose the transformation and variable that you wish to assign to the new column. Click the **OK** button. JMP places the re-expressed data in the new column.

COMMENTS

The log and square root re-expressions are found in the **Transcendental** menu of functions in the formula dialog.

MINITAB

To re-express a variable in MINITAB, choose **Calculator** from the **Calc** menu. In the Calculator dialog, specify a name for the new re-expressed variable. Use the **Functions List,** the calculator

buttons, and the **Variables list** box to build the expression. Click **OK.**

SPSS

To re-express a variable in SPSS, Choose **Compute** from the **Transform** menu. Enter a name in the Target Variable field. Use the calculator and Function List to build the expression. Move a

variable to be re-expressed from the source list to the Numeric Expression field. Click the **OK** button.

TI-NSPIRE

To re-express data, create a new list and enter the formula in the cell in the second row. For example, if one column has a list

named *time,* another list can be created using the formula *log(time).*

TI-89

To re-express data stored in a list, perform the re-expression on the whole list and store it in another list. For example, to use the common (base 10) logarithms of the data in list1, on the home screen, enter the command **log(list1)** STO▸ **list2.**

COMMENTS

▸ To find the log command, press CATALOG then ④ (L) arrow to log, and press ENTER.
▸ Natural logs are **LN** (press 2nd X).
▸ For square roots, press 2nd ×.

Chapter 11. Understanding Randomness

DATA DESK

Generate random numbers in Data Desk with the **Generate Random Numbers . . .** command in the **Manip** menu. A dialog guides you in specifying the number of variables to fill, the number of cases, and details about the values. For most simulations, generate random uniform values.

COMMENTS

Bernoulli Trials generate random values that are 0 or 1, with a specified chance of a 1.
Binomial Experiments automatically generate a specified number of Bernoulli trials and count the number of 1's.

EXCEL

The **RAND** function generates a random value between 0 and 1. You can multiply to scale it up to any range you like and use the INT function to turn the result into an integer.

COMMENTS

Published tests of Excel's random-number generation have declared it to be inadequate. However, for simple simulations, it should be OK. Don't trust it for important large simulations.

JMP

In a new column, in the **Cols** menu choose **Column Info...** In the dialog, click the **New Property** button, and choose **Formula** from the drop-down menu.

Click the **Edit Formula** button, and in the **Functions(grouped)** window click on **Random. Random Integer (10),** for example, will generate a random integer between 1 and 10.

MINITAB

In the **Calc** menu, choose **Random Data...** In the Random Data submenu, choose **Uniform...**

A dialog guides you in specifying details of range and number of columns to generate.

SPSS

The **RV.UNIFORM(min, max)** function returns a random value that is equally likely between the min and max limits.

TI-NSPIRE

To generate random integers, press (home), (1) for Calculator, (menu), (5) for Probability, (4) for Random, and (2) for Integer. Then type the range for the random integers, such as randInt(1,6).

To create a list of random integers, type the length of the list as the third value, such as randInt(1,6,10).

TI-89

To generate random numbers, move the cursor to highlight the name of a blank list. Use **5:RandInt** from the F4 (Calc) Probability menu. This command will produce any number of random integers in the specified range.

COMMENTS

Some examples:
RandInt(0,10) randomly chooses a 0 or a 1. This is an effective simulation of 10 coin tosses.
RandInt(1,6,2) randomly returns two integers between 1 and 6. This is a good way to simulate rolling two dice.
RandInt(0,56,3) produces three random integers between 0 and 56, a nice way to simulate the chapter's dorm room lottery.

Chapter 16. Random Variables

TI-NSPIRE

To compute the mean and standard deviation for a discrete random variable, enter the values in one named list and the probabilities in another. Then press (home), (1) for Calculator, (menu), (6) for

Statistics, (1) for Stat Calculations, and (1) for One-Variable Statistics. Enter 2 for the prompt for the number of lists, (tab) to OK, (enter), and complete the dialog box.

TI-89

To calculate the mean and standard deviation of a discrete random variable, enter the probability model in two lists:
- In one list (say, list1) enter the x-values of the variable.
- In a second list (say, list2) enter the associated probabilities $P(X = x)$.
- From the **STAT CALC** (F4) menu select **1-VarStats.** Use VAR-LINK to enter the list name list1 in the List box and list2 in the Freq box.

COMMENTS

You can enter the probabilities as fractions; the calculator will change them to decimals for you.
Notice that the calculator knows enough to compute only the standard deviation σ, but mistakenly uses $\bar{x}$ when it should say μ. Make sure you don't make that mistake!

Chapter 17. Probability Models

The only important differences among these functions are in what they are named and the order of their arguments. In these functions, pdf stands for "probability density function"—what we've been calling a probability model. The letters cdf stand for "cumulative distribution function," the technical term when we want to accumulate probabilities over a range of values. These technical terms show up in many of the function names. The term "cumulative" in a function name says that it corresponds to a cdf.

Generically, the four functions are as follows:

Geometric pdf (*prob, x*)	Finds the individual geometric probability of getting the first success on trial *x* when the probability of success is *prob*.	For example, the probability of finding the first Tiger Woods picture in the fifth cereal box is Geometric pdf(0.2, 5)
Geometric cdf (*prob, x*)	Finds the cumulative probability of getting the first success on or before trial *x*, when the probability of success is *prob*.	For example, the total probability of finding Tiger's picture in one of the first 4 boxes is Geometric cdf(0.2, 4)
Binomial pdf (*n, prob, x*)	Finds the probability of getting *x* successes in *n* trials when the probability of success is *prob*.	For example, Binomial pdf(5, 0.2, 2) is the probability of finding Tiger's picture exactly twice among 5 boxes of cereal.

DATA DESK

BinomDistr(*x, n, prob*) (pdf)
CumBinomDistr(*x, n, prob*) (cdf)

COMMENTS

Data Desk does not compute Geometric probabilities.
These functions work in derived variables or in scratchpads.

EXCEL

Binomdist(*x, n, prob, cumulative*)

COMMENTS

Set cumultive = *true* or for cdf, *false* for pdf.
Excel's function fails when *x* or *n* is large.
Possibly, it does not use the Normal approximation.
Excel does not compute Geometric probabilities.

JMP

Binomial Probability (*prob, n, x*) (pdf)
Binomial Distribution (*prob, n, x*) (cdf)

COMMENTS

JMP does not compute Geometric probabilities.

MINITAB

Choose **Probability Distributions** from the **Calc** menu.
Choose **Binomial** from the Probability Distributions submenu.
To calculate the probability of getting *x* successes in *n* trials, choose **Probability.**

To calculate the probability of getting *x* or fewer successes among *n* trials, choose **Cumulative Probability.**
For Geometric, choose **Geometric** from the Probability Distribution submenu.

SPSS

PDF.GEOM(*x, prob*)
CDF.GEOM(*x, prob*)

PDF.BINOM(*x, n, prob*)
CDF.BINOM(*x, n, prob*)

TI-NSPIRE

To compute geometric and binomial probabilities, press (menu), (5) for Probability, and (5) for Distributions. Select the menu item.

Pdf is for the probability distribution function; Cdf will display cumulative probabilities. Complete the dialog box.

TI-89

Find the commands under the F5 (Distributions) menu.

▸ F: **Geometric Pdf** will ask for *p* and *x*. It returns the probability of the first success occurring on the *x*th trial.
▸ G: **Geometric Cdf** will ask for *p* and the upper and lower values of interest, say *a* and *b*. It returns $P(a \leq X \leq b)$, the probability the first success occurs between the *a*th and *b*th trials, inclusive.
▸ A: **Binomial Pdf** asks for *n, p,* and *x*.
▸ B: **Binomial Cdf** asks for *n, p,* and the lower and upper values of interest.

COMMENTS

For Geometric variables, when finding $P(X \geq a)$ specify an upper value of infinity, 1EE99, or a very large number.
For Binomial variables, when finding $P(X \geq a)$, the upper value is *n*.

Chapter 19. Confidence Intervals for Proportions

DATA DESK

Data Desk does not offer built-in methods for inference with proportions.

COMMENTS

For summarized data, open a Scratchpad to compute the standard deviation and margin of error by typing the calculation. Then use **z-interval for individual μs.**

EXCEL

Inference methods for proportions are not part of the standard Excel tool set.

COMMENTS

For summarized data, type the calculation into any cell and evaluate it.

JMP

For a **categorical** variable that holds category labels, the **Distribution** platform includes tests and intervals for proportions. For summarized data, put the category names in one variable and the frequencies in an adjacent variable. Designate the frequency column to have the **role** of **frequency.** Then use the **Distribution** platform.

COMMENTS

JMP uses slightly different methods for proportion inferences than those discussed in this text. Your answers are likely to be slightly different, especially for small samples.

MINITAB

Choose **Basic Statistics** from the **Stat** menu.
- Choose **1Proportion** from the Basic Statistics submenu.
- If the data are category names in a variable, assign the variable from the variable list box to the **Samples in columns** box. If you have summarized data, click the **Summarized Data** button and fill in the number of trials and the number of successes.
- Click the **Options** button and specify the remaining details.

- If you have a large sample, check **Use test and interval based on normal distribution.** Click the **OK** button.

COMMENTS

When working from a variable that names categories, MINITAB treats the last category as the "success" category. You can specify how the categories should be ordered.

SPSS

SPSS does not find confidence intervals for proportions.

TI-NSPIRE

To compute a confidence interval for a population proportion, press ⌂, ① for Calculator, ⊞ (menu), ⑥ for Statistics, ⑥ for Confidence Intervals, and ⑤ for 1-Prop z-interval. Complete the

dialog box. Be sure to enter the number of successes, x, as a whole number, and the C level as a decimal, such as .99.

TI-89

To calculate a confidence interval for a population proportion:
- Go to the **Ints** menu (2nd F2) and select **5:1-PropZInt.**
- Enter the number of successes observed and the sample size.
- Specify a confidence level.
- Calculate the interval.

COMMENTS

Beware: When you enter the value of x, you need the count, not the percentage. The count must be a whole number. If the number of successes are given as a percentage, you must first multiply np and round the result.

Chapter 20. Testing Hypotheses About Proportions

DATA DESK

Data Desk does not offer built-in methods for inference with proportions. The **Replicate Y by X** command in the **Manip** menu will "reconstruct" summarized count data so that you can display it.

COMMENTS

For summarized data, open a Scratchpad to compute the standard deviation and margin of error by typing the calculation. Then perform the test with the **z-test for individual μs** found in the Test command.

EXCEL

Inference methods for proportions are not part of the standard Excel tool set.

COMMENTS

For summarized data, type the calculation into any cell and evaluate it.

JMP

For a **categorical** variable that holds category labels, the **Distribution** platform includes tests and intervals of proportions. For summarized data, put the category names in one variable and the frequencies in an adjacent variable. Designate the frequency column to have the **role** of **frequency.** Then use the **Distribution** platform.

COMMENTS

JMP uses slightly different methods for proportion inferences than those discussed in this text. Your answers are likely to be slightly different.

MINITAB

Choose **Basic Statistics** from the **Stat** menu.
▸ Choose **1Proportion** from the Basic Statistics submenu.
▸ If the data are category names in a variable, assign the variable from the variable list box to the **Samples in columns** box.
▸ If you have summarized data, click the **Summarized Data** button and fill in the number of trials and the number of successes.
▸ Click the **Options** button and specify the remaining details.

▸ If you have a large sample, check **Use test and interval based on Normal distribution.**
▸ Click the **OK** button.

COMMENTS

When working from a variable that names categories, MINITAB treats the last category as the "success" category. You can specify how the categories should be ordered.

SPSS

SPSS does not find hypothesis tests for proportions.

TI-NSPIRE

To compute a hypothesis test for a population proportion, press ⓐ, ① for Calculator, (menu), ⑥ for Statistics, ⑦ for Stat Tests,

and ⑤ for 1-Prop z-test. Complete the dialog box. Be sure to enter the number of successes, x, as a whole number.

TI-89

To do the mechanics of a hypothesis test for a proportion,
▸ Select **5:1-PropZTest** from the **STAT TESTS** [2nd][F1] menu.
▸ Specify the hypothesized proportion.
▸ Enter the observed value of x.
▸ Specify the sample size.
▸ Indicate what kind of test you want: one-tail lower tail, two-tail, or one-tail upper tail.

▸ Specify whether to calculate the result or draw the result (a normal curve with p-value area shaded.)

COMMENTS

Beware: When you enter the value of x, you need the *count,* not the percentage. The count must be a whole number. If the number of successes is given as a percent, you must first multiply np and round the result to obtain x.

Chapter 22. Comparing Two Proportions

DATA DESK

Data Desk does not offer built-in methods for inference with proportions. Use **Replicate Y by X** to construct data corresponding to given proportions and totals.

COMMENTS

For summarized data, open a Scratchpad to compute the standard deviations and margin of error by typing the calculation.

EXCEL

Inference methods for proportions are not part of the standard Excel tool set.

COMMENTS

For summarized data, type the calculation into any cell and evaluate it.

JMP

For a **categorical** variable that holds category labels, the **Distribution** platform includes tests and intervals of proportions. For summarized data, put the category names in one variable and the frequencies in an adjacent variable. Designate the frequency column to have the **role** of **frequency.** Then use the **Distribution** platform.

COMMENTS

JMP uses slightly different methods for proportion inferences than those discussed in this text. Your answers are likely to be slightly different.

MINITAB

To find a hypothesis test for a proportion, Choose **Basic Statistics** from the **Stat** menu.

Choose **2Proportions . . .** from the Basic Statistics submenu. If the data are organized as category names in one column and case IDs in another, assign the variables from the variable list box to the **Samples in one column** box. If the data are organized as two separate columns of responses, click on **Samples in different columns:** and assign the variables from the variable list box. If you have summarized data, click the **Summarized Data** button and fill in the number of trials and the number of successes for each group.

Click the **Options** button and specify the remaining details. Remember to click the **Use pooled estimate of p for test** box when testing the null hypothesis of no difference between proportions. Click the **OK** button.

COMMENTS

When working from a variable that names categories, MINITAB treats the last category as the "success" category. You can specify how the categories should be ordered.

SPSS

SPSS does not find hypothesis tests for proportions.

TI-NSPIRE

To compute a confidence interval for the difference between two population proportions, press ⌂, ① for Calculator, ⊞, ⑥ for Statistics, ⑥ for Confidence Intervals, and ⑥ for 2-Prop z-interval. Complete the dialog box. Be sure to enter each number of successes as a whole number, and the C level as a decimal, such as .99.

To compute a hypothesis test for the difference between two population proportions, press ⌂, ① for Calculator, ⊞, ⑥ for Statistics, ⑦ for Stat Tests, and ⑥ for 2-Prop z-test. Complete the dialog box. Be sure to enter each number of successes as a whole number.

TI-89

To calculate a confidence interval for the difference between two population proportions,

▸ Select **6:2-PropZInt** from the **STAT Ints** menu.
▸ Enter the observed counts and the sample sizes for both samples.
▸ Specify a confidence level.
▸ Calculate the interval.

To do the mechanics of a hypothesis test for equality of population proportions,

▸ Select **6:2-PropZTest** from the **STAT Tests** menu.
▸ Enter the observed counts and sample sizes.

▸ Indicate what kind of test you want: one-tail upper tail, lower tail, or two-tail.
▸ Specify whether results should simply be calculated or displayed with the area corresponding to the P-value of the test shaded.

COMMENTS

Beware: When you enter the value of x, you need the *count*, not the percentage. The count must be a whole number. If the number of successes is given as a percent, you must first multiply np and round the result to obtain x.

Chapter 23. Inferences About Means

DATA DESK

Select variables.

From the **Calc** menu, choose **Estimate** for confidence intervals or **Test** for hypothesis tests. Select the interval or test

from the drop-down menu and make other choices in the dialog.

EXCEL

Specify formulas. Find t^* with the TINV(alpha, df) function.

COMMENTS

Not really automatic. There's no easy way to find P-values in Excel.

JMP

From the **Analyze** menu, select **Distribution.** For a confidence interval, scroll down to the "Moments" section to find the interval limits. For a hypothesis test, click the red triangle next to the variable's name and choose **Test Mean** from the menu. Then fill in the resulting dialog.

COMMENTS

"Moment" is a fancy statistical term for means, standard deviations, and other related statistics.

MINITAB

From the **Stat** menu, choose the **Basic Statistics** submenu. From that menu, choose **1-sample t . . . .** Then fill in the dialog.

COMMENTS

The dialog offers a clear choice between confidence interval and test.

SPSS

From the **Analyze** menu, choose the **Compare Means** submenu. From that, choose the **One-Sample t-test** command.

COMMENTS

The commands suggest neither a single mean nor an interval. But the results provide both a test and an interval.

TI-NSPIRE

To compute a confidence interval for a population mean, press Ⓐ, ① for Calculator, (menu), ⑥ for Statistics, ⑥ for Confidence Intervals, and ② for *t*-interval. Select between Data and Stats, (tab) to OK, and press Ⓔ. Complete the dialog box. Be sure to enter the number of successes, *x*, as a whole number, and the C level as a decimal, such as .99.

To compute a hypothesis test for a population mean, press Ⓐ, ① for Calculator, (menu), ⑥ for Statistics, ⑦ for Stat Tests, and ② for *t*-test. Select between Data and Stats, (tab) to OK, and Ⓔ. Complete the dialog box.

TI-89

Finding a confidence interval:
In the **STAT Ints** menu, choose **2:TInterval.** Specify whether you are using data stored in a list or whether you will enter the mean, standard deviation, and sample size. You must also specify the desired level of confidence.
Testing a hypothesis:
In the **STAT Tests** menu, choose **2:T-Test.** You must specify whether you are using data stored in a list or whether you will

enter the mean, standard deviation, and size of your sample. You must also specify the hypothesized model mean and whether the test is to be two-tail, lower-tail, or upper-tail. Select whether the test is to be simply computed or whether to display the distribution curve and highlight the area corresponding to the P-value of the test.

Chapter 24. Comparing Means

There are two ways to organize data when we want to compare two independent groups. The data can be in two lists, as in the table at the start of this chapter. Each list can be thought of as a variable. In this method, the variables in the batteries example would be *Brand Name* and *Generic*. Graphing calculators usually prefer this form, and some computer programs can use it as well.

There's another way to think about the data. What is the response variable for the battery life experiment? It's the *Time* until the music stopped. But the values of this variable are in both columns, and actually there's an experiment factor here, too—namely, the *Brand* of the battery. So, we could put the data into two different columns, one with the *Times* in it and one with the *Brand*. Then the data would look as shown in the table to the right.

This way of organizing the data makes sense as well. Now the factor and the response variables are clearly visible. You'll have to see which method your program requires. Some packages even allow you to structure the data either way.

The commands to do inference for two independent groups on common statistics technology are not always found in obvious places. Here are some starting guidelines.

Time	Brand
194.0	Brand name
205.5	Brand name
199.2	Brand name
172.4	Brand name
184.0	Brand name
169.5	Brand name
190.7	Generic
203.5	Generic
203.5	Generic
206.5	Generic
222.5	Generic
209.4	Generic

DATA DESK

Select variables.
From the **Calc** menu, choose **Estimate** for confidence intervals or **Test** for hypothesis tests. Select the interval or test from the drop-down menu and make other choices in the dialog.

COMMENTS

Data Desk expects the two groups to be in separate variables.

EXCEL

From the Data Tab, Analysis Group, choose **Data Analysis.** Alternatively (if the Data Analysis Tool Pack is not installed), in the Formulas Tab, choose More functions > Statistical > TTEST, and specify Type=3 in the resulting dialog.
Fill in the cell ranges for the two groups, the hypothesized difference, and the alpha level.

COMMENTS

Excel expects the two groups to be in separate cell ranges. Notice that, contrary to Excel's wording, we do not need to assume that the variances are *not* equal; we simply choose not to assume that they *are* equal.

JMP

From the **Analyze** menu, select **Fit y by x.** Select variables: a **Y, Response** variable that holds the data and an **X, Factor** variable that holds the group names. JMP will make a dotplot. Click the **red triangle** in the dotplot title, and choose **Unequal variances.** The *t*-test is at the bottom of the resulting table. Find the P-value from the Prob>F section of the table (they are the same).

COMMENTS

JMP expects data in one variable and category names in the other. Don't be misled: There is no need for the variances to be unequal to use two-sample *t* methods.

MINITAB

From the **Stat** menu, choose the **Basic Statistics** submenu. From that menu, choose **2-sample t....** Then fill in the dialog.

COMMENTS

The dialog offers a choice of data in two variables, or data in one variable and category names in the other.

SPSS

From the **Analyze** menu, choose the **Compare Means** submenu. From that, choose the **Independent-Samples t-test** command. Specify the data variable and "group variable." Then type in the labels used in the group variable. SPSS offers both the two-sample and pooled-*t* results in the same table.

COMMENTS

SPSS expects the data in one variable and group names in the other. If there are more than two group names in the group variable, only the two that are named in the dialog box will be compared.

TI-NSPIRE

To compute a confidence interval for the difference between two population means, press ⌂, ① for Calculator, (menu), ⑥ for Statistics, ⑥ for Confidence Intervals, and ④ for 2-Sample *t*-interval. Select between Data and Stats, (tab) to OK, and ⏎. Complete the dialog box. Be sure to enter the C level as a decimal, such as .99.

To compute a hypothesis test for the difference between two population means, press ⌂, ① for Calculator, (menu), ⑥ for Statistics, ⑦ for Stat Tests, and ④ for 2-Sample *t*-test. Select between Data and Stats, (tab) to OK, and ⏎. Complete the dialog box.

TI-89

For a confidence interval: In the **STAT Ints** menu, choose **4:2-SampTInt.** You must specify if you are using data stored in two lists or if you will enter the means, standard deviations, and sizes of both samples. You must also indicate whether to pool the variances (when in doubt, say no) and specify the desired level of confidence.

To test a hypothesis: In the **STAT TESTS** menu, choose **4:2-SampTTest.** You must specify if you are using data stored in two lists or if you will enter the means, standard deviations, and sizes of both samples. You must also indicate whether to pool the variances (when in doubt, say no) and specify whether the test is to be two-tail, lower-tail, or upper-tail.

Chapter 25. Paired Samples and Blocks

DATA DESK

Select variables.
From the **Calc** menu, choose **Estimate** for confidence intervals or **Test** for hypothesis tests. Select the interval or test from the drop-down menu, and make other choices in the dialog.

COMMENTS

Data Desk expects the two groups to be in separate variables and in the same "Relation"—that is, about the same cases.

EXCEL

In Excel 2003 and earlier, select **Data Analysis** from the **Tools** menu. In Excel 2007, select **Data Analysis** from the **Analysis** Group on the **Data** Tab.
From the **Data Analysis** menu, choose **t-test: paired two-sample for Means.** Fill in the cell ranges for the two groups, the hypothesized difference, and the alpha level.

COMMENTS

Excel expects the two groups to be in separate cell ranges. **Warning:** Do not compute this test in Excel without checking for missing values. If there are any missing values (empty cells), Excel will usually give a wrong answer. Excel compacts each list, pushing values up to cover the missing cells, and then checks only that it has the same number of values in each list. The result is mismatched pairs and an entirely wrong analysis.

JMP

From the **Analyze** menu, select **Matched Pairs.** Specify the columns holding the two groups in the **Y Paired Response** Dialog. Click **OK.**

MINITAB

From the **Stat** menu, choose the **Basic Statistics** submenu. From that menu, choose **Paired t...** Then fill in the dialog.

COMMENTS

Minitab takes "First sample" minus "Second sample."

SPSS

From the **Analyze** menu, choose the **Compare Means** submenu. From that, choose the **Paired-Samples t-test** command. Select pairs of variables to compare, and click the arrow to add them to the selection box.

COMMENTS

You can compare several pairs of variables at once. Options include the choice to exclude cases missing in any pair from all tests.

TI-NSPIRE

For inference on a matched pair design, compute a third list of differences such as *diff = time2-time1.* Then construct the

confidence interval or conduct the hypothesis test in the same way as 1-sample procedures, using the list of differences.

TI-89

If the data are stored in two lists, say, list1 and list2, create a list of the differences: Move the cursor to the name of an empty list, and then use VAR-LINK to enter the command list1-list2. Press ENTER to perform the subtraction.

Since inference for paired differences uses one-sample *t*-procedures, select **2:T-Test** or **2:TInterval** from the **STAT Tests** or **Ints** menu. Specify as your data the list of differences you just created, and apply the procedure.

Chapter 26. Comparing Counts

DATA DESK

Select variables.
From the **Calc** menu, choose **Contingency Table.** From the table's HyperView menu, choose **Table Options.** (Or Choose **Calc > Calculation Options > Table Options.**) In the dialog, check the boxes for **Chi Square** and for **Standardized Residuals.** Data Desk will display the chi-square and its P-value below the table, and the standardized residuals within the table.

COMMENTS

Data Desk automatically treats variables selected for this command as categorical variables even if their elements are numerals.
The **Compute Counts** command in the table's HyperView menu will make variables that hold the table contents (as selected in the Table Options dialog), including the standardized residuals.

EXCEL

Excel offers the function
CHITEST(actual_range, expected_range), which computes a chi-square value for homogeneity. Both ranges are of the form UpperleftCell:LowerRightCell, specifying two rectangular tables that must hold counts (although Excel will not check for integer values). The two tables must be of the same size and shape.

COMMENTS

Excel's documentation claims this is a test for independence and labels the input ranges accordingly, but Excel offers no way to find expected counts, so the function is not particularly useful for testing independence. You can use this function only if you already know both tables of counts or are willing to program additional calculations.

JMP

From the **Analyze** menu, select **Fit Y by X.** Select variables: a Y, Response variable that holds responses for one variable, and an X, Factor variable that holds responses for the other. Both selected variables must be Nominal or Ordinal. JMP will make a plot and a contingency table. Below the contingency table, **JMP** offers a **Tests** panel. In that panel, the Chi Square for independence is called a **Pearson ChiSquare.** The table also offers the P-value.
Click on the Contingency Table title bar to drop down a menu that offers to include a **Deviation** and **Cell Chi square** in each cell of the table.

COMMENTS

JMP will choose a chi-square analysis for a **Fit Y by X** if both variables are nominal or ordinal (marked with an N or O), but not otherwise. Be sure the variables have the right type.
Deviations are the observed—expected differences in counts. Cell chi-squares are the squares of the standardized residuals. Refer to the deviations for the sign of the difference.
Look under **Distributions** in the **Analyze** menu to find a chi-square test for goodness-of-fit.

MINITAB

From the **Stat** menu, choose the **Tables** submenu. From that menu, choose **Chi Square Test . . . .** In the dialog, identify the columns that make up the table. Minitab will display the table and print the chi-square value and its P-value.

COMMENTS

Alternatively, select the **Cross Tabulation . . .** command to see more options for the table, including expected counts and standardized residuals.

SPSS

From the **Analyze** menu, choose the **Descriptive Statistics** submenu. From that submenu, choose **Crosstabs . . . .** In the Crosstabs dialog, assign the row and column variables from the variable list. Both variables must be categorical. Click the **Cells** button to specify that standardized residuals should be displayed. Click the **Statistics** button to specify a chi-square test.

COMMENTS

SPSS offers only variables that it knows to be categorical in the variable list for the Crosstabs dialog. If the variables you want are missing, check that they have the right type.

TI-NSPIRE

To conduct a χ^2 goodness of fit test, enter the observed and the expected values into two named lists. Then press (ⓐ), (1) for Calculator, (menu), (6) for Statistics, (7) for Stat Tests, and (7) for χ^2 GOF. Complete the dialog box.

To conduct a χ^2 test of independence or homogeneity, first enter the data into a matrix. Press (ctrl) (▦) and select the matrix icon.

Enter the dimensions and (tab) to OK, and (↵). Then type the data into the matrix. Then press ▸ to exit the matrix, press (ctrl) (↵) and a matrix name such as *ma* to store the matrix. To complete the test, press (ⓐ), (1) for Calculator, (menu), (6) for Statistics, (7) for Stat Tests, and (8) for χ^2 2-way Test. Complete the dialog box.

TI-89

To test goodness-of-fit, enter the observed counts in a list and the expected counts in another list. Expected counts can be entered as n*p, and the calculator will compute them for you. From the **STAT Tests** menu, select **7:Chi2 GOF.** Enter the list names using VAR-LINK and the degrees of freedom, $k - 1$, where k is the number of categories. Select whether to simply calculate or display the result with the area corresponding to the P-value highlighted.

To test a hypothesis of homogeneity or independence, you need to enter the data as a matrix. From the home screen, press [APPS] and select **6:Data/Matrix Editor,** then select **3:New.** Specify type as Matrix and name the matrix in the **Variable** box. Specify the number of rows and columns. Type the entries, pressing [ENTER] after each. Press [2nd] [ESC] to leave the editor. To do the test, choose **8:Chi2 2-way** from the **STAT Tests** menu.

Chapter 27. Inferences for Regression

DATA DESK

- ▸ Select *Y*- and *X*-variables.
- ▸ From the **Calc** menu, choose **Regression.**
- ▸ Data Desk displays the regression table.
- ▸ Select plots of residuals from the Regression table's HyperView menu.

COMMENTS

You can change the regression by dragging the icon of another variable over either the *Y*- or *X*-variable name in the table and dropping it there. The regression will recompute automatically.

EXCEL

- ▸ In Excel 2003 and earlier, select Data Analysis from the **Tools** menu. In Excel 2007, select Data Analysis from the **Analysis Group** on the Data Tab.
- ▸ Select Regression from the **Analysis Tools** list.
- ▸ Click the **OK** button.
- ▸ Enter the data range holding the Y-variable in the box labeled "Y-range".
- ▸ Enter the range of cells holding the X-variable in the box labeled "X-range."
- ▸ Select the **New Worksheet Ply** option.
- ▸ Select **Residuals** options. Click the **OK** button.

COMMENTS

The Y and X ranges do not need to be in the same rows of the spreadsheet, although they must cover the same number of cells. But it is a good idea to arrange your data in parallel columns as in a data table.

Although the dialog offers a Normal probability plot of the residuals, the data analysis add-in does not make a correct probability plot, so don't use this option.

JMP

- From the **Analyze** menu, select **Fit Y by X.**
- Select variables: a Y, Response variable, and an X, Factor variable. Both must be continuous (quantitative).
- JMP makes a scatterplot.
- Click on the red triangle beside the heading labeled **Bivariate Fit...** and choose **Fit Line.** JMP draws the least squares regression line on the scatterplot and displays the results of the regression in tables below the plot.
- The portion of the table labeled "Parameter Estimates" gives the coefficients and their standard errors, t-ratios, and P-values.

COMMENTS

JMP chooses a regression analysis when both variables are "Continuous." If you get a different analysis, check the variable types.

The Parameter table does not include the residual standard deviation s_e. You can find that as Root Mean Square Error in the Summary of Fit panel of the output.

MINITAB

- Choose **Regression** from the **Stat** menu.
- Choose **Regression...** from the **Regression** submenu.
- In the Regression dialog, assign the Y-variable to the Response box and assign the X-variable to the Predictors box.
- Click the **Graphs** button.
- In the Regression-Graphs dialog, select **Standardized residuals,** and check **Normal plot of residuals** and **Residuals versus fits.**

- Click the **OK** button to return to the Regression dialog.
- Click the **OK** button to compute the regression.

COMMENTS

You can also start by choosing a Fitted Line plot from the **Regression** submenu to see the scatterplot first—usually good practice.

SPSS

- Choose **Regression** from the **Analyze** menu.
- Choose **Linear** from the **Regression** submenu.
- In the Linear Regression dialog that appears, select the Y-variable and move it to the dependent target. Then move the X-variable to the independent target.
- Click the **Plots** button.

- In the Linear Regression Plots dialog, choose to plot the *SRESIDs against the *ZPRED values.
- Click the **Continue** button to return to the Linear Regression dialog.
- Click the **OK** button to compute the regression.

TI-NSPIRE

To compute a confidence interval for a population slope, first enter the data into two named lists. Then press (a), (1) for Calculator, (menu), (6) for Statistics, (6) for Confidence Intervals, and (7) for Linear Reg t-intervals. Select slope, (tab) to OK, and (enter). Complete the dialog box. Be sure to enter the C level as a decimal, such as .99.

To compute a hypothesis test for a population slope, first enter the data into two named lists. Then press (a), (1) for Calculator, (menu), (6) for Statistics, (7) for Stat Tests, and (A) for Linear Reg t-test. Complete the dialog box.

TI-89

Under **STAT Tests** choose **A:LinRegTTest.** Specify the two lists where the data are stored and (usually) choose the two-tail option. Select an equation name to store the resulting line. In addition to reporting the calculated value of t and the P-value, the calculator will tell you the coefficients of the regression equation (a and b), the values of r^2 and r, the value of s used in predic-

tion and confidence intervals, and the standard error of the slope. For 95% prediction and confidence intervals, choose **7:LinRegTint** from the **STAT Ints** menu. Specify the two lists where the data are stored, and select an equation name to store the resulting line. Select for an interval for the slope or for a response. If for a response, enter the x-value.

Chapter 28. Analysis of Variance

DATA DESK

- Select the response variable as Y and the factor variable as X.
- From the **Calc** menu, choose **ANOVA.**
- Data Desk displays the ANOVA table.
- Select plots of residuals from the ANOVA table's HyperView menu.

COMMENTS

Data Desk expects data in "stacked" format. You can change the ANOVA by dragging the icon of another variable over either the Y or X variable name in the table and dropping it there. The analysis will recompute automatically.

EXCEL

- ▸ In Excel 2003 and earlier, select **Data Analysis** from the Tools menu.
- ▸ In Excel 2007, select **Data Analysis** from the Analysis Group on the Data Tab.
- ▸ Select **Anova Single Factor** from the list of analysis tools.
- ▸ Click the **OK** button.
- ▸ Enter the data range in the box provided.
- ▸ Check the **Labels in First Row** box, if applicable.
- ▸ Enter an alpha level for the F-test in the box provided.
- ▸ Click the **OK** button.

COMMENTS

The data range should include two or more columns of data to compare. Unlike all other statistics packages, Excel expects each column of the data to represent a different level of the factor. However, it offers no way to label these levels. The columns need not have the same number of data values, but the selected cells must make up a rectangle large enough to hold the column with the most data values.

JMP

- ▸ From the **Analyze** menu select **Fit Y by X.**
- ▸ Select variables: a quantitative Y, Response variable, and a categorical X, Factor variable.
- ▸ JMP opens the **Oneway** window.
- ▸ Click on the red triangle beside the heading, select **Display Options**, and choose **Boxplots.**

- ▸ From the same menu choose the **Means/ANOVA.t-test** command.
- ▸ JMP opens the oneway ANOVA output.

COMMENTS

JMP expects data in "stacked" format with one response and one factor variable.

MINITAB

- ▸ Choose **ANOVA** from the Stat menu.
- ▸ Choose **One-way...** from the **ANOVA** submenu.
- ▸ In the One-way Anova dialog, assign a quantitative Y variable to the Response box and assign a categorical X variable to the Factor box.
- ▸ Check the **Store Residuals** check box.
- ▸ Click the **Graphs** button.
- ▸ In the ANOVA-Graphs dialog, select **Standardized residuals,** and check **Normal plot of residuals** and **Residuals versus fits.**

- ▸ Click the **OK** button to return to the Regression dialog.
- ▸ Click the **OK** button to compute the regression.

COMMENTS

If your data are in unstacked format, with separate columns for each treatment level, choose **One-way (unstacked)** from the **ANOVA** submenu.

SPSS

- ▸ Choose **Compare Means** from the **Analyze** menu.
- ▸ Choose **One-way ANOVA** from the **Compare Means** submenu.
- ▸ In the One-Way ANOVA dialog, select the Y-variable and move it to the dependent target. Then move the X-variable to the independent target.
- ▸ Click the **OK** button.

COMMENTS

SPSS expects data in stacked format. The **Contrasts** and **Post Hoc** buttons offer ways to test contrasts and perform multiple comparisons. See your SPSS manual for details.

TI-89

Under **STAT** Tests, choose **C:ANOVA**
- ▸ Specify the input method (Data or Stats) according to whether you have data entered as one list for each group or summary statistics for each group, and specify the number of groups. Press ÷.
- ▸ If Data, you will then be asked to supply the name of each list.
- ▸ If Stats, you will be asked for the stats for each group. Enter n, $\bar{x}$, and s for each group separated by commas and within curly braces ({and}).
- ▸ Press ÷ to perform the calculations.

COMMENTS

In addition to the ANOVA table output, the calculator creates three new lists—the means for each group (in the order specified) and *individual* 95% confidence interval upper and lower bounds.

Chapter 29. Multiple Regression

DATA DESK

- ▸ Select Y- and X-variable icons.
- ▸ From the **Calc** menu, choose **Regression.**
- ▸ Data Desk displays the regression table.
- ▸ Select plots of residuals from the Regression table's HyperView menu.

COMMENTS

You can change the regression by dragging the icon of another variable over either the Y- or an X-variable name in the table and dropping it there. You can add a predictor by dragging its icon into that part of the table. The regression will recompute automatically.

EXCEL

- ▸ In Excel 2003 and earlier, select **Data Analysis** from the **Tools** menu.
- ▸ In Excel 2007, select **Data Analysis** from the **Analysis Group** on the Data Tab.
- ▸ Select **Regression** from the **Analysis Tools** list.
- ▸ Click the **OK** button.
- ▸ Enter the data range holding the Y-variable in the box labeled "Y-range."
- ▸ Enter the range of cells holding the X-variables in the box labeled "X-range."
- ▸ Select the **New Worksheet Ply** option.
- ▸ Select **Residuals** options. Click the **OK** button.

COMMENTS

The Y and X ranges do not need to be in the same rows of the spreadsheet, although they must cover the same number of cells. But it is a good idea to arrange your data in parallel columns as in a data table. The X-variables must be in adjacent columns. No cells in the data range may hold non-numeric values.

Although the dialog offers a Normal probability plot of the residuals, the data analysis add-in does not make a correct probability plot, so don't use this option.

JMP

- ▸ From the **Analyze** menu select **Fit Model.**
- ▸ Specify the response, Y. Assign the predictors, X, in the **Construct Model Effects** dialog box.
- ▸ Click on **Run Model.**

COMMENTS

JMP chooses a regression analysis when the response variable is "Continuous." The predictors can be any combination of quantitative or categorical. If you get a different analysis, check the variable types.

MINITAB

- ▸ Choose **Regression** from the **Stat** menu.
- ▸ Choose **Regression . . .** from the **Regression** submenu.
- ▸ In the Regression dialog, assign the Y-variable to the Response box and assign the X-variables to the Predictors box.
- ▸ Click the **Graphs** button.

- ▸ In the Regression-Graphs dialog, select **Standardized residuals,** and check **Normal plot of residuals** and **Residuals versus fits.**
- ▸ Click the **OK** button to return to the Regression dialog.
- ▸ Click the **OK** button to compute the regression.

SPSS

- ▸ Choose **Regression** from the **Analyze** menu.
- ▸ Choose **Linear** from the **Regression** submenu.
- ▸ When the Linear Regression dialog appears, select the Y-variable and move it to the dependent target. Then move the X-variables to the independent target.
- ▸ Click the **Plots** button.

- ▸ In the Linear Regression Plots dialog, choose to plot the *SRESIDs against the *ZPRED values.
- ▸ Click the **Continue** button to return to the Linear Regression dialog.
- ▸ Click the **OK** button to compute the regression.

TI-89

Under **STAT Tests** choose **B:MultREg Tests**
- ▸ Specify the number of predictor variables, and which lists contain the response variable and predictor variables.
- ▸ Press ⊞ to perform the calculations.

COMMENTS

- ▸ The first portion of the output gives the F-statistic and its P-value as well as the values of R^2, $Adj'R^2$, the standard deviation of the residuals (s), and the Durbin-Watson statistic, which measures correlation among the residuals.

- ▸ The rest of the main output gives the components of the F-test, as well as values of the coefficients, their standard errors, and associated t-statistics along with P-values. You can use the right arrow to scroll through these lists (if desired).
- ▸ The calculator creates several new lists that can be used for assessing the model and its conditions: Yhatlist, resid, sresid (standardized residuals), leverage, and cookd, as well as lists of the coefficients, standard errors, t's, and P-values.

Answers

APPENDIX C

Here are the "answers" to the exercises for the chapters and the unit reviews. As we said in Chapter 1, the answers are outlines of the complete solution. Your solution should follow the model of the Step-By-Step examples, where appropriate. You should explain the context, show your reasoning and calculations, and draw conclusions. For some problems, what you decide to include in an argument may differ somewhat from the answers here. But, of course, the numerical part of your answer should match the numbers in the answers shown.

CHAPTER 2

1. Categorical
2. Categorical
3. Quantitative
4. Quantitative
5. Answers will vary.
6. Answers will vary.
7. *Who*—2500 cars
   What—Distance from car to bicycle
   Population—All cars passing bicyclists
8. *Who*—30 similar companies
   What—401(k) employee participation rates
   Population—All similar companies
9. *Who*—Coffee drinkers at a Newcastle University coffee station
   What—Amount of money contributed
   Population—All people in honor system payment situations
10. *Who*—120 first-run movies in 2005
    What—Length of the movie and profit
    Population—All first-run movies
11. *Who*—25,892 men aged 30 to 87
    What—Fitness level and cause of death
    Population—All men
12. *Who*—10 crankshafts at Cleveland Casting
    What—The pouring temperature of molten iron
    Population—All Cleveland Casting crankshafts
13. *Who*—54 bears
    Cases—Each bear is a case.
    What—Weight, neck size, length, and sex
    When—Not specified
    Where—Not specified
    Why—To estimate weight from easier-to-measure variables
    How—Researchers collected data on 54 bears they were able to catch.
    Variable—Weight
       Type—Quantitative
       Units—Not specified
    Variable—Neck size
       Type—Quantitative
       Units—Not specified
    Variable—Length
       Type—Quantitative
       Units—Not specified

   Variable—Sex
      Type—Categorical
14. *Who*—Students
    Cases—Each student is an individual.
    What—Age, race or ethnicity, number of absences, grade level, reading score, math score, and disabilities/special needs
    When—Current
    Where—Not specified
    Why—Keeping this information is a state requirement.
    How—The information is collected and stored as part of school records.
    Variable—Age
       Type—Quantitative
       Units—Not specified, probably years (perhaps years, months)
    Variable—Race or ethnicity
       Type—Categorical
    Variable—Days absent
       Type—Quantitative
       Units—Number of days
    Variable—Current grade level
       Type—Categorical (could be quantitative for some purposes)
    Variable—Standardized reading score
       Type—Quantitative
       Units—Not specified
    Variable—Standardized math score
       Type—Quantitative
       Units—Not specified
    Variable—Disability/special needs
       Type—Categorical
15. *Who*—Arby's sandwiches
    Cases—Each sandwich is a case.
    What—Type of meat, number of calories, and serving size
    When—Not specified
    Where—Arby's restaurants
    Why—To assess nutritional value of sandwiches
    How—Report by Arby's restaurants
    Variable—Type of meat
       Type—Categorical
    Variable—Number of calories
       Type—Quantitative
       Units—Calories
    Variable—Serving size
       Type—Quantitative
       Units—Ounces

16. *Who*—1180 Americans
    Cases—Each of the 1180 Americans surveyed is an individual in this poll.
    What—Region, age, political affiliation, and whether or not the person voted in the 2006 midterm congressional election
    When—First quarter of 2007
    Where—United States
    Why—The information was gathered for presentation in a Gallup public opinion poll.
    How—Phone survey
    Variable—Region
       Type—Categorical
    Variable—Age
       Type—Quantitative
       Units—Not specified (years?)
    Variable—Party affiliation
       Type—Categorical
    Variable—Voted in last election?
       Type—Categorical

17. *Who*—882 births
    Cases—Each of the 882 births is a case.
    What—Mother's age, length of pregnancy, type of birth, level of prenatal care, birth weight of baby, sex of baby, and baby's health problems
    When—1998–2000
    Where—Large city hospital
    Why—Researchers were investigating the impact of prenatal care on newborn health.
    How—Not specified exactly, but probably from hospital records
    Variable—Mother's age
       Type—Quantitative
       Units—Not specified; probably years
    Variable—Length of pregnancy
       Type—Quantitative
       Units—Weeks
    Variable—Birth weight of baby
       Type—Quantitative
       Units—Not specified, probably pounds and ounces
    Variable—Type of birth
       Type—Categorical
    Variable—Level of prenatal care
       Type—Categorical
    Variable—Sex
       Type—Categorical
    Variable—Baby's health problems
       Type—Categorical

18. *Who*—385 species of flowers
    Cases—Each of the 385 species at each of the 47 years is a case, for a total of 18,095 cases.
    What—Date of first flowering
    When—Not specified
    Where—Southern England
    Why—The researchers believe that early flowering indicates a warming of the overall climate.
    How—Not specified
    Variables—Date of first flowering
       Type—Quantitative
       Units—Days

19. *Who*—Experiment subjects
    Cases—Each subject is an individual.
    What—Treatment (herbal cold remedy or sugar solution) and cold severity
    When—Not specified
    Where—Not specified

    Why—To test efficacy of herbal remedy on common cold
    How—The scientists set up an experiment.
    Variable—Treatment
       Type—Categorical
    Variable—Cold severity rating
       Type—Quantitative (perhaps ordinal categorical)
       Units—Scale from 0 to 5
    Concerns—The severity of a cold seems subjective and difficult to quantify. Scientists may feel pressure to report negative findings of herbal product.

20. *Who*—American vineyards
    Cases—Each vineyard is a case.
    What—Size of vineyard, number of years in existence, state, varieties of grapes grown, average case price, gross sales, and percent profit
    When—Not specified
    Where—United States
    Why—To provide information for American grape growers
    How—Not specified
    Variable—Vineyard size
       Type—Quantitative
       Units—Acres
    Variable—Number of years in existence
       Type—Quantitative
       Units—Years
    Variable—State
       Type—Categorical
    Variable—Varieties of grapes grown
       Type—Categorical
    Variable—Average case price
       Type—Quantitative
       Units—Not specified (dollars?)
    Variable—Gross sales
       Type—Quantitative
       Units—Not specified (dollars?)
    Variables—Percent profit
       Type—Quantitative
       Units—Percent

21. *Who*—Streams
    Cases—Each stream is a case.
    What—Name of stream, substrate of the stream, acidity of the water, temperature, BCI
    When—Not specified
    Where—Upstate New York
    Why—To study ecology of streams
    How—Not specified
    Variable—Stream name
       Type—Identifier
    Variable—Substrate
       Type—Categorical
    Variable—Acidity of water
       Type—Quantitative
       Units—pH
    Variable—Temperature
       Type—Quantitative
       Units—Degrees Celsius
    Variable—BCI
       Type—Quantitative
       Units—Not specified

22. *Who*—Each model of automobile
    Cases—Each vehicle model is a case.
    What—Vehicle manufacturer, vehicle type, weight, horsepower, and gas mileage for city and highway driving
    When—Current
    Where—United States

Why—By the Environmental Protection Agency to track fuel economy of vehicles

How—The data are collected from the manufacturer of each model.

Variable—Manufacturer
  Type—Categorical
Variable—Vehicle type
  Type—Categorical
Variable—Weight
  Type—Quantitative
  Units—Not specified (pounds)
Variable—Horsepower
  Type—Quantitative
  Units—Not specified (horsepower)
Variable—Gas mileage, city
  Type—Quantitative
  Units—Miles per gallon
Variable—Gas mileage, highway
  Type—Quantitative
  Units—Miles per gallon
Concerns—Do manufacturers' ratings of their own vehicles' gas mileage reflect customer experience?

23. *Who*—41 refrigerator models
  Cases—Each of the 41 refrigerator models is a case.
What—Brand, cost, size, type, estimated annual energy cost, overall rating, and repair history
When—2006
Where—United States
Why—To provide information to the readers of *Consumer Reports*
How—Not specified
Variable—Brand
  Type—Categorical
Variable—Cost
  Type—Quantitative
  Units—Not specified (dollars)
Variable—Size
  Type—Quantitative
  Units—Cubic feet
Variable—Type
  Type—Categorical
Variable—Estimated annual energy cost
  Type—Quantitative
  Units—Not specified (dollars)
Variable—Overall rating
  Type—Categorical (ordinal)
Variable—Percent requiring repair in last 5 years
  Type—Quantitative
  Units—Percent

24. *Who*—32 volunteers
  Cases—Each volunteer is an individual.
What—Sex, height, handedness, distance walked, sideline crossed
When—Not specified
Where—Not specified
Why—To see if people naturally walk in circles
How—Collected during a test on a football field
Variable—Sex
  Type—Categorical
Variable—Height
  Type—Quantitative
  Units—Not specified (inches?)
Variable—Handedness
  Type—Categorical
Variable—Distance
  Type—Quantitative
  Units—Yards

Variable—Sideline crossed
  Type—Categorical

25. *Who*—Kentucky Derby races
What—Date, winner, margin, jockey, net proceed to winner, duration, track condition
When—1875 to 2008
Where—Churchill Downs, Louisville, Kentucky
Why—Not specified (To see trends in horse racing?)
How—Official statistics collected at race
Variable—Year
  Type—Quantitative
  Units—Day and year
Variable—Winner
  Type—Identifier
Variable—Margin
  Type—Quantitative
  Units—Horse lengths
Variable—Jockey
  Type—Categorical
Variable—Net proceeds to winner
  Type—Quantitative
  Units—Dollars
Variable—Duration
  Type—Quantitative
  Units—Minutes and seconds
Variable—Track condition
  Type—Categorical

26. *Who*—Indy 500 races
What—Year, winner, pole position, average speed, pole winner, pole speed
When—1911 to 2008
Where—Indianapolis
Why—Not specified (To see trends in auto racing?)
How—Official statistics at race
Variable—Year
  Type—Quantitative
  Units—Year
Variable—Winner
  Type—Categorical
Variable—Pole position
  Type—Categorical or possibly quantitative
  Units—Position number
Variable—Average speed
  Type—Quantitative
  Units—Miles per hour
Variable—Pole winner
  Type—Categorical
Variable—Average pole speed
  Type—Quantitative
  Units—Miles per hour

CHAPTER 3

1. Answers will vary.
2. Answers will vary.
3. Answers will vary.
4. Answers will vary.
5. a) Yes; each is categorized in a single genre.
  b) Thriller/Horror
6. a) Yes, each movie falls into only one category.
  b) PG-13
7. a) Comedy
  b) It is easier to tell from the bar chart; slices of the pie chart are too close in size.
8. a) G
  b) No. These displays show only the distribution of movies in these categories. They do not show any changes over time.

9. 1755 students applied for admission to the magnet schools program. 53% were accepted, 17% were wait-listed, and the other 30% were turned away.

10. Of the 1755 students who applied for admission to the magnet schools program, 29.5% were black or Hispanic, 16.6% were Asian, and 53.9% were white.

11. a) Yes. We can add because these categories do not overlap. (Each person is assigned only one cause of death.)
b) $100 - (27.2 + 23.1 + 6.3 + 5.1 + 4.7) = 33.6\%$
c) Either a bar chart or pie chart with "other" added would be appropriate. A bar chart is shown.

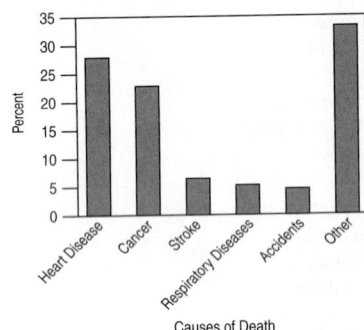

12. a) Yes, we can add because each crash was assigned to only one cause category.
b) $100 - (40 + 5 + 6 + 14 + 6) = 29\%$
c) Either a bar chart or pie chart with "other" added would be appropriate. A bar chart is shown.

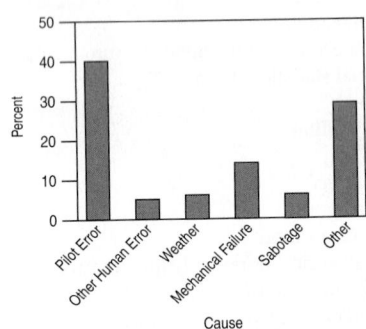

13. a) The bar chart shows that grounding and collision are the most frequent causes of oil spills. Very few have unknown causes.
b) A pie chart seems appropriate as well.

14. a) There are too many categories to make a meaningful bar chart or pie chart.
b) One way to reduce the number of categories is to collect many of the countries that didn't win many medals into an "other" category. Here we've put every country with fewer than 6 medals into "Other."

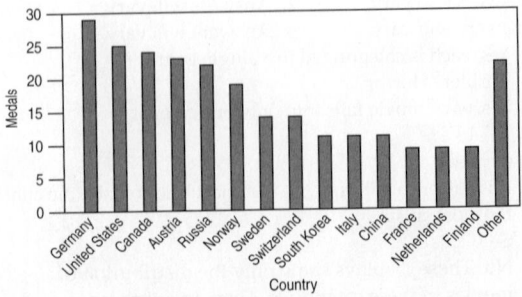

15. There's no title, the percentages total only 92%, and the three-dimensional display distorts the sizes of the regions.

16. a) The bars have false depth, which can be misleading. This is a bar chart, so the bars should have space between them. From a simple design standpoint, running the labels on the bars one way and the vertical axis label the other way is awkward.
b) The percentages sum to 100%. This is unlikely if the respondents were asked to name three methods each. For example, it would be possible for 80% of respondents to use ice at some time and another 75% to use electric stimulation. This is a case where adding to 100% seems wrong rather than correct.

17. In both the South and West, about 58% of the eighth-grade smokers preferred Marlboro. Newport was the next most popular brand, but was far more popular in the South than in the West, where Camel was cited nearly 3 times as often as in the South. Nearly twice as many smokers in the West as in the South indicated that they had no usual brand (12.9% to 6.7%).

18. Over 76% of guns collected in the buyback program were small-caliber handguns, a type of gun involved in only 20% of the homicides. Most homicides (55%) involved medium-caliber guns, yet these represented fewer than 20% of those collected. While large-caliber and other guns represented only 4% of those collected by the buyback program, such weapons were involved in 25% of the murders.

19. a) The column totals are 100%.
b) 31.7%
c) 60%
d) i. 35.7%; ii. can't tell; iii. 0%; iv. can't tell

20. a) The row and column margins are not 100%, but the table total is 100%.
b) PG-13 comedies make up 16.7% of all movies.
c) 12   d) 6
e) Yes. 46.7% + 31.7% = 78.4% of movies are rated PG-13 or R.

21. a) 82.5%   b) 12.9%   c) 11.1%
d) 13.4%   e) 85.7%

22. a) 59.9%   b) 14.1%   c) 18.3%   d) 10.9%

23. a) 73.9% 4-yr college, 13.4% 2-year college, 1.5% military, 5.2% employment, 6.0% other
b) 77.2% 4-yr college, 10.5% 2-year college, 1.8% military, 5.3% employment, 5.3% other
c) Many charts are possible. Here is a side-by-side bar chart.

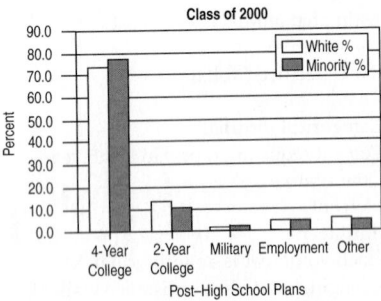

d) The white and minority students' plans are very similar. The small differences should be interpreted with caution because the total number of minority students is small. There is little evidence of an association between race and plans.

24. a) 45% "Liberal," 47% "Moderate," 8% "Conservative"
b) 43% "Liberal," 38% "Moderate," 18% "Conservative"
c)

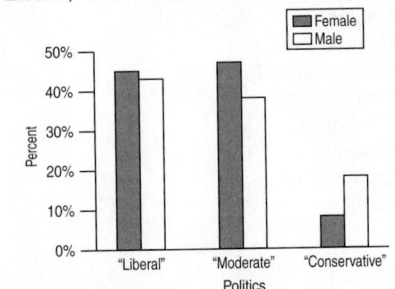

d) No. Although about the same fraction of males and females classified themselves as politically "Liberal," females were slightly more likely to be "Moderate." A much higher percentage of males (18%) considered themselves to be "Conservative" than did the females (8%).

25. a) 16.6%   b) 11.8%   c) 37.7%   d) 53.0%

26. a)

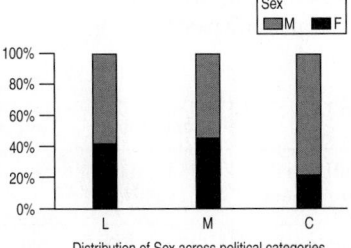

b) The overall percentage of females in the class is about 40%. That's true among "Liberals" and "Moderates", but the percentage of females among "Conservatives" is only about half that.

27. 1755 students applied for admission to the magnet schools program: 53% were accepted, 17% were wait-listed, and the other 30% were turned away. While the overall acceptance rate was 53%, 93.8% of blacks and Hispanics were accepted, compared to only 37.7% of Asians and 35.5% of whites. Overall, 29.5% of applicants were black or Hispanic, but only 6% of those turned away were. Asians accounted for 16.6% of all applicants, but 25.4% of those turned away. Whites were 54% of the applicants and 68.5% of those who were turned away. It appears that the admissions decisions were not independent of the applicant's ethnicity.

28. a) Foreign cars are defined as non-American. There are 45 + 102 = 147 non-American cars, or 147/359 = 40.95%.
b) There are 212 American cars, of which 107 or 107/212 = 50.47% were owned by students.
c) There are 195 students, of whom 107 or 107/195 = 54.87% owned American cars.
d)

		Totals	Percentages
Origin	American	212	59.05%
	European	45	12.53%
	Asian	102	28.41%

e) For students, the *Origin* distribution is:

		Driver	
		Student	Percentage
Origin	American	107	54.87%
	European	33	16.92%
	Asian	55	28.21%
	Totals	195	100.00%

For staff, it is:

		Driver	
		Staff	Percentages
Origin	American	105	64.02%
	European	12	7.32%
	Asian	47	28.66%
	Totals	164	100.00%

f) No, the marginal distributions look slightly different. A higher percentage of staff drive American cars, 64% compared to about 55% among students. Bar chars or pie charts could be used to compare.

29. a) 9.3%   b) 24.7%   c) 80.8%
d) No, there appears to be no association between weather and ability to forecast weather. On days it rained, his forecast was correct 79.4% of the time. When there was no rain, his forecast was correct 81.0% of the time.

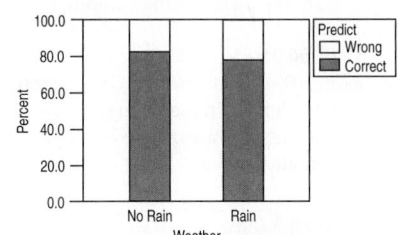

30. a) 22.7%   b) 52.9%   c) 39.7%
d)

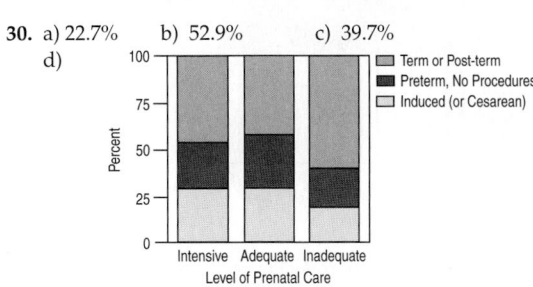

e) It appears that inadequate care leads to more term or post-term births and fewer induced or cesarean births.

31. a) Low 20.0%, Normal 48.9%, High 31.0%
b)

		Under 30	**30–49**	**Over 50**
Blood Pressure	Low	27.6%	20.7%	15.7%
	Normal	49.0%	50.8%	47.2%
	High	23.5%	28.5%	37.1%

c)

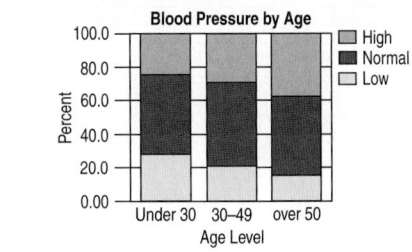

d) As age increases, the percent of adults with high blood pressure increases. By contrast, the percent of adults with low blood pressure decreases.
e) No, but it gives an indication that it might. There might be additional reasons that explain the differences in blood pressures.

32. a) For each classification of BMI (column), participants' self-reported exercise levels. These are column percentages. The percentages sum to 100% in each column, *not* across each row.
b)

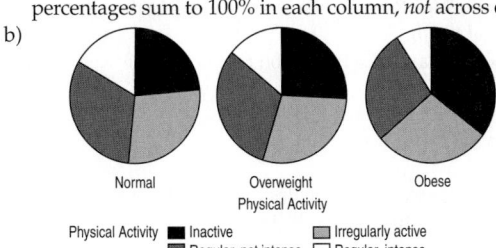

c) No. The graphical displays provide strong evidence that lack of exercise and BMI are not independent. While it may seem logical that lack of exercise causes obesity, the causality could run the other way, or a lurking variable may be the cause of the link between the variables.

33. No, there's no evidence that Prozac is effective. The relapse rates were nearly identical: 28.6% among the people treated with Prozac, compared to 27.3% among those who took the placebo.

34. Yes, the risk of bone fractures is about twice as high (10%) among the people who were taking SSRIs than among those who were not (5%).

35. a) 4.7%   b) 50.0%.
   c) There are about 50% of each sex in each age group, but it ranges from 48.8% female in the youngest group to 54.6% in the oldest. As the age increases, there is a slight increase in the percentage of female drivers.

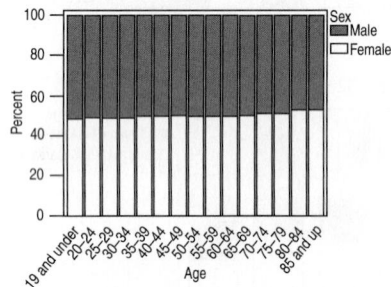

   d) There is a slight association. As the age increases, there is a small increase in the percentage of female drivers.

36. There seems to be a strong association between having a tattoo and contracting hepatitis C. There appears to be a much greater risk for tattoos done in commercial parlors, less for tattoos done elsewhere, and very little risk for those with no tattoos.

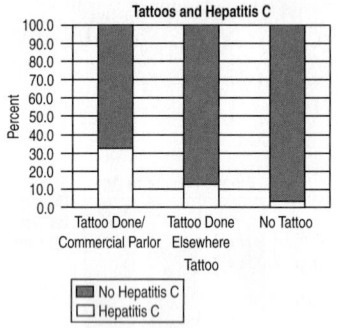

37. a) 160 of 1300, or 12.3%
   b) Yes. Major surgery: 15.3% vs. minor surgery: 6.7%
   c) Large hospital: 13%; small hospital: 10%
   d) Large hospital: Major 15% vs. minor 5%
      Small hospital: Major 20% vs. minor 8%
   e) No. Smaller hospitals have a higher rate for both kinds of surgery, even though it's lower "overall."
   f) The small hospital has a larger percentage of minor surgeries (83.3%) than the large hospital (20%). Minor surgeries have a lower delay rate, so the small hospital looks better "overall."

38. a) Overall, Pack Rats are late 5.6% of the time. Boxes R Us is late 6.0% of the time.
   b) Pack Rats have 3% late rate on regular and 16% on overnight. Boxes R Us have 2% late rate on regular and 7% on overnight, so they are better on both.
   c) Simpson's paradox

39. a) 42.6%
   b) A higher percentage of males than females were admitted: Males: 47.2% to females: 30.9%

c) Program 1: Males 61.9%, females 82.4%
   Program 2: Males 62.9%, females 68.0%
   Program 3: Males 33.7%, females 35.2%
   Program 4: Males 5.9%, females 7.0%

d) The comparisons in c) show that males have a lower admittance rate in every program, even though the overall rate shows males with a higher rate of admittance. This is an example of Simpson's paradox.

40. One possibility

	Company A	Company B
Full-time employees	40 local of 100 (40%)	90 of 200 (45%)
Part-time employees	170 of 200 (85%)	90 of 100 (90%)
Overall	210 of 300 (70%)	180 of 300 (60%)

CHAPTER 4

1. Answers will vary.   2. Answers will vary.
3. Answers will vary.   4. Answers will vary.
5. a) Unimodal (near 0) and skewed to the right. Many seniors will have 0 or 1 speeding tickets. Some may have several, and a few may have more than that.
   b) Probably unimodal and slightly skewed to the right. It is easier to score 15 strokes over the mean than 15 strokes under the mean.
   c) Probably unimodal and symmetric. Weights may be equally likely to be over or under the average.
   d) Probably bimodal. Men's and women's distributions may have different modes. It may also be skewed to the right, since it is possible to have very long hair, but hair length can't be negative.
6. a) Bimodal because you have players and parents. It may also be skewed to the right, since parents' ages can be higher than the mean more easily than lower.
   b) Unimodal and skewed to the right. There are probably many students with 0 or 1 sibling and some with 2 or more.
   c) Unimodal and symmetric. It will be unusual to have either very high or low pulse rates.
   d) Uniform. Each face of the die has the same chance of coming up, so the number of times should be about the same for each face.
7. a) Bimodal. Looks like two groups. Modes are near 6% and 46%. No real outliers.
   b) Looks like two groups of cereals, a low-sugar and a high-sugar group.
8. a) Looks bimodal, one mode near 65 inches, one near 71 inches. Could be two groups that look fairly symmetric within each group.
   b) Probably the mix of men and women in the group.
9. a) 78%
   b) Skewed to the right with at least one high outlier. Most of the vineyards are less than 90 acres with a few high ones. The mode is between 0 and 30 acres.
10. Skewed to the right, centered at around 30 to 31 minutes, with most observations between 29 to 32 minutes. Skewed to the right because it is easier to run much slower than usual, but hard to run the same amount *faster* than usual.
11. a) Because the distribution is skewed to the right, we expect the mean to be larger.

b) Bimodal and skewed to the right. Center mode near 8 days. Another mode at 1 day (may represent patients who didn't survive). Most of the patients stay between 1 and 15 days. There are some extremely high values above 25 days.

c) The median and IQR, because the distribution is strongly skewed.

12. a) Because the distribution is skewed to the right, we expect the mean to be larger.

b) Skewed to the right. Mode near 1. One outlier very far from the rest of data. Spread fairly small. Most of the data less than 5.

c) The median and IQR, because the distribution is strongly skewed.

13. a) 45 points   b) 37 points and 54 (or 55) points

c) In the Super Bowl teams typically score a total of about 45 points, with half the games totaling between 37 and 55 points. In only one fourth of the games have the teams scored fewer than 27 points, and they once totaled 75.

14. a) 14 points   b) 7 points and 21 (or 21.5) points

c) Winning Super Bowl teams typically outscore their opponents by 14 points, with half the margins of victory between 7 and 21 points. In only one fourth of the games has a team won by more than three touchdowns, although the greatest margin ever was 45 points.

15. a) The standard deviation will be larger for set 2, since the values are more spread out. SD(set 1) = 2.2, SD(set 2) = 3.2.

b) The standard deviation will be larger for set 2, since 11 and 19 are farther from 15 than are 14 and 16. Other numbers are the same. SD(set 1) = 3.6, SD(set 2) = 4.5.

c) The standard deviation will be the same for both sets, since the values in the second data set are just the values in the first data set + 80. The spread has not changed. SD(set 1) = 4.2, SD(set 2) = 4.2.

16. a) Set 2 will have a larger SD because 6 and 8 in set 2 are farther from 7 than 7 and 7 in set 1. The other numbers are the same. SD(set 1) = 2.1, SD(set 2) = 2.2.

b) The standard deviation will be the same for both sets, since the values in set 1 are just the values in set 2 + 90. The spread remains the same. SD(set 1) = 36.1, SD(set 2) = 36.1.

c) Set 2 has a wider range and will have a larger SD. SD(set 1) = 6.0, SD(set 2) = 7.2.

17. The mean and standard deviation because the distribution is unimodal and symmetric.

18. The mean and standard deviation because the distribution is unimodal and symmetric.

19. a) The mean is closest to $2.60 because that's the balancing point of the histogram.

b) The standard deviation is closest to $0.15 since that's a typical distance from the mean. There are no prices as far as $0.50 or $1.00 from the mean.

20. a) The mean is closest to 15 inches because that's the balancing point of the histogram.

b) The standard deviation is closest to 1 inch because a typical value lies about 1 inch from the mean. There are few points as far as 4 inches from the mean and none as far as 5 inches, so those are too large to be the standard deviation.

21. a) About 100 minutes

b) Yes, only 4 of these movies run that long.

c) The mean would be higher. The distribution is skewed high.

22. a) Unimodal and reasonably symmetric.

b) About 15%

c) $\bar{y} = 288.6$; any estimate between 285 and 290 is a good guess.

d) The median should be close to the mean because this distribution is symmetric. (In fact, it is 288.35.)

23. a) i. The middle 50% of movies ran between 97 and 119 minutes.

ii. On average, movie lengths varied from the mean run time by 19.6 minutes.

b) We should be cautious in using the standard deviation because the distribution of run times is skewed to the right.

24. a) i. The middle 50% of golfers had driving averages between 282 and 294 yards.

ii. The typical difference between golfers' driving distances and the mean was 9.3 yards.

b) The distribution is symmetric, so either is okay.

25. a) The median will probably be unaffected. The mean will be larger.

b) The range and standard deviation will increase; the IQR will be unaffected.

26. a) The mean will be smaller; the median will not be affected.

b) The range and standard deviation will be larger; the IQR won't change.

27. The publication is using the median; the watchdog group is using the mean, pulled higher by the several very expensive movies in the long right tail.

28. The company is using the mean of 7 days, while the union negotiators are using the median of 3 days. There may be high outliers, or a distribution of sick days that is generally skewed to the right.

29. a) Mean $525, median $450

b) 2 employees earn more than the mean.

c) The median because of the outlier.

d) The IQR will be least sensitive to the outlier of $1200, so it would be the best to report.

30. a) Median 66 inches, IQR 5 inches

b) Mean 67.12 inches, SD 3.79 inches

c)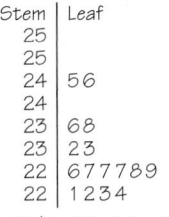

d) The distribution is centered near 67 in., with median 66 in. and mean 67.12 in. The IQR is 5 in., with the middle 50% of heights between 65 and 70 in. The distribution appears to be unimodal, but skewed slightly to the right.

31. a)

```
Stem | Leaf
  25 |
  25 |
  24 | 5 6
  24 |
  23 | 6 8
  23 | 2 3
  22 | 6 7 7 7 8 9
  22 | 1 2 3 4
```

22|1 = $2.21/gallon

b) The distribution of gas prices is unimodal and skewed to the right (upward), centered around $2.27, with most stations charging between $2.26 and $2.33 per gallon. The lowest and highest prices were $2.21 and $2.46.

c) There are two high prices separated from the other gas stations by a gap.

32. a)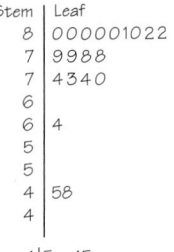

```
Stem | Leaf
   8 | 0 0 0 0 0 1 0 2 2
   7 | 9 9 8 8
   7 | 4 3 4 0
   6 |
   6 | 4
   5 |
   5 |
   4 | 5 8
   4 |
```

4|5 = 45 games

b) The distribution of games played by Gretzky is unimodal and skewed to the left.

c) Typically, Gretzky played about 80 games per season. The numbers of games played are tightly clustered in the upper 70s and low 80s.

d) Two seasons are low outliers, when Gretzky played fewer than 50 games. He might have been injured.

33. a) Since these data are strongly skewed to the right, the median and IQR are the best statistics to report.

b) The mean will be larger than the median because the data are skewed to the right.

c) The median is 4 million. The IQR is 4.5 million (Q3 = 6 million, Q1 = 1.5 million).

d) The distribution of populations of the states and Washington, DC, is unimodal and skewed to the right. The median population is 4 million. One state is an outlier, with a population of 34 million.

34. a) Median. The distribution is highly skewed to the left.

b) 79 games

c) The mean should be lower. There are two seasons when Gretzky played an unusually low number of games. Those will pull the mean down.

35. Skewed to the right, median at 36. Three low outliers, then a gap from 9 to 22.

36. a)

```
Stem │ Leaf
  22 │ 8
  21 │
  20 │ 66
  19 │
  18 │ 136
  17 │ 578
  16 │ 0022367
  15 │ 23367

15|2 = 152 Species
```

b) Skewed to the right; several high outliers. Center near 166.

37. a)

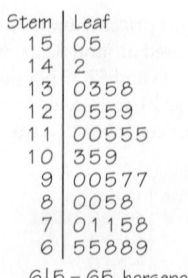

Hurricanes in Period 1944–2006

b) Slightly skewed to the right. Unimodal, mode near 2. Possibly a second mode near 5. No outliers.

38. This distribution is nearly uniform. Values range from 65 to 155 horsepower. The center is near 100.

```
Stem │ Leaf
  15 │ 05
  14 │ 2
  13 │ 0358
  12 │ 0559
  11 │ 00555
  10 │ 359
   9 │ 00577
   8 │ 0058
   7 │ 01158
   6 │ 55889

6|5 = 65 horsepower
```

39. a) This is not a histogram. The horizontal axis should split the number of home runs hit in each year into bins. The vertical axis should show the number of years in each bin.

b)

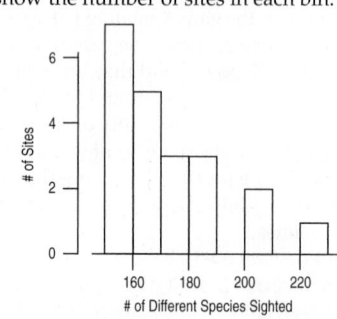

40. a) This is not a histogram. The horizontal axis should split the number of counts from each site into bins. The vertical axis should show the number of sites in each bin.

b)

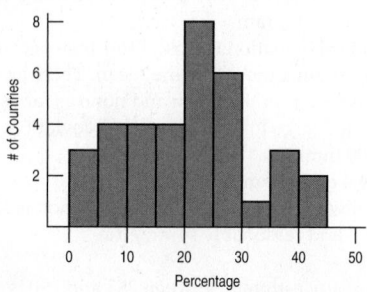

41. Skewed to the right, possibly bimodal with one fairly symmetric group near 4.4, another at 5.6. Two outliers in middle seem not to belong to either group.

```
Stem │ Leaf
  57 │ 8
  56 │ 27
  55 │ 1
  54 │
  53 │
  52 │ 9
  51 │
  50 │ 8
  49 │
  48 │ 2
  47 │ 3
  46 │ 034
  45 │ 267
  44 │ 015
  43 │ 0199
  42 │ 669
  41 │ 22

41|2 = 4.12 pH
```

42. Unimodal with center around 20%, but with a cluster of countries reporting over 35%.

43. Histogram bins are too wide to be useful.

44. a) Histogram bins are too narrow to be useful.

b) Skewed to the left, mode near 170, several outliers below 100. Fairly tightly clustered except for outliers.

45. Neither appropriate nor useful. Zip codes are categorical data, not quantitative. But they do contain *some* information. The leading

digit gives a rough East-to-West placement in the United States. So, we see that they have almost no customers in the Northeast, but a bar chart by leading digit would be more appropriate.

46. Not very much, since zip codes are categorical. However, there is *some* information in the first digit of zip codes. They indicate a general East (0–1) to West (8–9) direction. So, the distribution shows that a large portion of their sales occur in the West and another in the 32000 area. But a bar chart of the first digits might be a better display of this.

47) a) Median 239, IQR 9, Mean 237.6, SD 5.7
    b) Because it's skewed to the left, probably better to report Median and IQR.
    c) Skewed to the left; may be bimodal. The center is around 239. The middle 50% of states scored between 233 and 242. Alabama, Mississippi, and New Mexico scores were much lower than other states' scores.

48. a) Here is a histogram of the distribution:

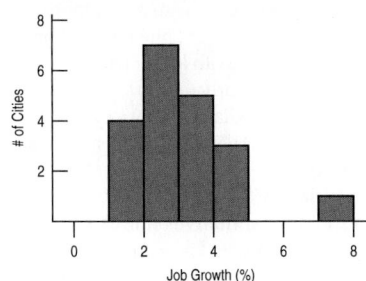

   b) Mean 3.07%, median 2.85%. Mean is higher because of the outlier.
   c) The median because of the outlier.
   d) IQR 1.1%, SD 1.37%
   e) The standard deviation is also influenced by the outlier. The better measure of spread is the IQR.
   f) Mean and median would be 1.2% lower. SD and IQR would not change.
   g) Median and IQR won't change very much. The middle value and the two quartiles will shift at most one data value. The mean and SD will decrease.
   h) The distribution of growth rates for the cities is unimodal and symmetric except for the one outlier, Las Vegas, at 7.5%. The median growth rate for these cities is 2.85%. The middle 50% of the cities had growth rates between 2.25% and 3.35%, for an interquartile range of 1.1%. The median and IQR are the best statistics to report when a distribution has outliers, but if the outlier were omitted, the average growth rate would be 2.84%, with a standard deviation of 0.91%.

49. In the year 2004, per capita gasoline use by state in the United States averaged around 500 gallons per person (mean 488.8, median 500.5). States varied in per capita consumption, with a standard deviation of 68.7 gallons. The only outlier is New York. The IQR of 96.9 gallons shows that 50% of the states had per capita consumption of between 447.5 and 544.4 gallons. The data appear to be bimodal, so the median and IQR are better choices of summary statistics.

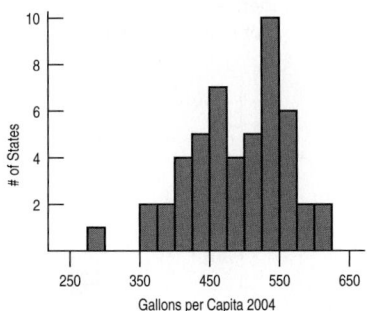

50. In 2005, the median increase in federal prison populations in 20 northeastern and midwestern states was 2.3%; only 4 of the 20 states showed a decrease. The distribution is unimodal and skewed to the right. The large IQR of 4.7% indicates much variability from state to state, with one fourth of these states experiencing prison population increases in excess of 5.5%.

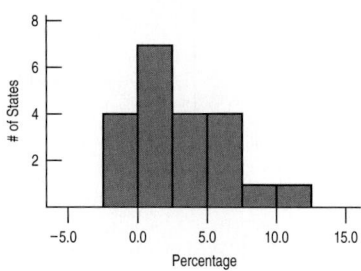

CHAPTER 5

1. Answers will vary.
2. Answers will vary.
3. Answers will vary.
4. Answers will vary.
5. a) Prices appear to be both higher on average and more variable in Baltimore than in the other three cities. Prices in Chicago may be slightly higher than in Dallas and Denver, but the difference is very small.
   b) There are outliers on the low end in Baltimore and Chicago and one high outlier in Dallas, but these do not affect the overall conclusions reached in part a).
6. a) Coffee, on average, is the most expensive.
   b) On average, newspapers are more expensive than public transportation rides, but there are cities in which a ride on public transportation costs more than a newspaper in other cities.
   c) There is one outlying city in each of the three boxplots, but removing this outlier will not affect the conclusions of part a) or b).
7. a) Essentially symmetric, very slightly skewed to the right with two high outliers at 36 and 48. Most victims are between the ages of 16 and 24.
   b) The slight increase between ages 22 and 24 is apparent in the histogram but not in the boxplot. It may be a second mode.
   c) The median would be the most appropriate measure of center because of the slight skew and the extreme outliers.
   d) The IQR would be the most appropriate measure of spread because of the slight skew and the extreme outliers.
8. a) Both plots show skewness to the high end.
   b) The histogram shows a bimodal distribution and skewness to the high end. Boxplots can't show multiple modes clearly.
   c) We'd prefer the median to the mean because of the skewness and possible outliers.
   d) We'd prefer the IQR to the standard deviation because of the skewness and possible outliers.
9. a) About 59%   b) Bimodal
   c) Some cereals are very sugary; others are healthier low-sugar brands.
   d) Yes
   e) Although the ranges appear to be comparable for both groups (about 28%), the IQR is larger for the adult cereals, indicating that there's more variability in the sugar content of the middle 50% of adult cereals.
10. a) Unimodal, symmetric   b) 3
    c) Results from the two procedures were markedly different.
    d) Biceps transfer
    e) No, one deltoid transfer was as good as at least 25% of the biceps transfers.
    f) Although the range in strength is approximately 2 units for both methods, the IQR is much smaller for the deltoid transfers, indicating more consistent results.

11. a)

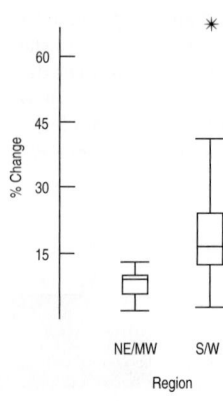

b) Growth rates in NE/MW states are tightly clustered near 5%. S/W states are more variable, and bimodal with modes near 14 and 22. The S/W states have an outlier as well. Around all the modes, the distributions are fairly symmetric.

12. a) The distribution is strongly skewed to the right, so use the median and IQR.
b) The IQR is 50, so the upper fence is the upper quartile +1.5 IQRs; that is, $78 + 75 = 153$. There appear to be 4 to 5 parks that should be considered as outliers with more than 153 camp sites.
c)

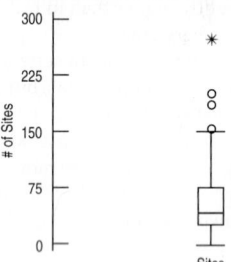

d) The distribution is unimodal with a strong skew to the right. There are several outliers past the $1.5 \times$ IQR upper fence of 153 camp sites. The median number of camp sites is 43.5 sites. The mean is 62.8 sites. The mean is larger than the median because it has been influenced by the strong skew and the outliers.

13. a) They should be put on the same scale, from 0 to 20 days.
b) Lengths of men's stays appear to vary more than for women. Men have a mode at 1 day and then taper off from there. Women have a mode near 5 days, with a sharp drop afterward.
c) A possible reason is childbirth.

14. a) Both are unimodal, skewed to the low end.
b) A higher proportion of blacks die between 25 and 74 years old. A higher proportion of whites die at ages older than 75.
c) Interval widths are not constant. Most bars are for 10-year intervals, but the first bars are only for ages 0–4, a 5-year span, and the last bars are for 85 and over.

15. a) Both girls have a median score of about 17 points per game, but Scyrine is much more consistent. Her IQR is about 2 points, while Alexandra's is over 10.
b) If the coach wants a consistent performer, she should take Scyrine. She'll almost certainly deliver somewhere between 15 and 20 points. But if she wants to take a chance and needs a "big game," she should take Alexandra. Alex scores over 24 points about a quarter of the time. (On the other hand, she scores under 11 points as often.)

16. a) Gas prices increased, on average, over the 3-year period, and the spread increased as well. The prices in 2002 were skewed to the left, with several low outliers. Since then, the distribution has been increasingly skewed to the right. There is a high outlier in 2004.
b) 2004 shows both the greatest range and the biggest IQR, so the prices varied a lot.

17. Women appear to marry about 3 years younger than men, but the two distributions are very similar in shape and spread.

18. Both fuel economy and its spread decrease from 4 to 6 to 8 cylinders (not enough data to compare 5-cylinder cars). The lower 75% of MPGs for the 8-cylinder cars corresponds roughly to the bottom 25% of MPGs for the 6-cylinder cars. All the 8-cylinder cars get less mileage than all the 4-cylinder cars, and their fuel economy is consistently low.

19. (*Note:* Numerical details may vary.) In general, fuel economy is higher in cars than in either SUVs or vans. There are numerous outliers on both ends for cars and a few high outliers for SUVs. The top 50% of cars gets higher fuel economy than 75% of SUVs and nearly all vans. On average, SUVs and vans get about the same fuel economy, although the distribution for vans shows less spread. The range for vans is about 40 mpg, while for SUVs it is nearly 30 mpg.

20. a) April   b) February   c) August
d) The median ozone level in June is slightly higher than in January, but June's readings are much more consistent. June does show two outliers, one low and one high.
e) Strong seasonal pattern with low consistent ozone concentrations in later summer/early fall and high variable concentrations in early spring. The medians follow a cyclic pattern, rising from January to April, then falling to October and rising again from October to December.

21. The class A is 1, class B is 2, and class C is 3.

22. No, boxplots are for quantitative data, and these are categorical, although coded as numbers. The numbers used for hair color and eye color are arbitrary, so the boxplot and any accompanying statistics for eye color make no sense.

23. a) Probably slightly left skewed. The mean is slightly below the median, and the 25th percentile is farther from the median than the 75th percentile.
b) No, all data are within the fences.
c)

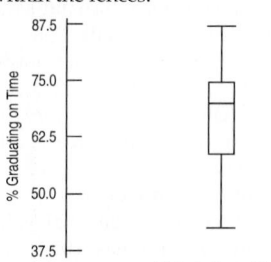

d) The 48 universities graduate, on average, about 68% of freshmen "on time," with percents ranging from 43% to 87%. The middle 50% of these universities graduate between 59% and 75% of their freshmen in 4 years.

24. a) Skewed to the right, since the mean is much larger than the median and the upper quartile is farther from the median than the lower quartile.
b) The IQR is about 36, so fences are below 0 and at 109. Since the range is 244 and the minimum is 6, the maximum is 250, which is certainly an outlier. Without knowing the data points, we are not sure of other outliers, but the standard deviation of 47.8 makes us suspect there are others.
c) We don't know if there are other outliers above the upper fence, but the boxplot may look something like this:

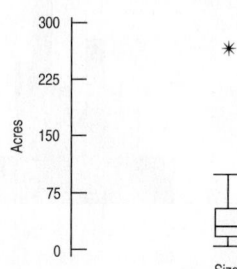

d) The vineyards range in size from 6 to 250 acres. The median size of the 36 vineyards is 33.5 acres, so half are larger and half are smaller. The middle 50% of the vineyards have sizes between 18.5 and 55 acres. The distribution of sizes is skewed to the right, with at least one outlier.

25. a) *Who:* Student volunteers
    What: Memory test
    Where, when: Not specified
    How: Students took memory test 2 hours after drinking caffeine-free, half-dose caffeine, or high-caffeine soda.
    Why: To see if caffeine makes you more alert and aids memory retention.

    b) Drink: categorical; Test score: quantitative.

    c)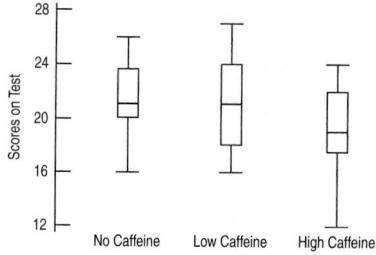

    d) The participants scored about the same with no caffeine and low caffeine. The medians for both were 21 points, with slightly more variation for the low-caffeine group. The high-caffeine group generally scored lower than the other two groups on all measures of the 5-number summary: min, lower quartile, median, upper quartile, and max.

26. a)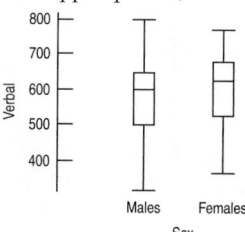

    b) Median score by females, at 625 points, is 25 points higher than that by males. Female mean is higher by 12. The middle 50% of females scored between 530 and 680, while the middle 50% of males scored between 515 and 650. The males did have a larger range, from 310 to 800, the highest score for both genders. Both distributions are slightly skewed to the left.

27. a) About 36 mph
    b) Q_1 about 35 mph and Q_3 about 37 mph
    c) The range appears to be about 7 mph, from about 31 to 38 mph. The IQR is about 2 mph.
    d) We can't know exactly, but the boxplot may look something like this:

    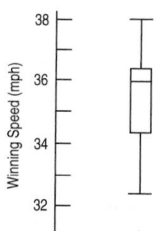

    e) The median winning speed has been about 36 mph, with a max of about 38 and a min of about 31 mph. Half have run between about 35 and 37 mph, for an IQR of 2 mph.

28. Distribution is essentially symmetric, with median near 225. The IQR is about 75 points. Extremes are about 80 and 380.

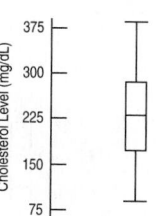

29. a) Boys   b) Boys   c) Girls
    d) The boys appeared to have more skew, as their scores were less symmetric between quartiles. The girls' quartiles are the same distance from the median, although the left tail stretches a bit farther to the left.
    e) Girls. Their median and upper quartiles are larger. The lower quartile is slightly lower, but close.
    f) $[14(4.2) + 11(4.6)]/25 = 4.38$

30. a) The median and IQR, because the means are much larger than the median and the SDs are much larger than the IQR, indicating either right skewness and/or outliers.
    b) Since the median rainfall for seeded clouds is more than 4 times that for unseeded clouds, it appears that seeding clouds may be effective.

31.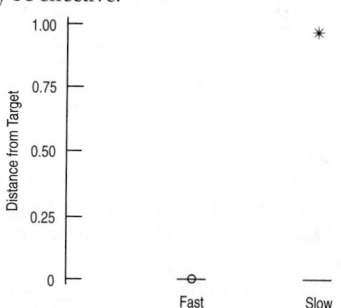

    There appears to be an outlier! This point should be investigated. We'll proceed by redoing the plots with the outlier omitted:

    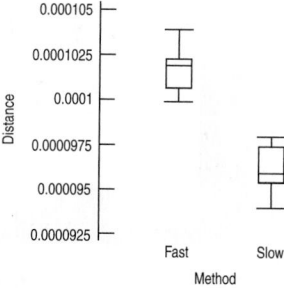

    It appears that slow speed provides much greater accuracy. But the outlier should be investigated. It is possible that slow speed can induce an infrequent very large distance.

32.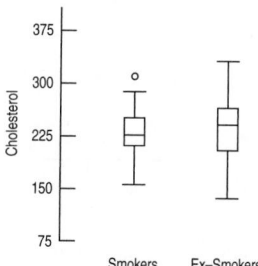

    Smoker's distribution has a lower median and is less spread out. There is an outlier in the smoker's group at 351. (Other answers are possible too. Could possibly say that ex-smoker's distribution is more symmetric if histogram or stem-and-leaf provided.)

33. a)

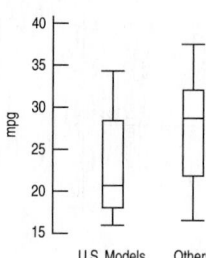

b) Mileage for U.S. models is typically lower, although the variability is about the same as for cars made elsewhere. The median for U.S. models is around 21 mpg, compared to 28 for the others. Half of U.S. models fall below the first quartile of others. (Other answers possible.)

34. a) (Other displays possible.)

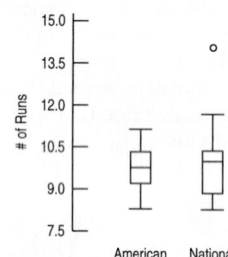

b) The National League scores slightly fewer runs in general, but the distribution is more spread out. The median for both leagues is about 9.5, but the American League's is slightly higher. There is a high outlier in the National League at about 14 runs.

c) Yes, it looks like an outlier. It is nominated by the boxplot rule and is well separated from all the other parks.

35. a) Day 16 (but any estimate near 20 is okay).
b) Day 65 (but anything around 60 is okay).
c) Around day 50

36. a)

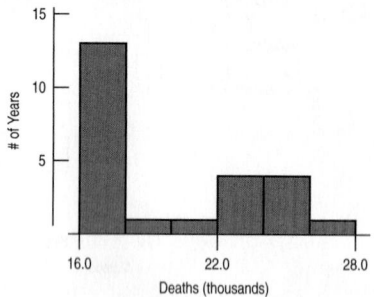

b)

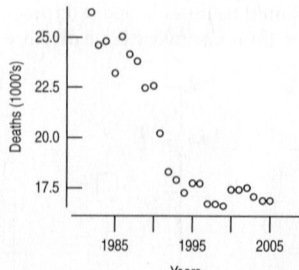

c) The histogram shows a bimodal distribution. The timeplot shows that drunk-driving deaths fell between 1980 and 1995, but after 1995, they have remained at about the same level.

37. a) Most of the data are found in the far left of this histogram. The distribution is very skewed to the right.

b) Re-expressing the data by, for example, logs or square roots might help make the distribution more nearly symmetric.

38. a) The data are extremely skewed to the right. That makes it difficult to determine a center. Our perception of spread will be affected by the very long tail.

b) Re-expressing the data by, for example, logs or square roots might help make the distribution more nearly symmetric.

39. a) The logarithm makes the histogram more symmetric. It is easy to see that the center is around 3.5 in log assets.
b) That has a value of around 2,500 million dollars.
c) That has a value of around 1,000 million dollars.

40. a) The numbers are manageable (each acre-foot is about 320,000 gallons).

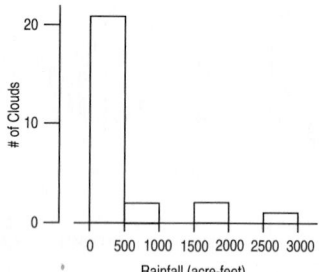

b) The distribution is very skewed to the right.
c)

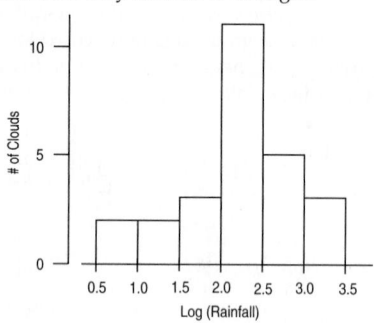

d) $\log_{10}$ is the exponent of 10, which results in the original value. For example, $10^3 = 1000$, so $\log_{10}$ of 1000 is 3.

41. a) Fusion time and group.
b) Fusion time is quantitative (units = seconds). Group is categorical.
c) Both distributions are skewed to the right with high outliers. The boxplot indicates that visual information may reduce fusion time. The median for the Verbal/Visual group seems to be about the same as the lower quartile of the No/Verbal group.

42. The analysis would probably be improved by using the log-transformed data. The distributions are more symmetric, and it's easier to compare the groups. The outliers are eliminated.

CHAPTER 6

1. a) 72 oz., 40 oz.   b) 4.5 lb, 2.5 lb
2. a) 264 sec, 138 sec   b) 240 sec, 138 sec
3. a) Skewed to the right; mean is higher than median.
b) $350 and $950.
c) Minimum $350. Mean $750. Median $550. Range $1200. IQR $600. Q1 $400. SD $400.
d) Minimum $330. Mean $770. Median $550. Range $1320. IQR $660. Q1 $385. SD $440.
4. a) Range 3.30 pounds. IQR 0.95 pounds.
b) Slightly skewed to the left because the mean is lower than the median and the first quartile is farther from the median than the third quartile.

c) Mean 96 oz. SD 10.4 oz. Q1 89.6 oz. Q3 104.8 oz. Median 99.2 oz. IQR 15.2 oz. Range 52.8 oz.

d) Mean 126 oz. SD 10.4 oz. Q1 119.6 oz. Q3 134.8 oz. Median 129.2 oz. IQR 15.2 oz. Range 52.8 oz.

e) Median, IQR.

5. Lowest score = 910. Mean = 1230. SD = 120. Q3 = 1350. Median = 1270. IQR = 240.

6. Maximum temperature = 51.8°F. Range = 59.4°F. Mean = 33.8°F. SD = 12.6°F. Median = 35.6°F. IQR = 28.8°F.

7. Your score was 2.2 standard deviations higher than the mean score in the class.

8. The boy's height is 1.88 standard deviations below the mean height of American children his age.

9. 65 **10.** 140 or above

11. In January, a high of 55 is not quite 2 standard deviations above the mean, whereas in July a high of 55 is more than 2 standard deviations lower than the mean. So it's less likely to happen in July.

12. In French she scored 1.25 standard deviations higher than the mean. On the math exam she scored 1.50 standard deviations higher than the mean, so she did "better" on the math exam.

13. The z-scores, which account for the difference in the distributions of the two tests, are 1.5 and 0 for Derrick and 0.5 and 2 for Julie. Derrick's total is 1.5, which is less than Julie's 2.5.

14. The z-scores, which account for the difference in the distributions of the two tests, are 0 and 1 for Reginald, for a total of 1.0. For Sara, they are 2.0 and −0.33, for a total of 1.67. While her raw total is lower, her z-score total is higher.

15. a) Megan b) Anna

16. a) To know something about their consistency and the chances they will last a certain amount of time.
   b) RockReady. Mean is larger and SD smaller.
   c) 16 hours is 2.5 SD higher than DuraTunes's mean, and 2.67 SD higher than RockReady's mean. So, although neither battery has much chance of lasting 16 hours, DuraTunes's chance is greater.

17. a) About 1.81 standard deviations below the mean.
   b) 1000 ($z = 1.81$) is more unusual than 1250 ($z = 1.17$).

18. a) The z-score for 20 mph is $z = (20 - 23.84)/3.56 = -1.08$. It is 1.08 SD below the mean speed.
   b) The z-score for 34 is $z = (34 - 23.84)/3.56 = 2.85$. For 10 mph, it's $z = (10 - 23.84)/3.56 = -3.89$. So, 10 mph is more unusual.

19. a) Mean = $1152 - 1000 = 152$ pounds; SD is unchanged at 84 pounds.
   b) Mean = $0.40(1152) = \$460.80$; SD = $0.40(84) = \$33.60$.

20. a) The new mean is 3.84 mph. The standard deviation is unchanged.
   b) The new mean is $23.84(1.609) = 38.359$ kph. (The speed limit is 32.18 kph.) The new standard deviation is 5.728 kph.

21. Min = $0.40(980) - 20 = \$372$; median = $0.40(1140) - 20 = \$436$; SD = $0.40(84) = \$33.60$; IQR = $0.40(102) = \$40.80$.

22. Mean = $100 + 10(8) = \$180$; SD = $10(2.4) = \$24$; max = $100 + 10(13) = \$230$; IQR = $10(3.2) = \$32$.

23. College professors can have between 0 and maybe 40 (or possibly 50) years' experience. A standard deviation of 1/2 year is impossible, because many professors would be 10 or 20 SDs away from the mean, whatever it is. An SD of 16 years would mean that 2 SDs on either side of the mean is plus or minus 32, for a range of 64 years. That's too high. So, the SD must be 6 years.

24. Probably 2000, but it could be 20,000 if they played to a few really large audiences (like sold-out stadiums).

25. a)
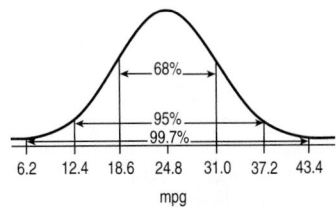
   b) 18.6 to 31.0 mpg c) 16%
   d) 13.5% e) less than 12.4 mpg

26. a)
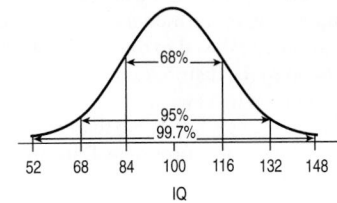
   b) 68 to 132 IQ points c) 16%
   d) 13.5% e) 2.5%

27. Any weight more than 2 standard deviations below the mean, or less than $1152 - 2(84) = 984$ pounds, is unusually low. We expect to see a steer below $1152 - 3(84) = 900$ pounds only rarely.

28. Any IQ more than 2 standard deviations above the mean, or $100 + 2(16) = 132$ points, is unusually high. We expect to find someone with an IQ over $100 + 3(16) = 148$ only rarely.

29. a)
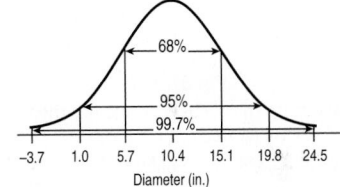
   b) Between 1.0 and 19.8 inches c) 2.5%
   d) 34% e) 16%

30. a)
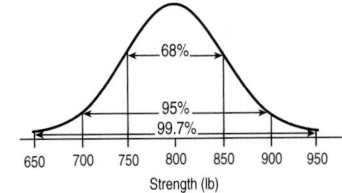
   b) Chances are 84% that the strength is at least 750 pounds. That means 16% of the time it will be below. Seems like much too high a percentage.
   c) 97.5%
   d) In the extreme case, I would want a very small probability of failure, maybe less than 1 out of a million. For this I'd need to be sure that 800 was 6 SDs higher than my limit. So I might not use the rivets if I needed them to have shear strength more than 500 pounds.

31. Since the histogram is not unimodal and symmetric, it is not wise to have faith in numbers from the Normal model.

32. The histogram is unimodal and roughly symmetric, and the normal probability plot looks quite straight, so a normal model is appropriate.

33. a) 16% b) 3.8%
   c) Because the Normal model doesn't fit well.
   d) Distribution is skewed to the right.

34. a) We know that 95% of the observations for a Normal model fall within 2 SDs of the mean. That corresponds to 23.84 − 2(3.56) = 16.72 mph and 23.84 + 2(3.56) = 30.96 mph.

b) The actual 97.5% and 2.5% tiles are 30.976 and 16.638, respectively. These are very close to the predicted values of 30.96 and 16.72 mph. The histogram is roughly unimodal and symmetric. It is very slightly right skewed and there is one outlier, but the Normal probability plot is quite straight. We should not be surprised that the approximation is good.

35. a) 2.5%
b) 2.5% of the receivers should gain less than −333 yards, but that's impossible, so the model doesn't fit well.
c) Data are strongly skewed to the right, not symmetric.

36. a) Median because the distribution is so skewed to the left.
b) IQR. Distribution is skewed.
c) 68%   **d)** More than 75%
e) Normal model is not appropriate. Data are strongly skewed.

37. a) 12.2%   **b)** 71.6% **c)** 23.3%
38. a) 89.4%   **b)** 26.6%   **c)** 20.4%
39. a) 1259.7 lb **b)** 1081.3 lb **c)** 1108 lb to 1196 lb
40. a) 126.3   **b)** 91.6   **c)** 79.5 to 120.5
41. a) 1130.7 lb **b)** 1347.4 lb **c)** 113.3 lb
42. a) 83.4   **b)** 132.9   **c)** 21.6
43. a)

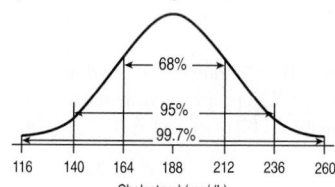

b) 30.85%   **c)** 17.00% **d)** 32 points   **e)** 212.9 points
44. a) No, that's more than 3 SDs above the mean.
b) 21.2%   **c)** 67.3%
d) Quartiles at 30,314 and 33,686 miles, so 3372 miles.
e) 27,623 miles
45. a) 11.1%   **b)** (35.9, 40.5) inches   **c)** 40.5 inches
46. a) Based on the Normal model, we expect 95% to be between 96.8 and 99.6°F.
b) 28.4%   **c)** 97.6°F
47. a) 5.3 grams   **b)** 6.4 grams
c) Younger because SD is smaller.
48. a) 3.26 grams **b)** 75.70 grams   **c)** 2.86 grams
d) The new tomatoes are more consistent in their weights.

PART I REVIEW

1. a)

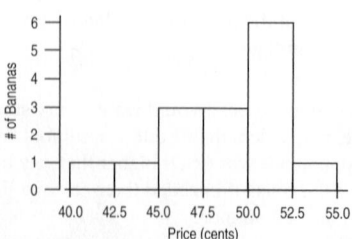

b) Median 49 cents, IQR 6 cents.
c) The distribution is unimodal and left skewed. The center is near 50 cents; values range from 42 cents to 53 cents.

2. a) It is (rounded to 1 decimal place), but there's no reason it should be unless the number of women receiving each type of care was roughly the same.
b) Yes, but they do not prove that adequate prenatal care is important for pregnant women. The mortality rate is quite a bit lower for women with adequate care than for other women, but there may be a lurking variable.

c) Intensive care is given for emergency conditions. The data do *not* suggest that the care is the *cause* of the higher mortality.
3. a) If enough sopranos have a height of 65 inches, this can happen.
b) The distribution of heights for each voice part is roughly symmetric. The basses are slightly taller than the tenors. The sopranos and altos have about the same median height. Heights of basses and sopranos are more consistent than those of altos and tenors.
4. With only 3 cases, probably no display is appropriate.
5. a) It means their heights are also more variable.
b) The z-score for women to qualify is 2.40, compared with 1.75 for men, so it is harder for women to qualify.
6. a) The distribution is unimodal and skewed to the right. The mode is near 100, and values range from 95 to 140.
b) The mean will be larger than the median, since the distribution is right skewed.
c) Create a boxplot with quartiles at 97 and 105.5, and median at 100. The IQR is 8.5, so the upper fence is at (1.5 × 8.5) + 105.5 = 118.25. There are several outliers to the right. There are no outliers to the left because the minimum at 95 lies well within the left fence at 97 − (1.5 × 8.5) = 84.25.
d) No, the Normal model is not appropriate for these data. They are unimodal but not symmetric.

7. a) *Who*—People who live near State University
   What—Age, attended college? Favorable opinion of State?
   When—Not stated
   Where—Region around State U.
   Why—To report to the university's directors
   How—Sampled and phoned 850 local residents
b) Age—Quantitative (years); attended college?—categorical; favorable opinion?—categorical.
c) The fact that the respondents know they are being interviewed by the university's staff may influence answers.

8. a)

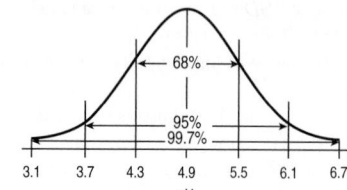

b) 3.3%   **c)** 6.7%
d) pH 4.40   **e)** pH 5.89
f) Quartiles at 4.50 and 5.30, so IQR is 0.80.
9. a) These are categorical data, so mean and standard deviation are meaningless.
b) Not appropriate. Even if it fits well, the Normal model is meaningless for categorical data.
10. a) Stream name—categorical; substrate—categorical; pH—quantitative; temperature—quantitative (°C); BCI—quantitative (units missing).
b) Bar chart or pie chart.
11. a)

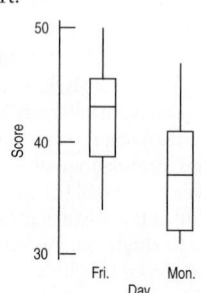

b) The scores on Friday were higher by about 5 points on average. This is a drop of more than 10% off the average score and shows that students fared worse on Monday after preparing for the test on Friday. The spreads are about the same, but the scores on Monday are a bit skewed to the right.

c)

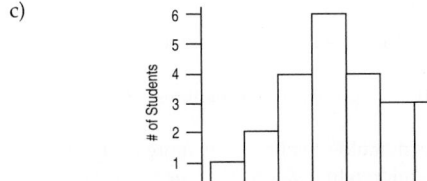

d) The changes (Friday–Monday) are unimodal and centered near 4 points, with a spread of about 5 (SD). They are fairly symmetric, but slightly skewed to the right. Only 3 students did better on Monday (had a negative difference).

12. No, you can't add these, since the groups are not disjoint.

13. a) Categorical
   b) Go fish. All you need to do is match the denomination. The denominations are not ordered. (Answers will vary.)
   c) Gin rummy. All cards are worth their value in points (face cards are 10 points). (Answers will vary.)

14. a)

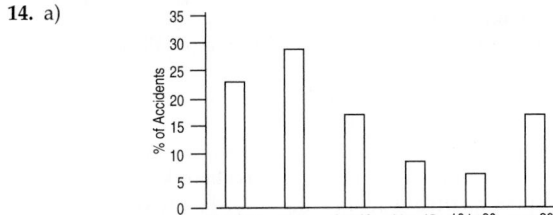

   b) We are given no information about how many miles are driven in each of these categories, so we have no idea how many accidents to *expect*. We also have no information about how many accidents were involved in compiling the data.

15. a) Annual mortality rate for males (quantitative) in deaths per 100,000 and water hardness (quantitative) in parts per million.
   b) Calcium is skewed right, possibly bimodal. There looks to be a mode down near 12 ppm that is the center of a fairly tight symmetric distribution and another mode near 62.5 ppm that is the center of a much more spread out, symmetric (almost uniform) distribution. Mortality, however, appears unimodal and symmetric with the mode near 1500 deaths per 100,000.

16. a) Overall mean is $(34 \times 1631.59 + 27 \times 1388.85)/(34 + 27) = 1524.15$ deaths per 100,000.
   b) Yes. Mortality for the Northern towns is generally higher than that for the South. Fully half of the towns in the South have mortality rates lower than the rates of *any* of the Northern towns. About 25% of Northern towns have rates higher than the rates of all of the Southern towns.

17. a) They are on different scales.
   b) January's values are lower and more spread out.
   c) Roughly symmetric but slightly skewed to the left. There are more low outliers than high ones. Center is around 40 degrees with an IQR of around 7.5 degrees.

18. Bimodal with modes around 50 and 80 minutes. Fairly symmetric around each mode.

19. a) Bimodal with modes near 2 and 4.5 minutes. Fairly symmetric around each mode.
   b) Because there are two modes, which probably correspond to two different groups of eruptions, an average might not make sense.
   c) The intervals between eruptions are longer for long eruptions. There is very little overlap. More than 75% of the short eruptions had intervals less than about an hour (62.5 minutes), while

more than 75% of the long eruptions had intervals longer than about 75 minutes. Perhaps the interval could even be used to predict whether the next eruption will be long or short.

20. The chance of an accident is not the same for different age groups. The distribution of ages for drivers involved in fatal crashes is not the same as that for other drivers. So, the two variables are not independent.

21. a)

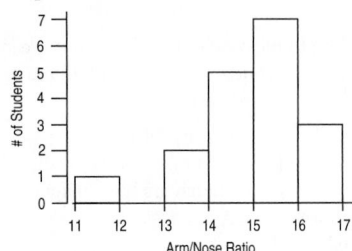

   The distribution is left skewed with a center of about 15. It has an outlier between 11 and 12.
   b) Even though the distribution is somewhat skewed, the mean and median are close. The mean is 15.0 and the SD is 1.25.
   c) Yes. 11.8 is already an outlier. 9.3 is more than 4.5 SDs below the mean. It is a very low outlier.

22. a) About 38%   b) 16%
   c) Data are bimodal, and the Normal model is inappropriate.

23. If we look only at the overall statistics, it appears that the follow-up group is insured at a much lower rate than those not traced (11.1% of the time compared with 16.6%). But most of the follow-up group were black, who have a lower rate of being insured. When broken down by race, the follow-up group actually has a higher rate of being insured for both blacks and whites. So the overall statistic is misleading and is attributable to the difference in race makeup of the two groups.

24. a) 3, 25.5, 36, 50.5, 70
   b) Because the IQR is so large, none are technically outliers, but the seasons with fewer than 20 home runs stand out as a separate group.
   c)

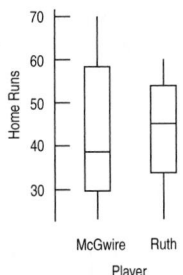

   d) Without the injured seasons, McGwire and Ruth's home run production distributions look similar. (Ruth's seasons as a pitcher were not included as well.) Ruth's median is a little higher, and he was a little more consistent (less spread), but McGwire had the two highest season totals.
   e)

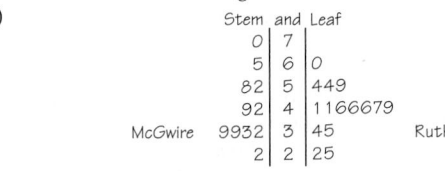

   Stem and Leaf

   7|0 = 70 home runs

   f) Now we can see how much more consistent Ruth was. Most of Ruth's seasons had home run totals in the 40s or 50s. McGwire's seasons are much more spread out (not even including the three at the bottom).

25. a)

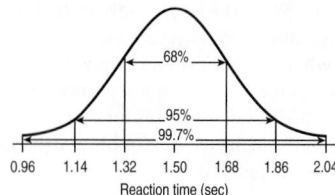

b) According to the model, reaction times are symmetric with center at 1.5 seconds. About 95% of all reaction times are between 1.14 and 1.86 seconds.

c) 8.2%   d) 24.1%

e) Quartiles are 1.38 and 1.62 seconds, so the IQR is 0.24 seconds.

f) The slowest 1/3 of all drivers have reaction times of 1.58 seconds or more.

26. a) *Who*—62 people
What—Number of objects correctly memorized
When—Not stated
Where—Not stated
Why—To see if music affects memorization ability
How—Randomized experiment

b) Type of music (categorical) and number of items remembered (quantitative).

c) Because we do not have all the data, we can't know exactly how the boxplots look, but we do know that the minimums are all within the fences and that two groups have at least one outlier on the high side.

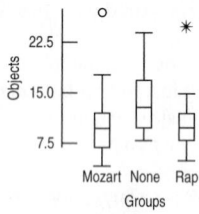

d) All three distributions are right skewed. Mozart and rap had very similar distributions. The scores for the None groups are, if anything, slightly higher than those for the other two groups. It is clear that groups listening to music did not score higher than those who heard none.

27. a)

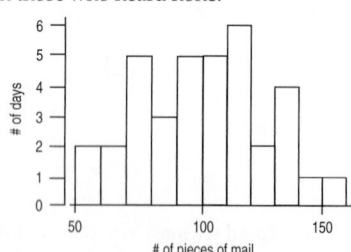

b) Mean 100.25, SD 25.54 pieces of mail.

c) The distribution is somewhat symmetric and unimodal, but the center is rather flat, almost uniform.

d) 64%. The Normal model seems to work reasonably well, since it predicts 68%.

28. a) 50.7%   b) 34.9%
c) 13.3%   d) 6.7%

29. a) *Who*—100 health food store customers
What—Have you taken a cold remedy?, and Effectiveness (scale 1 to 10)
When—Not stated
Where—Not stated
Why—Promotion of herbal medicine
How—In-person interviews

b) Have you taken a cold remedy?—categorical. Effectiveness—categorical or ordinal.

c) No. Customers are not necessarily representative, and the Council had an interest in promoting the herbal remedy.

30. a) Math/Sci—25.6%, Ag—41.7%, Humanities—19.3%, Other—13.5%.

b) Math/Sci—30.1%, Ag—46.0%, Humanities—13.3%, Other—10.6%.

c) Math/Sci—20.3%, Ag—39.1%, Humanities—24.6%, Other—15.9%.

d) No; it appears that oldest children are more likely than second-born children to major in Math or Science (30% to 20%), while second-born children are more likely than first-born to major in the Humanities (25% to 13%)

31. a) 38 cars

b) Possibly because the distribution is skewed to the right.

c) Center—median is 148.5 cubic inches. Spread—IQR is 126 cubic inches.

d) No. It's bigger than average, but smaller than more than 25% of cars. The upper quartile is at 231 inches.

e) No. 1.5 IQR is 189, and 105 − 189 is negative, so there can't be any low outliers. 231 + 189 = 420. There aren't any cars with engines bigger than this, since the maximum has to be at most 105 (the lower quartile) + 275 (the range) = 380.

f) Because the distribution is skewed to the right, this is probably not a good approximation.

g) Mean, median, range, quartiles, IQR, and SD all get multiplied by 16.4.

32. a) Fairly symmetric, almost uniform except for the right tail.

b) 47 horses

c) IQR is 47, so fences are at 7.5 and 195.5. Since the range is only 90, we know there can't be any observations lower than 35 (upper quartile − range) or greater than 168 (lower quartile + range), so no.

d) Distribution is very roughly uniform, not unimodal, so the Normal modal might not be reasonable.

e) 22 of 38, or about 58%.

f) Mean, median, and quartiles would increase by 10. SD, IQR, and range would not change.

33. a) 30.4%

b) If this were a random sample of all voters, yes.

c) 36.6%   d) 8.8%
e) 23.1%   f) 47.0%

34. Chief executives have a mean salary less than the median, so the distribution is likely to be skewed to the left. General and operations managers have a mean salary larger than the median, so their distribution is likely to be skewed to the right.

35. a) Republican—16,535, Democrat—17,183, Other— 20,666; or Republican—30.4%, Democrat—31.6%, Other—38.0%.

b)

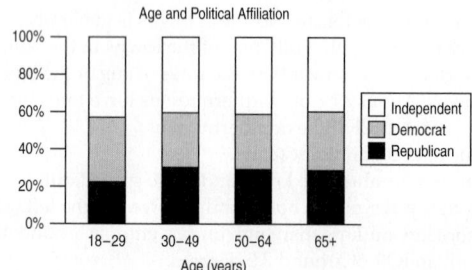

c) Among voters over 30, political affiliation appears to be largely unrelated to age. However there is some evidence that younger voters are less likely to be Republican

d) Voters who identified themselves as "Other" seem to be generally younger than Democrats or Republicans.

36. a) *Who*—Years from 1994 to 2003
What—Bicycle fatalities
When—1994–2003
Where—United States

Why—To study bicycle helmet safety
How—Bicycle Helmet Safety Institute Report

b)

Stem	Leaf
6	1
6	68
7	2
7	5569
8	12

6/1 = 610 − 619 fatalities

c)

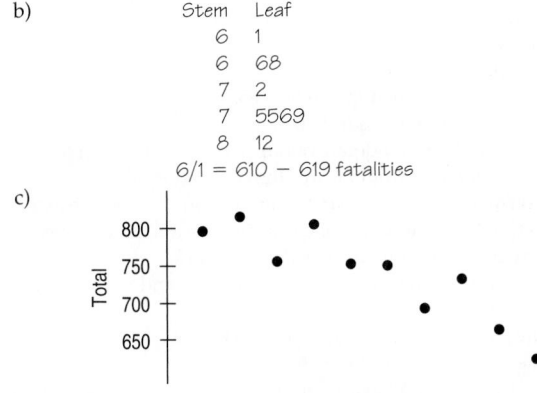

d) The stem-and-leaf display shows the distribution is skewed to the left. It also provides some idea about the center and spread of the annual fatalities.

e) The number of bicycle fatalities has tended to decrease over the 10-year period.

f) In the 10-year period from 1994 to 2003, reported bicycle fatalities decreased fairly steadily from about 800 per year to around 620 a year.

37. a) 0.43 hours.   b) 1.4 hours.
  c) 0.89 hours (or 53.4 minutes).
  d) Survey results vary, and the mean and the SD may have changed.

38. a) −9, 1, 4, 9, 25
  b)

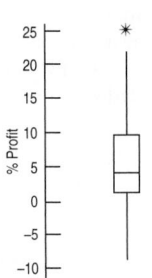

  c) Mean 4.72% of sales, SD 7.55% of sales.
  d) Fairly symmetric and unimodal, centered around 4% of sales. 50% of the companies report % profit between 1% and 9%. There are two outliers at 22% and 25% of sales.

CHAPTER 7

1. a) Weight in ounces: explanatory; Weight in grams: response. (Could be other way around.) To predict the weight in grams based on ounces. Scatterplot: positive, straight, strong (perfectly linear relationship).
  b) Circumference: explanatory. Weight: response. To predict the weight based on the circumference. Scatterplot: positive, linear, moderately strong.
  c) Shoe size: explanatory; GPA: response. To try to predict GPA from shoe size. Scatterplot: no direction, no form, very weak.
  d) Miles driven: explanatory; Gallons remaining: response. To predict the gallons remaining in the tank based on the miles driven since filling up. Scatterplot: negative, straight, moderate.

2. a) Price: explanatory; Number sold: response. To predict the number sold based on the price. Scatterplot: negative, straight, moderate.
  b) Depth: explanatory; Water pressure: response. To predict water pressure based on depth. Scatterplot: positive, straight, strong.

c) Visibility: explanatory; Depth: response. To predict depth based on visibility (although predicting visibility based on depth is also possible). Scatterplot: negative, straight(?), weak to moderate.
  d) Weight: explanatory; Reading score: response. To predict reading test scores based on weight. Scatterplot: positive, possibly straight, moderate.

3. a) Altitude: explanatory; Temperature: response. (Other way around possible as well.) To predict the temperature based on the altitude. Scatterplot: negative, possibly straight, weak to moderate.
  b) Ice cream cone sales: explanatory. Air-conditioner sales: response—although the other direction would work as well. To predict one from the other. Scatterplot: positive, straight, moderate.
  c) Age: explanatory; Grip strength: response. To predict the grip strength based on age. Scatterplot: curved down, moderate. Very young and elderly would have grip strength less than that of adults.
  d) Reaction time: explanatory; Blood alcohol level: response. To predict blood alcohol level from reaction time test. (Other way around is possible.) Scatterplot: positive, nonlinear, moderately strong.

4. a) Time: explanatory; Cost: response. To predict cost based on time. Scatterplot: positive, straight, strong.
  b) Time delay: explanatory; Distance: response. To predict the distance from the lightning based on the time delay of the thunder. Scatterplot: positive, straight, strong.
  c) Brightness: explanatory; Distance: response. To predict distance based on apparent brightness. Scatterplot: negative, curved, moderate.
  d) Weight of car: explanatory; Age of owner: response. To predict the age of the owner based on the weight of the car. (Or other way around.) Scatterplot: no direction, no shape, very weak.

5. a) None   b) 3 and 4   c) 2, 3, and 4
  d) 1 and 2   e) 3 and possibly 1

6. a) 1   b) 4   c) 2 and 4   d) 3   e) 2 and 4

7. There seems to be a very weak—or possibly no—relation between brain size and performance IQ.

8. Nonlinear form. Moderately strong. The rate has not been constant; the rate of increase from the beginning to about 1950 is steeper than from 1950 to the present. One winner in the early 1890s was quite slow.

9. a)

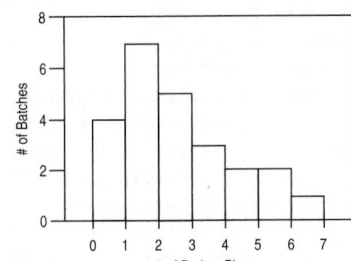

  b) Unimodal, skewed to the right. The skew.
  c) The positive, somewhat linear relation between batch number and broken pieces.

10. a)

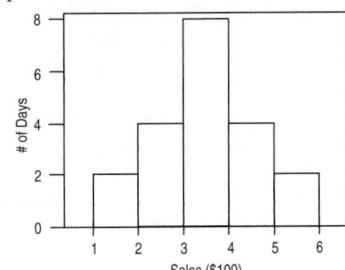

b) Positive linear relation, so sales are increasing.

c) Unimodal, symmetric sales; average sales around $350.

11. a) 0.006   b) 0.777   c) −0.923   d) −0.487

12. a) −0.977   b) 0.736   c) 0.951   d) −0.021

13. There may be an association, but not a correlation unless the variables are quantitative. There could be a correlation between average number of hours of TV watched per week per person and number of crimes committed per year. Even if there is a relationship, it doesn't mean one causes the other.

14. There may be an association, but not a correlation; type of car is not a quantitative variable.

15. a) Yes. It shows a linear form and no outliers.

b) There is a strong, positive, linear association between drop and speed; the greater the coaster's initial drop, the higher the top speed.

16. a) Yes. The plot shows a positive sign, linear form, and no outliers (though the last point on the right may be a bit out of line).

b) There's a strong, positive association. Experiments that showed a greater mean improvement among patients who took an antidepressant also showed a greater placebo effect.

17. The scatterplot is not linear; correlation is not appropriate.

18. The scatterplot is not linear, it has two high outliers (LA and NY), and there is a cluster of 8 cities that appear to be atypical. Correlation is not appropriate for such a relationship.

19. The correlation may be near 0. We expect nighttime temperatures to be low in January, increase through spring and into the summer months, then decrease again in the fall and winter. The relationship is not linear.

20. The relation might be nonlinear, resulting in a correlation close to 0 even though the association is strong, or there might be an outlier.

21. The correlation coefficient won't change, because it's based on z-scores. The z-scores of the prediction errors are the same whether they are expressed in nautical miles or miles.

22. The correlation coefficient won't change, because it's based on z-scores. The z-scores of the prediction errors are not changed by adding or subtracting a constant.

23. a) Assuming the relation is linear, a correlation of −0.772 shows a strong relation in a negative direction.

b) Continent is a categorical variable. Correlation does not apply.

24. a) Correlation cannot be greater than 1. There is an error.

b) Assuming the relation is linear, the strong correlation shows a relation, but it does not show *causality*.

25. a) Actually, yes, taller children will tend to have higher reading scores, but this doesn't imply causation.

b) Older children are generally both taller and are better readers. Age is the lurking variable.

26. a) No. It just means that in countries where cell phone use is high, the life expectancy tends to be high as well.

b) General economic conditions of the country could affect both cell phone use and life expectancy. That's the lurking variable.

27. a) No. We don't know this from the correlation alone. There may be a nonlinear relationship or outliers.

b) No. We can't tell from the correlation what the form of the relationship is.

c) No. We don't know from the correlation coefficient.

d) Yes, the correlation doesn't depend on the units used to measure the variables.

28. a) We don't know this from the correlation alone. There may be a nonlinear relationship or outliers.

b) No. We can't tell from the correlation what the form of the relationship is.

c) No. We don't know this from the correlation coefficient.

d) No. The correlation doesn't depend on the units used to measure the variables.

29. This is categorical data even though it is represented by numbers. The correlation is meaningless.

30. The source of data is categorical. The correlation is meaningless.

31. a) The association is positive, moderately strong, and roughly straight, with several states whose HCI seems high for their median income and one state whose HCI appears low given its median income.

b) The correlation would still be 0.65.

c) The correlation wouldn't change.

d) DC would be a moderate outlier whose HCI is high for its median income. It would lower the correlation slightly.

e) No. We can only say that higher median incomes are associated with higher housing costs, but we don't know why. There may be other economic variables at work.

32. a) The association is negative, quite strong, and fairly straight; there are no outliers.

b) The correlation would still be −0.84.

c) The correlation wouldn't change.

d) That year would have a very high mortgage amount for an interest rate that high. That would tend to weaken the correlation (bring it closer to 0).

e) No. We can only say that lower interest rates are associated with larger mortgage amounts, but we don't know why. There may be other economic variables at work.

33. a)

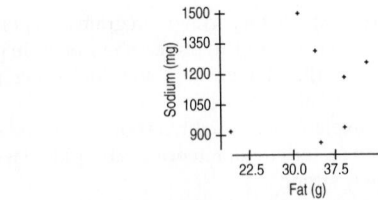

b) Negative, linear, strong.   c) −0.869

d) There is a strong linear relation in a negative direction between horsepower and highway gas mileage. Lower fuel efficiency is associated with higher horsepower.

34. a)

b) 0.934

c) There is a strong linear relation in a positive direction between the percent of teens using marijuana and the percent of teens using other drugs.

d) There is no indication of causality. There could be lurking variables.

35.

(Plot could have explanatory and predictor variables swapped.) Correlation is 0.199. There does not appear to be a relation between sodium and fat content in burgers, especially without the low-fat, low-sodium item. The correlation of 0.199 shows a weak relationship, even with the outlier included.

36. Correlation is 0.961. There appears to be a strong linear relation in a positive direction. The correlation of 0.961 supports the conclusion

of a strong relation. Even without the outlier at 410 calories and 19 grams of fat, the correlation is 0.837, still strong.

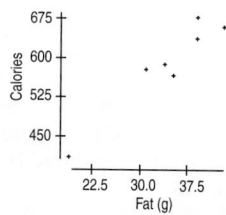

37. a) Yes, the scatterplot appears to be somewhat linear.
   b) As the number of runs increases, the attendance also increases.
   c) There is a positive association, but it does not *prove* that more fans will come if the number of runs increases. Association does not indicate causality.
38. a) There appears to be a moderately strong positive relationship between the number of wins and home attendance. The points show some significant scattering. The correlation of 0.697 is moderately strong.
   b) Winning. The correlation is only slightly higher for wins vs. attendance (0.697) than for runs vs. attendance (0.667).
   c) The correlation between runs and wins is 0.605.
39. A scatterplot shows a generally straight scattered pattern with no outliers. The correlation between *Drop* and *Duration* is 0.35, indicating that rides on coasters with greater initial drops generally last somewhat longer, but the association is weak.
40. a)

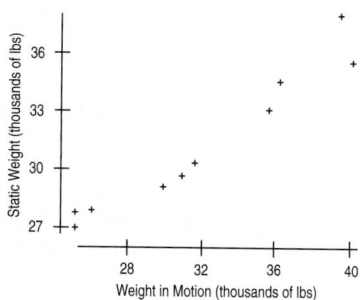

   b) Positive, linear, very strong.
   c) The new scale is able to predict the static weight fairly well, except possibly at the high end. It may be possible to predict the static weight from the new scale accurately enough to be useful. But the weight-in-motion measurements seem a bit too high.
   d) 0.965.
   e) The correlation would not change.
   f) At the higher end of the weight-in-motion scale, there is one point where the weight in motion is much higher than the static weight. The line does not seem to go through the points where x and y are equal. The new scale may have to be recalibrated.
41. a)

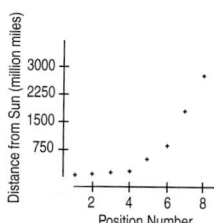

   The relation between position and distance is nonlinear, with a positive direction. There is very little scatter from the trend.
   b) The relation is not linear.

c)

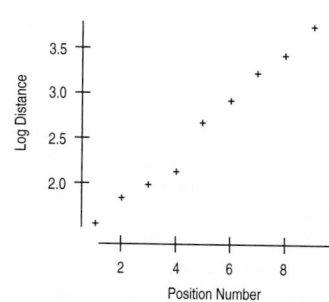

   The relation between position number and log of distance appears to be roughly linear.
42. a) 0.828
   b)

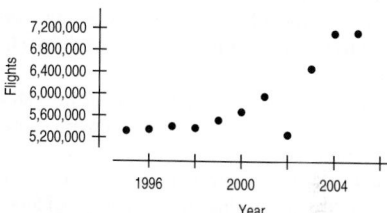

   The trend is positive and curved. There is a low outlier at 2002.
   c) The plot is not straight and has an outlier. Either violation would disqualify the correlation. It isn't unusual for growth in a business to be faster than linear. The outlier in 2002 is due to the drop in airline flights after the 9/11 attacks.

CHAPTER 8

1. 281 milligrams
2. 30.07 mpg
3. The potassium content is actually lower than the model predicts for a cereal with that much fiber.
4. The car gets better gas mileage than the model predicts for a car with that much horsepower.
5. The model predicts that cereals will have approximately 27 more milligrams of potassium for every additional gram of fiber.
6. The model estimates that cars lose an average of 0.84 miles per gallon for every additional 10 horsepower.
7. 81.5%   **8.** 75.5%
9. The true potassium contents of cereals vary from the predicted amounts with a standard deviation of 30.77 milligrams.
10. True fuel economy varies from the predicted amount with a standard deviation of 3.287 miles per gallon.
11. a) Model is appropriate.
    b) Model is not appropriate. Relationship is nonlinear.
    c) Model may not be appropriate. Spread is changing.
12. a) Model is not appropriate. Relationship is nonlinear.
    b) Model may not be appropriate. Spread is changing.
    c) Model is appropriate.
13. 300 pounds/foot. It's ridiculous to suggest an extra foot in length would add 3, 30, or 3000 pounds to a car's weight.
14. 1 ft/inch. A small increase of 1" in circumference wouldn't add 100' (or even 10') to the height of a tree, but surely more than 0.1 ft.
15. a) *Price* (in thousands of dollars) is y and *Size* (in square feet) is x.
    b) Slope is thousands of $ per square foot.
    c) Positive. Larger homes should cost more.
16. a) *Duration* (in seconds) is y and initial *Drop* (in feet) is x.
    b) Slope is seconds per foot.
    c) Positive. Bigger coasters probably provide longer rides.
17. A linear model on *Size* accounts for 71.4% of the variation in home *Price*.
18. A linear model on *Height* accounts for 12.4% of the variability in the *Duration* of the ride.

19. a) 0.845; + because larger homes cost more.
b) Price should be 0.845 SDs above the mean in price.
c) Price should be 1.690 SDs below the mean in price.

20. a) 0.352; + because bigger coasters probably give longer rides.
b) Duration should be about 0.352 SDs below the mean duration.
c) Duration should be about 1.056 SDs above the mean duration.

21. a) *Price* increases by about $0.061 × 1000, or $61.00, per additional sq ft.
b) 230.82 thousand, or $230,820.
c) $115,020; $6000 is the residual.

22. a) On average, rides last about 0.242 seconds longer per additional foot of initial drop.
b) 139.433 seconds
c) That's 7.333 seconds shorter than the model predicts, a negative residual.

23. a) R^2 does not tell whether the model is appropriate, but measures the strength of the linear relationship. High R^2 could also be due to an outlier.
b) Predictions based on a regression line are estimates of average values of y for a given x. The actual wingspan will vary around the prediction.

24. a) R^2 measures the amount of variation accounted for. It does not imply that x determines the values of y. Here, literacy rate *accounts for* 64% of the *observed* variation in life expectancy.
b) This interpretation is implying a causal relationship. Just because two variables are related does not mean one causes another. Here, one could say that an increase of 5% in literacy rate is associated with an average 2-year improvement in life expectancy.

25. a) Probably not. Your score is better than about 97.5% of people, assuming scores follow the Normal model. Your next score is likely to be closer to the mean.
b) The friend should probably retake the test. His score is better than only about 16% of people. His score is likely to be closer to the mean.

26. People on the cover are usually there for outstanding performances. Because they are so far from the mean, the performance in the next year is likely to be closer to the mean.

27. a) Probably. The residuals show some initially low points, but there is no clear curvature.
b) The linear model on *Tar* content accounts for 92.4% of the variability in *Nicotine*.

28. a) Yes, the relationship is not very strong, but it's reasonably straight.
b) The linear model on number of *Wins* accounts for 48.5% of the variation in *Attendance*.
c) The residuals spread out. There is more variation in *Attendance* as the number of wins increases.
d) The Yankees' attendance was about 13,000 fans more than we might expect given the number of wins.

29. a) $r = 0.961$
b) Nicotine should be 1.922 SDs below average.
c) Tar should be 0.961 SDs above average.

30. a) 0.696
b) Attendance should be 1.392 SDs above the average.
c) Number of games won should be 0.696 SDs below the average.

31. a) $\widehat{Nicotine} = 0.15403 + 0.065052\,Tar$
b) 0.414 mg
c) Predicted nicotine content increases by 0.065 mg of nicotine per additional milligram of tar.
d) We'd expect a cigarette with no tar to have 0.154 mg of nicotine.
e) 0.1094 mg

32. a) $\widehat{Attendance} = -14,364.5 + 538.915\,Wins$.
b) 12,581 people
c) Every win adds an average 538.915 people in attendance.
d) It means the team's average attendance is lower than the expected average for a team with as many wins.

33. e) 12,222.56 attendees. This means that the Cardinals averaged over 12,000 more attendees than one would predict for an AL team with 83 wins.

33. a) Yes. The relationship is straight enough, with a few outliers. The spread increases a bit for states with large median incomes, but we can still fit a regression line.
b) From summary statistics: $\widehat{HCI} = -156.50 + 0.0107\,MFI$; from original data: $\widehat{HCI} = -157.64 + 0.0107\,MFI$
c) From summary statistics: predicted HCI = 324.93; from original data: 324.87.
d) 223.09   e) $\widehat{z_{HCI}} = 0.65z_{MFI}$ f) $\widehat{z_{MFI}} = 0.65z_{HCI}$

34. a) Yes, the relationship is straight enough, both variables are quantitative, and the spread is roughly constant. There are no real outliers.
b) $\widehat{MortAmt} = 220.88 - 7.768\,IntRate$
($\widehat{MortAmt} = 220.89 - 7.775\,IntRate$ from original data)
c) $65.52 million ($65.39 from original data equation).
d) Yes, 20% is outside the range of the original x-variable. It may not be appropriate.
e) $\widehat{z_{Mort}} = -0.84\,z_{Int}$ f) $\widehat{z_{Int}} = -0.84\,z_{Mort}$

35. a) $\widehat{Total} = 539.803 + 1.103\,Age$
b) Yes. Both variables are quantitative; the plot is straight (although flat); there are no apparent outliers; the plot does not appear to change spread throughout the range of *Age*.
c) $559.65; $594.94
d) 0.14%
e) No. The plot is nearly flat. The model explains almost none of the variation in *Total Yearly Purchases*.

36. a) $\widehat{Total} = 28.73 + 0.0108\,Income$
b) Yes. Both variables are quantitative; the plot is straight enough; there are a few possible outliers, but none are extreme, and there are 500 points; the plot does not appear to change spread throughout the range of *Income*.
c) $244.76, $892.86   d) 52.1%
e) Yes, the model accounts for a good part of the variation in *Total Yearly Purchases*. The difference between the predicted amount of $208 and $922 is probably financially significant.

37. a) Moderately strong, fairly straight, and positive. Possibly some outliers (higher-than-expected math scores).
b) The student with 500 verbal and 800 math.
c) Positive, fairly strong linear relationship. 46.9% of variation in math scores is explained by verbal scores.
d) $\widehat{Math} = 217.7 + 0.662 \times Verbal$.
e) Every point of verbal score adds 0.662 points to the predicted average math score.
f) 548.5 points   g) 53.0 points

38. a) $\widehat{GPA} = -1.262 + 0.00214\,SAT$.
b) One could say that a person with a 0 SAT score would have a −1.262 GPA, but a score of 0 on the SAT is impossible, as is a negative GPA. So the y-intercept serves only to adjust the height of the line and is meaningless by itself.
c) The expected GPA is higher by 0.21 points for every additional hundred combined SAT points scored.
d) 3.23
e) *SAT* is probably only somewhat useful, since R^2 is only 0.221. There are many other factors that influence *GPA*.
f) Positive. My *GPA* would be higher than expected.

39. a) 0.685   b) $\widehat{Verbal} = 162.1 + 0.71 \times Math$.
c) The observed verbal score is higher than predicted from the math score
d) 516.7 points.   e) 559.6 points
f) Regression to the mean. Someone whose math score is below average is predicted to have a verbal score below average, but not as far (in SDs). So if we use *that* verbal score to predict math, they will be even closer to the mean in predicted math score than their observed math score. If we kept cycling back and forth, eventually we would predict the mean of each and stay there.

40. 1868.1 points

41. a)

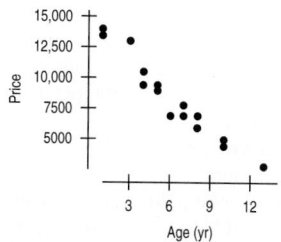

b) Negative, linear, strong.   c) Yes.   d) −0.972
e) *Age* accounts for 94.4% of the variation in *Advertised Price*.
f) Other factors contribute—options, condition, mileage, etc.

42. a) Yes. Plot shows positive, linear, fairly strong relationship.
b) Variation in marijuana use accounts for 87.3% of variation in use of other drugs.
c) $\widehat{Other\%} = -3.068 + 0.615\ Marijuana\%$.
d) Each additional percent of teens using marijuana adds 0.615 percent to the percentage of teens using other drugs.
e) The results do not *confirm* marijuana as a gateway drug. They do indicate an *association* between marijuana and other drug use, but that doesn't imply causation.

43. a) $\widehat{Price} = 14{,}286 - 959 \times Years$.
b) Every extra year of age decreases average value by $959.
c) The average new Corolla costs a predicted $14,286.
d) $7573
e) Negative residual. Its price is below the predicted value for its age.
f) −$1195
g) No. After age 14, the model predicts negative prices. The relationship is no longer linear.

44. a)

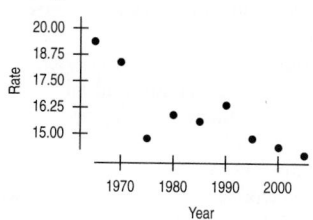

b) $\widehat{Birth\ Rate} = 25.345 - 0.11 \times Year$.
c)

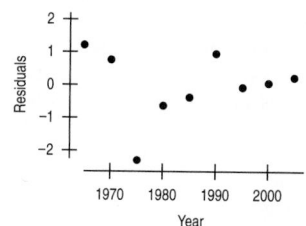

There's a low outlier for 1975; we might want to investigate.
d) The *Birth Rate* has been decreasing by 0.11 births per 1000 women per *Year*, on average.
e) 16.739 live births per 1000 women.
f) Model predicts 1.739 more live births per 1000 women.
g) 13.21 live births per 1000 women. Caution: We are extrapolating.
h) 11.55 live births per 1000 women. Extrapolating this far is dangerous.

45. a)

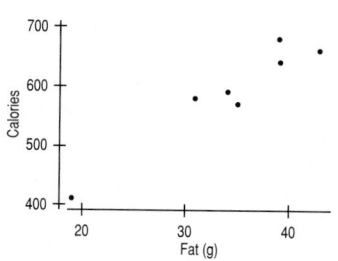

b) 92.3% of the variation in calories can be accounted for by the fat content.
c) $\widehat{Calories} = 211.0 + 11.06 \times Fat$.
d)

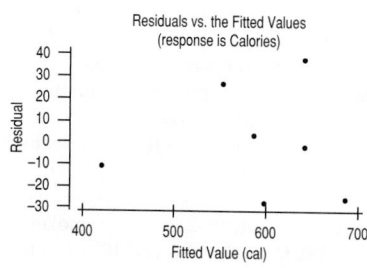

Residuals show no clear pattern, so the model seems appropriate.
e) Could say a fat-free burger still has 211.0 calories, but this is extrapolation (no data close to 0).
f) Every gram of fat adds 11.06 calories, on average.
g) 553.5 calories.

46. a) Yes.   b) Strong.
c) $\widehat{Calories} = 185.7 + 13.93 \times Fat$.
d) Every gram of fat adds 13.93 calories, on average.
e) A fat-free sandwich would have 185.7 calories, but this is extrapolation.
f) The sandwich has fewer calories than predicted for its grams of fat.

47. a) The regression was for predicting calories from fat, not the other way around.
b) $\widehat{Fat} = -15.0 + 0.083 \times Calories$. Predict 34.8 grams of fat.

48. a) The regression was for predicting calories from fat, not the other way around.
b) $\widehat{Fat} = -9.842 + 0.0644\ Calories$. Predict 15.9 grams of fat.

49. a) % Body Fat $= -27.4 + 0.25 \times Weight$.
b) Residuals look randomly scattered around 0, so conditions are satisfied.
c) *% Body Fat* increases, on average, by 0.25 percent per pound of *Weight*.
d) Reliable is relative. R^2 is 48.5%, but residuals have a standard deviation of 7%, so variation around the line is large.
e) 0.9 percent.

50. $\widehat{\%\ Body\ Fat} = -62.6 + 2.22 \times Waist$.
R^2 is 78.7%, and the standard deviation of the residuals is smaller (4.5%), so this model should be better able to predict.

51. a) $\widehat{HighJump} = 2.681 - 0.00671 \times 800mTime$. High-jump height is lower, on average, by 0.00671 meters per additional second of 800-m race time.
b) 16.4%
c) Yes, the slope is negative. Faster runners tend to jump higher.
d) There is a slight tendency for less variation in high-jump height among the slower runners than among the faster ones.
e) Not especially. The residual standard deviation is 0.060 meters, which is not much smaller than the SD of all high jumps (0.066 meters). The model doesn't appear to do a very good job of predicting.

52. a) $\widehat{LongJump} = 4.20053 + 1.10541 HighJump$. Every additional meter in the high jump is associated with an additional 1.1054 meters, on average, in the long jump.
b) 12.6%   c) Yes, the slope is positive.
d) The residuals plot is fairly patternless.
e) No. The residual SD is 0.196 meters, not much smaller than the SD of all long jumps (0.206), and R^2 is only 12.6%.

53. The sum of the squared vertical distances to any other line would be greater than 1790.

54. The sum of the squared vertical distances to any other line would be greater than 34,500.

CHAPTER 9

1. a) The trend appears to be somewhat linear up to about 1940, but from 1940 to about 1970 the trend appears to be nonlinear. From 1975 or so to the present, the trend appears to be linear.
   b) Relatively strong for certain periods.
   c) No, as a whole the graph is clearly nonlinear. Within certain periods (ex: 1975 to the present) the correlation is high.
   d) Overall, no. You could fit a linear model to the period from 1975 to 2003, but why? You don't need to interpolate, since every year is reported, and extrapolation seems dangerous.
2. a) The percent of men 18–24 who are smokers decreased dramatically between 1965 and 1990, but the trend has not been consistent since then.
   b) Stronger before 1990 than after.
   c) No. The relationship is not straight.
3. a) The relationship is not straight.
   b) It will be curved downward.
   c) No. The relationship will still be curved.
4. a) The relationship is not straight.
   b) It will be curved downward.
5. a) No. We need to see the scatterplot first to see if the conditions are satisfied, and models are always wrong.
   b) No, the linear model might not fit the data everywhere.
6. a) No, it just means that the model doesn't account for more than 13% of the variation of the response variable.
   b) She should look at s to judge the accuracy of predictions. R^2 is not a direct measure of accuracy.
7. a) Millions of dollars per minute of run time.
   b) Costs for movies increase at the same rate per minute.
   c) On average dramas cost about $20 million less for the same runtime.
8. a) The slopes are about the same, so the costs increase at about the same rate as movies get longer.
   b) Although the costs per minute are about the same, on average it costs about $20 million less to make an R-rated movie than a movie of the other rating types with the same running time.
   c) Omitting *King Kong* would make the slope for the PG-13 movies steeper. We'd conclude that the cost per minute of PG-13 movies was greater than the cost per minute of the other movies.
9. a) The use of the Oakland airport has been growing at about 59,700 passengers/year, starting from about 282,000 in 1990.
   b) 71% of the variation in passengers is accounted for by this model.
   c) Errors in predictions based on this model have a standard deviation of 104,330 passengers.
   d) No, that would extrapolate too far from the years we've observed.
   e) The negative residual is September 2001. Air traffic was artificially low following the attacks on 9/11.
10. a) Tracking errors averaged about 292 nautical miles in 1970 and have decreased an average of 5.23 nautical miles per year since then.
    b) Residuals based on this model have a standard deviation of 42.87 nautical miles.
    c) The model predicts an error of only 88.1 nautical miles in 2009. However, the NHC has *already* achieved better than 125-nautical-mile accuracy, so it seems safe to assume that they can maintain that level of success even if the decline in prediction errors doesn't continue.
    d) A tracking error of 90 nautical miles would be achieved if the trend fit by the regression model were to persist, but that is an extrapolation beyond the data.
    e) We should be cautious in assuming that the improvements in prediction will continue at the same rate.

11. a) 1) High leverage, small residual.
       2) No, not influential for the slope.
       3) Correlation would decrease because outlier has large z_x and z_y, increasing correlation.
       4) Slope wouldn't change much because the outlier is in line with other points.
    b) 1) High leverage, probably small residual.
       2) Yes, influential.
       3) Correlation would weaken, increasing toward zero.
       4) Slope would increase toward 0, since outlier makes it negative.
    c) 1) Some leverage, large residual.
       2) Yes, somewhat influential.
       3) Correlation would increase, since scatter would decrease.
       4) Slope would increase slightly.
    d) 1) Little leverage, large residual.
       2) No, not influential.
       3) Correlation would become stronger and become more negative because scatter would decrease.
       4) Slope would change very little.
12. a) 1) High leverage, makes large residual a bit smaller.
       2) Yes, influential.
       3) Correlation would increase because scatter would decrease.
       4) Slope would increase because outlier pulls the line toward 0.
    b) 1) High leverage, small residual.
       2) Yes, influential.
       3) Correlation would become weaker because outlier has high z_x and z_y values, increasing correlation.
       4) Slope would decrease because outlier pulls the line from nearly flat to a positive slope.
    c) 1) Little leverage, large residual.
       2) No, not influential.
       3) Correlation would become stronger because scatter would decrease.
       4) About the same.
    d) 1) High leverage, small residual.
       2) No, not influential.
       3) Correlation would become weaker and become less negative because outlier has large negative $z_x \times z_y$ value.
       4) Slope would stay about the same because outlier is consistent with slope determined by other points.
13. 1) e   2) d   3) c   4) b   5) a
14. 1) d   2) e   3) c   4) b   5) a
15. Perhaps high blood pressure causes high body fat, high body fat causes high blood pressure, or both could be caused by a lurking variable such as a genetic or lifestyle issue.
16. Perhaps playing computer games makes kids more violent, or more violent kids like to play computer games, or both behaviors could be influenced by a lurking variable such as the child's home life or a genetic predisposition to aggressiveness.
17. a) The graph shows that, on average, students progress at about one reading level per year. This graph shows averages for each grade. The linear trend has been enhanced by using averages.
    b) Very close to 1.
    c) The individual data points would show much more scatter, and the correlation would be lower.
    d) A slope of 1 would indicate that for each 1-year grade level increase, the average reading level is increasing by 1 year.
18. The graph has a positive trend, showing that increased *SAT* score is associated with increased *GPA*. However, the graph is based on averages, not individual data. While *SAT* may be a fair predictor of *GPA* on average, there may also be lurking variables.
19. a) *Cost* decreases by $2.13 per degree of average daily *Temp*. So warmer temperatures indicate lower costs.
    b) For an avg. monthly temperature of 0°F, the cost is predicted to be $133.

c) Too high; the residuals (observed − predicted) around 32°F are negative, showing that the model overestimates the costs.

d) $111.70   e) About $105.70

f) No, the residuals show a definite curved pattern. The data are probably not linear.

g) No, there would be no difference. The relationship does not depend on the units.

20. a) The slope indicates that *Fuel Efficiency* decreases by 0.1 mpg per mph of *Speed*; for every 10 mph of increase in speed, the fuel economy decreases by 1 mpg.

b) The intercept is the predicted fuel efficiency at 0 mph, which doesn't make any sense.

c) The residuals are negative, so the model is overestimating the mpg.

d) 27 mpg   e) 28.5 mpg

f) The association is strong but not linear.

g) No, the residuals indicate a clear nonlinear relationship.

21. a) 0.88

b) Interest rates during this period grew at about 0.25% per year, starting from an interest rate of about 0.64%.

c) Substituting 50 in the model yields a prediction of about 13%.

d) Not really. Extrapolating 20 years beyond the end of these data would be dangerous and unlikely to be accurate.

22. a) −0.8666

b) The *Age Difference* is decreasing by about 0.0166 years per *Year*.

c) About 1.61 years.

d) The latest data point is for the year 2003. Extrapolating to 2015 is risky because it depends on the assumption that the trend will continue in the same manner.

23. a) The two models fit comparably well, but they have very different slopes.

b) This model predicts the interest rate in 2000 to be 3.24%, much lower than the other model predicts.

c) We can trust the new predicted value because it is in the middle of the data used for the regression.

d) The best answer is "I can't predict that."

24. a) The data from the late 1800s to 1950 are high-leverage points. Since they generally follow the same linear trend as the 1975–1998 data, those data points increase the correlation and the R^2 value.

b) Probably. There is a slight bend, but it is difficult to tell whether it is enough to prevent using a linear model. We should proceed with caution.

c) The predicted *Age Difference* drops about 0.030 years per *Year*.

d) These data begin with $x = 75$. The y-intercept is the predicted difference in age at the year $x = 0$, quite an extrapolation.

25. a) Stronger. Both slope and correlation would increase.

b) Restricting the study to nonhuman animals would justify it.

c) Moderately strong.

d) For every year increase in life expectancy, the gestation period increases by about 15.5 days, on average.

e) About 270.5 days.

26. a) Only 1.3% of the variation in time is accounted for by the regression on year.

b) We can't say. The Outlier Condition is violated.

c) Probably not. This point doesn't have much leverage.

27. a) Removing hippos would make the association stronger, since hippos are more of a departure from the pattern.

b) Increase.

c) No, there must be a good reason for removing data points.

d) Yes, removing it lowered the slope from 15.5 to 11.6 days per year.

28. a) Errors of prediction should be smaller with the new model.

b) This regression accounts for only 4.1% of the variability in time, but it looks like swimmers are slowing down at the rate of about 6.2 minutes per year.

29. a) Answers may vary. Using the data for 1955–2000 results in a scatterplot that is relatively linear with some curvature. The residuals plot shows a definite trend, indicating that the data are not linear. If you used the line, for 2010 the predicted age is 26.07 years.

b) Not much, since the data are not truly linear and 2010 is 10 years from the last data point (extrapolating is risky).

c) No, that extrapolation of more than 50 years would be absurd. There's no reason to believe the trend from 1955 to 2000 will continue.

30. Answers may vary. There seem to be two periods. From 1980 to 1993 or so, there was a consistent increase of about 1% a year. A regression of those years gives:

$$\widehat{Percentage} = -1988.94 + 1.014\,Year$$

with an R^2 of 98.9%. The years from 1994 to 1998 are nearly flat. A model of those years yields:

$$\widehat{Percentage} = -87.28 + 0.060\,Year$$

with an R^2 of 17.3%.

How to use the data depends on the purpose of the study.

31.

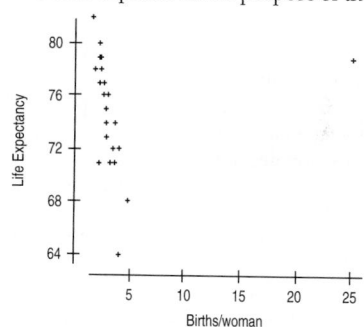

a) Except for the outlier, Costa Rica, the data appear to have a linear form in a negative direction.

b) The outlier is Costa Rica, whose data appear to be wrong, with 25 births per woman. That's impossible.

c) With Costa Rica, $r = 0.168$ and R-squared $= 2.8\%$, indicating that 2.8% of the variation in *Life Expectancy* is explained by the variation in *Births per Woman*. Without Costa Rica, $r = -0.796$ and R-squared $= 63.3\%$, indicating that 63.3% of the variation in *Life Expectancy* is explained by the variation in *Births/Woman*.

d) With Costa Rica, $\widehat{Life\ Expectancy} = 72.6 + 0.15\,Births$; without Costa Rica, $\widehat{Life\ Expectancy} = 84.5 - 4.44\,Births$.

e) The model with Costa Rica is not appropriate. The residuals plot shows a distinct outlier, which is Costa Rica. Removing Costa Rica gives a better residuals plot, suggesting that the linear equation is more appropriate.

f) With Costa Rica, the slope is near 0, suggesting that the linear model is not very useful. The y-intercept suggests that with no births, the life expectancy is about 72.6 years. Without Costa Rica, the slope is −4.44, indicating that an average increase of one child per woman predicts a lower life expectancy of 4.44 years, on average. The y-intercept indicates that a country with a birth rate of zero would have a life expectancy of 84.5 years. This is extrapolation.

g) While there is an association, there is no reason to expect causality. Lurking variables may be involved.

32. a) The relationship is positive (increasing speed over time), with a moderate amount of scatter. There is a general linear trend, with several deviations. There were no races from 1915 to 1918 or from 1940 to 1946 because of world wars. After each, speeds didn't recover for several years. The unexpectedly low speed for 2006 may be due to the absence of several leading riders over suspicions of drug use.

b) $\widehat{Avgspeed} = -272.745 + 0.156\,Year$

c) No. The relationship is not straight enough in the early part of the 20th century to fit a regression line. The variables are quantitative, and there are no real outliers.

33. a) The scatterplot is clearly nonlinear; however, the last few years—say, from 1970 on—do appear to be linear.

b) Using the data from 1970 to 2006 gives $r = 0.997$ and $\widehat{CPI} = -9052.42 + 4.61\ Year$. Predicted CPI in 2016 = 241.34 (an extrapolation of doubtful accuracy).

34. a) After an initial post-war increase, speeds leveled off in the 1960s. They began to climb again in the 1980s and have generally continued to do so. $\widehat{Avgspeed} = -205.8 + 0.123\ Year$

b) The slope says that average winning speed increases by about 0.123 kph per year.

c) Hinault's 1979 time has a residual of 2.79 kph. He raced much faster than the model would predict ($s_e = 1.34\ kph$). Armstrong's 2005 speed of 41.65 kph has a residual of -1.45 kph. Hinault's performance was more remarkable for its era.

CHAPTER 10

1. a) No re-expression needed.

b) Re-express to straighten the relationship.

c) Re-express to equalize spread.

2. a) Re-express to straighten the relationship.

b) Re-express to equalize spread.

c) No re-expression needed.

3. a) There's an annual pattern in when people fly, so the residuals cycle up and down.

b) No, this kind of pattern can't be helped by re-expression.

4. a) The plot shows a wavy pattern, likely part of an annual cycle that repeats each year.

b) Cyclic patterns like this cannot be straightened with a re-expression.

5. a) 16.44   b) 7.84   c) 0.36   d) 1.75   e) 27.59

6. a) 1.44   b) 630.96   c) 5.78   d) 1.67   e) 0.128

7. a) Fairly linear, negative, strong.

b) Gas mileage decreases an average 7.652 mpg for each thousand pounds of weight.

c) No. Residuals show a curved pattern.

8. a) The scatterplot shows upward curvature and decreasing spread.

b) Try plotting log (GDP) against Crowdedness.

9. a) Residuals are more randomly spread around 0, with some low outliers.

b) $\widehat{Fuel\ Consumption} = 0.625 + 1.178 \times Weight$.

c) For each additional 1000 pounds of *Weight*, an additional 1.178 gallons will be needed to drive 100 miles.

d) 21.06 miles per gallon.

10. a) No. The student has gone too far. We now see marked downward curvature.

b) A next step would be to try a "weaker" re-expression, like reciprocal square root or log (GDP).

11. a) Although more than 97% of the variation in GDP can be accounted for by this model, we should examine a scatterplot of the residuals to see if it's appropriate.

b) No. The residuals show clear curvature.

12. No, a trend that goes up and down erratically cannot be linearized by re-expression.

13. Yes, the pattern in the residuals is somewhat weaker.

14.

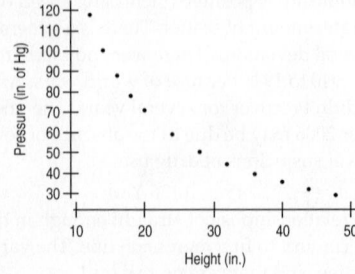

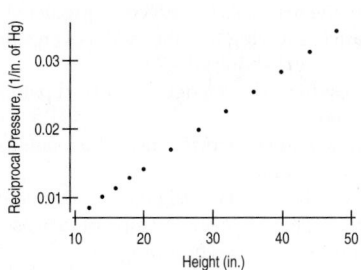

$1/Pressure$ is an exact relationship. R^2 is 100%.

$\widehat{reciprocal\ Pressure} = -0.000077 + 0.000713\ \widehat{Height}$

Predictor	Coeff	SE Coeff	T	P
Intercept	−0.00007670	0.00007813	−0.98	0.349
Height	0.00071307	0.00000260	274.30	0.000

$s = 0.0001057$ $R - Sq = 100.0\%$ $R - Sq(adj) = 100.0\%$

or $\widehat{\log(Pressure)} = 3.15 - 1.001\ \log(Height)$

15. a)

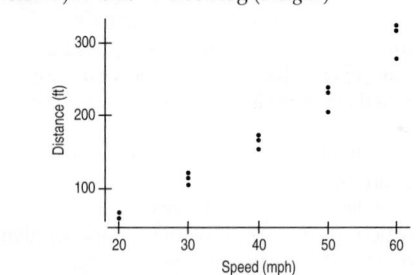

$\widehat{Distance} = -65.9 + 5.98\ Speed$.

But residuals have a curved shape, so linear model is not appropriate.

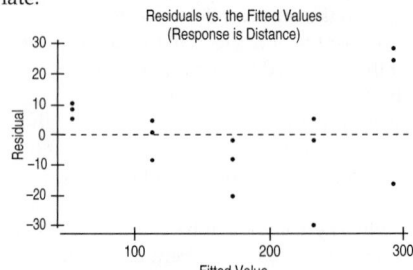

b)

$\sqrt{Distance}$ linearizes the plot.

c) $Predicted\ \sqrt{Distance} = 3.30 + 0.235 \times Speed$.

d) 263.4 feet.   e) 390.2 feet (an extrapolation)

f) Fairly confident, since $R^2 = 98.4\%$, and s is small.

16. a)

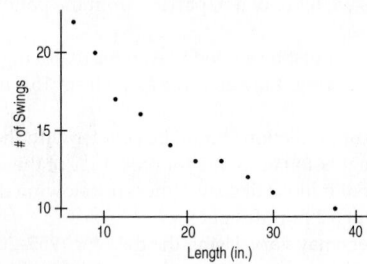

Relationship is curved, not straight.

b)

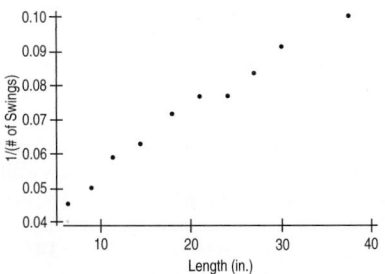

1/Swings makes the plot more linear.

c) $\widehat{1/(Swings)} = 0.0367 + 0.00176 \times Length.$
   or $\widehat{\log(Swings)} = 1.721 - 0.453 \log(Length)$

d) 22.9 swings.   e) 8.3 swings.

f) Fairly confident. These values are outside the data range, but not by a lot. Also, $R^2 = 98.1\%$.

17. a) The plot looks fairly straight. (It is okay to see a bend in the plot; there's one there.)

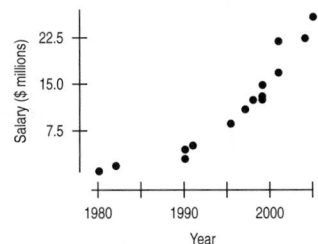

b) $\widehat{Salary} = -1913.88 + 0.965\,Year$

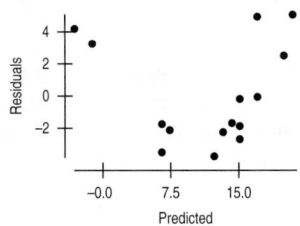

The residuals plot shows a strong bend.

c) *log(Salary)* works well.

d) $\log\widehat{(Salary)} = -109.133 + 0.05516\,Year$

18. a)

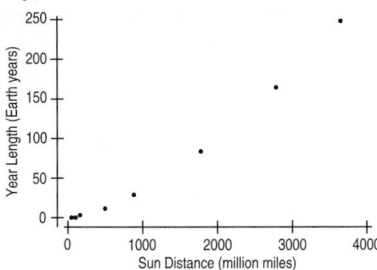

Concave upward curve.

b)

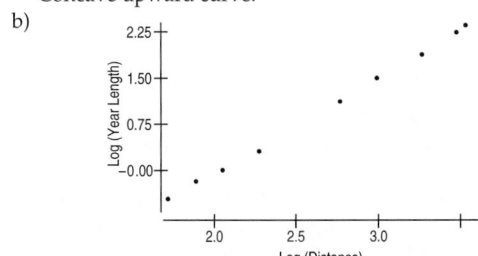

Taking the log of both variables linearizes the plot.

$\widehat{\log(Length)} = -2.95 + 1.5 \log(Distance).$

c) Fits perfectly. $R^2 = 100\%$.

19. a)

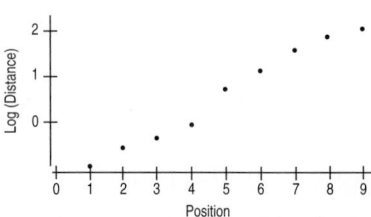

Log(*Distance*) against position works pretty well.

$\widehat{\log(Distance)} = 1.245 + 0.271 \times Position\ number.$

b) Pluto's residual is not especially larger in the log scale. However, a model without Pluto predicts the 9th planet should be 5741 million miles. Pluto, at "only" 3707 million miles, doesn't fit very well, giving support to the argument that Pluto doesn't behave like a planet.

20. $\widehat{\log(Distance)} = 1.32 + 0.23 \times New\ Position$, or

$\widehat{\log(Distance)} = 1.285 + 0.239\,New\ Position$ (without Pluto). The linear relationship is very strong ($R^2 = 99.5\%$). The new position works better for modeling log(*Distance*), since the other model had $R^2 = 98.2\%$.

21. The predicted log(*Distance*) of Eris is 3.685, corresponding to a distance of 4841 million miles. That's short of the actual average distance of 6300 million miles.

22. Both models have high R^2, but the Exercise 18 model has $R^2 = 1$. That indicates that perhaps we have found a physical law. If we find another system with the same pattern, it would add evidence for Titius-Bode "law"; otherwise, we would discount the law.

23. a)

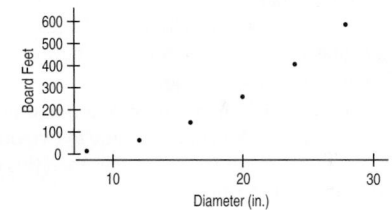

$\widehat{\sqrt{Bdft}} = -4 + diam$
The model is exact.

b) 36 board feet.   c) 1024 board feet.

24. a) $\widehat{Lift} = 164.97 + 2.56 Class.$

b)

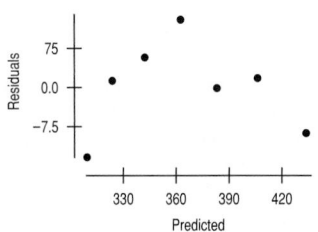

Residuals show a curved pattern; need a different model.

c) Re-expressing *Class* by reciprocal works well. (Other models are possible.)

$\widehat{Lift} = 568.727 - 15243(1/Class); R^2 = 99.4\%$

d) Less pattern in the residuals plot; higher R^2.

e) Asanidze's large negative residual indicates that he lifted less than expected.

25.

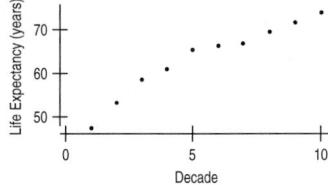

$\widehat{\log Life} = 1.685 + 0.18497 \log Decade$

26. a) $\widehat{Lift} = 573.257 - 15491.5(1/Class)$
b) This model predicts 391.0 kg for Asanidze.
c) He lifted 8.5 kg less than the model predicted.
27. The relationship cannot be made straight by the methods of this chapter.
28. $\dfrac{1}{\widehat{aveWeight}} = 1.603 + 0.00112(Oranges)$
29. a) $\widehat{\sqrt{Left}} = 8.465 - 0.06926(Age)$ b) 52.10 years
c) No; the residuals plot still shows a pattern.
30. a) The linear model:
$\widehat{Diameter} = 1.97 + 0.463 \times Age$ gives $R^2 = 98\%$, but there is a clear pattern in the residuals. The pattern oscillates, so it cannot be made straight with the methods of this chapter.
b) Less strong, since averages are less variable than individuals.

PART II REVIEW

1. % over 50, 0.69.
% under 20, −0.71.
% Graduating on time, −0.51.
% Full-time Faculty, 0.09
2. a) If no meals are eaten together, the *GPA* is 2.73.
b) For each increase of 1 meal, the *GPA* increases by 0.11.
c) 3.15
d) The predicted GPA was higher than the observed value.
e) There is no indication that there is a causal relation. There may be lurking variables.
3. a) There does not appear to be a linear relationship.
b) Nothing, there is no reason to believe that the results for the Finger Lakes region are representative of the vineyards of the world.
c) $\widehat{CasePrice} = 92.77 + 0.567 \times Years$.
d) Only 2.7 % of the variation in case price is accounted for by the ages of vineyards. Most of that is due to two outliers. We are better off using the mean price rather than this model.
4. a) No, no apparent relationship.
b) There is one very large leverage point.
c) Weaker. The cross point is extreme in the *x* direction and low on the *y*-axis, causing the correlation to weaken (become closer to 0).
d) The slope of the line would increase.
5. a) $\widehat{TwinBirths} = -5119590 + 2618.25 \times Year$.
b) Each year, the number of twins born in a year increases, on average, by approximately 2618.25.
c) 143,092.5 births. The scatterplot appears to be somewhat linear, but there is some curvature in the pattern. There is no reason to believe that the increase will continue to be linear 5 years beyond the data.
d) The residuals plot shows a definite curved pattern, so the relation is not linear.
6. a) 0.914
b) $\widehat{DJIA} = -2294.01 + 341.095(Year-1970)$.
c) Each year, the *DJIA* increased by about 341 points, on average.
d) The relationship does not appear to be linear, as the residuals have a definite pattern.
7. a) −0.520
b) Negative, not strong, somewhat linear, but with more variation as pH increases.
c) The BCI would also be average.
d) The predicted *BCI* will be 1.56 SDs of *BCI* below the mean *BCI*.
8. a) Number of motorboat registrations.
b) The association is fairly strong, linear, and has a positive direction.
c) 0.946
d) 89.5% of the variation in manatees killed is explained by the variation in powerboat registrations.
e) No, there is no proof of causality.

9. a) $\widehat{Manatee\,Deaths} = -45.67 \times 0.1315\,Powerboat\,Registrations$ (in 1000s).
b) According to the model, for each increase of 10,000 motorboat registrations, the number of manatees killed increases by approximately 1.315.
c) If there were 0 motorboat registrations, the number of manatee deaths would be −45.67. This is obviously a silly extrapolation.
d) The predicted number is 82.41 deaths. The actual number of deaths was 79. The residual is $79 - 82.41 = -3.41$. The model overestimated the number of deaths by 3.41.
e) Negative residuals would suggest that the actual number of deaths was lower than the predicted number.
f) Over time, the number of motorboat registrations has increased and the number of manatee kills has increased. The trend may continue. Extrapolation is risky, however, because the government may enact legislation to protect the manatee.
10. a) 73 points b) 7 points c) $r = 0.75$ d) 100 points
e) The regression equation is designed to predict final exam scores based on midterm exam scores. You would need to find the regression equation to predict midterm scores based on final exam scores.
f) −85 points
g) Increase. The point is unusual and has a high negative residual that would decrease the correlation; removing it would increase the correlation and the R^2 value.
h) Slope will increase.
11. a) −0.984 b) 96.9% c) 32.95 mph d) 1.66 mph
e) Slope will increase.
f) Correlation will weaken (become less negative).
g) Correlation is the same, regardless of units.
12. a) 0.473 b) A weak linear association in a positive direction.
c) The actual score was higher than the predicted value for Monday.
d) The predicted Monday score would be (0.473) SDs, or 2.37 points below the mean for Monday, or 34.9 points.
e) Predicted Monday Score $= 14.6 + 0.54 \times$ Friday Score.
f) 36.0 points
13. a) Weight (but unable to verify linearity).
b) As weight increases, mileage decreases.
c) *Weight* accounts for 81.5% of the variation in *Fuel Efficiency*.
14. a) *Displacement* and *Weight* (but unable to verify linearity).
b) No. Large engines (higher displacement) weigh more, so the engine itself may influence the weight somewhat. More likely, heavy cars are equipped with larger engines because their added weight needs a larger engine (higher displacement) to drive.
c) The change in units will not affect the correlation.
d) The predicted fuel economy will be 0.786 SDs below the mean fuel economy.
15. a) $\widehat{Horsepower} = 3.50 + 34.314 \times Weight$.
b) Thousands. For the equation to have predicted values between 60 and 160, the *X* values would have to be in thousands of pounds.
c) Yes. The residual plot does not show any pattern.
d) 115.0 horsepower.
16. *Sex* and *Colorblindness* are both categorical variables, not quantitative. Correlation is meaningless for them, but we can say that the variables are associated.
17. a) The scatterplot shows a fairly strong linear relation in a positive direction. There seem to be two distinct clusters of data.
b) $\widehat{Interval} = 33.967 \div 10.358 \times Duration$.
c) The time between eruptions increases by about 10.4 minutes per minute of *Duration* on average.
d) Since 77% of the variation in *Interval* is accounted for by *Duration* and the error standard deviation is 6.16 minutes, the prediction will be relatively accurate.
e) 75.4 minutes.
f) A residual is the observed value minus the predicted value. So the residual $= 79 - 75.4 = 3.6$ minutes, indicating that the model underestimated the interval in this case.

18. a) Yes, the R^2 values indicate that 97.2% of the Indian crocodile length and 98% of the Australian crocodile length are explained by the head size.
  b) The slopes of the regression equations are similar, as are the R^2 values.
  c) The two models have different y-intercepts. It means that the Indian crocodile is smaller.
  d) Predicted body length for the Indian crocodile is 389.4 cm but is 458.2 cm for the Australian croc. The skeleton was probably from an Indian crocodile.

19. a) $r = 0.888$. Although r is high, you must look at the scatterplot and verify that the relation is linear in form.
  b)

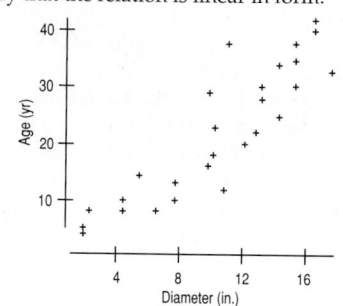

The association between diameter and age appears to be strong, somewhat linear, and positive.
  c) $\widehat{Age} = -0.97 + 2.21 \times Diameter$.
  d)

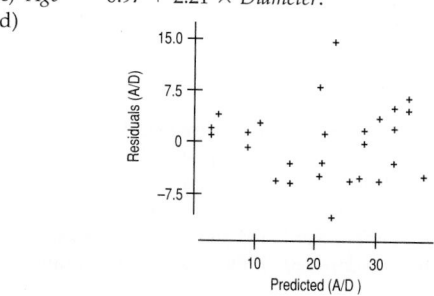

The residuals show a curved pattern (and two outliers).
  e) The residuals for five of the seven largest trees (15 in. or larger) are positive, indicating that the predicted values underestimate the age.

20. a) Yes.

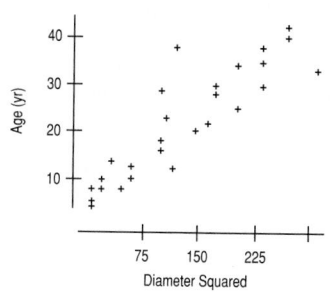

  b) $\widehat{Age} = 7.24 + 0.11 \times Diameter^2$.

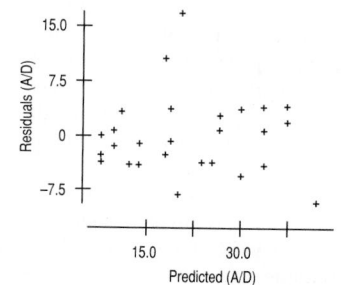

  c) The residuals appear to be more randomly scattered and have less of a pattern. But there are still some points with large residuals.
  d) 43.9 years old.

21. Most houses have areas between 1000 and 5000 square feet. Increasing 1000 square feet would result in either $1000(.008) = 8$ thousand dollars, $1000(.08) = 80$ thousand dollars, $1000(.8) = 800$ thousand dollars, or $1000(8) = 8000$ thousand dollars. Only \$80,000 is reasonable, so the slope must be 0.08.

22.

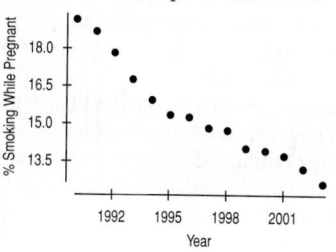

  a) The association is very strong, straight, and negative. However, there is some curvature to the scatterplot.
  b) -0.972
  c) Averaging increases the strength of the correlation, making it more negative.
  d) $\widehat{Percentage} = 946.92 - 0.4662 \times Year$. The model suggests that each year the average percent of women smoking while pregnant has decreased by about 0.47%. (Up-and-down curvature can't be straightened.)

23. a) The model predicts % smoking from year, not the other way around.
  b) $\widehat{Year} = 2027.91 - 202.74 \times \%\ Smoking$.
  c) The smallest % smoking given is 12.7, and an extrapolation to $x = 0$ is probably too far from the given data. The prediction is not very reliable in spite of the strong correlation.

24. a) There is a very weak linear relationship between quality of service and tip size.
  b) $R^2 = 1.21\%$, indicating that the variation in quality of service accounts for only about 1% of the variation in tip percentages.

25. The relation shows a negative direction, with a somewhat linear form, but perhaps with some slight curvature. There are several model outliers.

26. a) Latitude, since the correlation -0.848 is stronger (but unable to verify linearity).
  b) The same.
  c) The same. Changes in units do not change the shape of the association nor correlation of the data.
  d) The predicted average January temperature would be 0.738 SDs below the mean average January temperature.

27. a) 71.9%
  b) As latitude increases, the January temperature decreases.
  c) $\widehat{January\ Temperature} = 108.80 - 2.111 \times Latitude$.
  d) As the latitude increases by 1 degree, the average January temperature drops by about 2.11 degrees, on average.
  e) The y-intercept would indicate that the average January temperature is 108.8 when the latitude is 0. However, this is extrapolation and may not be meaningful.
  f) 24.4 degrees.
  g) The equation underestimates the average January temperature.

28. No. First, the R^2 indicates that only 4.6% of the variation in *Depression* is accounted for by *Internet Use*. That is a very low percentage. Even if it were higher, there may be lurking variables that explain the increase in both.

29. a) The scatterplot shows a strong, linear, positive association.
  b) There is an association, but it is likely that training and technique have increased over time and affected both jump performances.

c) Neither; the change in units does not affect the correlation.

d) The long-jumper would jump 0.925 SDs above the mean long jump, on average.

30. a) $\widehat{High\ Jump} = -20.76 + 0.331 \times Long\ Jump$.

b) As the long jump increases by 1 inch, the high jump increases by 0.331 inches, on average, according to the model.

c) 95.09 inches.

d) This equation is designed to predict high jump based on long jump. To predict the long jump, you would need the equation predicting long jump based on high jump.

e) $\widehat{Long\ Jump} = 99.26 + 2.5853 \times High\ Jump$. (99.055 + 2.588 High Jump with technology)

31. a) No relation; the correlation would probably be close to 0.

b) The relation would have a positive direction and the correlation would be strong, assuming that students were studying French in each grade level. Otherwise, no correlation.

c) No relation; correlation close to 0.

d) The relation would have a positive direction and the correlation would be strong, since vocabulary would increase with each grade level.

32. a) There is a strong, fairly linear, positive trend for the data; as the year of birth increases, the preterm birth rate increases.

b) The highest preterm birth rate is for mothers receiving adequate prenatal care utilization, and the lowest preterm birth rate is for mothers receiving inadequate prenatal care utilization. The slope is about the same for these relations. Mothers receiving intensive prenatal care utilization had more preterm births than mothers receiving inadequate prenatal care and fewer preterm births than mothers receiving adequate prenatal care; however, their rate of increase is higher.

c) No, there is undoubtedly an underlying variable explaining these differences. Perhaps the level of prenatal care was determined by the complications the mothers experienced during the pregnancy.

33. $\widehat{Calories} = 560.7 - 3.08 \times Time$.
Each minute extra at the table results in 3.08 fewer calories being consumed, on average. Perhaps the hungry children eat fast and eat more.

34. a) $\widehat{Gallons} = 12451.2 - 6.2 \times Year$.

b) The model is designed to predict gallons based on the year. That question asks for a prediction of the year based on the gallons.

c) $\widehat{Year} = 2008.2 - 0.16 \times Gallons$ (in 1000s). Prediction is 2008.

d) The prediction in the wrong direction is close because the relationship is so strong. The line that minimizes the y-residuals also does a pretty good job at minimizing the x-residuals.

35. There seems to be a strong, positive, linear relationship with one high-leverage point (Northern Ireland) that makes the overall R^2 quite low. Without that point, the R^2 increases to 61.5%. Of course, these data are averaged across thousands of households, so the correlation appears to be higher than it would be for individuals. Any conclusions about individuals would be suspect.

36. a) $\widehat{Weight} = -1971.26 | 1.091 \times Year$.

b) The scatterplot appears to have a linear trend in a positive direction; however, since the weights are averages, the linear trend would be stronger than for individual players.

c) 214.75 pounds. Possibly, but it is a 10-year extrapolation from a range of only 20 years.

d) 323.88 pounds. No, this is an extrapolation of 110 years, and the average weight seems absurd.

e) 1306.11 pounds. Absolutely ridiculous to have an average weight of over 1000 pounds. The prediction is an extrapolation of 1010 years.

37. a) 3.842   b) 501.187   c) 4.0

38. a) $\widehat{Weight} = -1121.66 + 0.673 \times Year$.

b) 2032.

c) No. That seems to be too much extrapolation for the model.

39. a) 30,818 pounds.

b) 1302 pounds.

c) 31,187.6 pounds.

d) I would be concerned about using this relation if we needed accuracy closer than 1000 pounds or so, as the residuals are more than ±1000 pounds.

e) Negative residuals will be more of a problem, as the predicted weight would overestimate the weight of the truck; trucking companies might be inclined to take the ticket to court.

40. a) Symmetric distributions of x and y may help ensure that the residuals around a line through the data will be more symmetric and normally distributed. The re-expressed data will not have extreme outliers as well.

b) Yes. The scatterplot shows a linear trend in a positive direction. It has some moderate outliers in the x direction. The residuals plot shows some moderate residuals in the negative direction and some curvature, but the model using re-expressed data is surely better than the model using the original data.

c) $\widehat{\log(Profit)} = -0.11 + 0.648 \times \log(Sales)$.   d) $124.43 million.

41. The original data are nonlinear, with a significant curvature. Using reciprocal square root of diameter gave a scatterplot that is nearly linear:

$$\widehat{1/\sqrt{Drain\ Time}} = 0.0024 + 0.219\ Diameter.$$

42. The scatterplot of *Cost per Chip* vs. *Chips Produced* has a significant curvature. Taking the log of *Chips Produced* and the log of *Cost per Chip* provided a nearly linear scatterplot.

$$\widehat{\log Cost\ per\ Chip} = 2.67 - 0.502 \times \log Chips\ Produced.$$

CHAPTER 11

1. Yes. You cannot predict the outcome beforehand.

2. The outcomes cannot be predicted beforehand. Each outcome, numbers 00 through 36, should be equally likely.

3. A machine pops up numbered balls. If it were truly random, the outcome could not be predicted and the outcomes would be equally likely. It is random only if the balls generate numbers in equal frequencies.

4. Answers may vary. **Rolling one die or two dice:** If the dice are fair, then each outcome, 1 through 6, should be equally likely. **Spinning a spinner:** Each outcome should be equally likely, but the spinner might be more likely to land on one outcome than another due to friction or design. **Shuffling cards and dealing a hand:** If the cards are shuffled adequately (7 times for riffle shuffling), the cards will be approximately equally likely.

5. Use two-digit numbers 00–99; let 00–02 = defect, 03–99 = no defect

6. Use random digits 0–9; let 0 = colorblind, 1–9 = no defect

7. a) 45, 10   b) 17, 22

8. 43, 9, 50, 13, 27

9. If the lottery is random, it doesn't matter which number you play; all are equally likely to win.

10. If the lottery is random, it doesn't matter which number you play; all are equally likely to win.

11. a) The outcomes are not equally likely; for example, tossing 5 heads does not have the same probability as tossing 0 or 9 heads, but the simulation assumes they are equally likely.

b) The even-odd assignment assumes that the player is equally likely to score or miss the shot. In reality, the likelihood of making the shot depends on the player's skill.

c) The likelihood for the first ace in the hand is not the same as for the second or third or fourth. But with this simulation, the likelihood is the same for each. (And it allows you to get 5 aces, which could get you in trouble in a real poker game!)

12. a) The numbers would represent the sums, but the sums are not all equally likely, as this simulation assumes.

b) The numbers of boys in a family of 5 children are not equally likely; for example, having a total of 5 boys is less likely than having 3 boys out of 5 children. The simulation assigns the same likelihood to each event.

c) The likelihoods for out, single, double, triple, and home run are not the same, but the simulation assumes they are.

13. The conclusion should indicate that the simulation *suggests* that the average length of the line would be 3.2 people. Future results might not match the simulated results exactly.

14. The simulation *suggests* that 24% of the people might contract the disease. The simulation does not represent what happened, but what might have happened.

15. a) The component is one voter voting. An outcome is a vote for our candidate or not. Use two random digits, giving 00–54 a vote for your candidate and 55–99 for the underdog.

b) A trial is 100 votes. Examine 100 two-digit random numbers, and count how many people voted for each candidate. Whoever gets the majority of votes wins that trial.

c) The response variable is whether the underdog wins or not.

16. a) The component is picking a single card. An outcome is the suit and denomination of the card. You could use the digits 01–52 for the 52 different cards, ignoring 00 and 53–99, or you could use a single digit 1, 2, 3, or 4 for the suit and then 01–13 for the denomination (ignoring 0, 5–9 for suits, and 00, 14–99 for denominations).

b) A trial is a single five-card hand. Use five sets of random numbers, ignoring repeated cards.

c) As response variable, we'll record whether the hand had two pairs, three of a kind, or neither.

17. Answers will vary, but average answer will be about 51%.

18. Answers will vary, but average answer will be about 5 boxes.

19. Answers will vary, but average answer will be about 26%.

20. Answers will vary, but few simulations will have any runs getting all 6 correct, leading us to conclude that the probability is very small. (The true probability is 0.00024.)

21. a) Answers will vary, but you should win about 10% of the time.
    b) You should win at the same rate with any number.

22. Answers will vary, but you should win about 10% of the time.

23. Answers will vary, but you should win about 10% of the time.

24. Answers will vary, but you should win about 10% of the time.

25. Answers will vary, but average answer will be about 1.9 tests.

26. Answers will vary, but average answer will be about 18%.

27. Answers will vary, but average answer will be about 1.24 points.

28. Answers will vary, but average answer will be 6.8 people.

29. Do the simulation in two steps. First simulate the payoffs. Then count until $500 is reached. Answers will vary, but average should be near 10.2 customers.

30. Answers will vary, but average answer will be about $37.

31. Answers will vary, but average answer will be about 3 children.

32. Answers will vary, but average answer will be slightly less than 6 children.

33. Answers will vary, but average answer will be about 7.5 rolls.

34. Answers will vary, but average answer will be about 2.8 rolls.

35. No, it will happen about 40% of the time.

36. Answers will vary, but average answer will be about 39%.

37. Answers will vary, but average answer will be about 37.5%.

38. Answers will vary, but average answer will be about 37.5%.

39. Three women will be selected about 7.8% of the time.

40. Answers will vary, but if cell phone usage is only 12%, you should find four or more cell phone users among 10 drivers only about 2% of the time. Because that's what you saw while waiting for your bus, you'd suspect that the legislator's claim of 12% is probably too low.

CHAPTER 12

1. a) No. It would be nearly impossible to get exactly 500 males and 500 females from every country by random chance.
   b) A stratified sample, stratified by whether the respondent is male or female.

2. a) No. It would be nearly impossible to get exactly 50 from each class by random chance.
   b) A stratified sample, stratified by the respondents' class.

3. a) Voluntary response.
   b) We have no confidence at all in estimates from such studies.

4. a) A cluster sample, with teams being the clusters.
   b) It is a reasonable solution to the problem of randomly sampling players because you can sample an entire team at once relatively easily but couldn't efficiently draw a random sample of all players on the same day.

5. a) The population of interest is all adults in the United States aged 18 and older.
   b) The sampling frame is U.S. adults with telephones.
   c) Some members of the population (e.g, many college students) don't have landline phones, which could create a bias.

6. a) They are using a stratified design in which countries are strata, and a random sample is drawn within each stratum. They don't specify how the random samples are drawn.
   b) The difference in population size has no effect whatever on the precision of estimates from these surveys. Only the sample size matters.

7. a) Population—All U.S. adults.
   b) Parameter—Proportion who have used and benefited from alternative medicine.
   c) Sampling Frame—All Consumers Union subscribers.
   d) Sample—Those who responded.
   e) Method—Questionnaire to all (nonrandom).
   f) Bias—Nonresponse. Those who respond may have strong feelings one way or another.

8. a) Population—U.S. adults?
   b) Parameter—Proportion who feel marijuana should be legalized for medicinal purposes.
   c) Sampling Frame—None given; potentially all people with access to Web site.
   d) Sample—Those visiting the Web site who responded.
   e) Method—Voluntary response (no randomization employed).
   f) Bias—Voluntary response sample. Those who visit the Web site and respond may be predisposed to a particular answer.

9. a) Population—Adults.
   b) Parameter—Proportion who think drinking and driving is a serious problem.
   c) Sampling Frame—Bar patrons.
   d) Sample—Every 10th person leaving the bar.
   e) Method—Systematic sampling (may be random).
   f) Bias—Those interviewed had just left a bar. They may think drinking and driving is less of a problem than do other adults.

10. a) Population—City voters.
    b) Parameter—Not clear; percentages of voters favoring issues?
    c) Sampling Frame—All city residents.
    d) Sample—Stratified sample; one block from each district.
    e) Method—Convenience sample within each stratum.
    f) Bias—Parameter(s) of interest not clear. Sampling within clusters is not random and may bias results.

11. a) Population—Soil around a former waste dump.
    b) Parameter—Concentrations of toxic chemicals.
    c) Sampling Frame—Accessible soil around the dump.
    d) Sample—16 soil samples.
    e) Method—Not clear.
    f) Bias—Don't know if soil samples were randomly chosen. If not, may be biased toward more or less polluted soil.

12. a) Population—Cars.
    b) Parameter—Proportion with up-to-date registration, insurance, and safety inspections.
    c) Sampling Frame—All cars on that road.
    d) Sample—Those actually stopped by roadblock.
    e) Method—Cluster sample of location; census within cluster.
    f) Bias—Time of day and location may not be representative of all cars.

13. a) Population—Snack food bags.
    b) Parameter—Weight of bags, proportion passing inspection.
    c) Sampling Frame—All bags produced each day.
    d) Sample—Bags in 10 randomly selected cases, 1 bag from each case for inspection.
    e) Method—Multistage random sampling.
    f) Bias—Should be unbiased.
14. a) Population—Dairy farms.
    b) Parameter—Proportion passing inspection.
    c) Sampling Frame—All dairy farms?
    d) Sample—Not clear. Perhaps a random sample of farms and then milk within each farm?
    e) Method—Multistage sampling.
    f) Bias—Should be unbiased if farms and milk at each farm are randomly selected.
15. Bias. Only people watching the news will respond, and their preference may differ from that of other voters. The sampling method may systematically produce samples that don't represent the population of interest.
16. Sampling error. The description of the sampling method suggests that samples should be representative of the voting population. Nonetheless, random chance in selecting the individuals who were polled means that sample statistics will vary from the population parameter, perhaps by quite a bit.
17. a) Voluntary response. Only those who see the ad, have Internet access, *and* feel strongly enough will respond.
    b) Cluster sampling. One school may not be typical of all.
    c) Attempted census. Will have nonresponse bias.
    d) Stratified sampling with follow-up. Should be unbiased.
18. a) Voluntary response. Only those who see the show *and* feel strongly will call.
    b) Possibly more representative than part a, but only strongly motivated parents go to PTA meetings (voluntary response).
    c) Multistage sampling, with cluster sample within each school. Probably a good design if most of the parents respond.
    d) Systematic sampling. Probably a reasonable design.
19. a) This is a multistage design, with a cluster sample at the first stage and a simple random sample for each cluster.
    b) If any of the three churches you pick at random is not representative of all churches, then you'll introduce sampling error by the choice of that church.
20. They will get responses only from people who come to the park to use the playground. Parents who are dissatisfied with the playground may not come.
21. a) This is a systematic sample.
    b) The sampling frame is patrons willing to wait for the roller coaster on that day at that time. It should be representative of the people in line, but not of all people at the amusement park.
    c) It is likely to be representative of those waiting for the roller coaster. Indeed, it may do quite well if those at the front of the line respond differently (after their long wait) than those at the back of the line.
22. The first sentence points out problems the respondent may not have noticed and might lead them to feel they should agree. The last phrase mentions higher fees, though, which could make people reject improvements to the playground.
23. a) Answers will definitely differ. Question 1 will probably get many "No" answers, while Question 2 will get many "Yes" answers. This is response bias.
    b) "Do you think standardized tests are appropriate for deciding whether a student should be promoted to the next grade?" (Other answers will vary.)
24. a) This is a voluntary response survey. Even a large sample size can't make it representative.
    b) The wording seems fair enough. It states the facts and gives voice to both sides of the issue.
    c) The sampling frame is, at best, those who visit this particular site and even then depends on their volunteering to respond to the question.
    d) This is a true statement.

25. a) Biased toward yes because of "pollute." "Should companies be responsible for any costs of environmental cleanup?"
    b) Biased toward no because of "old enough to serve in the military." "Do you think the drinking age should be lowered from 21?"
26. a) Seems neutral.
    b) Biased toward yes because of "great tradition." Better to ask, "Do you favor continued funding for the space program?"
27. a) Not everyone has an equal chance. Misses people with unlisted numbers, or without landline phones, or at work.
    b) Generate random numbers and call at random times.
    c) Under the original plan, those families in which one person stays home are more likely to be included. Under the second plan, many more are included. People without landline phones are still excluded.
    d) It improves the chance of selected households being included.
    e) This takes care of phone numbers. Time of day may be an issue. People without landline phones are still excluded.
28. Cell phones are more likely to be used by higher income individuals. This will cause an undercoverage bias. As cell phones grow in use, this problem will be lessened.
29. a) Answers will vary.
    b) Your own arm length. Parameter is your own arm length; population is all possible measurements of it.
    c) Population is now the arm lengths of you and your friends. The average estimates the mean of these lengths.
    d) Probably not. Friends are likely to be of the same age and not very diverse or representative of the larger population.
30. a) Mean gas mileage for the last six fill-ups.
    b) Mean gas mileage for the vehicle.
    c) Recent driving conditions may not be typical.
    d) Mean gas mileage for all cars of this make and model.
31. a) Assign numbers 001 to 120 to each order. Use random numbers to select 10 transactions to examine.
    b) Sample proportionately within each type. (Do a stratified random sample.)
32. a) Most likely that all laborers are selected, no managers, and few foremen. Bias may be introduced because the company itself is conducting the survey.
    b) Assign a number from 001 to 439 to each employee. Use a random-number table or software to select the sample.
    c) Still heavily favors the laborers.
    d) Stratify by job type (proportionately to the members in each).
    e) Answers will vary. Assign numbers 01 to 14 to each person; use a random-number table or software to do the selection.
33. a) Select three cases at random; then select one jar randomly from each case.
    b) Use random numbers to choose 3 cases from numbers 61 through 80; then use random numbers between 1 and 12 to select the jar from each case.
    c) No. Multistage sampling.
34. What conclusions they may be able to make will depend on whether fish with discolored scales are equally likely to be caught as those without. It also depends on the level of compliance by fishermen. If fish are not equally likely to be caught, or fishermen are more disposed to bringing discolored fish, the results will be biased.
35. a) Depends on the Yellow Page listings used. If from regular (line) listings, this is fair if all doctors are listed. If from ads, probably not, as those doctors may not be typical.
    b) Not appropriate. This cluster sample will probably contain listings for only one or two business types.
36. a) Petition may bias people to say they support the playground. Also, many may not be home on Saturday afternoon.
    b) If the food at the largest cafeteria is representative, this should be OK. However, those who really don't like the food won't be eating there.

CHAPTER 13

1. a) No. There are no manipulated factors. Observational study.
   b) There may be lurking variables that are associated with both parental income and performance on the SAT.
2. a) No, there are no manipulated factors. It is a retrospective study.
   b) No, there may be lurking variables, such as patient's age.
3. a) This is a retrospective observational study.
   b) That's appropriate because MS is a relatively rare disease.
   c) The subjects were U.S. military personnel, some of whom had developed MS.
   d) The variables were the vitamin D blood levels and whether or not the subject developed MS.
4. a) This is a stratified sample. The question was about population values (the proportions of men and women who look forward more to the commercials). No treatment was applied, so it is not an experiment.
   b) Yes.
5. a) This was a randomized, placebo-controlled experiment.
   b) Yes, such an experiment is the right way to determine whether black cohosh has an effect.
   c) 351 women aged 45 to 55 who reported at least two hot flashes a day.
   d) The treatments were black cohosh, a multiherb supplement, a multiherb supplement plus advice, estrogen, and a placebo. The response was the women's symptoms (presumably frequency of hot flashes).
6. a) This is an experiment. The picture is the controlled factor. Randomization may have been used in deciding which days each picture appeared.
   b) The treatment was the picture behind the coffee station. The response was the average contribution.
   c) The differences in money collected were larger than could be reasonably attributed to usual day-to-day variation.
7. a) Experiment.
   b) Bipolar disorder patients.
   c) Omega-3 fats from fish oil, two levels.
   d) 2 treatments.
   e) Improvement (fewer symptoms?).
   f) Design not specified.
   g) Blind (due to placebo), unknown if double-blind.
   h) Individuals with bipolar disease improve with high-dose omega-3 fats from fish oil.
8. a) Observational study.
   b) Prospective.
   c) Disabled women aged 65 and older with and without a vitamin B_{12} deficiency, unknown selection process.
   d) Suffering severe depression.
   e) As there is no random assignment, there is no way to know that the deficiency caused the severe depression.
9. a) Observational study.
   b) Prospective.
   c) Men and women with moderately high blood pressure and normal blood pressure, unknown selection process.
   d) Memory and reaction time.
   e) As there is no random assignment, there is no way to know that high blood pressure *caused* subjects to do worse on memory and reaction-time tests. A lurking variable may also be the cause.
10. a) Experiment.
    b) People suffering from insomnia.
    c) 2 factors: desserts and exercise (2 levels each).
    d) 4 treatments.
    e) Improvement in ability to sleep.
    f) Completely randomized.
    g) Not blind.
    h) Insomniacs who exercise and refrain from desserts will experience improved ability to sleep.

11. a) Experiment.
    b) Postmenopausal women.
    c) Alcohol—2 levels; blocking variable—estrogen supplements (2 levels).
    d) 1 factor (alcohol) at 2 levels = 2 treatments.
    e) Increase in estrogen levels.
    f) Blocked.
    g) Not blind.
    h) Indicates that alcohol consumption *for those taking estrogen supplements* may increase estrogen levels.
12. a) Observational study.
    b) Retrospective.
    c) Women exposed to dioxin from an industrial accident.
    d) Risk of breast cancer.
    e) As there is no random assignment, there is no way to know that the dioxin levels caused the increase in breast cancer; there may have been lurking variables that were not identified.
13. a) Observational study.
    b) Retrospective.
    c) Women in Finland, unknown selection process with data from church records.
    d) Women's lifespans.
    e) As there is no random assignment, there is no way to know that having sons or daughters shortens or lengthens the lifespan of mothers.
14. a) Experiment.
    b) People exposed to cold virus.
    c) 1 factor: herbal treatment (2 levels).
    d) 2 treatments.
    e) Severity of cold symptoms.
    f) No discussion of randomness.
    g) Blind, as subjects did not know if they received the herbal treatment or the placebo. Not clear if it was double-blind.
    h) There is no indication that the herbal treatment is effective.
15. a) Observational study.
    b) Prospective.
    c) People with or without depression, unknown selection process.
    d) Frequency of crying in response to sad situations.
    e) There is no apparent difference in crying response (to sad movies) for depressed and nondepressed groups.
16. a) Experiment.
    b) Racing greyhounds.
    c) 1 factor, diet with 3 levels.
    d) 3 treatments.
    e) Speed.
    f) Random assignment to order of diets; matched design before and after diet.
    g) No blinding.
    h) Greyhounds who eat diets high in vitamin C seem to run more slowly.
17. a) Experiment.
    b) People experiencing migraines.
    c) 2 factors (pain reliever and water temperature), 2 levels each.
    d) 4 treatments.
    e) Level of pain relief.
    f) Completely randomized over 2 factors.
    g) Blind, as subjects did not know if they received the pain medication or the placebo, but not blind, as the subjects will know if they are drinking regular or ice water.
    h) It may indicate whether pain reliever alone or in combination with ice water gives pain relief, but patients are not blinded to ice water, so placebo effect may also be the cause of any relief seen caused by ice water.
18. a) Experiment.
    b) Inactive dogs.
    c) 1 factor: dog food (assuming amount of food to be determined by weight or size of dog) (2 levels).

d) 2 treatments.

e) Weight.

f) Blocked by size of breed.

g) Blinded, assuming dog owners do not know which food the dog is receiving.

h) Assuming the dog owners followed the prescribed feeding levels, there could be a conclusion as to whether or not the dog food helped maintain healthy weight.

19. a) Experiment.

b) Athletes with hamstring injuries.

c) 1 factor: type of exercise program (2 levels).

d) 2 treatments.

e) Time to return to sports.

f) Completely randomized.

g) No blinding—subjects must know what kind of exercise they do.

h) Can determine which of the two exercise programs is more effective.

20. a) Observational study.

b) Prospective.

c) General public; two random samples.

d) The purpose of the study was to identify variables on which there was a difference, so no response variable(s) could be identified at the start of the study.

e) Identify differences between people who can be reached by regular polling methods and those who cannot.

21. They need to compare omega-3 results to something. Perhaps bipolarity is seasonal and would have improved during the experiment anyway.

22. They need a basis for comparison. Perhaps insomnia is related to the amount of daylight, and that changed during the time the experiment was conducted.

23. a) Subjects' responses might be related to many other factors (diet, exercise, genetics, etc). Randomization should equalize the two groups with respect to unknown factors.

b) More subjects would minimize the impact of individual variability in the responses, but the experiment would become more costly and time consuming.

24. a) Subjects' responses might be related to many other factors (diet, medications, genetics, etc.). Randomization should equalize the two groups with respect to unknown factors.

b) More subjects would minimize the impact of individual variability in the responses, but the experiment would become more costly and time consuming.

25. People who engage in regular exercise might differ from others with respect to bipolar disorder, and that additional variability could obscure the effectiveness of this treatment.

26. People who are overweight might have different sleep patterns, and that additional variability could obscure the effectiveness of this treatment.

27. Answers may vary. Use a random-number generator to randomly select 24 numbers from 01 to 24 without replication. Assign the first 8 numbers to the first group, the second 8 numbers to the second group, and the third 8 numbers to the third group.

28. Answers may vary. Use a random-number generator to randomly select 24 numbers from 01 to 24 without replication. Assign the first group of 4 numbers to the first treatment (no fertilizer, natural watering), the second group of 4 numbers to the second treatment (no fertilizer, daily water), the third group of 4 numbers to the third treatment (half fertilizer, natural watering), and so on to the sixth treatment.

29. a) First, they are using athletes who have a vested interest in the success of the shoe by virtue of their sponsorship. They should choose other athletes. Second, they should randomize the order of the runs, not run all the races with their shoes second. They should blind the athletes by disguising the shoes

if possible, so they don't know which is which. The timers shouldn't know which athletes are running with which shoes, either. Finally, they should replicate several times, since times will vary under both shoe conditions.

b) Because of the problems in (a), the results they obtain may favor their shoes. In addition, the results obtained for Olympic athletes may not be the same as for the general runner.

30. The "control" in this experiment is not the same for all swimmers. We don't know what "their old swimsuit" is. They should compare their new swimsuit to the same suit design. The order in which the swims are performed should be randomized. There may be a systematic effect from one swim to the next. Finally, there is no way to blind this test. The swimmers will know which kind of suit they have on, and this may bias their performance.

31. a) Allowing athletes to self-select treatments could confound the results. Other issues such as severity of injury, diet, age, etc., could also affect time to heal; randomization should equalize the treatment groups with respect to any such variables.

b) A control group could have revealed whether either exercise program was better (or worse) than just letting the injury heal.

c) Doctors who evaluated the athletes to approve their return to sports should not know which treatment the subject had.

d) It's hard to tell. The difference of 15 days seems large, but the standard deviations indicate that there was a great deal of variability in the times.

32. a) Self-selection could result in groups that were very different at the start of the experiment, making it impossible to attribute differences in the results to their diet.

b) This assured that the diets were followed and that all subjects in each group received comparable treatments.

c) The researchers can compare the change in blood pressure observed in the DASH group to the control group. They need to rule out the possibility that external variables (the season, news events, etc.) affected everyone's blood pressure.

d) We'd like to know the standard deviation of the changes, too. If it's very small, then 6.7 points would seem to be significant.

33. a) The differences among the Mozart and quiet groups were more than would have been expected from sampling variation.

b)

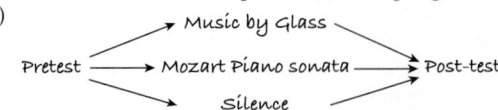

c) The Mozart group seems to have the smallest median difference and thus the *least* improvement, but there does not appear to be a significant difference.

d) No, if anything, there is less improvement, but the difference does not seem significant compared with the usual variation.

34. Use a retrospective observational study. For example, collect records from a random selection of police and emergency room logs for the past 3 years. Find the number of cases for the days when there is a full moon, when there is a waxing moon, when there is a waning moon, and when the moon is nearly dark. Compare the numbers for each group.

35. a) Observational, prospective study.

b) The supposed relation between health and wine consumption might be explained by the confounding variables of income and education.

c) None of these. While the variables have a relation, there is no causality indicated for the relation.

36. a) The difference in the depression rates for the two groups is greater than would be expected by natural sampling variation.

b) Observational study. There was no experimental treatment.

c) The difference could be explained by lurking variables. Perhaps swimmers are more affluent (can afford a membership at the Y or have access to a pool), or perhaps depressed people tend to swim less.

   d) Answers may vary. Give the subjects a test to measure depression. Then randomly assign the 120 subjects to one of three groups: the control group (no exercise program), the anaerobic exercise group, and the aerobic exercise group. Monitor subjects' exercise (have them report to a particular gym or pool). At the end of 12 weeks, administer the depression test again. Compare the post-exercise and pre-exercise depression scores.

37. a) Arrange the 20 containers in 20 separate locations. Use a random-number generator to identify the 10 containers that should be filled with water.

   b) Guessing, the dowser should be correct about 50% of the time. A record of 60% (12 out of 20) does not appear to be significantly different.

   c) Answers may vary. You would need to see a high level of success—say, 90% to 100%, that is, 18 to 20 correct.

38. Answers may vary. Randomly select half of the patients who agree to the study to get large doses of vitamin E after surgery. Give the other half a similar-looking placebo pill. Monitor their progress, recording the time until they have reached an easily agreed upon level of healing. Have the evaluating doctor blinded to whether the patient received the placebo or not. Compare the number of days until recovery of the two groups.

39. Randomly assign half the reading teachers in the district to use each method. Students should be randomly assigned to teachers as well. Make sure to block both by school and grade (or control grade by using only one grade). Construct an appropriate reading test to be used at the end of the year, and compare scores.

40.

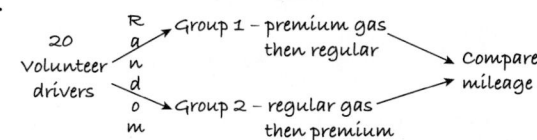

Answers may vary. This experiment has 1 factor (type of gasoline), at 2 levels (premium and regular), resulting in 2 treatments. The response variable is gas mileage. An experiment diagram for a matched design appears above. Have each of the volunteers use each kind of gas for a month. Randomly assign 10 of them to use regular first, the other 10 to use premium first. Ask them to keep driving logs (the number of miles driven and the gallons of gasoline) for each month. Compare the differences in the fuel economy for the two kinds of gasoline.

41. a) They mean that the difference is higher than they would expect from normal sampling variability.

   b) An observational study.

   c) No. Perhaps the differences are attributable to some confounding variable (e.g., people are more likely to engage in riskier behaviors on the weekend) rather than the day of admission.

   d) Perhaps people have more serious accidents and traumas on weekends and are thus more likely to die as a result.

42. a) Answers may vary. Randomly assign the eight patients to either the current medication or the new medication. Have nurses assess the degree of shingles involvement for the patient. Ask patients to rate their pain levels. Administer the medications for a prescribed time. Have nurses reassess the degree of shingles involvement. Ask patients to rate their pain levels post-medication. Compare the improvement levels.

   b) Let A = 1, B = 2 . . . H = 8.
Assign the first four randomly selected to the first group, the remainder to the second. So Group 1 is D, A, H, C, and Group 2 is B, E, F, G.

   c) Assuming that the ointments look alike, it would be possible to blind the experiment for the subject and for the administrator of the treatment.

   d) A block design with factors for gender and for ointment would be appropriate. Subjects would be randomly assigned to each treatment group in the blocked design.

43. Answers may vary. This experiment has 1 factor (pesticide), at 3 levels (pesticide A, pesticide B, no pesticide), resulting in 3 treatments. The response variable is the number of beetle larvae found on each plant. Randomly select a third of the plots to be sprayed with pesticide A, a third with pesticide B, and a third with no pesticide (since the researcher also wants to know whether the pesticides even work at all). To control the experiment, the plots of land should be as similar as possible with regard to amount of sunlight, water, proximity to other plants, etc. If not, plots with similar characteristics should be blocked together. If possible, use some inert substance as a placebo pesticide on the control group, and do not tell the counters of the beetle larvae which plants have been treated with pesticides. After a given period of time, count the number of beetle larvae on each plant and compare the results.

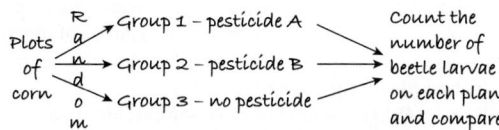

44. a) The students were not randomly assigned. Those who signed up for the prep course may be a special group whose scores would have improved anyway.

   b) Answers may vary. Find a group of volunteers who are willing to participate. Give all volunteers the SAT. Randomly assign the subjects to the review or no review group. Give the tutoring to the one group. After a reasonable time, retest both groups. See if the tutored group had a significant improvement in scores when compared with the no-review group.

   c) After the volunteers have taken the first SAT, separate the volunteers into blocks of low, average, and high SAT score performance. Now assign half of each block to the review and half to the no-review groups. Give the tutoring. Now retest all groups. Compare the differences between treatments for each block.

45. Answers may vary. Find a group of volunteers. Each volunteer will be required to shut off the machine with his or her left hand and right hand. Randomly assign the left or right hand to be used first. Complete the first attempt for the whole group. Now repeat the experiment with the alternate hand. Check the differences in time for the left and right hands.

46. Answers may vary. There are two factors: temperature of the water and wash cycle. Since each factor has 2 levels, there are 4 treatment groups (hot-reg., cold-reg., hot-del., cold-del.). It would be nice to have 32 shirts, but equal numbers of shirts in each group is not necessary. Randomly assign shirts to each of the 4 treatment groups. Rate the level of cleaning for the grass-stained shirts. Compare the 4 groups and determine the best use of the product.

47. a) Jumping with or without a parachute.

   b) Volunteer skydivers (the dimwitted ones).

   c) A parachute that looks real but doesn't work.

   d) A good parachute and a placebo parachute.

   e) Whether parachutist survives the jump (or extent of injuries).

   f) All should jump from the same altitude in similar weather conditions and land on similar surfaces.

   g) Randomly assign people the parachutes.

   h) The skydivers (and the people involved in distributing the parachute packs) shouldn't know who got a working chute. And the people evaluating the subjects after the jumps should not be told who had a real parachute either!

PART III REVIEW

1. Observational prospective study. Indications of behavior differences can be seen in the two groups. May show a link between premature birth and behavior, but there may be lurking variables involved.

2. Observational retrospective study. Can conclude there *may* be a link between tea drinking and survival after a heart attack. But lurking variables may also be involved.

3. Experiment, matched by gender and weight, randomization within blocks of two pups of same gender and weight. Factor: type of diet. Treatments: low-calorie diet and allowing the dog to eat all it wants. Response variable: length of life. Can conclude that, on average, dogs with a lower-calorie diet live longer.

4. Random sample. Population is all homes on the property tax list. Parameter is level of radon contamination. Procedure is probably a stratified random sample. May make inference about radon levels from sample to general population.

5. Observational prospective study. Indicates folate *may* help in reducing colon cancer for those with family histories of the disease.

6. Observational retrospective study. Indicates that during past 47 years first flowering has gotten earlier.

7. Sampling. Probably a simple random sample, although may be stratified by type of firework. Population is all fireworks produced each day. Parameter is proportion of duds. Can determine if the day's production is ready for sale.

8. Experiment, though there is no indication of a control group or differing treatments. Cosmetics *may* damage male reproductive systems in laboratory animals. Extrapolating to humans is risky.

9. Observational retrospective study. Living near strong electromagnetic fields may be associated with more leukemia than normal. May be lurking variables, such as socioeconomic level.

10. Observational retrospective study. Indicates no long-term risk of prostate cancer because of vasectomy.

11. Experiment. Blocked by sex of rat. Randomization is not specified. Factor is type of hormone given. Treatments are leptin and insulin. Response variable is lost weight. Can conclude that hormones can help suppress appetites in rats, and the type of hormone varies by gender.

12. Experiment. Factors are glaze type and temperature. Treatments are combinations of 4 glaze types and 3 temperatures. Response is age appearance of the pottery. Assuming an unbiased evaluator, can make a conclusion about best combination of glaze and temperature.

13. Experiment. Factor is gene therapy. Hamsters were randomized to treatments. Treatments were gene therapy or not. Response variable is heart muscle condition. Can conclude that gene therapy is beneficial (at least in hamsters).

14. Observational study (neither retrospective nor prospective). There seems to be a relationship between eye size and time of singing.

15. Sampling. Population is all oranges on the truck. Parameter is proportion of unsuitable oranges. Procedure is probably simple random sampling. Can conclude whether or not to accept the truckload.

16. Sampling. Population is all bottle-cap seals. Parameter is proportion with inadequate seals. Procedure is probably simple random sampling. Used to decide whether the bottling process is all right or needs to be adjusted.

17. Observational prospective study. Physically fit men may have a lower risk of death from cancer.

18. Experiment. Subjects blocked by calculus or not for analysis. Probably no randomization (students self-select into sections). Factor is software usage. Treatments are use of software or not. Response variable is final exam score. Can decide whether computer software is beneficial, and if so, is there a difference between the calculus and the noncalculus groups.

19. Answers will vary. This is a simulation problem. Using a random digits table or software, call 0–4 a loss and 5–9 a win for the gambler on a game. Use blocks of 5 digits to simulate a week's pick.

20. a) Answers will vary. This is a simulation problem. Using a random-digits table or software, obtain 3 numbers between 01 and 20 for your pick, then 5 others between 01 and 20. Count how many matches.
    b) With more numbers to choose from, the odds of winning go down dramatically.

21. Answers will vary.

22. a) Experiment. Treatments were actively imposed, rats were randomized, there was control (rats with no radio wave exposure), and they used many rats.
    b) The differences in brain tumors were small enough that they could be explained by sampling variability.
    c) May cause bias of some sort, intended or not.

23. a) Experiment. Actively manipulated candy giving, diners were randomly assigned treatments, control group was those with no candy, lots of dining parties.
    b) It depends on when the decision was made. If early in the meal, the server may give better treatment to those who will receive candy—biasing the results.
    c) A difference in response so large it cannot be attributed to natural sampling variability.

24. a)

    b) One factor (candy) with 4 levels (none, one, two, or one then another).
    c) 4.   d) Tip percentage.
    e) Probably blinded (diners probably weren't aware they were in an experiment), but not double-blinded.
    f) If "randomization" is done early in the meal, she may treat some better than others.

25. a) Voluntary response. Only those who feel strongly will pay for the 900 phone call.
    b) "If it would help future generations live a longer, healthier life, would you be in favor of human cloning?"

26. a) Water (quality and temperature) and material can vary. Results may be influenced by these confounding variables.
    b) Unrealistic conditions. This won't say how SparkleKleen works in normal situations.
    c) Might work, but if all the swatches were stained at the same time, the stains on the later swatches will have more time to "set in," causing bias against SparkleKleen.
    Other variables (changes in water temperature or pressure) won't be randomized.
    d) No guarantee that conditions are comparable.

27. a) Simulation results will vary. Average will be around 5.8 points.
    b) Simulation results will vary. Average will also be around 5.8 points.
    c) Answers will vary.

28. a) 10.6% chance to break at 900 pounds or less.
    b) Simulation results will vary. Use groups of three digits to simulate each rivet. For every one that is less than "106," denote as a failed rivet. Count how many rivets you need until 3 failures are reached.

29. a) Yes.
    b) No. Residences without phones are excluded. Residences with more than one phone had a higher chance.

c) No. People who respond to the survey may be of age but not registered voters.

d) No. Households who answered the phone may be more likely to have someone at home when the phone call was generated. These may not be representative of all households.

30. a) A difference in response so large it cannot be attributed to natural sampling variability.

b) More likely, younger looking individuals are more sexually active than older ones. We have no means of comparison (different levels of sexual activity in people of the same age, for example).

31. a) Does not prove it. There may be other confounding variables. Only way to prove this would be to do a controlled experiment.

b) Alzheimer's usually shows up late in life. Perhaps smokers have died of other causes before Alzheimer's can be seen.

c) An experiment would be unethical. One could design a prospective study in which groups of smokers and non-smokers are followed for many years and the incidence of Alzheimer's is tracked.

32. a) Randomized block experiment.

b) No—in an experiment we are looking for differences in response to treatments, not trying to generalize to all adults.

c) A sugar pill may affect digestion, confounding the experiment.

33.

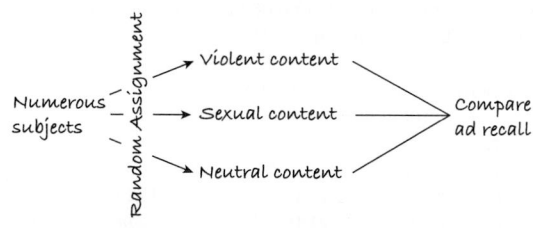

Numerous subjects will be randomly assigned to see shows with violent, sexual, or neutral content. They will see the same commercials. After the show, they will be interviewed for their recall of brand names in the commercials.

34. a) *Who*—900 Englishmen. *What*—reasons they go to a pub. *Why*—Kaliber alcohol-free beer wants to know. *When*—not stated. *Where*—England. *How*—survey (sampling method not specified).

b) Englishmen.

c) How those interviewed were chosen.

d) Probably convenience samples of those in pubs.

e) The study was financed by an alcohol-free beer.

35. a) May have been a simple random sample, but given the relative equality in age groups, may have been stratified.

b) 35.1%.

c) We don't know. Perhaps cell phones or unlisted numbers were excluded, and Democrats have more (or fewer) of those. Probably OK, though.

d) Do party affiliations differ for different age groups?

36. a) Nonresponse bias by conservatives.

b) U.S. adults.   c) Results do not total 100%.

d) Observed differences are of the same size one might expect from natural sampling variability.

37. The factor in the experiment will be type of bird control. I will have three treatments: scarecrow, netting, and no control. I will randomly assign several different areas in the vineyard to one of the treatments, taking care that there is sufficient separation that the possible effect of the scarecrow will not be confounded. At the end of the season, the response variable will be the proportion of bird-damaged grapes.

38. Since players vary in their ability to hit the ball, I will have each batter hit with both types of bats several times in a randomly chosen order. For each batter, calculate the average difference in distance, with metal or wood as the response variable.

39. a) We want all subjects treated as alike as possible. If there were no "placebo surgery," subjects would know this and perhaps behave differently.

b) The experiment looked for a difference in the effectiveness of the two treatments. (If we wanted to generalize, we would need to assume that the results for these volunteers are the same as on all patients who might need this operation.)

c) "Not statistically significant" means the difference in results were small enough that it could be explained by natural sampling variability.

40. Results will vary, but should be around 24%.

(Use a random-number table or software to do the simulation. Numbers 1 to 11 will represent the "worst" team, 12 to 21 the second worst, 22 to 30 your third-place team, 31 to 66 the others. Obtain random numbers in the range, ignoring any duplications of the team that gets to pick first. Count the number of times the third-place team picks first or second out of the total number of trials.)

41. a) Use stratified sampling to select 2 first-class passengers and 12 from coach.

b) Number passengers alphabetically, 01 = Bergman to 20 = Testut. Read in blocks of two, ignoring any numbers more than 20. This gives 65, 43, 67, 11 (selects Fontana), 27, 04 (selects Castillo).

c) Number passengers alphabetically from 001 to 120. Use the random-number table to find three-digit numbers in this range until 12 different values have been selected.

42. Simulation results will vary. Theoretically, the chance is about 3%. (Select four numbers between 01 and 20 to represent the Middle Easterners. Count the number of times in several repetitions where both to be searched are among the selected four numbers.)

43. Simulation results will vary.

(Use integers 00 to 99 as a basis. Use integers 00 to 69 to represent a tee shot on the fairway. If on the fairway, use digits 00 to 79 to represent on the green. If off the fairway, use 00 to 39 to represent getting on the green. If not on the green, use digits 00 to 89 to represent landing on the green. For the first putt, use digits 00 to 19 to represent making the shot. For subsequent putts, use digits 00 to 89 to represent making the shot.)

44. a) Answers may vary. A component in this simulation is a shot. Use pairs of random digits 00 to 99 to represent a shot. The way in which this component is simulated depends on the type of shot. For the first shot, let pairs of digits 01 to 80 represent hitting the green, and let pairs of digits 81 to 99, and 00, represent not hitting the green. If the first simulated shot misses the green, let 01 to 90 represent landing on the green, and 91 to 99, and 00, represent not landing on the green. Keep simulating shots until a shot lands on the green. Once on the green, let 01 to 20 represent sinking the putt on the first putt, and let 21 to 99, and 00, represent not sinking the putt on the first putt. If second putts are required, continue simulating putts until a putt goes in, with 01 to 90 representing making the putt, and 91 to 99, and 00, representing not making the putt. A run consists of following the guidelines above until the final putt is made. The response variable is the number of shots required until the final putt is made. The simulated average score on the hole is the total number of shots required divided by the total number of runs. According to 20 runs of this simulation, a pretty good golfer can be expected to average about 3.7 strokes per hole. Your simulation results may vary.

b) Answers may vary. The simulation is set up identically to part a, with the exception of the second shot. Now, let 01 to 10 represent hitting the green, and let 11 to 99, and 00, represent not hitting the green. According to 20 runs of this simulation, a pretty good golfer can be expected to average about 5.3 strokes per hole. Your simulation results may vary.

c) Answers may vary.

CHAPTER 14

1. a) S = {HH, HT, TH, TT}, equally likely.
   b) S = {0, 1, 2, 3}, not equally likely.
   c) S = {H, TH, TTH, TTT}, not equally likely.
   d) S = {1, 2, 3, 4, 5, 6}, not equally likely.
2. a) S = {2, 3, 4, 5, 6, 7, 8, 9, 10, 11, 12}, not equally likely.
   b) S = {BBB, BBG, BGB, BGG, GBB, GBG, GGB, GGG}, equally likely.
   c) S = {0, 1, 2, 3, 4}, not equally likely.
   d) S = {0, 1, 2, 3, 4, 5, 6, 7, 8, 9, 10}, not equally likely.
3. In this context "truly random" should mean that every number is equally likely to occur.
4. In circumstances "like this," rain occurs 25% of the time.
5. There is no "Law of Averages." She would be wrong to think that they are "due" for a harsh winter.
6. He's referring to the "Law of Averages," which doesn't exist. If rain in the fall and winter are independent of each other, a nice fall will have no bearing on winter rains.
7. There is no "Law of Averages." If at bats are independent, his chance for a hit does not change based on recent successes or failures.
8. a) There is no "Law of Averages." If crashes are independent, it makes no difference. If crashes were due to problems with the aircraft, another crash may be more likely; however, increased maintenance vigilance may *lessen* the chance of another crash.
   b) A good safety record may be due to competence rather than luck. In any even, if events are independent, the probability of a crash has not changed.
9. a) There is some chance you would have to pay out much more than the $300.
   b) Many customers pay for insurance. The small risk for any one customer is spread among all.
10. a) Almost all people make bets without winning the jackpot.
    b) This creates publicity—more people may be attracted to play. (And, of course, it's not a loss. We assume that they take in more than they pay out.)
11. a) Legitimate.   b) Legitimate.
    c) Not legitimate (sum more than 1).   d) Legitimate.
    e) Not legitimate (can't have negatives or values more than 1).
12. a) Not legitimate (sum less than 1).
    b) Not legitimate (sum greater than 1).   c) Legitimate.
    d) Not legitimate (can't have negatives).   e) Legitimate.
13. A family may own both a car and an SUV. The events are not disjoint, so the Addition Rule does not apply.
14. A home may have both a garage and a pool. The events are not disjoint, so the Addition Rule does not apply.
15. When cars are traveling close together, their speeds are not independent, so the Multiplication Rule does not apply.
16. There may be a genetic factor making the handedness of siblings not independent, so the Multiplication Rule does not apply.
17. a) He has multiplied the two probabilities.
    b) He assumes that being accepted at the colleges are independent events.
    c) No. Colleges use similar criteria for acceptance, so the decisions are not independent.
18. a) He has added the three probabilities.
    b) He is assuming that the events are disjoint.
    c) No. Many students get into more than one of the three, so the events are not disjoint and the probabilities cannot simply be added together.
19. a) 0.72   b) 0.89   c) 0.28
20. a) 0.13   b) 0.45   c) 0.87
21. a) 0.5184   b) 0.0784   c) 0.4816
22. a) 0.3025   b) 0.2025   c) 0.2431
23. a) Repair needs for the two cars must be independent.
    b) Maybe not. An owner may treat the two cars similarly, taking good (or poor) care of both. This may decrease (or increase) the likelihood that each needs to be repaired.

24. a) The Calculus backgrounds of the students must be independent.
    b) Yes. The professor assigned students to the groups at random.
25. a) 342/1005 = 0.340.
    b) 30/1005 + 50/1005 = 80/1005 = 0.080.
26. a) 950/2020 = 0.47.
    b) 424/2020 + 566/2020 = 990/2020 = 0.49.
27. a) 0.195   b) 0.913
    c) Responses are independent.
    d) People were polled at random.
28. a) 0.044   b) 0.624   c) 0.332
    d) Responses are independent.
    e) People were polled at random.
29. a) 0.2888   b) 0.7112
    c) (1 − 0.76) + 0.76(1 − 0.38) or 1 − (0.76)(0.38)
30. a) 0.2888   b) More likely in 1997. (0.4002)
31. a) 1) 0.30   2) 0.30   3) 0.90   4) 0.0
    b) 1) 0.027   2) 0.128   3) 0.512   4) 0.271
32. a) 1) 0.04   2) 0.51   3) 0.55
    b) 1) 0.041   2) 0.849   3) 0.974   4) 0.373
33. a) Disjoint (can't be both red and orange).
    b) Independent (unless you're drawing from a small bag).
    c) No. Once you know that one of a pair of disjoint events has occurred, the other is impossible.
34. a) Disjoint.
    b) Independent (unless they are related).
    c) No. Once you know that one of a pair of disjoint events has occurred, the other is impossible.
35. a) 0.0046   b) 0.125   c) 0.296   d) 0.421   e) 0.995
36. a) 0.027   b) 0.125   c) 0.001   d) 0.729   e) 0.784
37. a) 0.027   b) 0.063   c) 0.973   d) 0.014
38. a) 0.0225   b) 0.092   c) 0.00008   d) 0.556
39. a) 0.024   b) 0.250   c) 0.543
40. a) 0.148   b) 0.600   c) 0.344
41. 0.078   42. 0.469
43. a) For any day with a valid three-digit date, the chance is 0.001, or 1 in 1000. For many dates in October through December, the probability is 0. (No three digits will make 10/15, for example.)
    b) There are 65 days when the chance to match is 0. (Oct. 10–31, Nov. 10–30, and Dec. 10–31.) The chance for no matches on the remaining 300 days is 0.741
    c) 0.259   d) 0.049
44. a) Yes. There are 42 cards left in the deck: 26 black and only 16 red.
    b) No. There is no "long run." You'll see the whole deck after 52 cards, and you know there will be 26 of each color then.

CHAPTER 15

1. a) 0.68   b) 0.32   c) 0.04
2. a) 0.14   b) 0.23   c) 0.77
3. a) 0.31   b) 0.48   c) 0.31
4. a) 0.06   b) 0.50   c) 0.94
5. a) 0.2025   b) 0.6965   c) 0.2404   d) 0.0402
6. a) 0.193   b) 0.507   c) 0.067   d) 0.632
7. a) 0.50   b) 1.00   c) 0.077   d) 0.333
8. a) 0.333   b) 0.429   c) 0.667
9. a) 0.11   b) 0.27   c) 0.407   d) 0.344
10. a) i) 0.62   ii) 0.867   iii) 0.194   b) 0.66
11. a) 0.011   b) 0.222   c) 0.054   d) 0.337   e) 0.436
12. a) 0.103   b) 0.404   c) 0.209   d) 0.460   e) 0.559
13. 0.21   14. 0.08
15. a) 0.145   b) 0.118   c) 0.414   d) 0.217
16. a) 0.783   b) 0.013   c) 0.127   d) 0.586
17. a) 0.318   b) 0.955   c) 0.071   d) 0.009
18. a) 0.632   b) 0.140   c) 0:003   d) 0.624
19. a) 32%   b) 0.135

c) No, 7% of juniors have taken both.

d) No, the probability that a junior has taken a computer course is 0.23. The probability that a junior has taken a computer course *given* he or she has taken a Statistics course is 0.135.

20. a) 0.25   b) 0.875

c) No, 68% of all workers have health insurance. Of those with retirement plans, 87.5% have health insurance. These are not equal.

d) No, 49% of the people have both.

21. a) 0.266

b) No, 26.6% of homes with garages have pools; 21% of homes overall have pools.

c) No, 17% of homes have both.

22. a) 0.444

b) No, 4% of U.S. residents have been to both.

c) No, 18% of U.S. residents have been to Canada; 44.4% of U.S. residents who have been to Mexico have been to Canada. If independent, these would be equal.

23. Yes, $P(\text{Ace}) = 4/52$. $P(\text{Ace} \mid \text{any suit}) = 1/13$.

24. Yes, 1/3 of the dogs are male and 1/3 of the cats are male.

25. a) 0.17

b) No; 13% of the chickens had both contaminants.

c) No; $P(C \mid S) = 0.87 \neq P(C)$. If a chicken is contaminated with salmonella, it's more likely also to have campylobacter.

26. a) Yes; a student can enroll in only one college.

b) No. Nearly 42% of all students enrolled in the Agriculture college, but none of the Human Ecology students did.

c) No. Almost 7% of these students were both firstborn and enrolled in Human Ecology.

d) No. Over 19% of all students enrolled in Human Ecology, but only 13% of the firstborn students did.

27. No, only 32% of all men have high cholesterol, but 40.7% of those with high blood pressure do.

28. No, 86.7% of Republicans favor the death penalty, but only 62% of all voters do.

29. a) 95.6%

b) Probably. 95.4% of people with cell phones had landlines, and 95.6% of all people did.

30. a) 13.7%

b) No. 39.3% of people over 30 snored, but only 36.8% of all people did.

31. No. Only 34% of men were Democrats, but over 41% of all voters were.

32. No. Although only 12.5% of all cars were of European origin, about 17% of students drive European cars.

33. a) No, the probability that the luggage arrives on time depends on whether the flight is on time. The probability is 95% if the flight is on time and only 65% if not.

b) 0.695

34. a) Yes, the percentage who graduate depends on what kind of school they attend. The graduation rate is 75% for public school students and 90% for others.

b) 79.5%

35. 0.975   **36.** 66.0%

37. a) No, the probability of missing work for day-shift employees is 0.01. It is 0.02 for night-shift employees. The probability depends on whether they work day or night shift.

b) 1.4%

38. a) Rates of the lung condition are different for smokers and nonsmokers, 57% and 13%, respectively. If independent, these would be the same.

b) 0.231

39. 57.1%   **40.** 0.567

41. a) 0.20   b) 0.272   c) 0.353   d) 0.033

42. a) 15%   b) 0.20

c) No. Advance-purchase customers are less likely to be no-shows (5% vs. 30% among regular-fare customers.)

43. 0.563   **44.** 0.467   **45.** Over 0.999   **46.** 0.814

CHAPTER 16

1. a) 19   b) 4.2

2. a) 1.2   b) 280

3. a)

Amount won	$0	$5	$10	$30
P(Amount won)	$\frac{26}{52}$	$\frac{13}{52}$	$\frac{12}{52}$	$\frac{1}{52}$

b) $4.13   c) $4 or less (answers may vary)

4. a)

Amount won	$100	$50	$0
P(Amount won)	$\frac{1}{6}$	$\frac{5}{36}$	$\frac{25}{36}$

b) $23.61   c) $23 or less (answers may vary)

5. a)

Children	1	2	3
P(Children)	0.5	0.25	0.25

b) 1.75 children   c) 0.875 boys

Boys	0	1	2	3
P(Boys)	0.5	0.25	0.125	0.125

6. a)

Number of darts	1	2	3	4
P(Number of darts)	0.10	0.09	0.081	0.729

Amount won	$95	$90	$85	$80	−$20
P(Amount won)	0.10	0.09	0.081	0.073	0.656

b) 3.44 darts   c) $17.20

7. $27,000

8. $10,600

9. a) 7   b) 1.89

10. a) 0.75   b) 87.18

11. $5.44

12. $38.16

13. 0.83

14. $51.48

15. a) 1.7   b) 0.9

16. a) 2.25   b) 1.26

17. $\mu = 0.64, \sigma = 0.93$

18. a)

Profit	$100	−$9900	−$2900
P(Profit)	0.9975	0.0005	0.002

b) $89.00   c) $260.54

19. a) $50   b) $100

20. a) Yes, the expected value is $300.

b) $100   c) $360.56

21. a) No. The probability of winning the second depends on the outcome of the first.

b) 0.42   c) 0.08

d)

Games won	0	1	2
P(Games won)	0.42	0.50	0.08

e) $\mu = 0.66, \sigma = 0.62$

22. a) No. The chance to get the second contract depends on whether your company got the first.

b) 0.16   c) 0.14

d)

Contracts won	0	1	2
P(Contracts won)	0.14	0.70	0.16

e) $\mu = 1.02$, $\sigma = 0.55$

23. a)

Number good	0	1	2
P(Number good)	0.067	0.467	0.467

b) 1.40   c) 0.61

24. a)

Number males	0	1	2
P(Number males)	0.143	0.571	0.286

b) 1.14   c) 0.64

25. a) $\mu = 30$, $\sigma = 6$ b) $\mu = 26$, $\sigma = 5$ c) $\mu = 30$, $\sigma = 5.39$
d) $\mu = -10$, $\sigma = 5.39$ e) $\mu = 20$, $\sigma = 2.83$

26. a) $\mu = 60$, $\sigma = 12$ b) $\mu = 6$, $\sigma = 1.50$
c) $\mu = 92$, $\sigma = 12.37$ d) $\mu = 68$, $\sigma = 12.37$
e) $\mu = 24$, $\sigma = 4.24$

27. a) $\mu = 240$, $\sigma = 12.80$ b) $\mu = 140$, $\sigma = 24$
c) $\mu = 720$, $\sigma = 34.18$ d) $\mu = 60$, $\sigma = 39.40$
e) $\mu = 600$, $\sigma = 22.63$

28. a) $\mu = 44$, $\sigma = 6$ b) $\mu = 240$, $\sigma = 36$
c) $\mu = 32$, $\sigma = 4.24$ d) $\mu = 20$, $\sigma = 19.21$
e) $\mu = 240$, $\sigma = 20.78$

29. a) 1.8   b) 0.87
c) Cartons are independent of each other.

30. a) 10   b) 2.68
c) Packets are independent of each other. OK if different types of seeds; if all the same type (and lot), assumption would probably not be valid.

31. $\mu = 13.6$, $\sigma = 2.55$ (assuming the hours are independent of each other).

32. $\mu = 11.25$, $\sigma = 2.82$

33. a) $\mu = 23.4$, $\sigma = 2.97$
b) We assume each truck gets tickets independently.

34. a) $\mu = \$3840$, $\sigma = \$591.54$
b) We assume pledges are independent among callers.

35. a) There will be many gains of $150 with a few large losses.
b) $\mu = \$300$, $\sigma = \$8485.28$
c) $\mu = \$1,500,000$, $\sigma = \$600,000$
d) Yes. $0 is 2.5 SDs below the mean for 10,000 policies.
e) Losses are independent of each other. A major catastrophe with many policies in an area would violate the assumption.

36. a) Gamblers lose a relatively small amount most of the time, but there are a few large payouts.
b) $\mu = \$0.40$, $\sigma = \$268.33$
c) $\mu = \$80.00$, $\sigma = \$3794.73$
d) If the machine is played only 1000 times a day, we can't guarantee that it will be profitable, since $80 is about 0.02 SDs above 0. But if the casino has many slot machines, the chances of being profitable will go up.

37. a) 1 oz   b) 0.5 oz   c) 0.023
d) $\mu = 4$ oz, $\sigma = 0.5$ oz
e) 0.159
f) $\mu = 12.3$ oz, $\sigma = 0.54$ oz

38. a) $-\$20$ b) $46.10   c) 0.332
d) Costs for pets living together may not be independent.

39. a) 12.2 oz   b) 0.51 oz   c) 0.058

40. a) $D = $ cost for a dog; $C = $ cost for a cat;
total costs $= D_1 + D_2 + C$
b) Normal, $\mu = \$320$, $\sigma = \$55$ c) 0.073

41. a) $\mu = 200.57$ sec, $\sigma = 0.46$ sec
b) No, $z = \dfrac{199.48 - 200.57}{0.461} = -2.36$. There is only 0.009 probability of swimming that fast or faster.

42. a) $\mu = 37.6$ min, $\sigma = 3.7$ min
b) No, 30 min is more than 2 SDs below the mean.

43. a) $A = $ price of a pound of apples; $P = $ price of a pound of potatoes; Profit $= 100A + 50P - 2$
b) $63.00   c) $20.62
d) Mean—no; SD—yes (independent sales prices).

44. a) $B = $ number basic; $D = $ number deluxe;
Net $= 120B + 150D - 200$
b) $928.00   c) $187.45
d) Mean—no; SD—yes (sales are independent).

45. a) $\mu = 1920$, $\sigma = 48.99$; $P(T > 2000) = 0.051$
b) $\mu = \$220$, $\sigma = 11.09$; No—$300 is more than 7 SDs above the mean.
c) $P(D - \frac{1}{2}C > 0) \approx 0.26$

46. a) $\mu = 60$ lb, $\sigma = 0.346$ lb; $P(W > 60.5) \approx 0.074$
b) $\mu = \$35$, $\sigma = \$0.187$
c) $P(|D|) > 0.25 \approx 0.377$

CHAPTER 17

1. a) No. More than two outcomes are possible.
b) Yes, assuming the people are unrelated to each other.
c) No. The chance of a heart changes as cards are dealt so the trials are not independent.
d) No, 500 is more than 10% of 3000.
e) If packages in a case are independent of each other, yes.

2. a) Yes. Outcomes are {getting a 6} and {not getting a 6}.
b) No. More than two outcomes are possible.
c) If the dolls were manufactured independently of each other, yes.
d) No. The chance of a Democrat (or Republican) changes depending on who has already been picked so the trials are not independent.
e) Yes, assuming responses (and cheating) are independent among the students.

3. a) Use single random digits. Let $0, 1 = $ Tiger. Count the number of random numbers until a 0 or 1 occurs.
c) Results will vary.
d)

x	1	2	3	4	5	6	7	8	≥9
P(x)	0.2	0.16	0.128	0.102	0.082	0.066	0.052	0.042	0.168

4. a) Use random digits 1–6, ignoring 0, 7–9. Count the number of random numbers until a 1 occurs.
c) Results will vary.
d)

x	1	2	3	4	5	6	7	8	≥9
P(x)	0.167	0.139	0.116	0.096	0.080	0.067	0.056	0.047	0.232

5. a) Use single random digits. Let $0, 1 = $ Tiger. Examine random digits in groups of five, counting the number of 0's and 1's.
c) Results will vary.
d)

x	0	1	2	3	4	5
P(x)	0.33	0.41	0.20	0.05	0.01	0.0

6. a) Use pairs of random digits. Let 00–74 represent that a seatbelt is worn, 75–99 that it is not. Examine random digits in groups of five, counting the number of seatbelts.
c) Results will vary.
d)

x	0	1	2	3	4	5
P(x)	0.0	0.01	0.09	0.26	0.40	0.24

7. Departures from the same airport during a 2-hour interval may not be independent. All could be delayed by weather, for example.

8. What happens to pieces of luggage checked together and all traveling on the same flight probably aren't independent.

9. a) 0.0819   b) 0.0064   c) 0.992
10. a) 0.0184   b) 0.183
11. 5   12. 50   13. 20 calls   14. 100 donors
15. a) 25   b) 0.185   c) 0.217   d) 0.693
16. a) 12.5   b) 0.716   c) 0.0527   d) 0.528
17. a) 0.0745   b) 0.502   c) 0.211
    d) 0.0166   e) 0.0179   f) 0.9987
18. a) 0.032   b) 0.738   c) 0.00768
    d) 0.246   e) 0.901   f) 0.345
19. a) 0.65   b) 0.75   c) 7.69 picks
20. a) 4.8   b) 0.98   c) 1.25 shots
21. a) $\mu = 10.44, \sigma = 1.16$
    b) i) 0.812   ii) 0.475   iii) 0.00193   iv) 0.998
22. a) $\mu = 8, \sigma = 1.26$
    b) i) 0.107   ii) 0.624   iii) 0.302   iv) 0.967
23. $\mu = 20.28, \sigma = 4.22$
24. $\mu = 2.4, \sigma = 1.5$
25. a) 0.118   b) 0.324   c) 0.744   d) 0.580
26. a) 0.201   b) 0.453   c) 0.171   d) 0.989
27. a) $\mu = 56, \sigma = 4.10$
    b) Yes, $np = 56 \geq 10$, $nq = 24 \geq 10$, serves are independent.
    c) In a match with 80 serves, approximately 68% of the time she will have between 51.9 and 60.1 good serves, approximately 95% of the time she will have between 47.8 and 64.2 good serves, and approximately 99.7% of the time she will have between 43.7 and 68.3 good serves.
    d) Normal, approx.: 0.014; Binomial, exact: 0.016
28. a) $\mu = 160, \sigma = 5.66$
    b) Yes, $np = 160 \geq 10$, $nq = 40 \geq 10$.
    c) In matches with 200 arrows, about 68% of the time she will have between 154.34 and 165.66 bull's-eyes, about 95% of the time between 148.68 and 171.32 bull's-eyes, and about 99.7% of the time between 143.02 and 176.98 bull's-eyes.
    d) Yes, that's more than 3 SDs below the mean. The probability this happens is less than 0.0015.
29. a) Assuming apples fall and become blemished independently of each other, Binom(300, 0.06) is appropriate. Since $np \geq 10$ and $nq \geq 10$, $N(18, 4.11)$ is also appropriate.
    b) Normal, approx.: 0.072; Binomial, exact: 0.085
    c) No, 50 is 7.8 SDs above the mean.
30. a) $\mu = 18.75, \sigma = 4.05$
    b) Yes, $np = 18.75 \geq 10$, $nq = 131.25 \geq 10$.
    c) No, 22 is only 0.8 SD above the mean; this is likely to happen by natural sampling variability.
31. Normal, approx.: 0.053; Binomial, exact: 0.061
32. Normal, approx.: 0.094; Binomial, exact: 0.116
33. The mean number of sales should be 24 with SD 4.60. Ten sales is more than 3.0 SDs below the mean. He was probably misled.
34. If the coin is fair, expect 125 heads, with SD 7.91; 140 is 1.9 SDs above the mean. From the Normal approximation, the chance of 140 or more heads is 2.9%. That's pretty unlikely to happen.
35. a) 5   b) 0.066   c) 0.107   d) $\mu = 24, \sigma = 2.19$
    e) Normal, approx.: 0.819; Binomial, exact: 0.848
36. a) 0.042   b) 0.107   c) 5   d) $\mu = 10, \sigma = 2.83$
    e) Normal, approx.: 0.025; Binomial, exact: 0.027
37. $\mu = 20, \sigma = 4$. I'd want *at least* 32 (3 SDs above the mean). (Answers will vary.)
38. $\mu = 25, \sigma = 3.54$. I'd want *at least* 36 (3 SDs above the mean). (Answers will vary.)
39. Probably not. There's a more than 9% chance that he could hit 4 shots in a row, so he can expect this to happen nearly once in every 10 sets of 4 shots he takes. That does not seem unusual.
40. No. If she were to shoot several flights of 6 arrows, an archer of her ability could be expected to get all bull's-eyes about 26% of the time. That's not an unusual result.
41. Yes. We'd expect him to make 22 shots, with a standard deviation of 3.15 shots. 32 shots is more than 3 standard deviations above the expected value, an unusually high rate of success.

42. No. We'd expect her to hit the bull's-eye with 40 of the 50 arrows, with a standard deviation of 2.83 bull's-eyes. She got 45, less than 2 standard deviations above the mean. Good shooting, but probably not unusual for her.

PART IV REVIEW

1. a) 0.34   b) 0.27   c) 0.069
   d) No, 2% of cars have both types of defects.
   e) Of all cars with cosmetic defects, 6.9% have functional defects. Overall, 7.0% of cars have functional defects. The probabilities here are estimates, so these are probably close enough to say the defects are independent.
2. a) i) 0.60   ii) 0.90   iii) 0.615   iv) 0.80
   b) No. Fewer than half of the managers are female, but 60% of all workers are female.
3. a) C = Price to China; F = Price to France; Total = $3C + 5F$
   b) $\mu = \$5500, \sigma = \$672.68$ c) $\mu = \$500, \sigma = \$180.28$
   d) Means—no. Standard deviations—yes; ticket prices must be independent of each other for different countries, but all tickets to the same country are at the same price.
4. $np = 100 \geq 10$ and $nq = 9900 \geq 10$, so Success/Failure Condition is verified. Also, we assume that cases are independent. $\mu = 100, \sigma = 9.95$. Over 200 would be more than 10 SDs above the mean. Probability is essentially 0.
5. a) $\mu = -\$0.20, \sigma = \$1.89$ b) $\mu = -\$0.40, \sigma = \$2.67$
6. a) 3%
   b) No; 62% of those who can do it with their right hand can do it with their left, but 83.3% of those who can't do it with their right hand can do it with their left.
   c) No; 51% can use either hand.
7. a) 0.106   b) 0.651   c) 0.442
8. Expected (extra) cost of the cheaper policy with the deductible is $2.50, much less than the $12 for the no-deductible surcharge, so on average, she will save money by going with the deductible. But the standard deviation ($35.27) is evidence of risk. Value of the car shouldn't influence the decision.
9. a) 0.590   b) 0.328   c) 0.00856
10. a) $\mu = 6.775, \sigma = 2.60$ b) No, since $np = 6.78 < 10$
11. a) $\mu = 15.2, \sigma = 3.70$ b) Yes, $np \geq 10$ and $nq \geq 10$
    c) Normal, approx.: 0.080; Binomial, exact: 0.097
12. a)

Spaces	5	10	20
P(Spaces)	0.5	0.25	0.25

    b) $\mu = 10, \sigma = 6.12$
    c)

Spaces	0	1	2	3	4
P(Spaces)	1/3	1/6	1/6	1/6	1/6

    d) $\mu = 1.67, \sigma = 1.49$ e) $\mu = 11.67, \sigma = 6.30$
13. a) 0.0173   b) 0.591
    c) Left: 960; right: 120; both: 120
    d) $\mu = 120, \sigma = 10.39$
    e) About 68% chance of between 110 and 130; about 95% between 99 and 141; about 99.7% between 89 and 151.
14. $\mu = 8.33, \sigma = 6.30$
15. a) Men's heights are more variable than women's.
    b) Men (1.75 SD vs 2.4 SD for women)
    c) M = Man's height; W = Woman's height; $M - W$ is how much taller the man is.
    d) 5.1"   e) 3.75"   f) 0.913
    g) If independent, it should be about 91.3%. We are told 92%. This difference seems small and may be due to natural sampling variability.
16. a) 0.389   b) 0.284   c) 0.793   d) 0.896
17. a) The chance is 1.6×10^{-7}.   b) 0.952   c) 0.063
18. a) No, this does not confirm the advice. If you follow the advice, it seems there's only a 75% chance it goes up in the 3rd year.

b) It actually has risen in 73% of all years. Not much difference from their strategy.

19. $240

20. a) 0.240   b) 0.050   c) 0.383

21. a) 0.717   b) 0.588

22. a) 36   b) 5.02
    c) Because both $np = 36 \geq 10$ and $nq = 84 \geq 10$.
    d) There is a 68% chance between 30.98 and 41.02 (31 and 41 students); 95% chance between 25.96 and 46.04 (26 and 46 students); 99.7% chance between 20.94 and 51.06 (21 and 51 students).

23. a) $\mu = 100, \sigma = 8$ b) $\mu = 1000, \sigma = 60$
    c) $\mu = 100, \sigma = 8.54$ d) $\mu = -50, \sigma = 10$
    e) $\mu = 100, \sigma = 11.31$

24. Assuming policies are independent, we add the profit variances. The resulting standard deviation of total profit is less than the sum of the SDs of individual profits. This means the profit for the large company will be less variable than the total of profits for the smaller companies.

25. a) Many do both, so the two categories can total more than 100%.
    b) No. They can't be disjoint. If they were, the total would be 100% or less.
    c) No. Probabilities are different for boys and girls.
    d) 0.0524

26. a) $\mu = \$27.00, \sigma = \$9.90$
    b) Spending on different days is independent. This might not be reasonable, since a student may be more likely to spend less on a day after he had spent a lot.
    c) $\mu = \$94.50, \sigma = \$18.52$
    d) No. $50 is 2.4 SDs below the mean. We assumed independence of costs each day.

27. a) 21 days   b) 1649.73 som
    c) 3300 som extra. About 157-som "cushion" each day.

28. a) 4 lb   b) 3.20 lb   c) 0.894

29. No, you'd expect 541.2 homeowners, with an SD of 13.56. 523 is 1.34 SDs below the mean; not unusual.

30. $\mu = \$2.54, \sigma = \$0.94$

31. a) 0.018   b) 0.300   c) 0.26

32. a) 0.0156   b) 0.0039
    c) Answer b would become 0.0026.

33. a) 6   b) 15   c) 0.402

34. a) 3
    b) Expect to lose 1/6 of your current score.   c) 18
    d) Roll until you score 18 points, then stop.

35. a) 34%   b) 35%   c) 31.4%
    d) 31.4% of classes that used calculators used computer assignments, while in classes that didn't use calculators, 30.6% used computer assignments. These are close enough to think the choice is probably independent.

36. a) 0.5929   b) 0.407   c) 0.053   d) 0.0025

37. a) 1/11   b) 7/22   c) 5/11   d) 0   e) 19/66

38. a) $\mu = 18, \sigma = 3$ b) 6   c) No, σ is now 5.   d) 10 or more
    e) What appears "surprising" in the short run becomes expected in a large number of trials.

39. a) Expected number of stars with planets.
    b) Expected number of planets with intelligent life.
    c) Probability of a planet with a suitable environment having intelligent life.
    d) f_l: If a planet has a suitable environment, the probability that life develops.
       f_i: If a planet develops life, the probability that the life evolves intelligence.
       f_c: If a planet has intelligent life, the probability that it develops radio communication.

40. a) 0.017   b) 0.824   41. 0.991

42. a) 0.01   b) 0.0098   c) 0.366   d) First
    e) The chance of winning is 0.01 anywhere in line, so position does not matter.

CHAPTER 18

1. All the histograms are centered near 0.05. As n gets larger, the histograms approach the Normal shape, and the variability in the sample proportions decreases.

2. All the histograms are centered near 0.85. As n gets larger, the histograms approach the Normal shape, and the variability in the sample proportions decreases.

3. a)

n	Observed mean	Theoretical mean	Observed st. dev.	Theoretical st. dev.
20	0.0497	0.05	0.0479	0.0487
50	0.0516	0.05	0.0309	0.0308
100	0.0497	0.05	0.0215	0.0218
200	0.0501	0.05	0.0152	0.0154

b) They are all quite close to what we expect from the theory.
c) The histogram is unimodal and symmetric for $n = 200$.
d) The success/failure condition says that np and nq should both be at least 10, which is not satisfied until $n = 200$ for $p = 0.05$. The theory predicted my choice.

4. a)

n	Observed mean	Theoretical mean	Observed st. dev.	Theoretical st. dev.
20	0.8481	0.85	0.0803	0.0798
50	0.8507	0.85	0.0509	0.0505
75	0.8481	0.85	0.0406	0.0412
100	0.8488	0.85	0.0354	0.0357

b) They are all quite close to what we expect from the theory.
c) The histogram is unimodal and symmetric for $n = 75$.
d) The success/failure condition says that np and nq should both be at least 10, which, for $p = 0.85$ and $q = 0.15$, would be satisfied at a sample size of about 67. So my choice of $n = 75$ is reasonable.

5. a) Symmetric, because probability of heads and tails is equal.
   b) 0.5   c) 0.125   d) $np = 8 < 10$

6. a) Probability skewed right.
   b) No. $np = 5 < 10$ c) 0.10   d) 0.042

7. a) About 68% should have proportions between 0.4 and 0.6, about 95% between 0.3 and 0.7, and about 99.7% between 0.2 and 0.8.
   b) $np = 12.5, nq = 12.5$; both are ≥ 10.
   c)

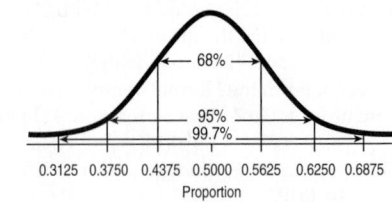

$np = nq = 32$; both are ≥ 10.
   d) Becomes narrower (less spread around 0.5).

8. a) $np = 20, np = 180$; both are ≥ 10.
   b) About 68% will be between 7.9% and 12.1%, about 95% will be between 5.8% and 14.2%, and about 99.7% should be between 3.7% and 16.3%.
   c) Same center, less spread.

9. This is a fairly unusual result: about 2.26 SDs below the mean. The probability of that is about 0.012. So, in a class of 100 this is certainly a reasonable possibility.

10. No. It's only 1.49 SDs above the mean.

11. a)

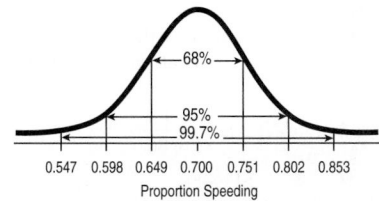

b) Both $np = 56$ and $nq = 24 \geq 10$. Drivers *may* be independent of each other, but if flow of traffic is very fast, they may not be. Or weather conditions may affect all drivers. In these cases they may get more or fewer speeders than they expect.

12. There is a 68% chance that between 20.2% and 32.6% are smokers, 95% chance between 14% and 38.8%, and 99.7% chance between 7.8% and 45%. The 50 adults would have to be a random sample whose choice to smoke is independent of the others in the sample. The Success/Failure Condition is satisfied, $np = 13.2$, and $nq = 36.8$. Both are ≥ 10.

13. a) Assume that these children are typical of the population. They represent fewer than 10% of all children. We expect 20.4 nearsighted and 149.6 not; both are at least 10.

b)

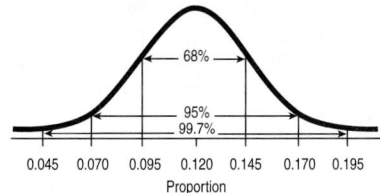

c) Probably between 12 and 29.

14. a) Assume that this bank's mortgages are typical of the population. They're fewer than 10% of all mortgages. We expect 27.7 foreclosures and 1703.3 "failures"; both are at least 10.

b)

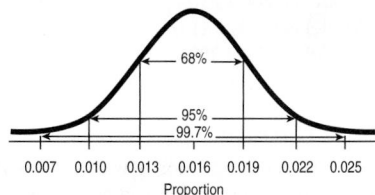

c) Probably between 17 and 38.

15. a) $\mu = 7\%$, $\sigma = 1.8\%$
b) Assume that clients pay independently of each other, that we have a random sample of all possible clients, and that these represent less than 10% of all possible clients. $np = 14$ and $nq = 186$ are both at least 10.
c) 0.048

16. a) Normal, $\mu = 30\%$, $\sigma = 4.6\%$; SRS, $np = 30$, $nq = 70 \geq 10$.
b) 0.234

17.

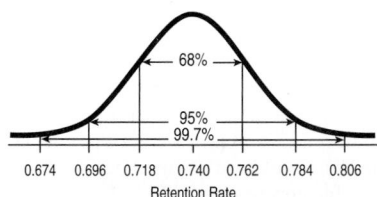

These are not random samples, and not all colleges may be typical (representative). $np = 296$, $nq = 104$ are both at least 10.

18.

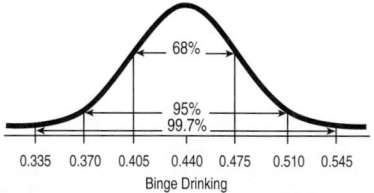

Conditions are met: It's a random sample, with $np = 88$ and $nq = 112$, both at least 10.

19. Yes; if their students were typical, a retention rate of $522/603 = 86.6\%$ would be over 7 standard deviations above the expected rate of 74%.

20. Probably not; his sample proportion of 39.3% is less than 1.5 standard deviations below the national result of 44%.

21. 0.212. Reasonable that those polled are independent of each other and represent less than 10% of all potential voters. We assume the sample was selected at random. Success/Failure Condition met: $np = 208$, $nq = 192$. Both ≥ 10.

22. 0.081. Assume the seeds are a random sample, germinate independently of each other, and represent less than 10% of all seeds of this type. Since the seeds are in the same pack, they may not be independent of each other, or be a random sample. Success/Failure Condition met: $np = 147.2$, $nq = 12.8$. Both ≥ 10.

23. 0.088 using $N(0.08, 0.022)$ model.
24. 0.960 using $N(0.04, 0.007)$ model.
25. Answers will vary. Using $\mu + 3\sigma$ for "very sure," the restaurant should have 89 nonsmoking seats. Assumes customers at any time are independent of each other, a random sample, and represent less than 10% of all potential customers. $np = 72$, $nq = 48$, so Normal model is reasonable ($\mu = 0.60$, $\sigma = 0.045$).
26. Answers will vary. Using $\mu + 2\sigma$ for "pretty sure," he should plan on having at least 47 steak specials ($\mu = 0.20$, $\sigma = 0.030$).
27. a) Normal, center at μ, standard deviation $\sigma/\sqrt{n}$.
b) Standard deviation will be smaller. Center will remain the same.
28. a) Skewed left, center at μ, standard deviation $\sigma/\sqrt{n}$.
b) Becomes more Normal, center at μ, standard deviation $\sigma/\sqrt{n}$.
c) Becomes more like the population—skewed to the left.
29. a) The histogram is unimodal and slightly skewed to the right, centered at 36 inches with a standard deviation near 4 inches.
b) All the histograms are centered near 36 inches. As n gets larger, the histograms approach the Normal shape and the variability in the sample means decreases. The histograms are fairly normal by the time the sample reaches size 5.
30. a) The histogram is unimodal, but strongly skewed to the right with several large outliers.
b) All the histograms are centered near 10,000 ($1000). As n gets larger, the variability in sample means decreases and histograms approach a Normal shape; however, they are still visibly skewed, with the possible exception of the last one at $n = 200$.
c) The rule of thumb doesn't seem to account for highly skewed distributions.
31. a)

n	Observed mean	Theoretical mean	Observed st. dev.	Theoretical st. dev.
2	36.314	36.33	2.855	2.842
5	36.314	36.33	1.805	1.797
10	36.341	36.33	1.276	1.271
20	36.339	36.33	0.895	0.899

b) They are all very close to what we would expect.
c) For samples as small as 5, the sampling distribution of sample means is unimodal and very symmetric.

d) The distribution of the original data is nearly unimodal and symmetric, so it doesn't take a very large sample size for the distribution of sample means to be approximately Normal.

32. a)

n	Observed mean	Theoretical mean	Observed st. dev.	Theoretical st. dev.
30	10,251.73	10,307.31	3359.64	3279.88
50	10,343.93	10,307.31	2483.84	2540.58
100	10,329.94	10,307.31	1779.18	1796.46
200	10,340.37	10,307.31	1260.79	1270.29

b) They are all reasonably close to what we would expect.
c) All the sampling distributions are still quite skewed, with the possible exception of the sampling distribution for $n = 200$, which is still somewhat skewed.
d) The distribution of the data is strongly skewed, so it will take a very large sample size before the sampling distribution of the mean is approximately normal.

33.

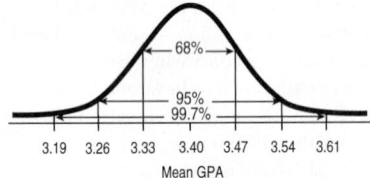

68%
95%
99.7%

3.19 3.26 3.33 3.40 3.47 3.54 3.61
Mean GPA

Normal, $\mu = 3.4$, $\sigma = 0.07$. We assume that the students are randomly assigned to the seminars and represent less than 10% of all possible students, and that individual's GPAs are independent of one another.

34.

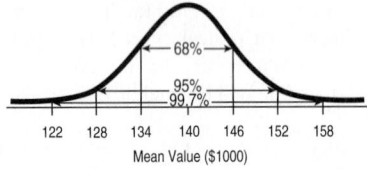

68%
95%
99.7%

122 128 134 140 146 152 158
Mean Value ($1000)

35. a) As the CLT predicts, there is more variability in the smaller outlets.
b) If the lottery is random, all outlets are equally likely to sell winning tickets.
36. The standard deviation of the sampling distribution model for the mean is $\dfrac{\sigma}{\sqrt{n}}$. So cities in which the average is based on a smaller number of drivers will have greater variation in their averages and will be more likely to be both safest and least safe.
37. a) 21.1%   b) 276.8 days or more
c) $N(266, 2.07)$ d) 0.002
38. a) 13.7%   b) 31.9"
c) $N(35.4, 2.1)$ d) 0.005
39. a) There are more premature births than very long pregnancies. Modern practice of medicine stops pregnancies at about 2 weeks past normal due date.
b) Parts (a) and (b)—yes—we can't use Normal model if it's very skewed. Part (c)—no—CLT guarantees a Normal model for this large sample size.
40. a) Some people work far longer than the mean +2 or 3 SDs.
b) The CLT says $\bar{y}$ is approximately Normal for large sample sizes, but not for samples of size 1 (individuals).
41. a) $\mu = \$2.00$, $\sigma = \$3.61$
b) $\mu = \$4.00$, $\sigma = \$5.10$
c) 0.191. Model is $N(80, 22.83)$.
42. a)

Amount won	40	0	−10
P(amount won)	1/6	5/36	25/36

b) $\mu = -\$0.28$, $\sigma = \$18.33$ c) $\mu = -\$1.40$, $\sigma = \$40.99$
d) 0.56

43. a) $\mu = 2.859$, $\sigma = 1.324$
b) No. The score distribution in the sample should resemble that in the population, somewhat uniform for scores 1–4 and about half as many 5's.
c) Approximately $N\left(2.859, \dfrac{1.324}{\sqrt{40}}\right)$.
44. a) $\mu = \$137.50$, $\sigma = \$148.56$
b) No. These 50 members will probably make donations typical of the current member population, and that's skewed to the right.
c) Approximately $N\left(137.5, \dfrac{148.56}{\sqrt{50}}\right)$.
45. About 20%, based on $N(2.859, 0.167)$.
46. Nearly 99%, based on $N(137.5, 16.61)$.
47. a) $N(2.9, 0.045)$ b) 0.0131   c) 2.97 gm/mi
48. a) 0.0478   b) 0.863   c) 0.0019   d) Essentially 0
49. a) Can't use a Normal model to estimate probabilities. The distribution is skewed right—not Normal.
b) 4 is probably not a large enough sample to say the average follows the Normal model.
c) No. This is 3.16 SDs above the mean.
50. a) Can't use a Normal model to estimate probabilities. The distribution is skewed right—not Normal.
b) Probably not. 10 is not a large sample. It depends on the amount of skewness of the distribution.
c) No. With 50 customers, $\bar{y}$ is $N(32, 2.83)$. The probability is 0.0023.
51. a) 0.0003. Model is $N(384, 34.15)$.   b) $427.77 or more.
52. a) 0.482. Model is $N(9984, 353.27)$.   b) $9531.27 or less.
53. a) 0.734
b) 0.652. Model is $N(10, 12.81)$.
c) 0.193. Model is $N(120, 5.774)$.
d) 0.751. Model is $N(10, 7.394)$.
54. a) 0.309
b) 0.696. Model is $N(4, 7.810)$.
c) 0.037. Model is $N(43, 1.118)$.
d) 0.325. Model is $N(4, 2.202)$.

CHAPTER 19

1. She believes the true proportion is within 4% of her estimate, with some (probably 95%) degree of confidence.
2. He believes the true proportion is within 3% of his estimate, with some (probably 95%) degree of confidence.
3. a) Population—all cars; sample—those actually stopped at the checkpoint; p—proportion of all cars with safety problems; $\hat{p}$—proportion actually seen with safety problems (10.4%); if sample (a cluster sample) is representative, then the methods of this chapter will apply.
b) Population—general public; sample—those who logged onto the Web site; p—population proportion of those who favor prayer in school; $\hat{p}$—proportion of those who voted in the poll who favored prayer in school (81.1%); can't use methods of this chapter—sample is biased and nonrandom.
c) Population—parents at the school; sample—those who returned the questionnaire; p—proportion of all parents who favor uniforms; $\hat{p}$—proportion of respondents who favor uniforms (60%); should not use methods of this chapter, since not SRS (possible non-response bias).
d) Population—students at the college; sample—the 1632 students who entered that year; p—proportion of all students who will graduate on time; $\hat{p}$—proportion of that year's students who graduate on time (85.0%); can use methods of this chapter if that year's students (a cluster sample) are viewed as a representative sample of all possible students at the school.
4. a) Population—people who recently bought new cars; sample—167 people surveyed; p—proportion of all new car buyers who

are dissatisfied with the salesperson; $\hat{p}$—proportion of those surveyed who are unsatisfied (3%); can't use methods of this chapter because only 5 people were dissatisfied.

b) Population—college students; sample—2883 students who were asked about cell phones; p—proportion of college students with cell phones; $\hat{p}$—proportion of those asked who have cell phones (8.4%); use methods of this chapter with caution—students entering the football stadium may not reflect all students.

c) Population—potato plants in the United States; sample—240 plants checked; p—proportion of all plants in the field with blight; $\hat{p}$—proportion of the 240 plants checked with blight (2.9%); can't use methods of this chapter—fewer than 10 "successes."

d) Population—employees at the company; sample—all employees that year; p—proportion of all employees who will have an injury on the job in a year; $\hat{p}$—proportion of that year's employees with an injury (3.9%); can use methods of this chapter if that year's employees are viewed as a random sample of all possible employees.

5. a) Not correct. This implies certainty.
   b) Not correct. Different samples will give different results. Many fewer than 95% will have 88% on-time orders.
   c) Not correct. The interval is about the population proportion, not the sample proportion in different samples.
   d) Not correct. In this sample, we *know* 88% arrived on time.
   e) Not correct. The interval is about the parameter, not the days.

6. a) Not correct. The interval is about the proportion of heads, not individual coins.
   b) Not correct. We know in this sample that 56% of the spins were heads.
   c) Correct.
   d) Not correct. The interval is about the proportion of heads, not about individual spins.
   e) Not correct. The interval is about the proportion of heads, not about the percentage of euros.

7. a) False   b) True   c) True   d) False
8. a) True   b) True   c) True   d) True
9. On the basis of this sample, we are 90% confident that the proportion of Japanese cars is between 29.9% and 47.0%.
10. On the basis of this sample, we are 95% confident that the proportion of paroles granted by the Nebraska Board of Parole is between 56.1% and 62.5%.
11. a) (0.798, 0.863)
    b) We're 95% confident that between 80% and 86% of all broiler chicken sold in U.S. food stores is infected with *Campylobacter*.
    c) The size of the population is irrelevant. If *Consumer Reports* had a random sample, 95% of intervals generated by studies like this will capture the true contamination level.
12. a) It's not clear how the sample was chosen, but coming from many stores in 23 states, it may be representative of all broiler chicken sold. And it's important that the researchers kept the chicken samples separated. The sample was far less than 10% of the population. There were 79 successes and 446 failures, both at least 10.
    b) (0.12, 0.18)
    c) We're 95% confident that between 12% and 18% of all broiler chicken sold in U.S. food stores is infected with *Salmonella*.
13. a) 0.025
    b) We're 90% confident that this poll's estimate is within ±2.5% of the true proportion of people who are baseball fans.
    c) Larger. To be more certain, we must be less precise.
    d) 0.039   e) less confidence
    f) No evidence of change; given the margin of error, 0.37 is a plausible value for 2007 as well.
14. a) 1.9% (0.019)
    b) The pollsters are 95% confident that the true proportion of adults who approve of attempts to clone humans is within 1.9% of the estimated 11%.

c) Smaller. Less confidence allows a narrower interval.
d) 1.6%   e) Larger
15. a) (0.0465, 0.0491). The assumptions and conditions for constructing a confidence interval are satisfied.
    b) The confidence interval gives the set of plausible values (with 95% confidence). Since 0.05 is outside the interval, that seems to be a bit too optimistic.
16. a) (0.0223, 0.0250). The assumptions and conditions for constructing a confidence interval are satisfied.
    b) Since 0.02 is below the confidence interval, it isn't one of the plausible values. Proceed with the plan.
17. a) (12.7%, 18.6%)
    b) We are 95% confident, based on this sample, that the proportion of all auto accidents that involve teenage drivers is between 12.7% and 18.6%.
    c) About 95% of all random samples will produce confidence intervals that contain the true population proportion.
    d) Contradicts. The interval is completely below 20%.
18. a) (10.6%, 14.0%)
    b) We are 90% confident, based on this sample, that the proportion of people contacted who may buy something is between 10.6% and 14.0%.
    c) About 90% of all random samples will produce confidence intervals that contain the true population proportion.
    d) Do the mass mailing. The interval is considerably above 5%.
19. Probably nothing. Those who bothered to fill out the survey may be a biased sample.
20. Nothing. Those who showed up for the meeting are probably a biased group. In addition, a show-of-hands vote may influence people, making the votes nonindependent.
21. a) Response bias (wording)   b) (54%, 60%)
    c) Smaller—the sample size was larger.
22. a) The class's interval will be larger, since its sample size is smaller.
    b) Newspaper: (50.2%, 55.8%) class: (49.4%, 58.6%).
    c) Students because 50% is in their interval.
23. a) (18.2%, 21.8%)
    b) We are 98% confident, based on the sample, that between 18.2% and 21.8% of English children are deficient in vitamin D.
    c) About 98% of all random samples will produce a confidence interval that contains the true proportion of children deficient in vitamin D.
24. a) (18.8%, 28.5%)
    b) We are 90% confident, based on this sample, that between 18.8% and 28.5% of women under 40 who are helped at this clinic will become pregnant and give birth.
    c) About 90% of all random samples will produce intervals that contain the true proportion.
    d) No, 25% is in the interval.
25. a) Wider. The sample size is probably about one-fourth of the sample size for all adults, so we'd expect the confidence interval to be about twice as wide.
    b) Smaller. The second poll used a slightly larger sample size.
26. a) The larger sample of private college students will produce a narrower confidence interval.
    b) Smaller; the overall sample size is bigger.
27. a) (15.5%, 26.3%)   b) 612
    c) Sample may not be random or representative. Deer that are legally hunted may not represent all sexes and ages.
28. a) 828 women
    b) This sample might be larger than 10% of the population and is far larger than the clinic's reported client load.
29. a) 141   b) 318   c) 564
30. a) 542, assuming $p = 0.50$.   b) 1504
    c) Would need 13,560 employers.
31. 1801   32. 1037   33. 384 total, using $p = 0.15$
34. 581   35. 90%
36. a) 50% is in the interval.   b) 98%

CHAPTER 20

1. a) $H_0: p = 0.30; H_A: p < 0.30$
   b) $H_0: p = 0.50; H_A: p \neq 0.50$
   c) $H_0: p = 0.20; H_A: p > 0.20$
2. a) $H_0: p = 0.40; H_A: p \neq 0.40$
   b) $H_0: p = 0.20; H_A: p < 0.20$
   c) $H_0: p = 0.60; H_A: p > 0.60$
3. Statement d is correct.
4. Statement d is correct.
5. No, we can say only that there is a 27% chance of seeing the observed effectiveness just from natural sampling variation. There is no *evidence* that the new formula is more effective, but we can't conclude that they are equally effective.
6. Yes. If there is no difference, there's only a 1.7% chance of seeing such a high sample proportion just from sampling variation.
7. a) No. There's a 25% chance of losing twice in a row. That's not unusual.
   b) 0.125   c) No, we expect that to happen 1 time in 8.
   d) Maybe 5? The chance of 5 losses in a row is only 1 in 32, which seems unusual.
8. a) 0.091
   b) Perhaps. If so, we'd get 3 vanillas in a row about 9% of the time.
   c) No. The probability of getting 4 vanillas in a row is only about 3%; that's unlikely to have happened by chance.
9. 1) Use p, not $\hat{p}$, in hypotheses.
   2) The question was about failing to meet the goal, so H_A should be $p < 0.96$.
   3) Did not check $0.04(200) = 8$. Since $nq < 10$, the Success/Failure Condition is violated. Didn't check 10% Condition.
   4) $188/200 = 0.94; SD(\hat{p}) = \sqrt{\dfrac{(0.96)(0.04)}{200}} = 0.014$
   5) z is incorrect; should be $z = \dfrac{0.94 - 0.96}{0.014} = -1.43$
   6) $P = P(z < -1.43) = 0.076$
   7) There is only weak evidence that the new instructions do not work.
10. 1) Use p in hypotheses, not $\hat{p}$.
    2) The question asks "*not accurate*," so H_A should be two sided: $p \neq 0.9$.
    3) The correct conditions are SRS, $750 < 10\%$ of county population; $(0.9)(750) \geq 10$, and $(0.10)(750) \geq 10$.
    4) $\hat{p} = 657/750 = 0.876; SD(\hat{p}) = \sqrt{\dfrac{(0.9)(0.1)}{750}} = 0.011$
    5) z is incorrect; should be $z = \dfrac{0.876 - 0.9}{0.011} = -2.18$
    6) $P = 2P(z < -2.18) = 0.029$
    7) There is only a 2.9% chance of observing a $\hat{p}$ as far from 0.90 by sampling variation, so we believe that the proportion of adults who drink milk here is different from the claimed 90%.
11. a) $H_0: p = 0.30; H_A: p > 0.30$
    b) Possibly an SRS; we don't know if the sample is less than 10% of his customers, but it could be viewed as less than 10% of all possible customers; $(0.3)(80) \geq 10$ and $(0.7)(80) \geq 10$. Wells are independent only if customers don't have farms on the same underground springs.
    c) $z = 0.73$; P-value $= 0.232$
    d) If his dowsing is no different from standard methods, there is more than a 23% chance of seeing results as good as those of the dowser's, or better, by natural sampling variation.
    e) These data provide no evidence that the dowser's chance of finding water is any better than normal drilling.
12. a) $H_0: p = 0.05; H_A: p > 0.05$
    b) SRS (not clear from information provided), $< 10\%$ of all children, $(0.05)(384) \geq 10$, and $(0.95)(384) \geq 10$.

c) $z = 6.28, P = 2 \times 10^{-10}$.
d) If the abnormality rate has not increased, the chance of observing at least 46 children with abnormalities in a sample of 384 is 2×10^{-10} (almost 0).
e) Reject H_0. These data show that the rate of abnormalities is now more than 5%.
f) We do not know that chemicals cause abnormalities, only that the rate is higher now than in the past.

13. a) $H_0: p_{2000} = 0.34; H_A: p_{2000} \neq 0.34$
    b) Students were randomly sampled and should be independent. 34% and 66% of 8302 are greater than 10. 8302 students is less than 10% of the entire student population of the United States.
    c) P $= 0.058$
    d) With such a small P-value, I reject H_0. There has been a statistically significant change in the proportion of students who have no absences.
    e) No. A difference this small, although statistically significant, is not meaningful. We might look at new data in a few years.
14. a) $H_0: p_{2000} = 0.31; H_A: p_{2000} \neq 0.31$
    b) Students are randomly sampled and should be independent; 31% and 69% of 8368 are both at least 10. 8368 students is less than 10% of the entire student population of the U.S.
    c) P $= 0.048$
    d) With a P-value so low, I reject H_0. There is a statistically significant difference in the proportion of college-educated mothers.
    e) A difference this small, although statistically significant, may not be meaningful.
15. a) $H_0: p = 0.05$ vs. $H_A: p < 0.05$
    b) We assume the whole mailing list has over 1,000,000 names. This is a random sample, and we expect 5000 successes and 95,000 failures.
    c) $z = -3.178$; P-value $= 0.00074$, so we reject H_0; there is strong evidence that the donation rate would be below 5%.
16. a) $H_0: p = 0.02$ vs. $H_A: p \neq 0.02$
    b) We assume the company has over 500,000 cardholders. This is a random sample, and we expect 1000 successes and 49,000 failures.
    c) $z = 5.878$; the P-value is less than 0.0001, so we reject H_0, concluding that there is strong evidence that the success rate will be more than 2%.
17. a) $H_0: p = 0.63, H_A: p > 0.63$
    b) The sample is representative. $240 < 10\%$ of all law school applicants. We expect $240(0.63) = 151.2$ to be admitted and $240(0.37) = 88.8$ not to be, both at least 10. $z = 1.58$; P-value $= 0.057$
    c) Although the evidence is weak, there is some indication that the program may be successful. Candidates should decide whether they can afford the time and expense.
18. a) $H_0: p = 0.46; H_A: p < 0.46$
    b) Assume the sample is representative of all applicants from this college. $180 < 10\%$ of all medical school applicants. We expect $180(0.46) = 82.8$ to be admitted and $180(0.54) = 97.2$ not to be, both at least 10. $z = -0.87$; P-value $= 0.19$
    c) The high P-value says this isn't an unusual result; it may be just year-to-year variation, as the president says.
19. $H_0: p = 0.20; H_A: p > 0.20$. SRS (not clear from information provided); 22 is more than 10% of the population of 150; $(0.20)(22) < 10$. Do not proceed with a test.
20. $H_0: p = 0.02; H_A: p > 0.02$. SRS; less than 10% of all washers and dryers made by the company; $(0.02)(60) < 10$. Do not proceed with a test.
21. $H_0: p = 0.03; p \neq 0.03$. $\hat{p} = 0.015$. One mother having twins will not affect another, so observations are independent; not an SRS; sample is less than 10% of all births. However, the mothers at this hospital may not be representative of all teenagers; $(0.03)(469) = 14.07 \geq 10; (0.97)(469) \geq 10$. $z = -1.91$; P-value $= 0.0556$. With a P-value this low, reject H_0. These data

show some evidence that the rate of twins born to teenage girls at this hospital is less than the national rate of 3%. It is not clear whether this can be generalized to all teenagers.

22. $H_0: p = 0.50; H_A: p > 0.50$. Results of one game should not affect another, so games are independent; data are all results for one season, which should be representative of all seasons; sample is less than 10% of all games; $(0.50)(240) \geq 10$; $(0.50)(240) \geq 10$. $z = 2.07$; P-value $= 0.02$. With a P-value this low, reject H_0. These data show strong evidence that the home team does have an advantage; they win more than 50% of games at home.

23. $H_0: p = 0.25; H_A: p > 0.25$. SRS; sample is less than 10% of all potential subscribers; $(0.25)(500) \geq 10$; $(0.75)(500) \geq 10$. $z = 1.24$; P-value $= 0.1076$. The P-value is high, so do not reject H_0. These data do not show that more than 25% of current readers would subscribe; the company should not go ahead with the WebZine on the basis of these data.

24. $H_0: p = 0.92; H_A: p < 0.92$. Seeds in a single packet may not be independent of each other. This is a cluster sample of all seeds in the packet. We may view this cluster as representative of all year-old seeds, in which case the sample is less than 10% of all seeds; $(0.92)(200) \geq 10$; $(0.08)(200) \geq 10$. $z = -3.39$; P-value $= 0.0004$ Because the P-value is very low, we reject H_0. There is strong evidence that these seeds have lost viability; their germination rate is less than 92%.

25. $H_0: p = 0.40; H_A: p < 0.40$. Data are for all executives in this company and may not be able to be generalized to all companies; $(0.40)(43) \geq 10$; $(0.60)(43) \geq 10$. $z = -1.31$; P-value $= 0.0955$. Because the P-value is high, we fail to reject H_0. These data do not show that the proportion of women executives is less than the 40% of women in the company in general.

26. $H_0: p = 0.19; H_A: p < 0.19$. $\hat{p} = 0.125$. $z = -1.41$; P-value $= 0.0793$. Because the P-value is high, we fail to reject H_0. These data do not show convincing evidence that Hispanics are represented in the jury pool at less than their 19% proportion in the population in general, but the data do indicate that the percentage is below what was expected.

27. $H_0: p = 0.103; H_A: p > 0.103$. $\hat{p} = 0.118; z = 2.06$; P-value $= 0.02$. Because the P-value is low, we reject H_0. These data provide evidence that the dropout rate has increased.

28. $H_0: p = 0.15; H_A: p > 0.15$. $\hat{p} = 0.25; z = 2.80$; P-value $= 0.0026$. The 95% confidence interval is $(0.165, 0.335)$. We must assume that the trees sampled are an SRS of the trees in the area and are less than 10% of all trees in the forest. The results are generalizable only to the Hopkins forest (or nearby if the forest is viewed as representative). Because the P-value is so low, we reject H_0. There is strong evidence that the proportion of trees damaged by acid rain in the Hopkins forest is higher than the 15% average for the Northeast.

29. $H_0: p = 0.90; H_A: p < 0.90$. $\hat{p} = 0.844; z = -2.05$; P-value $= 0.0201$. Because the P-value is so low, we reject H_0. There is strong evidence that the actual rate at which passengers with lost luggage are reunited with it within 24 hours is less than the 90% claimed by the airline.

30. $H_0: p = 0.40; H_A: p > 0.40$. $\hat{p} = 0.431; z = 1.29$; P-value $= 0.0977$ Because the P-value is high, we fail to reject H_0. These data do not show that at least 40% of the public recognizes the brand; I would not recommend they continue to advertise during Super Bowls on the basis of these data.

31. a) Yes; assuming this sample to be a typical group of people, $P = 0.0008$. This cancer rate is very unusual.
    b) No, this group of people may be atypical for reasons that have nothing to do with the radiation.

32. No; the P-value of 0.47 indicates that her school's results could be explained by random variation (sampling error).

CHAPTER 21

1. a) Two sided. Let p be the percentage of students who prefer Diet Pepsi. $H_0: p = 0.5$ vs. $H_A: p \neq 0.5$
   b) One sided. Let p be the percentage of teenagers who prefer the new formulation. $H_0: p = 0.5$ vs. $H_A: p > 0.5$
   c) One sided. Let p be the percentage of people who intend to vote for the override. $H_0: p = 2/3$ vs. $H_A: p > 2/3$.
   d) Two sided. Let p be the percentage of days that the market goes up. $H_0: p = 0.5$ vs. $H_A: p \neq 0.5$

2. a) Two sided. Let p be the percentage of students who prefer plastic. $H_0: p = 0.5$ vs. $H_A: p \neq 0.5$
   b) Two sided. Let p be the percentage of juniors applying for study abroad. $H_0: p = 0.1$ vs. $H_A: p \neq 0.1$
   c) One sided. Let p be the percentage of people who experience relief. $H_0: p = 0.22$ vs. $H_A: p > 0.22$
   d) One sided. Let p be the percentage of hard drives that pass the test the first time. $H_0: p = 0.6$ vs. $H_A: p > 0.6$

3. If there is no difference in effectiveness, the chance of seeing an observed difference this large or larger is 4.7% by natural sampling variation.

4. If harsher penalties and ad campaigns have made no difference in seat-belt use, there is a 17% chance of seeing an observed difference this large or larger by natural sampling variation.

5. $\alpha = 0.10$: Yes. The P-value is less than 0.05, so it's less than 0.10. But to reject H_0 at $\alpha = 0.01$, the P-value must be below 0.01, which isn't necessarily the case.

6. At $\alpha = 0.10$, they reach the same decision only if the P-value is > 0.10. We know only that it's > 0.05. But we know that it's > 0.01, so at $\alpha = 0.01$, they reach the same decision.

7. a) There is only a 1.1% chance of seeing a sample proportion as low as 89.4% vaccinated by natural sampling variation if 90% have really been vaccinated.
   b) We conclude that p is below 0.9, but a 95% confidence interval would suggest that the true proportion is between $(0.889, 0.899)$. Most likely, a decrease from 90% to 89.9% would not be considered important. On the other hand, with 1,000,000 children a year vaccinated, even 0.1% represents about 1000 kids—so this may very well be important.

8. a) There is only a 2.3% chance of seeing a sample proportion of 15.1% (or less) of students not attaining grade level by natural sampling variation if 15.9% is the true population value.
   b) Under old methods, 1352 students would not be expected to read at grade level. With the new program, 1284 did not. This is only a decrease of 68 students. It would depend on the costs of switching to the new program.

9. a) $(1.9\%, 4.1\%)$
   b) Because 5% is not in the interval, there is strong evidence that fewer than 5% of all men use work as their primary measure of success.
   c) $\alpha = 0.01$; it's a lower-tail test based on a 98% confidence interval.

10. a) $(0.498, 0.622)$; we are 95% confident that the true proportion of heads is between 49.8% and 62.2%.
    b) No; 0.50 is within the confidence interval, so it's a plausible value for the proportion of heads.
    c) $\alpha = 0.05$; it's a two-tailed test based on a 95% confidence interval.

11. a) $(0.274, 0.327)$
    b) Since 0.27 is not in the confidence interval, we reject the hypothesis that $p = 0.27$

12. a) $(19.0\%, 25.1\%)$
    b) Since the confidence interval extends well below 25%, we can't be sure that over 25% of men are stay-at-home dads. The company should not buy the ads.
    c) Yes, 25% is in the interval. It's a plausible value, but we can never prove that the null hypothesis is true.

13. a) The Success/Failure Condition is violated: only 5 pups had dysplasia.
    b) We are 95% confident that between 5% and 26% of puppies will show signs of hip dysplasia at the age of 6 months.
14. a) The Success/Failure Condition is violated: only 8 people weren't home team fans.
    b) We can be 95% confident that between 81% and 95% of the people attending the game are fans of the home team.
15. a) Type II error   b) Type I error
    c) By making it easier to get the loan, the bank has reduced the alpha level.
    d) The risk of a Type I error is decreased and the risk of a Type II error is increased.
16. a) Type II error   b) Type I error
    c) Lower alpha: It takes more evidence to classify the e-mail as spam.
    d) The risk of a Type I error is decreased and the risk of a Type II error is increased.
17. a) Power is the probability that the bank denies a loan that would not have been repaid.
    b) Raise the cutoff score.
    c) A larger number of trustworthy people would be denied credit, and the bank would miss the opportunity to collect interest on those loans.
18. a) Power is the probability the filter spots spam.
    b) Lower the cutoff score.
    c) More real messages would end up in the junk mailbox.
19. a) The null is that the level of home ownership remains the same. The alternative is that it rises.
    b) The city concludes that home ownership is on the rise, but in fact the tax breaks don't help.
    c) The city abandons the tax breaks, but they were helping.
    d) A Type I error causes the city to forego tax revenue, while a Type II error withdraws help from those who might have otherwise been able to buy a home.
    e) The power of the test is the city's ability to detect an actual increase in home ownership.
20. a) The null hypothesis is that a person is healthy (there is no parameter here). The alternative is that the person has Alzheimer's disease.
    b) A Type I error is a false positive—deciding a person has Alzheimer's disease when he or she doesn't.
    c) A Type II error is a false negative—failing to diagnose Alzheimer's disease when the person has it.
    d) A Type I error would require more testing, resulting in time and money lost. A Type II error would mean that the person did not receive the treatment he or she needed. A Type II error is much worse.
    e) The power is $1 - P(\text{Type II error}) = 1 - 0.08 = 0.92$
21. a) It is decided that the shop is not meeting standards when it is.
    b) The shop is certified as meeting standards when it is not.
    c) Type I   d) Type II
22. a) Deciding there has been an increase in defective items when there has not.
    b) Deciding the number of defectives is small when it has increased.
    c) Probably Type II, depending on the costs of shutting the line down. Generally, because of warranty costs and lost customer loyalty, defects that are caught in the factory are much cheaper to fix than defects found after items are sold.
    d) Type II
23. a) The probability of detecting a shop that is not meeting standards.
    b) 40 cars. Larger n.   c) 10%. More chance to reject H_0.
    d) A lot. Larger differences are easier to detect.
24. a) The probability of deciding there are too many defectives when this is true.

    b) Advantage: More power. Disadvantage: More work testing.
    c) Advantage: Smaller Type I error chance. Disadvantage: Less power.
    d) Increases because the effect size increases.
25. a) One-tailed. The company wouldn't be sued if "too many" minorities were hired.
    b) Deciding the company is discriminating when it is not.
    c) Deciding the company is not discriminating when it is.
    d) The probability of correctly detecting actual discrimination.
    e) Increases power.   f) Lower, since n is smaller.
26. a) One-tailed because we are interested only in whether the new design is more visible. If it's less visible, we don't care how much less visible it is.
    b) Deciding that new signs are more visible when they are not.
    c) Failing to decide that new signs are more visible when they are.
    d) Ability (probability) to detect that a more visible sign works.
    e) It will decrease power, because you'll need more evidence to reject H_0.
    f) It will decrease power because it will make the SD larger.
27. a) One-tailed. Software is supposed to decrease the dropout rate.
    b) $H_0: p = 0.13; H_A: p < 0.13$
    c) He buys the software when it doesn't help students.
    d) He doesn't buy the software when it does help students.
    e) The probability of correctly deciding the software is helpful.
28. a) $H_0: p = 0.20; H_A: p > 0.20$
    b) The company wants more "proof" the ad is effective before deciding it is.
    c) The probability of correctly deciding that more than 20% have heard the ad and recognize the product when it's true.
    d) 10%. More chance to reject H_0.
    e) Lower. Larger n has more power and smaller Type II error probability.
29. a) $z = -3.21, p = 0.0007$. The change is statistically significant. A 95% confidence interval is (2.3%, 8.5%). This is clearly lower than 13%. If the cost of the software justifies it, the professor should consider buying the software.
    b) The chance of observing 11 or fewer dropouts in a class of 203 is only 0.07% if the dropout rate is really 13%.
30. a) $z = 1.33, p = 0.0923$. The company should not renew the contract.
    b) There is a 9.23% chance of having 133 or more of 600 people in a random sample remember the ad if in fact no more than 20% of people do.
31. a) $H_A: p = 0.30$, where p is the probability of heads
    b) Reject the null hypothesis if the coin comes up tails—otherwise fail to reject.
    c) $P(\text{tails given the null hypothesis}) = 0.1 = \alpha$.
    d) $P(\text{tails given the alternative hypothesis}) = \text{power} = 0.70$
    e) Spin the coin more than once and base the decision on the sample proportion of heads.
32. a) The null hypothesis is that the drive is good $(p = 0.96)$. The alternative is that it is bad; $(p = 0.35)$, where p is the probability of passing the test.
    b) Reject the null hypothesis if the computer fails the test—otherwise fail to reject.
    c) $P(\text{fails test given the null hypothesis}) = 0.04 = \alpha$.
    d) $P(\text{fails test given the alternative hypothesis}) = \text{power} = 0.65$
33. a) 0.0464   b) Type I   c) 37.6%
    d) Increase the number of shots. Or keep the number of shots at 10, but increase alpha by declaring that 8, 9, or 10 will be deemed as having improved.
34. a) 0.0464   b) Type I   c) 0.6242
    d) Increase the number of pieces tested, or keep the number of pieces tested at 10, but increase the number of "acceptable broken pieces" from 1 to 2.

CHAPTER 22

1. It's very unlikely that samples would show an observed difference this large if in fact there is no real difference in the proportions of boys and girls who have used online social networks.

2. If in fact there is no difference in the proportion of the population who get science news from the Internet and from TV, then it's not unusual to observe a difference of 1% by sampling.

3. The ads may be working. If there had been no real change in name recognition, there'd be only about a 3% chance the percentage of voters who heard of this candidate would be at least this much higher in a different sample.

4. There's no evidence of a change. Even if there's been no real change in public opinion, there's a 37% chance we'd see this much difference—or more—from one sample to another.

5. The responses are not from two independent groups, but are from the same individuals.

6. The 790 parents are a subset of the 935 parents, so the two groups are not independent. That violates the independence assumption.

7. a) Stratified   b) 6% higher among males   c) 4%
   d)

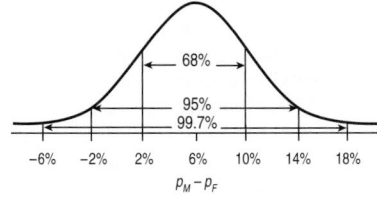

$p_M - p_F$

   e) Yes; a poll result showing little difference is only 1–2 standard deviations below the expected outcome.

8. a) Stratified   b) 2% higher among owners of Japanese cars
   c) 2.8%
   d)

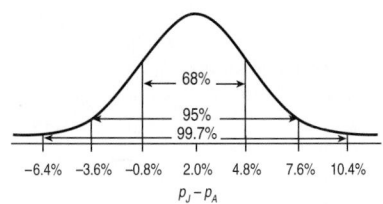

$p_J - p_A$

   e) Yes; a poll result showing greater satisfaction among owners of American cars (a negative difference) is less than one standard deviation below the expected outcome.

9. a) Yes. Random sample; less than 10% of the population; samples are independent; more than 10 successes and failures in each sample.
   b) (0.055, 0.140)
   c) We are 95% confident, based on these samples, that the proportion of American women age 65 and older who suffer from arthritis is between 5.5% and 14.0% more than the proportion of American men of the same age who suffer from arthritis.
   d) Yes; the entire interval lies above 0.

10. a) Yes. Random sample; less than 10% of the population; samples are independent; more than 10 successes and failures in each.
    b) (0.024, 0.040)
    c) We are 95% confident, based on these samples, that the proportion of 24-year-old American women who have graduated from high school is between 2.4% and 4.0% more than that of 24-year-old American men.
    d) Yes; the entire interval lies above 0.

11. a) 0.035   b) (0.356, 0.495)
    c) We are 95% confident, based on these data, that the proportion of pets with a malignant lymphoma in homes where herbicides are used is between 35.6% and 49.5% higher than the proportion of pets with lymphoma in homes where no pesticides are used.

12. a) 0.068   b) (0.116, 0.384)
    c) We are 95% confident, based on these data, that the proportion of patients who show improvement in carpal tunnel syndrome with surgery is between 11.6% and 38.4% more than the proportion who show improvement with wrist splints.

13. a) Yes, subjects were randomly divided into independent groups, and more than 10 successes and failures were observed in each group.
    b) (4.7%, 8.9%)
    c) Yes, we're 95% confident that the rate of infection is 5–9 percentage points lower. That's a meaningful reduction, considering the 20% infection rate among the unvaccinated kids.

14. a) Yes, subjects were randomly divided into independent groups, and more than 10 successes and failures were observed in each group.
    b) (−20%, 17%)
    c) No, because 0 is in the confidence interval. There's no evidence of any difference between Prozac and the placebo.

15. a) $H_0: p_V - p_{NV} = 0, H_A: p_V - p_{NV} < 0.$
    b) Because 0 is not in the confidence interval, reject the null. There's evidence that the vaccine reduces the rate of ear infections.
    c) 2.5%   d) Type I
    e) Babies would be given ineffective vaccinations.

16. a) $H_0: p_{prozac} - p_{placebo} = 0, H_A: p_{prozac} - p_{placebo} > 0$
    b) Because 0 is in the confidence interval, fail to reject the null. The data provide no evidence that Prozac is an effective treatment for anorexia.
    c) 2.5%   d) Type II
    e) We'd overlook a potentially helpful treatment.

17. a) Prospective study
    b) $H_0: p_1 - p_2 = 0; H_A: p_1 - p_2 \neq 0$ where p_1 is the proportion of students whose parents disapproved of smoking who became smokers and p_2 is the proportion of students whose parents are lenient about smoking who became smokers.
    c) Yes. We assume the students were randomly selected; they are less than 10% of the population; samples are independent; at least 10 successes and failures in each sample.
    d) $z = -1.17$, P-value $= 0.2422$. These samples do not show evidence that parental attitudes influence teens' decisions to smoke.
    e) If there is no difference in the proportions, there is about a 24% chance of seeing the observed difference or larger by natural sampling variation.
    f) Type II

18. a) Prospective study
    b) $H_0: p_1 - p_2 = 0; H_A: p_1 - p_2 < 0$, where p_1 is the proportion of people without depression who died within the 4 years and p_2 is the proportion of depressed people who died within the same period.
    c) Yes. We assume the patients were representative; they are less than 10% of the population; samples are independent; at least 10 successes and failures in each sample.
    d) $z = -2.22$, P-value $= 0.0131$. With a P-value this low, we reject H_0. This study indicates that the death rate for patients with heart disease who die within 4 years is less in nondepressed patients than in depressed patients.
    e) If there is no difference in the proportions, we will see an observed difference this large or larger only about 1.3% of the time by natural sampling variation.
    f) Type I

19. a) (−0.065, 0.221)
    b) We are 95% confident that the proportion of teens whose parents disapprove of smoking who will eventually smoke is between 22.1% less and 6.5% more than for teens with parents who are lenient about smoking.
    c) 95% of all random samples will produce intervals that contain the true difference.

20. a) (0.004, 0.209)
    b) We are 95% confident, based on these data, that the proportion of heart disease patients who die within 4 years is between 0.4% and 20.9% higher for depressed patients than for nondepressed patients.
    c) 95% of all random samples will produce intervals that contain the true value.
21. a) No; subjects weren't assigned to treatment groups. It's an observational study.
    b) $H_0: p_1 - p_2 = 0; H_A: p_1 - p_2 \neq 0. z = 3.56$, P-value = 0.0004. With a P-value this low, we reject H_0. There is a significant difference in the clinic's effectiveness. Younger mothers have a higher birth rate than older mothers. Note that the Success/Failure Condition is met based on the pooled estimate of p.
    c) We are 95% confident, based on these data, that the proportion of successful live births at the clinic is between 10.0% and 27.8% higher for mothers under 38 than in those 38 and older. However, the Success/Failure Condition is not met for the older women, since # Successes < 10. We should be cautious in trusting this confidence interval.
22. a) No; subjects weren't assigned to treatment groups. It's an observational study.
    b) $H_0: p_1 - p_2 = 0; H_A: p_1 - p_2 > 0. z = 2.56$, P-value = 0.005 (with technology, $z = 2.71$, P-value = 0.003). With a P-value this low we reject H_0. This study shows a significantly higher rate of low-weight babies born to mothers exposed to high levels of air pollution during pregnancy.
    c) We are 90% confident that the proportion of low-birth weight babies will be between 0.8% and 7.6% higher for mothers exposed to high levels of air pollution than those who were not.
23. a) $H_0: p_1 - p_2 = 0; H_A: p_1 - p_2 > 0. z = 1.18$, P-value = 0.118. With P-value this high, we fail to reject H_0. These data do not show evidence of a decrease in the voter support for the candidate.
    b) Type II
24. a) $H_0: p_1 - p_2 = 0; H_A: p_1 - p_2 > 0. z = 0.78$, P-value = 0.2166. With a P-value this high, we fail to reject H_0. There is no evidence, based on this information, that men are more likely than women to make online purchases of books.
    b) Type II   c) $(-0.04, 0.11)$
    d) With 95% confidence we estimate that the proportion of men who buy books online could be 4 percentage points lower than women, or up to 11 percentage points higher.
25. a) $H_0: p_1 - p_2 = 0; H_A: p_1 - p_2 \neq 0. z = -0.39$, P-value = 0.6951. With a P-value this high, we fail to reject H_0. There is no evidence of racial differences in the likelihood of multiple births, based on these data.
    b) Type II
26. a) $H_0: p_1 - p_2 = 0; H_A: p_1 - p_2 > 0. z = 0.24$, P-value = 0.4068. With a P-value this high, we fail to reject H_0. These data do not suggest that mammograms are effective in reducing breast cancer deaths.
    b) Type II
27. a) We are 95% confident, that between 67.0% and 83.0% of patients with joint pain will find medication A effective.
    b) We are 95% confident, that between 51.9% and 70.3% of patients with joint pain will find medication B effective.
    c) Yes, they overlap. This might indicate no difference in the effectiveness of the medications. (Not a proper test.)
    d) We are 95% confident that the proportion of patients with joint pain who will find medication A effective is between 1.7% and 26.1% higher than the proportion who will find medication B effective.
    e) No. There is a difference in the effectiveness of the medications.
    f) To estimate the variability in the difference of proportions, we must add variances. The two one-sample intervals do not. The two-sample method is the correct approach.

28. a) We are 95% confident, based on the data, that between 47.5% and 56.5% of male voters will vote for the candidate.
    b) We are 95% confident, based on the data, that between 40.7% and 49.3% of female voters will vote for the candidate.
    c) Yes, they overlap. There appears to be no discernible gender gap, but this is not the proper approach.
    d) We are 95% confident, that the proportion of men who will vote for the candidate is between 0.8% and 13.2% higher than the proportion of female voters who will vote for him.
    e) No. There is a gender gap.
    f) To estimate the variability in the difference of proportions, we must add variances. The two one-sample intervals do not. The two-sample method is the correct approach.
29. The conditions are satisfied to test $H_0: p_{young} = p_{old}$ against $H_A: p_{young} > p_{old}$. The one-sided P-value is 0.0619, so we may reject the null hypothesis. Although the evidence is not strong, *Time* may be justified in saying that younger men are more comfortable discussing personal problems.
30. The conditions are met to test the hypothesis that public and private colleges have the same retention rates. The two-sided P-value is 0.1996, too high to reject the null. We don't have evidence of an overall difference.
31. Yes. With a low P-value of 0.003, reject the null hypothesis of no difference. There's evidence of an increase in the proportion of parents checking the Web sites visited by their teens.
32. No. Based on a high P-value of 0.11, fail to reject the hypothesis that there's no age-based difference in computer gaming among teenage boys.

PART V REVIEW

1. H_0: There is no difference in cancer rates, $p_1 - p_2 = 0$. H_A: The cancer rate in those who use the herb is higher, $p_1 - p_2 > 0$.
2. a) Yes, $0.08 \times 325 = 26$, so we expect more than 10 successes and more than 10 failures.
   b) $\mu = 0.08, \sigma = 0.015$
   c)

   d) There is about a 68% chance of observing between 6.5% and 9.5% of colorblind males in a class of this size and a 95% chance of having between 5% and 11% colorblind males. Almost all classes of this size will have a percentage between 3.5% and 12.5% colorblind males.
3. a) 10.29
   b) Not really. The z-score is −1.11. Not any evidence to suggest that the proportion for Monday is low.
   c) Yes. The z-score is 2.26 with a P-value of 0.024 (two-sided).
   d) Some births are scheduled for the convenience of the doctor and/or the mother.
4. a) No. There is always natural sampling variation.
   b) 47.6% and 53.8% (using 1.96 SDs).
   c) We can't use a Normal model; we expect only 4 "successes".
   d) Less. Proportions farther from 0.5 have smaller standard deviations.
5. a) $H_0: p_1 = 0.40; H_A: p_1 < 0.40$
   b) Random sample; less than 10% of all California gas stations, $0.4(27) = 10.8, 0.6(27) = 16.2$. Assumptions and conditions are met.

c) $z = -1.49$, P-value $= 0.0677$

d) With a P-value this high, we fail to reject H_0. These data do not provide evidence that the proportion of leaking gas tanks is less than 40% (or that the new program is effective in decreasing the proportion).

e) Yes, Type II.

f) Increase α, increase the sample size.

g) Increasing α—increases power, lowers chance of Type II error, but increases chance of Type I error.
Increasing sample size—increases power, costs more time and money.

6. a) An experiment.

b) H_0: There is no difference, $p_1 - p_2 = 0$.
H_A: Patients with carbolic acid are more likely to live, $p_1 - p_2 > 0$.
$z = 2.91$, P-value $= 0.0018$; with a P-value so low, we reject H_0. These data show that carbolic acid is effective in increasing the chances of surviving an amputation.

c) We are not told whether the patients were randomized to the treatments. We are not told whether the experiment was double-blinded or even blinded at all. This could have biased the results toward a more favorable outcome.

7. a) The researcher believes that the true proportion of "A's" is within 10% of the estimated 54%, namely, between 44% and 64%.

b) Small sample   c) No, 63% is contained in the interval.

8. 0.647; $\bar{y} \approx N(3.5, 0.538)$. Rolls are independent.

9. a) Pew believes that the true proportion is within 3% of the 33% from the sample; that is, between 30% and 36%.

b) Larger, since it's a smaller sample.

c) We are 95% confident that the proportion of active traders who rely on the Internet for investment information is between 38.7% and 51.3%, based on this sample.

d) Larger, since it's a smaller sample.

10. a) 42.7% to 51.1%

b) Since the interval extends above 50%, it is possible that the death penalty does have majority support.

c) About 3382 people

11. a) Bimodal!

b) μ, the population mean. Sample size does not matter.

c) $\sigma/\sqrt{n}$; sample size does matter.

d) It becomes closer to a Normal model and narrower as the sample size increases.

12. a) Individuals were probably independent of one another, they are less than 10% of all African-American women, and there were clearly at least 10 successes and 10 failures. Is the sample a random sample from the population of African-American women?

b) 39.5% to 44.5%

c) We are 95% confident that between 39.5% and 44.5% of African-American women have a vitamin D deficiency.

d) 95% of all such random samples will produce intervals that contain the true proportion.

13. a) $\mu = 0.80$, $\sigma = 0.028$

b) Yes. $0.8(200) = 160$, $0.2(200) = 40$. Both ≥ 10.

c)

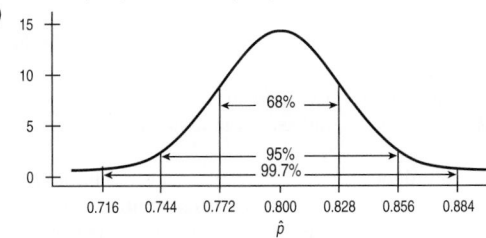

d) 0.039

14. a) $(-4.5\%, 6.5\%)$

b) No. The interval for the difference includes 0.

15. H_0: There is no difference, $p_1 - p_2 = 0$. H_A: Early births have increased, $p_1 - p_2 < 0$. $z = -0.729$, P-value $= 0.2329$. Because the P-value is so high, we do not reject H_0. These data do not show an increase in the incidence of early birth of twins.

16. a) We are 95% confident that the increase in the proportion of women who may develop side effects from magnesium sulfide is between 18.1% and 20.8%.

b) H_0: There is no difference, $p_1 - p_2 = 0$. H_A: Treatment prevents eclampsia, $p_1 - p_2 < 0$. $z = -4.84$, P-value $= 6.4 \times 10^{-7}$. Because the P-value is so low, we reject H_0. This study shows evidence that treatment with magnesium sulfide is effective in preventing eclampsia.

17. a) H_0: There is no difference, $p_1 - p_2 \geq 0$. H_A: Treatment prevents deaths from eclampsia, $p_1 - p_2 < 0$.

b) Samples are random and independent; less than 10% of all pregnancies (or eclampsia cases); more than 10 successes and failures in each group.

c) 0.8008

d) There is insufficient evidence to conclude that magnesium sulfide is effective in preventing eclampsia deaths.

e) Type II   f) Increase the sample size, increase α.

g) Increasing sample size: decreases variation in the sampling distribution, is costly. Increasing α: Increases likelihood of rejecting H_0, increases chance of Type I error.

18. a) 33.7%   b) 0.073

c)

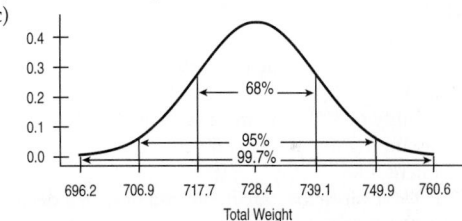

19. a) It is not clear what the pollster asked. Otherwise they did fine.

b) Stratified sampling.   c) 4%

d) 95%   e) Smaller sample size.

f) Wording and order of questions (response bias).

20. About 1800

21. a) H_0: There is no difference, $p = 0.143$. H_A: The fatal accident rate is lower in girls, $p < 0.143$. $z = -1.67$, P-value $= 0.0479$. Because the P-value is low, we reject H_0. These data give some evidence that the fatal accident rate is lower for girls than for teens in general.

b) If the proportion is really 14.3%, we will see the observed proportion (11.3%) or lower 4.8% of the time by sampling variation.

22. a) H_0: There is no difference, $p_1 - p_2 = 0$. H_A: There is a difference, $p_1 - p_2 \neq 0$.

b) There is a difference in the proportion of students with perfect pitch; more Asians are likely to have it.

c) If there is no difference, the observed 25% difference will be seen by sampling variation less than about 1 time in 10,000.

d) The data do not "prove" anything about genetic differences causing differences in perfect pitch—merely that Asian students are more likely to have it.

23. a) One would expect many small fish, with a few large ones.

b) We don't know the exact distribution, but we know it's not Normal.

c) Probably not. With a skewed distribution, a sample size of five is not a large enough sample to say the sampling model for the mean is approximately Normal.

d) 0.961

24. a) Random sample; less than 10% of all high-school students; many more than 10 successes and 10 failures. 90% confidence interval for proportion who cheat: 72.9% to 75.1%.
   b) Based on this information, we are 90% confident that the proportion of all high-school students who have cheated at least once on a test is between 72.9% and 75.1%.
   c) 90% of all such random samples will produce confidence intervals that contain the true proportion of students who cheat.
   d) Wider. More confidence means a larger margin of error.
25. a) Yes. $0.8(60) = 48, 0.2(60) = 12$. Both are ≥ 10.
   b) 0.834
   c) Higher. Bigger sample means smaller standard deviation for $\hat{p}$.
   d) Answers will vary. For $n = 500$, the probability is 0.997.
26. a) H_0: Progress is on track, $p = 0.20$. H_A: Progress is off track, $p > 0.20$.
   b) Random samples; less than 10% of all high-school students; many more than 10 failures and successes.
   c) 0.0008
   d) If the proportion were 20%, the probability of seeing a value as high as (or higher than) 23% in a random sample this large is about 0.0008.
   e) By 2006, the rate of cigarette smoking in high-school students was higher than the target of 20%.
   f) Type I
27. a) 54.4 to 62.5%
   b) Based on this study, with 95% confidence the proportion of Crohn's disease patients who will respond favorable to infliximab is between 54.4% and 62.5%.
   c) 95% of all such random samples will produce confidence intervals that contain the true proportion of patients who respond favorably.
28. No. The probability of 30% or more is very small: 6.1×10^{-5}.
29. At least 423, assuming that p is near 50%.
30. a) An experiment
   b) A one-sided test, since they are interested only in a decrease in percentage needing repairs.
   c) Deciding the additive reduces the number of repairs needed when there really is no difference.
   d) Deciding the additive makes no difference when it really does reduce the number of repairs needed.
   e) Type II
   f) Given that the two groups received roughly the same use and care, yes. They can't necessarily claim it will work for all cars, only cars similar to their fleet.
31. a) Random sample (?); certainly less than 10% of all preemies and normal babies; more than 10 failures and successes in each group. 1.7% to 16.3% greater for normal-birth weight children.
   b) Since 0 is not in the interval, there is evidence that preemies have a lower high school graduation rate than children of normal birth weight.
   c) Type I, since we rejected the null hypothesis.
32. a) We are 95% confident that between 11.6% and 16.4% of Texas children wear helmets when biking, roller skating, or skateboarding, based on these data.
   b) The data might not be a random sample.
   c) About 408, using the previous 14% as $\hat{p}$.
33. a) H_0: The computer is undamaged. H_A: The computer is damaged.
   b) 20% of good PCs will be classified as damaged (bad), while all damaged PCs will be detected (good).
   c) 3 or more.   d) 20%
   e) By switching to two or more as the rejection criterion, 7% of the good PCs will be misclassified, but only 10% of the bad ones will, increasing the power from 20% to 90%.
34. a) Increase   b) Decrease

35. The null hypothesis is that Bush's disapproval proportion is 66%—the Nixon benchmark. The one-tailed test has a z-value of -2.00, so the P-value is 0.0228. It looks like Bush's May 2007 ratings were better than the Nixon benchmark low.
36. The null hypothesis is that the percentage of students who attain a GPA of at least 3.5 remained 20% in 2000. The sample proportion of 25% is more than four standard deviations above the hypothesized rate, strong evidence the results are not due to chance. This may be an indication of grade inflation.
37. a) The company is interested only in confirming that the athlete is well known.
   b) Type I: the company concludes that the athlete is well known, but that's not true. It offers an endorsement contract to someone who lacks name recognition. Type II: the company overlooks a well-known athlete, missing the opportunity to sign a potentially effective spokesperson.
   c) Type I would be more likely, Type II less likely.
38. a) Although 27% of the people polled could identify her, her name recognition rate in the whole population could be less than the required 25%.
   b) Type II.   c) Higher.
39. I am 95% confident that the proportion of U.S. adults who favor nuclear energy is between 7 and 19 percentage points higher than the proportion who would accept a nuclear plant near their area.
40. We're 95% confident that between 46% and 60% of anorexia patients will drop out of treatment programs. However, this wasn't a random sample of all patients; they were assigned a treatment rather than choosing one on their own, and they may have had different experiences if they were not part of an experiment.

CHAPTER 23

1. a) 1.74   b) 2.37   c) 0.0524   d) 0.0889
2. a) 2.36   b) 2.62   c) 0.9829   d) 0.0381
3. Shape becomes closer to Normal; center does not change; spread becomes narrower.
4. The critical value becomes smaller, approaching 1.96.
5. a) The confidence interval is for the population mean, not the individual cows in the study.
   b) The confidence interval is not for individual cows.
   c) We *know* the average gain in this study was 56 pounds!
   d) The average weight gain of all cows does not vary. It's what we're trying to estimate.
   e) No. There is not a 95% chance for another sample to have an average weight gain between 45 and 67 pounds. There is a 95% chance that another sample will have its average weight gain within two standard errors of the true mean.
6. a) Nine out of 10 intervals will contain the true mean salary; different samples will produce different intervals.
   b) This is correct.
   c) The interval is for the population mean, not individual teachers.
   d) The interval is for the mean, not individual teachers.
   e) The interval addresses only Nevada teachers, not the entire country.
7. a) No. A confidence interval is not about individuals in the population.
   b) No. It's not about individuals in the sample, either.
   c) No. We know the mean cost for students in the sample was $1196.
   d) No. A confidence interval is not about other sample means.
   e) Yes. A confidence interval estimates a population parameter.
8. a) No. The confidence interval is not about the years in the sample.
   b) No. The confidence interval does not predict what will happen in any one year.

c) No. The confidence interval was not based on a sample of days.
d) Yes. The confidence interval estimates the true mean.
e) No. We know that the mean annual snowfall for the last century was 23".

 9. a) Based on this sample, we can say, with 95% confidence, that the mean pulse rate of adults is between 70.9 and 74.5 beats per minute.
b) 1.8 beats per minute
c) Larger

10. a) Based on this sample, we can say, with 95% confidence, that the mean age at which babies begin to crawl is between 29.2 and 31.8 weeks.
b) 1.3   c) Smaller

11. The assumptions and conditions for a t-interval are not met. The distribution is highly skewed to the right and there is a large outlier.

12. The assumptions and conditions for a t-interval are not met. There is one cardholder who spent over $3,000,000 on his card. This made the standard deviation, and hence the SE, huge, resulting in a t-interval too wide to be useful.

13. a) Yes. Randomly selected group; less than 10% of the population; the histogram is not unimodal and symmetric, but it is not highly skewed and there are no outliers, so with a sample size of 52, the CLT says $\bar{y}$ is approximately Normal.
b) (98.06, 98.51) degrees F
c) We are 98% confident, based on the data, that the average body temperature for an adult is between 98.06°F and 98.51°F.
d) 98% of all such random samples will produce intervals containing the true mean temperature.
e) These data suggest that the true normal temperature is somewhat less than 98.6°F.

14. a) The data are a representative sample of less than 10% of all days; $n = 44$ is a large sample.
b) ($122.20, $129.80)
c) We are 90% confident the mean daily income for the parking garage is between $122.20 and $129.80, based on this sample.
d) 90% of all such samples will produce intervals that contain the true mean daily fee.
e) No. The interval is below $130.

15. a) Narrower. A smaller margin of error, so less confident.
b) Advantage: more chance of including the true value. Disadvantage: wider interval.
c) Narrower; due to the larger sample, the SE will be smaller.
d) About 252

16. a) More chance the interval contains the true mean.
b) The interval would be wider.
c) More data would result in a narrower interval at same confidence level.
d) About 99

17. a) (709.90, 802.54)
b) With 95% confidence, based on these data, the speed of light is between 299,709.9 and 299,802.5 km/sec.
c) Normal model for the distribution, independent measurements. These seem reasonable here, but it would be nice to see if the Nearly Normal Condition held for the data.

18. a) 7.9 km/sec
b) Should be narrower. Different mean will change the center of the interval. Larger sample size and smaller standard deviation will reduce the SE. Since the t-critical value is also smaller, the margin of error will be smaller.
c) New interval: (836.72, 868.08) km/sec. This experiment is worse than the first, which included 710.5. This interval is considerably above that value. There may have been a bias in this experiment's measurements.

19. a) Given no time trend, the monthly on-time departure rates should be independent. Though not a random sample, these

months should be representative, and they're fewer than 10% of all months. The histogram looks unimodal, but slightly left-skewed; not a concern with this large sample.
b) $80.57 < \mu$(OT Departure%) < 81.80
c) We can be 90% confident that the interval from 80.57% to 81.80% holds the true mean monthly percentage of on-time flight departures.

20. a) Given no time trend, the monthly late-arrival rates should be independent. Though not a random sample, these months should be representative, and they're fewer than 10% of all months. The histogram looks unimodal and symmetric.
b) $19.19 < \mu$(Late Arrival%) < 20.97
c) We can be 99% confident that the interval from 19.19% to 21.0% holds the true mean monthly percentage of late flight arrivals.

21. The 95% confidence interval lies entirely above the 0.08 ppm limit, evidence that mirex contamination is too high and consistent with rejecting the null. We used an upper-tail test, so the P-value should therefore be smaller than $\frac{1}{2}(1 - 0.95) = 0.025$, and it was.

22. The 90% confidence interval contains the 325 mg limit, so they can't assert that the mean sodium content is less than 325—consistent with not rejecting the null. They used a lower-tail test, so we'd expect the P-value to be more than $\frac{1}{2}(1 - 0.90) = 0.05$, which it was.

23. If in fact the mean cholesterol of pizza eaters does not indicate a health risk, then only 7 of every 100 samples would have mean cholesterol levels as high (or higher) as observed in this sample.

24. If in fact this ball meets the velocity standard, then 34% of all samples of this size would have mean speeds at least as high as was recorded in this sample.

25. a) Upper-tail. We want to show it will hold 500 pounds (or more) easily.
b) They will decide the stands are safe when they're not.
c) They will decide the stands are unsafe when they are in fact safe.

26. a) Two-sided. If they're too big, they won't fit through the vein. If they're too small, they probably won't work well.
b) The catheters are rejected when in fact the diameters are fine, and the manufacturing process is needlessly stopped.
c) Catheters that do not meet specifications are allowed to be produced and sold.

27. a) Decrease α. This means a smaller chance of declaring the stands safe if they are not.
b) The probability of correctly detecting that the stands are capable of holding more than 500 pounds.
c) Decrease the standard deviation—probably costly. Increase the sample size—takes more time for testing and is costly. Increase α—more Type I errors. Increase the "design load" to be well above 500 pounds—again, costly.

28. a) Increase
b) The probability of correctly detecting deviations from 2 mm in diameter.
c) Increase   d) Increase the sample size or increase α.

29. a) $H_0: \mu = 23.3; H_A: \mu > 23.3$
b) We have a random sample of the population. Population may not be normally distributed, as it would be easier to have a few much older men at their first marriage than some very young men. However, with a sample size of 40, $\bar{y}$ should be approximately Normal. We should check the histogram for severity of skewness and possible outliers.
c) $(\bar{y} - 23.3)/(s/\sqrt{40}) \sim t_{39}$ d) 0.1447
e) If the average age at first marriage is still 23.3 years, there is a 14.5% chance of getting a sample mean of 24.2 years or older simply from natural sampling variation.
f) We lack evidence that the average age at first marriage has increased from the mean of 23.3 years.

30. a) $H_0: \mu = 26; H_A: \mu < 26$

b) We have a representative sample, fewer than 10% of all trips, and a large enough sample that skewness should not be a problem.

c) $(\bar{y} - 26)/(4.83/\sqrt{50}) \sim t_{49}$ d) 0.0789

e) If the average fuel economy is 26 mpg, the chance of obtaining a sample mean of 25.02 or less by natural sampling variation is 8%.

f) Since $0.05 < P < 0.10$, there is some evidence that the company may not be achieving the fuel economy goal.

31. a) Probably a representative sample; the Nearly Normal Condition seems reasonable. (Show a Normal probability plot or histogram.) The histogram is nearly uniform, with no outliers or skewness.

b) $\bar{y} = 28.78, s = 0.40$ c) (28.36, 29.21) grams

d) Based on this sample, we are 95% confident the average weight of the content of Ruffles bags is between 28.36 and 29.21 grams.

e) The company is erring on the safe side, as it appears that, on average, it is putting in slightly more chips than stated.

32. a) Probably a representative sample; the Nearly Normal Condition seems reasonable. (Show a Normal probability plot or histogram.) The data are fairly symmetric with no apparent outliers.

b) $\bar{y} = 28.98, s = 0.36$ c) (28.61, 29.36) grams

d) With 95% confidence, the average weight of the content of Doritos bags is between 28.61 and 29.36 grams.

e) The company is erring on the safe side, as it appears that, on average, it is putting in slightly more chips than stated.

33. a) Type I; he mistakenly rejected the null hypothesis that $p = 0.10$ (or worse).

b) Yes. These are a random sample of bags and the Nearly Normal Condition is met (Show a Normal probability plot or histogram.); $t = -2.51$ with 7 df for a one-sided P-value of 0.0203.

34. a) He would have made a Type II error.

b) No. We'll consider these races a representative sample, and the Nearly Normal Condition is met (Show your plot.); $t = -0.7646$ with 7 df for a one-sided P-value of 0.2347.

35. a) Random sample; the Nearly Normal Condition seems reasonable from a Normal probability plot. The histogram is roughly unimodal and symmetric with no outliers. (Show plot.)

b) (1187.9, 1288.4) chips

c) Based on this sample, the mean number of chips in an 18-ounce bag is between 1187.9 and 1288.4, with 95% confidence. The *mean* number of chips is clearly greater than 1000. However, if the claim is about individual bags, then it's not necessarily true. If the mean is 1188 and the SD deviation is near 94, then 2.5% of the bags will have fewer than 1000 chips, using the Normal model. If in fact the mean is 1288, the proportion below 1000 will be less than 0.1%, but the claim is still false.

36. a) Random sample. Nearly Normal Condition is reasonable by examining a Normal probability plot. The histogram is roughly unimodal (although somewhat uniform) and symmetric with no outliers. (Show your plot.)

b) (132.0, 183.7) calories

c) The mean number of calories in a serving of vanilla yogurt is between 132 and 183.7, with 95% confidence. We conclude that the diet guide's claim of 120 calories is too low.

37. a) The Normal probability plot is relatively straight, with one outlier at 93.8 sec. Without the outlier, the conditions seem to be met. The histogram is roughly unimodal and symmetric with no other outliers. (Show your plot.)

b) $t = -2.63$, P-value = 0.0160. With the outlier included, we might conclude that the mean completion time for the maze is not 60 seconds; in fact, it is less.

c) $t = -4.46$, P-value = 0.0003. Because the P-value is so small, we reject H_0. Without the outlier, we see strong evidence that

the average completion time for the maze is less than 60 seconds. The outlier here did not change the conclusion.

d) The maze does not meet the "one-minute average" requirement. Both tests rejected a null hypothesis of a mean of 60 seconds.

38. The data value of 102 feet is an outlier. When this is removed, the Normal probability plot is relatively straight. The Nearly Normal Condition seems satisfied. With the outlier removed, the histogram is roughly unimodal and symmetric with no other outliers. $H_0: \mu = 125; H_A: \mu > 125$ feet. With the outlier eliminated, $\bar{y} = 128.89, t = 3.29$, P-value = 0.01. With a P-value this low, we reject H_0. There is strong evidence to suggest that the tires will not bring the car to a complete stop within 125 feet. On the basis of these data, the company should not adopt the new tread pattern. Only 2 out of the 10 data values were less than the desired 125 feet, and 1 of these was an outlier.

39. a) $287.3 < \mu$(Drive Distance) < 289.9

b) These data are not a random sample of golfers. The top professionals are (unfortunately) not representative and were not selected at random. We might consider the 2006 data to represent the population of all professional golfers, past, present, and future.

c) The data are means for each golfer, so they are less variable than if we looked at all the separate drives.

40. a) The timeplot shows no pattern, so it seems that the measurements are independent. Although this is not a random sample, an entire year is measured, so it is likely that we have representative values. We certainly have fewer than 10% of all possible wind readings. Both the histogram and Normal probability plot suggest near normality.

b) Testing $H_0: \mu = 8$ mph vs. $H_A: \mu > 8$ mph with 1113 df gives $t = 0.1663$ for a P-value of about 0.44. Even though the observed mean wind speed is over 8 mph, I can't be confident that the true annual mean wind speed exceeds 8 mph. I would not recommend building a turbine at this site.

CHAPTER 24

1. Yes. The high P-value means that we lack evidence of a difference, so 0 is a possible value for $\mu_{Meat} - \mu_{Beef}$.

2. Yes. The high P-value means that we lack evidence of a difference, so 0 is a possible value for $\mu_{Meat} - \mu_{Beef}$.

3. a) Plausible values of $\mu_{Meat} - \mu_{Beef}$ are all negative, so the mean fat content is probably higher for beef hot dogs.

b) The difference is significant. c) 10%

4. a) Plausible values of $\mu_{Top} - \mu_{Front}$ are all negative, so the mean cycle time is probably higher for front-loading machines.

b) The difference is significant. c) 2%

5. a) False. The confidence interval is about means, not about individual hot dogs.

b) False. The confidence interval is about means, not about individual hot dogs.

c) True.

d) False. CI's based on other samples will also try to estimate the true difference in population means; there's no reason to expect other samples to conform to this result.

e) True.

6. a) False. The confidence interval is about means, not about individual machines.

b) False. The confidence interval is about means, not about individual machines.

c) False. CI's based on other samples will also try to estimate the true difference in population means; there's no reason to expect other samples to conform to this result.

d) True. e) True.

7. a) 2.927 b) Larger

c) Based on this sample, we are 95% confident that students who learn Math using the CPMP method will score, on average, between 5.57 and 11.43 points better on a test solving applied Algebra problems with a calculator than students who learn by traditional methods.

   d) Yes; 0 is not in the interval.

8. a) Based on this sample, we are 90% confident that people who receive either no or only verbal information about the image in a stereogram will take between 0.55 and 5.47 seconds longer, on average, to report they saw the image than people who receive both verbal and visual information.

   b) Yes, since 0 is not in the interval.   c) 2.46

   d) 90% of all random samples will produce intervals that contain the true value of the mean difference between the times for these two groups.

   e) Wider. More confidence means less precision.

   f) Possibly. The wider interval may contain 0.

9. a) $H_0: \mu_C - \mu_T = 0$ vs. $H_A: \mu_C - \mu_T \neq 0$

   b) Yes. Groups are independent, though we don't know if students were randomly assigned to the programs. Sample sizes are large, so CLT applies.

   c) If the means for the two programs are really equal, there is less than a 1 in 10,000 chance of seeing a difference as large as or larger than the observed difference just from natural sampling variation.

   d) On average, students who learn with the CPMP method do significantly worse on Algebra tests that do not allow them to use calculators than students who learn by traditional methods.

10. $H_0: \mu_C - \mu_T = 0$; $H_A: \mu_C - \mu_T \neq 0$. $t = 1.406$, df = 590.05, P-value = 0.1602. Because of the large P-value, we fail to reject H_0. Based on these samples, there is no evidence of a difference in mean scores on a test of word problems, whether students learned with CPMP or traditional methods.

11. a) (1.36, 4.64)

   b) No; 5 minutes is beyond the high end of the interval.

12. a) The mean rates are roughly equal, but females are more variable.

   b) Yes; boxplots look symmetric.

   c) $(-3.025, 3.275)$ d) Yes, 0 is in the interval.

13.

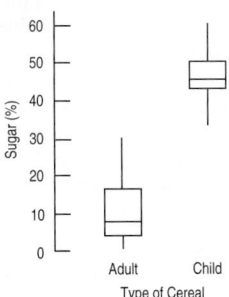

Random sample—questionable, but probably representative, independent samples, less than 10% of all cereals; boxplot shows no outliers—not exactly symmetric, but these are reasonable sample sizes. Based on these samples, with 95% confidence, children's cereals average between 32.49% and 40.80% more sugar content than adult's cereals.

14. a) Random sample (we assume), independent samples, histograms look unimodal and symmetric.

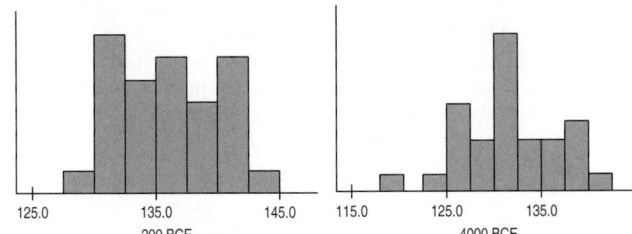

   b) (1.88, 6.66) mm

   c) These data provide evidence that mean maximum skull breadth in Egyptians in 200 B.C.E. was between 1.88 and 6.66 mm larger than that in 4000 B.C.E.

15. $H_0: \mu_N - \mu_C = 0$ vs. $H_A: \mu_N - \mu_C > 0$; $t = 2.207$; P-value = 0.0168; df = 33.4. Because of the small P-value, we reject H_0. These data do suggest that new activities are better. The mean reading comprehension score for the group with new activities is significantly (at $\alpha = 0.05$) higher than the mean score for the control group.

16. a) $H_0: \mu_L - \mu_S = 0$ vs. $H_A: \mu_L - \mu_S \neq 0$

   b) Don't know if the streams were a random sample or whether they are less than 10% of all Adirondack streams. Boxplots show outliers and shale may be skewed (median is equal to Q1 or Q3), but samples are large.

   c) Based on these data, it appears that water flowing over limestone is less acidic, on average, than water flowing over shale.

17. a)

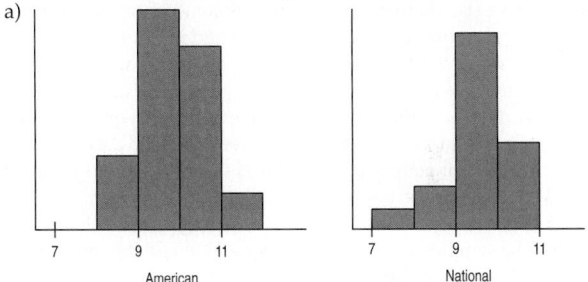

Both are unimodal and reasonably symmetric.

   b) Based on these data, the average number of runs in an American League stadium is between 9.36 and 10.23, with 95% confidence.

   c) No. The boxplot indicates it isn't an outlier.

   d) We want to work directly with the average difference. The two separate confidence intervals do not answer questions about the difference. The difference has a different standard deviation, found by adding variances.

18. a) Males: (18.67, 20.11) pegs; females: (16.95, 18.87) pegs.

   b) It suggests that there is no evidence of a difference, but the method is incorrect.

   c) (0.29, 2.67) pegs

   d) There is evidence of a difference: Males do better, on average, by between 0.29 and 2.67 pegs (with 95% confidence).

   e) Two-sample inference

   f) If we examine the gap between the means using two separate confidence intervals, we are essentially adding the margins of error, which are based on standard deviations. To be correct, we must instead add variances. That's exactly what happens when we create the confidence interval for the difference in means—the proper approach.

19. a) $(-0.18, 0.89)$

   b) Based on these data, with 95% confidence, American League stadiums average between 0.18 fewer runs and 0.89 more runs per game than National League stadiums.

   c) No; 0 is in the interval.

20. a) $H_0: \mu_N - \mu_S = 0$ vs. $H_A: \mu_N - \mu_S \neq 0$. $t = 6.47$, df = 53.49, P-value = 3.2×10^{-8}. Because the P-value is low, we reject H_0. On the basis of these data, there is clear evidence that mortality rates are different. The mean rate in the north is significantly higher.

   b) It will raise $\bar{y}$ for the north, but from looking at the boxplots and the fact that the mean and median are nearly the same, it probably will not change the conclusion of the test.

21. These are not two independent samples. These are before and after scores for the same individuals.

22. These are before and after scores for the same individuals, not independent samples, so we cannot proceed.

23. a) These data do not provide evidence of a difference in ad recall between shows with sexual content and violent content.

b) $H_0: \mu_S - \mu_N = 0$ vs. $H_A: \mu_S - \mu_N \neq 0$. $t = -6.08$, df = 213.99, P-value = 5.5×10^{-9}. Because the P-value is low, we reject H_0. These data suggest that ad recall between shows with sexual and neutral content is different; those who saw shows with neutral content had higher average recall.

24. a) With 95% confidence, those who watch shows with violent content remembered an average of between 0.6 and 1.6 fewer brand names than those who saw shows with neutral content.

b) If they want viewers to remember their brand, advise that they consider advertising in shows with neutral content in preference to those with violent content. Of course, costs of the ad should be considered.

25. a) $H_0: \mu_V - \mu_N = 0$ vs. $H_A: \mu_V - \mu_N \neq 0$. $t = -7.21$, df = 201.96, P-value = 1.1×10^{-11}. Because of the very small P-value, we reject H_0. There is a significant difference in mean ad recall between shows with violent content and neutral content; viewers of shows with neutral content remember more brand names, on average.

b) With 95% confidence, the average number of brand names remembered 24 hours later is between 1.45 and 2.41 higher for viewers of neutral content shows than for viewers of sexual content shows, based on these data.

26. a) He would conclude that recall is higher 24 hours later.

b) The samples are not independent. They are the same people asked at two different periods.

c) The first inquiry might influence people.

d) Use two different groups for each type of show. Interview one group immediately, the other 24 hours later.

27. $H_0: \mu_{big} - \mu_{small} = 0$ vs. $H_A: \mu_{big} - \mu_{small} \neq 0$; bowl size was assigned randomly; amount scooped by individuals and by the two groups should be independent. With 34.3 df, $t = 2.104$ and P-value = 0.0428. The low P-value leads us to reject the null hypothesis. There is evidence of a difference in the average amount of ice cream that people scoop when given a bigger bowl.

28. $H_0: \mu_{tumbler} - \mu_{highball} = 0$ vs. $H_A: \mu_{tumbler} - \mu_{highball} \neq 0$; glass size was assigned randomly; amount poured by individuals and by the two groups should be independent. With 194 df, $t = 7.71$; P-value < 0.001. The low P-value leads us to reject the null hypothesis and conclude that there is a difference in the amount of liquid that people pour, on average, when given a small wide tumbler as opposed to a tall narrow highball glass.

29. a) The 95% confidence interval for the difference is (0.61, 5.39). 0 is not in the interval, so scores in 1996 were significantly higher. (Or the t, with more than 7500 df, is 2.459 for a P-value of 0.0070.)

b) Since both samples were very large, there shouldn't be a difference in how certain you are, assuming conditions are met.

30. a) The observed differences are too large to attribute to chance or natural sampling variation.

b) Type I   c) No. There may be many other factors.

31. Independent Groups Assumption: The runners are different women, so the groups are independent. The Randomization Condition is satisfied since the runners are selected at random for these heats.

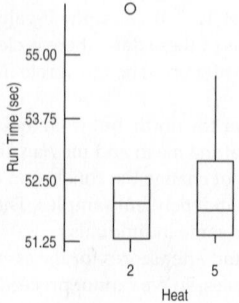

Nearly Normal Condition: The boxplots show an outlier, but we will proceed and then redo the analysis with the outlier deleted. When we include the outlier, $t = 0.035$ with a two-sided P-value of 0.97. With the outlier deleted, $t = -1.14$, with P = 0.2837. Either P-value is so large that we fail to reject the null hypothesis of equal means and conclude that there is no evidence of a difference in the mean times for runners in unseeded heats.

32. Independent Groups Assumption: The swimmers in the two heats are different women, so the groups are independent. The histograms indicate the Nearly Normal Condition is satisfied.

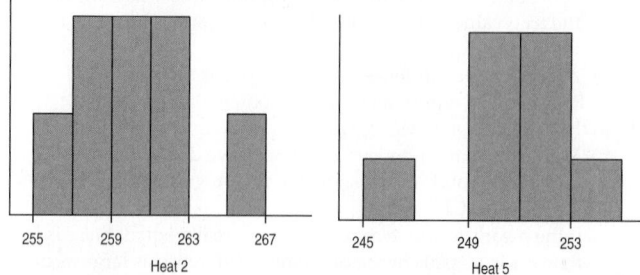

The Randomization Condition is not strictly satisfied since the swimmers are not selected at random, but if we can consider these heats to be representative of seeded heats, we may be able to generalize the results. With $t = 7.19$ and a P-value less than 0.001, we reject the null hypothesis of equal means. There is strong evidence to suggest that seeded swimming heats have different mean times. It appears that on average times were faster in heat 5.

33. With $t = -4.57$ and a very low P-value of 0.0013, we reject the null hypothesis of equal mean velocities. There is strong evidence that golf balls hit off Stinger tees will have a higher mean initial velocity.

34. With $t = -9.70$ and a P-value less than 0.0001, we reject the null hypothesis of equal mean distances. There is strong evidence that the mean distance traveled will be greater for golf balls hit off Stinger tees. We can be 95% confident that they'll travel an average of between 10.7 and 17.0 yards farther.

35. a) We can be 95% confident that the interval 74.8 ± 178.05 minutes includes the true difference in mean crossing times between men and women. Because the interval includes zero, we cannot be confident that there is any difference at all.

b) Independence Assumption: There is no reason to believe that the swims are not independent or that the two groups are not independent of each other.

Randomization Condition: The swimmers are not a random sample from any identifiable population, but they may be representative of swimmers who tackle challenges such as this.

Nearly Normal Condition: the boxplots show no outliers. The histograms are unimodal; the histogram for men is somewhat skewed to the right. (Show your graphs.)

36. a) $H_0: \mu_M - \mu_R = 0$ vs. $H_A: \mu_M - \mu_R > 0$. $t = -0.70$, df = 45.88, P-value = 0.7563. Because the P-value is so large, we do not reject H_0. These data provide no evidence that listening to Mozart while studying is better than listening to rap.

b) With 90% confidence, the average difference in score is between 0.189 and 5.351 objects more for those who listen to no music while studying, based on these samples.

37. a) $H_0: \mu_R - \mu_N = 0$ vs. $H_A: \mu_R - \mu_N < 0$. $t = -1.36$, df = 20.00, P-value = 0.0945. Because the P-value is large, we fail to reject H_0. These data show no evidence of a difference in mean number of objects recalled between listening to rap or no music at all.

b) Didn't conclude any difference.

38.

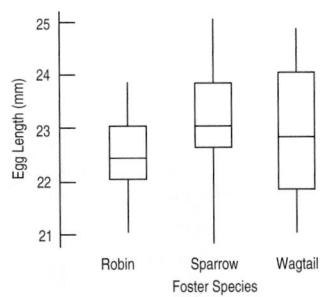

$H_0: \mu_S - \mu_R = 0$ vs. $H_A: \mu_S - \mu_R \neq 0$.
$t = 1.641$, df $= 21.60$, P-value $= 0.115$.

Since $P > 0.05$, fail to reject H_0. There is no evidence of a difference in mean cuckoo egg length between robin and sparrow foster parents.

$H_0: \mu_S - \mu_W = 0$ vs. $H_A: \mu_S - \mu_W \neq 0$.
$t = 0.549$, df $= 26.86$, P-value $= 0.587$.

Since $P > 0.05$, fail to reject H_0. There is no evidence of a difference in mean cuckoo egg length between sparrow and wagtail foster parents.

$H_0: \mu_R - \mu_W = 0$ vs. $H_A: \mu_R - \mu_W \neq 0$.
$t = -1.012$, df $= 23.60$, P-value $= 0.322$.

Since $P > 0.05$, fail to reject H_0. There is no evidence of a difference in mean cuckoo egg length between robin and wagtail foster parents.

In general, we should be wary of doing three t-tests on the same data. Our Type I error rate is not the same for doing three tests as it is for doing one test. However, because none of the tests showed significant differences, this is less of a concern here.

CHAPTER 25

1. a) Randomly assign 50 hens to each of the two kinds of feed. Compare production at the end of the month.
   b) Give all 100 hens the new feed for 2 weeks and the old food for 2 weeks, randomly selecting which feed the hens get first. Analyze the differences in production for all 100 hens.
   c) Matched pairs. Because hens vary in egg production, the matched-pairs design will control for that.
2. a) Randomly assign half the volunteers to do the puzzles in a quiet room and half to do them with MTV on. Compare times.
   b) Randomly assign half the volunteers to do a puzzle in a quiet room and half to do a puzzle with MTV on. Then have each do a puzzle under the other condition. Look at the differences in completion times.
   c) Matched pairs. People vary in their ability to do crossword puzzles.
3. a) Show the same people ads with and without sexual images, and record how many products they remember in each group. Randomly decide which ads a person sees first. Examine the differences for each person.
   b) Randomly divide volunteers into two groups. Show one group ads with sexual images and the other group ads without. Compare how many products each group remembers.
4. a) Select a random sample of freshmen. Weigh them when college starts in the fall and again when they leave for home in the spring. Examine the difference in weights.
   b) Weigh a random sample of freshmen as they enter college in the fall to determine their average weight. Repeat with a new sample of students at the end of the spring semester. Compare the mean weights of the two groups.

5. a) Matched pairs—same cities in different periods.
   b) There is a significant difference (P-value $= 0.0244$) in the labor force participation rate for women in these cities; women's participation seems to have increased between 1968 and 1972.
6. a) Two-sample. Clouds are independent of one another.
   b) Based on these data, there is some evidence of a difference (P-value $= 0.0538$) in the amount of rain between seeded and unseeded clouds.
7. a) Use the paired t-test because we have pairs of Fridays in 5 different months. Data from adjacent Fridays within a month may be more similar than data from randomly chosen Fridays.
   b) We conclude that there is evidence (P-value 0.0212) that the mean number of cars found on the M25 motorway on Friday the 13th is less than on the previous Friday.
   c) We don't know if these Friday pairs were selected at random. If these are the Fridays with the largest differences, this will affect our conclusion. The Nearly Normal Condition appears to be met by the differences, but the sample size is small.
8. a) The paired t-test is appropriate since we have pairs of Fridays in 6 different months. Data from adjacent Fridays within a month may be more similar than data from randomly chosen Fridays.
   b) We conclude that there is evidence (P-value 0.0211) that the mean number of admissions to hospitals found on Friday the 13th is more than on the previous Friday.
   c) We don't know if these Friday pairs were selected at random. Obviously, if these are the Fridays with the largest differences, this will affect our conclusion. The Nearly Normal Condition appears to be met by the differences, but sample size is small.
9. Adding variances requires that the variables be independent. These price quotes are for the same cars, so they are paired. Drivers quoted high insurance premiums by the local company will be likely to get a high rate from the online company, too.
10. Adding variances requires that the variables be independent. The wind speeds were recorded at nearby sites, so they are likely to be both high or both low at the same time.
11. a) The histogram—we care about differences in price.
    b) Insurance cost is based on risk, so drivers are likely to see similar quotes from each company, making the differences relatively smaller.
    c) The price quotes are paired; they were for a random sample of fewer than 10% of the agent's customers; the histogram of differences looks approximately Normal.
12. a) The outliers are particularly windy days, but they were windy at both sites, making the difference in wind speeds less unusual.
    b) The histogram and summaries of the differences are more appropriate because these are paired observations and all we care about is which site was more windy.
    c) The wind measurements at the same times at two nearby sites are paired. We should be concerned that there might be a lack of independence from one time to the next, but the times were 6 hours apart and the *differences* in speeds are likely to be independent. Although the sample is not random, we can regard a sample this large as generally representative of wind speeds at these sites. The histogram of differences is unimodal, symmetric, and bell-shaped.
13. $H_0: \mu(Local - Online) = 0$ vs. $H_A: \mu(Local - Online) > 0$; with 9 df, $t = 0.83$. With a high P-value of 0.215, we don't reject the null hypothesis. These data don't provide evidence that online premiums are lower, on average.
14. $H_0: \mu(2 - 4) = 0$ vs. $H_A: \mu(2 - 4) \neq 0$; $t = 2.667$ with 1114 df. The P-value of 0.008 is very low, so we reject the null. There's strong evidence that the average wind speed is higher at site 2.

15.

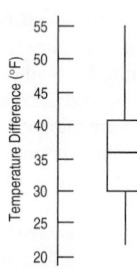

Data are paired for each city; cities are independent of each other; boxplot shows the temperature differences are reasonably symmetric, with no outliers. This is probably not a random sample, so we might be wary of inferring that this difference applies to all European cities. Based on these data, we are 90% confident that the average temperature in European cities in July is between 32.3°F and 41.3°F higher than in January.

16. We are 90% confident that the average women's winning marathon time is between 16.25 and 17.45 minutes higher than the men's based on this sample.

17. Based on these data, we are 90% confident that boys, on average, can do between 1.6 and 13.0 more push-ups than girls (independent samples—not paired).

18. a) $H_0: \mu_{NC} - \mu_C = 0$ vs. $H_A: \mu_{NC} - \mu_C \neq 0$; μ_{NC} is the mean for nonconfined inmates, μ_C is the mean for inmates confined to solitary.
b) Groups are independent of each other, not paired; random assignment to groups, less than 10% of all inmates, boxplot shows no outliers in either group.

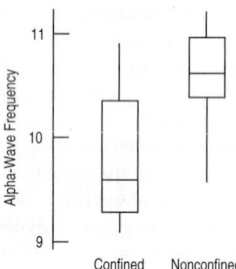

c) Two-sample t-test statistic: 3.357, df = 19.6, P-value = 0.0038.
d) Because the P-value is so small, we reject H_0. Solitary confinement makes a difference in mean alpha-wave frequencies; it seems those subjected to confinement have lower frequencies.

19. a) Paired sample test. Data are before/after for the same workers; workers randomly selected; assume fewer than 10% of all this company's workers; boxplot of differences shows them to be symmetric, with no outliers.

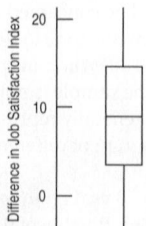

b) $H_0: \mu_D = 0$ vs. $H_A: \mu_D > 0$. $t = 3.60$, P-value = 0.0029. Because P < 0.01, reject H_0. These data show evidence that average job satisfaction has increased after implementation of the program.
c) Type I

20. a) $H_0: \mu_D = 0$ vs. $H_A: \mu_D > 0$. $t = 1.75$, df = 5, P-value = 0.0699. Because P > 0.05, we fail to reject H_0. These data do not provide enough evidence to conclude that the summer school program is worthwhile, at $\alpha = 0.05$.
b) Type II

21. $H_0: \mu_D = 0$ vs. $H_A: \mu_D \neq 0$. Data are paired by brand; brands are independent of each other; fewer than 10% of all yogurts (questionable); boxplot of differences shows an outlier (100) for Great Value:

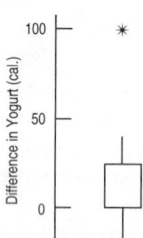

With the outlier included, the mean difference (Strawberry – Vanilla) is 12.5 calories with a t-stat of 1.332, with 11 df, for a P-value of 0.2098. Deleting the outlier, the difference is even smaller, 4.55 calories with a t-stat of only 0.833 and a P-value of 0.4241. With P-values so large, we do not reject H_0. We conclude that the data do not provide evidence of a difference in mean calories.

22. a) $H_0: \mu_D = 0$ vs. $H_A: \mu_D > 0$. $t = 4.47$, df = 9, P-value = 0.0008. Because of the very small P-value, we reject H_0. These data provide strong evidence that cars get significantly better mileage, on average, with premium than with regular gasoline.
b) (1.18, 2.82) miles per gallon
c) Premium gasoline costs more than regular.
d) $t = 1.25$, df = 17.89, P-value is 0.1144. Would have decided no difference. The variation in the cars' performances is larger than the differences.

23. a) Cars were probably not a simple random sample, but may be representative in terms of stopping distance; boxplot does not show outliers, but does indicate right skewness. A 95% confidence interval for the mean stopping distance on dry pavement is (131.8, 145.6) feet.
b) Data are paired by car; cars were probably not randomly chosen, but representative; boxplot shows an outlier (car 4) with a difference of 12. With deletion of that car, a Normal probability plot of the differences is relatively straight.

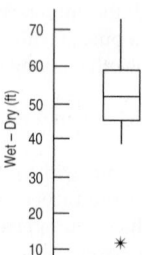

Retaining the outlier, we estimate with 95% confidence that the average braking distance is between 38.8 and 62.6 feet more on wet pavement than on dry, based on this sample. (Without the outlier, the confidence interval is 47.2 to 62.8 feet.)

24. a) Not a simple random sample, but most likely representative; stops most likely independent of each other; boxplot is symmetric with no outliers.

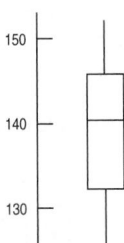

Based on these data, with 95% confidence, the average braking distance for these tires on dry pavement is between 133.6 and 145.2 feet.

b) Not simple random samples, but most likely representative; stops most likely independent of each other; less than 10% of all possible wet stops; Normal probability plots are relatively straight. Based on these data, with 95% confidence, the average increase in distance for these tires on wet pavement is between 51.4 and 74.6 feet

25. a) Paired Data Assumption: Data are paired by college. Randomization Condition: This was a random sample of public colleges and universities. 10% Condition: these are fewer than 10% of all public colleges and universities.

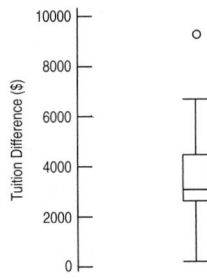

Normal Population Assumption: U.C. Irvine seems to be an outlier; we might consider removing it.

b) Having deleted the observation for U.C.-Irvine, whose difference of $9300 was an outlier, we are 90% confident, based on the remaining data, that nonresidents pay, on average, between $2615.31 and $3918.02 more than residents. If we retain the outlier, the interval is ($2759, $4409).

c) Assertion is reasonable; with or without the outlier, $3500 is in the confidence interval.

26. Using a t-test for paired differences, $t = -0.86$ and two-tailed $P = 0.396$. With a P-value so high, we fail to reject the null hypothesis of no mean difference. There is no evidence that sexual images in ads affects people's ability to remember the product being advertised.

27. a) 60% is 30 strikes; $H_0: \mu = 30$ vs. $H_A: \mu > 30$. $t = 6.07$, P-value $= 3.92 \times 10^{-6}$. With a very small P-value, we reject H_0. There is very strong evidence that players can throw more than 60% strikes after training, based on this sample.

b) $H_0: \mu_D = 0$ vs. $H_A: \mu_D > 0$. $t = 0.135$, P-value $= 0.4472$. With such a high P-value, we do not reject H_0. These data provide no evidence that the program has improved pitching in these Little League players.

28. If this group is representative of all students, we can be 95% confident that freshmen gain a mean of between 1.40 and 2.43 pounds during their first 12 weeks at college. That's strong evidence of a weight gain, but it's unlikely that it amounts to 15 pounds for the whole first year.

PART VI REVIEW

1. a) $H_0: \mu_{Jan} - \mu_{Jul} = 0$; $H_A: \mu_{Jan} - \mu_{Jul} \neq 0$. $t = -1.94$, df $= 43.68$, P-value $= 0.0590$. Since P < 0.10, reject the null.

These data show a significant difference in mean age to crawl between January and July babies.

b) $H_0: \mu_{Apr} - \mu_{Oct} = 0$; $H_A: \mu_{Apr} - \mu_{Oct} \neq 0$. $t = -0.92$; df $= 59.40$; P-value $= 0.3610$. Since P > 0.10, do not reject the null; these data do not show a significant difference between April and October with regard to the mean age at which crawling begins.

c) These results are not consistent with the claim.

2. $H_0: \mu_D = 0$; $H_A: \mu_D > 0$. $t = 1.36$; df $= 20$; P-value $= 0.0949$. Because the P-value is high, we do not reject H_0. These data do not show that floral scent improved the average maze completion time between scented and unscented.

3. $H_0: p = 0.26$; $H_A: p \neq 0.26$. $z = 0.946$; P-value $= 0.3443$. Because the P-value is high, we do not reject H_0. These data do not show that the Denver-area rate is different from the national rate in the proportion of businesses with women owners.

4. a) We are 95% confident the average savings in Canada for prescription drugs is between $77.57 and $174.43.

b) We are 95% confident that the average savings in Canada for prescription drugs is between 40.1% and 64.2%.

c) Using percents makes the histogram more unimodal and symmetric.

d) Probably would change. The pharmacy may have listed only the 12 drugs with the "best" savings.

5. Based on these data, we are 95% confident that the mean difference in aluminum oxide content is between -3.37 and 1.65. Since the interval contains 0, the means in aluminum oxide content of the pottery made at the two sites could reasonably be the same.

6. We are 95% confident that the proportion of streams in the Adirondacks with shale substrates is between 32.8% and 47.4%.

7. a) $H_0: p_{ALS} - p_{Other} = 0$; $H_A: p_{ALS} - p_{Other} > 0$. $z = 2.52$; P-value $= 0.0058$. With such a low P-value, we reject H_0. This is strong evidence that there is a higher proportion of varsity athletes among ALS patients than those with other disorders.

b) Observational retrospective study. To make the inference, one must assume the patients studied are representative.

8. Paired samples; boxplot shows no strong skewness or outliers. One might wonder how the individuals in the study were selected. We are 95% confident that average percentage of 15-year-old males who have been drunk is between 4.5% and 11.4% more than 15-year-old females for these countries. We cannot infer that these percentages are true for other countries.

9. $H_0: \mu = 7.41$; $H_A: \mu \neq 7.41$. $t = 2.18$; df $= 111$; P-value $= 0.0313$. With such a low P-value, we reject H_0. Assuming that Missouri babies fairly represent the United States, these data suggest that American babies are different from Australian babies in birth weight; it appears American babies are heavier, on average.

10. a) 88.6%   b) 82.2%

c) The petition would be certified when there are not enough valid signatures.

d) A correct petition is not certified.

e) $H_0: p = 0.822$; $H_A: p > 0.822$. $z = 7.48$; P-value $= 3.64 \times 10^{-14}$. With such a low P-value, we reject H_0. This sample provides sufficient evidence for certification of the petition.

f) Increase sample size.

11. a) If there is no difference in the average fish sizes, the chance of seeing an observed difference this large just by natural sampling variation is less than 0.1%.

b) If cost justified, feed them a natural diet.   c) Type I

12. We have two independent samples, but we don't know how these vehicles were chosen. Even if we consider them representative, the samples are small, and the data for SUVs are skewed to the right. These data are not appropriate for inferences.

13. a) Assuming the conditions are met, from these data we are 95% confident that patients with cardiac disease average between 3.39 and 5.01 years older than those without cardiac disease.

b) Older patients are at greater risk from a variety of other health issues, and perhaps more depressed.

14. a) We are 95% confident that the proportion of smokers is between 3.67% and 12.93% more in patients with heart disease than in those without heart disease, based on this information.

b) Yes. The interval lies completely above 0.

c) Could be a confounding variable. The smokers are at even greater risk for heart attack.

15. a) Stratified sample survey.

b) We are 95% confident that the proportion of boys who play computer games is between 7.0 and 17.0 percentage points higher than among girls.

c) Yes. The entire interval lies above 0.

16. a) Based on the data, we are 95% confident that the difference in recruiting is between −4.2% and 8.2%.

b) No. The interval contains 0 and negative values—the new strategy may actually lower acceptance rates.

17. Based on the data, we are 95% confident that the mean difference in words misunderstood is between −3.76 and 3.10. Because 0 is in the confidence interval, we would conclude that the two tapes could be equivalent.

18. $H_0: p_{VT} - p_{NH} = 0; H_A: p_{VT} - p_{NH} \neq 0. z = -0.59;$ P-value = 0.5563. With such a high P-value, we do not reject H_0. These data show no evidence of a difference in the rates of cesarean deliveries between Vermont and New Hampshire.

19. a)

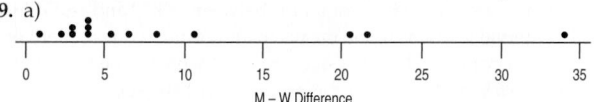

M − W Difference

The countries that appear to be outliers are Spain, Italy, and Portugal. They are all Mediterranean countries.

b) $H_0: \mu_D = 0; H_A: \mu_D > 0.$
$t = 5.56; df = 10;$ P-value = 0.0001. With such a low P-value,

we reject H_0. These data show that European men are more likely than women to read newspapers.

20. $H_0: \mu = \$10; H_A: \mu > \$10. t = 0.66; df = 13.$ P-value = 0.26. With such a high P-value, we do not reject H_0. These data do not provide evidence that he is likely to overspend his budget of $10 per day.

21. We are 95% confident that the proportion of American adults who would agree with the statement is between 57.0% and 63.0%.

22. $H_0: \mu_{Cert} - \mu_{UC} = 0; H_A: \mu_{Cert} - \mu_{UC} > 0. t = 1.57; df = 86;$ P-value = 0.0598. The P-value is low enough to reject H_0. These data provide evidence that students of certified teachers achieve higher mean reading scores than students of uncertified teachers.

23. Data are matched pairs (before and after for the same rooms); less than 10% of all rooms in a large hotel; uncertain how these rooms were selected (are they representative?). Histogram shows that differences are roughly unimodal and symmetric, with no outliers. A 95% confidence interval for the difference, before − after, is (0.58, 2.65) counts. Since the entire interval is above 0, these data suggest that the new air-conditioning system was effective in reducing average bacteria counts.

24. There is a significant difference in Math performance; students of certified teachers seem to do better. There is a significant difference in language performance at $\alpha = 0.05$ between students of certified teachers and those of uncertified teachers.

25. a) We are 95% confident that between 19.77% and 38.66% of children with bipolar symptoms will be helped with medication and psychotherapy, based on this study.

b) 221 children

26. a) $H_0: \mu_D = 0; H_A: \mu_D \neq 0. t = 0.381; df = 19;$ P-value = 0.7078. With a high P-value, we do not reject H_0. There is little evidence that the tests differ in mean difficulty.

b) $H_0: \mu_D = 0; H_A: \mu_D \neq 0.$ The boxplot of the differences shows three outliers, but they are symmetric. A paired t-test with

them included shows $t = -0.2531$ with 19 df and a P-value of 0.8029, showing no evidence of a difference in mean scores. With the three outliers removed, the t-test shows $t = -1.1793$ with 16 df and a P-value of 0.2555—still no evidence of a difference in means. We conclude the testing environment does not affect mean score.

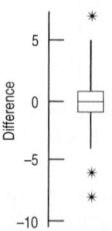

27. a) From this histogram, about 115 loaves or more. (Not Normal.) This assumes the last 100 days are typical.

b) Large sample size; CLT says $\bar{y}$ will be approximately Normal.

c) From the data, we are 95% confident that on average the bakery will sell between 101.2 and 104.8 loaves of bread a day.

d) 25

e) Yes, 100 loaves per day is too low—the entire confidence interval is above that.

28. a)

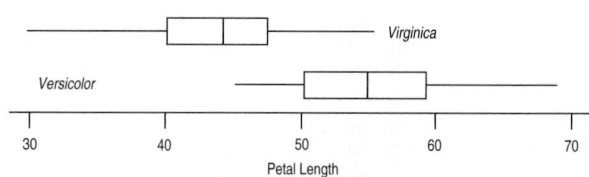

Petal Length

b) *Versicolor* generally has longer petals than *virginica*.

c) (10.14, 14.46)

d) We are 95% confident, based on the information given, that the average petal length for *versicolor* irises is between 10.14 and 14.46 mm longer than that of *virginica* irises.

e) Yes. The entire interval is above 0.

29. a) $H_0: p_{High} - p_{Low} = 0; H_A: p_{High} - p_{Low} \neq 0. z = -3.57;$ P-value = 0.0004. Because the P-value is so low, we reject H_0. These data suggest the IRS risk is different in the two groups; it appears people who consume dairy products often have a lower risk, on average.

b) Doesn't indicate causality; this is not an experiment.

30. a) $H_0: p = 0.16; H_A: p > 0.16. z = 4.42;$ P-value = 5×10^{-6}. Because the P-value is so low, we reject H_0. These data suggest that the proportion of tickets given to blacks on this section of the New Jersey Turnpike is unusually high.

b) Doesn't prove it; there may be other factors.

c) Answers will vary. Possibly what proportion of drivers on this highway is black?

31. Based on these data, we are 95% confident that seeded clouds will produce an average of between −4.76 and 559.56 more acrefeet of rain than unseeded clouds. Since the interval contains negative values, it may be that seeding is unproductive.

32. a) $H_0: \mu = 35.4; H_A: \mu < 35.4$

b) There is one high value, but it is technically not an outlier. Otherwise the histogram appears roughly unimodal and symmetric (only 6 values).

c) $t = 0.726;$ P-value = 0.7497. With such a large P-value, we do not reject H_0. The data do not show that the average weight is less than claimed.

d) $t = -0.53;$ P-value = 0.3107. Without the high outlier, the data still do not show that the average weight is significantly less than claimed.

e) We lack evidence to question the stated weight.

33. a) Randomizing order of the tasks helps avoid bias and memory effects. Randomizing the cards helps avoid bias as well.

b) $H_0: \mu_D = 0$; $H_A: \mu_D \neq 0$

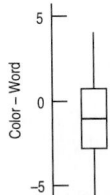

Boxplot of the differences looks symmetric with no outliers.

$t = -1.70$; P-value $= 0.0999$; do not reject H_0, because $P > 0.05$. The data do not provide evidence that the color or written word dominates.

34. a) $373.50
    b) They are 95% confident that the average loss in a home burglary is between $1644 and $2391, based on their sample.
    c) 95% of all random samples will produce confidence intervals that contain the true mean loss.
35. a) Different samples give different means; this is a fairly small sample. The difference may be due to natural sampling variation.
    b) $H_0: \mu = 100$; $H_A: \mu < 100$
    c) Batteries selected are a SRS (representative); fewer than 10% of the company's batteries; lifetimes are approximately Normal.
    d) $t = -1.0$; P-value $= 0.1666$; do not reject H_0. This sample does not show that the average life of the batteries is significantly less than 100 hours.
    e) Type II.
36. a) Based on these data, we are 90% confident that the average hamster litter will have between 7.11 and 8.33 babies.
    b) Larger—to be more confident, we need a wider interval.
    c) About 27 (based on t_{24}^*).

CHAPTER 26

1. a) Chi-square test of independence. We have one sample and two variables. We want to see if the variable *Account Type* is independent of the variable *Trade Type*.
   b) Other test. *Account Size* is quantitative, not counts.
   c) Chi-square test of homogeneity. We want to see if the distribution of one variable, *Courses*, is the same for two groups (resident and nonresident students).
2. a) Chi-square goodness-of-fit test. We want to see if the distribution of defects is uniform over the work days.
   b) Other test. *Cholesterol Level* is quantitative, not counts.
   c) Chi-square test of independence. We have data on two variables, *Political Leaning* and *Major*, for one group of students.
3. a) 10   b) Goodness-of-fit
   c) H_0: The die is fair (all faces have $p = 1/6$).
      H_A: The die is not fair.
   d) Count data; rolls are random and independent; expected frequencies are all bigger than 5.
   e) 5   f) $\chi^2 = 5.600$, P-value $= 0.3471$
   g) Because the P-value is high, do not reject H_0. The data show no evidence that the die is unfair.
4. a) Yellow, red: 21.2; orange, blue, green: 10.6; brown; 31.8
   b) Goodness-of-fit
   c) H_0: The distribution is as specified by the company.
      H_A: The distribution is not as specified.
   d) Count data; bag may not be a random sample, but most likely representative; expected counts are all bigger than 5.
   e) 5   f) $\chi^2 = 9.315$, P-value $= 0.0972$
   g) Because the P-value is high, do not reject H_0. These data do not provide evidence that the distribution is other than specified.
5. a) Weights are quantitative, not counts.
   b) Count the number of each kind of nut, assuming the company's percentages are based on counts rather than weights.

6. Data are averages, not counts.
7. H_0: The police force represents the population (29.2% white, 28.2% black, etc.). H_A: The police force is not representative of the population. $\chi^2 = 16516.88$, df $= 4$, P-value $= 0.0000$. Because the P-value is so low, we reject H_0. These data show that the police force is not representative of the population. In particular, there are too many white officers in relationship to their membership in the community.
8. H_0: Murders among women have the same distribution of weapons as all murders (63.4% guns, etc.). H_A: Murders among women have a different distribution of weapons than all murders. $\chi^2 = 389.54$, df $= 3$, P-value < 0.0001. Because the P-value is so low, we reject H_0. Women's murders do not have the same distribution of weapons as all murders nationwide. Women are much less likely to be killed by other weapons and more likely to be killed by personal attack.
9. a) $\chi^2 = 5.671$, df $= 3$, P-value $= 0.1288$. With a P-value this high, we fail to reject H_0. Yes, these data are consistent with those predicted by genetic theory.
   b) $\chi^2 = 11.342$, df $= 3$, P-value $= 0.0100$. Because of the low P-value, we reject H_0. These data provide evidence that the distribution is not as specified by genetic theory.
   c) With small samples, many more data sets will be consistent with the null hypothesis. With larger samples, small discrepancies will show evidence against the null hypothesis.
10. H_0: Digits are all equally likely (all occur with frequency 1/10). H_A: Digits are not all equally likely. $\chi^2 = 5.509$, df $= 9$, P-value $= 0.7879$. Because the P-value is large, we do not reject H_0. These data provide no evidence that the digits in pi are not all equally likely.
11. a) $96/16 = 6$ b) Goodness of Fit
    c) H_0: The number of large hurricanes remains constant over decades.
       H_A: The number of large hurricanes has changed.
    d) 15   e) P-value $= 0.63$
    f) The very high P-value means these data offer no evidence that the numbers of large hurricanes has changed.
    g) The final period is only 6 years rather than 10 and already 7 large hurricanes have been observed. Perhaps this decade will have an unusually large number of such hurricanes.
12. a) Goodness of Fit   b) $655/49 = 13.367$
    c) H_0: All numbers are equally likely;
       H_A: Some numbers are more likely than others.
    d) 48   e) 0.93
    f) The very high P-value means these data offer no evidence that some numbers are more likely to come up than others.
13. a) Independence
    b) H_0: Breastfeeding success is independent of having an epidural.
       H_A: There's an association between breastfeeding success and having an epidural.
14. a) Homogeneity
    b) H_0: The same proportion of articles used statistics in the three time periods surveyed. H_A: The proportion of articles using statistics changed over time.
15. a) 1   b) 159.34
    c) Breastfeeding behavior should be independent for these babies. They are fewer than 10% of all babies; we assume they are representative. We have counts, and all the expected counts are at least 5.
16. a) 2   b) 21.85
    c) These are counted data. One article shouldn't affect another (except perhaps for the rare article based on a previous one in an earlier year cohort included in this study). We can regard the selected years as representative of those nearby, and the authors (judging by their title) seem to want to regard these articles as representative of those appearing in other similar-quality medical journals, so they're fewer than 10% of all articles. All expected counts are at least 5.

17. a) 5.90 b) P-value < 0.005
    c) The P-value is very low, so reject the null. There's evidence of an association between having an epidural and subsequent success in breastfeeding.
18. a) 2.82 b) P-value < 0.001
    c) The P-value is very low, so reject the null. There's evidence that the percentage of medical journal articles that include statistics has changed over time.
19. a) $\dfrac{(190 - 159.34)}{\sqrt{159.34}} = 2.43$
    b) It appears that babies whose mothers had epidurals during childbirth are much less likely to be breastfeeding 6 months later.
20. a) $\dfrac{(14 - 21.85)}{\sqrt{21.85}} = -1.68$
    b) The residuals for No stats are decreasing and those for Stats are increasing over time, indicating that, over time, fewer articles are appearing without statistics.
21. These factors would not be mutually exclusive. There would be yes or no responses for every baby for each.
22. The methods would not have been mutually exclusive. Articles might use more than one statistical method.
23. a) 40.2% b) 8.1% c) 62.2% d) 285.48
    e) H_0: Survival was independent of status on the ship.
    H_A: Survival depended on the status.
    f) 3
    g) We reject the null hypothesis. Survival depended on status. We can see that first-class passengers were more likely to survive than passengers of any other class.
24. a) 15.0% b) 13.0% c) 730.4
    d) Independence
    e) H_0: *Rank* is independent of *Sex*. H_A: *Rank* and *Sex* are not independent.
    f) Count data; not a random sample, but all NYPD officers; expected counts all greater than 5.
    g) 5
    h) Because the P-value is so low, we reject H_0. *Sex* and *Rank* in the NYPD are not independent.
25. First class passengers were most likely to survive, while 3rd-class passengers and crew were under-represented among the survivors.
26. Women are overrepresented at the lower ranks and underrepresented at every rank from sergeant up.
27. a) Experiment—actively imposed treatments (different drinks)
    b) Homogeneity
    c) H_0: The rate of urinary tract infection is the same for all three groups. H_A: The rate of urinary tract infection is different among the groups.
    d) Count data; random assignment to treatments; all expected frequencies larger than 5.
    e) 2 f) $\chi^2 = 7.776$, P-value $= 0.020$.
    g) With a P-value this low, we reject H_0. These data provide reasonably strong evidence that there is a difference in urinary tract infection rates between cranberry juice drinkers, lactobacillus drinkers, and the control group.
    h) The standardized residuals are

	Cranberry	Lactobacillus	Control
Infection	-1.87276	1.19176	0.68100
No Infection	1.24550	-0.79259	-0.45291

From the standardized residuals (and the sign of the residuals), it appears those who drank cranberry juice were less likely to develop urinary tract infections; those who drank lactobacillus were more likely to have infections.

28. a) Homogeneity
    b) H_0: The distribution of *Car Origin* is the same for students and staff. H_A: The distribution of *Car Origin* is different for students and staff.
    c) Count data; random survey of cars in lots (probably can't generalize to other universities); expected frequencies greater than 5.
    d) $\chi^2 = 7.828$, df $= 2$, P-value $= 0.020$.
    e) With a P-value this low, we reject H_0. The distribution of car origins differs between students and staff. Examination of the residuals shows that students are more likely than staff to drive European cars and less likely than staff to drive American cars.
29. a) Independence
    b) H_0: *Political Affiliation* is independent of *Sex*.
    H_A: There is a relationship between *Political Affiliation* and *Sex*.
    c) Counted data; probably a random sample, but can't extend results to other states; all expected frequencies greater than 5.
    d) $\chi^2 = 4.851$, df $= 2$, P-value $= 0.0884$.
    e) Because of the high P-value, we do not reject H_0. These data do not provide evidence of a relationship between *Political Affiliation* and *Sex*.
30. a) Prospective study; individuals were selected and then subsequently followed.
    b) Independence
    c) $\chi^2 = 3.677$, df $= 3$, P-value $= 0.2985$. Because of the high P-value, we do not reject H_0. These data do not provide evidence of a relationship between the amount of fish in the diet and prostate cancer. (Data are for totals and cancer, not non-cancer and cancer.)
    d) No. There may be many other factors involved.
31. H_0: *Political Affiliation* is independent of *Region*. H_A: There is a relationship between *Political Affiliation* and *Region*. $\chi^2 = 13.849$, df $= 4$, P-value $= 0.0078$. With a P-value this low, we reject H_0. *Political Affiliation* and *Region* are related. Examination of the residuals shows that those in the West are more likely to be Democrat than Republican; those in the Northeast are more likely to be Republican than Democrat.
32. a) Survey b) Homogeneity
    c) $\chi^2 = 4.030$, df $= 4$, P-value $= 0.4019$. Because the P-value is so high, we fail to reject H_0. These data do not show evidence of a change in attitudes about the ideal family between 1991 and 2001.
33. a) Homogeneity
    b) H_0: The grade distribution is the same for both professors.
    H_A: The grade distributions are different.
    c)

	Dr. Alpha	Dr. Beta
A	6.667	5.333
B	12.778	10.222
C	12.222	9.778
D	6.111	4.889
F	2.222	1.778

Three cells have expected frequencies less than 5.

34. a) Homogeneity
    b) H_0: The crime distribution is the same for both phases of the moon.
    H_A: The crime distributions are different for the different moon phases.
    c)

	Full moon	Not full
Violent	2.558	2.442
Property	19.442	18.558
Drugs/alcohol	23.535	22.465
Domestic abuse	12.791	12.209
Other offenses	7.674	7.326

Two cells have expected counts less than 5.

35. a)

	Dr. Alpha	Dr. Beta
A	6.667	5.333
B	12.778	10.222
C	12.222	9.778
Below C	8.333	6.667

All expected frequencies are now larger than 5.
b) Decreased from 4 to 3.
c) $\chi^2 = 9.306$, P-value $= 0.0255$. Because the P-value is so low, we reject H_0. The grade distributions for the two professors are different. Dr. Alpha gives fewer A's and more grades below C than Dr. Beta.

36. a) Combine Violent and Domestic abuse into one category (or Violent and Other).
b) $\chi^2 = 2.877$, df $= 3$, P-value $= 0.4109$. Because the P-value is so high, we do not reject H_0. These data do not provide any evidence of a difference in type of illegal activity in different moon phases.

37. $\chi^2 = 14.058$, df $= 1$, P-value $= 0.0002$. With a P-value this low, we reject H_0. There is evidence of racial steering. Blacks are much less likely to rent in Section A than Section B.

38. $\chi^2 = 453.476$, df $= 1$, P-value < 0.0001. With a P-value this low, we reject H_0. Survival was related to gender. Females were much more likely than males to survive (72.98% to 21.20%).

39. a) $z = 3.74936$, $z^2 = 14.058$.
b) P-value $(z) = 0.0002$ (same as in Exercise 25).

40. a) $z = 21.29498$.   b) $z^2 = 453.476$.
c) Both P-values are near 0.

41. $\chi^2 = 5.89$, df $= 3$, P $= 0.117$. Because the P-value is >0.05, these data show no evidence of an association between the mother's age group and the outcome of the pregnancy.

42. $\chi^2 = 178.453$, df $= 12$, P-value < 0.0001, Because the P-value is so low, we reject H_0. There is a difference is education level among the different age groups. The largest component indicates those 35 to 44 were less likely to have only a high-school diploma. The second-largest indicates the same age group is more likely than other age groups to have at least four years of college.

CHAPTER 27

1. a) $\widehat{Error} = 453.22 - 8.37\,YearSince1970$; according to the model, the error made in predicting a hurricane's path was about 453 nautical miles, on average, in 1970. It has been declining at a rate of about 8.37 nautical miles per year.
b) $H_0: \beta_1 = 0$; there has been no change in prediction accuracy. $H_A: \beta_1 \neq 0$; there has been a change in prediction accuracy.
c) With a P-value < 0.001, I reject the null hypothesis and conclude that prediction accuracies have in fact been changing during this period.
d) 58.5% of the variation in hurricane prediction accuracy is accounted for by this linear model on time.

2. a) $\widehat{\%Other} = -3.068 + 0.615\%\,Marijuana$. The percentage of ninth graders in these countries who have used other drugs is estimated to have increased 0.615% for each 1% increase in the percentage of ninth graders who have used marijuana.
b) H_0: There is no (linear) relationship between use of marijuana and other drugs, $\beta_1 = 0$. H_A: There is a relationship, $\beta_1 \neq 0$.
c) $t = 7.85$, P-value $= 0.0001$. With such a low P-value, we reject H_0. Percentage of teens using other drugs seems to be positively related to percentage using marijuana.
d) Percentage using marijuana accounts for 87.3% of the variation in other drug usage for ninth graders in these countries.
e) The use of other drugs is associated with marijuana use, but there is no proof of causality. There may be lurking variables.

3. a) $\widehat{Budget} = -31.387 + 0.714\,RunTime$. The model suggests that movies cost about $714,000 per minute to make.
b) A negative starting value makes no sense, but the P-value of 0.07 indicates that we can't discern a difference between our estimated value and zero. The statement that a movie of zero length should cost $0 makes sense.
c) Amounts by which movie costs differ from predictions made by this model vary, with a standard deviation of about $33 million.
d) 0.154 $m/min
e) If we constructed other models based on different samples of movies, we'd expect the slopes of the regression lines to vary, with a standard deviation of about $154,000 per minute.

4. a) $\widehat{Price} = -0.312 + 94.5\,Size$. The model suggests that Saratoga houses cost about $94.5 per square foot.
b) The P-value for the intercept is 0.50. That means that we cannot discern a difference between the intercept value and zero. A value of $0 for a house of zero size makes sense.
c) Amounts by which house prices differ from predictions made by this model vary, with a standard deviation of about $54,000.
d) $2.393/sq. ft.
e) If we constructed other models based on different samples of houses, we'd expect the slopes of the regression lines to vary, with a standard deviation of about $2.39 per square foot.

5. a) The scatterplot looks straight enough, the residuals look random and nearly normal, and the residuals don't display any clear change in variability.
b) I'm 95% confident that the cost of making longer movies increases at a rate of between 0.41 and 1.02 million dollars per additional minute.

6. a) The scatterplot looks straight enough, the residuals look random and nearly normal, and the residuals don't display any clear change in variability.
b) I'm 95% confident that Saratoga housing costs increase at a rate of between $89.8 and $99.2 per square foot.

7. a) $H_0: \beta_1 = 0$; there's no association between calories and sodium content in all-beef hot dogs. $H_A: \beta_1 \neq 0$: there is an association.
b) Based on the low P-value (0.0018), I reject the null. There is evidence of an association between the number of calories in all-beef hot dogs and their sodium contents.

8. a) H_0: There is no linear relationship between *Age* and *Cholesterol Level*, $\beta_1 = 0$. H_A: Cholesterol levels change with age, $\beta_1 \neq 0$
b) $t = 3.00$, P-value $= 0.0056$. Because the P-value is so small, we reject H_0. These data show a significant positive relationship between *Age* and *Cholesterol Level*.

9. a) Among all-beef hot dogs with the same number of calories, the sodium content varies, with a standard deviation of about 60 mg.
b) 0.561 mg/cal
c) If we tested many other samples of all-beef hot dogs, the slopes of the resulting regression lines would be expected to vary, with a standard deviation of about 0.56 mg of sodium per calorie.

10. a) Among adults of the same age, cholesterol levels vary with a standard deviation of about 46 points.
b) 0.2574 pts/yr
c) If we tested many other samples of adults, the slopes of the resulting regression lines would be expected to vary with a standard deviation of 0.26 cholesterol points per year of age.

11. I'm 95% confident that for every additional calorie, all-beef hot dogs have, on average, between 1.07 and 3.53 mg more sodium.

12. I'm 95% confident that, on average, adult cholesterol levels increase between 0.27 and 1.28 points per year of age.

13. a) H_0: Difference in age at first marriage has not been changing, $\beta_1 = 0$. H_A: Difference in age at first marriage has been changing, $\beta_1 \neq 0$.
b) Residual plot shows no obvious pattern; histogram is not particularly Normal, but shows no obvious skewness or outliers.

c) $t = -7.04$, P-value < 0.0001. With such a low P-value, we reject H_0. These data show evidence that difference in age at first marriage is decreasing.

14. a)

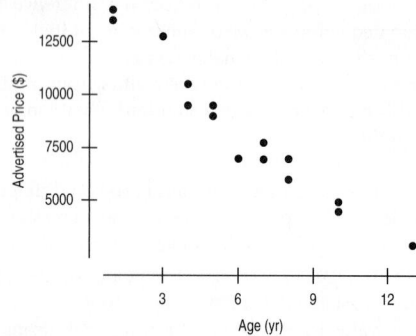

b) Yes, the plot seems linear.

c) $\widehat{Advertised\ Price} = 14286 - 959 \times Age$

d)

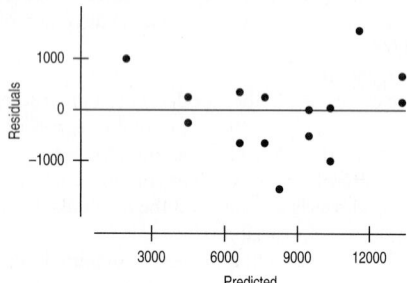

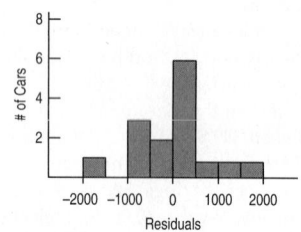

The residual plot shows some possible curvature. Inference may not be valid here, but we will proceed (with caution).

15. Based on these data, we are 95% confident that the average difference in age at first marriage is decreasing at a rate between 0.039 and 0.021 years per year.

16. Based on these data, we are 95% confident that a used car's *Price* decreases between $819.50 and $1099 per year.

17. a) H_0: *Fuel Economy* and *Weight* are not (linearly) related, $\beta_1 = 0$. H_A: *Fuel Economy* changes with *Weight*, $\beta_1 \neq 0$. P-value < 0.0001, indicating strong evidence of an association.

b) Yes, the conditions seem satisfied. Histogram of residuals is unimodal and symmetric; residual plot looks OK, but some "thickening" of the plot with increasing values.

c) $t = -12.2$, P-value < 0.0001. These data show evidence that *Fuel Economy* decreases with the *Weight* of the car.

18. a) H_0: There is no (linear) relationship between *SAT Verbal* and *SAT Math* scores, $\beta_1 = 0$. H_A: There is a relationship, $\beta_1 \neq 0$.

b) Assumptions seem reasonable, since conditions are satisfied. The scatterplot suggests a positive linear relationship. Residual plot shows no patterns (one outlier); histogram is unimodal and roughly symmetric.

c) $t = 11.9$; P-value < 0.0001. These data show evidence of a positive relationship between *SAT Verbal* and *SAT Math* scores.

19. a) $(-9.57, -6.86)$ mpg per 1000 pounds.

b) Based on these data, we are 95% confident that *Fuel Efficiency* decreases between 6.86 and 9.57 miles per gallon, on average, for each additional 1000 pounds of *Weight*.

20. a) $(0.581, 0.769)$

b) Based on the sample, we are 90% confident that average *SAT Math* scores increase between 0.58 and 0.77 points for each additional point scored on the *SAT Verbal* test.

21. a) We are 95% confident that 2500-pound cars will average between 27.34 and 29.07 miles per gallon.

b) Based on the regression, a 3450-pound car will get between 15.44 and 25.36 miles per gallon, with 95% confidence.

22. a) We are 90% confident that the mean Math score for students with a Verbal score of 500 is between 534.0 and 560.18.

b) Based on the regression, we are 90% confident that the Math score for the valedictorian will be between 569 and 800 (SAT scores cannot be higher than 800).

23. a) Yes. $t = 2.73$, P-value $= 0.0079$. With a P-value so low, we reject H_0. There is a positive relationship between *Calories* and *Sodium* content.

b) No. $R^2 = 9\%$ and s appears to be large, although without seeing the data, it is a bit hard to tell.

24. a) H_0: No linear relationship between *Brain Size* and *IQ*, $\beta_1 = 0$. H_A: There is evidence of a relationship, $\beta_1 \neq 0$. $t = 1.12$; this will not be significant.

b) With $R^2 = 6.5\%$, the relationship is very weak. There seem to be three students with large brains who also scored high. Without them, there seems to be no association at all.

25. Plot of *Calories* against *Fiber* does not look linear; the residuals plot also shows increasing variance as predicted values get large. The histogram of residuals is right skewed.

26. Scatterplot of *Temperature* against *Latitude* shows curvature (downward). Histogram of residuals is right skewed; residual plot shows decreasing variance as predicted values increase.

27. a) H_0: No (linear) relationship between *BCI* and *pH*, $\beta_1 = 0$. H_A: There seems to be a relationship, $\beta_1 \neq 0$.

b) $t = -7.73$ with 161 df; P-value < 0.0001

c) There seems to be a negative relationship; *BCI* decreases as *pH* increases at an average of 197.7 *BCI* units per increase of 1 *pH*.

28. a) $\widehat{Temp} = 15.31 + 0.004 \times CO_2$

b) Yes; $t = 4.44$, P-value $= 8.5 \times 10^{-5}$

c) The standard deviation of the residuals is 0.0809°C, so we don't expect the model to predict to an accuracy greater than about ±0.16°C.

29. a) H_0: No linear relationship between *Population* and *Ozone*, $\beta_1 = 0$. H_A: *Ozone* increases with *Population*, $\beta_1 > 0$. $t = 3.48$, P-value $= 0.0018$. With a P-value so low, we reject H_0. These data show evidence that *Ozone* increases with *Population*.

b) Yes, *Population* accounts for 84% of the variability in *Ozone* level, and s is just over 5 parts per million.

30. a) H_0: No linear relationship between *Sales* and *Profit*, $\beta_1 = 0$. H_A: There is a relationship, $\beta_1 \neq 0$. $t = 12.33$; P-value ~ 0. There seems to be a relationship between *Sales* and *Profit*.

b) Somewhat. *Sales* accounts for 66% of the variability in *Profits*; although s is nearly a half million dollars, the mean *Profit* for these companies is over $200 million.

31. a) Based on this regression, each additional million residents corresponds to an increase in average ozone level of between 3.29 and 10.01 ppm, with 90% confidence.

b) The mean *Ozone* level for cities with 600,000 people is between 18.47 and 27.29 ppm, with 90% confidence.

32. a) Based on this regression, average *Profits* increase between 0.078 and 0.107 million dollars ($78,000 and $107,000) for every million dollars in sales, with 95% confidence.

b) Based on the regression, the profit for Eli Lilly, Inc., should be between −281.1 and 1592.8 million dollars (−$281,100,000 and $1,592,800,000), with 95% confidence. This interval is so wide that it's really useless.

33. a) 33 batteries.

b) Yes. The scatterplot is roughly linear with lots of scatter; plot of residuals vs. predicted values shows no overt patterns; Normal probability plot of residuals is reasonably straight.

c) H_0: No linear relationship between *Cost* and *Cranking Amps*, $\beta_1 = 0$. H_A: *Cranking Amps* increase with cost, $\beta_1 > 0$. $t = 3.23$; P-value $= \frac{1}{2}(0.0029) = 0.00145$. With a P-value so low, we reject H_0. These data provide evidence that more expensive batteries do have more cranking amps.

d) No. $R^2 = 25.2\%$ and $s = 116$ amps. Since the range of amperage is only about 400 amps, an s of 116 is not very useful.

e) $\widehat{Cranking\ amps} = 384.59 + 4.15 \times Cost$.

f) $(1.97, 6.32)$ cold cranking amps per dollar.

g) *Cranking amps* increase, on average, between 1.97 and 6.32 per dollar of battery *Cost* increase, with 90% confidence.

34.

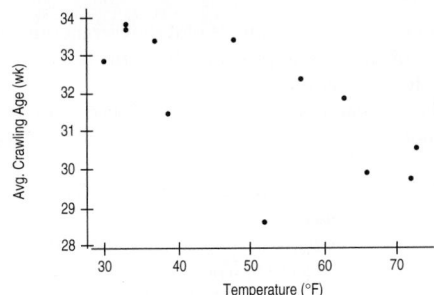

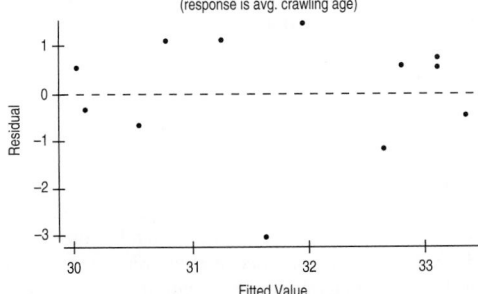

Residuals vs. the fitted values
(response is avg. crawling age)

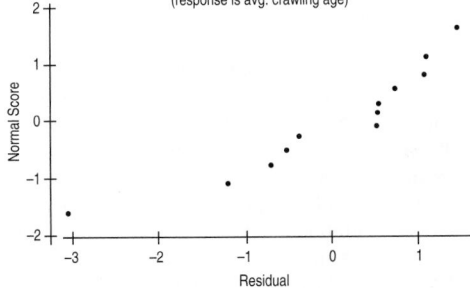

Normal Probability Plot of the Residuals
(response is avg. crawling age)

a) Weaker. Individuals are more variable than averages.

b) The residuals plot shows a large negative outlier; the Normal probability plot is not very straight, but the sample size is small. We will proceed with the inference with caution. H_0: No linear relationship between *6-Month Temperature* and *Crawling Age*, $\beta_1 = 0$. H_A: Linear relationship between *6-Month Temperature* and *Crawling Age*, $\beta_1 \neq 0$. t with 10 df is -3.10, with P-value 0.0113. With a P-value so low, we reject H_0. There's evidence of a negative relationship between temperature and crawling age.

c) $(-0.134, -0.022)$ weeks per °F.

35. a) H_0: No linear relationship between *Waist* size and *%Body Fat*, $\beta_1 = 0$. H_A: *%Body Fat* changes with *Waist* size, $\beta_1 \neq 0$. $t = 8.14$; P-value < 0.0001. There's evidence that *%Body Fat* seems to increase with *Waist* size.

b) With 95% confidence, mean *%Body Fat* for people with 40-inch waists is between 23.58 and 29.02, based on this regression.

36. a) $(0.145, 0.355)$ % per pound.

b) Based on the regression, average *%Body Fat* increases between 0.145% and 0.355% for each pound of weight.

c) With 95% confidence, a person weighing 165 pounds will have between 0% (-1.61%) and 29.32% body fat.

37. a) The regression model is $\widehat{Midterm2} = 12.005 + 0.721\ Midterm1$

	Estimate	Std Error	t-ratio	P-value
Intercept	12.00543	15.9553	0.752442	0.454633
Slope	0.72099	0.183716	3.924477	0.000221

RSquare	0.198982
s	16.78107
n	64

b) The scatterplot shows a weak, somewhat linear, positive relationship. There are several outlying points, but removing them only makes the relationship slightly stronger. There is no obvious pattern in the residual plot. The regression model appears appropriate. The small P-value for the slope shows that the slope is statistically distinguishable from 0 even though the R^2 value of 0.199 suggests that the overall relationship is weak.

c) No. The R^2 value is only 0.199 and the value of s of 16.8 points indicates that she would not be able to predict performance on *Midterm2* very accurately.

38. a) The regression model is $\widehat{MT\ total} = 46.062 + 1.580\ Homework$

	Estimate	Std Error	t-ratio	P-value
Intercept	46.0619317	14.4608874	3.18527698	0.00226414
Slope	1.5800598	0.197715	7.99160308	4.0922e-11

RSquare	0.507412
s	18.30433
n	64

b) The scatterplot shows a strong, mostly linear, positive relationship between midterm total and homework scores. There is a model outlier and a high influence point, but the model is not significantly changed by deleting both points. There is no obvious pattern in the residual plot. The regression model appears appropriate. The small P-value for the slope shows that the slope is statistically distinguishable from 0.

c) The R^2 value of 0.507 suggests that the overall relationship is fairly strong. However, this does not mean that midterm total is accurately predicted from homework scores. The error standard deviation of 18.30 points indicates that a prediction of midterm total could easily be off by 20 to 30 points or more. If this is a significant number of points for deciding grades, then homework score alone will not suffice.

39. H_0: Slope of *Effectiveness* vs *Initial Ability* $= 0$; H_A: Slope $\neq 0$

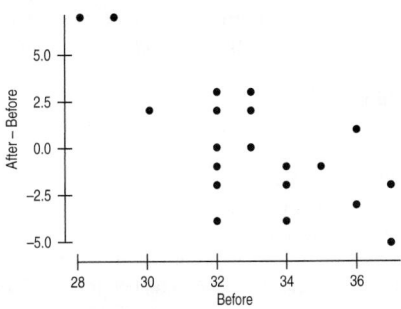

Scatterplot is straight enough. Regression conditions appear to be met. $t = -4.34$, df $= 19$, P-value $= 0.004$. With a P-value this small, we reject the null hypothesis. There is strong evidence that the effectiveness of the video depends on the player's initial ability. The negative slope observed that the method is more effective for those whose initial performance was poorest and less so for those whose initial performance was better. This looks like a case of regression to the mean. Those who were above average initially tended to be worse after training. Those who were below average initially tended to improve.

40. a) We'd like to know if there is a linear association between *Price* and *Fuel Efficiency* in cars. We have data on 2004 model-year cars giving their highway mpg and retail price. $H_0: \beta_1 = 0$ (no linear relationship); $H_A: \beta_1 \neq 0$.

b) The scatterplot fails the Straight Enough Condition. It shows a bend and it has an outlier. There is also some spreading from right to left, which would violate the Plot Thickens Condition. We cannot continue the analysis.

c) The conditions are not met; the regression equation should not be interpreted.

41. a) Data plot looks linear; no overt pattern in residuals; histogram of residuals roughly symmetric and unimodal.

b) H_0: No linear relationship between *Education* and *Mortality*, $\beta_1 = 0$. $H_A: \beta_1 \neq 0$. $t = -6.24$; P-value < 0.001. There is evidence that cities in which the mean education level is higher also tend to have a lower mortality rate.

c) No. Data are on cities, not individuals. Also, these are observational data. We cannot predict causal consequences from them.

d) $(-65.95, -33.89)$ deaths per 100,000 people.

e) *Mortality* decreases, on average, between 33.89 and 65.95 deaths per 100,000 for each extra year of average *Education*.

f) Based on the regression, the average *Mortality* for cities with an average of 12 years of *Education* will be between 874.239 and 914.196 deaths per 100,000 people.

42. a) Data plot looks linear; no obvious pattern in residuals; Normal plot of residuals is roughly linear.

b) H_0: No linear relationship between *Square Footage* and *Assessed Value*, $\beta_1 = 0$. H_A: Larger houses have higher assessed values, $\beta_1 > 0$. $t = 2.77$; P-value $= 0.0068$. With the very low P-value, we reject H_0. There is evidence that larger houses do have higher assessed values.

c) 32.5%

d) I am 90% confident that average *Assessed Value* increases between $4.58 and $19.22 for each square foot.

e) No. These are observational data, not an experiment. We cannot conclude anything about the consequences of changes. Also, the data are for one city and may not apply to others.

f) No. A 95% confidence interval for the assessed value of his 2100-square-foot house is from $45,619 to $78,572.

PART VII REVIEW

1. H_0: The proportions are as specified by the ratio 1:3:3:9; H_A: The proportions are not as stated. $\chi^2 = 5.01$; df $= 3$; P-value $= 0.1711$. Since $P > 0.05$, we fail to reject H_0. These data do not provide evidence to indicate that the proportions are other than 1:3:3:9.

2. a) 59 products b) 84.5% c) $(2.21, 2.78)$ dollars per minute

d) Average price increases between $2.21 and $2.78 for each extra minute of polishing time, with 95% confidence.

3. a) H_0: *Mortality* and *calcium concentration* in water are not linearly related, $\beta_1 = 0$; H_A: They are linearly related, $\beta_1 \neq 0$.

b) $t = -6.73$; P-value < 0.0001. There is a significant negative relationship between calcium in drinking water and mortality.

c) $(-4.19, -2.27)$ deaths per 100,000 for each ppm calcium.

d) Based on the regression, we are 95% confident that mortality (deaths per 100,000) decreases, on average, between 2.27 and 4.19 for each part per million of calcium in drinking water.

4. a) Based on these data, with 95% confidence 5-year yields are between 3.15 and 5.93% higher than 3-year yields, on average (paired data).

b) Yes (at least for this data set). The regression line is $\widehat{5\text{-}year} = 6.93 + 0.719\,\widehat{3\text{-}year}$. $H_0: \beta_1 = 0$ against $H_A: \beta_1 \neq 0$ has $t = 4.27$; P-value $= 0.0009$. Since P is so small, we reject H_0. There is evidence of an association. (But we don't know that this was an SRS or even a representative sample of large cap funds.)

5. 404 checks

6. a) $(36.21, 60.51)$ feet. b) Yes, 40 is in the interval.

c) Wider; we'd need to have a wider interval to be more confident.

d) Roughly 4 times as big—44 flights.

7. H_0: *Income* and *Party* are independent. H_A: *Income* and *Party* are not independent. $\chi^2 = 17.19$; P-value $= 0.0018$. With such a small P-value, we reject H_0. These data show evidence that income level and party are not independent. Examination of components suggests Democrats are most likely to have low incomes; Independents are most likely to have middle incomes, and Republicans are most likely to have high incomes.

8. a) 38

b) Yes. Data plot looks linear; residuals plot shows random scatter; histogram of residuals approximately Normal.

c) $(0.131, 0.177)$ foals per adult.

d) Based on this regression, we have 95% confidence that for every adult horse, an average of between 0.131 and 0.177 foals will be born.

e) A herd with 80 adult horses will have between 2.27 and 19.21 foals, with 90% confidence.

9. $H_0: p_L - p_R = 0$; $H_A: p_L - p_R \neq 0$. $z = 1.38$; P-value $= 0.1683$. Since $P > 0.05$, we do not reject H_0. These data do not provide evidence of a difference in musical abilities between right- and left-handed people.

10. a) Combining boys and girls, df $= 4$, $\chi^2 = 42.80$; P-value < 0.0001. Reject H_0. These data show evidence that the distribution of scores at this school is significantly different from national results.

b) Pooling scores of 2 or 1, df $= 3$, $\chi^2 = 5.59$, P-value $= 0.1336$. There is no significant difference in the distribution of scores for boys and girls.

11. a) $H_0: \mu_D = 0$; $H_A: \mu_D \neq 0$.
Boxplot of the differences indicates a strong outlier (1958). With the outlier kept in, the t-stat is 0, with a P-value of 1.00 (two sided). There is no evidence of a difference (on averag of actual and that predicted by Gallup. With the outlier taken out, the t-stat is still only -0.8525 with a P-value of 0.4106, so the conclusion is the same.

b) H_0: There is no (linear) relationship between predicted and actual number of Democratic seats won ($\beta_1 = 0$). H_A: There is a relationship ($\beta_1 \neq 0$). The relationship is very strong, with an R^2 of 97.7%. The t-stat is 22.56. Even with only 12 df, this is clearly significant (P-value < 0.0001). There is an outlying residual (1958), but without it, the regression is even stronger.

12. Conditions are met; df $= 4$; $\chi^2 = 6.14$; P-value $= 0.1887$. Since $P > 0.05$, we do not reject H_0. We do not have evidence of an association between duration of pregnancy and level of care.

13. Conditions are met; df $= 4$; $\chi^2 = 0.69$; P-value $= 0.9526$. Since $P > 0.05$, we do not reject H_0. We do not have evidence that the way the hospital deals with twin pregnancies has changed.

14. Conditions are met. $H_0: p_P - p_N = 0$; $H_A: p_P - p_N > 0$. $z = 1.96$; P-value $= 0.0249$. With such a small P-value, we reject H_0. These data provide moderately strong evidence to suggest that preemies are more likely than babies of normal weight to be of "subnormal height" as adults.

15. a) Based on these data, the average annual rainfall in LA is between 11.65 and 17.39 inches, with 90% confidence.

b) About 46 years

c) No. The regression equation is $\widehat{Rain} = -51.684 + 0.033 \times Year$. $R^2 = 0.1\%$. For the slope, $t = 0.12$ with P-value $= 0.9029$.

16. a) Independence—individuals classified on two variables.

b) H_0: No relationship between *Age* and *Party*; H_A: There is a relationship. df $= 6$, $\chi^2 = 16.66$; P-value $= 0.0106$. With such a small P-value, we reject H_0. These data indicate a relationship between *Age* and *Party*.

c) There seems to be a relationship between age and party: The "largest" standardized residual is for Republicans 18 to 29 (-2.07). The next-largest is for Independents in the same age group (1.79). These data indicate that young people are much more likely to be Independents than Republicans. We also see that the oldest group is more likely to be Democrat (1.57).

17. a) Linear regression is meaningless—the data are categorical.
    b) This is a two-way table that is appropriate. H_0: *Eye* and *Hair* color are independent. H_A: *Eye* and *Hair* color are not independent. However, four cells have expected counts less than 5, so the χ^2 analysis is not valid unless cells are merged. However, with a χ^2 value of 223.6 with 16 df and a P-value < 0.0001, the results are not likely to change if we merge appropriate eye colors.

18. a) H_0: There is no (linear) relationship between *Depression* and *Internet usage*. $\beta_1 = 0$. H_A: There is a linear relationship. $\beta_1 \neq 0$. $t = 2.76$; P-value $= 0.0064$. With such a small P-value, we reject H_0. These data provide strong evidence of a relationship between Internet use and depression.
    b) The study says nothing about causality. Many other factors may be involved.
    c) H_0: $\mu_D = 0$; H_A: $\mu_D \neq 0$. $t = -2.73$; P-value $= 0.014$. With such a small P-value, we reject H_0. These data suggest that mean depression level actually got better (decreased) during the study.

19. a) H_0: $p_Y - p_O = 0$; H_A: $p_Y - p_O \neq 0$. $z = 3.56$; P-value $= 0.0004$. With such a small P-value, we reject H_0. We conclude there is evidence of a difference in effectiveness; it appears the methods are not as good for older women.
    b) $\chi^2 = 12.70$; P-value $= 0.0004$. Same conclusion.
    c) The P-values are the same; $z^2 = (3.563944)^2 = 12.70 = \chi^2$.

20. a) Answers will vary. Stacked bar charts or pie charts would work well.
    b) Chi-square test of independence. One person's response should not influence another's. We have counts, an SRS of less than 10% of the population, and all expected cell frequencies are much larger than 5. With 8 df, $\chi^2 = 190.96$. Because the P-value < 0.001, reject the null. There is strong evidence to suggest that responses are not independent of age.

21. a) Positive direction, generally linear trend; moderate scatter.
    b) H_0: There is no linear relationship between *Interval* and *Duration*. $\beta_1 = 0$. H_A: There is a linear relationship, $\beta_1 \neq 0$.
    c) Yes; histogram is unimodal and roughly symmetric; residuals plot shows random scatter.
    d) $t = 27.1$; P-value ≤ 0.001. With such a small P-value, we reject H_0. There is evidence of a positive linear relationship between duration and time to next eruption of Old Faithful.
    e) The average time to next eruption after a 2-minute eruption is between 53.24 and 56.12 minutes, with 95% confidence.
    f) Based on this regression, we will have to wait between 63.23 and 87.57 minutes after a 4-minute eruption, with 95% confidence.

22. a) H_0: $\beta_1 = 0$; H_A: $\beta_1 \neq 0$. $t = 4.16$, P-value < 0.0001. These data show evidence of a positive relationship between number of meals eaten together and grades.
    b) No. R^2 is small and $s = 0.66$ points. So we could predict only to within 1.32 grade points at best.
    c) No. The slope is clearly not 0, but that doesn't mean the relationship is strong or the predictions are useful.

23. a) $t = 1.42$, df $= 459.3$, P-value $= 0.1574$. Since P > 0.05, we do not reject H_0. There's no evidence the two groups differed in ability at the start of the study.

b) $t = 15.11$; P-value < 0.0001. The group taught using the accelerated Math program showed a significant improvement.
    c) $t = 9.24$; P-value < 0.0001. The control group showed a significant improvement in test scores.
    d) $t = 5.78$; P-value < 0.0001. The Accelerated Math group had significantly higher gains than the control group.

24. H_0: $p = 0.512$; H_A: $p < 0.512$. $z = -3.35$; P-value $= 0.0004$. With such a small P-value, we reject H_0. These data provide evidence that exposure to dioxin reduces the rate of male births.

25. a) The regression—he wanted to know about association.
    b) There is a moderate relationship between cottage cheese and ice cream sales; for every million pounds of cottage cheese, 1.19 million pounds of ice cream are sold, on average.
    c) Testing if the mean difference is 0 (matched t-test). Regression won't answer this question.
    d) The company sells more cottage cheese than ice cream, on average.
    e) part (a)—linear relationship; residuals have a Normal distribution; residuals are independent with equal variation about the line. (c)—Observations are independent; differences are approximately Normal; less than 10% of all possible months' data.
    f) About 71.32 million pounds.   g) (0.09, 2.29)
    h) From this regression, every million pounds of cottage cheese sold is associated with an increase in ice cream sales of between 0.09 and 2.29 million pounds.

26. $\chi^2 = 14.96$; P-value $= 0.0006$. With such a small P-value, we reject H_0. These data provide evidence that continued treatment with inflixamab is of value.

27. Based on these data, the average weight loss for the clinic is between 8.24 and 10.06 pounds, with 95% confidence. The clinic's claim is plausible.

28. a) Yes. Data plot is linear; residuals plot shows random scatter; histogram of residuals is roughly normal.
    b) Yes; $t = 5.19$; P-value ≤ 0.0001.
    c) Weaker. Individuals are more variable than averages.
    d) Based on this regression, every extra year of median education in a city is associated with an increase of between $1500.48 and $3389.10 in median income.
    e) Based on the regression, median income for cities where residents spent an average of 11 years in school is between $32,198.81 and $33,526.67, with 90% confidence.

29. $\chi^2 = 8.23$; P-value $= 0.0414$. There is evidence of an association between *cracker type* and *bloating*. Standardized residuals for the gum cracker are -1.32 and 1.58. Prospects for marketing this cracker are not good.

30. a) H_0: $\mu_1 - \mu_2 = 0$; H_A: $\mu_1 - \mu_2 > 0$. $t = 0.98$; P-value $= 0.1864$. Since P > 0.05, we do not reject H_0. Weeklong study scores were not significantly higher.
    b) H_0: $p_1 - p_2 = 0$; H_A: $p_1 - p_2 \neq 0$. $z = -3.10$, P-value $= 0.0019$. With such a small P-value, we reject H_0. There is evidence of a difference in proportion for passing on Friday; it appears that cramming may be more effective.
    c) H_0: $\mu_D = 0$; H_A: $\mu_D > 0$. $t = 5.17$, P-value < 0.0001. These data show evidence that learning does not last for 3 days because mean score declined.
    d) Based on these data, the average number of words forgotten by crammers is between 3.03 and 7.05, with 95% confidence.
    e) Yes. Regression equation is $\widehat{Monday} = 14.6 + 0.536 \times Friday$. $t = 2.57$; P-value $= 0.0170$

Photo Acknowledgments

Note: Page numbers in **boldface** indicate chapter-level topics; page numbers in *italics* indicate definitions; FE indicates For Example references.

Index of TI Tips

Tables

Row					TABLE OF RANDOM DIGITS					
1	96299	07196	98642	20639	23185	56282	69929	14125	38872	94168
2	71622	35940	81807	59225	18192	08710	80777	84395	69563	86280
3	03272	41230	81739	74797	70406	18564	69273	72532	78340	36699
4	46376	58596	14365	63685	56555	42974	72944	96463	63533	24152
5	47352	42853	42903	97504	56655	70355	88606	61406	38757	70657
6	20064	04266	74017	79319	70170	96572	08523	56025	89077	57678
7	73184	95907	05179	51002	83374	52297	07769	99792	78365	93487
8	72753	36216	07230	35793	71907	65571	66784	25548	91861	15725
9	03939	30763	06138	80062	02537	23561	93136	61260	77935	93159
10	75998	37203	07959	38264	78120	77525	86481	54986	33042	70648
11	94435	97441	90998	25104	49761	14967	70724	67030	53887	81293
12	04362	40989	69167	38894	00172	02999	97377	33305	60782	29810
13	89059	43528	10547	40115	82234	86902	04121	83889	76208	31076
14	87736	04666	75145	49175	76754	07884	92564	80793	22573	67902
15	76488	88899	15860	07370	13431	84041	69202	18912	83173	11983
16	36460	53772	66634	25045	79007	78518	73580	14191	50353	32064
17	13205	69237	21820	20952	16635	58867	97650	82983	64865	93298
18	51242	12215	90739	36812	00436	31609	80333	96606	30430	31803
19	67819	00354	91439	91073	49258	15992	41277	75111	67496	68430
20	09875	08990	27656	15871	23637	00952	97818	64234	50199	05715
21	18192	95308	72975	01191	29958	09275	89141	19558	50524	32041
22	02763	33701	66188	50226	35813	72951	11638	01876	93664	37001
23	13349	46328	01856	29935	80563	03742	49470	67749	08578	21956
24	69238	92878	80067	80807	45096	22936	64325	19265	37755	69794
25	92207	63527	59398	29818	24789	94309	88380	57000	50171	17891
26	66679	99100	37072	30593	29665	84286	44458	60180	81451	58273
27	31087	42430	60322	34765	15757	53300	97392	98035	05228	68970
28	84432	04916	52949	78533	31666	62350	20584	56367	19701	60584
29	72042	12287	21081	48426	44321	58765	41760	43304	13399	02043
30	94534	73559	82135	70260	87936	85162	11937	18263	54138	69564
31	63971	97198	40974	45301	60177	35604	21580	68107	25184	42810
32	11227	58474	17272	37619	69517	62964	67962	34510	12607	52255
33	28541	02029	08068	96656	17795	21484	57722	76511	27849	61738
34	11282	43632	49531	78981	81980	08530	08629	32279	29478	50228
35	42907	15137	21918	13248	39129	49559	94540	24070	88151	36782
36	47119	76651	21732	32364	58545	50277	57558	30390	18771	72703
37	11232	99884	05087	76839	65142	19994	91397	29350	83852	04905
38	64725	06719	86262	53356	57999	50193	79936	97230	52073	94467
39	77007	26962	55466	12521	48125	12280	54985	26239	76044	54398
40	18375	19310	59796	89832	59417	18553	17238	05474	33259	50595

Table Z
Areas under the
standard normal curve

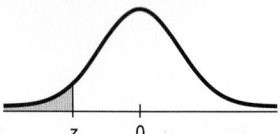

0.09	0.08	0.07	0.06	0.05	0.04	0.03	0.02	0.01	0.00	z
						Second decimal place in z				
0.0001	0.0001	0.0001	0.0001	0.0001	0.0001	0.0001	0.0001	0.0001	0.0001	−3.8
0.0001	0.0001	0.0001	0.0001	0.0001	0.0001	0.0001	0.0001	0.0001	0.0001	−3.7
0.0001	0.0001	0.0001	0.0001	0.0001	0.0001	0.0001	0.0001	0.0002	0.0002	−3.6
0.0002	0.0002	0.0002	0.0002	0.0002	0.0002	0.0002	0.0002	0.0002	0.0002	−3.5
0.0002	0.0003	0.0003	0.0003	0.0003	0.0003	0.0003	0.0003	0.0003	0.0003	−3.4
0.0003	0.0004	0.0004	0.0004	0.0004	0.0004	0.0004	0.0005	0.0005	0.0005	−3.3
0.0005	0.0005	0.0005	0.0006	0.0006	0.0006	0.0006	0.0006	0.0007	0.0007	−3.2
0.0007	0.0007	0.0008	0.0008	0.0008	0.0008	0.0009	0.0009	0.0009	0.0010	−3.1
0.0010	0.0010	0.0011	0.0011	0.0011	0.0012	0.0012	0.0013	0.0013	0.0013	−3.0
0.0014	0.0014	0.0015	0.0015	0.0016	0.0016	0.0017	0.0018	0.0018	0.0019	−2.9
0.0019	0.0020	0.0021	0.0021	0.0022	0.0023	0.0023	0.0024	0.0025	0.0026	−2.8
0.0026	0.0027	0.0028	0.0029	0.0030	0.0031	0.0032	0.0033	0.0034	0.0035	−2.7
0.0036	0.0037	0.0038	0.0039	0.0040	0.0041	0.0043	0.0044	0.0045	0.0047	−2.6
0.0048	0.0049	0.0051	0.0052	0.0054	0.0055	0.0057	0.0059	0.0060	0.0062	−2.5
0.0064	0.0066	0.0068	0.0069	0.0071	0.0073	0.0075	0.0078	0.0080	0.0082	−2.4
0.0084	0.0087	0.0089	0.0091	0.0094	0.0096	0.0099	0.0102	0.0104	0.0107	−2.3
0.0110	0.0113	0.0116	0.0119	0.0122	0.0125	0.0129	0.0132	0.0136	0.0139	−2.2
0.0143	0.0146	0.0150	0.0154	0.0158	0.0162	0.0166	0.0170	0.0174	0.0179	−2.1
0.0183	0.0188	0.0192	0.0197	0.0202	0.0207	0.0212	0.0217	0.0222	0.0228	−2.0
0.0233	0.0239	0.0244	0.0250	0.0256	0.0262	0.0268	0.0274	0.0281	0.0287	−1.9
0.0294	0.0301	0.0307	0.0314	0.0322	0.0329	0.0336	0.0344	0.0351	0.0359	−1.8
0.0367	0.0375	0.0384	0.0392	0.0401	0.0409	0.0418	0.0427	0.0436	0.0446	−1.7
0.0455	0.0465	0.0475	0.0485	0.0495	0.0505	0.0516	0.0526	0.0537	0.0548	−1.6
0.0559	0.0571	0.0582	0.0594	0.0606	0.0618	0.0630	0.0643	0.0655	0.0668	−1.5
0.0681	0.0694	0.0708	0.0721	0.0735	0.0749	0.0764	0.0778	0.0793	0.0808	−1.4
0.0823	0.0838	0.0853	0.0869	0.0885	0.0901	0.0918	0.0934	0.0951	0.0968	−1.3
0.0985	0.1003	0.1020	0.1038	0.1056	0.1075	0.1093	0.1112	0.1131	0.1151	−1.2
0.1170	0.1190	0.1210	0.1230	0.1251	0.1271	0.1292	0.1314	0.1335	0.1357	−1.1
0.1379	0.1401	0.1423	0.1446	0.1469	0.1492	0.1515	0.1539	0.1562	0.1587	−1.0
0.1611	0.1635	0.1660	0.1685	0.1711	0.1736	0.1762	0.1788	0.1814	0.1841	−0.9
0.1867	0.1894	0.1922	0.1949	0.1977	0.2005	0.2033	0.2061	0.2090	0.2119	−0.8
0.2148	0.2177	0.2206	0.2236	0.2266	0.2296	0.2327	0.2358	0.2389	0.2420	−0.7
0.2451	0.2483	0.2514	0.2546	0.2578	0.2611	0.2643	0.2676	0.2709	0.2743	−0.6
0.2776	0.2810	0.2843	0.2877	0.2912	0.2946	0.2981	0.3015	0.3050	0.3085	−0.5
0.3121	0.3156	0.3192	0.3228	0.3264	0.3300	0.3336	0.3372	0.3409	0.3446	−0.4
0.3483	0.3520	0.3557	0.3594	0.3632	0.3669	0.3707	0.3745	0.3783	0.3821	−0.3
0.3859	0.3897	0.3936	0.3974	0.4013	0.4052	0.4090	0.4129	0.4168	0.4207	−0.2
0.4247	0.4286	0.4325	0.4364	0.4404	0.4443	0.4483	0.4522	0.4562	0.4602	−0.1
0.4641	0.4681	0.4721	0.4761	0.4801	0.4840	0.4880	0.4920	0.4960	0.5000	−0.0

For $z \leq -3.90$, the areas are 0.0000 to four decimal places.

Table Z (cont.)
Areas under the
standard normal curve

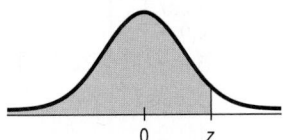

z	Second decimal place in z									
	0.00	0.01	0.02	0.03	0.04	0.05	0.06	0.07	0.08	0.09
0.0	0.5000	0.5040	0.5080	0.5120	0.5160	0.5199	0.5239	0.5279	0.5319	0.5359
0.1	0.5398	0.5438	0.5478	0.5517	0.5557	0.5596	0.5636	0.5675	0.5714	0.5753
0.2	0.5793	0.5832	0.5871	0.5910	0.5948	0.5987	0.6026	0.6064	0.6103	0.6141
0.3	0.6179	0.6217	0.6255	0.6293	0.6331	0.6368	0.6406	0.6443	0.6480	0.6517
0.4	0.6554	0.6591	0.6628	0.6664	0.6700	0.6736	0.6772	0.6808	0.6844	0.6879
0.5	0.6915	0.6950	0.6985	0.7019	0.7054	0.7088	0.7123	0.7157	0.7190	0.7224
0.6	0.7257	0.7291	0.7324	0.7357	0.7389	0.7422	0.7454	0.7486	0.7517	0.7549
0.7	0.7580	0.7611	0.7642	0.7673	0.7704	0.7734	0.7764	0.7794	0.7823	0.7852
0.8	0.7881	0.7910	0.7939	0.7967	0.7995	0.8023	0.8051	0.8078	0.8106	0.8133
0.9	0.8159	0.8186	0.8212	0.8238	0.8264	0.8289	0.8315	0.8340	0.8365	0.8389
1.0	0.8413	0.8438	0.8461	0.8485	0.8508	0.8531	0.8554	0.8577	0.8599	0.8621
1.1	0.8643	0.8665	0.8686	0.8708	0.8729	0.8749	0.8770	0.8790	0.8810	0.8830
1.2	0.8849	0.8869	0.8888	0.8907	0.8925	0.8944	0.8962	0.8980	0.8997	0.9015
1.3	0.9032	0.9049	0.9066	0.9082	0.9099	0.9115	0.9131	0.9147	0.9162	0.9177
1.4	0.9192	0.9207	0.9222	0.9236	0.9251	0.9265	0.9279	0.9292	0.9306	0.9319
1.5	0.9332	0.9345	0.9357	0.9370	0.9382	0.9394	0.9406	0.9418	0.9429	0.9441
1.6	0.9452	0.9463	0.9474	0.9484	0.9495	0.9505	0.9515	0.9525	0.9535	0.9545
1.7	0.9554	0.9564	0.9573	0.9582	0.9591	0.9599	0.9608	0.9616	0.9625	0.9633
1.8	0.9641	0.9649	0.9656	0.9664	0.9671	0.9678	0.9686	0.9693	0.9699	0.9706
1.9	0.9713	0.9719	0.9726	0.9732	0.9738	0.9744	0.9750	0.9756	0.9761	0.9767
2.0	0.9772	0.9778	0.9783	0.9788	0.9793	0.9798	0.9803	0.9808	0.9812	0.9817
2.1	0.9821	0.9826	0.9830	0.9834	0.9838	0.9842	0.9846	0.9850	0.9854	0.9857
2.2	0.9861	0.9864	0.9868	0.9871	0.9875	0.9878	0.9881	0.9884	0.9887	0.9890
2.3	0.9893	0.9896	0.9898	0.9901	0.9904	0.9906	0.9909	0.9911	0.9913	0.9916
2.4	0.9918	0.9920	0.9922	0.9925	0.9927	0.9929	0.9931	0.9932	0.9934	0.9936
2.5	0.9938	0.9940	0.9941	0.9943	0.9945	0.9946	0.9948	0.9949	0.9951	0.9952
2.6	0.9953	0.9955	0.9956	0.9957	0.9959	0.9960	0.9961	0.9962	0.9963	0.9964
2.7	0.9965	0.9966	0.9967	0.9968	0.9969	0.9970	0.9971	0.9972	0.9973	0.9974
2.8	0.9974	0.9975	0.9976	0.9977	0.9977	0.9978	0.9979	0.9979	0.9980	0.9981
2.9	0.9981	0.9982	0.9982	0.9983	0.9984	0.9984	0.9985	0.9985	0.9986	0.9986
3.0	0.9987	0.9987	0.9987	0.9988	0.9988	0.9989	0.9989	0.9989	0.9990	0.9990
3.1	0.9990	0.9991	0.9991	0.9991	0.9992	0.9992	0.9992	0.9992	0.9993	0.9993
3.2	0.9993	0.9993	0.9994	0.9994	0.9994	0.9994	0.9994	0.9995	0.9995	0.9995
3.3	0.9995	0.9995	0.9995	0.9996	0.9996	0.9996	0.9996	0.9996	0.9996	0.9997
3.4	0.9997	0.9997	0.9997	0.9997	0.9997	0.9997	0.9997	0.9997	0.9997	0.9998
3.5	0.9998	0.9998	0.9998	0.9998	0.9998	0.9998	0.9998	0.9998	0.9998	0.9998
3.6	0.9998	0.9998	0.9999	0.9999	0.9999	0.9999	0.9999	0.9999	0.9999	0.9999
3.7	0.9999	0.9999	0.9999	0.9999	0.9999	0.9999	0.9999	0.9999	0.9999	0.9999
3.8	0.9999	0.9999	0.9999	0.9999	0.9999	0.9999	0.9999	0.9999	0.9999	0.9999

For $z \geq 3.90$, the areas are 1.0000 to four decimal places.

Two tail probability	0.20	0.10	0.05	0.02	0.01	
One tail probability	0.10	0.05	0.025	0.01	0.005	
Table T	df						df
Values of t_α	1	3.078	6.314	12.706	31.821	63.657	1
	2	1.886	2.920	4.303	6.965	9.925	2
	3	1.638	2.353	3.182	4.541	5.841	3
	4	1.533	2.132	2.776	3.747	4.604	4
	5	1.476	2.015	2.571	3.365	4.032	5
	6	1.440	1.943	2.447	3.143	3.707	6
	7	1.415	1.895	2.365	2.998	3.499	7
	8	1.397	1.860	2.306	2.896	3.355	8
	9	1.383	1.833	2.262	2.821	3.250	9
	10	1.372	1.812	2.228	2.764	3.169	10
	11	1.363	1.796	2.201	2.718	3.106	11
	12	1.356	1.782	2.179	2.681	3.055	12
	13	1.350	1.771	2.160	2.650	3.012	13
	14	1.345	1.761	2.145	2.624	2.977	14
	15	1.341	1.753	2.131	2.602	2.947	15
	16	1.337	1.746	2.120	2.583	2.921	16
	17	1.333	1.740	2.110	2.567	2.898	17
	18	1.330	1.734	2.101	2.552	2.878	18
	19	1.328	1.729	2.093	2.539	2.861	19
	20	1.325	1.725	2.086	2.528	2.845	20
	21	1.323	1.721	2.080	2.518	2.831	21
	22	1.321	1.717	2.074	2.508	2.819	22
	23	1.319	1.714	2.069	2.500	2.807	23
	24	1.318	1.711	2.064	2.492	2.797	24
	25	1.316	1.708	2.060	2.485	2.787	25
	26	1.315	1.706	2.056	2.479	2.779	26
	27	1.314	1.703	2.052	2.473	2.771	27
	28	1.313	1.701	2.048	2.467	2.763	28
	29	1.311	1.699	2.045	2.462	2.756	29
	30	1.310	1.697	2.042	2.457	2.750	30
	32	1.309	1.694	2.037	2.449	2.738	32
	35	1.306	1.690	2.030	2.438	2.725	35
	40	1.303	1.684	2.021	2.423	2.704	40
	45	1.301	1.679	2.014	2.412	2.690	45
	50	1.299	1.676	2.009	2.403	2.678	50
	60	1.296	1.671	2.000	2.390	2.660	60
	75	1.293	1.665	1.992	2.377	2.643	75
	100	1.290	1.660	1.984	2.364	2.626	100
	120	1.289	1.658	1.980	2.358	2.617	120
	140	1.288	1.656	1.977	2.353	2.611	140
	180	1.286	1.653	1.973	2.347	2.603	180
	250	1.285	1.651	1.969	2.341	2.596	250
	400	1.284	1.649	1.966	2.336	2.588	400
	1000	1.282	1.646	1.962	2.330	2.581	1000
	∞	1.282	1.645	1.960	2.326	2.576	∞
Confidence levels		80%	90%	95%	98%	99%	

Two tails

One tail

Right tail probability		0.10	0.05	0.025	0.01	0.005
Table χ	df					
Values of χ_α^2	1	2.706	3.841	5.024	6.635	7.879
	2	4.605	5.991	7.378	9.210	10.597
	3	6.251	7.815	9.348	11.345	12.838
	4	7.779	9.488	11.143	13.277	14.860
	5	9.236	11.070	12.833	15.086	16.750
	6	10.645	12.592	14.449	16.812	18.548
	7	12.017	14.067	16.013	18.475	20.278
	8	13.362	15.507	17.535	20.090	21.955
	9	14.684	16.919	19.023	21.666	23.589
	10	15.987	18.307	20.483	23.209	25.188
	11	17.275	19.675	21.920	24.725	26.757
	12	18.549	21.026	23.337	26.217	28.300
	13	19.812	22.362	24.736	27.688	29.819
	14	21.064	23.685	26.119	29.141	31.319
	15	22.307	24.996	27.488	30.578	32.801
	16	23.542	26.296	28.845	32.000	34.267
	17	24.769	27.587	30.191	33.409	35.718
	18	25.989	28.869	31.526	34.805	37.156
	19	27.204	30.143	32.852	36.191	38.582
	20	28.412	31.410	34.170	37.566	39.997
	21	29.615	32.671	35.479	38.932	41.401
	22	30.813	33.924	36.781	40.290	42.796
	23	32.007	35.172	38.076	41.638	44.181
	24	33.196	36.415	39.364	42.980	45.559
	25	34.382	37.653	40.647	44.314	46.928
	26	35.563	38.885	41.923	45.642	48.290
	27	36.741	40.113	43.195	46.963	49.645
	28	37.916	41.337	44.461	48.278	50.994
	29	39.087	42.557	45.722	59.588	52.336
	30	40.256	43.773	46.979	50.892	53.672
	40	51.805	55.759	59.342	63.691	66.767
	50	63.167	67.505	71.420	76.154	79.490
	60	74.397	79.082	83.298	88.381	91.955
	70	85.527	90.531	95.023	100.424	104.213
	80	96.578	101.879	106.628	112.328	116.320
	90	107.565	113.145	118.135	124.115	128.296
	100	118.499	124.343	129.563	135.811	140.177

SO-CFM-372

Points to Stress. Under this heading, you will find key concepts and ideas to emphasize in your class lectures. Many of these annotations provide additional information and examples to help you clarify the point for your students.

POINT TO STRESS
A professor may use multiple teaching techniques such as provocative statements, questions, transparencies, and video clips within one lecture. These multiple methods of presenting the material increase the students' attention. Not knowing exactly what to expect next requires us to pay closer attention, thus taking in more information and being ready for surprise questions and other changes in format.

Certain characteristics of objects influence the degree to which these objects capture our attention. The intensity of a stimulus, its contrast with other stimuli in the environment, its familiarity or novelty are all factors that affect our attention. These factors often counterbalance each other and combine in complex ways.

Intensity refers to the overall impact that the external object has on the sensing person. Intense objects are more likely to capture one's attention. For example, although shouting at someone may not be polite, it is probably more likely to capture the person's attention than speaking in a normal tone of voice. Why do you suppose a bright red light is used for traffic "stop" signals? It is because this color has been shown to make the biggest impact physiologically on our visual receptors and hence is most likely to be noticed. Bigger things are also more likely to catch your eye than smaller things, because they excite a greater number of your visual receptors.

Teaching Notes. These sections provide class discussion questions.

TEACHING NOTE
What is the difference between memory and comprehension? Can you memorize something and not comprehend it? Can you comprehend something but not fully remember it?

Finally, *memory*, Nunnally's last dimension of mental ability, is the ability to recall material that was mentally processed in the past. Rote memory may be tested with lists like the one described in Figure 4-1. Memory may also be tested by asking a subject to read a paragraph and then answer questions about its content. Visual memory may be tested by asking subjects to recall images or pictures they have viewed.

Another View. Additional theoretical approaches useful in stimulating class discussion.

ANOTHER VIEW
Failure to support a theory may suggest the theory is wrong, but it is also possible that the design of the research was faulty. Inappropriate measures or inadequate controls may muddy interpretations of data. For example, it is often argued that low organizational commitment and low job satisfaction are antecedents of high absenteeism, but empirical evidence supporting this theory is weak. Recently, investigators have focused on defining absenteeism more precisely. Involuntary absenteeism (illness of self or child) would not be related to commitment or satisfaction whereas voluntary absenteeism (simply not wanting to go to work) would.

it is true. If there is very little correspondence between the hypothesized results and the actual findings, the theory must be rejected. The process must begin all over again, with the generation of a new theory. If there is only some correspondence between the projected and actual findings, the theory may need to be changed in some way so as to be consistent with the data. If there is almost complete correspondence between the hypothesized results and the actual findings, we may be tempted to claim that we have proven that the theory is true. Such a conclusion would not be warranted, however, unless we could establish that all other possible explanations for the results (explanations not accounted for by the theory) had been eliminated. Because this is almost never possible, we usually refer to data that correspond closely with a hypothesis as "supporting" rather than "proving" the theory.

Thus the scientific process relies on the constant interplay of theory and data. One without the other is no better than neither at all. For example, in the Middle Ages, an elaborate theory of body chemistry led early physicians to practice bloodletting, in which they applied leeches to various parts of the human body in an effort to heal various illnesses. The problem was that no one ever tested the theory to see if those who were treated in this fashion actually were better off than those who were not. Had such data been collected, people would have realized that the theory was false. Scientific knowledge requires both theory and data, and great mistakes can be made when one is divorced from the other.

Applying the Diagnostic Model. A cross-reference between the text and end-of-chapter Diagnostic Questions, this annotation is helpful in reminding students to think about how they would apply text concepts to work-life situations.

script A schema that involves well-known sequences of action

APPLYING THE DIAGNOSTIC MODEL
Diagnostic Question 5: What are some of the major events in your organization that you conceive in terms of scripts? How might your version of these scripts differ from others views of the same scripts?

and your friend chunked a 14-bit sequence into a one-word schema: *date*. Schemas that involve sequences of actions are called **scripts**, for the very good reason that they resemble the material from movies or plays.

Assume that your friend went on to report that he had dinner at his parents' house on Thursday, studied with friends on Wednesday, went to a college basketball game on Tuesday, and washed and waxed his car on Monday. You can probably envision scripts for all five activities that your friend engaged in Monday through Friday and in doing so chunk over a hundred bits of information into five tight pieces. This kind of efficiency makes scripts highly useful from an information-processing point of view.

ABOUT THE AUTHORS

John A. Wagner III is Associate Professor of Management in the College of Business at Michigan State University. Professor Wagner received his Ph.D. degree in Business Administration from the University of Illinois at Urbana-Champaign in 1982. At Illinois and at Michigan State, Professor Wagner has taught both undergraduate and graduate courses in management, organizational behavior, and organization theory and design.

A member of the editorial board of *Administrative Science Quarterly*, Professor Wagner also reviews frequently for such journals as the *Academy of Management Review*, *Organizational Behavior and Human Decision Processes*, and the *Journal of Applied Psychology*.

Professor Wagner is a member of the Academy of Management, the Midwest Academy of Management, the Strategic Management Society, and the Decision Sciences Institute.

In 1989, the Personnel/Human Resources Management Division of the Academy of Management presented Professor Wagner with its Scholarly Achievement Award.

Professor Wagner's primary research interests are in the fields of organization theory and organizational behavior. Some of his principal publications have examined the efficacy of participatory decision making, the long-term effects of incentive payment on group productivity, the influence of size on the performance of groups and organizations, and the meaning and measurement of organizational collectivism.

Professor Wagner has served as a consultant to several state and federal government agencies on projects aimed at downsizing organizational operations and streamlining procedures.

John R. Hollenbeck is Associate Professor of Management in the College of Business at Michigan State University. Professor Hollenbeck, who received his Ph.D. in Business Administration from New York University in 1984, teaches at both the undergraduate and graduate level, specializing in organizational behavior, human resource management, and organizational research methods.

In 1987 Professor Hollenbeck was presented with the Michigan State University Teacher Scholar Award for undergraduate teaching.

A member of the Executive Committee of the Organizational Behavior Division of the Academy of Management, Professor Hollenbeck serves on the Awards Committee of the Academy's Personnel/Human Resources Division. He is a member of Division 14 (Industrial/Organizational Psychology) of the American Psychological Association, where he has served on the Scientific Affairs Committee.

Professor Hollenbeck is a member of the editorial boards of three journals: *The Academy of Management Journal*, *Organizational Behavior and Human Decision Processes*, and *Personnel Psychology*. In addition, he reviews frequently for the *Journal of Applied Psychology* and the *Academy of Management Review*.

Professor Hollenbeck has published widely in the areas of organizational behavior and human resource management. Currently, his primary research interests are in the areas of goal setting and motivation theory, personnel selection and placement, work attitudes and turnover, and team decision-making under stress. His research has been funded by the Office of Naval Research, the Michigan Department of Public Health, and several private organizations.

PREFACE

Organizational behavior is an interdisciplinary field concerned with understanding and managing people at work. By definition, it is both research and application oriented. Wagner and Hollenbeck's *Management of Organizational Behavior* is intended to help present and future managers learn more about people and work. The authors have translated abstract theory and research results into straightforward and concise explanations integrating many real-world examples. This text is designed to be a complete teaching and learning tool that captures the reader's interest while it provides useful knowledge and develops critical thinking skills.

ANNOTATED INSTRUCTOR'S EDITION

This Annotated Instructor's edition (AIE) has been designed for users of Wagner and Hollenbeck's *Management of Organizational Behavior*. The purpose of the AIE is to provide the instructor with additional examples, ideas, and materials to enhance lectures and with questions to stimulate class discussion. The annotations which appear in the margins are of five types:

Points to Stress: Under this heading you will find key concepts and ideas that merit special emphasis. Where necessary, additional information or examples are provided to enhance students' understanding.

Teaching Notes: These annotations provide questions to stimulate class discussion and to help students apply the material to real-world situations.

Examples: These notes provide additional examples of how and where the text materials apply. A concerted effort has been made to provide examples of companies and events with which students are familiar.

Applying the Diagnostic Model: These brief notations refer you to the diagnostic question listed at the end of the chapter. By referring to these notations, you will be able to draw students' attention to the questions the material poses for assessing behavior in organizations.

Another View: Here you will find additional theoretical approaches and examples that differ from the proposed theoretical models. These can also be very useful in stimulating class discussion.

USE OF THIS TEXT

The philosophy behind the development of this text was to provide a balance between presenting theory and research and application of the material. For students to leave an organizational behavior class with a solid understanding of the complexities of explaining, predicting, and managing behavior in organizations, they must receive a grounding in theory and be directed in ways these theories apply in the real world. In addition to the examples drawn from publications such as *Business Week*, the *Wall Street Journal*, and *The New York Times* about how theories are put into practice, Wagner and Hollenbeck use cases, exercises and diagnostic questions to give students practice in making use of the theory. Students are presented theory and led through the process of diagnosing behavior in organizations, and developing programs to make adjustments in problem areas.

KEY FEATURES

Diagnostic Questions The diagnostic questions that conclude each chapter review significant chapter topics and show the student how the theories and concepts discussed can be applied to the problems of managing people in organizations.

International Boxes bring home the importance of understanding cultural differences when crossing national borders to do business. Examples of how companies adapt their behaviors to accommodate different values, beliefs, and norms are presented.

Management Issues Boxes integrate ethics and business environment issues into the organizational behavior context. Students begin to understand, through these examples, that organizations and the ways in which they function are not static. Understanding organizational behavior helps students realize that organizations must evolve with their environments or cease to exist.

In Practice Boxes present clear examples directly related to the material. Students can see how specific questions have been answered by different companies and how the answers direct management to different behaviors.

Cases and Exercises accompany each chapter. They are designed to give students practice in applying the material to real-world situations. A key learning tool is to be able to practice what has just been studied. It is through this process that students will gain an appreciation for the relevance of the material and begin developing skills in applying it.

ADDITIONAL SUPPLEMENTS

An *Instructor's Manual* contains expanded outlines of each chapter and thorough teaching notes for all in-text activities.

Management Live! The Video Book contains exercises, readings, and self-assessment materials on eighteen topics of interest to managers and people studying to become managers. These materials are enhanced by video clips contained in the package. *Management Live!* is available at a reduced price shrinkwrapped to the text. The videos are available upon adoption of the shrinkwrapped package. As you will see below, the Applications Pack provides teaching notes for tying these videos to Management of Organizational Behavior.

The *Applications Pack* provides additional exercises and cases which the instructor may hand out to the students. Comprehensive teaching notes are also provided to aid the instructor in integrating these additional materials into his/her program. The Applications Pack also provides the Instructor with teaching notes for thoroughly integrating the *ABC News/Prentice Hall Video Library* and the *Management Live!* Video/Book Collection with *Management of Organizational Behavior*. Also included in the Applications Pack are teaching notes for using *Acumen*.

Test Item File, a comprehensive bank of test questions, is comprised of 2,000 multiple choice, true/false and essay questions. Each question is referenced as to its type (factual or applied) and its level of difficulty.

Prentice Hall Test Manager is a sophisticated computerized version of the Test Item File described above. The Test Manager system allows you to add and edit questions as well as to assemble and save tests. Tests may be created either manually or randomly. Two test scrambling options provide a virtually infinite number of different versions of your tests. Test Manager is available in both 5.25″ and 3.5″ IBM formats.

Transparency Masters provide reproductions of all text figures in *Management of Organizational Behavior*. These may be reproduced as overhead transparencies or as student handouts.

Acumen is a powerful computerized managerial assessment development program. Shrinkwrapped to *Management of Organizational Behavior*, it costs only a few dollars more than the text price. *Acumen* is available in both 3.5″ and 5.25″ versions.

ABC News and Prentice Hall

The media age has established video as a dominant influence in American life. Video is one of the most dynamic and effective means of communication you can use to enhance learning in the classroom. But the quality of the video material and how well it relates to your course can make all the difference.

Prentice Hall and ABC News have brought together their talents in academic publishing and global reporting and are proud to present the most comprehensive video Teaching Package available in the college market today.

ABC News/PH Video Library for Management of Organizational Behavior

Prominent and respected anchors, such as David Brinkley, Ted Koppel, and Peter Jennings, bring their insights in organizational behavior into your classroom. ABC and Prentice Hall offer your students a resource of these feature and documentary-style videos, which relate directly to the concepts and applications in **Management of Organizational Behavior.**

The ABC News PH Video Library pulls together critically acclaimed selections from Nightline, Business World, On Business, This Week with David Brinkley, and World News Tonight. The programs are of extremely high production quality, present substantial content, and are hosted by well-versed, well-known anchors.

ABC News/PH Video Library for Management of Organizational Behavior offers selected programs such as:

Legalities of Integrity Tests in Job Interviews (World News Tonight)
Executive Adventures (On Business)
Behind the Scenes of "A Chorus Line" (Nightline)
Akio Morita of Sony (Nightline)
Telecommuting (Business World)

In the Annotated Instructor's Edition of **Management of Organizational Behavior,** the ABC News Icon will appear in the margins to indicate the availability of an appropriate ABC News Video. Further teaching notes are included in the Applications Pack, including discussion questions with suggested answers, which integrate the ABC News Videos with topics covered in **Management of Organizational Behavior** on a chapter-by-chapter basis.

ABC News and Prentice Hall . . . Moving Images, Making Lasting Impressions

The New York Times

and

PRENTICE HALL

present

MANAGEMENT and ORGANIZATIONAL BEHAVIOR
A Contemporary View

A Contemporary View

The New York Times and Prentice Hall, leading publishers in academia and world news are proud to cosponsor A CONTEMPORARY VIEW, a program designed to enhance student access to current and relevant information in the world of finance.

Your students will receive a 16-page dodger—a student version of *The New York Times* containing articles to be used in conjunction with *Management of Organizational Behavior*. The stories in the dodger are actual articles that appeared in current issues of *The New York Times* and relate specifically to the world of organizations. The selected articles include topics such as leadership styles, multicultural management, ethics and the environment, motivation, and emerging organizational structures. An index incorporated with the dodger cross-references each story to the appropriate discussion in the text for easy integration.

The Value of *The New York Times*

Knowledge of world events is invaluable. Reading a premier news publication such as *The New York Times* establishes a practice of staying abreast of the events happening in today's society. Students who deepen their appreciation of print in the learning environment will remain devoted to the medium throughout their professional and personal lives.

Service Is Our Cornerstone

Prentice Hall continues to lead the way in classroom innovation. The key to the Contemporary View Program is relevance. Therefore, a new collection of articles will be available each year. As a professor using *Management of Organizational Behavior*, you are eligible to receive a complimentary one-semester subscription of *The New York Times* for classroom use. Your students may subscribe at a special reduced rate in deliverable areas. To order *The New York Times* for your class, or for more information, call toll free:
1-800-631-1222
Prentice Hall and *The New York Times*. Together, we offer the most comprehensive and beneficial teaching package available.

NOTES

NOTES

NOTES

NOTES

NOTES

NOTES

MANAGEMENT
OF
ORGANIZATIONAL
BEHAVIOR

John A. Wagner III
John R. Hollenbeck

Michigan State University

Prentice Hall, Inc.
Englewood Cliffs, New Jersey 07632

Library of Congress Cataloging-in-Publication Data

WAGNER, JOHN A.,
 Management of organizational behavior/John A. Wagner III, John
R. Hollenbeck.
   p.   cm.
 Includes index.
 ISBN 0-13-556648-7
 1. Organizational behavior   2. Organizational behavior—Case
studies.   I. Hollenbeck, John R.   II. Title.
HD58.7.W24 1992
658.3—dc20   91–34368
                                                CIP

To Mary Jane, Allison, and Jillian Elizabeth Wagner
and
Harold J. Hollenbeck

Acquisition Editor: *Alison Reeves*
Development Editor: *Virginia Otis Locke*
Production Editor: *Esther S. Koehn*
Copy Editor: *Shirley Stone*
Designer: *Lorraine Mullaney*
Cover Designer: *Butler/Udell Design*
Photo Research: *Anita Duncan* and *Teri Stratford*
Page Layout: *Martin J. Behan*
Prepress Buyer: *Trudy Pisciotti*
Manufacturer Buyer: *Robert Anderson*
Supplements Editor: *David M. Scholder*
Marketing Manager: *Sandra Steiner*

© 1992 by Prentice-Hall, Inc.
A Simon & Schuster Company
Englewood Cliffs, New Jersey 07632

Printed in the United States of America
10 9 8 7 6 5 4 3 2 1

ISBN 0-13-556648-7

Prentice-Hall International (UK) Limited, *London*
Prentice-Hall of Australia Pty. Limited, *Sydney*
Prentice-Hall Canada Inc., *Toronto*
Prentice-Hall Hispanoamericana, S.A., *Mexico City*
Prentice-Hall of India Private Limited, *New Delhi*
Prentice-Hall of Japan, Inc., *Tokyo*
Simon & Schuster Asia Pte. Ltd., *Singapore*
Editora Prentice-Hall do Brasil, Ltda., *Rio de Janeiro*

BRIEF CONTENTS

CONTENTS

PART II
INDIVIDUALS IN ORGANIZATIONS

PART III
INTERPERSONAL, GROUP, AND INTERGROUP RELATIONS

10 INTERPERSONAL PROCESSES AND COMMUNICATION 329

PART IV
THE ORGANIZATIONAL CONTEXT

PART V
SUMMARY AND CAPSTONE

PREFACE

Organizational behavior is the study of human behavior in the workplace. This real-world focus makes OB a discipline that embraces both research and the practical application of research findings. Some OB texts tend to emphasize research; some focus on practical applications. Still others try to cover both areas by dotting their explorations of theory and research with applications material. Professors of organizational behavior have long struggled with the issue of whether theory and practice can be presented together effectively.

We think they can. The value of OB research is measured by its usefulness in the real world of business and industry. And what makes OB research useful is having the right tools to apply its theories and concepts. When we started writing this book we set ourselves two primary aims: first, to cover the field of OB with complete accuracy and second, to offer you, the student, solid guidance and practice in using the theories and concepts that we discuss. With these two goals before us, we created a set of tools that will help you acquire the skills and expertise you'll need to be effective managers in the complex world of work.

SMART THINKING: LINKING THEORY AND PRACTICE

Learning a handful of theories in rote fashion might not be too difficult, but it would probably be tedious, and it definitely would not be very useful. Learning theories and how to apply them may be more challenging at the outset but ultimately it is much more worthwhile. The key to the successful application of theory is critical thinking, smart thinking. To help you think critically about the material in this book, both now, as you study it, and later, as you manage organizational behavior in the workplace, we have incorporated several special features into our text: diagnostic issues and questions, diagnostic frameworks, exercises and cases, and in-text examples.

Diagnostic Issues and Questions

The first three chapters of this book are designed to orient you to the field of organizational behavior and to our particular focus on learning how to manage human behavior in organizations. With Chapter 4, we move into the core part of our book, in which each chapter starts out with a section called "Diagnostic Issues." These issues, like traditional learning objectives, will alert you to the significant concerns of each chapter and will guide you in applying the diagnostic approach we outline in Chapter 1, helping you target and solve real-world problems.

At the end of each chapter you will find a set of "Diagnostic Questions" that have several functions. First, they supplement the chapter's review questions in helping you to summarize the important issues covered in the chapter. Second, they supplement the study questions at the ends of cases, guiding you through the process of assembling the facts of a case, diagnosing the problem, and prescribing appropriate action. Finally, these questions offer you valuable guidelines that you'll use throughout your career in analyzing problems/situations and prescribing and implementing effective solutions.

Problem-solving is a big part of a manager's daily job. The tools you'll acquire in studying this book and in participating in classwork will help you build the foundation for a successful management career.

Diagnostic Frameworks

One problem students of organizational behavior often confront is having to learn many different theories of a given phenomenon without any guidance as to what to accept or reject. To eliminate this source of confusion we've created what we call diagnostic frameworks, or informal models that show you how each of several equally valid theories fit together. Even though a theory may fail to explain something completely, it may still have something useful to contribute when combined with other theories. We think you will find that these frameworks will help you organize complex material, learn it more efficiently, and remember it better. Chapters that present such frameworks include Chapter 7, "Motivation and Performance," Chapter 12, "Leadership," and Chapter 16, "Organization Design."

Exercises and Cases

This book's cases and exercises offer students solid practice in applying each chapter's concepts and theories. The cases will help you think about what you would actually do in a particular situation, based on what you've learned. They will also give you a chance to practice using some of the diagnostic questions in working toward solutions of the problems they pose.

The exercises we've selected for this book, which are designed to be performed in the classroom, are realistic models of problems that managers encounter on the job. For this reason they'll give you meaningful practice in applying your knowledge.

Examples

As you read the chapters of our book, you will find a wealth of examples of real people and of companies that are familiar to you. These examples help make the text's theories and concepts more concrete. We present some of these examples in a sentence or two in the course of explaining a particular aspect of a theory. More complex examples are discussed at greater length, in the stories that open each of our chapters or in the "In Practice" series of boxes. And you'll find other examples in the captions that describe the book's photographs of people on the job.

Integrating Ethical and International Dimensions

Because both ethics in business and industry and the international aspects of management are so important, we have integrated our coverage of these topics into the text. Throughout the book, "Management Issues" boxes discuss such subjects as responsibility for the *Exxon Valdez* oil spill, surveillance of employees without their knowledge, and the complex notion of comparable worth. For example, Chapter 9 discusses Canadian employers' struggle to figure out how to compensate jobs that are of equal worth but that are paid widely varying wages.

In many chapters you will find an "International OB" box that in some cases introduces a concept expanded on in the last chapter of the book. These boxes explore such issues as the way work groups are formed in different national cultures, the Japanese notion of *karoshi*, or death from overwork, and the rise of innovation in Taiwan. A box in Chapter 6, for example, traces the evolution of Taiwanese industry from copying foreign designs to creating new ones.

A Structure That Emphasizes Applications

We have structured our book to provide more than one level of application. As we've said, each of our chapters uses diagnostic issues and questions, in-text examples, opening stories, and end-of-chapter cases and exercises to help you apply what you've learned. In addition, in each major part of the book a concluding chapter expands on the applications already described and introduces you to a formal, applied OB field. Chapter 9, on managing individuals, discusses the applied field of human resources management (HRM). Chapter 14, on managing groups, discusses the applied field of organization development (OD). Chapter 18, on managing the organization, discusses the applied field of strategic management (SM) and returns to the topic of OD, this time at the organization-wide level. The last part of the book presents a single chapter in capstone fashion, exploring the challenge of applying OB concepts throughout the world.

THE DEVELOPMENT PROCESS

When we set out to write this book, we wanted to balance theory with practice. We wanted to cover micro and macro OB equally. We wanted the book to be completely accurate. And we wanted our book to be useful, employing a diagnostic perspective throughout. Prentice Hall's market research and development divisions helped us in many ways to achieve our goals. Initially, we worked with market research in creating a detailed questionnaire to survey OB instructors' thoughts on the distinctive features we planned for our book as well as on other general issues.

Out of the nearly 1500 professors of organizational behavior who received this questionnaire, to our great delight, over a third responded. Prentice Hall's market researchers fed this material into a database that tabulated the information and compared the data across variables. This feedback helped us refine our first draft of the book.

Next, a full-time in-house development editor was assigned to our project. This editor's job was to read our manuscript as if she were a student taking a course in OB for the first time. Combining the results of this experience with her editing skills, the editor helped us present the material clearly and interestingly.

Intensive rounds of reviews and reviewer conferences further advanced the development process. During the development of our first two drafts, Prentice Hall obtained over 50 reviews of our manuscript by peers and colleagues. At each stage, several reviewers met with us face to face to discuss issues and ideas. The insights and comments of these reviewers were invaluable in producing the final draft.

The long, intensive process of development has been well worth the effort. We are proud that, with the help of Prentice Hall and our many reviewers, we have provided you with an excellent, clearly written textbook. We are proud to have realized our vision of an accurate and engaging book that will help present and future managers think smart and work effectively.

SUPPLEMENTS PACKAGE

The tools that we've included in this textbook are those we think the most critical for students to acquire. However, a number of highly effective supplements are available to those who want to add a particular emphasis to one or another aspect of the course. Items that are designed strictly for the professor, such as the Test Item File, the Instructor's Manual, and Transparency Masters, are discussed in the preface to the Annotated Instructor's Edition. Following are the items that are of interest to both professors and students.

ABC News/Prentice Hall Video Library for Management of Organizational Behavior. Carefully selected videos from ABC News's award-winning programs bring the real world to life. This library provides a video for each chapter of the text, emphasizing the application of OB concepts. Teaching notes for integrating the videos with the text are included in the Applications Pack that we'll describe shortly.

Management Live! The Video Book; The Video Collection. Developed by Bob Marx, Peter Frost, and Todd Jick, this creative, video-based experiential workbook is available at a discounted price when shrinkwrapped to *Management of Organizational Behavior*. The companion *Video Collection* is available on adoption of the shrinkwrapped package (one per department). It contains exercises, readings, and self-assessment materials on eighteen topics of interest to managers and people studying to become managers. These materials are enhanced by video clips contained in *The Video Collection*. An *Instructor's Manual* to accompany the *Book/Video Collection* is available.

Acumen, Educational Version 2.0. Designed for use on an IBM-compatible personal computer, *Acumen* is a managerial assessment and development program that enables you to evaluate your managerial strengths and weaknesses by responding to 120 statements. The program then compares your graphic, personalized profile with a cross-section of professional managers and gives you a detailed report that suggests ways of improving your skills. Like *Management Live!*, *Acumen: Educational Version* is available at a discounted price when ordered shrinkwrapped to this text.

Applications Pack with Video Guide. The applications pack includes additional exercises and cases that can be used with *Management of Organizational Behavior* as well as teaching notes for both. It also includes teaching notes for the ABC/PH video library, the *Management Live!* video collection and book, and *Acumen*.

Management and Organizational Behavior: A Contemporary View, sponsored by The New York Times and Prentice Hall. This collection of timely newspaper articles gives students current information about topics discussed in the text. The articles come from one of the world's most distinguished newspapers, *The New York Times*, and demonstrate the vital connection between what you learn in the classroom and what is happening in the world about you.

ACKNOWLEDGMENTS

Our book has been influenced by the ideas and suggestions of many people. First, we would like to thank the instructors who reviewed and commented on initial drafts and helped us refine our ideas: Murray R. Baruch, University of Iowa; Robert A. Bolda, University of Michigan-Dearborn; Robert Bontempo, Columbia University; Joel Brockner, Columbia University; Donald Conlon, University of Delaware; Gerald R. Ferris, University of Illinois; Douglas M. Fox, Western Connecticut State University; Terry L. Gaston, Southern Oregon State College; Barrie Gibbs, Simon Fraser University; Stephen G. Green, Purdue University; James L. Hall, Santa Clara University; Nell Hartley, Robert Morris College; Diane Hoadley, University of South Dakota; Russell E. Johannesson, Temple University; Ralph Katerberg, University of Cincinnati; Kenneth A. Kovach, George Mason University; Charles Kuehl, University of Missouri, St. Louis; Vicki LaFarge, Bentley College; Edwin A. Locke, University of Maryland; Gail H. McKee, Roanoke College; Linda L. Neider, University of Miami; Aaron Nurick, Bentley College; Daniel Ondrack, University of Toronto; Christine Pearson, University of Southern California; Gary N. Powell, University of Connecticut; Gerald L. Rose, University of Iowa; Joseph G. Rosse, University of Colorado, Boulder; Carol Sales, Brock University, Ontario; Mel E. Schnake, Valdosta State College; Randall G. Sleeth; E. M. Teagarden, Dakota State University; Lucian Spataro, Ohio University; Gary L. Whaley, Norfolk State University; and David G. Williams, West Virginia University.

Second, we would like to acknowledge the special input we received from the reviewers who examined our text in depth and, meeting with us twice, helped us fine-tune its chapters: Hrach Bedrosian, New York University; Jeannette Davy, Arizona State University; Howard E. Mitchell, Wharton School, University of Pennsylvania; Ronald R. Sims, College of William and Mary; Roger Volkema, American University; Deborah L. Wells,

Creighton University; and Wayne M. Wormley, Drexel University. All of these individuals deserve special thanks for the time and effort they devoted to their task.

Third, we would like to thank the people at Prentice Hall whose patience and guidance enabled us to complete the task of writing a college textbook: Alison Reeves, assistant vice president and executive editor, who introduced us to Prentice Hall and managed our project from beginning to end, and her assistant, Diane Peirano; Garret White, publisher of the business and economics team; Virginia Otis Locke, senior editor, who helped us turn dry, wordy prose into clear, engaging text and forced us to live up to the promises made in this preface; Raymond Mullaney, vice president and editor in chief, who managed our book's development, and his assistant, Asha Rohra; Esther S. Koehn, supervisory production editor, who oversaw the entire production process and brought all of the words, diagrams, and pictures together into a meaningful whole; Sandra M. Steiner, marketing manager, and her assistant, Elizabeth Gamboa; Lorraine Mullaney, designer, who created the book's beautiful design; William Ethridge, editorial director; Jeanne Hoeting, production manager; Frances Russello, managing editor; Christine Wolf, design supervisor; Lorinda Morris-Nantz, director of photo archives; Anita Duncan and Teri Stratford, photo researchers; David Scholder, supplements editor; Bob Anderson, manufacturing buyer; Trudy Pisciotti, pre-press buyer; Elizabeth Robertson, scheduler; Kama Siegel, assistant editor; Lourdes Brun and Frances Falk, former assistants; and Dennis Hogan, former publisher of the business and economics team, who took a chance and signed a contract with first-time textbook authors.

Finally, we owe special thanks to our families, who put up with our occasional absences and our constant preoccupation with the task of writing this book. Without their support and understanding the book would not exist.

We conclude with a special invitation to you, our newest student. We want to know how you like our book and how you feel about the field of organizational behavior. We encourage you to contact us with your ideas, especially your suggestions for making future editions of our book even better. Please write to us at:

Michigan State University
Graduate School of Business Administration
Department of Management
East Lansing, Michigan 48824–1121

John A. Wagner III
John R. Hollenbeck

ACKNOWLEDGMENT OF ILLUSTRATIONS

CHAPTER 15

545, 546 Alen MacWeeney/Onyx; **555** Bartholomew/Gamma-Liaison; **561** Red Morgan; **563** © R. Ian Lloyd, Singapore; **571** Ann States/SABA.

CHAPTER 16

589, 590 Courtesy of General Motors; **594** Gerry Gropp/Sipa Press; **607** Charles Archambault; **615** Richard Howard; **619** Les Stone/Sygma.

CHAPTER 17

637, 638 Courtesy of National Bevpak, a subsidiary of National Beverage Corp.–Ann States/SABA; **650** National Bicycle Industrial Co.; **661** Courtesy of Volvo Cars of North America; **665** Courtesy of Chad Industries, Orange, California.

CHAPTER 18

681, 682 Courtesy of USX Corporation; **686** Robert Holmgren; **692** Steve Winter/Gamma-Liaison; **698** Steven Pumphrey; **702** © Celestial Seasonings Inc., Boulder, Colorado.

CHAPTER 19

731, 732 Caroline Parsons; **735** Nikolai Ignatiev/Matrix; **746** Karen Kasmauski; **754** Bill Gentile/Sipa Press.

C H A P T E R   1

ORGANIZATIONAL BEHAVIOR

Restoring American competitiveness is one of the goals of General Motor's new Saturn division, which aims to build a small car not only as good as Japanese cars but better. GM's poor reputation in the small-car field began to spread in 1970 with the Chevrolet Vega, which was poorly engineered and subject to breakdowns, and moved steadily downhill into the 1990s, when other automakers outdistanced the company in the marketplace. The Saturn models, like the sports touring sedan seen here, have got off to a slow start largely because of the company's determination to ensure both workers' motivation and a high quality product. Such things as teaching employees to inspect their own work and changing engine-mount specifications in mid-production caused delays, but for Saturn president Richard LeFauvre, building a reputation for excellence is worth it: "You absolutely have to bite the bullet on quality."

Although there's no free lunch, one thing comes awfully close: productivity. When it's growing, businesses can do the impossible. Companies can hand out raises, slash prices, and increase profits—sometimes all at once. . . . But for all its potential, productivity has not been living up to its promise lately. Output per worker has been growing, on average, less than 1 percent a year since 1973, compared with a rate of more than 2 percent in the 1960s. . . . [This lag] is clearly a culprit in America's declining competitiveness. U.S. trade rivals are scoring faster productivity growth, and America now ranks near the bottom among industrialized countries.[1]

This quote from a 1987 issue of *Business Week* is as true today as it was a few years ago. Companies throughout North America continue to face the choice of improving productivity or losing ground to aggressive competitors. Imagine yourself as a manager in such a company. Initial assessments indicate that lagging productivity is due to poor employee motivation, and your boss tells you to solve this problem. Your future with the company—and possibly the future of the company itself—may depend on whether you can find a way to improve employee motivation.

To help you decide what to do, you call in four highly recommended management consultants. After analyzing your company's situation, the first consultant states that many of today's jobs are so simple, monotonous, and uninteresting that they thwart employee motivation and fulfillment. As a result, employees become so bored and resentful that productivity falls off. The consultant recommends that you redesign your firm's jobs, making them more complex, stimulating, and fulfilling.

Consultant number two performs her own assessment of your company. As she reviews her findings, she agrees that monotonous work can reduce employee motivation. She says, however, that the absence of clear, challenging goals is an even greater threat to motivation and productivity. She goes on to say that such goals provide performance targets that draw attention to the work to be done and focus employee effort on successful performance. Therefore, the second consultant advises that you solve your company's productivity problem by implementing a program of formal goal setting.

Next, the third consultant conducts an investigation and concedes that both job design and goal setting can improve employee motivation. She suggests, however, that you consider a contingent payment program instead. She explains that contingent payment means paying employees according to their performance instead of giving them fixed salaries or hourly wages. For instance, salespeople may be paid commissions on their sales, production employees may be paid piece-rate wages according to their productivity, or executives may be paid bonuses according to their firm's profitability. The consultant's formal report points out that contingent payment programs change the way wages are *distributed* but not necessarily the *amount* of wages paid to the work force as a whole.

Finally, the fourth consultant examines your situation and agrees that any of the three approaches might work but describes another technique that is often used to deal with motivational problems—allowing employees to participate in decision making. He suggests that such participation gives employees a sense of belongingness or ownership that energizes productivity, and he recites an impressive list of companies—among them, General Motors, IBM, General Electric—that have recently established participatory programs.

[1] Joan Berger, "Productivity: Why It's the Number One Underachiever," *Business Week*, April 20, 1987, pp. 54–55.

In 1989, 58,000 Boeing machinists went out on strike, convinced by their union that wage increases were more desirable than the company's traditional year-end bonuses. Although bonuses don't contribute to benefits like pensions, organizational behavior research indicates that pay for performance is probably the most effective motivator in the workplace. How would you reconcile this finding with union demands for regular, guaranteed pay increases? Source: "Bonus Battles," Fortune, November 6, 1989, p. 9.

Later, alone in your office, you consider the four consultants' reports and conclude that you should probably recommend all four alternatives—just in case one or two of the consultants are wrong. However, you also realize that your company can afford the time and money needed to implement only one of the four approaches. What do you do? Which alternative should you choose?

According to a recent review of research comparing the effectiveness of these alternatives, if you chose the first one, job redesign, productivity would probably rise by about 9 percent.[2] An increase of this size would save your job, keep your company afloat, and probably earn you the company president's eternal gratitude. If you chose the second alternative, goal setting, productivity would probably increase by around 16 percent. This outcome would save your job and your company and might even put you in the running for a promotion. If you chose the third alternative, contingent payment, you could dust off that vice president nameplate hidden in your desk drawer and prepare for a bigger office. Productivity could be expected to increase by approximately 30 percent, ensuring you an executive position with the company until retirement.[3]

But what about the fourth alternative, employee participation in decision making? How might this approach affect productivity where low performance is attributable to poor motivation? Given that contemporary managers are increasingly choosing participatory programs to solve motivation problems, it stands to reason that this alternative should work at least as well as the other three. Surprisingly, however, participation usually has virtually no effect on productivity. It is likely to improve performance only when combined with one or more of the other three alternatives.[4] Consequently, as the manager in our story, if you chose participation you might soon be looking for a new job.

[2] Edwin A. Locke, Dena B. Feren, Vickie M. McCaleb, Karyll N. Shaw, and Anne T. Denny, "The Relative Effectiveness of Four Methods of Motivating Employee Performance," in Changes in Working Life, ed. K. D. Duncan, Michael M. Gruneberg, and D. Wallis, (Chichester, England: John Wiley, 1980), pp. 363–88.

[3] One study suggests that productivity might rise even more dramatically, increasing by more than 100 percent under certain situations and over extended periods of time. See John A. Wagner III, Paul A. Rubin, and Thomas J. Callahan, "Incentive Payment and Nonmanagerial Productivity: An Interrupted Time Series Analysis of Magnitude and Trend," Organizational Behavior and Human Decision Processes 42 (1988), 47–74.

[4] Locke et al., "Relative Effectiveness"; John A. Wagner III and Richard Z. Gooding, "Shared Influence and Organizational Behavior: A Meta-Analysis of Situational Variables Expected to Moderate Participation-Outcome Relationships," Academy of Management Journal 30 (1987), 524–41. For an explanation of why research findings do not support the popular idea that participation is effective, see Wagner and Gooding, "Effects of Societal Trends on Participation Research," Administrative Science Quarterly 32 (1987), 241–62.

IN PRACTICE

OB Gains in Leading Business Schools

Boxes like this one appear throughout this book to illustrate the significance of OB in today's management world. In this vein, if we haven't yet convinced you of the importance of studying OB, consider changes that have recently taken place at the University of Chicago. Long a haven for statistical business approaches, Chicago is developing courses on such OB topics as communication and team building. For Chicago, the change is the result of a soul searching that started when a group of students began challenging the school's numbers-oriented curriculum.

According to the dean of Chicago's business school, John P. Gould, faculty and alumni as well as students became concerned when Chicago rated only eleventh in a national survey of top business schools. A subsequent internal study revealed that Chicago's program had underemphasized the skills needed to manage people. To correct this deficiency, the study recommended new courses in areas such as leadership and management.

Chicago is not alone in making such changes. Dean John H. McArthur recently recruited leading OB scholars from other universities to redirect Harvard Business School toward a new emphasis on the management of people at work. Dean Donald Jacobs of Northwestern University's J. L. Kellogg School saw his MBA program ascend to the top in a national poll after he built a strong OB curriculum. If trends in business school curricula are reliable indicators, expertise in OB is critically important in today's business world.*

* David Greising, "Chicago's B-School Goes Touchy-Feely," *Business Week*, November 27, 1989, p. 140; Bruce Nussbaum and Alex Beam, "Remaking the Harvard B-School, *Business Week*, March 24, 1986, pp. 54–58; and John A. Byrne, "The Best B-Schools," *Business Week*, November 28, 1988, pp. 76–80.

How realistic is this story? Is the predicament it portrays an everyday problem? Echoing the quotation that opened this chapter, experts in the United States have recently bemoaned the fact that American firms' productivity levels are declining compared with foreign competitors. These writers point to "people problems" as an important cause of this situation.[5]

As the "In Practice" box indicates, top U.S. business schools are taking ideas like this seriously. U.S. politicians running for office have even recommended laws or policies that would mandate workplace efforts to improve employee motivation and performance.[6]

[5] Norman Jonas, "No Pain, No Gain: How America Can Grow Again," *Business Week*, April 20, 1987, pp. 68–69; William J. Hampton, "Why Image Counts: A Tale of Two Industries," *Business Week*, June 8, 1987, pp. 138–40; John A. Byrne, "How the Best Get Better," *Business Week*, September 14, 1987, pp. 98–99; Bruce Nussbaum, "Needed: Human Capital," *Business Week*, September 19, 1988, pp. 100–103; and Christopher Farrell and John Hoerr, "ESOPs: Are They Good for You?" *Business Week*, May 15, 1989, pp. 116–23.

[6] See Richard Fly, Douglas Harbrecht, Howard Gleckman, and Lee Walczak, "The Duke and the Democrats," *Business Week*, July 25, 1988, pp. 22–34.

American managers' increasing concern with employee satisfaction and performance owes a great deal to research in organizational behavior and management. A recent M.I.T. study, for example, suggests that better "people management" practices have enabled Japanese firms to sell higher-quality products at lower prices. Toshiba's annual corporate baseball tournament, by reinforcing company spirit, enhances employees' loyalty and commitment.
Source: *"Pacific Rim," Fortune,* 1989, p. 15.

INTERNATIONAL OB

Motivation in Western Europe

Most of the theories and concepts presented in this book are based on research conducted in North America. As a result, these ideas should be used as management tools only in the U.S., Canada, and other countries with similar cultural backgrounds. In other regions throughout the world, local cultural values encourage forms of organizational behavior that differ from the American norm. Chapter 19 examines many of these differences in detail and discusses how to translate material from this book for use abroad. If you are especially interested in international management, we suggest that you read Chapter 19 before proceeding to the central chapters of this book.

In addition to Chapter 19, we have included boxes like this one throughout the book to help you contrast American practices with the way things are done in other countries. These boxes indicate important ways in which organizational behavior varies throughout the world, reminding you of significant international differences.

One such difference concerns the effects of money and participation on work motivation. Although participation has little effect on motivation in the U.S., experience at Saab-Scania, Volvo, and other western European companies indicates that participation *does* have a strong, positive effect on performance in Sweden, Norway, and Germany. Conversely, money has a strong effect on motivation in North America, but workers in Sweden are hardly influenced by incentive payment. How can these differences be explained?

In part they seem to be the consequence of national laws throughout Europe that require that workers be allowed to participate in running their place of work. Owing to these laws, European workers have been participating in decision making for years. They perceive participation as an interesting, rewarding part of their jobs. In contrast, the U.S. does not have any major laws requiring workplace participation, and American workers often have little experience in employee participation, Thus many American employees really don't care whether they participate in making decisions. In sum, American and European workers value participation differently.

The different effects of money on motivation in America

For these workers in Dresden, Germany, participation in decision making is an important motivating factor, but apparently it is not for many United States workers. Money aside, how would you reconcile this apparent lack of interest in making decisions on the job with such American traditions as self-government and the rights of the individual?

and western Europe may also reflect the high income taxes collected in countries like Sweden. Offering Swedish employees more money for greater productivity is not much of an incentive, because they must pay up to 70 percent of any additional income in national income taxes. Little is left over as a reward for working harder. For Americans, however, only about 30 percent of incentive wages are consumed by national income taxes. Consequently, monetary rewards are valued differently by American and European employees.*

---

* Jonathan Kapstein and John Hoerr, "Volvo's Radical New Plant: 'The Death of the Assembly Line'?" *Business Week*, August 28, 1989, pp. 92–93; and Fred E. Emery and Einar Thorsrud, *Democracy at Work* (Leiden, Netherlands: Kroese, 1976).

Thus poor employee motivation is a widespread problem. American managers are painfully aware of the need to improve motivation at work and are struggling to find a way to do it. Our opening story is indeed realistic.

Without the kind of information that the study of organizational behavior provides, managers have no basis for accepting any one consultant's advice or for choosing one particular way to solve people problems instead of another. As a result, managers often make unsound decisions and wait for changes in productivity that never materialize. Fortunately, however, expertise in the field of organizational behavior can provide the insight necessary to avoid making these sorts of mistakes. It is the purpose of this book to provide you with this expertise, introducing you to the field of organizational behavior and helping you to develop the skills you will need to perform as an informed, effective manager. The theories and models we will discuss are meant primarily for application in North America. However, as the "International OB" box indicates, we will also discuss how things are done in other parts of the world so as to give you basic knowledge about international differences in the management of organizational behavior.

WHAT IS ORGANIZATIONAL BEHAVIOR?

organizational behavior (OB) A field of study that endeavors to understand, explain, predict, and change human behavior as it occurs in the organizational context.

Organizational behavior (OB) is a field of study that endeavors to understand, explain, predict, and change human behavior as it occurs in the organizational context. This definition has three corollaries:

1. OB focuses on observable behaviors, such as talking with coworkers, running equipment, or preparing a report. It also deals with internal states — thinking, perceiving, deciding, and similar hidden processes that accompany visible actions.
2. OB studies the behavior of people both as individuals and as members of groups and organizations.
3. OB also analyzes the "behavior" of groups and organizations per se. Neither groups nor organizations "behave" in the same sense that people do. In the organizational context, however, some events occur that cannot be explained in terms of individual behavior. These events must be examined in terms of group or organizational variables.

POINT TO STRESS
OB studies three determinants of behavior in organizations: individuals, groups, and the organizational context.

The approach this book takes to the topic of OB is indicated by its title— *Management of Organizational Behavior.* As the second half of the title, *Organizational Behavior,* suggests, we will introduce and explain the key theories and concepts that make up the field of OB. However, consistent with the first half of the title, *Management of,* we will also focus on the managerial uses of OB, that is, on making OB's concepts and theories useful to those of you who are or will soon be managers.

A MODEL OF SKILL ACQUISITION

Can a textbook really help managers learn how to use OB theories? Can classroom experiences actually help managers become better at their jobs? Managing is something that can be learned only by doing, isn't it? Many managers and potential managers doubt that books and classes on management can help people learn how to manage. Perhaps you share this doubt. However, the five-stage **skill-acquisition model** shown in Table 1–1 indicates how books and classroom activities *can* contribute to the development of effective management skills.[7]

---

[7] Based on information presented in Hubert L. Dreyfus and Stuart E. Dreyfus with Tom Athanasiou, *Mind over Machine: The Power of Human Intuition and Expertise in the Era of the Computer* (New York: Free Press, 1986). For an extended managerial example based on this model, see Robert E. Quinn, *Beyond Rational Management: Mastering the Paradoxes and Competing Demands of High Performance* (San Francisco: Jossey-Bass, 1988).

T A B L E 1-1		
How Managers Learn Management Skills		
STAGE	**GENERAL DESCRIPTION**	**MANAGEMENT ANALOGY**
Novice	Behaving mechanically. Following elementary rules and procedures.	Practicing applying textbook theories in case analyses and experiential exercises.
Advanced Beginner	Memorizing elementary rules. Using rules under various conditions. Developing circumstantial rules.	Discovering workplace cues that indicate applicability of textbook theories.
Competence	Developing "rules of thumb" and the ability to focus on important information.	Developing personalized models of how to solve management problems effectively.
Proficiency	Developing the ability to read situations unconsciously and respond intuitively.	Developing the ability to respond intuitively to particular management problems.
Expertise	Developing the ability to respond intuitively in a wide variety of situations.	Developing the ability to respond intuitively to a wide range of management problems.

Stage 1: Novice

novice   The stage of skill development in which people learn elementary rules and procedures that, followed mechanically, result in actions resembling skilled behaviors.

Developing skills of any kind begins with a **novice** stage. Novices are beginners who learn rules and procedures that when followed consciously and mechanically result in actions resembling skilled behaviors. For example, novice drivers learn a step-by-step procedure to pass slower cars. Pull up behind the car, check your rearview mirror to be sure no one is passing you, edge out to check for oncoming traffic, accelerate around the slower car, pull back into the lane only after you can see in your rearview mirror the whole front end of the car you have passed. By following this procedure carefully, a new driver can mimic the behavior of experienced drivers and pass slower cars without causing a wreck.

Similarly, novice managers and OB students like you can learn basic rules and procedures by studying textbook theories. You can discover how to use these rules and procedures as you complete case analyses or experiential exercises. A case is a narrative depicting an organizational situation. To perform a *case analysis*, you use textbook theories to identify problems in the case and formulate solutions. For example, you may read a chapter on work motivation in this book and use the theories you learn to analyze a case illustrating motivation problems. In performing case analyses, you learn how to use OB theories as tools to solve real-world problems.

An *experiential exercise* is an activity—often completed in a classroom—in which you learn by doing. For instance, after reading a chapter on motivation, you might complete a questionnaire assessing your own needs and motives. You might also work with classmates to determine how to award incentive payment to the members of a fictitious work group. Experiential exercises like these reinforce textbook theories and provide personal insight into how to use them.

Case analyses and experiential exercises give novice managers the chance to practice using textbook information about OB and management much as behind-the-wheel training enables novice drivers to practice textbook driving skills. When students work on cases and exercises in groups they can share what they know and help one another out. In these ways novices learn that theories provide "what-to-do" guidance that can help them cope with management problems in a skillful—though at this point mechanical—manner.

POINT TO STRESS
Cases and experiential exercises provide the novice with real-world situations. By applying textbook material, novices begin to see how theories apply to real-world problems and begin developing a core of experience.

Stage 2: Advanced Beginner

advanced beginner The stage of skill development in which people learn to base behaviors on an expanded set of rules that include both the elementary rules of novices and circumstantial rules discovered through experience.

As novices continue to learn theories and practice applying them, repetition reinforces the rules of skillful behavior, and it becomes easier to follow these rules. At the same time, as they apply rules in different circumstances, novices learn that they cannot follow the same rules in every situation. As they enter the **advanced beginner** stage, they learn to base behaviors on an expanded set of rules that now include both textbook rules and the circumstantial rules that experience has suggested.

By practicing under different road conditions, advanced beginner drivers learn to allow greater distance between cars when, for example, passing at higher speeds or driving in stormy weather. Similarly, entry-level managers advance their skills by learning that certain workplace cues indicate when particular textbook theories should be used as managerial tools and when they should not. For instance, after a year or two of experience a manager might discover that a company's low productivity is caused by a variety of factors besides poor motivation—faulty equipment, inadequate supervision, defective raw materials. He could then conclude that motivation theories would not help him solve productivity problems caused by these other factors.

competence The stage of skill development in which people replace basic rules with advanced rules of thumb that can be altered to fit a wide range of circumstances.

AN EXAMPLE: SOURCE PERRIER S.A.
Some people think that the way Source Perrier S.A. handled the discovery of traces of benzene, a possible carcinogen, in its sparkling water was less than competent. Although the company recalled its famous green bottles around the world, it was initially evasive and vacillating in its statements to the press about the cause of the problem. As a result, although the company finally diagnosed the problem as one of dirty filters and prescribed more frequent cleaning and checking of the filters, Perrier was still struggling a year later to regain its former leading role in the market. ("You Can Lead a Restaurateur to Perrier, But . . . ," *Business Week*, June 25, 1990, pp. 25–26; Alix M. Freedman, "Perrier Finds Mystique Hard to Restore," *The Wall Street Journal*, December 12, 1990, p. B1.)

Stage 3: Competence

As advanced beginners continue to gain experience and learn additional circumstantial rules, the number of rules they must remember becomes potentially overwhelming. At this point, advanced beginners develop **competence**. Instead of mechanically following rules, they formulate complex rules of thumb that can be altered to fit a broad range of circumstances. For example, instead of having a different rule for each driving condition, competent drivers develop personal theories about how to pass other cars that can be quickly adjusted to fit varying circumstances. Similarly, managers develop competence by relying less on the mechanical application of textbook theories and more on experimenting with new combinations of theories and new ways of assessing their applicability. Thus a competent manager might develop a personalized model of how to manage productivity by combining several OB theories of motivation with first-hand knowledge about factory automation and advice from others about the effectiveness of different supervisory styles.

In gaining competence, people learn how to focus on important information. In part, this ability comes from personal trial-and-error learning. Advice and guidance from other experienced individuals can also help sharpen competence. Textbook theories contribute to the development of competence, because they continue to influence the makeup of competent individuals' personal theories. Lessons learned as a novice are not forgotten, even as people grow more skilled.

Stage 4: Proficiency

proficiency The stage of skill development in which people learn how to read situations instinctively and respond to familiar circumstances intuitively, deliberating consciously only in unusual situations.

Competence, which is characterized by conscious thought and deliberate reasoning, develops into **proficiency** when people learn how to read situations instinctively and respond to familiar circumstances intuitively. Proficiency comes from experiencing the same situation so many times that the behaviors required to deal with it become automatic. Once proficiency is achieved, conscious deliberation is required only in unusual situations.

Proficient drivers approaching slower vehicles on a rainy day know intuitively if they are driving too fast and reduce their speed without thinking. However,

By the proficiency stage, many aspects of the job become habit. A worker doesn't think about tightening a drill bit, just as most of us don't think about locking the door or turning off the coffee pot. Managers, while becoming proficient, must be careful about what becomes habit. When acknowledging employees' efforts becomes habit, sincerity may be lost and the praise may lose its positive effect.

expertise The stage of skill development in which individuals develop the ability to act intuitively in a wide variety of situations, rarely needing to deliberate consciously.

a proficient driver experienced only in warm-weather driving must think about what to do when encountering snowy conditions for the first time. Proficiency reverts to conscious rule following until the driver learns how to drive on snow. Similarly, proficient managers do not have to think about which theories to use to solve familiar productivity problems—they know what to do. If poor productivity is caused by an unknown or unfamiliar factor, however, even the most proficient manager must consciously consider different ways of diagnosing the situation and attacking the problem. In doing this he may need to go back to textbook rules and theories.

Stage 5: Expertise

Expertise, the final stage of skill development, involves the ability to act intuitively in a wider variety of situations than the proficient manager can handle. Proficient individuals develop expertise as they become accustomed to so many different situations that few if any are novel enough to trigger conscious deliberation. Expert drivers often have little conscious awareness of their vehicles or driving conditions but retain complete intuitive awareness of the "feel" of the road. They can maintain a steady speed or react to changing conditions without needing to think about what to do. Expert managers develop a similar intuitive understanding that replaces conscious thought. They do not have to think specifically about motivation, supervision, or other such factors to solve a problem with, say, productivity. Instead, their gut instincts tell them that only certain factors cause the sort of low productivity they confront and that only a particular type of program will lead to acceptable improvement. Expert managers know what needs to be done without having to think about it.

THE DIAGNOSTIC MODEL

Expertise and the intuition it involves are based on a combination of personal experience and advice from others. But expertise is also anchored in the textbook theories learned as a novice. An important implication of this statement is that becoming an expert manager can begin with classroom training in which you learn and apply textbook theories to experiential exercises and case analyses. For this reason, exercises and cases are featured throughout this book. Particularly useful to you in analyzing cases will be the **diagnostic model** that is the foundation of our book and that we discuss next (see also Table 1–2).[8]

diagnostic model A four-step model that describes how managers perceive and solve problems and that is both a learning tool and an on-the-job guide.

Description

description Collecting information about a situation without attempting to explain either the cause of the situation or the motives of the people involved in it.

Description is the simple collection of information about a situation. No attempt is made at this point to explain either the cause of the situation or the motives of the people involved in it. Managers can obtain descriptive information in several ways. For instance, they can make personal observations. A manager can walk through a plant to estimate the amount of inventory on hand, observe production speed, and count the number of employees at work. Observation provides quick, first-hand information, but it cannot furnish data about widespread conditions

[8] Our diagnostic model is partly based on information presented in Judith R. Gordon, *A Diagnostic Approach to Organizational Behavior*, 2nd ed. (Boston: Allyn & Bacon, 1987); both models are derived from the action research model presented in Wendell French, "Organization Development Objectives, Assumptions, and Strategies," *California Management Review* 12 (Winter 1969), 23–34. We discuss the action research model in greater detail in Chapter 14.

		CASE ANALYSIS
STEP	MANAGERIAL ACTION	COUNTERPART
Description	Collection information about a situation without explaining causes or motives.	Reading a case and identifying relevant information.
Diagnosis	Determinining basic causes; identifying and stating the major problem.	Using classroom theories and concepts to identify a problem in the case.
Prescription	Specifying an appropriate solution.	Using the same theories and concepts to develop a case solution.
Action	Implementing the solution, evaluating its consequences.	Specifying the actions needed to implement the proposed solution.

TABLE 1-2
Using the Diagnostic Model in Case Analysis

or general situations. To acquire information about what employees think and feel about their work, managers can also distribute questionnaires throughout their firms. The major strength of this method is that it allows access to a wider range of opinions than can be collected through personal observation. However, questionnaires lack flexibility, because people can respond to only a limited number of items. In a third method, the face-to-face interview, a manager can ask questions as they come to mind, and interviewees can ask questions of their own. The major limitation of interviewing is the amount of time it requires. Managers must spend a lot of time interviewing different individuals to get a sense of what is going on in an organization, and employees must stop working at their normal jobs to participate in interview sessions. Fourth and last, written documents such as annual reports, office memoranda, personnel files, and production records can both provide new data and verify information acquired through the other three sources.

As you read this book, you can sharpen your descriptive skills by studying the cases carefully and teasing out the facts of each case. Remember that not all of the information presented in a case is relevant to the solution of a particular problem. Just as managers must separate relevant facts from irrelevant details, in performing a case analysis you must begin by deciding the relevance of different pieces of information and determining what to do where information is unavailable.

Diagnosis

diagnosis Looking for the causes of a troublesome situation and summarizing them in a problem statement.

In the step called **diagnosis**, managers look for the causes of the situation described and attempt to summarize them in a *problem statement*. Depending on their level of skill, managers rely on a mix of textbook theories, experience, and intuition as they diagnose the situation before them.

To begin developing your own diagnostic skills, you should study each theory presented in this book to develop a basic understanding of the variables and relationships it describes. Next, practice using the theory as a tool to help define the problem in a case. Then, as you become comfortable applying the theory, try combining it with several others to develop the type of expanded theory used by advanced beginners. For example, you might combine theories of employee

Management often involves hard decisions with no obvious right or wrong answers. Some of these decisions pose ethical dilemmas in which social values conflict with personal well-being or organizational interests. Consider Perrier's decision to pull potentially tainted Perrier water off the market rather than maintain the short-term profitability of the company and its retailers. Other decisions involve issues of social responsibility in which a company must identify its obligations as a "citizen" in society and decide how to fulfill them. For example, should a business firm leave unemployment programs to the government or try to help society by hiring and training the hard-core unemployed? Other issues concern the practice of management itself. For example, what right do managers have to tell others what to do?

To get you thinking about these kinds of issues, we have included boxes like this one throughout the book. We hope you will consider the issues they raise with care and discuss them with your classmates and instructor.

One basic issue is the idea of social responsibility itself. Should businesses even concern themselves with social responsibility? For one manager's answer to this question, consider remarks made by Henry Ford II while he was the head of Ford Motor Company:

Like governments and universities and other institutions, business is much better at some tasks than at others. Business is especially good at all the tasks that are necessary for economic growth and development. To the extent that the problems of society can be solved by providing more and better jobs, higher incomes of more people and a larger supply of goods and services, the problems can best be solved by relying heavily on business.

On the other hand, business has no special competence in solving many other urgent problems. Businesspeople, for example, know little about the problems involved in improving the education of ghetto children, the quality of ghetto family life, the relations between police and minority citizens or the administration of justice. Solutions to problems such as these will be more effective if they are left to political, educational, and social agencies. In short, our society will be served best if each of its specialized institutions concentrates on doing what it does best, and refuses either to waste its time or to meddle in tasks it is poorly qualified to handle.*

* Henry Ford II, *The Human Environment and Business* (New York: Weybright & Talley, 1970), pp. 30–31.

Many corporate leaders today disagree with Henry Ford about the role of business in dealing with society's problems. Almost all of the 206 CEOs of Fortune and Service 500 companies polled recently by Fortune *magazine believe that corporations must help educate and train workers, and many companies have already developed such programs.*
Sources: *Andrew Erdman, "How to Make Workers Better,"* Fortune, *October 22, 1990.*

Ford is saying that businesses have an obligation to make and use profits to promote economic well-being, and that government agencies should not interfere in this regard. Do you agree with this opinion? From Ford's viewpoint, who is responsible for equal opportunity and affirmative action—businesses that need talented employees of all races or government agencies charged with protecting the rights of all citizens? What do you think about Ford's assertion that different types of agencies and organizations should specialize in dealing with different kinds of social problems?†

† For additional information about this issue, see "The 'Responsible' Corporation: Benefactor or Monopolist?" *Fortune,* November 1973, pp. 56–58; Eli Goldston, "New Prospects for American Business," *Daedalus* 15 (Winter 1969), 78–79; and Milton Friedman, *Capitalism and Freedom* (Chicago: University of Chicago Press, 1962).

motivation, leadership, and job design to try to explain the causes of poor employee performance. Such personalized theories will help you to develop your skills further and to understand the complexity of many organizational problems.

prescription Developing a solution responding to a problem statement identified through diagnosis.

Prescription

Prescription involves developing a solution to the problem one has identified through diagnosis. Organizational problems are often multifaceted, and there is usually more than one way to solve a given problem. Therefore, successful managers usually consider several reasonable alternatives before choosing one to deal with a troublesome situation. Whether they review these alternatives mechanically, consciously, or intuitively depends, of course, on the level of skill they have achieved.

For the student, prescription involves following the theories she has applied during problem definition through to their logical conclusions. For instance, the same theory of employee motivation you use to diagnose a productivity problem may also suggest ways to reduce or eliminate the problem. The more theories you apply during diagnosis, the more comprehensive your final solution will be. Thus as you develop into an advanced beginner, the solutions you devise are likely to become increasingly thorough and more effective.

Action

Action involves implementing a proposed solution. In this step, managers must stipulate the specific actions needed to solve a particular problem. This step will require you, as a student, to indicate the actions required to implement and assess your proposed solution. Your action plan should specify a sequence of steps that indicate what needs to be done, who will do it, when it will be done, and how its effectiveness will be measured.

Planning actions is a value-laden process. Which actions are morally correct? Which ones are not? Managers considering various actions must constantly raise such questions of social responsibility. As the "Management Issues" box indicates, different managers have different ideas about what is right and what is wrong, what is responsible and what is irresponsible. We urge you to think about the ethical implications of the actions you propose, both now as a student of organizational behavior and in the future as a manager.

In sum, the five-stage skill-acquisition model shows you how you can shape your management skills by studying the contents of this book, and the four-step diagnostic model shows you how you can apply your knowledge to case analyses and managerial decision making. These models will help you get the most out of this book, not only as you read it today but as you refer back to it on the job.

Using the Diagnostic Approach

As you study the core chapters of this book, you will find yourself most of the time in one phase or another of our diagnostic model. As Chapter 3 points out, the steps of the model parallel the stages of the scientific method of research, and both model and method inform our approach to the topics of this book. First and foremost, they compel us to ask questions. To describe we need to ask, What are the basic data? To diagnose we need to ask, How are different factors related to each other? To prescribe we need to ask, How will these factors interact in different circumstances? And to take action we need to implement a plan and be ready to evaluate its success—by asking more questions.

To lead you into this information-seeking mode, we have begun each of our core chapters with a "Diagnostic Issues" section that previews for you the questions the chapter undertakes to answer. When you have finished the chapter you will be able to answer not only these initial questions but the more detailed and specific "Diagnostic Questions" that precede the chapter's exercise and cases. The diagnostic questions flesh out the earlier ones and help you apply what you have learned to analyzing the cases. For those of you who will become managers, these questions will also be invaluable guides to on-the-job problem solving.

THE CONTINGENCY PERSPECTIVE

contingency perspective The view that no single theory, procedure, or set of rules is useful in every situation and that each situation determines the usefulness of different management approaches.

Both the skill acquisition model and the diagnostic model suggest that managers draw on a variety of theories in order to cope with different situations. They also propose that managers have to determine which theory to apply in each specific situation. Thus these models both embody the **contingency perspective**, the view that no single theory, procedure, or set of rules is useful in every situation. Instead, according to the contingency perspective, the usefulness of a particular management approach depends on the situation being managed.

For the student and the beginning manager this view is sometimes more confusing than helpful. Too often, textbooks present a long list of theories and then leave the student without guidance in choosing a theory to follow in one or another managerial problem situation. We have tried in our book not only to describe the important theories and concepts of the field but to show you how to diagnose the surrounding situation and choose which specific theory to apply.

CHARTING THE FIELD OF ORGANIZATIONAL BEHAVIOR

Given that OB theories can be useful to managers, what exactly is the field of organizational behavior all about? How is it similar to the field of management? How do OB and management differ? OB traces its origins to the late 1940s, when a group of researchers in psychology, sociology, and other social sciences decided to work together to develop a new, comprehensive body of organizational research.[9] Despite the intentions of its founders, however, the field of OB has resisted unification. It is now divided into two distinct subfields: micro organizational behavior, deriving from psychology and the behavioral sciences, and macro organizational behavior, deriving largely from economics, sociology, and political science.

Micro Organizational Behavior

micro organizational behavior The subfield of OB concerned with understanding the behaviors of individuals working alone or in small groups.

Micro organizational behavior is concerned mainly with understanding the behaviors of individuals working alone or in small groups.[10] Four subfields of psychology were the principal contributors to micro OB. *Experimental psychology* provided theories of learning, motivation, perception, and stress. *Clinical psychology* furnished models of personality and human development. *Industrial psychology* offered theories of employee selection, workplace attitudes, and performance assessment. And *social psychology* supplied theories of socialization, leadership, and group dynamics. Owing to this heritage, micro OB has a distinctly psychological orientation.[11] Among the questions it examines are, What motivates employees to perform their jobs? What effects do differences in ability have on employee performance? Why do some employees feel satisfied with their jobs while others experience stress? What makes leaders effective? How can group performance be improved?

[9] Larry L. Greiner, "A Recent History of Organizational Behavior," in *Organizational Behavior*, ed. Steven Kerr (Columbus, Ohio: Grid Publishing, 1979) pp. 3–14.

[10] Larry L. Cummings, "Toward Organizational Behavior," *Academy of Management Review* 3 (1978), 90–98.

[11] Ibid.

Macro Organizational Behavior

macro organizational behavior The subfield of OB that focuses on understanding the actions of a group or an organization as a whole.

Macro organizational behavior focuses on understanding the "behaviors" of groups and organizations. The origins of macro OB can be traced to four principal disciplines: *Sociology* provided theories of structure, social status, and institutional relations. *Political science* offered theories of power, conflict, bargaining, and control. *Anthropology* contributed theories of symbolism, cultural influence, and comparative analysis. And *economics* furnished theories of competition and efficiency. Research on macro OB considers questions such as, How is power acquired and retained? How can conflicts be resolved? What mechanisms can be used to coordinate work activities? Why do we have different forms of organizational structure? How should an organization be structured in order to cope with surrounding circumstances?[12]

Related Domains of Research

Besides the two OB subfields, three other domains of research also focus on organizational topics. In general, these additional domains of human resource management, organization development, and strategic management arose from the same scientific disciplines as OB. Their origins, however, can also be traced to the field of management. Thus in contrast to OB's theoretical orientation, they emphasize *doing* things to shape behaviors in organizations, such as developing programs, facilitating change, and assessing results.

human resource management A domain of organizational research that focuses on devising practical, effective ways to manage employee behaviors.

Human Resource Management. Human resource management (HRM) is similar to micro OB; they share a common focus on the behavior of individuals. In fact, HRM studies often incorporate theories of micro OB. For example, incentive-payment programs devised in HRM are often based on micro OB theories of employee motivation. However, the two domains differ in one important respect. In contrast to micro OB's emphasis on theory development, HRM focuses mainly on devising practical, effective ways to manage employee behaviors. HRM researchers study different ways to select employees, to train and evaluate them, and to compensate them for their performance.[13]

organization development An area of organizational research that develops techniques to instill cooperation and manage change.

Organization Development. Organization development (OD) is an area of practical research that develops techniques to instill cooperation and manage change.[14] Some OD techniques focus on the personal development of individual organization members. A few others are concerned with the effectiveness of entire organizations. Most OD techniques, however, emphasize group and intergroup relations, highlighting ways in which cooperation can be strengthened and conflict resolved within and between groups. Thus there is a great deal of overlap between OD and micro OB research on behavior in small groups. In addition, there is significant overlap between OD and macro OB studies of group behavior. Yet

[12] Robert H. Miles, *Macro Organizational Behavior* (Santa Monica, Calif: Goodyear, 1980); Richard L. Daft and Richard M. Steers, *Organizations: A Micro/Macro Approach* (Glenview, Ill: Scott, Foresman, 1986).

[13] Wayne F. Cascio, *Applied Psychology in Personnel Management*, 2nd ed. (Englewood Cliffs, N.J.: Prentice Hall, 1989); and George T. Milkovich, *Personnel and Human Resource Management: A Diagnostic Approach* (Plano, Texas: Business Publications, 1985).

[14] Richard Beckhart, *Organizational Development: Strategies and Models* (Reading, Mass.: Addison-Wesley, 1969); and Wendell L. French and Cecil H. Bell, Jr., *Organization Development: Behavioral Science Interventions for Organization Improvement*, 4th ed. (Englewood Cliffs, N.J.: Prentice Hall, 1990).

OD's emphasis on the development of effective management techniques differs from the accent on theory development that characterizes both subfields of OB.

strategic management A domain of organizational research concerned with defining an organization's purpose and planning how to achieve organizational objectives.

Strategic Management. Like macro OB, **strategic management (SM)** focuses more on organizations than on individuals. However, in contrast to macro OB's concern with developing theories that explain organizational events, SM addresses the development of techniques to manage organizations.[15] For example, SM research provides advice about how to set organizational goals, determine competitive postures, and choose which business opportunities to pursue. Thus SM is the practical counterpart of macro OB.

Fitting the Domains Together

We have now identified five relatively distinct domains of current organizational research. Separating these domains are differences in *level of analysis* and *primary orientation*, as you can see in Figure 1-1. The vertical dimension in the figure, **level of analysis**, differentiates among the domains according to whether their primary focus is on the *individual*, the *group*, or the *organization*. The horizontal dimension, **primary orientation**, classifies the domains according to whether their main focus is on *theory* or *application*.

level of analysis A dimension that classifies the five areas of organizational research according to whether their primary focus is on the behaviors of individuals, of groups, or of organizations.

primary orientation A dimension that classifies the five areas of organizational research according to whether their main focus is on abstract theories or practical techniques.

Some commentators have suggested that the five domains pictured in Figure 1-1 should be permitted—or even required—to remain separated from one another.[16] For organizational researchers intent on developing in-depth expertise,

[15] H. Igor Ansoff, *Corporate Strategy* (New York: McGraw-Hill, 1965); Kenneth R. Andrews, *The Concept of Corporate Strategy* (Homewood, Ill: Richard D. Irwin, 1980); and Arthur A. Thompson, Jr., and A. J. Strickland III, *Strategic Management: Concepts and Cases*, 4th ed. (Plano, Texas: Business Publications, 1987).

[16] Cummings, "Toward Organizational Behavior," p. 90.

FIGURE 1-1

Subfields of Organizational Behavior and Domains of Related Research

The field of organizational behavior is divided into two subfields, micro organizational behavior and macro organizational behavior. Closely associated with these two subfields are three other domains of organizational research: human resource management, organization development, and strategic management.

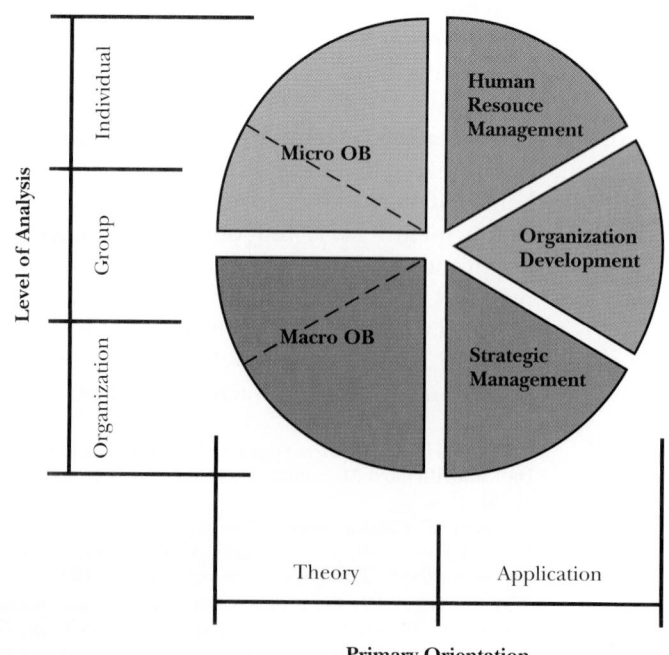

this suggestion makes sense; keeping the five domains separate makes it easier to specialize in a narrow area of knowledge. For present or future managers, however, knowing how to cope with a wide variety of situations is usually more important than being a specialist in any one area. From this perspective, separating the five domains and specializing in only one or two of them means that useful information may be omitted. Thus although this is primarily a book on OB, we also include special chapters on HRM, OD, and SM to provide the breadth of practical knowledge required to develop effective management skills.

OVERVIEW OF THE BOOK

Besides charting the five domains of organizational research, Figure 1-1 also diagrams the contents of this book. The following chapters strike a balance between the two orientations indicated in the figure, both theory and application. They also span all three levels of analysis shown in the figure by considering issues pertaining to individuals, groups, and organizations.

Part I consists of three introductory chapters, including this one, that overview the field of organizational behavior, survey the general topic of management, and briefly outline research methods in organizational behavior. You may read one, two, or all three of these chapters, depending on the focus of your course and the teaching goals of your instructor. Part II, on individuals in organizations, is composed of five chapters that introduce you to various theories about people's behavior in the work setting and a sixth chapter that focuses on the application of these theories to work-related issues. As you can see, this part of the book encompasses the upper left and upper right segments of Figure 1-1. In Part III of the book, on groups in organizations, four chapters present concepts of interpersonal, group, and intergroup behavior and a fifth chapter shows you how these concepts can be applied in managing groups in the organizational setting. Thus this part covers the two middle segments in Figure 1-1. Part IV, on the organizational context, is made up of three chapters that focus on the structure and design of organizations and the jobs that comprise them and a fourth chapter that applies this material to the management of entire organizations. This part, which focuses on change in the organizational context, encompasses the lower left and lower right segments of Figure 1-1. Part V consists of a single, capstone chapter that examines the challenges of applying OB theories throughout the world.

Summary

Organizational behavior is a field of research that helps predict, explain, and understand behaviors occurring in and among organizations. Learning how to use knowledge about OB for managerial purposes begins by studying textbook theories and using them to perform case analyses and experiential exercises. The five-stage *skill acquisition* model and the four-step *diagnostic model* provide helpful guidance for students engaged in this process of initial learning. The latter model in particular, through its steps of *description*, *diagnosis*, *prescription*, and *action*, facilitates the analysis of case material in this book and of real-world, on-the-job managerial problems. In common, both models share grounding in the *contingency perspective*, which suggests that the usefulness of a particular theory or concept depends on the situation being managed.

Organizational behavior's two subfields, *micro organizational behavior* and *macro organizational behavior*, reflect differences among the scientific disciplines that contributed to the founding of this field of study. Micro OB is concerned primarily with individual behavior and the behaviors of people in small groups. Macro OB focuses on the behaviors of people in larger groups and on the "behaviors" of organizations as entities.

Closely related to these subfields of OB are three other domains of organizational study, each of which is primarily concerned with the application of knowledge derived from OB subfields. *Human resources management* uses data from micro OB studies. *Strategic management* makes use of data from macro OB studies. *Organization development* uses data from both micro and macro OB studies.

The *primary orientation* of micro and macro OB is theoretical, whereas human resources management, strategic management, and organization development are applications oriented. The *levels of analysis* at which the latter three fields operate are the individual, the group, and the organization, respectively.

REVIEW QUESTIONS

1. What is the field of organizational behavior? What kinds of behavior does it examine? Why does it include examination of the "behaviors" of groups and organizations?

2. What are the five stages of skill development? What role do textbook theories and concepts play during each of these stages? How will the information you learn from this book affect your development as a manager?

3. What are the four steps of the diagnostic model? How can using this model help you develop managerial skills? Why is it important for you to refer to textbook theories at each step? How can you use your creativity in performing the diagnostic procedure?

4. What is the contingency perspective of management? In what ways is this perspective related to the skill-development process and diagnostic model?

5. What are the two subfields of OB? What are each of them about? Why have they developed separately? Why is it important for you to know about both of them?

6. Name the three domains of organizational research that are closely linked to OB. How is each of these domains linked to one or both subfields of OB? How do they differ from one another?

7. Explain how the two dimensions, level of analysis and primary orientation, map the five areas of organizational research examined in this book. Why is it important to have chapters on practical management—HRM, OD, SM—in a book on OB?

MANAGEMENT AND MANAGERS

According to Lee Iacocca, his mother—who came to the United States through Ellis Island— taught him to revere the Statue as a symbol of the American way of life. Iacocca himself has become a symbol of expert management in American business, although the company he rescued from bankruptcy in the late 1970s continues to be beset by problems. Recently, Iacocca initiated an extensive cost-cutting program and reorganized product development into vehicle rather than component-part teams. He also created the Bet Your Check program in which 2000 top managers put as much as 10 percent of their salaries in escrow on the assurance of doubling their money if they delivered budget cuts on time. They did, and the company paid out $15.4 million.
Source: Alex Taylor III, "Can Iacocca Fix Chrysler—Again?" Fortune, April 8, 1991, pp. 50–54.

Chrysler Corporation is hitting the road with a six-city promotional tour in which Chairman Lee A. Iacocca will tout his cars and his company. . . The number three U. S. auto maker will show off current and future products during the tour, which begins on February 21. . . In each city—New York, Chicago, Atlanta, Dallas, Los Angeles, and Washington, D.C.—Mr. Iacocca will preside over news conferences and receptions for financial analysts, money managers, dealers, and other invited guests . . . [who] will hobnob with Mr. Iacocca and other top company officials, "plus a smattering of vice presidents, group directors, and managers," said a company spokesperson. . . This will be the first time in several years that Mr. Iacocca has made so public a pitch for the company he heads. . . According to the spokesperson, Chrysler has been planning the tour "literally for months." "People are raising questions about Chrysler's viability," he said. "We want to give a positive answer to that question, and we're going to bring out our top salesman to do it."[1]

Chrysler's Lee Iacocca is living proof that successful managers sometimes become "executive celebrities."[2] Designer of the Ford Mustang—the first and most successful "muscle car" of the 1960s—Iacocca joined Chrysler in 1978, when the firm was on the brink of failure. Iacocca soon won public attention by negotiating $1.5 billion worth of federal loan guarantees and secured his reputation by repaying these loans well before the 1990 deadline. Recognizing Iacocca's management expertise and growing worldwide reputation, President Ronald Reagan asked him to head the Statue of Liberty-Ellis Island Centennial Commission. Leading Democrats promoted Iacocca as a potential presidential candidate in 1984 and 1988, but he declined to run. In perhaps the ultimate expression of executive celebrity, Iacocca appeared in a 1985 episode of "Miami Vice." Back in the business world, Iacocca's image as a triumphant manager was clinched when Chrysler purchased Jeep/American Motors from the French auto maker Renault and, with Japan's Mitsubishi Corporation, began the Diamond-Star joint venture to build Plymouth Lasers and Mitsubishi Eclipses.[3]

Although the general public has come to admire Lee Iacocca, few people who are not managers know much about the topic of management. Could you tell someone what management is? How current management practices have developed? What managers do? Probably not. Yet modern societies depend on the well-being of thousands of organizations ranging from industrial giants like Chrysler to local businesses like the corner grocery store. All of these organizations depend on competent management. Ignorance about management would doom a society like ours to failure. Therefore, it is critical that you know what management is and what managers do. This chapter is a primer on management thought. It begins by defining the concept of management and then considers several schools of thought about management and managers. The chapter con-

[1] Melinda Grenier Guiles, "Iacocca Roadshow Will Try to Make Point Chrysler Isn't Down and Out Despite Woes," *Wall Street Journal*, January 24, 1990, p. B4.

[2] Judith H. Dobrzynski and Jo Ellen Davis, "Business Celebrities," *Business Week*, June 23, 1986, pp. 100–105.

[3] Lee Iacocca and William Novak, *Iacocca: An Autobiography* (New York: Bantam Books, 1984); William J. Hampton, "Chrysler's Next Act," *Business Week*, November 3, 1986, pp. 66–72; and Dobrzynski and Davis, "Business Celebrities."

cludes by discussing what managers actually do, focusing on the skills the
and the roles they fill as they perform their jobs.

DEFINING MANAGEMENT

What is management? Stated simply, it is a process of influencing people's behaviors in organizations. To define management in greater detail, let's begin by considering a closely related question: What is an organization?

Three Attributes of Organizations

An **organization** is an arrangement of people and materials brought together to accomplish a purpose that would be beyond the means of individuals working alone. Three attributes enable an organization to achieve this feat: a mission, division of labor, and distribution of authority.

Mission. Each organization works toward a specific **mission**, which is its purpose or reason for being. As you can see in Table 2-1, a statement of mission identifies the primary goods or services the organization produces and the markets it hopes to serve. An organization's mission helps hold it together by giving members a shared sense of direction.

TEACHING NOTE
Ask students to define what they think management is. Return to this list as you develop the textbook definition. Where possible, weave in student definitions.

organization An assembly of people and materials brought together to accomplish a purpose that would be beyond the means of individuals working alone.

mission An organization's purpose or reason for being.

TABLE 2-1 Some Statements of Mission	
COMPANY	**MISSION**
Hershey Foods Corp.	Hershey Foods Corporation's basic business mission is to become a major, diversified food company. . . . A basic principle which Hershey will continue to embrace is to attract and hold customers with products and services of consistently superior quality and value.
Polaroid Corp.	Polaroid designs, manufactures, and markets worldwide a variety of products based on its inventions, primarily in the photographic field. These include instant photographic cameras and films, light polarizing filters and lenses, and diversified chemical, optical, and commercial products. The principal products of the company are used in amateur and professional photography, industry, science, medicine, and education.
MCI Communications, Inc.	MCI's mission is leadership in the global telecommunications services industry. Profitable growth is fundamental to that mission, so that we may serve the interests of our stockholders and our customers.
Litton Industries, Inc.	Litton is a technology-based company applying advanced electronics products and services to business opportunities in defense, industrial automation, and geophysical markets. Research and product engineering emphasis is on developing advanced products which the company manufactures and supplies worldwide to commercial, industrial, and government customers.

Source: Excerpted from recent annual stockholder reports.

division of labor The process and result of breaking difficult work into smaller tasks.

Division of Labor. In every organization, difficult work is broken into smaller tasks. Such **division of labor** can enhance *efficiency* because it simplifies tasks and makes them easier to perform. You can see a classic example of this effect in an analysis of the task of pin making by the eighteenth-century economist Adam Smith:

> One man draws out the wire, another straightens it, a third cuts it, a fourth points it, a fifth grinds it at the top for receiving a head. To make the head requires two or three more operations. [Using a division of labor such as this,] ten persons could make among them upward of forty-eight thousand pins a day. But if they had all wrought separately and independently they certainly could not each of them have made twenty; perhaps not one pin in a day.[4]

The division of labor enables organized groups of people to accomplish tasks that would be beyond their physical or mental capacities as individuals. Few people, for example, can build a car by themselves, but companies like Chrysler turn out thousands of cars each year by dividing the complex job of building a car into simple assembly-line tasks.

hierarchy of authority A pyramidal distribution of authority in which managers higher in the pyramid can tell managers in lower positions what to do.

Hierarchy of Authority. The **hierarchy of authority** is a third attribute of every organization. In very small organizations, the authority to issue commands, make decisions, and enforce obedience may be shared equally among all members. More often, however, authority is distributed among members in a hierarchical pattern such as the one shown in Figure 2-1. At the top of this hierarchy, the chief executive officer (CEO) has the authority to issue orders to every other member of the organization and to expect these orders to be obeyed. At successively lower levels, managers direct the activities of people beneath them but are constrained by the authority of managers above them.

[4] Adam Smith, *An Inquiry into the Nature and Causes of the Wealth of Nations*, 5th ed. (Edinburgh: Adam and Charles Black, 1859), p. 3.

Ford Motor Co. long wanted to enter the luxury-car market dominated by German companies and jumped at the opportunity to acquire Britain's prestigious Jaguar, which had fallen on hard times. Jaguar (left) had been spending an incredible 100 man-hours to produce a single car and was forced early in 1990 to recall most of the cars sold over the past two years because of defects that could cause fire to break out. Ford plans to improve the Jaguar to compete with the highly successful BMW and Daimler-Benz Mercedes (right), which are made on automated assembly lines.
Sources: Fortune, *January 29, 1990, p. 96; Paul Ingrassia and Jacqueline Mitchell, "Jaguar Recalling Majority of Cars Sold Over 2 Years," The Wall Street Journal, February 21, 1990.*

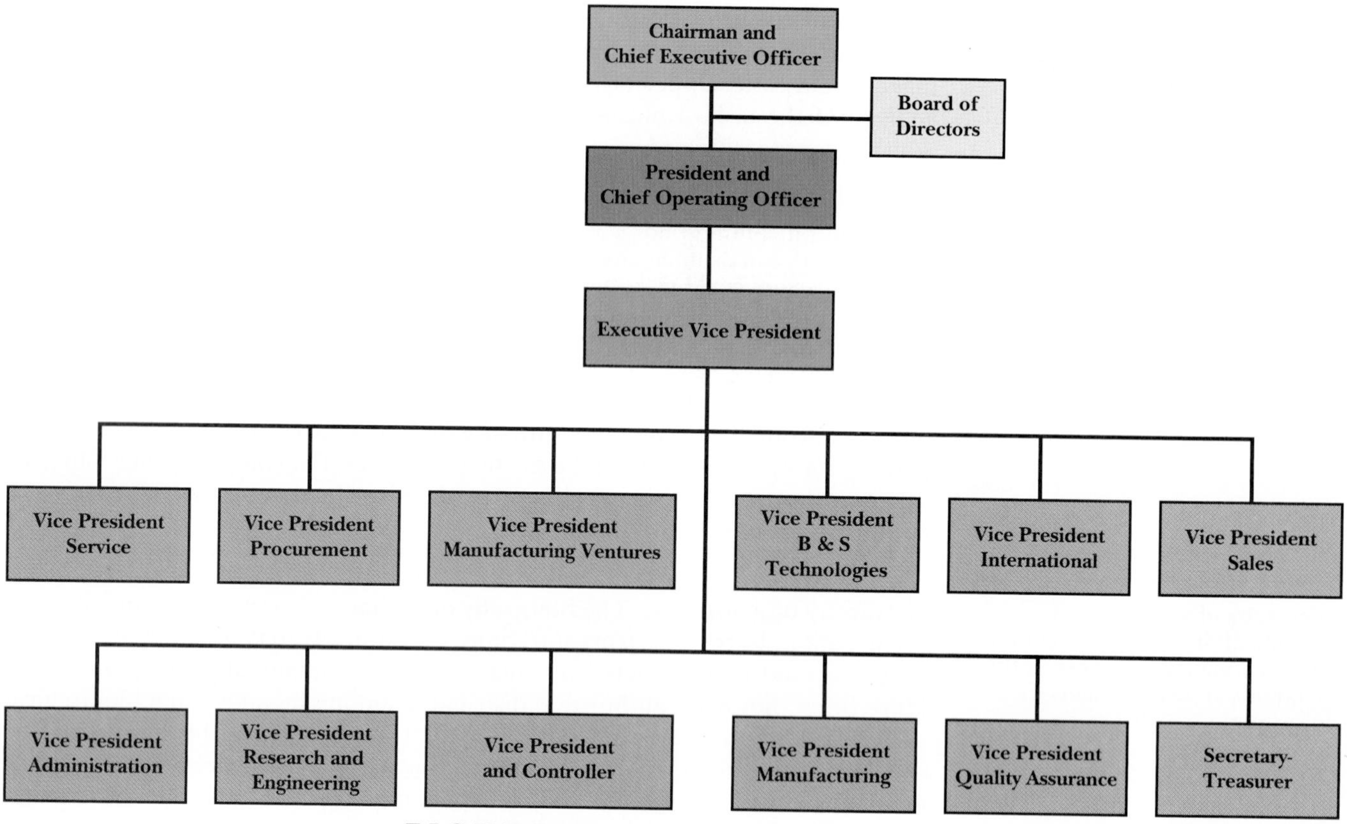

```
                    ┌──────────────────────┐
                    │ Chairman and         │
                    │ Chief Executive Officer│
                    └──────────────────────┘          ┌──────────────┐
                                 │──────────────────────│ Board of     │
                                 │                       │ Directors    │
                    ┌──────────────────────┐            └──────────────┘
                    │ President and        │
                    │ Chief Operating Officer│
                    └──────────────────────┘
                                 │
                    ┌──────────────────────┐
                    │ Executive Vice President│
                    └──────────────────────┘
```

Vice President Service	Vice President Procurement	Vice President Manufacturing Ventures	Vice President B & S Technologies	Vice President International	Vice President Sales

Vice President Administration	Vice President Research and Engineering	Vice President and Controller	Vice President Manufacturing	Vice President Quality Assurance	Secretary-Treasurer

FIGURE 2-1

Briggs-Stratton Organization Chart

An organization chart is a graphic representation of a firm's hierarchy of authority. The organization chart in this figure shows the top and middle management of Briggs-Stratton, a manufacturer of small gasoline engines used in lawn mowers, snow blowers, and similar equipment. Note that the company is divided *horizontally* into various functional departments—such as manufacturing and sales—whose efforts are unified through authority relations that extend *vertically* between vice presidents and the CEO.

Source: The 1987 annual report of the Briggs-Stratton Company.

A Functional Definition

management A process of planning, organizing, directing, and controlling organizational behaviors in order to accomplish a mission through the division of labor.

The three attributes of organizations that we have discussed help clarify the role of management in organizational life. In a sense, the first two are in conflict. The mission assumes the integration of effort, whereas the division of labor produces a differentiation of effort. As a result, an organization's members are simultaneously pushed together and pulled apart. It is managerial authority that reconciles this conflict by balancing the two opposing attributes. This balancing act is what managers do and what management is all about.

Management is thus a process of planning, organizing, directing, and controlling organizational behaviors in order to accomplish a mission through the division of labor. This definition incorporates several important ideas. First, management is a process, an ongoing flow of activities. It is not something that can be accomplished once and for all. Second, managerial activities affect the behaviors of both an organization's members and the organization itself. Third, to accom-

24

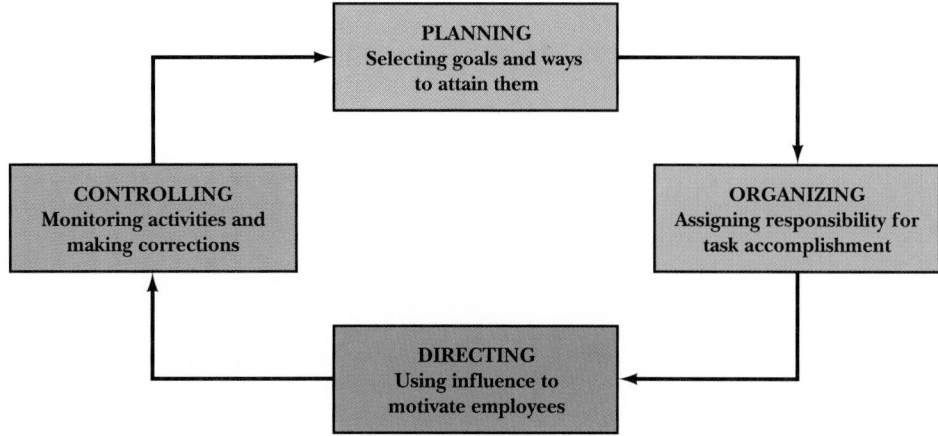

FIGURE 2-2
The Four Management Functions

plish a firm's mission requires organization. If the mission could be accomplished by individuals working alone, neither the firm nor its management would be necessary. Let us look next at the four functions of management: planning, organizing, directing, and controlling (see also Figure 2-2).

planning The management function of deciding what to do in the future; setting goals and establishing the means to attain them.

TEACHING NOTE
It is estimated that 80 percent of mergers do not meet financial expectations and that 50 percent fail completely. Of the failures, 75 percent have been attributed to unanticipated and poorly handled people problems. What do these facts suggest about the management of mergers? Discuss each component: planning, organizing, directing, controlling. (*Personnel Administrator*, 8/89, pp. 84–90)

POINT TO STRESS
Organizations exist to accomplish a variety of goals. The degree of goal accomplishment is a measure of management effectiveness. Goals imply purposefulness, rationality, and achievement.

organizing The management function of developing a structure of interrelated tasks and allocating people and resources within this structure.

Planning. **Planning** is a forward-looking process of deciding what to do. Managers who plan try to anticipate the future. They set goals and objectives for a firm's performance and identify the actions required to attain these goals and objectives. For example, managers at Sears planned a policy of increasing business by offering "everyday low prices." General Electric's managers were planning when they decided to have Black and Decker make GE steam irons.

In planning, managers set three types of goals and objectives:

1. *Strategic goals* are the outcomes that the organization as a whole expects to achieve in the pursuit of its mission.
2. *Functional* or *divisional objectives* are the outcomes that units within the firm are expected to achieve.
3. *Operational objectives* are the specific, measurable results that the members of an organizational unit are expected to accomplish.[5]

As shown in Figure 2-3, these three types of goals and objectives fit into an interdependent hierarchy. The focus of lower-order objectives is shaped by the content of higher-level goals, and, achieving higher-level goals depends on the fulfillment of lower-level objectives.

Goals and objectives are targets for organizational behavior. They help managers plan and implement a sequence of actions that will lead to goal attainment. For example, the sales objectives on which Sears based its pricing policy became performance targets for local stores and regional operations. Goals and objectives also serve as benchmarks of the success or failure of organizational behavior. When they review past performance, managers can judge effectiveness by assessing goal achievement. Thus GE's managers assessed the success of their decision to outsource steam iron production by comparing actual revenue and cost data with the profitability goals they had set.

Organizing. In **organizing**, managers develop a structure of interrelated tasks and allocate people and resources within this structure. Organizing begins with dividing an organization's labor and designing tasks that will lead to the achieve-

[5] Herbert Simon, *Administrative Behavior: A Study of Decision Making Processes in Administrative Organization* 3rd. ed. (New York: Free Press, 1976), pp. 257–78.

DEFINING MANAGEMENT

25

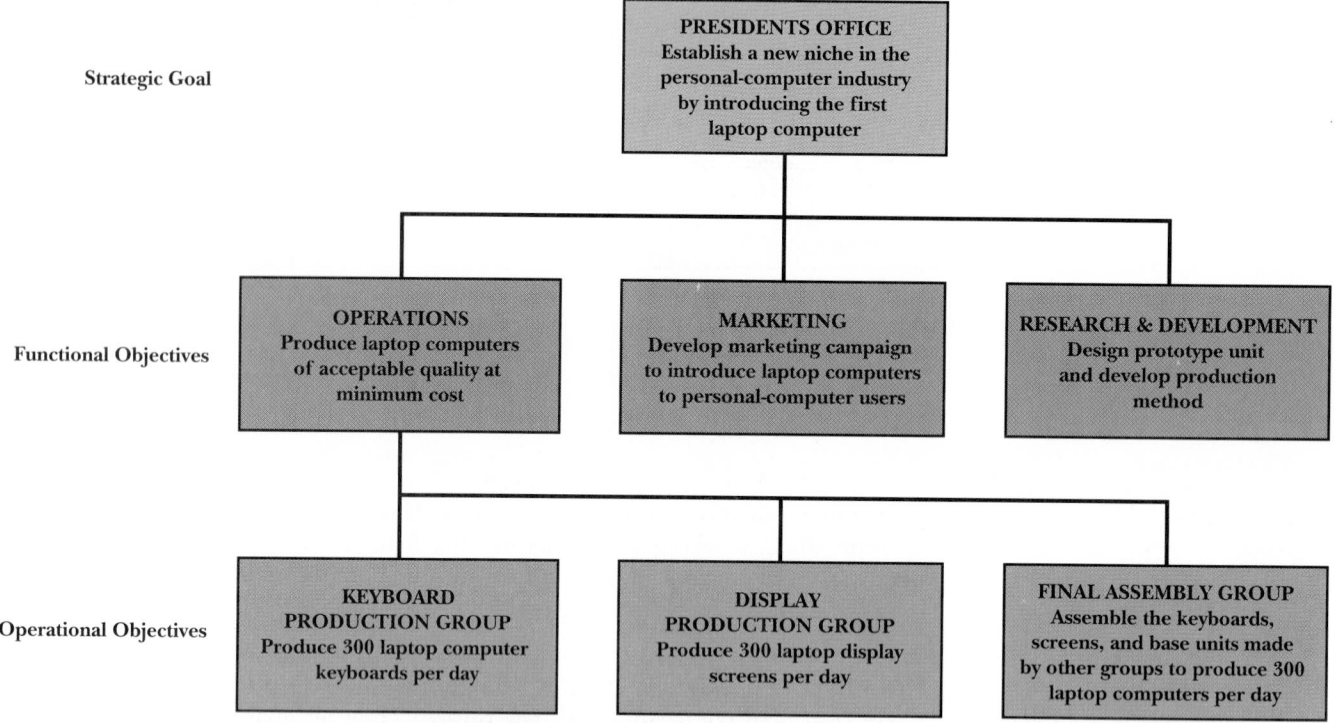

Strategic Goal

PRESIDENTS OFFICE
Establish a new niche in the
personal-computer industry
by introducing the first
laptop computer

Functional Objectives

OPERATIONS
Produce laptop computers
of acceptable quality at
minimum cost

MARKETING
Develop marketing campaign
to introduce laptop computers
to personal-computer users

RESEARCH & DEVELOPMENT
Design prototype unit
and develop production
method

Operational Objectives

**KEYBOARD
PRODUCTION GROUP**
Produce 300 laptop computer
keyboards per day

**DISPLAY
PRODUCTION GROUP**
Produce 300 laptop display
screens per day

FINAL ASSEMBLY GROUP
Assemble the keyboards,
screens, and base units made
by other groups to produce 300
laptop computers per day

FIGURE 2-3

The Hierarchy of Goals and Objectives

An organization's strategic goals set boundaries within which functional objectives are established. In turn, functional objectives shape the objectives of operational units. Thus accomplishing operational objectives contributes to the attainment of functional objectives and strategic goals.

POINT TO STRESS
Organizing a firm is necessary to develop a system that, when all its interrelated parts function well, is effective in accomplishing its goals and objectives. Compare to the human body, which is composed of thousands of parts working together. If any one part is not functioning correctly, the whole body suffers and is less effective in what the person tries to do. Ask students to imagine how hard it would be to do well on a test if they were suffering with a headache.

organizational unit A recognizable group of employees responsible for completing its own particular functional and/or operational objectives.

ment of organizational goals and objectives. In companies like General Motors, Boeing, and IBM, assembly lines are designed and built during this phase. Managers must next determine who will perform these tasks. But first they must analyze the tasks to determine the knowledge, skills, and abilities needed to perform them successfully. They can then select qualified employees or train other employees who lack the necessary qualifications.

Grouping tasks and the people who perform them into **organizational units** is another step in the organizing process. One type of organizational unit, a *department*, is formed by people who perform the same type of work. For instance, all employees who market an organization's goods or services can form a marketing department. Another type of unit, a *division*, is formed by people who do the company's work in the same geographic territory, who work with similar kinds of clients, or who make or provide the same type of good or service. For example, the Ford Motor Company has a European division that does business in Europe. General Dynamics has a division that specializes in military contracts. General Electric's consumer electronics division markets only household appliances.

directing The management function of encouraging and guiding employees' efforts toward the attainment of organizational goals and objectives.

Directing. Directing encourages member effort and guides it toward the attainment of organizational goals and objectives. Directing is partly a process of communicating goals and objectives to members. Managers must announce, clarify, and promote targets toward which effort should be directed. For example, Steve Jobs is directing when he meets with the employees of Next Computer Company to publicize sales objectives. Directing is also a process of learning

Is Management the Same Thing as Domination?

Many experts have proposed that organization is best understood as a process of domination. Since the beginning of recorded time, by banding together for political, military, or other purposes, groups of people have succeeded in bending others to their will.

Where there is organization there is usually management. Do managers then dominate the people they supervise? Certainly the earliest managers wielded absolute power over those who worked under them. Think, for example, of the Egyptian rulers under whose direction the pyramids were built. The planning, organization, direction, and control that were required to supervise the construction of the Great Pyramid at Giza seem almost to overshadow any modern effort at managing the work of others. More important, the slaves and laborers who performed the actual work were pressed into service and probably received little more than food and lodging in return for their Herculean efforts.

But surely today, you may say, workers can choose whether or not they wish to work for a particular organization, under a particular kind of management. Unions have greatly improved the working conditions and pay of many employees. Workers are increasingly given a voice in their own management. In some cases small groups of workers actually function without a direct supervisor. This does not sound as if workers are being dominated.

Critics suggest, however, that management relations that result in the majority working in the interests of the few are the hallmark of domination. They point out that over the course of history whether it was the building of pyramids or the maintenance of an army or the running of a multinational corporation—sometimes even the management of a family business—the lives and hard labor of many people have been used to serve the interests of a privileged few.*

Do you agree or disagree with the idea that management is a process of dominating others and controlling their behaviors? Why do you agree? Disagree? Is dominating or controlling others necessarily bad? Inherently good? What implications do your answers have for your future behavior as a manager?

* Gareth Morgan, *Images of Organization* (Beverly Hills, CA: Sage, 1986); Stewart Clegg and David Dukevley, *Organization, Class, and Control* (London, England: Routledge & Kegan Paul, 1980); and Loven Baritz, *The Servants of Power* (Middletown, Conn.: Wesleyan University Press, 1960).

POINT TO STRESS

Directing is getting employees to learn, accept, and move toward an organization's goals and objectives. The different roles/jobs employees carry out must be considered. For example, the director of a play must tell a story. Each actor has a role different from all the others and must not only learn her own role but understand how that role fits into the story. For the story to be told effectively, the director must guide all the actors in cooperating with each other. Both manager and director aid the learning and understanding processes through direction.

controlling The management function of evaluating the performance of an organization or organizational unit to determine whether it is progressing in the desired direction.

employees' desires and interests and of ensuring that these desires and interests are satisfied in return for successful goal-oriented performance. When Sam Walton of Wal-Mart chats with employees about their jobs and the products they're selling, he is being more than a friendly person—he's directing. Directing may also require managers to use personal expertise or charisma to inspire employees to overcome obstacles that might appear insurmountable. Lee Iacocca is the perfect example of a manager performing this sort of directing. In all instances, directing is a process in which managers *lead* their subordinates, influencing them to work together to achieve organizational goals and related objectives. It is this influence-using aspect of management direction that has sometimes led critics to remark that management is the same thing as domination (see "Management Issues" box).

Controlling. **Controlling** involves evaluating the performance of an organization and its units to see whether the firm is progressing in the desired direction. In typical performance evaluation, managers compare an organization's actual results with the desired results described in the organization's goals and objectives. For example, GE might compare its actual profitability with the profitability objectives set for outsourcing. To make this sort of evaluation, members of the organization must collect and assess performance information. A firm's accounting personnel first gather data about the costs and revenues of organizational activities. Marketing representatives may provide additional data about sales volume or the organization's position in the marketplace. Finance specialists then appraise organizational performance by determining whether the ratio of costs to revenues meets or surpasses a target level.

If evaluation reveals a significant difference between goals and actual performance, the control process enters a phase of *correction*. In this phase, managers return to the planning stage, redeveloping goals and objectives, and indicating how differences between goals and outcomes can be reduced. Then the cycle begins again, with new efforts at planning, organizing, directing, and controlling.

SCHOOLS OF MANAGEMENT THOUGHT

Our definition of management is based on a wealth of management thoughts and practices, many of which are thousands of years old. Consider the following:

1. As early as 3000 B.C., the Sumerians formulated missions and goals for government and commercial enterprises.
2. Between 3000 and 1000 B.C, the Egyptians successfully organized the efforts of thousands of workers to build the Pyramids.
3. Between 800 B.C and about 300 A.D. the Romans perfected the use of hierarchical authority.
4. Between 450 A.D. and the late 1400s, Venetian merchants developed commercial laws and invented double-entry bookkeeping.
5. In the early 1500s, at the request of an Italian prince, Niccolo Machiavelli prepared an analysis of power that is still widely read.
6. At about the same time, the Catholic Church perfected a governance structure involving the use of standardized procedures.

Truly modern management practices did not begin to develop, however, until the industrial revolution of the 1700s and 1800s. Inventions like James Watt's steam engine and Eli Whitney's cotton gin created new forms of mass production that made existing modes of administration obsolete. The field of industrial engineering—which arose to invent and improve workplace machinery—began to address the selection, instruction, and coordination of industrial employees. Toward the end of the industrial revolution, managers and engineers throughout North America and Europe focused on developing general theories of management.

1890–1940: The Scientific Management School

Management theories initially took the form of *management principles* intended to provide managers with practical advice about managing their firms. Most of these principles were written by practicing managers or others closely associated with the management profession. Among the first principles to be widely read were those of the **scientific management school**.

scientific management school
The school of management thought that focuses on increasing the efficiency of production processes in order to enhance organizational profitability.

All principles of scientific management reflected the idea that through proper management an organization could achieve profitability and long-term survival in the competitive world of business. Thus, theorists in the scientific management school devoted their attention to describing proper management and determining the best way to achieve it.

Frederick W. Taylor. The founder of scientific management, Frederick W. Taylor (1856–1915) developed his principles of scientific management as he rose from laborer to chief engineer at the Midvale Steel Works in Philadelphia, Pennsylvania. Shown in Table 2-2, these principles focused on increasing the efficiency of the workplace by differentiating managers from nonsupervisory workers and systematizing the jobs of both.

TABLE 2-2
Frederick W. Taylor's Principles of Scientific Management

1. *Assign all responsibility for the organization of work to managers rather than workers.* Managers should do all the thinking related to the planning and design of work, leaving workers the task of carrying it out.

2. *Use scientific methods to determine the one best way of performing each task.* Managers should design each worker's job accordingly, specifying a set of standard methods for completing the task in the right way.

3. *Select the person most suited to each job to perform that job.* Managers should match the abilities of each worker to the demands of each job.

4. *Train the worker to perform the job correctly.* Managers should train workers in how to use the standard methods devised for their jobs.

5. *Monitor work performance to ensure that specified work procedures are followed correctly and that appropriate results are achieved.* Managers should exercise the control necessary to guarantee that workers under their supervision always perform their jobs in the one best way.

6. *Provide further support by planning work assignments and eliminating interruptions.* Managers can help their workers continue to produce at a high level by shielding them from things that interfere with job performance.

Source: Based on Frederick W. Taylor, *The Principles of Scientific Management* (New York: Norton, 1911) pp. 34–40.

AN EXAMPLE: TOYOTA, HEWLETT-PACKARD
Today many companies like Toyota and Hewlett-Packard use a management approach according to which there are many ways to do a job, employees are responsible for monitoring productivity, quality, and work procedures, and each employee is cross-trained to do several jobs. How does this approach compare with that of scientific management?

According to Taylor, the profitability of an organization could be assured only by finding the "one best way" to perform each job. Managers could teach workers this technique and use a system of rewards and punishments to encourage its use. Consider Taylor's work to improve the productivity of coal shovelers at the Bethlehem Steel Company. As he observed these workers, Taylor discovered that a shovel load of coal could range from 4 to 30 pounds depending on the density of the coal being carried. By experimenting with a group of workers, Taylor discovered that shovelers could move the most coal in a day without suffering undue fatigue if each load of coal weighed 21 pounds. He then developed a variety of different shovels, each of which would hold approximately 21 pounds of coal of a particular density. After Taylor taught workers how to use these shovels, each shoveler's daily yield rose from 16 tons to 59. Moreover, the average wage per worker increased from $1.15 to $1.88 per day. Bethlehem Steel was able to reduce the number of shovelers in its yard from about 500 to 150, saving the firm about $80,000 per year.

Taylor's ideas influenced management around the world. In a 1918 article for the newspaper *Pravda*, Russian Communist Party founder Lenin even recommended that Taylor's scientific management be used throughout the USSR. In the U.S., Taylor's principles had such a dramatic effect on management that in 1912 he was called to testify before a special committee of the House of Representatives. Union employees and employers all objected to Taylor's idea that employers and employees should share the economic gains of scientific management and wanted Congress to do something about it. Nevertheless, with the newspaper publicity he gained from his appearance, Taylor found even wider support for his ideas and was soon joined in his work by other specialists.

Other Contributors. The husband-and-wife team of Frank (1868–1924) and Lillian Gilbreth (1878–1972), followed in Taylor's footsteps in pursuing the "one best way" to do any job. The Gilbreths are probably best known for their invention of "motion study," a procedure in which jobs are reduced to their most basic movements. A sample listing of these basic movements, each of which is called a *therblig* (Gilbreth spelled backward without inverting the *th*), is shown in Table 2-3. They also invented the microchronometer, a clock with a hand capable of

T A B L E 2-3
Therblig Motions

Search	Transport loaded	Dissemble	Transport empty
Find	Position	Inspect	Rest to relieve fatigue
Select	Assemble	Pre-position	Other unavoidable delay
Grasp	Use	Release load	Avoidable delay
Plan			

measuring time to the 1/2000 of a second. Using this instrument, analysts could precisely determine the time required by each of the movements needed to perform a job.

Another contributor to scientific management, Henry Gantt (1861–1919) developed a task-and-bonus wage plan that paid workers a bonus besides their regular wages if they completed their work in an assigned amount of time. Gantt's plan also provided bonuses for supervisors. Each supervisor's bonus was determined by the number of subordinates who met deadlines. If all subordinates finished on time, the supervisor received an additional bonus.[6] Gantt also invented the Gantt chart, a bar chart used by managers to compare actual with planned performance.[7] Present-day scheduling methods such as the program evaluation and review technique (PERT) are based on this invention.

Harrington Emerson (1853–1931), a third contributor to scientific management, applied his own list of twelve principles to the railroad industry in the early 1900s.[8] Among Emerson's principles were recommendations to establish clear objectives, seek advice from competent individuals, manage with justice and fairness, standardize procedures, reduce waste, and reward workers for efficiency. Late in his life, Emerson became interested in the selection and training of employees, stressing the importance of explaining scientific management to employees during their initial training. Emerson reasoned that sound management practices could succeed only if every member of the firm understood them.

1900–1950: The Administrative Principles School

At about the same time that Taylor and his colleagues were formulating their principles of scientific management, another group of theorists was developing the **administrative principles school**. In contrast to scientific management's emphasis on reducing the costs of production activities, this second school focused on increasing the efficiency of administrative procedures.

Henri Fayol. Considered the father of modern management thought, Henri Fayol (1841–1925) developed his principles of administration in the early 1900s while serving as chief executive of a French mining and metallurgy firm, Commentry-Fourchambault-Decazeville, known as "Comambault." Fayol was the first to identify the four functions of management we have already discussed: planning, organizing, directing, and controlling.[9] He also formulated the fourteen principles shown in Table 2-4 to help administrators perform their jobs.

administrative principles school The school of management thought that deals with streamlining administrative procedures in order to encourage internal stability and efficiency.

[6] Henry L. Gantt, "A bonus system of rewarding labor, *ASME Transactions* 23 (1901), 341–72; and *Work, Wages, and Profits* (New York: Engineering Magazine Company, 1910), pp. 18–29.

[7] Henry L. Gantt, *Organizing for Work* (New York: Harcourt, Brace, and Howe, 1919), pp. 74–97.

[8] Harrington Emerson, *The Twelve Principles of Efficiency* (New York: Engineering Magazine Company, 1912), pp. 59–367.

[9] Henri Fayol, *General and Industrial Management*, trans. Constance Storrs (London: Sir Isaac Pitman & Sons, 1949), pp. 19–43.

TABLE 2-4
Fayol's Fourteen Principles of Management

PRINCIPLE	DESCRIPTION
Division of work	A firm's work should be divided into specialized, simplified tasks. Matching task demands with work force skills and abilities will improve productivity. The management of work should be separated from its performance.
Authority and responsibility	Authority is the right to give orders, and responsibility is the obligation to accept the consequences of using authority. No one should possess one without the other.
Discipline	Discipline is performing a task with obedience and dedication. It can be expected only when a firm's managers and subordinates agree on the specific behaviors that subordinates will perform.
Unity of command	Each subordinate should receive orders from only one hierarchical superior. The confusion of having two or more superiors would undermine authority, discipline, order, and stability.
Unity of direction	Each group of activities directed toward the same objective should have only one manager and only one plan.
Individual versus general interests	The interests of individuals and the whole organization must be treated with equal respect. Neither can be allowed to supersede the other.
Remuneration of personnel	The pay received by employees must be fair and satisfactory to both them and the firm. Pay should be in proportion to personal performance, but employees' general welfare must not be threatened by unfair incentive-payment schemes.
Centralization	Centralization is the retention of authority by managers. It should be used when managers desire greater control. Decentralization should be used, however, if subordinates' opinions, counsel, and experience are needed.
Scalar chain	The scalar chain is a hierarchical string extending from the uppermost manager to the lowest subordinate. The line of authority follows this chain and is the proper route for organizational communications.
Order	Order, or "everything in its place," should be instilled whenever possible because it reduces wasted materials and efforts. Jobs should be designed and staffed with order in mind.
Equity	Equity means enforcing established rules with a sense of fair play, kindliness, and justice. Equity should be guaranteed by management, because it increases members' loyalty, devotion, and satisfaction.
Stability of tenure	Properly selected employees should be given the time needed to learn and adjust to their jobs. The absence of such stability undermines organizational performance.
Initiative	Members should be allowed the opportunity to think for themselves because this motivates performance and adds to the organization's pool of talent.
Esprit de corps	Managers should harmonize the interests of members by resisting the urge to split up successful teams. They should rely on face-to-face communication to detect and correct misunderstandings immediately.

Fayol believed that the number of management principles that might help improve an organization's operation is potentially limitless. He considered his principles to be flexible and adaptable, labeling them principles rather than laws or rules in order to avoid any idea of rigidity, as there is nothing rigid or absolute in [management] matters; everything is a question of degree. The same principle is hardly ever applied twice in exactly the same way, because we have to allow for different and changing circumstances, for human beings who are equally different and changeable, and for many other variable elements. The principles, too, are flexible, and can be adapted to meet every need; it is just a question of knowing how to use them.[10]

For Fayol, management was more than mechanical rule following. It required the sort of intuition and skillful application we have discussed in Chapter 1.

Max Weber. Max Weber (1864–1920) was a German sociologist who, though neither manager nor management consultant, had a major effect on twentieth-century management thought. Like Fayol, Weber was interested in the efficiency of different kinds of administrative arrangements. To figure out what makes organizations efficient, Weber analyzed the Egyptian empire, the Prussian army, the Roman Catholic church, and other large organizations that had functioned efficiently over long periods of time. Based on these analyses, Weber developed his model of **bureaucracy**, an idealized description of an efficient organization.

According to Weber, bureaucratic organizations are characterized by six important traits. First, employees of these organizations are selected and promoted solely on the basis of *technical competence*. Managers are appointed rather than elected and can be removed if they lack the expertise to perform their jobs correctly.

Second, bureaucratic organizations link superiors and subordinates through a *hierarchy of authority* in order to facilitate managerial control. The holder of each lower position in the hierarchy reports to, and is controlled by, the person in the position directly above.

Third, the tasks, responsibilities, and authority of the employees of bureaucratic organizations are defined by specific *rules and regulations* that give employees consistent, impartial guidance about how to behave on the job.

Fourth, bureaucratic organizations use a systematic *division of labor*. Employees carry out the work specified in their job descriptions, and other employees do not interfere so long as the work is performed in accordance with relevant job specifications.

Fifth, all administrative decisions, activities, and rules are set down in *written documentation*. Bureaucratic record keeping makes rules and decisions accessible to all employees. It also facilitates evaluation of existing rules and provides a source of information that can guide future decision making.

The sixth and last characteristic of bureaucratic organizations is *ownership or control by each organization* of its property, positions, and affairs. The rights and control associated with the positions in the organization's hierarchy belong to the organization and not to the employees occupying the positions.[11]

Weber's bureaucratic model, summarized in Table 2-5, provides for both the differentiation (through the division of labor and task specialization) and the integration (by the hierarchy of authority and written rules and regulations) necessary to get a specific job done. Weber believed that any organization with

bureaucracy An idealized description of an efficient organization based on clearly defined authority, formal record keeping, and standardized procedures.

AN EXAMPLE: FORD, CHRYSLER

We may identify organizations like Ford and Chrysler as bureaucratic, but close examination reveals variations of Weber's model. First, many promotions, especially at lower levels of the organization, are based on seniority rather than technical competence. Second, through unionization there is a recognition that employees have ownership rights to their jobs. Employees can organize and force management to share decision making power. We seldom see any of the management theories presented here in pure form. They are often modified to address organizational needs, social pressure, and government regulation.

[10] Henri Fayol, *Industrial and General Administration*, trans. J. A. Coubrough (Geneva, Switzerland: International Management Institute, 1930), p. 19.
[11] H. H. Gerth and C. Wright Mills, trans., *From Max Weber: Essays in Sociology* (New York: Oxford University Press, 1946); Nicos P. Mouzelis, *Organisation and Bureaucracy: An Analysis of Modern Theories* (Chicago: Aldine 1967); and Talcott Parsons, trans., *Max Weber: The Theory of Social and Economic Organization* (New York: Free Press, 1947).

MANAGEMENT AND MANAGERS

TABLE 2-5
Features of Bureaucratic Organizations

FEATURE	DESCRIPTION
Selection and promotion criteria	Expertise is the primary criterion. Friendship or other favoritism is explicitly rejected.
Hierarchy of authority	Superiors have the authority to direct subordinates' actions. They are responsible for ensuring that these actions are in the bureaucracy's best interests.
Rules and regulations	Unchanging regulations provide the bureaucracy's members with consistent, impartial guidance.
Division of labor	Work is divided into tasks that can be performed by the bureaucracy's members in an efficient, productive manner.
Written documentation	Records provide consistency and a basis for evaluation of bureaucratic procedures.
Separate ownership	Members cannot gain unfair or undeserved advantage through ownership.

Source: Based on information presented in H. H. Gerth and C. Wright Mills, trans., *From Max Weber: Essays in Sociology* (New York: Oxford University Press, 1946).

bureaucratic characteristics would be efficient. He also noted, though, that work in a bureaucracy could become so simple and undemanding that employees might grow dissatisfied and, as a result, less productive.

Other Contributors. A number of other management experts have contributed to the administrative principles school. James Mooney (1884–1957), was vice-president and director of General Motors and president of General Motors Overseas Corporation during the late 1920s when he created his principles of organization.[12] Mooney's *coordinative principle* highlighted the importance of organizing the tasks and functions in a firm into a coordinated whole. He defined coordination as the orderly arrangement of group effort to provide unity of action in the pursuit of a common mission. Mooney's *scalar principle* identified the importance of scalar, or hierarchical, chains of superiors and subordinates as a means of integrating the work of different employees. Finally, Mooney's *functional principle* stressed the importance of functional differences, such as marketing, manufacturing, and accounting. He noted how work in each functional area both differs from and interlocks with the work of other areas.

Lyndall Urwick (1891–1983), another writer in the administrative principles school, was a British military officer and director of the International Management Institute in Geneva, Switzerland. Urwick made his mark by consolidating the ideas of Fayol and Mooney with those of Taylor.[13] From Taylor, Urwick adopted the idea that systematic, rigorous investigation should inform and support the management of employees. Urwick also used Fayol's fourteen principles to guide managerial planning and control and Mooney's three principles of organization to structure his discussion of organizing. Urwick's synthesis thus bridged Taylor's scientific management and the administrative principles approach, and integrated the work of others within the framework of the four functions of management with which you are now familiar.

---

[12] James D. Mooney and Alan C. Reiley, *Onward Industry: The Principles of Organization and Their Significance to Modern Industry* (New York: Harper & Brothers, 1931); revised and published as James D. Mooney, *The Principles of Organization* (New York: Harper & Brothers, 1947).

[13] Lyndall Urwick, *The Elements of Administration* (New York: Harper & Brothers, 1944).

Mary Parker Follett (1868–1933), who became interested in industrial management in the 1920s, was among the first proponents of what became known as *industrial democracy*. Follett proposed that, to promote cooperation and attention to a company's overall mission and goals, every employee should have an ownership interest in the company.[14] As you will see, Follett's work contributed to the human relations school, which we describe next. Follett, who advanced a number of administrative principles, suggested that because organizational problems typically stem from a variety of interdependent factors, they tend to resist simple solutions. Here again she anticipated later theorists, contributing to the contingency perspective that we introduced in Chapter 1 and discuss in this chapter in more detail.

1930–1970: The Human Relations School

Although members of both the scientific management and administrative principles schools advocated the scientific study of management, they rarely evaluated their ideas in any formal way. This situation changed in the middle 1920s when university researchers began to use scientific methods to test existing management thought.

The Hawthorne Studies. The now-famous *Hawthorne studies*, performed in 1924 at Western Electric's Hawthorne plant near Chicago, were among the earliest attempts to use scientific techniques to examine human behavior at work.[15] This three-stage series of experiments was undertaken to assess the effects of varying physical conditions and management practices on workplace efficiency. The design of these studies reflected the Hawthorne research team's grounding in the scientific management school of thought.[16] The first-stage experiment tested the effects of workplace lighting on productivity. One group of employees (the test group) was exposed to varying intensities of illumination. A second group (the control group) worked under a constant degree of illumination. The researchers expected to find that among the workers exposed to varying lighting conditions productivity would also vary and that productivity would peak when an optimal level of lighting was reached. They also predicted that the control group's performance would not change because their lighting conditions remained constant. Unexpectedly, the productivity of *both* groups rose. Only after workplace illumination was reduced to the level of moonlight did the test group's productivity drop significantly! These results led the researchers to conclude that it was not illumination alone that affected worker productivity. Reviewing their findings, they hypothesized that changes in social conditions, that is, the increased attention the workers received from researchers, might underlie the observed rise in productivity in both groups.

To investigate this possibility, the Hawthorne researchers conducted three more experiments. First, over a two-year period, they observed the performance of a group of women who assembled telephone relays. In this group, tasks were

---

[14] Henry C. Metcalf and Lyndall Urwick, eds., *Dynamic Administration: The Collected Papers of Mary Parker Follett* (New York: Harper & Row, 1940). Also see Judith Garwood, "A Review of *Dynamic Administration: The Collected Papers of Mary Parker Follett*," *New Management* 2 (1984), 61–62.

[15] Alex Carey, "The Hawthorne Studies: A Radical Criticism," *American Sociological Review* 33 (1967), 403–16.

[16] L. J. Henderson, T. N. Whitehead, and Elton Mayo, "The Effects of Social Environment," in *Papers on the Science of Administration*, ed. Luther Gulick and L. Urwick (New York: Institute of Public Administration, 1937), pp. 143–58; Elton Mayo, *The Human Problems of an Industrial Civilization* (Cambridge, Mass.: Harvard University Press, 1933); F. J. Roethlisberger and William J. Dickson, *Management and the Worker* (Cambridge, Mass.: Harvard University Press, 1939).

simplified, working hours were shortened, rest breaks were permitted, friendlier supervision was instituted, and an incentive-payment system was introduced. Next, the researchers convened a second group of five women, also assembling relays, and put them on the same simplified tasks and incentive-payment plan, but without the special working hours, rest periods, and friendly supervision given the first group. Finally, a third group of women who performed a mica-splitting task were exposed to the same working hours, rest periods, and friendly supervision given the first group of relay assemblers but were not put on the incentive-payment system.

Reviewing the results of this second group of experiments, the researchers concluded that the productivity increases they observed probably resulted from a change in the workers' attitudes owing to the new, more friendly supervisory style instituted in the first and third experimental groups. From a series of interviews with the subjects of the second group of experiments, the researchers learned that each group had apparently influenced the productivity of its members by enforcing informal agreements about what a fair day's work should be.

Finally, the Hawthorne researchers conducted an experiment that focused on the productivity of a group of nine men who assembled terminal banks for telephone exchanges. After implementing an incentive system in which each man was paid according to the number of pieces of work he completed, the researchers found that the group developed agreements about what constituted a day's work. Based on these agreements, both underproducing "chiselers" and overproducing "rate busters" were punished by social isolation, even occasional physical violence. The researchers concluded that social factors—specifically, workers' adherence to group norms to ensure the continued satisfaction of social needs—explained the results observed across all the Hawthorne studies.

Interestingly, much later reanalyses of the Hawthorne data not only found weaknesses in the studies' methods and techniques but suggested yet another possible cause of the results obtained. As shown in Table 2-6, these later investigators discovered that incentive pay rather than social factors improved productivity in the relay assembly experiments. They also concluded that the net

TABLE 2-6
The Hawthorne Studies

EXPERIMENT	MAJOR CHANGES	RESULTS
Stage I		
Illumination study	Lighting conditions	Improved productivity at nearly all levels of illumination
Stage II		
First relay-assembly test	Job simplification, shorter work hours, rest breaks, friendly supervision, incentive pay	Thirty percent productivity improvement
Second relay-assembly test	Incentive pay	Twelve percent productivity improvement
Mica-splitting test	Shorter work hours, rest breaks, friendly supervision	Fifteen percent productivity improvement
Stage III		
Interview program	—	Discovery of presence of informal productivity norms
Bank-wiring-room test	Incentive pay	Emergence of productivity norms

effects of social factors could not even be correctly determined because other important elements of the work situation (the tasks being performed, rest periods, working hours) had been modified at the same time that social factors had been changed.[17] Consequently, conclusions reached by the original Hawthorne researchers are subject to doubt.

Douglas McGregor. Even though they were flawed in important respects, the Hawthorne studies raised serious questions about the efficiency-oriented focus of the scientific management and administrative principles schools. Most important, they stimulated debate about the importance of human satisfaction and personal development at work. The **human relations school** of management thought grew out of this debate, redirecting attention away from improving efficiency and toward increasing employee growth, development, and satisfaction.[18]

In 1960, Douglas McGregor (1906–1964), one spokesperson for the human relations school, contrasted the philosophy of the human relations approach with the efficiency orientation of the scientific management and administrative principles schools of management.[19] McGregor's **Theory X** incorporated the key assumptions about human nature shown in Table 2-7. McGregor suggested that theorists and managers holding these assumptions would describe management as follows:

1. Managers are responsible for organizing the elements of productive enterprise—money, materials, equipment, people—solely in the interest of economic efficiency.

2. The manager's function is to motivate workers, direct their efforts, control their actions, and modify their behavior to fit the organization's needs.

3. Without such active intervention by managers, people would be passive or even resistant to organizational needs; they must be persuaded, rewarded, and punished for the good of the organization.[20]

According to McGregor, the scientific management and administrative principles schools promoted a "hard" version of Theory X. They favored overcoming employees' resistance to organizational needs with strict discipline and economic rewards or sanctions. McGregor added that a "soft" version of Theory X seemed to underlie the Hawthorne studies. The Hawthorne researchers appeared to regard satisfaction and social relations mainly as rewards to employees who followed orders.

Theory Y, a contrasting philosophy of management that McGregor attributed to theorists, researchers, and managers from the human relations school, is based on the second set of assumptions shown in Table 2-7. McGregor indicated

---

human relations school The school of management thought that emphasizes increasing employee growth, development, and satisfaction.

Theory X A managerial point of view that assumes that nonmanagerial employees have little interest in attaining organizational goals and must therefore be motivated to fit the needs of the organization.

TEACHING NOTE
Theory X assumes people are not self-motivated and, as a result, must be coerced through rewards and punishment meted out by managers to work toward organizational goals. How can Theory X explain the existence of managers? How do these people become different from the rest?

Theory Y A managerial point of view that assumes that nonmanagerial employees will readily direct behavior toward organizational goals if given the opportunity to do so.

---

[17] Carey, "Hawthorne Studies"; R. H. Franke and J. D. Kaul, "The Hawthorne Experiments: First Statistical Interpretation," *American Sociological Review* 43 (1978), 623–43; and A. J. M. Sykes, "Economic Interests and the Hawthorne Researchers," *Human Relations* 18 (1965), 253–63.

[18] Examples from the body of research stimulated by the Hawthorne studies include Lester Coch and John R. P. French, Jr., "Overcoming Resistance to Change," *Human Relations* 1 (1948), 512–33; Leonard Berkowitz, "Group Standards, Cohesiveness, and Productivity," *Human Relations* 7 (1954), 509–14; and Stanley E. Seashore, *Group Cohesiveness in the Industrial Work Group* (Ann Arbor: University of Michigan Survey Research Center, 1954).

[19] Douglas McGregor, "The Human Side of Enterprise," *Management Review* 56 (1957), 22–28 and 88–92; Douglas McGregor, *The Human Side of Enterprise* (New York: McGraw-Hill, 1960).

[20] Adapted from McGregor, "The Human Side of Enterprise," p. 23.

T A B L E 2-7
Theory X and Theory Y Assumptions

Theory X assumptions
1. The average human being has an inherent dislike of work and will avoid it if possible.
2. Because they dislike work, most people must be coerced, controlled, directed, or threatened with punishment before they will put forth effort toward the achievement of organizational objectives.
3. The average human being prefers to be directed, wishes to avoid responsibility, has relatively little ambition, and wants security above all.

Theory Y Assumptions
1. Expending physical and mental effort at work is as natural as play and rest. The average human being does not inherently dislike work.
2. External control and the threat of punishment are not the only means to direct effort toward organizational objectives. People will exercise self-direction and self-control in the service of objectives to which they feel committed.
3. Commitment to objectives is a function of the rewards associated with their achievement. The most significant rewards—the satisfaction of ego and self-actualization needs—can be direct products of effort directed toward organizational objectives.
4. Avoidance of responsibility, lack of ambition, and emphasis on security are not inherent human characteristics. Under proper conditions, the average human being learns not only to accept but to seek responsibility.
5. Imagination, ingenuity, creativity, and the ability to use these qualities to solve organizational problems are widely distributed among people.

Source: Based on information presented in Douglas McGregor, *The Human Side of Enterprise* (New York: McGraw-Hill, 1960), pp. 33–34 and 47–48.

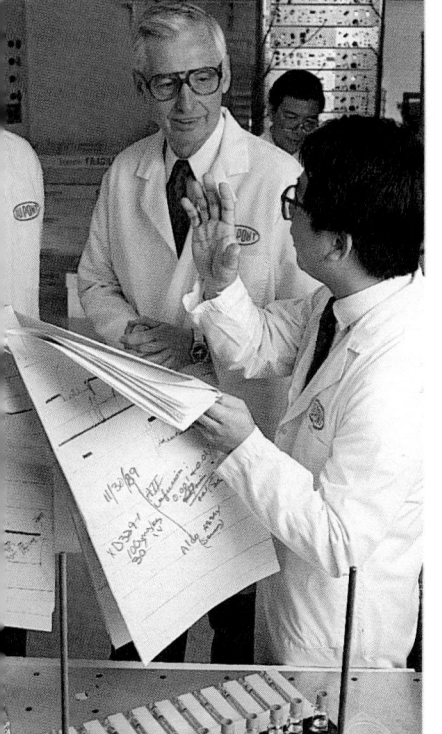

Summarizing Theory Y some 30 years after McGregor first proposed it, Dupont's CEO Edgar Woolard (center) says, "Employees have been underestimated. You have to start with the premise that people at all levels want to contribute and make the business a success." Woolard initiated the Adopt a Customer program in which blue-collar workers visit customers once a month, learn their needs, and literally serve as their representatives on the factory floor. Source: *Brian Dumaine, "Creating a New Company Culture,"* Fortune, *January 15, 1990.*

that individuals holding Theory Y assumptions would view the task of management as follows:

1. Managers are responsible for organizing the elements of productive enterprise—money, materials, equipment, people—in the interest of economic ends.

2. Because people are motivated to perform, have potential for development, can assume responsibility, and are willing to work toward organizational goals, managers are responsible for enabling people to recognize and develop these basic capacities.

3. The essential task of management is to arrange organizational conditions and methods of operation so that working toward organizational objectives is also the best way for people to achieve their own personal goals.[21]

Thus, unlike Theory X managers, who try to control their employees, Theory Y managers try to help employees learn how to manage themselves.

Other Contributors. Many management theorists, including Abraham Maslow (Chapter 7) and Frederick Herzberg (Chapter 17) embraced McGregor's Theory Y perspective in their work, discussing how personal autonomy and group participation might encourage employee growth, development, and satisfaction at work. We will discuss the ideas of these contributors to the human relations school later in the book.

1960–Present: The Open Systems School

The concern with employee satisfaction and development broadened to include a focus on organizational growth and survival with the emergence in the 1960s

[21] Adapted from McGregor, "The Human Side of Enterprise," pp. 88–89.

open systems school The school of management thought that characterizes every organization as a system that is open to the influence of the surrounding environment.

AN EXAMPLE: LINCOLN SAVINGS & LOAN ASSOCIATION

Savings and loans failures have been attributed to their managers' ignoring external influences like government regulations, investor expectations, and economic changes. Many S&L executives seemed to adopt a closed-system mentality that allowed them to think they could do business any way they wanted and were responsible to no one. This belief was undoubtedly encouraged by the initial failure of government regulators and examiners to spot such things as the ploy that allowed Charles Keating Jr. to siphon off $94 million from the Lincoln Savings & Loan Association. When external forces investigated and made demands for performance, the S&Ls fell apart. (Brooks Jackson, "How Regulatory Error Led to the Disaster at Lincoln Savings," *The Wall Street Journal*, 11/20/89.)

of the **open systems school**. According to this school of thought, every organization is a *system*—a unified structure of interrelated subsystems—and it is *open*, or subject to the influence of the surrounding environment. Together, these two ideas form the central tenet of the open systems approach. Organizations whose subsystems can cope with the surrounding environment can continue to do business, whereas organizations whose subsystems cannot cope do not survive. John Deere exemplifies a company whose subsystems—financial management, marketing and sales, human resource development—helped it survive in the competitive heavy-equipment industry of the 1980s. International Harvester is a company from the same industry whose subsystems failed to cope effectively; it no longer exists.

Daniel Katz and Robert L. Kahn. In one of the founding works of the open systems school, Daniel Katz and Robert Kahn identified the process shown in Figure 2-4 as essential to organizational growth and survival.[22] This process consists of the following sequence of events:

1. Every organization imports *inputs*, such as raw materials, production equipment, human resources, and technical know-how from the surrounding environment. For instance, Shell Oil Company hires employees and, from sources around the world, acquires unrefined oil, refinery equipment, and knowledge about how to refine petroleum products.

2. Some of these inputs are used to transform other inputs during a process of *throughput*. At Shell, employees use refinery equipment and know-how to transform unrefined oil into petroleum products like gasoline, kerosene, and diesel fuel.

3. The transformed resources are exported as *outputs*—saleable goods or services—to the environment. Petroleum products from Shell's refineries are loaded into tankers and transported to service stations throughout North America.

4. Outputs are exchanged for new inputs, and the cycle repeats. Shell sells its products and uses the resulting revenues to pay its employees and purchase additional oil, equipment, and know-how.

FIGURE 2-4
The Open Systems View of Organizations

[22] Daniel Katz and Robert L. Kahn, *The Social Phychology of Organizations* (New York: Wiley, 1966).

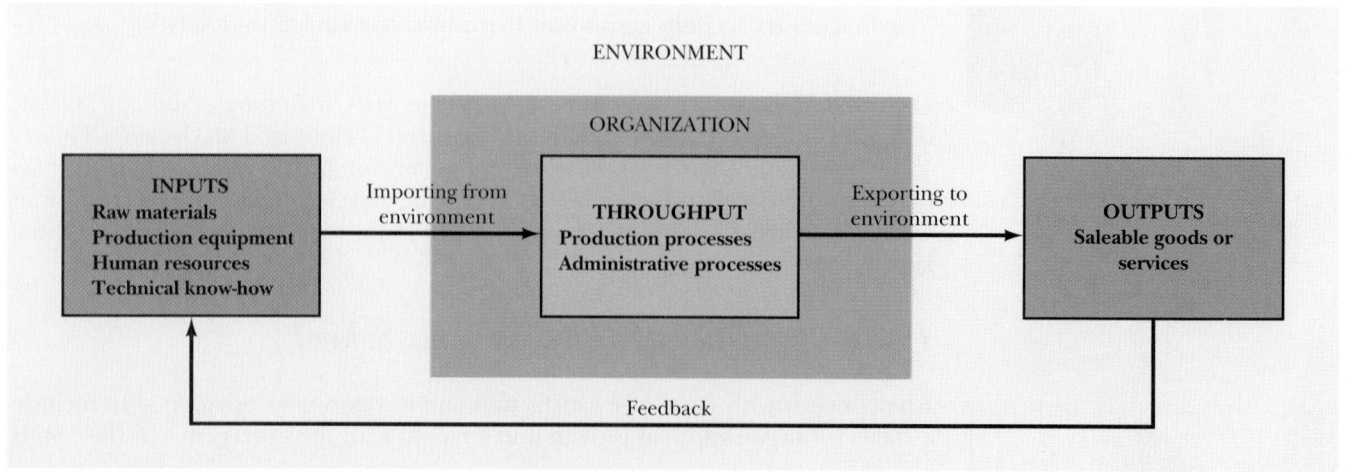

According to Katz and Kahn, organizations continue to grow and survive only as long as they import more inputs from the environment than they expend in producing the outputs exported back to the environment. *Information inputs* that signal how the environment and organization are functioning can help determine whether the organization will continue to survive; *negative feedback* indicates potential failure and the need to change the way things are being done.

Fred Emery and Eric Trist. In Katz and Kahn's model, the environment surrounding an organization is both the origin of needed resources and the recipient of transformed products. Accordingly, organizational survival depends on sensing the environment and adjusting to its demands. Describing environments and the demands they make so as to improve this sensing and adjustment was the goal of Fred Emery and Eric Trist, two early theorists of the open systems school.[23]

After noting that every organization's environment is itself composed of a collection of more or less interconnected organizations—supplier companies, competitors, and customer firms—Emery and Trist proposed four basic kinds of environments. The first kind, which they labeled *placid random environments*, are loosely interconnected and relatively unchanging. Organizations in such environments operate independently of each other, and one firm's decision to change the way it does business has little effect on the others. They are usually small—for example, landscape maintenance companies, construction firms, and industrial job shops—and can usually ignore each other and still stay in business by catering to local customers.

Placid clustered environments are more tightly interconnected. Here firms are grouped together into stable industries. Environments of this sort require organizations to cope with the actions of a *market*—a fairly constant group of suppliers, competitors, and customers. As a result, companies in these environments develop strategic moves and countermoves in response to competitors' actions. Grocery stores in the same geographic region often do business in this type of environment, using coupon discounts, in-store specials, and similar promotions to lure customers away from each other.

Disturbed reactive environments are as tightly interconnected as placid clustered environments but are less stable. Changes that occur in the environment itself have forceful effects on every organization. For instance, new competitors from overseas, increasing automation, and changing consumer tastes in the U.S. automobile market revolutionized the American auto industry in the 1970s and 1980s. As a result, GM, Ford, and Chrysler had to change their way of doing business, and a fourth long-time manufacturer, American Motors, ceased to exist. In such circumstances, organizations must respond not only to competitors' actions but to changes in the environment itself. It is very difficult to plan how to respond to these changes because their very nature makes future situations and actions hard to predict.

Turbulent fields are extremely complex and changeful. Companies operate in multiple markets. Public and governmental actions can alter the nature of an industry virtually overnight. Technologies advance at lightning speed. Finally, the amount of information needed to stay abreast of industrial trends is overwhelming. It is virtually impossible for organizations facing such uncertainty to do business in any consistent way. Instead, they must remain flexible, ready to adapt themselves to whatever circumstances unfold. Today's computer and communications industries exemplify this sort of environment.

[23] Fred E. Emery and Eric Trist, "The Causal Texture of Organizational Environments." *Human Relations* 18 (1965), 21–32; and *Towards a Social Ecology* (London: Plenum, 1973).

Other Contributors. As you can see, Emery and Trist hold that organizations must respond in different ways to different environmental conditions. Tighter environmental interconnections require greater awareness about environmental conditions, and more sweeping environmental change requires greater flexibility and adaptability. In Chapter 16, when we discuss designing organizations to fit environmental conditions, we will explore the ideas of open systems theorists including Paul Lawrence, Robert Duncan, and Jay Galbraith, who have stressed the need for organizations to adjust to their environments.

A Contingency Framework

As we pointed out in Chapter 1, the contingency perspective suggests that no single theory, perspective, or school of thought is entirely correct. There is no single best way to manage or even to think about the process of management. From this viewpoint, none of the four schools of management thought tells the whole story about management. Instead, as shown in Figure 2-5, each contributes valuable insights that supplement the others' contributions. The scientific management school focuses on making a profit in the *external* world by increasing the *efficiency* of production activities. The administrative principles school emphasizes improving *internal* operations by increasing the *efficiency* of administration. The human relations school concerns developing the *flexibility* to respond to the individual needs of members *inside* the organization. The open systems school focuses on developing the *flexibility* to respond to changes in the *external* environment.[24]

Parallels are evident among the schools. The scientific management and administrative behavior schools both promote attention to efficiency and stability. The human relations and open systems schools share a common emphasis on

[24] Our classification scheme is based on research conducted by Robert E. Quinn and associates. See, for example, Robert E. Quinn and John Rohrbaugh, "A Spatial Model of Effectiveness Criteria: Towards a Competing Values Approach to Organizational Analysis," *Management Science* 29 (1983), 363–77; Quinn, *Beyond Rational Management: Mastering the Paradoxes and Competing Demands of High Performance* (San Francisco: Jossey-Bass, 1988), pp. 50–54; and Quinn, Sue R. Faerman, Michael P. Thompson, and Michael R. McGrath, *Becoming a Master Manager: A Competency Framework* (New York: John Wiley, 1990), pp. 2–12.

FIGURE 2-5

The Four Schools of Management Thought

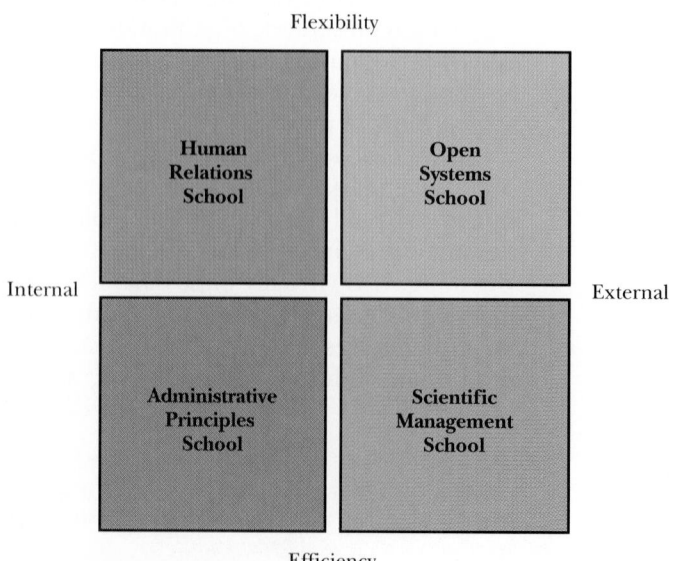

flexibility and change. The administrative principles and human relations schools focus on characteristics and procedures within the organization. The open systems and scientific management schools emphasize the importance of dealing with demands on the organization from external sources.

Each school of thought also has an opposite. The human relations school, with its emphasis on human growth and satisfaction, stands in stark contrast to the scientific management school's emphasis on employee efficiency and task simplification. The focus of the open systems school on adapting to environmental circumstances contrasts sharply with the administrative principles school's concern with developing stable, internally efficient operations. These differences reflect dilemmas that managers face every day. Shall we stimulate task performance or employee satisfaction? Shall we structure the organization to promote efficiency or flexibility? Shall we design jobs to encourage satisfaction or maximize profitability? We will discuss choice points like these throughout this book. For now, we conclude our discussion of management thought by repeating that in dealing with management dilemmas, no single approach is either always right or always wrong. Managers must make hard choices, but the perspectives offered by all four schools can help them weigh the alternatives and decide what to do.

WHAT DO MANAGERS DO?

Managers are people who plan, organize, direct, and control in order to manage organizations and organizational units. They are found in virtually every organization, from Exxon to the Girl Scouts (see "In Practice" box). Managers establish the directions to be pursued, allocate people and resources among tasks, supervise individual, group, and organizational performance, and assess progress toward goals and objectives. In order to fulfill these functions, they perform specific jobs, use a variety of skills, and play particular roles.

Managerial Jobs

Although all managers are responsible for the same functions, not all of them perform exactly the same jobs. Instead, as Figure 2-6 on page 4B shows, most organizations have three kinds of managers: top managers, middle managers, and supervisory managers. Figure 2-7 on page 43 illustrates the specific mix of planning, organizing, directing, and controlling performed by each of the three types of managers.[25]

Top Managers. *Top managers,* who are responsible for managing the entire organization, include *chairperson, president, chief executive officer, executive vice president,* and *chief operating officer.* The job of these managers consists mainly of performing the planning activities needed to develop the organization's mission and strategic goals. Top managers also perform organizing and controlling activities resulting from strategic planning. In controlling, they assess progress toward the attainment of strategic goals by monitoring information about activities occurring both within the firm and in its surrounding environment. Top management's responsibilities include making adjustments in the organization's overall direction on the basis of information reviewed in controlling procedures. Because strategic planning, organizing, and controlling require a great deal of time, top managers have little

manager A person who is responsible for planning, organizing, directing, and controlling behavior in organizations. *Top managers* are responsible for the entire firm; *middle managers* manage an organizational unit; *supervisory managers* manage the employees who do the firm's basic work.

AN EXAMPLE: DOMINO'S PIZZA
Tom Monaghan, who founded Domino's Pizza, takes a somewhat unusual approach to high-level management. Monaghan encourages his top managers to visit stores on a regular basis, and at each weekly executive team meeting, he and his managers tally the number of stores they have visited. Monaghan himself tries to visit two stores a week and enjoys flipping a few pizzas while on site. Domino's managers believe they enhance employee motivation by keeping in close touch with their employees in this way, and the success of the chain suggests they may be right. (*Fortune*, March 13, 1989, p. 38.)

[25] Luis R. Gomez-Mejia, Joseph E. McCann, and Ronald C. Page, "The Structure of Managerial Behaviors and Rewards," *Industrial Relations* 24 (1985), 147–54.

Not all successful managers are concerned about making a profit. Frances Hesselbein, former national executive director of Girl Scouts of the U.S.A., had other goals in mind as she reframed and pursued her organization's mission: to help each scout reach her highest potential.

Hesselbein, who was first lured into the Girl Scouts in the 1950s as a volunteer troop leader, moved up gradually to a full-time, paid position as a local Council director and then, in 1976, to national executive director of the organization.

A formidable challenge greeted Hesselbein. The Girl Scouts had seemingly lost touch with the society it served. Largely a white, middle-class group, it had failed to take note of important changes in its environment. Teenagers were losing interest in the Scouts, and adult volunteers were hard to find. Hesselbein undertook a major reexamination of the Girl Scouts' mission. "We kept asking ourselves very simple questions," she says. "What is our business? Who is the customer? And what does the customer consider value? Whether you're the Girl Scouts, IBM, or AT&T, you have to manage for a mission. . . . When you are clear about your mission, corporate goals and operating objectives flow from it."

Hesselbein gave the business more focus. She installed a planning system, reorganized the national staff, and introduced management training for both paid staff and volunteers. In keeping with her mission, she undertook market studies to determine how to help today's young girls to fulfill their best potential. She found that the Scouts needed to emphasize science, the environment, and business. "Math Whiz" and "Computer Fun" badges took the place of the traditional awards for household skills.

She championed "equal access" for all girls, and as a result the percentage of Girl Scouts from minority groups has tripled in the past ten years. Emphasizing again the mission of the organization, she told her staff that every girl in the country should be able to identify with the scout epitomized in the scouting handbook. "If I'm a Navajo child on a reservation, a newly arrived Vietnamese child, or a young girl in rural Appalachia, I have to be able to open that book and find myself. That's a very powerful message that 'I'm not an outsider,' that 'I can be part of something big.' "

Hesselbein's belief that everyone in the organization—scouts, volunteers, employees—must be free to develop to her fullest potential is exemplified in her innovative organization chart. Instead of the traditional pyramid with layers of increasing hierarchical power and influence, she designed a cir-

Reemphasizing the mission and purpose of the Girl Scouts has led to a revitalization of the organization and a new interest in the Scouts on the part of girls from all groups of society. Badges are now awarded for achievement in scientific and technological subjects, reflecting the entry of women into fields long dominated by men, and the percentage of Scouts who are from minority groups has tripled.

cular management structure with herself at the hub of the wheel-like diagram. "We don't talk about moving up or down," she says. "We move across. It's how we liberate the creative spirits of people."

Since leaving the Scouts, Hesselbein has become a leading speaker on management. Ranked as one of today's foremost managers by experts on leadership, she has made a videotape for management training at IBM and Motorola and also addresses classes at the Harvard Business School. As indicated by Hesselbein's accomplishments, good management practices—planning, organizing, directing, and controlling—can contribute to the success of non-profit firms just as they advance the fortunes of profit-oriented companies.*

* John A. Byrne, "Profiting from the Nonprofits: Much Can Be Learned from Some of the Best-Run Organizations Around," *Business Week*, March 26, 1990, pp. 66–74; Sally Helgesen, "The Pyramid and the Web," *The New York Times*, May 27, 1990, p. 13.

time to spend in directing subordinates' behavior. Typically, they delegate responsibility for directing such behavior to managers lower in the hierarchy of authority.

Middle Managers. *Middle managers* are usually responsible for the performance of a particular organizational unit and for implementing top managers' strategic

Types of Managers

Top managers occupy positions at the top of their organization's hierarchy of authority and are responsible for managing the entire organization. *Middle managers*, found immediately below, oversee a department or division. *Supervisory managers* are at the base of the hierarchy of authority, where they manage non-supervisory employees.

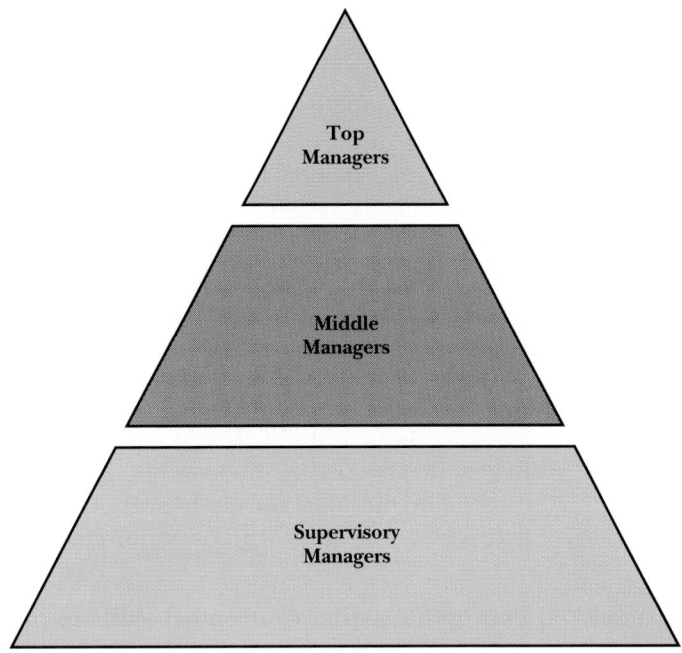

plans. During strategic implementation, each middle manager helps establish functional or divisional objectives that will guide unit performance toward targets established by strategic plans. Middle managers may also set operational objectives to be pursued at lower levels in the unit. They organize and direct unit members toward accomplishing functional or divisional objectives, and they control unit behavior by determining whether objectives are successfully attained. Labels such as *vice president*, *director* or *manager* are usually a part of a middle manager's title—for example, vice president of finance, director of personnel relations, manager of consumer affairs.

Supervisory Managers. *Supervisory managers*, often called *superintendents*, *supervisors*, or *foremen*, are charged with overseeing the nonsupervisory employees who perform the organization's basic work. Of the three types of managers, supervisory

F I G U R E 2-7

Managerial Functions and Types of Managers

Planning is the most important function of top managers; middle managers fulfill all four management functions about equally; directing is the most important function of supervisory managers.

Source: Based on information presented in Luis Gomez-Mejia, Joseph E. McCann, and Ronald C. Page, "The Structure of Managerial Behaviors and Rewards," Industrial Relations 24 (1985), 147–54.

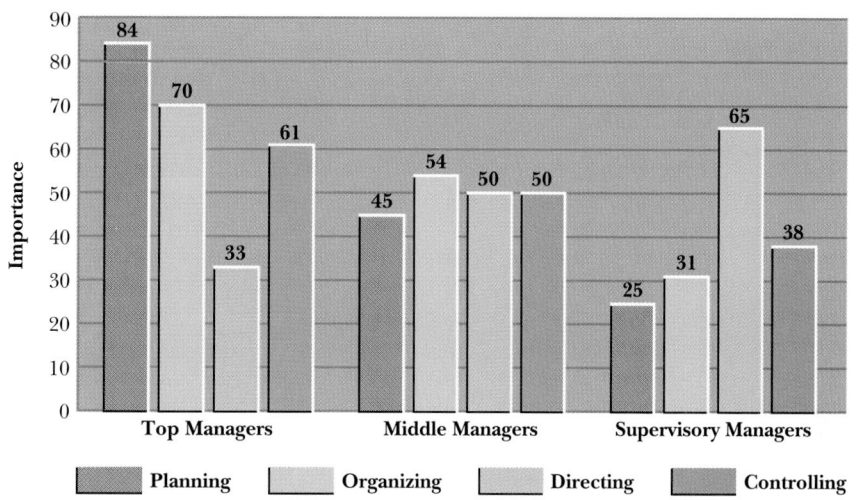

managers spend the greatest amount of time directing employees. Except for small, on-the-job adjustments, supervisory managers seldom perform planning and organizing activities. These activities are normally performed by the middle manager who has authority over the supervisory manager. Supervisory managers initiate the upward flow of information that middle and top managers use to control organizational behavior. They may also distribute many of the rewards or punishments used to control nonsupervisory employees' behaviors. Their ability to control subordinates' activities is limited, however, to the authority delegated to them by middle management.

Managerial Skills

Not surprisingly, the skills that managers must have to succeed in their jobs are largely determined by the combination of planning, organizing, directing, and controlling that they must perform. As diagrammed in Figure 2-8, each level of management has its own skill requirements.[26]

conceptual skills Management skills involving the ability to perceive an organization or organizational unit as a whole, to understand how its labor is divided into tasks and reintegrated by the pursuit of common goals or objectives, and to recognize important relationships between the organization or unit and its environment.

human skills Management skills involving the ability to work effectively as a group member and to build cooperation among the members of an organization or unit.

Conceptual Skills. Conceptual skills include the ability to perceive an organization or organizational unit as a whole, to understand how its labor is divided into tasks and reintegrated by the pursuit of common goals or objectives, and to recognize important relationships between the organization or unit and the environment that surrounds it. Conceptual skills involve a manager's ability to *think* and are most closely associated with planning and organizing. Thus, these skills are used quite frequently by top managers, who are responsible for organizationwide strategic endeavors.

Human Skills. Human skills include the ability to work effectively as a group member and to build cooperation among the members of an organization or unit. Managers with well-developed human skills can create an atmosphere of trust and security in which people can express themselves without fear of punishment or humiliation. Such managers, who are adept at sensing the desires,

[26] Robert L. Katz, "Skills of an Effective Administrator," *Harvard Business Review* 52 (1974), 90–102.

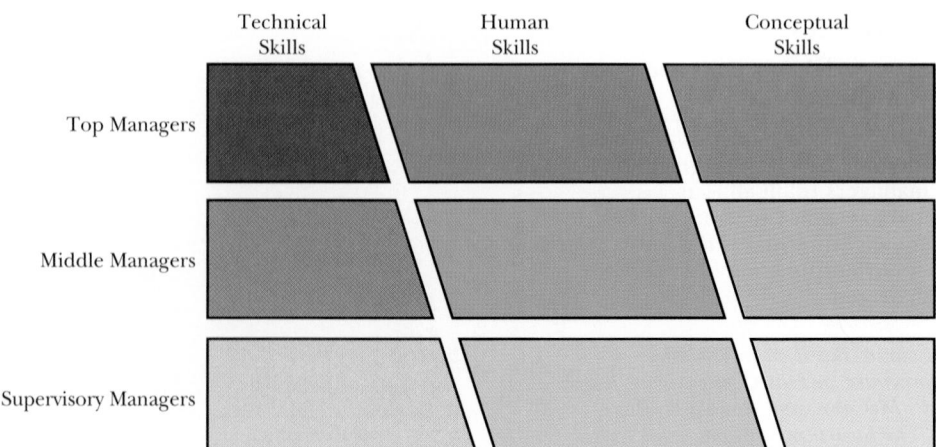

interests, and viewpoints of others, can often foresee others' likely reactions to prospective courses of action. Because all management functions require that managers interact with other employees to acquire information, make decisions, implement changes, and assess results, top, middle, and supervisory managers all need strong human skills.

technical skills Management skills involving an understanding of the specific knowledge, procedures, and tools used to make the goods or services produced by an organization or unit.

TEACHING NOTE
Preview Chapter 12's discussion of the relation between leadership and management. Have students consider whether a manager is a leader? A leader a manager? Are there differences? If not, why not have one term? If so, what are they?

managerial role Behaviors expected of managers in performing their jobs. Managers promote good interpersonal relations in the *interpersonal* role, receive and send information to others in the *informational* role, and determine the firm's direction in the *decisional* role.

Technical Skills. Technical skills involve an understanding of the specific knowledge, procedures, and tools used to make the goods or services produced by an organization or unit. Of the three managerial skills, technical skills are the most practical and down to earth. For example, a marketing manager must have skills in selling. An accounting supervisor must have bookkeeping skills. A supervisor of maintenance mechanics may need to have welding skills. To managers at the top or middle of an organization's hierarchy of authority, these skills are the least important. Technical skills are critical to managerial success only for supervisory managers overseeing employees who use technical skills in their work.

Managerial Roles

Like skills, **managerial roles** vary from one manager to another. As shown in Table 2-8, these roles cluster together in three general categories: interpersonal, informational, and decisional roles.[27]

[27] The list of ten roles described in this section is adapted from Henry Mintzberg, *The Nature of Managerial Work* (New York: Harper & Row, 1973). Other researchers who have described similar managerial roles include Sune Carlson, *Executive Behavior* (Stockholm: Stromsberg, 1951), and Rosemary Stewart, *Managers and Their Jobs* (London: MacMillan, 1967).

T A B L E 2-8
The Ten Roles of Managers

ROLE	DESCRIPTION
Interpersonal Roles	
Figurehead	Representing the organization or unit in ceremonial and symbolic activities
Leader	Guiding and motivating employee performance
Liaison	Linking the organization or unit with others
Informational Roles	
Monitor	Scanning the environment for information that can enhance organizational or unit performance
Disseminator	Providing information to subordinates
Spokesperson	Distributing information to people outside of the organization or unit
Decisional Roles	
Entrepreneur	Initiating changes that improve the organization or unit
Disturbance handler	Adapting the organization or unit to changing conditions
Resource allocator	Distributing resources within the organization or unit
Negotiator	Bargaining or negotiating to sustain organizational or unit survival

Source: Based on information presented in Henry Mintzberg, *The Nature of Managerial Work* (New York: Harper & Row, Publishers, 1973).

Interpersonal Roles. In three distinct *interpersonal roles*, managers create and maintain interpersonal relations to ensure the well-being of their organizations or units. Managers represent their organizations or units to other people in the *figurehead role*, which involves such ceremonial and symbolic activities as greeting visitors, attending awards banquets, and cutting ribbons to open new facilities. In the *leader role*, they motivate and guide employees by issuing orders, setting performance goals, and training subordinates. Managers create and maintain links between their organizations or units and others in the *liaison role*. For example, a company president may meet with the presidents of other companies at an industrial association conference.

Informational Roles. Because they serve as the primary authority figures of the organizations or units they supervise, managers have unique access to both internal and external information networks. In *informational roles* managers receive and transmit information within these networks. In the *monitor role*, managers scan the environment surrounding their organizations or units, seeking information to enhance performance. Activities in this role can range from reading periodicals and reports to trading rumors with managers in other firms or units. In the *disseminator role*, managers pass information to subordinates who would otherwise have no access to it. To disseminate information, managers may hold meetings with subordinates, write them memoranda, or telephone them. In the *spokesperson role*, managers distribute information to people outside of their organizations or units through annual stockholder reports, speeches, memos, and various other means.

Decisional Roles. In *decisional roles*, managers determine the direction to be taken by their organizations or units. In the *entrepreneur role*, managers must make decisions about improvements in the organizations or units for which they are responsible. Such decisions often entail initiating change. For example, a manager who hears about a new product opportunity may decide to commit the firm to producing it. She may also decide to delegate the responsibility for managing the resulting project to others. The *disturbance handler role* also involves change-oriented decisions. Managers acting in this role must often try to adapt to change beyond their personal control. For example, they may have to handle such problems as conflicts among subordinates, the loss of an important customer, or damage to the firm's building or plant.

In the *resource allocator role*, managers decide what specific resources will be acquired and who will receive them. Such decisions often involve important tradeoffs. For instance, if a manager decides to acquire personal computers for sales clerks, he may have to deny manufacturing department employees a piece of equipment. As part of the resource allocation process, priorities may be set, budgets established, and schedules devised. Finally, in the *negotiator role*, managers engage in formal bargaining or negotiations to acquire the resources needed for the survival of their organizations or units. In this role, for example, managers may negotiate with suppliers about delivery dates or bargain with union representatives about employee wages.

Differences Among Managers. Just as the mix of functions managers perform and the skills they use differ from one managerial job to another, so do the roles

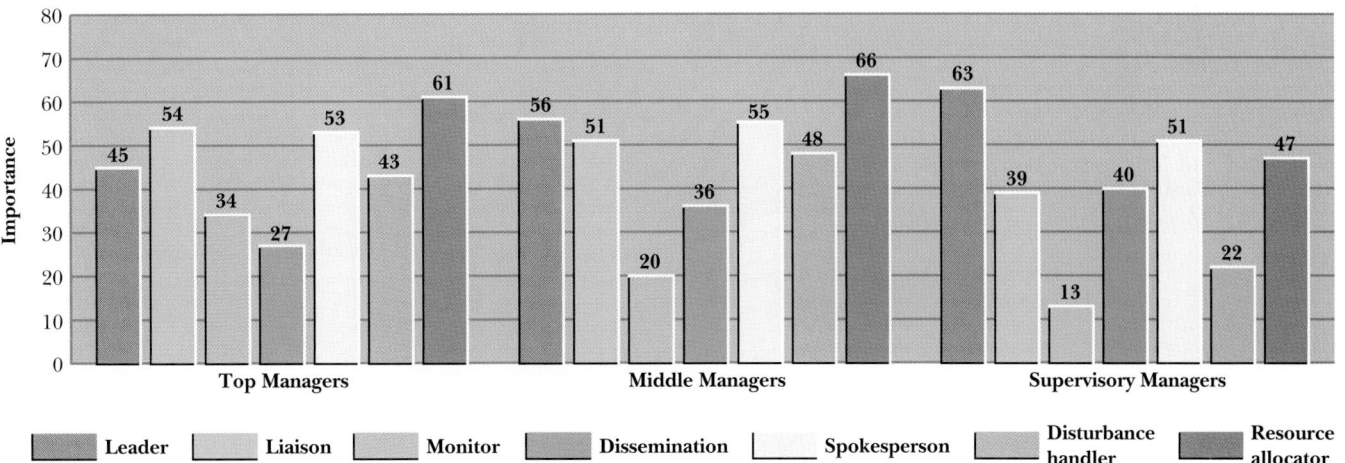

FIGURE 2-9

Managers' Jobs and the Roles They Fill

When researchers asked top, middle, and supervisory managers about the importance of the roles they perform, their answers revealed the data illustrated here. Note that the roles of figurehead, entrepreneur, and negotiator were not included in this survey.

Source: Based on information presented in Allen I. Kraut, Patricia R. Pedigo, D. Douglas McKenna, and Marvin D. Dunnette, "The Role of the Manager: What's Really Important in Different Management Jobs," Academy of Management Executive *3(1989), 286–93.*

managers fill. Figure 2-9 shows, for example, that the roles of liaison, spokesperson, and resource allocator are most important in the jobs of top managers, reflecting top management's responsibilities for planning, organizing, and controlling the strategic direction of the firm. In addition, monitor activities are more important for top managers than for others because they must scan the environment for pertinent information.

Among middle managers, leader, liaison, disturbance handler, and resource allocator roles are the most important. These roles are in keeping with middle management's job of organizing, directing, and controlling the functional or divisional units of the firm. The role of disseminator is also important in middle managers' jobs because they are responsible for explaining and implementing the strategic plans of top management.

For supervisory managers, the leader role is the most important. They spend most of their time directing non-supervisory personnel. They also act as spokespersons who disseminate information within their groups and serve as liasons that connect their groups with the rest of the organization. In addition, they acquire and distribute the resources their groups need to do their jobs.

THE NATURE OF MANAGERIAL WORK

To further analyze the classification of managerial roles we've just discussed, Henry Mintzberg observed a group of top managers at work for several weeks. Listing major activities and the time it took to perform

FIGURE 2-10

The Manager's Day

Source: Reproduced with the author's permission from Henry Mintzberg, The Nature of Managerial Work *(New York: Harper & Row, Publishers 1973), p. 39.*

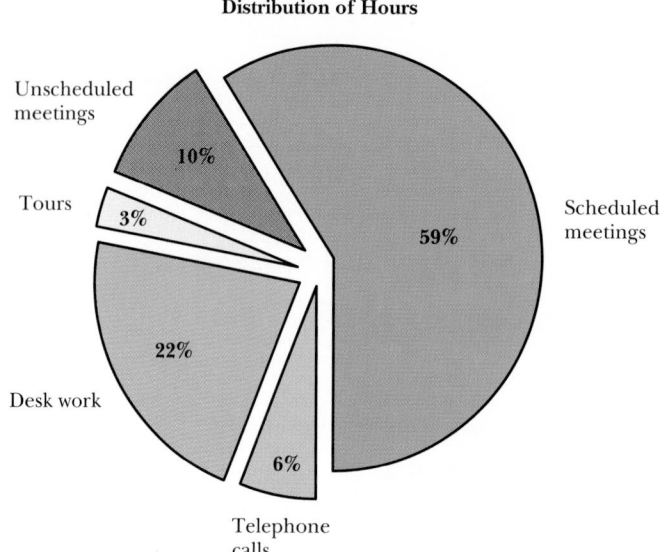

Distribution of Hours

Unscheduled meetings 10%

Tours 3%

Desk work 22%

Telephone calls 6%

Scheduled meetings 59%

AN EXAMPLE: COMPAQ COMPUTER
An article in Fortune *states managers need the "courage and foresight to concentrate on the critical tasks rather than trying to do it all." The founder of Compaq Computer had to learn this lesson. When the company started, Rod Canion knew what everyone else was doing every hour of the day. Now, with 9,000 employees, he has to prioritize, delegating less important decisions, responsibilities, and activities to lower levels of management.*

them, Mintzberg found that his managers spent by far the most time in scheduled meetings. When combined with unscheduled meetings, this activity accounted for almost 70 percent of the managers' time. As Figure 2-10 shows, the managers were left with barely a fifth of the day for desk work, and about a tenth for telephone calls and company tours.

Mintzberg also recorded the amount of time consumed by each instance of each activity. Scheduled meetings averaged a little over an hour in length and ranged from under 10 minutes to over 2 hours. Unscheduled meetings were normally shorter, lasting from a few minutes to about an hour and averaging approximately 12 minutes each. Periods of desk work and tours to inspect the company averaged around 11 to 15 minutes each and were fitted in between scheduled meetings and unscheduled interruptions. Telephone calls were almost always quite short, averaging about 6 minutes apiece.

Based on his observations, Mintzberg concluded that managers work in short bursts rather than in long, uninterrupted sessions. They frequently lack the time to complete rigorous planning, organizing, directing, and controlling. Instead, managing is often more a process of making erratic, *incremental adjustments* than one of following a routine, well-planned course of action.[28] Managing is a fast-paced, active profession.

SUMMARY

Management is a process of *planning*, *organizing*, *directing*, and *controlling* the behavior of others that makes it possible for an *organization* using a *division of labor* to accomplish a *mission* beyond the means of individuals working alone.

Theorists from four schools of thought have attempted to explain or improve management practices. Members of the *scientific management school* have tried to increase the efficiency of production processes in order to enhance marketplace

[28] James Brian Quinn, *Strategies for Change: Logical Incrementalism* (Homewood, Ill: Richard D. Irwin, 1980), p. 18.

profitability. Proponents of the *administrative principles school* have focused on enhancing the efficiency of administrative procedures. Researchers in the *human relations school* have put the emphasis on nurturing the growth and satisfaction of organization members. Theorists in the *open systems school* have highlighted the importance of coping with the surrounding environment.

Managers differ in terms of where they fit in the *hierarchy of authority*. These differences affect managers' jobs, influencing the *skills* they use, the behaviors they engage in, and the *roles* they fill. The job of manager is fast paced and allows the manager little time to devote to any single activity.

REVIEW QUESTIONS

1. How does an organization enable its members to accomplish a purpose beyond the means of individuals working alone? Why aren't organizations formed to achieve purposes that people can accomplish individually?

2. What is an organization's mission? Its division of labor? Its hierarchy of authority? How do these three organizational attributes clarify the role of management?

3. Describe the work of a manager in performing each of the four basic management functions: planning, organizing, directing, and controlling. How does planning affect organizing? How does organizing affect directing? How does directing affect controlling? How does controlling affect later planning?

4. What is the central idea underlying work in the scientific management school? What advice would an expert in this school give managers? Give an example of the kind of change an expert in scientific management might recommend if he or she were asked to improve the efficiency of your class.

5. How does the administrative principles school differ from the scientific management school? What is a management principle? How does it differ from a law or rule? What is a bureaucracy? Give an example of an organization that is extremely bureaucratic. Of an organization that is not very bureaucratic.

6. What does the human relations school focus attention on? According to Douglas McGregor, what sort of perspective do members of this school have on management? The Hawthorne researchers differed from members of the human relations school in what important respect?

7. What are the two key ideas underlying the open systems school? What central tenet do they support? Explain the cycle of events described by Katz and Kahn. According to your description, why is it important for managers to be able to diagnose environmental conditions and adapt their organizations to environmental changes as they occur?

8. Explain the contingency model formed by the four schools of management thought described in this chapter. If you were a manager having problems with employee satisfaction, which school of thought would you consult for advice? If you were concerned about efficiency, which schools of thought could probably help you out?

9. What do managers do? What are the three kinds of managers? How do hierarchical differences affect the job of being a manager? How do the management jobs at various levels fit together, connecting the whole organization with each of its parts?

10. What kinds of skills do all managers use to perform their jobs? Which of these skills are you learning about as you read this book? Which skill becomes

more important as managers move up the hierarchy of authority? Which one stays about the same? Which one loses importance?

11. What roles do managers perform? What effect do hierarchical differences have on the roles of managers?

12. Describe the job of being a manager. How realistic is the television portrayal of managers as doing little more than sitting behind desks, making decisions, and telling people what to do? How might this portrayal be made more realistic?

CHAPTER 3

THINKING CRITICALLY ABOUT ORGANIZATIONS: RESEARCH METHODS

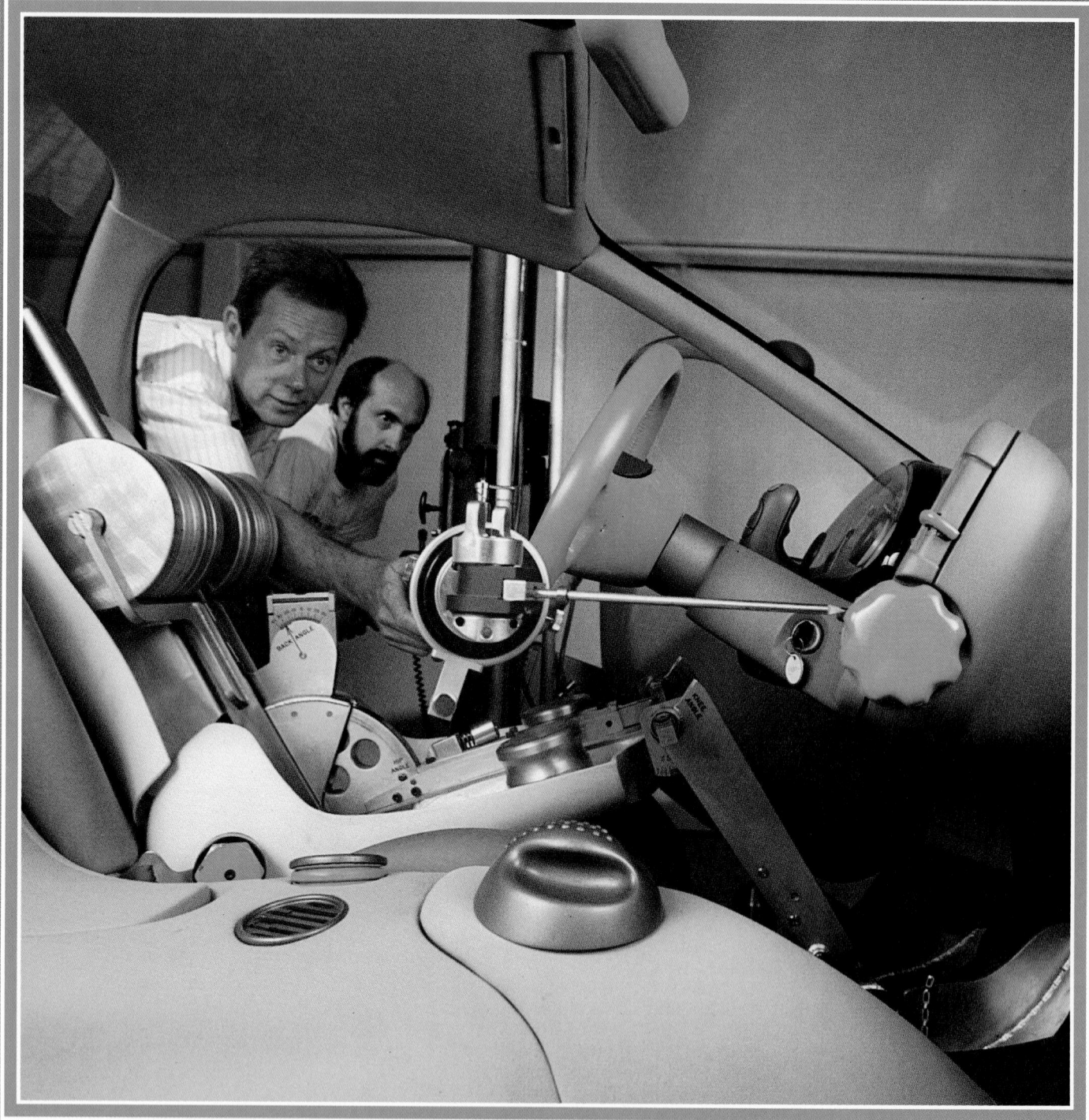

Is the United States number one? Is it in danger of losing its economic leadership? According to J. D. Power & Associates Inc., U.S. auto makers may not yet realize that their arch rival, Japan, is into a "second phase of quality," researching refinements of design and function that Japanese firms hope will win consumer approval. At Nissan, for example, engineers turn the results of anthropological research on what people want in cars into features that give cars a comfortable feel, such as computer-driven hydraulics that cushion jolts and level a car during quick stops and sharp cornering.

Source: "A New Era for Auto Quality," Business Week, October 22, 1990.

I know it may sound un-American, but I don't believe that we were born to be number one. America won't be number one forever as a genetic or historic right. I think that it is entirely conceivable that the United States could lose its world economic leadership. It is no longer unthinkable that we could just fade away. Remember that Britain kept saying that productivity stagnation was unacceptable even as her position steadily deteriorated. I think that unless this nation decides to do something about the problems that have generated our productivity stagnation, we will fade away.[1]

This statement on productivity in America paints a grim picture of our current economic situation. It also underlines our need for more and better knowledge about the effective management of organizations and the people who form them. The complexity and volatility of the modern world have raised the stakes associated with managerial success or failure as they have increased the challenge to organizations and entire societies.

The performance of a firm depends to a great extent on the competence of its managers, and the manager's job is essentially one of directing the behavior of others. This fact is recognized by the increasing tendency of firms to evaluate managers' performance on the basis of their success in handling behavioral problems with the firm. General Electric, for example, now assesses its managers' performance with an employee relations index in which all eight indicators of success or failure deal with employee behaviors like performance on the job, absenteeism, and the initiation of formal grievance procedures.[2]

RESEARCH IN ORGANIZATIONAL BEHAVIOR

Clearly, knowledge about organizational behavior has become increasingly critical to a manager's performance and to her long-term career success. Given this need for knowledge, it is not surprising that writers in this field often claim to have the information that managers need to excel at their jobs. Mitroff and Mohrman comment that "in this environment U.S. businesses often fall prey to every new management fad promising a painless solution, especially when it is presented in a neat, bright package."[3] Indeed, the demand for this knowledge has created a veritable cottage industry of "pop" management books; consider the "excellence" books—the best-selling *In Search of Excellence*, by Tom Peters and Robert Waterman, and its sequel, *A Passion for Excellence*, by Peters and Nancy Austin.[4]

Initially, Peters and Waterman outlined eight principles that they said distinguished excellent firms from others. Peters and Austin reduced this list to four. Since the appearance of these two books, many business firms have reportedly been attempting to conform their practices to one or the other set of principles.

If these and other similar books have already provided managers with all the answers they need, why do we devote an entire chapter in this textbook to

---

[1] Reprinted from P. Galagan, "Staying Alive: Jack Grayson on the American Productivity Crisis," *Training and Development Journal* (1984), 59–62.

[2] H. F. Merrihue and R. A. Katzell, "ERI—Yardstick of Employee Relations," *Harvard Business Review* 33 (1955), 91–99.

[3] I. I. Mitroff and S. A. Mohrman, "The Slack Is Gone: How the United States Lost Its Competitive Edge in the World Economy," *Academy of Management Executives* (1987), 69.

[4] Tom J. Peters and Robert Waterman, *In Search of Excellence* (New York: Harper and Row, 1982); and Tom J. Peters and Nancy Austin, *A Passion for Excellence* (New York: Random House, 1985).

research in organizational behavior? Let's take a closer look at some of the answers provided by Peters and his colleagues.

The widespread acceptance of the excellence books led researchers Michael Hitt and Duane Ireland, to attempt to replicate Peters' work.[5] Employing more rigorous research techniques and a sample of 185 firms (which included all those studied by Peters and Waterman as well as many others), Hitt and Ireland came to some startling conclusions. First, comparing market returns, they found no significant differences between the so-called excellent firms and others. Second, Hitt and Ireland discovered that the firms designated excellent were no more likely to report following the excellence principles than were other firms. Finally, these researchers found that whether or not firms were designated excellent, there was no relationship between their market performance and their adherence to the principles formulated by Peters and his associates.

Concluding their report, Hitt and Ireland comment that "excellent firms identified by Peters and Waterman may have not been excellent performers, and they may have not applied the excellence principles to any greater extent than did the general population of firms. Additionally, the data call into question whether these excellence principles are in fact related to performance."[6] Hitt and Ireland note that ultimately the problem with the "excellence" books is not so much that their message is wrong—even their authors did not intend that managers should follow their guidelines blindly. Rather, the problem lies in the fact that practicing managers are so intent in their search for answers that they often uncritically adopt the first useful-looking material they find, failing to survey and evaluate the wide range of research data available.

To avoid this quick-fix mentality, Hitt and Ireland offer several suggestions. First, they urge managers to keep current with the literature in the field of management and to pay particular attention to journal articles that translate research findings into practical guidelines. Second, they warn managers to be skeptical when simple solutions are offered, analyzing such solutions (and their supposed evidence) thoroughly. Third, Hitt and Ireland suggest that managers make sure that the concepts they apply are based on science rather than advocacy. They also urge managers to experiment with new solutions themselves whenever possible.

The purpose of this chapter is to help dispel the quick-fix mentality by focusing on Hitt and Ireland's three recommendations. We will begin by examining the nature of the scientific process in order to show you not only how you can use others' research findings but how you can conduct your own experiments. Next, we will show you how to be skeptical in evaluating the claims made by those who propose solutions to your problems. Finally, we will describe some of the scientific sources you can turn to in seeking answers to your managerial questions.

People who teach organizational behavior and management skills often lament the fact that there is not enough dialogue between practicing managers and researchers. This kind of dialogue can develop only when managers and researchers understand each other's work and appreciate its value for their own efforts. Practicing managers need to know what organizational behavior researchers do and why they do it the way they do. Researchers need to know what practitioners' most pressing problems are so that they can study issues that managers view as significant. It is in large part because we feel it is so important to create and encourage this kind of ongoing practitioner–researcher dialogue that we have included this chapter on research methods in our book.

---

[5] M. A. Hitt and R. D. Ireland, "Peters and Waterman Revisited: The Unended Quest for Excellence," *Academy of Management Executives* 2 (1987), 91–98.

[6] Ibid., p. 95.

EXPLORING THE SCIENTIFIC PROCESS

As part of their solution for restoring the United States's competitive edge, Mitroff and Mohrman note that what managers need is not simple solutions to complex problems but "a method for helping [them] debate and thereby assess the proposed attributes of excellence."[7] The purpose of this section is to describe one such method—the scientific method. Let us begin by comparing the scientific method to other ways of discovering truth.

Ways of Knowing

How do we come to know things? When we say we know that there are nine planets in our solar system, how do we know this is true? When we say that dropping atomic bombs on Nagasaki and Hiroshima caused the end of World War II in the Pacific, how do we know this is true? When we say we know that providing workers with specific and difficult goals leads them to perform better than just telling them to do their best, how do we know this is true? Finally, when we say we know that an organization's structure must match its technology and its environment, how do we know this is true?

Traditional Ways of Knowing. Philosophers of science have explored many ways of arriving at knowledge.[8] Figure 3-1 suggests that some ways of acquiring knowledge are more reliable than others. Thus among the traditional ways of knowing, personal experience and rationalism are more satisfactory than tenacity, intuition, and authority, and science is the most reliable of all. Let's look first at the traditional sources of knowledge.

By *tenacity* we mean believing that something is true simply because we have always believed it to be true. Many superstitions and prejudices are based on this method of knowing. *Intuition* means arriving at knowledge without relying on either reason or inference. This method is based on an appeal to propositions

---

[7] Mitroff and Mohrman, "Slack is Gone," p. 69.

[8] See, e.g., J. Buchler, *Philosophical Writings of Peirce* (New York: Dover, 1955); M. Cohen and E. Nagel, *An Introduction to Logic and the Scientific Method* (New York: Harcourt, 1954); M. Polyani, *Personal Knowledge* (Chicago: University of Chicago Press, 1958); L. B. Christenson, *Experimental Methodology* (Boston; Allyn & Bacon, 1977).

FIGURE 3-1

Ways of Knowing

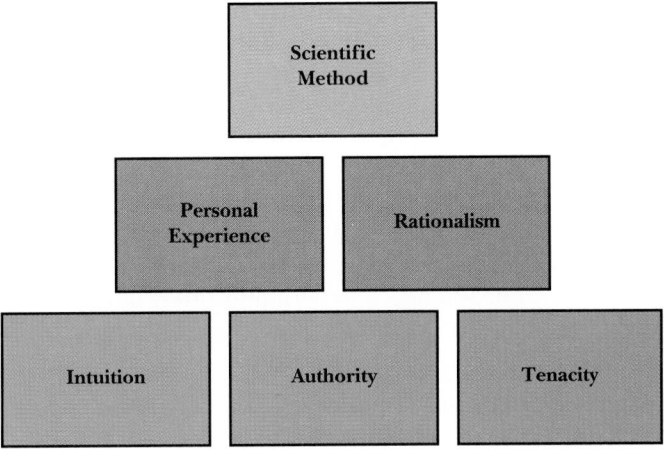

felt to be obvious and indisputable. Recall, for example, Thomas Jefferson's assertion in the introduction to the Declaration of Independence that "we hold these truths to be self-evident." By the method of *authority*, we may believe a statement or proposition to be true because it is made by a person whom we respect or who we feel is an expert on the topic. For example, if a doctor tells a pregnant woman her baby must be delivered by Caesarean section, the woman is likely to believe what he says.

Although most of us base some of our knowledge on these three methods, clearly they are not always reliable ways of coming to know something.[9] Two people or two groups often hold tenaciously to opposite beliefs; self-evident truths are often more evident to some than to others; authorities often give conflicting advice on the same issue. The main problem with these three methods is that they do not allow for reconciling the disagreements that inevitably arise.

We find the same problem with the fourth traditional way of knowing—*personal experience*. This is probably the most widespread source of knowledge for most of us. Most people tend to believe information they acquire through interacting with other people and the world at large and to conclude that their experience reflects truth. Again, different people may have different experiences that point to different truths. Even more problematic, people's perceptions and memories of their experiences are often biased, inaccurate, or distorted over time. Finally, even if we disregard inaccuracies of perception or memory, the fact remains that any one person can experience only a tiny fraction of all possible situations: thus the knowledge she acquires by personal experience will necessarily be extremely limited.

The last traditional method of attaining knowledge is *rationalism*, or the notion that we can acquire knowledge if we use the correct reasoning procedure. Logical deductions form the basis for such knowledge. For example, in mathematics, if A is greater than B and B is greater than C, logic tell us that A is greater than C. Once we step outside the world of mathematics, however, this kind of logic breaks down. Suppose, for example, that a baseball team—we'll call it Team A—beats another team, Team B, and Team B beats Team C; can we say that therefore Team A will beat Team C? No, because there are too many factors and too much variability over time in events like a baseball game or a company's advertising campaign. Such events involve many different individuals and many different conditions.

scientific method An objective method of expanding knowledge characterized by an endless cycle of theory building, hypothesis formation, data collection, empirical hypothesis testing, and theoretical modification.

objectivity In science, the degree to which a set of scientific findings are independent of any one person's opinion about them.

Scientific Method. The problems and pitfalls of these more traditional ways of knowing, and in particular their subjectivity, led to the development of the **scientific method**. As Charles Sanders Peirce has stated, "To satisfy our doubts . . . it is necessary that a method should be found by which our beliefs may be determined by nothing human, but by some external permanency. . . . The method must be such that the ultimate conclusion of every man shall be the same. Such is the method of science."[10]

Thus **objectivity**, which measures the degree to which scientific findings are independent of any one person's opinion about them, stands as the major difference between the scientific approach to knowledge and the other approaches described so far. Science as an enterprise is *public* in the sense that methods and results obtained by one scientist are shared with others. It is also *self-correcting*, in the sense that erroneous findings can be isolated through the replication of one scientist's work by another scientist. And it is *cumulative*, in the sense that one scientist's experiment often builds on the work of another. These features

---

[9] E. F. Stone, *Research Methods in Organizational Behavior* (Santa Monica, Calif.: Goodyear, 1978).

[10] Buchler, *Philosophical Writings of Peirce*, p. 18.

of the scientific method make it ideal as a means of generating reliable knowledge, and it is no coincidence that the physical, natural, and social sciences receive so much emphasis in today's colleges and universities. For all these reasons, it will be useful for us, in this chapter, to explore the nature of the scientific process more closely. We will look first at the major goals or purposes of science and then at how the scientific method is structured to achieve these objectives.

THE PURPOSES OF SCIENCE

The basic goal of science is to help us understand the world around us. Science defines the understanding it seeks as the ability to describe, explain, predict, and control the subjects of its inquiry. We will examine each of these objectives.

Description. The purpose of some research is simply *description*, that is, drawing an accurate picture of a particular phenomenon or event. In Chapter 2, for example, we presented data from Mintzberg's study of managerial roles (see Table 2-8). The purpose of Mintzberg's research was simply to find out what managers actually do on the job on a daily basis.[11] In Chapter 4, we will review descriptive research that attempts to describe the major dimensions of intelligence, or cognitive ability. In Chapter 17, we will look at descriptive research that seeks the dimensions best suited to describe the nature of jobs. The development of scientific knowledge usually begins with descriptive work. The ultimate criterion for evaluating all descriptive research is the fidelity with which it reflects the real world.

Explanation. The ultimate goal of science is *explanation*—stating why some relationship exists. Some might argue that as long as we can describe, predict and control things, why go any further? For example, if a manager in the insurance business knows that people with college degrees sell more life insurance than people with high school degrees, why does she need to know more than this? Why not just hire all college graduates for sales positions? Well, if researchers can uncover the reason for college graduates' greater success, the manager may be able to bring about the desired outcome (selling more insurance) in a more efficient or cost-effective way.

For example, suppose that college-educated salespeople outperform those without higher education not because of their years of study per se but because, on average, they are more self-confident, and self-confidence sells. If this were the case, the manager might be able to get the same high success rate by hiring, at lower salaries, high-school-educated people who are high in self-confidence or by hiring such persons and training them to boost their self-confidence. You can see that if we know the exact reason why something occurs we can usually explain and control it much more efficiently.

Prediction. *Prediction*, or stating what will happen in the future, is the primary goal of many scientific studies. Prediction requires that we know the relationships between certain conditions and outcomes. For example, in Chapter 8, we will look at research that attempts to predict who will leave organizations and who will stay. In Chapter 10, we will review studies that predict when decisions are made best by groups and when they are best left to individuals. In Chapter 18,

[11] H. A. Mintzberg, "Structured Observation as a Method to Study Managerial Work," *Journal of Management Studies* 7 (1970), 87–104.

we will discuss studies that attempt to predict strategic choices of various kinds of organizations. When we cannot accurately predict what will happen in a given situation, we have generally failed to understand it.

Control. Studies that focus on prediction often lead to further research in which the goal is to *control* the situation. Predictive studies often uncover relationships between antecedents and outcomes, and if it is possible to manipulate the antecedents, it may be possible to control the outcomes. In Chapter 8, for example, we will review studies that show that by manipulating pay practices, one can also manipulate how hard individuals will work. In Chapter 10, we will discuss research that shows how group communication patterns can be controlled by manipulating how chairs are arranged around a table. In Chapter 17, we will show how changing the design of work will lead to changes in worker attitudes. It is in the area of control that the interests of scientists and practitioners most clearly converge.

As we have seen already, managers in organizations are responsible for controlling the behaviors of others. Thus the more information a study provides on how control can be achieved, the more useful the study is to practicing managers. Indeed, research guided by the other three objectives is often perceived by managers as "academic" and not worthwhile. But as you have seen, studies dealing with control often are the by-products of earlier descriptive, explanatory and predictive studies. Without good descriptive, explanatory and predictive research, we would probably never do much successful research aimed at control.

The four objectives of science are often pursued by researchers in the order given above, that is, starting out with description and proceeding through control. Despite this fact, the four goals do not explicitly build on each other in any tight hierarchical fashion. That is, you can sometimes predict things that you cannot explain. For example, before Copernicus, people could predict the passage of day to night even though they could not explain why this occurred. Today, someone with little background in auto mechanics or chemistry may be able to control engine knocking by using higher octane fuel but not have any idea why this works.

The Interplay of Theory and Data

We have concluded that the scientific method is the best available means for arriving at knowledge, and we have laid out the goals, or purposes, of scientific inquiry. Now we need to consider precisely what scientific method entails. Figure 3-2 represents our conception of scientific inquiry and depicts science as a con-

F I G U R E 3-2

The Nature of the Scientific Process

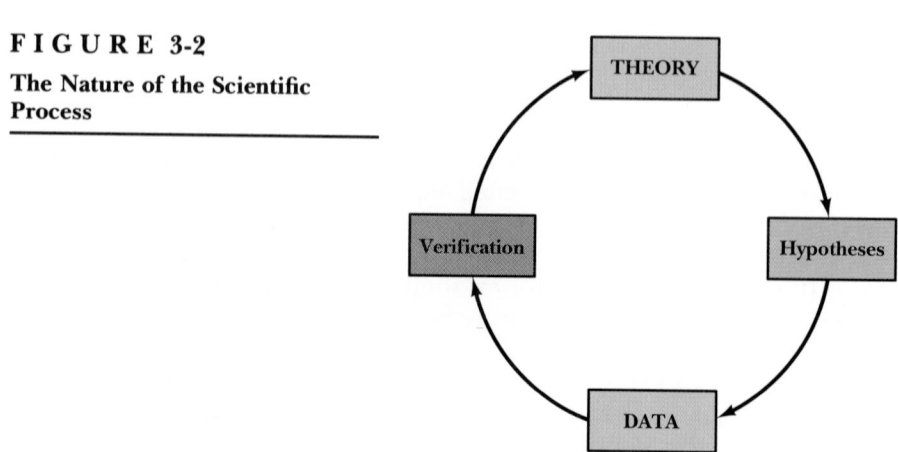

tinuous process that links theory, which resides in the world of abstract ideas, with data, which reside in the world of concrete facts. A theory is translated into real-world terms by the process of stipulating hypotheses, and real-world data are translated back into the realm of ideas through the process of verification. These processes then form a chain, and any one scientific study is only as strong as its weakest link. Moreover, if one link breaks apart (e.g., if the theory is not appropriately expressed in the hypotheses) then the whole process breaks down.

Kerlinger defines a **theory** as "a set of interrelated constructs, definitions, and propositions that present a systematic view of a phenomenon by specifying relations among variables."[12] With your understanding of the purposes of science and this definition of theory, can you see why theory plays such a central role in the scientific process? A good theory, through its constructs and definitions, should provide us with a clear description of a part of the real world. Moreover, by specifying relations among variables, a theory facilitates both prediction and control. Finally, a theory's systematic nature allows us to explain the relationships described. The remaining chapters of this book are filled with theories that attempt to help you understand how to manage the behavior of people in organizations.

Theory is only half of the scientific process, however. The other half deals with *data*, or real-world facts. Theories, to have any practical utility, cannot remain solely in the world of ideas. The scientific process requires that theories prove themselves in the world of data. Through a process of deduction, **hypotheses**, or specific predictions about the relationships between certain conditions in the real world, are generated. These hypotheses are related to the theory in the sense that if the theory is correct, then what the hypotheses predict should be found in the real world.

Here is where data enter the scientific process. Once hypotheses are formulated, data can be collected, and the hypothesized and actual patterns of results can be compared. Then, through the process of **verification**, this comparison can be used to check the accuracy of the theory—to judge the extent to which it is true. If there is very little correspondence between the hypothesized results and the actual findings, the theory must be rejected. The process must begin all over again, with the generation of a new theory. If there is only some correspondence between the projected and actual findings, the theory may need to be changed in some way so as to be consistent with the data. If there is almost complete correspondence between the hypothesized results and the actual findings, we may be tempted to claim that we have proven that the theory is true. Such a conclusion would not be warranted, however, unless we could establish that all other possible explanations for the results (explanations not accounted for by the theory) had been eliminated. Because this is almost never possible, we usually refer to data that correspond closely with a hypothesis as "supporting" rather than "proving" the theory.

Thus the scientific process relies on the constant interplay of theory and data. One without the other is no better than neither at all. For example, in the Middle Ages, an elaborate theory of body chemistry led early physicians to practice bloodletting, in which they applied leeches to various parts of the human body in an effort to heal various illnesses. The problem was that no one ever tested the theory to see if those who were treated in this fashion actually were better off than those who were not. Had such data been collected, people would have realized that the theory was false. Scientific knowledge requires both theory and data, and great mistakes can be made when one is divorced from the other.

Good Theories. One doesn't have to be a scientist to have a theory. Indeed, in our daily lives, we all develop informal or **implicit theories** about the world

[12] F. N. Kerlinger, *Foundations of Behavioral Research* (New York: Holt, Rinehart & Winston, 1986), p. 9.

explicit theories Internally consistent, formal theories that are subject to empirical test.

around us. We arrive at these theories through our personal experience and are often unaware that they exist. Many of these implicit theories can be lumped together under the general heading of common sense. Scientific theories are usually developed more formally. We will refer to these as **explicit theories** in order to distinguish them from implicit theories. As you will see throughout this book, much of what we are trying to do when we discuss the diagnostic stage of our overall model is to get you to replace your implicit theories with explicit theories that have been supported by research. However, explicit theories are not always better than implicit theories. Moreover, there are often multiple explicit theories that deal with a given subject, and some may be better than others. How do we judge whether a theory is good?

John B. Miner has offered several good criteria for judging the worth of theories.[13] First and foremost, a theory should contribute to the objectives of science: it should be useful in describing, explaining, predicting, or controlling important things. Most theories, whether implicit or explicit, meet this test.

Second, a theory must be logically consistent within itself. Here is where many implicit theories (and some explicit ones) fall short. For example, common sense tells us that "fortune favors the brave." On the other hand, common sense also tells us that "fools rush in where angels fear to tread," which has the opposite implication. Again common sense tells us that "two heads are better than one," but it also tells us that "too many cooks spoil the broth." As you can see, common sense, and many of the implicit theories on which it is based, is not good theory at all because it contradicts itself.

Third, it is also important that a theory be consistent with known facts. For example, many people have an implicit theory that "a happy worker is a productive worker." In fact, a vast amount of research shows that satisfaction and performance are actually unrelated.[14] Any theory that stated directly or implied that these two variables are related would not be a good theory.

A fourth criterion by which to evaluate a theory is its consistency with respect to future events. The critical notion here is not simply that the theory predicts but that it makes *testable* predictions. A prediction is testable if it can be refuted by data. A theory that predicts all possible outcomes says nothing at all. For example, if a theory states that a particular leadership style can increase, decrease, or leave employee performance unchanged, it has really said nothing about the relationship between that leadership style and worker performance.

Finally, simplicity is a desirable characteristic of a theory. Highly complex and involved theories are not only more difficult to test but more difficult to apply. Therefore, a theory that uses only a few concepts to predict and explain some outcome is preferable to one that does the same thing with more concepts. Theoretical simplicity, however, is hard to maintain. Theories are by their nature oversimplifications of the real world. The inductive nature of the scientific process, which requires that a theory be consistent with real-world data, inevitably pushes simple theories toward increasing complexity over time. A good theory is one that can walk the fine line between being too simple (when it will fail to predict events with any accuracy) and being too complex (when it is no longer testable or useful for any purpose).

Good Data. Because science deals with the interplay of theory and data, good data are just as important as good theory. Most of the data for testing theories are gathered through measures of the theory's important concepts. There are

---

[13] J. B. Miner, *Theories of Organizational Behavior* (Hinsdale, Ill.: Dryden Press, 1980).

[14] M. T. Iaffaldono and P. M. Muchinsky, "Job Satisfaction and Performance: A Meta-Analysis," *Psychological Bulletin* 97 (1985), 251–73.

RESEARCH METHODS

several characteristics that make some measures, and therefore some data, better than others.

First, the measures of the theoretical concepts that we are interested in must possess **reliability**; that is, they must be free of random errors and thus present a consistent, stable reflection of the underlying concept. Suppose, for example, that you were applying for graduate school and the person who was interviewing you was interested in your scholastic aptitude because he felt this measure was predictive of success in graduate school. Imagine, then, that to assess your aptitude the interviewer handed you two dice and asked you to toss them, at the same time suggesting that a high score would mean you had high aptitude and a low score that you had low aptitude. At this point, you would probably start wondering about the aptitude of the interviewer, as dice tossing is obviously a very poor measure of scholastic aptitude. Aside from this fact, one of the main problems with dice as a measure is that they generate completely random, unreliable numbers. That is, you could get a high score today, and tomorrow you could get a low score. There would be little consistency or agreement in your scholastic aptitude from one measurement to the next. Thus the unreliability of this measure makes it virtually worthless.

This is an extreme and obvious case, but consider the following. It was once believed that interviewers, after talking to job applicants in an unstructured way for about 30 minutes, could provide ratings reflecting the suitability of these people for the jobs for which they were being considered. Research showed, however, that these ratings were about as unreliable as the dice-tossing example cited above.[15] That is, an interviewer would rate an applicant high one day and then after some passage of time rate the same applicant differently another day. In making important decisions like admitting an applicant to graduate school, most institutions rely heavily on test scores like the Graduate Record Exam (GRE), the Graduate Management Admissions Test (GMAT), and the Law School Admissions Test (LSAT). Although these tests are not perfectly reliable (i.e., students taking them repeatedly will not get the exact same score each time), they do exhibit a high degree of consistency on retesting.

[15] R. Arvey and M. Campion, "The Employment Interview: A Summary and Review of Recent Research," *Personnel Psychology* 34 (1982), 281–322.

When people are conducting applied research, the speed with which data can be collected is often very important. Information needs to be brought to bear on problems before it's too late. When Frito-Lay issued hand-held Fujitsu computers to all its salespeople, the quality of the information about the company's products and its competitors' that fed into Frito-Lay's Dallas headquarters skyrocketed. In two days, for example, the company was able to revitalize a sales promotion that was lagging in some stores of a Von's chain in Los Angeles. Under its old system, the company might not have noticed the difference between store sales for weeks, too late to do anything about the problem.
Source: *"Managing,"* Fortune, *September 24, 1990, pp. 116–18.*

validity The degree to which a measure of an individual, group, organizational, or environmental attribute does what it is intended to do.

criterion-related validation Establishing validity by showing that a measure predicts some variable that, based on theory, it should predict.

content validation Establishing validity by showing that, according to expert judges, the measure samples the appropriate material.

construct validation Establishing validity by showing that a measure of a concept is congruent with the theory and data that support the concept.

POINT TO STRESS
Validity asks the question, are we measuring what we think we are measuring? Valid measures used to make hiring decisions must be job related.

POINT TO STRESS
You can have reliability and not have validity, but you cannot have validity without reliability.

standardization In the context of scientific measurement, the practice of ensuring that all people measure the same variables with the same instruments applied in the same manner.

Second, the measures of a theory's concepts must possess **validity**; that is, they must assess what they were meant to assess. To see if the GMAT is valid, for example, we might want to test whether those who perform better on the test actually perform better in graduate school. This means of testing validity is called **criterion-related validation**, because we test whether the measure predicts some criterion (e.g., grade point average) that it is supposed to be able to predict. We can also assess validity by having recognized experts on the concept the test is designed to measure rate the extent to which the test items actually represent that concept. This is called **content validation**, because we test whether the content of the test is appropriate according to experts on the subject.

In **construct validation** we assess the degree to which the measure actually taps some abstract concept. The most complex form of validation, construct validation requires all the steps associated with criterion-related and content validation as well as other tests. Among the latter tests, one of the most important is designed to show that the test is not contaminated, that is, that it does not measure something that it is *not* supposed to measure. For example, if we were trying to measure scholastic aptitude with a handwritten essay test and could show that students with poor handwriting consistently got lower grades than students with good handwriting (especially if we knew that the content of all the students' answers was the same) we would conclude that this measure was contaminated. It lacks construct validity because it is measuring something (i.e., handwriting ability) that it is not supposed to be measuring.

Reliability and validity are closely related. Reliability is necessary for validity, but it is not sufficient, because we could develop highly reliable measures that might not prove valid. For example, we could probably measure people's height reliably, but this measure would have little validity as a measure of scholastic aptitude (i.e., it could not predict who would do well in graduate school). Reliability is necessary for validity though, because an unreliable measure cannot pass any of the tests necessary for establishing validity. An unreliable measure does not relate well even to itself. Thus it is virtually impossible for such a measure to relate well to anything else. So the next time you see "Lottery Dice," which generate random numbers in the hope of allowing you to predict winning lottery numbers (numbers that are generated randomly), save your money.

A third desirable property of the measures of a theory's concepts is **standardization**. Standardization in the context of measurement means that everyone who measures the construct uses the same instrument in the same way. That is, the measures used to obtain data should ideally be common and the procedures for administering the measure well established. Holding reliability and validity constant, data obtained from standardized procedures are preferable to data obtained from procedures or measures that are unique to a particular situation. Jum Nunnally notes that because of the time and effort required to develop measures that are reliable and valid, a great deal of *efficiency* can be achieved by using existing standardized measures.[16]

Standardized measures provide two other advantages. First, they are far more likely than other measures to achieve *objectivity*. Because everyone uses the same procedures, the results of measurement are much less likely to be affected by who happens to be doing the assessing. Second, standardized measures facilitate the *communication* and *comparison* of results across situations. You could construct a scale to measure job satisfaction in your own company, but even if you did succeed in developing a reliable and valid measure (a difficult task), you could not compare the satisfaction level in your company to that in other companies. The Job Descriptive Index (JDI), for example, which is discussed in detail in

---

[16] Jum C. Nunnally, *Psychometric Theory* (New York: McGraw-Hill, 1978), p. 4.

Chapter 8, is a standardized measure of job satisfaction that has been used in hundreds of companies. For most standardized measures, a wealth of existing data allow you to compare your company to other companies that have measured the concept in which you are interested in the same way.

For these and other reasons, you would be foolish to try to develop your own measures for every situation. At worst, you would come up with measures that lack reliability and validity, and at best you would be reinventing the wheel. Moreover, even if the new wheel you developed were reliable and valid, it would not be comparable to the wheel that everyone else was using. For this reason, throughout this book we will provide you either with specific, standardized measures of various concepts or with the sources from which such measures can be obtained. It is possible, of course, that on some occasion you will need to test new concepts or develop measures that are unique to your situation. Such cases. however, will be the exception rather than the rule. Many of the measures you will need are already available.

CAUSAL INFERENCES

From an applied perspective, the beauty of the scientific method is its ability to isolate the probable causes for various outcomes. Once causes are identified, they can often be manipulated to bring about the specific outcomes we want. Good theory and good data take us a long way toward this objective, but they are not sufficient for making causal inferences. Making causal inferences depends not only on how the data are obtained but also on when the data are obtained and on what is done with the data once they are collected.

Criteria for Inferring Cause

According to philosopher John Stuart Mill (1806–1873), in order to establish that one thing causes another, we must be able to establish three things: the temporal precedence of one thing over the other; the covariation, or relationship between, the two things in question; and the absence of alternative explanations for the observed results.

temporal precedence The degree to which any measured cause actually precedes an effect in time.

Temporal Precedence. The first step is establishing **temporal precedence**, which simply means that the cause must occur before the effect. That is relatively easy to determine, although sometimes you can fool yourself if you are not precise. For example, some professors think that giving tests causes students to study. This notion is technically incorrect, for it has the effect (studying) occurring *before* the proposed cause (giving the test). More precisely, it is probably the fear of failing the test that causes studying, not the test itself.

covariation The degree to which two variables are associated with each other; the degree to which changes in one are related to changes in the other.

Covariation. Mill's second criterion for inferring cause is **covariation**, which simply means that the cause and effect are related. For example, if we believe that providing day care for employees' children causes less worker absenteeism, then there should be a relationship between company day care services and low employee absenteeism.

There are several ways to assess covariation, all of which rely on statistical methods. As this is not a statistics book, we will limit our discussion here to two simple but widely applicable statistical techniques. The first, known as a test of

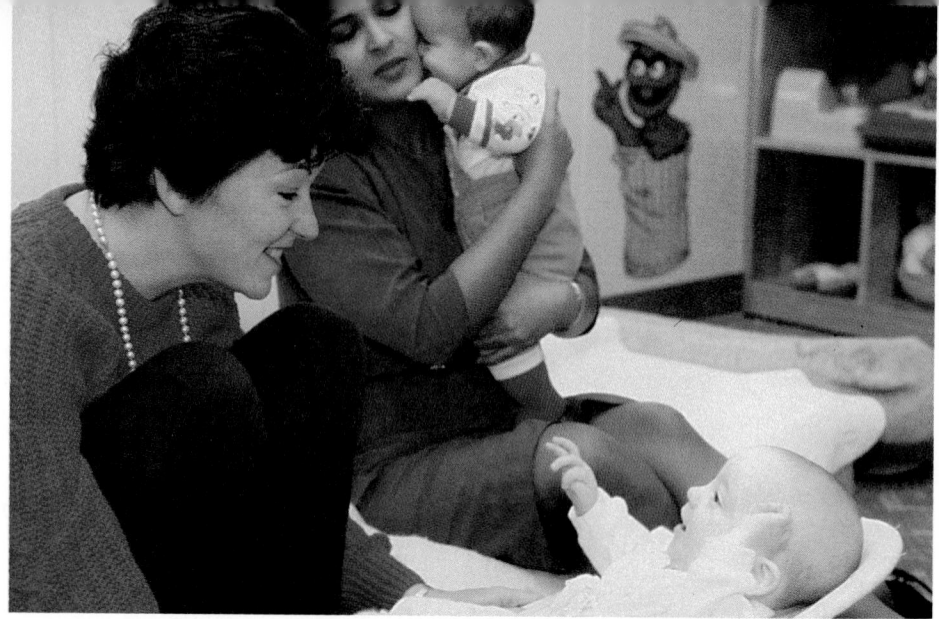

As the example in this chapter shows, it is very difficult to establish beyond a reasonable doubt that company-sponsored day care services decrease absenteeism. Nevertheless, Campbell Soup Company, one of the first corporations to provide on-site child care for employees, is convinced not only that its program has reduced absenteeism and turnover but that it has improved employee morale and enhanced the company image. Company representatives also say that the program has increased productivity in part, perhaps, by reducing anxiety on the part of employees whose children are in the program. Between 1982 and 1988, the number of companies that offered child-care benefits ranging from financial assistance to actual on-site care rose by more than 80 percent, and by 1989 some 200 corporations had established their own centers. Source: *Lena Williams, "Child Care at Job Site: Easing Fears,"* The New York Times, *March 16, 1989.*

AN EXAMPLE

The rising sun seems to have temporal precedence over the setting sun and both covary. So far in history, the correlation between these two events is a perfect 1.0. Can we then argue the rising of the sun causes the sun to set?

correlation coefficient A statistic that assesses the degree of relationship between two variables.

mean differences, compares the mean, or average, scores of two groups on a measure of the proposed causal factor. Table 3-1 presents data on absenteeism for two groups of workers: Ten work in Plant A, where there is an in-house day-care center, and ten work in Plant B, where there are no on-site provisions for day care. As you can see, the level of absenteeism is much higher for Plant B than Plant A. This simple analysis of mean differences suggests that day-care provision and absenteeism are in fact related. As we will see, however, more rigorous approaches will enable us to refine our conclusions. We might also test for mean differences between numbers of absences at Plant A before and after the establishment of the day care center (see Table 3-2). If the average absenteeism rates were higher before putting in the day-care center than they were after it was in place, we might again conclude (before engaging in more rigorous analyses) that there is a relationship between providing day care and lower absenteeism.

A second means of establishing covariation is through the use of the **correlation coefficient**. This statistic, a number that ranges from $+1.0$ to -1.0, is an expression of the relationship between two things. A $+1.0$ correlation means that there is a perfect positive relationship between the two measures in question. That is, as the value of one increases, the value of the other increases to the same

---

T A B L E 3-1
Absence Data at Two Hypothetical Plants

EMPLOYEE	NUMBER OF ABSENCES	
	Plant A (with day care)	Plant B (without day care)
01	10	12
02	11	11
03	8	13
04	11	8
05	3	16
06	4	14
07	3	10
08	2	4
09	1	2
10	5	3
Average	5.8	9.3

RESEARCH METHODS

T A B L E 3-2
Absence Data for One Hypothetical Plant at Two Different Times

	NUMBER OF ABSENCES AT PLANT A	
EMPLOYEE	Before Day Care	After Day Care
01	12	10
02	14	11
03	10	8
04	12	11
05	6	3
06	8	4
07	4	3
08	2	2
09	1	1
10	6	5
Average	7.5	5.8

relative degree. A correlation of -1.0 reflects a perfect negative relationship between the two measures in question. Here, as the value of one increases, the value of the other decreases, again to the same relative degree. Finally, a correlation of .00 indicates that there is no relationship whatever between the measures, so that as the value of one increases, the value of the other can be anything—high, medium, or low. To give you a feel for other values of the correlation coefficient, Figure 3-3 shows plots of 5 different correlation values, $+1.0$, $+.70$, $+.20$, .00 and $-.50$. As you can see, the sign of the correlation reveals whether the relationship is positive or negative, and the absolute value of the correlation reveals the magnitude of the relationship.

Let's go back to our employees at Plants A and B. In Table 3-3, in addition to the data on day care and rates of absenteeism we also show data on the ages of all the workers. We could use the correlation coefficient to answer the question, Is there a relationship between age and absenteeism? In fact, the correlation between age and absenteeism for these data is $-.50$, indicating that older workers are absent less often than younger ones. If one were to plot these data on a graph, where x is the horizontal axis and y the vertical axis, it would look just like the figure shown in Figure 3-3e.

One important question that must be answered at this point is how big a difference in means or how big a correlation is big enough to conclude that there is a relationship. Even if there is no relationship between day care and absenteeism, it is unlikely that the means for the two different groups will be exactly the same. Similarly, even if age is not related to absenteeism, it is unlikely that the observed correlation is exactly .00. What if this correlation is .10? Does that mean we have established a relationship? What about correlations of .40 or .70?

Common sense probably tells you that .10 is not going to be big enough, and that .70 probably is big enough. In fact, as is often the case, common sense may let you down here. The fact is that in some cases .10 may be big enough and in other cases .70 may *not* be big enough. The precise answer to this question is generally provided by tests of **statistical significance**.

statistical significance A numerical index of the probability that a relationship detected between two variables could be explained by luck or chance.

A relationship is said to be statistically significant when there is a very small probability (often set at 5 percent, or 1 in 20) that a relationship that size could be attained through chance alone. For example, you know already that dice tossing generates a random set of numbers. You should also be aware now that random numbers do not correlate well with anything, including other sets of random numbers. Yet if you were to toss dice ten times today and ten times tomorrow and then calculate the correlation, you would find that it was not .00 exactly.

FIGURE 3-3

Plots Depicting Various Levels of Correlation between Variables

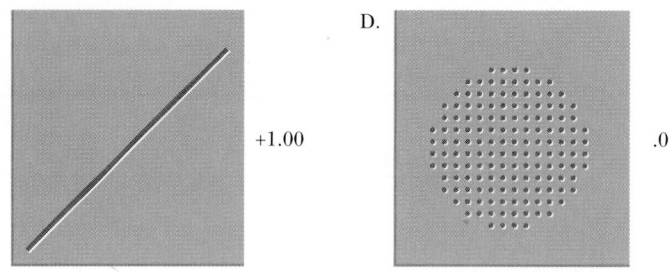

A.   +1.00   D.   .00

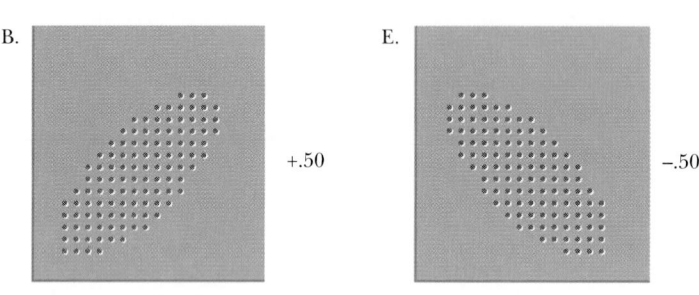

B.   +.50   E.   –.50

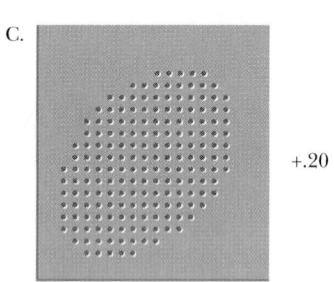

C.   +.20

TABLE 3-3
Absence and Age Data at Two Hypothetical Plants

EMPLOYEE	PLANT A (DAY CARE) No. of Absences	Age	PLANT B (NO DAY CARE) No. of Absences	Age
01	10	27	12	27
02	11	31	11	34
03	8	30	13	31
04	11	26	8	25
05	3	40	16	33
06	4	61	14	35
07	3	52	10	25
08	2	47	4	40
09	1	46	2	52
10	5	41	3	46
Average	5.8	40.1	9.3	34.6

Rather, it would probably be some small number like .02, .11 or −.13 just by chance alone. In fact, if you were to try this experiment for 100 days you would get 100 correlations that, though they would have a mean of zero, would not all be zero.

Of these 100 correlations, how many of them would be as large as .10? The answer is many of them—many more than 5 percent. Many of them would also be as large as .40, and again, more than 5 percent. On the other hand, few and less than 5 percent would be as large as .70. Thus one would conclude that the .70 is a statistically significant correlation but that the .10 and .40 are not. So far, common sense is performing well; statistical significance is a function of the size of the relationship.

There is one additional consideration beyond the size of the relationship that is critical to statistical significance, however, and that is *sample size*. If instead of performing our 100–trial experiment by tossing the dice ten times each day we tossed the dice ten thousand times each day we would observe quite a different set of correlations. Now almost all of the correlations would be .00. Few would be as large as .10, and none would be as large as .40 or .70. With this many tosses, all the chance elements that went into the previous nonzero correlations would be canceled out over the ten thousand trials in a way that could not happen in just ten tosses. Thus here even a correlation as small as .10 would be statistically significant. Table 3-4 shows the size of the correlation needed to achieve statistical significance for various sample sizes. We have already shown that .10 is large enough for a sample of 10,000. One thing this table shows is that .70 is not large enough if the sample is only 5. The moral of this story is that it is hard to detect statistically significant relationships when sample sizes are small, and this makes it hard to establish covariation in certain kinds of contexts. We have used a small sample in our day-care example to make it easier to present some basic concepts. Real experiments with this small number of subjects, however, are generally uninformative.

We must stress that both covariation and temporal precedence must be shown to infer cause—one without the other is insufficient. For example, suppose data on "supervisors' consideration for employees" were collected at Plants A and B at the *same time* that the data on absenteeism were collected. We might very well find a negative correlation between consideration and absenteeism and be tempted to conclude that lack of consideration on the part of supervisors causes high absenteeism. Here we have covariation but not temporal precedence. Under these conditions, we cannot conclude that lack of consideration caused high absenteeism; for all we know, the high rates of absenteeism could have caused supervisors to be less considerate. In general, when all data are collected at the same time, it is virtually impossible to establish causation. About all one can do under these circumstances is establish covariation.

T A B L E 3-4
Magnitude of Correlation Needed to Achieve Statistical Significance

NUMBER OF SUBJECTS	CORRELATION NEEDED
5	.75
10	.58
20	.42
40	.30
100	.19

There is an alternative explanation for the relation between sunrise and sunset. A third factor, the rotation of the earth around the sun, causes both events. Temporal precedence and covariation are necessary for causal inference, but not sufficient. Asking a question differently and improving research methods can invalidate earlier assumptions or findings.

selection threat A threat to validity created when experimental and control groups differ from each other before an experimental manipulation.

Eliminating Alternative Explanations. Once we have established both covariation and temporal precedence, it would seem that we are only one step away from establishing that something actually caused something else. Unfortunately, Mill's third criterion for establishing cause, the *elimination of alternative explanations*, is more like "one giant leap" than like "one small step." In our continuing example, if we are to infer that providing day care caused lower absenteeism, we must also show that there was not some other factor that actually caused the low rates of absenteeism. Most real-world situations are so complex, however, that it is often very difficult to rule out other possible explanations for the results one obtains. Indeed, this one problem, more than any other, is what makes it so much more difficult to conduct research in the social sciences than in the physical sciences. In the physical sciences, experimenters can use things like lead shields and vacuum chambers to protect the variables they are examining from outside influences.

This kind of tight control is harder to achieve in social science research, and indeed, some alternative explanations are so common that they are given special names. For example, the **selection threat** is the danger that we may claim that A caused B because of mean differences between groups when what really happened was that the groups were not the same to begin with.[17] Returning to our continuing example, if we had only the data on absenteeism in the two plants (the data in Table 3-1), because of the lower mean rate of absenteeism in the plant with day care we might have concluded that providing the day care caused the lower absenteeism. We have additional data, however, that show that age was negatively related to absenteeism, and it happens that workers in Plant A are older than those in Plant B. In fact, if we were to control for age by comparing only workers who were the same age (e.g., in the 35–65 range), we would find that for people in the same age groups there were no differences in absenteeism between the plants. Thus because the two groups selected for study were not the same to begin with, it would have been incorrect to conclude that providing day care led to low absenteeism. In fact, that was not the case at all. Rather, older workers are absent less often; the workers in Plant A are older than those in Plant B; and therefore the difference in absenteeism across the plants is apparently not due to the day care but to age.

At this point you may be saying "So what—what difference does it make?" It makes a huge difference if based on your incorrect judgment as to what caused what, your company invested a lot of money in providing day-care facilities on a corporationwide basis. This large investment would be based on your conclusion that day care would pay for itself through lower absenteeism. But because day care was actually irrelevant to absenteeism, this investment would be completely lost, and many people would be left wondering what happened?

The selection threat will be a problem any time you are comparing two groups that differed from each other even before you started out. Any time you compare two groups you are subject also to the **mortality threat**. In this situation, the groups may be comparable at the start, but because over time different kinds of people may leave each group, at the end the groups are no longer comparable. Suppose, for example, that you were on the staff of a college that wanted to set up a program to improve the chances that on graduation, undergraduate business majors would be accepted into the country's top MBA programs. You create two groups that are equal from the outset in terms of "commitment to pursue the MBA degree" and put one group of 50 students into the proposed program but do nothing with the other group of 50 students. Finally, the program that you and your colleagues devise is a long and difficult one that includes several advanced courses and requires students to write many term papers and to give a number of oral presentations.

mortality threat A threat to validity created when subjects who drop out of an experimental group differ on some significant characteristic or characteristics from those who drop out of the control group.

[17] T. D. Cook and D. T. Campbell, *Quasi-Experimentation: Design Analysis Issues for Field Settings* (Chicago: Rand McNally, 1979), p. 53.

You might very well find that because of the difficulty of the program, only 25 of the original 50 students actually complete it. If you were then to compare these 25 to the original, control group of 50 students who did not enter the program, you would probably find that those who completed the program were more frequently admitted to the best graduate schools than those who were in the control group. However, you would be incorrect to assume that it was the program that was the cause. If the 25 students who dropped out of the program were low in "commitment to pursue the MBA degree," the experimental and control groups were not equal in the end, because only those high in commitment remained in the experimental group, while people of high, medium, and low commitment remained in the control group.

The **history threat** is common in before-and-after studies. In this situation, the real cause is not the change you made but something else that happened at the same time. Returning to our old example, what if we had no data from Plant B and simply compared the mean number of absences for Plant A *before* day care with the mean number of absences *after* day-care (see Table 3-2)? We might see a lower average rate of absenteeism after we put in a day-care center than before and be tempted to infer that the center caused lower absenteeism. Accordingly, we might recommend that the program be extended to all other plants in the corporation, despite its high cost.

Suppose, however, that we obtained the "before" measure during the summer months and the "after" measure during the winter. It's possible that people find more reasons to be absent in the summer than in the winter; thus it could be the weather rather than the day-care center that caused the difference in absenteeism rates. Again, if we were to extend the day-care program throughout the corporation, we would find that it would not reduce absenteeism and would be left wondering why.

The history threat will be a problem any time you make before and after comparisons in the absence of a control group. Another common threat in this situation, the **instrumentation threat**, is the danger that a change from before to after will reflect not what you did but a change in the measurement instrument itself. For example, assume that you are the manager of a real estate sales agency that in 1972 initiated a strategy of advertising on television. You notice that sales volume in terms of absolute dollars increases steadily for the next three years. In 1975, though, television advertising prices get too high, and you switch to advertising in newspapers. You then notice that sales volume in terms of absolute dollars decreases steadily for the next two years. Detecting a pattern, in 1977 you reinstate television advertising and are happy to find that sales volume increases steadily for the next three years.

Who among us would not be tempted, given this experience, to conclude that television advertising causes increased sales in the real estate industry? We could all be wrong, however. Figure 3-4 shows the Consumer Price Index (CPI) for the years 1972 to 1980. The CPI is a measure of inflation, that is, the relative value of the U.S. dollar. You can see that the value of the dollar itself changes from one year to the next. For this reason, sales volume measured in dollars may change despite unchanging levels of real estate sales. Thus it could be that the form of advertising has no impact whatsoever on real estate sales and that the changing value of our measure of it (which is based on dollars) led us to the wrong inference.

Designing Observations to Infer Cause

Earlier we noted that obtaining good data is an issue not only of how but of when. Both the timing and the frequency of data collection affect our ability to

---

history threat A threat to validity created when some important variable other than the one manipulated experimentally changes during an experiment.

instrumentation threat A threat to validity created by artificial changes in the measurement device used to assess an experimental effect.

AN EXAMPLE: LORAL DEFENSE SYSTEMS AND GOODYEAR AEROSPACE
A study of Loral Defense System's acquisition of Goodyear Aerospace was complicated by the history threat. While attempting to measure the effects of the acquisition on employees, the researcher had to take careful note of the changes that occurred in the organization and why those changes occurred. For example, layoffs occurred shortly after the acquisition and the employees blamed the layoffs on the acquisition. But, at the same time the Gram-Rudman Act took effect, requiring a balanced budget. As a result, many defense contracts were cancelled or put on hold, resulting in a loss of jobs. Determining exactly what portion of changes were attributable to internal changes and what were attributable to external changes became a nearly impossible task.

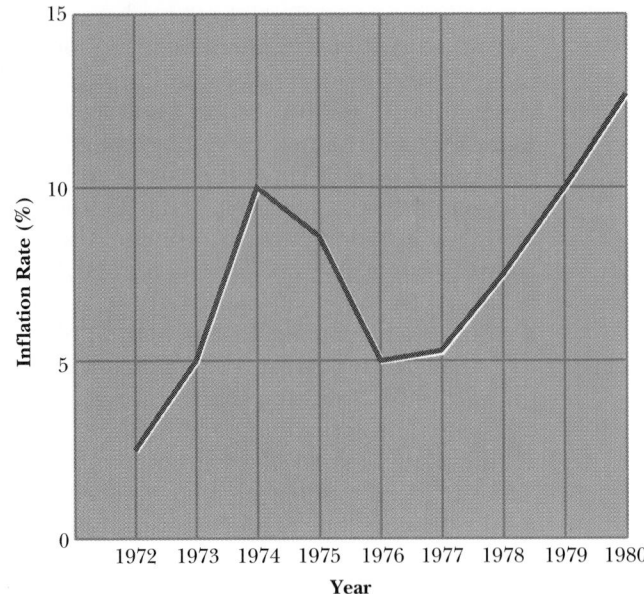

make causal interpretations. Deciding on the timing of measurement is a large part of research design.

Faulty Designs. Consider the two *faulty designs* shown in Figure 3-5. In the One Group Before-After design (Fig 3-5A), data are collected both before and after some event or treatment. If the after score ($CD2$) is different from the before score ($CD1$), it is assumed that the treatment (represented by Δ) caused the difference. The reason this is a faulty design is that both the history threat and the instrumentation threat are possible alternative explanations for the results. If in our day-care example, for instance, we collected data from only one plant, once in the summer and once in the winter, we would be using this type of faulty design. Were this the case, we would be open to the two threats.

F I G U R E 3-5

Two Faulty Research Designs

A. One Group Before-After

CD_1	Δ	CD_2
Collect data at Time 1	Change situation	Collect data at Time 2

B. After Only With Unequal Groups

Change situation for A	Collect data from A

Δ	CD_A
	CD_B

Do Not change situation for B	Collect data from B

TEACHING NOTE
Seeing that the productivity of new employees was low, a manager decided to train all new hires. After the new employees were trained, their productivity rose. Discuss how this design can be improved. For example, did the performance of the trained workers differ from that of workers who did not receive training?

In the After Only with Unequal Groups design (Fig. 3-5B), data are collected from two different groups, one of which receives some experimental treatment while the other does not. The reason this design is faulty is that we do not know that the groups were equal before the treatment or during the treatment, and thus both the selection and mortality threats are alternative explanations for the results. If in our day-care example, we collected data from both Plant A and Plant B without making sure that the people in those plants were similar (e.g., were the same age on average), we would have this kind of faulty design. Our research design would thus be subject to the two threats to validity.

Improved Designs. There are several ways to design studies that can help eliminate some of these threats. Let's take the One Group Before-After design, where our major threats are history and instrumentation. Here we are somewhat better off if we add a control group, turning the design into the Two Group Before-After design shown in Figure 3-6A. This design allows us to test whether the two groups were equal to begin with by comparing CD_{1A} with CD_{1B}. That is, in our day-care example, was the rate of absenteeism in Plants A and B similar before the treatment—the day-care center—was put in place? This design also allows us to test whether some historical factor other than the treatment could have caused the results. That is, if the real cause was time of the year (summer versus winter), we could expect a decrease in absenteeism in Plant B as we moved from Time 1 to Time 2, even though no day-care center was established there.

random assignment A method of increasing the validity of a study by ensuring that each subject has an equal probability of being assigned to any one experimental condition. Random assignment eliminates the *selection threat.*

The Two Group After-Only with Randomization model shown in Figure 3-6B is an even better design. It is just like the one shown in Figure 3-5B with one major exception: subjects are randomly assigned to groups. **Random assignment** of subjects to conditions means that each person has an equal chance of being placed in either the experimental or control group. This random arrangement can be achieved by pulling names out of a hat, flipping coins, tossing dice or using a random numbers table from a book on statistics. For example,

FIGURE 3-6
Two Improved Research Designs

A. Two Group Before-After

Collect data from Group A at Time 1	Change situation for A	Collect data from at Time 2
CD_{1A} CD_{1B}	Δ	CD_{2A} CD_{2B}
Collect data from Group B at Time 1	Do not change situation for B	Collect data from B at Time 2

B. Two Group After Only With Randomization

Randomly assign subjects to Groups A and B	Change situation for Group A	Collect data from Group A
R	Δ	CD_A CD_B
	Do not change situation for Group B	Collect data from Group B

in our day-care study, if at the outset we could simply have assembled the twenty workers at the two plants and then tossed a coin to see who would go to which plant, the odds are that when we were finished the two resulting groups would have been equal in age, that is, each group would have had roughly the same number of people of a given age.

In fact, the real value of randomization is that it not only equates groups on factors (like age) that we expect to influence our results, but it equates groups on virtually all factors. Thus in our day-care study, if we randomized the groups at the outset, we could be fairly confident that they would be equated not only on age but on other things, such as height and weight. You might not think that a person's height or weight would relate to absenteeism, but some research actually has found a relationship between absenteeism and weight.[18] Even if we were unaware of this relationship when we started the day-care study, it is nice to know that randomization solved a potential problem for us. Because of randomization's ability to rule out both anticipated and unanticipated selection threats, one should randomly assign subjects to treatments whenever possible.

Because randomization is not always possible, we often need other tools to rule out selection threats. For example, suppose that when we start our day-care experiment we know that workers at the two plants are not evenly distributed in terms of age, and we also know that age affects absenteeism. In the real world, we could not randomly move people from plant to plant; we would have to work with existing groups.

What, then, can we do to rule out age as the alternative explanation for our results? We have several choices. First, we could use *homogeneous groups*, that is, groups that do not differ on age. For example, we might compare absenteeism in the two plants but only among workers in the 25–35-year-old bracket. Thus, as you can see from Table 3-3, we would compare Subjects 1, 2, 3, and 4 in Plant A with Subjects 1, 2, 3, 4, 5, 6, and 7 in Plant B. With this sample, if we still found lower absenteeism in Plant A than in Plant B, we could not attribute the difference to age because all subjects were roughly the same age.

We could also equate groups by *matching subjects*. For example, we might use only the subjects in Plant A for whom there are corresponding subjects in Plant B, or subjects who are within two years of each other in age. Thus, looking again at Table 3-3, we could match subjects 1, 3, 5, 7, and 9 in Plant A with Subjects 1, 4, 8, 9, and 10 in Plant B. Again, if we found that absenteeism was lower in one plant than another we could not attribute this result to age because we equated the groups on this factor.

Finally, we could also *build the threat into the design*. By this we mean that we could simply treat age as another possible factor affecting rate of absenteeism and examine its effect at the same time that we examine the effect of day care. One advantage of building alternative explanations into your design is that it allows you to test for **interactions**. An interaction exists when the relationship between the treatment (e.g., the day-care center) and the outcome (e.g., absenteeism) depends upon some other variable (e.g., age). Figure 3-7 shows what we might find if we built the alternative explanation of age into our day-care study. As you can see, among the younger group day care does lower absenteeism somewhat, but among the older group it has no effect at all. Thus the relation between day care and absenteeism depends on the factor of age. Day care lowers absenteeism among younger workers who are likely to have young children but not among older workers, whose children are grown.

At this point you can see how many factors must be considered in designing studies that allow us to infer causality. Clearly the more variables we can control

interaction An experimental outcome in which the relationship between two variables changes depending on the presence or absence of some third variable.

---

[18] K. R. Parkes, "Relative Weight, Smoking and Mental Health as Predictors of Sickness and Absence from Work," *Journal of Applied Psychology* 72 (1987), 275–87.

FIGURE 3-7

The Effect of Age and Day-Care
Facilities on Absenteeism

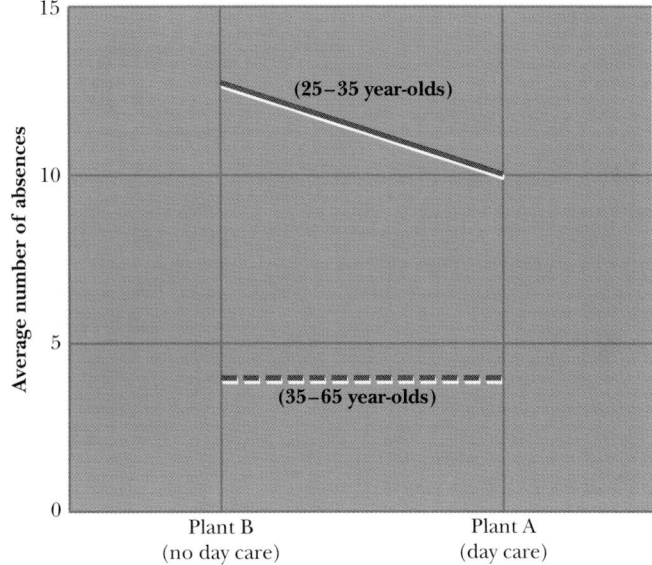

the tighter our research design and the more likely it is that our results will be significant. As the "Management Issues" box shows, however, the best designed studies sometimes manipulate people in ways that are open to attack on ethical grounds.

GENERALIZING RESEARCH RESULTS

Research is usually conducted with one sample, in one setting, across one time period. Often, however, we wish to know whether the results obtained in a unique sample-setting-time period would be the same had the study been conducted in some other sample-setting-time period. Such **generalizability** is sometimes of interest when we are conducting research, but it is always of interest when we evaluate research findings to see whether what worked for the investigators can be applied in a real-world setting.

generalizability The degree to which the result of a study conducted in one sample-setting-time configuration can be replicated in other sample-setting-time configurations.

AN EXAMPLE
Members of different races consistently obtain scores that differ substantially in tests of cognitive ability. Some people, who believe that these differences are caused by biases in the tests, have suggested that the tests be revised. The true problem, however, is a societal one. Minorities have fewer educational opportunities. Blaming testing for race differences is analogous to blaming a thermometer for the weather.

Sample, Setting, and Time

Our day-care example provides a good illustration of how results might not generalize across different samples. Recall that the results of our study eventually showed that day care reduced absenteeism among workers in the 25–35-year-old group but not in the older group. An astute manager who studied our results would be unlikely to recommend establishing a day-care center for her company without further investigation, because the data indicate that it lowers absenteeism rates in only one small group of workers.

Suppose, however, that we had used a design where we homogenized our subjects on age (i.e., used only those in the 25–35 bracket). In this case, we would have reported simply that providing day care reduced absenteeism. A manager who read these results and then, based on our experiment, instituted a day-care center in a company where most of the workers were between 35 and 65 years old would find that the results of our work did not generalize to his situation.

MANAGEMENT ISSUES

Ethical Issues in Organizational Behavior Research

Are hiring decisions based solely on an objective examination of applicants' credentials, or are these decisions biased by the race of the applicant? This is a question of critical importance for both organizations and the larger society. You will not be surprised to find, therefore, that this question has been examined in scientific research.

Let's look at a study in which resumés of non-existent "paper people" were sent to the personnel managers of 240 organizations.* These bogus resumés portrayed the educational attainments and work experiences of two applicants who were exactly the same on every dimension presented in the resumes *except race*. One half of the organizations were randomly selected to receive the resumé describing an African-American job applicant. The other half were sent the resumé describing a white candidate.

The addressees, who were responsible for sending out interview invitations to applicants submitting resumés, were asked to send any correspondence to an address that actually belonged to the research team. The researchers, Newman and Krzystofiak, collected the correspondence and checked to see if black and white candidates with identical resumés had an equal probability of receiving an invitation to interview with the companies in their sample.

This was a well-designed, tightly-controlled study that was well suited to ask the research question posed at the outset of this box. Given that it was a good research design, should we ask any other questions about it? Yes. We now need to ask, was this study ethical?

Consider, first, the personnel managers of the organizations selected for study. These are busy people who have enough work to do without pouring over resumés and sending correspondence to people who do not actually exist. Moreover, the managers did not ask to be in this research study, and in fact were unaware that they were even participating in research. Is this the research equivalent of a crank phone call?

Consider, too, other real applicants for the position in question. Most organizations go through employee selection in steps. First one group of applicants, let's say 10, are invited in for interviews based on their resumes. Perhaps five of these are called back for additional interviews. Finally, one applicant is selected. Suppose the bogus resumé was rated higher than the legitimate resumé of a real person? Then that individual, who otherwise would have obtained an interview, would have been eliminated from further consideration because of the research study.

Does the scientific-social value of this study outweigh its costs to personnel managers? Does the value of the study to the *researcher's career* outweigh its potential negative effects on some *job applicant's career*?

Before you pronounce your verdict consider an additional twist. Newman and Krzystofiak actually told some of the managers that they were part of a study. This allowed the researchers to compare managers who knew they were involved in research with those who did not. The results of this study were striking. African-American applicants were treated either equally or preferentially by managers who were aware of the research but were rated lower by managers who were not aware that they were being studied.

This study highlights some of the intricate issues involved in making decisions about the ethics of organizational research. In general, professional associations of researchers like the Academy of Management or the Society of Industrial/Organizational Psychology strongly recommend that all research subjects give informed consent to participation in research. In the study in question, however, informed consent from all subjects—the personnel managers—could change the "truth" of the findings.

Similar issues arise in studies where subjects know they are involved in research but are deceived as to its true nature. Professional guidelines clearly stipulate that honesty is the best policy when dealing with subjects. It does not seem to make sense that the first step some researchers take in their quest for "truth" is to deceive the very people asked to join in the project. Still, an abundance of sound empirical research shows that subjects who are aware of the exact nature of studies respond differently from subjects who are naive.†

It appears very difficult, if not impossible, to devise hard and fast rules for resolving ethical dilemmas like these. But professional researchers do lean on some general prescriptions. First, deception or lack of informed consent in research should be absolute last resorts. Second, researchers must demonstrate that (a) the problem under study is of critical importance, (b) there is a high probability that without the use of deception the study's validity will be compromised, (c) the subjects are not likely to be emotionally upset nor to sustain any long-term harm when the deception is revealed, (d) the researcher takes responsibility for removing any detrimental side effects of the study, and (e) the researcher fairly compensates the subjects for their time and efforts. Most universities have "Human Subject Committees" who review all faculty research proposals. For researchers not affiliated with academic institutions, however, there may be no such checks and balances.

Do you think the professional guidelines now in place are adequate to ensure the ethical nature of behavioral and social science research? If not, what other rules might be imposed? What if anything can we do about the independent, nonuniversity-affiliated researcher?

* J. Newman and F. Krzystofiak, "Self-Reports Versus Unobtrusive Measures: Balancing Method Variance and Ethical Concerns in Employment Discrimination Research," *Journal of Applied Psychology* 64 (1979), 82–85.

† J. H. Resnick and T. Schwartz, "Ethical Standards as an Independent Variable in Psychological Research," *American Psychologist* 28 (1973), 134–39.

This is one of the major drawbacks of making groups homogeneous; it limits one's ability to generalize results across other types of samples.

We may also be concerned about generalizing research results across settings. For example, suppose that the plants in our original study were both located in rural settings. Assume further that it is more difficult to obtain high-quality day care in rural settings than in urban settings. Someone reading our study who manages a plant in an urban area might establish a day-care center in her plant only to find that because child care is not a problem for her workers, the center has no effect on absenteeism. Here again, our results would not generalize to another setting.

Finally, we might also be concerned about whether our results would generalize across time. For example, suppose that we had conducted our study at a time when there was a huge labor shortage; that is, when many more jobs were available than there were people to fill them. At such a time, unemployment rates would be low, both parents might well be working, and many people who might in other circumstances serve as day-care providers would very likely be working at different and perhaps higher-paying jobs. Thus when we conducted our study, there may have been a great demand for day-care services but a small supply. By providing our own day-care services, we solved a major problem for our workers with small children, and this ultimately led to lower absenteeism rates.

Now let's move forward ten years, say, to a time when there is a labor surplus. Unemployment is high, there is a good chance that one parent is not working, and anyone capable of setting up a day-care center is open for business. In this situation, because the demand for day care is small and the supply of day-care services large, company-sponsored day care does not provide a needed service to employees, and so there is no relationship between providing day care and lowering absenteeism. Here our results do not generalize across time.

Facilitating Generalization

You may be wondering if there are any findings that are generalizable given the many factors that might differ from one unique sample-setting-time to another. From a researcher's perspective, can anything be done to increase the ability to generalize? The answer to both of these questions is yes. Technically, one can safely generalize from one sample to another if the original sample of people we study is *randomly selected* from the larger population of people to which we wish to generalize.

As an example of random selection, you may have noticed that in presidential elections the television networks usually declare a winner when less than 10 percent of the actual results are available. Making the wrong call here might be very embarrassing, so why are the networks taking such a big risk? They are not. The key to their success is that when they poll people who have just finished voting, they do so randomly. In this way they ensure that the small percentage of people they poll are by all odds very similar to the larger group of voters. In fact, this is a case in which the researchers are so sure that their results will generalize that they have no fear whatsoever in publicly declaring a winner way in advance of the results, even though there are huge costs associated with being wrong. Just as the ability to generalize to other people can be assured with random selection of subjects, generalizing from one setting or time to another can be assured only if we randomly sample settings and time periods.

Although random selection is the only way to ensure the ability to generalize, from a practical perspective it is often very difficult to achieve. Studies that employ

The research that has been going into the development of the new Globex system, being tested here by Reuters employees, may ultimately lead to major changes in the way traders in the world's futures markets do business. This new electronic trading system and others like it are making it possible to trade around the clock on a worldwide basis. Rapid technological changes like these highlight the issue of generalizability. Will research-based techniques developed for jobs in this century be equally useful in 21st-century jobs? Source: *David Zigas and Gary Weiss with Ted Holden and Richard A. Melcher, "A Trading Floor on Every Screen,"* Business Week, *November 5, 1990.*

random selection are usually huge in scale, requiring a large number of investigators and a great deal of money. More often, in the real world of research, the ability to generalize a finding is achieved not by one big experiment but by many small experiments, using the same measures, in which results are replicated over and over again in a host of different sample-setting-time configurations.[19] For example, in Chapter 7 we will discuss some research results that generalize very well—the repeated finding that high performance is more likely to result from setting specific and difficult goals than from offering vague goals like "do your best."

As we have noted, although generalizing results to other samples-settings-times is always of interest in evaluating research, it is not always interesting to the original researchers. Often research is conducted strictly to test or build theories. Here investigators may be less interested in what *does* happen than in what *can* happen.[20] For example, in Chapter 8, when we discuss stress, we will look at research that shows people can learn to control some of their own physiological processes, such as heart rate and blood pressure, when hooked up to special devices that give them feedback on these processes. One can think of few real-world samples-settings-times that would correspond to the situation in which subjects in this kind of research find themselves. That is not the point of this research, however. Its purpose is to test a theory that states that human beings can voluntarily control supposedly involuntary physiological responses when provided with the right feedback. There is nothing inherent in this theory that suggests it would not work with college sophomores in a laboratory setting at some specific time period. Thus, if the results fail to support the theory, the theory must be either rejected or modified and retested, and the fact that neither subjects, settings, nor times were randomized is completely irrelevant. With this kind of research, the ultimate aim is not to make the laboratory setting more like the real world but to make the real world more like the lab—that is, to change the real world in ways that benefit us all.

LINKING OB SCIENCE AND PRACTICE

Unless you intend to work as a staff member in a research capacity for some organization, you will probably have only a few opportunities to do much real experimentation yourself. As a practicing manager, how-

---

[19] Cook and Campbell, *Quasi-Experimentation.*
[20] D. G. Mook, "In Defense of External Invalidity," *American Psychologist* 38 (1983), 379–87.

RESEARCH METHODS

TABLE 3-5
**The Ten Most Influential Journals
in Organizational Behavior**

Micro Organizational behavior
1. *Journal of Applied Psychology*
2. *Organizational Behavior and Human Decision Processes*
3. *Personnel Psychology*
4. *Journal of Vocational Behavior*
5. *Journal of Occupational Psychology*

Macro Organizational behavior
1. *Administrative Science Quarterly*
2. *Academy of Management Journal*
3. *Academy of Management Review*
4. *Human Relations*
5. *Administration and Society*

Source: G. R. Salancik, "An Index of Subgroup Influence in Dependent Networks," *Administrative Science Quarterly 31* (1986), 207–11.

ever, you should know that there is a wealth of research just waiting to be discovered. Table 3-5 provides a list of the major scientific journals that publish theory and research related to topics in this book. These journals are rank ordered, within the categories of micro and macro organizational behavior (see Chapter 1 for a refresher on this distinction), by their influence in the field. A high ranking means that the findings reported in that journal tend to be more widely cited in the organizational sciences than are those in lower ranked journals.

A great deal of the research conducted in this area is performed by people working in university settings. Thus you may also be able to uncover research on the topics that interest you by contacting university faculty who publish a good deal of research on topics related to organizational behavior. Table 3-6 lists universities where OB researchers are particularly interested in micro organi-

TABLE 3-6
Research Productivity: The Top Forty Institutions in Micro Organizational Behavior

1. University of Houston	21. University of Pennsylvania
2. Ohio State University	22. University of South Carolina
3. University of Washington	23. University of Tennessee
4. Pennsylvania State University	24. University of California at Los Angeles (UCLA)
5. Michigan State University	25. Wayne State University
6. University of Illinois (Champaign)	26. Texas Christian University
7. University of Maryland	27. Bowling Green University
8. Purdue University	28. Florida International University
9. University of Georgia	29. Baruch College
10. University of Alberta	30. Colorado State University
11. University of California (Berkeley)	31. New York University
12. University of Kansas	32. Temple University
13. Virginia Polytechnic Institute	33. Old Dominion University
14. Iowa State University	34. Kansas State University
15. University of Michigan	35. University of Western Ontario
16. Northwestern University	36. Case Western Reserve
17. University of Minnesota	37. University of Texas (Austin)
18. University of Akron	38. Dartmouth College
19. Illinois State University	39. Flinders University
20. George Washington University	40. Howard University

Source: G. S. Howard, S. E. Maxwell, S. M. Berra, and M. E. Sernitzke, "Institutional Research Productivity in Industrial/ Organization Psychology," *Journal of Applied Psychology 70* (1985), 233–36.

T A B L E 3-7
Research Productivity: The Top 40 Institutions in Macro and Micro Organizational Behavior

1. Harvard University	21. University of California at Los Angeles
2. Columbia University	22. University of Minnesota
3. Massachusetts Institute of Technology	23. Northwestern University
4. University of California (Berkeley)	24. University of Maryland
5. Stanford University	25. Michigan State University
6. New York University	26. Boston University
7. Ohio State University	27. Arizona State University
8. Indiana University at Bloomington	28. University of South Carolina
9. Texas A&M University	29. Carnegie-Mellon University
10. University of Illinois (Champaign)	30. University of Chicago
11. University of Pennsylvania	31. Georgia Institute of Technology
12. Pennsylvania State University	32. University of Georgia
13. University of Houston	33. University of North Carolina
14. University of Michigan (Ann Arbor)	34. University of Pittsburgh
15. University of Washington	35. University of Cincinnati
16. University of Southern California	36. University of Nebraska (Lincoln)
17. Cornell University	37. University of Missouri (Columbia)
18. Purdue University	38. University of Alabama (Birmingham)
19. University of Wisconsin (Madison)	39. Auburn University
20. Virginia Polytechnic Institute	40. University of Illinois (Chicago)

Source: M. J. Stahl, T. L. Leap, and Z. Z. Wei, "Publication of Leading Management Journals as a Measure of Institutional Research Productivity," *Academy of Management Journal 31* (1988), 707–19.

zational behavior; Table 3-7 lists universities where investigators do both micro and macro OB research.

The Scientific and Diagnostic Models

In Chapter 1, we outlined the diagnostic model that we follow in Chapters 4 through 19 of this book. In this chapter, we have described the scientific model, which produced much of the information we discuss in the remaining chapters of this text. The two models can be compared by matching the four stages of the diagnostic model with the four objectives of the scientific method. The first objective of science and the first stage of the diagnostic model match exactly: *description*. In describing a situation, which is the first stage in diagnostic problem solving, it is generally a good idea to use methods of data collection that are reliable, valid, and standardized. In the model's second stage, diagnosis, we are concerned with determining the underlying causes for some situation. Diagnosis corresponds directly with the scientific goal of *understanding*. Here is where theories can be most useful, by separating key factors from irrelevant ones. The third diagnostic stage, *prescription*, corresponds to the scientific objective of *prediction* in that both require us to predict what may happen in the future. Finally, the *action* stage of the diagnostic model corresponds to the scientific goal of *control*: If our prescriptions and predictions are correct, we should be able to manipulate our environment in order to bring about the ends we seek. In implementing our actions, however, we must keep in mind the various threats to our ability to make valid causal inferences and, as far as possible, arrange our observations to eliminate these threats. This is critical, because if we mistakenly assume that a change in some outcome was due to our actions when in reality the change was due to something else, we may try this action again in the future only to find out how wrong we were.

The Scientist-Practitioner Model

We hope that by this point we have convinced you of the need to think about managing organizational behavior from a scientific point of view. Even though you may not actually conduct research yourself, you will find it invaluable to familiarize yourself with the wealth of scientific evidence available on topics that will be crucial to you, your employer, and your employees. Although this research may not provide you with all the answers, it will most assuredly give you something to think about and perhaps provide new slants on old problems. In addition, as a manager, you will be constantly bombarded by people claiming to be able to solve your problems (for a sizeable fee, of course). It is important that you be able to examine their claims with a critical eye, so that you won't be the victim of every fad that comes across your desk. The critical perspective provided from thinking scientifically will make you less likely to fall for this year's "quick fix of the century."

SUMMARY

It is important to generate reliable and valid knowledge in the area of organizational behavior. Traditional ways of ascertaining knowledge, such as rationalism, personal experience and reliance on authorities have many limitations. The advantage of science relative to these more traditional means is its *objectivity*, and science as an enterprise tends to be public, self-correcting and cumulative. The major goals of science are the description, explanation, prediction and control of various phenemona. These goals are achieved through an interplay of *theory* and *data*, whereby ideas contained in theories are expressed in testable *hypotheses*, which are then compared to actual data. The correspondence (or lack thereof) between the hypothesized results and the actual results are then used to verify, refute, or modify the theory. There are many similarities between the scientific method of inquiry and the diagnostic model laid out in Chapter 1.

Good theories are characterized by simplicity, self-consistency, and consistency with known facts, and they should contribute to the objectives of science. To be useful, data for testing theories should be *reliable* and *valid*, and there are many advantages to using established *standardized* measures.

At the core of many theories is the idea of establishing causes. Cause can be inferred only when one establishes *temporal precedence* and *covariation* and when all *alternative explanations* have been eliminated. This last is often the most troublesome aspect of research in the social sciences, and some kinds of threats like *selection*, *mortality*, *instrumentation*, and *history threats* are especially problematic. These threats can be partially ameliorated through research designs that use control groups and make these comparable to experimental groups through *randomization*, *matching* or *homogenization*.

To *generalize* the findings from one study—with a specific sample, in a specific setting, and during a specific time period—to another context it is necessary to randomly select samples, settings, and time periods. This is rarely achieved in the social sciences. However, if over time experimental results have repeatedly been confirmed in different samples and settings and at different times, it may be possible to generalize such findings.

REVIEW QUESTIONS

1. Many theories can be shown to follow a similar pattern. They start out simply, grow increasingly complex as empirical tests on the theory proliferate, and then die out or are replaced by new theories. Look back at the criteria for

a good theory and discuss why this pattern is so common. In your discussion, specify possible conflicts or inconsistencies among the "criteria for a good theory."

2.   Objectivity is one of the hallmarks of scientific inquiry. Yet all scientists can be shown to have their own subjective beliefs and biases surrounding the phenomena they study. Indeed, some scientists are motivated to do their work because of passionate beliefs about these phenomena. Discuss whether this kind of passion is an asset or a liability to the scientist. Discuss further how science can be an objective exercise when everyone who practices it can be shown to have biases. What prevents a passionate scientist from cheating or distorting results in favor of his beliefs?

3.   Experiments in organizations usually involve people other than the experimenters, that is, managers or employees. What are some of the ethical responsibilities of an experimenter with respect to these people? Is it ethical, for example, for an experimenter to use one group of employees as a control group when she strongly suspects that the treatment given to the experimental group will enhance their chances for success, promotion, or satisfaction? If the experimenter is afraid that explaining the nature of the experiment will cause people to act differently than they would otherwise (and hence ruin the experiment), is it ethical for her to deceive them about the study's true purpose?

4.   Philosopher of science Murray S. Davis once remarked that "the truth of a theory has very little to do with its impact." (See his 1978 article, "That's Interesting! Towards a Phenomenology of Sociology and a Sociology of Phenomenology," *Philosophy of the Social Sciences* 1, 309–44.) History, according to Davis, shows that the impact of a theory depends more on how interesting the theory is perceived to be by practitioners and scientists than on how much truth it holds. We listed criteria for good theories in this chapter; list what you think are criteria for "interesting" theories. Where do these two lists seem to conflict most, and what can be done by scientists and the practitioners they serve to generate theories that are both interesting and truthful?

C H A P T E R 4

ABILITY AND PERSONALITY

*When Toyota Motor Corporation decided to establish manufacturing operations in the United
States, it paid careful attention to individual differences in hiring its workers. With the slogan,
"Investing in the individual," it set about building good will among Americans through its
support of diverse aspects of the U.S. culture and society. The Toyota USA Foundation granted
some $87,000 to the Chesapeake Bay Foundation to support student trips and teacher training in
the biology and environment of the Bay watershed. Toyota has also "adopted" a high school in
Torrance, California, and will support its students and programs by donating funds to the school
and by encouraging company employees to volunteer their time in tutoring students and giving
them career guidance.*
Sources: *Toyota Motor Corporation and Chesapeake Bay Foundation, 1991;* Fortune, *March
25, 1991.*

In 1987, Toyota Motor Corporation became one of the first Japanese-owned firms to open up manufacturing operations in the United States. Many American businesses were amazed by the intensity of the selection procedures used by Toyota in staffing their new plant in Lexington, Kentucky. Applicants for even the lowest paying job on the shop floor went through a minimum of 14 hours of assessment procedures. Other applicants endured over 25 hours of testing for individual differences. William Osos, a regional official for the United Auto Workers, contrasted this procedure with his own experience in applying for work at an American-owned plant, where "you wrote down what work you had done before, and that was the end of it. If you knew somebody who worked in the plant and put in a good word for you that helped."[1]

DIAGNOSTIC ISSUES

In Chapter 1, when we introduced the diagnostic model we said we would begin each core chapter by reintroducing this model and by showing you how the questions it generates can help you to learn the material in the chapter. Like the sections that open all the other chapters, this first one raises questions that arise out of each of the model's stages. When you finish the chapter, you will be able not only to answer these questions but to use the more detailed "Diagnostic Questions" at the end of the chapter to analyze the book's cases and solve real problems on the job.

What sorts of questions about human abilities and personality characteristics does our diagnostic model lead us to raise? To begin with, How can we *describe* the major dimensions of human abilities? How do these abilities affect performance on various tasks? What are some of the important differences between people in terms of mental ability? How do these differences relate to performance on particular tasks? How can we *diagnose* the effects on a situation of people's varying social traits? Can a person's traits change because of the job she holds? Why are there sometimes big differences between a person's maximum performance level and the level she typically displays? How can we *prescribe* who should be assigned tasks that must be learned very quickly? How can we predict who will improve on the job and who will not? Can we prescribe what sorts of jobs different people are best suited for on the basis of measures of their personality characteristics? Finally, what *actions* can we take to measure human abilities and personality characteristics most effectively? How can we use information on these qualities to improve performance and enhance satisfaction?

THE MIRROR IMAGE FALLACY

Ralph Waldo Emerson once wrote that "the wise man shows his wisdom in separation, in gradation, and his scale of creatures and of merits is as wide as nature. . . . The foolish have no range in their scale, but suppose that every man is as every other man."

This passage gets to the heart of our chapter. Although most of us would readily acknowledge the truth of Emerson's statement, many people have a persistent tendency to assume that people are basically alike. This belief that the whole world is "just like me," called the **mirror image fallacy**, is attractive because

mirror image fallacy The false belief that all people are alike or that others share one's own abilities, beliefs, motives, or predispositions.

[1] R. Koenig, "Toyota Takes Pains and Time, Filling Jobs at Its Kentucky Plant," *The Wall Street Journal*, December 1, 1987, p. 21.

Attention to individual differences may become more and more important as companies around the world become increasingly globalized. Fitting into a particular corporate culture in one's own society is often hard enough (see Chapter 18). Succeeding in an organizational culture that is based on the customs of a different country can be twice as difficult. According to William Ahern (left), a vice president at Mitsui Taiyo Kobe Bank in New York City, American employees of Japanese firms must be culturally aware and "keep their egos in check." Scott Whitlock, executive vice president of American Honda Motors agrees. Honda is a company for team players, he says, one where people achieve with others, not by themselves.
Source: *"Managing,"* Fortune, *December 3, 1990.*

ANOTHER VIEW
Sometimes when we look in a mirror we see that people are different from us. Men see women as different; people of one race see those of another as different. These assumptions are often based on physical characteristics. Affirmative action laws (see Chapter 9) are designed to make us look deeper, beyond outside appearance.

personnel selection The process by which an organization decides who will and who will not be allowed to work for an organization.

it makes the world seem much easier to manage. For example, if an owner of a firm believes that everyone in her company shares her abilities, interests, beliefs, and values, she will consider it an easy task to organize and encourage her employees to pursue a common goal. Because the mirror image fallacy *is* a fallacy, however, she will soon find that the myriad differences among the people she employs will make her task far from easy.

As we will see, when the mirror image belief is allowed expression in the real world of management it can become very dangerous. First and foremost, it often leads people to make incorrect diagnoses, attributing to other factors many problems that are in fact due to individual differences. For example, if two people do not seem to be getting along, a manager who holds the mirror image belief may assume that some misperception or miscommunication has caused the problem and that clearing the communication channels will solve the problem. Often, however, each person in a conflict has a very clear picture of the other's position but simply has different values and interests. Just making it easier for two such people to communicate would have little problem-solving potential, and could even inflame matters. For example, in 1988, when Geraldo Rivera tried to bring members of the Congress of Racial Equality together with the Skinheads of the National Resistance and other extremist groups, a melee ensured in which Rivera's nose was broken.

Suppose a highly competent engineer with degrees from several of the best engineering schools is promoted to a position that requires him to manage engineers with somewhat less aptitude and education. If the managing engineer holds the mirror image fallacy, he may attribute poor performance among his subordinates to a lack of either effort or motivation rather than to a lack of ability (a problem he personally never faced). As a result, he might promise subordinates large raises for improving the quality of their work or threaten them with termination. These methods might well increase subordinates' desire to improve, but if their problems stem from lack of ability rather than lack of effort the manager's efforts will meet with failure.

The mirror image fallacy can also cause a failure to anticipate problems. The people responsible for the security of the marine barracks in Lebanon that was hit by a devastating terrorist bomb in 1983 plainly admitted that there was no provision for defending against that type of attack. The thought of such a suicidal mission was so utterly foreign to Western decisionmakers that it was unforeseen.

Foreign competition has heightened this need to anticipate problems. Clearly, Toyota executives are not assuming that all American workers are alike or that American workers are like Japanese workers. To realize the full potential of a labor force, managers must ensure that the right people are in the right places. The person who is best for one job will not necessarily be best for another. The increasing complexity of the world around us has forced businesses to make a greater effort to assess the unique abilities and personality characteristics needed to fill today's specialized jobs. Needless to say, matching the right people to the right jobs takes time, and it is difficult for firms that don't spend this time to compete with firms that do. Competition becomes extremely keen when firms share the same labor pool; some firms take the best employees, and others must make do with what's left.

SELECTION AND PLACEMENT

Organizations that take advantage of individual differences do so through either selection or placement. **Selection** is the process of choosing some applicants and rejecting others. It controls individual differences by

ABILITY AND PERSONALITY

personnel placement The process by which an organization assigns new employees to specific jobs.

AN EXAMPLE
Over the past 20 years we have come to realize that even the mentally handicapped can become assets to organizations when placed in jobs consistent with their abilities. Through proper job placement, they can become contributing members to society.

determining who enters the organization. **Placement** is the process of assigning individuals who have been selected to jobs. The distinction between these two processes is quite important. First, organizations in tight labor markets—that is, where there are few applicants for many positions—may not be able to reject many applicants. The only way to take advantage of individual differences, then, is through placement opportunities. The "baby bust" of the early 1970s has led experts to predict that this kind of tight labor market will confront most firms in the year 2000.

Second, although selection decisions provide competitive advantages to individual companies, accurate selection does not necessarily help the economy as a whole. That is, unless there is widespread unemployment, even the poorest applicants will be selected by some firm, and their lack of productivity will be reflected in the general economy. On the other hand, if effective placement would allow these individuals to be placed in jobs they could do well, the overall health of the economy would improve.[2]

HUMAN ABILITIES

Differences among individuals have been recognized for many thousands of years. But the systematic study of such variation and its application to matters of social concern did not begin until the nineteenth century.

Darwin and Galton

The study of individual differences can be traced directly to the work of the English naturalist Charles Darwin (1809–1882). Darwin's theory of evolution, which suggested that human beings are animals, legitimized the scientific study of the human being. Before Darwin, human beings were thought of as qualitatively different from animals and thus as subject to philosophic and theologic, but not scientific, inquiry.

In addition, Darwin's theory of natural selection and his notion of the "survival of the fittest" illustrated that even small, within-species differences can have important implications. For example, Darwin showed that although to the casual observer one finch may look just like any other, specific characteristics of different varieties of finch serve to adapt these birds to particular environments. As a result, when Darwin switched the customary environments of finches with long, curved bills (for eating out of flowers) and finches with short, stout bills (for eating off the hard ground), the mismatched birds died rapidly.

Expanding on Darwin's work, Sir Francis Galton (1822–1911) proposed that there might be important within-species differences in human beings that could have major implications for survival. Galton devoted his life to measuring and studying these differences, giving particular attention to their possible genetic origin. At the time that he conducted his research, the then new field of psychology was focusing largely on the mental makeup of the "average person." Galton's work ensured that the new area of study would not ignore individual differences.

physical ability Ability to perform a task involving body movement, strength, endurance, dexterity, force, or speed.

Physical and Psychomotor Abilities

Nonmental tests of individual differences are usually of two types.[3] One type measures general **physical abilities**, such as the ability to lift a weight. The other

[2] G. F. Dreher and P. R. Sackett, *Perspectives on Employee Status and Selection* (Homewood, Ill.: Irwin, 1984), p. 23.
[3] M. D. Dunnette, "Aptitudes, Abilities, and Skills" in *Handbook of Industrial and Organizational Psychology*, ed. Dunnette (Chicago: Rand McNally, 1976).

AN EXAMPLE

Contrary to popular belief, police departments do not have one set of minimum physical abilities necessary to do the job. Instead they have several standards. First, there are two basic standards, one for men and one for women. Then the standards are adjusted for age groups. For example, men 18 to 30 must run the mile in eight minutes while men 31 to 35 have 12 minutes. How can we argue that these individuals are equally able, physically, to protect the public? Why don't we have one standard that everyone must meet?

TABLE 4-1
Fleishman's Major Physical Abilities

1. **Static strength**	Maximizes force that can be exerted against external objects (e.g., lifting weights).
2. **Dynamic strength**	Muscular endurance in exerting force continuously or repeatedly (e.g., pull-ups).
3. **Explosive strength**	Ability to mobilize energy effectively in bursts (e.g., standing broad jump).
4. **Trunk strength**	Dynamic strength limited to leg muscles (e.g., leg lifts).
5. **Dynamic flexibility**	Ability to stretch trunk and muscles (e.g., twist-and-touch floor test).
6. **Gross body coordination**	Ability to coordinate action of several parts of the body while the body is in motion (e.g., jump rope test).
7. **Gross body equilibrium**	Ability to maintain balance (e.g., walking a balance beam).
8. **Stamina**	Capacity to sustain maximum effort requiring cardiovascular exertion (e.g., a one-mile run).

Source: E. A. Fleishman, *The Structure and Measurement of Physical Fitness* (Englewood Cliffs, N.J.: Prentice-Hall, 1964). pp. 31–42.

psychomotor ability Ability to perform a task involving coordination between physical and mental functions.

type assesses **psychomotor abilities**, or abilities that require the precise coordination of sensory information, such as sight or sound, with physical movements, such as threading a needle. Edwin Fleishman, a leading researcher in the area of human abilities, has conducted a number of studies of physical and psychomotor abilities and has been able to reduce a long list to 17 primary dimensions (see Tables 4-1 and 4-2).

Although a thorough analysis of a job is needed to determine whether it requires a particular physical capacity, in general, the abilities listed in Table 4-1 are most frequently needed in jobs of three types: the protective services, such as municipal police, fire, and prison-corrections departments; construction and other physically demanding industries; and professional athletics.

If police, fire, or correctional personnel lack the necessary physical abilities, they or the persons they seek to protect may be injured. Testing for the kinds of physical abilities listed in Table 4-1 is much more common now than in the past, when height and weight criteria substituted for specific abilities. Because height and weight measures are considered to discriminate unfairly against women and the members of some minority groups they are rarely used today.

Physical ability tests are also used to select employees for work such as construction, where jobs require physical strength and agility. Such tests can predict not only job performance but job-related injuries. For example, research by Chaffin has shown that the incidence of lower-back injury can be predicted by tests of physical strength.[4] This is a significant finding, since back-related disability claims have been rising at 14 times the rate of the population over the last ten years. Because employers are increasingly picking up the bill for employees' medical costs (and because lower-back pain is such a widespread and recurring affliction), tests that allow one to predict health problems for a job applicant are extremely cost effective. General Dynamics Corporation's electric boat unit has actually developed specific physical examinations designed to screen out individuals at risk for back-pain problems.

AN EXAMPLE: COCA-COLA

The rising costs of employee health care have led many companies not only to screen job applicants for physical problems but to develop programs to help people, once hired, to maintain their fitness. At Coca-Cola, CEO Brian Dyson trains regularly for triathlons, setting an example for the company's employees. "As long as you're fit you can do anything you want," he says. Dyson, 55 years old, regularly finishes among the top 20 in his age group; his best time for the 1500m swim, 40k bike, and 10k run is 2 hours 35 minutes. ("The CEO 1000," *Business Week*, October 19, 1990, p. 44.)

[4] D. B. Chaffin, "Human Strength Capability and Low Back Pain," *Journal of Occupational Medicine* 16 (1974), 248–54.

ABILITY AND PERSONALITY

APPLYING THE DIAGNOSTIC MODEL

Diagnostic Question 1: Are there any special features of this job that call for high levels of specific physical or motor abilities? Does the jobholder possess these abilities?

TABLE 4-2
Fleishman's Major Psychomotor Abilities

1. **Control precision**	Tasks requiring finely controlled muscular adjustments (e.g., moving a lever to a precise setting).
2. **Multi-limb coordination**	The ability to coordinate the movements of limbs simultaneously (e.g., packing a box with both hands).
3. **Response orientation**	The ability to make correct and accurate movements in relation to a stimulus under highly accelerated conditions (e.g., reaching and flicking a switch when a warning horn sounds).
4. **Reaction time**	The speed of a person's response when a stimulus appears (e.g., pressing a key in response to a bell).
5. **Speed of arm movement**	The speed of gross arm movements where accuracy is not required (e.g., gathering trash and throwing it into a large pile).
6. **Rate control**	The ability to make continuous motor adjustments relative to a moving target that changes in speed and direction (e.g., holding a rod on a moving rotor).
7. **Manual dexterity**	Skillful arm and hand movements in handling rather large objects under speeded conditions (e.g., rapidly placing blocks of different shapes into the correct holes of a special board).
8. **Finger dexterity**	Skillful manipulations of small objects with the fingers (e.g., attaching nuts and bolts).
9. **Arm-hand steadiness**	The ability to make precise arm-hand positioning movements that do not require strength (e.g., threading a needle).

Source: E. A. Fleishman, *The Structure and Measurement of Physical Fitness* (Englewood Cliffs, N.J.: Prentice-Hall, 1964). pp. 67–78.

POINT TO STRESS

Although the assessment of physical abilities is discussed in much more depth in Human Resource Management classes, it is important for all students to realize that job ability levels can no longer be set arbitrarily. Numerous court cases, like *Griggs vs Duke Power Company* and the Uniform Guidelines require most employers to do an in-depth job analysis listing skill requirements. Employers must also show a significant relation between required skill level and subsequent job performance.

Finally, tests of physical and motor abilities are used in professional athletics, where performance is highly dependent on physical characteristics. In this field of work, people over a certain age may have difficulty developing some of the necessary skills. Moreover, when training is provided, the gap between individuals who differ in some physical ability tends to get even larger.[5] Highly standardized tests (e.g., the 40-yard dash) are used in these areas, because it is often necessary to compare individuals from widely different geographic regions.

The motor ability tests shown in Table 4-2 have been most useful for predicting success in three broad types of occupation: vehicle operators (e.g., truck drivers, fork-lift operators); industrial employees (e.g., assembly-line workers, packagers); and people engaged in crafts and trades (e.g., carpenters, plumbers, electricians, mechanics).[6]

Cognitive Abilities

Binet and Spearman. Although Galton referred to his tests as mental tests, psychologists today classify most of his measurements under physical or pyschomotor abilities. It was actually Alfred Binet (1857–1911) who in 1905 created the first test of intelligence, or cognitive ability. Binet's first test attempted to distinguish between poor students who could benefit from remedial education and poor

APPLYING THE DIAGNOSTIC MODEL

Diagnostic Question 2: Are there any features of this job that require specific facets of mental ability? Does the jobholder possess these abilities?

[5] I. L. Goldstein, *Training: Program Development and Evaluation* (Belmont, Calif.: Brooks, Cole, 1986), p. 20.
[6] E. E. Ghiselli, *The Validity of Occupational Aptitude Tests* (New York: John Wiley, 1966), p. 38.

In the service industry, the need to meet an increasing variety of customer needs has heightened job complexity and led companies to alter their employee selection programs. In the past, answering telephone inquiries was not considered that demanding a job, but today companies are raising their requirements for these positions. For example, to staff its highly regarded telephone answer center, General Electric now recruits college graduates with good general cognitive ability, interpersonal skills, and sales experience. The company puts the recruits through six weeks of training in which they learn precisely how GE's machines function. Taking apart a washing machine and putting it back together can help a GE Answer Center phone representative explain to a caller how to operate the appliance or to make minor repairs. Source: "Selling," Fortune, December 5, 1990, pp. 43–44.

students who could not. Today, virtually all students in higher education have taken tests of cognitive ability, and many of these modern tests bear a close resemblance to Binet's initial effort. Tests of cognitive ability are also used increasingly in the business sector, as industries search for people who can learn quickly and who can solve complex problems.

Whereas Binet's work was highly practical, the British psychologist Charles Spearman (1863–1945) took a more scholarly approach, submitting many of the things that Binet took for granted to scientific test. For example, Binet assumed that intelligence was one dimensional, that is, that all his tests and items measured the same thing. As a result, Binet always reported intelligence test results in the form of one score.

Spearman set out to test this presumption of unidimensionality, and after years of collecting data and building mathematical models he concluded that all tests of intelligence were made up of two dimensions. The first, a generalized intelligence, he called "g"; the second was a unique dimension associated with each test. Today it is widely accepted that intellectual ability is multifaceted. We are more likely to characterize someone as "good with words, but bad with numbers," than "of average intelligence." This kind of distinction is especially critical in today's business climate, where increasing specialization demands a high level of one or two specific skills for most jobs rather than an average level of a broad set of abilities.

general cognitive ability The totality of an individual's mental capacity, summing across specific mental abilities such as verbal comprehension, quantitative aptitude, reasoning ability, and deductive ability.

General Cognitive Ability. Although mental abilities are not one dimensional, we do generally find positive relationships among people's performance on different kinds of mental tests. Thus scores across different types of tests are often summed and treated as an index of general intelligence. Specialists often substitute the term **general cognitive ability** for *intelligence* because the former term is more precise and because it conjures up less controversy over such issues as the role of genetic factors in mental ability. The term *intelligence* is used imprecisely in the lay community where the high social value placed on it complicates discussions of things like age, sex, and racial differences.

Specific Dimensions of Cognitive Ability. Although there is a common core among the many different kinds of mental tests, several dimensions of mental ability are sufficiently unique that they are worth assessing in their own right.

ABILITY AND PERSONALITY

Cognitive Ability Tests

APPLYING THE DIAGNOSTIC MODEL

Diagnostic Question 3: How much cognitive ability does the jobholder have? How will this influence the speed with which he will learn the job?

Diagnostic Question 4: How much improvising will the person have to do on this job? Does the jobholder have enough general cognitive ability to adapt to changing circumstances?

Diagnostic Question 5: If the jobholder lacks general cognitive ability, does he have sufficient work experience to make up for this lack?

Diagnostic Question 6: How complex is this job? If it is highly complex, does the jobholder have enough cognitive ability to handle this complexity?

The usefulness of mental ability tests in predicting task performance has been investigated in both academic and organizational contexts.

General Tests. In academic settings, researchers have found high correlations between tests like the Scholastic Aptitude Test (SAT) and first-year-college grade point average, or GPA (correlations in the .50s), as well as overall rank in class (correlations in the .60s).[8] These tests are more predictive for students in the physical sciences or math than in the humanities or the social sciences. They are less predictive of success in graduate school (correlations in the .30s), because most applicants for graduate school score relatively high in mental ability and are therefore a somewhat homogeneous group.

There is a great deal of evidence to suggest that general cognitive ability is also predictive of success in the world of work.[9] Research by Jack Hunter has shown that in virtually any job where planning, judgment, and memory are used in day-to-day performance, individuals high in cognitive ability will generally outperform those who are low in this ability. Other research has shown that the relationship between general mental ability and job performance increases as the job gets more complex in terms of decision making, planning, problem solving, and analyzing information.[10] General cognitive ability is important even for jobs not characterized by such complexity if these jobs require the person to learn something new. Individuals high in general cognitive ability will learn the job more quickly than others. In low-complexity jobs, experience over time often wipes out this initial difference between high- and low-ability individuals, as Figure 4-2 demonstrates.[11] That is, as months on the job increase, the initial performance differences attributable to differences in ability decrease. Thus general cognitive ability is important in two respects. It relates both to learning the job

---

[8] A. R. Jenson, *Bias in Mental Testing* (New York: Free Press, 1980), p. 313.

[9] J. E. Hunter, "Cognitive Ability, Cognitive Aptitudes, Job Knowledge, and Job Performance," *Journal of Vocational Behavior* 29 (1986), 340–62.

[10] R. L. Gutenberg, R. D. Arvey, H. G. Osburn, and R. P. Jeanneret, "Moderating Effects of Decision-Making/Information Processing Dimensions on Test Validities," *Journal of Applied Psychology* 68 (1983), 600–608.

[11] F. L. Schmidt, J. E. Hunter, A. N. Outerbridge, and S. Goff, "Joint Relation of Experience and Ability With Job Performance: Test of Three Hypotheses," *Journal of Applied Psychology* 73 (1988), 46–57.

F I G U R E   4-2

General Cognitive Ability and Experience on the Job as Determinants of Performance

and to performing the job when the job requires the person to deal continually with new situations.

Specific Tests. In certain specific jobs, tests of specific mental ability can add greatly to the predictive power of tests of general intelligence.[12] For example, spatial visualization is critical for draftsmen and personnel in technical positions, and it is an important component in jobs that require mechanical skills, such as machinist, forklift operator, and warehouse worker. Perceptual ability is important in positions such as accountant, bookkeeper, stenographer, proofreader, typist, and general office clerk. Verbal ability and reasoning are critical to success in executive, administrative, and professional positions. Numerical ability is important in jobs such as accountant, payroll clerk, and salesperson and in many types of supervisory jobs.

At the outset of this chapter we distinguished between selection and placement. The use of general mental ability tests is most appropriate in selection because of the wide variety of jobs to which this ability is relevant. All applicants can be rank ordered in terms of general cognitive ability and then assigned to jobs, with the assumption that "more is better" regardless of the nature of the job. The use of specific ability tests facilitates placement. The key here is to place individuals with different strengths in jobs requiring different specific abilities. The assumption is that there are few people who cannot do some job well, and thus a high level of overall firm performance can be obtained from the correct matching procedure.

Figure 4-3 illustrates how various kinds of ability combine to affect task performance. General cognitive ability is the main influence on general job knowledge. It influences both how fast a person can learn the job, and how readily the person can adapt to changing circumstances when on the job. Job complexity affects the relationship between general cognitive ability and general job knowledge because the more complex the job in terms of decision making, planning, and judgment, the more learning and improvising the job requires. In simple jobs, the impact of general cognitive ability is less, and experience on the job can often substitute for lesser mental ability.

[12] G. K. Bennett, H. G. Seashore, and A. G. Wesman, *Administrators Handbook for the Differential Aptitude Test*, Psychological Corporation (San Antonio, Texas: Harcourt, Brace, Jovanovich, 1982), p. 55.

FIGURE 4-3

Specific facets of mental, physical, and motor ability in general relate to specific aspects of a task. Whereas general cognitive ability will almost always be relevant, the importance of specific abilities can be determined only by a detailed job analysis. Experience is shown in Figure 4-3 as a moderator of the relationship between specific abilities and specific job skills because initial individual differences on these characteristics tend to increase as people develop more experience or are further trained.[13] Specific job skills combine with general job knowledge to create the individual's total job capability.

HUMAN PERSONALITY

Whereas abilities deal with the things an individual can or cannot do, personality deals with what a person is like. It is useful to distinguish between these two broad aspects of individual differences for several reasons. First, the two have relatively distinct histories. Second, there is more controversy over personality measurement methods than over methods of measuring ability. Finally, the evidence for the usefulness of ability measures as predictors of performance is clearer than the corresponding evidence for personality measures.

Freud and Guilford

The great impact of Darwin's theories was felt not only by researchers on human abilities but by early psychologists who were interested in the study of personality. In the wake of Darwin's theory of evolution, which implied a great deal of common ground between human beings and animals, it was not surprising that many early approaches to the study of personality described consistencies in human behavior as instincts. Unfortunately, just as unique dimensions piled up in research on intelligence, "instincts" piled up in the area of personality research. In fact, by the 1920s, over 800 different classes of instincts had been proposed![14] Clearly, such a major violation of the principle of theoretical simplicity (see Chapter 3) required a more economical approach. For many, this approach was Sigmund Freud's.

Freud, an Austrian physician and psychiatrist and the founder of psychoanalysis, proposed that all behavior is powered by two unconscious drives, or instincts: the *life instinct* and the *death instinct*. The former represented a striving for creative and constructive aims, and the latter represented a striving toward destructive outcomes. Freud accounted for individual differences in personality by suggesting that different people dealt with their fundamental drives in different ways. He postulated that the *ego* mediates between these two opposing forces, constantly trying to strike a compromise between them.

According to Freud, because many of the compromises arrived at fail to satisfy either drive, and because many of the desires stemming from both instincts are socially unacceptable, people use **defense mechanisms** to deflect this inner turmoil (see Table 4-3). Unhealthy, or neurotic, people, he said, expend so much energy erecting defense mechanisms that they have little left for productive work or satisfying relationships. Freud conceived of a healthy person as one who could successfully "love and work."

POINT TO STRESS
Many employers seem more interested in personality than in skills, despite the fact that personality is much harder to measure and quantify.

TEACHING NOTE
Is Freud's definition of health flawed? What about people who use work as a defense mechanism—"workaholics." High productivity may be a defense against other perceived or real inadequacies.

defense mechanism In Freudian psychology, a kind of mental operation by which individuals rechannel the energies linked to socially unacceptable urges.

---

[13] L. S. Gottfredson, "Societal Consequences of the g Factor in Employment," *Journal of Vocational Behavior* 29 (1986), 379–411.
[14] L. L. Berbard, *Instincts* (New York: Holt, 1924), p. 91.

TABLE 4-3	
Major Ego Defense Mechanisms	
Denial	Protecting the self from unpleasant realities by refusing to perceive them.
Repression	Preventing painful or dangerous thoughts from entering consciousness.
Displacement	Releasing pent-up feelings, usually of hostility, and directing them toward persons who are perceived as less dangerous than the individuals toward whom the feelings were originally directed.
Reaction formation	Preventing dangerous desires from being expressed by exaggerating attitudes and behaviors that are in direct opposition to such desires.
Sublimation	Substituting nonsexual, socially acceptable, usually creative activities for the expression of frustrated sexual desires.
Regression	Reverting to behaviors that are typical of an earlier stage of life, when one has fewer responsibilities and, generally, a lower level of aspiration.

Freudian theory has been widely criticized and not without reason. First, because it proposes unconscious states and drives, it is almost impossible to test empirically. Whatever supporting evidence there is for the theory comes out of psychotherapy sessions and thus relies on the memory of a therapist who is not objective and who could misinterpret subjects' responses. These criticisms notwithstanding, Freud left an indelible mark on the study of personality. Many current theories still employ the concept of unconscious causes of behavior, and many of our present methods of personality measurement claim to tap unconscious drives. In addition, although most people believe that Freud overemphasized the role of sexual factors, the role of early childhood experiences in personality development is now widely recognized.

Many recent approaches to personality owe less to Freud, however, than they do to J. P. Guilford and his associates.[15] By the 1950s, when Guilford began his research, the notion of instincts had given way to the notion of traits. Personality traits were similar to instincts in that they reflected simple behavioral consistencies (e.g., if a person exhibited a tendency to be curious across many situations, the trait of curiosity was invented as a means to describe this characteristic). No claim was made, however, that traits were strictly inherited. Guilford, like Spearman, used mathematical models to uncover the dimensions that underlay the many different characteristics that others had described. Table 4-4 describes ten of the major dimensions of personality identified by Guilford.

As we noted before, whereas ability refers to what individuals can and cannot do, personality refers to what people are like. The fact that many personality characteristics are described in everyday language—for example, aggressiveness, sociability, impulsiveness—is both good news and bad news. It is good news because most people can readily perceive individual differences in these qualities and can see how such variations might affect particular situations. It is bad news because terms adopted from everyday language are usually not very precise. This can create considerable difficulty in understanding, communicating, and using information obtained from scientific measures of personality. In the next three sections of this chapter, we will try to help clarify this point by describing some

[15] J. P. Guilford, *Personality* (New York: McGraw-Hill, 1959); R. B. Cattell, *The Scientific Analysis of Personality* (Baltimore: Penguin, 1965); and H. J. Eysenck, *The Structure of Personality* (London: Methuen, 1960).

TABLE 4-4
Guilford's Major Dimensions of Personality

	CHARACTERISTICS OF PERSONS SCORING HIGH ON DIMENSION
General activity	Need to be doing something all the time. Possess seemingly endless energy.
Restraint	Seem overcontrolled, stiff, and lacking in spontaneity. Very deliberate in actions; not impulsive.
Ascendance	Self-assured and ambitious. Like leadership and control over other people.
Sociability	Enjoy face-to-face dealings with others. Confident in social situations.
Emotional stability	Even tempered and not easily upset. Low scorers are seen as moody and over-emotional.
Objectivity	Possess a realistic view of self and others. Not easily hurt by others' remarks and sometimes insensitive to others' feelings.
Friendliness	Want to please others and avoid confrontation. Cooperative, agreeable, and easy to be with.
Thoughtfulness	Introspective and reflective about self. Prefer tasks that involve deep, prolonged, analytic thinking.
Personal relations	Tolerant of other people and accepting of rules and customs.
Masculinity	Behave in stereotypically male fashion. Low scorers are considered motherly, protective, and sensitive.

Source: Adapted from J. P. Guilford, *Personality* (New York: McGraw-Hill Book Co., 1959), pp. 51–53.

general and specific characteristics of personality, some ways of measuring these characteristics scientifically, and some of the existing evidence for the usefulness of measuring these characteristics in organizational contexts.

A Classification System

Given the vast number of personality characteristics that are described in the scientific literature, we need some type of classification scheme in order to understand both the characteristics themselves and their interrelationships. Nunnally assigns all the various facets of personality to five categories: (a) social traits, (b) motives, (c) personal conceptions, (d) emotional adjustment, and (e) personality dynamics.[16] In the following sections, we will describe each of these classes. We will also show specific examples of measures used in the real world to measure specific characteristics. If you see how a characteristic is actually measured in the real world, you will get a better idea of what it means.

social traits Behavior patterns that an individual typically displays when interacting with others in social contexts.

APPLYING THE DIAGNOSTIC MODEL
Diagnostic Question 7: Are there any features of this job that require certain types of social traits? Does the holder of this job possess those traits?

Social Traits. **Social traits** are behavior patterns that people typically manifest when interacting with others in social contexts. Social traits, which often represent the surface layer of personality, reflect the way a person appears to others.

One widely recognized approach to individual differences, based on the work of Carl Jung and translated into measures by Katherine Myers and Isabel Briggs, focuses on personal style.[17] Jung identified two basic personality types; introverts, who are shy and withdrawn, and extroverts, who are outgoing, ag-

[16] J. C. Nunnally, *Psychometric Theory* (New York: McGraw Hill, 1978), p. 546.
[17] D. B. Myers and K. C. Briggs, *Myers-Briggs Type Indicators* (Princeton, N.J.: Educational Testing Service, 1962).

gressive, and dominant. These personality types are matched by four problem-solving orientations or personal styles.

On one dimension, Myers and Briggs distinguish *feeling-type* individuals from *thinking-type* individuals. Feeling types are characterized as "sensitive to other peoples' feelings," "sympathetic," "disliking confrontation," and "warm and personable." Thinking types are characterized as "unemotional and hard-hearted," "logical and analytical," and "uninterested in relating" to other, nonthinking type individuals. As an example of this distinction, note the self-description of ex-college professor, millionaire real-estate developer Clay Hamner, who states, "I don't feel, I think. It wounds never occur to me to ask someone how they feel. I'm a tough touch for people who don't know me."[18]

Myers and Briggs also differentiate between *sensor types* and *intuitive types.* Sensor-type people are characterized by "empasizing action and wanting to get things done," "being highly precise in action and language," and "hard-charging." Intuitive types "emphasize conceptions and theories," "engage in long-range planning," and "are oriented to problem solving but not necessarily to application." Clay Hamner lined up a $105 million purchase of The Pantry in two days. He bought options on 1,500 acres near Atlanta for the Gwinnett Progress Center, over the telephone, sight unseen. He bought and then sold the Hotel Europa in Chapel Hill within 24 hours.

Table 4-5 shows some of the occupations for which Myers and Briggs feel these four types are best suited. Note how their classification system clearly predicts *both* occupations of a personality like Clay Hamner.

A social trait measure recently developed specifically for the work context is the Employee Reliability Scale.[19] According to this scale, low-reliability employees are characterized by (a) hostility toward rules, (b) thrill-seeking impulsiveness, (c) social insensitivity, and (d) alienation, or a feeling of detachment from others. Studies using this scale have found that high scores on this measure are related to supervisory ratings such as good attitudes, teamwork, punctuality, adaptability, and sales productivity. Low scores are related to turnover, high absenteeism rates, high equipment-repair records, and a greater incidence of reported injuries.

Finally, companies are becoming increasingly interested in the honesty of their prospective employees. As the Management Issues Box indicates, there has been an increase in the use of paper and pencil measures of honesty, particularly since polygraph, or lie detector, tests were outlawed.

[18] "Feats of Clay," *Business/North Carolina*, August 1989, pp. 23–31.
[19] J. Hogan and R. Hogan, "How to Measure Employee Reliability," *Journal of Applied Psychology* 74 (1988), 273–79.

T A B L E 4-5
Myers-Briggs' Personal Styles and Compatible Occupations

TYPE	OCCUPATION
Thinking	Lawyer, engineer, teacher, computer programmer
Feeling	Writer, nurse, social worker, psychologist
Sensation	Accountant, pilot, salesperson, physician, land developer
Intuitive	Scientist, artist, corporate planner, inventor

Source: Based on D. B. Myers and K. C. Briggs, *Myers-Briggs Type Indicators* (Princeton, N.J.: Educational Testing Service, 1962), pp. 15–35.

Jerry Pardue, vice-president of loss prevention for Super D Drugs, Inc., a Southeastern chain, noticed with horror a few years ago that shrinkage—retailing jargon for stolen goods—was rising dramatically. Mr. Pardue wanted to see if paper-and-pencil honesty tests could screen out thieves among job applicants.

He had quite a surprise. By his own calculations the tests have helped Super D save about $400,000 a year by reducing shrinkage. Absenteeism rates, substance abuse, lateness, and other forms of counterproductive behavior are down too. "As soon as we started using these tests to screen employees, the integrity of the work force improved all around" Pardue said.*

The passage of the Employee Polygraph Protection Act in 1988 prohibits the use of physiological lie detectors in most instances. Since then, many employers, like Super D Drugs, have turned to paper-and-pencil honesty tests as a means of ensuring integrity in the work force. An indication of this trend can be seen in the growth rate of the $25 million testing industry, roughly 20 percent annually.

Can these kinds of tests actually weed out thieves and other undesirable elements from an employer's work force? Although there is certainly a lot of anecdotal evidence like our story about Super D Drugs, there is very little independent research to support either advocates or critics of this kind of testing. The evidence we have is generally positive, but it is based on studies performed by the testing companies themselves, and thus obviously subject to bias. Independent studies

are difficult to perform because of the proprietary nature of the answer keys that go along with the tests. Setting this problem aside, researchers who have taken test results at face value, have found the validity of these tests is not sufficiently high to meet the "reasonable doubt" standards of our justice system.† Thus one could call into question the ethics of using these tests on job applicants who have been accused of no crime.

Another drawback is that the tests have relatively high rejection rates; 40 to 70 percent of applicants fail the tests. In many of the industries in which these kinds of tests are popular, such as retailing and the restaurant business, there are few applicants relative to the number of available positions. Moreover, the laws of many states (e.g., Massachusetts and Rhode Island) limit the use of these tests. Although no federal laws prohibit these tests as yet, the fact that Congress has appropriated funds to study "integrity" tests suggests that federal legislation is under consideration.

Ironically, some reviewers have noted that "it is not uncommon to encounter questionable, if not blatantly deceptive, sales tactics"‡ in the very aggressive marketing of some integrity tests. Thus organizations considering the use of integrity tests should not take the integrity of the testers for granted. This is one area in which you may need to do your own research to see what works and does not work for your own company.

* C. H. Deutsch, "Pen-and-Pencil Integrity Tests," *New York Times*, February 11, 1990, p. D1.

† K. R. Murphy, "Detecting Infrequent Deception," *Journal of Applied Psychology* 72 (1987), 611–14.

‡ P. R. Sackett, L. R. Burris and C. Callahan, "Integrity Testing in Personnel Selection: An Update," *Personnel Psychology* 42 (1989), 491–528.

---

motive A reflection of an individual's underlying drives, needs, and values

POINT TO STRESS
Motives direct behavior and, to a great extent, account for persistence in behavior. Without drives, needs, and values, there would be little or no behavior.
APPLYING THE DIAGNOSTIC MODEL
Diagnostic Question 8: What motives seem to drive this person? How do these motives influence the amount of energy that she brings to the task?

Motives. Motives reflect an individual's underlying drives, needs, and values. In general, motives are below the surface; that is, they reside within the person. Because this class of personality variables lies at the core of many theories of motivation, we will explore the specific topic of drives and needs in Chapter 7. Here we will look at some common measures of personal values.

The Minnesota Importance Questionnaire (MIQ), widely used in the work context, assesses personal values. It measures the importance that an individual places on various features of the work situation, classified into the following six categories:

1. *Autonomy*: Authority and responsibility
2. *Pleasant working conditions*: Being in a work situation that includes variety and activity
3. *Achievement*: The opportunity to use valued abilities and to be promoted
4. *Recognition*: Enjoying high social status and the respect of others
5. *Altruism*: Doing work that is consistent with moral values and providing needed social services
6. *Management responsibilities*: Managing other people and implementing company policies and practices

The MIQ is typically used in conjunction with another measure called the Occupational Reinforcer Pattern (ORP). The ORP, which is administered to current jobholders, asks which of the characteristics assessed by the MIQ actually describe the job for which applicants who have taken the test are being considered. Thus the two instruments are used together in an effort to match the applicants who favor a particular set of values with the job that most closely fits those values. Research has shown that when there is a close fit between values and job, job satisfaction will be greater and the risk of turnover will be less.[20]

personal conceptions A person's thoughts, attitudes, and beliefs about his social and physical environment.

locus of control The extent to which an individual believes that his own actions influence the environment.

APPLYING THE DIAGNOSTIC MODEL
Diagnostic Question 9: What are the major personal conceptions that guide this person? How do they influence the way he directs his energy?

Personal Conceptions. Personal conceptions reflect the way a person thinks about the social and physical environment. They reflect the person's major beliefs and her personal outlook on a variety of issues.

One widely used formulation is Julian Rotter's **locus of control**. Locus of control refers to the extent to which a person believes that his own actions influence his environment and the events that affect him. *Internals* believe that they control their own destiny. *Externals* believe that their lives are controlled by fate and other forces beyond their control. Some sample items from Rotter's measure of locus of control are shown in Table 4-6.

A major review of the organizational behavior literature on locus of control draws several conclusions about individuals who differ on this personality characteristic.[21] First, internals tend to be more easily motivated than externals, because they seek out and conform to rules that tie rewards to high performance. Thus internals work much better than externals under incentive systems. Second, internals tend to be self-motivated and to prefer participative approaches by supervisors, whereas externals want more directive supervision. This difference seems to spill over to other situations; for example, in leadership roles, internals

[20] E. Betz, "Need Reinforcer Correspondence as a Predictor of Satisfaction," *Personnel and Guidance Journal* 47 (1969), 878–83.

[21] P. E. Spector, "Behavior in Organizations as a Function of Employee Locus of Control," *Psychological Bulletin* 91 (1982), 482–97.

T A B L E 4-6
Items From the Rotter Internal-External Locus of Control Scale

1. a. Many of the unhappy things in people's lives are partly due to bad luck.
   b. People's misfortunes result from the mistakes they make.
2. a. In the long run, people get the respect they deserve in this world.
   b. Unfortunately, an individual's worth often passes unrecognized no matter how hard he tries.
3. a. No matter how hard you try, some people just won't like you.
   b. People who can't get others to like them don't understand how to get along with others.
4. a. Becoming a success is a matter of hard work; luck has little or nothing to do with it.
   b. Getting a good job depends mainly on being in the right place at the right time.
5. a. The average citizen can have an influence in governmental decisions.
   b. This world is run by the few people in power, and there is not much the little guy can do about it.
6. a. Many times I feel that I have little influence over the things that happen to me.
   b. It is impossible for me to believe that chance or luck plays an important role in my life.

Note: For each numbered item the respondent must choose the statement that best describes him or her. You can see how an internal-external pattern can emerge.
Source: Based on Julian Rotter, "Generalized Expectancies For Internal versus External Locus of Control," *Psychological Monographs* 80 (1966), 1–44.

ABILITY AND PERSONALITY

authoritarianism A set of personality characteristics that include ethnocentrism and strong tendencies to overvalue authority, to stereotype others, and to be suspicious and distrustful of people in general.

use more democratic approaches than do externals. Finally, internals seem better at collecting and processing information. Thus they seem to perform better on complex tasks even when general mental ability is held constant.

Research on another personal conception, **authoritarianism**, began in the aftermath of World War II as social scientists tried to understand the psychological forces that made the creation of the Nazi political system possible.[22] In particular, investigators wanted to know whether certain personality types were especially susceptible to Fascist, antidemocratic propaganda. The result of their efforts was the F scale (F stood for "fascism"), which identifies the "authoritarian" personality type, the person who is susceptible to this kind of influence. People who have high scores on the F scale can be characterized as follows:

> They believe that obedience and respect for authority should be the individual's primary values.
> They are ethnocentric: they believe in their own superiority and the superiority of the subgroup to which they belong. They distrust outsiders.
> They are superstitious. They often stereotype others and reject introspection and analysis.
> They feel that they live in a hostile environment full of threatening people.

People who score high on the F scale tend to be rigid in their approach to problem solving and generally score low on measures of creativity. They tend to follow written rules and procedures very closely and to be highly conformist. In general, they follow superiors' orders without question, and they are very apt to use punishment when dealing with subordinates. In the light of the recent emphasis on ethical issues in the business world, authoritarian people are undesirable candidates for key positions in any organization. By their nature, they are too susceptible to the lures of acting unethically.

emotional adjustment A class of personality variables that deal with the extent to which a person experiences affective distress or engages in socially unacceptable behaviors.

Type A behavior pattern A set of personality characteristics that include aggressiveness, competitiveness, and the tendency to work under self-induced time pressures.

Emotional Adjustment. **Emotional adjustment** deals with the extent to which a person experiences emotional distress or engages in socially unacceptable behavior. A person is said to be emotionally *maladjusted* if such distress prevents her from functioning properly in the environment or if it affects her health in a deleterious way.

Commonly seen in the business setting, especially among managers, is what has been called the **Type A behavior pattern**.[23] People with a Type A personality are characterized as aggressive and competitive. They set high standards for themselves and others and put themselves under constant time pressure.[24] Type B's, on the other hand, are free of such feelings of urgency. We discuss the Type A behavior pattern here as a maladjustment, because a tremendous amount of evidence links this characteristic to coronary heart disease. For example, one study of over 3,000 people found that individuals classified as Type A were twice as likely to suffer a fatal heart attack as those classified as Type B.[25] Another study showed that over 70 percent of coronary heart disease sufferers were Type A's.[26]

APPLYING THE DIAGNOSTIC MODEL
Diagnostic Question 10: Are there any signs of personality maladjustments that may be hindering this person's ability to concentrate on the task?

[22] T. W. Adorno, E. Frenkel-Brunswick, D. J. Levinson, and B. J. Sanford, *The Authoritarian Personality* (New York: Harper, 1950).

[23] J. H. Howard, D. A. Cunningham, and P. A. Rechnitzer, "Health Patterns Associated with Type A Behavior: A Managerial Population," *Journal of Human Stress* 2 (1976), 24–31.

[24] K. A. Mathews, "Psychological Perspectives on the Type A Behavior Pattern," *Psychological Bulletin* 91 (1982), 293–323.

[25] R. Rosenman and M. Friedman, "The Central Nervous System and Coronary Heart Disease," *Hospital Practice* 6 (1971), 87–97.

[26] C. D. Jenkins, "Psychologic and Social Precursors of Coronary Disease," *New England Journal of Medicine* 284 (1971), 244–55; 307–17.

Dealing with Type A behavior becomes a complex matter when we add one more ingredient to the mix. That is, Type A's tend to outperform Type B's on most tasks that require persistence and endurance. It would seem, then, that attempts to eliminate Type A personalities from an organization would risk destroying the firm's ability to accomplish its mission.

We are learning more, however, about this personality characteristic. A recent study of college professors with Type A personalities, for example, found that job performance was most influenced by just one part of the behavior pattern, that dealing with goal setting and time pressures. The negative emotional consequences of the behavior pattern sprang out of its aggressive and competitive components.[27] Thus the Type A pattern is a kind of two-edged sword; we might wield this sword to the advantage of the organization if we could just find individuals who were strong in some of its aspects but weak in others.

One of the most widely used personality assessment devices, the Minnesota Multiphasic Personality Inventory (MMPI) is designed to differentiate among various types of psychological disturbance, such as depression and paranoid behavior. Some of these dimensions are shown in Table 4-7. As a manager in a business organization, you would rarely encounter anyone whose scores on the MMPI were "clinically elevated," or high enough to indicate serious mental disorder and the need for specific treatment. However, research suggests that only mildly elevated levels of these kinds of characteristics can affect performance on complex tasks that demand a great deal of attention.[28] The reason is that even at low levels, these kinds of distress tend to distract a person from any task.

personality dynamics   A class of personality characteristics that deal with the integration and organization of traits, motives, personal conceptions, and adjustment.

self-esteem   the degree to which a person believes that she is a worthwhile and deserving individual.

TEACHING NOTE
The measurement of self-esteem is an excellent example of the difficulty of assessing personality. According to a (recent) survey, of the more than 100 self-esteem scales in use, many have little relationship to each other. For example, the Ghiselli and Rosenberg scales correlate at only .10. Which scale really measures self-esteem? Or is self-esteem multi-faceted? If so, what are these facets? Which facets do each of these scales measure? And which facets are the most important?

Personality Dynamics.   **Personality dynamics** constitute the most complex group of personality characteristics. This group deals with the principles by which all the other four classes are integrated and organized. Personality dynamics explain how all the other types of characteristics are put together to form the "whole person."

One such personality dynamic is referred to as the self-concept, and one of the most important aspects of the self-concept is self-esteem. **Self-esteem** (SE) describes the extent to which the person believes that he is a worthwhile and deserving individual. People with high self-esteem differ from people with low self-esteem in terms of all four sets of personality characteristics we have discussed—social traits, motives, personal conceptions, and emotional adjustment.[29]

[27] M. S. Taylor, E. A. Locke, C. Lee, and M. E. Gist, "Type A Behavior and Faculty Research Productivity: What are the Mechanisms?" *Organizational Behavior and Human Performance* 34 (1984), 402–18.

[28] Ruth Kanfer, Personal Communication, April 17, 1989.

[29] J. Brockner, *Self-Esteem at Work* (Lexington, Mass.: Lexington Books, 1988), p. 144.

T A B L E 4-7
Some Dimensions of Maladjustment Measured by the MMPI

Depression	Sadness, despondency, feelings of worthlessness.
Paranoia	Extreme suspiciousness and distrust of others; belief that others intend one harm.
Hypomania	Hyperactivity; inability to concentrate on things for any length of time.
Psychasthenia	Being subject to strong fears and compulsions.
Psychopathic deviance	Lacking a conscience, having little regard for the feelings of others, and getting into trouble frequently.

ABILITY AND PERSONALITY

With regard to social traits, high-SE individuals are less interested in seeking approval from others, less likely to model others, and less affected by others' feedback. Also, because they accept and feel good about themselves, high SE's tend to be more accepting of others. With regard to motives, high SE's are driven to meet their own self-set goals, which tend to be high and resistant to change. With regard to personal conceptions, high SE's tend to have an internal locus of control and to expect to succeed, regardless of the context. Finally, high self esteem tends to be correlated with emotional adjustment.

Interestingly, research does not always show high-self-esteem people to be better performers. In fact, they are less competent performers under some conditions, as when a little self-doubt is needed to spur the search for additional information.[30] Most studies suggest, however, that people with high self-esteem choose occupations that are compatible with their needs, and self-esteem is generally positively related to job satisfaction.[31]

To understand how the five sets of personality characteristics we have discussed work together, we might picture the personality as a ship. The visible social traits determine what the ship looks like above the waterline. The hidden motives form the ship's propeller, making it move. Personal conceptions form the compass. They locate the ship in its surroundings. Personality dynamics, which can be seen as the ship's rudder, determine the direction in which the ship will move.

What about emotional adjustment? When it's good, the ship sails smoothly. Maladjustment, however, can be seen as a crack in the hull. Atttention must be shifted from the ship's purpose—reaching its destination—to repairing the damage. In terms of the human personality, when people experience serious emotional distress they are likely to divert their energies from productive activity to concern with and attempts to solve the problems that beset them. If the crack in the ship's hull cannot be repaired—if a person cannot solve his problems—the entire enterprise is in trouble.

Measuring Personality Characteristics

We did not need a separate section on measurement when discussing human abilities because, for the most part, testing in this area is relatively straight forward and uncontroversial. There is much less consensus on the best way to measure personality characteristics. In fact, there continues to be substantial debate about the many measurement techniques that have been developed. In this section, we will look briefly at the types of measures employed in this area and at the advantages and disadvantages of each.

self-inventory A measure of personality characteristics that asks the individual to describe herself by means of standardized responses to questionnaire items.

Self-Inventories. By far the most frequently employed personality measure is the **self-inventory**, a paper-and-pencil test in which people are asked to describe themselves by answering questions. Most such inventories present people with a series of descriptions of personality characteristics and ask them to indicate, in a yes-no or agree-disagree format, whether each description applies to them. Although many such measures are developed by using the same mathematical models used by Spearman, these tests differ fundamentally from ability tests in that the items on personality measures have no right or wrong answers.

---

[30] H. M. Weiss and P. A. Knight, "The Utility of Humility: Self-Esteem, Information Search, and Problem Solving Efficiency," *Organizational Behavior and Human Performance* 25 (1980), 216–23.

[31] E. A. Locke, "The Nature and Causes of Job Satisfaction," in *Organizational and Industrial Psychology,* ed. M. D. Dunnette, (Chicago: Rand McNally, 1976), pp. 515–612.

The main advantage of the personality inventory is that it is quick and efficient—it can measure a large number of characteristics with a minimum of expense. Moreover, experts in the area of personality assessment generally agree that it is the best general approach to measuring personality. There are some drawbacks, however, to self-inventories.[32]

The main problem, referred to as **social desirability bias**, is that a person's responses may reflect not what she is actually like but what she would like you to think she is like. For example, in one study a researcher asked people to rate 140 items on the extent to which the items were socially desirable, that is, to indicate whether most people would think that having a given characteristic was good rather than bad.[33] The same researcher then asked another group of people to respond to the items and found a very high (.87) correlation between the rated desirability of the item and the likelihood that people would say yes when asked if it applied to them. This result could come about in several ways. Of course, it could be that most people actually have the desirable qualities they attribute to themselves. Most researchers think, however, that people who respond in this manner are just not being frank. It could also be that people simply do not know themselves very well and think they are providing factual responses when they are not. Both lack of self-knowledge and lack of frankness will present an inaccurate picture of an individual's personality.

Note that this type of bias cannot affect ability tests. That is, even though a person might wish to pass himself off as having strong mathematical ability, if he is actually weak in this ability, he will be unable to make a good showing. The social-desirability problem, probably more than any other, has fueled the perceived need for other approaches to measuring personality.

Observational Techniques. The observational technique asks one person to describe another. This kind of test may be conducted in a contrived setting, as when an interviewer talks with a person he has never met before and then makes a judgment as to what the person is like. Or observations may be made in everyday situations: a researcher may ask people who should know a particular person well (e.g., best friends, former boss, co-workers) to describe him. This method runs into trouble, however, because observers' judgments depend on their own abilities or personality characteristics, and different observers may make different judgments about the same person. This is especially likely when the observer has had only limited opportunity to interact with the person being rated. Suppose, for example, that a university professor is asked by a student's prospective employer to describe the student's personality, when the student has never participated in class discussions or met with the teacher outside of class. Under these conditions, the person doing the rating hasn't the vaguest idea of what the person being rated is truly like.

The employment interview is a good example of an observational technique for measuring personality. For example, at Mobil Oil Company, interviewers are trained to look for ten characteristics—among them initiative and flexibility—that signal a good fit with the company culture. Thomas Padden, a manager of the heavy-products staff at Mobil suggests that in looking for initiative the interviewer does not simply check off the number of extra-curricular activities the person reports, but tries to uncover the nature of those activities. For example, the interviewer might ask someone who listed "member of student government" whether he typed the minutes of the group's meetings or negotiated with college

[32] Nunnally, *Psychometric Theory*.

[33] A. L. Edwards, *The Measurement of Personality Traits by Scales and Inventories* (New York: Holt, 1970).

trustees. Either of these jobs might have been part of student government, but only one of them has implications for initiative.[34]

Projective Tests. You will recall that a significant difference between Freud's and Guilford's approaches to personality lay in Freud's assumption that important aspects of personality reside in the unconscious and that as a result the person is not aware of them. If Freud was right, the self-inventory is not a valid method of personality measurement. In addition, if people approach the world from behind the shield of the defense mechanisms described in Table 4-3, it is unlikely that untrained observers can offer us very helpful information about what a person is really like. In fact, although they reject many of Freud's other ideas, some personality researchers insist that most people have little consciousness, or awareness, of their true personalities. Thus because this belief casts doubt on the utility of self-inventories and untrained observations as measurement methods, some other technique must be used. For these researchers, projective measures of personality are the solution.

Projective measurement techniques are based on the notion that when people are presented with an ambiguous situation, one that is open to many different interpretations, they will project aspects of their own personalities into such situations. Thus in asking people to provide meaning to a vague stimulus of some sort, the **projective test** theoretically induces them to reveal themselves to a trained observer. Among the several types of projective measures, the Rorschach ink-blot test is probably the most famous. In this test, a person is shown a series of patterns (literally created by spreading ink on a surface and then blotting it) and asked to tell the interviewer what she sees in these patterns, or what images they call to her mind. In the Thematic Apperception Test (TAT), another widely used projective device, people are asked to write stories about pictures of people in different types of social situations. The stories are then judged in terms of the presence of several themes, such as the need for power, achievement, or affiliation.

Although there are scattered findings that particular techniques are useful for particular purposes, the evidence in general suggests that projective techniques do not provide valid measures of personality.[35] Because most of these techniques depend heavily on the subjective opinion of the test administrator, few projective devices are well standardized. Indeed, some investigators have suggested that the interpretations of these tests reveal more about the examiner than about the examinee. For example, one study showed that examiners who were rated as hostile by their colleagues were more likely than others to interpret subjects' ink-blot responses as reflecting hostility.[36] This kind of problem severely reduces the reliability of judgments across different measurements. For example, different judges may evaluate the same person in different ways.

In view of the many weaknesses in the measurement techniques we have discussed, how should we go about measuring personality? Our classification of personality characteristics can help us fit measurement methods to a number of the characteristic we are interested in. For example, social traits, which often represent the surface layer of personality, reflect the way a person appears to others. This class of variables therefore might be best assessed through observations by other people.

projective test A measure of personality in which individuals are asked to assign meaning to an ambiguous stimulus. Unconscious aspects of the personality are inferred from the person's responses.

---

[34] "The New Job Interview: Show Thyself," *Wall Street Journal*, December 4, 1989, p. B4.

[35] D. C. McClelland and R. E. Boyatzis, "Leadership Motive Patterns and Long-Term Success," *Journal of Applied Psychology* 67 (1982), 737–43.

[36] J. Masling, "The Influence of Personal and Situational Variables in Projective Testing," *Psychological Bulletin* 57 (1960), 65–85.

Personal conceptions represent an individual's beliefs, and these are probably best assessed by self-reports. Self-reports may also be appropriate to measure motives, which reflect an individual's underlying drives and needs. However, to guard against social desirability bias, we might want to use projective measures in addition to self-inventories. It is generally a good idea, in attempting to assess personality, to use multiple methods. If the judgments derived from multiple measures are not in agreement, you may be hard pressed to determine which is correct. On the other hand, if such judgments are in agreement, you can be reasonably confident that the picture painted of the individual truly reflects that individual rather than the painter, the brush, or the canvas.

Usefulness of Personality Testing

POINT TO STRESS
Considerable research indicates situation variables (e.g., time pressure, job structure) account for far more variance in explaining human behavior than do personality variables. These findings have led many researchers to ask, why measure personality at all?

Many companies, including General Motors, American Cyanamid, J. C. Penney, and Westinghouse, rely heavily on personality assessment programs to evaluate and promote employees. Many other firms use such programs as screens for initial hiring. For example, American Multi Cinema (AMC), the third largest theater chain in America, looks for individuals with "kinetic energy, emotional maturity, and the ability to deal with large numbers of people in a fairly chaotic situation."[37] Despite their widespread use in industry, however, the usefulness of personality measures in the explanation and prediction of human behavior has been criticized on several counts.

In Chapter 3 we discussed the difference between construct validity and criterion-related validity. Many of the measures of personality characteristics that we have examined have been attacked on the grounds of not achieving either type of validity. In terms of the constructs themselves, Walter Mischel has stated that "the assumption of massive behavioral similarity across diverse situations is untenable."[38] Rather, according to Mischel, people act differently in different situations. If someone could exhibit an internal locus of control when dealing with friends and family problems but exhibit an external locus of control when dealing with problems at work, does it even make sense to talk about a generalized notion of this concept?

With regard to criterion-related validity, after reviewing studies of the use of personality measures in personnel selection, Guion and Gottier concluded that "it is difficult, in the face of this summary, to advocate with a clear conscience the use of personality measures in most situations as a basis for making employment decisions about people."[39] This same conclusion was reached more recently by Landy.[40] The low criterion validity of personality measures is particularly noticeable when we compare the validity of these measures with tests of cognitive or physical abilities needed to perform a particular function. Even as predictors of work attitudes, personality variables have served mainly as accessories to other theories and have rarely accounted for a great deal of variance in employee reactions to work.[41] As a whole, especially in organizational contexts, personality variables have simply not had a great history of success in terms of predicting the kinds of outcomes managers are most interested in.

Despite these criticisms, others have held that the problem with research on personality in organizational contexts is not with the conceptions of personality

---

[37] "Can You Pass the Job Test?" *Newsweek*, May 5, 1986, pp. 46–51.
[38] Walter Mischel, *Personality and Assessment* (New York: John Wiley, 1968), p. 295.
[39] R. M. Guion and R. F. Gottier, "Validity of Personality Measures in Personnel Selection," *Personnel Psychology* 18 (1965), 135–64.
[40] F. J. Landy, *The Psychology of Work Behavior*, 4th ed. (New York: Free Press, 1985), p. 186.
[41] T. R. Mitchell, "Organizational Behavior," in *Annual Review of Psychology*, vol. 30, ed. M. R. Rosenzweig and L. W. Porter (Palo Alto, Calif.: Annual Reviews Institute, 1979), p. 70–119.

characteristics themselves. The problem, they have argued, lies in the way theory and data have been used. In the first place, the fact that most organizational research is conducted at one level of a single organization influences the degree to which individual differences in personality can be observed. According to these investigators, the recruitment-selection-training-retention cycle characteristic of most organizations tends to produce a group of employees who are rather similar in personal style. People who are widely different from the status quo, according to this perspective, either (a) are not attracted to the organization in the first place, (b) are never hired, (c) conform if they do get in, or (d) quit or are fired if they fail to conform. This restriction in range precludes the strong manifestation of differences attributable to personality.[42] A similar problem exists in laboratory studies where participants are often students of the same age and social class, come from the same geographic area, and may be majoring in the same subject. Some research suggests that personality characteristics serve as better predictors when the person being observed is exposed to different tasks and contexts, and all of his responses are summed across all observations.[43]

Other defenders of the role of personality characteristics have criticized the methodology of organizational personality research. Howard Weiss and Seymour Adler, for example, have noted that most research is either cross-sectional (i.e., all the data are collected at one time) or of relatively short duration.[44] The subtle influences of personality on behavior are probably best captured in longitudinal research that follows people over relatively long periods of time. Indeed, one study that runs counter to the vast majority of the literature on personality has shown that predictability on the basis of personality characteristics increases over time.[45] In some rare cases, successful predictions on the basis of such characteristics have been made as far as twenty years into the future.[46] Weiss and Adler also note that the theoretical rationale for including most personality variables in studies is weaker than it is for including situational variables such as goals or incentives.

Thus there is no consensus on how to measure personality characteristics. Moreover, the evidence for the predictive value of many of these measures is not impressive. This poor showing is probably best attributed to a lack of conceptual development regarding which characterisitics are likely to predict which outcomes and in what ways.

The five-fold breakdown we use here to classify different approaches to personality may help us bring everything together. First, in terms of the sheer volume of personality characteristics, this breakdown may make it easier for you to evaluate proposals for new characteristics. It may also help you to appreciate the relationship of a new personality characteristic to established ones.

Figure 4-4 offers a conceptual base from which to understand personality's effect on work outcomes. Personality dynamics, especially as revealed through motives and personal conceptions, probably have their greatest impact on motivation, in that they provide and direct the person's available energy for performing the task. Certain social traits, on the other hand, act more like specific facets of mental ability and physical or motor skills in that they are requirements for specific aspects of the job. For example, extroversion and an outgoing per-

---

[42] B. Schneider, "Interactional Psychology and Organizational Behavior," in *Research in Organizational Behavior*, ed. B. M. Staw and L. L. Cummings (Greenwich, Conn.: JAI Press, 1983), pp. 49–81.

[43] Seymour Adler and Howard M. Weiss, "Criterion Aggregation in Personality Research: Self-Esteem and Goal Setting" (Paper presented at the Ninetieth Annual Convention of the American Psychological Association, Washington, D.C., 1983).

[44] H. M. Weiss and S. Adler, "Personality and Organizational Behavior," in *Research in Organizational Behavior*, pp. 191–236.

[45] J. R. Hinrichs, "An Eight Year Follow-Up of a Management Assessment Center," *Journal of Applied Psychology* 63 (1978), 596–601.

[46] McClelland and Boyatzis, "Leadership Motive Patterns," p. 742.

FIGURE 4-4

How Personality Characteristics Combine to Affect Motivation, Skills, and Concentration

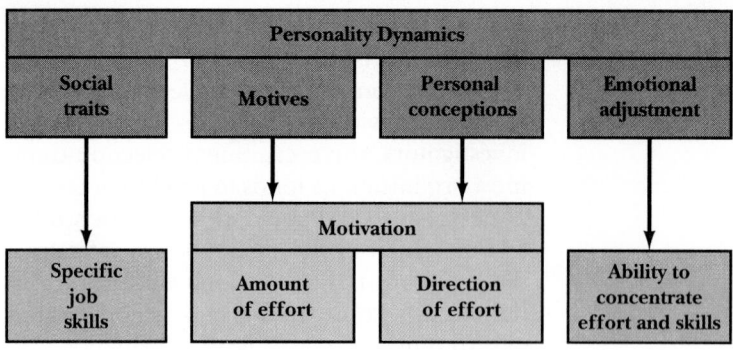

sonality may have a direct positive effect on how well a salesperson can do her job. Finally, the degree to which a person can concentrate her attention on the task at hand will be a function of her adjustment. Even a mild degree of maladjustment may distract attention from the task and thus from performance, especially in complex jobs.

ABILITY AND PERSONALITY IN TASK PERFORMANCE

Performance is a function of the interaction between ability and motivation, particularly as the latter is influenced by personality factors (see Figure 4-5). The relation between ability and motivation seems quite clear. Even if you have the ability to do a particular job you may not do it well unless you are motivated to do it. Conversely, you may have all the motivation in the world but if you lack the ability you will not do well. The relations between these two factors and the many varying personality characteristics that we have been able only to sample here—for instance, thinking versus intuiting or having high or low self-esteem—are much more complicated and often difficult to assess.

There is certainly anecdotal evidence to support the notion that personality characteristics interact with motives and abilities to affect performance. See, for example, the "In Practice" box, which describes the role of the commander of

FIGURE 4-5

How Abilities and Personality Characteristics Combine to Affect Job Performance

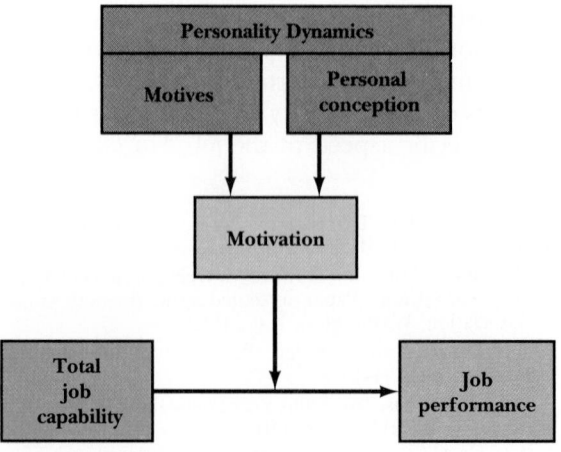

106

ABILITY AND PERSONALITY

The General's Resume

It was perhaps one of the most complex jobs ever designed. The head of the Allied Forces assembled to retake Kuwait had to orchestrate a war machine made up of forces from 28 different nations. This included 675,000 troops, hundreds of ships, and thousands of airplanes and tanks that had to work in a tightly coordinated fashion. What were the abilities, experiences, and personality characteristics of the man assigned the task, General H. Norman Schwarzkopf, and how did these match the task's requirements?

First and foremost, the leader of this effort needed a high degree of cognitive ability. Schwarzkopf's IQ is over 170 and he graduated in the top 10 percent of his class at West Point. When he was head of the U.S. Central Command in North Africa Schwarzkopf foresaw the Arab aggression in the Middle East. In fact, by 1983 he had already designed contingency plans (later called Operation Desert Storm) for a campaign to retake a Gulf state from an invading force. A mere five days before Saddam Hussein launched his attack in August 1990, Schwarzkopf and his staff were actually running an exercise predicated on the possibility that Iraq might overrun Kuwait. "Initially," said one British Commander, "we were taken aback by his gung-ho appearance, but in a very short time we came to realize that here was a highly intelligent soldier—a skilled planner."

Managing the diversity of the allied coalition also required interpersonal skills and sensitivity of the highest order. Schwarzkopf had to manage the four branches of the U.S. military, which are often rivals, as well as the forces of 28 foreign nations. Many of these countries, like Syria and Britain, were not on the best of terms with each other. Schwarzkopf's early years in Iran, where his father worked for the Central Intelligence Agency, gave him a sensitivity to Middle Eastern traditions and values. And when his family later moved to Europe, he attended foreign schools and learned German and French. These early experiences, along with his own later military assignments abroad helped prepare him for managing the multinational forces arrayed against Iraq.

Finally, Schwarzkopf's high self-esteem and need for achievement fitted him well for command. He decided to become a general when he was ten years old and friends from his West Point days still marvel at his single-minded ambition. According to a former roommate, retired General Leroy Suddath, Schwarzkopf "saw himself as a successor to Alexander the Great, and we didn't laugh when he said it."* Indeed, Schwarzkopf's abiding certitude and bristling self-assurance,

* Stormin' Norman on Top, *Time*, February 4, 1991, pp. 28–30.

General Norman Schwarzkopf brought to his role as Allied commander not only a mix of high intelligence, strong interpersonal skills, sensitivity to others, and a broad range of experience but an ability to obsess long enough to plan an effective strategy tempered by the knowledge of when enough is enough. "I thought about this plan every waking and sleeping moment," he says, and he made repeated modifications in his strategy. "But there comes a point where. . . . You say, OK, that's it." The effectiveness of Schwarzkopf's strategy and his implementation of it brought the war in the Persian Gulf to a close with a very small number of casualties. A miracle, he said. "but it will never be miraculous for the families of those people." Source: *Tom Mathews with C. S. Manegold and Thomas M. DeFrank, "A Soldier of Conscience,"* Newsweek, *March 11, 1991.*

in combination with his commanding physical presence—6'4", 245 lbs.—immediately strikes observers.

Schwarzkopf, a student of military history, has focused particularly on the Battle of Cannae, where in 216 Hannibal crushed the forces of Rome in the first real war of annihilation. History may record Operation Desert Storm as a mismatch of forces in the same class as that at Cannae. This battlefield mismatch, however, was also due in part to the excellent match between Schwarzkopf's abilities, experiences, personality, and the job he was assigned.

the Allied Forces in the Persian Gulf War. But we also have a considerable amount of empirical evidence.[47] Consider the findings of one study, diagrammed in Figure 4-6, which examined the relative success of salespersons who varied in terms of both ability and self-esteem. As you can see, salespeople who were not very skilled made low commissions regardless of whether their self-esteem was low or high.

[47] J. R. Hollenbeck and E. M. Whitener, "Reclaiming Personality Traits for Personnel Selection: Self-

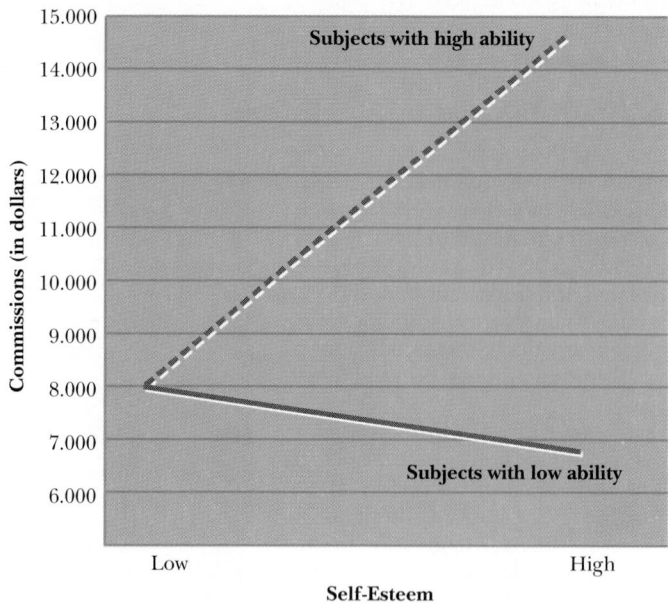

For the salespeople who were extremely skilled, however, self-esteem made a big difference. High-ability people with low self-esteem made no more in sales commissions than the low-ability salespeople did. On the other hand, high-ability people with high self-esteem earned nearly twice as much in commissions as those with low self-esteem.

Here again ability is the bottom line. If you lack the ability to do the job you won't do well. But if you have the ability to do the job, other factors may affect how well you do it. Together with other research that has supported the interactive relationship between ability and personality,[48] these studies suggest that measures of personality characteristics can indeed be useful if they are used in conjunction with measures of ability.

A good example of how ability and motivation interact can be seen in the increasing "foreignization" of U.S. schools of engineering. Becoming an engineer, especially at the Ph.D. level, takes both a high degree of cognitive ability and a lot of motivation, for years of training are required. Moreover, because yearly stipends for graduate training are so low ($12,000–$15,000) and opportunities for engineers with four-year degrees are numerous, most high-ability U.S. students opt for jobs in private industry, foregoing future training. High-ability foreigners who make it to the U.S. are much more motivated to pursue the Ph.D., despite poor funding. As a result, according to Iowa State University president, Gordon Eaton, soon 75 to 93 percent of the faculties of U.S. engineering schools will be foreign born.[49]

Interestingly, recent research has suggested that it is useful to distinguish between measures of a person's maximum level of performance and that person's typical level of performance. For example, in a study of cash-register operators, Paul Sackett and his colleagues obtained both a measure of typical performance (an objective measure of average daily processing speed) and a measure of max-

Esteem as an Illustrative Case," *Journal of Management* 14 (1988), 81–91 and J. R. Hollenbeck, A. P. Brief, E. M. Whitener and K. Pauli, "A Note on the Interaction of Personality and Aptitude in Personnel Selection," *Journal of Management* 14 (1988), 441–50.

[48] E. G. French, "The Interaction of Achievement Motivation and Ability in Problem Solving Success," *Journal of Abnormal and Social Psychology* 7 (1958), 306–9; and D. Kipnis, "A non-cognitive correlate of performance among Lower Aptitude Men," *Journal of Applied Psychology* 46 (1962), 76–80.

[49] *Time*, "Wanted: Fresh, Homegrown Talent," January 11, 1987, p. 65.

After reviewing performance data for several months, a manager finds that her data-entry people frequently performed at high-levels but that their typical performance was one-third less than their best. Ask, how would you diagnose these differences? Why didn't these people work at their maximum all the time? Are the differences attributable to motivation, physical abilities, work flow problems, fatigue? Prescribing and taking actions to move these people closer to their maximum performance depends on answers to these questions.

imum performance (a timed test of speed based on a standardized package of 25 items). When the data were analyzed, these researchers found that there was in fact very little relationship between an operator's maximum and typical performance.

This distinction may help untangle the effects of personality and ability on performance. Sackett and colleagues note that "it is reasonable to view the maximum performance measure as highly ability loaded, whereas motivational factors play a larger role in long term typical performance."[50]

Figure 4-7, which combines and expands on Figures 4-3, 4-4, and 4-5, summarizes what we have discussed in this chapter. The diagram shows that both maximum and typical performance levels are a function of one's motivation and total job capability. When both motivation and capability are high, typical performance will come very close to maximum performance, but when motivation is low, typical performance will be something substantially less than maximum performance. Typical performance will also be significantly less than maximum performance if emotional maladjustment detracts from the individual's ability to concentrate on the task. Finally, the figure reveals that other aspects of personality—social traits, motives, and personal conceptions—can influence the in-

[50] P. R. Sackett, S. Zedeck and L. Fogli, "Relations Between Measures of Typical and Maximum Performance," *Journal of Applied Psychology* 73 (1988), 482–86.

F I G U R E 4-7

How Individual Differences Combine to Affect Performance

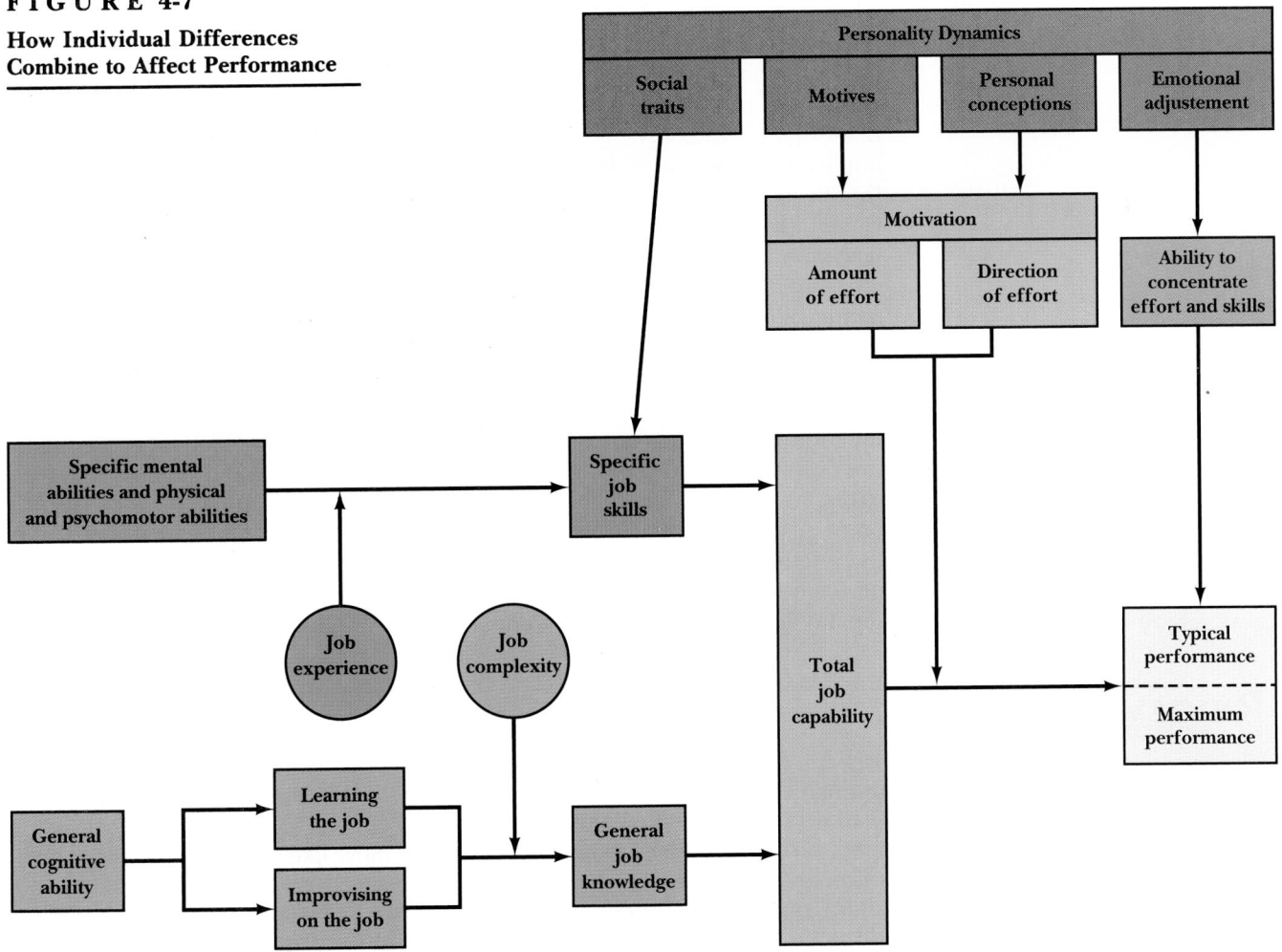

APPLYING THE DIAGNOSTIC MODEL
Diagnostic Question 12: How do we know what abilities and personality characteristics this jobholder possesses? How should we go about measuring these if we are unsure?

dividual's motivation in terms of both the person's level of effort and the way he will expend that effort.

An individual's total job capability is a function of her general job knowledge and her specific job skills. Specific job skills are primarily a function of specific facets of mental ability as well as specific physical and motor abilities that may be needed for particular jobs. Although personality plays a smaller role here, certain social traits may be required for certain jobs.

Finally, general cognitive ability (see the lower left-hand portion of Figure 4-7) influences general job knowledge by helping the individual to learn the job and to improvise on the job when nonroutine circumstances crop up. Job complexity moderates the relationship between general cognitive ability and job knowledge. As the job becomes more complex in terms of planning, decision making, and judgment, general cognitive ability becomes more important. For jobs that are low in complexity, general cognitive ability is less important. As the figure shows, under these conditions job experience becomes a more important influence on job knowledge.

Thus individual differences affect the behaviors of people in many ways. The manager who ignores these differences or assumes that everyone else is just like him is making a very big mistake that could be costly not only to him but to his company and his co-workers.

SUMMARY

Individuals differ on a number of different dimensions, and successfully taking advantage of this fact is essential to the effective control of organizational behavior. For example, individuals differ in their *physical* and *psychomotor abilities*, and there are many job situations in which people who lack the necessary abilities perform poorly and put themselves and others at risk for injuries. *General cognitive ability* is another characteristic on which individuals differ and one that has important implications for a much wider variety of jobs. Indeed, this characteristic is relevant for any job that requires planning and complex decision making on a daily basis. General cognitive ability also relates to both learning the job and adapting to new situations. Specific facets of cognitive ability, such as *verbal ability*, *quantitative ability*, *deduction*, *reasoning*, *spatial skills*, and *perceptual ability*, are important supplements to general cognitive ability for certain types of jobs.

Individuals also differ in personality characteristics. These differences in *social traits*, *motives*, *personal conceptions*, *emotional adjustment*, and *personality dynamics* often spill over into job-performance differences and have important implications for coordination and control. There is controversy, however, about how to measure personality, and the evidence for the utility of this kind of measure is not as strong as the evidence for the usefulness of ability measures. Further research on the use of personality measures in conjunction with ability measures and on the interaction between personality characteristics and typical versus maximum performance may provide stronger evidence for the usefulness of assessing personality in the organizational context.

REVIEW QUESTIONS

1.   In what ways are the underlying beliefs of companies that use rigorous testing and selection processes much like the underlying beliefs of early researchers like Darwin, Galton, and Binet?

2.   Do you think the mirror image fallacy is more likely to affect our assessments of others' abilities or of their personalities? Are there particular dimensions of ability or classes of personality characteristics that are especially susceptible to this kind of mistaken perception? Explain your answer.

3.   Think of someone you know who was highly successful in her chosen field. What were the important personal characteristics that led to this person's success? Now think of what would have happened if that person had chosen a different line of work. Do you think the person would have been successful no matter what she ventured into, or can you imagine lines of work for which she was poorly suited? How does your answer to this question relate to the selection-versus-placement distinction?

4.   Think of someone who is turned down for a job and told that he was rejected because of (a) his performance on a paper-and-pencil cognitive ability test, (b) an interviewer's assessment of his intelligence and self-esteem, (c) a projective test of his need for affiliation and need for achievement, (d) his responses to a personality inventory, or (e) his answers on a paper-and-pencil honesty test. What differential reactions would you expect from this person? Explain your answer.

5.   One of the authors knows of a firm that hands out several different kinds of tests to prospective employees but never actually scores them before making employment decisions. What message is indirectly being sent to applicants by firms that employ rigorous selection testing, and how might the intention to send such a message explain this company's behavior?

6.   Some nations, for instance Japan, have federally controlled education, which results in educational standardization. They also promote a clear hierarchy of primary schools, secondary schools, and universities, arranged in order of their prestige and quality. Other nations, including the United States, leave control over education to state and local authorities. How might this difference in education policies lead to differential needs on the part of employers in the area of personnel testing?

DIAGNOSTIC QUESTIONS

When attempting to come to grips with individual differences in resolving organizational problems, it may help to ask the following diagnostic questions.

1.   Are there any special features of this job that call for high levels of specific physical or motor abilities? Does the jobholder possess these abilities?

2.   Are there any features of this job that require specific facets of mental ability? Does the jobholder possess these abilities?

3.   How much cognitive ability does the jobholder have? How will this influence the speed with which he will learn the job?

4.   How much improvising will the person have to do on this job? Does the jobholder have enough general cognitive ability to adapt to changing circumstances?

5.   If the jobholder lacks general cognitive ability, does he have sufficient work experience to make up for this deficiency?

6.   How complex is this job? If it is highly complex, does the jobholder have enough cognitive ability to handle this complexity?

7.   Are there any features of this job that require certain types of social traits? Does the holder of this job possess those traits?

8.   What motives seem to drive this person? How do these motives influence the amount of energy that she brings to the task?

9.   What are the major personal conceptions that guide this person? How do they influence the way he directs his energy?

10.   Are there any signs of personality maladjustments that may be hindering this person's ability to concentrate on the task?

11.   What is the relation between the person's typical and maximum level of performance?

12.   How do we know what abilities and personality characteristics this jobholder possesses? How should we go about measuring these if we are unsure?

EXERCISE 4-1

MEASURING PERSONALITY DIFFERENCES[1]

JOHN E. OLIVER, *Valdosa State College*
GARY B. ROBERTS, *Kennesaw College*

Individual differences in personality characteristics such as social traits and personal conceptions affect the way people think, feel, and behave. The behavior of people in organizations is particularly likely to be affected by the four dimensions of *uncertainty avoidance*, *empathy-aggression*, *individualism-collectivism*, and *power distance*. If you have already read Chapter 19, you will recognize these as the dimensions Hofstede uses to differentiate organizational behaviors across cultures. In the exercise for that chapter we will take another look at these dimensions of human behavior.

In this exercise, after assessing yourself in terms of these four dimensions, you'll meet with class members who have assessed themselves as similar to you. You'll share your self-perceptions and learn how the four characteristics can affect people's satisfaction and performance in organizations. Then you'll meet with students who have assessed themselves as different from you and learn how individual differences can affect the way people work together.

STEP ONE: PRE-CLASS PREPARATION

To prepare for class, read the entire exercise. Next, find out where you stand on the four behavioral dimensions by completing the Personality Dimensions Questionnaire (Exhibit 4-1). As you respond to the questions, keep in mind that *there are no right or wrong answers*. This is a measure of individual differences, not a test of ability or knowledge. Just respond as honestly as you can. When you've completed the questionnaire, score your answers using the key at the end of this exercise.

STEP TWO: DISCUSSIONS IN SIMILAR GROUPS

The class session will begin by dividing into groups of people who have similar scores on the Personality Dimensions Questionnaire. Your instructor will set up four types of groups—one type to represent each of the four dimensions. You'll join a group for the dimension on which you get the highest total. Thus if your highest score is on power distance, you'll join a PD group. Suppose, however, that you have two or more scores—say

empathy-aggression and power distance—that are equally high. In this case, you would join either a PD or an EA group, depending on which one has the fewest members. The total number of groups will vary, of course, depending on the size of your class, but in general each group should have between four and six members.

Do not be concerned about your actual scores on the four dimensions. It's not necessarily good to be high on one, or low on another. What we're interested in here is how people who are *either alike* or *different from* each other interact and work together.

The task of every group is to answer the following questions:

1. Given that we have all rated ourselves similarly on X dimension, what do we think about this predisposition that we share? Are there good things about it? Are there bad things about it?
2. How does having this predisposition and the corresponding strengths and weaknesses that we've defined affect our behavior in the group? Our satisfaction with the group interaction?

Important: If your class will do Exercise 19-1 later in the term, be sure to make a list of the members of your group and to keep notes on your discussions in this step of Exercise 4-1. Later, in Exercise 19-1 you'll reconvene in the same groups to continue these discussions.

STEP THREE: DISCUSSIONS IN DISSIMILAR GROUPS

Now the class should divide into new groups of four to six members in which each of the four dimensions is represented by at least one high-scoring member. These

[1] The text of this exercise is adapted with the authors' permission from an exercise entitled "Personality Traits and Organizational Cultures: Two one-hour experiential learning exercises." The Personality Dimensions Questionnaire appearing in this exercise was developed by John Wagner and is based on dimensions described in Geert Hofstede, *Culture's Consequences: International Differences in Work-Related Values* (Beverly Hills, CA: Sage, 1980).

EXHIBIT 4-1
Personality Dimensions Questionnaire

Indicate whether you agree with each of the following statements by circling the appropriate number: 1 = Strongly Disagree, 2 = Disagree, 3 = Slightly Disagree, 4 = Neither Agree Nor Disagree, 5 = Slightly Agree, 6 = Agree, 7 = Strongly Agree.

1. I enjoy going to new and unfamiliar places.

   [1] [2] [3] [4] [5] [6] [7]

2. In an organization, subordinates should be allowed to participate in decision making.

   [1] [2] [3] [4] [5] [6] [7]

3. People are more likely to succeed in life if they are dominant and aggressive.

   [1] [2] [3] [4] [5] [6] [7]

4. Working alone is better than working in a group.

   [1] [2] [3] [4] [5] [6] [7]

5. People are more likely to succeed in life if they are compassionate and understanding.

   [1] [2] [3] [4] [5] [6] [7]

6. People in a group should do their best to cooperate with each other instead of trying to work things out on their own.

   [1] [2] [3] [4] [5] [6] [7]

7. People with power should have more status and prestige.

   [1] [2] [3] [4] [5] [6] [7]

8. Safety and security are the most important things in life.

   [1] [2] [3] [4] [5] [6] [7]

9. People in a group should be willing to make sacrifices for the sake of the group's welfare.

   [1] [2] [3] [4] [5] [6] [7]

10. I sympathize with people who are victims.

   [1] [2] [3] [4] [5] [6] [7]

11. Not knowing what is going to happen next makes me feel anxious and uncomfortable.

   [1] [2] [3] [4] [5] [6] [7]

12. Employees should be able to express disagreement with their managers.

   [1] [2] [3] [4] [5] [6] [7]

13. I admire people who are achievers.

   [1] [2] [3] [4] [5] [6] [7]

14. People like me want managers to make all the decisions.

   [1] [2] [3] [4] [5] [6] [7]

15. A group is more productive when its members do what they personally want to do rather than what the group wants them to do.

   [1] [2] [3] [4] [5] [6] [7]

16. I like not knowing what will happen tomorrow.

   [1] [2] [3] [4] [5] [6] [7]

---

groups should begin by having each member report the results of the Step Two discussions. After everyone in the group understands all four dimensions and their corresponding advantages and disadvantages, the group should answer the following questions:

1. Are there areas of unavoidable conflict among the four dimensions? How might such conflicts be managed?
2. Do the four dimensions complement each other in any way? Do some dimensions' advantages make up for other dimensions' disadvantages?

Then the group should select a spokesperson to present a five-minute summary of these answers to the rest of the class.

STEP FOUR: GROUP REPORTS AND CLASS DISCUSSION

In this last step, the spokespersons will present their summaries, and the class should look for similarities and differences among the conclusions that the various Step Three groups have reached. Your instructor will then share his or her personal observations of how the members of the similar and dissimilar groups worked together. Finally, the class should discuss how the individual differences examined in this exercise may affect organizational behavior. Here are some questions to consider in this discussion:

1. Did you see the advantages and disadvantages you associated with the different dimensions actually

influence work in the groups created in Step Three? What positive things happened? What negative things occurred?

2. What might be done to help people with different characteristics, attitudes, and beliefs work together? In what ways might the four personality dimensions examined here influence the management of organizational behavior?

CONCLUSION

Knowledge of individual differences such as those examined in this exercise can inform personnel selection and placement practices. Intelligently applied, it can help reduce absenteeism and turnover, and it can strengthen communication, motivation, and cooperation in organizations. Remember, though, that before we can use our knowledge about individual differences we must measure these differences properly, and our measures must be both reliable and valid. Otherwise, our personnel decisions may be inaccurate and subject to criticism or even legal action. Finally, it is crucial to keep in mind that many factors influence attitudes and performance at work: general and specific abilities, intelligence, job knowledge, personality characteristics,

and motivation (see Chapter 7) also have important effects on organizational behavior.

SCORING KEY

1. For questionnaire items 1, 2, 4, 5, 10, 12, 15, 16, subtract the number you circled from 8; the result is your score on each of these items. For example, if you circled the number (5) on item 1, your score for that item is 3.

    For questionnaire items 3, 6, 7, 8, 9, 11, 13, and 14, the number you circled is your item score.

2. Add your scores for items 1, 8, 11, and 16 and write the total here: _____. This is your score on *uncertainty avoidance* (UA).

3. Add your scores together for items 3, 5, 10, and 13 and write the total here: _____. This is your score on *empathy-aggression* (EA).

4. Add your scores together for items 4, 6, 9, and 15 and write the total here: _____. This is your score on *individualism-collectivism* (IC).

5. Add your scores together for items 2, 7, 12, and 14 and write the total here: _____. This is your score on *power distance* (PD).

DIAGNOSING ORGANIZATIONAL BEHAVIOR

John Matthews was a young executive at the divisional level of a large corporation. John, like a number of other young men in business throughout the United States, had been selected by higher-ups in his firm to attend a two-and-a-half-week executive development retreat.

The retreat was held at a remote camp in northern Minnesota. Although all of the necessary facilities for an enjoyable vacation were present, the structure and demands of the retreat left little time for relaxing and enjoying the surroundings. John was among 60 male executives who were registered to attend the retreat. They would spend 15 days living, working, and competing with one another.

ORGANIZATION AND ACTIVITIES

The 60 participants were broken down into 5 groups of 12. Each group was provided with a group leader. This leader was a senior corporate executive who had

CASE 4-1

EXECUTIVE RETREAT: A CASE OF GROUP FAILURE*

DONALD D. WHITE, *University of Arkansas*

previously attended the retreat. For 15 days, the men were involved in a variety of academic and athletic activities.

Selected sessions of the retreat were designated for "educational activities." The men participated in seminar sessions designed to deepen their understanding of central management decision making. These sessions involved a limited amount of lecture by either the group leader or a visitor. However, the majority of time devoted to academic pursuits was spent in case studies and a business game. Athletically, a good deal of the men's time was spent in physical fitness training and athletic competition. Finally, a few sessions were conducted along the lines of sensitivity training. (Chapter 14 discusses this method of group intervention).

Although a considerable amount of time was spent in intra-group activities, inter-group competition was

* Reprinted with the author's permission.

also fostered. In particular, groups competed athletically and through the business game.

The remaining portions of this case represent the reflections of John Matthews on his experiences at the executive retreat.

FIRST IMPRESSIONS

It is hard to express the emotions or thoughts that were going through my mind, let alone the minds of others, when I first met the members of my group. Until now, I had been working with business acquaintances in my company's San Francisco office. When I learned of my selection for the retreat and the manner in which it would be conducted, I wondered what my new associates would be like. Would we all remain for the full two and one-half weeks of the retreat? Would I be able to take the criticisms of others? How would our group do athletically and academically? And would the other members of my group resent the fact that I could not participate in the sporting events due to an old knee injury? Subconsciously, I had been establishing the criteria by which I would accept others and they would accept me.

During the first group meeting, I tried to learn the backgrounds of others who were with me. I went through the following process. I tried to find out where the others were from, what their education was, and the kind of experience they had accumulated. I discovered that the level at which one had worked within a firm together with whether or not he had held down a "home office" job were important because they created identification and solidarity between individuals; i.e., financial officers interacted with other financial officers, production managers with production managers, marketing people with others from marketing departments, and so forth. Our group leader made sure that he allowed enough time for all of us to meet each other before he walked in the door.

The group leader was the faculty member who had over-all responsibility for administrative functions in the group. He also graded papers and presentations, conducted all of our counseling sessions and was the all-around nursemaid for the group. Our leader, Mark, was a top-level corporate executive out of New York. This posed an immediate threat to some in the group when they first met him. After a few minutes of informal chitchat, Mark called everyone into a seminar room.

Mark made a low-key introduction of himself and the retreat. He emphasized that to be a success individually at "the camp," everyone had to cooperate and function as a group. He explained that no group always dominated intellectually or athletically. He related that his last group was not especially great in academics or

athletics yet their cumulative scores both in tests and games enabled them to become the top group at the retreat. This allowed certain privileges over other groups. The point Mark kept trying to make was that the men could no longer think of themselves as individuals. "The school theme" he said, "is 'Think—Communicate—Cooperate' and I suggest that you too adopt it as your guiding principle while you are here."

GROUP MEMBERS

The following are my recollections of the other members in our group.

Wally was an older member of the group and became the group student leader. He was a middle-level manager in a large company and had no formal technical training. This may have made him reluctant to assume a leadership role in the group. He appeared to be afraid of hurting other people's feelings even though his actions usually were justified.

Dave also did not have formal technical training; however, he was one of the few who had experience as a corporate president. He was an average student and speaker and above average in his writing ability. He appeared to be obsessed with sex. He called his wife every night, "studied" *The Art of Sensual Response*, and occasionally sniffed some musk oil which he had bought for his wife.

Jim was a financial analyst. He was one of two bachelors and was considered to be the playboy of the group. His goal was just to finish the retreat and get back to his home office.

Bob was a manager of production and operations for a leading producer of men's apparel. He, too, was a bachelor and considered himself to be a "ladykiller." To most of us he appeared to be conceited and boisterous. He claimed to be an authority on most subjects. He was also suspected of cheating in the 25-Mile Jogging Club (cheating on anything was strictly forbidden).

Larry was a director of public relations for a major steel producer. He had a liberal arts background and turned out to be our only distinguished graduate. Although he participated in everything, he never really assumed a leadership role and his contribution to the section was minimal. He was the only one (with the exception of me) who was not able to run a mile and a half in 12 minutes. He was a good speaker but a below average writer.

Rich was an internal financial consultant. He attended Harvard Business School and was later to be considered as one of the better executive prospects at the retreat. He was a good speaker and writer. Although he was very outspoken, he did make a lot of sense. He assumed the leader's role in two major exercises; however, he never did maintain his hold as leader over the group.

Wayne was a personnel director. He was an average student, writer, and speaker. He never did assume a leader's role, possibly because he was the most naive member of the group.

Ollie was a marketing manager and was considered the "country boy" of the group. He was an average student, good speaker, and good writer. He performed many odd jobs for us and was successful in leading us to two victories in athletics.

Gary was an executive vice president for a pipeline supplier whom I thought, at the beginning, would emerge as the leader of the group. He was poor academically, an average writer, and a good speaker. His additional duty was that of athletic chairman. Although he encouraged everyone to run 25 miles (25-Mile Club) during this period, he himself failed to achieve this goal.

Burrell was a personnel and public relations manager. He also was considered to be among the more promising men at the retreat. He was a fair speaker and an average writer. His additional duties were academic chairman and basketball coach. He was the type of guy that, if something were to go wrong, he would be in the middle of it.

Paul was manager for engineering for an electronics manufacturer. He had to spend three days of the first week of the retreat in the infirmary with a virus. This may have been one of the reasons why he was always trying to promote group functions when he got back. One thing I remember in particular about Paul is that he was always complaining about the "developmental rotation" program in his company. The program placed technically trained managers in functional areas other than their own for up to six months to provide them with career broadening. He saw the program as a threat to his own career but failed to see it as a threat to "general managers with no technical expertise." Over-all, Paul was an average speaker, writer, and student.

I, John, did not have formal education for my job as division director of industrial relations. I was an average student and writer and above average speaker. I considered myself to be a harmonizer of the group. I was the only member who was excused from sports because of an injury. Although I disagreed many times with decisions that were made, I usually went along with the group to the end.

The group members lived in three locations during the school. Living in Cabin I were Dave, Wally, Jim, Bob, John, and Larry. Wayne, Burrell, and Paul lived in Cabin II, while Ollie, Rich, and Gary lived in Cabin III. Bob and John generally walked to seminar sessions together as did Wally and Larry, Wayne, Burrell, and Paul and Ollie, Rich, and Gary. Dave and Jim walked separately to the sessions. Rich and Burrell studied together regularly. Larry, Rich, and Paul generally studied together.

GROUP ORGANIZATION AND ACTIVITIES

For convenience, Mark arranged the seating alphabetically around the table (see Exhibit 4-2). There was only one exception; Wally, the designated leader by virtue of age and experience, sat near the front. Following some brief introductions and a few administrative actions, goals of the group were established.

After much haggling about the goals, which ranged from totally idealistic to extremely pragmatic, the group decided on the following goals:

1. Everyone in the group would strive to complete the program and would seek to assure that our group was ranked first among various competing groups.
2. We would strive to be the best in sports.
3. Everyone would run at least 25 miles.
4. We would strive to maintain a harmonious atmosphere in the group.

Of immediate importance to the group was developing athletically rather than academically. (In final group ratings, athletics ranked a very close second to academics in total possible points that could be scored.) In fact, it wasn't until the latter part of the school that the section would come together in academics.

A couple of incidents that occurred during the retreat illustrated the extent of the group's success.

Toward the end of the first week, an entire afternoon was set aside for self-evaluation. The session resembled a sensitivity-training group session. Most groups had lunch followed with a little beer drinking to "loosen things up." After our loosening up we started our discussion. Several comments were made that should have provoked a fiery discussion, but for some reason they never did.

I don't believe we were open that afternoon. We looked at our leadership in academics, but none of us was willing to tell Burrell that he had a weak academic program. None of use would tell Gary that our athletics program was bad and that our group looked worse than most other groups with whom we competed. We all knew these things, but were unwilling to place the blame on anyone. Our group leader must have been totally frustrated at the end of the day. How could a group,

EXHIBIT 4-2
Seating Arrangement, First Day

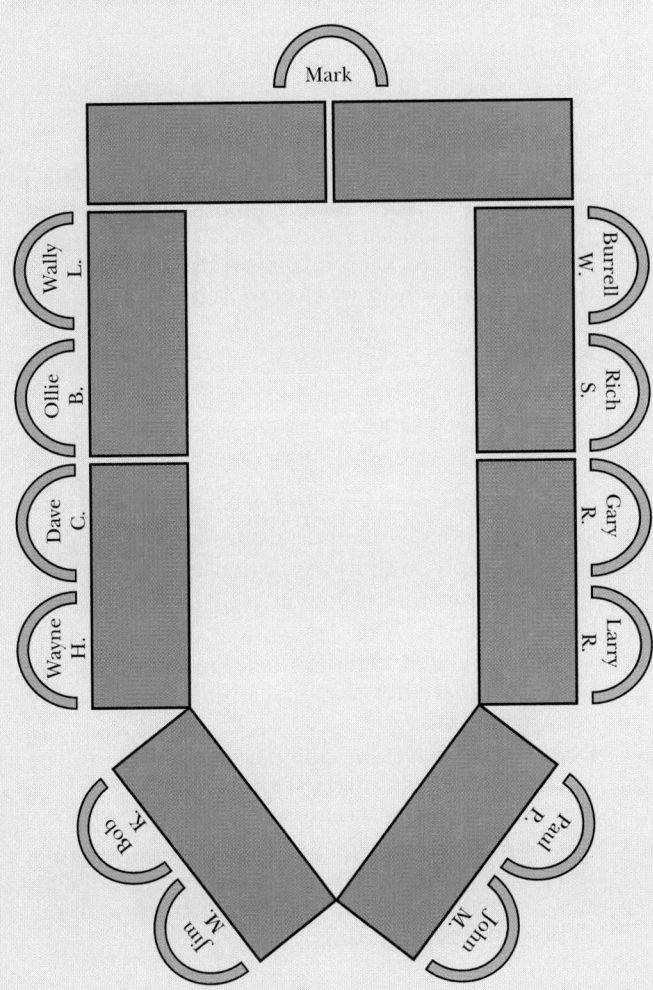

individual specialities in our own organizations, but couldn't muster up the same vitality and enthusiasm to carry forth this synthetically designed group toward goal achievement. Although we wouldn't openly admit it, we were not committed to our goals. Yet, even though we lacked this commitment, we still maintained the goals. Going back to the afternoon encounter session, someone suggested that we revise our goals in light of our successes and failures to date. Even though it was impossible to achieve the original goals set, they still were unchanged!

The seating as depicted in Exhibit 4-3 was the arrangement for the last few days.

EXHIBIT 4-3
Seating Arrangement, Last Few Days

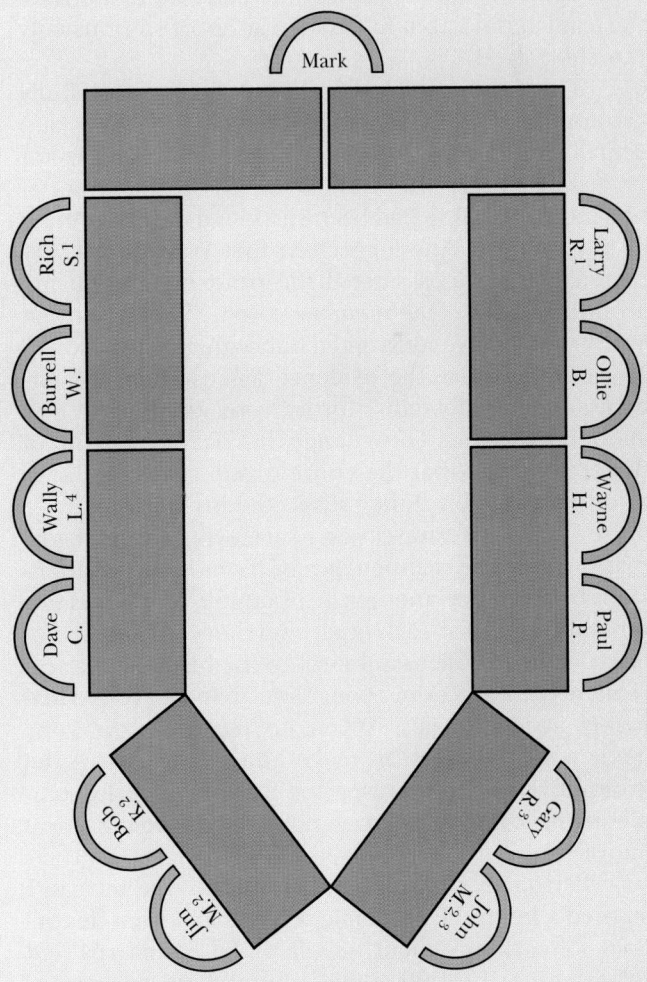

which had such high goals and such mediocre results, have allowed such an opportunity to pass by?

A few days later another group project was scheduled. An obstacle course, intriguingly called "Project X," consisted of a series of tasks to be performed by six people at a time. It was supposed to test the group's ability to recognize the problem, decide on a solution, and carry it out in a 15-minute period. During the break, we tallied our score, 0 for 5. Mark seemed very upset. It was the first time he got upset with the entire group. Larry commented on the episode:

> We didn't see "Project X" or even the rest of the retreat as a life or death situation. In a retreat where no one fails to graduate, it can hardly be considered as a threat to anyone's career if these group goals go unaccomplished.

Personally, I saw us as a group of individuals in search of a real leader. We were all strong in our own

1. These individuals finished in top third of class.
2. These individuals never changed their seats.
3. Gary and John sat next to each other during the last seven days.
4. Wally never did assume the leader's position at the end of the table except when he led the two seminars.

Our group finished the two-and-one-half-week session having accomplished the following: Out of 12 individuals, one finished as a distinguished graduate and a total of three finished in the top third of the class. Ollie made the observation, "Lacking strong leadership in education, we each went our separate ways in trying to wade through all the material."

Our second goal also shared defeat. At the beginning of the program it was felt that our group had a chance to do well in sports. During practice sessions, we appeared to be relatively good. However, practice sessions reflected one characteristic of the group. Generally, we were disorganized, and there was always a lot of joking going on. I believed this carried over to the games and resulted in less than full commitment to winning. Gary would get frustrated and try to motivate the team at times, but his sudden surge of spirit usually was short-lived.

The third goal also fell short of being successfully accomplished. Only six of the members of our group actually finished the 25 miles. Another important factor regarding the 25-Mile Club centered on the ethics of one individual. Bob had been suspected of not running all the miles that he logged. At first Wally and Larry had suspected this, as later all the group members living around Bob did. One member noted, "We all felt that Wally should have confronted him with our suspicions." However, because the evidence against Bob was circumstantial, Wally didn't formally say anything to Bob about the incident. One change that did develop out of the episode was that the entire group ceased to listen to or trust Bob once they suspected his cheating.

The group came closest to achieving the final goal that involved the maintenance of harmonious relationships between one another. An example of this was our mutual respect for each other's territory. As one member stated it, "When Paul went to the hospital, his seat remained vacant even though we didn't have permanently assigned seats. When he returned, everyone made a special effort to make him feel a part of the group. Even when we suspected Bob of not really completing his running, we tried not to make too big a deal out of it."

Personally, I think I got a lot out of the retreat. I learned a lot in the academic sessions and even discovered some things about myself that I hadn't realized before. But, truthfully, I never figured out our group. Sometimes I think it was a near disaster.

When you have read this case, look back at the chapter's diagnostic questions and choose the ones that apply to the case. Then use those questions with the ones that follow in your case analysis.

1. What abilities did each member of John Matthews' group have? How did group members complement one another in terms of their abilities? How did they clash?

2. What abilities did the group's tasks require? Did any of the members have these abilities? Could members without these abilities develop them while working on the group tasks?

3. Did the group members' personalities help or hurt the group's performance? If, on the basis of their personality characteristics, you were to choose which people to keep in the group and which to dismiss, who would you keep? Why?

CASE 4-2

FREIDA MAE JONES

Read Chapter 5's Case 5-1, "Freida Mae Jones." Next, look back at Chapter 4's diagnostic questions and choose the ones that apply to that case. Then use those questions with the ones that follow in your case analysis.

1. What job-related abilities does Freida Mae Jones have? What are her personal strengths? What are her weaknesses?

2. Does Freida have the abilities needed to do the work currently assigned to Paul Koehn? What abilities does such work require?

3. What can a company do to ensure that promotions and job assignments are based on job-relevant abilities? Do you think the Industrialist World Bank of Boston has done this?

CASE 4-3

THE LORDSTOWN PLANT OF GENERAL MOTORS

Read Chapter 17's Case 17-1, "The Lordstown Plant of General Motors." Next, look back at Chapter 4's diagnostic questions and choose the ones that apply to that case. Then use those questions with the ones that follow in this first analysis of the case.

1. How did the backgrounds of the Lordstown employees differ from those of workers in other GM plants? What abilities distinguished Lordstown workers from the others?

2. What abilities did the jobs at Lordstown require? Did these match the abilities of the work force? What were the results of the match or mismatch?

3. Do you think GM used the same management practices at Lordstown as at other assembly plants? If it did, should it have instead adapted its approach to local conditions at Lordstown?

C H A P T E R 5

PERCEPTION AND JUDGMENT

Even before its invasion of Kuwait, Iraq was a former ally of the Soviet Union and certainly was not a pro-western nation. Nevertheless, James Guerin, former CEO of International Signal and Control Group PLC, apparently helped Iraq acquire cluster bombs, bombs that could later have been deployed against the Allied Forces in the Persian Gulf War. When this and other questionable business transactions were revealed, Guerin, who had generally been perceived as an outstanding citizen, was regarded quite differently by people who thought they knew him. How would you reconcile two such different perceptions of the same person? How would you determine which is correct?

TABLE 5-1
Supervisors' and Subordinates' Views of Supervisors' Praise for Good Work

FORM OF RECOGNITION	FREQUENCY WITH WHICH SUPERVISORS SAY THEY GIVE RECOGNITION "VERY OFTEN"	FREQUENCY WITH WHICH EMPLOYEES SAY SUPERVISORS GIVE RECOGNITION "VERY OFTEN"
"A pat on the back"	82%	13%
Sincere and thorough praise	80	14
Training for better job	64	9
Special privileges	52	14
More interesting work	51	5
Added responsibility	48	10

Source: Adapted from R. Likert, *New Patterns in Management* (New York: McGraw-Hill Book Co., 1961), p. 71.

APPLYING THE DIAGNOSTIC MODEL
Diagnostic Question 1: On whose perceptions do you rely to describe and diagnose organizational problems? How might the choice of this person affect the description and diagnosis provided?

TEACHING NOTE
In an effort to reduce the effects of faulty perceptions on employment decisions, the courts are requiring that performance appraisals be based on objective, measurable, and valid information. Can everything that is important be measured quantitatively? Could we just feed information into a computer and let it make the decisions? Point out that this chapter will discuss specific ways of improving perceptual accuracy.

APPLYING THE DIAGNOSTIC MODEL
Diagnostic Question 2: Which of the four types of accuracy is required for specific organizational decisions that must be made based on perceptual data? Such decisions include promotion decisions and decisions about group or individual training.

elevation accuracy The degree to which a rater's assessment of an entire group of people, across a number of different dimensions, reflects the group's true standing on those dimensions.

partly on their sex perceived themselves as less qualified and their performance as lower than it actually was.[5] Finally, what *action* can managers take when those they supervise contest and reject managerial perceptions? For example, in the five years between 1982 and 1987, over eight thousand lawsuits were brought by terminated employees who disagreed with managerial assessments of their performance. In over 80 percent of these cases, the ex-employees were vindicated and received settlements that averaged over a half-million dollars.[6] What actions can managers take to avoid these kinds of settlements and convince others—judges or juries—that these managers' perceptions were valid?

PERCEPTUAL ACCURACY

Judging other people's performance is one of the primary functions of a manager. In learning how to make such judgments as accurately as possible, it is useful to distinguish among different kinds of accuracy because some are more difficult to achieve than others, and each is needed for a different purpose. Four kinds of accuracy can be identified: elevation accuracy, differential elevation accuracy, stereotype accuracy, and differential accuracy.[7]

Table 5-2 shows the "true" scores (i.e., the objective level of performance) of five hypothetical employees on five characteristics. A score of 1 represents a very low standing on the characteristic, and a score of 9 reflects a very high standing. Table 5-3 presents a supervisor's evaluations of each employee on each dimension. Comparing the true, or objective, scores shown in Table 5-2 with the supervisor's ratings shown in Table 5-3—which are, of course, based on the supervisor's perceptions—will help us illustrate the four types of perceptual accuracy.

Elevation accuracy, deals with how high or low the ratings are in general, (hence the term *elevation*).

[5] M. C. Heilman and D. O. Repper, "Intentionally Favored, Unintentionally Harmed? Impact of Sex-Based Preferential Selection on Self-Perceptions and Self-Evaluations," *Journal of Applied Psychology* 72 (1987), 62–68.

[6] J. B. Copeland, "The Revenge of the Fired," *Newsweek*, February 16, 1987, p. 49.

[7] L. J. Cronbach, "Processes Affecting Scores on 'Understanding of Others' and 'Assumed Similarity,' " *Psychological Bulletin* 52 (1955), 177–93.

TABLE 5-2

TABLE 5-2
"True" Scores of Five Workers on Five Different Characteristics

PERSON	TRAITS					TOTAL SCORE FOR PERSON
	Job Knowledge	Interpersonal	Initiative	Creativity	Loyalty	
Bob	7	2	8	5	7	29
John	8	3	7	6	3	27
Steve	9	1	8	4	5	27
Carol	8	3	9	5	4	29
Dean	8	1	8	5	6	28
Average rating for group on characteristic	8	2	8	5	5	

TOTAL SCORE FOR GROUP
140

TABLE 5-3
Supervisor's Ratings of Five Workers on Five Different Characteristics

PERSON	TRAITS					TOTAL RATING FOR PERSON
	Job Knowledge	Interpersonal	Initiative	Creativity	Loyalty	
Bob	9	2	7	6	7	31
John	6	2	8	7	2	25
Steve	7	3	4	2	7	23
Carol	4	2	5	3	2	16
Dean	8	1	8	5	6	28
Average rating for group on characteristic	6.8	2	6.4	4.6	4.8	

TOTAL RATING FOR GROUP
123

This measure reflects the degree to which the sum of all true values added up across both employees and dimensions (this sum is shown in the triangle in Table 5-2) matches the corresponding sum arrived at by the supervisor (shown in the triangle in Table 5-3). This could be thought of as the group's total score. When the two values differ substantially, a type of inaccuracy called leniency-severity may be operating. A leniency bias occurs when a supervisor rates all members of the group on all dimensions higher than they deserve. A severity bias reflects the opposite kind of error. In the data provided here, the supervisor appears to show severity rather than leniency, for the summed ratings in Table 5-3 are lower than the summed true scores in Table 5-2.

Elevation accuracy is particularly important when we need to compare individuals or groups rated by different supervisors. Suppose, for example, that two groups of employees have equal total true scores (in the triangle) but that the very lenient supervisor of one group rates its members significantly higher on the target characteristics, whereas the very tough supervisor of the other group rates its members as quite low. Imagine now that the organization in question has a pay-for-performance plan that rewards people based on their supervisors' ratings. The implementation of this plan might be perceived as quite unfair by those working under the severe supervisor, and over time this inaccuracy could lead to substantial conflict.

differential elevation accuracy The extent to which a rater's assessment of one individual, across a number of dimensions, reflects that person's true standing on those dimensions.

The second type of accuracy, **differential elevation accuracy**, reflects the degree to which the supervisor's ratings are accurate with respect to how high or low individuals are across all dimensions (hence the term *differential elevation*). This determination has implications for the accuracy of rank orders. This type of accuracy deals with how well the sum of each employee's true scores, shown in circles in the last column of Table 5-2, matches the same value as judged by the supervisor, shown in circles in the last column of Table 5-3.

Differential elevation accuracy is important in distinguishing among employees within the same group. Suppose, for example, that Bob, John, Steve, Carol, and Dean, whose scores and ratings are shown in Tables 5-2 and 5-3, were all candidates for promotion to a single, new position. Suppose further that the five dimensions assessed in the tables were considered equally important. Based on the supervisor's ratings, Bob would get the new job. Based on the true scores, however, Bob and Carol are equally qualified and, in fact, all five are very close on the five dimensions measured.

One rating error that has occurred here may be the similar-to-me bias. Our tendency to be attracted to people who are similar to us sometimes colors our ratings of others. Supervisors may unknowingly overestimate well-liked subordinates and, across all dimensions, assign excessively high ratings. Suppose, for example, that the supervisor whose ratings are shown in Table 5-3 is a male. The fact that his ratings of Carol are much lower than her true scores, might lead to charges of sexism and could severely damage the supervisor's credibility.

stereotype accuracy The extent to which a rater's assessment of a group of people, on a single dimension, reflects the group's true standing on that dimension.

The third type of accuracy, **stereotype accuracy**, deals with how accurately the supervisor describes the group as a whole in terms of its strengths and weaknesses.

The term *stereotype* has negative connotations because many stereotypes based on sex or race are innaccurate. However, we use the term here more broadly, recognizing that some group stereotypes may be accurate. For example, it is generally accurate to describe offensive linemen in football as bigger but slower than wide receivers.

Moreover, in the present context, the term *stereotype* focuses attention on general characteristics of a group as a whole rather than any individual within the group. The degree to which the group's true, or average, score on a given dimension (see squares at the bottom of each column in Table 5-2) matches the average rating on that dimension assigned the group by the supervisor (see squares at the bottom of each column in Table 5-3) gives us an index of stereotype accuracy. In this example, the supervisor assesses interpersonal skills accurately but is way off the mark in evaluating job knowledge and initiative.

This kind of accuracy is important when we are trying to diagnose group characteristics. Suppose that we were considering adopting some form of group training. The group described in Table 5-2 is strong in the areas of job knowledge and initiative (scores of 8 out of a possible 9), but weak in interpersonal skills (score of only 2). As we've already noted, the supervisor has accurately perceived this weakness, but he has underestimated the group's strength in initiative. Such a misperception could lead us to devise group training that would waste time focusing on developing skills the group already possessed.

TABLE 5-4		
The Four Kinds of Perceptual Accuracy		
	SUBJECT OF RATING	
RANGE OF RATING	Group	Individual
Across several dimensions	Elevation	Differential elevation
On one dimension	Stereotype	Differential

differential accuracy The extent to which a rater's assessment of one individual, on one single dimension, is reflective of the person's true standing on that one dimension.

The most difficult type of accuracy to achieve, **differential accuracy**, is the extent to which the supervisor correctly assesses different employees on different dimensions (hence the term *differential* accuracy). As you can see from Tables 5-2 and 5-3, this fourth type of accuracy is the degree to which the true score for each individual on each specific dimension matches the supervisor's corresponding rating (as examples, see the numbers in the diamonds in columns 2 and 3 of Tables 5-2 and 5-3). This kind of accuracy is important in diagnosing an individual employee's strengths and weaknesses, as we would need to do if we were considering entering him or her in a career development program.

For example, suppose Steve's grasp of his assignment, willingness to take on new projects, and consistent hard work have come to the attention of a company executive and she has suggested that he be considered for career development. As you can see, Steve's score on job knowledge is the highest of anyone's, and his score on initiative is also very high. He is one of the lowest, however, on interpersonal skills, a dimension on which his supervisor overrates him. Moreover, the supervisor seriously underestimates Steve on initiative and underestimates his job knowledge as well. Steve probably will not be recommended for the career development program if a decision is based on his supervisor's ratings, in particular, of low initiative. If the decision is based on Steve's true ratings, however, the company might decide to give him some help in improving his interpersonal skills and then move him into the program.

Differential accuracy requires a great deal of perceptual skill on the part of the supervisor, and for reasons we will discuss later in this chapter, it is very difficult to achieve. Fortunately, for some decisions this type of accuracy is not always necessary. For example, if the dimensions being assessed are equally important and the ratings are to be used for merit-pay-raise decisions, differential elevation accuracy is all that is needed. If decisions are to be made about which group or team in the company is performing the best and should serve as a model for others, making this identification requires only elevation accuracy. If some type of team development is being planned, then only stereotype accuracy is needed to ensure that the right kind of training program is selected. But if decisions are being made about the strengths and weaknesses of an individual employee, differential accuracy must be achieved. Table 5-4 shows how the four types of accuracy differ in terms of the level at which the rating is made, and the dimensional specificity of the rating.

THE PERCEPTUAL PROCESS

In the next five sections of this chapter we describe a model of the perceptual process that is adapted from the work of Robert Lord.[8] In examining each stage in this process, we will identify and describe the types of factors that

[8] Robert Lord, "An Information Processing Approach to Social Perceptions, Leadership and Behavioral Measurement in Organizations," in *Research in Organizational Behavior*, vol. 7, ed. B. M. Staw and L. L. Cummings (Greenwich, Conn.: JAI Press, 1985), pp. 87–128.

bias accuracy. We will also suggest ways of combatting these biases in order to maximize perceptual accuracy.

Before we explore our perceptual framework and learn what kinds of factors get in the way of perceptual accuracy, let us consider for a moment how human beings process information from the environment. Basically, we process such information in one of two ways, by controlled or by automatic processing. In **controlled processing**, perceivers are aware of the fact that they are processing information. **Automatic processing**, on the other hand, is characterized by a lack of awareness on the part of the perceiver. For example, an interviewer actively listening to a job applicant's answers to specific questions and taking notes is processing information in a controlled mode. However, at the same time that she is listening and taking notes, the interviewer may perceive certain nonverbal behaviors of the applicant (such as not maintaining eye contact, profuse sweating, or other signs of nervousness). Even if the interviewer isn't conscious of these behaviors at the time and makes no formal notes about them, they may still enter into her mind through automatic processing. These impressions, although unrecorded formally, may affect the interviewer's judgment of the job applicant later, when she must make a formal decision about that person. Moreover, because the original perceptions were processed automatically, the interviewer may not be aware of all the information that is affecting the judgment she makes. We will see that in many different stages of the perceptual process, automatic processing greatly increases the likelihood of perceptual inaccuracies.

Figure 5-3 presents a model of the perception and judgment process that will guide our discussion in the next few sections. As the figure shows, there is a stimulus in the objective environment that must be subjectively processed by the observer. After this information is processed, a judgment is eventually reported, and, as the dotted lines indicate, the subjective judgment and the environmental stimulus match if the perception is accurate. Decisions and behaviors based on accurate perceptions are likely to be effective in dealing with the stimulus. Behaviors and decisions that are inaccurate, however, are usually ineffective.

But how do we get to an accurate judgment, one that can yield useful choices and actions? That is the topic of this chapter. As Figure 5-3 shows, before information reaches the stage at which a final judgment is made, it moves through four distinct stages of the perceptual process: attention, organization, interpretation, and retrieval. Let's start with the attention stage. Our five senses are constantly bombarded with so much information that we cannot process even a fraction of it. In the *attention stage* all available information is filtered, with the result that only a tiny portion of it is actually allowed inside the system for further processing. Despite this filtering process, the complexity of the incoming information is still so great that in the second, *organization stage*, it must be further simplified. Specific bits of information are grouped into more meaningful chunks, which reduces processing requirements. In the third, *interpretation stage*, the per-

controlled processing A manner of information processing in which the perceiver is aware that he is processing information.

automatic processing A manner of information processing in which the perceiver is not aware that he is processing information.

AN EXAMPLE:
PHILADELPHIA
INTERNATIONAL AIRPORT
Although one way to heighten the accuracy of our perceptions is to operate solely in the controlled processing mode, maintaining constant vigilance can be very fatiguing. According to a study by the Federal Aviation Administration, vigilance tasks are also stressful, producing irritation in people and decreasing their subsequent attentiveness. In addition, many experiments have demonstrated a "vigilance decrement" that usually occurs within the first 20 to 35 minutes of a vigil. At Philadelphia International Airport, air traffic controllers are not allowed to work for more than an hour without taking a break. (*Psychology Today*, April 1986, p. 48.)

FIGURE 5-3

A Five-stage Model of the Perception and Judgment Process

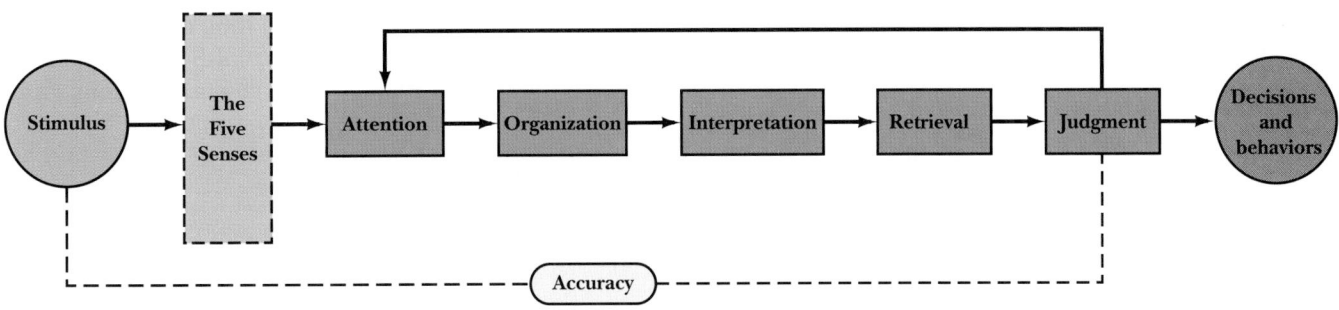

THE PERCEPTUAL PROCESS

ceiver assigns meaning to the information and attempts to determine its implications. In the *retrieval stage*, the fourth stage of the perceptual process, information must be retrieved from memory. Because the input of information and its recall for specific use may sometimes be separated by weeks, months, or even years, much information is probably lost. In the *judgment stage*, whatever information can be retrieved is used to make needed decisions.

ATTENTION STAGE

As we have said earlier, in the **attention stage** the incredible amount of available information is filtered so that some enters the system and some does not. Of all the information that does enter the system, the observer can attend consciously to only a subset. Thus the controlled mode of processing can be used for only a small fraction of the incoming information. As a result, some of what enters the system does so automatically, so that the person may not be aware of its presence.

Although information that is not processed in a controlled manner usually has little effect on subsequent evaluations, this is not always the case. For example, in the initial phase of one experiment, subjects were given a mild electric shock whenever a certain word appeared on a screen placed before them. Over time the subjects began to associate this word with the shock. In the second phase of the same experiment, words were flashed on the screen so rapidly that they could not be read. Despite the fact that subjects could not tell what the words were, they gave evidence of fear as measured by physiological arousal (increased heart rate, amount of perspiration) whenever the threatening word appeared.[9] Perceptions like this that enter the system in an automatic rather than a controlled fashion are often referred to as **subliminal perceptions**.

Although we cannot overlook the effects of subliminal perceptions, controlled perceptions dominate judgment. Moreover, because perceptions cannot be processed in a controlled mode without conscious attention to them, the *external* and *internal* factors that draw attention to certain objects are worth noting.

External Factors

Certain characteristics of objects influence the degree to which these objects capture our attention. The intensity of a stimulus, its contrast with other stimuli in the environment, its familiarity or novelty are all factors that affect our attention. These factors often counterbalance each other and combine in complex ways.

Intensity refers to the overall impact that the external object has on the sensing person. Intense objects are more likely to capture one's attention. For example, although shouting at someone may not be polite, it is probably more likely to capture the person's attention than speaking in a normal tone of voice. Why do you suppose a bright red light is used for traffic "stop" signals? It is because this color has been shown to make the biggest impact physiologically on our visual receptors and hence is most likely to be noticed. Bigger things are also more likely to catch your eye than smaller things, because they excite a greater number of your visual receptors.

[9] R. S. Corteen and B. Wood, "Autonomic Responses to Shock-Associated Words in an Unattended Channel," *Journal of Experimental Psychology* 94 (1972), 308–13.

attention stage The stage in the information processing cycle in which the individual decides what will be processed and what will be ignored.

TEACHING NOTE
Driving requires attention to traffic lights and signs, traffic flow, pedestrians, road hazards. Ask several students about their trip from home to school: How many lights do they pass? What are the speed limit signs? Can they describe their last trip in detail— numbers of cars and pedestrians, number of times they stopped? This is all information to which we are exposed but give little attention unless something out of the ordinary happens.

subliminal perception
Information that is enco[c] perceiver without his or [l] awareness.

POINT TO STRESS
A professor may use multiple teaching techniques such as provocative statements, questions, transparencies, and video clips within one lecture. These multiple methods of presenting the material increase the students' attention. Not knowing exactly what to expect next requires us to pay closer attention, thus taking in more information and being ready for surprise questions and other changes in format.

On the 1988 presidential campaign trail, George Bush answered questions about new taxes so many times he sought an answer that would really drive home his point: "Read my lips—no new taxes." The words "read my lips" have become so familiar that people now use it whenever they want to get the attention of others. Here a supporter of the 1990 Civil Rights Act is using the perceptual principle of frequency to try to get President Bush's attention.

APPLYING THE DIAGNOSTIC MODEL
Diagnostic Question 3: What characteristics of persons or objects that you have to evaluate "pull" your attention? How do these characteristics affect your judgments regarding these persons or objects?

AN EXAMPLE: BARTLES AND JAYMES
When the Bartles and Jaymes Wine Cooler television commercials appeared they were a novel approach to selling a wine product. The two individuals portrayed were down-home rather than elegant, they were dressed in work clothes rather than fine suits, and they were older rather than young and sophisticated. These novelties attracted our attention. We saw their wine cooler as refreshing, fun, and for everyone. These commercials met with great success in developing the market for Bartles and Jaymes.

APPLYING THE DIAGNOSTIC MODEL
Diagnostic Question 4: What kinds of expectations do you have for persons or objects that you need to evaluate? How might these expectations affect your judgment?

Counterbalanced against the notion of intensity, however, is the notion of *contrast*. As one television commercial correctly noted, "If you want to capture someone's attention, just whisper." Similarly, if you looked first at this sentence when you turned to this page, you have just demonstrated that contrast sometimes wins out over size or intensity in drawing one's attention. Thus in a sea of intense stimuli, a less intense stimulus can sometimes succeed in capturing our attention.

In general, the greater the *frequency* with which people encounter a given stimulus or the greater its familiarity, the more likely it is that they will attend to it. Thus the more often students complain about a particular aspect of a course, the more likely the teacher is to pay attention to their concerns. The more often a particular recording is played on the radio, the more likely it is that people will become aware of the tune.

Counterbalanced against frequency, is the notion of novelty. We may encounter a given stimulus so often that we begin to ignore it. Something that is totally new to our experience, on the other hand, may arouse our interest and draw our attention. Thus *novelty*, or the extent to which a stimulus is seen as new and different, is also important. Advertisers, who are professional attention getters, try to mix frequency and novelty together by repeating a fundamental theme with a minor innovation. During the very solemn lighting of the Olympic torch at the 1988 Summer Games, the producer of ABC television's Olympics coverage was heard to say: "Fine, but I wanted a Bud Light." Clearly this one advertising campaign's use of repetition (always ending with the same line) in combination with novelty (the same line capping many different setups) had captured the attention of at least one viewer.

Internal Factors

The perceiver's *expectations* of an object will often affect his evaluation of that object. One reason for this reaction is that attention is more easily drawn to objects that confirm our expectations. A good example can be seen in Eden and Shani's study of tank crews in the Israeli army.[10] In this study, the researchers told one set of tank commanders that test data indicated that some members of the crews assigned to them had exceptional ability. The tank commanders were also told

---

[10] D. Eden and A. B. Shani, "Pygmalian Goes to Boot Camp: Expectancy, Leadership and Trainee Performance," *Journal of Applied Psychology* 67 (1982), 194–99.

that some of the soldiers assigned to them were, according to the tests, only average. In reality, the soldiers were assigned to commanders randomly so that the two groups were equally able. Nevertheless, when asked later to rate the performance of their men, the commanders reported that the performance of the "exceptional" soldiers was better than the performance of soldiers who were said to be "average." The researchers explained their findings by noting that the commanders "naturally lavished more attention on individuals for whom they harbored more positive expectancies."[10] Thus expectations become a self-fulfilling prophecy in that they help bring about exactly what they predict.

The *needs and interests* of perceivers themselves can also influence attention. For example, when you are reading a newspaper, a story about a robbery that occurred on your street is much more likely to grab your attention than a story on overall trends in the U.S. crime rate. A news story on a surge in interest rates that many readers might ignore might attract a great deal of attention among workers in construction. For construction firms, high interest rates mean low demand for services and can even cause firms to go out of business.

One way in which needs and interests influence perceptions is by **perceptual defense**, in which a person completely blocks out a stimulus that is highly threatening to him. He simply does not perceive it. Studies have shown, for example, that when words are presented under speeded viewing conditions, subjects have a harder time recognizing threatening words (e.g., *rape, kill*) than nonthreatening words (e.g., *rake* or *mill*).[11]

Earlier we discussed the study by Heilman on the effects of granting preferential treatment to women and minorities on the perceptions of those who were favored. Given the influence of needs and interests on perceptions, it is not hard to anticipate the effect of such practices on those who are placed at a disadvantage.

perceptual defense The process by which an individual avoids processing information that is potentially threatening.

[11] E. McGinnies, "Emotionality and Perceptual Defense," *Psychological Review* 56 (1949), 244–51.

T A B L E 5-5
Differences in the Views of Black and White Americans on Opportunities for Blacks

		BLACKS	WHITES
Perceptual Differences			
	Do black Americans have the same opportunities as whites?		
1	In housing?		
	Yes	22%	48%
	No	75	47
2	In education?		
	Yes	38	73
	No	59	24
3	In employment?		
	Yes	26	59
	No	71	37
Judgmental Differences			
4	Should colleges admit some students whose records would not normally qualify them for admission?		
	Yes	33	15
5	Should businesses set a goal of hiring a minimum number of black employees?		
	Yes	62	32

Source: Adapted from "Attitudes in Black and White," *Time*, February 2, 1987, p. 37.

PERCEPTION AND JUDGMENT

Consider the results of an opinion survey of blacks and whites shown in Table 5-5. As you can see, there were marked differences in the way blacks and whites perceived and judged current opportunities for African Americans in the United States. These differences are probably best explained by the fact that each group is attending to different things. Given these differences in perceptual input, it is not surprising that the two sides come to markedly different judgments with respect to questions 4 and 5 in the Table. Working out these kinds of perceptual differences may be essential before true progress can be made in the area of civil rights.

Making Attention More Efficient

Clearly there are many ways a human observer can fail to portray her environment accurately. Before we look at some methods of improving our powers of attention, we should note one thing. There is no free lunch when it comes to increasing the accuracy of perceptions. Most of the solutions we will discuss require a controlled mode of information processing. As we noted at the outset, human observers can process only one thing at a time in this mode, and therefore time devoted to increasing accuracy of perceptions subtracts from time available for other aspects of the observer's job. Thus if we expect supervisors to provide detailed and accurate judgments of a subordinate's effectiveness, we must also understand that this responsibility will take time away from other dimensions of their jobs. Moreover, we should keep the four types of perceptual accuracy in mind. The kind of accuracy needed for one type of decision may be easier to achieve than the accuracy needed for another kind of decision.

Because one of the major problems at the attention stage is simply the amount of information available for processing, one way to improve accuracy is by *increased frequency of observations*. That is, we can increase the observer's exposure to the thing being observed. By making more observations, an observer may gather more information and thus heighten the accuracy of her perceptions.[12]

Because a second major problem at the attention stage is the representativeness of the information, the manner in which observations are obtained should also be considered. If we obtain observations by *random sampling* (see Chapter 3), we increase the probability that these observations will be accurate. If a supervisor observes a group of workers only at a given time on a given day or makes observations only when problems develop, the behaviors she observes may not truly reflect what is happening in this group.

The opportunity to observe employee work behaviors frequently, randomly, and secretively has increased rapidly with technological developments in the field of surveillance. In fact, this opportunity has increased so dramatically that some are beginning to raise ethical issues with regard to monitoring practices. Some of these practices and the ethical issues involved are described in the "Management Issues" box.

Finally, because observers have a tendency to ignore information that is inconsistent with their expectations, it is often good advice to actively *seek information that is inconsistent* with, or that disconfirms, one's current beliefs. For example, if in the past you have judged a particular worker as someone who "can't do anything right," approach the observation of that person from the opposite direction. That is, systematically search for things that the person *does* do right. You might surprise yourself.

---

[12] W. C. Borman, "Exploring the Upper Limits of Reliability and Validity of Performance Ratings," *Journal of Applied Psychology* 63 (1978), 135–44.

MANAGEMENT ISSUES

The Brave New World of Employee Surveillance

Nurses at Holy Cross Hospital in Silver Springs, Maryland, were surprised when they discovered that the silver box with red lights hanging on their locker room wall was actually a video camera. They were shocked when they learned that the pictures taken by this camera, which captured the nurses in various stage of undress, were being broadcast over the hospital's closed-circuit TV network. Although the hospital later claimed that the pictures were viewed only by the hospital's (male) chief of security, this did little to alleviate the nurses' outrage. Moreover, few outside observers of this incident felt that the hospital's stated purpose of trying to track disappearing narcotics justified its behavior.

Incidents like this are becoming widespread. The DuPont Company, for example, now uses hidden long-distance cameras to monitor its loading docks. At Delta Airlines, computers track which salespeople write the most reservations. At Management Recruiters, Inc., in Chicago, supervisors surreptitiously watch computerized schedules to see which interviewers talk to the most job candidates. Supervisors at the Internal Revenue Service can "tap into" telephone conversations between IRS agents and taxpayers calling for information.

Firms are not reluctant to use the data captured by these electronic devices in making decisions. Safeway Stores, for example, has installed dashboard computers on its 800 trucks. The computers record driving speed, idling time, and when and how long a truck is stopped. According to George Sveum, secretary of Teamsters Local 350 in Martinez, California, Safeway tries to suspend or discharge up to 20 drivers a year using this computerized data.

The increased use of computerized employee monitoring has been a product of two forces. First, there has always been a need to observe employees' work behaviors, and recent developments in surveillance technology have simply made this easier and less obtrusive. The second force is the increasing number of court cases dealing with "negligent hiring," where employers are held liable for the mistakes or crimes of employees. These two developments have led to a serious erosion of employee rights to privacy.

Finding the right balance between these rights and the employer's right and responsibility to monitor workers will be

Pilferage at loading docks, whether inland or on the waterfront, has always been a difficult problem to control. Now, with hidden cameras that have telephoto lenses and that can make continuous and constant observations, it is very difficult to get away with such theft. But for many people this means of surveillance raises ethical issues. Has the worker a right to privacy on the job? Is there a differnce between having a foreman watching you constantly or an executive vice president strolling by your desk several times a day and having your every action recorded by an electronic eye?

a difficult process. Undoubtedly it will play itself out in the courts and legislatures. Although current federal and state laws permit most forms of eavesdropping and electronic monitoring in the workplace, growing public awareness of such practices may lead to remedial legislation that provides employees with some assurance of privacy.*

* J. Rothfeder, M. Galen and L. Driscoll, "Is Your Boss Spying on You?: High Tech Snooping in 'the Electronic Sweatshop,'" *Business Week*, January 15, 1990, pp. 74–75.

ORGANIZATION STAGE

organization stage The stage in the information processing cycle in which many discrete bits of information are chunked into higher-level, abstract concepts.

Even though much information is automatically filtered out at the attention stage, the remaining information is still too abundant and too complex to be easily understood and stored. Because human perceivers can process only a few bits of information at a time in a controlled fashion, in the **organization stage** we further simplify and organize these incoming data. One method is to "chunk" several discrete pieces of information into a single piece of information that can be processed more easily.

PERCEPTION AND JUDGMENT

TEACHING NOTE
Have several students compare their notes for the last 15 minutes of lecture, reading portions of their notes verbatim. Point out how each person captured somewhat different information and organized it in a unique fashion. Although 30 people may be exposed to the same information, each person may hear it differently.

schema   Cognitive structures that group discrete bits of perceptual information in an organized fashion. (The term *schema* is used for both singular and plural forms.)

script   A schema that involves well-known sequences of action

APPLYING THE DIAGNOSTIC MODEL
Diagnostic Question 5: What are some of the major events in your organization that you conceive in terms of scripts? How might your version of these scripts differ from others views of the same scripts?

To show you how effective this kind of chunking can be, imagine your reaction if someone were to ask you to memorize a string of 40 numbers. You might very well doubt your capacity to memorize this many numbers regardless of how much time you were given. Your doubts are probably misplaced, however, because if asked, you could probably write down (a) your social security number, (b) your telephone number with area code, (c) your license plate number, (d) the month, date, and year of your birth, (e) your current zip code, and (f) your height and weight. You might say, "Well, yes, but these are only six numbers," but note that *a* and *b* have 9 digits each, *c* and *d* have 6, and *e* and *f* probably have 5; this comes to a grand total of 40 digits! The fact that we think of these as six numbers rather than forty shows how we mentally chunk things together. The fact that by the chunking process we can memorize many more than forty numbers (think of all the telephone numbers, zip codes, birthdays, and so on that you can recall) attests to the efficiency of this type of organizing process.

When we do this kind of chunking with non-numerical information we refer to the chunks as schema. **Schema** are cognitive structures that group discrete bits of perceptual information in an organized fashion.[13] In general, two types of schema are particularly important to understanding the processing of social-interpersonal information: scripts and prototypes.

Scripts

Suppose you are chatting with a friend and ask him, "So what did you do Friday night?" Imagine, then, that he replies, "I put on my best suit. I got in my car. I drove downtown and picked up a young woman at her apartment. We drove to an expensive restaurant. We had drinks. We made polite conversation. We ordered something to eat. We ate our dinner. We got back in my car. I drove the young woman back to her apartment. I walked her to her door. I stole a kiss. I got back in my car. I drove back home." You might say, "In other words, you went out on a date," and he might respond "Yes." Note how in the last two exchanges you and your friend chunked a 14-bit sequence into a one-word schema: *date*. Schemas that involve sequences of actions are called **scripts**, for the very good reason that they resemble the material from movies or plays.

Assume that your friend went on to report that he had dinner at his parents' house on Thursday, studied with friends on Wednesday, went to a college basketball game on Tuesday, and washed and waxed his car on Monday. You can probably envision scripts for all five activities that your friend engaged in Monday through Friday and in doing so chunk over a hundred bits of information into five tight pieces. This kind of efficiency makes scripts highly useful from an information-processing point of view.

Now imagine that from your friend's report of his date last night, you form the judgment that he probably does not have the money to go to a movie tonight. To your surprise, he says he can indeed go, because "last night didn't cost me a dime, my date paid for the dinner." Herein lies the problem with scripts, namely, that a given script may have different meanings for different people. Note that nowhere in your friend's 14-point description of his date did he mention how the bill was paid. You simply assumed that he paid because that is part of *your* script for a date—one that some might call sexist and outdated!

Thus while the kind of simplification that is granted by scripts is vital for efficient information processing, one should not lose sight of the fact that in scripting such sequences we may be adding things to the event that never took place or deleting things that did happen. Clearly there are numerous events in

---

[13] U. Neisser, *Cognition and Reality* (San Francisco: W. H. Freeman, 1976), p. 112.

organizations that prompt different scripts in different people's minds. Things such as taking a client to lunch, preparing a written report, or disciplining a subordinate, may involve sequences of behavior for some organizational members that are not the same as the sequences envisioned by others. Clarifying these scripts is essential if perceptual accuracy is the goal.

Prototypes

prototype One type of schema that involves a unified configuration of personal characteristics that are used to classify persons into "types."

Just as there are schema for simplifying events, there are also schema for simplifying the description of persons. **Prototypes** are schema that enable us to chunk information about people's characteristics.

For example, if you asked your precise friend what the person he dated was like, he might report that she was spirited, exuberant, outgoing, boisterous, and warm. "You mean she's an extrovert," you say. Here again we see multiple bits of information chunked into one word that is meant to carry a detailed description of a person. Like scripts, however, prototypes sometimes carry excess baggage and thus may not reflect the person accurately. You may recall how during the 1988 presidential election George Bush repeatedly labeled Michael Dukakis an "old-fashioned, liberal democrat" and a "card-carrying member of the American Civil Liberties Union." The purpose of this strategy was to get the voters to categorize Dukakis as a certain type of person, one whom many people would find unsuitable as a president of the United States.

In the area of organizational behavior, the leader prototype is an important one. Most managers want others to perceive them as leaders. What characteristics are likely to cause people to categorize someone in this way? According to research conducted by Robert Lord, the leader prototype is made up of the 12 characteristics shown in Table 5-6 and listed in descending order of importance. People who exhibit a majority of these characteristics will be seen as leaders. Moreover, according to Lord, people generally assume that anyone who is a leader must have these characteristics.

With a concept as broad as *leader*, there is a tendency to make distinctions among various kinds of leaders. Lord and his colleagues have found that there are specific prototypes for many kinds of leaders, including military, religious, business, political, labor, and minority leaders, to name just a few.

Prototypes can be negative. For example, the expression "empty suit" emerged in the business world in the late 1980s as a label for a particular type of manager. Specifically, an "empty suit" was a manager with a great deal of style and a "dress-for-success" image but little in the way of substance, skill, or deeply rooted values.[14]

[14] W. Kiechel, "How to Spot an Empty Suit," *Fortune*, November 20, 1989, p. 22.

APPLYING THE DIAGNOSTIC MODEL
Diagnostic Question 6: What are some of the major prototypes that exist for persons in your organization? How accurate are these prototypes?

TEACHING NOTE
People are likely to perceive the president of the United States as a leader particularly in his roles of head of the U.S. government in international meetings and negotiations and commander in chief of the country's armed forces. When President George Bush visited the troops in Saudi Arabia on Thanksgiving Day in 1990, he was given a particularly warm welcome. Ask students to study the qualities of the leader prototype proposed in Table 5-5 and then to say whether, by this measure, they would describe George Bush as a leader. If so, why? If not, why not? ("A Question of Time," *Newsweek*, December 3, 1990.)

T A B L E 5-6
Major Characteristics of the "Leader" Prototype (in descending order of importance)

1. Intelligent	5. Aggressive	9. Decisive
2. Outgoing	6. Determined	10. Dedicated
3. Understanding	7. Industrious	11. Educated
4. Articulate	8. Caring	12. Well dressed

Source: R. G. Lord, R. J. Foti, and D. DeVader. "A Test of Leadership Categorization Theory: Internal Structure, Information Processing, and Leadership Perceptions," *Organizational Behavior and Human Performance* 34 (1984), 343–78.

AN EXAMPLE

Many people perceive older workers as less productive in spite of research showing the opposite. The Age Discrimination Act was passed based on findings that older workers were just as productive as younger ones in terms of both quantity and quality of work. Older workers have fewer accidents and are often a stabilizing force for younger workers, helping them adjust to and understand changes.

AN EXAMPLE

Attractive men are more likely to be promoted in management than are less attractive men. Conversely, more attractive women are *less* likely to be promoted. Attractiveness in women apparently connotes femininity, emotionality and less stability. As a result, attractive women are thought to be less able to make tough management decisions.

halo error A rating error wherein a rater's judgment about a specific behavior is colored by her overall evaluation of the person she is rating.

POINT TO STRESS
The halo effect is also referred to as the *halo-horn effect*. The horn effect describes a negative bias in ratings on specific dimensions caused by an overall negative impression of an individual.

Stereotyping. Not all prototypes are useful. A *stereotype* is a widely held generalization about a group of people. Often it is a prototype organized around a person's race, sex, age, ethnic origin, socioeconomic group, or other sociocultural characteristics: for example, African-Americans, women, the elderly, Hispanic-Americans, blue-collar workers, homosexuals. In one study business students displayed a clear stereotype of the elderly.[15] Among other things, students described this group as less creative, less able to do physically demanding work, less able to change or be innovative. These perceptions led the students to make other negative judgments about elderly workers. For instance, they expressed the belief that these workers would be less likely than younger workers to benefit from training and development. Given the increasing age of our national work force, such stereotypes need to be examined closely.

Stereotypes are not confined to students. Many people feel that businesses systematically discriminate against older workers by forcing them into early retirement programs. For example, Richard E. Wilson, a former vice-president of Monarch Paper Company was demoted to a warehouse-maintenance job that included clean-up duty and other menial tasks, after failing to accept an early retirement "offer." A federal jury awarded Wilson $3.2 million and judged that the company was guilty of age bias. The jury suggested that the company's heavy-handed offer was part of a plan to eliminate older managers and replace them with younger ones.[16]

In some cases, workers are actually punished for not living up to widely shared stereotypes. For example, despite the fact that she had won a $34-million government contract, Ann B. Hopkins was not promoted to partner in the prestigious accounting firm of Price Waterhouse. Male colleagues criticized Hopkins for being "too macho" and suggested that she needed "a course at charm school."[17] The U.S. Supreme Court found this type of sex stereotyping a violation of the Civil Rights Act and awarded Hopkins a large settlement.

Halo error. Another type of perceptual bias, the **halo error**, occurs when a rater's judgment about specific facets of behavior are colored by his overall evaluation of the person he is rating.

For example, Figure 5-4 reproduces an instructor evaluation form in which students are asked to answer one general question and ten specific questions about an instructor's performance. Presumably, the ten dimensions tapped in the specific questions are unrelated. That is, it is assumed that a person might be very enthusiastic but not well organized. Similarly it is assumed that an instructor may be able to outline the direction of the course adequately but may not be very good at stimulating class discussion. Despite these assumptions by persons who constructed the rating form, actual ratings obtained from students often make it look as if the ten factors were highly interrelated.

For example, in filling out the questionnaire, Student A gave the instructor a "superior" ("S") on all eleven items. This instance of halo error suggests that the student's overall impression of his instructor has created a positive bias in his ratings of the specific dimensions of performance. That is, the student has developed a prototype of what a good instructor is, judges this person to be a good instructor, and therefore rates the person high on every single dimension. Clearly, if raters make no distinctions with respect to the specific dimensions in which we are interested, we have a major threat to differential accuracy. In this case, the

---

[15] B. Rosen and T. H. Jerdee, "The Influence of Age Stereotypes on Managerial Decisions," *Journal of Applied Psychology* 61 (1976), 428–32.

[16] E. G. Olson, "The Workplace Is High on the High Court's Docket," *Business Week*, October 10, 1988, pp. 88–89.

[17] Ibid.

QUESTIONNAIRE ITEMS

Overall Evaluation

1. How would you rate this instructor overall?

Specific Evaluation

2. Instructor's enthusiasm
3. Instructor's interest in teaching
4. Instructor's organization of the course material
5. Instructor's stimulation of class discussion
6. Adequacy of the course outline
7. Instructor's use of examples
8. Ease of taking notes during instructor's presentation
9. Instructor's concern for students' learning
10. Appropriateness of instructor's pace
11. Instructor's receptiveness to students' ideas

STUDENTS EVALUATIONS

	Student A	Student B	Student C
1.	● AA AV BA I	S AA ● BA I	S AA AV ● I
2.	● AA AV BA I	S ● AV BA I	S ● AV BA I
3.	● AA AV BA I	S ● AV BA I	S ● AV BA I
4.	● AA AV BA I	S ● AV BA I	S ● AV BA I
5.	● AA AV BA I	S ● AV BA I	S ● AV BA I
6.	● AA AV BA I	S ● AV BA I	S ● AV BA I
7.	● AA AV BA I	S ● AV BA I	S ● AV BA I
8.	● AA AV BA I	S ● AV BA I	S ● AV BA I
9.	● AA AV BA I	S AA AV ● I	S AA AV ● I
10.	● AA AV BA I	S AA AV ● I	S AA AV ● I
11.	● AA AV BA I	S AA AV ● I	S AA AV ● I

KEY:
S   Superior–exceptionally good course or instructor
AA   Above Average–better than the typical course or instructor
AV   Average–typical course or instructor
BA   Below Average–not as good as the typical course or instructor
I   Inferior–exceptionally poor course or instructor

FIGURE 5-4

Hypothetical Student Evaluations of Instructors

halo error undermines any attempt to help instructors improve in areas in which they are weak.

What makes the halo error such an intractable problem, however, is that we often do not know how much true relationship there is among the dimensions in which we are interested. Consider a recent study of major league baseball players.[18] In this study, several dimensions of player performance (e.g., batting average, runs batted in, stolen bases, fielding errors) were measured objectively. Subjects were then asked to rate several well-known players on these dimen-

[18] S. W. J. Kozlowski and M. P. Kirsch, "The Systematic Distortion Hypothesis, Halo, and Accuracy: An Individual Level Analysis," *Journal of Applied Psychology* 72 (1987), 252–61.

sions. The ratings did indicate a relationship among the dimensions. If a player was rated high on one dimension (e.g., hitting), for example, he tended to be rated high on others (e.g., fielding and running) even though the dimensions appear to tap different abilities.

Do these results reflect a halo error? No, because the true relationship among the dimensions was even greater than the ratings indicated. The raters did not show *enough* halo. Whereas hitting and fielding may *seem* to be different things, the objective data show that these two aspects of performance actually are related to each other. Separating illusory halo from true halo is not an easy matter. Before securing ratings one needs to think seriously about how and when one dimension of performance may be related to another.

Improving Perceptual Organization

The main problem at the organization stage is oversimplification of the information that we attend to. Given the limited information-processing skills that human beings have, it is unrealistic to expect people to completely give up the use of schemas. It is a good idea, on the other hand, to *elaborate prototypes and stereotypes*, that is, to make people aware of the prototypes and stereotypes they hold of other people. It is also a good idea to get observers to *abandon particular prototypes or scripts* that seem to lead them astray, and perhaps to replace these with other scripts and prototypes that will be accurate and still helpful in simplifying data.

When a person must work with social groups that differ from her own, it may be useful for her to *increase her exposure to different social groups*. This approach may go a long way toward helping the person develop more accurate prototypes. Research on expert-novice differences in all kinds of domains shows that experts differ from novices not because they ignore developing prototypes but because they develop more complex, detailed prototypes that are more accurate.[19] Thus as people develop experience with people and situations that are unfamiliar to them, their processes of organization become more complex and better able to reflect the underlying reality. Actively *searching for disconfirming prototype information*, is particularly useful in this regard.

POINT TO STRESS
As our economy becomes more dependent on international relationships, we are becoming more and more aware of our need to understand different cultures— their laws, their norms, their values—to work effectively with their individual citizens and their organizations (see also Chapter 19).

interpretation stage The stage in the information processing cycle in which meaning is attached to the relation among abstract concepts.

INTERPRETATION STAGE

Organizing discrete observations into scripts and prototypes is a first step toward making sense of what one perceives. However, the processes that occur during organization merely set the stage with identifiable actors and actions. Still further processing must occur to make complete sense of the incoming data, and much of it occurs at the **interpretation stage**. Here the perceiver tries to go beyond merely identifying "who is doing what with whom" and tries to uncover the reasons behind the actions. Often the same objective behavior can lead to quite different judgments depending on how the perceiver answers the question, "Why?" For example, the behavior of working late at the office each night could be interpreted in different ways by different people. One observer might see such behavior as a sign of an employee's drive and ambition, but another observer might see it as a sign that the person cannot keep up with the work that is assigned. It is easy to see how these two people who view the same behavior can come to quite different conclusions.

[19] L. T. DeFong and C. J. Ferguson-Hessler, "Information Processing Differences in Experts and Novices," *Journal of Applied Social Psychology*, 21 19–27.

Projection

projection A bias in the interpretation of information wherein the perceiver assumes that his own motivations explain the behaviors of others.

APPLYING THE DIAGNOSTIC MODEL
Diagnostic Question 8: Are your reasons for interpreting other people's actions self-serving? Do you project any of your own socially undesirable characteristics onto others?

AN EXAMPLE
Projection can be a source of conflict. For example, if two people working together have different beliefs about why people behave as they do and each tends to project his own beliefs, the two may be in trouble. John may be self-focused, working only for personal gain, whereas Allen views work as a team effort in which all gain. If he offers John assistance, Allen expects cooperation and possibly a thank you in return. John questions the motives behind the help, tends not to trust Allen, and is reluctant to cooperate. Conflict will arise and productivity will drop for these two people quite quickly.

One common way in which observers interpret the behaviors of others is to use themselves as a point of reference. In this context particularly you can see that interpretation can involve both controlled and automatic processing. In **projection**, people project their own thoughts and feelings onto other people, often unconsciously. That is, they assume, often incorrectly, that others share their own feelings or motivations.

A good example of the projection process can be found in Richard Mowday's study of government employees. Mowday was interested in organizational turnover as viewed by employees who stayed with an organization. Reminding employees that several of their coworkers had left the organization in recent years, he asked them why the other people had left. The subjects of the study were given the choice of three explanations. The first was that the former employees left because they were dissatisfied with their jobs. The second was that they were not dissatisfied but left to take better jobs. The third was they left both because they were dissatisfied and because they received better offers.

The results of this study indicated that employees who liked their own jobs and their organization were much less likely to believe that turnover resulted from others' dissatisfaction. Those who were themselves dissatisfied with their jobs and the organization, however, embraced dissatisfaction as an explanation for turnover and tended to downplay the possibility that better offers were the reason.

Thus projection can become a means of self-justification. Satisfied workers can defend their decisions to stay with the organization by viewing it as a good place to work and believing that those who left did so for reasons other than job dissatisfaction. The dissatisfied workers, on the other hand, can cite this turnover as proof that there is something wrong with their jobs or the organization.

Projection can also act as a means of self-protection. More often than not it is their socially undesirable characteristics that observers project onto others. For example, a person who is overly cynical may be quick to point to the moral failures of famous religious or political leaders in an effort to quell the anxieties she has about her own moral stance. Projecting negative traits in this fashion has a way of dragging one's perceptions of the world to a very low level. Everyone is seen in terms of his basest impulses. When combined with the potential for self-fulfilling prophecies, projection can seriously distort one's organization of information and can make one's entire world seem a desolate place.

Attribution

attribution The process whereby observers decide what caused the behavior of another person.

Attributions are the causal factors that observers use to explain the behaviors of other people. Attribution theory tries to explain how people decide among possible explanations for various outcomes such as high or low task performance. The explanations that observers arrive at have a great impact on subsequent evaluations and judgments in that even a successful performance can be turned into a loss if a certain kind of attribution is made.

Types of attributions. In general, those who have studied attribution have focused on four possible explanations for a person's success or failure on a task, depending on the *stability* of the cause (i.e., either stable or unstable) and the *locus of causality* (i.e., internal or external to the person.)[20] As shown in Figure

[20] B. Weiner, I. Freize, A. Kukla, L. Reed, S. Rest, and R. M. Rosenbaum, "Perceiving the Causes of Success and Failure" in *Attribution: Perceiving the Causes of Behavior*, ed. E. Jones, D. Kanouse, H. Kelley, R. Nesbitt, S. Valins, and B. Weiner (Morristown, N.J.: General Learning Press, 1971), pp. 45–61.

FIGURE 5-5

**Four Types of Attributions
of Task Success or Failure**

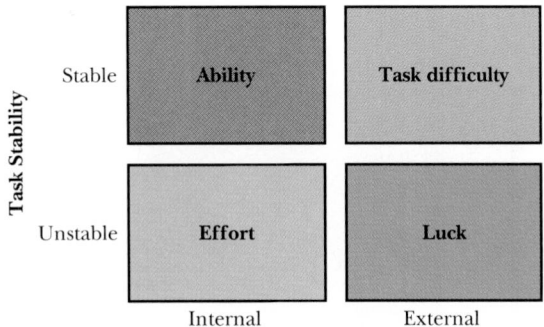

POINT TO STRESS
Attributions are influenced by
personality characteristics. For
example, low self-esteem people
tend to attribute their failures to
stable, internal causes like ability.
High self-esteem people are more
likely to attribute their failures to
unstable causes like bad luck or
lack of effort.

5-5, these four types of attributions are ability, effort, task difficulty, and luck.
Ability is a potential cause of performance that is stable (it is not expected to
change much over time) and internal (it is a characteristic of the performer).
Effort is also a characteristic that is internal to the performer, but unlike ability,
it may fluctuate from time to time. External causes include task difficulty (e.g.,
the performer was successful because it was an easy task) and luck. Task difficulty
tends to be a stable dimension, whereas luck is unstable.

Results of attribution. For any given instance of success or failure, the explanation that the observer chooses from these four possibilities will greatly influence
the judgments he eventually renders. At the end of the 1987 professional football
season, the coach of the rarely successful Cincinnati Bengals, Sam Wyche, found
many sportswriters calling for his resignation. Wyche did not resign, and in the
following year he lead the Bengals to the top and earned the team a spot in the
Super Bowl.

You might think that such an outstanding performance would have been
enough to get the critics off Wyche's back, but you would be wrong. Despite this
objective achievement, some sportswriters still held fast to their earlier published
judgments about the coach's lack of ability. How could they possibly do this?
Attribution theory explains this nicely. Some writers pointed to the fact that 1988
was an off year for most teams in the league and that winning in that year was
thus a relatively easy task. Others said that the Bengals had a lot of lucky breaks
in a lot of close ball games. Note that by explaining the Bengals' success in terms
of external factors (task difficulty and luck) that were beyond Wyche's control, these
observers were able to question his ability despite the objective fact that his team
posted more wins than the other 27 teams in the league.

But attributions can also work the other way. Someone who fails miserably
at a task can still be held in high regard if her failure is attributed to external
factors. In general, it is only when internal attributions are made that there is a
strong relationship between a performer's success or failure on the one hand and
a rater's opinion of that performer on the other. Some research even suggests
that some types of internal attributions are better than others. For example, success
that is perceived to be a function of effort is viewed more positively than success
that is seen as due to ability, because the former is more directly under the person's
control. The complex relationships between success and failure and ratings as a
function of attributions are illustrated in Figure 5-6.

These attributions are important from a diagnostic point of view because
they affect decisions regarding problem solving. If failure is attributed to lack of
ability, then selecting different applicants or retraining the current jobholders is
the solution. If lack of effort is the explanation, selecting different applicants or

INTERPRETATION STAGE

141

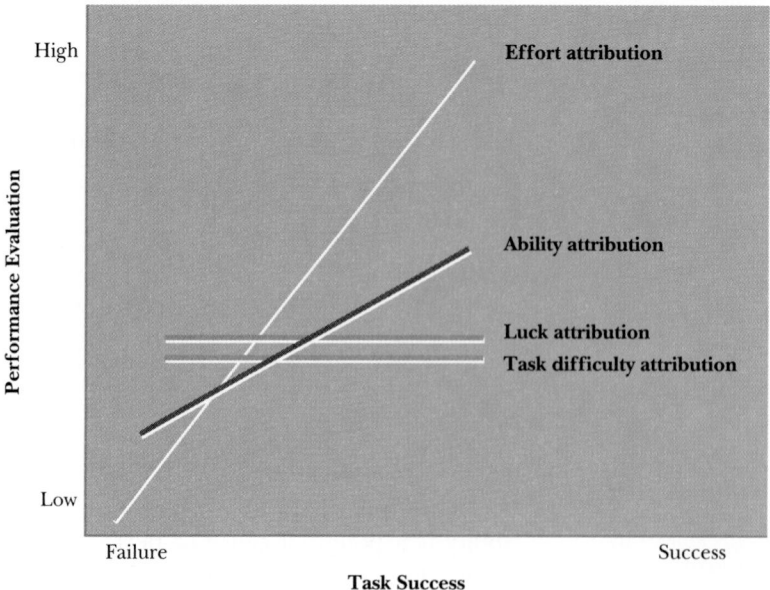

increasing jobholders' motivation is the answer. If failure is due to task difficulty, redesigning the task may be the best option.

Choosing attributions. What are some of the factors that lead observers to choose one type of attribution over another? Observers sometimes arrive at attributional decisions as a result of *self-serving biases*. That is, observers may choose the attributions that present the observers themselves in the best light. When rating their own performance, observers tend to attribute their successes to internal causes and their failures to external causes. Some interesting research by Bettman and Weitz illustrates this tendency. In their study of over 150 corporate annual reports, these investigators found that unfavorable organizational outcomes were likely to be attributed to external factors (e.g., market slowdowns, high interest rates, bad weather, foreign competition), whereas favorable outcomes were usually explained in terms of factors for which the firm was responsible (e.g., new marketing campaigns, new-product development, introduction of new technology).[21]

Self-serving biases influence attributions people make with regard to others' behavior in just the opposite way. Others' successes are explained by situational factors, while others' failures are attributed to internal factors.[22] Given these tendencies, it is easy to see how there can be a great deal of controversy in organizations over one person's merit as compared with another's, even when there is agreement on the objective facts of success and failure.

Causal determination is a complex process, and self-serving interests are not the only determinant of attributional choice. Other important factors include consensus, consistency, and distinctiveness.[23] *Consensus* refers to the extent to which everyone else experiences the same outcome (or engages in the same behavior) as the person being observed. If everybody who ever takes on the task fails, observers are much more prone to blame the task than the task performer.

[21] J. R. Bettman and B. A. Weitz, "Attributions in the Board Room: Causal Reasoning in Corporate Annual Reports," *Administrative Science Quarterly* 28 (1983), 165–83.

[22] H. H. Kelley and J. L. Michela, "Attribution Theory and Research," *Annual Review of Psychology* 12 (1980), 457–501.

[23] Kelley, "Attribution in Social Interaction," (Morristown, N.J.: General Learning Press, 1971), p. 71.

Who Was Responsible for the Exxon Valdez Disaster?

The grounding of the *Exxon Valdez* and resulting oil spill resulted in the devastation of 117 miles of coastline in Prince William Sound and along the Gulf of Alaska. The spill had a terrible effect on both plant and animal life (over 30,000 seagulls alone were killed), and severely damaged the economy of the surrounding areas. The spill was also costly for Exxon, which eventually paid over $2 billion for the cleanup. Last, but certainly not least, was the effect the event had on the ship's captain, Joseph Hazelwood, who was faced with the possibility of seven years in prison and fines of over $50,000. Answering the question of why this disaster took place, as well as determining what can be done to prevent another such disaster is critical, yet difficult.

According to Hazelwood's lawyers, the Coast Guard, which has responsibility to monitor ships in Prince William Sound, failed to warn the *Exxon Valdez* before it went aground. Hazelwood's attorneys also noted that Third Mate Gregory Cousins, who was in charge of the ship when it crashed, was qualified to run the vessel in Hazelwood's absence. Lastly, the widely held notion that the captain was drunk at the time of the accident was also disputed by Hazelwood's lawyers.*

Let us look at what two other people have had to say, both about the cause of the disaster and about what actions should be taken to prevent a reoccurrence. The first speaker is Oliver Holmes, Exxon tanker captain, retired after 43 years of service:

The Cause: "It is a wrong against his many friends and former colleagues at Exxon to portray him [Hazelwood] as victim rather than cause of that tragedy. Overlooked in many news accounts is the master's complex responsibility for his ship's safety at all times. If Captain Hazelwood had done what he was duty bound to do, there would have been no grounding. . . . The questions about double hulled ships, crew fatigue or shorthandedness; about the failure of the U.S. Coast Guard; . . . about the master's paperwork; even the questions about alcohol are either irrelevant or subordinate to the real cause of the grounding.

* P. A. Witteman, "Fall Guy or Villain?" *Time*, February 26, 1990, p. 34.

The Remedy: "We haven't reduced people at the lower level . . . but somewhere between the top of the house and the bottom there are employees who need more training, as well as managers who have to do a better job of evaluating people. What's motivating these people on the docks and in the ships? Are they upset? Is there too much time pressure? Maybe we'll have an industrial psychologist talk to them. We're not rushing people when they're moving oil. We want them to slow down."

Lawrence Rawl, chairman of the board at Exxon, was asked what he felt was the cause of the oil spill and, in particular, whether Exxon's huge personnel cutbacks in the 1980s put company safety and maintenance in jeopardy:

The Cause: "I think, in the end, the Alaska oil spill was caused by compounded human failure. In Louisiana [where an Exxon refinery exploded], that was legitimately an act of God. . . . As for Arthur Kill [site of 500,000-gallon spill off New York City], that was an act of God ripping that pipeline, but the way it was handled afterward was human error. . . . [As to cutbacks] . . . I don't have the answer, but I'm dissatisfied with sitting tight and hoping the bad luck goes away, because if you've got bad luck, you've missed something somewhere."

The Remedy: "We haven't reduced people at the lower level . . . but somewhere between the top of the house and the bottom there are employees who need more training, as well as managers who have to do a better job of evaluating people. What's motivating these people on the docks and in the ships? Are they upset? Is there too much time pressure? Maybe we'll have an industrial psychologist talk to them. We're not rushing people when they're moving oil. We want them to slow down."

The central question in the trial of Joseph Hazelwood was whether one man could be singled out for blame in the worst oil spill in U.S. history. In the end, a grand jury ruled he could not, and Captain Joseph Hazelwood was acquitted.

† O. Holmes, "Captain Hazelwood: The Safety Device that Failed," *The Wall Street Journal*, January 12, 1990, p. 1.

Consistency is the extent to which the person being observed obtains the same outcome (or engages in the same behavior) at different times. If the performer either succeeds or fails every time she attempts the task, then observers are likely to make internal attributions. Finally, *distinctiveness* is the extent to which the person achieves the same result on widely different tasks. If the person being rated succeeds or fails on whatever assignment is handed down, observers are much more likely to make internal attributions. Imagine how difficult it would be to criticize Sam Wyche if his success in 1988 had been followed up with similar success in 1989 and then in 1990—especially if he had switched teams.

In many cases, however, an incident occurs only one time, greatly complicating the process of determining the real cause. The "In Practice" box, for example, shows how three different people generated different attributions in order to account for the cause of the *Exxon Valdez* oil spill.

Increasing the Objectivity of Interpretations

Many judgments that observers must make can influence their self-perceptions. For this reason, an insecure, overly sensitive rater is not a reliable witness. An observer who has a high degree of *self-acceptance*—who knows and accepts her strengths and weaknesses—will have less psychological need to project negative characteristics onto others. The need to employ self-serving attributional biases will also be less in a rater who has high self-acceptance than in one who is low on this characteristic.

Because interpretation is so much "in the eye of the beholder," it is also a good idea to compare your interpretation of a person or a situation with others' interpretations. This process, called *reality testing*, is most useful when the people whose interpretations you are comparing with your own have attended to different information and perhaps developed prototypes that differ from yours.[24] Also, if other raters' needs and yours differ, the potential for self-serving biases is reduced even further. Following this reasoning, many firms insist on obtaining not only supervisor's ratings of an employee but peers', or coworkers', ratings as well.[25]

RETRIEVAL STAGE

retrieval stage The stage of the information processing cycle in which the observer tries to recall information about past events.

You can probably see by the time we reach the **retrieval stage,** that there may be little resemblance between the information that originally was available in the objective environment and that which is moving through the subjective perceptual system. Furthermore, in the real world, there is almost always a considerable delay between entering information into memory and retrieving it. Indeed for information to be retrieved, it must have been originally memorized. Memorizing information is typically a controlled rather than an automatic process, and therefore not all that is interpreted will necessarily be recalled. For example, think of how easy it is to *interpret* what is written on this page, compared to how difficult it is to *memorize* what is on it.

Moreover, not all that is stored will always be retrievable, there is a tendency for memory to decay overtime. A critical question is whether memory decay is selective or random. Most research suggests that such decay is quite selective. For example, studies by James Phillips and his colleagues show that people are likely to forget information that is inconsistent with the scripts or prototypes that were used to organize the information.[26] Look back at Table 5-5 where the prototypical traits of a leader are presented. If someone viewed as a leader possesses some characteristics that are unleaderlike, there is a good chance that after the passage of time, people will forget these characteristics. Or, a person who became a leader despite a low level of education might be recalled as having had a more extensive educational background than she did. Indeed, the most

TEACHING NOTE
Discuss memory in the elderly. Why can older people, often with great accuracy, recall events of 50 years ago whereas they may not remember what happened yesterday or what they had for breakfast a few hours ago? Is this selective memory loss? Is it some defect of attention, organization, or interpretation? Is such a defect absolute or may it be mediated by time, that is, may these processes simply operate more slowly?

[24] M. K. Johnson and C. L. Raye, "Reality Monitoring," *Psychological Review* 88 (1981), 67–85.
[25] D. Cederbloom and J. W. Lounsbury, "An Investigation of User Acceptance of Peer Evaluations," *Personnel Psychology* 33 (1980), 567–79.
[26] J. S. Phillips and R. G. Lord, "Schematic Information Processing and Perceptions of Leadership in Problem Solving Groups," *Journal of Applied Psychology* 67 (1982), 486–92.

worrisome aspect of Phillips's research is his finding that people were as likely to "recall" prototypical traits that were never there as they were to recall traits that were there.[27]

Once established in someone's memory, a prototype can persist for a long time. For example, Fred Malek, the president of Northwest Airlines, has struggled for 20 years with a mistake he made in 1971. Working for an embattled and, some say, paranoid Richard Nixon, Malek was asked to generate a list of Jews in the Bureau of Labor Statistics. Nixon apparently believed that a consortium of Jewish economists was twisting economic data to embarrass him. Although he ignored the order at first, Malek eventually complied. This fact first came out in the 1981 book *The Final Days* by Bob Woodward and Carl Bernstein, but it was in 1988 when the *Washington Post* put the story on the front page, that Malek was forced to resign as head of the Republican National Committee. Even though the Anti-Defamation League has formally accepted Malek's apologies, and its national director, Abraham Foxman, has said "enough is enough," the memory of that earlier event still seems to haunt Malek's career.

Point of View

APPLYING THE DIAGNOSTIC MODEL
Diagnostic Question 10: Do you take the point of view of a judge or a coach when evaluating others? How might each of these views affect your judgments?

APPLYING THE DIAGNOSTIC MODEL
Diagnostic Question 11: How are your present judgments influenced by judgments that you or others have made in the past?

One indication of the selectivity of retrieval can be found in studies that have asked subjects to take different points of view in recalling past information. For example, in one study subjects were asked to read a description of a house and then, at the retrieval stage, to recall as much as possible about the house. One set of subjects was asked to do this from the point of view of a prospective house buyer, whereas the other group were asked to take on the perspective of a burglar. Different details of the house were recalled depending on which perspective the subject took.[28] This shows clearly how memory involves a backward search from a current perspective to a desired piece of information.

Similar point-of-view effects can be found within organizations in the area of performance appraisal. Performance information is collected for different purposes and the purpose of an evaluation seems to affect the kind of information that managers are able to recall. In one study, managers who took the point of view of a coach and were asked to do an evaluation to aid employee development were able to recall much more specific behavioral information than managers who were asked to take the point of view of a judge whose ratings would determine merit raises.[29] Another study that showed the same kind of effect suggests that stereotypes have a greater influence on ratings when they are used in making administrative decisions than when the information is collected for other reasons. In this study, raters who held traditional stereotypes about women were much more likely to show bias in their ratings of female employees when the decision dealt with pay raises than when the ratings were collected for "research purposes."[30]

[27] J. S. Phillips, "The Accuracy of Leadership Ratings: A Categorization Perspective," *Organizational Behavior and Human Performance* 33 (1984), 125–38.

[28] R. C. Anderson and J. W. Pichert, "Recall of Previously Unrecallable Information Following a Shift in Perspective," *Journal of Verbal Learning and Verbal Behavior* 17 (1981), 1–12.

[29] S. Zedeck and W. F. Cascio, "Performance Appraisal Decisions as a Function of Rater Training and Purpose of the Appraisal," *Journal of Applied Psychology* 67 (1982), 752–58.

[30] G. H. Dobbins, R. L. Cardy and D. M. Truxillo, "The Effects of Purpose of Appraisal and Individual Differences in Stereotypes of Women on Sex Differences in Performance Ratings: A Laboratory and Field Study," *Journal of Applied Psychology* 73 (1988), 551–58.

Performance Cues

Another indication of the selectivity of the retrieval process comes from research on the "performance-cue effect."[31] In these studies, subjects are placed in groups and assigned a task. After performing the task for some period of time, the groups are given feedback on how well they have performed in comparison with other groups. Half of the groups are then selected randomly and told they performed very well, and the rest are told they performed poorly. The groups are then asked to describe group processes and the effectiveness of their leaders. Despite the fact that the two sets of groups actually performed at the same levels on similar tasks, group members invariably recall different processes, depending on the feedback they received. Groups who are told they did well recall positive aspects of leadership and group processes, whereas just the opposite occurs in groups who are told they did poorly.

This type of bias shows how dangerous it is to rely on retrospective reports of what certain individuals or companies did to achieve their success (recall our discussion of the "excellence" books in Chapter 3). In looking back on their success, individuals may recall things selectively, deleting things that may have been important and adding things that may never have occurred. Knowledge of the outcome may create serious biases in terms of what is recalled and what is forgotten. Thus the trustworthiness of the retrieval process should not go unquestioned. As we said in Chapter 3, it is important to collect information from multiple sources, using a number of different methods.

Increasing the Accuracy of Retrieval

APPLYING THE DIAGNOSTIC MODEL
Diagnostic Question 9: Do you keep external records of important past events, or do you rely exclusively on your own memory for important information?

One way to reduce demands on one's ability to recall information accurately is to maintain physical records of incidents or behaviors. Such logs, or *behavioral diaries*, can then be consulted whenever important judgments need to be made. In this way we can short-circuit our tendency to selectively recall some incidents and not others. For example, the results of one study on diary keeping found that raters who regularly recorded critical incidents of employee effectiveness showed much less evidence in their ratings of leniency and halo bias. The level of agreement among diary-keeping raters was also higher than the level of agreement among raters who relied on memory alone.[32]

Because recalling information from different points of view or for different purposes seems to elicit different information, having observers take *multiple perspectives* can increase the overall amount of information that is available for making judgments. Thus a rater might try to take the point of view of the person being rated, of her coworkers, or of her clients. Just going through this process will force the rater to consider the fact that other people's opinions may differ from hers and may force her to recall information that though perhaps buried, is yet retrievable.

JUDGMENT STAGE

In the final **judgment stage** of the process, the information available for use—the data that have survived to this stage in the perceptual process— must be condensed into one or more judgments. Turn back to Figure 5-4,

judgment stage The stage of the information processing cycle in which recalled information is weighted and aggregated to come up with a single overall judgment.

---

[31] M. C. Rush and L. L. Beauvais, "A Critical Analysis of Format Induced Versus Subject Imposed Bias in Leadership Ratings," *Journal of Applied Psychology* 66 (1981), 722–27.

[32] H. J. Bernardin and C. S. Walters, "Effects of Rater Training and Diary-Keeping on Psychometric Error in Rating," *Journal of Applied Psychology* 62 (1977), 64–69.

which shows three students' evaluations of an instructor on ten specific dimensions of performance and one overall judgment. Note that Student B's overall judgment is not the average of her ten specific ratings. Rather, it is more negative than her average rating and can probably be traced to the fact that this student thought that some dimensions were more important than others. This weighing process may differ for different people. For example, Student C, whose ratings of the instructor on the ten specific dimensions were identical to Student B's ratings came up with a different overall rating. It might be a little discouraging to find that even though Students B and C agreed on all ten specific points, they rendered different overall judgments when summarizing their perceptions. Thus, perceptual differences can result when raters apply different degrees of importance to various dimensions.

Not only can different raters give different weights to various pieces of information, but the same individual may apply different weights in rating different people. Suppose that Students B and C are actually the same person, and that B represents the student's rating of an instructor in a quantitative course where, for example, "organization" is weighted more heavily than "stimulation of discussion," whereas C represents a rating of a more qualitative course to which the student applies an opposite pattern of emphasis. Or, consider the possibility that the same person could apply different patterns of emphasis at different points in time. For example,"organization" might be viewed as more important for a freshman class, and "stimulation of class discussion" more important in an upper-level, elective course.

Effects of Earlier Judgments

If you look back at Figure 5-3 you will see that it shows an arrow going from the judgment stage back to the beginning of the perceptual process. This arrow represents the dynamic nature of the entire perception and judgment cycle. That is, not only do perceptions give rise to judgments, but judgments influence subsequent perceptions. As we will see, the **assimilation effect** is the tendency for present judgments to be biased in the direction of past judgments. Two of the ways in which assimilation can occur are priming and the confirmation bias.

Priming. Research shows that the ease with which information can be recalled affects the degree to which it is used in forming judgments.[33] **Priming** raters simply means forcing them to recall one set of events and then very soon afterward asking them to make a judgment to which the information just retrieved may be relevant. Because the primed information was so recently recalled, it is easier to retrieve than other information and thus is likely to be used in forming the required judgment.

For example, suppose that a sales manager has just received a call from an irate customer who is upset about a salesperson's behavior. Further, suppose that the salesperson's annual performance appraisal was conducted that same day. The manager may give this very recent, or primed, information a lot more weight in the annual review than she will give experiences with this salesperson that accrued earlier in the year. Past judgments may act the same way in influencing future perceptions and judgments. Indeed, one of the first things a rater often thinks of when making an evaluation is, How did I evaluate this in the past?

---

[33] A. Tversky and D. Kahneman, "Availability: A Heuristic for Judging Frequency and Probability," *Cognitive Psychology* 5 (1973), 207–32.

Confirmation bias. Assimilation can also be brought about through **confirmation bias**, the tendency for raters to seek out information that reaffirms their earlier judgments while discounting evidence that runs counter to earlier impressions.[34] This kind of bias seems particularly prevalent when ratings are made public, because raters often feel compelled to confirm past judgments so as to appear to be consistent.[35] Thus whereas priming seems to affect primarily the retrieval part of the perception process, confirmation bias seems largely to affect what observers attend to and how they interpret it.

Making Better Judgments

One way to improve the observers' accuracy is to *simplify and reduce the number of judgments* that are required, so that there is less subjective processing. Asking the rater to be a witness of behavior, rather than a judge of others' intentions or personality characteristics simplifies the task considerably. It reduces the need to make complex inferences and interpretations.

Also, emphasis should be placed on asking raters to form judgments only on those behaviors or dimensions that they have had *ample opportunity to observe*. Asking an interviewer to infer the intelligence of a job applicant after talking to him for 30 minutes is asking a lot. There are better means of collecting this kind of information, and it is more useful to focus on things that can be observed in the context of an interview, such as the degree to which a person can express herself verbally or the ease with which she handles a somewhat stressful situation.

Third, the overall context in which the evaluation is made should not be overlooked. People will make more accurate judgments in a *supportive climate*, that is, when they are given positive feedback for making accurate judgments, especially if those judgments are harsh and likely to engender negative reactions on the part of the person being rated. For example, a study by Margaret Padgett showed that in one organization, 70 percent of the supervisors intentionally inflated their ratings of subordinates, particularly when a rater felt that subordinates would react negatively to poor ratings. Raters did not feel free, in these circumstances, to be honest.[36] Such intentional biasing of ratings is most trou-

---

[34] J. Darley and R. Fazio, "Expectancy Confirmation Processes in the Social Interaction Sequence," *American Psychologist* 35 (1980), 867–81.

[35] M. Bazerman, R. Beekun and F. Schoorman, "Performance Evaluation in a Dynamic Context: A Laboratory Study of the Impact of Prior Commitment to the Ratee," *Journal of Applied Psychology* 67 (1982), 873–76.

[36] M. Padgett, "Performance Appraisal in Context: Motivational Influences on Performance Ratings" (Paper Presented at the Annual Meeting of the National Academy of Management, Washington, D.C., August 1989, p. 11).

One way to improve the accuracy of human judgment is to simplify the perceptual task. Advertising agencies like Romann and Tannenholz enlist the help of consumers in testing brands of various products. This group is testing the tastes of different toothpastes. By focusing their attention on a single aspect of a specific product, testers may be able to provide more accurate ratings of a number of brands.
Source: Fortune, *December 3, 1990, p. 40.*

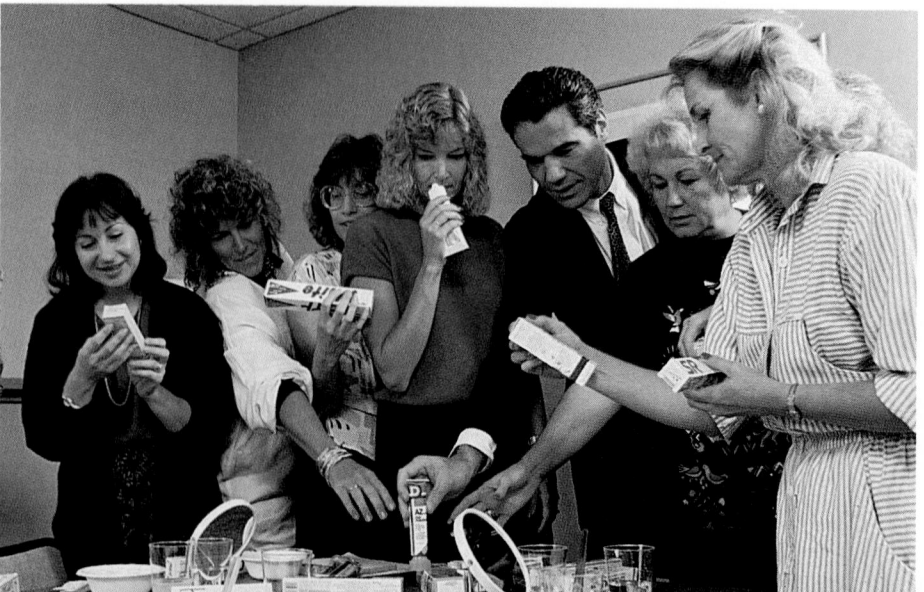

blesome, for it is hard enough to get accurate ratings when everyone is trying his best. If raters feel that accuracy is not in their own best interests, how can we strive to improve perceptual accuracy?

SUMMARY

APPLYING THE DIAGNOSTIC MODEL

Diagnostic Question 12: Which of the many ways to increase perceptual accuracy are most practical for your organization? How might different means of enhancing accuracy be more or less appropriate for different organizational needs (e.g., interviewing, performance appraisal, problem diagnosis)?

A thorough understanding of the perceptual process, that is, the process by which people encode and make sense out of the complex world around them, is critical to those who would manage organizational behavior. The very existence of perceptual illusions documents the fact that what we perceive is not always a very close approximation of objective reality. Accurately perceiving one's environment enhances the ability to both describe and diagnose current organizational problems. It also helps to ensure that the prescriptions and actions taken to solve those problems are directed toward the correct source. Accuracy in the perception of persons is especially critical. This type of perceptual accuracy takes four forms: *elevation accuracy*, *differential elevation accuracy*, *stereotype accuracy*, and *differential accuracy*. Different kinds of organizational problems require different kinds of accuracy. It is important to recognize this because some types of accuracy are easier than others to achieve.

Perceptual processing is either *controlled* or *automatic* and moves through five stages: *attention*, *organization*, *interpretation*, *retrieval* and *judgment*. At the *attention stage*, a small subset of all the information that is available to the five senses is selected for processing. The degree to which any stimulus attracts attention is a complex function of characteristics of the object and of the perceiver. At the *organization stage*, information is simplified. Complex behavioral sequences are converted into *scripts*, and people are represented by *prototypes*. A number of biases, including *stereotyping* and the *halo error*, can creep into this complex process. In the *interpretation stage* the condensed information is then interpreted, as perceivers employ *projection* or make *attributions* that allow them to understand why some object or person is as it is. If the information processed is not used immediately, it is stored in memory and later recovered in the *retrieval stage*. There is a tendency to lose information that is inconsistent with scripts and prototypes. Finally, at the *judgment stage*, the information processed in the four prior stages is used to come up with an evaluation of an object, person, or event. This evaluation, once made, often affects decisions, behaviors, and subsequent perceptions.

REVIEW QUESTIONS

1.   Which of the four types of perceptual accuracy require the greatest amount of controlled information processing? Which might be achieved through less energy-consuming, automatic information-processing strategies?

2.   List a set of traits that would make up the prototype for a yuppie, a hippie, an absent-minded professor, and a card-carrying member of the American Civil Liberties Union. Recalling Chapter 4, is your list dominated by ability or personality characteristics? What category of abilities (physical, psychomotor, cognitive) or personality characteristics (social traits, motives, personal conceptions, emotional adjustment, personality dynamics) is most heavily represented? What does this tell you about how prototypes are developed and in what ways they are most likely to be accurate?

3.   Sometimes the same behavioral episode in an organization—for example a fight among coworkers, a botched work assignment, or an ineffective meeting—

can be organized perceptually along the lines of a script or a prototype. How might the decision to use one or the other of these schemas affect things that occur later in the judgment process, such as attribution and memory recall?

4.   Think of four kinds of perceptual biases. Do each of these biases affect the diagnostic approach to problem solving (description-diagnosis-prescription-action) in the same way, or are some more problematic at certain stages than others?

5.   Subliminal advertising is restricted by law. There are limits on the number of images that a television viewer can be exposed to over certain specified time intervals. What is the basis for this legislation? In what ways is this kind of advertising categorically different from other subtle types of influence that are not restricted?

DIAGNOSTIC QUESTIONS

The following questions attempt to capture some of the major points of this chapter. They are designed to help you to diagnose perceptual problems in analyzing both case studies in this book and on-the-job problems now or in the future.

1.   On whose perceptions do you rely to describe and diagnose organizational problems? How might the choice of this person affect the description and diagnosis provided?

2.   Which of the four types of accuracy is required for specific organizational decisions that must be made based on perceptual data? Such decisions include promotion decisions and decisions about group or individual training.

3.   What characteristics of persons or objects that you have to evaluate "pull" your attention? How do these characteristics affect your judgments regarding these persons or objects?

4.   What kinds of expectations do you have for persons or objects that you need to evaluate? How might these expectations affect your judgment?

5.   What are some of the major events in your organization that you conceive in terms of scripts? How might your version of these scripts differ from others' views of the same scripts?

6.   What are some of the major prototypes that exist for persons in your organization? How accurate are these prototypes?

7.   What stereotypes do you hold of persons from particular social groups in your organization? Are these stereotypes accurate?

8.   Are your reasons for interpreting other people's actions self-serving? Do you project any of your own socially undesirable characteristics onto others?

9.   Do you keep external records of important past events, or do you rely exclusively on your own memory for important information?

10.   Do you take the point of view of a judge or a coach when evaluating others? How might each of these views affect your judgments?

11.   How are your present judgments influenced by judgments that you or others have made in the past?

12.   Which of the many ways to increase perceptual accuracy are most practical for your organization? How might different means of enhancing accuracy be more or less appropriate for different organizational needs (e.g., interviewing, performance appraisal, problem diagnosis)?

EXERCISE 5-1

SIX STYLES OF THINKING[1]

JANET MILLS, *Boise State University*

Perception and judgment processes involve many different styles of thinking, some more logical and analytical, others emotional or speculative. According to Edward deBono,[2] when we try to organize and interpret sensory data, different modes of thinking—for example, reasoning, intuiting, questioning observing—may crowd in all at once. To prevent the confusion that can result, deBono proposes that we teach people to do one kind of thinking at a time. To help you develop your thinking skills, this exercise introduces you to deBono's six styles of thinking (see Exhibit 5-1) and lets you practice using one style of thinking at a time.

[1] Adapted with the author's permission from an exercise entitled "Six Thinking Hats: An Exercise to Combat Confusion and Develop Thinking Skills."

[2] Edward deBono, *Six Thinking Hats* (Boston, MA: Little, Brown and Company, 1985).

EXHIBIT 5-1
Six Styles of Thinking

STYLE OF THINKING	WHAT IT IS	WHAT IT ISN'T
Factual-Literal	Neutral Objective Rational Mechanical	Argumentative Interpreting Extrapolating Opinion giving Persuading
Emotional-Intuitive	Emotional Based on gut reactions, hunches, intuition, personal taste, aesthetic preference	Justifying Explaining Accounting for Logical Consistent
Logical-Negative	Looking for problems Fault finding Objecting to alternatives Criticizing Focusing on risks and dangers Confronting factual-literal thinking	Emotional Argumentative Subjective
Speculative-Positive	Looking for opportunities Constructive Optimistic Focusing on benefits Emphasizing practical value Using dreams and visions Probing, exploring Finding logical support Making things happen	Emotional Intuitive Naive Unrealistic
Innovative-Creative	Developmental Fertile, nurturing Generating alternatives Getting new ideas Approaching old problems in new ways Being absurd, humorous, playful	Logical Judgmental Negative
Controlled-Regulated	Managing thinking Planning and organizing Monitoring and controlling Giving all types of thinking their turn Formally structuring thinking	Persuading Advocating Criticizing

Step One: Pre-Class Preparation

In class you will use deBono's thinking styles to make an important decision. To prepare for this class work, study Exhibit 5-1 to familiarize yourself with the six styles of thinking. Then read the rest of the exercise so you will know what to do.

Step Two: Thinking in a Single Style

Your instructor will divide your class into groups of four to six members each and will assign each group one of deBono's thinking styles. (Any number of groups may be formed, so long as the total number of groups is a multiple of six.) Each group should then review the thinking style it has been assigned, describing the style in detail and giving examples of thinking in that style. Next, the group should try thinking in its assigned style as it undertakes the following assignment:

> This term your class will perform exercises allowing you to experience personally many of the challenges of being a manager. Although you created temporary groups to perform the exercise in Chapter 4, you should now divide into permanent groups of four to six members each that will remain together for the rest of the term. The decision you must make today is, How should these groups be formed? At the end of today's session, your decision will actually be implemented, and permanent groups will be organized.

Before leaving this step, every member of each group should be completely familiar with the group's thinking style and be able to illustrate its use to others.

Step Three: Thinking in Multiple Styles

Your instructor will now reorganize the class into new groups made up of one person from each of the six different kinds of thinking-style groups. In these mixed-style groups you will again reach a decision about how to form permanent groups. It is important that as you complete this step you use the thinking style you learned in Step One. Once your group has made a decision, choose a spokesperson to report it to the class and to describe the way it was made. The spokesperson's presentation should last about five minutes.

Step Four: Class Discussion and Permanent Group Formation

As each group's spokesperson reports on its decision, all decisions should be listed on a chalk board, flip chart, or overhead projector transparency. After every group has presented its decision, the class as a whole must reach a final decision and form its permanent groups.

During this final judgment process, you should discuss questions like the following:

1. In what ways is each of the six thinking styles useful? What drawbacks are associated with each style?
2. Were any of the styles more powerful than others in influencing the decisions reached in Step Three? Were any of the styles virtually ignored?
3. Did any of the styles conflict with others? How did the groups deal with such conflicts?
4. What challenges do differences among the six styles create for managers engaged in decision making? How can the benefits of the various thinking styles be balanced against their costs?

Conclusion

Managing organizational behavior involves thinking processes that begin with perception and culminate in judgment and decision making. As this exercise illustrates, no single style of thinking is likely to be adequate during the entire process. Managers must be able to use different thinking styles, each to its best advantage. As Edward deBono suggests, the way to do this may be to use one style of thinking at a time.

DIAGNOSING ORGANIZATIONAL BEHAVIOR

Freida Mae Jones was born in her grandmother's Georgia farmhouse on June 1, 1949. She was the sixth of George and Ella Jones's ten children. Mr. and Mrs. Jones moved to New York City when Freida was four because they felt that the educational and career opportunities for their children would be better in the North. With the help of some cousins, they settled in a five-room

CASE 5-1

FREIDA MAE JONES*

MARTIN R. MOSER, *University of Lowell*

PERCEPTION AND JUDGMENT

apartment in the Bronx. George worked as a janitor at Lincoln Memorial Hospital, and Ella was a part-time housekeeper in a nearby neighborhood. George and Ella were conservative, strict parents. They kept a close watch on their children's activities and demanded they be home by a certain hour. The Jones believed that because they were black, the children would have to perform and behave better than their peers to be successful. They believed that their children's education would be the most important factor in their success as adults.

Freida entered Memorial High School, a racially integrated public school, in September 1963. Seventy percent of the student body was Caucasian, 20 percent black, and 10 percent Hispanic. About 60 percent of the graduates went on to college. Of this 60 percent, 4 percent were black and Hispanic and all were male. In the middle of her senior year, Freida was the top student in her class. Following school regulations, Freida met with her guidance counselor to discuss her plans upon graduation. The counselor advised her to consider training in a "practical" field such as housekeeping, cooking, or sewing, so that she could find a job.

George and Ella Jones were furious when Freida told them what the counselor had advised. Ella said, "Don't they see what they are doing? Freida is the top-rated student in her whole class and they are telling her to become a manual worker. She showed that she has a fine mind and can work better than any of her classmates and still she is told not to become anybody in this world. It's really not any different in the North than back home in Georgia, except that they don't try to hide it down South. They want her to throw away her fine mind because she is a black girl and not a white boy. I'm going to go up to her school tomorrow and talk to the principal."

As a result of Mrs. Jones's visit to the principal, Freida was assisted in applying to ten Eastern colleges, each of which offered her full scholarships. In September 1966, Freida entered Werbley College, an exclusive private women's college in Massachusetts. In 1970, Freida graduated summa cum laude in history. She decided to return to New York to teach grade school in the city's public school system. Freida was unable to obtain a full-time position, so she substituted. She also enrolled as a part-time student in Columbia University's Graduate School of Education. In 1975 she had attained her Master of Arts degree in Teaching from Columbia but could not find a permanent teaching job. New York City was laying off teachers and had instituted a hiring freeze because of the city's financial problems.

Feeling frustrated about her future as a teacher, Freida decided to get an MBA. She thought that there was more opportunity in business than in education. Churchill Business School, a small, prestigious school located in upstate New York, accepted Freida into its MBA program.

Freida completed her MBA in 1977 and accepted an entry-level position at the Industrialist World Bank of Boston in a fast-track management development program. The three-year program introduced her to all facets of bank operations, from telling to loan training and operations management. She was rotated to branch offices throughout New England. After completing the program she became an assistant manager for branch operations in the West Springfield branch office.

During her second year in the program, Freida had met James Walker, a black doctoral student in business administration at the University of Massachusetts. Her assignment to West Springfield precipitated their decision to get married. They originally anticipated that they would marry when James finished his doctorate and could move to Boston. Instead, they decided he would pursue a job in the Springfield-Hartford area.

Freida was not only the first black but also the first woman to hold an executive position in the West Springfield branch office. Throughout the training program Freida felt somewhat uneasy although she did very well. There were six other blacks in the program, five men and one woman, and she found support and comfort in sharing her feelings with them. The group spent much of their free time together. Freida had hoped that she would be located near one or more of the group when she went out into the "real world." She felt that although she was able to share her feelings about work with James, he did not have the full appreciation or understanding of her co-workers. However, the nearest group member was located one hundred miles away.

Freida's boss in Springfield was Stan Luboda, a fifty-five-year-old native New Englander. Freida felt that he treated her differently than he did the other trainees. He always tried to help her and took a lot of time (too much, according to Freida) explaining things to her. Freida felt that he was treating her like a child and not like an intelligent and able professional.

"I'm really getting frustrated and angry about what is happening at the bank," Freida said to her husband. "The people don't even realize it, but their prejudice comes through all the time. I feel as if I have to fight all the time just to start off even. Luboda gives Paul Koehn more responsibility than me and we both started at the same time, with the same amount of training. He's meeting customers alone and Luboda has accompanied me to each meeting I've had with a customer."

"I run into the same thing at school," said James. "The people don't even know that they are doing it. The other day I met with a professor on my dissertation committee. I've known and worked with him for over three years. He said he wanted to talk with me about a memo he had received. I asked him what it was about and he said that the records office wanted to know about my absence during the spring semester. He said that I had to sign some forms. He had me confused with Martin Jordan, another black student. Then he realized that it wasn't me, but Jordan he wanted. All I could think was that we all must look alike to him. I was angry. Maybe it was an honest mistake on his part, but whenever something like that happens, and it happens often, it gets me really angry."

"Something like that happened to me," said Freida. "I was using the copy machine, and Luboda's secretary was talking to someone in the hall. She had just gotten a haircut and was saying that her hair was now like Freida's—short and kinky—and that she would have to talk to me about how to take care of it. Luckly, my back was to her. I bit my lip and went on with my business. Maybe she was trying to be cute, because I know she saw me standing there, but comments like that are not cute, they are racist."

"I don't know what to do," said James. "I try to keep things in perspective. Unless people interfere with my progress, I try to let it slide. I only have so much energy and it doesn't make sense to waste it on people who don't matter. But that doesn't make it any easier to function in a racist environment. People don't realize that they are being racist. But a lot of times their expectations of black people or women, or whatever, are different because of skin color or gender. They expect you to be different, although if you were to ask them they would say that they don't. In fact, they would be highly offended if you implied that they were racist or sexist. They don't see themselves that way."

"Luboda is interfering with my progress," said Freida. "The kinds of experiences I have now will have a direct effect on my career advancement. If decisions are being made because I am black or a woman, then they are racially and sexually biased. It's the same kind of attitude that the guidance counselor had when I was in high school, although not as blatant."

In September 1980, Freida decided to speak to Luboda about his treatment of her. She met with him in his office. "Mr. Luboda, there is something that I would like to discuss with you, and I feel a little uncomfortable because I'm not sure how you will respond to what I am going to say."

"I want you to feel that you can trust me," said Luboda. "I am anxious to help you in any way I can."

"I feel that you treat me differently than you treat the other people around here," said Freida. "I feel that you are overcautious with me, that you always try to help me, and never let me do anything on my own."

"I always try to help the new people around here," answered Luboda. "I'm not treating you any differently than I treat any other person. I think that you are being

a little too sensitive. Do you think that I treat you differently because you are black?"

"The thought had occurred to me," said Freida. "Paul Koehn started here the same time that I did and he has much more responsibility than I do." (Koehn was already handling accounts on his own, while Freida had not yet been given that responsibility.)

"Freida, I know you are not a naive person," said Luboda. "You know the way the world works. There are some things which need to be taken more slowly than others. There are some assignments for which Koehn has been given more responsibility than you, and there are some assignments for which you are given more responsibility than Koehn. I try to put you where you do the most good."

"What you are saying is that Koehn gets the more visible, customer contact assignments and I get the behind-the-scenes running of the operations assignments," said Freida. "I'm not naive, but I'm also not stupid either. Your decisions are unfair. Koehn's career will advance more quickly than mine because of the assignments that he gets."

"Freida, that is not true," said Luboda. "Your career will not be hurt because you are getting different responsibilities than Koehn. You both need the different kinds of experiences you are getting. And you have to face the reality of the banking business. We are in a conservative business. When we speak to customers we need to gain their confidence, and we put the best people for the job in the positions to achieve that end. If we don't get their confidence they can go down the street to our competitors and do business with them. Their services are no different than ours. It's a competitive business in which you need every edge you have. It's going to take time for people to change some of their attitudes about whom they borrow money from or where they put their money. I can't change the way people feel. I am running a business, but believe me I won't make any decisions that are detrimental to you or to the bank. There is an important place for you here at the bank. Remember, you have to use your skills to the best advantage of the bank as well as your career."

"So what you are saying is that all things being equal, except my gender and my race, Koehn will get different treatment than me in terms of assignments," said Freida.

"You're making it sound like I am making a racist and sexist decision," said Luboda. "I'm making a business decision utilizing the resources at my disposal and the market situation in which I must operate. You know exactly what I am talking about. What would you do if you were in my position?"

When you have read this case, look back at the chapter's diagnostic questions and choose the ones that apply to the case. Then use those questions with the ones that follow in your case analysis.

1. Were Stan Luboda's perceptions of Freida Mae Jones accurate? What perceptual bias might explain the way Stan treated Freida? Might this same sort of bias explain his secretary's comments?

2. How might Stan's perceptual bias have affected his judgment of Freida's ability to do the same work as Paul Koehn? How did that judgment affect Freida's ability to put her knowledge and skills to the fullest use?

3. How can biases like those illustrated in this case undermine the performance of an organization? What can a firm do to improve its managers' perceptual accuracy? Why do such actions make good sense, both socially and financially?

CASE 5-2

WORLD INTERNATIONAL AIRLINES, INC.*

P. D. JIMERSON, *General Mills, Inc.*
DAVID L. FORD, *University of Texas at Dallas*

BACKGROUND

World International Airlines is a foreign-based multinational commercial air carrier. The corporate offices for its western hemisphere operations are located in New York City, New York. The company employs many hundreds of multilingual and multicultural employees, since its operations maintain World International terminals in South America, Central America, Mexico, the United States and Canada, all of which comprise the western hemisphere territory. In all of the continents the district managers, whose territory may involve several countries or, in the case of the United States, several states, are usually multilingual Europeans. The assistant managers are usually nationals of the country. The general manager of the western hemisphere is a native Spaniard while the personnel manager is a Spanish-American. Both the general manager and personnel manager are multilingual and multicultural. While many air carriers in the western hemisphere have experienced strikes and work stoppages in the past, there is no history of strikes having ever occurred at World International Airlines.

The present general manager is a man in his mid-fifties. He has worked his way up to the top of the western hemisphere's organization. He has hand-picked all of the men who are district managers in each of the

* Reprinted with the authors' permission.

aforementioned countries. He knows over 90 percent of the company's employees in the New York offices on a personal basis and is well liked by all of his subordinates. On one occasion when he was away attending a National Training Laboratories-sponsored workshop in California, the employees from the New York offices surprised him with a huge birthday cake, complete with the decorations and even champagne. His job performance has earned him influence and power in Spain. In addition, the present general manager's educational background is more along the classical line typical of many Spaniards in his socio-economic class (i.e., law, engineering, etc.) as opposed to a more applied business and management education background.

CHANGE IN SENIOR PERSONNEL CREATES PROBLEMS

The company is in a state of flux. The present general manager, John Nepia, is scheduled to be transferred to Barcelona, Spain and a new man, Stephen Esterant, has been sent to replace him. The present general manager has had little or no input into the selection of the new general manager. However, the incoming general manager has an outstanding record in the eastern hemisphere, and it is rumored that he is being groomed for something big. Coupled with this impending change in senior personnel is the fact that international flights are currently in a state of flux since a review committee is currently deliberating on a new rate structure.

John Nepia was scheduled to depart New York on April 30. Stephen Esterant arrived March 20. There was to be at least a 30-day transition period before the departure of Mr. Nepia. Problems on the setting of the international rate structure became acute on or about April 20. Mr. Nepia's departure was delayed and termed indefinite, since he was actively participating in the rate-setting negotiation with the FAA (Federal Aviation Administration).

During the transition period Stephen was to make himself acquainted with all of the district general managers as well as the rank and file employees. Stephen visited all of the district offices; he met and talked with district managers, sales personnel and operations personnel, and he made comments wherever he felt that company policy was not being followed. He seldom found anything worthy of praise if it did not comply with established company policy.

The first sign of difficulty came when the corporate chauffeur in New York asked to speak to John Nepia, the outgoing general manager. The outgoing general manager had maintained a policy of being accessible to any of the company's employees. The driver explained the following:

> I been with this company for five years now. I like my work and I like my job. But I don't believe that I should get less respect because I'm a driver. I don't like the idea of having to drive Mr. Esterant's wife and her friends around on a shopping trip in downtown New York. I don't think it's part of my responsibility to walk his dog or carry his wife's packages. I realize that I *work* for him, but I refuse to be treated as though I were his *servant*. I decided that if this treatment continues on his part, I will have to find out what grievance procedure is available and file an official complaint.

John Nepia was quick to assure the driver that the matter would be looked into and he would get in touch with him as soon as he knew more about the situation.

Other rumblings came from the operations employees. They contended that Stephen Esterant was thoughtless, unappreciative and distant in his interactions with them. They further believed that he felt and acted "too damn superior." For example, one of the operative employees related the following story. "Once Stephen visited the baggage-handling area at Chicago where a new computerized routing system was being tested. He had worked with a similar system before and immediately spotted some procedures which would increase the efficiency of operation. He proceeded to tell the employees that they didn't know what they were doing and questioned their intelligence."

The most recent sign of major discord came when the district managers sent a plea to John Nepia begging him to implore Spain to recall the new general manager. In their opinion, morale had suffered greatly, and Stephen was the direct cause.

The outgoing general manager and Jason Du-Bryne, the director of personnel, were good and long-time friends. They had survived many crises together. Therefore, John called in his trusted friend to seek advice and ponder their problem and their possible courses of action.

Jason was considered by many of the employees to be a firm, but fair administrator. He often prided himself on the fact that he was always available to talk to and help his people. He was often consulted by members of the firm concerning interpersonal matters. These consultations often concerned private as well as corporate issues.

During their meeting, the personnel manager acknowledged to John that the situation was indeed grave; however, at no point in the conversation did he indicate what his personal beliefs were concerning the problem. He stated that he did not believe that the heir apparent was technically incompetent. He also suggested the possibility that the heir apparent just did not understand the way of doing business in the western hemisphere. The meeting ended with John Nepia deciding that a conversation with Stephen Esterant was needed.

The New General Manager's Viewpoint

Stephen Esterant was named to head the western hemispheric operations of World International Airlines, Inc. as a reward for his outstanding service as a district manager in Spain. He was told that he was selected because he had been able to bring district offices into compliance with company operations policy and to maintain or increase sales volume at the same time.

Stephen had served in five other district posts prior to receiving this promotion. He was 32 years old and married to a lovely woman who was a member of a wealthy and influential family in Spain. In fact, Stephen's wife's family was one of a few wealthy families owning a substantial portion of World International Airlines stock. Stephen was a man who knew what he wanted and he knew how to get it. He moved briskly about his affairs asking no favors *from* anyone and giving no favors *to* anyone. He appeared to be the coming star in the organization.

Stephen Esterant received a memo from John Nepia requesting that he meet with him and Jason DuBryne, the personnel director, about a matter of apparent great importance. En route to the meeting he pondered over what would possibly be discussed. He, of course, had a few items on his own agenda. Since coming to New York, Stephen had become aware of several problems involved in his becoming the new general manager. He was displeased by the apparent lack of respect given to him by his subordinates as well as the "cocky" attitude of the hourly employees. He was sure that John and Jason were aware of the attitude problems and yet he could not understand why they had not dealt with these matters sooner and in a stronger manner. From Stephen's point of view, there was a need to run a tight ship, as he had done in the eastern hemisphere. He obviously had a distaste for the hourly employees' practice of calling managers by their first names and a lack of deference to those in authority, as was often done not only in the New York offices, but also throughout the rest of the western hemisphere operations. He also wanted to tell John and Jason that he needed to have them run less interference for him. Since he was soon to be general manager, he believed that he should start to handle inter-group conflict and decide about policy disputes so that the organization could easily recognize its new boss and leader. He resolved that if the opportunity arose in the meeting, he would raise these issues with John and Jason.

As he reached the door to John's office, Stephen turned the knob and jauntily entered the office to meet with John Nepia and Jason DuBryne, not really knowing what to expect.

When you have read this case, look back at the chapter's diagnostic questions and choose the ones that apply to the case. Then use those questions with the ones that follow in your case analysis.

1. How accurate were Stephen Esterant's perceptions of western hemisphere employees? What perceptual bias did he display? How did this bias affect his behavior?

2. Was Stephen the only person in this case whose perceptions were faulty? How accurate were the western-hemisphere employees' perceptions of Stephen? Did they display any perceptual biases?

3. What short-term action should John Nepia have taken to resolve the problems developing between Stephen and other airline employees? What long-term actions should the company have taken to prevent similar future problems?

CASE 5-3

PRECISION MACHINE TOOL

Read Chapter 6's Case 6-1, "Precision Machine Tool." Next, look back at Chapter 5's diagnostic questions and choose the ones that apply to that case. Then use those questions with the ones that follow in this first analysis of the case.

1. John Garner and Tom Avery reacted very differently to the prospect of selling their company to Ako Wang. Based on what you have learned in Chapter 5, explain why.

2. In what ways were each man's perceptions accurate? What inaccuracies, if any, can you identify? Did they show any perceptual biases? Which of the two men's perceptions do you think were the least biased? Why?

3. John and Tom had to evaluate alternatives, decide what to do, and then contact Ako Wang to inform him of their decision. As they went through these steps, how best could they have controlled for perceptual biases?

C H A P T E R 6

DECISION MAKING

According to Donald Regan, President Reagan's former chief of staff, Nancy Reagan's insistence on consulting astrologer Joan Quigley before any important decisions were made about the president's activities began "to interfere with the normal conduct of the presidency." Regan says in his memoir that for a time he kept a calendar on his desk with "good" days highlighted in green, "bad" in red, and "iffy" in yellow. That way he could remember when "it was propitious to move the President of the U.S. from one place to another, or schedule him to speak in public, or commence negotiations with a foreign power."

Source: *Barret Seaman, "Good Heavens!" Time, May 16, 1988, 25; Donald T. Regan, "For the Record," Time, May 16, 1988, pp. 26–36.*

In May of 1988, much of the world was shocked to learn that for some years decisions about the schedule of the President of the United States had been made at least partially in consultation with San Francisco astrologer Joan Quigley. In 1981 Quigley showed Ronald Reagan's wife, Nancy, that astrological charts could have predicted extreme danger for the president on March 30 of that year—the day that John Hinckley, Jr., attempted to assassinate the president. Obsessed with her fear for her husband's safety and convinced of Quigley's power to protect him, Nancy Reagan resolved that from that day on, no presidential public appearance would be confirmed without Quigley's seal of approval. Over just one four-month interval, Quigley forbade the president's public exposure on 47 days. Moreover, she characterized three of those four months as generally "bad."[1]

Many insiders consider the president's schedule the single most potent tool of the White House because it determines what the most powerful person in the Western world is going to do, with whom, and when. The fact that such important decisions as the timing of the 1987 Geneva summit talks were being influenced by an astrologer immediately captured the attention of the entire world. Most people have little respect for astrological interpretations and predictions, seeing them as baseless advice found in some newspapers—usually next to the comics. In a letter to the editor of *Time* magazine one person declared, "No wonder our government is in such confusion. It is ruled by a man who is ruled by his wife who is ruled by a friend who is ruled by the stars."[2]

Interestingly, in all the ensuing uproar, there was very little criticism of the actual decisions that were made. For example, no one ever claimed that the time ultimately selected for the Geneva summit was bad in any way. Rather, it was the nonrational way in which decisions were being reached that produced the controversy. Astrology was seen as totally irrational, and people felt that if any decisions ought to be based on rationality, certainly decisions about the schedule of the president of the United States should be among them.

DIAGNOSTIC ISSUES

In this chapter you will see that our diagnostic model brings to light many questions about the process of decision making. One of the most important is, What key factors enable us to *describe* and distinguish effective decisions from ineffective decisions? What factors differentiate between effective and ineffective decision-making processes? How can we *diagnose* a situation in which different ways of seeing a problem lead to different prospective solutions? Can we explain why a decision maker sticks to a failing course of action in the face of consistently negative feedback?

How can we *prescribe* whether an organizational decision maker should rely on past practices or whether he should opt for more innovative choices? Can we determine when a decision maker should lean toward risky decisions and when he should be cautious? Finally, what *actions* can a manager take to control biases in the decision-making process and to enhance the quality of decisions made? What actions can create a climate that is conducive to creative decision making?

[1] Adapted from "For the Record," *Time Magazine*, May 16, 1988, p 29.
[2] *Time*, June 5, 1989, p 5.

EVALUATING DECISION MAKING

aking decisions is an essential part of the job of any manager or executive. Although the nature of the decision depends on the manager's area of expertise, such as marketing or production, the responsibility for choosing among different courses of action is something that all managers share. Managerial decisions often have a great impact not only on an organization's future but on the future of all the people who depend on the organization or are in some way affected by its actions—customers, suppliers, competitors, even the general public. Thus the decisions made by key people in many organizations are often closely scrutinized.

Certainly the outcome of a decision is the ultimate basis for judging its effectiveness. However, many factors can affect that outcome, and it is often a long time before the outcome is known. Thus the decision-making process itself is often of great interest.

POINT TO STRESS
An individual may make a good decision at a given time, but in the implementation phase the environment may change suddenly. The quality of the outcome may then be poor, not because the decision was bad but because of events that could not be predicted or controlled.

APPLYING THE DIAGNOSTIC MODEL
Diagnostic Question 1: Are ineffective decisions actually poor-quality decisions? Or is the ineffectiveness a result of poor timing and low acceptance of decisions by those needed to carry them out?

distributive justice An individual's perception of the fairness of his reward in comparison with the rewards given others.

procedural justice Perceived fairness of the process by which reward allocations have been made.

APPLYING THE DIAGNOSTIC MODEL
Diagnostic Question 2: Is the process by which decisions are made likely to be scrutinized? What can be done to convince people that the process is not arbitrary?

Outcomes

Let's look first at some of the factors that affect our evaluation of a decision's ultimate *outcome*. First and perhaps most obvious is *decision quality*, or the degree to which the chosen course of action is ultimately seen as the best among all those available. Also of considerable importance is *decision timeliness*. The right decision at the wrong time is no better than a wrong decision. In addition, because in most organizational contexts decisions are implemented by people other than the decision makers, *decision acceptance* by others is one of the important criteria by which decisions come to be judged. A manager who makes high-quality decisions but who cannot demonstrate the quality of the decision to the people who must put its results into effect will be no more successful than a manager who makes low-quality decisions.

Processes

In evaluating decision quality, people are influenced by the *decision-making process*—by how just they feel a decision is and by how fairly they feel the decision was arrived at. Theorists in the area of decision making distinguish between two kinds of perceived justice; distributive justice and procedural justice. **Distributive justice** is the fairness of the amount of a reward as perceived by its recipient and as compared with rewards given to others.

Procedural justice, on the other hand, is the fairness of the way in which rewards are allocated as perceived by the recipients of the rewards.

Some research suggests that procedural justice is more important than distributive justice. For example, a study of 2,800 federal employees found that level of job satisfaction was more strongly related to employees' concerns about *the way their salaries were determined* than to their actual pay levels.[3]

Perceptions of procedural justice are critically important when distributive justice is seen as low. In one study, workers who were underpaid in comparison with other employees reacted negatively—for example, they displayed poor work attitudes and were often absent—when pay decisions seemed based on arbitrary

---

[3] S. Alexander and M. Ruderman, "The Role of Procedural and Distributive Justice in Organizational Behavior," *Social Justice Research* 3, (1987), 117–98.

DECISION MAKING

reasons.[4] On the other hand, when these same workers were led to believe that pay differentials were based on differences in performance, no such negative reactions took place.

These and other studies suggest that the public's reaction to the manner in which decisions were made about President Reagan's schedule was not an isolated phenomenon. On the contrary, it reflected a real concern that decisions affecting individuals and groups in our society be based on rational processes rather than on irrational or arbitrary judgments. In this chapter we examine the positive and negative features of a rational decision-making model and consider some practical examples of how this model can be used. After exploring several factors that interfere with our ability to make purely rational decisions, we examine a decision-making model that recognizes the difficulty—sometimes the impossibility—of making rational decisions. We conclude with a look at some processes that fuel creative decision making.

rational decision-making model A model in which decisions are made systematically and based consistently on the principle of economic rationality.

economic rationality The belief underlying rational decision-making models that people attempt to maximize their individual economic outcomes.

RATIONAL DECISION MAKING

Because of its ties to classic economic theories of behavior, the **rational decision-making model** is often referred to as the rational-economic model. One of the most important of this model's key assumptions is that of **economic rationality**, or the notion that people attempt to maximize their individual economic outcomes. The model also assumes that people will consistently, logically, and in a precise, mathematical way pursue this goal of economic maximization through the five steps we describe next (see also Figure 6-1).

Setting Objectives

The first step in the rational decision-making process involves *setting objectives*, or goals. For example, the manager of a dairy farm may set a goal of increasing sales by 10 percent over the coming year. Similarly, a student may set a goal of attaining a 3.0 grade point average (GPA) over the next semester. The rational model assumes that it is possible to specify objectives in terms of outcomes that can be easily observed and measured.[5]

The objectives established in the first stage of the decision-making process not only determine the target one is aiming for but provide standards with which the objective situation can be compared. As we will see in the next section, decisions can be reassessed and adjustments made whenever the situation appears to deviate from these preset standards.

Monitoring the Environment

Although setting a goal is a decision in itself, from the perspective of the rational decision-making model, a substantive decision is not made until some problem or prospect related to the established goal appears. Thus, in the second stage of the decision-making process *monitoring the environment* is necessary to see how

FIGURE 6-1

The Rational Decision-Making Process

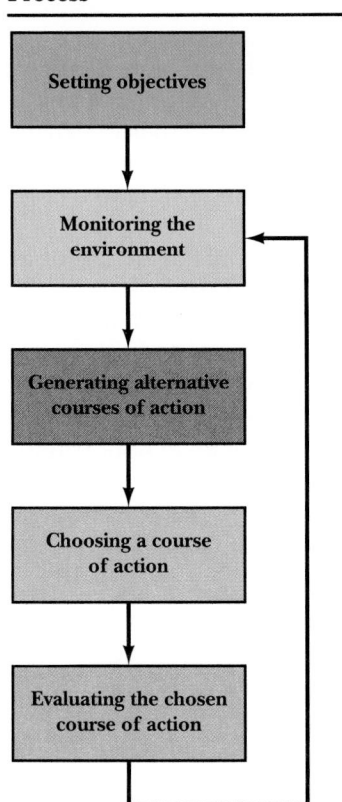

- Setting objectives
- Monitoring the environment
- Generating alternative courses of action
- Choosing a course of action
- Evaluating the chosen course of action

---

[4] J. Greenberg, "Reactions to Procedural Injustice in Payment Distributions: Do the Ends Justify the Means?" *Journal of Applied Psychology* 72 (1988), 55–61.

[5] L. S. Baird, R. W. Beatty and C. E. Schneier, *The Performance Appraisal Sourcebook* (Amherst, Mass.: Human Resource Development Press, 1982), p. 37.

When he became chairman of American Airlines, Bob Crandall set two major goals for the company: to become the leading air carrier in the U.S. and to derive 30 percent of its revenues from international traffic by 2000. By 1990 American had achieved the first goal and was well on its way to the next: the international share of its traffic had grown from 8 to 20 percent in just 8 years. Crucial to American's rise in both domestic and overseas markets was cost cutting. By emphasizing internal growth it avoided the high cost of labor that accompanies mergers and acquisitions. By flying into smaller cities on its new international routes, it has saved both itself and its passengers money and the hassle of crowds and delays. Source: *Kenneth Labich, "American Takes on the World,"* Fortune, *September 24, 1990, pp. 40–48.*

APPLYING THE DIAGNOSTIC MODEL
Diagnostic Question 4: Can the organization detect threats and opportunities in the environment accurately? Does it tend to see things in terms of one rather than the other of the phenomena?

well the individual or group responsible for meeting the goal is progressing toward it. This scanning can lead to several different outcomes. First, matters could proceed pretty much as the decision maker anticipated, and thus no new decisions would be required. In this case, the decision maker would simply "stay the course," continuing to engage in the behaviors and practices that are already in place.

A second possibility is that a problem, or *threat*, may arise in the environment, making the goal impossible to reach. The manager of the dairy farm, for example, may notice that a competing farm has started to advertise, and that his own farm's sales have decreased by 20 percent during the first quarter of the year. Or our student may discover that one of her courses is much more difficult than she thought it would be when she receives a failing grade on the first quiz. Such unanticipated events make staying the course an illogical reaction, and in each situation some different course of action must be decided upon.

Monitoring the environment may also reveal an unanticipated prospect of a favorable nature, or an *opportunity*, that would enhance the probability of goal attainment. For example, the dairy farm manager may have an opportunity to supply the local elementary schools and may have to decide whether to expand his operations in this direction. The student may learn that she can get extra credit in one of her classes if she participates in a research project and thus may have to decide whether to invest time in this activity.

Interestingly, research indicates that most real-world managers are more sensitive to perceived threats than to perceived opportunities. In fact, managers often infer a threat when information is ambiguous, and they sometimes ignore information that suggests an opportunity.[6] For example, when other small companies began developing clones or replicas of IBM personal computers (PCs), IBM first saw the competition from these numerous, tiny companies as a threat. Over time, however, IBM executives came to see the situation as an opportunity. An environment in which many smaller companies were producing IBM-compatible equipment with a considerable time lag would clearly be preferable to one in which major competitors like Apple were doing their own thing.

Generating Alternatives

The rational decision-making model requires that all potential solutions to a problem be identified before a decision is made. Without *generating alternatives*,

[6] S. E. Jackson and J. E. Dutton, "Discerning Threats and Opportunities," *Administrative Science Quarterly* 33 (1988), 370–87.

or examining all possible courses of action, we can never be sure we are choosing the best alternative. According to Paul Nutt, managers can choose possible actions in three major ways: they can use historical models, off-the-shelf processes, or nova techniques.

Nutt's research revealed that **historical decision models**, or actions taken in the past, are the most commonly used sources for generating alternatives. As Table 6-1 shows, he found that managers consult such sources almost half the time. There are several kinds of historical model. Very often we select from among *provincial models*; that is, we consider doing almost exactly what some other single firm or individual has already done. For example, when the material management system (a system for making purchases and managing raw-material inventories) of one of Nutt's subject companies failed, the company chose simply to make a carbon copy of the system used by its most successful competitor.

In the *enriched model*, a variation of the historical model, decision makers may visit several different plants or facilities that have dealt with the same problem and use what they learn from these site visits in generating alternatives. Thus if a university is thinking about restructuring its MBA program, it may send representatives to a number of different schools to explore the ways their programs are structured.

A third category of the historical model is the *pet idea*. In this variation, managers who have dealt at other times or in other places with the problem that now confronts them renew the alternatives they considered before. For example, Frank Lorenzo, former chairman of Texas Air, Continental Air, and Eastern Airlines, on more than one occasion chose bankruptcy in order to void or renegotiate noncompetitive union labor contracts. Each time he faced the same problem he used the same, highly controversial tactic.

In the second most widely used method of generating alternatives, the **off-the-shelf decision model**, decision makers use outside consultants who provide ready-made solutions. Sometimes a firm may seek only one proposal, but in other instances it may invite outside firms to bid competitively for the opportunity to solve the problem. For example, a firm with high turnover in its managerial ranks may fear that its executive salaries are no longer competitive. Rather than prepare its own wage-and-salary survey, the company may turn to consulting firms like

---

T A B L E 6-1
Methods Managers Use to Generate Alternative Courses of Action

METHOD		FREQUENCY OF USE (%)
Historical models		41
Provincial	20	
Enriched	6	
Pet idea	15	
Off-the-shelf processes		30
Extended (multiple proposals)	7	
Truncated (single proposal)	23	
Nova techniques		15
Internal staff	8	
External consultants	7	
Unclassified		14

Source: P. C. Nutt, "Types of Organizational Decision Processes," *Administrative Science Quarterly* 29 (1984), 14–50.

nova technique A method of generating alternatives by seeking new and innovative solutions.

AN EXAMPLE
Organizations have had to deal with change for generations, but it is only recently that managers have begun to realize that the impact of change on employees can cause serious organizational problems. The numerous mergers and acquisitions of the 1980s forced companies to find new ways of helping employees adjust to change. Many firms developed transition teams of employees from all levels of the organization to gather information about problem areas and generate alternative solutions. (*Personnel Administrator*, 8/89, pp. 84–90)

utility maximization A process by which a decision maker selects the one alternative that leads to the highest possible payoff.

APPLYING THE DIAGNOSTIC MODEL
Diagnostic Question 5: When deciding on a course of action, does the organization tend to rely more on past practice or on experimentation?

APPLYING THE DIAGNOSTIC MODEL
Diagnostic Question 6: Is the organization's environment best characterized by risk or uncertainty? How does this environmental characteristic affect decision-making processes?

expected value The projected value of an outcome that has less than a 100 percent probability of occurring. The expected value is derived mathematically by multiplying each possible outcome of a particular course of action by the probability that that outcome will occur.

the Hay Group, Management Compensation Services, or Growth Resources, Inc. Because making salary surveys is the consulting firm's bread and butter, these firms may be able to do a better job at less cost than the firm could do on its own.

When a problem has never been confronted before, historical and off-the-shelf methods may be inappropriate, and managers may use **nova techniques** to generate new ideas. Nova techniques seek ideas from many sources: the organization's own staff, outside consultants, or managers' colleagues in other firms. The key is to generate innovative ideas without reference to the practices of others.

Nutt's research, as reflected in Table 6-1, clearly suggests that managers tend to copy the ideas of others or to search for ready-made solutions to problems rather than generate new ideas. This tendency can probably be attributed to the speed and certainty associated with the first two approaches. Developing innovative approaches usually takes more time, and the outcome is harder to predict.

Choosing a Course of Action

The fourth step proposed by the rational decision-making model involves comparing all the alternatives that one has selected and choosing the one that seems most likely to lead to goal accomplishment. The final choice is determined by a cost-benefit, or **utility maximization**, approach, that is, the decision maker selects the alternative that leads to the maximum payoff.

Utility maximization is complicated by the fact that one cannot always be certain of the outcome associated with a particular choice. That is, there is not always a 100 percent probability that a given choice will lead to a given outcome. One's degree of certainty about the outcome of an alternative can be seen as falling somewhere along a continuum that ranges from *certainty* through a condition of *risk* to *uncertainty* (see Figure 6-2).

In some situations, you can be confident that a particular choice will lead to a particular outcome. For example, you can be assured that if you place $1,000 in a guaranteed savings account, the value of your money will grow each year by the rate of interest offered you by the savings institution.

In other cases, you cannot be sure of an outcome but you can try to assess the risk involved. Suppose you were to bet your $1,000 on a racehorse that gave odds of ten to one. Although you could not predict the precise outcome, you could be sure that one of two things would happen. Either the horse would win the race and you would win $9,000 ($10,000 less your investment of $1,000) or the horse would lose the race and you would lose your original bet.

An important characteristic of decisions made under the condition of risk is that we can compute **expected values** that permit us to compare alternatives with each other.

An expected value is derived by multiplying each possible outcome of a particular alternative by the probability that that outcome will occur. Table 6-2 shows the calculation of the expected values associated with each of three alternative investment opportunities. Clearly, Alternative 3 is the best overall investment in terms of expected value.

FIGURE 6-2

A Continuum of Uncertainty in Decision-Making Situations

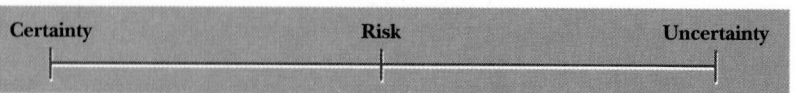

Certainty · Risk · Uncertainty

TABLE 6-2
Calculating Expected Values and Comparing Alternatives

Alternative 1
$1,000 investment with

50% chance of returning $0
50% chance of returning $3,000

Expected value = (.50)(0) + (.50)(3,000)
= $1,500

Alternative 2
$1,000 investment with

33.3% chance of returning $0
33.3% chance of returning $1,000
33.3% chance of returning $3,000

Expected value = (.333)(0) + (.333)(1,000) + (.333)(3,000)
= $1,332

Alternative 3
$1,000 investment with

25% chance of returning $0
25% chance of returning $500
25% chance of returning $1,000
25% chance of returning $5,000

Expected value = (.25)(0) + (.25)(500) + (.25)(1,000) + (.25)(5,000)
= $1,625

AN EXAMPLE: NORTHERN TRUST CORPORATION
Chicago's Northern Trust Corporation had solid expertise in managing assets for wealthy clients, but it decided to try its hand at real estate lending and other new ventures. After it was outgunned by bigger and more experienced competitors in these fields, the company figured the expected values of several strategies and decided to refocus its efforts on its core skills. It expanded into other states and opened up an office in London, where it has become a leader in cash management and custodial services for internal pension funds. Recognizing that its key asset is people, Northern has beefed up its management training programs and, calculating expected value far down the line, it has donated $1 million to a Chicago community group to help improve preschool, elementary, and secondary education.
Source: "Managing," *Fortune*, December 31, 1990, p. 72.

POINT TO STRESS
Loss aversion and other forms of decision bias underline the fact that people do not generally think in calculative, mathematical terms.

Evaluating Actions

According to the rational decision-making model, once a decision is selected and implemented, we must check periodically to see if the results of the decision are in line with the original goals. If the feedback from the environment suggests that the goals are being reached, the decision maker will stay with his chosen course of action. If results deviate from what is expected, however, the entire decision-making process may begin anew. Thus decision making according to this model is a dynamic process designed to be self-correcting over time.

THREATS TO RATIONAL DECISION MAKING

Rationality in decision making is neither easily nor commonly achieved. In Chapter 5 we used perceptual illusions to show that perception is not nearly as straightforward as it seems. Here we will use what we might call decision-making illusions to show why rationality is not so easy to achieve. Credit for documenting the existence of these illusions goes to Kahneman and Tversky, who first documented the existence of several of the illusions we examine here: loss aversion, availability bias, base rate bias, and regression to the mean.

Loss Aversion

As a prelude to this discussion, read the paragraph in Table 6-3 and decide what strategy you would choose if you were the sales executive faced with the situation described. If you decided to go with Strategy 1 and save the 200 accounts for sure, you are not alone. In fact, research with managers and nonmanagers alike

APPLYING THE DIAGNOSTIC MODEL
Diagnostic Question 7: Does the organization employ standard mathematical solutions to routine or programmable problems where they are appropriate?

Diagnostic Question 8: Do factors such as loss aversion or availability bias influence the organization's decisions adversely?

shows that this choice is preferred roughly three to one over the more risky Strategy 2.

Now turn to a similar decision situation, shown in Table 6-4, and decide what strategy you would use under these circumstances. If this time you chose Strategy 2, again you are not alone. Research shows that this choice is preferred roughly four to one over Strategy 1.[7]

The surprising thing about these results, though, is that the problems described are virtually identical. Read the paragraphs in Tables 6-3 and 6-4 once more. You can see that Strategy 1 is the same in both tables. The only difference is that in Table 6-3, it is expressed in terms of accounts *saved* (200 out of 600) whereas in Table 6-4 it is expressed in terms of accounts *lost* (400 out of 600). Clearly, if 200 are saved, 400 are lost and vice versa.

Strategy 2 is also the same in both situations, expressed in terms of accounts saved in one table and accounts lost in the other. If you compute expected values for all four strategies (refer to Table 6-2 for help in doing this), you will confirm that both situations have the same expected values. Why then is Strategy 1 preferred when the situation is described as it is in Table 6-3 and Strategy 2 preferred when the situation is outlined as in Table 6-4?

Because the expected values in these problem situations are all the same, none of the choice strategies presented are bad. Thus let's consider the decision situations outlined in Table 6-5. Most people show the same pattern of preference when presented with these decision choices, even though in each case, in terms of expected value one choice is clearly superior to the other. In Decision 1, most people choose the sure gain (a), even though the expected value for the risky gain (b) is clearly higher. Similarly, in Decision 2, most people take the risk on the larger loss (d), even though the expected value associated with the sure loss (c) is higher.

---

[7] A. Twersky and D. Kahneman, "The Framing of Decisions and the Psychology of Choice," *Science* 211 (1981), 453–58.

loss aversion bias The tendency of most decision makers to weigh losses more heavily than gains, even when the absolute value of each is equal.

AN EXAMPLE
Stocks are generally considered risky investments. While you may gain you also stand the chance of losing big. As a result, the average person either avoids the stock market or invests very conservatively. Yet, over the past two years large numbers of average people have lost large sums in the junk bond market. Marketing campaigns focusing on large and quick gains convinced many that there were no risks or that the risks were very small. Statistics and testimonials were often presented to support these claims. The long-term risk and the reasons certain investments were "junk" were not broached. As a result, normally risk-averse, conservative people invested heavily and lost heavily.

availability bias The tendency in decision makers to judge the likelihood that something will happen by the ease with which they can recall examples of it.

AN EXAMPLE: CHEVRON
Chevron, attempting to draw its credit card customers back, dropped the "discount for cash" in 1991 and began to offer the same price for credit or cash. Thus they rewarded customers for coming back and using their credit cards.

Research indicates that, in general, people have a slight preference for sure outcomes over risky ones. Studies also indicate, however, that people hate losing, and this **loss aversion bias** affects their decision making even more strongly than their preference for nonrisky situations. Thus, when given a choice between a sure gain and a risky gain, most people will take the sure thing and avoid the risk (showing the minor aversion to risk). When given a choice between a sure loss and a risky loss, however, most people will avoid the sure loss and take a chance on not losing anything (showing the major aversion to loss).

We can see this preference for risk over loss in many real-world situations. For example, in the early 1980s oil companies like Shell Oil and Amoco Oil found themselves faced with the increasing costs of managing credit-card accounts. To make up for the added cost of managing these accounts, these firms began to charge customers differentially depending on whether they paid cash for gasoline or used their credit cards. Initially, some companies described this extra charge as a credit surcharge (a sure loss). Their customers were outraged. Other companies advertised a discount for cash (a sure gain), and their customers thought this was great! Needless to say, nobody talks about credit surcharges anymore, and everyone offers discounts for cash.

Availability Bias

Let's explore another kind of bias in decision making. In a typical passage of English prose, does the letter *k* occur more often as the first or the third letter in a word? Twice as many people confronted with this problem choose first over third, although the fact is that *k* appears in the third spot almost twice as often as in the first. This phenomenon can be explained in terms of the **availability bias**, which means that people have a tendency to judge the likelihood that something will happen by the ease with which they can call examples to mind. Most people assume that *k* is more common at the beginning of words simply because it is easier to remember words beginning with *k* than words that have *k* as their third letter.

You can see the availability bias at work in the way people think about death, illness, and disasters. In general, people vastly overestimate the numbers of deaths caused by vividly imaginable events like an airplane crash and underestimate deaths caused by illness like emphysema or stroke. Deaths caused by sudden disasters are more easily called to mind because they are so vivid and so public, often making the first page of newspapers across the country. Death caused by illness, on the other hand, is generally private and thus less likely to be recalled.

Companies that employ risky technologies, such as nuclear power plants, must deal with the availability bias continuously. Ironically, some things they do

AN EXAMPLE

It is quite common for people to indicate they are afraid of flying; they are really afraid of crashing and being killed or injured. These same people do not report a similar fear of driving and yet tens of thousands of people are killed or injured in cars each year. The number of deaths and injuries resulting from plane crashes each year is in the hundreds. But plane crashes, even small ones, make the headlines while most car accidents are never even mentioned. A second factor accounting for this fear is the amount of control or decision-making power we think we have. In a plane we are powerless. We can't see the cockpit nor influence the pilot. In a car we feel we have control, as the driver, or at least influence, as the passenger.

APPLYING THE DIAGNOSTIC MODEL

Diagnostic Question 9: Do organizational members understand the concepts of base rate and regression to the mean? Do they take account of these influences making decisions?

to allay people's fears actually make things worse. For example, going over disaster scenarios and detailing what would be done in case of a nuclear accident actually makes residents of a community more fearful. Indeed, research shows that the more detail presented, the more vivid the picture becomes, the greater the probability of a disaster appears, and the more strongly people resist having such a plant in their community.[8]

The power of either the loss aversion bias or the availability bias is great in isolation, but when the two are combined, many decision makers find them impossible to resist. The insurance industry, for example, is financed almost solely through the compelling nature of this combination. Suppose you pay $300 for home owners' insurance that will pay your beneficiary $100,000 in case of a disaster. This is a terrible bet, because the odds that such a disaster will occur are less than .003 (i.e., 300/100,000). Nevertheless, because we often see fires on television news broadcasts (vividness plus frequency), and because one's home, if owned, represents such a huge investment (loss of great proportions), people are more than happy to make such bets.

Base Rate Bias

To understand a third type of decision bias, consider the decision-making problem described in Table 6-6. By now you are probably ready to guess that the answer that seems obviously correct—that the cab was a Blue—is actually wrong. You are right. In fact, the odds are much better that the cab was a Green. Figure 6-3 shows you mathematically why that is so.

If there were 100 cabs in the city, 85 would be Greens and 15 would be Blues. This is the base rate, that is, the initial probability given no other piece of information. On the premise—established in Table 6-6—that the witness (who provides an additional piece of information over and above the base rate) would be right 80 percent of the time, let us see what would happen in each possible scenario. If the cab in the accident was actually a Blue, the witness would identify it correctly as a Blue 12 times (.80 × 15 = 12) and would incorrectly identify it as a Green 3 times (.20 × 15 = 3). If the car was a Green, however, the witness would correctly identify it as a Green 68 times (.80 x 85) and misidentify it as a Blue 17 times (.20 x 85). Thus the odds are much greater that the witness's identification of the cab as a Blue was a misidentification of a Green cab (which

[8] P. Slovic, B. Fischhoff and S. C. Lichtenstein, "Cognitive Processes and Societal Risk Taking," in *ORI Research Bulletin*, vol. 16, (Eugene: Oregon Research Institute, 1974), pp. 16–76.

T A B L E 6-6

Identifying a Hit-and-Run Driver

A cab was involved in a hit-and-run accident.

Two taxicab companies serve the city: the Green Company operates 85 percent of the cabs, and the Blue Company operates the remaining 15 percent.

A witness identifies the hit-and-run cab as blue. When the court tests the witness's reliability under circumstances similar to those on the night of the accident, he correctly identifies the color of a cab 80 percent of the time and misidentifies it 20 percent of the time.

Which cab company was most probably involved in the hit-and-run accident?

Source: A. Tversky and D. Kahneman, "The Framing of Decisions and the Psychology of Choice," *Science* 211 (1981), 453–58.

FIGURE 6-3

How to Test the Witness's Accuracy in the Hit-and-Run-Driver Case

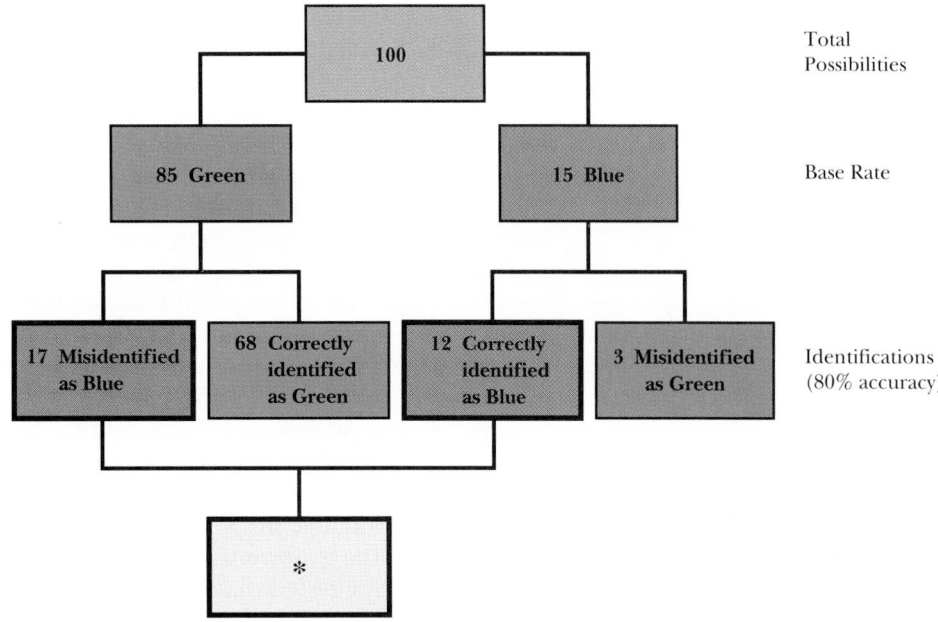

*The probability is greater that the cab was a misidentified Blue (17 out of 100) than a correctly identified Blue (12 out of 100).

happens 17 out of 100 times) than a correct identification of a Blue cab (which happens only 12 out of 100 times).

The reason why virtually everyone who approaches this problem naively gets it wrong is that we put too much weight on the evidence provided by the witness and not enough weight on the evidence provided by the base rate. That is what is meant by the term **base rate bias**. People tend to ignore the background information in this sort of case and to feel that they are dealing with something unique. In one example, decision makers will discount the evidence of how few cars are actually blue, and instead put more confidence in human judgment about the color of the car. Ignoring the base rate leads to misplaced confidence in decision making.

The problem of misplaced confidence is particularly pronounced when more than one probabilistic event is involved. As you might guess, this is more often than not the case in actual business ventures. Suppose, for example, that a house builder contracts to have a house completed by the end of the year. Assume further that her chances of accomplishing four specific tasks in time to meet this deadline are as follows:

base rate bias The tendency in decision makers to ignore the underlying objective probability or base rate, that a particular outcome will follow a particular course of action.

TASK	PROBABILITY
Get permits	excellent (90%)
Get financing	very good (80%)
Get materials	excellent (90%)
Get subcontractors to complete their work	very good (80%)

Reviewing these data, our builder might well conclude that there is a good to excellent chance that she can complete the project in the time specified in the contract. In fact, the odds that this will happen are only about 50-50. Multiplying the 4 probabilities together (.9 x .8 x .9 x .8 = .52) gives us just over 50 percent, which is hardly a good to excellent chance. The axiom known as Murphy's Law states that anything that can go wrong will go wrong. This may be a tad pessimistic,

THREATS TO RATIONAL DECISION MAKING

171

TEACHING NOTE
State-run lotteries also play on the loss aversion bias. States use lotteries to generate funds because voters are more likely to support a lottery than an increase in taxes. Taxes represent a sure loss. The lottery's sure loss, however, appears very small in contrast to its enormous potential gain. Discuss why millions of people who would vote against a tax are avid lottery players. They still pay the "tax" but in a way that gives them a (very slim) chance to win large sums of money.

AN EXAMPLE
Compulsive gamblers provide an excellent example of individuals who fail to understand regression to the mean. These people may hit a winning streak, and as that streak continues they become convinced it will last forever. Then they lose and lose repeatedly. Now they continue playing because they know their luck will change and they will be big winners again. They fail to understand that luck is random. If they gamble over a long period, they will at best break even. Extraordinary events that led to winning will be balanced by extraordinary events that lead to losing. The problem is that most end up on the losing side because they don't have sufficient resources to continue once they start losing.

regression toward the mean   The phenomenon whereby in a series of events that is influenced by complex factors, any single extraordinary event is almost sure to be followed by a more ordinary event.

but in a long series of events, the odds are quite good that any one event *will* go wrong, and sometimes it takes just one mishap to destroy an entire venture. Any business executive who is putting together a deal where the ultimate outcome depends on a series of discrete events, none of which are sure things, must keep this fact in mind.

Again, combining two biases can have a particularly powerful effect. The lottery industry gets people to make bad bets by combining the tendency to ignore the base rate with the availability bias. People's tendency to think that their chances of winning are greater than they are is fueled by marketing that provides vivid, graphic examples of the joy of winning the lottery. For example, you may have seen publicity pictures of lottery winners holding up ridiculous, four-foot checks. Moreover, winners are announced every day or every weekend, making it appear as if winning happens all the time. Publishing the names of losers, of course, would require producing a document the size of an encyclopedia every day.

The scope of such operations (state-run lotteries generate over $17 billion a year nationally)[9] shows how intelligent, hard-working people can be tricked into making poor decisions. Indeed, in some communities people have questioned the ethics of the government's involvement in what is actually gambling. Other people have suggested that lotteries are in effect a regressive tax on people who do not understand probability theory. This latter criticism is understandably linked with the observation that, seen as a fixed-rate tax, the purchase of a weekly lottery ticket hurts the poor much more than it hurts people of middle- to-upper income levels.

Regression toward the Mean

Table 6-7 presents another decision problem that confounds almost all naive subjects. Read over this exhibit and decide which stock you would like to purchase. By this time, you are probably so sensitized to being misled that you do not even want to be in the same room with the stock that has gone up five years in a row, let alone hold it. In fact, one study has shown that if over a fifty-year span, an investor continually bought only the stocks that had declined the most in value over each preceding five-year period he would have earned 30 percent above the average market return in each succeeding five-year period.[10] Nevertheless, most naive subjects jump on the continually rising stock because they fail to recognize a phenomenon called **regression toward the mean**.

To understand this phenomenon, you must realize that in any series of events that is influenced by a complex array of factors, any single extraordinary event is almost sure to be followed by a more ordinary event. That is, extraordinary events are more often than not preceded by events that are closer (i.e., they regress) to average (i.e., the mean). Because most people fail to appreciate this

[9] Tom Callahan, "Did Pete Do It? What Are the Odds?" *Time*, July 26, 1989, p. 92.

[10] L. Thaler and W. De Bront, "Regression to the Mean on the New York Stock Exchange," working paper, University of Wisconsin at Madison.

T A B L E 6-7
Regression toward the Mean at the New York Stock Exchange

Two stocks exist, one of which has decreased in value five consecutive years, and one that has increased in value for five consecutive years. Assuming you had to purchase and hold the stock for the next five years, which would you choose?

Source: A. Tversky and D. Kahneman, "The Framing of Decisions and the Psychology of Choice," *Science* 211 (1981), 453–58.

---

phenomenon, their decision making is biased. One of the authors once saw an insurance company almost abandon what was actually an effective incentive system—providing travel prizes to the top salesperson—because experience revealed that the winner's performance was always poorer in the following year. Given the complex number of factors that go into a final sales figure, this kind of variation is to be expected. Undoubtedly some of the lucky breaks that help make an incredibly successful year simply do not repeat themselves the next year. This is also the reason why in sports like baseball, repeating as the champion is considered one of the most difficult feats to accomplish.

APPLYING THE DIAGNOSTIC MODEL
Diagnostic Question 10: Do organizational decision makers stick to a single course of action too long (as in escalation of commitment)? Or do decision makers switch from one decision to another too quickly, making it impossible to learn in "noisy environments?"

POINT TO STRESS
Contrary to the assumption of the rational model that decisions are self-correcting, we tend to stick with a course of action in the face of information indicating the need for change.

escalation of commitment
Investing additional resources in failing courses of action that are not justified by any foreseeable payoff.

Escalation of Commitment

For a final demonstration of biased decision making, imagine the following scenario. You are on the ground floor of a building and have two minutes to get to a job interview on the fourth floor. You can either take the elevator, which can whisk you to the fourth floor in a matter of seconds, or you can take the stairs, which can get you to your destination in a few minutes. The elevator is the obvious choice, but this is a twenty-story building, and you don't know where the elevator is at the moment. You push the button, and nothing happens. You could immediately take to the stairs, but you decide to wait. After a few seconds you again look at the stairs and consider giving up on the elevator. But the elevator will surely arrive the moment you head up the stairs, so you continue to wait. Still no elevator appears, and you realize that you probably should have taken the stairs in the first place. By now, however, if you are going to have a chance of arriving on time you must continue to wait for the elevator. On the other hand, being a little late is better than being quite late. You reconsider the stairs. You conclude, "What the heck, I've waited this long, what's a little more time," and continue to wait. The elevator does not appear, and in a fit of disgust, you rush to the stairs, sprinting the four flights while thinking up excuses for why you are late and wondering, "Was that the elevator I just heard?"

If you can imagine your frustration in this situation, you have a feel for another threat to rationality in decision making called **escalation of commitment**. Escalation of commitment is a process in which people invest more and more heavily in an apparently losing course of action in order to justify their earlier decisions. Usually the investments that are made once this process gets started are disproportionate to any gain that could conceivably be realized. Consider the following memo sent by Undersecretary of State George Ball to President Lyndon Johnson prior to U.S. involvement in the Vietnam War:

> The decision you face now is crucial. Once large numbers of U.S. troops are committed to direct combat, they will begin to take heavy casualties in a war they are ill-equipped to fight in a non-cooperative if not downright hostile countryside. Once we suffer large casualties, we will have started a well-nigh irreversible process. Our involvement will be so great that we cannot—without national humiliation—stop short of achieving our complete objectives. Of the two possibilities, I think humiliation would be more likely than achievement of our objectives—even after we have paid terrible costs.[11]

Even in the face of evidence that costs are actually outstripping benefits, a decision maker may feel many different kinds of pressure to continue to act in accord with a particular decision.[12] For psychological reasons, the decision maker

---

[11] B. M. Staw, "The Escalation of Commitment to a Failing Course of Action," *Academy of Management Review* 6 (1981), 577–87.

[12] B. M. Staw and J. Ross, "Behavior in Escalation Situations," in *Research in Organizational Behavior*, ed. B. M. Staw and L. L. Cummings (Greenwich, Conn.: JAI Press, 1987), pp. 12–47.

may not want to appear inconsistent by changing course. He may not want to admit that he has made a mistake. Moreover, particularly where feedback is ambiguous or complex, perceptual distortions like the expectation effect we discussed in Chapter 5 can make the picture appear more hopeful than it really is. Because one cannot make perfect predictions regarding future outcomes, there is always the hope that staying the course will pay off. Moreover, the decision maker may have been rewarded in past situations for sticking it out. Although rare, such experiences are usually quite memorable (availability bias). The experience of giving up when it is appropriate is often not rewarded, at least in the short run, and thus is something people like to forget. Finally, sometimes a decision maker throws out cost-benefit analyses altogether, and develops a win-at-any-cost mentality. The quest to prove himself completely takes over, and the decision maker comes to resemble Captain Ahab in his obsessive pursuit of Moby Dick.

Factors That Prevent Rationality

The illusions we have discussed make rational decision making difficult but not impossible. Often, however, the complexity of real-world decision situations literally makes rationality impossible. Our exploration of the factors that prevent rational decision making will lead us to our discussion, in the next section, of a decision making model that acknowledges these barriers to rationality.

Nobel laureate Herbert A. Simon has remarked that "the capacity of the human mind for formulating and solving complex problems is very small compared with the size of the problems whose solution is required for objectively rational behavior in the real world."[13]

Simon's comment on the limits of human intelligence is not so much a condemnation of human beings as it is an acknowledgement of the complexity of the environment in which human beings must operate. Indeed, according to Simon and to others who have followed his lead, the complexity of the real world will typically overwhelm the decision maker at each step of the rational decision making process, making rationality completely impossible.

APPLYING THE DIAGNOSTIC MODEL
Diagnostic Question 3: Does the organization have clear goals and objectives, so that analytical solutions can be developed? Or is the organization characterized by a lack of goal consensus, thus requiring a bargaining solution?

Lack of Goal Consensus

A problem at the outset of the decision-making process is that rational decisions can be made only when there is general agreement on the goals to be pursued. In large organizations, such goal consensus is very hard to achieve. Neither individuals nor organizations consistently rank desired outcomes in the same way, and in complex organizations the only shared goals are those that are so vague as to be almost meaningless. For example, the public goal of a firm may be the "benefit of humankind through the development of technology." This is a fine goal around which everyone in the firm may rally. However, it is vague and provides no guidance on exactly how it is to be achieved. In running the day-to-day operations of the firm, the goal is worthless, and if there are disagreements among organizational members about specific objectives there is no way to proceed in any rational manner.

Let's return for a moment to our story about the president's astrologer. Every member of the president's cabinet probably has a different idea of how

[13] J. G. March and H. A. Simon, *Organizations* (New York: John Wiley, 1958), p. 10.

There was goal consensus among
political and military leaders of
Desert Storm: the goal was to get
Iraq out of Kuwait. The media
and general public often discussed
a second possible goal, to destroy
both Saddam Hussein and Iraq.
Because not all members of the
multinational coalition would have
subscribed to this secondary goal,
it never became formalized.
Without such consensus, no
rational decision could be taken.

TEACHING NOTE
Ask students to list some goals for
finance, marketing, production,
and human resources. Are these
goals in conflict? How can an
organization make decisions and
address these divergent goals?

the president should spend his time, as does probably every White House staffer
and every member of Congress. Indeed, probably no single agreed-upon rank
ordering could ever be established for what the president of the United States
should be trying to accomplish on a given day. Lack of goal consensus implies
that there is no one best solution, and some problems that the President has to
face will engender harsh negative reactions regardless of the stance he takes. For
example, whether he supports or opposes abortion, he will enrage one group of
people. Either the right-to-life group or the pro-choice group will picket the
White House. In such a no-win situation, perhaps the president's best alternative
is to keep a low profile and focus his attention instead on an issue that may
engender more consensus. For example, most people cringe at the thought of
burning the American flag (goal consensus). As a result, public-relations-minded
staff tried to direct President Bush toward a constitutional amendment to ban
flag burning.

In business organizations, too, different departments and different individuals have different ideas about what goals should be pursued. Because this
lack of goal consensus characterizes most organizations, the political processes
of bargaining and compromise are more commonly used to reach decisions than
are analytical processes in which people try to maximize their gains by the use
of techniques like linear programming. Although this kind of mathematical formulation is appropriate for some well-defined organizational problems, like scheduling routine purchases, it doesn't work for the kind of complex, often ill-defined
problems that top managers continually must face.

Increasingly, the curricula of business schools, which train future managers,
are being criticized for an excessive emphasis on quantitative methods. Experienced executives complain that graduates must be more than flesh-and-blood
calculators, and many schools have responded. For example, the University of
Chicago's Graduate School of Business, which we mentioned in Chapter 2, has
recently overhauled its MBA programs, eliminating the numbers-oriented focus.
Although the dean of the school, John Guild, insists that "we're going to keep
our pocket protectors," he also has insisted on instituting seminars focusing on
the "softer skills" of negotiation tactics and team building.[14]

Problems in Monitoring the Environment

Monitoring the environment, although crucial to identifying threats and opportunities, is not easy. *Time constraints*, for example, may interfere with accurate
monitoring. Returning again to our opening example, many problems cross the
president's desk that demand immediate action. In these cases, there is no time
to engage in surveys to determine where the public-at-large stands on the issue.
In addition, one may get *equivocal feedback*: Some evidence may suggest that a
threat or opportunity is developing, whereas other information may be contradictory. Soviet initiatives on nuclear arms talks, for example, were perceived by
some of the president's staff as genuine attempts to move toward world peace.
Others, however, pointed to Soviet support of leftist groups throughout Latin
America as contradicting that impression.

Means-End Relations

In rational decision making we encounter a problem when we try to generate
possible courses of action and then try to select the most promising one. Managers

[14] D. Greising, "Chicago's B-School Goes Touchy Feely," *Business Week*, February 12, 1990, p. 61.

In monitoring the organizational environment, it helps to have what a recent Fortune *article called the "laser-like focus" of Home Depot. Bernard Marcus (left), CEO and cofounder of the chain of do-it-yourself warehouse stores, says he wants to be the power retailer in any of his product categories. When the Swedish home furnishings chain IKEA bounded into the United States market with five stores, Home Depot had to make a fast and tough decision. On average, IKEA stores devote 200,000 square feet to furniture, whereas Home Depot warehouses gave their unfinished wood furniture departments only 2000 to 7000 square feet. With a much greater selection, IKEA could sell more and keep its prices even lower than Home Depot's. Home Depot decided to close its furniture section and to use the space to expand its selection of floor tiles and wallpaper.*
Source: *Susan Caminiti, "The New Champs of Retailing,"* Fortune, *September 24, 1990, pp. 85–86.*

AN EXAMPLE

In the late 1970s the Republicans blamed Democratic spending for runaway inflation. Shortly after the Republicans won the 1980 presidential election inflation began to dissipate, and this improvement was attributed to Reaganomics. In reality, both the inflation and its decline owed a lot to our dependence on OPEC oil. OPEC nations increased prices and reduced supplies in the 1970s, but reduced prices in the early 1980s when an oil glut developed. Through all the placing of blame and back slapping for success, little mention was made of our economy's dependence on what OPEC did.

bounded discretion The recognition that the alternatives offered to a decision maker are bounded by social, legal, moral, and organizational restrictions.

often cannot anticipate what actions will lead to what consequences. Because, as Simon points out, most real-world decisions are characterized by uncertainty, managers cannot even speculate on the odds. Under these conditions, one cannot compute expected values, and thus there is no common metric with which to compare various alternatives. This problem occurs especially with nonprogrammed decisions, that is, decisions that are called for only infrequently. In making these kinds of decisions, no one ever develops enough experience to assess the probabilities associated with any alternative easily. For example, how was the president able to decide what he should do in response to the democratization of the Soviet Union? This was an unprecedented event in history for which there was no clear knowledge of what actions might lead to what results.

Although the necessity to make nonprogrammed decisions sometimes creates paralysis, programmed decisions, on the other hand, often become so institutionalized that one wonders if any rational thought is ever given to them. For example, the president routinely makes time in his schedule to call the winning coach of the Super Bowl game, and he meets yearly with college athletes who win national championships in major sports. Nobody sits down each year and compares the time spent doing this with every other possible thing the president could be doing at these times. These are simply practices that have been carried out as long as most people can remember, and tradition rather than analytical thought dictates the decision.

You should also note that managers are not free to choose among all the choices they may generate. The term **bounded discretion**, first suggested by Herbert Simon, refers to the fact that the list of alternatives that any decision maker generates is restricted by social, legal, and moral norms. As Figure 6-4 indicates, the discretionary area within which acceptable choices can be made is bounded on all sides by the limits imposed by (a) unofficial social norms, (b) organizational rules and policies, (c) moral and ethical norms, and (d) legal re-

DECISION MAKING

FIGURE 6-4

The Concept of Bounded Discretion

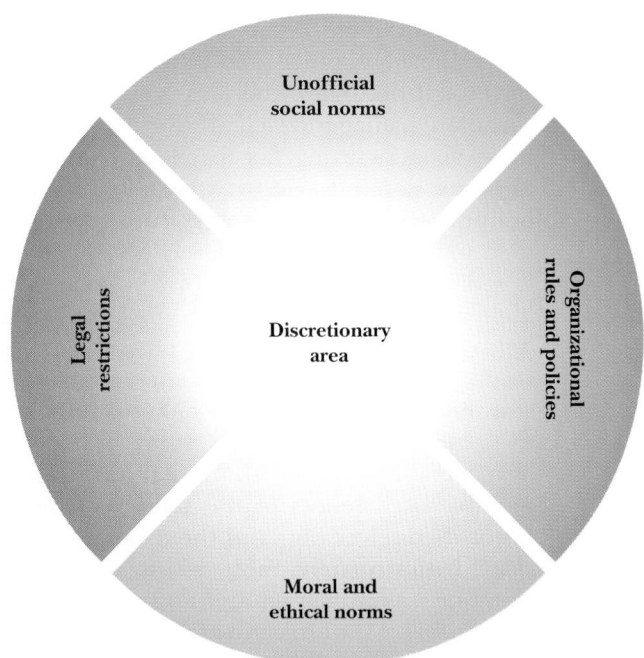

Unofficial social norms

Organizational rules and policies

Discretionary area

Legal restrictions

Moral and ethical norms

strictions. The boundaries between each of these sets of limitations and the discretionary area are not clear-cut, and as a result decision makers do not always know whether an alternative is in or out of bounds.

Moreover, recent litigation has blurred these boundaries even further. One recent phenomenon that has certainly caught the eyes of executives is personal liability for organizational decisions. Such liability may even extend to include criminal wrongdoing. For example, a New York jury found two executives at Pymm Thermometer Company guilty of assault and reckless endangerment for exposing a worker (who later developed brain damage) to poisonous metal. Similarly, in Illinois, Cook County attorneys successfully prosecuted three senior executives of Film Recovery Systems, Inc., on murder and reckless conduct charges after one of their workers died from inhaling cyanide fumes. These examples make it clear that managers must conform their decisions to societal norms, and that failure to do so may result in very negative consequences both for them and for their organizations.[15]

AN EXAMPLE: QUAKER OATS COMPANY

Trying to monitor several noisy environments can be too great a task for an organization. When competitors began to attack the Quaker Oats Company's Fisher-Price preschool-age toy division in 1988, Quaker went on the offensive with more sophisticated, electronic toys. It misjudged the ferocity of the competition, however, and within two years it lost more than $100 million in profits and was forced to cut Fisher-Price loose. Trying to predict outcomes in another noisy environment, Quaker ran into trouble again when it added the profitable Gaines pet food business to its own Ken-L-Ration and Kibbles'n Bits divisions. Heightened competition from Ralston and other companies forced Quaker's sales of pet food to decline by 14.5 percent by 1989. Although William Smithburg, Quaker's CEO, feels the merger with Gaines will pay off eventually, he admits it didn't accomplish what the company wanted. (Bill Saporito, "How Quaker Oats Got Rolled," *Fortune*, October 8, 1990, pp. 129–138.)

Noisy Environments

Finally, the idea proposed by the rational decision-making model that once a course of action has been initiated, it can be evaluated by checking its outcome against the original objective must be questioned. In complex environments, where many factors other than the chosen alternative can influence the outcome, making what seems to be the right choice may not lead invariably to the desired outcome. In fact, in *noisy environments*, where the link between actions and outcomes is tenuous, decision makers often place too much value on what happens in any given instance. This can lead them to assign too much importance to their own actions in bringing about the observed results and thus may inhibit their ability to learn from their experience.

[15] S. B. Garland, "This Safety Ruling Could Be Hazardous to Employers' Health," *Business Week*, February 12, 1989, p. 34.

For example, let us assume that in 2000 there will be a fax machine in every home and office, and the demand for overnight mail delivery will be close to zero. If one knew this was going to be the case, then a 1993 decision by a corporate manager to sell off overnight-mail-delivery firms and instead invest in firms manufacturing fax machines would be a good decision. It would be the right decision *in the long run* but 1993 might be too soon to put it in place. For example, the transition process from a time when relatively few have fax machines to the day when they are in every home and office may not be smooth. Moreover, fax sales for 1994 might be off for some unstable, unforeseen reason (such as high interest rates that limit business expansions). If overnight-delivery revenues hold constant that year, the corporate decision maker may very well question the wisdom of her earlier decision and be tempted to reverse it—a decision that could cost millions if the world in 2000 turned out to be as described above.

In noisy environments, one can make sense of action-outcome links only by making many observations of the same outcomes after the same actions. After multiple occurrences of the same sequence of events, the random influences factor themselves out, and the true nature of the action-outcome link becomes clear. Unfortunately, most decision makers in noisy environments fail to stick with one action long enough to sort out the effects of the chosen action on the outcome from the effects of random influences. This lack of consistency in decision making makes people move from one action to another without ever learning much about the action-outcome link associated with any one specific action.[16]

ADMINISTRATIVE DECISION MAKING

There are many contexts in organizations where the use of the rational decision-making model serves as a valuable decision aid. It is particularly useful in routine decision making where everyone agrees on the desired outcomes and the best methods for attaining those outcomes. But because of the many factors that make the rational decision-making model useless in some contexts, alternatives to the model have been developed. We will look next at one of the most widely cited alternatives.

Simon's **administrative decision-making model**, as outlined in Figure 6-5, a move away from the rational decision-making model, was an attempt to paint a more realistic picture of the way organizations make decisions.[17] According to Simon, although the rational decision-making model may be useful in outlining what organizations *should* do if strict rationality were possible, the administrative model provides a better picture of what effective organizations *actually* do when strict rationality is impossible. Simon's model differs from the rational model in four fundamental ways.

Satisficing versus Optimizing

According to Simon, optimal solutions require that the decision arrived at be better than all other possible decisions. For all the reasons we have discussed, that is simply not possible most of the time. So instead of striving for this impossible goal, organizations try to find **satisficing** solutions to the problems they confront.

administrative decision-making model A model in which decisions pursuant to negotiated goals are made based on satisficing rather than maximizing outcomes, through a sequential consideration of alternatives.

satisficing Settling for a decision alternative that meets some minimum level of acceptability, as opposed to trying to maximize utility by considering all possible alternatives.

[16] B. Brehmer, "Response Consistency in Probabilistic Inference Tasks," *Organizational Behavior and Human Performance* 22 (1978), 103–15.

[17] March and Simon, *Organizations*, 1958, pp. 10–12.

FIGURE 6-5

The Administrative Decision-Making Model

Source: Based on J. G. March and H. A. Simon, Organizations *(New York: John Wiley and Sons, 1958).*

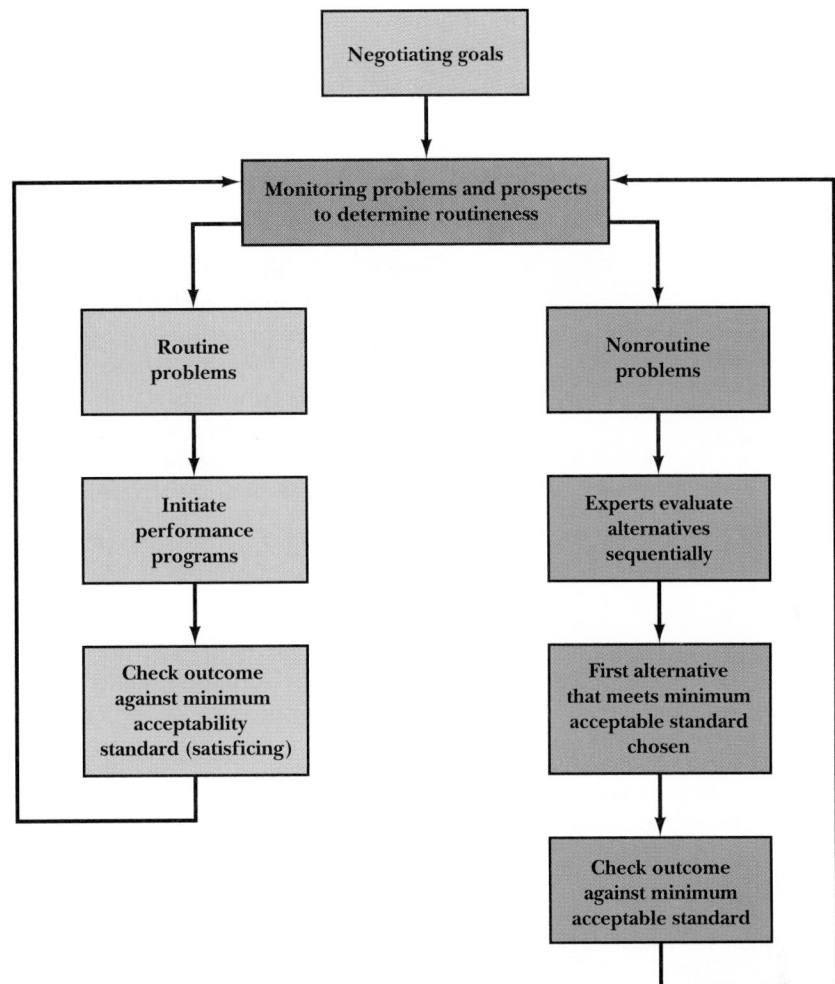

Satisficing means settling for an alternative that meets some minimum level of acceptability. Needless to say, it is much easier to achieve this goal than to try for an optimal solution, which reaquires that the alternative chosen be superior to all possible alternatives. Indeed, Simon likens it to the comparison between finding *a* needle in a haystack (satisficing) and finding the *biggest, sharpest* needle in the haystack (optimizing).

Considering Alternatives Sequentially

In searching for satisficing solutions, organizations further simplify the process by *considering alternatives sequentially* rather than simultaneously. That is, instead of first generating all possible alternatives and then comparing and contrasting them, each alternative is evaluated one at a time against the criteria for a satisficing outcome. The first alternative generated that is satisfactory is chosen, and the organization moves on to other problems.

For example, a firm needing to downsize (reduce its total number of employees) is faced with over a dozen means of bringing about this outcome. Rather than try to compare the expected results for every possible means with every other possible means (a step that may even be impossible if the firm has never tried to downsize before), the firm may just look at an early retirement program and think, that may do the trick. If the company puts such a program in place and it achieves the desired results, no further alternatives need be considered. If the plan does *not* work, some other reasonable alternative, like a hiring freeze, may be implemented. If this course of action fails, it may be followed by yet another attempt, such as closing down antiquated plants.

Performance Programs

performance programs Scripts that detail exactly what actions are to be taken by a job incumbent when confronted with a standard problem or situation.

Because many of the problems that an organization encounters are routine, new decisions need not be made each time these problems arise. Instead, organizations develop **performance programs**, or scripts that detail exactly what is to be done in a given situation. Typically these programs are set in motion without any conscious thought as to whether they are the best means of accomplishing the objective. For example, when a fire gong rings at a fire station, it immediately triggers a great deal of smoothly coordinated activity, and there is no need for any one firefighter to spend much time thinking about what he should do next.

Developing Experts

discretion An area of latitude wherein the decision maker can use her own judgment in developing and deciding among alternative decisions.

Even though we can develop programs that help simplify a task for an employee, uncertainty in the environment makes it impossible to develop perfectly detailed scripts that will be applicable everywhere. Thus there is still a need for **discretion**, or individual authority to change or modify performance programs while they are being carried out. The range of discretion tends to be limited to tightly defined areas, and experts are developed who become the decision makers for such areas. The advantage of using experts in decision making is that people with special expertise in an area can devise more accurate and more detailed scripts.

APPLYING THE DIAGNOSTIC MODEL
Diagnostic Question 11: Does the organization employ experts? Does it use these forces successfully, through loose coupling, integrating these people with its own staff?

In this way complexity is handled by breaking it up into discrete, more manageable chunks that can be handled by a single individual. Thus as we will discuss in Chapters 15–17, an organization may be divided into departments (e.g., personnel) which are divided into units (e.g., personnel selection) which are divided into areas (first-line managerial selection) which are divided into job categories (college recruiter) which are divided into individual jobs (western regional interviewer). The occupant of an individual job typically focuses in on one very narrow area of organizational problem solving.

Failing to differentiate and delegate in this fashion can lead to a situation where one person is simply trying to do much and, in the process, not accomplishing anything. For example, many considered the disastrous $1.2 billion losses recorded by the Bank of New England Corporation (BNE) in the fourth quarter of 1989 to be just this kind of failure. Many claimed that CEO Walter Connolly became involved in many decisions that most bankers would delegate to low-level loan specialists. As the former president of a BNE subsidiary noted, "Walter Connolly could very successfully run a $6 billion institution. . . . When he tried to keep his hands on everything in a $30 billion bank, nothing got done."[18]

As the "In Practice" box illustrates, the combination of expertise and autonomy is often the perfect solution to managing complexity—even the complexity inherent in the management of diversified businesses within a large corporation.

Coupling Programs and Experts

If we break up jobs into small parts, we reduce the burden on any one individual, but then we must integrate each person's contribution with everyone else's.

[18] L. Jereski, "A Stomach for the Bank That Ate New England," *Business Week*, February 5, 1990, p. 68.

DECISION MAKING

Revolutions in Management from RPM Inc.

In the late 1980s many firms grew through acquisition strategies. That is, rather than grow through internal development, they simply bought out smaller firms and added them to their corporate portfolios. Most corporations employing this strategy found that buying these companies was a lot easier than managing them. The tremendous complexity of managing multiple businesses that were sometimes unrelated to each other often overwhelmed corporate leadership. This happened particularly often when corporate leadership had the financial expertise to put the deal together but lacked the technical expertise necessary to run the business that was purchased. More recently, however, corporations growing by acquisitions have recognized the need for technical expertise and have obtained it from an obvious, but long overlooked source: former owners.

The traditional approach was to "clean house." Former owners departed, and the acquiring corporation brought in a totally new management team. Increasingly, however, corporations are employing former owners as experts and giving them considerable decision-making autonomy.

RPM Inc. of Medina, Ohio is one holding company that has employed this practice successfully for a long time. In fact, in just over two decades Chairman Thomas Sullivan persuaded 25 business owners to sell their companies to him and then to run them as independent divisions of RPM. Retaining former owners and keeping them happy and productive is not an easy task, yet RPM has managed this time and time again.

RPM's formula for success is quite straightforward. Rather than merge or relocate the firms he purchases, Sullivan almost always leaves them in place. Owners are retained and given a share of the firms' earnings. Decision making is highly decentralized; each owner produces an annual plan that is evaluated and refined by RPM top management. The former owners and corporate staff then meet face-to-face and iron out any differences in the original and revised plans. The resulting plans are an effective combination of the former owners' technical expertise and RPM's financial discipline. The plans are often more daring than those typically produced by a large, unwieldy corporation yet more in line with market realities than they might be if drawn up by free-wheeling entrepreneurs.

RPM's strategy has reaped huge benefits. At year end 1989, the firm reported earnings of $24.2 million, nearly five times the $5.1 million earned ten years earlier. Moreover, despite numerous economic swings, RPM's earnings and sales have increased for 24 years in a row. This huge growth has been achieved with the addition of only two headquarters staff members since 1976. This latter fact alone stands as a testament to the effectiveness of managing complexity through the use of decentralized decision-making experts.*

* *Source*: M. Selz, "RPM Bases Success on Keeping Owners of the Firms It Buys," *Wall Street Journal*, June 19, 1990, p. B1.

---

loosely coupling Managing interrelatedness across different functional areas by not allowing the actions or decisions of one functional unit to have an overly large or immediate impact on the actions or decisions of other functional units.

Chunking does not change the fact that organization members are interrelated, and it is unrealistic to think that one expert can operate unaffected by others or that one set of programs can be activated independently of others. In integrating groups, the complexity of planning is greatly simplified by **loosely coupling** the different parts, that is, weakening the effect that one subgroup has on another so that each can plan and operate almost as if the other were not there.

For example, a production department's work is greatly facilitated by having steady operations, that is, operations that do not fluctuate a great deal over time. Yet the sales department may be subject to wild swings in consumer demand that need to be met quickly. These two departments can make decisions more easily if inventories can be used as buffers. That is, by letting products accumulate in inventories, the production department can go on operating (for a while at least) as if there is a steady demand for the product, even if sales are down. Later, when demand may exceed production capacity, the sales group can sell out of inventories without making rush demands on the production process.

Comparing Models

Figure 6-5 makes it possible to compare the administrative decison-making model with the rational decision-making model. As the figure shows, although the administrative model is also a goal-driven model, it recognizes that day-to-day operational goals are rarely givens and instead come about only through negotiation

and compromise. Once goals are set, the decision maker monitors the environment and analyzes any problems that arise, to determine if they are programmed or nonprogrammed in nature. If a problem is routine, a well-established performance program is executed to deal with it. If a satisficing solution is found the process is complete. If the problem is nonroutine, however, possible solutions are analyzed sequentially, and sometimes experts are called in to help evaluate alternatives. The first reasonable alternative is then implemented, and if it results in a satisficing solution, the process is complete. Otherwise, a second reasonable alternative is attempted, and then the process continues until a satisficing solution is reached.

CREATIVITY IN DECISION MAKING

One elusive quality essential to organizational decision making is creativity. We will define **creative decisions** as choices that are new and unusual but effective. Neither the rational nor the administrative decision-making process gets at the issue of how creative decisions are produced. Indeed, there are elements in both models that make the generation of creative solutions to problems less, rather than more, likely. For example, strictly following the rational decision-making model's historical method or off-the-shelf methods of generating alternatives (See Table 6-1) will rarely result in innovation. In fact, these methods will probably lead to innovation only by chance or by mishap.[19]

The need for creative solutions to problems becomes especially important as the environment changes more and more rapidly. In such environments, old solutions to problems become outdated relatively quickly, and firms that can shift most readily to new and effective operations gain a huge competitive advantage. The "International OB" box illustrates the lengths to which some countries will go to encourage this kind of activity. In this section we will emphasize the creativity process and how organizations can enhance creativity by selecting appropriate people and altering the workplace environment.

The Creative Process

Studies of people engaged in the creative process or of the decision-making processes of people who were famous for their creativity suggest that a discernible pattern of events leads up to most innovative solutions. Most creative episodes can be broken down into the four distinct stages shown in Figure 6-6: preparation, incubation, insight, and verification.

Contrary to what most people think, creative solutions to problems rarely come out of the blue. Indeed, anyone waiting around for an idea to pop into her head more than likely will be waiting a long time. Although creative ideas do sometimes come about serendipitously, this seems to be the exception more than the rule. More often than not, innovations are first sparked by a problem or perceived need. Because creative decision making is in this way like other decision-making processes,[20] it should not surprise you to know that **preparation**, the first stage in the creative process, involves assembling materials. Analogous to the rational model's "generating alternatives" stage, preparation is character-

Archer Daniels Midland (ADM) has gained a reputation for innovation. It keeps turning out an expanding variety of products made from three basic resources: corn, wheat, and oilseeds. Currently the company makes vegetable oil, animal feed, vodka, flour, caramel coloring, sorbitol (used in making Vitamin C supplements), corn syrup, and ethanol. In an impressive show of creativity, the company married the low-margin product ethanol with an increasingly profitable one, high fructose corn syrup (HFCS), producing both with the same milling process. This process turns out ethanol in the winter, when some cities already require the use of this clean-burning but still costly motor fuel. In the summer, when people consume more soft drinks, the mill switches to high fructose corn syrup (HFCS), which is fast becoming the sweetener of choice in the soft drink industry.
Source: *Ronald Henkoff, "Oh, How the Money Grows at ADM," Fortune, October 8, 1990, pp. 105–116.*

preparation A stage in the creative decision making process in which the person accumulates information needed to solve a problem.

---

[19] T. S. Kuhn, *The Structure of Scientific Revolutions* (Chicago: University of Chicago Press, 1962), pp. 22–39.

[20] D. G. Marquis, "The Anatomy of Successful Innovations," *Managing Advancing Technology* 1 (1972), 34–48.

From Imitators to Innovators: The Taiwan Conversion

For many years, the Taiwanese were more than happy to copy products that had already been developed in the West or in Japan. A large supply of cheap labor and a relatively weak currency allowed the Taiwanese to make a comfortable living simply by producing inexpensive clones of others' toys, electronics and athletic equipment. Today, however, this practice is a less viable option. Since 1986, the cost of labor has increased almost 40 percent, and the value of the New Taiwan dollar has risen to as high as 50 percent of the U.S. dollar. No longer able to compete on the basis of producing at low cost, Taiwanese firms have had to develop other avenues of competition. Long-standing practices have been abandoned, and creativity and innovation have become the pillars of the future Taiwan economy.

The Taiwan government has played an important role in moving the country from imitation to innovation, and the country's new methods of success are studied and copied by other Pacific Rim nations such as South Korea and Indonesia. One of Taiwan's first steps was to increase expenditures for research and development (R&D). In fact, this type of investment has doubled in just the last five years. Typically, R&D funding goes to projects that can be shared by many small companies.

Second, through tax laws, the Taiwanese government encourages companies to substitute machines for workers. Quite purposefully, it is encouraging labor-intensive businesses to move overseas. Low-tech factories are being closed, and instead Taiwan firms, with government aid, are purchasing small Silicon Valley enterprises. The new thinking is also evidenced by Taiwan's stepped-up foreign investments, including roughly $1 billion in mainland China, which is still officially at war with Taiwan.

When they cannot buy entire businesses, Taiwan companies are not afraid to pirate key personnel. For example, Bobo Wang, a young and rising star in the Xerox Corporation was recently lured to Taiwan. The government set Wang up in a Silicon Valley-like industrial park, ideally situated between the capital city and two of the area's finest engineering schools. The park had cheap rent, clean water, and affordable electricity. The payoff came in 1990, when Microtek, the company founded by Wang, hit the market with a new machine that feeds not only black-and-white but color photographs into a computer for processing. The machine was highly coveted by desktop publishers, and in six months little Microtek sold 11,000 of them at almost $3,000 each.

Besides adding relocated Asian-Americans to their staffs, Taiwanese companies maintain close ties with U.S. markets by such means as sending key personnel to U.S. seminars and workshops. Microtek and many other firms keep a staff of technical people in the United States to monitor trends. These companies encourage staff members to visit other countries for exposure to different ideas.

Most Taiwan companies are, like Microtek, small, entrepreneurial, and able to move quickly into new areas. As Robert Hsieh, Microtek's head of office automation notes, "We move fast. . . . Japan has committees, committees, and committees to decide something. We can jump right on things because all I need is one signature." Although small size and lack of brand-name familiarity are handicaps, it looks as though creative solutions to these problems will be forthcoming soon.*

* Source: D. Darlin, "Unlikely Leader: Taiwan, Long Noted for Cheap Imitations, Becomes an Innovator," Wall Street Journal, June 1, 1990, p. 1.

---

incubation A stage in the creative decision-making process in which the person apparently stops attending to the problem at hand.

ized by plain, old-fashioned hard work. In attempting to solve the problem, the creative person immerses himself in all existing solutions to the problem, usually to the point of saturation.

The second stage, **incubation**, differs greatly from steps in other decision-making models. Rather than come to a decision immediately after assembling

FIGURE 6-6

Steps in the Creative Decision-Making Process

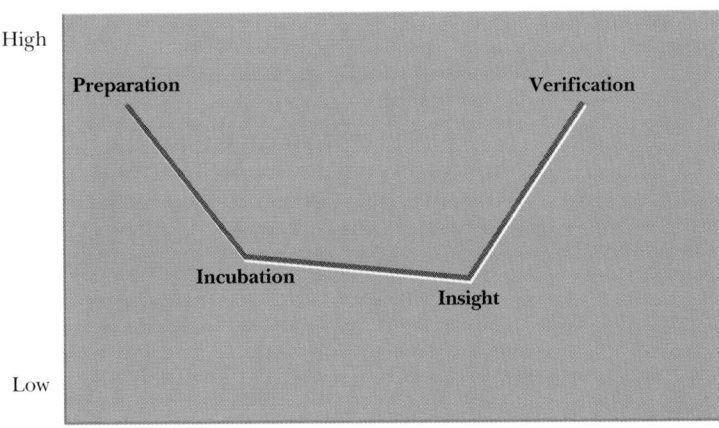

and evaluating the relevant material, creative decision makers enter a period during which they seem not to expend any visible effort on the problem. Sometimes, out of frustration or sheer exhaustion they may give up working on the problem temporarily and turn to other things.

It has been suggested that when people seem to give up on a problem in this way they are actually continuing to work on it but on an unconscious level. Allusions to "unconscious" processes like this, however, are in large part signs that although research has been able to identify the stage of incubation, it has been unsuccessful so far in explaining what really goes on during this stage.

After a person spends some time in the incubation stage, the solution to the problem typically manifests itself in a flash of inspiration, or **insight**. Usually, the person is engaged in some other task when this insight comes to her. This third stage in the creative decision-making process is a very delicate one because insights tend to be ephemeral. They can disappear as quickly as they appear. It is because insights can come at any time that many truly creative people have a habit of carrying a notebook or of jotting ideas down on whatever piece of paper is available to make sure an idea is not lost. Arthur Laffer, for example, is supposed to have sketched the basic ideas of supply-side economics on the back of a cocktail napkin at a Washington, D.C., bar. This theory served as the intellectual underpinning for "Reaganomics" which dominated public policy decisions in the 1980s.

The fourth stage of the creative decision-making process is **solution verification**. Here, the solution formulated in the insight stage is tested more rigorously to determine its usefulness for solving the problem. This stage in creative decision making is very much like the rational decision-making model's stage of evaluating action.

Typically, the verification process takes a lot of time. In fact, it resembles the preparation stage in the amount of hard work it requires. This is primarily because people, particularly if they have a lot invested in traditional ideas and methods, resist change. They have to be convinced to try a new approach, and we can rarely convince them without independently verifying the new findings.

Creative People

Certain characteristics of individuals seem to be associated with creative endeavor. First, there seems to be a modest relationship between creativity and general cognitive ability (see Chapter 4) and the specific capacities of reasoning and deduction. According to J.P. Guilford, some minimum threshold of intelligence seems to be necessary for creative work. As shown in Figure 6-7, however, once we get above that minimum threshold, general intelligence becomes less critical. In terms of the creative process, a severe lack of cognitive ability would probably impair one's ability to get immersed in information about the problem. Once the threshold is reached, however, hard work is probably more important than raw intelligence for creative achievement.

It appears that such personality characteristics as interests, attitudes, and motivation are more important than intelligence in distinguishing creative people from the general population. One common characteristic of creative people is that they set high goals for themselves which may make them dissatisfied with the status quo and current solutions to problems.[21] High levels of aspiration may also explain why creative people often do not seem to feel any particular loyalty to an employer and are instead highly mobile, moving from company to com-

---

[21] D. W. MacKinnon, "Assessing Creative Persons," *Journal of Creative Behavior* 1 (1967), 303–4.

FIGURE 6-7

The Relationship between
Cognitive Ability and Creativity

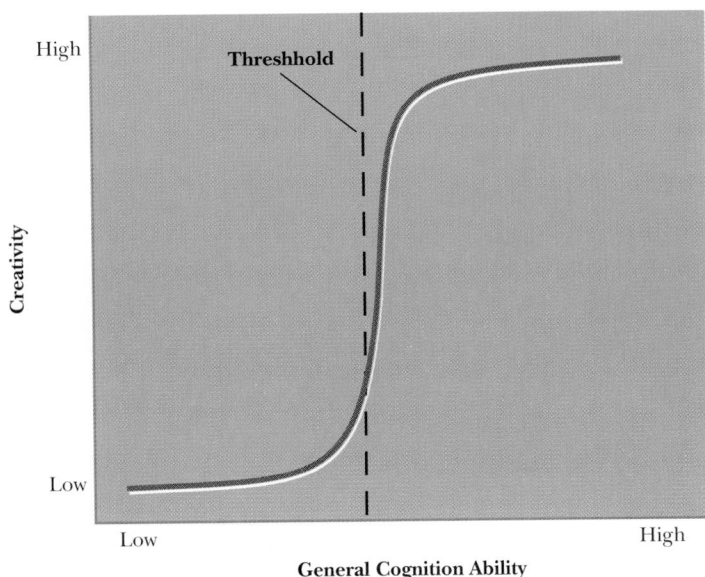

pany.[22] Like most valued commodities, creative talent is highly sought after. Thus it is often hard for a company to hold on to its creative people.

It has also been suggested that the creative person is *persistent* and has a high energy level.[23] These characteristics are probably particularly useful in the stages of preparation and verification, which demand hard work for long periods of time. Persistent people will stick with something despite obstacles and setbacks, and people with a lot of energy can continue to work hard for long periods of time. Both these qualities may help creative people to assemble more relevant information and to test their ideas more exhaustively.

Self-acceptance and independence also seem to relate to creativity.[24] These characteristics probably are needed most in the verification stage, where maintaining one's confidence in the face of criticism and even ridicule becomes important.

Finally, *age* seems to be related to creativity. In a seminal study of people recognized for their creativity, one consistent finding was that regardless of the field in which the person did his or her work (the fields studied included mathematics, physics, biology, chemistry, medicine, music, painting, and sculpture), creativity peaked between the ages of 30 and 40 (see Figure 6-8).[25]

Creativity-Inducing Situations

Selecting people who have characteristics that seem to be related to creativity is not the only option organizations have for increasing innovativeness. Providing *specific and difficult goals* and *firm deadlines* actually seems to stimulate creative achievement. Although they can be taken too far, in general, time pressures seem to aid rather than hinder the creative process. This also seems to be true of

[22] T. Rotundi, "Organizational Identification: Issues and Implications," *Organizational Behavior and Human Performance* 13 (1975), 95–109.

[23] E. Randsepp, "Are You a Creative Manager?" *Management Review* 58 (1978), 15–16.

[24] H. G. Gough, "Techniques for Identifying the Creative Research Scientist," *Conference on the Creative Person* (Berkeley: University of California, Institute of Personality Assessment, 1961).

[25] H. C. Lehman, *Age and Achievement* (Princeton, N.J.: Princeton University Press, 1953, pp. 50–61.

FIGURE 6-8

The Relationship between Age and Creativity

From J. P. Guilford, "Some Misperceptions Regarding Measurement of Creative Talents," Journal of Creative Behavior 5 *(1971), 86–99.*

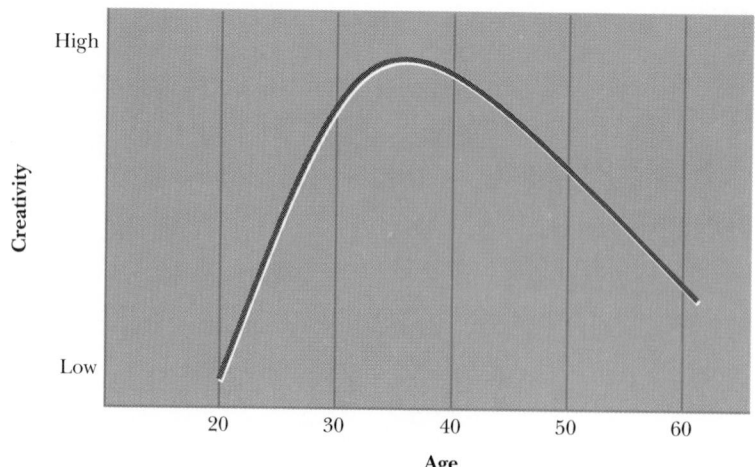

budgetary pressures. Perhaps one reason is that these kinds of constraints limit the use of historical options or "off-the-shelf" solutions. Consequently, the organization is forced to come up with innovative, nova-type techniques.[26]

Some firms even set goals for creativity. As an example, 3M company, based in Minneapolis, has set a goal that 25 percent of its total revenues should come from new products developed in the past five years. Currently those revenues are running closer to 30 percent, and 3M insists that nearly 70 percent of its annual $12 billion in sales comes from ideas that originated from the work force.[27]

Although a bit trite, there is some truth in the old saying that "necessity is the mother of invention," and creativity often springs from organizations that have their backs to the wall. For example, David Luther, senior vice-president and corporate director of quality at Corning notes that "desperation is a good motive. . . . Customers came to us and said if we didn't change, they'd go somewhere else."[28] This point of view can also be seen in the "International OB" box referred to earlier, which illustrates the way changes in the economy in Taiwan forced businesses to think along more creative lines.

Certain characteristics of what is called the *organizational culture* (see Chapter 18) may also be related to creativity. First, the degree to which organizations recognize and reward creativity is of paramount importance.[29] Although you might think otherwise, unfortunately, many organizations place more emphasis on following existing written rules and procedures than on experimenting with new procedures.

A culture that promotes creativity must ensure not only that innovativeness is reinforced but that experimentation leading to failure is not punished. Executives like James Burke, CEO of Johnson and Johnson, attempt to create a climate where the risks of innovation are minimized. Burke, in fact, has gone so far as to tell his employees, "We won't grow unless you take risks. . . . Any successful company is riddled with failures. There's just no other way to do it."[30] Although they needn't reward every failure, companies that want to encourage innovation might consider giving an award for "the best failed experiment." Recognition of this sort would drive home the point that the organization values risk taking and trying new approaches more than it fears mistakes.

[26] G. Zaltman, R. Duncan and J. Holbek, *Innovations and organizations* (New York: John Wiley, 1969), p. 191.

[27] T. Curry, J. E. Gallagher and W. McWhirter, "Let's Get Crazy," *Time,* June 11, 1990, pp. 41–42.

[28] Ibid.

[29] H. A. Shepard, "Innovation-Resisting and Innovation- Producing Organizations," *Journal of Business* 40 (1967), 470–77.

[30] W. Guzzardi, "The National Business Hall of Fame," *Fortune,* 121 (1990), pp. 17–29.

Finally, since much creativity comes out of collaborative efforts among different individuals, organizations should promote *internal diversity and exposure* among organizational members. If all the people in a group have the same interests, experiences, strengths, and weaknesses, they will be less likely to generate new ideas than if they are of divergent backgrounds and capabilities. We will have more to say about this in Chapter 11 when we discuss group decision making.

Moreover, because different organizations do different things in different places, exposing people to varying kinds of experiences, such as foreign assignments, professional development seminars, or extended leaves may help shake up overly routine decision-making processes. The notion that difference and variety encourage creative thinking receives some support from the finding that organizations that emphasize external recruiting seem to be more innovative than firms that promote from within.[31] This does not mean that organizations should completely abandon promote-from-within policies. However, the value of mixing new and long-tenured employees in fostering a climate of creativity may be considerable.

SUMMARY

Both the outcomes of decisions and the processes by which decisions are made are of interest to managers of organizational behavior. The *rational decision-making model* proceeds through five steps wherein the decision maker sets objectives, monitors the environment to determine progress toward those objectives, generates alternative courses of action when problems are detected, chooses a course of action, and finally, evaluates the chosen course of action for its ability to accomplish the original objectives. Many types of programmed decisions such as setting production standards and determining inventory levels, can be handled by this model. Other kinds of problems, however, may exceed our capacity for rational analysis. Moreover, illusions, or threats, like *loss aversion*, *availability bias*, and *escalation of commitment* can seriously interfere with rational decision making. For a whole host of nonprogrammed decisions, rational decision making simply is not possible. With nonprogrammed decisions, few of the assumptions required for deriving optimal solutions are met. Under these conditions, the *administrative decision-making model* provides a more accurate depiction of the decision-making process. The administrative model emphasizes *satisficing* rather than *optimizing* and also differs from the rational model by suggesting that alternatives be evaluated sequentially. The creative process in decision making proceeds through the four stages of *preparation*, *incubation*, *insight* and *verification of solutions*. Both personal (cognitive ability, persistence, energy level, age) and situational (organizational goals, culture, diversity) characteristics affect creativity in the workplace.

REVIEW QUESTIONS

1. Compare and contrast the decision-making process associated with rational decision making, administrative decision making, and creative decision making. At what points do these three models diverge most? What implications does this have for decision makers who employ one model when they should be employing the other?

---

[31] B. Schneider and N. Schmitt, *Staffing Organizations* (Glenview, Ill: Scott, Foresman, 1986), p. 71.

2.   Assume for a moment that over time, the price of a stock is not stationary, but ever increasing. Is a bias that ignores the base rate or regression to the mean really a bias in this context? Can you think of other instances in which what looks like a decision-making bias may actually facilitate decision making?

3.   Suppose that managerial jobs can be distinguished by the kinds of decision processes required. For example, some jobs call for rational decision making, others require administrative decision-making processes, and still others require creative decision processes. If you were a recruiter, what personal characteristics would you look for in staffing these three kinds of positions? Do you think it is possible to find one individual who would be comfortable with all three kinds of processes? If so, what characteristics would this person display?

4.   Escalation of commitment to a failing course of action has been widely researched and it is easy to call to mind many examples of this kind of mistake. The flip side of this mistake, however, is giving up too soon, which has not been studied much and for which it is hard to think of examples. Why can't we recall such events? How might researchers in this area be victims of "availability bias"?

5.   Recall the distinctions drawn in this chapter between certainty, risk, and uncertainty, and think of a decision you made recently that conformed to each of these conditions. Which of these conditions is most pervasive for managers? What weight in causing these conditions would you assign to (a) organizational level, that is, top, middle, or supervisory management; (b) the individual manager's experience or cognitive ability; (c) the nature of the organization's industry; and (d) characteristics of the manager's subordinates?

DIAGNOSTIC QUESTIONS

In analyzing an organization's decision-making process, the following set of diagnostic questions may be useful.

1.   Are ineffective decisions actually poor-quality decisions? Or is the ineffectiveness a result of poor timing and low acceptance of decisions by those needed to carry them out?

2.   Is the process by which decisions are made likely to be scrutinized? What can be done to convince people that the process is not arbitrary?

3.   Does the organization have clear goals and objectives, so that analytical solutions can be developed? Or is the organization characterized by a lack of goal consensus, thus requiring a bargaining solution?

4.   Can the organization detect threats and opportunities in the environment accurately? Does it tend to see things in terms of one rather than the other of these phenomena?

5.   When deciding on a course of action, does the organization tend to rely more on past practice or on experimentation?

6.   Is the organization's environment best characterized by risk or uncertainty? How does this environmental characteristic affect decision-making processes?

7.   Does the organization employ standard mathematical solutions to routine or programmable problems where they are appropriate?

8.   Do factors such as loss aversion or availability bias influence the organization's decisions adversely?

9.   Do organizational members understand the concepts of base rate and regression to the mean? Do they take account of these influences in making decisions?

10.   Do organizational decision makers stick to a single course of action too long (as in escalation of commitment)? Or do decision makers switch from one decision to another too quickly, making it impossible to learn in "noisy environments"?

11.   Does the organization employ experts? Does it use these forces successfully, through loose coupling, successfully integrating these people with its own staff?

12.   Does the organization seek creativity? What does it do to select people or design situations so as to enhance creativity?

EXERCISE 6-1

DECISION-MAKING HEURISTICS AND BIASES*

ARIEL S. LEVI, *Wayne State University*
LARRY E. MAINSTONE, *University of Michigan*

Managers often work under complex, changing circumstances that require rapid and accurate decision making. They must constantly process information, diagnose problems, think creatively, and develop solutions under conditions of substantial uncertainty. Technical aids that facilitate decision making like computerized information systems and diagnostic decision trees can be helpful, but they cannot substitute for human judgment.

In making judgments of all kinds, we often violate some fundamental laws of logic and probability. Why do we do this? Because we use *heuristics*, or mental strategies for processing the information we need to make a decision. Heuristics are really rules of thumb. For example, a common heuristic is, When it's cloudy, carry an umbrella. Heuristics can produce accurate results. If it rains, you'll be glad you have the umbrella. On the other hand, they can also bias decision making. If you carry an umbrella and it doesn't rain, you will be sorry. The biases that we have discussed in this chapter—loss aversion, availability, base rate, and regression to the mean—are all rooted in the use of heuristics.

This exercise focuses on those decision making processes in which people collect and mentally combine information to make estimates and evaluations of people, objects, and events. To make you more aware of heuristic biases and to stimulate class discussion about how to control their effects, we will show you how several heuristics work and how they influence judgment and decision making.

STEP ONE: PRE-CLASS PREPARATION

To prepare for class, read this entire exercise and then write down your responses to the four problems that follow. Try your hardest to come up with the best answer to each one. The first problem will be familiar to you, for we discussed it in this chapter.

1. A cab company was involved in a hit-and-run accident at night. Two cab companies, the Green and the Blue, operate in the city. You have the following information:
    A. 85 percent of the cabs in the city are Green and 15 percent are Blue.
    B. A witness identified the cab as a Blue cab. The court tested her ability to identify cabs under night visibility conditions. When presented with a sample of cabs (half Blue and half Green), the witness made correct identifications in 80 percent of the cases and erred in 20 percent.

    Question: The probability that a Blue cab was involved in the accident is _____ percent.

2. Bill is 34 years old. He is intelligent, compulsive, unimaginative, and sedentary. In school, he was strong in mathematics but weak in science and social studies.

    Rank order the following statements by their probability of being correct, using 1 for the most probable and 8 for the least probable.

    Ranking
    _____ A. Bill is a physician who plays poker for a hobby.
    _____ B. Bill is an architect.
    _____ C. Bill is an accountant.
    _____ D. Bill plays jazz for a hobby.
    _____ E. Bill surfs for a hobby.
    _____ F. Bill is a television reporter.
    _____ G. Bill is an accountant who plays jazz for a hobby.
    _____ H. Bill climbs mountains for a hobby

3. In an average year in the United States, do more people die of fire or drowning? Circle the letter that indicates your answer:
    A. Fire
    B. Drowning
    C. About the same

4. Many professional athletes believe that to be pictured on the cover of *Sports Illustrated* is bad luck because one's performance is likely to decline during the weeks and months afterward. Do you think there is any basis for this belief? If so, what is it?

    _____

    _____

    _____.

* Adapted with the authors' permission from an exercise entitled "A Group-Based Procedure for Revealing Judgmental Heuristics and Biases." Related information is reported in Ariel S. Levi and Larry E. Mainstone, "A Strategy for Teaching about Judgmental Bias and Methods for Improving Judgment," *Organizational Behavior Teaching Review*, 10 (1985), 9–24.

STEP TWO: COMPARING INDIVIDUAL AND GROUP ANSWERS

The class should divide into groups of four to six members each. If you formed permanent groups in Exercise 5-1, you should assemble in those same groups. In each group you should begin by discussing the four problems just presented and reach agreement on an answer for each one. Next, compare your group's answers with those you and other members reached working alone. Then select a spokesperson to compile this information and prepare a brief presentation for the entire class. The spokesperson should discuss the means and distributions of individual responses to each question, the group's response to each question, and the way the group reached agreement. For example, did the group take a majority vote or was it persuaded by one dominant individual?

STEP THREE: GROUP PRESENTATIONS AND DISCUSSION

The spokespersons should take no more than five minutes each to report their group's results to the rest of the class. The class should try to uncover reasons for any differences between individual responses and group agreements and for any differences among different groups' responses.

Next, your instructor will reveal the best answers for the four questions and will discuss the heuristics and biases that the problems illustrate.

STEP FOUR: GROUP JUDGMENT TASKS

Once everyone understands the heuristics and biases discussed in Step Three, the judgment groups should reconvene and choose the best answer to each of the following four problems.

1. As of 1985, approximately what percentage of the male working-age population in Japan was guaranteed lifetime employment? _____ percent.
2. Film critics have noted that sequels like "The Godfather III" or "Ghostbusters II" are usually of lower quality than their predecessor films. Sequels are usually financially less successful as well. Critics sometimes accuse the film studios of trying to "milk" a good idea. Do you think this is a valid criticism and, if so, how do you explain the phenomenon?

3. A panel of psychologists have interviewed and administered personality tests to 30 engineers and 70 lawyers, all successful in their fields. The following is one of the descriptions written about these 100 people, based on the results of the tests.

    > Joan is a 45-year-old woman. She is married and has four children. She is generally conservative, careful, and ambitious. She shows no interest in political or social issues and spends most of her time on her many hobbies which include camping, sailing, and mathematical puzzles.

    Question: The probability that Joan is an engineer is _____ percent.
4. Rank order the following events by their probability of occurrence this year, using 1 for the most probable and 4 for the least probable.
    Ranking
    _____ A. The industrial midwest will gain in economic strength.
    _____ B. The industrial midwest will decline in economic strength.
    _____ C. The fortunes of the major U.S. automakers will improve.
    _____ D. While the fortunes of the major U.S. automakers will improve, the industrial midwest will decline in economic strength because of the continued exodus of companies to the Sun Belt.

After reaching agreement on how to answer the four problems, each group should again appoint a spokesperson to report on the group's four answers and how they were reached.

STEP FIVE: GROUP REPORTS AND CLASS DISCUSSION

After each spokesperson has presented a report and the class has discussed similarities and differences between the groups' answers, your instructor again will reveal the best answers and discuss relevant heuristics and biases. The class should then discuss the problems raised by heuristics and biases, paying special attention to the following questions:

1. Some people won't admit they have biases. Others react defensively when their biases are pointed out. Are biases a sign of abnormality? Of low intelli-

gence? Of not trying hard enough to make the right decision?

2. Are group judgments generally more accurate than individual judgments? Or can groups impair accuracy? What factors might determine whether groups improve or impair accuracy?

3. What real-world organizational examples can you think of that illustrate each of the biases you learned about in this exercise? Do any of these examples show the effects of more than one bias?

4. NASA has made several monumental errors in recent years, including the decision to launch Challenger despite warnings from the space shuttle's designer and the decision to orbit the Hubble Space Telescope without adequately checking the focus of its mirrors. Based on what you have learned from this exercise, what advice would you give NASA to help it avoid similar mistakes in the future?

CONCLUSION

Managers need to learn about heuristics and biases for several reasons. First, heuristic-based biases are pervasive. They affect many organizational phenomena ranging from performance appraisal to strategic planning. Second, because these biases often operate unconsciously, neither intelligence, expertise, nor motivation to be accurate can protect us against them. Once we become aware of these biases, however, we can learn to step back each time and ask ourselves whether one or more of them have led us to misdiagnose a problem, make a poor decision, or fail to respond effectively to a crisis or opportunity. As organizational work and decision making become more complicated and challenging, bias-free managerial judgment becomes more and more important.

DIAGNOSING ORGANIZATIONAL BEHAVIOR

CASE 6-1

PRECISION MACHINE TOOL*

JANET BARNARD, *Rochester Institute of Technology*

John Garner, president of Precision Machine Tool, watched the elegantly tailored Mr. Wang leave the office after making his disturbing proposition. Of course, John had known that his own production people were working with Suzuki Machines on developing specifications for a machining center that would help solve Precision's nagging quality problems. Negotiations were winding down and the Japanese firm's price quotation was expected. Ako Wang's name on today's appointment calendar, therefore, was no surprise. From past dealings with Asian firms, John had expected the traditional old-world formalities that precede the closing of a sale for a major piece of capital equipment. In fact, he had braced himself for the usual rich combination of urbane courtesy and sharp technology. The United States machine tool industry has a strong bias to "buy American," and Precision was no exception.

This morning Wang had performed as expected, but this time Suzuki wasn't intent on making a sale. True, the proposal on John's desk contained a purchase document, but it wasn't a quotation for a $250,000 heavy-duty machining center. Instead, it was a formal invitation to discuss the purchase of Precision Machine Tool by Suzuki Machines.

"Lorraine," John spoke into the intercom on his desk, "see if you can find Tom and ask him to come up."

While waiting for his partner, John stood by the window that overlooked a big machining bay on the floor below. Even through the heavy insulated glass he could hear the ceaseless clamor of the big machines that were making high-precision parts for the lathes that Precision produced and sold to the automobile industry to use in their factories. John watched as an operator checked the control panel of a new cobalt-blue lathe that stood among the aging machines on the shop floor. The hulking lathe was state-of-the-art machine technology, precise and sophisticated—with a manufacturer's nameplate that read, "Suzuki/Made in Japan."

The machine tool industry is unique in that it uses its own products to make its own products, and much of Precision's old equipment was becoming dulled by decades of use. In a desperate move to stem customer complaints about quality, they had bought the computer-controlled Japanese lathe to use for making parts for the machines they produced. To buy foreign-made machinery went against the grain, but many domestic toolmakers were buying imported machines because they were more efficient to operate, gave a higher quality of output, and were cheaper than American-made equipment. For some toolmakers it was the alternative to joining the 20 percent of the nation's machine tool companies that had gone out of business recently.

John turned away from the window and took a sales printout from a desk drawer. His company was in better shape than many of the medium-sized toolmakers, but that wasn't saying a lot. Sales were down 30 percent. Booked orders were weak, and quality rejects due to the aging and long-used equipment ate into profits. Precision was a victim of recession in the automobile industry. Like many other machine toolmakers, it had never fully recovered. John looked up as his partner entered the office.

Tom Avery flung himself into the chair that had been occupied by Ako Wang. Precision's works manager was a big man, blunt and outspoken, and a first-class tool design engineer. Tom ran the manufacturing and materials management end of the business and, with John, had founded Precision Machine Tool. His reaction to John's news about Wang's proposition was expressed in a single word that was short, direct, and explosively negative.

Despite excess capacity in American plants, Japan's share of the American machine tool market was increasing and was hotly resented by the domestic industry. The fire was currently being fueled by Japan's determination to increase exports of cars to the United States now that the voluntary quota system had expired. There was no corresponding assurance that there would be any increase in the trickle of American goods that were allowed to enter Japan. As a result the decrease in American market share could be significant in an industry linked so closely with automobiles and steel. Employment would be hard hit as well, and Precision's 312 employees were down 22 percent.

"That was my first reaction, too, Tom. But I think we should think this through." John held up his hand to silence his partner as Tom leaned forward, scowling. "Let me go on for a minute. Our industry is in its worst crisis since the Depression. Sure, Precision's done better than some, but our sales are down to $16 million and you know what the reject rate is doing to costs. Orders have softened steadily, Tom, and that's what worries me most. We've had a reputation for top-quality machine tools from the time we opened our shop.

"Precision Machine Tool has always been synonymous with precision quality. We're losing that, Tom." John went over to the bay window. "Sixty percent of our equipment is old, some of it more than 20 years. Accuracy of these machines is unreliable, they're expensive to operate, and not worth any more rebuilding. We're in a spiral, Tom; without profits, we can't afford to modernize the plant. And with obsolete machines, we can't compete with foreign imports, not in price and not in quality."

The late 1970s were the apex of the domestic machine tool industry. There was a record backlog in orders that couldn't be filled because of inadequate production capacity. Industrial customers waited two years for machine tool orders they needed today. The domestic industry was too busy to notice that several years before, Japan had identified machine tools as a growth industry and started subsidizing modern factories. Now was the time to cash in. American manufacturers were turning overseas for fast delivery of high-quality, inexpensive machines and machine tools to use in their production processes. During the 1980–1981 recession in the automobile industry, American tool firms had little capital for investment. When the economy recovered, they were left behind in the marketplace. Lately, subsidiaries of big Japanese toolmakers began to appear in the United States, along with an occasional Japanese acquisition of a domestic firm.

"What are you telling me, John? That you want to sell out?" Tom's voice was tight. "This is the most exciting industry around right now. We've got the wonders of automation to sell these days, the futuristic manufacturing systems. You want them all to be Japanese, John? Or West German, or Korean, or everything but American? You want to get out of the race just when we've survived the cash crunch from buying the new equipment we do have? That Japanese lathe, for starters."

It had taken Tom a long time to accept the idea of using an imported machine in their own production process, but the harshness in his voice was gone as he said, "Listen, John. You're a financial expert, but I know that yesterday's production gives us yesterday's dollars. Why not get rid of this patch-and-mend philosophy and shop for some real capital to modernize the plant? The U.S. capabilities for producing the computer software that meshes the tools together is superior to anybody's. We've got access to that, John, and all Precision needs is modern machines to get the edge we need to stay in the race." Tom waited. He knew that John was a financial conservative, dedicated to financing capital improvements from profits.

John turned from the window and sat down in the chair across from his partner. Tom knew as well as he did that it wasn't a matter of catching up with the competition; it would be necessary to leapfrog over a moving target. A big capital investment meant a big debt, and interest rates would tend to be high for a firm in the troubled machine tool industry. Precision would become highly leveraged and could risk ruin. There were too many "ifs." What if there were a downturn in the economy . . . or if too many customers were irrevocably lost during the transition . . . or if foreign toolmakers slashed prices to protect their market share . . . or if software companies outside the machine tool industry won important orders in the area of software expertise where American toolmakers had the edge? It was ironic, but John knew that Tom would infinitely prefer bankruptcy to selling out to foreign competition.

"All right, Tom," John took a deep breath. "Look at *this* scenario, and think about it for a minute. If we wanted to go the retrenchment route, maybe it makes sense to sell a line of imports to help us finance some new equipment. I don't like the idea any better than you do, but it would be temporary, Tom, and it's profitable. They wouldn't have to be Asian. The dollar's strong, and we might be able to buy West German machine tools at a price that would give us a good markup." John stopped, expecting Tom's outburst that was as vehement as his response to the Wang proposition.

Both of the men, especially Tom, had always been severely critical of the strategy of some of the hard-up domestic toolmakers who acted as distributors of imported machines and machine tools, or who bought imported products and customized them for special-order customers. John had to agree with the tone of derision and contempt in Tom's words. Selling imported machine tools in direct competition with your own industry was quite different from using a couple of pieces of imported equipment to beef up your own production process in a crisis.

Not only could the practice spell doom for the domestic industry, but the knife cut a lot deeper. One of the opportunity costs of selling foreign goods is that while a manufacturer is doing it he tends not to improve his own technology and capabilities. The machine tool industry is at the core of modern manufacturing, and the country that controls state-of-the-art machines and machine tools has the advantage of being able to make better cars and aircraft and drilling equipment—and ballistic missles.

It was inconsistent that an industry that had spent so much time and resources trying to get the federal government to provide protection from foreign competition by limiting imports was itself buying those same imports. Buying foreign was repugnant, John knew that, but it was a trade-off. Was it worth it?

"What are our options, Tom?" John's voice was quiet. "Is it better to commit ourselves to a debt that could wipe us out? Is it better to fold? Whatever we do, we've both got to buy into it, right? We always have." Tom smiled, and John knew that they were both remembering the early days of Precision Machine Tool when they operated on a shoestring and sat down together every Thursday to decide what bills they could afford to pay.

Things seemed more complex now, and even more uncertain. During the life of Precision, its industry had experienced a revolutionary change in products, and the past ten years had been either feast or famine for

domestic machine tools. Now the race to build the factories of the future would be won by the nation that had the most efficient computerized operations to produce the cheapest, most reliable products. For the owners of Precision Machine Tool, the price paid for falling behind was high, and the risks in trying to stay in the race was great.

Precision's key executives had to take a number of complex variables into account in making their decision. Important economic and political factors impacted Precision's ability to compete in its industry. Foreign competition and foreign technologies posed a serious threat both to the machine tool industry and to the future of domestic manufacturing. Personal attitudes and values as well influenced John and Tom in their task.

Lorraine sent out for sandwiches and coffee, and as the afternoon passed the two men examined the future of their industry and their place in it. When they parted late that night, their decision had been made—and they agreed to meet in John's office the next morning at ten for his telephone call to Ako Wang.

When you have read this case, look back at the chapter's diagnostic questions and choose the ones that apply to the case. Then use those questions with the ones that follow in your case analysis.

1. At which step of the rational decision making process are John Garner and Tom Avery now? How would you characterize their decision making up to this point?

2. What threats to rational decision making are evident? Are there factors that make rational decision making impossible? What specific steps should John and Tom take to complete the process of deciding whether to sell or not?

3. If they follow your suggestion, what decision do you think the two will make. In the long run, how satisfactory do you expect this decision to be. Why?

CASE 6-2
BOB COLLINS*

RICHARD E. DUTTON, *University of South Florida*
RODNEY C. SHERMAN, *University of South Florida*

Bob Collins was employed by the Mansen Company, a division of Sanford, Barnes, Inc., a diversified company engaged mainly in the manufacture and sale of men's and women's apparel. The Miami plant of the Mansen Company is the largest of the 19 manufacturing loca-

* Reprinted with the authors' permission.

tions and has, in its organization, an industrial engineering unit.

As department head of industrial engineering in the Miami plant, Jim Douglas also has the responsibility for all industrial engineering functions in the Florida Region. This includes three smaller plants within a 275-mile radius of Miami. Jim reports to the Miami plant manager, Mr. Scott, for local projects and to the Florida regional manager, Mr. Glenn, for projects of a regional nature. Mr. Glenn has been regional manager for many years, but only for the previous 23 months had this been his sole responsibility. Prior to this time, he was also the manager of the Miami plant, and he was still a dominant personality in the plant, partially because of Mr. Scott's indecisiveness.

Assisting Jim in Miami are two other industrial engineers, Bob Collins and Mark Douglas (see Exhibit 6-1). Mark was hired in September, 1992, soon after his release from the army, and had been with Mansen about 27 months. Bob had been with the company for about 21 months since leaving his last position because of a conflict there regarding a heavy workload and a schedule requiring some night work. Bob had freely given this information during his preemployment interview, but no effort had been made to uncover the past employer's version of the situation.

Jim holds an associate degree in industrial engineering from a two-year technical school, while both Bob and Mark have bachelor's degrees, in history and

business, respectively. All three men are army veterans. Jim and Mark served as enlisted men for nine and two years, respectively, Jim becoming a staff sergeant and Mark, a sergeant. Bob served as an officer for four years, reaching the rank of captain. Bob had displayed a talent for creative and imaginative thinking in regard to mechanical development and was assigned a majority of the projects that delved into the creation, installation, and improvement of mechanical innovations and devices. In addition, he and the local head of mechanical development, Ned Larson, worked together on many of their own original ideas, both in the planning and development.

CURRENT SITUATION

One day Bob came upon an interoffice memo, in Jim's incoming mail box, containing a question from Mr. Glenn about a mechanical project on which Bob was working. Feeling that he could save time for Jim, he picked up the memo, read it, and proceeded to Mr. Glenn's office to answer the question. When he returned, he simply put the memo back in Jim's mail box and went on with other work.

Later in the day Jim returned to the office to answer his mail and came upon Mr. Glenn's memo. Jim sought out Bob and the following conversation ensued:

Jim: Bob, I've got a short note here from Mr. Glenn asking about the status of the cuff machine project I gave you. Where do we stand on that now?

Bob: Well, as I mentioned before, all we have to build is the automatic stacking device and then we should have the machine about ready to go. Some of the parts won't be in until the first of the next week, but it should only take a day after that to finish.

Jim: Okay, that's good. How about answering this memo to Mr. Glenn and we'll have him up-to-date on this thing?

Bob: I already have. I saw the memo earlier and went on in and brought him up on how the project stands.

Jim: Did you get a copy of this too?

Bob: No, I saw yours and decided to save you time so I went ahead and answered his question.

Jim: You mean you got this out of my box?

Bob: Yeah, I saw it as I was coming to my desk and decided to go ahead and get it out of the way.

Jim: Oh well, I'll just hold on to this for a while then.

Several days later, Jim and Bob were discussing one of Bob's new ideas, and the discussion became very

---

EXHIBIT 6-1
Partial Organization Chart—The Mansen Company

Regional Manager
Mr. Glenn

↓

Plant Manager
Mr. Scott

↓

Industrial Engineering
Department Head
Jim Douglas

↓

Bob Collins   Mark Douglas

DECISION MAKING

heated when Jim rejected the idea as too expensive, in both time and money.

Jim: And another thing, Bob, I don't want you going through my mail box again. What's in there is none of your business unless I assign it to you.

Bob: I was just trying to do you a favor and get the memo answered. If you don't want me to do that, then I won't.

Jim: You would have answered eventually, but I don't want you to do these things unless I tell you to. I'm in charge of the department, and I have to know what's going on. That reminds me of another thing. From now on, you tell me about all of the projects you're working on. I don't want any more secret projects being worked on without my knowing it. I feel pretty stupid when Mr. Scott or Mr. Glenn asks me a question about something I've never even heard of. From now on, you tell me about your ideas, and if we can work it into the schedule we will; otherwise it will have to wait until we can get to it. This also means not going to Mr. Glenn with your ideas first, and then telling me that he thinks it's a good idea and should be developed. I'll approve the ideas first, and then we'll check with him if necessary.

Bob: I know you're talking about the new sleeve hemming stacker, and I just happened to mention it to Mr. Glenn this morning at coffee break, and he wanted to know more about it. I had to tell him about it when he asked.

Jim: That's right. In that case you couldn't have done anything else, but from now on make sure you've cleared these ideas with me before going to him.

Mark came into the office, and the discussion was ended.

The following day, Bob and Mark were leaving the office together, and Bob told Mark about his discussion with Jim.

Bob: Mark, I'm so mad at Jim I'd like to quit and walk out of here right now. I know darn well I could make more money somewhere else and wouldn't have to put up with Jim. You know what really gets me down is the thousands of dollars that I can prove I've saved the company, and I can't get a decent raise. I know Jim is making about $36,000, and I feel I'm worth as much as he is, but I do realize that they have to pay him more because he's a department head. However, he's not worth the amount of difference in our salaries. I feel I should be able to get at least $1,000 a year more than I'm get-

ting now, but the "Book" won't allow that much of a raise at one time. And besides that, I'd feel more like putting out more for the company. As it is, I want to do my best, but it's hard to feel that way when you aren't fairly paid for your work.

Mark: You know what chances you've got of getting *that* kind of a raise! What started all of this anyway?

Bob: Well, I was telling Jim about my idea for the fronts presser and he turned it down, just like he's done most of my ideas.

Mark: Did he tell you *why* he turned it down?

Bob: Said it would be too expensive and would take too much time. Mark, it would save us a penny a dozen which would be about $5,000 a year; they're just time studies to try and satisfy some operator who doesn't really want to work.

Mark: I know. My projects are like that too, and he turned down my idea for revamping the boxing department. You know what a bottleneck that has been. My first estimate, which was conservative, was savings of $50,000 a year plus being able to get out our weekly production. We're not anywhere near that now and spending twice the amount of money we need to. This would also allow the warehouse to have half of the present boxing area. But Jim says it can't be done because there would have to be too much coordination between departments, and that it would take someone with more authority than we have to make it work. I told him if we were to work up the proposal and send it to Mr. Scott, he couldn't pass up those savings on a system that's workable. Of course, you know how Scott hates to make decisions, but if Mr. Glenn knew about it, it would be our main project until it was installed. You know how he likes those dollar signs.

Bob: Yeah, I know. Jim doesn't seem to understand that these little projects don't save us any money and yet he turns down ideas that will save us thousands of dollars a year. You know he doesn't know anything about mechanical development. And besides that, when you try to explain something to him and he doesn't understand it, he says it won't work. But I know that he takes some of these ideas and mentions them to Mr. Scott and Mr. Glenn and takes credit for them. I don't like that one little bit, and I'm going to tell him so one of these days. Then, after telling me my idea for the fronts presser wasn't any good, he chewed me out for going through his mail box. That happened a couple of days ago. When I tried to do him a favor by answering a question Mr. Glenn had asked in a memo, he got all upset. He didn't know anything about it anyway, so what's the difference?

Mark: Well, do you think it was right to go through his mail?

Bob: Well, I just happened to see the subject of the memo and I knew it was concerned with my project so I went ahead and answered the question. I didn't go through his mail; the memo was right on top, and I just happened to see it on my way to my desk.

Mark: Yes, but you *did* get into his personal mail box and went ahead without him knowing about it. Do you see what I mean, Bob? I mean he *is* the head of the department, and he needs to know what goes on within the department.

Bob: But he doesn't have to know *everything* I'm working on. It's none of his business. Most of the things Ned and I do are our own ideas, and he doesn't have a thing to do with them—he doesn't even understand them. Anyway, he told me he didn't want me working on any "secret" projects, that I was to tell him about all of my ideas before I did anything with them. Well, I'll tell you, I'm going ahead and do the projects he assigns me, but I'm *still* going to work on my own ideas whenever I get a chance. Here comes Mr. Scott, I'll see you later.

After closing hours that night and after Bob had left, Jim and Mark were still in the office.

Jim: Mark, did Bob tell you about our little discussion yesterday afternoon?

Mark: He said you had a few words.

Jim: Bob's just getting too big for his own britches. If he doesn't like something I say or do, he acts like a little child. Goes around pouting and gloomy for two or three days. He's just going to have to learn that he's not running the department, although I'm sure he feels he could do a better job than I'm doing. But the thing is that he can't take any criticism. Some of his ideas are good, but others are just too far out and we don't have the time for them. He's going to have to realize that we have other things to do besides mechanical development. I know a lot of our projects cost us more to carry out than can be saved in terms of dollars, but if we can show an operator what is being done is right—or if it's wrong—admit the error and correct the situation, then that can be worth as much as saving several thousand of dollars a year. Al-

though we are becoming increasingly automated, we have to remember that people are still our main source of production and that without their cooperation, we're out of business. Besides, mechanical development isn't even his job, but because he has had some good ideas, I've let him work with Ned on them. I know he's sensitive, and that he is worth a lot to the company because some of his ideas are worthwhile, but if he doesn't change his ways, I'm going to have to talk to Mr. Glenn about letting him go. I've got to run this department, and we can't do our best when he acts up like he does.

When you have read this case, look back at the chapter's diagnostic questions and choose the ones that apply to the case. Then use those questions with the ones that follow in your case analysis.

1. As the case suggests, Bob Collins has displayed a talent for creative and imaginative thinking. Has Jim Douglas taken full advantage of this talent?

2. Jim is thinking about disciplining Bob for going through Jim's mail. Might Bob's actions be responsive to Jim's failure to recognize Bob's creative abilities? What alternatives to discipline might Jim consider?

3. What should Jim do to manage Bob's creativity more effectively? In general, what special requirements does the task of managing creative employees like Bob place on managers like Jim?

CASE 6-3

NURSE ROSS

Read Chapter 10's Case 10-1, "Nurse Ross." Next, look back at Chapter 6's diagnostic questions and choose the ones that apply to that case. Then use those questions with the ones that follow in this first analysis of the case.

1. What factors in this case make rational decision making impossible? How can the management of Benton Hospital make a satisfactory decision as to how to integrate the ward and clinic facilities?

2. Compare the approach you have just recommended with the one actually followed. What differences can you detect? Did these differences contribute to any of the problems that confronted Nurse Ross?

3. Is there anything Dr. Peake could have done when he first arrived to eliminate or modify the factors that are blocking rational decision making? Would such actions have affected your recommendations? How?

DECISION MAKING

MOTIVATION AND PERFORMANCE

Although many retail apparel chains filed for bankruptcy in the late 1980s and early 1990s, Phillips-Van Heusen entered the retailing industry for the first time in 1987 and was soon doing better than ever. One reason was the extraordinary $1 million bonus plan for top managers that Larry Phillips, PVH chairman, dreamed up (see opening story). Phillips (left) and PVH president Bruce Klatsky (center) encouraged their senior executives to find new markets for PVH's shirts, sweaters, and casual shoes, such as the J. C. Penney Co., with whom they confer here. PVH also established its own Profiles stores, where a new sweater collection with PVH's own label was sold for the first time. PVH's incentive program has not only motivated top managers' performance but has encouraged their cooperation in such ventures as Profiles, thus sharpening the company's focus on its goals and objectives.
Source: *Christopher Knowlton, "11 Men's Million-Dollar Motivators,"* Fortune, *April 9, 1990, pp. 8–10.*

What would you do with a million dollars? This is a hypothetical question that we have probably all been asked at one time or another. For 11 executives at Phillips-Van Heusen (PVH), however, it was not a hypothetical question. The corresponding question, What would you do *for* a million dollars, was also very real to them. A million dollars was exactly what each of these executives would receive if the company's earnings per share grew at a 35 percent compound annual rate during the four years ending in January 1992. The first $500,000 came in installments for meeting four year-end goals; the next $500,000 would come as the kicker for meeting the final goal.

This unique incentive plan was the brainchild of Lawrence S. Phillips, PVH's chairman. In the words of one Wall Street analyst this program "has been critical to the success, evolution, and expanding growth of Phillips-Van Heusen." Like many innovative programs, this one was born out of necessity. One long-standing problem faced by PVH was that the demand for men's apparel was susceptible to wide, cyclical swings associated with the notoriously fickle retail industry. The need to solve this recurrent problem became immediate in 1987, when another problem emerged. In an attempt to ward off an outside takeover attempt from a Dallas-based investment firm, PVH was forced to take drastic defensive maneuvers. These maneuvers left the company with $150 million in debt and a devastated stock price of $8, down from $28 less than a year earlier. To solve these problems, Phillips decided that it was "important that we tackle, very aggressively, both the expansion of our third-party [wholesale] business and the creation of our own retail unit."

The only problem with this two-pronged strategy was that the wholesale group and the retail group became potential competitors with each other. Thus there was a strong need to foster camaraderie and cooperation among these units. This unique incentive plan, based upon *organization-wide*, rather than divisional results, was the solution.

The results were evident almost immediately. Soon after the announcement of the plan, the division heads began to meet more frequently and to discuss their operations more openly. Clever cooperative efforts emerged, such as a joint venture between the "sweater" and "shirt" groups to combine sales by offering color-coordinated combinations. A feeling of mutual dependency and team spirit began to emerge. For example, after failing to meet one of the interim goals because of poor returns from the retail group, Michael Blitzer, president of PVH Retail announced with conviction that "you can be damn sure those ten guys are not going to look me in the eye two years from now and tell me that I blew their million dollars." Thus the PVH plan showed not only that money still motivates but that when used creatively, it can reinforce a team-oriented culture that rivets attention on team goals.[1]

The concept of motivation is central to the understanding of human behavior. For example, in everyday life, homicide detectives conducting criminal investigations must establish not only that a suspect had the opportunity to commit a murder but that the person had a specific *motive* for the crime. Indeed, the failure to establish a suspect's motivation is often considered a fatal flaw in attempts to prosecute criminal cases. Our greatest actors have often said that the success of their performances on stage or on screen depends largely on their ability to

[1] Based on C. Knowlton, "11 Men's Million Dollar Motivator," *Fortune*, April 9, 1990, pp. 8–10.

understand the motivation of the characters they portray. In preparing for their roles, they often go to great lengths to research and read as much as they can about their characters, including their personality characteristics and the kinds of circumstances in which they find themselves.

Understanding the reasons behind behaviors and related attitudes is also critical to managerial success. Managing behaviors in the workplace effectively requires the ability to predict human reactions and to prescribe and take appropriate actions. Managing organizational behavior is thus greatly aided by theories that help us understand people's motives. The purpose of this chapter is to introduce and discuss several theories of motivation.

Given the managerial importance of understanding human motivation, it should come as no surprise that there are a large number of well-developed theories of motivation. In general, each of these theories defines **motivation** as the reason or reasons behind an individual's action. Although there are over 100 definitions of motivation in the literature, almost all hold that motivation deals with the factors that initiate, direct, and sustain human behavior over time.[2]

Unfortunately, the sheer number of definitions and theories of motivation can obscure understanding rather than encourage it. We will try to avoid this problem in three ways. First, we begin our discussion of motivation with a brief overview of several dimensions that reduce some of the dissimilarities among theories of motivation. Second, rather than try to cover every theory of motivation, we will focus our attention on just six; expectancy theory, need theory, reinforcement theory, self-efficacy theory, goal-setting theory, and equity theory. Third, we will develop a diagnostic model to make it clear how the theories interrelate. In doing this, we will show how certain theories are best at describing certain aspects of motivation.

motivation The factors that initiate, direct, and sustain human behavior over time.

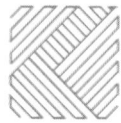

DIAGNOSTIC ISSUES

In the area of work motivation and performance, the diagnostic model evokes many critical questions. To begin with, how do we *describe* the many varying needs and values that characterize different people? How do we describe the ways in which people learn that certain behaviors will lead to certain outcomes? Can we *diagnose* the reasons why some people perform poorly no matter how hard they try? Can we explain why some people approach their jobs with confidence in their ability to perform well whereas others are plagued by fear and self-doubt?

Can we *prescribe* the kinds of rewards that will enhance motivation and avoid those that won't? Can we prescribe how people evaluate what is fair and what is unfair? Can we predict which employees will feel fairly treated and which will not? Finally, what *actions* can we take to enhance performance? Should we set performance goals? How should these goals be set?

CLASSIFYING THEORIES OF MOTIVATION

The theories we will discuss in this chapter can be differentiated in several ways. First, we can distinguish between content theories and process theories of motivation. **Content theories** like the need and goal-setting

content theories Theories of motivation that attempt to specify what sorts of events or outcomes motivate behavior.

[2] J. P. Campbell, M. P. Dunnette, E. E. Lawler, and K. E. Weick, *Managerial Behavior, Performance and Effectiveness* (New York: McGraw-Hill, 1970), p. 77; and F. J. Landy and W. S. Becker, "Motivation

theories try to answer the question of *what* motivates behavior—what sorts of events or outcomes cause people to behave as they do. **Process theories** like the equity and expectancy theories try to answer the question of *how* events and outcomes interact to motivate people's behavior.

The theories discussed in this chapter can also be differentiated by the degree to which their descriptions of motivation include *cognition* as an important element. Cognition here refers to the process of consciously thinking about personal actions and the probable outcomes associated with those actions. Cognitive theories like expectancy theory and self-efficacy theory suggest that motivation is heavily influenced by conscious thought and purposeful choice. Need and other "acognitive" theories emphasize unconscious drives. These theories suggest that people may not be aware of some of the factors that influence their behavior. Certain learning theories also downplay cognitive factors. These theories describe human behavior essentially as a consequence of the rewards and punishments an individual has received in the past. According to this view, human actions are more mechanical reactions than the product of conscious choice.

Finally, the *breadth* of a theory refers to its generality, its ability to explain different types of behavior. "Narrow" theories attempt to specify in great detail the antecedents of one or a few specific behaviors and attitudes. Goal-setting theory, for example, deals almost solely with performance. Self-efficacy theory deals primarily with persistence on a task. Other theories, such as the expectancy and equity theories, are "broad" theories because they can be used to explain a wide variety of behaviors including performance, persistence, job choice, work attitudes, absenteeism, and turnover.

A DIAGNOSTIC MODEL OF MOTIVATION AND PERFORMANCE

The diagnostic model of motivation developed in this chapter is an elaboration of Vroom's *expectancy theory*,[3] particularly as it was extended by Porter and Lawler and supplemented by several other theories.

Expectancy Theory

Expectancy theory is a broad, cognitive, process theory of motivation. It is broad because it explains a diverse set of personal outcomes, most notably, a person's desire to perform and the level of effort the person is willing to exert. The theory is cognitive because it places a strong emphasis on personal thoughts and judgment processes. Finally, it is a process theory because it focuses on how motivation occurs. These features, together with the considerable literature that supports the theory's predictions, make it an excellent base for an integrative approach.[4]

The three major components underlying expectancy theory are the concepts of valence, instrumentality, and expectancy. Sometimes the first letters of these words are used to form the term "VIE" theory.

Theory Reconsidered," in *Research in Organizational Behavior*, ed. L. L. Cummings and B. M. Straw (Greenwich, Conn.: JAI Press, 1988), p. 101.

[3] V. H. Vroom, *Work and Motivation* (New York: John Wiley, 1964), pp. 55–71; and L. W. Porter and E. E. Lawler, *Managerial Attitudes and Performance* (Homewood, Ill.: Richard D. Irwin, 1968), pp. 107–39.

[4] T. R. Mitchell, "Expectancy Models of Job Satisfaction, Occupational Choice, and Effort: A Theoretical, Methodological, and Empirical Appraisal," *Psychological Bulletin* 81 (1974), 1053–77.

valence The amount of satisfaction an individual anticipates receiving from a particular outcome.

POINT TO STRESS

Valences can change over time. As particular needs are met, the promise of things that will further meet those needs loses impact. For example, a person may turn down a promotion because his need for money, power, and responsibility have become less important than his need for time with family and friends. Keeping up with what motivates many people over time is a complex task.

Valence. The concept of **valence** is based on the assumption that at any given time, a person prefers certain outcomes to others. Valence is a measure of the attraction a given outcome holds for an individual, or the satisfaction she anticipates receiving from a particular outcome.

Outcomes can have positive, negative, or zero valence. An outcome is said to have a positive valence when a person would rather attain it than not attain it. For example, in our opening story, for a group of top-management executives in their early 40s, almost all of whom started at PVH in entry-level positions as merchandising assistants or store clerks, $1 million would surely have a very positive valence. If a person prefers *not* to attain the outcome, the outcome is said to have a negative valence. For example, not wanting to have to look other PVH executives in the eye after being responsible for their losing $1 million clearly had negative valence for the head of PVH's retail unit. When a person is indifferent to attaining an outcome, the outcome is assigned a valence of zero. PVH CEO Phillips has noted that "this company has no corporate baloney tied to it. Nobody here gets a car or a country club paid for. This is a very straight, down-to-earth company."[5] Clearly, Phillips is saying that these kinds of perks should have no valence for anyone who wants to work for him.

It is important to distinguish between valence and value. Valence refers to *anticipated* satisfaction. *Value* represents the *actual* satisfaction a person experiences from attaining a desired outcome. Vroom has noted that "An individual may desire an object but derive little satisfaction from its attainment—or he may strive to avoid an object which he later finds to be quite satisfying. At any given time there might be quite a discrepancy between the anticipated satisfaction from an outcome (i.e., its valence) and the actual satisfaction that it provides (i.e., its value)."[6] Thus when we are trying to understand someone's motivation, we must determine his valence for a particular outcome, not the value he will place on that outcome.

instrumentality A person's subjective belief about the relationship between performing a behavior and receiving an outcome.

Instrumentality. The second major component of expectancy theory is instrumentality. **Instrumentality** refers to a person's subjective belief about the relationship between performing a behavior and receiving an outcome. This belief is sometimes referred to as a performance-outcome expectation. Determining

[5] Knowlton, "Million-Dollar Motivator," p. 8.
[6] Vroom, *Work and Motivation*, p. 27.

A well motivated work force for whom work accomplishment has a high valence can make a big difference in bottom-line results. Recently, a team of Federal Express clerks spotted and solved a billing problem that was costing the company $2.1 million a year.
Source: *Brian Dumaine, "Who Needs a Boss?" Fortune, May 7, 1990, pp. 50–52.*

202

how a person perceives instrumentalities is important because a person's desire to perform a particular behavior is likely to be strong only when both valence and instrumentality are perceived as acceptably high. Thus we need to know more than the satisfaction an individual expects as the consequence of attaining a particular outcome. We also need to know what the person believes she must do in order to obtain that outcome. Thus, the key to the motivation of the PVH executives is not that a million dollars may be available to them but that this bonus will be forthcoming only if they accomplish their objectives.

Expectancy.   The third element of expectancy theory consists of the concept for which the theory is named: expectancy. **Expectancies** refer to beliefs regarding the link between making an effort and actually performing well, or effort-performance expectations. Whereas knowledge about valences and instrumentalities tells us what an individual *wants to do*, we cannot know what the individual will *try to do* without knowing his expectancies. According to Vroom, "Whenever an individual chooses between alternatives which involve uncertain outcomes, it seems clear that his behavior is affected not only by his preferences among these outcomes, but also by the degree to which he believes these outcomes to be probable."[7]

Money, prestige, and performing an enjoyable task may have high valence for most of us. Furthermore, we may see that playing basketball in the National Basketball Association is instrumentally related to achieving all three of these outcomes. Despite these high levels of valence and instrumentality, however, few of us are likely to try out for a professional team, because we doubt we could make the team.

In the same way, although executives at PVH wanted the $1 million bonus and clearly understood what they needed to do to get it, their motivation could still have been low if they felt that the goal was impossible because of factors outside their control. For example, although the PVH group met its first interim goal, it did not meet its second. A plague of bad luck and world events conspired against them. First, much of the manufacturing of sweaters had been outsourced to China, and the Tiananmen Square riots severely affected these operations. Other work was outsourced to Puerto Rico, where damage from Hurricane Hugo forced the closing of two other sweater mills. Finally, in the same disastrous year, the largest purchasers of PVH clothing, Federal Department Stores and Allied Department Stores, filed for bankruptcy. If PVH executives had seen these events as destroying all hope of reaching their objectives—they did not—the $1 million bonus would no longer have served as an incentive.

Thus expectancy theory defines motivation in terms of desire and effort, and sees the achievement of desired outcomes as an interactive function of valences, instrumentalities, and expectancies. Desire comes about only when both valence and instrumentality are high, and effort comes about only when all three are high.

Supplemental Theories

There are two primary reasons why we need to supplement expectancy theory with five other motivation theories. First, a number of other theories deal in much more detail with certain specific components of motivation. Therefore, they help to elaborate expectancy theory constructs. For example, need theories provide important insights into how valences are developed and how they can change over time. Learning theories explain how perceptions of instrumentality come

---

[7] Ibid., p. 17.

about. Self-efficacy theory describes the origin of effort-performance expectancies as well as the ways in which they are maintained.

Second, we need to extend expectancy theory to explain outcomes other than desire and effort. For example, to predict performance, expectancy theory needs information about human ability (see Chapter 4) and about how people set goals. In order to predict satisfaction levels, we need to know what equity theory tells us about social comparisons. Thus along with expectancy theory, our diagnostic model will also incorporate ideas from the need, learning, self-efficacy, goal-setting, and equity theories.

Overview of the Model

To develop our diagnostic model of motivation and performance we will use a 6–5–4 heuristic. That is, the model contains *six components* put together in *five steps* to explain *four outcomes*.

One of our six components, "abilities," has already been explained in Chapter 4, and will be only briefly touched on here. Three of the remaining five are valence, instrumentality, and expectancy. These have already been defined, but we will elaborate on each using need, learning, and self-efficacy theories. The remaining two components are accuracy of role perceptions, and equity, particularly as these are delineated in goal-setting theory and equity theory.

The model's five steps are the key places where components combine to influence outcomes. These steps build on each other progressively. For example, as we have already suggested, the components of valence and instrumentality combine to determine desire. Desire then combines with another component, expectancy, to determine effort.

The four outcomes of interest to us are desire, effort, performance, and satisfaction. Because we are also interested in how these outcomes might affect variables like valence and instrumentality, in our fifth step we will look at valences,

FIGURE 7-1

A Diagnostic Model of Motivation and Performance

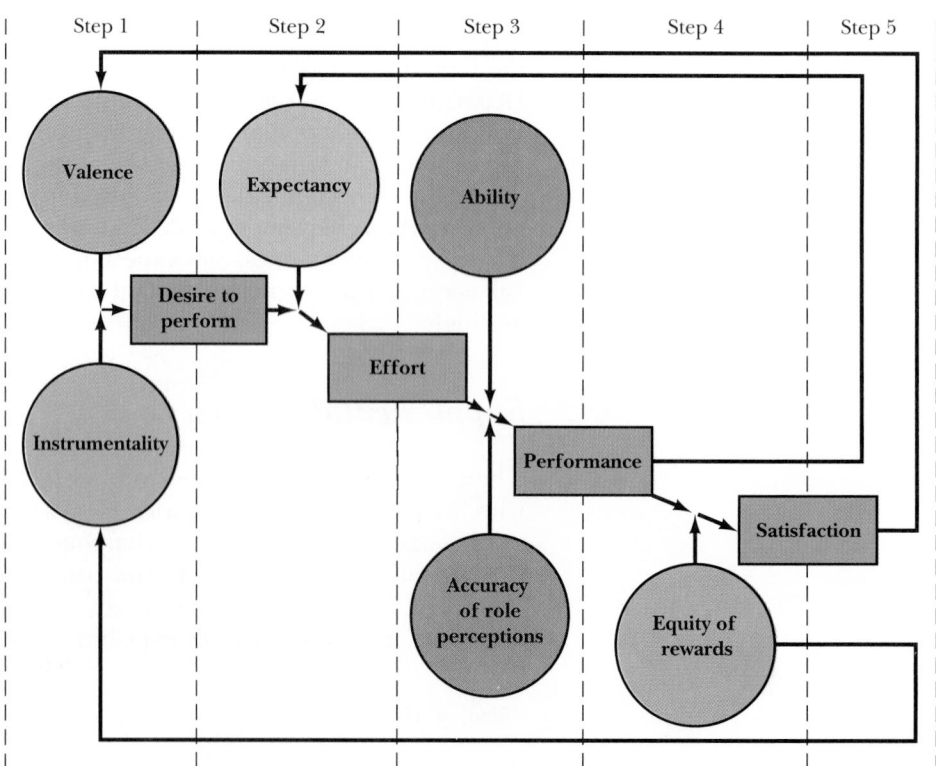

instrumentalities, and expectancies at Time 2 as a function of satisfaction levels at Time 1.

Figure 7-1 presents our model of motivation and performance graphically. As you can see, the six components are shown in circles, the four outcomes are shown in rectangles, and the five steps in which these are combined are indicated by vertical lines. In line with expectancy theory, the first two steps of the model suggest that valence and instrumentality combine to influence desire to perform, and desire and expectancy combine to determine effort. In the third step, effort combines with ability and accuracy of role perceptions to influence performance. In the fourth step, performance and equity of rewards combine to influence satisfaction. In the fifth and last step, the dynamics of the model become apparent. The levels of satisfaction, rewards, and performance feed back to influence valences, instrumentalities, and expectancies, respectively. With this overview in mind, we will describe and illustrate each component of the model and show why the components interact as we have suggested.

VALENCE: NEED THEORIES

APPLYING THE DIAGNOSTIC MODEL
Diagnostic Question 1: What are the most important needs of the person I am trying to motivate?

Given the central role played by valence in determining motivation, it seems critically important that we understand how an individual's valences are determined. As you know from your own personal experience, people differ greatly in their personal preferences. For example, one person may decide to be a missionary, another to become a stockbroker, and each may be quite satisfied with the choice. We would not expect Mother Teresa and Ivan Boesky to share the same valences. Need theories are especially helpful in understanding not only how valences originate but why they differ among people.

Maslow's Need Hierarchy

Abraham Maslow was a clinical psychologist and a pioneer in the development of *need theories*. Little existed in the way of empirical, scientific studies of motivation in Maslow's day. He based his own theory on 25 years of experience treating individuals of varying degrees of psychological health.

Maslow's need theory proposed the existence of five distinct types of needs: physiological, safety, love, esteem, and self-actualization. These needs, according to Maslow, are genetically based and characteristic of all human beings. Moreover, he argued, these five needs are arranged in the hierarchy shown in Figure 7-2 and influence motivation on the basis of need **prepotency**. Prepotency means that needs residing higher in the hierarchy can influence motivation only if needs that are lower are largely satisfied.

At the lowest level of Maslow's hierarchy are the *physiological needs*, such as hunger and thirst. According to Maslow, these physiological needs possess the greatest initial prepotence. If they are unfulfilled, no other needs can influence an individual's motivation. Maslow said that when a person's needs for things like food are unfulfilled,

> Capacities that are not useful for this purpose lie dormant, or are pushed into the background. The urge to write poetry, the desire to acquire an automobile, the interest in American history, the desire for a new pair of shoes are, in the extreme case, forgotten or become of secondary importance. For the [person] who is extremely and dangerously hungry, no other interests exist but food.[8]

Maslow's need theory A theory of motivation that suggests that behavior is driven by the urge to fulfill five fundamental needs: physiological and safety needs love, esteem, and self-actualization.

prepotency The notion arising from Maslow's theory that higher-order needs can influence motivation only if lower-order needs are largely satisfied.

[8] A. H. Maslow, "A Theory of Human Motivation," *Psychological Reports* 50 (1943), 370–96.

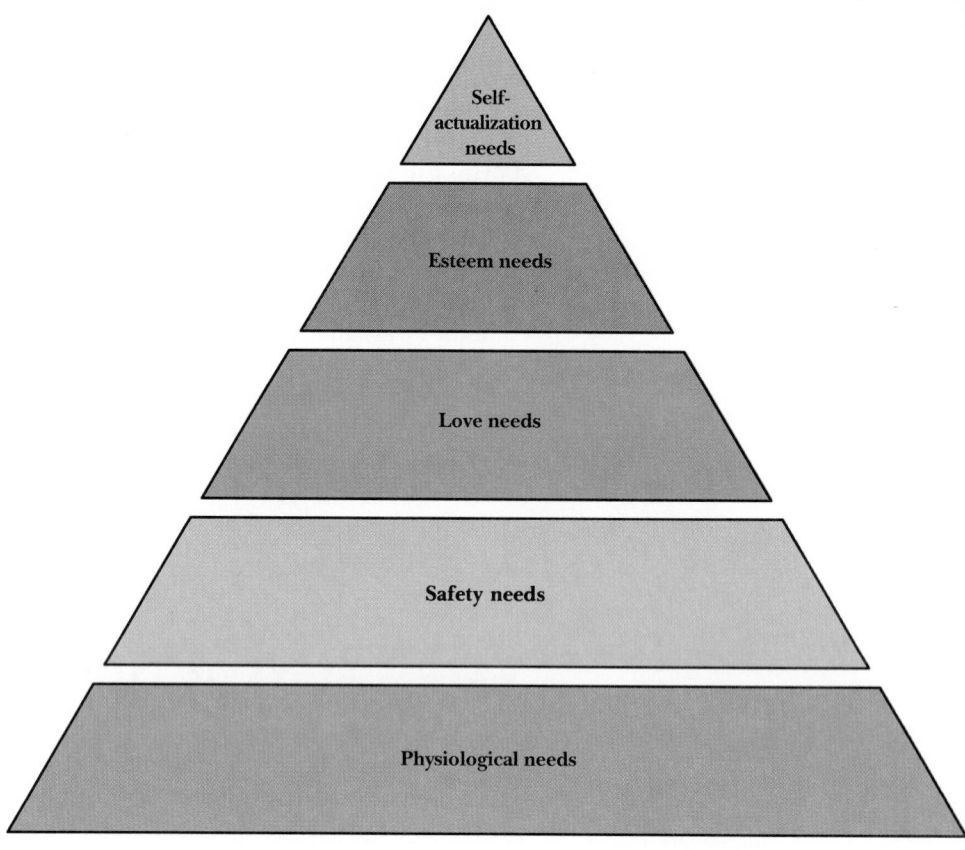

On the other hand, once physiological needs are mostly gratified, they no longer serve as strong motivating elements. Under these conditions, the second-level, *safety needs*, increase in strength. Safety needs have to do with acquiring objects and relationships that protect their possessor from future threats, especially threats to the person's ability to satisfy physiological needs.

If both physiological and safety needs are mostly fulfilled, *love needs* become prepotent. Maslow used the term *love* in a broad sense to refer to preferences for affection from others as well as a sense of community or belongingness. The need for friends, family, and colleagues falls within this category. Maslow classified sexual desires among the physiological needs.

At the fourth level of needs in Maslow's hierarchy are *esteem needs*. Maslow grouped two distinct kinds of esteem within this category. Social esteem consists of the respect, recognition, attention, and appreciation of others. Self-esteem reflects an individual's own feelings about personal adequacy. Consequently, esteem needs can be satisfied partly from sources outside an individual and partly from sources within.

The last set of needs, at the top of Maslow's hierarchy, are the *self-actualization needs*. Maslow felt that if all needs beneath self-actualization were fulfilled, a person could be considered generally satisfied. He also suggested that "since in our society, basically satisfied people are the exception, we do not know much about self-actualization, either experimentally or clinically." For this reason, perhaps, Maslow failed to define this category of needs precisely. In general, these needs seem to involve an individual's desire to realize her full potential. In Maslow's words, self-actualization ". . . might be phrased as the desire to become more and more what one is, to become everything that one is capable of becoming."[9]

[9] Ibid.

The way the need for self-actualization expresses itself varies among people. It may lead one individual to strive to be an ideal teacher. In another person, it may be expressed athletically. In still another, it may be expressed in artistic endeavors. Unlike all the other needs identified by Maslow, self-actualization needs can never be fully satisfied. Hence, the picture of human motivation drawn by Maslow is one of constant striving as well as constant deprivation of one sort or another.

Perhaps because of its simplicity, Maslow's theory has been widely accepted by managers and management educators. It is of interest because of its place in history as one of the earliest motivation models and as a precursor to more modern theories of motivation. Maslow, however, failed to provide researchers with clear-cut measures of his concepts, and his theory has not received much empirical support. In a comprehensive review of studies testing Maslow's theory, Wahba and Bridwell concluded that there was little evidence for (a) Maslow's hierarchical classification scheme of five different types of needs, (b) his deprivation-domination hypothesis, which proposed that unfulfilled needs should dominate others, or (c) his gratification-activation hypothesis, which asserted that gratified needs activate higher-order needs.[10]

Alderfer's ERG Theory

Largely in response to the lack of research support for Maslow's need hierarchy, Alderfer attempted to reformulate the theory.[11] Rather than five levels of needs, Alderfer proposed three: existence, relatedness, and growth. The first initial of each of these needs furnished the name for the **ERG theory** that resulted.

ERG theory A theory of motivation developed by Alderfer that suggests that behavior is dirven by the urge to fulfill three essential needs: existence, relatedness, and growth.

At the base of Alderfer's ERG theory are *existence needs*, which incorporate Maslow's physiological needs as well as those safety needs satisfied by the possession of material objects. In the middle of Alderfer's theory are *relatedness needs*, which include safety needs satisfied by the presence of other people as well as love needs and needs for social esteem. Finally, at the top of Alderfer's theory are *growth needs*, which encompass needs for self-esteem and self-actualization.

In proposing fewer levels of need, Alderfer's theory is more economical than Maslow's. It is also more flexible in that it allows for the possibility that several needs may affect motivation simultaneously. Alderfer agreed with Maslow that need importance progresses up the hierarchy one step at a time. He argued, however, that individuals frustrated by failed attempts to satisfy some higher-level need will often regress and reaffirm the importance of lower-level needs. Alderfer constructed the instrument shown in Table 7-1 to measure satisfaction with each of his three types of needs. Research has supported the hypothesized three dimensions of this instrument as well as its reliability and construct validity. Moreover, predictions based on Alderfer's theory have received greater research support than those based on Maslow's theory.[12]

---

[10] M. A. Wahba and L. G. Bridwell, "Maslow Reconsidered: A Review of Research on the Need Hierarchy," *Organizational Behavior and Human Performance* 15 (1976), 121–40.

[11] C. P. Alderfer, "An Empirical Test of a New Theory of Human Needs," *Organization Behavior and Human Performance* 4 (1969) 142–75; and *Existence, Relatedness and Growth: Human Needs in Organizational Settings*, (New York: Free Press, 1972), p. 27.

[12] C. P. Alderfer, R. E. Kaplan, and K. K. Smith, "The Effect of Variation in Relatedness Need Satisfaction on Relatedness Desires" *Administrative Science Quarterly* 19 (1974), 507–32; and J. P. Wanous and A. Zwany, "A Cross-Sectional Test of Need Hierarchy Theory," *Organizational Behavior and Human Performance* 18 (1977), 78–97.

Existence Needs
Pay:
1. Compared to the rates for similar work here my pay is good.
2. Compared to similar work in other places my pay is poor.
3. I do not make enough money from my job to live comfortably.
4. Compared to the rates for less demanding jobs my pay is poor.
5. My pay is adequate to provide for the basic things in life.
6. Considering the work required, the pay is what it should be.

Fringe Benefits:
1. Our fringe benefits do not cover many of the areas they should.
2. The fringe benefit program here gives nearly all the security I want.
3. The fringe benefit program here needs improvement.
4. Compared to other places, our fringe benefits are excellent.

Relatedness Needs
Respect from Superiors:
1. My boss will play one person against another.
2. My boss takes account of my wishes and desires.
3. My boss discourages people from making suggestions.
4. It's easy to talk with my boss about my job.
5. My boss does not let me know when I could improve my performance.
6. My boss gives me credit when I do good work.
7. My boss expects people to do things his way.

8. My boss keeps me informed about what is happening in the company.

Respect from Peers:
1. My co-workers are uncooperative unless it's to their advantage.
2. I can count on my co-workers to give me a hand when I need it.
3. I cannot speak my mind to my co-workers.
4. My co-workers welcome opinions different from their own.
5. My co-workers will not stick out their necks for me.

Belongingness (Love):
1. I have developed close friendships in my job.
2. I have an opportunity in my job to help my co-workers quite a lot.

Status (Esteem):
1. I have the feeling that my job is regarded as important by other people.
2. My job gives me status.

Growth
1. I seldom get the feeling of learning new things from my work.
2. I have an opportunity to use many of my skills at work.
3. In my job I have the same things to do over and over.
4. My job requires that a person use a wide range of abilities.
5. My job requires making one or more important decisions every day.
6. I do not have the opportunity to do challenging things at work.

Source: C. P. Alderfer, "An Empirical Test of a New Theory of Human Needs," *Organization Behavior and Human Performance* 4 (1969).

Murray's Theory of Manifest Needs

Henry Murray's *theory of manifest needs*, developed before Maslow's theory, defined needs as recurrent concerns for particular goals or end states.[13] Each need was made up of two components. The first dealt with the object toward which the need was directed. The second was concerned with the intensity or the strength of the need for that particular object. Murray proposed over twenty needs, several of which are described in Table 7-2.

Because Murray's needs are not arranged in any hierarchical fashion, the theory has considerable flexibility. Like Alderfer, Murray held that an individual could be motivated by more than one need at a time, and he also suggested that at times, needs could conflict with each other. Unlike Maslow, who viewed needs as innate and genetically determined, Murray regarded needs as learned.

[13] H. A. Murray, *Explorations in Personality* (New York: Oxford University Press, 1938).

TABLE 7-2
Some of Murray's Manifest Needs

Achievement	To do one's best, to be successful, to accomplish tasks requiring skill and effort, to be a recognized authority, to accomplish something important, to do a difficult job well
Deference	To get suggestions from others, to find out what others think, to follow instructions and do what is expected, to praise others, to accept leadership of others, to conform to custom
Order	To keep things neat and orderly, to make advance plans, to organize details of work, to have things arranged so they run smoothly without change
Autonomy	To be able to come and go as desired, to say what one thinks about things, to be independent of others in making decisions, to do things without regard for what others may think
Affiliation	To be loyal to friends, to participate in friendly groups, to form strong attachments, to share things with friends, to write letters to friends, to make as many friends as possible
Succorance	To have others provide help when in trouble, to seek encouragement from others, to have others be kindly and sympathetic, to receive a great deal of affection from others
Dominance	To argue for one's point of view, to be a leader in groups to which one belongs, to persuade and influence others, to supervise and direct the actions of others
Nurturance	To help friends when they are in trouble, to treat others with kindness and sympathy, to forgive others and do favors for them, to show affection and have others confide in one
Change	To do new and different things, to travel, to meet new people, to have novelty and change in daily routine, to try new and different jobs, to participate in new fads and fashions
Endurance	To keep at a job until it is finished, to work hard at a task, to work at a single job before taking on others, to stick at a problem even though no apparent progress is being made
Aggression	To attack contrary points of view, to tell others off, to get revenge for insults, to blame others when things go wrong, to criticize others publicly, to read accounts of violence.

Source: H. A. Murray, *Explorations in Personality* (New York: Oxford University Press, 1938), pp. 152–205.

To measure the needs proposed in his theory, Murray developed the Thematic Apperception Test (TAT). As you may recall from Chapter 4, the TAT is a projective test in which a person views an ambiguous picture and makes up a story about it. The person is asked to include in her story such information as, What is going on in the picture? Who is involved? What led up to the situation portrayed? What are the people pictured thinking and feeling? What will happen next? What will the outcome be? The projective test assumes that people will project their own thoughts and feelings onto an ambiguous stimulus, unconsciously revealing their own needs.

McClelland's Theory of Achievement Motivation

Just as Alderfer's theory grew out of Maslow's, Murray's work was extended and expanded upon by others. Most notably, David McClelland developed a theory of motivation that has focused particularly on the need for achievement (nAch).[14]

[14] D. C. McClelland, *The Achieving Society* (Princeton, N.J.: Van Nostrand Press, 1963).

How do you nurture achievement motivation, find innovative uses for natural products, and save the rain forests of the Amazon all in one fell swoop? If you're Cultural Survival Enterprises, you help indigenous people form cooperatives to harvest and sell rain forest products. CSE's biggest product is Rainforest Crunch, a confection made with cashews and Brazil nuts that's sold as a candy as well as an ingredient in Ben & Jerry's Rainforest Crunch Ice Cream. CSE puts the profit from this and other endeavors into the hands of forest residents, enabling them to resist selling out to miners, loggers, and cattle-ranchers.
Source: *"Ideas for 1991,"* Fortune, January 14, 1991, p 42.

According to McClelland certain types of situations are preferred by people with a high need for achievement and tend to elicit achievement striving from these people. These types of situations often offer the opportunity to take personal responsibility. They may also permit people to receive personal credit for the consequences of their actions.

People with a high need for achievement also prefer situations characterized by the availability of clear and unambiguous feedback about personal performance. Task difficulty is yet another situational characteristic important to high nAch individuals. According to McClelland, such people prefer tasks of intermediate difficulty—where the probability of success is close to 50-50—to tasks that are too easy or too difficult. Finally, situations that have a future orientation or permit the development of novel or innovative solutions are attractive to achievement-oriented people.

The four need theories we have discussed are just a few of the many theories that contribute to the complexity of the valence construct. Many things other than pay have valence for workers, a fact that is being increasingly recognized by both management and organized labor. Thus we see labor contracts pushed by the United Auto Workers that trade pay for job security. We see employers like the Park Plaza Hotel in Boston providing employer-financed housing aid for its workers. And we see many employers providing company-sponsored day care.[15]

INSTRUMENTALITY: LEARNING THEORIES

The understanding of valence contributed by need theories provides us with only one piece of the motivation puzzle—what human beings want. In order to understand behavior, we need to know not just what outcomes people want but what they believe will lead to the attainment of desired outcomes.

AN EXAMPLE
When trying to motivate workers to increase productivity, a company may offer an increase in pay. Employees evaluate the likelihood of receiving the pay if performance increases are achieved. But hidden outcomes, real and perceived, are also evaluated. If I increase my productivity, my co-workers may shun me, I will have to eat alone, and no one will talk to me. If I increase performance, I will improve my chances for training or promotion. If I increase my productivity, I will lose time with my family. The valence and instrumentality of each of these outcomes is assessed and a desired pay increase may become less desirable because of these other outcomes that may result. If this happens, productivity will not improve.

[15] P. C. Judge, "U.A.W. Faces Test at Mazda Plant," *New York Times*, March 27, 1990, p. D1; and A. R. Karr, "Housing Benefits for Workers Expected to Get Boost from Change in Labor Law," *Wall Street Journal*, April 9, 1990, p. 3.

MOTIVATION AND PERFORMANCE

These beliefs are referred to as instrumentalities in expectancy theory. Learning theories help clarify how relationships between behaviors and rewards come to be perceived. They also provide information that allows us to estimate the character, permanence, and strength of these relationships and are thus another useful supplement to expectancy theory.

The notion that human beings generally behave so as to maximize pleasure and minimize pain, first formulated by the ancient Greek philosophers as the concept of **hedonism**, is part of virtually all modern theories of motivation. It is especially conspicuous in learning theories, all of which attempt to explain behavior in terms of the associations people form between performance and the receipt of some pleasurable or painful outcome. We will look at three types of learning theory: classical conditioning; reinforcement theory, or operant conditioning; and social learning theory.

Classical Conditioning

Does the name Pavlov ring a bell? It might, because it was Ivan Pavlov, a Russian physiologist working early in this century, whose research on conditioning a dog to salivate at the sound of a bell led him to formulate the theory of classical conditioning. **Classical conditioning** creates associations between an **unconditioned stimulus**, something that is known to consistently produce a certain response, and a **conditioned stimulus**, something that has never produced that response. When a conditioned and an unconditioned stimulus are repeatedly presented to a subject at the same time, the subject eventually comes to make the same response to the conditioned stimulus that he originally made to the unconditioned stimulus.

For example, in his experiments at the Soviet Military Academy, Pavlov found that he could make hungry dogs salivate at the sound of a bell (the conditioned stimulus), if in prior conditioning processes, the bell was paired with the presentation of food (the unconditioned stimulus). Later, in the United States, John Watson demonstrated the role played by classical conditioning in human learning by consistently presenting a young child with white, furry objects at the same time as he made a frightening loud noise. As a result, the child became afraid of all white, furry objects.[16]

Classical conditioning provides insights into the way people form associations between different stimuli even when there is no outwardly obvious relationship between them. Managers need to be sensitive to this type of learning so they can avoid becoming aversive conditioned stimuli themselves. For example, a well-intentioned manager who takes up his boss's time only when a problem develops may unwittingly be teaching his boss to expect nothing but problems when the manager approaches. As management consultant Jane Halpert notes, "If every time you see a person, he's bringing you bad news, you tend to want to avoid him."[17] It would be a good idea for this manager to develop a relationship with his supervisor that entails something other than discussion of problem situations.

Reinforcement Theory

Reinforcement theory, or as it is also referred to, operant conditioning, grew out of E. L. Thorndike's work on the *law of effect*. According to Thorndike,

---

hedonism The belief that human beings generally behave so as to maximize pleasure and minimize pain.

classical conditioning Learning that occurs when a neutral stimulus, through repeated pairing with a stimulus that elicits a specific response, comes to elicit that same response.

unconditioned stimulus A stimulus that naturally and invariably produces a given response.

conditioned stimulus A stimulus that is initially neutral but when repeatedly paired with an unconditioned stimulus, elicits the response associated with the latter stimulus.

---

[16] J. B. Watson and R. Raynor, "Conditioned Emotional Reactions," *Journal of Experimental Psychology* 20 (1920), 1–14.

[17] W. Kiechel, "Breaking Bad News to the Boss," *Forbes*, April 9, 1990, pp. 70–71.

Diagnostic Question 2: What
contingencies has this person
learned over the course of his or
her reinforcement history?

reinforcement theory A theory
of motivation that suggests that
people are motivated to engage in
or avoid certain behaviors because
of past rewards and punishments
associated with those behaviors.

positive reinforcement The
increase in a response that occurs
when engaging in the response
leads to obtaining a pleasurable
stimulus.

POINT TO STRESS
Both operant conditioning and
reinforcement theory assume that
learning is not a cognitive process
but simply a process of connecting
rewards received to behavior
performed. The pigeon doesn't
think.

extinction The gradual
disappearance of a response that
occurs after the cessation of
positive reinforcement.

**APPLYING THE DIAGNOSTIC
MODEL**
Diagnostic Question 3: How can I
make the receipt of outcomes that
have positive valence for this
person contingent upon
performing at a high level?

Diagnostic Question 4: What
outcomes that have negative
valence for this person can I
remove, contingent upon
performance at a high level?

negative reinforcement The
increase in a response that occurs
when engaging in the response
leads to the removal of an aversive
stimulus.

punishment A decrease in a
response that occurs when
engaging in the response leads to
receiving an aversive stimulus.

Of several responses made to the same situation, those which are accompanied or closely followed by satisfaction to the animal will, other things being equal, be more firmly connected with the situation, so that, when it recurs, they will be more likely to recur; those that are accompanied or closely followed by discomfort to the animal will, other things being equal, have their connections with that situation weakened, so that when it recurs, they will be less likely to recur. The greater the satisfaction or discomfort, the greater is the strengthening or weakening of the bond."[18]

The learning process described in reinforcement theory differs substantially from the one depicted in classical conditioning. **Reinforcement theory** proposes that a subject comes to make a specific response because that response has been reinforced by a specific outcome. For example, if a pigeon receives a pellet of food every time it pecks a particular spot on the wall of its cage, in time it will peck that spot repeatedly in order to receive the **positive reinforcement** of food. Here, the unconditioned stimulus and the conditioned stimulus are essentially the same (food), and they both *follow* the conditioned response (pecking). In classical conditioning, however, the unconditioned stimulus and the conditioned stimulus are presented to the subject at the same time, and *precede* the conditioned response. Pavlov's bell, for example, became a sign of food to the dog and thus came to elicit a conditioned (salivary) response.

The term *operant conditioning*, more or less synonymous with reinforcement learning, got its name from the fact that in this type of learning the subject must perform some operation in order to receive the reinforcing outcome. Thus the subject plays an active role in responding. The pigeon must perform a specific behavior in order to receive his reward. In classical conditioning, however, a more or less automatic response comes to be made to a stimulus that never elicited that response before. Consider our manager's boss, for example, who begins to respond automatically with apprehension when the manager appears at his door.

Extinction occurs in both types of learning. For example, if our manager starts talking with his supervisor about such things as improvements in production and new ideas, as we suggested, his supervisor will form a new, positive association with him and the negative one will weaken and eventually extinguish. Suppose we stop giving our pigeon food pellets when he pecks at the accustomed spot. Eventually, the pecking response too will extinguish; the pigeon will give it up in the absence of reinforcement.

Companies are becoming increasingly interested in basing pay specifically on performance. They are recognizing the potential rewards of linking the reinforcer of pay directly with the response they want to reinforce, or successful job performance. One GTE spokesperson has noted that "our philosophy is reward for performance. . . . Inflation isn't a factor as it was a generation ago." Similarly, compensation specialists at Gantos Corporation note that they are "aware of the inflation rate" but believe that paying for performance is "the most equitable, sensible way."[19]

Negative reinforcement and punishment are two other ways to influence behavior. In **negative reinforcement**, the likelihood that a person will engage in a particular behavior is increased because the behavior is followed by the removal of something the person dislikes. In **punishment**, the likelihood of a given behavior is decreased because it is followed by something that the person dislikes. The distinctions among positive reinforcement, extinction, negative reinforcement, and punishment are shown in Figure 7-3. The figure shows reinforcement theory's ability to explain how to strengthen and weaken behaviors as well as its ability to predict the effects of rewards that have positive or negative valences.

[18] E. L. Thorndike, *The Elements of Psychology* (New York: Seiler, 1911), p. 244.
[19] "Labor Letter," *Wall Street Journal*, March 6, 1990, p. 1.

FIGURE 7-3

Effects of Methods of Reinforcement on Behavioral Response

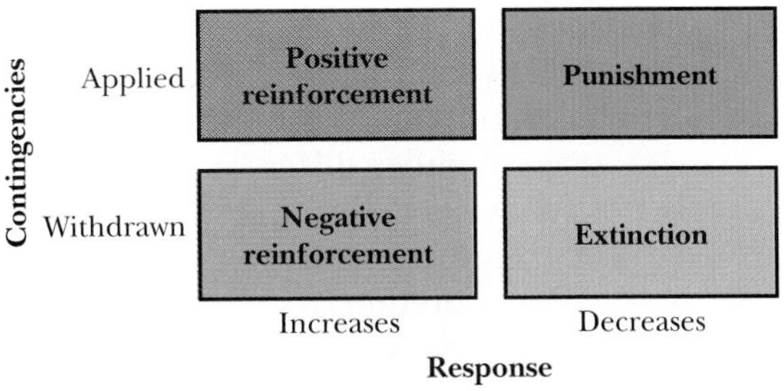

Managers in organizations sometimes contend that they cannot make use of reinforcement theory because they do not have enough resources to give positive reinforcements. For example, they cannot always raise salaries or award bonuses as they might like. What Figure 7-3 makes clear is that positive reinforcement is only one of a number of ways to increase the frequency of a desired behavior. Managers can also employ negative reinforcement to increase a response. They can find something about the job that people do not like and, when employees engage in desired behaviors, remove it. For example, a sales manager who wants to increase sales and who knows that salespeople hate to complete paperwork associated with their work might offer to shift the responsibility for completing paperwork to others if the salespeople increase their productivity. The sales force's enthusiasm for selling might increase noticeably as a result.

As Figure 7-3 clearly shows, punishment suppresses an incorrect response. Unfortunately, however, punishment does nothing to increase the frequency of correct responses. Moreover, although punishment may appear to eliminate an unwanted behavior, its effect is only temporary.[20] A final drawback in the use of punishment is that it often leads to undesirable side effects. Most commonly, these take the form of negative emotional reactions from those who have been punished. Negative reactions are especially likely to occur when punishment is rendered publicly rather than privately. We will have more to say about organizations' use of discipline programs in Chapter 9.

Another important factor in the usefulness of reinforcement theory is the *schedule of reinforcement* used to pace the delivery of reinforcing outcomes. There are five different kinds of schedules: continuous, fixed ratio, variable ratio, fixed interval, and variable interval. In a *fixed-ratio schedule*, reinforcement is given after the occurrence of a fixed number of target behaviors. For example, under a piece-rate payment plan a worker might earn one dollar for every ten products produced. *Continuous reinforcement* is a special case of fixed ratio reinforcement in which there is one-to-one correspondence between the occurrence of a behavior and the receipt of a reward (e.g., one dollar for every product produced). Under fixed-ratio or continuous schedules, initial learning tends to occur rapidly and the behavior, once learned, is typically exhibited quite often thereafter. On the negative side, however, behaviors reinforced in this fashion tend to extinguish quite rapidly if reinforcing outcomes are withheld.

Behaviors are much more resistant to extinction when reinforced on a *variable-ratio* schedule. This schedule is similar to the fixed-ratio schedule in that the receipt of an outcome is contingent on the number of behaviors exhibited. On a variable-ratio schedule, however, the number of responses required to obtain an outcome is not the same every time. Thus, an employee working under a variable ratio piece-rate payment system would earn one dollar for every 10 products produced on average, but his actual earnings might be contingent on producing a different number of items from one time to the next. The number of produced products needed to obtain a reward might be 8 in one case, 12 in

AN EXAMPLE

To work at all, punishment requires constant monitoring. Consider the repeat traffic offender. She pays numerous fines, has her licenses revoked, and maybe does jail time, yet she still refuses to obey the law. She gambles she won't get caught.

TEACHING NOTE

Continuous reinforcement, if used too long, can lose its potency as a motivator. Ask students how they would feel if every time they performed a specific behavior correctly they were told, "Good job, keep it up." After a while the reinforcer is no longer important. At some point, we begin to feel we are not trusted to behave correctly without being watched and motivation goes down.

---

[20] R. L. Solomon, "Punishment," *American Psychologist* 19 (1964), 239–53.

another, and 3 in yet another. Notably, slot machines produce rewards in this manner, and anyone who has observed people playing these machines will attest to their ability to motivate high rates of behavior despite little reinforcement.

Interval schedules differ from ratio schedules in that time rather than the number of behaviors determines rewards. On a *fixed-interval* schedule, individuals receive initial reinforcement after exhibiting a target behavior, and reinforcement continues after a predetermined time interval that does not vary from one reinforcement episode to the next. Because the receipt of outcomes is tied to the passage of time rather than to personal performance, the number of target behaviors encouraged by this type of schedule is typically quite low. Nevertheless, fixed-interval schedules are used quite often in business. Hourly pay reflects the passage of time but is unrelated to whether employees have behaved productively while at work.

A *variable-interval* schedule, which provides greater resistance to extinction, also gears the distribution of outcomes to the passage of time. However, unlike fixed-interval schedules, the amount of time between the receipt of outcomes differs from one reinforcement episode to the next. For example, a production supervisor who visits a remote work station sometimes once a week, sometimes once a month, sometimes twice a month, is using a variable-interval schedule. This schedule increases the probability that employees at the remote station will work at a constant rate; they never know when the supervisor might show up. Compare this situation to that of employees managed by a supervisor whose arrival can be predicted precisely because she uses a fixed-interval schedule.

Although they differ in important respects, all the reinforcement schedules we have discussed have one thing in common. They are initially activated by the occurrence of a particular behavior. When we want to encourage a complex behavior that might not occur on its own the process of shaping can be helpful.

shaping Bringing about a desired behavior by rewarding successive approximations to that behavior.

Shaping means rewarding successive approximations to a desired behavior. For example, it is virtually impossible for someone who has never played golf to pick up a club and execute a perfect drive the first time. Moreover, left on her own to try and try again with no instruction, a novice golfer is unlikely ever to exhibit the correct behavior. In shaping, rather than wait for the correct behavior to occur on its own, one begins by rewarding close approximations. Then over time, rewards are held back until the person gets closer and closer to the right behavior. Thus a golf instructor might at first reward a novice golfer for holding the club with the right grip. To obtain a second reward, the novice may be required not only to display the correct grip but to stand at the right distance from the ball. To obtain additional rewards, the novice may have to do both of these things and perform the appropriate backswing, and so on. In this way, initially simple behaviors are shaped into a complex desired behavior.

Social Learning

social-learning theory A theory of motivation originated by Bandura that suggests that behavior is often driven by the desire of an observer to model the behavior of some other person.

Social-learning theory, as proposed by Albert Bandura, is a theory of observational learning that holds that most people learn behaviors by observing others and then *modeling* the behaviors they perceive to be effective. This sort of observational learning contrasts markedly with the process of learning through direct reinforcement.

For example, suppose a worker observes a colleague who, after giving bad news to her manager, is punished. Strict reinforcement theory would suggest that when confronted with the same task, the observing worker will be neither more nor less prone to be the bearer of bad tidings because she has received no

TEACHING NOTE
Ask students to list things they
believe they learned through
reinforcement. Many items on the
list will be from early childhood
(e.g., tying their shoes, cleaning
their rooms). Then ask them to list
things they have learned through
observation. This list will probably
be much longer. What are the
advantages of learning by
observation rather than by
reinforcement?

AN EXAMPLE
Small children mimic other
children and adults. Oftentimes
what they mimic is inappropriate,
like using foul language. They are
not selective in what they mimic.
Teenagers tend to model their
behavior after their heroes and
popular individuals. Although
adults may question their
judgment, teenagers are more
selective about who and what they
copy.

direct reinforcement herself. Social-learning theory suggests otherwise, however. Despite the fact that the worker may never have directly experienced the fate of her colleague, she will nonetheless learn by observation that this boss "shoots the messenger." She will doubtless conclude that the best response in such situations is to keep quiet. Even though the manager would probably never agree that problems should be covered up, this may be the precise message sent by his behavior.

It is important to distinguish social learning, or modeling, from simple imitation or mimicry. When we imitate or mimic others, we simply copy their behaviors. When we model our behaviors on others, however, we are very selective. Before we decide to imitate a model, we consider such things as our own abilities and the effectiveness of the model's behavior in the situation surrounding her. For example, a study of pairs of lower-level supervisors and their direct subordinates showed that subordinates attempting to succeed on the job did not always imitate the values and behaviors of their direct supervisors. In fact they imitated only those supervisors whom they perceived as competent and successful. Also, subordinates who were low in self-esteem were much more likely to model supervisor behaviors than those who possessed higher self-esteem. Thus, rather than simply imitating their supervisors, the subordinates first tried to discern whether their supervisors were worth imitating. They also considered whether they could do better on their own.[21]

Besides its focus on learning by observation, social-learning theory proposes that people can reinforce or punish their own behaviors; that is, they can engage in *self-reinforcement*. According to Bandura, a self-reinforcing event occurs when (a) tangible rewards are readily available for the taking, (b) people deny themselves free access to those rewards, and (c) they allow themselves to acquire the rewards only after achieving difficult self-set goals.[22] Consider the behavior of many novelists. Once alone and seated at their typewriters, a considerable number of authors refuse to take a break until they have written a certain number of pages. Obviously, these people can get up and leave any time they wish. However, they deny themselves the reward of a rest until they have accomplished their self-set goal.[23] Research has indicated that self-reinforcement can be used to help people stop smoking, overcome drug addiction, cure obesity, improve study habits, enhance scholastic achievement, and reduce absenteeism.[24]

[21] H. M. Weiss, "Subordinate Imitation of Supervisory Behavior," *Organizational Behavior and Human Performance* 19 (1977), 89–105; "Social Learning of Work Values in Organizations," *Journal of Applied Psychology* 63 (1978), 711–18; Weiss and J. B. Shaw, "Social Influences on Judgments on Tasks," *Organizational Behavior and Human Performance* 24 (1979), 126–40; and T. L. Rakestraw and Weiss, "The Interaction of Social Influences and Task Experience on Goals, Performance and Performance Satisfaction," *Organizational Behavior and Human Performance* 27 (1981), 326–44.

[22] A. Bandura, "Self-Reinforcement: Theoretical and Methodological Considerations," *Behaviorism* 4 (1976), 135–55.

[23] I. Wallace, "Self-Control Techniques of Famous Novelists," *Journal of Applied Behavioral Analysis* 10 (1977), 515–25.

[24] F. H. Kanfer and J. S. Phillips, *Learning Foundations of Behavior Therapy* (New York: John Wiley, 1970), p. 59; Kanfer, "Self-Regulation: Research, Issues, and Speculation," in *Behavior Modification in Clinical Psychology* ed. C. Neuringer and J. Michael (New York: Appleton-Century-Crofts, 1974), pp. 178–220; M. J. Mahoney, N. G. Moura, and T. C. Wade, "The Relative Efficacy of Self-Reward, Self-Punishment, and Self-Monitoring Techniques for Weight Loss," *Journal of Consulting and Clinical Psychology* 40 (1973), 404–7; C. S. Richards, "When Self-Control Fails: Selective Bibliography of Research on the Maintenance Problems in Self-Control Treatment Programs," *JSAS: Catalog of Selected Documents in Psychology* 8 (1976), 67–68; E. L. Glynn, "Classroom Applications of Self-Determined Reinforcement," *Journal of Applied Behavioral Analysis* 3 (1970), 123–30; and C. A. Frayne and G. P. Latham, "Application of Social Learning Theory to Employee Self-Management of Attendance," *Journal of Applied Psychology* 72 (1987), 387–92.

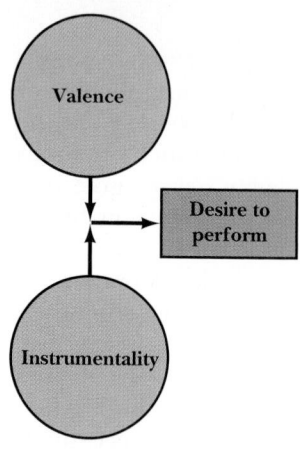

FIGURE 7-4

Step 1. The Desire to Perform as a Function of Valence and Instrumentality

STEP 1: DESIRE TO PERFORM AS A FUNCTION OF VALENCE AND INSTRUMENTALITY

Valence and instrumentality are the first two parts of our diagnostic model of motivation. As shown in Figure 7-4, these two concepts combine to influence the desire to perform. People will be motivated to perform at a high level so long as they perceive that receiving high-valence outcomes is contingent upon strong personal performance. Our understanding of the process depicted in Figure 7-4 is based in part on need theories, which help explain what outcomes individuals will perceive as having a positive valence. In addition, because reinforcement theories explain how people learn about contingencies, they also provide insight into the process that makes people want to perform. As the "In Practice" box suggests, when valence and instrumentality issues are not dealt with satisfactorily, employees' desire to perform their jobs may plummet to zero.

EXPECTANCY: SELF-EFFICACY THEORY

Although actually part of Bandura's social-learning theory, self-efficacy is an important topic in its own right.[25] **Self-efficacy** refers to the judgments people make about their ability to execute courses of action required to deal with prospective situations. People high in self-efficacy feel they can master, or have mastered, some specific task. As you can see, self-efficacy is a perception and may be accurate or faulty.

Self-efficacy differs from self-esteem in that it is usually more task specific or situation specific. For example, an artist may have a positive self-image generally but may feel little self-efficacy when confronted with the need to repair his car. Self-efficacy also differs from the concept of locus of control discussed in Chapter 5. An internal locus of control implies that a person perceives that his actions are responsible for the outcomes he receives. It does not, however, imply that the person perceives himself as having the ability to actually execute the actions required to obtain desired outcomes. To underscore this distinction, Bandura reserved the term *efficacy-based futility* for individuals who have both an internal locus of control *and* low self-efficacy.[26]

Self-Efficacy and Behavior

Self-efficacy perceptions strongly influence human behavior. In Bandura's words:

> People avoid activities that they believe exceed their coping capabilities, but they undertake and perform assuredly those that they judge themselves capable of managing. . . . Judgments of self-efficacy also determine how much effort people will expend and how long they will persist in the face of obstacles or aversive experiences. When beset with difficulties people who entertain serious doubts about their capabilities slacken their efforts or give up altogether, whereas those who have a strong sense of efficacy exert greater effort to master the challenges.[27]

[25] A. Bandura, "Self-Efficacy Mechanism in Human Behavior," *American Psychologist* 37 (1982), 122–47.

[26] Ibid.

[27] Ibid.

216

Valences, Instrumentalities, and the Boeing Strike

Most tasks required to construct a modern aircraft, such as riveting, wiring, metal binding, wing assembly, and cockpit installation, are done in large hangars. These hangars seem cavernous when they are absolutely still, as the Boeing hangars were in October, 1989, when 58,000 workers walked off their jobs.

The Boeing strike had implications for more than one employer or one industry. Indeed, the strike became symbolic of organized labor's crucial fight against bonuses as substitutes for pay raises. This practice was unheard of in the late 1970s. By 1990, however, 42 percent of all American workers covered by union agreements were receiving such bonuses. This early battle between union and management helps illustrate the usefulness of the valence and instrumentality concepts. It also showcases the process of social exchange in a context where foreign competition is intense and where failure to reach an agreement could mean economic suicide for one or both parties.

Valences deal with needs. The notion of substituting bonuses for raises first emerged to meet the needs of employers. Unlike raises, bonuses do not count toward pensions, vacations, sick pay, or overtime. And they do not compound or accumulate year after year. In the early 1980s a wave of deregulation and an influx of foreign competition put almost unbearable pressures on many U.S. manufacturing firms. Companies pleaded hardship and turned to their unions for relief. In many cases the unions agreed to forego yearly wage raises, in return for yearly bonuses.

Boeing employees were willing to agree to such provisions because it was obvious that the company had fallen on hard times. Moreover, the bonuses met the needs of many Boeing employees. In the early 1980s, Boeing's work force was one of the youngest in America and for them, a big year-end bonus (given out ten days before Christmas) was "like winning the lottery."

Times and needs change, however. In 1989, Boeing was no longer in danger as a company. In fact, when the strike occurred, Boeing was enjoying its most profitable period ever, having piled up orders to build 1,600 planes for $80 billion. Executive salaries, which were often tied to stock performance, increased dramatically. Moreover, in the intervening period, despite union concessions, Boeing had engaged in several layoffs that reduced the company's size at one point to less than a quarter of what it had been in the 1970s. This experience, coupled with the aging of Boeing's work force, made needs for security and long-term economic growth more im-

portant than the instant gratification associated with hefty bonuses. The economic impact of the bonuses was detailed in a union circular, which showed that since 1983, Boeing had paid bonuses equal to 31 percent of each worker's gross pay. The circular explained that if the same 31 percent had been in the form of wages, compounding (i.e., accumulation) alone would have given employees three times as much income.

Boeing's management resisted union demands. Their long-term plan was to move the entire work force toward a system that included not only bonuses but also annual profit-sharing payments. The latter would rise during good times, and disappear during the inevitable bad times. Management saw both these payments as "rewards for profitability and hard work." But while management was attempting to make performance instrumental to receiving high pay, the unions were interested in setting some instrumentalities of their own. In particular, the contingency they set for accepting this kind of flexible pay in place of raises was an agreement by management to forgo all future layoffs.

Thus needs, valences, and instrumentalities drive the behaviors of both employers and employees. Moreover, these factors change over time, so that a solution that was once viable may no longer be successful. Approaches to worker motivation that fail to take these facts into consideration are as empty as the hangars at Boeing in 1989.*

* Based on L. Uchitelle, "Boeing's Fight over Bonuses: Trend Threatens Traditional Raises," *New York Times*, October 12, 1989, p. D1.

APPLYING THE DIAGNOSTIC MODEL

Diagnostic Question 6: Does the person I am trying to motivate actually have the ability to accomplish the tasks that he or she is attempting to perform?

Sources of Self-Efficacy

Given the effect that feelings of self-efficacy can have on behavior, it is important to know where these feelings come from. In his research, Bandura has identified four different sources of self-efficacy beliefs. First, self-efficacy can be based on a person's *past accomplishments*. Past instances of successful behavior increase personal feelings of self-efficacy, especially when these successes seem attributable to unchanging factors such as personal ability or a manageable level of task difficulty. Conversely, past instances of failure tend to reduce personal feelings of self-efficacy. However, if these failures can be attributed to causes over which one has no control, such as bad luck, or that one has the power to change, such as effort, then relatively high self-efficacy beliefs can be maintained even in the face of failure.

The link between self-efficacy theory and social-learning theory is made clear in Bandura's second source of self-efficacy beliefs: *observation of others*. Merely watching someone else perform successfully on a task may increase an individual's sense of self-efficacy with respect to the same task. In his research, Bandura often cured children who had irrational fears of dogs with a treatment in which the fearful children repeatedly observed other children of the same age and sex approaching and petting dogs. After repeated observations, the self-efficacy of the fearful children increased to the point where they were able to approach and pet the dogs themselves.

It is important to note that characteristics of the observer and model can influence the effects of observation on feelings of self-efficacy. For instance, the observer must judge the model to be both credible and similar to the observer (in terms of personal characteristics like ability and experience) if observation is to influence efficacy perceptions.

A third source of self-efficacy is *verbal persuasion*. Convincing people that they can master a behavior will under some circumstances increase their perceptions of self-efficacy. The characteristics of the source and the target of the communication, however, can affect the influence that persuasion has on self-efficacy perceptions. Again, people who are perceived as credible and trustworthy are most able to influence others' self-efficacy perceptions through verbal persuasion.

Self-efficacy training is gaining wide acceptance in business and industry. For example, Dale Carnegie and Associates reports that companies enrolled over 170,000 people in their "you-can-do-it" training courses in the last decade alone. More and more blue-collar workers are enrolling in such courses. At a Ford Motor Company stamping plant in Buffalo, New York, for instance, a joint labor-management training center has as many as three Carnegie classes a day. Carnegie CEO Stewart Levine explains, "We focus on the critical skills of self-confidence, so that [people] on the assembly line can make decisions, even though they're thousands of miles away from the CEO."[28]

Logical verification is another source of self-efficacy perceptions. By logical verification, people can generate perceptions of self-efficacy at a new task if they can perceive a logical relationship between the new task and a task they have already mastered. Suppose, for example, that a highly competent secretary is worried she will not be able to master a word processor. If the secretary can be convinced that the word processor is nothing more than a glorified typewriter, which she knows very well how to handle, her perceived self-efficacy as a word processor may well increase.

STEP 2: EFFORT AS A FUNCTION OF DESIRE AND EXPECTANCY

In Step 1 we stated that the desire to perform is influenced by both valence and instrumentality. Now we can state that effort is a function of desire to perform and expectancy. In other words, the level of effort a person will

---

[28] A. Bernstein, "How to Work the Line and Influence People," *Business Week*, May 7, 1990, pp. 140–41.

AN EXAMPLE
A man confessed to a friend that he had not continued in school because he was afraid of math. Specifically, as a child, he could never do fractions. The friend, who had seen considerable evidence that the man handled fractions expertly, was surprised. The man was an excellent carpenter. He was continually making measurement conversions in his head, calculating fractions of inches for adjustments. The friend proceeded to point out numerous examples of the man's expertise. By the end of the day the man was working through fractions on paper with the same agility with which he had worked them through in his head.

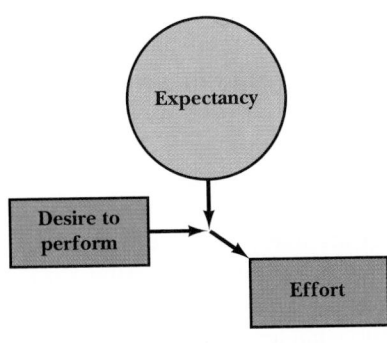

put forth is determined by three components of our model: valence, instrumentality, and expectancy.

Self-efficacy theory is particularly useful as an explanation of how expectancies are formed and how they can be changed. However, as Figure 7-5 suggests, a person's belief that she can perform possesses no motivational value unless she truly desires to excel. Similarly, simply wanting to excel will not bring about high levels of effort unless the person has some belief that it is possible to do so.

Our diagnostic model of motivation now consists of the three basic components of expectancy theory: valences, instrumentalities, and expectancies. This so-called VIE theory is quite useful in understanding how people are motivated to perform and how they decide what effort they will put forth. Porter and Lawler have argued, however, that to fully understand performance we must supplement this theory with additional components.[29] In the next section we will examine one such component: the effect of setting specific goals on the accuracy with which people perceive their tasks.

goal-setting theory A theory of motivation originated by Locke that suggests that behavior is driven by goals and aspirations, such that specific and difficult goals lead to higher levels of achievement.

ACCURACY OF ROLE PERCEPTIONS: GOAL-SETTING THEORY

Role perceptions are people's beliefs about what they are supposed to be accomplishing on the job and how. Role perceptions are accurate when people facing a task know what needs to be done, how much needs to be done, and who will have the responsibility to do it. Role accuracy of this sort guarantees that the energy devoted to task accomplishment will be directed toward the right activities and outcomes. At the same time, it decreases the amount of energy wasted on unimportant goals and activities. Goal-setting theory can help us understand how to enhance the accuracy of role perceptions.

Important Goal Attributes

Employees are often told, "Do your best." This axiom is a standard instruction intended to guide job performance in everyday situations. Yet, research by Edwin Locke, the leading advocate of **goal-setting theory**, and numerous other organizational scientists, has consistently demonstrated that vague instructions like this can actually undermine personal performance. In contrast, over 100 studies have provided evidence supporting the assertion that performance is enhanced by goals that are both specific and difficult. In summarizing these studies, Locke and several of his colleagues concluded that "the beneficial effect of goal setting on task performance is one of the most robust and replicable findings in the

[29] Porter and Lawler, "Managerial Attitudes."

Diagnostic Question 7: Does the person I am trying to motivate have specific difficult performance goals in mind?

Diagnostic Question 8: Is the person I am trying to motivate committed to goals?

psychological literature."[30] As you can see from Table 7-3, setting specific goals has improved performance in a wide variety of jobs.

Although initially Locke defined a goal as simply "what the individual is trying to do,"[31] he has long emphasized that specific and difficult goals lead to higher performance than vague or simple goals. According to Locke, specific and difficult goals seem to promote greater effort and to enhance persistence. Moreover, they are likely to encourage people to develop effective task strategies. Their primary virtue, however, is that they direct attention.

By directing our attention to specific desired results, goals clarify both what is important and what level of performance is needed. Goal setting is most effective when teamed with feedback so that progress can be monitored. Because goals clarify what needs to be done they lead to accurate role perceptions. For example, for the executives at Phillips-Van Heusen, any result could have been considered consistent with the goal of "do your best." However, these top managers were told that they had to move the stock price to $28 per share by January 1, 1992. This was a much more specific goal and open to much less interpretation. In addition, PVH provided clear feedback on progress toward this goal. The stock price was charted on a bulletin board placed in the reception area at corporate headquarters, where PVH executives could hardly fail to notice it.

Goal Commitment and Participation

Another factor that affects performance is the extent to which a person feels committed to a goal. Specific and difficult goals lead to increased performance only when there is high **goal commitment**. The requirement that people be committed to goals makes the managerial use of goal-setting programs somewhat exacting, because goals that are particularly difficult are typically met with less personal commitment. Taken to an extreme, there is a good chance that people will view a goal that is set excessively high as impossible and thus will reject it altogether.

Fortunately, research has examined several ways to increase commitment to difficult goals. One important factor is the degree to which the goals are public rather than private. For example, in one study, students for whom difficult GPA goals were made public (posted on a bulletin board) showed higher levels of

goal commitment A person's willingness to put forth effort in accomplishing goals and unwillingness to lower or abandon goals.

AN EXAMPLE: TRW
TRW, once a loosely knit group of some 80 different businesses, took a long, hard look at itself, its customers, and its environment. According to William Lawrence, executive vice president for planning, technology, and government affairs, the company decided that it needed focus—it needed to figure out and to build on what it did best. This Cleveland-based company then shed nearly half of its businesses. Fixing its sights on the growing automotive airbag industry, TRW developed a clear purpose and specific goals and made sure that all employees—from the chairman to hourly workers—knew exactly what their jobs were. With accurate role perceptions assured, the company was able to grab early leadership in a fast-moving market. (Ronald Henkoff, "How to Plan for 1995," *Fortune*, December 31, 1990, p. 70.)

---

[30] Edwin A. Locke, "Toward a Theory of Task Motivation and Incentives," *Organizational Behavior and Human Performance* 3 (1968), 145.

[31] Locke, "Task Motivation," p. 159. For a later review, see Locke, K. N. Shaw, L. Saari, and G. P. Latham, "Goal Setting and Task Performance: 1968–1980," *Psychological Bulletin* 80 (1981), 125–52. For research on the focusing affects of goals, see Locke and J. F. Bryan, "The Directing Function of Goals in Performance," *Organizational Behavior and Human Performance* 4 (1969), 35–42.

TABLE 7-3
Jobholders Who Have Improved Performance in Goal-Setting Programs

Telephone servicepersons	Marine recruits
Baggage handlers	Union bargaining representatives
Typists	Bank managers
Salespersons	Assembly-line workers
Truck loaders	Animal trappers
College students	Maintenance technicians
Sewing machine operators	Dock workers
Engineering research scientists	Die casters
Loggers	

commitment to those goals than students whose goals were kept private. So one way to commit yourself to a course of action is to tell people what goal you are working toward.

This study also found a significant positive relationship between need for achievement and goal commitment. Moreover, the relationship between need for achievement and commitment was especially strong when the goals were set by the subjects themselves, as opposed to when they were assigned to the subjects by an outside party. Figure 7-6 depicts this complex relationship between goal origin (self-set vs. assigned by others), need for achievement, and goal commitment.

The complex relationship found in this study is typical of research in the area of goal setting. Prior to studies like this one, it was widely assumed that participation in the goal-setting process would invariably enhance performance and commitment. That is, it was thought that the more input people had in establishing their goals, the more committed they would be to them. This common-sense notion has not held up well in scientific studies, however. Many contemporary studies have failed to find significant differences between participative goal-setting groups and assigned-goal groups in terms of either goal commitment or performance. Generally speaking, studies that have found positive effects for participation have been able to do so only within limited subsamples of research participants.[32] Thus although commitment is important in enhancing the effect of goals on performance, participation does not guarantee commitment.

Goals and Strategies

As Table 7-3 shows, goal setting has been used to increase performance on a variety of jobs. Yet most of the early research on goal setting consisted of studies that focused attention on relatively simple tasks. That was not an accident or an oversight; it was quite intentional. In characterizing his pioneering research on goal setting, Locke stated that "the research to be reported here involves pre-

[32] G. P. Latham, M. Erez, and E. A. Locke, "Resolving Scientific Disputes through the Joint Design of Crucial Experiments by the Antagonists: Application to the Erez-Latham Dispute Regarding Participation in Goal Setting," *Journal Of Applied Psychology* 73 (1978), 753–72.

AN EXAMPLE
A real-estate broker in a major metropolitan area set a goal to increase office sales by 20% in 1990. The agents in the brokerage regarded this as an unrealistic goal. Real-estate was depreciating, sales were declining, and major companies in the area were downsizing. People couldn't afford to sell or buy homes. As a result, the agents did not become committed to the goal. Many adopted a more realistic goal of maintaining 1989 sales levels. Others left the company to escape the pressure.

APPLYING THE DIAGNOSTIC MODEL
Diagnostic Question 9: Does the person I am trying to motivate know the best strategies for accomplishing these goals?

F I G U R E   7-6

Effect of Need for Achievement and Type of Goal on Goal Committment

High need for achievement leads to high goal commitment, but only when people set their own goals.

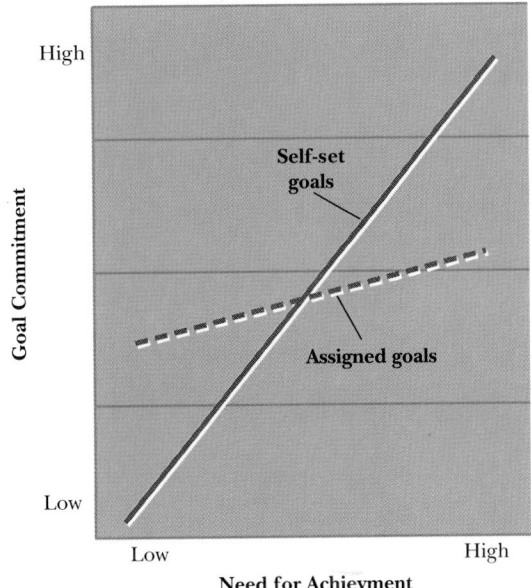

dominantly simple tasks in which learning complex new skills and making long-term plans and strategies is not necessary to achieve goals."[33]

More recent research has extended goal-setting theory into more complex task domains, however, and in these situations, the links between goals, effort, and performance are not so direct. A review of these later studies indicated that while goals have positive effects for all tasks, the magnitude of the effect is stronger for simple tasks than for complex tasks.[34] Figure 7-7 shows how the effect of goal difficulty on performance decreases as task complexity increases.

The primary reason why the goal-setting effect for complex tasks is not as straightforward as it is for simple tasks involves task strategies. With complex tasks, the *task strategies* or plans of action that people devise have a big impact on the outcome of their efforts. This impact can obscure and in some rare cases even wipe out goal-setting effects.

Research on goal setting and task strategies by Christopher Earley and his colleagues suggests that whereas setting specific and difficult goals may lead to increased strategy development, there is no guarantee that the resulting strategies will always be effective.[35] Moreover, because developing strategies consumes time that might otherwise be devoted to task performance, there may be situations where goals actually hinder performance. Accordingly, Earley has shown that specific and difficult goals may be least effective for "tasks in which (a) performance is primarily a function of strategy rather than performance, (b) there are many available strategies, (c) the optimal strategy is neither obvious nor readily identified, and (d) little opportunity to test hypotheses retrospectively exists."[36]

[33] Locke, "Task Motivation," p. 161.

[34] R. E. Wood, E. A. Locke, and A. J. Mento, "Task Complexity as a Moderator of Goal Effects: A Meta-Analysis," *Journal of Applied Psychology* 72 (1987), 416–25.

[35] P. C. Earley and B. C. Perry, "Work Plan Availability and Performance: An Assessment of Task Strategy Priming on Subsequent Task Completion," *Organizational Behavior and Human Decision Processes* 39 (1987), 279–302; and Earley, P. Wajnaroski, and W. Prest, "Task Planning and Energy Expended: Exploration of How Goals Influence Performance," *Journal of Applied Psychology* 72 (1987), 107–14.

[36] P. C. Earley, T. Connolly, and G. Ekegren, "Goals, Strategy Development and Task Performance: Some Limits on the Efficacy of Goal Setting," *Journal of Applied Psychology* 74 (1989), 24–33.

FIGURE 7-7

Goal Difficulty, Task Complexity, and Performance

When tasks are relatively simple, there is a close relation between the difficulty of the goal and the person's performance. That is, a person will work hard to achieve a difficult goal but may make less effort to reach an easy one. When tasks are complex, however, the relation between goal difficulty and performance changes. A person trying to achieve a difficult goal in a highly complex task may not be able to sustain the level of performance that he can on a simpler task.

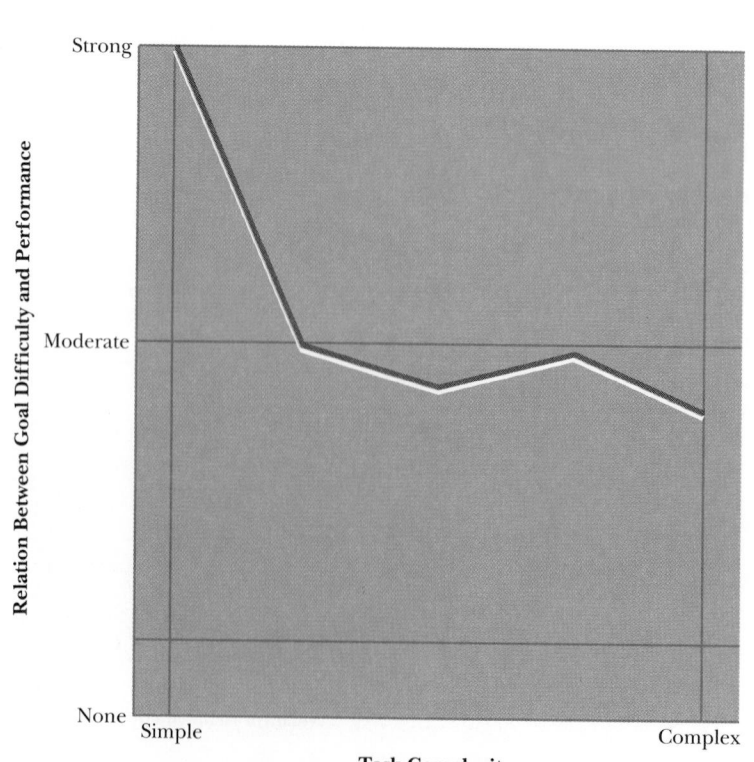

Although research on performance strategies has yielded findings that complicate the results of other goal-setting studies, this research is helpful in delineating the specific role-clarifying effect of goals. In simple tasks, where the *means* to perform a task are clear, specific and difficult goals lead to higher performance because they clarify the *ends* toward which task performance should be directed. In complex tasks, however, the *means* are not clear. Individuals performing such tasks do not know how to go about them in the best way, so merely clarifying the *ends* sought is unlikely to enhance performance. In sum, goals can clarify the ends to which role-related performance should be directed and may even promote strategy-development efforts. Goals alone offer no guidance about how to attain these ends, however, and thus need to be teamed with effective strategies.

ABILITY AND EXPERIENCE REVISITED

Predicting task performance from goals is also contingent upon the abilities of the jobholder. We discussed abilities at great length in Chapter 4, so here we will focus only on how such individual differences interact with goal setting and task strategies.

Three things are worth noting. First, it should be obvious that people lacking requisite abilities cannot perform a complex task even under the most favorable goal-related circumstances. Second, there are some subtle relations among goal setting, attention, and cognitive capacity that affect task performance. Recall that one of the ways in which goal setting affects performance is by directing attention to the kinds of results that are desired. Kanfer and Ackerman have developed a model that recognizes that different people have different amounts of cognitive ability to bring to bear on a task, and that this limits how much they can attend to at any one time.[37] Because it diverts attention from the task to the goal, goal setting may be particularly damaging to people who have low ability or who are still learning the task. Such people need to devote all their attention to the task, and goal setting for them is unlikely to enhance performance.

POINT TO STRESS
Goals must be consistent with a person's abilities.

Finally, it should be noted that individuals high in ability are more likely to develop effective strategies. For one thing, high-ability people, especially those high in reasoning and deduction, can often figure out good strategies prior to working on the task. Second, individuals high in cognitive ability learn more quickly and are therefore more likely to deduce effective strategies through trial-and-error.[38]

Thus, although motivation is critical to performance, the lessons learned in Chapter 4 about the importance of ability should not be forgotten. For all but the most simple of tasks, there is no substitute for ability.

STEP 3: PERFORMANCE AS A FUNCTION OF EFFORT, ACCURACY OF ROLE PERCEPTIONS, AND ABILITY

At last, in this third step, we can see how motivation and other factors combine to determine performance (see Figure 7-8). Specifically, performance will be high when a person (a) puts forth significant effort, (b) has the ability to exhibit the right behaviors and (c) focuses her effort properly.

[37] R. Kanfer and P. L. Ackerman, "Motivation and Cognitive Abilities: An Integrative/Aptitude-Treatment Interaction Approach to Skill Acquisition," *Journal of Applied Psychology* 74 (1989), 657–90.

[38] J. Shapiro, "Goal Setting, Cognitive Ability, and Task Strategy," Master's Thesis, Michigan State University, 1990, p. 79.

FIGURE 7-8

Step 3. Performance as a
Function of Effort, Accuracy
of Role Perceptions, and Ability

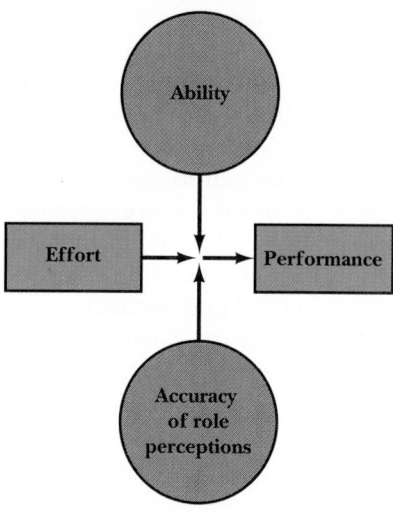

Proper focus is achieved by making role requirements clear, that is, by clarifying
the ends sought through goal setting and by clarifying the means to be used in
achieving these goals through the development of task strategies.

EVALUATION OF REWARDS: EQUITY THEORY

Motivation, we have seen, is a complex and dynamic process in which
valences, instrumentalities, expectancies, role perceptions, and abilities
interact to affect performance. Now we will add another component
to our diagnostic model—the effect of rewards for successful performance on
the individual's satisfaction. Equity theory seems particularly useful in examining
the intricate relations among motivation, performance, and rewards.

Equity, Equality, and Social Comparisons

equity theory A theory of
motivation originated by Adams
that suggests that behavior is
motivated by the desire to reduce
guilt or anger associated with
social exchanges that are perceived
to be unfair.

Equity theory is a theory of social exchange that describes the process by which
people determine whether they have received fair treatment. More specifically,
as shown in Figure 7-9, equity theory holds that people make judgments about
fairness by forming a ratio of their perceived investments (or inputs) and perceived
rewards (or outcomes). They then compare this ratio to a similar ratio reflecting
the perceived costs and benefits of some other person. If these ratios are not
equal, the situation is perceived as unfair. For example, is it fair for the CEO of
a savings and loan institution that lost $120 million to make $4.8 million in salary?
That is precisely what happened for CenTrust Bank of Miami chairman, David
Paul. He was ousted, however, after an investigation by federal regulators who
were given tips by disgruntled employees.[39]

[39] J. Fierman, "The People Who Set the CEO's Pay," *Fortune*, March 12, 1990, pp. 58–66.

FIGURE 7-9

Algabraic Expression of How
People Make Equity Comparisons

$$\frac{I_{person}}{O_{person}} = \frac{I_{Reference\ person}}{O_{Reference\ person}}$$

Diagnostic Question 10: Who is
the reference person for the
person I am trying to motivate?

Diagnostic Question 11: Are the
rewards I am giving equitable with
respect to this reference person?

Equity theory does not require that outcomes or inputs be equal for equity to exist. A person receiving fewer desirable outcomes than someone else may still feel fairly treated if he sees himself as contributing fewer inputs than the other person.

Adams identified a number of possible inputs and outcomes that might be incorporated in equity comparisons (see Table 7-4). Two crucial issues complicate the use of Adams's theory. First, equity judgments are based on individual perceptions of inputs and outcomes, and perceptions of the same inputs or outcomes may differ markedly from one person to the next. For instance, an employer may be less impressed with an employee's experience than the employee himself, and the employee may not see the status of his job the same way the employer does. Second, the importance associated with a particular input or outcome differs from one person to the next. Thus, one employee might consider pay an extremely important outcome whereas another might place greater emphasis on job security. Without even realizing it, an employer who was unaware of differences like these could easily create inequity by treating two people identically.

Responses to Inequity

We are much more tolerant of
positive inequity than negative
inequity. Ask students how many
of them would come to your office
if you subtracted 2 points that
they deserved from their exams.
Most will indicate they would
come in for their 2 points. Then
ask them who would come in if
you added 2 points they didn't
deserve—or 5 points—or 10
points. Have students discuss the
differences in their responses to
under and over reward and their
justifications for keeping
undeserved points.

According to Adams, after a person perceives inequity, she experiences an unpleasant emotional state of either anger or guilt. When people perceive themselves as receiving a greater share of outcomes than they deserve, they may feel guilty. Indeed, a significant amount of the early research on equity theory dealt with the guilt-bearing *over-compensation effect*. In these studies, some workers who were led to believe they were overpaid proceeded to produce at higher levels out of guilt.[40] About the $1 million bonus offer, one PVH executive stated flat out, "I can honestly say that I was embarrassed by it. I don't know that I deserve it."[41]

But results like these are not so common. Research has shown that perceived inequities associated with the *under-compensation effect* are far more frequent and more potent. Moreover, perceiving oneself as at the low end of the equity comparison results in anger, a much stronger reaction. With respect to the huge salaries of upper-level executives, for example, one middle-level financial analyst notes that "it disturbs me when someone on high dictates that no matter how hard you work or what you do, you're only going to get a 6% increase, and if

[40] J. S. Adams and W. B. Rosenbaum, "The Relationship of Worker Productivity to Cognitive Dissonance about Wage Inequities," *Journal of Applied Psychology* 46 (1962), 161–64.

[41] Knowlton, "Million-Dollar Motivator," p: 10.

T A B L E 7-4	
Inputs and Outcomes in Equity Theory	
INPUTS	**OUTCOMES**
Education	Pay
Intelligence	Satisfying supervision
Experience	Seniority benefits
Training	Fringe benefits
Skill	Status symbols
Social status	Job perquisites
Job effort	Working conditions
Personal appearance	
Health	
Possession of tools	

you don't like it you can take a hike. Yet whatever they have negotiated for themselves—10%, 20%, or 30%—is a different issue."[42]

Whether guilt or anger, the tension associated with inequity motivates the person to do something to reduce the inequity. Adams specified six possible reactions. First, the individual might *alter personal inputs*, either increasing them or decreasing them depending upon the type of inequity perceived. In situations characterized by the over-compensation effect described above, an increase in effort inputs might be expected in response to the receipt of an outcome that was greater than expected. The analogous response in situations of undercompensation would be a decrease in personal inputs. Someone feeling underpaid on a job might put forth less effort.

A second response to inequity is to try to *alter personal outcomes*. Individuals who feel they are relatively underpaid according to the market may demand raises and threaten to leave or strike. An area of increased activity recently is the renegotiation of so-called two-tier labor contracts. These contracts, like the substitution of bonus for pay raise we discussed in the "In Practice" box, were born in the recession-ridden early 1980s. In an effort to keep foundering employers alive, unions agreed that although current workers' wages would not be lowered, all newly hired employees would be brought in at a lower wage rate. These employees would be on a different pay scale that would leave them forever underpaid relative to current employees, who were on the higher tier. As one disgruntled, second-tier American Airlines pilot put it, "We like to describe it as mortgaging the unborn."[43] Interest in these kinds of plans peaked in 1985, when 11 percent of union employees were covered by such agreements. As the business climate improved, second-tier employees, who soon began to outnumber first-tier employees, fought to dismantle this inequitable solution. By 1990, this pressure brought the 11 percent figure down to 6 percent, and according to James Martin, a leading researcher in this area, such provisions "are going to be fairly rare" by 2000.[44]

A third way of responding to inequity is to use what is called *cognitive distortion*, that is, rationalizing the results of one's comparisons. For example, if a manager feels she is overpaid for her effort, rather than increase that effort she may rationalize the imbalance by overvaluing her inputs. For example, a relatively overpaid worker may exaggerate the importance of her level of education.

People can also distort their perceptions of outcomes. In one equity-theory study, subjects who were underpaid for a particular task justified this underpayment by stating that the task they were working on was more enjoyable than the task performed by research participants who were overpaid, even though their tasks were identical.

Yet another means of eliminating an inequity via cognitive distortion is to change the reference person. A salesperson who brings in less revenue than his peers may claim, "You can't compare me with them because I have a different territory." By this statement, he seeks to disqualify his peers as reference persons.

A fourth way to restore equity is to take some action that will *change the behavior of the reference person*. Workers who in the eyes of their peers perform too well on piece-rate systems often earn the derogatory title of *rate buster* (see Chapter 2). Research has shown that if name calling of this sort fails to constrain personal productivity, more direct tactics may be invoked. In one study, for instance, the researchers coined the term *binging* to refer to a practice in which

Suppose an employee seeks equity through a more challenging job challenge rather than through a higher salary? John Allegretti (center), a trainee at the Hyatt Regency hotel chain, couldn't wait for the eight years or more that it takes to become a manager in the hotel business. When Hyatt found out that Allegretti was considering a job with a waste-recycling company, it found a way to keep him on board. The company asked him to head a project to reduce waste at its Chicago hotel. Allegretti built this project into a new waste-consulting company, called International ReCycle Co. Inc., that now has 24 clients in 8 states. Source: *James E. Ellis, "Feeling Stuck at Hyatt? Create a New Business,"* Business Week, *December 10, 1990, p. 195.*

[42] A. Bennett, "Caught in the Middle: Managers Don't Mind that the CEO Makes a Lot of Money, but Raise Questions About Fairness," *The Wall Street Journal*, April 18, 1990, p. 9.

[43] R. Tomsho, "Employers and Unions Feeling Pressure to Eliminate Two-Tier Labor Contracts," *Wall Street Journal*, April 20, 1990, p. 25.

[44] Ibid.

INTERNATIONAL OB

Establishing Equity Overseas

If you were given a chance to work in any city in the world, what city would you choose? San Francisco? London? New York? San Diego? Paris? Hong Kong? How about Khartoum, in Sudan? Many Americans might hesitate at the thought of working in Khartoum, and after doing a little homework, they might refuse outright. In 1989, the United States State Department designated Khartoum the most difficult place in the world to live and work.*

What might North Americans find objectionable about Khartoum? To begin with, Sudan, of which Khartoum is the capital city, is a developing nation, and it stills lacks many of the technological advances that Americans have come to take for granted. In 1989, many phones in the city had not worked for years. Live power lines drooped over pot-holed roads and dirt paths. The city was also plagued by labor unrest. For example, in one month alone, bank workers, bus drivers, electricians, pilots, doctors, pharmacists, postal workers, engineers, and university staff were all on strike.

Sudan borders on two countries that have experienced considerable political unrest, Chad and Ethiopia. It also borders on Libya, whose leader, Muammar Qadaffi, is considered by many to be unstable and potentially dangerous. The Sudan government itself has had repeatedly to fend off rebel insurgents, and the re-emergence of civil war is an ever present threat.

Like several African countries, Sudan has oil and several ports on the Persian Gulf. American companies thus have a legitimate reason to do business in the country. So a very realistic question is, What would an American organization have to offer an employee to get him to work in Sudan? As large corporations continue to internationalize, compensation issues associated with overseas assignments, particularly to hardship posts, have become salient. Approaching these problems from an equity perspective highlights two of the major problems in this area: keeping expatriates whole, and keeping local country nationals satisfied.

Keeping expatriates whole means enabling United States personnel stationed abroad to maintain the same standard of living that they enjoy at home. Organizations try to do this by adding an incentive component and an equalization component to the job.† The *incentive component*, essentially a bonus, takes the form of a percentage increment in salary. The U.S. State Department, for example, recommends a 25 percent premium for Khartoum. The *equalization component* is an additional payment to adjust for differentials in buying power of an employee's base salary in one country versus another. Equalization payments typically include allowances, for housing, private school for children, and taxes (expatriates are taxed by both the United States and the host nation). It also includes a cost-of-living allowance to adjust for price differences on such necessities as food.

With the addition of both incentive and equalization components, compensation for overseas jobs can be quite high. This raises more problems: how to justify such high salaries for Americans when local country nationals (LCNs) are doing similar work yet being paid less. Further, many U.S. firms overseas employ third-country nationals (TCNs), or people who are neither U.S. citizens nor citizens of the country where they work. No incentive or equalization payments are made to TCNs. Among both LCNs and TCNs who compare themselves with U.S. expatriates, the possibility for perceptions of inequity and exploitation is great.‡

To minimize these problems, organizations employ two strategies. Sometimes they try to limit awareness of pay differences by keeping salary information secret or to mask the issue by paying people in different currencies. Or they may make an expatriate job a relatively short-term consulting assignment that includes training LCNs and TCNs. As organizations become increasingly experienced in internationalization, other more creative solutions to the inequity problems posed by expatriates, LCNs, and TCNs will probably evolve. A clear understanding of equity theory should help future managers to achieve such solutions.

---

* R. Winslow, "How Khartoum Won No. 1 Ranking as a Hardship Post," *The Wall Street Journal*, April 26, 1989, p. 1.

† G. T. Milkovich and J. M. Newman, *Compensation*, (Richard D. Irwin, Homewood IL, 1990), p. 545.

‡ D. Darlin, "When in Rome," *The Wall Street Journal*, May 3, 1988, p. 1.

---

workers periodically punched suspected rate busters in the arm until they reduced their level of effort.

Finally, if all else fails, equity can be secured by *leaving an inequitable situation* altogether. Turnover and absenteeism are common means of dealing with perceptions of unfairness in the workplace. Research emphasizes that it is the perception of unfairness, not absolute level of pay, that causes such escape behaviors. In many cases, workers receiving low pay in an absolute sense have very good attendance records because they also perceive their inputs as being low.[45]

---

[45] M. Patchen, *The Choice of Wage Comparisons* (Englewood Cliffs, N.J.: Prentice-Hall, 1961), p. 38.

Choosing among Responses. Given all these methods of reducing inequity, it is not surprising that equity theory is frequently criticized for its failure to specify how equity is most likely to be restored. As the "International OB" box points out, the increasing globalization of business and industry is putting ever greater pressure on us to figure out how to maintain equity among people from different countries and with different training and backgrounds.

Adams has offered some guidelines, however, as to how people behave in the face of inequity. First, an individual will attempt to maximize highly valued outcomes. Second, inputs that are costly or difficult to increase will rarely be altered. Third, the person will resist both distortions and real changes in inputs and outcomes that are central to his self-concept. Fourth, it is easier for a person to mentally distort the inputs and outcomes of others than his own inputs and outcomes. Fifth, people will leave a relationship only when the magnitude of the inequity is large and there are no other means of eliminating it. Sixth, the longer someone has used another as an object of comparison, the more resistant he will be to changing to another reference person.

STEP 4: SATISFACTION AS A FUNCTION OF PERFORMANCE AND EQUITY

Figure 7-10 incorporates the implications of equity theory into our model of human motivation. As the figure shows, the level of satisfaction experienced by the individual will be a function of both the person's performance and the perceived equity of the rewards she receives for that performance. The rewards will contribute to high levels of satisfaction only when the person perceives them as equitable. We will have more to say about job satisfaction in our next chapter.

STEP 5: MOTIVATIONAL COMPONENTS AS A FUNCTION OF PRIOR OUTCOMES

The fifth and final step needed to complete our model deals with the feedback loops that make the model dynamic over time. Let's look back at Figure 7-1, which shows three such feedback loops. First, there is a feedback loop going from level of satisfaction back to valence. Recall that valence, as a construct, deals with anticipated satisfaction, not realized satisfaction. The feedback loop allows for the possibility that an outcome a person thought might be valent might not bring him much real satisfaction when he actually receives it. His valence for such an outcome would then decrease relative to what it was at an earlier time.

Another feedback loop goes from rewards experienced after performance to instrumentalities. This loop is meant to suggest that the receipt of rewards at

FIGURE 7-10

Step 4. Satisfaction as a Function of Performance and Equity of Rewards

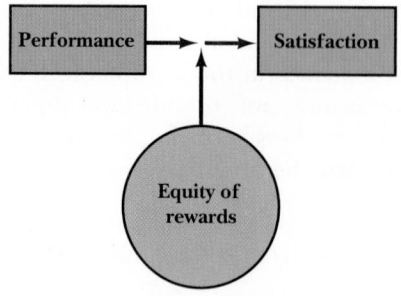

MOTIVATION AND PERFORMANCE

one time will affect the person's perceived instrumentalities at later times. If high performance is not followed by equitable rewards, extinction of the performance response could take place. High performance followed by equitable rewards, on the other hand, may be positively reinforcing.

APPLYING THE DIAGNOSTIC MODEL
Diagnostic Question 12: If the person I am trying to motivate was initially higher in motivation than he or she is at present, which of the three feedback loops in the diagnostic model accounts for the loss in motivation?

Finally, there is also a feedback loop from performance to expectancy. This loop affirms the fact that expectancies and self-efficacy are based at least partially on prior performance. All else equal, successful performance strengthens self-efficacy and leads to high expectancies. Failing at a task, however, generally leads to lower levels of self-efficacy.

These three feedback loops in our model of motivation create the possibility that motivation can change over time. For example, Figure 7-1 suggests that a highly motivated person might lose motivation for any of three reasons. First, a person starting out with high expectancies might discover during job performance that she cannot perform nearly as well as she thought. Decreased self-efficacy would reduce expectancy perceptions and lower motivation would probably result. Second, a person might discover that performing well on a job does not lead to the desirable outcomes he initially expected. His motivation could be expected to diminish as projected instrumentalities fail to materialize. Third, experience with the rewards received from performing a job might lead a person to discover faults with initial valences. That is, rewards expected to yield satisfaction do not do so. Motivation might drop owing to the absence of desirable rewards if no other performance-contingent rewards were available.

SUMMARY

Our diagnostic model of *motivation* and *performance* is based on *expectancy theory* and incorporates notions from five other theories of motivation including *need theory*, *learning theory*, *self-efficacy theory*, *goal-setting theory*, and *equity theory*. The model focuses on explaining four outcomes. The first, desire to perform, is a function of *valences* and *instrumentalities*. A person's desire to perform well will be high when valent rewards are associated with high performance. The second outcome, *effort*, is a function of desire to perform and *expectancy*. Effort will be forthcoming only when individuals want to perform well and when they believe they can do so. The third outcome, *performance*, is a function of effort, *accurate role perceptions*, and *ability*. Performance will be high only when individuals with the requisite abilities and knowledge of desired goals and strategies put forth their best effort. The model also shows how *satisfaction*, the fourth outcome, is a product of past performance and the perceived *equity of rewards* received for performing well. The dynamic nature of the motivation process is revealed in the way present levels of satisfaction, perceived equity of rewards, and performance affect future levels of valence, instrumentality, and expectancy.

REVIEW QUESTIONS

1. Recent research suggests that personality characteristics, like individual needs, may be determined more by genetic factors than we have thought. Take each of the four different need theories described in this chapter and discuss whether this new evidence supports, contradicts or is irrelevant to that theory.

2. Specific, difficult goals have been suggested to enhance performance, but researchers have also shown that performance will be high only when expectancies are high. We might think that as goals become increasingly difficult, expectations

for accomplishing these goals would decrease. Can you resolve this apparent contradiction between goal-setting theory and expectancy theory?

3.   Free agents in baseball are players whose contracts with current teams have expired. They put themselves on the "auction block" at the end of their option year (the last year of their current contract) and offer their services to the highest bidder. Some experts have suggested that, according to expectancy theory, a player's performance will be higher in the option year than in the year following his signing of a new contract. Others have suggested that, according to equity theory, performance will be higher after the new contract is signed. Why might these theories make different predictions in this case, and which do you feel is right?

4.   Imagine that you work for Phillips-Van Heusen and that your immediate supervisor is one of the 11 executives vying for the million-dollar bonus. How do you think this incentive would affect your boss's attitude toward your work? How would you feel if he got the bonus and you got your normal raise for that year? Analyze this situation using equity theory and describe what steps you think should be taken to reduce feelings of inequity among PVH employees other than the 11 top managers.

5.   Analyst Daniel Shore once called motivation researchers "servants of power" because their research was often used to manipulate lower-level workers. Is trying to motivate people necessarily exploitative? Are there any conditions under which providing external motivation might be exploitative? Which theories of motivation do you feel are exploitive? Which ones are not?

DIAGNOSTIC QUESTIONS

When you are trying to diagnose a situation in which motivation may be a problem, the following questions can guide your inquiry:

1.   What are the most important needs of the person I am trying to motivate?

2.   What contingencies has this person learned over the course of his or her reinforcement history?

3.   How can I make the receipt of outcomes that have positive valence for this person contingent upon performing at a high level?

4.   What outcomes that have negative valence for this person can I remove, contingent upon performance at a high level?

5.   Does the person I am trying to motivate believe that he or she can perform well? If not, what can I do to increase the person's self-efficacy perceptions?

6.   Does the person I am trying to motivate actually have the ability to accomplish the tasks that he or she is attempting to perform?

7.   Does the person I am trying to motivate have specific, difficult performance goals in mind?

8.   Is the person I am trying to motivate committed to goals?

9.   Does the person I am trying to motivate know the best strategies for accomplishing these goals?

10.   Who is the reference person for the person I am trying to motivate?

11.   Are the rewards I am giving equitable with respect to this reference person?

12.   If the person I am trying to motivate was initially higher in motivation than he or she is at present, which of the three feedback loops in the diagnostic model accounts for the loss in motivation?

EXERCISE 7-1
MANAGING MOTIVATION: AN EXERCISE IN POSITIVES AND NEGATIVES[1]

JANET MILLS, *Boise State University*
MELVIN MCKNIGHT, *Northern Arizona University*

What can managers do to improve workplace performance? One approach is to review the model and theories of motivation presented in this chapter and to identify some *positive* steps that one can take to improve employee motivation. Another, more unusual approach would be to review the models we've discussed, identify *negative* factors that discourage employee motivation, and take steps to eliminate or avoid these factors. In this exercise you will have a chance to experiment with both of these approaches as you consider situations involving motivational issues that you have experienced and think about how you might motivate employees who will be working for you in the future.

STEP ONE: PRE-CLASS PREPARATION

In class you are going to work together to develop lists of adjectives that describe what it is like to be motivated. To prepare for this class, think of a recent situation in which you were highly motivated and intensely involved in what you were doing. The situation can be from any context, including school, work, sports, hobbies, or travel. Try to remember the situation so vividly that you can "see" it in your mind. Now, list 10 to 15 words or phrases that describe how you felt in the situation. Such words or phrases might include "being challenged," "feeling stressed," or "wanting to prove I could do it."

_____

_____

_____

_____

_____.

Next, think of another recent situation in which you lacked motivation—in which you were bored, apathetic, turned off. Again, picture the situation vividly and list 10 to 15 words or phrases that describe your experience. These words or phrases might include "having no choice," "feeling resentful," or "feeling rejected."

_____

_____

_____

_____

_____.

STEP TWO: SHARING POSITIVE AND NEGATIVE EXPERIENCES

The class should divide into groups of 4 to 6 members each (if you have already formed permanent groups, reassemble in those same groups again). In each group, members should take turns discussing each of the highly motivated situations they have recalled and should share their lists of words and phrases that describe how they felt in these situations. A spokesperson should keep a master list of all of the words and phrases reported during this discussion. After everyone has had a turn, group members should take turns discussing the situations in which they lacked motivation and should share their lists of descriptive words and phrases. The spokesperson should again keep a master list.

STEP THREE: DEVELOPING MOTIVATIONAL PROGRAMS

The instructor should reconvene the class and ask each spokesperson to read the master lists developed during Step Two. As the spokespersons read their lists, your instructor will record the words and phrases on a chalk board or flipchart. He or she should make two lists: one, positive list for motivated situations and another negative list for unmotivated situations. After both lists are completed, your instructor will break the class up again into groups and will assign half of the groups Task A and the other half Task B:

Task A: Imagine that you manage a staff of 20. Productivity and satisfaction are at an acceptable level—neither especially high nor especially low. Your objective is to interfere with productivity and satisfaction by weakening employees' motivation.

[1] Adapted with the authors' permission from a paper entitled "Two Exercises for Teaching about Motivation."

You have one week to do this. Describe what specific actions you would take to accomplish this objective and explain how and why your actions would work.

Task B: Imagine that you manage a staff of 20. Productivity and satisfaction are at an acceptable level—neither especially high nor especially low. Your objective is to improve productivity and satisfaction by strengthening employees' motivation. You have one week to do this. Describe what specific actions you would take to accomplish this objective and explain how and why your actions would work.

All group members should record the group's recommendations, and a spokesperson should be appointed to report back to the class. After the two sets of groups have each completed their tasks (that is, one set of groups have performed Task A and the other Task B) and if time permits, your instructor will ask the groups to perform the second task (either Task B or Task A). As each group completes its second task its members should again record the group's recommendations, and a spokesperson should be ready to present a summary to the class.

STEP FOUR: CLASS DISCUSSION

The class should reconvene and each spokesperson should summarize the results of Step Three. As group summaries are given, the instructor and the class should work together to develop a master list of negative actions that would weaken employee motivation and positive actions that would strengthen it. After all groups have completed their reports, the class should discuss the results of this exercise, focusing on the following questions:

1. Which negative actions designed to weaken motivation would probably have the greatest effect? Why? Which negative actions would probably have the least effect? Why?

2. Which positive actions designed to strengthen motivation would probably have the greatest effect? Why? Which positive actions would probably have the least effect? Why?

3. Combine the strongest positive and negative actions together to form a list of motivational "dos" and "don'ts." How might managers use this list? What kinds of motivational programs does it suggest?

CONCLUSION

Motivating employees is a process of energizing them to perform their work in a productive, satisfying manner. On a daily basis, managers face the task of determining how to motivate employee performance in the most effective way. To accomplish this task, it is important that managers know not only what to do but also what *not* to do. Clearly, the axiom "one can know a thing best only by also knowing its opposite" applies to the issue of employee motivation.

DIAGNOSING ORGANIZATIONAL BEHAVIOR

CASE 7-1

THE PRODUCTION OF KCDE-TV*
C. PATRICK FLEENOR, *Seattle University*

KCDE-TV is one of two television stations in Tuttle, a city of 100,000 population, with a metropolitan area population of 175,000.

KCDE-TV (and radio) for some time had serious morale problems, especially in the television production department. KCDE employed 85 people in six departments: general office; data processing; news; engineering; radio; and television production. The television production group formed the single largest department, about 20 people. The functional areas of the production department are: announcing, directing, switching, camera operating, and video tape operating. See Exhibit 7-1 for description of these functions.

As is the case with many small to medium-sized stations, KCDE was looked upon as a training ground by many members of both management and staff. This was a reason offered by management on occasion for not granting a raise to an employee. It was suggested to the employee that if he wished to remain at KCDE he had better accept his present wage as the maximum for the foreseeable future. He then would find it necessary to move on to a bigger city if he expected to be paid more for the same job. The turnover, especially in the radio and production departments, was high.

Announcers are responsible for performing live commercials and programs, and for providing audio recordings for locally produced slide, film, and video tape commercials. Since the work load is variable, they typically have other duties, e.g., writing commercial copy, or reading news for the radio station.

The director is ostensibly the most creative member of the crew. He is responsible for the "on-air" presentation. He either recommends a set for a commercial or program, or approves an idea presented by some other member of the crew. During the actual broadcast or recording session, the director is in charge of all activities.

The switcher, sometimes referred to as the technical director, performs the physical operations at the control board required to put various video sources on the air, and to mix the sources at the director's command. He also is responsible for loading slides and film on the various projectors.

The video tape operator loads and "cues" video tapes on the video tape machines for the playback of commercials and programs on the air. he also sets up the machines for the recording of commercials and programs. The video tape machines are extremely complicated and quite difficult to operate, requiring a practiced touch for trouble-free operation.

The cameramen operate the large studio cameras, moving them on the director's cue and selecting the shots the director asks for. The cameramen do the actual construction of the sets and do most of the lighting, sometimes under the direct supervision of the director.

An additional member of the operating crew is an engineer, who is not a member of the production department. He is expected to provide technical advice to the director. His primary responsibility however, is the maintenance of the expensive, complicated, electronic gear.

Each employee negotiated his own salary with management since there was no union representation. There were no published salary ranges, but staff members knew that approximate ranges in 1990 were as follows:

Announcers	$1,650–$2,000/mo.
Directors	1,650–1,800
Switchers	1,200–1,300
Video Tape Operators	1,200–1,300
Cameramen	800–1,200

The salaries were based on a 48-hour, six day week. Much conversation among the crew members centered around what they all agreed was a low pay scale. As one of the crew members put it regularly in conversation: "No where else can you work a six-day week, a night shift, and virtually every holiday for such lousy money."

Benefits were another sore point. The company made group insurance available, but there was no retirement program. Though providing paid vacations, the company paid the vacationing employee for two 40-hour weeks. The two-week paycheck then was less by sixteen hours of overtime what the employee was acustomed to.

Working conditions with regard to physical comfort and safety were adequate and about average for the industry.

It was a common feeling among the crew members that they were being "used" to one degree or another by management. The men knew that many general office workers for the city's major private employers and the state government were making more money than they, working better hours and shorter weeks. Adding salt to the wound was the feeling that the television job required infinitely more creative ability then the general office workers needed or had. At the same time, most felt their jobs were intrinsically interesting, and far more challenging than office or administrative work.

Great animosity was directed toward the assistant general manager of the station. His previous post was chief engineer of the station, where he was tagged with the nickname "Overkill" by some members of the engineering department. This name was inspired by his tendency to over-react to situations. On one occasion he had fired an employee for smoking in the television control room. Though parts of the studio and control areas were posted against smoking, members of the staff looked upon this regulation as trivial. Care was taken not to smoke only when the assistant general manager was in the immediate area.

More than once, "Overkill" threatened to have a vital piece of equipment removed, ". . . unless you guys take better care of it." The threats were obviously hollow, since the station couldn't operate without the equipment. He had been heard to refer to the operating crew and the engineering department, or various members as "coolies."

The leader of the production department itself was not spared the crew's wrath. Every member of the crew looked upon Gary Brown, the production supervisor, as, as one of the switchers put it, "a miserable, two-timing s.o.b." More than one of the men had had the experience of making a request for a raise, only to find some weeks later that the production supervisor had "forgotten to take it up," or to be counseled that "this just isn't the right time to ask." It had been observed by everyone on the production staff that Gary often delivered different versions of a story to upper management than he gave to his subordinates. It was generally felt that he always sided with management, especially "Overkill," rather than backing his subordinates.

The general manager of the station, Gordon Frederick, was a retired military officer and an ex-mayor of the city. He was active in political causes and was out of town frequently, leaving the day-to-day operation of the station to the assistant general manager. Most of the staff members looked upon Frederick as being a slightly befuddled autocrat since he conducted regular "inspections" when in the building, and indulged a fetish for small detail, such as seeing that the flags were removed from the flagpole in front of the building promptly at sunset. He was responsible for, and for the most part the author of, a booklet of company rules and regulations called the Blue Book. In the Blue Book were voluminous descriptions of each job title within the organization, and page upon page of rules pertaining to coffee breaks, use of company telephones, and virtually every other activity within the building.

The Blue Book was treated with varying degrees of contempt by most staff members, and with utter contempt by the production department. Those who had been in the military service insisted parts of the Blue Book text were lifted wholesale from military manuals. It was felt that the book's only value was to management, in that some obscure regulation could be used to chastise an employee, while other rules were totally ignored. For example, the Blue Book stated that the company had a policy against members of the same family being employed. However, Overkill's son Steve worked as a full-time cameraman, one of the director's wives worked in the office, and the husband of the TV program director served as a technician.

The Blue Book also contained rules for communication between departments, the management feeling being that the rank and file of one department should not communicate directly with their counterparts in other departments in matters of operations. For example, if a newsman became upset at a cameraman,

director, or any other member of the production staff in connection with a newscast, he was to inform the news director, who would then take the matter up with the production supervisor. This rule was totally ignored.

Though the Blue Book delineated a very rigid chain of command, it was fairly common for orders to the production crew to come from "Overkill," the program director, or Brown, the production supervisor. On occasion, in the case of an equipment failure or similar emergency, these orders would conflict, resulting in confusion until the three decided upon a common plan.

Job security was felt to be nonexistent. Many of the workers felt directly threatened by "Overkill" and verbally expressed their fear of his capricious behavior.

Seemingly arbitrary changes of shift upset some of the men. In early Spring of 1990, one of the directors was moved to the position of video tape operator. Though his salary was left at its old level, this move involved a real loss of prestige. No explanation was given to members of the crew. A cameraman was promoted directly to the position of director, by-passing several switchers. Again, there was no explanation.

Sabotage, in the name of "games," became quite common among the operating crew. It was not too unusual for a film projector to be mis-threaded, causing the film to be torn to ribbons when the projector was started, resulting in program down-time. Program sets would occasionally topple over during a video taping session, or microphones would refuse to work. One favorite trick was the tripping of master light breakers for the control room areas. Another was pounding on the wall of an area where an announcer was on the air. One of the more ingenious acts involved the wiring of a prop telephone on the tv news set. The phone was then rung during a newscast, causing the newsman to "break up." Though members of management never appeared to suspect sabotage, its occurrence was by no means rare.

Also in the Spring of 1990, Ron E., the announcer, came to work for KCDE radio. The television and radio control areas were adjacent to one another, and some of the announcers worked both radio and television. There was a great deal of social contact between employees of both sides.

At the end of his first pay period, Ron became tremendously upset. His check totaled about $100 less for the two-week period than he thought it would be. According to Ron, the radio station manager had hired him at $1900 a month, but his first check was paid at the rate of $1700 a month. Ron promptly complained to his supervisor, and the matter was taken to the general manager. He informed Ron that the radio station manage did not have the authority to hire an announcer at such a salary as Ron had been promised. There was no offer to compromise on the salary. Frederick offered to pay Ron's moving expenses back to the city he had left just weeks before. Ron's answer was, "And what the hell am I supposed to do for a job if I do return?" Feeling he had no choice, Ron accepted the lower salary.

In May, about a month after the salary episode, Ron began questioning other employees about the possibility of unionizing the station. His idea was met with great enthusiasm by the members of the production department. More than one of them indicated that though they did not like unions, they liked the management of KCDE even less. The few holdouts expressed fear for their jobs, but no one expressed any pro-management thoughts.

Several meetings were held with union representatives and the union formally notified Mr. Frederick of their intention to organize the production department. This action was met with disbelief on the part of Frederick, followed soon by a meeting to stress to employees that "the door is always open, and you know we're interested in your problems." Union "horror" stories soon followed, accompanied by a frigid atmosphere and veiled threats by both sides. Rumor generation reached very high levels.

In early August, Ron E., was fired for "inattention to duties." He filed an unfair labor practices suit against the station management with the National Labor Relations Board. The filing of the suit served to freeze the unionization proceedings until the suit was resolved.

In the meantime, Frederick, Brown and "Overkill" turned to a well known management consulting firm for help in analysis of the organizational and personnel problems.

When you have read this case, look back at the chapter's diagnostic questions and choose the ones that apply to the case. Then use those questions with the ones that follow in your case analysis.

1. How does KCDE-TV's practice of paying low wages affect employees' equity perceptions? Does your answer to this question help explain the morale problems evident among the station's employees?

2. From the point of view of employee motivation, how have the station's managers made a bad situation even worse? In answering this question, review the actions of "Overkill," Gary Brown, and Gordon Frederick, and consider employee reactions of each of these people.

3. What does the station's treatment of Ron E. tell you about the way KCDE-TV manages its employees? If you were asked to join this company as a new general manager, what would you do first to improve employee motivation? What long-term actions would you initiate?

CASE 7-2

CONNORS FREIGHT LINES*

RICHARD PETERSON, *University of Washington*

Connors Freight Lines is a large, interstate trucking concern serving the north, central, and western states. Its head office is in Fargo, North Dakota, and it has forty-three terminals, with Chicago at one extreme and Los Angeles at the other. The La Crosse, Wisconsin, terminal has been in existence for sixty of the company's seventy-five years and enjoys a fair reputation competitively.

The technical organization of the La Crosse plant consists of a fleet of twenty-seven pickup trucks, ten town-tractors, two fork-lifts, and a cart line hookup track to facilitate loading. This branch is housed at a typical freight terminal which is superior to most other trucking firms, from the technical standpoint, although considered only "adequate" from the standpoint of its social organization. Since the plant is located well away from the business district, most drivers and dock workers bring their lunches, and a small lunchroom is provided for their convenience, where they can also buy coffee or milk. This lunchroom is furnished with long tables and benches, measures approximately 15' x 20' and will comfortably seat about twenty-four workers.

The formal organization of the company, which employs approximately 180 people, consists of the terminal manager, Ralph Preston, and his assistant terminal manager, Jason Hobbs. Exhibit 7-2 shows the company organizational structure.

Although the company appears successful enough in solving its external problems, the high rate of absenteeism, the generally low morale among the truckers, and their relatively short tenure of employment are puzzling internal problems that have vexed management for the past several years. The truck drivers at Connors are strongly union-oriented which results in feelings of mixed loyalties. To some extent, therefore, an undercurrent of conflict is felt in this area by the company as well as by the workers.

It is part of the company's policy to select its supervisors from the ranks of the drivers. This assures them of men who are experienced with the specific job and its problems. In theory, at least, this also rewards the employee with the advancement from a job at worker-level to a position within the organization. Upon becoming a supervisor, the employee is no longer a union member, so he works overtime without any pay

* Reprinted with the author's permission.

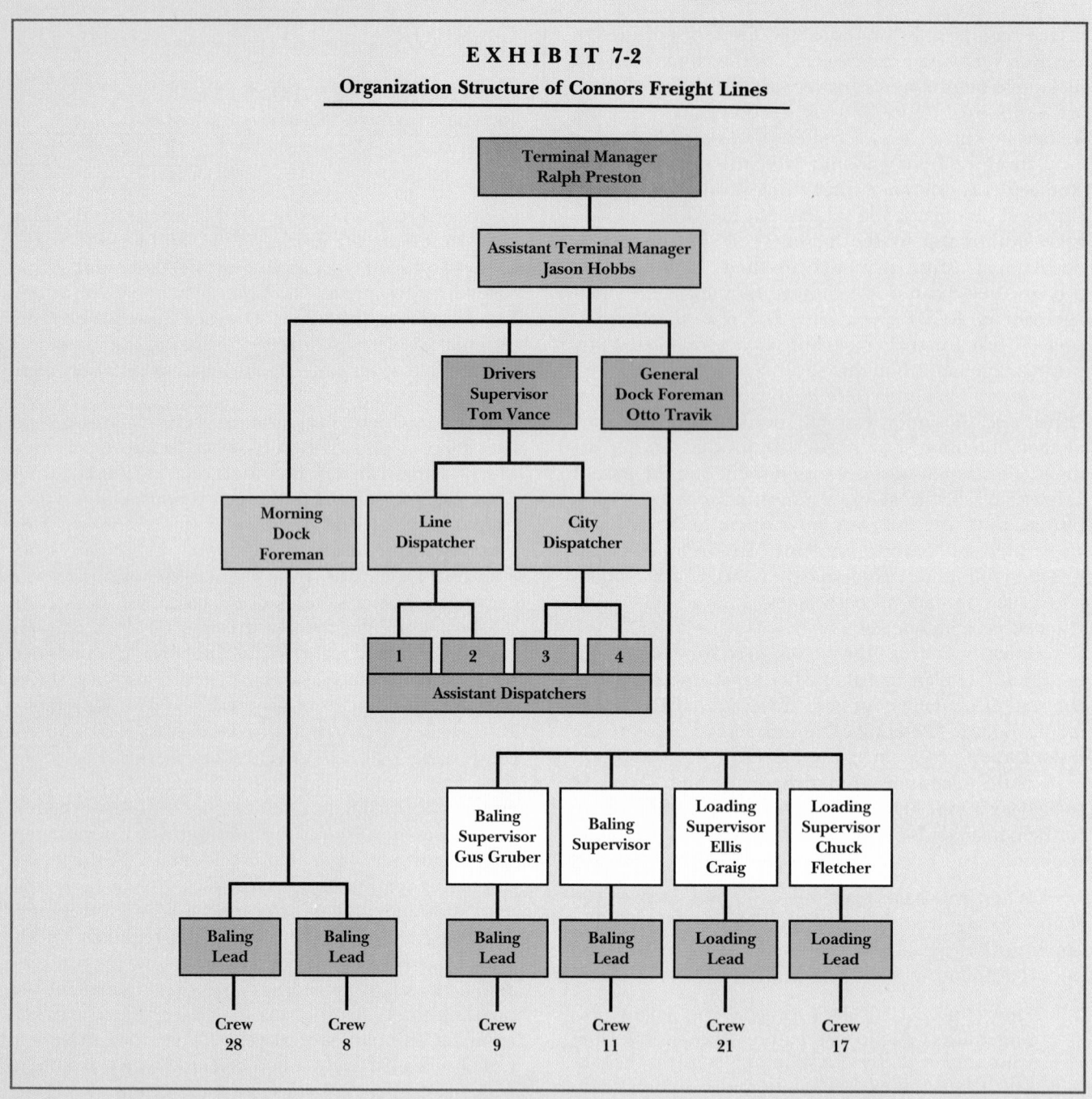

EXHIBIT 7-2

Organization Structure of Connors Freight Lines

Terminal Manager
Ralph Preston

Assistant Terminal Manager
Jason Hobbs

Drivers Supervisor Tom Vance

General Dock Foreman Otto Travik

Morning Dock Foreman

Line Dispatcher

City Dispatcher

1 2 3 4

Assistant Dispatchers

Baling Supervisor Gus Gruber

Baling Supervisor

Loading Supervisor Ellis Craig

Loading Supervisor Chuck Fletcher

Baling Lead

Baling Lead

Baling Lead

Baling Lead

Loading Lead

Loading Lead

Crew 28

Crew 8

Crew 9

Crew 11

Crew 21

Crew 17

and forfeits his seniority standing. The new supervisor is now salaried, with his base income slightly higher than that which he had earned as a driver on straight time. He no longer wears work clothes as his job is considered a "clean" one, and it would seem that some increase in prestige should also accompany this advancement.

Workers of supervisory caliber at Connors have not been too plentiful among the ranks of truck drivers, and the high rate of turnover aggravates this further. The drivers fall mainly into three categories (1) family

men in their 30s and 40s who are settled in their job and, because of various circumstances, have decided to make a career of it; (2) young fellows in their twenties who have a few years of college and must earn money in order to continue toward a degree; and (3) young fellows for whom it is simply "a job," and who in time will probably make up the bulk of the first group or drift on to other employment elsewhere.

Chuck Fletcher would belong in the second of these categories. At 25, he had three years of college

236

but was still uncertain about his field of interest so decided to work a while, keep his eyes open, and do some thinking. Son of parents who were both college graduates, he was well above average in intelligence. Other truckers on his shift both liked and respected him, as did the members of management that had come to know him. He had worked for Connors for two years, spending half his shift on the dock, loading, and the other half as a driver.

Because of the nature of the freighting business, there is a great need for overtime help, since prompt movement of freight is a large part of the "product" which the company sells. Chuck maintained a good attitude about this added work load which was lightened somewhat by his paychecks, reflecting the regular union demand for overtime pay. Not all workers showed as good an attitude toward this overtime work even with the added pay, and several among the top 5 percent in seniority who were given a choice as to overtime work, flatly refused any work other than the regular 8-hour shift, regardless of compensation.

There is a fairly intricate technique involved in the correct loading of trucks and trailers so that the weight is kept under the maximum allowed in highway regulations and is distributed evenly and correctly. The more fragile or perishable items must be given special attention and the merchandise loaded with logic in respect to the order of its being unloaded at the delivery point. Chuck caught on quickly to these loading techniques and soon attracted the attention of the dock foreman, Otto Travik. In a conversation with Chuck's loading supervisor, Ellis Craig, Travik suggested that Craig "keep his eye on" Chuck, with the idea of possibly bringing him into management in the future when an opening might occur within the organization. Craig, who was aware of Chuck's ability to get along well with other workers and knew him as a hard worker, agreed that he was "worth watching."

In January 1991 at one of the regular Wednesday morning staff meetings, Hobbs, the assistant terminal manager, mentioned that a sales job would soon open up, and that they planned to move Al Johnson into this spot. Johnson's move to Sales would then leave a supervisor's job open on the loading dock. Preston, the terminal manager, asked for suggestions as to who might best fill this spot. Thomas Vance, the drivers' supervisor, suggested Ford Wheeler, who had been working on day shift, was thirty-five years old, and a former school teacher. Preston agreed that Wheeler was a good prospect, but Travik and Craig suggested they consider Chuck Fletcher for the position. Qualifications of both men when then compared and discussed and the decision left open, pending more thought on the two candidates, and the actual job opening.

At about this time, Craig said to Fletcher one day:

"I know you like the shift you're on, but would you consider cancelling your bid on it and taking the St. Paul loading job that's driving me crazy? I know the late night shift is a crummy one, but this would be the chance you've wanted to learn to drive the heavy-duty trucks. Vance is our official qualifier, and he's on that shift, so I'm sure by coming to work early you could qualify inside of a month. I'll help expedite the whole thing if you'll help me out on this. After all, the annual bids come up in five months, so if you don't like the job, you can always re-bid. And keep this under your hat—there's going to be a supervisor's job open before too long, and if you make good on this job I think your chances for it would be excellent."

Chuck thought about the St. Paul loading job. He didn't like the night shift, and the job was an especially "dirty" one, but he took it for two reasons: he had been hoping for a chance at supervision, and he also had wanted time to learn to drive the heavy-duty rigs both for the increased pay diesel drivers drew and also because it offered him a challenge.

At a staff meeting in June, 1991, after other articles of business were out of the way, the conversation ran something like this:

Preston: There is now a definite need for a new supervisor, and I hope you have all been keeping this situation in mind and giving it earnest thought. What are your suggestions?

Vance: I still think Ford Wheeler is your boy.

Preston: Yes, I've been seriously considering him. Checked into his background from his job application and I was really quite impressed. Do the rest of the workers respect him? Would they work for him?

Craig: I don't feel he has any of the workers' respect! For one thing he's lazy, and we'll be setting the poor example that all you have to do to get ahead in the organization is do nothing. On the other hand, there's Fletcher who really is liked by the other workers, and respected too. In the time he has worked for me he has shown himself a hard worker, often doing more than is required. For a couple of months he's been coming to work an hour early to practice driving heavy-duty trucks on his own time and hoping to get qualified—and by the way, Vance, he's been ready for his test for some time—he shows a great desire to do a thorough job and accepts more than his share of responsibility. To me, this adds up to material for a good company man.

Vance: Hell! He's a union boy straight down the line. Don't you remember how he initiated two valid grievances against us last year? There has even been some talk of his being named shop steward. But Wheeler, now, has a college degree as a school teacher. He can hardly wait to shake loose from the union.

Preston remembered seeing Fletcher many times, and in all kinds of weather, out practicing driving the diesel rig in the loading zones before his shift time. In checking over Fletcher's job application he found it almost as good as Wheeler's, considering the differences in their ages.

After some further discussion Preston decided in favor of Fletcher. The supervisory position was offered to him on the usual 90-day provision which would allow either him or the company to terminate the arrangement at the end of that time, with no loss of status or seniority on his part. After a few days of consideration, Fletcher accepted the new assignment and went to work as a supervisor of the same crew with which he had been working previously.

Preston gave Fletcher a short "welcoming" talk before his first day in the new job, and in it he covered three main points. He first suggested that Fletcher try to be tactful in his initiation of ideas, maybe even to the point of making his cooperating supervisors sometimes think that the idea came from them. Secondly, he stressed the importance of demanding respect from his crew. "If the occasion demands a reprimand, see that it's done in some private place where others can't see or hear you. Otherwise you both lose face. It must be kept private between you and your man." That led to the third point, that of not being too familiar with the workers on your crew. "It's not really any of our business what you do off the job, but it'd be best if you don't hobnob with your men," Preston had said.

Soon afterwards, at Craig's invitation, Fletcher joined the bowling team of which both Preston and Vance were members. He and Craig, with whom he had always had a good relationship, were becoming close friends since he had been made supervisor. They'd usually go somewhere for a couple of beers together on the nights after bowling.

One night Fletcher was off on his game. After some ribbing from Vance and Preston, he mentioned to Craig how tired he was from so much overtime, "tired and disgusted" was the way he put it. He told Craig that earlier that day he had overheard Preston "chewing out old Gus Gruber" down in the steel bay, and the incident had both embarrassed and disappointed Fletcher. Craig, at that time, was planning a vacation to the East coast with his family and invited Fletcher to come with them. Chuck said he knew he needed a rest and a change

of scenery, but he was short of funds. Last year he had gone up to Canada with some of the men from his crew, but that was when he was getting paid for overtime.

Craig previously had included Fletcher on a couple of family outings and at one time, when the men were alone, they got to talking shop as usual. Fletcher seemed discouraged about the attitude of his crew. "We have always gotten along so well before, but now there's no more kidding. No one hardly cracks a smile and I keep getting jibes like 'Gettin' rich on your overtime, Chuck?' or 'How do you like your new raise by now?'

Part of his interest in taking the supervising job was that he hoped he'd find ways of easing the obvious friction between management and labor, and this development was really discouraging.

The weekly staff meetings, which were attended by all management personnel, were the occasions when affairs concerning the company's technical and social organization and welfare were brought up for discussion. A variety of issues were constantly being introduced, listened to, considered and many were acted upon promptly. At the third staff meeting Chuck himself made the suggestion that the company might see fit to supply work aprons to the workers, adding that these were not expensive items when bought in quantity, and that the gesture of goodwill on the part of the company might help to improve the rather poor relations between it and the workers. This suggestion was listened to but not acted upon.

Another suggestion was made by Fletcher within the next few weeks: that of switching of lunchrooms between the workers and supervisors. The lunchroom reserved for the twelve supervisors was twice the size of the one into which forty-five drivers were crowded at peak times to have their lunches, and this necessitated several of them sitting on the floor or standing to eat. This suggestion, too, was not acted upon, although Craig and Travik both thought it was "worth considering."

Within a short space of time, however, the company was gratified by the outcome of two other suggestions which Fletcher made in work operations. One of these resulted in substantial savings to the company, and the second, in a greater profit. The first was accomplished by Fletcher's simplifying of a certain loading procedure and organizing it in such a way that two men from another crew, ordinarily required for extra help at this time, were no longer needed. Dispensing with this usual short-handing of the second crew enabled it to finish its own loading on schedule, resulting in more efficient moving of freight on the two jobs and less need for overtime on both.

At about this time, two company officials from Fargo were visiting the terminal and Preston took them on a tour of the docks. As they passed the loading area,

Fletcher looked up and nodded. Farther on, as they circled the dock, one of the officials said, "Where is this new supervisor, Fletcher, that you've mentioned a few times lately?"

Preston replied, "Oh, he's the fellow in the red shirt we just passed down there at the north end. I'll introduce him to you at coffee break if he comes up."

The second instance in which Fletcher's suggestion worked to the company's advantage took place toward the end of the ninety-day trial period. It was not presented at the staff meeting but to Hobbs in person one day when he happened to be out on the dock as Fletcher came on shift. Fletcher's attitude, which had always been friendly and respectful, seemed to have changed in recent weeks, and Hobbs, who had become confident that Fletcher was proving to be good supervisory material, was a little puzzled by his apparently growing coolness.

Hobbs: Good morning, Fletcher.

Fletcher: Good morning, Mr. Hobbs. May I speak to you a minute when you have the time?

Hobbs: Sure. Right now is as good a time as any.

Fletcher: It's about the St. Paul run. While I was on the St. Paul-loading job I realized that on the last schedule to St. Paul at night there was always more freight moving from La Crosse to St. Paul than returned from St. Paul to La Crosse. We'd always need a 'double' heading east and then always have to haul back one 'empty.'

A college friend of mine is the son of a truck farmer over east of the Mississippi, and I was talking to him the other day. I think we could arrange for a full load of produce to be picked up near Winona. Then we'd have a full paying load in both directions.

I've already said something to Vance about it yesterday but didn't hear anything from him so thought I'd let you know. The arrangements could be made easily. I'll give you the farmer's name, if you're interested. It's all up to you, of course.

Hobbs thought Fletcher seemed a bit abrupt, but the suggestion pleased and impressed him as it did Preston when he was told about it later in the day.

On Thursday of the last week of the ninety-day trial period, Fletcher came into the office of the assistant manager and told Hobbs that he was going to exercise his option and ask to be returned to worker status the following Monday.

Hobbs was surprised by this unexpected turn of events, but said, "Well, we'll live up to our side of the bargain. I'm sure Mr. Preston will want to hear your reasons for this decision, though. He has some free time at two o'clock. Could you come in and see him then?"

In Preston's office later:

Preston: What's this I hear about your request to be returned to your old job as trucker?

Fletcher: Yes, I've decided to hang it up. The long hours are getting me down.

Preston: Well, you knew what the hours were like when you took the job. Aren't they the same ones you worked as a driver?

Fletcher: Yes. Well, I thought perhaps I could expedite a few things and maybe shorten them a little, I guess.

Preston: Well, are you sure you gave this "expediting" your best efforts?

Fletcher: Yes, sir I did. Until I got to the point where I felt I was knocking my head against a stone wall.

Preston: Are you sure it's the long hours, or are there some other reasons?

Fletcher: Well, the money and the shift, coupled with the long hours.

Preston: Well, we think you were doing a fine job, and we are already taking steps in these areas you're dissatisfied with. In two weeks we hope to add one new supervisor and possibly two to cut down hours worked by the supervisors.

Fletcher: Well, that would make it a lot easier all right . . .

Preston: And there will be an opening in Dispatch in the not-too-distant future that would be a day-shift job rather than a night one.

Fletcher: That would be a real improvement . . .

Preston: So that just leaves the problem of money. How much more would you say you'd need to make it worth your while to stay on?

Fletcher: A hundred dollars a month, sir.

Preston: Hmm, well, I think we can probably make some kind of arrangements that will make you more contented. I'll check with the head office and let you know on Monday.

Fletcher: All right. Fine. You can let me know.

After Fletcher left his office, Preston said to Hobbs, "I think the head office will go for an eighty-five dollar raise anyway, and that will probably be enough to hold Fletcher here. Put a note in his locker-box stating we will meet his demands, but don't state a definite amount of money. I have to talk to the head office first."

Hobbs: That's still a lot of money. Do you realize that would make him the sixth highest paid man in this terminal? He'll be jumping over eight men.

Preston: Fletcher has already saved the company more than he'll make in a year on that produce haul alone, and he has come up with quite a number of ideas that have helped us out. He's a thinking boy and that's the kind we need. He has a bright future and I want to keep him with us.

The Monday morning mail contained the following note addressed to Preston:

Dear Mr. Preston:

Please accept this notice of my resignation from the company, to be effective in two weeks.

I have thought it over carefully and this is the only possible solution.

Thank you for your generous offer, even though I cannot accept it.

Sincerely,
Charles S. Fletcher

When you have read this case, look back at the chapter's diagnostic questions and choose the ones that apply to the case. Then use those questions with the ones that follow in your case analysis.

1. What motivated Chuck Fletcher when he was doing his initial job at Connors Freight Lines, on Otto Travik's loading dock? What motivated him to perform so well on the St. Paul loading job? What changed when Chuck moved into the supervisory job?

2. Explain Chuck's letter of resignation. Why did he decide to quit instead of continuing in his position as supervisor? Is there anything the company could do to get him to change his mind? If there is, can the company take such action and should it?

3. What should Connors do to ensure that the kinds of problems described in this case do not happen again in the future?

CASE 7-3

CHANCELLOR STATE UNIVERSITY

Read Chapter 9's Case 9-2, "Chancellor State University." Next, look back at Chapter 7's diagnostic questions and choose the ones that apply to that case. Then use those questions with the ones that follow in this first analysis of the case.

1. What effect is the salary compression problem at Chancellor State University likely to have on the motivation of faculty members?

2. University teaching is a profession in which it is fairly easy to move from one institution to another. Which faculty members would be the most likely to become so dissatisfied with the situation at Chancellor State as to leave? Which would be least likely to leave? How might the university's salary compression problem and its effect on faculty members affect students?

3. What can Fred Kennedy do to cope with his department's morale problems? What rewards other than salary could he use to motivate his faculty's performance? How should such rewards be administered?

CHAPTER 8

SATISFACTION AND STRESS

Despite its ongoing battle with its Seattle employees' union and a recent loss in earnings, Nordstrom made ambitious plans for the five-year period beginning in 1990. This one-time family shoe store with a legendary reputation for service expects to add 20 new stores in 13 states with the goal of 80 stores and sales of $5 billion by 1995. The retailer's owners, all members of the Nordstrom family, say they aren't worried about complaints from disgruntled employees or union threats to picket new stores. They are concerned, however, about the chain's earnings. Although Nordstrom carries less debt than many of its competitors, management says it will cut back the number of planned openings if earnings don't improve.
Source: Dori Jones Yang and Laurie Zinn, "Will 'The Nordstrom Way' Travel Well?" Business Week, September 3, 1990, pp. 82–83.

She had worked weeks without a day off and pulled 15-hour shifts without a break. She had stockpiled service awards and customer thank-you notes. Yet she thought she was going to be fired, and she wasn't alone. In fact, according to employment counselor Alice Snyder, she was "the fourth person from that store that I've seen this week." "That store" was Nordstrom, the Seattle-based clothing store, and the strung-out workers who marched through Snyder's office suffering from ulcers, colitis, hives, and hand tremors were among a growing army of "Nervous Nordies." Worn down by the relentless pressure associated with working at Nordstrom, workers "were dropping like flies," according to Patty Bemis, a one-time top salesperson.

According to disgruntled employees, the sources of stress at Nordstrom were many. There were constant threats of job loss for failing to meet quotas. For example, the manager of a cosmetics department told employees that if the goals she had set for them were not met within 60 days, they would be fired. Moreover, the numbers of hours that employees were expected to put in created problems away from the job. "It reminds me of a cult the way they program you to devote your life to Nordstrom," noted Cheri Validi, a four-year salesperson in women's clothing. Some employees felt deprived by their work requirements of basic needs like sleep and nutrition. Lori Lucas, who developed an ulcer while working for Nordstrom said, "I'd be up til 3 A.M. doing my letters and doing my manager's books. Before you know it, your whole life is Nordstrom. But you couldn't complain because . . . the next thing you know, you're out the door."

Pressure at Nordstrom reportedly extended into such areas as dress and demeanor. Becoming a Nordie involved getting "a certain look," and managers were said to have strong-armed new salesclerks into purchasing expensive new wardrobes—all from Nordstrom. Management was also accused of using "Secret Shoppers"—people dressed to look like customers—to determine whether salespersons were smiling enough.

According to employees, the atmosphere at Nordstrom created a cut-throat mentality. Some salespeople engaged in "sharking" activities. They might hog the cash register, taking all the "walk-up" business for themselves. Or they would cut deals with temporary, non-commission salespeople, getting the latter to generate false sales records. Some even went so far as to use rival employees' ID numbers when customers returned merchandise.

In early 1990, workers' complaints began to have severe negative consequences for Nordstrom. First, tips from disgruntled workers led to an investigation by the Washington State Department of Labor and Industry. The investigation found that the firm had systematically violated state law by failing to pay employees for a variety of off-the-clock duties. The agency ordered Nordstrom to pay back wages—as much as $30 to $40 million—or face legal action.

Lawsuits brought by individuals also cost the firm large sums of money. Cindy Nelson was fired reportedly because five of her coworkers accused her of sharking. She sued for "wrongful discharge" and "failure to receive due process," and a King County Superior Court jury made Nordstrom offer her a $180,000 award for damages.

All of these developments led to a deterioration in the firm's image, and Nordstrom stock tumbled $3.25 (from roughly $33.00 to $29.75) in one week alone. Profits fell too. In fact, for 1989, the firm reported a decline in profits for the first time since it went public in 1971.[1]

---

[1] S. C. Faludi, "At Nordstrom Stores, Service Comes First— But At a Price," *Wall Street Journal*, February 20, 1990, p. 1.

Because most organizations are not in the job-satisfaction business, it is sometimes difficult to make employers see the importance of recognizing workers' attitudes and feelings about their work. Establishing this recognition is made more difficult by research that fails to support a clear relationship between job satisfaction and job performance.[2] Working in organizations has several important effects on organizational members that cannot be overlooked—even by those interested only in financial profits. Some managers who would not think of failing to maintain their physical equipment and resources give little thought to their human resources, and the consequences can be disastrous.

This chapter focuses on the attitudes and emotions that people experience in the workplace. We will start by defining job satisfaction and job stress, and we will then show how cognitive appraisal theory ties these two constructs together. Then, to convince you of the importance of attitudes and emotions at work, we will examine the consequences of dissatisfaction and stress. We will examine the cost of such problems both in human terms and in terms of dollars and cents. Next, we will review the major sources of dissatisfaction and stress in work environments. Finally, we will discuss methods and techniques to eliminate dissatisfaction and stress or to help employees cope with these unpleasant phenomena.

DIAGNOSTIC ISSUES

As in preceding chapters, the diagnostic model calls attention to a host of important issues in the area of job satisfaction and job stress. Can we *describe* the kinds of situations in which people experience too much stress? Are there common symptoms that tell us when people may be having problems with stress? How can we *diagnose* why some people seem quite satisfied with their work while others seem dissatisfied? What explains the fact that some workers are generally satisfied but others aren't, no matter what kinds of jobs they hold? How can we explain why some jobs that seem quite simple to perform actually cause high levels of dissatisfaction and stress?

Can we *prescribe* which individuals should be assigned to jobs that are stress-inducing and which individuals should not? Are there measures of job satisfaction that can help predict rates of absenteeism and turnover? Finally, what *actions* can managers take to eliminate dissatisfaction and stress? If they can't remove all the stressful aspects of a job, what can they do to help people cope with stress? What can employees do for themselves, when away from work to cope with work-related stress?

DEFINING SATISFACTION AND STRESS

Satisfaction and stress are closely related but separate concepts. In this chapter, we will show what they have in common as well as what is unique to each.

Satisfaction

job satisfaction The perception that one's job enables one to fulfill important job values.

Job satisfaction is a pleasurable feeling that "results from the perception that one's job fulfills or allows for the fulfillment of one's important job values."[3] There

[2] M. T. Iaffaldono and P. M. Muchinsky, "Job Satisfaction and Job Performance: A Meta-Analysis," *Psychological Bulletin* 97 (1985), 251–73.

[3] E. A. Locke, "The Nature and Causes of Job Dissatisfaction," in *Handbook of Industrial/Organizational Psychology*, ed. M. D. Dunnette (Chicago: Rand McNally, 1976), pp. 901–69.

is a considerable amount of research on the topic. Reviewing this literature in 1976, Edwin Locke estimated that there were over 3,000 articles or studies dealing with job satisfaction. More recent reviews show that this interest is still alive and well.[4]

Key Components. There are three key components of our definition of job satisfaction: values, importance of values, and perception. First, job satisfaction is a function of values. In his 1976 review, Locke defined *values* as "what a person consciously or unconsciously desires to obtain."[5]

Locke, however, distinguished between values and needs. Needs, he said, are best thought of as "objective requirements" of the body that are essential for maintaining life, such as the needs for oxygen and for water. Values, on the other hand, are "subjective requirements" that exist in the person's mind. Needs are innate, and all people have the same needs. People learn values, however, and thus their values differ widely. As you can see, Locke's values include many of the higher-order needs we discussed in Chapter 7, such as the need for self-actualization.

The second important component of our job satisfaction is *importance*. People differ not only in the values they hold but in the importance they place on those values, and these differences are critical in determining the degree of their job satisfaction. One person may value job security above all else. Another may be most concerned with the opportunity to travel. Yet another person may be most interested in doing work that is fun or that helps others.

The last important component of our definition of job satisfaction is *perception*. Satisfaction is based on our perception of the present situation and our values. That is, will the job help me obtain what I want? Remember that perceptions may not be completely accurate reflections of objective reality. When they are not, we must look at the person's perception of the situation—not the actual situation—to understand her reactions.

Thus the three components of values, importance, and perception help us define job satisfaction. A person will be satisfied with a job when her *perception* of what the job offers exceeds her values, and the more *important* those values are to her the more intense her satisfaction will be.

Conceptualizing job satisfaction in terms of the value-perception-importance triad reveals the many different ways that people can become satisfied or dissatisfied with work. Table 8-1 shows three sets of examples. In the first example, Bill and Susan, both of whom work on the same job and perceive the job in the same way, wind up having different levels of satisfaction. That's because they have different values and place importance on different things. The second example pairs Bill with Sherry, who works on a different job. Although both these people share the same values and importance ratings, their jobs make their perceptions different and thus they have different levels of satisfaction. Finally, in the third example, we compare two people who differ on all three dimensions but are equally satisfied in their work.

Measuring Job Satisfaction. Most attempts to measure workers' satisfaction levels rely on self-reports. Some measures, like the Job Descriptive Index (JDI), emphasize aspects of work, such as pay, work itself, supervision, coworkers. Other measures, like the Faces Scale, emphasize overall satisfaction. Table 8-2 shows several items from these two measures.

[4] M. T. Iaffaldono and P. M. Muchinsky, "Job Satisfaction and Job Performance."
[5] Locke, "Job Dissatisfaction."

TABLE 8-1
Hypothetical Examples of Varying Levels of Satisfaction

EXAMPLE 1	PERCEPTION	VALUE	IMPORTANCE	SATISFACTION
Bill	Job provides opportunity to travel.	Likes to travel.	Considered important.	Very satisfied.
Susan	Job provides opportunity to travel.	Dislikes travel.	Considered unimportant.	Not satisfied.
EXAMPLE 2				
Bill	Job provides opportunity to travel.	Likes to travel.	Considered important.	Very satisfied.
Sherry	Job does not allow one to travel.	Likes to travel.	Considered important.	Not satisfied.
EXAMPLE 3				
Bill	Job provides opportunity to travel.	Likes to travel.	Considered important.	Very satisfied.
John	Job provides high job security.	Likes high job security.	Considered important.	Very satisfied.

The JDI and the Faces Scale are useful measures for the manager who wants to assess the satisfaction levels of all his employees. They are easy to use across the board because the JDI requires minimal reading skills and the Faces Scale requires none. The tests' reliability and validity have been supported by many studies.[6] Because there is a wealth of data on the use of these measures, it is easy to compare their results in one firm with their results at another.

Stress

stress An unpleasurable emotional state resulting from the perception that a situational demand exceeds one's capacity and that it is very important to meet the demand.

The term stress has been used widely and with varying meanings over the years. Indeed, researchers and practitioners alike have lamented the imprecision with which this term is used.[7] We will adopt the definition provided by Joseph McGrath, a leading researcher in this area, who defined **stress** as an unpleasurable emotional state resulting from the perception that a situational demand one feels it is important to meet exceeds one's capacity. As in the case of satisfaction, we will find it easier to understand the nature of stress if we break this definition into three key components. We'll look first at perception of the demand, then at importance, and finally at perception of one's capacity.

The first component, *perception of the demand*, emphasizes that stress involves the interaction between the person and his environment. It is the person's perception that something is happening "out there," not objective reality, that creates

[6] S. M. Johnson, P. C. Smith and C. M. Tucker, "Response Format of the Job Descriptive Index: Assessment of Reliability and Validity by the Multitrait-Multimethod Matrix," *Journal of Applied Psychology* 67 (1982), 500–505.

[7] R. L. Kahn, "Some Propositions toward a Researchable Conceptualization of Stress," in *Social and Psychological Factors in Stress*, ed. J. E. McGrath (New York: Holt, Rinehart & Winston, 1970), pp. 19–37.

the potential for stress. For example, unfounded rumors about
will create stress, even though no real threat exists. On the othe
agement actually is planning to close a factory but keeps its mee
its plan secret, workers will experience no stress.

The second component, *importance*, is critical for the same rea
to our definition of satisfaction. Unless a demand threatens some in
it will not cause stress. The rumored plant closing may not crea
worker who is about to retire in two weeks. Interestingly, a deman
perceived as negative as to create stress. Stress can also be associated
that have positive consequences (i.e., opportunities). This kind of s
times referred to as **eustress**. A manager notified that she is bein
for a promotion is likely to feel more stressed than she did before
of the opportunity. Even if she knows that at worst, she will be rig
is now, the possibility of failing to secure a potential gain can invol

Finally, the third component, *perception of one's capacity to mee*
highlights the notion that the person must interpret the demand in
perceived ability to handle it. Clearly, if a person perceives that he can
with a demand, he feels no stress. But what if the person is overwhel
demand, seeing no possible chance that he can meet it? If we assur
stress will be very high we will be wrong. Research shows that stress
highest when the perceived difficulty of the demand closely matches th
perceived capacity to meet the demand. The reason is that as the diffi
and the ability level get closer and closer, the outcome becomes in
uncertain. It is this uncertainty that creates the stress, not the fear of a
outcome.

eustress A particular kind of
stress created when an individual
is confronted with an opportunity.

cognitive appraisal theory and show how it distinguishes satisfaction from stress and relates these concepts to each other. We will also show how the theory relates these emotions to physiological, cognitive, and behavioral responses.

Overview of Cognitive Appraisal Theory. In Chapter 5, we saw how important perception is in determining peoples' understanding of objective reality. As we have seen in this chapter, perceptions are key components in our definitions of satisfaction and stress. Two leading psychologists, Arnold and Lazarus have emphasized the role of perception in the formation of emotional reactions.[9] The cognitive appraisal theory of satisfaction and stress summarized in Figure 8-1 is based on the work of these two researchers. As we will see, cognitive appraisal theory emphasizes the interplay between cognitive, emotional, behavioral, and physiological responses. It highlights emotions like satisfaction and stress as complex, mind-body interactions.

The first step shown in Figure 8-1 is *perception*. The person perceives a demand in the external environment. In the next step, of **primary appraisal**, the person judges whether the demand is good or bad, an opportunity or a threat. A positive appraisal will lead to *satisfaction*, and processing will stop. But a negative appraisal will lead to *dissatisfaction*, an unpleasant state that the person wants to escape. He will then make a **secondary appraisal**, in which he assesses his *ability to cope* with the environmental demand. If he judges himself able to cope and succeeds in doing so, the appraisal process stops. If the person is uncertain about

[8] J. E. McGrath, "Stress and Behavior in Organizations," in *Handbook of Industrial/Organizational Psychology*, pp. 1315–73.

[9] M. B. Arnold, *Emotion and Personality* (New York: Columbia University Press, 1960); and R. S. Lazarus, "Emotions and Adaptation: Conceptual and Empirical Relations," in *Nebraska Symposium on Motivation*, ed. W. J. Arnold (Lincoln: University of Nebraska Press, 1968).

...aisal   In cognitive
...ry, the first stage in
...essment of the
... in which she judges
... object in the
... is good or bad,
... harmful, an
... or a threat.

...ppraisal   In cognitive
...eory, the second stage
...'s assessment of the
...t in which he judges
... to cope with perceived
... opportunities in the
...nt.

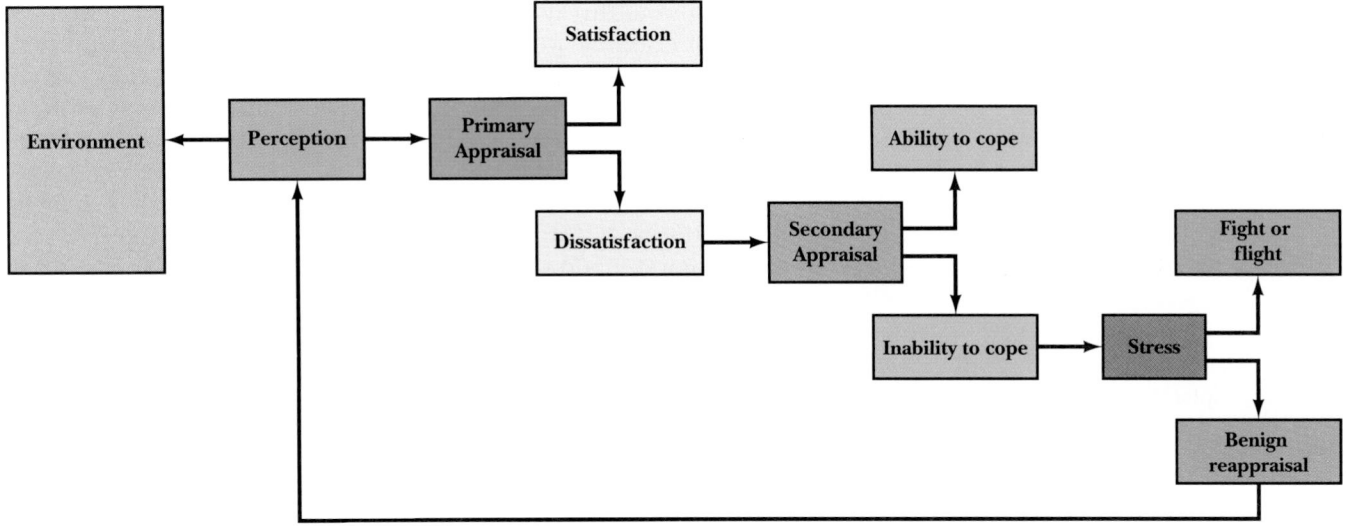

FIGURE 8-1

Satisfaction and Stress in the Cognitive Appraisal Theory of Emotion

fight or flight response A response to stress in which a person confronts and overcomes a stressful demand or escapes it by leaving the scene.

benign reappraisal A response to stress in which the person reassesses an apparently threatening environmental demand and modifies his original perception of it.

AN EXAMPLE
Two people faced with the same stimulus may have quite different emotional responses. For example, two men have just been told their wives are pregnant. One is thrilled and the other frightened. The first individual places a high value on children and views having a child as important. He also feels he has the capacity to be a good father. The second man may not value children, doesn't want children, or may not feel he has the capacity to be a good father.

his *ability to cope*, however, he will experience *stress*. Stress is even more unpleasant than dissatisfaction, and to get rid of it the person will try to eliminate the environmental demand. In primarily behavioral actions, he can **fight**, or remove the stressor (e.g., a manager can fire an insubordinate employee), or he can take **flight**, simply leaving the scene (e.g., a person can walk away from someone who is annoying her). Alternatively, in **benign appraisal**, a person can use cognitive means to change his goals or values so as to adjust his perception of the environmental demand.

All strong emotions are accompanied by some sort of physiological response. A student who's just been admitted to her first-choice medical school may feel her heart racing as she sprints to the phone to call her family. In general, however, it is the negative emotions that are accompanied by the most intense physiological reactions. As we will see, it is particularly important to understand the relations between such reactions and our experience of stress. In our model, this experience reflects a more complex judgment of an environmental demand than do either satisfaction or dissatisfaction. When we feel stressed we are judging not only that a demand is threatening but that we can't deal with it. Our emotional response is also more complex, typically including such unpleasant feelings as anger and fear. In the next three sections, we will explore the links between emotions and physiological response, emotions and cognitive-behavioral responses, and the increasingly common syndrome of job burnout.

Emotions and Physiological Responses: The General Adaptation Syndrome. As Darwin's theory of evolution (see Chapter 4) made clear, there are many similarities between human beings and lower animals. You may not be surprised to learn that at the biological level, the physiological changes that occur in a deer when it realizes a mountain lion is closing in are not very different from the changes that occur in a human being who is similarly threatened.

Indeed, the body's physiological reaction to any threat is an adaptive process that probably once had great survival value. When threatened, the body produces chemicals that cause the blood pressure to rise and that divert the blood from the skin and digestive organs to the muscles. Blood fats are then released to provide a burst of energy and to enhance blood clotting in case of injury. These changes are adaptive in the sense that they ready the person either to physically fight or to flee some threat.[10] Unfortunately, these same physiological changes

[10] D. H. Funkenstein, "The Physiology of Fear and Anger," *Scientific American* 192 (1955), 74–80.

occur today in response to threats regardless of whether increased physical capacity is adaptive. For example, workers who hold jobs with many demands over which they have little control (e.g., middle managers) are three times more likely to suffer from high blood pressure than other workers. The increased physical capacity purchased with high blood pressure is not going to help these workers cope with the demands they face.[11]

When the threat facing the individual is prolonged over time, other changes begin that prepare the body for a long battle. The body begins to conserve resources by retaining water and salts. Extra gastric acid is produced to increase the efficiency of digestion in the absence of blood (which has been diverted away from internal organs). The body begins to act as if it is under seige.[12]

Hans Selye, a prominent physician and researcher, developed the theory of the **general adaptation syndrome**, which has been useful in explicating the relationship between stress and physical-physiological or psychosomatic symptoms. According to Selye, the body's reaction to prolonged stress occurs in three stages: alarm. resistance, and exhaustion (see Figure 8-2).

In the first, *alarm stage*, the person identifies the threat. Whether this threat is physical (e.g., a threat of bodily injury) or psychological (e.g., a threat to self-esteem), the same physiological changes ensue. You have probably noticed that people suffering from diverse illnesses often share such common symptoms as headache, fatigue, aching muscles, and loss of appetite.

In the *resistance stage* the organism seems to become resilient to the pressures created by the original stressor. All the symptoms that occurred in the alarm stage disappear, even though the stressor itself is still in place. This resistance seems to be accomplished through increased levels of hormones secreted by the pituitary gland and the adrenal cortex.

If exposure to the threatening stressor continues, the person reaches the *exhaustion stage*, when he can no longer maintain resistance. Pituitary gland and adrenal cortex activity slows down and the person can no longer adapt to the continuing stress. Many of the physiological symptoms that originally appeared

general adaptation syndrome
The theory developed by Hans Selye that the body's response to stress occurs in three distinct stages: alarm, resistance and exhaustion.

[11] R. Winslow, "Study Uncovers New Evidence Linking Strain on the Job and High Blood Pressure," *Wall Street Journal*, April 11, 1990, p. B18.

[12] D. Foley, "How to Avoid 'the Perfect Day for a Heart Attack,' " *Prevention*, Sept. 6, 1986, pp. 54–58.

FIGURE 8-2

The General Adaptation Syndrome

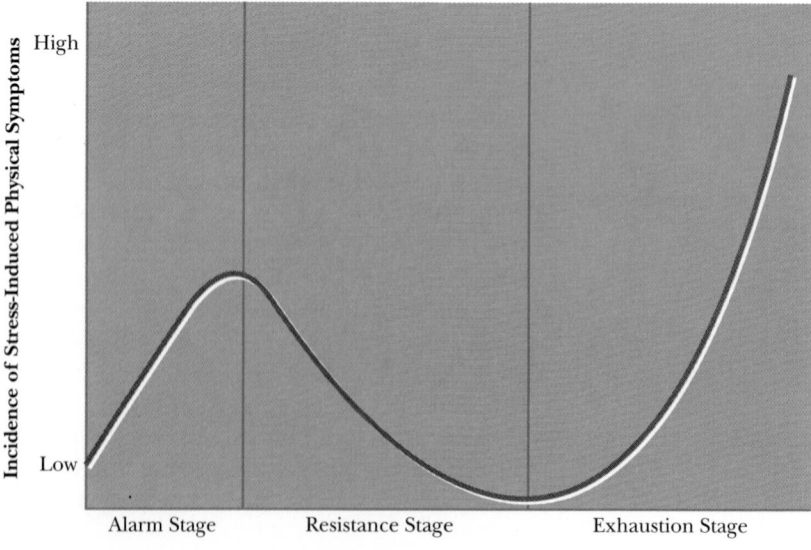

SATISFACTION AND STRESS

Now manager of a franchise owned by ProForma Business Products, Keith Beck left his position as manager of executive placement and development at Mervyn's, a nationwide department store chain because he had "mentally resigned from the company." Burnout in Beck's case came after he was passed over for a promotion and then had to help lay off hundreds of employees when the company, facing intense retailing competition, closed a regional base in Dallas and then fired 600 employees from the company's California headquarters.
Source: "Executive Life," Fortune, December 17, 1990, pp. 52–56.

job burnout A condition of emotional, physical, and mental exhaustion resulting from prolonged exposure to intense, job-related stress.

in the alarm stage now reappear. The stage of exhaustion is rarely reached, and in fact, stress continued beyond the resistance stage can lead to severe physical damage, even to death.

Emotions and Behavioral-Cognitive Responses. Let's look again at the final stage of our cognitive appraisal theory of satisfaction and stress, where the person confronts two basic alternatives. In a primarily behavioral response, she can fight or she can flee from the stressor. Or, in a primarily cognitive response, she can reappraise the situation and modify her original perceptions.

For example, in a fight response, a Nordstrom salesperson who is frightened of losing her job may begin to work 15 hours a day, seven days a week to meet the pressing demands of her sales quota. Or, she may choose a flight response; she may remove herself from the situation by applying for work with another employer. In general, people try behavioral means of eliminating threats first because they actually change an unpleasant environment. In some cases, however, behavioral responses may not be possible. For example, the person may not be able to work any harder or may not have other employment opportunities.

We have called the cognitive mode of response to a threatening stimulus *benign reappraisal* because in this mode the person simply rearranges her values and perceptions. Although the original threat persists, the stress goes away. For example, another Nordstrom employee reappraised the consequences of losing her job: "I remember thinking I'm making less than $20,000 a year, why am I killing myself? Nordstrom was the most unfair place I ever worked."[13] This worker simply reappraised the thought of being fired from Nordstrom and concluded that it might not be that negative after all.

Job Burnout. In some cases, benign reappraisal may be as difficult as fighting or fleeing. One extreme form of job stress has been labeled **job burnout**. Burnout is a condition of emotional, physical, and mental exhaustion resulting from prolonged exposure to intense, job-related stress. The term was coined to refer to a particular pattern of stress development that occurs in the human services sector of the economy.[14] Burnout is prevalent in occupations that require a great deal of contact with people who are in serious need of help. Candidates for burnout include social workers, nurses, teachers, public defenders, doctors, and police officers.

Two defining characteristics of a burned-out worker are a feeling of *low personal accomplishment* from not being able to meet all the needs of the people one serves, and *emotional exhaustion*, which stems from the never-ending demands of those in need. The unique problem faced by human services workers, however, is their inability, in the face of unsuccessful fight-flight responses, to benignly reappraise their situation. It is difficult for a physician trained in the art of healing others to simply dismiss the death of a longtime patient with a "win some, lose some" attitude.

The third and perhaps most revealing characteristic of a burned-out worker is *depersonalization*. In depersonalization the worker begins to treat persons as objects. For example, a doctor may refer to a patient as "the bleeding ulcer in room 305." Depersonalization seems to become particularly pronounced when the person in question feels she is not being supported by colleagues or supervisors.[15]

[13] Ibid.

[14] H. J. Freudenberger, "Staff Burnout," *Journal of Social Issues*, 30 (1974), 159–64.

[15] S. E. Jackson, R. L. Schwab and R. S. Schuler, "Toward an Understanding of the Burnout Phenomenon," *Journal of Applied Psychology* 71 (1986), 630–40; and D. W. Russell, E. Altmaier and V. Van Velzen, "Job-Related Stress, Social Support and Burnout Among Classroom Teachers," *Journal of Applied Psychology* 72 (1987), 269–79.

TEACHING NOTE
Discuss the nature of the
relationship between satisfaction
and stress. It has been proposed
not only that dissatisfaction causes
stress but that as stress continues,
dissatisfaction increases (Scheck,
Kinicki, and Davy, ongoing study).

APPLYING THE DIAGNOSTIC
MODEL
Diagnostic Question 1: Why might
the organization be concerned
about dissatisfaction and stress?
What costly health-related or
behavioral problems linked to
stress are evident?

Diagnostic Question 2: Are the
values that the organization
expects its member to uphold
consistent with the members' own
values and needs?

Stress and Need-Value Conflicts

One last point should be made about cognitive appraisal theory. The model as outlined in Figure 8–1 implies a relatively strong, negative relationship between job satisfaction and job stress. Considerable research has borne out this prediction, and for the most part, it is accurate to say that chronic dissatisfaction leads to stress.[16] Is it ever possible for someone who is basically satisfied with his work to develop stress and stress-related symptoms? The answer is yes.

Earlier we distinguished needs (objective, bodily requirements) from values (subjective, mental desires and wishes). There is usually a close correspondence between one's needs and one's values, but that is not always the case. Pursuing values that are misaligned with needs may lead to momentary satisfaction but ultimately damage a person. Thus an executive who is burning the candle at both ends may love her job but may still feel stressed. Moreover, the job she loves so much could very well make her emotionally sick in the long run. In this example, the value for achieving more and more at work may be at variance with the body's need for nutrition or sleep. Such a person might show symptoms of stress without necessarily feeling dissatisfied with the job itself.

COSTS OF DISSATISFACTION AND STRESS

As we have noted already, because the explicit goals of most firms are financial or consumer oriented rather than worker oriented, it is sometimes difficult to convince managers of the need to monitor and act on problems of dissatisfaction and stress. In this section we will examine the costs, in both financial and human terms, of neglecting this critical area of management.

Health Problems

As we have already seen, work-related attitudes and emotions have a great impact on workers' health and well-being. A fact of organizational life in the 1990s is that employing organizations bear much of the cost for employee health care. A Department of Labor survey, for example, has shown that 96 percent of medium and large firms provide health insurance for all employees.[17] The term *fringe benefit* hardly applies to health insurance today. It is virtually impossible to attract good talent to one's organization if such insurance is not part of the package. Although wages have risen over the last 30 years, the spiraling costs of medical fees of doctors, hospitals, and other health care personnel and facilities have increased the cost of patient insurance three times as much as wage increases.[18]

Besides paying for general health insurance, employers are increasingly finding themselves held liable for specific incidents of stress-related illness. The Occupational Safety and Health Act of 1970 (OSHA) and many state laws hold employing organizations accountable "for all diseases arising out of and in the course of employment."[19] Because research has shown a strong link between stress and mental disorders, it was possible for an overworked advertising ex-

---

[16] C. Cooper and R. Payne, *Stress at Work* (London: John Wiley, 1978), p. 39.

[17] U.S. Department of Labor, *Employee Benefits 1985* (Washington, D.C.: Chamber of Commerce, 1984).

[18] D. W. Belcher and T. J. Atchison, *Compensation Administration* (Englewood Cliffs, N.J.: Prentice Hall, 1987), p. 57.

[19] *Analysis of Workers' Compensation Laws* (Washington, D.C.: U.S. Chamber of Commerce, 1985), p. 3.

IN PRACTICE

Employer Health Care: Asset or Liability?

It is becoming quite clear that business managers need to do something to bring the rising costs of health care under control. However, the traditional means of cost cutting used by employers is coming under closer scrutiny, and many are now asking whether the cure for high costs might not be worse than the disease.

Typically, to reduce business-related health-care costs firms have hired outside insurers. These insurers build custom-designed health plans in which a selected group of "preferred" doctors work for lower fees. These health maintenance organizations (HMOs) or preferred provider organizations (PPOs) have been effective in the short-term management of costs, but some observers are beginning to question their long-term effectiveness. These observers' concerns center on two issues: employer liability for physician malpractice, and the long-term ineffectiveness of inexpensive treatments for mental health problems.

There has always been a concern that cost cutting through HMOs and PPOs might diminish the quality of health care. The fact, however, that businesses themselves can be held liable for problems arising out of this cost cutting is a recent phenomenon. That is, if the company's health plan is deficient or if a preferred provider engages in malpractice, the company can be sued. Employers are particularly vulnerable when they offer huge discounts to employees who opt for the HMO or PPO and pass on their costs to employees who prefer to stay with plans that allow them to choose their own physicians. The Employee Retirement Income Securities Act of 1974 created a fiduciary responsibility for anyone who offers employee benefits, including health care. The interpretation of this act's provision that the employer must act as a "prudent person" forms the core of many lawsuits against executives who act as fiduciaries in selecting health-care providers.

Two other traditional cost-cutting practices that can lead to trouble deal with utilization reviews and physician incentives. *Utilization review* is a term that really means stingier use of medical care. For example, bed rest at home may be substituted for a long stay in the hospital. If these general rules deny the patient the care he needs, problems arise. For example, in 1986, after Lois Wickline's doctors performed car-diovascular surgery on her leg they advised her to stay in the hospital eight days to recuperate. The state of California's medical program said four was enough. Wickline's leg became infected, however, and subsequently had to be amputated. One state court awarded her $500,000, and another stated bluntly that "third-party payers can be held accountable when cost-limitation programs corrupt medical judgments."*

Physician incentives are programs that give bonuses to doctors contingent upon reducing costs. One way a physician can reduce patient costs is to neglect to refer him to expensive specialists. Here too, where such incentives promote malpractice, those offering the incentives can be held liable.

Traditional cost cutting programs are being attacked from a different angle in the area of mental health. Specifically, some observers are now questioning whether short-term cost-cutting maneuvers may actually increase long-term costs for mental health problems. Treatment for such afflictions as depression, job-related stress, and drug and alcohol abuse account for up to 20 percent of employers' health costs, and that makes these types of treatment a prime target for cost cutters. But a recent study conducted at McDonnell-Douglas indicates that the best financial returns are obtained when people suffering from these kinds of disorders are given the best possible service.

For example, a crucial feature of McDonnell-Douglas's program is the requirement that the whole family be included in any treatment program. In almost all cases, this arrangement results in higher first-year costs, but it turns out to be one of the most important aspects of long-term recovery and cost management. For example, when families were included in chemical dependency treatment programs for McDonnell-Douglas employees, treatment costs were on average $8,400 less in each of the second through the fifth years of treatment than they were for families who were not involved in first-year treatment. For psychiatric patients, the cost difference in follow-up years was $11,000. Thus the message from studies like these is that quality health care and cost-effective health care may not be mutually exclusive alternatives.

* M. Galen, "Are Companies Cutting Too Close to the Bone?" *Business Week*, October 30, 1989, pp. 143–44.

---

ecutive who was the victim of a nervous breakdown to successfully sue his employer.[20] Indeed, stress-induced mental disorders are the fastest rising category of occupational disease, and the number of lawsuits involving organizations and allegedly stress-damaged employees is increasing at a rapid rate.

Studies have also revealed a link between job stress and actual physical disorders like coronary heart disease, which is one of the major causes of death in this country. As employers are held increasingly liable for the onset of this disease and others such as hypertension and ulcer, the financial cost of stress becomes a major problem. As the "In Practice" box indicates, although there is

[20] R. Poe, "Does Your Job Make You Sick?" *Across the Board* 9 (1987), 34–43.

often a tradeoff between employee health benefits and company liability, some companies are finding ways to provide quality health care that is also cost effective.

Absenteeism and Turnover

Dissatisfaction and stress not only create direct costs for organizations in terms of health care programs. They also are the source of indirect costs, most notably in the form of absenteeism and turnover. Dissatisfaction is one of the major reasons for absenteeism, a very costly organizational problem. In the early 1980s, for example, executives at General Motors announced that absenteeism was costing the company a billion dollars every year. Among the 500,000 members of the United Auto Workers (UAW) employed by GM, casual absenteeism—the failure of employees to report to work as scheduled—was 5 percent. Thus on any given day, 25,000 people were absent. With 250 scheduled workdays in a year, this added up to 6,250,000 lost workdays, or 50 million hours. The average wage at GM was roughly $10 an hour plus $5 an hour in fringe benefits that were paid whether or not an employee showed up for work. Each hour lost cost GM $5 in fringe benefits and $15 for a temporary replacement—a total of $20 an hour. Multiply this $20 figure by the 50 million lost hours, and you will see why GM considered casual absenteeism a billion-dollar problem.[21] Replacing workers who leave the organization voluntarily is a costly undertaking. One high-tech company, Hewlett-Packard, estimates that the cost of replacing one middle-level manager is $40,000.[22] And replacement costs are not the only issue here. If people who leave an organization are better performers than those who stay, turnover lowers the productivity of the remaining work force. This kind of "negative employee flow" is most likely to affect complex jobs that take a long time to learn.[23] Under these conditions, companies lose the investment they have made in employee development. In the worst cases, disgruntled, experienced employees take jobs with competitors. A company's investment in employee development not only is lost but actually winds up as an investment in a competing firm, that gains access to a lot of knowledge of the first firm's operations.

Dissatisfaction is also a major cause of declining organizational commitment. **Organizational commitment** is the degree to which people identify with the organization that employs them. Commitment implies a willingness to put forth a great deal of effort on the organization's behalf and an intention to stay with the organization for a long time.

The subject of organizational commitment has been attracting a great deal of attention of late. Many employers fear that staffing policies they pursued in the 1980s may have killed company loyalty in the 1990s. A decade ago, to cope with ferocious global competition, deregulation, hostile takeovers, and unprecedented levels of corporate debt, many firms were forced to slash labor costs through massive layoffs. According to the Department of Commerce, 4.7 million workers who had held their jobs for more than three years have been dismissed since 1983. For example, in 1989, during one two-month period, Chrysler, Kodak, Campbell's Soup, Sears, and RJR Nabisco let 13,000 workers go. Between 1980 and 1989, General Motors fired over 150,000 workers. As chief economist for the AFL-CIO Rudy Oswald notes, "Workers have a right to be upset and angry.

organizational commitment
Identification with one's employer that includes the willingness to work hard on behalf of the organization and the intention to remain with the organization for an extended period of time.

ANOTHER VIEW
Although satisfaction and commitment are not perfectly related, recent studies by Williams and Hazer and by Davy, Kinicki, and Scheck demonstrate a strong causal link between the two constructs. Job satisfaction is an immediate antecedent of commitment, and it mediates the effects of work characteristics on commitment. (*Journal of Vocational Behavior*, 6/91)

[21] C. R. Deitsch and D. A. Ditts, "Getting Absent Workers Back on the Job: The Case at General Motors," *Business Horizons* 11 (1981), 52–53.

[22] W. R. Wilhelm, "Helping Workers to Self-Manage Their Careers," *Personnel Administrator* 28 (1983), 83–89.

[23] J. W. Boudreau and C. J. Berger, "Decision-Theoretic Utility Analysis Applied to Employee Separations and Acquisitions," *Journal of Applied Psychology* 70 (1985), 581–619.

SATISFACTION AND STRESS

AN EXAMPLE

A contract specialist at a major defense company said he had been permanently laid off from jobs three times since 1981. He expected to be laid off from his current job by November 1991. His wife had also seen two of her past employers go out of business during the same time period. The business she was currently working for was up for sale. This kind of history makes it very difficult for individuals to commit themselves to organizations. This couple began to view each job as only temporary. As a result, for their own protection, they are always scanning the job market and keeping their resumes in circulation. This couple is not an isolated case.

AN EXAMPLE

A recent study of an acquisition reports that the acquired company experienced nearly a 200 percent increase in voluntary turnover in the first two years following the acquisition. This increase was attributed to high levels of stress and reduced levels of satisfaction and commitment. Many changes, like increased performance standards, cuts in pay and benefits, and layoffs, had caused the increased stress and reduced satisfaction and commitment. The turnover data indicate that many chose to end the stress through flight (*Personnel Administrator*, August, 1989, pp. 84–90).

They have been bought and sold and have seen their friends and relations fired and laid off in large numbers. There is little bond between employers and workers anymore."[24] Evidence provided by surveys of American workers bolster this claim. When asked, "Compared with ten years ago, are employees today more loyal or less loyal to their companies?" 63 percent said less, and only 22 percent said more. A full 50 percent of those responding said it was likely that they would change employers in the next five years. Thus, just when American businesses are trying to inculcate a new sense of worker participation and involvement, many of their employees are looking to reduce their levels of commitment and dependency.

Sample items from the most frequently used measure of organizational commitment are shown in Table 8-3. Job satisfaction and organizational commitment are positively, but not perfectly, related.[25] A person can be high on one, yet low on the other. For example, professional workers like lawyers, teachers, or journalists are particularly likely to say they love their work, but would be just as happy doing it for one employer as for another.

Performance Failures

Severe levels of stress can affect a person's concentration on his job. At very high levels of stress a person simply stops concentrating on the task and focuses instead on the stress. In jobs requiring attention to detail, performance may suffer as stress goes beyond a tolerable level. Interestingly, very low levels of stress can have similar effects on task concentration. When a job is so undemanding that workers become bored, they are likely to daydream and to focus on nonjob factors, which also lowers performance. The inverted-U relationship between stress and job performance is depicted in Figure 8-3.

Widespread dissatisfaction among a firm's employees can diminish the firm's reputation in the labor market. The best applicants (those with the most ability)

[24] J. Castro, "Where Did All the Gung-Ho Go?" *Time*, September 11, 1989, pp. 52–55.
[25] L. J. Williams and J. T. Hazer, "Antecedents and Consequences of Satisfaction and Commitment on Turnover Models: A Reanalysis Using Structural Equations Modeling," *Journal of Applied Psychology* 71 (1986), 219–27.

T A B L E 8-3
Items Measuring Organizational Commitment

I find that my values and this organization's values are very similar.	<u>Agree</u> or Disagree
I am proud to tell others that I work for this organization.	<u>Agree</u> or Disagree
I could just as well be working for a different organization as long as the type of work was similar.	Agree or <u>Disagree</u>
This organization really inspires the very best in me in terms of job performance.	<u>Agree</u> or Disagree
It would take very little change in my present circumstances to cause me to leave this organization.	Agree or <u>Disagree</u>
I am extremely glad that I chose this organization to work for over others I was considering at the time I joined.	<u>Agree</u> or Disagree

Note: Underlined responses indicate a committed employee.
Source: Based on Mowday, Steers, and Porter.

FIGURE 8-3

The Inverted-U Relationship between Stress and Performance

Moderate amounts of stress tend to have an energizing effect. As you can see, performance is generally highest when people are under some degree of pressure. But where there's no stress at all or too much of it performance breaks down, as people become bored and inattentive or completely overwhelmed.

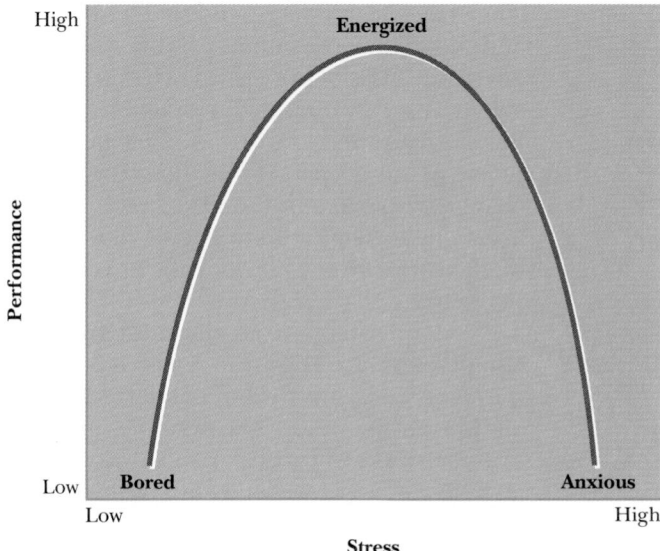

gravitate toward the most desirable employers. Employers perceived as undesirable end up with labor-pool "leftovers." For example, one of the best sources of recruits for an employer in terms of subsequent performance and turnover are *walk-ins*.[26] Walk-ins are people who go out of their way to apply to an organization even though no formal openings have been advertised. These walk-ins usually do their homework and know both the company and themselves very well. This applicant self-selection leads to a good person-organization fit. For similar reasons, another excellent source of recruits is referrals by current, satisfied employees. Thus firms with poor reputations as employers find it difficult, if not impossible, to recruit the best applicants. Indeed, sometimes, the only way such firms can staff their positions is to pay higher wages than the competition. Competition for job applicants will likely become even more intense in the next 20 years, as the United States goes through an impending labor shortage. So few firms can afford to have a bad reputation in the labor pool.

Human Costs

POINT TO STRESS
A major spouse-abuse intervention program in the Twin Cities has reported steadily increasing case loads that have been traced back to job-related stress in the abuser (e.g., fear of losing a job or having lost a job).

It would be quite unfair to imply that the only reason organizations care about employee attitudes and emotions is their need to save money. Some managers feel morally responsible for maintaining a reasonably high level of satisfaction among workers. One executive has put it this way: "I have over 500 people who spend 60 percent of their waking hours in my plant five, six, sometimes seven days a week, and I don't know how much time traveling to and from work and thinking about their jobs. If they're basically unhappy with their work, that means I'm responsible for one helluva lot of human misery, particularly if they go home and take it out on their families."[27]

Indeed, research has borne out this executive's worst fears. A significant correlation has been found between job stress and domestic violence such as spouse or child abuse. In particular, wife abuse has been related to sex-related

[26] P. J. Decker and E. T. Cornelius, "A Note of Recruiting Sources and Job Survival Rates," *Journal of Applied Psychology* 64 (1974), 463–64; and D. P. Schwab, "Recruiting and Organizational Participation," in *Personnel Management*, ed. K. M. Rowland and G. R. Ferris (Boston: Allyn & Bacon, 1982), pp. 103–28.
[27] H. J. Reitz, *Behavior in Organizations* (Homewood, Ill.: Irwin-Dorsey, 1981), p. 202.

SATISFACTION AND STRESS

status inconsistency. That is, abuse is most likely to occur with underachieving husbands who hold jobs that are lower in socioeconomic status than the jobs held by their wives. Wife abuse is also common among men who have lost their jobs.[28]

Alcohol and drug abuse have also been related to job-induced stress.[29] Thus many employers develop employee assistance programs (EAPs) to help workers with personal problems that have often been triggered or made worse by events or conditions in the workplace. Whether one wishes to approach the topic from a financial or a moral perspective, there is simply no escaping the importance of work attitudes and job-related stress.

Identifying Symptoms of Dissatisfaction and Stress

Because of the varied and important costs associated with employee dissatisfaction and stress, the identification of such problems should be a major part of the job description of every manager. In some cases, employees themselves report problems in these areas. Often, however, employees are afraid to admit that they cannot meet some of the demands their jobs impose on them. Similarly, workers dissatisfied with some facet of their job may censor themselves to avoid sounding like chronic complainers. Finally, the attitudes of some workers may have got so bad that they may see reporting dissatisfaction as a waste of time.

For this reason, it is critical for managers to monitor the kinds of physiological, cognitive, or behavioral responses that we discussed earlier for clues to underlying levels of dissatisfaction and stress. It is also a good idea for managers to be aware of well-known sources of dissatisfaction and stress. As we will see in the next section, many of these sources have been documented empirically by organizational researchers.

SOURCES OF DISSATISFACTION AND STRESS

There are many areas within organizations from which dissatisfaction and stress can arise. According to the conceptual scheme depicted in Figure 8-4, behavior in organizations (*ABC*) can be thought of as the interaction of three separate systems. First, there is the *physical and technological environment* (A) in which behavior takes place. Second, there is the *social environment* (B), or interpersonal relations among organizational members. Third, there is the *person* (C) whose behaviors and reactions are of interest to us.

Dissatisfaction and stress can originate in any of these three systems but more commonly arise in the overlapping areas shown in the figure. The *behavior settings* (AB), where the physical and social environments overlap, are the physical surroundings as they affect workers. Hence, they deal with factors such as crowding or privacy. In the *organization tasks* (AC) person and physical environment come together. The task is simply what the person's job is, that is, his formal function in carrying out the organization's mission. Finally, the *organization role*, (BC), which involves an interaction between the person and the social environment, includes the behavioral expectations that other people in the organization

[28] J. Barling and A. Rosenbloom, "Work Stressors and Wife Abuse," *Journal of Applied Psychology* 71 (1986), 346–50; C. A. Hornung, B. C. McCullough, and T. Sugimoto, "Status Relationships in Marriage: Risk Factors in Spouse Abuse," *Journal of Marriage and Family* 43 (1981), 675–92; and M. A. Straus, R. J. Gelles and S. K. Steinmetz, *Behind Closed Doors* (New York: Anchor, 1981), p. 21.

[29] H. Peyser, "Stress and Alcohol," in *Handbook of Stress*, ed. L. Goldberger and S. Breznitz (New York: Free Press, 1982), pp. 135–171.

FIGURE 8-4

**Sources of Dissatisfaction
and Stress**

*Adapted from J. E. McGrath, "Stress
and Behavior in Organizations," in
Handbook of Industrial/
Organizational Psychology, ed. M. D.
Dunnette (Chicago: Rand McNally,
1976).*

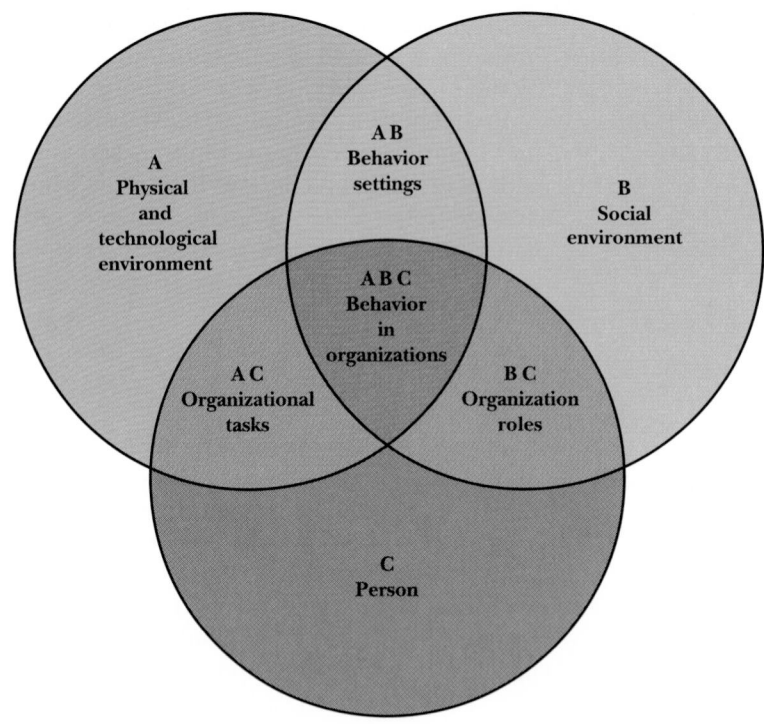

have for the person. These expectations include not only the things specified in the organization's formal definition of the job but many things that are not. For example, a production supervisor's subordinates may come to expect her consistently to show an interest in their families and to advise them on personal problems, even though the supervisor's job description requires her only to deal with problems that arise on the job. Expectations for role behaviors develop gradually. They are subject to negotiation between the person and other organization members who have a stake in how the person performs the job.[30] With this discussion as our framework, let us look more closely at each of the six areas where dissatisfaction and stress can originate.

Physical-Technological Environment

APPLYING THE DIAGNOSTIC MODEL
Diagnostic Question 3: Are any aspects of the physical environment (such as noise, darkness, hazards) causing stress among organizational members?

Although the Hawthorne research discussed in Chapter 2 had the effect of moving investigators away from studying the physical environment, we have documented evidence that some physical factors can engender negative emotional reactions in workers. For example, studies have shown that *extremes in temperatures* can affect job attitudes as well as performance and decision making. Research has also shown that there are different optimal *lighting requirements* for different tasks, and perceived darkness has been found to correlate significantly with job dissatisfaction.[31]

[30] D. R. Ilgen and J. R. Hollenbeck, "The Structure of Work: Job Design and Roles," in *Handbook of Industrial/Organizational Psychology*, 2nd ed., ed. M. D. Dunnette (San Diego: Contemporary Psychology Press).

[31] G. B. Meese, M. I. Lewis, D. P. Wyon, and R. Kok, "A Laboratory Study of the Effects of Thermal Stress on the Performance of Factory Workers," *Ergonomics* 27 (1982), 19–43; H. D. Ellis, "The Effect of Cold on Performance of Serial Choice Reaction Time and Various Discrete Tasks," *Human Factors* 24 (1982), 589–98; D. G. Hayward, "Psychological Factors in the Use of Light and Lighting in Buildings," in *Designing for Human Behavior: Architecture and the Behavioral Sciences*, ed. J. Lang, C. Burnette, W. Moleski and D. Vachon (Stroudsburg, Pa.: Dowden, Hutchinson, and Ross, 1974), pp. 120–29; and G. R. Oldham and N. L. Rotchford, "Relationships Between Office Characteristics and Employee Reactions: A Study of the Physical Environment," *Administrative Science Quarterly* 28 (1983), 542–56.

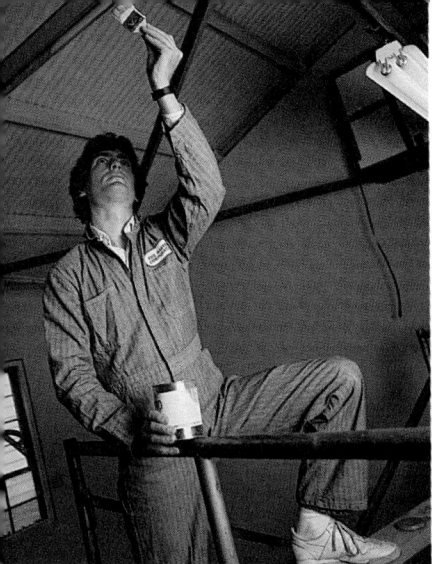

Can sick-building syndrome be prevented or cured? Architectural designer Paul Bierman-Lytle of New Canaan, Connecticut, believes it can, and he searches all over the world for building and decorating materials that are nonhazardous, such as formaldehyde-free plywood from Finland. Bierman-Lytle believes that manufacturers of paint and other materials will eventually develop environmentally safe products, but in the meantime he plans to open a chain of stores that will carry products he has judged safe.
Source: *"Environment,"* Fortune, *July 2, 1990, p. 88.*

functional attraction Satisfaction with other persons in the workplace that comes about because these other people help one attain valued work outcomes.

entity attraction Satisfaction with other persons in the workplace that comes about because these people share one's fundamental values, attitudes or philosophy.

APPLYING THE DIAGNOSTIC MODEL
Diagnostic Question 4: Are any aspects of the social environment (such as hostile coworkers or supervisors) causing stress among organizational members?

Moreover, research on how people perceive tasks shows that physical features of the environment such as *cleanliness*, *working outdoors*, and *health hazards* are very important in the way people perceive their tasks.[32] These factors often have more important effects on attitudes at work than do the more psychologically based factors that we will discuss later.[33] The fact that many jobs are seen as physically unattractive is reflected in the "compensating differentials," or extra pay, that employers must often give to workers who hold such jobs.

More recent research has focused on some very subtle characteristics of the physical environment. In fact, researchers have coined the term *sick building syndrome* to refer to physical structures whose indoor air is contaminated by invisible pollutants. Ironically, as buildings have become better insulated, this syndrome has become more common. With today's office technology, fumes from copier machine liquids, carbonless paper, paint, rugs, synthetic draperies, wall paneling, and cleaning solvents are held in a building's inside air and endlessly recycled. In addition, to conserve energy, many firms turn off air conditioning and heating systems on weekends. The stagnant air provides fertile breeding grounds for mold and bacteria, which are spewed forth into the work area come Monday morning.

Sometimes problems like these get so bad that workers take matters into their own hands. In June of 1988, 70 workers picketed the headquarters of the Environmental Protection Agency in Washington, charging that the air inside the building was so contaminated that it caused burning eyes, fatigue, dizziness, and breathing difficulty. In California, one worker won $600,000 in a lawsuit that claimed that formaldehyde fumes in a new office building caused him to lose consciousness and suffer permanent brain damage. Restoring the air purity in contaminated structures can be costly. At the Veterinary Teaching Hospital at the University of Florida, for example, it cost $6 million to clean up a brand-new building that cost only $10 million to build.[34]

Social Environment

Two primary sets of people in the organization serve as potential sources of satisfaction or frustration for the employee: supervisors and coworkers. There are two major ways in which these people can engender positive or negative reactions in a worker. First, the employee may be satisfied with her supervisor or coworkers because these people help her attain some valued outcome. This attitude is referred to as **functional attraction.**

On the other hand, a person may also be attracted to others because their values, attitudes, or philosophy are fundamentally similar to his. This attraction is referred to as **entity attraction**. The fact that people can make such a distinction is evident when people say they like their manager as a supervisor but not as a person. The greatest degree of satisfaction with supervisors and coworkers will be found where both kinds of attraction are at work. Thus, many organizations try to foster a culture of shared values among employees. We will discuss organizational culture in Chapter 18. For now we will simply note that although generating a strong unitary culture throughout an entire organization is difficult, significant increases in satisfaction can be achieved even if only direct supervisors and subordinates come to share some values.[35] Managers must take care, however,

[32] E. F. Stone and H. G. Gueutal, "An Empirical Derivation of the Dimensions along Which Characteristics of Jobs are Perceived," *Academy of Management Journal* 28 (1985), 376–96.

[33] S. J. Zacarro and E. F. Stone, "Incremental Validity of an Empirically Based Measure of Job Characteristics," *Journal of Applied Psychology* 73 (1988), 245–52.

[34] A. Toufexis, "Got That Stuffy, Run-Down Feeling," *Time*, June 6, 1988, p.

[35] B. M. Meglino, E. C. Ravlin, and C. L. Adkins, "A Work Values Approach to Corporate Culture: A Field Test of the Value Congruence Process and Its Relationship to Individual Outcomes," *Journal of Applied Psychology* 74 (1989), 424–33.

social support A surrounding environment in which people are sympathetic and caring.

buffering The notion that certain positive factors in the person's environment can limit the capacity of other factors to create dissatisfaction and stress.

lest workers think that management is encouraging the development of entity attraction among a firm's employees simply because it wants something from them.

Social support is the active provision to a person, by other people in her environment, of sympathy and caring. Many writers have suggested that social support from supervisors and coworkers can buffer employees from stress. The notion behind **buffering** is illustrated in Figure 8-5 where, as you can see, the presence of people who are supportive can lower the incidence of stress- related symptoms under conditions of high stress. Our evidence for this effect has come largely from research in medical contexts that shows that recovery and rehabilitation from illness proceed better when the patient is surrounded by caring friends and family.[36]

The concept of buffering, however, is somewhat controversial. For example, although one organizational behavior study showed that student nurses who received social support were much better able to perform their jobs in the face of stress than nurses who received little support,[37] a second study found just the opposite. In the latter study, social support actually made the effects of some stressors more powerful.[38] For example, nurses who had strong social support actually found role conflict more stressful than did nurses who lacked social support. This could have been because the nurses' families, friends, and other supporters placed additional role-related demands on them. A third study, this time of workers in the construction industry, found that social support had little effect on stress one way or the other.[39] Despite the conflicting evidence regarding social support as a buffer, one important thing on which all three studies agreed is that social support is an independent predictor of stress and dissatisfaction.

[36] R. E. Mitchell, A. G. Billings, and R. M. Moos, "Social Support and Well-Being: Implications for Prevention Programs," *Journal of Primary Prevention* 11 (1982), 77–98.

[37] K. R. Parkes, "Occupational Stress Among Student Nurses: A Natural Experiment," *Journal of Applied Psychology* 67 (1982), 784–96.

[38] G. M. Kaufman and T. A. Beerh, "Interactions between Job Stressors and Social Support: Some Counter-Intuitive Results," *Journal of Applied Psychology* 71 (1986), 522–30.

[39] G. C. Ganster, M. R. Fusilier, and B. T. Mayes, "Role of Social Support in the Experience of Stress at Work," *Journal of Applied Psychology* 71 (1986), 102–11.

FIGURE 8-5

How Social Support May Buffer Stress

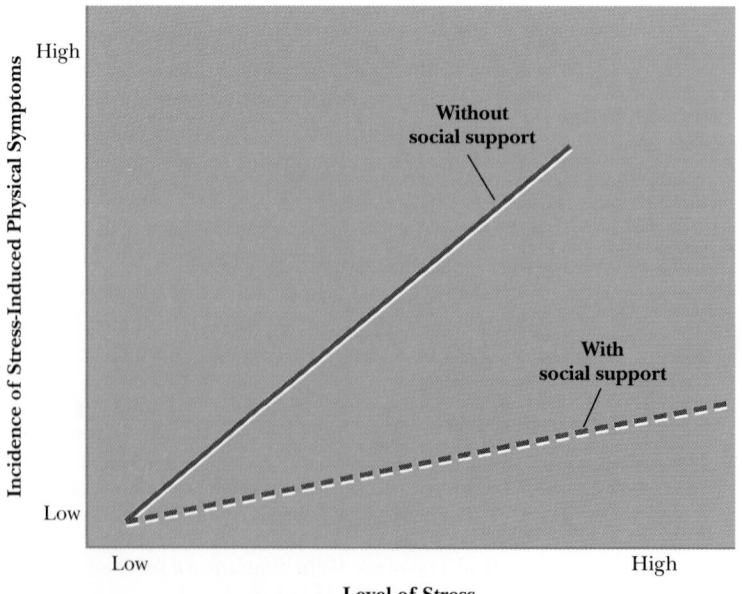

SATISFACTION AND STRESS

That is, holding type of stressor constant, people receiving support were less stressed and dissatisfied. Social support may be a good thing in and of itself, but it may not be the remedy for stress that many have proposed.

Behavior Settings

social density An index of crowding, typically calculated as the number of people occupying an area divided by the number of square feet in that area.

privacy The freedom to work unobserved by others and without undue interruption.

Looking again at Figure 8-4 we see that at the intersection of physical and social environments is the behavior setting. Two important and interrelated aspects of the behavior setting are social density and privacy. **Social density** is a measure of crowding. It is calculated by dividing the number of people in a given area by the number of square feet in that area. **Privacy**, on the other hand, is the freedom to work without either observation by others or unnecessary interruption. Research with clerical workers has shown that job satisfaction decreases as social density increases.[40] Moreover, social density is a particular problem when it is compounded by lack of privacy, as when workers' stations are not enclosed by walls or partitions. Research by Greg Oldham and Yitzak Fried found that neither crowding nor lack of partitions alone predicted turnover in university office staff, but when workers were both crowded and without privacy, turnover was exceptionally high.[41]

open-office plan A physical work environment that minimizes interior walls and partitions.

APPLYING THE DIAGNOSTIC MODEL
Diagnostic Question 5: Are any aspects of the behavioral settings (such as crowding or lack of privacy) causing stress among organizational members?

TEACHING NOTE
Many banks still use open office designs. Ask students to discuss their feelings when having to do very personal business in an environment where everyone can see and hear you. How would they feel if they had to work in such an environment all the time?

These findings have significant implications for organizations with **open-office plans**. In the late 1960s, open-office plans enjoyed enormous popularity among design professionals in Western countries. The open-office plan is characterized by an absence of the interior walls and partitions that more conventional designs used to define private work spaces. Typically in open-office plans, all office personnel, from clerks to managers, are located in one large open space. Advocates of this approach hoped that open designs would increase communication and thus improve work efficiency and lower operating costs.

Research on open-office designs suggests that these hopes have gone largely unrealized, primarily because of problems with crowding and lack of privacy. Firms that have moved away from open offices and have returned to more conventional designs have increased work satisfaction either by giving workers more space and thus decreasing social density or by installing partitions and thus providing real privacy.[42] Moreover, crowding and privacy are not the sole issues. Other research shows that the egalitarian nature of open offices leads some employees, particularly managerial and professional staff, to resent the loss of perceived status attached to having a private office.[43]

The Person

negative affectivity A person's tendency to often experience feelings of subjective distress such as anger, contempt, disgust, guilt, fear, and nervousness.

Because both stress and dissatisfaction ultimately reside within a person, it is not surprising that many who have studied these outcomes have focused on individual difference variables. The term **negative affectivity**, for example, describes a dimension of subjective distress that includes such unpleasant mood states as

---

[40] R. I. Sutton and A. Rafaeli, "Characteristics of Work Stations as Potential Occupational Stressors," *Academy of Management Journal* 30 (1987), 260–76.

[41] G. R. Oldham and Y. Fried, "Employee Reactions to Workspace Characteristics," *Journal of Applied Psychology* 72 (1987), 75–84.

[42] G. R. Oldham, "Effects of Change in Workspace Partitions and Spatial Density on Employee Reactions: A Quasi-Experiment," *Journal of Applied Psychology* 73 (1988), 253–60.

[43] M. D. Zalesny and R. V. Farace, "Traditional versus Open Offices: A Comparison of Sociotechnical, Social Relations, and Symbolic Meaning Perspectives," *Academy of Management Journal* 30 (1987), 240–59.

anger, contempt, disgust, guilt, fear, and nervousness.[44] Table 8-4 shows some items that are used to assess individual differences on some aspects of negative affectivity.

People who are generally high in negative affectivity tend to focus on both their own negative qualities and those of others. Such people are also more likely to experience significantly higher levels of distress than are individuals who are low on this dimension. Being familiar with the concept of negative affectivity is important for two reasons. First, this notion highlights the fact that some people bring stress and dissatisfaction with them to work. Such people may be relatively dissatisfied regardless of what steps are taken by the organization or the manager. For example, research by Barry Staw and his colleagues showed that degree of negative affectivity in early adolescence predicted overall job satisfaction in adulthood.[45] These investigators also found significant correlations between work attitudes measured over a five-year period even when workers changed employers or occupations.[46] This too points to an underlying personal predisposition to negative affectivity.

Second, negative affectivity influences both a person's perception of a situation (e.g., a task, a supervisor) and her perception of her level of stress. Thus one needs to be cautious in interpreting the relation between situation and stress when both factors are measured by the employee's perceptions. A study by Arthur Brief and his colleagues found that what appeared to be a strong relationship between job stress and health complaints, was in reality quite weak.[47] As Figure

[44] D. Watson, L. A. Clark, and A. Tellegen, "Development and Validation of Brief Measures of Positive and Negative Affect: The PANAS Scales," *Journal of Personality and Social Psychology* 54 (1988), 1063–70.

[45] B. M. Staw, N. E. Bell, and J. A. Clausen, "The Dispositional Approach to Job Attitudes: A Lifetime Longitudinal Test," *Administrative Science Quarterly* 31 (1986), 56–78.

[46] Staw and J. Ross, "Stability in the Midst of Change: A Dispositional Approach to Job Attitudes," *Journal of Applied Psychology* 70 (1985), 469–80.

[47] A. P. Brief, M. J. Burke, J. M. George, B. S. Robinson, and J. Webster, "Should Negative Affectivity Remain an Unmeasured Variable in the Study of Job Stress," *Journal of Applied Psychology* 73 (1988), 193–200.

TABLE 8-4
Items from a Measure of Negative Affectivity

This scale consists of a number of words that describe different feelings and emotions. Please indicate to what extent you have the following feelings _____ .*

Read each item in the second and third columns and then write the number from the scale in the first column that corresponds to your answer in the space next to that word.

1	Very slightly or not at all	_____ interested	_____ irritable
		_____ distressed	_____ alert
		_____ excited	_____ ashamed
2	A little	_____ upset	_____ inspired
		_____ strong	_____ nervous
3	Moderately	_____ guilty	_____ determined
		_____ scared	_____ attentive
4	Quite a bit	_____ hostile	_____ jittery
		_____ enthusiastic	_____ active
5	Extremely	_____ proud	_____ afraid

* By filling in the blank with such instructions as "at this moment," "today," or "generally," we can use this scale to measure feeling states at particular times or over different time periods.

FIGURE 8-6

Influence of Negative Affectivity on Both Self-Reports of Stress and Health Complaints

The relation between employees' reports of stress and their reports of health problems is tenuous, as the broken line in B indicates. Negative affectivity, or the tendency to experience frequent feelings of distress, may be the cause of both reported stress and reported health problems.

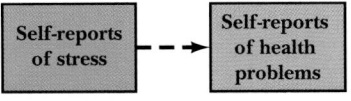

A Apparent Causal Relationship

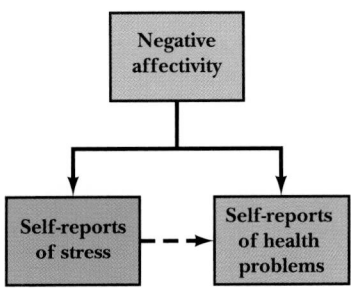

B Actual Causal Relationship

APPLYING THE DIAGNOSTIC MODEL
Diagnostic Question 6: Do certain characteristics of organization members contribute to stress problems at work (e.g., negative affectivity or the type A behavior pattern)?

POINT TO STRESS
Research has shown stress is additive. People bring stress at home to work. Work stress is added on. By the end of the day, both home stress and work stress go home. When stress remains high for a prolonged period it can become difficult for individuals to separate out sources of stress.

8-6 indicates, negative affectivity caused both the perceptions of being stressed *and* the health complaints, thus inflating the stressor–health complaint relationship.

Although the origins of negative affectivity are not completely known, interesting research by Richard Arvey suggests that a genetic component may be involved.[48] Arvey studied identical twins who had been separated at birth and reared apart. Thirty-four pairs of such twins were located and assessed on general job satisfaction as well as satisfaction with intrinsic and extrinsic aspects of their jobs. Amazingly, there was a significant correlation between twin pairs' ratings on general satisfaction and intrinsic satisfaction. Despite the fact that these twins were reared apart and were working in different jobs, the work attitudes they expressed were very similar.

A second critical individual-difference variable, one we looked at in Chapter 4, is the Type A behavior pattern. The unrealistic expectations of the impatient, ambitious, and overly aggressive Type A person makes him particularly susceptible to dissatisfaction and stress. This susceptibility may also account for the Type A person's 2-to-1 risk ratio, compared to the Type B person, for developing coronary heart disease. Research suggests that the aggressiveness and hostility of the Type A pattern are its most damaging aspects.[49]

Although the focus of this chapter is on stress induced by the job, one must recognize that stress can originate in events outside of work. This nonjob-related stress cannot simply be turned off when people enter the plant gates or office doors. Table 8-5 presents a list of various life-change events that are sources of stress. On this scale, 100 represents the most stressful event, and 11 the least stressful. Typically, subjects are asked to check off any event that has occurred in their lives in the preceding 12 months and to sum the corresponding values. In one of the original studies using this scale, people with scores between 0 and 150 usually reported good health in the year that followed. People who scored over 300, however, had a 70 percent chance of contracting a major illness the following year. It must be admitted that subsequent research has not always documented strong relations between scores on this scale and subsequent illness.

[48] R. D. Arvey, T. J. Bouchard, N. L. Segal, and L. M. Abraham, "Job Satisfaction: Genetic and Environmental Components," *Journal of Applied Psychology* 74 (1989), 187–93.
[49] R. B. Williams, "Type A Behavior Pattern and Coronary Heart Disease: Something Old and Something New," *Behavioral Medicine Update* 6 (1984), 29–33.

TABLE 8-5
The Stress-Inducing Impact of Some Work and Nonwork Related Events

LIFE EVENT	SCALE VALUE	LIFE EVENT	SCALE VALUE
Death of spouse	100	Change in responsibilities at work	29
Divorce	73	Son or daughter leaving home	29
Marital separation	65	Trouble with in-laws	29
Jail term	63	Outstanding personal achievement	28
Death of a close family member	63	Spouse begins or stops work	26
Major personal injury or illness	53	Begin or end school	26
Marriage	50	Change in living conditions	25
Fired from work	47	Revision of personal habits	24
Marital reconciliation	45	Trouble with boss	23
Retirement	45	Change in work hours or conditions	20
Major change in health of family		Change in residence	20
member	44	Change in schools	20
Pregnancy	40	Change in recreation	19
Sex difficulties	39	Change in church activities	19
Gain of a new family member	39	Change in social activities	18
Business readjustment	39	Mortgage or loan less than $10,000	17
Change in financial state	38	Change in sleeping habits	16
Death of a close friend	37	Change in number of family	
Change to a different line of work	36	get-togethers	15
Change in number of arguments		Change in eating habits	15
with spouse	35	Vacation	13
Mortgage over $10,000	31	Christmas	12
Foreclosure of mortgage or loan	30	Minor violations of the law	11

Source: L. O. Ruch and T. H. Holmes, "Scaling of Life Change: Comparison of Direct and Indirect Methods," *Journal of Psychosomatic Research* (June 1971), p. 213.

Moreover, some hardy individuals can experience a whole host of these life events without showing any signs of stress.[50] Interestingly, as you have probably noticed, the life event scale lists a number of events that seem quite positive, such as marriage, vacation, and outstanding personal achievement. Why do you suppose such events cause stress?

Organizational Tasks

In spite of the influence of dispositional levels, nothing predicts a person's level of satisfaction or stress better than the nature of her job.[51] Table 8-6 shows a list of some of the most and least stressful jobs. Innumerable aspects of tasks have been linked to dissatisfaction and stress. Moreover, as you will see in Chapter 14, some elaborate theories relating task characteristics to worker reactions have been formulated and extensively tested. In general, the key factors that determine satisfaction and stress are task complexity, physical strain, and task meaningfulness.

[50] T. H. Holmes and R. H. Rahe, "The Social Readjustment Rating Scale," *Journal of Psychosomatic Research* 40 (1967), 213– 18; D. V. Perkins, "The Assessment of Stress Using Life Change Scales," in *Handbook of Stress*, ed. L. Goldberger and S. Breznitz (New York: Free Press, 1982), pp. 320–31; and S. C. Kobasa, "Stressful Life Events, Personality, and Health: An Inquiry into Hardiness," *Journal of Personality and Social Psychology* 35 (1983), 1–11.

[51] B. A. Gerhart, "How Important Are Dispositional Factors as Determinants of Job Satisfaction? Implications for Job Design and Other Personnel Programs," *Journal of Applied Psychology* 72 (1987), 493–502.

TABLE 8-6	
Jobs Characterized as High and Low in Stress	
HIGH-STRESS JOBS	**LOW-STRESS JOBS**
Manager	Farm laborer
Foreman	Craft worker
Nurse	Stock handler
Waitress	College professor
Air traffic controller	Heavy-equipment operator

AN EXAMPLE: NEW ENGLAND POWER POOL
As the summer of 1989 began, the New England Power Pool still didn't know whether the various sources of power it was counting on to increase its electricity generation would come through. This uncertainty could only add to the tension felt by workers in the control room, who know they must make quick and effective decisions in emergency situations. If they don't, an entire city can suffer a brownout or, worse, a complete blackout like the one that engulfed New York City in the summer of 1977. This tension added to the boredom that these jobs induce because they are so uneventful most of the time can cause workers to feel frustrated, even stressed out. ("Energy," *Fortune*, June 5, 1989, pp. 122–23.

Task Complexity. Although some tasks are too complex, it is very common to find a strong positive relationship between *task complexity* and satisfaction. The boredom generated by simple, repetitive jobs that are not mentally challenging leads to frustration for the worker. This frustration manifests itself in the form of dissatisfaction, stress and, ultimately, tardiness, absenteeism, and turnover.[52]

Boredom created by lack of task complexity can also hinder performance on certain types of jobs. For example, airport security personnel, air traffic controllers, operators in nuclear power stations, medical technicians, and inspectors on production floors all belong in a class of jobs that require *vigilance*. Workers on these jobs must continually monitor equipment and be prepared to respond to critical events. However, because such events are so rare these jobs are exceedingly boring, and boredom results in poor concentration. Ultimately, this inattention results in performance breakdowns of often serious dimensions.

For example, in 1985, TWA flight 847 from Athens to Rome was hijacked and forced to land in Beirut, beginning an ordeal that lasted 17 days. The hijackers had been able to pass a dozen hand grenades and handguns through airport security. This performance lapse was attributed by experts to a breakdown in vigilance on the part of security personnel, who up until that day had never personally encountered a smuggling incident.

Interestingly, some research indicates that on tasks requiring vigilance, some of what are typically referred to as "disabilities" can often enhance performance. Some basic research has shown that blind subjects do better than sighted people on auditory vigilance tasks, and that deaf people do better than hearing subjects on visual vigilance tasks. Judging from these results, businesses might serve themselves well by hiring the "disabled" for jobs that entail certain types of vigilance.[53]

The Department of Labor (DOL), which keeps vital statistics on thousands of jobs in the United States, rates all jobs in terms of three dimensions of complexity: interaction with other people, use of data, and actions. Table 8-7 shows the numerical rating system used by the DOL in rating jobs, and Table 8-8 shows some actual DOL ratings of jobs with which you may be familiar. An examination of these two tables should give you a feel for at least one system of defining complexity.

APPLYING THE DIAGNOSTIC MODEL
Diagnostic Question 7: How can the jobs that need to be performed by organization members be described in terms of complexity, meaning, and physical demand? How might characteristics of the jobs relate to dissatisfaction and stress?

Physical Strain. Another important determinant of work satisfaction is how much physical strain and exertion the job involves.[54] This factor is sometimes overlooked in this age of technology, where much of the physical strain associated with jobs has been removed by automation. Indeed, the very fact that technology

---

[52] L. W. Porter and R. M. Steers, "Organizational, Work and Personal Factors in Employee Absenteeism and Turnover," *Psychological Bulletin* 80 (1973), 151–76.

[53] J. S. Warm and W. N. Dember, "Awake at the Switch," *Psychology Today*, April 1986, pp. 46–50.

[54] Locke, "Job Dissatisfaction."

SOURCES OF DISSATISFACTION AND STRESS

TABLE 8-7
U.S. Department of Labor's Rating System for Evaluating Task Complexity*

PEOPLE	DATA	ACTIONS
0 Mentor	0 Synthesize	0 Set up
1 Negotiate	1 Coordinate	1 Precision work
2 Instruct	2 Analyze	2 Operate/Control
3 Supervise	3 Compile	3 Drive
4 Divert	4 Compute	4 Manipulate
5 Persuade	5 Copy	5 Tend
6 Speak/Signal	6 Compare	6 Feed
7 Serve		7 Handle
8 Take instruction		

* Lower numbers indicate greater complexity.

continues to advance highlights the degree to which physical strain is universally considered an undesirable work characteristic. Many jobs, however, can still be characterized as physically demanding.

Task Meaningfulness. Finally, it is also important for the worker to believe that his work has value. The Peace Corps recruits applicants by describing its work as "the toughest job you will ever love." Similar recruiting advertisements for Catholic priests note that "the pay is low but the rewards are infinite." Indeed, there are over one million volunteer workers in the U.S. alone who perform their jobs almost exclusively because of the meaning attached to the work. Some of these jobs are low in complexity and high in physical exertion. People who do them, however, view themselves as performing a worthwhile service. This perception overrides the other two factors and ultimately contributes to high levels of satisfaction.

Organization Roles

organization role The total set of expectations that people who interact with an organizational member have for that person and his performance of his job.

Look back at Figure 8-4. As you can see, the **organization role** occurs at the intersection of the social environment and the person (BC). The person's role in the organization can be defined as the total set of expected behaviors that both the person and other people who make up the social environment have for the

TABLE 8-8
U.S. Department of Labor Ratings of Job Complexity for Some Common Jobs

	PEOPLE	DATA	ACTIONS	AVERAGE RATING
Optometrist	0	1	1	1
Nuclear engineer	6	0	1	2
Psychological counselor	0	1	7	3
Photo journalist	6	0	2	3
Airplane pilot	6	2	2	3
Elementary school teacher	2	2	7	4
Auto mechanic	8	3	1	4
Food concession manager	6	1	7	5
Welder	8	4	2	5
Bicycle assembler	8	2	7	5
Tree pruner	8	6	4	6
Ticket taker	8	6	7	7

SATISFACTION AND STRESS

role incumbent.[55] As we noted earlier, these behaviors include all the formal aspects of the job as well as the expectations of coworkers, supervisors, clients, or customers. These expectations have a great impact on how the person responds to the work. Three of the most researched aspects of roles are role ambiguity, role conflict, and role scope.

role ambiguity Lack of clarity about the expectations of a person's role in an organization.

APPLYING THE DIAGNOSTIC MODEL
Diagnostic Question 8: How clear and unambiguous are the role expectations that are being sent to various organization members? How might the ambiguity of some expectations relate to stress?

Role Ambiguity. Role ambiguity refers to the level of uncertainty or lack of clarity surrounding expectations about the person's role in the organization. It is an indication that the person in the role does not have enough information about what is expected of her. What should she do? How should she do it? Role ambiguity can also stem from a lack of information about the rewards for performing well and the punishments for failing to do the right thing or for doing things the wrong way. For example, imagine that you were in a class where an instructor assigned a term paper but neglected to tell you (a) what topics were pertinent, (b) how long the paper should be, (c) when it was due, (d) how it would be evaluated, and (e) how much it was worth toward the final course grade. Would you feel stress under these circumstances?

role conflict Conflict or incompatibility between the demands facing a person who occupies a particular role.

Role Conflict. Role conflict is the recognition of incompatible or contradictory demands that face the person who occupies the role. Role conflict can occur in many different forms. *Intersender role conflict* occurs when two or more people in the social environment convey mutually exclusive expectations. For example, a middle manager may find that upper management wants severe reprimands for worker absenteeism but that the workers themselves expect consideration of their needs and personal problems.

Intrasender role conflict occurs when one person in the social environment holds two competing expectations. A research assistant for a magazine editor may be asked to write a brief but detailed summary of a complex and lengthy article from another source. In trying to accomplish this task, the assistant may experience considerable distress over what to include and what to leave out of the summary.

A third form of role conflict is called *interrole conflict*. Most of us occupy multiple roles, and the expectations for our different roles may conflict. A parent who has a business trip scheduled during his daughter's first piano recital will feel torn between the demands of two roles.

Finally, *person-role conflict* arises when the role occupant's own expectations for the role conflict with the expectations of others in her role set. For example, a new college instructor who values research but is told to disregard this aspect of his job and to concentrate solely on teaching might experience this type of conflict. Sometimes organizations can avoid such conflict through selection procedures. In fact, some colleges recruit only professors who have little interest in research, actively discouraging research activity.

TEACHING NOTE
Many graduate students are experts imparting knowledge in their role as teachers. But in their role as students they are no longer the experts. How do you shift roles so quickly?

role scope The total number of expectations that exist for the person occupying a particular role.

Role Scope. Role scope refers to the absolute number of expectations that exist for the person occupying the role. In role overload, too many expectations or demands are placed on the role occupant, and in role underload we have the opposite problem. Because researchers have focused primarily on jobs with high role scope, they have tended to look at the negative consequences of jobs that are too challenging. Jobs that are too high in role scope also demand a tremendous amount of time from incumbents, and as you can see from the "Management Issues" box, there is considerable danger in this.

---

[55] S. E. Jackson and R. S. Schuler, "A Meta-Analysis and Conceptual Critique of Research on Role Ambiguity and Role Conflict in Work Settings," *Organizational Behavior and Human Decision Processes* 36 (1987), 16–78.

INTERNATIONAL OB

Karoshi—One Japanese Import that Americans May Not Want

Other than nationality, what do the heads of a major Japanese robotics firm, a large Japanese publishing house, and Japan's leading communications empire have in common? According to the Japanese press, all three died from *karoshi*. To the millions of Japanese upper- and middle-level managers who work in perpetual overdrive, *karoshi* is a well-known term for "death by overwork."* On the average, Japanese managers work over 500 hours more a year than those in West Germany, and over 250 hours more than American managers. Yet just at a time when American firms are importing more and more types of Japanese management practices, the widespread problems with *karoshi*, the dark side of the Japanese economic miracle, seem to question the wisdom of adopting *all* Japanese business practices.

Is *karoshi* being imported into the United States? Surveys seem to indicate yes. In a recent poll, 77 percent of 206 CEOs from the Fortune 500 and Service 500 companies indicated that "large U.S. companies will have to push their managers harder if we are to compete successfully with the Japanese." Moreover, whereas 62 percent of the CEOs agreed that managers are working longer hours today than ten years ago, a whopping 91 percent disagreed with the notion that "companies are pushing managers too hard." Texaco CEO James Kinnear summarized the feelings of the respondents, when he stated that "heads of companies must set objectives and monitor employee performance—and if that leads to longer hours, then so be it!" Winston Wallin, CEO of Medtronics, one of the few dissenters, noted that "executives who work 80 hours a week are not likely to have the breadth of knowledge that they ought to have. Managers are likely to be more creative if they have a little balance in their lives."†

Back in Japan, however, the tide may be turning Wallin's way. The government recently announced a $2 million study of *karoshi*, and some of the major firms are forcing their man-

agers to take time off. The Sony Corporation announced in 1990 that all employees would be required to take a vacation of a week or two, whether they wanted it or not. Moreover, as part of a nationwide drive toward a five-day week, Japanese banks now close down on Saturdays. Of course, old ways die hard, and several prominent institutions have spread the word that employees are expected to make up the lost time by tacking on more hours during the week.

The willingness of the younger Japanese managers to adhere to such expectations, however, is decreasing. Executives say that new hires are shying away from working on Sundays, and that young Japanese managers insist on having time to be with their families. These managers are more likely to agree with the sentiments expressed in a book entitled *Karoshi*—supposedly a set of reflections written in the appointment calendar of an advertising agency executive who died of this syndrome. In one passage, he notes that "people become inured to the ease of a slavelike existence. They are bought by money. They are bound by time. The slaves of the past most likely had time to eat with their own families."

* D. E. Sanger, "Tokyo Tries to Find Out if 'Salarymen' Are Working Themselves to Death," *New York Times*, March 19, 1990, p. A8.
† S. Solo, "Stop Whining and Get Back to Work," *Fortune*, March 12, 1990, pp. 49–51.

Assessing the problems that people experience with regard to their roles on the job (see Table 8-9) is only the first step. Getting rid of the sources of stress or finding ways to cope with them is the next task and, as we will see, it is often a challenging one.

ELIMINATING OR COPING WITH DISSATISFACTION AND STRESS

Given the huge direct and indirect costs associated with dissatisfaction and stress in organizations, it is not surprising that a great many ways to deal with stress have been proposed. Some have received more research sup-

TABLE 8-9
Some Items That Measure Role Problems

Role conflict
1. Having to work under conflicting guidelines
2. Having to work with two or more groups of people who expect different things from you
3. Having conflicting demands from people at work
4. Not knowing exactly what your responsibilities are
5. Having to do things you feel should be done differently
6. Being uncertain about how much authority you have

Role ambiguity
7. Not knowing how you must perform to do your work well
8. Being uncertain whether you have divided your time properly between the work you have to do
9. Doing work where you can't always be certain what's expected of you
10. Doing work where it's hard to get all the necessary information, resources, or materials
11. Feeling that you don't have all the necessary skills to do your job well

Role underload
12. Working on tasks that could be done by someone less qualified
13. Doing work that is repetitive or boring
14. Working on things you feel are not absolutely necessary or helpful

Role overload
15. Working under continuous time pressure
16. Working hard to meet deadlines

APPLYING THE DIAGNOSTIC MODEL
Diagnostic Question 9: Is the organization most interested in eliminating the sources of stress, or is it willing (or forced) to deal only with stress-related symptoms?

port than others. The 12 specific approaches we discuss are organized according to whether they attempt to eliminate stress and dissatisfaction at the source or merely deal with the physiological and behavioral symptoms associated with stress. Clearly, interventions aimed at the source are preferable to those aimed only at the symptoms. Because it is not always possible to eliminate stressors, however, research on symptom-based approaches is valuable. Figure 8-7 shows the 12 interventions associated with the stages of the overall dissatisfaction-stress model that constitute their targets: primary appraisal, secondary appraisal, stress, fight-flight, and benign reappraisal.

Targeting the Primary Appraisal

Organizational interventions that aim at the primary appraisal focus on the stressors themselves. These methods attempt to change characteristics of either tasks or roles.

FIGURE 8-7

Stress-Reduction Interventions that Target Stages of the Cognitive Appraisal Theory of Emotion

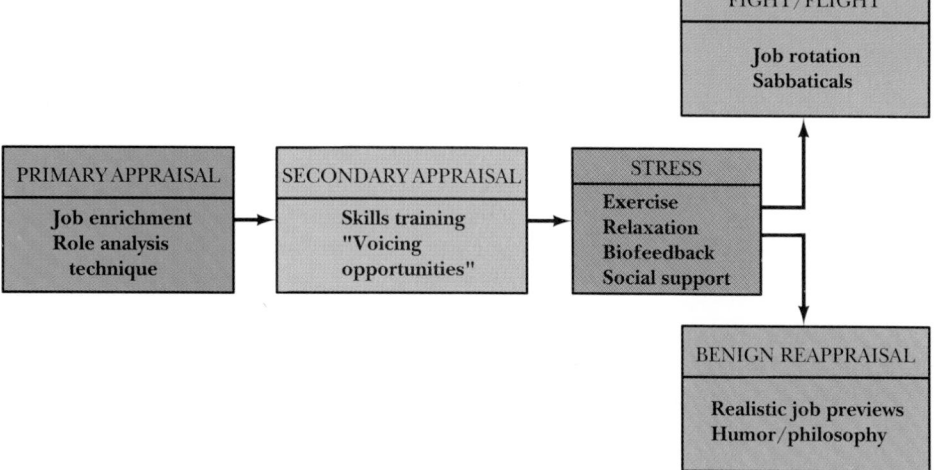

Job enrichment. Because the nature of the task is such a strong influence on dissatisfaction and stress, some of the most effective means of reducing negative reactions to work focus on the task. *Job enrichment* methods include many techniques designed to add complexity and meaning to a person's work. As the term *enrichment* suggests, this kind of intervention is directed at jobs that are boring because of their repetitive nature or low scope. Although enrichment is not universally successful in bringing about improved employee reactions to work, it can be very useful. In Chapter 17 we will discuss a number of job enrichment techniques.

APPLYING THE DIAGNOSTIC MODEL
Diagnostic Question 10: How can overly simple jobs in the organization be enriched? How can jobs characterized by too many conflicting role requirements be simplified?

Role Analysis Technique. Role problems rank right behind job problems in creating distress. *Role analysis technique* is designed to clarify role expectations for a jobholder by improving communication between her and her *role set*, or the supervisors, coworkers, subordinates, and other employees with whom she interacts regularly.

In role analysis, both jobholder and role set members are asked to write down their expectations. These people are then gathered together to review their lists. All expectations are written down so that ambiguities can be removed and conflicts identified. Where there are conflicts, the group as a whole tries to decide how these conflicts should be resolved. When this kind of analysis is done throughout an organization, instances of overload and underload may be discovered, and role requirements may be traded off, so that more-balanced roles can be developed. Compared to job enrichment, there has not been a great deal of research on role analysis. What little research there is suggests that this technique may be a useful means for reducing role pressures. We will discuss role analysis technique in greater detail in Chapter 14.

Targeting the Secondary Appraisal

Interventions that target the secondary appraisal aim at teaching the person how to cope with the demands that are creating the stress or dissatisfaction.

Skills Training. The key to the secondary appraisal process is to remove the person's self-doubt about his ability to do what he must to eliminate the stressor. *Skills training* is one way of accomplishing this end. For example, training in time management and goal prioritization has been successful in reducing managers' physiological stress symptoms such as rapid pulse rate and high blood pressure.[56] Subjects in one study first decided on their most important work values. They were then taught how to pinpoint goals, how to identify roadblocks to successful goal accomplishment, and how to seek the collaboration of coworkers in achieving these goals. Other research points to the importance of good job skills in overcoming stress. Lois Tetrick and James LaRocco found that the greater job incumbents' ability to predict, understand, and control events occurring on the job, the less stress they experienced. Moreover, being able to understand and control these events weakened the effect of perceived stress on job satisfaction.[57] Thus increasing a person's management and technical skills increases his capacity for resolving stressful situations.

[56] N. S. Bruning and D. R. Frew, "Effects of Exercise, Relaxation, and Management Skills Training on Physiological Stress Indicators: A Field Experiment," *Journal of Applied Psychology* 72 (1987), 515–21.
[57] L. E. Tetrick and J. M. LaRocca, "Understanding, Prediction and Control as Moderators of the Relationship Between Perceived Stress, Satisfaction and Pyschological Well-Being," *Journal of Applied Psychology* 72 (1987), 538–48.

voice The formal opportunity to complain to the organization about one's work situation.

APPLYING THE DIAGNOSTIC MODEL
Diagnostic Question 11: Do employees have outlets for registering complaints? Do they have any influence in decisions that affect how they conduct their jobs?

AN EXAMPLE
Police departments are increasingly faced with highly stressed officers. Some become so stressed they develop emotional problems that make it unsafe and unwise to put them on patrol. Many departments make an effort to give these individuals desk jobs. One major metropolitan department refers to these individuals as the rubber gun brigade. The problem is there are no longer enough desk jobs to meet the increasing demand. Police departments need to develop programs that will focus on reducing the amount of stress to which any individual is exposed and helping individuals develop healthy coping methods.

Voicing and Participating in Decision Making. A person's ability to handle dissatisfying or stressful work experiences is also enhanced when he feels he has an opportunity to air his problems. The formal opportunity to complain to the organization about one's work situation has been referred to as **voice**.[58] Work by Dan Farrell and his colleagues has shown that voicing provides employees with an active, constructive outlet for their work frustrations.[59] Research with nurses shows that the provision of such voice mechanisms as grievance procedures, employee attitude surveys, and question-and-answer sessions between employees and management all lead to better worker attitudes and less turnover.[60]

One step beyond voicing opinions is the chance to take action or make decisions based on one's opinions. *Participation in decision making* (PDM) provides opportunities for workers to provide input into important organizational decisions that involve their work. In a field experiment that randomly assigned subjects to PDM and non-PDM conditions, Susan Jackson found that PDM in the form of bimonthly information-sharing meetings among nursing staff and nursing supervisors resulted in reduced role conflict and ambiguity. In turn there was less emotional stress and absenteeism and fewer nurses resigned. Figure 8-8 depicts the process suggested by this study. Many organizations that use PDM have adopted *quality circles* as an approach to improving productivity and quality of work life. Quality circles (which will be discussed in more detail in Chapter 17) take advantage of the extensive knowledge of a company's operations that lower-level workers have often developed. Typically they are groups of three to thirty workers who meet for about an hour every week or two on company time to discuss production problems or problems with product quality. Although these groups are performance oriented, one hoped-for side effect is the enhancement of employee satisfaction.

Targeting the Symptoms of Stress

In some situations, neither roles, tasks, nor individual capacities can be altered sufficiently to reduce dissatisfaction and stress. Here we must aim our interventions at the symptoms of stress. Although clearly not as desirable as eliminating the stressors themselves, eliminating the symptoms is better than nothing. Some interventions that fall in this category focus exclusively on physiological reactions to stress.

[58] A. O. Hirshman, *Exit Voice and Loyalty* (Cambridge, Mass.: Harvard University Press, 1970), p. 51.

[59] D. Farrell, "Exit, Voice, Loyalty and Neglect as Responses to Job Dissatisfaction: A Multidimensional Scaling Study," *Academy of Management Journal* 26 (1983), 596–607.

[60] D. G. Spencer, "Employee Voice and Employee Retention," *Academy of Management Journal* 29 (1986), 488–502.

FIGURE 8-8

How Participation in Decision Making Affects Stress and Withdrawal from Work

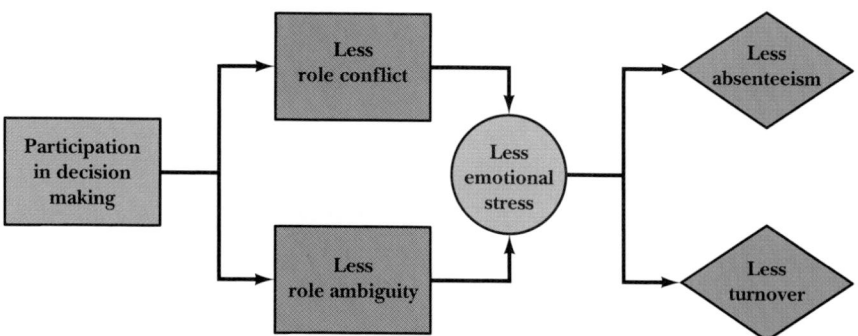

APPLYING THE DIAGNOSTIC MODEL
Diagnostic Question 12: What types of programs would be most useful in handling stress-related physiological symptoms that arise in this organization?

Exercise Programs. Physical conditioning, particularly in the form of *aerobic exercise*, helps make a person more resistant to the physiological changes, such as high blood pressure, that accompany stress reactions. CTI, a Knoxville-based manufacturer of medical equipment, and Southwestern Bell are but two of the many companies that hold aerobic exercise classes for their employees. Other firms organize group hikes and cross-country ski trips. Tenneco, a diversified manufacturing company even provides its employees, free of charge, with a gym that occupies 25,000 square feet. Here one can find basketball and racquetball courts, a workout area, a glass-enclosed running track with piped-in music and $200,000 worth of exercise and body-building equipment. Tenneco chairman, James Ketelsen, feels that the $3 million it takes to run the center is well worth it. He comments: "Our testing process discovered problems that could have been fatal. I'm sure we've saved some lives. How do you put a value on that?"[61] Research strongly supports the fact that this kind of program can be successful in reducing stress-related symptoms.[62] Other programs not only focus on encouraging positive behaviors, like exercise but also aim at discouraging negative behaviors that make the person less resistant to stress, like smoking or over eating.[63] Research by Katherine Parkes shows that both these factors are significantly associated with sickness and absenteeism and increase the negative impact of existing stressors on absenteeism and stress reactions.[64]

Relaxation Programs. Another approach to treating stress symptoms is to employ *relaxation techniques*. When under a severe amount of stress (as when preparing a fight or flight response), many of the body's muscle systems tighten. Relaxation programs focus on eliminating tenseness in most of the major muscle groups, including the hand, forearm, back, neck, face, foot, and ankle. Relaxing all these major muscle groups lowers blood pressure and pulse rate and reduces other physiological stress manifestations. One experiment that dealt with relaxation therapy in a social service agency found that subjects who were randomly assigned to such therapy reported less anxiety than did control subjects.[65]

Some forms of relaxation therapy, such as *meditation*, include mental concentration as well as tension reduction. In one popular form of meditation, transcendental meditation (TM), the person sits comfortably, relaxing all muscle groups, concentrating on a single thought, and repeating a special sound (a *mantra*) for 20 minutes. Advocates of TM have made claims that it can cure everything from physical disease to male-pattern baldness. Although many of these claims must be taken lightly, there is evidence from organizational research that TM can reduce heart rate, blood pressure, and oxygen consumption.[66]

Biofeedback. It was once thought that people had no voluntary control over physiological responses. **Biofeedback** machines that allow a person to monitor her own physiological reactions have changed all that.[67] Indeed, with the right

biofeedback A technique that uses machines to monitor bodily functions thought to be involuntary, such as heart beat and blood pressure, so that a person can learn to regulate these functions.

[61] M. Freudenheim, "Assessing the Corporate Fitness Craze," *New York Times*, March 18, 1990, p. D1.

[62] Bruning and Frew, "Exercise, Training, and Management Skills Training."

[63] K. D. Brownell, A. J. Stunkard, and P. E. McKeon, "Weight Reduction at the Worksite: A Promise Partially Fulfilled," *American Journal of Psychiatry* 142 (1985), 47–52.

[64] K. R. Parkes, "Relative Weight, Smoking and Mental Health as Predictors of Sickness and Absence from Work," *Journal of Applied Psychology* 72 (1987), 275–86.

[65] D. C. Ganster, B. T. Mayes, W. E. Sime, and G. D. Tharp, "Managing Organizational Stress: A Field Experiment," *Journal of Applied Psychology* 67 (1982), 533–42.

[66] D. Kuna, "Meditation at Work," *Vocational Guidance Quarterly* 12 (1975), 342–46.

[67] N. E. Miller, "Learning of Visceral and Glandular Responses," *Science* 163 (1969), 1271–78.

feedback, some people can learn to control brain waves, muscle tension, heart rate, and even body temperature. Biofeedback training teaches people to recognize when these physiological reactions are taking place as well as how to lower the levels of these symptoms when under stress. A biofeedback program set up by Equitable Life Insurance, for example, led to an 80 percent reduction in visits to the company's health center for stress-related problems.[68]

Social Support Groups. Because, as we have seen, a supportive environment can reduce stress, many organizations encourage team sports both at work and in off hours. The hope behind softball and bowling leagues is that group cohesiveness and support for individual group members will be increased through socializing and team effort. Although management certainly cannot ensure that every stressed employee will develop friends, it can make it easier for employees to interact.

Targeting Flight Reactions

Other means of coping with stress that cannot be eliminated at the source focus on flight reactions. (Remember, in the fight reaction the person attacks the source of the stress directly.) These interventions allow a person time away from the stressful environment.

Job Rotation. Does it surprise you to learn that air traffic controllers at Chicago's O'Hare airport are restricted to 90–minute work periods? They are required to rest for a period of time before returning to their stations. Many employers employ **job rotation** in much the same way. Although one may not feel capable of handling the stress or putting up with the dissatisfying aspects of a particular job indefinitely, it is often possible to do so temporarily. Rotation is supported by the resistance stage notion proposed in Selye's general adaptation syndrome, which we discussed earlier. Thus, during the Vietnam War, soldiers would "rotate stateside," or return to the U.S., after a tour of duty lasting for a specified time period. The idea behind this policy was that the stress of combat would be easier to manage if the soldier knew it would be over in the foreseeable future.

Job rotation in more conventional organizations can do even more than simply spread out the stressful aspects of a particular job. It can also increase the complexity of the work and provide valuable cross-training in jobs so that any one person eventually comes to understand many different jobs. This makes for a more flexible work force and increases the workers' appreciation for the other tasks that must get done in order for the organization to complete its mission. We will discuss job rotation and its consequences again in Chapter 17.

Sabbaticals. At the upper executive level, more and more organizations are also providing *sabbaticals*. These are extended periods of time away from work, lasting anywhere from six months to a year. The purpose of these leaves is to reenergize a person and provide a new perspective on life and work that can be achieved only when the whole process is viewed from a distance. The chairman of Apple Computers, John Sculley, took off for six months. Sculley claims that while building a barn he came up with an entirely new vision of the strategic direction his company should take over the next ten years.

---

job rotation A process whereby an individual is systematically moved from one job to another over the course of time.

POINT TO STRESS
The Japanese have used many of the techniques discussed here. Participative decision making, job rotation, and formal exercise programs are important components of human resource management in Japan. The Japanese believe that these programs allow more effective use of people. Much of Japan's high productivity has been attributed to the more effective use of human resources.

---

[68] J. S. Manuso, "Executive Stress Management," *Personnel Administrator* 24 (1979), 23–26.

Targeting Benign Reappraisal

Finally, there are also interventions that focus on making cognitive adjustments to stressful environments.

Realistic Job Previews. If the negative aspects of a job cannot be changed, managers should be up front with prospective jobholders about the nature of the work. Many companies are hesitant to admit to the undesirable aspects of a job when trying to recruit workers, for fear that nobody will take the job. Fooling someone into taking a job, however, is not good for the company or the person. Similarly, a fancy job title, like executive assistant, for a job that is more accurately described as typist is self-defeating. These practices raise peoples' expectations unjustly. The firm winds up attracting people to the job who would not be at all interested in joining the company if they knew what was truly involved in the work. The ultimate result is increased turnover. *Realistic job previews* (RJPs) lower expectations and are likely to attract workers whose values will more closely match the actual job situation (see also Chapter 14). When these people do go on the job and begin to experience its negative aspects, they are likely to see them as not so bad, saying, "I knew the job was difficult when I took it."

Steve Premack and John Wanous reviewed 21 separate realistic job preview experiments and found that although previews have not been 100 percent effective, there is certainly evidence that they can reduce subsequent turnover. Audiovisual presentations in these previews make them especially useful.[69]

Humor and Philosophical Approaches. Cognitively reappraising a situation through the use of *humor* is also an effective means of making an environment seem less threatening. A number of professional comedians have come from family backgrounds that were filled with stress and uncertainty. Humor for them was a means of diffusing and minimizing such stress.

Humor is also useful for managing one's image. For example, President Reagan had a full-time joke writer, Landon Parvin, on his staff. According to Parvin, for every Reagan speech he would "put together three to six pages of one-liners" that "relied mostly on self-deprecatory humor." Because it was thought that voters might be afraid that the President was too old, his age was the frequent target of Parvin's jokes. For example, speaking to the Washington Press Corps, Reagan noted that the Press Corps was founded in 1919 and quipped, "It feels just like yesterday." When giving a talk to the 105th annual meeting of the American Bar Association, Reagan said, "It isn't true that I attended the first meeting."[70] Obviously, the tactic here is to get people laughing about age, a potentially worrisome issue. If we can laugh about it, how serious an issue can it really be? This is benign reappraisal in its purest form.

Reappraisal can also take the form of changes in one's *personal philosophy* on life. A manager whose career seems to have come to a dead end may simply adopt the stance that "family should come first anyway." By devaluing career goals, he feels less troubled by his inability to accomplish those goals. Indeed, the notion of being born again to a new set of values, whether they be religious or otherwise, is a way of untying oneself from a dissatisfying past life. If this past life brought only dissatisfaction and stress, the substitution of a new set of values can certainly do no harm.

---

[69] S. L. Premack and J. P. Wanous, "A Meta-Analysis of Realistic Job Preview Experiments," *Journal of Applied Psychology* 70 (1985), 706–19.

[70] E. M. Miller, "Working Hard for the Last Laugh, *Time*, August 15, 1983, p. 16.

SUMMARY

Among the great variety of attitudes and emotions generated in the workplace, the most important constructs are *job satisfaction* and occupational *stress*. Job satisfaction is a pleasurable emotional state resulting from the perception that a job helps one attain valued outcomes. Occupational stress, an unpleasant emotional state, comes from the perceived uncertainty that one can meet the demands of a job when attaining important, valued outcomes is at stake. *Cognitive appraisal theory* is useful in comparing and contrasting these two constructs. Satisfaction is a product of the *primary appraisal*, whereas stress is a product of the *secondary appraisal*. There are multiple responses to stress, including physiological responses (high blood pressure, rapid pulse rates), behavioral responses (e.g., *fight or flight*), and cognitive reactions (e.g., *benign reappraisal*). These stress reactions have important consequences for organizations, particularly in the financial costs of health care, absenteeism and turnover, and performance failure. There are six discrete sources from which dissatisfaction and stress originate: These sources include the *physical-technical environment*, the *social environment*, the *behavior setting*, the *person*, the *organizational task*, and the *organizational role*. A dozen different intervention programs are aimed at eliminating the stress-inducing event, enabling the person to eliminate or cope with the stressor or, failing these efforts, eliminating the symptoms of stress. Some of these are *job enrichment*, *skills training*, *biofeedback*, *job rotation*, and *realistic job previews*.

REVIEW QUESTIONS

1. Recall from Chapter 2 some of the many roles a manager must play in her organization. Which of these roles do you think create the most stress, and which are probably the least stressful? Which role do you think most managers derive their greatest satisfaction from? Compare your answers to these three questions and speculate on the relationship between satisfaction and stress for managerial employees.

2. Reexamine the scale of life-change events presented in Table 8–5. Do you think that on a year-to-year basis more stress is created by on-the-job or off-the-job events? Do you think that more stress is generated by positive events or negative events? Can you think of particular years in a person's life when both positive and negative sources of stress are likely to be exceedingly high?

3. Organizational turnover is generally considered a negative outcome, and many organizations spend a great deal of time, money, and effort trying to reduce turnover. Can you think of any situations where an increase in turnover might be just what an organization needs? What are some steps that organizations might take to enhance functional types of turnover? Do you think mass firings of ineffective workers are likely to enhance overall organization effectiveness, or do you think that they would have deleterious effects on the firm's ability to recruit the most desirable applicants?

4. We saw in this chapter that characteristics like negative affectivity and the Type A behavior pattern are associated with aversive emotional states including dissatisfaction and stress. Do you think these tendencies are learned, or genetically determined? If these tendencies are learned, from reinforcement theory perspective, what reinforcers might sustain the behaviors associated with these characteristics? Although by their nature these characteristics seem associated with aversive outcomes, from an operant perspective, something must be reinforcing them for them to persist over time.

5. If off-the-job stress begins to spill over and create on-the-job problems, what do you think are the rights and responsibilities of the manager in helping the employee overcome these problems? If employees are engaged in unhealthy, off-the-job behavior patterns such as smoking, overeating, or alcohol abuse, what

are the rights and responsibilities of the employer to change these behaviors? Are intrusions into such areas an invasion of privacy? Is it a benevolent-paternalistic move on the part of the employer? Or is it simply a prudent financial step taken to protect the firm's investment?

DIAGNOSTIC QUESTIONS

In evaluating an organization to determine where and when dissatisfaction and stress may be problems and how to go about resolving these problems, the following diagnostic questions may prove a useful start.

1. Why might the organization be concerned about dissatisfaction and stress? What costly health-related or behavioral problems linked to stress are evident?

2. Are the values that the organization expects its members to uphold consistent with the members' own values and needs?

3. Are any aspects of the physical environment (such as noise, darkness, hazards) causing stress among organizational members?

4. Are any aspects of the social environment (such as hostile coworkers or supervisors) causing stress among organizational members?

5. Are any aspects of the behavioral settings (such as crowding or lack of privacy) causing stress among organizational members?

6. Do certain characteristics of organization members contribute to stress problems at work (e.g., negative affectivity or the Type A behavior pattern)?

7. How can the jobs that need to be performed by organization members be described in terms of complexity, meaning, and physical demand? How might characteristics of the jobs relate to dissatisfaction and stress?

8. How clear and unambiguous are the role expectations that are being sent to various organization members? How might the ambiguity of some expectations relate to stress?

9. Is the organization most interested in eliminating the sources of stress, or is it willing (or forced) to deal only with stress-related symptoms?

10. How can overly simple jobs in the organization be enriched? How can jobs characterized by too many conflicting role requirements be simplified?

11. Do employees have outlets for registering complaints? Do they have any influence in decisions that affect how they conduct their jobs?

12. What types of programs would be most useful in handling stress-related physiological symptoms that arise in this organization?

LEARNING THROUGH EXPERIENCE

EXERCISE 8-1
ROLES: UNDERSTANDING SOURCES OF STRESS*
Patrick Doyle, St. Lawrence College

All of us fulfill various roles in our lives. Each of these roles is a set of expectations about good or appropriate behaviors that people hold for us because we occupy a specific social position. When people's expectations for us in one of our roles conflict with the way we see ourselves in that role, or when expectations associated with one role conflict with expectations associated with another that we must fulfill, we tend to experience stress.

To reduce stress we must first recognize its causes. In order to do this we must identify the different roles we hold and recognize the expectations associated with each one. Suppose, for example, that you are a student in the business school, and that you are married and have several children. You are treasurer of your school's management club, and you are also a part-time employee in the campus bookstore. All of these roles and the people with expectations about your role performance could be diagrammed as shown in Exhibit 8-1. Starting at the upper left of this diagram, your instructors will expect you to attend class and do your home-

work. Your boss at the bookstore will expect you to be at your job on time and to perform the tasks you are assigned. Your parents will expect you to telephone them occasionally and to do well in school. Your spouse will expect you to be a companion and help support the family. The members of your club will expect you to keep accurate records of club revenues and expenses. And your children will expect you to spend time with them and to show an interest in their activities.

You can imagine potential conflicts among these varying role expectations. One such conflict could be expressed as shown in the illustration at the top of the next page.

Group discussions can often help identify ways of resolving such conflicts. For example, a family discussion might lead to the suggestion that you plan to take part in family activities earlier in the term, when school-

*Adapted from: J. William Pfeiffer, *The 1986 Annual: Developing Human Resources* (San Diego, CA: University Associates, Inc., 1986). Used with permission.

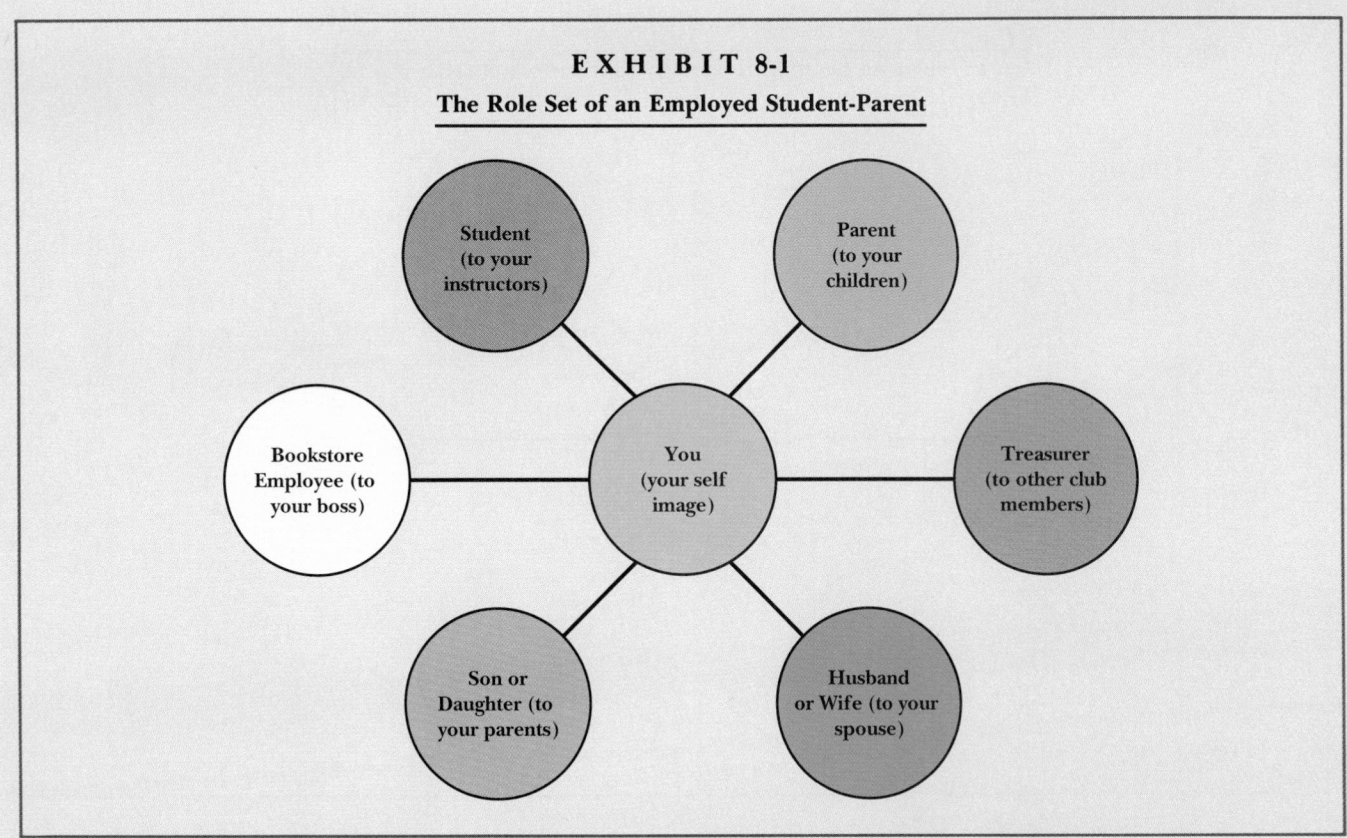

EXHIBIT 8-1
The Role Set of an Employed Student-Parent

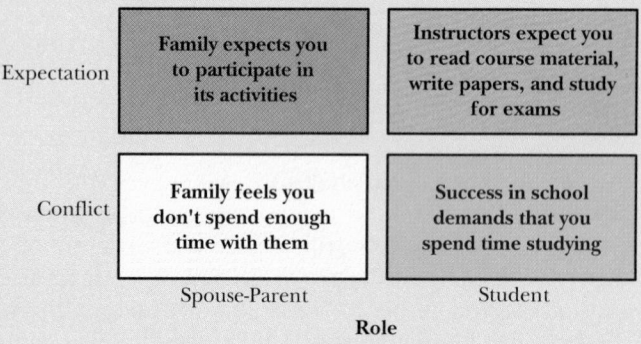

	Spouse-Parent	Student
Expectation	Family expects you to participate in its activities	Instructors expect you to read course material, write papers, and study for exams
Conflict	Family feels you don't spend enough time with them	Success in school demands that you spend time studying

Role

work demands are somewhat lighter, and focus more energy on coursework as the term progresses and you have projects to complete and final exams to take.

Reducing role-related stress requires you not only to identify the causes of stress but to modify role expectations and resolve conflicts among your different roles. This exercise gives you the opportunity to develop these kinds of skills so that you can use them to manage the stress in your life.

STEP ONE: PRE-CLASS PREPARATION

In class you are going to discuss the roles you actually occupy and consider how to reduce any stress that you feel as a result of conflicting role expectations. To prepare for this discussion, begin by completing the Roles Diversity Sheet in Exhibit 8-2. Be sure to identify *at least* four different roles. Next, fill in the Roles Characteristics Sheet in Exhibit 8–3, listing each of the roles you identified in the Role Diversity Sheet and, for each role, identifying the people who provide you with role expectations and the expectations themselves. Then, think about which pairs of expectations you have listed might conflict and list these, one on each side of the top portion of the Role Conflict-Resolution Sheet in Exhibit 8-4. In the bottom portion of the sheet, list any ways you can think of that you might cope with these conflicts or resolve them.

STEP TWO: PRE-CLASS PREPARATION

The class should divide into groups of four to six members each (if you formed permanent groups in an earlier

---

EXHIBIT 8-2
Roles Diversity Sheet

Instructions: Use this sheet to identify four to six roles that you fill.

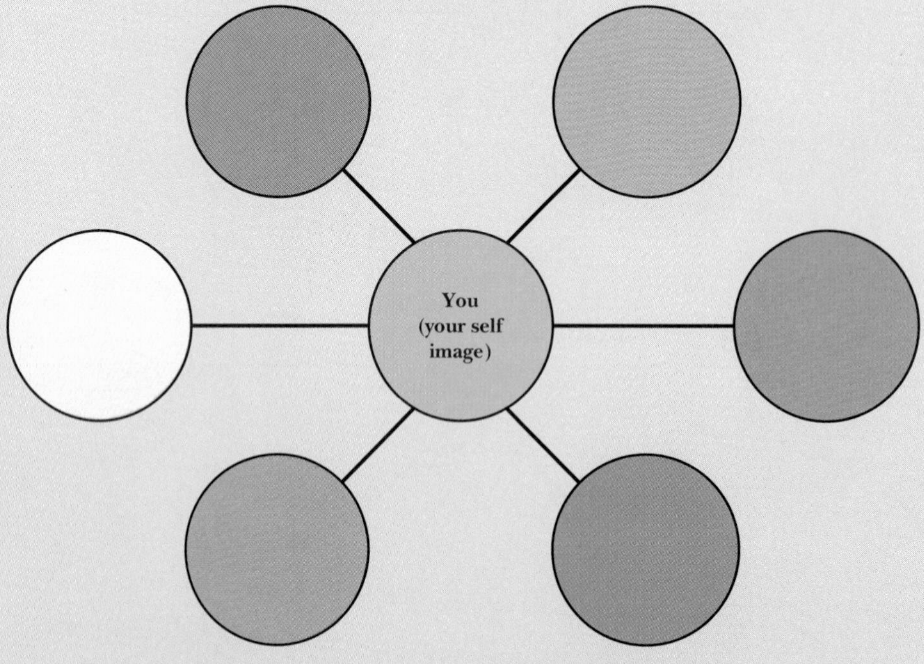

You
(your self image)

278

EXHIBIT 8-3
Roles Characteristics Sheet

List the roles you have identified as follows: First, list these roles in descending order, according to how comfortable you feel in them; the most comfortable first, the least comfortable last. Then, for each role, list the people who provide you with expectations for that role and describe those expectations.

ROLES	EXPECTATIONS

EXHIBIT 8-4

Roles Conflict-Resolution Sheet

Given the roles that you have listed, what expectations are likely to conflict?

EXPECTATIONS THAT CONFLICT

What are some potential ways to resolve these conflicts?

POTENTIAL RESOLUTIONS

SATISFACTION AND STRESS

exercise, reassemble in those same groups). Each group member should begin by describing one of the conflicts identified in Step One, being sure to include descriptions of the roles involved, the people holding expectations for those roles, and the expectations themselves.

After every group member has described a conflict, the group should discuss potential ways to reduce the stress caused by each of the conflicts identified. To begin this discussion, members should refer to potential solutions identified during Step One and should provide each other with feedback about those solutions. The group should then search for about additional ways to solve the role conflicts of its members.

If time permits, group members may describe additional conflicts and ask the group to consider how to solve them. Throughout Step Two a spokesperson should take notes and be prepared to summarize the group's discussion for the class as a whole.

STEP THREE: CLASS DISCUSSION

The instructor should convene the entire class. As each group spokesperson summarizes a group's discussion, the instructor should list all role conflicts and potential solutions the group finds acceptable on a blackboard, flipchart, or transparency projector. The class should look for similarities and contrasts among the results of Step Two, being sure to discuss the following questions:

1. What is the most common type of role-related stress experienced by members of this class? The second most common type? The third most common type? What expectations are causing each kind of stress? From whom do these expectations originate?

2. For each type of stress, what seem to be the most effective methods of stress reduction? Did any approaches seem promising initially but fail ultimately to be useful?

3. What kind of support might you need from others around you to implement the stress reduction approaches you've identified? What obstacles stand in your way? What are the benefits of adopting one or more of these approaches? How might they change your life?

CONCLUSION

Stress originating in conflicts among role expectations is ever present in modern organizations and everyday life. It is important, therefore, that you have the skills needed to identify role conflicts and to decide what to do about them. The process of charting roles, recognizing expectations, and identifying potential conflicts and solutions might seem overly complicated and time consuming. However, the costs of failing to deal with role conflict are even greater: prolonged stress can cause severe physical and psychological illness, limiting your ability to lead a satisfying and productive life.

DIAGNOSING ORGANIZATIONAL BEHAVIOR

CASE 8-1
NO RESPONSE FROM MONITOR TWENTY-THREE*
ROBERT D. JOYCE, *Innovative Management*

Loudspeaker: IGNITION MINUS 45 MINUTES.

Paul Keller tripped the sequence switches at control monitor 23 in accordance with the countdown instruction book just to his left. All hydraulic systems were functioning normally in the second stage of the spacecraft booster at checkpoint 1 minus 45. Keller automatically snapped his master control switch to GREEN and knew that his electronic impulse along with hundreds of others from similar consoles within the Cape Kennedy complex signaled continuation of the countdown.

Free momentarily from data input, Keller leaned back in his chair, stretched his arms above his head, and

then rubbed the back of his neck. The monitor lights on console 23 glowed routinely.

It used to be an incredible challenge, fantastically interesting work at the very fringe of man's knowledge about himself and his universe. Keller recalled his first day in Brevard County, Florida, with his wife and young daughter. How happy they were that day. Here was the future, the good life . . . forever. And Keller was going to be part of the fantastic, utopian future.

Loudspeaker: IGNITION MINUS 35 MINUTES.

* Reprinted with the publisher's permission from Robert D. Joyce, *Encounters in Organizational Behavior* (New York: Pergamon Press, 1972), pp. 168–72.

Keller panicked! His mind had wandered momentarily, and he lost his place in the countdown instructions. Seconds later he found the correct place and tripped the proper sequence of switches for checkpoint 1 minus 35. No problem. Keller snapped master control to GREEN and wiped his brow. He knew he was late reporting and would hear about it later.

Damn! he thought, I used to know countdown cold for seven systems monitors without countdown instructions. But now . . . you're slipping Keller . . . you're slipping, he thought. Shaking his head, Keller reassured himself that he was overly tired today . . . just tired.

Loudspeaker: IGNITION MINUS 30 MINUTES.

Keller completed the reporting sequence for checkpoint 1 minus 30, took one long last drag on his cigarette, and squashed it out in the crowded ashtray. Utopia! Hell! It was one big rat race and getting bigger all the time. Keller recalled how he once naively felt that his problems with Naomi would disappear after they left Minneapolis and came to the Cape with the space program. Now, 10,000 arguments later, Keller knew that there was no escape.

Only one can of beer left, Naomi? One stinking lousy can of beer, cold lunchmeat, and potato salad? Is that all a man gets after 12 hours of mental exhaustion?

Oh, shut up, Paul! I'm so sick of you playing Mr. Important. You get leftovers because I never know when you're coming home . . . your daughter hardly knows you . . . and you treat us like nobodies . . . incidental to your great personal contribution to the Space Age.

Don't knock it, Naomi. That job is plenty important to me, to the Team, and it gets you everything you've ever wanted . . . more! Between this house and the boat, we're up to our ears in debt.

Now don't try to pin our money problems on me, Paul Keller. You're the one who has to have all the same goodies as the scientists earning twice your salary. Face it, Paul. You're just a button-pushing technician regardless of how fancy a title they give you. You can be replaced, Paul. You can be replaced by any S.O.B. who can read and punch buttons.

Loudspeaker: IGNITION MINUS 25 MINUTES.

A red light blinked ominously indicating a potential hydraulic fluid leak in subsystem seven of stage two. Keller felt his heartbeat and pulse rate increase. Rule 1 . . . report malfunction immediately and stop the count. Keller punched POTENTIAL ABORT on the master control.

Loudspeaker: THE COUNT IS STOPPED AT IGNITION MINUS 24 MINUTES 17 SECONDS.

Keller fumbled with the countdown instructions. Any POTENTIAL ABORT required a cross-check to separate an actual malfunction from sporadic signal error. Keller began to perspire nervously as he initiated standard cross-check procedures.

"Monitor 23, this is Control. Have you got an actual abort, Paul?" The voice in the headset was cool, but impatient, "Decision required in 30 seconds."

"I know, I know," Keller mumbled, "I'm cross-checking right now."

Keller felt the silence closing in around him. Cross-check one proved inconclusive. Keller automatically followed detailed instructions for cross-check two.

"Do you need help, Keller?" asked the voice in the headset.

"No, I'm O.K."

"Decision required," demanded the voice in the headset. "Dependent systems must be deactivated in 15 seconds."

Keller read and reread the console data. It looked like a sporadic error signal . . . the system appeared to be in order.

"Decision required," demanded the voice in the headset.

"Continue count," blurted Keller at last. "Subsystem seven fully operational." Keller slumped back in his chair.

Loudspeaker: THE COUNT IS RESUMED AT IGNITION MINUS 24 MINUTES 17 SECONDS.

Keller knew that within an hour after lift off, Barksdale would call him in for a personal conference. "What's wrong lately, Paul?" he would say. "Is there anything I can help with? You seem so tense lately." But he wouldn't really want to listen. Barksdale was the kind of person who read weakness into any personal problems and demanded that they be purged from the mind the moment his men checked out their consoles.

More likely Barksdale would demand that Keller make endless practice runs on cross-check procedures while he stood nearby . . . watching and noting any errors . . . while the pressure grew and grew.

Today's performance was surely the kiss of death for any wage increase too. That was another of Barksdale's methods of obtaining flawless performance . . . which would surely lead to another scene with Naomi . . . and another sleepless night . . . and more of those nagging stomach pains . . . and yet another imperfect performance for Barksdale.

Loudspeaker: IGNITION MINUS 20 MINUTES.

The monitor lights at console 23 blinked routinely.

"Keller," said the voice in the earphone. "Report, please."

"Control, this is Wallace at monitor 24. I don't believe Keller is feeling well. Better send someone to cover fast!"

Loudspeaker: THE COUNT IS STOPPED AT IGNITION MINUS 19 MINUTES 33 SECONDS.

"This is Control, Wallace. Assistance has been dispatched and the count is on temporary hold. What seems to be wrong with Keller?"

"Control, this is Wallace, I don't know. His eyes are open and fixed on the monitor, but he won't respond to my questions. It could be a seizure or . . . a stroke."

When you have read this case, look back at the chapter's diagnostic questions and choose the ones that apply to the case. Then use those questions with the ones that follow in your case analysis.

1. Explain Paul Keller's behavior at monitor 23. Why did he have trouble remembering and following countdown instructions? Why did he fail to answer Wallace's questions?
2. How were Keller's problems at home related to his behavior at work? Why did he have trouble sleeping? What might explain his stomach pains?
3. What should Barksdale do about Keller and his problems? Can an organization like NASA do anything to identify employees like Paul and to help them cope with their problems before they become serious?

CASE 8-2

CAMERAN MUTUAL INSURANCE COMPANY*

ROBERT J. COX, *Salt Lake Community College*

Cameran Mutual Insurance Company is a large national insurance company that has been in business since the early 1900s. The company is best known for its loss prevention service and for its workers compensation policies. The company also takes pride in personal sales and service to its industrial insurance accounts.

Recently, Mrs. Kay was referred to a local insurance office in Salt Lake City, Utah, where she had applied for a job. She had a college degree in sociology, some business background, five years' experience with public relations-type jobs, and an excellent reputation at her previous jobs for being reliable, dependable, and a hard worker. She was well qualified for the job except

* Reprinted with the author's permission.

for her lack of technical knowledge about the insurance business. However, she typed up a resume and made an appointment for a job interview.

The job interview was long and intense. First, the potential supervisor, Mrs. Perry, interviewed her for 30 minutes. Afterwards, she was required to fill out an application form. Upon completion of the form she was called back into the office to talk with the district sales manager, Mr. Landers. At the conclusion of this interview she was asked to fill out a more in-depth questionnaire that required far more detailed information. Finally, she was asked to come into the office for a third interview. While Mrs. Kay was a bit overwhelmed by the length of interviews and the personal data required on the forms, she was nevertheless flattered by the personal attention and felt that the extra time and depth of concern was a good omen for her chances of securing the job for which she applied. In this final interview, both Mrs. Perry and Mr. Landers asked, "What is your major goal in life? If you had a chance to do anything over again in your life what would it be? Do you have any objections to working with people who smoke? If you get this job, what do you think you would dislike the most?" Such questions caught her off guard, but somehow she felt she responded favorably in the eyes of the interrogators.

The job was described in the interview as being the "right-hand" assistant to the sales manager, Mr. Landers. It would be Mrs. Kay's job to fill in whenever he was out of the office: to prepare rates as specified by the underwriters, prepare reports, collect data for policies, handle phone calls, file, type, and perform other duties assigned by the sales manager and the supervisor of sales assistance, and even to handle some duties assigned by the district sales manager and underwriters. Since she'd need to be licensed by the state to sell policies, the company would pay for the on-the-job training and also the cost of the license fee.

Early that afternoon, Mrs. Kay was informed that, provided her references and other job information checked out, she would have the job and would start her job training the following Monday. She would be working very closely with the woman whom she would be replacing, Mrs. Mone. Mrs. Mone had agreed to stay on for the next three weeks to help with the orientation process.

Mrs. Kay arrived at work early Monday morning so she could become oriented to the office. She met Mr. Johnson, the Claims Manager, Mr. Metts, the Loss Prevention Manager, and several other workers in the office. When Mr. Landers arrived, Mrs. Kay was asked to fill out additional legal and administrative forms, including government licensing forms, a bonding contract, and insurance forms. She was then oriented to many of the company policies and benefits. She learned that raises were to be based upon the quality and quantity of work she performed, not on seniority. There was a mandatory probation period of three months, and then she would be eligible for insurance and sick leave benefits. There were also educational benefits that included full financial reimbursement for all classes dealing with insurance and reimbursement for half the price of the books used. There were also many additional benefits offered.

After a rather formal introduction to most of the workers in the office, Mrs. Kay was shown to her desk and told to start "on-the-job reading" of manuals in a prescribed manner. She started her studies in insurance with an introduction and description of the loss prevention program. Because this particular office served Utah, Wyoming, and Montana, it was necessary for some part of the staff to be out of the office much of the time, leaving a large amount of clerical work to be done by those who stayed in the office, including Mrs. Mone (who handled both technical assistance and routine clerical responsibilities).

During the course of the day there was time for Mrs. Kay to observe office functions and procedures. She also watched, with growing interest, the relationship between Mr. Landers and Mrs. Mone. Mrs. Kay was rather surprised at the behavior of Mr. Landers. Without any apparent provocation, except for a minor mistake, Landers burst into a fit of rage belittling Mrs. Mone in front of all the other workers. Mrs. Mone was apparently accustomed to her supervisor's behavior because she put up with his temper tantrum and did not get upset over the incident.

Later that day, Mrs. Kay talked briefly with Mrs. Mone about Mr. Landers. She said, "Well, you see, everybody in the company below the level of Executive Vice President has two or more bosses. For example, you will have Mr. Landers and Mrs. Perry as your main supervisors. Later on, you'll learn that the handling of many of your accounts and your bosses' accounts will be subject to the judgment and releases of the underwriting department. In a sense, you'll be taking on a third boss." She talked further about Mr. Landers. "Mr. Landers tends to get angry without regard to who is at fault. You'll also find out that there will be occasions where Mrs. Perry will direct you to do one thing, and Mr. Landers will tell you to do almost the complete opposite. There will also be occasions when they will both direct you to do the same chore, but they will use different terminology for the specific tasks they want you to accomplish. I've tried to find assigned tasks in the procedural manuals, but many times I've found that I've had to ask either Mr. Landers or Mrs. Perry for directions to complete the task, only to find that Mr. Landers gets angry, and Mrs. Perry is out of her office. Just don't let him bully you around. If you are right

(and you'd better be sure you are), stick to your guns and you'll come out o.k."

Reflection on the day's activities caused Mrs. Kay to feel good about the people she'd be working with, but she still felt a little apprehensive about Mr. Landers. Getting into her car, she immediately sensed the day's accumulation of the cigarette smoke that had adhered to her clothes and hair. It was soon apparent that the smoking of the other employees made her physically ill. This surprised her a bit since she had smoked up until four years before.

The next day, accustomed to arriving a few minutes early for work, Mrs. Kay arrived at 8:15 A.M., just fifteen minutes before work was to begin. She straightened out her desk and then pondered over the events of the previous day. When the clock reached 8:30 A.M. she started immediately into her studies. She later talked to the Claims Manager, Mr. Johnson, who informed her of his departmental functions. This helped her grasp how she fit into the picture of this office.

Mr. Landers instructed Mrs. Mone to spend at least an hour a day teaching Mrs. Kay the clerical duties. No specific hour was mentioned; however, and at three o'clock that afternoon, Mr. Landers again verbally assaulted Mrs. Mone because she hadn't instructed Mrs. Kay on the clerical duties. Twenty other people in the office looked on and listened to the argument. Mrs. Kay felt sympathetic toward Mrs. Mone and wondered if this was the way that she would be treated by Mr. Landers. She talked with another technical-clerical person, Sherry Olsen, who told Mrs. Kay that every time that Mrs. Mone stood up to Mr. Landers she felt like applauding. She added, "If he ever yells at you, don't feel embarrassed because all of the staff knows how he reacts and we're all used to it."

At the end of the day, Mrs. Kay put on her coat and prepared to go home. The instant she sat in the car, the nauseating smell of stale tobacco recaptured her attention. The odor became very strong, and almost overwhelmed her. This added to the anxiety that she already felt. There were several heavy smokers in the office, and the office had very little ventilation. It was quite easy to accumulate smoke in her clothes. Arriving home, supressed feelings rose to the surface and she became very tempermental, even hostile. This was contrary to her nature, but she supposed that the anxieties in acquiring a new job, the irritating physical effects of the office smoke, and the problems of working with a very temperamental boss had finally taken their toll. Family members unfortunately were most convenient and subject to the venting of her frustrations.

That night she pondered over the events of the past two days. She tried to weigh the benefits and drawbacks of her new job. Could she do it? She would receive a fair salary with good benefits, on-the-job training, and have opportunities for advancement. However, there were obvious complications. Much would be expected of her: a much heavier work load than most new employees carry, the "pool hall" working conditions, and a very temperamental boss worried her. Could she expect to have much impact on working conditions or her boss? Would there be a way of implementing new office procedures or practices?

When you have read this case, look back at the chapter's diagnostic questions and choose the ones that apply to the case. Then use those questions with the ones that follow in your case analysis.

1. Why did Mrs. Kay find the smell of cigarette smoke on her clothing so disagreeable? Why did she become temperamental at home after her second day at Cameran Mutual Insurance Company? What attitudes about work might be symptomized by these feelings and behaviors?

2. How might having two or three bosses affect an employee's attitudes toward her work? Her ability to perform satisfactorily? How would you advise Cameran management on this issue?

3. Under the circumstances described, how long would you expect Mrs. Kay to remain with Cameran Insurance? Why? What improvements, if any, would you recommend in the way the company's management treats employees?

CASE 8-3

CONNORS FREIGHT LINES

Review this case, which appears in Chapter 7. Next, look back at Chapter 8's diagnostic questions and choose the ones that apply to the case. Then use those questions with the ones that follow in your case analysis.

1. Was Chuck Fletcher dissatisfied with his supervisory job? Why? What symptoms do you see in this case that seem to signal his dissatisfaction? What factors might be causing this dissatisfaction?

2. Was Chuck's supervisory job stressful? What clues suggest that Chuck was experiencing stress? What sources of stress can you identify?

3. What actions might Connors Freight Lines take to reduce the dissatisfaction and stress experienced by supervisors like Chuck? Are there things the company currently does that it should stop doing? Are there things it doesn't do that it should start doing?

MANAGING INDIVIDUALS IN ORGANIZATIONS

Whether you are a line manager and supervise employees who do the organization's work or a staff manager in human resources, selecting and placing applicants in the company's jobs, knowing some simple rules of the game can save you and your company a lot of heartache. One of these is, Know and respect one another's expertise and call on it when you need to. If Larry Buck's supervisor (see opening story) had referred his caller to human resources rather than trying to handle the inquiry himself, Hall & Co. might not have been out $2 million.

Larry Buck's boss, an executive at Frank B. Hall and Company, an insurance firm in Houston, worked 50 hours a week keeping up to speed on insurance matters and had little time to keep track of what was happening in human resources. He certainly could not be expected to have the expertise in this area that members of the company's personnel department doubtless possessed. Thus, what happened one day in 1987 is not so surprising. The recently dismissed Buck, angry about his treatment by Hall and Company, hired a private investigator who telephoned Buck's former boss and represented himself as an executive of another insurance company. The private investigator pretended that he was interested in hiring Buck. Asked by the investigator what kind of person Buck was, the Hall and Company manager told the detective exactly what he thought: "Buck was a Jekyll and Hyde person, a classic sociopath."

An outraged Buck sued his former employer for malicious slander and libel and won before the U.S Supreme Court. Hall and Company had to pay Buck over $600,000 in lost wages and a $1,300,000 penalty, for a grand total of close to two million dollars. Local or long distance, this is a high price for a three-minute telephone call! Moreover, it was an unnecessary expense, for the litigation could have been avoided if Buck's former manager had known what was standard operating procedure for his company's personnel department.[1] There is an increasing tendency among former employees to bring suits against past employers who provide damaging references. The old rule "if you can't say anything nice about a person, don't say anything at all" is common procedure for today's human resource staff. A company staff member responding to a referral request should share only information about the duties of the job the person in question was performing. If compelled to describe the person, the staff member should share only information that is objectively verifiable.

APPLYING THE DIAGNOSTIC MODEL
Diagnostic Question 1: In this organization, who are the experts in P/HR? What is the best way to contact these people to discuss issues of mutual concern?

Organizations often distinguish between line managers and staff managers. Line managers oversee the creation, production, and distribution of organizational products or services. Staff managers provide advice, assistance, and guidance to line managers in certain functional specialty areas (e.g., accounting and finance). Although executives and line managers need to be informed in all the functional areas of staff expertise, in the area of organizational behavior, the link between managers and the human resources, or personnel, department is of paramount importance.

The functional specialty area of personnel/human resources (P/HR) deals with enhancing the effectiveness of the organization's employees.[2] P/HR departments handle employee recruitment, employee selection, training of employees, worker surveys, employee compensation, and collective bargaining. The increasingly litigious nature of our society has made it necessary also for P/HR specialists to keep up to date on legislative and litigation issues in the area of human resources. Most public or private organizations that employ at least 150 people find that the complexity of the problems in this area are sufficient to create a separate department to carry out these activities.[3]

---

[1] Based on "The Revenge of the Fired," *Newsweek*, February 16, 1987, pp. 46–47.
[2] G. T. Milkovich and J. W. Boudreau, *Personnel/Human Resource Management: A Diagnostic Approach* (Plano, Texas: Business Publications, 1988), pp. 3–4.
[3] H. G. Heneman, D. P. Schwab, J. A. Fossum, and L. D. Dyer, *Personnel/Human Resource Management* (Homewood, Ill.: Richard Irwin, 1986), p. 17.

289

TABLE 9-1
Areas of Overlap among Personnel Directors and Executives*

	MANAGERS			PERSONNEL DIRECTORS		
	Setting Policy	Advising	Controlling	Setting Policy	Advising	Controlling
Employment testing	55%	81%	42%	50%	50%	46%
Performance appraisal	60	83	48	51	56	44
Affirmative action programs	71	61	65	76	79	63
Developing pay structures	56	51	42	36	26	29
Incentive systems	45	63	36	53	71	46
Training	53	71	43	52	56	40
Punishment and discipline	59	85	37	58	57	44
Career planning	62	73	44	65	70	51
Average percent	58	71	45	55	58	45

* Entries represent percentage of respondents who deal regularly with the area represented.
Source: H. C. White and M. N. Wolfe, "The Role Desired for Personnel Administration," *Personnel Administrator* 25 (1980), 90–91.

Because this book is addressed to current and future executives and line managers and not to personnel specialists, it would be inappropriate to go into the details of P/HR here. There is, however, a great deal of overlap in the interests and needs of line managers and personnel specialists. Thus effective understanding and communication between them is critical in managing a company's human resources. The Hall and Company experience is a good example. Their P/HR group surely knew the dangers of providing negative information about a former employee. Better communication between P/HR and Larry Buck's manager could have saved the company two million dollars that day in Houston.

Research conducted with groups of managers and P/HR specialists suggests that eight primary areas of OB and P/HR overlap. These eight areas are listed in Table 9-1, which shows the nature and level of involvement for each group in each of the eight areas. Clearly, line managers and executives are heavily involved in all areas.

As you will see, the practices of P/HR specialists are often based on the theories of organizational behavior described in this section of the book. Thus this chapter serves a dual purpose. It shows you how organizational behavior and human resources management interrelate, and it further illustrates the way many of the concepts and theories we have explored in the preceding five chapters find application in industry. Table 9-2 also fulfills two purposes. It tells you which earlier chapter each of the major sections in this chapter relates to, and in so doing, it provides you with an outline of this chapter.

TABLE 9-2
How Organizational Behavior and Human Resource Management Mesh

ORGANIZATIONAL BEHAVIOR	HUMAN RESOURCE MANAGEMENT
Human attributes (Chapter 4)	Employment testing
Perception and judgment (Chapter 5)	Performance appraisal
Administrative decision making (Chapter 6)	Affirmative action programs
Equity theory (Chapter 7)	Developing pay structures
Expectancy theory (Chapter 7)	Incentive systems
Reinforcement theory (Chapter 7)	Training and discipline programs
Cognitive appraisal theory (Chapter 8)	Career development

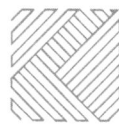

DIAGNOSTIC ISSUES

Applying the diagnostic model to human resources management highlights several important issues for managers. For starters, can we *describe* areas of organizational behavior that are of interest both to line managers and to the human resources staff? Can we develop systems for analyzing and describing jobs that will make recruitment and selection of employees more efficient? What *diagnostic* techniques can help us explain why selection and promotion procedures that seem fair sometimes fail to create a racially integrated workforce? How can we distinguish between situations in which merit raises and profit-sharing will be useful and situations in which pay-for-performance plans may prove disastrous?

How can we *prescribe* which job applicants will perform well and which will struggle unsuccessfully? What can we do to ensure that employees will respond positively to performance appraisals? What *actions* can we take to create a pay structure that will be viewed as fair? What can we do to change the behaviors of workers who engage in counterproductive activities?

TESTING EMPLOYEES' ABILITIES

One constantly recurring problem in organizations is deciding whom to hire for various job openings when there are many more applicants than positions. Without a crystal ball, one cannot look into the future and know which applicants will become high performers and which will fail. Using certain kinds of tests, however, employers can forecast these outcomes with a fair degree of accuracy and thereby increase organizational effectiveness. Moreover, effective testing and screening programs help individual job applicants—even those who are rejected. If applicants who are turned away would have been unsuccessful if hired, steering them into other occupations for which they are better qualified contributes to their own development.[4] Testing may also be useful in encouraging unsuccessful applicants to go back and improve their skills and qualifications through further education.[5] Thus for both employer and applicant, predicting job success accurately is important.

We will now look at two models of testing that are widely used in predicting job success—aptitude testing and skill testing. Both forms of testing are based on the theories of human attributes that we discussed in Chapter 4. As you study this section, you may want to refer back to that discussion.

Aptitude Testing

In Chapter 4, we noted that there were basically eight major physical abilities, nine major psychomotor abilities, eight major mental abilities, and five categories of personality characteristics (see Tables 4-4, 4-5, and 4-6 and Figure 4-4). **Aptitude testing** is illustrated in Figure 9-1. First, we perform a job analysis to determine which human attributes are most relevant to good performance on the job in which we are interested. Next we purchase any one of a number of existing tests that measure these particular human attributes. Then in our initial "test of the tests," we administer them to a group of job applicants. We keep the scores a secret (so as not to create expectation effects—see Chapter 5) from those who

aptitude testing Measuring broad, general abilities of job applicants.

APPLYING THE DIAGNOSTIC MODEL

Diagnostic Question 2: What are some of the major attributes of the important jobs in this organization? What kinds of aptitude tests might predict success in these jobs?

---

[4] G. F. Dreher and P. R. Sackett, *Perspectives of Employee Staffing* (Homewood, Ill.: Richard Irwin, 1984), p. 57.

[5] B. Schneider and N. Schmitt, *Staffing Organizations* (Glenview, Ill.: Scott, Foresman, 1986), p. 91.

FIGURE 9-1

The Process of Aptitude Testing

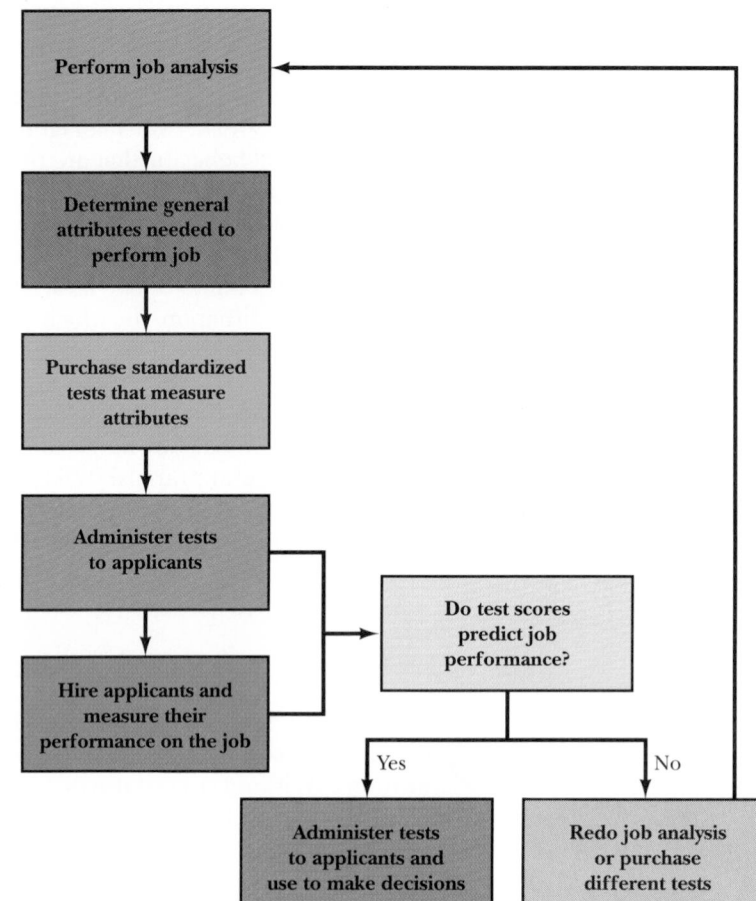

APPLYING THE DIAGNOSTIC MODEL
Diagnostic Question 3: For what jobs in this organization could skill tests be developed? What kinds of work sample tests would be most appropriate for these jobs?

The Aetna Institute for Corporate Education, headed by Badi Foster, former Harvard education professor, teaches Hartford inner-city residents to do basic math, write an acceptable memo, and compose a business letter. To be admitted to this program, would-be job applicants must pass an aptitude test, be drug-free, and have positive self-esteem. Foster admits his program can't handle all the inner-city educational problems. But it has made a small dent: two years after Aetna Life launched this program 47 people had completed the course and were hired by the company.
Source: Joel Dreyfuss, "Get Ready for the New Work Force," Fortune, April 23, 1990.

will be appraising the actual performance of whoever gets the job. In this trial period, the tests are not generally used for making selection decisions, because we have not yet proven their predictive value.

After a period of time, we look for a relationship between success on the tests and success on the job. That is usually done by calculating the correlation coefficient (see Chapter 3) between the original test scores and scores derived from a performance appraisal discussed later in this chapter. If there is a relationship, we can conclude that the tests are valid. (You may recall from Chapter 3 that this is referred to as criterion-related validity.) Once we have validated our tests in this fashion we can start to use them to help us decide which applicants to accept and which to reject for a particular job. Of course, if there is no re-

AN EXAMPLE
Often, we need to validate tests for jobs in which only a few people are hired each year. In these situations it would take far too long to develop a large enough sample to test validity. In these cases we can give the tests to be validated to current holders of the job we're hiring for and correlate their scores with performance ratings. High test scores should be related to high performance scores. This is a weaker method of validation, but it can be used when the situation warrants it.

lationship between the test scores and on-the-job performance, we may have to reanalyze the job or obtain new tests and go through the procedure again.

An example of aptitude testing is the Graduate Management Admissions Test (GMAT), which most prestigious MBA programs require applicants to take. An analysis of the task that confronts an MBA student clearly indicates the need for verbal and quantitative ability, two broad facets of ability tapped by this test. Because individuals who perform well on the test have the abilities required for the position, it should come as no surprise that the test predicts success in MBA programs.[6]

Many employers also use tests of verbal and quantitative ability, such as the Differential Aptitude Test or the Employee Aptitude Survey.[7] This kind of testing is called aptitude testing because the same attribute (and even the same test) may be used to select people for different jobs. That is, general mental ability may be needed for a whole host of jobs in an organization, and anyone applying for any of these jobs may be tested with the same instrument. Indeed, tests of general mental aptitude have been successfully validated across a wide variety of job categories.[8]

Skill Testing

skill testing Measuring narrow, job-specific abilities of job applicants.

Skill testing ties the test specifically to the job in question. One test or set of tests is designed to fit each position. The process underlying this method, shown in Figure 9-2, rests on the assumption that the best predictor of future behavior is

[6] A. Jenson, *Bias in Mental Testing* (New York: Free Press, 1980), p. 239.

[7] G. K. Bennett, H. G. Seashore and A. G. Wesman, *Differential Aptitude Tests: Administrator's Handbook* (New York: Psychological Corporation, 1982), p. 55; and F. L. Ruch and W. W. Ruch, *Employee Aptitude Survey: Technical Report* (Los Angeles: Psychological Services, 1980), p. 2.

[8] J. E. Hunter, "Cognitive Abilities, Cognitive Aptitudes, Job Knowledge, and Performance," *Journal of Vocational Performance* 29 (1986), 411–14.

FIGURE 9-2
The Process of Skill Testing

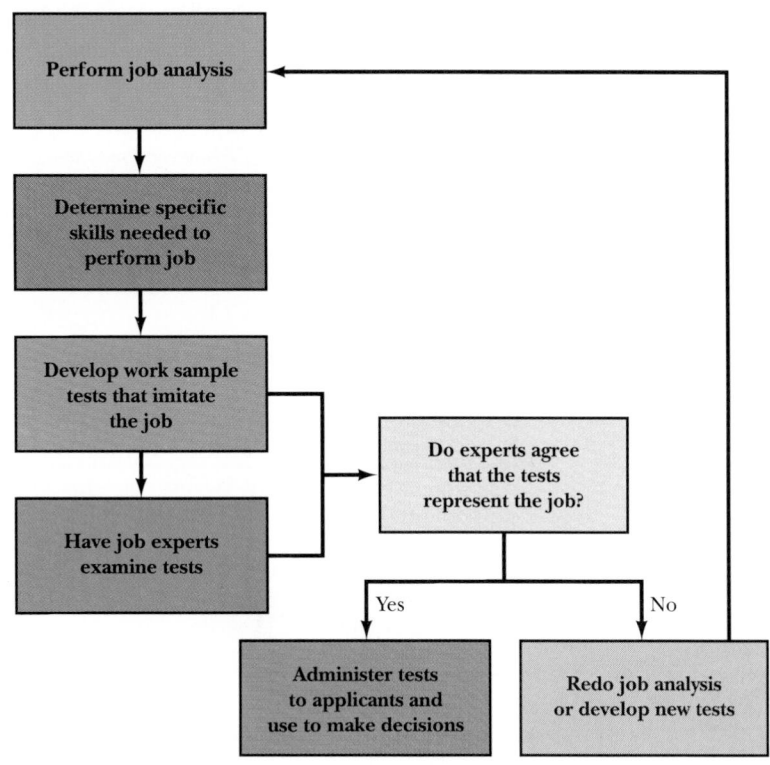

past behavior. The key differences between aptitude testing and skills testing are specificity and focus. Aptitude tests are generic and focus on abstract human attributes. Skill tests are specific and focus on concrete human behaviors. Aptitude tests tend to be purchased "off the rack" whereas skill tests are usually developed on site.

The key in skill testing is to develop tests that imitate or mimic the job in miniature form by *sampling* the specific behaviors the job requires. Because the behaviors tested are sampled from the job, these tests are often referred to as **work sample tests**. If the applicant cannot perform well on the "miniaturized job," it is unlikely that she will be able to perform well on the real job.

Work sample tests come in great variety. Look at Table 9-3, for example. This is a question from a **situational interview** test given to someone applying for a sales position in the Zales chain of jewelry stores. In the interview, the applicant must respond off the top of his head, and his response is scored for appropriateness on the scale shown in the table.

Another type of work sample test is the *business simulation*. In one well-known simulation, six applicants must work together as a group to operate a manufacturing company. This test calls for the job applicants to purchase raw materials, manufacture the product, develop a marketing plan, and then distribute and sell the product. IBM found that performance on this simulated manufacturing problem accurately predicted the number of promotions obtained by 94 middle managers three years into the future.[9]

There are also exercises called *in-basket tests* that confront the applicant with the kinds of problems she will encounter on the job. The term *in-basket* derives from the fact that the problems are often actual company memos or letters presented to the applicant as if they had just arrived in her "in" basket. As in the case of other tailored tests, it is assumed that if a job applicant can handle the simulated problem presented to her in a test, she will be able to handle similar problems on the job. Figure 9-3 shows an array of items confronting applicants in one in-basket test.

The most common way of validating a work sample test is to employ expert judgments about how well the test actually imitates the job. If experts agree that the content of the test accurately represents the content of the job, the test can be considered valid (this is content validity, as you may recall from Chapter 3), and we can use it to screen applicants.[10] Again, however, if the experts do not

---

[9] W. F. Cascio, *Managing Human Resources: Productivity, Quality of Worklife, Profits* (New York: McGraw-Hill, 1989), p. 211.

[10] G. F. Dreher and P. R. Sackett, "Some Problems with Applying Content Validity Evidence to Assessment Center Ratings," *Academy of Management Review* 6 (1981), 551–60.

work sample tests Tests that present job applicants with realistic simulations of actual job problems and ask them to indicate how they would handle them.

situational interview A work sample test in which the applicant is asked to respond orally to hypothetical problems that might confront him while working on the job.

TEACHING NOTE
In-basket tests and business simulation are commonly included in an extensive series of tests known as "assessment centers." These centers go to great trouble to replicate the work situation. They are quite expensive, but many companies report their benefits far outweigh their costs. They are hiring the right people.

AN EXAMPLE: APPLE, MOTOROLA
Companies are realizing the large costs associated with hiring the wrong person. Many companies like Apple and Motorola realize they need people who can deal with ambiguity and are participative and customer oriented. For these reasons, they are defining job requirements more sharply. They are also formalizing and improving testing. A key component in their new testing is observing how an individual functions in a simulated job situation. If you want to know how people will work, don't ask them, watch them (*Industry Week*, March 7, 1988, pp. 31–34).

TABLE 9-3

A Question from a Zales Work Sample Test for Potential Jewelry Store Salespersons

A customer comes into the store to pick up a watch he left for repair. The repair was supposed to have been completed a week ago, but the watch is not back yet from the repair shop. The customer becomes very angry. How would you handle this situation?

1 • Tell the customer it isn't back yet and ask him or her to check back with you later.
2 • Apologize, tell the customer that you will check into the problem and call him or her back later.
3 • Put the customer at ease and call the repair shop while the customer waits.

Adapted from J. A. Weekley and J. A. Gier, Reliability and validity of the situational interview for a sales position. *Journal of Applied Psychology* 72 1987, pp. 484–87.

FIGURE 9-3

**Items from a Typical
In-Basket Test**

*Source: N. Fredericksen, "Factors in
In-Basket Performance,"
Psychological Monograph 76
(1962), 22–41.*

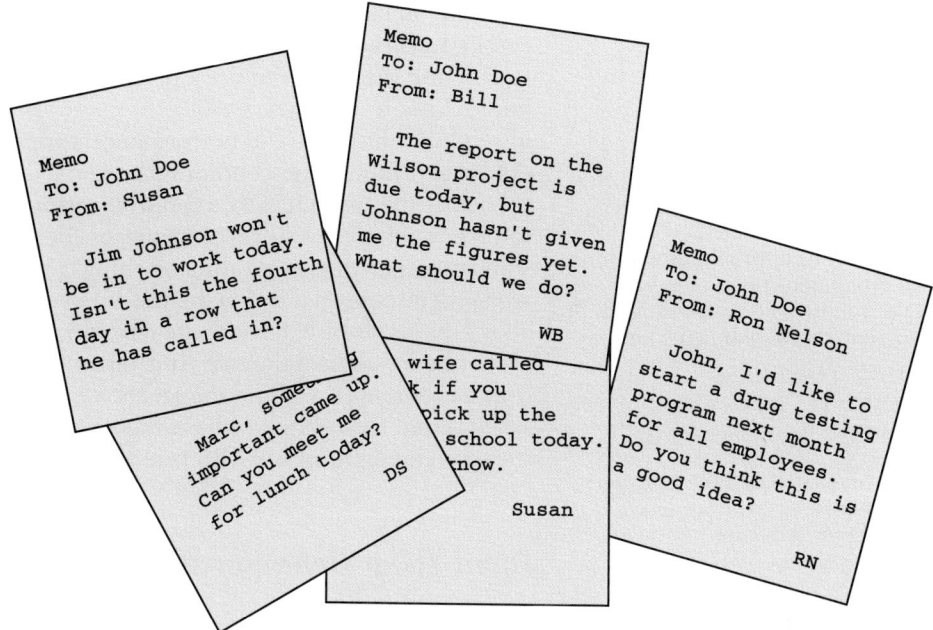

agree that our test samples the job in question accurately, we must reanalyze the job or design new tests.

APPRAISING PERFORMANCE BEHAVIORALLY

In Chapter 5, we discussed a number of barriers to perceptual accuracy. In fact, there are so many ways for bias to find its way into the subjective judgment process that many people question whether supervisors can evaluate their subordinates accurately. In one influential Supreme Court case, the court found that a General Motors foreman's judgments of workers' "ability, merit, and capacity" were too subjective to be fairly used in making promotion decisions. In other cases, supervisors' judgments of such qualities of their subordinates as leadership, resourcefulness, capacity for growth, loyalty, mental alertness, and personal conduct have been cited as "susceptible to partiality and to the personal taste, whim or fancy of the evaluator."[11]

APPLYING THE DIAGNOSTIC MODEL
Diagnostic Question 4: How free from bias are the subjective performance ratings of supervisors in this firm? If objective measures are used, how much control do employees have over the functions that these measures assess?

If we cannot rely on subjective measures, what about objective measures? Sales volume, products produced per hour, customer repeat business, or departmental financial performance all seem at first glance like attractive alternatives to subjective judgments. A little scrutiny, however, usually reveals several very unattractive features in measures of this sort. The most serious problem is that many things that influence such seemingly objective indicators are not under the control of the employee.[12] An insurance agent's sales volume, for example, depends on the general level of the economy, the amount and nature of advertising, competitors' policies, territory served, and many other factors over which the

---

[11] R. D. Arvey and R. H. Faley, *Fairness in Selecting Employees* (Reading, Mass.: Addison-Wesley, 1988), p. 37.
[12] F. L. Landy and J. L. Farr, *The Measurement of Work Performance: Methods, Theory and Applications* (New York: Academic Press, 1983), p. 111.

salesperson has no control. Similarly, the number of products produced per hour can be influenced by such things as machinery breakdowns, quality of raw materials, and the quality of work performed on the product earlier on the assembly line.

One solution to the performance appraisal dilemma that originated out of theories of human perception is the behaviorally based appraisal. One such approach is the **behaviorally anchored rating scale (BARS).** The BARS attempts to take much of the subjectivity out of the performance appraisal process while still basing performance rating on things the employee can control.[13] Subjectivity is lessened somewhat by having the supervisor make simple, objective judgments about the kinds of behaviors the employee has exhibited. Focusing on behaviors also enhances perceptions of controllability. The worker has much more control over behaviors (e.g., how she treats a customer) than over the outcomes (e.g., sales volume) associated with most objective measures. Let's look next at the four-step development of a BARS (see also Figure 9-4).

Identifying Behavioral Dimensions

In the first step, a group of experts, which may include supervisors, high-performing job holders, the company's owners, and some of the company's clients, are asked to *identify the important dimensions of performance* on the job in question. For example, in a hospital setting, an expert panel might decide that the important dimensions of a nurse's job are (a) demonstrating organizational ability, (b) showing compassion for patients, (c) working well with physicians, and (d) fulfilling technical responsibilities competently.

---

[13] H. J. Bernardin and P. C. Smith, "A Clarification of Some Issues Regarding the Development and Use of Behaviorally Anchored Rating Scales," *Journal of Applied Psychology* 66 (1981), 458–63.

F I G U R E 9-4

Steps in Developing Behaviorally Anchored Rating Scales

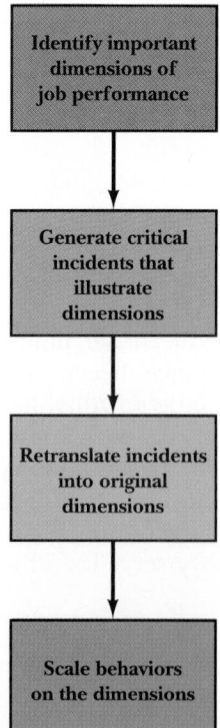

Generating Critical Behaviors

When the basic dimensions of the job in question have been established, a second group of experts is assembled to *generate behaviors that illustrate the performance dimensions*. It is this group's job to illustrate both effective and ineffective behaviors on the dimension with accounts of events that have actually occurred on the job. For example, for the dimension of organizational ability, this group may generate stories about effective nurses who "request early trays for patients who may take longer to eat," or "check medication orders for each day and arrange a schedule of distribution." Stories about ineffective nurses might include "make several trips to the supply closet to get a footboard" or "frequently leave important work undone in order to leave on time."

Retranslating Behaviors

When both the performance dimensions and the critical behaviors have been agreed on, a third group of job experts is convened (this group can be the same as the first one if there are few experts available). The job of this group is to *retranslate the behaviors back into the dimensions* they were chosen to represent. This process is called *retranslation*, because it is rather like translating one language into another. Here, one person translates from the original language to the second language, and then a different person translates that result back into the original language. If the original and the retranslation agree, one can conclude that everyone is "speaking the same language." Retranslation serves as a check on judges' accuracy. It is the basic reason for using multiple groups in the BARS development process.

For example, if one of the behaviors that was supposed to represent organizational ability is perceived by the retranslation group as representing working well with physicians, the description is eliminated from the scale for organizational ability. Conversely, if the retranslation group correctly identifies the behaviors, it is clear that everyone agrees.

Scaling Behaviors

In the last phase of the BARS building process, the behaviors that survive the retranslation process must be rated on a scale of effectiveness that ranges from 0, for completely ineffective, to 10, for very effective. In *scaling behaviors on a dimension*, we determine the mean, or average, value of each of the eleven ratings assigned to the scale by the members of a fourth group of experts (the second group if there are insufficient numbers of experts). Behaviors can also be dropped from the scale if there is substantial disagreement on rating—for example, if one person assigns a particular behavior a 2 on the scale but another assigns it an 8. This kind of disagreement suggests that the groups agree that the behavior reflects the dimension, but that the groups cannot agree on the extent to which the behavior is effective or ineffective.[14] Figure 9-5 shows what the final BARS might look like for the dimension of organizational ability in a nurse's position.

Research evidence has sometimes, though not always, found behaviorally anchored rating scales to generate more reliable ratings of performance than traditional measures. If too much time elapses between observation and rating, the BARS sometimes performs much like a subjective rating of personality char-

---

[14] R. Jacobs, D. Kafry and S. Zedeck, "Expectations of Behaviorally Anchored Rating Scales," *Personnel Psychology* 33 (1980), 595–640.

FIGURE 9-5

A Behaviorally Anchored Rating Scale for the Dimension of Organizational Ability for a Nursing Job

Raters are instructed to circle the number that best exemplifies the typical behavior of the person they are rating.

Reprinted from: J. Sheridan, "Measurement of Job Performance in Nursing Homes," report submitted to Health Resources Administration, 1980, p. 77.

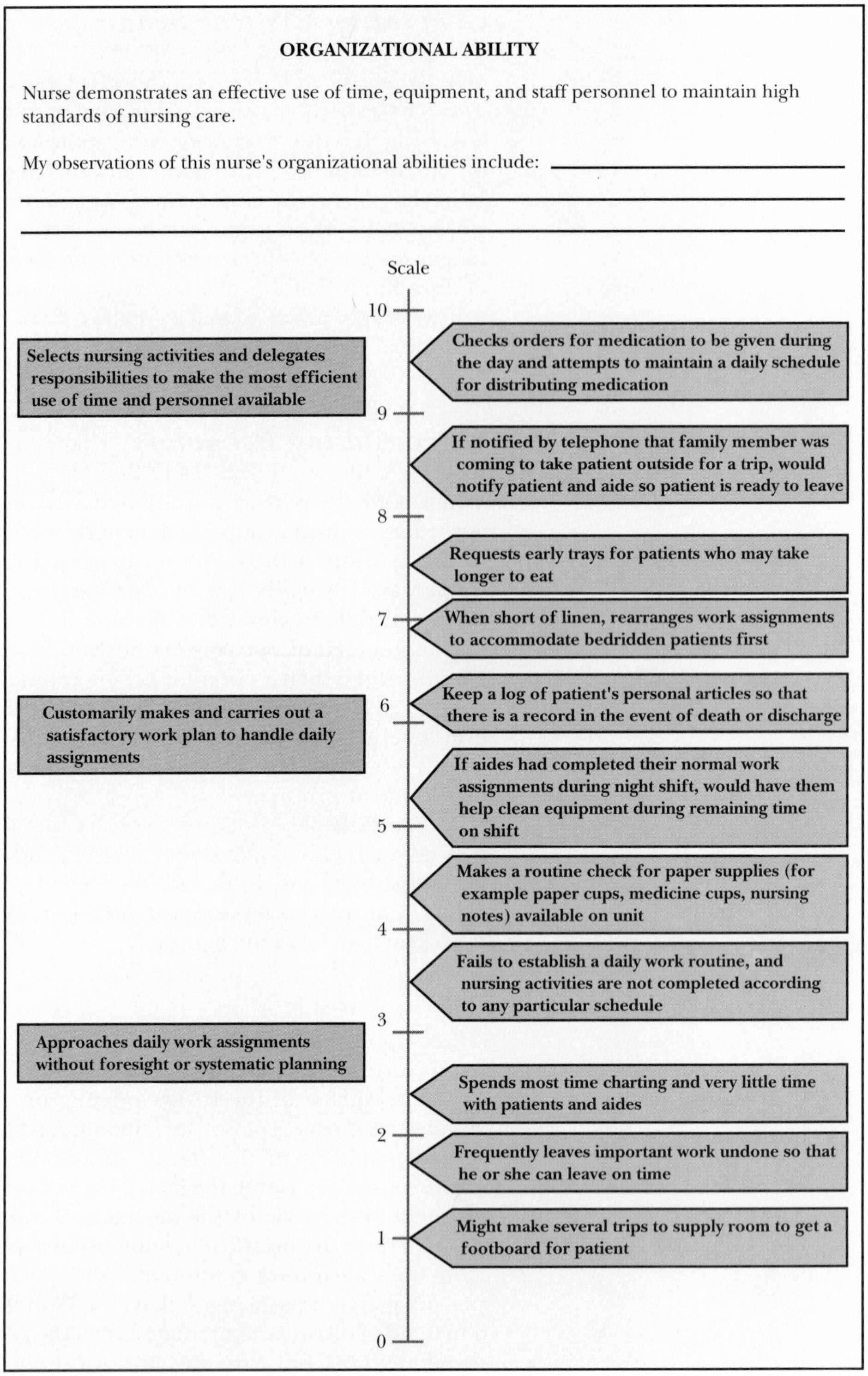

ORGANIZATIONAL ABILITY

Nurse demonstrates an effective use of time, equipment, and staff personnel to maintain high standards of nursing care.

My observations of this nurse's organizational abilities include: _____

_____

_____

Scale

10 — Checks orders for medication to be given during the day and attempts to maintain a daily schedule for distributing medication

Selects nursing activities and delegates responsibilities to make the most efficient use of time and personnel available

9 — If notified by telephone that family member was coming to take patient outside for a trip, would notify patient and aide so patient is ready to leave

8 — Requests early trays for patients who may take longer to eat

7 — When short of linen, rearranges work assignments to accommodate bedridden patients first

6 — Keep a log of patient's personal articles so that there is a record in the event of death or discharge

Customarily makes and carries out a satisfactory work plan to handle daily assignments

If aides had completed their normal work assignments during night shift, would have them help clean equipment during remaining time on shift

5 — Makes a routine check for paper supplies (for example paper cups, medicine cups, nursing notes) available on unit

4 — Fails to establish a daily work routine, and nursing activities are not completed according to any particular schedule

3 — Approaches daily work assignments without foresight or systematic planning

Spends most time charting and very little time with patients and aides

2 — Frequently leaves important work undone so that he or she can leave on time

1 — Might make several trips to supply room to get a footboard for patient

0 —

AN EXAMPLE: AT&T, GENERAL MOTORS

Affirmative Action laws were intended to move us past making decisions based on erroneous stereotyping. Chusmir and Durand argue that we need research and training programs aimed at differentiating between perception and reality. For example, it is generally thought that women have higher turnover rates than men. In reality, turnover rates are higher in lower-level jobs, where most women are. When turnover rates for men in these jobs are compared to the rates for women there are no differences. AT&T and General Motors report women in upper-level jobs actually quit less frequently than men. Thus, the instability of women on the job appears to be a myth. (*Training and Development Journal*, August, 1987, pp. 32–37)

acteristics.[15] Therefore, most experts agree that use of the BARS should be augmented with behavioral diaries (see Chapter 5 for a refresher on this).

The principal advantages of the BARS are that it is rooted in actual behaviors that are controlled by the employee, and that it requires relatively straightforward judgments of things the supervisor can observe directly. In addition, the common

[15] K. R. Murphy, C. Martin and M. Garcia, "Do Behavioral Observation Scales Measure Observation?" *Journal of Applied Psychology* 67 (1982), 562–67.

practice of involving subordinates in developing the scale increases their understanding of the process and hence their commitment to it and acceptance of it. Finally, the BARS provides a common frame of reference for supervisors and subordinates. It not only provides clear and specific feedback to poor performers but minimizes opportunities for disagreement and conflict.

AFFIRMATIVE ACTION PROGRAMS

In Chapter 6, on decision making, we noted that although there are a lot of pressures on organizations to make decisions rationally, many of the problems that confront decision makers are too complex to be solved by mathematical algorithms. One such problem that confronts most businesses today is managing affirmative action programs. These programs aim to increase the selection and promotion rates of employees who are members of various socially defined subgroups (for example, African-Americans, Hispanic-Americans, women, the elderly, the handicapped). Of the four factors prohibiting rational decision making discussed in that chapter, two stand out in the context of affirmative action planning: lack of goal consensus and bounded discretion. In the next two sections we will elaborate on the issues that complicate affirmative action programs, and show one widely used, but controversial solution.

Constraints on Decision Making and Actions

Managing affirmative action programs is characterized by *lack of goal consensus*. Planning in this context sometimes puts the goals of maximizing productivity and integrating the work force at odds with each other. It can also pit two American values against each other. The notion that people should be rewarded on the basis of achievement, or merit, can conflict with the notion of equal civil rights. The problem becomes, What is an organization to do when test scores or performance evaluations result in the selection or promotion of more white males than minority group members or women? This problem comes to the fore in two areas—tests of mental ability, which usually show large racial differences, and tests of certain physical abilities (like upper-body strength), which generate sex differences.[16] Should test scores or performance evaluations be ignored so that the work force can be integrated? Or should occupational segregation be ignored so as to promote those with the highest scores? In the United States, we have wrestled with this problem for years and still have not solved it.[17]

Managing affirmative action programs also involves the problem of *bounded discretion*, which means that the decision maker's list of alternatives is restricted by the social environment. In the context of affirmative action, bounds are often set for organizational decisions by social forces outside the organization itself.

In order to make personnel decisions, organizations must discriminate between employees on one basis or another every day. Yet laws like the Civil Rights Act of 1964, the Age Discrimination Act of 1967, and the Vocational Rehabilitation Act of 1978 prohibit discrimination if it relates to race, sex, religion, national origin, age, or handicapped status. Groups specified in this legislation are often referred to as **protected groups**.[18]

[16] Jenson, *Bias in Mental Testing*, p. 241; and Arvey and Faley, *Fairness in Selecting Employees*, p. 41.
[17] "Bigots in the Ivory Tower," *Time*, January 23, 1989, p. 22.
[18] K. J. McCulloch, *Selecting Employees Safely Under the Law* (Englewood Cliffs, N.J.: Prentice-Hall, 1981), p. 191.

disparate treatment An illegal practice in personnel selection wherein a person from one subgroup is asked to respond to questions, take tests, or display skills that are not asked of applicants from other groups.

disparate impact The tendency of a particular personnel selection practice to result in the hiring of a smaller percentage of the members of a particular group of job applicants than of other groups of applicants.

top-down selection within subgroups Personnel selection practice in which scores on a test are arrayed by subgroup and a flexible goal for representation from each group is decided upon. Selection proceeds from the highest to lowest score in each subgroup until all positions are filled and all subgroups are represented.

In 1990 President George Bush signed into law the Americans with Disabilities Act, which bars discrimination based on age, sex, religious, ethnic or racial origin, or physical or mental disability virtually across the board. Organizations throughout the United States in both the public and private sectors must henceforth hire without prejudice and must provide special facilities for the physically handicapped. Failure to comply with the new law may incur formal injunctions against violators and require back pay for victims of discrimination. Source: Stephen A. Holmes, "House Approves Bill Establishing Broad Rights for Disabled People," The New York Times, *May 23, 1990, p.1.*

A review of the litigation in this area reveals that employers are bounded by two primary rules. First, it is illegal to engage in disparate treatment. **Disparate treatment** means that members of protected groups are subjected to employment practices that differ from those used with other groups. In a selection interview, a fire department manager cannot ask a woman, "Does your husband approve of your working?" unless he asks a man, "Does your wife approve of your working?" The apparent silliness of the latter question illustrates the sexist nature of the first question.

Second, any practice that, although it appears neutral, is shown to have adverse or disparate impact on any group must be shown to be a strict business necessity. **Disparate impact** is the harmful effect on the employment opportunities of a protected group of a particular practice, even if the practice is universally applied. For example, both male and female applicants for the job of firefighter may be subjected to a test of upper-body strength. This neutral-appearing practice usually results in the rejection of many more women than men. If this happens, the fire department must be able to prove that the test is a valid predictor of performance on the job. That is, it must establish job necessity.

Most people accept the arguments associated with the disparate-treatment restriction, yet there is quite a bit of controversy associated with the disparate-impact restriction. To crystallize this controversy for you, we will take the example of an employer who uses a validated test of general cognitive ability in managerial selection for both African-American and white applicants. Keep three facts in mind. First, roughly a tenth of the United States population is African-American. Second, because of cultural disadvantages and other factors, the scores of African-Americans on tests of cognitive ability are typically lower than those of whites. Third, across a wide variety of jobs, there is a weak but reliable relationship between general cognitive ability and performance. Clearly, if organizations are concerned only with hiring people who get the top scores on their tests, they will hire proportionately fewer minority applicants who, in general, achieve lower scores on such tests.

Top-Down Selection within Subgroups

One solution to the problem posed by affirmative action has been the development of the concept of top-down selection within subgroups.[19] To understand how this type of selection process works, let's look first at a process in which goal consensus maximizes performance and there are no bounds on decision making. This situation is often referred to as *pure top-down selection*.

Suppose we have 55 open positions and 20 applicants, and we are using a test of general cognitive ability for selection. The data could very well turn out as in Table 9-4. If we ignore the goal of integrating the work force, we can simply take each individual's score on our best-predictor-of-performance test, rank order the applicants, and hire the top scorers (those whose test scores are above the cutoff score). At this cutoff, however, almost all African-American applicants are rejected.

In a context where we place value on the goal of integrating the work force and equal civil rights, this outcome is unacceptable. The alternative, therefore, is **top-down selection within subgroups**.[20] With this method, one draws two separate cutoffs, one for each group (see Table 9-5). This way, the five jobs are filled with each group's top scorers (circled). In the example in Table 9-5, this

[19] L. J. Cronbach, E. Yalow, and G. Schaeffer, "A Mathematical Structure for Analyzing Fairness in Selection," *Personnel Psychology* 33 (1980), 692–703.

[20] Ibid., Cronbach et. al.

TABLE 9-4
Pure Top-Down Selection

APPLICANT NUMBER	TEST SCORE	
1	100	
2	98	
3	97	
4	94	Select
5	94	
		------Cut-Off Point
6	92	
7	90	Reject
8	89	
9	86	
10	85	
11	85	
12	85	
13	84	
14	81	
15	78	
16	74	
17	73	
18	68	
19	65	
20	57	

procedure would result in our hiring one minority member, one woman, and three other people.

Top-down selection is not the same thing as a quota. Quotas in personnel selection set aside a specific number of positions for protected group members irrespective of their standing on selection tests. Although top-down-within-subgroups strategy is a less controversial technique than the quota, many still object to its use because of its ability to put nonminority candidates at a disadvantage.

TABLE 9-5
Top-Down Selection within Subgroups

MINORITY GROUP NUMBERS		WOMEN		OTHERS	
Applicant No.	Test Score	Applicant No.	Test Score	Applicant No.	Test Score
(7)	90	(8)	89	(1)	100
11	85	10	85	(2)	98
13	84	15	78	(3)	97
14	81			4	94
				5	94
				6	92
				9	86
				12	85
				16	74
				17	73
				18	68
				19	65
				20	57

Gearing Up for Workforce 2000

Consider the plight of Harold Epps, a plant manager at Digital Electronics Corporation in Boston. Epps has 350 workers in his plant assembling computer keyboards. These 350 people come from 44 different countries and speak 19 different languages. It is routine practice in this firm to issue plant announcements printed in English, Chinese, French, Spanish, Portuguese, Vietnamese, and Haitian Creole.

Workforce 2000 may sound like a futuristic, science fiction thriller to some, but for many managers like Epps, the future is now. *Workforce 2000* is actually the title of a study performed by the Hudson Institute on demographic changes in the United States and the implications of these changes for managing businesses in the years ahead. For employers the two most striking findings of this study are the type of work that will be required in the twenty-first century and who will be available to do it.

With respect to the second finding, it is clear that the work force will grow very slowly throughout the next decade, as ageing baby boomers begin to retire, and there are fewer workers available to take their places. Moreover, the workers who will be available will have more diverse backgrounds. Today, 47 percent of new entrants into the U.S. labor pool are native white males. In the year 2000, this group will constitute only 15 percent of new entrants.

As for the type of work to be done, it is also clear that new jobs in the year 2000 will demand much higher skill levels than the jobs of today. There will be few jobs for those who cannot read or perform basic mathematics. The trends in skill requirements, coupled with the trends in the nature of the work force, have made "managing diversity" a key issue.* Companies from Goodyear to Hewlett-Packard to Procter and Gamble are creating positions to handle this function—for example, "Diversity Manager" or "Vice President for Diversity." Jill Kanin-Lovers, vice-president at Towers Perrin, a management consulting firm, notes that "for the first time at many companies, human resources is a strategic issue." Companies are making a whole host of maneuvers to survive the population shifts that will confront them in the years ahead.

Some of the push comes in the area of recruiting. Procter and Gamble, for example, makes a concerted effort to recruit a higher percentage of women and minority engineers than we find with undergraduate degrees in engineering (10 percent for women and 15 percent for minorities). McDonald's provides summer corporate internships and year-round res-

taurant management programs for minority college students. Xerox gives minority prospects reprints of an article in *Black Enterprise* magazine that rates the company as one of the best places for African-Americans to work. Hewlett-Packard employs the same strategy with a copy of *Working Women* that rates HP as one of the best places for women to work.

There is also a push to increase the general level of sensitivity to cultural differences in the work place. For example at the DEC keyboard plant in Boston, problems developed when a young man newly arrived from a foreign country tried to give a present to a female colleague who had come to the United States from a third country. In the man's culture, accepting such a gift signaled that the woman was romantically interested. The woman held no such feelings, however. Rather, in her culture, it was considered inappropriate ever to turn down a gift. Cultural clashes like this one can only be averted by training workers to recognize and understand the cultures of their coworkers.

Pushing in yet another direction, firms have begun to reach out and provide remedial training to workers who otherwise could not meet their qualifications. Word-processing jobs at Aetna Life Insurance, for example, require basic skills in reading and math. When applications for these jobs went down 40 percent, Aetna launched classes in inner-city areas teaching residents basic office skills in intensive 14-week courses. Aetna has hired all 47 graduates of this program and is delighted with their performance levels.

Finally, firms are also pushing to retain women and minority workers once they are hired and trained. Corning found, for example, that between 1980 and 1987, African-American and women professionals left the company at roughly twice the rate of white men. A diversity-training program for managers and professionals helped reduce attrition rates for both groups. Avon encourages employees to organize into African-American, Hispanic, and Asian networks by granting them official recognition and providing a senior mentor to act as an advisor to the group. The cosmetics company once had a similar network for women but disbanded it after the program achieved its objective (today women hold 79 percent of the management positions at Avon). Avon's chief of human resources told a reporter that "My objective is to create an organization where people don't feel a need for a black network, a Hispanic network, or an Asian network, just as women decided they didn't need their network."†

* J. Dreyfus, "Get Ready for the New Workforce," *Fortune*, April 23, 1990, p. 12.

† Ibid.

Another problem is that the technique has sometimes been used without informing those involved. For example, the United States Employment Service (USES) used this type of scoring procedure with its General Aptitude Test. The USES would give prospective employers applicants' scores that were adjusted for race without informing employers of the adjustment. When word was leaked to the press, officials at the Departments of Labor and Justice saw to it that the practice was stopped.

Interestingly, while the country vigorously debates affirmative action issues, large-scale changes in the nature of the U.S. labor pool are making such debates increasingly irrelevant. As the "In Practice" box shows, changes in the size and nature of the work force are giving "diversity" issues a level of urgency that affirmative action has never experienced.

DEVELOPING EQUITABLE PAY STRUCTURES

Most complex organizations employ many different people who have many different skills and perform many different jobs. Paying employees would be a simple matter if everyone were happy receiving the same pay. Unfortunately that is not the case. For many reasons, occupants of some jobs must be paid more than others. Some jobs may be more dangerous than others. Some jobs may require much more education and training. Some jobs require more of a person's time. Organizations are stuck with the problem of having to pay different wages for different jobs. **Pay structure** refers to the organization's hierarchical arrangement of jobs in terms of pay differentials.

The pay differentials created by the pay structure can pose a problem, however. Those who receive lower wages may perceive the differentials as unfair. As we made clear in our discussion of equity theory (see Chapter 7), the fact that some people are getting more rewards than others could have detrimental effects on the others' job performance and on their attitudes toward the company.

So developing a pay structure that determines the pay for each job relative to all other jobs is often a sensitive and problematic organizational task. As is clear from equity theory, differentials in pay will be tolerated only if there are perceived differentials in input as well. The solution to this problem for some firms is job evaluation.

In **job evaluation**, an organization investigates its jobs and the duties and responsibilities of each job in order to devise a hierarchy of value or worth on which a pay scale can be based. There are many ways of performing job evaluation, but the most widespread method is the point system. Although point plans vary, all involve three basic elements—determining compensable factors, rating jobs in terms of these factors, and using these ratings to determine the worth of each job relative to the others in the firm.

pay structure An organization's hierarchical arrangement of jobs expressed in terms of pay differentials.

job evaluation A process by which the pay differentials for jobs throughout the organization are established based on differences in job requirements.

San Antonio's USAA, an insurance and investment management company, gets high marks from clients for prompt and satisfactory settlement of claims. Under CEO Robert McDermott's leadership, the company has expanded its assets from $200 million to $19 billion over just two decades. With a 69-year record of no layoffs, a 4-day, 38-hour work week, and an extensive benefits program, the firm has significantly reduced turnover and absenteeism despite a pay structure *that includes no fewer than 30 distinct pay grades. Employees are assessed both individually and as team members on such things as number of policies sold and pleasantness with clients, and every month, sales teams with the best scores are publicly commended.*
Source: "Managing," Fortune, *February 25, 1991.*

Deriving Compensable Factors

compensable factors Those aspects of a job for which an organization is willing to pay a premium.

APPLYING THE DIAGNOSTIC MODEL
Diagnostic Question 7: What compensable factors does this organization value? How does the organization translate these factors into a pay structure that reflects job-based pay differentials?

Compensable factors are those aspects of a job for which the organization is willing to pay a premium. They are dimensions that differentiate jobs and that are important enough to warrant differences in pay. In equity theory terms, these factors are the official "inputs" that can be used to justify pay differentials. There are several criteria for designating a compensable factor. It must be work related, it must be consistent with the organization's values, and it must be accepted by all parties involved. Let's look at some real-world examples of compensable factors.

The federal government is one of the nation's largest employers. Keeping government wages and salaries equitable is a major task, particularly because for many jobs—for example, the job of a federal judge—there are no direct analogs in the private sector. Table 9-6 describes the four compensable factors used by the government's *General Schedule System* to differentiate jobs for the purpose of devising pay scales. These factors are knowledge required, supervisory controls, job guidelines, and job complexity. Any pay differences between jobs in the federal government can be explained by reference to these four factors. A high-paying job, like that of a federal judge, requires more knowledge, has less supervisory control and fewer written guidelines and is more complex than a lower-paying job, like that of night watchman at a federal courthouse.

In the private sector, the *Hay system* is probably the most widely known method of job evaluation. It has been used in over 5,000 companies, including 130 of the nation's 500 largest corporations. Table 9-4 also describes this system's three compensable factors: know-how, problem solving, and accountability.[21] In the Hay system, higher-paying jobs are characterized as requiring more know-how, more problem solving, and more accountability than lower-paying jobs. A CEO of a major corporation will make more money under this system than a worker on the janitorial staff owing to these kinds of differences in their jobs.

---

[21] G. T. Milkovich and J. M. Newman, *Compensation* (Plano, Texas: Business Publications, 1987), p. 311.

TABLE 9-6
Some Widely Used Compensable Factors

The United States Government's General Schedule System
1. Knowledge required—The nature and extent of information or facts that the worker must understand to do the work.
2. Supervisory controls—The extent of direct or indirect controls exercised by the supervisor and the level of responsibility on the part of the employee.
3. Job guidelines—The nature of the job guidelines and level of judgment needed to apply them.
4. Job complexity—The number, nature, variety, and intricacy of tasks performed; the difficulty of the job and the originality required.

The Hay System
1. Know-how—The sum total of every kind of skill required for acceptable job performance.
2. Problem solving—The original thinking required by the job in analyzing, evaluating, creating, reasoning, and reaching conclusions.
3. Accountability—Answerability for actions and for the consequences thereof; the effect or end results.

Scaling Jobs and Weighting Compensable Factors

Once compensable factors have been chosen, we need to *scale*, or rate, them as they apply to a group of specific jobs and then weight the factors in terms of their relative importance in these jobs. Let's look at just two of the skills that might be required by the staff of a management consulting firm: mathematical skills and interpersonal skills. The first step is to determine the mathematic functions required by the jobs included in the analysis and order them in terms of complexity, or difficulty. Suppose we come up with the five levels of mathematical skill listed in Table 9-7, ranging from no math used at all to the use of fairly advanced techniques. We then rate these levels of skill from 1 to 5, as the table shows. By a similar process, we establish levels of interpersonal skills and rate those also from 1 to 5.

The next step in the point-plan method of job analysis is to weight the compensable factors by importance in the jobs under consideration. Sometimes this is done by means of a particular kind of statistical analysis, but it can also be done by having a committee make a collective judgment as to the factors' relative importance. Let's assume that a committee has decided that in the management consulting jobs we're analyzing, interpersonal skills are twice as important as mathematical skills. This means that in figuring points for different levels of math skills we must use a weight factor of 1 and for levels of interpersonal skills a weight factor of 2. As Table 9-7 shows, points for math skill levels are the same as the ratings for these levels, but for interpersonal skills, points are twice the ratings (because we multiply each of the latter ratings by 2 to get its point value).

TABLE 9-7
Job-Evaluation Scaling Format for Two Skills: Mathematical and Interpersonal

MATHEMATICAL SKILLS

Points	Rating	Description
1	1	No mathematics used at all.
2	2	Simple arithmetic computations involving addition, subtraction, multiplication, or division.
3	3	Computations involving decimals, percentages, fractions, and/or basic statistics.
4	4	Computations involving algebra, (i.e., solving for an unknown) or geometry (i.e., calculating areas, volumes).
5	5	Computations involving the use of trigonometry, logarithms, exponents, or advanced statistics.

INTERPERSONAL SKILLS

Points	Rating	Description
2	1	No interactions with any other persons.
4	2	Interactions with others that involve providing or receiving information or documents.
6	3	Interactions with others that require explanation or interpretation of information.
8	4	Interactions that involve discussions with stakeholders on issues regarding policies or programs. Impact limited to one department.
10	5	Interactions that involve implementation decisions that will impact on the entire organization.

Now, to differentiate jobs in terms of the salary each merits, we need only to determine how much of each type of skill a particular job requires. Suppose that we rate the job of vice president for human resources 2 on math skills and 4 on interpersonal skills. This job would have a total point score of 10 (2 + 8). But the job of accountant-statistician, rated 4 on math skills and 2 on interpersonal skills, would have a total point score of only 8 (4 + 4). Thus even though each of these jobs requires one skill rated 2 and another skill rated 4, because the skills are weighted differently, one—the vice-president's job—will receive a higher salary. In the next section, we'll see how specific wages are determined.

Converting Points into Wages

Once we have a set of compensable factors that have been assigned points we need a way of translating these points into actual wages. Typically, organizations use benchmark jobs to help make this conversion. **Benchmark jobs** are jobs that are similar to jobs in other organizations. Because there is usually ample information on the going market rate for such jobs, it is a fairly easy task to determine the relationship between scores on compensable factors and market wages for particular jobs. In practice, as you can see from Figure 9-6, this relationship is depicted as a scatterplot.

Once the relationship between points and wages are known for benchmark jobs, pay rates for nonbenchmark jobs can be determined by using the figure's diagonal line that captures the relationship between total points and pay for benchmark jobs. We can simply locate the point on the horizontal axis that corresponds to the point value of the job in question, draw a line perpendicular to the horizontal that intersects the diagonal line, and then by drawing a line perpendicular to the vertical axis, locate the appropriate wage per hour on that axis. Thus in Figure 9-6, a job worth 30 points will be paid $6.85 an hour; a job worth 45 points will be paid $8.15 an hour.[22]

What job evaluation does, then, from an equity theory perspective, is to justify outcome differentials in pay by documenting input differentials in the form of compensable factors. Even though some employees are paid less than others, they feel equitably treated because it can be shown that these more highly paid others contribute more to the company.

benchmark jobs Common jobs for which market rates are readily available and which are used, along with point plans, to determine salaries for uncommon jobs.

TEACHING NOTE
Ask students what factors their employers are compensating them for. What do they base their beliefs on? One of the biggest problems in compensation is communicating clearly to the employee what the company considers compensable factors. If employees don't know what these factors are, they may feel they have been paid inequitably.

[22] P. A. Katz, "Specific Job Evaluation Systems: White-Collar Jobs in the Federal Civil Service," in *Handbook of Wage and Salary Administration*, ed. M. Rock (New York: McGraw-Hill, 1984) pp. 14/1–14/10.

FIGURE 9-6

Determining Pay for Nonbenchmark Jobs Based on Pay and Points for Benchmark Jobs

In this graph, Xs represent benchmark jobs for which job evaluation points and market wage rates are known. The dashed line O_1 represents a nonbenchmark job whose 30 job evaluation points enable us to determine its wage of $6.85 per hour. The dashed line O_2 represents another nonbenchmark job, worth 45 job evaluation points, that merits a wage of $8.15 per hour.

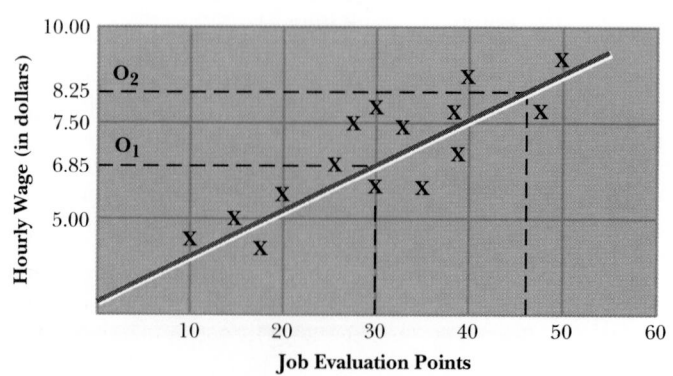

306

MANAGEMENT ISSUES

Comparable Worth Becomes a Reality in Ontario

It is well accepted throughout the United States and Canada that men and women should be paid equally for equal work. This is guaranteed in both countries by legislative acts. But what happens when men and women perform different jobs? In 1990, female workers earned only 65 cents for every dollar earned by male workers. Very little of this pay differential is caused by a difference in the pay of women and men working in the same jobs. Most of it is attributable to occupational segregation. Men hold "men's jobs" and women hold "women's jobs," and many argue that women's jobs pay less than they are worth. *Comparable worth* is a controversial remedy to this kind of discrimination. It refers to the practice of giving equal pay for work of comparable value when jobs are not the same.

For example, in Denver, Colorado, nurses (predominantly women) felt that their jobs were at least of equal value to the work performed by "tree trimmers" (predominantly men), yet they were paid significantly less. A study commissioned by the state of Washington showed that the jobs of truck driver (predominantly male) and laundry worker (predominantly female) were equal in skills, mental demand, accountability, and working conditions. Nevertheless, the drivers earned $1,574 a month compared to only $1,114 for the laundry workers. At the University of Washington, the predominantly female faculty of one department claimed that they were underpaid relative to faculty in departments that were predominantly male. These claims were all rebuffed by the courts. In the United States, the practice of paying market wages even where the result is pay discrimination against women appears legitimate.*

In Canada, on the other hand, the province of Ontario is moving full speed ahead with the notion of comparable worth. Because of a new Ontario law, employers in the province are struggling with the problem of how to compare jobs as diverse as secretary and warehouse worker, janitor and telephone operator, librarian and construction worker. Ontario is the first big jurisdiction anywhere to make private firms adopt comparable-worth plans. All businesses with more than 500 workers must do their own job evaluations to determine where pay discrimination is taking place. They must establish point values for each job and then pay a salary for that job that is based on the point values established.

At the Toronto Sun Publishing Corporation, for example, a job-evaluation system was put in place that rated jobs

on over 30 compensable factors (including working conditions, required education, and deadline pressures). This job-evaluation plan uncovered many disparities. The same number of points were given to the jobs of switchboard operator (held mostly by women) and night cleaning staffperson (held mostly by men), yet the operators were paid $25 a week less than the cleaners. To redress this disparity, the operators were given catch-up raises. Librarians got the biggest boost—almost $6 an hour—to make their pay comparable to that of entry-level engineers.

The law and the plans that it stimulates have triggered disputes, however. The Energy and Chemical Workers Union wanted Consumer's Gas Company to have one provincewide job-evaluation system. The alternative was one system for each different operating system in each different geographic region. A total provincewide plan would push wages in small towns up toward the high levels of wages in Toronto. The Pay Equity Commission, an administrative body set up to settle such disputes ruled against the union. It allowed Consumers Gas to adopt three different regional plans. Thus even in Ontario, prevailing markets, while not afforded the same weight as in the United States, are still given some deference.†

† L. Kilpatrick, "In Ontario, 'Equal; Pay for Equal Work' Becomes a Reality, but Not Very Easily," *Wall Street Journal*, March 9, 1990, p. B1.

* G. T. Milkovich and J. M. Newman, *Compensation* (Homewood, Ill.: BPI/Irwin, 1990), p. 000.

---

comparable worth Theory that sex differences in wages are attributable to discrimination and that such discrimination can be eliminated through job evaluation.

Some people have suggested that job evaluation might be used to fight pay discrimination against women. There is considerable debate on the notion of **comparable worth**—the idea that pay for jobs typically held by men (such as heavy equipment operator) and jobs typically held by women (such as kindergarten teacher) should be equalized. As the "Management Issues" box shows, this idea is more widely accepted in Canada than in the United States.

PAY-FOR-PERFORMANCE PROGRAMS

In theory, one of the least controversial statements one can make about work compensation is that it is important to tie pay to job performance, so that the better the worker performs, the higher he is paid. However, the actual implementation of programs to bring about such a relationship is often quite difficult. To get a feel for this difficulty, consider the following issues that arise when one tries to pay for performance.

Should pay increases be based on outcomes that occur at the individual level (worker performance) or the group level (work group or organizational performance)? At the individual level, the organization may create competition among co-workers, destroying team morale. At the group level, individuals may have a hard time seeing how their own performance relates to group performance and outcomes. In expectancy-theory terms, these kinds of conditions sever the performance-outcome relationship, or instrumentality (see Chapter 7).

If the firm decides to stay at the individual level, should the firm set up the rules for payment in advance, tying future pay to the eventual level of objective productivity? This sounds like a good idea, but the firm will be unable to forecast its labor costs. Moreover, because the price of the product sold or service rendered cannot be known in advance, the firm also cannot anticipate revenue. If the firm waits till the end of the year to see how much money is available for merit pay, people will not know how their performance relates to their pay. Moreover, if the firm, like most firms, engages in pay secrecy to protect people's privacy, how can anyone actually know how fair the merit system is?

If the firm decides to keep incentives at an organizational level, should they be based on cost savings and distributed yearly, or on profits and distributed on a deferred basis? The calculations and accounting procedures required by cost-savings plans are enormous and complex, but rewards are distributed quickly. Profit-sharing plans are much easier to handle from an accounting perspective. Because their rewards are distributed on a deferred basis, however, they are less motivating than cost-savings plans.

Asking all these questions illustrates the complexity inherent in putting into practice the seemingly simple concept of paying for performance. Covering all the complexities of these issues is well beyond the scope of this chapter. However, we will examine the distinguishing features of four different kinds of pay-for-performance programs: merit-based plans, incentive plans, cost-savings plans and profit-sharing plans.

Merit Pay and Incentive Systems

Individual pay-for-performance plans base pay, at least partially, on the accomplishments of individual workers. There are two types of individual programs, those based on merit and those based on incentives.

Merit-based pay plans are by far the easiest to administer and control. In these programs, performance is assessed at the end of the fiscal year by either subjective ratings or ratings on behaviorally based scales. Also at the end of the year, a fixed sum of money is allocated to wage increases. This sum is distributed to individuals in amounts proportional to their performance ratings.

In designing merit-based programs there are three major considerations. First, what will the average performer receive? Many firms try to make sure that average performers are at least able to keep up with inflation. As a result, the midpoint of the rating scales used is often tied to the yearly consumer price index (CPI).

AN EXAMPLE

Incentive systems may be bad for business. Several studies show that tangible rewards can actually lower levels of performance, particularly in jobs requiring creativity. These studies have shown that intrinsic interest in a task declines when an individual is given an external reason for doing it. It is argued that "using money as a motivator leads to progressive degradation in the quality of everything produced." (*Inc.*, January, 1988, pp. 93–94)

---

merit-based pay plans Basing pay increases on subjective ratings of performance made at year end and allocating increases as a percentage of available funds based on these ratings.

Diagnostic Question 8: How does
the firm reward employees for
their performance? If it uses pay
as a reward, is a merit or incentive
plan most appropriate?

Second, what will a poor performer receive? Companies rarely decrease an employee's wages because of performance deficiencies. In terms of buying power, however, raises that fail to cover the CPI are actually wage decreases. Is it in the best interest of the firm to hurt poor performers intentionally? If so, how much damage does the firm wish to inflict? How replaceable are these people if they are prompted to leave the firm?

Finally, how much will high performers receive? Will high performers at the top of a pay grade receive the same raise as those at the bottom? In discussing job evaluation, we noted that some jobs needed to be above other jobs in the pay hierarchy because of the different kinds of inputs that go into each. Paying for performance could cause top performers in a job lower in the hierarchy to surpass (through yearly raises) low performers in upper-level jobs. To prevent this kind of problem, most firms scale raises so that the raise level is a function of both performance and position in the pay grade for the job in question (see Table 9-8). A top performer at the top of a pay grade receives a smaller percentage raise than a top performer who is at the bottom of the grade. Another desirable outcome of scaling raises as a function of both performance and position in the pay grade is that the absolute dollar values of the raises for top performers are more equal. If those at the upper and lower ends of the grade were given the same percentage raise, the result would be a much greater dollar value for people at the top.

Although the performance ratings that determine merit pay have traditionally come from supervisors, this practice is starting to change. In the service sector of the economy, many companies trying to enhance customer service have eliminated the middle man (the supervisor). These companies tie merit pay raises to customer ratings obtained from surveys. For example, at GTE customer ratings are weighted 35 percent in annual merit pay decisions for certain managerial groups.[23]

incentive systems A process by which future pay is made contingent on individual performance based on objective performance indicators and using established quantitative rules.

Incentive systems differ from merit systems in two ways. First, incentive programs stipulate the rules by which payment will be made in advance, so that the worker can calculate exactly how much money she will earn if she achieves a certain level of performance. Second, rewards in an incentive program are based on objective measures of performance. Let's look at two plans: piece-work plans and standard-hour plans.

Incentive systems have a limited
effect on increasing productivity.
Workers tend to increase
performance to a point they feel is
safe, one that ensures a consistent
increase in pay over their base rate
but that is not enough for
management to increase standards
of performance.

In simple *piece-work plans*, a standard of productivity per time interval is set, and any productivity beyond that standard is rewarded with a set amount per unit. This type of plan is easy for the worker to understand, and it creates an obvious performance-outcome expectancy. On the other hand, the standard must often be adjusted. If the standard is initially set too low, labor costs can get out

[23] S. Phillips, A. Dunkin, J. B. Treece, and K. H. Hammonds, "King Customer: At Companies That Listen Hard and Respond Fast, Bottom Lines Thrive," *Business Week*, March 12, 1990, pp. 88–94.

TABLE 9-8
Percent of Pay Increase as a Function of Both Performance and Position in a Job's Pay Range

POSITION IN RANGE	PERFORMANCE RATING				
	Unsatisfactory	Improvement Needed	Competent	Commendable	Superior
Fourth quartile	0%	0%	4%	5%	6%
Third quartile	0%	0%	5%	6%	7%
Second quartile	0%	0%	6%	7%	8%
First quartile	0%	2%	7%	8%	9%
Below minimum	0%	3%	8%	9%	10%

of hand. If it is set too high, workers will reject it when they find that even though they are trying harder, the standard cannot be reached (and pay is the same as before). But if the standard is flexible, gradual raises in the standard will be viewed as a management trick, and decreases will cause some workers to try to manipulate the system by lowering output. Furthermore, without built-in safeguards, these programs also lead workers to achieve quantity at the price of quality.[24]

The *standard-hour plan* stipulates normal time requirements and pay rates for certain tasks. For example, a heating and cooling company may tell its workers that the normal time required to remove an old furnace and replace it with a new one is six hours and that a worker will be paid $120 for this task. Highly skilled and motivated mechanics may be able to complete the job in less time, however. Standard-hour plans pay the worker the set amount even when the work takes less time than normal. If the skilled and hard-working mechanic can install two furnaces in eight hours, his pay for the day will be $240, compared to $120 for the novice or leisurely worker who can install only one furnace in eight hours. Standard-hour plans are more suitable than piece-work plans for complex, nonrepetitive tasks that require numerous skills for completion.[25]

Profit Sharing and Cost Savings

profit-sharing plans Fringe benefit programs in which profits are calculated at year end and distributed to employees, typically in a deferred fashion.

APPLYING THE DIAGNOSTIC MODEL
Diagnostic Question 9: How does the organization reward people for organizational performance? If it uses pay, would a cost-saving or a profit-sharing approach be most appropriate?

cost-saving plans Fringe benefit programs in which organizations pay workers year-end bonuses out of money saved through employee suggestions, increased efficiency, or increased productivity.

TEACHING NOTE
Ask students how employees might react if they have worked hard and then find there are no profits to share. One of the problems with profit sharing is that many factors that neither employees nor management can control result in their being no profits to be distributed.

As the name suggests, **profit-sharing plans** distribute organizational profits to employees. According to recent estimates, some 20 percent of United States firms have such plans in place, and these plans are becoming increasingly popular. *Cash distribution plans* provide full payment soon after profits are determined (annually or quarterly). Because of tax advantages, however, most plans—indeed as many as 80 percent—are deferred.[26] In these plans, current profits accumulate in employee accounts, and a cash payment is made only when a worker becomes disabled, leaves the firm, retires, or dies. Of course, not all profits are redistributed. Research suggests that the percent of profits distributed may range from a low of 14 percent to a high of 33 percent.[27]

Employees often find it hard to see the connection between their activities and their company's profits. When multiple businesses are involved, they may find it even harder to see the link between their efforts and corporate profits. For this reason, some firms adopt **cost-saving plans**, that pay workers bonuses out of the money the company has saved as a result of the efficient performance of its workers. Workers often have more control over the costs of doing business than over profit making. Thus they can easily see the connection between their own work and cost reduction.

One type of cost-saving plan, called the *Scanlon Plan*, is designed to reduce labor costs. Incentives are calculated as a function of labor costs relative to the sales value of production (SVP). SVP is the revenue that would be obtained from sales of all the goods or services produced in a time unit. Say that in one year, $100,000 worth of labor is needed to generate $500,000 in goods and services. In the next year, however, that same amount of goods and services can be produced with $70,000 worth of labor. A portion of the $30,000 saved is distributed to the workers. A typical practice is to return 50 percent of such savings to the workers, retain 25 percent for the firm, and place 25 percent in a fund to cover future years where a "negative bonus" might occur.

[24] C. W. Hamner, "How to Ruin Motivation with Pay," *Compensation Review* 21 (1975), 88–98.
[25] R. I. Henderson, *Compensation Management* (Englewood Cliffs, N.J.: Prentice-Hall, 1984), p. 75.
[26] Bureau of National Affairs, "Incentive Pay Schemes Seen as a Result of Economic Employee Relation Change," *BNA Daily Report*, October 9, 1984, p. 1.
[27] R. McCaffery, *Managing the Employee Benefits Process* (New York: AMACOM, 1983), p. 17.

MANAGING INDIVIDUALS IN ORGANIZATIONS

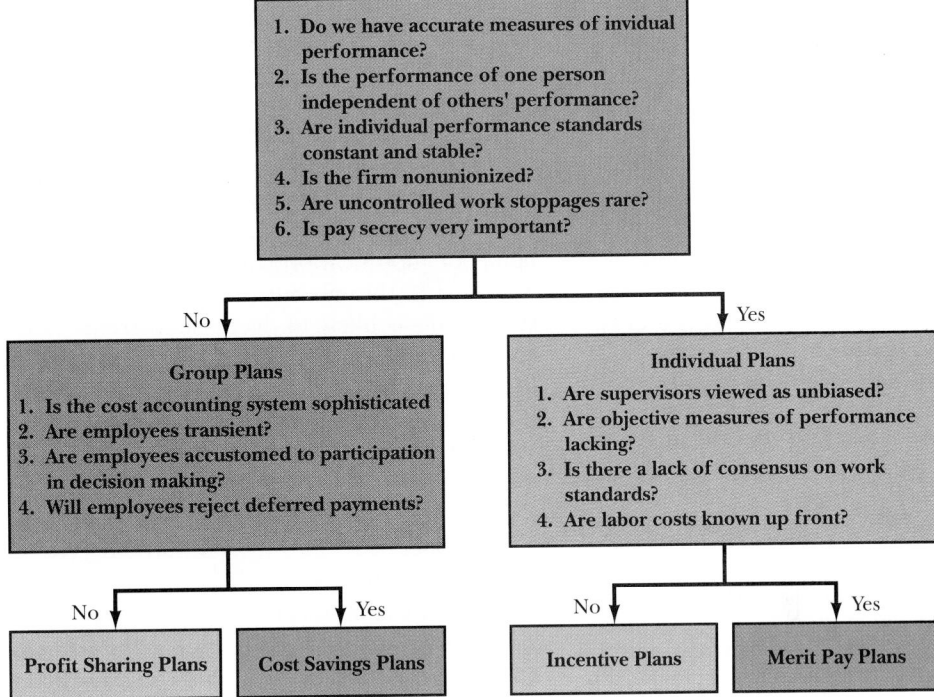

1. Do we have accurate measures of invidual performance?
2. Is the performance of one person independent of others' performance?
3. Are individual performance standards constant and stable?
4. Is the firm nonunionized?
5. Are uncontrolled work stoppages rare?
6. Is pay secrecy very important?

No / Yes

Group Plans
1. Is the cost accounting system sophisticated
2. Are employees transient?
3. Are employees accustomed to participation in decision making?
4. Will employees reject deferred payments?

Individual Plans
1. Are supervisors viewed as unbiased?
2. Are objective measures of performance lacking?
3. Is there a lack of consensus on work standards?
4. Are labor costs known up front?

No / Yes   No / Yes

Profit Sharing Plans | Cost Savings Plans | Incentive Plans | Merit Pay Plans

Although we have sampled only a few of the many pay-for-performance programs currently in use, by now you should have some feel for the kinds of issues raised by such programs. Figure 9-7 provides some guidance on choosing a suitable plan. It tells you under what circumstances an individual or a group plan is appropriate and in what situations specific individual or group plans are most effective.

TRAINING AND DISCIPLINE

New employees rarely know everything they need to know to perform a specific job in a specific context. Learning and applying what is learned to the job context lies at the core of all training programs. Although there are many kinds of training programs, we will focus on behavior modification inasmuch as it is a direct extension of the reinforcement theories of motivation we discussed in Chapter 7. Because modern behavior modification programs have their roots in B. F. Skinner's work on operant conditioning, they often emphasize positive reinforcement rather than punishment as a behavior change strategy. We will follow our discussion of behavior modification with an explanation of training that emphasizes punishment and discipline.

Behavior Modification

behavior modification
Application of contingent rewards in order to bring about behavioral change.

Behavior modification programs, though they vary slightly depending on the nature of the firm and the job, generally proceed in three distinct stages:

1. Identifying key behaviors
2. Monitoring key behaviors
3. Establishing positive contingencies between key behaviors and valued rewards

Diagnostic Question 10: What are some of the critical behaviors that need to be displayed in a job? How might a behavior modification program be used to increase the frequency of the needed behaviors?

AN EXAMPLE: SUPER VALU STORES

Super Valu Stores CEO Michael Wright notes that in a growing labor shortage his company must compete for its work force and train employees to maintain the company's standard of service to customers. At the firms's Cub Food markets in Denver, managers train workers to smile, maintain good posture, and dress properly for work. The company also encourages its employees to pursue their educations and improve their skills. Using a computer system to match worker shifts with customer traffic, managers can arrange for an employee to work for an hour, leave to take a chemistry exam, and return to work the same day. Such Cub market policies have kept the stores' turnover at less than half the rate for the supermarket industry. (Patricia Sellers, "What Customers Really Want," *Fortune*, June 4, 1990, pp. 58–64.)

In the first stage we use job analysis to *identify a few critical behaviors* that should occur on the job in question. Only a few behaviors can be isolated because it would be impossible to monitor the hundreds that occur daily. At the same time, it is hoped that the 5-to-10 percent of all behaviors identified at this stage will actually represent 80-to-90 percent of the day-to-day activities that characterize the performance of the job in question. Other criteria for the behaviors selected are observability and ease of measurement.

This first step in identifying behaviors is important, because if anything, behavior modification can sometimes work too well. That is, any behavior that is critical to the job but not identified at this stage—and thus not monitored or rewarded—is likely to decrease in frequency. Unfortunately, supervisors do not always realize what they actually want until they start rewarding what they think they want. For example, a supervisor may think she wants productivity but may find that needed behaviors once taken for granted (like those dealing with quality, safety, teamwork, social relations, or general maintenance of facilities) start to disappear. Supervisors may also fall into the trap of selecting behaviors because they are easy to measure rather than because they are related to performance. These supervisors may then find themselves in a position of hoping for one thing while rewarding another.[28]

After the behaviors to be reinforced are selected, the next step is to *monitor the key behaviors*, that is, to observe the frequency with which these behaviors are already being exhibited. (This could be zero if the behavior desired has never been exhibited before.) Often tally sheets are developed to record observations, and behaviors are aggregated and displayed in a chart form. The degree to which the behavior occurs prior to the initiation of reinforcement is called baseline data.

Finally, after some period of collecting this baseline data, valued rewards are provided contingent upon the frequency with which workers display the critical behaviors. This step is referred to as *establishing positive contingencies*, and the nature of the rewards can vary. A reward may be money, time off from work, or simply praise from a supervisor. Following the application of these contingencies, critical behaviors (which are still being recorded) should increase in frequency relative to the baseline.

Behavior modification programs have been used in industry for some time. A substantial amount of evidence substantiates their effectiveness in a wide variety of jobs. Table 9-9 describes programs and results in five organizations.

[28] S. Kerr, "On the Folly of Rewarding A While Hoping for B," *Academy of Management Journal* 18 (1975), 769–83.

TABLE 9-9
Successful Applications of Behavior Modification Programs

COMPANY	NUMBER OF EMPLOYEES	BEHAVIOR TARGETED	REINFORCER USED	RESULT
Emery Air Freight	500	Low productivity and poor quality	Praise and recognition	Cost savings
Connecticut General Life Insurance	3,000	Absenteeism	Earned time off	Drastic reduction in absenteeism
City of Detroit garbage collectors	1,100	Poor productivity per man-hour	Pay bonus	Savings of $51.6 million
B. F. Goodrich Chemical Company	100	Failure to meet production schedules	Freedom to choose job activities	Production increases of 300%
ACDC Electronics	350	96% attendance	Praise	98% attendance

Behavior modification programs emphasize the positive rather than the negative control of behavior that is fueled by fear of punishment. (You may want to look at Chapter 7 for a refresher on the distinctions among punishment, negative reinforcement, positive reinforcement, and extinction.) Many behavior modification proponents believe that although punishment is often effective in reducing the frequency of an undesirable behavior, it does nothing to promote a desirable behavior. In addition, the punished employee often has negative emotional reactions to the person who punishes her and, by extension, to the firm itself. Let's look next at the use of punishment and disciplinary tactics in modern industry.

Punishment and Discipline Programs

AN EXAMPLE
There are some situations in which disciplinary actions are necessary for a legal defense. For example, in sexual harassment cases a company must demonstrate that it has a disciplinary policy in place and that this policy is implemented when sexual harassment complaints are made.

There are several reasons why punishment is still widely practiced in many organizations despite the objections of behavior modification experts. First, not everyone is aware of the benefits of positive reinforcement. Managers and others need to become more knowledgeable about the methods we have just discussed. But also, there are some behaviors that are so damaging to the firm or to the employee himself that stopping them is crucial, even if it does nothing to promote positive behaviors. Moreover, failure to take immediate disciplinary action may sometimes imply acceptance or even approval of the offending behavior.

Several steps can be taken to improve the effectiveness of punishment programs. First, punishment should be *progressive*. It should proceed from a simple oral warning to a written formal notice to actual disciplinary action and, if the behavior persists, to termination of employment. Second, punishment should be *immediate*, rather than delayed. This maximizes the perceived contingency between the offending behavior and the punishment. It also minimizes any perceptions that the offending behavior is being used as a pretext to punish the person for something else. Third, punishment should be *consistent* so that no matter who commits the offense or in what circumstances, the punishment is the same. Fourth, punishment should be *impersonal* and directed at the offense itself, not at the person committing the offense. Fifth and last, punishment should be *documented*, so that a paper trail of evidence supports the fact that punishment was progressive, immediate, consistent, and impersonal. Figure 9-8 shows the sort of documentation that might support managerial action with respect to an employee's persistent absenteeism.

Sometimes an offense calls for a face-to-face meeting between supervisor and employee. Table 9-10 lists nine rules for conducting an effective *disciplinary meeting*.[29] When you are conducting a disciplinary meeting with a union member, a tenth rule must be observed. The offender has the right to have a witness at the meeting. Failure to allow a witness to be present has been interpreted by the Supreme Court as a violation of the National Labor Relations Act.[30]

Some firms that employ work teams try to get coworkers involved in disciplining their colleagues. According to Ralph Stoyer, CEO of Johnsonville Foods, discipline is best administered by the people who have to work with the individual. If you came to work late, which would you prefer, a lecture from your boss or protests from your seven coworkers who feel you let them down?[31]

Of course, the "capital punishment" of organizational life is *termination of employment*. Union workers are pretty well protected from this form of punishment,

[29] Cascio, *Managing Human Resources*, p. 55.
[30] D. Israel, "The Weingarten Case Sets Precedent for Co-Employee Representation," *Personnel Administrator* 28 (1983), 23–26.
[31] W. Kiechel, "How to Discipline in the Modern Age," *Fortune*, May 7, 1990, pp. 31–32.

FIGURE 9-8

A Written Reprimand for a Worker Who Is Repeatedly Absent

```
Memo
To: Jack Levitt
From: John Doe
Re: Absenteeism

   Friday, March 10th, 19X9 marked your
third absence in two weeks.

   I now find it necessary to tell you
in writing that this kind of absenteeism
will not be condoned.

   If you persist in this kind of action,
which is clearly a violation of Section 6 of
the work rules, I may be forced to take
disciplinary action, up to and including
dismissal.


                              John Doe
```

employment at will   A provision that either party in the employment relationship can terminate the relationship at any time, even without reason.

except perhaps in the case of blatant and outrageous offenses. For many workers not represented by a union, however, termination is an ever-present danger. An **employment at will** doctrine is created whenever someone agrees to work for an employer for an unspecified length of time. Until recently, such an arrangement could be terminated at the whim of either party.

Increasingly, however, ex-employees are fighting back and suing for "wrongful discharge." Moreover, unlike discrimination cases involving age, sex, or race, in which employees can sue only for back wages, wrongful discharge suits can obtain punitive damages for plaintiffs as well. The fact that these cases are often

---

TABLE 9-10
Rules for Conducting an Effective Disciplinary Meeting

1. Conduct the interview in a quiet, private place.
2. Come prepared with the employee's personnel file to document past digressions or exemplary actions.
3. Do not aggressively prosecute the employee. Merely state facts in a straightforward fashion.
4. Clarify the exact work rule violated and its importance to the firm.
5. Allow the employee to defend or explain his or her actions without interruption.
6. Stay unemotional, and treat the employee with respect. Never use foul language, and never touch the employee.
7. If the problem was a result of a misunderstanding, simply admit it, and close the case.
8. If the problem was an honest mistake, take steps to prevent its reoccurrence, and communicate the expectation that it will not happen again.
9. Focus on the future. After discipline is administered, forget the past and express confidence in the employee's potential.

---

heard by a jury may explain why plaintiffs are so frequently vindicated. Juries tend to identify more readily with individual victims than with large, impersonal corporations, and welcome the chance to punish a company. Research indicates that juries find for the plaintiff in 75 percent of cases, awarding them, on average, over $500,000.[32]

In some instances, these cases are brought up under charges of fraud. For example, Ian Dowie, a former office-products executive at IBM, was lured to Exxon in 1979 with promises of a division presidency, an eventual $100,000-a-year contract, and hefty bonuses from a company profit-sharing plan. When none of these promises were honored and Dowie complained, he was fired. Dowie sued for fraud and breach of contract, and in 1986 a New York jury awarded him $10.1 million in damages—$9 million of which was to punish Exxon.[33]

Managers can take several steps to avoid problems of this sort. First, the rules we have laid out for discipline programs should be followed closely, especially the rules dealing with documentation. Second, in recruiting people for jobs, no promises of job security should be made unless one intends to keep them without exception. Third, employee handbooks or manuals should be developed that clearly spell out the rights and responsibilities of employees. These handbooks should specifically outline the disciplinary process and the actions or behaviors (drug use, theft, excessive absenteeism or tardiness) that will initiate disciplinary action. If these handbooks are distributed to people before they are hired, they are often treated as implied contracts. Indeed, in many cases where handbooks are not available, employees may actually demand written guarantees. Gerald Simmons, president of Handy Associates, an executive recruiting firm, notes: "People are holding out for letters specifying what they would get if the company lessens their responsibilities, moves them to another location, or uses another of the typical methods corporations use when they want to harass someone into resigning."[34]

Finally, employers must make frank performance appraisals if they are to assemble the documentation necessary to support a pattern of misbehavior or inadequate performance. A supervisor who inflates a subordinate's ratings for ten years in order to avoid conflict and then finally gets fed up and fires the person is going to have to do a lot of explaining to a jury.

APPLYING THE DIAGNOSTIC MODEL
Diagnostic Question 11: What are some damaging behaviors that occur on the job? How might a discipline program be developed to decrease the frequency of those behaviors?

AN EXAMPLE
Patrick McConnell argues that employee discipline should be positive in nature and that managers should avoid jumping to conclusions. Every effort should be made to solve problems through training or other appropriate measures. Thus, discipline for poor performance should be structured around what the individual and the company can do to improve performance (*Personnel Journal*, 3/86, pp. 64–71)

APPLYING THE DIAGNOSTIC MODEL
Diagnostic Question 12: What kinds of career development programs might be helpful in creating a good fit between the changing needs of individuals and the changing needs of the organization?

TEACHING NOTE
Many organizations, downsizing their labor force, are providing career counseling, or *outplacement*, programs to help employees find new jobs. Have students discuss whose responsibility careers are: ours, our employer's, or both.

CAREER DEVELOPMENT PROGRAMS

Cognitive appraisal theory deals with the process by which a person decides if his job conforms to his work-related values. As we discussed in Chapter 8, job dissatisfaction and a lack of organizational commitment stem from the perception that the rewards offered by one's work do not satisfy one's needs and values. Moreover, when a person is uncertain about whether he can ever make a job meet his needs, he experiences stress, which in turn increases the probability that he will resign.

Two things increase the difficulty of matching an individual's values to job rewards. First, different people have different values. Second, a person's values tend to change over time. As individuals' needs and values change, so do the needs of organizations, especially in these days of mergers, acquisitions, and downsizing. Organizations are trying to be "leaner and meaner" and, often, to

---

[32] Firing Line: Legal Challenges Force Firms to Revamp Ways They Dismiss Workers," *Wall Street Journal*, September 13, 1983, p. 1.

[33] M. Geyelin, "Fired Managers Winning More Lawsuits: Raising Stakes, Many Now Seek Punitive Awards," *Wall Street Journal*, September 7, 1989, p. 1.

[34] C. H. Deutsch, "When a Handshake Isn't Enough," *New York Times*, February 4, 1990, p. D29.

do more things with fewer people. In the highly competitive environment in which many organizations now operate, a firm can afford neither an excess of human resources nor an absence of required talent. At the same time, companies cannot dismiss all their old employees and hire new ones every time there is a shift in strategy or a shock wave in the economy. How can organizations maintain flexibility and adaptability while encouraging commitment among talented people? For some firms, career-development programs are the answer.

A *career* has been defined as an individually perceived sequence of attitudes and behaviors associated with work experiences over the span of a person's life.[35] A career typically unfolds in stages. A *career stage* is a period marked by a similarity of experiences that set one period apart from another. These experiences can be role transitions, crises, or other turning points. Figure 9-9 shows some of the major career stages that researchers have identified and the approximate ages at which they occur.

Career-development programs are created by organizations to help improve the fit between changing individual needs and values and changing organizational realities. Career-development activities in an organization attempt to provide for the effective, long-term utilization and development of human talent. Although each firm's career-development program may have idiosyncratic features, many include career counseling, mentoring, encouragement to women and minorities, and assistance to dual-career couples.

Career Counseling

Career counseling is designed to help individual employees understand and assess their own needs and values. It also helps them learn about opportunities in the firm that may help them satisfy these personal imperatives. In many companies, employees can participate in programs that focus on self-analysis. They can attend

[35] D. T. Hall, *Careers in Organizations* (Glenview, Ill.: Scott, Foresman, 1976), p. 112.

FIGURE 9-9

Major Career Stages in a Person's Life Span

In the stage of *exploration*, children and young people explore a wide range of jobs, seeking a personal identity. In the *trial* stage, the young adult may have a series of jobs and employers as he tries to match his identity with a work role. In the *establishment* stage, the person commits himself to one work role and develops as a specialized and involved worker. Negotiating the *midcareer transition*, the person reassesses the match between identity and work role, choosing to grow in the job, to maintain the same performance level, or to slow down. Finally, in the *exit* stage, the person becomes less involved in his work role, preparatory to eventual retirement.
Source: Adapted from D. T. Hall Careers in Organizations *(Glenview, Ill.: Scott, Foresman, 1976), 57.*

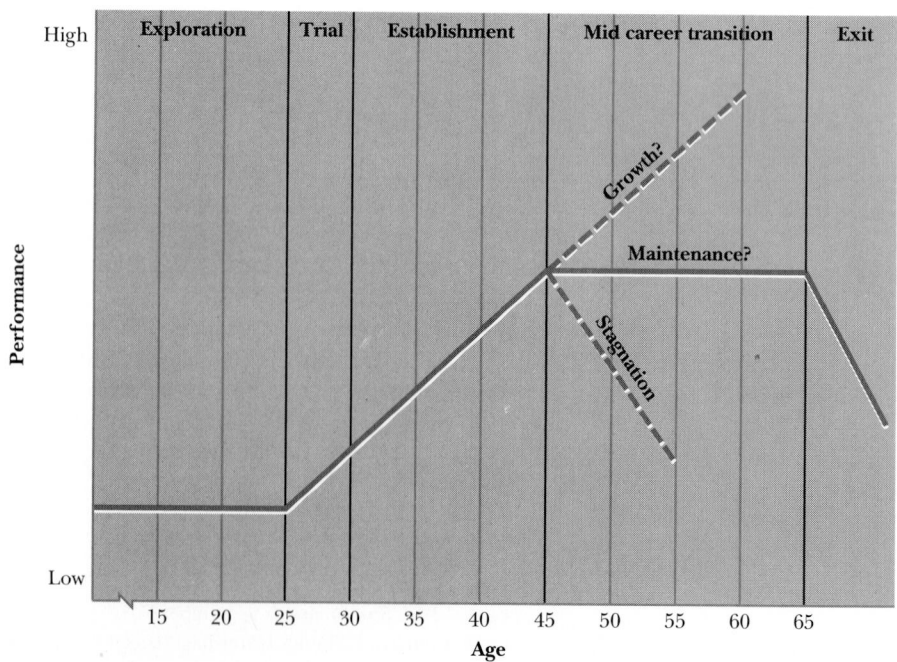

workshops on life and career planning or on interpersonal relationships. They can stop in at assessment centers for aptitude and interest evaluations.[36]

Mentor-Protégé Relationships

mentoring relationship A partnership between a senior and a junior colleague in which the senior partner promotes the development of the junior partner.

A **mentoring relationship** is a partnership between a senior and a junior colleague in which the senior partner promotes the development of the junior partner.[37] The senior partner might engage in *career advancement* activities such as coaching, providing visibility, protecting his protégé from threats, or providing her with challenging assignments. The mentor might also offer his protégé *personal support* by extending friendship, enhancing the junior person's confidence and self-concept.

It's not hard to see how these kinds of activities can help a younger person setting out on a career path. Perhaps it's less obvious just how these activities benefit the mentor. Typically, the mentor is established in her career and at the stage where she can experience growth, maintenance, or stagnation. Working with young, energetic junior people, who will be guiding the company long after the mentor is gone, is an excellent way of maintaining one's own motivation and desire for growth. At the same time, it allows one to pass on to a younger generation the unique knowledge about the firm and its business that can be achieved only by spending years in the trenches.

Organizations can promote mentoring relationships by providing opportunities for exchanges between senior and junior colleagues. They can provide training to mentors on how to handle these special interpersonal relationships. They can also reward managers who faithfully discharge their mentoring responsibilities, especially where they do so at some cost to their own careers.

Women and Minorities

Research shows that career-planning programs are especially needed for women and minority-group members, who generally have less access than white males to career-enhancing experiences.[38] Even in the 1990s, a majority of upper-level professional and managerial positions are still occupied by white males. Those

[36] J. W. Walker and T. G. Gutterridge, *Career Planning Practices* (New York: AMACOM, 1979), p. 36.

[37] K. E. Kram, *Mentoring at Work* (Glenview, Ill.: Scott, Foresman, 1984), p. 80.

[38] F. S. Hall and M. H. Albrecht, *The Management of Affirmative Action* (Santa Monica, Calif.: Goodyear, 1979), p. 101.

"He's an unlimited knowledge base," Jeffrey Parker, undergraduate student at Atlanta's Morehouse University, says of his mentor, Philip M. Butterfield. Butterfield, a Citibank vice-president, says he volunteered for the bank's Fellows Program because he felt mentoring was the best way to "make a change in someone's life." As a result of his mentoring relationship, which included two summer internships at Citibank, Parker has chosen a business major and plans to go on to law school after graduation. Source: Michel Marriott, "Matching Those Who Need Guidance with Those Who Have Been There," The New York Times, June 27, 1990.

executive tend to develop mentor-protégé relationships with other white males unless they are encouraged to do otherwise. Some organizations, such as Baxter Health Care Corporation and the Federal National Mortgage Association, go out of their way to make sure managers develop their female or minority employees. Both companies tie managerial bonuses to objective indicators of female and minority success. These programs achieve results. The Federal National Mortgage Association points to the fact that the representation of females at senior levels of management went from 4 percent to 25 percent between 1981 and 1990. Such programs fight stereotypical views about women and minorities in the hope that members of these groups will not be shunted into unchallenging, low-visibility, career-retarding job assignments.[39]

Finally, although managers are generally reluctant to give negative feedback to any subordinate, they are especially self-conscious about giving it to people who are unlike themselves for fear of being perceived as bigoted or sexist. Yet frank and open negative feedback is vital to new employees who are still learning the ropes and need proper guidance about the do's and don'ts of organizational life.

Organizational career-development programs have to overcome these obstacles by fostering good relations between minority protégés and non-minority mentors. That is the way to dispel stereotypes. Organizations also need to encourage company-wide networks of women and minority-group members so that they can exchange experiences and learn from each other. Moreover, if affirmative action programs do bring in women and minorities who might not otherwise have been hired, it is imperative to provide additional training and developmental experiences for these groups. Some organizations ignore scores on valid predictors of future job success to meet hiring quotas and provide no remedial training, then have trouble retaining "protected groups." Such companies have to consider whether they are helping or hurting the cause of integrating the work force.

Dual-Career Couples

Another issue addressed by career-development programs is the dual-career couple. If it is challenging to manage a single career, it can be exceedingly difficult to manage two careers in the same household at the same time. Almost invariably, the needs of one career will sometimes conflict with the needs of the other. For example, one person may be offered a desirable position in another geographical area, and the spouse may be forced to take a less desirable job in order to make the move. On top of this, if a dual-career couple have children, the role conflict (see Chapter 8) generated by the many competing demands on the couple, their family, and the organization may seem an almost unsurmountable obstacle.

Organizations can do several things to help meet this challenge. First, *flexible working arrangements* at least give the couple a fighting chance at managing the many demands on their time. For example, at Colgate Palmolive, some workers are on flexible hours, with a core time (10:00 AM–2:00 PM) when they must be at the office. The remaining part of their eight-hour day can be made up by coming in as early as 6:00 AM or staying as late as 7:00 PM.

Company-sponsored day care can help solve one of the major problems faced by dual-career couples. Indeed, the number of companies that offered some form of child-care assistance grew from 2,500 to 5,400 from 1986 to 1990 alone. The list of such companies includes some of the nation's most prestigious employers— IBM, Time-Warner, American Express, DuPont, Honeywell and Levi-Strauss.

[39] C. Trost, "Firms Heed Women Employee's Needs," *Wall Street Journal*, November 22, 1989, p. B1.

Some companies, such as Stride Rite, even offer programs for employees' aging relatives. Care for the older generation is an increasing concern for baby-boomers.[40] Stride Rite spent $700,000 to convert 8,500 square feet of its head-quarters into an elder-care center. The company estimates that it costs $140 dollars a week to care for an ageing relative.

Company *relocation programs* that help find suitable employment for the spouse of a newly promoted or transferred employee are another useful resource. Finally, *minimizing travel requirements* can benefit dual-career couples. Today, with fax machines in most offices and the technology for holding conferences and seminars by satellite hook-up widely available, it is possible to cut down on business travel.

One controversial proposal is the development of separate career paths for employees who prefer to put family before work. These so-called mommy-tracks have been advocated by some who insist that women who are as committed to their families as to their careers should be given more flexible roles in the work-place, and eased off the "fast track."[41] Critics of these plans maintain that mommy tracks reinforce sexual stereotypes and assume that the burden of child-rearing responsibilities rests solely on women. Companies like Johnson and Johnson try to spare workers such devil's bargains by providing *family-care leaves*. Under these programs, workers can leave the job for one year with full benefits and guarantees of reemployment and career opportunities upon reinstatement.[42]

SUMMARY

There are eight areas of overlap between organizational behavior and human resource management. Employment testing programs, both *aptitude-based* and *skill-based*, are applications of theories of human attributes. Theories of perception and judgment inform the general process of performance appraisal as well as the specific development of *behaviorally anchored rating scales*. Affirmative action hiring and promotion programs make use of the administrative decision-making model. The *top-down-selection-within-subgroups* approach is one solution to the problem of integrating the work force when valid selection practices result in *disparate impact* on members of protected groups. Two theories of motivation, equity theory and expectancy theory, are useful in justifying pay differentials both across and within jobs. *Job evaluation* is a process for establishing equitable pay differentials across jobs. Merit and incentive systems are used to establish equitable pay differentials among employees working on the same job. *Profit-sharing plans* and *cost-saving plans* distribute rewards to all employees based on performance of the organization as a whole. Reinforcement theory offers potential solutions to problems regarding training and organizational discipline. Career-development programs offer many ways of enhancing employee commitment, satisfaction, and performance.

[40] C. Lawson, "Hope for the Working Parent: Company Care Plans Spread Slowly," *New York Times*, March 15, 1990, p. C1 and K. Teltsch, "For Younger and Older, Workplace Day Care," *New York Times*, March 10, 1990, p. A1.

[41] J. Solomon, "Schwartz of 'Mommy Track' Notoriety Prods Firms to Address Women's Needs," *Wall Street Journal*, December 11, 1989, p. B13.

[42] C. H. Deutsch, "Saying No to the 'Mommy Track': Some Companies Don't Require Part-Time Professional to Sacrifice Their Careers," *New York Times*, January 28, 1990, p. 29.

REVIEW QUESTIONS

1.   Although employment tests are used to meet the needs of employers, why might the nature of tests selected be important to job applicants? That is, what does the type of selection instrument that a firm uses tell the applicant about the firm? What type of selection program, aptitude or skill, is likely to leave the best impression on the job applicant? Why?

2.   Imagine two different pharmaceutical companies that employ the same job categories yet differ in their business strategy. One is trying to increase market share through innovation (developing new and better drugs). The other sticks to established products and tries to increase market share by lowering costs. Why might the two firms wind up with dramatically different pay structures even if both go through the process of job evaluation described in this chapter?

3.   We discussed four types of pay-for-performance systems in this chapter: merit based, incentives, profit sharing, and cost saving. Compare each to the process used to develop a behavior modification program, and rank order the four in terms of how well each conforms to behavior modification principles.

4.   Assume your organization wants to establish either an individually based or organizationally based pay-for-performance program. Which do you think would be most suitable and why? Assume further that your organization wants to use *both* individual and organizationally based rewards. Which pair (incentive + profit sharing, incentive + cost saving, merit + profit sharing, merit + cost saving) seems the best and the worst match in your opinion? Why?

5.   We have drawn a distinction between line personnel and staff personnel. Many managers have the job of supervising staff personnel. In what way might this kind of task be more or less difficult than managing nonstaff personnel? Think about the task of managing human resource specialists in particular. In what way might this kind of task be more or less difficult than managing other types of staff personnel?

DIAGNOSTIC QUESTIONS

 Few line managers and executives ever become specialists in personnel-human resources. Those with supervisory responsibilities, however, must have some working knowledge of the key issues to avoid costly mistakes. Moreover, this knowledge will facilitate communication and interaction between managers and P/HR specialists. This integration may be focused by concentrating on the following questions:

1.   In this organization, who are the experts in P/HR? What is the best way to contact these people to discuss issues of mutual concern?

2.   What are some of the major attributes of the important jobs in this organization? What kinds of aptitude tests might predict success in these jobs?

3.   For what jobs in this organization could skill tests be developed? What kinds of work sample tests would be most appropriate for these jobs?

4.   How free from bias are the subjective performance ratings of supervisors in this firm? If objective measures are used, how much control do employees have over the functions that these measures assess?

5.   Are there any jobs in this firm for which a BARS might usefully be developed?

6.   How does this firm manage the dual goals of selecting and promoting on the basis of merit and of integrating the work force?

7.   What compensable factors does this organization value? How does the organization translate these factors into a pay structure that reflects job-based pay differentials?

MANAGING INDIVIDUALS IN ORGANIZATIONS

8. How does the firm reward employees for their performance? If it uses pay as a reward, is a merit or incentive plan most appropriate?

9. How does the organization reward people for organizational performance? If it uses pay, would a cost-saving or a profit-sharing approach be most appropriate?

10. What are some of the critical behaviors that need to be displayed in a job? How might a behavior modification program be used to increase the frequency of the needed behaviors?

11. What are some damaging behaviors that occur on the job? How might a discipline program be developed to decrease the frequency of those behaviors?

12. What kinds of career-development programs might be helpful in creating a good fit between the changing needs of individuals and the changing needs of the organization?

EXERCISE 9-1
MOTIVATION AND PAY RAISE
ALLOCATION*

EDWARD E. LAWLER III, *University of Southern California*

Although piecerate, commission, or other continuous reinforcement plans provide rewards to employees that are highly motivating, many organizations can only pay employees fixed salaries and make yearly salary adjustments. Thus an annual pay raise is about the only indicator the employees of such firms have of how the organization views their performance. Making salary decisions under these circumstances is a critically important but immensely difficult task. In this exercise you will get to experience the demands of this task firsthand.

STEP ONE: PRE-CLASS PREPARATION

Read the instructions to the "Employee Profile Sheet" below and then decide on a pay increase for each of the eight employees. Be prepared to explain your decisions in class.

STEP TWO: GROUP DISSCUSION

The class should divide into groups of four to six members each (if you have already formed permanent groups, reassemble in those same groups). Each member should share the recommendations he or she made in Step One and explain his or her reasons for them. After all members have reported, the group should analyze and try to explain differences among everyone's recommendations. The group should then develop a set of recommendations that it can agree on. A spokesperson should be appointed to present the group's recommendations to the class.

STEP THREE: CLASS DISCUSSION

Your instructor will reassemble the class so that group spokespersons can present their reports. As each group's recommendations are presented the instructor will record them on a blackboard, flipchart, or overhead projector. After the reports are completed, the class

should look for similarities and differences among the recommendations of different groups and consider the following questions:

1. What kinds of differences could you detect among the eight managers described on the Employee Profile Sheet? Which of these differences served as factors that affected your pay raise decisions?
2. What were the reasons for basing pay raises on each of these factors? Why did you choose to concentrate on some differences among the eight supervisors and to ignore others?
3. If there are differences among the recommendations of different groups, how can you explain them? If there are similarities, what caused them?
4. What would probably happen if you implemented the recommendations made by your class? How would each of the managers react? How would these reactions affect the performance of the company?

CONCLUSION

Using rewards to motivate employees requires that the receipt of rewards be tied as closely as possible to instances of successful performance. Performance-reward contingency is certainly the strongest when piecerate or commission plans are implemented. If companies have no choice but to pay yearly salaries, a minimal degree of contingency can still be established if yearly raises are tied directly to employee performance.

As you have learned in this exercise, maintaining such a tight connection between employee performance and yearly raises can be more difficult than it sounds, sometimes requiring painful decisions. At some point in your future role as manager, it may be your job to make such decisions and to explain them to your subordinates. Base your decisions on sound information about employee performance, and make sure that your explanations are clearly understood.

* The Employee Profile Sheet is reprinted with the author's permission.

Employee Profile Sheet

You must make salary increase recommendations for eight managers whom you supervise. They have just completed their first year with the company and are now to be considered for their first annual raise. Keep in mind that you may be setting precedents that will shape future expectations and that you must stay within your salary budget. Otherwise, there are no formal company policies to restrict you as you decide how to allocate raises. Write the raise you would give each manager in the space to the left of each name. You have a total of $26,000 in your budget for pay raises.

$_____ A. J. Adams. Adams is not, as far as you can tell, a good performer. You have discussed your opinion with others and they agree completely. However, you know that Adams has one of the toughest work groups to manage. Adams' subordinates have low skill levels and the work is dirty and hard. If you lose Adams, you are not sure that you could find an adequate replacement. Current salary: $30,000.

$_____ B. K. Berger. Berger is single and seems to lead the life of a carefree swinger. In general, you feel that Berger's job performance is not up to par, and some of Berger's "goofs" are well known to other employees. Current salary: $33,750.

$_____ C. C. Carter. You consider Carter to be one of your best subordinates. However, it is quite apparent that other people don't agree. Carter has married into wealth and, as far as you know, doesn't need any more money. Current salary: $37,000.

$_____ D. Davis. You happen to know from your personal relationship that Davis badly needs more money because of certain personal problems. Davis also happens to be one of your best managers. For some reason, your enthusiasm is not shared by your other subordinates and you have heard them make joking remarks about Davis's performance. Current salary: $34,000.

$_____ E. J. Ellis. Your opinion is that Ellis just isn't cutting the mustard. Surprisingly enough, however, when you check with others to see how they feel you find that Ellis is very highly regarded. You also know that Ellis badly needs a raise. Ellis was recently divorced and is finding it extremely difficult to support a young family of four as a single parent. Current salary: $30,750.

$_____ F. M. Foster. Foster has turned out to be a very pleasant surprise, has done an excellent job, and is seen by peers as one of the best people in your group of managers. This surprises you because Foster is generally frivolous and doesn't seem to care very much about money or promotions. Current salary: $32,700.

$_____ G. K. Gomez. Gomez has been very successful so far. You are particularly impressed by this because Gomez's is one of the hardest jobs in your company. Gomez needs money more than many of your other subordinates and is respected for good performance. Current salary: $35,250.

$_____ H. A. Hunt. You know Hunt personally. This employee seems to squander money continually. Hunt has a fairly easy job assignment, and your own view is that Hunt doesn't do it especially well. You are thus surprised to find that several of the other new managers think that Hunt is the best of the new group. Current salary: $31,500.

CASE 9-1
DENVER DEPARTMENT STORES*
J. B. RITCHIE, *Brigham Young University*
PAUL H. THOMPSON, *Brigham Young University*

In the early spring of 1991 Jim Barton was evaluating the decline in sales volume experienced by the four departments he supervised in the main store of Denver Department Stores, a Colorado retail chain. Barton was at a loss as to how to improve sales. He attributed the slowdown in sales to the current economic downturn affecting the entire nation. However, Barton's supervisor, Mr. Cornwall, pointed out that some of the other departments in the store had experienced a 15 percent gain over the previous year. Cornwall added that Barton was expected to have his departments up to par with the others in a short period of time.

BACKGROUND

Jim Barton had been supervisor of the sporting goods, hardware, housewares, and toy departments in the main store of Denver Department Stores for three of the ten years he had worked for the chain. The four departments were situated adjacent to each other on the ground floor of the store. Each department had a head sales clerk who reported to Mr. Barton on merchandise storage and presentation, special orders, and general department upkeep. The head sales clerks were all full-time, long-term employees of Denver Department Stores, having an average of about eight years' experience with the chain. The head clerks were also expected to train the people in the department they supervised. The rest of the staff in each department was made up of part-time employees who lived in or near Denver. Most of the part-time people were students at nearby universities who worked to finance their education. In addition, there were two or three housewives who worked about ten hours a week in the evenings.

All sales personnel at Denver Department Stores were paid strictly on an hourly basis. Beginning pay was just slightly over the minimum wage and raises were given based on length of employment and work performance evaluations. The salespeople in the housewares and sporting goods departments were paid about forty cents an hour more than the clerks in the other departments because it was thought that more sales ability and experience were needed in dealing with the people who shopped for items found in those departments.

As a general rule the head sales clerk in each department did not actively sell, but kept the department well stocked and presentable, and trained and evaluated sales personnel. The part-time employees did most of the clerk and sales work. The role of the sales clerk was seen as one of answering customer questions and ringing up the sale rather than actively selling the merchandise except in the two departments previously mentioned where a little more active selling was done.

The sales clerks in Barton's departments seemed to get along well with each other. The four department heads usually ate lunch together. If business was brisk in one department and slow in another, the sales people in the slower area would assist in the busy department. Male clerks often helped female clerks unload heavy merchandise carts. Store procedure was that whenever a cash register was low on change a clerk would go to a master till in the stationery department to get more. Barton's departments, however, usually supplied each other with change, thus avoiding the longer walk to the master till.

Barton's immediate supervisor, Mr. Cornwall, had the reputation of being a skilled merchandiser and in the past had initiated many ideas to increase the sales volume of the store. Some of the longer-term employees said that Mr. Cornwall was very impatient and that he sometimes was rude to his subordinates while discussing merchandising problems with them.

The store manager, Mr. Blanding, had been with Denver Department Stores for twenty years and would be retiring in a few years. Earlier in his career Mr. Blanding had taken an active part in the merchandising aspect of the store, but recently he had delegated most of the merchandising and sales responsibilities to Mr. Cornwall.

SITUATION

Because of Mr. Cornwall's concern, Barton consulted with his department supervisors about the reason for the declining sales volume. The consensus reached was that the level of customer traffic had not been adequate to allow the departments to achieve a high sales volume. When Barton presented his problem to Mr. Cornwall, Cornwall concluded that since customer traffic could

* Reprinted with the publisher's permission from pages 148–52 of *Organizations and People: Readings, Cases, and Exercises in Organizational Behavior*, 3rd ed., J. B. Ritchie and Paul H. Thompson, copyright © 1984 by West Publishing Company. All rights reserved.

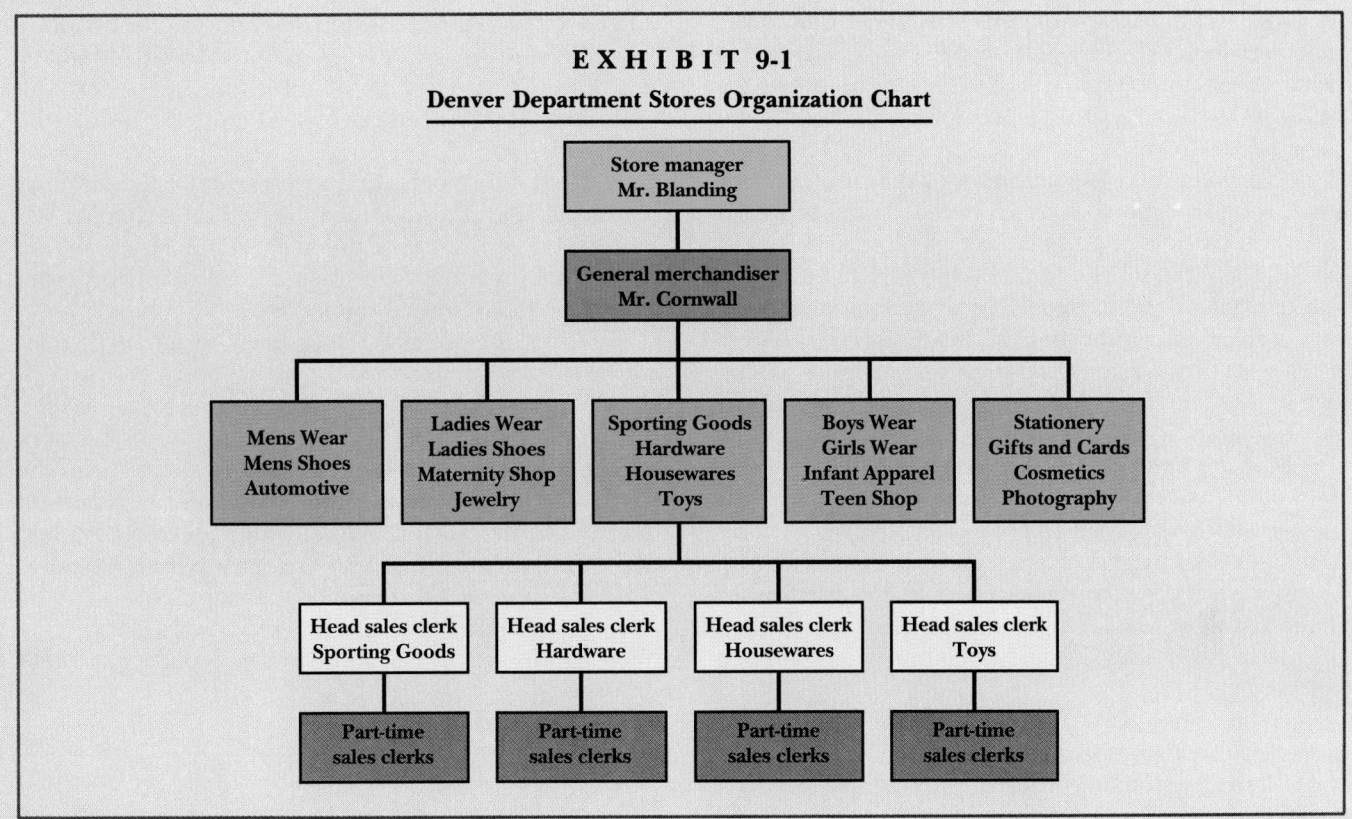

EXHIBIT 9-1

Denver Department Stores Organization Chart

Store manager
Mr. Blanding

General merchandiser
Mr. Cornwall

| Mens Wear Mens Shoes Automotive | Ladies Wear Ladies Shoes Maternity Shop Jewelry | Sporting Goods Hardware Housewares Toys | Boys Wear Girls Wear Infant Apparel Teen Shop | Stationery Gifts and Cards Cosmetics Photography |

Head sales clerk
Sporting Goods

Head sales clerk
Hardware

Head sales clerk
Housewares

Head sales clerk
Toys

Part-time sales clerks

Part-time sales clerks

Part-time sales clerks

Part-time sales clerks

not be controlled and since the departments had been adequately stocked throughout the year, the improvement in sales would have to be a result of increased effort on the part of the clerks in each department. Cornwall added that if sales didn't improve soon the hours of both the full- and part-time sales clerks would have to be cut back. Later Barton found out that Cornwall had sent a letter around to each department informing employees of the possibility of fewer hours if sales didn't improve.

A few days after Barton received the assignment to increase sales in his department, Mr. Cornwall called him into his office again and suggested that each sales person carry a personal tally card to record daily sales. Each clerk would record his or her sales and at the end of the day the personal sales tally card would be totaled. Cornwall said that by reviewing the cards over a period of time he would be able to determine who were the "deadwood" and who were the real producers. The clerks were to be told about the purpose of the tally card and that those clerks who had low sales tallies would have their hours cut back.

Barton told Cornwall he wanted to consider this program and also discuss it with the head salespeople before implementing it. He told Mr. Cornwall that the next day was his day off but that when he returned to work the day after he would discuss this proposal with the head sales clerks.

Upon returning to the store after his day off, Mr. Barton was surprised to see each of his salespeople carrying a daily tally sheet. When he asked Mr. Cornwall why the program had been adopted so quickly, Cornwall replied that when it came to improvement of sales, no delay could be tolerated. Barton wondered what effect the new program would have on the personnel in each of his departments.

When Mr. Cornwall issued the tally cards to Barton's salespeople, the head sales clerks failed to fill them out. Two of the head clerks had lost their tally cards when Cornwall came by later in the day to see how the program was progressing. Cornwall issued the two head clerks new cards and told them that if they didn't "shape up" he would see some "new faces" in the departments.

The part-time salespeople filled out the cards completely, writing down every sale. The rumor that those clerks who had low sales tallies would have their hours cut spread rapidly. Soon the clerks became much more active and aggressive in their sales efforts. Customers were often approached more than once by different clerks in each department. One elderly lady complained that while making her way to the restroom in the back of the hardware department she was asked by four clerks if she needed assistance in making a selection.

When Barton returned the day after the institution of the program, the head sales clerks asked him about the new program. Barton replied that they had no alternative but to follow Cornwall's orders or quit. Later that afternoon the head clerks were seen discussing the situation on their regular break. After the break the head clerks began waiting on customers and filling out their sales tally cards.

Not long after the adoption of the program, the stock rooms began to look cluttered. Unloaded carts lined the aisles of the stock room. The shelves on the sales floor were slowly emptied and remained poorly stocked. Sales of items that had a large retail value were especially sought after and the head sales clerks were often seen dusting and rearranging these more expensive items. The head clerks' tally sheets always had the greatest amount of sales when the clerks compared sheets at the end of each day. (Barton collected them daily and delivered them to Cornwall.) The friendly conversations among salespeople and between clerks and customers were shortened and sales were rung up on the cash register and completed in a much shorter time. Breaks were no longer taken as groups and when they were taken they seemed to be much shorter than before.

When sales activity was slow in one department, clerks would migrate to other departments where there were more customers. Sometimes conflicts between clerks arose because of competition for sales. In one instance the head clerk of the hardware department interrupted a part-time clerk from the toy department who was demonstrating a large and expensive table saw to a customer. The head clerk of the hardware department introduced himself as the hardware specialist and sent the toy clerk back to his own department.

Often customers asked for items which were not on the shelves of the sales floor. When the clerk looked for the item it was found on the carts which jammed the stock room aisles. Some customers were told the item they desired wasn't in stock and later the clerk would find it on a cart in the stock room.

When Barton reported his observations of the foregoing situations to Mr. Cornwall, he was told that it was a result of the clerks' adjusting to the new program and to not worry about it. Cornwall pointed out, however, that sales volume had still not improved. He further noted that the sum of all sales reported on the tally sheets was often $500 to $600 more than total department sales according to the cash register.

A few weeks after the instigation of the tally card system Cornwall walked through the hardware department and stopped beside three carts of merchandise left in the aisle of the stock room from the morning of the day before. He talked to the head clerk in an impatient tone and asked him why the carts weren't unloaded. The clerk replied that if Mr. Cornwall had any questions about the department he should ask Mr. Barton. Cornwall picked up the telephone and angrily dialed Barton's office. Barton told him that the handling of merchandise had been preempted by the emphasis on the tally card system of recording sales. Cornwall slammed down the receiver and stormed out of the department.

That afternoon, at Barton's request, Blanding, Cornwall, and Barton visited the four departments. After talking with some of the salespeople, Mr. Blanding sent a memo announcing that the tally card program would be discontinued immediately.

After the program had been terminated, sales clerks still took their breaks separately and conversations seemed to be limited to only the essential topics needed to run the department. Barton and the head sales clerk didn't talk as freely as they had before and some of the head clerks said that Mr. Barton had failed to represent their best interests to Cornwall. Some of the clerks said they thought the tally card system was Barton's idea. The part-time people resumed the major portion of the sales and clerking jobs and the head clerks returned to merchandising. Sales volume in the department didn't improve.

When you have read this case, look back at the chapter's diagnostic questions and choose the ones that apply to the case. Then use those questions with the ones that follow in your case analysis.

1. Do you agree with Mr. Cornwall that keeping sales records and terminating clerks who fail to perform acceptably should stimulate greater effort among the Denver department store's salesforce? Why? Or why not?

2. Describe the steps that the management of the Denver Department Stores should take to set up a more effective performance appraisal program than Mr. Cornwall's. Why would your program work better?

3. Suppose the Denver stores decided to implement a pay-for-performance program to motivate salesforce performance. What kind of program would you recommend? Why? Describe how you would implement your program.

CASE 9-2
CHANCELLOR STATE UNIVERSITY*
THOMAS R. MILLER, *Memphis State University*

THE SETTING

Chancellor State University is a large, urban university in the Midwest. Although the University had experienced rapid growth for several years, overall enrollment had stabilized. The School of Business Administration, however, had continued to grow, drawing students away from programs in the School of Education and the College of Arts and Sciences as well as attracting new students concerned with future vocational opportunities. The faculty and administration of the business school were pleased to see the enrollment growth as it signaled acceptance of their degree programs, but the enrollment expansion also created strong pressure to expand the business faculty.

Under normal circumstances, faculty expansion would simply have meant an active recruitment effort by school administrators. But the situation at Chancellor State was representative of a national phenomenon of enrollment growth in business schools that had resulted in a strong demand for doctorally qualified faculty in the face of a relatively short supply. Thus, faculty recruitment at many business schools had become a priority activity, rather than merely one of the many administrative responsibilities of deans and department heads.

At Chancellor State, Fred Kennedy, Chairman of the Management Department, had been actively seeking new faculty members for his staff, which had the heaviest course load in the school. As is often customary in academia, the faculty in the Department of Management participated in recruitment, spending considerable time meeting with the faculty candidates in an effort to evaluate their candidacy for a faculty position. Faculty members could then make recommendations as to whether or not the prospect should be tendered an offer to join the staff.

THE CONFERENCE

It was late in February, and several prospective faculty members had visited Chancellor State for campus job interviews. Early one Friday morning, Kennedy was in his office reviewing the job files of prospective faculty members. He looked up as he heard the voice of Larry Gordon, an assistant professor of management who was now in his third year at Chancellor State.

"Good morning, Fred," said Larry, as he walked into the Department office. "Do you have a couple of minutes? I want to talk with you about something."

Fred gestured to him to come into his office.

"Sure, Larry, what's on your mind?"

After entering Fred's office, Larry closed the door, indicating to Fred that this was not to be just a casual, friendly conversation.

"Fred," Larry began, "I was wondering what you thought about the prospective faculty member we had in here for an interview last week. I've been talking with a couple other faculty members about him, and they're not really all that impressed. He seems to be OK, I guess, but we may be able to do better. Are we going to make him an offer? If we do, he's sure not worth top dollar in my opinion."

"Well, I've received some of the written evaluations back from the faculty, and they seem to be fairly positive," replied Fred. "They're not as favorable as they could be, but the other faculty seem to think that he would be acceptable and that he could work out pretty well on our staff. His academic credentials are not bad, and he has had some good experience. Given the state of the market for business faculty in his specialty, I expect that we'll extend an offer to him. By the way, I know that he already has a couple of offers in hand from our competition."

Fred could readily see that Larry was not pleased to hear all of this. From their earlier conversations, Fred could anticipate Larry's next comment.

"Yeah, O.K., I can see that we could use him, but what kind of money are we offering in these new positions?" questioned Larry. "I don't mean to pry into somebody else's business, but what sort of salary is the department offering our new faculty?"

Fred winced at this question. He had in the past made no secret about general salary ranges for new faculty members. In fact, this information was generally known throughout the school. But this had become a very sensitive issue in the last few years, given the rapid increases in starting salaries for new business faculty members.

"Well, Larry, I guess you know that we're paying competitively for our new faculty. With our enrollment increase we've got to increase our teaching staff, and to do that we're probably going to have to meet the market," Fred responded.

Larry was obviously not satisfied with this response and was becoming irritated with the conversation. "Fred, I assume that by 'meeting the market' you mean that we're going to offer this guy two or three thousand dollars more than some of us who have been here for several years are now making. This new guy has not yet finished his doctorate, has very little teaching experience, has no publications, and, in my opinion, is not as good as a lot of our current faculty. How much can you justify paying for an unknown quantity? I think it's just unfair to the present faculty to offer him more money than many of us are making. When is somebody going to do something for us? Fred, I'm not unhappy here in this department, but I'm sure going to keep my eyes open for other opportunities. I feel sure that I could move to another school at a higher rank and increase my salary significantly. You may think I'm wrong and maybe I shouldn't feel this way, but this situation is just not fair!"

Fred sighed and tried to calm Larry down. "Larry,

I know what you're concerned about, and I'm certainly sympathetic to the problem. After all, this salary compression issue affects me in the same way it does you. I can assure you that I have reservations about paying the kind of money we are for new faculty in light of our existing faculty salaries, but I don't believe that we can attract the kind of faculty we want by paying less than competitive rates. Although this seems to create some internal inequities, I hope that we'll have sufficient salary increase money to make some adjustments to reduce these discrepancies. Certainly I want to be able to reward and retain our productive people . . ."

Larry, feeling a little embarrassed by his earlier emotional statement, interjected: "I know you've got other problems, Fred, and I didn't mean to lash out at you. I know it's not really your fault, but a lot of the other faculty are talking about this salary issue. It surely doesn't help morale any when a new, inexperienced assistant professor is hired for more than some of the associate professors are making."

"Yes, I'm well aware of this, Larry, and I'm making the Dean aware of it as well. We're certainly going to do what we can to try to resolve this salary compression problem," Fred responded.

As Larry moved toward the door, he continued to make his point: "Well, I hope you can do something soon because it's most inequitable at the present time. People are pretty upset about it, and it's likely to cause the department some turnover problems in the future. No one likes to be treated unfairly. I'll see you later, Fred. I've got to run to class. Maybe we can talk about it again later."

As Larry walked out of his office Fred reflected on their conversation. It reminded him of other discussions he had had previously with several other faculty members. In fact, Larry had hinted at his dissatisfaction before, but had not been so outspoken about it. Yes, the salary compression problem was reaching a crisis. No longer was it a matter of the "new hires" nearing the salaries of some present faculty; it was a matter of their exceeding them. Never in his experience had Fred recalled a labor market for faculty that was this chaotic.

Fred had puzzled over this dilemma before, but he had not been able to come up with a solution for the problem. He wondered if, in fact, there was a solution that would enable him to hire the new personnel he wanted without offending some of the present staff. Maybe it's just one of those "no win" administrative

situations, he mused. Perhaps this was something that could be discussed with the other department chairmen and the Dean as some of them had basically the same problem. Maybe then, he would have a better idea of how to deal with the situation. He certainly hoped so!

When you have read this case, look back at the chapter's diagnostic questions and choose the ones that apply to the case. Then use those questions with the ones that follow in your case analysis.

1. How might a job analysis help Fred Kennedy restore equity to the pay structure of the management department at Chancellor State? How could he get the management faculty to cooperate?

2. What kind of pay-for-performance program is being used now in the department? Critique this program, pointing out specific strengths and weaknesses in Chancellor State's approach.

3. How do you think the department's current salary structure would affect faculty motivation? How could this situation be improved? What steps should Fred take to implement your suggestion?

CASE 9-3

THE SLAB YARD SLOWDOWN

Read Chapter 11's Case 11-2, "The Slab Yard Slowdown." Next, look back at Chapter 9's diagnostic questions and choose the ones that apply to that case. Then use those questions with the ones that follow in this analysis of the case.

1. What kind of pay-for-performance program is being used to determine the pay of Midland Steel's scarfers? Why isn't this program working out?

2. What steps would you take to modify the scarfers' practice of altering torch tips? Would punishment play a role in your approach? How? Why? What results would you expect from the steps you would implement?

3. If you suceeded in eliminating the scarfers' tip altering practice, do you think the company's pay-for-performance program would begin to work? What additional steps, if any, would the company have to take to make its program succeed?

C H A P T E R   10

INTERPERSONAL PROCESSES AND COMMUNICATION

Some members of the Prentice Hall Management of Organizational Behavior *team. Left to right, standing: Martha Coffman, senior copywriter; Sandra M. Steiner, senior marketing manager; Lori Cowen, advertising manager; Robert Anderson, manufacturing buyer; Frances Russello, managing editor; Liz Robertson, scheduler; Trudy Pisciotti, pre-press buyer; Kris Ann Cappelluti, supplements production manager. Seated: Diane Peirano, editorial assistant; Alison Reeves, assistant vice president and executive editor; David Scholder, supplements editor; Esther S. Koehn, supervisory production editor; Virginia Otis Locke, senior editor; Ruta Kysilewskyj, senior advertising designer.*

eople in organizations seldom work alone. It is more usual for them to interact with other people as they perform their jobs. For example, the book you are reading is the product of many people working together. The book's authors wrote initial drafts of each chapter and redrafted these chapters several times. A staff of market researchers at Prentice Hall, the book's publisher, gathered information about organizational behavior courses to guide the development of the book. A panel of reviewers—college instructors, like yours—offered opinions to help fine-tune the book's contents. A development editor analyzed reviewers' responses and made additional suggestions to make the book more readable. An art director engaged a designer to plan the book's layout and its cover. Photo researchers located the photographs that open each chapter and illustrate discussions. A pre-press buyer contracted with a compositor to set the book in type. A paper buyer purchased the quantities of paper required to print the desired number of copies. A manufacturing buyer contracted with a printer to do the actual printing of the book. A production manager worked with all of these people in overseeing the book's progress to completion.

At the same time, a marketing manager planned the sales and advertising campaigns for the book and informed the publisher's sales representatives about the book. The sales representatives brought the book to the attention of your instructor and helped to make sure that your bookstore stocked it in time for you to purchase it. Throughout the entire process, the executive editor who had contracted the authors initially coordinated activities at Prentice Hall and continued to keep in touch with the authors.

Group activities like the ones that produced this book are more the rule than the exception in organizational life. As we indicated in Chapter 2, organizations exist to produce goods and services that individuals working alone could not produce. So although it is important for managers to know how the individual-level characteristics we've discussed in Chapters 4–9 shape organizational behavior, they must also understand how group factors affect the way people behave on the job. Therefore, in this chapter we identify and explore key group processes. First we discuss several reasons why people band together and characterize the various types of interdependence that connect people. Then we examine the process of making and taking roles, in which people's behaviors are shaped by the expectations of others. Next we look at communication and the function it plays in linking people together. We conclude with a discussion of how socialization processes help shape and maintain relations among people at work.

DIAGNOSTIC ISSUES

everal diagnostic questions will guide us through this chapter. First, can we *describe* the various ways in which people come to depend on each other in work environments? What kinds of communication networks facilitate job performance and enhance job satisfaction? What *diagnostic* information can help us understand the situation in which a person develops a role for himself that goes beyond his written job description? Can we explain how some people become informal group leaders and amass power that exceeds the power of people higher up in the organizational hierarchy?

Can we *prescribe* how people should be socialized early in their employment so as to ensure successful performance and satisfaction? What can we tell a group to do when it must cope with an overwhelming amount of information? Finally, what *actions* can implement the most suitable communication medium—oral, written, nonverbal—for delivering a particular message? What can we do to ensure that organization members will conform to assigned work roles?

GROUP INTERACTION AND INTERDEPENDENCE

People in organizations have a rich variety of interconnections. Their work may require them to associate with each other as a regular part of job performance. They may belong to the same group in their organization's structure, although as the example that opened this chapter indicated, people form important interpersonal relationships not only with individuals in other departments but with people in outside organizations. People may band together to share resources, such as access to valuable equipment or pools of money. In addition, many employees may form friendships and get together away from work as well as on the job. Connections like these make interpersonal interactions a very important fact of organizational life.

Why Do People Interact with One Another?

Whether in organizations or in the societies that surround them, people form and maintain relations with others for many reasons. Let's look at five major forces that propel people into such relations—evolutionary adaptation, need satisfaction, interpersonal attraction, shared goals, and group activities (see Figure 10-1).

Evolutionary Adaptation. The tendency for people to associate with each other probably first arose from the benefits our ancestors derived from forming groups. Prehistoric humans undoubtedly discovered that living with others offered advantages for survival that were not available to the solitary individual. Banding together helped protect people from faster, physically stronger predators. People found they could hunt and farm more successfully together than alone. Gathering

FIGURE 10-1

Why People Form Groups

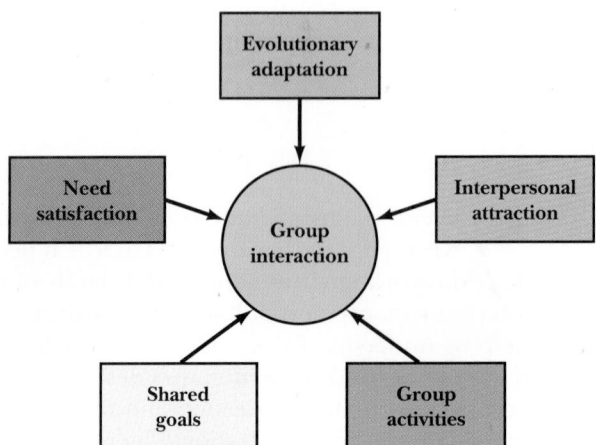

into groups also facilitated adaptive innovation. One person could pass his discovery on to others, who could refine it over time. Finally, mutual protection and nurturance sheltered reproduction and child-rearing activities. For all of these reasons, social behavior became adaptive and a part of what it is to be a human being.[1]

Need Satisfaction. Although it was the necessity of satisfying the lower-order needs such as food and safety that first promoted group formation, socializing eventually fulfilled other, higher-order needs as well. In particular, people now interact to fulfill needs for affiliation or social satisfaction. Furthermore, relations with others who are seen as prestigious can enhance a person's perceived self-worth. Thus other people can help fulfill one's desires for esteem and shape one's self-concept.[2]

Interpersonal Attraction. People may also be drawn to others by their attractiveness. Interpersonal attraction tends to be high when people have *similar attitudes and beliefs*. Interaction with others of like mind can reinforce one's own world views.[3] Interpersonal attraction can also be a function of the *dissimilarity of needs and abilities*. For instance, a person with a high need for power will tend to be attracted to people who have complementary, submissive needs. Finally, *proximity* plays an important part in determining the strength of interpersonal attraction (see Figure 10-2). Unless they are near one another, people generally do not have the opportunity to discover their similarities and dissimilarities.

Goal Pursuit. People may also band together to pursue mutual goals that they can achieve more effectively by working together. Mothers Against Drunk Driving (MADD) is a good example of a close-knit group of people who associate with each other to pursue a common goal—to protect society from the danger posed by drunk drivers. On the other hand, the members of a local school board may have very different goal-related reasons for associating with one another. One person may join the board because he is concerned about his children's education. Another may join out of her wish to control property tax rates. Many organizations—particularly business concerns—are held together by their members' interests in accomplishing specific economic goals. In a business, some members' goals may be shared and others may be quite distinctive. They may all advocate good wages and benefits, but some may be interested in power, and others may want to do a particular kind of work. Thus goals need not be shared for people

[1] F. L. Ruch and P. G. Zimbardo, *Psychology and Life* (Glenview, Ill.: Scott, Foresman, 1971), p. 32.
[2] R. Brown and J. Williams, "Group Identification: The Same Thing to All People?" *Human Relations* 43 (1984), 547–60.
[3] W. H. Whyte, *The Organization Man* (New York: Simon & Schuster, 1956), p. 55.

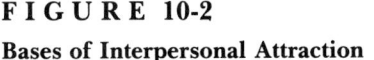

FIGURE 10-2

Bases of Interpersonal Attraction

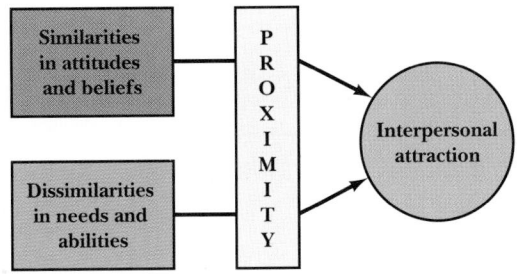

to band together. It is only necessary that the association helps people satisfy goals they value personally.

Attractive Activities. Finally, some people join groups to take part in activities that appeal to them. For example, some employees may form an organizational softball team because there is a mutual interest in playing this sport on nice spring days.

Types of Interdependence

APPLYING THE DIAGNOSTIC MODEL
Diagnostic Question 1: What type of interdependence unites you with other employees? What sorts of problems might this interdependence cause?

pooled interdependence A type of interaction where individuals draw off a common resource pool but do not interact with each other in any other way.

As they work with each other, people form patterns of *interdependence*. They come to depend on each other for information, raw materials, social support, help in performing a task, and other equally important resources. This interdependence typically takes one of four forms: pooled, sequential, reciprocal, or team interdependence (see Figure 10-3).[4]

Pooled Interdependence. **Pooled interdependence**, as Figure 10-3A suggests, occurs among people who draw resources from a shared pool but have little else in common. For example, a college golf team is made up of individuals with pooled interdependence. The players share the course, the coach, and other resources, but when it comes to a match, they are on their own. One player cannot hit the ball for another, carry another's clubs, line up others' putts, or do much of anything to assist a teammate. In the end, the score for the team is simply the sum of the individual players' scores. Similarly, in an organization like Metropolitan Life Insurance, individual data-entry specialists draw off a common pool of work that needs to be entered into the firm's computers. Yet each data-entry person works alone in entering information. As with the golf team, the total amount of work accomplished by the group is simply the sum of all the individual accomplishments.

Pooled interdependence, the simplest form of interdependence, requires little or no interpersonal interaction. For example, although data-entry personnel

[4] J. D. Thompson, *Organizations in Action* (New York: McGraw-Hill, 1967), p. 41; and A. H. Van de Ven, A. L. Delbecq, and R. Koenig, Jr., "Determinants of Coordination Modes within Organizations," *American Sociological Review* 41 (1976), 322–38.

FIGURE 10-3

Types of Interdependence

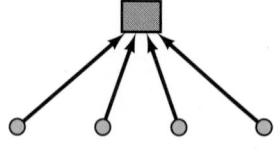

A Pooled Interdependence

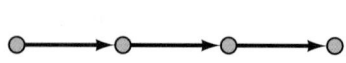

B Sequential Interdependence

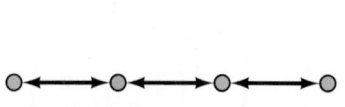

C Reciprocal Interdependence

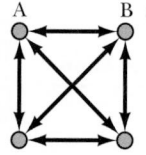

D Team Interdependence

at Metropolitan Life may be sitting right next to each other, there is no necessity for them to interact.

sequential interdependence A type of interaction in which individuals are arrayed in a chain of one-way links.

AN EXAMPLE: COMPAQ COMPUTER
Sequential interdependence can be a very effective way of coordinating workflow. Receiving a steady flow of U.S.-made parts, workers on the assembly line in Compaq Computer's Scotland plant ·vere able to produce at a rate that increased the company's overseas sales by 102 percent in 1988. (Edward Prewitt, "America's Biggest Exporters," *Fortune*, July 17, 1989, p. 50.)

Sequential Interdependence. **Sequential interdependence** (see Figure 10-3B) is a chain of one-way interactions in which people later in the chain depend on those who precede them. People earlier in the chain, however, are independent of those who follow them. Thus sequentially interdependent relationships are said to be *asymmetric*. Some people depend on others who do not depend on them. People on an assembly line in a company like RCA or Navistar are connected by sequential interdependence. Workers earlier in the line produce partial assemblies that workers later in the line complete. By its nature, sequential interdependence prevents people at the end of the chain from performing their jobs unless people at the head of the chain have performed theirs. People at the head of the chain, however, can complete their tasks no matter what people at the end do.

Sequential interdependence usually involves some form of direct interaction. For example, people on an assembly line sometimes talk with each other to pass on information about work coming along the line. Although sequential interdependence is more complex than pooled interdependence, its one-way asymmetry makes it less complex than the types of interdependence we discuss next.

reciprocal interdependence A type of interaction in which there are two-way links among individuals.

Reciprocal Interdependence. In **reciprocal interdependence** (see Figure 10-3C), a network of two-way relationships tie a collection of people together. A good example of this kind of interdependence is the relationship between a sales force and a clerical staff. Sales representatives rely on clerks to complete invoices and process credit card receipts, and clerks depend on salespeople to generate sales. Reciprocal interdependence also occurs among the members of a hospital staff. Doctors depend on nurses to check patients periodically, administer medications, and report alarming symptoms. Nurses depend on doctors to prescribe medications and to specify what symptoms to look for.

Reciprocal interdependence always involves direct interaction of one sort or another, such as face-to-face communication, telephone conversations, or written instructions As a result, people who are reciprocally interdependent are more tightly interconnected than are individuals who are interconnected by either pooled or sequential interdependence. In addition, reciprocal interdependence is significantly more complex than either pooled or sequential interdependence. It incorporates symmetric, two-way interactions in which each person depends on the person who depends on her.

team interdependence A type of group interaction in which every group member depends on every other.

Team Interdependence. **Team interdependence**, depicted in Figure 10-3D, develops in a tight network of reciprocal interdependence. What makes team interdependence the most complex form of interdependence is that all members of the team or group are reciprocally interdependent on each other. As in reciprocal interdependence, people who depend on each other interact directly. In team interdependence, however, interactions tend to be more frequent, more intense, and of greater duration than in any other type of interdependence.

For example, in the brand-management groups that oversee the development of new products at firms like Colgate-Palmolive and Procter and Gamble, product designers, market researchers, production engineers, and sales representatives are linked by a completely connected network of two-way relationships. The product designers interact with the market researchers, product engineers, and sales representatives. The market researchers also interact with both the product engineers and sales staff, who in turn interact with each other. Similarly, the teams of engineers and scientists who design NASA spacecraft and satellites are linked by team interdependence.

AN EXAMPLE: THE
CHALLENGER DISASTER
The results of conflict among team
members are exemplified by the
1986 Challenger disaster. Would
the O-rings hold at sub-freezing
temperatures? A few team
members felt safety was the issue
and any doubt should delay the
launch, but others felt the
evidence wasn't compelling and
they needed to get the shuttle
program back on schedule. After
the disaster, it was revealed that
some team members had known
about the O-ring problem for
some time but kept it secret. The
conflict arose from a lack of
consensus as to three goals: safety,
schedule, and business. The
business issue (keep the problem
quiet) probably caused the last-
minute conflict between safety and
schedule. The results of this
conflict are recorded forever.

role The typical and expected
behaviors that characterize an
individual's position in some social
context.

established task elements The
components of work roles that are
contained in written job
descriptions and formally
recognized in the organization.

The type of interdependence that characterizes the group has important
managerial implications. First, there is a greater potential for conflict as the
complexity of the interdependence increases from pooled to team situations.
Second, turnover has more of an influence on the group when the group is
characterized by team interdependence. One person's absence affects a large
number of interactions. In some teams, the loss of a single player can make all
the rest of the players perform below par. Third, teams tend to be more flexible
and can adapt more quickly to changing environments than groups unified by
less complex forms of interdependence.

ROLES AND GROUP INTERACTION

Within the networks of interdependence that characterize group interactions, people come to expect each other to behave in particular ways.
Taxi drivers expect passengers to pay them when they reach their
destinations. Instructors expect students to complete required assignments before
coming to class. Spectators expect sports stars to exhibit skill and perseverance
as they compete. Expectations such as these—and the behaviors they presup-
pose—make up roles that connect individuals interpersonally.

A **role** consists of the typical behaviors that characterize a person's position
in a social context.[5]

You'll recall that in Chapter 8 we discussed the concept of *organization role*,
defining it as the set of behaviors that both the employee and the people with
whom she regularly interacts expect her to perform in doing her job. We pointed
out that in addition to the formal expectations of a jobholder, which are generally
determined by a company's top management, many additional, informal expectations evolve over time. Table 10-1 suggests some ways in which formal jobs
and informal work roles differ.

According to Ilgen and Hollenbeck, work roles are comprised of two kinds
of tasks.[6] First, there are the **established task elements** that make up the job.
The job is a formal position and comes with a written statement of the tasks it
entails. *Job descriptions* are generally prepared by managers or others at the upper
levels of an organization's hierarchy. As a result, there is a fair amount of agreement at the outset as to what constitutes the established task elements of the job.
Because job descriptions are prepared before the fact by people who do not

[5] B. J. Biddle, *Role Theory: Expectations, Identities and Behaviors* (New York: Academic Press, 1979),
p. 20.
[6] D. R. Ilgen and J. R. Hollenbeck, "The Structure of Work: Job Design and Roles," in *Handbook of
Industrial/Organizational Psychology*, ed. M. Dunnette (Houston: Consulting Psychologist Press, in press).

TABLE 10-1
The Job versus Work Role Distinction

JOB	WORK ROLE
1. Created by the owners of the organi-zation or their agents independently of the role occupant	1. Created by everyone who has a stake in how the role is performed, includ-ing the role incumbent
2. Has elements that are objective, for-mally documented, and about which there is considerable consensus	2. Has elements that are subjective, not formally documented, and open to negotiation
3. Static and relatively constant	3. Constantly changing and developing

In teaching, emergent task elements often seem to overwhelm established task elements. One school superintendent listed 52 nonacademic issues that his teachers must now deal with, ranging from day care to suicide prevention. At a 1990 conference, students, educators, business leaders, and politicians discussed ways that businesses can help overburdened schools. Here James Smith and Amanda Beliveau of New Hampshire's Thayer High School discuss needs with teacher Jean Kennedy and principal Dennis Littky. Typical of the programs discussed is the Arizona Business Coalition's task force, which has reshaped high school curricula, trained teachers, and provided technology, money, and employee volunteers to help students become productive members of the work force.
Source: Nancy J. Perry, "Schools: Tackling the Tough Issues," Fortune, December 17, 1990, pp. 143–56.

---

emergent task elements The components of work roles that are not formally recognized by the organization but arise out of expectations held by others for the role incumbent.

APPLYING THE DIAGNOSTIC MODEL
Diagnostic Question 2: What important components of your unique work role are not part of your job description?

TEACHING NOTE
Job descriptions help develop boundaries around jobs, clarifying individuals' roles. As tasks emerge, however, those boundaries often become fuzzy and individuals may overstep into other people's roles. This can cause conflict and result in tasks not being completed. How do we develop boundaries and clarify roles while allowing flexibility for jobs to change?

actually perform the job they are usually incomplete. Job descriptions usually take no account of job incumbents or of the complex and dynamic environments in which jobs must be performed.

As a person begins to do a job, it becomes clear to her and all those around her that tasks never detailed in the written job description need to be performed for the role to be successfully played. These added-on tasks are referred to as **emergent task elements**. For example, secretarial workers are increasingly being called on nowadays to perform a variety of duties other than typing, filing, and answering telephones. As business has grown more complex and executives' time has become more precious, some secretaries have expanded their roles. "Today's executive secretaries have started to assume many of the burdens of middle management," according to Nancy Shuman, vice president of a New York placement firm called Career Blazers. For example, Kay Kilpatrick, assistant to Richard Smith, chairman of General Cinema Corporation, finds herself doing tasks that were never part of her job description. She runs the firm's employee matching gift program and charitable corporate gift program. She also evaluates stock portfolios and handles distributions from the company's corporate assets.[7]

As you can see from Figure 10-4, established and emergent task elements can combine in different ways. At one extreme is the *bureaucratic prototype*. This jobholder performs no duties other than those written in the job description. That is, his work role is made up entirely of established task elements. Many low-level jobs in automated, assembly-line factories are this type.

Almost completely opposite is the *loose-cannon prototype*, in which the few established elements are greatly outnumbered by emergent elements. For example, when General Motors hired the flamboyant Ross Perot, ex-CEO of Electronic Data Systems, as a general organizational trouble shooter, it gave Perot wide latitude in his role. He decided to tour GM plants and criticize what he thought were inefficient management practices. Eventually GM leadership tired of the role Perot developed and wound up paying over $3 million just to get rid of him. Similarly, ex-army colonel Oliver North was pretty much allowed to do whatever he wanted in his role in President Reagan's National Security Council. One of the tasks he set for himself was the ill-fated "arms for hostages" deal with Iran, which generated a great deal of controversy in the final years of Reagan's presidency.

Finally, Figure 10-4C also shows the *job-similarity–role-difference prototype*. Here two individuals have the same job, but special characteristics of the incum-

---

[7] D. Fanning, "Calling on Secretaries to Fill in the Gaps," *New York Times*, March 11, 1990, p. A12.

FIGURE 10-4
**Job versus Work Role:
Established and Emergent
Elements**

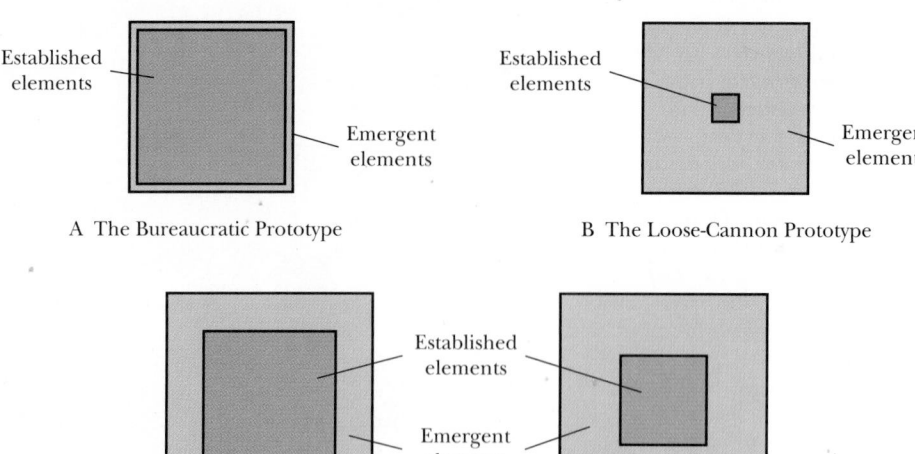

A The Bureaucratic Prototype

B The Loose-Cannon Prototype

C The Job-Similarity–Role-Difference Prototype

bents lead to the development of many emergent elements in one job but few in the other. On a football team, for example, an outside linebacker has established duties to contain the outside run and cover running backs in passing situations. A rookie linebacker will typically perform just those duties. An eight-year veteran, because of his experience, may have an expanded role with many emergent elements, such as serving as a team leader, calling defensive formations, and making decisions about whether to accept or decline penalties.

As Figure 10-4 suggests, a work role often includes a whole lot more than the job itself. It is important for managers to distinguish between jobs and more broadly defined work roles for several reasons. First, the manager must identify and reward individuals who are performing expanded roles to reinforce the behavior. At the same time, it is important to make sure that people do not expand their roles in directions that conflict with other people's formal responsibilities. Finally, when the established elements of a job are not well documented, it is critical that a manager anticipate the direction that a particular job applicant might take in the job before hiring him.

Dimensions of Role Specialization

**APPLYING THE DIAGNOSTIC
MODEL**
Diagnostic Question 3: What are
some of the important functional,
hierarchical, and inclusionary
dimensions in your organization?
Sketch an inverted-cone diagram
to represent these dimensions in
your firm.

Edger Schein has developed a conceptual framework that is particularly useful in understanding organizations as a set of interrelated roles.[8] Schein's model differentiates organizational roles from each other and from nonorganizational roles along three dimensions: functional, hierarchical, and inclusionary.

Functional. The *functional* dimension refers to the various tasks performed by members of the organization. Figure 10-5A shows the typical functional dimension of a business organization and a university: marketing, production, accounting, human resources, research and development, and finance. Similarly, the functional dimensions common to many universities are shown in Figure 10-5B. They include the schools of business, social sciences, arts and letters, medicine, engineering, and law. The roles performed in each of these dimensions are quite distinct because the jobholders are trying to accomplish different aspects of the organization's overall mission.

[8] E. H. Schein, *Organizational Psychology* (Englewood Cliffs, N.J.: Prentice Hall, 1970), pp. 111–33.

FIGURE 10-5
The Functional Dimension
of Organizations

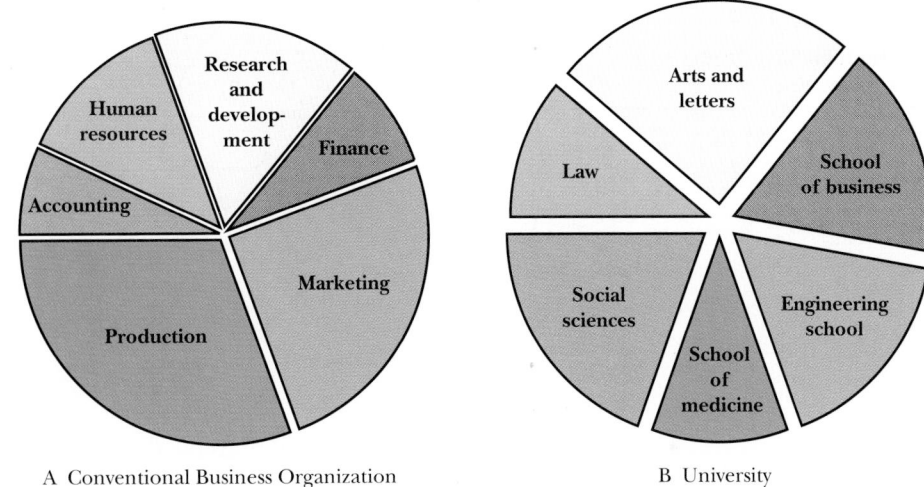

A Conventional Business Organization

B University

POINT TO STRESS
More and more organizations are
finding their hierarchies
cumbersome, expensive, and
inefficient. As a result, middle
levels of management are being
cut and companies reorganized.

Hierarchical. Schein's *hierarchical* dimension concerns the distribution of rank, or the official lines of supervisory authority. As you will recall from Chapter 2, hierarchy has to do with who is officially responsible for the actions of whom. In traditional organizations, this dimension takes a triangular shape, in which the highest ranks are held by relatively few people. The roles performed by people higher in the pyramid differ from the roles of individuals lower in the pyramid largely in that the former have greater authority and power. In a highly centralized organization like the army, this triangle is often rather steep (see Figure 10-6A).

A decentralized organization has fewer levels of authority and looks like a flattened pyramid. Despite their militaristic nature, most city police departments

FIGURE 10-6

**The Hierarchical Dimension
of Organizations**

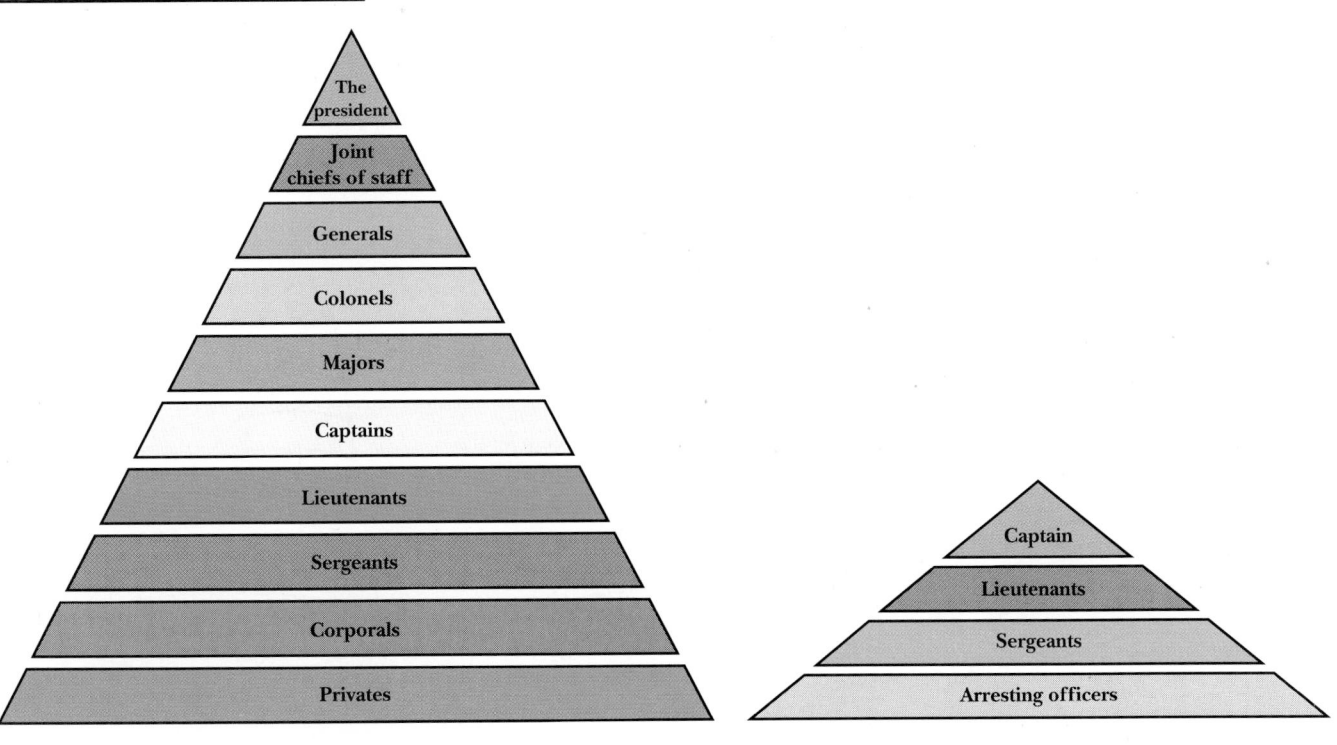

A Military Organization

B City Police Department

FIGURE 10-7

The Inclusionary Dimension of Organizations

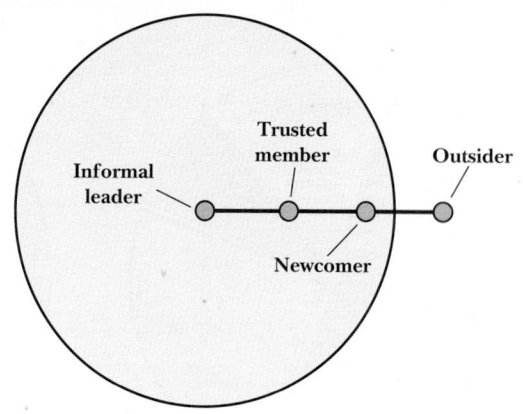

AN EXAMPLE

Youth gangs provide an excellent example of testing newcomers before welcoming them in the group. Gangs often have extreme initiation rites that must be passed. These rites focus on individuals demonstrating that they subscribe to the values and motives of the gang and are willing to do anything necessary for the gang. Rites of passage into these gangs include robbery, burglary, assault, and, in some cases today, murder. If you can't perform the assigned task, you will not be allowed to wear the gang colors.

have fewer levels of hierarchy than the army. Most employees are arresting officers, the highest rank is captain, and there are only two genuine levels of hierarchy between the top and bottom (see Figure 10-6B).

Inclusionary. The third dimension of Schein's model, the *inclusionary* dimension, reflects the degree to which an employee of an organization finds herself at the center or on the periphery of things. As you can see from the circular diagram in Figure 10-7, a person may move from being an outsider, beyond the organization's periphery, to being an informal leader, at the center of the organization. A job applicant, or outsider, joins the organization and becomes a newcomer, just inside the periphery. To move further along the radial dimension shown in the figure the newcomer must become accepted by others as a full member of the organization. This move can be accomplished only by proving that one shares the same assumptions as others about what is important and what is not. Usually, newcomers must first be tested—formally or informally—as to their abilities, motives, and values before being granted inclusionary rights and privileges.

Putting all three of Schein's dimensions together lets us depict an organization as a three-dimensional inverted cone, as shown in Figure 10-8. The entire organization and all the individual roles that comprise it can be conceptualized in terms of the three dimensions of function, hierarchy and inclusion. This conception helps us integrate individual and organizational issues through the concept of role. Look first at Figure 10-8A. This represents a military operation, with its tall hierarchy and small number of functional units. As the "Management

FIGURE 10-8

The Inverted Cone, Three Dimensional Model of Organizational Roles

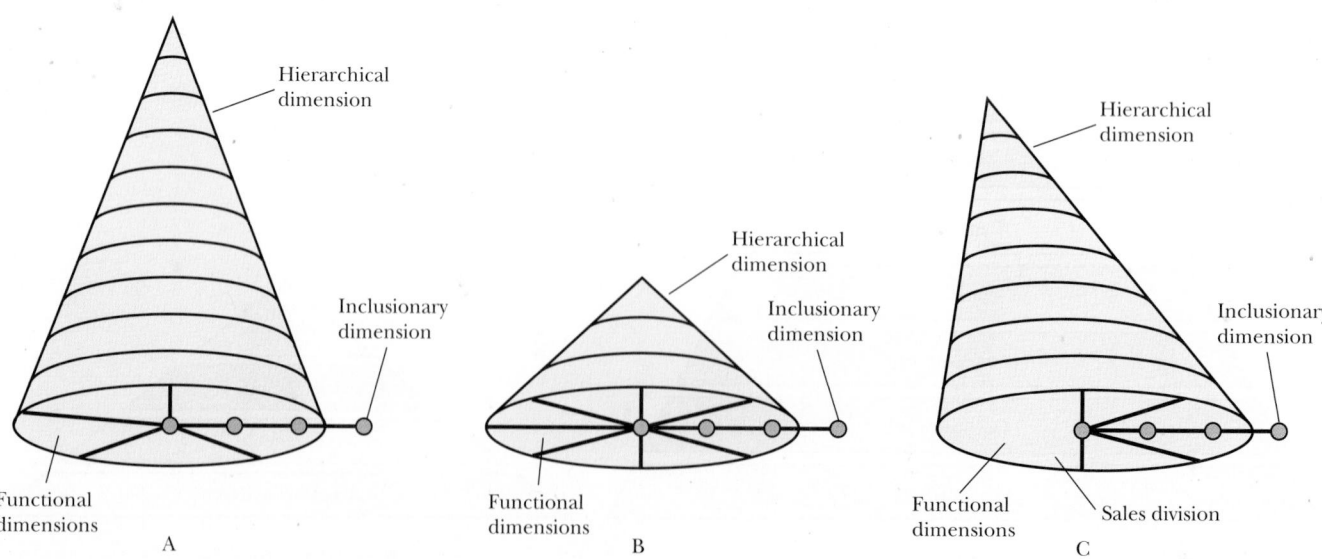

MANAGEMENT ISSUES

Too Many Chiefs and Not Enough Indians?

Many critics of American enterprise have suggested that organizations in this country have become top heavy. Richard Rosencrance, author of *America's Economic Resurgence: A Bold Strategy*, notes several disturbing statistics. First, more than half the employees of the modern American business corporation are not involved directly with production or service to consumers. For example, at General Motors, 77 percent of employees fill administrative and office positions, whereas only 23 percent handle operational, production jobs. The percentages of white-collar employees at Mobil, General Electric, and Du Pont all run over 50 percent. On the other hand, some more profitable companies, such as Ford, have only 37 percent administrative workers. Compare these data with the 8 to 10 percent administrative staff typical of similar large corporations in Japan.

This proclivity to generate too many chiefs and not enough Indians is not restricted to business ventures. Top heaviness is also present in the military and education sectors of the economy. For example, the number of senior military officers (three-star officers and full generals and admirals) is considerably greater now than it was at the peak of World War II, despite the fact that these officers command forces that are only a sixth as large as they were during that war. At both the high-school and college levels, administrators are now beginning to outnumber faculty. School districts in Los Angeles, New York, Philadelphia, and Denver all have more administrators and support staff than actual teachers.

To counter this trend, many have argued that corporate, government, military and educational bureaucracies need to be ruthlessly pruned. Executives in firms of the future need to be fewer in number but broader in talents. Workers need to be greater in number and also more willing to assume responsibility for decision making. In fact, in the future, the

very distinction between white- and blue-collar workers may disappear altogether. In the interim, the retraining of both those at the top and those at the bottom of the corporate ladder must be a high priority. Tall narrow cones are easily toppled. Shorter, wider cones are more resilient.

Source: R. Rosecrance, "Too Many Bosses, Too Few Workers," *New York Times*, July 15, 1990, p. 11.

Issues" box suggests, many critics of U.S. management practices feel that this military-style model is too common in American organizations.

In Figure 10-8B we see a research-and-development firm with a relatively flat hierarchy but a large number of functional departments. Finally, Figure 10-8C shows a real estate sales organization with a slightly skewed, or tilted, cone. The tilt of the cone indicates that the most important function of the firm is sales and that almost all positions at the top of the hierarchy are filled with people who moved up from the sales department. In this firm, a newcomer in the personnel department might be more of an outsider than a newcomer in sales. The probability that the former will ever rise to be CEO is almost zero given the role structure depicted in the figure.

Taking and Making Roles

Defining organizations as systems of roles highlights the fact that organizations are structured in terms of role behaviors, not in terms of the unique acts of specific individuals. Indeed, one of the strengths of formal organizations is their constancy under conditions of persistent turnover of personnel. For example, the New York Yankees are perceived to be a single, unified entity, despite the fact that within any two- to three-year period, 50 percent of the team members change. Thus roles are of crucial importance to organizations. The process by which they are developed is a central concern for the study of organizational behavior. In this section we will examine the model of the role-taking process shown in Figure 10-9.

The Role Set. A **role set** is a group of people who must interact with a role occupant either formally or informally. Typically a role set includes such people as an employee's supervisor and subordinates, other members of the employee's functional unit, and members of adjacent functional units that share tasks, clients, or customers.

Because they have something at stake in the role occupant's performance, members of the role set develop role expectations or **norms**. Norms are strong beliefs about how the role should be performed. There are norms for both the formal requirements of the job, or its established task elements, and its generally agreed upon informal rules, or emergent task elements.[9]

Norms may evolve out of a number of sources, as indicated in Table 10-2. Sometimes *precedents* that are established in early exchanges simply persist over time and become traditions. For example, students take certain seats on the first day of class and, even though the instructor establishes no formal seating arrangement, they tend over time to keep the same seats. Norms may also be *carryovers* from other situations. That is, people may generalize from what they have done in the past in other, similar situations. A person may stand when called on to make a presentation at a meeting because he was trained to stand in these circumstances. Sometimes norms reflect *explicit statements from others*. A part-time summer worker, for instance, may be told by more experienced workers to "slow down and save some work for tomorrow." Finally, some *critical historical event* may influence norms. Suppose, for example, that a secretary leaks important company secrets to a competitor. In response to this incident, an unwritten norm may evolve that requires that all sensitive information be typed personally, not delegated. In Chapter 11 we will discuss the different kinds of norms that often develop in organizations.

Of course, expectations and norms would have little effect on organizational behavior if they stayed inside the heads of role-set members. But they do not. As shown in Figure 10-9, norms, or role expectations, are "sent" to the role occupant. Some of the messages transmitted in *role sending* are informational and tell the focal person what is going on. Others are attempts to influence the role occupant in one way or the other (e.g., by letting her know what rewards or punishments will follow adherence to or disregard of particular norms). Some influence attempts may be directed toward accomplishing organizational objectives. Others may be unrelated to, or even contrary to, official requirements ("Don't

[9] J. R. Hackman, "Toward Understanding the Role of Tasks in Behavioral Research," *Acta Psychologica* 31 (1979), 97–128.

role set   The entire group of individuals who have an interest in and expectations about the way a role occupant performs his job.

norms   A strong set of expectations that members of a role set have for the role occupant.

APPLYING THE DIAGNOSTIC MODEL
Diagnostic Question 4: Who are the members of your role set? What are their expectations for the occupant of your role? How do they communicate these expectations to you?

POINT TO STRESS
Established norms tell us informally what an organization considers appropriate behavior and how the organization will treat us. When we accept such norms we form a kind of psychological contract that often more clearly defines expectations and becomes more important than formal written contracts, job descriptions, or policy handbooks.

FIGURE 10-9

The Role-Taking Process

Adapted from D. Katz and R. L. Kahn, The Social Psychology of Organizations (New York: John Wiley, 1978), p. 112.

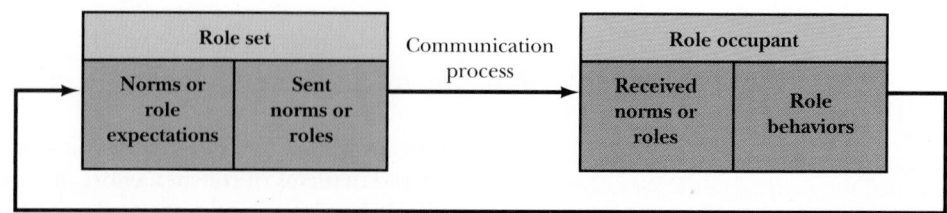

INTERPERSONAL PROCESSES AND COMMUNICATION

TABLE 10-2
Sources of Group Norms

1. Precedents set by earlier interactions in a specific group
2. Carryovers from earlier interactions that group members had with people in other groups they belonged to
3. Explicit statements and agreements among group members
4. Critical events in the group's history

Source: Adapted from D. C. Feldman, "The Development and Enforcement of Group Norms," *Academy of Management Review* 9 (1984), 47–53.

APPLYING THE DIAGNOSTIC MODEL

Diagnostic Question 5: What are your own expectations for your role? How do these expectations compare with your role set's expectations, and what should be done about any discrepancies?

Diagnostic Question 6: What are the primary reasons that people in your organization conform to their role expectations—compliance, identification, or internalization? What do these choices imply for monitoring employees' behavior.

role occupant   The current incumbent of an existing organizational work role.

waste your time filling out that form every time you have a complaint from a customer").

As long as the role occupant complies with the expectations of the role senders, the senders will attend to their own jobs. However, if the role occupant starts to deviate from the sent expectations, the role senders, their expectations, and their means of enforcing compliance will become quite visible. The part-time summer worker who fails to heed the warnings of more experienced personnel may soon become the victim of derision, practical jokes, or isolation.

As we will see in the next section of this chapter, communication can be a very complex process. Role sending is a form of communication and, as such, its success depends on a number of factors that we will examine shortly. For now, we need simply know that messages about role expectations may not always be transmitted as clearly as possible, and they may not always be accurately understood by recipients.

The Role Occupant.   Although it is through the sent role that the members of the organization communicate the do's and don'ts associated with a role, it is the "received" role that is the immediate influence on the behavior of a **role occupant**. As we will see, a number of factors may distort a message or cause it to be misunderstood. But even when messages are communicated effectively, senders' role expectations are often not met by role occupants. Several types of role conflict that we discussed in Chapter 8 can prevent a role receiver from meeting the expectations of a sender.

First of all, *inter-sender role conflict* may place competing, mutually exclusive demands on the role occupant. If the person meets one sender's expectations she may violate the expectations of another.

In addition, the role occupant may experience *person-role conflict*. He may have some ideas about how the role should be performed that conflict with role-sender demands. Finally, inter-role conflict can not only interfere with the performance of one or more roles (in Chapter 8 we contrasted the roles of parent and employee), it can in some cases lead an employee to flagrantly violate the expectations of his role set. For example, because news of the latest developments at Apple Computer was valuable currency in the social milieu of Silicon Valley, many Apple employees leaked sensitive information to friends. Clearly their desire to appear in the know conflicted with the organization's need for secrecy. To stop the practice, Apple came up with buttons for all employees that read, "I know a lot but I can keep a secret."[10]

After sorting through the various sent roles, the role occupant typically makes a decision about which role behaviors to perform. His reason for choosing one expectation over another affects both the stability of his behavior and the

[10] G. P. Zachary, "At Apple Computer Proper Office Attire Includes a Muzzle," *Wall Street Journal*, October 6, 1989, p. 1.

compliance Behaving in accord with norms out of fear of punishment or hope of reward.

degree to which the role sender must monitor his behavior. Let us look at three primary reasons for conforming to sent role expectations—compliance, identification, and internalization.[11]

In response to a role sender's communication, a role incumbent may choose **compliance** because the role sender has the power to reward or punish. Although this sounds like a good way of making sure that role behavior coincides with the sent role, as we will see, compliance is actually the least stable method of achieving conformity. To produce true compliance in a role occupant, a role sender must have three characteristics. First, she must be able to monitor the role occupant closely. Unless you know what a person is doing, you can hardly reward or punish him appropriately.

Second, the role sender must be able to deliver the rewards and punishments she has promised. If threats of punishments and pledges of rewards begin to be perceived as bluffs and empty promises, compliance will stop.

Third, the role sender's power must be stable. If there is a good chance that the role sender's power to reward and punish will soon diminish, compliance may not be the best means of achieving conformity. The moment the power balance changes, conformity will cease.

identification Behaving in accord with norms out of respect and admiration for one or more members of the role set.

A role incumbent may also conform to a role sender's expectations through **identification**. She may be attracted to the role sender; she may want to be like the role sender; and she may want the role sender's approval. Role behaviors are often a product of this kind of process. Role occupants may find someone they admire and then adopt the behaviors and attitudes of this person. Indeed, people whose actions and attitudes are imitated and adopted by others are referred to as *role models*.

TEACHING NOTE
Ask students to differentiate between identification and mimicking a role.

Although more stable than compliance, identification is also hard to achieve over extended periods of time. First, role models are only human themselves. Often the more one learns about them, the less heroic or romantic they seem. When sordid details of the lives of the evangelical religious leaders Jimmy Swaggert and James Bakker became public, many of their followers were forced into some serious soul-searching. Second, as individuals develop, they often outgrow one role model and move on to another. Or they may stop modeling the behavior of others entirely, having developed a strong sense of their own roles.

internalization Behaving in accord with norms that are consistent with one's own value system.

Finally, a person may conform to a role sender's expectations because the sent role is consistent with the role occupant's own personal value system. Sometimes called **internalization**, this method of achieving conformity is also referred to as "winning the hearts and minds." Mary Kay Cosmetics, for example, which relies on a sales force that cannot be monitored, tries to equate selling success with personal growth and development. The company fosters the belief among its sales representatives that selling cosmetics will increase their interpersonal skills and their self-confidence. As a result, generating sales revenues is important to Mary Kay sales staff for more than financial reasons. (Just in case this pitch fails, the organization also provides sales commissions.) Theoretically, then, the sales representatives are motivated to perform their role because it leads to self-development and self-enhancement.

Because people's value systems are quite stable, (see Chapter 4), internalization is the most reliable method of achieving conformity. It also requires minimal monitoring, for role occupants come to monitor themselves. Indeed, some role behaviors, like a soldier's falling on a grenade to save his comrades, can be achieved only through internalization.

The Role Episode. A single iteration of the role expectation—sent roles—received role—role behavior sequence is called a *role episode*. As Figure 10-9 shows,

[11] H. C. Kelman, "Processes of Opinion Change," *Public Opinion Quarterly* 25 (1961), 57–87.

the role-taking process is cyclical. The way the role occupant conforms to the role sender's expectations on a particular occasion will affect the role sender's expectations at a later point in time.

An initial influence attempt that results in conformity reinforces expectations. If the same initial influence attempt is met with a defensive counterattack, however, role senders may quickly modify their expectations. If the role occupant conforms partially when placed under a little pressure, the pressure may be increased in the next cycle. If the role occupant is obviously overwhelmed by the role, the role senders may agree to "lay off" until the person has developed a little further. Thus, when viewed as a system of roles, the organization is seen not as an immutable, objective entity but as a flexible, negotiated set of expectations and behaviors.

COMMUNICATION AND ROLES

We have suggested that an organization is best conceptualized as a system of negotiated roles in which both the organization and the roles are characterized by dimensions of function, rank, and inclusion. We have also suggested that at the heart of the role-taking process is the exchange of ideas between members of the role set (the role senders) and role occupants.

If roles are conceived of as the bricks of an organization, communication is the cement that holds those bricks together. **Communication** is the exchange of ideas through a common system of symbols. It is the only means by which people can transmit messages to one another about such things as role expectations and norms. We need to explore next how communication takes place and what factors may enhance or impede it.

communication The exchange of information between people through a common set of symbols.

The Communication Process

In Figure 10-9, we simply drew a straight line between the sent role and the received role to denote the transmission of ideas between members of the role set and the role occupant. Now we need to expand on that oversimplification. A much better model envisions the communication process in three stages:

1. Encoding information into a message
2. Transmitting the message by some medium
3. Decoding the information from the message[12]

encoding The process by which a communicator's abstract idea is translated into the symbols of language for transmission to someone else.

Encoding the Message. Encoding is the process by which a communicator's abstract idea is translated into the symbols of language and thus into a message that can be transmitted to someone else. The idea is subjective and known only to the communicator. The message, because it employs a common set of symbols, can be understood by other people who know the communicator's language.

APPLYING THE DIAGNOSTIC MODEL
Diagnostic Question 7: What means of communication (oral, written, or nonverbal) do you use for different organizational tasks? Are these the optimal choices for the purposes you seek to achieve? Why?

Communication Media. The *medium* is the carrier of the message and is objectively observable. That is, it exists outside the communicator's mind and can be perceived by others. We can further characterize the media of organizational communication by the human senses they use. Oral speech uses hearing. Written

[12] C. Shannon and W. Weaver, *The Mathematical Theory of Communication*, (Urbana: University of Illinois Press, 1948), p. 17.

AN EXAMPLE: J. C. PENNEY COMPANY

Using direct satellite broadcasts from their Dallas headquarters, J. C. Penney executives have their improved communications with buyers for individual stores. When a merchandise manager senses that a particular item of clothing is going to be a big seller, he can display a few samples via satellite to buyers, who can then order from headquarters or from suppliers. This electronic method of communicating with role occupants greatly increases the accuracy of the message. It also makes it possible for Penney buyers to react more quickly to trends, it reduces the time they must spend traveling to see manufacturers' offerings, and, of course, it cuts costs. ("Selling," *Fortune*, September 24, 1990, pp. 94–98.)

AN EXAMPLE

Union contracts present an excellent example of attempts to be precise and accurate in communicating in writing the roles and responsibilities of management and union members. Although weeks and sometimes months are spent negotiating and writing these contracts, most formal grievance procedures reflect two different interpretations of the contract. (*Journal of Labor Research*, forthcoming 1992)

TEACHING NOTE

Have students discuss messages that may be lost when communicators are not face to face. Is the telephone an efficient way to communicate a message? Is teaching over television effective?

decoding The process by which a transmitted message is converted into an abstract idea in the mind of the person to which the communication is directed.

noise A collective term for a number of factors that can distort a message as it is transmitted from one person to another.

---

documentation uses vision or touch (Braille). Nonverbal communication may use at least four of the five basic senses.

Oral communication relies predominantly on the sense of hearing; its symbols are based on sounds. Face-to-face conversations, meetings, and telephone calls are the most commonly used forms of communication in organizations. Look back at Figure 2-10, where you will see that as much as 75 percent of a manager's time is devoted to meetings and telephone calls.[13] Oral communications are fast. One can encode information quickly, and the feedback cycle is rapid. If the receiver is unclear about the message, she can immediately ask for clarification. Presenting a proposal orally, for example, provides much more opportunity for clarifying questions than preparing a written report. Therefore, oral messages are generally efficient in handling the day-to-day problems that arise in organizations.

Although much oral communication in organizations takes place in meetings, the business meeting is generally perceived as an inefficient, unproductive activity. Can anything be done to overcome this problem? Yes, indeed. Many organizations have taken specific remedial steps. For example, Hercules, Inc., a chemical and synthetics manufacturer in Wilmington, Delaware mounts huge clocks at the back of every conference room to make it easier for meeting chairs to halt endless deliberations.[14] It is also a good idea to invite to a meeting only those people who have something to contribute or something at stake. Participants should know what the content of the meeting will be at least a day in advance, and an agenda and any required reading should be distributed before the meeting. Finally, the person leading the meeting should make sure that the shy get heard—even if this requires direct questioning—and that domineering people are held in check.

Sometimes *written communication* is preferred over oral communication. Although written messages are more slowly encoded, they allow the communicator to use more precise language. A sentence in a labor contract, for example, may be rewritten five or six times to make certain that everyone involved knows exactly what it means. The aim is to minimize the possibility of any future confusion or argument over interpretation. Written materials also provide a "hard copy" of the communication that can be stored and retrieved for later purposes. For example, a supervisor may write a formal memo to an employee informing her that she has been late for work 10 of the last 11 days and if she does not begin coming in on time she will be fired. If the behavior continues, the supervisor has documented evidence that the employee received fair warning.

Nonverbal communication covers a variety of transmission modes that rely on something other than the written and spoken word.[15] Table 10-3 describes several forms of nonverbal communication. It is easy to underestimate the impact that messages relayed this way have on others' perceptions, in part because this form of communication often reaches the listener at relatively low levels of awareness.

Decoding the Message. To complete the communication process, the message sent must be subjected to **decoding**, or translated in the mind of the receiver. When all works well, the resulting idea or mental image corresponds closely to the sender's idea or mental image.

Unfortunately, there is no shortage of things that can go wrong, making communication ineffective. The term **noise** refers to the factors that can distort a message. Noise can occur at any stage of the process. For example, a person may not encode exactly what he means to say. A surgeon may tell a nurse that "the patient has a cancerous tumor on his hand and it needs to be removed,"

---

[13] H. Mintzberg, *The Nature of Managerial Work* (New York: Harper & Row, 1973), p. 22.

[14] S. Hsu, "Dull Meeting? Turn the Tables," *Bergen Record*, July 30, 1990, p. C1.

[15] F. Williams, *The New Communications* (Belmont, Calif.: Wadsworth, 1989), p. 45.

1. **Paralinguistics** A form of language in which meaning is conveyed through variations in speech qualities, such as loudness, pitch, rate, and number of hesitations
2. **Kinesics** The use of gestures, facial expressions, eye movements, and body postures in communicating emotions
3. **Haptics** The use of touch in communicating, as in a handshake, a pat on the back, or an arm around the shoulder
4. **Chronemics** Communicating status through the use of time. For example, making people wait or allowing some people to go ahead of others
5. **Iconics** The use of physical objects or office designs to communicate status or culture, such as the display of trophies or diplomas
6. **Dress** Communicating values and expectations through clothing and other dimensions of physical appearance

meaning that the tumor must be removed. The nurse, however, prepares the patient for an amputation of his hand.

Another communication error may lie in the selection of the medium for a message. Suppose you write a memo giving a colleague information about the date and time of an important meeting and a clear outline of the meeting agenda. You leave the memo in your colleague's mailbox. Unfortunately, your colleague, who is a sales representative, does not come into the office that day to collect her mail, so no communication takes place. A telephone call instead might have avoided this problem. Some executives feel the apparently growing need to communicate quickly with others, which explains the skyrocketing popularity of electronic telephone pagers.

Finally, problems can occur at the receiving end of communication. Because nonverbal language is as culture specific as any other form of language, you must choose your gestures and interpret others' gestures with caution. Putting your arm around an employee, for example, may be your way of saying, "We're all in this thing together, and we'll help one another out," but an employee may interpret your behavior as a sexual advance. In the next section, we will examine the great variety of noise factors that act as barriers to effective communication. We will use the role-taking model in Figure 10-9 to put these factors in a single framework.

Barriers to Effective Communication

Several organizational, interpersonal, and personal factors can either help or hinder communication within organizations, depending on how they are handled. These barriers need to be removed if effective communication is to take place. We will examine spatial arrangements of people and offices; characteristics of the communicators, both role sender and role occupant; and interpersonal differences in language and experience.

Organizational Factors: Spatial Arrangements. The nature of the physical space occupied by jobholders inevitably affects patterns of communication. If an organization wants to promote the development of interpersonal relations, for example, it must place people in close physical proximity (although not too close, as we saw in Chapter 8). All else equal, people who work closely together have more opportunities to interact and are more likely to form lasting relationships than people who are physically distant. Apparently, whether you are a clerk, a

Partitions like these used at Xerox Corp. make it possible for people to exchange information directly with each other. But do they provide enough privacy and personal space? Do you think companies choose the type of arrangement pictured primarily to encourage worker interaction or to save money on costs of heating, air conditioning, and construction?

college professor, or a member of a bomber crew, the closer you work to other people, the more often you will communicate with them.[16]

Sometimes it is useful to distinguish between actual physical proximity and psychological proximity. For example, architectural arrangements can create psychological barriers to communication that can discourage interaction. On the other hand, arrangements that channel the flow of people who are moving about toward a common area, such as a reception area, a water fountain, and a bank of elevators, can create opportunities for spontaneous interaction. You'll remember that in Chapter 8 we discussed the open-office concept developed in an effort to increase the amount and quality of interaction among organizational members. As we learned, although this kind of design does increase interaction and communication, it does not always lead to greater job satisfaction. Indeed, offices can get too open, forcing people into too much interaction. People also have needs for privacy and personal space. Office designs that fail to recognize these needs may increase interaction and communication at the expense of satisfaction and productivity.

Role-Set Factors. Whether the purpose of the communication is to inform or persuade, the *credibility* of the source will largely determine whether the message is internalized by the role occupant. *Credibility* refers to the degree to which the information provided by the source is believable. Credibility is a function of two factors. The first factor is expertise, or the source's knowledge of the topic at hand. The second is trustworthiness, or the degree to which the recipient believes the communicator has no hidden motives. Thus a new manager may view an older, more experienced executive whose job is secure as a credible source of information and advice. The same manager, however, may take a skeptical view of advice from a fellow newcomer who may be competing with him for the same promotion six months down the road. The latter probably doesn't know any more than the former and may in fact have something to gain from his failure.

Protecting one's credibility is vital to one's long-term success in any company. If you lose your credibility you lose just about any influence you might have on others in your role set, and you may find yourself isolated and vulnerable. Much like the boy who cried wolf too often, organizational members who have lost credibility in the eyes of other role-set members may find themselves alone in

AN EXAMPLE
Managers who, after a merger, tell employees that there will be no rapid major changes lose credibility. Employees have heard and read about the horrors that follow a merger, they see changes already happening, and they do not believe the management line.

[16] J. T. Gullahorn, "Distance and Friendship as Factors in the Gross Interaction Matrix," *Sociometry* 15 (1952), 123–34.

INTERPERSONAL PROCESSES AND COMMUNICATION

APPLYING THE DIAGNOSTIC MODEL

Diagnostic Question 9: How do others in your group perceive your influence and credibility? How do these perceptions affect the communications others send you and the communications you send to them?

confronting crisis situations. As the "In Practice" box shows, there is some evidence that employees' trust of top management has deteriorated recently. Many companies are going to great lengths to win back this trust.

A *power imbalance* between a role sender and a role occupant may also impede communication. Although we will discuss the concept of power at some length in Chapter 13, at this point we will consider just two kinds of power. One kind of power is based on legitimate authority, that is, the organizationally sanctioned ability of one person to reward or punish another. We will refer to the other kind of power as *status*, or the degree of prestige associated with a person's social position. Status is a function or both formal and informal authority.

The inverted-cone conception of organizations is useful in showing that communication can move in many directions. Legitimate authority affects the communication of messages that move along the vertical dimension of the inverted cone. A *downward communication* moves from a member of higher legitimate authority to a member of lower authority. Although today many successful executives do spend time on the shop floor talking with line workers, traditionally this type of communication has moved down the hierarchy one step at a time. The president of a large company would communicate with a vice-president, perhaps, who would speak to a line manager, who would issue directions to workers. As you may imagine, in this situation there are many opportunities for a mistranslation somewhere along the chain. It is a good idea for the initiator of a message to check the way it is being received.

POINT TO STRESS

The more people a message must pass through the more it is distorted. Each person interprets the message and filters out part of it before sending it on. Compare this to the game of telephone that children play.

Upward communication flows from people low in the organizational hierarchy to people above them. Because people at upper levels of the hierarchy have a great deal of power to reward and punish employees at lower levels, the latter are sometimes inhibited in their upward communication. Insecure lower-level workers may have a tendency to forget about the losses and exaggerate the wins when reporting information upward, leaving those at upper levels with a distorted sense of reality. Similarly, lower-level employees who are unsure about how to do their jobs may be reluctant to ask for assistance, fearing to appear unknowledgeable.[17] Here too, upper-level managers may get a distorted view of the competencies and capabilities of those who serve under them.

Finally, distortion can also occur in *radial communication*, which moves between a relatively peripheral member, say a newcomer, and a more central member, say an informal leader. As in the case of upward communication, the newcomer's reluctance to reveal ignorance to informal leaders may be a barrier to communication. Moreover, long-tenured, central members may share knowledge and language that newcomers find difficult to get a handle on, resulting in miscommunication and misunderstandings.

APPLYING THE DIAGNOSTIC MODEL

Diagnostic Question 10: What are some of your major values, beliefs, and frames of reference? How might these affect the way you interpret communications you receive from others in the group?

Role-Occupant Factors. As we saw in Chapter 5, people are bombarded with much more information than they can possibly attend to. Therefore, incoming information has to be filtered. The same thing holds true for communication. All the factors that affect perceptions affect communication as well.

A role occupant's *beliefs and values* will shape the way she interprets a message from a role sender. If she is anxious, for example, about how she is performing, she may place a great deal of weight on any message sent that reflects positive feedback. A person's strongly held beliefs can also affect his interpretation of messages sent. Perceivers tend to discount information that is not in accord with their beliefs. Messages that contradict those beliefs may not be correctly under-

---

[17] A. S. Tsui, "A Role Set Analysis of Managerial Reputation," *Organizational Behavior and Human Decision Processes* 34 (1984), 64–96.

IN PRACTICE

Closing the Credibility Gap

Most top managers are quick to declare, "People are our most important asset." Yet all too often in recent years, this kind of declaration has been followed almost in the next breath with layoffs. Similarly, although many top managers stress the importance of quality, workers often see these same people as evaluating them solely in terms of the number of products pushed out the door. According to pollster Ilene Gochman of Opinion Research, "The days when management could say, 'Trust us, this is for your own good,' are over. Employees have seen that if the company steams off on some new strategic tack and it doesn't work, employees lose their jobs, not management."*

The credibility gap between what top managers say and what they do is matched by the gap in lifestyles between top managers and the rank-and-file employees. After Time, Inc., acquired Warner Communications, Time's CEO, J. Richard Munro stood to make $12 million on the deal. When asked about this, he responded, "That sounds like a lot of money unless you live in New York and live in the world I live in." Similarly, Citicorp chairman, John Reed, summarily dismissed an aide who had a problem with Reed's "Can't talk now—gotta meet [tennis star] Jimmy Connors." Finally, when asked if he ever shared his vision of the company's future with his employees, a head of an insurance company remarked in a surprised tone, "You mean, sit with them in little red plastic chairs and drink coffee out of styrofoam cups?" Comments like these have tended to dispel the myth that top managers and employees share something in the way of common goals or problems. The fact that top-executive salaries in the U.S. are commonly 100 times that of the average employee (this

* Based on Farnum A. "The Trust Gap," *Fortune*, December 4, 1989, pp. 56–78.

Herb Kelleher, Southwest Airlines CEO, with some of the air carrier's staff.

rate is only 15 to 1 in Europe and Japan) adds fuel to this already raging fire.

Executives at many U.S. companies are working hard to close the credibility gap. Their efforts point to a number of positive, concrete behaviors that others could model to increase the perception of shared goals by the top and bottom halves of the organization.

One way to close the gap is to go through the bad times together as well as the good times. Herb Kelleher, CEO of Southwest Airlines, said, "If there's going to be a downside you should share it. When we were experiencing hard times two years ago, I went to the board and told them I wanted to

---

AN EXAMPLE

Constriction of communication channels often occurs after a merger or acquisition. Usually so many changes occur so fast that crisis may follow crisis. Managers tend to go into a "war room" mode of management. They cluster together trying to interpret what is happening and develop strategies to deal with each crisis. Little information filters out of these meetings, leaving everyone wondering what is happening and why. Without this information, employees begin to generate their own answers, which are often incorrect.

stood. Take the example of a sexist supervisor who believes that women are incapable of performing in managerial roles. He may distort or ignore messages that suggest a female manager is doing a good job but dutifully write down any evidence of her shortcomings.

The *frame of reference* of the receiver will also determine how well communication takes place. In Chapter 5, we noted that research by Tjosvold showed that the manager's frame of reference affects his decisions.[18] That is, managers made quite different decisions about the same situation depending on whether they interpreted the situation as a crisis or as a challenge. In this study, manager's interpretations also affected their communications. As shown in Table 10-4, when managers interpreted a situation as a crisis, they were less likely to ask questions, less knowledgeable about others' arguments, and less interested in hearing additional arguments. A crisis mentality appears to lead to a constriction of communication channels. The crisis frame of reference is particularly likely to crop up when communications must occur under heavy time pressures.

[18] D. Tjosvold, "Effects of Crisis Orientation on Managers' Approach to Controversy in Decision Making," *Academy of Management Journal* 27, 130–38.

cut my salary. I cut all the officers' bonuses 10 percent, mine 20 percent."

Herman Miller, one of the U.S.'s top furniture manufacturers makes sure that the ratio of executive salaries to that of the average worker never exceeds 20 to 1. Vermont ice-cream maker, Ben and Jerry's Homemade, goes even further, putting a cap on at a 5 to 1 ratio. Both these firms firmly believe that disproportionate top-management salaries disrupt teamwork.†

In addition to clamping down on salaries, some firms have moved away from executive perks, like country club memberships and chauffeurs. These so-called benefits create a psychological distance between the ranks and in doing so create more problems than they solve. For example, Union Carbide's headquarters on Park Avenue was at one time the most hierarchical, class-conscious office environment in New York City. When the firm moved its operations to Danbury, Connecticut, it took the move as the perfect opportunity to invoke change. According to Jim Barton, Carbide's director of general services, "In terms of amenities, everybody had the same stuff. It was an egalitarian approach: 2,350 private offices, all the same size. No executive parking. No executive dining room." These changes led to both increased productivity and satisfaction among employees.

There is no substitute in the pursuit of reduced psychological distance for reducing physical distance. Many top executives find that exposure to rank-and-file jobs promotes trust and understanding. For example, Darryl Hartley-

Leonard, president of Hyatt Hotels, put his entire headquarters staff to work for a day changing sheets, pouring coffee, and running elevators. The president himself worked as a doorman alongside veteran porter Bill Kurvers. When asked what the president got from his experience other than tips, Kurvers noted quickly: "He got respect."

At Lincoln Electric, MBA's—even those from the most prestigious schools—spend eight weeks on a welding line. "We want them to understand the difficulty of the factory environment and have respect for people out there," says President Don Hastings. He adds, "These MBA's have got a big target on their backs. People have to see that they are not just traders coming in from the financial world."

Finally, although less dramatic, the use of regular employee surveys for increasing trust and cooperation should not be overlooked. For example, Preston Trucking, a Maryland-based carrier with nearly $600 million in yearly revenue, solicits its workers' ideas and opinions on a regular basis. By attending to workers' input and acting on it, the employees come to believe that management respects and trusts them.

This mutual respect can come in handy, especially during a crisis. Chuck Dunlop, manager at Preston's dock in Kearney, New Jersey, needed to save money by closing down the dock on the Friday before Christmas in 1988. However, a contract with the Teamsters stipulated that Friday was an official holiday. Dunlop could close the plant, but 35 workers would have to be paid anyway. Shop steward Carl Conoscenti told the drivers that they were entitled to the money, that no one would think less of them for taking it, and that no one would even know if they took it. None took it. Dunlop commented: "These are teamsters . . . This is *New Jersey*." One wonders how employees working for Time's Munro or Citicorp's John Reed might have responded to such a crisis.

† J. Greenwald, Advice to Bosses: Try a Little Kindness. *Time*, September 11, 1989, p. 56

Differences in frames of reference often loom large in horizontal communication, or communication between functional departments of an organization. When such communications do occur, the radically different frames of reference adopted by members of each department can seriously complicate the communication process. For example, members of a sales department, who are close to consumers, may frame their communications in terms of revenue. Mem-

TABLE 10-4

Framing a Problem as a Crisis or a Challenge: The Effect on Communication

	MEAN RESPONSE*	
	Crisis	Challenge
1. Number of questions asked	2.77	3.84
2. Knowledge of others' arguments	2.25	3.30
3. Interest in hearing more arguments	4.64	6.46

* Responses were measured on a scale that ranged from 1, "few" or "very little" to 7, "many" or "a great deal."

bers of a production group, on the other hand, who are closer to the manufacturing end, may frame messages in terms of costs. These different frames of reference cause each group to focus on different kinds of information and they may spend much of their time talking past each other.

Interpersonal Differences.   Not only do different functional units have different frames of reference—they often speak different languages. Most specialized units develop their own **jargon**. Jargon is extremely useful. It maximizes information exchange with a minimum of time and symbols by taking advantage of the shared training and experience of its users. A coach of a football team, for example, may tell his quarterback to tell the team, "Left 41 out on three." This simple five-word message conveys a wealth of information. It provides detailed instructions to 11 people about complex behavioral sequences they are expected to perform. Jargon also may prevent others from understanding what is being communicated, which may be desirable. For example, a quarterback may have to change a play at the line of scrimmage in full earshot of the opposing players.

On the other hand, because jargon is likely to confuse anyone lacking the same training and experience, it can be a barrier to communication between groups. Often technical specialists get to the point where they use jargon unconsciously and indeed have a hard time expressing themselves in any other terms. This can become a permanent disability, greatly reducing people's career opportunities outside their own small groups.

SOCIALIZATION, ROLES, AND COMMUNICATION

We have pointed out that people choose to work together to accomplish things they cannot accomplish alone. In so doing, they develop and take on specialized roles. We have also noted how these roles are formed and maintained. Group members develop communication processes so that they can exchange information. In the final section of this chapter, we will focus on two special facets of communication. Acknowledging that communication is part of all roles, we will examine some special roles in which communication is the key purpose of the role. We will also discuss a special kind of communication, called *socialization*, which takes place when a person first assumes a new role.

Communication Roles

Many problems can result from poor communication across roles. So it is not surprising that sometimes we create special roles whose sole purpose is to improve the effectiveness of communication. We will discuss three special communication roles in the organizational setting—gatekeeper, cosmopolitan, and opinion leader.[19]

Gatekeepers.   One special communication role is that of the gatekeeper. The **gatekeeper**, as illustrated in Figure 10-10A, is responsible for controlling the messages sent through a particular communication channel. Gatekeepers are essential to prevent information overload from reducing the effectiveness of those in upper-level roles.

[19] E. M. Rogers and R. A. Rogers, *Communications in Organizations* (New York: Free Press, 1978), p. 31.

---

jargon   Idiosyncratic use of language that is often useful among specialists but that inhibits their ability to communicate with nonspecialists.

APPLYING THE DIAGNOSTIC MODEL
Diagnostic Question 8: How would you characterize the communication network in your group? How does this network deal with or affect information overload, communication speed, and communication accuracy?

APPLYING THE DIAGNOSTIC MODEL
Diagnostic Question 11: Which people in your organization occupy communication roles (gatekeepers, cosmopolitans, opinion leaders)? How can their special expertise help you in the performance of your role?

gatekeeper   A person responsible for controlling messages sent through a particular communication channel.

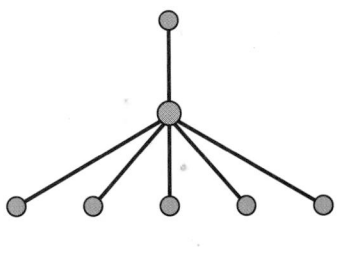

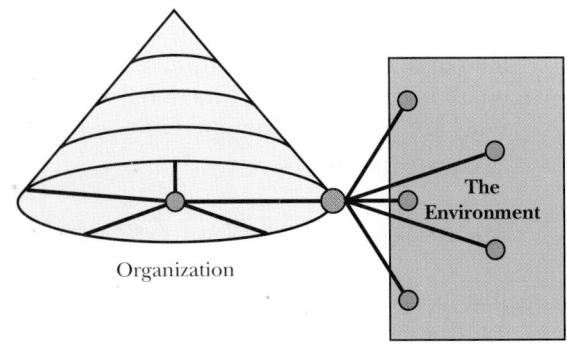

The Environment

Organization

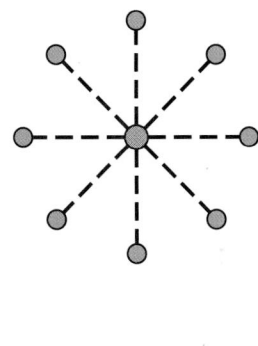

A Gatekeeper   B Cosmopolitan   C Opinion Leader

FIGURE 10-10
Special Communication Roles

cosmopolitan A person who has many important contacts outside the organization and develops special knowledge from these contacts.

For example, the role of chief of staff to the United States president is to shield the president from the multitude of people who want to influence him, seek his advice, or ask for favors. This gatekeeper must make important decisions regarding who will and will not have access to the president. Control over access to people in positions of power and authority gives a gatekeeper himself a great deal of power.

Cosmopolitans. Cosmopolitans are people with many contacts outside the organization who have special knowledge that the firm needs (see Figure 10-10B). Such people may be affiliated with professional groups that allow them to keep up to date on specific subjects, or they may have former associations that give them special expertise. For example, many defense contractors hire ex-Pentagon officials as consultants. The latter are very knowledgeable about the inner workings of defense contract decision processes and have a good understanding of the Pentagon's needs.

The ease with which some cosmopolitan workers move between organizational boundaries can sometimes be a problem. They develop a host of contacts outside the firm and many times decide to set up shop on their own. Rather than restrict their movement, some organizations have tried to give these employees their freedom in return for service. In fact, some companies trying to downsize set up consulting arrangements with departing employees. They are reducing the payroll and overhead while at the same time retaining the service of people who are familiar with the companies. Paul Kiczec, a systems analyst at Marine Midland Bank, developed a software program that helped commercial banks keep track of sales. The more he talked with people from other banks while developing the project, the more convinced he became that he had a marketable item on his hands. Kiczec got Marine Midland to give him the rights to the software in return for a royalty. Then he formed his own business. His biggest client right now? Marine Midland, although he hopes to reduce Marine's share to 10 to 20 percent in a few years.[20]

opinion leader A person who has special access to an organization's informal channels of communication and therefore has enhanced ability to influence others.

Opinion Leaders. A third communication role is that of the **opinion leader** (see Figure 10-10C). This person, has special access to an organization's informal channels of communication, often referred to as the grapevine. He has a greater than average ability to communicate with others in the organization and influence them. For example, an ex-foreman who was respected by both management and employees may still have many informal ties to both workers and management that enable him to shape organizational policies.

[20] C. H. Deutsch, "Turning a Boss into a Client," *New York Times*, July 8, 1990, p. 23.

Opinion leaders occupy a central position in the organization. For this reason they are closer both to the sources of legitimate authority and to opinion leaders in other functional units. The effective use of opinion leaders and their informal lines of communication is often what separates firms that are effectively integrated on all hierarchical and functional dimensions from firms in which these dimensions are poorly coordinated.

Socialization of New Group Members

Just as there are certain situations in organizations in which a particular communication function is essential, there are certain occasions on which a very specific kind of communication seems especially critical. One of these is a person's entry into a new role within the organization. **Organizational socialization** is the process by which an individual acquires the social knowledge and skills necessary to assume an organizational role.[21] It is the process of learning the ropes and entails much more than simply learning the technical requirements associated with one's job. It deals with learning about the organization, its values, its culture, its past history, its potential, and where the newly admitted member fits in.

Socialization occurs any time an individual moves along any of the three dimensions of organizations (hierarchical, functional, or inclusionary). It is particularly intense when the target person is crossing all three organizational boundaries at once. When a person joins a new firm, she crosses the inclusionary boundary as she moves from nonmember to member status, and she crosses functional and hierarchical boundaries as she joins a particular function unit, such as the advertising department, at a specific hierarchical level, such as account executive. It is at this time that the organization has the most instructing and persuading to accomplish. It is also the time when a person is most susceptible to being taught and influenced. Being the new kid on the block typically causes a person to be anxious and self-conscious, and so the faster one can learn the ropes, the better.

Desired Goals. Although instruction and persuasion are part of all socialization programs, different firms may seek different end goals in this process. Some organizations may pursue a **custodianship** response. Here the newcomer takes a caretaker's stance toward the means and ends associated with his role. He is not to question the status quo; he is merely to conform to it. A popular expression in the U.S. Marine Corps, paraphrased from Tennyson's "Charge of the Light Brigade," is "Ours is not to question why; ours is but to do or die."

When an organization hopes that the new member will change either the means by which her role is performed or the ends sought by the role, it may have as a goal **role innovation**. For example, in the medical community a subtle effort is being made—largely by health maintenance organizations, (HMOs)—to change the role of health-care provider. Instead of simply treating the sick, providers would promote physical and mental health and work to prevent illness. Of course, custodianship and role innovation are two ends of a continuum, and many firms seek something in the middle.

Socialization Tactics. Firms use several tactics in socializing new members. The chosen strategy will have a considerable impact on how the new recruit turns

[21] J. Van Maanen and E. H. Schein (1979). Toward a theory of organizational socialization. *Research in Organizational Behavior*. eds. B. Staw and E. L. Cummings, JAI Press: Greenwich, Conn., pp. 209–64.

APPLYING THE DIAGNOSTIC MODEL
Diagnostic Question 12: What do you wish to accomplish when socializing new group members (custodianship or innovation)? How should your socialization program be designed to promote your goal?

organizational socialization The process by which a person acquires the social knowledge and skills necessary to assume an organizational role.

POINT TO STRESS
Research has shown that most people leave a company within the first year of joining it. Much of this turnover has been attributed to poor socialization of newcomers into the organization.

custodianship A product of socialization in which a new group member adopts the means and ends associated with the role unquestioningly.

role innovation A product of socialization in which a new group member is expected to improve on both the goals for her job and the means of achieving them.

INTERPERSONAL PROCESSES AND COMMUNICATION

out. There are four critical dimensions of socialization strategies: collective-individual, sequential-random, serial-disjunctive, and divestiture-investiture (see Figure 10-11). The first technique in each of these pairs indoctrinates the newcomer in the custodianship role. The second member of each pair leads the newcomer into role innovation.

In *collective socialization*, a group of new recruits may be put through a particular set of experiences together. This method is characteristic of army boot camps, fraternities, sororities, and management-training courses. In collective processes much of the socialization is accomplished by the recruits themselves. For example, Marine Corps recruits may abuse each other verbally or even physically, something the corps itself would not do.

In contrast, in *individual socialization*, recruits are taken one at a time and put through unique experiences. This treatment is characteristic of apprenticeship programs or on-the-job learning. It has much more variable results than collective socialization does and success depends a great deal on the qualities of the recruit herself.

The second dimension of socialization strategies is called *sequential socialization*. This technique takes recruits through a given sequence of discrete and identifiable steps leading to the target role. A physician's training, for example, includes several observable steps; the undergraduate premed program, medical school, an internship, and a residency. A person must complete all of these before taking specialist board examinations. Usually, in sequential processes, each stage builds on the prior stage. Moreover, there is a tendency on the part of those doing the socializing to suggest that all the other steps are easier than the current one. The algebra teacher socializing the student to the world of math notes that geometry will be easy if one understands algebra. The geometry teacher notes that trigonometry will be painless if one appreciates geometry. This type of presentation helps recruits keep focused on the current stage. It minimizes the discouragement that comes with the knowledge that they are a long way from where they need to be. Sometimes successful completion of a step is rewarded with a ceremony, such as a graduation, which reinforces the person's feeling of accomplishment.

At the other end of this dimension are *random socialization* processes, in which there is no rhyme or reason to learning experiences. The steps are unknown, ambiguous, or continually changing. Training for a general manager, for example, tends to be much less rigorously specified than that for a medical professional. Some managers rise from lower ranks, some come from other occupations, some come straight from MBA programs.

Socialization strategies also differ on the amount of help and guidance they give newcomers in learning their new roles. In *serial socialization*, experienced members of the organization teach the newcomers about the roles they are about to assume. These experienced members become role models or mentors for the new members. In police departments, for example, rookies are assigned as partners to older, veteran officers. Some observers have suggested that this practice creates a remarkable degree of inter-generational stability in the behaviors of

FIGURE 10-11

The Custodianship-Innovation Continuum and Its Socialization Techniques

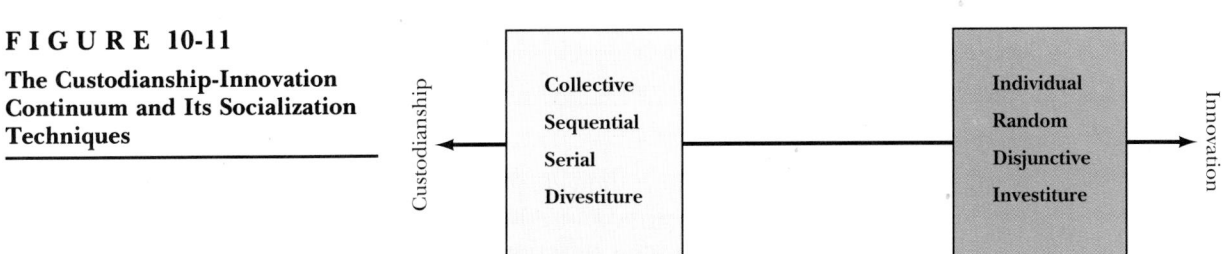

POINT TO STRESS
Women and minorities often lack role models in moving up the organizational hierarchy. Their uncertainty as to how to behave and interact may be complicated by the fact that no one in the organization has had experience dealing with these groups.

police officers. This method of socialization allows the recruit to see into the future in that he can expect to wind up much like the role model. This works well with the right role model. If, however, the picture of the future is not very flattering, the newcomer may never become committed to the organization.

In *disjunctive socialization*, on the other hand, a recruit must learn by herself how to handle a new role. For example, the first woman partner in a conservative law firm may find few people if any on the scene who have faced her unique problems. She may be completely on her own in coping with these problems. Disjunctive socialization is sometimes brought on by organizations who "clean house," that is, sweep out all the older members of the organization and replace them with new personnel. Such a shakeup causes almost all employees of the firm to relearn their role. Typically the organization hopes that the result will be more creativity in problem solving.

The fourth dimension of socialization deals with the degree to which a socialization process confirms or disconfirms the value of a newcomer's personal identity. *Divestiture socialization* ignores or denies the value of the newcomer's personal characteristics. The organization wants to tear the person down to nothing and then rebuild him as a completely new and different individual. Some organizations require either explicitly or implicitly that the recruit sever old relationships, undergo intense harassment from experienced members, and engage for long periods of time in doing the dirty work of the trade (work that is associated with low pay and low status). The organization promotes these ordeals in the belief that those who emerge from them will be strongly committed to the group's goal and identity.

In contrast, *investiture socialization* affirms the value to the organization of the recruit's particular personal characteristics. The organization says, in effect, "We like you just the way you are." It implies that rather than change the recruit, the firm hopes the recruit will change the organization. Under these conditions, the organization may try to make the recruit's transition process as smooth and painless as possible.

Designing Socialization Programs. The strategy you choose in designing your socialization program depends on your goal: custodianship or innovation. If you want to generate a custodianship response, you will want to use collective, sequential, serial, and divestiture socialization. In this way, you will give every recruit the same experiences in the same order. You will promote his acceptance of the status quo by replacing his own beliefs and values with those of experienced members. If, on the other hand, you want to promote innovation in the role, you will use just the opposite tactic. You will want a unique and individualized program for each recruit. The program must value and support the newcomer's particular personality characteristics and style. Figure 10-11 summarizes these two approaches.

By designing the appropriate socialization programs, organizations provide a link to the past and a blueprint for their future. The decisions reached about what new members will experience will profoundly affect the way the history and culture of an organization is preserved. Decisions should also promote innovation where necessary so that the organization can master the challenges of the future.

SUMMARY

Joining into groups allows individuals to accomplish goals and satisfy needs that could not be met if they acted alone. Individuals within groups become dependent upon each other in several different ways. The simplest type of relationship involves *pooled interdependence*; the most complex form involves *team interdependence*. Interdependencies among individuals lead to the evolution of *roles*. These roles capture the expectations that members of a person's *role set* have for the person occupying a given work role. Roles can be differentiated on *functional*, *hierarchical* and *inclusionary* dimensions. Whereas roles are the building blocks of organizations, communication is the cement that hold these blocks together. *Communication* involves the encoding, transmission, and decoding of information sent from one person to another via any one of a number of media. There are several barriers to communication effectiveness such as spatial arrangements, status differentials and jargon. Special organizational roles such as those of *gatekeeper*, *cosmopolitan*, and *opinion leader* deal almost exclusively with communication issues. Certain recurring organizational problems, like *socialization* of new members, require special attention. In particular, depending upon the goal of one's socialization program, different communicators, using different tactics, might be required for different positions.

REVIEW QUESTIONS

1.   Of the four types of interdependence discussed in this chapter, which type do you think is most adversely affected by turnover among organizational members? Which type of interdependence is most adversely affected by turnover in group leadership? How might the nature of the turnover process affect the kind of interdependence one builds into groups?

2.   Two trends in organizations are a move toward more employee involvement (through employee ownership) and, at the same time, a move towards greater use of temporary workers. What is likely to be the strongest force affecting employee owners and temporary workers as far as compliance, identification, and internalization with group norms are concerned? If you specify different forces for these two groups, why might the use of both tactics be on the rise at the same time?

3.   Socialization refers to the impact that the group or organization has on the individual. We noted that this impact tends to be greatest when the individual is moving through more than one dimension at a time (e.g., functional and hierarchical). In contrast, when is the individual most likely to have the greatest impact on the organization? (Are there honeymoon periods? Do lame ducks have any influence?) How might your answer depend on the tactics of socialization employed when bringing the individual into the new group?

4.   What role do ceremonies play in the socialization process of someone crossing an important organizational boundary? If one looks at the three kinds of boundaries that one can traverse, where are ceremonies most frequently encountered, and why? What role do ceremonies play in the motivation of group members who are not crossing a boundary but are merely observers at the affair?

5.   In communication, it has been said that "the medium is the message." What are the factors that one should consider when choosing a medium for one's communication? Some of the greatest leaders of all time actually wrote very little. What might explain why people who are perceived as strong leaders avoid leaving a paper trail of writings? When might writing be used to enhance leadership?

DIAGNOSTIC QUESTIONS

When you are confronting problems of communication or socialization or attempting to deal with difficulties that arise out of interpersonal problems, the following questions may prove useful.

1.   What type of interdependence unites you with other employees? What sorts of problems might this interdependence cause?

2. What important components of your unique work role are not part of your job description?

3. What are some of the important functional, hierarchical and inclusionary dimensions in your organization? Sketch an inverted-cone diagram to represent these dimensions in your firm.

4. Who are the members of your role set? What are their expectations for the occupant of your role? How do they communicate these expectations to you?

5. What are your own expectations for your role? How do these expectations compare with your role set's expectations, and what should be done about any discrepancies?

6. What are the primary reasons that people in your organization conform to their role expectations—compliance, identification, or internalization? What do these choices imply for monitoring employees' behavior?

7. What means of communication (oral, written, or nonverbal) do you use for different organizational tasks? Are these the optimal choices for the purposes you seek to achieve? Why?

8. How would you characterize the communication network in your group? How does this network deal with or affect information overload, communication speed, and communication accuracy?

9. How do others in your group perceive your influence and credibility? How do these perceptions affect the communications others send you and the communications you send to them?

10. What are some of your major values, beliefs, and frames of reference? How might these affect the way you interpret communications you receive from others in the group?

11. Which people in your organization occupy communication roles (gatekeepers, cosmopolitans, opinion leaders)? How can their special expertise help you in the performance of your role?

12. What do you wish to accomplish when socializing new group members (custodianship or innovation)? How should your socialization program be designed to promote your goal?

EXERCISE 10-1
THE ADVERTISING FIRM: A GROUP FEEDBACK ACTIVITY*

JEANNE LINDHOLM, *College of William Mary*

Working successfully with others requires that people establish and maintain effective patterns of interdependence among each other. To develop such interdependence coworkers must be aware of how they work together, and they must take the time to figure out how to improve the way they do things. In this exercise you will learn how periodic feedback can help groups improve their effectiveness in performing a task. You will also develop a greater awareness of how you work with others.

STEP ONE: PRE-CLASS PREPARATION

Before class, read the entire exercise and familiarize yourself with the tasks described below.

STEP TWO: GROUP WORK—NAMING THE PRODUCT

The class should divide into groups of 4 to 6 members each (if you have formed permanent groups in an earlier exercise, reassemble in those same groups). Your group should then complete the following task:

> You and your fellow group members are a project team for an advertising firm. A candy company has just created a new candy bar made of chocolate, caramel, and pecans and has employed your advertising firm to *name the bar* and to *design a one-minute radio commercial* for it. In this step of the exercise your team will be given five minutes to name the new bar. Later you will spend about thirty minutes designing the radio commercial, and then presenting it to the class.

STEP THREE: PROCESS FEEDBACK—NAMING THE PRODUCT

Your instructor will terminate Step Two after five minutes and ask all group members to remain seated together. The instructor will then lead a class discussion in which the members of each group should respond to the following questions:

1. What name did the group come up with? How pleased are each of the group members with the group's final choice?

2. How was the final choice made? How pleased are each of the group members with the way this choice was made?

3. Is there anything the group might do differently to improve its effectiveness in its next task, of working on the radio commercial?

STEP FOUR: GROUP WORK—CREATING A COMMERCIAL

The class should resume working in groups and should now begin work on the radio commercial advertising the new candy bar. You will need to pace yourselves, keeping in mind that you will have about 30 minutes to complete this task.

After fifteen minutes the instructor will interrupt the groups and will ask you to discuss the following questions:

1. As you work on the radio commercial, to what extent are you using your greatest assets as a member of your group? How do you feel about this? What changes need to be made?

2. How satisfied are you with the way in which ideas are being shared in your group? How satisfied are you with the way in which the task is being completed? What improvements should be made?

3. In what ways are the members of your group encouraging one another to contribute? In what ways are you discouraging one another? Are these actions helping or hurting the performance of your group?

4. When you compare your group's work in Step Four with its work in Step Two, what improvements can you see? What improvements still need to be made?

5. In view of your answers to these questions, what will you do differently during the next 15 minutes as you complete the radio commercial?

Your instructor will guide all groups in discussing these questions openly and and will encourage you to consider each question carefully. Then all groups will return to the task of developing the radio commercial.

---

* Adapted from J. William Pfeiffer, *The 1988 Annual: Developing Human Resources* (San Diego, CA: University Associates, Inc., 1988). Used with permission.

STEP FIVE: GROUP PRESENTATIONS

After 15 minutes more the instructor will reconvene the class and will ask each group to present its commercial. After all of the commercials have been presented the instructor will lead the class in a discussion of the following questions:

1. How pleased are you with the quality of your group's commercial? How pleased are you with the quality of your group's presentation?
2. How did the group decide who would present the commercial? How pleased are you with the process that was used to make this decision?
3. On a scale of 1 = low to 10 = high, rate your group on the following dimensions of effective teamwork:
    _____ Participation (giving everyone the opportunity to contribute)
    _____ Sensitivity (taking one another's feelings into account)
    _____ Openness (being able to say what you really think)
    _____ Flexibility (being able to change and correct problems)
    _____ Commitment (feeling responsible for completing the group's task)
    _____ Risk taking (trying things out that might seem outlandish at first)
    In the light of your ratings, what can you say about the way your group works together? What changes still need to be made? What will you personally do in the future to improve the effectiveness of your group?
4. What have you learned about the role of periodic feedback in improving a group's effectiveness?

CONCLUSION

To work together effectively, interdependent individuals must receive periodic feedback that lets them know how they're doing and whether changes need to be made. In the absence of such feedback, people working together cannot tell whether they are fulfilling their roles effectively and may perform poorly. At present as you work with your classmates and throughout your managerial career, you should seek out periodic feedback and provide feedback in return to those with whom you are working.

DIAGNOSING ORGANIZATIONAL BEHAVIOR

CASE 10-1

NURSE ROSS*

WILLIAM M. FOX, *University of Florida*

The following situation was reported by Miss Jackson, who had known Miss Evelyn Ross for several years and had also worked in some of the same hospitals as Miss Ross on different occasions.

Miss Ross, a registered nurse, began working at Benton Hospital when she was 31 years old. This hospital was an industrial hospital in a fairly large city on the West Coast. The bed capacity of the hospital was about 150, but 50 to 100 patients received treatment daily through the hospital's clinic facilities. The hospital was built and operated by a large shipbuilding concern. All the employees of the company's shipyards and their dependents could receive medical care through the company's hospitalization plan.

The nursing staff was headed by a director of nurses who had two assistants. One was in charge of nursing services in the hospital, and the other was in charge of the clinic nursing services. However, the two departments operated as a coordinated unit, and personnel were exchanged between them in the event the work load became too heavy in either place.

The medical director of the hospital, Dr. Peake, was energetic and his manner was usually quite brusque. Although he was a stickler for discipline and efficiency, he was fair in his treatment of the staff and they respected him and cooperated well. Dr. Peake had many progressive ideas and had helped to build the hospital up from 75 to 150 beds. The new ideas he had were discussed in staff conferences. Any persons or heads of departments who might be affected by proposed changes participated in these conferences.

Miss Ross worked as a head nurse, both in the hospital and in the clinic, during her employment there. (Miss Jackson at that time was employed as assistant head nurse in the clinic.) Miss Ross resigned her position to enter the Army Nurse Corps as a first lieutenant. She served in the Army for two and a half years, most of which was duty in the South Pacific. During the time she was overseas, she was promoted to captain. She was transferred to reserve status upon leaving the corps. Shortly after this she took a three-month course in operating room supervision.

In the meantime, Miss Jackson had moved to the East Coast and was employed at Hughes Hospital, a large industrial hospital in a relatively small New England city. They had corresponded during this time and Miss Jackson wrote that the position of operating room supervisor would soon be open at the hospital and thought Miss Ross had a good chance of getting the position if she wanted to move to the East Coast. Miss Ross applied to the director of nursing at the hospital and was accepted for the position. She began working soon thereafter at Hughes.

Hughes Hospital was set up much like Benton Hospital. It took care of the medical needs of most of the community in addition to serving the employees of the Hughes Steel Company, the city's principal employer. It had clinic facilities for emergency and outpatient care. The bed capacity was 250 and the clinic staff treated well over 100 patients daily, although often a complete record of the number of patients was not kept.

The organization of the nursing department was quite similar to that of Benton Hospital with one important exception: the hospital department and the clinic department were operated as two completely separate units. The clinic was in a building separate from the hospital building; thus the problem of moving a stretcher case from the clinic to the hospital was an extreme ordeal. Besides the lack of proper equipment for moving patients, there was a shortage of male orderlies, and nurses' aides had to be utilized for this arduous task. This shortage of personnel and equipment was especially acute when emergency cases and accident victims came into the clinic and had to be moved to the hospital with a minimum loss of time and disturbance.

The director of nurses, Miss McHaffey, was about 45 years old; she had been at the hospital three years. Miss Linden had been the hospital supervisor for six months, and Miss Hartman had been employed as a clinic supervisor for over a year. There were 24 graduate nurses employed in the hospital wards, 30 aides, and 10 maids. The staff under Miss Hartman in the clinic consisted of five graduate nurses, four aides, and two maids. The orderly personnel numbered only six for all three shifts. One was utilized throughout the hospital on the evening shift, one on the night shift, and during the day shift one worked in the clinic, one in the operating room (O.R.), and one for each of the two men's wards in the hospital. Miss Ross, as supervisor of the O.R., had a staff of four nurses, three aides, and the one orderly. The nurses in the O.R. rotated turns, being "on call" each night for any emergency surgery cases.

Miss Ross found that the work was quite strenuous and often entailed long hours, but she was deeply interested in it and never seemed to object. She frequently stayed to help in emergency surgery cases, as a number of rather serious accidents occurred from time to time in the steel plants that the hospital served. Miss McHaffey praised her highly for increasing the efficiency and cleanliness in the operating rooms.

Dr. McMillan, the medical director of the hospital, was nearly 65 years old. He had been employed as a company doctor for the Hughes Steel Company for over 20 years. Dr. McMillan would usually arrive at his offices in the hospital about nine in the morning, would dictate answers to his correspondence, make sporadic rounds of some of the hospital wards (very rarely did he put in an appearance at the clinic), leave for lunch promptly at noon, and, only two or three times a week return to the hospital for a few hours after lunch. On his occasional ward rounds, he would stop at the floor nurse's desk, inquire if everything was going all right, then say, "Fine! Fine!" and go on his way.

When Dr. McMillan suffered a heart attack severe enough to prevent him from retaining his position at the hospital, a new medical director had to be found. The president of the steel company was familiar with the West-Coast shipbuilding concern and knew Dr. Peake had been at Benton. He contacted Dr. Peake to see if he would be interested in the position as the hospital medical director. Dr. Peake accepted. He entered the new situation with his usual brusque and energetic manner and made complete daily rounds in the clinic and hospital. He often spent considerable time talking to patients, nurses, aids, and the staff physicians.

After nearly a month of concentrated observation of the clinic and hospital routines, Dr. Peake had a conference with Miss McHaffey and the nursing supervisors. He criticized the "unprofessional attitude" of several of the nurses, and said he had had complaints from many of the patients about the care they were receiving. He asked why so many of the nurses seemed to be away from their wards when he made morning rounds. Miss McHaffey said the nurses were permitted to leave the wards at intervals between nine and eleven to have coffee in the hospital dining room. The time for this was not rigidly enforced. Dr. Peake also talked to Dr. Albright, the staff physician in charge of the clinic, and to the clinic nurses to ascertain why the clinic patients often had to wait so long to see a doctor in the clinic. (The gist of these conferences was given by Miss Jackson, who was assistant supervisor of the clinic.) The clinic staff agreed that there was definitely a "bottleneck" in the clinic, but they felt that it was due primarily to a shortage of personnel when needed most, the inconvenience of having to transport the patients the distance to the hospital, and the lack of satisfactory laboratory facilities in the clinic itself. Dr. Peake told the staff that the new additions being built onto the hospital were going to be utilized for clinic facilities. In the meantime, he said he would try to help them find some way to ease the situation.

During the second week of August of that year, Miss McHaffey asked Miss Ross to come into her office.

Miss McHaffey: Miss Ross, Dr. Peake tells me that you worked with him at Benton Hospital. I knew that he had been at Benton at one time, but didn't realize that it was during the same time you were there. He said that you are familiar with the clinic-hospital arrangement there and told me to relieve you of your present position so that you may help to coordinate the clinic and hospital units here.

Miss Ross: I'm sorry to hear that. I have been very happy with my present position. Will I be working in the clinic or in the hospital?

Miss McHaffey: Both. I want you to know that I consider Miss Linden a very capable supervisor and I don't want her to be hurt in this new arrangement. Also, I want to know everything that is going on down there. I expect you to report to me at least once a day. I don't know what Dr. Peake expects you to do that hasn't already been done. He should hire more people if he expects to make this a model hospital. He comes in here and all he does is criticize.

Miss Ross: I'll do the best I can. I am familiar with the setup that Dr. Peake had at Benton. Maybe I can help put it into operation here.

A few hours later Dr. Peake entered Miss Ross' office in the O.R. unit.

Dr. Peake: Hello, Rossie, I have a new job for you.

Miss Ross: Miss McHaffey has told me about it.

Dr. Peake: You know how things were at Benton. I want the units to be set up in exactly that way here. During the past few months I have arranged for another physician to help out in the clinic during their busy hours and we've hired a couple more aides, but there doesn't seem to be too much improvement. Maybe you can help me find out what

the trouble is there. Our new building program has been started and when it is finished I want the two units to be operating as one integrated unit. I don't like to take you away from the surgery—you've been doing a fine job here—but I feel you can help me get the clinic and hospital units functioning better together.

Miss Ross: I can try, Dr. Peake.

Dr. Peake: Good! Now I don't want you to go through anybody—if you have any problems, come right to me!

Miss Ross—knowing the strained relationship between Dr. Peake and Miss McHaffey—was especially dubious about bypassing her immediate supervisor, the director of nurses. She decided at that time it would be best to observe the regular channels of communications.

Miss Ross reported for her new job and discussed Dr. Peake's plans and ideas for integrating the two units with both Miss Linden and Miss Hartman. She also told them that the reason he picked her for the job was because she had worked at Benton under him. They had known that both she and Miss Jackson had worked at Benton for a time while Dr. Peake was there. Neither of the supervisors seemed very surprised. Miss Linden remarked that it sounded like another of Dr. Peake's "wild ideas." Both Miss Linden and Miss Hartman seemed concerned over the shortage of an adequate staff and said that any changes that would improve the situation would be welcomed.

Personnel problems were especially acute in the hospital at that time. Several staff members were off duty because of illness and there were more patients than usual. The clinic was open Saturday and Sunday for emergencies only. One nurse and two aides were on duty weekends but were not too busy. Miss Ross arranged to transfer the two aides to the hospital for the weekends. Miss Linden was elated with the additional help. On the following Wednesday, the clinic was far behind in its work because of an emergency that had arisen. Miss Ross went to Miss Linden to see if someone could go over for the afternoon to help. The following conversation ensued:

Miss Ross: Miss Hartman is swamped. She had an emergency to take care of and the other patients are not being seen. Have you anyone you can send to help?

Miss Linden: I am not going to send anyone to that clinic. They have enough help! We are too short here.

362

Miss Ross went over to one of the wards and found two of the aides in the ward kitchen drinking coffee. She asked if they were slack right then.

One of them said, "Oh, sure. We haven't had very much to do all afternoon."

Miss Ross returned to Miss Linden and told her of the episode. She asked that one of them be sent to help out in the clinic. Miss Linden complied reluctantly.

Shortly after this Miss Linden went on a vacation for two weeks. Miss McHaffey asked Miss Ross to take charge of the hospital unit until her return. Thus Miss Ross was faced with the problem of making out time schedules for all the nurses, aides, orderlies, and maids employed in the hospital unit. Dr. Peake had also asked her to initiate a study to determine the personnel needs in the various hospital wards and the clinic departments, and to help with the plans for the layout of new equipment in the building additions. During the two weeks of Miss Linden's absence, Miss Ross found that 1) one ward had more nurses than another one, although the work loads were the same, and 2) maids were not doing the cleaning assigned to them and some were not even aware of what their duties were. With the cooperation of Miss Hartman and the approval and permission of Miss McHaffey, Miss Ross arranged to reallocate the nursing personnel so that all wards would have equal coverage in relation to their work loads. She made out schedules to provide available clinic help as relief in the hospital on weekends and instructed the maids as to their duties.

There seemed to be a gradual improvement in the amount and quality of patient care and most of the employees seemed to be more satisfied when they were placed in jobs where they were kept busy and understood their duties. Several patients commented on the improved care they received after the changes had been made. Dr. Peake praised Miss Ross and Miss McHaffey for the success of the new program.

Two days after Miss Linden returned from her vacation Miss Ross was called the office of the director of nurses.

Miss McHaffey: Miss Ross, Miss Linden has requested a transfer to the operating room, because she doesn't think you and she will get along. She is doing a good job in the hospital and I don't want to lose her. Hereafter, you will not interfere with the operation of the hospital unit and its personnel. Miss Linden will take care of everything over there.

Miss Ross: I don't understand, Miss McHaffey. Do you mean that my job is finished?

Miss McHaffey: No. You are to continue working in the clinic and help set up new de-partments there as the building program continues. I really don't know what made Dr. Peake think you would be able to do anything to improve the situation. He will just have to realize that we haven't sufficient personnel.

Miss Ross left the interview feeling very confused as to her exact status because she knew Dr. Peake would expect her to continue to try to coordinate the two units.

When you have read this case, look back at the chapter's diagnostic questions and choose the ones that apply to the case. Then use those questions with the ones that follow in your case analysis.

1. Why must Benton Hospital coordinate its hospital wards with its clinic? What kind of interdependence now links the hospital and the clinic and how has it affected relations between the units? How might this mode of interdependence be changed so as to coordinate the two units more effectively and improve the relations between them?

2. Examine the role that Dr. Peake expects Nurse Ross to play in linking the hospital with its clinic. What do other members of Nurse Ross's role set expect of her? Is it realistic to expect Nurse Ross to succeed in filling the role Dr. Peake has assigned her?

3. Describe the character and quality of the communication among the people involved in this case. Has it promoted mutual understanding or hindered it? Do all of the people in the case appear to be adequately socialized? What might Benton Hospital do to help Nurse Ross accomplish her task? How could it improve its communication networks and socialization procedures?

C A S E 10-2

CAMERAN MUTUAL INSURANCE COMPANY

Review this case, which appears in Chapter 8. Next, look back at Chapter 10's diagnostic questions and choose the ones that apply to the case. Then use those questions with the ones that follow in your case analysis.

1. Describe the socialization process that Mrs. Kay went through as she began to work at Cameran Insurance. What are the strengths and weaknesses of this sort of socialization?

2. Many people believe that most interpersonal problems boil down to faulty communication. How

might improved communications affect the working relationship between Mrs. Kay and Mr. Landers? What good things might result? What negative outcomes might occur?

3. If you were asked to advise Cameran Insurance about how to manage the process of introducing new employees to the organization, what changes would you recommend? Why? What things would you want the company to keep doing the same way?

C A S E 10-3

BETA BUREAU

Read Chapter 17's Case 17–2, "Beta Bureau." Next, look back at Chapter 10's diagnostic questions and choose the ones that apply to that case. Then use those questions with the ones that follow in this analysis of the case.

1. What type of interdependence would you expect to find in the modules formed in Beta Bureau? What kinds of problems does this type of interdependence stimulate? What are its strengths?

2. Suppose you were assigned the task of socializing new module members. What sort of program would you develop? What would be its aim? How successful would it be in achieving this aim?

3. Why did many supervisors at Beta Bureau react negatively to the changes caused by modularization? If you were responsible for making supervisory jobs more attractive to traditional managers, what would you do?

GROUP FORMATION AND PERFORMANCE

GE's conversion of its Salisbury plant to team-based production, though ultimately highly successful, met with resistance at first. Some workers were reluctant to accept more responsibility and to move constantly from job to job and they quit, mirroring GE's own initial reluctance to commit itself to the teamwork concept. By 1989, however, nearly 20 percent of GE's 120,000 employees throughout the United States were members of work teams, and the corporation's goal was to raise that to 35 percent by the end of the year.
Source: John Hoerr, "The Payoff from Teamwork," Business Week, July 10, 1989, pp. 58–59.

EXERCISE 11-1
WILDERNESS SURVIVAL: A CONSENSUS-SEEKING TASK*

DONALD T. SIMPSON, *Eastman Kodak Company*

In *consensus acceptance*, all members of a group agree to support a group decision. Everyone in the group is actively involved in reaching consensus acceptance. Everyone discusses the issues, and all members' ideas are incorporated into the group's ultimate decision. This method of making a decision pools the knowledge and experience of all a group's members and gains each individual's personal support.

Consensus, however, is difficult to attain. Moreover, achieving it takes more time than other methods of decision making such as majority rule or autocratic imposition of one or only a few members. To achieve consensus agreement in a group, all members of the group must do the following:

1. Before meeting with the group they must consider and be ready to state their personal positions to the group. They must also keep in mind that the decision making process is incomplete until everyone has explained his own ideas and the group as a whole has reached a decision.
2. They must recognize their obligation to explain their own opinions fully, so that the group can benefit from each member's thinking.
3. They must accept the obligation to listen to the opinions of all other group members, and they must be able to modify their own positions on the basis of logic and improved understanding.

4. Realizing that differences of opinion are normal and helpful, they must not resort to conflict-reducing techniques like voting, compromising, or giving in to others. They must believe that in exploring differences, they will arrive at the best course of action.

In this exercise you will first work alone, deciding what you would do to survive in the wilderness. You will then join a group and reach a consensus about the best approach to wilderness survival. As you work together, remember that consensus is difficult to attain and that not every decision made during the wilderness survival task may meet with everyone's unqualified approval. However, there should be a general feeling of support from all members before a group decision is finalized. Take the time to listen, to consider *all* members' views, and to make your own view known. Be reasonable in arriving at a group decision.

STEP ONE: PRE-CLASS PREPARATION

Read the entire exercise, then fill in your answers to the questions on the Wilderness Survival Work Sheet in Exhibit 11-1. Be sure to bring your answers to class.

* Adapted from J. William Pfeiffer and John E. Jones (eds.), *The 1976 Annual Handbook for Group Facilitators* (San Diego, CA: University Associates, Inc., 1976). Used with permission.

EXHIBIT 11-1
Wilderness Survival Work Sheet

Here are twelve questions concerning personal survival in a wilderness situation. Your first task is *individually* to select the best of the three alternatives given for each item. Try to imagine yourself in the situation depicted. Assume that you are alone and have a minimum of equipment, except where specified. The season is fall. The days are warm and dry, but the nights are cold.

After you have completed this task individually,

you will again consider each question as a member of a small group. Your group will have the task of deciding, *by consensus*, the best alternative for each question. Do not change your individual answers, even if you change your mind in the group discussion. Both the individual and group solutions will later be compared with the answers provided by a group of experienced naturalists who conduct classes in woodland survival.

	Your Answer	Your Group's Answer

1. You have strayed from your party in trackless timber. You have no special signaling equipment. The best way to attempt to contact your friends is to:
   a. call "help" loudly but in a low register.
   b. yell or scream as loud as you can.
   c. whistle loudly and shrilly.

2. You are in "snake country." Your best action to avoid snakes is to:
   a. make a lot of noise with your feet.
   b. walk softly and quietly.
   c. travel at night.

3. You are hungry and lost in wild country. The best rule for determining which plants are safe to eat (those you do not recognize) is to:
   a. try anything you see the birds eat.
   b. eat anything except plants with bright red berries.
   c. put a bit of the plant on your lower lip for five minutes; if it seems all right, try a little.

4. The day becomes dry and hot. You have a full canteen of water (about one liter) with you. You should:
   a. ration it—about a cupful a day.
   b. not drink until you stop for the night, then drink what you think you need.
   c. drink as much as you think you need when you need it.

5. Your water is gone; you become very thirsty. You finally come to a dried-up watercourse. Your best chance of finding water is to:
   a. dig anywhere in the stream bed.
   b. dig up plant and tree roots near the bank.
   c. dig in the stream bed at the outside of a bend.

6. You decide to walk out of the wild country by following a series of ravines where a water supply is available. Night is coming on. The best place to make camp is:
   a. next to the water supply in the ravine.
   b. high on a ridge.
   c. midway up the slope.

7. Your flashlight glows dimly as you are about to make your way back to your campsite after a brief foraging trip. Darkness comes quickly in the woods and the surroundings seem unfamiliar. You should:
   a. head back at once, keeping the light on, hoping the light will glow enough for you to make out landmarks.
   b. put the batteries under your armpits to warm them, and then replace them in the flashlight.
   c. shine your light for a few seconds, try to get the scene in mind, move out in the darkness, and repeat the process.

8. An early snow confines you to your small tent. You doze with your small stove going. There is danger if the flame is:
   a. yellow.
   b. blue.
   c. red.

9. You must ford a river that has a strong current, large rocks, and some white water. After carefully selecting your crossing spot, you should:

a. leave your boots and pack on.

b. take your boots and pack off.

c. take off your pack, but leave your boots on.

10. In waist-deep water with a strong current, when crossing the stream, you should face: _____ _____

a. upstream.

b. across the stream.

c. downstream.

11. You find yourself rimrocked: your only route is up. The way is mossy, slippery rock. You should try it: _____ _____

a. barefoot.

b. with boots on.

c. in stocking feet.

12. Unarmed and unsuspecting, you surprise a large bear prowling around your campsite. As the bear rears up about ten meters from you, you should: _____ _____

a. run.

b. climb the nearest tree.

c. freeze, but be ready to back away slowly.

STEP TWO: GROUP CONSENSUS SEEKING

The class should divide into groups of four to six members each (if you have formed permanent groups, reassemble in those groups). Your group should then seek consensus answers to each of the questions in the Wilderness Survival Work Sheet. You should appoint a spokesperson to report on the results of Steps One and Two to the entire class.

STEP THREE: CLASS DISCUSSION

Your instructor will reconvene the class, asking the members of each group to remain seated together. The instructor will then inform everyone of the right answers to the questions. Group members should then determine how many questions they answered correctly and give this information to their group spokesperson. The group spokesperson should also determine how many questions the group answered correctly.

Next, the instructor will call on each spokesperson and ask for the information needed to complete the tabulation below for each group:

After your instructor has completed this table on a blackboard, flipchart, or overhead projector, class members should discuss the consensus-seeking procedures they used in their groups and should consider the following questions:

1. What specific behaviors promoted productivity in your group? What behaviors hindered group productivity? What rules could be set up to encourage helpful behaviors and discourage harmful ones?

2. Did groups with wider ranges of individual scores have a harder time reaching consensus than groups whose individual scores were more similar? What effects, if any, did differences among the opinions of individual members have on your group's ability to reach consensus?

3. How might a group use socialization to improve its ability to reach group consensus? How might having clearly defined roles help the members of a group reach consensus more rapidly?

OUTCOME	GROUP 1	GROUP 2	GROUP 3 . . .
Range of individual scores (low-high)			
Average of individual scores			
Group score			

4. How might factors such as size, member motivation, cohesiveness, and the group's communication structure affect a group's ability to reach consensus? Could group judgment errors occur during the process of reaching consensus? What could be done to avoid these errors?

CONCLUSION

Requiring consensus acceptance increases the time it takes to make decisions and solve problems. However, it also enables every member to have an effect on the decisions that are made and encourages acceptance of those decisions. Whether consensus should be sought depends on a variety of factors. The need for speedy decisions will argue against it. If decision acceptance and commitment are important or if group members have uncommon abilities or insights that can be shared in consensus-building discussions, it would seem worthwhile to pursue consensus. The size and communication structure of the group will also affect the decision to seek consensus. Consensus can be very useful, but it can also be impractical.

DIAGNOSING ORGANIZATIONAL BEHAVIOR

CASE 11-1
HOVEY AND BEARD COMPANY*

GEORGE STRAUSS, *University of California, Berkeley*
ALEX BAVELAS, *Massachusetts Institute of Technology*

PART 1

The Hovey and Beard Company manufactured wooden toys of various kinds: wooden animals, pull toys, and the like. One part of the manufacturing process involved spraying paint on the partially assembled toys. The operation was staffed entirely by women.

The toys were cut, sanded, and partially assembled in the wood room. Then they were dipped into shellac, following which they were painted. The toys were predominantly two-colored; a few were made in more than two colors. Each color required an additional trip through the paint room.

For a number of years, production of these toys had been entirely handwork. However, to meet tremendously increased demand, the painting operation had recently been re-engineered so that the eight women who did the painting sat in a line by an endless chain of hooks. These hooks were in continuous motion, past the line of women and into a long horizontal oven. Each woman sat at her own painting booth, so designed as to carry away fumes and to backstop excess paint. She would take a toy from the tray beside her, position it in a jig inside the painting cubicle, spray on the color according to a pattern, then release the toy and hang it on the hook passing by. The rate at which the hooks moved had been calculated by the engineers so that each woman, when fully trained, would be able to hang a painted toy on each hook before it passed beyond her reach.

The women working in the paint room were on a group bonus plan. Since the operation was new to them, they were receiving a learning bonus which decreased by regular amounts each month. The learning bonus was scheduled to vanish in six months, by which time it was expected that they would be on their own— that is, able to meet the standard and to earn a group bonus when they exceeded it.

PART 2

By the second month of the training period, trouble had developed. The women learned more slowly than had been anticipated, and it began to look as though their production would stabilize far below what was planned for. Many of the hooks were going by empty. The women complained that the hooks were going by too fast, and that the time-study man had set the rates wrong. A few women quit and had to be replaced with new workers, which further aggravated the learning problem. The team spirit that the management had expected to develop automatically through the group bonus was not in evidence except as an expression of what the engineers called "resistance." One woman whom the group regarded as its leader (and the management regarded as the ringleader) was outspoken in making the various complaints of the group to the foreman: The job was a messy one, the hooks moved too fast, the incentive pay was not being correctly calculated, and it was too hot working so close to the drying oven.

* Exerpt from William Foote Whyte, *Money and Motivation* (New York: Harper & Row, Publishers, Inc., 1955). Reprinted by permission of the publisher.

PART 3

A consultant who was brought into this picture worked entirely with and through the foreman. After many conversations with him, the foreman felt that the first step should be to get the workers together for a general discussion of the working conditions. He took this step with some hesitation, but he took it on his own volition.

The first meeting, held immediately after the shift was over at 4:00 in the afternoon, was attended by all eight women. They voiced the same complaints again: The hooks went by too fast, the job was too dirty, the room was hot and poorly ventilated. For some reason, it was the last item that they complained of most. The foreman promised to discuss the problem of ventilation and temperature with the engineers, and he scheduled a second meeting to report back to the workers. In the next few days the foreman had several talks with the engineers. They and the superintendent felt that this was really a trumped-up complaint, and that the expense of any effective corrective measure would be prohibitively high.

The foreman came to the second meeting with some apprehension. The women, however, did not seem to be much put out, perhaps because they had a proposal of their own to make. They felt that if several large fans were set up so as to circulate the air around their feet, they would be much more comfortable. After some discussion, the foreman agreed that the idea might be tried out. The foreman and the consultant discussed the question of the fans with the superintendent, and three large propeller-type fans were purchased.

PART 4

The fans were brought in. The women were jubilant. For several days the fans were moved about in various positions until they were placed to the satisfaction of the group. The workers seemed completely satisfied with the results, and relations between them and the foreman improved visibly.

The foreman, after this encouraging episode, decided that further meetings might also be profitable. He asked the women if they would like to meet and discuss other aspects of the work situation. The women were eager to do this. The meeting was held, and the discussion quickly centered on the speed of the hooks. The women maintained that the time-study expert had set the hooks at an unreasonably fast speed and that they would never be able to reach the goal of filling enough of them to make a bonus.

The turning point of the discussion came when the group's leader frankly explained that the point wasn't that they couldn't work fast enough to keep up the hooks, but they couldn't work at that pace all day long. The foreman explored the point. The women were unanimous in their opinion that they could keep up with the belt for short periods if they wanted to. But they didn't want to because if they showed they could do this for short periods, they would be expected to do it all day long. The meeting ended with an unprecedented request: "Let us adjust the speed of the belt faster or slower, depending on how we feel." The foreman agreed to discuss this with the superintendent and the engineers.

The reaction of the engineers to the suggestion was negative. However, after several meetings, it was granted that there was some latitude within which variations in the speed of the hooks would not affect the finished product. After considerable argument with the engineers, it was agreed to try out the workers' ideas.

With misgivings, the foreman had a control with a dial marked "low, medium, fast" installed at the booth of the group leader; she could now adjust the speed of the belt anywhere between the lower and upper limits that the engineers had set.

PART 5

The women were delighted, and spent many lunch hours deciding how the speed of the belt should be varied from hour to hour throughout the day. Within a week the pattern had settled down to one in which the first half hour of the shift was run on what the women called medium speed (a dial setting slightly above the point marked "medium"). The next two and one-half hours were run at high speed; the half hour before lunch and the half hour after lunch were run at low speed. The rest of the afternoon was run at high speed with the exception of the last 45 minutes of the shift, which was run at medium.

In view of the women's reports of satisfaction and ease in their work, it is interesting to note that the constant speed at which the engineers had originally set the belt was slightly below medium on the dial of the control that had been given the women. The average speed at which the women were running the belt was on the high side of the dial. Few, if any, empty hooks entered the oven, and inspection showed no increase of rejects from the paint room.

Production increased, and within three weeks (some two months before the scheduled ending of the learning bonus) the women were operating at 30 to 50 percent above the level that had been expected under the original arrangement. They were collecting their base pay, a considerable piece-rate bonus, and the learning bonus which, it will be remembered, had been set to decrease with time and not as a function of current productivity. The women were earning more now than many skilled workers in other parts of the plant.

PART 6

Management was besieged by demands that this inequity be taken care of. With growing irritation between superintendent and foreman, engineers and foreman, superintendent and engineers, the situation came to a head when the superintendent revoked the learning bonus and returned the painting operation to its original status. The hooks moved again at their constant, time-studied designated speed; production dropped again; and within a month, all but two of the eight workers had quit. The foreman himself stayed on for several months but, feeling aggrieved, then left for another job.

When you have read this case, look back at the chapter's diagnostic questions and choose the ones that apply to the case. Then use those questions with the ones that follow in your case analysis.

1. At what stage of development was the group of toy painters at Hovey and Beard? What does this tell you about the group's roles and norms? What kind of task was the group performing? Did the group's members have the required abilities? How did the group's roles, norms, and task affect its performance?

2. Explain why the workers were upset with the line speed established by management but then established an average line speed that was even higher. Why did productivity drop when the line was returned to its original speed?

3. Why did most of the workers quit their jobs? What made the company take actions that had such dire consequences? Is there anything the company could have done that would have restored equity with other parts of the plant without precipitating dissatisfaction among the painters?

CASE 11-2

THE SLAB YARD SLOWDOWN*

HAL B. GREGERSEN, *Brigham Young University*
PAUL H. THOMPSON, *Brigham Young University*

Even the tremendous roar of the nearby open-hearth furnace would not have drowned out Bob Flint's

screams when he was handed the latest slab yard payroll report. Flint, a division superintendent at Midland Steel's Dayton, Ohio plant, was outraged to discover that a crew of "scarfers" had reported a 410 percent incentive pay performance level on last week's day shift. To earn the $81 hourly wage reported by the scarfers would have required a physically impossible work pace.

THE OVERALL STEEL PRODUCTION PROCESS

Raw iron ore is melted down in large blast furnaces to form basic pig-iron soup. The soup is transported to open-hearth furnaces where certain alloys are added to produce a molten steel mixture of malleability and strength, tailored to unique specifications. This mixture is reheated, then poured into ingot molds and transported to a mill for further processing.

In the mill, the ingot molds are removed and the ingots rolled and chopped into steel slabs of predetermined dimensions. After the slabs cool to 1,000 degrees, jet streams of water are turned on them to further reduce their temperature. At this point, the scarfers mount the steel slabs and use their torches to burn off any cracks, scabs, or blemishes on the slab surface. When the scarfers have completed their work, the slabs are transferred to a rolling mill, reheated, and rolled into either plates or coils, depending on customer specifications.

THE SCARFING PROCESS

The scarfers were specifically responsible for burning all cuts, cracks, scabs, and blemishes off the surface of the steel slabs before they were reheated and rolled into plates or coils in the rolling mill. Their function was vital, since failure to remove defects in the steel slabs could have resulted in scrapping much of the metal during the rolling process. The defects could ruin the metal in two ways: (1) impurities could prevent the steel from reaching the required level of malleability during reheat, causing the metal to snap during the coiling process; or (2) even a small blemish could become greatly enlarged during the rolling process; e.g.; a two-inch-wide blemish on the surface of a six-inch-thick slab of steel would be stretched out over the face of over 100 feet of finished product when the slab was rolled into a plate one-tenth of an inch thick.

The scarfers' task was achieved by standing on top of the steel slabs in thick, wooden-soled boots and cutting paths along the steel surface with a heavy blowtorch. Obviously, the working conditions were not comfortable. The workers endured the discomfort of extreme heat, bulky clothing and protective goggles, moisture from the water jets used to cool the slabs, and an immediate environment full of acid fumes from molten steel.

This harsh environment dictated that a scarfer be in top physical condition. Some of the workers were very large physically while others were not. Regardless of his size, each scarfer was exceptionally strong. The majority of the scarfers were either junior high or high school dropouts. Most scarfers would spend their entire working lives in the slab yard, working until well into their sixties; consequently, many had worked together for years. Over time, they had developed good friendships and did many things together after work.

Added to the physical discomforts of the environment and the monotony of the task were the inherent dangers of working in an enclosed area among eight-foot stacks of hot steel slabs. The slab yard was the division with the highest accident rate in the corporation. According to the shift foremen, a high percentage of Midland's scarfers had died on the job from accidents or from heart attacks caused by overexertion.

THE CURRENT INCENTIVE PLAN

Because scarfing was one of the most difficult, dangerous, and dirty tasks at Midland Steel, the job was more financially attractive than others. Under the current incentive plan, scarfing had become the highest-paying blue-collar position in the steel plant. In fact, good scarfers would make over $80,000 a year; consequently, they were considered by themselves and others to be the "elite" of steelworkers. However, the large pay differential between scarfers and other blue-collar workers created tension between the two groups.

The scarfers' incentive system was based on measures of production output, i.e., actual number of slabs and amount of square inches scarfed. Measurement of square inches and slab counts were performed by inspectors who were also responsible for slab quality. Markers carried out another function, that of marking areas on the slabs that needed scarfing. Markers, as well as inspectors, received compensation based on the number of slabs scarfed.

In addition to the piece-rate incentive system, management provided scarfers with a base pay which was calculated by the dollar amount of rolled or coiled steel which would have been wasted in absence of the scarfing process. To increase their variable incentive pay, scarfers had often skipped their hourly heat breaks.

When the bargaining union had first negotiated the incentive plan in 1966, management had expected scarfers to average about 150 percent of their base pay, but scarfers saw the incentive plan as a much larger money-making opportunity. Throughout the twenty-five years of the plan's existence, scarfers had averaged 262 percent of base pay.

THE PROBLEM

Under Midland Steel's quantity-oriented incentive system, division superintendent Flint knew that the scarfers had been cheating for years by altering their blowtorch tips so that they could scarf more steel slabs per shift, thereby earning a higher wage. The workers would simply drill larger holes in the tip of their blowtorch so that the flame broadened, enabling them to burn a larger path with one pass of the torch over the steel slab. The larger torch path enabled the scarfer to finish more slabs during a shift, thereby increasing his income. However, by broadening the torch flame, the worker also decreased the flame's intensity such that the burn into the steel would not penetrate deeply enough to lift out the defects and blemishes from the steel. Rather, the weaker flame would simply cover the defect with molten steel, hiding it from the inspector's visual check. These hidden defects resulted in costly scrap in the final steel making process.

In addition, markers would often mark slabs for scarfing that did not need treatment in order to increase slab count and total square inches scarfed. Since quantity incentives were also offered to inspectors, they typically qualified inferior steel as acceptable in order to increase their own and scarfers' pay.

The only real contact between the scarfers and management was via the shift foremen. The shift foremen were thought of as the bridge between white- and blue-collar workers (each shift had two foremen—one in each slab yard). Their function was to ensure that the scarfers worked safely, kept on schedule, and did quality work. This was not an easy task considering the strong-willed and independent-minded attitudes of the scarfers.

The shift foremen also had the task of inspecting the blowtorches. Because some of the scarfers had altered their tips, a large number of defects had caused the rejection of significant quantities of finished steel products. In order to decrease their rejection rate, the shift foremen conducted periodic inspections to ensure that the tips met regulations. Managers at any level higher than the shift foremen were prohibited by union contract from conducting these inspections themselves.

THE RAID

Management, including the shift foremen, believed that greed was the motivating force behind the scarfers' altering their blowtorch tips. Management also thought that the incentive plan was a problem in that it rewarded output but not quality. The newer scarfers particularly were seen as guilty because they had been under pressure to keep up with older employees. The tip alteration problem had been going on for a long time, and the finished product rejection rate at the Dayton plant was twice the corporate rate. With the obvious abuse of the incentive plan (410 percent by one crew for one period), management felt that it was time to make an example of the violators and to correct the abuses.

Blowtorch tip checks had been conducted before on the initiative of the shift foremen to discover and replace altered torch tips and to mildly reprimand the men, but Flint felt the whole thing had gone too far this time. Now was the time for more drastic action. Flint felt that the recent 410 percent incentive pay performance level provided him with an opportunity to slow the scarfers who was boss. It would also serve as an excuse for him to implement changes in the types of torch tips the scarfers used and in their incentive system.

Consequently, Bob Flint ordered his shift foremen to conduct an immediate torch tip check on all scarfing crews, telling them that any workers found using altered tips were, in his words, "to be dealt with." When Dee Colton, the shift foreman at the time, conducted a jobsite tip check in the slab yard, he found 50 percent of the scarfers using altered tips. Following past procedure, he issued each guilty worker a reprimand requiring two days off without pay. When Colton reported his actions to Flint, however, Flint was incensed. He demanded that Colton fire the guilty men immediately and that he fire any other scarfers found cheating in the next two shifts. Despite the fact that the first shift had warned the next shift—swing shift—that a tip check was on, three men supposed that the action taken would not be any different from that of the usual inspections and therefore they did not bother to change their tips. They were subsequently fired. By the time the graveyard shift shuffled into the slab yard that night, the word had gotten around that top management had its hands in this crackdown. No altered tips were found during the graveyard shift.

Two days later the scarfers began a wildcat slow-down which created a bottleneck potentially costing the company millions of dollars.

When you have read this case, look back at the chapter's diagnostic questions and choose the ones that apply to the case. Then use those questions with the ones that follow in your case analysis.

1. Midland Steel's scarfers all seemed to produce at about the same rate. Can you explain this? What specific factors contributed to the level of productivity accepted by the scarfers as a "fair day's work"?

2. Why didn't the scarfers expect to be fired even if caught with an altered torch tip? What caused the wildcat slowdown? What steps should Midland Steel's management have taken to deal with the slowdown?

3. What long-term changes need to be made to maintain productivity in the slab yard? What must management do to get the group of scarfers to change the way they perform the slab yard work?

CASE 11-3

EXECUTIVE RETREAT: A CASE OF GROUP FAILURE

Review this case, which you read in Chapter 4. Next, look back at Chapter 11's diagnostic questions and choose the ones that apply to the case. Then use those questions with the ones that follow in your case analysis.

1. Through what stages of development did John Matthews' group progress during the fifteen days it spent together? Did the group's development affect its ability to perform effectively? How?

2. What kind of task was the group performing? What abilities does such a group task require? Did the group have these abilities?

3. Explain why the group failed to accomplish most of its goals. If it were to reconvene and try to accomplish the same goals again, what advice would you give the group to improve its chances for success?

GROUP FORMATION AND PERFORMANCE

CHAPTER 12

ERSHIP

DEF

ent

SHIP

A DIAGNOSTIC

ADERSHIP

One of the most charismatic leaders in history, Martin Luther King, Jr. had a dream of what society could be, a dream that millions of Americans made their own. Winner of the Nobel Peace Prize in 1964, King dedicated his life to achieving rights for all citizens by nonviolent methods.

And so, in May 1941, Hitler stood master of Europe. It was an incredible achievement. Less than ten years before, he had tricked and blundered his way into the leadership of a penniless and disarmed nation. Now, from the Pyrenees to the Arctic Circle, from Brittany to Warsaw to Crete, this ex-corporal ruled virtually unchallenged over more of Europe than any man that had governed since the days of the Roman Empire. And his friends and allies ruled in Moscow, Tokyo, Rome, Madrid. His only remaining enemy, Britain, was badly mauled and begging the United States for supplies.[1]

From the mid-1950s until the late 1960s, Martin Luther King was the most important leader of a non-violent civil rights movement that transformed the politics of America and inspired oppressed people throughout the world. During this period, black Americans attained more progress than in the previous century. The system of *de jure* segregation was overturned in the South and essential legislation was enacted, enabling blacks to make significant strides toward resolving . . . the conflict between the nation's democratic ideals of freedom and equality and its practice of denying basic rights to black citizens.[2]

As we noted in Chapter 11, few important tasks or goals can be accomplished by one person working alone. It is for this reason that there are so many organizations in our society. On the other hand, few groups or organizations can accomplish much without the help of a single individual acting as a leader. Leadership is the force that energizes and directs groups. Some leaders, like Hitler, move groups in ways that greatly hinder their ultimate survival. Others, like Martin Luther King Jr., move groups in directions that vastly improve their ultimate welfare. Given the centrality of leadership to the behavior of people in groups and to organizational achievement, whether good or evil, it is of paramount importance that we understand how leaders emerge and what makes them effective.

DIAGNOSTIC ISSUES

Our diagnostic model highlights several important questions in the area of leadership. Can we *describe* the characteristics of people who develop into strong leaders? How do differences among groups affect the kinds of leaders who emerge? Can we *diagnose* different leadership styles? Can we explain why some people who are capable managers never become capable leaders? Can we *prescribe* which behaviors a leader should engage in to stimulate both productivity and satisfaction? Can we predict which kinds of situations will enhance the effectiveness of a particular style of leadership? Finally, what *actions* can organizations take to screen people for leadership positions? What sorts of abilities and personality characteristics should we look for? When should leaders use group decision-making processes and when should they make decisions on their own?

[1] "World War II: The Desperate Years," *Time*, September 4, 1989, p. 34.
[2] J. A. Colaiaco, *Martin Luther King, Jr.* (New York: St. Martin's Press, 1988), p. 37.

Defining Leadership

Supreme Court Justice Potter Stewart was once asked to define pornography so as to distinguish it from more conventional art. He replied that although he could not define pornography, he certainly knew what it was when he saw it. You get a similar feel for people's conceptions of leadership. Most people have a hard time explaining what they mean by it, and those who do offer definitions disagree on some of the major dimensions. At the same time, when people are asked to name strong leaders throughout history, they respond in a remarkably consistent way. Table 12-1 lists a number of people who are almost always cited as strong leaders.

Looking at the list in Table 12-1 may give you an idea of how difficult it is to come up with a definition of leadership that is specific enough to be useful yet broad enough to include people that differ so greatly from each other. Try to think of characteristics of these people or of their followers or of the situations they faced that were common to all. Thinking about each of these historic figures and the contexts within which they served as leaders will help dramatize the complexity of the leadership concept for you. As you study this chapter it will help you understand why there has been so little consensus among theories of leadership.

Components of Leadership

One concept that you may have found in common among the people listed in Table 12-1 is their *ability to influence* others. The notion of influence certainly should be paramount in any definition of leadership. Yet would we consider a mugger who enters a subway train and induces the group gathered there to hand over their personal belongings a leader? Most people would recognize this person's influence but they would not consider his act one of leadership.

A leader's influence must to some degree be *sanctioned by followers*. In some situations a person may be compelled by her followers to lead, and in other situations a leader may be merely tolerated for the time being. Still, the idea that followers voluntarily surrender control over their own behavior to someone else seems to be an integral part of any definition of leadership. Finally, a complete definition of leadership must include the nature of the context in which leadership occurs. Leadership occurs in *goal-oriented* group contexts.

Leadership versus Management

It is important to distinguish between leadership and management, two concepts that are frequently confused. To begin with, in Mintzberg's conception of management (see Table 2-8 in Chapter 2), the role of leader is just one of ten roles

ANOTHER VIEW
Many situations exist in which, it would appear, followers did not voluntarily surrender control. Dictatorships and military governments found in Central America are examples. Followers have not been given a choice. Have students discuss how these types of leaders stay in power. What methods do they use? Are they really leaders?

TABLE 12-1
Conventional Examples of Strong Leaders

Adolf Hitler	Martin Luther King, Jr.
Mahatma Gandhi	Napoleon Bonaparte
Mao Tse-Tung	Moses
Franklin D. Roosevelt	Abraham Lincoln
Winston Churchill	Golda Meir

As Chairman of the Joint Chiefs of Staff and leader of the country's military forces, General Colin Powell symbolizes the strength and purpose of the United States both to the men and women he leads and to the rest of the world.

leadership The use of noncoercive influence to direct and coordinate the activities of the members of an organized group toward the accomplishment of group objectives.

commonly ascribed to managers. Leadership, according to Mintzberg, deals explicitly with guiding and motivating employees.[3] From this point of view leadership is a managerial task, albeit one that many believe today's managers fulfill rather poorly.

The literature offers almost innumerable definitions of leadership. However, Arthur Jago's is the one that best captures the essence of the three components we have enumerated and that maintains the manager-leader distinction. Jago has defined **leadership** as the use of noncoercive influence to direct and coordinate the activities of the members of an organized group toward the accomplishment of group objectives.[4]

THE IMPORTANCE OF LEADERSHIP

Leaders perform several essential functions for the groups they serve.[5] They are responsible for generating and maintaining the required *level of effort* needed from individual group members. Leaders are also responsible for *directing the effort* of group members in ways that promote group survival and goal accomplishment. One important aspect of directing groups is ensuring the *coordination of effort* among group members. Finally, leaders *facilitate group membership* by attracting people to the group and its mission and by meeting the needs of group members. In sum, leaders help move a group in directions consistent with its mission and at the same time hold the group together.

In addition to these four goal-directed functions, leaders serve an important symbolic function for both group members and outsiders. It is virtually impossible for every employee in an organization to understand everything that goes on in the firm, especially one surrounded by a complex, dynamic environment. As we saw in Chapter 5, when the complexity of a stimulus exceeds a person's cognitive capacity, the person attempts to simplify the stimulus. In the organizational context, the leader provides the means for much of this simplification.

The leader offers a logically compelling and emotionally satisfying focal point for people who are trying to understand the causes and consequences of organized activity. Many of these causes and consequences of what organizations

[3] Henry Mintzberg, *The Nature of Managerial Work*. (New York: Harper and Row, 1973).
[4] A. Jago, "Leadership: Perspectives in Theory and Research," *Management Sciences 28* (1982), 315–36.
[5] D. Katz and R. Kahn, *The Social Psychology of Organizations* (New York: John Wiley, 1978), p. 125–37.

TEACHING NOTE
Leaders are often the scapegoats
when things go wrong. They carry
the blame and often suffer the
consequences. So why do some
individuals strive to be leaders?

do are obscure, uncertain, and perhaps even objectionable. Focusing on the leader reduces these complexities to simple terms that people can easily understand and communicate. Sometimes, of course, this leads to misguided actions. For example, when a baseball team has a losing season, the owner and fans take comfort in firing the manager, even when what might truly be needed is to fire all the players. According to Meindl and Ehrlich, "As an explanatory concept, leadership has assumed a special status. Not merely a prosaic alternative that people dispassionately consider on an equal footing with other explanations, it has achieved a heroic, larger-than-life value."[6]

In this chapter we will put leadership into perspective. We will integrate the many theories and lines of research on leadership into a single unifying framework. With an appreciation of the major themes and the significant results of the work that has been done on leadership, you will be in a position to apply this knowledge in becoming an effective leader yourself.

THE TRANSACTIONAL MODEL: A DIAGNOSTIC FRAMEWORK

To make it easier for you to understand the dozen or so different theories of leadership we explore in this chapter, we will start with a conceptual framework that encompasses each theory. With this framework in place, we can examine each approach and fit it into the overall scheme. This single framework will also allow us, in the last section of the chapter, to build one integrated model of leadership that incorporates significant features of each theory we discuss.

Edward Hollander has convincingly suggested that the leadership process is best understood as the occurrence of *mutually satisfying transactions* among leaders and followers within a particular situational context.[7] As Figure 12-1 indicates, the *locus of leadership* is found where these three forces—*leaders*, *followers*, and *situations*—come together. In Hollander's view, one can understand leadership only by gaining an appreciation of the important characteristics of these three forces and of the ways in which they interact.

[6] J. R. Meindl and S. B. Ehrlich, "The Romance of Leadership and the Evaluation of Organizational Performance." *Academy of Management Journal 30* (1987), 91–109.
[7] Edward P. Hollander, *Leadership Dynamics* (New York: Free Press, 1978).

FIGURE 12-1

The Transactional Model of Leadership

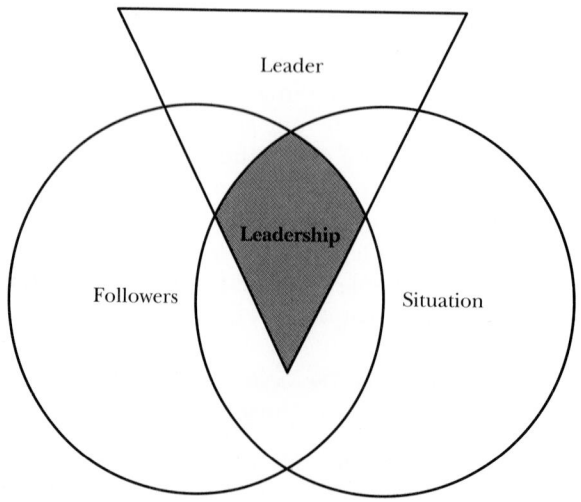

412

The importance of followers is being recognized by companies like AT&T. As baby boomers enter their 40s, they expect to move into leadership positions. The problem is that there are not enough leadership positions to meet those expectations. As a result, AT&T is emphasizing the importance of followers and their contributions. Nothing would get done if everyone were a leader. Hitler never would have conquered Europe if no one followed him. An important function of followers is to assess the quality of leadership rather than follow blindly. In the United States we have mechanisms in our laws and government regulations to affect our leadership.

AN EXAMPLE: THE SOVIET UNION

As the Soviet Union inches toward democracy, we begin to see attempts at changing leadership. In June of 1991 the Soviets *elected* a president for the first time in history. The Soviet people, as well as the rest of the world, must wait and see what this will mean. What the president's role will be versus that of Prime Minister Gorbachev is not yet defined. Experts feel that economic conditions within Russia will, over time, be a critical determinant of what type of leadership will prevail.

Leaders often recognize the important role played by their followers. At a 1936 speech in Nuremberg, for example, Hitler declared to 200,000 followers, "The wonder of this age is that you have found me—an unknown man among millions."[8] One nameless soldier-follower of Hitler's later remarked that "We, like the soldiers of other countries, were trained to obedience. We had not been brought up free to demonstrate our opposition under the protection of a liberal constitution. We had the same sensitivities that all humans have, but during a time of difficult decisions, we lacked political vision." The role of the follower has also been pointed to by Martin Luther King Jr.'s biographers. Ella Baker writes that "the movement made Martin rather than Martin making the movement."[9]

To appreciate the importance to a leader of followers, ask yourself the following questions. Could a person with Hitler's characteristics rise to power in a post-Vietnam war United States, where opposing almost any government act is close to a national pastime? Could King's peaceful, patient approach to civil rights have won the hearts of Shiite Muslims in Iran after the fall of the shah? Can anyone establish a position of leadership with people, such as a group of intellectuals, who reject the very idea that they need to be led?[10]

Turning to the characteristics of the situation, would Mahatma Gandhi's program of civil disobedience have been as successful had he been opposing the Nazis instead of the British? In 1989, could Tadeusz Mazowiecki have risen as the first non-Communist leader of Poland in forty years if the Polish economy had not been in total shambles? At Apple Computer, John Sculley reportedly believed in "borderline anarchy" and held that "a good company should constantly be stretching." Would Sculley's practice of almost annual reorganization in an industry where creativity is king have been effective if he had been the manager of a ball-bearing plant? These questions underline the complex nature of leadership and the important contribution to successful leadership of the three distinct elements of leader, follower, and situation.

Not all theoretical approaches to leadership emphasize the three-dimensional character of the leadership process proposed by Hollander. In fact, as we will see, universal approaches generally focus on one dimension, interaction approaches on two, and only a few, comprehensive approaches consider all three dimensions. We will examine each of these approaches in turn.

UNIVERSAL APPROACHES TO LEADERSHIP

Some of the earliest probes into the nature of leadership focused almost uniformly on leader characteristics. These *universal theories* emphasized personality traits, abilities, typical behaviors, and decision-making styles. These first approaches were followed by others that also focused on just one aspect of leadership—either the follower or the situation. Born out of the failure of leader-focused studies to explain the richness of the leadership process, these newer, anti-leadership theories discounted the leader almost entirely.

Qualities of the Leader

The earliest approaches to leadership, often referred to as the *great man theories of leadership*, held that leaders were born, not made. Sir Francis Galton, whose

[8] "World War II: Blitzkrieg," *Time*, August 28, 1989, p. 39.
[9] D. J. Garrow, *Bearing the Cross* (New York: Random House, 1986), p. 2.
[10] Hedley Donovan, "Managing Your Intellectuals," *Fortune*, October 29, 1989, pp. 177–80.

studies of individual differences we discussed in Chapter 4, argued in 1869 that the qualities found in great leaders were inherited. Later, researchers influenced by behavioral schools of thought discarded this idea, suggesting instead that the characteristics associated with successful leadership could be learned.

Physical and Mental Abilities. Studies of the physical characteristics of leaders have yielded rather weak but consistent relationships between a person's *energy level* and her ability to rise to a position of leadership.[11] Still weaker and less consistent results have been found for characteristics like height. Oddly, we tend to think of leaders as tall people even though many are not. Consider the leaders listed in Table 12-1. No more than half of them could have been considered tall or physically imposing people.

Research on mental abilities has produced few substantial predictors of leadership quality and effectiveness, although again some consistent findings have been reported. *General cognitive ability* (intelligence—see Chapter 4) seems to be one of the best overall predictors of leadership ability.[12] Specific *technical skills* or *knowledge about a group's task* also show modest relationships with success in leadership.[13]

Personality Characteristics. Researchers have related the development and effectiveness of leadership skills to almost all of the five classes of personality characteristics that we examined in Chapter 4. For example, there is evidence that leaders tend to exhibit the social trait of *dominance*, and that leadership potential is associated with the motives of *need for achievement* and *need for power*.[14] In addition, the personality dynamic of *self-esteem* (or self-confidence or self-assurance) seems to be related to leadership across a wide variety of situations and followers.[15]

These findings notwithstanding, for every personality characteristic that does appear to be related to leadership potential, skill, or effectiveness there are probably ten others for which no such evidence exists. As the "Management Issues" box indicates, even the basic moral character of the leader does not seem to relate in any simple manner to success. Moreover, the relations that have been found are modest at best. It was this failure to find significant relationships between leadership and individual personal qualities that led researchers to explore other approaches to understanding this important concept.

Leader Behaviors

University of Michigan Studies. In the late 1950s, Rensis Likert and other researchers at the University of Michigan began a series of leadership studies. They set out to identify aspects of leaders' behavior that might differentiate those who performed well from those who did not. Interviewing supervisors and clerical workers at the Prudential Insurance Company, these investigators concluded that there were two general classes of supervisory behavior.[16] **Employee-centered**

APPLYING THE DIAGNOSTIC MODEL
Diagnostic Question 3: What leadership-related abilities (e.g., general cognitive ability, task knowledge, supervisory skills) do you or your managers possess? How does this relate to the five components of motivation?

APPLYING THE DIAGNOSTIC MODEL
Diagnostic Question 4: What leadership related personality characteristics (e.g., need for power, self-esteem, charisma) do you or your managers possess? How does this relate to the five components of motivation?

ANOTHER VIEW
A recent study by Kirkpatrick and Locke proposes several key traits that characterize leaders. These traits are drive, leadership motivation, honesty and integrity, self-confidence, cognitive ability, and knowledge of work. The researchers argue that while possession of these traits does not guarantee one will become a leader, they do help the individual develop necessary skills, formulate an organizational vision and an effective plan to pursue it, and take the necessary steps to implement the vision and make it reality. They conclude that leaders are not like other people; it takes a special kind of person to master the challenges of opportunity. (*The Executive*, May 1991, pp. 48–60)

employee-centered behaviors
Leadership behaviors designed to meet the social and emotional needs of group members.

[11] G. Yukl, *Leadership in Organizations* (Englewood Cliffs, NJ: Prentice Hall, 1981), p. 71.

[12] R. M. Stodgill, *Handbook of Leadership* (New York: Free Press, 1974), p. 112.

[13] R. Katz, "Skills of an Effective Administrator," *Harvard Business Review 72* (1974), pp. 90–101.

[14] E. Constantini and K. H. Craik, "Personality and Politicians: California Party Leaders, 1960–1976," *Journal of Personality and Social Psychology 38* (1980), 641–61; and E. G. Ghiselli, *Explorations in Managerial Talent* (1966), p. 92.

[15] Jago, "Leadership," p. 319.

[16] Rensis Likert, *New Patterns of Management* (New York: McGraw-Hill, 1961), p. 36.

MANAGEMENT ISSUES

Managers and Moral Leadership: Is Pursuing Self-Interest Enough?

The predominant business ideology in the United States holds that when firms and individuals pursue their own self-interests, market forces bring about the most efficient use of resources and result in the greatest satisfaction of people's needs. Apparent breakdowns in this theory could be seen, for example, in the days of the Industrial Revolution when sweatshops employed ten-year-olds for seventy-hour work weeks. Although using cheap child labor was in a firm's self-interest, it was quite questionable whether this was in the best interests of society as a whole. Similar debates have appeared more recently on *greenmail*, *golden parachutes*, *downsizing*, *insider trading*, and *acid rain*.* In each of these areas, one could argue that the interests of the few are at odds with the interests of the many.

Organizational leaders have to balance the needs of executives, employees, consumers, stockholders, suppliers, distributors, and the general public, all the while keeping their companies financially secure. With all this to manage, is it fair to ask them to be moral leaders of the nation and their communities as well? This is a volatile issue that is likely to generate considerable debate in the years ahead.

Clearly when one looks at contemporary business leaders it seems that what is traditionally defined as a strong moral character is neither a requisite of nor a deterrent to leadership success. Roy Vagelos is the CEO of Merck and Company, a large and highly successful pharmaceutical firm. Several years ago, Merck developed a drug named ivermectin that cured river blindness. River blindness is caused by the bite of a particular variety of fly that deposits a parasite in a person. The parasite can grow to two feet in length. If and when it reaches the victim's eye, blindness results. The disease has afflicted an estimated 40 million people in Africa, the Middle East, and Latin America. Unfortunately, the drug was expensive to produce, and most people who needed it could not afford to pay for it. Vagelos directed Merck to donate the drug to the World Health Organization (WHO) despite the substantial expense associated with this giveaway.†

James Burke, CEO of Johnson and Johnson has also been recognized as a moral business leader for his handling of the Tylenol tampering incident. In this incident, seven people died when an anonymous person placed cyanide in capsules that were already on store shelves. Even though Johnson and Johnson was clearly not responsible for the poisonings, Burke directed the company to recall all unused Tylenol, at a cost of over $100 million. Burke went on many television talk shows in a one man crusade to save Johnson and Johnson's image. Today, both Tylenol and Johnson and Johnson are stronger than ever, and Burke has become recognized as a business hero. In his own estimation, the reason the Tylenol rescue succeeded was "not that we did anything clever, but just that we are a company that tries to do the right thing."‡

Other business leaders have built financially successful enterprises despite being "at the margin" on ethical issues. For example, Thomas Jones, the longtime controversial CEO of Northrup, has been pursued almost constantly for potentially unethical or illegal activity. Northrup is a large aviation defense contractor that produces, among other things, the F-20 Tigershark fighter, the Stealth bomber, and the guidance system for the MX missile. In the 1960s Jones established a secret fund in Paris to launder illegal political contributions in the United States. In the early 1970s Northrup was again involved in scandal when it gave a bribe of $450,000 to two Saudi Arabian generals. Later, in 1974, Jones pleaded guilty to illegally contributing $150,000 to Richard Nixon's presidential campaign. Then, in 1983, Northrup was charged with influence peddling involving a bribe of $55 million made to a Korean official to help get his government to purchase a large number of F-20s. In 1988, Jones fought several suits brought by Northrup's own employees. They claimed that the company had billed the government $400 million in false labor charges for work supposedly done on the secret Stealth bomber.

Yet despite all this turbulence, Jones has engineered a successful financial performance for Northrup. One ex-Northrup executive commented, "Jones did a phenomenal job with the company. When I joined in 1963, the company had broken the $300 million mark. Now it sells more than $3 billion. I do think he tends to cut corners a bit ethically sometimes. . . . He'll do whatever is required for his company."§

Although one could debate forever the utility or disutility of a strong moral character as a requisite for leadership, what should not be overlooked in this debate is that leadership inherently involves moral philosophy. In the process of selecting and pursuing economic or social goals and policies, the leader of an organization is guided by his values and moral principles. Many times the values behind the leader's behavior are not even well thought out or articulated. Nevertheless, the leader's actions are value based. Moreover, these values are communicated, often subliminally, to followers and thus affect their behaviors. As one ethics scholar has commented, "The challenge for the manager is not whether to include ethical theory and criteria in strategic choice, but rather when and how."‖

* G. F. Cavanagh, *American Business Values* (Englewood Cliffs, NJ: Prentice Hall, 1990).
† M. Waldoz, "Merck to Donate Drug for "River Blindness," *The Wall Street Journal*, October 22, 1987, p. 38.

‡ L. Shames, *The Big Time: Harvard Business School's Most Successful Class and How It Shaped America* (New York: Mentor, 1986), p. 159.
§ R. Nader and W. Taylor, *The Big Boys* (New York: Pantheon Press, 1986), p 210.
‖ E. A. Murray, "Ethics and Corporate Strategy," in *Corporations and the Common Good*, ed. Robert B. Dickie and Leroy S. Rouner (Notre Dame: University of Notre Dame Press, 1986), p 115.

job-oriented behaviors
Leadership behaviors that focus on careful supervision of employees' work methods and performance level.

AN EXAMPLE
Drill instructors in military boot camps are job oriented, or focused on initiating structure. Their objective is to establish discipline and, within a short period of time, teach recruits to effectively fight and defend themselves and their comrades under fire. They are not concerned with the social needs, but rather, the survival needs of these men and women.

consideration leader behavior aimed at meeting the social and emotional needs of workers such as helping them, doing them favors, looking out for their best interests, and explaining decisions

initiating structure leader behaviors aimed at meeting the group's task requirements, such as getting workers to follow rules, monitoring performance standards, clarifying roles, and setting goals

The employee-centered, circular management structure initiated by Frances Hesselbein when she was executive director of the Girl Scouts of America does away with the traditional rising layers of management. According to Hesselbein, this type of management enables people to move "across" the organizational structure rather than up or down and fosters innovation and creativity.
Source: John A. Byrne, "Profiting from the Nonprofits," Business Week, March 26, 1990, pp. 66–74.

behavior aimed at meeting the social and emotional needs of group members. **Job-oriented behavior** focused on careful supervision of employees' work methods and task accomplishment. These two orientations were seen as mutually exclusive. A leader could display one pattern or the other but not both.

The first group of studies at Michigan indicated that work attitudes were better and productivity was higher in the groups led by supervisors who displayed employee-centered behaviors. These studies, however, measured both the independent and the dependent variables at the same time. As a result, we cannot tell whether supervisors' personal concern caused the high productivity and good attitudes or whether these positive employee behaviors attracted supervisory attention.

To clarify this point, Morse and Reimer undertook a follow-up field study in which they trained some supervisors to use job-centered behaviors and others to use employee-centered behaviors in interacting with employees.[17] The results of this study supported one of the earlier findings but not the other. Leaders' employee-centered behavior did appear to cause more positive attitudes among workers. However, productivity was higher among workers supervised by leaders who used a job-centered approach.

Ohio State University Studies. While Likert and his colleagues were exploring the effects of different leader behaviors, Edwin Fleishman was conducting similar research at Ohio State University. Analyzing workers' responses to a questionnaire by means of a sophisticated statistical procedure, the Ohio State group concluded that most supervisory behaviors could be assigned to either one of two dimensions: **consideration** or **initiating structure**.[18] Table 12-2 shows some items from the Leadership Behavior Description Questionnaire (LBDQ) that evolved out of the original Ohio State studies. As you can see, the consideration dimension resembles the Michigan group's employee-centered orientation. Both dimensions address individual and social needs of workers. Similarly, the initiating structure dimension is like the job-centered orientation identified by Likert and his associates. Both of these dimensions focus on issues of supervision and task accomplishment.

Despite their similarities, the Michigan and Ohio State conceptions differ in an important way. The former sees the two dimensions of leader behavior as mutually exclusive, whereas the latter sees these dimensions as coexisting. Thus, as Figure 12-2 suggests, a leader might be high on both consideration and initiating structure, high on one dimension and low on the other, low on both, moderately high on one and low on the other—or almost any combination of degrees on the two dimensions. Subsequent research with the scales has suggested that the relationship between the two dimensions is neither perfectly negative (as the Michigan studies imply) or perfectly zero (as the Ohio State studies imply). Rather, in roughly half the studies that used both sets of scales, a weak, positive relationship between the two was apparent.[19] In other words, there is a slight tendency for leaders who are considerate to initiate structure.

Initially it was thought that the most effective leader is one who engages in both consideration and initiating-structure types of behavior. Research at Navistar (formerly International Harvester) failed to indicate any clear pattern, however. Consideration seemed to relate positively to worker attitudes but not

[17] N. C. Morse and E. Reimer, "The Experimental Change of a Major Organizational Variable," *Journal of Abnormal and Social Psychology 52* (1956), 120–29.

[18] R. M. Stodgill and A. E. Coons, *Leader Behavior: Its Description and Measurement* (Columbus, OH: Ohio State University, Bureau of Business Research, 1957), p. 75.

[19] P. Weissenberg and M. H. Kavanaugh, "The Independence of Initiating Structure and Consideration: A Review of the Evidence, *Personnel Psychology 25* (1972), 119–30.

APPLYING THE DIAGNOSTIC MODEL

Diagnostic Question 5: How readily do you think your abilities and characteristics or the abilities and characteristics of your managers can be changed?

Diagnostic Question 6: What leader behaviors (e.g., consideration, initiating structure, contingent rewarding/punishing) do you or your managers typically employ? How might this relate to the five components of motivation?

T A B L E 12-2

Items Similar to Items in the Leader Behavior Description Questionnaire

Consideration Items
1. Is easy to get along with
2. Puts ideas generated by the group into operation
3. Treats everyone the same
4. Lets followers know of upcoming changes
5. Explains actions to all group members

Initiation of Structure Items
1. Tells group members what is expected
2. Promotes the use of standardized procedures
3. Makes decisions about work methods
4. Clarifies role relationship among group members
5. Sets specific goals and monitors performance closely

---

to productivity. Initiating structure was not clearly related to either attitudes or productivity.[20]

The Leadership Grid®. Despite the conflicting results of the Michigan and Ohio State studies on the behavioral dimensions of leadership, Blake and Mouton developed the notion of the managerial grid, republished in 1991 as the **leadership grid figure** by Blake and McCanse.[21] Based on their own research, Blake and Mouton identified two attitudinal dimensions, as distinguished from the behavioral dimensions identified at Ohio and Michigan. As Figure 12-3 shows, the leadership grid proposes five different styles of leadership, based on the interaction between *concern for people* and *concern for production*. Each of these two dimensions is measured on a scale of 1 (low) to 9 (high). Blake and Mouton have

leadership grid figure A two-dimensional representation of leadership behaviors in which concern for people and concern for production combine to produce five behavioral styles.

[20] Stodgill and Coons, *Leader Behavior*, p. 112.

[21] R. Blake and J. S. Mouton, *The Managerial Grid III: The Key to Leadership Excellence* (Houston: Gulf Publishing Co., 1985); and R. Blake and A. A. McCanse, *Leadership Dilemmas—Grid Solutions* (Houston: Gulf Publishing Co, 1991).

F I G U R E 12-2

Initiating Structure and Consideration in the Ohio State Studies

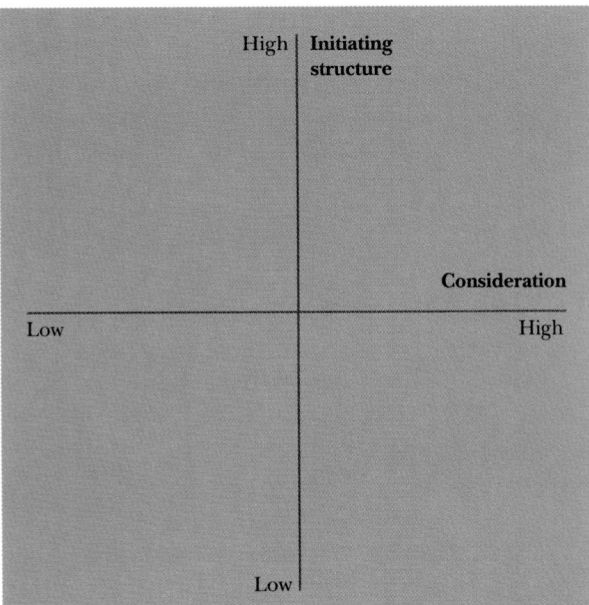

FIGURE 12-3

The Leadership Grid® Figure

Source: The Leadership Grid®
Figure from Leadership Dilemmas—Grid
Solutions, by Robert R. Blake and
Anne Adams McCanse. Houston: Gulf
Publishing Company, p. 29. Copyright
© 1991, by Scientific Methods, Inc.
Reproduced by permission of the
owners.

TEACHING NOTE

Have students place the leaders listed in Table 12-1 into the Leadership Grid. If they are not familiar enough with these individuals, have them try to classify contemporary leaders like Saddam Hussein, Gorbachev, and Bush. Ask them how readily these individuals could be trained to use different leadership styles.

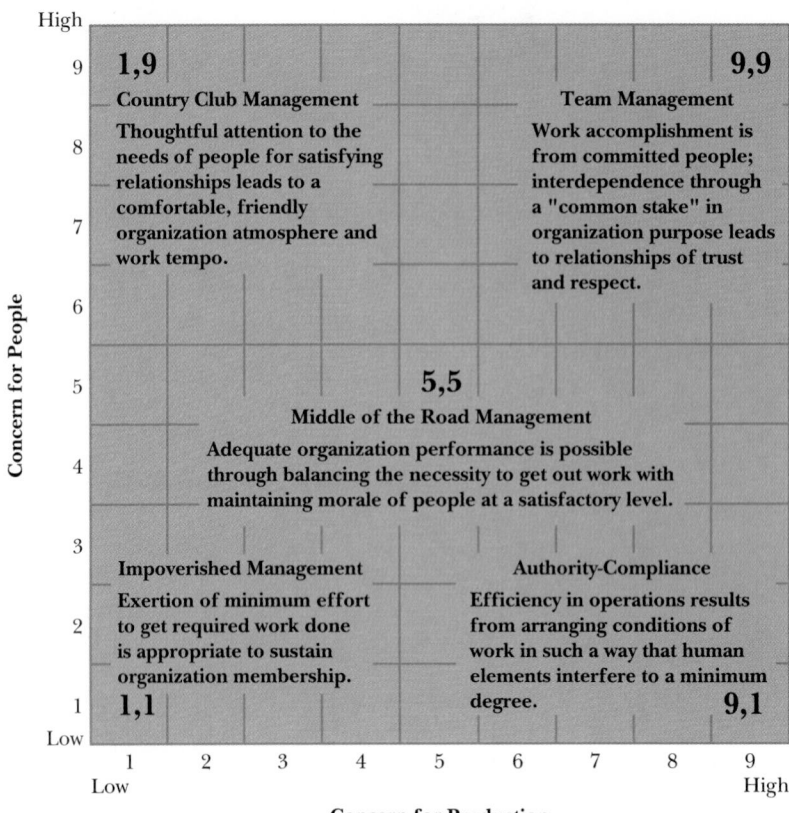

explicitly stated that the 9,9 Team Management style (see the upper righthand corner of the grid) is "the one best way" to lead, and they have developed an elaborate training program to move managers in that direction. Managers find the program appealing because it points to two specific sets of behaviors—consideration and initiating structure—that they can engage in to enhance the attitudes and performance of their group.

Despite its appeal, however, the managerial grid approach lacks empirical support from rigorous scientific studies, either in the lab or in the field. In fact, some investigators have gone so far as to label the whole 9,9 idea a myth.[22] In response to such criticisms, Blake and Mouton have offered only a conceptual defense, arguing that their theory *should* work.[23] In the absence of sound, supportive empirical data from independent researchers, it is hard to see the managerial grid as the final word on leadership effectiveness. Indeed, a good deal of research argues against the notion that there is any "one best way" of leading, irrespective of followers and situations.

Leaders' Decision-Making Styles

At the time of the Michigan and Ohio State studies, a third line of research on universal approaches to leadership was well under way under the direction of Kurt Lewin at the University of Iowa. The Iowa group studied the leader's manner

---

[22] L. L. Larson, J. G. Hunt, and R. Osburn, "The Great Hi-Hi Leader Myth: A Lesson from Occam's Razor," *Academy of Management Journal 19* (1976), 628–41.

[23] R. Blake and J. S. Mouton, "A Comparative Analysis of Situationalism and 9,9 Management by Principle," *Organizational Dynamics 24* (1982), 21.

When Rosetta Riley took over the job of quality chief at Cadillac in 1985, customers were complaining that the company's cars were less powerful and smaller than they used to be. Using a democratic style of leadership, Riley worked with fellow executives and local dealers to create a new responsiveness to customers. She encouraged dealers to alert headquarters to problems and urged executives to begin calling recent buyers to ask how they liked their cars. Soon, market research reports were ranking Cadillac first among U.S. cars in terms of overall customer satisfaction.
Source: *"Commissar of Quality,"* Fortune, *March 25, 1991, p. 131.*

authoritarian leader   A leader who makes almost all decisions by herself, minimizing the input of subordinates.

democratic leader   A leader who works to ensure that all subordinates have a voice in making decisions.

laissez-faire leader   A leader who lets a group run itself, with minimal intervention from upper levels of the organizational hierarchy.

APPLYING THE DIAGNOSTIC MODEL
Diagnostic Question 7: Do you and your managers tailor your behaviors to different followers?

of making decisions and the effect that varying decision styles had on subordinates' rates of productivity and general satisfaction.

Lewin and his colleagues looked at three different decision-making styles: authoritarian, democratic, and laissez faire. The **authoritarian leader** made virtually all decisions by himself. The **democratic leader** worked with the group to help members come to their own decisions. The **laissez-faire leader** did just what this French term means—he left the group alone to do whatever it wanted.

The first Iowa study examined these leadership styles in groups of ten-year-old boys who were members of a hobby club. The investigators found that almost every group preferred a democratic leader best. Members of groups led by an authoritarian leader were either extremely submissive or extremely aggressive in interacting with each other. They were the most likely of any club members to quit the organization. Authoritarian groups were the most productive but only when members were closely supervised. When left alone, these groups tended to stop working. The results of this decision-style research were interesting and provocative. Like the personal characteristic and behavioral approaches discussed earlier, however, the Iowa studies revealed only rather modest relationships between leader style and follower behavior.

These three streams of early leadership research focused almost exclusively on the qualities of the leader. Although this early work has been abandoned, it continues to influence current theoretical and experimental work. Almost all of the comprehensive theories that we will discuss shortly include the three principal ways of characterizing a leader that we have explored: abilities and personal characteristics, behavioral styles, and decision-making styles (Figure 12-4). The value of this initial work lies also in its having pointed the way to the phenomena we look at next—the situation in which the leader finds himself and the followers who surround him.

Anti-Leadership Approaches

The primary problem of all three of the approaches we have discussed is that they specify one best way to lead regardless of the characteristics of followers and situations. Subsequent research has cast doubt on this notion. For example, although leadership requires intelligence, one study indicated that if a leader is far more intelligent than her followers, she may have difficulty communicating with them. As a result she may be less effective than a leader of more modest

FIGURE 12-4
Three Leadership Dimensions

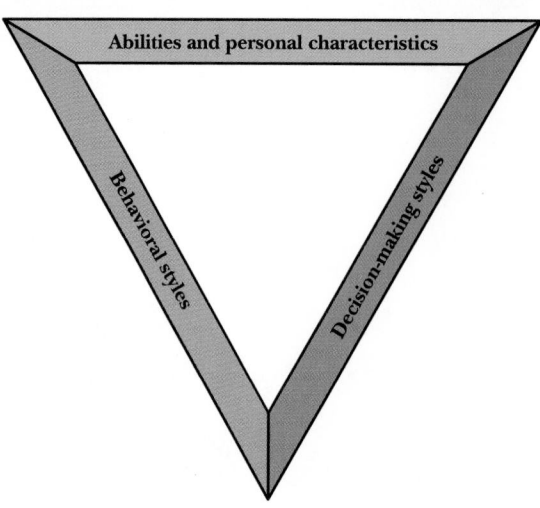

intelligence.[24] Another study that looked at decision-making styles among consumer-loan officers suggested that some followers actually prefer authoritarian leaders.[25] Finally, a number of studies indicate that initiating structure is more likely to be associated with increased productivity when tasks are complex than when they are relatively simple.[26] Findings like these led some to the conclusion that leadership per se did not really matter that much, and several *anti-leadership theories* began to take shape.

Situations and Leader Irrelevance. Jeffrey Pfeffer has argued that leadership is actually quite irrelevant to most organizational outcomes.[27] Pfeffer stresses a situation-based approach to understanding leadership. First, he says, *factors outside the leader's control* tend to affect profits and other critical elements in the business context more than anything a leader might do. Consider the recent plight of Lockheed Airlines. Much of this firm's revenue comes from defense contracts. When the federal government's 1990 budget slashed defense spending dramatically, CEO Daniel Tellep found himself almost powerless to prevent massive losses.

Second, even leaders at relatively high levels tend to have *unilateral control over only a few resources*. Moreover, the capricious use of any set of resources is constrained by the leader's obligation to account for her behavior to other people both within and outside the organization. Even the CEO of a major corporation has to answer to shareholders, consumers, and government regulators.

Finally, the *selection process* through which all leaders must go filters people in such a way that those in leadership positions tend to act in similar ways. For example, the process used to select the president of the United States makes it impossible for some types of people (for example, illiterates, introverts, radicals of either the left or right, or people with past sexual indiscretions) to rise to that position. The people who make it through screening procedures of this kind tend to be alike in more ways than they are different. According to Pfeffer, "homogenizing" leaders in this way reduces the impact that any change in leadership has on an organization's outcomes. For all these reasons, Pfeffer has argued,

AN EXAMPLE
What factors or resources does the President of the United States control? The president has very little authority to make unilateral decisions. Congress, the courts, and, to a lesser extent, the general public control the decisions of the president. For example, he can promise a tax cut, but only Congress can decide if a tax cut or increase will take place.

---

[24] Stodgill, *Handbook of Leadership*, p. 70.

[25] H. Tosi, "Effect of the Interaction of Leader Behavior and Subordinate Authoritarianism," Proceedings of the Annual Convention of the American Psychological Association, 1971, 473–74.

[26] A. K. Korman, "Consideration, Initiating Structure, and Organizational Criteria: A Review," *Personnel Psychology 19* (1966), 349–61.

[27] J. Pfeffer, "The Ambiguity of Leadership," *Academy of Management Review 2* (1977), 104–12.

situations are much more important determinants of events than leader characteristics.

Followers and Attribution Theory. Other anti-leader approaches emphasize follower perceptions as the key component of leadership. *Attribution theory* (see also Chapter 5) starts out with the relatively straightforward notion that people need to be able to make sense of the world around them. In particular, they need to be able to infer causes for important things that happen in their lives.[28]

In determining the causes of events, however, people have a built-in tendency to give too much credit to other people or to place too much blame on them.[29] Therefore, even though leaders may have little impact on their organizations, it is important for people both inside and outside the organization to think that the leader's impact is great. Attributing organizational outcomes to leaders makes the follower's world seem more predictable and manageable. For example, if the local economy is failing, people need to figure out why. In drawing conclusions about the causes of this situation, people are much more likely to direct their attention to the mayor's office than to esoteric notions of business cycles. People feel they can exercise some control over the mayor's actions because they can defeat him in the next election or even impeach him. They feel powerless, however, in the face of business cycles. People need heroes and scapegoats, and the leader, according to the attributional approach, fulfills this need.

In this framework, the successful leader is sensitive to the causal attributions of followers and can *manipulate followers' causal perceptions* in a favorable manner. Successful leaders can separate themselves from organizational failures and associate themselves with organizational successes. A president facing a large budget deficit might blame the shortfall on previous administrations and at the same time take credit for low unemployment rates, which could just as easily be attributed to past administrations.

Critique of Anti-Leadership Theories. Like the universal theories of leadership, the theories that emphasize either situations or followers seem overly simplistic. For example, in support of his leadership irrelevance conception, Pfeffer interpreted a study by Lieberson and O'Connor as concluding that "compared to other factors, administration had limited effects on organizational outcomes." If one examines this study closely, however, one finds that its authors actually concluded that "leadership influence on profit margins exceeds that for either industry or company effects."[30] It would seem that leadership can affect even bottom-line figures.

Although followers' perceptions are certainly central to leadership, even attributional models show how these perceptions are influenced from the very outset by leader behaviors. Each of the universal approaches we have discussed lays out an important part of the leadership transaction process—the characteristics of either the leader, the situation, or the follower. Because of their single-variable focus, however, none of these approaches is adequate.

Substitutes for Leadership. Kerr and Jermier's substitutes for leadership theory can be seen as a conservative version of the various anti-leadership approaches. Although this theory emphasizes characteristics of situations and followers, it

AN EXAMPLE: PRESIDENT JIMMY CARTER
In the late 1970s President Carter was blamed for high inflation and for not being able to end the hostage crisis in Iran. Much of the general public was unable to see or understand that inflation was being caused by oil prices set by OPEC nations. In addition, evidence is now emerging that the hostage crisis was taken out of Carter's control by political game-playing. Iran perceived that it could influence the presidential election and cut a better deal for return of the hostages. Thus, Carter took the blame and lost the election for issues he had little or no control over. Had he been better at manipulating people's perceptions, he might have been able to maintain a better image.

---

[28] B. J. Calder, "An Attribution Theory of Leadership," in *New Directions in Organizational Behavior*, ed. B. Staw and G. Salancik (Chicago: St. Clair Press, 1976).

[29] Fritz Heider, *The Psychology of Interpersonal Relations*. (New York: John Wiley, 1958), pp. 15–21.

[30] S. Leiberson, and J. F. O'Connor, "Leadership and Organizational Performance: A Study of Large Corporations," *American Sociological Review 37* (1972), 117–30.

Succession Planning: The Leader's Final Test

When Raymond Danner, founder and CEO of Shoney's, Inc., appointed J. Mitchell Boyd to succeed him, he took his successor into his office and showed him a plaque on the wall. Engraved there was the statement, "The final test of greatness in a CEO is how well he chooses a successor." Six months later, however, Danner and his associates on Shoney's board of directors pressured Boyd into resigning his new post. Shoney's stock dropped from $13\frac{1}{8}$ to $12\frac{1}{8}$ the day that Boyd resigned, and it appeared that Ray Danner was about to flunk that final test of greatness.

According to David K. Wachtel, president of Charley O's restaurant chain and former CEO of Shoney's, this was "the classic case of the founder not being able to get over his ego and let go." Clearly, Danner could not have been upset with the financial performance of Shoney's under Boyd's reign. With Boyd at the helm, the company experienced earnings of $115 million on sales of $680 million—$10 million more in profits than Danner himself had projected.

The problems with Boyd seemed to be over differences in corporate direction and personal style. Danner's focus was on day-to-day operations, holding down costs and sticking to Shoney's short but experience-tested menu. Danner, the son of a poor Louisville paperhanger, had a rough-and-ready leadership style—he often terrorized employees. For example, Danner once shoved a cook's arm into a large vat of soup to drive home the point that the soup was not hot enough to serve. Boyd on the other hand was smooth and polished, an intellectual with a prestigious MBA background. He spent much more money on marketing and advertising and was continuously experimenting with the restaurant's menu.*

Problems associated with leadership succession are hardly limited to Shoney's. Since Perry Ellis died in 1986, the fashion company that bears his name has faced nothing but adversity. After Ellis's death, the firm lost many key employees and went through several unsuccessful managerial changes. Despite attempts to restructure, the firm has struggled to maintain market share in the face of rising criticism that the clothes produced lack quality and appeal.

Although the importance of leadership-succession planning has long been recognized (in fact, no less a figure than God told Moses to instruct his successor, Joshua, on how to carry on after Moses left) there are built-in problems associated with shifting power that make it a difficult task nonetheless. To begin with, truly exceptional leaders are rare. As we have seen, legendary leaders are usually the result of a unique, one-time alignment of leader, follower, and situational characteristics. In addition, these kinds of leaders are rarely able to teach others the same vision or inspirational skills that made them so successful. In fact, they may shine so brightly and be so imposing that they drive away or stifle others who might otherwise be effective leaders.

Nowhere is this more evident than in entrepreneurial, family-run businesses. The child of such a family rarely feels free to be his own person. For example, Frederick Wang was never able to manage Wang Laboratories as successfully as his father, An Wang, the founder of the company. In fact, things got so bad that the father had to force his son to resign as president and CEO after the firm suffered a string of losses and problems in 1988.†

To avoid some of the pitfalls in succession, firms can

* W. Konrad, "Shoney's Needs a Recipe for Succession," *Business Week*, December 25, 1990, p. 52.

† G. Fuchsburg, "Loss of a Star Can Cast Shadow Over Firm," *The Wall Street Journal*, April 9, 1990, p. B1.

---

substitute for leadership
Someone or something in the leader's environment that affects workers' attitudes, perceptions, or behaviors in such a way that the leader's role is made superfluous.

APPLYING THE DIAGNOSTIC MODEL
Diagnostic Question 11: What are some of the major characteristics of the followers or the situation that might substitute or neutralize leader behavior?

reintroduces leader behaviors to the process. In many ways, the substitutes approach solves many of the shortcomings of the other anti-leader approaches. It even borders on what we will classify later as comprehensive theories of leadership. For these reasons it will serve as a good bridge to the interactive and comprehensive approaches to leadership that we will discuss next.

The **substitutes for leadership** theory argues that traditional leader behaviors such as initiating structure and consideration are often made irrelevant by certain characteristics of followers or situations.[31] According to the substitutes model, the success of a particular leadership behavior depends on characteristics of followers or of situations that can act to *substitute* for that particular behavior. Figure 12-5 illustrates the effect of a substitute. Here, consideration on the part of the leader leads to follower satisfaction when boring tasks must be performed. When tasks are intrinsically satisfying, however, leader consideration has no effect. It is not necessary because the satisfying nature of the task substitutes for the leader behavior.

[31] S. Kerr and J. M. Jermier, "Substitutes for Leadership: Their Meaning and Measurement" *Organizational Behavior and Human Decision Processes* 22 (1978), 375–403.

take several steps. First, the leader should choose her successor long before the moment of transition. One advantage of this procedure is that the successor can acquire the kind of experience and knowledge that will be necessary when she takes hold of the reins. It also assures that succession takes place quickly and smoothly and reinforces the image that what is happening is not an unforeseen emergency. Finally, it helps decrease some of the inevitable intragroup competition that can take place when power is suddenly up for grabs. In a battle for power, the firm is so gutted by strife and hard feelings that it is impossible for whoever emerges from the pack to lead effectively.

For example, Malcolm Forbes, Jr. held a press conference at Forbes headquarters the day after his father's death. Malcolm Jr., had already become president and deputy editor-in-chief at *Forbes* magazine, but he felt it necessary to assure financial writers at the conference that he was ready, willing, and able to guide Forbes through the next century.

Second, firms should explicitly remind both employees and the public at large that there is a lot more to their organization than just one person. This is true even at firms that have flaunted a strong leader in the past. For example, Fidelity Investments, Inc. always sold itself to investors by plugging the record of Peter Lynch—the stock analyzer who turned the firm's Magellan Fund into the nation's largest mutual fund. The day after Lynch made public his intention to leave Fidelity, the firm held a press conference. At this conference, they announced his successor, Morris Smith, and emphasized that Magellan made up only 11 percent of the firm's assets. Fidelity

vice-president George Vanderheiden reassured investors that Fidelity "is a well-oiled machine,"—an analogy that explicitly depersonalized the reasons behind the firm's past success.‡

Third, while tempering the glorious history of the past leader, firms must also be careful to manage expectations about the new leader. The unique alignment of leader, follower, and situation that created the legendary leader may not be present for the new leader. The successor's credentials had better be able to speak for themselves, and excessive praise of the new leader at this point can often make matters worse. Lyman Wood, president of Brennan College Services, recalls, "My worst memory was having my predecessor trumpet my arrival out of all proportion—and there I was with no choice but to try to live up to these unrealistic expectations." Unrealistic expectations often wind up in anger and depression for both the new leader and the followers.§

Fourth, firms and ex-leaders should avoid making too many overly restrictive commitments that serve to bind the new leader. Confining her in this way will stifle her capacity for independent thought and action. It is natural for the ex-leader to hope that every small detail of his vision will continue indefinitely into the future—but that is unrealistic. The ex-leader's final acts are to groom a successor, develop a strong, multifaceted organization, and prepare the followers for the changeover. Then, all that is left is the hard part—getting out of the way.

‡ Ibid.
§ M. Feinberg, "Secrets of Successful Succession Planning," *The Wall Street Journal*, November 12, 1990, p. B1.

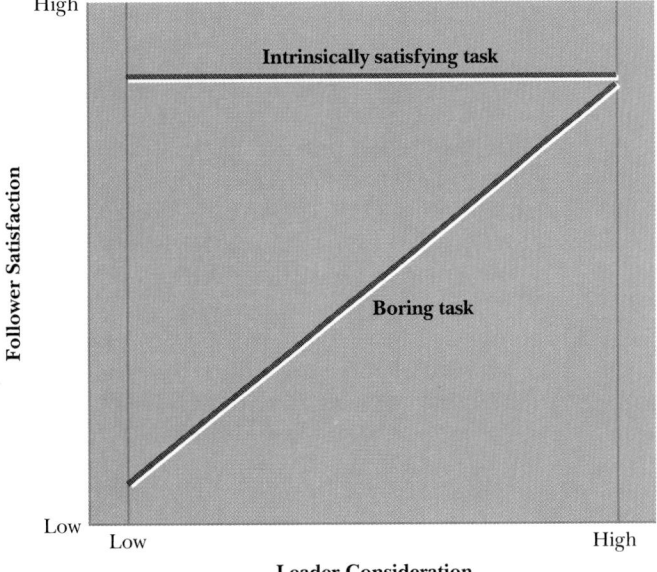

FIGURE 12-5

How a Situational Characteristic Can Substitute for Leader Behavior

By far the most rigorous and comprehensive study of the substitutes concept has been conducted by Philip Podsokoff and his colleagues at Indiana University.[32] With a sample of over 600 people from three large firms, Podsokoff explored the degree to which 13 different variables moderated the relationship between seven different kinds of leader behaviors and eight possible outcomes. The comprehensiveness of this study is reflected in the fact that it allowed for a test of 728 ($13 \times 7 \times 8$) possible substitute effects.

Only 44 of the possible 728 interactions were statistically significant, and only 19 of these were in the predicted direction. Inasmuch as these numbers are close to what one would expect to find by chance, the results of this study cast doubt on the substitutes for leadership approach.

The substitutes for leadership theory did solve some of the problems with the universal approaches to leadership by incorporating the three components of leader, follower, and situation. It also rejected the narrow focus and the one-best-way philosophy. However, the substitutes approach has a major shortcoming in its lack of strong theoretical links explaining when and why various aspects of leadership and characteristics of followers or situation interact. The theories discussed in the next two sections retain many of the virtues of the substitutes approach—its broader scope and notion of contingency—but tighten the theory relating the leader to the situation and follower. Moreover, as the "In Practice" box on pages 422–23 shows, nothing can substitute completely for a leader. Leader replacement and succession are often difficult issues to manage, but they are critical to firms in a state of transition.

INTERACTION THEORIES OF LEADERSHIP

Interaction theories of leadership see the leadership process as evolving out of an interaction between two dimensions. For example, the influence of one dimension, such as leader behavior, may be contingent upon the nature of a second dimension, such as the situation. Although certainly more complex than the universal theories, these later approaches still neglect one major component of the transactional framework.

The Leadership Motivation Pattern

The theory behind the leadership motivation pattern grew out of David McClelland's research on characteristics of the leader.[33] McClelland has proposed that leaders must either have a high need for achievement (see Chapter 7) or display what he has called the **leadership motivation pattern (LMP)**. The leadership motivation pattern is a composite of three specific characteristics: a high need for power, a low need for affiliation with others, and a high degree of self-control.

McClelland also argues that there are two types of leadership situations. The *entrepreneurial situation* is found in small organizations or in small technical units in large organizations where a few key people do most of the work them-

leadership motivation pattern (LMP) A composite behavior pattern composed of a high need for power, a low need for affiliation, and a high degree of self control, that predicts success in bureaucratic leadership situations.

[32] P. M. Podsokoff, B. P. Niehoff, S. B. McKenzie, and M. L. Williams, *Organizational Behavior and Human Decision Processes*, in press.

[33] David McClelland, *Power: The Inner Experience* (New York: Irvington, 1975).

selves. The *bureaucratic situation* is found in the context of large, formalized, tightly structured organizations.

McClelland suggests that need for achievement is critical to leaders in entrepreneurial situations and that the leadership motivation pattern is essential for success in bureaucratic situations. According to McClelland, people high in need for achievement are primarily interested in their own progress and much less interested in influencing and encouraging others. As a result, although the need for achievement is useful in small groups or technical groups where one person's progress readily spills over into group progress, it is not that critical for leadership success in large organizations.

In large, bureaucratic organizations, the three-characteristic configuration of the leadership motivation pattern is much more useful. A person who has a strong need for power also has an interest in influencing and controlling others, a prerequisite for leading a group of people. A low need for affiliation enables a leader to make difficult decisions without worrying excessively about being unpopular. Finally, high self-control makes it possible for a person to use his power to get things done within the organizational rules of the game.

How do these predictions work out in the real world? One study of 246 AT&T managers, hired between 1956 and 1960 and followed up 16 years later, tested McClelland's theory.[34] The study predicted that high LMP scorers would be successful in nontechnical areas, which were generally bureaucratic in organizations, but not in technical ones, which tended to be more entrepreneurial. The results, plotted in Figure 12-6, show that this was exactly what happened. Nontechnical managers who had high LMP scores had a 75 percent rate of promotions. However, managers in technical units whose LMP scores were high had only a 25 percent promotion rate. The key to success for the latter managers was need for achievement. In fact, the correlation between need for achievement and success was twice as high in technical areas as in nontechnical ones.

These data strongly suggest that the effect of a set of leader characteristics on leaders' success depends on the situation in which the leader is performing. Entrepreneurial situations call for leaders high in need for achievement, but bureaucratic situations call for a leader high in LMP.

[34] David C. McClelland and R. E. Boyatzis, "Leadership Motive Patterns and Long-Term Success in Management," *Journal of Applied Psychology* 67 (1982), pp. 737–43.

FIGURE 12-6

Relation between LMP Scores and Success among AT&T Managers

Source: David C. McClelland and R. E. Boyatzis, "Leadership Motive Pattern and Long-Term Success in Management," Journal of Applied Psychology 67 *(1982), pp. 737-43.*

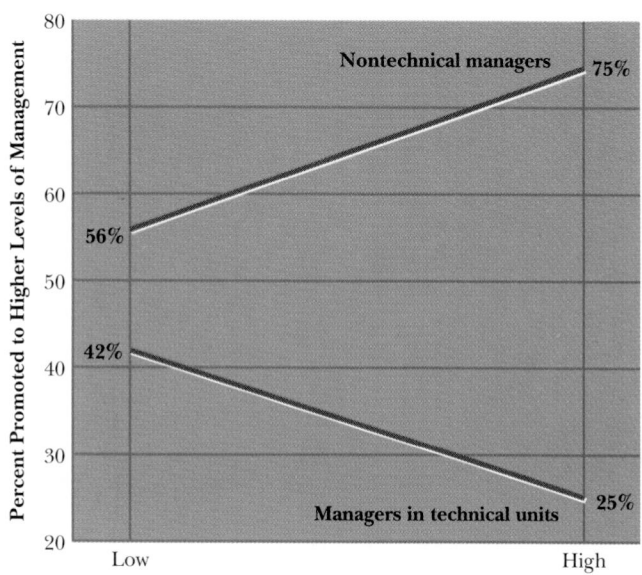

Vertical Dyad Linkage

vertical dyad Two persons who are related hierarchically, such as a supervisor-subordinate pair.

Another interaction approach to leadership examines the relations between the leader behavioral styles of consideration and initiating structure and certain follower characteristics. In George Graen's *vertical dyad linkage (VDL)* theory of leadership, a **vertical dyad** consists of two persons who are linked hierarchically, for example, a supervisor and a subordinate.[35] Most research based on the Ohio State studies measures leader consideration or initiating structure by averaging subordinates' ratings of leaders. VDL proponents, however, focus on the ratings of single followers. They argue that there is no such thing as an "average" leadership score. Instead, they insist, each supervisor-subordinate relationship is unique. A supervisor may be considerate toward one person but not another. Similarly, the leader may initiate structure for some workers but not others.

The importance of distinguishing dyadic from average scores has been supported by subsequent research. For example, Figure 12-7 compares correlations between (a) leader consideration and follower satisfaction and (b) leader initiating structure and follower role clarity as measured both by dyadic scores and average scores.[36] As you can see, the correlations obtained using dyadic scores were almost twice as high as those obtained with average scores. This suggests that leaders do behave differently with different subordinates and that these differences spill over into worker reactions.

The VDL approach also suggests that leaders tend to classify subordinates into in-group members and out-group members. According to Graen, *in-group members* are not only capable of doing more than the tasks outlined in a formal job description but are willing to do so. Once a leader has identified these people she generally gives them more and more latitude, authority, and consideration so that they become informal assistants. *Out-group members*, on the other hand, either cannot or will not expand their roles beyond formal requirements. Leaders assign these individuals more routine tasks, give them less consideration, and

[35] George Graen, "Role-Making Processes within Complex Organizations," in *Handbook of Industrial/ Organizational Psychology*, ed. M. D. Dunnette (Chicago: Rand McNally, 1976), pp. 1210–59.
[36] R. Katerberg and P. Hom, "Effects of Within-Group and Between-Groups Variation in Leadership," *Journal of Applied Psychology 66* (1981), 218–23.

FIGURE 12-7

Measuring the Relations between Leader Behaviors and Follower Outcomes by Dyadic Ratings and Average Group Ratings

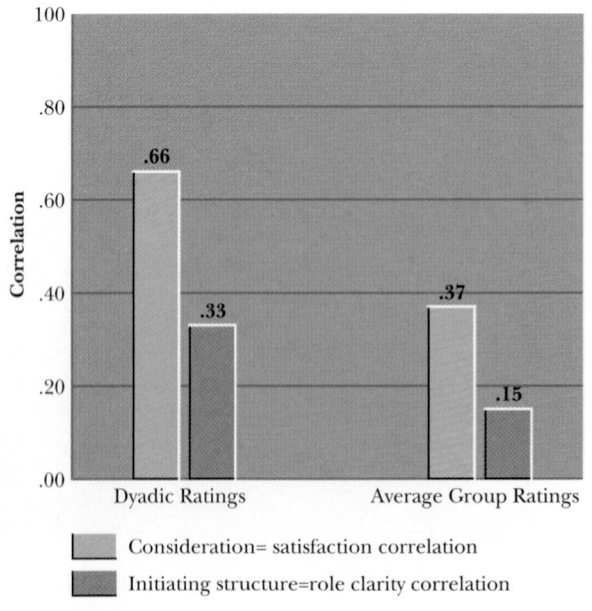

□ Consideration= satisfaction correlation
■ Initiating structure=role clarity correlation

communicate less often with them so that they become little more than hired hands. The Leader-Member Exchange Questionnaire, a self-report answered by subordinates, is often used to measure in-group versus out-group status.[37] Table 12-3 shows some items from this questionnaire.

Whether distinguishing among subordinates in this manner improves a leader's effectiveness depends on the leader's reasons for placing some people in the in group and others in the out group. Graen has suggested that a supervisor will "tend to select members who are compatible with him in terms of work competence and interpersonal skills . . . people he can trust to do the right thing."[38] This sounds like an effective leadership style. Graen has also noted, however, that "in some cases the selection probably is based upon the supervisor's prejudices concerning race, religion, or ethnic background."[39] Separating members into in groups and out groups based on information that is not related to performance could interfere with leader effectiveness. Highly competent and committed workers may differ from their supervisors but may excel if given in-group status and support. Unfortunately, the evidence suggests that in-group selections are often made capriciously. Demographic similarity and things like mutual interests outside of work have been found to predict in-group membership.[40]

Vertical dyad linkage theory suggests that as a potential leader you may find it worthwhile to behave in different ways with different followers. It also points out that you should base your distinctions among subordinates on their performance capabilities, not on the ways in which they are either similar to or different from you.

APPLYING THE DIAGNOSTIC MODEL
Diagnostic Question 8: How do you or your managers determine in-group vs. out-group status for group members? For example, do you base your selection on competence or on demographic similarity?
POINT TO STRESS
While in-groups and out-groups exist, how those groups are formed and how members are treated can have legal ramifications under the affirmative action laws.
TEACHING NOTE
Have students discuss issues regarding leaders/managers treating individuals differently and the laws requiring equal treatment. To what does equal treatment apply? What restrictions does equal treatment place on treating people differently?

[37] G. Graen, R. Liden, and W. Hoel, "Role of Leadership in the Employee Withdrawal Process," *Journal of Applied Psychology 67* (1982), 868–72.

[38] Graen, "Role-making Processes," p. 1242.

[39] Graen, et al., "Role of Leadership," p. 869.

[40] D. Duchon, S. G. Green, and T. D. Taber, "Vertical Dyad Linkage: A Longitudinal Assessment of Antecedents, Measures, and Consequences," *Journal of Applied Psychology 71* (1986), 56–60; A. S. Tsui and C. A. O'Reilly, "Beyond Simple Demographic Effects: The Importance of Relational Demography in Superior-Subordinate Dyads," *Academy of Management Journal 32* (1989), 402–23; and A. Crouch and P. Yetton, "Manager-Subordinate Dyads: Relationships Among Task and Social Contract, Manager Friendliness and Subordinate Performance in Management Groups," *Organizational Behavior and Human Decision Processes 41* (1988), 65–82.

T A B L E 12-3
Items That Assess Leader-Member Exchange

1. How flexible do you believe your supervisor is about evolving change in *your* job? 4 = Supervisor is enthused about change; 3 = Supervisor is lukewarm to change; 2 = Supervisor sees little need to change; 1 = Supervisor sees no need for change.

2. Regardless of how much formal organizational authority your supervisor has built into his/her position, what are the chances that he/she would be personally inclined to use his/her power to help you solve problems in your work? 4 = He certainly would; 3 = Probably would; 2 = Might or might not; 1 = No.

3. To what extent can *you* count on your supervisor to "bail you out," at his/her expense, when *you* really need him/her? 4 = Certainly would; 3 = Probably; 2 = Might or might not; 1 = No.

4. How often do you take suggestions regarding your work to your supervisor? 4 = Almost always; 3 = Usually; 2 = Seldom; 1 = Never.

5. How would *you* characterize *your* working relationship with your supervisor? 4 = Extremely effective; 3 = Better than average; 2 = About average; 1 = Less than average.

The five items are summed for each participant, resulting in a possible range of scores from 5 to 20.

Life Cycle Model

According to the *life cycle model* of Paul Hersey and Kenneth Blanchard, the effectiveness of a leader's decision style depends very largely on his followers' level of maturity, or their job experience and emotional maturity.[41] The life cycle model proposes two basic decision styles: *task orientation* and *relationship orientation*. As you might guess, these concepts were derived from the initiating structure and consideration dimensions of the early Ohio State studies.

The life cycle model suggests that there are four types of decision styles: telling, selling, participating, and delegating. In the *telling style*, which is characterized by high task orientation and low relationship orientation, the leader simply tells the follower what to do. In the *selling style*, characterized by both high task and high relationship orientations, the leader tries to sell his ideas to subordinates, to convince them that her decision is appropriate. The *participating style* is marked by a high relationship orientation but a low task orientation. The leader who uses this style of decision making includes subordinates in discussions so that decisions are made by consensus. Finally, in the *delegating style*, which is low on both task and relationship orientations, the leader actually turns things over to followers and lets them make their own decisions.

Like all interaction theories, the life cycle model proposes that there is more than one way to lead. The type of decision style that a leader should adopt depends on the level of maturity of his followers. By maturity is meant both job maturity,

[41] Paul Hersey and Kenneth Blanchard, *Management of Organizational Behavior* (Englewood Cliffs, NJ: Prentice Hall, 1977), p. 11–35.

FIGURE 12-8

The Life Cycle Model of Leadership in Four Dimensions

Three of this model's four dimensions are easily seen. *Relationship orientation* may be low (bottom half of the rectangular box model) or high (top half). *Task orientation* may also be low (left half of the model) or high (right half). *Follower maturity* ranges from very low (front of the model) to very high (back). The fourth dimension, *leader effectiveness*, is represented by the highlighted cell at each follower maturity level. For example, at the high maturity level, the highlighting of the cell for the participating leader style—which is high on relationship orientation and low on task orientation—indicates that at this level this style is the most effective.

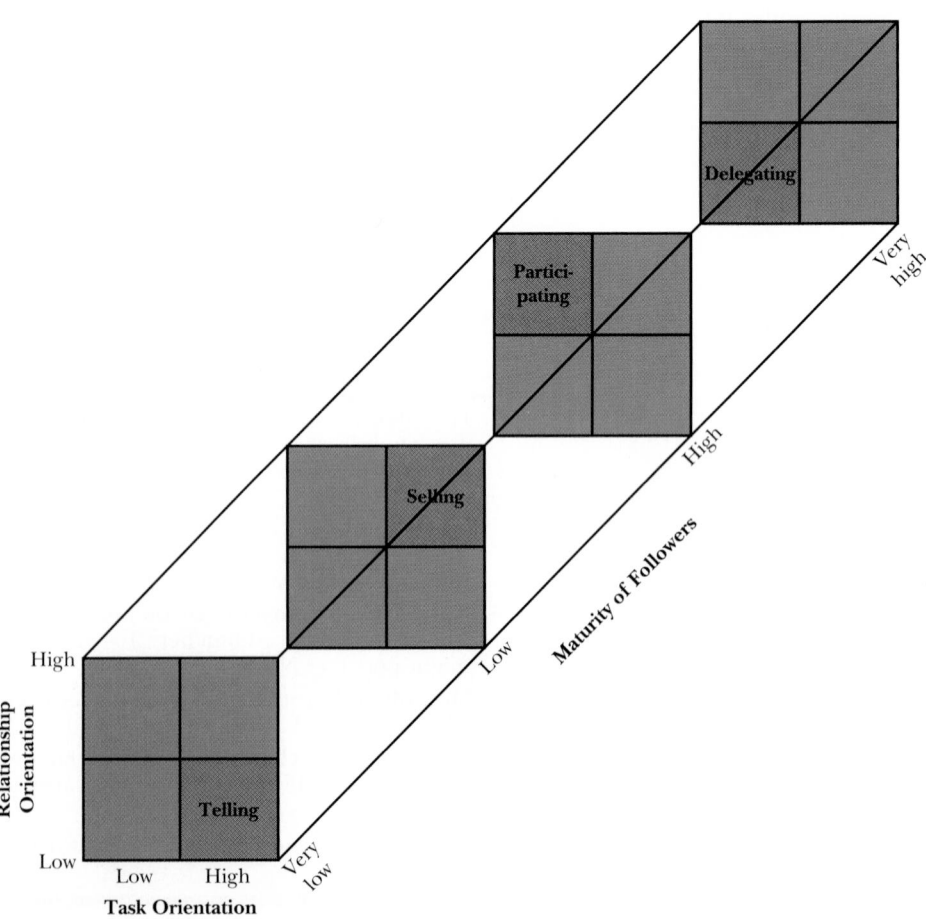

Diagnostic Question 9: What kind
of decision-styles (autocratic,
participating, delegating) do you
or your managers typically
employ?

Diagnostic Question 10: How
might the typical decision-style
that you or your managers employ
need to be tailored to your
followers—in terms, for example,
of their maturity or their
commitment to goals—or to the
situation?

or task experience and skill, and psychological maturity, or feelings of self-worth and self-acceptance. As Graeff has correctly pointed out, the life cycle model has four dimensions: task orientation, relationship orientation, maturity, and effectiveness.[42] Figure 12-8 illustrates the four dimensions of this model.

As you can see, this model suggests that for followers at very low levels of maturity, telling is the most effective leadership decision style. As followers move from very low to moderately low levels of maturity, a selling style becomes more effective. When followers show a moderately high level of maturity, participating is the most effective style, and at the very highest levels of follower maturity, the delegating style leaves followers essentially on their own.

Like the managerial grid, the life cycle approach has been well received by practitioners because of its intuitive appeal. However, empirical research has not supported it. The most rigorous study of the approach to date suggests two conclusions, neither of which is encouraging. First, *satisfaction* with supervision is no higher in situations where leader style matches follower maturity level in the way the model prescribes than in situations where there is a mismatch. Second, the notion that *performance* will be higher in matched situations is supported at only one level of maturity, the lowest.[43] Thus, the most we can say at this point is that with workers at low levels of maturity, the telling style is slightly more effective in eliciting good performance than the other styles.

COMPREHENSIVE THEORIES OF LEADERSHIP

Comprehensive leadership theories are interaction theories, like the ones we have just explored. Unlike the latter, however, comprehensive theories incorporate all three of the elements of the transactional approach to leadership—leader, follower, and situation. The three theories that we will examine next differ only in that each tends to focus on a particular leader characteristic—either a personal characteristic, a behavioral orientation, or a decision style.

Fiedler's Contingency Model

Think for a moment of someone with whom you have really had difficulty working. In fact, try to choose from all the people you have ever worked with the one you worked with least well. Now rate this person on the qualities listed in the *Least Preferred Coworker Scale* shown in Table 12-4. Put a checkmark in the blank that best represents your judgment of the person you have in mind. For example, 1 and 8 suggest *extremely* unpleasant or pleasant, 2 and 7 *quite* unpleasant or pleasant, 3 and 6 *moderately* unpleasant or pleasant, and 4 and 5 *somewhat* unpleasant or pleasant. Then total your answers to get a final score.

A low score on this scale (between 16 and 64) will indicate that you described your *least preferred coworker* in relatively harsh terms and that you would take a *task orientation* to leadership. Task oriented leaders, according to Fred Fiedler, the author of this theory, emphasize completing tasks successfully, even at the

---

[42] G. L. Graeff, "The Situational Leadership Theory: A Critical Review," *Academy of Management Review* 7 (1983), 285–91.

[43] R. P. Vecchio, "Situational Leadership Theory: An Examination of a Prescriptive Theory," *Journal of Applied Psychology* 72 (1988), 444–51.

T A B L E 12-4
Items from the Least Preferred Coworker Scale

Pleasant	8	7	6	5	4	3	2	1	Unpleasant
Friendly	8	7	6	5	4	3	2	1	Unfriendly
Rejecting	1	2	3	4	5	6	7	8	Accepting
Helpful	8	7	6	5	4	3	2	1	Frustrating
Unenthusiastic	1	2	3	4	5	6	7	8	Enthusiastic
Tense	1	2	3	4	5	6	7	8	Relaxed
Distant	1	2	3	4	5	6	7	8	Close
Cold	1	2	3	4	5	6	7	8	Warm
Cooperative	8	7	6	5	4	3	2	1	Uncooperative
Supportive	8	7	6	5	4	3	2	1	Hostile
Boring	1	2	3	4	5	6	7	8	Interesting
Quarrelsome	1	2	3	4	5	6	7	8	Harmonious
Self-assured	8	7	6	5	4	3	2	1	Hesitant
Efficient	8	7	6	5	4	3	2	1	Inefficient
Gloomy	1	2	3	4	5	6	7	8	Cheerful
Open	8	7	6	5	4	3	2	1	Guarded

Note: LPC score is the sum of the answers to these 16 questions. High scores indicate a relationship orientation; low scores, a task orientation.
Source: From *Leadership and effective management* by Fred E. Fiedler and Martin M. Chomers. Copyright © 1974 by Scott, Foresman & Co. Reprinted by permission.

expense of interpersonal relations. A low score on the scale reflects a leader's inability to overlook the negative traits of a poorly performing subordinate.

On the other hand, a high score (between 80 and 128) will indicate that you described your least preferred coworker in relatively positive terms and that you would take a *relationship orientation* to leadership. Relationship-oriented leaders, according to Fiedler, are permissive, considerate leaders who can maintain good interpersonal relationships even with workers who are not contributing to group accomplishment. The leader's orientation toward either tasks or relationships is the central piece in the complex and controversial theory of leadership that Fiedler has proposed.

Fiedler's model is called a contingency theory of leadership because it holds that the effectiveness of a leader's orientation depends both on the leader's followers and on the situation in which she is functioning. A leadership context can be placed on a continuum of favorability, in which the interaction among three factors defines eight positions of varying favorability (see Figure 12-9). The three factors are leader-follower relations, task structure, and leader position power.

Leader-follower relations are good if followers trust and respect the leader and poor if they don't. Good relations are more favorable for leader effectiveness than poor relations. **Leader task structure** is high when a group has clear goals and clear means for achieving these goals. High task structure is more favorable

leader-follower relations A component of Fiedler's contingency theory that describes the level of trust and respect between leader and follower.

leader task structure A component of Fiedler's contingency theory that describes the clarity of goals and of means—end relationships in a group's task.

FIGURE 12-9

How Situation Favorability Is Determined by Leader-Follower Relations, Leader Task Structure, and Leader Position Power

Source: Adapted from Fred E. Fiedler. A Theory of Leadership Effectiveness (New York: McGraw-Hill, 1967). Reprinted by permission.

Leader-follower relations	Good				Poor			
Leader task structure	High		Low		High		Low	
Leader position power	Strong	Weak	Strong	Weak	Strong	Weak	Strong	Weak
Situations	I	II	III	IV	V	VI	VII	VIII

Very favorable ←——————→ Very unfavorable

---

leader position power A component of Fiedler's contingency theory that describes the degree to which the leader can administer significant rewards and punishments to followers.

for the leader than low task structure. Finally, **leader position power** is the ability to reward or punish subordinates for their behavior. Clearly, the more power a leader has, the more favorable the situation is from the leader's perspective.

As examples, Fiedler has suggested that the respected leader of a bomber crew might fit into situation I in Figure 12-9 (highly favorable); the disliked chairman of a volunteer committee asked to plan an office picnic on a nice Sunday afternoon might fit into situation VIII (very unfavorable); and the elected director of a food cooperative might fit into situation IV (moderately favorable).[44]

Fiedler's analysis of a number of studies that used the Least Preferred Coworker Scale suggested to him that task-oriented leaders are most effective in situations that are either extremely favorable or extremely unfavorable (I, II, VII, and VIII in Figure 12-9). Relationship-oriented leaders, he found, were most successful in situations of moderate favorability (III, IV, V, and VI in Figure 12-9).

Evidence for Validity. Over the last 25 years this theory has aroused considerable interest. One review described 24 studies that tested the model's predictions directly.[45] Figure 12-10 shows the empirical support for the model. For each of the eight situations described by the model, average correlations and the range of correlations between a leader's LPC score and his effectiveness are plotted.

These data suggest that the model works quite well at the extremes. For example, both the average correlation and the range of correlations between an LPC score and leader effectiveness are negative in situations I and VIII, as the model predicts. Similarly, in situations IV and V, the average correlation and the range of correlations are both positive, again as the model predicts. However, LPC scores fail to predict leader effectiveness in situations II, III, VI, and VII.

Leader Match Training. One offshoot of this theory is the leader training program called LEADER MATCH, which attempts to translate OB theory into managerial practice. Fiedler has commented that it is easier to change almost anything in the job situation than personality and leadership style, and this belief in the immutability of human beings is reflected in LEADER MATCH training.[46] This self-paced, programmed text tells leaders not to change their styles but instead to try to manipulate a situation. For example, if leader-follower relations are poor, the leader may try to raise morale by giving bonuses or time off. If a task is unstructured, a leader may break it down into simpler subtasks. If position

TEACHING NOTE
Given the earlier discussion which argued leaders are often constrained by the situation, what hope does a leader have to effectively change a situation to match his style? Would it be better to match leaders to the situation?

---

[44] Fred E. Fiedler, *A Theory of Leadership Effectiveness* (New York: McGraw-Hill, 1967), pp. 120–37.

[45] L. H. Peters, D. D. Hartke, J. T. Pohlmann, "Fiedler's Contingency Theory of Leadership: An Application of the Meta-analysis Procedures of Schmidt and Hunter," *Psychological Bulletin 97* (1985), 274–85.

[46] Fred E. Fiedler, "Engineering the Job to Fit the Manager," *Harvard Business Review 43* (1965) 115–22.

FIGURE 12-10

Evidence for Fiedler's Contingency Theory of Leadership

In Situations I-III and VIII, the correlations between Least Preferred Coworker (LPC) Scale scores and effectiveness are negative, as the theory predicts. In Situations IV-VII, again as the theory predicts, the correlations between LPC scores and effectiveness are positive.

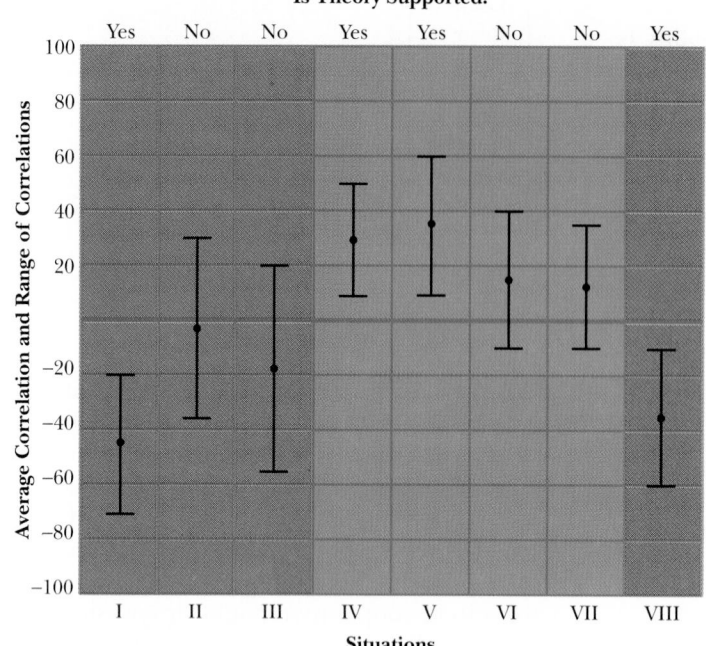

APPLYING THE DIAGNOSTIC MODEL

Diagnostic Question 12: How favorable is the situation for the leader in terms of task structure, position power, and leader-member relations? What can be done to make this situation more favorable?

power is low, a leader may try to increase her authority by seeing that all information is channeled through her.

Critique of Fiedler's Model. Both contingency theory and LEADER MATCH training have been subject to considerable criticism. The theory has been criticized as "too data driven." According to his critics, Fiedler started with a set of results that he tried to explain, rather than with a logical, deductive theory. Moreover, there is continuing controversy over why low LPC leaders should be best in situations that are either extremely good or extremely bad but not in situations of moderate favorability. It is somewhat disconcerting that a theory with a history as long as Fiedler's has not yet been able to specify why its predictions turn out to be correct. The LPC measure itself has aroused controversy. Critics have questioned what the scale actually measures and how well it measures this variable.[47]

LEADER MATCH training has been criticized for using questionable measures of performance and for failing to control for rater expectation biases (see Chapter 5) and the so-called Hawthorne effect (see Chapter 2).[48] In addition, some people have noted that the classifications of situations in the LEADER MATCH booklet are not what the original theory would predict.[49] Indeed, critics have suggested that mismatches could total 60 percent.[50] Finally, as Figure 12-10 shows, current evidence suggests for as many as half of the situations that

[47] A. K. Korman, "Contingency Approaches to Leadership: An Overview," in *Contingency Approaches to Leadership*, ed. J. G. Hunt and L. L. Larson (Carbondale: Southern Illinois Press, 1974), p. 24; and C. A. Schriesheim, B. D. Bannister, and W. H. Money, "Psychometric Properties of the LPC Scale: An Extension of Rice's Review," *Academy of Management Review* 4 (1979), 287–90.

[48] B. Kabanoff, "A Critique of LEADER MATCH and Its Implications for Leadership Research," *Personnel Psychology* 34 (1981),749–64.

[49] A. G. Jago and J. W. Ragan, "The Trouble with LEADER MATCH Is That It Doesn't Match Fiedler's Contingency Model," *Journal of Applied Psychology* 71 (1986), 555–59.

[50] A. G. Jago and J. W. Ragan, "Some Assumptions Are More Troubling Than Others: Rejoinder to Chemers and Fiedler," *Journal of Applied Psychology* 71 (1986), 564–65.

might be encountered, LPC scores show no consistent relationship with performance.[51]

Despite all these problems, however, there certainly seem to be some situations (I, II, VII, and VIII) where the theory works quite well. In particular, assigning task oriented, low LPC leaders to groups in situations of extreme favorability or unfavorability seems to be a defendable prescription. Moreover, Fiedler's contribution and that of his followers cannot be denied. They have proposed the first well developed leadership theory that includes all three aspects of the transactional model.

Vroom-Yetton Decision Tree Model

The first comprehensive model we looked at focused on personality characteristics of the leader. Now we will examine a model that centers on leader decision styles. The *decision tree model of leadership* originated by Victor Vroom and his colleagues emphasizes the fact that leaders achieve success through effective decision making. The model describes effective decisions as of *high quality, well-accepted* by followers, and made in a *timely* fashion.[52] According to this theory, leaders whose decisions do not meet all three criteria will ultimately fail.

Vroom recognizes four general classes of leadership style: *authoritarian, consultative, delegation,* and *group based.* He then breaks these down into seven specific decision styles. Three of these are appropriate to all decisions, two are appropriate only to decisions regarding individual followers, and two are appropriate only to decisions regarding an entire group of followers (see Table 12-5). We will focus our attention on the processes that involve groups.

Like all comprehensive theories of leadership, the decision-tree model proposes that the most effective leadership style depends on characteristics of both the situation and the followers. Specifically, the model asks eight questions—three about the situation and five about the followers—in order to determine which of the seven leadership styles outlined in Table 12-5 is best. The decision tree presented in Figure 12-11 on page 435 makes the question-and-answer process easy.[53] Responding to questions A through H will lead you to one of 18 answers, each of which identifies one or more decision styles that are appropriate to the problem you confront. To choose among two or more styles, the leader must decide whether she wishes to maximize the speed of decision making or the personal development of subordinates. Autocratic approaches favor speed, whereas consultative or group approaches favor employee growth.

Using the Decision Tree. The model may seem a little intimidating at first, but going through a few simple examples will show you that it is actually quite easy to follow. Suppose you are a camp counselor trying to get a dozen 12-year-olds to march two miles to a campsite. You have to decide between two routes. The marked path is four miles long, but if you go off the path—into the woods, over a large hill, and through rough terrain—the trek will be only one mile. How can you decide on a route? Let's try using the decision tree model in Figure 12-11.

[51] M. M. Chemers and F. E. Fiedler, "The Trouble with Assumptions: A Reply to Jago and Ragan," *Journal of Applied Psychology* 71 (1986), 560–663.

[52] Victor H. Vroom, "Leadership," in *Handbook of Industrial/Organizational Psychology*, ed. M.D. Dunnette (Chicago: Rand-McNally, 1976), p. 912.

[53] V. H. Vroom and A. G. Jago, "Decision Making as a Social Process: Normative and Descriptive Models of Leader Behavior," *Decision Sciences* 5 (1974), 743–69.

TABLE 12-5
The Seven Decision Styles in the Vroom-Yetton Decision-Tree Model of Leadership

For All Problems

AI   You solve the problem or make the decision yourself, using information available to you at the time.

AII   You obtain any necessary information from subordinates, then decide on the solution to the problem yourself. You may or may not tell subordinates what the problem is, in getting the information from them. The role played by your subordinates in making the decision is clearly one of providing specific information which you request, rather than generating or evaluating solutions.

CI   You share the problem with the relevant subordinates individually, getting their ideas and suggestions without bringing them together as a group. Then *you* make the decision. This decision may or may not reflect your subordinates' influence.

For Individual Problems

GI   You share the problem with one of your subordinates and together you analyze the problem and arrive at a mutually satisfactory solution in an atmosphere of free and open exchange of information and ideas. You both contribute to the resolution of the problem with the relative contribution of each being dependent on knowledge rather than formal authority.

DI   You delegate the problem to one of your subordinates, providing him with any relevant information that you possess, but giving him responsibility for solving the problem by himself. Any solution which the person reaches will receive your support.

For Group Problems

CII   You share the problem with your subordinates in a group meeting. In this meeting you obtain their ideas and suggestions. Then, *you* make the decision which may or may not reflect your subordinates' influence.

GII   You share the problem with your subordinates as a group. Together you generate and evaluate alternatives and attempt to reach agreement (consensus) on a solution. Your role is much like that of chairman, coordinating the discussion, keeping it focused on the problem, and making sure that the critical issues are discussed. You do not try to influence the group to adopt "your" solution and are willing to accept and implement any solution which has the support of the entire group.

Note: A stands for *authoritarian*, C for *consultative*, D for *delegation*, and G for *group-based* decision styles.

Question A: Yes. If the children can follow the unmarked route, it is the shorter way.

Question B: No. You don't know the children's climbing skills.

Question C: Yes. What needs to be done going either way is clear, that is, you need to get the children home safely.

Question D: Yes. You must deliver all twelve children to the campsite, so refusal by even one is unacceptable.

Question E: Yes. You are older, more experienced, and the children respect you.

Question F: No. They would all be happy to stop and camp where they are.

You have now arrived at answer number 10. According to this model, a wholly group-based decision such as GII in Table 12-5 is out of the question. If it is getting dark and time is of the essence, the autocratic style of AII will be most appropriate. If it is early in the morning, however, and you want the children to learn from the experience, a more consultative style, like CII, may be appropriate.

Here's another example. A corporate vice president has just been given the responsibility for starting up a new plant in a developing country, and she must choose a plant manager. Should it be one of her five current and highly experienced plant managers? Should it be someone from outside the firm who has had experience working overseas? Should it be a citizen of the target country?

A. Is there a quality requirement such that one solution is likely to be more rational than another? (Situation)
B. Do I have sufficient information to make a high quality decision? (Situation)
C. Is the problem structured? (Situation)
D. Is acceptance of decision by subordinates critical to effective implementation? (Followers)
E. If I were to make the decision by myself, is it reasonably certain that it would be accepted by my subordinates? (Followers)
F. Do subordinates share the organizational goals to be attained in solving this problem? (Followers)
G. Is conflict among subordinates likely in preferred solutions? (This question is irrelevant to individual problems.) (Followers)
H. Do subordinates have sufficient information to make a high quality decision? (Followers)

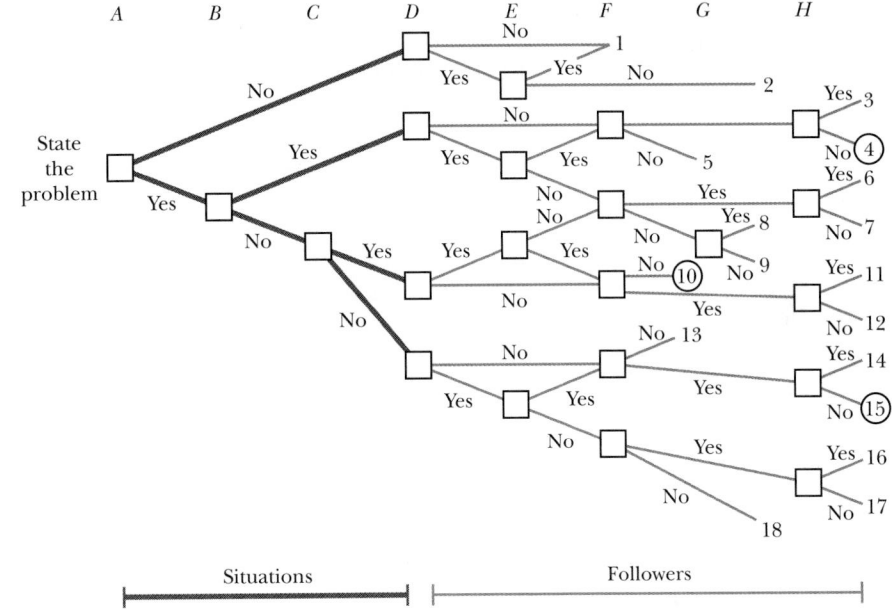

Answers and Appropriate Leadership Styles (see also Table 12-5)

Answer Number	Individual Problems	Group Problems	Answer Number	Individual Problems	Group Problems
1	AI, AII, CI, DI, GI	AI, AII, CI, CII, GII	10	AII, CI	AII, CI, CII
2	DI, GI	GII	11	AII, CI, DI, GI	AII, CI, CII, GII
3	AI, AII, CI, DI, GI	AI, AII, CI, GII, GII	12	AII, CI, GI	AII, CI, CII, GII
4	AI, AII, CI, GI	AI, AII, CI, CII, GII	13	CI	CII
5	AI, AII, CI	AI, AII, CI, CII	14	DI, CI, GI	CII, GII
6	DI, GI	GII	15	CI, GI	CII, GII
7	GI	GII	16	DI, GI	GII
8	CI, GI	CII	17	GI	GII
9	CI, GI	CI, CII	18	CI, GI	CII

The vice president might move through the decision tree as follows:

Question A: Yes. Some managers may be better suited than others.
Question B: No. The vice president may not know all the details of the candidates' personal lives, interests, or past experience that would be relevant to the assignment.
Question C: No. This is a new problem for the company, and thus there are no clear guidelines dictating what steps to take.
Question D: Yes. Any one of the vice president's current managers could find good jobs with other firms in their own country if they refused the overseas job.

Question E: No. The decision will have too large an impact on subordinates' lives.

Question F: Yes. They have been with the company a long time and are committed to the organization.

Question H: No. There are many details of the assignment that only the vice president knows.

The no response to question H leads to answer number 17. This answer, applied to a group problem, eliminates both autocratic and consultative styles, and recommends the GII, group-based decision style.

Evidence for Validity. Early studies of the model's usefulness asked managers to think about past decisions that were effective or ineffective and had them trace their decision processes back to see if they had followed the model's prescriptions. When the managers' decision processes were consistent with the model, 68 percent of decisions were effective, compared to only 22 percent when decisions violated the model.[54] Support for the model has also been obtained in studies that did not rely strictly on retrospective self-reports. In addition, when outside observers have been asked to rate decisions arrived at by groups following the model's rules, they have tended to rate these decisions as more effective than those taken by groups that violated the model's rules.[55]

Research also indicates that most managers' natural decision processes seem to violate the model's prescriptions. In particular, in both recalled and standardized problem sets, managers tend to overuse the group-based style CII and to underutilize the autocratic AI and group-based GII styles.[56] In addition, sex, experience, and organizational level may affect choice of style. Women tend to be more participative than men, students are more participative than experienced executives, and upper-level executives seem to be more participative than lower-level executives.[57]

Recent Extensions of the Model. Recently the decision tree model has been extended in several ways. First, it has become clear that the model should consider other leader characteristics besides decision style. One study, for example, has shown that using a group-based decision style when there is likely to be conflict among subordinates works only for managers who are skilled in conflict management. The study concludes that leaders who have trouble managing conflict should ignore model rules that dictate a group solution because they may not be able to manage the ensuing debate.[58]

The model has also been extended to include answers to questions that do not present simple yes or no choices. That is, the new model allows answers that are probability estimates or ratings on scales that include several numbers. This added complexity has been embedded in a computer software program that guides the manager through the decision-making process.[59] Evidence on the validity or

---

[54] V. H. Vroom and P. W. Yetton, *Leadership and Decision Making* (University of Pittsburgh Press, 1973), p. 12.

[55] R. H. Field, "A Test of the Vroom-Yetton Normative Model of Leadership," *Journal of Applied Psychology 67* (1982), 523–32.

[56] Vroom and Yetton, *Leadership*, p. 13.

[57] R. M. Steers, "Individual Differences in Participative Decision Making," *Human Relations 30* (1977), 837–47; A. G. Jago and V. H. Vroom, "Predicting Leader Behavior from a Measure of Behavioral Intent," *Academy of Management Journal 21* (1978), 715–21; and A. G. Jago, "Hierarchical Level Determinants of Participative Leader Behavior" (Ph.D. dissertation, Yale University, 1977).

[58] A. Crouch and P. Yetton, "Manager Behavior, Leadership Style, and Subordinate Performance: An Empirical Extension of the Vroom-Yetton Conflict Rule," *Organizational Behavior and Human Decision Processes 39* (1987), 384–96.

[59] V. H. Vroom and A. G. Jago, *The New Leadership: Cases and Manuals for Use in Leadership Training* (New Haven, Conn.: 1987). Authors retain all rights to decision trees, cases, and computer software.

batting average of the new model is unavailable, and many managers may be unwilling to turn their decision making over to a computer. As a result, it is hard to draw any conclusion about these extensions of the decision tree model.

Path Goal Theory

By far the most comprehensive theory of leadership to date and the one that best exemplifies all the aspects of the transactional model, is the *path-goal theory of leadership* originated by Martin Evans and Robert House.[60] At the core of this theory is the notion that the primary purpose of the leader is to motivate followers. Because they saw motivation as essential to the leader role, Evans and House built their theory on a base of expectancy theory. You will remember that we looked closely at expectancy theory in Chapter 7. Figure 12-12 is a condensed version of our treatment of that theory that emphasizes (a) followers' performance and satisfaction of followers as the primary outcomes of interest and (b) the five motivational variables that leaders may be able to influence through their behaviors or decision styles: valences, instrumentalities, expectancies, accuracy of role perceptions, and equity of rewards.

Manipulating Motivation. The job of the leader, according to path-goal theory, is to manipulate these five factors in desirable ways. The theory's proponents recommend that leaders

> *Manipulate follower valences* by recognizing or arousing needs for outcomes that the leader can control
> *Manipulate follower instrumentalities* by ensuring that high performance results in satisfying outcomes for followers
> *Manipulate follower expectancies* by reducing frustrating barriers to performance
> *Manipulate the accuracy of role perceptions* by making the paths to effective performance clear through coaching and direction
> *Manipulate equity of rewards* by increasing the amount and types of rewards available when followers perform well.[61]

[60] Martin G. Evans, "The Effect of Supervisory Behavior on the Path-Goal Relationship," *Organizational Behavior and Human Decision Processes 5* (1970), 277–98; and R. J. House, "A Path-Goal Theory of Leadership Effectiveness," *Administrative Science Quarterly 16* (1971), 321–38.

[61] R. J. House and T. R. Mitchell, "Path-Goal Theory of Leadership," *Journal of Contemporary Business 3* (1974), 81–97.

APPLYING THE DIAGNOSTIC MODEL
Diagnostic Question 2: Which of the five components of follower motivation (i.e., valence, instrumentality, expectancy, accurancy of role perceptions, or rewards) seems most lacking in your setting?

POINT TO STRESS
The complexity of these manipulations is apparent when recalling the motivation model. For example, each of the followers will place a different valance on outcomes. In addition, valences are not stable, but change over time. Thus, to manipulate valences, the leader must first know each individual's need level for the outcome and must monitor changes in need levels.

FIGURE 12-12

A Condensed Version of Our Model of Motivation and Performance

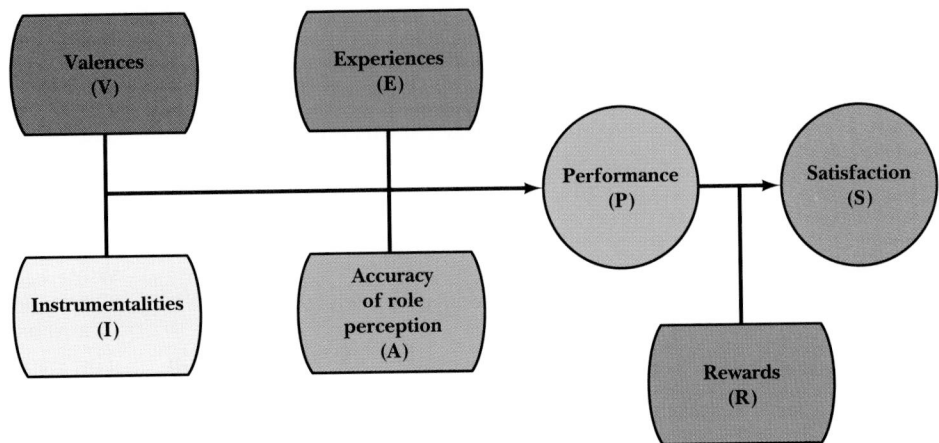

Behavioral Styles. Path-goal theory proposes four behavioral styles that can enable leaders to manipulate these five expectancy theory variables: directive, supportive, participative, and achievement-oriented leadership. As you can see from Table 12-6, these styles are composed of both behaviors, like initiating structure, and decision styles, like the authoritarian approach.

Note that the leader's effectiveness in each case will depend on follower and situation characteristics that may also affect these components. Much like the substitutes for leadership approach, path-goal theory recognizes that follower or situational characteristics may make leader behavior unnecessary or impossible.[62]

The full complexity of the path-goal model can be appreciated by looking at Figure 12-13, which combines Table 12-6 and Figures 12-1 and 12-12 to show how each one of the five aspects of motivation can be influenced by leader behavior styles, characteristics of the followers, and characteristics of the situation, as well as by interactions among variable across categories.

Evidence for Validity. The sheer complexity of the model makes a single comprehensive test of it impossible. Instead, researchers have tested small parts of it. Some of their findings are as follows:

> *Leader* participative behavior results in satisfaction in *situations* where the task is nonroutine but only for *followers* who are nonauthoritarian.[63]
>
> *Leader* directive behavior produces high satisfaction and high performance but only among *followers* who have high needs for clarity.[64]
>
> *Leader* supportive behavior results in follower satisfaction but only in *situations* where the task is highly structured.[65]
>
> *Leader* achievement-oriented behavior results in improved performance but only when *followers* are committed to goals.[66]

[62] A. C. Filley, R. J. House, and S. Kerr, *Managerial Processes and Organizational Behavior* (Glenview, Ill.: Scott, Foresman, 1976), p. 91.

[63] R. T. Keller, "A Test of the Path-Goal Theory of Leadership with Need for Clarity as a Moderator in Research and Development Organizations," *Journal of Applied Psychology 74* (1989), 208–12.

[64] Ibid.

[65] J. E. Stinson and T. W. Johnson, "A Path-Goal Theory of Leadership: A Partial Test and Suggested Refinements," *Academy of Management Journal 18* (1975), 242–52.

[66] M. Erez and I. Zidon, "Effect of Goal Acceptance on the Relationship between Goal Difficulty and Performance," *Journal of Applied Psychology 69* (1984), 69–78.

T A B L E 12-6
Path-Goal Theory's Four Behavioral Styles

Directive leadership	The leader is authoritarian. Subordinates know exactly what is expected of them, and the leader gives specific directions. Subordinates do not participate in decision making.
Supportive leadership	The leader is friendly and approachable and shows a genuine concern for subordinates.
Participative leadership	The leader asks for and uses suggestions from subordinates but still makes the decisions.
Achievement-oriented leadership	The leader sets challenging goals for subordinates and shows confidence that they will attain these goals.

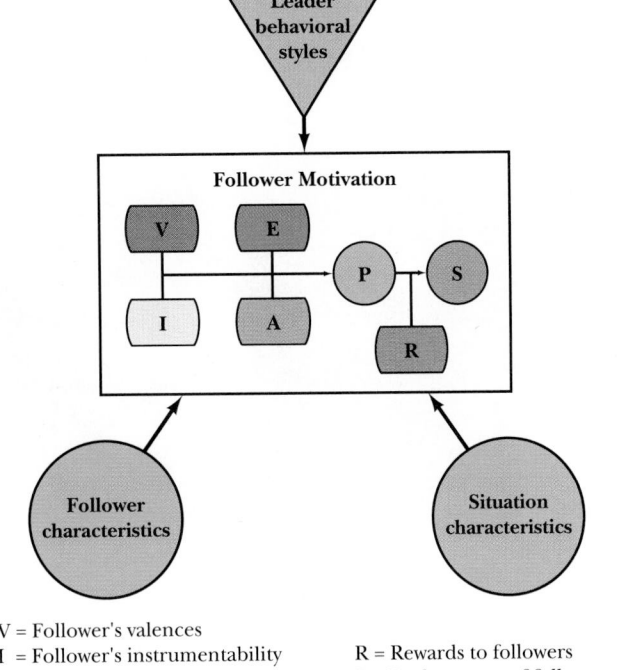

V = Follower's valences
I = Follower's instrumentability
E = Follower's expectancy
A = Accuracy of role perceptions

R = Rewards to followers
P = Performance of followers
S = Satisfaction of followers

Perhaps because the theory is so complex, no one has yet mounted a comprehensive study of path goal theory in which every variable is tested, like the Podsokoff study of substitutes for leadership theory. The theoretical framework provided by path-goal theory, however, is an excellent one for generating, testing, and understanding the complexities of the leadership process. Moreover, its tie to the expectancy theory of motivation makes it particularly suitable for leadership as conceptualized by Mintzberg, that is, the leader as a group motivator.

NEO-UNIVERSAL THEORIES

Although the increasing complexity of leadership theories presents an exciting challenge to theorists and researchers, it has left many people dissatisfied with what appears to be a lack of progress. In addition, it has created something of a backlash. Investigators have begun to propose what we may call *neo-universal theories of leadership* that, like some of the universal models, focus on a particular characteristic of the leader and exclude followers and situations.

Charismatic Leadership

charismatic leadership
Creating a new vision of an organization and getting group members to commit themselves enthusiastically to the new mission, structure, and culture embodied in the vision. Encouraging members to transcend self-interests on behalf of the organization as a whole.

For example, theories of **charismatic leadership** emphasize the ability of the leader to communicate new visions of an organization to followers.[67] Charismatic leaders or, as they are sometimes called, transformational leaders, raise followers' awareness of the importance and value of group goals, often getting people to

[67] J. M. Burns, *Leadership* (New York: Harper & Row, 1978), p. 52.

CEO Wayne Calloway had a vision that Pepsico could become the best consumer products company in the world. He envisioned a day when pizza, tacos, and chicken would be as convenient and easy to get as a bag of potato chips and set about changing the way the company's Pizza Hut, Taco Bell, and KFC (formerly Kentucky Fried Chicken) offered their services to the public. The company's delivery and carryout services are earning twice what the eat-in restaurants do, and already you can buy Pizza Hut's pizza from vendors at football stadiums and basketball arenas, in school cafeterias, and at airport shops.
Source: Patricia Sellers, "Pepsi Keeps on Going After No. 1," Fortune, March 11, 1991, pp. 62–70.

AN EXAMPLE
President John F. Kennedy is often characterized as a charismatic leader. In the early 1960s he provided a vision for the nation, a direction for all to move. This vision was one of vibrance and advancement. We would be the best educated and most innovative country in the world. We would be explorers, putting a man on the moon by the end of the decade. This dynamic, youthful, goal-oriented vision captured the country.

transcend their own interests. Charismatic leaders "raise the stakes" of organizational performance by convincing subordinates of the importance of the leader's vision. It is vision that distinguishes top performing managers from more ordinary managers.[68]

The demand for leaders with vision has never been higher. Stephen Garrison, chairman of Ward Howell International, an executive search firm, said in an interview, "In a third or a half of our searches these days, our client companies specify that they want a high level of vision."[69] Vision has become increasingly necessary since most businesses have finished (or are about to finish) major restructuring and now need to find new avenues to increased profitability. Visionaries are not just those who can predict or paint a picture of the future but those who can draw a map showing where the organization is now and where it should be in the imagined future. Vision is what distinguishes leaders from mere futurologists. Leaders like Apple Computer's Stephen Jobs, People's Express Airline's Donald Burr, and McDonald's Ray Kroc not only had vision—a computer in every home, no-frills air travel for everyone, inexpensive short-menu restaurants for every small town—but got others to help them make their vision reality.

Sometimes the leader whose vision created a successful enterprise has trouble relinquishing control of his creation to a successor. As the "In Practice" box suggests, there are a number of things a leader can do to ensure a smooth transition to a new leader and perhaps to new goals and new ways of achieving them.

Operant Model of Leadership

The *operant theory of leadership*, proposed by Judith Komacki and her colleagues at Purdue University, is also a throwback to universal approaches.[70] This approach, which is based on learning theory (see Chapter 7), focuses on specific leader behaviors: monitors and consequences. **Monitors** are behaviors associated with collecting performance information. For example, leaders may spotcheck subordinates' work or ask others to report on it. **Consequences** are behaviors in

monitors Leader behaviors that involve collecting performance information on subordinates.

consequences Leader behaviors that involve administering rewards and punishments contingent upon subordinate performance.

[68] J. J. Hater and B. M. Bass, "Superiors' Evaluations and Subordinates' Perceptions of Transformational Leadership," *Journal of Applied Psychology* 73 (1988), 695–702.
[69] W. Keichel, "A Hard Look at Executive Vision," *Fortune*, October 23, 1989, p. 44.
[70] J. L. Komacki and M. L. Desselles, *Supervision Reexamined: The Role of Monitors and Consequences* (Boston: Allyn & Bacon, in press).

which supervisors demonstrate their knowledge of employees' performance by taking action that responds to it. For example, a leader may give feedback on a worker's performance, praising a good job and criticizing a poor one; in short, he metes out rewards and punishment.

In several studies conducted by Komacki and her colleagues, the extent to which supervisors engage in these two types of behaviors has been found to relate to group performance.[71]

These neo-universal approaches have identified several important features of leader characteristics and behaviors. It seems likely, however, that they will ultimately have to incorporate qualities of followers and situations. No matter how charismatic the leader, a cynical audience may defeat his efforts to transform them. Similarly, overmonitoring a competent, high-self-esteem professional may engender little but resentment and dissatisfaction.

THE TRANSACTIONAL MODEL REVISITED

We started this chapter with a discussion of Mintzberg's and Hollander's conceptions of leadership. A relatively simple approach, Mintzberg's is defined in motivational terms. Hollander, on the other hand, sees leadership as a complex transaction involving characteristics of the leader, the followers, and the situation. These ideas provided a framework for our discussion of twelve theories and approaches to leadership that vary in the breadth of their view, or the number of variables they consider, and in the emphasis they place on different leader characteristics (see Figure 12-14).

The dynamic relationships among elements of these several theories as they fit together in an integrated *transactional model of leadership* are depicted in Figure 12-15. At the core of this model is the notion that leaders exist in order to meet the performance and satisfaction needs of individual group members. Through

[71] J. L. Komacki, M. L. Desselles, and E. D. Bowman, "Definitely Not a Breeze: Extending an Operant Model of Effective Supervision to Teams," *Journal of Applied Psychology* 74 (1989), 522–29.

FIGURE 12-14

Theories and Approaches to Leadership

		Leader Characteristics		
		Trait	Behavior	Decision Style
Breadth of Approach	Universal	Trait approaches	Ohio State & Michigan studies: Consideration & initiating structure	Iowa studies: Autocratic Democratic Laissez-faire
	Interactive	Trait and situation: Leadership motivation pattern	Behavior and follower: Vertical dyad linkage	Decision style and follower: Life cycle approach
	Comprehensive	Fiedler's contingency theory	Path-goal theory	Decision tree model

FIGURE 12-15

The Fully Articulated Transactional Model of Leadership

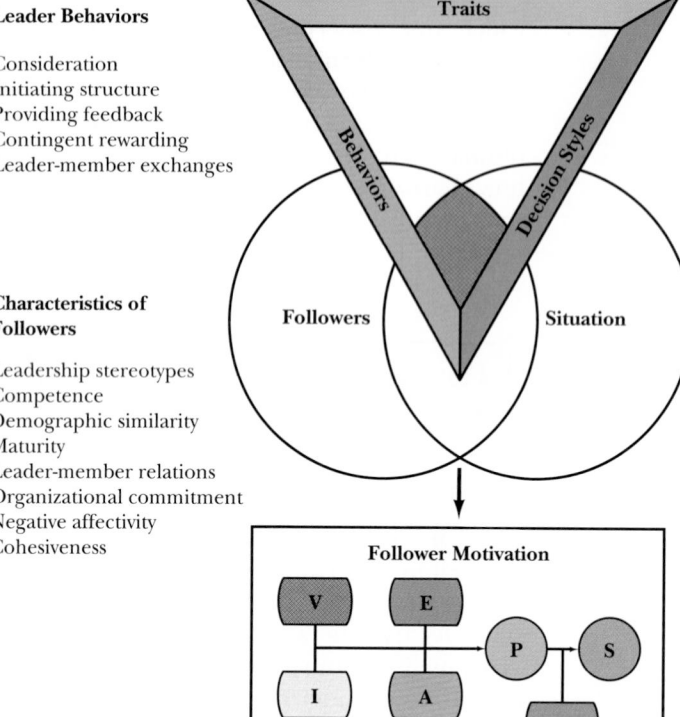

Leader Traits

Energy level	Supervisoty ability	Charisma
Cognitive ability	Dominance	LPC
Task knowledge	Self-confidence	LMP

Leader Behaviors

Consideration
Initiating structure
Providing feedback
Contingent rewarding
Leader-member exchanges

Leader Decision Styles

Autocratic
Laissez-faire
Participative
Delegative

Characteristics of Followers

Leadership stereotypes
Competence
Demographic similarity
Maturity
Leader-member relations
Organizational commitment
Negative affectivity
Cohesiveness

Characteristics of Situations

Economic conditions
Selection systems
Technical nature of tasks
Task structure
Position power
Contextual factors
Organizational formalizatio
Spatial distance

their abilities and personality characteristics, their behaviors, and their decision styles, leaders must affect their followers' valences, instrumentalities, expectancies, role perceptions, and outcomes, or rewards.

At the same time, leaders must recognize that these phenomena do not exist in a vacuum and are affected by a variety of follower and situational characteristics. Thus a characteristic that works well with one unique configuration of situation and followers is unlikely to work well with another configuration. The initial match between the leader, the situation, and the follower is critical for leader emergence. The leader's ability to adapt to changing situations and followers will determine her staying power over time.

If our model looks complex, it is because the phenomenon it attempts to describe is complex. If leadership were easy, everyone would be doing it.

SUMMARY

A great deal of theory and research has examined the topic of leadership. Leadership differs from management in that leading is one of the tasks of managerial work. The emergence and continued success of a *leader* is a complex function of his characteristics, characteristics of his *followers*, and characteristics of the *situation*.

Some of the more important personal characteristics of a leader seem to be high intelligence, need for power, energy level, and charisma and low need for affiliation. These characteristics are typically manifested in particular leader behaviors or decision styles.

The more important dimensions of leader behavior include *consideration* of employee needs, sometimes referred to as *relationship orientation* or concern for people; *initiating structure*, sometimes referred to as *task orientation*, or concern for production; and leader-member exchange behaviors that separate subordinates into in-groups and out-groups. Two other critical leader behaviors are *monitors*, or keeping track of employee performance, and *consequences*, or administering rewards and punishments contingent on performance.

Leader characteristics lead to different leader decision styles. Some *authoritarian* leaders make all decisions for their followers, whereas others take a *laissez faire* approach and leave followers to do as they please. Still others take a *democratic* approach, working actively with followers to ensure that all group members have a chance to contribute to a task. According to the *transactional model of leadership*, the effectiveness of these different behaviors and decision styles is contingent on characteristics of the followers and of the situation.

Followers differ along several important dimensions. They may be highly knowledgeable, mature, professional, and committed to the organization and its mission or they may be quite the opposite. Different leadership styles will be required to work effectively with followers with these different characteristics.

The situation in which leader and followers find themselves also affects the relationship between leader characteristics, behaviors, and decision styles on the one hand and leader effectiveness on the other. Where the leader has great *position power*, goals are clear, *leader task structure* is high, and *leader-follower relations* are characterized by trust and respect, one set of behaviors or decision styles may be required. In situations where the opposite conditions hold, a completely different kind of leader or leader style may be needed.

REVIEW QUESTIONS

1.   Theories of leadership differ in terms of how adaptable they suggest the leader can be. Of the theories charted in Figure 12-14, choose two that suggest the leader is immutable and two that suggest the leader is readily adaptable. Which of these two conflicting perspectives seems most likely to be true? Are leaders born or are they made?

2.   Most of the early research on leadership was done with leaders who were almost exclusively white and male. In the "Management Issues" box in Chapter 9 we discussed the report "Workforce 2000," which suggests that few of the new entrants in the labor force in the year 2000 will be white males. Which theories of leadership may need to be seriously reexamined because of this change, and which do you feel will generalize well to the new work force?

3.   We discussed the Least Preferred Coworker Scale in this chapter. Although no such instrument exists, what if there were a Least Preferred Leader Scale? Who would be your least preferred leader, and why do you object so strongly to this person? Can you think of followers other than yourself, or situations other than the one you were in, where this person might be an excellent leader?

4.   Although one can think of a few exceptions, in general people who achieve preeminence as leaders in business organizations do not achieve success as political leaders. What are some characteristics of leaders, followers, or situations that make this kind of transition difficult?

5.   The list of often-cited leaders in Table 12-1 clearly includes both saints and sinners. Why is it that the general moral character of the leader apparently plays no consistent role in a leader's emergence or continuation in power?

DIAGNOSTIC QUESTIONS

We have attempted to develop and describe a model of the leadership process that will help you diagnose organizational problems attributed to poor leadership. The following questions should help you apply the transactional model in performing diagnostic analyses.

1. Compared to other aspects of managerial work, how important to you is the specific process of leadership?

2. Which of the five components of follower motivation—valence, instrumentality, expectancy, accuracy of role perceptions, or rewards—seems most lacking in your setting?

3. What abilities related to leadership (e.g., general cognitive ability, task knowledge, supervisory skills) do you or your managers have? How do these abilities relate to the five components of motivation?

4. What personality characteristics that are related to leadership (e.g., self-esteem, need for power, charisma) do you or your managers have? How do these characteristics relate to the five components of motivation?

5. How readily do you think your abilities and characteristics or the abilities and characteristics of your managers can be changed?

6. What leader behaviors (e.g., consideration, initiating structure, contingent rewarding and punishing) do you or you managers typically employ? How do these behaviors relate to the five components of motivation?

7. Do you and your managers tailor your behaviors to different followers?

8. How do you or your managers determine in-group versus out-group status for group members? For example, do you base your selection on competence or on demographic similarity?

9. What kind of decision styles (autocratic, participating, delegating) do you or your managers typically employ?

10. How should this decision style be tailored to your followers? Should you fit it to their level of maturity? their commitment to goals? or to the situation?

11. What are some of the major characteristics of the followers or of the situation you confront that might substitute for leader behaviors or neutralize them?

12. How favorable is the situation for the leader in terms of leader task structure, position power, and leader-member relations? What can be done to make this situation more favorable?

EXERCISE 12-1
EXECUTIVE PIE: SHARING LEADERSHIP*

STEPHAN H. PUTNAM, *University of North Carolina at Chapel Hill*

We tend to think of leaders as using their influence to help people achieve stated goals. Leaders, however, may choose to share their influence with others in pursuing a given mission or purpose. As you'll recall, in this chapter we discussed the Vroom-Yetton model, which proposed seven distinct leadership decision styles (see Table 12-5). From these seven styles we can distill four basic modes of handling influence and power in decision making. When a leader chooses the *autocratic* mode, she decides what to do without sharing any influence at all. If she chooses the *consultative* mode, she will share a modest amount of influence by asking others for their opinions before making a decision herself. If she chooses the *group-based* mode she will share her influence, encouraging others in the group to participate equally in the decision making. And if she chooses the *delegation* mode she will transfer all of her influence to others. This exercise will give you the opportunity to consider tradeoffs among these four leader decision styles as you decide which to use in dealing with a common management problem.

STEP ONE: PRE-CLASS PREPARATION

In the class before the one in which you will perform this exercise, your instructor will divide the class into groups of four to six people (if your class has formed permanent groups, you should reassemble in those groups). The instructor will then assign each group member the role of representative of one of six departments of a manufacturing business. Thus you will be a representative of sales and marketing, accounting and finance, manufacturing and operations, purchasing and legal affairs, research and development, or human resource management. Write the name of your role in the space provided in the "Task Role and Agenda" in Exhibit 12-1. Then read the task role and agenda description and prepare to follow it in the next class.

STEP TWO: GROUP DECISION MAKING

At the beginning of class, each group should meet and immediately elect or appoint a leader. In each group, every member should write his role on a piece of paper and place it in front of him so that everyone else will know what department he represents. When the groups have completed these tasks, the instructor will give each group leader a large paper circle. This "pie" represents all of the decision making power in the group, and it can be divided and distributed among the group's mem-

---

* Adapted from J. William Pfeiffer and John E. Jones, eds., *The 1977 Annual Handbook for Group Facilitators* (San Diego, Calif.: University Associates, Inc., 1977). Used with permission.

EXHIBIT 12-1
Task Role and Agenda Sheet

You are a member of a central steering committee composed of the heads of the departments in an industrial plant. The departments represented are sales and marketing, accounting and finance, manufacturing and operations, purchasing and legal affairs, research and development, and human resources management. Leadership of your committee rotates among the department representatives every six months.

Today's meeting has been called to allocate $100,000 which has come from company headquarters. This money must be spent within the next 60 days or be lost to taxes. The only restriction on your decision is that the money must be used to benefit the employees of your plant and to raise their morale. Your committee has 30 minutes to decide how the money will be spent.

You are the representative of the _____ department. Before the class in which your group will make its decision on this matter, you should think about your role and do whatever research is necessary to understand what functions are performed by departments like yours. You will be expected to fulfill your assigned role convincingly and with the expertise that one might expect of a department head.

bers as the leader sees fit. Once the leader gives the pie or any part of it to another person, power travels with that part by percentage. For instance, if the leader cuts the pie into four equal pieces and gives three to other individuals, keeping one for herself, each of those four members has equal decision-making power in the group and the remaining members have none.

After the pie has been distributed and everyone understands how decision-making power is to be shared in this task, each group will have 30 minutes to reach the decision outlined in Exhibit 12-1. As group members work on the task, the group can further subdivide pieces of the pie if this seems desirable. A spokesperson should be appointed to keep a record of group activities and to report back to the class afterward.

STEP THREE: CLASS DISCUSSION

After 30 minutes the instructor should assemble the entire class, and spokespersons should report on the activities of their group. The class should then discuss its experience in allocating influence among the members of a group. Be sure to address the following questions:

1. For each group, which of Vroom's decision-making modes was the group's original choice? Which other mode(s) did the group try out subsequently? Which mode seemed to work the best? Which mode seemed least useful? Why?

2. Was it easy or difficult to move from one mode to another? How might rough transitions be smoothed?

3. If you had it to do over again, would you want your group to use the same decision-making modes? What changes, if any, would you make? Which things would you want to stay the same?

4. What aspects of a decision and the situation in which it must be made are most likely to influence the usefulness of a particular mode? Discuss your answer in the context of Vroom's model as it's described in this chapter.

5. In a real organization, what mechanisms might take the place of the "executive pie" in the regulating influence sharing? In your future job, how much say do you expect to have over who will have influence and who won't? Do you think you'll find this acceptable?

CONCLUSION

The process of sharing influence and reaching a decision is complicated and requires great managerial insight. Leading a group through decision-making procedures can involve as much attention to *how* to decide as to *what* to decide. No single mode of decision making is suitable for every situation. Instead, to perform effectively, managers must adopt the contingency approach, diagnosing each situation to determine which decision mode to employ.

DIAGNOSING ORGANIZATIONAL BEHAVIOR

CASE 12-1

SONIA HARRIS*

ARVA R. CLARK, *Simmons College*

In the spring of 1992 Sonia Harris was a financial analyst for Multinational Fiber Products, Inc., a multinational corporation based in San Francisco. Before enrolling in the Graduate Program in Management at Simmons College as a full-time student, Harris had been a computer programmer for eight years. In 1992 she was 30 years old.

Case Writer: What do you do as a financial analyst, Sonia?

S.H.: I'm in the earnings section of the financial reporting division within the comptroller's department. My group collects earnings information from all our regions worldwide and

compiles it for public reporting and government reporting.

When you collect this data, you have to be sure it's correct, so you're constantly checking and cross-checking. I'm the initial control person, in charge of seeing that all the earnings information comes in from all the regions I'm responsible for. I contact each of my regions every month and ask them for their net income—get a snapshot of what's going on. I also take care of any corporate adjustments to these numbers. Then I do an analysis of the numbers and prepare an explanation of changes over the last month or quarter or year.

After we get the numbers in, we as a group present worldwide earnings to the comptroller. I represent my regions in the presentations.

C.W.: How do you know the reasons for the monthly changes in your regions?

S.H.: The regions send in letters. If a letter is not complete I contact them. This process takes one day. My whole job is done in one day each month. The rest of the month is spent in preparing for that day, doing whatever comparisons and analyses can be done beforehand so that when the numbers come in, I'll be able to get everything done. Then, about 20 hours after the numbers actually come in, we make our presentation. It's a very tight time schedule.

At the end of each quarter, we put out a whole book of analyses. The book is solid numbers; there's much more data than at the end of the month. This usually takes three or four days to prepare and we work every night or an entire weekend. At year-end, we have ten times more data coming from the regions and we have to reconcile it all. That's really a panic time.

It's funny. We have a series of really tight time schedules that we know we're going to have. We know we're going to have a crisis every month, but I can't determine why. Why must all the monthly analysis be done within one day? Nobody knows, except that it's traditional. People get "up" for it. The adrenalin starts flowing. They like it. There's a sense of accomplishment, getting it done in such a short time. That's the biggest challenge we have. My theory is that these deadlines exist because they give people a sense of accomplishment.

C.W.: How is your earnings section structured?

S.H.: There are four of us—three women and one man—who report to the division leader, Pete Rinaldi. Another section, consisting of five people, does all the government reporting, and we often work with them. Then our two groups report to Aaron Rappaport, the head of the financial reporting division.

C.W.: How long have you been working in your present capacity?

S.H.: For a month. For five months before that I was an assistant to the woman who held my present job.

C.W.: Is that a typical way to come into the department?

S.H.: That's exactly the way it is done. It's considered a good way to see what's happening in the company. It's the most active section in the comptroller's department. Every month you get a snapshot of what the company is doing worldwide. It's considered excellent exposure.

C.W.: Have you worked in any other department in the company, or any other section in the comptroller's department?

S.H.: I was in the computer section within the comptroller's department for three months.

C.W.: Why did you leave?

S.H.: I wasn't happy there and I told people I wasn't happy. I never really had a boss there. The man in charge of that section and most of the rest of the staff travel constantly. So I had no direction.

C.W.: How did you let the company know you were unhappy there?

S.H.: I talked to a man from another department. I was doing a job for him but didn't report to him on a formal basis. He did talk to my boss and I assume he conveyed my discontent. After three months I was moved to the earnings reporting section.

C.W.: How did you choose this field, control?

S.H.: I was inclined towards the comptroller's area because it is operational. That means you get to know what a company is doing. We can call up anyone, anywhere in the company, and ask where the numbers come from. They have to explain—whether the markets have changed or expenses have gone up, whatever is going on. It's amazing the amount of information you can get from this position.

But, basically, all the comptroller's people do is collect numbers and reconcile them. If I stay in this area, I'll never get to use any finance, which I feel badly about. I love finance and did well in it at Simmons.

C.W.: Can you move from the comptroller's area to finance?

S.H.: I've asked about that. It's been done once, but it's very rare. Finance would be in the treasurer's department. They look at investments, the loan situation, the debt-equity situation, the purchasing of stock, taking care of stock options—things I'm interested in. I wasn't aware enough of the difference between the comptroller's department and the treasurer's department when I was interviewing for a job.

C.W.: What is the career progression for people in the comptroller's area?

S.H.: They go to the operating divisions. In corporate,

we get a surface picture. In the operational regions you actually have to deal with real-life problems. Usually, people relocate to different operational regions and move up the ladder in different areas. Then they'll get promoted to corporate to be group head, division head, department head, comptroller.

The company makes it very clear that you have to be free to relocate anywhere in the world. Virtually everyone on our floor has worked in South America, Africa, and Asia. They come back to corporate on higher levels after learning about different regions.

People relocate, move fast, climb up in the corporation. I see them as having no other lives except their careers. They've changed homes so many times that their nuclear families are the only thing they have to relate to except for the company. They are constantly moving. And they work very long hours. It's not clear to me *why* they work such long hours. That's another problem.

C.W.: Are long hours a part of the job as a way to show you're in earnest?

S.H.: Right. In my section, for example, I report to Claire Herzlinger, the supervisor who used to have my job, and to Pete Rinaldi, the head of the earnings section. Claire also reports to Pete Rinaldi. Claire is a workaholic. She gets in at 8 o'clock, an hour early, and works until 7 o'clock at night. Pete didn't always work that many hours but Claire told him he wasn't working hard enough. Since then he's been working longer hours. She really runs the whole floor.

C.W.: Does she expect long hours from you?

S.H.: Yes. For the last few weeks things have been really slow because we finally wrapped up our year-end report. When we're working on year-end, we work every night, every weekend, and we have no time off. I can understand that because we have a lot of work to do. But now people are still working late and there's not enough work to fill up my time during regular working hours. I've been putting in just regular working hours and Claire doesn't approve.

C.W.: How do you know that? Does she tell you?

S.H.: Once she told me. Otherwise she just glares. She works late every night. You know, you can always find something to do—material to read, people to talk to—if there isn't anything else to do.

Everyone watches to see what time you come in and when you leave. This morning I was fifteen minutes late and it was a mortal sin. In fact, it's a venial sin if you come in on time. If you come in a half hour early, then you're okay. And although you're supposed to come in early and stay late, it's okay to take a longer lunch hour than the alloted 45 minutes. It's perfectly okay to take an hour or an hour and a half. I don't understand that.

C.W.: What is your biggest problem here?

S.H.: Claire is my biggest problem. I'm terrified of her! I don't know anyone who isn't terrified of her.

C.W.: Both men and women?

S.H.: Even the comptroller! She has a terrible reputation.

C.W.: Why?

S.H.: She's so aggressive. Let me give you an example of what I'm going through right now. We're rewriting our instructions manual that tells the regions how to submit earnings reports. I was happy to see the improvements that were made. It looked great, a very professional job. So, since I report to Claire on things like that, I took it to her and said, "I'm reviewing the revised manual. They're revamped a lot of the troublesome areas. It really looks good. I want you to read it."

She came in the next morning and said, "Did you read this? This is awful! They tried to make these changes last year and I told them it wouldn't do! This is the same thing as last year!"

I guess some of the division people tried to make some changes last year and Claire absolutely forbade it. There was a lot of conflict between her and Francis McArthur, the person responsible for the manual, and they never revised it. So here I find I'm fighting her on the same issues. So I said, "Claire, what exactly don't you like about it?" She didn't like "this page" and "this page" and "this page." Finally, I said—trying to pin her down to something specific—"What is the thing you hate the most, Claire?" It was like handling fire. It's incredibly hard to get down to the nuts and bolts with her. She made me feel like I hadn't read the thing. She said, "Well, what do you *like* about it?" I went blank. I said, "I don't know anymore." She made me feel like dust, like dirt.

She does that to everybody. Francis McArthur, who did the revision, is a really strong man. He has very strong opinions. She made him back down in exactly the same way. She is very smart, and when she says something, she always has her back-up information. And she comes on so strong—her whole approach is overwhelming. She used to have my job, dealing with the regions, and she dealt with them in the same way. There were problems there.

C.W.: How much does she have to do with your career? Is she important for your promotions?

S.H.: I don't know. I think if she backs someone, that would carry a lot of weight. But if she puts someone down, that might not have as much weight because she's so offensive.

C.W.: Any other problems?

S.H.: Well, I came to this company with another woman from my class at Simmons, Marianna Perry, and we're as different as we can be. But because we went to Simmons together, we're thought of as twins. Our attitudes and our experiences and our abilities are completely different, but that doesn't matter. Marianna and Claire and I are the only professional women on this floor and we're all in the same section. We're all completely different, yet all of us are considered to be the same. We're not distinct. We're known as "Pete's harem."

C.W.: Is it usual or unusual to have that many women in one section?

S.H.: Unusual. This company has just started hiring women. I remember once I was issuing instructions to the regions and I signed my name as "Sonia A. Harris." One of the men in the department told me to sign "S. A. Harris," never to let the regions know I'm a woman.

Another problem is that when I came to this section, I think I was seen as somewhat uninterested in my job. Marianna and Claire really like the frenetic pace of the job. I would be happier if things would calm down and were better organized. Because I'm not like Claire or Marianna in that regard, some people saw me as uninterested.

C.W.: Who told you that—your boss, Pete Rinaldi?

S.H.: No, my peers. Claire and Marianna.

C.W.: How would you like people to see you?

S.H.: I would like to be seen as very competent and very organized, someone who can get the job done, someone who could take a situation where the energy might be frenetic and be able to calm it down. That's my personal goal, which is in tune with my personality.

C.W.: Are you saying that the company doesn't value that approach?

S.H.: Some people do, but not everyone. I was talking to a man from another section who told me about his experience as comptroller in our Chi-

cago office. He said that when he went there, people were running around, working late, but they couldn't get the work done. He went in, organized the work and stabilized the department. People didn't have to work overtime anymore and they weren't frantic at the end of every month. Then *his* boss came in and said, "What's wrong? This place used to be working overtime. People used to work all the time. What's wrong?" The comptroller said, "Well, I got it organized." His boss was critical; he didn't like the change. People in this department seem to lack perspective.

C.W.: Sonia, what about your future? If you want to move out of this department to the treasurer's department, who would you talk to about it?

S.H.: I've already talked to the personnel people about it, but they didn't like that at all. You're supposed to talk to your immediate boss. But I happen to know that there are some openings in the treasurer's department. I have some informal contracts over there.

C.W.: How do you know people in the treasurer's department?

S.H.: I have contacts all over the company. That's part of my job. They have to report earnings like everybody else.

C.W.: In general, how are you feeling about the company after nine months?

S.H.: I don't like it. I don't like the emphasis on long hours. I find that childish and unprofessional. I don't like the lack of challenge. The lack of direction on my first job—that was absolutely unacceptable. To go into a job and have no one to report to for three months—it's not the way to operate. They have no built-in procedure for taking care of new employees.

Let me tell you about the review procedure. They are supposed to review you once a year but they don't have to tell you about it. If they do tell you about it, they may or may not let you see the actual review papers. You're not allowed to express any differing opinions with your immediate supervisor on reviews. The type of savvy you're supposed to get, you have to get on your own and they'll respond to it by giving you promotions. But if you don't get a promotion, that's the only way you know that you're not doing well. It's very hard cornering managers to get a review. It's all up to the employee. You have to do it yourself. I don't like that at all.

I suspect I'm not ambitious enough to be at Multinational Fiber Products. When I was interviewing candidates for jobs here recently, one man in the department told me: "What we are looking for is people who are overachievers." You not only have to be brilliant—everybody here is brilliant—you have to be an overachiever. So you have all these brilliant overachievers doing their underchallenging jobs and what you end up with is an incredibly political situation because there is no place else to put your energy. I'd rather have something that is more intrinsically challenging because I'm no good at politics. So, frankly, I'm thinking of looking for a new job after I've been here for a year.

C.W.: Has this been a lost year for you then?

S.H.: Oh, no! This company has a fantastic reputation. Just to be hired by Multinational Fiber Products is desirable. It means you're a good person. They don't hire bad people; they're very picky. You really get a feeling for that when you're here and interviewing candidates.

C.W.: So how do you feel about this year?

S.H.: I've tried something new. This is entirely different from any other job I've had. It's what I wanted to do and I'm not at all disappointed that I tried it.

It's becoming clearer and clearer to me what I really want to do. As I think about this year and my past jobs, I realize that I'd really like to be a manager. The substance of the job isn't as important to me as the opportunity to manage people. Multinational Fiber wouldn't be so bad if I could see the prospect of being a manager somewhere down the road. I really don't care about job content when you come right down to it. If I could be a manager, I wouldn't care what the other responsibilities were.

If I can't be a manager, I'd like to get more into a financial area. A lot of people in San Francisco have their own tiny companies. I've run into some who need help in doing their books. That would be interesting. I'd be learning about people's businesses and they would be learning how to keep their books. I have the ability to teach people and it's really what I like to do, interacting with people. That's a possibility I'll be looking into.

When you have read this case, look back at the chapter's diagnostic questions and choose the ones that apply to the case. Then use those questions with the ones that follow in your case analysis.

1. What kind of a leader is Claire Herzlinger, Sonia Harris's supervisor? Do you think that other people at Multinational Fiber Products share Sonia's criticisms of Claire? What kinds of effects do leaders like Claire have on their subordinates' attitudes and performance? Is Claire affecting her subordinates in this way?

2. Do you agree with Sonia that the harried pace of work at Multinational is probably caused by the company's managers? May situational factors be contributing to the hectic work pace?

3. If Sonia achieves her goal of becoming a manager, what kind of a leader is she likely to be? Is her style of leadership likely to be consistent with the situational demands at Multinational? If she stays with the company, is Sonia likely to succeed as a manager and leader?

CASE 12-2

THE CASE OF DICK SPENCER*

MARGARET P. FENN, *University of Washington*

After the usual banter when old friends meet for cocktails, the conversion between a couple of university professors and Dick Spencer, a former student who was now a successful businessman, turned to Dick's life as a vice-president of a large manufacturing firm.

"I've made a lot of mistakes, most of which I could live with, but this one series of incidents was so frustrating that I could have cried at the time," Dick said in response to a question. "I really have to laugh at how ridiculous it is now, but at the time I blew my cork."

Spencer was plant manager of Modrow Company, a Canadian branch of the Tri-American Corporation. Tri-American was a major producer of primary aluminum, with integrated operations ranging from the mining of bauxite through the processing and fabrication of aluminum into a variety of products. The company also made and sold refractories and industrial chemicals. The parent company had wholly owned subsidiaries in five separate United States locations and had foreign affiliates in 15 different countries.

Tri-American mined bauxite in the Jamaican West Indies and shipped the raw material by commercial vessels to two plants in Louisiana where it was processed into alumina. The alumina was then shipped to reduction plants in one of three locations for conversion into primary aluminum. Most of the primary aluminum was then moved to the companies' fabricating plants for further processing. Fabricated aluminum items in-

* Reprinted with the author's permission.

cluded sheet, flat, coil, and corrugated products; siding; and roofing.

Tri-American employed approximately 22,000 employees in the total organization. The company was governed by a board of directors which included the chairman, vice-chairman, president, and twelve vice-presidents. However, each of the subsidiaries and branches functioned as independent units. The board set general policy, which was then interpreted and applied by the various plant managers. In a sense, the various plants competed with one another as though they were independent companies. This decentralization in organizational structure increased the freedom and authority of the plant managers, but also increased the pressure for profitability.

The Modrow branch was located in a border town in Canada. The total work force in Modrow was 1,000. This Canadian subsidiary was primarily a fabricating unit. Its main products were foil and building products such as roofing and siding. Aluminum products were gaining in importance in architectual plans, and increased sales were predicted for this branch. Its location and its stable work force were the most important advantages it possessed.

In anticipation of estimated increases in building product sales, Modrow had recently completed a modernization and expansion project. At the same time, their research and art departments combined talents in developing a series of twelve new patterns of siding which were being introduced to the market. Modernization and pattern development had been costly undertakings, but the expected return on investment made the project feasible. However, the plant manager, who was a Tri-American vice-president, had instituted a campaign to cut expenses wherever possible. In his introductory notice of the campaign, he emphasized that cost reduction would be the personal aim of every employee at Modrow.

Salesman

The plant manager of Modrow, Dick Spencer, was an American who had been transferred to this Canadian branch two years previously, after the start of the modernization plan. Dick had been with the Tri-American Company for 14 years, and his progress within the organization was considered spectacular by those who knew him well. Dick had received a Master's degree in Business Administration from a well-known university at the age of 22. Upon graduation he had accepted a job as salesman for Tri-American. During his first year as a salesman, he succeeded in landing a single, large contract which put him near the top of the sales-volume leaders. In discussing this phenomenal rise in the sales

volume, several of his fellow salesmen concluded that his looks, charm, and ability on the golf course contributed as much to his success as his knowledge of the business or his ability to sell the products.

The second year of his sales career, he continued to set a fast pace. Although his record set difficult goals for the other salesmen, he was considered a "regular guy" by them, and both he and they seemed to enjoy the few occasions when they socialized. However, by the end of the second year of constant traveling and selling, Dick began to experience some doubt about his future.

His constant involvement in business affairs disrupted his marital life, and his wife divorced him during the second year with Tri-American. Dick resented her action at first, but gradually seemed to recognize that his career at present depended on his freedom to travel unencumbered. During that second year, he ranged far and wide in his sales territory, and successfully closed several large contracts. None of them was as large as his first year's major sale, but in total volume he again was well up near the top of salesmen for the year. Dick's name became well known in the corporate headquarters, and he was spoken of as "the boy to watch."

Dick had met the president of Tri-American during his first year as a salesman at a company conference. After three days of golfing and socializing they developed a relaxed camaraderie considered unusual by those who observed the developing friendship. Although their contacts were infrequent after the conference, their easy relationship seemed to blossom the few times they did meet. Dick's friends kidded him about his ability to make use of his new friendship to promote himself in the company, but Dick brushed aside their jibes and insisted that he'd make it on his own abilities, not someone's coattail.

By the time he was 25, Dick began to suspect that he did not look forward to a life as a salesman for the rest of his career. He talked about his unrest with his friends, and they suggested that he groom himself for sales manager. "You won't make the kind of money you're making from commissions," he was told, "but you will have a foot in the door from an administrative standpoint, and you won't have to travel quite as much as you do now." Dick took their suggestions lightly, and continued to sell the product, but was aware that he felt dissatisfied and did not seem to get the satisfaction out of his job that he had once enjoyed.

By the end of his third year with the company, Dick was convinced that he wanted a change in direction. As usual, he and the president spent quite a bit of time on the golf course during the annual company sales conference. After their match one day, the president kidded Dick about his game. The conversion drifted back to business, and the president, who seemed to be in a jovial mood, started to kid Dick about his sales ability. In a joking way, he implied that anyone could sell a product as good as Tri-American's, but that it took real "guts and know-how" to make the products. The conversation drifted to other things, but the remark stuck with Dick.

Sometime later, Dick approached the president formally with a request for transfer out of the sales division. The president was surprised and hesitant about this change in career direction for Dick. He recognized the superior sales ability that Dick seemed to possess, but was unsure that Dick was willing or able to assume responsibility in any other division of the organization. Dick sensed the hesitancy, but continued to push his request. He later remarked that it seemed that the initial hesitancy of the president convinced Dick that he needed an opportunity to prove himself in a field other than sales.

Trouble-Shooter

Dick was finally transferred back to the home office of the organization and indoctrinated into production and administration roles in the company as a special assistant to the senior vice-president of production. As a special assistant, Dick was assigned several trouble-shooting jobs. He acquitted himself well in this role, but in the process succeeded in gaining a reputation as a ruthless head hunter among the branches where he had performed a series of amputations. His reputation as an amiable, genial, easy-going guy from the sales department was the antithesis of the reputation of a cold, calculating head hunter which he earned in his trouble-shooter role. The vice-president, who was Dick's boss, was aware of the reputation that Dick had earned but was pleased with the results that were obtained. The faltering departments that Dick had worked in seemed to bloom with new life and energy from Dick's recommended amputations. As a result, the vice president began to sing Dick's praises, and the president began to accept Dick in his new role in the company.

Management Responsibility

About three years after Dick's switch from sales, he was given an assignment as assistant plant manager of an English branch of the company. Dick, who had remarried, moved his wife and family to London, and they attempted to adapt to their new routine. The plant manager was English, as were most of the other employees. Dick and his family were accepted with reservations into the community life as well as into the plant life. The difference between British and American philosophy and performance within the plant was marked for Dick who was imbued with modern managerial concepts and methods. Dick's directives from headquarters were to update and upgrade performance in this branch. However, his power and authority were less than those of his superior, so he constantly found himself in the position of having to soft pedal or withhold suggestions that he would have liked to make, or innovations that he would have liked to introduce. After a frustrating year and a half, Dick was suddenly made plant manager of an old British company which had just been purchased by Tri-American. He left his first English assignment with mixed feelings and moved from London to Birmingham.

As the new plant manager, Dick operated much as he had in his trouble-shooting job for the first couple of years of his change from sales to administration. Training and reeducation programs were instituted for all supervisors and managers who survived the initial purge. Methods were studies and simplified or redesigned whenever possible, and new attention was directed toward production which better met the needs of the sales organization. A strong controller helped to straighten out the profit picture through stringent cost control; and, by the end of the third year, the company showed a small profit for the first time in many years. Because he felt that this battle was won, Dick requested transfer back to the United States. This request was partially granted when nine months later he was awarded a junior vice president title, and was made manager of a subsidiary Canadian plant, Modrow.

Modrow Manager

Prior to Dick's appointment as plant manager at Modrow, extensive plans for plant expansion and improvement had been approved and started. Although he had not been in on the original discussions and plans, he inherited all the problems that accompany large-scale changes in any organization. Construction was slower in completion than originally planned, equipment arrived before the building was finished, employees were upset about the extent of change expected in their work routines with the installation of additional machinery and, in general, morale was at a low ebb.

Various versions of Dick's former activities had preceded him, and on his arrival he was viewed with dubious eyes. The first few months after his arrival were spent in a frenzy of catching up. This entailed constant conferences and meetings, volumes of reading of past reports, becoming acquainted with the civic leaders of the area, and a plethora of dispatches to and from the home office. Costs continued to climb unabated.

By the end of his first year at Modrow, the building program had been completed, although behind schedule, the new equipment had been installed, and some revamping of cost procedures had been incorporated. The financial picture at this time showed a substantial loss, but since it had been budgeted as a loss, this was not surprising. All managers of the various divisions had worked closely with their supervisors and accountants in planning the budget for the following year, and Dick began to emphasize his personal interest in cost reduction.

As he worked through his first year as plant manager, Dick developed the habit of strolling around the organization. He was apt to leave his office and appear anywhere on the plant floor, in the design office, at the desk of a purchasing agent or accountant, in the plant cafeteria rather than the executive dining room, or wherever there was activity concerned with Modrow. During his strolls he looked, listened, and became acquainted. If he observed activities which he wanted to talk about, or heard remarks that gave him clues to future action, he did not reveal these at the time. Rather he had a nod, a wave, a smile, for the people near him, but a mental note to talk to his supervisors, managers, and foremen in the future. At first his presence disturbed those who noted him coming and going, but after several exposures to him without any noticeable effect, the workers came to accept his presence and continue their usual activities. Supervisors, managers, and foremen, however, did not feel as comfortable when they saw him in the area.

Their feelings were aptly expressed by the manager of the siding department one day when he was talking to one of his foremen: "I wish to hell he'd stay up in the front office where he belongs. Whoever heard of a plant manager who had time to wander around the plant all the time? Why doesn't he tend to his paper work and let us tend to our business?"

"Don't let him get you down," joked the foreman. "Nothing ever comes of his visits. Maybe he's just lonesome and looking for a friend. You know how these Americans are."

"Well, you may feel that nothing ever comes of his visits, but I don't. I've been called into his office three separate items within the last two months. The heat must really be on from the head office. You know these conferences we have every month where he reviews our financial progress, our building progress, our design progress, etc? Well, we're not really progressing as fast as we should be. If you ask me we're in for continuing trouble."

In recalling his first year at Modrow, Dick had felt constantly pressured and badgered. He always sensed that the Canadians he worked with resented his presence since he was brought in over the heads of the operating staff. At the same time he felt this subtle resistance from his Canadian work force, he believed that the president and his friends in the home office were constantly on the alert, waiting for Dick to prove himself or fall flat on his face. Because of the constant pressures and demands of the work, he had literally dumped his family into a new community and had withdrawn into the plant. In the process, he built up a wall of resistance toward the demands of his wife and children who, in turn, felt as though he was abandoning them.

During the course of the conversation with his university friends, he began to recall a series of incidents that probably had resulted from the conflicting pressures. When describing some of these incidents, he continued to emphasize the fact that his attempt to be relaxed and casual had backfired. Laughingly, Dick said, "As you know, both human relations and accounting were my weakest subjects during the Master's program, and yet they are two fields I felt I needed the most at Modrow at this time." He described some of the cost procedures that he would have liked to incorporate. However, without the support and knowledge furnished by his former controller, he busied himself with details that were unnecessary. One day, as he described it, he overheard a conversation between two of the accounting staff members with whom he had been working very closely. One of them commented to the other, "For a guy who's a vice-president, he sure spends a lot of time breathing down our necks. Why doesn't he simply tell us the kind of systems he would like to try, and let us do the experiments and work out the budget?" Without commenting on the conversation he overhead, Dick then described himself as attempting to spend less time and be less directive in the accounting department.

Another incident he described which apparently had real meaning for him was one in which he had called a staff conference with his top-level managers. They had been going "hammer and tongs" for better than an hour in his private office, and in the process of heated conversation had loosened ties, taken off coats, and really rolled up their sleeves. Dick himself had slipped out of his shoes. In the midst of this, his secretary reminded him of an appointment with public officials. Dick had rapidly finished up his conference with his managers, straightened his tie, donned his coat, and had wandered out into the main office in his stocking feet.

Dick fully described several incidents when he had disappointed, frustrated, or confused his wife and family by forgetting birthdays, appointments, dinner engagements, etc. He seemed to be describing a pattern of behavior which resulted from continuing pressure and frustration. He was setting the scene to describe his baffling and humiliating position in the siding department. In looking back and recalling his activities during this first year, Dick commented on the fact that his frequent wanderings throughout the plant had resulted in a nodding acquaintance with the workers, but probably had also resulted in foremen and supervisors spending more time getting ready for his visits and reading meaning into them afterwards than attending to their specific duties. His attempts to know in detail the accounting procedures being used required long hours of concentration and detailed conversations with the accounting staff, which were time-consuming and very frustrating for him, as well as for them. His lack of attention to his family life resulted in continued pressure from both his wife and family.

The Siding Department Incident

Siding was the product which had been budgeted as a large profit item of Modrow. Aluminum siding was popular among both architects and builders because of its possibilities in both decorative and practical uses. Panel sheets of siding were shipped in standard sizes on order; large sheets of the coated siding were cut to specifications in the trim department, packed, and shipped. The trim shop was located near the loading platforms, and Dick often cut through the trim shop on his wanderings through the plant. On one of his frequent trips through the area, he suddenly became aware of the fact that several workers responsible for the disposal function were spending countless hours at high-speed saws cutting scraps into specified lengths to fit into scrap barrels. The narrow bands of scrap which resulted from the trim process varied in length from 7 to 27 feet and had to be reduced in size to fit into the disposal barrels. Dick, in his concentration on cost reduction, picked up one of the thin strips, bent it several times and fitted it into the barrel. He tried this with another piece and it bent very easily. After assuring himself that bending was possible, he walked over to a worker at the saw and asked why he was using the saw when material could easily be bent and fitted into the barrels, resulting in saving time and equipment. The worker's response was "We've never done it that way, sir. We've always cut it."

Following his plan of not commenting or discussing matters on the floor, but distressed by the reply, Dick returned to his office and asked the manager of the siding department if he could speak to the foreman

of the scrap division. The manager said, "Of course, I'll send him up to you in just a minute."

After a short time, the foreman, very agitated at being called to the plant manager's office, appeared. Dick began questioning him about the scrap disposal process and received the standard answer: "We've always done it that way." Dick then proceeded to review cost-cutting objectives. He talked about the pliability of the strips of scrap. He called for a few pieces of scrap to demonstrate the ease with which it could be bent, and ended what he thought was a satisfactory conversation by requesting the foreman to order heavy-duty gloves for his workers and use the bending process for a trial period of two weeks to check the cost saving possibilities.

The foreman listened throughout the most of this hour's conference, offered several reasons why it wouldn't work, raised some questions about the record-keeping process for cost purposes, and finally left the office with the forced agreement to try the suggested new method of bending, rather than cutting, for disposal. Although he was immersed in many other problems, his request was forcibly brought home one day as he cut through the scrap area. The workers were using power saws to cut scraps. He called the manager of the siding department and questioned him about the process. The manager explained that each foreman was responsible for his own processes, and since Dick had already talked to the foreman, perhaps he had better talk to him again. When the foreman arrived, Dick began to question him. He received a series of excuses, and some explanations of the kind of problems they were meeting by attempting to bend the scrap metal. "I don't care what the problems are," Dick nearly shouted, "when I request a cost-reduction program instituted, I want to see it carried through."

Dick was furious. When the foreman left, he phoned the maintenance department and ordered the removal of the power saws from the scrap area immediately. A short time later the foreman of the scrap department knocked on Dick's door reporting his astonishment at having maintenance men step into his area and physically remove the saws. Dick reminded the foreman of his request for a trial at cost reduction to no avail, and ended the conversation by saying that the power saws were gone and would not be returned, and the foreman had damned well better learn to get along without them. After a stormy exit by the foreman, Dick congratulated himself on having solved a problem and turned his attention to other matters.

A few days later Dick cut through the trim department and literally stopped to stare. As he described it, he was completely nonplussed to discover gloved workmen using hand shears to cut each piece of scrap.

When you have read this case, look back at the chapter's diagnostic questions and choose the ones that apply to the case. Then use those questions with the ones that follow in your case analysis.

1. At the time Dick Spencer assumed his managerial job with the branch of Tri-American in Great Britain, what style of leadership did he exhibit? What kinds of outcomes did his leadership prompt?

2. What sort of a leader did Dick become when he took over the Modrow facility? How successful was he as a leader in the Canadian plant?

3. What does the siding department incident tell you about Dick's approach to leadership and management? Why didn't the foreman adopt his suggestions and direct subordinates to bend scrap pieces instead of cutting them? If you were Dick Spencer, what would you do when you discovered the workmen using shears to cut the scrap?

C A S E 12-3

BOB COLLINS

Review this case, which you read in Chapter 6. Next, look back at Chapter 12's diagnostic questions and choose the ones that apply to the case. Then use those questions with the ones that follow in your case analysis.

1. What sort of a leader was Jim Douglas? Was he doing a good job of leading his subordinates, Bob Collins and Mark Douglas?

2. Why did Bob become upset with Jim? What could Jim have done differently to avoid the problems that were growing between the two men?

3. If you were asked to advise Jim on matters of leadership style, what advice would you give him? Could he follow your advice? Why? Why not?

CHAPTER 13

POWER, POLITICS, AND CONFLICT

In 1985, the leader of the 100-year-old Aluminum Company of America (Alcoa) undertook an unsuccessful program of diversification through acquiring companies in other industries. It took three years, nearly $500 million, and some power plays on the part of board members to convince the company that it should stick to the business it knew best—the production of aluminum.

Shareholders of Aluminum Company of America (Alcoa) can be excused if they missed the unveiling of the company's new strategic plan. It wasn't exactly a media event. But if they had read their recent annual reports closely, the change was right there. In his understated and direct style, Paul H. O'Neill, named Alcoa chairperson a year ago, simply told shareholders to forget what they had been hearing about diversification and acquisitions. Alcoa's future, he explained, lies in the aluminum business. Period.[1]

As suggested by this excerpt from a 1988 *Business Week* article, Alcoa's then-emerging strategy of refocusing its efforts on its original business—marketing aluminum—differed markedly from the plans of Charles W. Parry, Alcoa's chairperson prior to O'Neill. After spending nearly $500 million under Parry's direction to acquire twelve companies outside of the aluminum industry, Alcoa was going to sell off its nonaluminum businesses under O'Neill and return to its roots. What explained this sudden reversal in strategic direction? Answer: a well-hidden but nonetheless decisive political struggle which had occurred between Parry and Alcoa's board of directors.

Trouble had begun brewing early in 1985 following Parry's announcement of his intention to diversify half of Alcoa's business. Alcoa was then a century-old company steeped in conservatism, and the company's directors favored careful planning and familiar ways of doing things. In contrast, Parry's approach would require Alcoa to take a variety of precarious risks, seeking out unfamiliar businesses and embracing unexpected opportunities. Especially appealing, in Parry's view, were acquisitions in the aerospace industry. He bought the TRE Corporation, a Los Angeles-based defense manufacturer, and attempted to buy both LTV Aerospace, a division of the subsequently bankrupt LTV Corporation, and Goodyear Tire and Rubber Company's aerospace unit. The Goodyear business was subsequently purchased by Loral Corp. Parry was also attracted to various high-tech companies including Allen-Bradley Company, the factory automation specialist, which Parry tried to acquire but lost in a bidding war to Rockwell International Corp.

Fearing the perilousness of Parry's plans, several of Alcoa's directors began voicing personal concerns about the company's long-term security. Among them was W. H. Krome George, Alcoa's chief executive prior to Parry, who formed a coalition to press for Parry's resignation and search for a replacement. The coalition began its task by attempting to acquire Cummins Engine Company, a diesel engine manufacturer. The secret plan was to bring Cummins' CEO, Henry B. Schacht, into Alcoa as the firm's new head. The Cummins deal fell through, though, when an unsuspecting Parry concluded that Alcoa and Cummins shared little in common and rejected the merger. The board was forced to seek other options. In February 1987, the board quietly asked O'Neill, one of its own members, to take the chairperson's job. Several weeks later, the coalition called Parry to a meeting at Manhattan's River Club and abruptly demanded his resignation. O'Neill, who had been deputy director of the federal government's Office of Management and Budget during the 1970s and president of International Paper, was then publicly named Alcoa's chief executive officer.

As part of his severance settlement, Parry agreed to stay on for two years to facilitate the change of command. By 1989, O'Neill was completely in charge and Alcoa was again the world's largest producer of aluminum. Yet, with the coalition's victory over Parry, Alcoa became a captive of cyclical changes in

---

[1] Michael Schroeder, "The Quiet Coup at Alcoa: How the Board Rejected a New Vision for the Aluminum Giant," *Business Week*, June 27, 1988, p. 58.

world demand for aluminum. In the early 1990s, the market for aluminum declined substantially, and Alcoa's earnings fell with it.[2]

As shown by events at Alcoa, power and politics can gain some people jobs and force others aside. Power can also increase productivity and efficiency or reduce them substantially. Power, politics, and conflict can even decide the existence and strategic direction of entire organizations such as Alcoa. Rather than being the exception, Alcoa was one of many companies that experienced internal strife during the 1980s. Similar "palace revolts" ousted other top executives, such as Alegis/United Air Lines' Richard J. Ferris and Apple Computer's Steve Jobs. Political processes continue to influence business organizations in the 1990s. Current restructuring, often stimulated as much by internal politics as by external economic conditions, is prompting executives to search out new strategic directions for their firms. In the process, political considerations are altering the careers of thousands of managers and nonmanagers, creating opportunities for some but costing many others their jobs.[3]

Understanding power, politics, and conflict is therefore critical to managerial success—and survival—in today's organizations. For this reason, we begin Chapter 13 with a discussion of the nature, sources, and consequences of using power. Then, we will consider organizational politics, the process through which people acquire power and use it to pursue personal gains. Finally, we will explore conflict among groups, tracing through the origins, processes, and results of intergroup confrontation in organizations.

DIAGNOSTIC ISSUES

Our diagnostic model helps us raise a number of important questions about power, politics, and conflict. To begin with, can we *describe* different sources from which organization members derive power? In which situation is conflict a healthy sign and in which does it indicate severe problems within the organization? Can we *diagnose* how the aftermath of a conflict sets the stage for future conflicts? Can we explain why political decision making and negotiation sometimes take the place of more rational methods for making decisions?

Can we *prescribe* a plan of action to deal with conflict before it manifests itself? Can we prescribe how to manage interdependence that might otherwise cause destructive conflict? Finally, what managerial *actions* can be taken to prevent employees from engaging in destructive politics? If such politicking is inevitable, what actions can managers take to protect the organization?

POWER IN ORGANIZATIONS

power The ability to influence the conduct of others and resist unwanted influence in return.

If someone asked you to define **power**, how would you respond? Many people might think of a powerful person like Alcoa's Krome George and define power as the ability to influence the behaviors of others, getting them to do

[2] Jonathan P. Hicks, "Is That a Dark Cloud or a Silver Lining for Aluminum?" *The New York Times,* July 22, 1990, p. 12F.
[3] John A. Byrne, Wendy Zeller, and Scott Ticer, "Caught in the Middle: Six Managers Speak Out on Corporate Life," *Business Week,* September 12, 1988, 80–88.

things they would otherwise avoid.[4] For other people, the image of a less powerful person like Alcoa's William Parry might come to mind, leading them to define power as the ability to avoid others' attempts to influence one's own behavior. In truth, both these views are correct because **power** can be formally defined as the ability to influence the conduct of others and resist unwanted influence in return.[5]

Why do people seek power over others? The work of David McClelland, a researcher interested in determining why some people succeed as managers while others do not, provides a clue.[6] McClelland deduced that people are driven to gain and use power by a need for power—*n Pow*—that develops during childhood and adolescence. McClelland thus suggested that experience teaches some people to seek and use power, contributing to the development of a high n Pow. Others, he said, learn to avoid its use and develop a low n Pow as a result. (See Chapter 7 for further discussion of McClelland's learned need theory).

The need for power can have several different effects on the way people behave. Generally speaking, people with high n Pow are competitive, aggressive, prestige conscious, action oriented, and prone to join groups. They are likely to be effective managers if, in addition to pursuing power, they also:

> Use power to accomplish organizational goals instead of using it to satisfy personal interests;
>
> Coach subordinates and use participatory management techniques rather than autocratic, authoritarian methods; and
>
> Remain aware of the importance of managing interpersonal relations but avoid developing close relationships with subordinates.[7]

According to McClelland, seeking power and using it to influence others are not activities in and of themselves to be shunned or avoided. In fact, the process of management *requires* that power be put to appropriate use—where appropriateness is determined on the basis of several competing ethical concerns (see the "Management Issues" box).

Interpersonal Sources of Power

If management requires the use of power, then how do people in organizations acquire the power needed to influence others' behaviors? That is, from where does a manager's power originate? In their pioneering work aimed at identifying different types of power in organizations, John French and Bertram Raven sought to answer such questions by identifying the major bases, or sources, of power in organizations.[8] The five sources and types of power they discovered are overviewed in Table 13-1.

---

[4] Robert A. Dahl, "The Concept of Power," *Behavioral Science*, 2 (1957), 201–15; Abraham Kaplan, "Power in Perspective," in *Power and Conflict in Organizations*, ed. Robert L. Kahn and Elise Boulding (London: Tavistock, 1964), pp. 11–32.

[5] V. V. McMurray, "Some Unanswered Questions on Organizational Conflict," *Organization and Administrative Sciences* 6 (1975), 35–53.

[6] David C. McClelland, *Power: The Inner Experience* (New York: Irvington Publishers, 1975) pp. 3–29; David C. McClelland and David H. Burnham, "Power Is the Great Motivator," *Harvard Business Review* 54 (1976), pp. 100–110.

[7] McClelland and Burnham, "Power is the Great Motivator."

[8] John R. P. French, Jr. and Bertram Raven, "The Bases of Social Power," *Studies in Social Power*, ed. Dorwin Cartwright (Ann Arbor: Institute for Social Research, University of Michigan, 1959), pp. 150–65.

MANAGEMENT ISSUES

The Ethics of Power

How should power holders determine whether the use of power is appropriate? One approach is to adopt the *utilitarianist* perspective and judge the appropriateness of the use of power in terms of the consequences of this use. Does using power provide the greatest good for the greatest number of people? If the answer to this question is yes, then, according to the utilitarian perspective, power is being used appropriately.

A second perspective, derived from the theory of *moral rights*, suggests that power is used appropriately only when no one's personal rights or freedoms are sacrificed. It is certainly possible for many people to derive great satisfaction from the use of power to accomplish some purpose, thus satisfying utilitarian criteria, and at the same for the rights of a few individuals to be abridged, an indication of inappropriateness according to the theory of moral rights. Power holders seeking to use their power appropriately must therefore respect the rights and interests of the minority as well as look after the well-being of the majority.

A third perspective, drawn from various theories of *social justice*, suggests that even having respect for the rights of everyone in an organization may not be enough to fully justify the use of power. In addition, those using power must treat people equitably, making sure that people who are similar in relevant respects are treated similarly and that people who are different in relevant respects are treated differently in proportion to the differences between them. Power holders must also be accountable for injuries caused by their use of power and must be prepared to provide compensation for these injuries in order for the use of power to be considered appropriate.

Obviously, the three perspectives offer conflicting criteria: there are no simple answers to questions concerning the appropriateness of using power.* Instead, power holders must seek a balance among concerns for efficiency, entitlement, and equity as they exercise influence over the behaviors of others.

* For an additional discussion of the three perspectives, see Gerald F. Cavanagh, Dennis Moberg, and Manuel Velasquez, "The Ethics of Organizational Politics," *Academy of Management Review*, 6 (1981), 363–74.

reward power   Interpersonal power based on the ability to control how desirable outcomes are distributed.

Reward Power.   The first type of power referred to in the table, **reward power**, is based on the ability to allocate rewarding outcomes—either the receipt of positive things or the elimination of negative things. If you are free to decide whether other people will receive rewarding outcomes, you can influence them by handing out rewards in return for their conformity to your demands. Praise, promotions, raises, desirable job assignments, and time off from work are outcomes that managers can often control. If they can control them, managers can use them to acquire and maintain reward power. Similarly, eliminating unwanted outcomes, such as unpleasant working conditions or mandatory overtime, can be used to reward employees. For instance, police officers given clerical support to help complete crime reports generally look at this reduction of paperwork as rewarding. Because company policies, union contracts, or similar constraints can

---

T A B L E   13-1
Five Types of Power and Their Sources

TYPE OF POWER	SOURCE OF POWER
Reward	Control over rewarding outcomes
Coercive	Control over punishing outcomes
Legitimate	Occupation of legitimate position of authority
Referent	Attractiveness; charisma
Expert	Expertise, knowledge, talent

Source: Based on John R. P. French, Jr., and Bertram Raven, "The Bases of Social Power," in *Studies in Social Power*, ed. Dorwin Cartwright (Ann Arbor, MI: Institute for Social Research, University of Michigan, 1959), pp. 150–65.

AN EXAMPLE
The fear that underlies coercive power can be readily seen under dictatorships and military government. People, whether officials or ordinary citizens, are afraid to speak out against these governments. Consider the changes that have occurred throughout eastern Europe—in Poland, for example—since the fall of totalitarian governments.

coercive power Interpersonal power based on the ability to control the distribution of undesirable outcomes.

legitimate power Interpersonal power based on holding a position of formal authority.

POINT TO STRESS
Authority is accorded to people whose positions in the organization give them formal, or legitimate, power.

referent power Interpersonal power based on the possession of attractive personal characteristics.

AN EXAMPLE
Recently, athletes have been criticized for promoting products. The strength of their referent power on poor, minority youth is great. Kids' need to identify with heroes is so strong that young people have been murdered for the shoes on their feet. Many people argue that high-profile athletes need to be more careful in how and what they endorse.

expert power Interpersonal power based on the possession of expertise, knowledge, and talent.

sometimes abridge managers' ability to control the distribution of rewards, however, managers cannot always count on having reward power.

Coercive Power. While reward power involves the allocation of desirable outcomes, **coercive power** is based on the distribution of undesirable outcomes—either the receipt of something negative or the removal of something positive. People who control undesirable outcomes can get others to conform to their wishes by threatening to penalize them in some way. Coercive power exploits fear. An employee may be afraid that if she resists direction she will be punished. To influence subordinates' behaviors, managers may resort to punishments such as public scoldings, assignment of undesirable tasks, or loss of pay. Taken to the extreme, managers may threaten employees with layoffs, demotions, or dismissals.

Legitimate Power. **Legitimate power** is based on norms, values, and beliefs which teach that particular people have the legitimate right to govern or influence others. From childhood we learn to accept the commands of authority figures, first parents and then teachers. It is this well-learned lesson that gives people with authority the power to influence other people's attitudes and behaviors. In most organizations, authority is distributed in the form of a hierarchy (see Chapter 2). People who hold positions of hierarchical authority are accorded legitimate power by virtue of the fact that they are office holders. So, the vice-president of marketing at a firm like Coca-Cola or Scott Paper issues orders and expects people in subordinate positions to obey them because of the clout that being a vice-president affords.

Referent Power. Have you ever admired a teacher, a student leader, or someone else whose personality, way of interacting with other people, values, goals, or other characteristics were exceptionally attractive? If so, you probably found yourself wanting to develop and maintain a close, continuing relationship with her. This desire can give her **referent power** over you. Because you hold her in such esteem, you are likely to be influenced by her attitudes and behaviors. In time you may identify with her to such an extent that you begin to think and act like her. Referent power is also called *charismatic power*.

Famous religious leaders and political figures often develop and use referent power. Mahatma Gandhi, John F. Kennedy, Martin Luther King, Jr., and Nelson Mandela are all examples of 20th-century people who have used personal charisma to profoundly influence the thoughts and behaviors of others. Referent power can also be put to more everyday use. Consider advertising's use of famous athletes and actors to help sell products. Athletic shoe manufacturers like Nike, Reebok and L.A. Gear, for example, employ sports celebrities like Michael Jordan of the Chicago Bulls basketball team and Joe Montana of the San Francisco Forty-Niners football team as spokespeople in an effort to influence consumers to buy their products. Similarly, movie makers try to ensure success of their films by getting stars like Jack Nicholson, Michelle Pfeiffer, and Tom Cruise to appear in them.

Expert Power. **Expert power** is based on the possession of expertise, knowledge, and talent. People who are seen as experts in a particular area can influence others in two ways. They can provide other people with knowledge that enables or causes them to change their attitudes or behavior, or they can demand conformity to their wishes as the price of the knowledge others need. Thus experts such as doctors, lawyers, and accountants provide advice that influences what their clients do. By expressing their own opinions, media critics shape people's attitudes about new books, movies, recordings, and television shows. Auto mechanics, plumbers, and electricians also exert a great deal of influence over customers who are not themselves talented craftspeople.

How People Respond to Interpersonal Power

POINT TO STRESS
Power implies a dependent relationship. People are dependent on the person holding power for rewards, removal of undesirable conditions, direction, values, and/or information or expertise.

How do employees respond when managers use the different kinds of power identified by French and Raven? According to Herbert Kelman, three distinctly different types of reactions are likely responses to attempts to influence behavior. As indicated in Chapter 10, these responses are compliance, identification, and internalization (see Table 13-2).[9]

Compliance. As you will recall from Chapter 10, compliance ensues when people conform to the wishes or directives of others to acquire favorable outcomes for themselves in return. They adopt new attitudes and behaviors not because the latter are agreeable or personally fulfilling but because they lead to specific rewards and approval or avoid specific punishments and disapproval. As indicated in Chapter 10, if people adopt attitudes and behaviors for these reasons, they are likely to continue to display them only as long as the receipt of favorable outcomes remains contingent on conformity.

APPLYING THE DIAGNOSTIC MODEL
Diagnostic Question 2: Which types of interpersonal power are currently in use? Are these types of interpersonal power likely to generate the compliance, identification, or internalization needed to energize appropriate behaviors? Might other types of interpersonal power be more effective?

Of the different types of power identified by French and Raven, which are most likely to stimulate compliance? The answer is reward and coercive power, which are based on linking employee performance with the receipt of positive or negative outcomes. Employees who work harder because a supervisor with reward power has promised them incentive payments are displaying compliance behavior. By choosing to work harder, they are essentially pursuing monetary rewards for their compliance with managerial desires for high productivity. They are likely to work harder only as long as incentive payments continue. Similarly, employees who work harder to avoid punishments administered by a supervisor with coercive power are likely to continue doing so only while the threat of punishment remains salient. In both of these examples, the same performance-outcome linkages that underlie reward and coercive power also encourage compliance, that is, conformity based on the receipt of rewards or the avoidance of punishments.

[9] Herbert C. Kelman, "Compliance, Identification, and Internalization: Three Processes of Attitude Change," *Journal of Conflict Resolution* 2 (1958), 51–60.

T A B L E 13-2
Three Responses to Interpersonal Power

LEVEL	DESCRIPTION
Compliance	Conformity based on desire to gain rewards or avoid punishment. Continues as long as rewards are received or punishment is withheld.
Identification	Conformity based on attractiveness of the influencer. Continues as long as a relationship with the influencer can be maintained.
Internalization	Conformity based on the intrinsically satisfying nature of adopted attitudes or behaviors. Continues as long as satisfaction continues.

Source: Based on Herbert C. Kelman, "Compliance, Identification, and Internalization: Three Processes of Attitude Change," *Journal of Conflict Resolution*, 2 (1958), 51–60.

Identification. Identification, you will recall from Chapter 10, takes place when people accept the direction or influence of other people because they want to establish or maintain satisfying relationships with these people. People come to believe in what they are doing because they identify with those who have asked them to do it, not necessarily because the specific nature of what they are asked to do is important to them. Referent power, discussed by French and Raven, is based on the same sort of personal attractiveness as identification. Consequently, referent power and identification are likely to be closely associated with each other. The use of referent power is likely to stimulate and be stimulated by identification. Charismatic leaders like Steven Jobs, now with Next Inc., or Chrysler's Lee Iacocca have power over others because of their own personal attractiveness. They are able to continue influencing other people's behaviors for as long as identification continues.

Internalization. Through internalization, people may adopt others' attitudes and behaviors because personal needs are satisfied or specific problems are solved. Another reason may be they find those attitudes and behaviors to be congruent with their own personal values. In either case, they accept the others' influence wholeheartedly. It follows that legitimate and expert power can stimulate internalization. Both forms of power rely on personal credibility—the extent to which a person is perceived as truly possessing authority or expertise. This credibility can be used to convince people of the intrinsic importance of the attitudes and behaviors they are being asked to adopt.

As we discussed in Chapter 10, internalization leads people to find newly adopted attitudes and behaviors personally rewarding and self-reinforcing. People who have internalized certain attitudes or behaviors will continue thinking or behaving in a particular way because they find these attitudes or behaviors satisfying in and of themselves. Therefore, a supervisor who can use her expertise to convince colleagues to use consultative leadership can expect the other managers to continue consulting with their subordinates long after she has withdrawn from the situation. In a related vein, a manager whose legitimate power lends credibility to the orders he issues can expect his subordinates to follow those orders even in the absence of rewards, punishments, or charismatic attraction.

A Model of Interpersonal Power: Assessment

French and Raven describe the different kinds of interpersonal power used in organizations, and Kelman's thoughts help identify how people respond to this use. Though valuable as a tool for understanding power and its consequences, the model summarizing their ideas, shown in Figure 13-1, is not entirely without fault. There is some question as to whether the five bases of power it describes are the separate, independent concepts that French and Raven propose or whether they are so closely interrelated as to be virtually indistinguishable from one another. The idea that reward, coercive, and legitimate power often derive from company policies and procedures has led some researchers to subsume these three types of power under **organizational power**. Because expert and referent power are based on personal expertise or charisma, they have sometimes been subsumed under **personal power**.

In fact, French and Raven's five bases of power may be even more closely interrelated than this categorization suggests. In their study of two paper mills, Charles Greene and Philip Podsakoff found that changing just one source of managerial power affected employees' perceptions of three other types of power.[10]

organizational power Types of interpersonal power (reward, coercive, and legitimate power) that often derive from company policies and procedures.

personal power Types of interpersonal power (expert and referent power) that are based on the possession of certain personal traits or characteristics.

[10] Charles N. Greene and Philip M. Podsakoff, "Effects of Withdrawal of a Performance-Contingent Reward on Supervisory Influence and Power," *Academy of Management Journal* 24 (1981), 527–42.

FIGURE 13-1

A Model of Interpersonal Power

Based on Herbert C. Kelman, "Compliance, Identification, and Internalization: Three Processes of Attitude Change," *Journal of Conflict Resolution* 2 (1958), 51–60; and Mario Sussmann and Robert P. Vecchio, "A Social Influence Interpretation of Worker Motivation," *Academy of Management Review* 7 (1982), 177–86.

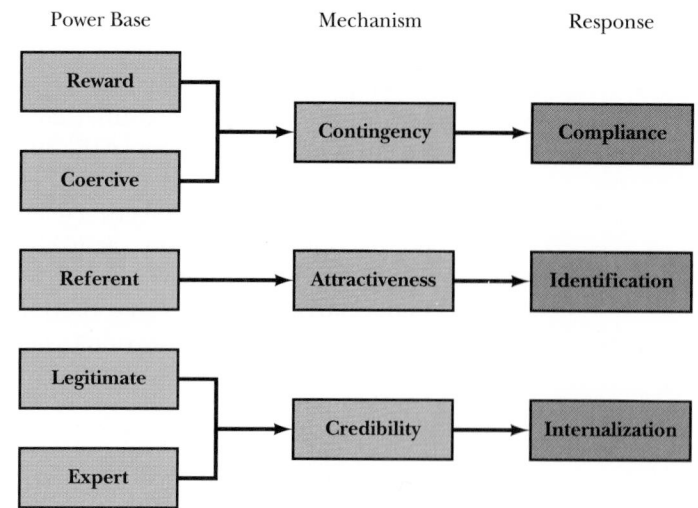

Initially, both paper mills used an incentive-payment plan in which employees' pay was determined by supervisors' monthly performance appraisals. At one mill, the incentive plan was changed to an hourly wage system in which seniority determined an employee's rate of pay. The existing incentive plan was left in place at the other mill. Following this change, the researchers found that employees at the first mill perceived their supervisors as having significantly less reward power—as we might expect—but they also saw significant changes in their supervisors' punishment, legitimate, and referent power. Specifically, they attributed a great deal more punishment power to their supervisors as well as a little less referent power and substantially less legitimate power (see Figure 13-2).

In contrast, employees in the second mill, where the incentive payment remained unchanged, reported no significant changes in their perceptions of their supervisors' reward, punishment, legitimate, and referent power. Because all other conditions were held constant in both mills, employees' changed per-

FIGURE 13-2

Effects of a Change in Method of Payment on Perceived Bases of Power

The change from incentive payment controlled by supervisory appraisals to hourly wages based on seniority eliminated supervisors' reward power and also affected subordinates' perceptions of their supervisor's legitimate, referent, and punishment power. Based on Charles N. Greene and Philip M. Podsakoff, "Effects of Withdrawal of a Performance-Contingent Reward on Supervisory Influence and Power," *Academy of Management Journal* 24 (1981), 527–42.

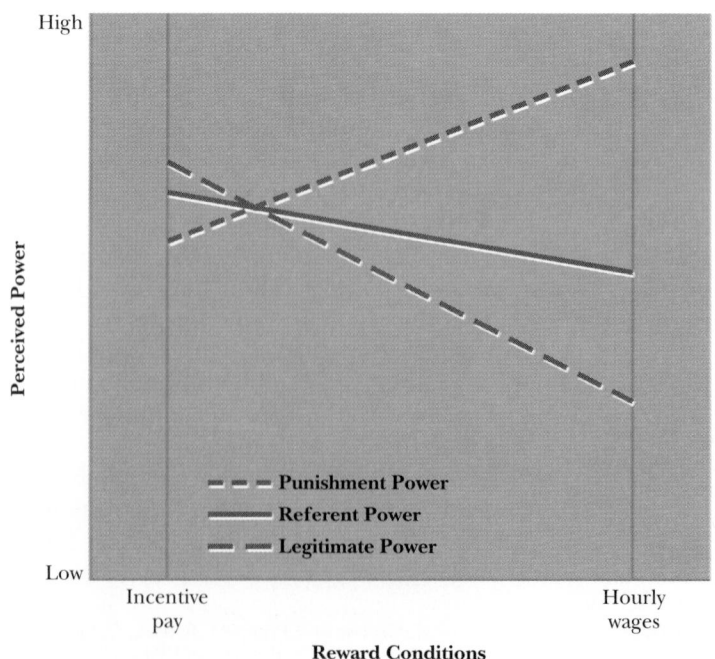

POWER, POLITICS, AND CONFLICT

FIGURE 13-3

The Critical Contingencies Model of Group Power

The ability to cope with uncertainty or to provide other critical contingencies combines with a group's centrality and substitutability to influence its power. For instance, a personnel department's ability to identify and hire skilled employees gives it power if other departments cannot attract employees on their own but know that the personnel department can do it for them. Based on David J. Hickson, C. Robin Hinings, Cynthia A. Lee, Rodney H. Schneck, and Johannes M. Pennings, "A Strategic Contingencies Theory of Intraorganizational Power," *Administrative Science Quarterly* 16 (1971), 216–229.

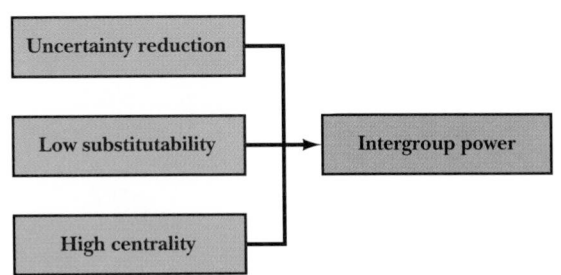

ceptions in the first mill could not be attributed to other unknown factors. Consequently, this study suggests that perceptions of reward, coercive, legitimate, and referent power are closely linked in the workplace. We can therefore conclude that four of the five types of power identified by French and Raven appear virtually indistinguishable to interested observers.[11] Despite its limitations, the model formed by joining French and Raven's classification scheme with Kelman's is useful in analyzing social influence and *interpersonal* power in organizations. Managers can use the model to help predict how subordinates will conform to directives based on a particular type of power. For example, how likely is it that the use of expertise will result in long-term changes in subordinates' behavior? Since the model shown previously in Figure 13-1 indicates that internalization is stimulated by the use of expert power, long-term behavioral changes are quite likely to occur. Alternatively, subordinates may find the model useful as a means of understanding—and perhaps influencing—the behaviors of their superiors. Can you explain why an employee interested in influencing his boss to permanently change her style of management is best advised to try using personal expertise?

Structural Sources of Power

In addition to the interpersonal sources discussed so far, power also originates in the *structure* of patterned work activities and flows of information found in every organization. In Chapter 15 we will examine the topic of organization structure in great detail, so we will limit our current discussion to those characteristics that shape power relations—uncertainty reduction, substitutability, and centrality. As depicted in Figure 13-3, these three variables combine to form the critical contingencies model of power.[12]

critical contingencies Events, activities, or objects that are required by an organization and its various parts to accomplish organizational goals and to ensure continued survival.

Uncertainty Reduction. Critical contingencies are things an organization and its various parts need in order to accomplish organizational goals and continue surviving. For example, the raw materials needed by a company to manufacture the goods it sells are critical contingencies. So, too, are the employees who make these goods, the customers who buy them, and the banks that provide loans to buy equipment. Information can also be a critical contingency. Consider the

[11] Another criticism of our model concerns problems with the measures and methods used to study the French and Raven classification scheme. For further information about these problems and their effects on power research see Gary A. Yukl, *Leadership in Organizations* (Englewood Cliffs, NJ: Prentice Hall, 1981), pp. 38–43; and Philip M. Podsakoff and Chester A. Schreisheim, "Field Studies of French and Raven's Bases of Power: Critique, Reanalysis, and Suggestions for Future Research," *Psychological Bulletin* 97 (1985), 387–411.

[12] David J. Hickson, C. Robin Hinings, Cynthia A. Lee, Rodney H. Schneck, and Johannes M. Pennings, "A Strategic Contingencies Theory of Intraorganizational Power," *Administrative Science Quarterly*, 16 (1971), 216–29; Jeffrey Pfeffer and Gerald R. Salancik, *The External Control of Organizations: A Resource Dependence Perspective* (New York: Harper & Row, 1978) p. 231; and Jeffrey Pfeffer, *Power in Organizations* (Marshfield, Mass.: Pitman, 1981), pp. 109–22.

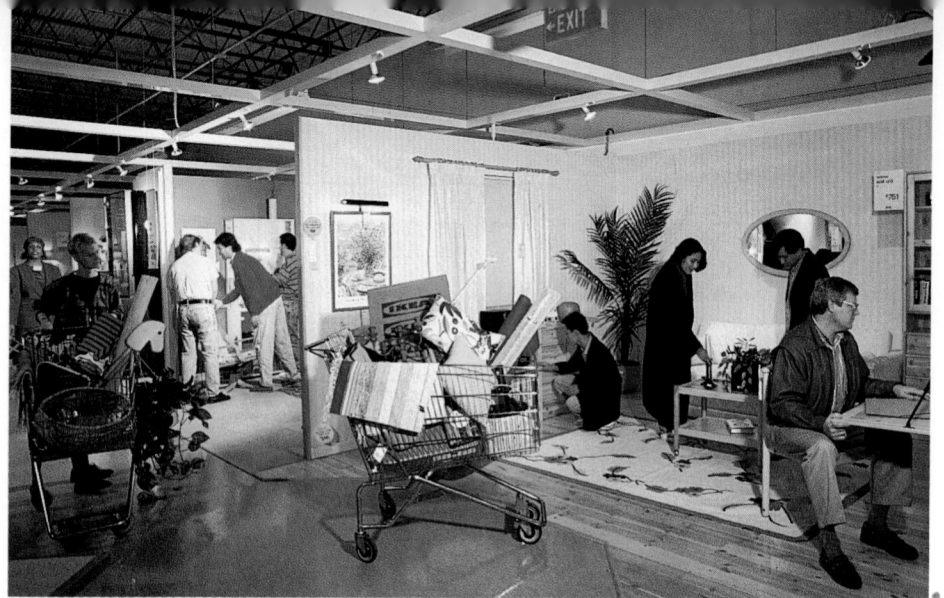

The power of the Swedish furniture and housewares seller IKEA to control two crucial resources—its goods and its customers' good will—enabled it to nearly double its sales between 1987 and 1991. IKEA, which has stores in 22 countries including the United States, regularly undersells its competitors by 20 to 40 percent. It employs fewer salespeople, substituting a catalog of prices and specifications and plenty of space and time for customers to roam about, gather information, and decide on a purchase. The store's low prices and its relaxed atmosphere as well as its free babysitting services keep its customers very satisfied.
Source: Bill Saporito, "IKEA's Got 'Em Lining Up," Fortune, March 11, 1991, p. 72.

AN EXAMPLE: SEARS
Sears has been able to maintain its prices and product quality through resource control. Sears contracts with suppliers who are dependent on Sears for as much as 90 percent of their business. As a result, Sears has considerable power in negotiating prices and enforcing quality standards.

financial data used by banks to decide whether to grant loans, or the mailing lists required by catalog merchandisers to locate and attract prospective customers.

Uncertainty about the continued availability of such critical contingencies threatens organizational well-being. For example, if a purchasing manager cannot be certain she can buy raw materials at reasonable prices, her organization's ability to start or continue productive work is compromised. Similarly, when a marketing department reports shifting consumer tastes, its firm's ability to sell what it has produced is threatened. Thus, as explained by Jerry Salancik and Jeffrey Pfeffer, the critical contingencies model of power is based on the principle that "those [individuals or groups] most able to cope with [their] organization's critical problems and uncertainties acquire power."[13] In other words, individuals or groups that can reduce uncertainty on behalf either of other groups or of the whole organization may be able to exert influence by trading uncertainty reduction for whatever they want in return.

One way to reduce uncertainty is to gain *resource control*, that is, to acquire and maintain access to those resources that are otherwise difficult to get.[14] A personnel department may be able to reduce an important source of uncertainty in an organization that has had problems attracting qualified employees if it can hire and retain an acceptable work force. Similarly, a purchasing department that can negotiate discounts on raw materials can help reduce uncertainty as to whether the firm can afford to continue to produce its line of goods. Finally, a sales group that helps maintain market share by keeping key customers satisfied is preserving another very important resource, or critical contingency. Each of these groups, by delivering crucial resources and thereby reducing success-threatening uncertainty, gains power.[15]

Information control offers another way of reducing uncertainty in organizations. Providing information about critical contingencies is particularly useful when such information can be used to predict or prevent threats to organizational operations.[16] Suppose, for example, that a telecommunication company's legal department learns of impending legislation that will restrict the company's ability to buy additional television stations unless it divests stations it already owns. By

[13] Gerald R. Salancik and Jeffrey Pfeffer, "Who Gets Power and How They Hold On to It: A Strategic-Contingency Model of Power," *Organizational Dynamics* 5 (1977), 3–4.

[14] Rosabeth Moss Kanter, "Power Failures in Management Circuits," *Harvard Business Review*, 57 (1979), 65–75.

[15] Robert H. Miles, *Macro Organizational Behavior* (Santa Monica, Calif.: Goodyear, 1980), pp. 171–72.

[16] Ibid., p. 171.

POWER, POLITICS, AND CONFLICT

alerting management and recommending ways to form subsidiary companies to allow continued growth, the firm's legal department may eliminate a lot of uncertainty for the firm. Or suppose an automobile manufacturer's market research group identifies an emerging consumer demand for safe cars. By providing important information about consumers' preferences, the market research department reduces uncertainty for the firm.

A third way to reduce uncertainty is to acquire *decision-making control*, that is, to have input into the initial decisions about what sorts of resources are going to be critical contingencies. At any time, events may conspire to give certain groups power over others, power that allows the former to determine the rules of the game or to decide such basic issues as what the company will produce, to whom it will market the product, and what kinds of materials, skills, and procedures are needed. In our opening story, Charles Parry apparently thought he had acquired decision-making control when he became Alcoa's chief executive. Since Alcoa's founding, the company's sole business had been the production and distribution of aluminum. Parry wanted to redefine Alcoa, however, as a company involved partly in the aluminum business but also in a variety of other business ventures. If the company had adopted this revised mission, aluminum production and marketing would have become less important. Other contingencies associated with conducting the new businesses would have emerged as critical sources of uncertainty.

Having the ability to create and impose definitions enables powerful groups to remain powerful. For instance, Alcoa's board members were knowledgeable about the aluminum industry and capable of reducing many of its critical uncertainties themselves. They retained power by imposing their definition of the company—"Alcoa is an aluminum business"—on Parry and other company officials. Having power can even enable those already in power to make the contingencies they manage more important to organizational well-being. Thus marketing research departments sometimes report the results of their research using advanced statistics that cannot be interpreted by other managers. Management therefore develops additional reliance on marketing researchers as interpreters of the reports they generate. In this manner, power can be used to acquire power of even greater magnitude—"the rich get richer."[17]

substitutability The extent to which other people or groups can grant access to the same critical contingencies provided by the focal person or group.

Substitutability. Whether individuals or groups gain power as a result of their success at reducing uncertainty depends partly on their **substitutability**. Simply put, if others can serve as substitutes and reduce the same sort of uncertainty, then individuals or departments who need help in coping with uncertainty can turn to a variety of sources and no single source is likely to acquire much power. For example, a personnel department's ability to attract potential employees does not help it gain power if other groups in the same organization are also able to bring in new workers. Similarly, a legal department's ability to interpret laws and regulations is unlikely to yield power for the department if legal specialists working in other departments can fulfill the same function. Owing to the presence of substitutes, other departments are able to ignore the pressures of any particular group, and so each group's ability to amass power is undermined.

If others, however, cannot get help in coping with uncertainty from any but the target person or group, this person or group is clearly in a position to bargain uncertainty reduction for desired outcomes. Alternatively, the person or group can withhold assistance as punishment for failure to conform. For example, a research and development group that is a company's sole source of new product ideas can threaten to reduce the flow of innovation if the firm does not provide

[17] Gerald R. Salancik and Jeffrey Pfeffer, "The Bases and Uses of Power in Organizational Decision Making," *Administrative Science Quarterly* 19 (1974), 470.

the resources it wants. Or a hospital's staff of physicians can raise or lower the number of patients it is willing to see depending on the treatment it receives from hospital administrators. As you can see, the less substitutability there is in a situation, the more likely it is that a particular person or group will be able to amass power.[18]

centrality The position of person or group within the flow of work in an organization.

AN EXAMPLE
For years personnel departments had little power because much of what they did (e.g., recruit applicants, process paperwork) could be done just as well by others in the organization. In recent years these departments have been gaining in power. They hold expertise regarding affirmative action and labor laws. They are now in a position to reduce uncertainty for the organization by ensuring that personnel policies and procedures comply with state and federal regulations. No other part of the organization can provide this expertise, and personnel experts have become better at letting the rest of the organization know of their abilities.

APPLYING THE DIAGNOSTIC MODEL
Diagnostic Question 3: If power inadequacies exist, are they traceable to limitations in the ability to reduce uncertainty? To high substitutability? To low centrality? What actions can be taken to correct these deficiencies?

Centrality. The ability of a person or a group to acquire power is also affected by its **centrality**, or its position within the flow of work in the organization. The ability to reduce uncertainty is not likely to affect a group's power if no one outside the group knows it has this ability and no one inside the group knows how important the ability is. Ignorance of both types is especially likely in the case of groups that (1) do not have a lot of connections with other groups and (2) have little or no effect on the flow of work through the firm. Thus, simply because few other people know of its existence, a clerical staff located on the periphery of a company is unlikely to be able to amass power even if its typing and filing activities bring it in direct contact with critically important information. Even when uncertainty emerges that the staff could help resolve, it is ignored because no one is aware of the knowledge and abilities the staff members possess.

The Critical Contingencies Model: Assessment

There is strong research support for the critical contingencies model's suggestion that power is a function of uncertainty reduction, substitutability, and centrality. For instance, an analysis of British manufacturing firms in business during the first half of the 20th century confirmed this idea. The analysis revealed that accounting departments dominated organizational decision making in the depression era preceding World War II because they kept costs down at a time when money was scarce.[19] Following the war, power shifted to purchasing departments as money became more readily available and strong consumer demand made access to plentiful supplies of raw materials more important. Then during the 1950s, demand dropped so precipitously that marketing became the most important problem facing British firms. As a result and as the model predicts, marketing and sales departments that succeeded in increasing company sales gained power over important decision making processes.

In another study, researchers examined 29 departments of the University of Illinois, looking at the departments' national reputations, teaching loads, and financial receipts from outside contracts and grants.[20] Results indicated that each department's ability to influence university decision making was directly related to its reputation, teaching load, and grant contributions. In addition, the amount of contract and grant money brought in from the outside had an especially strong effect on departmental power. Contracts and grants are sources of operating funds critical to the survival of a public institution like the University of Illinois. Thus, as predicted by the critical contingencies model, the power of each of the departments in the university was directly related to its ability to contribute to the management of critical contingencies.

An even more intriguing piece of evidence supporting the critical contingencies model was discovered by Michel Crozier, a French sociologist who studied

---

[18] Hickson et al., "A Strategic Contingencies Theory," p. 40.

[19] Henry A. Landsberger, "A Horizontal Dimension in Bureaucracy," *Administrative Science Quarterly*, 6 (1961), 299–332.

[20] Salancik and Pfeffer, "The Bases and Uses of Power." See also Jeffrey Pfeffer and Gerald R. Salancik, "Organizational Decision Making as a Political Process: The Case of a University Budget," *Administrative Science Quarterly* 19 (1974), 135–51.

POWER, POLITICS, AND CONFLICT

a government-owned tobacco company located just outside of Paris.[21] As described by Crozier, maintenance mechanics in the tobacco company sought control over their working lives by refusing to share knowledge needed to repair crucial production equipment. The mechanics memorized repair manuals and threw them away so that no one else could refer to them. In addition, they refused to let production employees or supervisors watch as they repaired the company's machines. They also trained their replacements in a closely guarded apprenticeship process so that outsiders could not learn what they knew. Some mechanics even altered equipment so completely that the original manufacturer could not figure out how it worked. In this manner, the tobacco company's maintenance mechanics retained absolute control over the information and skill required to repair production equipment. Because mechanical problems were the most critical form of uncertainty threatening the tobacco plant's productivity, the mechanics' ability to control machine stoppages gave them power over production workers and their supervisors. In essence, maintenance personnel ran the production facility as a result of the information they alone possessed about its equipment.

Crozier's account of the tobacco factory mechanics illustrates the usefulness of the critical contingencies model in explaining why people who have hierarchical authority and formal power sometimes lack the influence needed to manage workplace activities. If subordinates have knowledge, skills, or abilities required to manage critical contingencies, thereby reducing troublesome uncertainties, they may gain the power to refuse to obey hierarchical superiors. Correspondingly, as long as superiors must depend on subordinates to manage such contingencies, it will be the subordinates and not the superiors who determine which orders will be followed and which will be ignored.[22]

In sum, the critical contingencies model appears to depict the structural bases of power quite accurately. Its utility for contemporary managers lies in the observation that the roots of power lie in the ability to solve crucial organizational problems. It is important for managers to know about these roots because such knowledge can help them acquire and hold on to the power needed to do their jobs. We will note several tactics that can be used for these purposes as we discuss politics and political processes in organizations.

politics Activities in which individuals or groups acquire power and use it to advance their own interests.

POINT TO STRESS
It is important to note the positive aspects of politics. We generally discuss politics and politicians with disdain, but it is often politics that aid in bringing about necessary change or retaining needed stability. This occurs in organizations as well as governments.

ORGANIZATIONAL POLITICS

Politics can be defined as activities in which individuals or groups acquire power and use it to advance their own interests. Politics is power in action.[23]

In organizations, we can distinguish politics from other uses of power by emphasis on self-interest. Although political behavior may provide organizational benefit, often it is not intended to do so. Politics also differs from other uses of power in that it is often present outside the formally recognized network of

[21] Michel Crozier, *The Bureaucratic Phenomenon* (Chicago: University of Chicago Press, 1964), pp. 153–54.

[22] Chester I. Barnard, *The Functions of the Executive* (Cambridge, Mass.: Harvard University Press, 1938) p. 163; David Mechanic, "Sources of Power of Lower Participants in Complex Organizations," *Administrative Science Quarterly* 7 (1962), 349–64; Lyman W. Porter, Robert W. Allen, and H. L. Angle, "The Politics of Upward Influence in Organizations," in *Research in Organizational Behavior*, vol. 3, ed. Barry M. Staw and Larry L. Cummings (Greenwich, Conn: JAI Press, 1981) pp. 109–50; and Richard S. Blackburn, "Lower Participant Power: Toward a Conceptual Integration," *Academy of Management Review* 6 (1981), 127–31.

[23] Robert W. Allen, Dan L. Madison, Lyman W. Porter, Patricia A. Renwick, and Bronston T. Mayes, "Organizational Politics: Tactics and Characteristics of Its Actors," *California Management Review* 22 (1979), 77–83; Bronston T. Mayes and Robert W. Allen, "Toward a Definition of Organizational Politics," *Academy of Management Review* 2 (1977), 672–78; Victor Murray and Jeffrey Gandz, "Games Executives Play: Politics at Work," *Business Horizons* 23 (1980), 11–23; Pfeffer, *Power in Organizations*, p. 6.

hierarchical authority. It is the informal, unapproved face of power in organizations and may sometimes involve dishonesty or outright deception.

Political behavior is not necessarily bad. The unsanctioned, unanticipated changes wrought by political processes on outdated policies and procedures can, in fact, enhance organizational well-being. They do so by ridding companies of familiar but dysfunctional ways of doing things.[24] For instance, it was political in-fighting that led managers at Apple to reconsider an earlier decision to avoid any sort of compatibility with archrival IBM's personal computers. Subsequent hardware and software developments enabled Apple's Macintosh computers to read data files created by IBM PCs. Apple was then able to expand its market by convincing former IBM customers that they could convert to Macintosh machines without losing existing files.

Nonetheless, because politics has a negative connotation, political behavior is seldom discussed openly in organizations. Indeed, managers and employees may even deny that politics has any influence whatsoever on organizational activities. Research indicates, however, that politicking *does* occur and that it has measurable effects on organizational behavior.[25] We will now discuss many of the findings reported in research on politics and also introduce several of the management recommendations it suggests.

Personality and Politics

Why do people engage in politics? As with power in general, certain personal characteristics predispose people to exhibit political behaviors. For example, there is the need for power (n Pow) identified by McClelland and discussed previously. Just as n Pow drives people to seek out influence over others, it also motivates them to use this power for political gain. Other researchers have suggested that people who evidence the personality characteristic of Machiavellianism may also be inclined toward politics. **Machiavellianism** is defined as the tendency to seek to control other people through opportunistic, manipulative behaviors. Self-conscious people may be less likely than others to get involved in office politics because they fear being singled out as a focus of public attention and being evaluated

Machiavellianism A personality trait characterized by the tendency to seek to control other people through opportunistic, manipulative behaviors.

---

[24] Miles, *Macro Organizational Behavior*, p. 155.

[25] Allen et al., "Organizational Politics," p. 77; Murray and Gandz, "Games Executives Play;" and Abraham Zaleznik, "Power and Politics in Organizational Life," *Harvard Business Review* 48 (1970), 47–60.

According to several executive search firms, the 1990s will force CEOs to fulfill new assignments. Entrepreneurial skills will be less in demand, and the ability to raise capital and reduce debt will be at a premium. When he took over at General Dynamics in January 1991, William Anders declared that the defense industry faced a shrinking market and that it must become smaller, more productive, and focus more on profitability. Six months later he moved the company's headquarters from St. Louis to the Washington, D.C. area, clearly signaling the company's commitment to its principal customers at the Pentagon.
Source: Jennifer Reese, "CEOs: More Churn at the Top," Fortune, March 11, 1991, pp. 12–13; Richard W. Stevenson, "Mr. Anders Moves to Washington," The New York Times, June 23, 1991.

negatively for engaging in politics. This fear keeps them from seeking power and using it for personal gain.[26]

Antecedent Conditions and Politics

In addition to personality characteristics such as n Pow and Machiavellianism, certain antecedent conditions also encourage political activity in organizations (see Figure 13-4). One such antecedent condition is *uncertainty* of the sort that can be traced to ambiguity and change (see Table 13-3). Uncertainty reduces constraints on behaviors. If their behaviors are hidden or disguised by ambiguity and change, people find it possible to engage in self-interested activities that would otherwise be prohibited. Uncertainty also triggers political behavior because it gives people something to be political about. People want to find a way to reduce uncertainty and to benefit personally from its reduction.

As an example of how uncertainty can encourage politics, consider the following:

> In a budget presentation of all departments in a local community college, one head made a very pious and long presentation outlining the very deep cuts to be made in his area. The rest were moved by his presentation and a portion of his funds were restored. Following the meeting, the head asked for feedback on his "performance" from a small group of his staff who were also at the meeting. All were unanimous in awarding full marks for presentation, piety, general dishonesty, and success of the game plan.

In this example, uncertainty surrounding an important resource—the distribution of a community college's budget among its departments—provided the motivation for a department head to engage in politics. The head purposely deceived his colleagues to win a budget increase for himself and his department. Uncertainty surrounding budgetary matters also allowed the department head to keep information about his department's budget to himself and his staff. That enabled him to get away with his deception. Had the community college required that all departmental spending records be made public, others hearing the head's budgetary presentation, would have been aware of the deception and would probably have denied his appeals.

Besides uncertainty, some other antecedent conditions that may encourage political behavior are *organizational size, hierarchical level, membership heterogeneity,* and *decision importance.* Politicking is more prevalent in larger organizations than in smaller ones. The presence of a greater number of people is more likely to hide the behaviors of any one person, enabling him to engage in political behaviors with less fear of discovery. Politics is also more common among middle and upper

[26] David C. McClelland, "The Two Faces of Power," *Journal of International Affairs,* 24 (1970), 32–41; R. Christie and F. L. Geis, *Studies in Machiavellianism* (New York: Academic Press, 1970), pp. 1–9; and Gerald R. Ferris, Gail S. Russ, and Patricia M. Fandt, "Politics in Organizations," in *Impression Management in the Organization,* ed. R. A. Giacalone and P. Rosenfeld (Hillsdale, N.J.: Erlbaum, 1989), pp. 143–70.

APPLYING THE DIAGNOSTIC MODEL
Diagnostic Question 4: Is politics undermining satisfaction or performance in the organization? Can antecedent conditions promoting politics be eliminated?

FIGURE 13-4

A Model of the Emergence of Politics

Based on Gerald R. Ferris, Gail S. Russ, and Patricia M. Fandt, "Politics in organizations," in *Impression Management in the Organization* ed. R. A. Giacalone and P. Rosenfeld (Hillsdale, N.J.: Erlbaum 1989), pp. 143–70.

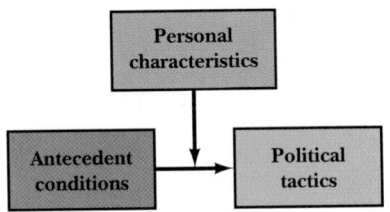

T A B L E 13-3
Types of Uncertainty That Encourage Politics

1. Interruptions in the availability of critical resources or of information about these resources.
2. Ambiguity (no clear meaning) or equivocality (more than one possible meaning) in the information that is available.
3. Goals, objectives, work roles, or performance measures that are not well defined.
4. Unclear decision rules for such things as who should make decisions, how decisions should be reached, or when decision making should occur.
5. Change of any type; for example, reorganization, budgetary reallocations, or procedural modifications.
6. Dependence on other individuals or groups, especially when such dependence is accompanied by competitiveness or hostility.

Source: Based on Don R. Beman and Thomas W. Sharkey, "The Use and Abuse of Corporate Politics," *Business Horizons* 30 (1987), 26–30; Anthony Raia, "Power, Politics, and the Human Resource Professional," *Human Resource Planning* 8 (1985), 198–209; and John P. Kotter, "Power, Dependence, and Effective Management," *Harvard Business Review* 53 (1977), 125–36.

AN EXAMPLE
Several members of a department at a major university used a combination of coalition formation and impression management to gain considerable power in department decisions. The coalition wanting power met behind closed doors and made decisions on how they would vote on issues. The vote was generally in the direction favored by the chair. Before long a definite in-group and out-group formed. Within three years the destruction of the department had begun. Productive faculty began to leave. This was compounded by an inability to hire replacements due partly to budget constraints and partly to the conflict in the department.

coalition A group that forms to allow its members to combine their political strength in order to pursue interests they hold in common.

managers than at other levels of the organizational hierarchy. That is because the power required to engage in self-interested influence attempts is usually concentrated at upper hierarchical layers. In heterogenous organizations, members share few interests and values and therefore see things very differently. Under such circumstances, political processes are likely to emerge as members compete to decide whose interests will be satisfied and whose will not. Finally, important decisions stimulate more politics than unimportant decisions do simply because less important issues merit less interest and attention.

Political Tactics

When personal characteristics and antecedent conditions are favorable, a variety of political tactics may surface. Each tactic is intended to increase the power of one person or group relative to others. When power increases, so does the likelihood that the person or group will be able to seek out and acquire self-interested gains.[27]

Acquiring Interpersonal Power: Forming Affiliations. Forming **coalitions** or political affiliations with each other is one important way for people to increase their power and pursue political gain.[28] By banding together, people can share any control they might have over rewards or punishments. They can also combine their expertise, legitimacy, and charisma. Pooling their power in this manner enables the members of a coalition to pursue mutually desirable outcomes that would be beyond their individual grasp. For instance, employees sometimes form unions to pursue economic rewards and improved working conditions. Collective bargaining enables union members to obtain wages and conditions far superior to those they could demand as individuals. On the other side of the coin, companies also form trade associations to exchange information about collective bargaining and union agreements. This enables employers to help each other bargain for affordable union contracts.

[27] Pfeffer, *Power in Organizations*; Richard L. Daft and Richard M. Steers, *Organizations: A Micro/Macro Approach* (Glenview, Ill.: Scott, Foresman and Company, 1986), pp. 488–89; and Allen et al., "Organizational Politics," pp. 77–83.
[28] William B. Stevenson, Jone B. Pearce, and Lyman W. Porter, "The Concept of Coalition in Organization Theory and Research," *Academy of Management Review* 10 (1985), 256–68.

POWER, POLITICS, AND CONFLICT

The Fine Art of Ingratiation

Suppose you want to improve your position with your boss. How do you do it? One way involves flattery, but through an indirect approach so as to be inconspicuous. Identify a co-worker of yours or a colleague of your boss who interacts with your boss on a regular basis, then praise your boss in front of this person. He or she will often pass your praise on. Another way to ingratiate yourself is to intentionally disagree with your boss, only to yield to your boss's persuasion at a later point in time. This approach enables you to avoid the stigma of looking like a yes man while still allowing you to gain your boss's trust by concurring with her way of seeing things. A third way is to present yourself as what you suspect is your boss's idea of a perfect employee. Being seen working late at the office—if your boss likes hard workers—is one way of managing the impression you make. *Not* being seen working late—if your boss likes employees who work smarter rather than harder—is another approach. In either case, impression management of this sort helps communicate the message that you accept the norms and values your boss espouses, thus, that you are worthy of your boss's respect and consideration.

Does "sucking up" of this sort pay off? Yes, according to several research studies. Employees who ingratiate themselves are more likely to have favored in-group status with those who can grant them power. They may be the recipients of important task assignments, delegated to them in order to prepare them for advancement. At the same time, ingratiators often come to believe in what they are doing, meaning that their attempts to play the role of perfect employee may, in fact, increase their sense of job commitment and improve their level of performance. A moderate amount of ingratiation may benefit everyone.*

* Edward E. Jones, *Ingratiation* (New York: Appleton-Century-Crofts, 1964); Camille B. Wortman and Joan A. W. Linsenmeier, "Interpersonal attraction and techniques of ingratiation in organizational settings," in *New Directions in Organizational Behavior*, eds. Barry M. Staw and Gerald R. Salancik, (Chicago: St. Clair Press, 1977), pp. 133–78; Rahul Jacob, "Sucking up pays off," *Fortune* December 18, 1989, p. 12.

As part of the process of forming political affiliations, doing favors for others is sometimes used to create a sense of indebtedness. Those receiving favors are obliged to reciprocate by doing favors in return. People who pursue this tactic can increase the dependence of others by building up a bank of favors that are owed them. In the U. S. Congress, for instance, members from industrial regions will vote for bills providing farm subsidies with the understanding that farm-state representatives will reciprocate by supporting bills that secure industrial assistance grants.

cooptation Making former adversaries into allies by involving them in planning and decision making-processes.

Besides exchanging favors, people engaging in politics sometimes use cooptation to preserve their interests in the face of adversity. In **cooptation** former adversaries are made into allies by involving them in planning and decision-making processes. Colleges and universities often use this tactic during periods of campus unrest, inviting student protestors to join university representatives on administrative committees. Making opponents part of the team often silences their objections. However, such a tactic risks major changes in plans and decisions.

ingratiation The use of praise and compliments to gain the favor or acceptance of others.

impression management Behaving in ways intended to build a positive public image.

Finally, ingratiation and impression management can be used to build and maintain political relationships. As described more fully in the "In Practice" box, **ingratiation** is the use of praise and compliments to gain the favor or acceptance of others. Similarly, **impression management** involves behaving in ways intended to build a positive image. Both ingratiation and impression management are used to increase personal attractiveness in order to increase the likelihood that others will seek a close relationship.

Acquiring Structural Power: Controlling Critical Resources. Another approach that can be used to amass power and further personal interests is to gain access to supplies of critical resources and control their distribution. As suggested by the critical contingencies model of organization power, controlling the supply of a critical resource gives people power over those whose success or survival depends on having access to that resource. A warehouse manager, for example, can decide which orders will be filled immediately and which will be delayed. He can acquire power by determining who will receive timely supplies of critical raw materials. As a political tool, power of this sort can be used to ensure that personal interests are satisfied.

Similarly, controlling access to information sources provides power over those who need information to reduce uncertainty. As part of this tactic, political players often attempt to control access to people who are sources of important information or expertise. It is not uncommon, for instance, for managers to shield from others in their firm staff specialists who advise them. Engineers working on new product development are often sequestered from other employees; cost accountants may be separated from others in a company's accounting department. Employees like these are an important resource because they possess critical information that is unavailable elsewhere.

To succeed as a political tactic, controlling access to important resources, information, or people, requires eliminating substitutes for these critical resources and discrediting alternative definitions of what is critical. The presence of substitutes counteracts attempts to win power by controlling critical resources because political efforts are neutralized. In addition, successful control of critical resources requires that people have at least the centrality needed to identify which resources are critical and which are not.

Negative Politics. If all else fails, the political upper hand can sometimes be gained over others by attacking or blaming them. This strategy reduces the power they can wield in their own defense. Blaming others for negative outcomes and making them **scapegoats** for failures is one way political players attempt to triumph over adversaries. Another way is to denigrate or belittle others' accomplishments. Either approach involves a direct attack on the interpersonal sources of power others might possess in an attempt to weaken their political position. Such an attack could create doubt about their ability to control rewards and punishments or reduce their credibility, legitimacy, or attractiveness. Negative politicking can also provide reasons for creating substitute sources of critical resources or information or reducing the degree of centrality enjoyed by a person or group. After all, who would want an incompetent individual or group in charge of something that is critically important to organizational survival?

Managing Destructive Politics

You can imagine some of the consequences when people band together, hoard resources, or belittle each other for no other reason than to get their own way. Morale may suffer; battle lines between contending individuals or groups may impede important interactions; energy that should go into productive activities may instead be spent on planning attacks and counterattacks if politicking is left uncontrolled. For this reason, controlling political behavior is a big part of every manager's job.[29]

Set an Example. One way to manage destructive politics is to set an example. Managers who do not tolerate deceit and dirty tricks and refuse to engage in politics themselves make it clear that political tactics are inappropriate. Subordinates are thus discouraged from engaging in destructive political activities. In

[29] The political management techniques described in this section are based on discussions in Robert P. Vecchio, *Organizational Behavior* (Chicago, Dryden Press, 1988), pp. 270–72; and Gregory Moorhead and Ricky W. Griffin, *Organizational Behavior* (Boston: Houghton Mifflin, 1989), pp. 377–78.

scapegoats People who are blamed, whether rightly or not, for the failures of groups or organizations.

476   POWER, POLITICS, AND CONFLICT

contrast, managers who engage in politics—blaming their mistakes on others, keeping critical information from others—convey the message that politics are acceptable. It is little wonder that subordinates in such situations are themselves prone to politicking.

AN EXAMPLE: PRESTON TRUCKING
Controlling political behavior increasingly means closing the trust gap between management and employees. When people trust each other there is no need for politicking. Preston Trucking regularly surveys its employees attitudes and solicits their suggestions for improvements in the company's operations. In one year alone the suggestion program brought in 4,412 money-making ideas. And interestingly, even though employees get no material reward for their ideas they continue to submit them because they feel that management respects them and is attentive to them. ("Managing," *Fortune*, December 4, 1989.)

APPLYING THE DIAGNOSTIC MODEL
Diagnostic Question 5: Are managers controlling politics by setting an example, encouraging open communications, managing coalitions, confronting game players, and anticipating future occurrences?

Communicate Openly. By sharing all relevant information with coworkers and colleagues you can alleviate destructive politics. Managers who communicate openly with their peers, superiors, and subordinates eliminate the political advantage of withholding information or blocking access to important people. Information that everyone already knows cannot be hoarded or hidden. In addition, open communication ensures that everyone understands and accepts resource allocations. Such understanding eliminates the attractiveness of political maneuvers intended to bias distribution procedures. Shrinking the potential benefits of destructive politicking acts to lessen the incidence of political behaviors.

Reduce Uncertainty. A third way to minimize destructive political behavior is to reduce uncertainty. Clarifying goals, tasks, and responsibilities makes it easier to assess people's behaviors and makes politics difficult to hide. Opening up decision making processes by consulting with subordinates or involving them in participatory decision processes (see Chapters 11 and 12) helps to make decisions understandable and discourages undercover politicking. In Chapter 14, we will discuss how explaining changes helps encourage people to accept them and reduces fears that can stimulate politics.

Manage Informal Coalitions and Cliques. Managing informal coalitions and cliques can also help reduce destructive politics. As you will see in Chapter 18, influencing the norms and beliefs that steer group behaviors can ensure that employees continue to serve organizational interests. When cliques resist less severe techniques, job reassignment becomes a viable option. Group politicking is abolished by eliminating the group.

Confront Political Game Players. A fifth approach to managing politics is to confront political game players. When people engage in politics despite initial attempts to discourage such activities, a private meeting between superior and subordinate may be enough to curb the subordinate's political pursuits. If not, it may be necessary to resort to disciplinary measures. Punishments such as a public reprimand or a period of layoff without pay ensure that the costs of politicking outweigh its benefits. If this does not work, managers having to cope with damaging politics may have no choice but to dismiss political game players.

Anticipate the Emergence of Damaging Politics. In any effort to control political behavior, awareness and anticipation are critical. If managers are aware that circumstances are conducive to politicking, they can try to prevent politics altogether. Detection of any of the personal characteristics or antecedent conditions discussed earlier should be interpreted as a signal indicating the need for management intervention *before* destructive politics crop up.

INTERGROUP CONFLICT

conflict A process of opposition and confrontation that can occur between either individuals or groups.

Conflict—a process of opposition and confrontation that can occur between either individuals or groups—is an inevitable feature of organizational life. As we pointed out in Chapter 8, in organizations we may see role

intergroup conflict A process of confrontation that occurs when one group obstructs the progress of one or more other groups.

conflicts *within people* that reflect differences between personal values and role expectations. In Chapter 10 we discussed the conflict *between people* that can arise out of contradictions among their varying role expectations. Now we turn to another important kind of conflict that occurs in organizations—conflict that occurs *between groups*.

Conflict between groups, or **intergroup conflict**, can be defined as a process of confrontation that occurs when one group obstructs the progress of one or more other groups.[30] Key to this definition is the idea that conflict between groups involves confrontation, that is, disputes among groups over clashing interests. Also important is the notion that conflict is a process—something that takes time to unfold, rather than an event that occurs in an instant and then disappears. Finally, to the extent that obstructing group progress threatens group effectiveness and organizational performance, our definition implies that group conflict is a problem that managers must be able to control. In this section, we will discuss the antecedent conditions that stimulate conflict and describe how conflict develops. We will also discuss how to manage conflict when it threatens group and organizational performance. Before considering these important points, though, we will first take a closer look at the basic nature of conflict.

Is Conflict Necessarily Bad?

POINT TO STRESS
It is important to point out that conflict, like politics, can be positive, introducing new ideas and bringing about positive change.

Intergroup conflict might seem inherently undesirable. In fact, many of the classic models of organization and management discussed in Chapter 2 support this view, suggesting that intergroup conflict is an indicator of failure or inadequacy in the formal design of an organization. Classic theorists often likened organizations to machines and therefore tended to see conflict as symptomatic of breakdown. Managers in the days of Henri Fayol and Frederick Taylor concerned themselves with discovering ways either to avoid conflict between groups or to suppress it as quickly and forcefully as possible.

Modern theorists, however, suggest that intergroup conflict is not necessarily bad.[31] To be sure, they say, *dysfunctional* intergroup conflict—confrontation between groups that hinders progress toward organizational goals—does occur. In the late 1980s, for example, a long period of labor-management conflict over wages at Hormel, a Minnesota meat packing firm, nearly caused the firm's failure. In 1989, a dispute between labor and management involving wages and working conditions helped push Eastern Airlines into bankruptcy.

Current research, however, suggests that intergroup conflict is sometimes *functional*, having positive effects such as the following:

1. Intergroup conflict can lessen social tensions, helping to stabilize and integrate relationships. If resolved in a way that allows the discussion and dissipation of disagreements between groups, it can serve as a safety valve that vents pressures built up over time.

2. It lets groups express rival claims and provides the opportunity to readjust inventories and allocations. Resource pools may thus be consumed more effectively due to conflict-induced changes.

3. It helps to maintain the level of stimulation or activation required to function innovatively. In so doing, intergroup conflict can serve as a source of motivation to seek adaptive change (see Chapter 14).

4. It supplies feedback about the state of interdependencies and power distributions that form the organization's structure. As a result, structural char-

---

[30] Miles, *Macro Organizational Behavior*, p. 122.

[31] Robert E. Quinn, *Beyond Rational Management: Mastering the Paradoxes and Competing Demands of High Performance* (San Francisco: Jossey-Bass, 1988), p. 2.

acteristics that promote coordinated effort are more visible and more readily understood (see Chapter 15).

5. Intergroup conflict can help provide a sense of group identity and purpose by clarifying differences and boundaries between groups. Outcomes of this sort are discussed in greater detail later in this chapter.[32]

At the very least, intergroup conflict can serve as a red flag or warning of the need for change. Believing that conflict can have positive effects, contemporary managers try to manage or resolve conflict, not simply avoid or suppress it.

Antecedent Conditions

In order for conflict to occur among groups, several antecedent conditions must be present—interdependence, political indeterminism, and diversity (see Figure 13-5).

Interdependence. Interdependence refers to relations among two or more groups in which the groups depend on each other for assistance, information, compliance, feedback, or other coordinative actions.[33] As you will recall from Chapter 10, four types of interdependence can link groups together—pooled, sequential, reciprocal, and team (see Figure 10-3). Any linkages between groups can become sources of conflict. For example, two groups that share a pooled source of funds may fight over who will receive money to buy new office furniture. Similarly, groups organized along a sequential assembly process may fight about the pace of work. In the absence of interdependence, however, groups have nothing to fight about and, in fact, may not even know of each other's existence.

Political Indeterminism. The emergence of intergroup conflict also requires that the political pecking order among groups be subject to question. If power relations among interdependent groups are unambiguous and stable and if they are accepted as valid by all, appeals to authority will replace conflict, and differences among groups then will be resolved in favor of the most powerful group. Only a group whose power is uncertain will gamble on getting its way through conflict rather than by using power and authority. For this reason, groups in a newly reorganized company are much more likely to engage in conflict than groups in an organization with a stable hierarchy of authority.

[32]Lewis Coser, *The Functions of Social Conflict* (New York: Free Press, 1956), p. 154; Miles, *Macro Organizational Behavior*, p. 123.

[33] Miles, *Macro Organizational Behavior*, p. 131.

AN EXAMPLE
Often it is unclear to employees what it takes to get a promotion. As a result, there may be a great deal of competition among employees, each trying their own political tactics to improve their chances. By clarifying in as much detail as possible what the requirements are to be successful in receiving a promotion, uncertainty is reduced. Appropriate behaviors are defined and contenders are advised to spend their energies on accomplishing those behaviors. This not only reduces benefits from politicking but also reduces the time a person has for this activity and gives him more time to do his job.

FIGURE 13-5

Conditions that Stimulate Intergroup Conflict

Conflict among groups can occur only when groups differ among themselves but are dependent on one another and when lines of authority are obscured by uncertain power relations. For instance, conflict between an engineering department and a research and development department is likely when they must work together to develop new products but neither has the authority to tell the other what to do.

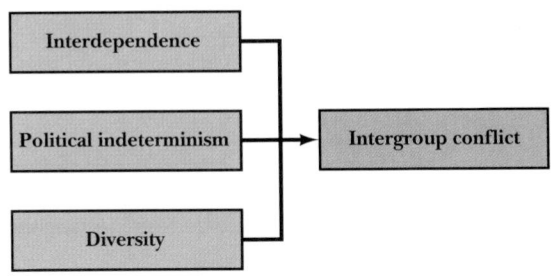

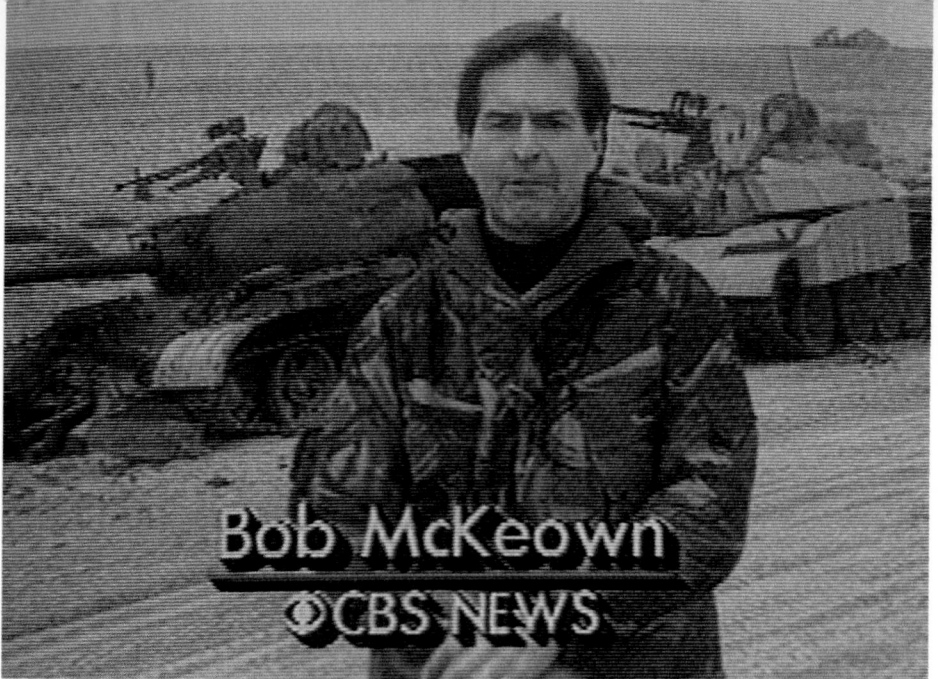

During the Persian Gulf War, the Pentagon attempted to manage a diversity of media representatives by means of a "pool": small groups of reporters were taken periodically to the field but closely watched by military public affairs officers. Some networks obeyed this rule, but others tried to circumvent it. CBS's Bob McKeown, for example, got a scoop by rushing to Kuwait City as the war was drawing to a close. Is there any way that the military could have reconciled its goal to maintain the security of its forces with the media's goal to get the best stories?
Source: Jonathan Alter, "Clippings from the Media War," Newsweek, *March 11, 1991, p. 52.*

APPLYING THE DIAGNOSTIC MODEL
Diagnostic Question 6: Do antecedent conditions—interdependence, political indeterminism, diversity—favor the emergence of intergroup conflict? Is there evidence that dysfunctional latent, perceived, or felt conflict is brewing? Can any of the antecedent conditions be modified to resolve the conflict before it manifests itself?

Diversity. Finally, in order for intergroup conflict to emerge there must be differences or disagreements among the groups that are worth fighting over. Let's look at some of the most common types of diversity that stimulate intergroup conflict.[34]

Owing to differences in the functions they perform, organizational groups may have varying *group goals*. Table 13-4 describes some differences in the goal orientations of marketing and manufacturing groups. As you can see, each group's approach reflects its particular orientation—marketing's focus on customer ser-

[34] Miles, Ibid., pp. 132–38; and John M. Ivancevich and Michael T. Matteson, *Organizational Behavior and Management*, 2nd ed. (Homewood, Ill.: BPI-Irwin, 1990), pp. 309–12.

T A B L E 13-4
Differences in Goal Orientations: Marketing and Manufacturing

GOAL FOCUS	MARKETING APPROACH	MANUFACTURING APPROACH
Product variety	Customers demand variety.	Variety causes short, often uneconomical production runs.
Capacity limits	Manufacturing capacity limits productivity.	Inaccurate sales forecasts limit productivity.
Product quality	Reasonable quality should be achievable at a cost that is affordable to customers.	Offering options that are difficult to manufacture undermines quality.
New products	New products are the firm's life blood.	Unnecessary design changes are costly.
Cost control	High cost undermines the firm's competitive position.	Broad variety, fast delivery, high quality, and rapid responsiveness are not possible at low cost.

Source: Based on information presented in B. S. Shapiro, "Can Marketing and Manufacturing Coexist?" *Harvard Business Review*, 55, 5 (September-October 1977), 104–14.

POWER, POLITICS, AND CONFLICT

AN EXAMPLE
One of the arguments against just-in-time management is that marketing departments do a poor job of forecasting demand. As a result, production is often asked to make changes more rapidly than it can. For smooth operation, the two groups have to recognize and appreciate their different time orientations. Production must realize that predicting customer preferences is not a precise art. In turn, marketing must accept that rapid production changes are as likely as stopping a train on a dime. This conflict can be reduced if top management supports work on improving and simplifying product designs and involving production people in product design. This tends to speed up later production changes. (*Attaining Manufacturing Excellence*, Hall, 1987).

vice, manufacturing's concern with smooth production runs. In such situations, conflicts may occur over whose goals to pursue and whose to ignore.

Groups also may have different *time orientations*. For example, tasks like making a sale to a regular customer require only short-term planning and can be initiated or altered quite easily. On the other hand, tasks like traditional assembly-line manufacturing operations necessitate a longer time frame because such activities require extensive preplanning and are not easy to change once they have begun. Certain tasks, such as the strategic planning activities that plot an organization's future, may even require time frames of several years. When differences between time orientations exist among groups in a firm, conflicts develop about which orientation should regulate task planning and performance.

Often, *resource allocations* among organizational groups are unequal. In state-supported universities, for example, business schools often receive more funding than liberal arts programs. Such differences usually stem from the fact that groups must compete with each other to get a share of their organization's resources. Generally someone wins and someone loses, laying the groundwork for additional rounds of conflict.

Another source of conflict may be the practices used to *evaluate* and *reward* groups and their members. Consider, for example, that manufacturing groups are often rewarded for efficiency, achieved by minimizing the quantity of raw materials consumed in production activities. Sales groups, on the other hand, are more likely to be rewarded for flexibility, which sacrifices efficiency. Conflict is likely to arise in such situations as each group tries to meet its own performance criteria or tries to force others to adopt the same criteria.

In addition, *status discrepancies* invite conflict over stature and position. Although a group's status is generally determined by its position in the organization's hierarchy of authority—groups higher in the hierarchy having higher status— sometimes other criteria also influence status.[35] A group might argue that its status should depend on the knowledge possessed by its members. Or a group might assert that status should be conferred on the basis of such factors as loyalty, seniority, or visibility.

Conflict can emerge in *jurisdictional disputes* when it is unclear who has responsibility for something. For example, if both the personnel and the employing departments interview a prospective employee, the two groups may get into a dispute over which has the ultimate right to offer employment and which must take the blame if mistakes are made.

Finally, groups can differ in the *values, assumptions*, and *general perceptions* that guide their performance. It is not unusual for a group to develop stereotypic perceptions of other groups that it deals with regularly. These perceptions *may* exaggerate otherwise minimal differences between groups in order to give each group a strong sense of identity. In turn, these exaggerated perceptions may cause conflict that centers on what might otherwise appear to be unimportant dissimilarities.

Stages of Conflict

As we mentioned previously, conflict is more appropriately thought of as a continuing process than as a discrete event. To simplify managerial analysis and action planning, the conflict process can be broken into the sequence of five developmental stages diagrammed in Figure 13-6.[36]

---

[35] David Ulrich and Jay B. Barney, "Perspectives on Organizations: Resource Dependence, Efficiency, and Population," *Academy of Management Review* 9 (1984), 471–81.

[36] Louis R. Pondy, "Organizational Conflict: Concepts and Models," *Administrative Science Quarterly* 12 (1967), 296–320.

FIGURE 13-6

A Model of the Conflict Process

Based on Louis R. Pondy, "Organizational Conflict: Concepts and Models," *Administrative Science Quarterly* 12 (1967), 296–320.

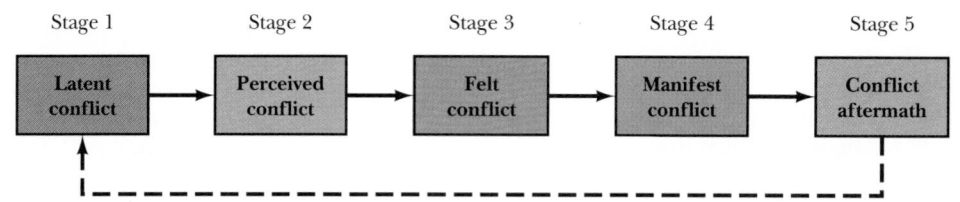

latent conflict The stage of conflict development at which dissension is only suspected or at best dimly perceived.

Latent Conflict. Conflict begins with antecedent conditions—interdependence, political indeterminism, and diversity—that provide the basis for argument, competition, and finally conflict. Sometimes these conditions stem from the aftermath of a preceding conflict. For example, managers in a firm such as Kodak may try tightening the linkages between the company's research and development group and its manufacturing group in an effort to deal with prior disagreements over the development of manufacturing techniques for new cameras. Unintentionally, though, tighter interdependence between the two groups may actually set the stage for additional episodes of intergroup conflict. At the stage of **latent conflict**, dissension is only suspected or at best dimly perceived. There is no open conflict at this stage.

perceived conflict The stage of conflict development at which problems are readily perceived and everyone involved in the conflict knows that it exists.

Perceived Conflict. In the second stage, **perceived conflict**, problems are readily perceived and everyone involved in the conflict knows it exists. During this stage, groups sometimes choose to ignore the problem. If they decide to act, however, perceptions can often be changed, misunderstandings reconciled, and the conflict resolved before it progresses any further. Thus, if the conflict between Kodak's research and manufacturing groups evolves to this stage, the members of both groups are aware that a problem exists. They usually can sit down together and iron out differences without attacking each other or causing further misunderstanding.

felt conflict The stage of conflict development at which people are not only aware of the conflict but feel tense, anxious, angry, or otherwise upset.

Felt Conflict. There is an important difference between perceived and **felt conflict**. When conflict is felt, people are not only aware of it but often feel tense, anxious, angry, or otherwise upset. These feelings emerge as group members involved in the developing conflict personalize the situation, internalizing and attempting to cope with the pressures building between their groups. Conflict can be defused at this stage only by providing a safety valve that allows conflictive feelings to be vented without detrimental effect. In the case of Kodak's research and manufacturing groups, the groups might be forced to confront each other in a carefully monitored meeting. Each group is allowed to verbally attack the other and react to the other's verbal attack in return. By letting groups work through their conflict verbally instead of confronting each other at work, meetings such as the intergroup mirroring intervention described in Chapter 14 help control the extent to which felt conflict undermines workplace performance.

manifest conflict The stage of conflict development at which people engage in behaviors that are clearly intended to frustrate or block their opponents.

Manifest Conflict. At the stage of **manifest conflict** people engage in behaviors that are clearly intended to frustrate or block their opponents. Such behaviors can range from refusals to cooperate to sabotage, verbal abuse, or even physical aggression. Kodak's research-and-development group might refuse to provide manufacturing personnel with the specifications needed to initiate trial production runs. In retaliation, the manufacturing group might sabotage the raw material inventories used by research-and-development scientists as they experiment with different ways of manufacturing cameras. At this stage, relations between conflicting groups are at greatest risk, and organizational performance is seriously

482

POWER, POLITICS, AND CONFLICT

jeopardized. If manifest conflict is not resolved quickly, group and organizational productivity is likely to suffer.

conflict aftermath The stage of conflict development at which conflict sets the stage for later situations and events.

Conflict Aftermath. In the last stage of conflict, **conflict aftermath**, a conflict that has already occurred sets the stage for later situations and events. If the conflict is resolved, the basis for more cooperative relations may be established and the likelihood of recurrences of the conflict reduced. If the conflict is merely suppressed or smoothed over, however, even more serious episodes of conflict may develop. Discord may persist until either differences between conflicting groups are finally resolved or relations between the groups are dissolved. The seeds of future conflicts are often sown by current conflicts.

Effects of Intergroup Conflict

Conflict, especially when manifest, affects relationships within and between groups in several ways. We will look first at changes that typically occur within conflicting groups and then at the changes that occur in the relations between such groups.[37]

APPLYING THE DIAGNOSTIC MODEL
Diagnostic Question 7: Is there evidence of manifest intergroup conflict? Do the dysfunctional, destructive effects of this conflict outweigh its functional benefits? Can any of the antecedent conditions be modified to resolve the conflict?

Changes Within Groups. Within groups engaged in manifest conflict, changes of four types are often observed. First, as we saw in Chapter 11, external threats such as intergroup conflict bring about *increased group cohesiveness*. As a result, groups engaged in intergroup conflict become more attractive and important to their members. Ongoing conflict also stimulates an *emphasis on task performance*. All efforts within each conflicting group are directed toward meeting the challenge posed by other groups, and concerns about individual members' satisfaction lose importance. A sense of urgency surrounds task performance; defeating the enemy becomes uppermost, and there is much less goofing off.

In addition, when a group faces conflict, its members will often submit to *autocratic leadership* to manage the crisis, perceiving participatory decision making as slow and weak. Strong, authoritarian leaders often emerge as a result of this shift. A group in such circumstances is also likely to place much more emphasis on standard procedures and centralized control. As a result, it becomes characterized by *structural rigidity*. By adhering to established rules and creating and strictly enforcing new ones, the group seeks to eliminate any conflicts that might exist among its members and ensure the repetition of task successes.

APPLYING THE DIAGNOSTIC MODEL
Diagnostic Question 10: Is there evidence of conflict aftermath remaining from a previous episode that might encourage additional conflict? Can this aftermath be altered before destructive conflict reemerges?

Changes Between Groups. In addition to these four changes within groups, four changes often occur in relations between conflicting groups. One such change concerns the *hostility* that conflict arouses between groups. Often this hostility surfaces in the form of hardened we-they attitudes. Each group sees itself as virtuous and other groups as enemies. Intense dislike often accompanies these negative attitudes. As attitudes within each group become more negative, group members develop *distorted perceptions* of other groups. They begin to emphasize negative traits and deemphasize positive ones. Negative stereotyping results, creating even greater differences between groups and further strengthening cohesiveness within each group.

---

[37] Muzafer Sherif and Carolyn W. Sherif, *Groups in Harmony and Tension* (New York: Harper, 1953) pp. 229–295; Andrew D. Szilagyi, Jr., and Marc J. Wallace, Jr., *Organizational Behavior and Performance*, 4th ed. (Glenview, Ill.: Scott, Foresman, 1987), p. 301; James L. Gibson, John M. Ivancevich, and James H. Donnelly, Jr., *Organizations: Behavior, Structure, Process* (Plano, Texas: Business Publications, 1988), pp. 314–16; Ivancevich and Matteson, *Organizational Behavior and Management*, pp. 313–15.

In time, negative attitudes and perceptions of group members are likely to fuel a *decrease in communication* among conflicting groups. The isolation that results only adds to the conflict, making resolution even more difficult. At the same time, however, conflicting groups often engage in *increased surveillance* intended to provide information about the attitudes, weaknesses, and likely behaviors of other groups. This covert monitoring of other groups' activities is considered essential to staying ahead of the others and winning the conflict. "Facts" that validate negative stereotypes are given preference to information that portrays intergroup relations in more accurate terms. Additional we-they thinking and further conflict are stimulated as a consequence.

Resolving Conflict through Restructuring or Bargaining and Negotiation

Situations in which conflict causes negative changes can be managed in a variety of ways. In general, conflict-management techniques can be clustered into the two categories shown in Figure 13-7. These are restructuring techniques that focus on changing the nature or meaning of relations among interdependent groups, and bargaining and negotiation procedures that are intended to reduce conflict-causing diversity among the interests of different groups.

APPLYING THE DIAGNOSTIC MODEL
Diagnostic Question 8: If interdependence or political indeterminism can be modified, which of the approaches to managing interdependence seems most suitable?

Restructuring Intergroup Relations. Conflict requires interdependence. Thus it is possible to manage or resolve intergroup conflict by restructuring the connections that tie conflicting groups together.[38] One way to do so involves *developing superordinate goals*, that is, identifying and pursuing a set of performance targets that conflicting groups can achieve only by working together. By requiring conflicting groups to work together to succeed, superordinate goals tighten interdependence. In turn, sharing a common fate requires the groups to look beyond their differences and learn how to cooperate with each other. In the automobile industry, for instance, unions and management fearing plant closures have forgone adversarial relations in order to strengthen the competitiveness of their firms. Teamwork has replaced conflict in the pursuit of the superordinate goal of producing high quality products for today's world markets.

AN EXAMPLE
During World War II President Roosevelt was able to eliminate union/management conflict for a while by getting the two sides to join together to accomplish a superordinate goal: to quickly produce the materials needed to win the war.

Another way to manage conflict by restructuring interdependence involves *clarifying hierarchical distinctions* and making the political position of each group readily apparent. If it is feasible, this political clarification affects interdependence between groups by strengthening the groups' understanding of how and why they are connected. This approach helps resolve conflict because it reduces the political indeterminism that must exist for conflict to occur. *Expanding the supply of critical resources* is yet a third way to restructure because it removes a major source of conflict between groups that draw from the same supply. Pools of critical

[38] Muzafer Sherif, "Superordinate Goals in the Reduction of Intergroup Conflict," *American Journal of Sociology* 63 (1958), 349–56; Jay R. Galbraith, "Organization Design: An Information Processing View," *Interfaces* 4 (1974), 28–36; and Pfeffer, *Power in Organizations*.

F I G U R E   13-7

Conflict Resolution Techniques

Intergroup conflict can be resolved by managing interdependence or by managing diversity. Both these approaches seek to reduce or eliminate the negative antecedent effects of diversity, political indeterminism, or interdependence.

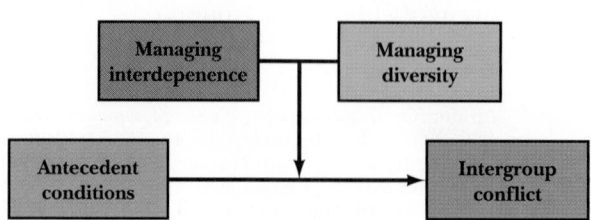

resources are not easily enlarged—which is what makes them critical to begin with. When this method is successful it decreases the amount of interdependence between groups by making groups compete less for available resources. For example, one way to eliminate interoffice conflicts over the availability of shared computers is to buy every department a network of personal computers. Therefore organizations, such as Ford Motor Company's Casting Division, purchase large quantities of used computers at reduced prices instead of a few new ones at full retail. Although the used machines may be slower and less powerful, they provide a greater number of employees with ready access to critical computer resources.

Conflicting groups can also be "decoupled," or separated from each other, by *implementing buffering devices*. In some coffee shops and restaurants, for example, you have probably seen a wheel or spindle on which waiters hang written food orders for the cooks. Eliminating direct communication between these two groups of employees substantially reduces the incidence of conflict.[39] Buffering devices like this one reduce interdependence substantially and they help to manage conflict.

A final approach to manipulate interdependence involves *designing self-contained groups*. Regrouping the members of conflicting groups into new groups that perform their work independently of others is effective. For instance, an engineering department and a computer-assisted drafting department may be involved in conflicts about responsibilities for setting deadlines, approving specifications, and so on. These departments can be reorganized into "cells" that consist of engineers and drafting technicians who complete the entire process of design and drafting without outside assistance. Interdependence between groups is completely eliminated by this approach.

Methods of conflict resolution like these are *structural* techniques because they affect the way an organization's work is divided into tasks and reintegrated into a meaningful whole. We will return to some of these techniques in Chapter 15, when we discuss the kinds of structural mechanisms that are used to coordinate intergroup relations.

Bargaining and Negotiation. In addition to employing structural techniques, managers attempting to resolve destructive conflict may try to work out the differences that generate conflicting interests and concerns. Bargaining and negotiation are two closely associated processes that are often used to accomplish this aim. **Bargaining** is a process that occurs between groups with conflicting interests in which offers, counteroffers, and concessions are exchanged as the groups search for some mutually acceptable resolution. **Negotiation**, in turn, is the process in which the groups decide what each will give and take in the exchange between them.[40]

In the business world, relations between management and labor are often the focus of bargaining and negotiation. Stories about negotiations between unions and local employers appear in newspapers and on news broadcasts on a regular basis. Usually unions are bargaining for better wages, working conditions, benefits, and job security, while management representatives are bargaining to hold down labor costs and ensure continued profitability. However, bargaining and negotiation also occur elsewhere in organizations as people and groups try to satisfy their own desires and control the extent to which they must sacrifice in order to satisfy others. In tight economies, groups of secretaries who are dependent on the same supply budget may have to bargain with each other to

bargaining A process in which offers, counteroffers, and concessions are exchanged as conflicting groups search for some mutually acceptable resolution.

negotiation A process in which groups with conflicting interests decide what each will give and take in the exchange between them.

---

[39] William Foote Whyte, *Human Relations in the Restaurant Industry* (New York: McGraw-Hill, 1948), pp. 67–70.
[40] J. Z. Rubin and B. R. Brown, *The Social Psychology of Bargaining and Negotiation* (New York: Academic Press, 1975), p. 3.

see who will get new computer equipment and who will have to make do with what is already available. A company's sales force may try to negotiate favorable delivery dates for their best clients by offering manufacturing personnel leeway in meeting deadlines for other customers' orders. Research-and-development specialists may offer to make their new discoveries easier to produce if only the manufacturing department will give them additional time.

In deciding whose conflicting interests will be satisfied and whose will not, groups engaged in bargaining and negotiation can choose the degree to which they will assert themselves and look after their own interests. They can also decide whether they will cooperate with their adversary and put its interests ahead of their own. As Figure 13-8 indicates, there are five general approaches to managing diverse interests that are characterized by different mixes of assertiveness and cooperativeness:[41]

1. **Competition** (assertive, uncooperative). This involves overpowering other groups in the conflict and promoting the concerns of one's own group at the expense of the other groups. One way to accomplish this aim is by resorting to authority to satisfy the concerns of one's own group. Thus the head of a group of account executives may appeal to the Director of Advertising to protect the group's turf from the intrusions by other account execs.

2. **Accommodation** (unassertive, cooperative). This is allowing other groups to satisfy their own concerns at the expense of one's own group. Differences are smoothed over to maintain superficial harmony. A purchasing department that fails to meet budgetary guidelines because it overspends on raw materials to satisfy the demands of production groups is trying to use accommodation to cope with latent conflict.

3. **Avoidance** (unassertive, uncooperative). This approach requires staying neutral at all costs. Or it may be refusing to take an active role in conflict resolution procedures. The finance department that sticks its head in the sand and hopes that dissension about budgetary allocations will blow over is exhibiting avoidance.

4. **Collaboration** (assertive, cooperative). This involves attempting to satisfy the concerns of all of the groups by working through differences and seeking

competition A conflict management technique that involves attempts to overpower other groups in the conflict and to promote the concerns of one's own group at the expense of the other groups.

accommodation A conflict management technique that involves allowing other groups to satisfy their own concerns at the expense of one's own group.

avoidance A conflict management technique that involves staying neutral at all costs and refusing to take an active role in conflict resolution procedures.

collaboration A conflict management technique that involves attempting to satisfy the concerns of all conflicting groups by working through differences and seeking out optimal solutions in which everyone gains.

[41] Kenneth W. Thomas, "Conflict and Conflict Management," *Handbook of Industrial and Organizational Psychology* ed. Marvin D. Dunnette (Chicago: Rand McNally, 1976), pp. 889–935; also see Kenneth W. Thomas, "Toward Multidimensional Values in Teaching: The Example of Conflict Behaviors," *Academy of Management Review* 2 (1977), 472–89.

FIGURE 13-8

Managing Diversity

Thomas L. Ruble and Kenneth Thomas, "Support for a Two-Dimensional Model of Conflict Behavior," *Organizational Behavior and Human Performance* 16 (1976), 143–55. Reprinted with permission.

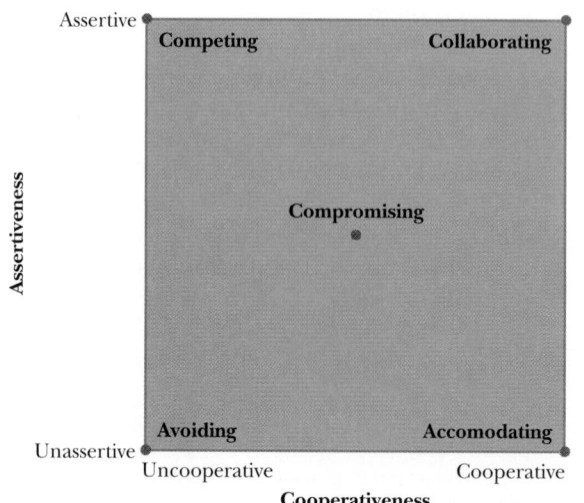

compromise A conflict management technique that involves seeking partial satisfaction of all conflicting groups through exchange and sacrifice.

solutions so that everyone gains as a result. A marketing department and a manufacturing department that meet on a regular basis to plan mutually acceptable production schedules are collaborating.

5. **Compromise** (mid-range assertiveness and cooperativeness). This approach seeks partial satisfaction of all groups through exchange and sacrifice, settling for acceptable rather than optimal resolution. Contract bargaining between union representatives and management typically involves significant compromise by both sides.

As indicated in Table 13-5, the appropriateness of each of these five approaches depends on the situation surrounding the conflict and, often, the time pressure for a negotiated settlement.

T A B L E   13-5
When Different Styles of Managing Diversity Should Be Applied

STYLE	APPLICATION
Competing	When quick, decisive action is required; to cope with crises. On important issues where unpopular solutions must be implemented, such as cost cutting or employee discipline. On issues vital to organizational welfare when your group is certain that its position is correct. Against groups who take advantage of noncompetitive behavior.
Accommodating	When your group is wrong and wants both to show reasonableness and to encourage the expression of a more appropriate view. When issues are more important to groups other than yours, to satisfy others and maintain cooperation. To build credits or bank favors for later issues. To minimize losses when your group is outmatched and losing. When harmony and stability are especially important.
Avoiding	When a conflict is trivial or more important conflicts are pressing. When there is no chance that your group will satisfy its own needs. When the costs of potential disruption outweigh the benefits of resolution. To let groups cool down and gain perspective. When others can resolve the conflict more effectively.
Collaborating	To find an integrative solution when conflicting concerns are too important to be compromised. When the most important objective is to learn. To combine the ideas of people with different perspectives. To gain commitment through the development of consensus. To work through conflicting feelings in individuals and between groups.
Compromising	When group concerns are important but not worth the disruption of more assertive styles. When equally powerful groups are committed to pursuing mutually exclusive concerns. To achieve temporary or transitional settlements. To arrive at expedient resolutions under time pressure. As a backup when neither competing nor problem-solving styles are successful.

Source: Adapted with permission from Kenneth W. Thomas, "Toward Multidimensional Values in Teaching: The Example of Conflict Behaviors," *Academy of Management Review*, 2 (1977), 487.

Beyond these styles of bargaining and negotiation, experts on organizational development have devised an assortment of conflict management techniques based on structured bargaining and negotiation. The next chapter describes several of these techniques in step-by-step detail (see Chapter 14's section on Intergroup Interventions).

SUMMARY

Power is the ability to influence others and to resist their influence in return. Compliance, identification, and internalization are outcomes that may result from the use of five types of interpersonal power—*reward, coercive, legitimate, referent,* or *expert*. In addition to interpersonal sources, power also grows out of uncertainty surrounding the continued availability of *critical contingencies*. It is thus based on the ability to reduce this uncertainty and is enhanced by low *substitutability* and high *centrality*.

Politics is a process in which power is acquired and used to advance self-interests. It is stimulated by a combination of personal characteristics and antecedent conditions and can involve a variety of tactics ranging from *controlling* supplies of critical *resources* to attacking or blaming others. There are several techniques to manage politicking. These include setting an example and confronting political game players.

Intergroup conflict is a process of opposition and confrontation that requires the presence of interdependence, political indeterminism, and diversity. It develops in a sequence of five stages—*latent conflict, perceived conflict, felt conflict, manifest conflict,* and *conflict aftermath*. Intergroup conflict can be resolved by restructuring intergroup relations or managed through bargaining and negotiation.

REVIEW QUESTIONS

1.   Is power being exercised when a manager orders a subordinate to do something the subordinate would do even without being ordered? When a subordinate successfully refuses to follow orders? When a manager's orders are followed despite the subordinate's reluctance?

2.   What is the difference between reward power and coercive power? What do these two types of power have in common? How are they similar to legitimate power? How do they differ from both expert and referent power?

3.   Why must uncertainty, centrality, and low substitutability *all* be present in order for power to be acquired. Explain how a group's power might be reduced by increasing substitutability.

4.   How does uncertainty encourage politics? What can managers do to control this antecedent condition?

5.   What is a coalition? What is gained by forming one? Explain how political tactics like impression management and doing favors for others can make forming a coalition easier.

6.   How can controlling information serve as a political tactic? What can managers do to guard against this tactic?

7.   Why does intergroup conflict require interdependence? How does political indeterminism influence whether this sort of conflict will occur? Based on your answers to these two questions, what can managers do to resolve intergroup conflicts without attempting to reduce diversity?

8.   Explain how one episode of conflict can set the stage for another. Based on what you know about the different stages of conflict, at which stage would you expect to find conflict the easiest to resolve? At which stage would you expect it to be most resistant to resolution?

9. Why is accommodation unlikely to succeed as a conflict management technique in most instances? Under what specific conditions is it most useful?

DIAGNOSTIC QUESTIONS

Considered together, power, politics, and conflict constitute a complex collection of political processes. If they take a dysfunctional turn, these processes can undermine the productivity and satisfaction of individuals and groups, jeopardizing organizational performance as a result. We suggest asking the following diagnostic questions to help manage the political face of organizations:

1. Do individuals and groups in the organization have the power required to function productively and interact effectively?

2. Which types of interpersonal power are currently in use? Are these types of interpersonal power likely to generate the compliance, identification, or internalization needed to energize appropriate behaviors? Might other types of interpersonal power be more effective?

3. If power inadequacies exist, are they traceable to limitations in the ability to reduce uncertainty? To high substitutability? To low centrality? What actions can be taken to correct these deficiencies?

4. Is politics undermining satisfaction or performance in the organization? Can antecedent conditions promoting politics be eliminated?

5. Are managers controlling politics by setting an example, encouraging open communications, managing coalitions, confronting game players, and anticipating future occurrences?

6. Do antecedent conditions—interdependence, political indeterminism, diversity—favor the emergence of intergroup conflict? Is there evidence that dysfunctional latent, perceived, or felt conflict is brewing? Can any of the antecedent conditions be modified to resolve the conflict before it manifests itself?

7. Is there evidence of manifest intergroup conflict? Do the dysfunctional, destructive effects of this conflict outweigh its functional benefits? Can any of the antecedent conditions be modified to resolve the conflict?

8. If interdependence or political indeterminism can be modified, which of the approaches to managing interdependence seems most suitable?

9. If diversity can be modified, which style of bargaining and negotiation best fits the conflict situation?

10. Is there evidence of conflict aftermath remaining from a previous episode that might encourage additional conflict? Can this aftermath be altered before destructive conflict reemerges?

EXERCISE 13-1
CONFLICT AND DISARMAMENT*
NORMAN BERKOWITZ, *Boston College*
HARVEY HORNSTEIN, *Columbia University*

Working in organizations means depending on others and having them depend on you. This interdependence can often stimulate conflict because individuals as well as organizational units have different needs and viewpoints. Although conflict is inescapable and sometimes harmful, you can learn to manage it. The purpose of this exercise is to give you a chance to observe and experience the feelings generated by conflict and confrontation and to examine strategies for developing collaboration among organizational units *before* conflicts arise.

STEP ONE: PRE-CLASS PREPARATION

To prepare for this exercise, label sixteen 3 × 5 index cards as follows:

1. On five of the cards, draw a large black "X" on one side only.
2. On three of the cards, write "Five Dollars" and "$5.00" on both sides.
3. On four of the cards, write "One Dollar" and "$1.00" on both sides.
4. On one of the cards, write "Fifty Cents" and "$.50" on both sides.
5. On two of the cards, write "Twenty Cents" and "$.20" on both sides.
6. On one of the cards, write "Ten Cents" and "$.10" on both sides.

In addition, familiarize yourself with the Disarmament Exercise Rules, described next, and briefly skim the rest of the exercise.

Disarmament Exercise Rules

The Disarmament Exercise is a game played by two teams. Each team can win money or lose it to a World Bank, which holds the funds for the game. Your objective, as a team member, will be to win as much money as you can. Each team will consist of 4 to 6 players. If there are more players on one team than the others, the extra players will assist the instructor and act as referees.

The Funds. Each player will distribute the $20 of "money" that he prepared before the exercise as follows:

1. He will put $15 in a team treasury. At the end of the game, or exercise, whatever is left in each team's treasury will be divided equally among the team members as a measure of the team's performance.
2. He will give $5 to the World Bank, which will be managed by a referee.

Your instructor will deposit an amount equal to the deposits of both teams in the World Bank. For example, for two teams of four players each, each team's treasury will contain $60 and the World Bank will hold $80 for that team—$20 from each team, plus a matching fund of $40 from the instructor.

Special Roles. At the start of Step Two, each team will have 15 minutes to review the instructions for the exercise and plan a team strategy. You must also select people to fill four special roles. No person can fill more than one role at the same time, and the roles can be reassigned at any time by a majority vote of the team. The roles are as follows:

1. Two *negotiators*: Their functions will be explained shortly.
2. A *team representative*: The representative will inform the referee of group decisions about such things as initiation and acceptance of negotiations, moves, and attacks. All such communications must be in writing. The referee will not acknowledge messages in any other form nor will he accept messages from any other team member.
3. One *recorder*: The recorder records team moves on the form provided in Exhibit 13-1. The record must include the action taken by the team in each move, and the team's weapon status at the end of each move. The recorder should also note who initiates decision making and how the team makes final decisions.

The Weapons. Members of each team should pool the cards they have marked with an "X" and then discard

* Adapted with the authors' permission from an exercise entitled "The Disarmament Game." This adaptation copyright © 1992 Prentice-Hall.

EXHIBIT 13-1

Record of Results

Move	SET 1 Actual Number of Armed Weapons	SET 1 Action Taken for This Move (Attack, Not Attack, Negotiate)	SET 2 Actual Number of Armed Weapons	SET 2 Action Taken for This Move (Attack, Not Attack, Negotiate)	SET 3 Actual Number of Armed Weapons	SET 3 Action Taken for This Move (Attack, Not Attack, Negotiate)	SET 4 Actual Number of Armed Weapons	SET 4 Action Taken for This Move (Attack, Not Attack, Negotiate)	SET 5 Actual Number of Armed Weapons	SET 5 Action Taken for This Move (Attack, Not Attack, Negotiate)	SET 6 Actual Number of Armed Weapons	SET 6 Action Taken for This Move (Attack, Not Attack, Negotiate)	
1.													
2.													
3.													
4.													
5.													
6.													
7.													
8.													
9.													
10.													
Ending no. of armed weapons of other team													Total of all sets
Line 1 $ paid to other team													
Line 2 $ paid to World Bank													
Line 3 $ received from other team													
Line 4 $ received from World Bank													
											Total Results (Lines 3 + 4-1-2)		

all but 20. Each of the 20 remaining cards is a weapon: when the card is turned so that the "X" is visible, the weapon is armed; when the card is turned so that the blank side is visible, the weapon is unarmed. To begin the exercise, place all twenty of your team's weapons, in the armed condition ("X" side up), where all your team members can see them. During the course of the exercise, your own team's weapons should remain in your possession and out of sight of the other team.

Exercise Procedure. The exercise will consist of several sets of moves and negotiations, completed according to the following instructions:

1. Sets
   A. As many sets as possible will be completed in the time available. Payments will be made after each set.

B. Each set will consist of no more than ten moves for each team. An attack following any move ends a set. If there is no attack, the set ends after the tenth move. Each team has two minutes to make a move. At the end of two minutes, you must have moved two, one, or none of your team's weapons from "armed" to "unarmed" status. You may not rearm weapons that have been disarmed. If you fail to decide on a move in the allotted time, the status quo will be counted as your move. In addition, during each move you must decide whether you want to attack or negotiate (see "Negotiations"). You must communicate your decision to the referee within fifteen seconds of the end of the move.

C. Each team may announce an attack on the other team following any two-minute move period except for the third, sixth, and ninth pe-

riods. You may not attack during negotiations.
  D. Once a set ends, you should begin a new set immediately with all weapons armed. Continue with as many sets as time permits.
2. Negotiations
  A. Between moves your team will have the opportunity to communicate with the other team through negotiations.
  B. You may call for negotiations during the fifteen seconds between move periods. The other team may accept or reject your request to negotiate. Negotiations can last no longer than two minutes.
  C. Following negotiations, the next two-minute move period will start immediately after the negotiators have rejoined their teams.
  D. Negotiators may say whatever is necessary to most benefit their team.
  E. The team is not necessarily bound by agreements made by its negotiators.
  F. Negotiators *must* meet after the third, sixth, and ninth moves.

The Payoff. Teams will make payments or receive payoffs at the end of each set according to the following rules:

1. If there is an attack, the set ends immediately. The team with the greater number of armed weapons wins $.50 per member for each armed weapon it has *over and above* the number of armed weapons held by the other team. This is paid directly from the treasury of the losing team to the treasury of the winning team. If both teams have the same number of armed weapons when there is an attack, both teams pay the World Bank $.50 per member.
2. If there is no attack after ten moves, the set ends. If your team has more disarmed weapons than armed weapons, the World Bank should pay it $.20 per member for each surplus disarmed weapon. If your team has fewer disarmed weapons than armed weapons, it must pay the World Bank $.20 per member for each excess armed weapon.

Notes to Referee. While managing the exercise keep the following tips in mind:

1. Keep the pairs of teams separated, so that each team cannot hear the other's conversations. If possible, arrange for each team pair to be in a separate room.
2. To time the length of moves, use a watch with a second hand (your own or a classmate's). An alarm clock that will ring accurately at two-minute intervals would be helpful.
3. Be sure to permit the teams only the specified times. You may need to be forceful in directing

the teams when to start and end moves in order to keep them on schedule.
4. You will be the sole manager of World Bank funds and should also check the accuracy of the teams' record keeping.
5. You must not assist either team.

STEP TWO: DISARMAMENT SETS

The class should divide into teams of four to six members each (if you have already formed permanent groups, reassemble in those groups). Your instructor will pair teams up and select extra members to serve as referees. Each pair of teams should then be assigned a referee and should immediately begin the Disarmament Exercise.

STEP THREE: CLASS DISCUSSION

Your instructor will notify all teams when the exercise is over. Each team should make sure its record of results (Exhibit 13-1) is complete and should calculate its profits or losses. A spokesperson should be elected to report briefly to the class on the team's activities. Funds remaining in the team's treasury should be divided equally among its members to further clarify the gains or losses of the team. The class should then reconvene and all team spokespersons should give their reports. Finally, the class should discuss the following questions:

1. What was each team's goal? Did the teams become aware of the need to collaborate with each other and the advantages of doing so? What part did trust between teams play in this exercise?
2. If you were to complete the exercise again, would you do anything differently?
3. If one of the "teams" in an analogous real-world situation were the manufacturing department and the other a sales department of an industrial company and the World Bank were the marketplace, what recommendations would you make to the two departments based on what you learned in this exercise?

CONCLUSION

As you have learned in this exercise, looking after the interests of one's own group and ignoring or even subverting the interests of another group can stimulate disagreements. These disagreements, when coupled with mistrust and opportunism, can ignite in intergroup conflict. Such conflict occurs when interdependent groups lack other means of resolving differences among themselves. It is thus important that groups who must work together develop trust and the ability to handle differences in constructive ways. You will learn more about how to accomplish these twin goals by completing the intergroup mirroring intervention in Exercise 14-1.

DIAGNOSING ORGANIZATIONAL BEHAVIOR

CASE 13-1

CITY NATIONAL BANK*

J. B. RITCHIE, *Brigham Young University*
PAUL H. THOMPSON, *Brigham Young University*

After having worked two months during the previous summer for City National Bank. I returned again this summer to work while on my break from school. Though I would only be there for four months, they hired me on as a full-time staff member replacing a woman who recently terminated. They also hired a woman just out of high school to help handle the extra work load our resort town gets throughout the vacation months. These additions brought our operations division up to seven women plus the assistant manager over our division (see Exhibit 13-2).

The same day I started, a new woman transferred up from a larger branch to our division to take over the note department. Marilyn, the new woman, was not very well liked by most of the workers in the branch because of some negative reports which had preceded her arrival and because her family "owned" the town we worked in.

City National Bank, like any other large bank with many branches, had standardized policies, procedures, and regulations. In order to protect customers, employees, and the corporation, these procedures have to be followed. The bank has auditors who come around periodically to check the books and operational procedures of the branches to assure maintenance of the high standards.

Our branch has relaxed several of the rules and has developed some policies unique to our branch. In part this stems from the informal and friendly relationships shared between customers and employees in our small town, where we know most of the customers on a first-name basis. Unlike our branch, the larger branch Marilyn previously worked in followed procedures strictly and supposedly did everything "according to the book." Marilyn let us know that we were inefficient and backward. Soon bad feelings developed and came to a head in the early part of August.

In August we were in the process of changing managers. Our assistant manager had also left on vacation, so we had a former auditor in management training and a newly promoted supervisor filling the man-

agement positions for two weeks. Their job was to sit in for people on vacations throughout our division until each was placed in his own branch. The new manager and assistant manager were upset with our lax attitude toward many rules, some of which we had never even heard of, and they set out to shape up our branch. Among the things we needed to reform were the opening and closing procedures, keeping our keys with us at *all* times, always locking our cash boxes, balancing procedures, check cashing policy, and several other regulations. Marilyn was happy with the new situation and told us "it was about time," but the rest of the branch was very defensive and uncooperative with the temporary management. During this time Marilyn changed several of her responsibilities with the temporary supervisor's permission, and when our assistant manager returned from vacation, there was a great deal of tension between them.

One afternoon while the assistant manager was out, Marilyn went around on her own and picked up three sets of keys that were lying on a table and turned them over to the other assistant manager, saying that the "women should be taught a lesson." This all occurred after the branch had closed, and there were some frantic minutes spent searching for the keys. Marilyn said nothing and we all kept looking until someone remembered seeing Marilyn in their work area. When confronted she merely told the women that Paul, the assistant manager, had the keys. When the whole story was put together, there was a lot of name-calling and derogatory comments, with Marilyn getting the silent treatment for almost a week.

Our management never took an official stand that was enforced. There was never a confrontation between the opposing sides. When Marilyn would approach them on a point, they would satisfy her by agreeing that she had a good point, and when the other side brought

* Reprinted with the publisher's permission from pages 46–48 of *Organizations and People: Readings, Cases, and Exercises in Organizational Behavior*, 3rd Ed. by J. B. Ritchie and Paul H. Thompson, copyright © 1984 by West Publishing Company. All rights reserved.

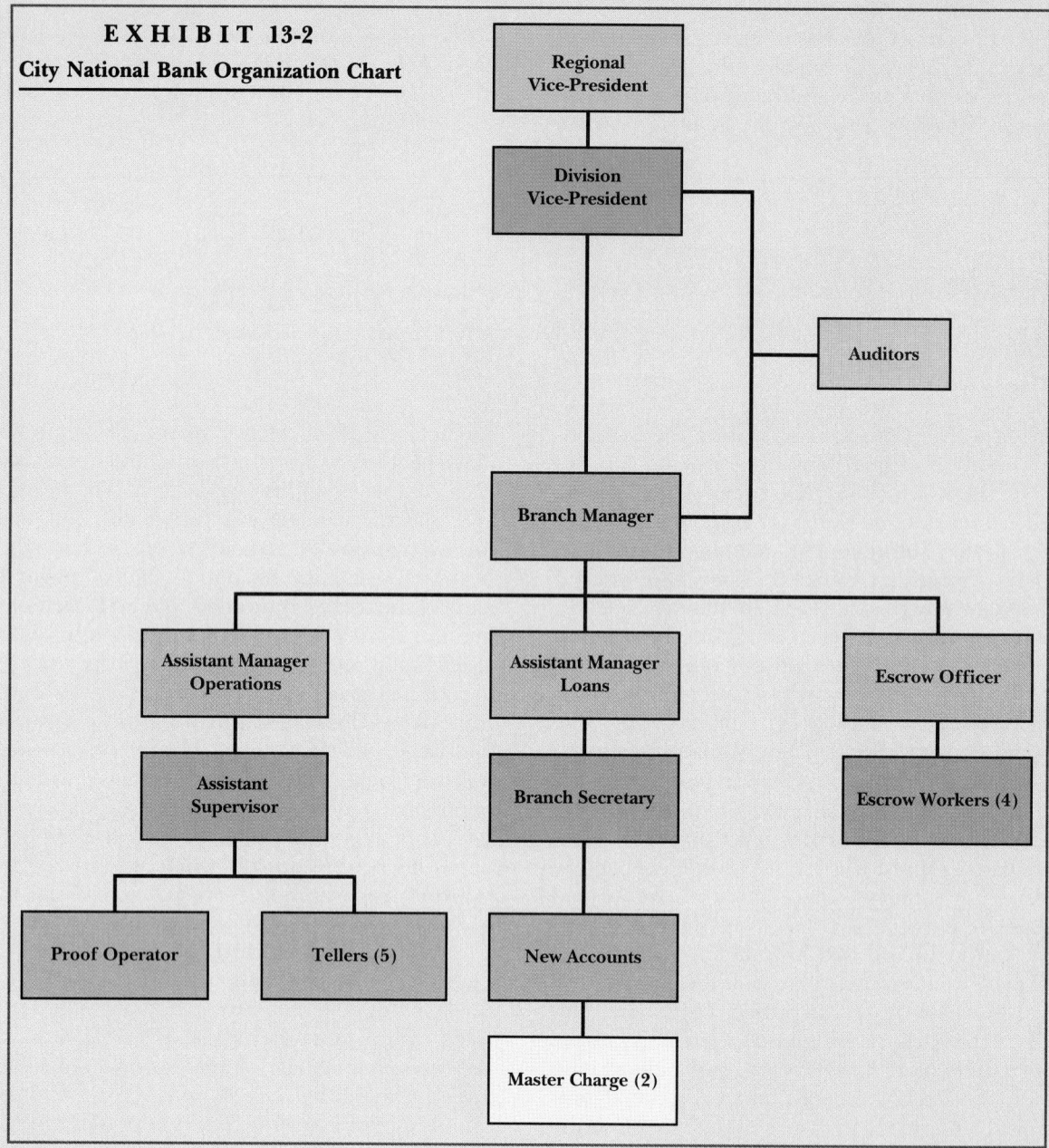

EXHIBIT 13-2
City National Bank Organization Chart

Regional Vice-President

Division Vice-President

Auditors

Branch Manager

Assistant Manager Operations

Assistant Manager Loans

Escrow Officer

Assistant Supervisor

Branch Secretary

Escrow Workers (4)

Proof Operator

Tellers (5)

New Accounts

Master Charge (2)

up complaints, they would agree with them, too. A harmony between practices and policies was never reached. As the summer ended, many problems were compounded because there was no consistent authority and so many things "had changed" according to the original employees. We never knew what was expected. By Labor Day, I left to return to school and four other women had quit or transferred out of my hometown branch of City National Bank.

When you have read this case, look back at the chapter's diagnostic questions and choose the ones that apply to the case. Then use those questions with the ones that follow in your case analysis.

1. What sort of power were City National Bank's new manager and assistant manager using as they tried to reform the procedures followed by the employees at their branch? How did the employees react? Why did they react this way?

2. What should management have done to control Marilyn's political behavior? What was the effect of what they did do?

3. What stage of development had the conflict reached when the writer left the bank to return to school? What do you think is likely to happen in the future if management makes no effort to manage this conflict? What steps should management take to deal with this situation?

CASE 13-2

RONDELL DATA CORPORATION*

JOHN A. SEEGER, *Bentley College*

"God damn it, he's done it again!"

Frank Forbus threw the stack of prints and specifications down on his desk in disgust. The Model 802 wide-band modulator, released for production the previous Thursday, had just come back to Frank's Engineering Services Department with a caustic note that began, "This one can't be produced, either. . . ." It was the fourth time Production had kicked the design back.

Frank Forbus, director of engineering for Rondell Data Corp., was normally a quiet man. But the Model 802 was stretching his patience; it was beginning to look just like other new products that had hit delays and problems in the transition from design to production during the eight months Frank had worked for Rondell. These problems were nothing new at the sprawling old Rondell factory; Frank's predecessor in the engineering job had run afoul of them too, and had finally been fired for protesting too vehemently about the other departments. But the Model 802 should have been different. Frank had met two months before (July 3, 1988) with the firm's president, Bill Hunt, and with factory superintendent Dave Schwab to smooth the way for the new modulator design. He thought back to the meeting. . . .

"Now we all know there's a tight deadline on the 802," Bill Hunt said, "and Frank's done well to ask us to talk about its introduction. I'm counting on both of you to find any snags in the system, and to work together to get that first production run out by October second. Can you do it?"

"We can do it in Production if we get a clean design two weeks from now, as scheduled," answered Dave Schwab, the grizzled factory superintendent. "Frank and I have already talked about that, of course. I'm setting aside time in the card room and the machine shop, and we'll be ready. If the design goes over schedule, though, I'll have to fill in with other runs, and it will cost us a bundle to break in for the 802. How does it look in Engineering, Frank?"

"I've just reviewed the design for the second time," Frank replied. "If Ron Porter can keep the salesmen out of our hair, and avoid any more last minute changes, we've got a shot. I've pulled the draftsmen off three other overdue jobs to get this one out. But, Dave, that means we can't spring engineers loose to confer with

your production people on manufacturing problems."

"Well, Frank, most of those problems are caused by the engineers, and we need them to resolve the difficulties. We've all agreed that production bugs come from both of us bowing to sales pressure, and putting equipment into production before the designs are really ready. That's just what we're trying to avoid on the 802. But I can't have 500 people sitting on their hands waiting for an answer from your people. We'll have to have *some* engineering support."

Bill Hunt broke in, "So long as you two can talk calmly about the problem I'm confident you can resolve it. What a relief it is, Frank, to hear the way you're approaching this. With Kilmann (the previous director of engineering) this conversation would have been a shouting match. Right, Dave?" Dave nodded and smiled.

"Now there's one other thing you should both be aware of," Hunt continued. "Doc Reeves and I talked last night about a new filtering technique, one that might improve the signal-to-noise ratio of the 802 by a factor of two. There's a chance Doc can come up with it before the 802 reaches production, and if it's possible, I'd like to use the new filters. That would give us a real jump on the competition."

Four days after that meeting, Frank found that two of his key people on the 802 design had been called to Production for emergency consultation on a bug found in final assembly: two halves of a new data transmission interface wouldn't fit together because recent changes in the front end required a different chassis design for the back end.

Another week later, Doc Reeves walked into Frank's office, proud as a new parent, with the new filter design. "This won't affect the other modules of the 802 much," Doc had said, "Look, it takes three new cards, a few connectors, some changes in the wiring harness, and some new shielding, and that's all."

Frank had tried to resist the last-minute design changes, but Bill Hunt had stood firm. With a lot of overtime by the engineers and draftsmen, Engineering Services should still be able to finish the prints in time.

Two engineers and three draftsmen went onto 12-hour days to get the 802 ready, but the prints were still five days late reaching Dave Schwab. Two days later, the prints came back to Frank, heavily annotated in red. Schwab had worked all day Saturday to review the job, and had found more than a dozen discrepancies in the prints—most of them caused by the new filter design and insufficient checking time before release. Correction of those design faults had brought on a new generation of discrepancies; Schwab's cover note on the second return of the prints indicated he'd had to release the machine capacity he'd been holding for the 802. On the third iteration, Schwab committed his photo and plating capacity to another rush job. The 802 would be

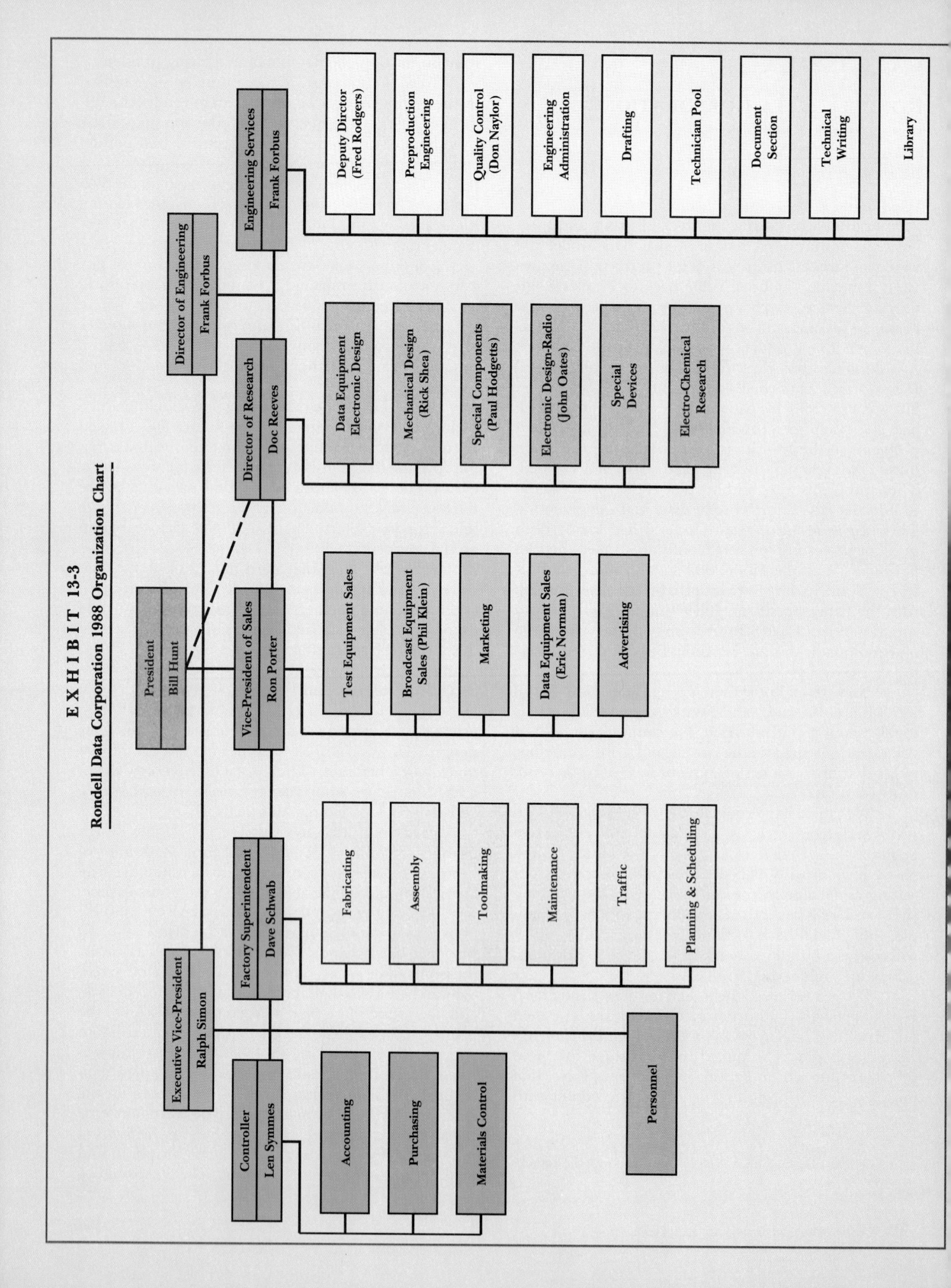

EXHIBIT 13-3
Rondell Data Corporation 1988 Organization Chart

at least one month late getting into production. Ron Porter, Vice President for Sales, was furious. His customer needed 100 units *NOW*, he said. Rondell was the customer's only late supplier.

"Here we go again," thought Frank Forbus.

COMPANY HISTORY

Rondell Data Corp. traced its lineage through several generations of electronics technology. Its original founder, Bob Rondell, had set the firm up in 1920 as "Rondell Equipment Co." to manufacture several electrical testing devices he had invented as an engineering faculty member at a large university. The firm branched into radio broadcasting equipment in 1947, and into data transmission equipment in the early 1960s. A well-established corps of direct sales people, mostly engineers, called on industrial, scientific, and government accounts, but concentrated heavily on original equipment manufacturers. In this market, Rondell had a long-standing reputation as a source of high-quality, innovative designs. The firm's salespeople fed a continual stream of challenging problems into the Engineering Department, where the creative genius of Ed "Doc" Reeves and several dozen other engineers "converted problems to solutions" (as the sales brochure bragged). Product design formed the spearhead of Rondell's growth.

By 1988, Rondell offered a wide range of products in its two major lines. Broadcast equipment sales had benefitted from the growth of UHF TV and FM radio; it now accounted for 35% of company sales. Data transmission had blossomed, and in this field an increasing number of orders called for unique specifications, ranging from specialized display panels to entirely untried designs.

The company had grown from 100 employees in 1947 to over 800 in 1988. (Exhibit 13-3 shows the current organization chart of key employees.) Bill Hunt, who had been a student of the company's founder, had presided over most of that growth, and took great pride in preserving the "family spirit" of the old organization. Informal relationships between Rondell's veteran employees formed the backbone of the firm's day-to-day operations; all the managers relied on personal contact, and Hunt often insisted that the absence of bureaucratic red tap was a key factor in recruiting outstanding engineering talent. The personal management approach extended throughout the factory. All exempt employees were paid on a straight salary plus a share of the profits. Rondell boasted an extremely loyal group of senior employees, and very low turnover in nearly all areas of the company.

The highest turnover job in the firm was Frank Forbus's. Frank had joined Rondell in January of 1988, replacing Jim Kilmann, who had been director of engineering for only 10 months. Kilmann, in turn, had replaced Tom MacLeod, a talented engineer who had made a promising start, but had taken to drink after a year in the job. MacLeod's predecessor had been a genial old timer who retired at 70 after 30 years in charge of engineering. (Doc Reeves had refused the directorship in each of the recent changes, saying, "Hell, that's no promotion for a bench man like me. I'm no administrator.")

For several years, the firm had experienced a steadily increasing number of disputes between research, engineering, sales, and production people—disputes generally centered on the problem of new product introduction. Quarrels between departments became more numerous under MacLeod, Kilmann, and Forbus. Some managers associated these disputes with the company's recent decline in profitability—a decline that, in spite of higher sales and gross revenues, was beginning to bother people in 1987. President Bill Hunt commented:

> Better cooperation, I'm sure, could increase our output by 5–10%. I'd hoped Kilmann could solve the problems, but pretty obviously he was too young, too arrogant. People like him—that conflict type of personality—bother me. I don't like strife, and with him it seemed I spent all my time smoothing out arguments. Kilmann tried to tell everyone else how to run their departments, without having his own house in order. That approach just wouldn't work, here at Rondell. Frank Forbus, now, seems much more in tune with our style of organization. I'm really hopeful now.
>
> Still, we have just as many problems now as we did last year. Maybe even more. I hope Frank can get a handle on Engineering Services soon. . .

THE ENGINEERING DEPARTMENT: RESEARCH

According to the organization chart (see Exhibit 13-3), Frank Forbus was in charge of both research (really the product development function) and engineering services (which provided engineering support). To Forbus, however, the relationship with research was not so clear-cut:

> Doc Reeves is one of the world's unique people, and none of us would have it any other way: He's a creative genius. Sure, the chart says he works for me, but we all know Doc does his own thing. He's not the least bit interested in management routines, and I can't count on him to take any responsibility in scheduling projects, or checking budgets, or what-have-

you. But as long as Doc is director of research, you can bet this company will keep on leading the field. He has more ideas per hour than most people have per year, and he keeps the whole engineering staff fired up. Everybody loves Doc—and you can count me in on that, too. In a way, he works for me, sure. But that's not what's important.

"Doc" Reeves—unhurried, contemplative, casual, and candid—tipped his stool back against the wall of his research cubicle and talked about what *was* important:

Development engineering. That's where the company's future rests. Either we have it there, or we don't have it.

There's no kidding ourselves that we're anything but a bunch of Rube Goldbergs here. But that's where the biggest kicks come from—from solving development problems, and dreaming up new ways of doing things. That's why I so look forward to the special contracts we get involved in. We accept them not for the revenue they represent, but because they subsidize the basic development work which goes into all our products.

This is a fantastic place to work. I have a great crew and they can really deliver when the chips are down. Why, Bill Hunt and I (he gestured toward the neighboring cubicle, where the president's name hung over the door) are likely to find as many people here at work at ten P.M. as at three in the afternoon. The important thing here is the relationships between people; they're based on mutual respect, not on policies and procedures. Administrative red tape is a pain. It takes away from development time.

Problems? Sure, there are problems now and then. There are power interests in production, where they sometimes resist change. But I'm not a fighting man; you know I suppose if I were, I might go in there and push my weight around a little. But I'm an engineer, and can do more for Rondell sitting right here, or working with my own people. That's what brings results.

Other members of the Research Department echoed Doc's views and added some additional sources of satisfaction with their work. They were proud of the personal contacts they had built up with customers' technical staffs—contacts that increasingly involved travel to the customers' factories to serve as expert advisors in preparation of overall system design specifications. The engineers were also delighted with the department's encouragement of their personal development, continuing education, and independence on the job.

But there were problems, too. Rick Shea, of the mechanical design section, noted,

In the old days I really enjoyed the work—and the people I worked with. But now there's a lot of irritation. I don't like someone breathing down my neck. You can be hurried into jeopardizing the design.

John Oates, head of the radio electronic design section, was another designer with definite views:

Production engineering is almost nonexistent in this company. Very little is done by the preproduction section in engineering services. Frank Forbus has been trying to get preproduction into the picture, but he won't succeed because you can't start from such an ambiguous position. There have been three directors of engineering in three years. Frank can't hold his own against the others in the company. Kilmann was too aggressive. Perhaps no amount of tact would have succeeded.

Paul Hodgetts was head of special components in the R & D department. Like the rest of the department he valued bench work. But he complained of engineering services.

The services don't do things we want them to do. Instead, they tell us what they're going to do. I should probably go to Frank, but I don't get any decisions there. I know I should go through Frank, but this holds things up, so I often go direct.

The Engineering Department: Enginering Services

The Engineering Services Department provided ancillary services to R & D, and served as liaison between engineering and the other Rondell departments. Among its main functions were drafting; management of the central technicians' pool; scheduling and expediting engineering products; documentation and publication of parts lists and engineering orders; preproduction engineering (consisting of the final integration of individual design components into mechanically compatible packages); and quality control (which included inspection of incoming parts and materials, and final inspection of subassemblies and finished equipment). Top management's description of the department included the line, "ESD is responsible for maintaining cooperation with other departments, providing services to the development engineers, and freeing more valuable people in R & D from essential activities which are diversions from and beneath their main competence."

Many of Frank Forbus's 75 employees were located in other departments. Quality control people were scattered through the manufacturing and receiving areas, and technicians worked primarily in the research area or the prototype fabrication room. The remaining ESD personnel were assigned to leftover nooks and crannies near production or engineering sections.

Frank Forbus described his position:

My biggest problem is getting acceptance from the people I work with. I've moved slowly rather than risk antagonism. I saw what happened to Kilmann, and I wanted to avoid that. But although his precipitate action had won over a few of the younger R & D people, he certainly didn't have the department's backing. Of course it was the resentment of other departments which eventually caused his discharge. People have been slow accepting me here. There's nothing really overt, but I get a negative reaction to my ideas.

My role in the company has never been well defined, really. It's complicated by Doc's unique position, of course, and also by the fact that ESD sort of grew by itself over the years, as the design engineers concentrated more and more on the creative parts of product development. I wish I could be more involved in the technical side. That's been my training, and it's a lot of fun. But in our setup, the technical side is the least necessary for me to be involved in.

Schwab (production head) is hard to get along with. Before I came and after Kilmann left, there were six months intervening when no one was really doing any scheduling. No work loads were figured, and unrealistic promises were made about releases. This puts us in an awkward position. We've been scheduling way beyond our capacity to manufacture or engineer.

Certain people within R & D, for instance John Oates, head of the radio electronic design section, understand scheduling well and meet project deadlines, but this is not generally true of the rest of the R & D department, especially the mechanical engineers who won't commit themselves. Most of the complaints come from sales and production department heads because items—like the 802—are going to production before they are fully developed, under pressure from sales to get out the unit, and this snags the whole process. Somehow, engineering services should be able to intervene and resolve these complaints, but I haven't made much headway so far.

I should be able to go to Hunt for help, but he's too busy most of the time, and his major interest is the design side of engineering, where he got his own start. Sometimes he talks as though he's the engineering director as well as president. I have to put my foot down; there are problems here that the front office just doesn't understand.

Sales people were often observed taking their problems directly to designers, while production frequently threw designs back at R & D, claiming they could not be produced and demanding the prompt attention of particular design engineers. The latter were frequently observed in conference with production supervisors on the assembly floor. Frank went on:

The designers seem to feel they're losing something when one of us tries to help. They feel it's a reflection on them to have someone take over what they've been doing. They seem to want to carry a project right through to the final stages, particularly the mechanical people. Consequently, engineering services

people are used below their capacity to contribute and our department is denied functions it should be performing. There's not as much use made of engineering services as there should be.

Frank Forbus's technician supervisor added his comments:

Production picks out the engineer who'll be the "bum of the month." They pick on every little detail instead of using their heads and making the minor changes that have to be made. The fifteen-to-twenty-year people shouldn't have to prove their ability any more, but they spend four hours defending themselves and four hours getting the job done. I have no one to go to when I need help. Frank Forbus is afraid. I'm trying to help him but he can't help me at this time. I'm responsible for fifty people and I've got to support them.

Fred Rodgers, whom Frank had brought with him to the company as an assistant, gave another view of the situation:

I try to get our people in preproduction to take responsibility but they're not used to it and people in other departments don't usually see them as best qualified to solve the problem. There's a real barrier for a newcomer here. Gaining people's confidence is hard. More and more, I'm wondering whether there really is a job for me here.

(Rodgers left Rondell a month later.) Another of Forbus's subordinates gave his view:

If Doc gets a new product idea you can't argue. But he's too optimistic. He judges that others can do what he does—but there's only one Doc Reeves. We've had 900 production change orders this year—they changed 2,500 drawings. If I were in Frank's shoes I'd put my foot down on all this new development. I'd look at the reworking we're doing and get production set up the way I wanted it. Kilmann was fired when he was doing a good job. He was getting some system in the company's operations. Of course, it hurt some people. There is no denying that Doc is the most important person in the company. What gets overlooked is that Hunt is a close second, not just politically but in terms of what he contributes technically and in customer relations.

This subordinate explained that he sometimes went out into the production department but that Schwab, the production head, resented this. Personnel in production said that Kilmann had failed to show respect for oldtimers and was always meddling in other departments' business. This was why he had been fired, they contended.

Don Taylor was in charge of quality control. He commented:

I am now much more concerned with administration and less with work. It is one of the evils you get into. There is tremendous detail in this job. I listen to everyone's opinion. Everybody is important. There shouldn't be distinctions—distinctions between people. I'm not sure whether Frank has to be a fireball like Kilmann. I think the real question is whether Frank is getting the job done. I know my job is essential. I want to supply service to the more talented people and give them information so they can do their jobs better.

THE SALES DEPARTMENT

Ron Porter was angry. His job was supposed to be selling, he said, but instead it had turned into settling disputes inside the plant and making excuses to waiting customers. He jabbed a finger toward his desk:

> You see that telephone? I'm actually afraid nowadays to hear it ring. Three times out of five, it will be a customer who's hurting because we've failed to deliver on schedule. The other two calls will be from production or ESD, telling me some schedule has slipped again.
>
> The Model 802 is typical. Absolutely typical. We padded the delivery date by six weeks, to allow for contingencies. Within two months the slack had evaporated. Now it looks like we'll be lucky to ship it before Christmas. (It was now November 28.) We're ruining our reputation in the market. Why, just last week one of our best customers—people we've worked with for 15 years—tried to hang a penalty clause on their latest order.
>
> We shouldn't have to be after the engineers all the time. They should be able to see what problems they create without our telling them.

Phil Klein, head of broadcast sales under Porter, noted that many sales decisions were made by top management. Sales was understaffed, he thought, and had never really been able to get on top of the job.

> We have grown further and further away from engineering. The director of engineering does not pass on the information that we give him. We need better relationships there. It is very difficult for us to talk to customers about development problems without technical help. We need each other. The whole of engineering is now too isolated from the outside world. The morale of ESD is very low. They're in a bad spot—they're not well organized.
>
> People don't take much to outsiders here. Much of this is because the expectation is built up by top management that jobs will be filled from the bottom. So it's really tough when an outsider like Frank comes in.

Eric Norman, order and pricing coordinator for data equipment, talked about his own relationships with the production department:

Actually, I get along with them fairly well. Oh, things could be better, of course, if they were more cooperative generally: They always seem to say, "It's my bat and my ball, and we're playing by my rules." People are afraid to make production mad; there's a lot of power in there. But you've got to understand that production has its own set of problems. And nobody in Rondell is working any harder than Dave Schwab to try to straighten things out.

THE PRODUCTION DEPARTMENT

Dave Schwab had joined Rondell just after the Korean War, in which he had seen combat duty (at the Yalu River) and intelligence duty at Pyong Yang. Both experiences had been useful in his first year of civilian employment at Rondell's: the wartime factory superintendent and several middle managers had been, apparently, indulging in highly questionable side deals with Rondell's suppliers. Dave Schwab had gathered evidence, revealed the situation to Bill Hunt, and had stood by the president in the ensuing unsavory situation. Seven months after joining the company, Dave was named Factory Superintendent.

His first move had been to replace the fallen managers with a new team from outside. This group did not share the traditional Rondell emphasis on informality and friendly personal relationships, and had worked long and hard to install systematic manufacturing methods and procedures. Before the reorganization, production had controlled purchasing, stock control, and final quality control (where final assembly of products in cabinets was accomplished). Because the wartime events, management decided on a check-and-balance system of organization and removed these three departments from production jurisdiction. The new production managers felt they had been unjustly penalized by this organization, particularly since they had uncovered the behavior that was detrimental to the company in the first place.

By 1988, the production department had grown to 500 employees, of whom 60% worked in the assembly area—an unusually pleasant environment that had been commended by *Factory* magazine for its colorful decoration, cleanliness, and low noise level. An additional 30% of the work force, mostly skilled machinists, staffed the finishing and fabrication department. About 60 others performed scheduling, supervisory, and maintenance duties. Production workers were nonunion, hourly-paid, and participated in both the liberal profit-sharing program and the stock purchase plan. Morale in production was traditionally high, and turnover was extremely low.

Dave Schwab commented:

To be efficient, production has to be a self-contained department. We have to control what comes into the department and what goes out. That's why purchasing, inventory control, and quality ought to run out of this office. We'd eliminate a lot of problems with better control there. Why, even Don Naylor in QC would rather work for me than for ESD; he's said so himself. We understand his problems better.

The other departments should be self-contained, too. That's why I always avoid the underlings, and go straight to the department heads with any questions. I always go down the line.

I have to protect my people from outside disturbances. Look what would happen if I let unfinished, half-baked designs in here—there'd be chaos. The bugs have to be found before the drawings go into the shop, and it seems I'm the one who has to find them. Look at the 802, for example. (Dave had spent most of Thanksgiving Day [it was now November 28] red-pencilling the latest set of prints.) ESD should have found every one of those discrepancies. They just don't check drawings properly. They change most of the things I flag, but then they fail to trace through the impact of those changes on the rest of the design. I shouldn't have to do that.

And those engineers are tolerance crazy. They want everything to a millionth of an inch. I'm the only one in the company who's had any experience with actually machining things to a millionth of an inch. We make sure that the things that engineers say on their drawings actually have to be that way and whether they're obtainable from the kind of raw material we buy.

That shouldn't be production's responsibility, but I have to do it. Accepting bad prints wouldn't let us ship the order any quicker. We'd only make a lot of junk that had to be reworked. And that would take even longer.

This way, I get to be known as the bad guy, but I guess that's just part of the job. (He paused with a wry smile.) Of course, what really gets them is that I don't even have a degree.

Dave had fewer bones to pick with the sales department because, he said, they trusted him.

When we give Ron Porter a shipping date, he knows the equipment will be shipped then.

You've got to recognize, though, that all of our new product problems stem from sales making absurd commitments on equipment that hasn't been fully developed. That always means trouble. Unfortunately, Hunt always backs sales up, even when they're wrong. He always favors them over us.

Ralph Simon, age 65, executive vice president of the company, had direct responsibility for Rondell's production department. He said:

There shouldn't really be a dividing of departments among top management in the company. The president should be czar over all. The production people ask me to do something for them, and I really can't do it. It creates bad feelings between engineering and production, this special attention that they (R & D) get from Bill. But then Hunt likes to dabble in design. Schwab feels that production is treated like a poor relation.

THE EXECUTIVE COMMITTEE

At the executive committee meeting of December 6, it was duly recorded that Dave Schwab had accepted the prints and specifications for the Model 802 modulator, and had set Friday, December 29, as the shipping date for the first 10 pieces. Bill Hunt, in the chairperson's role, shook his head and changed the subject quickly when Frank tried to open the agenda to a discussion of interdepartmental coordination.

The executive committee itself was a brainchild of Rondell's controller, Len Symmes, who was well aware of the disputes that plagued the company. Symmes had convinced Bill Hunt and Ralph Simon to meet every two weeks with their department heads, and the meetings were formalized with Hunt, Simon, Ron Porter, Dave Schwab, Frank Forbus, Doc Reeves, Symmes, and the personnel director attending. Symmes explained his intent and the results:

Doing things collectively and informally just doesn't work as well as it used to. Things have been gradually getting worse for at least two years now: We had to start thinking in terms of formal organization relationships. I did the first organization chart, and the executive committee was my idea too—but neither idea is contributing much help, I'm afraid. It takes top management to make an organization click. The rest of us can't act much differently until the top people see the need for us to change.

I had hoped the committee especially would help get the department managers into a constructive planning process. It hasn't worked out that way because Mr. Hunt really doesn't see the need for it. He uses the meetings as a place to pass on routine information.

MERRY CHRISTMAS

"Frank, I didn't know whether to tell you now, or after the holiday." It was Friday, December 22, and Frank Forbus was standing awkwardly in front of Bill Hunt's desk.

"But, I figured you'd work right through Christmas Day if we didn't have this talk, and that just wouldn't have been fair to you. I can't understand why we have such poor luck in the engineering director's job lately. And I don't think it's entirely your fault. But . . ."

Frank only heard half of Hunt's words, and said nothing in response. He'd be paid through February 28 . . . He should use the time for searching . . . Hunt would help all he could . . . Jim Kilmann was supposed to be doing well at his own new job, and might need more help . . .

Frank cleaned out his desk, and numbly started home. The electronic carillon near his house was playing a Christmas carol. Frank thought again of Hunt's rationale: conflict still plagued Rondell—and Frank had not made it go away. Maybe somebody else could do it.

"And what did Santa Claus bring you, Frankie?" he asked himself.

"The sack. Only the empty sack."

When you have read this case, look back at the chapter's diagnostic questions and choose the ones that apply to the case. Then use those questions with the ones that follow in your case analysis.

1. Why was Rondell Data unable to fill Frank Forbus's position successfully? What would you advise Frank to do to increase his chances of keeping his job?

2. What made Doc Reeves so powerful? What problems did Doc's power cause the rest of the company? What should be done to manage Doc and his relations with the rest of the engineering staff?

3. Describe the conflict between Rondell's departments. What is causing this conflict? How should it be managed? What specific steps would you advise Rondell's management to take?

CASE 13-3

NEWCOMER-WILLSON HOSPITAL

Read Chapter 16's Case 16-1, "Newcomer-Willson Hospital." Next, look back at Chapter 13's diagnostic questions and choose the ones that apply to that case. Then use those questions with the ones that follow in this first analysis of the case.

1. What is the source of the hospital's administrators' power? What is the source of the physicians' power? The source of the nurses' power? Which of these groups seems to have the most clout? The least?

2. What caused the conflict between the hospital administration and the staff doctors? Between the doctors and the staff nurses? Are these conflicts disruptive? In what ways?

3. How can the conflicts at Newcomer-Willson Hospital be resolved? Can a negotiated settlement be reached? Or are structural modifications needed?

MANAGING GROUP AND INTERGROUP RELATIONS: ORGANIZATION DEVELOPMENT I

NCR's new concurrent engineering *approach to product development, in which people in several different departments work side by side, offers both opportunities and challenges. Frequent exchange of information among team members keeps everyone better informed about the progress of work and can generate ideas about new ways of doing things. For example, a new product is a pen-and-tablet device intended to replace the keyboard on a personal computer. But the often intense interaction among team members can also lead to arguments and dissension, and managers using this approach need to be skilled in facilitating cooperation and in helping group members resolve conflicts.*

At its plant in Atlanta, NCR Corp. is trying a new approach to product development. Like most manufacturers, NCR used to develop products in a series of steps, starting with design and engineering, then letting contracts for various materials, parts, and services, then finally going into production. Each step was largely independent of the others, and changes made at any post-design stage caused major traumas. The late fixes would ripple back through a project, causing everything that had gone before to be reworked. That would delay the product and push costs through the ceiling. So NCR decided to test a new method. In 1988 it tore down the wall that separates most design and manufacturing departments. Now, all the plant's 100-odd engineers are located in a pool of identical cubicles, so the specialists involved in design, software, hardware, purchasing, manufacturing, and field support all work side by side and compare notes constantly. This makes for more synergy, curbs late fixes, and achieves what William R. Sprague, NCR's senior manufacturing engineer in Atlanta, calls the overriding factor—getting products out on time.[1]

NCR's new approach, called concurrent engineering, is catching on in companies in the United States and Canada and is being touted as a development that could help North America regain its competitive edge throughout the world.[2] Besides NCR, newly merged with American Telephone and Telegraph, John Deere and Company, Motorola, and Westinghouse's Electronic Systems Group are also using concurrent engineering to reduce the costs and time required to design new products and get them to the marketplace. These gains have their price, though. To work as planned, concurrent engineering requires intense cooperation within the teams formed to complete design assignments. Collaboration is also required between different design teams so that information about one team's innovations can filter outward to other teams and be used to reduce product-development times even further. Managers of concurrent engineering operations must therefore be experts in encouraging teamwork within and among groups.

Similar trends toward using teams at work are also noticeable in the manufacturing phase of company operations. For instance, in 1986 Caterpillar Inc., a United States manufacturer of tractors and other heavy equipment, adopted flexible manufacturing methods that made obsolete the old way of organizing work into individualized tasks. By 1990, many of the company's employees had been grouped into highly interdependent teams. Today, each of these teams operates a self-sufficient cell of robots and other computerized machinery that can be reconfigured quickly to adapt to new production requirements. This flexibility enables Caterpillar to be extremely responsive to changing customer demands.[3] However, it also requires that managers be adept at managing group and intergroup relations.

Just as NCR is not alone in its adoption of concurrent engineering, Caterpillar is but one of many companies that have turned to group manufacturing processes. Firms such as IBM, Procter and Gamble, and Monsanto have begun

---

[1] Reprinted from Otis Port, Zachary Schiller, and Resa W. King, "A Smarter Way to Manufacture," *Business Week*, April 30, 1990, p. 110.

[2] Ibid., pp. 110–17.

[3] Brian Bremmer, "Can Caterpillar Inch Its Way Back to Heftier Profits?" *Business Week*, September 25, 1989, pp. 75–78.

AN EXAMPLE: XEROX CORPORATION

Xerox rethought every facet of its business after losing a major portion of its market to Japanese competitors. A key problem was the length of time it took Xerox to get a new product from the design stage to the sales floor. Using a new complex network of product development teams, crisis teams, and problem-solving teams, Xerox was able to reduce the time it took to launch a new product from five years to two and a half years and cut manufacturing costs in half. (*Taking Charge of Manufacturing,* Ettlie, 1988).

to rely on the team approach.[4] General Motors has also experimented with team production at its NUMMI plant in California and is implementing what it has learned at its Saturn plant in Tennessee (see the "International OB" box). If this trend continues, by the beginning of the 21st century, most large companies in North America will employ team methods either to design or to manufacture their products. More than ever before, managers will have to be experts at helping employees as they form work teams, participate in group decision making, and cope with the transition from working alone to working closely with others. Managerial success will also require expertise in stabilizing groups and resolving occasional confrontations between group members or among groups.

As you know, we have already overviewed much of this expertise in the previous four chapters. By now you should be aware of the basic facts and theories you will need to function effectively as a manager of team operations. We have not spent much time, however, talking about specific actions managers can take and programs they can implement to manage relations in and among groups. In this chapter, we turn our attention to discussing such actions and programs in greater detail. As indicated in Chapter 1, it is the second of three chapters in this book that addresses what can actually be done to manage organizational behavior. We will explore the field of organization development, which is the study and practice of changing the way people interact in organizations. After defining the field of organization development, we will examine resistance to change and ways to overcome it. In the process, we will talk about factors that can act as forces for and against change in organizations. We will also discuss what it means to be a *change agent* charged with overseeing an organization-development intervention. We will conclude by describing a variety of such interventions that managers can use to change and manage interpersonal, group, and intergroup relations on the job.

DIAGNOSTIC ISSUES

Our diagnostic model points to several important issues in organization development and change. How do we *describe* an organization's current situation and the kinds of problems we are having with group and intergroup relations? Can we distinguish the forces or groups within the organ-

[4] James G. Ellis, "Monsanto Is Teaching Old Workers New Tricks." *Business Week,* August 21, 1989, p. 67.

ization that will favor change from those that will resist it? What *diagnostic* procedures can we use to decide which changes to make and how to make them? How can we distinguish the situation in which organization development efforts will lead to permanent change from the situation in which the effects of such efforts will not last?

How can we *prescribe* the most effective method for overcoming resistance to change? Can we predict whether the changes we need to make will be best achieved by someone inside the organization or by an outside consultant? Finally, in putting organization development plans into *action*, how do we decide who should evaluate the effectiveness of these plans? Who should have access to this information?

WHAT IS ORGANIZATION DEVELOPMENT?

organization development A planned approach to interpersonal, group, intergroup, and organization-wide change that is comprehensive and long term and under the guidance of a change agent.

intervention A particular organization development technique, such as counseling or team building, that is used to stimulate change in organizations.

Organization development, often referred to as *OD*, is a process of planning, implementing, and stabilizing the results of change in organizations. OD is also a field of research that specializes in developing and assessing specific **interventions** or change techniques.[5] As both a management process and a field of research, OD is characterized by five important features:

1. *Emphasis on planned change.* The field of organization development evolved out of the need for a systematic, planned approach to managing change in organizations. It is OD's emphasis on *planning* that distinguishes it from other kinds of organizational changes that are more spontaneous or less methodical.

2. *A social-psychological focus.* OD interventions can stimulate change at many different levels—interpersonal, group, intergroup, or organization wide. The field of OD is thus neither purely psychological (focused solely on individuals) nor purely sociological (focused solely on organizations) but instead incorporates a mixture of both orientations.

AN EXAMPLE
It has been estimated that the change period following a merger or acquisition is from 5 to 7 years. Often, it is assumed change ends when the papers are signed. This assumption can lead to failure of mergers and acquisitions (*The Human Side of Mergers and Acquisitions*, 1989, pp. 7–9).

change agent A person who manages the OD process, serving both as a catalyst for change and as a source of information about OD.

3. *Attention to comprehensive change.* Although every OD intervention focuses on a specific organizational target, planners also keep in mind the effects of change on the *total system.* No OD intervention is designed and implemented without considering its effects on the rest of the organization.

4. *Long-range orientation.* OD experts emphasize that change is a continuing process that can sometimes take months—or even years—to produce desired results. Although managers often face pressures for quick, short-term gains, the OD process is not intended to yield stopgap solutions.

5. *Guidance by a change agent.* OD interventions are designed, implemented, and assessed with the help of a **change agent**, who serves both as a catalyst for change and as a source of information about the OD process. Successful organization development does not grow out of an unguided, do-it-yourself approach to organizational change.[6]

---

[5] Gordon L. Lippitt, Petter Longseth, and Jack Mossop, *Implementing Organizational Change* (San Francisco: Jossey-Bass, 1985), p. 3; and Ellen Fagenson and W. Warner Burke, "The Current Activities and Skills of Organization Development Practitioners," *Academy of Management Proceedings*, August 13–16, 1989, 251.

[6] Alan C. Filley, Robert J. House, and Steven Kerr, *Managerial Process and Organizational Behavior*, 2nd ed. (Glenview, Ill.: Scott, Foresman, 1976), pp. 488–90; and Wendell L. French and Cecil H. Bell, Jr., *Organization Development: Behavioral Science Interventions for Organizational Improvement*, 4th ed. (Englewood Cliffs, N.J.: Prentice Hall, 1990), pp. 21–22.

Together, these five features suggest the following definition: Organization development is a planned approach to interpersonal, group, intergroup, and organization wide change that is comprehensive, long-term, and under the guidance of a change agent.

The Change Agent

As indicated in our definition of OD, a change agent is a person who takes responsibility for overseeing the process of organization development. Such a person might be from outside the **client organization**—the organization undergoing development. An **external change agent** of this sort may be the employee of a consulting firm such as Arthur Andersen, she may be a specialist who is also a university professor, or she may be a full-time independent consultant. In any case, she probably has one or more graduate degrees in specialties that focus on individual and group behavior in organizations—organization development, organizational behavior, management, industrial/organizational psychology, or related fields of social science. This background, combined with an outsider's somewhat greater objectivity, supports a professional perspective that enables the external change agent to manage the change process in an unbiased manner. On the downside, the external change agent is unlikely to know as much about organizational policies, practices, and politics as someone from inside the firm.

Sometimes the change agent may already work for the client organization. Often, but not always, this person will be an employee of the firm's human resource management department or, in large companies like General Motors or IBM, of a separate training-and-development group. Being employed by the organization gives the **internal change agent** insight into the details of day-to-day events and procedures.[7] The internal change agent, however, may not always have the training of an external consultant, and his objectivity may be considerably less than that of an outside observer.

As you can see, choosing between external and internal change agents involves tradeoffs between professional expertise and objectivity, on the one hand, and insider familiarity and insight, on the other.[8] Sometimes an organization can seek a middle road, combining professional objectivity with insider familiarity by using a team of external and internal change agents. Although this seems like a very sensible approach, it is not used often now, partly because it is quite expensive. It seems likely, however, that as the emphasis on group production methods and intergroup collaboration continues to grow, the external-internal change agent team may become more common.

Resistance to Change

Change, often both the impetus and the product of OD efforts, involves modifying the ways the people and groups in an organization normally work together. Table 14-1 outlines the three types of change that take place in organizations—adaptive, innovative, and revolutionary. Whenever we attempt to set any one of these kinds of changes in motion we can expect resistance, for people tend to resist what they perceive as a threat to the established way of doing things. Eliminating or greatly decreasing this resistance is critical to the success of all OD interventions or other change-oriented processes.

---

[7] Stephen C. Harper, "The Manager as Change Agent: 'Hell No' to the Status Quo," *Industrial Management* 3 (1989), 8–11.

[8] Manuel London, *Change Agents* (San Francisco: Jossey-Bass, 1988), p. 194.

client organization An organization involved in the process of organization development.

external change agent An OD change agent who is not a member of the client organization.

internal change agent An OD change agent who is a member of the client organization.

AN EXAMPLE
There is growing evidence that, with some types of organizational changes like a merger or major restructuring, an external change agent has distinct advantages. Changes causing tremendous uncertainty and job insecurity are often not handled well by internal change agents; they too are affected by the uncertainty and insecurity. Internal change agents succumb to the same pressures everyone else feels. The external change agent is better able to separate himself from the situation, stay objective, and make the necessary changes.

APPLYING THE DIAGNOSTIC MODEL
Diagnostic Question 1: What sort of change is being contemplated? What sources of resistance to this change exist in the organization? What might you do to overcome this resistance? How successful are you likely to be?

T A B L E 14-1
Types of Organizational Change

TYPE OF CHANGE	DESCRIPTION	EXAMPLE IN THE RETAIL SALES INDUSTRY	COMPLEXITY, COST, AND UNCERTAINTY	POTENTIAL RESISTANCE
Adaptive	Reintroducing a familiar practice—one used previously or used elsewhere in the organization	Adoption of longer business hours during holiday shopping season	Lowest	Lowest
Innovative	Introducing a practice that is new to the organization but used elsewhere in the industry	Use of point-of-sale computers connected to cash registers to keep running account of inventory	Moderate	Moderate
Revolutionary	Introducing a practice that is new to the organization and new to the industry	Use of automated salesclerk machines that allow customers to complete an entire sales transaction without human assistance	Highest	Highest

Source: Bases on Paul C. Nutt, "Tactics of Implementation," *Academy of Management Journal* 29 (1986), 230–61; and Manuel London and John Paul Mac-Duffie, "Technical Innovations: Case Examples and Guidelines," *Personnel* (1987), 26–38.

AN EXAMPLE
Following a merger, employees feel tremendous job insecurity. As a result, they take steps to preserve their own jobs. Cooperation and team spirit disappear and self-preservation rules (*Journal of Vocational Behavior*, 6/91).

POINT TO STRESS
Uncertainty results in a sense of loss of control. Individuals take steps to regain control. These steps often focus on returning to the old way (*Personnel Administrator*, 8/89, pp. 84–90).

Sources of Resistance. Resistance to change may take different forms. It may be physical, intellectual, or emotional. Some of the most common sources of this resistance are individual self-interest, fear of the unknown, general mistrust, fear of failure, differing perceptions and goals, possible loss of status, social disruption, pressure from peers, managerial tactlessness, poor timing in introducing changes, personality conflicts, and bureaucratic inertia.[9]

Change can threaten *self interest*. People may fear that change will make it difficult or impossible for them to continue to satisfy personal needs and desires at work. For example, employees accustomed to working alone may feel that a change to team operations costs them the autonomy to decide how to do their own jobs. Conversely, employees used to working on teams may mourn the loss of social interaction if new job assignments require them to work alone.

Uncertainty about what to expect, or *fear of the unknown*, can arouse anxiety and create resistance. Especially when innovative or radically different ideas are introduced without warning, employees likely to be affected become fearful of the implications of change. Consider how you would react if it were suddenly announced that your college was going to change from semesters to quarters (or vice versa). Most likely, your first response would be to resist this change. You would worry about how credits obtained under the old system would transfer into the new system, about the courses taught in the new system, and about whether you would still be able to graduate on time.

When people mistrust others' intentions and behavior, they are particularly apt to be suspicious of impending change. *General mistrust* encourages secrecy or even deception, either of which creates further doubts about the intentions underlying change. At Eastern Airlines during the period of Frank Lorenzo's ownership, mechanics kept secret diaries detailing alleged rushed repairs and unsafe working conditions. Later they turned the diaries over to union representatives

[9] Paul R. Lawrence, "How to Deal with Resistance to Change," in *Organizational Change and Development*, ed. G. W. Dalton, P. R. Lawrence, and L. E. Greiner, (Homewood, Ill.: Irwin-Dorsey, 1970), pp. 181–97; Rino J. Patty, "Organizational Resistance to Change: The View from Below," *Social Service Review* 48 (1974), 371–72; Gerald Zaltman and Robert Duncan, *Strategies for Planned Change* (New York: Wiley-Interscience, 1977), pp. 98–121; and Joseph Stanislao and Bettie C. Stanislao, "Dealing with Resistance to Change," *Business Horizons*, 26 (1983), 74–78.

and government officials. One of the reasons for the mechanics' secret activities was their mistrust of Lorenzo. He had a history of trying to eliminate unions in the companies he owned. The mechanics' activities, along with similar actions by pilots, cabin attendants, and other airline employees, eventually undermined cost-saving changes that Lorenzo was trying to implement and pushed Eastern into bankruptcy. Mistrust can thus doom to failure an otherwise well-conceived change program.

The challenge posed by change may cause some members of an organization to doubt their personal competence. The *fear of failure* stimulated by this self-doubt may increase peoples' reluctance to support efforts to change familiar practices. General Motors employees throughout the Midwest who were invited to move to the new Saturn plant in Tennessee in the late 1980s sometimes voiced personal concerns about their ability to succeed in Saturn's new, high-tech assembly operations. Many refused to move even though they faced likely layoffs if they remained in their old jobs.

POINT TO STRESS
Often goals are not explained to employees. As a result, employees don't know what the benefits of change are.

Because individuals and groups have *differing perceptions and goals*, what is good for one group may sometimes be bad for another. Members of different groups may often have legitimate disagreements about whether a particular change is necessary or about how it may affect them. For example, a company's accounting department may expect to gain through the adoption of a new computerized information system. They know it will provide up-to-the-minute information on work in progress, making bookkeeping and auditing activities easier to perform. Employees in the company's manufacturing department, however, may anticipate being hurt by the system, because it will require them to take time out from their production activities to enter data into remote computer terminals. Groups that do not benefit from a given change may choose to resist it. In our example, manufacturing employees would probably be resistant to attempts to implement the new computerized system.

Organizational change often threatens existing power distributions and may require the elimination of jobs. The *prospective loss of status* that such change may pose can build strong resistance among the individuals and groups affected. For instance, managers in a company moving toward participatory decision making may be apprehensive about their loss of power over subordinates if employees are allowed to make decisions for themselves. Or the members of a company's human resource management department may resist attempts to automate staffing and selection procedures because they are concerned that their own importance will be diminished.

Change often causes *social disruption*; that is, it disturbs existing traditions or relationships within and between groups. The threat that interpersonal and group dynamics will be thrown into disequilibrium can stimulate resistance. A seemingly harmless attempt to combine research specialists and product engineers in a new-product-development group may unintentionally stimulate extreme resistance within the company's existing research and engineering departments. In highly cohesive groups, people are likely to resist change simply out of fear that it will break the group apart.

Peer pressure may cause employees who are not directly affected by a particular change to resist it nevertheless, to protect the interests of friends or co-workers. For example, employees are sometimes laid off when automation enables fewer individuals to produce more. In such cases, workers facing possible layoff may pressure their colleagues to resist automation, despite the fact that most employees will remain on the job and perform the same basic tasks as before. Even people who might otherwise support change may be pressured into resistance. Secretaries already familiar with a new word processing program may resist its introduction if the other secretaries in their office voice strong opposition.

People may resist change that is introduced at an inopportune time or in an insensitive manner. *Tactlessness and poor timing* can undermine routine acceptance. Awarding executive bonuses at the same time that hourly wages are being cut exemplifies the sort of tactlessness that has got companies like GM into trouble from time to time. Choosing the December holiday season as the time to announce that plant closings will require employees to move their families from one location to another is another example of how poor timing can ruin what might otherwise be seen as a positive change.

Resistance can evolve out of *personality conflicts* between those who advocate change and the individuals and groups directly affected by a proposed change. Imagine yourself in the position of working for a boss you don't especially like. Are you more likely to follow your boss's lead and accept change, or might you instead resist your boss in order to assert your personal feelings? Indeed, not getting along with others in the organization can substantially limit an individual's ability to encourage successful change.

Finally, resistance may be built into the very structure of the organization in the form of *bureaucratic inertia*. Managers of large, bureaucratic organizations like Exxon, General Dynamics Corporation, or the Chase Manhattan Bank often complain that their employees lack initiative and flexibility. Yet when inflexible, bureaucratic rules and rigid, standardized procedures are used to manage organizational behavior, is it any wonder that flexible, adaptive behavior is rare?

Forces for Change. Opposing these sources of resistance to change are a number of forces that favor or promote change. These forces are found both within the firm and outside of it.

External pressures for change come from a number of different sources—changes in international markets, shifts in national business and industry, shifting economic conditions, new governmental laws and regulations, changing population trends, and technological advances.[10]

Engaging in *international trade* for the first time opens up new problems and opportunities for a company. Dealing with different national cultures, economies, and organizing styles often highlights limitations in the way things have been done in the firm and provides the initial impetus for change. For instance, when American car manufacturers decided to enter the domestic Japanese market they had to learn how to conduct the door-to-door visits that Japanese automobile salespeople make. The purpose of the visits is to form close relationships with their customers and cultivate long-term loyalty.

Changing customer tastes, the entry of new competitors, and the introduction of new goods and services that may replace or substitute for established products all may cause *industry shifts*, that is, changes in the pattern of interfirm relationships within an industry. In turn, they usually require internal changes in the way each firm in the industry does business. For example, yogurt makers' introduction of frozen yogurt into the ice cream market influenced the sales of ice-cream manufacturers, who responded by introducing their own frozen yogurt desserts.

Changes in interest rates, inflation, labor markets, and other *economic conditions* can affect a company's ability to do business. Thus it can strengthen the drive to change the way things are done in the organization. Just-in-time inventorying procedures, which substantially reduce the amount of inventory maintained by a company, were adopted by American manufacturers during the early 1980s. The aim was to counteract the effects of growing inventory costs. Firms

---

[10] Andrew D. Szilagyi, Jr., and Marc J. Wallace, Jr., *Organizational Behavior and Performance*, 4th ed. (Glenview, Ill.: Scott, Foresman, 1987), pp. 635–36.

Though unusual, war is an environmental force that can have a very sudden and intense impact on the activities within a firm. During the Persian Gulf war, production of General Dynamics' M-1 Abrams tank was high. The war ended early in 1991, and more than 300 assemblers stood to lose their jobs by the middle of the year. The company, its new CEO William Anders, and others began efforts to convince the U.S. government that it should not close down the country's tank industry and at the same time began trying to sell the tank on the international market. Source: Bill Saporito, "This War Doesn't Mean a Windfall: General Dynamics," Fortune, February 25, 1991, pp. 40–42.

AN EXAMPLE
As a result of the Baby Bust of the 60s and 70s, we are facing a labor shortage. Some companies may recruit older workers or hire inner-city people to fill their labor needs. Such tactics often require special training or remedial education for these people.

throughout the U.S. and Canada also sought out energy-efficient production equipment and manufacturing methods to counteract the soaring costs of petroleum products and electricity.

Changes in *government regulations* or laws that regulate business practices, such as antitrust regulations, employment laws, safety codes, and tax acts, are an important external force favoring change within a company. For instance, deregulation in the air transportation industry in the early 1980s led to the emergence of several "super carriers," such as Northwest and U.S. Air. Their growth by merger stimulated a multitude of changes in company operations.

Changing birthrates and lifespans are among the *population trends* that affect societal age groups and consequently, the different pools from which an organization draws employees and customers. Having to meet changing employee demands, consumer preferences, and client needs mandates new ways of doing business and thus organizational change. For example, as the baby-boom generation grows older, hospitals are putting more emphasis on geriatric medicine and less on other specialties.

Changes in *technology*—the machines, procedures, and know-how used to create goods and services—create new products, industries, and ways of organizing work. For instance, food processors who failed to redesign canned or frozen items for convenient use in microwave ovens saw their market position shrink considerably during the 1980s. To keep up with spreading technological change or to create their own new technologies, firms must emphasize innovation and successful adaptation.

In addition to external, environmental forces, internal pressures within an organization also act as forces for change. Among the most important internal forces promoting change are organizational crises such as shortages of raw materials, increased understanding of the need for change, a drop in production quality or quantity, changing viewpoints of organization members, and a gut feeling that change is needed.[11]

More than anything else, the sense of emergency produced by an *organizational crisis* can stimulate support for change. For example, companies that suddenly face shortages of critical tools or materials may be forced to find substitutes: Milk producers turned to plastic crates when metal ones became too

---

[11] Edgar Huse and James Bowditch, *Behavior in Organizations: A Systems Approach to Managing* (Reading, Mass.: Addison-Wesley, 1973), p. 391.

expensive to be affordable. The United States Mint began to manufacture copper-centered coins when the high price of silver made it impossible to produce solid dimes, quarters, and half dollars.

Increased knowledge about a problem motivates people to attempt to solve it by changing the way things are done. For instance, Chrysler employees willingly participated in a wage-reduction program during the early 1980s after learning that their company was on the verge of declaring bankruptcy. Since then, wage reduction has become fairly common among United States firms as a way to cope with adverse economic conditions and foreign competition.

A significant decrease in the *quality or quantity of production*, resulting in fewer saleable goods or services, will stimulate concerns about long-term survival and make people more receptive to the idea of change. Having fewer customers is likely to make restaurant employees more open to changes in the firm's menu and pricing habits. Similarly, losing the business of a long-standing industrial buyer is likely to make the employees of an electronic-components manufacturer like Motorola more receptive to designing products faster and manufacturing them more efficiently.

New information, education, or additional experience can bring about *changes in management or work-force viewpoints*. With changing perceptions may also come an increased desire to change the way things are done in the firm. For example, companies that fund continuing education programs typically find it necessary to make jobs more interesting and challenging, because better-educated people demand more stimulating work.

Finally, change is sometimes supported simply by the feeling that it is needed. No specific factor may underlie or explain the *felt need for change*. People may just feel that things have been the same for too long. A senior employee's gut instinct, developed through years of experience, may be the only clue that a problem exists.

forcefield analysis A diagnostic method that depicts the array of forces for and against a particular change in a graphic analysis; often used as a component of the OD process.

APPLYING THE DIAGNOSTIC MODEL
Diagnostic Question 3: Based on a forcefield analysis, what is the likely combined effect of the forces for and against change? Is it realistic to attempt to institute change or should the status quo be accepted instead?

POINT TO STRESS
Students should understand that resistance to change cannot be eliminated; it can only be reduced and its impact on the organization minimized.

Overcoming Resistance through Forcefield Analysis. Setting change in motion requires identifying and overcoming sources of resistance, on the one hand, and encouraging and strengthening sources of support, on the other. **Forcefield analysis** is a diagnostic method that depicts the array of forces for and against a particular change in a graphic analysis. It is a useful tool for managers and change agents who are attempting to envision the situation surrounding a prospective change. As Figure 14-1 shows, in a forcefield analysis, two lines are drawn, one representing the organization's present situation (the solid horizontal line) and the other the organization after the desired change has been put into effect (the dashed horizontal line). Next, forces identified as supporting change are depicted as arrows pushing in the direction of the desired change. Forces resisting change are drawn as arrows pushing in the opposite direction. The length of each arrow indicates the perceived strength of the force represented by the arrow relative to the other forces in the forcefield.

There is no universal, fail-safe way to overcome the resistant factors identified in a forcefield analysis. Of the many available options, the six that are used most often include:

1. *Education and communication.* Disseminating information about the need and rationale for a prospective change through one-on-one discussions, group meetings, and written memos or reports. This approach is best used where change is being undermined by a lack of information or where available information is inaccurate. Its strength is that once persuaded through education, people will often help with the implementation of change. Its primary weakness is that it can be quite time consuming if many people must be involved.

FIGURE 14-1

Forcefield Analysis

During the middle 1980s, IBM, Compaq, and other U.S. manufacturers faced the task of introducing a new line of computers to maintain position in an increasingly competitive world market for personal computers. Forces resisting this change included the following: (1) differing perceptions among the management of American companies about the need for new products (as opposed to modest improvements of existing lines), (2) employee concerns about the social disruption likely to occur as old work groups disbanded to staff new production facilities, (3) bureaucratic inertia stemming from the rules and procedures used to coordinate existing ways of doing things, and (4) employee fears about not being able to cope with the demands of new production technologies. Opposing these forces were others supporting change. Those forces included the following: (1) growing competition from Asian computer manufacturers, (2) a drive in American computer firms to introduce greater factory automation to cut costs and increase quality, (3) and a general sense of impending crisis throughout the American computer industry. In the end, forces supporting change won out with the introduction of IBM's micro-channel series and Compaq's Desqpro personal computers. Both lines proved to be quite successful in the marketplace.

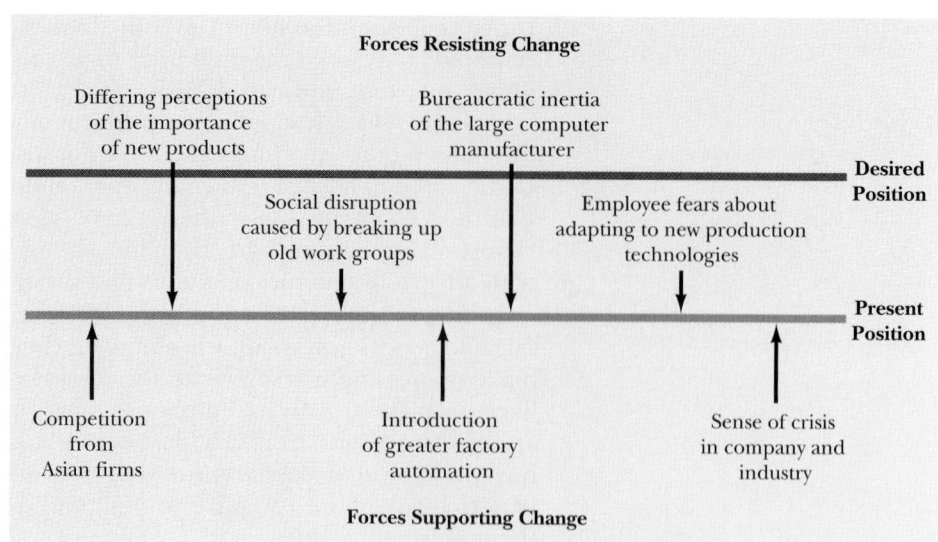

2. *Participation and involvement.* Involving those to be affected by a change in its design and implementation. Employees meet in special committees or task forces to participate in the decision making. There are two situations where this option is most effective. It works well when information required to manage change is dispersed among many people and where employees who have considerable power are likely to resist change if not involved themselves. It facilitates information exchange among people and breeds commitment among the people involved, but it can slow down the process if participants design an inappropriate change or stray from the task at hand.

3. *Facilitation and support.* Providing job training and emotional support through instructional meetings and counseling sessions for employees affected by a change. This method is most useful when people are resisting change because of problems with personal adjustment. No other method works as well with adjustment problems, but it can consume significant amounts of time and money and still fail.

4. *Bargaining and negotiation.* Working with resistant employees through bargaining and tradeoffs to provide them with incentives to change their minds. This technique is sometimes used if an individual or group with the power to block a change is likely to lose out if the change takes place. Negotiation can be a relatively easy way to avoid such resistance but can prove costly if it alerts other individuals and groups that they might be able to negotiate additional gains for themselves.

5. *Hidden persuasion.* Using covert efforts and providing information on a selective basis to get people to support desired changes. This approach is sometimes used when other tactics will not work or are too expensive. It can be a quick and inexpensive way to dissolve resistance. However, it can lead to future problems if people feel manipulated.

6. *Explicit and implicit coercion.* Using power and threats of negative consequences to change the minds of resistant individuals. It tends to be used when speed is essential and where those initiating change possess considerable power. It can overcome virtually any kind of resistance. Its weakness is that it can be risky if it leaves people angry.[12]

[12] John P. Kotter and Leonard A. Schlesinger, "Choosing Strategies for Change," *Harvard Business Review* 57 (1979), 102–21; and John M. Ivancevich and Michael T. Matteson, *Organizational Behavior and Management*, 2nd ed. (Homewood, Ill.: BPI-Irwin, 1990), pp. 621–22.

Values That Guide Change

When set in motion by organization-development processes, organizational change is guided by a set of fundamental values. These values direct the way OD interventions influence relations among people and groups and include the following:

1. *The needs and aspirations of human beings are the reasons for organized effort in society.* Out of this value grows a strong concern with enhancing the personal development and satisfaction of all the members of an organization. This concern then creates a self-fulfilling prophesy. The belief that you can grow and develop in personal and organizational competence tends to produce conditions conducive to such growth and development. When managers at American Steel and Wire Company, formerly a unit of USX, became concerned about the personal development of their employees, they began involving all employees in strategic-planning processes. They found that the resulting exchange of information created a team feeling that led to even greater employee involvement in carrying out strategic objectives. Today, American Steel is among the lowest cost, highest-quality wire and rod producers in the U.S., while retaining strong loyalty and a sense of enhanced personal worth among its employees (for more information about how American Steel encourages teamwork, see the "In Practice" box later in this chapter).[13]

2. *Openness is essential to working together effectively.* This value holds that work and life can be more worthwhile and organized effort more effective and enjoyable if people openly express their feelings and sentiments to each other. Openness is sometimes hazardous for employees, though, as might be the case if a subordinate were to be openly critical of her boss without receiving protection from possible retribution. It is also naive to expect two groups engaged in conflict to trust each other enough to communicate openly without the help of a neutral party or procedure. So it is unrealistic to believe that openness will develop in the workplace without first eliminating the negative effects of hierarchical and political barriers.

3. *Commitment to both action and research is required.* It is possible to get so caught up in the action of implementing change that instituting the careful design, controls, and other necessary elements of formal research go by the board. However, research into the nature of change processes and the effectiveness of different interventions is an indispensable part of the OD process. If we don't use scientific methods (see Chapter 3) to study what we do as we manage organizational change, we will never know why something works when it does or why it doesn't work when it fails.

4. *Democratization, or power sharing, in organizations is a valued outcome of the OD process.* Placing a high value on humanizing the workplace and building a participatory atmosphere does not mean that we want to reduce or neutralize the power of owners and managers. The goal of organization development is to increase everyone's power by encouraging the development of both technical and human relations competence in all employees.[14]

When put into practice, these four values promote the assumptions summarized in Table 14-2. As you can see from these assumptions, the OD framework emphasizes the necessity for organizations to help facilitate the personal growth of their members. Equally, it emphasizes the importance of employees' contributions to the continued well-being of the organization. All change stimulated

---

[13] Joan E. Rigdon, "Team Builders Shine in Perilous Waters," *Wall Street Journal*, October 29, 1990, p. B1.

[14] French and Bell, *Organization Development*, pp. 49–50.

TABLE 14-2

Assumptions Underlying the Practice of Organization Development

People

Have inherent needs for personal growth and development.

Want to contribute more to their organization than unchanged conditions often allow.

Desire to be accepted and interact cooperatively in at least one small group of peers.

Look to their work group as an important reference group that helps them form personal beliefs, values, and norms.

In groups

Members must assist each other with effective leadership and membership behaviors. Formal leaders cannot be expected to do all that is required to keep groups performing effectively.

Suppressed feelings and attitudes reduce problem solving, personal growth, and satisfaction.

The level of interpersonal trust, support, and cooperation is typically much lower than is either necessary or desirable.

Solutions to problems are only likely to succeed if all members alter their mutual relationships; doing things alone will not help.

In organizations

Groups linked together by work activities affect the attitudes and beliefs of each other's members.

Conflicts between people and groups in which one party wins and the other loses do not provide long-term solutions to organizational problems.

Changes stimulated by OD interventions must be reinforced and sustained by appropriate changes in the human resource management system—performance appraisal, compensation, training, staffing, task and communication processes and procedures.

Members place a high value on collaboration and cooperation, and seek to avoid exploitation or manipulation.

Source: Based on Wendell L. French and Cecil H. Bell, Jr., *Organization Development: Behavioral Science Interventions for Organizational Improvement*, 4th ed. (Englewood Cliffs, N.J.: Prentice Hall, 1990), pp. 44–51.

by OD is thus seen as having the dual goals of human fulfillment and organizational accomplishment.

THE ORGANIZATION-DEVELOPMENT PROCESS

Regardless of the type of change being pursued, organization development follows a multiple-step process. We can gain insight into the way the process of OD is conducted from the Lewin development model, the planned change model, and the action research model.

The Lewin Development Model

Lewin development model A three-step model of the development process that is followed in every successful OD intervention.

The **Lewin development model**, named for its creator, the social scientist Kurt Lewin, is a three-step model of the development process that takes place during every successful OD intervention. According to this model, OD progresses through stages of unfreezing, transforming, and refreezing.[15]

[15] Kurt Lewin, *Field Theory in Social Science* (New York: Harper & Row, 1951), pp. 228–29. See also Marvin W. Weisbord, *Productive Workplaces: Organizing and Managing for Dignity, Meaning, and Community* (San Francisco: Jossey-Bass, 1987), pp. 14–23.

How to Institutionalize Teamwork

To institutionalize the results of change means to stabilize them and make them permanent. Suppose you have completed a group intervention that resulted in a good team spirit and an ability to work together effectively. Now you want to institutionalize the outcome. What do you do? Tom Tyrrell, president of American Steel and Wire Company, has several answers. First you don't lay off employees if the economy slows down. Instead, you make what cuts you can through attrition, leaving the jobs of newly retired employees unfilled. You keep the rest of your employees busy rebuilding machinery and doing maintenance chores that have to be ignored during busier times. In so doing, you are helping to maintain the teams and team relations you have worked so hard to build. Second, you treat all your employees with the same sense of fairness. All employees at American Steel and Wire receive the same vacations and benefits and share the same formula for profit sharing. In this way, inequity that might otherwise undermine teamwork is avoided. Third, you give your em-

ployees greater control over their work and working lives. Process managers (foremen) at American Wire once made virtually every day-to-day decision about production activities, but today lower-level employees now make many of these decisions. In addition, they are encouraged to look beyond daily activities and to think about long-term cost cutting and quality assurance. This way, employees have the feeling that teams and team tasks are *their* creations instead of something forced on them by a distant management. Fourth, you hold social get-togethers to give employees the chance to relax and enjoy each other's company. American Steel and Wire recently had international dinner costing $25,000, but it was worth much more, according to Tyrrell. Getting everyone together gives people the chance to overcome anxieties and build camaraderie. Team effectiveness improves even more as a result.

Joan E. Rigdon, "Team Builders Shine in Perilous Waters," *Wall Street Journal*, October 29, 1990, p. B1.

unfreezing The first step in the Lewin development model; the step in which one tries to weaken old attitudes, values, and behaviors and to get people ready for change.

POINT TO STRESS
Generally businesses strive for satisfied employees. During the unfreezing step of introducing change, however, they may actually strive to generate dissatisfaction. People only seek and support change when the current situation is perceived as bad.

transforming The second step in the Lewin development model; the step in which change actually occurs.

Unfreezing. Unfreezing is a preparatory step in which one tries to weaken old attitudes, values, and behaviors and get people ready for change. New and different experiences or information that challenges routine perceptions facilitate this unfreezing process. The forcefield analysis procedure described earlier can be especially helpful at this stage, because it clarifies which forces and perceptions must be weakened and which should be encouraged.

To stimulate unfreezing, change agents use various OD interventions, for instance, the counseling, team building, and intergroup mirroring techniques described later in this chapter. Their purpose is to increase people's awareness of challenging information and encourage employees to question current behaviors and attitudes. This questioning can lead to greater readiness for change, because the less satisfied people are with the status quo, the more likely they are to feel that change is necessary.

Transforming. Transforming is the step in which change actually occurs. It takes place as organization members first identify with the change agent. They begin to internalize the values of organization development and to adopt new attitudes and behaviors at work. This process often requires (1) facilitation, in which the change agent helps members understand why change is necessary; and (2) training, in which employees learn how they will be affected by the change and what will be expected of them after the change has taken place. These techniques are an integral part of all OD interventions and help dispel most remaining resistance to change.

refreezing The third step in the Lewin development model; the change that took place during the transforming step becomes stable and permanent.

Refreezing. Refreezing focuses on institutionalizing change. In this step, the change that took place during the transforming step becomes stable and permanent. During refreezing, new attitudes, values, and behaviors are integrated into everyday organizational processes and procedures. For example, leaders become less directive as subordinates assume newly developed decision-making roles. Reward systems change so that they reinforce cooperation instead of competition. Managers and their subordinates meet regularly to encourage greater communication (see the "In Practice" box for additional examples). Refreezing does not imply rigidity or resistance to future change. Indeed, because of their experience with successful organization development, the members of an organization learn not to fear change but to welcome it instead.

The Planned Change Model

planned change model A model of the OD process that is an expansion of the Lewin development model. Tells how OD proceeds when an off-the-shelf intervention is implemented.

According to the Lewin model, we must prepare for organizational change and monitor its progress if we are to expect an OD intervention to produce lasting results. The **planned change model** is an expansion of Lewin's approach in which the basic steps of unfreezing, transforming, and refreezing have been elaborated to produce an organization-development action guide.[16] As Figure 14-2 shows, in this expanded model, Lewin's three steps have become seven stages.

Scouting. During the first stage of the planned change model, a change agent and the management of a client organization jointly explore a particular orga-

[16] R. Lippitt, J. Watson, and B. Westley, *The Dynamics of Planned Change* (New York: Harcourt, Brace, & World, 1958) pp. 129–44; and D. A. Kolb and A. H. Frohman, "An Organization Development Approach to Consulting," *Sloan Management Review* 12 (1970), 51–65.

FIGURE 14-2

The Planned Change Model Compared to the Lewin Development Model

From Edgar F. Huse, Organization Development and Change, *2nd ed. (St. Paul, Minn.: West, 1980), p. 87. Reprinted with permission of the publisher.*

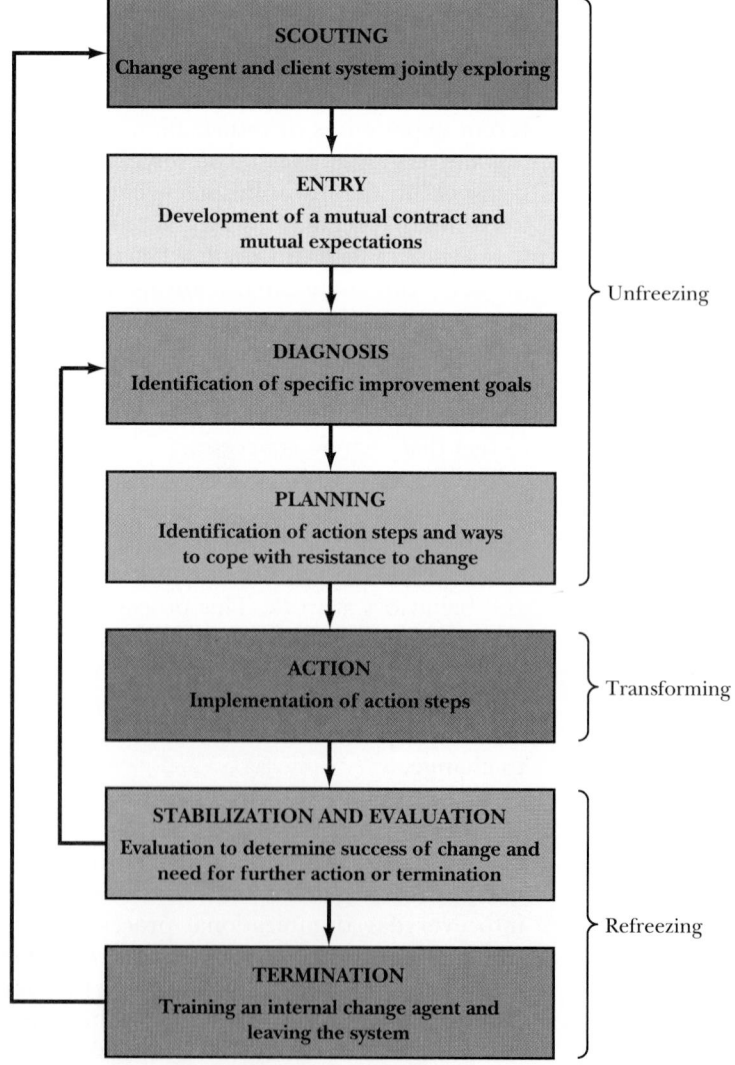

nizational problem or situation. If an external change agent has been called in, it is at this stage that she decides whether to become involved. Meanwhile, management checks out her expertise and ability. If an internal change agent is involved, scouting is an exercise in employee selection (see Chapter 9).

APPLYING THE DIAGNOSTIC MODEL
Diagnostic Question 4: If change is a realistic goal, do cost, availability, expertise, and similar considerations favor using an external change agent? An internal change agent? A team of external and internal change agents?

Entry. Together, the change agent and the client organization develop a contract and expectations as to how stages three to seven will be carried out. If an external change agent is involved, the contract may specify how long she will work with the client organization and whether she will train an internal agent to take over after her departure. The contract may also specify what results the client organization wants. An honest contract, however, will *not* guarantee these results. Just as medical doctors cannot guarantee successful treatment and lawyers cannot guarantee success in the courtroom, OD change agents cannot guarantee successful change.

Diagnosis. The change agent and the client identify general goals for improvement, specific problems to be addressed, and indicators of success to be used during postaction evaluation. They boil down the general expectations that were identified during the entry stage and turn them into concrete problem statements. Also they look for various off-the-shelf OD interventions that have been used to solve similar problems in other organizational settings.

Planning. The change agent and representatives of the client organization jointly develop a plan of action, including specific ways of dealing with potential resistance to planned change. They select for use in the client organization one of the OD interventions identified during diagnosis and make preparations for its implementation.

Action. The change agent oversees implementation of the plan of action. Members of the client organization participate in an OD intervention that leads them through the process of determining what specific changes to make and how to make them. Then they implement these changes.

Stabilization and Evaluation. The change agent and representatives of the client organization work together to assess the accomplishments of the intervention, using the success indicators identified during the diagnosis stage. If further action is required, the planned change process returns to the diagnosis stage. If not, the process moves on to the termination stage.

Termination. Termination ends the change agent-client organization relationship. If an external change agent is involved and contractual agreements specify it, the change agent trains others in the organization to maintain the change. Then she either leaves the organization or moves on to another assignment in the firm. If an internal change agent is involved, she will be assigned new duties.

The Action Research Model

action research model A model of the OD process that permits the development and assessment of original, innovative interventions.

The **action research model** is a bit more complicated than the planned change model, incorporating a recurrent cycle of data-based action planning (see Figure 14-3). Organization-development programs based on the planned change model usually involve the implementation of standard off-the-shelf interventions that

FIGURE 14-3

The Action Research Model

Based on Wendell French,
"Organization Development:
Objectives, Assumptions, and
Strategies," California Management
Review *12 (1969),* 26.

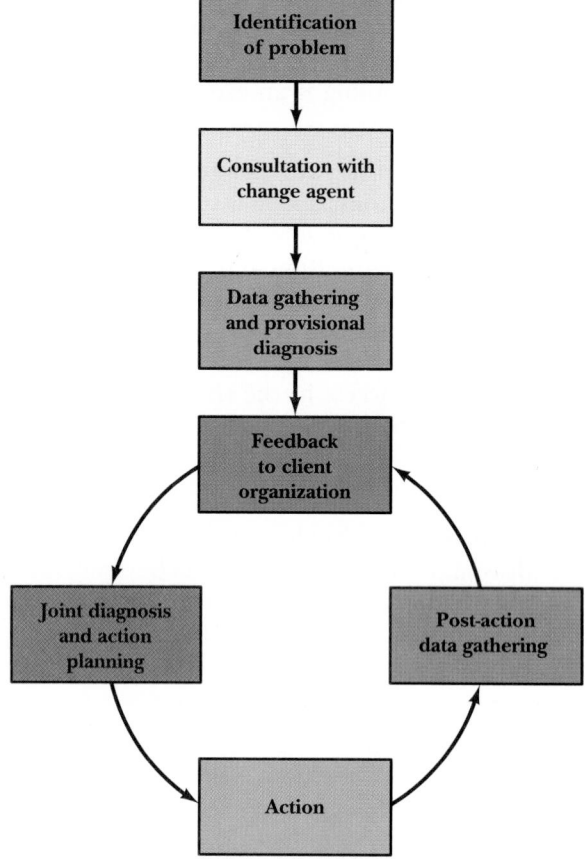

have been used before. The action research model, however, permits the development and assessment of original, innovative procedures.

APPLYING THE DIAGNOSTIC MODEL
Diagnostic Question 5: Once chosen, does the change agent behave in accordance with the basic values and assumptions of OD? If the aim is to use an existing OD intervention to solve an organizational problem, is the change agent adhering to the planned change model of OD? If the aim is to develop and assess a new type of intervention, do change agent behaviors conform to the action research model?

Problem Identification. In the initial stage of action research, someone in an organization, often a top manager, perceives problems that might be solved with the assistance of an organization-development change agent. Specific problem statements can usually be formulated at this stage. Sometimes, though, problem identification cannot progress beyond an uneasy feeling that something is wrong. Consultation with a change agent may then be required to crystallize the problems.

Consultation. In the second stage, the manager and a change agent clarify the perceived problems and consider ways of dealing with them. During this discussion, they assess the degree of fit between the organization's needs and the change agent's expertise. For example, if the organization is troubled by poor interpersonal relations, does the change agent know how to help people interrelate? Or if the organization is having problems with intergroup conflict, does the change agent have experience in conflict management? If the agent fits the situation, action research progresses to the next stage. If not, another change agent is called in and consultation begins anew.

Data Gathering and Provisional Diagnosis. The change agent initiates the diagnostic process by gathering data about the organization and its perceived problems. He observes, interviews, and questions employees and analyzes performance records. If an external change agent is employed, a member of the organization may assist during this process, facilitating the agent's entry into the firm and providing access to a significant amount of otherwise hidden or unavailable data. The agent concludes this stage by examining the data and performing a provisional analysis and diagnosis of the situation.

Feedback to the Client Organization. Next, the data and provisional diagnosis are submitted to the client organization's top management group. Informing top management early on of the OD process under way is absolutely necessary to secure the managerial support that any OD effort must have to succeed. The change agent is careful, during this presentation, to preserve the anonymity of people serving as sources of information. Identifying them could jeopardize their willingness to cooperate later on.

Joint Diagnosis and Action Planning. During the fifth stage of the action research model, the change agent and the top management group discuss the meaning of the data, their implications for organizational functioning, and any need for further data gathering and diagnosis. At this point, other people throughout the organization may also become involved in the diagnostic process. Sometimes employees meet in feedback groups and react to the results of top management's diagnostic activities. At other times, work groups elect representatives, who then get together to exchange views and report back to their coworkers. If the firm is unionized, union representatives may also be consulted. No matter which members of the organization are specifically involved in the change process, however, the important point to remember is that in action research, the change agent does not impose interventions on the client organization. Instead, members of the organization deliberate jointly with the change agent and work as a team to develop wholly new interventions and plan specific action steps.

Action. Next, the company puts plan into motion and executes its action steps. In addition to implementing the jointly designed intervention, action may involve such activities as additional data gathering, further analysis of the problem situation, and supplementary action planning.

Postaction Data Gathering and Evaluation. Because action research is a cyclical process, data are also gathered after actions have been taken. The purpose is to monitor and assess the effectiveness of an intervention. In evaluating the intervention, groups in the client organization review the data and decide whether they need to rediagnose the situation, perform further analyses of the situation, and develop new interventions. The change agent's role during this process is to serve as an expert on research methods as applied to the process of development and evaluation. In filling this role, he will probably perform data analyses, summarize the results of these analyses, guide subsequent rediagnoses, and position the organization for further intervention.

Managing Organization Development

What do the three models we have just described have to say about managing the process of organization development? The Lewin development model highlights the fact that successful OD efforts require adequate preparation and careful stabilization. The planned change model shows how problems can sometimes be solved with existing OD interventions like the ones we describe next. The action research model summarizes the process of inventing, implementing, and eval-

uating new OD interventions. Each of these models thus provides a working framework for change agents. Guided by a set of basic values, they can manage OD programs intended to bring about significant changes in organizational behavior.

ORGANIZATION-DEVELOPMENT INTERVENTIONS

You now know about the role of change agent and the values that underlie the field of organization development. You also understand the process of OD and how it incorporates existing interventions or leads to the creation of new ones. However, you have yet to learn about the kind of actions involved in working through a specific OD intervention. We turn our attention next to this final topic.

A Matrix of Interventions

There are a large number—perhaps hundreds—of different OD interventions. Many of these interventions are widely known and employed often by change agents pursuing the planned change approach. Some of them also serve as a source of ideas for change agents who are following the action research approach to create and evaluate new interventions. In this section, we examine nine well-known interventions that, as Table 14-3 shows, differ in terms of two principal factors—depth and target.

depth The degree or intensity of change that an organization development intervention is designed to stimulate.

Depth. The **depth** of an intervention is the degree or intensity of change that the intervention is designed to stimulate.[17] A *shallow* intervention is intended mainly to provide people with information or facilitate communication. Interpersonal counseling interventions, for instance, often involve little more than acquainting individuals with ideas they might not otherwise consider. This sort of exposure to new knowledge can trigger a modest amount of cognitive or behavioral change but is not intended to alter deeply held feelings or opinions. Little personal risk is involved in interventions of this sort.

[17] Roger Harrison, "Choosing the Depth of Organizational Intervention," *Journal of Applied Behavioral Science* 6 (1970), 181–202.

T A B L E 14-3
Organization-Development Interventions

TARGET	FOCAL PROBLEM	DEPTH		
		Shallow	Moderate	Deep
Interpersonal relations	Problem fitting in with others	Counseling	Role analysis technique	Sensitivity training
Group relations and leadership	Problem with working as a group	Process consultation	Team building	Team development
Intergroup relations	Problem with relations between groups	Third-party peacemaking	Intergroup mirroring	Intergroup team building
Organization-wide relations	Problem with functioning effectively	Interventions described in Chapter 18		

APPLYING THE DIAGNOSTIC
MODEL
Diagnostic Question 6: Do the
depth and target of the OD
intervention being implemented
seem to match the problem being
experienced by the organization?
Note that the shallowest
intervention that is likely to
stimulate the required amount of
change is the one that should be
selected for implementation.

POINT TO STRESS
It is important to note depth of
intervention must fit with the
change needed. If the intervention
is too shallow, change will not
happen. If too deep, more harm
than good may be done.

In sharp contrast, a *deep* intervention is intended to effect massive psychological and behavioral change. An intervention of this type and the OD change agent guiding it both attack basic beliefs, values, and norms in an attempt to bring about fundamental changes in the way people think, feel, and behave. Taken to the extreme, a deep intervention, such as interpersonal sensitivity training, can even resemble a session of brainwashing in which participants risk exposure to extreme psychological injury. Therefore, deep interventions must be approached with great caution and require the guidance of an expert change agent.

Between these two extremes, interventions of *moderate* depth seek to challenge existing attitudes and bring about changes in people's points of view without precipitating major psychological change. Often, interventions of this sort involve getting people with differing viewpoints together to discuss the way they perceive each other and the organization. For instance, the role analysis technique described below is a structured procedure of moderate depth that enables colleagues to trade opinions about each other's role responsibilities. They exchange more than the simple information of a shallow intervention, because participants in a role analysis session bargain with each other about personal roles and argue about interpersonal expectations. The sort of behavioral change caused by role analysis, however, falls far short of the more extreme psychological change aimed for in deeper interventions.

target The specific focus of an
OD intervention's change efforts.

Target. The **target** is what an intervention focuses on. In Chapter 9, we discussed the individual-level target of human resource management (career planning and development, employee mentoring, and so on.) In addition, interpersonal, group, intergroup, and organization-wide relations can serve as targets of OD interventions. We will wait until Chapter 18 to discuss organization-wide interventions, because we have not yet described the types of organizational problems that such interventions can solve. For now, we will concentrate on interventions aimed at solving the kinds of interpersonal, group, and intergroup problems discussed in Chapters 10–13.

*Some business leaders like Max
DePree believe that organization
development interventions must take
on a spiritual nature. DePree, who
heads Herman Miller, a leading
furniture design and manufacturing
firm, says that businesses must offer
workers the community and
psychological sustenance that
churches, families, and neighborhoods
no longer provide. DePree proposes a
"covenant" between company and
employee as the basis for superior
management.*
*Source: "Should Your Company Save
Your Soul?" Fortune, January 14,
1991, p. 31.*

Interpersonal Interventions

Interpersonal interventions focus on solving several of the problems that people sometimes have in fitting in with others at work. We highlighted some of those problems in Chapter 10. Depending on the particular intervention, attempts may be made to define personal roles, clarify social expectations, or strengthen sensitivity to others' needs and interests.

counseling A shallow interpersonal OD intervention in which a change agent meets, either one on one or in small groups, to provide helpful information to people who are having trouble relating with others.

Counseling. Counseling is a shallow interpersonal intervention in which an OD change agent meets, either one on one or in small groups, with people who are having trouble relating with others. In these meetings, the change agent's role is to suggest alternative ways of looking at things and provide information that otherwise might not be considered.[18] In the process, the change agent may act as a coach who encourages troubled people to find out how others see their behavior and also how others expect them to behave. In fulfilling the roles of coach and counselor, the change agent is as nondirective as possible. Changes and improvements in behavior, if any, are decided upon and implemented after counseling has run its course. The change agent neither advocates nor supervises any particular type of change.

realistic previewing A technique sometimes used during counseling interventions to help people form sensible expectations about workplace relationships.

During the counseling process, a technique called **realistic previewing** is sometimes used to help people form sensible expectations about workplace relationships. Realistic previewing was initially developed as an employee selection and socialization technique.[19] It was intended to reduce turnover by acquainting prospective employees with both the positive and negative aspects of a job prior to their decision to work for a company (see Chapter 9). This was in contrast to the positive-only orientations that new employees often received. It is now also used as an intervention aimed at solving interpersonal problems. In a realistic previewing intervention of this type, a counselor provides employees with information that they might not otherwise have about interpersonal relations. The counselor might talk about coworkers' opinions of their boss and her management style or about forming informal friendships on the job or even about off-the-job activities, such as company picnics or bowling leagues. In addition to the counselor, a group of coworkers are sometimes assembled to provide their own comments about working conditions and interpersonal relations. This kind of intervention helps employees form realistic impressions about their coworkers. Consequently, it can reduce the likelihood that unreasonable expectations will undermine interpersonal relations as they develop.

role analysis technique An interpersonal OD intervention of moderate depth intended to help people form and maintain effective working relationships by clarifying role expectations.

Role Analysis Technique. The **role analysis technique** (RAT), an interpersonal intervention of moderate depth, is also intended to help people form and maintain effective working relationships.[20] As we saw in Chapter 10, people at work fill specialized *roles* in which they are expected to engage in specific sorts of behavior. Often, however, employees lack a clear idea of what their roles entail, or they are overburdened by role demands. RAT, outlined in Figure 14-4, is designed to help reduce role ambiguity and conflict by clarifying interpersonal expectations and responsibilities.

To initiate a RAT intervention, the occupant of a troublesome role contacts a change agent about her problem and receives instruction from the agent on

---

[18] Edgar H. Schein, *Process Consultation: Its Role in Organization Development* (Reading, Mass.: Addison-Wesley, 1969), p. 115.

[19] John P. Wanous, *Organizational Entry: Recruitment, Selection, and Socialization of Newcomers* (Reading, Mass.: Addison-Wesley, 1980), p. 43.

[20] Ishwar Dayal and John M. Thomas, "Operation KPE: Developing a New Organization," *Journal of Applied Behavioral Science* 4 (1968), 473–506.

FIGURE 14-4

Steps in the Role Analysis
Technique

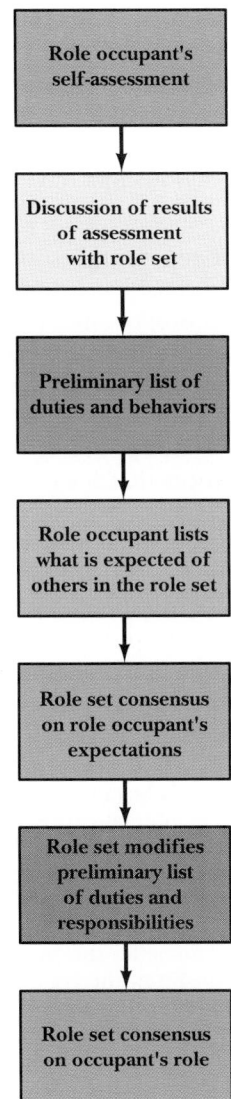

Role occupant's
self-assessment

Discussion of results
of assessment
with role set

Preliminary list of
duties and behaviors

Role occupant lists
what is expected of
others in the role set

Role set consensus
on role occupant's
expectations

Role set modifies
preliminary list
of duties and
responsibilities

Role set consensus
on occupant's role

the RAT procedure. Next, she works alone to analyze the rationale for the role as well as its place in the organizational network of interpersonal relations. She tries to learn how to use her role in meeting personal, group, and organizational goals. Then she discusses the results of her analysis in a meeting attended by everyone whose work is directly affected by her role. During this discussion, the change agent lists on a blackboard or flipchart the specific duties and behaviors of the role as identified by the role occupant. The rest of the group suggest corrections to this list. Behaviors are added or deleted until the role occupant is satisfied that the role as she performs it is defined accurately and completely.

Next, the change agent directs attention to the role occupant's expectations of others. To begin this step, the role occupant lists her expectations of the roles that are connected with her own. The group then discusses and modifies these expectations until everyone agrees on them. After this, all participants have the opportunity to modify their expectations about the person's role, in response to her expectations of them. So, as you can see, RAT is a process of negotiation. The person who is the focus of the intervention can ask others to do things for her, and others can ask her to do something for them.

In the final step of role analysis technique, the subject writes a summary or profile of her role as it has been defined. This profile specifies which behaviors are absolutely required and which are discretionary. It thereby constitutes a clearly defined listing of the role-related activities to be performed by the role occupant. The meeting continues, focusing on the roles of the other RAT participants, until all relevant interpersonal relationships have received adequate clarification.

Sensitivity Training. As a deep interpersonal intervention, **sensitivity training** focuses on developing greater sensitivity to oneself, to others, and to one's relations with others.[21] Designed to promote emotional growth and development, sensitivity training typically takes place in a closed session away from work. It may involve a collection of people who do not know each other, a group of people who are well acquainted, or a combination of both. A sensitivity-training session may last for as little as half a day or may go on for several days. It is begun by a change agent, who announces that his role is to serve solely as a nondirective resource. He then lapses into silence, leaving the participants with neither a leader nor an agenda to guide interpersonal activities. The purpose of putting people in such an ambiguous situation is to force them to structure relationships among themselves and, in the process, question long-held assumptions about themselves, about each other, and about interpersonal relations.[22]

Sensitivity-training participants take part in an intense exchange of ideas, opinions, beliefs, and personal philosophies as they struggle with the process of structuring interpersonal relations. Here is a description of one four-day session.

> The first evening discussion began with a rather neutral opening process which very soon led to strongly emotional expression of concern. . . . By the second day the participants had begun to express their feelings toward each other quite directly and frankly, something they had rarely done in their daily work. As the discussion progressed it became easier for them to accept criticism without becoming angry or wanting to strike back. As they began to express long-suppressed hostilities and anxieties the "unfreezing" of old attitudes, old values, and old approaches began. From the second day onward the discussion was spontaneous and uninhibited. From early morning to long past midnight the process of self-examination and confrontation continued. They raised questions they had never felt free to ask before. Politeness and superficiality yielded to openness and emotional expression and then to more objective analysis of themselves and their relationships at work. They faced up to many conflicts and spoke of their differences. There were tense moments, as suspicion, distrust, and personal antagonisms were aired, but more issues were worked out without acrimony.[23]

By completing this process, people learn more about their own personal feelings, inclinations, and prejudices and about what other people think of them.

A word of warning: Sensitivity training is a deep intervention that can initiate profound psychological change. It is not uncommon for participants to engage in intensely critical assessments of themselves and others that can be both difficult and painful. Therefore, the change agent overseeing sensitivity training *must* be a trained professional who can help participants deal with criticism in a con-

sensitivity training A deep interpersonal OD intervention that focuses on developing greater sensitivity to oneself, to others, and to one's relations with others through an intense, leaderless group experience.

ANOTHER VIEW
The effects of sensitivity training may not be long lasting. People who participate in such training undergo changes away from their normal surroundings. When they return to their jobs, they may find that their coworkers don't understand or trust their new ideas and behaviors and may feel pressured by their colleagues to revert to old ways. What actions might a company take to prevent this loss of learning and continued resistance to change?

[21] John P. Campbell and Marvin D. Dunnette, "Effectiveness of T-Group Experiences in Managerial Training and Development," *Psychological Bulletin* 65 (1968), 73–104.

[22] Elliot Aronson, "Communication in Sensitivity Training Groups," in *Organization Development: Theory, Practice, and Research,* ed. Wendell L. French, Cecil H. Bell, Jr., and Robert A. Zawacki (Plano, Texas: Business Publications, 1983), pp. 249–53.

[23] G. David, "Building Cooperation and Trust," in *Management by Participation,* ed. A. J. Marrow, D. G. Bowers, and S. E. Seashore (New York: Harper & Row, 1967), pp. 99–100.

structure manner. In the absence of expert help, participants could risk serious psychological trauma.[24]

Group Interventions

Group interventions are designed to solve many of the problems with group performance and leadership identified in Chapters 11 and 12. In general, such interventions focus on helping the members of a group learn how to work together to fulfill the group's task and maintenance requirements.

process consultation A shallow group-level OD intervention in which a change agent meets with a work group and helps its members examine group processes such as communication, leadership and followership, problem solving, and cooperation.

team diagnostic session A group OD intervention of moderate depth in which a change agent and a work group critique the group's performance and look for ways to improve it.

At The New England, an insurance company founded in 1835, management attempts to monitor changes in the economic environment and to help the company's associates respond to these changes. Here a group meets in a "process improvement" session, focusing on the ways insurance agents, agencies, and the home office can work together to generate new business. (Left to right) James Medeiros, Judith Precourt, Christopher Frachette, Linda Collins, Bonnie Mallin, Patrick Hanlon (standing), Karen Rosser, and Scott Andrews.
Source: "A Celebration of the American Worker," Fortune, September 24, 1990.

Process Consultation. **Process consultation** is a relatively shallow, group-level OD intervention. In a process consultation intervention, a change agent meets with a work group and helps its members examine group processes, such as communication, leadership and followership, problem solving, and cooperation. The specific approach taken during this exploration, which varies from one situation to another, may include:

1. Stimulus questions asked by the change agent that direct attention to relationships among group members. Ensuing discussions between group members may focus on ways to improve these relationships and on how such relationships can influence group productivity and effectiveness.
2. A process analysis session, during which the change agent watches the group as it works, followed by feedback sessions in which the change agent discusses his observations about how the group maintains itself and how it performs its task. There may also be supplementary feedback sessions to allow the change agent to clarify the events of earlier sessions for individual group members.
3. Suggestions made by the change agent, which may pertain to group membership, communication, and interaction patterns and the allocation of work duties, responsibilities, and authority.[25]

Whatever the change agent's approach in a given situation, his primary focus in process consultation is on making a group more effective by getting its members to pay more attention to important *process* issues. He wants them to focus on *how* things are done in the group, rather than on the issues of *what* is to be done, which normally dominate a group's attention. The ultimate goal of process consultation is to help the group improve its ability to solve its own problems by increasing the ability of members to identify and correct faulty group processes.[26]

Team Diagnostic Sessions. A **team diagnostic session** is a group intervention of moderate depth that normally takes place outside the work setting and may last from four to eight hours. In a team diagnostic session, a change agent and a work group critique the group's performance. During the session, each member of the group has the opportunity to exchange personal perceptions about group problems with every other member. Group members prepare separately for the

[24] Carl A. Bramlette and Jeffrey H. Tucker, "Encounter Groups: Positive Change or Deterioration," *Human Relations* 34 (1981), 303–14.
[25] Schein, *Process Consultation*, pp. 102–3; and Christian F. Paul and Albert C. Gross, "Increasing Productivity and Morale in a Municipality: Effects of Organization Development," *Journal of Applied Behavioral Science* 17 (1981), 59–78.
[26] Schein, *Process Consultation*, p. 135.

session by asking themselves such questions as, Where are we going? How are we doing? What opportunities should we take advantage of? and What problems do we have that we should work on? When the group assembles, the change agent uses one of the following techniques to make personal perceptions public:[27]

1. A whole-group discussion, in which every member makes personal contributions.
2. Subgroup discussions, in which the larger group is broken down into smaller groups for intensive explorations. Members then report the results of subgroup discussion back to the total group.
3. Pair discussions, in which two people discuss their ideas with each other and report the results back to the total group.

After group members have described their personal perceptions to the larger group, the session progresses through the third, fourth, fifth, and sixth steps shown in Figure 14-5. First, the group discusses the issues uncovered in the first two steps and tries to categorize them; for example, goal difficulty problems, role ambiguity problems, or leadership problems. Next, the group enters into an action planning phase in which each problem category is assigned to a subgroup, which develops a solution strategy. Subgroups then report their strategies back to the larger group, and the group decides whether to accept or reject recommended strategies.

[27] French and Bell, *Organization Development*, p. 129.

FIGURE 14-5

Steps in a Team Diagnostic Session

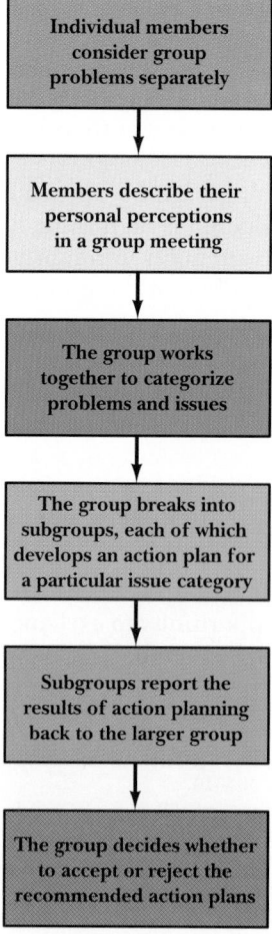

Depending upon the strategies they accept, group members may move on to other interventions. For instance, they may enter into an RAT intervention as part of a strategy developed to counteract role ambiguity. Or they may undertake a team-development session (discussed next) to help overcome problems with leadership and group functioning.

team development A deep group-level extension of interpersonal sensitivity training in which a group of people who work together on a daily basis meet over an extended period to assess and modify group processes.

Team Development. **Team development** is a deep, group-level extension of interpersonal sensitivity training. In a team-development intervention, a group of people who work together on a daily basis meet over an extended period of time to assess and modify group processes.[28] Throughout these meetings, participants focus their effort on achieving a balance of such basic components of teamwork as:

1. An understanding of, and commitment to, common goals
2. Involvement of as many group members as possible, in order to take advantage of the complete range of skills and abilities available to the group
3. Analysis and review of group processes on a regular basis to ensure that there are sufficient maintenance activities
4. Trust and openness in communication and relationships
5. A strong sense of belonging on the part of all members[29]

To begin team development, the group first engages in a lengthy diagnostic meeting in which a change agent helps members identify group problems and map out possible solutions. The change agent asks them to observe interpersonal and group processes and to be prepared to comment on what they see. Thus the group works on two basic issues. They look for solutions to problems of everyday functioning that have come up in the group, and they observe the way group members interact with each other during the meeting.

Based on the results of these efforts, team development then proceeds in two specific directions. First, the change agent and group implement the interventions chosen during diagnosis, to solve the problems the group is able to identify. Second, the change agent initiates group sensitivity training to uncover additional problems that might otherwise resist detection.

> As the group fails to get [the change agent] to occupy the traditional roles of teacher, seminar leader, or therapist, it will redouble its efforts until in desperation it will disown him and seek other leaders. When they too fail, they too will be disowned, often brutally. The group will then use its own brutality to try to get the [change agent] to change his task by eliciting his sympathy and care for those it has handled so roughly. If this manoeuver fails, and it never completely fails, the group will tend to throw up other leaders to express its concern for its members and project its brutality onto the consultant. As rival leaders emerge it is the job of the consultant, so far as he is able, to identify what the group is trying to do and explain it. His leadership is in task performance, and the task is to understand what the group is doing "now" and to explain why it is doing it.[30]

As you can see, group sensitivity training is really an interpersonal sensitivity training intervention conducted with an intact work group. It enables coworkers to critique and adjust interpersonal relations problems that are inevitable during the workday.

---

[28] Robert T. Golembiewski, *Approaches to Planned Change, Part 1: Orienting Perspectives and Micro-Level Interventions* (New York: Marcel Dekker, 1979), p. 301.

[29] Gordon L. Lippitt, *Organization Renewal* (New York: Appleton-Century-Crofts, 1969), pp. 107–13.

[30] A. K. Rice, *Learning for Leadership* (London: Tavistock Publications, 1965), pp. 65–66.

Intergroup Interventions

Intergroup interventions focus on solving the types of intergroup problems identified in Chapter 13. In general, these problems concern politicking, conflict, and associated breakdowns in intergroup coordination. Thus OD interventions developed to manage intergroup relations involve various open communication techniques and conflict resolution methods.

third party peacemaking A shallow OD intervention in which a change agent seeks to resolve intergroup misunderstandings by encouraging communication between or among groups.

Third-Party Peacemaking. **Third-party peacemaking** is a relatively shallow intervention in which a change agent seeks to resolve intergroup misunderstandings by encouraging communication between or among groups. The change agent, who is not herself a member of any of the groups and is referred to as a third party, guides a meeting between the groups. To be productive, the meeting must be characterized by the following attributes:

1. *Motivation*: All groups must be motivated to try to resolve their differences.
2. *Power*: A stable balance of power must be established between the groups.
3. *Timing*: Confrontations must be synchronized so that no one group can gain an information advantage over another.
4. *Emotional release*: People must be given the time to work through the negative thoughts and feelings that have built up between the groups. They need to recognize and express their positive feelings as well.
5. *Openness*: Conditions must favor openness in communication and mutual understanding.
6. *Stress*: There should be enough stress, enough pressure on group members to motivate them to give serious attention to the problem but not so much that the problem appears insoluble.[31]

The change agent facilitates communication between the groups both directly and indirectly. In a direct fashion, she may interview group members before an intergroup meeting, help to put together a meeting agenda, monitor the pace of communication between groups during the meeting, or actually referee the interaction. Acting in a more subtle, indirect way, she may schedule the meeting at a neutral site or establish time limits for intergroup interaction. The whole process can be as short as an afternoon but more likely as long as several months of weekly sessions. As the result of actions like these, the group members begin to learn things about each other and their relationship that can help them focus on common interests and begin to overcome conflictive tendencies.

intergroup mirroring An OD intervention of moderate depth in which representatives from several groups tell the members of a particular group with whom they interact how the people they represent perceive the host group.

host group A group that is experiencing difficulties in working with other groups and asks them to send representatives to an intergroup mirroring intervention.

Intergroup Mirroring. **Intergroup mirroring** is an intervention of moderate depth in which representatives from several groups tell the members of another group with whom they all interact how the people they represent perceive this other group.[32] This other group is the **host group**—a group that is having difficulty in working with the rest of the groups. To begin a mirroring intervention, the host group asks key members of the other groups to come to a meeting. Next, as Figure 14-6 indicates, a change agent interviews the people who will attend the meeting and gets a sense of the problems between the groups.

The mirroring meeting begins with a feedback session in which the change agent reports on the results of premeeting interviews. Following this session, the representatives from all groups except the host group break into small groups

---

[31] Richard E. Walton, *Interpersonal Peacemaking: Confrontation and Third Party Consultation* (Reading, Mass.: Addison-Wesley, 1969), pp. 94–115.

[32] French and Bell, *Organization Development*, pp. 147–48.

530 Managing Group and Intergroup Relations: Organization Development I

FIGURE 14-6

Steps in an Intergroup Mirroring Intervention

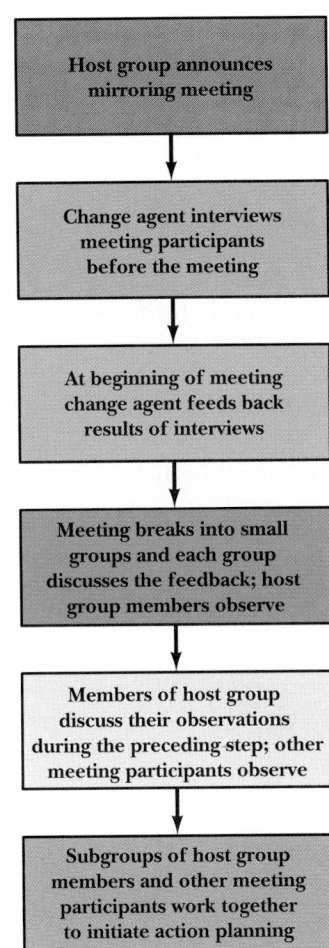

and discuss in greater detail the data provided by the change agent. During these discussions, members of the host group sit around the perimeter of each discussion group, forming a "fishbowl." They observe what goes on in order to learn about the other groups' feelings and perceptions. Next, the host group reconvenes and its members discuss what they have observed. This time, the representatives of the other groups form a fishbowl around the discussion and provide clarification as needed. They want to make sure the host group clearly understands how it is perceived.

The informational stage of the mirroring intervention ends at this point. Small groups composed of host- and other-group members begin to work on the key problems uncovered during the two fishbowl discussions, and on developing strategies to solve them. When they have finished, the larger group reconvenes to make a master list. This list then serves as the basis of an action plan that is devised to help the host group improve relations with the other groups. Once the action plan has been established, people are assigned to specific tasks, and target dates are set for task completion. At a follow-up meeting, held later, the large group assesses progress toward completion of the action plan and allows for the correction of faulty aspects of the plan.

intergroup team building A deep OD intervention intended to improve communication and interaction between work-related groups.

Intergroup Team Building. Intergroup team building is a deep intervention that has three primary aims. They are (1) to improve communication and interaction between work-related groups, (2) to decrease counterproductive competition between the groups, and (3) to replace group-centered perspectives with an orientation that recognizes the necessity for various groups to work together.[33] As indicated in Figure 14-7, during the first step of intergroup team building, two groups (or their leaders) meet with an OD change agent and discuss whether relations between the groups can be improved. If both groups agree that improvement is possible, the change agent asks the two groups to commit themselves to searching for ways of improving their relationship. Once they do so, they move to the second step. The two groups meet in separate rooms, and each makes two lists. The group's perceptions, thoughts, and attitudes toward the other group are on one list. Their thoughts about what the other group is likely to say about them are on the other. In the third step, the two groups reconvene and compare their lists. Each group can compare its view of the other group with the way the other group expects to be seen. Discrepancies uncovered during this comparison are discussed during the fourth step, when the groups meet separately. Each one reacts to what it has learned about itself and the other group. Then each group makes a list of important issues that need to be resolved between the two groups.

[33] Robert R. Blake, H. A. Shepard, and Jane S. Mouton, *Managing Intergroup Conflict in Industry* (Houston: Gulf, 1965), pp. 36–100; and Richard Beckhard, *Organization Development: Strategies and Models* (Reading, Mass.: Addison-Wesley, 1969), pp. 33–35.

FIGURE 14-7

Steps in an Intergroup Team-Building Intervention

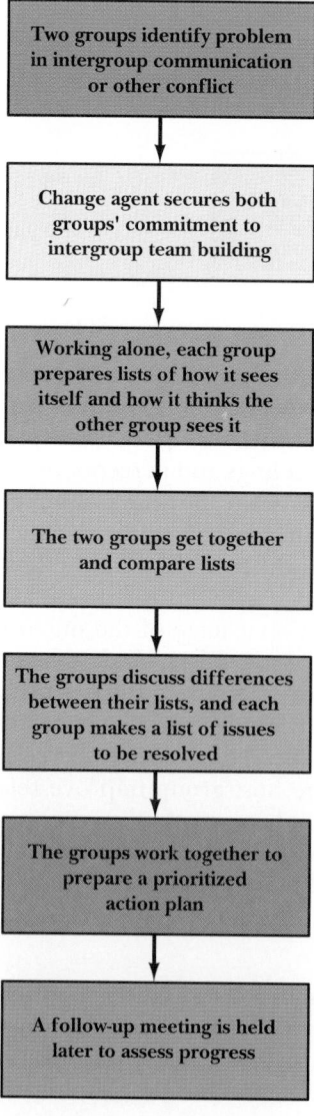

Two groups identify problem in intergroup communication or other conflict

Change agent secures both groups' commitment to intergroup team building

Working alone, each group prepares lists of how it sees itself and how it thinks the other group sees it

The two groups get together and compare lists

The groups discuss differences between their lists, and each group makes a list of issues to be resolved

The groups work together to prepare a prioritized action plan

A follow-up meeting is held later to assess progress

MANAGING GROUP AND INTERGROUP RELATIONS: ORGANIZATION DEVELOPMENT I

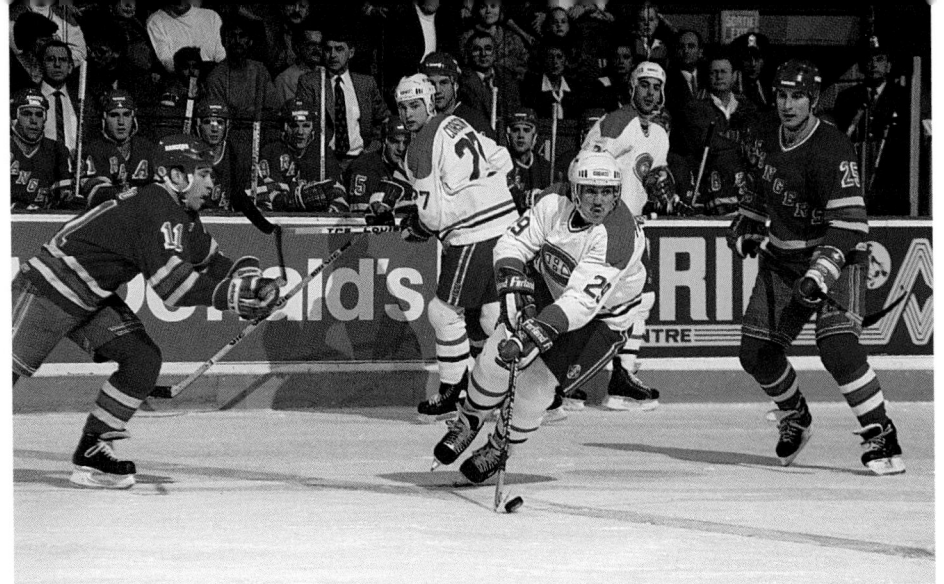

You may not think of the NHL's hockey teams as needing to engage in intergroup team building—as needing to improve their communication and develop a joint perspective. That is just what they must do, however, if the League is to expand its present 21 teams to 28 and make hockey as popular in the U.S. South as it is in Canada. To get the TV exposure necessary to such nationwide popularity the entire NHL will need to cooperate to get more large cities to offer homes to teams. Source: Bill Saporito, "The Hockey Puck as Globalizer," Fortune, January 14, 1991.

The two groups come back together during the fifth step and compare the lists of issues, setting priorities. Then they work together on a plan of action to resolve the issues in order of their priority. They assign individual responsibilities and target dates for completion. The final step is a follow-up meeting held later on to assess progress. At that time, additional actions are planned as required to make sure that intergroup cooperation will continue over the long run.

EVALUATING DEVELOPMENT AND CHANGE

No matter which organization-development intervention you decide to use, the concluding stage of the OD process always consists of an evaluation of effectiveness. Based on the results of this evaluation, efforts may be devoted to ensuring the permanence of newly developed attitudes, values, and behaviors. Alternatively, OD may begin anew, and additional interventions may be initiated to stimulate further change. Table 14-4 contains a checklist of questions that can be useful in deciding what criteria to use and how to measure them when evaluating the effectiveness of organization development.

As suggested by the checklist, resources are expended to acquire the outcomes generated by the OD process. So OD's effectiveness must be judged partly

APPLYING THE DIAGNOSTIC MODEL
Diagnostic Question 7: Is evaluation an integral part of the OD effort? As evaluation criteria and procedures are established, are serious attempts made to guard against bias and distortion?

Diagnostic Question 8: How likely is it that positive change will persist after the OD effort has ended? What can be done to ensure lasting change?

TABLE 14-4
Criteria for Evaluating Change Efforts

CRITERION	SUGGESTED QUESTIONS
Overall Results	
Desired outcomes	1. What were the intended outcomes of the intervention? How do they compare with the outcomes actually realized?
Guiding assumptions	2. How explicit were the assumptions that guided the intervention? Did experience prove them to be both valid and appropriate? Did everyone understand and agree with the intervention's purpose as a result?
	3. How consistent with current theories of OB and OD are these assumptions? Was everything currently known with regard to the intervention's focus and purpose incorporated in the intervention?

TABLE 14-4 Cont'd.

CRITERION	SUGGESTED QUESTIONS
Phase of intervention	
Initiation	4. What was the reason for starting the intervention? Who was initially involved? Was the intervention initiated because of a broadly felt need or a narrow set of special interests?
Entry	5. What activities were there at the start of the intervention process? Who was involved in them? Was the intervention implemented prematurely, without adequate diagnosis? Did unnecessary resistance arise as a result?
Diagnosis	6. What specific diagnostic activities took place? Were they carried out fully and effectively? What aspects of the organization were diagnosed to determine the target and depth of the intervention that was implemented?
Planning	7. How was the intervention planned, and who planned it? How were resources used in this effort? How explicit and detailed were the plans that resulted?
Action	8. What was actually done? When was it done? Who did it? How do the answers to these questions compare with the action plan as initially developed?
Evaluation	9. Was evaluation included from the outset as part of the intervention? Were deficiencies identified during evaluation corrected through a careful, planned modification to the intervention or its action plan?
External factors	
Work-force traits	10. Were the results of the intervention affected, either positively or negatively, by work-force characteristics, e.g., age, gender, education, unemployment level?
Economy	11. What was the state of the economy and the firm's market at the time of the intervention? Did economic factors affect the success of the intervention?
Environment	12. How much did the organization's environment change over the course of the intervention? Are the intended results of the intervention still desirable given the organization's current environment?
Internal factors	
Size	13. How large is the organization? Did its size permit access to the resources required for the intervention to succeed?
Technology	14. What is the organization's primary product, and what sort of technology is used to make it? Do the results of the intervention fit in or conflict with the requirements of this technology?
Structure	15. How mechanistic or organic is the organization's structure? Do the results of the intervention fit in or conflict with this structure?
Culture	16. What are the organization's prevailing norms and values concerning change? Concerning involvement in OD interventions?

Source: Based on Noel Tichy and Jay N. Nisberg, "When Does Work Restructuring Work? Organizational Innovations at Volvo and GM," *Organizational Dynamics*, (1976), 13–36; and Wendell L. French, "A Checklist for Organizing and Implementing an OD Effort," in *Organization Development: Theory, Practice, and Research*, rev. ed., ed. Wendell L. French, Cecil H. Bell, Jr., and Robert A. Zawacki (Plano, Texas: Business Publications, 1983) pp. 451–59.

in terms of its outcomes. In addition, measuring its effectiveness requires remembering why it was undertaken to begin with and assessing what took place during each stage of the OD process. This procedure guarantees that an OD effort labeled "effective" not only accomplished its intended purpose but did so in a manner that left everyone more informed about the process of change and how to manage it. Finally, the effects of external and internal factors—whether positive or negative—on the OD process must be examined and cataloged for subsequent reference. That way, the factors that support change can be called into play when needed again in the future, and the ones that are resistant can be anticipated and neutralized.

SUMMARY

Organization development is both a field of research and a collection of *interventions* intended to stimulate planned change in organizations. Associated with OD is a concern about managing resistance to change and strengthening forces that favor change. *Forcefield analysis* is a technique that can be used to aid in the pursuit of these complementary goals.

OD is based on a set of underlying values and assumptions that stress the importance of encouraging human growth and development in organizations. *External* or *internal change agents* are guided by these values and assumptions as they guide the process of implementing OD interventions. When already established interventions are used, the properly managed OD process unfolds as described by the *planned change model*. When new interventions are being invented and assessed for the first time, they follow the *action research model*.

OD interventions differ from each other in the *depth* of change they are intended to stimulate and the types of organizational behavior that are their *target*. No matter which intervention is used, all OD efforts should conclude with an evaluation of program effectiveness.

REVIEW QUESTIONS

1. What differences are there between organization development and other approaches to change in organizations? Are these differences important? Why or why not?

2. What is an external change agent? What is an internal change agent? What are the major advantages and disadvantages of each type of change agent? Why don't more companies use teams composed of both types?

3. Suppose you were given the assignment of developing a new grading system for your OB class. Draw a forcefield analysis showing the major forces for and against change that you would probably encounter while implementing your new grading system. What would you do to weaken the forces against change? How would you strengthen the forces for change? Is it likely that your change intervention would succeed?

4. In what ways is the Lewin development model similar to the planned change model? How do the two models differ from one another? What does the action research model tell you about OD that is not discussed in the other two?

5. Why is it important to avoid using an intervention that is deeper than needed to stimulate the required amount of change? How can you increase the likelihood that the intervention is focused on the appropriate target?

6. Which off-the-shelf intervention would you choose for each of the following situations: two groups that strongly disagree about both groups' task responsibilities; a person who understands her role in a group but can't seem to get along with her coworkers; a group of people who get along with each other but fail to be as productive as expected.

7. Why is it always important to evaluate the results of an OD intervention? What kinds of information should you collect and consider during an evaluation?

DIAGNOSTIC QUESTIONS

Organization development is a source of action steps that you can use to solve many of the interpersonal, group, and intergroup problems identified in this part of the book. The following diagnostic questions offer practical guidance as you go through this process.

1.   What sort of change is being contemplated? What sources of resistance to this change exist in the organization? What might you do to overcome this resistance? How successful are you likely to be?

2.   What are the forces in the organization that favor the change? How can you strengthen them? How strong can they be made?

3.   Based on a forcefield analysis, what is the likely combined effect of the forces for and against change? Is it realistic to attempt to institute change, or should the status quo be accepted instead?

4.   If change is a realistic goal, do cost, availability, expertise, and similar considerations favor using an external change agent? An internal change agent? A team of external and internal change agents?

5.   Once chosen, does the change agent behave in accordance with the basic values and assumptions of OD? If the aim is to use an existing OD intervention to solve an organizational problem, is the change agent adhering to the planned change model of OD? If the aim is to develop and assess a new type of intervention, do change agent behaviors conform to the action research model?

6.   Do the depth and target of the OD intervention being implemented seem to match the problem? Note that the shallowest intervention likely to stimulate the required amount of change is the one that should be selected for implementation.

7.   Is evaluation an integral part of the OD effort? As evaluation criteria and procedures are established, are serious attempts made to guard against bias and distortion?

8.   How likely is it that positive change will persist after the OD effort has ended? What can be done to ensure lasting change?

EXERCISE 14-1
INTERGROUP MIRRORING*

MARK S. PLOVNICK, *University of the Pacific*
RONALD E. FRY, *Case Western Reserve University*
W. WARNER BURKE, *Teachers College, Columbia University*

Intergroup mirroring is an organization development intervention used to help bring to the surface the root causes of conflict between two groups. This technique is also designed to create conditions under which a win-win attitude can prevail and mutual problem solving can occur. The mirroring process involves three major phases:

1. *Imagery.* Each group develops images of itself and of the other group. This first step elicits stereotypes and untested assumptions about "them" and "us." Often just correcting misperceptions of each other can bring groups closer together.

2. *Confrontation.* Each group acknowledges its uniqueness and its differences from the other group. The aim here is to specify differences that are accepted as valid but that are causing conflict. Without this clarification and labeling of just what is in conflict, meaningful resolution is unlikely. Smoothing over or avoiding the conflict is more apt to occur.

3. *Bonding.* Groups experiencing conflict that is apparently unreconcilable begin to collaborate and address problems mutually. They become more alike in their perceptions of the necessity of working together and of understanding one another.

In order to succeed, intergroup mirroring must occur between groups that are both motivated to improve the situation and relatively equal in the power they can bring to bear. If one group favors the current situation, it is rational for that group to avoid the changes otherwise stimulated by intergroup mirroring. And if one group can force the other to do things it wouldn't otherwise consider doing, true collaboration is unlikely.

Anyone who wants to use intergroup mirroring must understand some basic assumptions underlying its design. First, the technique assumes that both groups in conflict are acting with integrity and good intentions. If you believe that groups actually wish to harm, punish, or humiliate one another there can be no resolution of differences. Second, the intervention assumes that both groups are responsible for the conflict between them. This responsibility may be unequal, but it must be shared.

STEP ONE: PRE-CLASS PREPARATION

The upcoming class session will give you the opportunity to experience a mirroring intervention for yourself. Prepare for it by reading the entire exercise before coming to class.

STEP TWO: GENERATING IMAGES— HOMOGENOUS GROUPS

The class should divide into the same pairs of groups that completed the Disarmament Exercise in Exercise 13-1. If the class has not yet completed Exercise 13-1 it should do so before proceeding further.

To begin Step Two, each group should meet by itself and list as many responses as possible to the three statements that follow. In filling in the blanks with as many words or phrases as you can, think about your own group and the group that is paired with yours.

1. How we see the other group: "We see the other group as _____."

2. How we see ourselves: "We see ourselves as _____."

3. How we think the other group sees us: "The other group sees us as _____."

Brainstorming or going around the group at first will help elicit ideas. As ideas are presented they should be listed on a chalkboard or flipchart so that everyone in the group can see them. Try to reach a consensus on whatever you decide to include in your group's lists. If opinion is clearly split on an item, note this next to the item. A spokesperson should be appointed to prepare a presentation for Step Three.

STEP THREE: SHARING IMAGES— HETEROGENOUS GROUPS

Each pair of groups should form one discussion group. Spokespersons for each group should begin this step

* Adapted with the authors' permission from Mark S. Plovnick, Ronald E. Fry, and W. Warner Burke, *Organization Development: Exercises, Cases, and Readings* (Boston: Little, Brown, 1982), pp. 89–93. Copyright © 1982 by Mark S. Plovnick, Ronald E. Fry, and W. Warner Burke.

by reporting the results of Step Two. During these reports, listeners may ask for information providing further clarification or understanding, but they may not debate any aspect of either report.

STEP FOUR: IDENTIFYING DISCREPANCIES—HOMOGENOUS GROUPS

All groups should reform, and each group should examine what it said about itself and what was said about it. Make a list that everyone can see of as many discrepancies as you can identify between how the group views itself, how the other group views it, and how it thought the other group would view it.

STEP FIVE: SHARING AND PRIORITIZING DISCREPANCIES—HETEROGENOUS GROUPS

In this step each pair of groups should get together again and group spokespersons should report on the results of Step Four. Similarities in the discrepancies identified by the two groups should be noted and combined. Then each pair of groups should decide on two to four discrepancies that the pair would like to work on.

STEP SIX: PROBLEM-SOLVING—HETEROGENOUS SUBGROUPS

Mixed subgroups made up of members from each of the groups in the pairing should discuss how to manage, neutralize, or eliminate the two to four discrepancies identified at the end of Step Five. Each subgroup should then prepare some recommendations to be considered by the class as a whole in Step Seven. A spokesperson

should be appointed to report on the discrepancies examined and actions recommended by the subgroup.

STEP SEVEN: CLASS DISCUSSION

The entire class should assemble, and all spokespersons should give their reports. As actions are recommended, class members should discuss each action's practicality and likelihood of success. A master list of discrepancies and ways of overcoming their effects should then be created. The last part of the class should be devoted to discussing the usefulness of intergroup mirroring and should focus on the following questions:

1. Knowing what you do now, what would you do differently if you were required to complete the Disarmament Exercise again? Would you change any of your answers to the questions at the end of Exercise 13-1?

2. When would you recommend an intergroup mirroring intervention? When would you *not* recommend one?

3. How might intergroup mirroring help interdependent task groups (such as manufacturing employees and sales personnel) set realistic performance goals? What other benefits would mirroring provide such groups?

CONCLUSION

Intergroup mirroring can be especially helpful in eliciting and exploring general attitudes and feelings groups hold toward one another. Mirroring can also be useful for getting at specific procedural problems that have arisen between groups who need each other to get their work done. As you may have discovered during this exercise, it is usually easier to generate perceptions and even to identify discrepancies than to agree on how to change them. This is why an intervention like intergroup mirroring is an invaluable aid to managers who must cope with intergroup conflict and build a sense of intergroup team spirit.

DIAGNOSING ORGANIZATIONAL BEHAVIOR

It all started so positively. Three days after graduating with his degree in business administration, Mike Wilson started his first day at a prestigious insurance company—Consolidated Life. He worked in the Policy Issue Department. The work of the department was mostly

CASE 14-1
THE CONSOLIDATED LIFE CASE: CAUGHT BETWEEN CORPORATE CULTURES*

JOSEPH WEISS, *Bentley College*
MARK WAHLSTROM, *Bentley College*
EDWARD MARSHALL, *Bentley College*

* Reprinted by permission of the publisher from *The Journal of Management Case Studies*, 2, 238–43. Copyright © 1986 by Elsevier Science Publishing Co., Inc.

clerical and did not require a high degree of technical knowledge. Given the repetitive and mundane nature of the work, the successful worker had to be consistent and willing to grind out paperwork.

Rick Belkner was the division's vice-president, "the man in charge" at the time, an actuary by training and a technical professional described in the division as "the mirror of whomever was the strongest personality around him." It was also common knowledge that Belkner made $60,000 a year while he spent his time doing crossword puzzles.

Mike was hired as a management trainee and promised a supervisory assignment within a year. However, because of a management reorganization, it was only six weeks before he was placed in charge of an eight-person unit. The reorganization was intended to streamline workflow, upgrade and combine the clerical jobs, and make greater use of the computer system. It was a drastic departure from the old way of doing things and created a great deal of animosity and anxiety among the clerical staff.

Management realized that a flexible supervisory style was necessary to pull off the reorganization without immense turnover, so the firm gave its supervisors a free hand to run their units as they saw fit. Mike used this latitude to implement group meetings and training classes in his unit. In addition, he assured all members raises if they worked hard to attain them. By working long hours, participating in the mundane tasks with his unit, and being flexible in his management style, he was able to increase productivity, reduce errors, and reduce lost time. Things improved so dramatically that he was noticed by upper management and earned a reputation as a "superstar" despite being viewed as free-spirited and unorthodox. The feeling was that his loose, people-oriented management style could be tolerated because his results were excellent.

After a year, Mike received an offer from a different Consolidated Life division located across town. Mike was asked to manage an office in the marketing area. The pay was excellent and it offered an opportunity to turn around an office in disarray. The reorganization in his present division at Consolidated was almost complete, and most of his mentors and friends in management had moved on to other jobs. Mike decided to accept the offer. In his exit interview, he was assured that if he ever wanted to return, a position would be made for him. It was clear that he was held in high regard by management and staff alike. A huge party was thrown to send him off.

The new job was satisfying for a short time, but it became apparent to Mike that it did not have the long-term potential he was promised. After bringing on a new staff, computerizing the office, and auditing the books, he began looking for a position that would both

challenge him and give him the autonomy he needed to be successful.

Eventually, word got back to Rick Belkner that Mike was looking for another job. Rick offered Mike a position with the same pay he was now receiving and control over a 14-person unit in his old division. After considering other options, Mike decided to return to his old division, feeling that he would be able to progress steadily over the next several years.

Upon his return to Consolidated Like, Mike became aware of several changes that had taken place in the six months since his departure. The most important change was the hiring of a new divisional senior vice-president, Jack Greely. Greely had been given total authority to run the division. Rick Belkner now reported to Jack.

Belkner's reputation was now that he was tough but fair. It was necessary for people in Jack's division to do things his way and "get the work out." Mike also found himself reporting to one of his former peers, Kathy Miller, who had been promoted to manager during the reorganization. Mike had always "hit it off" with Miller and foresaw no problems in working with her.

After a week, Mike realized the extent of the changes that had occurred. Gone was the loose, casual atmosphere that had marked his first tour in the division. Now, a stricter, task-oriented management doctrine was practiced. Morale of the supervisory staff had decreased to an alarming level. Jack Greely was the major topic of conversation in and around the division. People joked that MBO now meant management by "oppression," not by "objectives."

Mike was greeted back with comments like "Welcome to prison" and "Why would you come back here? You must be desperate!" It seemed as if everyone was looking for new jobs or transfers. Their lack of desire was reflected in the poor quality of work being done.

Mike felt that a change in the management style of his boss was necessary in order to improve a frustrating situation. Realizing that it would be difficult to affect Greely's style directly, Mike requested permission from Belkner to form a Supervisors' Forum for all the managers on Mike's level in the division. Mike explained that the purpose would be to enhance the existing management-training program. The Forum would include weekly meetings, guest speakers, and discussions of topics relevant to the division and the industry. Mike thought the Forum would show Greely that he was serious about both his job and improving morale in the division. Belkner gave the O.K. for an initial meeting.

The meeting took place, and ten supervisors who were Mike's peers in the company eagerly took the opportunity to "Blue Sky" it. There was a euphoric attitude about the group as they drafted their statement of intent. It read as follows:

TO: Rick Belkner
FROM: New Issue Services Supervisors
SUBJECT: Supervisors' Forum

On Thursday, June 11, the Supervisors' Forum held its first meeting. The objective of the meeting was to identify common areas of concern among us and to determine topics that we might be interested in pursuing.

The first area addressed was the void that we perceived exists in the management-training program. As a result of conditions beyond anyone's control, many of us over the past year have held supervisory duties without the benefit of formal training or proper experience. Therefore, what we propose is that we utilize the Supervisors' Forum as a vehicle with which to enhance the existing management-training program. The areas that we hope to affect with this supplemental training are: (a) morale/job satisfaction, (b) quality of work and service, (c) productivity, and (d) management expertise as it relates to the life insurance industry. With these objectives in mind, we have outlined below a list of possible activities that we would like to pursue.

1. Further utilization of the existing "in-house" training programs provided for manager trainees and supervisors, i.e., Introduction to Supervision, E.E.O., and Coaching and Counseling.
2. A series of speakers from various sections in the company. This would help expose us to the technical aspects of their departments and their managerial style.
3. Invitations to outside speakers to address the Forum on management topics such as management development, organizational structure and behavior, business policy, and the insurance industry. Suggested speakers could be area college professors, consultants, and state insurance officials.
4. Outside training and visits to the field. This could include attendance at seminars concerning management theory and development relative to the insurance industry. Attached is a representative sample of a program we would like to have considered in the future.

In conclusion, we hope that this memo clearly illustrates what we are attempting to accomplish with this program. It is our hope that the above outline will be able to give the Forum credibility and establish it as an effective tool for all levels of management within New Issue. By supplementing our on-the-job training with a series of speakers and classes, we aim to develop prospective management's role in it. Also, we would like to extend an invitation to the underwriters to attend any programs at which the topic of the speaker might be of interest to them.

cc: J. Greely
    Managers

The group felt the memo accurately and diplomatically stated their dissatisfaction with the current situation. However, they pondered what the results of their actions would be and what else they could have done.

Shortly after the memo had been issued, an emergency management meeting was called by Rick Belkner at Jack Greely's request to address the "union" being formed by the supervisors. Four general managers, Rick Belkner, and Jack Greely were at that meeting. During the meeting, it was suggested the Forum be disbanded to "put them in their place." However, Rick Belkner felt that if "guided" in the proper direction the Forum could die from the lack of interest. His stance was adopted, but it was common knowledge that Jack Greely was strongly opposed to the group and wanted its founders dealt with. His comment was "It's not a democracy and they're not a union. If they don't like it here, then they can leave." A campaign was directed by the managers to determine who the main authors of the memo were so they could be dealt with.

About this time, Mike's unit had made a mistake on a case, which Jack Greely was embarrassed to admit to his boss. This embarrassment was more than Jack Greely cared to take from Mike Wilson. At the managers' staff meeting that day, Greely stormed in and declared that the next supervisor to "screw up" was out the door. He would permit no more embarrassments of his division and repeated his earlier statement about "people leaving if they didn't like it here." It was clear to Mike and everyone else present that Mike Wilson was a marked man.

Mike had always been a loose, amiable supervisor. The major reason his units had been successful was the attention he paid to each individual and how they interacted with the group. He had a reputation for fairness, was seen as an excellent judge of personnel for new positions, and was noted for his ability to turn around people who had been in trouble. He motivated people through a dynamic, personable style and was noted for his general lack of regard for rules. He treated rules as obstacles to management and usually used his own discretion as to what was important. His office had a sign saying "Any fool can manage by rules. It takes an uncommon man to manage without any." It was an approach that flew in the face of company policy, but it had been overlooked in the past because of his results. However, because of Mike's actions with the Supervisors' Forum, he was now regarded as a thorn in the side, not a superstar, and his oddball style only made things worse.

Faced with the fact that he was rumored to be out the door, Mike sat down to appraise the situation.

When you have read this case, look back at the chapter's diagnostic questions and choose the ones that apply to the case. Then use those questions with the ones that follow in your case analysis.

1. Why was Mike Wilson so successful in his first managerial position at Consolidated Life? Evaluate his approach in managing the animosity and anxiety that the company's reorganization had created among his subordinates.

2. Was the Supervisors' Forum a success? Why or why not? What was it intended to accomplish? What forces favored its success? What forces stood in its way? Is there anything that Mike and the other managers could have done to increase top management's acceptance of the forum?

3. Describe Jack Greely's attitude toward change. Given Greely's stance and Mike's situation after the forum, what would you advise Mike to do?

C A S E 14-2
L. J. SUMMERS COMPANY*

J. B. RITCHIE, *Brigham Young University*
PAUL H. THOMPSON, *Brigham Young University*

Jon Reese couldn't think of a time in the history of L. J. Summers Company when there had been as much anti-company sentiment among the workers as had emerged in the past few weeks. He knew that Mr. Summers would place the blame on him for the problems with the production workers because Jon was supposed to be helping Mr. Summer's son, Blaine, to become oriented to his new position. Blaine had only recently taken over as production manager of the company (see Exhibit 14-1). Blaine was unpopular with most of the workers, but the events of the past weeks had caused him to be resented even more. This resentment had

* Reprinted with the publisher's permission from pages 344–48 of *Organizations and People: Readings, Cases, and Exercises in Organizational Behavior*, 4th ed. by J. B. Ritchie and Paul H. Thompson, copyright © 1988 by West Publishing Company. All rights reserved.

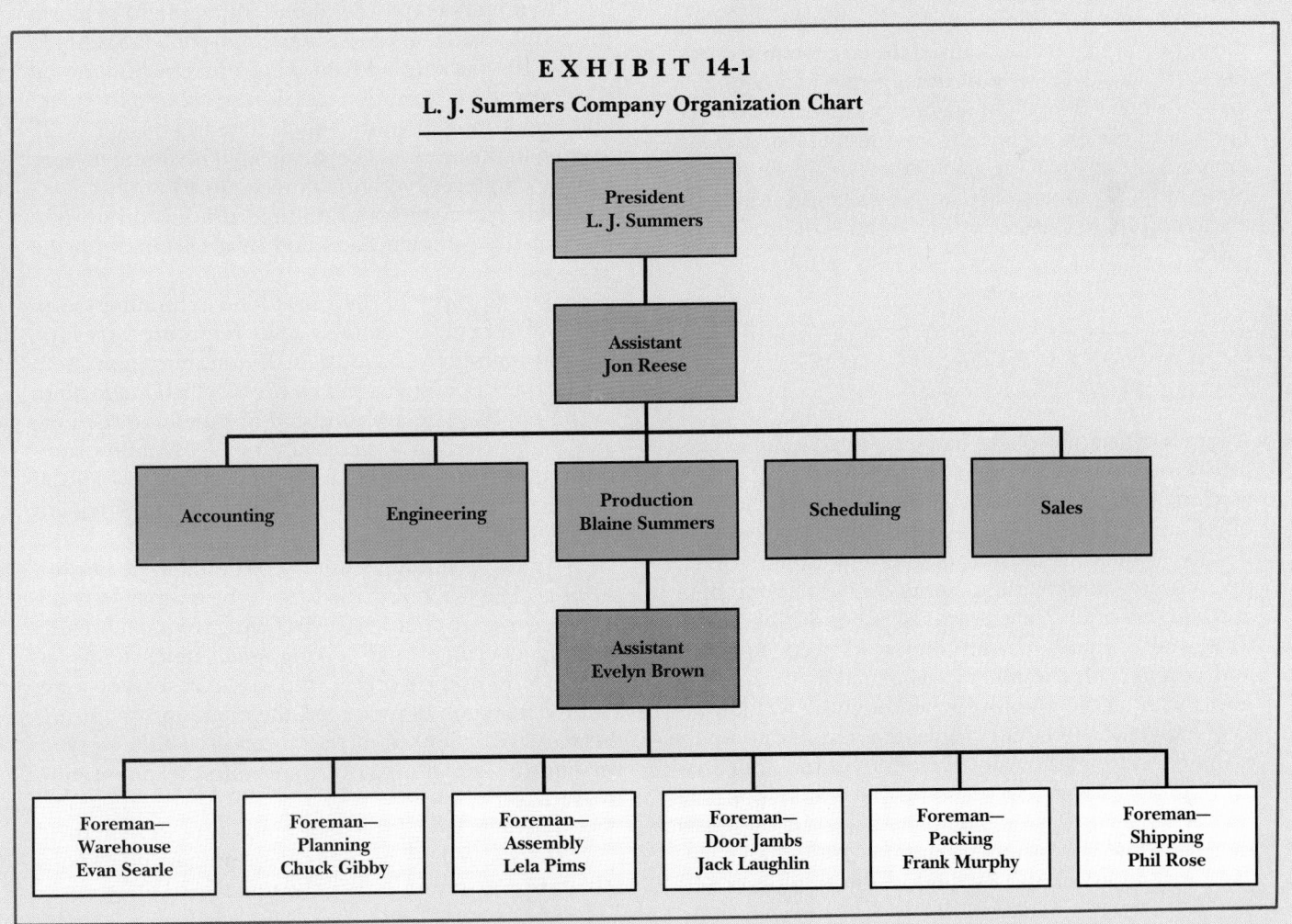

EXHIBIT 14-1
L. J. Summers Company Organization Chart

President
L. J. Summers

Assistant
Jon Reese

Accounting | Engineering | Production
Blaine Summers | Scheduling | Sales

Assistant
Evelyn Brown

Foreman—
Warehouse
Evan Searle

Foreman—
Planning
Chuck Gibby

Foreman—
Assembly
Lela Pims

Foreman—
Door Jambs
Jack Laughlin

Foreman—
Packing
Frank Murphy

Foreman—
Shipping
Phil Rose

increased to the point that several of the male workers had quit and all the women in the assembly department had refused to work.

The programs that had caused the resentment among the workers were instituted by Blaine to reduce waste and lower production costs, but they had produced completely opposite results. Jon knew that on Monday morning he would have to explain to Mr. Summers why the workers had reacted as they did and that he would have to present a plan to resolve the employee problems, reduce waste, and decrease production costs.

COMPANY HISTORY

L. J. Summers Company manufactured large sliding doors made of many narrow aluminum panels held together by thick rubber strips, which allowed the door to collapse as it was opened. Some of the doors were as high as 18 feet and were used in buildings to section off large areas. The company had grown rapidly in its early years due mainly to the expansion of the building program of the firm's major customer, which accounted for nearly 90 percent of Summers' business.

When L. J. Summers began the business, his was the only firm that manufactured the large sliding doors. Recently, however, several other firms had begun to market similar doors. One firm in particular had been bidding to obtain business from Summers' major customer. Fearing that the competitor might be able to underbid his company, Mr. Summers began urging his assistant, Jon, to increase efficiency and cut production costs.

CONDITIONS BEFORE THE COST REDUCTION PROGRAMS

A family-type atmosphere had existed at Summers before the cost reduction programs were instituted. There was little direct supervision of the workers from the front office, and no pressure was put on them to meet production standards. Several of the employees worked overtime regularly without supervision. The foremen and workers often played cards together during lunchtime, and company parties after work were common and popular. Mr. Summers was generally on friendly terms with all the employees, although he was known to get angry if something displeased him. He also participated freely in the daily operations of the company.

As Mr. Summers's assistant, Jon was responsible for seeing to it that the company achieved the goals established by Mr. Summers. Jon was considered hardworking and persuasive by most of the employees and

had a reputation of not giving in easily to employee complaints.

Blaine Summers had only recently become the production manager of Summers. He was in his early 20s, married, and had a good build. Several of the workers commented that Blaine liked to show off his strength in front of others. He was known to be very meticulous about keeping the shop orderly and neat, even to the point of making sure that packing crates were stacked "his way." It was often commented among the other employees how Blaine seemed to be trying to impress his father. Many workers voiced the opinion that the only reason Blaine was production manager was that his father owned the company. They also resented his using company employees and materials to build a swing set for his children and to repair his camper.

Blaine, commenting to Jon one day that the major problem with production was the workers, added that people of such caliber as the Summers' employees did not understand how important cost reduction was and that they would rather sit around and talk all day than work. Blaine rarely spoke to the workers but left most of the reprimanding and firing up to his assistant, Evelyn Brown.

Summers employed about 70 people to perform the warehousing, assembly, and door-jamb building, as well as the packing and shipping operations done on the doors. Each operation was supervised by a foreman, and crews ranged from 3 men in warehousing to 25 women in the assembly department. The foremen were usually employees with the most seniority and were responsible for quality and on-time production output. Most of the foremen had good relationships with the workers.

The majority of the work done at Summers consisted of repetitive assembly tasks requiring very little skill or training; for example, in the pinning department the workers operated a punch press, which made holes in the panels. The job consisted of punching the hole and then inserting a metal pin into it. Workers commented that it was very tiring and boring to stand at the press during the whole shift without frequent breaks.

Wages at Summers were considered to be low for the area. The workers griped about the low pay but said that they tried to compensate by taking frequent breaks, working overtime, and "taking small items home at night." Most of the workers who worked overtime were in the door-jamb department, the operation requiring the most skill. Several of these workers either worked very little or slept during overtime hours they reportedly worked.

The majority of the male employees were in their mid-20s; about half of them were unmarried. There was a great turnover among the unmarried male workers. The female employees were either young and single or older married women. The 25 women who worked in production were all in the assembly department under Lela Pims.

THE COST REDUCTION PROGRAMS

Shortly after Mr. Summers began stressing the need to reduce waste and increase production, Blaine called the foremen together and told them that they would be responsible for stricter discipline among the employees. Unless each foreman could reduce waste and improve production in his department, he would either be replaced or receive no pay increases.

The efforts of the foremen to make the workers eliminate wasteful activities and increase output brought immediate resistance and resentment. The employees' reactions were typified by the following comment: "What has gotten into Chuck lately? He's been chewing us out for the same old things we've always done. All he thinks about now is increasing production." Several of the foremen commented that they didn't like the front office making them the "bad guys" in the eyes of the workers. The workers didn't change their work habits as a result of the pressure put on them by the foremen, but a growing spirit of antagonism between the workers and the foremen was apparent.

After several weeks of no improvement in production, Jon called a meeting with the workers to announce that the plant would go on a 4-day, 10-hour-a-day work week in order to reduce operating costs. He stressed that the workers would enjoy having a three-day weekend. This was greeted with enthusiasm by some of the younger employees, but several of the older women complained that the schedule would be too tiring for them and that they would rather work five days a week. The proposal was voted on and passed by a two-to-one margin. Next Jon stated that there would be no more unsupervised overtime and that all overtime had to be approved in advance by Blaine. Overtime would be allowed only if some specific job had to be finished. Those who had been working overtime protested vigorously, saying that this would only result in lagging behind schedule, but Jon remained firm on this new rule.

Shortly after the meeting, several workers in the door-jamb department made plans to stage a work slowdown so that the department would fall behind schedule and they would have to work overtime to catch up. One of the workers, who had previously been the hardest working in the department said, "We will tell them that we are working as fast as possible and that we just can't do as much as we used to in a five-day week. The only thing they could do would be to fire us, and they would

never do that." Similar tactics were devised by workers in other departments. Some workers said that if they couldn't have overtime they would find a better paying job elsewhere.

Blaine, observing what was going on, told Jon, "They think I can't tell that they are staging a slowdown. Well, I simply won't approve any overtime, and after Jack's department gets way behind I'll let him have it for fouling up scheduling."

After a few weeks of continued slowdown, Blaine drew up a set of specific rules, which were posted on the company bulletin board early one Monday morning (see Exhibit 14-2). This brought immediate criticism from the workers. During the next week they continued to deliberately violate the posted rules. On Friday two of the male employees quit because they were penalized for arriving late to work and for "lounging around" during working hours. As they left they said they would be waiting for their foreman after work to get even with him for turning them in.

EXHIBIT 14-2
Production Shop Regulations

1. Anyone reporting late to work will lose one half hour's pay for each five minutes of lateness. The same applies to punching in after lunch.
2. No one is to leave the machine or post without the permission of the supervisor.
3. Anyone observed not working will be noted, and if sufficient occurrences are counted the employee will be dismissed.

That same day the entire assembly department (all women) staged a stoppage to protest an action taken against Myrtle King, an employee of the company since the beginning. The action resulted from a run-in she had with Lela Pims, foreman of the assembly department. Myrtle was about 60 years old and had been turned in by Lela for resting too much. She became furious, saying she couldn't work 10 hours a day. Several of her friends had organized the work stoppage after Myrtle had been sent home without pay credit for the day. The stoppage was also inspired by some talk among the workers of forming a union. The women seemed to favor this idea more than the men.

When Blaine found out about the incident he tried joking with the women and in jest threatened to fire them if they did not begin working again. When he saw he was getting nowhere he returned to the front office. One of the workers commented, "He thinks he can send us home and push us around and then all he has to do is tell us to go back to work and we will. Well, this place can't operate without us."

Jon soon appeared and called Lela into his office and began talking with her. Later he persuaded the women to go back to work and told them that there would be a meeting with all the female employees on Monday morning.

Jon wondered what steps he should take to solve the problems at L. J. Summers Company. The efforts of management to increase efficiency and reduce production costs had definitely caused resentment among the workers. Even more disappointing was the fact that the company accountant had just announced that waste and costs had increased since the new programs had been instituted, and the company scheduler reported that Summers was farther behind on shipments than ever before.

When you have read this case, look back at the chapter's diagnostic questions and choose the ones that apply to the case. Then use those questions with the ones that follow in your case analysis.

1. Critique Blaine Summers's behavior as a change agent. Why did the L. J. Summers Company's employees resent his actions? What should he do to curb this resentment?
2. Why did the door-jamb department employees stage a work slowdown? What do you think about Blaine's reaction to this slowdown? What would you advise Blaine to do instead?
3. What would you advise Jon Reese to do in order to correct the situation at Summers? What intervention would you implement if hired as an organization development change agent? What results would you expect this intervention to yield?

C A S E 14-3

SONIA HARRIS

Review this case, which appears at the end of Chapter 12. Next, look back at Chapter 14's diagnostic questions and choose the ones that apply to the case. Then use those questions with the ones that follow in your case analysis.

1. Might Sonia Harris benefit from a counseling intervention? What would you expect such an intervention to accomplish? What benefits would it have for the Multinational Fiber Company?
2. What kind of an organization development intervention would you recommend to improve relations between Claire Herzlinger and Multinational Fiber's other employees? What is the purpose of an intervention of this type? How would it improve interpersonal relations?
3. If you were given the job of revising the company's instruction manual and getting management and workers to accept the revision, how would you go about this task?

C H A P T E R 15

STRUCTURING THE ORGANIZATION

In 1987 John Sculley undertook a second reorganization of Apple Computer Inc., but by 1990 the company was in trouble again. Its domestic market share and rate of growth had declined, and its income had leveled off. According to Sculley, the firm was confused about where it was going. Dissension between the two largely autonomous divisions of marketing and product development kept both from being in touch with the market. For example, this structural problem may have caused Apple to miss a rare opportunity. Microsoft's OS/2, introduced in 1987 to give IBM PCs graphics capability, wasn't selling. Apple could have attracted frustrated OS/2 customers with lower-priced Macintoshes, but it elected to keep its prices high. Late in 1990 Apple's new Chief Operating Officer, Michael Spindler, announced detailed plans for regaining market share, while Sculley took over the research and development work on computers for the 21st century.

High on the list of Apple Computer's proficiencies is "event marketing"—turning corporate announcements into spectacles that reap extensive press coverage. Lately though, the press has been reporting events that the Cupertino, California, company would prefer not to publicize at all. From the end of May to the middle of June, Apple reorganized in a rush, fired 20 percent of its work force, and relieved Steven P. Jobs, Apple's 30-year-old cofounder and chairman, of all operating duty. John Sculley, 46, president and chief executive, noted with regret that Apple's moves were attracting as much attention as a popular television soap opera.

These changes are part of an overhaul Sculley has been working on since being wooed away from the presidency of Pepsi-Cola USA. From the beginning Sculley's mission was to improve Apple's responsiveness to retailers and customers. That required merging the company's nine highly decentralized divisions, most of which had broad responsibility for a product line, into an organization structured according to such business functions as engineering, manufacturing, and marketing.

Transforming Apple was more challenging than Sculley first imagined. Under Jobs the company had acquired a single-minded focus on products. The former chairman had often electrified Apple's employees with talk of "insanely great" new computers. But Sculley managed to consolidate Apple's divisions into just three—a sales division for all products, a division for the Apple II family of products, and one with the Macintosh as its focus.[1]

Shortly after this mid-1985 reorganization described in *Fortune* magazine, Apple's president, John Sculley, further transformed the firm by merging the Apple II and Macintosh divisions into a single department for product development and manufacturing. Sales and marketing remained combined in a second department. Sculley also implemented strict financial controls and formal reporting procedures within the scaled-down, streamlined version of Apple, Inc., increasing the efficiency of the firm's operations. In addition, the authority to make major decisions was taken away from middle management and pushed upward to the top of the organization. What remained after these changes bore little resemblance to the sprawling patchwork of nine loosely connected, autonomous divisions that Steve Jobs and Apple's other founder, Steve Wozniak, had created.[2]

Apple's transformation involved changes in the way the firm's employees were organized. Employees formerly grouped together according to the specific type of computer they produced and sold (Apple II, Lisa, Macintosh, among others) were regrouped according to the kind of work they performed (research and development, manufacturing, sales and marketing, and so on). The new methods of coordination—Sculley's formal procedures and controls—served to keep everyone in Apple's different groups working together toward the firm's overall objectives. In addition, there were changes in the hierarchical network of connections among the groups in Apple. Formerly each of the nine original divisions had been largely self-sufficient and able to succeed by working alone.

---

[1] Based on Bro Uttal, "Behind the Fall of Steve Jobs," *Fortune*, August 5, 1985, pp. 20–21.

[2] Katherine M. Hafner and Geoff Lewis, "Apple's Comeback," *Business Week*, January 19, 1987, pp. 84–89; Deborah C. Wise, "Can John Sculley Clean up the Mess at Apple?" *Business Week*, July 29, 1985, pp. 70–71; and Uttal, "Fall of Steve Jobs," pp. 20–24.

The groups created by Sculley's reorganization scheme became extremely interdependent and had to learn how to work together. Finally, Apple's reorganization involved critical modifications in decision-making procedures. By confining decision making to top management, Sculley gave himself and other top managers the authority to personally coordinate intergroup relations at Apple.

These changes constituted major modifications to Apple's *structure*. The composition of Apple's basic groups, the methods used to coordinate activities within those groups, the mechanisms used to link the groups together, and the distribution of decision making throughout the firm were all new. Like the steel framework of a building or the skeletal system of the human body, the structure of an organization like Apple separates its different parts from each other and also helps keep those parts interconnected. Specifically, an organization's structure groups its members together and separates the resulting groups from one another. In so doing, the structure also creates relationships of interdependence or interconnectedness within and among these groups. It is the structure of an organization that gives shape to the various interpersonal, group, and intergroup processes examined in Chapters 10–14. It defines the context that surrounds and influences the individual-level processes discussed in Chapters 4–9.

For this reason, an organization's structure has widespread effects on behavior and productivity within the firm. Managers like Apple's John Sculley must know how to structure their organizations in different ways. They must also know about the strengths and weaknesses of the different ways of structuring. In this chapter, we will examine the basic elements of an organization's structure—how the organization's work is divided and allocated among groups and the linking and coordinating mechanisms that facilitate the smooth flow of that work.

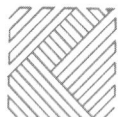

DIAGNOSTIC ISSUES

Our diagnostic model highlights several important questions about organization structure. What are some of the major dimensions that *describe* the way firms are structured? What different ways of coordinating work are used in different kinds of structures? How can we *diagnose* whether an organization's structure will promote efficiency or flexibility and adaptability? Can we explain the reason why employees in an organization are grouped into structural units?

Can we *prescribe* how to link the units in an organization together? Can we specify an adequate balance in a firm between hierarchical control and the need to keep work flowing? Finally, what kinds of *actions* should managers take to devise ways of coordinating work in the organization? What corresponding adjustments should they make to the way jobs are designed and responsibilities are assigned?

WHAT IS ORGANIZATION STRUCTURE?

organization structure The relatively stable network of interconnections or interdependencies among the people and tasks that make up an organization.

As you learned in Chapter 2, all organizations consist of people performing tasks that when combined, create goods or services whose production lies beyond the abilities of people working alone. Whether as well known as Apple or as anonymous as a locally owned convenience store, every organization is characterized by a pattern of interrelated tasks that are essential to its efficient functioning. This identifiable **organization structure** consists of a relatively stable

network of interconnections or interdependencies among the people and tasks that make up an organization.[3]

An organization's structure serves as a framework that both integrates the individuals and groups that constitute the organization and differentiates them from each other. Typically, the structure of an organization takes the form of a hierarchy such as the one diagrammed in the organization chart shown in Figure 2-1. Balancing structural *integration* and *differentiation* within structural hierarchies is an important challenge facing managers.[4] The ability to sense a workable balance and arrange an organization's structure accordingly can have a major effect on the firm's productivity.

FORMING AND COORDINATING STRUCTURAL UNITS

Managers first strive to balance integration and differentiation by determining the most effective way to group employees. This process is one of seeking acceptable answers to questions such as, "Who should work with whom?" "What mechanisms should be used to coordinate activities in each group?" and "How large should each group be?"

Unit Grouping

unit grouping The process of grouping the members of an organization into work groups or units.

Answering the first question, Who should work with whom? is neither simple nor straightforward. In general, however, the people in an organization are placed in work groups or units according to similarities either in what they do or in what they produce.[5] To illustrate these two types of **unit grouping**, let us imagine a company that makes wooden desks, bookshelves, and chairs. To produce each of these products, four basic activities are required. A receiver must unpack and stock the raw materials required for the product. A fabricator must shape and assemble the raw materials into a partially completed product. A finisher must complete the assembly operation by painting and packaging the product. Finally, a shipper must dispatch the finished products to the organization's customers. In summary, the manufacturing work force of the company consists of 12 employees organized into three assembly lines consisting of four employees each.

functional grouping People are grouped into units according to similarities in the functions they perform. Grouping word-processing typists into a word-processing pool is an example.

market grouping People are grouped into units according to similarities in the products they make or markets they serve. Grouping the members of an automobile assembly line into an assembly unit is an example.

What the management of this company must decide is whether to group its 12 employees by the tasks they perform, called **functional grouping**, or by the products they produce or markets they serve, **market grouping**. As you will see, each method has both advantages and disadvantages, so the choice presents managers with a tradeoff. Let's start by considering what functional grouping, or grouping by the means of production, will do for the firm. The upper panel of Figure 15-1 shows how the four tasks from each assembly line can be grouped together so that the four resulting work units consist of people with the same sets of abilities, knowledge, and skills. The receivers, the fabricators, the finishers, and the shippers thus form four *functional* work units.

[3] James G. March and Herbert A. Simon, *Organizations* (New York: John Wiley 1958) p. 4; James D. Thompson, *Organizations in Action* (New York: McGraw-Hill, 1967), p. 51; W. Richard Scott, *Organizations: Rational, Natural, and Open Systems*, 2nd ed. (Englewood Cliffs, N.J.: Prentice Hall, 1987), p. 15.

[4] Paul R. Lawrence and Jay W. Lorsch, "Differentiation and Integration in Complex Organizations," *Administrative Science Quarterly* 12 (1967), 1–47; and Lawrence and Lorsch, *Organization and Environment* (Homewood, Ill.: Richard D. Irwin, 1967), p. 7.

[5] Henry Mintzberg, *The Structuring of Organizations* (Englewood Cliffs, N.J.: Prentice Hall, 1979), pp. 108–29.

FIGURE 15-1

Unit Grouping

People are put into work units according to similarities in what they do or what they produce. The upper diagram shows an organization with its units formed around similarities in what their members do, that is, functional similarities. The lower diagram shows the same organization with its units formed around similarities in what their members produce, in other words, market similarities.

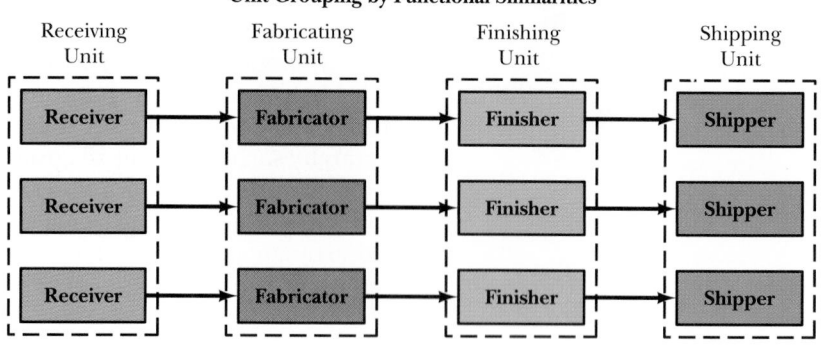

Unit Grouping by Functional Similarities

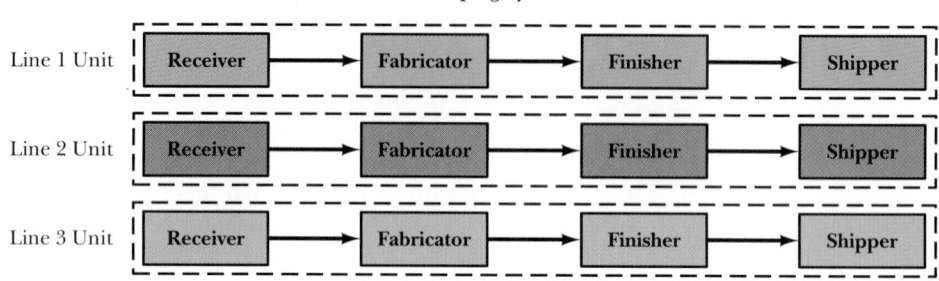

Unit Grouping by Market Similarities

APPLYING THE DIAGNOSTIC MODEL

Diagnostic Question 1: Are structural units grouped according to functional or market similarities? What does this tell you about the relative emphasis placed by the firm on efficiency? On adaptability?

Functional work units help integrate and coordinate employees who perform the same sorts of tasks. Employees in such units can exchange information about task procedures, sharpening their knowledge and skills. They can also help each other out when necessary. This sort of cooperation can greatly enhance unit productivity.

Functional grouping can also allow the organization to take advantage of certain other cost savings. Suppose that in our example, receivers for all three of the assembly lines typically need only five hours a day to complete their work and are idle for the remaining three hours. If receiving is handled in a single work unit, the firm can economize by employing two receivers instead of three. In a full eight-hour day, these two people should be able to complete the unit's work (3 workers × 5 hours each = 15 hours; 2 workers × 8 hours each = 16 hours), minimizing idle time and improving the firm's economic efficiency.

Functional grouping, however, separates people performing different tasks along the same work flow. This differentiation can encourage slowdowns that block the flow, reducing productivity as a result. For instance, suppose the finisher on the desk-assembly line has nothing to do and wants the desk fabricator to speed up in order to provide more work. Because of functional grouping, the two people are in different units, and there is no simple way for them to communicate with each other directly. Instead, the desk finisher must rely on hierarchical communication linkages between the fabricating and finishing units. The finisher must tell the supervisor of the finishing unit about the problem. The finishing supervisor must notify the superintendent overseeing all manufacturing operations. The manufacturing superintendent must talk with the supervisor of the fabricating unit. And the fabricating supervisor must tell the desk fabricator to work faster. Meanwhile, productivity suffers owing to the absence of direct communication along the flow of work.

Now let's consider what happens if the firm decides to create *market-based* work units. Such units can be formed on the basis of similarities in products, in customer groups, or in geographic regions served. We will assume that our company decides to create units based on its three product lines of desks, bookshelves, and chairs. The lower panel of Figure 15-1 illustrates this option. Now the company is grouped into three work units, and each unit completely contains one of the firm's product-assembly lines.

One of the primary strengths of market grouping is that it integrates activities along the flow of work. Each separate work flow is completely enclosed within a single unit. If employees who fill different functions along the assembly line need to coordinate with each other to maintain the flow of work, they can do so without difficulty. As you can imagine, in an organization grouped by market similarities, work tends to flow very smoothly.

Owing to its encouragement of work-flow integration, market grouping also enhances the organization's adaptability. Operations on any of the firm's three assembly lines can be halted or stopped without affecting the rest of the company. For example, suppose the desk-assembly line in the company is shut down because of poor sales. To simulate this situation, cover the upper assembly line in the bottom panel of Figure 15-1 with a piece of paper. You can see that neither of the remaining two units will be affected in any major way. Work on the bookcase and chair lines can continue without interruption. Under functional grouping, however, the firm would not have the same degree of flexibility. If you cover the upper assembly line in the top panel of Figure 15-1, you will note that all four of the units created by functional grouping would be affected if desk production were interrupted. Complete reorganization would be required by any long-term disturbance. Thus the adaptability allowed by the relative independence of different work flows in an organization grouped according to market similarities is not available to companies composed of functionally grouped units.

Despite its strengths, however, market grouping does not permit the scale economies offered by functional grouping. Under the market grouping arrangement, people who perform the same function (e.g., receiving) cannot help or substitute for one another. In addition, at times they will inevitably duplicate one another's work, adding to the firm's overall costs. Moreover, it becomes very difficult for people who perform the same task to trade information about such things as more efficient work procedures and how to improve task skills. So just as functional grouping does not allow the adaptability of market grouping, market grouping does not incorporate the economic efficiency of functional grouping.

Unit Coordination Mechanisms

Deciding how an organization's employees should be grouped is one of the most important structural decisions a manager can make. Assigning people to work units helps establish physically close relationships among the members of each unit, making it easier for them to engage in face-to-face communications. Grouping workers can also make it easier to carry out superior-subordinate interactions. Furthermore, people who work close together are more likely to develop common norms and behaviors. These similarities can make it easier to establish standardized work processes and outputs. All these outcomes encourage greater coordination.

Coordination is a process in which otherwise disorganized actions are integrated so as to produce a desired result. Different parts of our bodies, for example, work together to produce complex, coordinated behaviors. Your hands follow a trajectory plotted by your eyes in order to catch a ball. Your hands also manipulate your car's steering wheel at the same time that your foot depresses the accelerator pedal. It would be very difficult, if not impossible, to catch the ball if you could not see it. It would be dangerous to accelerate or even to move the car if you could not control its direction.

AN EXAMPLE: VOLVO
Rather than have its cars put together by workers on assembly lines, Volvo has for a long time used a market-grouping system of teamwork, in which groups of workers assemble individual cars. In 1990 worker absenteeism began to increase, however, and in early 1991 it reached 24 percent, the highest among automakers not only in Sweden but around the world. The prices of the company's cars rose to compensate for this rising worker cost. Ask students to consider what could have happened to cause this increase in no-shows in a system that has prided itself on teamwork and worker participation. (Jennifer Reese, "Saab and Volvo Hit a Pothole in U.S.," *Fortune*, March 25, 1991, p. 16.)

In similar fashion, the members of an organization by working together can accomplish outcomes that would be beyond the abilities of any one person working alone. By helping to mesh interdependent task activities, three **unit coordination mechanisms** sustain the structural interconnections in the units of an organization. These mechanisms are mutual adjustment, direct supervision, and standardization. As the primary means by which organizational activities are integrated, they are the glue that holds the organization together.[6]

Mutual Adjustment. By **mutual adjustment**, we mean coordination accomplished through interpersonal communication in which coworkers who occupy positions of similar hierarchical authority share job-related information.[7] Mutual adjustment, the simplest of the three unit coordination mechanisms, is the face-to-face exchange of information about how a job should be done and who should do it. A group of factory-maintenance mechanics who are examining service manuals and discussing how to fix a broken conveyor belt are coordinating by means of mutual adjustment.

Note that in mutual adjustment, information is exchanged by people who can exercise at least partial control over the tasks they perform. Clearly, unless the people doing the communicating possess this control, they cannot coordinate their activities in this way.

Direct Supervision. In **direct supervision**, one person takes responsibility for the work of a group of others.[8] As part of their responsibilities, direct supervisors may determine which tasks need to be performed, who will perform them, and how they will be linked together to produce the desired end result. They may then issue orders to subordinates under their jurisdiction, check to see that these orders have been executed, and redirect subordinates as needed to fulfill additional work responsibilities. The owner of a grocery store is functioning as a direct supervisor when having instructed an employee to restock the shelves, he finds that the clerk has completed the job and directs her next to change the signs advertising the week's specials. A shop-floor supervisor overseeing a potato chip production line is also coordinating by direct supervision when she orders workers to stop the line to adjust the heat in its drying oven.

Standardization. **Standardization** coordinates work by providing employees with carefully worked out standards and procedures that guide the performance of their tasks. This kind of coordination is achieved on the drawing board, before the work to be performed is actually undertaken.[9] So long as drawing-board plans are followed and the work situation remains essentially unchanged, interdependence is maintained.

There are four types of standardization—standardization of work processes, or behaviors, and standardization of outputs, skills, and norms. As you will see from the brief descriptions that follow, each kind of standardization draws on procedures discussed in other parts of the book. For example, work process standardization can grow out of the process of job design examined in Chapter 17. Output standardization can incorporate the sort of goal setting discussed in Chapter 7. Skill standardization can be based on human resource training pro-

---

[6] Ibid., pp. 2–3; March and Simon, *Organizations*, p. 160; and Jay R. Galbraith, *Designing Complex Organizations* (Reading, Mass.: Addison-Wesley, 1973), p. 4.

[7] Thompson, *Organizations in Action*, p. 62.

[8] Mintzberg, *Structuring of Organizations*, pp. 3–4. See also the discussion in Chapter 2 of Henri Fayol's management principles.

[9] Mintzberg, *Structuring of Organizations*, p. 5.

cedures identified in Chapter 9. And norm standardization can grow out of the interpersonal socialization described in Chapter 10.

Work process standardization is sometimes called *behavioral standardization*. It involves specifying the precise behaviors or actions employees must perform to accomplish their assigned tasks. Specified behaviors are a part of the process, or behavioral standards, for a particular job. They link that job with other jobs in the organization. For instance, the behavioral specifications for the worker who is responsible for filling soda-pop bottles may include step-by-step instructions for controlling the flow of the soda, checking for cracked bottles, positioning the bottles for filling, and placing filled bottles on a conveyor line. The behavioral specifications for the worker who is responsible for capping the bottles may include step-by-step instructions for checking that all bottles on the conveyor line have been filled, that none have cracked during filling and capping, that all caps are tightly secured, and that all properly capped bottles move forward along the line to the shipping department. In this example, the two people are connected by the conveyor line and are able to work together without any further coordination.

Output standardization is the formal designation of output targets, or performance goals. For example, a sales representative of a publishing company might be assigned the goal of getting university English departments to adopt 1,000 copies of a new English grammar textbook within a 12-month period. Alternatively, workers on an assembly line in a baseball glove manufacturer might be given the goal of producing 25 gloves per hour. Unlike employees working under behavioral standardization, people coordinated by output standardization are free to decide for themselves how to attain their goals. So long as everyone accomplishes his goal, work continues, unchanged, and no one needs to engage in further coordination.

Skill standardization involves specifying in advance the skills, knowledge, or abilities that people must have to perform a task competently. Because skill standardization is aimed at regulating characteristics of people rather than jobs, it is used most often in situations where neither work process standards nor output targets can be easily specified.

For example, few experts agree on the precise behaviors that high-school teachers should engage in while teaching. Moreover, there is a general consensus that the *output indicators* for the job of teaching, such as course grades and standardized test scores, have little validity as measures of teaching success. (Grades can be artificially inflated, and test scores can be undermined by pretest coaching.) On the other hand, almost all community school districts mandate that their teachers be certified by an agency of the state, and such certification often requires that teachers not only hold certain educational degrees but give evidence of having acquired specific knowledge and skills. Thus all teachers hired by a school district that requires state certification should possess a more or less standardized set of job qualifications or skills.

Skilled employees seldom need to communicate with each other to figure out what to do and can usually predict with reasonable accuracy what other similarly skilled employees will do on the job. Consequently, on jobs staffed by specially skilled employees, there may be much less need to coordinate work behaviors in other ways.

As you will recall from Chapters 10 and 11, norms describe desirable and expected forms of behavior. They are statements of what people should or should not do. For example, a norm often found in work organizations is that people should arrive at their jobs on time and should not leave before the end of the normal workday. *Norm standardization* is present when the members of a unit or organization share a set of beliefs about the acceptability of particular types of behavior and so tend to behave in ways that are generally approved. For instance, Ford Motor employees who believe that "quality is job one" work in ways that

AN EXAMPLE: MCDONALD'S
McDonald's provides step by step instructions of how to make an Egg McMuffin. These steps cover equipment to use, how long the egg should cook, and what should be done while waiting. The result is a product whose parts are ready at the same time for quick assembling and whose quality is consistent (standardized).

AN EXAMPLE
The nursing profession relies heavily on skill standardization. Nurses must past a common state board exam testing their knowledge and understanding of when to use various skills. Skills are generally retested periodically by the employing hospital. As a result, as an accident victim is wheeled into emergency the nursing staff functions as a team with minimal communication. Everyone knows his job. The doctor can give complex orders by using only a few words, and the orders are understood and carried out with great efficiency. If just one person lacks sufficient skills, an individual's life could be lost.

enhance product quality. They do not need to discuss the merits of this philosophy with each other or to be directed by a supervisor to produce quality products. Similarly, the employees of United Parcel Service who share a belief in the importance of on-time delivery tend to make their deliveries promptly or to see that their subordinates do so.

Choosing among the Mechanisms. Managers charged with designing an organization structure continually confront choices among the three unit coordination mechanisms we have discussed (these mechanisms are reviewed in Table 15-1). Most of the time, two or more of these mechanisms are used concurrently to integrate work activities in organizational units. In such instances, one of them serves as the primary mechanism used to solve most coordination problems. The others (if present) serve as secondary mechanisms that supplement the primary mechanism, backing it up in case it fails to provide enough integration. It is up to managers to determine which mechanism will serve as the primary means of coordination and which ones (if any) will act as secondary mechanisms. Besides cultural tendencies (see the "International OB" box), two factors govern such choices—first, the number of people whose efforts must be coordinated in order to ensure the successful performance of interdependent tasks; second, the relative stability of the situation in which the tasks to be coordinated must be performed.[10]

In small groups, of about 12 people or fewer, coordination is often accomplished by everyone's doing what comes naturally. Employees communicate face to face, using mutual adjustment to fit individual task behaviors into the group's overall network of interdependence. No other coordination mechanisms are needed, and none are used. Family farms and specialty restaurants are often organized around this type of coordination. Similarly, students working together on textbook exercises or case analyses often coordinate among themselves using mutual adjustment alone.

Suppose a group is made up of many more than 12 people—as many as 30, 40, or even 50. Now face-to-face mutual adjustment may fail to sustain purposeful interdependence. It may need to be replaced by some other unit coor-

---

[10] Mintzberg, *Structuring of Organizations*, pp. 7–9; and Mintzberg, "The Structuring of Organizations," in *The Strategy Process: Concepts, Contexts, and Cases*, ed. James Brian Quinn, Henry Mintzberg, and Robert M. James (Englewood Cliffs, N.J.: Prentice Hall, 1988), pp. 276–304.

APPLYING THE DIAGNOSTIC MODEL

Diagnostic Question 2: Which of the three unit coordination mechanisms are used in the organization's units? Which is used as the primary means of coordination, and what does this tell you about the importance the firm accords to efficiency? to flexibility?

APPLYING THE DIAGNOSTIC MODEL

Diagnostic Question 3: What secondary coordination mechanisms back up the primary mechanisms used throughout the firm? Are unit activities coordinated adequately?

TABLE 15-1
Unit-Coordination Mechanisms

MECHANISM	DEFINITION
Mutual Adjustment	Face-to-face communication in which hierarchical equals exchange information about work procedures
Direct Supervision	The direction and coordination of the work of a unit by one person who issues direct orders to the unit's members
Standardization	The planning and implementation of standards and procedures that regulate work performance
Work process/behavior standardization	The specification of sequences of task behaviors
Output standardization	Establishment of goals or desired end results of task performance
Skill standardization	Specification of the abilities, knowledge, and skills required by a particular task
Norm standardization	Inculcation of attitudes and beliefs that lead to specific desired behaviors

INTERNATIONAL OB

Effects of National Cultures on Organization Structures

Research has revealed four dimensions along which national cultures can be differentiated (see Chapter 19). Two of these dimensions, uncertainty avoidance and power distance, affect the emergence of different unit coordination mechanisms.*

The first dimension, uncertainty avoidance, concerns the extent to which people are comfortable with ambiguous situations and an unknown future. In national cultures with strong uncertainty avoidance, people try to make the present more comfortable by developing extensive rules and regulations to standardize behavior. There are far fewer rules in national cultures characterized by weak uncertainty avoidance.

The second dimension, power distance, is the degree to which people feel that power differences should be minimized and political equality encouraged. In national cultures characterized by small power distance, participatory processes are preferred over hierarchical procedures. In cultures with tendencies toward large power distance, the opposite is true, and hierarchy is the preferred way of coordinating activities.

What effects do these two international dimensions have on the choice of unit coordination mechanisms? In countries with strong tendencies toward uncertainty avoidance (Greece, Japan, France), rules and regulations developed to cope with uncertainty also push toward the emergence of standardization even before information-processing requirements make it necessary. In countries with weak uncertainty avoidance (U.S., Canada, Sweden), firms turn to standardization only when other methods of unit coordination fail to perform appropriately. In countries favoring large power distance (Mexico, India, Hong Kong), direct supervision appears early and dominates unit coordination even after the emergence of extensive standardization. In countries with small power distance (Israel, Sweden, Great Britain), mutual adjustment may be used even when process loss would seem to necessitate its

* Geert Hofstede, *Culture's Consequences: International Differences in Work-Related Values* (Beverly Hills, Calif.: Sage, 1980).

replacement. Thus, although the same unit coordination mechanisms are used in organizations throughout the world, cultural tendencies can alter the specific point at which one type of unit coordination mechanism is replaced by another.

---

dination mechanism. As you can see from Figure 15-2, the reason mutual adjustment alone cannot effect coordination among large numbers of people is that the number of needed communication links rises geometrically as the number of individuals rises arithmetically. Clearly, the members of larger groups would have to spend so much time communicating with each other that they would have very little time to actually complete their tasks. Process loss (see Chapter 11) would substantially undermine the group's productivity.

Thus, in larger groups, direct supervision takes the place of mutual adjustment as the primary means of coordinating group activities. Direct supervision reduces the number of linkages needed to coordinate tasks to one superior-subordinate linkage for each employee. In communicating information to her subordinates, the direct supervisor acts as a proxy for the group as a whole. To use an analogy, the direct supervisor functions like the sort of automated switching facility that routes telephone messages from callers to receivers. She originates direct orders and collects performance feedback while channeling information from one interdependent group member to another. In such situations, however,

FIGURE 15-2

Group Size and Mutual Adjustment Links

For each person added to a group, the number of links needed to coordinate through mutual adjustment rises geometrically. Thus, although 2 people need only 1 link, 3 people need 3, and 6 people need 15. In an organization of 30 people, how many links would be required?

Number of people	Number of Links	Group Configuration
2	1	
3	3	
4	6	
5	10	
6	15	

AN EXAMPLE

Employees often use mutual adjustment during periods of significant organization change. Standard procedures lose their usefulness and either supervisors are not sure of what is happening or employees don't believe what they are told. As a result, employees depend on each other for answers and clarifications.

mutual adjustment still continues as a supplementary coordination mechanism. When the direct supervisor is unavailable or does not know how to solve a particular problem, employees resort to face-to-face communication among themselves to try to figure out what to do.

Besides clarifying how direct supervision functions as a unit coordination mechanism, the telephone-switching analogy also helps to explain the failure of even direct supervision to coordinate the activities of members in very large groups. Just as a switching facility can be overloaded by an avalanche of telephone calls, in successively larger groups the direct supervisor is increasingly burdened by the need to obtain information and channel it to the right people. Ultimately, the direct supervisor must succumb to information overload. She is unable to keep up with the demands for coordination made by her subordinates.

At this point, the third unit coordination mechanism, standardization, replaces direct supervision as the primary means of sustaining coordinated task interdependence. As long as the conditions on which established standards have been based continue to prevail, coordination by standardization can prevent information overload. That is because it greatly reduces or eliminates the amount of communication needed for effective coordination. Workers are performing prespecified task behaviors, producing prespecified task outputs, using prespecified task skills, or conforming to prespecified workplace norms. Therefore, members of very large groups can complete complex, interdependent networks of task activities with little or no need for further coordination.

Where standardization is the primary means of coordination, direct supervision and mutual adjustment are still available for use as secondary coordination mechanisms. Direct supervision may be used to make sure that workers on the assembly line adhere to standards. Mutual adjustment may also be used on the assembly line to cope with machine breakdowns, power outages, or other temporary situations in which standard operating procedures lose effectiveness. In this example, the secondary mechanisms of direct supervision and mutual adjustment supplement the primary mechanism of standardization, filling in the holes left by momentary failures in the primary mechanism.

Standardization requires stability. If the conditions envisioned during the planning of a particular standardization program change, the usefulness of that program may be destroyed. For example, behavioral specifications that detail computerized check-in procedures are likely to be of little use to hotel-registration personnel facing a long line of guests and a dead computer screen. Similarly, output specifications requiring the sale of 50 pairs of denim jeans per day provide no useful guidance if denim goes out of fashion.

When changing conditions completely undermine coordination via standardization, mutual adjustment sometimes reemerges as the primary unit coordination device. When large groups or organizations face rapidly changing conditions that make standardization impossible, they rely heavily on face-to-face communication. The process loss associated with mutual adjustment in these situations is simply tolerated as a necessary cost of staying in business.

The three means of coordination, then, form a continuum, from mutual adjustment at one extreme, through direct supervision and standardization in the middle ranges, and back to mutual adjustment at the other extreme.[11] This continuum is depicted in Figure 15-3. It is important to remember that as coordination needs progress from left to right along the continuum, mechanisms to the left are not completely abandoned. So at the point represented all the way to the right of the continuum, standardization, direct supervision, and secondary mutual adjustment are available to supplement the mutual adjustment that serves as the primary means of coordination.

When you are choosing among coordination mechanisms, you must keep in mind that there is a critical tradeoff between the *costs* of using a particular mechanism and the *flexibility* it permits. The simplest coordination mechanism, mutual adjustment, requires neither extensive preplanning nor preexisting hierarchical differentiation. Therefore, it affords a high degree of flexibility. Changing circumstances can be readily accommodated. The links forged by mutual adjustment, however, cannot usually be banked for future use. Instead, each time mutual adjustment is used, it generates new coordination costs. These costs may take the form of time, effort, and similar resources that must be diverted away from task-related activities and directed toward face-to-face communications. The costs of each instance of mutual adjustment tend to be modest, but over time they add up and become quite significant.

In contrast, the initial costs of standardization are quite high. Planning standards and procedures often means contracting with specialists, and otherwise productive resources must be diverted to the planning process. Yet once a program of standardization has been designed and implemented, it no longer consumes resources of major significance. The coordination costs associated with standardization can therefore be amortized or spread over long periods of time and across long production runs. The result will be an extremely low coordination cost per unit of goods or services produced. Standardization is thus less costly than mutual adjustment. As mentioned earlier, however, standardization requires that the work situation remain essentially unchanged—changing conditions would render existing standards obsolete. So it lacks the flexibility of mutual adjustment.

The flexibility of direct supervision lies between the extremes of mutual adjustment and standardization. Because it presupposes a hierarchy of authority,

FIGURE 15-3

A Continuum of Coordination Mechanisms

The three basic means of coordinating unit activities—mutual adjustment, direct supervision, and standardization—fulfill increasingly demanding needs for coordination and come into use at varying points along a continuum of unit size and complexity.

[11] Mintzberg, *Structuring of Organizations*, p. 7.

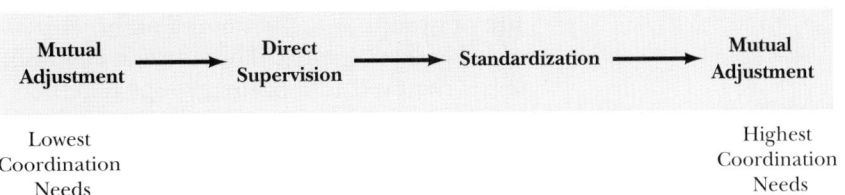

it lacks the spontaneity and fluidity of mutual adjustment. Yet because direct supervision requires much less planning than standardization, it is more flexible. Not surprisingly, the coordination costs of direct supervision also fall between those of mutual adjustment and standardization. Although supervision requires fewer costly communication links than mutual adjustment, new coordination costs are generated each time a supervisory action is taken.

Unit Size

Besides having major effects on the costs and quality of coordination, the different coordination devices used in a group also determine the appropriate **unit size**, or **span of control**, of the group. As we have implied, mutual adjustment is most effective in small units of about 2 to 12 people in which each member is able to talk with everyone else. In contrast, large units of 50 or more people can be coordinated effectively by means of standardization, which requires minimal face-to-face communication. Groups of moderate size can be handled by direct supervision, which requires less communication than mutual adjustment but more than standardization.

The optimal size of a particular work unit is also a function of the kinds of tasks its members and supervisor perform. Homogeneity of tasks reduces the variety of problems likely to arise and makes it easier for the unit's supervisor and members to solve the ones that do come up. For example, coping with the work problems of a group of 12 computer operators is easier than overseeing a group of 6 computer operators and 6 computer programmers. All 12 operators are likely to require help only with operating problems (following directions found in computer manuals, dealing with machine breakdowns, deciding the priority of different computer jobs). In the programmers' group, half will have one kind of problem, while the other half are likely to need help with a whole set of entirely different problems (plotting the basic logic of programs yet to be written, finding errors in newly written programs, determining how to update older programs). Consequently, the more similar the tasks performed by a group's members, the larger the group can be without overloading the supervisor or requiring extensive mutual adjustment.

In addition, optimal unit size depends on the extent to which employees in the unit must consult with the supervisor in order to perform their work. Although the demands on the supervisor's time in a small group can be reasonable, they may prove overwhelming in a larger group. The manageability of a particular unit size may also reflect the number of nonsupervisory duties the supervisor must perform. Such duties will reduce the time she has available to manage group activities.

To some degree, the size of a work unit is determined by its hierarchical position. Typically, groups found lower in a firm's hierarchy are larger than groups at higher levels. This tendency is generally due to the fact that the work performed at lower levels tends to be relatively less complex, therefore, less difficult to standardize and supervise than work performed at higher levels. It may also be caused by attempts to reduce distortion in the flow of information traveling up and down the hierarchy. As shown in Figure 15-4, increasing the size of groups at the bottom of an organization's hierarchy reduces the number of lower-level groups. Therefore, it can also reduce the number of hierarchical levels required to link the groups together. In turn, because fewer people are involved in the firm's vertical communication channels, cutting down on hierarchical levels lessens the chance that information transferred along vertical communication channels will be distorted.

FIGURE 15-4

Unit Size and Vertical
Information Flow

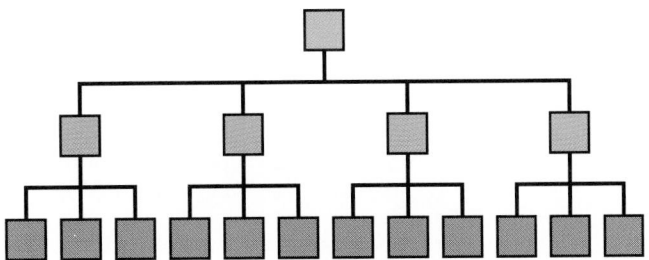

A Units of 3 with 3 Hierarchical Levels

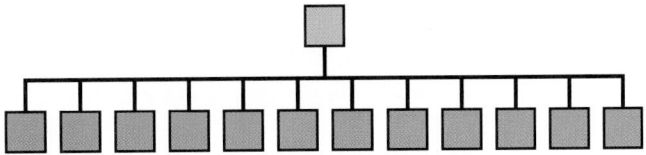

B 1 Unit of 12 with 2 Hierarchical Levels

Increasing the size of units at the bottom of an organization's hierarchy can decrease the number of levels in the hierarchy. The result is reduced distortion in information flowing up and down the hierarchy.

You can see that there is no magic number of members that will guarantee unit effectiveness. Instead, the size that is most appropriate for a particular group depends on the factors we have just mentioned. It is management's responsibility to assess these factors and adjust the group's size accordingly.

STRUCTURAL CORRELATES OF STANDARDIZATION

Besides affecting the type of coordination mechanism used to integrate activities within structural units, the decision to coordinate by standardization affects the nature of two other structural characteristics—formalization and specialization. They are *correlates* of standardization, meaning that they change in association with changes in standardization. We will discuss them more fully in this section.

Formalization

formalization The process of planning the regulations that control organizational behavior; also the written documentation produced by the planning process. Jobs, work flows, or general rules may be formalized.

Formalization is the process of planning the regulations that control organizational behavior. It also refers to the written documentation produced by the planning process. As you can see, there is a clear relationship between formalization and standardization. Formalization gives concrete form to the behavioral, output, skill, and norm standards used to coordinate by standardization. Thus you are likely to find extensive formalization only in organizations that make use of standardization as a coordination mechanism. In organizations that make little use of standardization, formalization will be almost nonexistent. Figure 15-5 shows a few items from an instrument that is useful in assessing the amount of formalization in a firm.[12]

[12] D. S. Pugh, D. J. Hickson, C. R. Hinings, and C. Turner, "Dimensions of Organization Structure," *Administrative Science Quarterly*, 13 (1968), 65–91; J. H. K. Inkson, Pugh, and Hickson, "Organization, Context, and Structure: An Abbreviated Replication," *Administrative Science Quarterly* 15 (1970), 318–29; and John Child, "Organization, Structure, and Strategies of Control: A Replication of the Aston Study," *Administrative Science Quarterly* 17 (1972), 163–77.

FIGURE 15-5

**A Questionnaire Measure
of Formalization**

Circle your response to each of the following items as they apply to the organization in question.

1. Written job descriptions are available for
   a. operative employees only.
   b. operative employees and first-line supervisors only.
   c. operative, first-line supervisory, and middle management personnel.
   d. operative, first-line supervisory, middle and upper-middle management personnel.
   e. all employees, including senior management.

2. Where written job descriptions exist, how closely are employees supervised to ensure compliance with standards set in the job description?
   a. very loose   d. close
   b. loose   e. very close
   c. moderately close

3. How much latitude are employees allowed from the standards?
   a. a great deal   d. very little
   b. a large amount   e. none
   c. a moderate amount

4. What percentage of nonmanagerial employees are given written operating instructions or procedures for their jobs?
   a. 0–20%   d. 61– 80%
   b. 21–40%   e. 81–100%
   c. 41–60%

5. Of those nonmanagerial employees given written instructions or procedures, to what extent are they followed?
   a. none   d. a great deal
   b. little   e. a very great deal
   c. some

6. To what extent are supervisors and middle managers free from rules, procedures, and policies when they make decisions?
   a. a very great deal   d. little
   b. a great deal   e. none
   c. some

7. What percentage of all the rules and procedures that exist within the organization are in writing?
   a. 1–20%   d. 61– 80%
   b. 21–40%   e. 81–100%
   c. 41–60%

Scoring: For all items, a = 1, b = 2, c = 3, d = 4, e = 5. Add up the score for all seven items. The sum of the item scores is the degree of formalization (out of a possible 35).

AN EXAMPLE
Universities have general rules in place to govern all faculty regardless of what college or department they are in. These rules include requirements to hold class at scheduled times, give exams at scheduled times, and keep regular office hours. This formalization is necessary. Without it, students taking classes in several colleges, like liberal arts and business, would be unable to organize their programs of study and would have no idea of what to expect of their professors.

Types of Formalization. There are three types of formalization—formalization by job, by work flow, and by rules (see Table 15-2). *Formalization by job* is used to set up coordination by work process standardization. It is the process of planning and documenting the sequence of steps employees must take to perform their jobs. For example, a fast-food restaurant may prepare procedures manuals that specify how long employees should cook each type of food they serve, what condiments they should use to flavor it, and how they should package it for the customer. Likewise, a bank may develop detailed job descriptions that specify steps tellers must follow to cash checks, deposit funds, accept loan payments, and so forth.

Formalization by workflow is the process of establishing standards, or goals, for the flow of work in a firm. It provides the underpinnings for output standardization. An example is the posted monthly sales goals that insurance sales representatives are expected to achieve. Another example would be a set of standards for display-screen brightness, keyboard responsiveness, and exterior appearance prepared for workers assembling laptop personal computers. They can use these standards to assess the quality of their output.

TABLE 15-2
The Three Types of Formalization

TYPES OF FORMALIZATION	DEFINITION
Formalization by job	Planning and documentation of the details of task performance, such as the specific steps to be taken and the sequence of those steps
Formalization by work flow	Planning and documentation of work-flow standards, such as quality specifications and daily output goals
Formalization by rule	Planning and documentation of general workplace rules and procedures

In an effort to build sales and market share, Burger King CEO Barry Gibbons streamlined the chain's management structure and changed the formal rules by which employees do their jobs—that is, the way they prepare and serve their customers. Responding to consumer demand for more nutritional foods, Gibbons changed some ingredients—french fries are now cooked in 100% vegetable oil—and created new dishes—a broiled-chicken sandwich. Source: Gail deGeorge, "Can Barry Gibbons Put the Sizzle Back in Burger King?" Business Week, October 22, 1990.

In *formalization by rules*, general rules and procedures are planned and documented to govern all the members of an organization irrespective of the specific jobs they perform. For instance, everyone in an organization might be subject to the rule that lunch breaks are to be taken between noon and 1:00 P.M. You can find other examples of this type of formalization in the work-rules sections of collective-bargaining contracts or in the policy manuals that many organizations prepare to guide employees' conduct.

Formalization by rules is closely associated with both work flow and output standardization. Because they help restrict the range of activities likely to occur in the workplace, general rules streamline the task of specifying appropriate task behaviors or outputs.

Problems and Substitutes. Despite the fact that formalization contributes to the reduction of coordination costs by making standardization possible, it can also lessen efficiency in several ways. For one thing, the very existence of rules can encourage the practice of following them to the letter. Some employees may interpret rules that were intended to describe *minimally* acceptable levels of performance as describing the *maximum* level of performance for which they should aim. As a result, their performance may suffer. For example, the rule that tables in a self-serve restaurant must be wiped clean at least once every hour may have the unintended consequence of curbing more frequent cleaning.

In addition, rigid adherence to rules and regulations can discourage initiative and creativity among workers, and the organization can lose its ability to adapt to new or changing conditions. For instance, rules requiring lengthy approval reviews for even minor design changes limited the ability of American firms to improve existing products or introduce new ones in the consumer-electronics market of the 1980s. As a result, American companies like General Electric and Sunbeam are no longer major manufacturers in markets for everything from hair-curling irons to stereo receivers.

Formalization can also undermine performance by narrowing the scope of workplace activities to the point where employees become bored. If boredom is unrelieved, it can lead to dissatisfaction and rebelliousness. Groups of workers may develop informal social structures in which low productivity is the norm. Employees may even turn to dangerous horseplay or costly sabotage to break up the monotony or to get even with a company they perceive as insensitive and uncaring.

Formalization can become an instrument of political maneuver. That is, it can be used to develop rules, regulations, and standards that favor the interests of some groups over others, thus maintaining the power of the favored groups. The individuals or groups who make an organization's rules can write rules that assure their power. They can require the continued use of resources or knowledge

561

that only they themselves can provide. Moreover, they can develop regulations that force the organization as a whole to move in the direction they prefer. To the extent that this kind of gamesmanship distracts from productivity, organizational efficiency is impaired.

There are several substitute measures that managers can take to avoid problems like the foregoing and yet coordinate by means of standardization. Those measures are professionalization, training, and socialization, all of which concern the possession of specific skills (see Table 15-3). In **professionalization**, managers hire people to perform certain work for which useful written specifications do not exist and in some cases cannot be prepared.[13] Such **professionals** are people who develop work-related knowledge, skills, and abilities in training programs conducted outside an employing organization. For example, teachers learn how to teach in schools of education, and medical doctors acquire their skills in medical schools. Similarly, business professionals develop their expertise in business programs in colleges and universities. Professionalization is one way of establishing skill standardization.

Professional skills are both portable and nonsubstitutable. That is, they can be employed in a variety of different organizational circumstances to perform tasks that lie beyond the abilities of people who are not themselves professionals. Professional skills can form the basis for the coordination of work, specifically by means of skill standardization. Because the professional learns the rules and standards of conduct needed to perform her job during the professional-training process, further formalization may not be required.

When the knowledge and skills needed to perform the work of an organization can be acquired within the organization itself, **training** (see Chapter 9) may be an effective substitute for formalization. Such training, provided by the employing organization, is purposely organization specific and often job specific. No attempt is made to teach the trainee the sort of generalized code of conduct that professionals learn. When training can be conducted on the job rather than in a formal program of instruction, it often does not require the use of written documentation. As with professionalization, training enables us to coordinate by means of skill standardization but without formalization by job, work flow, or rules.

Finally, organizations can use *socialization* (see Chapter 10) to teach employees, particularly newcomers, the norms of the organization. To the extent that these norms regulate behavior and activities required to coordinate the flow of work, coordination by norm standardization can be enacted without formalized written rules and procedures.

---

[13] Richard H. Hall, "Professionalism and Bureaucratization," *American Sociological Review* 33 (1968), 92–104; and Jerald Hage and Michael Aiken, "Relationship of Centralization to Other Structural Properties," *Administrative Science Quarterly* 12 (1967), 72–91.

professionalization The use of professionals to perform work for which useful written specifications do not exist and in some cases, cannot be prepared.

professionals People who develop work-related knowledge, skills, and abilities in training programs conducted outside an employing organization. Examples include teachers, lawyers, doctors, and managers trained in schools of business.

training The process of teaching organization-specific and, often, job-specific skills on the job or in a formal program sponsored by the employing organization.

APPLYING THE DIAGNOSTIC MODEL
Diagnostic Question 5: If standardization is present, is the appropriate type of formalization or substitute for formalization—professionalization, training, socialization—also present?

TABLE 15-3 Substitutes for Formalization	
SUBSTITUTE	**DEFINITION**
Professionalization	The use of professionally trained people whose abilities, knowledge, and skills equip them to perform work for which written specifications have not been developed
Training	Teaching an organization's employees skills needed to perform specific jobs within the organization
Socialization	Teaching new employees the norms of the organization

At Singapore Airlines, one of the world's youngest and most profitable air carriers, learning how to deal with emergencies is an important part of cabin attendants' skills training. Source: Louis Kraar, "Ten to Watch Outside Japan," Fortune, March 25, 1991, p. 26.

Specialization

specialization The division of an organization's work into specialist jobs of narrow scope and limited variability.

Just as formalization (or the substitutes discussed above) precedes standardization, **specialization** usually follows from its use. Specialization refers to the way an organization's work is divided into individualized tasks. In some organizations, everyone performs the same sort of *generalized* tasks. In others, employees perform *specialized* tasks that differ from each other in significant ways. The questionnaire measure shown in Figure 15-6 captures this distinction, asking you to indicate which functions in an organization are carried out by specialists.

Specialization can be seen in the scope or variety of activities included in the employees' jobs. The higher the degree of specialization, the narrower the scope of each job's activities.[14] Assembly-line work is extremely specialized, because each worker is responsible for only a small task—attaching a chrome strip to the door of a car, placing the label on a bottle of liquid detergent, putting an assembled calculator in a shipping box. At the opposite extreme, being the owner and sole employee of a small company is not a specialized task because the same

[14] Mintzberg, *Structuring of Organizations*, p. 69.

FIGURE 15-6

A Questionnaire Measure of Specialization

Check each of the following activities that is carried out by at least one full time person in your organization.

Strategic planning
Production planning
Inspection/quality control
Buying
Inventory control
Research and development
Maintenance
Computing
Transport
Industrial engineering
Operations research

for all 22 items. The sum of the item scores

horizontal specialization The type of specialization in which the work performed at a given hierarchical level is divided into specialized jobs. An example is to divide secretarial work into the jobs of typist, receptionist, and file clerk.

AN EXAMPLE: NCR, MOTOROLA

Personnel departments provide a range of specialization. Typically in smaller companies personnel specialists perform a broad range of tasks. A few people do recruiting, training, compensation, and the like. These people share the tasks and pitch in where needed. In large organizations like NCR and Motorola we see highly specialized personnel departments. They have compensation specialists who deal only with pay and benefits. Training specialists just train. Often times these departments get so specialized that they become fragmented, and the various functions don't share important information. For example, training doesn't get information from those responsible for performance evaluations as to the type of training needed and who specifically needs training.

vertical specialization The type of specialization in which the management of work is separated from the performance of that work. It establishes the number of levels of hierarchy in an organization.

APPLYING THE DIAGNOSTIC MODEL

Diagnostic Question 6: If standardization and formalization are evident in the firm, are jobs horizontally specialized? Is vertical specialization also present? Does each type of specialization seem balanced in relation to the other?

FIGURE 15-7
Horizontal Specialization

person is responsible for doing all the buying, bookkeeping, selling, and other jobs required to keep the company in business.

Types of Specialization. There are two types of specialization—horizontal and vertical. **Horizontal specialization** refers to the way the work to be performed in each hierarchical level (see Chapter 2), or horizontal "slice," of an organization is divided into discrete, individualized jobs. At one extreme, or in *low horizontal specialization*, the work at a particular hierarchical level is distributed among workers as generalist jobs. The first panel of Figure 15-7 depicts an office arrangement in which filing, typing, and telephone-answering duties are distributed equally among three generalist secretaries. All three secretarial jobs are virtually identical, and each person performing one of these jobs can readily substitute for any of the others.

In contrast, at the other extreme, or in *high horizontal specialization*, the work within a hierarchical level is distributed in the form of specialist jobs. This form of specialization is shown in the second panel of Figure 15-7, where each of three employees—a typist, a file clerk, and a receptionist who answers the phone—has entirely separate duties.

Although the same type of work can be performed in both situations depicted in Figure 15-7, higher horizontal specialization has the potential to produce greater productivity.[15] First, people performing horizontally specialized jobs complete the same task activities again and again. This repetition enables them to learn their jobs thoroughly and, over time, to sharpen job-related knowledge and skills. Second, high horizontal specialization substantially reduces the amount of time lost in switching from one task activity to another, because such specialized jobs consist of a limited number of different activities. Third, because it is easier to identify and analyze a smaller number of critical task activities, high horizontal specialization makes it easier to develop new methods and new equipment. It is the potential for high productivity associated with these three benefits that makes horizontal specialization attractive to managers.

The other type of specialization, **vertical specialization**, describes the division of an organization into hierarchical levels. As you can see in Figure 15-8, the higher the degree of specialization, the more vertical layers an organization's hierarchy of authority contains—and the greater the separation of the management of a task from its performance. The upper panel of Figure 15-8 illustrates low vertical specialization. With only a single managerial layer, much of the actual management of the organization's tasks rests with those who perform them. For instance, workers on the shop floor of a company like Whirlpool or Gaines Pet Food may design their own jobs and decide who will do them. They may also order their own raw materials, set their own work hours, and even hire and fire coworkers. In the second hierarchy shown in Figure 15-8, higher vertical specialization has produced several managerial layers. Here, hierarchical superiors rather than the people who actually perform the work generally handle man-

---

[15] For further support of this statement see the discussion of Adam Smith's ideas in Chapter 2.

E ORGANIZATION

FIGURE 15-8
Vertical Specialization

agement activities. Thus decision making about work methods, job assignments, hours of work, and such is taken away from shop-floor employees and made the task of management.

Problems with Specialization. As the level of horizontal specialization rises in an organization, so does the degree of vertical specialization. That is because the more specialized each job is, the less easily an employee working at the base of the company can see the big picture required to coordinate with others, let alone manage the organization as a whole. A first managerial level is added above specialized workers to handle coordination and management tasks. If this new level becomes so specialized that it can no longer develop and pursue organizational goals, a second, third, fourth, and perhaps even fifth managerial level may be added. You can see that achieving productivity through horizontal specialization incurs significant costs in the form of higher management payroll expenses.

Specialization also increases the costs of management in other significant ways. Vertical specialization increases the number of managers, who must then spend increasing amounts of time coordinating with each other to oversee organizational activities. This additional coordination raises costs. Furthermore, horizontal specialization increases task interdependence, because it creates networks of tasks that must be performed together in order to accomplish the desired end result. For example, the typist, file clerk, and telephone receptionist envisioned in Figure 15-7 must cooperate closely in order to complete the same work that any one of the three secretaries in the same figure could complete alone.

Specialization can also force an organization to pay for specialists who may be idle at a given moment but who are expected to be needed at some future time. These carrying costs can add up. Imagine a small hospital that must offer full-time employment to a highly skilled cardiovascular surgeon so that he will be available to perform the three or four operations per month that the hospital requires. The hospital has a choice. Either it must bear the costs of carrying an idle specialist or it must reduce the scope of the services it can offer its patients.

Finally, specialization can sometimes simplify an organization's jobs to the point that they become tiresome and unchallenging. If such oversimplification

-force motivation and, as
is problem in some depth

. It is up to managers to
enefits of standardization
andardization which gen-
alization is a difficult but

f each unit closer together,
rating different units from
e fabricators in our furniture
Then the interdependencies
, finisher, and shipper who
ndaries and cannot be easily
sk, bookcase, and chair units.
bricating, finishing, or ship-
ping—for the different product lines will now find it hard to confer with one
another. As suggested by these examples, the distance between units created by
unit grouping makes it difficult to coordinate interunit linkages with the same
types of mutual adjustment, direct supervision, and standardization used to coordinate activities within units. So although grouping people into units can facilitate coordination within the units, it decreases the ability to coordinate interdependencies among units.

Departmentation

In an effort to cope with interunit coordination problems, managers apply some
of the same principles to the job of designing the overall structure of an organization that are also applied to the task of grouping units together. The result
is two different types of **departmentation** schemes, each of which corresponds
to one of the kinds of unit grouping described earlier.[17]

To understand them, let's consider an organization that consists of four
functional areas—marketing, research, manufacturing, and accounting—and
three product lines—automobiles, trucks, and small gasoline engines. Figure 15-
9 shows the departmentation in this firm. Each box represents one of the four
functions, and each of the horizontal work flow depicts one of the three product
lines. The first type of departmentation, *functional departmentation*, is the equivalent
of performing functional grouping at the top of the firm. All marketing tasks
are combined into a single marketing department, all research tasks are combined
into a single research department, and so forth. As with functional unit grouping,
the result is a form of departmentation that is economically efficient but relatively
inflexible.

In contrast, the second type of departmentation, *divisional departmentation*,
is the same as using market unit grouping at the top of the organization. Instead

departmentation The process of grouping structural units into larger clusters. In functional departmentation, units are grouped into departments according to functional similarities—similarities in the work they do. In divisional departmentation, units are grouped into divisions according to market similarities—similarities in their products or in customers or geographical areas they serve.

POINT TO STRESS
By now it should be clear there is no way to structure a perfect organization. Either flexibility will be sacrificed for economic efficiency or economic efficiency will be sacrificed for flexibility. When a firm must accomplish complex tasks it is impossible to optimize on both counts.

[16] Glenn R. Carroll, "The Specialist Strategy," *California Management Review* 26 (1984), 126–37.

[17] Pradip N. Khandwalla, *The Design of Organizations* (New York: Harcourt, Brace, Jovanovich, 1977), pp. 489–97; and A. Walker and Jay Lorsch, "Organizational Choice: Product versus Function," *Harvard Business Review* 46 (1968), 129–38.

Types of Departmentation

A comparison between the diagrams in this figure and the ones shown in Figure 15-1 shows how unit grouping and departmentation involve the same sort of logic applied at different organizational levels.

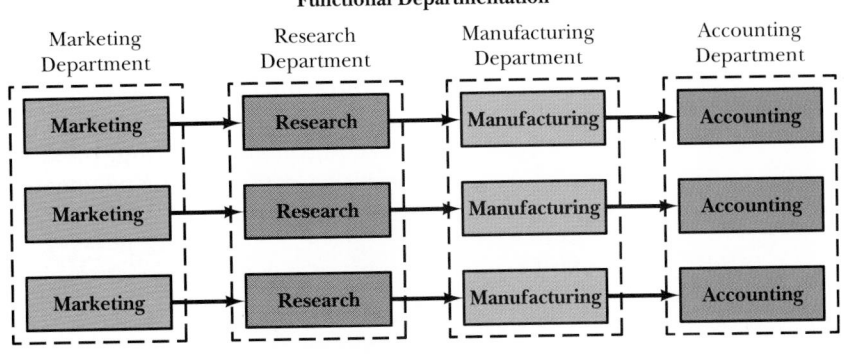

Functional Departmentation

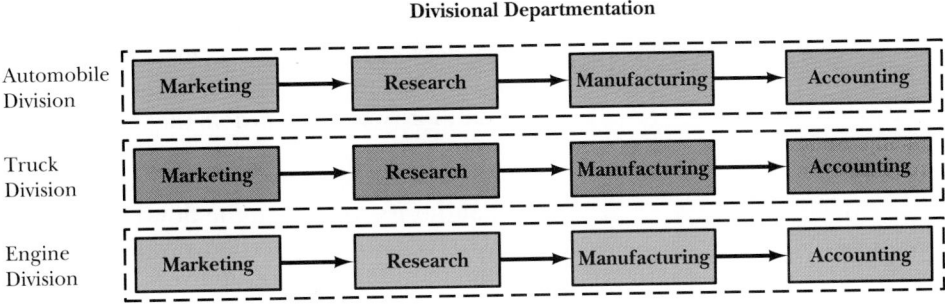

Divisional Departmentation

APPLYING THE DIAGNOSTIC MODEL

Diagnostic Question 7: Is the firm's structure based on functional or divisional departmentation? What does this tell you about the probable balance between efficiency and adaptability in the organization?

AN EXAMPLE: KODAK

It might seem risky to locate your research and development department thousands of miles from your headquarters. But Kodak, which is headquartered in Rochester, N.Y., found that by building a research center in the Tokyo suburbs it was able to keep better in touch with Japanese competitors' new developments and it could compete more successfully for scientific and engineering talent in Japanese universities and firms. Kodak's Tokyo facility is the center for some of its most advanced research, such as growing crystals for making computer microchips and developing a system of sending fax images from one computer to many locations simultaneously. ("Technology," *Fortune*, March 25, 1991, pp. 88–92.)

of being clustered into marketing, manufacturing, research, and accounting departments, the organization's activities are grouped into product divisions—an automobile division, a truck division, and a gasoline engine division. When an organization's clients differ more than its products, the organization's work may be grouped according to differences in the clients served. For instance, there might be a military contracts division, a wholesale distribution division, and an aftermarket parts division. In a third alternative, when an organization's operations are spread throughout the world, its parts may be geographically grouped into a North American division, an Asian division, and a European division. In any of these forms, the organization possesses division-by-division flexibility. Each division can tailor its response to the particular demands of its own market. For example, Ford's Lincoln-Mercury division can decide to redesign its luxury market automobiles to be more Mercedes-like without having to worry about Ford products and markets. The economic efficiency of functional departmentation is sacrificed, however, because effort is duplicated across the organization's three product lines. Lincoln-Mercury's product design studios duplicate Ford's, but the two divisions' studios cannot be consolidated without losing divisional flexibility. So as with unit grouping, managers making departmentation decisions must grapple with a tradeoff between economy and flexibility.

By clustering related groups together, departmentation accentuates similarities that facilitate the management of intergroup relations. Specifically, in an organization structured around functional departmentation, groups in the same department share the same specialized knowledge, language, and ways of looking at the company's business. For instance, all the members of a marketing department share the same general marketing know-how. They talk about things like market segmentation and market share and generally agree that the best way to ensure their company's success is by appealing to customer needs. A manager charged with coordinating different units in the marketing department can base her actions on this common knowledge, language, and viewpoint despite the fact that she is dealing with several different groups of employees. She can manage the different groups using the same basic management approach.

Similarly, in an organization structured around divisional departmentation, groups in the same division share interests in the same basic line of business. Thus all employees in the truck division of a company like General Motors or Ford are concerned about doing well in the truck industry. This commonality allows the manager of a division to treat groups performing different functions—marketing, manufacturing, research, and so forth—in much the same way. He doesn't have to tailor management practices to the functional specialty of each particular group.

Hierarchy and Centralization

By grouping units together, divisionalization creates clusters of units and a layer of managers having responsibility for the activities of particular clusters. Hierarchy (see Chapter 2) can then be used to control intergroup relations. Specifically, a manager having hierarchical authority over a particular cluster of units can use this authority to issue orders that, when followed, will help coordinate activities among those units. For instance, the manager having hierarchical authority over all the manufacturing operations of the furniture factory in Figure 15-1 can use that authority to smooth the flow of work among units if functional grouping is used to form them. Alternatively, the manufacturing manager can help facilitate communication among employees performing similar functions on different lines if market grouping is used. In turn, interdependencies that span different clusters of units can be coordinated by managers higher in the organization's hierarchy. For example, problems between the manufacturing department of our furniture company and other departments, such as sales, accounting, or personnel, can be dealt with by the executive responsible for overseeing the various department managers. Hierarchical authority, then, can be used to coordinate intergroup relations among units in much the same way that direct supervision is used to coordinate interpersonal relations within units.

The use of hierarchy as an intergroup coordination mechanism differs from one organization to the next as to the level of managers—top, middle, or supervisory—who have the ultimate authority to make the decisions and issue the orders that coordinate intergroup activities and direct the organization's overall progress. Left to their own devices, many top managers in the United States favor **centralization**, the concentration of authority and decision making at the top of a firm.[18] Centralization affords top managers a high degree of certainty. Because they alone make the decisions in centralized firms, they can be sure not only that decisions are made but that they are made in accordance with their own wishes. In addition, centralization can minimize the time needed to make decisions. That is because only an extremely limited number of people are involved in the decision-making processes. In Figure 15-10, a questionnaire measure of centralization illustrates the kind of authority and decision making that can be centralized. This measure can be used to assess the amount of centralization in an organization.

Despite centralization's appeal to top management, **decentralization**, is increasingly common in modern organizations (see the "In Practice" box). In decentralization, authority and decision making are dispersed downward and outward through the hierarchy. Several factors push otherwise reluctant top

centralization The concentration of authority and decision making at the top of an organization; the opposite of decentralization.

decentralization The dispersion of authority and decision making downward and outward through the hierarchy of an organization; the opposite of centralization.

[18]Pugh et al., "Dimensions of Organization Structure," p. 72; J. Hage and M. Aiken, "Relationship of Centralization to Other Structural Properties" *Administrative Science Quarterly* 12 (1967), 72–92; Peter Blau, "Decentralization in Bureaucracies," in *Power in Organizations*, ed. Mayer N. Zald (Nashville, Tenn.: Vanderbilt University Press, 1970), pp. 42–81; N. M. Carter and J.B. Cullen, "A Comparison of Centralization/Decentralization of Decision Making Concepts and Measures," *Journal of Management* 10 (1984), 259–68; and Roger Mansfield, "Bureaucracy and Centralization: An Examination of Organizational Structure," *Administrative Science Quarterly* 18 (1973), 477–78.

FIGURE 15-10

A Questionnaire Measure of Centralization

Circle your response to each of the following items as they apply to the organization in question.

1. How much direct involvement does top management have in gathering the information input that they will use in making decisions?
   a. none
   b. little
   c. some
   d. a great deal
   e. a very great deal
2. To what degree does top management participate in the interpretation of the information input?
   a. 0–20%
   b. 21–40%
   c. 41–60%
   d. 61– 80%
   e. 81–100%
3. To what degree does top management directly control the execution of the decision?
   a. 0–20%
   b. 21–40%
   c. 41–60%
   d. 61– 80%
   e. 81–100%

For questions 4 through 10, use the following responses:
   a. very great
   b. great
   c. some
   d. little
   e. none

4–10. How much discretion does the typical first-line supervisor have over
   4. Establishing his or her unit's budget?
   5. Determining how his or her unit's performance will be evaluated?
   6. Hiring and firing personnel?
   7. Personnel rewards (i.e., salary increases, promotions)?
   8. Purchasing of equipment and supplies?
   9. Establishing a new project or program?
   10. How work exceptions are to be handled?

Scoring: For all items, a = 1, b = 2, c = 3, d = 4, e = 5. Add up the score for all ten items. The sum of the item scores is the degree of centralization (out of a possible 50).

managers toward its implementation. First, some decisions require top managers to consider a great deal of information. The managers may become overloaded by the task of processing all this information and therefore find it useful to involve more people in the decision-making process. Second, decentralization may be stimulated by a need for flexibility. If local conditions require that different parts of an organization respond differently, managers of those organizational units must be empowered to make their own decisions. Third, decentralization may be useful in dealing with employee motivation problems if those problems can be solved by according employees control over workplace practices and conditions (we will discuss this further in Chapter 17). In any of these cases, the failure to decentralize can undermine attempts to coordinate intergroup relations.

APPLYING THE DIAGNOSTIC MODEL
Diagnostic Question 8: Are intergroup relations adequately coordinated by hierarchy and centralization, or are additional intergroup coordination mechanisms needed?

Intergroup Coordination Mechanisms

Even when an organization is suitably centralized or decentralized, sometimes it is impossible to coordinate all the interdependencies among an organization's units by means of departmentation and hierarchy alone. One way of dealing with the need for additional coordination among units is to reduce it. Decoupling mechanisms accomplish this by severing connections between groups. When the number of connections cannot be reduced, unit-linking mechanisms can be used instead to promote the exchange of information among interdependent work units.[19] Table 15-4 previews the mechanisms that we will discuss next.

[19] Galbraith, *Designing Complex Organizations*, 14–18.

Changes at IBM

Apple was not the only computer company to restructure itself during the 1980s. IBM, the world's leading computer manufacturer, also reorganized its 387,000 employees. Unlike Apple, however, IBM moved toward decentralization.

What triggered this change? IBM had lost sight of computer customers and their changing tastes. Where large mainframe computers—IBM's specialty—had once dominated the industry, business customers now wanted networks of smaller, more powerful desktop models. Sales were soaring in personal computers, a market segment IBM had joined only as a reluctant latecomer. The company had not even tried to develop leading products in the laptops and work-station markets, two areas of high future growth.

John Akers, IBM's chief executive officer, recognized that their problems arose from a common source, the company's tendency to make all important decisions in its Armonk, New York, corporate office. There, a corporate management board of 18 senior executives relied on experience and consensus to decide about everything from advertising campaigns to new directions in product development. This centralization blinded the company to many important changes in the computer industry. It also created a tendency to respond to those few changes the management board did learn about by in-sisting on doing things "the IBM way"—which meant remaining the same mainframe-producing giant.

In IBM's new structure, the firm is now composed of nearly autonomous business divisions. Each year, division managers negotiate business plans with Akers and the board and then go off to run their divisions. Where managers once took conflicts between divisions to the board, now they meet among themselves to solve interdivisional problems. Problems that resist resolution are referred only one level up. Peer negotiation has replaced direct orders. Moreover, the customer now has the company's undivided attention, and divisions can respond to customer demands without referring up the hierarchy to top management. As a result, the corporate management board is spending two-thirds less time with month-to-month management issues (leaving more time for the sort of long-term strategic management we will discuss in Chapter 18). Customers are once again getting products and services that meet their needs.* IBM is again returning to its former position of prominence.

* Joel Dreyfuss, "Reinventing IBM," *Fortune*, August 14, 1989, pp. 30–39.

TEACHING NOTE
Many companies have improved productivity by implementing just-in-time (JIT) programs that increase interdependence between work groups. Slack resources have been reduced or eliminated, lowering inventory carrying costs. Ask students what this suggests about coordinating work efforts (*Managing the Design-Manufacturing Process*, 1990).

TABLE 15-4
Intergroup Coordination Mechanisms

MECHANISM	ACTION
Decoupling Mechanisms	
Slack resources	Lessen the ability of one unit to affect the activities of another interdependent unit by creating buffer inventories.
Self-contained tasks	Redesign work formerly performed by two or more interdependent units and assign it to new groups composed of representatives from all the original units.
Unit-linking mechanisms	
Vertical information systems	Facilitate the transfer and processing of large amounts of information throughout an organization's hierarchy of authority.
Lateral linkage devices	Facilitate the use of unit-coordination mechanisms in handling intergroup dependencies.
Liaison position	Make one individual responsible for ensuring that communications flow freely and directly between interdependent groups.
Representative groups	Create task force or standing committee in which representatives from interdependent groups coordinate by means of mutual adjustment.
Integrating managers	Create a managerial position that spans interdependent groups and has the authority to issue orders and expect obedience.
Matrix structure	Combine functional and divisional departmentation; use only as a last resort.

decoupling mechanisms
Mechanisms that regulate intergroup coordination by making work units less dependent on each other. The use of decoupling mechanisms involves making adjustments to the relationships formed among units during departmentation.

slack resources The type of decoupling mechanism in which groups are separated from each other by buffer inventories.

APPLYING THE DIAGNOSTIC MODEL
Diagnostic Question 9: If additional coordination is needed, is decoupling feasible? If not, which unit linking mechanism or mechanisms should be employed?

self-contained tasks The type of decoupling mechanism formed by combining the work of two or more interdependent units and assigning their work to several independent work units. Those units are then staffed by people drawn from each of the original units.

unit-linking mechanisms
Mechanisms that regulate intergroup coordination by linking interdependent units more closely together.

Decoupling Mechanisms. Two **decoupling mechanisms**—slack resources and self-contained tasks—regulate intergroup coordination by making work units less dependent on each other. Both these approaches involve adjustments to the relationships formed among units during departmentation.

Slack resources help to decouple otherwise interconnected groups by creating buffers that lessen the ability of one group to affect the activities of another. For example, suppose one group assembles telephone handsets, and another group connects finished handsets to telephone bodies to form fully assembled units. The two groups are sequentially interdependent, because the second group's ability to perform its work is contingent on the first group's ability to complete its task (see Chapter 10 for a more complete definition of sequential interdependence). Work in the second group comes to a halt if the handset-assembly group stops producing. If, however, we create a buffer inventory—a supply of finished handsets—that the second group can draw on when the handset-assembly group is not producing anything, we have at least temporarily decoupled the two work groups. There are various ways to create and replenish this inventory: (1) delay the second group's startup, (2) have the handset-assembly group work extra hours, (3) form an additional group of handset assemblers from time to time, or (4) buy extra handsets from another source.

Another way of reducing interdependence among groups is to create **self-contained tasks** by combining the work of two or more interdependent units and assigning this work to several independent work units. Typically, such self-contained units are staffed by employees drawn from each of the original interdependent units. For example, engineering and drafting units might have problems coordinating engineering specifications and the drawings produced by the drafting unit. These two units might be regrouped into several independent engineering-drafting units. Each one can produce product specifications and drawings without outside assistance. After this regrouping, the original two units would no longer exist. Similarly, a marketing research unit and a sales unit might be merged into a single marketing unit to coordinate information about a firm's market more effectively. In both these examples, key interdependencies that lie outside the original units are contained within redesigned units. They can then be handled by one or more of the unit coordination mechanisms we have already discussed.

Unit-Linking Mechanisms. Sometimes concerns about minimizing inventory costs rule out the use of slack resources. Among U.S. manufacturers, for instance, the cost of carrying excessive inventory is a growing concern and has stimulated

IBM used to store its parts and equipment in some 21 warehouses of its own, but in 1989 it worked out an arrangement with Federal Express that is a good example of a self-contained task. *Now Federal Express stores all IBM's materials, and an IBM employee can call just one telephone number rather than several in order to locate the part she needs. This consolidation has reduced the need for workflow coordination without endangering crucial inventory levels.*
Source: Thomas A. Stewart, "There Are No Products—Only Services," Fortune, *January 14, 1991, p. 32.*

increasing interest in just-in-time (JIT) procedures. Using JIT, items are produced for use only when needed, eliminating the cost of having unused inventory lying around. In addition, work often cannot be divided into self-contained tasks. For example, the task of producing the parts required to make a car and assembling them into a final product is so immense that many groups (in fact, many companies) must be involved. In such cases, intergroup interdependencies can be coordinated instead by means of various **unit-linking mechanisms**. Vertical information systems facilitate information transmission among different hierarchical levels, and lateral linkage devices foster direct communication between interdependent units.

Vertical information systems usually consist of mainframe computers with remote terminals or networks of personal computers that can be used to input and exchange information about organizational performance. If you have taken courses in computer science, you have probably had experience with a computer network similar to those used in businesses as vertical information systems. Managers use such systems to communicate among themselves and store information for later review. These systems not only facilitate the transfer of large amounts of information up and down an organization's hierarchy of authority. They also make it easier to process this information centrally without overloading normal supervisory processes. Thus these systems make it possible for managers at higher levels of a firm's hierarchy to coordinate intergroup relations. Just as a direct supervisor issues orders to subordinates in order to coordinate activities within a work unit, so hierarchical supervisors of two or more units can use vertical information systems to issue orders and gather the performance feedback needed to coordinate activities among those units. The fact that many organizations have recently added the corporate position of chief information officer (CIO) reflects the growing use of vertical information systems to coordinate organizational activities.

There are times when decoupling mechanisms cannot be used, and centralized information processing is undermined by either the absence or overabundance of timely, relevant information. Even then, managers can use **lateral linkage devices** to foster direct communication between interdependent units. For example, an employee may be assigned a **liaison position** in which he is responsible for seeing that communications flow directly and freely between interdependent units. The liaison position is an alternative to hierarchical communication channels. It reduces both the time needed to communicate between units and the amount of information distortion likely to occur. The person occupying a liaison position has no authority to issue direct orders decreeing coordination but relies instead on negotiation, bargaining, and persuasion. He may also mediate between units in conflict, resolving differences and moving the units toward voluntary intergroup coordination.

The liaison position is the least costly of the lateral linkage devices. Because one person handles the task of coordination, a minimum of a firm's resources are diverted from the primary task of production. Moreover, because the position has no formal authority, it is also the least disruptive of normal hierarchical relationships.

Sometimes a liaison position is not strong enough to solve interunit coordination problems. Managers then have the option of turning to a second unit linking mechanism, **representative groups**, to coordinate activities among interdependent units. Representative groups consist of representatives from the interdependent units, who meet to coordinate intergroup activities. There are two kinds of representative groups. One, called a **task force**, is formed to complete a specific task or project and is then disbanded. By encompassing intergroup

---

vertical information system The type of unit-linking mechanism in which computer networks are used to facilitate managerial communication and information processing. Managers can deal with more coordination information than would otherwise be possible.

lateral linkage devices The types of unit-linking mechanism in which hierarchical links between interdependent units are supplemented by various avenues of mutual adjustment or direct supervision.

liaison position The type of unit-linking mechanism in which one person is made responsible for seeing that communication flows directly and freely between interdependent units. A liaison position has no authority, so its occupant must rely on negotiation, bargaining, and persuasion to move interdependent units toward voluntary coordination.

representative groups The type of unit-linking mechanism in which representatives of interdependent units meet to coordinate intergroup activities.

task force The type of representative group formed to complete a specific task or project and then disbanded.

interdependencies, the task force transforms intergroup coordination problems into intragroup problems so that the mechanism of mutual adjustment can be used temporarily. Companies like Colgate-Palmolive or Procter and Gamble form product task groups. They draw members from advertising, marketing, manufacturing, and product research departments to identify consumer needs, design new products that respond to these needs, and manage their market introduction. Once a new product is successfully launched, the task force responsible for its introduction is dissolved, and its members return to their former units.

standing committee The type of representative group formed to meet on a regular basis to discuss and resolve intergroup coordination problems. A standing committee has no specific task, nor is it expected to disband at any particular time.

The other type of representative group is a more-or-less permanent one. Like the members of the task force, the members of this group, called the **standing committee**, represent interdependent work units, but they meet on a regular basis to discuss and resolve intergroup coordination problems. No specific task is assigned to the standing committee, nor is the committee expected to disband at any particular time. An example of a standing committee is a factory's Monday morning production meeting. At that meeting, representatives from production control, purchasing, quality assurance, shipping, and the company's different assembly groups overview the week's production schedule and try to anticipate coordination problems.

Like task forces, standing committees make it possible to use intragroup mutual adjustment to coordinate intergroup relations. Despite their usefulness in this regard, both these linkage devices are more costly than the liaison position. The reason is that through process loss, group meetings inevitably consume otherwise productive resources. In addition, because representative groups (especially task forces) are sometimes designed to operate outside customary hierarchal channels, they can prove quite disruptive to normal operations.

Occasionally, neither liaison positions nor representative groups are enough to solve interunit coordination problems. The *integrating manager* is a third type of lateral linkage device. Like the liaison officer, the integrating manager mediates between interdependent units, but unlike the liaison officer, the integrating manager has the formal authority to issue orders and expect obedience. He can tell interdependent groups what to do to coordinate their work. Project managers at companies like Rockwell International and Lockheed fill the role of integrating manager. They oversee the progress of a project by making sure that the various planning, designing, assembling, and testing groups work together successfully.

Normally, an integrating manager issues orders only to the supervisors of the units she is coordinating. Giving orders to the people who report to these supervisors would violate the principle of unity of command (see Chapter 2) and would confuse employees. They would feel they were being asked to report to two supervisors. Because an integrating manager disrupts normal hierarchical relationships, shortcircuiting the relationship between unit supervisors and their usual superior, this device is used much less often than either the liaison position or representative groups.

matrix organization structure The type of lateral linkage mechanism that is an organization structure incorporating both functional and divisional departmentation.

Once in a while, even integrating managers cannot provide the guidance needed to coordinate activities among units. In these rare instances, a fourth type of lateral linkage device, called the **matrix organization structure**, is sometimes employed. Matrix structures are the most complicated of the mechanisms used to coordinate intergroup relations and are extremely costly to sustain. They incorporate both functional and divisional departmentation and coordinate intergroup linkages through a complex network of mutual adjustment. We will discuss the matrix organization structure in greater detail in Chapter 16, because it is both a coordination mechanism and a specific type of organization structure. Now we will just say that matrix structures are appropriate only when all other intergroup coordination mechanisms have proven ineffective.

SUMMARY

An *organization's structure* is a relatively stable network of interdependencies among the people and tasks that make up the organization. It is created, first, by the process of *unit grouping* in which the organization's members are grouped together into structural units on the basis of similarities in what they perform, called *functional grouping*, or similarities in the products they produce or markets they serve, called *market grouping*. Within these units, the *unit coordination mechanisms* of *mutual adjustment, direct supervision*, or *standardization* are used to manage interdependence. Along with several other factors, the type of coordination mechanism used in a unit affects the optimum size of the unit.

Using standardization to coordinate within units encourages the emergence of two other structural characteristics, *formalization* and *specialization*. Professionalization, training, or socialization sometimes substitute for formalization. Organizations are also structured by mechanisms that coordinate relations among units. Those are *departmentation* (either functional or divisional), *centralization*, and various *decoupling mechanisms* or *unit-linking mechanisms*.

APPLYING THE DIAGNOSTIC MODEL

Diagnostic Question 10: Across the array of structural characteristics that you have identified does a general profile stressing efficiency emerge or does a profile stressing flexibility and adaptability appear instead?

REVIEW QUESTIONS

1. Given that an organization's structure integrates and differentiates activities in the organization, tell which of the following structural characteristics provide integration and which produce differentiation: unit grouping, unit coordination mechanisms, formalization, specialization, departmentation, hierarchy and centralization.

2. With respect to unit grouping, why is market grouping more flexible than functional grouping? If your company sold pencils, pens, and notebook paper, which type of unit grouping would provide the greatest benefit? Why?

3. Explain why standardization requires stability. Why is mutual adjustment so much more flexible? How does direct supervision fit between the two extremes? What mechanism(s) would you use to coordinate a television-assembly unit of 50 employees? Six custom jewelry makers? A dozen door-to-door magazine salespeople? Why?

4. Suppose you want to increase the size of a work unit that is already having coordination problems. What can you do to ensure that the larger group will be adequately coordinated?

5. Draw and explain a diagram showing how standardization, formalization, and specialization are interrelated. If an organization decided to institute participatory decision making, thereby replacing standardization with mutual adjustment, what effects does your diagram suggest this plan would have on the organization's structure?

6. Explain how professionalization, training, and socialization can act as supplements or substitutes for formalization. Name some other purposes these three processes serve in organizations.

7. What kinds of departmentation can be used to cluster structural units together? How do departmentation and hierarchy work together to resolve coordination problems among structural units?

8. How do decoupling mechanisms reduce interunit coordination problems? How does this approach differ from that of unit-linking mechanisms?

DIAGNOSTIC QUESTIONS

You can use what you have learned in this chapter to identify the specific characteristics of an organization's structure and to gain insight into the probable strengths and weaknesses of that structure. The following questions are provided to assist this diagnostic process.

1. Are structural units grouped according to functional or market similarities? What does that tell you about the relative emphasis placed by the firm on efficiency? On adaptability?

2. Which of the three unit coordination mechanisms are used in the organization's units? Which is used as the primary means of coordination, and what does that tell you about the importance the firm accords to efficiency? To flexibility?

3. What secondary coordination mechanisms back up the primary mechanisms used throughout the firm? Are unit activities coordinated adequately?

4. Are the sizes of unit groups consistent with the types of unit coordination mechanisms being used? With the tasks of unit members and supervisors? With the units' positions in the organizational hierarchy? With needs for undistorted vertical information flow?

5. If standardization is present, is the appropriate type of formalization or substitute for formalization—professionalization, training, socialization—also present?

6. If standardization and formalization are evident in the firm, are jobs horizontally specialized? Is vertical specialization also present? Does each type of specialization seem balanced in relation to the other?

7. Is the firm's structure based on functional or divisional departmentation? What does that tell you about the probable balance between efficiency and adaptability in the organization?

8. Are intergroup relations adequately coordinated by hierarchy and centralization, or are additional intergroup coordination mechanisms needed?

9. If additional coordination is needed, is decoupling feasible? If not, which unit-linking mechanism or mechanisms should be employed?

10. Across the array of structural characteristics that you have identified does a general profile stressing efficiency emerge or does a profile stressing flexibility and adaptability appear instead?

EXERCISE 15-1

RESTRUCTURING A UNIVERSITY*

ERIC PANITZ, *University of Detroit*

How to group units into departments or divisions is a key issue facing managers as they structure an organization. As we have discussed in this chapter, such grouping is accomplished by combining units together according to either what they do or what they produce.

Concerns about administrative overhead can also influence choices about how to form units and combine them together. Administrative overhead is the cost of supervising the units and departments or divisions in a firm. It consists largely of the salaries paid to managers employed as supervisors. Sometimes it is possible to group units together in such a way that fewer managers are needed.

Like other organizations, universities have structures that consist of units (called departments) grouped together into large clusters (called colleges). Administrators who manage educational institutions like yours must choose what sorts of groups to form. In the process, they must often consider tradeoffs affecting administrative overhead, asking themselves which sort of grouping will minimize the number of people needed to manage organizational activities. In this exercise, you will experience some of the problems and make some of the decisions that the management of an educational institution must make in this process.

STEP ONE: PRE-CLASS PREPARATION

In class you will work with other class members to restructure a university. In preparation for this exercise, session, read the following description of Midwestern State University and then familiarize yourself with the rest of the exercise. Next, list two or more structural modifications that could be made to help the university save money:

1. _____

2. _____

3. _____

4. _____

5. _____

Be prepared to explain your suggestions and describe their benefits.

Midwestern State University has a student population of 14,500 undergraduates and 1,500 graduate students, divided among the university's colleges as shown in the table at the bottom of the page.

The university must reduce its budget. All of the "easy" actions such as eliminating travel budgets, leaving open positions unfilled, and trimming operating expenses have been implemented. Now the Vice President for Academic Affairs is considering restructuring the organization to save money. However, the Board of Regents which governs the university has made the following stipulations:

1. The number of currently employed teaching faculty must not be reduced.

2. No increases can be made in the number of administrative personnel or staff members.

3. No employee can be terminated, but employees' positions and responsibilities may be changed.

* Adapted with the author's permission from an exercise entitled "Restructuring the University—An Experiential Exercise." This adaptation copyright © 1992 Prentice Hall.

COLLEGE	UNDERGRADUATE STUDENTS	MASTERS STUDENTS	DOCTORAL STUDENTS
Business Administration	4500	600	29
Natural Sciences and Mathematics	1500	125	38
Humanities and Social Sciences	1750	25	25
Law Enforcement	2000	65	0
Allied Health and Nursing	750	40	3
Education	2800	405	42
Health and Recreation	1200	85	18

4. No existing undergraduate program may be modified so extensively that it would be put in a position of losing its accreditation.

5. Graduate programs are not to be considered during reorganization; they will be integrated into whatever organization results from restructuring the undergraduate programs.

The current structure of the university is shown in Exhibit 15-1. We will describe three colleges—the Colleges of Business Administration, Allied Health, and Natural Sciences—in some detail shortly. In addition to these three colleges, the university has a College of Humanities and Social Sciences, a College of Education, a College of Health and Recreation, a College of Law Enforcement (which also offers first aid courses) and a separate Graduate School. The latter school advises graduate students, approves theses and dissertations, and manages associated paperwork. Graduate courses are taught by faculty housed in the university's seven colleges.

Every college (including the Graduate School) is headed by a Dean who receives a salary of $90,000–$150,000 depending on seniority, job responsibilities, scholarly reputation, and other factors. Each college consists of several departments, and every department is headed by a chairperson. Department chairs receive a 20 percent salary supplement (added to the salary they would normally receive as members of the teaching faculty). In addition, to give them the time they need for their managerial duties, they are relieved of all teach-ing responsibilities. All department chairs are former faculty members who can return to teaching when their services as chairperson are no longer required.

The College of Business Administration

As shown in Exhibit 15-2, the College of Business Administration consists of 5 departments and 13 programs. These programs are summarized next. Not only business majors but many students in the Colleges of Allied Health and Natural Sciences take several business courses as part of their major or minor programs. For example, Health Record Administration majors must take the Principles of Management course.

Department of Business Administration and Marketing. This department includes the management, marketing, general business, and coal mining administration programs. There are a total of 17 faculty, 4 in management, 4 in marketing, 6 in general business, and 3 in coal mining administration.

Management offers three majors: Business Administration trains students to become mid-level managers. Industrial Relations develops skills in human resource management, organization development, and labor relations. Operations Management develops skills in production management, quality control, inventory administration, and operations research.

Marketing offers two majors: Marketing prepares students for positions in sales and sales management,

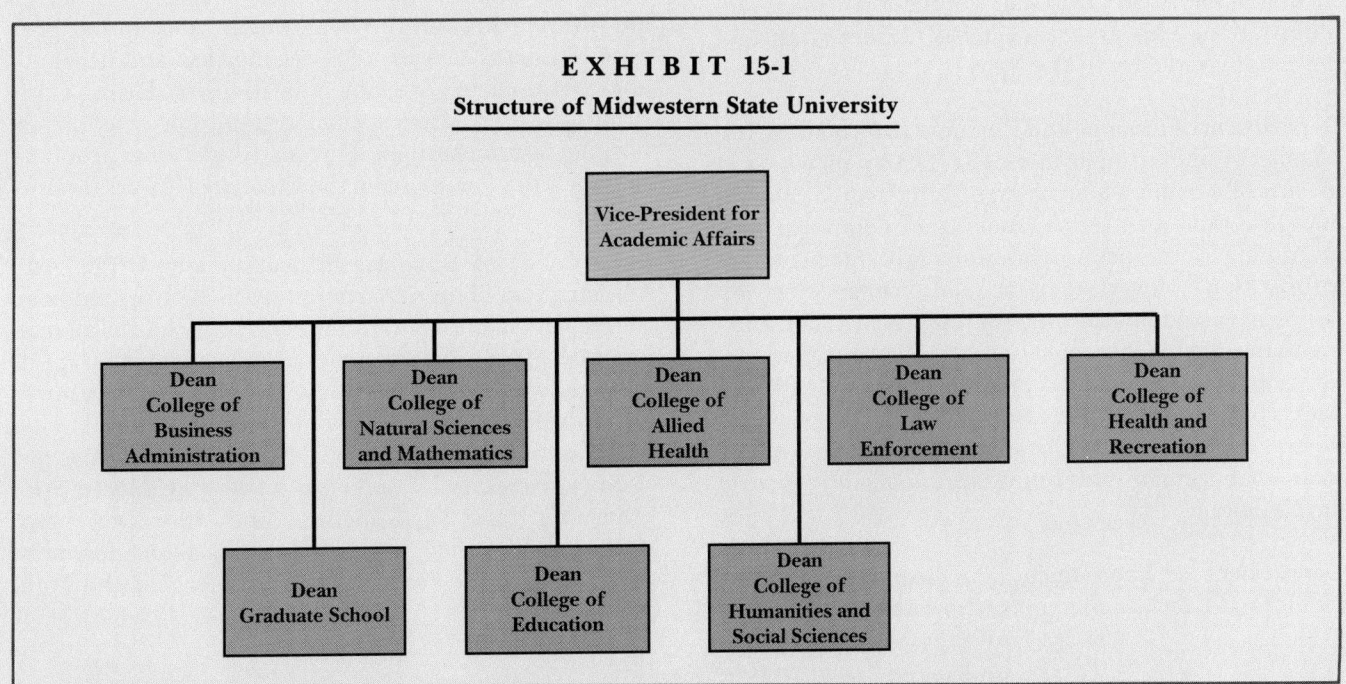

EXHIBIT 15-1
Structure of Midwestern State University

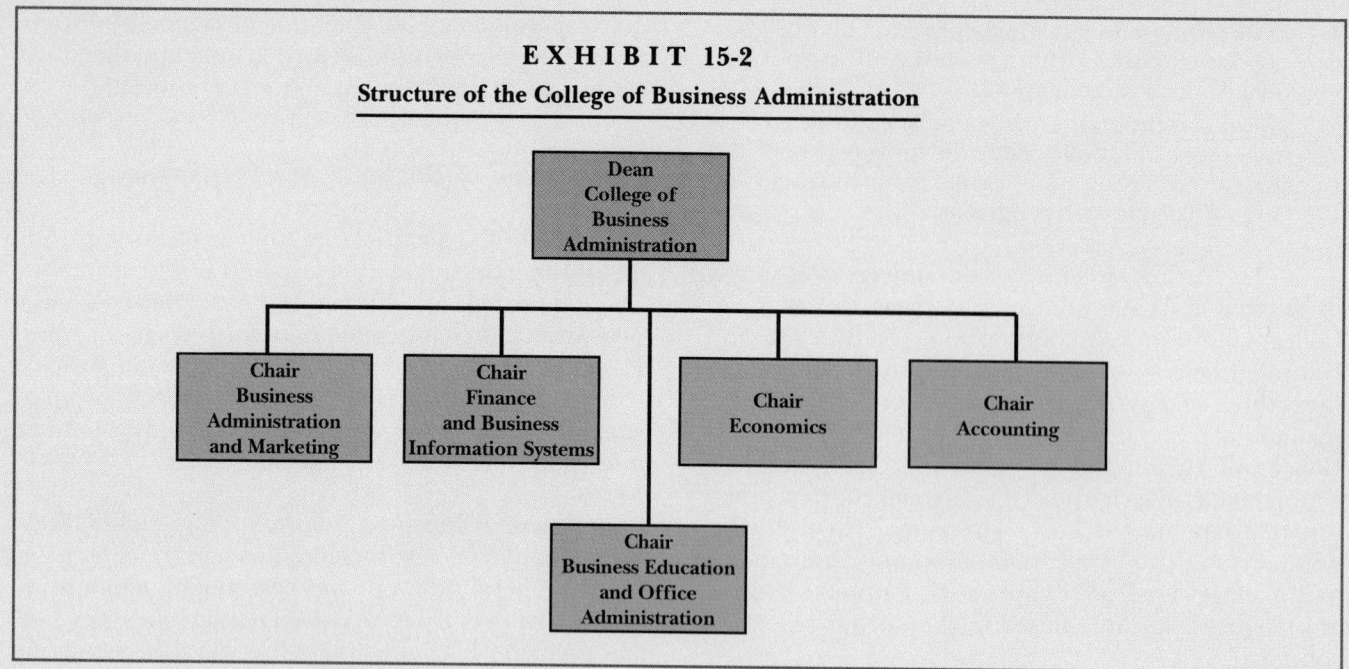

EXHIBIT 15-2

Structure of the College of Business Administration

retailing, market research, promotion, and advertising. Transportation qualifies students for jobs in the field of transportation and physical distribution.

General Business offers courses in law, the capstone policy course, and courses in small business administration. There is also a major in general business for students planning to attend professional or graduate schools.

Finally, *Coal Mining Administration* qualifies students to enter managerial positions in the coal industry such as mine supervision and occupational safety administration. This program reflects the importance of the mining industry in the state.

Department of Finance and Business Information Systems. This department houses 16 faculty members, 4 in each of 4 majors. Management Information Systems and Programming trains students in computer programming for business applications and in management information system development and management. Finance gives students the tools for financial decision making to needed for careers in corporate finance, banking, or investment. Insurance provides students with the background needed for careers in the insurance industry. Real Estate develops students' capabilities in real estate management, marketing, appraisal, and property development.

Department of Economics. The Economics Department has 11 faculty members who teach primarily in-

troductory economics and business statistics courses. Advanced courses are offered to all majors in the College of Business Administration. Economics majors are offered through both the College of Business and the College of Humanities.

Department of Accounting. The Accounting Department includes 11 faculty members who teach the introductory accounting course to all business majors as well as business minors from other colleges. Two majors are offered through the department. Accounting prepares students to seek CPA certification and accountancy positions in government or industry. Health Care Administration offers specific training for positions in hospital administration. This is a cooperative program that includes faculty from the College of Allied Health.

Department of Business Education and Office Administration. This department has 8 faculty members who teach business communications courses for all majors and office management, secretarial, and business education courses to students with the following majors. Office Administration provides the skills required to fill positions as executive secretaries and administrative assistants. Secretarial Training is a two-year degree program that trains legal, medical, and other specialized secretaries in office services. Business Education prepares students to teach business subjects at the high school level.

The College of Natural Sciences and Mathematics

The College of Natural Sciences and Mathematics, diagrammed in Exhibit 15-3, has 6 departments that offer 13 programs. Many of this college's courses are general education requirements that must be completed by all students in the university. In addition, many courses in the biology and chemistry departments serve as preparatory courses for students in the allied health programs (nursing, medical technology, environmental health).

Department of Biology. This department has 17 faculty members and offers 5 majors. General Biology provides an overview of the biological sciences and the component fields of ecology, botany, physiology, biostatistics, entomology, vertebrate and invertebrate biology, and cell biology. Microbiology emphasizes the study of pathogenic and nonpathogenic bacteria, fungi, virology, and parasitology in clinical and nonclinical settings. Wildlife Biology focuses on the management and health of terrestrial wildlife and its environments. Aquatic and Fisheries Biology concerns the management of fisheries and their habitats, including pollution control and other aspects of marine biology. Environmental Resources Biology offers a broad view of economic and environmental aspects of biological resources.

Department of Chemistry. This department has 10 faculty members who offer 2 majors. Chemistry contains coursework in analytical, physical, and organic chemistry and biochemistry. Chemical Technology is a two-year program that prepares students for positions as laboratory technicians.

Department of Geology. This department has 8 faculty members and offers 3 majors designed to prepare students for careers in the petroleum, coal mining, and other related industries, and for teaching assignments at the secondary level. The majors offered are Geology, Earth Science, and a two-year program in Geological Engineering.

Department of Mathematics and Computer Science. This is a 12-member department offering majors in Computer Science, Mathematics, and Statistics. Most of the courses in this department are part of the general education requirement that all students must fulfill. The department has few math majors and is trying to cope with the effects of the specialized courses in statistics offered by other colleges such as Business and Law Enforcement.

Department of Natural Resources. The Department of Natural Resources has 6 faculty members and offers science courses for non-majors who must take science to fulfill general education requirements. These courses

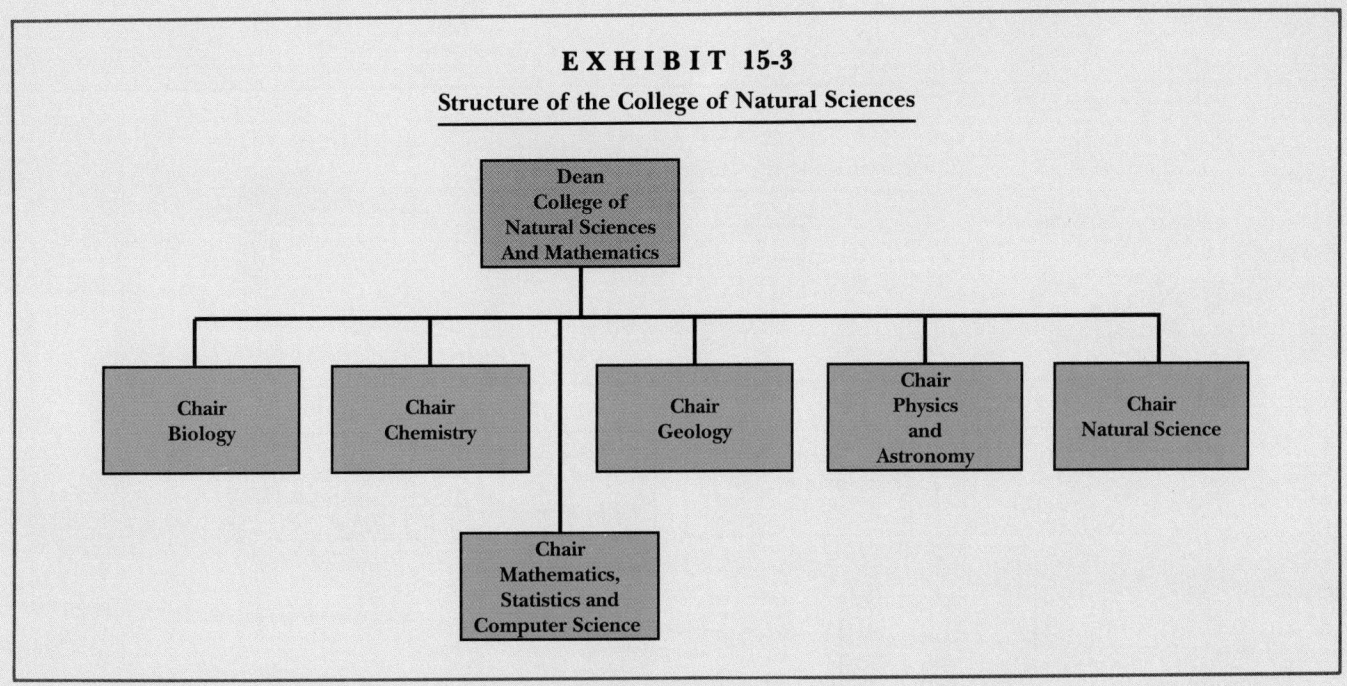

EXHIBIT 15-3

Structure of the College of Natural Sciences

Dean
College of
Natural Sciences
And Mathematics

Chair
Biology

Chair
Chemistry

Chair
Geology

Chair
Physics
and
Astronomy

Chair
Natural Science

Chair
Mathematics,
Statistics and
Computer Science

emphasize the historical development of scientific knowledge and its effects on present-day life.

Department of Physics and Astronomy. This 6-member department offers a physics major and teaches courses on physics and astronomy that are listed as general education requirements.

The College of Allied Health and Nursing

The College of Allied Health and Nursing, shown in Exhibit 15-4, has 37 faculty members and offers 14 programs in the departments of emergency medical technology, environmental health, health records administration, medical technology, occupational therapy, and nursing.

Department of Emergency Medical Technology (EMT). This three-member department offers one-year programs leading to certification in EMT and advanced EMT, and a two-year program leading to an EMT Services Management degree. Students are trained in techniques and management of ambulance services and accident management. Similar courses are offered in the College of Law Enforcement, but the Law program does not include a certification or degree program in this area.

Department of Environmental Health. The Department of Environmental Health has three full-time and one part-time faculty member. It offers a program in applied biology and chemistry with emphasis on public health aspects of pollution control, disease transmission, and waste disposal. Students are trained to manage related types of public health problems.

Department of Health Records Administration. This department has 3 faculty members and offers 4 programs from a one-year certification in medical transcription through a four-year degree in health record administration. The program is designed to train students in the effective management, storage, and retrieval of hospital records.

Department of Medical Technology. This department's seven faculty members offer a 2-year medical assistant degree and 4-year degrees in medical technology and medical laboratory technology. These programs are designed to train students to perform the medical testing required to support physician decision making, and to prepare students to attain the certification needed to work in hospital laboratories, clinics, or medical testing facilities. Areas of study are hematology, clinical chemistry, clinical microbiology, parasitology, and similar subjects.

Department of Occupational Therapy. The Department of Occupational Therapy offers a 4-year training program in physical therapy and has 7 faculty members.

Department of Nursing. The Department of Nursing is the largest department in the college, offering 2 pro-

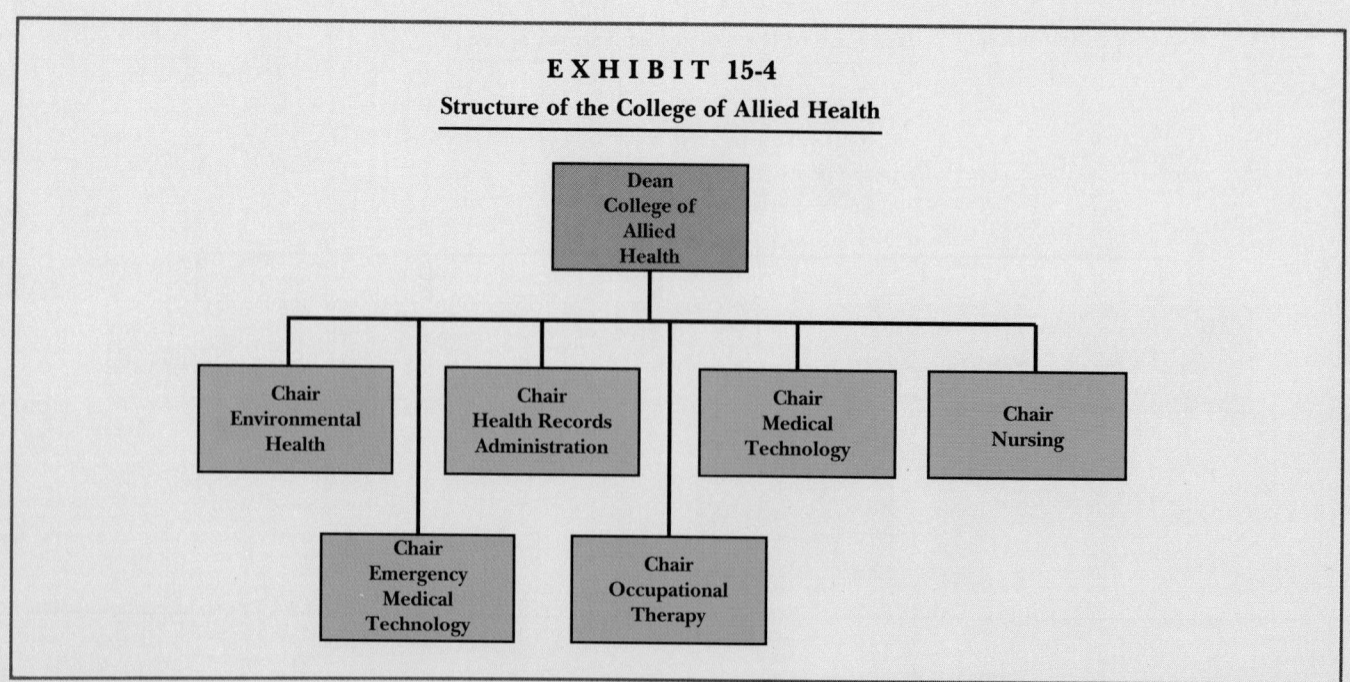

EXHIBIT 15-4
Structure of the College of Allied Health

Dean College of Allied Health

Chair Environmental Health

Chair Health Records Administration

Chair Medical Technology

Chair Nursing

Chair Emergency Medical Technology

Chair Occupational Therapy

STRUCTURING THE ORGANIZATION

grams leading to 2-year or 4-year nursing degrees along with state RN (registered nurse) or LPN (licensed practical nurse) certification.

STEP TWO: SHARING SUGGESTIONS

The class should divide into groups of about 4 to 6 members each. If your class has already formed permanent groups, these groups should reconvene. In each group, members should describe and explain their reasons for the structural modifications they recommend to solve Midwestern State's fiscal problems. If you think of additional modifications during these discussions, you should work together to describe them. Each group should combine all of the modifications it has thought of into a single list.

STEP THREE: DESIGNING A NEW STRUCTURE

After the group has listed all proposed modifications, members should work together to find the combination that will create a structure that saves as much money as possible yet remains consistent with the requirements of the Board of Regents. You should redraw Exhibits 15-1 through 15-4 as necessary to show the changes your group is suggesting. One group member should be appointed spokesperson to explain the group's proposal to the rest of the class.

STEP FOUR: GROUP REPORTS AND DISCUSSION

Each group spokesperson should summarize the results of the group's design efforts in a five-minute presentation to the rest of the class. The list of modifications considered by the group as well as its revised structural diagrams should be presented on a chalkboard, flipchart, or overhead transparencies.

STEP FIVE: CLASS DISCUSSION OF THE RESTRUCTURING PROCESS

The class should review the lists of modifications and structures developed by the groups, looking for similarities and differences among the ideas of different groups. If time permits, the class should try to reach consensus on one combination of modifications that will be consistent with the Regents' guidelines and achieve the greatest budget reduction. The class may discuss questions like the following during this process.

1. Which modifications seemed obvious at first but later proved to be inappropriate?
2. Which modifications came to the surface only after further consideration?
3. Which modifications, if any, conflicted with others? In the event of such conflicts, how did you decide which modification to choose?
4. In what ways is Midwestern State University similar to your school or educational program? What sort of structural modifications might the administrators of your school or program make if required to reduce costs?
5. In general, what should you do to an organization's structure to economize on administrative overhead?

CONCLUSION

Forming structural units and combining them into departments or divisions involves tradeoffs between the economy of functional grouping and the flexibility of divisional groupings. The costs of administrative overhead may also enter into decisions about how to form units and cluster them together. Subtle changes in structure can sometimes save an organization a lot of money without affecting its ability to accomplish its mission and goals.

DIAGNOSING ORGANIZATIONAL BEHAVIOR

CASE 15-1

O CANADA*

BONNIE J. LOVELACE, *University of Alberta*
ROYSTON GREENWOOD, *University of Alberta*

The Public Service Commission of Canada (PSC) is responsible for the provision of a comprehensive human resources management service to the fifty-two departments and agencies of the federal public service. Initially, the commission operated through six branches.

* Reprinted with the author's permission.

One of these, the Staff Development Branch (SDB), is the focus of the present case. The case examines the SDB subsequently as it began to experience problems of financial restraint. The SDB was responsible for the provision of:

1. regularly scheduled courses in a variety of professional, technical, and general subjects;
2. regular and special courses for senior and executive managers;
3. specialized, custom-designed courses on a consulting basis as needed;
4. a research and development service on federal adult educational needs.

Since its creation, the SDB had grown steadily. Its members were highly qualified professionals in their specific fields, and the SDB provided them with extensive training in adult education methods. The SDB served the federal government on a cost-recoverable basis. That is, it had to market courses and cover *all* of its costs, including overhead. Courses were sold to client departments at prices comparable to those charged for similar programs available on the open market. Before this case was written, SDB had enjoyed more business than it could handle. It had an excellent reputation, and there had been no lack of funds within departmental training budgets.

In prior years, the SDB had about 250 members and was organized as shown in Exhibit 15-5. Each of the five directorates was a cost center, responsible for forecasting its own revenues and costs. Although the SDB technically operated on a "branch break-even" basis, each directorate operated on the assumption that it should cover costs.

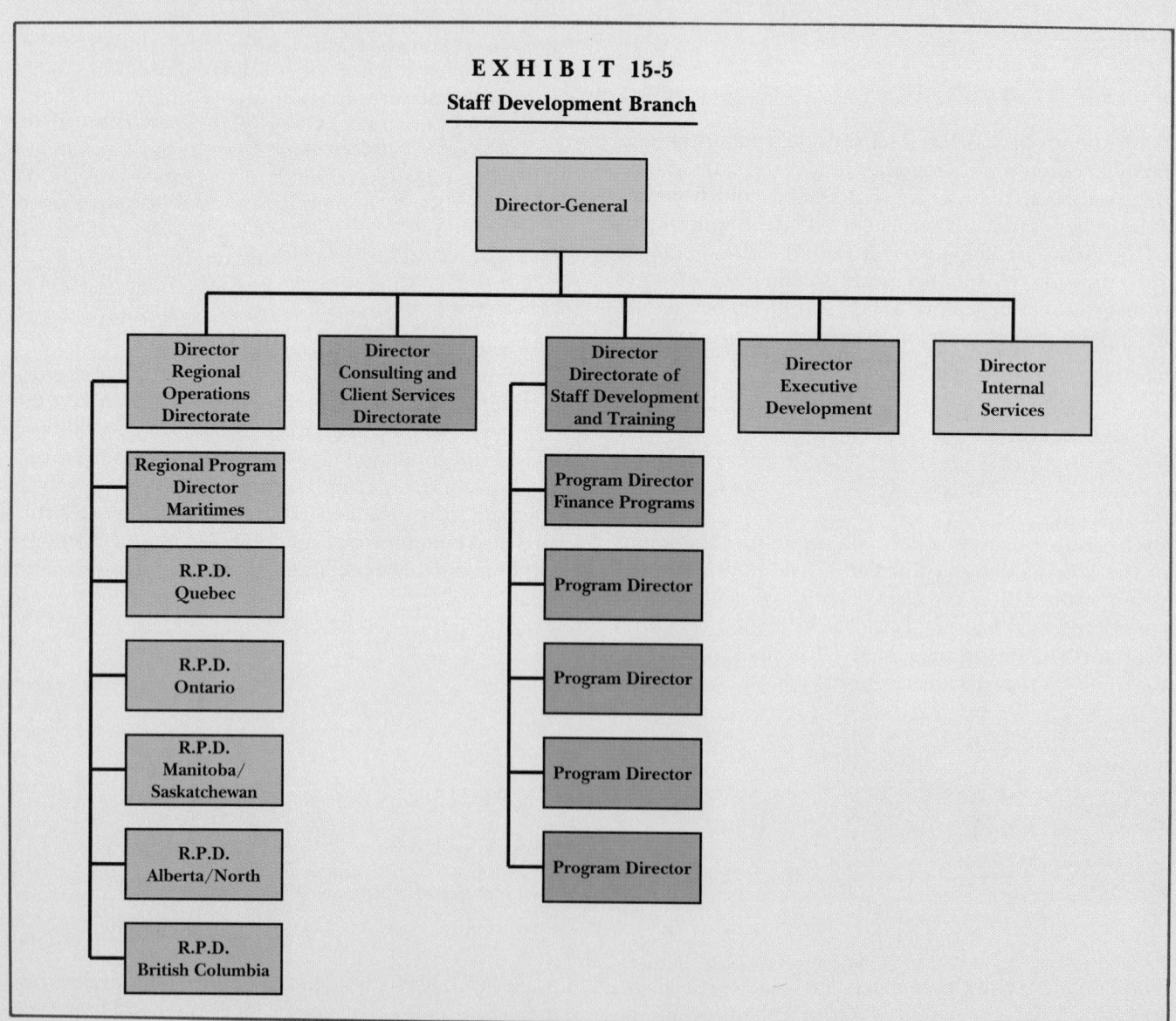

EXHIBIT 15-5
Staff Development Branch

The two largest directorates within the branch were the Directorate of Staff Development and Training (DSDT) and the Regional Operations Directorate (ROD). These provided the bulk of the regularly scheduled courses offered by the branch. The primary division of responsibility between the DSDT and the ROD was that the DSDT serviced the Ottawa region, where the vast bulk of the public service was located, and the ROD serviced the rest of Canada. The six regional units of the ROD and the DSDT operated the same courses, but in different locations.

DIRECTORATE OF STAFF DEVELOPMENT AND TRAINING

The DSDT was organized in terms of *six programs*, each headed by a program manager and staffed by up to fourteen people (including two clerks). Each program had its special field and provided a full range of courses within that field. Trainers within a program did most of their own course design and teaching and would hire outside consultants only for very special courses offered on a limited basis.

The client group of the DSDT included any public servant in the Ottawa region who was not a senior manager or an executive. The latter groups were serviced through the Executive Education Directorate. The DSDT trainers worked singly or in teams, depending on the course and their experience. Each trainer generally was responsible for one or two courses that would be taught ten to fifteen times a year.

Consulting and custom design work in the Ottawa region was not handled by the DSDT. Such work would be handled through the Consulting and Client Services Directorate (CCSD). If a client department wanted a regular course to be run in-house and for itself alone (as opposed to sending participants to the DSDT's courses), the DSDT would "sell" an appropriate trainer to the Consulting Directorate. Regional units of the ROD also could use (and be charged for) DSDT trainers.

Marketing and registration for Ottawa courses were handled through the Internal Services Directorate. Program units within the DSDT concentrated on the provision of high-quality, regularly scheduled courses in Ottawa, leasing out trainers to consulting or to regional operations when time permitted and as need demanded.

Essentially, the task facing the DSDT was reasonably straightforward: develop and teach courses in one city for a very large population. The directorate was large and operated through four levels of management providing a heavy schedule of repeated courses. Each of these levels of management had controls and pressures that affected the next.

REGIONAL OPERATIONS DIVISION

The ROD had a small headquarters group in Ottawa, headed by the director. The six regional offices were located in Halifax, Montreal, Toronto, Winnipeg, Edmonton, and Vancouver. Each regional office was headed by a regional program director and staffed by two or three full-time trainers, supported by a secretary, a registry clerk, and a student from Waterloo University who administered the Open Learning Systems Correspondence courses.

Regional offices catered to federal public servants in the regions and handled most SDB business within their area. The basic role of the ROD was to provide the same spectrum of courses for the regions as was offered in Ottawa by the DSDT. However, because of the lower volume of demand, regional trainers were generalists and were required to teach and manage a variety of courses that in Ottawa were divided between the six program areas. The regional trainers were responsible for all administrative support services. They would design and advertise courses, prepare necessary materials, set up the classroom, teach, and assess the course. In addition, the regional trainers would administer, but not teach, a wide range of other courses. These courses would be taught by local consultants or Ottawa trainers (from DSDT) hired by the regional trainer.

The director and the trainers in the regions spent a considerable amount of time visiting clients, advertising programs, and putting out newsletters. The registry clerk spent most of her time contacting departmental training officers, looking for course participants. She also ensured that administrative letters and details were put out on time by the trainers. In addition, the trainers and director actively sought out consulting work, which they set up and discharged themselves.

The regions carried high overhead and travel expenses and had smaller clients with smaller budgets. The trainers were conscious that every penny counted. At the same time, quality had to be maintained. In times of trouble, most rules were set aside, and people within the regional offices worked together to generate new ideas for courses. The regional offices were small enough to encourage considerable face-to-face interaction.

Relationships between the DSDT and the regional offices of ROD had always been good. Many of the regional people had worked in the DSDT earlier in their careers. Two regional directors had worked through the ranks of the Ottawa division. Minor skirmishes had often occurred over the years, generally relating to problems with a few DSDT trainers who tended to head for the regions and demand that everyone from the director down should cater to their every whim. These few were well known, however, and avoided when possible. The

Regional Operations Directorate, however, deliberately sought persons who preferred smaller working groups, diverse tasks, and a great deal of autonomy. The DSDT tended more towards individuals who had a particular specialty and taught it, leaving their senior managers to handle the "paperwork." There was no question that Ottawa trainers felt strong ownership of "their" courses and, given the opportunity, wanted a say in the regions. The regional people taught "everyone's" courses, depending on the schedule, and were just as happy to find local people who could do the others with a little guidance.

From Boom to Bust

In prior years, the SDB had enjoyed a booming business. There was no lack of funding in departmental training budgets, and the branch had all the business it could handle. Although the economy seemed to be slumping, it did not appear serious. Rumors, however, were circulating about cutbacks as the full force of the economic downturn began to make itself felt. The Treasury Board demanded thorough reviews of departmental budgets, and one of the first areas cut by most departments was training. The SDB, on full cost recovery, found its market suddenly less affluent.

In June, the regional directors were in Halifax for their semi-annual meeting. They usually met in one of the regions during September for a general meeting and again in January in Ottawa for a budget meeting. This year, however, they were meeting in June because a major educational conference was taking place for two days at Dalhousie University at which some of the top experts in the field were featured speakers. The regional directors had agreed with their boss, George Hudson, that they would work Sunday through Wednesday to handle regular business, leaving Thursday and Friday for the conference.

On Tuesday afternoon, the group was discussing what the ensuing months might hold. . . .

"I'm worried," mused Herb Aiken of Halifax. "My registrations are dropping off, and we're looking at cancelling courses. You know what that means; trainers sitting around on the overhead with nothing to do."

Sarah Wilson from Edmonton concurred. She had just received a telex from her office informing her that a three-day course set to start the next day had just suffered seven last-minute cancellations.

"That only leaves eight people; we can't do it, financially or pedagogically. And we've sunk training time and administrative costs into it. I'm going to have to

call and tell my staff to contact the other participants and try to postpone. This is very bad for business, though, and we can't keep it up."

She left to make her call. Thomas Russell from Vancouver picked up the ball:

"The funny thing is, our clients are willing to lose the one-third late cancellation penalty, rather than pay the whole course fee. Forecasting revenues is becoming impossible, and we're barely keeping our heads above water. Where is this taking us?"

George Hudson tried to soothe everyone's fears, saying everyone in Ottawa was still doing okay and was optimistic. The directors looked at one another, each silently thinking that it was always the regions that got hit first. It was easier for the Ottawa mandarins to make cuts where the pain wasn't staring them in the face every morning. At that moment, Sarah returned and told George there was an urgent phone call for him. He left, and the others continued to discuss the future. Hudson returned about ten minutes later, his face grim.

"There's very bad news," he said flatly. "Treasury Board issued a directive this morning stating that all nonessential training is to be reviewed and cancelled whenever possible. The phones are ringing off the wall and everything on our books is on hold until October or November."

The situation worsened during the summer. Regional trainers were out visiting their clients constantly, trying desperately to drum up business, selling a day's consulting here, working on a problem there. It was difficult. Many clients were in offices located significant distances from the regional centers. Regional directors, however, were on the rampage over travel costs and telephone bills. But, as one Toronto trainer said to her boss one day:

"A letter a day just won't do it! We need to talk to them, get them to spend whatever money they've got on our courses, rather than buying on the private market."

Alice Waters knew this was true, but she *had* to keep costs trimmed to the bone. The Treasury Board had told departments to trim training costs. Given their smaller budgets, many departments preferred to provide their own training or use consultants.

By late summer, a few courses were beginning to pick up registrants as people began to sort out their budgets. Some Ottawa courses were rescheduled, but there was still a lot of slack.

One morning, Sam Wisler of Winnipeg called Vancouver.

"I just had a long talk with Mike White, the Financial Management Program manager. He wants to negotiate with us about having his trainers do all the resourcing on our regional financial courses from now on. Did a lot of talking about how we should be saving branch funds by keeping the money inside wherever possible."

Thomas Russell, listening carefully, said:

"Well, in the past, we could never get their trainers, unless somebody wanted to visit his relatives and made a deal with us. All the regions hire local consultants for courses we don't teach ourselves. Saves all those travel costs. However, it's worth thinking about. What did you tell him?"

Wisler replied:

"Just that. We should all think about it. The way I see it, things are getting better, but we may never see those good times again. If we can get Programs to do some of our courses (which are the same ones being done in Ottawa), and for less than our local consultants can do it, we'll be helping each other. They've got a lot of trainers with expensive time on their hands, and we've got courses our own staff can't do, especially in EDP, Finance, and Personnel. Maybe we can help each other. I think I should talk it over with Hudson, and see about putting out a telex to all regions on it. We can discuss it on our next teleconference."

By November, both Ottawa and the regions had managed to reschedule most of their courses, but at drastically reduced registration levels. This meant costs were more or less the same, but revenues were way down. Even though the branch had an official policy that break-even was calculated on the branch level, everyone knew that cost centers losing money were vulnerable. And each program, each region, was a cost center. They closed ranks. People who had worked well together for years with colleagues in the other directorate suddenly discovered negative characteristics of which they had previously been unaware. ROD jealously defended its right to hire local resources; the DSDT stubbornly insisted that course manuals were their property. Each group saw the other as untrustworthy, and open communication virtually ended. This was on everyone's mind as the regional directors held a conference by phone one morning. George Hudson opened the discussion:

"I've been getting feedback from all of you by telex on progress with Programs. My assessment so far is that they want to sell you their trainers' time to cut their overhead, and you're willing to buy it as long as charges are comparable to what it costs when you resource these programs locally. However, it appears that what they want to charge exceeds your local costs. Not only that, but each of you is negotiating separate agreements."

Thomas Russell broke in angrily:

"You can say that again. Mike White wants to send me two trainers to do the four-day "Fundamentals of Budget Formulation and Control" course, and he wants a total of nineteen days of time plus travel costs to do it. But he offered to do it for Sarah with one trainer and fewer days of time. What's going on here?"

Sarah's reply was consistent with what everyone had been experiencing.

"The month my course is scheduled is one where most of Mike's trainers are booked. He gave me whatever time was left. It seems that they want to dump all their excess time on us. Well, our budgets won't take it."

Evelyn D'anjou in Montreal continued:

"We've got to negotiate standard charges. And they must be reasonable ones, or we'll go to local, as we've always done when we had to rely on ourselves."

George Hudson, sensing that feelings were heating up and deciding a teleconference was not the best medium for this discussion, told everyone to sit back. He promised to meet with the DSDT Director, Bob Smythe, and talk things over as soon as possible.

The next day a furious Alice Waters was on the phone to George Hudson.

"Things are getting totally out of hand. I phoned Mike this morning to tell him we couldn't accept the charges he wants for our next financial course, so I had hired the Jameson people to do it. He tells me that's just fine, but all those new regulations for budgeting are being worked into the course, and his people are the only ones who can do it. And he refuses to release the new course manual because he claims it's not in its final form. George, you know we can't do outdated courses in the regions. I have to have that course book. Those manuals are branch property, not DSDT property! The Programs develop them because that's part of their responsibility, but it's policy that they must be made available to the regions, because we have to offer the same course out here. Mike as much as hinted that we will all be having trouble getting manuals for the Programs from now on. He says when things were slack over the summer, they revamped many of our courses, but the changes are still being tested. We're being blackmailed!"

Alice stopped, having run out of breath. George questioned her, giving her time to cool off a bit, but he was concerned. Alice was one of his best managers, a skilled trainer herself, and one who was more than able to negotiate solid agreements with her colleagues. If her problem-solving skills were not helping, they were in trouble.

"Have you considered training some of your own staff to do the more specialized courses, Alice? Maybe we can reduce our dependence on the Programs that way."

Alice was not mollified:

"George, you know what our trainers do. Everything . . . teach, administer, market, consult, clean up classrooms, weekends in airports. They just don't have time for more. Besides, why train them to do a course that's only offered twice a year in their own region. . . . We have others that run frequently both on our regular schedule and on an in-house basis. But the Ottawa trainers only have their one or two little courses to think about. No marketing, no consulting. Even big training centers with everything done for them! They walk out of our classrooms on Friday night and don't clean up a thing! They say that's our job, not theirs. Well, we don't have big staffs catering to our small offices, and it's our weekend, too. But I'm getting off the topic. . . . What about those course books? I've already telexed the other regions to warn them about what's happening."

Inwardly, Hudson groaned. Every one of his directors would be up in arms by the end of the day. He promised Alice he'd go to Bob Smythe, the director of DSDT, to talk matters over, and hung up. Glancing at his telephone, he could see the lights coming on; it was starting already. Thankful it was Friday, he told his clerk to hold the calls and left to find Bob Smythe.

A half hour later, Hudson returned, and dictated a telex: everyone was to sit tight. Smythe was meeting with his managers Monday morning to discuss the matter.

The following Tuesday, Hudson picked up the teleconference phone to address his regions. He wondered how much he'd get through before the protests began.

"I just had a meeting with Bob Smythe. His managers claim we're doing outdated courses and that they should be given control of course content. They also believe we should hire their resources before any consultants, to help minimize branch downtime. Smythe agrees with them, and they're tabling the matter with the Director General at the next management committee meeting."

There was silence as the six listeners digested this news, each realizing the potential consequences. Then Sam Wisler in Winnipeg spoke angrily:

"This is incredible. They want to make money at our expense! Are we working for the same place or aren't we? What the hell is going on here? We won't let those bastards get away with this!"

Herb Aiken's language was much stronger, but the message was the same. Hudson listened to the chorus of angry voices for a while and then asked for everything in the way of financial ammunition, details of travel costs, local costs, and Programs changes. Then he ended the call.

The SDB Management Committee came to the conclusion that branch resources should be used whenever possible to teach branch courses. The regions and the Programs were instructed to work out standard charges to be used in the January budget exercise for the upcoming fiscal year. The point was noted that the regions had to provide up-to-date courses and, if that involved using the DSDT resources, that was the way things had to be.

In January, two of the regional directors came to Ottawa to meet two representatives from the DSDT. The objective was to settle standard charges for all courses. Preparation time, teaching time, travel time, and administrative responsibilities would be fixed. Ratios were to be agreed upon and used as formulas for all courses in the future. Alice Waters and Sarah Wilson had convassed the other regional directors on acceptable alternatives and had full authority from them to act. They had requested that the two DSDT representatives come with the same authority, as time was running out. The group met for a full day on the Monday and, by the end of it, the two regional directors were exhausted and frustrated. The DSDT representatives were demanding costly ratios, were not giving an inch, and had to take back any proposals to their own director for his approval. And he was away until Wednesday.

That night Alice and Sarah paid a late night visit to Hudson, venting their anger openly. The regions could not survive the charges being imposed by DSDT. It seemed that the SDB had some fundamental decisions to make about its internal affairs, decisions that were beyond the authority of Wilson and Waters. Those decisions had to be made before the new budgets were drafted.

Despite meeting again on Tuesday and Wednesday, the DSDT and ROD representatives failed to agree on standard charges. The matter was again put to the Management Committee. The committee reiterated its position that in-house resources had to be used and decided that the regions would have to live with the Program demands.

In March, the Regional Operations Directorate tabled its budget for the upcoming fiscal year. It showed a substantial projected loss. The DSDT tabled its budget, showing a substantial projected profit.

When you have read this case, look back at the chapter's diagnostic questions and choose the ones that apply to the case. Then use those questions with the ones that follow in your case analysis.

1. What type of unit grouping was used to form the directorates that make up the Public Service Commission's Staff Development Branch? What are the strengths and weaknesses of this type of structure?
2. What sorts of coordination mechanisms are being used to coordinate activities in the SDB's unit groups? Given the apparent importance of controlling costs in the SDB, how suitable do you think these mechanisms are? What are the costs and benefits of structuring an organization in this manner?
3. What triggered the breakdown between the Regional Operations Directorate and the Directorate of Staff Development and Training? What interunit mechanisms should the SBD consider implementing in order to deal with this situation? Prepare a diagram that shows how the SBD's structure would look after implementing your suggestions.

CASE 15-2

RONDELL DATA CORPORATION

Review this case, which appears in Chapter 13. Next, look back at Chapter 15's diagnostic questions and choose the ones that apply to the case. Then use those questions with the ones that follow in your case analysis.

1. What type of unit grouping was used to form the departments in the Rondell Data Corporation? What are the strengths and weaknesses of this approach?
2. How are Rondell's research and engineering departments interconnected? What relations do these departments have with the company's sales and production departments? What triggered the breakdown in coordination among Rondell's departments? What interunit mechanisms should Rondell's management consider implementing in order to improve matters? Draw a diagram of the kind of structure the company would have if it implemented your suggestions.
3. Did Frank Forbus fail because of personal deficiencies? Given Rondell's current structure, could someone else have performed his job successfully?

CASE 15-3

DUMAS PUBLIC LIBRARY

Read Chapter 18's Case 18-2, "Dumas Public Library." Next, look back at Chapter 15's diagnostic questions and choose the ones that apply to that case. Then use those questions with the ones that follow to begin your analysis of the case.

1. How were the unit groups formed in the city government of Kimball, New Mexico? What are the benefits of this method of unit grouping? What are its weaknesses?
2. What mechanisms were used to coordinate activities within the units of the Kimball government structure? Between or among these units? Do these mechanisms seem appropriate? What changes would you recommend?
3. Is the conflict between Debra Dickenson and Helen Hendricks simply a matter of incompatible personalities? Are the coordinating mechanisms used in the city government contributing to this conflict? If they are, what modifications would you make in order to resolve the conflict?

CHAPTER 16

ORGANIZATION DESIGN

When Roger Smith, former CEO of General Motors, reorganized the huge company in 1984, GM had already announced its design of the Saturn and its intention to create a separate company to produce it. As we go to press in mid-1991, the Saturn has yet to prove itself in the marketplace. But the company's employees seem to agree that two of Saturn's most important achievements are the new design of the organization that is producing it and the new roles that design has created for managers, workers, and union representatives.

At an upcoming meeting in New York, Roger B. Smith, chairman of General Motors Corporation, will ask the GM board of directors to approve a plan to reduce the company's five car divisions—Chevrolet, Pontiac, Oldsmobile, Buick and Cadillac—to two. The new Chevrolet-Pontiac division will produce and sell only GM's smaller cars, and the Buick-Oldsmobile-Cadillac division will handle the larger models. Beyond merging nameplates, Smith's plan will change General Motors in other important ways. From the company's beginning, design, engineering, and marketing activities have been directed by central staffs at the top corporate level. Under the new plan, these activities will be handled by separate staffs at the two divisions. In charge of these divisions will be two new executive vice presidents who will be strictly accountable for divisional profit and performance to GM's top management.[1]

The 1984 meeting announced in *Business Week* had been in the making for quite some time. Like Apple's John Sculley, the management of General Motors Corporation had struggled, beginning in the late 1960s, with the task of streamlining a sprawling corporation and controlling the costs of its operations. In an attempt to reduce the duplication of effort that was then increasing the cost of production, GM's top management had consolidated many of the assembly operations of the firm's five automotive divisions—Chevrolet, Pontiac, Buick, Oldsmobile, and Cadillac—into a single unit, the General Motors Assembly Division. Responsibility for major purchasing, product design, and engineering decisions had also migrated from the "nameplate" automotive divisions to GM's Detroit headquarters. By the early 1970s General Motors had moved toward a structure that resembled, in many important respects, the structure of Sculley's reorganized Apple (described at the beginning of Chapter 15).[2]

Unfortunately, however, whereas Sculley's Apple flourished, the centralized General Motors failed to perform as expected. Publicly, GM's management blamed this failure on conflicts over product development that erupted between the newly consolidated assembly division and the remnants of the five nameplate divisions which continued to handle product distribution.[3] Privately, detractors also suggested that the restructuring reduced GM's sensitivity to changes in the preferences of American car buyers, leading the company to produce cars that nobody wanted to purchase.[4] Whatever the cause, Roger Smith reacted by proposing his two-division reorganization plan in early 1984, and GM's board of directors gave its approval. Restructuring began later that year.

Why did the same kind of structural changes that put Apple back on its feet in the 1980s prove to be inappropriate for GM in the 1970s? Was it due to differences between the two firms' products—Apple's computers versus GM's cars? Or differences between the sizes of the two companies—the large Apple versus the mammoth General Motors? Or yet other differences? Without guidance, managers like Apple's Sculley and GM's Smith are unlikely to know the

[1] Tom Nicholson, James C. Jones, and Erik Ipsen, "GM Plans a Great Divide," *Business Week*, January 9, 1984, p. 68.

[2] Ibid., pp. 68–69.

[3] "Can GM Solve Its Identity Crisis?" *Business Week*, January 23, 1984, pp. 32–33; and William J. Hampton and James R. Norman, "General Motors: What Went Wrong?" *Business Week*, March 16, 1987, pp. 102–110.

[4] J. Patrick Wright, *On A Clear Day You Can See General Motors* (Grosse Pointe, Mich: Wright Enterprises, 1979), pp. 219–20.

answers to questions like these. Organizational well-being may be threatened as a result. Contemporary managers, whether they are maintaining existing structures or devising new ones, therefore need to know about various structural alternatives and their effects on performance. For this reason, in this chapter we will consider the several different types of contemporary organizational structure. We will look especially at how the structural characteristics discussed in Chapter 15—unit grouping and coordination mechanisms, formalization, specialization, interunit coordination mechanisms—combine to determine the strengths and weaknesses of each structural type. We will then examine the factors that influence the effectiveness of each type of structure and show how a particular structure's strengths and weaknesses make it appropriate for some situations but not others. In so doing, we will present a contingency model of organization design that provides guidance to managers engaged in structuring modern organizations.

DIAGNOSTIC ISSUES

Our diagnostic model highlights many critical questions in the area of organization design. For starters, we need to *describe* the structure of an organization and determine whether it is performing effectively. We must also be able to characterize the situation in which the organization must do its work. How do we *diagnose* what kind of technology is used by the firm and what its effects are on the firm's structure? How do we assess the environment surrounding an organization and determine whether the organization is suitably structured?

How do we *prescribe* the right type of coordinating mechanism for a particular type of environment? How can we determine what structure an organization should have in light of the organization's age and maturity? Finally, what *actions* can managers take to match structure to an organization's technology? What steps can managers take to improve the fit between an organization's structure and its environment?

A CONTINGENCY MODEL OF ORGANIZATION DESIGN

Is there a single *best* form of organizational structure? Our quick comparison between GM and Apple suggests that no one type of structure is suitable for all organizations. Instead, each type of organization structure has unique strengths and weaknesses that make it more appropriate for some situations than for others. Structuring an organization involves choices among various alternatives.

organization design The process of diagnosing the situation that confronts a particular organization and selecting and putting in place the organization structure most appropriate for that situation.

Organization design is the process of making these choices. That is, management diagnoses the situation confronting a particular organization and selects and puts in place the structure that seems most appropriate. It is guided by the contingency perspective, described in Chapter 1, that no single management approach—or structure—is useful in every situation and that the usefulness of a particular approach depends on the situation being managed. Organization design is thus based on the principle that the degree to which a particular type of *structure* will contribute to the *effectiveness* of an organization depends on *contingency factors* that impinge on the organization (see Figure 16-1).

FIGURE 16-1

The Contingency Model
of Organization Design

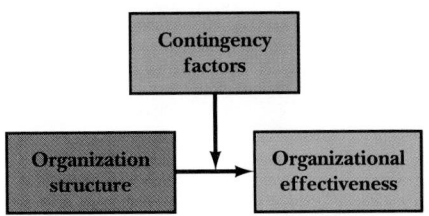

POINT TO STRESS
It is a rather new concept to
regard employees as a
constituency group. Managers are
now paying attention to the costs
and difficulties in replacing
employees.

organizational effectiveness The
degree to which an organization is
successful in achieving its goals
and objectives while at the same
time ensuring its continued
survival by satisfying the demands
of interested parties, such as
suppliers and customers.

constituency groups Groups
such as employees, customers, and
suppliers upon whom the survival
of an organization depends.
Constituency groups make
demands that they expect will be
fulfilled in return for their support
of the organization.

Organizational Effectiveness

Organizational effectiveness, the ultimate aim of organization design, is a measure of an organization's success in achieving its goals and objectives. In an effective organization, employee behaviors, group and intergroup processes, and
organizational factors all contribute toward accomplishing the goals and objectives
that serve as the firm's purpose. You will recall from Chapter 2 that goals and
objectives of this sort might include targets pertaining to profitability, growth,
market share, product quality, efficiency, stability, or similar outcomes.[5] An organization that fails to accomplish such goals is ineffective because it is not fulfilling
its purpose.

An effective organization must also satisfy the demands of the various **constituency groups** that provide it with the resources it needs to survive. For example, as Figure 16-2 suggests, if a company satisfies customers' demands for
desirable goods or services, it will probably continue to enjoy its customers' pa-

---

[5] James L. Price, "The Study of Organizational Effectiveness," *Sociological Quarterly* 13 (1972), 3–15;
Stephen Strasser, J. D. Eveland, Gaylord Cummings, O. Lynn Deniston, and John H. Romani,
"Conceptualizing the Goal and System Models of Organizational Effectiveness—Implications for
Comparative Evaluative Research," *Journal of Management Studies* 18 (1981), 321–40; and Y. K. Shetty,
"New Look at Corporate Goals," *California Management Review* 22 (1979), 71–79.

FIGURE 16-2

**Types of Constituency Groups
and Their Demands**

Circles surrounding the central
organization in this figure
represent a few of the many
constituency groups whose
demands must be satisfied to
ensure continued organizational
survival. Arrows pointing toward
the center indicate the sorts of
demands that the constituency
groups make on the organization.
Arrows pointing outward indicate
the kinds of responses that an
effective organization might make
in return.

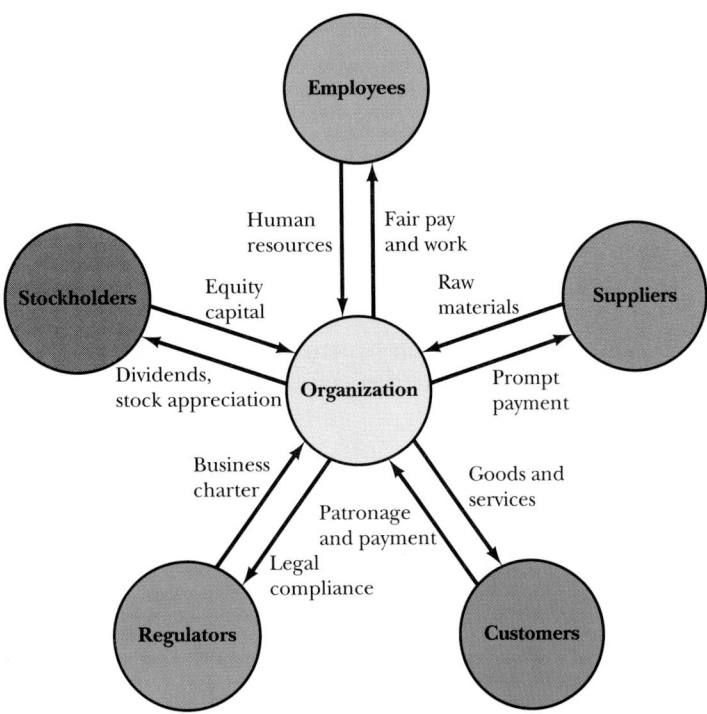

organizational productivity The amount of goods or services produced by an organization. Higher productivity means that more goods or services are produced.

organizational efficiency The ratio of outputs produced per unit of inputs consumed; minimizing the raw materials and energy consumed by the production of goods and services.

AN EXAMPLE: AMERICA WEST AIRLINES
In 1991 the survival of America West Airlines was threatened because of the carrier's inability to make timely payments to leasors supplying their planes and to bond holders for loans they made to the airline. These two groups have considerable power and could use that power to force a restructuring of the organization.

tronage. If it satisfies its suppliers' demands for payment in a timely manner, the suppliers will probably continue to provide it with needed raw materials. If it satisfies its employees' demands for fair pay and satisfying work, it will probably be able to retain its workers and recruit new employees. If it satisfies stockholder demands for profitability, it will probably have continued access to equity funding.[6] If a firm fails to satisfy any one of these demands, however, its effectiveness will be weakened, because the potential loss of needed resources, such as customers or employees, threatens its survival.

Effectiveness is not the same thing as **organizational productivity**. The latter concept does not take into account whether a firm is producing the *right* goods or services.[7] A modern company producing more buggy whips than ever before is certainly productive, but it is also ineffective, because few people need buggy whips in today's society. Effectiveness is not efficiency, either. **Organizational efficiency** means minimizing the raw materials and energy consumed by the production of goods and services. It is usually measured as the ratio of outputs produced per unit of inputs consumed.[8] Thus efficiency means *doing the job right*, whereas effectiveness means *doing the right job*—determining whether a company is producing what it ought to produce in light of the goals, objectives, and constituency demands that influence its performance and are its reason for being.

Structural Alternatives

The extent to which an organization is effective is strongly influenced by its structure. One type of structure enhances the ability to attain goals and satisfy constituencies, while others are likely to detract from these pursuits. In general, organization structures differ from one another along a dimension ranging from *mechanistic* to *organic*.[9]

[6] Constituency models of effectiveness and other examples of constituencies and their interests are discussed by Paul S. Goodman, Johannes M. Pennings and Associates, *New Perspectives on Organizational Effectiveness* (San Francisco: Jossey-Bass, 1977); John A Wagner III and Benjamin Schneider, "Legal Regulation and the Constraint of Constituent Satisfaction," *Journal of Management Studies* 24 (1987), 189–200; and Raymond F. Zammuto, "A Comparison of Multiple Constituency Models of Organizational Effectiveness," *Academy of Management Review* 9 (1984), 606–16.

[7] Richard Z. Gooding and John A Wagner III, "A Meta-Analytic Review of the Relationship between Size and Performance: The Productivity and Efficiency of Organizations and Their Subunits," *Administrative Science Quarterly* 30 (1985), 462–81.

[8] Ibid.

[9] Tom Burns and G. M. Stalker, *The Management of Innovation* (London: Tavistock Publications, 1961), pp. 119–22.

ORGANIZATION DESIGN

TABLE 16-1
Comparison of Mechanistic and Organic Structures

CHARACTERISTICS OF MECHANISTIC STRUCTURES	CHARACTERISTICS OF ORGANIC STRUCTURES
Tasks are highly specialized. It is often not clear to members how their tasks contribute to accomplishing organizational objectives.	Tasks are broad and interdependent. Relation of task performance to attainment of organizational objectives is emphasized.
Tasks remain rigidly defined unless formally altered by top management.	Tasks are continually modified and redefined by means of mutual adjustment among task holders.
Specific roles (rights, duties, technical methods) are defined for each member.	Generalized roles (acceptance of the responsibility for overall task accomplishment) are defined for each member.
Control and authority relationships are structured in a vertical hierarchy.	Control and authority relationships are structured in a network of both vertical and horizontal connections.
Communication is primarily vertical, between superiors and subordinates.	Communication is both vertical and horizontal depending on where needed information resides.
Communication is mainly in the form of instructions and decisions issued by superiors, performance feedback, and requests for decisions sent from subordinates.	Communication takes the form of information and advice.
Loyalty to the organization and obedience to superiors are insisted upon.	Commitment to organizational goals is more highly valued than loyalty or obedience.

Source: Based in part on Tom Burns and G. M. Stalker, *The Management of Innovation* (London: Tavistock Publications, 1961), pp. 120–122.

---

mechanistic structures Machine-like organization structures designed to enhance efficiency; characterized by large amounts of formalization, standardization, specialization, and centralization.

organic structures Organism-like organization structures designed to enhance flexibility and innovation; characterized by large amounts of mutual adjustment and decentralization.

AN EXAMPLE
Highly organic structures are often found in entrepreneurial organizations where emphasis is on flexibility and creativity rather than on efficiency. As the organization grows and matures, there is often a shift toward a mechanistic structure focusing more on efficiency.

structural contingency factors Characteristics of an organization and its surrounding circumstances that influence whether its structure will contribute to organizational effectiveness.

Wholly **mechanistic structures** are machinelike. They permit workers to complete routine, narrowly defined tasks in an efficient manner, but they lack flexibility. As Table 16-1 indicates, they are characterized by large amounts of formalization, standardization, and specialization. Therefore, they are much the same as the *bureaucratic* form of organization described by Max Weber nearly a century ago (see Chapter 2). Mechanistic structures are centralized and usually have *tall hierarchies* of vertical authority and communication relationships such as the one depicted in the upper panel of Figure 16-3.

Highly **organic structures** are more like living organisms in that they are innovative and can adapt to changing conditions. Owing to their flexibility, however, organic structures lack the stability or constancy that enables more mechanistic structures to perform routine work efficiently. Organizations with organic structures rely more on mutual adjustment than on formalization, standardization, and specialization. Their divisions, departments, and people are connected by a decentralized network. Often, though not always, this network takes the form of a *flat hierarchy*, such as the one shown in the lower panel of Figure 16-3. Its emphasis on horizontal relationships can help reduce the number of vertical layers required to process information and manage activities.

A particular organization's structure may fit at any point along the mechanistic-organic dimension. The more mechanistic the structure, the more efficient but less flexible it will be. The more organic the structure, the more flexible but less efficient it will be. These differences in efficiency and flexibility can be traced to the mechanisms used to coordinate work activities. As you will remember from Chapter 15, standardization embodies low long-term coordination costs and is thus the basis of mechanistic efficiency. Mutual adjustment, on the other hand, is quite flexible and is therefore the source of organic flexibility.

Structural Contingencies

Whether mechanistic efficiency or organic flexibility will lead to greater organizational effectiveness depends on the influence of a variety of factors. **Structural**

Tall Hierarchy

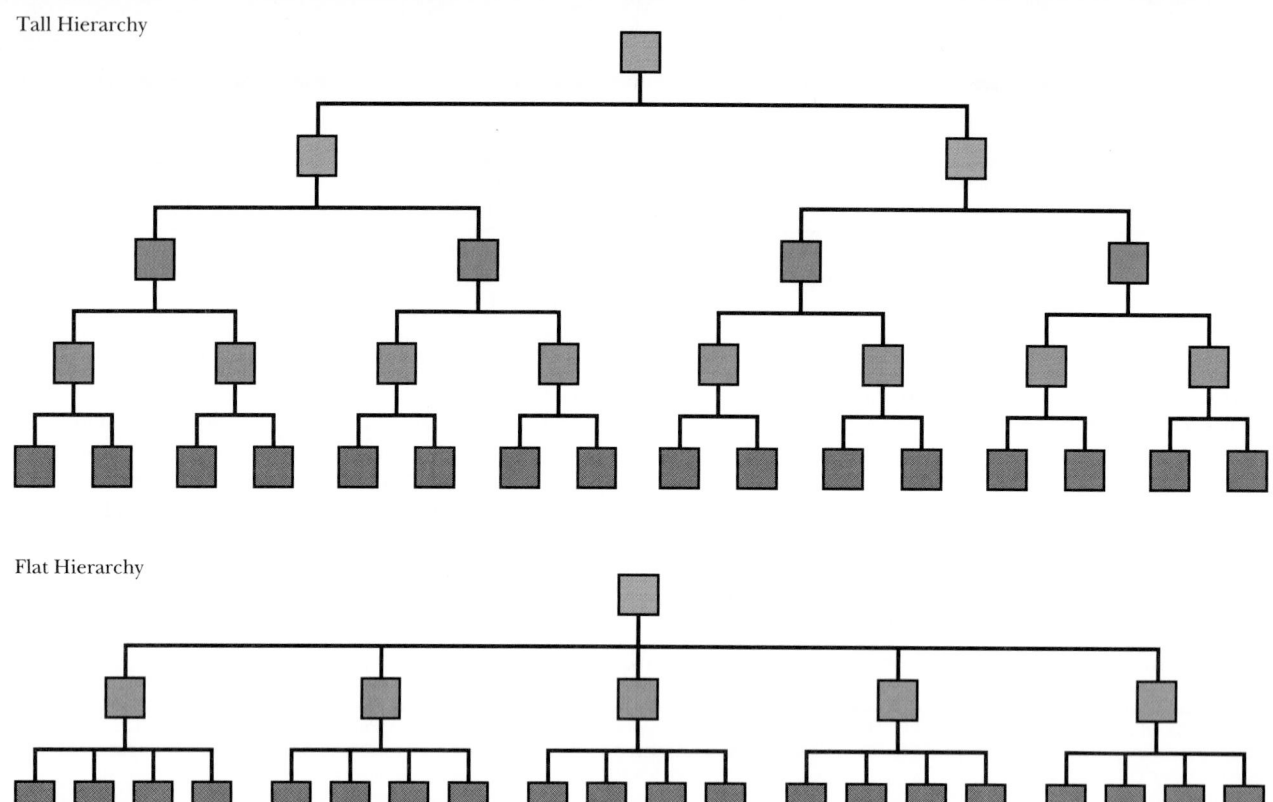

Flat Hierarchy

F I G U R E 16-3

Tall and Flat Organizational Hierarchies

closed system contingencies
Structural contingency factors that are characteristics of the organization. Size and technology are the most dominant of these factors.

open system contingencies
Structural contingency factors that are characteristics of the environment surrounding an organization.

contingency factors such as characteristics of an organization and the surrounding situation shape or constrain the firm's work. Two perspectives on organizational design—the closed system perspective and the open system perspective—help to identify several contingency factors that can influence the success of different organization structures.

The *closed system perspective* focuses on enhancing the efficiency of activities within organizations but tends to ignore the effects of outside factors.[10] So **closed system contingencies** are factors within the organization. Of these contingencies, the two most noteworthy ones are the organization's *size* and the *core technology* it employs to produce its goods or services. Closely related to the size of an organization are two additional contingencies, the firm's *age* and its stage of *life-cycle* development.

In contrast, you may recall from Chapter 2 that the *open system perspective* recognizes that organizations are shaped by the circumstances that surround them. To remain viable, companies must acquire energy, materials, and other resources from the external environment and offer finished goods or services in return.[11] **Open system contingencies** thus consist of features of the *environment* surrounding a firm.

Considered together, closed and open system contingencies constitute the situation that managers must diagnose correctly to design a structure that promotes organizational effectiveness. Later in this chapter, we will examine the contingency factors of size, core technology, and environment in greater detail and discuss their effects on organization design. First, however, we turn to an overview of the specific types of structures found in contemporary organizations.

[10] James D. Thompson, *Organizations in Action* (New York: McGraw-Hill, 1967), p. 4.

[11] Thompson, *Organizations in Action*; Daniel Katz and Robert L. Kahn, *The Social Psychology of Organizations*, 2nd ed. (New York: John Wiley, 1978), p. 3.

ORGANIZATION DESIGN

TYPES OF ORGANIZATION STRUCTURE

How do the various structural characteristics discussed in Chapter 15 combine to form fully developed organization structures? What kinds of structures are formed as a result? Which of these structures are more mechanistic and efficient, and which are more organic and flexible? The answers to questions like these are found in the "menu" of alternative structures overviewed next.

Simple Structures

simple structure An uncomplicated type of organization structure that relies on mutual adjustment or direct supervision to achieve coordination.

As suggested by their name, **simple structures** are the least complicated form of structure used in today's organizations. Their simplicity stems from their reliance on the relatively uncomplicated coordination mechanisms of mutual adjustment or direct supervision to integrate work activities. There are two types of simple structures—simple undifferentiated and simple differentiated structures.

simple undifferentiated structure A type of simple organization structure in which coordination is achieved solely by means of mutual adjustment.

Simple Undifferentiated Structure. In a **simple undifferentiated structure**, coordination is accomplished solely by mutual adjustment. As you will recall from Chapter 15, mutual adjustment is a process in which coworkers talk with each other to determine how to coordinate work among themselves. Because talking with other people is natural for most of us, mutual adjustment is easy to initiate and relatively simple to sustain. Thus simple undifferentiated structures can often be established and perpetuated fairly easily.

As Figure 16-4 suggests, there is no hierarchy of authority in a simple undifferentiated structure. Such a structure is nothing more than an organization of people who decide what to do by talking with each other as they work. No single individual has the authority to issue orders, and there are few if any written procedures to guide performance. A group of friends who decide to open a small restaurant, gift shop, or similar sort of business might, at the outset, adopt this type of structure for the business. Similarly, the kind of classroom discussion group you might form to analyze the cases or do the exercises in this book could be thought of as a small organization with a simple undifferentiated structure.

The primary strengths of simple undifferentiated structures are their simplicity and extreme flexibility. Networks of face-to-face conversations occur spontaneously and can be reconfigured almost instantly. For instance, adding another member to a small classroom discussion group is likely to cause only a momentary lapse in the group's activities. A major weakness of these structures, however, is their limitation to small organizations. Suppose you were a manager in an advertising firm composed of 25 or 30 people. Could you rely solely on mutual adjustment to ensure that the firm's accounts were properly handled? Probably you could not, because as you'll recall from Chapter 11, process loss undermines the usefulness of face-to-face coordination when it is applied to groups much larger than 12 people. Mutual adjustment requires so many links between people in large groups that valuable time and effort are lost in the exchange of needed

FIGURE 16-4

The Simple Undifferentiated Structure

The simple undifferentiated structure is the simplest of all structures, relying only on face-to-face mutual adjustment to coordinate work.

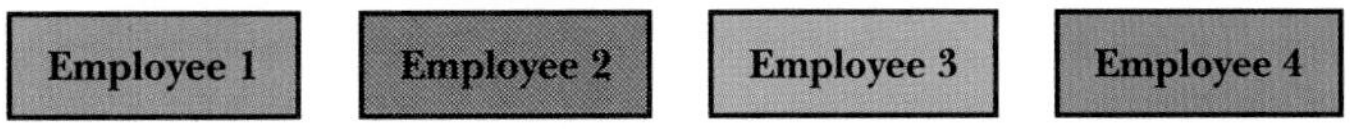

| Employee 1 | Employee 2 | Employee 3 | Employee 4 |

information. As a result, only a very small organization can function effectively with a simple undifferentiated structure.

Another important—and related—weakness is that simple undifferentiated structures cannot provide the coordination needed to accomplish complex tasks. It is difficult to imagine 12 or so people being able to assemble cars, but simple undifferentiated structures cannot coordinate the efforts of larger groups of people in an efficient manner. Complicated work requires a more complicated form of organization structure.

simple differentiated structure A type of simple structure in which coordination is achieved by means of direct supervision.

Simple Differentiated Structure. In the second type of organizational structure, the **simple differentiated structure**, direct supervision replaces mutual adjustment as the primary unit coordination mechanism. You probably encounter organizations with simple differentiated structures every day—a bookstore or clothing shop on your campus or the family-owned grocery store or gas station in your neighborhood. Look at the diagram of a simple differentiated structure in Figure 16-5. The relatively flat hierarchy pictured in this figure contains small but significant amounts of *vertical specialization* and *centralization*. One person (usually the firm's owner or the owner's management representative) retains the hierarchical authority needed to coordinate work activities by means of *direct supervision*. As a secondary mechanism, mutual adjustment is used to deal with coordination problems that direct supervision cannot resolve. For example, while the owner of a small insurance office is at the post office getting the morning mail, clerks in the insurance office may talk among themselves to decide who will answer the telephone and who will process paperwork until she returns.

The simple differentiated structure can coordinate larger numbers of people than can the simple differentiated structure. The reason is that, as indicated in Chapter 15, shifting to direct supervision eliminates much of the process loss associated with reliance on mutual adjustment alone. In addition, because its decision-making powers are centralized in the hands of single person, a simple differentiated structure can respond rapidly to changing conditions. At the same time, this structure retains a good deal of flexibility, because it avoids standardization. Its weaknesses, however, are its inability to coordinate the activities of more than about 50 people and its failure to provide the integration needed to accomplish complex tasks. It is just as unlikely that a group of people could organize themselves to produce cars by using a combination of direct supervision and mutual adjustment as it is that they might organize themselves for such a task using mutual adjustment alone. A single direct supervisor would soon be overwhelmed by the vast amount of information required to know which sort of cars to produce, what parts to order, whom to order them from, how to assemble them properly, and so forth.

APPLYING THE DIAGNOSTIC MODEL
Diagnostic Question 1: What kind of structure does the organization currently have? What is its primary means of coordination? Do the titles of its line vice presidents provide evidence of one particular form of structure?

Diagnostic Question 2: Is the organization's current structure the one dictated by its size? If the organization has a simple undifferentiated structure, does it have twelve or fewer employees? If the organization has a simple differentiated structure, does it have 50 or fewer employees? If the organization has a complex structure, is it substantially larger?

Complex Structures

Both kinds of simple structures are overwhelmed by the coordination requirements of complicated tasks. Standardization—of processes, outputs, skills, or

FIGURE 16-5

The Simple Differentiated Structure

The emergence of hierarchical, superior-subordinate differentiation distinguishes the simple differentiated structure from the simple undifferentiated structure.

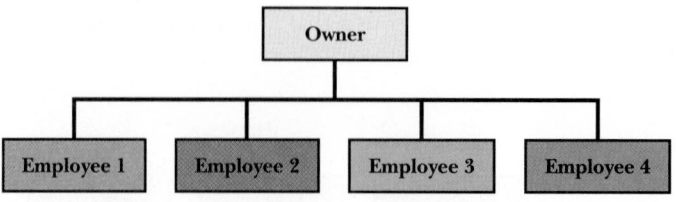

complex structure A type of organization structure that relies on standardization to achieve coordination; a complex structure is therefore characterized by noticeable formalization and specialization.

norms—becomes useful in this situation. It greatly reduces the amount of information that must be exchanged and the number of decisions that must be made as work is being performed. Such standardization is the hallmark of **complex structures**. In these structures, direct supervision and mutual adjustment are retained as secondary mechanisms that kick in when standardization fails to meet all coordination needs. This combination of unit coordination mechanisms allows organizations with complex structures to integrate the variety of jobs needed to perform complicated, demanding work. There are three major kinds of complex structures—functional structures, divisional structures, and matrix structures.

functional structure An efficient but inflexible type of complex structure characterized by functional departmentation and centralization.

AN EXAMPLE: AMERICA WEST AIRLINES
Organizational structures change as companies mature and expand. For example, when America West began flying in 1983 it relied on a simple differentiated structure. This allowed the airline considerable flexibility. Supervisors had the authority to coordinate work activities but were not tied to rigid guidelines or job descriptions imposed by formalization and specialization. Flight attendants were often assigned to work reservations and billing. Baggage handlers often cleaned the plane. As the airline grew from serving five cities to serving 60 cities, including some in Canada and Japan, the structure became complex with considerable standardization and centralization. Jobs became more clearly defined and limited in scope and the hierarchy grew rather tall.

Functional Structure. Functional structures are characterized by four key attributes. First, because they are complex structures, functional structures are based on coordination by *standardization*. As you'll recall from Chapter 15, standardization is preceded by *formalization* and contributes to *specialization*. So both of these characteristics are also features of functional structures. Second, functional structures are *centralized*. Most if not all important decisions are made by one or a few people at the tops of firms with functional structures—for instance, decisions leading to the formation of organizational goals, objectives, and mission statements. Third, owing to their centralization, functional structures require a great deal of vertical communication and tend to have *tall hierarchies*. Fourth, these structures are characterized by *functional departmentation*. That is, units within them are grouped into departments that are named for the functions their members perform, such as marketing, manufacturing, or accounting.

As Figure 16-6 suggests, one of the easiest ways to determine whether a particular firm has a functional structure is to examine the titles held by its vice-presidents. If the firm has a complex structure, and all of its vice-presidents have titles that indicate what their subordinates do—for example, vice president of manufacturing, vice president of marketing, vice president of research and development—the firm has a functional structure. If one or more vice-presidents have other sorts of titles, however, for instance, vice president of the consumer finance division or vice president of European operations—the firm has another type of structure.

Organizations like credit unions and locally owned banks, car dealerships, department stores, and many junior colleges, colleges, and universities have functional structures. The primary strength of this type of structure is its economic efficiency. Standardization minimizes the long-term cost of coordination. In addition, centralization makes it possible for workers to focus their attention on their work rather than having to take time out to make needed decisions. Functional structures, however, have a critical weakness. They lack flexibility. The standardization that provides so much efficiency not only takes lengthy planning and documentation (formalization) to set in place but requires that the same standards be followed again and again. This inflexibility reduces the functional structure's ability to cope with instability or change. Functional departmentation adds to this rigidity, because as you will recall from Chapter 15, changes to any work flow in a company organized by functional departmentation also affect the other work flows in the organization.

A functional structure can coordinate the work of an organization effectively if the firm limits itself to one type of product, produces this product in a single geographic location, and sells to only one general type of client. Many organizations produce more than one product, however, or do business in several different locations or seek to serve a variety of clients. Such diversity of products, locations, or clients injects variety into the information a firm needs to make managerial decisions. This variety overloads the centralized decision-making processes on which the functional structure is based. Let's look next at the kind of

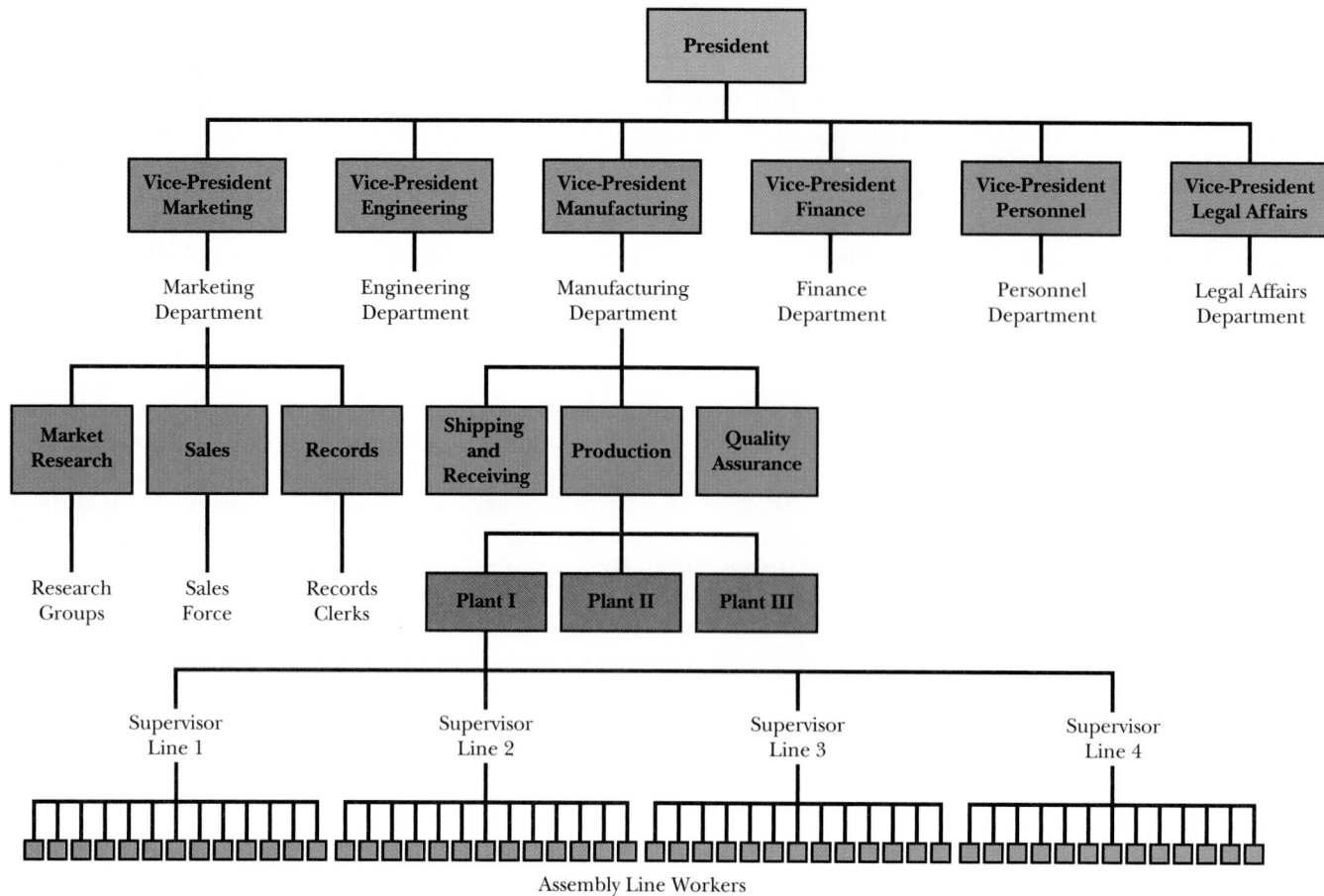

FIGURE 16-6

The Functional Structure

The functional structure uses a combination of functional grouping, centralization, and standardization to organize work activities. The marketing and manufacturing departments are shown in some detail to illustrate how the departments of a functional structure are composed of smaller units that facilitate coordination among employees who must work closely with each other.

divisional structure A flexible but inefficient type of complex structure characterized by divisional departmentation and moderate decentralization.

product structure A type of divisional structure formed by grouping units according to similarities in the products they make and sell.

geographic structure A type of divisional structure formed by grouping units according to similarities in their geographic location.

market structure A type of divisional structure formed by grouping units according to similarities in the clients or customers they serve.

structure that enables an organization to cope with this sort of diversity—the divisional structure.

Divisional Structure. Like functional structures, **divisional structures** are complex structures characterized by *standardization*, and, therefore, by *formalization* and *specialization*. Unlike functional structures, however, divisional structures are moderately *decentralized*. Decision making is pushed downward one or two hierarchical layers, so a company's vice-presidents and sometimes their immediate subordinates share in the process of digesting information and making key decisions. *Divisional departmentation* is another notable feature that distinguishes divisional structures from functional structures. Units in divisional structures are grouped together according to similarities in products, geographic locations, or clients. For this reason, divisional structures are also sometimes called **product structures**, **geographic structures**, or **market structures**.

Look at Figure 16-7, which depicts several different divisional structures. Can you see how they differ from the functional structure diagrammed in Figure 16-6? You're right: In each of the structures in Figure 16-7, vice-presidential titles include product, geographic, or client names. Note, though, that in these

ORGANIZATION DESIGN

A Product Structure

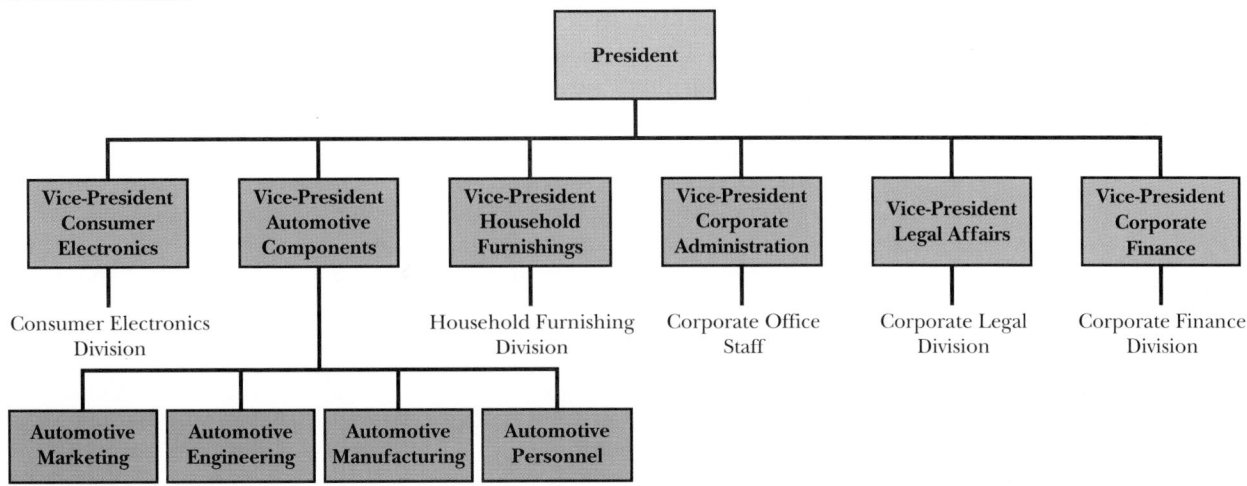

B Geographic Structure

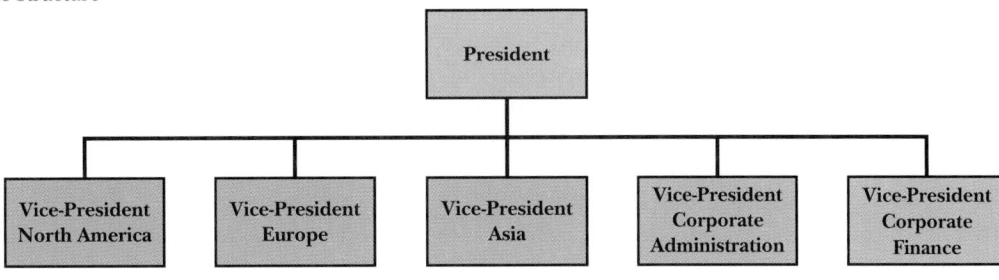

C Client Structure

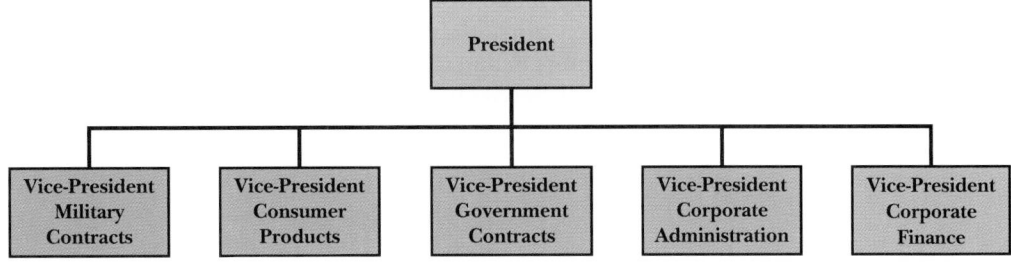

FIGURE 16-7

Divisional Structures

The divisional structure incorporates decentralization, standardization, and divisional departmentation. The diagram shows three divisional structures in which departmentation is based on (a) product similarities, (b) geographic similarities, and (c) client similarities. Note how the lower groupings of the automotive-components division shown in panel A resemble a small functional structure.

divisional structures, vice-presidents of *staff* units have titles that sound like functions; for example, vice-president of legal affairs and vice-president of corporate finance. (If you don't remember what staff means, review the discussion in Chapter 9 of the line-versus-staff distinction).

The divisional structure's departmentation and moderate decentralization give it a degree of adaptability not found in the functional structure. Each unit, or division, of such a structure can react to issues concerning its own product, geographic region, or client group fairly independently of other units. It must not, however, lose sight of the overall organization's goals and objectives. For example, the vice-president of consumer electronics, shown in the upper panel of Figure 16-7, can make decisions affecting the production and sales of clock radios and steam irons without consulting with the company's president or other vice-presidents. This degree of independence even allows a division to stop doing business without seriously interrupting the operations of the organization's other units. For example, the division of TRW that fulfills NASA space contracts could discontinue doing business without affecting work in the firm's credit information division. Remember, though, that each division in a divisional structure is itself organized like a functional structure—take another look at the product structure shown in Figure 16-7. As a result, a particular division cannot change products, locations, or clients without serious interruption to its own internal operations. For example, the decision at TRW to reduce reliance on military contracts would require that the division servicing such contracts be substantially reorganized.

The adaptability that is the main strength of divisional structures comes at the price of increased costs because of duplication of effort across divisions. For example, every division is likely to have separate sales forces even though that means that salespeople from several different divisions may repeatedly visit the same customer. So the primary weakness of divisional structures is the fact that they are, at best, only moderately efficient.

matrix structure An extremely flexible but also extremely costly type of complex structure characterized by both functional and divisional departmentation as well as high decentralization. Also called the simultaneous structure.

Matrix Structure. Matrix structures, like divisional structures, are adopted by organizations that must integrate work with a variety of products, locations, or customers. Firms that have matrix structures, however, need even more flexibility than divisional structures allow. They try to achieve this flexibility by reintegrating functional specialists across different product, location, or customer lines. Because matrix structures use functional and divisional departmentation *simultaneously* to group structural units together, they are also called *simultaneous structures*.

Figure 16-8 illustrates the matrix structure of a firm that has three divisions, each of which manufactures and sells a distinct product line. Each box, or cell, in the matrix is a distinct group composed of a small hierarchy of supervisors and one or more structural units having both functional and divisional responsibilities. For example, cell 1 in the figure is a consumer-electronics-marketing group composed of units that market televisions, radios, cellular telephones, and other electronic merchandise. Cell 2 is the automotive-components-engineering group consisting of engineering units that design automobile engines, suspensions, steering assemblies, and other such items. Cell 3 is a household-products-manufacturing group made up of facilities that produce furniture polish, floor wax, window cleaner, and other household supplies. Note that the figure also indicates that staff units in a matrix structure are often excluded from the matrix itself. The three staff departments shown in the diagram—personnel, finance,

FIGURE 16-8

The Matrix Structure

The matrix structures increase the flexibility of a large organization by incorporating both functional departmentation (shown horizontally in the diagram) and divisional departmentation (shown vertically). The groupings represented by each cell perform functional duties related to a particular type of product.

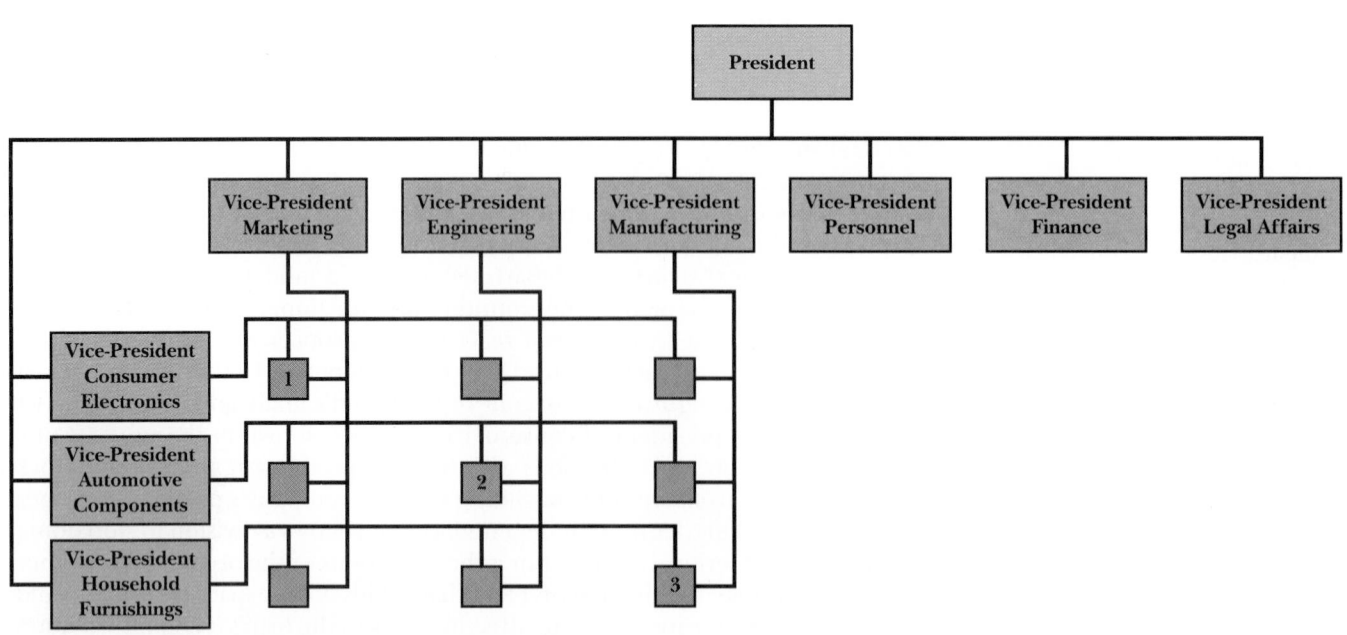

and legal affairs—provide advice to top management but are not part of the matrix.

You will recall from Chapter 15 that the matrix is both a structure and a lateral linkage device used to interconnect the units in a firm. *Mutual adjustment* is the primary means of coordination within the upper layers of a matrix structure, and decision making is *decentralized* among matrix managers. Both these characteristics allow top managers to reconfigure relationships among the cells in the matrix, promoting extreme flexibility. Because of their dual responsibilities, each matrix cell has two bosses—a functional boss and a divisional boss. This arrangement violates Fayol's principle of unity of command (see Chapter 2). Thus mutual adjustment must also be used in the upper layer of each cell to cope with conflicting orders from above.

Beneath the upper layer of each cell, however, formalization, standardization, and specialization are used to integrate work activities. Both direct supervision and lower-level mutual adjustment serve as supplementary mechanisms that coordinate cell activities. For instance, once managers at the top of the matrix structure shown in Figure 16-8 have decided to manufacture a new kind of floor wax, formalization is used to develop new standards. Standardization is then used to coordinate activities in the units in the household-products-manufacturing cell that make this new product. Direct supervisors help employees learn the new standards and also work to correct deficiencies in the standards as they become apparent. In addition, employees engage in mutual adjustment to cope with problems that their supervisor cannot resolve. Thus a matrix structure basically consists of a simple structure designed into the upper layers of a complex structure. The simple structure injects mutual adjustment to encourage communication, coordination, and flexibility among the managers who oversee organizational operations. Beneath this simple structure is a *tall hierarchy* of cells, cell supervisors, and nonsupervisory employees.

The primary strength of matrix structures is their extreme flexibility. They can adjust to changes that would overwhelm other complex structures. Why do you suppose, then, that matrix structures are extremely rare? The reason is that they have one primary and very crucial weakness. They are extremely costly to operate. In part, this costliness stems from the proliferation of managers in matrix firms. They need two complete sets of vice-presidents. Matrix structures also incorporate the same sort of duplication of effort—multiple sales forces, for instance—that make divisional structures so expensive to operate. Moreover, because employees near the top must deal with two bosses and often conflicting orders, working in a matrix is a stressful situation. It can lead to absenteeism, turnover, and ultimately to lowered productivity and higher human resource costs.

More important, however, matrix structures are economically inefficient, because they rely on mutual adjustment as their primary coordination mechanism despite extremely costly levels of process loss. Matrix structuring thus represents the decision to put up with costly coordination in order to secure high flexibility. Firms that choose matrix structures and function effectively thereafter are generally those that face radical change that would destroy them if they could not easily adapt to such change. Those firms include the Monsanto Company, Prudential Insurance, and the Chase Manhattan Bank. In effect, they are choosing the lesser of two evils—the inefficiency of a matrix rather than dissolution. Firms that try matrix organization but later abandon it do not face the degree of change required to justify the costs of the matrix approach. Among those firms have been Phillips Petroleum and Texas Instruments.

Strategic Business Unit Structure. **Strategic business unit (SBU) structures** are a recent invention that offer an attractive alternative to matrix structuring. These

strategic business unit structure An extremely complex structure consisting of two or more autonomous strategic business units (SBUs), which themselves have complete organization structures.

Diagnostic Question 4: Does the
organization have an SBU
structure? If so, perform separate
technology and environment
analyses—that is, answer the rest
of the diagnostic questions for
each SBU as though it were a
separate organization.

AN EXAMPLE: LORAL
DEFENSE SYSTEMS
Loral Defense Systems used an
SBU structure after buying
Goodyear Aerospace: Each
goodyear facility became a
division. This was a significant
transition for former Goodyear
managers who were used to
decisions being made in Akron.
They now have to make many
decisions they once only had to
implement.

FIGURE 16-9

**The Strategic Business Unit
(SBU) Structure**

The SBU structure resembles a
divisional structure in which each
"division"—an SBU—is itself an
autonomous firm with a
functional, divisional, or matrix
structure.

structures encourage adaptability by *deintegrating* divisions of a large organization rather than by integrating functional and divisional elements, as the matrix structure does. SBU structures redefine divisions as autonomous, fully independent structures, called strategic business units, or SBUs. They allow divisions to fend for themselves with little or no interference from the rest of the firm. (Note that *SBU structure* refers to a complete organization structure while *SBU* refers to a single business entity within an SBU structure.) Thus an SBU structure is actually a "structure of structures." Each SBU has its own relatively complete organization structure, and its chief reports to an upper-level manager in the hierarchy of the parent corporation. An SBU may have a simple or functional structure. or even a divisional structure if its parent is a very large organization. Divisional SBUs can be created by combining several divisions from the parent structure into the same strategic business unit.

Figure 16-9 shows an SBU structure. Such structures are quite *decentralized*. SBU managers three or four levels below the parent firm's CEO may have broad decision-making authority. At the same time, SBU structures are coordinated through *standardization*. Thus they are characterized by high degrees of bureaucratic *formalization* and *specialization*. They have extremely *tall hierarchies* and are organized around *divisional departmentation*. As we've already noted, however, each of an SBU structure's "divisions" is actually a self-sufficient business concern.

A major strength of SBU structures is their ability to provide the coordination required to integrate large or extremely large organizations without incurring the inefficiency of matrix structuring. Thus, prior to its mid-1980s breakup, Beatrice had an SBU structure. General Electric still does and is experimenting with ways to improve it. A weakness of SBU structures, however, is the requirement that top managers give up control over day-to-day operations within each SBU. They retain only the ability to sell laggard SBUs or to buy attractive ones from other companies. Although this might sound like a small price to pay for needed flexibility, it has discouraged the adoption of SBU structures in many instances. Another deterrent to the use of this type of organization is the fact that even SBU structures have some degree of inefficiency inasmuch as their divisional departmentation means substantial duplication of effort. A final drawback is that SBU structures are not useful when strong links are needed between the different parts of an organization. For example, it is difficult to imagine organizing a hospital as an SBU structure. Too many transfers are required among the units of a hospital to allow any of them to have the autonomy of an SBU.

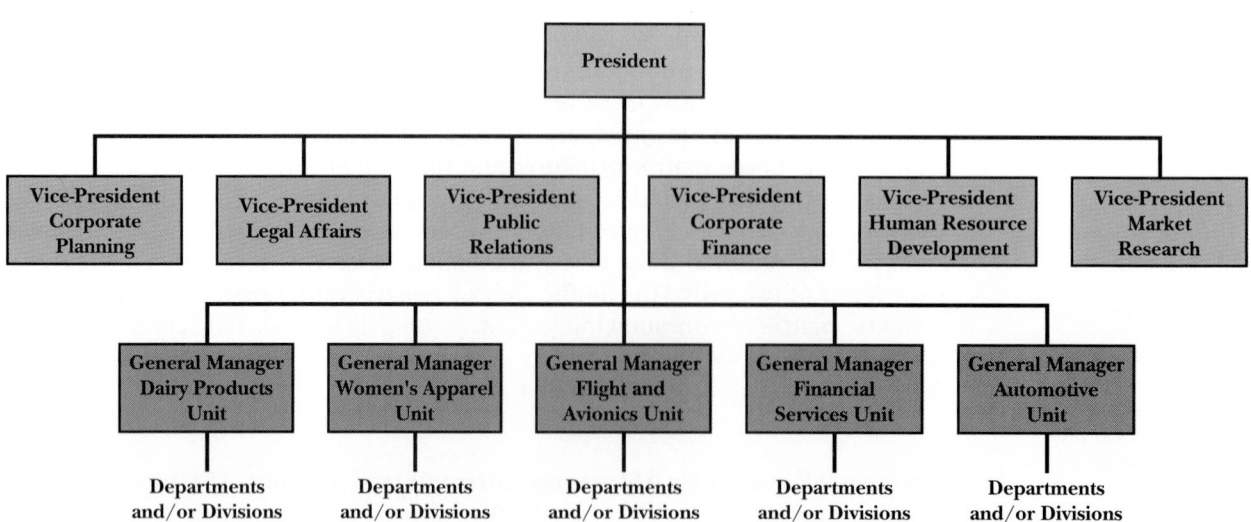

TABLE 16-2
Diagnostic Design Chart I: Organizational Structure

Structural Type	COORDINATION MECHANISMS			BUREAUCRATIC CORRELATES		OTHER CHARACTERISTICS			OUTCOMES	
	Mutual Adjustment	Direct Supervision	Standardization	Formalization	Specialization	Centralization	Hierarchy	Depart-mentation Scheme	Efficiency	Flexibility
Simple undifferentiated	Primary	None	None	Absent	Absent	Low	Flat	None	Low-high*	High
Simple differentiated	Secondary	Primary	None	Absent	Absent	High	Flat	None	Low-high*	High
Functional	Secondary	Secondary	Primary	Present	Present	High	Tall	Functional	High	Low
Divisional	Secondary	Secondary	Primary	Present	Present	Moderate	Tall	Divisional	Moderate	Moderate
Matrix	Primary and Secondary	Secondary	Secondary	Present	Present	Low	Tall	Simultaneous	Extremely low	High
SBU	Secondary	Secondary	Primary	Present	Present	Low	Tall	Divisional	Low	High

* Note: Simple structures are highly efficient at handling simple work but inefficient when complex work must be performed.

Diagnostic Design Chart I

Table 16-2 reviews the six organizational structures we have described, indicating how the different structural characteristics explained in Chapter 15 fit together to form each structural type. We said earlier that organizational structures vary along a scale ranging from mechanistic to organic. Which of these six structural types do you think is the most mechanistic? Which is the most organic? Where do the others fit into the mechanistic-organic dimension we have proposed?

The most straightforward way to address these questions is to look at Table 16-2 and review the type of coordination used in each of the six structures. Bear in mind that mechanistic structures rely mainly on standardization, while organic structures are based primarily on mutual adjustment.[12] Centralization is also a useful indicator, because mechanistic structures tend to be more centralized than organic ones. Let's review the different structural types in the order in which they appear on the continuum in Figure 16-10, moving from the most mechanistic—the functional structure—to the most organic—the simple undifferentiated structure.

Functional structures are the most mechanistic, because they are highly centralized and use standardization as their primary mechanism of unit coordination.

Divisional structures, which also use standardization as their primary unit

[12] Henry Mintzberg, *The Structuring of Organizations* (Englewood Cliffs, N.J.: Prentice-Hall, 1979), p. 86.

FIGURE 16-10

A Continuum of Organization Structures

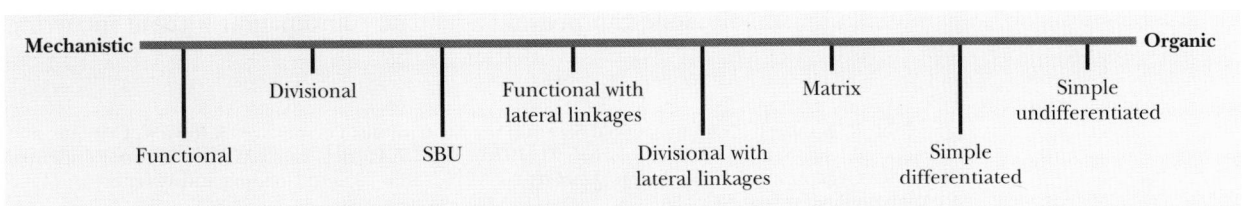

Mechanistic — Functional — Divisional — SBU — Functional with lateral linkages — Divisional with lateral linkages — Matrix — Simple differentiated — Simple undifferentiated — Organic

coordination mechanism, are basically mechanistic, but they are slightly more organic than functional structures because they are less centralized.

SBU structures are somewhat mechanistic, because they are based primarily on standardization. But because they are highly decentralized, they are more organic than divisional structures.

Matrix structures rely on mechanistic standardization beneath their upper layers. They are no more decentralized than SBU structures, but they are more organic, because they use mutual adjustment to coordinate activities at the top.

Simple differentiated structures are more organic than matrix structures because they do not use standardization. On the other hand, because they are centralized, they are not completely organic.

Simple undifferentiated structures are the most organic of all, because they rely solely on mutual adjustment and are completely decentralized.

Functional and divisional structures can be made more organic with the use of *lateral linkage devices* that coordinate interunit relations through mutual adjustment or direct supervision (these devices are discussed in Chapter 15). As shown in Figure 16-10, these modified structures then fit between the SBU and matrix structures on our continuum, yielding a total of eight structural alternatives. These eight alternatives form a menu of possibilities available for consideration during the process of organization design.

CONTINGENCY FACTORS

We turn our attention now to the three contingency factors that have the greatest influence on whether a particular form of structure will contribute to organizational effectiveness. They are organization size, core technology, and the external environment. We will begin by discussing the effects of organization size and, in the process, will also examine the related effects of the age of an organization and its stage of life-cycle development.

Organization Size

organization size The number of members in an organization; its volume of sales, clients, or profits; its physical capacity (e.g., a hospital's number of beds or a hotel's number of rooms); or the total financial assets that it controls.

Organization size can be defined as the number of members in an organization; its volume of sales, clients, or profits; its physical capacity (e.g., a hospital's number of beds or a hotel's number of rooms); or the total financial assets it controls.[13] In keeping with our focus on structural coordination, we will adopt the first definition. We will consider *size* to be the number of members or employees within the organization and thus the number of people whose activities must be integrated and coordinated.

The size of an organization affects its structure mainly by determining which of the three coordination mechanisms that we discussed in Chapter 15 is most appropriate as the primary means of unit coordination in the organization. The three mechanisms, you will recall, are mutual adjustment, direct supervision, and standardization. (Size also has direct effects on organizational performance. See the "In Practice" box.) In extremely small organizations of 12 or fewer people, mutual adjustment alone can provide adequate coordination without incurring

---

[13] John R. Kimberly, "Organizational Size and the Structuralist Perspective: A Review, Critique, and Proposal," *Administrative Science Quarterly* 21 (1976), 571–97; Patricia Yancy Martin, "Size in Residential Service Organizations," *Sociological Quarterly* 20 (1979), 569–79; and Gooding and Wagner, "A Meta-Analytic Review," p. 463.

Organization Size and Performance: Is Bigger Better?

When it comes to organizations, what size is "just right?" Experience at AT&T seems to suggest that smaller is better. The telephone company has downsized significantly in the last six years, shrinking by 92,000 to its current size of 281,000. Though still enormous, AT&T's smaller size has enabled its fewer remaining employees to make more decisions themselves and react more quickly to customer needs. Customer response appears positive. Other once-giant firms, ranging from Johnson & Johnson to Hewlett-Packard Company have similarly downsized or reorganized into smaller groups to foster flexibility and innovation.*

However, the success of other companies seems to suggest that bigger is better. The Illinois Tool Works, a manufacturer of nuts, bolts, nails, and screws, succeeds precisely because of its largeness. By reducing prices and maintaining high production volumes, the firm blocks other potential competitors from entering its midwestern fastener market.† More generally, the 500 largest companies in the United States ac-

count for one-third of the nation's gross national product and employ three-quarters of the scientists and engineers who work in industry (encouraging creativity and innovation). Companies like General Motors, USX, and General Electric have the resources to do things that smaller firms can't—incubating new-product ideas for years, surviving through periods of temporary adversity, rewarding successful employees with career promotions.‡

Although current trends seem to favor downsizing, research actually indicates that larger organizations produce more goods or services than smaller organizations. At the same time, though, they are *not* any more efficient than smaller firms. It takes 100 employees to do 100 employees' worth of work, regardless of the overall size of the organization. Thus, from the standpoint of overall performance, bigger appears to be better, but managers who expect some net gain in efficiency from size changes of any sort are likely to be disappointed.§

* Claudia H. Deutsch, "Less Is Becoming More at AT&T," *New York Times*, June 3, 1990, 25.

† Ronald Henkoff, "The Ultimate Nuts and Bolts Company," *Fortune*, July 16, 1990, pp. 70–73.

‡ John A. Byrne, "Is Your Company Too Big?" *Business Week*, March 27, 1989, pp. 84–94.

§ Gooding and Wagner, "A Meta-Analytic Review," p. 478.

overwhelming process loss. Thus small organizations of this sort can have simple undifferentiated structures and survive without additional structural mechanisms.

If more than about a dozen people try to coordinate by means of mutual adjustment alone, however, so much time, energy, and other resources are diverted away from productive activities and into coordination procedures that performance declines substantially. As we noted in Chapter 15, the activities of larger numbers of people—30, 40, or even 50—can be coordinated instead by direct supervision. This reduces the amount of time and effort that must be devoted to coordination. Simple differentiated structures are ideally suited for this slightly larger organization.

What happens when an organization exceeds 50 people? As you might guess, both simple undifferentiated structures and simple differentiated structures are

The U.S. Postal Service is such a huge organization—it has nearly 700,000 employees—that standardization *has to be its primary mode of coordination. Yet technological-environmental changes demand constant flexibility, something this highly bureaucratic organization does not have. Repeated raises in rates have led many large mailers to desert the Postal Service for lower-priced delivery services. But the powerful American Postal Workers Union has so far prevented the elimination of many jobs that new technology could replace. What do you think the postal service can do to improve its efficiency, cut costs, and keep its workers motivated and happy?*
Source: Susan B. Garland and Mark Lewyn, "Can Tony Frank Get Postal Workers to Cut Him Some Slack?" Business Week, *August 27, 1990.*

overwhelmed by coordination requirements. As the primary means of unit coordination, mutual adjustment becomes extremely expensive, and direct supervision is bogged down by growing information-processing needs. Standardization thus assumes the role of primary unit coordination mechanism, and the organization takes on a complex form of structure. Whether the most effective structure for a company will be a functional or divisional structure depends on the effects of the other contingency factors, which we will discuss below.

Finally, in extremely large organizations of thousands of employees, standardization sometimes proves inadequate as the primary means of unit coordination. The reason is the difficulty of formalizing the wide variety of activities that must go on during the normal course of business. In such "super organizations," standardization may be replaced by the mutual adjustment of matrix structuring if business activities have to remain integrated across different functions and markets. Alternatively, SBU structuring may be used to decouple parts of an extremely large organization, creating several smaller, more easily managed, strategic business units.

life-cycle model A model that proposes that organizational growth progresses through a series of stages, each of which has its own structural requirements.

Organization Age. Closely related to the size of an organization is its age. Increasing age tends to be positively related to greater size. Thus, a firm may require different structures as it grows older. This process of growing out of one structure and into another is captured quite explicitly in life-cycle models such as the one described next.

Organization Life Cycle. As proposed in the **life cycle model** developed by Robert Quinn and Kim Cameron, organizational growth progresses through a series of stages, each of which has its own structural requirements. In the *entrepreneur stage*, one person or a small group of people create an organization and identify the firm's initial purpose. As commitment to this purpose develops, initial planning and implementation bring the firm to life. There is little, if any, formal coordination. Usually mutual adjustment or direct supervision suffices as the primary means of unit coordination. So, at this stage, the firm can be organized as a *simple undifferentiated* or *simple differentiated structure*.

APPLYING THE DIAGNOSTIC MODEL
Diagnostic Question 3: Is the organization's current structure the one suggested by the firm's life cycle stage and age? If the firm is young and in the entrepreneur or collectivity stage, does it have a smaller, simple structure? If it is older and in the formalization or elaboration stage, does it have a larger, complex structure?

An organization experiences rapid growth in the *collectivity stage*. To cope with changes stemming from the growth of their firm and provide stability, members begin to routinize activities. At the same time, members spend long hours at work and develop a strong sense of identification with the organization and its mission. Communication and structural relations remain informal, and the organization may retain a simple structure, but the newly developing rules and procedures indicate the beginning of formalization.

The *formalization stage* is characterized by the division of work into different functional areas, the development of systematic evaluation and reward procedures, and formal planning and goal setting to determine the organization's direction. Professional managers may replace the firm's owners as the day-to-day bosses who run the company. Management emphasizes efficiency and stability, and work becomes even more routine. In the process, the organization adopts a *functional structure*.

TEACHING NOTE
Refer to the opening description of Apple. Have students trace Apple's development. What stage is it currently in? Which structure best describes Apple now?

To adapt to changing conditions and to pursue continued growth, a firm in the *elaboration stage* may seek out new product, location, or client opportunities. As the company's business diversifies, its functional structure loses the ability to coordinate work activities, and there is a need for divisional departmentation and greater decentralization. The organization may then adopt a *divisional structure*. If the firm continues to mature even further, continued growth and diversification may require additional structural elaboration. The company may then adopt a *matrix* or *SBU structure* to cope with its greater size or with its need for greater flexibility.

608   ORGANIZATION DESIGN

Diagnostic Design Chart II

As you can see, the larger an organization becomes, the more likely it is to develop increasingly complex structures. As it ages, then, it tends to become more formalized, standardized, and specialized. Age pushes firms toward the adoption of functional, divisional, matrix, or SBU structures. Youth may allow companies to perform effectively with simple undifferentiated or simple differentiated structures. The chart in Table 16-3 reviews the stages of development described in the Quinn-Cameron model. It puts into life-cycle terms the influence of changes in organizational size and age.

Core Technology

technology The knowledge, procedures, and equipment used in an organization to transform unprocessed resources into finished goods or services.

core technology The dominant technology used in performing work at the base of the organization.

An organization's **technology** consists of the *knowledge, procedures,* and *equipment* used to transform unprocessed resources into finished goods or services.[14] **Core technology** is a more specific term, pertaining to the dominant technology used in performing work at the base of the organization. You can find core technologies in the assembly lines on which cars are manufactured at GM, Ford, and Chrysler; in the kitchens where fast foods are prepared at McDonald's, Burger King, and Wendy's; in the offices in state employment agencies and job-training centers where job applicants are processed and in the reactor buildings where electricity is generated at nuclear power plants. In this section, we introduce two contingency models—the *Woodward manufacturing model* and the *Thompson service model.* They propose that core technology influences the effectiveness of an organization by placing certain coordination requirements on its structure.

Woodward's Manufacturing Technologies. Joan Woodward, a British researcher who started studying organizations in the early 1950s, was one of the

[14] Charles Perrow, "A Framework for the Comparative Analysis of Organizations," *American Sociological Review* 32 (1967), 194–208; and Denise Rousseau, "Assessment of Technology in Organizations: Closed versus Open System Approaches," *Academy of Management Review* 4 (1979), 531–42.

TABLE 16-3
Diagnostic Design Chart II: Stages in the Structural Life Cycle

STAGE	PRIMARY CHARACTERISTICS	STRUCTURAL TYPES
Entrepreneur	Determination of firm's purpose Growth of commitment Initial planning and implementation Reliance on mutual adjustment	Simple undifferentiated Simple differentiated
Collectivity	Rapid growth and change Development of routine activities Appearance of rules and procedures	Simple undifferentiated Simple differentiated
Formalization	Division of work into functions Systematic evaluation and rewards Formal planning and goal setting Entry of professional management Emphasis on efficiency, stability	Functional
Elaboration	Search for new opportunities Diversification, decentralization Maturation and continued growth	Divisional Matrix SBU structure

Source: Based on Robert E. Quinn and Kim Cameron, "Organizational Life Cycles and Shifting Criteria of Effectiveness: Some Preliminary Evidence," *Management Science* 29 (1983), 29–34.

---

first proponents of the view that an organization's technology can have tremendous impact on structural effectiveness.[15] She began her work by studying 100 British manufacturing firms, examining their organizational structures and their relative efficiency and success in the marketplace. Analyzing her data, she discovered that not all companies that had the same type of structure were equally effective. Hypothesizing that these differences in effectiveness might be traced to differences in core technologies, Woodward devised a classification scheme of three basic types of manufacturing technology—*small-batch production*, *mass production*, and *continuous process production*. When she tested her theory by reanalyzing data from the 100 firms she found evidence to support it.

small-batch production A type of manufacturing technology that involves the production of one-of-a-kind items or small quantities of goods designed to meet unique customer specifications. Also called unit production.

SMALL-BATCH PRODUCTION **Small-batch production** (also called *unit production*) is a technology that involves the manufacture of one-of-a-kind items or small quantities of goods designed to meet unique customer specifications. Such items range from specialized electronic instruments, weather satellites, and space shuttles to custom-tailored clothing and custom-made leather sandals. To make this kind of product, craftspeople work alone or in small, close-knit groups. Because customer specifications often change from one order to another, it is almost impossible to predict what will be required on the next job. Thus the work in firms using small-batch technologies varies in an unpredictable way.

It is this unpredictability that fuels the effect of small-batch technologies on organizational structures and effectiveness. Unpredictability impedes advance planning and therefore makes it difficult to coordinate by means of standardization. It is impossible to plan legitimate standards for use in a future that cannot be foreseen. Instead, employees must decide for themselves how to perform their jobs. When employees work alone, they are guided by their own skills and expertise and by customer specifications. When employees work in groups, they coordinate with one another by means of mutual adjustment.

APPLYING THE DIAGNOSTIC MODEL

Diagnostic Question 5: Is the organization's primary purpose to manufacture a tangible product? Does it use small batch production, continuous process production, or flexible cell production? If so, does the organization have an organic structure: simple or matrix, or functional or divisional with lateral linkages? or does it use mass production? Does the organization have a mechanistic structure: functional or divisional?

Woodward found that the important role played by mutual adjustment in coordinating small-batch production was pivotal. Her research showed that among organizations using this type of technology, firms with organic structures were significantly more likely to be successful than companies with mechanistic structures. This suggests that an organization employing small-batch technology would be wise to adopt one of the structures found on the right side of the continuum in Figure 16-10. A simple, matrix, or laterally linked functional or divisional structure may help a company with small-batch production technology be more effective. Similarly, because every SBU in an SBU structure is itself an organization with its own structure and technology, each SBU using small-batch technology should opt for an organic form of structure—again, one of the types found on the right side of Figure 16-10.

mass production A type of manufacturing technology in which the same product is produced repeatedly, either in large batches or in long production runs. Also called large-batch production.

MASS PRODUCTION In **mass production** (also referred to as *large-batch production*), the same product is produced repeatedly, either in large batches or in long production runs. For instance, rather than producing a few copies of this book each time an order was received, Prentice Hall initially printed thousands of copies at the same time and warehoused them to fill incoming orders. Other examples of mass production range from word-processing pools in which midterm examinations are consolidated and typed in large batches to car-manufacturing operations in which hundreds of thousands of Ford Escorts are made on an assembly line that remains virtually unchanged for years at a time.

As these examples suggest, work in mass production technologies is intentionally repetitive and remains so over the course of extended periods of time.

[15] Joan Woodward, *Management and Technology* (London: Her Majesty's Stationery Office, 1958). See also Woodward's *Industrial Organization: Theory and Practice* (London: Oxford University Press, 1975).

Employees perform the same jobs over and over. They know that the work they'll do tomorrow will be the same as the work done today. This stability and routineness facilitates planning and formalization. As a result, a company is likely to use standardization to reduce the long-term costs of coordination. Woodward's research thus revealed that mass-production firms with mechanistic structures were far more likely to be effective than those with organic structures. Therefore, structures on the left side of our continuum—functional or divisional structures—are more apt to enhance the effectiveness of firms or SBUs employing mass production than are structures on the right side of the continuum.

CONTINUOUS PROCESS PRODUCTION In **continuous process production**, automated equipment makes the same product in the same way for an indefinite period of time. For example, at Phillips Petroleum, one refinery makes nothing but gasoline, another refines motor oil, and a third produces only diesel fuel. The equipment used in this type of technology is designed to produce one product and cannot readily be used for any other. Moreover, there is no starting and stopping once the equipment has been installed. Machines in continuous process facilities perform the same tasks without interruption.

Of the three types of technology discussed so far, continuous process production involves the most routine work. Few changes, if any, occur in production processes even over the course of many years. You might expect, then, that organizations using continuous process production would be most effective if structured along mechanistic lines. Interestingly, however, closer examination reveals that few, if any, of the people involved in continuous process production perform routine, repetitive jobs. Machines perform these jobs instead. The people are technicians, who monitor production equipment—watching dials and gauges, checking machinery, inspecting finished goods—and who deal with the problems that arise when this equipment fails to function properly. Although some of these problems occur again and again and can be planned for in advance, a significant number are emergencies that have never happened before and cannot be anticipated. Some of the most critical work performed by people in continuous process production technologies is therefore highly unpredictable, and as a result, standardization is not feasible. Mutual adjustment, sometimes in conjunction with direct supervision, is the dominant mode of coordination. Technicians who oversee production equipment manage unusual events by conferring with each other and devising solutions to emergencies as they arise.

Therefore, Woodward's finding that firms using continuous process production technologies were most effective when structured organically is not surprising. We can conclude that structures on the right side of the continuum shown in Figure 16-10—simple, matrix, or laterally linked functional or divisional structures—are the ones most likely to encourage effectiveness in organizations or SBUs that use continuous process production.

FLEXIBLE CELL PRODUCTION: A RECENT DEVELOPMENT Since Woodward's studies, advances in computers, robotics, and automation have helped create another type of manufacturing technology, **flexible cell production**. In this system, a group, or cell, of computer-controlled production machines are connected by a flexible network of conveyors. These conveyors can be rapidly reconfigured to adapt the cell for different production tasks. This technology is used mainly to produce a wide variety of machined metal parts—pistons for car engines, hinges for the passenger entries of jet planes, parts for the lock on the front door of your house or apartment. Conceivably, though, it could be used to make virtually any kind of product.

As in continuous process production, work in flexible cells is performed by automated equipment. The only people involved are technicians who monitor the equipment and handle problems. But while continuous process production facilities can make only a single product, flexible cells can make many different things. In this respect, flexible cell production resembles small-batch production. It is an efficient method of producing one-of-a-kind items or small quantities of similar items built to satisfy unique customer specifications.

Inasmuch as Woodward found mutual adjustment the most effective coordination mechanism for both continuous process and small-batch production technologies, an organic structure would seem most suitable for a firm using flexible cell production. Indeed, a study of 110 manufacturing firms in New Jersey revealed a significant positive relationship between organic structuring and the effectiveness of organizations with flexible cells.[16] We can use this information to update Woodward's research. Referring to Figure 16-10, we will suggest that companies or SBUs employing flexible cell technologies are likely to be more effective if they adopt simple, matrix, or laterally linked functional or divisional structures.

Thompson's Service Technologies. Because Woodward focused her research solely on manufacturing firms, her contingency model is applicable only to technologies used to produce tangible goods. Today, however, firms that provide services—telephone communications, appliance repair, vacation planning—make up an increasingly critical element of the United States economy as well as the economies of other nations. Thus another contingency model, developed by James D. Thompson, is also quite useful, because it examines technologies often employed in service organizations. These technologies, diagrammed in Figure 16-11, are mediating technology, long-linked technology, and intensive technology.[17]

[16] Frank M. Hull and Paul D. Collins, "High Technology Batch Production Systems: Woodward's Missing Type," *Academy of Management Journal* 30 (1987), 786–97.

[17] Thompson, *Organizations in Action, pp. 15–18.*

FIGURE 16-11

Thompson's Service Technologies

This figure depicts the three service technologies identified by Thompson. Rectangles represent work groups or organizations, circles represent employees, and arrows represent work flows. These diagrams form a measure of core technology if preceded by instructions to respondents to choose the picture that best illustrates the way work is performed in their organizations.

A Mediating Technology

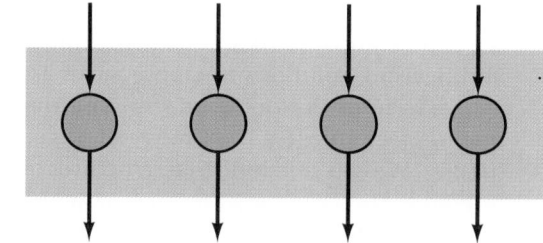

B Long-Linked Technology

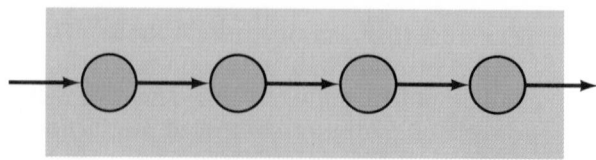

C Intensive Technology

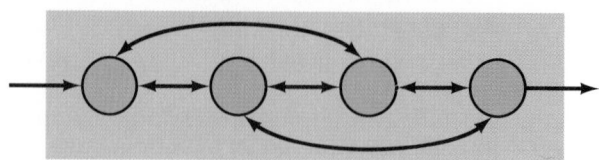

ORGANIZATION DESIGN

MEDIATING TECHNOLOGY A mediating technology provides services that link clients together. For example, banks connect depositors who have money to invest with borrowers who need loans; insurance companies enable their clients to pool risks, permitting one person's losses to be covered by joint investments; and telephone companies provide the equipment and technical assistance people need to talk with each other from separate locations.

When mediating technology is used to provide a service, employees usually serve each client individually. For instance, bank tellers serve customers one at a time. Consequently, as depicted in Figure 16-11A, bank tellers and workers in other mediating technologies normally perform their jobs without assistance from others in their organization. Assuming adequate training, a single bank teller can handle a deposit or withdrawal without requiring help from other tellers. She and other workers, however, may share equipment, such as the central computer that keeps track of all bank transactions.

APPLYING THE DIAGNOSTIC MODEL
Diagnostic Question 6: Is the organization's primary purpose instead to provide a service? Does it use intensive technology? If so, does the organization have any one of the organic structures named above? Or does it use mediating technology or long-linked technology? If so, does the organization have one of the mechanistic structures named above?

Although in the service organization individual employees work independently, many perform the same job. Coordination in such firms is needed to make sure that workers provide a consistently high quality of service and offer the same basic service to each client. Thus managers in service firms develop lists of the different types of clients their organization is likely to serve and devise a particular standard operating procedure to be followed while serving each type of client. For example, a bank teller will follow one procedure while serving a client who is making a savings account deposit, another procedure when waiting on a client who is making a loan payment, and yet a third when helping a client open a new checking account. This standardization of work processes or behaviors means that firms or SBUs using mediating technologies to coordinate worker-client relationships are most likely to be effective when structured mechanistically. Going back to our continuum in Figure 16-10, it would appear that either functional or divisional structures would be most suitable for such firms or SBUs.

LONG-LINKED TECHNOLOGY Thompson's long-linked technology is analogous to Woodward's mass production technology. Both refer to sequential chains of simplified tasks. A service-sector example of this type of technology is the state employment agency that requires all clients to follow the same lock-step procedures. Each client moves along an "assembly line" starting with registration and progressing through assessment, counseling, training, and placement activities. Another example of long-linked technology in a service organization consists of the in-person registration procedures used at some universities. If you haven't experienced this sort of registration yourself, imagine moving along a line of tables, picking up courses at one, having identification pictures taken at another, buying meal passes at a third, and making tuition payments at a fourth. The sort of sequential movement—from one table or station to the next—that characterizes long-linked technology is diagrammed in Figure 16-11B.

Like firms that use mass-production technology, those that use long-linked technology coordinate by means of standardization. According to Thompson, the effectiveness of a firm using long-linked technology is likely to be enhanced by mechanistic structuring. Our continuum suggests, therefore, that long-linked technology will be most effectively paired with functional or divisional structures.

INTENSIVE TECHNOLOGY Intensive technology consists of work processes whose configuration may change as employees receive feedback from the clients they serve. The specific assortment of services to be rendered to a particular patient in a hospital, for example, depends on the symptoms this particular patient exhibits. A patient entering the hospital's emergency room complaining of chest pains may be rushed to an operating room and then to a cardiac-care unit. A patient with a broken arm may be shuttled from the emergency room to the radiology lab for an x ray and then back to the emergency room for splinting. A third patient with uncertain symptoms may be checked into a room for further observation and testing (see Figure 16-11C).

To fit itself to the needs of each client, a firm using intensive technology must be able to reorganize itself again and again. So above all, it must have flexibility. And because the needs of future clients cannot be foretold, the work of such a firm is too unpredictable to be successfully formalized. Both flexibility and unpredictability require the use of mutual adjustment as a coordinating mechanism. Thus firms using intensive technology will be best suited by structures located toward the right side of our continuum. They will require simple or matrix structures or else, laterally linked functional or divisional structures.

Diagnostic Decision Chart III

The Woodward and Thompson technology models we have discussed will be important to you as a manager. They help to identify which organization structure is most likely to enhance the effectiveness of a firm using a specific type of technology. The main points of these models are summarized in the diagnostic design chart shown in Table 16-4. As indicated in this table, standardization and mechanistic structuring generally enhance the effectiveness of firms using technologies that are suited to more-routine work—mass production, mediating, and long-linked technologies. Mutual adjustment, on the other hand, promotes effectiveness in firms that use technologies suited to unpredictable, often rapidly changing requirements—mass production, mediating, and long-linked technologies.[18] As we would expect, and as Table 16-4 shows, the first group of technologies tend to be best served by the mechanistic type of structure (found on the left side of our Figure 16-10 continuum), the second group by the organic structural type (found on the right side of the continuum).

[18] Charles Perrow, "A Framework for the Comparative Analysis of Organizations," *American Sociological Review* 32 (1967), 194–208; Raymond G. Hunt, "Technology and Organization," *Academy of Management Journal* 13 (1970), 235–52; and William H. Starbuck, "Organizational Growth and Development," in *Handbook of Organizations*, ed. J. G. March (New York: Rand McNally, 1965), Chapter 11.

T A B L E 16-4
Diagnostic Design Chart III: Technological Contingencies

INDUSTRY TYPE	TECHNOLOGY	STRUCTURAL TYPE	SIZE	
			SMALLER STRUCTURE	LARGER STRUCTURE
Manufacturing	Small batch	Organic	Simple*	Matrix
			Laterally linked functional	Laterally linked divisional
	Mass	Mechanistic	Functional	Divisional
	Continuous process	Organic	Simple*	Matrix
			Laterally linked functional	Laterally linked divisional
	Flexible Cell	Organic	Simple*	Matrix
			Laterally linked functional	Laterally linked divisional
Service	Mediating	Mechanistic	Functional	Divisional
	Long Linked	Mechanistic	Functional	Divisional
	Intensive	Organic	Simple*	Matrix
			Laterally linked functional	Laterally linked divisional

* Note: Simple structures may be either simple undifferentiated or simple differentiated.

ORGANIZATION DESIGN

The External Environment

An organization's **environment** consists of everything outside the organization. Suppliers, customers, and competitors are part of an organization's environment as are the governmental bodies that regulate its business, the financial institutions and stockholders that supply it with funding, and the labor market that provides it with employees. In addition, general factors such as the economic, geographic, and political conditions that impinge on the firm are part of its environment. Central to this definition is the idea that the term *environment* refers to things external to the firm. The internal "environment" of a firm, more appropriately called the company's culture, will be discussed in Chapter 18.

As a structural contingency factor, an organization's environment influences the effectiveness of its structure by placing certain coordination and information-processing requirements on the firm. Five specific environmental characteristics influence structural effectiveness—environmental change, complexity, uncertainty, hostility, and diversity.

Environmental Change. **Environmental change** concerns the extent to which conditions in an organization's environment change unpredictably. At one extreme, an environment is stable if it does not change at all or if it changes only in a cyclical, predictable way. An example of such a stable environment is the one that surrounds many of the small firms in Amish communities throughout the midwestern United States. Amish religious beliefs require the rejection of modern conveniences, such as automobiles, televisions, and automated farm equipment. So Amish blacksmiths, dry-goods merchants, and livestock breeders have conducted business in much the same way for generations. Another stable environment is that surrounding firms that sell Christmas trees. The retail market for cut evergreen trees is predictably strong in November and December but absent at other times of the year.

At the other extreme, an environment is dynamic when it changes over time in an unpredictable manner. Because the style of dress changes so frequently in societies like ours, the environment surrounding companies in the fashion industry is quite dynamic. Similarly, the environment surrounding companies in the consumer-electronics industry has changed dramatically. New products, such as projection television, videotape machines, and Walkman radios, have created entirely new markets. Older products have been redesigned to incorporate computer microchips, digital displays, infrared remote controls, and similar technological breakthroughs.

Environmental change affects the structure of an organization by influencing the predictability of the firm's work and, therefore, the method of coordination

Videoconferencing may become an increasingly important way of coordinating an organization's workflow as firms continue to establish departments, divisions, and subsidiaries in locations far from headquarters offices. PictureTel of Peabody, Massachusetts, has developed a video system that can accommodate the starts, stops, and interruptions of normal conversation, even when people seem to be talking all at once. In addition, a "windowing feature" lets people see themselves as they appear to conference participants at the other end of the line.
Source: "Videoconferencing Gets Cheaper," Fortune, March 11, 1991, p. 76.

used to integrate work activities.[19] Stability allows managers to complete the planning needed to formalize organizational activities. One can predict and plan for variation when it is cyclical in nature. Firms operating in stable environments can and do use standardization as their primary coordination mechanism and will typically elect to do so to reduce long-term coordination costs. Mechanistic structures are the most likely to prove effective in such instances.

In addition, as with the technological contingencies discussed earlier, each SBU in an SBU structure has its own environment and should therefore be structured in accordance with that environment. Thus an SBU dealing with a stable environment should have a mechanistic structure.

In a dynamic environment, it is difficult to establish formal rules and procedures. In fact, it is useless for managers to try to plan for a future they cannot foresee. Members of an organization or SBU facing a dynamic environment must adapt to changing conditions instead of relying on inflexible, standardized operating procedures. Dynamism in the environment leaves management with little choice but to rely on mutual adjustment as a primary coordination mechanism. Organic structuring is therefore appropriate.

environmental complexity An environmental characteristic referring to the degree to which an organization's environment is complicated and therefore difficult to understand.

AN EXAMPLE
With cessation of the cold war, the defense industry has become much more dynamic. Defense spending is being cut and, as a result, fewer contracts are available. Even if a contract is won it may be cancelled long before its termination date. This situation has made it very difficult for many companies to plan and, in some cases, survive.

Environmental Complexity. **Environmental complexity** is the degree to which an organization's environment is complicated and therefore difficult to understand. A simple environment is composed of relatively few component parts— for example, suppliers, competitors, types of customers. So there is not much that can affect organizational performance. A local Amoco gas station does business in a relatively simple environment. It orders most of its supplies from a single petroleum distributor, does business almost exclusively with customers who want to buy gasoline or oil for their cars, and can limit its attention on the competitive activities of a fairly small number of nearby stations. On the other hand, a complex environment consists of a large number of component parts. The environments of aviation firms like Boeing and McDonnell Douglas are extremely complex, including an enormous number of suppliers, many different types of customers, and scores of foreign and domestic competitors.

Complexity influences structural effectiveness by affecting the amount of knowledge and information people must process to understand the environment and cope with its demands.[20] Consider an inexpensive digital watch. If you took this watch apart, you would probably not have much trouble putting it back together again, because it has very few parts—a computer chip programmed to keep time, a digital liquid-crystal face, a battery, and a case. With only a few minutes of practice or simple instructions, you could quickly learn to assemble this watch. Now suppose you had the pieces of a Rolex watch spread out before you. Could you reassemble the watch? Probably not, because it is made up of hundreds of springs, screws, gears, and other parts. Learning to assemble a Rolex properly would require intensive training.

Similarly, the organization facing a simple environment—one with few "parts"—can understand environmental events and meet the challenges they pose using a minimal amount of knowledge and processing little new information. For instance, a local lawn-care firm that is losing business can determine the reason for its plight simply by telephoning a few former customers and asking

---

[19] Burns and Stalker, *The Management of Innovation*; C. R. Hinings, D. J. Hickson, J. M. Pennings, and R. E. Schneck, "Structural Conditions of Intraorganizational Power," *Administrative Science Quarterly* 19 (1974), 22–44; Robert B. Duncan, "Multiple Decision-Making Structures in Adapting to Environmental Uncertainty: The Impact of Organizational Effectiveness," *Human Relations* 26 (1973), 273–91.

[20] Robert B. Duncan, "Characteristics of Organizational Environments and Perceived Environmental Uncertainty," *Administrative Science Quarterly* 17 (1972), 313–27; and Jay R. Galbraith, *Designing Complex Organizations* (Reading, Mass.: Addison-Wesley, 1973), pp. 4–6.

them why they stopped using the company's service. However, organizations in complex environments—environments with many "parts"—must draw on a considerable store of knowledge and process an overwhelming amount of information to understand environmental events. To find the reason for their loss of market share in the early 1980s and again in the late 1980s, Chrysler Corporation analyzed competitors' marketing strategies and performed extensive market studies of consumer preferences. To recapture market share, Chrysler also worked with hundreds of suppliers to increase the quality and reduce the cost of the parts used to produce its cars.

How does environmental complexity affect organizational structures? Simply stated, environmental complexity influences the suitability of centralization as an interunit coordination mechanism. As you will recall from Chapter 15, in centralization, decision making is limited to a select group of top managers. Centralization thus minimizes the number of people available to digest information and determine its meaning. Because simple environments require little information processing, organizations operating in such environments can be centralized and function quite effectively. However, because environmental complexity requires the ability to process and understand large amounts of information, centralized organizations in complex environments can suffer the effects of information overload. One way to cope with this information overload is to involve more individuals in information-processing activities. Thus organizations like Chrysler that are attempting to cope with complex environments often decentralize decision making. That way, they include more people—more brains—in the process of digesting and interpreting information.

environmental uncertainty An environmental characteristic formed by the combination of change and complexity that reflects the absence of information about environmental factors, activities, and events.

Environmental Uncertainty. In addition to pointing out distinctive environmental differences, the two environmental dimensions of change and complexity also combine to define yet another important environmental characteristic—**environmental uncertainty** (see Figure 16-12). Uncertainty reflects the absence of information about environmental factors, activities, and events.[21] It undermines

[21] Jay R. Galbraith, *Organization Design* (Reading, Mass.: Addison-Wesley, 1977), p. 4.

FIGURE 16-12

Environmental Uncertainty as a Function of Change and Complexity

The amount of change and complexity in an organization's environment affects the degree to which executives perceive the environment as uncertain. As this figure shows, greater change and increasing complexity contribute to higher levels of perceived uncertainty.
Based on Robert B. Duncan, "Characteristics of Perceived Environments and Perceived Environmental Uncertainty," Administrative Science Quarterly *17 (1972), 313–27.*

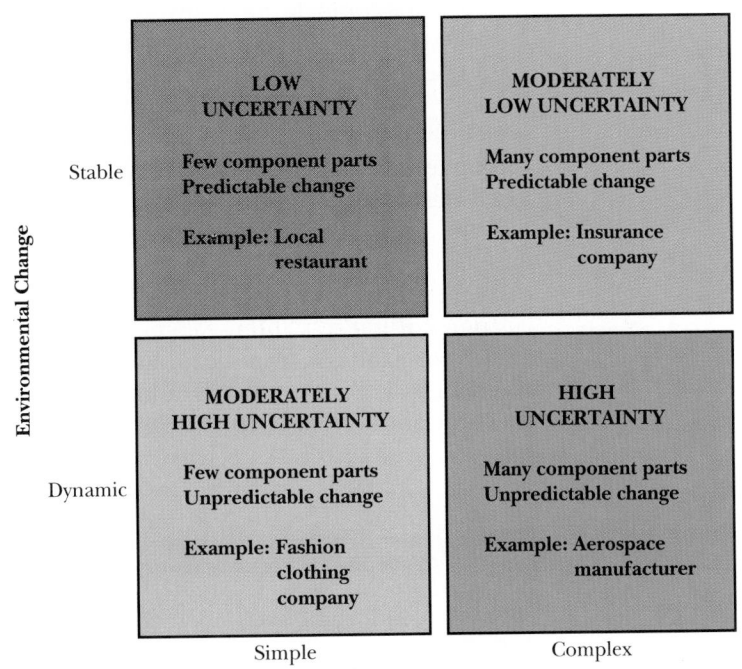

an organization's ability to manage current circumstances and plan for the future. To cope with uncertainty, organizations try to find better ways of acquiring information about the environment. This effort often involves the creation of boundary-spanning positions that can strengthen the information linkage between an organization and its environment.

A **boundary spanner** is a member or unit of an organization that interacts with people or firms in the organization's environment.[22] Salespeople, who have contact with customers, purchasing departments that deal with suppliers of raw materials, and top managers who in their figurehead roles represent the company to outsiders are all boundary spanners. In boundary-spanning roles, employees or units

Monitor the environment for information that is relevant to the organization

Serve as gatekeepers, simplifying incoming information and ensuring that it is routed to the appropriate people in the firm

Warn the organization of environmental threats and initiate activities that protect it from them

Represent the organization to other individuals or firms in its environment, providing them with information about the organization

Negotiate with other organizations to acquire raw materials and sell finished goods or services

Coordinate any other activities that require the cooperation of two or more firms.[23]

These activities enable boundary spanners to provide the organization with information about its environment that can help make change and complexity more understandable. As a result, the organization can adapt to its environment instead of being overwhelmed by unexpected environmental events.

Environmental Receptivity. **Environmental receptivity** is the degree to which an organization's environment supports the organization's progress toward fulfilling its purpose. In a munificent environment, a firm is able to acquire the raw materials, employees, technology, and capital resources needed to perform productively. In such an environment, the firm finds a receptive market for its products. Its competitors, if any, do not threaten its existence. Regulatory bodies do not try to impede its progress. Thus, for example, the environment surrounding the McDonald's fast-food chain at the time of its founding was munificent. Few other fast-food franchises existed, labor was fairly plentiful in the post-Korean war era, and a convenience-minded middle class was emerging throughout North America. Organizations involved in military contracting enjoyed this same munificent environment during the Reagan presidency.

In a hostile environment, the opposite situation obtains. An organization may have great difficulty acquiring, or may be unable to acquire, needed resources, employees, knowledge, or money. The firm's future may also be threatened by customer disinterest, intense competition, or severe regulation. During the early 1980s, for example, American auto producers faced an intensely hostile environment as U.S. consumers shunned Fords, Chryslers, and Chevys in favor of fuel-efficient Toyotas, Hondas, and Nissans. The Phillip-Morris Company and

[22] J. Stacy Adams, The Structure and Dynamics of Behavior in Organization Boundary Roles," in *Handbook of Industrial and Organizational Psychology*, ed. M. D. Dunnette (Chicago: Rand McNally, 1976), pp. 1175–99.

[23] Howard Aldrich and Diane Herker, "Boundary Spanning Roles and Organization Structure," *Academy of Management Review* 2 (1977), 217–39; Robert Miles, *Macro Organizational Behavior* (Santa Monica, Calif.: Goodyear, 1979), pp. 320–39; and Richard L. Daft and Richard M. Steers, *Organizations: A Micro/Macro Approach* (Glenview, Ill.: Scott, Foresman, 1986), p. 299.

MANAGEMENT ISSUES
The Ethics of Influencing the Environment

Although organizations are influenced by their environments, the opposite is equally true. Organizations often attempt to influence their environments. For example:

Facing increasing public resistance and threatened anti-smoking legislation in many states and communities, Phillip-Morris began a series of advertisements extolling the importance of constitutional rights. The implicit message of these advertisements, printed and televised in 1989 and 1990, was that smoking is a personal choice with which government agencies should not interfere.

The savings and loan (S&L) bailout of the 1990s resulted from the deregulation of the 1980s. This deregulation developed as the S&L industry lobbied congressional regulation committees to allow S&Ls greater leeway in making loans to real estate partnerships and other speculative business ventures. The same industry supposedly controlled by federal regulations had a strong hand in developing those regulations.

Are activities of this sort right or wrong? Why? Are all attempts by organizations to influence the surrounding environment necessarily good or bad? How might society determine which (if any) of such activities are improper and guard against them? What is sacrificed when it becomes necessary to be on guard?

The same environment that created S&L failures also caused trouble in the general banking industry.

other members of the tobacco industry had to cope with extreme hostility later in the same decade. C. Everett Koop, then surgeon general of the United States, set a goal of eliminating smoking throughout North America by the year 2000 (see the "Management Issues" box for a related discussion).

Environmental hostility, though normally temporary, represents a crisis that must be dealt with quickly and effectively if the firm is to survive. An organization facing such hostility either finds a way to deal with it—for example, by substituting one raw material for another, marketing a new product, or lobbying against threatening regulations—or it ceases to exist. Thus American automobile manufacturers quickly designed smaller, higher-efficiency cars, such as the Ford Escort and Plymouth Reliant, and began to market them in the mid-1980s. Phillip-Morris in 1989 began a massive advertising campaign centered on the U.S. Constitution and "the freedom to choose."

To deal with the crisis of a hostile environment, firms that are normally decentralized because of environmental complexity will centralize decision making for a limited period of time.[24] This temporary centralization facilitates crisis management. Because it reduces the number of people who must be consulted to make a decision the organization can respond to threatening conditions more quickly. It is important to emphasize that centralization established in response to a hostile environment should remain in effect only so long as the hostility persists. When munificence reappears, a firm dealing with a complex environment will perform effectively only if it reinstates decentralized decision making.

[24] Mintzberg, *Structuring of Organizations*, p. 281.

environmental diversity The degree to which an organization's environment is varied or heterogeneous in nature.

environmental domain A part or segment of the environment in which an organization does business.

Environmental Diversity. **Environmental diversity** refers to the number of distinct **environmental domains** served by an organization. A firm in a uniform environment serves a single type of customer, provides a single kind of product, and conducts its business in a single geographic location. Thus it serves only a single domain. A campus nightclub, for example, that caters to the entertainment needs of local college students operates in a uniform environment. So does a building-materials firm whose sole product is concrete, which it sells only to local contractors. In contrast, an organization in a *diverse* environment produces an assortment of products, serves various types of customers, or has offices or other facilities in several geographic locations. It does business in several different domains. IBM, for instance, sells computers to businesses, universities, and the general public. General Electric handles consumer electronics, financial services, jet engines, and diesel locomotives. Ford Motor Company markets cars in North America, South America, and Europe.

Environmental diversity affects an organization by influencing the amount of diversity that must be built into its structure.[25] In organizations with uniform environments managers can use *functional* departmentation to group units together. Because firms in uniform environments face only a single domain, they need concern themselves only with information about a single kind of environment, and they need react to only a single set of environmental events. Functional departmentation, which facilitates this sort of unified information processing and response, is therefore sufficient in such situations. The absence of environmental diversity permits the firms to operate effectively without significant internal diversification.

In organizations with diverse environments, however, management must use divisional departmentation so as to gather work associated with each product, customer, or location into its own self-contained division. Companies in diverse environments face a number of distinct domains and must acquire information about each in order to cope with its particular demands. Divisional departmentation allows these firms to keep track of each domain separately and respond to the demands of one domain independently of others. If managers didn't structure the firm this way, work on one product could get in the way of work on other products, services rendered to one type of customer could detract from services provided to other types of customers, or operations at one location could impede operations at other locations.

Environmental uniformity, then, favors functional departmentation, therefore a functional structure. Environmental diversity requires divisional departmentation and either a divisional, matrix, or SBU structure, depending on other contingency factors. It is important to note that organizational size affects the likelihood that a firm will have to deal with appreciable environmental diversity. It's only when an organization has grown from a relatively small firm into a larger organization that it begins to evidence product, customer, or location diversity. Moreover, because a small firm has a simple structure and uses neither functional nor divisional departmentation, environmental diversity is not an issue for such an organization.

Diagnostic Design Chart IV

Environments have five distinct characteristics—change, complexity, uncertainty, receptivity, and diversity. Therefore, diagnosing the nature of a firm's environment during the process of organization design requires that managers do five environmental analyses more or less simultaneously. To help you perform this

[25] Thompson, *Organizations in Action*, pp. 25–38.

ORGANIZATION DESIGN

sort of diagnosis, we've devised the *decision tree* shown in Figure 16-13. If you trace through the branches of this tree as you answer each of the four questions we suggest, you will be led to the most suitable organizational structure for the environment you are diagnosing.

Each question deals with one of the environmental characteristics we have just examined. Note that we need not ask a separate question about uncertainty. Because it is a combination of change and complexity, uncertainty is assessed by the answers to questions 1 and 2 (see Figure 16-12).

APPLYING THE DIAGNOSTIC MODEL
Diagnostic Question 9: Do the various contingencies seem to mandate the same type of structure? If not, and if there is no evidence of recent changes that might serve as an explanation, look for faulty diagnosis in one or more of your contingency analyses.

1. *Is the environment stable or dynamic?* The answer to this question identifies the amount of change in the environment and helps to determine whether standardization or mutual adjustment is likely to be the more effective co-ordination mechanism for the firm under analysis. Stable environments either do not change or change in a predictable, cyclical manner and thus allow the use of standardization. Dynamic environments change in unpredictable ways and require mutual adjustment as a result.

2. *Is the environment simple or complex?* Here the answer will be an assessment of environmental complexity and will indicate whether centralization or decentralization is more appropriate for the firm. Simple environments are easy to figure out and allow centralization. Complex environments require

FIGURE 16-13

Diagnostic Decision Chart IV: Environmental Contingencies

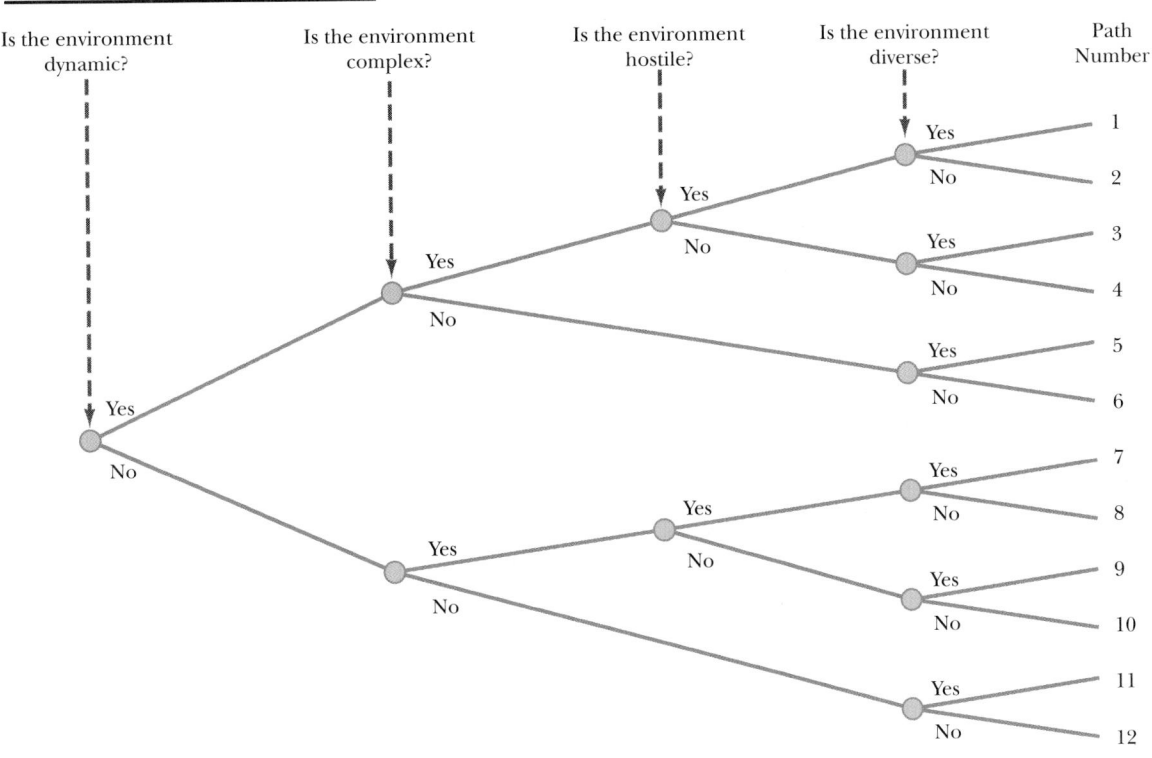

Path Number	Structural Alternatives
1	Divisional with lateral linkages or matrix; temporarily centralized; extensive boundary spanning
2	Functional with lateral linkages or matrix; temporary centralization; extensive boundary spanning
3	Divisional with lateral linkages or matrix; extensive boundary spanning
4	Functional with lateral linkages or matrix; extensive boundary spanning

Path Number	Structural Alternatives
5	Divisional with lateral linkages
6	Functional with lateral linkages
7	Divisional; temporary centralization
8	Functional
9	Divisional
10	Functional
11	Simple or divisional, depending on size
12	Simple or functional, depending on size

Diagnostic Question 10: If the
different contingencies *do* seem to
point toward the same type of
structure, is this structure the
same as the one the organization
has now? If not, the structure
recommended by your analysis
should be implemented; structural
redesign is needed. If, however,
the current structure is the one
recommended by your analysis,
structural deficiencies are not the
cause of the organization's
problems. Look for individual or
group-level problems instead.

a great deal of information processing and therefore exert pressure toward decentralization.

3. *Is the environment munificent or hostile?* This question is relevant only if an organization has a complex environment and is thus decentralized. How it is answered gauges environmental receptivity and indicates whether temporary centralization is necessary. Munificent environments are resource rich and allow continued decentralization, but hostile environments are resource poor and stimulate crises that mandate temporary centralization.

4. *Is the environment uniform or diverse?* In response to this question, a manager must evaluate environmental diversity so as to determine what form of departmentation to use. Environmental uniformity allows the structural uniformity of functional departmentation. Environmental diversity requires the structural diversity of divisional departmentation.

SUMMARY

Organization design is the process of structuring an organization so as to enhance *organizational effectiveness* in the light of the *contingency factors* it faces. *Functional* and *divisional* structures are *mechanistic*. *Matrix, simple differentiated,* and *simple undifferentiated* structures are *organic*. *Strategic business unit* structures can combine both mechanistic and organic parts.

Whether an organization should have a mechanistic structure or an organic structure depends on the effects of *closed system contingency* factors within the organization, and on the influence of *open system contingency* factors outside of it.

One closed system contingency, the *size* of an organization in terms of the number of members it has, influences the amount of coordination and information exchange needed to integrate work activities and thus the likely effectiveness of different coordination mechanisms. Size is often related to two other contingency factors, *life-cycle stage* and *age*. Organizations usually grow larger as they grow older and progress from earlier stages of development to later ones.

An additional closed system contingency factor, *technology*, affects the degree to which an organization's work is more or less routine and thus, again, the appropriateness of different coordination mechanisms. More routine technologies such as *mass production, mediating technology,* and *long-linked technology* can be coordinated with standardization. Less routine technologies such as *small-batch production, continuous process production, flexible cell production,* and *intensive technology* require the use of mutual adjustment. The organization's *environment* is an open system contingency factor that places coordination and information-processing requirements on the firm. These requirements stem from environmental *change, complexity, uncertainty, receptivity,* and *diversity*. In turn, these requirements influence the coordination, decision-making, boundary-spanning, and departmentation processes that give rise to the organization's structure.

REVIEW QUESTIONS

1. Name a specific business organization in your community and identify three of its most important constituency groups. What interests do each of the constituency groups expect the organization to fulfill? How does the organization's structure affect its ability to satisfy these interests? Is the company effective?

2. In what major way do simple undifferentiated structures differ from simple differentiated structures? What is the importance of this difference? What strengths do both kinds of simple structure share? What are the major weaknesses of each one?

3. How do complex structures differ from simple structures? Which complex structure is the most mechanistic? Why? Which is the most organic? Why? Compare the strengths and weaknesses of these two types of complex structure.

4. What typical effect does the age of an organization have on its size? According to the Quinn-Cameron life-cycle model, how does growing older and larger affect organizational structure?

5. In which of Woodward's and Thompson's technologies is work routine and predictable? In which of them is work nonroutine and unpredictable? What kinds of structures are most fitting for each of the two clusters of technologies you have identified? In general terms, how does the routineness and predictability of an organization's technology affect the appropriateness of different types of structure?

6. Explain why environmental change impedes an organization's ability to coordinate by means of standardization. What sort of coordination is used instead? Why does environmental complexity push toward decentralization? Given that environmental uncertainty is a combination of change and complexity, what effects besides increased boundary spanning would you expect it to have on the way an organization is structured?

7. Why is environmental receptivity an issue only for organizations facing complex environments? Under conditions of environmental complexity, why should the centralization stimulated by hostility be eliminated as the environment becomes munificent?

8. After reviewing the story that opened this chapter, use the concept of environmental diversity to explain the failure of General Motors' attempted use of functional departmentation to group its assembly units together. What had GM's management hoped to gain by using functional grouping? Why didn't things work out as expected?

DIAGNOSTIC QUESTIONS

The following questions should help you during the process of designing or redesigning an organization's structure by guiding you through the diagnosis of the contingency factors we have discussed.

1. What kind of structure does the organization currently have? What is its primary means of coordination? Do the titles of its line vice-presidents provide evidence of one particular form of structure?

2. Is the organization's current structure the one dictated by its size? If the organization has a simple undifferentiated structure, does it have no more than about 12 employees? If the organization has a simple differentiated structure, does it have no more than about 50 employees? If the organization has a complex structure, is it substantially larger?

3. Is the organization's current structure the one suggested by the firm's life-cycle stage and age? If the firm is young and in the entrepreneur or collectivity stage, does it have a smaller, simple structure? If it is older and in the formalization or elaboration stage, does it have a larger, complex structure?

4. Does the organization have an SBU structure? If so, perform separate technology and environment analyses. That is, answer the rest of the diagnostic questions for each SBU as though it were a separate organization.

5. Is the organization's primary purpose to manufacture a tangible product? Does it use small-batch production, continuous process production, or flexible cell production? If it uses one of these, does the organization have an organic structure? Is it simple or matrix, functional, or divisional with lateral linkages? Or does it use mass production? Does the organization have a mechanistic structure? Is it functional or divisional?

6. Is the organization's primary purpose instead to provide a service? Does it use intensive technology? If so, does the organization have one of the organic

structures named above? Or does it use mediating technology or long-linked technology? If one of these, does the organization have one of the mechanistic structures named above?

7.   Is the organization's external environment stable or dynamic? Does the organization's primary mode of coordination match the amount of change in its environment? Is the environment simple or complex? Does the degree of decentralization in the organization's structure match the amount of complexity in its environment?

8.   Is the organization's external environment uncertain? If so, is there evidence of significant boundary-spanning activities? Is the environment munificent or hostile? If hostility exists and the environment is also complex, is the organization temporarily centralized? Is the environment uniform or diverse? Does the type of departmentation used to structure the firm match the diversity of its environment?

9.   Do the various contingencies seem to mandate the same type of structure? If not, and if there is no evidence of recent changes that might serve as an explanation, look for faulty diagnosis in one or more of your contingency analyses.

10.   If the different contingencies *do* seem to point toward the same type of structure, is this structure the same as the one the organization has now? If not, the structure recommended by your analysis should be implemented; structural redesign is needed. If, however, the current structure is the one recommended by your analysis, structural deficiencies are not the cause of the organization's problems. Look for individual or group-level problems instead.

EXERCISE 16-1
OPEN SYSTEM PLANNING*
MARK S. PLOVNICK, *University of the Pacific*
RONALD E. FRY, *Case Western Reserve University*
W. WARNER BURKE, *Columbia University*

Open system planning (OSP) is a technique that can be used to clarify an organization's mission and plan how to achieve it in the face of demands and expectations originating in the environment. These demands come from such constituency groups as employees, customers, and raw material suppliers. OSP is an integral part of the process of strategic planning and a useful way to improve an organization's understanding of its environment (see Chapter 18).

An open system planning intervention begins with a discussion of the organization's basic goals, mission, and reason for being. The participants in an initial OSP session then identify the constituency groups that can have an impact on the organization's accomplishment of its goals and mission. Participants describe the current relations between the organization and each of its constituency groups. They assess these relations, deciding whether they satisfy both the organization and the constituency group. If assessment uncovers dissatisfaction, participants determine how relations ought to be to achieve a good balance between organizational effectiveness and constituency satisfaction.

OSP participants then assess the organization's current response to each constituency group by answering several questions. What does this type of constituency want from us? What are we currently doing in response to this demand? Is our current response moving us closer to where we want to be in relation to our organization's goals and purpose? Finally, OSP participants decide what actions, if any, must be taken to redirect the organization toward the desired state of affairs.

This exercise will give you the opportunity to experience the process of open system planning first-hand as you work with other class members to assess the environment and constituency groups of a real organization.

STEP ONE: PRE-CLASS PREPARATION

In class you will perform parts of each phase of the open system planning process. The focal organization will be the school or education program in which you are enrolled (unless your instructor specifies another organization instead). In preparation for class, read the entire exercise. Then think about what the basic mission

or purpose of the organization should be and write a mission statement here:

_____

_____

_____

Finally, identify five constituency groups that expect the organization to do something for them. List these groups and their demands here:

1. _____

2. _____

3. _____

4. _____

5. _____

STEP TWO: DEFINING A MISSION

The class should divide into planning groups of four to six members each. If you have already established permanent groups, reassemble in those groups. In each group, members should share the mission statements they developed before class and reach a consensus about the organization's mission. Next, your instructor will lead the entire class in developing a mission statement to be used for the rest of the exercise.

STEP THREE: IDENTIFYING CONSTITUENCY GROUPS

The class as a whole should agree on the key constituency groups in the environment that create demands on the organization. If the organization you have decided to focus on is your school, examples of these

* Source: Mark S. Plovnick, Ronald E. Fry, and W. Warner Burke, *Organization Development: Exercises, Cases, and Readings* (Boston: Little, Brown and Company, 1982), pp. 67–73. Copyright © 1982 by Mark S. Plovnick, Ronald E. Fry, and W. Warner Burke.

groups might include students, faculty, alumni, and employees. Each planning group should be assigned one constituency group to consider in the next step of this exercise (every group should be assigned a different constituency).

STEP FOUR: OPEN SYSTEM PLANNING IN GROUPS

In order to experience the OSP process, each planning group should complete the remaining phases of the process, focusing only on the constituency group assigned to it. Following is a description of the five phases you should complete.

1. *Identification of current demands.* What does the constituency group currently expect from or demand of the organization? (estimated time: 15 minutes)
2. *Current response.* For the one or two most important of these demands, what is the organization's current response? Is it a response of action or inaction? (estimated time: 10 minutes)
3. *Future demands.* Considering the current response and whatever changes or trends you think are likely to occur over the next two years, what are the key demands from your constituency group going to be two years from now? (estimated time: 10 minutes)
4. *Ideal state.* Imagine two years from now. What kinds of expectations or demands would the organization like to see coming from the constituency group you are analyzing? (estimated time: 10 minutes)
5. *Identifying gaps and planning action.* Compare and contrast the results of phases 3 and 4. What gaps exist between anticipated and ideal demands? Choose one of these gaps and suggest what the organization should do to alter its current course. (estimated time: 15 minutes)

Note that you have one hour to complete these five phases. You will have to manage your time carefully. A spokesperson should be ready to present a summary of the group's work in Step Five.

STEP FIVE: PLANNING GROUP REPORTS AND DISCUSSION

A spokesperson from each group should take no more than five minutes to summarize what the group discovered, discussed, and concluded about the constituency group it examined. Any suggestions for action should be listed on a chalkboard, overhead transparency, or flipchart.

STEP SIX: CLASS DISCUSSION OF THE OSP PROCESS

The class as a whole should now review the OSP process just completed and discuss the following questions:

1. Do the planning group reports reveal any potential conflicts between satisfying different constituency groups' demands? That is, will satisfying one constituency make it difficult to satisfy another? How might such conflicts be resolved?
2. Do the actions proposed by the planning groups fit together into a meaningful action plan? If so, describe the plan. If not, how can they be made to do so?
3. What changes, if any, would you recommend in the mission statement developed in Step Two? How does knowledge about an organization's environment affect perceptions of its mission?
4. How would you expect OSP as conducted in real organizations to differ from the process you have completed in class? In what ways might it be more complicated? Less complicated?
5. What kinds of organizations or situations do you see as likely candidates for OSP? Which ones should probably not use it?

CONCLUSION

Open system planning is not a panacea for all organizations. A significant investment of time and energy is required. In addition, the OSP process may involve a great deal of ambiguity and stress. However, OSP can enable organizations to manage their responses to important but often conflicting environmental demands. OSP can also be used as a method to help the members of organizations achieve consensus about organizational missions and goals. The resulting consensus can provide the coherent sense of direction needed to function effectively in complex, changing environments.

DIAGNOSING ORGANIZATIONAL BEHAVIOR

CASE 16-1

NEWCOMER-WILLSON HOSPITAL*

SAMUEL M. WILSON, *Temple University*

The administrative process in a hospital is complex. There are not many organizations with such cumbersome structures that still succeed. The administrative task is neither clear, definite, nor clean cut at any time. If it were, the doctors would not want it to be. So, we have to set up an elaborate framework of communications, especially with committee structures, in order to keep things moving.

This is the view of how hospitals are organized and run held by Mr. William Baker, the professionally trained and experienced hospital Director (administrator) of the Newcomer-Willson Hospital.

THE HOSPITAL BUSINESS— BACKGROUND AND CURRENT STATUS

In medieval Europe, hospitals were lodging places for travelers and were supported primarily by religious organizations. Later these places started taking care of old people and the homeless. Originally they were not institutions for the care and recovery of the sick. These charitable institutions provided for the needy, and doctors did provide some medical care mostly as a benevolence. Until about 1850 hospitals were usually very poor substitutes for home care. Living standards, the evolution of modern medicine, along with many scientific and technological developments associated with the detection and treatment of illnesses, have caused hospitals to become important centers for medical care and training of doctors and nurses. Today doctors usually have a primary interest in the establishment and operation of hospitals because their chief professional interest is medical care, i.e., the same objective as that of hospital management.

The purpose of all hospitals is to provide medical care. Most hospitals provide this care on a "nonprofit" basis from an accounting point of view. However, hospitals are business organizations and must conduct their long-run activities in such a manner that total revenue from all sources equals total expenditures, i.e., they must break even.

HOSPITAL ADMINISTRATION

The growth of hospitals in size and numbers during the first part of the twentieth century ushered in a new era in hospital administration. Traditionally hospital administrators were either doctors or nurses who devoted whatever time was needed to the administrative matters of the organization. In small hospitals, this was not a particularly time-consuming job especially when a full-time clerical assistant was used. The growth of hospitals (size and numbers) challenged this arrangement. Doctors and nurses were taken from their respective professional fields too much of the time, and they did not have the training or experience necessary for successful managers. As far back as 1938 the University of Chicago established the first program designed especially for hospital administration. Since that time many other universities have initiated such programs while many Schools of Business Administration feel that they prepare graduates who are qualified for this type of work.

THE NEWCOMER-WILLSON HOSPITAL

The Newcomer-Willson Hospital is located in a growing and prosperous suburban community. It has grown rapidly in the last 15 years, and with its recent expansion a total of 285 beds, 45 bassinets, and 700 rooms for different purposes are provided. It has a good rate (85%) of bed utilization as compared to the national average. There are 700 full-time employees, 400 medical staff members (of which 270 are courtesy members), and a nursing school of 160 students.

The policies and organization of the hospital were reviewed two years ago by Cresap, McCormick and Paget, Management Consultants. The Consultant's report covered all areas of the hospital in a rather broad way. It was generally favorable. It noted that over the past several years prime attention had been given to organization for administration, but that there were some problem areas needing additional attention.

AUTHORITY STRUCTURE AT NEWCOMER-WILLSON HOSPITAL: TOP MANAGEMENT

Exhibit 16-1 reveals the top and middle management organization at Newcomer-Willson. The top corporate body is the Board of Trustees, which is composed of

* Reprinted with the author's permission.

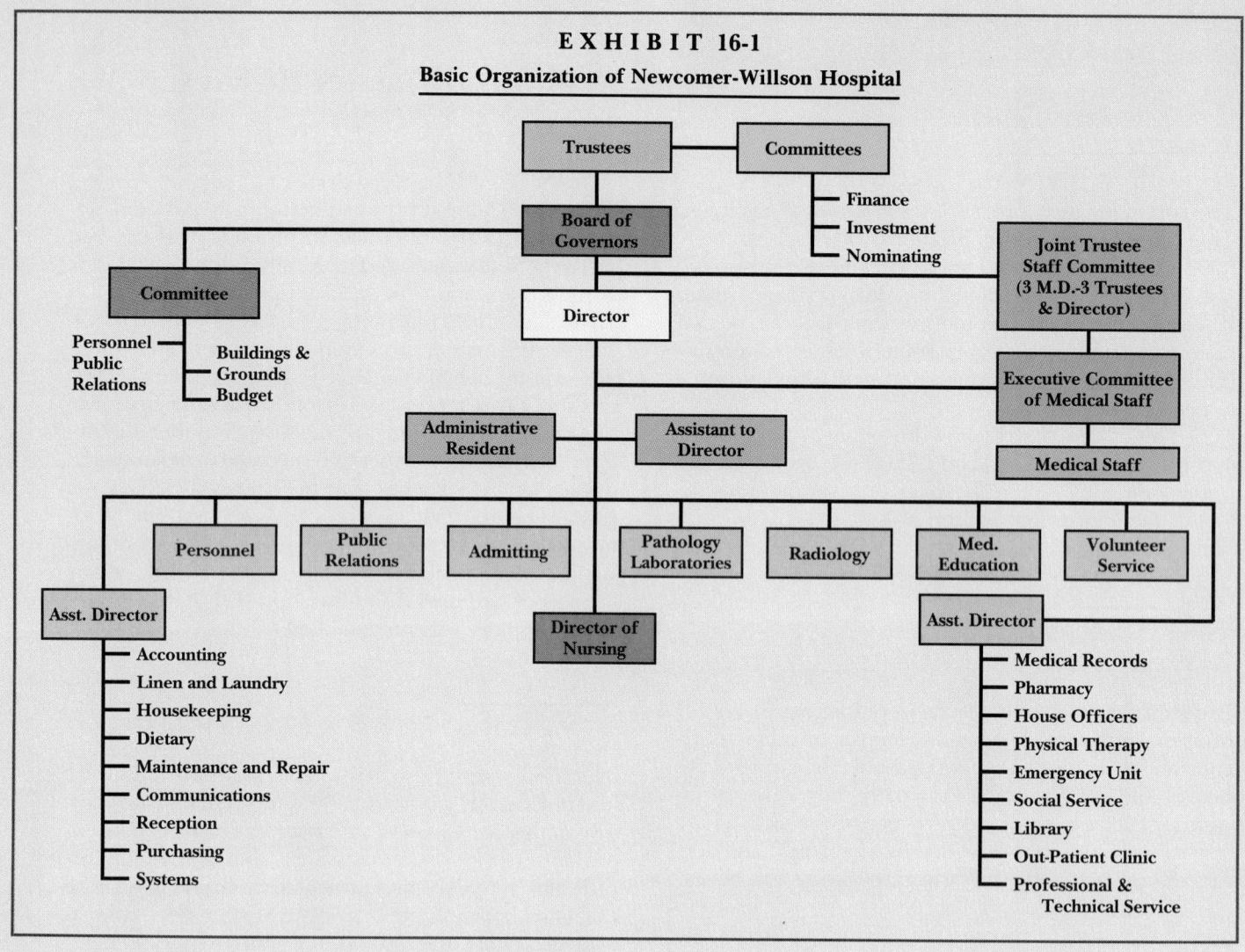

EXHIBIT 16-1

Basic Organization of Newcomer-Willson Hospital

about 70 Volunteers (nonmedical) from a variety of fields. Most of the trustees are local citizens of some standing in the community.

The most important body in the administrative process which deals with top level considerations and problems is the Board of Governors. This Board includes the four corporate officers from the Board of Trustees and nine elected trustees. The Board of Governors is responsible for the general administration of the organization, and it appoints all members of the Medical Staff and all other key people of the hospital.

The Medical Staff organization is headed by an Executive Committee. This committee operates within the framework of the By-laws of the Corporation and more specifically within its own By-laws which were approved by the Board of Governors. This committee may report to either the Director, Mr. Baker, or to the Joint Trustee Medical Staff Committee, depending on a variety of situations. In the past there has been a deliberate attempt at times to override the Director because of the

nature of the problems. This has not usually been achieved, however, because the Director is a member of the Joint Committee.

The medical staff is composed of doctors who have private practice in the surrounding community. They use the hospital when the need arises. They serve on committees of various types (see committees on Exhibit 16-2. These staff members are highly trained in their individual professional fields of medicine, and their primary concern is for their particular patients who are in the hospital. In fact, this concern causes some problems for Mr. Baker, the nursing staff, and others in supervisory positions because the hospital staff must think in terms of all of the patients, not just one or a few.

The chief full-time administrative position is that of the Director, currently held by Mr. Baker. Mr. Baker has been with the hospital for several years. He is a Fellow in hospital administration and has a great deal of administrative experience in several different positions of various organizations. He has been a prime

ORGANIZATION DESIGN

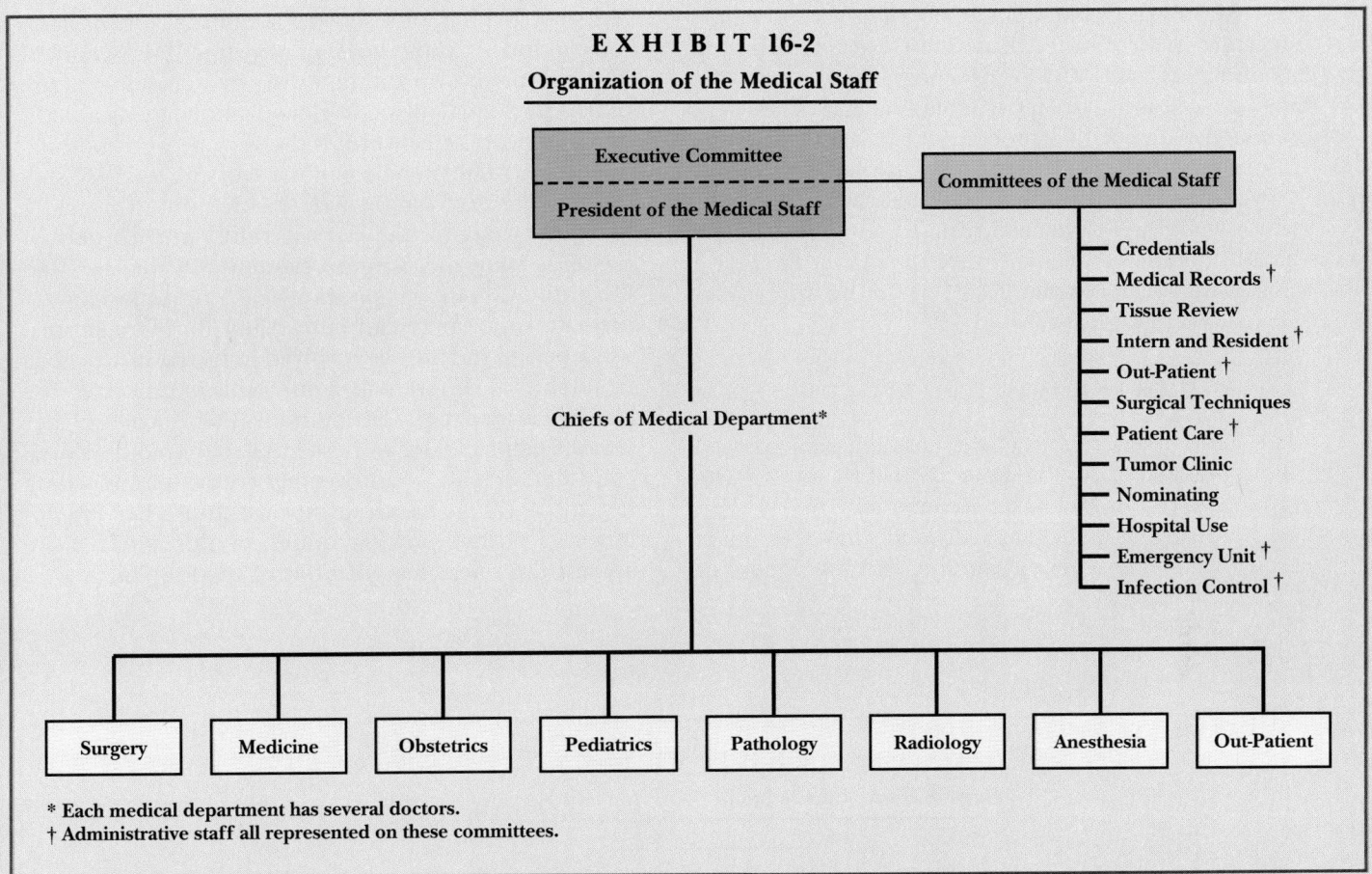

EXHIBIT 16-2

Organization of the Medical Staff

Executive Committee

President of the Medical Staff

Committees of the Medical Staff

- Credentials
- Medical Records †
- Tissue Review
- Intern and Resident †
- Out-Patient †
- Surgical Techniques
- Patient Care †
- Tumor Clinic
- Nominating
- Hospital Use
- Emergency Unit †
- Infection Control †

Chiefs of Medical Department*

| Surgery | Medicine | Obstetrics | Pediatrics | Pathology | Radiology | Anesthesia | Out-Patient |

* Each medical department has several doctors.
† Administrative staff all represented on these committees.

mover in establishing an improved administrative organization with the establishment of committees, by-laws for the medical staff, and written descriptions of duties and responsibilities for all managers and committees. He has an excellent rapport with the medical staff doctors, the members of the Board of Governors, and his subordinates. Although the administrative process seems to bog down at times, the network of communications through the organization and committee structure ultimately yields satisfactory results. Mr. Baker is a member of most of the administrative committees of the hospital and spends a great deal of time with committee meetings.

Mr. Baker emphasizes the fact that the administrator's job involves many problems and believes a study made by Charles Prall is reasonably representative. This study shows the percentage of administrators (by type) who reported one or more problems in several given categories. The summary report is presented in Exhibit 16-3.

EXHIBIT 16-3

Percentage of Hospital Administrators Reporting One or More Problems in Specific Categories—By Type of Administrator

PROBLEM AREAS	PERCENTAGE REPORTING ONE OR MORE PROBLEMS BY AREA		
	LAYMEN	DOCTORS	NURSES
Working with doctors	40	41	71
Improvement of Medical Care	50	63	90
Business and Finance	61	40	43
Public Relations	50	50	50
Physical Plant	33	25	50

Mr. Baker looks upon the extensive committee structure at the Trustee, Board of Governors, and top medical staff level with mixed feelings. On the one hand, they are release valves for troublesome issues and represent the democratic process which keeps everyone informed. On the other hand, they are very numerous, slow-acting, time-consuming, and reach few decisions which would not have been reached on a more timely basis by the Director working directly with the Executive Committee of the Medical Staff or the Board of Governors.

MIDDLE MANAGEMENT ORGANIZATION

The full-time operations of the hospital are organized and conducted under the supervision of 12 persons who report directly to Mr. Baker (see Exhibit 16-1).

Nursing activities are divided into two main groups; namely, Nursing Services and the School of Nursing. Exhibit 16-4 shows the internal organization for conducting these nursing activities. The School of Nursing has its own autonomy to a great extent because of its educational mission. Along with the medical staff, Nursing Services constitutes the very heart of hospital operations. The members of the Nursing staff are specialized and perform their duties on a round-the-clock basis every day of the year (by shifts) in such departments as Medicine, Surgery, Obstetrics, Pediatrics, and Operating Room. Theoretically each nurse has an immediate superior (Head Nurse) but during a normal work period she may be involved in taking instructions from several different persons. This is especially the case when dealing with the individual doctors of her various patients. Her work seems to run smoothly until "outsiders" create confusion and frustration by telling her to do things which do not constitute her job, interfere with her primary duties, or things which are against the rules or regulations of the hospital.

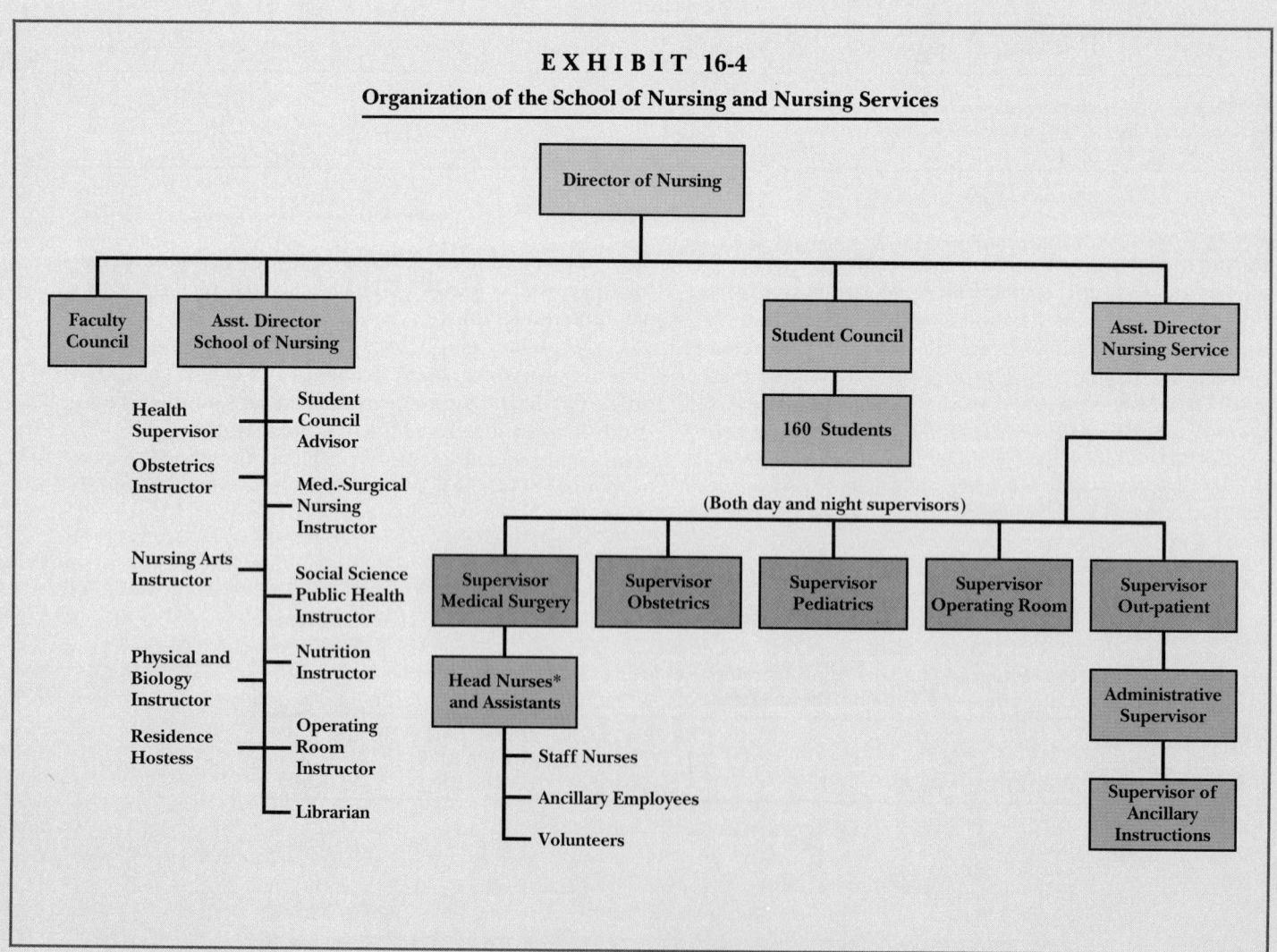

EXHIBIT 16-4

Organization of the School of Nursing and Nursing Services

ORGANIZATION DESIGN

The Medical Staff Organization includes the doctors who are associated with the hospital. (See Exhibit 16-2 for the internal organization of the Medical staff.) The hospital's existence depends on the requirements for patient care as determined by the individual medical staff members. Sometimes the individual doctors do not fully realize the "public utility"nature of the hospital. While the doctor is concerned with his patient, the hospital personnel are concerned with all patients. This leads to some difficulties in such activities as scheduling of operating rooms, proper use of precautionary methods, and administering the rules and regulations established by the Board of Governors and outside agencies.

ADMINISTRATIVE AND MEDICAL STAFF CONFLICT

Until about 15 tears ago, Newcomer-Willson had a doctor administrator. After a brief experience with a layman administrator, it went back to a doctor. About nine years ago it decided again to employ a nonmedical professional administrator. Since then, strides have been made in the administrative activities of the hospital. Basic problems do arise, however, which seem to indicate the need for further improvement in the organizational arrangement and/or the administrative processes. For example, the Head (physician) of one of the full-time departments recently demanded "individual professional status" which he thought the medical staff members had and which he did not have. The problem became so serious that the physician threatened to resign if he did not get the status desired. After considerable discussions with various people and in several committee meetings, the Board of Governors decided that the current status would not be changed substantially and the issue seemed to have been settled—with no resignations. A by-product of this action was the clear evidence that the Board of Governors is the "governing" body of the hospital.

Another incident reflects the type of problems which Mr. Baker and the administrative supervisors face. Only recently a patient had been placed on "precaution" by her doctor. Her physician, accompanied by several resident doctors, came into the patient's room without observing the precautionary rules. The staff nurse reported this incident immediately to her supervisor. The supervisor ordered all the doctors from the room, explaining the reasons to them. The physician took the patient off "precaution" on the spot and remained in the room. This upset the nursing staff; however, the next morning the physician placed the patient back on precaution. After discussing the problem with the Director of Nursing and the physician, Mr. Baker

had to decide what must be done in this instance and also in the future to reduce or eliminate such situations.

Mr. Baker said: "I suppose our problem is that the nature of hosptial operations makes it necessary to violate some of the essential characteristics of good organization which authorities like Urwick emphasized. Maybe Emerson was right when he said poor organization is the 'hook-worm disease' of industry. It is a disease we haven't completely cured. Everyone seems to have too many bosses, but somehow we do get the job done."

When you have read this case, look back at the chapter's diagnostic questions and choose the ones that apply to the case. Then use those questions with the ones that follow in your case analysis.

1. What constituencies must be satisfied if Newcomer-Willson Hospital is to operate effectively? What demands are made by these constituencies? How well is the hospital satisfying them?

2. What kind of structure does the hospital have now? What are the strengths and weaknesses of this type of structure? How does the hospital's structure affect its effectiveness?

3. What contingency factors have an impact on the hospital's effectiveness? In view of these factors, what sort of structure should the hospital have? What modifications must be made to the current structure?

CASE 16-2
CORRECTIONS CORPORATION OF AMERICA*
WINTHROP KNOWLTON, KENNEDY SCHOOL OF GOVERNMENT, *Harvard University*

Thomas W. Beasley and Doctor R. Crants, Jr., first met at West Point. After graduating in 1966, they served together in Vietnam where Beasley received the silver star and two bronze stars for valor.

The turn of the decade found both men back in school, Beasley at Vanderbilt University's School of Law and Crants enrolled in a joint Law and Business School program at Harvard.

After receiving his law degree in 1973, Beasley returned to his hometown of Dickson Falls, Tennessee, and began his own law practice. One observer said:

> He's really a country boy who likes farming. But he got bored with the practice of law and turned his attention to politics. He'd gotten a taste of that at Vanderbilt and in 1974 he ran Lamar Alexander's losing

* Reprinted with the author's permission.

campaign for the Tennessee governorship. Lamar lost bad, and from 1974 to 1978 the Democrats controlled the House, the Senate, and the State House. Tom became chairman of the Republican Party. Over the next four years he represented the party in all the important debates; he learned how to handle the press; and he developed a certain expertise under fire.

"Doc" Crants came to Tennessee in 1974 at the urging of his old friend Beasley and another West Point classmate then practicing law in Nashville. While Beasley operated in the world of politics, Crants pursued a career in real estate development, banking, and financial consulting. "Somewhere along the line," he says, "I became aware of the value of franchises—nursing homes and radio and television stations, for example. I discovered that the process of applying for licenses was exceedingly cumbersome—dealing with all that bureaucratic red tape—and not many people were willing to put up with it. I *was*. I didn't mind doing the dirty work, and occasionally I found that at the end of the process, I was the only bidder for a valuable property."

In 1978 Beasley ran Lamar Alexander's second, and successful, campaign for the governorship. Afterwards, he was charged with helping find candidates for the governor's cabinet. One job seemed unfillable: that of State Commissioner of Corrections. Beasley said:

> It was the one area where you couldn't find anyone with a good reputation. People would have a good reputation *for awhile* but they always seemed to lose it and get fired. There weren't any exceptions. It's a very difficult job to begin with, of course, but it always gets worse. The prison population grows, and legislatures are unwilling to provide money. So the institutions become more crowded. Indeed, they become inhumane. There are scandals and lawsuits. Many of our state correctional systems have been declared unconstitutional under the cruel and inhumane treatment of prisoners provision of the Constitution. The commissioner is the one who takes the rap.

While the new governor eventually filled the position, the new commissioner did not last, and during Alexander's first term Beasley kept wondering whether the public sector was really capable of running a decent corrections system. By 1982, with the Republicans in power in Washington, there was much talk in the air of reducing the activities of government in general and providing the private sector with a greater role in solving societal problems. It seemed to Beasley that the corrections area was one where this might be possible. He began taking small groups of Nashville businessmen to dinner at the Bellemeade Country Club to see if they could be persuaded to put money in a new venture that would try to run prisons for a profit. One businessman who was interested was his old West Point classmate, "Doc" Crants.

Not long after the two men agreed, late in 1982, to go into business together, they enjoyed two remarkable strokes of good fortune. It turned out that several other Nashville businessmen were considering the same kind of proposition. Jack Massey, the founder of American Hospital Supply of America, and his young colleague, Lucius E. Burch III, had formed a new venture-capital firm, Massey Burch, in June 1981, to provide a formal vehicle for venture-capital investments for a very limited number of large outside clients. Already highly successful in the privatization of hospitals, already committed to the concept of building strong franchises (as in the case of at least one other extraordinarily successful investment, Kentucky Fried Chicken Corporation), Massey Burch was now considering starting a company called Prisons Corporations of America. Hearing of Beasley's and Crants's efforts to launch a similar concern, Burch set up a meeting of the parties, and a deal was quickly struck. For an investment of $500,000 Massey Burch would obtain a 50 percent interest in the Beasley/Crants venture. It would be called Corrections Corporation of America. Beasley would be president and chairman of the Board of Directors; Crants, secretary, treasurer, and vice chairman; and Burch would become a director and senior adviser.

The second lucky stroke came in the form of T. Don Hutto, an experienced corrections professional who had just been replaced as head of Virginia's corrections system as a result of the election of a new governor, Democrat Charles Robb. Hutto was the chairman-elect of the American Correctional Association (the ACA), the watchdog trade association that had been struggling since 1977 to formulate humane operating standards for American prisons. The ACA was planning to hold its next convention in Nashville in early 1983. Beasley, Crants, and Hutto met, and Hutto was brought aboard as executive vice-president of the new firm.

THE U.S. DETENTION AND CORRECTIONS SYSTEM

The American prison system consist of two broad types of institutions (although, as indicated below, the two sometimes overlap): facilities for detention and for correction.

The most widespread of the former are local jails where individuals are held awaiting trial or, having been convicted, are held awaiting sentencing to a corrections institution. The most common corrections institutions are penal farms and workhouses, where individuals are sent to serve terms of less than six years and penitentiaries (state and federal) where they serve longer terms for more serious crimes. Because corrections institutions are often overcrowded, some prisoners who would ordinarily be sent to the latter are kept instead in local jails serving out their terms alongside other individuals who have yet to be found guilty of any crime or who have been convicted of relatively minor offenses. Sometimes individuals are "detained" in penal farms and workhouses, so that these serve both as "detention" and "corrections" facilities. In addition, there are a number of specialized institutions where illegal immigrants or juveniles are held for a variety of purposes.

Describing U.S. jails in a 1981 report, the ACA wrote:

Since the first jail built in the Jamestown Colony, Virginia, in 1608, American society has caged human beings like animals with the expectation that they would return to the community and become law-abiding citizens. Traditional jail procedures have changed very little since the beginning of the 17th century, continuing to reflect a punitive "lock 'em up" approach in facilities. Prisoners often languish in idleness and boredom, under primitive conditions. Jails were primarily designed then, and still are, to enable a small number of staff to confine securely a comparatively large number of prisoners.

From the small to the large, today's detention facilities are poorly equipped to handle their diverse populations ranging from homeless drunks and the mentally ill to individuals accused of every conceivable crime. When such human beings are imprisoned with little regard for individual needs and rights their further deterioration is virtually assured. At detention facilities, the intake point in the criminal justice system, petty offenders comingle with hard-core criminals. Contempt for the law as well as knowledge and skills for future criminal careers are readily available to the young and first offenders. Their exposure to and acceptance of such opportunity is reflected by their high rates of failure and recidivism.

Despite decades of failure and criticism, most jails have remained unchanged, continuing to function as human warehouses. The reasons for this phenomenon can be any combination of such factors as limited local financing, public indifference and apathy, emphasis on restraint, and public ambivalence regarding concepts of punishment and treatment. Jails are the least studied and least understood of all penal institutions and are often held in low esteem by the public and its officials. Supporting jail reform has become a politically unpopular cause, and the tendency to minimize the importance of detention facilities and their impact on society is growing. The rare existence of sound operational standards for detention facilities and the failure of the many states to enforce those that do exist further depress the situation.[1]

Justice Department data released in the spring of 1985 indicated:

- In 1974, the United States prison population in federal and state penal institutions totaled approximately 229,000. By 1984, this number had doubled to 464,000. Between 1980 and 1984 the prison population increased by more than 40 percent. These increases were spurred in part by the arrival at the prison-prone ages (20–29 years) of the "baby boom" generation.

- In addition to state and federal prison systems, there are approximately 3,400 local jails within the United States. On an given day, there are approximately 300,000 people residing in these jails. In 1984, a total of 11,500 sentenced prisoners were held in local jails because of state prison overcrowding, an increase of more than 40 percent over the previous year.

- The number of prison admissions from court for every 100 serious crimes reported to the police declined from 6.3 in 1960 to 2.3 in 1970. The ratio began to increase in 1981, reaching 4.0 in 1983. If the 1960 rate of prison admissions relative to crime had prevailed in 1983, the number of offenders sentenced that year would have been about 100,000 higher than the 173,000 who were actually incarcerated.

- The strain placed on prison systems by the rapid influx of prisoners in recent years has been accompanied by a series of court interventions charging that individual prisons or whole state systems are violating the constitutional provision prohibiting "cruel and unusual treatment" of prisoners. In 1984 the entire prison systems in eight states were operating under court order or consent decree. Thirty-three state correctional systems had been declared unconstitutional, either in whole or in part, by the federal courts or were currently defending litigation in that regard.

- Despite substantial increases in expenditures for new prison capacity in recent years, both federal and state systems were operating at over-capacity occupancy rates in 1984: federal institutions at 110–137 percent of capacity and states at 105–116 percent (depending on how capacity is measured). One in 10 prisoners was housed in a facility built before 1875, half of all prisoners in facilities 40 or more years old.[2]

Experts predict that inmate populations will level off from about 1985 to 1990 when they are again expected to increase. These projections are based on expected population age groupings and the fact that young people are the high-risk group and commit most crimes.

The Immigration and Naturalization Service (INS) reports that they arrested 1.2 million illegal aliens

---

[1] *Standard for Adult Local Detention Facilities*, 2nd ed. American Correctional Association, 1981.

[2] *Prisoners in 1984*, Bureau of Justice Statistics Bulletin.

in 1983. Because of jail overcrowding and despite the fact that INS operates over 3,000 beds, INS finds itself without places to detail aliens once they are arrested. It is not uncommon for local district offices to stop arresting aliens as early as 9 A.M. each day because there is no further place to detain them.

It is expected that the illegal alien problem will worsen over the next 10 years and that the need for detention of aliens will increase. Even if Congress passes a reform bill, most observers believe that this will increase the numbers in detention. This is based on the premise that declaring amnesty for present aliens within this country will only serve to encourage other aliens to enter illegally in the future.

CCA's Operating Strategy

The Company's managers and owners believed that the state of the nation's detention and corrections system provided it with a unique commercial opportunity. As Lucius Burch put it, in an early memorandum on CCA:

> It is evident that change in the corrections industry is necessary and imminent. Public entities clearly are not managing the prison systems effectively.
>
> While the American public has taken a "hard-line" attitude in regard to the suppression of crime, this attitude has not translated itself into support for tax initiatives to build additional facilities to handle the increased population. There is no political constituency for building and housing criminals. It is a low budget priority and will not receive proper attention absent private intervention.
>
> Further, as public treasuries grow leaner in the years ahead, experts believe that the role of private enterprise in criminal justice will increase.
>
> Prisons in the United States are aging and are in need of repair or replacement. Private entities can relieve municipalities (and ultimately taxpayers) from the burden of additional capital outlays and high operating expenses for these facilities. In fact, a corrections corporation is a tax-paying entity and will provide additional capital through payment of state and federal taxes.
>
> Most importantly, however, a private entity can manage correctional facilities more economically and efficiently than the governmental entities. A private entity can be competitive in its costs relative to governmental operations for several reasons:

- Personal economies can be achieved through careful attention to the design of the facility. In a twenty-four (24) hour operation, such economies are of major importance.
- Further economies can be achieved through mass purchasing, an advantage not available to small single jail operations.
- Private entities are not required to operate under cumbersome bureaucratic purchasing regulations which, inevitably,

increase the cost of supplies and materials. As a result, a private operation can trim an estimated 10–25 percent off the cost of running conventional facilities. The bidding procedures designed to maintain honesty, while resulting in low bids for the government, usually do not result in low prices.

With solid financial backing from Massey Burch, with a small cadre of managers experienced in politics (Beasley), business (Crants), and corrections management (Hutto), CCA determined that it would attempt to contract with local, state, and federal governmental bodies for the detention of persons for whom a minimum to medium level of security was required and who presented, in the management's opinion, "a relatively low risk of violence or other untoward behavior." The company would build, own (or lease), staff, and manage such correctional and detention facilities. It would contract with government bodies to "detain or incarcerate, provide food and other necessities for, and supervise persons in the correctional and detention facilities owned by government bodies."

Although this statement of purpose seemed relatively straightforward, the leaders of the company knew that implementation would be no simple matter. They understood that the process of "selling" their concept to local elective officials (sheriffs, county executives, and county commissions) would be time-consuming and difficult. Many local officials had a stake in the existing system. For them, employment opportunities in county jails represented a source of patronage. At the state and federal level there would be legislative oversight committees to deal with as well as elective and appointive officials in the executive branches. How would these public bodies feel about turning over the administration of corrections facilities to a for-profit body?

What standards would need to be established so that the public sector would feel at home delegating these kinds of responsibilities to private concerns? The ACA had recently set a standard of 70 square feet of cell space for each prisoner. The federal standard was 60 square feet. But it was estimated that only 1 percent of all facilities met these levels. There were numerous instances, in fact, where four prisoners were crowded into cells of less than 60 square feet, where there were no outdoor recreation areas, and where health, sanitation, and hygiene facilities were lamentable. Indeed, it was the public sector's inability or unwillingness to provide these standards that created the opportunity that CCA wished now to exploit. Should CCA strive to build and run facilities that met the ACA standards or some less stringent test that would still substantially improve upon existing practice?

There were a number of other questions on CCA's

managers' minds in 1983 as it began formulating its first bids:

- What role, if any, would public sector officials play in monitoring what went on inside a prison once CCA "owned" or managed it?
- How much competition did CCA want? It was important that the public sector have more than one firm bidding for this kind of business. On the other hand, nothing would damage the privatization concept faster than fly-by-night operators who failed to make good on their promises. Was there any way CCA could influence the competitive environment in which it operated? Should it even attempt to do so?
- What liabilities would the company be exposing itself to in the management of detention and corrections facilities? And how could these be contained?
- How would the company meet its staffing requirements—both for its own management and for the staff required within the facilities it operated—given the poor reputation of the public officials now in the business.
- What would the media make of all this? And the liberal academic intelligentsia? And how could *they* be "managed"? Or should the company even try?

THE FIRST CONTRACT

The company submitted its first proposal and bid to manage a detention facility on August 26, 1983, to the Immigration and Naturalization Service for the operation of an approximately 67,600-square-foot, 350-bed detention facility for illegal aliens in Houston, Texas. On October 6, 1983, the company received notification from the INS that its proposal and bid to operate the facility had been accepted and that the INS contract was effective from that date. On October 11, 1983, the company purchased an approximately 5.84-acre site in Houston for construction of the facility for a price of approximately $763,000. On October 21, 1983, the company entered into a construction contract with Trimble & Stephens Co. of Houston, Texas, for the construction of the facility. Construction commenced on October 26, 1983, and was completed on April 20, 1984. The company also incurred architectural and engineering fees in connection with the construction.

The initial term of the INS contract terminated on September 30, 1984. The INS renewed the INS contract to September 30, 1985 and subsequently sent the company a letter of intent to renew the contract until September 30, 1986. The INS had the option to renew

the INS contract for three consecutive one-year periods with adjustments in the amounts paid by the INS to the company during such renewal periods. These adjustments were designed, in part, to offset inflation. Unless the company operated the facility for at least five years, the company would likely lose a substantial amount of money on the project. The company was to receive $25.74 per day per person from the INS through September 1985.

When you have read this case, look back at the chapter's diagnostic questions and choose the ones that apply to the case. Then use those questions with the ones that follow in your case analysis.

1. What major constituencies are likely to make demands on the Corrections Corporation of America as it opens prison facilities throughout the United States? What demands are these constituencies apt to make? Do you believe that CCA will be able to satisfy these demands?

2. At what life-cycle stage is CCA? What type of structure should it have? Does it appear that the company does in fact have this type of structure?

3. How will CCA's expected growth affect its structure? What structural changes will have to be made as the company grows larger? What should CCA's current management do to prepare the company for these changes?

CASE 16-3

DUMAS PUBLIC LIBRARY

Read Chapter 18's Case 18-1, "Dumas Public Library," or review it if you have read it for Case 15-3. Next, look back at Chapter 16's diagnostic questions and choose the ones that apply to that case. Then use those questions with the ones that follow in your analysis of the case.

1. What constituency groups are making demands on the city government of Kimball, New Mexico? What are their demands? How successful has the city been in its efforts to meet those demands?

2. What kind of structure does the city government have? How has this structure contributed to the conflict between Debra Dickenson and Helen Hendricks? What structural changes should be made to minimize the reoccurrence of similar conflicts in future?

3. Diagram what you think would be the most effective structure for Kimball's city government. Where do the library board and city council belong in your diagram?

C H A P T E R 17

JOB DESIGN

At National Bevpak, a bottling and canning plant that is a subsidiary of the National Beverage Corp., this worker spends his day checking the flow of bottles along the conveyor belt. How can a company keep employees who perform boring jobs like this motivated and satisfied? It could teach workers other tasks, such as packing the bottles in boxes, so they could rotate from one job to another. Or it could try to invent a machine to do these simple, highly repetitive tasks. What other approaches might an employer take?

I stand in one spot, about a two- or three-feet area, all night. . . . We do about thirty-two [welding] jobs per car, per unit. Forty-eight units an hour, eight hours a day. Thirty-two times forty-eight times eight. Figure it out. That's how many times I push that button. . . . You dream, you think of things you've done. I drift back continuously to when I was a kid and what me and my brothers did. . . . [Y]ou're nothing more than a machine. They give better care to that machine than they will to you. They'll have more respect, give more attention to that machine. . . . Somehow you get the feeling that the machine is better than you are.[1]

*****

The other day when I was proofreading [insurance policy] endorsements I noticed some guy had insured his store for $165,000 against vandalism and $5,000 against fire. Now that's bound to be a mistake. They probably got it backwards. . . . I was just about to show it to [my supervisor] when I figured, wait a minute! I'm not supposed to read these forms. I'm just supposed to check one column against another. And they do check. . . . They don't explain this stuff to me. I'm not supposed to understand it. I'm just supposed to check one column against the other. . . . If they're gonna give me a robot's job to do, I'm gonna do it like a robot! Anyway, it just lowers my production record to get up and point out someone else's error.[2]

job design The process of deciding what specific tasks each job holder should perform in the context of the overall work that an organization must accomplish.

It's easy to understand why workers who perform monotonous, unchallenging jobs like these feel bored and frustrated. Yet almost all such jobs are the result of conscious, deliberate planning. Why do managers intentionally design jobs that are so unappealing? What do they expect to gain by simplifying work so drastically? Can anything be done to counteract the negative effects of oversimplifying work—effects like the welder's detached daydreaming and the insurance clerk's decision to overlook an obvious error? Or, can oversimplification be avoided altogether?

We will seek answers to questions such as these in this chapter as we examine theories and methods of **job design**, the process of deciding what specific tasks each jobholder should perform in the context of the overall work of an organization. In Chapters 15 and 16, we focused on the structural characteristics that help coordinate tasks and integrate the work of an organization. Here, we will consider the opposite side of the coin, so to speak, as we see how an organization's work is divided into jobs that can be performed by individual workers.

We will begin by overviewing one approach to job design, the efficiency perspective, which originated in work on scientific management discussed in Chapter 2. Today, it belongs to the field of industrial engineering. Next, we will turn our attention to another approach, the satisfaction perspective, which arose largely in reaction to problems with the efficiency perspective. It is based on ideas about motivation, satisfaction, and performance like those we have already discussed in Chapters 7 and 8. We will conclude by describing several recent developments that are changing the way work is organized and accomplished in modern organizations.

[1] Studs Terkel, *Working* (New York: Avon Books, 1972), pp. 221–23.
[2] Barbara Garson, *All the Livelong Day: The Meaning and Demeaning of Routine Work* (New York: Penguin Books, 1977), p. 171.

DIAGNOSTIC ISSUES

Before we start our discussion of job design, let's consider some issues our diagnostic model raises. To begin with, can we *describe* how the organization's work is divided into jobs and the effects of this division of labor on employees? In designing its jobs, is the organization primarily concerned with efficiency or with satisfaction? Can we *diagnose* whether a job needs to be simplified? What signals might indicate instead that a job has been made too simple?

On what basis do we *prescribe* that a job be redesigned? How can we predict who will react positively to enriched jobs and who won't? Finally, what *actions* can managers take to make jobs more challenging and rewarding? What can be done if such actions "overstretch" the work force?

THE EFFICIENCY PERSPECTIVE ON JOB DESIGN

To achieve *efficiency*, companies minimize the resources that are consumed providing a product or service. The **efficiency perspective** on job design is concerned with creating jobs that economize on time, human energy, raw materials, and other productive resources. It is the basis for the field of **industrial engineering**, which focuses on maximizing the efficiency of the methods, facilities, and materials used to produce commercial products. Industrial engineers design products—whether tangible goods or intangible services—in a way that simplifies production processes. They also develop standard procedures and materials to cut production costs, design and test production machinery to ensure proficient operation, and devise inspection procedures to guarantee product quality. Among the pioneers of this approach was Frederick Winslow Taylor (1856–1915), whose studies of efficiency in the workplace gave rise to the field of scientific management.

Scientific Management

As he rose from laborer to chief engineer at the Midvale Steel Company, Taylor formulated a set of principles of scientific management that today's managers continue to consult in designing and managing jobs in organizations (see Table 2-2). To illustrate the potential benefits of following these principles, Taylor often told a story about a worker named Schmidt who loaded pig iron, or ingots of iron, into railroad cars for the Bethlehem Steel Company. At Bethlehem, before Taylor analyzed the job of loading iron, each employee would pick up a pig of iron that weighed 92 pounds, carry it up a ramp to the door of a railroad car, and drop it into the car. In this manner, a worker could load about $12\frac{1}{2}$ long tons in a day (2,240 pounds = 1 long ton). Taylor set out to improve this rate of productivity by selecting a worker, Schmidt, and offering him $1.85 per day rather than the usual $1.15 per day to do the job of loading pig iron exactly as Taylor instructed. Following Taylor's directions (which involved loading each pig at a prescribed pace and taking periodic rest breaks to allow muscle recuperation), Schmidt was able to load $47\frac{1}{2}$ long tons per day. Considering wages paid and tons of pig iron loaded, Taylor's approach yielded a cost per ton of 3.9 cents versus the old way's cost of 9.2 cents.

Doubt surrounds the authenticity of many of the details recounted in Tay-

efficiency perspective An approach to job design that focuses on the creation of jobs that economize on time, human energy, raw materials, and other productive resources.

industrial engineering A branch of engineering that concerns itself with how to maximize the efficiency of the methods, facilities, and materials used to produce commercial products.

TEACHING NOTE
Ask students to give examples of unskilled workers being paid high wages. Are they being compensated for the boredom of the job?

JOB DESIGN

lor's pig iron story, suggesting that it be interpreted more as an illustrative allegory rather than as a factual account.[3] But whether allegory or fact, it shows that besides Taylor's trademark concern with efficiency and orderliness, an important feature of his approach was the idea of gainsharing—the sharing by both employers and employees of the economic gains that resulted from applying the principles of scientific management. In Taylor's era, this idea was considered a threat to the well-being of both unions and employers. Unions often used employees' wage dissatisfaction to gain support for unionization. Employers did not want to give up their claim to their firms' profits. Despite this initial resistance, Taylor's approach caught on, aided by the publicity he gained when he appeared before the United States House of Representatives to defend his principles. Others built on his ideas about engineering jobs and the methods used to perform them.

Methods Engineering

methods engineering An area of industrial engineering that attempts to improve the methods used to perform work.

One of the most important descendants of Taylor's pioneering work, **methods engineering**, is an area of industrial engineering that attempts to improve the methods used to perform work. It incorporates two related endeavors—process engineering and human factors engineering.

process engineering A type of methods engineering in which specialists study the sequence of tasks required to produce a particular good or service and examine how these tasks fit together into an integrated job.

Process Engineering. Process engineering studies the sequence of tasks required to produce a particular good or service and examines the way these tasks fit together into an integrated job. It also analyzes tasks to see which should be performed by human beings and which by machines and tries to determine how workers can perform their jobs most efficiently. Process engineers examine the good or service to be produced and decide what function, if any, human beings should serve in its production. They also determine the need for some employees to serve as managers, to direct and control the flow of work (see Chapter 2), and they differentiate the resulting managerial jobs from those of nonmanagerial workers. They specify the procedures for employees to follow, the equipment they should use, and the physical layout of offices, work stations, and materials-storage facilities.

AN EXAMPLE: HUGHES AIRCRAFT

At Hughes Aircraft, process engineering enabled employees to cut the time required to build a satellite in half and saved the company millions of dollars. Employees set to work and identified 131 steps in their procedures that needed modification. Small changes in just 30 of the steps that seemed most urgent—for example, moving a hole a quarter of an inch to one side so a testing probe could be inserted more easily—added up to the savings in time and money. As Joe Sanders, group vice president for operations said, "When you stretch it out on the wall . . . . you say, What the hell am I doing that for?" (Ronald Henkoff, "Cost Cutting: How to Do It Right," *Fortune*, April 9, 1990, p. 48.)

Consider the task of selling a sweater. The process chart in Figure 17-1 details the original job design. To sell one sweater, clerks regularly performed 46 different work activities and walked a distance of 318 feet. To redesign the job of selling sweaters, process engineers analyzed the salespersons' movements and actions guided by questions like the following:

1. Should all the work activities now included in the job actually be in this one job?
2. Does the job holder currently perform some of the work activities in the job in a random fashion when order and consistency might promote greater efficiency?
3. Can some of the work activities be batched, that is, performed in groups for several sales transactions, rather than separately for each one?
4. Can instructions for managing work activities be standardized?
5. Does the layout of the workplace (including equipment and supplies) facilitate the completion of work activities?

[3] See Charles D. Wrenge and Amedeo G. Perroni, "Taylor's Pig-Tale: A Historical Analysis of Frederick W. Taylor's Pig-Iron Experiments," *Academy of Management Journal* 17 (1974), 6–27.

FIGURE 17-1

Process Chart of Sweater Sale, Original Method

This chart depicts a method of selling a sweater before process engineering. The large circles denote operations performed at a fixed location. The small circles indicate movements by which the clerk moves toward an object or changes the location of an object. The inverted triangles signify delays during which the clerk is motionless. Thus the original process consists of 26 activities, 14 movements, and 6 delays and requires the clerk to walk a distance of 318 feet.

From Marvin E. Mundel, Motion and Time Study: Improving Productivity, *6th ed. (Englewood Cliffs, N.J.: Prentice Hall, 1985), pp. 591–92. Reprinted with the publisher's permission.*

BASIC CHART FORM

<u>Process chart-man</u> Type of chart	<u>383-#138</u> Department
<u>Original</u> Original or proposed	<u>S.W.</u> Chart by
<u>Sweater sale</u> Subject charted	<u>9/14</u> Date charted

DIST.	SYMBOL	EXPLANATION
	▽	Waits to greet customer (from counter)
	●	Greets customer, inquires and shows black cardigan
	●	Gets black pullover from display case
	●	Shows to customer
	●	Puts cardigan on to display; too small for customer
	●	Takes cardigan off
	●	Replaces cardigan in display case
90'	○	Walks to stock room
5'	○	Climbs ladder (splintery and unsafe)
	●	Gets box
	●	Takes out 1 cardigan of larger size
5'	○	Climbs down ladder
90'	○	Returns to customer
	●	Shows cardigan to customer
12'	○	Walks around counter
	●	Gets whisk broom
12'	●	Returns to customer
	●	Brushes lint off sweater
	●	Hands brush to customer at customers request
	▽	Waits while customer brushes sweater
	●	Accepts brush; lays brush on counter
	▽	Customer agrees to buy
	●	Asks if "charge or cash" (charge)
12'	○	Walks to end of sales counter
	●	Gets sales book
12'	○	Returns to customer
10'	○	Walks to shelf

6. Is the time and effort consumed by a particular work activity so great that it should be broken into a sequence of smaller activities?

7. Can some of these smaller work activities be eliminated by physical rearrangement of the workplace or by the use of different equipment (calculators rather than adding machines, word processors instead of typewriters)?

8. If some of the smaller work activities were performed in a different sequence, could any of them be eliminated or combined?

9. Could overall productivity be increased by redistributing work activities among a group of workers?[4]

[4] Adapted from Marvin E. Mundel, *Motion and Time Study: Improving Productivity*, 6th ed. (Englewood Cliffs, N.J.: Prentice Hall, 1985), p. 115.

FIGURE 17-1

Cont'd.

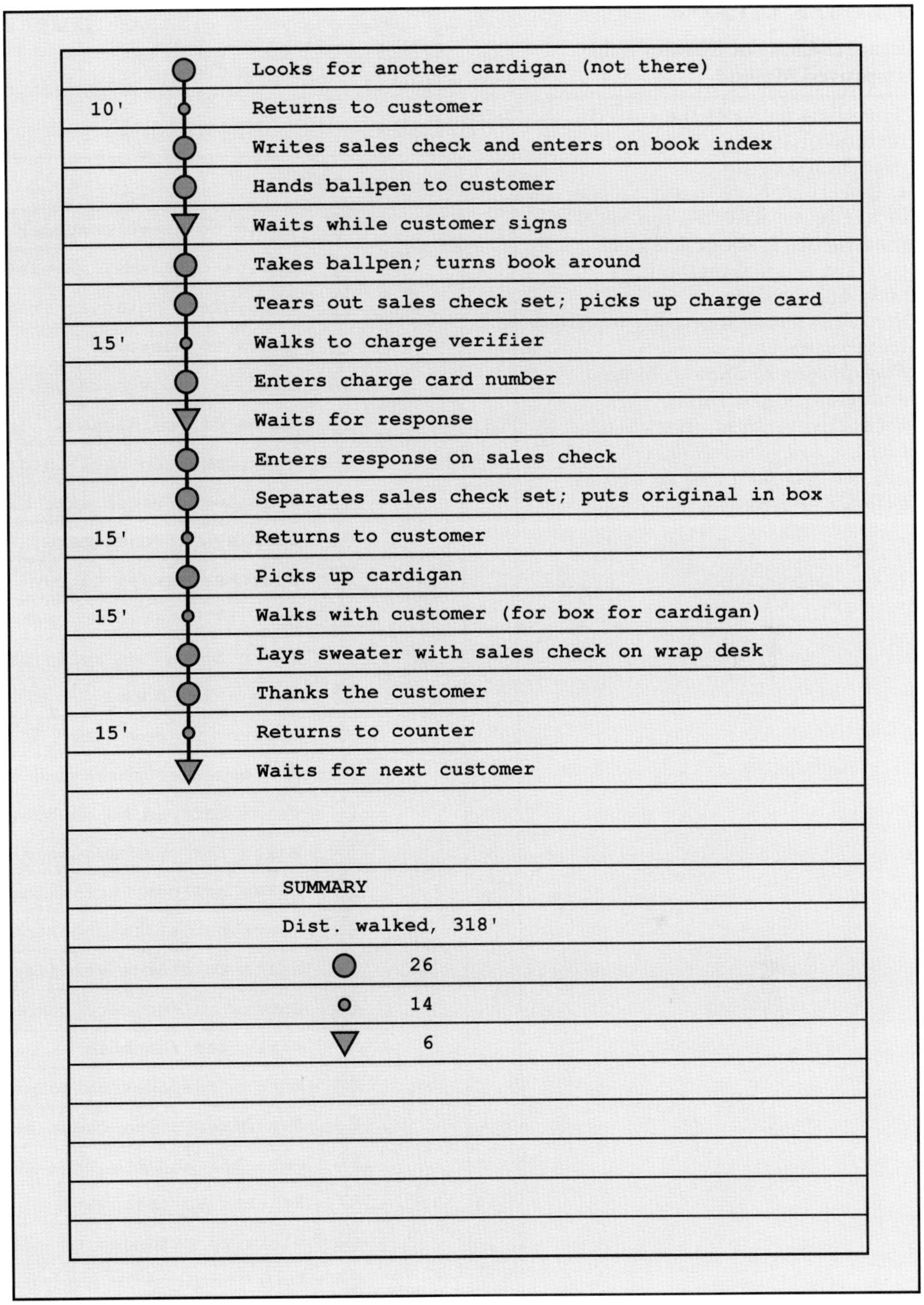

	●	Looks for another cardigan (not there)
10'	○	Returns to customer
	●	Writes sales check and enters on book index
	●	Hands ballpen to customer
	▽	Waits while customer signs
	●	Takes ballpen; turns book around
	●	Tears out sales check set; picks up charge card
15'	○	Walks to charge verifier
	●	Enters charge card number
	▽	Waits for response
	●	Enters response on sales check
	●	Separates sales check set; puts original in box
15'	○	Returns to customer
	●	Picks up cardigan
15'	○	Walks with customer (for box for cardigan)
	●	Lays sweater with sales check on wrap desk
	●	Thanks the customer
15'	○	Returns to counter
	▽	Waits for next customer

SUMMARY

Dist. walked, 318'

●	26
○	14
▽	6

Now look at Figure 17-2. As you can see, the improved method that the process engineers devised reduced the job of selling sweaters by 16 activities and saved clerks 264 feet of walking.

Human Factors Engineering. In **human factors engineering**, sometimes called **ergonomics**, experts design machines, operations, and work environments so that they match human capacities and limitations. Table 17-1 lists some of the concerns of human factors engineering. Note that this area of methods engineering differs noticeably from process engineering. Process engineers fit people to jobs; human factors engineers fit jobs to people.

When people make mistakes at work, human factors engineers ask, Is the

human factors engineering A type of methods engineering in which experts design machines, operations, and work environments so that they match human capacities and limitations.

ergonomics Another name for human factors engineering, a type of methods engineering that focuses on designing machines to match human capacities and limitations.

FIGURE 17-2

Process Chart of Sweater Sale, Improved Method

This chart indicates the correct method of selling a sweater, determined by process engineering. Note that it consists of five fewer activities, ten fewer movements, and one less delay than the original method. In addition, the distance walked by the sales clerk is shortened by 264 feet (83 percent).
From Marvin E. Mundel, Motion and Time Study: Improving Productivity, *6th ed. (Englewood Cliffs, N.J.: Prentice Hall, 1985), pp. 593–94. Reprinted with the publisher's permission.*

BASIC CHART FORM

Process chart-man — Type of chart 383-#138 — Department
Original — Original or proposed S.W. — Chart by
Sweater sale — Subject charted 9/14 — Date charted

DIST.	SYMBOL	EXPLANATION
	▽	Waits to greet customer (from counter)
	●	Greets customer, inquires, and shows black cardigan
	●	Gets black pullover from case
	●	Shows to customer
	●	Puts on black cardigan to display
	●	Takes off cardigan
	●	Puts cardigan back into case
	●	Takes out larger size black cardigan
	●	Displays to customer
	▽	Customer agrees to buy
	●	Asks if "charge or cash" (charge)
12'	○	Walks to end of sales counter
	●	Gets sales book
12'	○	Returns to customer
	●	Writes sales check and enters on book index
	●	Hands ballpen to customer
	▽	Waits for customer to sign
	●	Takes ballpen; turns book around
	●	Tears out sales check set; picks up charge card
15'	○	Walks to charge verifier
	●	Enters charge card number
	▽	Waits for response
	●	Enters response on sales check
	●	Separates sales check set; puts original in box
	●	Gets box and bag from under counter
15'	○	Returns to customer
	●	Picks up cardigan
	●	Puts cardigan in box; box in bag
	●	Hands to customer and thanks the customer
	▽	Waits for next customer

SUMMARY

Category	Original	Proposed	Saved
Distance walked	318'	54'	264'
●	26	21	5
○	14	4	10
▽	6	5	1

TABLE 17-1
Areas of Study in Human Factors Engineering

AREA OF STUDY	EXAMPLES
Physical aspects of the user-machine interface	Size, shape, color, texture, and method of operation of displays and controls for such things as cars, home appliances, and industrial and commercial equipment.
Cognitive aspects of the user-machine interface	Understanding of instructions and other information; style of dialogue between computer and user
Workplace design and workspace layout	Layout of offices, factories, home kitchens, and other places in which people work; detailed relationships between furniture and equipment and between different equipment components
Physical environment	Effects of climate, noise and vibration, illumination, and chemical/biological contaminants on human performance and health

Source: Adapted with the publisher's permission from I. A. R. Galer, *Applied Ergonomics Handbook* (London: Butterworth, 1987), p. 6.

equipment being used partially to blame for these mistakes? Are mistakes made when certain kinds of equipment are used rather than others? Is it possible to redesign equipment so as to minimize or even eliminate human error? More often than not, the effects of human fallibility and carelessness can be substantially decreased by minimizing error-provoking features of jobs and equipment. For example, shape-coded controls like those shown in Figure 17-3 can be used to reduce aircraft accidents caused by reaching for the wrong control.

Work Measurement: Motion and Time Studies

Besides engineering the methods used to perform jobs, industrial engineers sometimes also examine the motions and time required to complete each job. Such work can be traced to Frederick Winslow Taylor's principles of scientific management but is more directly the product of research by Frank and Lillian Gilbreth, who set out to find the "one best way" to do any job (see also Chapter 2). In the course of this pursuit, the Gilbreths developed motion study, a procedure that reduces jobs to their most basic movements. As you will recall from Chapter 2,

FIGURE 17-3

Shape-Coding to Reduce Flying Errors

The knobs shown in this figure are intended to help pilots differentiate among control levers without looking at them. Two general rules were followed during the design process: (1) The shape of a control should suggest its purpose, and (2) the shape should be distinguishable even when gloves are worn.
Adapted from C. T. Morgan, J. S. Cook, A. Chapanis, and M. W. Lund, Human Engineering Guide to Equipment Design *(New York: McGraw-Hill Book Company, 1963), p. 25. Reprinted with permission of McGraw-Hill, Inc.*

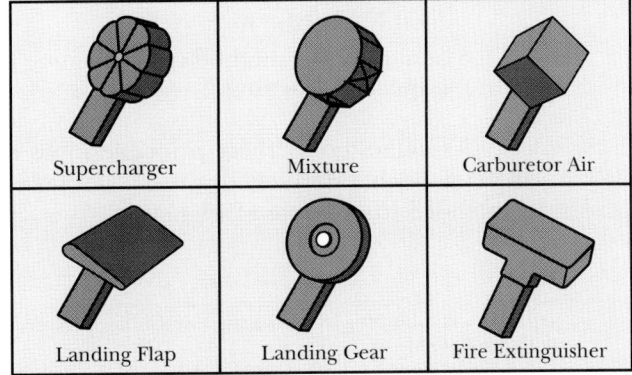

Supercharger   Mixture   Carburetor Air

Landing Flap   Landing Gear   Fire Extinguisher

each of these basic movements is called a therblig (a near reversal of *Gilbreth*) and consists of motions such as "search," "grasp," and "assemble." The Gilbreths also developed procedures to determine the time required by each of the movements needed to perform a job. Time-and-motion analysis was thus invented.

Like Taylor, the Gilbreths were fanatics about efficiency in all aspects of their lives. For example, through a lengthy analysis, Frank determined that 12 was the number of children that allowed the most efficient use of a family's resources, so the Gilbreths had 12 children. Among other projects aimed at increasing the efficiency of everyday living, they attempted to simplify the English alphabet, the typewriter, and spelling. Frank even went so far as to apply time-and-motion experimentation to the task of shaving:

> For a while he tried shaving with two razors, but finally gave it up. "I can save forty-four seconds," he grumbled, "but I wasted two minutes this morning putting a bandage on my throat." It wasn't the [nicked] throat that really bothered him. It was the two minutes.[5]

The Gilbreths developed their methods to eliminate unnecessary motion and effort and to set accurate job standards, or expected levels of job performance. These procedures constituted an indispensable addition to the efficiency perspective on job design. They gave rise to **work measurement**, an area of industrial engineering concerned with measuring the amount of work accomplished and developing standards for performing work of an acceptable quantity and quality. Work measurement includes micromotion analysis, memomotion analysis, and time study procedures.

work measurement An area of industrial engineering concerned with measuring the amount of work accomplished and developing standards for performing work of an acceptable quantity and quality.

micromotion analysis A type of work measurement in which industrial engineers analyze the hand and body movements required to do a job.

Micromotion Analysis. In **micromotion analysis**, industrial engineers analyze the hand and body movements required to do a job. This technique is a direct descendant of the motion study methods devised by the Gilbreths. Their therbligs continue to be used in current micromotion procedures. Industrial engineers usually perform micromotion analysis by using a slow-speed film or videotape of a person performing her job. They analyze the movements involved in the task and try to improve performance efficiency by means of principles like the following:

1. Try to have both hands doing the same thing at the same time or to balance the work of the two hands.
2. Try to avoid using the hands simply for holding. Use specialized jigs, vises, or clamps instead.
3. Keep all work inside a work area bounded by the worker's reach.
4. Relieve the hands of work wherever possible.
5. Eliminate as many therbligs or as much of a therblig as possible and combine therbligs when possible.
6. Arrange therbligs in the most convenient order. Each therblig should flow smoothly into the next.
7. Standardize the method of performing the job in the manner that promotes the quickest learning.[6]

As suggested by these principles, jobs designed by means of micromotion analysis are characterized by great economy of motion. The technique, of course, is intended to promote efficiency.

---

[5] Frank B. Gilbreth, Jr., and Ernestine Gilbreth Carey, *Cheaper by the Dozen* (New York: Thomas Y. Crowell, 1948), p. 2.
[6] Adapted from Mundel, *Motion and Time Study*, p. 309.

JOB DESIGN

memomotion analysis A type of work measurement in which industrial engineers examine longer activity sequences by using slow speed to film or videotape a person at work and then playing back the resulting film or tape at normal speed.

Memomotion Analysis. **Memomotion analysis** is used to analyze jobs that are less repetitive than most assembly-line jobs and that have longer activity sequences. This time, the analyst uses a slow speed to film or videotape a person at work and then plays back the resulting film or tape at normal speed. As you can imagine, job activities are sped up—one hour of activity may be viewed in as little as four minutes—and the analyst is able to identify and observe gross movements that normally occur over long periods of time.

The type of movements identified in this sort of analysis and some of its potential benefits are illustrated in Figure 17-4, which shows two alternative layouts of a dentist's office. In the upper diagram, a memomotion analysis has captured the travel paths of the dentist and an assistant during a typical patient visit. In the lower diagram, the analyst has again traced the travel paths of the dentist and his assistant during a similar patient visit but after the dentist's office has been rearranged in accordance with the recommendations of the memomotion analyst. You don't have to look very hard to see that many fewer motions are shown in the lower diagram and that they stretch over a smaller area in space.

Time-Study Techniques. Time-study techniques are generally used to measure the time consumed by job performance, but they are sometimes used also to specify the time that a particular job should take to complete. In **stopwatch time analysis**, an analyst uses a stopwatch to time the sequence of motions needed to complete a job. In **standard time analysis**, the analyst matches the results of micromotion analysis with standard time charts to determine the average time that should be required to perform a job. When combined with micromotion analyses, the results of either type of time analysis can be used to create job element descriptions such as the one shown in Figure 17-5.

stopwatch time analysis A time study technique in which an analyst uses a stopwatch to time the sequence of motions needed to complete a job.

standard time analysis A time study technique in which an analyst matches the results of micromotion analysis with standard time charts to determine the average time that should be required to perform a job.

F I G U R E 17-4

Memomotion Analysis of a Dentist's Office

Initial film viewings showed the travel paths sketched in (a). Memomotion analysis led to the redesigned office and resulting travel paths shown in (b).

From Marvin E. Mundel, Motion and Time Study: Improving Productivity, *6th ed. (Englewood Cliffs, N.J.: Prentice Hall, 1985), pp. 320–21. Reprinted with the publisher's permission.*

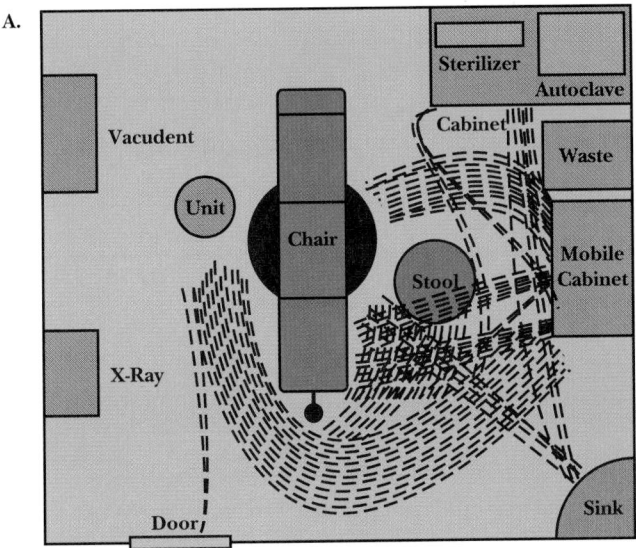

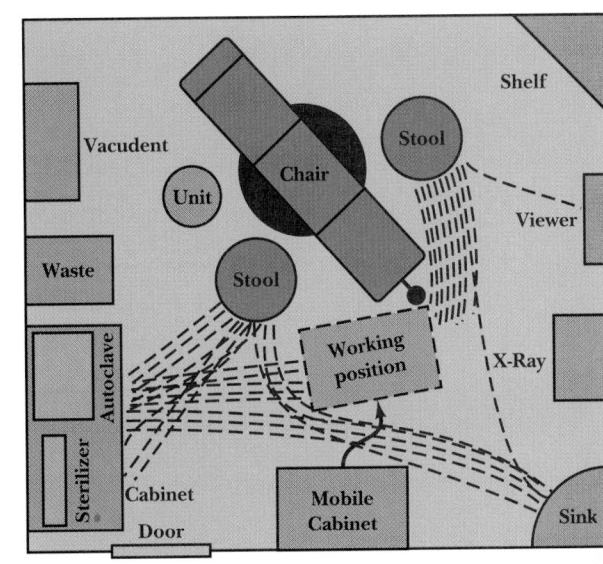

PRODUCTION STANDARD-- JOB SUMMARY SHEET

C.O.00.1120.0

Page No. __1__ Of __3__ Dept. __Process Ham__ Study Date __2-16-__ Study No. __3-6-111__

Operation __Bone 4½-5½ S.S. Picnics-Skin On__ Job __Bone Picnics__

Backing _ _ _ _ _ _ Machine No. _ _ _ _ _ _ Ave. Equip. __Job Symbol 150-h__

No.	DETAILED DESCRIPTION OF JOB ELEMENTS		Std. Allowed Time/Unit
	LEFT HAND	**RIGHT HAND**	
1.	Grasp picnic from conveyer or from pile at side of operator. Slide picnic on cutting board to position in front of operator - face side up.	Hold knife.	.0371
2.	Pick up steel from cutting board - Hold steel - aside steel to cutting board.	Slide knife blade across steel. (3 strokes)	.0408
3.	Smooth out skin around shank end - pull up on skin during cut - use skin as handle to roll picnic over so skin side is up. Roll picnic back so face side is up - aside skin to conveyer.	Slit skin on shank end - cut skin away from skin side of picnic complete skinning by cutting away skin from shank end on face side.	.2786
4.	Hold picnic down on cutting board.	Make horizontal cut at side of picnic length of the arm bone to the area in the center of the picnic near the bone.	.1390
5.	Hold picnic - grasp and hold bone at center area. Hold bone at blade end, until bone is free of picnic.	Guide knife along contour of arm bone, around center area of bone until entire center area is free of picnic. Cut away meat at blade end of arm bone. Cut away meat from shank end of arm bone until bone is completely free of picnic.	.5877
6.	Aside bone to top conveyer - hit counter on return motion to record number of picnics boned.	Aside picnic to bottom conveyer.	.0370

Converted to		Cwt.	Std. Allowed Time Min/Unit	1.1762
Basis Unit	**Wt. Per Piece**	4.83 Lbs.	Std. mos/ 100 Units	1.9603
Std. Allowed Time Min/Unit		.2435	Units/Mos.	51.0
Std. was 100/Units		.4058	Approved by	A.D.
Units/Mos.		246.4	Approved by	

FIGURE 17-5

Summary of the Job of Preparing Hams

This description summarizes the hand movements and time required to process a picnic ham. It brings together the results of micromotion analysis and standard time analysis.

From Marvin E. Mundel, Motion and Time Study: Improving Productivity, *6th ed. (Englewood Cliffs, N.J.: Prentice Hall, 1985), p. 502. Reprinted with permission of Patrick Cudahy, Inc.*

648

Health Risks of Repetition

Besides reducing satisfaction, oversimplified, repetitive tasks can also depress employee health. According to government sources, repetitive motion disorders accounted for 147,000 instances of job-related illness in 1989.* Such disorders result from repeating the same motions with arms and hands throughout the day. Also called cumulative trauma injuries, repetitive motion disorders include carpal tunnel syndrome, a painful wrist ailment found among people who use typewriter keyboards for extended periods of time, and white finger, an injury common among operators of pneumatic drills that is caused by intense vibration.

To combat the health effects of repetition, the U.S. Department of Labor's Occupational Safety and Health Administration (OSHA) has warned employers of the negative health effects of repetition and has developed plans with some to reduce the presence of repetitive tasks. For example, Chrysler Corporation has hired experts to study jobs at five of the company's assembly plants. It has also begun to rotate workers among tasks to break up repetition over the course of each working day. Working with human factors engineers, Chrysler has redesigned many jobs and developed special tools to reduce or eliminate repetitive motions at work. In this manner, the company expects to begin to bring the problem under control within three years.†

* "Repetitive Motion Disorders Lead Increase in Job Illnesses," *New York Times*, November 16, 1990, p. D7.

† "Chrysler Agrees to Curtail Repetitive Tasks for Workers," *Lansing State Journal*, November 3, 1989, p. 4B.

Evaluating Industrial Engineering and the Efficiency Perspective

APPLYING THE DIAGNOSTIC MODEL
Diagnostic Question 1: Does the design of the organization's current jobs seem to reflect the efficiency or the satisfaction perspective? Are most jobs simplified, or have attempts been made to alter job range or depth?

Diagnostic Question 2: If the efficiency perspective appears dominant, do productivity and satisfaction data support the idea that jobs have not been oversimplified? Or does faltering productivity and conspicuous dissatisfaction indicate that oversimplification may be a problem?

POINT TO STRESS
Viewing people mechanistically can result in a loss of efficiency. Individuals will participate in activities that make the job less boring. Such activities are often counter productive.

satisfaction perspective An approach to job design that suggests that fitting the characteristics of jobs to the needs and interests of the people who perform them provides the opportunity for satisfaction at work.

Consistent with the efficiency perspective that serves as their foundation, all the industrial engineering methods we have described attempt to enhance productivity by simplifying jobs. Often, industrial engineers using these methods can improve productivity dramatically. There is, however, a danger that simplification will be carried too far, leading to the creation of oversimplified jobs like those of the welder and clerk described at the beginning of this chapter. This danger looms particularly large in bureaucratic organizations, because the common use in such firms of standardization and formalization increases the pressure to simplify work (see Chapter 15).

Why worry about oversimplification? As we said in Chapters 7 and 8, workers performing oversimplified, routine jobs may become bored, resentful, and dissatisfied. They may even engage in sabotage because of the absence of challenge and interest in their work. Oversimplified work may also lead employees to be absent a lot of the time or to start looking for other jobs. Oversimplification can even have dire health consequences, as discussed in the "In Practice" box. So the same simplification intended to enhance the efficiency of work processes may actually reduce that efficiency if carried to an extreme.

THE SATISFACTION PERSPECTIVE ON JOB DESIGN

What can be done to counteract the effects of oversimplification or to make sure that jobs are not oversimplified to begin with? The answer to this question, offered initially by Lillian Gilbreth, is that jobs should be designed in such a way that performing them creates feelings of fulfillment in their holders.[7] This idea is the central tenet of the **satisfaction perspective**

[7] Lillian M. Gilbreth, *The Psychology of Management* (New York: MacMillan, 1921), p. 19.

TABLE 17-2
Two Perspectives on Job Design

EFFICIENCY PERSPECTIVE	SATISFACTION PERSPECTIVE
Tasks are shaped mainly by technology and organizational needs.	Tasks are shaped at least partly by workers' personal needs.
Tasks are repetitive and narrow.	Tasks are varied and complex.
Tasks require little or no skill and are easy to learn and perform.	Tasks require well-developed skills and are difficult to learn and perform.
The management and performance of work are separated into different jobs.	The management and performance of work are merged in the same job.
It is assumed that there is only one best way to do each job. Tools and methods are developed by staff specialists.	It is assumed that each job can be performed in several ways. Tools and methods are often developed by the people who use them.
Workers are an extension of their equipment and perform according to its requirements. Work is often machine paced.	Workers use equipment but are not regulated by it. The pace of work is set by people rather than machines.
Primarily extrinsic rewards (incentive wages) are used to motivate performance.	Intrinsic rewards (task achievements) are used with extrinsic rewards to motivate performance.
Social interaction is limited or discouraged.	Social interaction is encouraged, in some cases required.
Efficiency and productivity are the ultimate goals of job design.	Satisfaction and fulfillment are the ultimate goals of job design.

horizontal job enlargement A type of job design based on the idea that increasing the number of tasks a job holder performs will reduce the repetitive nature of the job and thus eliminate worker boredom.

job range The number of tasks a job holder performs to complete the job.

job extension A type of horizontal job enlargement in which several simplified jobs are combined to form a single new job.

One way to make workers proud of their work and customers happy with their purchases is to produce a unique product every time. At Matsushita's National Bicycle Industrial Co., employees work with computers and robots to make customized bicycles, each one tailored to the specifications of an individual customer. The 20-worker plant can produce any of 11,231,862 variations on 18 models of racing, road, and mountain bikes in 199 color patterns and about as many sizes as there are people. Source: Susan Moffat, "Japan's New Personalized Production," Fortune, October 22, 1990.

on job design. The satisfaction perspective suggests that fitting the characteristics of jobs to the needs and interests of the people who perform them provides the opportunity for satisfaction at work.[8] Table 17-2 contrasts this approach with the efficiency perspective discussed in the last section. Among the methods of job design developed with the satisfaction perspective in mind are horizontal job enlargement, vertical job enrichment, comprehensive job enrichment, and sociotechnical enrichment.

Horizontal Job Enlargement

To counteract oversimplification, managers sometimes attempt to boost the complexity of work by increasing the number of task activities a job entails. This approach, called **horizontal job enlargement**, is based on the idea that increasing **job range**, or the number of tasks a jobholder performs, will reduce the repetitive nature of the job and thus eliminate worker boredom.[9] Increasing job range in this manner is *horizontal* enlargement, because the job is created out of tasks from the same horizontal "slice" of an organization's hierarchy.

Job Extension. Some horizontal job enlargement programs involve **job extension**, an approach in which several simplified jobs are combined to form a single new job. For example, the job of our insurance company clerk, which consists solely of proofreading, might be extended by adding filing and telephone-an-

[8] Gerald R. Salancik and Jeffrey Pfeffer, "An Examination of Need-Satisfaction Models of Job Attitudes," *Administrative Science Quarterly* 22 (1977), 427–56.
[9] The classic piece on this approach to counteracting oversimplification is Charles R. Walker and Robert H. Guest, *The Man on the Assembly Line* (Cambridge, Mass.: Harvard University Press, 1952).

POINT TO STRESS
Combining several boring tasks does little to relieve boredom.

AN EXAMPLE
Job extension not only fails to make jobs more interesting, it can reduce efficiency and satisfaction. Workers may even get angry because they are distracted by new facets of the job. The proofreader, for example, may now be constantly interrupted by the phone and get little proofreading done.

job rotation A type of horizontal job enlargement in which workers are rotated among several jobs in a structured, predefined manner.

job depth The amount of discretion a job holder has to choose job activities and outcomes.

vertical job enrichment A type of job design based on the idea that giving job holders the discretion to choose job activities and outcomes will improve their satisfaction.

motivator factors Characteristics of the job that according to Frederick Herzberg, influence the amount of satisfaction experienced at work.

swering tasks. Similarly, the welder's job might be extended by adding other assembly operations to it.

Organizations as diverse as Maytag, AT&T, and the U.S. Civil Service have implemented job extension in one form or another. However, especially when a number of simple, easy-to-master tasks are combined, it is easy for workers to view job extension as giving them more of the same routine, boring work to do. Thus, although initial tests seemed promising, most research has suggested that job extension rarely succeeds in reversing oversimplification sufficiently to strengthen employee motivation and satisfaction.[10]

Job Rotation. In **job rotation**, workers are rotated among several jobs in a structured, predefined manner. Rotation of this sort creates horizontal enlargement without combining or otherwise redesigning a firm's jobs. For instance, a supermarket employee might run a checkout lane for a specific period of time and then, switching jobs with another employee, restock shelves for another set period of time.

As workers rotate, they perform a wider variety of tasks than they would if limited to a single job. Again, though, critics have observed that job rotation often achieves little more than having people perform several boring, routine jobs rather than one. As a result, although companies including Ford Motor Company and Western Electric have tried job rotation, it has generally failed to improve worker motivation or satisfaction.[11]

Vertical Job Enrichment: Herzberg's Two-Factor Theory

The failure of horizontal job enlargement to successfully counteract the undesirable effects of oversimplification has led many managers to try other approaches instead. Many such trials involve attempts to increase **job depth**, that is, the amount of discretion a jobholder has to choose job activities and outcomes. This approach, called **vertical job enrichment**, is based on the work of Frederick Herzberg, an industrial psychologist who studied the causes of employee satisfaction and dissatisfaction at work.[12]

Herzberg, who began his research in the mid-1950s, started out by interviewing 200 engineers and accountants in nine companies, asking them to describe incidents at work that had made them feel "exceptionally good" or "exceptionally bad" about their jobs. From these interviews, Herzberg concluded that satisfaction, or feeling good, and dissatisfaction, or feeling bad, should be thought of as independent concepts, not opposites on a single continuum, as traditional views had held. What this suggests is that a person might feel more satisfied with her job without feeling less dissatisfied, more dissatisfied without feeling less satisfied, and so forth. Figure 17-6 contrasts Herzberg's two-factor view with the traditional single-factor view of satisfaction in which higher satisfaction always accompanies lower dissatisfaction and vice versa.

As he dug further into his interview data, Herzberg also found that certain characteristics of the work situation seemed almost always to affect employee satisfaction. Quite different work characteristics appeared to be associated with employee dissatisfaction. **Motivator factors**, such as achievement or recognition,

[10] J. D. Kilbridge, "Reduced Costs through Job Enlargement: A Case," *Journal of Business* 33 (1960), 357–62; J. F. Biggane and P. A. Stewart, "Job Enlargement: A Case Study," in *Design of Jobs*, ed. Louis E. Davis and James C. Taylor, (New York: Penguin, 1972), pp. 264–76; and Gerald E. Susman, "Job Enlargement: Effects of Culture on Worker Responses," *Industrial Relations* 12 (1973), 1–15.

[11] Ricky W. Griffin, *Task Design: An Integrative Approach* (Glenview, Ill.: Scott, Foresman, 1982), p. 25.

[12] Frederick Herzberg, Bernard Mausner, and Barbara Bloch Snyderman, *The Motivation to Work* (New York: John Wiley, 1959).

FIGURE 17-6

Contrasting Views of Satisfaction and Dissatisfaction

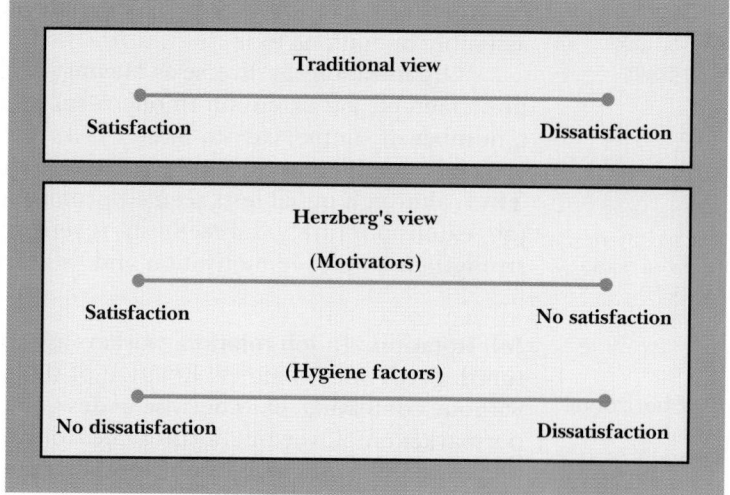

increased satisfaction. Their absence produced a lack of satisfaction but not active dissatisfaction. On the other hand, **hygiene factors**, such as company policy or employees' relationships with their supervisors, usually led to serious dissatisfaction and rarely contributed to a gain in satisfaction.

Armed with this distinction, Herzberg then noticed that only the motivator factors identified in his research seemed able to increase the incentive to work. Hygiene factors, he said, could help to maintain motivation but would more often contribute to a decrease in motivation. If you examine Figure 17-7, you will see that many of Herzberg's hygiene factors are the very job characteristics emphasized by the efficiency perspective on job design. You can undoubtedly anticipate Herzberg's argument. In fact, he contended that following the principles advocated by Taylor, the Gilbreths, and later specialists in industrial engineering would create oversimplified jobs that could only dissatisfy and demotivate workers. Thus, he suggested, managers should pay less attention to things like working conditions and salary and instead design jobs that incorporate opportunities for such positive outcomes as growth, achievement, and recognition.

Over the years, many critics have attacked Herzberg's ideas.[13] Among the most serious criticisms are the following:

1. The *critical-incident technique* that Herzberg used, in which he asked people to recall earlier feelings and experiences, is a questionable research method subject to errors in perception or memory and to subconscious biases. Thus the validity of his conclusions is questionable.

2. Herzberg's interviewees, engineers and accountants, were all members of professional, white-collar occupational groups and male (few women were engineers or accountants in Herzberg's day). Women, minorities, and members of other occupational groups, such as salespeople or industrial laborers, could be expected to answer Herzberg's questions differently.

hygiene factors Characteristics of the job that according to Frederick Herzberg, influence the amount of dissatisfaction experienced at work.

[13] For example, see Robert. J. House and Lawrence A. Wigdor, "Herzberg's Dual-Factor Theory of Job Satisfaction and Motivation: A Review of the Empirical Evidence and a Criticism," *Personnel Psychology* 20 (1967), 369–89; Marvin D. Dunnette, John P. Campbell, and Milton D. Hakel, "Factors Contributing to Job Dissatisfaction in Six Occupational Groups," *Organizational Behavior and Human Performance* 2 (1967), 146–64; Joseph Schneider and Edwin A. Locke, "A Critique of Herzberg's Classification System and a Suggested Revision," *Organizational Behavior and Human Performance* 6 (1971), 441–58; Donald P. Schwab and Larry L. Cummings, "Theories of Performance and Satisfaction: A Review," *Industrial Relations* 9 (1970), 408–30; and Richard J. Caston and Rita Braito, "A Specification Issue in Job Satisfaction Research," *Sociological Perspectives* 28 (1985), 175–97.

FIGURE 17-7

Herzberg's Motivator Factors
and Hygiene Factors

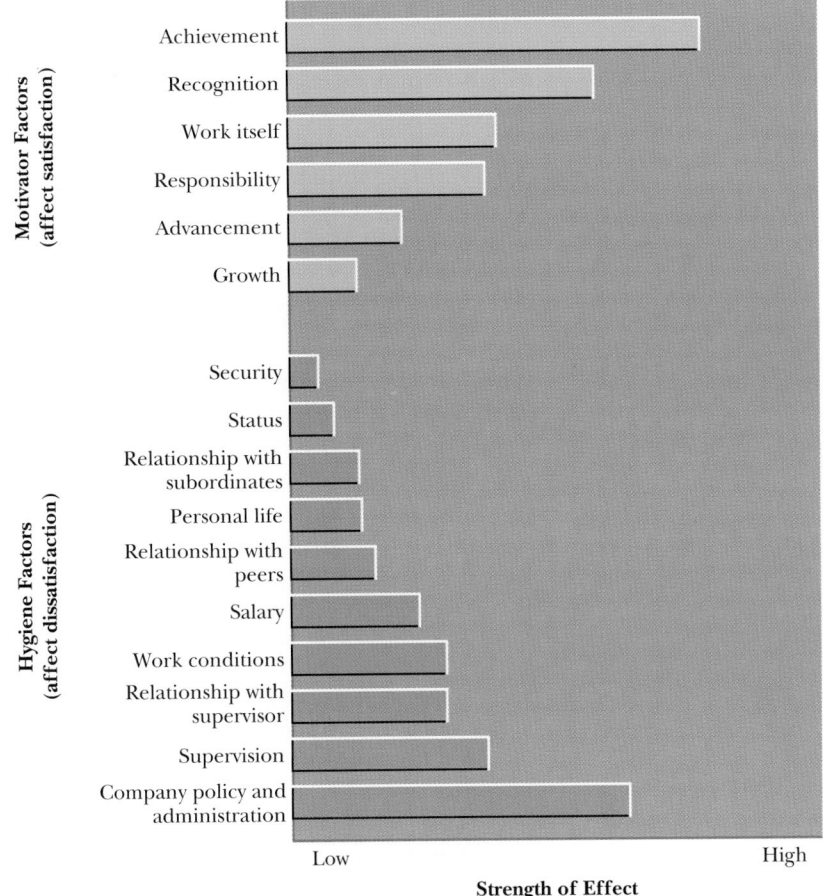

3. Other studies have failed to replicate Herzberg's results. You will recall from Chapter 3 that such failure casts grave doubts on the merits of research findings.

4. Job design programs based on Herzberg's model almost always fail to stimulate work-force satisfaction of lasting significance.

Despite these criticisms, Herzberg's theory is widely known among managers and continues to stimulate interest in questions of motivation, satisfaction, and job design. In addition, it has influenced more recent approaches to job design by highlighting the importance of designing jobs that satisfy human desires for growth, achievement, and recognition. Owing to questions about its validity, however, Herzberg's two-factor theory is not a useful guide for managerial actions.[14]

Comprehensive Job Enrichment: The Hackman-Oldham Job Characteristics Model

Although neither the horizontal loading of job enlargement nor the vertical loading of Herzberg's job enrichment is able to counteract oversimplification when used separately, **comprehensive job enrichment** programs that combine

comprehensive job enrichment A type of job design that combines both horizontal and vertical improvements to stimulate employee motivation and satisfaction.

[14] Griffin, *Task Design*; also see J. Richard Hackman, "On the Coming Demise of Job Enrichment," in *Man and Work in Society*, ed. Eugene Louis Cass and Frederick G. Zimmer (New York: Van Nostrand, 1975), pp. 45–63

both horizontal and vertical improvements are usually quite successful in stimulating motivation and satisfaction. Many such programs are based on the model of job design developed by J. Richard Hackman and Greg Oldham, which is shown in Figure 17–8.[15]

As the figure suggests, Hackman and Oldham proposed that five core job characteristics influence workers' experience of three critical psychological states. These critical states, they said, lead to a variety of work and personal outcomes. This process is affected by certain moderating factors, so that the model works as expected under some circumstances but not others. Let's look at the Hackman-Oldham model in greater detail.

Core Job Characteristics. According to Hackman and Oldham, jobs that in and of themselves are likely to motivate performance and contribute to employee satisfaction exhibit the following five **core job characteristics:**

core job characteristics Job characteristics identified in the Hackman-Oldham model that lead their holders to experience certain critical psychological states.

1. *Skill variety.* The degree to which a job holder must carry out a variety of different activities and use a number of different personal skills in performing the job.
2. *Task identity.* The degree to which performing a job results in the completion of a whole and identifiable piece of work and a visible outcome that can be recognized as the result of personal performance.
3. *Task significance.* The degree to which a job has a significant impact on the lives of other people, whether those people are coworkers in the same firm or other individuals in the surrounding environment.
4. *Autonomy.* The degree to which the job holder has the freedom, independence, and discretion necessary to schedule work and to decide what procedures to use in carrying it out.

[15] Hackman and Oldham, "Motivation through the Design of Work: Test of a Theory," *Organizational Behavior and Human Performance* 16 (1976), 250–79; Hackman and Oldham, *Work Redesign* (Reading, Mass.: Addison-Wesley, 1980); Karlene H. Roberts and William H. Glick, "The Job Characteristics Approach to Task Design: A Critical Review," *Journal of Applied Psychology* 86 (1981), 193–217; and Ramon J. Aldag, Steve H. Barr, and Arthur P. Brief, "Measurement of Perceived Task Characteristics," *Psychological Bulletin* 99 (1981), 415–31.

FIGURE 17-8

The Hackman-Oldham Job Characteristics Model

Adapted with the publisher's permission from J. Richard Hackman and Greg R. Oldham, "Motivation through the Design of Work: Test of a Theory," Organizational Behavior and Human Performance 16 (1976), 256.

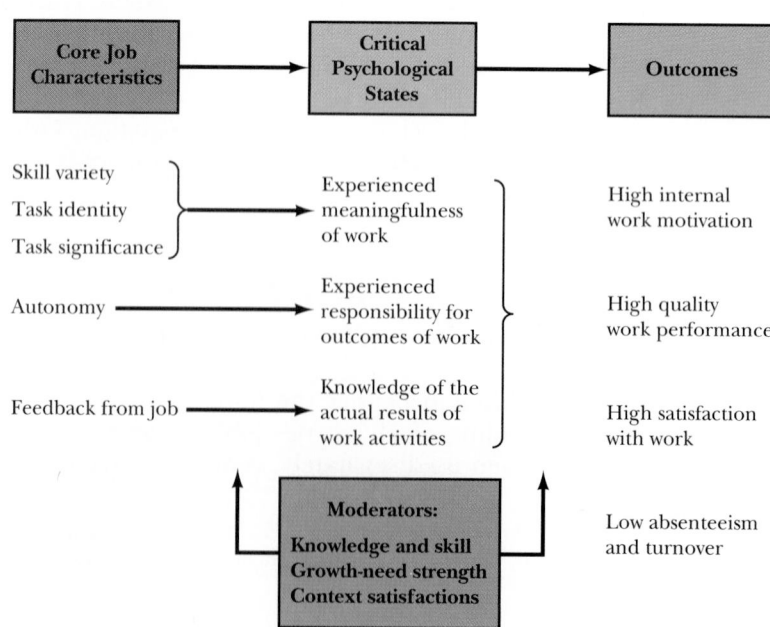

5. *Feedback*. The degree to which performing the activities required by the job provides the worker with direct and clear information about the effectiveness of her performance.

critical psychological states
Mental conditions identified in the Hackman-Oldham model as being triggered by the presence of certain core job characteristics.

Critical Psychological States. These five core job characteristics, in turn, influence the extent to which employees experience three **critical psychological states**:

1. *Experienced meaningfulness of work*. The degree to which a worker experiences her job as having an outcome that is useful and valuable to her, the company, and the surrounding environment.
2. *Experienced responsibility for work outcomes*. The degree to which the worker feels personally accountable and responsible for the results of her work.
3. *Knowledge of results*. The degree to which the worker maintains an awareness of the effectiveness of her work.[16]

As Figure 17-8 shows, each of the job characteristics influences one particular psychological state. Specifically, *skill variety*, *task identity*, and *task significance* are seen as affecting the *experienced meaningfulness of work*. Thus jobholders should experience their jobs as meaningful if they must use a variety of activities and skills to produce an identifiable piece of work that influences the lives of others. *Autonomy*, on the other hand, influences the job holder's *experienced responsibility for work outcomes*. This means that workers who have the discretion to determine work procedures and outcomes should feel responsible for the results of their work. Finally, *feedback* determines whether a worker will have *knowledge of the results of work*. Through information about performance effectiveness that comes from the job itself, the jobholder can maintain an awareness of how effectively she is performing.

Work and Personal Outcomes. According to the Hackman-Oldham model, if workers experience these three psychological states, several work and personal outcomes may result. First, workers may view their jobs as interesting, challenging, and important and may be motivated to perform them simply because they are so stimulating, challenging, and enjoyable. *High internal work motivation*, or being "turned on" to job performance by its personal consequences, is thus one possible outcome. Second, experiencing the three critical psychological states and the internal, or intrinsic, motivation they arouse can encourage *high-quality work performance* (and sometimes greater production quantity).[17] Third, workers who experience the three psychological states do so because their work allows them opportunities for personal learning, growth, and development. As we saw in Chapter 8, these kinds of experiences generally promote *high satisfaction with work*. Fourth, work that includes these kinds of experiences also tends to be associated with *lower absenteeism* and *turnover*.

APPLYING THE DIAGNOSTIC MODEL
Diagnostic Question 3: Can the firm's current technology be changed to the degree required by job enrichment methods? Are jobs mainly individualized, indicating the appropriateness of Hackman-Oldham enrichment? Or are jobs often performed by groups of people working together closely, suggesting the sociotechnical approach?

Moderating Factors. The Hackman-Oldham model proposes several moderating factors. They determine whether its core job characteristics will indeed trigger the critical psychological states leading to the work and personal outcomes we have identified. The first of these moderators is the worker's *knowledge and skill*.

[16] Hackman and Oldham, "Design of Work," pp. 256–57.
[17] Raymond A. Katzell, Penny Bienstock, and Paul H. Faerstein, *A Guide to Worker Productivity Experiments in the United States 1971-1975* (New York: New York University Press, 1977) p. 14; Edwin A. Locke, Dena B. Feren, Vickie M. McCaleb, Karyll N. Shaw, and Anne T. Denny, "The Relative Effectiveness of Four Methods of Motivating Employee Performance," in *Changes in Working Life*, ed. K. D. Duncan, Michael M. Gruneberg, and D. Wallis (London: John Wiley, 1980), pp. 363–88; and Richard E. Kopelman, "Job Redesign and Productivity: A Review of the Evidence," *National Productivity Review* 4 (1985), 237–55.

To succeed on a job with the five core job characteristics, a worker must have the knowledge and skill required to perform the job successfully. People who cannot perform a job because they lack the necessary knowledge or skill will only feel frustrated by their failure. The motivational aims of job enrichment will be thwarted.

Growth-need strength, the strength of a worker's need for personal growth, is the second factor that affects the operation of the Hackman-Oldham model. Workers who have strong growth needs are attracted to enriched work because it offers the opportunity for growth. On the other hand, workers whose need for growth is weak are likely to feel overburdened by the opportunities offered them. Therefore they will try to avoid enriched work.

Finally, certain *context satisfactions* can act as moderator factors that influence the Hackman-Oldham model's applicability. Hackman and Oldham identified them as satisfaction with pay, with job security, with coworkers, and with supervisors. Workers who feel exploited and dissatisfied because they are poorly paid, feel insecure about their jobs, or have abusive coworkers or unfair supervision are likely to view job enrichment as just one more type of exploitation. Context dissatisfaction can thus negate the expected benefits of Hackman-Oldham job enrichment.

The Job Diagnostic Survey. To put their model to use, Hackman and Oldham developed a questionnaire, the **Job Diagnostic Survey** (JDS), that measures workers' perceptions of the five core job characteristics, the three critical psychological states, and different moderating factors. (The Task Diagnostic Questionnaire that appears in Exercise 17-1 is based on many of the questions that appear in the JDS and can be substituted for it in the following procedure.) Figure 17-9 shows two ways to use the data on core job characteristics acquired with the JDS. On the left side of the figure, JDS scores on each of the five characteristics have been plotted for two jobs. As you can see, Job A rates well across the five characteristics, but Job B rates poorly overall. Its reasonably high level of task significance and moderate levels of skill variety and task identity are offset by rather poor feedback and very little autonomy. Clearly, enrichment of Job B will require changes in four of the five core job characteristics.

POINT TO STRESS
We often assume people want more responsibility and more opportunities to grow and advance. Data are accumulating indicating that this assumption is not always valid and thus may not be a good basis for developing programs and assessing individuals.

Job Diagnostic Survey A questionnaire that measures workers' perceptions of core job characteristics, critical psychological states, and different moderating factors.

FIGURE 17-9

JDS Profiles and MPS Comparison

From J. R. Hackman, "Work Design," in Improving Life at Work: Behavioral Science Approaches to Organizational Change, *ed. J. R. Hackman and J. L. Suttle (Santa Monica, Calif.: Goodyear Publishing Co., 1977), p. 135. Reprinted with the authors' permission.*

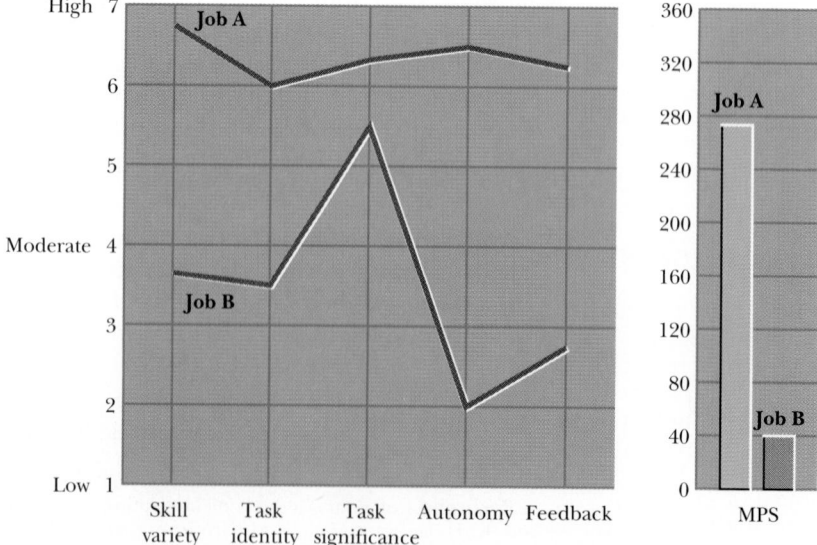

JOB DESIGN

motivating potential score A summary score calculated from data obtained with the Job Diagnostic Survey.

On the right side of Figure 17-9, a bar chart reflects the calculation for each of the two jobs of a **motivating potential score** (MPS), entering the JDS scores into the following formula:

$$\text{MPS} = \frac{(\text{Variety} + \text{Identity} + \text{Significance})}{3} \times \text{Autonomy} \times \text{Feedback}$$

This formula reflects the Hackman-Oldham model's conception of skill variety, task identity, and task significance as exerting a joint effect on experienced meaningfulness. These three characteristics are averaged before being combined with autonomy and feedback scores. (If you perform Exercise 17-1 you will use this same formula to compute a motivational summary score or MSS.) Some investigators are beginning to use a simpler additive formula to perform this combination instead of the multiplicative one suggested by Hackman and Oldham. Which one will ultimately prove to be the most useful is currently unclear.[18]

Examination of the bar chart suggests that the MPS facilitates gross comparisons among jobs and can help managers locate trouble spots quickly even if a large number of different jobs must be scrutinized. A lower MPS means a job needs enrichment while a higher MPS indicates that the job doesn't. However, the MPS does not tell managers how to enrich a particular job. To learn that, they must construct profiles like the ones in the left-hand part of Figure 17-9, which reveal a job's specific strengths and weaknesses.

There are a number of ways of correcting the deficiencies identified by a JDS analysis. To enhance skill variety and task identity, oversimplified jobs can be *combined* to form enlarged modules of work. For example, the production of a toaster could be redesigned so that the entire appliance is constructed by a single employee working alone rather than by a dozen people working on an assembly line. *Natural units of work* can be formed by clustering similar tasks into logical or inherently meaningful groups. For example, a data-entry clerk who formerly selected work orders randomly from a stack might be given sole responsibility for the work of an entire department or division. This intervention might strengthen both task identity and task significance for the clerk.

In an effort to increase task variety, autonomy, and feedback, a firm can give workers the responsibility for *establishing* and *managing client relationships*. In several General Motors plants, assembly workers can telephone people who have recently purchased a GM car to ask for feedback about product quality. Or to increase autonomy, managerial duties can be designed into a particular job through *vertical loading*. Finally, to increase feedback, *feedback channels* can be opened by adding to a job such things as quality-control duties and computerized feedback mechanisms.

Sociotechnical Enrichment: The Tavistock Model

The Hackman-Oldham model focuses on jobs as individualized units of work, each performed by a single employee. It is therefore not appropriate for jobs performed by closely interacting groups of workers. How can managers counteract

---

[18] Hugh J. Arnold and Robert J. House, "Methodological and Substantive Extensions of the Job Characteristics Model of Motivation," *Organizational Behavior and Human Performance* 25 (1980), 161–83; Arthur P. Brief, Marc J. Wallace, Jr., and Ramon J. Aldag, "Linear vs. Non-Linear Models of the Formation of Affective Reactions: The Case of Job Enlargement," *Decision Sciences* 7 (1976), 1–9; Gerald R. Ferris and David C. Gilmore, "A Methodological Note on Job Complexity Indices," *Journal of Applied Psychology* 70 (1985), 225–27; and Denis D. Umstot, Cecil H. Bell, Jr., and Terrence R. Mitchell, "Effects of Job Enrichment and Task Goals on Satisfaction and Productivity: Implications for Task Design," *Journal of Applied Psychology* 61 (1976), 379–94.

TEACHING NOTE
Have students discuss methods of enriching jobs on an assembly line in a bottling plant. What kinds of technological changes would be needed?

AN EXAMPLE
Enlarged work modules in which an individual produces a complete or nearly complete product resemble the old craft industries. A shoe-maker made a complete pair of shoes while the silversmith made a teapot. These individuals used a number of skills to complete the work and task identity was high. The work from raw materials to finished products was theirs. Many, like silversmiths, signed their work to make their identity with the object known.

AN EXAMPLE: PRATT & WHITNEY
In 1986 customers of jet-engine maker Pratt & Whitney were defecting in droves to archrival General Electric Co. because of poor service. With new methods of empowering workers and building teams, the company began to improve its competitive position, and by 1989 it had increased orders eightfold. Rather than stand at one workstation all day, P&W workers now move about the plant in doing their jobs. Workers are also encouraged to get close to their customers. They participate in "gripe" sessions with customers invited to the plant, and they take short-term assignments working for customers (P&W pays their salaries) to learn their clients' needs. (Todd Vogel, "Where 1990s-Style Management Is Already Hard at Work," *Business Week*, October 23, 1989, pp. 92–100.)

sociotechnical enrichment A type of job design that recognizes the importance of satisfying the needs of employees within the technical requirements of an organization's production system.

the negative effects of oversimplified *group* work? Answer: They can make use of some form of **sociotechnical enrichment**, an approach to designing group jobs that recognizes the importance of satisfying employees' personnel needs.

Sociotechnical Principles. Sociotechnical enrichment originated in the early 1950s when researchers from England's Tavistock Institute set out to correct faults in the processes used to mine coal in Great Britain.[19] Historically, coal had been mined by teams of miners working closely with each other to pool efforts, coordinate activities, and cope with the physical threats of mining. With the advent of powered coal-digging equipment in the 1930s and 1940s, however, coal mining changed drastically. Teams were split up, and miners often found themselves working alone along the long walls of exposed coal created by the equipment. Mining, normally a hazardous, physically demanding occupation anyway, grew even more unbearable owing to changes stimulated by the new technology. Miners expressed their dissatisfaction with these circumstances through disobedience, absence, and occasional violence.

The Tavistock researchers soon realized that the roots of the miners' dissatisfaction lay in the loss of the social interaction that mining teams had provided and that had made the dangerous, demanding job of mining more tolerable. It appeared to the researchers that technology had been allowed to supersede important social factors and that performance in the mine could be improved only if this balance were redressed.

Indeed, after small teams were formed to operate and provide support for clusters of powered equipment, production rose substantially. This experience led the Tavistock researchers to suggest that work-force productivity could be hurt when either social or technical factors alone were allowed to shape work processes. They further suggested that job designs that balanced social (socio) and technological (technical) factors—*sociotechnical designs*—encourage both performance and satisfaction.

In other words, employees should work in groups that allow them to talk with each other about their work as they do it. Those work groups should include the people whose frequent interaction is required by the production technology being used. For instance, salespeople, register clerks, and stock clerks who must often interact with each other to serve customers in a department store should be grouped together to facilitate communication about work. Salespeople and clerks from other departments should not be included in the group, because they do not share job-related interdependencies with the group's members.

In the course of performing their research, the Tavistock sociotechnical researchers identified the following psychological requirements as critical to worker motivation and satisfaction:

1. The content of each job must be reasonably demanding or challenging and provide some variety, although not necessarily novelty.
2. Performing the job must have perceivable, desirable consequences. Workers should be able to identify the products of their efforts.
3. Workers should be able to see how the lives of other people are affected by the production processes they use and the things they produce.
4. Workers must have decision-making authority in some areas.
5. Workers must be able to learn from the job and go on learning. This implies appropriate performance standards and adequate feedback.

---

[19] Eric L. Trist and K. W. Bamforth, "Some Social and Psychological Consequences of the Longwall Method of Coal-Getting," *Human Relations* 4 (1951), 3–38.

6. Workers need the opportunity to give and receive help and to have their work recognized by others in the work place.[20]

The Tavistock group, who worked mainly in England and Norway, developed this list of required job characteristics independently of Hackman and Oldham, who worked only in the United States. But as you can see, items 1 through 5 correspond loosely with the five core job characteristics of the Hackman-Oldham model. Item 6 highlights the emphasis placed by sociotechnical enrichment on interpersonal relations and social satisfaction in the workplace.

semiautonomous groups Groups that are subject to the management direction needed to ensure adherence to organizational policies but are otherwise responsible for managing themselves.

Semiautonomous Groups. Contemporary sociotechnical designs normally involve **semiautonomous groups**. These groups are subject to the management direction needed to ensure adherence to organizational policies but are otherwise responsible for managing group activities. Within each such group

> Individuals must move about within the group spontaneously and without being ordered to do so, because it is necessary to the efficient functioning of the [group]. . . . If we observe the group in action, we will see movements of individuals between different jobs. When an especially heavy load materializes at one work station and another is clear for the moment, we will see the person at the latter spontaneously move to help out at the former. . . . It is a natural and continuous give and take within a group of people, the object being to attain an established production target. . . . The group members are not merely carrying out a certain number of tasks. They are also working together, on a continuing basis, to coordinate different tasks, bearing responsibility, and taking whatever measures are necessary to cope with the work of the entire unit.[21]

As they work together in this manner, the members of a semiautonomous group are able to (1) rotate in and out of tasks to enhance skill variety; (2) work together on a group product that is a whole, identifiable piece of work; (3) influence the lives of other members of the group and the lives of those who consume the group's output; (4) decide as a group who will belong to the group and what tasks group members will perform; (5) obtain feedback from group members about task performance; and (6) count on the help and support of other group members if it is needed.

Shop-Floor Implications. Figure 17-10 contrasts a traditional assembly line with semiautonomous groups. As you can see, the decision to adopt sociotechnical design principles has important implications for shop-floor operations. In both panels of the figure, workers are assembling truck engines. In panel A, each worker performs a simplified job in which he takes a part from a storage bin and attaches it to a partially completed truck engine as it moves along a conveyor. In panel B, however, workers are grouped into semiautonomous groups, each of which removes a bare engine block from a conveyor loop, assembles a complete truck engine from parts in surrounding storage bins, and returns the finished engines to the conveyor loop for transportation to other truck-assembly operations. As suggested by this example, sociotechnical job designs normally eliminate traditional assembly-line operations. The "International OB" box describes another example of how the sociotechnical approach can affect shop-floor layout.

[20] Adapted from Fred Emery and Einar Thorsrud, *Democracy at Work: The Report of the Norwegian Industrial Democracy Program* (Leiden, The Netherlands: H. E. Stenfert Kroese, 1976), p. 14.

[21] David Jenkins, trans., *Job Reform in Sweden: Conclusions from 500 Shop Floor Projects* (Stockholm: Swedish Employers' Confederation, 1975), pp. 63–64.

FIGURE 17-10

Comparison of an Assembly Line and Semiautonomous Groups

Adapted with the publisher's permission from Jan-Peder Norstedt and Stefan Aguren, The Saab-Scania Report *(Stockholm: Swedish Employers' Confederation, 1973), pp. 35, 37.*

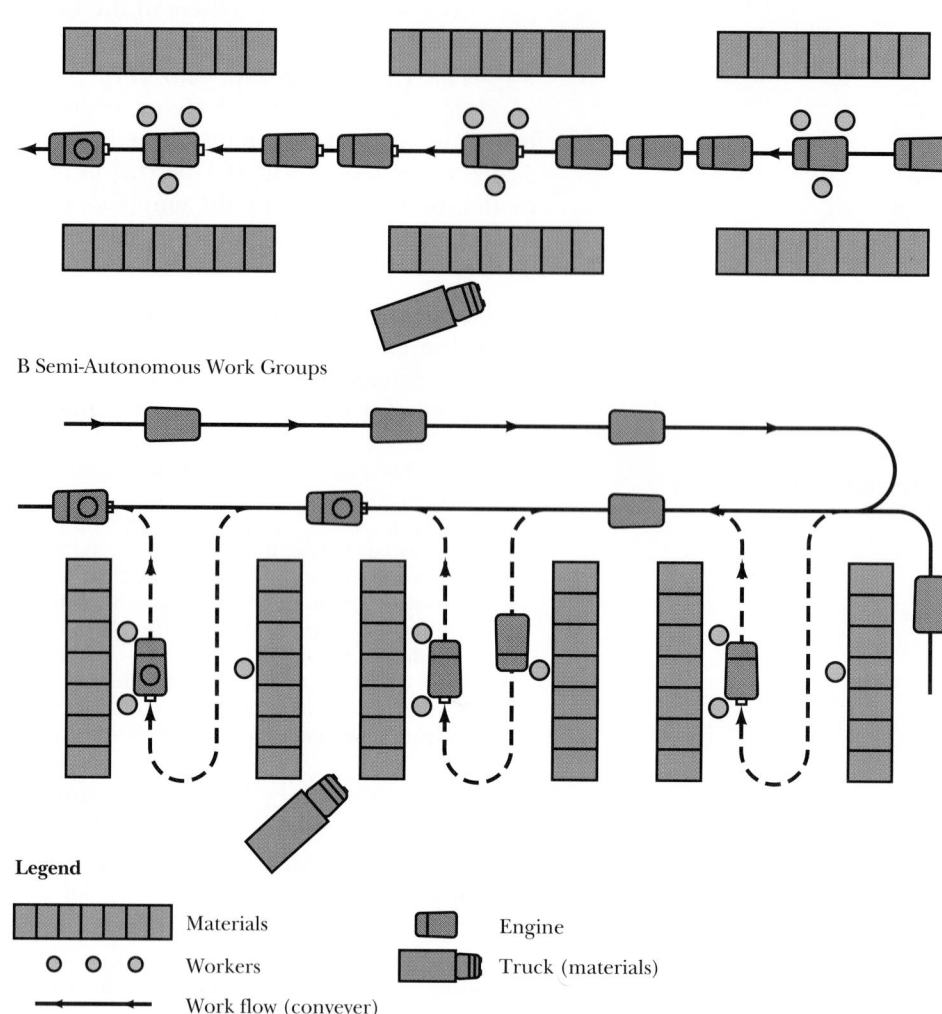

A Traditional Assembly Line

B Semi-Autonomous Work Groups

Legend

▦ Materials		▢ Engine	
○ ○ ○ Workers		▰ Truck (materials)	
◀— Work flow (conveyer)			

Evaluating Job Enrichment and the Satisfaction Perspective

Consistent with the satisfaction perspective that serves as their foundation, all enlargement and enrichment techniques are aimed at designing jobs that satisfy the needs and interests of their holders. As we have already indicated, methods that consist solely of horizontal enlargement or vertical enrichment have largely failed to achieve this goal. However, methods of job design that incorporate *both* horizontal enlargement and vertical enrichment have proven effective in stimulating work-force motivation and satisfaction in a wide variety of situations.

Research on the Hackman-Oldham model has sometimes failed to verify the existence of five distinct job characteristics.[22] It is also unclear whether JDS

[22] Studies that have confirmed the existence of five distinct characteristics include Ralph Katz, "Job Longevity as a Situational Factor in Job Satisfaction," *Administrative Science Quarterly* 23 (1978), 204–23; and R. Lee and A. R. Klein, "Structure of the Job Diagnostic Survey for Public Service Organizations," *Journal of Applied Psychology* 67 (1982), 515–19. Studies that have failed to reveal confirmatory evidence include Randall B. Dunham, "The Measurement and Dimensionality of Job Characteristics," *Journal of Applied Psychology* 61 (1976), 404–9; Jeannie Gaines and John M. Jermier, "Functional Exhaustion in a High Stress Organization," *Academy of Management Journal* 26 (1983), 567–86; Jon L. Pierce and Randall B. Dunham, "The Measurement of Perceived Job Characteristics: The Job Diagnostic Survey vs. the Job Characteristics Inventory," *Academy of Management Journal* 21 (1978),

The Sociotechnical Layout of Volvo's E Plant

One example of how sociotechnical job design can affect the physical layout of the workplace is depicted in Figure 17-11. In Volvo's E-shaped plant in Skovde, Sweden, the work of several semiautonomous groups is combined to produce one completed product. In the four "arms" of the plant, semi-autonomous groups complete machining operations that produce parts for gasoline engines. Assembly is done in the main plant, located in the body of the *E*. Besides assembly operations, the main plant also houses several groups that perform final quality inspections. The building is intended to separate the machining groups to give each its own identity, yet facilitate interactions among groups that occur as work flows from one group to another.

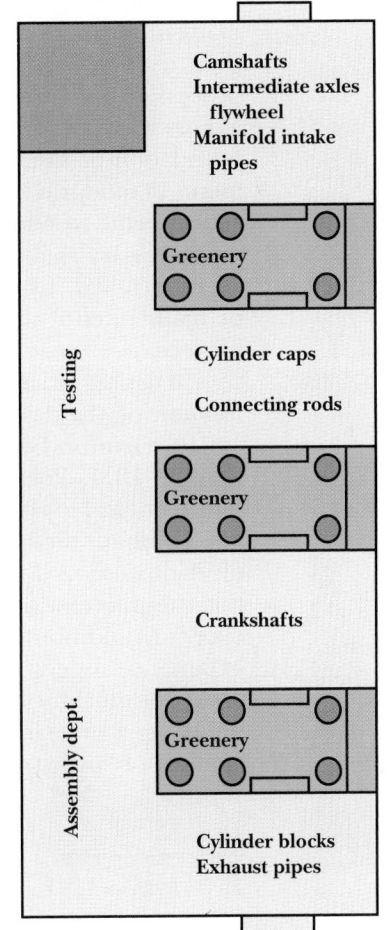

FIGURE 17-11
The Volvo Skovde Plant

From Stefan Aguren and Jan Edgren, New Factories: Job Design through Factory Planning in Sweden *(Stockholm: Swedish Employers' Confederation, 1980), p. 92. Reprinted with the publisher's permission.*

questionnaire items measure objective, stable job characteristics or only subjective, changing worker opinions.[23] Some researchers have even questioned whether job characteristics like those identified by Hackman and Oldham truly influence motivation and satisfaction. They suggest instead that employees' feelings about themselves and their work might be affected more by the opinions of others in

---

123–28; and Denise M. Rousseau, "Technological Differences in Job Characteristics, Job Satisfaction, and Motivation: A Synthesis of Job Design Research and Sociotechnical Systems Theory," *Organizational Behavior and Human Performance* 19 (1977), 18–42.

[23] Objectivity is suggested by studies such as Rickey W. Griffin, "A Longitudinal Investigation of Task Characteristics Relationships," *Academy of Management Journal* 42 (1981), 99–113; Eugene F. Stone and Lyman W. Porter, "Job Characteristics and Job Attitudes: A Multivariate Study," *Journal of Applied Psychology* 60 (1975), 57–64; and Carol T. Kulik, Greg R. Oldham, and Paul H. Langner, "Measurement

the surrounding social context.[24] We will return to this idea and discuss it in detail as we talk about *social information processing* and organizational culture in Chapter 18. Finally, some disagreement exists as to whether the moderators identified by Hackman and Oldham actually influence the model's applicability.[25]

Nonetheless, the Hackman-Oldham model has served as the basis of successful job design programs at Texas Instruments, AT&T, Motorola, Xerox, and many other firms of similar size and reputation. Such programs are not without their drawbacks. They are usually incompatible with assembly-line production processes. To enrich jobs using the Hackman-Oldham approach, a firm must almost always abandon the sort of simplified, repetitive tasks that serve as the foundation of assembly lines. Consequently, companies with substantial investments in modernized assembly lines are often reluctant to try Hackman-Oldham enrichment. In addition, because some 5 to 15 percent of the work force lack the necessary skills, growth needs, or context satisfactions, they are likely to be "overstretched" by enriched work. Therefore, a cluster of unenriched jobs must be maintained if the firm wants to avoid losing a significant number of its employees.

Turning to sociotechnical enrichment, this approach started out in Europe, influencing the design of jobs in firms such as Norsk Hydro, Volvo, Saab-Scania, and the Orrefors Glass Works. Now American companies such as Xerox, Cummins Engine, IBM, Polaroid, and General Electric have begun to experiment with sociotechnical job design.[26] Virtually the same outcomes stimulated by the Hackman-Oldham method are produced by the sociotechnical approach. Sociotechnical job designs do not always improve productivity, but most evidence indicates that they increase motivation and satisfaction and decrease absenteeism and turnover.[27] In addition, as is true for programs based on the Hackman-Oldham model, experience suggests that a small but significant number of workers are likely to resist sociotechnical enrichment. Consequently, either a few jobs must be left unchanged or managers must be prepared to deal with a small but significant amount of "overstretching."

EMERGING METHODS

Within the last two decades, several additional approaches to job design have emerged as researchers and managers have sought new ways to improve workplace productivity and satisfaction. Of these emerging methods, quality circles, alternative work schedules, and automation and robotics have had the greatest effects on the way work is now designed.

APPLYING THE DIAGNOSTIC MODEL

Diagnostic Question 5: If the satisfaction perspective appears dominant, do productivity and satisfaction data suggest that jobs have been enriched without creating work that is overdemanding? Or do falling productivity and satisfaction indicate that workers are being "overstretched" and asked to do more than they can?

Diagnostic Question 6: If you are facing an overenrichment problem, do the results of work measurement procedures suggest ways to simplify jobs enough to facilitate successful performance while still retaining opportunities for growth, achievement, and recognition?

of Job Characteristics: Comparison of the Original and the Revised Job Diagnostic Survey," *Journal of Applied Psychology* 73 (1988), 462–66. Other studies that seem to support the subjectivity side of the argument include Arthur P. Brief and Ramon J. Aldag, "The Job Characteristic Inventory: An Examination," *Academy of Management Journal* 21 (1978), 659–70; and Philip H. Birnbaum, Jiing-Lih Farh, and Gilbert Y. Y. Wong, "The Job Characteristics Model in Hong Kong," *Journal of Applied Psychology* 71 (1986), 598–605.

[24] Gerald R. Salancik and Jeffrey Pfeffer, "A Social Information Processing Approach to Job Attitudes and Task Design," *Administrative Science Quarterly* 23 (1978), 224–53.

[25] Arthur P. Brief and Raymon J. Aldag, "Employee Reactions to Job Characteristics: A Constructive Replication," *Journal of Applied Psychology* 60 (1975), 182–86; and Henry P. Sims and Andrew D. Szilagyi, "Job Characteristic Relationships: Individual and Structural Moderators," *Organizational Behavior and Human Performance* 17 (1976), 211–30.

[26] Richard E. Walton, "From Control to Commitment in the Workplace," *Harvard Business Review* 63 (1985), 76–84.

[27] For instance, see Thomas G. Cummings and Edmond S. Molloy, *Strategies for Improving Productivity and the Quality of Work Life* (New York: Praeger, 1977), pp. 38–49.

Quality Circles

quality circles Small groups of employees who meet on company time to identify and resolve job-related problems.

APPLYING THE DIAGNOSTIC MODEL
Diagnostic Question 4: If technological considerations prohibit job enrichment, might quality circles provide enough relief to restore motivation and satisfaction? If not, can you eliminate the troublesome jobs through automation?

Many companies suffer the negative consequences of job oversimplification. They are unable or unwilling, however, to modify production equipment or methods to the extent required by the Hackman-Oldham and sociotechnical models. In some of these firms, managers are trying to use **quality circles** to counteract oversimplification.

Quality circles (QCs) are small groups of employees, ranging in size from about 3 to 30 members, who meet on company time to identify and resolve job-related problems. Although usually thought of as a Japanese management technique, QCs were actually invented in the United States and exported to Japan by American quality experts Edward Demming and J. M. Juran in the years following World War II.[28] In North America, companies such as Lockheed, Westinghouse, Eastman Kodak, Procter and Gamble, General Motors, Ford, and Chrysler have implemented quality circles. Sometimes they call them quality teams, work teams, or productivity teams. The objectives of all such groups include:

Reducing assembly errors and enhancing product quality
Inspiring more effective teamwork and cooperation in everyday work groups
Promoting a greater sense of job involvement and commitment
Increasing employee motivation
Creating greater decision-making capacity
Substituting problem prevention for problem solving
Improving communication in and between work groups
Developing harmonious relations between management and employees
Promoting leadership development among nonmanagerial employees[29]

POINT TO STRESS
Many organizations had false starts with quality circles because they were unwilling to relinquish control. QCs, not allowed to deal with substantive issues, quickly began to distrust management and, as a result, many of these groups dissolved.

Ordinarily, QC membership is voluntary and stable over time. The amount of time spent in QC activities may range from an hour a month to a few hours every week. Topics of discussion may include quality control, cost reduction, improvement of production techniques, production planning, and even long-term product design.[30] Over the course of many meetings, the activities of a typical QC proceed through a series of steps:

1. Initially, members of the QC identify concerns they have about their work and workplace in a group discussion coordinated by their supervisor or a specially trained facilitator. Often, the facilitator is an internal change agent with expertise in many of the group-level organization-development interventions described in Chapter 14.

2. QC members next examine their concerns and look for ways to collapse or integrate them into specific projects. For instance, concerns about production speed and raw-material quality may be grouped together in a production-methods project. Concerns about workplace safety and worker health may be put into a work environment project.

3. Members perform initial analyses of their QC's projects using various group decision-making techniques and tools, including data gathering, graphs, checklists, or charts.

4. The QC then reaches consensus decisions about the feasibility and impor-

---

[28] William L. Mohr and Harriet Mohr, *Quality Circles: Changing Images of People at Work* (Reading, Mass.: Addison-Wesley, 1983), p. 13.

[29] Donald L. Dewar, *The Quality Circle Handbook* (Red Bluff, Calif.: Quality Circle Institute, 1980), pp. 17–104.

[30] Gerald R. Ferris and John A Wagner III, "Quality Circles in the United States: A Conceptual Reevaluation," *Journal of Applied Behavioral Science* 21 (1985), 155–167.

tance of different projects, deciding which ones to abandon and which ones to pursue.

5. Representatives from the QC make a presentation or recommendation to management that summarizes the work of their group.

6. Management reviews the recommendation and makes a decision. Often, the decision is that QC members will have the opportunity to implement their own recommendations.[31]

QCs fight oversimplification by giving employees the opportunity to participate in the management of their jobs rather than by modifying existing work technologies. For example, employees who work on an assembly line for 39 hours each week might meet as a QC group during the last hour to evaluate the assembly line's performance and prepare for the following week's work. They might also meet in an extended session once a month to discuss more-complicated issues and resolve more-difficult problems. These monthly sessions offer an opportunity for more managerial activity, group autonomy, and information exchange than the regular QC meetings allow. To the extent that QC meetings focus workers' attention on the outputs of the whole assembly line, they may reinforce task identity and task significance.

Research evidence on the effects of QCs as a form of job enrichment is sketchy. What information we do have suggests that QCs have little effect on productivity but can enhance feelings of satisfaction and involvement significantly.[32] The magnitude of such effects is usually smaller than from job enrichment programs based on the Hackman-Oldham model or the Tavistock sociotechnical model. That is understandable, because workers who participate in QCs must still perform unenriched jobs during most of the time spent at work.

Alternative Work Schedules

Besides reshaping jobs themselves, concerns about employee satisfaction have also led managers and OB specialists to reshape the way work is scheduled. This reshaping leads to such innovations as flexible-hour programs, compressed work weeks, and job sharing. Virtually all such plans are intended to stimulate satisfaction by helping employees balance the time and scheduling demands of their work with those of their nonworking lives.

Some firms offer employees the chance to participate in **flexible-hour programs**, also known as *flextime*. These programs generally specify a block of core hours during which everyone must be on the job but allow each employee to choose when she will start and quit work. Core hours are generally the period between 10:00 A.M. and 3:00 P.M. One employee might choose to come to work as late as 10:00 A.M. and remain until 6:00 P.M. Another might arrive at 7:00 A.M. and leave at 3:00 P.M. Flexible-hour programs make it easier for employees to coordinate work responsibilities with nonwork obligations. They allow parents to start work later in the morning if child care is unavailable earlier, for instance. Consequently, flextime can help decrease absenteeism and turnover.[33]

flexible-hour programs
Alternative work schedules that specify a block of core hours during which everyone must be on the job but allow each employee to choose when to start work and when to go home. Also called flextime.

TEACHING NOTE
In Phoenix, Arizona, flextime is supported as a means of reducing air pollution by reducing traffic congestion. Flextime allows travel to be spread over several hours, keeping freeways open and making travel faster.

[31] Barbara Rae Lee, "Organization Development and Group Perceptions: A Study of Quality Circles," Ph.D. dissertation, University of Minnesota, 1982; and Mike Robson, *Quality Circles: A Practical Guide*, 2nd ed. (Hants, England: Gower, 1988), pp. 47–62.

[32] Robert P. Steel and Russell F. Lloyd, "Cognitive, Affective, and Behavioral Outcomes of Participation in Quality Circles: Conceptual and Empirical Findings," *Journal of Applied Behavioral Science* 24 (1988), 1–17; Howard H. Greenbaum, Ira T. Kaplan, and William Metlay, "Evaluation of Problem Solving Groups: The Case of Quality Circle Programs," *Group and Organization Studies* 13 (1988), 133–47; and Kimberly Buch and Raymond Spangler, "The Effects of Quality Circles on Performance and Promotions," *Human Relations* 43 (1990), 573–82.

[33] Robert Golembiewski and Carl W. Proehl, Jr., "A Survey of the Empirical Literature on Flexible Workhours: Character and Consequences of a Major Innovation," *Academy of Management Review* 3 (1978), 837–55.

JOB DESIGN

compressed work week An alternative work schedule that allows employees to work a 40-hour week in fewer than the normal five days.

TEACHING NOTE
Some hospitals have gone to a 3/36 schedule for nurses (working three days a week, twelve hours each day). Discuss some of the problems that can occur as a result of a 12-hour shift in this environment. Why would hospitals choose this system? What might the risks be in compressing the work weeks in this type of job?

job sharing An alternative work schedule in which two or more part-time employees are allowed to share a full-time job.

APPLYING THE DIAGNOSTIC MODEL
Diagnostic Question 7: If work measurement fails to reveal a remedy, can methods engineering be used to create new jobs that are both performable and capable of adequate enrichment? If not, can you eliminate the troublesome jobs through automation?

industrial robots Machines that can be programmed to repeat the same sequence of work movements over and over again.

Some firms also offer a work schedule known as the **compressed work week**, enabling employees to work a 40-hour week in fewer than the normal five days. Most programs of this sort are based on the 4/40 work week in which employees work for ten hours on each of four days. Nearly 2000 organizations, ranging from IBM to the U.S. Army Corps of Engineers, have experimented with 4/40 scheduling. Many have reported that it has helped reduce costs, improve efficiency, reduce absenteeism, and improve satisfaction.[34]

Job sharing is another scheduling innovation in which companies attempt to reconcile the desire of employees to work part time with the organization's need to staff jobs on a full-time basis. Two or more part-time employees are allowed to share a full-time job. For example, one worker might perform a job from 8:00 A.M. to noon, and another might perform the same job from 1:00 P.M. to 5:00 P.M. Job sharing can be beneficial to both companies and employees, because it allows firms to employ capable people who might otherwise be unable to work. However, the effects of this type of alternative work schedule on such outcomes as productivity, satisfaction, and absenteeism have yet to be carefully documented.

Automation and Robotics

Automation offers managers who are designing jobs yet another alternative. For years, automation in the form of assembly-line manufacturing created many of the most oversimplified, demotivating, dissatisfying jobs in industry. Today, however, with the invention of automated technologies that can totally replace people in production processes, automation is sometimes an attractive option.

In some firms, much of a traditional assembly operation is left unchanged, but the most repetitive or physically demanding tasks are turned over to automated equipment. Frequently, this equipment consists of **industrial robots**, or machines that can be programmed to repeat the same sequence of work movements over and over again. Robots have been introduced throughout the automotive industry, taking over various painting and installation jobs. In fact, the welding job described at the beginning of this chapter is currently performed by robots on many North American auto-assembly lines.

Robots are not without their flaws. At General Motors, for example, employees regularly tell stories of one robot busily smashing windshields installed

[34] Paul Dickson, *The Future of the Workplace* (New York: Weybright and Talley, 1975), p. 219.

When Steve Jobs left Apple Computer and started Next Inc. he wanted to design a manufacturing process as advanced as the new computer it would produce. Next's engineers and software designers came up with a process that used some of the most advanced industrial robots around, and the company was soon producing about $100 million of hardware in a year. This automated system not only relieves workers of boring, repetitive tasks but does these tasks more accurately than people can. For example, Next circuitboards that are soldered by a robot have a solder-joint defect rate that is 1/10 the defect rate typical in the industry.
Source: Mark Alpert, "The Ultimate Computer Factory," Fortune, February 26, 1990, pp. 75–79.

by another or of a group of robots painting each other instead of the cars passing them by on the assembly line. Proper programming is obviously an essential aspect of introducing robots into the workplace. Careful planning, implementation, and adjustment is thus essential. Experience has also shown that building a robot capable of performing anything more than the simplest of jobs is often cost prohibitive. Consequently, the American population of robots is not the hundreds of thousands once predicted but about 37,000, according to the Robotic Industries Association of Ann Arbor, Michigan. Nonetheless, robots are an effective way to cope with many repetitive jobs that people don't want or are not very good at.[35]

Computer-integrated manufacturing in the form of *flexible manufacturing cells* is another type of automated technology but one that focuses primary attention on adaptability instead of robotic repetitiveness (see Chapter 16). Products made in such cells include gear boxes, cylinder heads, brake components, and similar machined-metal components used in the automotive, aviation, and construction-equipment industries. Companies throughout Europe, Japan, and North America are also experimenting with using flexible manufacturing cells to manufacture items out of sheet metal.[36]

Each flexible manufacturing cell consists of a collection of automated production machines that cut, shape, drill, and fasten metal components together. These machines are connected with each other by convertible conveyor grids that allow quick rerouting to accommodate changes from one product to another. It is possible, for instance, to produce a small batch of automotive door locks, then switch over to machine and finish a separate batch of crankshafts for automotive-air-conditioner compressors. It is simply a matter of turning some machines on and others off, then activating those conveyors that interconnect the machines that are in use. In this manner, the same collection of machines can make more than 100 different products without having to alter the cell substantially.[37]

Workers in a flexible manufacturing cell need never touch the product being produced, nor do they perform simple, repetitive production tasks. Instead, their jobs consist of the surveillance and decision making required to change the cell from one product configuration to another and to oversee equipment operations. Often, a cell's work force forms a semiautonomous group to accommodate the sizeable amount of mutual adjustment that must occur to keep production flowing smoothly. Employees in a flexible manufacturing cell thus exercise expertise in teamwork, problem solving, and self-management as they work.[38]

At its core, automation of this sort represents a return to the efficiency perspective of industrial engineering. Some jobs resist enrichment, and it is more effective to turn them over to machines than attempt to convert them into interesting, enjoyable work. But with the efficiency perspective, the danger that human satisfaction will be ignored also returns. So once workplace automation is established, managers must find a way to ensure that the worker-held jobs that remain offer each person the opportunity to experience sufficient levels of motivation and satisfaction.

[35] Peter T. Kilborn, "Brave New World Seen for Robots Appears Stalled by Quirks and Costs," *New York Times*, July 1, 1990, p. C7.

[36] Robert B. Kurtz, *Toward a New Era in U. S. Manufacturing* (Washington, D.C.: National Academy Press, 1986), p. 3.

[37] Ramchandran Jaikumar, "Postindustrial manufacturing," *Harvard Business Review* 44 (1986), 69–76.

[38] Peter Senker, *Towards the Automatic Factory: The Need for Training* (New York: Springer-Verlag, 1986), pp. 27–43.

SUMMARY

Contemporary *job design* began with Frederick Taylor's pioneering work on scientific management. Other experts, notably Frank and Lillian Gilbreth, refined Taylor's ideas and founded the field of *industrial engineering*. Frederick Herzberg differentiated between *motivator* and *hygiene factors*, and joined other specialists in introducing early models of *horizontal job enlargement* and *vertical job enrichment*. Eventually, two perspectives on job design emerged—the *efficiency perspective* and the *satisfaction perspective*.

Today's industrial engineering methods continue to reflect the efficiency perspective. *Methods engineering* attempts to improve the methods used to perform work. *Work measurement* examines the motions and time required to complete each job. *Comprehensive job enrichment* programs are guided by the *Job Diagnostic Survey* and, like *sociotechnical job enrichment* methods, are based on the satisfaction perspective. Other methods of designing jobs that have evolved in recent years include *quality circles, flexible-hour programs, job sharing*, and *robotics*.

REVIEW QUESTIONS

1. Explain how following Taylor's principles of scientific management can simplify the jobs in an organization. What are some of the positive effects of this simplification? What negative effects might occur?

2. What do the fields of process engineering and human factors engineering share in common? How do they differ from one another? Are they more likely to enhance satisfaction or efficiency? Why?

3. What effects do motion and time studies have on the design of jobs? What type of work measurement would you use to analyze the job of installing engines on an automobile assembly line? What type would you use to analyze the job of sorting and shelving library books?

4. Why do horizontal job-enlargement programs like job extension and job rotation often fail to stimulate employee satisfaction?

5. How do job design programs based on the Hackman-Oldham job characteristics model differ from programs based on Herzberg's motivator-hygiene model? Of the two types of programs, which are most likely to lead to significant improvements in employee motivation and satisfaction?

6. In what ways is the sociotechnical model of job design similar to the Hackman-Oldham model? In what ways do the two models differ? Which would you use if you were designing the job of a postal carrier? Which would you use to design the job of a surgical team?

7. Which of the following job design methods are products of the efficiency perspective, and which are products of the satisfaction perspective—quality circles, alternative work schedules, automation and robotics. Which of these approaches would you select to enrich jobs in a newly built assembly line? Which would you use to design jobs that resist all attempts at enrichment?

DIAGNOSTIC QUESTIONS

The number of different job design methods available today invites confusion. They require managers to consider carefully which of the various methods to use in solving the specific job design problems faced by their organizations. To help alleviate this confusion, the following diagnostic questions are provided.

1. Does the design of the organization's current jobs seem to reflect the effi-

ciency or the satisfaction perspective? Are most jobs simplified, or have attempts been made to alter job range or depth?

2.   If the efficiency perspective appears dominant, do productivity and satisfaction data support the idea that jobs have not been oversimplified? Or does faltering productivity and conspicuous dissatisfaction indicate that oversimplification may be a problem?

3.   Can the firm's current technology be changed to the degree required by job enrichment methods? Are jobs mainly individualized, indicating the appropriateness of Hackman-Oldham enrichment? Or are jobs often performed by groups of people working together closely, suggesting the sociotechnical approach?

4.   If technological considerations prohibit job enrichment, might quality circles provide enough relief to restore motivation and satisfaction? If not, can you eliminate the troublesome jobs through automation?

5.   If the satisfaction perspective appears dominant, do productivity and satisfaction data suggest that jobs have been enriched without creating work that is overdemanding? Or do falling productivity and satisfaction indicate that workers are being "overstretched" and asked to do more than they can?

6.   If you are facing an overenrichment problem, do the results of work measurement procedures suggest ways to simplify jobs enough to facilitate successful performance while still retaining opportunities for growth, achievement, and recognition?

7.   If work measurement fails to reveal a remedy, can methods engineering be used to create new jobs that are both doable and capable of adequate enrichment? If not, can you eliminate the troublesome jobs through automation?

EXERCISE 17-1
REDESIGNING A SIMPLIFIED JOB*
MARK S. PLOVNICK, *University of the Pacific*
RONALD E. FRY, *Case Western Reserve University*
W. WARNER BURKE, *Columbia University*

The drive toward efficiency in modern organizations sometimes results in the creation of oversimplified jobs that fail to challenge or involve the worker. People in these jobs often express strong dissatisfaction as product quality falls and both absenteeism and turnover increase. Comprehensive job enrichment is intended to counteract this situation. This technique changes job content in several ways. It combines tasks and forms natural work units so as to increase skill variety, task identity, and task significance. It gives each worker more autonomy on the job. And it opens up channels for immediate feedback to the worker. In this exercise you will diagnose a job and suggest how it might be enriched, thus experiencing for yourself the initial steps of job redesign.

STEP ONE: PRE-CLASS PREPARATION

In Step Two of this exercise you will work in groups to diagnose one of the following jobs:

1. A job your group has observed and discussed before class. If your instructor assigns you this option you should meet as a group, decide what job you want to observe, verify with your instructor that your choice is appropriate, and meet on site and observe the jobholder as he performs the job for _____. Be sure to take notes to refresh your memory as you perform the rest of the exercise.
2. A job that has been performed by a member of your class. If your instructor assigns you this option, he or she will ask for volunteers from class to serve as interviewees. In this role, they will describe in depth a job they have performed. Depending on the number of volunteers, each group may have a volunteer to interview, several groups may have to combine together to interview the same volunteer, or the class as a whole may be required to conduct a single interview. Depending on your instructor's preference, interviews may be conducted before or at the beginning of class.
3. A videotaped job. If your instructor assigns this option he or she will show a videotape of a job being performed and you will diagnose the job based on what you see. Your instructor will supply the videotape and will show it at the beginning of class.

Now read the remainder of the exercise and prepare for the next step as you need to, depending on the option your instructor has chosen.

STEP TWO: COMPLETING THE TASK DIAGNOSTIC QUESTIONNAIRE (TDQ)

The class should divide into groups of four to six members each. If you have already formed permanent groups, you should reassemble in those groups. If your instructor has assigned you the task of observing a job before coming to class, the groups you form for this step of the exercise should be the same as the ones you formed to observe the job. The members of each group should work together to fill out the Task Diagnostic Questionnaire (TDQ) shown in Exhibit 17-1. The group should reach consensus on its response to each item. Any significant disagreements should be noted next to the relevant item on the TDQ.

STEP THREE: SCORING AND DISCUSSING THE TDQ

Here is how you score your group's evaluations, on the Task Diagnostic Questionnaire, of the job you're examining:

1. For items 2, 4, 6, 8, 10, and 12, subtract the number you circled from 8. The result is your score for each of these items. For items 1, 3, 5, 7, 9, 11, 13, 14, and 15, the number you circled is the correct score.
2. Add your scores for items 1, 4, and 13, divide by 3, and write the result here: _____. This is the job's score on *skill variety*.
3. Add your scores for items 2, 7, and 12, divide by 3, and write the result here: _____. This is the job's score on *task identity*.
4. Add your scores for items 5, 10, and 14, divide by

* Exercise adapted with the authors' permission from Mark S. Plovnick, Ronald E. Fry, and W. Warner Burke, *Organization Development: Exercises, Cases, and Readings* (Boston: Little, Brown, 1982), pp. 94–105. Copyright © 1982 by Mark S. Plovnick, Ronald E. Fry, and W. Warner Burke. Questionnaire adapted with the authors' permission from the Job Diagnostic Survey developed by J. Richard Hackman and Greg R. Oldham.

EXHIBIT 17-1

Task Diagnostic Questionnaire

This questionnaire contains statements with which you may agree or disagree. It is intended to reveal how people perceive different kinds of jobs. It is not intended to measure how much someone likes or dislikes a job but to elicit as accurate and objective a description as possible. To be useful, the questionnaire must be answered honestly.

In the blank next to each statement, write the number that represents how accurate you think the statement is in describing the job: 1 = very inaccurate, 2 = mostly inaccurate, 3 = slightly inaccurate, 4 = uncertain, 5 = slightly accurate, 6 = mostly accurate, 7 = very accurate.

_____ 1. The job requires the worker to use a number of complex, high-level skills.

_____ 2. The way the job is structured, the worker does not have the opportunity to do a complete piece of work from beginning to end.

_____ 3. Just doing the job provides the worker with many chances to figure out how well she is doing.

_____ 4. The job is quite simple and repetitive.

_____ 5. In this job, a lot of other people can be affected by how well the work is done.

_____ 6. The job does not give the worker any chance to use his personal initiative or judgment in carrying out the work.

_____ 7. The job allows the worker to completely finish every piece of work she begins.

_____ 8. The job itself provides very few clues about whether the worker is performing well.

_____ 9. The job gives the worker considerable independence and freedom in the way he does the work.

_____ 10. The job is not very significant or important in the overall work of the organization.

_____ 11. The job permits the worker to decide for herself what needs to be done.

_____ 12. The job is only a small part of an overall piece of work that is finished by other people or by automated machines.

_____ 13. The job requires the worker to do many different things, using a variety of skills and talents.

_____ 14. The results of the job have a significant effect on the lives and well-being of other people.

_____ 15. The job itself provides clues about how well the worker is doing; feedback from coworkers or supervisors is nor needed.

3, and write the result here: _____. This is the job's score on *task significance*.

5. Add your scores for items 6, 9, and 11, divide by 3, and write the result here: _____. This is the job's score on *autonomy*.

6. Add your scores for items 3, 8, and 15, divide by 3, and write the result here: _____. This is the job's score on *feedback*.

Next, using these group consensus scores, draw a job profile for your job like the one shown in Figure 17-9 in this chapter. Then calculate a motivational summary score, or MSS (similar to the motivation potential score discussed in this chapter), using the formula that follows.

MSS

$$= \frac{(\text{Skill Variety} + \text{Task Identity} + \text{Task Significance})}{3}$$

$$\times \text{ Autonomy} \times \text{ Feedback}$$

Your group should discuss the meaning of the scores it has derived and the results of its diagnosis. Then appoint a spokesperson to present a report to the class. The report should include information about what kind of job the group diagnosed, the group's scores on each of the five dimensions, significant disagreements among group members, and the MSS calculated by the group. The spokesperson should also show a diagram of the job profile developed by the group on a blackboard, flipchart, or overhead transparency.

STEP FOUR: DEVELOPING COMPREHENSIVE JOB ENRICHMENT STRATEGIES

The class should convene and all spokespersons should give their reports. After discussing the different jobs the class has analyzed, members should return to their groups and each group should develop a strategy to enrich the job it has diagnosed. The following points should be considered during this step:

1. Which specific job characteristics need enrichment? Which, if any, are good enough as is? What specific actions should be taken to enrich the job along the dimensions that need further help?

2. Who in the organization will be responsible for developing the strategy if additional refinement is needed? Who will be responsible for implementing it? How will the effectiveness of the implementation be measured? Who will perform this evaluation?

3. Before redesigning the job, what additional data should be collected? From whom? By whom? To whom should the data be fed back? What will this person or persons do with it?

4. What are some likely sources of resistance to the strategy you have developed? How should they be dealt with?

The group spokesperson should prepare an overview of the group's strategy for presentation to the class.

STEP FIVE: STRATEGY REPORTS AND CLASS DISCUSSION

The class should reconvene and each spokesperson should report on the results of Step Four. Class members should ask questions of clarification as needed to understand each strategic plan. The total class should then compare, contrast, and critique the strategies developed by the groups, being sure to address the following points.

1. To what extent did each strategy emphasize employee involvement? Changes in the job itself? Changes in the context surrounding the job? Does the strategy appear workable?

2. What consequences would the strategy have for the structure of the organization? For current policies? For the distribution of power in the firm?

3. For each strategy, what positive and negative indirect effects might it have on those individuals who are not directly involved in it? How might these individuals act as forces for or against change?

CONCLUSION

This exercise has introduced you to some of the factors underlying the nature of work and to the complexity of issues involved in trying to redesign jobs. Job redesign programs can be applied to jobs in the consumer products industry, jobs in service industries, white collar jobs, blue collar jobs—any place where work has been oversimplified to the point of reducing worker satisfaction and productivity.

DIAGNOSING ORGANIZATIONAL BEHAVIOR

Although the events we're about to describe took place more than twenty years ago, this case remains a classic example of certain kinds of problems that managers and employees face in modern industrial workplaces. Today, companies throughout the United States wrestle with exactly the same predicaments as those that confronted General Motors in its Lordstown plant in the early 1970s.

INTRODUCTION

In December 1971, the management of the Lordstown Plant was very much concerned with an unusually high rate of defective Vegas coming off the assembly line.

CASE 17-1
THE LORDSTOWN PLANT OF GENERAL MOTORS*
HAK-CHONG LEE, YONSEI UNIVERSITY†

For the previous several weeks, the lot with a capacity of 2,000 cars had been filled with Vegas which were waiting for rework before they could be shipped out to the dealers around the country.

The management was particularly disturbed by the fact that many of the defects were not the kinds of quality deficiency normally expected in an assembly production of automobiles. There was a countless number of Vegas with their windshields broken, upholstery

* Reprinted with the author's permission.

† This case was developed for instructional purposes from published sources and interviews with the General Motors Assembly Division officials in Warren, Michigan and Lordstown, Ohio. The Public Relations Office of GMAD read the case and made minor corrections in it. However, the author is solely responsible for the content of the case. The author appreciates the cooperation of General Motors.

slashed, ignition keys broken, signal levers bent, rearview mirrors broken, or carburetors clogged with washers. There were cases in which, as the Plant Manager put it, "the whole engine blocks passed by 40 men without any work done on them."

Since then, the incident in the Lordstown Plant has been much publicized in news media, drawing public interest. It has also been frequently discussed in the classroom and in the academic circles. While some people viewed the event as "young worker revolt," others reacted to it as a simple "labor problem." Some viewed it as "worker sabotage," and others called it "industrial Woodstock."

This case describes some background and important incidents leading to this much publicized and discussed industrial event.

The General Motors Corporation is the nation's largest manufacturer. The Company is a leading example among many industrial organizations which have achieved organizational growth and success through decentralization. The philosophy of decentralization has been one of the most valued traditions in General Motors from the days of Alfred Sloan in the 1930s through Charles Wilson and Harlow Curtice in the 1950s and up to recent years.

Under decentralized management, each of the company's car divisions, Cadillac, Buick, Oldsmobile, Pontiac and Chevrolet, was given maximum autonomy in the management of its manufacturing and marketing operations. The assembly operations were no exception, each division managing its own assembly work. The car bodies built by Fisher Body were assembled in various locations under maximum control and coordination between Fisher Body and each car division.

In the mid-1960s, however, the decentralization in divisional assembly operations was subject to a critical review. At the divisional level, the company was experiencing serious problems of worker absenteeism and increasing cost with declines in quality and productivity. They were reflected in the overall profit margins which were declining from 10% to 7% in the late 1960s. The autonomy in the divided management in body manufacturing and assembly operations, in separate locations in many cases, became questionable under the declining profit situation.

In light of these developments, General Motors began to consolidate in some instances the divided management of body and chassis assembly operations into a single management under the already existing General Motors Assembly Division (GMAD) in order to better coordinate the two operations. The GMAD was given an overall responsibility to integrate the two operations in these instances and see that the numerous parts and components going into car assembly got to the right places in the right amounts at the right times.

THE GENERAL MOTORS ASSEMBLY DIVISION (GMAD)

The GMAD was originally established in the mid 1930s, when the company needed an additional assembly plant to meet the increasing demands for Buick, Oldsmobile, and Pontiac automobiles. The demand for these cars was growing so much beyond the available capacity at the time that the company began, for the first time, to build an assembly plant on the west coast which could turn out all three lines of cars rather than an individual line. As this novel approach became successful, similar plants turning out a multiple line of cars were built in seven other locations in the east, south and midwest. In the 1960s, the demand for Chevrolet production also increased, and some Buick-Oldsmobile-Pontiac plants began to assemble Chevrolet products. Accordingly, the name of the division was changed to GMAD in 1965.

In order to improve quality and productivity, the GMAD increased its control over the operations of body manufacturing and assembly. It reorganized jobs, launched programs to improve efficiency, and reduced the causes of defects which required repairs and rework. With many positive results attained under the GMAD management, the company extended the single management concept to six more assembly locations in 1968 which had been run by the Fisher Body and Chevrolet Divisions. In 1971, GM further extended the concept to four additional Chevrolet-Fisher Body assembly facilities, consolidating the separate management under which the body and chassis assembly had been operating. One of these plants was the Lordstown Plant.

The series of consolidations brought to eighteen the number of assembly plants operated by the GMAD. In terms of total production, they were producing about 75% of all cars and 67% of trucks built by GM. Also in 1971, one of the plants under the GMAD administration began building certain Cadillac models, thus involving GMAD in production of automobiles for each of the GM's five domestic car divisions as well as trucks for Chevrolet, GMC truck, and GM's Coach Division.

THE LORDSTOWN COMPLEX

The Lordstown complex is located in Trumbull County in Ohio, about 15 miles west of Youngstown and 30 miles east of Akron. It consists of the Vega assembly plant, the van-truck assembly plant, and Fisher Body metal fabricating plant, occupying about 1,000 acres of land. GMAD, which operates the Vega and van-truck assembly plants, is also located in the Lordstown complex. The three plants are in the heart of the heavy industrial triangle of Youngstown, Akron and Cleveland. With Youngstown as a center of steel production,

Akron the home of rubber industries, and Cleveland as a major center for heavy manufacturing, the Lordstown complex commands a good strategic and logistic location for automobile assembly.

The original assembly plant was built in 1964–1966 to assembly Chevrolet Impalas. But in 1970 it was converted into Vega assembly through extensive redesign. The van-truck assembly plant was constructed in 1969, and the Fisher Body metal fabricating plant was further added in 1970 to carry out stamping operations to produce sheet metal components used in Vega and van assemblies. In October 1971, the Chevrolet Vega and van-assembly plants and Fisher Body Vega assembly plants which had been operating under separate management were merged into a single jurisdiction of the GMAD.

WORK FORCE AT THE LORDSTOWN PLANT

There are over 11,400 employees working in the Lordstown Plant (as of 1973). Approximately 6,000 people, of whom 5,500 are on hourly payroll, work in the Vega assembly plant. About 2,600 workers, 2,100 of them paid hourly, work in van-truck assembly. As members of the United Auto Workers Union, Local 1112, the workers command good wages and benefits. They start out on the line at about $5.00 an hour, get a 10¢ an hour increase within 30 days, and another 10¢ after 90 days. Benefits come to $2.50 an hour. The supplemental unemployment benefits virtually guarantee the worker's wages throughout the year. If the worker is laid off, he gets more than 90% of his wages for 52 weeks. He is also eligible for up to six weeks for holidays, excused absence or bereavement, and up to four weeks vacation.

The work force at the plant is almost entirely made up of local people with 92% coming from the immediate area of a 20-mile radius. Lordstown itself is a small rural town of about 500 residents. A sizable city closest to the plant is Warren, 5 miles away, which together with Youngstown supplies about two-thirds of the work force. The majority of the workers (57.5%) are married, 7.6% are home owners, and 20.2% are buying their homes. Of those who do not own their homes (72%), over one-half are still living with their parents. The rest live in rented houses or apartments.

The workers in the plant are generally young. Although various news media reported the average worker age as 24 years old, and in some parts of the plant as 22 years, the company records show that the overall average worker age was somewhat above 29 years as of 1971–72. The national average is 42. The work force at Lordstown is the second youngest among GM's 25 assembly plants around the country. The fact that the Lordstown plant is GM's newest assembly plant may partly explain the relatively young work force.

The educational profile of the Lordstown workers indicates that only 22.2% have less than a high school education. Nearly two-thirds or 62% are high school graduates, and 16% are either college graduates or have attended college. Another 26% have attended trade school. The average education of 13.2 years makes the Lordstown workers among the best educated in GM's assembly plants.

THE VEGA ASSEMBLY LINE

Conceived as a major competitive product against the increasing influx of foreign cars which were being produced at as low as one-fourth the labor rate in this country, the Vega was specifically designed with a maximum production efficiency and economy in mind. From the initial stages of planning, the Vega was designed by a special task team with most sophisticated techniques, using computers in designing the outer skin of the car and making the tapes that form the dies. Computers were also used to match up parts, measure the stack tolerances, measure safety performance under head-on collision, and make all necessary corrections before the first 1971 model car was ever built. The 2300-cubic-centimeter all-aluminum, 4-cylinder engine, was designed to give gas economy comparable to the foreign imports.

The Vega was also designed with the plant and the people in mind. As the GM's newest plant, the Vega assembly plant was known as the "super plant" with the most modern and sophisticated designs to maximize efficiency. It featured the newest engineering techniques and a variety of new power tools and automatic devices to eliminate much of the heavy lifting and physical labor. The line gave the workers an easier access to the car body, reducing the amount of bending and crawling in and out, as in other plants around the country. The unitized body in large components like prefab housing made the assembly easier and lighter with greater body integrity. Most difficult and tedious tasks were eliminated or simplified, on-line variations of the job were minimized, and the most modern tooling and mechanization was used to the highest possible degree of reliability.

It was also the fastest moving assembly line in the industry. The average time per assembly job was 36 seconds with a maximum of 100 cars rolling off the assembly line per hour for a daily production of 1,600 cars from two shift operations. The time cycle per job in other assembly plants averaged about 55 seconds. Although the high speed of the line did not necessarily imply greater work load or job requirement, it was a part of GM's attempt to maximize economy in Vega assembly. The fact that the Vega was designed to have 43% fewer parts than a full-size car also helped the high-speed line and economy.

IMPACT OF GMAD AND REORGANIZATION IN THE LORDSTOWN PLANT

As stated previously, the assembly operations at Lordstown had originally been run by Fisher Body and Chevrolet as two plants. There were two organizations, two plant managers, two unions, and two service organizations. The consolidation of the two organizations into a single operating system under the GMAD in October 1971 required a difficult task of reorganization and dealing with the consequences of manpower reduction such as a work slowdown, worker discipline, grievances, etc.

As duplicate units such as production, maintenance, inspection, and personnel were consolidated, there was a problem of selecting the personnel to manage the new organization. There were chief inspectors, personnel directors and production superintendents as well as production and service workers to be displaced or reassigned. Unions which had been representing their respective plants also had to go through reorganization. Union elections were held to merge the separate union committees at Fisher Body and Chevrolet in a single union bargaining committee. This eliminated one full local union shop committee.

At the same time, GMAD launched an effort to improve production efficiency more in line with that in other assembly plants. It included increasing job efficiency through reorganization and better coordination between the body and chassis assembly, and improving controls over product quality and worker absenteeism. This effort coincided with adjustments in line balance and work methods. Like other assembly plants, the Vega assembly plant was going through an initial period of diseconomy caused by suboptimal operations, imbalance in the assembly line, and a somewhat redundant work force. According to management, line adjustment and work changes were a normal process in accelerating the assembly operation to the peak performance the plant had been designed for after the initial break-in and start-up period.

As for job efficiency, the GMAD initiated changes in those work sequences and work methods which were not well coordinated under the divided managements of body and chassis assembly. For example, previous to the GMAD, Fisher Body had been delivering the car body complete with interior trim to the final assembly lines, where oftentimes the workers soiled the front seats as they did further assembly operations. GMAD changed this practice so that the seats were installed as one of the last operations in building the car. Fisher Body also had been delivering the car body with complete panel instrument frame which made it extremely difficult for the assembly workers to reach behind the frame in installing the instrument panels. The GMAD improved the job method so that the box containing the entire instrument panel was installed on the assembly line. Such improvements in job sequences and job methods resulted in savings in time and the number of workers required. Consequently, there were some jobs where the assembly time was cut down and/or the number of workers was reduced.

GMAD also put more strict control over worker absenteeism and the causes for defective work; the reduction in absenteeism was expected to require fewer relief men, and the improvement in quality and less repair work were to require fewer repairmen. In implementing these changes, the GMAD instituted a strong policy of dealing with worker slowdowns via strict disciplinary measures including dismissal. It was rumored that the inspectors and foremen passing defective cars would be fired on the spot.

Many workers were laid off as a result of the reorganization and job changes. The union was claiming that as many as 700 workers were laid off. Management, on the other hand, put the layoff figure at 375 to which the union later conceded. Although management claimed that the changes in job sequence and method in some assembly work did not bring a substantial change in the overall speed or pace of the assembly line, the workers perceived the job change as "tightening" the assembly line. The union charged that the GMAD brought a return of an old-fashioned line speedup and a "sweatshop style" of management reminiscent of the 1930s, making the men do more work at the same pay. The workers were blaming the "tightened" assembly line for the drastic increase in quality defects. As one worker commented, "That's the fastest line in the world. We have about 40 seconds to do our job. The company adds one more thing and it can kill us. We can't get the stuff done on time and a car goes by. The company then blames us for sabotage and shoddy work."

The number of worker grievances also increased drastically. Before GMAD took over, there were about 100 grievances in the plant. After GMAD's entry, grievances increased to 5,000, 1,000 of which were related to the charge that too much work had been added to each job. Worker resentment was particularly great in the "towveyor" assembly and seat sub-assembly areas. The "towveyor" is the area where engines and transmissions are assembled. Like seat sub-assembly there is a large concentration of workers working together in close proximity. Also, these jobs are typically performed by beginning assemblers who are younger and better educated.

The workers in the plant were particularly resentful of the company's strict policy in implementing the changes. They stated that the tougher the company became, the more they would stiffen their resistance even though other jobs were scarce in the market. One worker said, "In some of the other plants where the GMAD did the same thing, the workers were older and they took this. But, I've got 25 years ahead of me in this plant." Another worker commented, "I saw a woman running to keep pace with the fast line. I'm not going to run for anybody. There ain't anyone in that plant that is going to tell me to run." One foreman said, "The problem with the workers here is not so much that they don't want to work, but that they just don't want to take orders. They don't believe in any kind of authority."

While the workers were resisting management orders, there were some indications that the first-line supervisors had not been adequately trained to perform satisfactory supervisory roles. The average supervisor at the time had less than 3 years of experience, and 20% of the supervisors had less than 1 year's experience. Typically, they were young, somewhat lacking in knowledge of the provisions of the union contract and other supervisory duties, and less than adequately trained to handle the workers in the threatening and hostile environment which was developing.

Another significant fact was that the strong reactions of the workers were not entirely from the organizational and job changes brought about by the GMAD alone. Management felt that the intense resentment was particularly due to the nature of the work force in Lordstown. The plant was not only made up of young people, but also the work force reflected the characteristics of "tough labor" in steel, coal and rubber industries in the surrounding communities. Many of the workers in fact came from families who made their living working in these industries. Management also noted that the worker resistance had been much greater in the Lordstown Plant than in other plants where similar changes had been made.

A good part of the young workers' resentment also seemed to be related to the unskilled and repetitive nature of the assembly work. One management official admitted that the company was facing a difficult task in getting workers to "take pride" in the product they were assembling. Many of them were participating in the company's tuition assistance plan which was supporting their college education in the evening. With this educated background, obviously assembly work was not fulfilling their high work expectations. Also, the job market was tight at the time, and they could neither find any meaningful jobs elsewhere nor, even if found, they could not afford to give up the good money and fringe benefits they were earning on their assembly-line jobs. This frustrated them, according to company officials.

Many industrial engineers were questioning whether management could continue simplifying assembly line work. As the jobs became easier, simpler, and repetitive, requiring less physical effort, there were fewer and fewer traces of skill and increased monotony. Worker unrest indicated that employees not only wanted to go back to the work pace prior to the "speedup" (pre-October pace), but also wanted the company to do something about the boring and meaningless assembly work. One worker commented, "The company has got to do something to change the job so that a guy can take an interest in the job. A guy can't do the same thing 8 hours a day year after year. And it's got to be more than the company just saying to a guy, 'Okay, instead of 6 spots on the weld, you'll do 5 spots.'"

As the worker resentment mounted, the UAW Local 1112 decided in early January 1972 to consider possible authorization for a strike against the Lordstown Plant in a fight against the job changes. In the meantime, the union and management bargaining teams worked hard on worker grievances; they reduced the number of grievances from 5,000 to a few hundred; management even indicated that it would restore some of the eliminated jobs. However, the bargaining failed to produce accord on the issues of seniority rights and shift preference, which were related to wider issues of job changes and layoff.

A vote was held in early February 1972. Nearly 90% of the workers came out to vote in the heaviest turnout in the history of the Local. With 97% of the votes supporting, the workers went out on strike in early March.

In March 1972, with the strike in effect, the management of the Lordstown Plant was assessing the impact of the GMAD and the resultant strike in the Plant. It was estimated that the work disruption because of the worker resentment and slowdown had already cost the company 12,000 Vegas and 4,000 trucks amounting to $45 million. There had been repeated closedowns of assembly lines since December 1971, because of the worker slowdowns and the cars passing down the line without all necessary operations performed on them. The car lot was full with 2,000 cars waiting for repair work.

There had also been an amazing number of complaints from Chevrolet dealers, 6,000 complaints in November alone, about the quality of the Vegas shipped to them. This was more than the combined complaints from the other assembly plants.

The strike in the Lordstown Plant was expected to affect other plants. The plants at Tonawanda, New York and Buffalo, New York were supplying parts for

Vega. Despite the costly impact of the worker resistance and the strike, management felt that the job changes and cost reductions were essential if the Vega were to return a profit to the company. The plant had to be operating at about 90% capacity to break even because its highly automated features cost twice as much as had been estimated.

While the company had to do something to increase the production efficiency in the Lordstown Plant, the management was wondering whether it couldn't have planned and implemented the organizational and job changes differently in view of the costly disruption of operations that the Plant had been experiencing.

When you have read this case, look back at the chapter's diagnostic questions and choose the ones that apply to the case. Then use those questions with the ones that follow in your case analysis.

1. What kinds of jobs are performed in the General Motors Lordstown plant? Would further simplification reduce workers' complaints? Should the jobs be enriched instead?

2. Does it make sense to recommend that major changes be made to Lordstown's assembly lines? Why, or why not?

3. Which job design approach seems best suited to the situation at Lordstown? What steps should be taken to implement this approach? How successful is it likely to be in improving workforce attitudes throughout the Lordstown plant?

CASE 17-2

BETA BUREAU*
DONALD AUSTIN WOOLF, UNIVERSITY OF OKLAHOMA

PART A

The Sigma Agency is a large division of the Epsilon Department, a cabinet-level department of the federal government. It has primary responsibility for the administration of a law providing a variety of services to a large number of citizens. In general terms, the Agency is organized in terms of operating bureaus, an administrative and staff services bureau, and a bureau providing support services. Each of the operating bureaus administers or assists in administering a separate portion of the law. Beta Bureau operates regional claims processing centers for the Sigma Agency.

Claims are filed by applicants at widely dispersed branch offices which are administered by a branch office bureau. Those claims which are strictly routine in nature

may be authorized by representatives at the branch offices. All others—about half of the total work load—are forwarded to Beta Bureau processing centers along with the record of actions taken on those claims authorized at the branch office. The processing center reviews all actions taken at the branch level, processes initial claims not authorized by branches, reviews or authorizes changes in eligibility of existing claimants (post-entitlement), and initiates recovery action in cases where claimants have received services in excess of that permitted by law, rule, or regulation. Claim files are physically maintained at and by the processing centers. Finally, information on all actions taken is transmitted to the central data storage and retrieval system, located at Headquarters near Washington, D.C. Central Data operates as a separate, service bureau, and provides this service to all operating bureaus. Each of the Centers employs about 2,000 people.

Most of the bureaus—including Beta Bureau—have been organized along "functional" lines, that is, relatively large sections of people in which all do the same or very similar work. Accordingly, processing centers have had an intake unit, which receives and sends correspondence, records, claims, and files, a records unit, and two kinds of claims units, each constituting several sections, which are devoted to initial claims and post-entitlement claims, respectively. Records assembles various documents relating to a given case, places them in a folder, and routes the folder to the appropriate section. Accordingly, queueing occurs at records, and at each of the subsequent sections to which the file is sent. Because of the magnitude of records, they formerly were moved from place to place in large canvas "tubs" mounted on casters, with about 11,000 folders in each tub. Folders frequently failed to have all information necessary to complete processing, so were rerouted to other sections or even other centers, where queueing occurred again. In the past, it has taken from one to two weeks for a given case to move through a queue. As queues multiplied, the time necessary for processing sometimes extended to several months.

Authorizers processing initial claims held the most prestigious professional jobs, post-entitlement authorizers holding a lesser grade and pay status. The sections were relatively large, having as many as sixty kinds of cases. This specialization was formal in some instances and informal in others. Since all authorizers were evaluated in terms of number of cases processed and the accuracy thereof, there was a tendency for difficult cases to be rerouted, ostensibly for more documentation, or because another authorizer was deemed to have superior expertise in the problem associated with the case. Given the queueing phenomenon associated with functional organization, Beta Bureau experienced a chronic problem with aged cases. Unsurprisingly, claimants were likely to file complaints, sometimes with Sigma Agency, frequently with Congressmen, and occasionally with the Executive Office of the President. Sigma Agency maintained a special headquarters unit to process these complaints and to continue communication with the claimant, the elected official referring the complaint, and the bureau responsible for the claim. Meanwhile, the claim, most likely in transit, could prove exceedingly difficult to track down. Accordingly, processing centers developed "freeze lists" of aged claims. All claims on the freeze list were to be assigned highest priority until located and processed.

By the late 1980s processing time, error rates, employee morale, turnover, and service to clients had reached unacceptable levels. Documentation of cases became critical in several areas, especially in "unassociated material," i.e., documents needed to complete a claims case which, for one reason or another, never found their way to the claims folder for that case.

Efforts were made to improve the existing system through upgrading the data storage and retrieval system, and through tightening controls. A somewhat higher proportion of new claims were authorized at the branch office, enabling the same official to follow through from start to finish on a new claim. Numerous additional changes in equipment and procedure were authorized to expedite processing and to improve control. Although there is little hard evidence to suggest that firmer discipline was exercised, awards for exceptional service were increased and publicized.

Results of these efforts were disappointing, serving mainly to slow the decline in service, rather than to reverse it. Accordingly, bureau top management decided that the basic structure itself would have to be revised. A special staff was authorized to design and experiment with organizational structures to identify a system which would enhance service to the clientele and improve case control as well as productivity. Among things to be considered were job enlargement and enrichment, physical layout, composition of the work group and supervision.

PART B

Becaused of increasing problems of administration under the existing structure, top management of Beta Bureau of the Sigma Agency initiated a study of alternative forms of organization utilizing its in-house special staff. After initial research and planning at headquarters, special staff conducted field research at a processing center. Interviews and meetings were held with all levels of management, professional and support personnel, and union representatives. Results of the research indicated that it would be appropriate to conduct a pilot study to determine further the feasibility of work units organized along lines different from the traditional functional organization.

The Bureau Director, working with special staff and relevant line managers, proposed a concept of a "processing center within a processing center." Accordingly, the kinds of work to be done in such a unit, optimum size, positions, support staff, equipment, and facilities had to be determined as well as the appropriate grade and pay for new positions created. After some discussion, it was decided to call the new type of work group a "module," and the concept, "modular organization."

A number of combinations and variations in size, composition, workflow support equipment, span of control, chain of command, and support services were tried. Experimentation with and evaluation of two pilot work units over a period of two years produced a viable structure, although not one considered by staff to be "optimum." Top management determined that the problems leading to the experimentation were of such urgency that further study and experimentation were precluded. Also, in the interim, a number of new buildings had been built to house existing centers and it was felt that moving from existing facilities to the new ones could be combined with the change in organizational structure.

Matters were further complicated by a number of major, new amendments to the law relating to the programs being administered. New positions were created, necessitating authorization by the Civil Service Commission, which proved more difficult than had been anticipated. All of these events placed a massive burden on the existing training staff as well as on management from the first-line level to the top. Rank-and-file employees were also obviously affected by the rapid rate of change. Concurrent adoption of new technologies of case handling further complicated matters. The result was a kind of "future shock" felt by all concerned. Finally, during the latter stages of phasing in modularization, a massive reorganization of top management took place following the retirement of the bureau director who had initiated the original study.

Interviews with employees produced responses such as, "I wish the world would just stop for about six months so I could catch up," and, "We might just be able to do a workmanlike job on this program if Congress would quit making special exceptions for left-handed Eskimo veterans of the Korean War," or, "If management *really* knew what it was doing, we wouldn't have procedural changes every fifteen minutes."

Modules which emerged from this process contained about fifty employees each, supervised by a module manager and two assistant managers. A technical adviser served as a resource for professionals in the module. The latter position tended to be of a "rotating" nature in some locations, i.e., different rank-and-file professionals were temporarily assigned to the position rather than having it as a permanent assignment. Case records were specifically assigned to and physically located in each module. Accordingly, at the time a case was assigned an identification number it was determined which of the over two hundred modules would have virtually absolute responsibility for any future claim related to the case. Each time a folder was moved, information as to location change was fed to a central computer through a network terminal. For the most part, individual authorizers followed through on a single case until it was completed. Queueing was reduced to a minimum.

Two years after initial installation, a "faculty fellow" from a state university was assigned to attempt evaluation. Most of his efforts were directed toward job satisfaction. Initially, productivity was determined to be extremely difficult to measure because of the massive changes in the law, increased mechanization of some activities, difficulty in evaluating increased complexities in the program and resulting impact on processing, and a number of changes in data bases. Realignment of the work load between branch offices, processing centers and headquarters combined to make precise evaluation of productivity unobtainable. Nevertheless, some useful base data were obtained.

Two years later, a follow-up study was commissioned almost immediately after the conversion to modular organization was completed. In addition to job satisfaction, aggregate measures of productivity, turnover, absenteeism, processing time, control and relative cost were obtained. From the initiation of modular organization, massive change was a continuous phenomenon. In addition to the change in the form of organization, substantial revision of relevant legislation was passed, creating a number of new programs. In general, the new laws tended to make all but purely routine cases more difficult to process. Estimates of the level of increased difficulty ranged around fifteen percent. During the second study, the bureau was engaged in a project to upgrade data processing equipment such that each module had complete access to the master data file at headquarters. Accordingly, given the variety of changes, it was difficult to measure the precise impact of reorganization.

Nevertheless, a variety of findings were demonstrable. In comparison with other operating bureaus doing comparable work both absenteeism and turnover declined. Cost of administration as a proportion of total cost declined. "Freeze lists" declined by 85 percent or more demonstrating a marked improvement in control. Average processing time declined slightly where the new remote terminals had not been installed, and declined markedly where they were installed. Job satisfaction for the modules studied improved in four out of five categories measured; however, there were significant differences in perceived job satisfaction among different classes of employees. In general, lower-level clericals liked the change to modular organization, while professionals and first-line supervisors exhibited considerable variance in their opinion. Of the total number surveyed, about 80 percent preferred modules to functional organization. Few, if any, changes in quality control were demonstrated, but a decline attributable to the massive change in procedures as well as the law would not have been unexpected. Such a decline did not take place.

In the early stages, union representatives expressed substantial reservations about the proposed reorganization. This initial reluctance declined in most of the processing centers, although one local continued to maintain an official attitude of opposition. The attitude change was attributable in part to a modest net increase in pay-grade level resulting from the reorganization.

Summing up results of the studies, there appear to be few, if any, problems with the structural configuration of modules. They will work in this kind of service operation. Nevertheless, problems still remain with the operation of the centers as a whole. A few of these are structural, but most appear procedural or managerial.

In the process of abolishing sections of people all of whom were doing about the same thing and substituting modules of people doing different things, a number of one-of-a-kind positions were created. For example, some of the specialized sections merged into modules had only a couple of dozen people in them. This resulted in each module having only a single specialist in that category after reorganization. Accordingly, if the incumbent were promoted or left the job, the function served went uncovered. Although other professionals could be temporarily assigned to the function, they did not like it, and such occurrences were disruptive. Meanwhile, support functions such as recruiting and training were geared to the old-style sections, and would allow vacancies to accumulate in substantial numbers before selecting new candidates.

Job enlargement has been found also to be a mixed blessing. Professionals, and indeed, entire sections had tended to specialize in particular kinds of cases under the functional pattern of organization. With modularization, it became necessary to become proficient in all kinds. This latter kind of problem was intensified for module managers. Previously, they needed only to have detailed technical knowledge about a single phase of the processing operation, and how it interfaced with other parts; under the new scheme of organization it became necessary to be familiar with all parts of it.

Among the most visible results of the latter observation was a substantial exodus of former section heads destined to become—or who had become—module managers. Although headway was being made, the highest proportionate number of vacancies in modules was at the managerial level, with up to one-third of the modules, system-wide, operating either with "acting" (temporary) managers or without the usual complement of a manager and two assistant managers in each module. Estimates are that up to one-third of the former supervisors retired or transferred to other jobs in the federal government. On a brighter note, the remaining managers plus a number of newer appointees appear to be more flexible, more knowledgeable, and to consider the new position more of a challenge.

A continuing problem is that of substantial variation in productivity between individuals and between modules. Moreover, marginal personnel are at least benignly tolerated to the effect that an employee could be producing up to four times the amount of his neighbor doing the same work, and getting identical pay. The result of this amounted to rewarding poor performance rather than excellence. Predictably, a lack of consistent application of policy appeared because of relatively poor intermodular communication in some instances. In viewing the physical arrangement of modules, it was noteworthy that barriers, such as files, tables, and coatracks were placed so as to impede movement between the modules. In part, this was done so that anyone entering or leaving a module would have to pass in view of the manager, a form of control. On the other hand, it discouraged professionals and others from seeking counsel from their fellows in other modules. Professionals also aired complaints about inequitable distribution of the work load.

A substantial minority of professionals perceived a loss of status in that they were physically located with lower-level personnel. By contrast, the overwhelming majority of all personnel approved of the opportunity to observe all phases of the work as well as the integration of it. Good producers were very pleased with the increased accountability found in the modules, but not as pleased with what they viewed as inadequate management response to poor work.

In summation, the movement of an organization of this size from a traditional, functional mode to a form not previously tried on a large scale in service organizations during a period marked by a new construction, new law, and new procedures was an accomplishment of considerable magnitude in itself. Improvements in relative cost, case control, processing time (which means improved service to clientele), absenteeism, turnover,

and job satisfaction were observed. Remaining problems include interfacing staff support and service to the modular structure, intermodular communication, consistency of application of policy, and improving performance of some low producers. These would not appear to be insurmountable.

When you have read this case, look back at the chapter's diagnostic questions and choose the ones that apply to the case. Then use those questions with the ones that follow in your case analysis.

1. What kinds of jobs were being performed in the Beta Bureau processing centers during the time period covered in Part A of this case? How did employees react to these jobs? What other effects of the design of Beta Bureau's jobs were evident? What do you think caused employees' reactions and the other effects of job design in the bureau's processing centers?

2. What kind of job design approach did management take in creating the jobs described in Part B of the case? Was this approach successful? What effects did it have?

3. What problems were caused by the job redesign program at Beta Bureau? How would you deal with these problems?

C A S E 17-3

THE PRODUCTION DEPARTMENT AT KCDE-TV

Review this case, which appears in Chapter 7. Next, look back at Chapter 17's diagnostic questions and choose the ones that apply to the case. Then use those questions with the ones that follow in your case analysis.

1. How did the employees of KCDE's production department feel about their jobs? What effects did these attitudes have on job performance?

2. How might jobs in the production department be redesigned to improve employee attitudes and performance? What specific approach to job design should be used?

3. What steps should the management of KCDE-TV take to implement your program of job redesign? To what extent should employees be involved in the process of designing and implementing job changes at KCDE?

CHAPTER 18

MANAGING THE ORGANIZATION: STRATEGY, CULTURE, AND ORGANIZATION DEVELOPMENT II

In February of 1991, just 19 months after he became chairman of USX, Charles Corry succeeded in defusing two of the steel and energy giant's major problems. Satisfying the demands of investors, Corry issued a new class of stock representing the firm's steel assets but kept the steel unit within the organization, thus retaining this highly profitable business. At the same time, Corry negotiated a tentative labor settlement in bargaining that United Steel Workers union president Lynn Williams called "open, communicative, and realistic."
Source: Clare Ansberry, "USX Posts Rise of 9% in Profit for 4th Quarter," The Wall Street Journal, January 30, 1991, p. A9; Michael Schroeder, "How Charles Corry Became a Dragon Slayer at USX," Business Week, February 18, 1991, p. 35.

When Charles A. Corry was named chief of USX Corp. in June, some Wall Streeters wondered if he could fill the shoes of his charismatic predecessor, David M. Roderick. But on October 2, Corry made it clear that he aims to put his own mark on USX. Repudiating Roderick's troubled, $3 billion purchase of natural gas producer Texas Oil and Gas Corp. (TXO), Corry announced that he's putting TXO's energy reserves up for sale.

USX says it plans to use any proceeds to buy back stock and pay off debt. Analysts expect the bulk of the $1.5 billion to $2 billion raised from the sale of reserves will go to a massive stock buyback. And, ultimately, observers and former USX executives who know Corry well expect him to proceed with delayed plans to spin off at least 20% of the $5.8 billion steel unit to the public. That would still leave USX with its giant Marathon Oil Co. unit, which accounts for 53% of sales.[1]

As you know from reading this 1989 report, USX is a dominant player in the international petroleum industry. What you may not know, though, is that USX was once called U.S. Steel and until the early 1980s was the largest steel company in the world. What could possibly explain the company's metamorphosis from steel producer to petroleum refiner? Answer: from 1982 onward, USX's top managers intentionally redirected the firm into the increasingly profitable petroleum industry and out of the weakening world market for steel.

U.S. Steel—the world's first billion-dollar business—was born in 1901 when J. P. Morgan merged ten steel companies to overtake a competing steel empire owned by Andrew Carnegie. From then until the early 1980s, the firm maintained a leading role in the North American steel industry, serving as a major supplier for the automotive and heavy manufacturing industries. All that changed in 1982, however, when David M. Roderick, U.S. Steel's chair at the time, acquired Marathon Oil Company for $5.9 billion and began to transform it into a business that was more oil producer than steel manufacturer. Later in the same decade, the company changed its name to USX and strengthened its position in the petroleum industry by acquiring Husky Oil in 1984 and Texas Oil and Gas Corporation in 1985.[2]

As U.S. Steel ventured out of steel and into petroleum, the firm's business situation changed radically. From the relatively simple, stable steel industry where up to the end of the 1970s, it had held a comfortable position, the company now found itself moving into the more complex, dynamic environment of the petroleum industry of the 1980s. With this move came changes in the company's jobs and technology, away from the routine work and large-batch techniques of its foundry operations toward the nonroutine tasks and continuous process methods of its oil refineries. All these new conditions required that flexibility replace bureaucracy and that members of the firm learn to cope with change. USX managers thus found themselves wrestling with the task of refocusing the beliefs,

[1] Gregory L. Miles and Mark Ivey, "A New Iron Man Recasts USX," *Business Week*, October 16, 1989, p. 37.

[2] William C. Symonds, Gregory L. Miles, Mark Ivey, and Steven Prokesch, "The Toughest Job in Business: How They're Remaking U.S. Steel," *Business Week*, February 25, 1985, p. 50–56; Miles, Cheryl Debes, Richard A. Melcher, and James R. Norman, "It's USX vs. Everybody: Even If the Raiders Back Off, the Company Must Restructure Drastically," *Business Week*, October 6, 1986, pp. 26–27; and Miles and Ivey, "Iron Man Recasts USX."

values, and norms that had guided the company since the early 1900s. It was necessary to deemphasize norms and values that supported adherence to time-tested procedures and encourage creativity and innovation.[3] By the time Charles Corry replaced Roderick in 1989, USX was primarily an oil company and was struggling to separate itself from its bureaucratic past.

You will recall from Chapter 1 that this chapter is the last of three in our book (the others are Chapters 9 and 14) that focus attention on application issues—things that managers *do* to manage organizational behavior. It addresses the task of managing the organization as a whole. As is evident from our brief analysis of USX, managing the organizational context is an extremely complex task. Managers must plot the strategic direction of their firm. They must also design an organization structure of interdependent jobs to meet the demands of their company's work. In addition, they must shape the norms, values, and ways of thinking that influence behavior throughout the organization. In other words, managing the organizational context involves the management of *strategy*, *structure* and *job design*, and *culture*.

In keeping with what we have learned by examining management activities at USX, we will discuss the topics of strategy and culture in this chapter. Continuing the discussion begun in Chapter 14, we will also examine several organization-development interventions that can be used to manage organization-wide changes such as those that took place in USX throughout the 1980s. In addition, we will note how strategy, culture, and change at the organization level are related to the topics of structure and job design discussed in Chapters 15–17. After completing this chapter, you will be familiar with the broad range of activities involved in managing the organization as a whole.

DIAGNOSTIC ISSUES

When we talk about managing the entire organization, applying our diagnostic model raises some critical questions. Can we *describe* an organization's strengths and weaknesses and the way they affect its ability to survive? Can we identify the competitive forces arrayed against a firm? Can we *diagnose* and understand a situation in which an organizational culture persists in spite of the need for change? How can we *prescribe* which technique to use to overcome such resistance and stimulate needed change? Can we prescribe whether a business should grow and how fast? Finally, what *actions* can managers take to define the mission and strategy of their firm? What can they do to encourage changes that will enhance organizational effectiveness?

STRATEGIC MANAGEMENT AND THE FORMAL ORGANIZATION

A **strategy** is a plan of action that states an organization's goals and outlines the resources and activities required to achieve them.[4] **Strategic management** is thus a process of setting organizational goals and directing the organization toward goal achievement. During the process of strategic management, managers make decisions and implement changes that shape the **formal**

strategy   A plan of action that states an organization's goals and outlines the resources and activities required to achieve them.

strategic management   A process of setting organizational goals and directing the organization toward goal achievement.

---

[3] Symonds et al. "Toughest Job in Business."

[4] Roger Evered, "So What *Is* Strategy?" *Long Range Planning* 16 (1983), 57–72; and Ari Ginsberg, "Operationalizing Organizational Strategy: Toward an Integrative Framework," *Academy of Management Review* 9 (1984), 548–57.

formal organization Those aspects of an organization that are officially sanctioned, including intentionally designed structures and jobs.

organization of intentionally designed structures and jobs. New jobs are created, old jobs are eliminated, and structural arrangements are altered to fit the demands of the organization's strategy.

For example, in our opening story, David Roderick and his colleagues at U.S. Steel were engaged in strategic management as they formulated the strategy of diversifying into the oil industry. They were also performing strategic management as they put their strategy into action by acquiring Marathon Oil and other petroleum companies. With this change in U.S. Steel's business came new markets and technologies, therefore, pressures to create a new organization structure and new jobs. The formal organization moved toward change in response to the needs of the company's strategy. Later, Charles Corry and his staff were involved in strategic management when they sold off many of the assets acquired under Roderick's leadership. Again, as the company's business changed, pressure mounted to redesign its structure and the jobs performed by its employees. In the case of U.S. Steel, later USX, strategic management and the formal organization proved to be closely interrelated.

TEACHING NOTE
A large company with a long history, like USX (formerly U.S. Steel), has a tremendous amount of inertia to overcome. How can managers combat that inertia? How can managers change direction?

The Strategic Management Process

What do managers like Roderick and Corry do to develop and manage the strategy of a company? As shown in Figure 18-1, the process of strategic management involves five key phases—defining the organization's mission and strategic goals, analyzing its current situation, formulating a strategic plan, implementing the plan, and evaluating strategic performance.

Defining the Mission and Goals of the Organization. Strategic management begins with the definition of the mission of the firm. As you will recall from Chapter 2, an organization's mission is its purpose, or reason for being. Often the mission of a company can be expressed in the form of a statement that identifies the primary goods or services the company produces and the markets it hopes to serve. In conjunction with the mission, managers also identify strategic goals to be achieved as the organization progresses toward accomplishing its mission. Such goals might include gaining control over 20 percent of the market served by the company or doubling the company's level of profitability in five years. Together, the mission and goals developed during the initial phase of strategic management provide the organization with a sense of direction and serve as an indication of what must be accomplished for the organization to function successfully.

APPLYING THE DIAGNOSTIC MODEL
Diagnostic Question 1: What are the organization's current strengths and weaknesses? What opportunities and threats are present in the environment? What might be done to capitalize on current strengths and opportunities? How might current weaknesses and threats be avoided or even converted into strengths and opportunities?

Performing a Situation Analysis. The aim of any strategy is to capitalize on what the organization does well and avoid what the organization does poorly. Thus the second phase of the strategic management process involves a careful

FIGURE 18-1

The Strategic Management Process

Defining the mission and goals of the organization → Performing a situation analysis → Formulating a strategic plan → Implementing the strategic plan → Evaluating strategic performance

Feedback from evaluation

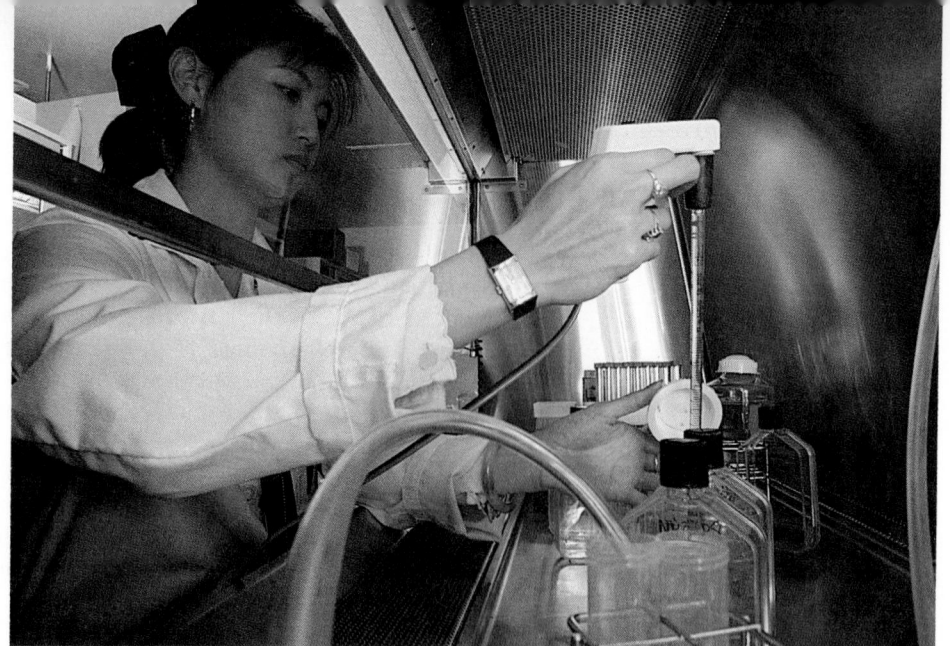

In 1980, Robert A. Swanson founded Genentech Inc. with a mission of building an independent pharmaceutical company that would do research in genetic engineering and develop products that would cure cancer, heart disease, and other major illnesses. But its first major product, the heart-attack drug known as TPA (tissue plasminogen activator), produced only half the sales the company had projected, and it became increasingly difficult to plow enough money back into research to keep the company's many projects going. In 1990 Swanson sold a controlling interest in the company to Roche Holding Ltd., the Swiss parent of Hoffmann-La Roche Co. Genentech's mission could change further if Roche takes over, as its stock options would allow it to do.
Source: Joan C. Hamilton et al., "Why Genentech Ditched the Dream of Independence," Business Week, February 19, 1990, pp. 36–37.

situation analysis An analysis of internal strengths and weaknesses and external opportunities and threats that is performed during the process of strategic planning.

AN EXAMPLE: USX
Many would argue that USX, formerly U.S. Steel, moved to oil not so much because of a weakening steel market, but because they couldn't compete in steel. Steel mills had become obsolete. Little had been invested in updating the mills to make them efficient. As a result, U.S. Steel couldn't compete with new providers like Japan, who was using state of the art equipment which allowed them to produce a better product at a cheaper price. On the other hand, by the end of 1990 USX's steel unit was the most profitable major U.S. steelmaker, producing tubular pipe for the oil industry (Clare Ansberry, "USX Posts Rise of 9% in Profit for 4th Quarter," *The Wall Street Journal*, January 30, 1991, p. A9).

situation analysis (see Figure 18-1).[5] To perform a situation analysis, top management begins by looking at the company's *strengths*, which are characteristics of the company that can help it achieve its strategic goals. Examples of such strengths include the United Parcel Service's reputation for providing fast, dependable delivery and IBM's reputation for providing some of the best client services of any computer manufacturer. In both these instances, the company's reputation is a strength that helps attract customers. Other strengths might include having experienced, knowledgeable managers and a skilled work force, holding extensive financial reserves, and owning the rights to important information or equipment.

Next is an assessment of the company's *weaknesses*. Those are the characteristics of the company that can block its progress toward goal achievement. Lacking a clear strategic direction is a notable weakness, as are operating out of obsolete facilities, failing to perform adequate research and development, and lacking essential managerial talent.

The situation analysis then focuses on conditions outside the organization. Managers first assess *opportunities* or characteristics of the environment surrounding the company that might help it meet or exceed its strategic goals. A healthy economy, readily available raw materials, and an expanding market are all environmental opportunities. The situation analysis concludes with an examination of *threats*. These threats are environmental factors that may prevent the organization from attaining its strategic goals. Examples are adverse government policies, growing competitive pressures, scarce labor resources, and changing consumer tastes.

Formulating A Strategic Plan. During the third phase of strategic management, managers work to develop a strategic plan that will take advantage of the strengths and opportunities uncovered during situation analysis and minimize the negative effects of weaknesses and threats. Strategies formulated during this phase are of the three types depicted in Figure 18-2—corporate, business, and functional.[6]

[5] Arthur A. Thompson, Jr., and A. J. Strickland III, *Strategic Management: Concepts and Cases*, 4th ed. (Plano, Texas: Business Publications, 1987), p. 97.

[6] Dan E. Schendel and Charles Hofer, eds., *Strategic Management: A New View of Business Policy and Planning* (Boston: Little, Brown, 1979), pp. 11–14; Milton Leontiades, *Strategies for Diversification and Change* (Boston: Little, Brown, 1980), p. 63; and Robert H. Hayes and Steven C. Wheelwright, *Restoring Our Competitive Edge: Competing through Manufacturing* (New York: John Wiley, 1984), p. 28.

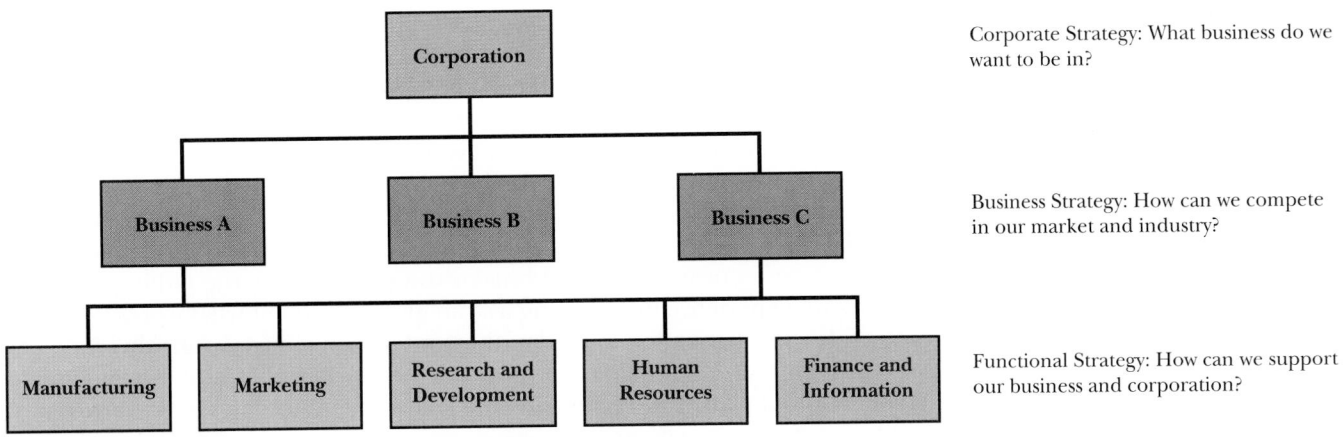

Corporate Strategy: What business do we want to be in?

Business Strategy: How can we compete in our market and industry?

Functional Strategy: How can we support our business and corporation?

FIGURE 18-2

Three Types of Strategies

Organizations consist of a corporate body often made up of several different businesses, each of which is composed of a number of functional units. *Corporate* strategies concern the mix of businesses constituting the organization. *Business* strategies specify the goals and actions of each business. *Functional* strategies indicate how each of the functions in a particular business will contribute to the success of the business and the company that owns it.

AN EXAMPLE: SOUTHERN CALIFORNIA EDISON

Sometimes companies map out alternative strategies years in advance in an effort to prepare for different contingencies. When Southern California Edison, which provides electricity to nearly 4 million customers, found that unexpected events such as nuclear accidents or new restrictions on sulfur emissions made every long-range plan it had constructed useless it adopted a technique known as scenario planning. The company developed 12 possible versions of its situation ten years hence, including such events as a Middle East oil crisis, expanded environmentalism, and an economic boom. It also built flexibility into its system to enable it to power or depower generating plants quickly and to buy electricity from other utilities.

They specify the *means* or procedures to be used to accomplish the *ends* or objectives that are in the organization's strategic goals. Later in this chapter, following a description of the strategic management process, we will examine the three types of strategies shown in Figure 18-2.

Implementing the Strategic Plan. Implementation involves putting the strategy into place and getting individuals and groups in the organization to execute their part of the strategic plan. It focuses on activities such as

Developing an organization structure capable of coordinating the jobs that must be performed to fulfill the company's mission

Formulating budgets and resource allocation procedures that will help the company attain strategic goals

Inspiring employee commitment to the company's strategy and mission

Linking employee motivation and the organization's reward system to the achievement of strategic goals

Creating a culture of norms and values that supports the strategic mission and successful goal attainment

Developing an information system to track and control the process of strategic implementation

Exerting the leadership needed to drive strategic implementation forward and to stimulate continual improvement in the execution of the company's strategy.[7]

Implementing an organization's strategy means evaluating and sometimes changing a firm's structure, technology, information flows, and leadership patterns to achieve the strategic goals of the firm. Management must examine the formal organization to determine what strategy-supporting features are required and then do what must be done to put these features in place.

[7] Thompson and Strickland, *Strategic Management*, p. 11.

Evaluating Strategic Performance. The final phase of the process of strategic management is an evaluation of the organization's performance to determine whether strategic goals are being met. It is not unusual for evaluation procedures to uncover operational shortcomings or changing conditions that require adjusting the strategic implementation or even the strategic plan itself. For instance, strategic adjustments may be required by changes in the internal strengths and weaknesses of the company that occur as some employees leave and others take their place. Or external opportunities and threats may shift as new competitors enter the market or technological breakthroughs influence the industry. Simply building up experience in executing a strategy can pinpoint what works and what doesn't. It stimulates additional planning and a change in strategic direction. As shown in Figure 18-1, evaluation feeds back to earlier phases of strategic management, influencing future missions and goals, situation analyses, strategic plans, and implementation efforts.

Types of Strategies

The direction ultimately pursued by an organization is the product of several types of strategies: (1) a corporate strategy that determines what businesses the firm is in, (2) different business strategies that influence the direction of each of the businesses in the firm, and (3) various functional strategies that help coordinate the activities of the different functional areas in each of the firm's businesses. Let's take a closer look at these three types of strategies.

corporate strategy A strategic plan that specifies the desired mix of businesses in a firm and how resources should be allocated among them.

Corporate Strategies. Corporate strategies address the question, *What business or businesses do we want to be in?* When a firm is involved in only a single business, the answer to this question may seem relatively obvious. For instance, the managers of a college bookstore probably have little difficulty determining that the business of their firm is selling college texts and that because there is a continuing demand for textbooks from one term to the next, they want to stay in that business.

When a company is involved in several different businesses, however, top management may find it hard to answer questions about the firm's present and prospective composition. The company's businesses may be as diverse as fashion apparel, consumer electronics, and petroleum production. To visualize the task of managing a firm consisting of several businesses, take a look at the **BCG Matrix** diagrammed in Figure 18-3. This matrix, developed by the Boston Consulting Group (BCG), is a tool that can be used to track the **portfolio** of businesses owned by an organization.[8]

BCG Matrix A management tool that depicts the mix of businesses owned by a firm.

portfolio The mix of different businesses owned by an organization.

Businesses that appear in the upper left-hand quadrant of the BCG Matrix are called *stars*, because they are the most desirable of all the businesses in the matrix. Stars are very valuable, because each controls a large share of the market in a growing industry. Firms normally try to hold on to the stars they already own as they try to find new stars for their portfolios.

In the lower-left-hand quadrant of the BCG Matrix, *cash cows* control large market shares of stagnant, less desirable industries. Cash cows generate a lot of money because of their market position. Management, however, has little reason to spend this money on developing them further, because their industries are so weak. Instead, they become sources of funds for maintaining existing stars or developing new ones.

[8] Bruce D. Henderson, *The Experience Curve Reviewed, IV* (Boston: Boston Consulting Group, 1973), p. 3; and Barry Hedley, "A Fundamental Approach to Strategic Development," *Long Range Planning,* 9 (1976), 2–11.

F I G U R E 18-3

The BCG Matrix

The vertical axis of the BCG Matrix maps industry growth rates for all businesses in the firm's portfolio. The horizontal axis charts each business's market share. Each business is represented by a circle whose size indicates the percentage of corporate revenues that business generates. The position of each circle within the matrix indicates the market share that the business it represents holds as well as the growth rate of the industry in which the business functions.

Based on Bruce D. Henderson, The Experience Curve Reviewed, IV: The Growth Share Matrix of the Product Portfolio *(Boston: Boston Consulting Group, 1973) p. 3; and Barry Hedley, "Strategy and the Business Portfolio," Long Range Planning 10 (1977), 2–11.*

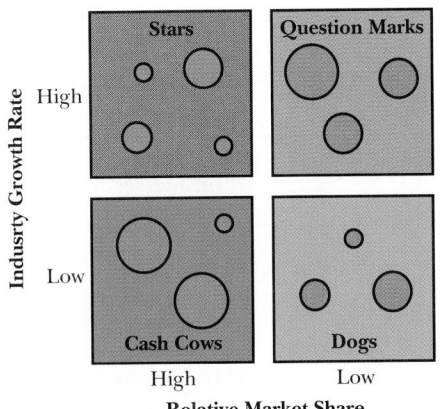

AN EXAMPLE: FORD AEROSPACE, LORAL DEFENSE SYSTEMS

Many businesses centered on defense have moved to the question mark quadrant of the BCG matrix. This has occurred as a result of defense spending cuts. Only companies that can meet a specific need, do it well, and underbid the competitors repeatedly will survive. A recent example was the sale of Ford Aerospace to Loral Defense Systems. Ford decided to get out of the shrinking, highly competitive aerospace business.

convergent growth strategies A class of corporate strategies that involve starting, acquiring, or maintaining businesses in the market and industry already served by the firm. Included in this class are the concentration strategy and the horizontal integration strategy.

evolutionary growth strategies A class of corporate strategies in which the firm develops products, markets, or businesses that are similar though not identical to the ones already associated with the company. Product development, market development, and concentric diversification are all evolutionary growth strategies.

Businesses in the upper-right-hand quadrant of the BCG Matrix are called *question marks*. The future of a question mark, which has only a small market share in a healthy, growing industry, is uncertain. If its market position cannot be improved, such a business may consume more resources than it produces and become a costly burden to its owner. On the other hand, by reallocating money from a cash cow or even a star, it may be possible to transform a question mark into a star for less money than it would cost to start a successful business from scratch or to buy one from another company.

Businesses in the lower-right-hand quadrant of the BCG Matrix, called *dogs*, occupy small market positions in stagnant industries and are the least desirable type of business to own. A firm will usually try to sell the dogs in its portfolio. If it cannot find buyers, the firm may choose to close the dogs and absorb the resulting short-term loss rather than endure a long-term drain on company resources.

The BCG Matrix demonstrates that managing a company's portfolio involves the acquisition of desirable businesses, the spin off of undesirable businesses, and the transfer of resources from lucrative businesses to those in need of additional support. Let's look next at some of the corporate strategies (see Table 18-1) that companies follow in performing these activities.[9]

A company uses **convergent growth strategies** to start, acquire, or maintain businesses in the market and industry it already serves. In the strategy known as *concentration*, a firm attempts to strengthen the market position of a business it already owns by using the firm's resources to sell the business's product more effectively in its current market. For example, in 1991 Apple Computer initiated a concentration strategy by introducing lower-priced versions of its most popular products and kicking off an aggressive marketing campaign. In *horizontal integration*, a firm tries to increase its market share by starting or acquiring other businesses that make the same product it produces itself. For instance, in 1990 GM opened its new Saturn division, reflecting the pursuit of a strategy of horizontal integration in the automotive industry.

Using **evolutionary growth strategies**, an organization may develop products, markets, or businesses that are similar though not identical to the ones already associated with the company. In *product development*, a firm tries to strengthen its position by selling a new product in its current market. For example, during the 1980s, Anheuser-Busch introduced a line of low-alcohol beverages to

---

[9] The strategies discussed in this section are based on a typology presented in Richard Z. Gooding, "Structuring Strategic Problems: Antecedents and Consequences of Alternative Decision Frames" (Ph.D. dissertation, Michigan State University, 1989). See also J. A. Pearce and R. B. Robinson, Jr., *Strategic Management: Strategy Formulation and Implementation* (Homewood, Ill.: Richard D. Irwin, 1985); and Thompson and Strickland, *Strategic Management.*

AN EXAMPLE: MOTOROLA

In the early 1980s Japanese companies flooded the market with cellular phones and pagers and took the lead in the world's production of semiconductors. Adopting a product development strategy, Motorola decided to out-Japan Japan and upgraded quality, pared costs, and poured billions into research, development, and capital improvements. By the end of 1989 the company had regained considerable market share in semiconductors. It had produced the MicroTac cellular phone, which was sleeker and lighter than its closest rival—made by Matsushita—and for which backorders began to pile up quickly. And Motorola had developed a wristwatch pager that, according to one customer, "scooped the world by a year or two." (Susan Moffat, "Japan's New Personalized Production," *Fortune*, October 22, 1990.)

AN EXAMPLE: GOODYEAR AEROSPACE

Goodyear Aerospace had the opportunity to diversify concentrically but turned it down. For years Goodyear made blimps for the military. In the 1970s private firms like Sea World approached Goodyear to make blimps for commercial use. Goodyear refused, leaving the door open for competitors to appear and fill that need. As a result, Goodyear lost its military contracts to a competitor in 1987.

revolutionary growth strategies A class of corporate strategies in which the firm develops or acquires businesses that are quite different from businesses it already owns. Included in this class are forward integration, backward integration, and conglomerate diversification.

adaptive decline strategies A class of corporate strategies in which the firm reduces its size by either downsizing or eliminating one or more businesses. Retrenchment, divestiture, and liquidation are all adaptive decline strategies.

TABLE 18-1
Types of Corporate Strategies

STRATEGIC GOAL	DESCRIPTION	SPECIFIC STRATEGIES
Convergent growth	Acquisition or maintenance of businesses in the firm's current market and industry	Concentration Horizontal integration
Evolutionary growth	Acquisition of businesses in markets and industries that are similar but not identical to those currently served by the firm	Product development Market development Concentric Diversification
Revolutionary growth	Acquisition of businesses in markets and industries that differ from those currently served by the firm	Forward integration Backward integration Conglomerate diversification
Adaptive decline	Downsizing, or elimination of some of the firm's current businesses	Retrenchment Divestiture Liquidation

---

take further advantage of the distribution network of supermarkets and convenience stores that the firm's brewery business had already established. Following a strategy of *market development*, a firm sells its current product to new markets. The decision of CBS Records to start a mail-order club to sell CDs and tapes exemplifies the market-development strategy, because it enabled the firm to branch out beyond the network of retailers that normally sell sound recordings. In *concentric diversification*, an organization establishes a business that is similar but not identical to its current businesses in products, markets, and technologies. This sort of strategy is exemplified by Boeing Aircraft's decision to enter the market for civilian aircraft when orders for military planes and armament declined at the end of World War II.

As their name suggests, **revolutionary growth strategies** call for a firm to develop or acquire businesses that are quite different from businesses it already owns. In *forward integration*, a company grows by acquiring one or more businesses in the distribution channel that connects it with the final consumers of its products. For example, Benetton, a manufacturer of sweaters and other clothing, acquired warehousing facilities and retail stores so that it could retail its own products. Conversely, in *backward integration*, a firm acquires businesses in the channels that connect it with the raw materials it needs. For instance, the Goodyear Tire and Rubber Company owns plantations throughout the world that it bought to ensure an uninterrupted supply of rubber for its tires and other products. In *conglomerate diversification*, a company establishes businesses in areas totally unrelated to its current undertakings. General Electric's move into financial services, including consumer credit operations and venture capital financing, exemplifies this strategy.

When none of the corporate growth strategies we have discussed are successful in improving or at least maintaining the profitability of an organization's portfolio of businesses, the organization may be forced to choose an **adaptive decline strategy**. It either downsizes or eliminates one or more businesses. Sometimes a firm will cut back on the operating levels of a business to reduce current costs. By such temporary *retrenchment*, the firm hopes to give the business a chance to deal with its problems and reemerge in a stronger position. American Motors attempted to survive in the mid-1980s by scaling back other products and focusing on the production and sales of Jeeps. When its retrenchment strategy failed to

MANAGEMENT ISSUES

Company for Sale: Should Employees Be Let In on the News?

Some managers argue that employees should be the last to know the details about a pending sale of their business unit. What they don't know won't hurt their productivity or morale. Other managers argue just the opposite, suggesting that employees who are let in on the news from the beginning are more able to cope with the changes that occur once the sale actually takes place. Consider the following story as you think about which position you agree with.

In December 1989, the Chrysler Corporation publicized its plan to sell its Gulfstream Aerospace subsidiary. Within a day, Allen E. Paulson, Gulfstream's founder and chief executive, announced that he would attempt to buy the company back from Chrysler. Why was Paulson so quick to act?

In 1985, the year Paulson had sold Gulfstream to Chrysler, the company was healthy and productive. "We were working hard, everyone had a piece of the action, we all felt good," recalled James Swindells, Gulfstream's director of completion center operations. Then came the announcement that the company was up for sale. One rumor (false) had it that Paulson was leaving Gulfstream. Another (false) was that layoffs were just ahead. A third (true) was that the new owner knew little about aerospace. "Everyone was insecure, handing his résumé to anyone who came in to have an airplane serviced," Swindells said. Things got even worse when employees discovered that Chrysler was going to be the new owner. "Workers were wor-

rying they'd be moved to Detroit or asked to build cars in Savannah, and managers were worrying if they'd have jobs at all," said Preston Blackwelder, Gulfstream's director of personnel.

This experience taught Paulson that keeping employees in the dark about the sale of their business can do more harm than good. It is certainly true that publicizing a pending sale can stimulate a bidding war that undermines the original buyer's position and can create a sense of worrisome uncertainty among employees. Keeping things out in the open, however, gives management the chance to deal with prospective issues before they turn into problems. In the Gulfstream of 1989, managers held weekly meetings to field employee questions and dispel rumors. One particularly troublesome rumor was that Mitsubishi was negotiating for the company. Management responded immediately by posting a notice that Mitsubishi was looking to buy one of Chrysler's auto plants, not Gulfstream. Workers clearly appreciated such responses. Unlike five years ago, no one quit for fear of ownership. "I've faced uncertainties in aviation before, and I've always made calls right away," says service mechanic Kenneth Farris. "This time I'm waiting patiently."*

* Claudia H. Deutsch, "Letting Employees in on The News," *New York Times*, March 4, 1990, p. F37.

TEACHING NOTE
More and more organizations are resorting to decline strategies. Why? What environmental factors contribute to this development?

APPLYING THE DIAGNOSTIC MODEL
Diagnostic Question 3: For each business, what competitive forces—existing market rivals, potential new entrants, suppliers, sources of substitute products, customers—threaten continued success? Should the business pursue cost leadership? Differentiation? Focus?

business strategy A strategic plan that specifies the way a business plans to compete in its market and industry. Overall cost leadership, differentiation, and focus are all business strategies.

work, however, the company was purchased by Chrysler. Sometimes reduction becomes permanent elimination. In *divestiture* a firm sells its entire interest in one of its businesses to another firm, as exemplified by Chrysler's sale of its military tank and armament business to General Dynamics in the early 1980s. In *liquidation*, a firm breaks up a failing business, salvaging whatever it can by selling one or more parts of the otherwise unsaleable business or merging parts of the failed business with other businesses it owns. Texas Instruments liquidated its personal-computer division in the mid-1980s after IBM became the dominant player in the personal-computer industry.

Decline strategies present the special problem of managing workers who face the prospect of losing their jobs. Should the managers of the business that is about to be divested tell employees about upcoming changes? Or should management keep news about the sale to itself? The "Management Issues" box considers some of the problems for employees when a business is put up for sale.

Business Strategies. For each business in the corporate portfolio, management formulates a **business strategy** in an attempt to ensure the business's successful performance. Business strategies address the question, How can we compete most effectively in our market and industry? In searching for an answer, managers assess the effects of five competitive forces. Those forces are the business's current rivals in the marketplace, its potential rivals in the form of new entrants into the market, firms that make products that might substitute for the business's own, the business's suppliers of raw materials, and the business's customers (see Figure 18-4). On the basis of this assessment, managers typically choose one of three

FIGURE 18-4

The Five-Forces Model of Business Competition

According to Michael Porter, managers planning a business strategy must consider five factors when choosing how their business will compete: rivals in the business's market, who compete with the business for the same customers; potential new entrants, who might become additional rivals; suppliers, who can affect business activities by altering the price or availability of raw materials; sources of substitute products, which might draw customers away from the business's market; and the business's customers, who might choose to buy from another firm. Based on Michael E. Porter, *Competitive Strategy: Techniques for Analyzing Industries and Competitors* (New York: Free Press, 1980).

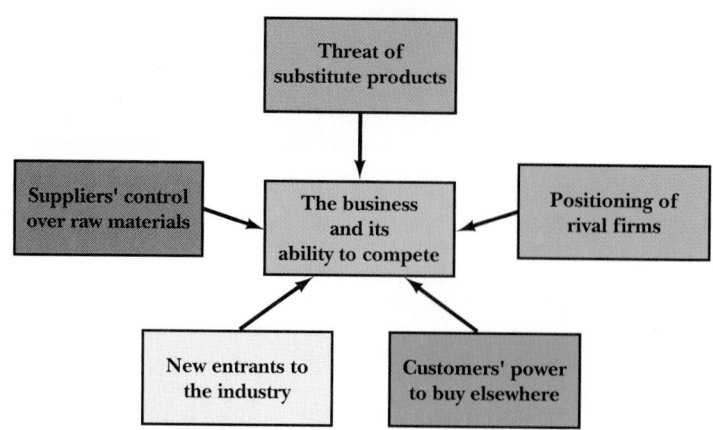

The Campbell Soup Co., facing declining profits, attempted to move into other product lines but stretched its resources too thin. When David Johnson took over as CEO in 1990 he concluded that the company had failed to capitalize on its strongest product. Campbell soups continue to command almost two-thirds of the U.S. market for soup. In a differentiation strategy, the company brought out a new and unique version of its current product: a line of soups just for children. In a further effort to reinforce brand loyalty, the company brought back its famous "Campbell Kids," featuring them, with new friends, on the labels of the new product.
Source: "From Soup to Nuts and Back to Soup," Business Week, *November 5, 1990, p. 114.*

types of competitive strategies—overall cost leadership, differentiation, or focus.[10]

The competitive thrust in the strategy of *overall cost leadership* is on cutting operating costs and selling at prices below those of competitors. In the retailing industry, for example, Wal-Mart has used this strategy to compete with K Mart, Woolco, and other discount chains. Clearly, in order to sell goods at the lowest market prices, a firm must organize with the utmost efficiency. It must be able to exert tight control over the costs of research and development, production, sales, and similar functional activities as well as over general overhead expenses. Careful cost accounting, close financial control, and investment in labor-saving equipment are all hallmarks of the cost-leadership strategy.

Cost leadership not only helps a business compete on the basis of price. It enables the business to defend itself against price-war conditions and earn acceptable profits in markets dominated by price competition. This strategy also makes it easy to use price cutting as a defense against new entrants into the market as well as against companies that offer substitute products. A highly efficient operation can absorb increasing costs of raw materials without forcing a significant rise in selling prices. Of course, building a reputation for consistently offering goods or services at lower prices than any competitor's also goes a long way toward ensuring customer loyalty.

Using the strategy of *differentiation*, a firm tries to distinguish its product or service. It creates something that differs in important ways from other competitive products—something that will be perceived as unique. Differentiation can be accomplished in many ways. Mercedes Benz and BMW, for example, use a com-

---

[10] Michael E. Porter, *Competitive Strategy: Techniques for Analyzing Industries and Competitors* (New York: Free Press, 1980), p. 35.

bination of design innovations and upscale advertising to build images of exceptional quality, durability, and luxury. Sony Electronics and Compaq Computer set themselves apart from competitors by staying on the cutting edge of technological advance. Caterpillar Tractor plays up its extensive dealer and service network as distinguishing it from its competitors.

The purpose of the differentiation strategy, however it is accomplished, is to create brand loyalty among a business's customers and in that way, insulate the business from the rivalry of both current and potential competitors. Brand loyalty, in turn, can provide something of a cushion in a situation where the firm is forced to raise selling prices. The loyalty stimulated by differentiation may reduce customers' sensitivity to price. Differentiation also allows the increased profit margins needed for protection against increasing raw materials costs.

In the *focus* business strategy, a firm decides to concentrate on serving a particular, or target, market. This strategic decision enables the firm to contain costs by limiting the scope of its research and development, advertising, marketing, and sales efforts. By tailoring its product to the special needs of its target group of customers, the firm can more easily differentiate itself from competitors. It does not have to meet the needs of a broad array of potential customers. Consider the strategy of the Fort Howard Paper Company, which targets industrial suppliers of paper goods and avoids the retail market. To these suppliers of factories, schools, government offices, and similar facilities, Fort Howard sells washroom paper products at prices that competitors like Kimberly-Clark or Scott Paper — which serve a wide consumer market—cannot beat. As you can see, the focus strategy lets a firm use elements of both the overall cost leadership and the differentiation strategies. It can maintain low costs by limiting itself to a specific market segment and distinguish itself from other competitors by focusing on that segment alone.

Functional Strategies

As you will recall from Chapters 15 and 16, businesses are organized in structural units responsible for specific *functions*. Some such units that are common to many firms are the departments for manufacturing, marketing, materials management, research and development, human resources, and financial-information systems.[11] **Functional strategies** address the question, How can we support our firm's corporate and business strategies most effectively? They help ensure that unit activities play a role in meeting corporate and business goals. Such strategies can not only build new strengths in a firm; they can also turn a company's weaknesses into strengths. For instance, in the 1970s, various food-processing businesses overhauled their manufacturing and marketing operations and then introduced generic food products. This move transformed money-losing products into profitable lines of overall cost leadership. A decade later, the Nippon Electric Company engaged in aggressive research and development and added a unique product—laptop computers—to an aging line of desktop computers. The company then initiated a new and successful differentiation strategy.

How Strategy Shapes the Formal Organization

The three types of strategies we have described exert strong influence over one another. Successful functional strategies increase the likelihood of success in

[11] Charles W. L. Hill and Gareth R. Jones, *Strategic Management: An Integrated Approach* (Boston: Houghton Mifflin, 1989), p. 230.

business operations, and successful business strategies improve the position of the corporation as a whole. Conversely, an effective corporate strategy also improves the resource positions of the firm's businesses, and a successful business strategy underwrites continued activity in the business's functional units.

In addition, as we said at the beginning of this chapter, strategic management influences a firm's formal organization of intentionally designed jobs and structural arrangements. One way it does so is by affecting the contingencies that determine structural effectiveness.[12] Consider the following points:

1. The environmental diversity surrounding a firm depends largely on whether the firm chooses to concentrate on a single business or develop a portfolio of different businesses. As long as an organization consists of a single business, it will probably face a uniform environment. If, however, it has a portfolio of different businesses, its management can expect to encounter significant environmental diversity.

2. The levels of complexity, stability, uncertainty, and hostility of the firm's environment are also affected by strategic decisions as to which businesses to acquire, which to keep, and which to sell. For instance, when Honeywell Bull acquired Zenith Data Systems, it left the relative stability of the mainframe computer industry for the much more volatile personal-computer market. Similarly, when R. J. Reynolds acquired Nabisco, the firm encountered a less hostile environment in the prepared-foods industry than the one it had previously dealt with as a tobacco manufacturer.

3. The size of a firm is a direct consequence of strategic choice. Adding additional businesses to the corporate portfolio and pursuing growth strategies in existing businesses increases organizational size. In contrast, reducing the size of the corporate portfolio or pursuing adaptive decline strategies reduces the size of an organization.

4. The technology employed in a business is often the consequence of choosing one business strategy over another. For instance, choosing to pursue an overall cost leadership strategy in a manufacturing firm pushes toward the adoption of an efficient technology—most often, mass production or continuous operations. In contrast, a differentiation strategy may necessitate the custom-building adaptability of unit or flexible-cell technologies.[13]

AN EXAMPLE: GENERAL MOTORS

General Motors' ads have attempted to differentiate its products from those of Ford, based on quality. While Ford tells us "quality is job one," GM demonstrates quality through longer warranty periods. While GM offers 3-year or 50,000 mile warranties, Ford still offers the standard 12 months or 12,000 miles. If quality weren't good, GM wouldn't run the risk of offering extended warranties and have to absorb the cost of repairs for an additional two years. Thus, the longer warranty should create an image of greater quality and suggest that Ford is afraid to stand behind its quality claims.

Strategy also influences the formal organization by affecting the design of jobs. Just as a strategy of overall cost leadership favors the adoption of efficient technologies, it calls for efficiency-oriented approaches to job design, such as work measurement and job simplification. American automobile companies often compete on the basis of cost, as shown by the countless rebates and discount programs they have offered. And their assembly lines consist of simplified jobs, such as the welding task described at the beginning of Chapter 17. On the other hand, a differentiation strategy that emphasizes product quality may favor the adoption of job enrichment methods. Experiences at Volvo and Saab indicate that the comprehensive enrichment afforded by sociotechnical job design can have a substantial positive effect on product quality.

In sum, the process of strategic management creates many of the contingencies and conditions that influence managers as they design organizational structures and the jobs within them. In turn, the structures and jobs that managers design define the nature of the organizations, businesses, and functional units that are the focus of further strategic management activities. Strategic management is therefore an integral part of the process of managing the formal organization.

[12] Alfred D. Chandler, Jr., *Strategy and Structure* (Garden City, N.Y.: Anchor Books, 1966), pp. 7–17; and Danny Miller, "Relating Porter's Business Strategies to Environment and Structure: Analysis and Performance Implications," *Academy of Management Journal* 31 (1988), 280–308.

[13] Porter, *Competitive Strategy*, pp. 35–40.

TEACHING NOTE
Ask students to consider: If employees realize in advance that business is down, work orders are down, and if things don't pick up they will be let go, why keep a layoff a secret until the last minute? What is to be gained? What might be gained if early notice of layoffs was given?

informal organization The unofficial rules, procedures, and interconnections that develop as employees make spontaneous, unauthorized changes in the way things are done.

grapevine The unofficial communication network within the informal organization through which employees trade gossip and rumors about their jobs, their coworkers, and the organization.

culture The shared attitudes and perceptions in an organization that are based on a set of fundamental norms and values and help members understand the organization.

Beneath every formal organization of official jobs and structural relationships lies an **informal organization** of unofficial rules, procedures, and interconnections that develops as employees make spontaneous, unauthorized changes in the way things are done. An important part of the informal organization is the **grapevine**, an unofficial communication network through which employees trade gossip and rumors about their jobs, their coworkers, and the organization. In many organizations, hearing from the grapevine that the company's business is off often precedes formal management announcements of layoffs or other cutbacks. Similarly, it is not uncommon for rumors that a favorite boss is about to retire to predate the formal announcement by several weeks. Although the information transmitted in a company's grapevine lacks official authorization, it does not necessarily lack validity.

The workings of the informal organization are also illustrated by the employee-created performance quotas that sometimes take the place of formal production goals. When employees perceive official standards as unfair or overly demanding, they may develop unofficial, lower standards of productivity. They may also develop a complex social structure of informal rewards and punishments around these standards. If employees catch peers working too hard, they may subject them to verbal threats, social isolation, and even physical abuse.[14]

In both these examples, the informal organization arises as day-to-day adjustments to the formal way of doing things create a **culture** of attitudes and understandings that are shared among coworkers. An organization's culture is

> A pattern of basic assumptions—invented, discovered, or developed [by a firm's members] to cope with problems of external adaptation and internal integration—that has worked well enough to be considered valid and, therefore, to be taught to new members as the correct way to perceive, think, and feel in relation to those problems.[15]

Culture is thus an informal, shared way of perceiving life and membership in an organization that binds members together and influences what they think about themselves and their work.

In the process of helping to create a mutual understanding of organizational life, organizational culture fulfills four basic functions:

1. It gives members an organizational identity. Sharing norms, values, and perceptions gives people a sense of togetherness that helps promote a feeling of common purpose.

2. It facilitates collective commitment. The common purpose that grows out of a shared culture tends to elicit strong commitment from all those who accept the culture as their own.

3. It promotes system stability. By encouraging a shared sense of identity and commitment, culture encourages lasting integration and cooperation among the members of an organization.

4. It shapes behavior by helping members make sense of their surroundings. An organization's culture serves as a source of shared meanings that explain why things occur the way they do.[16]

---

[14] Fritz J. Roethlisberger and William J. Dickson, *Management and the Worker* (Cambridge, Mass.: Harvard University Press, 1939), p. 523.

[15] Edgar H. Schein, *Organizational Culture and Leadership* (San Francisco: Jossey-Bass, 1985), p. 9.

[16] Robert Kreitner and Angelo Kinicki, *Organizational Behavior* (Homewood, Ill.: BPI-Irwin, 1989), pp. 649–50. See also Linda Smircich, "Concepts of Culture and Organizational Analysis," *Administrative Science Quarterly* 28 (1983), 339–58.

By fulfilling these four basic functions, the culture of an organization serves as a sort of social glue that helps reinforce persistent, coordinated behaviors at work.

Elements of Organizational Culture

APPLYING THE DIAGNOSTIC MODEL
Diagnostic Question 5: What cultural norms and values guide behaviors and understandings in the firm? Do surface elements reinforce these deeper elements? Do differences between surface elements and cultural norms and values suggest, ongoing cultural change? Is this change desirable?

Deep within the culture of every organization are a collection of fundamental norms and values that shape members' behaviors and help them understand the surrounding organization. In some companies, for example, Polaroid, 3M, and DuPont Chemical, cultural norms and values emphasize the importance of discovering new materials or technologies and developing them into new products. In other companies, such as AT&T and Maytag Appliances, cultural norms and values focus on high product quality.[17] Fundamental norms and values like these are the ultimate source of the shared perceptions, thoughts, and feelings that constitute the culture of an organization.

How are these fundamental norms and values expressed? How are they passed from one person to another? Certain surface elements of the culture help employees interpret everyday events in the organization and are the principal means by which cultural norms and values are communicated from one person to another. *Ceremonies, rites,* and *rituals* reinforce particular norms and values by demonstrating their worth through special events. *Stories* and *myths* exemplify norms and values by encompassing them in memorable narratives. In the exploits of *heroes* and *superstars,* organization members see the fruits of adhering to the firm's norms and values. *Symbols* and special *language* are constant reminders of important norms and values (see Table 18-2).

[17] Terrence E. Deal and Allan A. Kennedy, *Corporate Cultures: The Rites and Rituals of Corporate Life* (Reading, Mass.: Addison-Wesley, 1982), p. 15.

TABLE 18-2
Surface Elements of Organization Cultures

ELEMENT	DESCRIPTION
Ceremonies	Special events in which organization members celebrate the myths, heroes, and symbols of their firm
Rites	Ceremonial activities meant to communicate specific ideas or accomplish particular purposes
Rituals	Actions that are repeated regularly to reinforce cultural norms and values
Stories	Accounts of past events that illustrate and transmit deeper cultural norms and values
Myths	Fictional stories that help explain activities or events that might otherwise be puzzling
Heroes	Successful people who embody the values and character of the organization and its culture
Superstars	Extraordinary individuals who personify the upper limits of attainment in the organization and its culture
Symbols	Objects, actions, or events that have special meanings and enable organization members to exchange complex ideas and emotional messages
Language	A collection of verbal symbols that often reflect the organization's particular culture

ceremonies Special events in which the members of an organization celebrate the myths, heroes, and symbols of their culture.

rite A ceremonial activity meant to communicate particular messages or accomplish specific purposes.

AN EXAMPLE: GOODYEAR
Goodyear reinforced the values of company loyalty and years of service by presenting each 25-year employee a mantle clock engraved with Goodyear, the employee's name and length of service. After the sale of Goodyear Aerospace, many employees, approaching the 25-year mark, voiced concern over whether the clock would be given and what company name would appear on it.

ritual A ceremonial event that occurs repeatedly and continues to reinforce key norms and values.

Ceremonies, Rites, and Rituals. Ceremonies are special events in which the members of a company celebrate the myths, heros, and symbols of their culture.[18] Ceremonies thus exemplify and reinforce important cultural norms and values. In sales organizations like Mary Kay or Amway, annual ceremonies are held to recognize and reward outstanding sales representatives. Part of the reason for holding these ceremonies is to inspire sales representatives who have been less effective to adopt the norms and values of their successful colleagues. In personifying the "Mary Kay approach" and the "Amway philosophy," the people who are recognized and rewarded in these ceremonies greatly enhance the attractiveness of their companies' special philosophies.

Often, organizational ceremonies incorporate various **rites**—ceremonial activities meant to send particular messages or accomplish specific purposes. Let's look next at some of the most common kinds of organizational rites.[19]

Rites of passage are used to initiate new members and can convey important aspects of the culture to them. An elaborate example of this phenomenon is military boot camp. The entire experience is designed to inculcate a particular military culture into recruits. In some business firms, new recruits are required to spend considerable time talking with veteran employees and learning about cultural norms and values by listening to stories about their experiences at work. In other companies, however, the rite of passage is merely a brief talk about company rules and regulations delivered by a human resources staff member to newcomers during their first day at work. It is little more than a formal welcoming and doesn't really help newcomers learn about the culture of the firm.

When employees are transferred, demoted, or fired because of low productivity, incompatible values, or other personal failings, *rites of degradation* may draw the attention of others to the limits of acceptable behavior. Today, rites of degradation are generally deemphasized, involving little more than quiet reassignment, but they have on occasion been quite dramatic. In the early days of NCR, executives would learn that they had lost their jobs by discovering their desks burning on the lawn in front of corporate headquarters.

Rites of enhancement also emphasize the limits of appropriate behavior but in a positive way. They recognize increasing status or position in a firm and may range from simple promotion announcements to intricate recognition ceremonies, such as the Mary Kay and Amway ceremonies we have already described.

In *rites of integration*, members of an organization become aware of the common feelings that bond them together. Often in rites of this sort, official titles and hierarchical differences are intentionally ignored so that members can get to know each other as people rather than as managers, staff specialists, clerks, or laborers. At Tandem Computer, for example, a Friday "TGIF" is held each week, giving employees the opportunity to chat informally over pizza and drinks. Company picnics, golf outings, softball games, and holiday parties can also serve as rites of integration.

A rite that is repeated on a regular basis becomes a **ritual**, a ceremonial event that continually reinforces key norms and values. The morning coffee break is a ritual that strengthens important workplace relationships. So, too, is the annual stockholder meeting held by management to convey cultural norms and values to company shareholders. Just as routine coffee breaks enable coworkers to gossip among themselves and reaffirm important interpersonal relationships, annual stockholder meetings give the company the opportunity to strengthen connections between it and people who would otherwise have little more than a limited financial interest in its continued well-being.

---

[18] Deal and Kennedy, *Corporate Cultures*, p. 63.
[19] Janice M. Beyer and Harrison M. Trice, "How an Organization's Rites Reveal Its Culture," *Organizational Dynamics* 15 (1987), 3–21.

CEO Robert Paluck believes that Convex Computer's annual summer picnic inspires enthusiasm among his employees. For this rite of integration in 1990 the company had 200 tons of sand trucked in to its headquarters in Richardson, Texas, and, at the suggestion of the company's 1200 employees, filled a pool with 72 gallons of iced raspberry Jell-O. Five vice presidents followed Paluck himself in splashing down into the goo. Paluck uses other rites and rituals to reinforce team spirit and feels it pays off. In 1989, the company's earnings on its supercomputers, which sell for up to $2.5 million each, rose by 93%. Source: Mark M. Colodny, "High-Tech CEO Splashes Down," Fortune, August 27, 1990, p. 104.

story An account of past events that all employees are familiar with and that serves as a reminder of cultural understandings.

Stories and Myths. Stories are accounts of past events that all employees are familiar with and that serve as reminders of cultural values.

> Bill Hewlett and Dave Packard are "legends in their own time" to the employees of Hewlett-Packard. New employees learn from a slide presentation shown when they first arrive that "Bill and Dave" started the company in Bill's garage and made some of the first products using the Hewlett kitchen oven. They hear informally from many employees stories about how Bill and Dave expect employees to address them by their first names. Stories emphasize and legitimate the management philosophy to avoid long-term debt and . . . layoffs. Stories also help define, in a way mere statements can't, what the "HP way" is.[20]

As organization members tell stories and think about the messages the stories convey, concrete examples facilitate their later recall of the concepts presented. It is easier, for instance, to remember that Bill and Dave want to be called by their first names than to memorize an abstract rule stating that undue formality is discouraged at Hewlett-Packard. Stories also provide information about historical events in the development of a company that can improve employees' understanding of the present.

> In one organization, employees tell a story about how the company avoided a mass layoff when almost every other company in the industry . . . felt forced to lay off employees in large numbers. The company . . . managed to avoid a layoff of 10% of their employees by having everyone in the company take a 10% cut in salary and come to work only 9 out of 10 days. This company experience is thus called the "nine day fortnight."[21]

The story of the nine-day fortnight vividly captures the cultural value that looking after employees' well-being is the right thing to do. Present-day employees continue to tell the story among themselves as a reminder that the company will avoid layoffs as much as possible during economic downturns.

myth A story that provides a fictional but plausible explanation for something that might otherwise seem puzzling.

A **myth** is a special type of story that provides a fictional but likely explanation for an event or thing that might otherwise seem puzzling or mysterious. Ancient civilizations often invented myths involving gods and other supernatural forces

[20] Alan L. Wilkins, "Organizational Stories as Symbols Which Control the Organization," in *Organizational Symbolism*, ed. Louis R. Pondy, Peter J. Frost, Gareth Morgan, and Thomas C. Dandridge (Greenwich, Conn.: JAI Press, 1983), pp. 81–92.

[21] Ibid.

to explain natural occurrences such as the rising and setting of the sun, the phases of the moon, and the formation of thunderstorms. Similarly, the members of an organization sometimes develop fictionalized accounts of the company's founders, origins, or historical development to provide a framework for explaining current activities in their firm. In many instances, organizational myths actually contain at least a grain of truth. For example, myths told throughout General Motors about the management prowess of Alfred P. Sloan, one of the company's earliest chief executives, are based in part on a study of GM's structure and procedures that Sloan performed in 1919–20. It is this bit of truthful information that make myths sound completely true.

Heroes and Superstars. Heroes are people who embody the values of an organization and its culture. They serve as role models, illustrating personal performance that is not only desirable but attainable.

heroes People who embody the values of an organization's culture and serve as role models for other members in the organization.

> Richard A. Drew, a banjo-playing college dropout working in 3M's research lab during the 1920s, [helped] some colleagues solve a problem they had with masking tape. Soon thereafter, DuPont came out with cellophane. Drew decided he could go DuPont one better and coated the cellophane with a colorless adhesive to bind things together—and Scotch tape was born. In the 3M tradition, Drew carried the ball himself by managing the development and initial production of his invention. Moving up through the ranks, he went on to become technical director of the company and showed other employees just how they could succeed in similar fashion at 3M.[22]

The deeds of heroes are out of the ordinary but not so far out as to be beyond the capabilities of other employees. In contrast, organizational **superstars** are so extraordinary that they rise above their peers and sometimes even become symbols for an entire industry. For example, in the early 1900s, Thomas Edison personified the then astounding advance of commercial electricity. More recently, Drexel Burnham Lambert's Michael Milken has become a symbol—a negative one—of the excesses of the investment industry of the late 1980s.

superstars Cultural figures who are so extraordinary that they rise above their peers and sometimes even become symbols for an entire industry.

Symbols and Language. Symbols are objects, actions, or events to which people have assigned special meanings. Company logos, flags, and trade names are symbols that come readily to mind. Mercedes's three-point star logo is synonymous with quality in most people's minds, and even the youngest children know that the McDonald's golden arches mark the locations of certain fast-food restaurants. In organizations, symbols may also include official titles, such as chief operating officer. Or special eating facilities, official automobiles, or airplanes may be given symbolic status. Sometimes even the size of an employee's office or its placement or furnishings have special symbolic value.[23]

symbol An object, action, or event to which people have assigned special meaning.

Symbols mean more than might seem immediately apparent. For instance, despite the fact that a reserved parking space is just a few square feet of asphalt, it may symbolize its holder's superior hierarchical status or clout. It is the ability to convey a complex message in the efficient, economical manner that makes symbols so useful and important.

> When two people shake hands, the action symbolizes their coming together. The handshake may also be rich in other kinds of symbolic significance. Between free-masons it reaffirms a bond of brotherhood, and loyalty to the order to which they belong. Between politicians it is often used to symbolize an intention to cooperate and work together. To members of the counter-culture of the 1960s and early 1970s, their special hand clasp and a cry of "Right On!" affirmed a set

[22] Deal and Kennedy, *Corporate Cultures*, pp. 40–41.

[23] Jeffrey Pfeffer, *Power in Organizations* (Marshfield, Mass.: Pitman, 1981), p. 50.

of divergent values and opposition to the system. The handshake is more than just a shaking of hands. It symbolizes a particular kind of relationship between those involved.[24]

Clearly, we need symbols. They are able to convey complex ideas in a simple manner, and they enable people to convey emotional messages that cannot easily be put into words. Without symbols, many of the fundamental norms and values of an organization's culture could not be communicated among the members.

Language, too, is a means for sharing cultural ideas and understandings. In many organizations, the language members use is itself a reflection of the organization's particular culture. At Microsoft, for example, a young-techie vocabulary has developed, largely because of the youth of the firms' founder, Bill Gates (mid-30s) and its work force (median age, about 31). A confusing situation is called random. Bandwidth refers to the amount of information one can absorb. Things that go right are labeled radical, cool, or super.[25] Whatever the source of a common vocabulary, the fact that such a vocabulary exists attests to the presence and acceptance of a shared set of norms and values.

language A system of shared symbols that the members of an organization use to communicate cultural ideas and understandings.

Managing Organizational Culture

APPLYING THE DIAGNOSTIC MODEL
Diagnostic Question 6: Does the culture help hold the organization together in a way that supports the formal organization? Does it provide social information that is consistent with the firm's purpose, strategic direction, and general well-being?

social information Information growing out of cultural norms, values, and shared opinions that shapes the way people perceive themselves, their jobs, and the organization.

Organizational culture grows out of the informal organization of unofficial ways of doing things. It influences the formal organization by shaping the way employees perceive and react to formally defined jobs and structural arrangements. Put another way, cultural norms and values provide **social information** that helps employees determine the meaning of their work and the organization around them. For example, in a company that promotes the "Protestant work ethic"— the idea that working hard is the way to get ahead in life—employees are led to view their jobs as critical to personal success and therefore as important, interesting, challenging, and in other ways worthwhile. By encouraging employees to perceive success as something to be valued and pursued, these norms also encourage the development of a need for achievement (see Chapter 7) and motivate hard work and high productivity. As Figure 18-5 indicates, cultural norms and

[24] Gareth Morgan, Peter J. Frost, and Louis R. Pondy, "Organizational Symbolism," in *Organizational Symbolism*, pp. 3–38.

[25] Richard Brandt, "The Billion-Dollar Whiz Kid," *Business Week*, April 13, 1987, pp. 68–76.

F I G U R E 18-5

Cultural Elements as Social Information

Cultural norms and values are a source of basic social information that influences how members perceive the formal organization, their own needs and interests, and their work. Work behaviors that are based on these perceptions can stimulate cultural change.
Based on Gerald R. Salancik and Jeffrey Pfeffer, "A Social Information Processing Approach to Job Attitudes and Task Design," Administrative Science Quarterly *23 (1978), 224–53.*

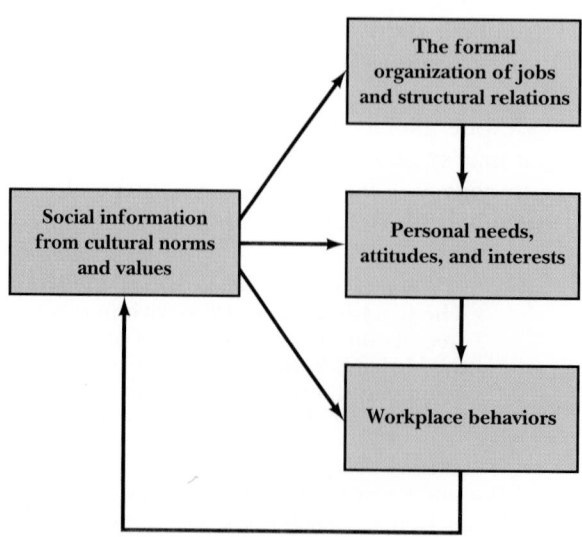

values convey social information that can influence the way people choose to behave on the job. They do so by affecting the way employees perceive themselves, their work, and the organization.

Can organizational culture be managed? You might be tempted to answer no to this question for the following reasons:

1. Cultures are so spontaneous, elusive, and hidden that they cannot be accurately diagnosed or intentionally changed.
2. Considerable experience and deep personal insight are required to truly understand an organization's culture, making cultural management infeasible in most instances.
3. There may be several subcultures in a single organizational culture, complicating the task of managing organizational culture to the point where it becomes impossible to accomplish.
4. Cultures provide organization members with continuity and stability. Therefore, members are likely to resist even modest efforts at cultural management or change because of concerns about discontinuity and instability.[26]

Most management experts disagree with these arguments, however, and suggest that organizational cultures can be managed by using the two approaches we discuss next.

symbolic management A process in which managers attempt to influence deep cultural norms and values by shaping the surface cultural elements that organization members use to express and transmit cultural understandings.

Symbolic Management. In **symbolic management**, managers attempt to influence deep cultural norms and values by shaping the surface cultural elements—such as symbols, stories, and ceremonies—that people use to express and transmit cultural understandings.[27] Managers can accomplish shaping of this sort in a number of ways. They can issue public statements about their vision for the future of the company. They can recount stories about themselves and the company. They can use and enrich the shared company language. In this way, managers not only communicate the company's central norms and key values but devise new ways of expressing them.

Managers who practice symbolic management realize that every managerial behavior broadcasts a symbolic message to employees about the norms and values of the organization. They consciously choose to do specific things that will symbolize and strengthen a desirable culture. For example, deciding to promote from within and avoid hiring people from outside the firm sends employees the message that strong performance is rewarded by career advancement. This message reinforces cultural norms and values that favor hard work. Filling positions by hiring from other organizations gives precisely the opposite message—hard work may *not* be rewarded by promotion—and undermines cultural norms and values that suggest otherwise.

The fact that symbolic management involves the manipulation of symbols is apt to lead some managers to play down its importance. Telling stories, performing ceremonies, and anointing heros might seem soft-headed or a waste of time to managers who do not understand the importance of managing cultural understandings. However, underestimating the importance of symbolic management can have disastrous consequences. A good illustration was the failure by Kraft Foods to identify and manage cultural inconsistencies between itself and Celestial Seasonings, a producer of specialty teas that it acquired in the mid-1980s

[26] John B. Miner, *Organizational Behavior: Performance and Productivity* (New York: Random House, 1988), p. 571. See also Harrison M. Trice and Janice M. Beyer, "Using Six Organizational Rites to Change Culture," in *Gaining Control of the Corporate Culture*, ed. R. H. Kilmann, M. J. Saxon, and R. Serpa (San Francisco: Jossey-Bass, 1985), pp. 370–99.

[27] Jeffrey Pfeffer, "Management as Symbolic Action: The Creation and Maintenance of Organizational Paradigms," in *Research in Organizational Behavior*, vol. 3, ed. L. L. Cummings and B. M. Staw (Greenwich, Conn.: JAI Press, 1981), pp. 1–52.

Celestial Seasonings Wasn't Kraft's Cup of Tea

Bicycling is an article of faith at Celestial Seasonings, Inc., the herbal tea company based in Boulder, Colorado. The sport appeals to the same sort of health-conscious consumers who are likely to be herbal tea drinkers. Celestial thus hosts a top pro cycle race, the Red Zinger, named after one of its teas. To drive home the company's feelings, President Barnet M. Feinblum often pedals a racer of his own.

No wonder trouble started brewing when Celestial's corporate owner, Kraft, Inc., ordered the company to cut its ties with cycling. Instead, Feinblum signed a deal with bicycling legend Greg LeMond to ride on Celestial's team. Relations between Kraft's board-room executives and the more casual Feinblum deteriorated further when Kraft, responding to anonymous letters charging drug use at Celestial, slipped an undercover agent into the company.

Such culture clashes helped push Kraft to put Celestial up for sale. Certainly the tea producer was quirky to the point of having trouble fitting into any large, bureaucratic firm. However, Kraft, too, was at fault. In its effort to create a mass market for herbal teas, Kraft lost sight of Celestial's wholesome image and frolicsome spirit. Celestial will certainly miss Kraft's deep pockets: New-product development and marketing may prove increasingly difficult without Kraft's resources. However, Feinblum believes he wrested his company away from Kraft in the nick of time. On the eve of Celestial's parting, Philip Morris bought Kraft. "It was bad enough being part of Velveeta's company," says Feinblum. "Can you imagine Celestial Seasonings being owned by a tobacco company?"*

* Based on Sandra D. Atchison, "Why Celestial Seasonings Wasn't Kraft's Cup of Tea," *Business Week*, May 8, 1989, p. 76.

(see the "In Practice" box). Managers at companies ranging from Disney to DuPont agree that managing symbols is a critical part of their job.[28]

Organization Development. Another way of managing the culture of an organization is to use organization-development interventions like those discussed in Chapter 14 and also later in this chapter. OD interventions can contribute to cultural management by helping the members of an organization progress through the following steps:

1. *Identifying current norms and values.* OD interventions typically require people to list the norms and values that guide their attitudes and behaviors at work. This kind of list gives members insight into the organization culture.

2. *Plotting new directions.* OD interventions often make it possible for the members of an organization to evaluate present personal, group, and organization goals and consider whether these goals represent the objectives they really want to achieve. Evaluation of this sort often points out the need to plot new directions.

3. *Identifying new norms and values.* Those OD interventions that stimulate think-

[28] Brian Dumaine, "Creating a New Company Culture," *Fortune*, January 15, 1990, 127–31.

ing about new directions also provide organization members with the opportunity to develop new norms and values that will promote a move toward new goals.

4. *Identifying culture gaps.* To the extent that current (step 1) and desired (step 3) norms and values are articulated, the OD process enables organization members to identify as culture gaps the differences between the current and the desired.

5. *Closing culture gaps.* OD gives people the opportunity to forge agreements that new norms and values replace old ones and that every employee will take responsibility for managing and reinforcing these changes.[29]

When people engage in behaviors that are consistent with the new norms and values developed in an OD intervention, they reduce culture gaps and, in effect, change the organization's culture.

CHANGING THE ORGANIZATIONAL CONTEXT: ORGANIZATION DEVELOPMENT II

Besides the interpersonal, group, and intergroup interventions discussed in Chapter 14, OD experts have also designed a variety of organization-level interventions. As you will recall from Chapter 14, the general purpose of any OD intervention is to provide a planned, systematic way of introducing and managing long-term change in organizations. The purpose of organization-level OD interventions is thus to introduce and manage planned change throughout the entire organization.

Organization-Level Change

APPLYING THE DIAGNOSTIC MODEL
Diagnostic Question 7: Do organization-wide problems suggest a need for organization-level OD? Would the information sharing of survey feedback provide adequate help? Do management problems seem to require the more intensive analysis of organizational confrontation? Or do problems with the organization's environment mandate the outward orientation of open system planning?

What sorts of organization-wide change might be stimulated by organization-level OD interventions? Four of the most important kinds of change are (1) shaping the informal organization, (2) redesigning the formal organization, (3) improving the fit between the informal and formal organizations, and (4) planning and implementing strategies.

Shaping Culture and the Informal Organization. Suppose you are David Roderick, chair of U.S. Steel during the firm's transformation into USX. You are the head of a company with a culture that stresses risk avoidance, stability, and originality. Changes underway in your company's business, however, make it imperative for your employees to adopt a culture of norms and values that favor creativity, innovation, and flexibility. What do you do?

Managers in this predicament often turn to organization-level OD interventions owing to the ability of these interventions to shape the informal organization. As is true for any OD intervention, the ones aimed at organization-wide change can help create an awareness of existing norms and values, facilitate evaluation and redirection of the organization and its culture, and encourage efforts to identify and eliminate cultural gaps. Compared to less-expansive OD techniques, organization-level interventions have the added advantage of stimulating change that pervades the entire organization. Such interventions are therefore especially useful when broad-based cultural change is the goal.

[29] Miner, *Organizational Behavior*, pp. 574–75. See also Ralph H. Kilmann, *Beyond the Quick Fix* (San Francisco: Jossey-Bass, 1984), pp. 105–23.

Redesigning Structures, Jobs, and the Formal Organization. Managers who must alter the formal organization to match contingency conditions face the daunting task of actually implementing structural change. Imagine, for instance, the job of dividing a centralized, functionally structured firm into a decentralized company of several divisions. That chore, you will recall from Chapter 15, was precisely what Roger Smith undertook at General Motors during the 1980s. How does one decide which specific structural relations to change and which to preserve? What jobs should be redesigned, and which approach to job design should be used? How can employees' views on structural change and job modification be collected and assessed? Is there any way to stimulate acceptance of such sweeping, eventful changes?

Organization-level interventions, like every other type of OD intervention, have a grounding in the developmental process described in Chapter 14. So they are of use to managers who want to initiate participatory diagnosis and problem-solving procedures. As we noted in Chapter 14, such procedures give employees input into the change process and stimulate employee acceptance of the changes. Consequently, organization-level OD interventions are quite useful in obtaining opinions from every part of an organization, involving all employees in design procedures, and obtaining total commitment to the structural changes and job modifications that result.

POINT TO STRESS
The formation of "transition teams" made up of employees from all parts of a business, has been suggested to facilitate change and encourage employee involvement (*Personnel Administrator*, 8/89, pp. 84–90).

Improving the Informal-Formal Fit. Sometimes the informal organization and its culture of norms and values support the formal organization of structural relations and jobs. You will find that kind of support, for instance, at Next Computer, where norms and values favoring creativity and flexibility flourish in an adaptable, organic organization structure. Frequently, though, the informal and formal organizations do not fit together that well. Recall that David Roderick faced both cultural rigidity and the need for structural flexibility at U.S. Steel in the early 1980s. What can be done in such cases to improve the degree of fit?

Here, too, organization-level OD provides a way of dealing with an important organizational problem. Through it, the company can make employees aware of the organization's formal structural arrangements, goals, and strategic direction. Such awareness can help them develop, and become committed to, a culture of norms and values consistent with the formal organization. Conversely, organization-level OD interventions can also improve the fit between the informal and formal organizations by calling attention to existing norms and values and strengthening employee commitment to supportive structural relations.

Planning and Implementing Strategies. Finally, organization-level OD interventions can play a major role in strategic planning and implementation. If they are used during strategic planning, their participatory nature can help inspire widespread commitment to the mission and strategic goals of the firm. In addition, they can prove helpful in initiating the development of norms and values that reinforce the strategy of the firm. Such interventions can also make the task of fitting the firm's structure and jobs to its strategy more manageable. OD does this by lowering resistance to change and encouraging employee involvement throughout the process of structural and job redesign.

Organization-Level OD Interventions

What specific organization-level OD interventions can be used for the purposes just described? To answer this question, we will now look at three OD interventions that differ in depth but have the same target of organization-wide change. They are survey feedback, organizational confrontation, and open system planning. Notably, these three interventions fit within the unfilled cells of the matrix of OD interventions shown in Table 14-3. To conclude our overview of organization-level OD, we will discuss how several interventions are sometimes combined to form comprehensive programs intended to enhance the quality of working life.

survey feedback A shallow organization-level OD intervention intended to stimulate information sharing throughout the entire organization.

Survey Feedback. The main purpose of **Survey feedback** is to stimulate information sharing throughout the entire organization. Planning and implementing change are of secondary importance. It is, then, a relatively shallow organization-level intervention. The survey feedback procedure normally proceeds in four stages.[30] First, under the guidance of a trained change agent, top management engages in preliminary planning, deciding such questions as who should be surveyed and what questions should be asked. Other organization members may also participate in this first stage if their expertise or opinions are needed. Second, the change agent and his staff administer the survey questionnaire to all organization members. Depending on the kinds of questions to be asked and issues to be probed, any of the questionnaire items we have included in this book may be included in the survey questionnaire. Third, the change agent categorizes and summarizes the data. After presenting it to management he holds group meetings to let everyone who responded to the questionnaire know the results. Fourth, the groups that received the feedback information hold meetings to discuss the survey. The group leaders—perhaps a foreman or an assistant vice-president— take the groups through an interpretation of the data, helping them to diagnose the results and identify specific problems, make plans for constructive changes, and prepare to report on the data and proposed changes with groups at the next lower hierarchical level. The change agent usually acts as a process consultant during these discussions so as to ensure that all group members get to contribute their opinions.

Survey feedback, as you can see from Figure 18-6, is very different from the traditional questionnaire method of gathering information. In survey feed-

[30] Floyd C. Mann, "Studying and Creating Change," in *The Planning of Change*, ed. W. G. Bennis, K. D. Benne, and R. Chin (New York: Holt, Rinehart & Winston, 1961), pp. 605–13.

POINT TO STRESS
Often surveys are done, but their results are neither shared nor discussed, and no changes suggested by survey results are implemented. Such inaction can do more harm than good. Trust is destroyed when the expectation that up-to-date input will improve conditions is violated.

FIGURE 18-6

Two Approaches to Data Collection by Questionnaire

Survey feedback differs from more traditional uses of questionnaires in its emphasis on influence sharing and participatory decision making. During a survey feedback intervention, all organization members likely to be affected by subsequent changes are involved in providing and analyzing data. *Adapted with the publisher's permission from Wendell L. French and Cecil H. Bell, Jr.,* Organization Development: Behavioral Science Interventions for Organization Improvement, *4th ed. (Englewood Cliffs, N.J.: Prentice Hall, 1990), p. 170.*

	Traditional Approach	Survey Feedback or OD Approach
Data collected from:	Workers and maybe foreman	Everyone in the system or subsystem
Data reported to:	Top management, department heads, and perhaps to employees through newspaper	Everyone who participated
Implications of data are worked on by:	Top management (maybe)	Everyone in work teams, with workshops starting at the top (all superiors with their subordinates)
Third-party intervention strategy:	Design and administration of questionnaire, development of report	Obtaining concurrence on total strategy, design and administration of questionnaire, design of workshops, appropriate interventions in workshops
Action planning done by:	Top management only	Teams at all levels
Probable extent of change and improvement:	Low	High

back, not only are data collected from everyone, from the highest to the lowest level of the hierarchy, but everyone in the organization participates in analyzing the data and in planning appropriate actions. These key characteristics of survey feedback reflect OD's basic values, which stress the critical importance of participation as a means of encouraging commitment to the organization's goals and stimulating personal growth and development.

organizational confrontation An organization-level OD intervention of moderate depth that enables managers to assess the internal workings of the organization and plan corrective actions as needed.

Organizational Confrontation. Organizational confrontation is an organization-level OD intervention of moderate depth. Its purpose is to provide a setting in which the managers of a firm can assess the internal workings of the company and plan corrective actions as needed. To conduct organizational confrontation, a management meeting is called. The organization's top manager—sometimes with the help of a change agent—guides the group through the six-step process shown in Figure 18-7.[31]

In the first step, *climate setting*, the manager states the goals of the confrontation. She stresses the need for free and open discussion of issues and problems and pledges that participants will not be punished for voicing their opinions. The change agent, if included in the meeting, may give a brief presentation on the importance of open communication and cooperative problem solving.

[31] Richard Beckhard, "The Confrontation Meeting," *Harvard Business Review* 45 (1967), 149–55.

FIGURE 18-7

Steps in an Organization Confrontation Intervention

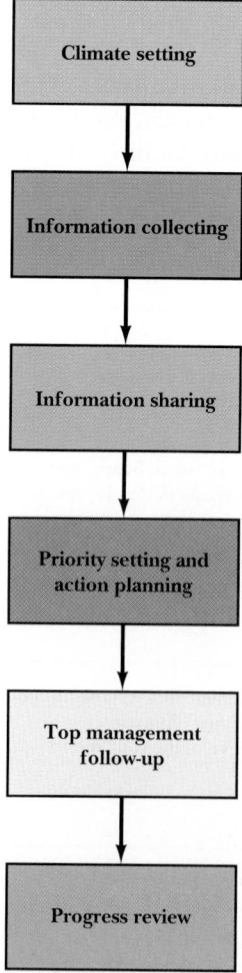

- Climate setting
- Information collecting
- Information sharing
- Priority setting and action planning
- Top management follow-up
- Progress review

The meeting then breaks up into small, heterogeneous groups of managers from different functional areas and hierarchical levels, reflecting a cross-section of problems and interests. In this *information-collecting* step, superiors are not put in the same group with direct subordinates, and top managers meet as a separate group. All groups are instructed as follows:

> Think of yourself as an individual with needs and goals. Also think as a person concerned about the total organization. What are the obstacles, "demotivators," poor procedures or policies, unclear goals, or poor attitudes that exist today? What different conditions, if any, would make the organization more effective and make life in the organization better?[32]

The groups work on this task for about an hour, and in each group a spokesperson records the results of the discussions.

In the third step, of *information sharing*, the groups reassemble into the larger group, and each spokesperson reports his group's findings to the meeting. As they are presented, these findings are listed on a blackboard, flipchart, or similar display. After everyone has completed his presentation, the resulting list of items is categorized into a few major topics. These topics reflect different types of problems (communication, motivation problems, job design problems), relationships (union-management relations, hierarchical connections, coworker affiliations), and functional matters (personnel management issues, accounting problems, marketing matters).

To get started on *priority setting and action planning*, the fourth step, the meeting coordinator reviews the list of categories and items. Then participants form functional work teams reflecting the way they are normally organized, for example, accountants in one group and salespeople in another. Each group is headed by its manager, and all groups are asked to do four things: (1) identify and prioritize problems that seem related to work in the group's functional area, (2) plan preliminary action steps to solve these problems, (3) identify any remaining problems and priority issues that require top management's attention, and (4) determine how to communicate the results of the confrontation meeting to their subordinates.

In the fifth step, *top management follow-up*, the top-management group meets alone to review the action plans developed in the fourth step by the functional teams and determine what additional actions, if any, should be taken. The resulting follow-up action plan is communicated to all managers within a few days.

The sixth and last step is a *progress review*—an appraisal session in which all confrontation participants meet four to six weeks later to review progress and, where necessary, develop supplementary action plans. This step is crucial for top management because it must demonstrate its commitment to follow through on suggestions made during the confrontation meeting.

The six steps of organizational confrontation provide a forum allowing the quick, accurate diagnosis of organizational problems and the development of constructive solutions. The intervention enhances hierarchical communication, increases involvement, and encourages managerial commitment throughout the organization.

open system planning A deep organization-level intervention that helps the members of an organization devise ways to accomplish the mission of their firm in light of the demands of environmental constituency groups.

Open System Planning. Open system planning is a fairly deep organization-level intervention. It is distinguished by its focus on the organization as a system open to its surrounding environment. That is, it sees the organization as a configuration of work processes that depend for their good function on external situations and events that impinge on them. The primary purpose of open system

[32] Beckhard, "Confrontation Meeting," p. 154.

planning is to help the members of an organization devise ways to accomplish the mission of their firm in the light of demands and constraints that originate with constituency groups in the organization's environment. These groups include raw-material suppliers, potential employees, and customers (see Chapter 15). The intervention consists of the following five steps (see Figure 18-8):

1. *Identification of core mission or purpose.* The members of the organization meet and, through open discussion, define the firm's basic goals, purpose, and reason for being.

2. *Identification of important constituency groups.* Then participants identify the environmental constituencies that can affect the firm's ability to accomplish its goals and purpose.

3. *"Is" and "ought" planning.* Next, participants describe current relations between the organization and its constituencies. They consider each constituency separately, focusing on the importance and duration of the relationship. Other factors are the frequency with which the parties are in contact and the organization's ability to sense and react to changes in the constituency group. Then participants determine how satisfactory the relationship *is* to both organization and constituency. If this assessment uncovers deficiencies, participants then specify what the relationship *ought* to be to be satisfactory to both sides.

4. *Current responses to constituency groups.* Participants then assess the organization's current response to each constituency group by answering these questions: What does this constituency want from us? What are we currently doing in response to this demand? Is our current response moving us closer to where we want to be in relation to our company's goals and purpose?

5. *Action planning.* If the current situation is not what it ought to be, and if the organization's current response to its constituency groups is not adequate, participants face the final task of deciding how to redirect the firm's behavior. In planning corrective action, firm members usually consider these questions: What actions should be taken, and who should take them? What

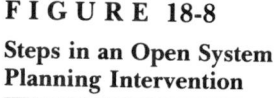

FIGURE 18-8

Steps in an Open System Planning Intervention

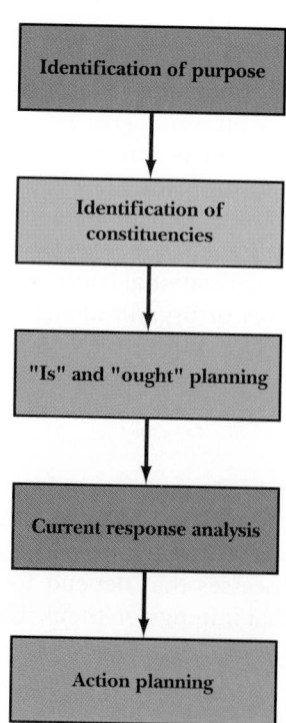

STRATEGY, CULTURE AND ORGANIZATION DEVELOPMENT II

resource allocations are necessary? What time table? When should each action start and finish? Who will prepare a progress report, and when will it be due? How will actions be evaluated to ensure progress in the desired direction?[33]

Unlike other OD interventions, open system planning directs primary attention to factors *outside* the organization that can influence organizational performance. It is especially useful during strategic planning, providing a structured yet participatory way to establish a firm's mission and set the strategic goals required to accomplish this mission. Open system planning can also help identify critical environmental contingencies during the process of organization design. Identifying them encourages the development of a better fit between an organization's structure and its environment.

quality of work life The degree to which work and membership in an organization facilitates the satisfaction of important personal needs and interests.

Quality-of-Work-Life Programs. Quality-of-work-life (QWL) programs often use one or more of the organization development techniques we have discussed to improve the "degree to which members of a work organization are able to satisfy important personal needs through their experiences in the organization."[34] In addition, QWL programs incorporate changes contributing to (1) adequate and fair compensation, (2) a safe and healthy work environment, (3) the immediate opportunity to use and develop human capacities, (4) future opportunities for continued growth and security, and (5) social integration in the work organization.[35] Such changes often involve the introduction of participatory decision making, payment according to skill development, and various work-environment improvements.

Improving the quality of working life became a major concern in the United States in the early 1970s in response to the publication of *Work in America*, a government report. It claimed that American workers were overwhelmingly dissatisfied with their jobs and employers. Although much of this report was subsequently discredited, it aroused considerable interest in improving the working lives of employees throughout North America.[36] It is this interest, combined with a growing concern about the competitiveness of American industry in the world marketplace, that continues to focus managerial attention on the importance of managing the organizational context in a way that enhances both productivity and satisfaction.

SUMMARY

Strategic management is a process of setting organizational goals and directing progress toward goal attainment. It consists of five steps: defining the mission and goals of the organization, performing a *situation analysis*, formulating a stra-

---

[33] William G. Dyer, *Strategies for Managing Change* (Reading, Mass.: Addison-Wesley, 1984), pp. 149–50.

[34] J. Lloyd Suttle, "Improving Life at Work—Problems and Perspectives," in *Improving Life at Work: Behavioral Science Approaches to Organizational Change*, J. Richard Hackman and J. Lloyd Suttle (Santa Monica, Calif.: Goodyear Press, 1976), p. 4.

[35] Richard E. Walton, "Quality of Working Life—What Is It?" *Sloan Management Review* 15 (1973), 11–21.

[36] James O'Toole, Elizabeth Hansot, William Herman, Neal Herrick, Elliot Liebow, Bruce Lusignan, Harold Richman, Harold Sheppard, Ben Stephansky, and James Wright, *Work in America: Report of a Special Task Force to the Secretary of Health, Education, and Welfare* (Cambridge, Mass.: MIT Press, 1973). For another point of view, see John B. Miner, *Theories of Organizational Behavior* (Hinsdale, Ill.: Dryden Press, 1980), p. 263.

tegic plan, implementing the plan, and evaluating the resulting performance. The direction pursued by the company is a function of three types of strategies: first a *corporate strategy* of *convergent growth*, *evolutionary growth*, *revolutionary growth*, or *adaptive decline*; second, one or more *business strategies* of overall cost leadership, differentiation, or focus; and third, various *functional strategies* that support the organization's corporate and business strategies. Strategic management influences the process of designing an organization's structure and jobs by influencing environmental, organizational, and technological contingencies. Thus strategic management is inseparable from the structure and design issues discussed in Chapters 15–17.

Besides the *formal organization* of structures and jobs, every firm also has an *informal organization* of deep-seated cultural norms and values as well as surface expressions of these norms and values. The latter include *ceremonies, rites, rituals, stories, myths, heroes, superstars, symbols*, and *special language*. A firm's *culture* is a cohesive force that also influences the way the firm's members perceive the formal organization, their behaviors, and themselves. *Symbolic management* and OD interventions can be used to manage the culture of an organization.

Organization-level OD interventions can also help manage change in the formal organization and integrate the formal and informal aspects of a firm. One organization-level intervention, *survey feedback*, is a questionnaire-based technique used to gather and analyze organization-wide opinions and data. Another intervention, *organizational confrontation*, helps management examine and improve the organization as a whole. *Open system planning*, a third technique, directs attention to the organization's environment and is especially helpful during strategic planning and structural contingency analyses. *Quality-of-working-life* programs integrate such techniques with the aim of improving employee development and satisfaction.

REVIEW QUESTIONS

1.   Why do managers perform strategic management? What would happen to a firm if the process of strategic management were not performed? What would happen if strategic plans were ignored? If the results of strategic implementation were not evaluated?

2.   What is the difference between a strength and an opportunity? Between a threat and a weakness? Why are both internal and external conditions assessed during a situation analysis?

3.   How does a company's corporate strategy affect the business strategies it pursues? What effects do its business strategies have on the firm's corporate strategy? How are the organization's business and functional strategies interrelated?

4.   As a manager, you face the task of reversing cultural norms that currently favor low performance. What do you do to accomplish this task? What role do the surface elements of culture play in your plan?

5.   How do cultural norms and values act as social information? What effects does this information have on organizational behavior? Why is it important for managers to take social information into account when designing jobs and structuring the organization?

6.   How are the strategies, structure and jobs, and culture of an organization interrelated? Explain how organization-level OD can be used to manage these interrelationships.

DIAGNOSTIC QUESTIONS

Managing the organization context—devising strategies, shaping cultures, and fitting the formal and informal organizations together—means asking a lot of questions and gathering a lot of data. The diagnostic questions that follow will get you started on this challenging task.

1. What are the organization's current strengths and weaknesses? What opportunities and threats are present in the environment? What might be done to capitalize on current strengths and opportunities? How might current weaknesses and threats be avoided or even converted into strengths and opportunities?

2. What business or businesses comprise the organization? If the organization owns more than one business, which should be kept and which should be sold? Should the firm grow by moving into new businesses? Should it instead consider adaptive decline? How should resources be distributed among the firm's businesses to ensure overall success?

3. For each business, what competitive forces—existing market rivals, potential new entrants, suppliers, sources of substitute products, customers—threaten continued success? Should the business pursue cost leadership? Differentiation? Focus?

4. What functional needs are stimulated by the firm's corporate and business strategies? What functional strategies are required to ensure that these needs can be satisfied?

5. What cultural norms and values guide behaviors and understandings in the firm? Do surface elements reinforce these deeper elements? Do differences between surface elements and cultural norms and values suggest ongoing cultural change? Is this change desirable?

6. Does the culture help hold the organization together in a way that supports the formal organization? Does it provide social information that is consistent with the firm's purpose, strategic direction, and general well-being?

7. Do organization-wide problems suggest a need for organization-level OD? Would the information sharing of survey feedback provide adequate help? Do management problems seem to require the more intensive analysis of organizational confrontation? Or do problems with the organization's environment mandate the outward orientation of open system planning?

EXERCISE 18-1
THINKING ABOUT ORGANIZATION CLIMATES*
BENJAMIN SCHNEIDER, *University of Maryland*

An *organization climate* is the general perception that people have of the organization in which they work. In this chapter we have defined organizational culture as the collection of norms and values that shape the behavior of people in an organization. Climate is the "feel" of the organization that grows out of its culture. The ebb and flow of everyday activities, events, practices, and procedures make up an organization's climate.

How do we assess the climate of an organization? We can have employees answer a series of questions about their personal impressions. Or we can have employees engage in a group discussion and keep track of what they say. You will have the opportunity to experience both of these approaches in this exercise as you examine the climate of your classroom and your educational institution.

STEP ONE: PRE-CLASS PREPARATION

To prepare for class, read the entire exercise. Next, thinking about your classroom as an organization, complete the Organization Climate Questionnaire shown in Exhibit 18-1.

STEP TWO: STRATEGIC PLANNING IN GROUPS

The class should divide into groups of 4 to 6 members each (if you have formed permanent groups, reassemble in those groups). In each group, members should share their responses to each questionnaire item and arrive at a group consensus. You should note significant disagreements, and the group should try to determine the reasons for such disagreements. After the group has reached agreement on its answers, members should examine any gaps the group has uncovered between the way things actually occur in the classroom and the way they would occur in an ideal situation. For each gap, the group should develop a strategy that could reduce the difference between actual and ideal circumstances. Each strategy should specify the actions to take, who should take them and when, and how to identify successful change. A spokesperson should be appointed to

report to the class on the group's discussion and strategic plan.

STEP THREE: ACTION PLANNING

Group spokespersons should report the results of Step Two to the class. As presentations are made, the class should look for similarities in the gaps identified by different groups and in the strategies developed to deal with them. After all spokespersons have completed their reports the class should develop an action plan for implementing the strategies it feels are the most likely to have a positive impact on the classroom climate. The instructor should list relevant action steps on a blackboard, overhead transparency, or flipchart, and specific class members should be assigned responsibility for overseeing their implementation.

STEP FOUR: GROUP DISCUSSION

The class should break into the same small groups. Each group's members should then describe the "dos" and "don'ts" that create the climate experienced by students at your school. It helps to recall your days as a new student and to think about what you had to learn to "fit in" to the activities around you. A spokesperson should prepare a brief presentation to be given to the class.

STEP FIVE: CLASS DISCUSSION

Spokespersons should present their reports. The class should then develop a comprehensive listing of the cultural norms it has identified and consider the following questions:

1. In what ways has the climate of your educational institution influenced the climate of your classroom? What aspects of your classroom's climate

* Adapted with the author's permission. The questionnaire appearing in this exercise was developed by Benjamin Schneider and C. J. Bartlett and is reprinted with permission. Those interested in a version useful for research should contact Benjamin Schneider, Department of Psychology, University of Maryland, College Park, MD 20742.

EXHIBIT 18-1
Organization Climate Questionnaire

Using the spaces in the left-hand column, describe the ideal practices and procedures you would like to see in this organization (your classroom). Next, using the spaces in the right-hand column, describe what you believe actually happens now in this organization. To record your answers, use the following scale: 1 = almost never, 2 = infrequently, 3 = sometimes, 4 = frequently, 5 = very frequently.

Ideal   Actual

_____   1. This organization takes care of the people who work in it.   _____

_____   2. Members keep up with current events outside the organization.   _____

_____   3. People in this organization ask each other how they are doing in reaching their goals.   _____

_____   4. Management effectively balances people problems and production problems.   _____

_____   5. There are definite "in" and "out" groups in the organization.   _____

_____   6. This organization encourages employees to exercise their own initiative.   _____

_____   7. This organization takes an active interest in the progress of its members.   _____

_____   8. Members of this organization have a wide range of interests.   _____

_____   9. More experienced members of this organization take time to help newer members.   _____

_____   10. Neither people problems nor production problems receive undue management attention.   _____

_____   11. Members of this organization always have complaints no matter what is done to correct them.   _____

_____   12. This organization willingly accepts members' ideas for change.   _____

_____   13. This organization recognizes that its life depends upon its members.   _____

_____   14. Members keep themselves informed on many topics besides their immediate job-related activities.   _____

_____   15. People in this organization speak openly about each other's shortcomings.   _____

_____   16. There is a sense of purpose and direction in this organization.   _____

_____   17. Members are prone to overstate and exaggerate their accomplishments.   _____

_____   18. Management does not exercise unnecessary control over members' activities.   _____

To complete pre-class preparation, score your responses using the following guide:

SCALE	ADD ITEMS	IDEAL TOTAL	ACTUAL TOTAL	GAP (IDEAL—ACTUAL)
Support	1, 7, 13	_____	_____	_____
Quality	2, 8, 14	_____	_____	_____
Openness	3, 9, 15	_____	_____	_____
Leadership	4, 10, 16	_____	_____	_____
Conflict	5, 11, 17	_____	_____	_____
Autonomy	6, 12, 18	_____	_____	_____

seem unaffected by the surrounding institution? Can you think of any ways in which your classroom's climate might affect the climate of your school?

2. How easy would it be to implement the action plan you developed in Step Three? How likely is it that your plan would have lasting effects on the climate of your classroom? What could you do to increase this likelihood?

3. Suppose you were asked to develop an action plan to change the climate of your educational institution. What additional complications would you have to deal with? How does the size of the organization affect the ability of its management to identify and modify features of its climate?

4. How are an organization's climate and culture interrelated? How do the two differ? How might managing the climate of an organization affect the organization's culture? How might managing the culture affect the climate?

CONCLUSION

Climate is an umbrella concept that summarizes numerous detailed perceptions in a small number of general dimensions. It describes how people think and feel about the organizations around them, and how they react to the norms and beliefs they encounter in these organizations. As you have probably discovered in this exercise, climates are difficult to measure—people in an organization often differ in their perceptions of what the organization is all about—and hard to change. Nevertheless, managers need to be aware of an organization's climate and must be able to intercede if it appears likely to threaten the effectiveness of the firm.

DIAGNOSING ORGANIZATIONAL BEHAVIOR

CASE 18-1
HARVARD UNIVERSITY STAFF SURVEY*
HILLARY BALLANTYNE, *Boston University*
FRED K. FOULKES, *Boston University*

INTRODUCTION

Daniel D. Cantor was in his ninth year as Director of Personnel for Harvard University. Cantor had joined the university in 1976 after many years of personnel work in industry and in the Peace Corps.

Cantor viewed the decade of his tenure as an eventful and productive one for the personnel function. Of the programs enacted under his leadership, he thought the recently conducted 1985 attitude survey was one of his significant accomplishments. Although attitude surveys were common tools of the personnel trade in corporations, among universities only Stanford was known to have conducted a staff survey when Cantor had introduced the idea in 1984. In addition, Harvard had been experiencing a unionization drive, which subjected any move on the administration's part to scrutiny. As Cantor faced the difficult administrative task of collecting, assimilating and acting on the results of the attitude survey in an organization that was highly decentralized, he, nevertheless, was optimistic about the work that needed to be done in the weeks and months ahead.

STRUCTURE OF HARVARD UNIVERSITY

Founded sixteen years after the arrival of the Pilgrims at Plymouth, Harvard University had grown from twelve students with a single master to an enrollment of some 16,000 degree candidates. This included students in the undergraduate college, in ten graduate and professional schools, and in an extension school, taught by a faculty of over 3,000.

The university had two governing boards. The Harvard Corporation, consisting of the president and Fellows of Harvard College, was the university's executive board. This seven-member board was responsible for the day-to-day management of the university's finances and business affairs. Significant matters of educational and institutional policy were also brought before the president and Fellows.

The Board of Overseers consisted of thirty members who were elected at large by graduates of Harvard. Through standing and visiting committees, the overseers learned about educational and administrative policies and practices of the university, provided advice to the corporation, and approved important actions of that body. Both the corporation and overseers had to approve major teaching and administrative appointments.

The expression "every tub on its own bottom" was often used to describe the decentralized organization and financial arrangement of the ten faculties overseeing Harvard's separate schools and colleges. Each faculty was headed by a dean, appointed by the president,

* Copyright © 1988 by the Human Resources Policy Institute, School of Management, Boston University. Reprinted with permission.

and approved by the board of overseers and Harvard Corporation. Each was directly responsible for its own academic programs, finances and organization. President Derek Bok directly controlled approximately 10% of the university's $700 million annual budget. Reporting to President Bok were five vice presidents (Administration, Finance, General Counsel, Development, and Government, Public and Community Affairs) and ten Deans.

Harvard's endowment and other funds were valued at more than $2.8 billion in 1985. Strong alumni support, through the recently completed Harvard Campaign, enabled the Faculty of Arts and Sciences, which is responsible for the education of more than half of Harvard's students, to renovate classroom buildings and the residential houses, strengthen the excellence of its faculty, and maintain its commitment to provide adequate financial aid for qualified students.

The university encompassed over 400 buildings spread across a radius of several miles. While most of Harvard was located in Cambridge, both the medically oriented schools and the business school were located in Boston. The business school was just across the Charles River from Harvard College, the undergraduate houses, and the Kennedy School of Government and the other schools and administrative offices. The schools of medicine, public health and dental medicine, however, were in Boston, approximately three miles from the main campus.

THE HARVARD STAFF

In 1985, Harvard employed nearly 12,000 people, 9,000 of whom held staff positions. The fifth largest employer in the Commonwealth of Massachusetts, Harvard employed 10% of the people working in the city of Cambridge, Massachusetts. The staff was occupied in a wide variety of jobs, from grounds maintenance to skilled laboratory work to clerical, library and administrative positions.

Fourteen hundred of Harvard's staff belonged to seven unions. The unionized employees had jobs in food and custodial services, skilled trades, security, the print shop and the cogeneration plant. Of the 7,500 non-union staff employees, about one quarter were occupied in unskilled or semi-skilled work. Another quarter made up the bulk of the technical, secretarial and clerical workforce that supported the teaching, research, professional and administrative functions at the university. Approximately 82% of these non-exempt employees were women. Roughly 3,500 exempt professional and administrative positions made up the rest of Harvard's staff. The exempt staff included all supervisors and managers who were not faculty members.

THE PERSONNEL FUNCTION

Harvard's central personnel function in 1985 consisted of six department heads and eighty staff, half of whom were professionals. Each of the schools at Harvard had its own personnel officer who reported to the administrative dean with "dotted line" responsibility to the central organization.

The relationships between central personnel and the school's personnel offices varied, according to Dan Cantor, from "close and comfortable to we-don't-need-you!" Benefits were administered entirely by central personnel, from distribution of information to medical form processing. Central personnel issued wage guidelines by job grade to the schools. Posting for all open positions at the university was also done centrally, although actual recruitment was done both centrally and by the schools themselves.

UNION ORGANIZATION CAMPAIGN

In 1984 Dan Cantor had been aware of the potential results to be gleaned from employee surveys, and for years had felt that such a project would be beneficial for Harvard. But two major obstacles existed: the union organizing campaign and the decentralized structure of the university. Cantor described these dilemmas:

> The idea of doing an attitude survey was long thought of. We were so spread out in terms of how we govern that there was no coordinated way of getting feedback from staff. The union organizing efforts impacted the project both positively and negatively. It increased concern about how people feel but also hindered a survey project that might have been construed as an unfair labor practice.

Organizing efforts among technical and clerical workers had been going on since the mid-seventies, focused on the three schools in the medical area. But staff members there had voted against bargaining units twice in the past eight years. In 1984 the National Labor Relations Board handed down a decision requiring the union to treat the entire Harvard technical and clerical force as a single potential bargaining unit, forcing the organizing to go campus-wide.

BACKGROUND OF HARVARD'S STAFF MEMBER SURVEY

Dan Cantor felt that the 1984 NLRB decision enlarging the bargaining unit to the entire university created an opportunity for conducting an attitude survey among staff. He introduced the idea at that time to the Personnel Policy Council, a group that consisted of nine administrative deans and three of Harvard's vice presidents. Although Stanford was the only university known to have surveyed its people, the council favored the idea. Cantor had suggested a 100% sample of one-third of the schools every three years. To take the project forward, during the summer of 1984, the council appointed a committee headed by Robert Scott, Vice President of Administration. Scott was seen as best suited to chair the committee because of his extensive administrative responsibilities and his knowledge of computers and data analysis.

The committee was composed of administrators and personnel officers representing a cross-section of the university. They met more than a half dozen times over five months beginning in the fall. Dan Cantor also consulted with faculty experts from the School of Education and the Business School about how best to proceed.

The committee received proposals for developing the survey from three consulting firms. The criteria used to make the selection were:

- The consultants' willingness and ability to understand the unique nature of Harvard's project
- The availability of a large database and the ability to use it
- Price
- Competence in developing and using data

The committee ultimately chose Opinion Research Corporation (ORC), a division of Arthur D. Little, to conduct the survey, although ORC's bid was not the lowest.

The allocation of the survey's cost was an issue because of Harvard's decentralized organization. The committee decided that the cost of the survey would be charged to each of the "tubs" on a per capita basis. The charge was higher for the units of the central organization than for the schools, but averaged under $10 per staff member. The decision was also made to survey the entire university rather than one-third at a time.

ORC developed a 24-page survey of over 100 items that varied in the number and type of possible responses. The main subjects of inquiry were compensation, performance evaluation, working conditions, career development and training, communications, and productivity. The survey was intended for the approximately 7,000 Harvard staff members, which excluded the faculty and members of bargaining units.

After the committee and other concerned individuals and groups had reviewed and approved the survey, it was mailed to employees at their homes in April 1985, seven weeks after the arrival of a letter to each employee from President Bok advising them of the survey and asking for their cooperation.

The questionnaire was sent to employees, with the covering letter signed by a vice president of ORC. All questionnaires were to be returned by April 19. So that ORC could analyze the results of the survey by various groupings of employees, respondents were asked to check off the Harvard unit where they worked; whether they were in academic/research, administration, or the library, the number of years they had worked for the university; the number of positions they had held at Harvard; their age, sex, and race; whether their immediate supervisor was a faculty member or a non-faculty member; their level of education; and whether their employment status was full or part time. Respondents were asked not to sign their names and they were assured that ". . . there are always enough people in any [employee] grouping so that no individual can be identified" and that if there were not 10 people in a group, the results of that group would not be released but instead would be combined with another group of employees.

Fifty-nine percent of the administrative/professional staff and 46% of the support staff completed the survey, which represented an overall response rate of 55% of Harvard staff members. During May and June ORC prepared a report that contained the principal results of the survey. ORC also prepared reports for each of Harvard's principal units. ORC delivered the results of the survey during the early part of the summer. Only President Bok and Messrs. Steiner and Cantor would see the complete results, including a comparison of responses by school.

With both President Bok and Mr. Steiner on vacation, Dan Cantor studied the results carefully. He was scheduled to go on vacation in a week, and the beginning of the fall term was just six weeks away. Cantor knew, however, that a well thought out action plan was needed by the end of August or, at the latest, immediately after the Labor Day weekend. Cantor also recalled that in President Bok's February 21 letter to each Harvard staff member, he had pledged that ". . . you will receive results from the survey, and an opportunity to discuss the results."

When you have read this case, look back at the chapter's diagnostic questions and choose the ones that apply to the case. Then use those questions with the ones that follow in your case analysis.

1. What type of organization development intervention does Harvard's employee attitude survey represent? Why did the university decide to undertake this survey? What benefits did Dan Cantor and the committee headed by Robert Scott expect the survey to yield? Do you agree with the committee's criteria for selecting the consultants who conducted the survey? Why or why not?

716

2. Do you think the committee will invite employees to participate in interpreting the results of the survey? Should they? What are the strengths of the survey approach? The weaknesses?

3. Who should be involved in deciding how to deal with problems identified by the survey? How participatory should this process be? Why?

CASE 18-2

DUMAS PUBLIC LIBRARY*

MARK HAMMER, *University of San Diego*
GARY WHITNEY, *University of San Deigo*

It came as a surprise when Jeff Mallet learned of the conflict between Debra Dickenson and Helen Hendricks because he knew them both personally and regarded them both as competent administrators. Debra Dickenson, 38, was the youngest mayor in the state when she was elected three years ago, and was the first female mayor in Kimball's history. She was widely recognized for her high levels of energy and dedication. Helen Hendricks, 62, had been the head librarian at Dumas Public Library for 15 years and was widely acknowledged among Kimball citizens as being primarily responsible for the high quality of the library services to the community.

Dumas Public Library serves the citizens of Kimball, New Mexico, a town of 20,000 people in rural Eastern New Mexico State. Kimball is dominated by the 16,000-student state university located there and this university presence creates a rather unique clientele for the public library. The library has enjoyed a history of solid citizen support and has until recently benefitted from cordial relations between the library staff and the city's administration.

The library is housed in a modern, air-conditioned structure with carpeted floors and attractive furnishings. Approximately 35,000 volumes are on the shelves. The 1988 budget, including payroll, acquisition of new books, and building maintenance, was $195,000.

The library has no formal organization. Helen Hendricks has reporting to her five fulltime employees, three of whom are professional librarians. Completing the staff are ten halftime permanent employees, ten to twelve unpaid volunteers, and an occasional intern from the university.

The city is governed by an elected city council and mayor. Day-to-day administration is the responsibility of Ralph Riesen, the City Supervisor, who is a permanent employee of the city.

Jeff Mallet, Professor of Management, first learned about the existence of strained relationships between the library and the city administration from Linda Turner, Adult Services Librarian. According to Linda, feelings of distrust and animosity toward City Hall had been growing recently among the library staff. Linda was concerned about the unhealthy climate that this hostility was creating at the library.

Several weeks later Jeff had an opportunity to talk with Debra Dickenson and Ralph Riesen. Jeff said he had heard that relations between City Hall and the library were not good. Debra and Ralph confirmed that relations between the two groups had reached an intolerably low level, and they agreed something would have to be done about it. Debra and Ralph expressed bewilderment about what could be done to improve the situation. "If you have any ideas or suggestions. I'd certainly like to hear them," Debra said.

Jeff suggested that it might prove helpful to have an outsider interview members of both groups to provide some independent perspective. He volunteered his services for this purpose. Debra and Ralph readily agreed to Jeff's offer.

The next day Jeff was talking to Paul Everest, a fellow business faculty member and consultant, about the situation at the library. Jeff invited Paul to join him on the case and Paul accepted.

Next week Jeff made a series of personal visits and phone calls to the key staff members from City Hall and the library. An agreement was reached to have Jeff and Paul interview both groups and make recommendations. Appointments were made for an interview with Debra Dickenson and Ralph Riesen at City Hall, followed by one with Helen Hendricks, Linda Turner, and Maude Richardson [Children's Librarian] at the library.

THE VIEW FROM CITY HALL

Debra: I'm really concerned about the way things have developed between us here at City Hall and the library staff. There is animosity between these two groups, and the situation has been worse over the past few months. There's not nearly the level of cooperation that there should be.

I'll be eager to consider any suggestions that you (professors) might have for how to improve the situation. I know that something has to be done, and I'm willing to devote some time and effort to working on it.

The problem at the library is that I no longer have administrative control over their operations. In the past the library has reported to the mayor through the city supervisor and that has worked reasonably well. Recently however, we discovered that legislation passed back in the 1930's makes it very clear that the Library Board of Trustees has the legal authority for the conduct of the day-to-day operations of the library.

My concern is that since the library is a part of the city administration, the city is legally responsible for its operations. I'm talking specifically about legal liability for such things as personnel selection, equal employment opportunity regulations, purchasing guidelines, and budgeting procedures set down by the state. In the case of lawsuits and budget overruns it seems clear to me that the city will be liable and hence we need to have administrative control over these matters. Also it just makes good common sense for us to coordinate certain administrative functions from City Hall, such as personnel selection and budgeting. Basically the library staff agrees with us on this, and we have been doing many of these functions at City Hall.

One of the things that irks me most about Helen Hendricks (Head Librarian) and her staff is that they continue to insist on politicizing the budget making process, even when they know or should know that this is an extremely disruptive and unfair practice. I have made it pretty clear to all the department heads within the city that the budget making process should be one where budget requests are submitted to the city administration and to the City Council along with the implications of funding increases or decreases. Based on that input, the City Council then decides on the services that it wants in a non-emotional manner. The City Council represents the citizens and that is a perfectly democratic procedure.

Prior to the recent budget preparation period the City Council gave budget directives to all city departments. The Library Board chose to ignore these directives and submitted their own budget. Subsequently the library staff started a big political campaign to pack the council chambers at all the budget hearings with patrons of the library and other citizens who supported the library's request for more funding.

I have tried to point out to Helen how disruptive and unfair this is. The fact is that almost every city department serves some consistency and could if they were so inclined rally citizen support from among their clients or constituents to bring political pressure to bear on the City Council and other members of the city administration to fund their individual projects. It seems obvious to me that this is a chaotic way to try to prepare a city budget. Special interest politics has no place in the preparation of the city budget which is fair to all parties concerned. Only people who have looked at the entire city budget and have considered the total revenues available to the city and the cost and benefit trade-offs made by each one of the city departments are in any position to judge whether any particular department is reasonably funded or not. The fact is that there are prime financial needs in all of the city departments and the library is not alone.

I support the library wholeheartedly; we all do. I'm just not one bit impressed when the librarians campaign to have a flock of citizens pack the council chambers to stand there and tell us that they support the library. That is not a helpful input to the budget making process. Everybody supports the library.

Following one occurrence of inappropriate political lobbying last fall, I expressed my annoyance to Walter Roy [chairperson of the Library Board of Trustees]. Subsequently Helen was told by the Board to cease her lobbying activities. I think she got the message, but I know the lobbying did not stop. That tells me that the Trustees do not have control over the library staff.

Don't get me wrong. Helen Hendricks has done a marvelous job down there at the library, but things just haven't been the same since her husband died unexpectedly two years ago. She seems to have retreated into a womb or something. I think she uses the library staff as a personal support group. I don't know who is running the library anymore, but it certainly isn't Helen. I think the staff is running the library to tell you the truth.

Ralph: I too have noticed the worsening relations between us and the library staff. Part of the problem may be the physical isolation of the library and the fact that they don't interact much with other city personnel. [The library is three blocks from City Hall.]

If you ask me I think there is a case of paranoia down there at the library. Some of them seem to believe that I'm out to get them. In fact, I have a definite feeling that several of the library staff members think that I'm some sort of an ogre.

I think many of the problems that the library staff think they have are more imaginary than real. I remember once I talked to Helen and she was complaining about some things. I asked her to make me a list of grievances that they had, ways in which they had less money or things that weren't satisfactory. Do you know, I've never gotten any list from Helen. I really don't think they have any substantial problems that aren't of their own making.

Debra: I get the impression that the library staff feels that they are picked on and mistreated. The fact is that the library has the best working conditions of almost any other department in the city. Not only are their working conditions congenial and agreeable, but the clientele they serve are all happy and supportive of the library. It's a totally positive environment. That's quite a bit different from the city engineer's department where they have to talk to irate contractors and home owners, or the police who have to deal with drug offenders and unhappy traffic violators.

I'm still very confused about the proper roles of

the library administration, the Library Board of Trustees, and the city administration.

Ralph: Lynn King [the city finance director] is another player in this scenario. Lynn probably has more interaction with the library staff on a day-to-day basis than anybody else here in City Hall. She deals with them on matters of auditing, purchasing procedures, and employee selection procedures. There have been disagreements and friction generated over a number of these issues. Lynn really distrusts Helen as an administrator.

Debra: I really would like to hear from the library staff on their perceptions of what our problems are. I don't really know what they think.

One of the areas that Helen and I have had disagreements about has been that of Helen's classification within the city administrative system. Helen seems to think that she should be classified as a department director. The trouble is that Helen's responsibilities are simply not equivalent to those of other department directors within the city. Each of the other directors has at least two major administrative functions reporting to him or her. For example, the Director of Public Safety has both Police and Fire reporting to him.

When we reorganized the city administration recently, we changed it so that Helen was reporting to the mayor through the director of public services, Jack Feldner. Helen got all bent out of shape that she wasn't reporting directly to the mayor and that she had to report through someone else. She made such a fuss about it that we finally agreed to her request and Ralph issued a memo of understanding to Helen to the effect that she still had direct access to us here at City Hall and that we would interact with her on a direct basis.

One of the City Council members introduced a proposal to classify Helen as a department head recently, but this proposal was withdrawn at my request. I'm afraid that as a result some people are getting the impression that I am not really supportive of the library. I really am, but my concern in this matter is with equity—all the other department directors have considerably more administrative responsibility than Helen does and they wouldn't consider it fair to have Helen classified as a department director.

Ralph: Helen keeps raising the issue of her salary level. I'm convinced that Helen is fairly paid in relation to other city employees. The trouble is that all city employees are underpaid compared to university salaries and we're *never* going to catch up. Dissatisfaction with pay is just one of those things that we have to accept and live with.

Despite what Helen says, I don't think salary is that big a problem. I remember from the supervision class that you (Jeff) taught that according to Herzberg, pay is a hygiene factor. I don't think that we're going to solve any big problems down at the library by working on hygiene factors.

Debra: An incident that happened recently will illustrate what I consider to be totally unprofessional conduct on the part of the library staff. As you know, I recently refused to reappoint Cecil Hockman to the Library Board of Trustees after his first term expired. Now as the mayor, I have the duty and obligation to the citizens of Kimball to appoint people to boards that I think are best qualified to do the jobs. I had my reasons for not reappointing Cecil; reasons which I consider to be good. Because we are making agreements with the Trustees about the administration of the library I want trustees who will work with us to try to reach a compromise. Cecil has never agreed to any compromise action and would stop library cooperative efforts.

What happened was that somebody down at the library called a reporter and told them about my refusal to reappoint Cecil Hockman. They apparently said that I had a vendetta going against Cecil and that a reporter should look into this. The reporter did check with Mr. Hockman and got a bunch of quotes from him concerning my nonsupport for library programs. Then the reporter called me and asked me if I wanted to respond to the charges. *I was furious.* I told her, "No, I do not want to respond." I did explain my duties and responsibilities as mayor to the reporter and she subsequently decided that there was no story.

Sometimes I feel like calling Helen up here on the carpet and telling her to shape up her act or get out. It becomes clearer to me all the time that, whatever else she is, Helen is not a competent administrator.

If the problems we're having with administration at the library can't be solved we are going to be forced to look at the issue of regionalization of this library, that is, having the city library join the county system along with the library in Morton. However, it is apparent to me that the idea of regionalization is extremely threatening to everybody down at the library. This showed up recently when the Capital Expenditures Committee recommended, among other things in its report to City Council, that the feasibility of regionalization of the city's library, cemetery, and health care facilities be studied. You wouldn't believe how upset the librarians became over that recommendation. They got a City Council member to make a motion that the recommendation be deleted from the Committee's report, and unfortunately it passed. The librarians clearly didn't even want the issue studied!

THE VIEW FROM THE LIBRARY

Helen: I'm surprised and delighted to hear you (professors) report that Debra Dickenson and Ralph Riesen are really interested in improving relations with us here at the library. I feel that we have been wasting a lot of time down here because of the poor relations we have with City Hall, and I wasn't at all sure how concerned they felt about it up there.

One of the main problems that I see between us and the city administration is their general resentment toward anything involving political pressure. I sense that Debra and Ralph get upset when the community voices opinions which are contrary to their views. I sometimes get the feeling that they would like to run the city without interference from citizens. However, that's the very nature of the political process. The mayor's job is inherently a political one. You shouldn't be in that position and expect to be immune from public pressure. So, I don't think it's appropriate that Debra gets upset when the citizens rally to support a program that they want.

During the recent budget hearings we have had lots of good people come to our defense. The Library Board of Trustees have been very supportive. The AAUW (American Association of University Women) has several members who have been strong supporters. These friends have been instrumental in helping us make the case to the mayor and the City Council that the community really supports a quality program here at the library.

Linda: We don't seem to have any problems of misunderstanding or nonsupport from either the Library of Board of Trustees or the City Council. I feel good about our relations with both of these groups. When we have gone to the City Council with our recommendations and proposals, they have been sympathetic and supportive. In the budget hearings both the Library Board and the City Council supported our proposed budget over the objections of Debra and Ralph. In effect, we bypassed the city administration and we came out better than if we had gone to them first, as they apparently wanted us to do.

One example of a way in which we have felt "under attack" by City Hall has been the way they have acted in regard to the appointment of members of the Library Board of Trustees.

Helen: That's right. You probably heard that just recently Debra refused to reappoint Cecil Hockman to the Board for a second term. Now Cecil has been a strong, energetic supporter of the library. He has given a great deal of his time and dedication to public service on the Library Board. Mr. Hockman's first term on the Board has just recently expired, and for no apparent reason Debra has declined to reappoint him, even though it has been customary in the past that members serve for two terms. So, Cecil Hockman is not only eligible for reappointment, but he has demonstrated in his first term that he is a dedicated and concerned public citizen.

It seems apparent to us that Debra resents anyone who supports the library as strongly as Cecil Hockman did. You see, Cecil initiated some legal research which determined that the Library Board of Trustees has the ultimate legislative authority for the administration of the library. Furthermore, Cecil Hockman took the initiative to argue our budget proposals before the City Council. Debra did not appreciate either of these, I am sure, and now it seems that she is out to get him.

In the past, I have always participated with the mayor when selecting candidates for the Library Board. The mayor has always been glad to have my input and opinion on which citizens would be good for the Library Board. None of that consultation has gone on between Debra and me recently; I just find out about her Board appointments by reading the newspaper.

Linda: Another way that we have felt attacked by the city administration has been the way we were treated in the recent reorganization of the city administrative hierarchy.

Helen: What they did was to demote the library by changing the reporting patterns so that instead of reporting directly to the city supervisor, I was directed to report through Jack Feldner, the director of public services.

This reassignment of the library was a serious downgrading of our status within the city. I was really upset when I learned that they expected me to report *through* Jack Feldner. Why, I have more education than Jack does. I have longer service to the City of Kimball than he does, and I supervise a *lot* of people here at the library. The very idea that the library with its staff of professionals should be considered subordinate to someone whose main concern is parks and recreation was an appalling idea to us over here. You see, that demotes us from one of the major functional units within the city administration to merely one of the concerns of the Parks and Recreation Department. I don't have anything against Jack Feldner, but I don't think it's right to have the city library subordinate to him and his department.

I was told that in the reorganization of the city administration I was not considered an administrator (Department Director level) because I supervise so few people. However, Lynn King [Finance Director] only supervises a few people, and she doesn't have the education I do either.

Maude: I don't think that they regard us as professionals over here, but we *are* professionals. Each one of us has had five years of college plus additional professional training, and yet we continually get treated as if we were mere clerks.

Linda: An incident which illustrates the library's diminished status was City Hall's insistence that Helen could not retain the title of "Library Director." The title

"Library Director" is common among librarians having similar jobs to Helen's. Among the staff here at the library, it seems the logical choice of position titles. And yet the city administration insisted that Helen could not be called a "Director." So they suggested that we call her the "Library Supervisor." Of course, "supervisor" denotes someone just above the clerical level; someone who is supervising a bunch of clerks. That seems natural to them, but the idea is appalling over here. We hassled back and forth over different possible titles for Helen's position and finally settled on "City Librarian." This title is less descriptive than "Library Director" and reflects Helen's lowered status in the city.

Maude: I don't think Helen is regarded as an administrator by the city administration. I don't think they really know how many people she has reporting to her, or how much leadership it takes to coordinate all the volunteer help we have. Helen has a substantial administrative job to keep this library running smoothly.

Helen: Going along with that is their resistance to paying me a salary reflecting my abilities and contribution. My salary is simply not in line with the requirements of this job, my education, and the experience I have with the City of Kimball. I know that I'm paid less than many other people in the city who have less education and less experience than I do. The city administration simply refuses to recognize the importance of my job.

Jeff: How would your salary compare, Helen, to other library directors having similar jobs around the state?

Helen: Well, I would have to say that my salary today reflects some very significant adjustments upward which were made during the 1960s. At that time the university was under heavy pressure to equalize the salaries of its female professionals, and the City of Kimball also upgraded their women's salaries at the same time. So I shared with some other women in some impressive gains during the 1960's.

If you looked just at the figures, my salary wouldn't look that far off relative to other city librarians. However, the figures don't reflect the quality of education I have received, the length of my service to the City of Kimball, and the contributions that I have made to the development of this library today.

Jeff: Could you give us an example or two of specific ways that the library's effectiveness has been impaired by the actions of members of the city administration?

Helen: Certainly. One good example would be the copier incident. That's a long story. Sometime ago we experienced an equipment failure with the copier which we had for patrons to use. Therefore, I asked permission from the Board of Trustees to allocate Kimball Fund [donated] money to purchase a new copier, and they

approved. I went ahead with procedures to order a new copier. The next thing that I learned was that Debra had disallowed the purchase. She said that I should have checked with her first.

I was flabbergasted. I had never felt that I had to check with the mayor on decisions like that. Furthermore, I was angry because she had ruled on the decision without checking into what the reasons for it were. I felt "zapped" by Debra, like I have in several other situations.

It seems to me that I did the right thing by checking with my Board of Trustees on the decision I made. As you know, by legislation they have the responsibility for the administrative functions of the library. When they have approved a decision like this, what basis does the mayor have for interfering in our decision?

Another way that Debra has demonstrated her lack of support for the library is by advancing the idea that the library should be regionalized to become a part of the county system. Anybody who knows anything about the library regards this as a preposterous idea.

In the first place, to seriously consider the idea of regionalization you would have to undertake a rather comprehensive study of the consequences. That in itself would be a major, expensive undertaking, which I don't think Debra is ready to shoulder. It is clear to me if such a study were done, the result would overwhelmingly favor the present organizational arrangement. We have very little in common with the Morton Library, and nothing at all to be gained by being put in the county system. Kimball is a unique community with citizens who have very different expectations from those in the remainder of the county, which is largely rural. The whole idea of regionalization is so preposterous that it seems to me to be irresponsible to even advance the idea.

I get the feeling that Debra is accumulating a check-list against me. I have had a fear for sometime now that Debra could at any time try to have me fired. I get the feeling in talking to them that I'm not getting straight messages from them.

At least there's one thing to be grateful for—I just passed my sixty-second birthday and can't be deprived of my pension if I am fired or forced to resign. I would like to stay on until I am sixty-five, but the way things are going between Debra and me I never know.

I get to feeling sad and hopeless and despairing when I think about the way I'm regarded at City Hall. I think it's tragic when someone like me has given many dedicated years of service and has made major contributions to building a strong program, and then finds themselves spending their last few years in an atmosphere of distrust and unappreciation. I think I deserve better.

Linda: The distrust in our relationship shows itself

practically every time we have an interaction. Recently I have taken on the duties of Adult Services Librarian and have been out visiting members of other city departments discussing ways that the library could be of service to them. I have had really warm and friendly receptions from everybody I have visited, with the exception of Ralph Riesen. When I talked to him in the same way that I had the other people, I felt like I got a cold shoulder. He seemed very uninterested. What I would most like would be to talk straight to Debra and Ralph and get straight answers in return.

Helen: We shouldn't overlook the fact that there have been some positive developments recently. For example, the new personnel officer, Joyce Gardner, came down and visited us last week. She was very understanding and very sympathetic about our problems. I am rather optimistic that many of our problems concerning selection, advertising, and interviewing will be better now that Joyce is here.

Linda: The recent hiring of two part-time people with Joyce's advice and help is an example of how well things *can* be done and how we and the city administration can work together. We should find more ways to use our separate expertise cooperatively!

Helen: Also, I am encouraged by the cooperation I have been getting from Jack Feldner. He recently responded favorably to my request for a crew to come over here and help with moving books away from an area where we had a leaking roof. I haven't always felt that I've had Jack's complete support and cooperation, but lately I've been feeling better about that.

One example of an item I'll bet is on Debra's checklist against me is the fact that the library is over its budget this year. Now the reason for this is that since the budgeting processes have been centralized in City Hall, I simply haven't had access to the kind of information I need to keep track of the budget. I'm afraid that I'm going to be unjustifiably blamed for this situation. This is an example of the kind of information I should not have to ask for—they should automatically give it to me.

Linda: I *am* concerned about the way that these crises with the city affect our morale and productivity. I have observed that when these crises come up we of the staff cease to care about our work as much, we spend *much* time rehashing incidents to reassure ourselves, and we do not do as good a job because we do not feel secure or appreciated. I am amazed to see myself doing this, as I like my job, but I do find myself lowering the quality of my work when I feel threatened, and I see others doing it too. So, continued bad feelings are counterproductive and inefficient.

Maude: One indicator of the kind of relationship which Debra has with us down here in the library is the reaction she gets when she comes down here. I remember a time when she was down here recently. We were

all very nervous and very alert. It was like we all suspected that she was up to no good being down here, and we had to watch her every step.

INITIAL MEETINGS

After reviewing what they had learned in the meetings with City Hall and the library staff, Jeff and Paul decided to recommend a series of four two-hour meetings. They formulated tentative meeting agendas and sent copies to each of the five prospective participants. After informal checks had established the agreement of each of the five to the proposed meetings, the consultants sent a confirming memo to each, announcing the time and place for each of the four meetings.

Meeting 1, March 19

The agenda presented by the consultants for the first meeting included a brief introduction by the consultants, an expectations check, a sharing appreciations exercise, and a closing process check.

Following the introduction, the participants were asked to participate in an expectations check. For the first half of this exercise each person was asked to write on two separate sheets of paper 1) their hopes, and 2) their fears for the upcoming series of meetings. In the second half of the exercise these hopes and fears were shared, posted on newsprint, and discussed. This exercise activity took about 40 minutes.

The "sharing appreciations" exercise contained four steps. In the first step each of the participants were given 3 × 5 cards and asked to write appreciation messages to other participants. Each message was to be addressed to another person on a separate card and was to be unsigned. A format suggested was, "I appreciate _____ about you." Each person was asked to write at least one such message to each of the other four participants present.

In step two of the appreciations exercise the cards were collected and sorted and then read by one facilitator while the other wrote the appreciations on newsprint. The result was one large newsprint sheet of appreciation messages for each of the five participants.

In step three each person was instructed to add to their individual sheets other things for which they would like to be appreciated, or for which they felt they deserved appreciation.

Step four consisted of a series of one-on-one conferences where each participant met individually with each of the other four participants for five minutes each. During these conferences each member of the pair was asked to *acknowledge* to the other person the appreciations which had been contributed by other participants, and further to acknowledge the appreciations which he or she had contributed or agreed with.

The sharing appreciations exercise took about 30 minutes.

The final activity for Meeting 1 was a process check, where participants were invited to share their feelings about the activities of the first meeting and about the upcoming meetings.

The expectations check generated a list of hopes and fears which was posted on two large sheets of newsprint. The main themes reflected in the "hopes" list included desires to improve working relations and communications between the library and the city, to clarify reporting patterns, to know others as individuals, to develop a more relaxed atmosphere among group members, to confront differences, to reduce felt threats, and to restore library staff confidence.

The list of fears included the following: that the library would become even more committed to single issue political activity; that the meetings would result in "unpleasant repercussions" for some; that information shared in the meetings would get out and be damaging or embarrassing; that the meetings would be a waste of time; that the library would move further away from the rest of the city and become more entrenched; and that Debra and Ralph would become too busy to attend one or more of the meetings.

The general mood during the meeting was one of cautiousness. Jeff and Paul noted that the appreciations shared were quite general and that some uneasiness was sensed during the appreciation sharing exercise. The process check at the end of the meeting revealed mildly positive reactions. Ralph seemed cool and reserved; he said that there were no dramatic gains but that he was willing to continue. Linda seconded Ralph's sentiment. Debra and Maude seemed to be more positive and appeared to feel reassured. Helen appeared to have very positive feelings about the meeting; she expressed reduced apprehensions about the meetings and increased comfort with the other participants.

Meeting 2, March 21

The meeting began with a brief introduction to the planned activities by Paul. He also apologized for having to leave early that day. Instructions were then given for the first phase of an "image exchange" exercise. Participants were told that each group was to meet in a separate room and prepare two lists. The first list was to summarize their own group's images of the other group, including thoughts, attitudes, feelings, perceptions, and behavior. The second list was to predict what the other group's images recorded in their first list would be.

After approximately 30 minutes of list preparation time, the two groups were reconvened to share the lists. During the list-sharing period a ground rule was enforced with disallowed debate and discussion but which allowed questions for clarification.

The librarians were invited to share their list of images of the city administration first. As they did so Jeff (Paul had gone) summarized the entries on newsprint. Next the city administration's images of the library were shared and posted. Time was allowed for clarification questions after each list had been aired.

Next the two groups shared their predictions of the other group's list with the librarians again going first. The time required for the sharing of the four lists was approximately 40 minutes. These four lists are reproduced in Exhibit 18-2.

Following the image exchange period the groups were again sent to separate rooms. This time each group was instructed to create a prioritized list of issues needing resolution. Twenty minutes was allocated for this activity.

The final activity for Meeting 2 was the sharing of the two lists of priority issues. Exhibit 18-3 shows the priority issues which were generated in this activity. This sharing and posting used up the remainder of the meeting time available.

At the conclusion of the meeting, Jeff's impression was that there was a general sense of tension relief that this long-repressed animosity was finally out in the open. Debra appeared to feel particularly good about the meeting when she left. Jeff was impressed by the casualness and informality with which Ralph engaged in musing conversation concerning the meeting with the three librarians for fifteen minutes after the meeting. This was the first time that Jeff could remember Ralph's being relaxed and at ease in any of the meetings concerning the library. Jeff guessed that Ralph might have felt good that some real progress had been made during this meeting.

Two days after this meeting, Linda reported to Jeff that the librarians left the meeting feeling quite discouraged.

Consultants' Meeting, March 22

Jeff Mallet and Paul Everest met at Paul's house to compare notes on the progress of the meetings so far, and to discuss strategy for the upcoming meetings.

When Paul saw the two priority lists of issues for resolution which had been generated by the two groups, he had an immediate reaction. Paul noted that the items listed by the librarians appeared to reflect a willingness to compromise, collaborate or negotiate; whereas, those

EXHIBIT 18-2
Image Exchange Data From Second Meeting of City Hall and Library Administration

I. Library Administration views of City Hall
1. They are suspicious of the library.
2. They are well-intentioned but inept.
3. They are uninterested in the library program.
4. They are protective of their own power.
5. They are unfriendly.
6. They want the library to accept administrative changes from City Hall, but are unwilling to accept administrative changes made by the Library Board.
7. They don't really want public input.
8. They are very willing to put library staff (esp. Helen) between power play of City Hall and the Library Board.
9. They are personally against Helen.

II. City Hall views of Library Administration
1. They have limited or no respect for the administrative abilities of City Hall.
2. "Massive paranoia" exists among the library staff.
3. The librarians have been operating a propaganda organ:
   • Internally with library staff
   • Externally with City Council and the public
4. The library staff has used the Library Board as a separate political support group.
5. There has been a concerted program by the librarians to establish a separate political base and become invulnerable.
6. Library personnel operate a tight "clique."
7. Library personnel distrust (and dislike and despise . . .) City Hall.
8. Library personnel wish to do their own thing without coordination.
9. Library personnel don't readily accept administrative assistance.

III. Library Administration's predictions of City Hall views of Library Administration
1. They think we are paranoid.
2. They think we are snobbish & isolated.
3. They think we are spreading our views of the problem among staff & public.
4. They think we are overprotective of the library.
5. They think we are inappropriately political.
6. They think we are encouraging the Library Board to move away from City Hall.

IV. City Hall's predictions of Library views of City Hall
1. They think that we believe the library is not a critical service; it is dispensable, or first to go in a crunch.
2. They think we are non-supportive of the library.
3. They think we discriminate against the library.
4. They think we impose unreasonable guidelines.
5. They think we have a vendetta against the library.
6. They think we are uncaring and unhelpful.
7. They think that the library gets the short end of resource allocations.
8. They think that we are fast to control and restrict, but seldom volunteer assistance.

items listed by Debra and Ralph appeared to reflect the expectation that it was the library which should do the changing. Jeff and Paul wondered if this was a pattern. They recalled other times when they had vague feelings that perhaps Debra or Ralph or both regarded the meetings as an opportunity to get the library to shape up. Following the meeting, Jeff had the feeling that the three librarians had seemed to take the instructions and the sessions more seriously than did Debra and Ralph. Jeff had hoped that the period for sharing the four lists would leave everyone in an introspective mood. This seemed to take place for the librarians, but not for Debra and Ralph.

After reflecting on the outcomes from Meeting 2,

EXHIBIT 18-3

Priority Issues for Resolution: Second Meeting, City Hall and Library Administration

I. Priorities of Library Staff
   1. Clarify the role of the Library Board of Trustees:
      a) State-wide
      b) City-wide
      c) Vis-a-vis the library staff
   2. Clarify the roles of the library staff, library administration, and City Hall.
   3. Reach agreement regarding appropriate political activity for the library.
   4. Develop mutual respect for one anothers' administrative abilities.

II. Priorities of City Hall
   1. (Debra) Inappropriate political activity.
   2. (Ralph) Resolve the perception that City Hall is doing something "bad" to Helen, i.e., perceived vendetta.
   3. Library's impression that City Hall is uninterested in the library program.
   4. Library's impression that members of City Hall are being protective of their own power.

Jeff reported feeling overwhelmed by the pervasiveness of the issue concerning appropriate political activity. His review had led him to the conclusion that this issue was so fundamental to all the problems being experienced between the library and City Hall that it was likely to be futile to work on any specific issues before addressing this major one.

As Jeff saw it, there were two major questions which needed to be resolved. First, what is the relationship of the Library Board of Trustees to City Hall? And secondly, how are the diametrically opposed views expressed by the library and City Hall concerning appropriate political activity going to be resolved? It seemed to Jeff that neither of these issues could be settled by the group which had been meeting with Jeff and Paul. Instead, it seemed more plausible that these issued needed to be referred to either the Library Board or to the City Council.

Paul agreed that there were no instant solutions in sight, and that the appropriate strategy for where to go with the present group was not at all apparent.

After some discussion, Paul and Jeff agreed on the prognosis that until the overriding issue of political activity was dealt with, administrative issues would probably be resistant to solution. They further agreed that it seemed unlikely that solutions to the political activity question could be generated from within the present group, and that action strategies to address this issue probably would have to come from the City Council or the Library Board.

Concerning strategy for Meeting 3, Paul and Jeff agreed to begin it by reviewing for the participants the consultants' interpretations of the outcome of Meeting 2 and to invite them to join in a problem-solving session

concerning appropriate action strategies. Paul and Jeff could think of two strategies which might prove fruitful:

1. Refer the issue of appropriateness of political activity to the City Council with a request for a definitive guideline on what activities are appropriate.

2. Have Debra and Helen get together, with or without a process consultant, to work out an agreement concerning political activity.

Jeff and Paul discussed whether the issue of the newspaper reporter being called should be brought up and dealt with at the next meeting. They agreed that Debra had stored up much resentment over this issue, and that if it came out it could be a "heavy" confrontation. Jeff and Paul were very uncertain about whether the issue could be constructively dealt with in one meeting. The uncertainties concerning the outcome of such a confrontation led Paul and Jeff to agree that they should probably try to avoid confronting this issue at the next meeting.

FURTHER DISCUSSIONS

Meeting 3, March 24

As Jeff and Paul arrived at the Savings Bank Community Room for Meeting 3, they exchanged the sentiment, "God knows what's going to happen today!"

As participants entered the meeting room, they were given a three-page handout summarizing the previous meeting's outcomes. This handout contained the data generated in Meeting 2 from the image exchange exercise and the priority issues for resolution list (Exhibits 18-3 and 18-4).

Jeff began by sharing some of his and Paul's reflections concerning the pervasiveness of the political issue. He raised the question about whether administration concerns could be addressed while the political issue remained unsolved. He further voiced some skepticism concerning whether the present group was the appropriate one to settle the political issue, or whether it could.

At this time, Jeff spent some time reflecting on the nature of the conflict over political activities. He tried to summarize the position of each of the two parties to the conflict. In doing this Jeff emphasized his understanding that each of the parties had a position which

was logically defensible, internally consistent, and supportable by others.

Jeff concluded by inviting the group members to comment on the consultants' diagnosis of the problem, and to join in a problem-solving session to identify reasonable options which could be taken. The remainder of the meeting time was used for unstructured discussion, with the exception of a brief process check at the end of the meeting.

Paul served in a process observation role during this meeting. During the time that Jeff was giving an overview of the problem situation, Paul noted the reactions of the five participants. Linda, Helen, and Ralph all seemed quite attentive. Debra and Maude were observed to be staring intently at their handouts for long periods of time. This was particularly true for Maude who hardly shifted her gaze from her handout for al-

---

EXHIBIT 18-4
City Hall and Library Administration Action List

ISSUE	ACTION
• Calling reporter anonymously	• Announcement at staff meeting (Helen will do. OK to break confidentially.)
• Offer by Debra to spend time in library	• Helen will schedule with staff and Debra.
• Reporting relations	• Helen will draft memo to Library Board by May 2 asking them for direction or clarification on the following issues: • Legal liability; errors & omissions • Property • Maintenance • Reporting relations • Political activity • Debra and Ralph will review memo • Ralph and Helen will attend May 2 meeting of Library Board.
• Maintaining good relations	• Debra, Ralph, Helen, Linda, and Maude will meet for brown bag luncheons. • First luncheon: Tuesday, Apr. 24 12:00 to 1:00 in Ralph's office; Linda will facilitate. Participants to begin with "check-in" concerning problem issues and good news. • Facilitator and location will rotate for subsequent luncheons. • Brown bag discussion item: exchanging of staff people
• Perception that City Hall is going to do something bad to Helen	• Brown bag luncheon "check-in" item
• Perception that City Hall is "inept" • Perception that library staff is "incompetent"	• All such evaluative stereotypes were declared inoperative by Jeff, who banned their use in thought and speech.

most twenty minutes. Paul noted that Maude looked dejected, and that she was avoiding eye contact with others present. Because the meeting room was chilly, Maude (along with most of the others) was feeling physically cold. Maude had also mentioned that she was coming down with a cold.

After Jeff had finished his introductory remarks. Debra abruptly initiated a discussion of political activity on the part of the library staff. Debra's remarks may be paraphrased as follows:

> Politics is a fact of life now. The library staff has started something that will be very hard to stop. They have politicized the budgeting process and it will be very hard to go back to a non-political procedure. What I need to know from the library staff is whether these activities are going to continue. If they are, there are going to be unpleasant repercussions which the library staff should understand.
>
> There are two things that are really bothering me; first, the fact that someone from the library called a newspaper reporter to ask that my "vendetta" against Cecil Hockman and the library be investigated. When I got that telephone call from the reporter, I felt 'angry, betrayed, and nonplussed.' Second is the issue of political activity by library staff members aimed at packing the City Council chambers with citizens supporting the library. That represents a clear violation of instructions from the Library Board, and leads me to wonder, "Who's running the library, anyway?"

When Debra made the point that the library staff had disregarded instructions concerning political activity, Helen pointed out that the library staff did not perceive that they had received any such instructions. Following Helen's point, discussion proceeded in another direction, with no overt evidence that Helen's comment was heard or understood.

Following Debra's expression of her feelings about the telephone call, Helen and Linda expressed consternation that the telephone call had been made. Both made it very clear that they thought such a telephone call was inappropriate. Linda said, "I didn't realize we had sunk to that low a level," and Helen seconded Linda's sentiment. During this conversation Maude was noticeably quite, and was avoiding eye contact.

Ralph said, "When I come in the library, I feel hostility all around me." When Ralph had said this, Paul intervened and asked Ralph to focus on his personal feelings when he was in this situation. Ralph's responses generally depicted his impressions of library staff members' attitudes. Paul pursued the issue by asking Ralph two more times to focus on and report his own feelings in this situation. After Ralph's responses again did not describe his own feelings, Jeff probed him by asking if he might have been feeling hurt, or disliked, or disrespected. In response to this prompting Ralph acknowledged that some of these guesses were accurate.

At this point Paul intervened with a few observations designed to set the stage for the librarians to air some of their feelings. With a few minor exceptions, the librarians did not divulge their feelings on issues.

At one point in the conversation Debra offered "to spend a week working in the library," if that would help to resolve some of the problems. Helen responded to this offer with apparent guardedness, citing the difficulties of time scheduling and the requirements of attending the human understanding workshops currently being conducted for all city employees. Debra seemed annoyed that Helen's reaction to her offer was not totally positive. At this point Maude made a pointed observation to Debra: "I have to tell you that there are some people in the library who will be pretty hostile toward you."

The question of whether the library should regionalize by joining the Morton County system was raised. Debra expressed dismay that the Library staff, the Library Board of Trustees, and several others had reacted so vehemently to the proposal that regionalization should be studied. The librarians responded to Debra's sentiment by assertively pointing out that the proposal [which had been part of a report to the City Council by the Capital Expenditures Committee] did not call for a study but called for *implementation* which was to occur by January 1, 1990. Both Debra and Ralph replied that they were sure that the wording of the Capital Expenditures Committee report was that the January 1, 1990 date was the deadline for *completion of a study*. The librarians were equally certain that their interpretation of the report was correct. Members of both groups vowed to get a copy of the committee report to bring to the next meeting.

Discussion of the regionalization issue continued. Helen referred to a previous study concerning regionalization which had been conducted by the League of Women Voters. This study had gathered some utilization data. Helen felt that the study supported her opinion that regionalization would be most unwise. Debra said that she had not seen or heard of the League study, and was very interested: "That's the kind of information I need to know."

At this point one of the librarians volunteered that they had prepared a "fact sheet" concerning the regionalization issue. Debra expressed surprise at hearing about the fact sheet. Paul noted that Debra seemed annoyed about learning about the fact sheet, and that Ralph gave the librarians a dirty look during this time.

The librarians at this point explained that the fact sheet was prepared in response to a request by an individual City Council member.

Lively discussion of substantive issues was continuing when Jeff interrupted at a few minutes before the end of the meeting time to ask for a process check. During this check the general sentiment expressed was, "Whew! we really got into it today!" Linda said that she thought a lot had been accomplished, and nods of agreement from other participants were noted. Ralph acknowledged some real accomplishment for the first time. Paul and Jeff shared both surprise and relief that the issue concerning the reporter had been successfully dealt with and largely defused. In fact, they expressed the view that the whole issue of political activity had been defused at least somewhat.

Meeting 4, March 25

The meeting began with Paul and Jeff suggesting a review of the "Priority Issues for Resolution" list generated in Meeting 2. The consultants suggested that the group make an "action/no action" decision for each of the priority issues. This was to provide some closure for this last of four scheduled meetings.

During the last half of the meeting Paul started an action list on newsprint, and he and Jeff pressed the participants for specific action commitments as the discussion approached agreement.

The last ten minutes of the meeting were spent reviewing the list of hopes and fears generated at the beginning of Meeting 1.

The action list that Paul constructed on newsprint during the last half of the meeting is shown in Exhibit 18-4. The last issue on the action list, i.e., the perceptions of "ineptness" and "incompetence," still had not been discussed as the end of the meeting time approached. Jeff called attention to the issue, and shared the perception of the consultants that the range of specific behaviors which each group found upsetting in the other group seemed quite small, too small to support the "inept" or "incompetent" generalizations. He pointed out that feedback on specific behaviors had been constructively shared during the four meetings, but that feedback on broad evaluative generalization was hard to respond to constructively. Jeff urged each participant to consciously avoid lapsing into the use of such evaluative stereotypes, and instead to concentrate on specific behaviors.

During the review of the hopes and fears lists the general feeling was that most of the hopes had been either partially or fully realized, and that most of the fears had dissipated. Concerning the fear that the meetings might prove to be a waste of time, Ralph said, "that remains to be seen." Concerning the hope that better working relations would be developed, all participants seemed to agree that this had been accomplished.

FOLLOW-UP

A survey instrument called the Intergroup Profile was used by the consultants to measure the climate existing between City Hall and the library staff. This instrument has eight Likert-type questions concerning relationships existing between two groups. Measurements were taken in March before the first intergroup meeting, and in May, six weeks after the last meeting. Parallel measurements were obtained from nine separate control organizations.

Data analysis revealed that the Library/City Hall climate prior to the meetings was considerably worse than that existing in any of the nine control organizations ($p < .0001$). Following the meetings the Library/City Hall climate scores had improved substantially ($p = .001$), but were still lower than the scores of any control organization.

In early August, four months after meeting four, a two-page written evaluation form was filled out by each of the five meeting participants. Their responses reflected general agreement that, as a result of the meetings, the climate between City Hall and the library had improved, but not dramatically.

Ralph Riesen commented, "We achieved a better understanding of positions, but no real resolution of conflicts. The conflicts that exist are political rather than personal."

Debra Dickenson noted that the meetings had provided ". . . a good chance to share concerns," and that they resulted in ". . . better feelings for the individuals involved." She continued,

> There is a period of transition that is required—just plain time to see how we all deal with the next 'challenge to authority.' Political changes have an effect. I don't feel the library personnel understand the scope of City demands and needs any better than before. In my opinion they just feel we are being nicer to them. Their anxieties are relieved a bit so the climate is improved. There is a value to that without a doubt.

Helen Hendricks noted three specific changes which had resulted from the meetings:

1. The Librarian is aware that her personal situation cannot improve but she is not threatened by further deterioration of her position.

2. The administrative reporting pattern between Library Administration, Library Board, and City Supervisor has improved.

3. The Library Staff are more united and supportive than ever.

Additional comments made by Helen included the following:

I believe the Library's fears and concerns were substantiated by the meetings but it was good to bring them into the open. The Librarian's and City Supervisor's personal contacts are slightly improved.

The problems at the Library stemmed from the City Administration decision to regroup the City program with the resultant down-grading of the library service and personal demotion of the Librarian—the view of the Library. The City Administration did not recognize this as the cause.

Linda Turner reported that the meetings ". . . relieved the Mayor's mind by allowing her a chance to 'let off steam.' Coming from the library, I [now] feel more relaxed in talking with the Mayor and City Supervisor—though not totally relaxed. The City Librarian and City Supervisor can now talk to each other—this is by far the most important result."

Maude Richardson concurred with Linda and Helen that the relationship between the City Librarian and the City Supervisor was much more comfortable. She also observed that "foul-ups at City Hall are no longer seen as personally directed at the library."

When you have read this case, look back at the chapter's diagnostic questions and choose the ones that apply to the case. Then use those questions with the ones that follow in your case analysis.

1. What kinds of problems in the Dumas library and the Kimball city government did Jeff Mallet and Paul Everest uncover in their interviews? What expectations did they create by their interviewing process? Were these expectations fulfilled by later actions and events?

2. What kind of organization development intervention did Jeff and Paul implement? Why did they

choose this particular type of intervention? Did they implement it properly? Was it the best choice, or should they have used another intervention instead?

3. Critique Jeff's and Paul's overall effectiveness as OD change agents. What would you have done differently? Would you hire them to manage change in your organization?

CASE 18-3

O CANADA

Review this case, which appears in Chapter 15. Next, look back at Chapter 18's diagnostic questions and choose the ones that apply to the case. Then use those questions with the ones that follow in your case analysis.

1. What organization development interventions could the Public Service Commission use to identify and diagnose the problems confronting the Directorate of Staff Development and Training and the Regional Operations Directorate? Which intervention would probably work best? Why? What steps should be taken to implement this intervention?

2. What OD interventions could be used to deal with the kinds of problems likely to be identified by your initial intervention? Describe a specific program of interventions and implementation steps that you would recommend.

3. How would your program benefit the Public Service Commission and its divisions? What obstacles might it encounter? How would you overcome such resistance?

C H A P T E R   19

INTERNATIONAL DIMENSIONS: ORGANIZATIONAL BEHAVIOR ABROAD

THE GRADUATION CEREMONY OF
INTERNATIONAL UNIVERSITY OF JAPAN
国際大学大学院修了式

With help from Dartmouth College's Amos Tuck School of Business, the International University of Japan opened its International Management MBA Program in 1988 and graduated its first class two years later. Students in the initial program, most of whom were supported by their companies, found the "giant" textbooks somewhat daunting and the pressure to offer their opinions in class disconcerting, particularly because all instruction in the new program is offered in English. The students liked their U.S. professors' openness to them, however, and even learned to evaluate them at the end of the term. One student said that he had never asked "Why?" in his business career but was now very quick to ask this question. His wife was worried, however, lest he be "kicked out [of his firm] for insulting the existing way of doing things."
Source: Patricia A. Langan, "Trying to Clone U.S.-Style MBAs," Fortune, October 8, 1990.

Backed in most cases by corporate cash and charged with the mission of deciphering the often inscrutable American culture, Japanese "salarymen" are coming to America. They have attended American business schools in small numbers since the 1960s, even though corporate leaders in Japan have preferred to teach business the old-fashioned way—on the job. But Japanese students now account for more than five percent of the total enrollment at America's top business schools. To them, the MBA has come to mean Mastering Being in America.

More than three fourths of Japanese business students in the United States are sponsored by their employers, and many of them expect to manage Japanese subsidiaries in the U.S. someday. Naoki Yamamori is typical. Formerly a bank branch manager for Dai-Ichi Kangyo in Tokyo, he came to MIT's Sloan School of Management nearly two years ago. Though he hopes to learn more about finance, Yamamori's foremost goals are "to have an American friend and understand how they think." Jinei Yamaguchi, a Mitsubishi Bank employee attending Sloan, echoes that notion. "Business school is more to learn about America," he says, "than to learn about net present value." Yamaguchi says his experience at Sloan should pay off later if he has to communicate with branch managers of Mitsubishi's American subsidiaries.[1]

POINT TO STRESS
It is important for managers to understand cultural and national differences if they are to succeed in other countries. The old days of expecting others to conform to "our" way are over. Organizations are searching for ways to manage that fit the organization's goals and objectives and the norms and values of the country and culture they are operating in.

In the world economy of the 1990s, it is becoming increasingly difficult to find a company that doesn't conduct business across national boundaries. Indeed, "American" companies like McDonald's, Coca-Cola, and General Motors and "Japanese" companies like Sony, Honda, and Mitsubishi do so much business outside their original homelands that they are more accurately described as global enterprises. As discussed in the "In Practice" box, multinational companies and stateless corporations spanning several countries are quite common in today's business world.

How seriously must the managers of such organizations take differences in nationality? How significantly do international differences affect the management of organizational behavior? In response to these questions, consider that Japanese companies are paying top dollar for younger salarymen (managers) to learn about the American way of life so that the Japanese approach to management can be fitted to the task of managing American employees. As suggested by the actions of the Japanese—who have amassed a strong record of success in multinational management—contemporary managers *must* take international differences seriously and adapt familiar management practices to compensate for these differences. Otherwise, they risk losing out in today's competitive world markets.[2]

The purpose of this chapter is to discuss important international differences. How do we fit the management practices we have examined in this book—practices that are overwhelmingly American in origin and focus—to the job of managing people and organizations in other national cultures? We will begin by examining the ways the cultures of nations differ from one another, identifying four dimensions that delineate key international differences. Next, we will take a closer look at how these differences affect organizations and organizational behavior.

[1] Todd Barrett, "Mastering Being in America: Japanese Are Flocking to U.S. Business Schools," *Newsweek*, February 5, 1990, p. 64.

[2] For a discussion of how the Japanese have adapted familiar management practices for use abroad, see Peter B. Smith, "The Effectiveness of Japanese Styles of Management: A Review and Critique," *Journal of Occupational Psychology* 57 (1984), 121–36.

The Rise of the Stateless Corporation

Today, dozens of American industrial companies, including IBM, Gillette, Xerox, Dow Chemical, and Hewlett-Packard sell more of their products outside the U.S. than they do at home, and American service firms such as McDonald's, Time-Warner, Disney, and American Express do at least 20 percent of their business in foreign markets. Many of these companies are *multinationals*—globe-spanning organizations that treat foreign operations as outposts that produce products designed and engineered back home. Others, however, are *stateless corporations*, or firms that customize products to regional tastes and are totally localized in the many sites they own.

Stateless is a relatively new strategy. How is it working out? Otis Elevator's decision to design its newest high-speed elevator in five different countries—systems integration in the U.S., motor drives in Japan, door systems in France, electronics in Germany, and gear components in Spain—saved more than $10 million in design costs and two years of product-devel-

opment time. Canada's Northern Telecom Ltd. moved many of its operations into the U.S. and became eligible to bid for Japanese contracts as an American company. This eligibility is critical to Northern Telecom. Japan now favors U.S. tele-communications companies over Canadian firms because of the politically sensitive U.S.-Japanese trade gap. Honda Motor Company is able to circumvent anti-Japanese trade barriers in Israel, Taiwan, and South Korea by shipping American-made Accords and Civics to these destinations. Statelessness thus appears to offer important competitive advantages to companies that are doing business in the international markets of the 1990s.

Adapted from William J. Holstein, Stanley Reed, Jonathan Kapstein, Todd Vogel, and Joseph Weber, "The Stateless Corporation: Forget Multinationals—Today's Giants Are Really Leaping Boundaries," *Business Week*, May 14, 1990, pp. 98–105.

We will narrow our focus to two of the four dimensions as we compare management practices in Scandinavian, Japanese, and Israeli kibbutz organizations. We will conclude by discussing the task of adapting familiar management techniques to the demands of different national cultures.

DIAGNOSTIC ISSUES

In this final chapter, we will again use our diagnostic model to call attention to several important considerations. In examining the international context we need to ask how we can *describe* how nations and their cultures differ from one another. What dimensions can we use to capture and understand such differences? How do specific countries differ on these dimensions? How can we *diagnose* whether a particular theory we have discussed will apply in a particular national culture? Is there any way to assess whether a management practice developed in one nation's culture will work in another?

Can we *prescribe* what sorts of structure, jobs, and leadership will work best in a particular country? Can we predict what sorts of efforts to increase motivation will be appropriate? Finally, what *actions* can managers take to adapt management practices that work in one country for use in another? How can managers adapt their own behavioral styles to living and working overseas?

NATIONAL CULTURES AND ORGANIZATIONAL BEHAVIOR

We have seen that organizational behavior is influenced by social information that originates in the organization's culture of norms and values (Chapter 18). Behavior in organizations is also shaped by social information residing in the **national culture** of societal norms and values that is part of the organization's environment. Sometimes the norms and values in an

national culture The collection of societal norms and values in the environment surrounding an organization.

Changing an organization's culture can be a very difficult task, but turning around an entire nation's norms and values is a job of overwhelming proportions. The Soviet Union's new efforts at entrepreneurism have touched only a very small percent of society. Because the new ventures have not yet become fully and legally independent of the state, other organizations may not want to follow the same path, and the fledgling companies themselves may not survive. Meanwhile, many workers like these Russian women still labor for low pay on state-owned collectives.
Source: Richard I. Kirkland Jr., "Can Capitalism Save Perestroika?" Fortune, July 30, 1990.

AN EXAMPLE: McDONALD'S
Cultural differences can arise when dealing with customers. McDonald's had difficulty serving customers in England. As all North Americans know, when you go into McDonald's you stand in line and when you get to the counter you give your complete order. The English weren't used to this. When they reached the counter, they ordered one item and then stood in another line for another item. McDonald's had to educate their new market in the idea of fast, efficient service.

organization's culture seem not only to mirror but to actually intensify characteristics of the surrounding national culture.[3] For example, promptness, which is viewed favorably in the U.S. national culture, is valued so highly in some American firms that missing a deadline by an insignificant amount of time is considered grounds for dismissal. In other instances, norms and values from the national culture can even overwhelm those residing in the organization's culture. Whistle blowing—in which one person reports organizational wrongdoing to others in authority—is an example. A whistle blower must violate organizational values favoring loyalty or secrecy to honor societal values that endorse honesty and truthfulness.

Frequently, however, the norms and values of an organization's culture are imported directly from, and are therefore consistent with, the national culture. For example, norms in American organizations assert that "a fair day's work should receive a fair day's pay." Those norms are based on societal norms of fairness and reciprocity that lie at the core of the American culture. Similarly, the value placed on deferring to hierarchical superiors in Japanese organizations is based on societal values that emphasize the importance of obedience and harmony. In cases like these, societal norms and values are themselves sources of social information that have direct effects on organizational behavior.

DIMENSIONS OF CROSS-CULTURAL DIFFERENCES

How do societal norms and values differ from one national culture to another? In what ways are the cultures of different countries similar? What effects do these similarities and differences have on people's attitudes and behaviors? Geert Hofstede set out to answer these questions by surveying employees of IBM offices located in 40 countries throughout the world. As he examined the data from 116,000 questionnaires, Hofstede discovered that most differences among national cultures were described by four dimensions—*uncertainty avoidance, masculinity-femininity, individualism-collectivism,* and *power distance.*[4]

[3] Andre Laurant, "The Cultural Diversity of Western Conceptions of Management," *International Studies of Management and Organization* 13 (1983), 75–96.

[4] Hofstede, "Motivation, Leadership, and Organization: Do American Theories Apply Abroad?" *Organizational Dynamics* 9 (1980), 42–63; and *Culture's Consequences: International Differences in Work-Related Values* (Beverly Hills, Calif.: Sage, 1984), pp. 153–212.

Uncertainty Avoidance

uncertainty avoidance A cross-cultural dimension that refers to the degree to which people are comfortable with ambiguous situations and with the inability to predict future events with assurance.

Uncertainty avoidance concerns the degree to which people are comfortable with ambiguous situations and with the inability to predict future events with assurance. People with weak uncertainty avoidance feel comfortable even though they are unsure about current activities or future events. Their attitudes are expressed in the following statements:

Life is inherently uncertain and is most easily dealt with if taken one day at a time.

It is appropriate to take risks in life.

Deviation from the norm is not threatening; tolerance of differences is essential.

Conflict and competition can be managed and used constructively.

There should be as few rules as possible, and rules that cannot be kept should be changed or eliminated.[5]

In contrast, people characterized by strong uncertainty avoidance are uncomfortable when they are unsure what to expect. Their attitudes about uncertainty and associated issues can be stated as follows:

The uncertainty inherent in life is threatening and must be fought continually.

Having a stable, secure life is extremely important.

Deviant persons and ideas are dangerous; neither should be tolerated.

Conflict and competition can unleash aggression and must be avoided.

There is a need for written rules and regulations; if people do not adhere to them it is because of human frailty, not defects in the rules and regulations themselves.[6]

In cultures characterized by high uncertainty avoidance, behavior is motivated at least partly by people's fear of the unknown and their need to cope with this fear. In addition, people in such cultures try to reduce or avoid uncertainty by establishing extensive formal rules. For instance, having extensive laws about marriage and divorce reduces uncertainty about the structure and longevity of family relationships. If uncertainty proves unavoidable, people with a cultural aversion to uncertainty may instead engage in *ritualistic* activities that help them cope with the anxiety that uncertainty arouses. Such activities may include the development of extensive planning systems that are designed to speculate about the future and make it seem more understandable and predictable. They dispel some of the anxiety even if the resulting plans prove to be completely useless when put into action. Seeking to cope with uncertainty, people may also hire "experts" who seem to have the ability to apply knowledge, insight, or skill to the task of making something uncertain into something understandable. Like the sorcerers or witch doctors of primitive cultures, these experts need not actually accomplish anything so long as they are perceived as understanding what others do not.

Masculinity-Femininity

Hofstede used the term masculinity to refer to the degree to which a culture is founded on values that emphasize independence, aggressiveness, dominance,

[5] Hofstede, "Motivation, Leadership, and Organization," p. 47.
[6] Ibid.

and physical strength. According to Hofstede, people in a national culture tilted toward masculinity hold beliefs such as the following:

> Sex roles in society should be clearly differentiated; men are intended to lead and women to follow.
> Independent performance and visible accomplishments are what counts in life.
> People live in order to work.
> Ambition and assertiveness provide the motivation behind behavior.
> People admire the successful achiever.[7]

Femininity, according to Hofstede, describes a society's tendency to favor such values as interdependence, compassion, empathy, and emotional openness. People in a national culture oriented toward femininity hold such beliefs as the following:

> Sex roles in society should be fluid and flexible; sexual equality is desirable.
> The quality of life is more important than personal performance and visible accomplishments.
> People work in order to live.
> Helping others provides the motivation behind behavior.
> People sympathize with the unfortunate victim.[8]

masculinity-femininity A cross-cultural dimension that refers to the degree to which a culture is founded on values that emphasize independence, aggressiveness, dominance, and physical strength, on the one hand, or interdependence, compassion, empathy, and emotional openness, on the other.

Together, masculinity and femininity form the dimension of **masculinity-femininity** in Hofstede's model of cross-cultural differences. One important effect of the differences mapped by this dimension is the way a nation's work is divided into jobs and distributed among its citizens. In masculine national cultures, women are forced to work at lower-level jobs. Managerial work is seen as the province of men who have the ambition and independence of thought required to succeed at decision making and problem solving. Women also receive less pay and recognition for their work than their male counterparts. Only in "feminine" occupations—teacher or nurse or supporting roles such as secretary or clerk—are women allowed to manage themselves. Even then, female supervisors are often required to imitate their male bosses in order to be accepted as managers.

In contrast, equality between the sexes is the norm in national cultures labeled feminine. Neither men nor women are considered to be better managers, and no particular occupation is seen as masculine or feminine. Both sexes are equally recognized for their work, and neither is required to mimic the behaviors of the other for the sake of acceptance in the workplace.

Individualism-Collectivism

individualism-collectivism A cross-cultural dimension that refers to two opposing points of view on the norms and values of a national culture—whether they should place greater emphasis on satisfying personal interests or on looking after group needs.

Individualism-collectivism, according to Hofstede, is the tendency of a culture's norms and values to emphasize either satisfying personal needs or looking after the needs of the group. From the point of view of individualism, pursuing personal interests is more important, and succeeding in the pursuit of these interests is critical to both personal and societal well-being. If each person takes care of personal interests, then everyone will be well off. Consistent with this perspective, the members of individualistic national cultures espouse the following attitudes:

> "I" is more important than "we."
> People are identified by their personal traits.

[7] Ibid. p. 49.
[8] Ibid.

Success is a personal achievement. People function most productively when working alone.

People should be free to seek autonomy, pleasure, and security through their own personal efforts.

Every member of society should take care of his personal well being and the well being of immediate family members.[9]

In contrast, the collectivist perspective emphasizes that group welfare is more important than personal interests. If you hold this view, you believe that only by belonging to a group and looking after its interests can people secure their own well-being and that of the broader society. The members of collectivistic national cultures are thus inclined to make personal sacrifices for the sake of their groups—ensuring group well-being even if personal hardships must be endured occasionally. They agree that:

"We" is more important than "I."

People are identified by the characteristics of the groups they belong to.

Success is a group achievement. People contribute to group performance, but groups alone function productively.

People can achieve order and security and fulfill their duty to society only through group membership.

Every member of society should belong to a group that will secure members' well being in exchange for loyalty and occasional self-sacrifice.[10]

APPLYING THE DIAGNOSTIC MODEL
Diagnostic Question 1: What are the characteristics of your own national culture? Is it more individualistic or collectivistic? Are its values consistent with more or less power distance? Is it characterized more by masculinity or feminity? Does it favor strong or weak uncertainty avoidance?

TEACHING NOTE
Is the United States an individualistic or collectivistic society? How would you explain unions? Are we changing?

power distance   A cross-cultural dimension that refers to the degree to which the members of a society accept differences in power and status among themselves.

How do these two very different views of the individual in society affect people's attitudes and behaviors? Principally, they determine the degree to which people feel independent, on the one hand, or interdependent, on the other. In individualistic national cultures, membership in a group is something that can be initiated and terminated whenever convenient. The individualist does not necessarily have a strong feeling of commitment to any of the groups to which she belongs. In collectivistic national cultures, however, changes in membership status can be traumatic. Joining and leaving a group can be like finding and then losing one's sense of identity. The collectivist feels a very strong sense of commitment to his group.

Power Distance

Power distance refers to the degree to which the members of a society accept differences in power and status among themselves. In national cultures that tolerate only a small degree of power distance, norms and values specify that power differences should be minimal and political equality encouraged. People in these cultures show a strong preference for participatory decision making, and tend to distrust autocratic, hierarchical types of governance. Such people hold the following beliefs:

Superiors should consider subordinates "people just like me," and subordinates should regard superiors in the same way.

Superiors should be readily accessible to subordinates.

Using power is neither inherently good nor inherently evil; whether power is good or evil depends on the purposes for, and consequences of, its use.

Everyone in a society has equal rights, and these rights should be universally enforced.[11]

[9] Ibid., p. 48.
[10] Ibid.
[11] Ibid., p. 46.

In contrast, national cultures characterized by a large degree of power distance are based on norms and values that stipulate that power should be distributed hierarchically. People in these cultures advocate the use of autocratic authority and supervision to coordinate individual efforts. They hold the following beliefs:

Superiors and subordinates should consider each other to be different kinds of people.
Superiors should be inaccessible to subordinates.
Power is a basic fact of society; notions of good and evil are irrelevant.
Power holders are entitled to special rights and privileges.[12]

Power distance affects attitudes and behaviors by affecting the integration of a society's various parts. When only a small degree of power distance is favored, government hierarchies are noticeably flatter, and societal decision making tends to be decentralized and participatory. In contrast, where societal norms and values favor larger power distance, decision making tends to be centralized, and governmental hierarchies are taller. Authoritarian, autocratic government is the hallmark of larger power distance.

THREE CULTURAL EXAMPLES

Hofstede's four-dimensional model is not without its critics. For instance, a study that used the same four dimensions to assess the societal values of American, Japanese, and Taiwanese managers in Taiwan revealed problems with measurement validity and reliability (see Chapter 3 for a discussion of these kinds of problems).[13] Nonetheless, the Hofstede model is the most comprehensive cross-cultural framework currently available, and it provides useful insights into key international differences. Table 19-1 summarizes the results of Hofstede's analysis, indicating average scores on the four dimensions for each of the 40 countries included in the study.

What effects do the differences found by Hofstede's research have on the management of organizational behavior? To answer this question, let's look at two of Hofstede's dimensions—individualism-collectivism and power distance. We will examine the ways these two dimensions of international differences influence the management of structure, motivation, leadership, and job design.

The United States national culture is extremely individualistic. It is also oriented toward larger degrees of power distance than many of the other cultures included in Hofstede's study (see Table 19-1). As a result, the American models and practices described throughout this book are attuned to individualism and large power distance. How would the models and practices described in this book look if shaped instead by a national culture of individualism and small power distance? We will answer this question shortly when we look at how things are done in the Scandinavian countries of Norway and Sweden. As Table 19-1 shows, the cultures of both these countries are individualistic (though not to the same extreme as the U.S. culture) and oriented toward relatively small power distance.

How would the models and practices we have described in this book appear if attuned instead to collectivism and a large level of power distance? We will address this second question by analyzing the workings of Japanese organizations. As shown in Table 19-1, Japan's national culture is fairly similar to that of the United States in terms of power distance but is noticeably more collectivistic.

[12] Ibid.
[13] Rhysong Yeh, "Values of American, Japanese, and Taiwanese Managers in Taiwan: A Test of Hofstede's Framework," *Academy of Management Proceedings 1988*, pp. 106–110.

TABLE 19-1
A Comparison of Cultural Characteristics

NATIONAL CULTURE	UNCERTAINTY AVOIDANCE	MASCULINITY-FEMININITY	INDIVIDUALISM-COLLECTIVISM	POWER DISTANCE
Argentina	86	56	46	49
Australia	51	61	90	36
Austria	70	79	55	11
Belgium	94	54	75	65
Brazil	76	49	38	69
Canada	48	52	80	39
Chile	86	28	23	63
Colombia	80	64	13	67
Denmark	23	16	74	18
Finland	59	26	63	33
France	86	43	71	68
Great Britain	35	66	89	35
Germany	65	66	67	35
Greece	112	57	35	60
Hong Kong	29	57	25	68
India	40	56	48	77
Iran	59	43	41	58
Ireland	35	68	70	28
Israel	**81**	**47**	**54**	**13**
Italy	75	70	76	50
Japan	**92**	**95**	**46**	**54**
Mexico	82	69	30	81
Netherlands	53	14	80	38
Norway	**50**	**8**	**69**	**31**
New Zealand	49	58	79	22
Pakistan	70	50	14	55
Peru	87	42	16	64
Philippines	44	64	32	94
Portugal	104	31	27	63
South Africa	49	63	65	49
Singapore	8	48	20	74
Spain	86	42	51	57
Sweden	**29**	**5**	**71**	**31**
Switzerland	58	70	68	34
Taiwan	69	45	17	58
Thailand	64	34	20	64
Turkey	85	45	37	66
United States	46	62	91	40
Venezuela	76	73	12	81
Yugoslavia	88	21	27	76

Note: Larger numbers signify greater amounts of uncertainty avoidance, masculinity, individualism, and power distance.
Source: Based on Geert Hofstede, "Motivation, Leadership, and Organization: Do American Theories Apply Abroad?" *Organizational Dynamics* 9 (1980), 42-63.

How would the models and practices described in this book look if fitted to a culture that is characterized by collectivism and small power distance—a culture exactly opposite to the U.S. national culture? We will approach this third question by looking at practices in Israeli kibbutz organizations. The Israeli national culture is characterized by collectivism and very small power distance, and kibbutz work organizations are even more collectivistic than the Israeli norm shown in Table 19-1.

To summarize, we will begin by describing structure, motivation, leadership, and job design in Scandinavian, Japanese, and Israeli kibbutz organizations. That will give us an idea of how cross-cultural differences in individualism-collectivism and power distance affect organizational behavior. We will then conclude by comparing these descriptions with some of the ideas about management and organizational behavior that we introduced in earlier chapters.

Scandinavian Industrial Democracies

industrial democracies
Industrial organizations that are required by law to permit their members to govern themselves.

In Norway and Sweden, companies like Volvo, Saab, Hunsfos Paper, and Norsk Hydro include many participatory features in their management practices. Most Scandinavian firms are **industrial democracies**, organizations required by law to permit their members to govern themselves. In Norway, the first companies to become industrial democracies were government-owned firms. That was in the mid-1930s, after the formal establishment of national labor-management relationships. In the 1970s, many private firms followed suit and began to experiment with various types of shop-floor participation.[14] The movement began even earlier in Sweden, where labor-management cooperation was mandated by the government in 1928. Later laws led to the creation of works councils in many Swedish industrial firms. Those councils soon became participatory forums in which organizational policies related to working conditions, productivity, employee facilities, training, and corporate expansion programs could be developed through negotiations between labor and management representatives. Various types of shop-floor participation grew out of the policies formulated by these works councils.[15]

Structure. Because of their emphasis on worker participation, Norwegian and Swedish industrial democracies are structured differently than the typical American industrial firm. An American organization is usually built around a single, more-or-less bureaucratic hierarchy. Scandinavian firms, on the other hand, often incorporate a **collateral structure**—a second hierarchy of groups and committees that parallels and sometimes takes the place of the primary managerial hierarchy.[16] This collateral structure, a portion of which is shown in Figure 19-1, is made of three types of groups—works councils, special-interest committees, and semiautonomous work groups.[17] Let's look for a moment at what these groups do.

collateral structure A second hierarchy of groups and committees that parallels and sometimes takes the place of the primary managerial hierarchy in Scandinavian industrial democracies.

works council A committee of worker representatives who are elected by their peers and management representatives who are appointed by top management. It oversees policy formulation in Scandinavian industrial democracies.

The **works council** is composed of worker representatives who are elected by their peers and management representatives who are appointed by top management. There is usually only one works council in an organization. It is the council's responsibility to develop overall organizational policies and procedures.

---

[14] Fred E. Emery and Einar Thorsrud, *Form and Content in Industrial Democracy* (London: Tavistock, 1964), p. 24; *Democracy at Work* (Leiden, the Netherlands: Kroese, 1976), p. 46; and Bjorn Gustavson and Gerry Hunnius, *New Patterns of Work Reform: The Case of Norway* (Oslo: Universitetsforlaget, 1981), p. 37.

[15] *Job Reform in Sweden* (Stockholm: Swedish Employers' Confederation [SAF], 1975), p. 6.

[16] Dale E. Zand, "Collateral Organization: A New Change Strategy," *Journal of Applied Behavioral Science* 10 (1974), 63–89.

[17] H. Lindestad and G. Rosander, *The Scan Vast Report* (Stockholm: Swedish Employers' Confederation [SAF], 1977), pp. 3–12; Fred E. Emery and Einar Thorsrud, *Democracy at Work* (Leiden, the Netherlands: Kroese, 1976), pp. 27–32; and Joep F. Bolweg, *Job Design and Industrial Democracy: The Case of Norway* (Leiden, the Netherlands: Martinus Nijhoff, 1976), pp. 98–109.

FIGURE 19-1

The Structure of Advisory Committees in a Scandinavian Firm

The collateral structure of Kockums, Malmo, a shipbuilder, has two layers—general advisory committees clustered around the company's works council and special advisory committees charged with providing input about specific topics and issues. *Adapted with the publisher's permission from H. G. Jones,* Planning and Productivity in Sweden *(London: Croom Helm, 1976), p. 111.*

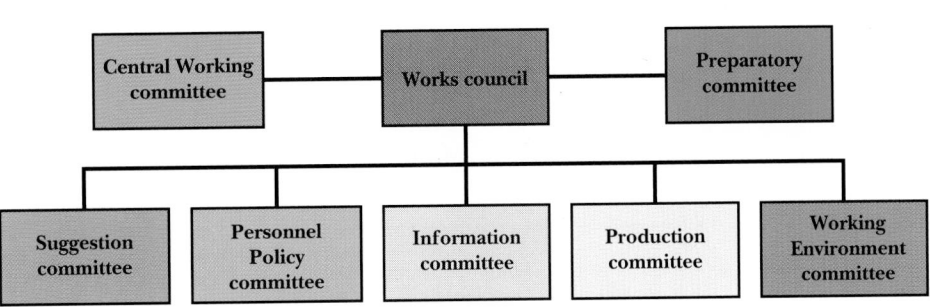

special-interest committees
Committees in Scandinavian industrial democracies composed of worker and manager representatives. They combine with middle management to produce yearly reports that assist works councils with the task of formulating company policies.

Works councils have little or no direct decision-making power, but they provide a forum in which worker representatives can express their opinions and thereby be instrumental in shaping the mission and strategic direction of the firm.

Special-interest committees, which are also composed of worker and manager representatives, provide the works council with advice on specific issues, such as job design, plant sanitation, personnel practices, and environmental safety. These committees combine with middle management to produce yearly reports that assist works councils with the task of formulating company policies. Such reports might include an analysis of water and air pollution caused by the company, a set of guidelines for curbing absenteeism, or a proposal about how to reduce the amount of inventory kept on hand.

Semiautonomous work groups consist of groups of employees who are given the responsibility for completing a particular job. As described in Chapter 17, in each group members negotiate the breakdown of work responsibilities into individualized tasks and decide who will perform these tasks. They may also have a say in hiring and firing decisions affecting group membership.

In addition to these three types of groups, works councils are typically supported by several general advisory committees located lower in the organization hierarchy. Such advisory groups may include suggestion committees, personnel-policy committees, or information committees. As a whole, the collateral system shown in Figure 19-1 provides a structure in which employees and management can work together to influence company policies and procedures.

Motivation. In Scandinavian industrial democracies, national welfare programs meet most lower-order needs for food, clothing, shelter, and health care. Methods of motivating employees, therefore, are aimed at satisfying higher-order needs for autonomy, growth, and development. The semiautonomous work groups we have just described are a good example of this approach, for these groups give employees the opportunity to control their own working lives—to manage themselves. Scandinavian managers consider semiautonomous groups an important means of encouraging productivity as well as combating turnover and absenteeism.

In most Scandinavian firms, wages consist of a base payment designed to further ensure employees' basic welfare, an incentive payment based on the productivity of the semiautonomous work group an employee interacts with, and a second incentive payment based on the number of job skills a worker has.[18] This system of payment reflects and reinforces the group-oriented structure of the Scandinavian industrial firm and also encourages continued personal development.

Leadership. As you might expect, Scandinavian managers often do not supervise employees directly, nor are they always required to issue direct orders to coordinate work activities. Instead, they usually function as boundary spanners who facilitate relations within and between groups while allowing employees to handle many coordination responsibilities themselves. Top managers span organizational boundaries. They gather information about the organization's environment, such as competitors' actions, changing customer tastes, and raw-material availability, and relay it to the works council. Then, with input from the company works council, top management charts the long-term policies that establish the direction and purpose of the firm.

Middle-level managers span the boundaries between top management and the shop floor. They serve as advisors to special interest groups or join these

---

[18] *Pay Reform in Sweden* (Stockholm: Swedish Employers' Confederation [SAF], 1977) p. 18; and H. Lindestadt and J. P. Norstedt, *Autonomous Groups and Payment by Result* (Stockholm: Swedish Employers' Confederation [SAF], 1973).

groups as managerial representatives. Lower-level managers span boundaries between semiautonomous work groups. They coordinate intergroup relations on the shop floor by helping to distribute raw materials and forward completed work. They also serve as interpersonal facilitators within semiautonomous groups, resolving conflicts and helping group members interact in the course of participatory decision making.

Job Design. In Scandinavian industrial democracies, jobs are frequently designed according to the sociotechnical approach described in Chapter 17. You will recall that this approach is based on the notion that individualized, standardized, simplified jobs not only prevent workers from satisfying social, esteem, and actualization needs but subject them to boring and repetitive work.[19] In an effort to guard against these negative effects of job simplification, Scandinavian industrial democracies divide work to be done into job clusters that are distributed to the semiautonomous work groups we've described. Group members then have the freedom to determine task assignments, work procedures, and personnel policies in their own groups.[20]

Large Japanese Corporations

Unlike Scandinavian industrial democracies, large Japanese corporations are based on hierarchical management procedures. However, they also support and depend on close interpersonal relations.[21] The Japanese organization is a product of many influences. Most important was the cultural seclusion of the country between 1634 and 1868 and the feudal and kinship traditions that developed during this era. The period of seclusion encouraged the development of a homogeneous culture based on feudal master-servant obligations that permeated the life of the family as well as societal life. These obligations created a kind of permanent state of dependence within families and between families and their feudal lords. Children were obligated to follow the wishes of their parents, who were required to obey their feudal master. Feudal *zaibatsu* organizations were built around these chains of vertical obligation, passing from children through father to master. The *zaibatsu* were large companies spanning many businesses and owned by wealthy, powerful clans.

[21] Peter B. Smith, "The Effectiveness of Japanese Styles of Management: A Review and Critique," *Journal of Occupational Psychology* 57 (1984), 121–36.

The period of Allied occupation that followed World War II also had a profound influence on Japanese organizations. During the occupation, in an effort to westernize Japanese businesses, American military commanders forcibly disbanded the *zaibatsu* and required companies to permit the formation of workers' unions. In addition, massive capital restoration programs funded by the U.S. and its allies enabled the Japanese to build modern, sophisticated production facilities. Subsequently, Japanese businesses reorganized into six industrial conglomerates (Mitsui, Mitsubishi, Sumitomo, Fuyo, Sanwa, and Dai-Ichi Kangyo) and a number of independent firms, such as Nissan, Toyota, and Sony. The six conglomerates were essentially the original *zaibatsu*, now centered in banks rather than feudal clans. In addition, a collection of smaller, "satellite" companies developed to supply the large corporations with raw materials, parts, and supplementary labor. These companies often started out under the management of a retiree from one of the large organizations.

[19] P. G. Herbst, *Alternatives to Hierarchies* (Leiden, the Netherlands: Kroese, 1976); and G. I. Susman, *Autonomy at Work: A Sociotechnical Analysis of Participative Management* (New York: Praeger, 1976).
[20] Ulla Ressner and Evy Gunnarsson, *Group Organised Work in the Automated Office* (Brookfield, Vt.: Gower, 1986).

Structure. Outwardly, the structures of most large Japanese corporations resemble the functional or divisional structures of large American companies. In fact, Japanese organization charts often display the same hierarchy of vertical relations that characterize an American firm's organization chart. In Japanese firms, however, these vertical relationships are often patterned after the parent-child (*oyabun-kobun*) relations of traditional Japanese families. In the organizational *oyabun-kobun* relationship, a subordinate is encouraged to feel loyal and obligated to his superior as well as dependent on him. This feeling of dependence in turn encourages—in fact, requires—acquiescence to autocratic, hierarchy-bound management policies and practices.[22]

Subordinates' loyalty and dependence is tested regularly by *ringi* decision making. In this process, decision making is initiated by subordinates, and possible decisions are circulated upward for superiors' approval. (The Japanese word *ringi* derives from *ringiseido*, which means "a system of reverential inquiry about a superior's intentions.") Though seemingly participatory, this form of decision making is often little more than an exercise in anticipating superiors' wishes. Subordinates try to make only those suggestions that their superiors will approve in order to avoid embarrassment.[23] Japanese aversion to standing out in a crowd and concerns about saving face would make such embarrassment intolerable.

Although at first glance, Japanese and American organization charts look quite similar, there is a critical difference between the Japanese and American approaches to structuring large organizations. In American organizations, the vertical lines of command that appear in organization charts are meant to be the only formal channels of communication. To communicate with someone at the same horizontal level, an employee is expected to pass a message up the hierarchy to a superior who then sends it downward to the final recipient. In Japanese corporations, however, whether they are conglomerates or independents, certain formally designated *horizontal* relationships are accorded the same degree of importance as the vertical relationships depicted in the organization chart. These horizontal relationships, which allow communication to flow across the hierarchy rather than having to go up and down, connect managers who entered the company at the same time and are encouraged by several socialization processes.

More specifically, the managerial hierarchy in most large Japanese firms is discontinuous. Lower-level managers are recruited from the ranks of shop-floor workers but cannot rise above the rank of foreman. Middle-level managers are recruited from Japanese universities and are able to rise to the top of their company. When they first join a large Japanese firm, candidates for middle-management positions are trained in groups, and throughout their careers with the company, they are encouraged to maintain contact with other members of their "entering class." Most white-collar employees in large corporations remain with the same firm from college graduation until retirement. During the period of training, a Japanese manager rotates from one functional area to another, often becoming more of a generalist than a specialist and cultivating a collection of horizontal linkages that unite him with management peers across functional boundaries.[24]

latticework hierarchy The structure of vertical and horizontal relationships found in many large Japanese corporations.

This **latticework hierarchy** of vertical and horizontal relationships, which connects Japanese superiors and subordinates in a single managerial unit, is

---

[22] Ronald Dore, *British Factory—Japanese Factory* (Berkeley: University of California Press, 1973); Peter F. Drucker, *Management* (New York: Harper & Row, 1974); and Nina Hatvany and C. V. Pucik, "Japanese Management Practices and Productivity," *Organizational Dynamics* 9 (1981), 5–21.

[23] For further discussion on this topic see S. Prakash Sethi, Nobuaki Namiki, and Carl L. Swanson, *The False Promise of the Japanese Miracle: Illusions and Realities of the Japanese Management System* (Boston: Pitman, 1984), pp. 34–41.

[24] Robert E. Cole, *Japanese Blue Collar* (Berkeley: University of California Press, 1971) p. 122; and R. J. Samuels, "Looking behind Japan Inc.," *Technology Review* 83 (1981), 43–46.

analogous to a comprehensive lateral linkage mechanism. Continuing relations among management peers from different functional areas, such as marketing and manufacturing, help stimulate harmony and coordination between functional groups. Do you think, then, that the Japanese latticework structure resembles the American matrix structure? Both types of structure involve simultaneous functional and divisional grouping. Note, however, that the latticework structure falls short of being a true matrix, because in large Japanese firms, decision-making authority remains highly centralized. Thus the Japanese latticework structure is unique.

Motivation. Managing motivation in large Japanese corporations is primarily a matter of reinforcing each employee's sense of loyalty, obligation, and dependence on superiors and coworkers. The widespread practice among Japanese firms of offering lifetime employment (to age 56) to their permanent employees goes a long way toward encouraging workers' loyalty. Japanese employees find it difficult to behave disloyally toward a firm that is willing to commit itself to them up to their retirement.[25]

nenko system A Japanese system of payment in which the pay an employee receives is determined by a basic wage plus merit supplements and job-level allowances.

Loyalty is also encouraged by the **nenko system** of wage payment that is used throughout Japan. Under the *nenko* system, the employee's pay is composed of a basic wage plus merit supplements and job-level allowances. The basic wage, which constitutes about 55 percent of total pay, consists of the employee's starting wage plus yearly increases. Those increases are determined (in order of importance) by seniority, or length of service with the company; age; and supervisory ratings on such qualities as seriousness, attendance, performance, and cooperativeness.[26]

Merit supplements make up an additional 15 percent of the employee's pay and are based on supervisory assessments of specific job behaviors. They are meant, in principle, to reward exemplary performance. In fact, merit supplements are heavily influenced by seniority, because they are calculated as a percentage of the basic wage. Moreover, junior employees' performance is typically rated below senior employees' work regardless of real differences between the two.[27] Clearly, Japanese merit supplements reward loyalty and longevity with the company.

Job-level allowances, which account for about 30 percent of each worker's total pay, are based on the importance of each worker's job in relation to the other jobs in the organization. Job-level allowances may sound similar to pay increments that result from American job-evaluation procedures. In Japan, however, each employee's position in the hierarchy of jobs—which affects his job-level allowance—is more directly influenced by seniority than by skill.[28]

Thus, as you can see, seniority is the single most important factor in determining a Japanese worker's compensation. It affects the basic wage, it affects merit supplements, and it affects the job-level allowances. The large Japanese firm is like a family in the sense that its employees spend their lives in a stable social setting and receive positions of increasing importance as they grow older.[29] Along with the *nenko* method of compensation, this familylike system rewards its members for loyalty to the company over everything else. Employees' decisions

[25] Cole, *Japanese Blue Collar*, pp. 72–100; and Dore, *British Factory—Japanese Factory*, pp. 74–113. It is important to note that *nenko* employment occurs only in large firms and applies to no more than a third of Japan's labor force; for further information, see T. K. Oh, "Japanese Management: A Critical Review," *Academy of Management Review* 1 (1976), 14–25.

[26] Cole, *Japanese Blue Collar*, p. 75.

[27] Dore, *British Factory—Japanese Factory*, p. 112.

[28] Cole, *Japanese Blue Collar*, p. 79; and Dore, *British Factory—Japanese Factory*, p. 390. Review Chapter 9 for further information on job evaluation in the United States.

[29] Rodney Clark, *The Japanese Company* (New Haven, Conn.: Yale University Press, 1979), p. 38.

to attend work and to perform productively grow out of this sense of loyalty and obligation to the familylike firm.

Leadership. The primary leadership task in large Japanese corporations is to guarantee the continued existence of vertical dependence—the acceptance of hierarchical relationships and obedience to superiors. At the lowest levels of management, shop-floor foremen (Japanese supervisors are still virtually all male) wear two hats. By day, they perform the same types of jobs as their subordinates. Indeed, to the casual observer, superior and subordinate seem almost indistinguishable at work. Away from work, however, foremen inquire at the homes of sick employees about their health and the welfare of their families. From time to time, they also do such things as giving subordinates small gifts and hosting social events for them and their families. In return, subordinates are led, by their sense of *giri*, or "obligation," to perform as their supervisors direct. Subordinates cannot fully repay their obligations by returning visits, reciprocating gift giving, helping sponsor social events, or even following orders. Japanese *giri* requires unending obedience. Lower-level management's primary leadership function, then, is to maintain subordinates' followership. This system enables management in large Japanese firms to operate autocratically.[30]

Leaders in middle and upper management ensure similar followership among their subordinates by immersing them in a program of intensive socialization during organizational entry. Initially, groups of new members learn the history of the organization, its mission, and its values. Then, as we have indicated, these "entering classes" spend several years together rotating among functional departments to develop horizontal ties and acquire a generalist's understanding of the organization. During this period, management leaders emphasize vertical relations by teaching and repeatedly encouraging the use of the *ringi* decision-making process that we have already discussed. By the time they have completed the process of socialization, managerial employees have learned the importance of followership, called *tsukiai*, and typically behave accordingly.[31]

Job Design. In the United States, jobs are usually individualized and allocated among the members of an organization on a one-to-one basis. Because many Japanese corporations adopted American industrial engineering practices during the postwar occupation, Japanese assembly lines today are often composed of similarly individualized tasks.[32] However, the interpretation of such jobs is de-

---

[30] Dore, *British Factory—Japanese Factory*, p. 228.

[31] Reiko Atsumi, "Tsukiai—Obligatory Personal Relationships of Japanese White-Collar Company Employees," *Human Organization* 38 (1979), 63–70.

[32] Koji Matsumoto, *Organizing for Higher Productivity: An Analysis of Japanese Systems and Practices* (Tokyo: Asian Productivity Organization, 1982), pp. 27–31.

In 1990 the Japanese Ministry of International Trade and Industry (MITI) issued an industrial policy paper in which, against tradition, it put human concerns before strictly business issues. MITI came out strongly for longer vacations for employees and better child-care facilities for them. One result was the establishment in Tokyo of a combined nursery school and home for the aged. Some of the older people help to care for the young ones, and all gain from companionship with the children. There are a number of such combined facilities in the United States, but formal child or elder care has lagged in Japan largely because families in which both parents work outside the home are rare and the extended family system usually provides such care.
Source: "Looking Ahead," Fortune, September 24, 1990.

746

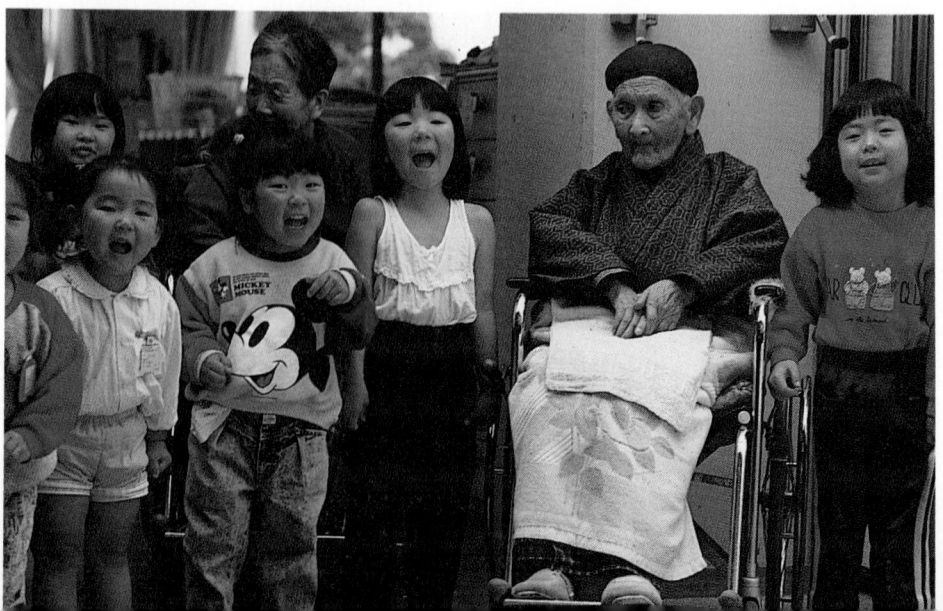

cidedly Japanese. Japanese workers consider the tasks they perform parts of a larger, group job. Consistent with this interpretation, they do not wish to be publicly recognized for good personal performance, and they prefer to share the blame for poor productivity with others, regardless of personal fault.[33] Although the tasks themselves may be individualized, task performance is seen as a group activity.

Away from assembly lines, work is more often divided into group tasks and assigned to groups of employees. Quality Circles (QCs), which we discussed in Chapter 17, are representative of the kinds of groups often formed among shop-floor employees to encourage working together. On the surface, a QC might seem similar to a semiautonomous work group formed by the sociotechnical method of job design employed in Scandinavian industrial democracies. However, semiautonomous groups in Scandinavian firms are formed by means of participatory decision making in works councils and other advisory committees, whereas groups in Japanese corporations are formed by edict from above. Furthermore, Scandinavian groups are intended to allow workers to participate directly in managing their working lives. Participation of this type is rare in Japanese work groups. In Japan, the family, or parent-child, model rules the work group as it does the larger society, and superiors retain control over subordinates' actions at all times.

Israeli Kibbutz Industries

Israeli kibbutzim combine the member commitment found in large Japanese corporations with the participatory aspect of Scandinavian industrial democracies. The **kibbutz** (*kibbutzim* is the plural form) is a close-knit community of people organized on the principles of collective ownership and direct participation in self-governance. The kibbutzim, which range in size from 40 to 2,000 members, were started early in the 20th century by young European Jews who moved to Palestine to escape anti-Semitic persecution. Consistent with the philosophies of the European youth movements of the day, kibbutz settlers favored Marxist economics and a rural, naturalistic lifestyle. These beliefs led them to establish a number of small agricultural communities in which they shared everything from basic foodstuffs to civic obligations.

Newly formed kibbutzim were immediately threatened by Arab neighbors, who held conflicting religious beliefs and whose lifestyles were based on open-land herding. In addition, most *kibbutzniks* (kibbutz members) had little farming experience and were quite poor. In order to survive, members had to share the few possessions they had brought with them as well as the wages they received for working outside the kibbutz. Of course, that fit in with their Marxist beliefs. After several years of hardship, the success that followed reinforced kibbutzniks' belief in the importance of sharing and helping each other out. Consequently, contemporary kibbutzim are still organized around principles that stress sharing both work responsibilities and the material results.[34]

Structure. In most kibbutzim, even clothing is owned collectively, and it is cleaned, mended, and distributed in central facilities. Children spend time with their parents each day but are raised in communal quarters. Individuals and families alike eat many of their meals in central dining halls. Kibbutzniks are

kibbutz A close-knit community of people located in Israel and organized on the principles of collective ownership and direct participation in self-governance.

[33] William G. Ouchi, *Theory Z: How American Business Can Meet the Japanese Challenge* (Reading, Mass.: Addison-Wesley, 1981), pp. 48–49.

[34] H. Darin-Drabkin, *The Other Society* (Harcourt, Brace, & World, 1963) pp. 66–70.

allowed only a limited number of personal possessions, most of which are received as gifts from nonkibbutz relatives.[35] Because the material goods of each kibbutz are collectively owned, decisions about their acquisition and distribution must be reached through public consensus. Thus weekly assembly meetings are held in which voting members participate in determining kibbutz policies and procedures.

In these meetings, interested parties raise issues that need to be decided, and after open discussion, participants are polled and the decision is made. The secretariat—an administrative board consisting of elected officials—is empowered only to implement policies approved by the kibbutz assembly. No official is permitted to act outside assembly mandates. Thus, as Figure 19-2 shows, the structure of an Israeli kibbutz is inverted, unlike the structures of American, Scandinavian, and Japanese firms. The officers of the kibbutz occupy positions of lower status than the assembly, which consists of the kibbutz membership as a whole.

The kibbutz depicted in Figure 19-2, like many kibbutzim, operates one business—a farm. Some kibbutzim, however, have added industrial branches to supplement their agricultural activities. These branches are organized around industrial assemblies and internal governance processes that reflect those of the larger kibbutz. Each industrial branch's officials are elected by its work force and are charged with implementing decisions reached by the branch's assembly. Decisions that might affect the larger kibbutz—major decisions about industrial operations, for example—are taken by the full kibbutz assembly. Officials of the kibbutz secretariat oversee industrial operations to make sure they conform with

---

[35] Melford Spiro, *Kibbutz: Venture in Utopia* (Cambridge, Mass.: Harvard University Press, 1958), p. 207; Yosef Criden and Saadia Gelb, *The Kibbutz Experience: Dialogue in Kfar Blum* (New York: Herzl, 1974), p. 100; Joseph Blasi, *The Communal Experience of the Kibbutz* (New Brunswick, N.J.: Transaction Books, 1986).

FIGURE 19-2

The Structure of an Agricultural Kibbutz

Amitai Etzioni, *The Organizational Structure of the Kibbutz* (New York: Arno Press, 1980), p. 266. Reprinted with the author's permission.

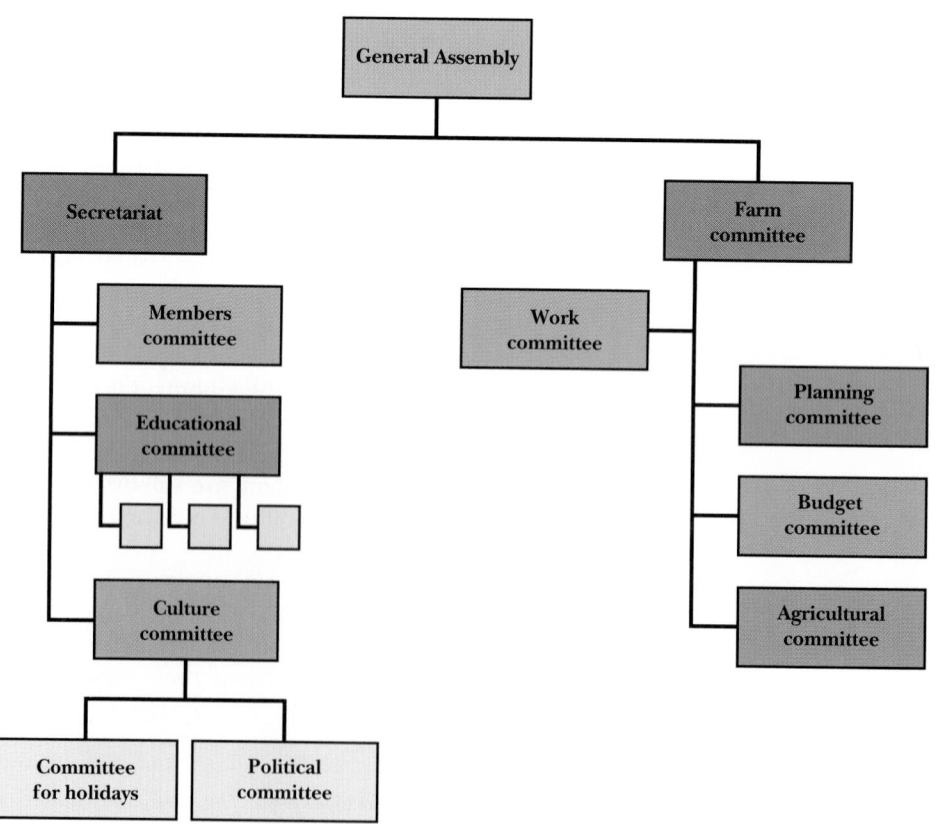

kibbutz policies. Nevertheless, kibbutz industrial branches are the most decentralized, participatory form of industrial organization in the world.[36]

Motivation. Kibbutzniks are motivated by concerns about the security of their community and by values that encourage hard work. The founders of the first kibbutzim believed that people realized their true worth only by working hard, and motivation in the kibbutz has continued to derive from the notion that physical labor is an intrinsically important endeavor. Simcha Ronen compared the job satisfaction of workers in a kibbutz industrial branch with the satisfaction of industrial employees in Israel's private sector. Ronen found that kibbutzniks' satisfaction was strongly influenced by the tasks they performed. In contrast, private-sector workers in Ronen's study based job satisfaction on such things as power, prestige, and wealth. Their degree of satisfaction was strongly related to the wages and benefits they received, not to the tasks they performed.[37]

Kibbutznik motivation also grows out of concerns for community well-being. Within the kibbutz, every member derives equal benefit from kibbutz resources. For example, everyone gets the same amount and kind of food, everyone gets the same clothing to wear, everyone lives in a similar kind of structure. This system of wageless compensation based on egalitarian sharing—except for cash allowances for travel outside the kibbutz and small amounts of pocket money for miscellaneous purposes—guarantees the well-being of all kibbutz members. Because members need not concern themselves with personal welfare, they are free to direct their full attention to improving the state of the kibbutz as a whole. Kibbutzniks identify this freedom from concern for personal security a valuable reward of kibbutz membership. Desires to preserve this freedom are an important motivational force underlying kibbutznik performance.[38]

Leadership. Leadership is shared by all members of each kibbutz, being vested in the kibbutz assembly rather than in the hands of a small management group. Kibbutz officials, as we have already noted, occupy positions *beneath* the general assembly in the inverted kibbutz hierarchy. They are elected by the assembly, serve fixed terms of office, and cannot hold the same office for more than one consecutive term. As a result, the ability of any officeholder to amass the power needed to function as an autonomous leader is strictly limited. On occasion, a small number of members may rotate among the various kibbutz offices, allowing the possible emergence of a dominant group. However, since participation in the kibbutz assembly is nearly universal—all members who are kibbutz-born and 19 or older or who have completed a one-year naturalization program can vote in assembly meetings—officials are under the watchful eyes of every other member. It is virtually impossible to try to usurp the assembly's power without being caught.

Similarly, leadership in kibbutz industrial branches is vested in branch assemblies. Officials in each branch are elected by the branch's assembly and are required to rotate out of office after completing their term. As in the wider kibbutz, ethical codes and member sentiment both oppose the emergence of officeholder-leaders.[39]

[36] A. S. Tannenbaum, B. Kavcic, M. Rosner, M. Vianello, and G. Weiser, *Hierarchy in Organizations* (San Francisco: Jossey-Bass, 1974), p. 34.

[37] Ronen, "Personal Values: A Basis for Work Motivational Set and Work Attitudes," *Organizational Behavior and Human Performance* 21 (1978) 80–107.

[38] Criden and Gelb, *The Kibbutz Experience*, p. 33.

[39] Criden and Gelb, *The Kibbutz Experience*, pp. 37–57; and Blasi, *The Communal Experience of the Kibbutz*, p. 112.

Job Design. All kibbutz jobs are shared, and all kibbutzniks work at them from childhood onward. Adult members of the kibbutz feel that introduction to the community's work at a young age is tremendously important, because it teaches children about the relationship between personal productivity and the well-being of the overall community.[40] Reflecting this primary concern with the welfare of the kibbutz, jobs are designed and assigned so as to enhance the society's welfare. Personal abilities and interests are not often given much attention. In practice, if a kibbutz member has skills that qualify her for a particular job, and if that job is open, she will be assigned to it. However, members who fail to make such a match are assigned to a general labor pool and may be drafted into any kibbutz branch that is short of workers.

Jobs are assigned by branch officials operating according to assembly mandates. All workers rotate in and out of less desirable jobs so as to share the burden of disagreeable work. For example, the lowest-level kitchen jobs are filled in this manner, although a permanent staff of kitchen workers is also maintained. So although kibbutz jobs are designed and assigned as individualized tasks, because members rotate among jobs and never get attached to any one, they tend to see their jobs as just a small part of the overall work of the community.

A CROSS-CULTURAL COMPARISON

We have briefly described several important features of Scandinavian industrial democracies, large Japanese businesses, and Israeli kibbutz industries. How do these three kinds of organizations compare with one another and with the models and practices described throughout this book? What cultural effects are uncovered by such comparisons?

To respond to these questions, recall how most American business firms are structured. They are hierarchies in which coordination is achieved mainly by standardization implemented from above and secondarily by direct supervision (Chapter 15). Employee attendance and performance are motivated by rewards that satisfy personal needs and are received in proportion to personal performance (Chapter 7). Leadership in American firms is largely a task of coordinating the work of individuals and looking after their personal well-being (Chapter 12). Jobs in the United States are designed and assigned as individualized tasks (Chapter 17).

These features of American business firms would seem to reflect the individualism and large power distance of the American national culture. Let's isolate the effects of individualism by looking for similarities in the way things are done in U.S. business firms and Scandinavian industrial democracies. In both, people are motivated to do their jobs by the desire to satisfy personal needs, and achievement on the job is seen as attributable to individual effort. In Scandinavian countries, however, specific tasks and procedures are often performed by groups and group incentives are sometimes used to motivate a high level of performance.

To examine some of the effects of collectivism on organizational behavior, let's look for similarities in the workings of large Japanese businesses and Israeli kibbutzim, both are surrounded by collectivistic national cultures. One similarity is that work motivation grows out of a sense of loyalty and a concern for the well-being of everyone in the organization. Another similarity is that both Japanese and Israeli employees view job performance as something people can accomplish only by working together.

Based on these comparisons, it appears that the amount of individualism or collectivism in the national culture has major effects on employee motivation

---

[40] Milford Spiro, *Children of the Kibbutz* (Cambridge, Mass.: Harvard University Press, 1975), p. 266.

FIGURE 19-3

A Two-Dimensional Comparative Analysis

Individualism

American business firms	Scandinavian industrial democracies
Large Japanese corporations	Israeli kibbutz industries

Larger Power Distance

Smaller Power Distance

Collectivism

and job design. Cultural individualism focuses attention on the performance and satisfaction of the individual. Cultural collectivism directs attention to the performance and satisfaction of the group.

What about the effects of power distance? Let's begin by examining the workings of American business organizations and large Japanese businesses, both of which are located in national cultures oriented toward large power distance. Structurally, these firms strongly emphasize coordination processes that are based on a combination of rules, procedures, and direct orders issued by hierarchical superiors. They use direction rather than participation to integrate work activities. Leadership activities reflect this directive orientation, focusing on supervising subordinates and making sure that superiors' orders are carried out.

However, in Scandinavian industrial democracies and Israeli kibbutz industries, both of which are surrounded by national cultures with small power distance, structural relationships are far more participatory. People who occupy positions lower in the structural hierarchy are able to advise hierarchical superiors or even become directly involved in decision making. Leadership is more a process of facilitating communication and mutual adjustment than one of ensuring that subordinates follow the directives of superiors.

From this pattern, we can conclude that the power distance of the surrounding national culture has strong effects on organizational structure and leadership processes. Large power distance stimulates directive structural relationships and leadership that is focused on supervision. Small power distance supports participatory structural relationships and leadership that concentrates on interpersonal facilitation.

In sum, the cross-cultural comparison depicted in Figure 19-3 suggests that features of the surrounding national culture can have significant effects on the characteristics of an organization and on the management of organizational behavior. With this in mind, we will now see how to fit management procedures to the demands of different national cultures.

MANAGING IN A MULTICULTURAL WORLD

Although the cross-cultural differences we have discussed are still very much in evidence, it is sometimes suggested that management practices throughout the world are becoming more alike as nations grow modern and prosperous.[41] Consistent with this suggestion, practices developed in one culture are occasionally borrowed for use in another. For instance, sociotechnical interventions patterned after Scandinavian programs and quality circles resembling Japanese QC groups are so prevalent in contemporary American companies that they are considered part of the U.S. approach to job design. You will recall that we discussed both of these interventions in Chapter 17.

Instances of cross-cultural borrowing like these seem to support the **convergence hypothesis**, that organizations and management practices throughout the world are growing more alike.[42] In a review of studies that have examined this hypothesis, John Child found some evidence for convergence but also some evidence for divergence, or continued cross-cultural differences. Interestingly, Child found that the studies that supported convergence typically focused on organizational variables, such as structure and technology, whereas the studies that revealed divergence in practices usually concerned employee attitudes, beliefs, and behaviors.[43] He concluded that organizations themselves may be becoming more alike throughout the world but that people in these organizations are maintaining their cultural distinctiveness. Therefore, the cultural convergence hypothesis is not entirely valid: management in a multicultural world currently requires an understanding of cultural differences and will continue to do so for quite some time.

Understanding Behavior in Other Cultures

How can the information in this chapter help managers understand cultural differences? You can find out by putting yourself into the managerial role and applying the Hofstede model to diagnose several intercultural differences. In the following examples, try using the four dimensions of the model to explain the differences described.

1. *Feelings about progress.* Being modern and future oriented is highly valued in many of today's national cultures. In others, tradition, the status quo, and the past are highly revered. From the modernist perspective, something that has been around for a while may seem old-fashioned or obsolete. To a traditionalist, the same thing may be perceived as trustworthy, proven, and worthwhile.

2. *Tendencies toward confrontation or consensus.* In some national cultures, it is important to smooth over differences to preserve agreement. Emphasis is placed on building consensus among coworkers and avoiding personal confrontation. In other national cultures, conflict and confrontation are accepted or even encouraged. In such cultures, conflict is often perceived as a signal of the need for change.

3. *Locus of control.* Some national cultures instill a sense of personal responsibility for the outcomes of individual behaviors. Others focus on external, social causes to explain similar outcomes. Although rewarding people for personal performance is logical in a culture that values personal respon-

convergence hypothesis A theoretical assertion that organizations and management practices throughout the world are growing more alike.

AN EXAMPLE: EUROTUNNEL AND TRANSMANCHE LINK
The building of the Chunnel, the tunnel under the English Channel that will link England and France by 1993, was plagued by cost overruns and bickering. Transmanche Link, the Franco-British construction consortium doing the work for the Chunnel's owner, Eurotunnel, needed a project manager who could work with people from different countries. They recruited Jack K. Lemley, an American who was then consulting on a hydroelectric project near Katmandu, Nepal, the latest in his long list of construction jobs around the world. After a few months on the job Lemley won high praise from both Eurotunnel and Transmanche for greatly improving coordination between the French and British construction crews. According to Lemley, "It's an advantage to be a third nationality . . . . I can't be accused of taking sides." (Shawn Tully, "The Hunt for the Global Manager," *Fortune*, May 21, 1990, p. 142.)

[41] Clark Kerr, John T. Dunlop, Frederick H. Harbison, and Charles A. Meyers, *Industrialism and Industrial Man* (Cambridge, Mass.: Harvard University Press, 1960), pp. 282–88; John Kenneth Galbraith, *The New Industrial State* (Boston: Houghton Mifflin, 1967), pp. 11–21; and F. Harbison, "Management in Japan," in *Management in the Industrial World: An International Analysis*, ed. F. Harbison and C. A. Meyers (New York: McGraw-Hill, 1959), pp. 249–64.

[42] Peter J. Dowling and Randall S. Schuler, *International Dimensions of Human Resource Management* (Boston: PWS-Kent, 1990), pp. 163–64.

[43] Child, "Culture, Contingency, and Capitalism in the Cross-National Study of Organizations," in *Research in Organizational Behavior*, ed. Larry L. Cummings and Barry M. Staw (Greenwich, Conn.: JAI Press, 1981), pp. 303–56.

sibility, it may not be understood in a culture that believes that behaviors are caused by outside forces.

4. *Status and social position.* In some national cultures, status is accorded on the basis of family, class, ethnicity, or even accent. High-status people in such cultures are often allowed to impose their will on people of lower status even when such people are as knowledgeable and competent as they. In other cultures, where status is earned through personal achievement, shared governance—majority rule or participatory decision making—is more apt to be valued over personal fiat. In such cultures, expertise usually outranks social position in determining who will be involved in decision-making procedures.[44]

Were you able to explain the first example? As you probably figured out, differing feelings about progress are produced by cross-cultural differences in uncertainty avoidance. Cultures with aversion to uncertainty honor tradition and feel threatened by new ways of doing things. Cultures with tolerance for uncertainty more readily embrace modern ways.

The second example focuses on conflict avoidance, a cultural tendency that is also closely associated with uncertainty avoidance. Simply put, conflict creates uncertainty, and cultures that cannot deal with uncertainty cannot handle the competition and aggression that conflict unleashes. In contrast, cultures that can tolerate uncertainty are also able to cope with conflict.

The third example concerns locus of control, and as you have probably surmised, arises out of cross-cultural differences in individualism and collectivism. On the one hand, the sense of personal responsibility stimulated by believing that the locus of control for personal behavior lies inside the individual is consistent with the norms and values of an individualistic national culture. On the other hand, the focus on social causes for behaviors that is prompted by an external locus of control is compatible with cultural collectivism.

The fourth example shows how cultural differences in power distance can show up as differences in the way status and social position are accorded and perceived. Cultures in which status and position are birthrights—the special-privilege approach—also tend to be oriented toward large power distance. In contrast, cultures in which status and position are awarded according to personal abilities—the equal opportunity approach—are more inclined toward smaller power distance.

APPLYING THE DIAGNOSTIC MODEL

Diagnostic Question 3: Based on your diagnosis of cultural differences, what adjustments should you make to familiar management practices to fit them to the current cultural situations?

Diagnostic Question 4: Are there colleagues within the organization who can help you with the necessary adaptation? Should others outside of the organization be asked to help instead?

Diagnostic Question 5: Might other new entrants in the future benefit from what you learn as you cope with cultural adjustment in the present? Will informal mentoring provide future entrants with enough guidance? Should a formal adjustment program be designed and implemented instead?

Managing Cross-Cultural Differences

Understanding international differences such as the ones depicted in our four cross-cultural examples is important for managers. It can help them determine whether they need to alter familiar management practices before using them abroad. Looking beyond our four examples, a glance back at Table 19-1 indicates that the national cultures of the United States and Canada are approximately equal in terms of power distance, uncertainty avoidance, masculinity, and individualism. Owing to this similarity, U.S. managers can expect to succeed in Canada and Canadian managers can anticipate working effectively in the United States without making major adjustments to customary management practices.

According to Hofstede's findings, however, the level of uncertainty avoidance in Denmark is about half that found in the United States and Canada. Therefore, American managers are likely to find it necessary to tailor their normal way of doing things if required to work in Denmark. More generally, managers

[44] For additional examples, see Leonard Sayles, "A 'Primer' on Cultural Dimensions," *Issues and Observations of the Center for Creative Leadership* 9 (1989), 8–9.

How Ugly Are Today's Americans?

The year 1958 marked the publication of *The Ugly American*, a novel in which William J. Lederer and Eugene Burdick criticized the United States foreign policy in Southeast Asia. The late 1950s were a time of unrest throughout the area as the French fought Communist armies in Vietnam and Russian agents spread Communist doctrine in the neighboring countries of Laos, Cambodia, and Burma. Through a sequence of stories about events in the fictional country of Sarkhan, Lederer and Burdick described how the Russians learned the local language, traveled the countryside, and provided basic necessities that made the lives of farming peasants somewhat easier. In contrast, according to Lederer and Burdick, American ambassadors assigned to duty in Southeast Asia often refused to learn the language, rarely if ever ventured out of their embassy compounds, and were interested only in building highways and dams that had high political visibility but lacked useful function. It was thus no wonder that the Russians were more successful in winning the hearts and minds of the local population. By bringing in American ways and expecting the locals to adapt, the Americans were devaluing the surrounding national culture and offending the very people they had been sent to assist.

Lederer and Burdick's book was a huge success in the U.S. and the term *ugly American* was soon applied to anyone who went abroad and refused to respect and abide by local customs. Discussions in schools, businesses, and churches focused on the importance of developing greater sensitivity and respect for the many different cultures throughout the world. Americans agonized over their cultural "ugliness" and sought ways to eliminate it.

More than three decades later, late in 1990, a military coalition that included a large contingent of American soldiers was gathering in Saudi Arabia to shield against an Iraqi attack from neighboring Kuwait. The dominant religion in Saudi Arabia, Wahhabism (a puritanical strain of Sunni Islam), forbade women to drive, to travel unaccompanied, or to expose anything other than their hands and eyes in public. At the request of Saudi officials, American military commanders instructed all female soldiers to wear long sleeves (despite the desert heat), refrain from driving in Saudi towns, and travel in groups that included men.*

* William Dowell, Dean Fischer, and Christopher Ogden, "Lifting the Veil: A Secretive and Deeply Conservative Realm, Saudi Arabia Suddenly Finds Itself on the Sword Edge of Change," *Time*, September 24, 1990, pp. 38–44.

These instructions caused an immediate uproar in the U.S. In speeches before Congress, representatives likened the Saudi's treatment of women to the treatment of blacks under South Africa's system of apartheid. Women's-rights organizations roundly criticized Saudi practices and the American military's hasty compliance. Breaking ranks with the military, some female soldiers even complained during television interviews that the restrictions were unfair.

The controversy in Saudi Arabia grew out of a clash between the norms and values of the increasingly feminine U.S. national culture and those of the strongly masculine Saudi national culture. Who was right in this situation—those Americans who argued that ugliness should be avoided and that female soldiers should observe local customs, or those Americans who argued instead that treating women differently from men can never be justified and that Saudi traditions should be overturned? If you had been the military commander charged with deciding what to do, what decision would you have made? What facts or opinions would you use to justify your decision? Would you be an ugly American?

working in national cultures characterized by *weaker uncertainty avoidance* must learn to cope with higher levels of anxiety and stress while reducing their reliance on planning, rule making, and other familiar ways of absorbing uncertainty. How to deal with this sort of situation? One way is to develop off-the-job interests and activities. You can take weekend trips to places of historical interest, study the culture's art and music, participate in sports activities. On the other side of the coin are managers working in cultures with *stronger uncertainty avoidance* than they are accustomed to. They must learn not only to accept but to participate in the development of seemingly meaningless rules and unnecessary planning in order to help other organization members cope with stressful uncertainty. Rituals that at first glance might appear useless or even irrational may serve the very important function of diminishing what is perceived to be an otherwise intolerable level of uncertainty.

Let us next look at the dimension of masculinity-femininity. Female managers working in cultures that display *more cultural masculinity* than their own face the prospect of receiving less respect at work than they feel they deserve (see the "Management Issues" box for a related discussion). To cope with gender discrimination of this sort, a female manager may have to seek out male mentors in senior management to secure her place in the organization. Conversely, male managers in national cultures that display *more cultural femininity* than their own must control aggressive tendencies and learn to treat members of both sexes with equal dignity and respect. Acting as mentors, female managers can help men accomplish transformations of this sort by demonstrating that women are as adept at their jobs as men.

The next dimension is individualism-collectivism. Managers who must work in national cultures that are more *individualistic* than their own must first learn to cope with the sense of rootlessness that comes from the absence of close-knit group relationships. They must also learn not to be embarrassed by personal compliments, despite their belief that success stems from group effort. At work, they must develop an appreciation of the importance of rewarding individuals equitably and must adjust to the idea that organizational membership is impermanent. Conversely, managers attempting to work in cultures that are more *collectivistic* than their own must learn to deal with demands for self-sacrifice in support of group well-being. They must also learn to accept equal sharing in lieu of equity and exchange at work. Consequently, they must refrain from paying individual employees compliments and instead praise group performance. Managers adjusting to collectivistic national cultures must also learn to understand that belonging to an organization in such cultures is more than just a temporary association. It is an important basis of each employee's personal identity.

Finally, managers working in cultures that favor *less power distance* must cope with initial discomfort stemming from an unfamiliar decentralization of authority and from feeling less in control. They must also learn to be less autocratic and more participatory in their work with others. On the other hand, managers facing cultural tendencies toward *more power distance* must accept the role that centralization and tall hierarchies play in maintaining what is deemed to be an acceptable level of control. They must also adopt a more authoritarian, autocratic style of management, and they may even find that subordinates, if asked to participate in decision making, will refuse on the grounds that decision making is management's rightful job.

SUMMARY

Whether comparisons are made within a single *national culture* or across different national cultures, no two organizations are exactly alike. And no two people in the world hold exactly the same beliefs and values. Thus our discussions in this chapter have necessarily involved generalization. Not every Japanese organization has a fully developed *latticework hierarchy*, and not every kibbutznik is completely collectivistic. Nevertheless, firms in a particular national culture are more like each other than they are like organizations in other national cultures. Moreover, people in the same national culture tend to think and act more similarly than people from different cultures. Cross-cultural differences exist and can have significant effects on the management of organizational behavior.

Many of these cross-cultural differences are captured by four dimensions: *uncertainty avoidance*, *masculinity-femininity*, *individualism-collectivism*, and *power distance*. The latter two of these dimensions explain many of the differences between American, Scandinavian, Japanese, and Israeli *kibbutz* management practices. Compared to American firms Scandinavian companies are far more participatory, Japanese organizations involve stronger feelings of obligation and dependence, and Israeli kibbutzim require greater member commitment to community well-being. Considered together, the four dimensions are helpful in understanding why people in a particular national culture behave the way they do and can be useful to managers as they adapt familiar management practices for use in unfamiliar cultures.

REVIEW QUESTIONS

1. Compared to the national culture of Sweden, what level of uncertainty avoidance characterizes the national culture of the United States? Of the two countries, in which one would you expect to find greater evidence of ritualistic behavior? Why? How would your answers to these questions change if you were asked to compare the U.S. with Greece?

2. Hofstede's findings indicate that the U.S. national culture at the time of his research was more masculine than many of the other national cultures he examined. In your opinion, is the American culture still as masculine as Hofstede's research suggests? Why?

3. According to Hofstede's research, the four most individualistic national cultures are found in the United States, Australia, and Great Britain. Can you think of a reason why these four countries share this cultural characteristic? Does your answer also explain the relatively strong individualism of the Canadian national culture?

4. Would you expect the structures of organizations in Denmark to be taller or flatter than those of organizations in the United States? Why? How are organization structures in Mexico likely to compare to those in Denmark and the U.S.?

5. Comparing U.S. business firms with Scandinavian industrial democracies, large Japanese businesses, and Israeli kibbutz industries, which of the four has the most equitable payment system? Which has the most egalitarian payment system? What explains this difference?

6. If you had to adapt the theories and models described in this book for use in Chile, what kinds of changes would you think about making? Explain your answer using Hofstede's four-dimensional model.

DIAGNOSTIC QUESTIONS

As we have shown, national cultures can have major effects on organizations and the behaviors of their members. Thus it is critical that you, as a manager, are able to assess the characteristics of the national culture surrounding your organization and understand the effects of that culture on organizational behavior. The following diagnostic questions are provided to facilitate this process.

1. What are the characteristics of your own national culture? Is it more individualistic or collectivistic? Are its values consistent with more or less power distance? Is it characterized more by masculinity or femininity? Does it favor strong or weak uncertainty avoidance?

2. What are the characteristics of the national culture surrounding your or-

ganization? How do they compare with the characteristics of your own national culture?

3.	Based on your diagnosis of cultural differences, what adjustments should you make to familiar management practices to fit them to the current cultural situation?

4.	Are there colleagues in the organization who can help you with the necessary adaptation? Should others outside the organization be asked to help instead?

5.	Might other new entrants in the future benefit from what you learn as you cope with cultural adjustment in the present? Will informal mentoring provide future entrants with enough guidance? Should a formal adjustment program be designed and implemented instead?

EXERCISE 19-1

DEGREES OF CULTURAL CONSISTENCY: SOCIETAL AND ORGANIZATIONAL*

JOHN E. OLIVER, *Valdosta State College*
GARY B. ROBERTS, *Kennesaw College*

The cultures of organizations and the societies that surround them consist of fundamental norms and values that are manifested in many ways such as taking part in ceremonies, telling stories about people and events, and honoring heroes. In both organizations and societies, norms and values are shared among members and give rise to common attitudes and understandings. Culture is thus a source of similarity and commonality that helps people communicate with one another and make sense of their world.

Can a society's culture affect the cultures of its organizations? Can the norms and values in an organization's societal environment shape the norms and values that influence what its employees think and do at work? After reading Chapters 18 and 19 you probably realize that the answer to both of these questions is yes. However, you have yet to examine first-hand how one cultural domain can affect another. This exercise gives you the opportunity to examine differing degrees of cultural consistency and to explore their potential consequences.

STEP ONE: PRE-CLASS PREPARATION

If your class completed Exercise 4-1 earlier this term, you should review your answers to the Personality Dimensions Questionnaire (Exhibit 4-1). If your class did *not* complete Exercise 4-1, you should fill out its questionnaire now and score your answers according to the instructions in Step One of Exercise 4-1. When you have either reviewed your earlier answers or completed the questionnaire, you should read the rest of this exercise to prepare for class activities.

STEP TWO: FORMING HOMOGENOUS GROUPS

If your class completed Exercise 4-1 earlier you should reassemble in the same homogenous groups you formed for Step Two of that exercise. When each group gets together it should review its notes from the discussions that took place during Exercise 4-1.

If your class did not complete Exercise 4-1 you should instead divide into groups as directed in Step Two of that exercise (if you have been assembling in

permanent groups throughout this term you should *not* do so this time). Once you have formed groups you should briefly discuss the two questions in Step Two of Exercise 4-1 and be sure that everyone in your group understands the nature and potential effects of the similarities shared among group members.

STEP THREE: EXAMINING CULTURAL CONSISTENCIES

After groups have formed and considered their similarities, everyone should read the Organization Culture Profiles shown in Exhibit 19-1. Next, each group should discuss the following issues and arrive at a consensus on the group's response:

1. Which organization profile is most consistent with the attitudes and interests that are shared among the members of your group? Explain your choice.

2. What inconsistencies can you find between each of the remaining three profiles and the "culture" of your group? What effects might these inconsistencies have on the attitudes and behaviors of people like you?

3. Which of the societal cultures listed in Table 19-1 in this chapter resemble your group in terms of their most pronounced characteristic? How would you expect people in these societal cultures to think and behave?

4. What kind of culture would you expect to find in organizations located in the societal cultures you have identified? Why? Which organizational culture would you be least likely to find? Why?

STEP FOUR: HETEROGENOUS GROUP DISCUSSIONS

Next the class should divide into groups consisting of one individual from each of the four kinds of homogenous groups. Members of these new heterogenous

* Adapted with the authors' permission from an exercise entitled "Personality Traits and Organizational Cultures: Two One-Hour Experiential Learning Exercises." The organization culture profiles included in this exercise are based on information presented in Terrence E. Deal and Allen A. Kennedy, *Corporate Cultures: The Rites and Rituals of Corporate Life* (Reading, MA: Addison-Wesley, 1982).

EXHIBIT 19-1
Organizational Culture Profiles

	TYPE OF CULTURE			
	COMPETITIVE	ENTHUSIASTIC	PROFICIENT	METHODICAL
Type of risks assumed	High	Low	High	Low
Type of feedback from decisions	Fast	Fast	Slow	Slow
The ways survivors and heroes behave in this culture	They have a tough attitude. They are individualistic. They can tolerate all-or-nothing risks.	They are super salespeople. They often are friendly, hail-fellow-well-met types. They use a team approach to problem solving. They are nonsuperstititious.	They can endure long-term ambiguity. They always double check their decisions. They are technically competent. They have a strong respect for authority.	They are very cautious and protective of their own flank. They are orderly and punctual. They are good at attending to detail. They always follow established procedures.
Strengths of the people who best match this culture	They can get things done in short order.	They are able to produce a high volume of work quickly.	They can generate high quality inventions and major scientific breakthroughs.	They bring order and system to the workplace.
Weaknesses of the people who best match this culture	They do not learn from past mistakes. Everything tends to be short-term in orientation. The virtues of cooperation are ignored.	They look for quick-fix solutions. They have a short-term time perspective. They are more committed to action than to problem solving.	They are extremely slow in getting things done. Their organizations are vulnerable to short-term economic fluctuations. Their organizations often face cash-flow problems.	There is lots of red tape. Initiative is downplayed. They work long hours.

groups should share the results of Step Three with one another and work to develop a diagram or table that matches the four dimensions (uncertainty avoidance, empathy-aggression, individualism-collectivism, power distance) with the four types of organizational culture. This diagram or table should also indicate which societal cultures reflect dominant amounts of each of the four dimensions. Note: there need not necessarily be a one-to-one correspondence between the four dimensions and the four cultures. Each group should appoint a spokesperson to report to the class on the results of its discussion.

STEP FIVE: CLASS DISCUSSION

As the group spokespersons give their report the rest of the class should ask whatever questions are needed to clarify each presentation. The instructor should summarize the presentations on a blackboard, flipchart, or overhead transparency, and the class should look for similarities and differences among the different group reports. All differences should be discussed and resolved by consensus and the class should consider the following questions:

1. With which of the four dimensions is each organizational culture most compatible? Least compatible? What does this imply about the relations between societal cultures and organizational cultures?

2. What kind of organizational culture seems most compatible with the Mexican societal culture? Why? With the Greek societal culture? Why?

3. What implications do the results of this exercise have for the task of international management? Why can't managers working outside their native culture simply ignore cultural differences?

CONCLUSION

Seldom do people or organizations fall neatly into any one category. Organizational and societal cultures are similarly difficult to categorize. Every organizational culture exhibits some of the characteristics of each of the cultural profiles shown in Exhibit 19-1. Moreover, members of the same societal culture may differ from one another almost as much as they differ from the members of other cultures. Overgeneralization is thus unwise and can artificially simplify what is actually quite complex. Nevertheless, limited use of the dimensions and profiles we have examined can help managers reduce complexity to a tolerable level and perform effectively in a variety of different cultures and contexts.

DIAGNOSING ORGANIZATIONAL BEHAVIOR

CASE 19-1
LANGLEY INTERNATIONAL GROWERS, INC.*

RAE ANDRE, *Northeastern University*

David Langley is the 58-year-old president of Langley International Growers, Inc., a Connecticut-based firm with annual sales close to $4 million. Like his father and grandfather before him, Langley grows flowers for distribution to wholesale markets along the eastern coast of the United States. Domestically, he runs about 12 acres of greenhouses, putting the company among the top ten greenhouse operations in the United States. In 1980, Langley started a subsidiary in Santa Nueva, an island republic in the Caribbean. The following account is based on discussions with Langley in December of 1982.

PART I

We first became interested in going abroad when we heard about Santa Nueva from one of our competitors. Because of the fuel situation we had decided it might be a good idea to hedge our bets and move to a climate where there's no fuel requirement. So we went down to Santa Nueva to see the competitor's operation, and we thought he was doing a good job. Sixty percent of the flowers used in the United States today are imported. Those flowers are grown with labor that costs $2 a day, versus what we have to pay at minimum wage, around $4.25 per hour. We thought there was money in it.

But establishing yourself in a foreign country is not easy. There are pitfalls. The laws of these governments look like they welcome business coming in. It's all on the books. The laws are there to help you. The government itself wants you there in a lot of these countries. What they spend nationally on oil alone exceeds the money they get from exports, which constantly puts them in the doghouse internationally. They can't buy anything outside, so they're constantly in a state of devaluation relative to everybody else. In addition to that, they have an enormous birthrate which constantly keeps their poverty in place. (You go down there with the idea you're going to help them out from that standpoint—forget it, because they're not going to let you do it.) There's a lot of money to be made if you know how. You go down there and start spending your money and then you find out that the laws have to be administered by people, and the people are where the hangups come because they don't obey the law. They circumvent it to their own benefit. In other words, they make it difficult for you for various reasons. The political aims start to disappear in the bureaucracy.

For example, we have to import a lot of things into the country because they don't have them. Well, they can let that stuff sit down at that dock until some guy clears it. They let our crates of greenhouses sit there for two to three *months*. It threw us way back, cost us thousands of dollars. We had importers down there who knew their business. All the paperwork was right, but all the guy says is "I don't think this is right," and it gets kicked back and forth. One problem is that a lot of the government income is taxes on imports, so they're very strict, particularly if it's an American company that's shipping. This holdup means that everybody's benefiting except the poor guy who has to cut the flowers, because they've made work for the guy who's pushing that paper all around, they've made work for the guy down at the dock, they've made work for the phone operator. You're constantly checking, checking, checking. It's a make-work scheme, in any sense of the word, and they're masters at it. A lot of these foreign countries, they don't operate, they make work.

For example, down where we are, the bureaucracy has increased 50 percent in four years—50 percent more government employees. Going through the airports, there's a guy who puts the tag on your thing and there's a guy who takes it off, five feet away. You can tell them your problem, but they don't understand the problem of business having to get that money moving that's sitting there. They don't realize that they have to collect their taxes and build their country from business. I wanted a land map. It took us three hours to get that

* Reprinted with the author's permission.

map and it had to go through three different people, and I had to sign a paper that I would get my lawyer to state the reason I wanted that map. Even answering the phone: the conversations are long, the conversations are flowery. They don't get to the point, and this, of course, is frustrating if you're not used to it. And especially if you're paying for a long distance phone call. It takes them a half hour to say good morning!

Back about six months ago, we needed this particular type of spreader for an insecticide and I wanted to make sure it was there. We wanted it the following morning and I wanted it delivered that day. And my secretary is on the phone talking to this guy twenty-five miles away. He kept saying, *"Mañana"* ("tomorrow"), and I kept saying, *"Ayer"* ("yesterday"). The secretary kept saying, *"Mañana"* and I kept saying, *"Ayer."* Finally, with negotiations back and forth, I got it.

So you have to tighten things. They don't respect you if they know they're getting away with it, because everybody's watching everybody else. We made that mistake. We were too easy. Of course, these people are very hungry. Their unemployment rate is tremendous. The established rate is 40 percent, but they don't count everybody. If they counted everybody, its around 80 percent. You learn as you go, and you learn from talking to people. They don't respect softness and, yet, they don't respect anybody who's going around shouting and yelling either. You've got to have them understand who's boss. We had a guy who was coming in late all the time, so we gave him a written notice. After you hire him for three months, the government says you own him: it costs you money to let him go. If he's there three months, you might have to give him another three months. After a year, you might have to give him another six months. The guy was late. We gave him the notice. He still was late, so we had to let him go. It didn't cost us anything. See, if he breaks the company rules and they're allowable rules according to the law, then you can get rid of him without any pay. But we had to get tougher and tougher and tougher. It's so easy to be easy because the labor's cheap. But you have to realize that any time you're not making money, it's coming out of capital, so labor's not cheap then.

We weren't knowledgeable about the culture. We assumed that they're like us—sort of like us—if you have the language. That's the mistake you make. You can hire a Neuvan to run the place if you've got the language yourself, but you have to know the language so well that you get the innuendoes, and that's something none of us have. Very few Americans have that. You could hire a Cuban, Mexican, or a Puerto Rican, but even they do not think like we do. They're more apt to identify with the person instead of identifying with the problem. They identify with their emotions and they think, "Well, poor guy." We hadn't taken one dime out of there yet,

and we were asked for a raise. They'll say, "Look at these poor people here. Don't you think they ought to be given hope?" Well, we hired them and they had 60 percent unemployment in the town, and yet I didn't give them hope?

They're big on expectations and poor on execution down there. You have to have an American boss, period. All those guys underneath can be Nuevans, but you gotta have an American boss because you have to teach those Nuevans how you want it. If you go down there and take their way of doing it, you've lost everything you've ever had.

When we went into the village there the only means of transportation was the truck that we had bought. Now, almost everybody rides up in a new Honda motorcycle. The standard of living has gone up.

There was a lot of petty thievery when we built the place. A bar down the end of the street was built in the last year since we built. The owner didn't have a thing before. He's the guy that plowed our property and worked the field before we put our greenhouses up. Right after we built, he was able to build himself a bar and a dancehall from similar materials that were used to construct Langley Greenhouses. I call it Langley's bar. It's right at the end of our road. I often stop in for a *cerveza* ("beer").

All the foreigners do better than we Americans do. Number one, they can bribe the governments. We put strings on our businessmen that are absolutely abominable and then holler that we can't export anything. Another thing is the gringo approach. The Japanese are a new face in there and they operate a little differently than we do. They always say, "Yes." We say, "No," but they say "Yes" and don't mean it, so it doesn't hurt as much. It's a different approach and they've sold one hell of a lot of cars. In fact, I have never driven an American car down there. If you want to go buy an American truck, just forget it. We have a little two-cylinder Japanese truck that's running up and down those hills for fifty kilometers per gallon.

PART II

Our manager there came originally from Puerto Rico. I hired him when he was fifteen or sixteen years old. He's worked for me for seventeen years. We usually have about 75 percent Puerto Ricans working for us up North. I am like a father to him. He had a child and named it after me.

I sent him down and he's been as happy as a lark, but that's Jorge. That's not everybody. You can't generalize on Jorge. He's doing a good job. Of course, he lost his first wife because she thought the girls were a little too loose in Santa Nueva. She wan't wrong. She walked out on him. I don't know whether he got divorced, but he got married again. To give you just a brief insight into it, we won't allow him to hire any woman under forty. We don't want him to be passing his favors out. There's another reason for that. They'll come and work for you, but if they get pregnant, you have to give them at least a month off with pay. Sure enough, they'll come when they're already pregnant. First thing you know, you have five young ladies pregnant, nobody to do the work and you're paying for it. We hired the first secretary, thought she could speak English, took me nine months to get rid of her. I fired her. She couldn't speak a word of English. "Yes," or "No." She had the books all fouled up. Pregnant, too. She lied to us about it. She came back and had the kid here. Now he's an American citizen.

We have three managers and twenty employees. We have an office manager, a pack and ship manager, and an overall manager. These people can all speak Spanish. Then we have about twelve men and eight women. The women do the bunching. The men do the cutting, are night watchmen, and all kinds of things. You look in that packing shed down there and it's probably identical to this one up here, only there are no conveyers. I won't put a conveyer down there because it'll only work about three hours. And that's when I'm working. The more we check up on them, the more controls we put on them in equipment, the more apt they are to say, "This doesn't work now." It's so simple to break a computer. You spit at it or push the wrong damn buttons and it's done. We sent down one of the finest little power mowers you can buy. We started it up before it went down. It worked perfectly. It was eight or nine months later and three mechanical overhauls before we got that thing working. Last year I arrived and found ten or fifteen men cutting the fields with machetes. I'm still not sure that isn't the cheapest. If you hire them for $2 a day, they're telling you something. They really are telling you something. You can hire their people, on certain jobs anyway, cheaper than you can use the damned equipment. You won't see a lot of bookkeeping machines in Santa Nueva. They use people and they'll get it right. They'll have a calculator, but that's about the extent of it. You might in a very big American company, but not generally.

We have parameters for the manager: checklists for his rounds, a checklist for his maintenance, a checklist for his nightman. You must be specific. You don't just walk out and say "clean." You've got to say, "Clean this table, clean that table, clean this." Write it down and give it to him. If you don't do that some of it will be forgotten, some of it just won't be done. And then you can't come in and holler, because the guy will say he didn't hear you. They're really sharp this way. You have to be specific. You have to draw it step by step or

they just won't do it. If they have a package of cigarettes, the empty packs will go onto the floor, until you tell them, "The next time you do that. . . . out. We are not going to have that. This is not the way we're going to be." You go to the company next door to ours where he never enforced these things and it's a dump. Not that he doesn't make money, but it's a dump, a literal dump. It's terrible.

We have the manager take videotape pictures around the plant every week so we see what the plants look like, see what the surroundings look like, see what the housekeeping looks like. We also have him send all the bills, the bank balances, and the payroll up each week.

We have a problem with visitors, too. We have to keep them out. They'll just drop in and say, "Can I see the place?" and they'll take up the manager's time and they'll take up the office time. When they get to talking, they'll talk about their grandfather, their father, their brothers, their sisters, and it's on your time. So we had to discourage that. We had to fence the place to keep the horses out, the cows out, and the people out. Just so you can keep control of the flowers. I don't know if we stopped it. If you have a fence, you have to say, "Don't crawl over the fence."

A lot of growers don't do things the way we do them, even up North. I like it written down. I hate verbal orders, unless it's just a day order. If it's a long-term deal it should be written down and put in the policy. "This is what we do in this way at this particular time." We're known to have the best place in Santa Nueva and there are a lot of flower growers. In the town, we're known as operating a very tight ship.

PART III

The hotels are owned by the government and they're rented from the government. It's amazing. These beautiful hotels rent for two or three hundred dollars a month. And you should see the way they keep it. Terrible. You can't swim in the swimming pool. It's green. It's a beautiful swimming pool and I know how to tell them to keep it, but they won't. If I were going to be down there a lot, I'd take my own chlorine and fix it. It would only cost $100 to use the pool the whole time I was there. Probably, they'd give me free drinks out of it, they'd make it up. They just don't know how to do things. They fool around.

When I was robbed at the hotel, I went to the police station and gave them a list just because I wanted it for the insurance company. Nothing happened. Nobody found anything. I didn't eat in that hotel for the next two months, the next two times I was there. I wouldn't go in their dining room, because I knew those guys knew who did it. I was there alone. Somebody had to be watching and the town is too small not to know the thief and I knew the police knew. I found out the hotel was responsible, but you can't get blood out of a stone, so I said "All right, I want a 10 percent discount rate until this is paid off on my hotel room," which they went along with. After that, they put a guard on me. Every night I have a guard—a private guard. They give him a peso. He's sitting right outside my door. I've never felt physically afraid.

I went down to town one night trying to negotiate for this land. Downtown at night looks like a country road. The house lights are on, but there are no street lights. I went down there negotiating with this family right in their house. (The guy who said that he owned the land really made me mad. I had it all negotiated and later found out we couldn't get a clear title.) Anyway, I'm sitting there in this house with this family. Nobody can speak English, and I can't speak Spanish, but we're negotiating. It was this guy and his son, who could speak English a little, and the whole family—his wife, and relatives. They all come in to look at me. Everybody was just staring. All of a sudden, I started to wiggle my ears, and I'll tell you, they had a hilarious time. My wife was up in the hotel. She was worried I'd disappeared in the middle of the night down in a strange country. I didn't get home till one or two o'clock in the morning.

But it's gorgeous. It's a paradise. You couldn't believe it until you see it. Everything grows. You can have a terrific amount of flowers; I love to go there. I'm getting homesick for it. I would say they've treated us very well. After all, it's their country. It's not up to them to change . . . we're trying to take a profit out of it.

PART IV

The United States has rules, too. It's just a new ballgame and you should detail it right from the beginning. We should have had notebooks, which was my fault. We should have had everything detailed—the duties, the laws of the country, the work rules. If I were to do it now, I would have all this stuff researched and if we ever expand again, we'll know what we're doing. And there'll be no problem. I spent quite a bit of time down there last June when we were planting, but I should have spent two months down there. I did spend practically that much time down there off and on, but I should have been right there and taken over the job of doing it. It's not the Nuevans' fault at all. I might lose it if I don't get down there more. If I were going to do it over again, I wouldn't invest down there, but if I had to do it, I'd still pick that country. I didn't make a mistake in the country. We did not do that. It's probably the best of the lot. They're more democratic than most. The

problem is that poverty does strange things. Poverty will turn those people into almost anything if they don't get it straightened out. That birth rate should be zero right now, but the population is going to double in the next ten years. It takes somebody to say you can't have any more children or to teach them birth control. You have to instill that over three or four generations. And this is where we've lost them. We get insurance against that. Our government insures us if we're taken over down there because of riot or insurrection or government acquisition. Otherwise, you couldn't get any loans. You'll see people with jobs there and you wouldn't believe it. Take the waiters in the hotels. You'll go down there today and five years from now and they're practically working for nothing. There's nobody in the hotels from one day to the next. Yet, they'll be there. They have no place to go. There's no place to go except the United States and there are 500,000 Nuevans working in the United States. You literally can't get a plane reservation back to the States during the first two weeks of January.

When you have read this case, look back at the chapter's diagnostic questions and choose the ones that apply to the case. Then use those questions with the ones that follow in your case analysis.

1. Using Hofstede's dimensions, characterize the societal culture of Santa Nueva. What kinds of adaptations would a manager trained in United States methods have to make to succeed in this culture? Has David Langley made any of these adaptations? Is Langley an "ugly American"?

2. How did Langley learn about the culture of Santa Nueva? You're read his account of his experiences, and you now know something about managing organizational behavior in the United States and in other cultures. If you had to start doing business in Santa Nueva and learn first-hand about Santa Nuevan ways, would you take Langley's approach? Why or why not?

3. How likely is it that Langley's company will be successful in the long run? What lessons does this case teach about the task of managing in another culture?

CASE 19-2

WORLD INTERNATIONAL AIRLINES

Review this case, which appears in Chapter 5. Next, look back at Chapter 19's diagnostic questions and choose the ones that apply to the case. Then use those questions with the ones that follow in your case analysis.

1. On the basis of Hofstede's dimensions and the fact that Stephen Esterant is Spanish, how would you

expect Stephen to behave toward subordinates? Does he behave in the manner you would predict?

2. What are the cultural differences between Stephen's native Spain and the United States location of World International Airlines? Do these differences explain any of the problems that Stephen is having with his American subordinates?

3. If you had to train Stephen in how to interact with American employees, how would you do it? Would he be likely to follow your advice? Describe specific steps you would consider taking to increase this likelihood.

CASE 19-3

PRECISION MACHINE TOOL

Review this case, which appears in Chapter 6. Next, look back at Chapter 19's diagnostic questions and choose the ones that apply to the case. Then use those questions with the ones that follow in your case analysis.

1. What cultural differences existed between the American managers of Precision Machine Tool and Ako Wang, a Japanese businessman? How did these differences affect the deliberations described in the case?

2. If you were a consultant hired by John Garner and Tom Avery, what would you do to make them aware of how cultural factors are affecting their decision making? What modifications would you suggest they make?

3. What sort of response do you expect John and Tom to communicate to Ako Wang? Why? Would this be your response if you were in John's and Tom's position?

GLOSSARY

accommodation A conflict management technique that involves allowing other groups to satisfy their own concerns at the expense of one's own group (p. 486).

action The diagnostic model step in which one stipulates the specific actions needed to implement a prescribed solution (p. 13).

action research model A model of the organization development process that permits the development and assessment of original, innovative interventions (p. 519).

adaptive decline strategies A class of corporate strategies in which the firm reduces its size by either downsizing or eliminating one or more businesses. Retrenchment, divestiture, and liquidation are all adaptive decline strategies (p. 690).

additive task A group task in which the accomplishments of each member of the group are added to those of other members (p. 374).

administrative decision-making model A model in which decisions pursuant to negotiated goals are made based on satisficing rather than maximizing outcomes, through a sequential consideration of alternatives (p. 178).

administrative principles school The school of management thought that deals with streamlining administrative procedures in order to encourage internal stability and efficiency (p. 30).

advanced beginner The stage of skill development in which people learn to base behaviors on an expanded set of rules that include both the elementary rules of novices and circumstantial rules discovered through experience (p. 9).

aptitude testing Measuring broad, general abilities of job applicants (p. 291).

assimilation effect The tendency for present judgments to be biased in the direction of past judgments (p. 147).

attention stage The stage in the information processing cycle in which the individual decides what will be processed and what will be ignored (p. 130).

attribution The process whereby observers decide what caused the behavior of another person (p. 140).

authoritarianism A set of personality characteristics that include ethnocentrism and strong tendencies to overvalue authority, to stereotype others, and to be suspicious and distrustful of people in general (p. 99).

authoritarian leader A leader who makes almost all decisions by herself, minimizing the input of subordinates (p. 418).

automatic processing A type of information processing in which the perceiver is not aware that he is processing information (p. 129).

availability bias The tendency in decision makers to judge the likelihood that something will happen by the ease with which they can recall examples of it (p. 169).

avoidance A conflict management technique that involves staying neutral at all costs and refusing to take an active role in conflict resolution procedures (p. 486).

bargaining A process in which offers, counteroffers, and concessions are exchanged as conflicting groups search for some mutually acceptable resolution (p. 485).

base rate bias The tendency in decision makers to ignore the underlying objective probability, or base rate, that a particular outcome will follow a particular course of action (p. 171).

BCG Matrix A management tool that depicts the mix of businesses owned by a firm (p. 688).

behavior modification Application of contingent rewards in order to bring about behavioral change (p. 311).

behavioral masking A phenomenon whereby the simple presence of other group members masks, or hides, the behaviors of one member (p. 381).

behaviorally anchored rating scale (BARS) A method of performance appraisal in which each judgment point along the scale is illustrated with examples of concrete, on-the-job behaviors (p. 296).

benchmark jobs Common jobs for which market rates are readily available and which are used, along with point plans, to determine salaries for uncommon jobs (p. 306).

benign reappraisal A response to stress in which the person reassesses an apparently threatening environmental demand and modifies his original perception of it (p. 249).

biofeedback A technique that uses machines to monitor bodily functions thought to be involuntary, such as heart beat and blood pressure, so that a person can learn to regulate these functions (p. 272).

boundary spanner A member or unit of an organization that interacts with individuals or firms in the organization's environment (p. 618).

bounded discretion The recognition that the alternatives available to a decision maker are bounded by social, legal, moral, and organizational restrictions (p. 176).

brainstorming A group decision-making process based on a set of rules intended to encourage idea generation (p. 393).

buffering The notion that certain positive factors in the person's environment can limit the capacity of other factors to create dissatisfaction and stress (p. 260).

bureaucracy An idealized description of an efficient organization based on clearly defined authority, formal record keeping, and standardized procedures (p. 32).

business strategy A strategic plan that specifies the way a business plans to compete in its market and industry. Overall cost leadership, differentiation, and focus are all business strategies (p. 691).

centrality The position of a person or group within the flow of work in an organization (p. 470).

centralization The concentration of authority and decision making at the top of an organization; the opposite of decentralization (p. 568).

ceremonies Special events in which the members of an organization celebrate the myths, heroes, and symbols of their culture (p. 697).

change agent A person who manages the organization development process, serving both as a catalyst for change and as a source of information about OD (p. 507).

charismatic leadership Creating a new vision of an organization and getting group members to commit themselves enthusiastically to the new mission, structure, and culture embodied in the vision. Encouraging them to transcend self-interests on behalf of the organization as a whole (p. 439).

choice shift The tendency for groups to make decisions that appear more extreme than the decisions group members would make on their own. In *risky shift* group decisions appear more risky than decisions made by individuals, and in *cautious shift* group decisions appear more cautious than decisions made by individuals (p. 389).

classical conditioning Learning that occurs when a neutral stimulus, through repeated pairing with a stimulus that elicits a specific response, comes to elicit that same response (p. 211).

client organization An organization involved in the process of organization development (p. 508).

closed system contingencies Structural contingency factors that are characteristics of the organization. Size and technology are the most dominant of these factors (p. 596).

coalition A group that forms to allow its members to combine their

765

political strength in order to pursue interests they hold in common (*p. 474*).

coercive power   Interpersonal power based on the ability to control the distribution of undesirable outcomes (*p. 463*).

cohesiveness   A measure of the interpersonal attraction among members of a group and their attraction to the group as a whole (*p. 384*).

collaboration   A conflict management technique that involves attempting to satisfy the concerns of all conflicting groups by working through differences and seeking out optimal solutions in which everyone gains (*p. 486*).

collateral structure   A second hierarchy of groups and committees that parallels and sometimes takes the place of the primary managerial hierarchy in Scandinavian industrial democracies (*p. 741*).

communication   The exchange of information between people through a common set of symbols (*p. 345*).

communication structure   The pattern of interactions by which group members share information. In the *wheel*, a central hub member communicates with all other members, who communicate only with her. In the *Y*, the members of two pairs can communicate with each other and with the hub but the pairs cannot communicate directly. The *chain* links members sequentially so that some can communicate with two people but others with only one. In the *circle*, each member can communicate with two others. In the *completely connected network* each group member can communicate directly with every other (*p. 386*).

comparable worth   Theory that sex differences in wages are attributable to discrimination and that such discrimination can be eliminated through job evaluation (*p. 307*).

compensable factors   Those aspects of a job for which an organization is willing to pay a premium (*p. 304*).

competence   The stage of skill development in which people replace basic rules with advanced rules of thumb that can be altered to fit a wide range of circumstances (*p. 9*).

competition   A conflict management technique that involves attempts to overpower other groups in the conflict and to promote the concerns of one's own group at the expense of the other groups (*p. 486*).

competitive group rewards   Group rewards distributed in such a way that members receive equitable rewards in exchange for successful performance as individuals in a group (*p. 383*).

complex structure   A type of organization structure that relies on standardization to achieve coordination. A complex structure is therefore characterized by noticeable formalization and specialization (*p. 599*).

compliance   Behaving in accord with norms out of fear of punishment or hope of reward (*p. 344*).

comprehensive job enrichment   A type of job design that combines both horizontal and vertical improvements to stimulate employee motivation and satisfaction (*p. 653*).

compressed work week   An alternative work schedule that allows employees to work a 40-hour week in fewer than the normal five days (*p. 665*).

compromise   A conflict management technique that involves seeking partial satisfaction of all conflicting groups through exchange and sacrifice (*p. 487*).

conceptual skills   Management skills involving the ability to perceive an organization or organizational unit as a whole, to understand how its labor is divided into tasks and reintegrated by the pursuit of common goals or objectives, and to recognize important relationships between the organization or unit and its environment (*p. 44*).

conditioned stimulus   A stimulus that is initially neutral but when repeatedly paired with an unconditioned stimulus, elicits the response associated with the latter stimulus (*p. 211*).

confirmation bias   The tendency for raters to seek out information that supports and reaffirms their earlier judgments (*p. 148*).

conflict   A process of opposition and confrontation that can occur between either individuals or groups (*p. 477*).

conflict aftermath   The stage of conflict development at which conflict sets the stage for later situations and events (*p. 483*).

conformity   Loyal but uncreative adherence by group members to both pivotal and peripheral norms (*p. 377*).

conjunctive task   A group task in which all group members must contribute to task performance (*p. 374*).

consequences   Leader behaviors that involve administering rewards and punishments contingent upon subordinate performance (*p. 440*).

consideration   Leader behavior aimed at meeting the social and emotional needs of workers such as helping them, doing them favors, looking out for their best interests, and explaining decisions (*p. 416*).

constituency groups   Groups such as employees, customers, and suppliers upon whom the survival of an organization depends. Constituency groups make demands that they expect will be fulfilled in return for their support of the organization (*p. 593*).

construct validation   Establishing validity by showing that a measure of a concept is congruent with the theory and data that support the concept (*p. 62*).

content theories   Theories of motivation that attempt to specify what sorts of events or outcomes motivate behavior (*p. 200*).

content validation   Establishing validity by showing that, according to expert judges, the measure samples the appropriate material (*p. 62*).

contingency perspective   The view that no single theory, procedure, or set of rules is useful in every situation and that each situation determines the usefulness of different management approaches (*p. 14*).

continuous process production   A type of manufacturing technology in which automated equipment makes the same product in the same way for an indefinite period of time (*p. 611*).

controlled processing   A manner of information processing in which the perceiver is aware that he is processing information (*p. 129*).

controlling   The management function of evaluating the performance of an organization or organizational unit to determine whether it is progressing in the desired direction (*p. 27*).

convergence hypothesis   A theoretical assertion that organizations and management practices throughout the world are growing more alike (*p. 752*).

convergent growth strategies   A class of corporate strategies that involve starting, acquiring, or maintaining businesses in the market and industry already served by the firm. Included in this class are the concentration strategy and the horizontal integration strategy (*p. 689*).

cooperative group rewards   Group rewards distributed in such a way that each member receives an equal reward in exchange for the successful performance of the group (*p. 382*).

cooptation   Making former adversaries into allies by involving them in planning and decision-making processes (*p. 475*).

core job characteristics   Job characteristics identified in the Hackman-Oldham model that lead their holders to experience certain critical psychological states (*p. 654*).

core technology   The dominant technology used in performing the organization's basic work (*p. 609*).

corporate strategy   A strategic plan that specifies the desired mix of businesses in a firm and the way resources should be allocated among them (*p. 688*).

correlation coefficient   A statistic that assesses the degree of relationship between two variables (*p. 64*).

cosmopolitan   A person who has many important contacts outside the organization and develops special knowledge from these contacts (*p. 353*).

cost-saving plans   Fringe benefit programs in which organizations pay workers year-end bonuses out of money saved through employee suggestions, increased efficiency, or increased productivity (*p. 310*).

counseling   A shallow, interpersonal organization development intervention in which a change agent meets either one-on-one or small groups to provide helpful information to people who are having trouble relating with others (*p. 524*).

covariation   The degree to which two variables are associated with each other; the degree to which changes in one are related to changes in the other (*p. 63*).

creative individualism   Acceptance by group members of pivotal norms and rejection of peripheral ones (*p. 377*).

criterion-related validation   Establishing validity by showing that a measure predicts some variable that, based on theory, it should predict (*p. 62*).

critical contingencies   Events, activities, or objects that are required by by an organization and its various parts to accomplish organizational goals and to ensure continued survival (*p. 467*).

critical psychological states Mental conditions identified in the Hackman-Oldham model as being triggered by the presence of certain core job characteristics (*p. 655*).

culture The shared attitudes and perceptions in an organization that are based on a set of fundamental norms and values and that help members understand the organization (*p. 695*).

custodianship A product of socialization in which a new group member adopts the means and ends associated with the role unquestioningly (*p. 354*).

decentralization The dispersion of authority and decision making downward and outward through the hierarchy of an organization; the opposite of centralization (*p. 568*).

decision bias The theory that group decisions are not affected by the loss aversion bias that affects individual decision making (*p. 390*).

decoding The process by which a transmitted message is converted into an abstract idea in the mind of the person to whom the communication is directed (*p. 346*).

decoupling mechanisms Mechanisms that regulate intergroup coordination by making work units less dependent on each other. The use of decoupling mechanisms involves making adjustments to the relationships formed among units during departmentation (*p. 571*).

deductive ability An individual's capacity to use logic and to evaluate the implications of various arguments (*p. 89*).

defense mechanism In Freudian psychology, a kind of mental operation by which individuals rechannel the energies linked to socially unacceptable urges (*p. 93*).

Delphi technique A group decision-making process in which the group never meets in person but instead corresponds with a central leader who initiates activities and receives all the resulting information (*p. 394*).

democratic leader A leader who works to ensure that all subordinates have a voice in making decisions (*p. 419*).

departmentation The process of grouping structural units into larger clusters. In functional departmentation units are grouped into departments according to functional similarities—similarities in the work they do. In divisional departmentation, units are grouped into divisions according to market similarities—similarities in their products, their customers, or the geographical areas they serve (*p. 566*).

depth In organization development, the degree or intensity of change that an intervention is designed to stimulate (*p. 522*).

description The diagnostic model step in which information about a situation without attempting to explain either the cause of the situation or the motives of the people involved in it (*p. 10*).

diagnosis The diagnostic model step in which one looks for the causes of a troublesome situation and summarizes them in a problem statement (*p. 11*).

diagnostic model A four-step model that describes how managers perceive and solve problems. The model is both a learning tool and an on-the-job guide (*p. 10*).

differential accuracy The extent to which a rater's assessment of one individual, on one single dimension, is reflective of the person's true standing on that one dimension (*p. 128*).

differential elevation accuracy The extent to which a rater's assessment of one individual, across a number of dimensions, reflects that person's true standing on those dimensions (*p. 127*).

differentiation The second stage of group development, characterized by conflicts that erupt as members seek agreement on the purpose, goals, and objectives of the group and the roles of its members (*p. 372*).

diffusion of responsibility The sense among group members that responsibility is shared broadly rather than shouldered personally (*p. 381*).

directing The management function of encouraging and guiding employees' efforts toward the attainment of organizational goals and objectives (*p. 26*).

direct supervision A unit coordination mechanism in which one person takes responsibility for the work of a group of others. She determines which tasks need to be performed, who will perform

them, and how they will be linked together to produce the desired end result (*p. 552*).

discretion An area of latitude wherein the decision maker can use her own judgment in developing and deciding among alternative decisions (*p. 180*).

disjunctive task A group task that can be completed by single group members working alone (*p. 374*).

disparate impact The tendency of a particular personnel selection practice to result in the hiring of a smaller percentage of the members of a particular group of job applicants than of other groups of applicants (*p. 300*).

disparate treatment An illegal practice in personnel selection wherein a person from one subgroup is asked to respond to questions, take tests, or display skills that are not asked of applicants from other groups (*p. 300*).

distributive justice An individual's perception of the fairness of his reward in comparison with the rewards given others (*p. 162*).

divisional structure A flexible but inefficient type of complex structure characterized by divisional departmentation and moderate decentralization (*p. 600*).

division of labor The process and result of breaking difficult work into smaller tasks (*p. 23*).

economic rationality The belief underlying rational decision-making models that people attempt to maximize their individual economic outcomes (*p. 163*).

efficiency perspective An approach to job design that focuses on the creation of jobs that economize on time, human energy, raw materials, and other productive resources (*p. 640*).

elevation accuracy The degree to which a rater's assessment of an entire group of people, across a number of different dimensions, reflects the group's true standing on those dimensions (*p. 125*).

emergent task elements The components of work roles that are not formally recognized by the organization but arise out of expectations held by others for the role incumbent (*p. 337*).

emotional adjustment A class of personality variables that deal with the extent to which a person experiences affective distress or engages in socially unacceptable behaviors (*p. 99*).

employee-centered behaviors Leadership behaviors designed to meet the social and emotional needs of group members (*p. 414*).

employment at will A provision that either party in the employment relationship can terminate the relationship at any time, even without reason (*p. 314*).

encoding The process by which a communicator's abstract idea is translated into the symbols of language for transmission to someone else (*p. 345*).

entity attraction Satisfaction with other persons in the workplace that comes about because these people share one's fundamental values, attitudes or philosophy (*p. 259*).

environment The context surrounding an organization, consisting of economic, geographic, and political conditions that impinge on the firm (*p. 615*).

environmental change A measure of the extent to which conditions in an organization's environment change unpredictably (*p. 615*).

environmental complexity A measure of the degree to which an organization's environment is complicated and therefore difficult to understand (*p. 616*).

environmental diversity A measure of the degree to which an organization's environment is varied or heterogeneous in nature (*p. 620*).

environmental domain A part or segment of the environment in which an organization does business (*p. 620*).

environmental receptivity The degree to which an organization's environment supports the organization's progress toward fulfilling its purpose (*p. 618*).

environmental uncertainty An environmental characteristic formed by the combination of change and complexity that reflects the absence of information about environmental factors, activities, and events (*p. 617*).

equity theory A theory of motivation originated by Adams that suggests that behavior is motivated by the desire to reduce guilt or anger associated with social exchanges that are perceived to be unfair (*p. 224*).

ERG theory   A theory of motivation developed by Alderfer that suggests that behavior is driven by the urge to fulfill three essential needs: existence, relatedness, and growth (*p. 207*).

ergonomics   Another name for human factors engineering, a type of methods engineering that focuses on designing machines to match human capacities and limitations (*p. 643*).

escalation of commitment   Investing additional resources in failing courses of action that are not justified by any foreseeable payoff (*p. 173*).

established task elements   The components of work roles that are contained in written job descriptions and formally recognized in the organization (*p. 336*).

eustress   A particular kind of stress created when an individual is confronted with an opportunity (*p. 247*).

evolutionary growth strategies   A class of corporate strategies in which the firm develops products, markets, or businesses that are similar though not identical to the ones already associated with the company. Product development, market development, and concentric diversification are all evolutionary growth strategies (*p. 689*).

expectancy   A person's beliefs regarding the link between his efforts and his performance (*p. 203*).

expectancy theory   A broad, cognitive, process theory of motivation that explains behavior as a function of expectancies, instrumentalities, and valences (*p. 201*).

expected value   The projected value of an outcome that has less than a 100 percent probability of occurring. The expected value is derived mathematically by multiplying each possible outcome of a particular course of action by the probability that that outcome will occur (*p. 166*).

expertise   The stage of skill development in which individuals develop the ability to act intuitively in a wide variety of situations, rarely needing to deliberate consciously (*p. 10*).

expert power   Interpersonal power based on the possession of expertise, knowledge, and talent (*p. 463*).

explicit theories   Internally consistent, formal theories that are subject to empirical test (*p. 60*).

external change agent   An organization development change agent who is not a member of the client organization (*p. 508*).

extinction   The gradual disappearance of a response that occurs after the cessation of positive reinforcement (*p. 212*).

felt conflict   The stage of conflict development at which people are not only aware of the conflict but feel tense, anxious, angry, or otherwise upset (*p. 482*).

fight or flight response   A response to stress in which a person confronts and overcomes a stressful demand or escapes it by leaving the scene (*p. 249*).

flexible cell production   A type of manufacturing technology in which computer-controlled production machines are connected together in a group, or cell, by a flexible network of conveyors that can be rapidly reconfigured for different production tasks (*p. 611*).

flexible-hour programs   Alternative work schedules that specify a block of core hours during which everyone must be on the job but allow each employee to choose when to start and quit work. Also called flextime (*p. 664*).

forcefield analysis   A diagnostic method that depicts the array of forces for and against a particular change in a graphic analysis; often used as a component of the OD process (*p. 513*).

formal groups   Groups that serve specific organizational purposes. In *work groups* employees work together to produce their firm's goods or services. Higher-level managers and the managers they supervise work together in *management teams*. *Temporary groups* are formed to accomplish a specific task. *Intermittent groups* are composed of people who do not work with each other but meet regularly to exchange work-related information (*p. 370*).

formal organization   Those aspects of an organization that are officially sanctioned, including intentionally designed structures and jobs (*p. 685*).

formalization   The process of planning the regulations that control organizational behavior; also the written documentation produced by the planning process. Jobs, work flows, or general rules may be formalized (*p. 559*).

formation   The initial stage of group development, characterized by

uncertainty and anxiety. People try to determine which behaviors will be appropriate and what contributions members should be expected to make to the group (*p. 372*).

functional attraction   Satisfaction with other persons in the workplace that comes about because these other people help one attain valued work outcomes (*p. 259*).

functional grouping   People are grouped into units according to similarities in the functions they perform. Grouping word-processing typists into a word-processing pool is an example (*p. 549*).

functional strategy   A strategic plan that specifies how a functional unit will contribute to the attainment of corporate and business goals (*p. 693*).

functional structure   An efficient but inflexible type of complex structure characterized by functional departmentation and centralization (*p. 599*).

gatekeeper   A person responsible for controlling messages sent through a particular communication channel (*p. 352*).

general adaptation syndrome   The theory developed by Hans Selye that the body's response to stress occurs in three distinct stages: alarm, resistance and exhaustion (*p. 250*).

general cognitive ability   The totality of an individual's mental capacity, summing across specific mental abilities such as verbal comprehension, quantitative aptitude, reasoning ability, and deductive ability (*p. 88*).

generalizability   The degree to which the result of a study conducted in one sample-setting-time configuration can be replicated in other sample-setting-time configurations (*p. 73*).

geographic structure   A type of divisional structure formed by grouping units according to similarities in their geographic location (*p. 600*).

goal commitment   A person's willingness to put forth effort in accomplishing goals and unwillingness to lower or abandon goals (*p. 220*).

goal-setting theory   A theory of motivation originated by Locke that suggests that behavior is driven by goals and aspirations, such that specific and difficult goals lead to higher levels of achievement (*p. 219*).

grapevine   The unofficial communication network within the informal organization through which employees trade gossip and rumors about their jobs, their coworkers, and the organization (*p. 695*).

group decision making   A group task in which the ultimate aim is to solve a problem or make a decision (*p. 388*).

group effectiveness   An assessment of the extent to which a group is accomplishing its task in the most productive and satisfactory manner (*p. 368*).

groupthink   A threat to the effective performance of groups that develops in highly cohesive groups whenever strivings for harmony and unanimity override efforts to appraise group judgments realistically (*p. 390*).

halo error   A rating error wherein a rater's judgment about a specific behavior is colored by her overall evaluation of the person she is rating (*p. 137*).

hedonism   The belief that human beings generally behave so as to maximize pleasure and minimize pain (*p. 211*).

heroes   People who embody the values of an organization's culture and serve as role models for other members in the organization (*p. 699*).

hierarchy of authority   A pyramidal distribution of authority in which managers higher in the pyramid can tell managers in lower positions what to do (*p. 23*).

historical decision model   A method of generating alternatives for current decisions by reviewing processes that were used in the past (*p. 165*).

history threat   A threat to validity created when some important variable other than the one manipulated experimentally changes during an experiment (*p. 69*).

horizontal job enlargement   A type of job design based on the idea that increasing the number of tasks a job holder performs will reduce the repetitive nature of the job and thus eliminate worker boredom (*p. 650*).

horizontal specialization The type of specialization in which the work performed at a given hierarchical level is divided into specialized jobs. An example is to divide secretarial work into the jobs of typist, receptionist, and file clerk (*p. 564*).

host group A group that is experiencing difficulties in working with other groups and that asks those groups to send representatives to an intergroup mirroring intervention (*p. 530*).

human factors engineering A type of methods engineering in which experts design machines, operations, and work environments so that they match human capacities and limitations (*p. 643*).

human relations school The school of management thought that emphasizes increasing employee growth, development, and satisfaction (*p. 36*).

human resource management A domain of organizational research that focuses on devising practical, effective ways to manage employee behaviors (*p. 15*).

human skills Management skills involving the ability to work effectively as a group member and to build cooperation among the members of an organization or unit (*p. 44*).

hygiene factors Characteristics of the job that according to Frederick Herzberg, influence the amount of dissatisfaction experienced at work (*p. 652*).

hypothesis A specific, testable prediction, derived typically from a theory, about the relationship between two variables (*p. 59*).

identification Behaving in accord with norms out of respect and admiration for one or more members of the role set (*p. 344*).

implicit theories Loose, informal theories about phenomena that people rarely test in a rigorous, empirical fashion (*p. 59*).

impression management Behaving in ways intended to build a positive public image (*p. 475*).

incentive systems A process by which future pay is made contingent on individual performance based on objective performance indicators and using established quantitative rules (*p. 309*).

incubation A stage in the creative decision-making process in which the person apparently stops attending to the problem at hand (*p. 183*).

individualism-collectivism A cross-cultural dimension that refers to two opposing points of view on the norms and values of a national culture—whether they should place greater emphasis on satisfying personal interests or on looking after group needs (*p. 737*).

individual roles In groups, roles that focus on the satisfaction of members' personal needs and interests even when they conflict with the well-being of the group (*p. 376*).

industrial democracies Industrial organizations, common in Scandinavian countries, that are required by law to permit their members to govern themselves (*p. 741*).

industrial engineering A branch of engineering that concerns itself with how to maximize the efficiency of the methods, facilities, and materials used to produce commercial products (*p. 640*).

industrial robots Machines that can be programmed to repeat the same sequence of work movements over and over again (*p. 665*).

informal groups Groups that satisfy personal needs of their members. *Friendship groups* form among people who like being with each other. *Interest groups* develop among people who want to achieve some mutually beneficial objective (*p. 369*).

informal organization The unofficial rules, procedures, and interconnections that develop as employees make spontaneous, unauthorized changes in the way things are done. (*p. 695*).

information overload A condition in which a person is presented with more information than he can possibly process (*p. 387*).

ingratiation The use of praise and compliments to gain the favor or acceptance of others (*p. 475*).

initiating structure Leader behaviors aimed at meeting the group's task requirements, such as getting workers to follow rules, monitoring performance standards, clarifying roles, and setting goals (*p. 416*).

insight A stage in the creative decision-making process in which the solution to a problem manifests itself in a flash of inspiration (*p. 184*).

instrumentality A person's subjective belief about the relationship between performing a behavior and receiving an outcome (*p. 202*).

instrumentation threat A threat to validity created by artificial changes in the measurement device used to assess an experimental effect (*p. 69*).

integration The third stage of group development, which is focused on reestablishing the central purpose of the group in light of the structure of roles developed during differentiation (*p. 373*).

interaction An experimental outcome in which the relationship between two variables changes depending on the presence or absence of some third variable (*p. 72*).

interactive group A group whose members interact in unstructured, face-to-face relationships like those that take place during ordinary conversations (*p. 388*).

intergroup conflict A process of confrontation that occurs when one group obstructs the progress of one or more other groups (*p. 478*).

intergroup mirroring An organization development intervention of moderate depth in which representatives from several groups tell the members of a particular group with whom they interact how the people they represent perceive the host group (*p. 530*).

intergroup team building A deep OD intervention intended to improve communication and interaction between work-related groups (*p. 531*).

internal change agent An OD change agent who is a member of the client organization (*p. 508*).

internalization Behaving in accord with norms that are consistent with one's own value system (*p. 344*).

interpretation stage The stage in the information processing cycle in which meaning is attached to the relation among abstract concepts (*p. 139*).

intervention An organization development technique, such as counseling or team building, that is used to stimulate change in organizations (*p. 507*).

jargon Idiosyncratic use of language that is often useful among specialists but that inhibits their ability to communicate with nonspecialists (*p. 352*).

job burnout A condition of emotional, physical, and mental exhaustion resulting from prolonged exposure to intense, job-related stress (*p. 251*).

job depth The amount of discretion a jobholder has to choose job activities and outcomes (*p. 651*).

job design The process of deciding what specific tasks each jobholder should perform in the context of the overall work that an organization must accomplish (*p. 639*).

Job Diagnostic Survey A questionnaire that measures workers' perceptions of core job characteristics, critical psychological states, and different moderating factors (*p. 656*).

job evaluation A process by which the pay differentials for jobs throughout the organization are established based on differences in job requirements (*p. 303*).

job extension A type of horizontal job enlargement in which several simplified jobs are combined to form a single new job (*p. 650*).

job-oriented behaviors Leadership behaviors that focus on careful supervision of employees' work methods and performance level (*p. 416*).

job range The number of tasks a jobholder performs to complete the job (*p. 650*).

job rotation The process by which employees are moved periodically from one type of job to another in order to increase their job satisfaction. In formal job rotation programs, workers are rotated among specific jobs in a systematic fashion (*pp. 273, 651*).

job satisfaction The perception that one's job enables one to fulfill important job values (*p. 244*).

job sharing An alternative work schedule in which two or more part-time employees are allowed to share a full-time job (*p. 665*).

judgment stage The stage of the information processing cycle in which recalled information is weighted and aggregated to come up with a single overall judgment (*p. 146*).

kibbutz A close-knit community of people located in Israel and organized on the principles of collective ownership and direct participation in self-governance (*p. 747*).

laissez-faire leader A leader who lets a group run itself, with minimal intervention from upper levels of the organizational hierarchy (*p. 419*).

language A system of shared symbols that the members of an organization use to communicate cultural ideas and understandings (*p. 700*).

latent conflict The stage of conflict development at which dissension is only suspected or at best dimly perceived (*p. 482*).

lateral linkage devices The types of unit-linking mechanism in which hierarchical links between interdependent units are supplemented by various avenues of mutual adjustment (*p. 572*).

latticework hierarchy The structure of vertical and horizontal relationships found in many large Japanese corporations (*p. 744*).

leader-follower relations A component of Fiedler's contingency theory that describes the level of trust and respect between leader and follower (*p. 430*).

leader position power A component of Fiedler's contingency theory that describes the degree to which the leader can administer significant rewards and punishments to followers (*p. 430*).

leadership The use of noncoercive influence to direct and coordinate the activities of the members of an organized group toward the accomplishment of group objectives (*p. 411*).

leadership grid figure A two-dimensional representation of leadership behaviors in which concern for people and concern for production combine to produce five behavioral styles (*p. 417*).

leadership motivation pattern (LMP) A composite behavior pattern composed of a high need for power, a low need for affiliation, and a high degree of self control, that predicts success in bureaucratic leadership situations (*p. 424*).

leader task structure A component of Fiedler's contingency theory that describes the clarity of goals and of means-end relationships in a group's task (*p. 430*).

legitimate power Interpersonal power based on holding a position of formal authority (*p. 463*).

level of analysis A dimension that classifies the five areas of organizational research according to whether their primary focus is on the behaviors of individuals, of groups, or of organizations (*p. 16*).

Lewin development model A three-step model of the development process that is followed in every successful OD intervention (*p. 516*).

liaison position The type of unit-linking mechanism in which one person is made responsible for seeing that communication flows directly and freely between interdependent units. A liaison position has no authority, so its occupant must rely on negotiation, bargaining, and persuasion to move interdependent units toward voluntary coordination (*p. 572*).

life-cycle model A model that proposes that organizational growth progresses through a series of stages, each of which has its own structural requirements (*p. 608*).

locus of control The extent to which an individual believes that his own actions influence the environment (*p. 98*).

loosely coupling Managing interrelatedness across different functional areas by not allowing the actions or decisions of one functional unit to have an overly large or immediate impact on the actions or decisions of other functional units (*p. 181*).

loss aversion bias The tendency of most decision makers to weigh losses more heavily than gains, even when the absolute value of each is equal (*p. 169*).

Machiavellianism A personality trait characterized by the tendency to seek to control other people through opportunistic, manipulative behaviors (*p. 472*).

macro organizational behavior The subfield of organizational behavior that focuses on understanding the actions of a group or an organization as a whole (*p. 15*).

maintenance roles Group roles that help ensure a group's continued existence by building and preserving strong interpersonal relations among its members (*p. 375*).

management A process of planning, organizing, directing, and controlling organizational behaviors in order to accomplish a mission through the division of labor (*p. 24*).

manager A person who is responsible for planning, organizing, directing, and controlling behavior in organizations. *Top managers* are responsible for the entire firm; *middle managers* manage an organizational unit; *supervisory managers* manage the employees who do the firm's basic work (*p. 41*).

managerial role Behaviors expected of managers in performing their jobs. Managers promote good interpersonal relations in the *interpersonal* role, receive and send information to others in the *informational* role, and determine the firm's direction in the *decisional* role (*p. 45*).

manifest conflict The stage of conflict development at which people engage in behaviors that are clearly intended to frustrate or block their opponents (*p. 482*).

market grouping A grouping of people into units according to similarities in the products they make or markets they serve. Grouping the members of an automobile assembly line into an assembly unit is an example (*p. 549*).

market structure A type of divisional structure formed by grouping units according to similarities in the clients or customers they serve (*p. 600*).

masculinity-femininity A cross-cultural dimension that refers to the degree to which a culture is founded on values that emphasize independence, aggressiveness, dominance, and physical strength, on the one hand, or interdependence, compassion, empathy, and emotional openness, on the other (*p. 737*).

Maslow's need theory A theory of motivation that suggests that behavior is driven by the urge to fulfill five fundamental needs: physiological and safety needs love, esteem, and self-actualization (*p. 205*).

mass production A type of manufacturing technology in which the same product is produced repeatedly, either in large batches or in long production runs. Also called large-batch production (*p. 610*).

matrix organization structure The type of lateral linkage mechanism that has an organization structure incorporating both functional and divisional departmentation. It is a complex network of mutual adjustment through which intergroup relations are coordinated (*p. 573*).

matrix structure An extremely flexible but also extremely costly type of complex structure characterized by both functional and divisional departmentation as well as high decentralization. Also called the simultaneous structure (*p. 602*).

maturity The fourth and final stage of group development, in which members begin to fulfill their prescribed roles and work toward attaining group goals. Many of the agreements reached about goals, roles, and norms are formalized, or preserved in written documentation during this stage (*p. 373*).

mechanistic structures Machine like organization structures designed to enhance efficiency; characterized by large amounts of formalization, standardization, specialization, and centralization (*p. 595*).

membership groups Formal and informal groups to which people belong. (*p. 369*).

memomotion analysis A type of work measurement in which industrial engineers examine longer activity sequences by using slow speed to film or videotape a person at work and then playing back the resulting film or tape at normal speed (*p. 647*).

mentoring relationship A partnership between a senior and a junior colleague in which the senior partner promotes the development of the junior partner (*p. 317*).

merit-based pay plans Basing pay increases on subjective ratings of performance made at year end and allocating increases as a percentage of available funds based on these ratings (*p. 308*).

methods engineering An area of industrial engineering that attempts to improve the methods used to perform work (*p. 641*).

micromotion analysis A type of work measurement in which industrial engineers analyze the hand and body movements required to do a job (*p. 646*).

micro organizational behavior The subfield of organizational behavior concerned with understanding the behaviors of individuals working alone or in small groups (*p. 14*).

mirror image fallacy The false belief that all people are alike or that others share one's own abilities, beliefs, motives, or predispositions (*p. 83*).

mission An organization's purpose or reason for being (*p. 22*).

monitors Leader behaviors that involve collecting performance information on subordinates (*p. 440*).

mortality threat A threat to validity created when subjects who drop out of an experimental group differ on some significant characteristic or characteristics from those who drop out of the control group (*p. 68*).

motivating potential score A summary score calculated from data obtained with the Job Diagnostic Survey (*p. 657*).

motivation The factors that initiate, direct, and sustain human behavior over time (*p. 200*).

motivator factors Characteristics of the job that according to Frederick Herzberg, influence the amount of satisfaction experienced at work (*p. 651*).

motive A reflection of an individual's underlying drives, needs, and values (*p. 97*).

mutual adjustment A unit coordination mechanism in which coordination is accomplished via face-to-face communications. Coworkers who occupy positions of similar hierarchical authority exchange information about how a job should be done and who should do it (*p. 552*).

myth A story that provides a fictional but plausible explanation for something that might otherwise seem puzzling (*p. 698*).

national culture The collection of societal norms and values in the environment surrounding an organization (*p. 734*).

negative affectivity A person's tendency to often experience feelings of subjective distress such as anger, contempt, disgust, guilt, fear, and nervousness (*p. 261*).

negative reinforcement The increase in a response that occurs when engaging in the response leads to the removal of an aversive stimulus (*p. 212*).

negotiation A process in which groups with conflicting interests decide what each will give and take in the exchange between them (*p. 485*).

nenko system A Japanese system of payment in which the pay an employee receives is determined by a basic wage plus merit supplements and job-level allowances (*p. 745*).

noise A collective term for a number of factors that can distort a message as it is transmitted from one person to another (*p. 346*).

nominal group technique A group decision-making process in which face-to-face interaction of group members is limited (*p. 394*).

norms A strong set of expectations that members of a role set have for the role occupant (*p. 342*).

nova technique A method of generating alternatives by seeking new and innovative solutions (*p. 166*).

novice The stage of skill development in which people learn elementary rules and procedures that, followed mechanically, result in actions resembling skilled behaviors (*p. 8*).

objectivity In science, the degree to which a set of scientific findings are independent of any one person's opinion about them (*p. 56*).

off-the-shelf decision model A method of generating alternatives for current decisions by consulting agents external to the organization that have standardized, ready-made alternatives (*p. 165*).

open-office plan A physical work environment that minimizes interior walls and partitions (*p. 261*).

open revolution Rejection by group members of both pivotal and peripheral norms (*p. 377*).

open system contingencies Structural contingency factors that are characteristics of the environment surrounding an organization (*p. 596*).

open system planning A deep organization-level intervention that helps the members of an organization devise ways to accomplish the mission of their firm in light of the demands of environmental constituency groups (*p. 707*).

open systems school The school of management thought that characterizes every organization as a system that is open to the influence of the surrounding environment (*p. 38*).

opinion leader A person who has special access to an organization's informal channels of communication and therefore has enhanced ability to influence others (*p. 353*).

organic structures Organism-like organization structures designed to enhance flexibility and innovation; characterized by large amounts of mutual adjustment and decentralization (*p. 595*).

organization An assembly of people and materials brought together to accomplish a purpose that would be beyond the means of individuals working alone (*p. 22*).

organization design The process of diagnosing the situation that confronts a particular organization and selecting and putting in place the organization structure most appropriate for that situation (*p. 592*).

organization development (OD) A planned approach to interpersonal, group, intergroup, and organization-wide change that is comprehensive and long term and that is guided by a change agent. Organization development research develops techniques to encourage cooperation and to manage change (*pp. 15, 507*).

organization role The total set of expectations that people who interact with an organizational member have for that person and his performance of his job (*p. 260*).

organization size The number of members in an organization; its volume of sales, clients, or profits; its physical capacity (e.g.,a hospital's number of beds or a hotel's number of rooms); or the total financial assets that it controls (*p. 606*).

organization stage The stage in the information processing cycle in which many discrete bits of information are chunked into higher-level, abstract concepts (*p. 134*).

organization structure The relatively stable network of interconnections or interdependencies among the people and tasks that make up an organization (*p. 548*).

organizational behavior (OB) A field of study that endeavors to understand, explain, predict, and change human behavior as it occurs in the organizational context (*p. 7*).

organizational commitment Identification with one's employer that includes the willingness to work hard on behalf of the organization and the intention to remain with the organization for an extended period of time (*p. 254*).

organizational confrontation An organization-level OD intervention of moderate depth that enables managers to assess the internal workings of the organization and plan corrective actions as needed (*p. 706*).

organizational effectiveness The degree to which an organization is successful in achieving its goals and objectives while at the same time ensuring its continued survival by satisfying the demands of interested parties, such as suppliers and customers (*p. 593*).

organizational efficiency The ratio of outputs produced per unit of inputs consumed; minimizing the raw materials and energy consumed by the production of goods and services (*p. 594*).

organizational power Types of interpersonal power (reward, coercive, and legitimate power) that often derive from company policies and procedures (*p. 465*).

organizational productivity The amount of goods or services produced by an organization. Higher productivity means that more goods or services are produced (*p. 594*).

organizational socialization The process by which a person acquires the social knowledge and skills necessary to assume an organizational role (*p. 354*).

organizational unit A recognizable group of employees responsible for completing its own particular functional and/or operational objectives (*p. 26*).

organizing The management function of developing a structure of interrelated tasks and allocating people and resources within this structure (*p. 25*).

participatory management A management style in which managers and nonmanagers work together to make decisions about what products to produce, which raw materials to purchase, what production processes to use, and similar issues (*p. 367*).

pay structure An organization's hierarchical arrangement of jobs expressed in terms of pay differentials (*p. 303*).

perceived conflict The stage of conflict development at which problems are readily perceived and everyone involved in the conflict knows that it exists (*p. 482*).

perceptual ability An individual's capacity to quickly and accurately recognize visual details (*p. 89*).

perceptual defense The process by which an individual avoids processing information that is potentially threatening (*p. 132*).

performance programs Scripts that detail exactly what actions are to be taken by a job incumbent when confronted with a standard problem or situation (*p. 180*).

peripheral norms Group norms for which adherence is desirable but not essential (*p. 376*).

personal conceptions A person's thoughts, attitudes, and beliefs about his social and physical environment (*p. 98*).

personal power Types of interpersonal power (expert and referent power) that are based on the possession of certain personal traits or characteristics (*p. 465*).

personality dynamics A class of personality characteristics that deal with the integration and organization of traits, motives, personal conceptions, and adjustment (*p. 100*).

personnel placement The process by which an organization assigns new employees to specific jobs (*p. 85*).

personnel selection The process by which an organization decides who will and who will not be allowed to work for an organization (*p. 84*).

persuasive argumentation The theory that when group discussions uncover arguments favoring extreme positions, moderate group members may switch to more extreme choices (*p. 390*).

phenomenal absolutism The belief that one's perceptions reflect reality perfectly (*p. 121*).

physical ability Ability to perform a task involving body movement, strength, endurance, dexterity, force, or speed (*p. 85*).

pivotal norms Group norms for which adherence is an absolute requirement of continued group membership (*p. 376*).

planned change model A model of the organization development process that is an expansion of the Lewin development model and that describes the implementation of an off-the-shelf intervention (*p. 518*).

planning The management function of deciding what to do in the future; setting goals and establishing the means to attain them (*p. 25*).

politics Activities in which individuals or groups acquire power and use it to advance their own interests (*p. 471*).

pooled interdependence A type of interaction where individuals draw off a common resource pool but do not interact with each other in any other way (*p. 334*).

portfolio The mix of different businesses owned by an organization (*p. 688*).

positive reinforcement The increase in a response that occurs when engaging in the response leads to obtaining a pleasurable stimulus (*p. 212*).

power The ability to influence the conduct of others and resist unwanted influence in return (*p. 460*).

power distance A cross-cultural dimension that refers to the degree to which the members of a society accept differences in power and status among themselves (*p. 738*).

preparation A stage in the creative decision-making process in which the person accumulates information needed to solve a problem (*p. 182*).

prepotency The notion arising from Maslow's theory that higher-order needs can influence motivation only if lower-order needs are largely satisfied (*p. 205*).

prescription The diagnostic model step of developing a solution to a problem statement that has been identified through diagnosis (*p. 12*).

primary appraisal In cognitive appraisal theory, the first stage in one's assessment of the environment in which one judges whether some object in the environment is good or bad, beneficial or harmful, an opportunity or a threat (*p. 248*).

primary orientation A dimension that classifies the five areas of organizational research according to whether their main focus is on abstract theories or practical techniques (*p. 16*).

priming Forcing raters to recall a specific set of events so that subsequent judgments will be biased by what is recalled (*p. 147*).

privacy The freedom to work unobserved by others and without undue interruption (*p. 261*).

procedural justice Perceived fairness of the process by which reward allocations have been made (*p. 162*).

process consultation A shallow group-level OD intervention in which a change agent meets with a work group and helps its members examine group processes such as communication, leadership and followership, problem solving, and cooperation (*p. 527*).

process engineering A type of methods engineering in which specialists study the sequence of tasks required to produce a particular good or service and examine how these tasks fit together into an integrated job (*p. 641*).

process loss The difference between what is produced by a group of individuals and what would be produced by the same people working alone (*p. 379*).

process theories Theories of motivation that attempt to specify how different kinds of events or outcomes motivate behavior (*p. 201*).

product structure A type of divisional structure formed by grouping units according to similarities in the products they make and sell (*p. 600*).

production blocking The negative effect on productivity caused by people getting in each other's way as they try to perform a group task (*p. 379*).

professionalization The use of professionals to perform work for which useful written specifications do not exist and in some cases, cannot be prepared (*p. 562*).

professionals People who develop work-related knowledge, skills, and abilities in training programs conducted outside an employing organization. Examples include teachers, lawyers, doctors, and managers trained in schools of business (*p. 562*).

proficiency The stage of skill development in which people learn how to read situations instinctively and respond to familiar circumstances intuitively, deliberating consciously only in unusual situations (*p. 9*).

profit-sharing plans Fringe benefit programs in which profits are calculated at year end and distributed to employees, typically in a deferred fashion (*p. 310*).

projection A bias in the interpretation of information wherein the perceiver assumes that his own motivations explain the behaviors of others (*p. 140*).

projective test A measure of personality in which individuals are asked to assign meaning to an ambiguous stimulus. Unconscious aspects of the personality are inferred from the person's responses (*p. 103*).

protected groups Groups of people defined in Civil Rights legislation who warrant special consideration in personnel selection, placement, and other procedures (*p. 299*).

prototype One type of schema that involves a unified configuration of personal characteristics that are used to classify persons into "types" (*p. 136*).

psychomotor ability Ability to perform a task involving coordination between physical and mental functions (*p. 86*).

punishment A decrease in a response that occurs when engaging in the response leads to receiving an aversive stimulus (*p. 212*).

quality circles Committees or small groups of employees charged with identifying and solving productivity problems on the job. Quality circles typically use a participatory approach (*pp. 367, 663*).

quality of work life The degree to which work and membership in an organization facilitates the satisfaction of important personal needs and interests (*p. 709*).

quantitative ability A specific form of cognitive ability that deals with the understanding and application of mathematical rules and operations (*p. 89*).

random assignment A method of increasing the validity of a study by ensuring that each subject has an equal probability of being assigned to any one experimental condition. Random assignment eliminates the *selection threat* (*p. 71*).

rational decision-making model A model in which decisions are made systematically and based consistently on the principle of economic rationality (*p. 163*).

realistic previewing A technique sometimes used during counseling interventions to help people form sensible expectations about workplace relationships (*p. 524*).

reasoning ability An individual's capacity to invent solutions to many different types of problems (*p. 89*).

reciprocal interdependence A type of interaction in which there are two-way links among individuals (*p. 335*).

reference groups Groups of people with whom individuals compare themselves in order to assess their own personal attitudes or behavior (*p. 369*).

referent power Interpersonal power based on the possession of attractive personal characteristics (*p. 463*).

refreezing The third step in the Lewin development model in which the change that took place during the transforming step becomes stable and permanent (*p. 517*).

regression toward the mean The phenomenon whereby in a series of events that is influenced by complex factors, any single extraordinary event is almost sure to be followed by a more ordinary event (*p. 172*).

reinforcement theory A theory of motivation that suggests that people are motivated to engage in or avoid certain behaviors because of past rewards and punishments associated with those behaviors (*p. 212*).

reliability The degree to which a measure of an individual, group, organizational, or environmental attribute is free from random error and thus replicable (*p. 61*).

representative groups The type of unit-linking mechanism in which representatives of interdependent units meet to coordinate intergroup activities (*p. 572*).

retrieval stage The stage of the information processing cycle in which the observer tries to recall information about past events (*p. 144*).

revolutionary growth strategies A class of corporate strategies in which the firm develops or acquires businesses that are quite different from businesses it already owns. Included in this class are forward integration, backward integration, and conglomerate diversification (*p. 690*).

reward power Interpersonal power based on the ability to control how desirable outcomes are distributed (*p. 462*).

rite A ceremonial activity meant to communicate particular messages or accomplish specific purposes (*p. 697*).

ritual A ceremonial event that occurs repeatedly and continues to reinforce key norms and values (*p. 697*).

role The typical and expected behaviors that characterize an individual's position in some social context (*p. 336*).

role ambiguity Lack of clarity about the expectations of a person's role in an organization (*p. 267*).

role analysis technique An interpersonal OD intervention of moderate depth intended to help people form and maintain effective working relationships by clarifying role expectations (*p. 524*).

role conflict Conflict or incompatibility between the demands facing a person who occupies a particular role (*p. 267*).

role innovation A product of socialization in which a new group member is expected to improve on both the goals for her job and the means of achieving them (*p. 354*).

role occupant The current incumbent of an existing organizational work role (*p. 343*).

role scope The total number of expectations that exist for the person occupying a particular role (*p. 267*).

role set The entire group of individuals who have an interest in and expectations about the way a role occupant performs his job (*p. 342*).

satisfaction perspective An approach to job design that suggests that fitting the characteristics of jobs to the needs and interests of the people who perform them provides the opportunity for satisfaction at work (*p. 649*).

satisficing Settling for a decision alternative that meets some minimum level of acceptability, as opposed to trying to maximize utility by considering all possible alternatives (*p. 178*).

scapegoats People who are blamed, whether rightly or not, for the failures of groups or organizations (*p. 476*).

schema Cognitive structures that group discrete bits of perceptual information in an organized fashion. (The term *schema* is used for both singular and plural forms) (*p. 135*).

scientific management school The school of management thought that focuses on increasing the efficiency of production processes in order to enhance organizational profitability (*p. 28*).

scientific method An objective method of expanding knowledge characterized by an endless cycle of theory building, hypothesis formation, data collection, empirical hypothesis testing, and theoretical modification (*p. 56*).

script A schema that involves well-known sequences of action (*p. 135*).

secondary appraisal In cognitive appraisal theory, the second stage in a person's assessment of the environment in which he judges his capacity to cope with perceived threats or opportunities in the environment (*p. 248*).

selection threat A threat to validity created when experimental and control groups differ from each other before an experimental manipulation (*p. 68*).

self-contained tasks The type of decoupling mechanism formed by combining the work of two or more interdependent units and assigning their work to several independent work units. Those units are then staffed by people drawn from each of the original units (*p. 571*).

self-efficacy The judgments people make about their ability to execute courses of action required to deal with prospective situations (*p. 216*).

self-esteem The degree to which a person believes that she is a worthwhile and deserving individual (*p. 100*).

self-inventory A measure of personality characteristics that asks the individual to describe herself by means of standardized responses to questionnaire items (*p. 101*).

semiautonomous groups Groups that are subject to the management direction needed to ensure adherence to organizational policies but are otherwise responsible for managing group activities (*p. 659*).

sensitivity training A deep interpersonal OD intervention that focuses on developing greater sensitivity to oneself, to others, and to one's relations with others through an intense, leaderless group experience (*p. 526*).

sequential interdependence A type of interaction in which individuals are arrayed in a chain of one-way links (*p. 335*).

shaping Bringing about a desired behavior by rewarding successive approximations to that behavior (*p. 214*).

simple differentiated structure A type of simple structure in which coordination is achieved by means of direct supervision (*p. 598*).

simple structure An uncomplicated type of organization structure that relies on mutual adjustment or direct supervision to achieve coordination (*p. 597*).

simple undifferentiated structure A type of simple organization structure in which coordination is achieved solely by means of mutual adjustment (*p. 597*).

situation analysis An analysis of internal strengths and weaknesses and external opportunities and threats that is performed during the process of strategic planning (*p. 686*).

situational interview A work sample test in which the applicant is asked to respond orally to hypothetical problems that might confront him while working on the job (*p. 294*).

skill-acquisition model A five-stage model of the process of developing expertise in a particular behavior (*p. 8*).

skill testing Measuring narrow, job-specific abilities of job applicants (*p. 293*).

slack resources The type of decoupling mechanism in which groups are separated from each other by buffer inventories (*p. 571*).

small-batch production A type of manufacturing technology that involves the production of one-of-a-kind items or small quantities of goods designed to meet unique customer specifications. Also called unit production (*p. 610*).

social comparison The theory that when people in groups hear others voicing extreme positions they often abandon their cautious choices and revert to their initial extreme positions (*p. 390*).

social density An index of crowding, typically calculated as the number of people occupying an area divided by the number of square feet in that area (*p. 261*).

social desirability bias The tendency for individuals responding to self-inventories to describe themselves in socially flattering ways (*p. 102*).

social information Information growing out of cultural norms, values, and shared opinions that shapes the way people perceive themselves, their jobs, and the organization (*p. 700*).

social-learning theory A theory of motivation originated by Bandura that suggests that behavior is often driven by the desire of an observer to model the behavior of some other person (*p. 214*).

social loafing The choice by some group members to take advantage of others by doing less work, working more slowly, or in other ways contributing less to group productivity (*p. 379*).

social support A surrounding environment in which people are sympathetic and caring (*p. 260*).

social traits Behavior patterns that an individual typically displays when interacting with others in social contexts (*p. 95*).

sociotechnical enrichment A type of job design that recognizes the importance of satisfying the needs of employees within the technical requirements of an organization's production system (*p. 658*).

solution verification A stage in the creative decision-making process wherein the person tests the efficacy of a proposed novel solution (*p. 184*).

span of control Another name for unit size; the number of people under the supervision of a single direct supervisor (*p. 558*).

spatial visualization An individual's capacity to mentally manipulate objects in space and time (*p. 89*).

special-interest committees Committees in Scandinavian industrial democracies composed of worker and manager representatives. They combine with middle management to produce yearly reports that assist works councils with the task of formulating company policies (*p. 742*).

specialization The division of an organization's work into specialist jobs of narrow scope and limited variability (*p. 563*).

standard time analysis A time study technique in which an analyst matches the results of micromotion analysis with standard time charts to determine the average time that should be required to perform a job (*p. 647*).

standardization In scientific measurement, the practice of ensuring that all people measure the same variables with the same instruments applied in the same manner. In structural design, a unit coordination mechanism in which work is coordinated by providing employees with carefully worked out standards and procedures that guide the performance of their tasks. It is coordination achieved on the drawing board, before the work is actually undertaken, and may involve standardization of work processes and behaviors, outputs, skills, or norms. (*p. 552*).

standing committee The type of representative group formed to meet on a regular basis to discuss and resolve intergroup coordination problems. A standing committee has no specific task, nor is it expected to disband at any particular time (*p. 573*).

statistical significance A numerical index of the probability that a relationship detected between two variables could be explained by luck or chance (*p. 65*).

stereotype accuracy The extent to which a rater's assessment of a group of people, on a single dimension, reflects the group's true standing on that dimension (*p. 127*).

stopwatch time analysis A time study technique in which an analyst uses a stopwatch to time the sequence of motions needed to complete a job (*p. 647*).

story An account of past events that all employees are familiar with and that serves as a reminder of cultural understandings (*p. 698*).

strategic business unit structure An extremely complex structure consisting of two or more autonomous strategic business units (SBUs), which themselves have complete organization structures (*p. 603*).

strategic management A process of setting organizational goals and directing the organization toward goal achievement. Strategic management research is concerned with defining an organization's purpose and planning how to achieve organizational objectives (*pp. 16, 684*).

strategy A plan of action that states an organization's goals and outlines the resources and activities required to achieve them (*p. 684*).

stress An unpleasurable emotional state resulting from the perception that a situational demand exceeds one's capacity and that it is very important to meet the demand (*p. 246*).

structural contingency factors Characteristics of an organization and its surrounding circumstances that influence whether its structure will contribute to organizational effectiveness (*p. 595*).

subliminal perception Information that is encoded by a perceiver without his or her awareness (*p. 130*).

substitutability The extent to which other people or groups can grant access to the same critical contingencies provided by the focal person or group (*p. 469*).

substitute for leadership Someone or something in the leader's environment that affects workers' attitudes, perceptions, or behaviors in such a way that the leader's role is made superfluous (*p. 422*).

subversive rebellion Acceptance by group members of peripheral norms but rejection of pivotal ones (*p. 377*).

superstars Cultural figures who are so extraordinary that they rise above their peers and sometimes even become symbols for an entire industry (*p. 699*).

survey feedback A shallow organization-level OD intervention intended to stimulate information sharing throughout the entire organization (*p. 705*).

symbol An object, action, or event to which people have assigned special meaning (*p. 699*).

symbolic management A process in which managers attempt to influence deep cultural norms and values by shaping the surface cultural elements that organization members use to express and transmit cultural understandings (*p. 701*).

target The specific focus of an OD intervention's change efforts (*p. 523*).

task force A type of representative group that is formed to complete a specific task or project and then disbanded (*p. 572*).

task-oriented roles Group roles that focus on making a contribution to successful task performance and accomplishing the group's task (*p. 375*).

team development A deep group-level extension of interpersonal sensitivity training in which a group of people who work together on a daily basis meet over an extended period to assess and modify group processes (*p. 529*).

team diagnostic session A group OD intervention of moderate depth in which a change agent and a work group critique the group's performance and look for ways to improve it (*p. 527*).

team interdependence A type of group interaction in which every group member depends on every other (*p. 335*).

technical skills Management skills involving an understanding of the specific knowledge, procedures, and tools used to make the goods or services produced by an organization or unit (*p. 45*).

technology The knowledge, procedures, and equipment used in an organization to transform unprocessed resources into finished goods or services (*p. 609*).

temporal precedence The degree to which any measured cause actually precedes an effect in time (*p. 63*).

theory A set of interrelated constructs, definitions, and propositions that present a systematic view of phenomena by specifying relations among variables (*p. 59*).

Theory X A managerial point of view that assumes that nonmanagerial employees have little interest in attaining organizational goals and must therefore be motivated to fit the needs of the organization (*p. 36*).

Theory Y A managerial point of view that assumes that nonmanagerial employees will readily direct behavior toward organizational goals if given the opportunity to do so (*p. 36*).

third party peacemaking A shallow OD intervention in which a change agent seeks to resolve intergroup misunderstandings by encouraging communication between or among groups (*p. 530*).

top-down selection within subgroups Personnel selection practice in which scores on a test are arrayed by subgroup and a flexible goal for representation from each group is decided upon. Selection proceeds from the highest to lowest score in each subgroup until all positions are filled and all subgroups are represented (*p. 300*).

training The process of teaching organization-specific and, often, job-specific skills on the job or in a formal program sponsored by the employing organization (*p. 562*).

transforming In the Lewin development model, the step in which change actually occurs (*p. 517*).

Type A behavior pattern A set of personality characteristics that include aggressiveness, competitiveness, and the tendency to work under self-induced time pressures (*p. 99*).

uncertainty avoidance A cross-cultural dimension that refers to the degree to which people are comfortable with ambiguous situations and with the inability to predict future events with assurance (*p. 736*).

unconditioned stimulus A stimulus that naturally and invariably produces a given response (*p. 211*).

unfreezing In the Lewin development model, the step in which one tries to weaken old attitudes, values, and behaviors and to get people ready for change (*p. 517*).

unit coordination mechanism A mechanism that sustains structural interconnections in structural units by helping to mesh interdependent task activities (*p. 552*).

unit grouping The process of grouping the members of an organization into work groups or units (*p. 549*).

unit linking mechanisms Mechanisms that regulate intergroup coordination by linking interdependent units more closely together (*p. 571*).

unit size The number of people who belong to a structural unit (*p. 558*).

utility maximization A process by which a decision maker selects the one alternative that leads to the highest possible payoff (*p. 166*).

valence The amount of satisfaction an individual anticipates receiving from a particular outcome (*p. 202*).

validity The degree to which a measure of an individual, group, organizational, or environmental attribute does what it is intended to do (*p. 62*).

verbal ability A specific type of cognitive ability that deals with the comprehension and use of language (*p. 89*).

verification A stage in the scientific process in which scientists assess the degree to which hypotheses based on theories match empirical data (*p. 59*).

vertical dyad Two persons who are related hierarchically, such as a supervisor-subordinate pair (*p. 426*).

vertical information system The type of unit linking mechanism in which computer networks are used to facilitate managerial communication and information processing. Managers can deal with more coordination information than would otherwise be possible (*p. 572*).

vertical job enrichment A type of job design based on the idea that giving job holders the discretion to choose job activities and outcomes will improve their satisfaction. (*p. 651*).

vertical specialization The type of specialization in which the management of work is separated from the performance of that work. It establishes the number of levels of hierarchy in an organization (*p. 564*).

voice The formal opportunity to complain to the organization about one's work situation (*p. 271*).

work measurement An area of industrial engineering concerned with measuring the amount of work accomplished and developing standards for performing work of an acceptable quantity and quality (*p. 646*).

work sample tests Tests that present job applicants with realistic simulations of actual job problems and ask them to indicate how they would handle them (*p. 294*).

works council A committee of worker representatives who are elected by their peers and management representatives who are appointed by top management. It oversees policy formulation in Scandinavian industrial democracies (*p. 741*).

INDEXES

NAME

COMPANY

SUBJECT

Abilities, human, 83, 85–93, (*see also* Individual differences):
cognitive, 87–93, 110, 184 (*see also* Cognitive abilities)
deductive, 89
and goal setting, 223
and group formation, 333
memory, 89–90
and motivation relationship, 106–10, 204–5, 215
perceptual, 89–90
and performance, 106–10, 223
and personality relationship, 106–10
physical, 85–87, 110
psychomotor, 86–87, 110
quantitative, 89–90
reasoning, 89–90
spatial visualization, 89–90
tests/testing, 90–93, 291–94
verbal, 89–90
Absenteeism:
and dissatisfaction, 254–55
as inequity response, 227
and job design, 655, 658, 664–65
and stress, 254–55, 265, 271, 603
and weight relationship, 72
Academy of Management, 74
Accommodation, and conflict management, 486
Acquisition strategies, 181
Action-outcome link, in decision making, 178
Action research model, 10, 519–21
Action stage, and OD, 519, 521
Action stage, diagnostic model, 11, 13, 78, 161
in motivation, 200
in perception, 125
Adaptation (*see also* Organization design; Structure, organization):
in environments, 38–40
Adaptive decline strategy, 690–91
Administrative decision-making model, 178–82
considering alternatives sequentially, 179
developing experts, 180
discretion in, 180
loosely coupling, 180–81
performance programs, 180
satisficing, 178–79
Administrative principles school, 30–34, 36, 40–41
Adopt a customer program, 37
Advanced beginner stage, management skills learning, 8–9
Affiliations, and power, 474–75
Affirmative action programs:
constraints on decision making, 299–300
general, 318
and perception, 124–25
After only with unequal groups research design, 71
Age Discrimination Act, 299
Agreements:
and work loads, 35
Alarm stage, in General adaptation syndrome, 250
Alderfer's ERG theory, 207–8
Alternatives:
elimination of, in research, 68–69
explanations, in causal inference, 63–69, 72
in problem solving, 13
Alternative work schedules, 664–65
Americans with Disabilities Act, 300
Antecedents, in predictive studies, 58
Anthropology, 15
Anti-leadership theories, 420–24
Applicants, and hiring decisions, 74, 105
Aptitude testing, 291–94
Aspirations, in goal setting, 219
Assessment centers, 294
Assessment procedures:
of applicants (*see* Selection)
of personality, 102, 104
Assimilation effect, 147
Assumptions:
about human nature, 37 (*see also* Theory X)
Asymmetric relationships, 335
Atmosphere (*see* Climate; Culture; Environment)
Attention stage, in perceptual process, 129–33
external factors, 130–31
internal factors, 131–33
and subliminal perceptions, 130

Attitudes:
entity attraction, 259
functional attraction, 259
and group formation, 333
Hawthorne studies, 34–36
negative affectivity, 261–63
and physical environment, 258–59
and stress, 244
and voicing opinions, 271
Attractive activities, and group formation, 334
Attribution theory, and leadership, 421
Attributions, in interpretation stage of perceptual process, 140–44
Authoritarianism, as personality characteristic, 99
Authority:
hierarchy of, 23, 28, 32, 42–43, 481, 558, 564, 572, 597, 603–4
managerial, 24
and power, 463, 471
principle of management, 31n
as way of knowing, 56
Autocratic leadership, 483, 746
Automatic processing, 129
Automation, and job design, 665–66
Autonomy, and job design, 654–55, 657
Availability bias, as threat to decision making, 169–70
Avoidance, and conflict management, 486, 753
Avoiders, as type of individual role, 376

Backward integration strategy, 690
Bargaining:
and change, 514
in conflict resolution, 485
in decision making, 175
Base rate bias, 170–72
BCG Matrix, 688
Behavior, organizational (*see* Organizational behavior)
Behavior modification programs, 311–13
Behavior patterns, 95
conformity, to norms, 377
creative individualism, and norms, 377
in norm adjustment, 376–77
open revolution, and norms, 377
subversive rebellion, and norms, 377
Type A, 99–100, 263
Type B, 99–100, 263
Behavioral diaries, 146, 298
Behavioral masking, 381
Behavioral responses, to stress, 249, 251
Behaviorally anchored rating scale (BARS), 296–99
Beliefs:
and change, 523
in communication, 349
and power, 463
Benchmark jobs, 306
Benign appraisal/reappraisal, 249, 251, 274
Bet Your Check program, 20
Bias, social desirability, 102
Binging, 226
Biofeedback, 272–73
Blockers, as type of individual role, 376
Bonus plans (*see also* Compensation; Wages):
motivating effect, 3, 4
task-and-bonus wage plan, 30
Boundary spanner, 618, 742
Bounded discretion, 176–77, 299–300
Brainstorming, 393–94
Breadth, of theories, 201
Buffering:
and conflict, 485
from stress, 260
Building threat, into research design, 72
Bureaucracy (*see also* Mechanistic structure):
definition, 32
features of, 32–33
model of, 32–33
organizational traits, 32, 595
weaknesses of, 33
Bureaucratic inertia, and change, 511
Bureaucratic prototype, work role, 337
Bureaucratic situation, and leadership, 425
Burnout, job, 251
Business simulation test, 294
Business strategies, 691–93

Career:
counseling, 316–17
definition, 316
development programs, 315–19
dual careers, 318–19
mentoring relationship, 317–18
paths, 319
stages, 316
Carryovers, in norm formation, 342
Case(s):
analyses, 8, 11
definition, 8
and diagnostic model, 10
Cash cows, and BCG Matrix, 688
Cash distribution plans, 310
Categories, personality, 95–101
Causal inference, 63–73
eliminating alternative explanations, 68–69
criteria for inferring cause, 63–69
designing observations, 69–73
threats to, 68–69
Cautious shift, in decision making, 390
Centrality, and power, 470 (*see also* Power)
Centralization (*see also* Decentralization):
principle of management, 31n
and structure, 568, 599, 605–6, 617, 619, 621–22, 755
Ceremonies, organizational, and culture, 697
Certainty, in decision making, 166, 169
Chain, communication structure, 386
Change, organizational, 15 (*see also* Change agent; Organization development):
and culture, 703–4
evaluating, 533
and forcefield analysis, 513–14
forces for change, 511–13
overcoming resistance, 513–14
resistance to, 508–14
sources of resistance, 509–11
types, 508–9n
and values, 515–16
Change agent, 506–8, 520, 524, 526–27, 529–30, 532, 705–6 (*see also* Organization development)
external, 508, 519
internal, 508
Characteristics of good data, 61–62
Charisma, to inspire, 27
Charismatic leadership, 439–40
Charismatic power, 463
Chief executive officer, in hierarchy of authority, 23
Child care, 64
Choice shift, in decision making, 389–90
Chronemics, as form of nonverbal communication, 347n
Circle, communication structure, 386
Civil Rights Act, 299
Classical conditioning, 211–12
Classification system, of personality, 95–101
Client organization, 508
Climate (*see also* Culture; Environment):
cut-throat, 243
organizational, 44
supportive, in making judgments, 148
Climate setting, and OD, 706
Clinical psychology, 14
Coalitions (*see also* Group):
and politics, 477
and power, 474
Coercion, and change, 514
Coercive power, 463, 466–67 (*see also* Power)
Cognition, in motivation, 201
Cognitive abilities, 87–93
dimensions, 89–90
general, 88, 184
and goal setting, 223
and job complexity, 110
and leadership, 414
and performance, 300
specific, 88–90
specific tests, 92–93
tests, 90–93
Cognitive appraisal theory, 248–52, 315
Cognitive distortion, 226
Cognitive responses, to stress, 251, 274
Cognitive theories of motivation, 201
Cohesiveness:
of groups (*see* Group, cohesiveness)
and productivity, 385

Followers, as type of maintenance role, 376
Forcefield analysis, and change, 513–14, 517
Formal groups, 370–71
Formal organization, 684–85
Formalization, and structure, 559–62, 595, 599, 600, 603–4, 609, 649
Formalization stage, in organization life cycle, 608
Formation stage of group development, 372
Forward integration strategy, 690
Frame of reference in communication, 350–52
Free riding, 380
Frequency, external factor in attention stage, 131
Frequency of data collection, 69–73
Friendship groups, 369–70
Functional attraction, 259
Functional departmentation, 566–67, 599, 602, 620 (*see also* Functional grouping; Functional structure)
Functional dimension, of organizations, 338, 354
Functional grouping, and structure, 549–50
Functional objectives, 25
Functional principle of organization, 33
Functional strategies, 693–94
Functional structure, 549–51, 599–600, 604, 606, 608, 611, 613–14, 744 (*see also* Functional departmentation)

Gainsharing, 641
Gantt chart, 30
Gatekeeper, as communication role, 352
Gender discrimination, 755
General adaptation syndrome, 249–51, 273
General Aptitude Test, 302
General intelligence (*see* Intelligence)
General Schedule System, for pay scales, 304
Generalizability, of research, 73–76
Geographic structures (*see* Divisional structure)
GMAT, 293
Goal commitment, 220
    and need for achievement, 221
Goal consensus:
    in affirmative action, 299
    in decision making, 174–75
Goal origin, 221
Goal-setting theory, 219–23
Goals (*see also* Objectives):
    and abilities, 223
    assigned, 221
    as benchmarks, 25
    challenging, 3
    and change, 510
    and conflict, 484
    in decision making, 174–75
    definition, 220
    difficult, 219–20, 222
    and group formation, 333–34
    and industrial democracy, 34
    means-ends relationship, 223–24
    organizational, 733
    as part of planning, 25
    and performance, 222, 381–82
    personal, 37
    private, 220–21
    public, 220–21
    research, 76
    of science, 57–58
    self-set, 221
    setting, 46, 76, 100
    shared, 174
    simple, 220, 222
    specific, 219–20, 222
    strategic goals, 25
    and strategy, 222, 684, 685
    and structure, 560
    superordinate, 484
    and task complexity, 222
    understanding, as scientific goal, 78
    vague, 220
Government regulations, and change, 512
Graduate Management Admissions Test, 293
Grapevine, 695 (*see also* Communication)
Gratification-activation hypothesis, 207
Group (*see also* Intergroup conflict; Intergroup coordination mechanism; Intergroup interventions; Intergroup mirroring; Intergroup team buliding; Satisfaction, of group members; Team development; Team interdependence; Teams; Teamwork; Work group):
    characteristics of, 368
    cohesiveness, 373, 384–86, 390–91, 483, 510

communication structure, 386–88
and conflict, 485
consensus, 373
constituency, 593–94
control, in research, 71
cultural differences, 371*n*
decision making, 388–97
definition, 368
development of, 371–77
effectiveness, 368
equating, in research, 72
experimental, in research, 71
goals, 373
homogeneous, 72
interaction, 332–45, 388
interventions, and OD, 527–29
maintenance activities, 379
matching of, 72
motivation, 381
norms, 35, 373, 376–77, 385
and organizational behavior, 7
participation, 37
productivity, 368, 375, 386–88
products of development, 373–77
purposes, 369–71
random assignment, 71–72
and research designs, 70–73
rewards, 382–83
roles, 375–76
self-contained, 485
semiautonomous groups, 659, 666, 742–43, 747
size, 380–81
stages of development, 372–73
and structure, organization, 549
tasks, 373–75
and threats to validity, 68–69
types, 369–71
why groups form, 332–34
Group observers, as type of maintenance role, 376
Groupthink, in decision making, 390–92
Growth:
    of employees, 36
    of organizations, 38
Growth needs, 207, 742
Growth need strength, 656

Hackman-Oldham Job Characteristics Model, 653–57, 659, 662, 664
Halo error:
    in organization stage of perceptual process, 137–39
    in retrieval stage of perceptual process, 146
Haptics, as form of nonverbal communication, 347*n*
Harmonizers, as type of maintenance role, 376
Hawthorne:
    effect, 432
    environment, 36
    studies, 34–36, 258
    work situation, 36
Hay system, in job evaluation, 304
Health, and job design, 649*n*
Health plans, 253
Hedonism, 211
Heroes, organizational, and culture, 699
Herzberg's Two-Factor Theory, 651–53
Heterogeneity, and power, 473–74
Hierarchy:
    of authority, 23, 28, 32, 42–43, 481, 558, 564, 572, 597, 603–4
    as coordination mechanism, 568–69
    of goals and objectives, 25–26
    latticework, 744–45
    organizational, 24, 339
    and power, 473–74
    and scalar chain, 31, 33
    and structure, 558
    and unity of command principle, 31*n*, 573, 603
Hiring decision, 74, 105
Historical decision model, 165
History threat to validity, 68, 70
Homogeneous groups, 72, 75
Honesty tests, 97
Horizontal integration strategy, 689
Horizontal job enlargement, 650–51, 660
Horizontal specialization, 564
Host group, in intergroup interventions and OD, 530
Human behavior:
    and Hawthorne studies, 34–36
    and Theory X, 36–37
Human factors engineering, 643–45

Human relations school:
    assumptions, 36–37, 40–41
    definition, 36
    Hawthorne studies, 34–36
    Mary Parker Follett, 34
    scientific methods, 34
Human resource management (*see also* Personnel):
    definition, 15, 18
    and micro OB, 15
Human skills, 44–45
Human Subjects Committee, 74
Humor as coping mechanism, 274
Hygiene factors, 652
Hypothesis and scientific process, 59

Iconics as form of nonverbal communication, 347*n*
Idea evaluation in decision making, 392
Idea generation in decision making, 392
Identification:
    as response to power, 465
    and role conforming, 344
Illumination, Hawthorne studies, 34
Illusion (*see* Perception)
Imitation in learning, 215
Implicit theories, and scientific process, 59–60
Impression management, and power, 475
In Search of Excellence, 53–54
In-basket tests, 294
Incentive, in motivation, 203
Incentive payment:
    and Hawthorne studies, 35
    and locus of control, 98
    programs, 15, 227, 308
Incentive systems, 309–10
Inclusionary dimension of organizations, 339, 354
Incremental adjustments of time, 48
Incubation stage in creative decision making process, 183–84
Individual adjustment to norms, 376
Individual differences (*see also* Abilities; Personality; Selection), 94, 110
    in communication, 348, 352
    in hiring, 82
    and mirror image fallacy, 84
    social traits, 95, 105, 110
    study of, 85
Individual roles, 376
Individual versus general interests, principle of management, 31*n*
Individualism-collectivism dimension, of culture, 737–39, 750–51, 753, 755
Induction (*see* Scientific process)
Industrial democracy, 34, 741
Industrial engineering, 28, 640–43, 646, 649
Industrial psychology, 14
Industrial revolution:
    and management practices, 28
Industrial robots, 665
Inequity (*see* Equity theory)
Informal groups, 369–70
Informal organization, 695–703
Information:
    acquisition of, 10–11, 13
    and environment, 39
    evaluating, 27
    inputs, and open systems, 39
    methods of acquiring, 10–11
    networks, 46
    overload, 387, 617
    social, 700
    upward, 42
Information, social, 700, 734
Information-collecting step, and OD, 707
Information givers as type of task-oriented role, 375
Information seekers as type of task-oriented role, 375
Information sharing step, and OD, 707
Informational roles, 45–46
Informed consent in research, 74
Ingratiation, and power, 475
In-group members:
    and leadership, 426–27
    and power, 475*n*
Initiating structure, 416–17, 426
Initiative principle of management, 31*n*
Initiators as type of task-oriented role, 375
Innovation in decision making, 182, 185
Inputs and open systems, 38
Insight stage in creative decision-making process, 184
Instincts, 93–94